Organic Chemistry

EIGHTH EDITION

Paula Yurkanis Bruice

University Of California
Santa Barbara

PEARSON

Editor in Chief: Jeanne Zalesky
Senior Acquisitions Editor: Chris Hess
Product Marketing Manager: Elizabeth Ellsworth
Project Manager: Elisa Mandelbaum
Program Manager: Lisa Pierce
Editorial Assistant: Fran Falk
Marketing Assistant: Megan Riley
Executive Content Producer: Kristin Mayo
Media Producer: Lauren Layn
Director of Development: Jennifer Hart
Development Editor: Matt Walker
Team Lead, Program Management: Kristen Flathman
Team Lead, Project Management: David Zielonka
Production Management: GEX Publishing Services
Compositor: GEX Publishing Services
Art Specialist: Wynne Au Yeung
Illustrator: Imagineering
Text and Image Lead: Maya Gomez
Text and Image Researcher: Amanda Larkin
Design Manager: Derek Bacchus
Interior and Cover Designer: Tamara Newnam
Operations Specialist: Maura Zaldivar-Garcia
Cover Image Credit: Olga Yakovenko/Shutterstock

Library of Congress Cataloging-in-Publication Data

Bruice, Paula Yurkanis
Organic chemistry / Paula Yurkanis Bruice, University of California,
   Santa Barbara.
Eighth edition. | Upper Saddle River, NJ: Pearson Education,
   Inc., 2015. | Includes index.
LCCN 2015038746 | ISBN 9780134042282 | ISBN 013404228X
LCSH: Chemistry, Organic—Textbooks.
LCC QD251.3 .B78 2015 | DDC 547--dc23
   LC record available at http://lccn.loc.gov/2015038746

ISBN 10: 0-13-404228-X; ISBN 13: 978-0-13-404228-2 (Student edition)
ISBN 10: 0-13-406659-6; ISBN 13: 978-0-13-406659-2 (Instructor's Review Copy)

8 2021

www.pearsonhighered.com

To Meghan, Kenton, and Alec
with love and immense respect
and to Tom, my best friend

Brief Table of Contents

Complete List of In-Chapter Connection Features

Medical Connections

Fosamax Prevents Bones from Being Nibbled Away (2.8)
Aspirin Must Be in its Basic Form to be Physiologically Active (2.10)
Blood: A Buffered Solution (2.11)
Drugs Bind to Their Receptors (3.9)
Cholesterol and Heart Disease (3.16)
How High Cholesterol is Treated Clinically (3.16)
The Enantiomers of Thalidomide (4.17)
Synthetic Alkynes Are Used to Treat Parkinson's Disease (7.0)
Synthetic Alkynes Are Used for Birth Control (7.1)
The Inability to Perform an S_N2 Reaction Causes a Severe Clinical Disorder (10.3)
Treating Alcoholism with Antabuse (10.5)
Methanol Poisoning (10.5)
Anesthetics (10.6)
Alkylating Agents as Cancer Drugs (10.11)
S-Adenosylmethionine: A Natural Antidepressant (10.12)
Artificial Blood (12.12)
Nature's Sleeping Pill (15.1)
Penicillin and Drug Resistance (15.12)
Dissolving Sutures (15.13)
Cancer Chemotherapy (16.17)
Breast Cancer and Aromatase Inhibitors (17.12)
Thyroxine (18.3)
A New Cancer-Fighting Drug (18.20)
Atropine (19.2)
Porphyrin, Bilirubin, and Jaundice (19.7)
Measuring the Blood Glucose Levels in Diabetes (20.8)
Galactosemia (20.15)
Why the Dentist is Right (20.16)
Resistance to Antibiotics (20.17)
Heparin–A Natural Anticoagulant (20.17)
Amino Acids and Disease (21.2)
Diabetes (21.8)
Diseases Caused by a Misfolded Protein (21.15)
How Tamiflu Works (22.11)
Assessing the Damage After a Heart Attack (23.5)
Cancer Drugs and Side Effects (23.7)
Anticoagulants (23.8)
Phenylketonuria (PKU): An Inborn Error of Metabolism (24.8)
Alcaptonuria (24.8)
Multiple Sclerosis and the Myelin Sheath (25.5)
How Statins Lower Cholesterol Levels (25.8)
One Drug—Two Effects (25.10)
Sickle Cell Anemia (26.9)
Antibiotics That Act by Inhibiting Translation (26.9)
Antibiotics Act by a Common Mechanism (26.10)
Health Concerns: Bisphenol A and Phthalates (27.11)

Biological Connections

Poisonous Amines (2.3)
Cell Membranes (3.10)
How a Banana Slug Knows What to Eat (7.2)
Electron Delocalization Affects the Three-Dimensional Shape of Proteins (8.4)

Naturally Occurring Alkyl Halides That Defend Against Predators (9.5)
Biological Dehydrations (10.4)
Alkaloids (10.9)
Dalmatians: Do Not Fool with Mother Nature (15.11)
A Semisynthetic Penicillin (15.12)
Preserving Biological Specimens (16.9)
A Biological Friedel-Crafts Alkylation (18.7)
A Toxic Disaccharide (20.15)
Controlling Fleas (20.16)
Primary Structure and Taxonomic Relationship (21.12)
Competitive Inhibitors (23.7)
Whales and Echolocation (25.3)
Snake Venom (25.5)
Cyclic AMP (26.1)
There Are More Than Four Bases in DNA (26.7)

Chemical Connections

Natural versus Synthetic Organic Compounds (1.0)
Diamond, Graphite, Graphene, and Fullerenes: Substances that Contain Only Carbon Atoms (1.8)
Water—A Unique Compound (1.12)
Acid Rain (2.2)
Derivation of the Henderson-Hasselbalch Equation (2.10)
Bad-Smelling Compounds (3.7)
Von Baeyer, Barbituric Acid, and Blue Jeans (3.12)
Starch and Cellulose—Axial and Equatorial (3.14)
Cis-Trans Interconversion in Vision (4.1)
The Difference between $\Delta G^{\ddagger}$ and E_a (5.11)
Calculating Kinetic Parameters (End of Ch 05)
Borane and Diborane (6.8)
Cyclic Alkenes (6.13)
Chiral Catalysts (6.15)
Sodium Amide and Sodium in Ammonia (7.10)
Buckyballs (8.18)
Why Are Living Organisms Composed of Carbon Instead of Silicon? (9.2)
Solvation Effects (9.14)
The Lucas Test (10.1)
Crown Ethers—Another Example of Molecular Recognition (10.7)
Crown Ethers Can be Used to Catalyze S_N2 Reactions (10.7)
Eradicating Termites (10.12)
Cyclopropane (12.9)
What Makes Blueberries Blue and Strawberries Red? (13.22)
Nerve Impulses, Paralysis, and Insecticides (15.19)
Enzyme-Catalyzed Carbonyl Additions (16.4)
Carbohydrates (16.9)
β-Carotene (16.13)
Synthesizing Organic Compounds (16.14)
Enzyme-Catalyzed Cis-Trans Interconversion (16.16)
Incipient Primary Carbocations (18.7)
Hair: Straight or Curly? (21.8)
Right-Handed and Left-Handed Helices (21.14)
β-Peptides: An Attempt to Improve on Nature (21.14)
Why Did Nature Choose Phosphates? (24.1)
Protein Prenylation (25.8)
Bioluminescence (28.6)

Contents

MasteringChemistry®

for Organic Chemistry

MasteringChemistry tutorials guide you through the toughest topics in chemistry with self-paced tutorials that provide individualized coaching. These assignable, in-depth tutorials are designed to coach you with hints and feedback specific to your individual misconceptions. For additional practice on Acids and Bases, go to MasteringChemistry, where the following tutorials are available:

- Acids and Bases: Definitions
- Acids and Bases: Factors That Influence Acid Strength
- Acids and Bases: Base Strength and the Effect of pH on Structure
- Acids and Bases: Predicting the Position of Equilibrium

MasteringChemistry®

for Organic Chemistry

Mastering Chemistry tutorials guide you through the toughest topics in chemistry with self-paced tutorials that provide individualized coaching. These assignable, in-depth tutorials are designed to coach you with hints and feedback specific to your individual misconceptions. For additional practice on Molecular Models, go to MasteringChemistry where the following tutorials are available:

- Basics of Model Building
- Building and Recognizing Chiral Molecules
- Recognizing Chirality in Cyclic Molecules

PART TWO Electrophilic Addition Reactions, Stereochemistry, and Electron Delocalization 141

Using the *E,Z* system to name alkenes was moved to Chapter 4, so now it appears immediately after using cis and trans to distinguish alkene stereoisomers.

MasteringChemistry®

for Organic Chemistry

MasteringChemistry tutorials guide you through the toughest topics in chemistry with self-paced tutorials that provide individualized coaching. These assignable, in-depth tutorials are designed to coach you with hints and feedback specific to your individual misconceptions. For additional practice on Interconverting Structural Representations, go to MasteringChemistry where the following tutorials are available:

- Interconverting Fischer Projections and Perspective Formulas
- Interconverting Perspective Formulas, Fischer Projections, and Skeletal Structures
- Interconverting Perspective Formulas, Fischer Projections, and Newman Projections

5 Alkenes: Structure, Nomenclature, and an Introduction to Reactivity • Thermodynamics and Kinetics 190

6 The Reactions of Alkenes • The Stereochemistry of Addition Reactions 235

Catalytic hydrogenation and relative stabilities of alkenes were moved from Chapter 6 to Chapter 5 (thermodynamics), so they can be used to illustrate how $\Delta H°$ values can be used to determine relative stabilities.

MasteringChemistry®

for Organic Chemistry

MasteringChemistry tutorials guide you through the toughest topics in chemistry with self-paced tutorials that provide individualized coaching. These assignable, in-depth tutorials are designed to coach you with hints and feedback specific to your individual misconceptions. For additional practice on Drawing Curved Arrows: Pushing Electrons, go to MasteringChemistry where the following tutorials are available:

- An Exercise in Drawing Curved Arrows: Pushing Electrons
- An Exercise in Drawing Curved Arrows: Predicting Electron Movement
- An Exercise in Drawing Curved Arrows: Interpreting Electron Movement

All the reactions in Chapter 6 follow the same mechanism the first step is always addition of the electrophile to the sp^2 carbon bonded to the most hydrogens.

7 The Reactions of Alkynes • An Introduction to Multistep Synthesis 288

8 Delocalized Electrons: Their Effect on Stability, pK_a, and the Products of a Reaction • Aromaticity and Electronic Effects: An Introduction to the Reactions of Benzene 318

Chapter 8 starts by discussing the structure of benzene because it is the ideal compound to use to explain delocalized electrons. This chapter also includes a discussion of aromaticity, so a short introduction to electrophilic aromatic substitution reactions is now included. This allows students to see how aromaticity causes benzene to undergo electrophilic substitution rather than electrophilic addition—the reactions they have just finished studying.

Traditionally, electronic effects are taught so students can understand the directing effects of substituents on benzene rings. Now that most of the chemistry of benzene follows carbonyl chemistry, students need to know about electronic effects before they get to benzene chemistry (so they are better prepared for spectroscopy and carbonyl chemistry). Therefore, electronic effects are now discussed in Chapter 8 and used to teach students how substituents affect the pK_a values of phenols, benzoic acids, and anilinium ions. Electronic effects are then reviewed in the chapter on benzene.

MasteringChemistry®

for Organic Chemistry

MasteringChemistry tutorials guide you through the toughest topics in chemistry with self-paced tutorials that provide individualized coaching. These assignable, in-depth tutorials are designed to coach you with hints and feedback specific to your individual misconceptions. For additional practice on Drawing Resonance Contributors, go to MasteringChemistry where the following tutorials are available:

- Drawing Resonance Contributors: Moving π Electrons
- Drawing Resonance Contributors: Predicting Aromaticity
- Drawing Resonance Contributors: Substituted Benzene Rings

PART THREE   Substitution and Elimination Reactions   390

9   Substitution and Elimination Reactions of Alkyl Halides   391

The two chapters in the previous edition on substitution and elimination reactions of alkenes have been combined into one chapter. The recent compelling evidence showing that secondary alkyl halides do not undergo S_N1 solvolysis reactions has allowed this material to be greatly simplified, so now it fits nicely into one chapter.

10   Reactions of Alcohols, Ethers, Epoxides, Amines, and Sulfur-Containing Compounds   458

11 Organometallic Compounds **508**

The discussion of palladium-catalyzed coupling reactions has been expanded, and the cyclic catalytic mechanisms are shown.

12 Radicals **532**

MasteringChemistry®

for Organic Chemistry

MasteringChemistry tutorials guide you through the toughest topics in chemistry with self-paced tutorials that provide individualized coaching. These assignable, in-depth tutorials are designed to coach you with hints and feedback specific to your individual misconceptions. For additional practice on Drawing Curved Arrows in Radical Systems, go to MasteringChemistry where the following tutorials are available:

- Curved Arrows in Radical Systems: Interpreting Curved Arrows
- Curved Arrows in Radical Systems: Drawing Curved Arrows
- Curved Arrows in Radical Systems: Drawing Resonance Contributors

In addition to the more than 170 spectroscopy problems in Chapters 13 and 14, there are 60 additional spectroscopy problems in the *Study Guide and Solutions Manual*.

Chapters 13 and 14 are modular, so they can be covered at any time.

PART FIVE Carbonyl Compounds 685

15 Reactions of Carboxylic Acids and Carboxylic Acid Derivatives 686

The focus of the first chapter on carbonyl chemistry is all about how a tetrahedral intermediate partitions. If students understand this, then carbonyl chemistry becomes pretty straightforward. I found that the lipid materil that had been put into this chapter in the last edition detracted from the main message of the chapter. Therefore, the lipid material was removed and put into a new chapter exclusively about lipids.

16 Reactions of Aldehydes and Ketones • More Reactions of Carboxylic Acid Derivatives 739

17 Reactions at the α-Carbon 801

This chapter was reorganized and rewritten for ease of understanding.

New art adds clarity.

22 Catalysis in Organic Reactions and in Enzymatic Reactions   1030

23 The Organic Chemistry of the Coenzymes, Compounds Derived from Vitamins   1063

Increased emphasis on the connection between the reactions that occur in the laboratory and those that occur in cells.

24 The Organic Chemistry of the Metabolic Pathways   1099

25 The Organic Chemistry of Lipids 1127

The lipid material previously in the chapter on carboxylic acids and their derivatives has been moved into this new chapter. The discussion of terpenes from the metabolism chapter has also been moved into this chapter, along with some new material.

26 The Chemistry of the Nucleic Acids 1155

Preface

The guiding principle behind this book is to present organic chemistry as an exciting and vitally important science. To counter the impression that the study of organic chemistry consists primarily of memorizing a multitude of facts, I have organized this book around shared features and unifying concepts, while emphasizing principles that can be applied again and again. I want students to apply what they have learned to new settings and to learn how to reason their way to solutions. I also want them to see that organic chemistry is a fascinating discipline that is integral to their daily lives.

Preparing Students for Future Study in a Variety of Scientific Disciplines

This book organizes the functional groups around mechanistic similarities. When students see their first reaction (other than an acid–base reaction), they are told that all organic compounds can be divided into families and that all members of a family react *in the same way*. And to make things even easier, each family can be put into one of four groups, and all the families in a group react *in similar ways*.

"Organizing What We Know About Organic Chemistry" is a feature based on these statements. It lets students see where they have been and where they are going as they proceed through each of the four groups. It also encourages them to remember the fundamental reason behind the reactions of all organic compounds: *electrophiles react with nucleophiles*. When students finish studying a particular group, they are given the opportunity to review the group and understand why the families came to be members of that particular group. The four groups are covered in the following order. (However, the book is written to be modular, so they could be covered in any order.)

- **Group I: Compounds with carbon-carbon double and triple bonds**. These compounds are nucleophiles and, therefore, react with electrophiles—undergoing electrophilic addition reactions.
- **Group II: Compounds with electron-withdrawing atoms or groups attached to sp^3 carbons**. These compounds are electrophiles and, therefore, react with nucleophiles—undergoing nucleophilic substitution and elimination reactions.
- **Group III: Carbonyl compounds**. These compounds are electrophiles and, therefore, react with nucleophiles—undergoing nucleophilic acyl substitution, nucleophilic addition, and nucleophilic addition-elimination reactions. Because of the "acidity" of the α-carbon, a carbonyl compound can become a nucleophile and, therefore, react with electrophiles.
- **Group IV: Aromatic compounds**. Some aromatic compounds are nucleophiles and, therefore, react with electrophiles—undergoing electrophilic aromatic substitution reactions. Other aromatic compounds are electrophiles and, therefore, react with nucleophiles—undergoing nucleophilic aromatic substitution reactions.

The organization discourages rote memorization and allows students to learn reactions based on their pattern of reactivity. It is only after these patterns of reactivity are understood that a deep understanding of organic chemistry can begin. As a result, students achieve the predictive capacity that is the beauty of studying science. A course that teaches students to analyze, classify, explain, and predict gives them a strong foundation to bring to their subsequent study of science, regardless of the discipline.

As students proceed through the book, they come across ~200 interest boxes that connect what they are studying to real life. Students don't have to be preparing for a career in medicine to appreciate a box on the experimental drug used to treat Ebola, and they don't have to be preparing for a career in engineering to appreciate a box on the properties that a polymer used for dental impressions must have.

The Organization Ties Together Reactivity and Synthesis

Many organic chemistry textbooks discuss the synthesis of a functional group and the reactivity of that group sequentially, but these two groups of reactions generally have little to do with one another. Instead, when I discuss a functional group's reactivity, I cover the synthesis of compounds that are formed as a result of that reactivity, often by having students design syntheses. In Chapter 6, for example, students learn about the *reactions* of alkenes, but they *do not* learn about the *synthesis* of alkenes. Instead, they learn about the synthesis of alkyl halides, alcohols, ethers, epoxides, alkanes, etc.—the compounds formed when alkenes react. The synthesis of alkenes is not covered until the reactions of alkyl halides and alcohols are discussed—compounds whose reactions lead to the synthesis of alkenes.

This strategy of tying together the reactivity of a functional group and the synthesis of compounds resulting from its reactivity prevents the student from having to memorize lists of unrelated reactions. It also results in a certain economy of presentation, allowing more material to be covered in less time.

Although memorizing different ways a particular functional group can be prepared can be counterproductive to enjoying organic chemistry, it is useful to have such a compilation of reactions when designing multistep syntheses. For this reason, lists of reactions that yield a particular functional group are compiled in Appendix III. In the course of learning how to design syntheses, students come to appreciate the importance of reactions that change the carbon skeleton of a molecule; these reactions are compiled in Appendix IV.

Helping Students Learn and Study Organic Chemistry

As each student generation evolves and becomes increasingly diverse, we are challenged as teachers to support the unique ways students acquire knowledge, study, practice, and master a subject. In order to support contemporary students who are often visual learners, with preferences for interactivity and small "bites" of information, I have revisited this edition to make it more compatible with their learning style by streamlining the narrative and using organizing bullets and subheads. This will allow them to study more efficiently with the text.

The book is written much like a tutorial. Each section ends with a set of problems that students need to work through to find out if they are ready to go on to the next section, or if they need to review the section they thought they had just mastered. This allows the book to work well in a "flipped classroom."

For those who teach organic chemistry after one semester of general chemistry, Chapter 5 and Appendix II contain material on thermodynamics and kinetics, so those topics can be taught in the organic course.

An enhanced art program with new and expanded annotations provides key information to students so that they can review important parts of the chapter with the support of the visual program. Margin notes throughout the book succinctly repeat key points and help students review important material at a glance.

Tutorials follow relevant chapters to help students master essential skills:
- Acids and Bases
- Using Molecular Models
- Interconverting Structural Representations
- Drawing Curved Arrows
- Drawing Resonance Contributors
- Drawing Curved Arrows in Radical Systems
- Synthesis and Retrosynthetic analysis

MasteringChemistry includes additional online tutorials on each of these topics that can be assigned as homework or for test preparation.

Organizational Changes

Using the *E,Z* system to distinguish alkene stereoisomers was moved to Chapter 4, so now it appears immediately after using cis and trans to distinguish alkene stereoisomers.

Catalytic hydrogenation and the relative stabilities of alkenes was moved from Chapter 6 to Chapter 5 (thermodynamics), so it can be used to illustrate how $\Delta H°$ values can be used to determine relative stabilities. Moving this has another advantage—because catalytic hydrogenation is the only reaction of alkenes that does not have a well-defined mechanism, all the remaining reactions in Chapter 6 now have well-defined mechanisms, all following the general rule that applies to all electrophilic addition reactions: the first step is always the addition of the electrophile to the sp^2 carbon bonded to the most hydrogens.

Chapter 8 starts by discussing the structure of benzene because it is the ideal compound to use to explain delocalized electrons. This chapter also includes a discussion on aromaticity, so a short introduction to electrophilic aromatic substitution reactions is now included. This allows students to see how aromaticity causes benzene to undergo electrophilic substitution rather than electrophilic addition—the reactions they just finished studying.

Traditionally, electronic effects are taught so students can understand the activating and directing effects of substituents on benzene rings. Now that most of the chemistry of benzene follows carbonyl chemistry, students need to know about electronic effects before they get to benzene chemistry (so they are better prepared for spectroscopy and carbonyl chemistry). Therefore, in this edition electronic effects are discussed in Chapter 8 and used to teach students how substituents affect the pK_a values of phenols, benzoic acids, and anilinium ions. Electronic effects are then reviewed in the chapter on benzene.

The two chapters in the previous edition that covered the substitution and elimination reactions of alkyl halides have been combined into one chapter (Chapter 9). The recent compelling evidence showing that alkyl halides do not undergo S_N1 solvolysis reactions has allowed this material to be greatly simplified, so now it fits nicely into one chapter.

I have found that teaching carbonyl chemistry before the chemistry of aromatic compounds (a change made in the last edition) has worked well for my students. Carbonyl compounds are probably the most important organic compounds, and moving them forward gives them the prominence they should have. In addition, the current location of the chemistry of benzene allows it and the chemistry of aromatic heterocyclic compounds to be taught sequentially.

The focus of the first chapter on carbonyl chemistry should be all about how a tetrahedral intermediate partitions. If students understand this, then carbonyl chemistry becomes relatively easy. I found that the lipid material that had been put into this chapter detracted from the main message of the chapter. Therefore, the lipid material was removed and put into a new chapter: The Organic Chemistry of Lipids. The discussion of terpenes from the metabolism chapter has also been moved into this chapter, and some some new material has been included.

Modularity/Spectroscopy

The book is designed to be modular, so the four groups (Group I—Chapters 6, 7, 8; Group II—Chapters 9 and 10; Group III—Chapters 15, 16, 17; Group IV—Chapters 18 and 19) can be covered in any order.

Sixty spectroscopy problems and their answers—in addition to ~170 spectroscopy problems in the text—can be found in the *Study Guide and Solutions Manual*. The spectroscopy chapters (Chapters 13 and 14) are written so that they can be covered at any time during the course. For those who prefer to teach spectroscopy before all the functional groups have been introduced—or in a separate laboratory course—there is a table of functional groups at the beginning of Chapter 13.

An Early and Consistent Emphasis on Organic Synthesis

Students are introduced to synthetic chemistry and retrosynthetic analysis early in the book (Chapters 6 and 7, respectively), so they can start designing multistep syntheses early in the course. Seven special sections on synthesis design, each with a different focus, are introduced at appropriate intervals. There is also a tutorial on synthesis and retrosynthetic analysis that includes some examples of complicated multistep syntheses from the literature.

Example 2

Starting with ethyne, how could you make 2-bromopentane?

$$HC\equiv CH \xrightarrow{\ ?\ } CH_3CH_2CH_2CHCH_3$$

ethyne

Br

2-bromopentane

2-Bromopentane can be prepared from 1-pentene, which can be prepared from 1-pentyne. 1-Pentyne can be prepared from ethyne and an alkyl halide with three carbons.

Problems, Solved Problems, and Problem-Solving Strategies

The book contains more than 2,000 problems, many with multiple parts. This edition has many new problems, both in-chapter and end-of-chapter. They include new solved problems, new problem-solving strategies, and new problems incorporating information from more than one chapter. I keep a list of questions my students have when they come to office hours. Many of the new problems were created as a result of these questions.

The answers (and explanations, when needed) to all the problems are in the accompanying *Study Guide/Solutions Manual*, which I authored to ensure consistency in language with the text. The problems within each chapter are primarily drill problems. They appear at the end of each section, so they allow students to test themselves on material just covered before moving on to the next section. Short answers provided at the end of the book for problems marked with a diamond give students immediate feedback concerning their mastery of a skill or concept.

Selected problems are accompanied by worked-out solutions to provide insight into problem-solving techniques, and the many Problem-Solving Strategies teach students how to approach various kinds of problems. These skill-teaching problems are indicated by LEARN THE STRATEGY in the margin. These strategies are followed by one or more problems that give students the opportunity to use the strategy just learned. These problems, or the first of a group of such problems, are indicated in the margin by USE THE STRATEGY.

The *Study Guide/Solutions Manual* has a practice test at the end of each chapter and contains Special Topics Sections on molecular orbital theory and how to solve problems on pH, pK_a, and buffer solutions.

Powerpoint

All the art in the text is available on PowerPoint slides. I created the PowerPoint lectures so they would be consistent with the language and philosophy of the text.

Students Interested in The Biological Sciences and MCAT[2015]

I have long believed that students who take organic chemistry also should be exposed to bioorganic chemistry—the organic chemistry that occurs in biological systems. Students leave their organic chemistry course with a solid appreciation of organic mechanism and synthesis. But when they take biochemistry, they will never hear about Claisen condensations, S_N2 reactions, nucleophilic acyl substitution reactions, etc., although these are extremely important reactions in cells. Why are students required to take organic chemistry if they are not going to be taught how the organic chemistry they learn repeats itself in the biological world?

Now that the MCAT is focusing almost exclusively on the organic chemistry of living systems, it is even more important that we provide our students with the "bioorganic bridge"—the material that provides the bridge between organic chemistry and biochemistry. Students should see that the organic reactions that chemists carry out in the laboratory are in many ways the same as those performed by nature inside a cell.

The seven chapters (Chapters 20–26) that focus primarily on the organic chemistry of living systems emphasize the connection between the organic reactions that occur in the laboratory and those that occur in cells.

> *Each organic reaction that occurs in a cell is explicitly compared*
> *to the organic reaction with which the student is already familiar.*

For example, the first step in glycolysis is an S_N2 reaction, the second step is identical to the enediol rearrangement that students learn when they study carbohydrate chemistry, the third step is another S_N2 reaction, the fourth step is a reverse aldol addition, and so on. The first step in the citric acid cycle is an aldol addition followed by a nucleophilic acyl substitution reaction, the second step is an E2 dehydration followed by the conjugate addition of water, the third step is oxidation of a secondary alcohol followed by decarboxylation of a 3-oxocarboxylate ion, and so on.

We teach students about halide and sulfonate leaving groups. Adding phosphate leaving groups takes little additional time but introduces the students to valuable information if they are going on to study biochemistry.

Students who study organic chemistry learn about tautomerization and imine hydrolysis, and students who study biochemistry learn that DNA has thymine bases in place of the uracil bases in RNA. But how many of these students are ever told that the reason for the difference in the bases in DNA and RNA is tautomerization and imine hydrolysis?

Colleagues have asked how they can find time to fit the "bioorganic bridge" into their organic chemistry courses. I found that tying together reactivity and synthesis (see p. xxiii) frees up a lot of time. (This is the organization I adopted many years ago when I was trying to figure out how to incorporate the bioorganic bridge into my course.) And if you find that this still does not give you enough time, I have organized the book in a way that allows some "traditional" chapters to be omitted (Chapters 12, 18, 19, and 28), so students can be prepared for biochemistry and/or the MCAT without sacrificing the rigor of the organic course.

The Bioorganic Bridge

Bioorganic chemistry is found throughout the text to show students that organic chemistry and biochemistry are not separate entities but rather are closely related on a continuum of knowledge. Once students learn how, for example, electron delocalization, leaving-group propensity, electrophilicity, and nucleophilicity affect the reactions of simple organic compounds, they can appreciate how these same factors influence the reactions of organic compounds in cells.

In Chapters 1–19, the bioorganic material is limited mostly to "interest boxes" and to the last sections of the chapters. Thus, the material is available to the curious student without requiring the instructor to introduce bioorganic topics into the course. For example, after hydrogen bonding is introduced in Chapter 3, hydrogen boding in proteins in DNA is discussed; after catalysis is introduced in Chapter 5, catalysis by enzymes is discussed; after the stereochemistry of organic reactions is presented in Chapter 6, the stereochemistry of enzyme-catalyzed reactions is discussed; after sulfonium ions are discussed in Chapter 10, a biological methylation reaction using a sulfonium ion is examined and the reason for the use of different methylating agents by chemists and cells is explained; after the methods chemists use to activate carboxylic acids are presented (by giving them halide or anhydride leaving groups) in Chapter 15, the methods cells use to activate these same acids are explained (by giving them phosphoanhydride, pyrophosphate, or thiol leaving groups); and after condensation reactions are discussed in Chapter 17, the mechanisms of some biological condensation reactions are shown.

In addition, seven chapters in the last part of the book (Chapters 20–26) focus on the organic chemistry of living systems. These chapters have the unique distinction of containing more chemistry than is typically found in the corresponding parts of a biochemistry text. Chapter 22 (Catalysis in Organic Reactions and in Enzymatic Reactions), for example, explains the various modes of catalysis employed in organic reactions and then shows that they are identical to the modes of catalysis found in reactions catalyzed by enzymes. All of this is presented in a way that allows students to understand the lightning-fast rates of enzymatic reactions. Chapter 23 (The Organic Chemistry of the Coenzymes, Compounds Derived from Vitamins) emphasizes the role of vitamin B_1 in electron delocalization, vitamin K as a strong base, vitamin B_{12} as a radical initiator, biotin as a compound that transfers a carboxyl group by means of a nucleophilic acyl substitution reaction, and describes how the many different reactions of vitamin B_6 have common mechanisms—with the first step always being imine formation. Chapter 24 (The Organic Chemistry of Metabolic Pathways) explains the chemical function of ATP and shows students that the reactions encountered in metabolism are just additional examples of reactions that they already have mastered. In Chapter 26 (The Chemistry of the Nucleic Acids), students learn that $2'$-OH group on the ribose molecules in RNA catalyzes its hydrolysis and that is why DNA, which has to stay intact for the life of the cell, does not have $2'$-OH groups. Students also see that the synthesis of proteins in cells is just another example of a nucleophilic acyl substitution reaction. Thus, these chapters do not replicate what will be covered in a biochemistry course; they provide a bridge between the two disciplines, allowing students to see how the organic chemistry that they have learned is repeated in the biological world.

ENGAGING MIXED SCIENCE MAJORS IN ORGANIC CHEMISTRY

Students better understand the relevance of what they're studying by seeing the connections between the reactions of organic compounds that occur in the laboratory and those that occur in a cell. Changes throughout this edition provide students with this much-needed "bioorganic bridge," while maintaining the rigor of the traditional organic course.

For example, we teach students about halide and sulfonate leaving groups. Adding phosphate leaving groups takes little additional time, but it introduces students to valuable information, particularly if they are taking organic chemistry because of an interest in the biological sciences. Students who are studying organic chemistry learn about tautomerization and imine hydrolysis, and students studying biochemistry learn that DNA has thymine bases in place of the uracil bases in RNA. But how many of these students are ever told that the reason for the difference in the bases in DNA and RNA is tautomerization and imine hydrolysis?

The NADP$^+$ formed in this reaction has to be reduced back to NADPH by NADH. Every NADH formed in a cell can result in the formation of 2.5 ATPs (Section 24.10). Therefore, reducing dihydrofolate comes at the expense of ATP. This means that the synthesis of thymine is energetically expensive, so there must be a good reason for DNA to contain thymine instead of uracil.

The presence of thymine instead of uracil in DNA prevents potentially lethal mutations. Cytosine can tautomerize to form an imine (Section 17.2), which can be hydrolyzed to uracil (Section 16.8). The overall reaction is called a **deamination** because it removes an amino group.

More Applications Than Any Other Organic Text

NEW! and Updated Application boxes connect the discussion to medical, environmental, biological, pharmaceutical, nutritional, chemical, industrial, historical, and general applications and allow students to relate the material to real life and to potential future careers.

Using Genetic Engineering to Treat the Ebola Virus

Plants have long been a source of drugs—morphine, ephedrine, and codeine are just a few examples (Section 10.9). Now scientists are attempting to obtain drugs from plants by biopharming. Biopharming uses genetic engineering techniques to produce drugs in crops such as corn, rice, tomatoes, and tobacco. To date, the only biopharmed drug approved by the Food and Drug Administration (FDA) is one that is manufactured in carrots and used to treat Gaucher's disease.

An experimental drug that was used to treat a handful of patients with Ebola, the virus that was spreading throughout West Africa, was obtained from genetically engineered tobacco plants. The tobacco plants were infected with three genetically engineered plant viruses that are harmless to humans and animals but have structures similar to that of the Ebola virus. As a result of being infected, the plants produced antibodies to the viruses. The antibodies were isolated from the plants, purified, and then used to treat the patients with Ebola.

The experimental drug had been tested in 18 monkeys who had been exposed to a lethal dose of Ebola. All 18 monkeys survived, whereas the three monkeys in the control group died. Typically, drugs go through rigorous testing on healthy humans prior to being administered to infected patients (see page 290). In the Ebola case, the FDA made an exception because it feared that the drug might be these patients' only hope. Five of the seven people given the drug survived. Currently, it takes about 50 kilograms of tobacco leaves and about 4 to 6 months to produce enough drug to treat one patient.

tobacco plants

The Birth of the Environmental Movement

Alkyl halides have been used as insecticides since 1939, when it was discovered that DDT (first synthesized in 1874) has a high toxicity to insects and a relatively low toxicity to mammals. DDT was used widely in World War II to control typhus and malaria in both the military and civilian populations. It saved millions of lives, but no one realized at that time that, because it is a very stable compound, it is resistant to biodegradation. In addition, DDT and DDE, a compound formed as a result of elimination of HCl from DDT, are not water soluble. Therefore, they accumulate in the fatty tissues of birds and fish and can be passed up the food chain. Most older adults have a low concentration of DDT or DDE in their bodies.

In 1962, Rachel Carson, a marine biologist, published *Silent Spring*, where she pointed out the environmental impacts of the widespread use of DDT. The book was widely read, so it brought the problem of environmental pollution to the attention of the general public for the first time. Consequently, its publication was an important event in the birth of the environmental movement. Because of the concern it raised, DDT was banned in the United States in 1972. In 2004, the Stockholm Convention banned the worldwide use of DDT except for the control of malaria in countries where the disease is a major health problem.

In Section 12.12, we will look at the environmental effects caused by synthetic alkyl halides known as chlorofluorohydrocarbons (CFCs).

GUIDED APPROACH TO PROBLEM SOLVING

Essential Skill Tutorials

These tutorials guide students through some of the topics in organic chemistry that they typically find to be the most challenging. They provide concise explanations, related problem-solving opportunities, and answers for self-check. The print tutorials are paired with MasteringChemistry online tutorials. These are additional problem sets that can be assigned as homework or as test preparation.

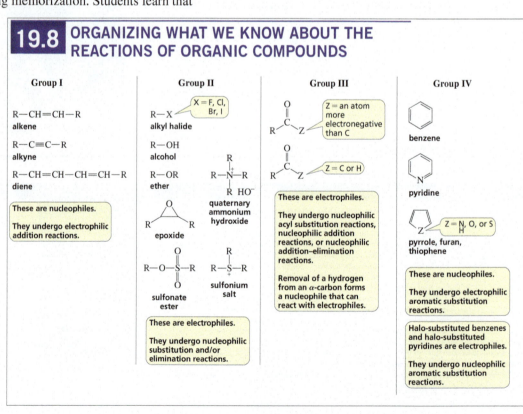

Organizing What We Know About the Reactivity of Organic Compounds

This organization emphasizes the unifying principles of reactivity and offers an economy of presentation while discouraging memorization. Students learn that

- organic compounds can be classified into *families* and that all members of a family react in the same way.
- the families can be put into one of four *groups* and that all the family members in a group react in similar ways.

The Organizing What We Know table builds as students work sequentially through the four groups.

Group I: electrophilic addition reactions

Group II: nucleophilic substitution reactions and elimination reactions

Group III: nucleophilic acyl substitution reactions, nucleophilic addition reactions, and nucleophilic addition–elimination reactions

Group IV: electrophilic (and nucleophilic) aromatic substitution reactions

19.8 ORGANIZING WHAT WE KNOW ABOUT THE REACTIONS OF ORGANIC COMPOUNDS

Group I

R—CH=CH—R
alkene

R—C≡C—R
alkyne

R—CH=CH—CH=CH—R
diene

These are nucleophiles.

They undergo electrophilic addition reactions.

Group II

R—X   X = F, Cl, Br, I
alkyl halide

R—OH
alcohol

R—OR
ether

R—N⁺R with R and HO⁻
quaternary ammonium hydroxide

epoxide

R—O—S—R
sulfonate ester

R—S⁺—R
sulfonium salt

These are electrophiles.

They undergo nucleophilic substitution and/or elimination reactions.

Group III

Z = an atom more electronegative than C

Z = C or H

These are electrophiles.

They undergo nucleophilic acyl substitution reactions, nucleophilic addition reactions, or nucleophilic addition–elimination reactions.

Removal of a hydrogen from an α-carbon forms a nucleophile that can react with electrophiles.

Group IV

benzene

pyridine

Z = N, O, or S

pyrrole, furan, thiophene

These are nucleophiles.

They undergo electrophilic aromatic substitution reactions.

Halo-substituted benzenes and halo-substituted pyridines are electrophiles.

They undergo nucleophilic aromatic substitution reactions.

Emphasis on the Strategies Needed to Solve Problems and Master Content

Passages explaining important problem-solving strategies are clearly labeled with a **LEARN THE STRATEGY** label. Follow-up problems that require students to apply the just-learned strategy are labeled with a **USE THE STRATEGY** label. These labels, which are implemented throughout the text, allow students to easily find important content and practice its use.

LEARN THE STRATEGY

PROBLEM-SOLVING STRATEGY

Using Basicity to Predict the Outcome of a Nucleophilic Acyl Substitution Reaction

What is the product of the reaction of acetyl chloride with CH_3O^-? The pK_a of HCl is −7; the pK_a of CH_3OH is 15.5.

To identify the product of the reaction, we need to compare the basicities of the two groups in the tetrahedral intermediate so we can determine which one will be eliminated. Because HCl is a stronger acid than CH_3OH, Cl^- is a weaker base than CH_3O^-. Therefore, Cl^- is eliminated from the tetrahedral intermediate and methyl acetate is the product of the reaction.

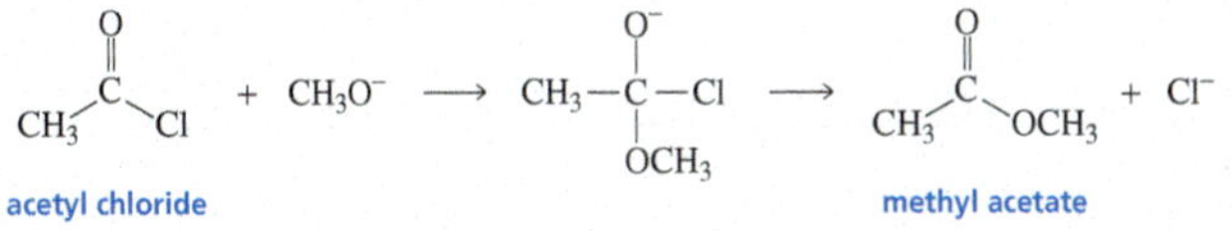

USE THE STRATEGY

PROBLEM 7 ◆

a. What is the product of the reaction of acetyl chloride with HO^-? The pK_a of HCl is −7; the pK_a of H_2O is 15.7.
b. What is the product of the reaction of acetamide with HO^-? The pK_a of NH_3 is 36; the pK_a of H_2O is 15.7.

PROBLEM 8 ◆

What is the product of an acyl substitution reaction—a new carboxylic acid derivative, a mixture of two carboxylic acid derivatives, or no reaction—if the new group in the tetrahedral intermediate is the following?

a. a stronger base than the substituent that is attached to the acyl group
b. a weaker base than the substituent that is attached to the acyl group
c. similar in basicity to the substituent that is attached to the acyl group

---

Designing a Synthesis

This recurring feature helps students learn to design multi-step syntheses and facilitates the development of complex problem-solving skills. Many problems include the synthesis of well-known compounds such as Novocain®, Valium®, and Ketoprofen®.

DESIGNING A SYNTHESIS V

17.20 MAKING NEW CARBON–CARBON BONDS

When planning the synthesis of a compound that requires the formation of a new carbon–carbon bond:

- locate the new bond that needs to be made and perform a disconnection—that is, break the bond to produce two fragments.
- determine which of the atoms that will form the new bond should be the electrophile and which should be the nucleophile.
- choose a compound with the desired electrophilic and nucleophilic groups.

Example 1

The new bond that needs to be made in the synthesis of the following β-diketone is the one that makes the second five-membered ring:

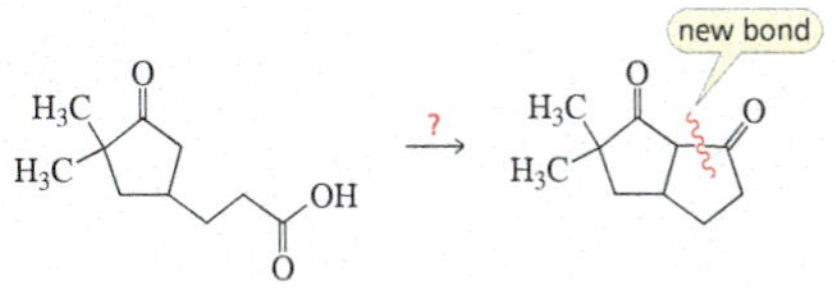

It is easy to choose between the two possibilities for the electrophile and nucleophile because we know that a carbonyl carbon is an electrophile.

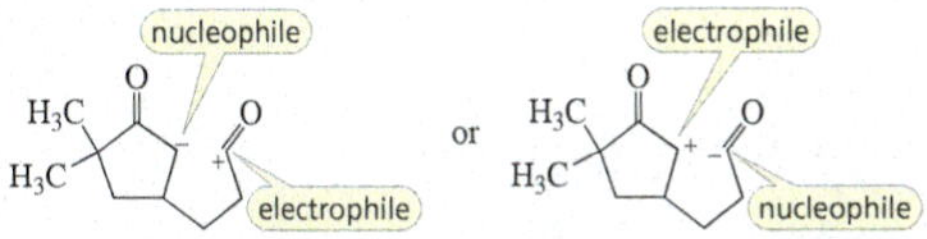

If we know what the starting material is, we can use it as a clue to arrive at the desired compound. For example, an ester carbonyl group would be a good electrophile for this synthesis because it has a group that would be eliminated. Moreover, the α-hydrogens of the ketone are more acidic than the α-hydrogens of the ester, so the desired nucleophile would be easy to obtain. Thus, converting the starting material to an ester (Section 15.18), followed by an intramolecular condensation, forms the target molecule.

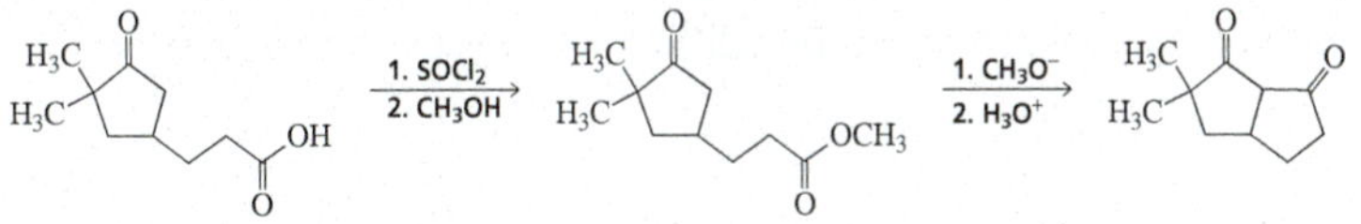

MasteringChemistry®

www.masteringchemistry.com

MasteringChemistry motivates student to learn outside of class and arrive prepared for lecture. The text works with MasteringChemistry to guide students on what they need to know before testing them on the content. The third edition continually engages students through pre-lecture, during-lecture, and post-lecture activities that all include real-life applications.

DYNAMIC STUDY MODULES

Help Students Learn Chemistry Quickly!

Now assignable, Dynamic Study Modules enable your students to study on their own and be better prepared for class. The modules cover content and skills needed to succeed in organic chemistry: fundamental concepts from general chemistry; practice with nomenclature, functional groups, and key mechanisms; and problem-solving skills. For students who want to study on the go, a mobile app that records student results to the MasteringChemistry gradebook is available for iOS and Android devices.

Spectroscopy Simulations

NEW! Six NMR/IR Spectroscopy simulations (a partnership with ACD labs) allow professors and students access to limitless spectral analysis with guided activities that can be used in the lab, in the classroom, or after class to study and explore spectra virtually. Activities authored by Mike Huggins, University of West Florida, prompt students to utilize the spectral simulator and walk them through different analyses and possible conclusions.

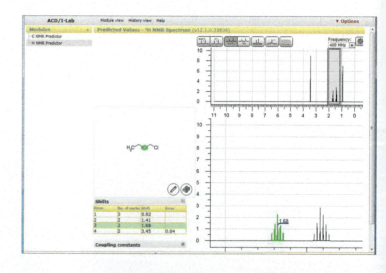

RESOURCES IN PRINT AND ONLINE

Supplement	Available in Print?	Available Online?	Instructor or Student Supplement	Description
MasteringChemistry® (isbn: 0134019202)		✓	Instructor and Student Supplement	MasteringChemistry® from Pearson has been designed and refined with a single purpose in mind: to help educators create that moment of understanding with their students. The Mastering platform delivers engaging, dynamic learning opportunities—focused on your course objectives and responsive to each student's progress—that are proven to help students absorb course material and understand difficult concepts.
Test Bank (isbn: 013406657X)		✓	Instructor Supplement	This test bank contains over 2500 multiple-choice, true/false, and matching questions. It is available in print format, in the TestGen program, and in Word format, and is included in the item library of MasteringChemistry®.
Instructor Resource Materials (isbn: 0134066596)		✓	Instructor Supplement	This provides an integrated collection of online resources to help instructors make efficient and effective use of their time. It includes all artwork from the text, including figures and tables in PDF format for high-resolution printing, as well as pre-built PowerPoint™ presentations. The first presentation contains the images embedded within PowerPoint slides. The second includes a complete lecture outline that is modifiable by the user. Powerpoints of the in-chapter worked examples are also included.
Study Guide and Solutions Manual (isbn: 0134066588)	✓		Instructor and Student Supplement	This manual for students, written by Paula Bruice, contains exercises and all key terms used in each chapter. In addition, you will find additional spectroscopy problems. This Solutions Manual provides detailed solutions to all in-chapter, as well as end-of-chapter, exercises in the text.

ACKNOWLEDGMENTS

It gives me great pleasure to acknowledge the dedicated efforts of many good friends who made this book a reality. In particular, a loud thanks goes to Richard Morrison of the University of Georgia, who read every page, made critically important suggestions, checked every answer in the *Solutions Manual*, and created many new end-of-chapter problems—and to Jordan Fantini of Dennison University, who created the interactive digital modules designed to engage students in the fundamental principles of reactivity while advancing their visualization skills. I am also deeply grateful to Laura Sessions of Valencia College who checked every inch of the book for accuracy; David Yerzley, M.D., for his assistance with the section on MRI; Warren Hehre of Wavefunction, Inc., and Alan Shusterman of Reed College for their advice on the electrostatic potential maps; and to my student, Jeremy Davis, who created the dog cartoons that appear on page 144. I also thank my many students, who pointed out sections that needed clarification, searched for errors, and whose questions guided the creation of new problems.

The following reviewers have played an enormously important role in the development of this book.

Eighth Edition Contributors

Richard Morrison, *University of Georgia*
Jordan Fantini, *Denison University*

Eighth Edition Accuracy Reviewers

David Boyajian, *Palomar College*
Gayane Godjoian, *Los Angeles Mission College*
Laura B. Sessions, *Valencia College*

Eighth Edition Reviewers

Ardeshir Azadnia, *Michigan State University*
Christopher Beaudry, *Oregon State University*
Thomas Bertolini, *University of Southern California*
Adam Braunschweig, *University of Miami*
Alexei Demchenko, *University of Missouri–St. Louis*
Christina DeMeo, *Southern Illinois University*
Steve Samuel, *SUNY Old Westbury*
Susan Schelble, *Metropolitan State University*

Seventh Edition Reviewers

Jason P. Anderson, *Monroe Community College*
Gabriele Backes, *Portland Community College*
Michael A. G. Berg, *Virginia Tech*
Thomas Bertolini, *University of Southern California*
Daniel Blanchard, *Kutztown University*
Ned Bowden, *University of Iowa*
Nancy Christensen, *Waubonsee Community College*
Veronica Curtin-Palmer, *Northeastern University*

Benjamin W. Gung, *Miami University—Oxford Ohio*
Matthew E. Hart, *Grand Valley State University*
Donna K. Howell, *Park University*
Tim Humphry, *Gonzaga University*
Frederick A. Luzzio, *University of Louisville*
Robert C. Mebane, *University of Tennessee—Chattanooga*
Delbert Howard Miles, *University of Central Florida*
Richard J. Mullins, *Xavier University*
Feliz Ngasse, *Grand Valley State University*
Anne B. Padias, *University of Arizona*
Matt A. Peterson, *Brigham Young University*
Christine Ann Prius, *Arizona State University*
Michael Pollastri, *Northeastern University*
Michael Rathke, *Michigan State University*
Harold R. Rodgers, *California State University Fullerton*
Webster Santos, *Virginia Tech*
Jacob D. Schroeder, *Clemson University*
Edward B. Skibo, *Arizona State University*
David Spivak, *Louisiana State University*
Zhaohui Sunny Zhou, *Northeastern University*

Seventh Edition Accuracy Reviewers

Jordan Fantini, *Denison University*
Malcolm D.E. Forbes, *University of North Carolina*
Stephen Miller, *University of Florida*
Christopher Roy, *Duke University*
Chad Snyder, *Western Kentucky University*

Many people made this book possible, but at the top of the list is my editor, Jeanne Zalesky, who has been involved and supportive at every stage of its creation and whose many talents guided the book to make it as good as it could be. I am also extremely grateful to have had the opportunity to work with Matt Walker, the development editor. His insights into how today's students learn and his creative art development skills have had a huge effect on this edition. I am also grateful to Elisa Mandelbaum, the project editor, whose attention to detail and creation of manageable deadlines made the book actually happen. And I want to thank the other talented and dedicated people at Pearson whose contributions made this book a reality:

I particularly want to thank the many wonderful and talented students I have had over the years, who inspired me, challenged me, and who taught me how to be a teacher. And I want to thank my children, from whom I may have learned the most.

To make this textbook as user friendly as possible, I would appreciate any comments that will help me achieve this goal in future editions. If you find sections that could be clarified or expanded, or examples that could be added, please let me know. Finally, this edition has been painstakingly combed for typographical errors. Any that remain are my responsibility. If you find any, please send me a quick email so they can be corrected in future printings of this edition.

Paula Yurkanis Bruice
University of California, Santa Barbara
pybruice@chem.ucsb.edu

About the Author

Paula Bruice with Zeus, Bacchus, and Abigail

Paula Yurkanis Bruice was raised primarily in Massachusetts. After graduating from the Girls' Latin School in Boston, she earned an A.B. from Mount Holyoke College and a Ph.D. in chemistry from the University of Virginia. She then received an NIH postdoctoral fellowship for study in the Department of Biochemistry at the University of Virginia Medical School and held a postdoctoral appointment in the Department of Pharmacology at the Yale School of Medicine.

Paula has been a member of the faculty at the University of California, Santa Barbara since 1972, where she has received the Associated Students Teacher of the Year Award, the Academic Senate Distinguished Teaching Award, two Mortar Board Professor of the Year Awards, and the UCSB Alumni Association Teaching Award. Her research interests center on the mechanism and catalysis of organic reactions, particularly those of biological significance. Paula has a daughter and a son who are physicians and a son who is a lawyer. Her main hobbies are reading mysteries and biographies and enjoying her pets (three dogs, two cats, and two parrots).

Organic Chemistry

PART ONE

An Introduction to the Study of Organic Chemistry

The first three chapters of this textbook cover a variety of topics with which you need to be familiar to start your study of the reactions and synthesis of organic compounds.

Chapter 1 Remembering General Chemistry: Electronic Structure and Bonding

Chapter 1 reviews the topics from general chemistry that are important to your study of organic chemistry. The chapter starts with a description of the structure of atoms and then proceeds to a description of the structure of molecules. Molecular orbital theory is introduced.

Chapter 2 Acids and Bases: Central to Understanding Organic Chemistry

Chapter 2 discusses acid–base chemistry, a topic that is central to understanding many organic reactions. You will see how the structure of a molecule affects its acidity and how the acidity of a solution affects molecular structure.

Chapter 3 An Introduction to Organic Compounds: Nomenclature, Physical Properties, and Representation of Structure

To discuss organic compounds, you must know how to name them and be able to visualize their structures when you read or hear their names. In **Chapter 3,** you will learn how to name five different families of organic compounds. This will give you a good understanding of the basic rules for naming compounds. Because the compounds examined in the chapter are the reactants or the products of many of the reactions presented in the first third of the book, you will have numerous opportunities to review the nomenclature of these compounds as you proceed through these chapters. Chapter 3 also compares and contrasts the structures and physical properties of these compounds, which makes learning about them a little easier than if the structure and physical properties of each family were presented separately. Because organic chemistry is a study of compounds that contain carbon, the last part of Chapter 3 discusses the spatial arrangement of the atoms in both chains and rings of carbon atoms.

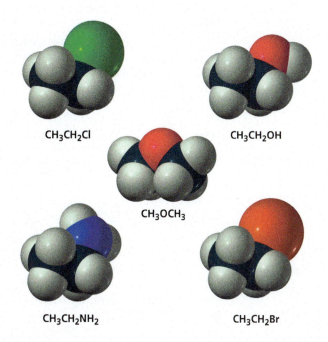

CH_3CH_2Cl

CH_3CH_2OH

CH_3OCH_3

$CH_3CH_2NH_2$

CH_3CH_2Br

1 Remembering General Chemistry: Electronic Structure and Bonding

NOTE TO THE STUDENT
- Biographies of the scientists mentioned in this text book can be found on the book's Website.

To stay alive, early humans must have been able to distinguish between different kinds of materials in their world. "You can live on roots and berries," they might have said, "but you can't eat dirt. You can stay warm by burning tree branches, but you can't burn rocks."

By the early nineteenth century, scientists thought they had grasped the nature of that difference, and in 1807, Jöns Jakob Berzelius gave names to the two kinds of materials. Compounds derived from living organisms were believed to contain an immeasurable vital force—the essence of life. These he called "organic." Compounds derived from minerals—those lacking the vital force—were "inorganic."

Because chemists could not create life in the laboratory, they assumed they could not create compounds that have a vital force. You can imagine their surprise when, in 1828, Friedrich Wöhler produced urea—a compound excreted by mammals—by heating ammonium cyanate, an inorganic mineral.

$$\overset{+}{N}H_4 \ \overset{-}{O}CN \xrightarrow{\text{heat}} \underset{H_2N \qquad NH_2}{\overset{O}{\underset{\|}{C}}}$$

ammonium cyanate
inorganic mineral

urea
"organic" compound

For the first time, an "organic" compound had been obtained from something other than a living organism and, therefore, did not contain a vital force. Thus, chemists needed a new definition for "organic compounds." **Organic compounds** are now defined as *compounds that are based on carbon.*

Organic compounds are compounds that are based on carbon.

Why is an entire branch of chemistry devoted to the study of carbon-containing compounds? We study organic chemistry because just about all of the compounds that make life possible and that make us who we are—proteins, enzymes, vitamins, lipids, carbohydrates, DNA, RNA—are organic compounds. Thus, the chemical reactions that take place in living systems, including our own bodies, are reactions of organic compounds. Most of the compounds found in nature—those that we rely on for food, clothing (cotton, wool, silk), and energy (natural gas, petroleum)—are organic compounds as well.

Organic compounds are not limited to those found in nature. Chemists have learned how to synthesize millions of organic compounds not found in nature, including synthetic fabrics, plastics, synthetic rubber, and even things such as compact discs and Super Glue. And most importantly, almost all commonly prescribed drugs are synthetic organic compounds.

Some synthetic organic compounds prevent shortages of naturally occurring compounds. For example, it has been estimated that if synthetic materials—nylon, polyester, Lycra—were not available for clothing, all of the arable land in the United States would have to be used for the production of cotton and wool just to provide enough material to clothe us. Other synthetic organic compounds provide us with materials we would not have—Teflon, Plexiglas, Kevlar—if we had only naturally occurring organic compounds. Currently, there are about 16 million known organic compounds, and many more are possible that we cannot even imagine today.

Why are there so many carbon-containing compounds? The answer lies in carbon's position in the periodic table. Carbon is in the center of the second row of elements. We will see that the atoms to the left of carbon have a tendency to give up electrons, whereas the atoms to the right have a tendency to accept electrons (Section 1.3).

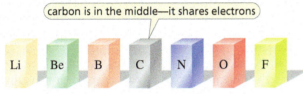

carbon is in the middle—it shares electrons

Li Be B C N O F

the second row of the periodic table

Because carbon is in the middle, it neither readily gives up nor readily accepts electrons. Instead, it shares electrons. Carbon can share electrons with several kinds of atoms as well as with other carbon atoms. Consequently, carbon forms millions of stable compounds with a wide range of chemical properties simply by sharing electrons.

Natural Versus Synthetic Organic Compounds

It is a popular belief that natural substances—those made in nature—are superior to synthetic ones—those made in the laboratory. Yet when a chemist synthesizes a compound, such as penicillin or morphine, the compound is the same in all respects as the compound synthesized in nature. Sometimes chemists can even improve on nature. For example, chemists have synthesized analogues of penicillin—compounds with structures similar to that of penicillin—that do not produce the allergic responses that a significant fraction of the population experiences from naturally produced penicillin or that do not have the bacterial resistance of the naturally produced antibiotic (Section 15.11).

Chemists have also synthesized analogues of morphine that have the same pain-killing effects but, unlike morphine, are not habit-forming. Most commercial morphine is obtained from opium, the juice extracted from the species of poppy shown in the photo. Morphine is the starting material for the synthesis of heroin. One of the side products formed in the synthesis has an extremely pungent odor; dogs used by drug enforcement agencies are trained to recognize this odor (Section 15.16). Nearly three-quarters of the world's supply of heroin comes from the poppy fields of Afghanistan.

a field of poppies in Afghanistan

When we study organic chemistry, we learn how organic compounds react. Organic compounds consist of atoms held together by covalent bonds. When an organic compound reacts, some of these covalent bonds break and some new covalent bonds form.

Covalent bonds form when two atoms share electrons,
and they break when two atoms no longer share electrons.

How easily a covalent bond forms or breaks depends on the electrons that are shared, which, in turn, depends on the atoms to which the electrons belong. So if we are going to start our study of organic chemistry at the beginning, we must start with an understanding of the structure of an atom—what electrons an atom has and where they are located.

1.1 THE STRUCTURE OF AN ATOM

An atom consists of a tiny dense nucleus surrounded by electrons that are spread throughout a relatively large volume of space around the nucleus called an electron cloud. The nucleus contains **positively charged protons** and **uncharged neutrons,** so it is positively charged. The **electrons** are **negatively charged.** The amount of positive charge on a proton equals the amount of negative charge on an electron. Therefore, the number of protons and the number of electrons in an uncharged atom must be the same.

Electrons move continuously. Like anything that moves, electrons have kinetic energy, and this energy counteracts the attractive force of the positively charged protons that pull the negatively charged electrons toward the nucleus.

Protons and neutrons have approximately the same mass and are about 1800 times more massive than an electron. Most of the *mass* of an atom, therefore, is in its nucleus. Most of the *volume* of an atom, however, is occupied by its electron cloud. This is where our focus will be because it is the electrons that form chemical bonds.

The **atomic number** of an atom is the number of protons in its nucleus. The atomic number is unique to a particular element. For example, the atomic number of carbon is 6, which means that all uncharged carbon atoms have six protons and six electrons. Although atoms can gain electrons and become negatively charged or lose electrons and become positively charged, the number of protons in an atom of a particular element never changes.

The **mass number** of an atom is the sum of its protons and neutrons. Although all carbon atoms have the same *atomic number*, they do not all have the same *mass number*. Why? Because carbon atoms can have varying numbers of neutrons. For example, 98.89% of all carbon atoms have six neutrons—giving them a mass number of 12—and 1.11% have seven neutrons—giving them a mass number of 13. These two different kinds of carbon atoms (^{12}C and ^{13}C) are called **isotopes.**

> The nucleus contains positively charged protons and uncharged neutrons.
>
> The electrons are negatively charged.

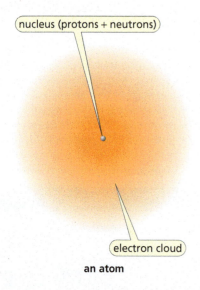

nucleus (protons + neutrons)

electron cloud

an atom

> atomic number = the number of protons in the nucleus
>
> mass number = the number of protons + the number of neutrons

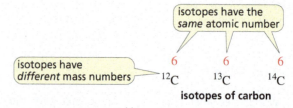

isotopes have the *same* atomic number

isotopes have *different* mass numbers

^{12}C ^{13}C ^{14}C

isotopes of carbon

Carbon also contains a trace amount of ^{14}C, which has six protons and eight neutrons. This isotope of carbon is radioactive, decaying with a half-life of 5730 years. (The *half-life* is the time it takes for one-half of the nuclei to decay.) As long as a plant or an animal is alive, the ^{14}C that is lost through exhalation or excretion is constantly replenished. When it dies, however, it no longer ingests ^{14}C. Consequently, its ^{14}C is slowly lost through radioactive decay. Therefore, the age of a substance derived from a living organism can be determined by its ^{14}C content.

The **atomic mass** is the weighted average of the isotopes in the element. Because an *atomic mass unit (amu)* is defined as exactly 1/12 of the mass of ^{12}C, the mass of ^{12}C is 12.0000 amu; the mass of ^{13}C is 13.0035 amu. Therefore, the atomic mass of carbon is 12.011 amu because $(0.9889 \times 12.0000) + (0.0111 \times 13.0035) = 12.011$. The **molecular mass** is the sum of the atomic masses of all the atoms in the molecule.

> atomic mass = the weighted average mass of the isotopes in the element
>
> molecular mass = the sum of the atomic masses of all the atoms in the molecule
>
> A molecule is a group of two or more atoms held together by bonds.

PROBLEM 1 ◆
Oxygen has three isotopes, ^{16}O, ^{17}O, and ^{18}O. The atomic number of oxygen is 8. How many protons and neutrons does each of the isotopes have?

PROBLEM 2 ◆
a. How many protons do the following species have? (See the periodic table inside the back cover of this book.)
b. How many electrons does each have?
  1. Na^+ **2.** Ar   **3.** Cl^-

PROBLEM 3 ◆
Chlorine has two isotopes, ^{35}Cl and ^{37}Cl; 75.77% of chlorine is ^{35}Cl, and 24.23% is ^{37}Cl. The atomic mass of ^{35}Cl is 34.969 amu, and the atomic mass of ^{37}Cl is 36.966 amu. What is the atomic weight of chlorine?

Albert Einstein

The bronze sculpture of Albert Einstein, on the grounds of the National Academy of Sciences in Washington, D.C., measures 21 feet from the top of the head to the tip of the feet and weighs 7000 pounds. In his left hand, Einstein holds the mathematical equations that represent his three most important contributions to science: the photoelectric effect, the equivalency of energy and matter, and the theory of relativity. At his feet is a map of the sky.

1.2 HOW THE ELECTRONS IN AN ATOM ARE DISTRIBUTED

For a long time, electrons were perceived to be particles—infinitesimal "planets" that orbit the nucleus of an atom. In 1924, however, Louis de Broglie, a French physicist, showed that electrons also have wave-like properties. He did this by combining a formula developed by Albert Einstein relating mass and energy with a formula developed by Max Planck relating frequency and energy. The realization that electrons have wave-like properties spurred physicists to propose a mathematical concept known as quantum mechanics to describe the motion of an electron around a nucleus.

Quantum mechanics uses the same mathematical equations that describe the wave motion of a guitar string to characterize the motion of an electron around a nucleus. The version of quantum mechanics most useful to chemists was proposed by Erwin Schrödinger in 1926.

According to Schrödinger, the electrons in an atom can be thought of as occupying a set of concentric shells that surround the nucleus (Table 1.1).

Table 1.1 Distribution of Electrons in the First Four Shells				
	First shell	**Second shell**	**Third shell**	**Fourth shell**
Atomic orbitals	s	s, p	s, p, d	s, p, d, f
Number of atomic orbitals	1	1, 3	1, 3, 5	1, 3, 5, 7
Maximum number of electrons	2	8	18	32

- The first shell is the one closest to the nucleus. The second shell lies farther from the nucleus. The third and higher numbered shells lie even farther out.

- The shells contain subshells known as **atomic orbitals.** We will see that an atomic orbital has a characteristic shape and energy and occupies a characteristic volume of space (Section 1.5).

- Each shell contains one s atomic orbital. Each second and higher shell—in addition to its s atomic orbital—contains three *degenerate p* atomic orbitals. **Degenerate orbitals** are orbitals that have the same energy. The third and higher shells—in addition to their s and p atomic orbitals—contain five degenerate d atomic orbitals, and the fourth and higher shells also contain seven degenerate f atomic orbitals.

- A maximum of two electrons can coexist in an atomic orbital. (See the Pauli exclusion principle on p. 6.) Therefore, the first four shells, with 1, 4, 9, and 16 atomic orbitals, respectively, can contain a maximum of 2, 8, 18, and 32 electrons.

Degenerate orbitals are orbitals that have the same energy.

In our study of organic chemistry, we will be concerned primarily with atoms that have electrons only in the first two shells.

Ground-State Electronic Configuration

The **ground-state electronic configuration** of an atom describes the atomic orbitals occupied by the atom's electrons when they are all in the available orbitals with the lowest energy. If energy is applied to an atom in the ground state, one or more electrons can jump into a higher-energy orbital. The atom then would be in an excited state and have an **excited-state electronic configuration.**

The ground-state electronic configurations of the smallest atoms are shown in Table 1.2. (Each arrow—whether pointing up or down—represents one electron.)

Atom	Name of element	Atomic number	$1s$	$2s$	$2p_x$	$2p_y$	$2p_z$	$3s$
H	Hydrogen	1	↑					
He	Helium	2	↑↓					
Li	Lithium	3	↑↓	↑				
Be	Beryllium	4	↑↓	↑↓				
B	Boron	5	↑↓	↑↓	↑			
C	Carbon	6	↑↓	↑↓	↑	↑		
N	Nitrogen	7	↑↓	↑↓	↑	↑	↑	
O	Oxygen	8	↑↓	↑↓	↑↓	↑	↑	
F	Fluorine	9	↑↓	↑↓	↑↓	↑↓	↑	
Ne	Neon	10	↑↓	↑↓	↑↓	↑↓	↑↓	
Na	Sodium	11	↑↓	↑↓	↑↓	↑↓	↑↓	↑

Table 1.2 The Electronic Configurations of the Smallest Atoms

Three rules specify which atomic orbitals an atom's electrons occupy:

1. The **aufbau principle** (*aufbau* is German for "building up") states that

    an electron always goes into the available orbital with the lowest energy.

When using the aufbau principle rule, it is important to remember that the closer the atomic orbital is to the nucleus, the lower is its energy. Because the $1s$ orbital is closer to the nucleus, it is lower in energy than the $2s$ orbital, which is lower in energy—and closer to the nucleus—than the $3s$ orbital. When comparing atomic orbitals in the same shell, we see that an s orbital is lower in energy than a p orbital, and a p orbital is lower in energy than a d orbital.

relative energies of atomic orbitals

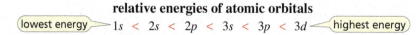
lowest energy $\quad 1s \; < \; 2s \; < \; 2p \; < \; 3s \; < \; 3p \; < \; 3d \quad$ highest energy

2. The **Pauli exclusion principle** states that

    no more than two electrons can occupy each atomic orbital, and the two electrons must be of opposite spin.

This is called an exclusion principle because it limits the number of electrons that can occupy an atomic orbital and, therefore, any particular shell. (Notice in Table 1.2 that opposite spins are designated by ↑ and ↓.)

These first two rules allow us to assign electrons to atomic orbitals for atoms that contain one, two, three, four, or five electrons.

- The single electron of a hydrogen atom occupies a $1s$ orbital.
- The second electron of a helium atom fills the $1s$ orbital.
- The third electron of a lithium atom occupies a $2s$ orbital.
- The fourth electron of a beryllium atom fills the $2s$ orbital.
- The fifth electron of a boron atom occupies one of the $2p$ orbitals. (The subscripts x, y, and z distinguish the three $2p$ orbitals.) Because the three p orbitals are degenerate, the electron can be put into any one of them.

Before we can discuss atoms containing six or more electrons, we need the third rule.

3. Hund's rule states that

> *when there are two or more atomic orbitals with the same energy,*
> *an electron will occupy an empty orbital before it will pair up with another electron.*

In this way, electron repulsion is minimized.

- Therefore, the sixth electron of a carbon atom goes into an empty $2p$ orbital, rather than pairing up with the electron already occupying a $2p$ orbital (see Table 1.2).
- There is one more empty $2p$ orbital, so that is where nitrogen's seventh electron goes.
- The eighth electron of an oxygen atom pairs up with an electron occupying a $2p$ orbital rather than going into the higher-energy $3s$ orbital.

The locations of the electrons in the remaining elements can be assigned using these three rules.

Valence and Core Electrons

The major factor that determines the chemical behavior of an element is the number of valence electrons it has. **Valence electrons** are electrons in an atom's outermost shell. Electrons in inner shells (below the outermost shell) are called **core electrons.** For example, carbon has four valence electrons and two core electrons (Table 1.2). Valence electrons participate in chemical bonding; core electrons do not.

Elements in the same column of the periodic table have similar chemical properties because they have the same number of valence electrons. If you examine the periodic table inside the back cover of this book, you will see that lithium and sodium, which have similar chemical properties, are in the same column because each has one valence electron.

> Valence electrons are electrons in the outermost shell.
>
> Core electrons are electrons in inner shells.
>
> The chemical behavior of an element depends on its electronic configuration.

PROBLEM 4 ♦

How many valence electrons do the following atoms have?

a. boron   **b.** nitrogen   **c.** oxygen   **d.** fluorine

PROBLEM 5 ♦

a. Write the ground-state electronic configuration for chlorine (atomic number 17), bromine (atomic number 35), and iodine (atomic number 53).

b. How many valence electrons do chlorine, bromine, and iodine have?

PROBLEM 6

Look at the relative positions of each pair of atoms listed here in the periodic table. How many core electrons does each have? How many valence electrons does each have?

a. carbon and silicon   **c.** nitrogen and phosphorus
b. oxygen and sulfur   **d.** magnesium and calcium

1.3 COVALENT BONDS

Now that you know about the electronic configuration of atoms, let's now look at why atoms come together to form bonds. In explaining why atoms form bonds, G. N. Lewis proposed that

> *an atom is most stable if its outer shell is either filled or contains eight electrons,*
> *and it has no electrons of higher energy.*

According to Lewis's theory, an atom will give up, accept, or share electrons to achieve a filled outer shell or an outer shell that contains eight electrons. This theory has come to be called the **octet rule** (even though hydrogen needs only two electrons to achieve a filled outer shell).

Achieving a Filled Outer Shell by Losing or Gaining Electrons

Lithium (Li) has a single electron in its $2s$ orbital. If it loses this electron, lithium ends up with a filled outer shell—a stable configuration. Lithium, therefore, loses an electron relatively easily. Sodium (Na) has a single electron in its $3s$ orbital; so it, too, loses an electron easily.

Because only valence electrons are used in bonding, only valence electrons are shown in the following equations. (The symbol for the element represents the protons, neutrons, and core electrons.) Each valence electron is shown as a dot. When the single valence electron of lithium or sodium is removed, the species that is formed is called an ion because it carries a charge.

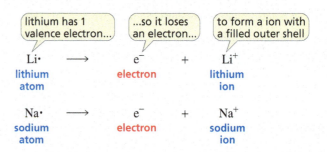

Each of the elements in the first column of the periodic table readily loses an electron because each has a single electron in its outermost shell.

Fluorine has seven valence electrons (Table 1.2). Therefore, it readily acquires an electron to fill its outer shell. Gaining the electron forms F⁻, a fluoride ion.

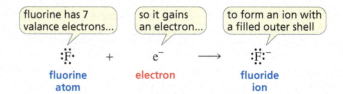

Elements in the same column of the periodic table as fluorine (such as chlorine, bromine, and iodine) also need only one electron to have an outer shell of eight; so they, too, readily acquire an electron.

$$:\ddot{\text{Cl}}\cdot \quad + \quad e^- \quad \longrightarrow \quad :\ddot{\text{Cl}}:^-$$

chlorine atom · electron · chloride ion

Elements (such as fluorine and chlorine) that readily acquire an electron are said to be **electronegative.**

A hydrogen atom has one valence electron. Therefore, it can achieve a completely empty shell by losing an electron, or a filled outer shell by gaining an electron.

$$\text{H}\cdot \quad \longrightarrow \quad e^- \quad + \quad \text{H}^+$$

hydrogen atom · electron · proton — stable because its outer shell is empty

$$\text{H}\cdot \quad + \quad e^- \quad \longrightarrow \quad \text{H}:^-$$

hydrogen atom · electron · hydride ion — stable because its outer shell is filled

Loss of its sole electron results in a positively charged **hydrogen ion.** A positively charged hydrogen ion is called a **proton** because when a hydrogen atom loses its valence electron, only the hydrogen nucleus—which consists of a single proton—remains. When a hydrogen atom gains an electron, a negatively charged hydrogen ion—called a **hydride ion**—is formed.

PROBLEM 7 ◆

a. Find potassium (K) in the periodic table and predict how many valence electrons it has.
b. What orbital does the unpaired electron occupy?

Achieving a Filled Outer Shell by Sharing Electrons

Instead of giving up or acquiring electrons to achieve a filled outer shell, an atom can achieve a filled outer shell by sharing a pair of electrons. For example, two fluorine atoms can each attain a filled second shell by sharing their unpaired valence electrons. A bond formed as a result of sharing electrons between two nuclei is called a **covalent bond.** A covalent bond is commonly shown by a solid line rather than by a pair of dots.

> each fluorine shares 1 of its 7 valence eletrons...

> ...to form a covalent bond...

> which can be denoted by a solid line

$$:\ddot{F}\cdot \ + \ \cdot\ddot{F}: \ \longrightarrow \ :\ddot{F}:\ddot{F}: \ \ or \ \ :\ddot{F}-\ddot{F}:$$

> each fluorine is surrounded by 8 electrons

A covalent bond is formed when two atoms share a pair of electrons.

Two hydrogen atoms can form a covalent bond by sharing electrons. As a result of covalent bonding, each hydrogen acquires a stable, filled first shell.

> each hydrogen shares its valence electron...

> ...to form a covalent bond

$$H\cdot \ + \ \cdot H \ \longrightarrow \ H:H \ \ or \ \ H-H$$

> each hydrogen is surrounded by 2 electrons

Similarly, hydrogen and chlorine can form a covalent bond by sharing electrons. In doing so, hydrogen fills its only shell, and chlorine achieves an outer shell of eight electrons.

$$H\cdot \ + \ \cdot\ddot{Cl}: \ \longrightarrow \ H:\ddot{Cl}: \ \ or \ \ H-\ddot{Cl}:$$

> H is surrounded by 2 electrons
> Cl is surrounded by 8 electrons

Let's now look at the bonds formed by oxygen, nitrogen, and carbon. Oxygen has six valence electrons, so it needs to form two covalent bonds to achieve an outer shell of eight electrons. Nitrogen, with five valence electrons, must form three covalent bonds, and carbon, with four valence electrons, must form four covalent bonds. Notice that all the atoms in water, ammonia, and methane have filled outer shells.

$$2\ H\cdot \ + \ \cdot\ddot{O}: \ \longrightarrow \ H-\underset{\underset{H}{|}}{\ddot{O}}:$$

> oxygen forms 2 covalent bonds

water

$$3\ H\cdot \ + \ \cdot\ddot{N}\cdot \ \longrightarrow \ H-\underset{\underset{H}{|}}{\overset{\cdot\cdot}{N}}-H$$

> nitrogen forms 3 covalent bonds

ammonia

$$4\ H\cdot \ + \ \cdot\overset{\cdot}{\underset{\cdot}{C}}\cdot \ \longrightarrow \ H-\underset{\underset{H}{|}}{\overset{\overset{H}{|}}{C}}-H$$

> carbon forms 4 covalent bonds

methane

> notice that each O, N, C is surrounded by 8 electrons, and each H is surrounded by 2 electrons

Nonpolar and Polar Covalent Bonds

A nonpolar covalent bond is a covalent bond between atoms with essentially the same electronegativity.

A polar covalent bond is a covalent bond between atoms with different electronegativities.

Covalent bonds are classified as nonpolar or polar depending on the difference in the electronegativities of the atoms that share the electrons. **Electronegativity** is a measure of the ability of an atom to pull electrons toward itself. The electronegativities of some of the elements are shown in Table 1.3. Notice that electronegativity increases from left to right across a row of the periodic table and from bottom to top in any of the columns.

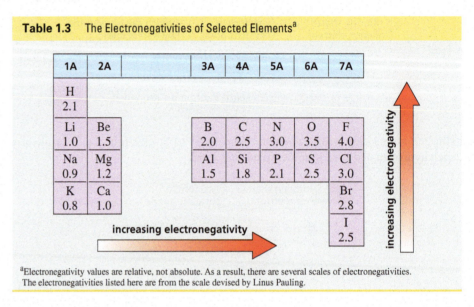

Table 1.3   The Electronegativities of Selected Elements[a]

1A	2A		3A	4A	5A	6A	7A
H 2.1							
Li 1.0	Be 1.5		B 2.0	C 2.5	N 3.0	O 3.5	F 4.0
Na 0.9	Mg 1.2		Al 1.5	Si 1.8	P 2.1	S 2.5	Cl 3.0
K 0.8	Ca 1.0						Br 2.8
							I 2.5

increasing electronegativity →

↑ increasing electronegativity

[a]Electronegativity values are relative, not absolute. As a result, there are several scales of electronegativities. The electronegativities listed here are from the scale devised by Linus Pauling.

If the electronegativity difference between the bonded atoms *is less than 0.5*, then the bond is a **nonpolar covalent bond.** That is, the atoms share the bonding electrons equally—the electrons represented by the bond are symmetrically distributed around each atom. Examples of nonpolar covalent bonds are shown below.

$$H-H \quad F-F \quad C-C \quad C-H$$

If the electronegativity difference between the bonded atoms *is between 0.5 and about 1.9*, then the bond is a **polar covalent bond.** The bonding electrons are unsymmetrically distributed because the bonding atoms have sufficiently different electronegativities; one end of a polar covalent bond has a partial negative charge (δ^-), and one end has a partial positive charge (δ^+). The negative end of the bond is the end with the more electronegative atom. The greater the difference in electronegativity between the bonded atoms, the more polar the bond.

sodium chloride crystals (table salt)

Salar de Uyuni in Bolivia—the largest deposit of natural lithium in the world

Lithium salts are used clinically. Lithium chloride (LiCl) is an antidepressant, lithium bromide (LiBr) is a sedative, and lithium carbonate (Li_2CO_3) is used to stabilize mood swings in people who suffer from bipolar disorder. Scientists do not yet know why lithium salts have these therapeutic effects.

The direction of bond polarity can be indicated with an arrow. By convention, chemists draw the arrow in the direction the electrons are pulled. Thus, the head of the arrow is at the negative end of the bond; a short perpendicular line near the tail of the arrow marks the positive end of the bond. (Physicists draw the arrow in the opposite direction.)

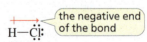

the negative end of the bond

If the electronegativity difference between the atoms *is greater than 1.9*, the atoms do not share their electrons. One of the atoms transfers an electron to the other, and the resulting ions (for example, Na^+ and Cl^-) are held together by **electrostatic attraction**—an attractive force between opposite charges. Sodium chloride is an example of an **ionic compound** (also known as a **salt**). Ionic compounds are formed when an element on the left side of the periodic table *transfers* one or more electrons to an element on the right side of the periodic table.

You can think of nonpolar covalent bonds—where the electrons are shared equally—being at one end of the continuum of bond types and ionic compounds—where no electrons are shared—at the other end. Polar covalent bonds fall somewhere in between.

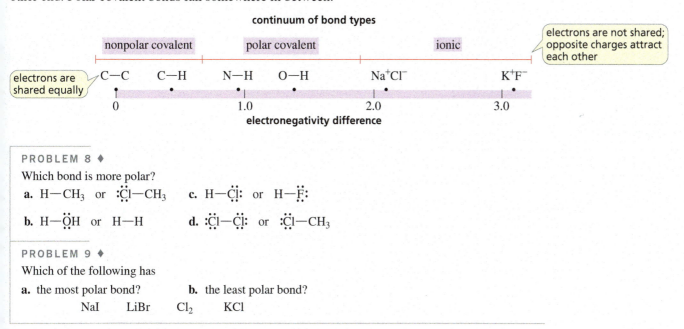

PROBLEM 8 ♦

Which bond is more polar?

a. H—CH₃ or :C̈l—CH₃   **c.** H—C̈l: or H—F̈:

b. H—ÖH or H—H   **d.** :C̈l—C̈l: or :C̈l—CH₃

PROBLEM 9 ♦

Which of the following has

a. the most polar bond?   **b.** the least polar bond?

NaI   LiBr   Cl₂   KCl

Dipole Moments of Bonds

A polar covalent bond has a **dipole**—that is, it has a negative end and a positive end. The size of the dipole is indicated by the dipole moment. The **dipole moment** of a bond is equal to the magnitude of the charge on either atom (because the partial positive charge and the partial negative charge have the same magnitude) times the distance between the two charges:

dipole moment of a bond = size of the charge × the distance between the charges

A dipole moment is reported in a unit called a **debye (D)** (pronounced de-bye). Because the charge on an electron is 4.80×10^{-10} electrostatic units (esu) and the distance between charges in a polar covalent bond has units of 10^{-8} cm, the product of charge and distance has units of 10^{-18} esu cm; so 1.0 D $= 1.0 \times 10^{-18}$ esu cm. Thus, a dipole moment of 1.5×10^{-18} esu cm is simply stated as 1.5 D. The dipole moments of bonds commonly found in organic compounds are listed in Table 1.4.

Table 1.4 The Dipole Moments of Some Common Bonds

Bond	Dipole moment (D)	Bond	Dipole moment (D)
H—C	0.4	C—C	0
H—N	1.3	C—N	0.2
H—O	1.5	C—O	0.7
H—F	1.7	C—F	1.6
H—Cl	1.1	C—Cl	1.5
H—Br	0.8	C—Br	1.4
H—I	0.4	C—I	1.2

When a molecule has only one covalent bond, its dipole moment is identical to the dipole moment of the bond. For example, the dipole moment of hydrogen chloride (HCl) is 1.1 D because the dipole moment of the H—Cl bond is 1.1 D.

The dipole moment of a molecule with more than one covalent bond depends on the dipole moments of all the bonds in the molecule and the geometry of the molecule. We will look at this in Section 1.16, after you learn about the geometry of molecules.

LEARN THE STRATEGY

PROBLEM 10 SOLVED

Use the symbols δ^+ and δ^- to show the direction of the polarity of the bond between carbon and oxygen.

$$H_3C—OH$$

According to Table 1.3, the electronegativity of carbon is 2.5 and the electronegativity of oxygen is 3.5. Because oxygen is more electronegative than carbon, oxygen has a partial negative charge and carbon has a partial positive charge.

$$\overset{\delta^+\ \ \ \delta^-}{H_3C—OH}$$

USE THE STRATEGY

PROBLEM 11 ◆

Use the symbols δ^+ and δ^- to show the direction of the polarity of the indicated bond in each of the following compounds:

a. HO—H   c. $H_3C—NH_2$ e. HO—Br   g. I—Cl

b. F—Br   d. $H_3C—Cl$ f. $H_3C—Li$ h. $H_2N—OH$

PROBLEM 12 SOLVED

Determine the partial negative charge on the fluorine atom in a C—F bond. The bond length is 1.39 Å*, and the bond dipole moment is 1.60 D. The charge on an electron is 4.80×10^{-10} esu.

SOLUTION If fluorine had a full negative charge, the dipole moment would be

$$(4.80 \times 10^{-10} \text{ esu})(1.39 \times 10^{-8} \text{ cm}) = 6.67 \times 10^{-18} \text{ esu cm} = 6.67 \text{ D}$$

Knowing that the dipole moment is 1.60 D and that the dipole moment would be 6.67 D if fluorine had a full negative charge, we can calculate that the partial negative charge on the fluorine atom is 0.24 of a full charge:

$$\frac{1.60 \text{ D}}{6.67 \text{ D}} = 0.24$$

PROBLEM 13

Explain why HCl has a smaller dipole moment than HF, even though the H—Cl bond is longer than the H—F bond.

Electrostatic Potential Maps

Understanding bond polarity is critical to understanding how organic reactions occur, because a central rule governing the reactivity of organic compounds is that *electron-rich atoms or molecules are attracted to electron-deficient atoms or molecules* (Section 5.5). **Electrostatic potential maps** (often called simply potential maps) are models that show how charge is distributed in the molecule under the map. The potential maps for LiH, H_2, and HF are shown here.

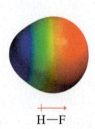

Li—H   H—H   H—F

The colors on a potential map indicate the relative charge distribution in the molecule and, therefore, the degree to which the molecule (or an atom in the molecule) attracts other species. Red, signifying the most negative electrostatic potential, is used for regions that attract electron-deficient species most strongly. Blue is used for areas with the most positive electrostatic

---

*The angstrom (Å) is not a Système International (SI) unit. Those who prefer SI units can convert Å into picometers (pm). 1 Å = 100 pm. Because the angstrom continues to be used by many organic chemists, we will use angstroms in this text book. Dipole moment calculations require the bond length to be in centimeters: $1 \text{ Å} = 10^{-8}$ cm.

potential—regions that attract electron-rich species most strongly. Other colors indicate intermediate levels of attraction.

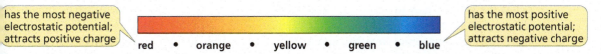

| has the most negative electrostatic potential; attracts positive charge | | has the most positive electrostatic potential; attracts negative charge |

red • orange • yellow • green • blue

The potential map for LiH shows that the hydrogen atom (red) is more electron-rich than the lithium atom (blue). By comparing the three maps, we can tell that the hydrogen in LiH is more electron-
rich than a hydrogen in H_2, whereas the hydrogen in HF is less electron-rich than a hydrogen in H_2.

Because a potential map roughly marks the "edge" of the molecule's electron cloud, the map tells
us something about the relative size and shape of the molecule. A particular atom can have different
sizes in different molecules, because the size of an atom in a potential map depends on its electron
density. For example, the negatively charged hydrogen in LiH is bigger than a neutral hydrogen in
H_2, which is bigger than the positively charged hydrogen in HF.

PROBLEM 14 ♦

After examining the potential maps for LiH, HF, and H_2, answer the following questions:

a. Which compounds are polar?
b. Why does LiH have the largest hydrogen?
c. Which compound has the hydrogen that would be most apt to attract a negatively charged molecule?

1.4 HOW THE STRUCTURE OF A COMPOUND IS REPRESENTED

First, we will see how compounds are represented using Lewis structures. Then we will look at the
kinds of structures that are used more commonly for organic compounds.

Lewis Structures

The chemical symbols we have been using, in which the valence electrons are represented as dots or
solid lines, are called **Lewis structures.** Lewis structures show us which atoms are bonded together
and tell us whether any atoms possess *lone-pair electrons* or have a *formal charge*, two concepts
described below. The Lewis structures for H_2O, H_3O^+, HO^-, and H_2O_2 are shown here.

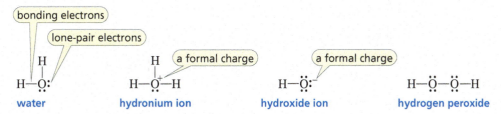

water   hydronium ion   hydroxide ion   hydrogen peroxide

Notice that the atoms in Lewis structures are always lined up linearly or at right angles. Therefore,
they do not tell us anything about the bond angles in the actual molecule.

Lone-Pair Electrons

When you draw a Lewis structure, make sure hydrogen atoms are surrounded by two electrons and
C, O, N, and halogen (F, Cl, Br, I) atoms are surrounded by eight electrons, in accordance with
the octet rule. Valence electrons not used in bonding are called **nonbonding electrons, lone-pair
electrons,** or simply, **lone pairs.**

> Lone-pair electrons are valence electrons that do not form bonds.

Formal Charge

Once the atoms and the electrons are in place, you must examine each atom to see whether a formal
charge should be assigned to it. A **formal charge** is the *difference* between the number of valence
electrons an atom has when it is not bonded to other atoms and the number it "owns" when it is
bonded. An atom "owns" all of its lone-pair electrons and half of its bonding (shared) electrons.
(Notice that half the bonding electrons is the same as the number of bonds.)

$$\begin{array}{c} \textbf{formal} \\ \textbf{charge} \end{array} = \begin{array}{c} \textbf{the number of} \\ \textbf{valence electrons} \end{array} - \left(\begin{array}{c} \textbf{the number of} \\ \textbf{lone-pair electrons} \end{array} + \begin{array}{c} \textbf{the number of} \\ \textbf{bonds} \end{array} \right)$$

Now we will see how to apply the equation for formal charge to three of the oxygen-containing species on p. 13. An oxygen atom has six valence electrons (Table 1.2). In water (H_2O), oxygen "owns" six electrons—four lone-pair electrons and half of the four bonding electrons. (Notice that one-half of the bonding electrons is the same as the number of bonds.) Because the number of electrons it "owns" is equal to the number of its valence electrons ($6 - 6 = 0$), the oxygen atom in water does not have a formal charge.

The oxygen atom in the hydronium ion (H_3O^+) "owns" five electrons: two lone-pair electrons plus three (half of six) bonding electrons. Because the number of electrons oxygen "owns" is one less than the number of its valence electrons ($6 - 5 = 1$), its formal charge is $+1$.

The oxygen atom in the hydroxide ion (HO^-) "owns" seven electrons: six lone-pair electrons plus one (half of two) bonding electron. Because oxygen "owns" one more electron than the number of its valence electrons ($6 - 7 = -1$), its formal charge is -1.

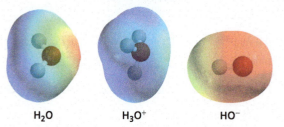

H_2O H_3O^+ HO^-

PROBLEM 15 ◆

An atom with a formal charge does not necessarily have more or less electron density than the atoms in the molecule without formal charges. We can see this by examining the potential maps for H_2O, H_3O^+, and HO^-.

a. Which atom bears the formal negative charge in the hydroxide ion?
b. Which atom has the greater electron density in the hydroxide ion?
c. Which atom bears the formal positive charge in the hydronium ion?
d. Which atom has the least electron density in the hydronium ion?

Drawing Lewis Structures

Nitrogen has five valence electrons (Table 1.2). Prove to yourself that the appropriate formal charges have been assigned to the nitrogen atoms in the following Lewis structures:

$$\text{H}-\overset{\overset{\displaystyle\cdot\cdot}{|}}{\underset{|}{\text{N}}}-\text{H} \qquad \text{H}-\overset{\overset{\displaystyle\text{H}}{|}}{\underset{|}{\overset{+}{\text{N}}}}-\text{H} \qquad \text{H}-\overset{\overset{\displaystyle\cdot\cdot}{|}}{\underset{|}{\text{N}}}:^{-} \qquad \text{H}-\overset{\displaystyle\cdot\cdot}{\underset{|}{\text{N}}}-\overset{\displaystyle\cdot\cdot}{\underset{|}{\text{N}}}-\text{H}$$

$$\text{H} \qquad\qquad\quad \text{H} \qquad\qquad\quad \text{H} \qquad\qquad \text{H} \quad \text{H}$$

ammonia   ammonium ion   amide anion   hydrazine

Carbon has four valence electrons. Take a moment to make sure you understand why the carbon atoms in the following Lewis structures have the indicated formal charges:

〔a carbocation〕 〔a carbanion〕 〔a radical〕

$$\text{H}-\overset{\overset{\displaystyle\text{H}}{|}}{\underset{\underset{\displaystyle\text{H}}{|}}{\text{C}}}-\text{H} \qquad \text{H}-\overset{\overset{\displaystyle\text{H}}{|}}{\underset{\underset{\displaystyle\text{H}}{|}}{\overset{+}{\text{C}}}} \qquad \text{H}-\overset{\overset{\displaystyle\text{H}}{|}}{\underset{\underset{\displaystyle\text{H}}{|}}{\text{C}}}:^{-} \qquad \text{H}-\overset{\overset{\displaystyle\text{H}}{|}}{\underset{\underset{\displaystyle\text{H}}{|}}{\text{C}}}\cdot \qquad \text{H}-\overset{\overset{\displaystyle\text{H}\;\text{H}}{|\;\;|}}{\underset{\underset{\displaystyle\text{H}\;\text{H}}{|\;\;|}}{\text{C}-\text{C}}}-\text{H}$$

methane   methyl cation   methyl anion   methyl radical   ethane

A carbocation is a species that contains a positively charged carbon.

A species containing a positively charged carbon is called a **carbocation,** and a species containing a negatively charged carbon is called a **carbanion.** (Recall that a *cation* is a positively charged ion and an *anion* is a negatively charged ion.) A species containing an atom with a single unpaired electron is called a **radical** (often called a **free radical**).

A carbanion is a species that contains a negatively charged carbon.

Hydrogen has one valence electron, and each halogen (F, Cl, Br, I) has seven valence electrons; so the following species have the indicated formal charges:

A radical is a species that contains an atom with an unpaired electron.

$$\text{H}^+ \qquad \text{H}:^{-} \qquad \text{H}\cdot \qquad :\overset{\cdot\cdot}{\underset{\cdot\cdot}{\text{Br}}}:^{-} \qquad :\overset{\cdot\cdot}{\underset{\cdot\cdot}{\text{Br}}}\cdot \qquad :\overset{\cdot\cdot}{\underset{\cdot\cdot}{\text{Br}}}-\overset{\cdot\cdot}{\underset{\cdot\cdot}{\text{Br}}}: \qquad :\overset{\cdot\cdot}{\underset{\cdot\cdot}{\text{Cl}}}-\overset{\cdot\cdot}{\underset{\cdot\cdot}{\text{Cl}}}:$$

hydrogen ion   hydride ion   hydrogen radical   bromide ion   bromine radical   bromine   chlorine

PROBLEM 16

Give each atom the appropriate formal charge:

a. $CH_3-\overset{\displaystyle .\,.}{\underset{\displaystyle |}{O}}-CH_3$ $\quad$ b. $H-\overset{\displaystyle .\,.}{\underset{\displaystyle |}{C}}-H$ $\quad$ c. $CH_3-\overset{\displaystyle CH_3}{\underset{\displaystyle CH_3}{\overset{|}{\underset{|}{N}}}}-CH_3$ $\quad$ d. $H-\overset{\displaystyle H\;\;\;H}{\underset{\displaystyle H\;\;\;H}{\overset{|\;\;\;\;|}{\underset{|\;\;\;\;|}{N-B}}}}-H$

While studying the molecules in this section notice that when the atoms do not bear a formal charge or an unpaired electron,

- carbon always forms 4 covalent bonds and has no lone pairs.
- nitrogen always forms 3 covalent bonds and has 1 lone pair.
- oxygen always forms 2 covalent bonds and has 2 lone pairs.
- a halogen always forms 1 covalent bond and has 3 lone pairs.
- hydrogen always forms 1 covalent bond and has no lone pairs.

$-\overset{\displaystyle |}{\underset{\displaystyle |}{C}}-$	$-\overset{\displaystyle .\,.}{\underset{\displaystyle |}{N}}-$	$:\overset{\displaystyle .\,.}{\underset{\displaystyle |}{O}}-$	$:\overset{..}{\underset{..}{F}}-$ $:\overset{..}{\underset{..}{Br}}-$ $:\overset{..}{\underset{..}{Cl}}-$ $:\overset{..}{\underset{..}{I}}-$	$H-$

$\underset{\text{# of bonds +}\atop\text{# of lone pairs}\atop\text{always equals 4}}{\quad}$

4 bonds	3 bonds	2 bonds	1 bond	1 bond
$\underline{0 \text{ lone pairs}}$	$\underline{1 \text{ lone pair}}$	$\underline{2 \text{ lone pairs}}$	$\underline{3 \text{ lone pairs}}$	
4	4	4	4	

When it is neutral:
C forms 4 bonds
N forms 3 bonds
O forms 2 bonds
a halogen forms 1 bond
H forms 1 bond

Notice that to have a complete octet, the number of bonds and the number of lone pairs must total four (except for hydrogen, which requires only two electrons to fill its shell).

These numbers are very important to remember when you are drawing structures of organic compounds because they provide a quick way to recognize when you have made a mistake. Atoms with more bonds or fewer bonds than is required for a neutral atom must have either a formal charge or an unpaired electron.

Each atom in the following Lewis structures has a filled outer shell. Notice that because none of the molecules has a formal charge or an unpaired electron, H forms 1 bond, C forms 4 bonds, N forms 3 bonds, O forms 2 bonds, and Br forms 1 bond. Notice, too, that each N has 1 lone pair, each O has 2 lone pairs, and Br has 3 lone pairs.

2 covalent bonds holding 2 atoms together is called a double bond

$H-\overset{\displaystyle H}{\underset{\displaystyle H}{\overset{|}{\underset{|}{C}}}}-\overset{..}{\underset{..}{Br}}:$ $\quad$ $H-\overset{\displaystyle H}{\underset{\displaystyle H}{\overset{|}{\underset{|}{C}}}}-\overset{..}{\underset{..}{O}}-\overset{\displaystyle H}{\underset{\displaystyle H}{\overset{|}{\underset{|}{C}}}}-H$ $\quad$ $H-\overset{\displaystyle :O:}{\overset{\displaystyle \|}{C}}-\overset{..}{\underset{..}{O}}-H$ $\quad$ $H-\overset{\displaystyle H}{\overset{|}{C}}-\overset{..}{\underset{\displaystyle H\;\;H}{\overset{|}{N}}}-H$ $\quad$ $:N\equiv N:$

3 covalent bonds holding 2 atoms together is called a triple bond

PROBLEM-SOLVING STRATEGY

Drawing Lewis Structures $\qquad$ **LEARN THE STRATEGY**

a. Draw the Lewis structure for CH_4O. $\quad$ b. Draw the Lewis structure for HNO_2.

a. 1. **Determine the total number of valence electrons** (4 for C, 1 for each H, and 6 for O adds up to $4 + 4 + 6 = 14$ valence electrons).

2. **Distribute the atoms,** remembering that C forms 4 bonds, O forms 2 bonds, and each H forms 1 bond. Always put the hydrogens on the outside of the molecule because H can form only 1 bond.

$H-\overset{\displaystyle H}{\underset{\displaystyle H}{\overset{|}{\underset{|}{C}}}}-O-H$

3. **Form bonds and fill octets with lone-pair electrons**, using the number of valence electrons determined in **1.**

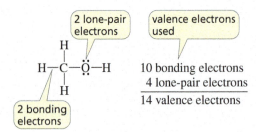

10 bonding electrons
4 lone-pair electrons
─────────────────────
14 valence electrons

4. **Assign a formal charge** to any atom whose number of valence electrons is not equal to the number of its lone-pair electrons plus the number of bonds. (None of the atoms in CH_4O has a formal charge.)

b. **1. Determine the total number of valence electrons** (1 for H, 5 for N, and 6 for each O adds up to $1 + 5 + 12 = 18$ valence electrons).

2. Distribute the atoms, putting the hydrogen on the outside of the molecule. If a species has two or more oxygen atoms, avoid oxygen–oxygen single bonds. These are weak bonds, and few compounds have them.

$$H—O—N—O$$

3. **Form bonds and fill octets with lone-pair electrons,** using the number of valence electrons determined in **1.**

$$H—\overset{..}{\underset{..}{O}}—\overset{..}{N}—\overset{..}{\underset{..}{O}}:$$

6 bonding electrons
12 lone-pair electrons
─────────────────────
18 valence electrons

If after all the electrons have been assigned, an atom (other than hydrogen) does not have a complete octet, use a lone pair on an adjacent atom to form a double bond to the electron-deficient atom.

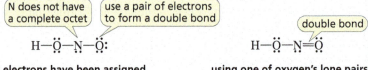

18 electrons have been assigned

using one of oxygen's lone pairs to form a double bond gives N a complete octet

4. **Assign a formal charge** to any atom whose number of valence electrons is not equal to the number of its lone-pair electrons plus the number of bonds. (None of the atoms in HNO_2 has a formal charge.)

USE THE STRATEGY

PROBLEM 17 SOLVED

Draw the Lewis structure for each of the following:

a. NO_3^- c. $^-C_2H_5$ e. $CH_3\overset{+}{N}H_3$ g. HCO_3^-

b. NO_2^+ d. $^+C_2H_5$ f. NaOH   h. HCO_2^-

SOLUTION TO 17 a. The total number of valence electrons is 23 (5 for N and 6 for each of the three Os). Because the species has one negative charge, we must add 1 to the number of valence electrons, for a total of 24. The only way we can arrange one N and three Os and avoid O—O single bonds is to place the three Os around the N. We then use the 24 electrons to form bonds and fill octets with lone-pair electrons.

$$:\overset{..}{\underset{}{O}}:$$
$$|$$
$$:\overset{..}{\underset{..}{O}}—N—\overset{..}{\underset{..}{O}}:$$

incomplete octet

All 24 electrons have been assigned, but N does not have a complete octet. We complete N's octet by using one of oxygen's lone pairs to form a double bond. (It does not make a difference which oxygen atom we choose.) When we check each atom to see whether it has a formal charge, we find that two of the Os are negatively charged and that the N is positively charged, for an overall charge of −1.

$$\overset{\displaystyle :\!\ddot{O}:}{\underset{+}{\overset{\|}{:\ddot{\underset{..}{O}}-N-\ddot{\underset{..}{O}}:}}^{-}}$$

SOLUTION TO 17 b. The total number of valence electrons is 17 (5 for N and 6 for each of the two Os). Because the species has one positive charge, we must subtract 1 from the number of valence electrons, for a total of 16. The 16 electrons are used to form bonds and fill octets with lone-pair electrons.

(incomplete octet)

$$:\ddot{\underset{..}{O}}-N-\ddot{\underset{..}{O}}:$$

Two double bonds are necessary to complete N's octet. We find that the N has a formal charge of +1.

$$\ddot{\underset{..}{O}}=\overset{+}{N}=\ddot{\underset{..}{O}}$$

PROBLEM 18 ♦

a. Draw two Lewis structures for C_2H_6O.   **b.** Draw three Lewis structures for C_3H_8O.

(*Hint:* The two Lewis structures in part **a** are **constitutional isomers**—molecules that have the same atoms but differ in the way the atoms are connected. The three Lewis structures in part **b** are also constitutional isomers.)

Kekulé Structures

Kekulé structures are like Lewis structures except that lone pairs are normally omitted.

Kekulé structures

$$\underset{\overset{|}{H}}{\overset{\overset{|}{H}}{H-C-Br}} \qquad \underset{\overset{|}{H}\;\;\overset{|}{H}}{\overset{\overset{|}{H}\;\;\overset{|}{H}}{H-C-O-C-H}} \qquad \overset{\overset{O}{\|}}{H-C-O-H} \qquad \underset{\overset{|}{H}\;\overset{|}{H}}{\overset{\overset{|}{H}}{H-C-N-H}} \qquad N\equiv N$$

Condensed Structures

Condensed structures omit some (or all) of the covalent bonds and list atoms bonded to a particular carbon (or nitrogen or oxygen) next to it (with a subscript if there is more than one of a particular atom). Lone pairs are usually not shown, unless they are needed to draw attention to some chemical property of the molecule.

condensed structures CH_3Br CH_3OCH_3 HCO_2H CH_3NH_2 N_2

(Although lone pairs are not shown, you should remember that neutral nitrogen, oxygen, and halogen atoms always have them: one pair for nitrogen, two pairs for oxygen, and three pairs for a halogen.)

   You can find examples of Kekulé and condensed structures in Table 1.5. Notice that because none of the molecules in Table 1.5 has a formal charge or an unpaired electron, each C has four bonds, each N has three bonds, each O has two bonds, and each H or halogen has one bond.

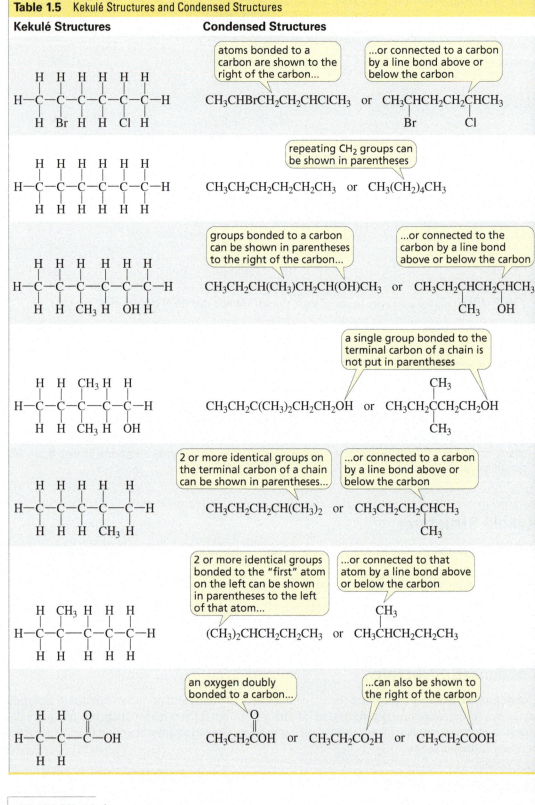

Table 1.5 Kekulé Structures and Condensed Structures

Kekulé Structures   **Condensed Structures**

atoms bonded to a carbon are shown to the right of the carbon...

...or connected to a carbon by a line bond above or below the carbon

$CH_3CHBrCH_2CH_2CHClCH_3$ or $CH_3CHCH_2CH_2CHCH_3$

Br   Cl

repeating CH_2 groups can be shown in parentheses

$CH_3CH_2CH_2CH_2CH_2CH_3$ or $CH_3(CH_2)_4CH_3$

groups bonded to a carbon can be shown in parentheses to the right of the carbon...

...or connected to the carbon by a line bond above or below the carbon

$CH_3CH_2CH(CH_3)CH_2CH(OH)CH_3$ or $CH_3CH_2CHCH_2CHCH_3$

CH$_3$ OH

a single group bonded to the terminal carbon of a chain is not put in parentheses

$CH_3CH_2C(CH_3)_2CH_2CH_2OH$ or $CH_3CH_2CCH_2CH_2OH$

CH$_3$ / CH$_3$

2 or more identical groups on the terminal carbon of a chain can be shown in parentheses...

...or connected to a carbon by a line bond above or below the carbon

$CH_3CH_2CH_2CH(CH_3)_2$ or $CH_3CH_2CH_2CHCH_3$

CH$_3$

2 or more identical groups bonded to the "first" atom on the left can be shown in parentheses to the left of that atom...

...or connected to that atom by a line bond above or below the carbon

$(CH_3)_2CHCH_2CH_2CH_3$ or $CH_3CHCH_2CH_2CH_3$

CH$_3$

an oxygen doubly bonded to a carbon...

...can also be shown to the right of the carbon

CH_3CH_2COH or $CH_3CH_2CO_2H$ or CH_3CH_2COOH

PROBLEM 19 ♦

Draw the lone-pair electrons that are not shown in the following condensed structures:

a. $CH_3CH_2NH_2$ **c.** CH_3CH_2OH **e.** CH_3CH_2Cl

b. CH_3NHCH_3 **d.** CH_3OCH_3 **f.** $HONH_2$

PROBLEM 20 ♦

Draw condensed structures for the compounds represented by the following models (black = C, gray = H, red = O, blue = N, and green = Cl):

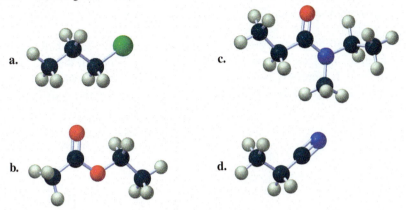

a.

c.

b.

d.

PROBLEM 21 ♦

Which of the atoms in the molecular models in Problem 20 have

a. three lone pairs?   **b.** two lone pairs?   **c.** one lone pair?   **d.** no lone pairs?

PROBLEM 22

Change the following condensed structures to Kekulé structures:

a. $CH_3NH(CH_2)_2CH_3$ **b.** $(CH_3)_2CHCl$ **c.** $(CH_3)_3CBr$ **d.** $(CH_3)_3C(CH_2)_3CHO$

Skeletal Structures

Skeletal structures show the carbon–carbon bonds as lines, but do not show the carbons or the hydrogens that are bonded to the carbons. Each vertex in a skeletal structure represents a carbon, and each carbon is understood to be bonded to the appropriate number of hydrogens to give the carbon four bonds. Atoms other than carbons are shown, and hydrogens bonded to atoms other than carbon are also shown.

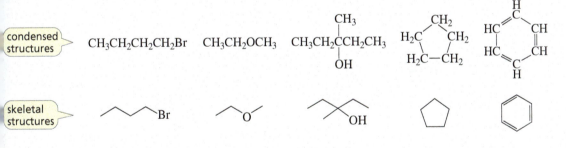

PROBLEM 23

Convert the models in Problem 20 to skeletal structures.

1.5 ATOMIC ORBITALS

We saw that electrons are distributed into different atomic orbitals (Table 1.2). An atomic orbital is a three-dimensional region around the nucleus where an electron is most likely to be found. Because the **Heisenberg uncertainty principle** states that both the precise location and the exact momentum of an atomic particle cannot be simultaneously determined, we can never say precisely where an electron is—we can only describe its probable location.

s Atomic Orbitals

Mathematical calculations indicate that an *s* atomic orbital is a sphere with the nucleus at its center, and experimental evidence supports this theory. Thus, when we say that an electron occupies a 1*s* orbital, we mean that there is a greater than 90% probability that the electron is in the space defined by the sphere.

Because the second shell lies farther from the nucleus than does the first shell (Section 1.2), the average distance from the nucleus is greater for an electron in a 2*s* orbital than it is for an electron in a 1*s* orbital. A 2*s* orbital, therefore, is represented by a larger sphere. Because of the greater size of a 2*s* orbital, its average electron density is less than the average electron density of a 1*s* orbital.

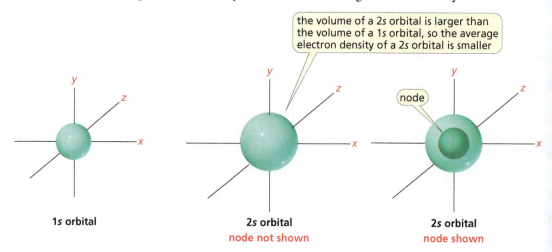

the volume of a 2*s* orbital is larger than the volume of a 1*s* orbital, so the average electron density of a 2*s* orbital is smaller

node

1s orbital

2s orbital
node not shown

2s orbital
node shown

An electron in a 1*s* orbital can be anywhere within the orbital, but a 2*s* orbital has a region where the probability of finding an electron falls to zero. This is called a **node.** Nodes occur because electrons have both particle-like and wave-like properties. A node is a consequence of the wave-like properties of an electron.

There are two types of waves: traveling waves and standing waves. Traveling waves move through space. Light is an example of a traveling wave. A standing wave, on the other hand, is confined to a limited space. The vibrating string of a guitar is a standing wave—the string moves up and down but does not travel through space. The amplitude is (+) in the region above where the guitar string is at rest, and it is (−) in the region below where the guitar string is at rest—the two regions are said to have opposite phases. The region where the guitar string has no transverse displacement (zero amplitude) is a node.

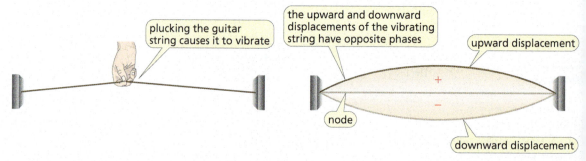

plucking the guitar string causes it to vibrate

the upward and downward displacements of the vibrating string have opposite phases

upward displacement

node

downward displacement

An electron behaves like a standing wave, but unlike the wave created by a vibrating guitar string, it is three-dimensional. This means that the node of a 2*s* orbital is actually a spherical surface within the 2*s* orbital. Because the electron wave has zero amplitude at the node, there is zero probability of finding an electron at the node.

p Atomic Orbitals

Unlike *s* orbitals that resemble spheres, *p* orbitals have two lobes. Generally, the lobes are depicted as teardrop shaped, but computer-generated representations reveal that they are shaped more like doorknobs (as shown on the right on the top of p. 21). Like the vibrating guitar string, the lobes have opposite phases, which can be designated by plus (+) and minus (−) or by two different colors.

(Notice that in this context, + and − indicate the phase of the orbital; they do not indicate charge.) The node of the *p* orbital is a plane—called a **nodal plane**—that passes through the center of the nucleus, between its two lobes. There is zero probability of finding an electron in the nodal plane of the *p* orbital.

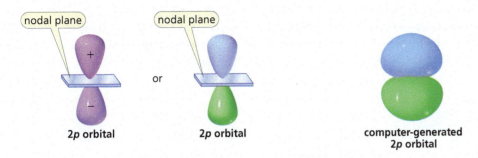

| 2*p* orbital | or   2*p* orbital | computer-generated 2*p* orbital |

In Section 1.2, we saw that the second and higher numbered shells each contain three degenerate *p* atomic orbitals. The p_x orbital is symmetrical about the *x*-axis, the p_y orbital is symmetrical about the *y*-axis, and the p_z orbital is symmetrical about the *z*-axis. This means that each *p* orbital is perpendicular to the other two *p* orbitals. The energy of a 2*p* orbital is slightly greater than that of a 2*s* orbital because the average location of an electron in a 2*p* orbital is farther away from the nucleus.

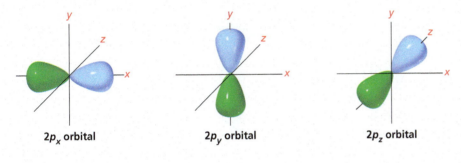

| 2p_x orbital | 2p_y orbital | 2p_z orbital |

PROBLEM 24

Draw the following orbitals:

a. 3*s* orbital   **b.** 4*s* orbital   **c.** 3*p* orbital

1.6 AN INTRODUCTION TO MOLECULAR ORBITAL THEORY

How do atoms form covalent bonds in order to form molecules? The Lewis model, which shows atoms attaining a complete octet by sharing electrons, tells only part of the story. A drawback of the model is that it treats electrons like particles and does not take into account their wave-like properties.

Molecular orbital (MO) theory combines the tendency of atoms to fill their octets by sharing electrons (the Lewis model) with their wave-like properties, assigning electrons to a volume of space called an orbital. According to MO theory, covalent bonds result when atomic orbitals combine to form *molecular orbitals*. Like an atomic orbital, which describes the volume of space around an atom's nucleus where an electron is likely to be found, a **molecular orbital** describes the volume of space around a molecule where an electron is likely to be found. And like atomic orbitals, molecular orbitals, too, have specific sizes, shapes, and energies.

> *An atomic orbital surrounds an atom.*
> *A molecular orbital surrounds a molecule.*

Forming a sigma (σ) Bond

Let's look first at the bonding in a hydrogen molecule (H_2). Imagine a meeting of two separate H atoms. As the $1s$ atomic orbital of one hydrogen atom approaches the $1s$ atomic orbital of the other hydrogen atom, the orbitals begin to overlap. The atoms continue to move closer, and the amount of overlap increases until the orbitals combine to form a molecular orbital. The covalent bond that is formed when the two s orbitals overlap is called a **sigma (σ) bond**. A σ bond is cylindrically symmetrical—the electrons in the bond are symmetrically distributed about an imaginary line connecting the nuclei of the two atoms joined by the bond.

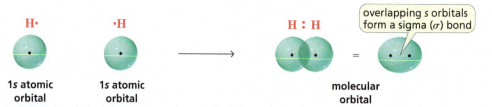

As the two orbitals begin to overlap, energy is released because the electron in each atom is attracted to its own nucleus *and* to the nucleus of the other atom (Figure 1.1).

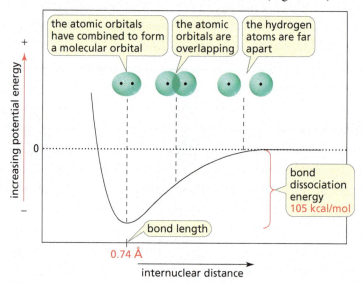

▶ **Figure 1.1**
The change in energy that occurs as two hydrogen atoms approach each other. The internuclear distance at minimum potential energy is the length of the H—H covalent bond.

Minimum energy corresponds to maximum stability.

The attraction of the two negatively charged electrons for the two positively charged nuclei is what holds the two H atoms together. The more the orbitals overlap, the more the energy decreases, until the atoms are so close that their positively charged nuclei begin to repel each other. This repulsion causes a large increase in energy. Figure 1.1 shows that minimum energy (maximum stability) is achieved when the nuclei are a particular distance apart. This distance is the **bond length** of the new covalent bond; the bond length of the H—H bond is 0.74 Å.

As Figure 1.1 shows, energy is released when a covalent bond forms. We see that when the H—H bond forms, 105 kcal/mol of energy is released. Breaking the bond requires precisely the same amount of energy. Thus, the **bond dissociation energy**—a measure of bond strength—is the energy required to break a bond or the energy released when a bond is formed. Every covalent bond has a characteristic bond length and bond dissociation energy.

Bonding and Antibonding Molecular Orbitals

Orbitals are conserved. In other words, the number of molecular orbitals formed must equal the number of atomic orbitals combined. In describing the formation of an H—H bond, we combined two atomic orbitals but discussed only one molecular orbital. Where is the other molecular orbital? As you will see shortly, it is there; it just doesn't contain any electrons.

It is the wave-like properties of the electrons that cause two atomic orbitals to form two molecular orbitals. The two atomic orbitals can combine in an additive (constructive) manner, just as two light waves or two sound waves can reinforce each other (Figure 1.2a). The constructive combination of two s atomic orbitals is called a σ **(sigma) bonding molecular orbital.**

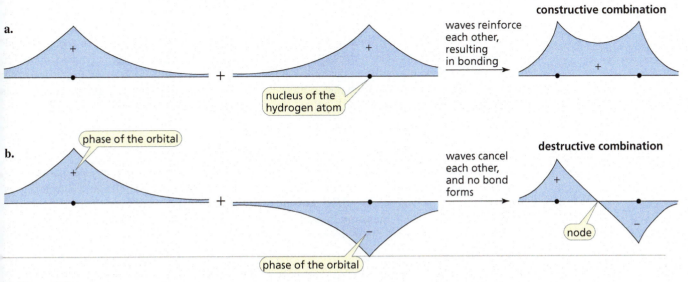

▲ **Figure 1.2**
(a) Waves with the same phase (+/+) overlap constructively, reinforcing each other and resulting in bonding.
(b) Waves with opposite phases (+/−) overlap destructively, canceling each other and forming a node.

The two atomic orbitals can also combine in a destructive way, canceling each other. The cancellation is similar to the darkness that results when two light waves cancel each other or to the silence that results when two sound waves cancel each other (Figure 1.2b). The destructive combination of two *s* atomic orbitals is called a **σ^* antibonding molecular orbital.** An antibonding orbital is indicated by an asterisk (*), which chemists read as "star." Thus, σ^* is read as "sigma star."

The σ bonding molecular orbital and the σ^* antibonding molecular orbital are shown in the molecular orbital (MO) diagram in Figure 1.3. In an MO diagram, the energies of both the atomic orbitals and the molecular orbitals are represented as horizontal lines, with the bottom line being the lowest energy level and the top line the highest energy level.

When two atomic orbitals overlap, two molecular orbitals are formed—
one lower in energy and one higher in energy than the atomic orbitals.

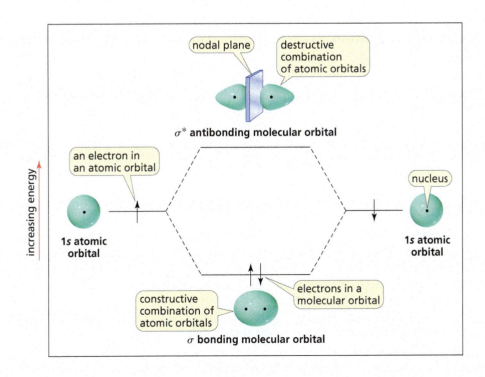

◀ **Figure 1.3**
Atomic orbitals of H· and molecular orbitals of H_2. Before covalent bond formation, each electron is in an atomic orbital. After covalent bond formation, both electrons are in the bonding MO. The antibonding MO is empty.

The electrons in the bonding molecular orbital are most likely to be found between the nuclei, where they can more easily attract both nuclei simultaneously. This increased electron density between the nuclei is what binds the atoms together (Figure 1.2a). Electrons in the antibonding molecular orbital, however, are most likely to be found anywhere except between the nuclei, because a nodal plane lies between the nuclei (Figure 1.2b). As a result, electrons in the antibonding orbital leave the positively charged nuclei more exposed to one another. Therefore, electrons in the antibonding orbital detract from, rather than assist in, the formation of a bond.

Electrons in a bonding MO assist in bonding.
Electrons in an antibonding MO detract from bonding.

The MO diagram shows that the bonding molecular orbital is lower in energy and is, therefore, more stable than the individual atomic orbitals. This is because the more nuclei an electron senses, the more stable it is. The antibonding molecular orbital, with less electron density between the nuclei, is less stable—and, therefore, higher in energy—than the atomic orbitals.

Electrons are assigned to the molecular orbitals using the same rules used to assign electrons to atomic orbitals:

- electrons always occupy available orbitals with the lowest energy (the aufbau principle).
- no more than two electrons can occupy a molecular orbital (the Pauli exclusion principle).

Thus, the two electrons of the H—H bond occupy the lower-energy bonding molecular orbital (the σ bonding MO in Figure 1.3), where they are attracted to both positively charged nuclei. It is this electrostatic attraction that gives a covalent bond its strength. We can conclude, therefore, that the strength of the covalent bond increases as the overlap of the atomic orbitals increases.

Covalent bond strength increases as atomic orbital overlap increases.

The MO diagram in Figure 1.3 allows us to predict that H_2^+ would not be as stable as H_2 because H_2^+ has only one electron in the bonding molecular orbital. Using the same diagram, we can also predict that He_2 does not exist, because the four electrons of He_2 (two from each He atom) would fill the lower energy bonding MO *and* the higher energy antibonding MO. *The two electrons in the antibonding MO would cancel the advantage to bonding that is gained by the two electrons in the bonding MO.*

PROBLEM 25 ◆
Predict whether He_2^+ exists.

Forming a pi (π) Bond

When two *p* atomic orbitals overlap, the side of one orbital overlaps the side of the other. The side-to-side overlap of two parallel *p* orbitals forms a bond that is called a **pi (π) bond.**

Side-to-side overlap of two in-phase *p* atomic orbitals (blue lobes overlap blue lobes and green lobes overlap green lobes) is a constructive overlap and forms a π bonding molecular orbital, whereas side-to-side overlap of two out-of-phase *p* orbitals (blue lobes overlap green lobes) is a destructive overlap and forms a π^* (read "pi star") antibonding molecular orbital (Figure 1.4). The π bonding MO has one node—a nodal plane that passes through both nuclei. The π^* antibonding MO has two nodal planes.

The VSEPR Model

You can find a more extensive discussion of molecular orbital theory in Special Topic II in the *Study Guide and Solutions Manual*.

To learn even more about the bonds in a molecule, organic chemists turn to the **valence-shell electron-pair repulsion (VSEPR) model**—a model for predicting molecular geometry based on the minimization of electron repulsion between regions of electron density around an atom. In other words, atoms share electrons by overlapping their atomic orbitals and, because electron pairs repel each other, the bonding electrons and lone-pair electrons around an atom are positioned as far apart as possible. Thus, a Lewis structure gives us a first approximation of the structure of a simple molecule, and VSEPR gives us a first glance at the shape of the molecule.

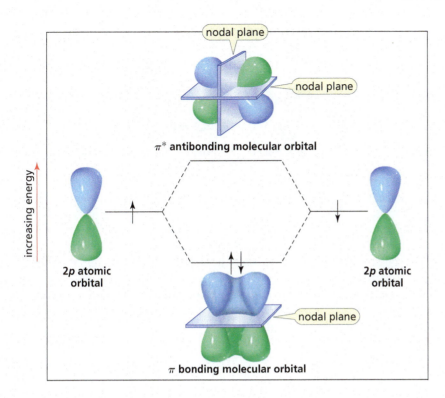

increasing energy →

π* antibonding molecular orbital

nodal plane

nodal plane

2p atomic orbital

2p atomic orbital

nodal plane

π bonding molecular orbital

Side-to-side overlap of two p orbitals forms a π bond. All other covalent bonds in organic molecules are σ bonds.

In-phase overlap forms a bonding MO; out-of-phase overlap forms an antibonding MO.

◄ **Figure 1.4**
Side-to-side overlap of two parallel p atomic orbitals forms a π bonding molecular orbital and a π* antibonding molecular orbital.

Because organic chemists generally think of chemical reactions in terms of the changes that occur in the bonds of the reacting molecules, the VSEPR model often provides the easiest way to visualize chemical change. However, the model is inadequate for some molecules because it does not allow for antibonding molecular orbitals. We will use both the MO and the VSEPR models in this book. Our choice will depend on which model provides the best description of the molecule under discussion. We will use the VSEPR model in Sections 1.7–1.13.

PROBLEM 26 ♦

Indicate the kind of molecular orbital (σ, σ*, π, or π*) that results when the two atomic orbitals are combined:

a. 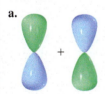 +

b. +

c.  +

d.  +

1.7 HOW SINGLE BONDS ARE FORMED IN ORGANIC COMPOUNDS

We will begin our discussion of bonding in organic compounds by looking at the bonding in methane, a compound with only one carbon. Then we will examine the bonding in ethane, a compound with two carbons attached by a carbon–carbon single bond.

The Bonds in Methane

Methane (CH_4) has four covalent C—H bonds. Because all four bonds have the same length (1.10 Å) and all the bond angles are the same (109.5°), we can conclude that the four C—H bonds in methane are identical. Four different ways to represent a methane molecule are shown here.

Representations of Methane

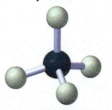

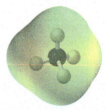

perspective formula   ball-and-stick model   space-filling model   electrostatic potential map

In a **perspective formula**, bonds in the plane of the paper are drawn as solid lines (and they must be adjacent to one another), a bond protruding out of the plane of the paper toward the viewer is drawn as a solid wedge, and one projecting back from the plane of the paper away from the viewer is drawn as a hatched wedge.

The potential map of methane shows that neither carbon nor hydrogen carries much of a charge: there are neither red areas, representing partially negatively charged atoms, nor blue areas, representing partially positively charged atoms. (Compare this map with the potential map for water on p. 37.) The absence of partially charged atoms can be explained by the similar electronegativities of carbon and hydrogen, which cause them to share their bonding electrons relatively equally (see p. 11). Methane, therefore, is a **nonpolar molecule.**

You might be wondering how carbon can form four covalent bonds when it has only two unpaired valence electrons (Table 1.2). Carbon has to form four covalent bonds otherwise it would not complete its octet. We need, therefore, to come up with an explanation that accounts for the fact that carbon forms four covalent bonds when it has only two valence electrons.

If one of the electrons in carbon's $2s$ orbital is promoted into its empty $2p$ orbital, then carbon will have four unpaired valence electrons (and then four covalent bonds can be formed).

The blue colors of Uranus and Neptune are caused by the presence of methane, a colorless and odorless gas, in their atmospheres. Natural gas—called a fossil fuel because it is formed from the decomposition of plant and animal material in the Earth's crust—is approximately 75% methane.

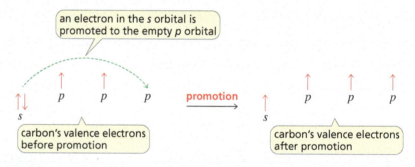

However, we know that the four C—H bonds in methane are identical. How can they be identical if carbon uses an s orbital and three p orbitals to form these four bonds? Wouldn't the bond formed with the s orbital be different from the three bonds formed with p orbitals? The four C—H bonds are identical because carbon uses hybrid atomic orbitals.

Hybrid Orbitals

Hybrid orbitals result from combining atomic orbitals.

Hybrid orbitals are mixed orbitals that result from combining atomic orbitals. The concept of combining atomic orbitals, called **hybridization,** was first proposed by Linus Pauling in 1931.

If the one s and three p orbitals of the second shell are all combined and then apportioned into four equal orbitals, each of the four resulting orbitals will be one part s and three parts p. This type of mixed orbital is called an sp^3 (read "s-p-three," not "s-p-cubed") orbital. The superscript 3 means that three p orbitals were mixed with one s orbital—the superscript 1 on the s is implied—to form the four hybrid orbitals. Each sp^3 orbital has 25% s character and 75% p character. The four sp^3 orbitals are degenerate—that is, they all have the same energy.

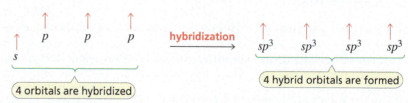

Like a *p* orbital, an *sp*³ orbital has two lobes. The lobes differ in size, however, because the *s* orbital adds to one lobe of the *p* orbital and subtracts from the other lobe (Figure 1.5).

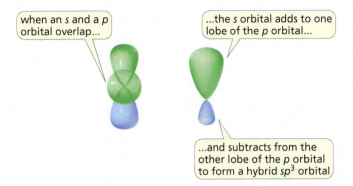

◄ Figure 1.5
The *s* orbital adds to one lobe of the *p* orbital and subtracts from the other. The result is a hybrid orbital with two lobes that differ in size.

The larger lobe of the *sp*³ orbital is used to form covalent bonds. The stability of an *sp*³ orbital reflects its composition; it is more stable than a *p* orbital, but not as stable as an *s* orbital (Figure 1.6). (To simplify the orbital dipictions that follow, the phases of the orbitals will not be shown.)

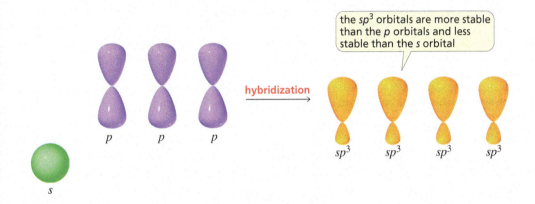

◄ Figure 1.6
An *s* orbital and three *p* orbitals hybridize to form four *sp*³ orbitals. An *sp*³ orbital is more stable (lower in energy) than a *p* orbital but less stable (higher in energy) than an *s* orbital.

Tetrahedral Carbon; Tetrahedral Bond Angle

The four *sp*³ orbitals adopt a spatial arrangement that keeps them as far away from each other as possible. They do this because electrons repel each other, and moving as far from each other as possible minimizes the repulsion. (See the description of the VSEPR model on p. 24.)

When four *sp*³ orbitals move as far from each other as possible, they point toward the corners of a regular tetrahedron—a pyramid with four faces, each an equilateral triangle (Figure 1.7a). Each of the four C—H bonds in methane is formed from the overlap of an *sp*³ orbital of carbon with the *s* orbital of a hydrogen (Figure 1.7b). This explains why the four C—H bonds are identical.

Electron pairs stay as far from each other as possible.

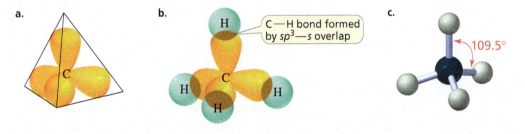

◄ Figure 1.7
For clarity, the smaller lobes of the *sp*³ orbitals are not shown.

(a) The four *sp*³ orbitals are directed toward the corners of a tetrahedron, causing each bond angle to be 109.5°. This arrangement allows the four orbitals to be as far apart as possible.

(b) An orbital picture of methane, showing the overlap of each *sp*³ orbital of carbon with the *s* orbital of a hydrogen.

(c) The tetrahedral bond angle is 109.5°.

The angle between any two lines that point from the center to the corners of a tetrahedron is 109.5° (Figure 1.7c). Therefore, the bond angles in methane are 109.5°. This is called a **tetrahedral bond angle.** A carbon, such as the one in methane, that forms covalent bonds using four equivalent *sp*³ orbitals is called a **tetrahedral carbon.**

If you are thinking that hybrid orbital theory appears to have been contrived just to make things fit, then you are right. Nevertheless, it gives us a very good picture of the bonding in organic compounds.

The Bonds in Ethane

Each carbon in ethane (CH_3CH_3) is bonded to four other atoms.

$$\begin{array}{ccc} & H & H \\ & | & | \\ H- & C- & C-H \\ & | & | \\ & H & H \end{array}$$

ethane

One bond connecting two atoms is called a **single bond.** All the bonds in ethane are single bonds.

To bond to four atoms, each carbon uses four sp^3 orbitals as they do in methane. (Figure 1.8). One sp^3 orbital of one carbon of ethane overlaps an sp^3 orbital of the other carbon to form the C—C bond.

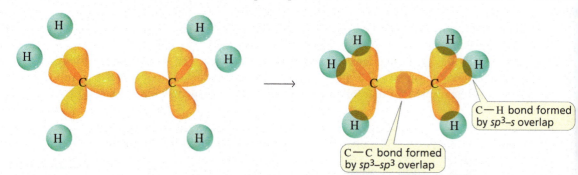

C—H bond formed by sp^3–s overlap

C—C bond formed by sp^3–sp^3 overlap

▲ **Figure 1.8**
An orbital picture of ethane. The C—C bond is formed by sp^3–sp^3 overlap, and each C—H bond is formed by sp^3–s overlap. (The smaller lobes of the sp^3 orbitals are not shown.) As a result, both carbons are tetrahedral and all bond angles are ~109.5°.

The three remaining sp^3 orbitals of each carbon overlap the s orbital of a hydrogen to form a C—H bond. Thus, the C—C bond is formed by sp^3–sp^3 overlap, and each C—H bond is formed by sp^3–s overlap.

Because both carbons are tetrahedral, each of the bond angles in ethane is nearly the tetrahedral bond angle of 109.5°. The length of the C—C bond is 1.54 Å. The potential map shows that ethane, like methane, is a nonpolar molecule.

Representations of Ethane

NOTE TO THE STUDENT

• It is important to understand what molecules look like in three dimensions. Therefore, be sure to visit the MasteringChemistry Study Area and look at the three-dimensional representations of molecules that can be found in the molecule gallery prepared for each chapter.

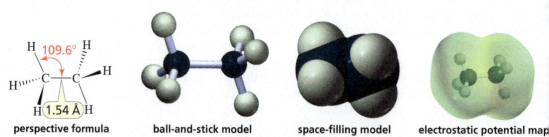

perspective formula   ball-and-stick model   space-filling model   electrostatic potential map

The MO diagram illustrating the overlap of an sp^3 orbital of one carbon with an sp^3 orbital of another carbon shows that the two sp^3 orbitals overlap end-on (Figure 1.9). End-on overlap forms a cylindrically symmetrical bond—a sigma (σ) bond (Section 1.6). *All single bonds in organic compounds are sigma bonds.* Thus, all the bonds in methane and ethane are sigma (σ) bonds.

Notice in Figure 1.9 that the electron density of the σ bonding MO is concentrated between the nuclei. This causes the back lobes (the nonoverlapping green lobes) to be quite small.

PROBLEM 27 ◆

What orbitals are used to form the 10 sigma bonds in propane ($CH_3CH_2CH_3$)?

PROBLEM 28

Explain why a σ bond formed by overlap of an s orbital with an sp^3 orbital of carbon is stronger than a σ bond formed by overlap of an s orbital with a p orbital of carbon.

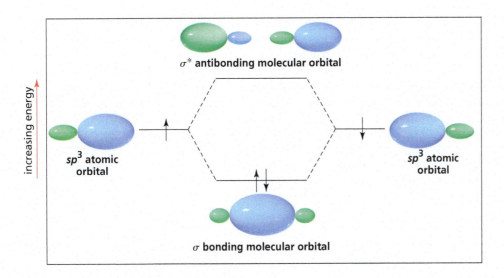

σ* antibonding molecular orbital

*sp*³ atomic orbital

*sp*³ atomic orbital

σ bonding molecular orbital

increasing energy

Sigma bonds are cylindrically symmetrical.

All single bonds in organic compounds are sigma bonds.

◄ **Figure 1.9**
End-on overlap of two *sp*³ orbitals forms a σ bonding molecular orbital and a σ* antibonding molecular orbital.

1.8 HOW A DOUBLE BOND IS FORMED: THE BONDS IN ETHENE

The carbons in ethene (also called ethylene) form two bonds with each other. Two bonds connecting two atoms is called a **double bond.** Each of the carbons forms four bonds, but each carbon is bonded to only three atoms.

$$\underset{H}{\overset{H}{C}}=\underset{H}{\overset{H}{C}}$$

ethene
ethylene

To bond to three atoms, each carbon hybridizes three atomic orbitals: an *s* orbital and two of the *p* orbitals. Because three orbitals are hybridized, three hybrid orbitals are formed. These are called *sp*² orbitals. After hybridization, each carbon atom has three degenerate *sp*² orbitals and one unhybridized *p* orbital:

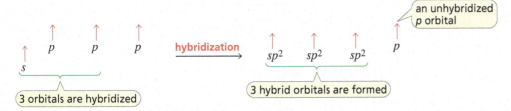

an unhybridized *p* orbital

hybridization

3 orbitals are hybridized

3 hybrid orbitals are formed

To minimize electron repulsion, the three *sp*² orbitals need to get as far from each other as possible. Therefore, the axes of the three orbitals lie in a plane, directed toward the corners of an equilateral triangle with the carbon nucleus at the center. As a result, the bond angles are all close to 120° (Figure 1.10a).

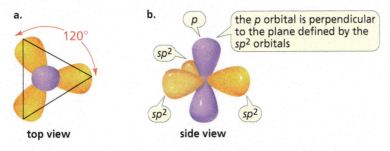

a.

120°

top view

b.

p

the *p* orbital is perpendicular to the plane defined by the *sp*² orbitals

*sp*²

*sp*²   *sp*²

side view

◄ **Figure 1.10**
(a) The three degenerate *sp*² orbitals lie in a plane, oriented 120° from each other. (The smaller lobes of the *sp*² orbitals are not shown.)
(b) The unhybridized *p* orbital is perpendicular to this plane.

Because an sp^2 hybridized carbon is bonded to three atoms and three points define a plane, it is called a **trigonal planar carbon.** The unhybridized p orbital is perpendicular to the plane defined by the axes of the sp^2 orbitals (Figure 1.10b).

The two bonds in the double bond are not identical. One bond results from the overlap of an sp^2 orbital of one carbon with an sp^2 orbital of the other carbon; this is a sigma (σ) bond because it is cylindrically symmetrical (Figure 1.11a). Each carbon uses its other two sp^2 orbitals to overlap the s orbital of a hydrogen to form the C—H bonds.

The second carbon–carbon bond results from side-to-side overlap of the two unhybridized p orbitals. Side-to-side overlap of p orbitals forms a pi (π) bond (Figure 1.11b). Thus, one of the bonds in a double bond is a σ bond, and the other is a π bond. All the C—H bonds are σ bonds. (Remember that all single bonds in organic compounds are σ bonds.)

A double bond consists of one σ bond and one π bond.

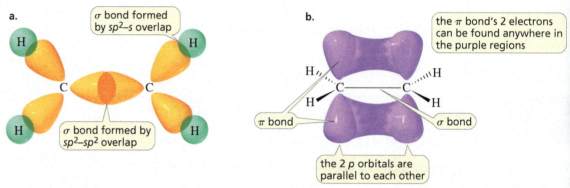

▲ **Figure 1.11**
(a) One C—C bond in ethene is a σ bond formed by sp^2–sp^2 overlap, and the C—H bonds are σ bonds formed by sp^2–s overlap.
(b) The second C—C bond is a π bond formed by side-to-side overlap of a p orbital of one carbon with a p orbital of the other carbon. The two p orbitals are parallel to each other.

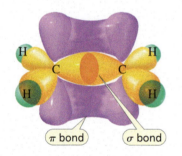

▲ **Figure 1.12**
The two carbons and four hydrogens lie in the same plane. Perpendicular to that plane are the two parallel p orbitals. This results in an accumulation of electron density above and below the plane containing the two carbons and four hydrogens.

To achieve maximum overlap, the two p orbitals that overlap to form the π bond must be parallel to each other (Figure 1.11b). This forces the triangle formed by one carbon and two hydrogens to lie in the same plane as the triangle formed by the other carbon and two hydrogens. As a result, all six atoms of ethene lie in the same plane, and the electrons in the p orbitals occupy a volume of space above and below the plane (Figure 1.12).

The potential map for ethene shows that it is a nonpolar molecule with a slight accumulation of negative charge (the pale orange area) above the two carbons. (If you could turn the potential map over, you would find a similar accumulation of negative charge on the other side.)

Representations of Ethene

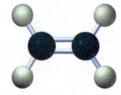

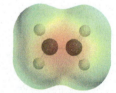

a double bond consists of one σ bond and one π bond

perspective formula   ball-and-stick model   space-filling model   electrostatic potential map

Four electrons hold the carbons together in a carbon–carbon double bond, but only two electrons hold the carbons together in a carbon–carbon single bond. Therefore, a carbon–carbon double bond is stronger (174 kcal/mol) and shorter (1.33 Å) than a carbon–carbon single bond (90 kcal/mol and 1.54 Å).

Diamond, Graphite, Graphene, and Fullerenes: Substances That Contain Only Carbon Atoms

The difference that hybridization can make is illustrated by diamond and graphite. Diamond is the hardest of all substances, whereas graphite is a slippery, soft solid most familiar to us as the lead in pencils. Both materials, in spite of their very different physical properties, contain only carbon atoms. The two substances differ solely in the hybridization of the carbon atoms.

Diamond consists of a rigid three-dimensional network of carbon atoms, with each carbon bonded to four others via sp^3 orbitals.

The carbon atoms in graphite, on the other hand, are sp^2 hybridized, so each bonds to only three other carbons. This trigonal planar arrangement causes the atoms in graphite to lie in flat, layered sheets. Because there are no covalent bonds between the sheets, they can shear off from neighboring sheets.

Diamond and graphite have been known since ancient times—but a third substance found in nature that contains only carbon atoms was discovered in 2004. Graphene is a one-atom-thick planar sheet of graphite. It is the thinnest and lightest material known. It is transparent and can be bent, stacked, or rolled. It is harder than diamond, and it conducts electricity better than copper. In 2010, the Nobel Prize in Physics was given to Andre Geim and Konstantin Novoselov of the University of Manchester for their ground-breaking experiments on graphene.

Fullerenes are also naturally occurring compounds that contain only carbon. Like graphite and graphene, fullerenes consist solely of sp^2 carbons, but instead of forming planar sheets, the carbons join to form spherical structures. (Fullerenes are discussed in Section 8.18.)

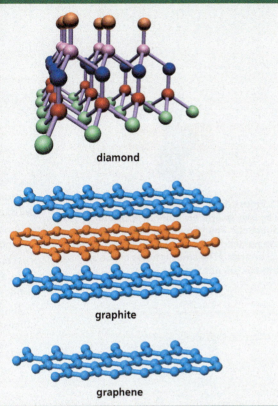

diamond

graphite

graphene

PROBLEM 29 SOLVED

Do the sp^2 carbons and the indicated sp^3 carbons have to lie in the same plane?

a.

$$\underset{\text{CH}_3}{\overset{\text{H}}{\diagdown}}\text{C}=\text{C}\underset{\text{H}}{\overset{\text{CH}_3}{\diagup}}$$

b.

$$\underset{\text{CH}_3}{\overset{\text{H}}{\diagdown}}\text{C}=\text{C}\underset{\text{H}}{\overset{\text{CH}_2\text{CH}_3}{\diagup}}$$

SOLUTION The two sp^2 carbons and the atoms that are bonded to each of the sp^2 carbons all lie in the same plane. The other atoms in the molecule will not necessarily lie in the same plane as these six atoms. By putting stars on the six atoms that do lie in the same plane, you will be able to see whether the indicated atoms lie in the same plane. They are in the same plane in part **a,** but they are not necessarily in the same plane in part **b.**

a.

$$\underset{\text{CH}_3}{\overset{\text{H}}{\diagdown}}\text{C}=\text{C}\underset{\text{H}}{\overset{\text{CH}_3}{\diagup}}$$

b.

$$\underset{\text{CH}_3}{\overset{\text{H}}{\diagdown}}\text{C}=\text{C}\underset{\text{H}}{\overset{\text{CH}_2\text{CH}_3}{\diagup}}$$

1.9   HOW A TRIPLE BOND IS FORMED: THE BONDS IN ETHYNE

The two carbons in ethyne (also called acetylene) are held together by three bonds. Three bonds connecting two atoms is called a **triple bond.** Each of the carbons forms four bonds, but each carbon is bonded to only two atoms—a hydrogen and another carbon:

$$\text{H}-\text{C}\equiv\text{C}-\text{H}$$

ethyne
acetylene

Oxyacetylene torches are used to weld and cut metals. The torch uses acetylene and mixes it with oxygen to increase the temperature of the flame. An acetylene/oxygen flame burns at ~3,500 °C.

To bond to two atoms, each carbon hybridizes two atomic orbitals—an s and a p. Two degenerate sp orbitals result.

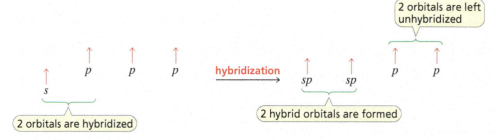

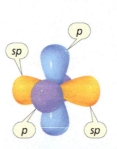

▲ **Figure 1.13**
The two *sp* orbitals point in opposite directions. The two unhybridized *p* orbitals are perpendicular to each other and to the *sp* orbitals. (The smaller lobes of the *sp* orbitals are not shown.)

Each carbon atom in ethyne, therefore, has two sp orbitals and two unhybridized p orbitals. To minimize electron repulsion, the two sp orbitals point in opposite directions. The two unhybridized p orbitals are perpendicular to each other and are perpendicular to the sp orbitals (Figure 1.13).

One of the sp orbitals of one carbon in ethyne overlaps an sp orbital of the other carbon to form a carbon–carbon σ bond. The other sp orbital of each carbon overlaps the s orbital of a hydrogen to form a C—H σ bond (Figure 1.14a). Because the two sp orbitals point in opposite directions, the bond angles are 180°.

Each of the unhybridized p orbitals engages in side-to-side overlap with a parallel p orbital on the other carbon, resulting in the formation of two π bonds (Figure 1.14b).

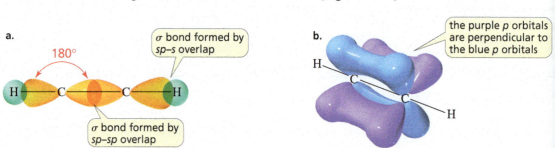

▲ **Figure 1.14**
(a) The C—C σ bond in ethyne is formed by *sp–sp* overlap, and the C—H bonds are formed by *sp–s* overlap. The carbon atoms and the atoms bonded to them form a straight line.
(b) The two carbon–carbon π bonds are formed by side-to-side overlap of the two *p* orbitals of one carbon with the two *p* orbitals of the other carbon.

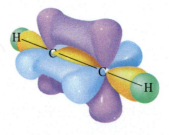

▲ **Figure 1.15**
The triple bond has an electron-dense region above and below and in front of and in back of the internuclear axis of the molecule.

Thus, a triple bond consists of one σ bond and two π bonds. Because the two unhybridized p orbitals on each carbon are perpendicular to each other, they create regions of high electron density above and below *and* in front of and in back of the internuclear axis of the molecule (Figure 1.15).

The overall result can be seen in the potential map for ethyne—the negative charge accumulates in a cylinder that wraps around the egg-shaped molecule.

Representations of Ethyne

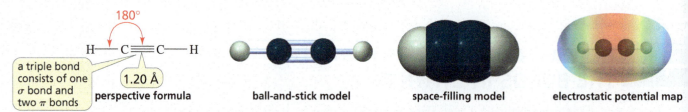

a triple bond consists of one σ bond and two π bonds

1.20 Å

perspective formula   ball-and-stick model   space-filling model   electrostatic potential map

A triple bond consists of one σ bond and two π bonds.

The two carbon atoms in a triple bond are held together by six electrons, so a triple bond is stronger (231 kcal/mol) and shorter (1.20 Å) than a double bond (174 kcal/mol and 1.33 Å).

PROBLEM 30

Put a number in each of the blanks:

a. ___ s orbital and ___ p orbitals form ___ sp^3 orbitals.

b. ___ s orbital and ___ p orbitals form ___ sp^2 orbitals.

c. ___ s orbital and ___ p orbitals form ___ sp orbitals.

PROBLEM 31 SOLVED

a. Draw the Lewis structure of H_2CO.

b. Describe the orbitals used by the carbon atom in bonding and indicate the approximate bond angles.

LEARN THE STRATEGY

SOLUTION TO 31 a. Our first attempt at a Lewis structure (drawing the atoms with the hydrogens on the outside of the molecule) shows that carbon is the only atom that does not form the needed number of bonds.

$$H-C-O-H$$

If we place a double bond between carbon and oxygen and move the H from O to C (which still keeps it on the outside of the molecule), then all the atoms end up with the correct number of bonds. Lone-pair electrons are used to give oxygen a filled outer shell. When we check to see if any atom needs to be assigned a formal charge, we find that none of them does.

$$\overset{:O:}{\underset{H-C-H}{\|}}$$

SOLUTION TO 31 b. Because carbon forms a double bond, we know that it uses sp^2 orbitals (as it does in ethene) to bond to the two hydrogens and the oxygen. It uses its "leftover" p orbital to form the second bond to oxygen. Because carbon is sp^2 hybridized, the bond angles are approximately 120°.

PROBLEM 32

For each of the given species:

a. Draw its Lewis structure.

b. Describe the orbitals used by each carbon atom in bonding and indicate the approximate bond angles.

   1. H_2CO_2 **2.** HCN   **3.** CCl_4 **4.** H_2CO_3

USE THE STRATEGY

1.10 THE BONDS IN THE METHYL CATION, THE METHYL RADICAL, AND THE METHYL ANION

Not all carbon atoms form four bonds. A carbon with a positive charge, a negative charge, or an unpaired electron forms only three bonds. Now we will see the orbitals that carbon uses when it forms three bonds.

The Methyl Cation ($^+CH_3$)

The positively charged carbon in the methyl cation is bonded to three atoms, so it hybridizes three orbitals—an s orbital and two p orbitals. Therefore, it forms its three covalent bonds using sp^2 orbitals. Its unhybridized p orbital remains empty. The positively charged carbon (and the three atoms bonded to it) lie in a plane. The unhybridized p orbital stands perpendicular to the plane.

The carbon in $^+CH_3$ is sp^2 hybridized.

Representations of Methyl Cation

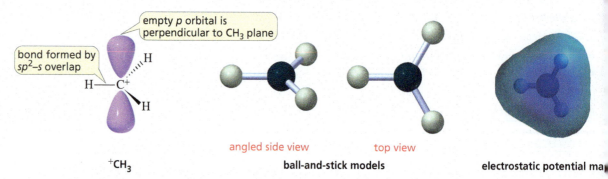

angled side view top view

$^+CH_3$ ball-and-stick models electrostatic potential map

The Methyl Radical (·CH₃)

The carbon in ·CH₃ is sp² hybridized.

The carbon atom in the methyl radical is also sp^2 hybridized. The methyl radical, though, has one more electron than the methyl cation. That electron is unpaired, and it resides in the p orbital, with half of the electron density in each lobe. Although the methyl cation and the methyl radical have similar ball-and-stick models, the potential maps are quite different because of the additional electron in the methyl radical.

Representations of the Methyl Radical

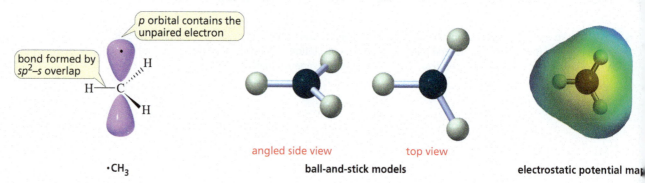

angled side view top view

·CH₃   ball-and-stick models electrostatic potential map

The Methyl Anion (⁻CH₃)

The carbon in ⁻CH₃ is sp³ hybridized.

The negatively charged carbon in the methyl anion has three pairs of bonding electrons and one lone pair. Four pairs of electrons are farthest apart when the four orbitals containing the bonding and lone-pair electrons point toward the corners of a tetrahedron. Thus, the negatively charged carbon in the methyl anion is sp^3 hybridized. In the methyl anion, three of carbon's sp^3 orbitals each overlap the s orbital of a hydrogen, and the fourth sp^3 orbital holds the lone pair.

Representations of the Methyl Anion

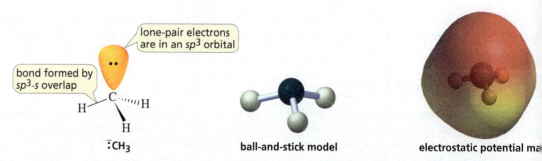

⁻CH₃   ball-and-stick model electrostatic potential map

Take a moment to compare the potential maps for the methyl cation, the methyl radical, and the methyl anion.

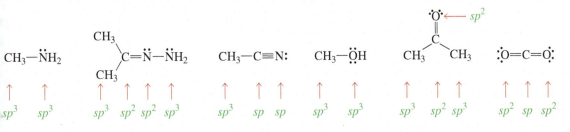

| one σ bond | one σ bond / one π bond | one σ bond / two π bonds |

Bond order describes the number of covalent bonds shared by two atoms. A single bond has a bond order of one, a double bond has a bond order of two, and a triple bond has a bond order of three.

Hybridization

The easiest way to determine the hybridization of carbon, nitrogen, or oxygen is to count the number of π bonds it forms:

The hybridization of a C, N, or O is $sp^{(3 \text{ minus the number of } \pi \text{ bonds})}$.

- If it forms no π bonds, it is sp^3 hybridized.
- If it forms one π bond, it is sp^2 hybridized.
- If it forms two π bonds, it is sp hybridized.

The exceptions are carbocations and carbon radicals, which are sp^2 hybridized—not because they form a π bond, but because they have an empty or a half-filled p orbital (Section 1.10).

$CH_3-\ddot{N}H_2$ $(CH_3)_2 C=\ddot{N}-\ddot{N}H_2$ $CH_3-C{\equiv}N{:}$ $CH_3-\ddot{O}H$ $(CH_3)_2C{=}\ddot{O} \leftarrow sp^2$ ${:}\ddot{O}{=}C{=}\ddot{O}{:}$

sp^3 sp^3 sp^3 sp^2 sp^2 sp^3 sp^3 sp sp sp^3 sp^3 sp^3 sp^2 sp^2 sp^2 sp sp^2

PROBLEM 40 SOLVED

In what orbitals are the lone pairs in each of the following molecules?

a. $CH_3\ddot{O}H$ **b.** (structure: C with =Ö above, CH₃ and CH₃ below)   **c.** $CH_3C{\equiv}N{:}$

SOLUTION

a. Oxygen forms only single bonds in this compound, so it is sp^3 hybridized. It uses two of its four sp^3 orbitals to form σ bonds (one to C and one to H) and the other two for its lone pairs.

a. $CH_3\ddot{O}H$ (sp^3)   **b.** ($sp^2 \rightarrow \ddot{O} \leftarrow sp^2$)   **c.** $CH_3C{\equiv}N{:}$ (sp)

b. Oxygen forms one π bond in this compound, so it is sp^2 hybridized. It uses one of its three sp^2 orbitals to form the σ bond to C and the other two for its lone pairs.

c. Nitrogen forms two π bonds in this compound, so it is sp hybridized. It uses one of the sp orbitals to form the σ bond to C and the other one for its lone pair.

Bond Length and Bond Strength

The shorter the bond, the stronger it is. In comparing the lengths and strengths of carbon–carbon single, double, and triple bonds, we see that the carbon–carbon bond gets shorter and stronger as the number of bonds holding the two carbon atoms together increases (Table 1.7). As a result, triple bonds are shorter and stronger than double bonds, which are shorter and stronger than single bonds.

The greater the electron density in the region of overlap, the stronger the bond.

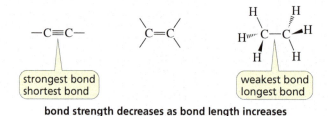

strongest bond
shortest bond

weakest bond
longest bond

bond strength decreases as bond length increases

The shorter the bond, the stronger it is.

Table 1.7   Comparison of the Bond Angles and the Lengths and Strengths of the Carbon–Carbon and Carbon–Hydrogen Bonds in Ethane, Ethene, and Ethyne

Molecule	Hybridization of carbon	Bond angles	Length of C—C bond (Å)	Strength of C—C bond (kcal/mol)	Length of C—H bond (Å)	Strength of C—H bond (kcal/mol)
ethane	sp^3	109.5°	1.54	90.2	1.10	101.1
ethene	sp^2	120°	1.33	174.5	1.08	110.7
ethyne	sp	180°	1.20	230.4	1.06	133.3

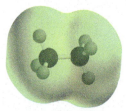

Ethane

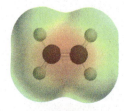

Ethene

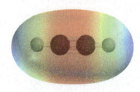

Ethyne

A C—H σ bond is shorter than a C—C σ bond (Table 1.7) because the *s* orbital of hydrogen is closer to the nucleus than is the sp^3 orbital of carbon. Consequently, the nuclei are closer together in a bond formed by sp^3–*s* overlap than they are in a bond formed by sp^3–sp^3 overlap. In addition to being shorter, a C—H σ bond is stronger than a C—C σ bond because there is greater electron density in the region of overlap of an sp^3 orbital with an *s* orbital than in the region of overlap of an sp^3 orbital with an sp^3 orbital.

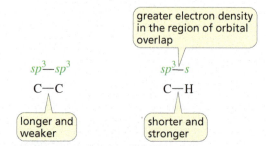

greater electron density in the region of orbital overlap

sp^3—sp^3
C—C

sp^3—s
C—H

longer and weaker

shorter and stronger

the greater the electron density in the region of orbital overlap, the stronger and shorter the bond

Hybridization Affects Bond Length and Bond Strength

The length and strength of a C—H bond depend on the hybridization of the carbon. The more *s* character in the orbital used by carbon to form the bond, the shorter and stronger the bond— *again because an s orbital is closer to the nucleus than is a p orbital.* Thus, a C—H bond formed by an *sp* carbon (50% *s*) is shorter and stronger than a C—H bond formed by an sp^2 carbon (33.3% *s*), which, in turn, is shorter and stronger than a C—H bond formed by an sp^3 carbon (25% *s*).

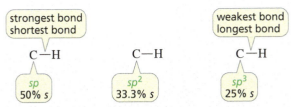

bond strength increases as bond length decreases

A double bond (a σ bond plus a π bond) is stronger (174 kcal/mol) than a single bond (a σ bond; 90 kcal/mol), but it is not twice as strong. Therefore, we can conclude that the π bond of a double bond is weaker than the σ bond. The π bond is weaker because the side-to-side overlap that forms a π bond is less effective for bonding than is the end-on overlap that forms a σ bond (Section 1.6).

The strength of a C—C σ bond given in Table 1.7 (90 kcal/mol) is for a bond formed by sp^3–sp^3 overlap. A C—C σ bond formed by sp^2–sp^2 overlap is expected to be stronger, however, because of the greater s character in the overlapping sp^2 orbitals; it has been estimated to be ~112 kcal/mol. We can conclude then that the strength of the π bond of ethene is about 62 kcal/mol (174 − 112 = 62).

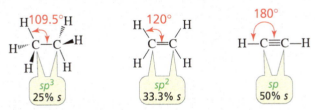

strength of the double bond = 174 kcal/mol
strength of the sp^2—sp^2 σ bond = −112 kcal/mol
strength of the π bond = 62 kcal/mol

a π bond is weaker than a σ bond

Hybridization Affects Bond Angles

We saw that bond angles depend on the orbital used by carbon to form the bond. The greater the amount of s character in the orbital, the larger the bond angle. For example, sp^3 carbons have bond angles of 109.5°, sp^2 carbons have bond angles of 120°, and sp carbons have bond angles of 180°.

The more s character in the orbital, the shorter the bond.

The more s character in the orbital, the stronger the bond.

The more s character in the orbital, the larger the bond angle.

A π bond is weaker than a σ bond.

bond angle increases as s character in the orbital increases

PROBLEM 41 ◆

Which of the bonds in a carbon–oxygen double bond has more effective orbital–orbital overlap: the σ bond or the π bond?

PROBLEM 42 ◆

Would you expect a C—C σ bond formed by sp^2–sp^2 overlap to be stronger or weaker than a C—C σ bond formed by sp^3–sp^3 overlap?

PROBLEM 43

Caffeine is a natural insecticide found in the seeds and leaves of certain plants, where it kills insects that feed on the plant. Caffeine is extracted for human consumption from beans of the coffee plant, from Kola nuts, and from the leaves of tea plants. Because it stimulates the central nervous system, it temporarily prevents drowsiness. Add caffeine's missing lone pairs to its structure.

caffeine

coffee beans

PROBLEM 44

a. What is the hybridization of each of the carbon atoms in the following compound?

$$CH_3CHCH=CHCH_2C\equiv CCH_3$$
$$|$$
$$CH_3$$

b. What is the hybridization of each of the C, N, and O atoms in the following compounds:

PROBLEM-SOLVING STRATEGY

LEARN THE STRATEGY

Predicting Bond Angles

Predict the approximate bond angle of the C—N—H bond in $(CH_3)_2NH$.

First, we need to determine the hybridization of the central atom (the N). Because the nitrogen atom forms only single bonds, we know that it is sp^3 hybridized. Next, we look to see if there are lone pairs that affect the bond angle. An uncharged nitrogen has one lone pair. Based on these observations, we can predict that the C—N—H bond angle is about 107.3°. This is the same as the H—N—H bond angle in NH_3, which is another compound with an sp^3 nitrogen and one lone pair.

USE THE STRATEGY

PROBLEM 45 ◆

Predict the approximate bond angles for

a. the C—N—C bond angle in $(CH_3)_2\overset{+}{N}H_2$.

b. the C—N—H bond angle in $CH_3CH_2NH_2$.

c. the H—C—N bond angle in $(CH_3)_2NH$.

d. the H—C—O bond angle in CH_3OCH_3.

1.16 DIPOLE MOMENTS OF MOLECULES

In Section 1.3, we saw that if a molecule has one covalent bond, then the dipole moment of the molecule is identical to the dipole moment of the bond. When molecules have more than one covalent bond, the geometry of the molecule must be taken into account because both the *magnitude* and the *direction* of the individual bond dipole moments (the vector sum) determine the overall dipole moment of the molecule.

The dipole moment depends on
the magnitude and direction of the individual bond dipoles.

Because the direction of the bond dipoles must be taken into account, symmetrical molecules have no dipole moment. In carbon dioxide (CO_2), for example, the carbon is bonded to two oxygens; so it uses sp orbitals to form the two C—O σ bonds. The remaining two p orbitals on the carbon form the two C—O π bonds. The sp orbitals form a bond angle of 180°, which causes the individual carbon–oxygen bond dipole moments to cancel each other. Carbon dioxide, therefore, has a dipole moment (symbolized by the Greek letter μ) of 0 D.

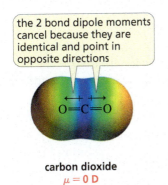

the 2 bond dipole moments cancel because they are identical and point in opposite directions

carbon dioxide
$\mu = 0$ D

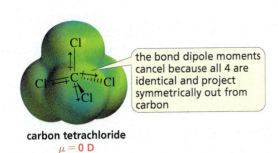

the bond dipole moments cancel because all 4 are identical and project symmetrically out from carbon

carbon tetrachloride
$\mu = 0$ D

Carbon tetrachloride (CCl_4) is another symmetrical molecule. The four atoms bonded to the p^3 carbon atom are identical and project symmetrically out from the carbon atom. Thus, as with CO_2, the symmetry of the molecule causes the bond dipole moments to cancel. Therefore, carbon tetrachloride also has no dipole moment.

The dipole moment of chloromethane (CH_3Cl) is greater (1.87 D) than the dipole moment of its C—Cl bond (1.5 D) because the C—H dipoles are oriented so that they reinforce the dipole of the C—Cl bond. In other words, all the electrons are pulled in the same general direction.

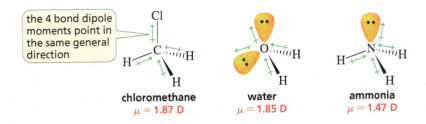

the 4 bond dipole moments point in the same general direction

chloromethane
$\mu = 1.87$ D

water
$\mu = 1.85$ D

ammonia
$\mu = 1.47$ D

The dipole moment of water (1.85 D) is greater than the dipole moment of a single O—H bond (1.5 D) because the dipoles of the two O—H bonds reinforce each other; the lone-pair electrons also contribute to the dipole moment. Similarly, the dipole moment of ammonia (1.47 D) is greater than the dipole moment of a single N—H bond (1.3 D).

PROBLEM 46 ◆

Which of the following molecules would you expect to have a dipole moment of zero? To answer parts **g** and **h**, you may need to review the Problem Solving Strategy on p. 39.

a. CH_3CH_3 **c.** CH_2Cl_2 **e.** $H_2C{=}CH_2$ **g.** $BeCl_2$
b. $H_2C{=}O$ **d.** NH_3 **f.** $H_2C{=}CHBr$ **h.** BF_3

PROBLEM 47

Account for the difference in the shape and color of the potential maps for ammonia and the ammonium ion in Section 1.11.

PROBLEM 48 ◆

If the dipole moment of CH_3F is 1.847 D and the dipole moment of CD_3F is 1.858 D, which is more electronegative: hydrogen or deuterium?

ESSENTIAL CONCEPTS

Section 1.0

- **Organic compounds** are compounds that are based on carbon.

Section 1.1

- The **atomic number** of an atom is the number of protons in its nucleus.
- The **mass number** of an atom is the sum of its protons and neutrons.
- **Isotopes** have the same atomic number but different mass numbers.
- **Atomic mass** is the weighted average of the isotopes in the element.
- **Molecular weight** is the sum of the atomic masses of all the atoms in the molecule.

Section 1.2

- The closer the atomic orbital is to the nucleus, the lower its energy.
- **Degenerate** orbitals have the same energy.
- The **electronic configuration** of an atom describes the atomic orbitals occupied by the atom's electrons.
- Electrons are assigned to orbitals (atomic or molecular) following the **aufbau principle,** the **Pauli exclusion principle,** and **Hund's rule.**
- **Core electrons** are electrons in inner shells. **Valence electrons** are electrons in the outermost shell.

Section 1.3

- An atom is most stable if its outer shell is either filled or contains eight electrons and it has no electrons of higher energy.
- The **octet rule** states that an atom will give up, accept, or share electrons in order to fill its outer shell or attain an outer shell with eight electrons.
- **Electronegative** elements readily acquire electrons.
- **Electrostatic attractions** are attractive forces between opposite charges.
- A **covalent bond** is formed when two atoms share a pair of electrons.
- A **polar covalent bond** is a covalent bond between atoms with different **electronegativities.** A nonpolar covalent bond is a bond between atoms with similar electronegativities.
- The greater the difference in electronegativity between the atoms forming the bond, the closer the bond is to the ionic end of the continuum.
- A polar covalent bond has a **dipole** (a positive end and a negative end), measured by a **dipole moment.**
- The **dipole moment** of a bond is equal to the size of the charge times the distance between the charges.

Section 1.4

- **Lone-pair electrons** are valence electrons that are not used in bonding.
- A **proton** is a positively charged hydrogen ion; a **hydride ion** is a negatively charged hydrogen ion.

- **formal charge** = the number of valence electrons − the number of electrons the atom has to itself (all the lone-pair electrons and one-half of the bonding electrons)
- **Lewis structures** indicate which atoms are bonded together and show **lone pairs** and **formal charges.**
- When the atom is neutral, C forms 4 bonds, N forms 3 bonds, O forms 2 bonds, and H or a halogen forms 1 bond.
- When the atom is neutral, C has no lone pairs, N has 1 lone pair, O has 2 lone pairs, and a halogen has 3 lone pairs.
- A **carbocation** has a positively charged carbon, a **carbanion** has a negatively charged carbon, and a **radical** has an unpaired electron.

Section 1.5

- An **atomic orbital** tells us the volume of space around the nucleus where an electron is most likely to be found.
- There is zero probability of finding an electron at a **node.**

Section 1.6

- According to **molecular orbital (MO) theory,** covalent bonds result when atomic orbitals combine to form **molecular orbitals.**
- **Minimum energy** corresponds to **maximum stability.**
- Two atomic orbitals combine to give a **bonding MO** and a higher energy **antibonding MO.**
- Electrons in a bonding MO assist in bonding. Electrons in an antibonding MO detract from bonding.
- Cylindrically symmetrical bonds are called **sigma** (σ) bonds; side-to-side overlap of parallel p orbitals forms a **pi** (π) **bond.**
- Bond strength is measured by the **bond dissociation energy.**

Section 1.7

- To be able to form four bonds, carbon must promote an electron from an s orbital to an empty p orbital.
- Carbon forms bonds using **hybrid orbitals.**
- Bonding pairs and lone-pair electrons around an atom stay as far apart as possible.
- All **single bonds** in organic compounds are σ bonds.

Section 1.8

- A **double bond** consists of one σ bond and one π bond; the doubly bonded carbon is sp^2 hybridized.

Section 1.9

- A **triple bond** consists of one σ bond and two π bonds; the triply bonded carbon is sp hybridized.

Section 1.10

- Carbocations and carbon radicals form bonds using sp^2 orbitals; carbanions form bonds using sp^3 orbitals.

Sections 1.11–1.13

- Nitrogen, oxygen, and the halogens form bonds using **hybrid orbitals.**

- Lone pairs decrease the bond angle because they are more diffuse than bonding electrons.

Section 1.14

- **Molecular geometry** is **determined by hybridization:** sp^3 is tetrahedral, sp^2 is trigonal planar, and sp is linear.

Section 1.15

- The **hybridization** of C, N, or O depends on the number of π bonds the atom forms: no π bonds = sp^3, one π bond = sp^2, and two π bonds = sp. Exceptions are carbocations and carbon radicals, which are sp^2.

- The greater the electron density in the region of orbital overlap, the shorter and stronger the bond.

- The more s character in the orbital used to form a bond, the shorter and stronger the bond and the larger the bond angle.
- The shorter the bond, the stronger it is.
- Triple bonds are shorter and stronger than double bonds, which are shorter and stronger than single bonds.
- A π bond is weaker than a σ bond.

Section 1.16

- The **dipole moment** of a molecule depends on the magnitude and direction of all the individual bond dipoles.

GLOSSARY

The definitions of the key words used in each chapter can be found at the beginning of each pertinent chapter in the *Study Guide/Solutions Manual*. The definitions of all the key words used in this book also can be found in the Glossary on p. G-1.

PROBLEMS

49. Draw a Lewis structure for each of the following:

   a. N_2H_4 **b.** CO_3^{2-} **c.** N_2H_2 **d.** CO_2 **e.** HOCl

50. a. Which of the following has a polar covalent bond?
   b. Which of the following has a bond closest to the ionic end of the bond spectrum?

   CH_3NH_2 CH_3CH_3 CH_3F CH_3OH

51. What is the hybridization of all the atoms (other than hydrogen) in each of the following? What are the bond angles around each atom?

   a. NH_3 **c.** $^-CH_3$ **e.** $^+NH_4$ **g.** HCN   **i.** H_3O^+
   b. BH_3 **d.** ·CH3   **f.** $^+CH_3$ **h.** $C(CH_3)_4$ **j.** $H_2C{=}O$

52. Draw the condensed structure of a compound that contains only carbon and hydrogen atoms and that has

   a. three sp^3 hybridized carbons.
   b. one sp^3 hybridized carbon and two sp^2 hybridized carbons.
   c. two sp^3 hybridized carbons and two sp hybridized carbons.

53. Predict the approximate bond angles:

   a. the C—N—H bond angle in $(CH_3)_2\overset{+}{N}H_2$ **c.** the C—N—H bond angle in $(CH_3)_2NH$
   b. the C—O—H bond angle in CH_3OH **d.** the C—N—C bond angle in $(CH_3)_2NH$

54. Draw the ground-state electronic configuration for each of the following:

   a. Mg   **b.** Ca^{2+} **c.** Ar   **d.** Mg^{2+}

55. Draw a Lewis structure for each of the following:

   a. CH_3NH_2 **b.** HNO_2 **c.** $NaNH_2$ **d.** NH_2O^-

56. What is the hybridization of each of the carbon and oxygen atoms in the following compound?

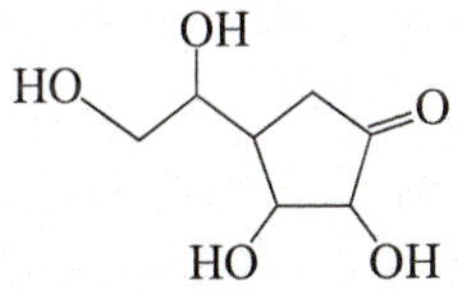

57. Rank the bonds from most polar to least polar.

    a. $C-O, C-F, C-N$ **b.** $C-Cl, C-I, C-Br$ **c.** $H-O, H-N, H-C$ **d.** $C-H, C-C, C-N$

58. Draw a Lewis structure for each of the following:

    a. CH_3CHO **b.** CH_3OCH_3 **c.** CH_3COOH

59. Draw a skeletal structure for each of the compounds in Problem 58.

60. What is the hybridization of the indicated atom in each of the following?

    a. $CH_3CH{=}CH_2$ **b.** $CH_3\overset{O}{\underset{}{C}}CH_3$ **c.** CH_3CH_2OH **d.** $CH_3C{\equiv}N$ **e.** $CH_3CH{=}NCH_3$ **f.** $CH_3OCH_2CH_3$

61. Predict the approximate bond angles for the following:

    a. the $C-N-H$ bond angle in $CH_3CH_2NH_2$ **c.** the $C-C-N$ bond angle in $CH_3C{\equiv}N$
    b. the $F-B-F$ bond angle in $^-BF_4$ **d.** the $C-C-N$ bond angle in $CH_3CH_2NH_2$

62. Show the direction of the dipole moment in each of the following bonds (use the electronegativities given in Table 1.3):

    a. H_3C-Br **b.** H_3C-Li **c.** $HO-NH_2$ **d.** $I-Br$ **e.** H_3C-OH **f.** $(CH_3)_2N-H$

63. Draw the missing lone-pair electrons and assign the missing formal charges for the following:

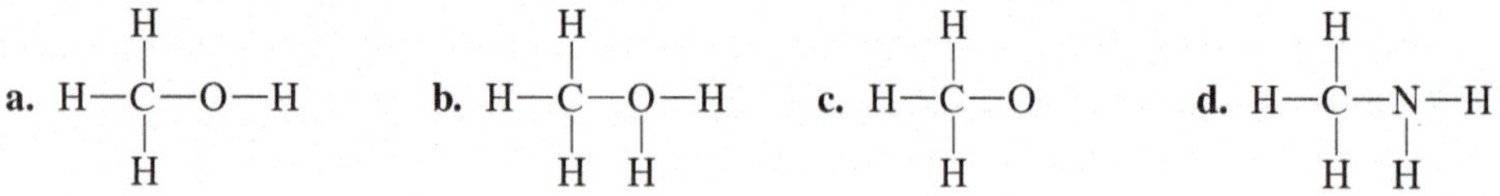

    a. $H-\overset{H}{\underset{H}{C}}-O-H$ **b.** $H-\overset{H}{\underset{H}{C}}-O-H$ **c.** $H-\overset{H}{\underset{H}{C}}-O$ **d.** $H-\overset{H}{\underset{H}{C}}-N-H$

64. a. Which of the indicated bonds in each molecule is shorter?
    b. Indicate the hybridization of the C, O, and N atoms in each of the molecules.

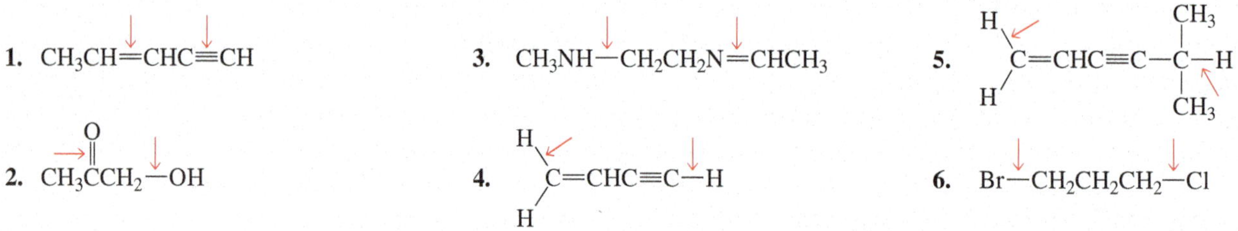

    1. $CH_3CH{=}CHC{\equiv}CH$ **3.** $CH_3NH-CH_2CH_2N{=}CHCH_3$ **5.** $\overset{H}{\underset{H}{}}C{=}CHC{\equiv}C-\overset{CH_3}{\underset{CH_3}{C}}-H$

    2. $CH_3\overset{O}{\underset{}{C}}CH_2-OH$ **4.** $\overset{H}{\underset{H}{}}C{=}CHC{\equiv}C-H$ **6.** $Br-CH_2CH_2CH_2-Cl$

65. For each of the following molecules, indicate the hybridization of each carbon and give the approximate values of all the bond angles:

    a. $CH_3C{\equiv}CH$ **b.** $CH_3CH{=}CH_2$ **c.** $CH_3CH_2CH_3$ **d.** $CH_2{=}CH-CH{=}CH_2$

66. Draw a Lewis structure for each of the following:

    a. $(CH_3)_3COH$ **b.** $CH_3CH(OH)CH_2CN$ **c.** $(CH_3)_2CHCH(CH_3)CH_2C(CH_3)_3$

67. Draw a skeletal structure for each of the compounds in Problem 66.

68. Rank the following compounds from highest dipole moment to lowest dipole moment:

69. In which orbitals are the lone pairs in nicotine?

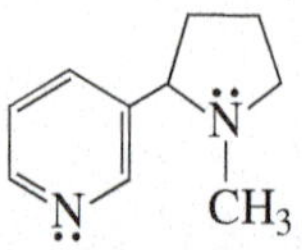

70. Indicate the formal charge on each carbon that has one. All lone pairs are shown.

$$H-\overset{\overset{\displaystyle H}{|}}{\underset{\underset{\displaystyle H}{|}}{C}}: \qquad H-\overset{\overset{\displaystyle H}{|}}{\underset{\underset{\displaystyle H}{|}}{C} \qquad H-\overset{\overset{\displaystyle H}{|}}{\underset{\underset{\displaystyle H}{|}}{C}\cdot \qquad H-\ddot{C}-H \qquad CH_2=\ddot{C}H \qquad CH_2=CH \qquad CH_2=\dot{C}H$$

71. Do the sp^2 carbons and the indicated sp^3 carbons lie in the same plane?

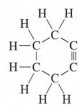

72. a. Which of the species have bond angles of 109.5°?
  b. Which of the species have bond angles of 120°?

$H_2O \qquad H_3O^+ \qquad {}^+CH_3 \qquad BF_3 \qquad NH_3 \qquad {}^+NH_4 \qquad {}^-CH_3$

73. Which compound has a longer C—Cl bond?

$$CH_3CH_2Cl$$
at one time, it was used as a refrigerant, an anesthetic, and a propellant for aerosol sprays

$$CH_2=CHCl$$
used as the starting material for the synthesis of a plastic that is used to make bottles, flooring, and clear packaging for food

74. Which compound has a larger dipole moment: CH_3Cl or CH_2Cl_2?

75. Explain why the following compound is not stable:

$$\begin{array}{c}
\text{H}\quad\text{H} \\
\text{H}\quad\overset{\displaystyle |}{\underset{\displaystyle |}{\text{C}}} \\
\text{H}-\text{C}\qquad\text{C} \\
\text{H}-\text{C}\qquad\text{C} \\
\overset{\displaystyle |}{\text{H}}\quad\text{C} \\
\text{H}\quad\text{H}
\end{array}$$

76. Explain why CH_3Cl has a greater dipole moment than CH_3F even though F is more electronegative than Cl.

77. a. Draw a Lewis structure for each of the following:

  1. $CH_3N_2^+$ **2.** HNO_3 **3.** N_3^- **4.** CH_3CONH_2 **5.** O_3

  b. Draw a structure for each of the species that shows approximate bond angles.
  c. Which species have no dipole moment?

78. There are three isomers with molecular formula $C_2H_2Cl_2$. Draw their structures. Which one does not have a dipole moment?

2 Acids and Bases: Central to Understanding Organic Chemistry

Decades of acid rain have devastated the Norway spruce trees near Hora Svatého Šebestiána in the Czech Republic.

The chemistry in this chapter explains such things as the cause of acid rain and why it destroys monuments and plants, why exercise increases the rate of breathing, how Fosamax prevents bones from being nibbled away, and why blood has to be buffered and how that buffering is accomplished. Acids and bases play an important role in organic chemistry. What you learn about them in this chapter will reappear in almost every other chapter in the book in one form or another. The importance of organic acids and bases will become particularly clear when you learn how and why organic compounds react.

It is hard to believe now, but at one time, chemists characterized compounds by tasting them. Early chemists called any compound that tasted sour an acid (from *acidus*, Latin for "sour"). Some familiar acids are citric acid (found in lemons and other citrus fruits), acetic acid (found in vinegar), and hydrochloric acid (found in stomach acid—the sour taste associated with vomiting).

Compounds that neutralize acids, thereby destroying their acidic properties, were called bases, or alkaline compounds. Glass cleaners and solutions designed to unclog drains are familiar alkaline solutions.

2.1 AN INTRODUCTION TO ACIDS AND BASES

We will look at two definitions for the terms *acid* and *base,* the Brønsted–Lowry definitions and the Lewis definitions. We will begin with the Brønsted–Lowry definitions.

According to Brønsted and Lowry, an **acid** is a species that loses a proton, and a **base** is a species that gains a proton. (Remember that positively charged hydrogen ions are called protons.) For example, in the reaction shown below, hydrogen chloride (HCl) is an acid because it loses a proton, and water is a base because it gains a proton. The reaction of an acid with a base is called an **acid–base reaction** or a **proton transfer reaction.** Notice that the reverse of an acid–base reaction is also an acid–base reaction. In the reverse reaction, H_3O^+ is an acid because it loses a proton, and Cl^- is a base because it gains a proton.

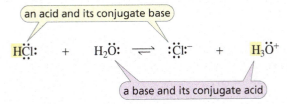

Water can accept a proton because it has two lone pairs, either of which can form a covalent bond with the proton, and Cl^- can accept a proton because any one of its four lone pairs can form a covalent bond with a proton. Thus, according to the Brønsted–Lowry definitions:

> *Any species that has a hydrogen can potentially act as an acid.*
> *Any species that has a lone pair can potentially act as a base.*

Both an acid and a base must be present in an acid–base reaction, because an acid cannot lose a proton unless a base is present to accept it. Most *acid–base reactions are* **reversible.** Two half-headed arrows are used to designate reversible reactions.

a reversible reaction:
A and B form C and D.
C and D form A and B.

$$A + B \rightleftharpoons C + D$$

an irreversible reaction:
A and B form C and D.
C and D *do not* form A and B.

$$A + B \longrightarrow C + D$$

Most acid–base reactions are reversible.

When an acid loses a proton, the resulting species without the proton is called the **conjugate base** of the acid. Thus, Cl^- is the conjugate base of HCl, and H_2O is the conjugate base of H_3O^+. When a base gains a proton, the resulting species with the proton is called the **conjugate acid** of the base. Thus, HCl is the conjugate acid of Cl^-, and H_3O^+ is the conjugate acid of H_2O.

an acid and its conjugate base

$$HCl: \;\; + \;\; H_2O: \;\; \rightleftharpoons \;\; :Cl:^- \;\; + \;\; H_3O^+$$

a base and its conjugate acid

A conjugate base is formed by removing a proton from an acid.

A conjugate acid is formed by adding a proton to a base.

Another example of an acid–base reaction is the reaction between ammonia and water: ammonia (NH_3) is a base because it gains a proton, and water is an acid because it loses a proton. In the reverse reaction, ammonium ion $(^+NH_4)$ is an acid because it loses a proton, and hydroxide ion (HO^-) is a base because it gains a proton. (In Section 2.5, you will learn why the arrows in these acid–base reactions are not the same length.)

a base and its conjugate acid

$$NH_3 \;\; + \;\; H_2O: \;\; \rightleftharpoons \;\; ^+NH_4 \;\; + \;\; HO:^-$$

an acid and its conjugate base

- $^+NH_4$ is conjugate acid of NH_3.
- HO^- is conjugate base of H_2O.
- H_2O is conjugate acid of HO^-.
- NH_3 is conjugate base of $^+NH_4$.

Notice in the first of these two reactions that water is a base and in the second that water is an acid. Water can behave as a base because it has a lone pair, and it can behave as an acid because it

A strong base has a high affinity for a proton.

A weak base has a low affinity for a proton.

has a proton that it can lose. In Section 2.4, we will see how we can predict that water is a base in the first reaction and is an acid in the second reaction.

Acidity is a measure of the tendency of a compound to lose a proton, whereas **basicity** is a measure of a compound's affinity for a proton. A strong acid has a strong tendency to lose a proton. This means that its conjugate base must be weak because it has little affinity for the proton. A weak acid has little tendency to lose its proton, indicating that its conjugate base is strong because it has a high affinity for the proton. Thus, the following important relationship exists between an acid and its conjugate base:

The stronger the acid, the weaker its conjugate base.

For example, HBr is a stronger acid than HCl, so Br^- is a weaker base than Cl^-.

PROBLEM 1 ◆

Which of the following are *not* acids?

CH_3COOH　　CO_2　　HNO_2　　$HCOOH$　　CCl_4

PROBLEM 2 ◆

Consider the following reaction:

$$HBr \ + \ {}^-C\equiv N \ \rightleftharpoons \ Br^- \ + \ HC\equiv N$$

a. What is the acid on the left side of the equation?　**e.** What is the acid on the right side of the equation?
b. What is the base on the left side of the equation?　**f.** What is the base on the right side of the equation?
c. What is the conjugate base of the acid on the left?　**g.** What is the conjugate base of the acid on the right?
d. What is the conjugate acid of the base on the left?　**h.** What is the conjugate acid of the base on the right?

PROBLEM 3 ◆

Draw the products of the acid–base reaction when

a. HCl is the acid and NH_3 is the base.　　　**b.** H_2O is the acid and $^-NH_2$ is the base.

PROBLEM 4 ◆

a. What is the conjugate acid of each of the following?
　1. NH_3　　　**2.** Cl^-　　　**3.** HO^-　　　**4.** H_2O

b. What is the conjugate base of each of the following?
　1. NH_3　　　**2.** HBr　　　**3.** HNO_3　　　**4.** H_2O

2.2　pK_a AND pH

When a strong acid such as hydrogen chloride is dissolved in water, almost all the molecules dissociate (break into ions), which means that the *products* are favored at equilibrium—the equilibrium lies to the right. When a much weaker acid, such as acetic acid, is dissolved in water, very few molecules dissociate, so the *reactants* are favored at equilibrium—the equilibrium lies to the left. A longer arrow is drawn toward the species favored at equilibrium.

$$H\ddot{C}l: \ + \ H_2\ddot{O}: \ \rightleftharpoons \ H_3\ddot{O}^+ \ + \ :\ddot{\underset{..}{C}l}:$$

hydrogen chloride

$$\underset{CH_3}{\overset{\displaystyle :O:\atop \displaystyle \|\;\;}{C}}\!\!-\!\ddot{O}H \ + \ H_2\ddot{O}: \ \rightleftharpoons \ H_3\ddot{O}^+ \ + \ \underset{CH_3}{\overset{\displaystyle :O:\atop \displaystyle \|\;\;}{C}}\!\!-\!\ddot{O}:$$

acetic acid

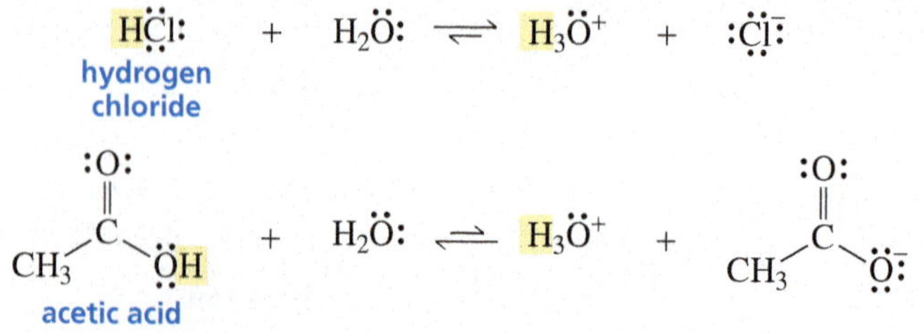

Defining K_{eq}

The degree to which an acid (HA) dissociates in an aqueous solution is indicated by the **equilibrium constant** of the reaction, K_{eq}. Brackets are used to indicate the concentrations of the reactants and products (in moles/liter).

$$HA + H_2O \rightleftharpoons H_3O^+ + A^-$$

$$K_{eq} = \frac{[H_3O^+][A^-]}{[HA][H_2O]}$$

Defining K_a

The degree to which an acid (HA) dissociates is normally determined in a dilute solution, so the concentration of water remains essentially constant. Combining the two constants (K_{eq} and H_2O) allows the equilibrium expression to be rewritten using a new equilibrium constant, K_a, called the **acid dissociation constant.**

$$K_a = \frac{[H_3O^+][A^-]}{[HA]} = K_{eq}[H_2O]$$

Thus, the acid dissociation constant is the equilibrium constant multiplied by the molar concentration of water (55.5 M).

Defining pK_a

The larger the acid dissociation constant, the stronger the acid—that is, the greater its tendency to lose a proton. Hydrogen chloride, with an acid dissociation constant of 10^7, is a stronger acid than acetic acid, with an acid dissociation constant of 1.74×10^{-5}. For convenience, the strength of an acid is generally indicated by its **pK_a** value rather than its K_a value, where

The stronger the acid, the more readily it loses a proton.

$$pK_a = -\log K_a$$

The pK_a of hydrogen chloride is -7, and the pK_a of acetic acid, a much weaker acid, is 4.76. Notice that the stronger the acid, the smaller its pK_a value.

The stronger the acid, the smaller its pK_a value.

very strong acids	$pK_a < 1$
moderately strong acids	$pK_a = 1-3$
weak acids	$pK_a = 3-5$
very weak acids	$pK_a = 5-15$
extremely weak acids	$pK_a > 15$

Unless otherwise stated, the pK_a values given in this text indicate the strength of the acid *in water*. Later (in Section 9.14), you will see how the pK_a value of an acid is affected when the solvent is changed.

Defining pH

The concentration of protons in a solution is indicated by **pH**. This concentration is written as either $[H^+]$ or, because a proton in water is solvated, as $[H_3O^+]$.

$$pH = -\log[H^+]$$

The pH values of some commonly encountered solutions are shown in the margin. Because pH values decrease as the acidity of the solution increases, we see that lemon juice is more acidic than coffee and that rain is more acidic than milk. Solutions with pH values less than 7 are acidic, whereas those with pH values greater than 7 are basic. The pH of a solution can be changed simply by adding acid or base to the solution.

Do not confuse pH and pK_a. The pH scale is used to describe the acidity of a *solution*, whereas the pK_a indicates the tendency of a compound to lose its proton. Thus, the pK_a is characteristic of a particular compound, much like a melting point or a boiling point.

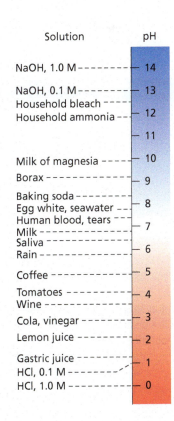

Solution	pH
NaOH, 1.0 M	14
NaOH, 0.1 M	13
Household bleach	
Household ammonia	12
	11
Milk of magnesia	10
Borax	9
Baking soda	
Egg white, seawater	8
Human blood, tears	
Milk	7
Saliva	
Rain	6
Coffee	5
Tomatoes	4
Wine	
Cola, vinegar	3
Lemon juice	2
Gastric juice	1
HCl, 0.1 M	
HCl, 1.0 M	0

PROBLEM 5 ♦

a. Which is a stronger acid: one with a pK_a of 5.2 or one with a pK_a of 5.8?
b. Which is a stronger acid: one with an acid dissociation constant of 3.4×10^{-3} or one with an acid dissociation constant of 2.1×10^{-4}?

PROBLEM 6 ♦

An acid has a K_a of 4.53×10^{-6} in water. What is its K_{eq} for reaction with water in a dilute solution? ($[H_2O] = 55.5$ M)

PROBLEM-SOLVING STRATEGY

LEARN THE STRATEGY

Determining K_a from pK_a

Vitamin C has a pK_a value of 4.17. What is its K_a value?

You will need a calculator to answer this question. Remember that $pK_a = -\log K_a$. Press the key labeled 10^x; then enter the negative value of the pK_a and press =. You should find that vitamin C has a K_a value of 6.8×10^{-5}.

USE THE STRATEGY

PROBLEM 7

Butyric acid, the compound responsible for the unpleasant odor and taste of sour milk, has a pK_a value of 4.82. What is its K_a value? Is it a stronger acid or a weaker acid than vitamin C?

PROBLEM 8

Antacids are compounds that neutralize stomach acid. Write the equations that show how Milk of Magnesia, Alka-Seltzer, and Tums remove excess acid.

a. Milk of Magnesia: $Mg(OH)_2$ **b.** Alka-Seltzer: $KHCO_3$ and $NaHCO_3$ **c.** Tums: $CaCO_3$

PROBLEM 9 ♦

Are the following body fluids acidic or basic?

a. bile (pH = 8.4)   **b.** urine (pH = 5.9)   **c.** spinal fluid (pH = 7.4)

Acid Rain

Rain is mildly acidic (pH = 5.5) because water reacts with the CO_2 in the air to form carbonic acid (a weak acid with a pK_a value of 6.4).

$$CO_2 + H_2O \rightleftharpoons H_2CO_3$$
carbonic acid

In some parts of the world, rain has been found to be much more acidic (pH values as low as 4.3). This so-called acid rain is formed where sulfur dioxide and nitrogen oxides are produced, because water reacts with these gases to form strong acids—sulfuric acid ($pK_a = -5.0$) and nitric acid ($pK_a = -1.3$). Burning fossil fuels for the generation of electric power is the factor most responsible for forming these acid-producing gases.

Acid rain has many deleterious effects. It can destroy aquatic life in lakes and streams; it can make soil so acidic that crops cannot grow and forests are destroyed (see p. 50); and it can cause the deterioration of paint and building materials, including monuments and statues that are part of our cultural heritage. Marble—a form of calcium carbonate—decays because protons react with CO_3^{2-} in the marble to form carbonic acid, which decomposes to CO_2 and H_2O (the reverse of the reaction shown above on the left).

photo taken in 1935

photo taken in 1994

Statue of George Washington in Washington Square Park in Greenwich Village, New York.

$$CO_3^{2-} \underset{}{\overset{H^+}{\rightleftharpoons}} HCO_3^{-} \underset{}{\overset{H^+}{\rightleftharpoons}} H_2CO_3 \rightleftharpoons CO_2 + H_2O$$

2.3 ORGANIC ACIDS AND BASES

Carboxylic Acids

The most common organic acids are carboxylic acids—compounds that have a COOH group. Acetic acid and formic acid are examples of carboxylic acids. Carboxylic acids have pK_a values ranging from about 3 to 5, so they are weak acids. The pK_a values of a wide variety of organic compounds are listed in Appendix I.

$$CH_3-\overset{\overset{\displaystyle O}{\|}}{C}-OH \qquad H-\overset{\overset{\displaystyle O}{\|}}{C}-OH$$

acetic acid   formic acid
$pK_a = 4.76$ $pK_a = 3.75$

Alcohols

Alcohols—compounds that have an OH group—are much weaker acids than carboxylic acids, with pK_a values close to 16. Methyl alcohol and ethyl alcohol are examples of alcohols. We will see why carboxylic acids are stronger acids than alcohols in Section 2.8.

$$CH_3OH \qquad CH_3CH_2OH$$

methyl alcohol   ethyl alcohol
$pK_a = 15.5$ $pK_a = 15.9$

Amines

Amines are compounds that result from replacing one or more of the hydrogens bonded to ammonia with a carbon-containing substituent. Amines and ammonia have such high pK_a values that they rarely behave as acids—they are more likely to act as bases. In fact, they are the most common organic bases. We will see why alcohols are stronger acids than amines in Section 2.6.

$$CH_3NH_2 \qquad NH_3$$

methylamine   ammonia
$pK_a = 40$ $pK_a = 36$

Protonated Compounds

We can assess the strength of a base by considering the strength of its conjugate acid—remembering that *the stronger the acid, the weaker its conjugate base*. For example, based on their pK_a values, protonated methylamine (10.7) is a stronger acid than protonated ethylamine (11.0), which means that methylamine is a weaker base than ethylamine. (A protonated compound is a compound that has gained an additional proton.) Notice that the pK_a values of protonated amines are about 11.

$$CH_3\overset{+}{N}H_3 \qquad CH_3CH_2\overset{+}{N}H_3$$

protonated methylamine   protonated ethylamine
$pK_a = 10.7$ $pK_a = 11.0$

Protonated alcohols and protonated carboxylic acids are very strong acids, with pK_a values < 0.

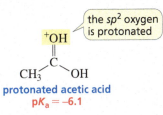

the sp^2 oxygen is protonated

$$CH_3\overset{+}{\underset{H}{O}}H \qquad CH_3CH_2\overset{+}{\underset{H}{O}}H \qquad CH_3-\overset{\overset{\displaystyle +OH}{\|}}{C}-OH$$

protonated methyl alcohol   protonated ethyl alcohol   protonated acetic acid
$pK_a = -2.5$ $pK_a = -2.4$ $pK_a = -6.1$

Notice that it is the doubly bonded oxygen of the carboxylic acid that is protonated (meaning that it acquires the proton). You will see why this is so when you read the Problem-Solving Strategy on p. 68.

Poisonous Amines

Exposure to poisonous plants is responsible for an average of 63,000 calls each year to poison control centers. Hemlock is an example of a plant known for its toxicity. It contains eight different poisonous amines—the most abundant is coniine, a neurotoxin that disrupts the central nervous system. Ingesting even a small amount can be fatal because it causes respiratory paralysis, which results in oxygen deprivation to the brain and heart. A poisoned person can recover if artificial respiration is applied until the drug is flushed from the system. A drink made of hemlock was used to put Socrates to death in 399 B.C.; he was condemned for failing to acknowledge the gods that the natives of the city of Athens worshipped.

hemlock

coniine

Alcohols, Carboxylic Acids, and Amines are Acids and Bases

We saw in Section 2.1 that water can behave as both an acid and a base. An alcohol, too, can behave as an acid and lose a proton, or it can behave as a base and gain a proton.

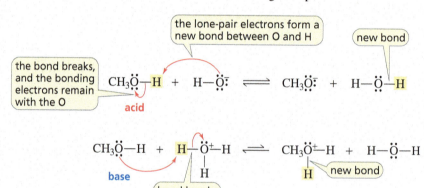

A curved arrow points **from** *the electron donor* **to** *the electron acceptor.*

Chemists frequently use curved arrows to indicate the bonds that are broken and formed as reactants are converted into products. They are called *curved arrows* (and are always red in this book) to distinguish them from the *straight* arrows used to link reactants with products in the equation for a chemical reaction. Each curved arrow with a two-barbed arrowhead signifies the movement of two electrons. The arrow always points *from* the electron donor *to* the electron acceptor.

In an acid–base reaction, one of the arrows is drawn *from* a lone pair on the base *to* the proton of the acid. A second arrow is drawn *from* the electrons that the proton shared *to* the atom on which they are left behind. As a result, the curved arrows let you follow the electrons to see what bond is broken and what bond is formed in the reaction.

A carboxylic acid also can behave as an acid (lose a proton) or as a base (gain a proton).

Similarly, an amine can behave as an acid (lose a proton) or as a base (gain a proton).

$$CH_3\overset{\cdot\cdot}{N}H + H-\overset{\cdot\cdot}{\underset{\cdot\cdot}{O}}{:}^- \rightleftharpoons CH_3\overset{\cdot\cdot}{\underset{\cdot\cdot}{N}}H + H-\overset{\cdot\cdot}{\underset{\cdot\cdot}{O}}-H$$

bond breaks | H | acid | new bond

$$CH_3\overset{\cdot\cdot}{N}H + H-\overset{+}{\underset{\cdot\cdot}{O}}-H \rightleftharpoons CH_3\overset{+}{N}H + H-\overset{\cdot\cdot}{\underset{\cdot\cdot}{O}}-H$$

base | bond breaks | new bond

It is important to know the approximate pK_a values of the various classes of compounds we have looked at. An easy way to remember them is in increments of five, as shown in Table 2.1. (R is used when the particular carboxylic acid, alcohol, or amine is not specified.)

- Protonated alcohols, protonated carboxylic acids, and protonated water have pK_a values less than 0.
- Carboxylic acids have pK_a values of ~5.
- Protonated amines have pK_a values of ~10.
- Alcohols and water have pK_a values of ~15.

Table 2.1 Approximate pK_a Values

pK_a < 0	pK_a ~ 5	pK_a ~ 10	pK_a ~ 15
$R\overset{+}{O}H_2$ protonated alcohol	$R\overset{O}{\underset{}{C}}OH$ carboxylic acid	$R\overset{+}{N}H_3$ protonated amine	ROH alcohol
$\overset{+OH}{R-C-OH}$ protonated carboxylic acid			H_2O water
H_3O^+ protonated water			

NOTE TO THE STUDENT
• You need to remember these approximate pK_a values because they will be very important when you learn about the reactions of organic compounds. These values are also listed inside the back cover of this book for easy reference.

PROBLEM 10 ◆

Draw the conjugate acid of each of the following:

a. CH_3CH_2OH b. $CH_3CH_2O^-$ c. $CH_3\overset{O}{\underset{}{C}}O^-$ d. $CH_3CH_2NH_2$ e. $CH_3CH_2\overset{O}{\underset{}{C}}OH$

PROBLEM 11

a. Write an equation showing CH_3OH reacting as an acid with NH_3 and an equation showing it reacting as a base with HCl.

b. Write an equation showing NH_3 reacting as an acid with CH_3O^- and an equation showing it reacting as a base with HBr.

PROBLEM 12 ◆

Estimate the pK_a values of the following compounds:

a. $CH_3CH_2CH_2NH_2$ b. $CH_3CH_2CH_2OH$ c. CH_3CH_2COOH d. $CH_3CH_2CH_2\overset{+}{N}H_3$

LEARN THE STRATEGY

PROBLEM-SOLVING STRATEGY

Determining the Most Basic Atom in a Compound

Which atom of the following compound is more apt to be protonated when an acid is added to a solution of the compound?

$$HOCH_2CH_2CH_2NH_2$$

An easy way to solve this problem is to look at the pK_a values of the conjugate acids of the groups, remembering that the weaker acid has the stronger conjugate base. The stronger base is the one more apt to be protonated.

$pK_a < 0$ $pK_a \sim 10$ group more apt to be protonated

$$\overset{+}{H}OCH_2CH_2CH_2\overset{+}{N}H_3 \quad\quad HOCH_2CH_2CH_2NH_2 \xrightleftharpoons{HCl} HOCH_2CH_2\overset{+}{N}H_3 \; Cl^-$$
H

The conjugate acids have pK_a values of ~0 and ~10. Because the $^+NH_3$ group is the weaker acid, the NH_2 group is the stronger base, so it is the group more apt to be protonated.

USE THE STRATEGY

PROBLEM 13 ◆

a. Which is a stronger base: CH_3COO^- or $HCOO^-$? (The pK_a of CH_3COOH is 4.8; the pK_a of $HCOOH$ is 3.8.)
b. Which is a stronger base: HO^- or $^-NH_2$? (The pK_a of H_2O is 15.7; the pK_a of NH_3 is 36.)
c. Which is a stronger base: H_2O or CH_3OH? (The pK_a of H_3O^+ is -1.7; the pK_a of $CH_3\overset{+}{O}H_2$ is -2.5.)

PROBLEM 14 ◆

Using the pK_a values in Section 2.3, rank the following species from strongest base to weakest base:

$$CH_3NH_2 \quad CH_3\bar{N}H \quad CH_3OH \quad CH_3O^- \quad CH_3\overset{O}{\overset{\|}{C}}O^-$$

2.4 HOW TO PREDICT THE OUTCOME OF AN ACID–BASE REACTION

Now let's see how to predict that water behaves as a base when it reacts with HCl (the first reaction in Section 2.1) and as an acid when it reacts with NH_3 (the second reaction in Section 2.1). To determine which of two reactants is the acid, we need to compare their pK_a values.

Reaction of H_2O with HCl

The reactants are water ($pK_a = 15.7$) and HCl ($pK_a = -7$). Because HCl is the stronger acid (it has the lower pK_a value), it is the reactant that loses a proton. Therefore, HCl is the acid and water is the base in this reaction.

this is the acid   reactants   reactants   this is the acid
$$HCl \; + \; H_2O \quad\quad\quad NH_3 \; + \; H_2O$$
$pK_a = -7$ $pK_a = 15.7$ $pK_a = 36$ $pK_a = 15.7$

Reaction of H_2O with NH_3

The reactants are water ($pK_a = 15.7$) and NH_3 ($pK_a = 36$). Because water is the stronger acid (it has the lower pK_a value), it is the reactant that loses a proton. Therefore, water is the acid and ammonia is the base in this reaction.

PROBLEM 15 ◆

Does methanol behave as an acid or a base when it reacts with methylamine? (*Hint:* See page 55 for the structures of methanol and methylamine.)

2.5 HOW TO DETERMINE THE POSITION OF EQUILIBRIUM

To determine the position of equilibrium for an acid–base reaction (that is, to determine whether reactants or products are favored), we need to compare the pK_a value of the acid on the left of the equilibrium arrows with the pK_a value of the acid on the right of the arrows. The equilibrium favors *formation* of the weaker acid (the one with the higher pK_a value). In other words, the equilibrium lies toward the weaker acid.

products are favored because the weaker acid is a product

$$CH_3\overset{O}{\overset{\|}{C}}OH \quad + \quad NH_3 \quad \rightleftharpoons \quad CH_3\overset{O}{\overset{\|}{C}}O^- \quad + \quad \overset{+}{N}H_4$$

stronger acid
$pK_a = 4.8$

weaker acid
$pK_a = 9.4$

In an acid–base reaction, the equilibrium favors formation of the weaker acid.

reactants are favored because the weaker acid is a reactant

$$CH_3CH_2OH \quad + \quad CH_3NH_2 \quad \rightleftharpoons \quad CH_3CH_2O^- \quad + \quad CH_3\overset{+}{N}H_3$$

weaker acid
$pK_a = 15.9$

stronger acid
$pK_a = 10.7$

Because the equilibrium favors formation of the weaker acid, we can say that

> an acid–base reaction will favor products
> if the conjugate acid of the base that gains the proton
> is a weaker acid than the acid that loses a proton in the first place.

The precise value of the equilibrium constant can be calculated from the following equation:

$$pK_{eq} = pK_a \text{ (reactant acid)} - pK_a \text{ (product acid)}$$

Thus, the equilibrium constant for the reaction of acetic acid with ammonia is 4.0×10^4.

$$pK_{eq} = 4.8 - 9.4 = -4.6$$
$$K_{eq} = 10^{4.6} = 4.0 \times 10^4$$

And the equilibrium constant for the reaction of ethyl alcohol with methylamine is 6.3×10^{-6}.

$$pK_{eq} = 15.9 - 10.7 = 5.2$$
$$K_{eq} = 10^{-5.2} = 6.3 \times 10^{-6}$$

PROBLEM 16

a. For each of the acid–base reactions in Section 2.3, compare the pK_a values of the acids on either side of the equilibrium arrows to prove that the equilibrium lies in the direction indicated. (The pK_a values you need are found in Section 2.3 or in Problem 13.)

b. Do the same for the acid–base reactions in Section 2.1.

PROBLEM 17

Ethyne (HC≡CH) has a pK_a value of 25, water has a pK_a value of 15.7, and ammonia (NH_3) has a pK_a value of 36. Draw the equation, showing equilibrium arrows that indicate whether reactants or products are favored, for the acid–base reaction of ethyne with

a. HO^-.   **b.** $^-NH_2$.

c. Which would be a better base to use if you wanted to remove a proton from ethyne, HO^- or $^-NH_2$?

PROBLEM 18 ◆

Which of the following bases can remove a proton from acetic acid in a reaction that favors products?

$$HO^- \qquad CH_3NH_2 \qquad HC{\equiv}C^- \qquad CH_3OH \qquad H_2O \qquad Cl^-$$

PROBLEM 19 ◆

Calculate the equilibrium constant for the acid–base reaction between the reactants in each of the following pairs:

a. $HCl + H_2O$ **b.** $CH_3COOH + H_2O$ **c.** $CH_3NH_2 + H_2O$ **d.** $CH_3\overset{+}{N}H_3 + H_2O$

2.6 HOW THE STRUCTURE OF AN ACID AFFECTS ITS pK_a VALUE

Stable bases are weak bases.

The strength of an acid is determined by the stability of the conjugate base that forms when the acid loses its proton: the more stable the conjugate base, the stronger the acid. (The reason for this is explained in Section 5.7.)

The more stable the base, the stronger its conjugate acid.

A stable base readily bears the electrons it formerly shared with a proton. In other words, stable bases are weak bases—they do not share their electrons well. Thus, we can say:

The weaker the base, the stronger its conjugate acid.

The weaker the base, the stronger its conjugate acid.

or

The more stable the base, the stronger its conjugate acid.

Now let's look at two factors that affect the stability of a base—its electronegativity and its size.

Electronegativity

The atoms in the second row of the periodic table are all *similar in size*, but they have very *different electronegativities*, which increase across the row from left to right. Of the atoms shown, carbon is the least electronegative and fluorine is the most electronegative.

relative electronegativities

C < N < O < F

most
electronegative

If we look at the acids formed by attaching hydrogens to these elements, we see that the most acidic compound is the one that has its hydrogen attached to the most electronegative atom. Thus, HF is the strongest acid and methane is the weakest acid.

> *When atoms are similar in size, the strongest acid has its hydrogen attached to the most electronegative atom.*

When atoms are similar in size, the strongest acid has its hydrogen attached to the most electronegative atom.

relative acidities

CH_4 < NH_3 < H_2O < HF

strongest
acid

If we look at the stabilities of the conjugate bases of these acids, we find that they, too, increase from left to right, because the more electronegative the atom, the better it bears its negative charge. Thus, *the strongest acid has the most stable (weakest) conjugate base.*

relative stabilities

$^-CH_3$ < $^-NH_2$ < HO^- < F^-

most
stable

The effect that the electronegativity of the atom bonded to a hydrogen has on the compound's acidity can be appreciated when the pK_a values of alcohols and amines are compared. Because oxygen is more electronegative than nitrogen, an alcohol is more acidic than an amine.

$$CH_3\underline{OH}$$
methyl alcohol
pK_a = 15.5

$$CH_3\underline{NH_2}$$
methylamine
pK_a = 40

Again, because oxygen is more electronegative than nitrogen, a protonated alcohol is more acidic than a protonated amine.

$$CH_3\overset{+}{O}H_2$$
protonated methyl alcohol
pK_a = -2.5

$$CH_3\overset{+}{N}H_3$$
protonated methylamine
pK_a = 10.7

PROBLEM 20 ◆

Rank the ions ($^-CH_3$, $^-NH_2$, HO^-, and F^-) from most basic to least basic.

Hybridization

Because hybridization affects electronegativity and electronegativity affects acidity, the hybridization of an atom affects the acidity of the hydrogen bonded to it. An *sp* hybridized atom is more electronegative than the same atom that is sp^2 hybridized, which is more electronegative than the same atom that is sp^3 hybridized.

relative electronegativities

most electronegative — sp > sp^2 > sp^3 — least electronegative

An sp carbon is more electronegative than an sp^2 carbon, which is more electronegative than an sp^3 carbon.

Therefore, ethyne is a stronger acid than ethene and ethene is a stronger acid than ethane, because the most acidic compound is the one with its hydrogen attached to the most electronegative atom.

	sp		*sp²*		*sp³*	
strongest acid	HC≡CH	>	H₂C=CH₂	>	CH₃CH₃	weakest acid
	ethyne		ethene		ethane	
	pK_a = 25		pK_a = 44		pK_a > 60	

Why does the hybridization of the atom affect its electronegativity? Electronegativity is a measure of the ability of an atom to pull the bonding electrons toward itself. Thus, the most electronegative atom is the one with its bonding electrons closest to the nucleus. The average distance of a 2*s* electron from the nucleus is less than the average distance of a 2*p* electron from the nucleus. Therefore, an *sp* hybridized atom with 50% *s* character is the most electronegative, an sp^2 hybridized atom (33.3% *s* character) is next, and an sp^3 hybridized atom (25% *s* character) is the least electronegative.

Pulling the electrons closer to the nucleus stabilizes the carbanion. Once again we see that the stronger the acid, the more stable (the weaker) its conjugate base. Notice that the electrostatic potential maps show that the strongest base (the least stable) is the most electron-rich (the most red).

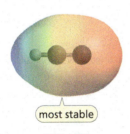

most stable

HC≡C⁻

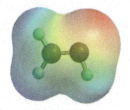

H₂C=C̄H

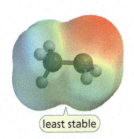

least stable

CH₃C̄H₂

PROBLEM 21 ◆

Rank the carbanions shown in the margin from most basic to least basic.

PROBLEM 22 ◆

Which is a stronger acid?

$$CH_3\overset{+}{\underset{H}{O}}CH_3 \quad \text{or} \quad \overset{\overset{+}{O}H}{\underset{CH_3}{\overset{\|}{C}}}CH_3$$

PROBLEM 23

a. Draw the products of the following reactions:

A $HC\equiv CH$ + $CH_3\bar{C}H_2$ $\rightleftharpoons$

B $H_2C=CH_2$ + $HC\equiv C^-$ $\rightleftharpoons$

C CH_3CH_3 + $H_2C=\bar{C}H$ $\rightleftharpoons$

b. Which of the reactions favor formation of the products?

PROBLEM 24

Which reaction in Problem 23 has the smallest equilibrium constant?

Size

> **Size overrides electronegativity when determining relative acidities.**

When comparing atoms that are very different in size, the *size* of the atom is more important than its *electronegativity* in determining how well it bears its negative charge. For example, as we proceed down a column in the periodic table, the atoms get larger and the *stability* of the anions *increases* even though the electronegativity of the atoms *decreases*. Because the stability of the bases increases going down the column, the strength of their conjugate acids *increases*. Thus, HI is the strongest acid of the hydrogen halides (that is, I⁻ is the weakest, most stable base), even though iodine is the least electronegative of the halogens (Table 2.2).

When atoms are very different in size,
the strongest acid has its hydrogen attached to the largest atom.

> **When atoms are very different in size, the strongest acid has its hydrogen attached to the largest atom.**

relative size

F^- < Cl^- < Br^- < I^-

(largest)

relative acidities

HF < HCl < HBr < HI

(strongest acid)

HF

HCl

HBr

HI

Why does the size of an atom have such a significant effect on stability that it more than overcomes any difference in electronegativity?

The valence electrons of F⁻ are in a $2sp^3$ orbital, the valence electrons of Cl⁻ are in a $3sp^3$ orbital, those of Br⁻ are in a $4sp^3$ orbital, and those of I⁻ are in a $5sp^3$ orbital. The volume of space occupied by a $3sp^3$ orbital is significantly larger than the volume of space occupied by a $2sp^3$ orbital because a $3sp^3$ orbital extends out farther from the nucleus. Because its negative charge is spread over a larger volume of space, Cl⁻ is more stable than F⁻.

Thus, as a halide ion increases in size (going down the column of the periodic table), its stability increases because its negative charge is spread over a larger volume of space. As a result, HI is the strongest acid of the hydrogen halides because I⁻ is the most stable halide ion. The potential maps shown in the margin illustrate the large difference in size of the hydrogen halides.

Table 2.2 The pK_a Values of Some Simple Acids

CH$_4$	NH$_3$	H$_2$O	HF
pK_a = 60	pK_a = 36	pK_a = 15.7	pK_a = 3.2
		H$_2$S	HCl
		pK_a = 7.0	pK_a = −7
			HBr
			pK_a = −9
			HI
			pK_a = −10

In summary:

- atomic size does not change much as we move from left to right across a row of the periodic table, so the atoms' orbitals have approximately the same volume. Thus, electronegativity determines the stability of the base and, therefore, the acidity of its conjugate acid.

- atomic size increases as we move down a column of the periodic table, so the volume of the orbitals increases. The volume of an orbital is more important than electronegativity in determining the stability of a base and, therefore, the acidity of its conjugate acid.

LEARN THE STRATEGY

PROBLEM 25 ♦

Rank the halide ions (F$^-$, Cl$^-$, Br$^-$, and I$^-$) from strongest base to weakest base.

USE THE STRATEGY

PROBLEM 26 ♦

a. Which is more electronegative, oxygen or sulfur?

b. Which is a stronger acid, H$_2$O or H$_2$S?

c. Which is a stronger acid, CH$_3$OH or CH$_3$SH?

PROBLEM 27 ♦

Which is a stronger acid?

a. HCl   or   HBr

b. CH$_3$CH$_2$CH$_2$\overset{+}{N}H_3$ or   CH$_3$CH$_2$CH$_2$\overset{+}{O}H_2$

c.    or

black = C
gray = H
blue = N
red = O

d. CH$_3$CH$_2$CH$_2$OH   or   CH$_3$CH$_2$CH$_2$SH

PROBLEM 28 ♦

a. Which of the halide ions (F$^-$, Cl$^-$, Br$^-$, and I$^-$) is the most stable base?

b. Which is the least stable base?

PROBLEM 29 ♦

Which is a stronger base? (The potential maps in the margin can help you answer part **a**.)

a. CH$_3$O$^-$ or   CH$_3$S$^-$ **b.** H$_2$O   or   HO$^-$ **c.** H$_2$O   or   NH$_3$ **d.** CH$_3\overset{\text{O}}{\overset{||}{\text{C}}}O^-$ or   CH$_3$O$^-$

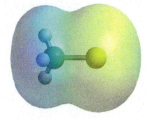

CH$_3$O$^-$

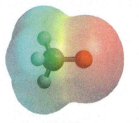

CH$_3$S$^-$

2.7 HOW SUBSTITUENTS AFFECT THE STRENGTH OF AN ACID

Although the acidic proton of each of the following carboxylic acids is attached to the same atom (an oxygen), the four compounds have different pK_a values:

Inductive electron withdrawal increases the strength of an acid.

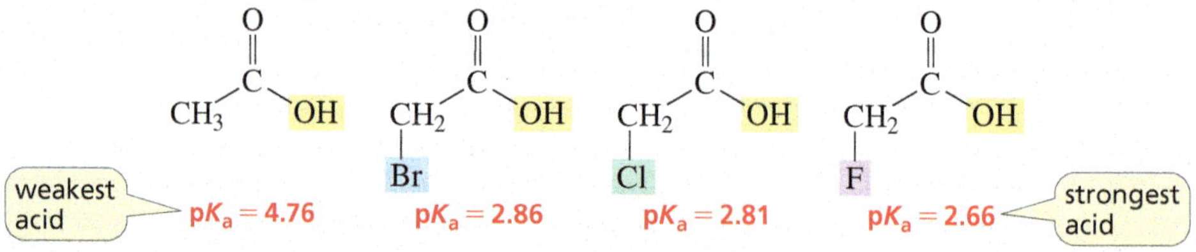

The different pK_a values indicate that there must be another factor that affects acidity other than the atom to which the hydrogen is bonded.

From the pK_a values of the four carboxylic acids, we see that replacing one of the hydrogens of the CH₃ group with a halogen increases the acidity of the compound. (The term for replacing an atom in a compound is *substitution*, and the new atom is called a *substituent*.) The halogen is more electronegative than the hydrogen it has replaced, so the halogen pulls the bonding electrons toward itself more than a hydrogen would. Pulling electrons through sigma (σ) bonds is called **inductive electron withdrawal**.

If we look at the conjugate base of a carboxylic acid, we see that inductive electron withdrawal *decreases the electron density* about the oxygen that bears the negative charge, thereby stabilizing it. And we know that stabilizing a base increases the acidity of its conjugate acid.

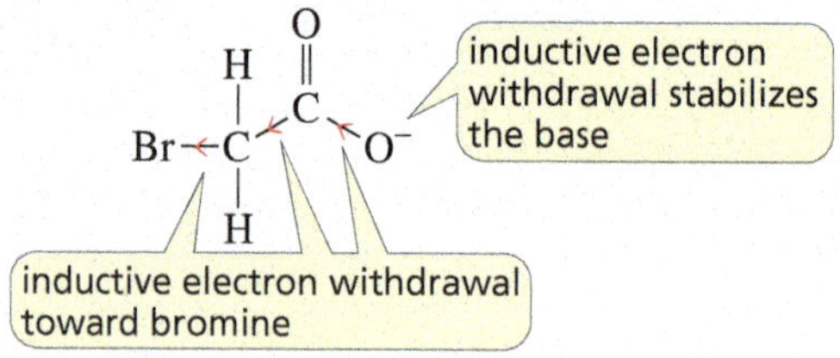

The pK_a values of the four carboxylic acids shown above decrease (become more acidic) as the electron-withdrawing ability (electronegativity) of the halogen increases. Thus, the fluoro-substituted compound is the strongest acid because its conjugate base is the most stabilized by inductive electron withdrawal.

As shown below, the effect a substituent has on the acidity of a compound decreases as the distance between the substituent and the acidic proton increases.

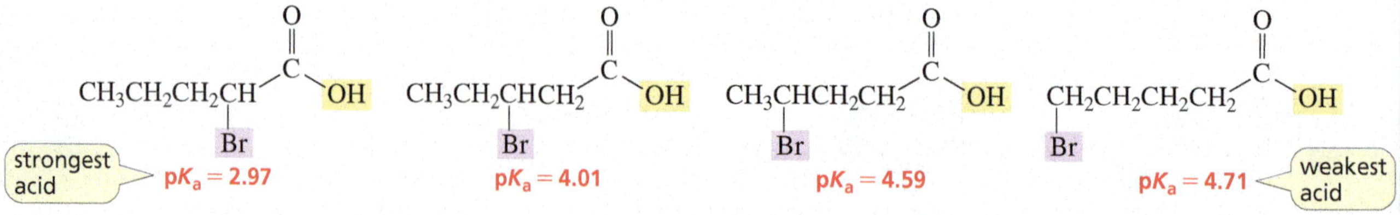

PROBLEM-SOLVING STRATEGY

LEARN THE STRATEGY

Determining Relative Acid Strength

a. Which is a stronger acid?

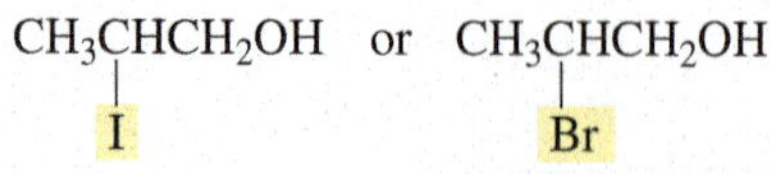

$$CH_3CHCH_2OH \quad \text{or} \quad CH_3CHCH_2OH$$
$$I \phantom{\quad \text{or} \quad CH_3CHCH_2}Br$$

When you are asked to compare two items, pay attention to where they differ and ignore where they are the same. These two compounds differ only in the halogen that is attached to the middle carbon. Because bromine is more electronegative than iodine, there is greater inductive electron withdrawal from oxygen in the brominated compound. The brominated compound, therefore, has the more stable conjugate base, so it is the stronger acid.

b. Which is a stronger acid?

$$\overset{\text{F}}{\underset{\text{F}}{\overset{|}{\underset{|}{\text{CH}_3\text{CCH}_2\text{OH}}}}} \quad \text{or} \quad \overset{\text{F}}{\underset{\text{F}}{\overset{|}{\underset{|}{\text{CH}_2\text{CHCH}_2\text{OH}}}}}$$

These two compounds differ in the location of one of the fluorines. Because the second fluorine in the compound on the left is closer to the O—H bond than is the second fluorine in the compound on the right, the compound on the left is more effective at withdrawing electrons from the oxygen. Thus, the compound on the left has the more stable conjugate base, so it is the stronger acid.

PROBLEM 30 ◆

USE THE STRATEGY

Which is a stronger acid?

a. $CH_3OCH_2CH_2OH$ or $CH_3CH_2CH_2CH_2OH$

b. $CH_3CH_2CF_2CH_2\overset{+}{N}H_3$ or $CH_3CH_2CF_2CH_2\overset{+}{O}H_2$

c. $CH_3OCH_2CH_2CH_2OH$ or $CH_3CH_2OCH_2CH_2OH$

d. $CH_3\overset{\text{O}}{\overset{||}{C}}CH_2OH$ or $CH_3CH_2\overset{\text{O}}{\overset{||}{C}}OH$

PROBLEM 31 ◆

Rank the following compounds from strongest acid to weakest acid:

$$\underset{\text{F} \quad \text{F}}{\overset{|\quad|}{\text{CH}_2\text{CHCH}_2\text{COOH}}} \qquad \text{CH}_3\text{CH}_2\text{CH}_2\text{COOH} \qquad \underset{\text{F}}{\overset{|}{\text{CH}_2\text{CH}_2\text{CH}_2\text{COOH}}} \qquad \underset{\text{F}}{\overset{|}{\text{CH}_3\text{CHCH}_2\text{COOH}}}$$

PROBLEM 32 ◆

Which is a stronger base?

a. $CH_3\overset{\text{O}}{\overset{||}{\underset{\underset{\text{Br}}{|}}{\text{CH}}}\text{CO}^-}$ or $CH_3\overset{\text{O}}{\overset{||}{\underset{\underset{\text{F}}{|}}{\text{CH}}}\text{CO}^-}$

c. $BrCH_2CH_2\overset{\text{O}}{\overset{||}{C}}O^-$ or $CH_3CH_2\overset{\text{O}}{\overset{||}{C}}O^-$

b. $CH_3\overset{}{\underset{\underset{\text{Cl}}{|}}{\text{CH}}}CH_2\overset{\text{O}}{\overset{||}{C}}O^-$ or $CH_3CH_2\overset{}{\underset{\underset{\text{Cl}}{|}}{\text{CH}}}\overset{\text{O}}{\overset{||}{C}}O^-$

d. $CH_3\overset{\text{O}}{\overset{||}{C}}CH_2CH_2O^-$ or $CH_3CH_2\overset{\text{O}}{\overset{||}{C}}CH_2O^-$

PROBLEM 33 SOLVED

If HCl is a weaker acid than HBr, why is $ClCH_2COOH$ a stronger acid than $BrCH_2COOH$?

SOLUTION To compare the acidities of HCl and HBr, we need to compare the stabilities of their conjugate bases, Cl^- and Br^-. (Notice that an H—Cl bond breaks in one compound and an H—Br bond breaks in the other.) Because we know that the size of the atom to which the hydrogen is attached is the most important factor in determining its stability, we know that Br^- is more stable than Cl^-. Therefore, HBr is a stronger acid than HCl.

In comparing the acidities of the two carboxylic acids, we again need to compare the stabilities of their conjugate bases, $ClCH_2COO^-$ and $BrCH_2COO^-$. (Notice that an O—H bond breaks in both compounds.) The only way the conjugate bases differ is in the electronegativity of the atom that is drawing electrons away from the negatively charged oxygen. Because Cl is more electronegative than Br, Cl exerts greater inductive electron withdrawal. Thus, it has a greater stabilizing effect on the base that is formed when the proton leaves, so the chloro-substituted compound is the stronger acid.

2.8 AN INTRODUCTION TO DELOCALIZED ELECTRONS

We have seen that a carboxylic acid has a pK_a value of about 5, whereas the pK_a value of an alcohol is about 15. Because a carboxylic acid is a much stronger acid than an alcohol, we know that the conjugate base of a carboxylic acid is considerably more stable than the conjugate base of an alcohol.

$$CH_3CH_2O\text{—}H$$
$$pK_a = 15.9$$

$$CH_3\quad O\text{—}H$$
$$pK_a = 4.76$$

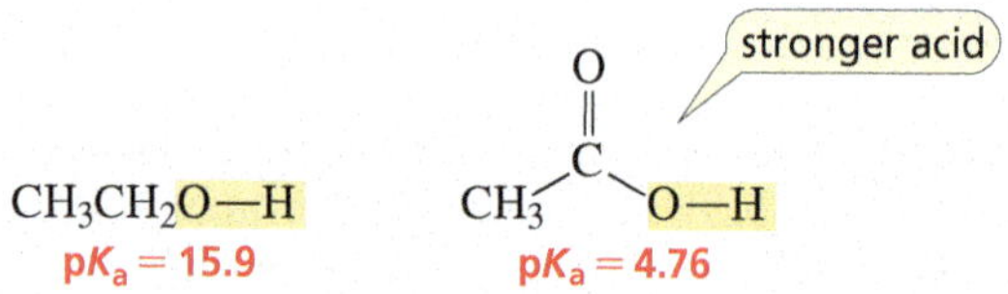

Two factors cause the conjugate base of a carboxylic acid to be more stable than the conjugate base of an alcohol.

Inductive Electron Withdrawal

The conjugate base of a carboxylic acid has a doubly bonded oxygen where the conjugate base of an alcohol has two hydrogens. Inductive electron withdrawal by this electronegative oxygen decreases the electron density of the negatively charged oxygen, thereby stabilizing it and increasing the acidity of the conjugate acid.

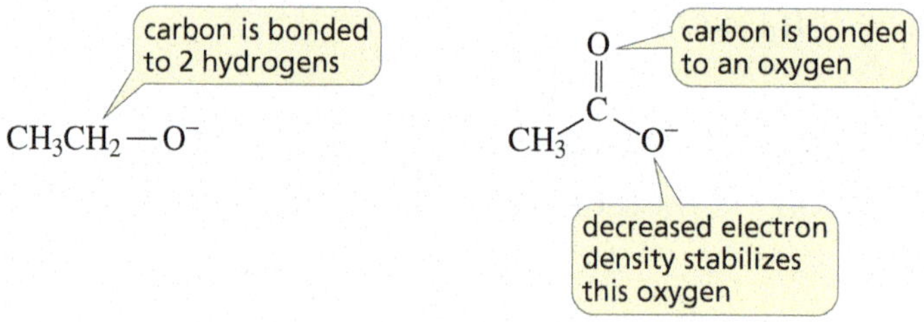

$$CH_3CH_2\text{—}O^-$$

Delocalized Electrons

When an alcohol loses a proton, the electrons left behind reside on its single oxygen atom. These electrons are said to be *localized* because they belong to only *one* atom. The negative charge, therefore, resides only on the oxygen atom.

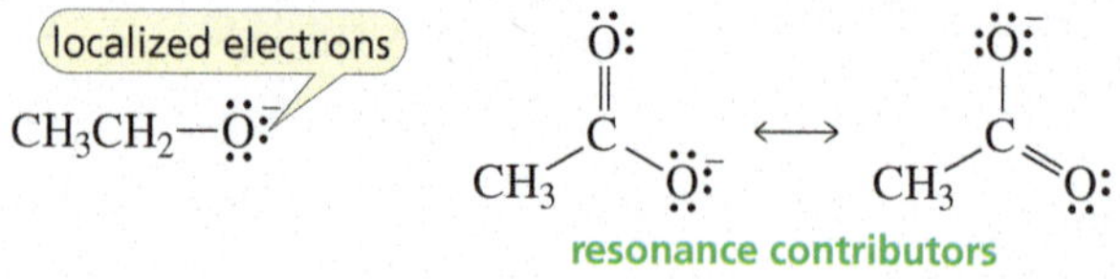

resonance contributors

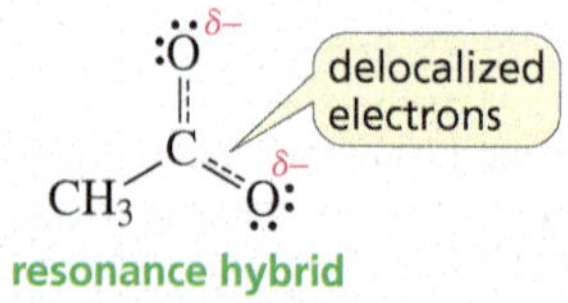

resonance hybrid

In contrast, when a carboxylic acid loses a proton, the electrons left behind are shared by three atoms—two oxygens and a carbon. These electrons are *delocalized* because they belong to more than two atoms. The negative charge, therefore, is shared by both oxygens. As a result, the conjugate base is stabilized because, as we have seen, decreasing the electron density of an atom stabilizes it.

The two structures shown for the conjugate base of the carboxylic acid are called **resonance contributors**. Neither resonance contributor alone represents the actual structure of the conjugate base—they only approximate the actual structure. Instead, the actual structure—called a **resonance hybrid**—is a composite of the two resonance contributors. The double-headed arrow between the two resonance contributors is used to indicate that the actual structure is a hybrid.

Notice that the second resonance contributor is obtained by moving a lone pair on an atom of the first resonance contributor toward an sp^2 carbon and breaking the π bond. Thus, the two resonance contributors differ only in the location of their π electrons and lone-pair electrons—all the atoms stay in the same place.

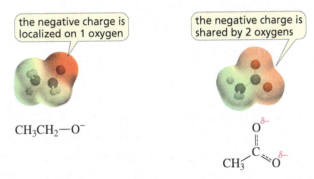

The negative charge is shared equally by the two oxygens, and both carbon–oxygen bonds are the same length—they are not as long as a single bond, but they are longer than a double bond. The resonance hybrid is drawn using dotted lines to show the delocalized electrons. We see that in the resonance hybrid, an electron pair is spread over *two oxygens and a carbon.*

the negative charge is localized on 1 oxygen

the negative charge is shared by 2 oxygens

$CH_3CH_2-O^-$

Delocalized electrons are shared by more than two atoms.

Thus, the combination of inductive electron withdrawal and the ability of two atoms to share the negative charge makes the conjugate base of the carboxylic acid more stable than the conjugate base of the alcohol.

Delocalized electrons are very important in organic chemistry—so important that all of Chapter 8 is devoted to them. By that time, you will be thoroughly comfortable with compounds that have only localized electrons, and we can then further explore how to recognize when a compound has delocalized electrons and how delocalized electrons affect the stability, reactivity, and pK_a values of organic compounds.

Fosamax Prevents Bones from Being Nibbled Away

Fosamax is used to treat osteoporosis, a condition characterized by decreased bone density. Under normal conditions, the rate of bone formation and the rate of bone resorption (breakdown) are carefully matched. In osteoporosis, resorption is faster than formation, so bone is nibbled away, causing bones to become fragile (they actually start to resemble honeycombs). Fosamax goes specifically to the sites of bone resorption and inhibits the activity of cells responsible for resorption. Studies have shown that normal bone is then formed on top of Fosamax, and the rate of bone formation becomes faster than the rate of its breakdown. (Trade name labels in this book are green.)

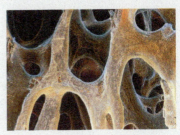

normal bone

bone with osteoporosis

Fosamax®

LEARN THE STRATEGY

PROBLEM-SOLVING STRATEGY

Determining the Site of Protonation in a Compound with Delocalized Electrons

Which atom in the following compound is more apt to be protonated when an acid is added to a solution of the compound?

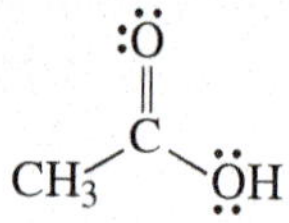

In the Problem-Solving Strategy on p. 58, you learned that when a compound has two basic atoms, the one that is the stronger base is the one more apt to be protonated in an acidic solution. When a compound has delocalized electrons, you need to draw its resonance contributors to determine which atom is the stronger base.

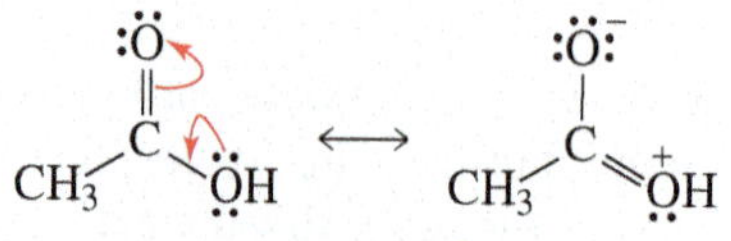

resonance contributors

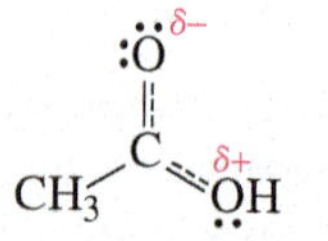

resonance hybrid

One of the resonance contributors has a negative charge on one of the oxygens and a positive charge on the other. Therefore, those oxygens in the resonance hybrid have a partial negative charge and a partial positive charge, respectively. The oxygen with the partial negative charge is the stronger base, so it is the oxygen that is more apt to be protonated in an acidic solution.

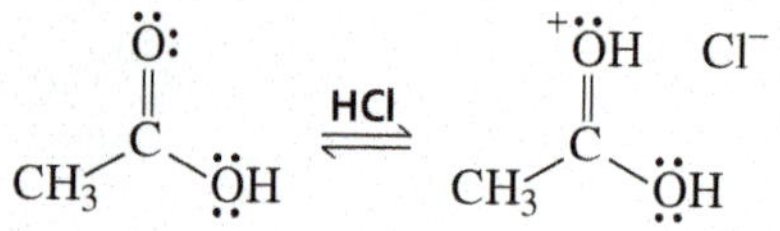

USE THE STRATEGY

PROBLEM 34

For each of the following compounds, indicate the atom that is protonated when an acid is added to a solution of the compound.

a. 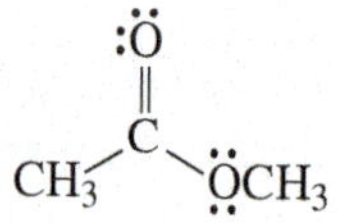
b. 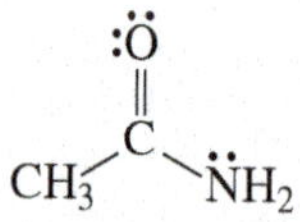
c. 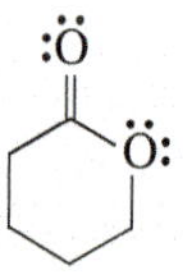
d. 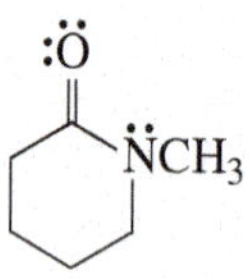

PROBLEM 35 ♦

Draw resonance contributors for each of the following:

a. 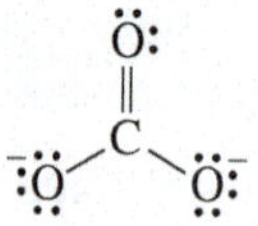
b. 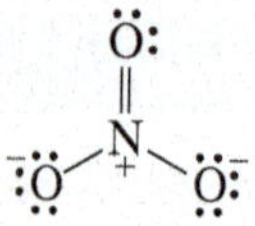

PROBLEM 36 ♦

Which is a stronger acid? Why?

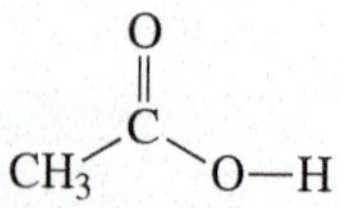 or 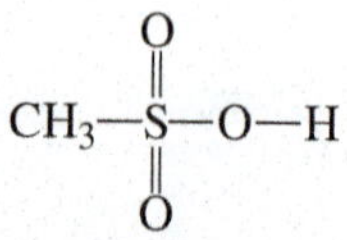

PROBLEM 37

Fosamax (shown on the previous page) has six acidic groups. The active form of the drug, which has lost two of its acidic protons, is shown in the box. (Notice that the phosphorus atom in Fosamax and the sulfur atom in Problem 36 can be surrounded by more than eight electrons because P and S are below the second row of the periodic table.)

a. Why are the OH groups bonded to phosphorus the strongest acids of the six groups?
b. Which of the remaining four groups is the weakest acid?

2.9 A SUMMARY OF THE FACTORS THAT DETERMINE ACID STRENGTH

We have seen that the strength of an acid depends on five factors: the *size* of the atom to which the hydrogen is attached, the *electronegativity* of the atom to which the hydrogen is attached, the *hybridization* of the atom to which the hydrogen is attached, *inductive electron withdrawal*, and *electron delocalization*. All five factors affect acidity by affecting the stability of the conjugate base.

1. **Size:** As the atom attached to the hydrogen increases in size (going down a column of the periodic table), the strength of the acid increases.

2. **Electronegativity:** As the atom attached to the hydrogen increases in electronegativity (going from left to right across a row of the periodic table), the strength of the acid increases.

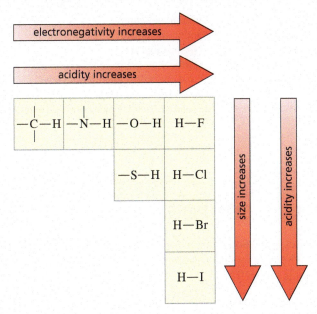

3. **Hybridization:** The electronegativity of an atom changes with hybridization as follows: $sp > sp^2 > sp^3$. Because an sp carbon is the most electronegative, a hydrogen attached to an sp carbon is the most acidic, and a hydrogen attached to an sp^3 carbon is the least acidic.

strongest acid $HC \equiv CH$ > $H_2C = CH_2$ > CH_3CH_3 **weakest acid**

sp sp^2 sp^3

4. **Inductive electron withdrawal:** An electron-withdrawing group increases the strength of an acid. As the electronegativity of the electron-withdrawing group increases or as it moves closer to the acidic hydrogen, the strength of the acid increases.

strongest acid
$$\underset{F}{CH_3CHCH_2} \overset{O}{\underset{}{\overset{\|}{C}}} OH \; > \; \underset{Cl}{CH_3CHCH_2} \overset{O}{\underset{}{\overset{\|}{C}}} OH \; > \; \underset{Br}{CH_3CHCH_2} \overset{O}{\underset{}{\overset{\|}{C}}} OH \; > \; CH_3CH_2CH_2 \overset{O}{\underset{}{\overset{\|}{C}}} OH$$
weakest acid

strongest acid
$$\underset{F}{CH_3CHCH_2} \overset{O}{\underset{}{\overset{\|}{C}}} OH \; > \; \underset{F}{CH_2CH_2CH_2} \overset{O}{\underset{}{\overset{\|}{C}}} OH \; > \; CH_2CH_2CH_2 \overset{O}{\underset{}{\overset{\|}{C}}} OH$$
weakest acid

5. **Electron delocalization:** An acid whose conjugate base has delocalized electrons is more acidic than a similar acid whose conjugate base has only localized electrons.

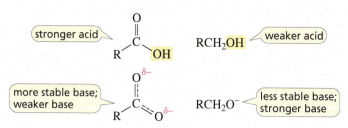

PROBLEM 38 ♦

Using the table of pK_a values given in Appendix I, answer the following:

a. Which is the most acidic organic compound in the table?
b. Which is the least acidic organic compound in the table?
c. Which is the most acidic carboxylic acid in the table?
d. Which is more electronegative: an sp^3 oxygen or an sp^2 oxygen? (*Hint:* Pick a compound in Appendix I with a hydrogen attached to an sp^2 oxygen and one with a hydrogen attached to an sp^3 oxygen and compare their pK_a values.)
e. Which compounds demonstrate that the relative electronegativities of a hybridized nitrogen atom are $sp > sp^2 > sp^3$?

2.10 HOW pH AFFECTS THE STRUCTURE OF AN ORGANIC COMPOUND

Whether a given acid loses a proton in an aqueous solution depends on both the pK_a of the acid and the pH of the solution. The relationship between the two is given by the **Henderson–Hasselbalch equation.** (Its derivation is on pp. 72–73.)

the Henderson–Hasselbalch equation

$$pK_a = pH + \log \frac{[HA]}{[A^-]}$$

This is an extremely useful equation because it tells us whether a compound exists in its acidic form (with its proton retained) or in its basic form (with its proton removed) at a particular pH. Knowing this will be important when we are assessing the reactivity of organic compounds.

acidic form		basic form
RCOOH	$\rightleftharpoons$	RCOO$^-$ + H$^+$
ROH	$\rightleftharpoons$	RO$^-$ + H$^+$
R$\overset{+}{N}$H$_3$	$\rightleftharpoons$	RNH$_2$ + H$^+$

The Henderson–Hasselbalch equation tells us

- when the pH of a solution equals the pK_a of the compound that undergoes dissociation, the concentration of the compound in its acidic form (HA) equals the concentration of the compound in its basic form (A$^-$) (because log 1 = 0).

- when the pH of the solution is less than the pK_a of the compound, the compound exists primarily in its acidic form.

- when the pH of the solution is greater than the pK_a of the compound, the compound exists primarily in its basic form.

In other words,

compounds exist primarily in their acidic forms in solutions that are more acidic than their pK_a values and primarily in their basic forms in solutions that are more basic than their pK_a values.

A compound exists primarily in its acidic form (HA) when the pH of the solution is less than the compound's pK_a value.

A compound exists primarily in its basic form (A$^-$) when the pH of the solution is greater than the compound's pK_a value.

PROBLEM-SOLVING STRATEGY

Determining the Structure at a Particular pH

Write the form of the compound that predominates in a solution with a pH = 5.5.

a. CH_3CH_2OH ($pK_a = 15.9$)   **b.** $CH_3CH_2\overset{+}{O}H_2$ ($pK_a = -2.5$)   **c.** $CH_3\overset{+}{N}H_3$ ($pK_a = 11.0$)

To answer this question, we need to compare the pH of the solution with the pK_a value of the compound's dissociable proton.

a. The pH of the solution is more acidic (5.5) than the pK_a value of the compound (15.9). Therefore, the compound exists primarily as CH_3CH_2OH (with its proton).

b. The pH of the solution is more basic (5.5) than the pK_a value of the compound (-2.5). Therefore, the compound exists primarily as CH_3CH_2OH (without its proton).

c. The pH of the solution is more acidic (5.5) than the pK_a value of the compound (11.0). Therefore, the compound exists primarily as $CH_3\overset{+}{N}H_3$ (with its proton).

PROBLEM 39 ◆

For each of the following compounds (here shown in their acidic forms), write the form that predominates in a solution with a pH = 5.5:

a. CH_3COOH ($pK_a = 4.76$)   **d.** HBr ($pK_a = -9$)   **g.** HNO_2 ($pK_a = 3.4$)

b. $CH_3CH_2\overset{+}{N}H_3$ ($pK_a = 11.0$)   **e.** $^+NH_4$ ($pK_a = 9.4$)   **h.** HNO_3 ($pK_a = -1.3$)

c. H_3O^+ ($pK_a = -1.7$)   **f.** $HC\equiv N$ ($pK_a = 9.1$)   **i.** $HO\overset{+}{N}H_3$ ($pK_a = 6.0$)

PROBLEM 40 ◆

As long as the pH is not less than _____, at least 50% of a protonated amine with a pK_a value of 10.4 will be in its neutral, nonprotonated form.

PROBLEM 41 SOLVED

Indicate whether a carboxylic acid (RCOOH) with a pK_a value of 4.5 has more charged molecules or more neutral molecules in a solution with the following pH:

1. pH = 1   **2.** pH = 3   **3.** pH = 5   **4.** pH = 7   **5.** pH = 10   **6.** pH = 13

SOLUTION First determine whether the compound is charged or neutral in its acidic form and charged or neutral in its basic form: a carboxylic acid is neutral in its acidic form (RCOOH) and charged in its basic form ($RCOO^-$). Then compare the pH and pK_a values and remember that when the pH of the solution is less than the pK_a value of the compound, then more molecules are in the acidic form, but when the pH is greater than the pK_a value of the compound, then more molecules are in the basic form. Therefore, at pH = 1 and 3, there are more neutral molecules in the solution, and at pH = 5, 7, 10, and 13, there are more charged molecules in the solution.

PROBLEM 42 ◆

a. Indicate whether a protonated amine ($R\overset{+}{N}H_3$) with a pK_a value of 9 has more charged or more neutral molecules in a solution with the pH values given in Problem 41.

b. Indicate whether an alcohol (ROH) with a pK_a value of 15 has more charged or more neutral molecules in a solution with the pH values given in Problem 41.

PROBLEM 43

A naturally occurring amino acid such as alanine has a group that is a carboxylic acid and a group that is a protonated amine. The pK_a values of the two groups are shown.

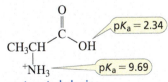

protonated alanine
a protonated amino acid

a. If the pK_a value of a carboxylic acid such as acetic acid is about 5 (see Table 2.1), then why is the pK_a value of the carboxylic acid group of alanine so much lower?

b. Draw the structure of alanine in a solution at pH = 0.

c. Draw the structure of alanine in a solution at physiological pH (pH 7.4).

> **d.** Draw the structure of alanine in a solution at pH = 12.
> **e.** Is there a pH at which alanine is uncharged (that is, neither group has a charge)?
> **f.** At what pH does alanine have no net charge (that is, the amount of negative charge is the same as the amount of positive charge)?

If we know the pH of the solution and the pK_a of the compound, the Henderson–Hasselbalch equation makes it possible to calculate precisely how much of the compound is in its acidic form and how much is in its basic form.

For example, when a compound with a pK_a of 5.2 is in a solution of pH 5.2, half the compound is in the acidic form and the other half is in the basic form (Figure 2.1). When the pH is one unit less than the pK_a of the compound (pH = 4.2), there is 10 times more compound present in the acidic form than in the basic form (because log 10 = 1).

$$5.2 \ = \ 4.2 \ + \ \log \frac{[\text{HA}]}{[\text{A}^-]}$$

$$1.0 \ = \ \log \frac{[\text{HA}]}{[\text{A}^-]} \ = \ \log \frac{10}{1}$$

When the pH is two units less than the pK_a of the compound (pH = 3.2), there is 100 times more compound present in the acidic form than in the basic form (because log 100 = 2).

Now consider pH values that are greater than the pK_a value: when the pH is 6.2, there is 10 times more compound in the basic form than in the acidic form, and at pH = 7.2, there is 100 times more compound present in the basic form than in the acidic form.

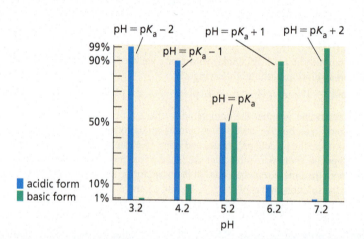

▶ **Figure 2.1**
The relative amounts of a compound with a pK_a of 5.2 in the acidic and basic forms at different pH values.

Derivation of the Henderson–Hasselbalch Equation

The Henderson–Hasselbalch equation can be derived from the expression that defines the acid dissociation constant:

$$K_a = \frac{[\text{H}_3\text{O}^+][\text{A}^-]}{[\text{HA}]}$$

Take the logarithms of both sides of the equation remembering that when expressions are multiplied, their logs are added. Thus, we obtain

$$\log K_a = \log [\text{H}_3\text{O}^+] \ + \ \log \frac{[\text{A}^-]}{[\text{HA}]}$$

Multiplying both sides of the equation by −1 gives us

$$- \log K_a \ = \ - \log [\text{H}_3\text{O}^+] - \log \frac{[\text{A}^-]}{[\text{HA}]}$$

Substituting pK_a for $-\log K_a$, substituting pH for $-\log [H_3O^+]$, and inverting the fraction (which means the sign of its log changes), we get

$$pK_a = pH + \log \frac{[HA]}{[A^-]}$$

LEARN THE STRATEGY

PROBLEM 44 SOLVED

a. At what pH is the concentration of a compound, with a $pK_a = 8.4$, 100 times greater in its basic form than in its acidic form?

b. At what pH is the concentration of a compound, with a $pK_a = 3.7$, 10 times greater in its acidic form than in its basic form?

SOLUTION 44 a. If the concentration in the basic form is 100 times greater than the concentration in the acidic form, then the Henderson–Hasselbalch equation becomes

$$pK_a = pH + \log 1/100$$
$$8.4 = pH + \log 0.01$$
$$8.4 = pH - 2.0$$
$$pH = 10.4$$

There is a faster way to get the answer: if 100 times more compound is present in the basic form than in the acidic form, then the pH will be two units more basic than the pK_a value. Thus, pH = 8.4 + 2.0 = 10.4.

SOLUTION 44 b. If 10 times more compound is present in the acidic form than in the basic form, then the pH will be one unit more acidic than the pK_a value. Thus, pH = 3.7 − 1.0 = 2.7.

USE THE STRATEGY

PROBLEM 45 ◆

a. At what pH is the concentration of a compound, with a $pK_a = 8.4$, 100 times greater in its acidic form than in its basic form?

b. At what pH is 50% of a compound, with a $pK_a = 7.3$, in its basic form?

c. At what pH is the concentration of a compound, with a $pK_a = 4.6$, 10 times greater in its basic form than in its acidic form?

PROBLEM 46 ◆

For each of the following compounds, indicate the pH at which

a. 50% of the compound is in a form that possesses a charge.

b. more than 99% of the compound is in a form that possesses a charge.

1. CH_3CH_2COOH ($pK_a = 4.9$)   **2.** $CH_3\overset{+}{N}H_3$ ($pK_a = 10.7$)

The Henderson–Hasselbalch equation is very useful in the laboratory for separating compounds in a mixture. Water and diethyl ether are barely soluble in each other, so they form two layers when combined. Diethyl ether is less dense than water, so the ether layer lies above the water layer.

- Charged compounds are more soluble in water than in diethyl ether.
- Uncharged compounds are more soluble in diethyl ether than in water (Section 3.10).

Two compounds, such as a carboxylic acid (RCOOH) with a $pK_a = 5.0$ and a protonated amine ($R\overset{+}{N}H_3$) with a $pK_a = 10.0$, dissolved in a mixture of water and diethyl ether, can be separated by adjusting the pH of the water layer. For example, if the pH of the water layer is 2, then the carboxylic acid and the protonated amine are both in their acidic forms because the pH of the water is less than the pK_a values of both compounds. The acidic form of a carboxylic acid is not charged, whereas the acidic form of an amine is charged. Therefore, the uncharged carboxylic acid dissolves in the ether layer, and the positively charged protonated amine dissolves in the water layer.

For the most effective separation, the pH of the water layer should be at least two units away from the pK_a values of the compounds being separated. Then the relative amounts of the compounds in their acidic and basic forms will be at least 100:1 (Figure 2.1).

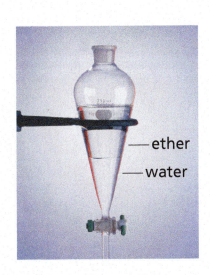
—ether
—water

LEARN THE STRATEGY

PROBLEM 47 SOLVED

Given that $C_6H_{11}COOH$ has a $pK_a = 4.8$ and $C_6H_{11}\overset{+}{N}H_3$ has a $pK_a = 10.7$, what pH would you make the water layer to cause both compounds to dissolve in it?

SOLUTION The compounds must be charged to dissolve in the water layer. The carboxylic acid is charged in its basic form—it is a carboxylate ion. For > 99% of the carboxylic acid to be in its basic form, the pH must be two units *greater* than the pK_a of the compound. Thus, the water should have a pH > 6.8. The amine is charged in its acidic form—it will be an ammonium ion. For > 99% of the amine to be in its acidic form, the pH must be two units *less* than the pK_a value of the ammonium ion. Thus, the water should have a pH < 8.7. Both compounds will dissolve in the water layer if its pH is 6.8–8.7. A pH in the middle of the range (for example, pH = 7.7) is a good choice.

USE THE STRATEGY

PROBLEM 48 ◆

Given the data in Problem 47:

a. What pH would you make the water layer to cause the carboxylic acid to dissolve in the water layer and the amine to dissolve in the ether layer?

b. What pH would you make the water layer to cause the carboxylic acid to dissolve in the ether layer and the amine to dissolve in the water layer?

Aspirin Must Be in Its Basic Form to Be Physiologically Active

Aspirin has been used to treat fever, mild pain, and inflammation since it first became commercially available in 1899. It was the first drug to be tested clinically before it was marketed (Section 7.0). Currently one of the most widely used drugs in the world, aspirin is one of a group of over-the-counter drugs known as NSAIDs (nonsteroidal anti-inflammatory drugs).

Aspirin is a carboxylic acid. When we look at the reaction responsible for its fever-reducing, pain-reducing, and anti-inflammatory properties in Section 15.9, we will see that the carboxylic acid group must be in its basic form to be physiologically active.

The carboxylic acid group has a pK_a value of ~5. Therefore, it is in its acidic form while it is in the stomach (pH = 1–2.5). The uncharged acidic form can pass through membranes easily, whereas the negatively charged basic form cannot. Once the drug is in the cell (pH 7.4), it is in its active basic form and, therefore, is able to carry out the reaction that reduces fever, pain, and inflammation.

The undesirable side effects of aspirin (ulcers, stomach bleeding) led to the development of other NSAIDs (p. 114). Aspirin also has been linked to the development of Reye's syndrome, a rare but serious disease that affects children who are recovering from a viral infection such as a cold, the flu, or chicken pox. Therefore, it is now recommended that aspirin not be given to anyone under the age of 16 who has a fever-producing illness.

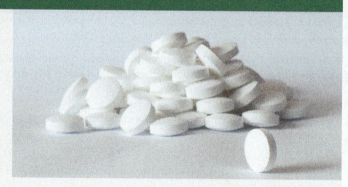

acidic form   **Aspirin**   basic form

2.11 BUFFER SOLUTIONS

A solution of a weak acid (HA) and its conjugate base (A⁻) is called a **buffer solution.** The components of three different buffer solutions are shown here.

acetic acid/sodium acetate	formic acid/sodium formate	methylammonium chloride/methylamine
CH_3COOH	$HCOOH$	$CH_3\overset{+}{N}H_3\ Cl^-$
$CH_3COO^-\ Na^+$	$HCOO^-\ Na^+$	CH_3NH_2

A buffer solution maintains nearly constant pH (in a range of one pH unit in either side of the pK_a of the conjugate acid) when small amounts of acid or base are added to it, because the weak acid can give a proton to any HO^- added to the solution and its conjugate base can accept any H^+ that is added to the solution.

can give a proton to HO^-

$$HA \ + \ HO^- \longrightarrow A^- \ + \ H_2O$$

$$A^- \ + \ H_3O^+ \longrightarrow HA \ + \ H_2O$$

can accept a proton from H_3O^+

NOTE TO THE STUDENT

• Buffer solutions are discussed in detail in Special Topic I in the *Study Guide and Solutions Manual*. By working the problems you will find there, you will see just how useful the Henderson–Hasselbalch equation is for dealing with buffer solutions.

PROBLEM 49 ◆

Write the equation that shows how a buffer made by dissolving CH_3COOH and $CH_3COO^-Na^+$ in water prevents the pH of a solution from changing appreciably when

a. a small amount of H^+ is added to the solution.
b. a small amount of HO^- is added to the solution.

PROBLEM 50 SOLVED

You are planning to carry out a reaction that produces hydroxide ion. For the reaction to take place at a constant pH, it will be buffered at pH = 4.2. Would it be better to use a formic acid/formate buffer or an acetic acid/acetate buffer? (Note: The pK_a of formic acid = 3.75 and the pK_a of acetic acid = 4.76.)

SOLUTION Constant pH is maintained because the hydroxide ion produced in the reaction removes a proton from the acidic form of the buffer. Thus, the better choice of buffer is the one that has the highest concentration of buffer in the acidic form at pH = 4.2. Because formic acid's pK_a is 3.75, the majority of the buffer is in the basic form at pH = 4.2. Acetic acid, with pK_a = 4.76, has more buffer in the acidic form than in the basic form. Therefore, it is better to use acetic acid/acetate buffer for your reaction.

Blood: A Buffered Solution

Blood is the fluid that transports oxygen to all the cells of the human body. The normal pH of human blood is ~7.4. Death results if this pH decreases to less than ~6.8 or increases to greater than ~8.0 for even a few seconds.

Oxygen is carried to cells by a protein in the blood called hemoglobin (HbH^+). When hemoglobin binds O_2, hemoglobin loses a proton, which would make the blood more acidic if it did not contain a buffer to maintain its pH.

$$HbH^+ \ + \ O_2 \ \rightleftharpoons \ HbO_2 \ + \ H^+$$

A carbonic acid/bicarbonate (H_2CO_3/HCO_3^-) buffer controls the pH of blood. An important feature of this buffer is that carbonic acid decomposes to CO_2 and H_2O, as shown below:

$$\underset{\text{bicarbonate}}{HCO_3^-} \ + \ H^+ \ \rightleftharpoons \ \underset{\text{carbonic acid}}{H_2CO_3} \ \rightleftharpoons \ CO_2 \ + \ H_2O$$

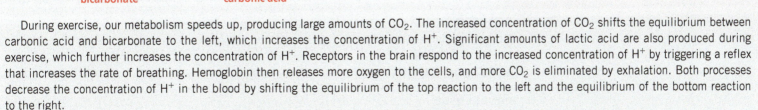

During exercise, our metabolism speeds up, producing large amounts of CO_2. The increased concentration of CO_2 shifts the equilibrium between carbonic acid and bicarbonate to the left, which increases the concentration of H^+. Significant amounts of lactic acid are also produced during exercise, which further increases the concentration of H^+. Receptors in the brain respond to the increased concentration of H^+ by triggering a reflex that increases the rate of breathing. Hemoglobin then releases more oxygen to the cells, and more CO_2 is eliminated by exhalation. Both processes decrease the concentration of H^+ in the blood by shifting the equilibrium of the top reaction to the left and the equilibrium of the bottom reaction to the right.

Thus, any disorder that decreases the rate and depth of ventilation, such as emphysema, decreases the pH of the blood—a condition called acidosis. In contrast, any excessive increase in the rate and depth of ventilation, as with hyperventilation due to anxiety, increases the pH of blood—a condition called alkalosis.

2.12 LEWIS ACIDS AND BASES

In 1923, G. N. Lewis offered new definitions for the terms acid and base. He defined

an acid as a species that accepts a share in an electron pair and
a base as a species that donates a share in an electron pair.

All Brønsted-Lowry (proton-donating) acids fit the Lewis definition because all proton-donating acids lose a proton and the proton accepts a share in an electron pair.

Remember that curved arrows show where the electrons start from and where they end up.

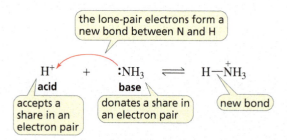

the lone-pair electrons form a new bond between N and H

$$H^+ \quad + \quad :NH_3 \quad \rightleftharpoons \quad H-\overset{+}{N}H_3$$

acid — accepts a share in an electron pair

base — donates a share in an electron pair

new bond

Lewis acids, however, are not limited to compounds that lose protons. Compounds such as aluminum chloride ($AlCl_3$), ferric bromide ($FeBr_3$), and borane (BH_3) are acids according to the Lewis definition because they have unfilled valence orbitals that can accept a share in an electron pair. These compounds react with a compound that has a lone pair, just as a proton reacts with a compound that has a lone pair.

Lewis acid: Need two from you.

Lewis base: Have pair, will share.

$$\begin{array}{c} Cl \\ | \\ Cl-Al \\ | \\ Cl \end{array} \quad + \quad CH_3\overset{..}{O}CH_3 \quad \rightleftharpoons \quad \begin{array}{cc} Cl & \\ | & \overset{new\ bond}{} \\ Cl-Al\overset{-}{-}\overset{+}{\overset{..}{O}}-CH_3 \\ | & | \\ Cl & CH_3 \end{array}$$

aluminum chloride
Lewis acid

dimethyl ether
Lewis base

$$\begin{array}{c} Br \\ | \\ Br-Fe \\ | \\ Br \end{array} \quad + \quad \begin{array}{c} H \\ | \\ :N-H \\ | \\ H \end{array} \quad \rightleftharpoons \quad \begin{array}{cc} Br & H \\ | & | \\ Br-\overset{-}{Fe}-\overset{+}{N}-H \\ | & | \\ Br & H \end{array}$$

ferric bromide
Lewis acid

ammonia
Lewis base

Thus, the Lewis definition of an acid includes all proton-donating compounds and some additional compounds that do not have protons. Throughout this text, the term *acid* is used to mean a proton-donating acid, and the term **Lewis acid** is used to refer to non-proton-donating acids such as $AlCl_3$ and BF_3.

All bases are **Lewis bases** because they all have a pair of electrons that they can share, either with a proton or with an atom such as aluminum, boron, or iron.

PROBLEM 51

Draw the products of the following reactions. Use curved arrows to show where the pair of electrons starts and where it ends up.

a. $ZnCl_2$ + $CH_3\overset{..}{\underset{..}{O}}H$ $\rightleftharpoons$ **b.** $FeBr_3$ + $:\overset{..}{\underset{..}{Br}}:^-$ $\rightleftharpoons$ **c.** $AlCl_3$ + $:\overset{..}{\underset{..}{Cl}}:^-$ $\rightleftharpoons$

PROBLEM 52

What products are formed when each of the following reacts with HO^-?

a. CH_3OH **c.** $CH_3\overset{+}{N}H_3$ **e.** $^+CH_3$ **g.** $AlCl_3$
b. $^+NH_4$ **d.** BF_3 **f.** $FeBr_3$ **h.** CH_3COOH

ESSENTIAL CONCEPTS

Section 2.1

- An **acid** is a species that loses a proton; a **base** is a species that gains a proton.
- **Acidity** is a measure of the tendency of a compound to lose a proton.
- **Basicity** is a measure of a compound's affinity for a proton.
- A strong base has a high affinity for a proton; a weak base has a low affinity for a proton.
- The stronger the acid, the weaker its conjugate base.

Section 2.2

- The strength of an acid is given by the **acid dissociation constant** (K_a) or by its pK_a value.
- The stronger the acid, the smaller its pK_a value.
- The **pH** of a solution indicates the concentration of protons in the solution; the smaller the pH, the more acidic the solution.

Section 2.3

- Approximate pK_a values are as follows: protonated alcohols, protonated carboxylic acids, and protonated water < 0; carboxylic acids ~ 5; protonated amines ~ 10; alcohols and water ~ 15.
- Curved arrows indicate the bonds that are broken and formed as reactants are converted into products.

Section 2.4

- The more acidic of two reactants is the one that loses a proton in an **acid–base reaction.**

Section 2.5

- In an acid–base reaction, the equilibrium favors formation of the weaker acid.

Section 2.6

- The strength of an acid is determined by the stability of its conjugate base: the more stable (weaker) the base, the stronger its conjugate acid.

- When atoms are similar in size, the strongest acid has its hydrogen attached to the most electronegative atom.
- Hybridization affects acidity because an sp hybridized atom is more electronegative than an sp^2 hybridized atom, which is more electronegative than an sp^3 hybridized atom.
- When atoms are very different in size, the strongest acid has its hydrogen attached to the largest atom.

Section 2.7

- **Inductive electron withdrawal** increases acidity: the more electronegative the electron-withdrawing group and the closer it is to the acidic hydrogen, the stronger the acid.

Section 2.8

- **Delocalized electrons** (electrons that are shared by more than two atoms) stabilize a compound.
- A **resonance hybrid** is a composite of the **resonance contributors**—structures that differ only in the location of their π electrons and lone-pair electrons.

Section 2.10

- The **Henderson–Hasselbalch equation** gives the relationship between pK_a and pH: a compound exists primarily in its acidic form (with its proton) in solutions more acidic than its pK_a value and primarily in its basic form (without its proton) in solutions more basic than its pK_a value.

Section 2.11

- A **buffer solution** contains both a weak acid and its conjugate base.

Section 2.12

- A **Lewis acid** is a species that accepts a share in an electron pair; a **Lewis base** is a species that donates a share in an electron pair.
- In this text, the term *acid* is used to mean a proton-donating acid; the term **Lewis acid** is used to refer to non-proton-donating acids such as $AlCl_3$ or BF_3.

PROBLEMS

53. Which is a stronger base?

- **a.** HS^- or HO^-
- **b.** CH_3O^- or $CH_3\overset{..}{N}H^-$
- **c.** CH_3OH or CH_3O^-
- **d.** Cl^- or Br^-
- **e.** CH_3COO^- or CF_3COO^-
- **f.** $CH_3CHClCOO^-$ or $CH_3CHBrCOO^-$

54. Draw curved arrows to show where the electrons start and where they end in the following reactions:

a. $\overset{..}{N}H_3 + H-\overset{..}{\underset{..}{C}l}: \rightleftharpoons\ ^+NH_4 + :\overset{..}{\underset{..}{C}l}:^-$

b. $H_2\overset{..}{O}: + FeBr_3 \rightleftharpoons H_2\overset{..}{O}{}^+\!\!-\!\bar{F}eBr_3$

c.

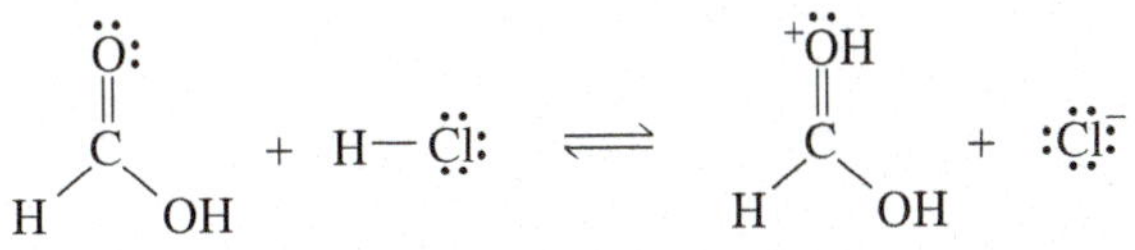

55. a. Rank the following alcohols from strongest to weakest acid:

$$CCl_3CH_2OH \qquad CH_2ClCH_2OH \qquad CHCl_2CH_2OH$$
$$K_a = 5.75 \times 10^{-13} \qquad K_a = 1.29 \times 10^{-13} \qquad K_a = 4.90 \times 10^{-13}$$

b. Explain the relative acidities.

56. a. Rank the following carboxylic acids from strongest to weakest acid:

$$CH_3CH_2CH_2COOH \qquad CH_3CH_2CHCOOH \qquad ClCH_2CH_2CH_2COOH \qquad CH_3CHCH_2COOH$$
$$K_a = 1.52 \times 10^{-5} \qquad\qquad\quad | \qquad\qquad\qquad K_a = 2.96 \times 10^{-5} \qquad\qquad\quad |$$
$$Cl \qquad\qquad\qquad\qquad\qquad\qquad\qquad\qquad Cl$$
$$K_a = 1.39 \times 10^{-3} \qquad\qquad\qquad\qquad\qquad\qquad\qquad K_a = 8.9 \times 10^{-5}$$

b. How does the presence of an electronegative substituent such as Cl affect the acidity of a carboxylic acid?
c. How does the location of the substituent affect the acidity of the carboxylic acid?

57. Draw the products of the following reactions:

a. $CH_3\ddot{O}CH_3 + BF_3 \longrightarrow$ **b.** $CH_3\ddot{O}CH_3 + H{-}Cl \longrightarrow$ **c.** $CH_3\ddot{N}H_2 + AlCl_3 \longrightarrow$

58. For the following compound,
a. draw its conjugate acid.   **b.** draw its conjugate base.

$$HOCH_2CH_2CH_2NH_2$$

59. Rank the following compounds from strongest to weakest acid:

$$CH_3CH_2OH \qquad CH_3CH_2NH_2 \qquad CH_3CH_2SH \qquad CH_3CH_2CH_3$$

60. For each of the following compounds, draw the form that predominates at pH $= 3$, pH $= 6$, pH $= 10$, and pH $= 14$:

a. CH_3COOH **b.** $CH_3CH_2\overset{+}{N}H_3$ **c.** CF_3CH_2OH
   $pK_a = 4.8$ $pK_a = 11.0$ $pK_a = 12.4$

61. Give the products of the following acid–base reactions and indicate whether reactants or products are favored at equilibrium. (Use the pK_a values that are given in Section 2.3.)

         O   O
         ‖   ‖

a. $CH_3\overset{\text{O}}{\overset{‖}{C}}OH + CH_3O^- \rightleftharpoons$ **c.** $CH_3\overset{\text{O}}{\overset{‖}{C}}OH + CH_3NH_2 \rightleftharpoons$
b. $CH_3CH_2OH + \ ^-NH_2 \rightleftharpoons$ **d.** $CH_3CH_2OH + HCl \rightleftharpoons$

62. a. Rank the following alcohols from strongest to weakest acid.
b. Explain the relative acidities.

$$CH_2{=}CHCH_2OH \qquad CH_3CH_2CH_2OH \qquad HC{\equiv}CCH_2OH$$

63. A single bond between two carbons with different hybridizations has a small dipole. What is the direction of the dipole in the indicated bonds?

a. $CH_3\overset{\downarrow}{-}CH{=}CH_2$ **b.** $CH_3\overset{\downarrow}{-}C{\equiv}CH$

64. For each compound, indicate the atom that is most apt to be protonated.

                                   CH_3 CH_3

a. $CH_3{-}\underset{\overset{|}{OH}}{CH}{-}CH_2NH_2$ **b.** $CH_3{-}\underset{\overset{|}{NH_2}}{\overset{\overset{|}{C}}{C}}{-}OH$ **c.** $CH_3{-}\underset{\overset{|}{NH_2}}{\overset{\overset{|}{C}}{C}}{-}CH_2OH$

65. a. Given the K_a values, estimate the pK_a value of each of the following acids without using a calculator (that is, is it between 3 and 4, between 9 and 10, and so on?):

    1. nitrous acid (HNO_2), $K_a = 4.0 \times 10^{-4}$ **3.** bicarbonate (HCO_3^-), $K_a = 6.3 \times 10^{-11}$ **5.** formic acid ($HCOOH$), $K_a = 2.0 \times 10^{-4}$
    2. nitric acid (HNO_3), $K_a = 22$ **4.** hydrogen cyanide (HCN), $K_a = 7.9 \times 10^{-10}$ **6.** phosphoric acid (H_3PO_4), $K_a = 2.1$

b. Determine the exact pK_a values, using a calculator.
c. Which is the strongest acid?

66. Tenormin, a member of the group of drugs known as beta-blockers, is used to treat high blood pressure and improve survival after a heart attack. It works by slowing down the heart to reduce its workload. Which atom in Tenormin is the most basic?

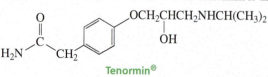

Tenormin®
atenolol

67. From which of the following compounds can HO⁻ remove a proton in a reaction that favors product formation?

$$CH_3COOH \qquad CH_3CH_2NH_2 \qquad CH_3CH_2\overset{+}{N}H_3 \qquad CH_3C\equiv CH$$
$$\text{A} \qquad\qquad \text{B} \qquad\qquad\quad \text{C} \qquad\qquad\quad \text{D}$$

68. a. For each of the following pairs of reactions, indicate which one has the more favorable equilibrium constant (that is, which one most favors products):

1. $CH_3CH_2OH + NH_3 \rightleftharpoons CH_3CH_2O^- + \overset{+}{N}H_4$
   or
   $CH_3OH + NH_3 \rightleftharpoons CH_3O^- + \overset{+}{N}H_4$

2. $CH_3CH_2OH + NH_3 \rightleftharpoons CH_3CH_2O^- + \overset{+}{N}H_4$
   or
   $CH_3CH_2OH + CH_3NH_2 \rightleftharpoons CH_3CH_2O^- + CH_3\overset{+}{N}H_3$

b. Which of the four reactions has the most favorable equilibrium constant?

69. You are planning to carry out a reaction that produces protons. The reaction will be buffered at pH = 10.5. Would it be better to use a protonated methylamine/methylamine buffer or a protonated ethylamine/ethylamine buffer? (pK_a of protonated methylamine = 10.7; pK_a of protonated ethylamine = 11.0)

70. Which is a stronger acid?

a. $CH_2=CHCOOH$ or CH_3CH_2COOH

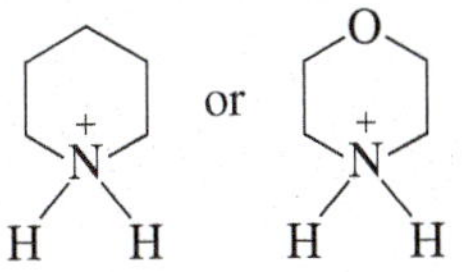

c. $CH_2=CHCOOH$ or $HC\equiv CCOOH$

b.

or

d.

or

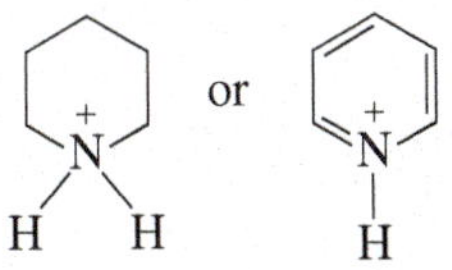

71. a. Without using a calculator, estimate the pH of each of the following solutions:
   1. $[HO^-] = 3.2 \times 10^{-5}$ 2. $[H_3O^+] = 8.3 \times 10^{-1}$ 3. $[H_3O^+] = 1.7 \times 10^{-3}$

b. Determine the exact pH, using a calculator.

72. Citrus fruits are rich in citric acid, a compound with three COOH groups. Explain the following:

a. The first pK_a (for the COOH group in the center of the molecule) is lower than the pK_a of acetic acid.
b. The third pK_a is greater than the pK_a of acetic acid.

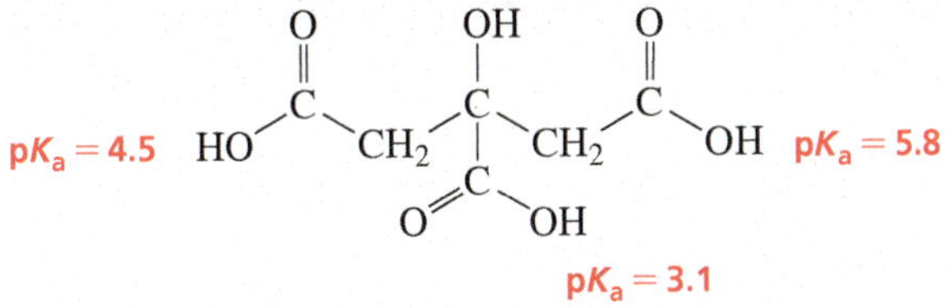

73. Given that pH + pOH = 14 and that the concentration of water in a solution of water is 55.5 M, show that the pK_a of water is 15.7.
(*Hint:* pOH = −log [HO⁻])

74. How could you separate a mixture of the following compounds? The reagents available to you are water, ether, 1.0 M HCl, and 1.0 M NaOH.
(*Hint:* See Problem 47.)

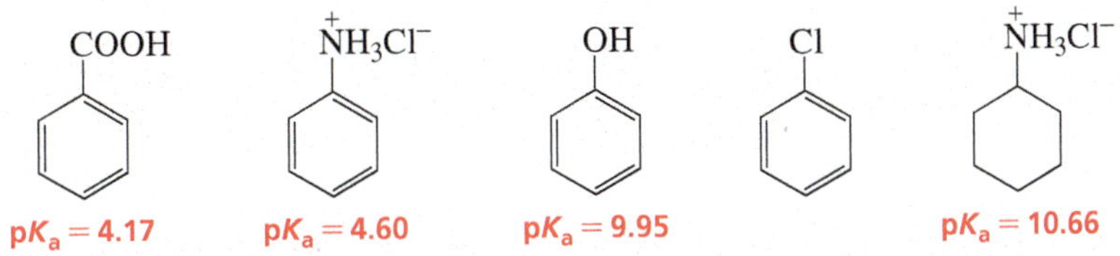

75. Carbonic acid has a pK_a of 6.1 at physiological temperature. Is the carbonic acid/bicarbonate buffer system that maintains the pH of the blood at 7.4 better at neutralizing excess acid or excess base?

76. a. If an acid with a pK_a of 5.3 is in an aqueous solution of pH 5.7, what percentage of the acid is present in its acidic form?
b. At what pH does 80% of the acid exist in its acidic form?

77. Calculate the pH values of the following solutions: (*Hint:* See Special Topic I in the *Study Guide and Solutions Manual*.)

a. a 1.0 M solution of acetic acid ($pK_a = 4.76$)
b. a 0.1 M solution of protonated methylamine ($pK_a = 10.7$)
c. a solution containing 0.3 M HCOOH and 0.1 M HCOO⁻ (pK_a of HCOOH = 3.76)

ESSENTIAL SKILL
TUTORIAL

Enhanced by MasteringChemistry®

TUTORIAL ACIDS AND BASES

This tutorial is designed to give you practice working problems based on some of the concepts you learned in Chapter 2. Most of the concepts are given here without explanation because full explanations can be found in Chapter 2.

An Acid and Its Conjugate Base

An acid is a species that can lose a proton (the Brønsted–Lowry definition). When an acid loses a proton (H^+), it forms its conjugate base. When the proton comes off the acid, the conjugate base retains the electron pair that had attached the proton to the acid.

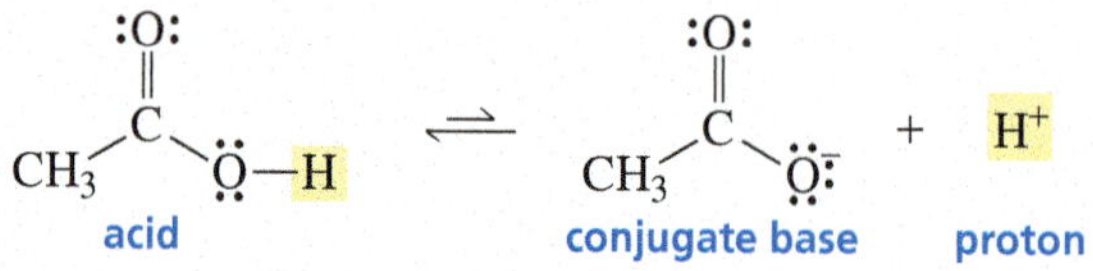

Often, the lone pairs and bonding electrons are not shown.

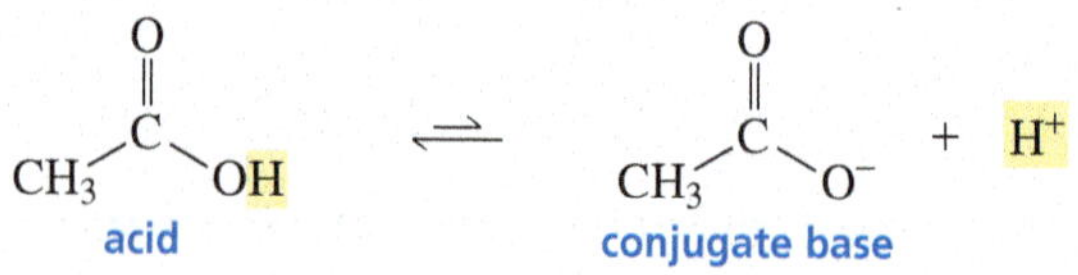

$$CH_3\overset{+}{O}H_2 \rightleftharpoons CH_3OH + H^+$$

acid — conjugate base

Notice that a neutral acid forms a negatively charged conjugate base, whereas a positively charged acid forms a neutral conjugate base. (The difference in charge *decreases* by one because the acid *loses* H^+.)

PROBLEM 1 Draw the conjugate base of each of the following acids:

a. CH_3OH **b.** $CH_3\overset{+}{N}H_3$ **c.** CH_3NH_2 **d.** H_3O^+ **e.** H_2O

A Base and Its Conjugate Acid

A base is a species that can gain a proton (the Brønsted–Lowry definition). When a base gains a proton (H^+), it forms its conjugate acid. To gain a proton, a base must have a lone pair that it can use to form a new bond with the proton.

$$CH_3\ddot{\underset{\cdot\cdot}{O}}{:}^- + H^+ \rightleftharpoons CH_3\ddot{\underset{\cdot\cdot}{O}}{-}H$$

base — conjugate acid

Often, the lone pairs and bonding electrons are not shown.

$$CH_3O^- + H^+ \rightleftharpoons CH_3OH$$

base — conjugate acid

$$CH_3NH_2 + H^+ \rightleftharpoons CH_3\overset{+}{N}H_3$$

base — conjugate acid

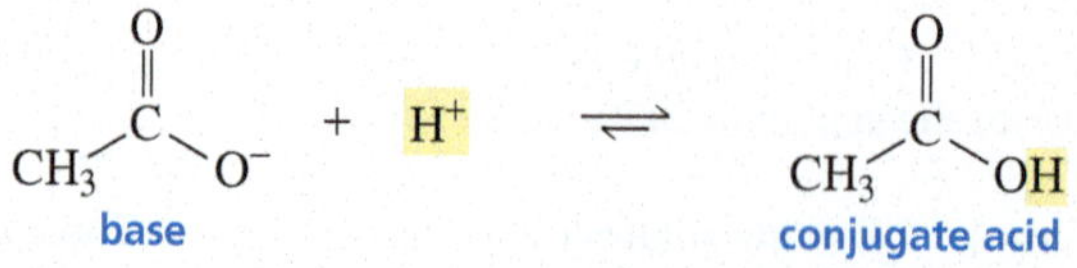

Notice that a negatively charged base forms a neutral conjugate acid, whereas a neutral base forms a positively charged conjugate acid. (The difference in charge *increases* by one because the compound *gains* H^+.)

PROBLEM 2 Draw the conjugate acid of each of the following bases:

a. H_2O **b.** HO^- **c.** CH_3OH **d.** NH_3 **e.** Cl^-

Acid–Base Reactions

An acid cannot lose a proton unless a base is present to accept the proton. Therefore, an acid always reacts with a base. The reaction of an acid with a base is called an acid–base reaction or a proton transfer reaction. Most acid–base reactions are reversible reactions.

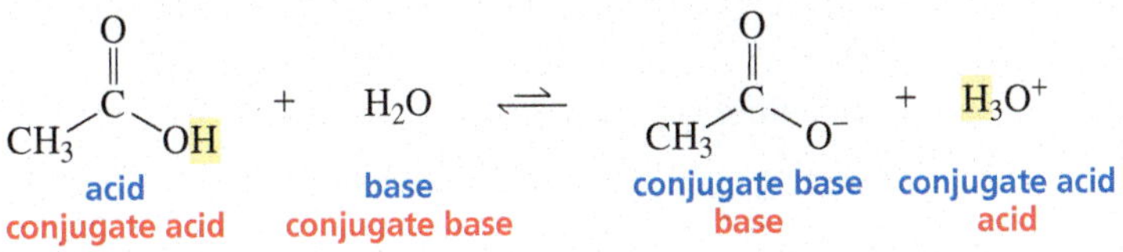

Notice that an acid reacts with a base in the forward direction (blue labels) and an acid reacts with a base in the reverse direction (red labels).

The Products of an Acid–Base Reaction

Both CH_3COOH and H_2O in the preceding reaction have protons that can be lost (that is, both can act as acids), and both have lone pairs that can form a bond with a proton (that is, both can act as bases). How do we know which reactant loses a proton and which gains a proton? We can determine this by comparing the pK_a values of the two reactants; these values are 4.8 for CH_3COOH and 15.7 for H_2O. The stronger acid (the one with the lower pK_a value) is the one that acts as an acid (it loses a proton). The other reactant acts as a base (it gains a proton).

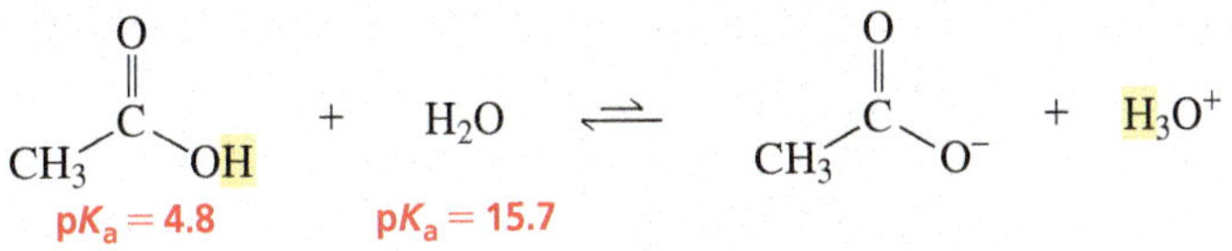

PROBLEM 3 Draw the products of the following acid–base reactions:

a. $CH_3\overset{+}{N}H_3$ + H_2O **c.** $CH_3\overset{+}{N}H_3$ + HO^-

b. HBr + CH_3OH **d.** CH_3NH_2 + CH_3OH

The Position of Equilibrium

Whether an acid–base reaction favors formation of the products or formation of the reactants can be determined by comparing the pK_a value of the acid that loses a proton in the forward direction with the pK_a value of the acid that loses a proton in the reverse direction. The equilibrium favors the reaction of the stronger acid to form the weaker acid. The following reaction favors formation of the reactants, because $CH_3\overset{+}{O}H_2$ is a stronger acid than CH_3COOH.

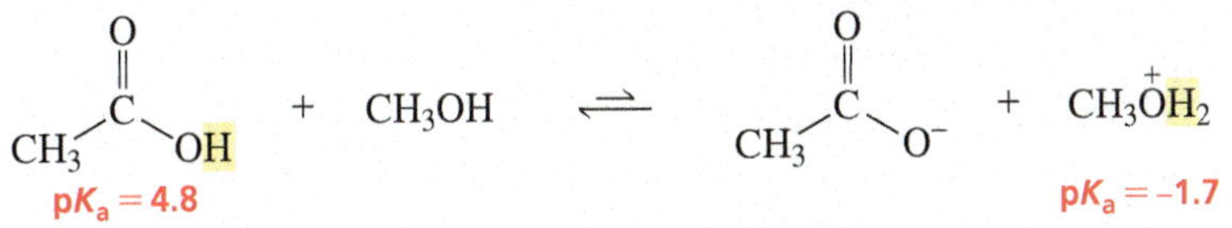

The next reaction favors formation of the products, because HCl is a stronger acid than $CH_3\overset{+}{N}H_3$.

$$HCl \;+\; CH_3NH_2 \;\rightleftharpoons\; Cl^- \;+\; CH_3\overset{+}{N}H_3$$
$$\text{p}K_a = -7 \qquad\qquad\qquad\qquad\qquad\qquad \text{p}K_a = 10.7$$

PROBLEM 4 Which of the reactions in Problem 3 favor formation of the reactants, and which favor formation of the products? (The pK_a values can be found in Sections 2.3 and 2.6.)

Relative Acid Strengths When the Proton Is Attached to Atoms Similar in Size

The atoms in the second row of the periodic table are similar in size, but they have different electronegativities.

relative electronegativities

$$C \;<\; N \;<\; O \;<\; F \;\text{—(most electronegative)}$$

When acids have protons attached to atoms similar in size, the strongest acid is the one with the proton attached to the more electronegative atom. The relative acid strengths are as follows:

$$\text{(strongest acid)—}HF \;>\; H_2O \;>\; NH_3 \;>\; CH_4\text{—(weakest acid)}$$

A positively charged atom is more electronegative than the same atom when it is neutral. Therefore,

$$CH_3\overset{+}{N}H_3 \quad\text{is more acidic than}\quad CH_3NH_2$$

$$CH_3\overset{+}{O}H_2 \quad\text{is more acidic than}\quad CH_3OH$$

When the relative strengths of two acids are determined by comparing the electronegativities of the atoms to which the protons are attached, both acids must possess the same charge. Therefore,

$$CH_3\overset{+}{O}H_2 \quad\text{is more acidic than}\quad CH_3\overset{+}{N}H_3$$

$$CH_3OH \quad\text{is more acidic than}\quad CH_3NH_2$$

PROBLEM 5 Which is the stronger acid?

a. CH_3OH or CH_3CH_3 c. CH_3NH_2 or HF
b. CH_3OH or HF   d. CH_3NH_2 or CH_3OH

The Effect of Hybridization on Acidity

The electronegativity of an atom depends on its hybridization.

$$\text{(most electronegative)—}sp \;>\; sp^2 \;>\; sp^3$$

Once again, the strongest acid has its proton attached to the most electronegative atom. Thus, the relative acid strengths are as follows:

$$\text{(strongest acid)—}HC\equiv CH \;>\; H_2C=CH_2 \;>\; CH_3CH_3\text{—(weakest acid)}$$
$$\qquad\qquad\quad sp \qquad\qquad\quad sp^2 \qquad\qquad sp^3$$

PROBLEM 6 Which is the stronger acid?

a. CH_3CH_3 or $HC\equiv CH$ **b.** $H_2C=CH_2$ or $HC\equiv CH$ **c.** $H_2C=CH_2$ or CH_3CH_3

Relative Acid Strengths When the Proton Is Attached to Atoms Very Different in Size

The atoms in a column of the periodic table become considerably larger as you go down the column.

largest halide ion $\quad I^- > Br^- > Cl^- > F^-\quad$ smallest halide ion

When comparing two acids with protons attached to atoms that are very different in size, the stronger acid is the one attached to the larger atom. Thus, the relative acid strengths are as follows:

strongest acid $\quad HI > HBr > HCl > HF\quad$ weakest acid

PROBLEM 7 ◆ Which is the stronger acid? (*Hint*: You can use the periodic table at the back of this book.)

a. HCl or HBr **b.** CH_3OH or CH_3SH **c.** HF or HCl **d.** H_2S or H_2O

The Effect of Inductive Electron Withdrawal on Acidity

Replacing a hydrogen with an electronegative substituent—one that pulls bonding electrons toward itself—increases the strength of the acid.

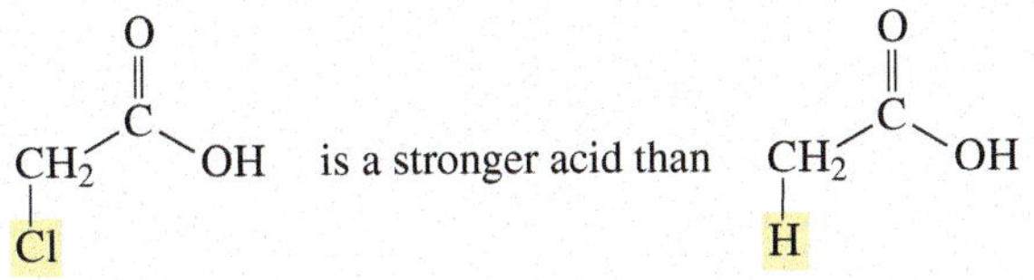

is a stronger acid than

The halogens have the following relative electronegativities:

most electronegative $\quad F > Cl > Br > I\quad$ least electronegative

The more electronegative the substituent that replaces a hydrogen, the stronger the acid. Thus, the relative acid strengths are as follows:

strongest acid $\quad CH_3\underset{F}{CH}COOH > CH_3\underset{Cl}{CH}COOH > CH_3\underset{Br}{CH}COOH > CH_3\underset{I}{CH}COOH\quad$ weakest acid

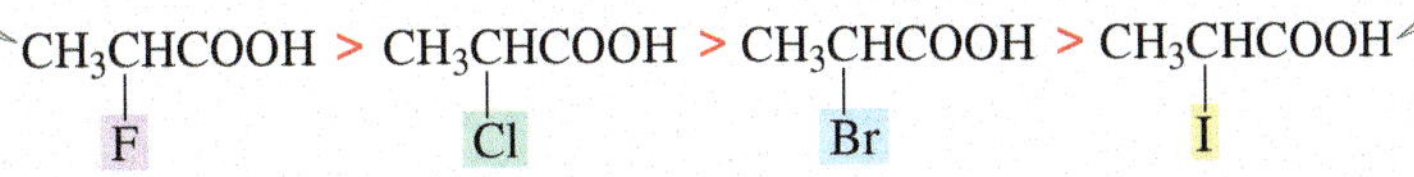

The closer the electronegative substituent is to the group that loses a proton, the stronger the acid is. Thus, the relative acid strengths are as follows:

strongest acid $\quad CH_3CH_2\underset{Cl}{CH}COOH > CH_3\underset{Cl}{CH}CH_2COOH > \underset{Cl}{CH_2}CH_2CH_2COOH\quad$ weakest acid

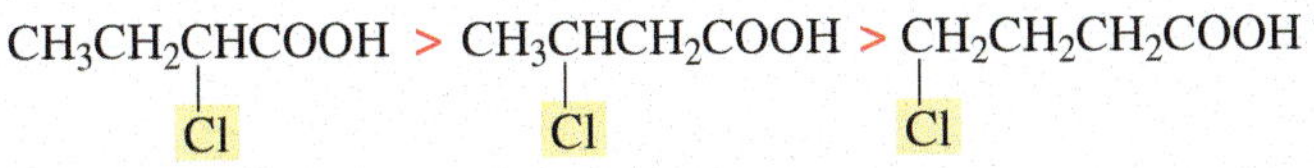

PROBLEM 8 Which is the stronger acid?

a. $ClCH_2CH_2OH$ or FCH_2CH_2OH

c. $CH_3CH_2OCH_2OH$ or $CH_3OCH_2CH_2OH$

b. $\underset{Br}{\overset{Br}{CH_2}}CHCH_2OH$ or $CH_3\underset{Br}{\overset{Br}{C}}CH_2OH$

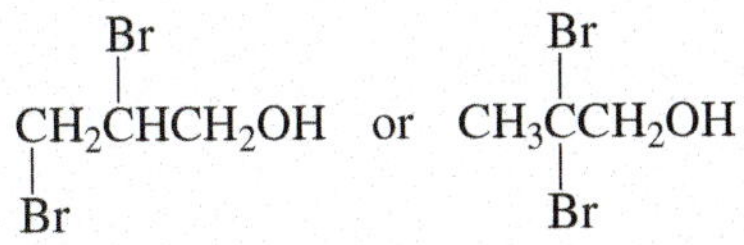

d. $CH_3\overset{O}{\overset{\|}{C}}CH_2CH_2OH$ or $CH_3CH_2\overset{O}{\overset{\|}{C}}CH_2OH$

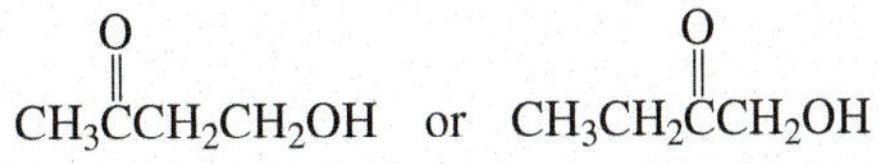

Relative Base Strengths

Strong bases readily share their electrons with a proton. In other words, the conjugate acid of a *strong* base is a *weak* acid because it does not readily lose a proton. This allows us to say, *the stronger the base, the weaker its conjugate acid* (or *the stronger the acid, the weaker its conjugate base*).

For example, which is the stronger base?

a. CH_3O^- or $CH_3\overset{-}{N}H$ **b.** $HC\equiv C^-$ or $CH_3\overset{-}{C}H_2$

To answer the question, first compare their conjugate acids:

a. CH_3OH is a stronger acid than CH_3NH_2 (because O is more electronegative than N). Because the stronger acid has the weaker conjugate base, $CH_3\overset{-}{N}H$ is a stronger base than CH_3O^-.
b. $HC\equiv CH$ is a stronger acid than CH_3CH_3 (an sp hybridized carbon is more electronegative than an sp^3 hybridized carbon). Therefore, $CH_3\overset{-}{C}H_2$ is a stronger base.

PROBLEM 9 Which is the stronger base?

a. Br^- or I^- **d.** $H_2C=\overset{-}{C}H$ or $HC\equiv C^-$
b. CH_3O^- or CH_3S^- **e.** $FCH_2CH_2COO^-$ or $BrCH_2CH_2COO^-$
c. $CH_3CH_2O^-$ or CH_3COO^- **f.** $ClCH_2CH_2O^-$ or $Cl_2CHCH_2O^-$

Weak Bases Are Stable Bases

Weak bases are stable bases because they readily bear the electrons they formerly shared with a proton. Therefore, we can say, *the weaker the base, the more stable it is.* We can also say, *the stronger the acid, the more stable (the weaker) its conjugate base.*

For example, which is a more stable base: Cl^- or Br^-?

To determine this, first compare their conjugate acids:

HBr is a stronger acid than HCl (because Br is larger than Cl). Therefore, Br^- is a more stable (weaker) base.

PROBLEM 10 Which is the more stable base?

a. Br^- or I^- **d.** $H_2C=\overset{-}{C}H$ or $HC\equiv C^-$
b. CH_3O^- or CH_3S^- **e.** $FCH_2CH_2COO^-$ or $BrCH_2CH_2COO^-$
c. $CH_3CH_2O^-$ or CH_3COO^- **f.** $ClCH_2CH_2O^-$ or $Cl_2CHCH_2O^-$

Electron Delocalization Stabilizes a Base

If a base has localized electrons, then the negative charge that results when the base's conjugate acid loses a proton belongs to one atom. On the other hand, if a base has delocalized electrons, then the negative charge that results when the base's conjugate acid loses a proton is shared by two (or more) atoms. A base with delocalized electrons is more stable than a similar base with localized electrons.

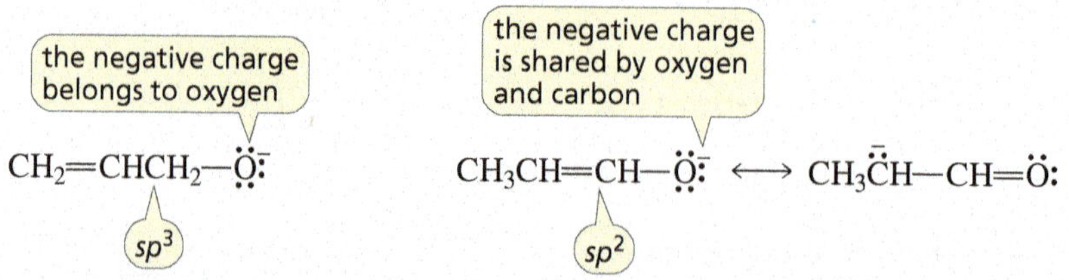

To determine if a base has delocalized electrons, we simply need to check the electrons left behind when the base's conjugate acid loses a proton.

- If these electrons are on an atom bonded to an sp^3 carbon, then the electrons belong to only one atom—that is, the electrons are localized.

- If these electrons are on an atom bonded to an sp^2 carbon, then the electrons are delocalized.

PROBLEM 11   Which is a more stable base?

a.

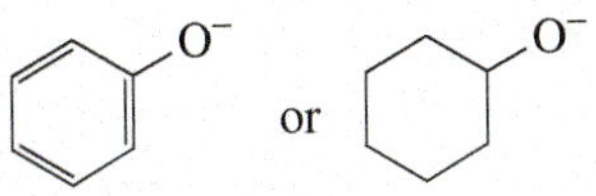

or

b.

or

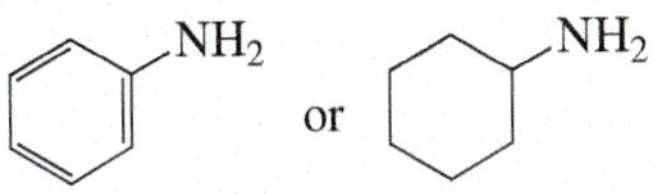

Remembering that the more stable (weaker) base has the stronger conjugate acid, solve Problem 12.

PROBLEM 12   Which is the stronger acid?

a.

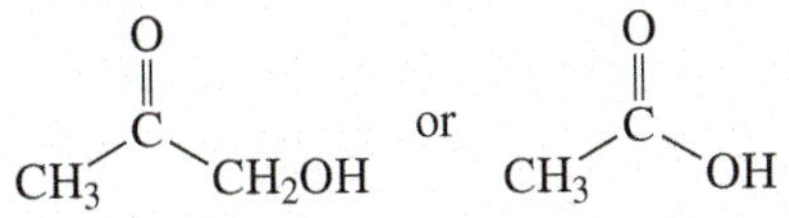

or

b.

or

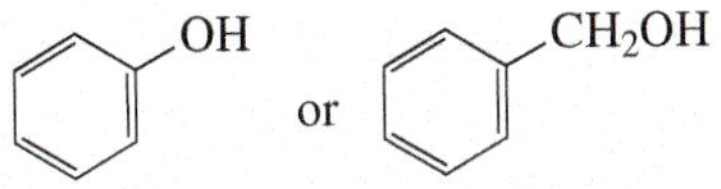

Compounds With More Than One Acidic Group

If a compound has two acidic groups, then a base will remove a proton from the more acidic of the two groups first. If a second equivalent of base is added, then the base will remove a proton from the less acidic group.

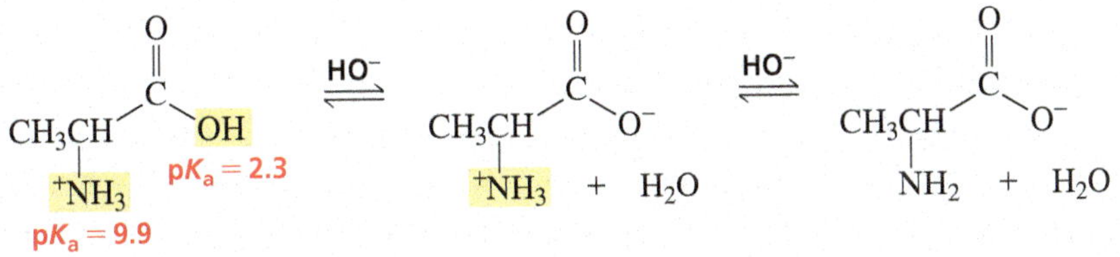

Similarly, if a compound has two basic groups, then an acid will protonate the more basic of the two groups first. If a second equivalent of acid is added, then the acid will protonate the less basic group.

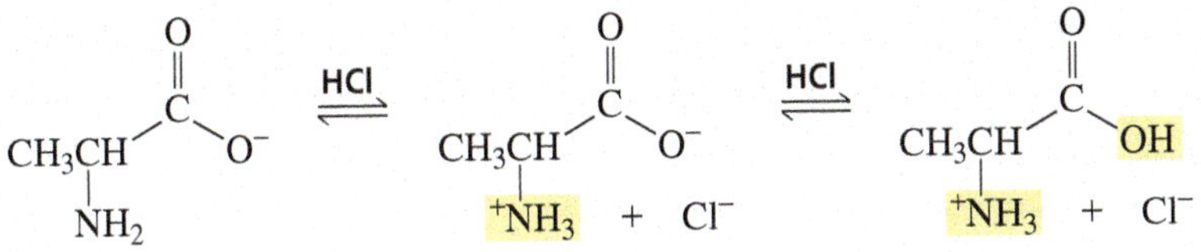

PROBLEM 13

a. What species is formed when one equivalent of HCl is added to $HOCH_2CH_2NH_2$?

b. Does the following compound exist?

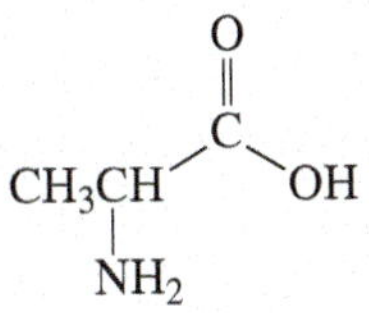

The Effect of pH on Structure

Whether an acid is in its acidic form (with its proton) or its basic form (without its proton) depends on the pK_a value of the acid and the pH of the solution:

- When $pH < pK_a$, the compound exists primarily in its acidic form.
- When $pH > pK_a$, the compound exists primarily in its basic form.

In other words, when the solution is more acidic than the pK_a value of the acid, the compound is in its acidic form. But when the solution is more basic than the pK_a value of the acid, the compound is in its basic form.

PROBLEM 14

a. Draw the structure of CH_3COOH ($pK_a = 4.7$) at pH = 2, pH = 7, and pH = 10.

b. Draw the structure of CH_3OH ($pK_a = 15.5$) at pH = 2, pH = 7, and pH = 10.

c. Draw the structure of $CH_3\overset{+}{N}H_3$ ($pK_a = 10.7$) at pH = 2, pH = 7, and pH = 14.

ANSWERS TO PROBLEMS ON ACIDS AND BASES

PROBLEM 1 SOLVED

a. CH_3O^- b. CH_3NH_2 c. $CH_3\bar{N}H$ d. H_2O e. HO^-

PROBLEM 2 SOLVED

a. H_3O^+ b. H_2O c. $CH_3\overset{+}{O}H_2$ d. $^+NH_4$ e. HCl

PROBLEM 3 SOLVED

a. $CH_3\overset{+}{N}H_3$ + H_2O $\rightleftharpoons$ CH_3NH_2 + H_3O^+

b. HBr + CH_3OH $\rightleftharpoons$ Br^- + $CH_3\overset{+}{O}H_2$

c. $CH_3\overset{+}{N}H_3$ + HO^- $\rightleftharpoons$ CH_3NH_2 + H_2O

d. CH_3NH_2 + CH_3OH $\rightleftharpoons$ $CH_3\overset{+}{N}H_3$ + CH_3O^-

PROBLEM 4 SOLVED

a. reactants   b. products   c. products   d. reactants

PROBLEM 5 SOLVED

a. CH_3OH b. HF c. HF d. CH_3OH

PROBLEM 6 SOLVED

a. $HC\equiv CH$ b. $HC\equiv CH$ c. $H_2C=CH_2$

PROBLEM 7 SOLVED

a. HBr b. CH_3SH c. HCl d. H_2S

PROBLEM 8 SOLVED

a. FCH_2CH_2OH b. $CH_3\overset{\displaystyle Br}{\underset{\displaystyle Br}{C}}CH_2OH$ c. $CH_3CH_2OCH_2OH$ d. $CH_3CH_2\overset{\displaystyle O}{\overset{\|}{C}}CH_2OH$

PROBLEM 9 SOLVED

a. Br^- c. $CH_3CH_2O^-$ e. $BrCH_2CH_2COO^-$

b. CH_3O^- d. $H_2C=\bar{C}H$ f. $ClCH_2CH_2O^-$

PROBLEM 10 SOLVED

a. I^- c. CH_3COO^- e. $FCH_2CH_2COO^-$

b. CH_3S^- d. $HC\equiv C^-$ f. $Cl_2CHCH_2O^-$

PROBLEM 11 **SOLVED**

a.   **b.**

PROBLEM 12 **SOLVED**

a. ![structure of acetic acid with O double bonded to C, CH₃ and OH attached] **b.** ![structure of phenol with OH attached to benzene ring]

PROBLEM 13 **SOLVED**

a. $HOCH_2CH_2\overset{+}{N}H_3$

b. The compound does not exist. For it to be formed, a base would have to be able to remove a proton from a group with a $pK_a = 9.9$ more readily than it would remove a proton from a group with a $pK_a = 2.3$. This is not possible, because the lower the pK_a, the stronger the acid—that is, the more readily the group loses a proton. In other words, a weak acid cannot lose a proton more readily than a strong acid can.

PROBLEM 14 **SOLVED**

a. CH_3COOH at pH = 2, because pH < pK_a
    CH_3COO^- at pH = 7 and 10, because pH > pK_a

b. CH_3OH at pH = 2, 7, and 10, because pH < pK_a

c. $CH_3\overset{+}{N}H_3$ at pH = 2 and 7, because pH < pK_a
    CH_3NH_2 at pH = 14, because pH > pK_a

3

An Introduction to Organic Compounds

Nomenclature, Physical Properties, and Structure

A solar eclipse occurs when the Moon aligns with and obscures the Sun. Eclipsed conformers align the same way.

The material in this chapter explains why drugs with similar physiological effects often have similar structures, how high cholesterol is treated clinically, why fish is served with lemon, how the octane number of gasoline is determined, and why starch (a component of many of the foods we eat) and cellulose (the structural material of plants) have such different physical properties even though both are composed only of glucose.

The presentation of organic chemistry in this book is organized according to how organic compounds react. When a compound undergoes a reaction, a new compound is synthesized. In other words, while you are learning how organic compounds react, you are simultaneously learning how to synthesize organic compounds.

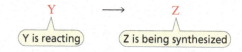

The compounds that are synthesized by the reactions we will study in Chapters 5–12 are primarily alkanes, alkyl halides, ethers, alcohols, and amines. Later in this chapter, we will look at the structures and physical properties of compounds in these five families. As we learn about these compounds, we must be able to refer to them by name. Therefore, we will begin our study of organic chemistry by learning how to name these five families of compounds.

Alkanes

Alkanes are composed of only carbon atoms and hydrogen atoms and contain only *single bonds*. Compounds that contain only carbon and hydrogen are called **hydrocarbons.** Thus, an alkane is a hydrocarbon that has only single bonds.

Alkanes in which the carbons form a continuous chain with no branches are called **straight-chain alkanes.** The names of the four smallest straight-chain alkanes have historical roots, but the others are based on Greek numbers. It is important that you learn the names of at least the first 10 straight-chain alkanes in Table 3.1.

Table 3.1 Nomenclature and Physical Properties of Straight-Chain Alkanes

Number of carbons	Molecular formula	Name	Condensed structure	Boiling point (°C)	Melting point (°C)	Density[a] (g/mL)
1	CH_4	methane	CH_4	−167.7	−182.5	
2	C_2H_6	ethane	CH_3CH_3	−88.6	−183.3	
3	C_3H_8	propane	$CH_3CH_2CH_3$	−42.1	−187.7	
4	C_4H_{10}	butane	$CH_3CH_2CH_2CH_3$	−0.5	−138.3	
5	C_5H_{12}	pentane	$CH_3(CH_2)_3CH_3$	36.1	−129.8	0.5572
6	C_6H_{14}	hexane	$CH_3(CH_2)_4CH_3$	68.7	−95.3	0.6603
7	C_7H_{16}	heptane	$CH_3(CH_2)_5CH_3$	98.4	−90.6	0.6837
8	C_8H_{18}	octane	$CH_3(CH_2)_6CH_3$	125.7	−56.8	0.7026
9	C_9H_{20}	nonane	$CH_3(CH_2)_7CH_3$	150.8	−53.5	0.7177
10	$C_{10}H_{22}$	decane	$CH_3(CH_2)_8CH_3$	174.0	−29.7	0.7299
11	$C_{11}H_{24}$	undecane	$CH_3(CH_2)_9CH_3$	195.8	−25.6	0.7402
12	$C_{12}H_{26}$	dodecane	$CH_3(CH_2)_{10}CH_3$	216.3	−9.6	0.7487
13	$C_{13}H_{28}$	tridecane	$CH_3(CH_2)_{11}CH_3$	235.4	−5.5	0.7546
⋮	⋮	⋮	⋮	⋮	⋮	⋮
20	$C_{20}H_{42}$	eicosane	$CH_3(CH_2)_{18}CH_3$	343.0	36.8	0.7886
21	$C_{21}H_{44}$	heneicosane	$CH_3(CH_2)_{19}CH_3$	356.5	40.5	0.7917
⋮	⋮	⋮	⋮	⋮	⋮	⋮
30	$C_{30}H_{62}$	triacontane	$CH_3(CH_2)_{28}CH_3$	449.7	65.8	0.8097

[a]Density is temperature-dependent. The densities given are those determined at 20 °C ($d^{20°}$).

The family of alkanes shown in the table is an example of a homologous series. A **homologous series** (*homos* is Greek for "the same as") is a family of compounds in which each member differs from the one before it in the series by one **methylene (CH_2) group.** The members of a homologous series are called **homologues.**

homologues differ by one CH_2 group

$CH_3CH_2CH_3$ $CH_3CH_2CH_2CH_3$

The relative numbers of carbons and hydrogens in the alkanes in Table 3.1 show that the general molecular formula for an alkane is C_nH_{2n+2}, where n is any positive integer. So if an alkane has one carbon, it must have four hydrogens; if it has two carbons, it must have six hydrogens; and so on.

Methane, Ethane, and Propane

We saw that carbon forms four covalent bonds and hydrogen forms only one covalent bond (Section 1.4). This means that there is only one possible structure for an alkane with molecular formula CH_4 (methane) and only one possible structure for an alkane with molecular formula C_2H_6 (ethane). We examined the structures of these compounds in Section 1.7. There is also only one possible structure for an alkane with molecular formula C_3H_8 (propane).

name	molecular formula	Kekulé structure	condensed structure	ball-and-stick model
methane	CH_4	$\begin{array}{c}H\\ \mid \\ H-C-H \\ \mid \\ H\end{array}$	CH_4	
ethane	C_2H_6	$\begin{array}{cc}H & H\\ \mid & \mid \\ H-C-C-H \\ \mid & \mid \\ H & H\end{array}$	CH_3CH_3	
propane	C_3H_8	$\begin{array}{ccc}H & H & H\\ \mid & \mid & \mid \\ H-C-C-C-H \\ \mid & \mid & \mid \\ H & H & H\end{array}$	$CH_3CH_2CH_3$	
butane	C_4H_{10}	$\begin{array}{cccc}H & H & H & H\\ \mid & \mid & \mid & \mid \\ H-C-C-C-C-H \\ \mid & \mid & \mid & \mid \\ H & H & H & H\end{array}$	$CH_3CH_2CH_2CH_3$	

Butane

There are, however, two possible structures for an alkane with molecular formula C_4H_{10}—a straight-chain alkane called butane and a branched alkane called isobutane. Both of these structures fulfill the requirement that each carbon forms four bonds and each hydrogen forms one bond.

$CH_3CH_2CH_2CH_3$ *butane*

$\underset{\underset{CH_3}{|}}{CH_3CHCH_3}$ *isobutane*

constitutional isomers

$\underset{\underset{CH_3}{|}}{CH_3CH-}$ a carbon bonded to 1 H and 2 CH_3 groups is called an "iso" structural unit

Compounds such as butane and isobutane that have the same molecular formula but differ in the way the atoms are connected are called **constitutional isomers**—their molecules have different constitutions. In fact, isobutane got its name because it is an *iso*mer of butane. The structural unit consisting of *a carbon bonded to a hydrogen and two CH_3 groups*, which occurs in isobutane, has come to be called "iso." Thus, the name *isobutane* tells you that the compound is a four-carbon alkane with an iso structural unit.

Pentane

There are three alkanes with molecular formula C_5H_{12}. You have already learned how to name two of them. Pentane is the straight-chain alkane. Isopentane, as its name indicates, has an iso structural unit and five carbons. We cannot name the other branched-chain alkane without defining a name for a new structural unit. (For now, ignore the names written in blue.)

$CH_3CH_2CH_2CH_2CH_3$ pentane

$\underset{\underset{CH_3}{|}}{CH_3CHCH_2CH_3}$ isopentane

$\underset{\underset{CH_3}{|}}{\overset{\overset{CH_3}{|}}{CH_3CCH_3}}$ 2,2-dimethylpropane

Hexane

There are five constitutional isomers with molecular formula C_6H_{14}. Again, we are able to name only two of them, unless we define new structural units.

$$CH_3CH_2CH_2CH_2CH_2CH_3 \qquad CH_3CHCH_2CH_2CH_3 \qquad \begin{array}{c} CH_3 \\ | \\ CH_3CCH_2CH_3 \\ | \\ CH_3 \end{array}$$

common name: **hexane**

systematic name: **hexane**

CH_3 (below second structure)

isohexane
2-methylpentane

2,2-dimethylbutane

$$CH_3CH_2CHCH_2CH_3 \qquad CH_3CH-CHCH_3$$
$$| \qquad\qquad\qquad\qquad | \quad | $$
$$CH_3 \qquad\qquad\qquad CH_3 \ CH_3$$

3-methylpentane   **2,3-dimethylbutane**

Heptane

There are nine alkanes with molecular formula C_7H_{16}. We can name only two of them (heptane and isoheptane).

$$CH_3CH_2CH_2CH_2CH_2CH_2CH_3 \qquad CH_3CHCH_2CH_2CH_2CH_3$$

common name: **heptane**

systematic name: **heptane**

CH_3

isoheptane
2-methylhexane

$$CH_3CH_2CHCH_2CH_2CH_3 \quad CH_3CH-CHCH_2CH_3 \quad CH_3CHCH_2CHCH_3 \quad CH_3CH_2CHCH_2CH_3$$
$$| \qquad\qquad\qquad | \ \ | \qquad\qquad | \ \ | \qquad\qquad | $$
$$CH_3 \qquad\qquad\quad CH_3 \ CH_3 \qquad CH_3 \ CH_3 \qquad\quad CH_2CH_3$$

3-methylhexane   **2,3-dimethylpentane**   **2,4-dimethylpentane**   **3-ethylpentane**

$$\begin{array}{c} CH_3 \\ | \\ CH_3CCH_2CH_2CH_3 \\ | \\ CH_3 \end{array} \qquad \begin{array}{c} CH_3 \\ | \\ CH_3CH_2CCH_2CH_3 \\ | \\ CH_3 \end{array} \qquad \begin{array}{c} CH_3 \ CH_3 \\ | \quad | \\ CH_3C-CHCH_3 \\ | \\ CH_3 \end{array}$$

2,2-dimethylpentane   **3,3-dimethylpentane**   **2,2,3-trimethylbutane**

Systematic/IUPAC Nomenclature

The number of constitutional isomers increases rapidly as the number of carbons in an alkane increases. For example, there are 75 alkanes with molecular formula $C_{10}H_{22}$ and 4347 alkanes with molecular formula $C_{15}H_{32}$. To avoid having to memorize the names of thousands of structural units, chemists have devised rules for creating systematic names that describe the compound's structure. That way, only the rules must be learned. Because the name describes the structure, these rules make it possible to deduce the structure of a compound from its name.

This method of nomenclature is called **systematic nomenclature.** It is also called **IUPAC nomenclature** because it was designed by a commission of the International Union of Pure and Applied Chemistry (abbreviated IUPAC and pronounced "eye-you-pack") in 1892.

The IUPAC rules have been continually revised by the commission since then. A name such as *isobutane*—a nonsystematic name—is called a **common name.** When both names are shown in this book, common names are shown in red and systematic (IUPAC) names in blue. Before we can understand how a systematic name for an alkane is constructed, we must learn how to name alkyl groups.

PROBLEM 1 ♦

a. How many hydrogens does an alkane with 17 carbons have?

b. How many carbons does an alkane with 74 hydrogens have?

PROBLEM 2

Draw the structures of octane and isooctane.

3.1 ALKYL GROUPS

Removing a hydrogen from an alkane results in an **alkyl group** (or an **alkyl substituent**). Alky groups are named by replacing the "ane" ending of the alkane with "yl." The letter "R" is used to indicate any alkyl group.

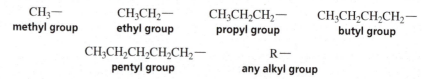

CH₃—	CH₃CH₂—	CH₃CH₂CH₂—	CH₃CH₂CH₂CH₂—
methyl group	ethyl group	propyl group	butyl group

CH₃CH₂CH₂CH₂CH₂—	R—
pentyl group	any alkyl group

If a hydrogen in an alkane is replaced by an OH, the compound becomes an **alcohol;** if it is replaced by an NH_2, the compound becomes an **amine;** if it is replaced by a halogen, the compound becomes an **alkyl halide;** and if it is replaced by an OR, the compound becomes an **ether.**

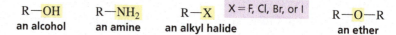

R—OH	R—NH₂	R—X	X = F, Cl, Br, or I	R—O—R
an alcohol	an amine	an alkyl halide		an ether

The alkyl group name followed by the name of the class of the compound (alcohol, amine, and so on) yields the common name of the compound. The two alkyl groups in ethers are listed in alphabetical order. The following examples show how alkyl group names are used to build common names:

CH₃OH	CH₃CH₂NH₂	CH₃CH₂CH₂Br	CH₃CH₂CH₂CH₂Cl
methyl alcohol	ethylamine	propyl bromide	butyl chloride

CH₃I	CH₃CH₂OH	CH₃CH₂CH₂NH₂	CH₃CH₂OCH₃
methyl iodide	ethyl alcohol	propylamine	ethyl methyl ether

Notice that for most compounds, there is a space between the name of the alkyl group and the name of the class of compound. For amines, however, the entire name is written as one word.

methyl alcohol

methyl chloride

methylamine

PROBLEM 3 ◆

Name each of the following:

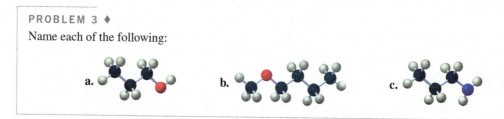

a.   b.   c.

Three-Carbon Alkyl Groups

There are two alkyl groups—the propyl group and the isopropyl group—that contain three carbons.

- A propyl group is obtained when a hydrogen is removed from *a primary carbon* of propane. A **primary carbon** is a carbon bonded to only one other carbon.

- An isopropyl group is obtained when a hydrogen is removed from the *secondary carbon* of propane. A **secondary carbon** is a carbon bonded to two other carbons.

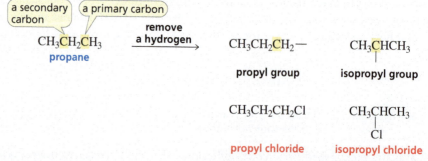

a secondary carbon ⟩ ⟨ a primary carbon		
CH₃CH₂CH₃	CH₃CH₂CH₂—	CH₃CHCH₃
propane	propyl group	isopropyl group

	CH₃CH₂CH₂Cl	CH₃CHCH₃
		Cl
	propyl chloride	isopropyl chloride

remove a hydrogen

Notice that an isopropyl group, as its name indicates, has three carbon atoms arranged as an iso structural unit—that is, a carbon bonded to a hydrogen and to two CH₃ groups.

Molecular structures can be drawn in different ways. For example, isopropyl chloride is drawn below in two ways. Both representations depict the same compound. Although the two-dimensional representations may appear at first to be different (the methyl groups are placed at opposite ends in one structure and at right angles in the other), the structures are identical because carbon is tetrahedral. The four groups bonded to the central carbon—a hydrogen, a chlorine, and two methyl groups—point to the corners of a tetrahedron. If you rotate the three-dimensional model on the right 90° in a clockwise direction, you should be able to see that the two models are the same.

two ways to draw isopropyl chloride

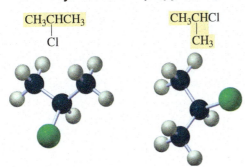

$$CH_3CHCH_3$$
$$|$$
$$Cl$$

$$CH_3CHCl$$
$$|$$
$$CH_3$$

NOTE TO THE STUDENT
• Build models of the two representations of isopropyl chloride to see that they represent the same compound.

Four-Carbon Alkyl Groups

There are four alkyl groups that contain four carbons. Two of them, the butyl and isobutyl groups, have a hydrogen removed from a primary carbon. A *sec*-butyl group has a hydrogen removed from a secondary carbon (*sec*-, sometimes abbreviated *s*-, stands for secondary), and a *tert*-butyl group has a hydrogen removed from a tertiary carbon (*tert*-, often abbreviated *t*-, stands for tertiary). A **tertiary carbon** is bonded to three other carbons. Notice that the isobutyl group is the only one with an iso structural unit.

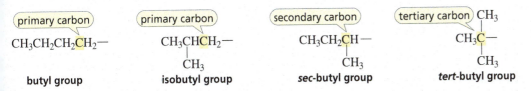

primary carbon	primary carbon	secondary carbon	tertiary carbon
$CH_3CH_2CH_2CH_2-$	CH_3CHCH_2- $\|$ CH_3	CH_3CH_2CH- $\|$ CH_3	CH_3 $\|$ CH_3C- $\|$ CH_3
butyl group	**isobutyl group**	**sec-butyl group**	**tert-butyl group**

A primary carbon is bonded to one carbon, a secondary carbon is bonded to two carbons, and a tertiary carbon is bonded to three carbons.

The names of straight-chain alkyl groups often have the prefix "*n*" (for "normal") to emphasize that the carbons are in an unbranched chain.

$$CH_3CH_2CH_2CH_2Br$$

butyl bromide
or
n-**butyl bromide**

$$CH_3CH_2CH_2CH_2CH_2F$$

pentyl fluoride
or
n-**pentyl fluoride**

Like the carbons, the hydrogens in a molecule are also referred to as primary, secondary, and tertiary. **Primary hydrogens** are attached to a primary carbon, **secondary hydrogens** are attached to a secondary carbon, and **tertiary hydrogens** are attached to a tertiary carbon.

primary hydrogens	secondary hydrogens	tertiary hydrogen
$CH_3CH_2CH_2CH_2OH$	CH_3CH_2CHOH $\|$ CH_3	CH_3CHCH_2OH $\|$ CH_3
primary carbon	secondary carbon	tertiary carbon

Primary hydrogens are attached to a primary carbon, secondary hydrogens to a secondary carbon, and tertiary hydrogens to a tertiary carbon.

A chemical name must specify one compound only. The prefix "*sec*," therefore, can be used only for *sec*-butyl compounds. The name "*sec*-pentyl" cannot be used because pentane has two different secondary carbons. Thus, removing a hydrogen from a secondary carbon of pentane produces one of two different alkyl groups, depending on which hydrogen is removed. As a result, *sec*-pentyl chloride would specify two different alkyl chlorides, so it is *not* a correct name.

A name must specify one compound only.

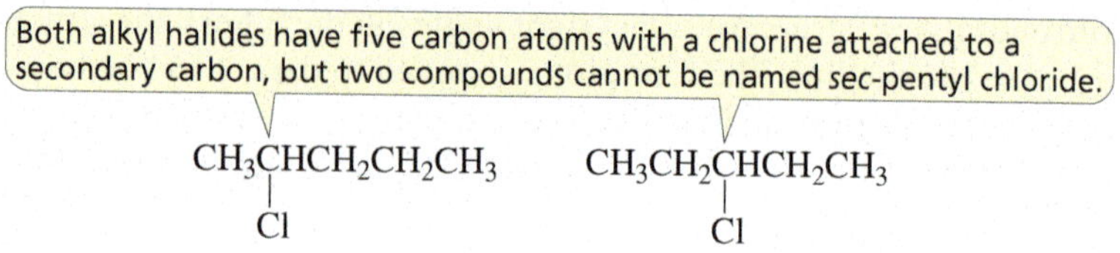

> Both alkyl halides have five carbon atoms with a chlorine attached to a secondary carbon, but two compounds cannot be named *sec*-pentyl chloride.

$$CH_3CHCH_2CH_2CH_3 \qquad CH_3CH_2CHCH_2CH_3$$
$$\quad | \qquad\qquad\qquad\qquad\qquad |$$
$$\quad Cl \qquad\qquad\qquad\qquad\qquad Cl$$

The prefix "*tert*" can be used for both *tert*-butyl and *tert*-pentyl compounds because each of these substituent names describes only one alkyl group. The name "*tert*-hexyl" cannot be used because it describes two different alkyl groups.

$$CH_3 \qquad\qquad CH_3 \qquad\qquad CH_2CH_3 \qquad\qquad CH_3$$
$$| \qquad\qquad\quad | \qquad\qquad\qquad | \qquad\qquad\qquad |$$
$$CH_3C{-}Br \qquad CH_3C{-}Br \qquad CH_3CH_2C{-}Br \qquad CH_3CH_2CH_2C{-}Br$$
$$| \qquad\qquad\quad | \qquad\qquad\qquad | \qquad\qquad\qquad |$$
$$CH_3 \qquad\qquad CH_2CH_3 \qquad\qquad CH_3 \qquad\qquad\qquad CH_3$$

   tert-butyl bromide   *tert*-pentyl bromide

> Both alkyl bromides have six carbon atoms with a bromine attached to a tertiary carbon, but two different compounds cannot be named *tert*-hexyl bromide.

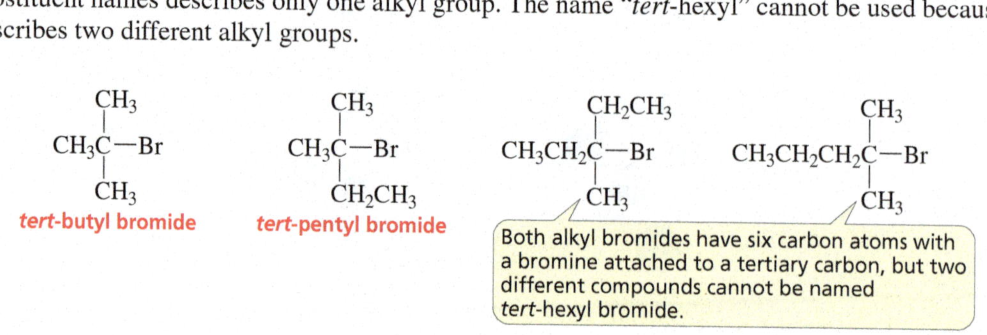

Notice in the following structures that whenever the prefix "iso" is used, the iso structural unit is at one end of the molecule and the group replacing a hydrogen is at the other end:

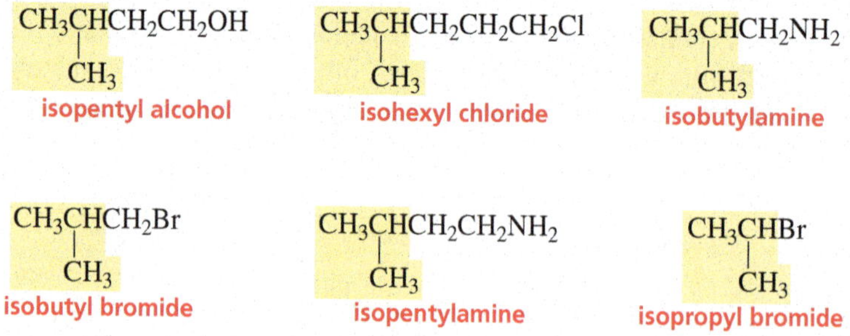

$$CH_3CHCH_2CH_2OH \qquad CH_3CHCH_2CH_2CH_2Cl \qquad CH_3CHCH_2NH_2$$
$$| \qquad\qquad\qquad\qquad | \qquad\qquad\qquad\qquad |$$
$$CH_3 \qquad\qquad\qquad\qquad CH_3 \qquad\qquad\qquad\qquad CH_3$$

  isopentyl alcohol   isohexyl chloride   isobutylamine

$$CH_3CHCH_2Br \qquad CH_3CHCH_2CH_2NH_2 \qquad CH_3CHBr$$
$$| \qquad\qquad\qquad\qquad | \qquad\qquad\qquad\qquad |$$
$$CH_3 \qquad\qquad\qquad\qquad CH_3 \qquad\qquad\qquad\qquad CH_3$$

  isobutyl bromide   isopentylamine   isopropyl bromide

Notice that an iso group has a methyl group on the next-to-the-last carbon in the chain. Also notice that all isoalkyl compounds have the substituent (OH, Cl, NH$_2$, and so on) on a primary carbon, except for isopropyl, which has the substituent on a secondary carbon. Thus, the isopropyl group could have been called a *sec*-propyl group. Either name would have been appropriate because the group has an iso structural unit and a hydrogen has been removed from a secondary carbon. Chemists decided to call it isopropyl, however, which means that "*sec*" is used only for *sec*-butyl.

Alkyl group names are used so frequently that you need to learn them. Some of the most common alkyl group names are compiled in Table 3.2.

Table 3.2   Names of Some Common Alkyl Groups

methyl	$CH_3{-}$	isobutyl	$CH_3CHCH_2{-}$ $\quad\vert$ $\quad CH_3$	pentyl	$CH_3CH_2CH_2CH_2CH_2{-}$
ethyl	$CH_3CH_2{-}$			isopentyl	$CH_3CHCH_2CH_2{-}$ $\quad\vert$ $\quad CH_3$
propyl	$CH_3CH_2CH_2{-}$	*sec*-butyl	$CH_3CH_2CH{-}$ $\qquad\quad\vert$ $\qquad\quad CH_3$		
isopropyl	$CH_3CH{-}$ $\quad\vert$ $\quad CH_3$			hexyl	$CH_3CH_2CH_2CH_2CH_2CH_2{-}$
butyl	$CH_3CH_2CH_2CH_2{-}$	*tert*-butyl	$CH_3C{-}$ $\quad\vert$ $\quad CH_3$	isohexyl	$CH_3CHCH_2CH_2CH_2{-}$ $\quad\vert$ $\quad CH_3$

PROBLEM 4 ◆

Draw the structure of a compound with molecular formula C_5H_{12} that has

a. one tertiary carbon.   **b.** no secondary carbons.

PROBLEM 5 ◆

Draw the structures and name the four constitutional isomers with molecular formula C_4H_9Br.

PROBLEM 6 ◆

Which of the following statements can be used to prove that carbon is tetrahedral?

a. CH_3Br does not have constitutional isomers.
b. CBr_4 does not have a dipole moment.
c. CH_2Br_2 does not have constitutional isomers.

PROBLEM 7 ◆

Draw the structure for each of the following:

a. isopropyl alcohol   **c.** *sec*-butyl iodide   **e.** *tert*-butylamine
b. isopentyl fluoride   **d.** *tert*-pentyl alcohol   **f.** *n*-octyl bromide

PROBLEM 8 ◆

Name the following compounds:

a. $CH_3OCH_2CH_3$

b. $CH_3OCH_2CH_2CH_3$

c. 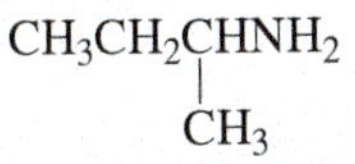 $CH_3CH_2\overset{\overset{\displaystyle CH_3}{|}}{C}HNH_2$

d. $CH_3CH_2CH_2CH_2OH$

e. 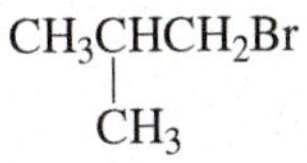 $CH_3\overset{\overset{\displaystyle CH_3}{|}}{C}HCH_2Br$

f. 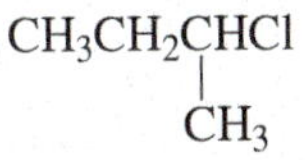 $CH_3CH_2\overset{\overset{\displaystyle CH_3}{|}}{C}HCl$

3.2   THE NOMENCLATURE OF ALKANES

The systematic name of an alkane is obtained using the following rules:

1. Determine the number of carbons in the longest continuous carbon chain. The longest continuous chain is not always in a straight line; sometimes you must "turn a corner" to obtain the longest continuous chain. This chain is called the **parent hydrocarbon.** The name that indicates the number of carbons in the parent hydrocarbon becomes the alkane's "last name." For example, a parent hydrocarbon with eight carbons is called *octane*.

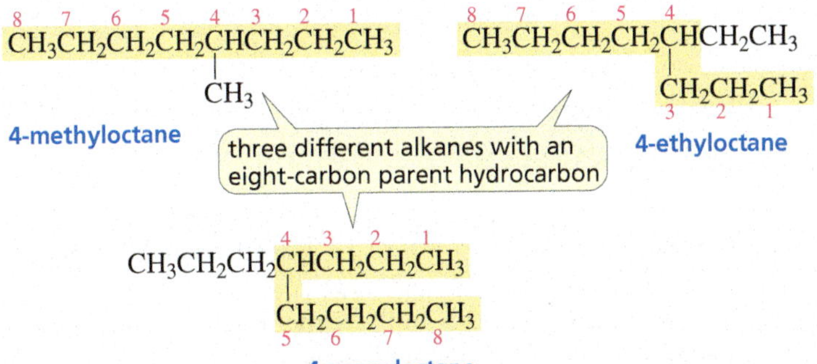

LEARN THE STRATEGY

$$\overset{8}{C}H_3\overset{7}{C}H_2\overset{6}{C}H_2\overset{5}{C}H_2\overset{4}{C}H\overset{3}{C}H_2\overset{2}{C}H_2\overset{1}{C}H_3$$
$$\underset{CH_3}{|}$$

4-methyloctane

$$\overset{8}{C}H_3\overset{7}{C}H_2\overset{6}{C}H_2\overset{5}{C}H_2\overset{4}{C}HCH_2CH_3$$
$$\underset{\underset{3\quad2\quad1}{CH_2CH_2CH_3}}{|}$$

4-ethyloctane

three different alkanes with an eight-carbon parent hydrocarbon

$$CH_3CH_2CH_2\overset{4}{C}H\overset{3}{C}H_2\overset{2}{C}H_2\overset{1}{C}H_3$$
$$\underset{\underset{5\quad6\quad7\quad8}{CH_2CH_2CH_2CH_3}}{|}$$

4-propyloctane

First determine the number of carbons in the longest continuous chain.

2. The name of any alkyl substituent that is attached to the parent hydrocarbon is placed in front of the name of the parent hydrocarbon, together with a number to designate the carbon to which the alkyl substituent is attached. The carbons in the parent chain are numbered in the direction that gives the substituent as low a number as possible. The substituent's name and the name of the parent hydrocarbon are joined into one word, preceded by a hyphen that connects the substituent's number with its name.

$$\overset{1}{C}H_3\overset{2}{C}H\overset{3}{C}H_2\overset{4}{C}H_2\overset{5}{C}H_3 \qquad \overset{6}{C}H_3\overset{5}{C}H_2\overset{4}{C}H_2\overset{3}{C}H\overset{2}{C}H_2\overset{1}{C}H_3 \qquad \overset{1}{C}H_3\overset{2}{C}H_2\overset{3}{C}H_2\overset{4}{C}H\overset{5}{C}H_2\overset{6}{C}H_2\overset{7}{C}H_2\overset{8}{C}H_3$$

Number the chain in the direction that gives the substituent as low a number as possible.

CH3	CH2CH3	CH2CH2CH3
2-methylpentane	3-ethylhexane	4-propyloctane
not	not	not
4-methylpentane	4-ethylhexane	5-propyloctane

Only systematic names have numbers; common names *never* contain numbers.

$$\begin{array}{c} CH_3 \\ | \\ CH_3CHCH_2CH_2CH_3 \end{array}$$

Numbers are used only for systematic names, never for common names.

common name: isohexane
systematic name: 2-methylpentane

3. If more than one substituent is attached to the parent hydrocarbon, the chain is numbered in the direction that produces a name containing the lowest of the possible numbers. The substituents are listed in alphabetical order, with each substituent preceded by the appropriate number. In the following example, the correct name contains a 3 as its lowest number, whereas the incorrect name contains a 4 as its lowest number:

$$\begin{array}{c} CH_3CH_2CHCH_2CHCH_2CH_2CH_3 \\ | \qquad | \\ CH_3 \quad CH_2CH_3 \end{array}$$

Substituents are listed in alphabetical order.

5-ethyl-3-methyloctane
not
4-ethyl-6-methyloctane
because 3 < 4

If two or more substituents are the same, the prefixes "di," "tri," and "tetra" are used to indicate how many identical substituents the compound has. The numbers indicating the locations of the identical substituents are listed together, separated by commas. There are no spaces on either side of a comma. There must be as many numbers in a name as there are substituents. The prefixes "di," "tri," "tetra," "*sec*," and "*tert*" are ignored in alphabetizing substituents.

A number and a word are separated by a hyphen; numbers are separated by a comma.

"di," "tri," "tetra," "*sec*," and "*tert*" are ignored in alphabetizing substituents.

$$\begin{array}{c} CH_3CH_2CHCH_2CHCH_3 \\ | \qquad | \\ CH_3 \quad CH_3 \end{array} \qquad \begin{array}{c} \qquad\qquad CH_2CH_3 \\ \qquad\qquad | \\ CH_3CH_2CCH_2CH_2CHCH_3 \\ | \qquad\qquad | \\ CH_3 \qquad CH_3 \end{array}$$

2,4-dimethylhexane   5-ethyl-2,5-dimethylheptane

> numbers are separated by a comma;
> a number and a word are separated by a hyphen

$$\begin{array}{c} CH_2CH_3 \quad CH_3 \\ | \qquad\quad | \\ CH_3CH_2CCH_2CH_2CHCHCH_2CH_2CH_3 \\ | \qquad\qquad | \\ CH_2CH_3 \; CH_2CH_3 \end{array} \qquad \begin{array}{c} \qquad\qquad\qquad CH_3 \\ \qquad\qquad\qquad | \\ CH_3CH_2CH_2CHCH_2CH_2CHCH_3 \\ \qquad\qquad | \\ CH_3CH_2 \end{array}$$

3,3,6-triethyl-7-methyldecane   5-ethyl-2-methyloctane

4. When numbering in either direction leads to the same lowest number for one of the substituents, the chain is numbered in the direction that gives the lowest possible number to one of the remaining substituents.

$$\begin{array}{c} CH_3 \\ | \\ CH_3CCH_2CHCH_3 \\ | \quad | \\ CH_3 \; CH_3 \end{array} \qquad \begin{array}{c} CH_3 \quad CH_2CH_3 \\ | \qquad | \\ CH_3CH_2CHCHCH_2CHCH_2CH_3 \\ | \\ CH_3 \end{array}$$

2,2,4-trimethylpentane   6-ethyl-3,4-dimethyloctane
not   not
2,4,4-trimethylpentane   3-ethyl-5,6-dimethyloctane
because 2 < 4   because 4 < 5

5. If the same substituent numbers are obtained in both directions, the first group listed receives the lower number.

$$CH_3CH_2\underset{\underset{\underset{CH_3}{|}}{\underset{|}{CHCH_2}}CHCH_2CH_3}{\overset{\overset{CH_2CH_3}{|}}{}}$$

3-ethyl-5-methylheptane
not
5-ethyl-3-methylheptane

Only if the same set of numbers is obtained in both directions does the first group listed get the lower number.

6. Systematic names for branched substituents are obtained by numbering the alkyl substituent starting at the carbon attached to the parent hydrocarbon. This means that the carbon attached to the parent hydrocarbon is always the number-1 carbon of the substituent. In a compound such as 4-(1-methylethyl)octane, the substituent name is in parentheses; the number inside the parentheses indicates a position on the substituent, whereas the number outside the parentheses indicates a position on the parent hydrocarbon. (If a prefix such as "di" is part of a branch name, it *is* included in the alphabetization.)

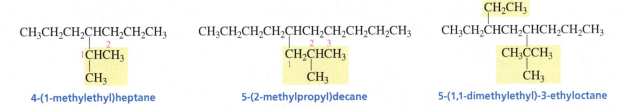

4-(1-methylethyl)heptane   **5-(2-methylpropyl)decane**   **5-(1,1-dimethylethyl)-3-ethyloctane**

If the substituent has a common name, the common name can be used instead of the parenthetical name.

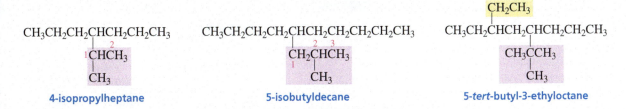

4-isopropylheptane   **5-isobutyldecane**   **5-*tert*-butyl-3-ethyloctane**

If the substituent does not have a common name, the parenthetical name must be used.

$$CH_3CH_2CH_2CH_2\underset{\underset{CH_3CHCHCH_3}{|}}{CHCH_2}\underset{\underset{CH_3}{|}}{CHCH_2CH_3}$$

6-(1,2-dimethylpropyl)-4-propyldecane   **2,3-dimethyl-5-(2-methylbutyl)decane**

7. If a compound has two or more chains of the same length, the parent hydrocarbon is the chain with the greatest number of substituents.

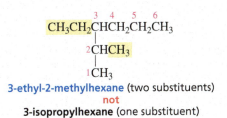

3-ethyl-2-methylhexane (two substituents)
not
3-isopropylhexane (one substituent)

In the case of two hydrocarbon chains with the same number of carbons, choose the one with the most substituents.

These rules allow you to name thousands of alkanes, and eventually you will learn th[e] additional rules necessary to name many other kinds of compounds. The rules are importan[t] for looking up a compound in the scientific literature, because it usually is listed by its system[m] atic name. Nevertheless, you must also learn common names because they are so entrenched i[n] chemists' vocabularies that they are widely used in scientific conversation and are often foun[d] in the literature.

Look at the systematic names (the ones written in blue) for the isomeric hexanes and isomeri[c] heptanes shown on p. 91 to make sure you understand how they are constructed.

USE THE STRATEGY

PROBLEM 9 ◆

What is each compound's systematic name?

a. $\underset{\underset{CH_3}{|}}{CH_3CH_2CH}\underset{\underset{CH_3}{|}}{CH_2C}CH_3$ (with CH_3 substituents)

b. $CH_3CH_2C(CH_3)_3$

c. $CH_3CH_2CH_2CHCH_2CH_2CH_3$
      $\overset{|}{CH_3CHCH_2CH_3}$

d. $\underset{\underset{CH_3}{|}}{CH_3CH}CH_2CH_2\underset{\underset{CH_3}{|}}{C}CH_3$

e. $CH_3CH_2C(CH_2CH_3)_2CH(CH_3)CH(CH_2CH_2CH_3)_2$

f. $\underset{\underset{CH_2CH_2CH_3}{|}}{CH_3C}\overset{\overset{CH_3\ CH_2CH_2CH_3}{|}}{-}CHCH_2CH_3$

g. $CH_3CH_2C(CH_2CH_3)_2CH_2CH_2CH_3$

h. $CH_3CH_2CH_2CH_2CHCH_2CH_2CH_3$
      $\overset{|}{CH(CH_3)_2}$

i. $\underset{\underset{CH_2CH_3}{|}}{CH_3CH}CH_2CH_2\underset{\underset{CH_3}{|}}{CH}CH_3$

How Is the Octane Number of Gasoline Determined?

The gasoline engines used in most cars operate by creating a series of carefully timed, controlled explosions. Fuel is mixed with air in the engine cyclinders, compressed, and then ignited by a spark. If the fuel ignites too easily, the heat of compression can initiate combustion before the spark plug fires. The result is a pinging or knocking sound and a loss of power.

Higher-quality fuels are less likely to knock. The quality of a fuel is indicated by its octane number. Straight-chain hydrocarbons have low octane numbers and make poor fuels. Heptane, for example, with an arbitrarily assigned octane number of 0, causes engines to knock badly. Branched-chain alkanes have more hydrogens bonded to primary carbons. These are the bonds that require the most energy to break and, therefore, make combustion more difficult to initiate, thereby reducing knocking. For example, 2,2,4-trimethylpentane does not cause knocking and has arbitrarily been assigned an octane number of 100.

$CH_3CH_2CH_2CH_2CH_2CH_2CH_3$
heptane
octane number = 0

$\underset{\underset{CH_3}{|}}{CH_3C}\overset{\overset{CH_3\ CH_3}{|\ \ \ |}}{CH_2}CHCH_3$
2,2,4-trimethylpentane
octane number = 100

The octane number of a gasoline is determined by comparing its knocking with the knocking of mixtures of heptane and 2,2,4-trimethylpentane. The octane number given to the gasoline corresponds to the percent of 2,2,4-trimethylpentane in the matching mixture. Thus, a gasoline with an octane rating of 91 has the same "knocking" property as a mixture of 91% 2,2,4-trimethylpentane and 9% heptane. The term *octane number* originated from the fact that 2,2,4-trimethylpentane contains eight carbons. Because slightly different methods are used to determine the octane number, gasoline in Canada and the United States has an octane number that is 4 to 5 points less than the same gasoline in Europe and Australia.

PROBLEM 10 SOLVED

Draw the structure for 2,2-dimethyl-4-propyloctane.

SOLUTION The name of the parent hydrocarbon is *octane*, so the longest continuous chain has eight carbons. Now draw the parent hydrocarbon and number it.

$$\overset{1}{C}-\overset{2}{C}-\overset{3}{C}-\overset{4}{C}-\overset{5}{C}-\overset{6}{C}-\overset{7}{C}-\overset{8}{C}$$

Put the substituents (two methyl groups and a propyl group) on the appropriate carbons.

$$\begin{array}{ccccccc} & CH_3 & & CH_2CH_2CH_3 & & & \\ & | & & | & & & \\ C-&C-&C-&C-&C-C-C-C \\ & | & & & & & \\ & CH_3 & & & & & \end{array}$$

Add the appropriate number of hydrogens so that each carbon is bonded to four atoms.

$$\begin{array}{ccccccc} & CH_3 & & CH_2CH_2CH_3 & & \\ & | & & | & & \\ CH_3-&C-&CH_2-&CH-&CH_2-CH_2-CH_2-CH_3 \\ & | & & & & \\ & CH_3 & & & & \end{array}$$

PROBLEM 11 ◆

Draw the structure for each of the following:

a. 2,2-dimethyl-4-isopropyloctane
b. 2,3-dimethylhexane
c. 4,4-diethyldecane

d. 2,4,5-trimethyl-4-(1-methylethyl)heptane
e. 2,5-dimethyl-4-(2-methylpropyl)octane
f. 4-(1,1-dimethylethyl)octane

PROBLEM 12 SOLVED

a. Draw the 18 isomeric octanes.
b. Give each isomer its systematic name.
c. How many isomers have common names?

d. Which isomers contain an isopropyl group?
e. Which isomers contain a *sec*-butyl group?
f. Which isomers contain a *tert*-butyl group?

SOLUTION TO 12 a. Start with the isomer with an eight-carbon continuous chain. Then draw isomers with a seven-carbon continuous chain plus one methyl group. Next, draw isomers with a six-carbon continuous chain plus two methyl groups or one ethyl group. Then draw isomers with a five-carbon continuous chain plus three methyl groups or one methyl group and one ethyl group. Finally, draw a four-carbon continuous chain with four methyl groups. (Your answers to Problem 12 b. will tell you whether you have drawn duplicate structures, because if two structures have the same systematic name, they represent the same compound.)

PROBLEM 13

Give each substituent on the ten-carbon chain a common name and a parenthetical name.

a.   b.   c.   d.

PROBLEM 14 ◆

Draw the structure and give the systematic name for a compound with molecular formula C_5H_{12} that has

a. only primary and secondary hydrogens.
b. only primary hydrogens.

c. one tertiary hydrogen.
d. two secondary hydrogens.

3.3 THE NOMENCLATURE OF CYCLOALKANES

Cycloalkanes are alkanes with their carbon atoms arranged in a ring. Cycloalkanes are named by adding the prefix "cyclo" to the alkane name that signifies the number of carbons in the ring.

cyclopropane   cyclobutane   cyclopentane   cyclohexane

Cycloalkanes are generally written as skeletal structures. Because of the ring, a cycloalkane has two fewer hydrogens than an acyclic (noncyclic) alkane with the same number of carbons. This means that the general molecular formula for a cycloalkane is C_nH_{2n}.

The rules for naming cycloalkanes resemble the rules for naming acyclic (noncyclic) alkanes:

LEARN THE STRATEGY

1. In a cycloalkane with an attached alkyl substituent, the ring is the parent hydrocarbon unless the substituent has more carbons than the ring. In that case, the substituent is the parent hydrocarbon and the ring is named as a substituent. *There is no need to number the position of a single substituent on a ring.*

If there is only one substituent on a ring, do not give that substituent a number.

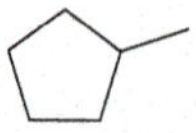

the substituent has more carbons than the ring

methylcyclopentane 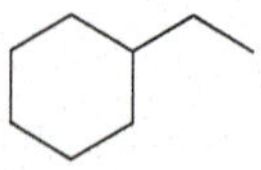   ethylcyclohexane 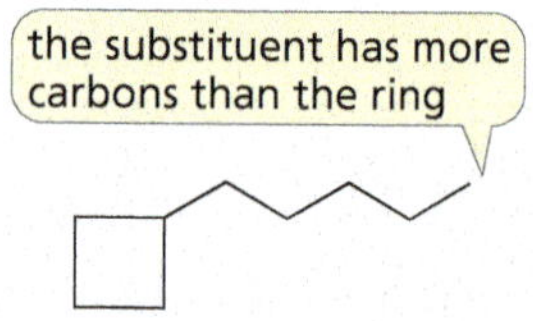   1-cyclobutylpentane

2. If the ring has two different substituents, they are listed in *alphabetical order* and the number-1 position is given to the substituent listed first.

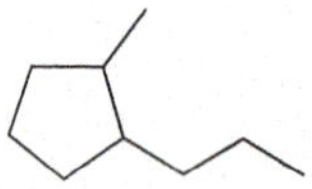

1-methyl-2-propylcyclopentane 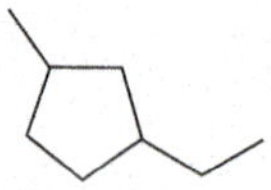   1-ethyl-3-methylcyclopentane 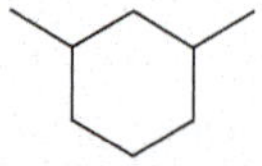   1,3-dimethylcyclohexane

3. If there are more than two substituents on the ring, they are listed in alphabetical order, and the substituent given the number-1 position is the one that results in a second substituent getting as low a number as possible. If two substituents have the same low numbers, the ring is numbered—either clockwise or counterclockwise—in the direction that gives the third substituent the lowest possible number.

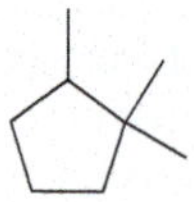

1,1,2-trimethylcyclopentane
not
1,2,2-trimethylcyclopentane
because 1 < 2
not
1,1,5-trimethylcyclopentane
because 2 < 5

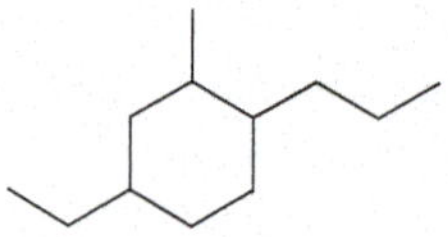

4-ethyl-2-methyl-1-propylcyclohexane
not
1-ethyl-3-methyl-4-propylcyclohexane
because 2 < 3
not
5-ethyl-1-methyl-2-propylcyclohexane
because 4 < 5

USE THE STRATEGY

PROBLEM 15 ◆

What is each compound's systematic name?

a.

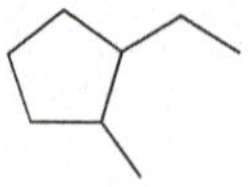

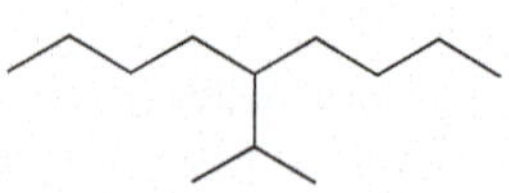

b.

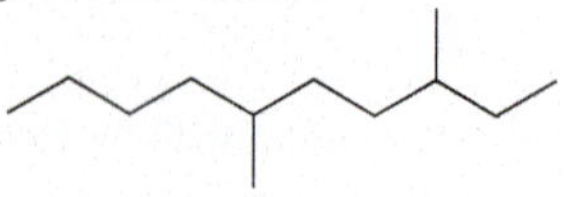

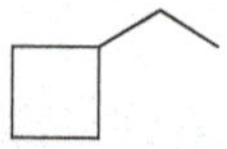

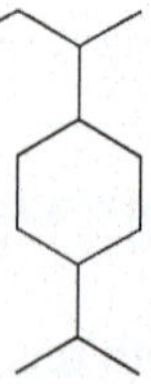

c.

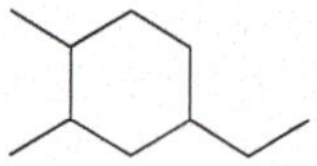

d.

e.

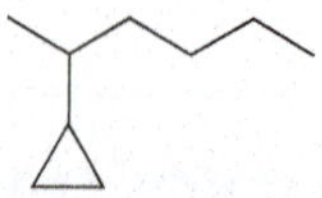

f.

g.

h.

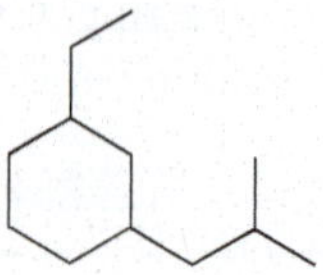

i.

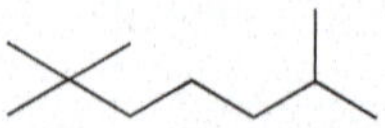

PROBLEM-SOLVING STRATEGY

Interpreting a Skeletal Structure

How many hydrogens are attached to each of the indicated carbons in cholesterol?

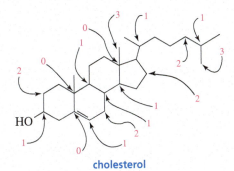

cholesterol

None of the carbons in the compound has a charge, so each needs to be bonded to four atoms. Thus, if the carbon has only one bond showing, it must be attached to three hydrogens that are not shown; if the carbon has two bonds showing, it must be attached to two hydrogens that are not shown; and so on. Check each of the answers (shown in red) to see that this is so.

PROBLEM 16

How many hydrogens are attached to each of the indicated carbons in morphine?

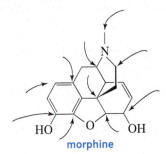

morphine

PROBLEM 17 ♦

Convert the following condensed structures into skeletal structures:

a. $CH_3CH_2CH_2CH_2CH_2CH_2OH$

b. $CH_3CH_2CH_2CH_2CH_2CH_3$

c. $CH_3CH_2CHCH_2CHCH_2CH_3$ with CH_3 and CH_3 branches

d. $CH_3CH_2CH_2CH_2OCH_3$

e. $CH_3CH_2NHCH_2CH_2CH_3$

f. $CH_3CHCH_2CH_2CHCH_3$ with CH_3 and Br branches

Condensed structures show atoms but show few, if any, bonds, whereas skeletal structures show bonds but show few, if any, atoms.

PROBLEM 18

Convert the structures in Problem 9 into skeletal structures.

PROBLEM 19

Draw a condensed and a skeletal structure for each of the following:

a. 3,4-diethyl-2-methylheptane

b. 2,2,5-trimethylhexane

3.4 THE NOMENCLATURE OF ALKYL HALIDES

An **alkyl halide** is a compound in which a hydrogen of an alkane has been replaced by a halogen. Alkyl halides are classified as primary, secondary, or tertiary, depending on the carbon to which the halogen is attached (Section 3.1).

- A **primary alkyl halide** has a halogen attached to a primary carbon.
- A **secondary alkyl halide** has a halogen attached to a secondary carbon.
- A **tertiary alkyl halide** has a halogen attached to a tertiary carbon.

The carbon to which the halogen is attached determines whether an alkyl halide is primary, secondary, or tertiary.

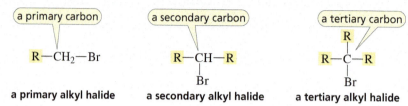

a primary carbon   a secondary carbon   a tertiary carbon

$R—CH_2—Br$ $R—CH—R$ $R—C—R$
                                     | |
                                     Br   Br

a primary alkyl halide   a secondary alkyl halide   a tertiary alkyl halide

The lone-pair electrons on the halogens are generally not shown unless they are needed to draw your attention to some chemical property of the atom.

Common Names

The common names of alkyl halides consist of the name of the alkyl group, followed by the name of the halogen—with the "ine" ending of the halogen name (fluorine, chlorine, bromine, and iodine) replaced by "ide" (fluoride, chloride, bromide, and iodide). Notice that although a name must specify only one compound, a compound can have more than one name.

A compound can have more than one name, but a name must specify only one compound.

Remember that numbers are used only for systematic names, never for common names.

	CH_3Cl	CH_3CH_2F	CH_3CHI	CH_3CH_2CHBr
			$\|$	$\|$
			CH_3	CH_3
common name:	methyl chloride	ethyl fluoride		
systematic name:	chloromethane	fluoroethane		
			isopropyl iodide	sec-butyl bromide
			2-iodopropane	2-bromobutane

Systematic Names

In the IUPAC system, alkyl halides are named as substituted alkanes. The prefixes for the halogens end with "o" (that is, fluoro, chloro, bromo, and iodo). Therefore, alkyl halides are also called haloalkanes.

CH_3F
methyl fluoride

CH_3
|
$CH_3CH_2CHCH_2CHCH_3$
|
Br

2-bromo-4-methylhexane

1-chloro-6,6-dimethylheptane

1-ethyl-2-iodocyclopentane

CH_3Cl
methyl chloride

(1-bromoethyl)cyclopentane

4-bromo-2-chloro-1-methylcyclohexane

CH_3Br
methyl bromide

PROBLEM 20 ◆

Give two names for each of the following alkyl halides and indicate whether each is primary, secondary, or tertiary:

a. $CH_3CH_2CHCH_3$
              |
              Cl

b. (cyclohexane with Br)

c. $CH_3CHCH_2CH_2CH_2Cl$
              |
              CH_3

d. (isopropyl with F)

PROBLEM 21

Draw the structures and provide systematic names for parts **a**, **b**, and **c** by substituting a chlorine for a hydrogen of methylcyclohexane:

a. a primary alkyl halide   **b.** a tertiary alkyl halide   **c.** three secondary alkyl halides

CH_3I
methyl iodide

3.5 THE NOMENCLATURE OF ETHERS

A compound with an oxygen attached to two alkyl groups is called an **ether.**

- In a **symmetrical ether,** the two alkyl groups are the same (R and R).
- In an **unsymmetrical ether,** the two alkyl groups are different (R and R′).

R—O—R
a symmetrical ether

R—O—R′
an unsymmetrical ether

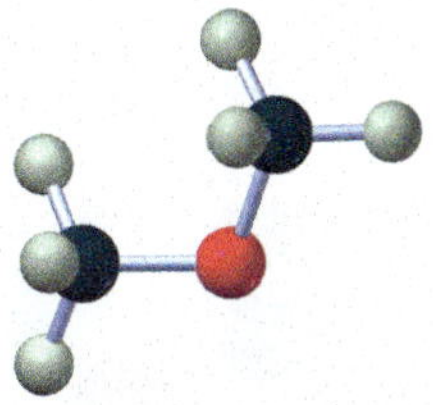
dimethyl ether

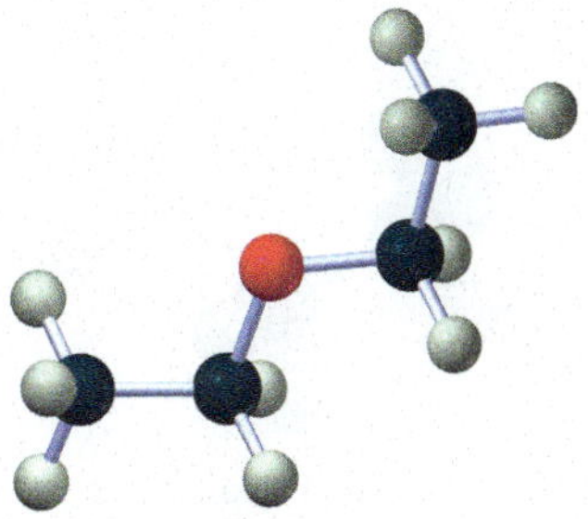
diethyl ether

Chemists sometimes neglect the prefix "di" when they name symmetrical ethers. Try not to do this.

Common Names

The common name of an ether consists of the names of the two alkyl substituents (in alphabetical order), followed by the word "ether." The smallest ethers are almost always named by their common names.

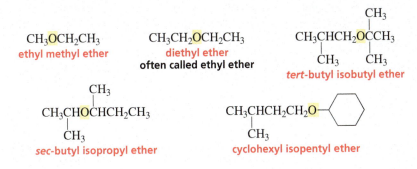

$CH_3OCH_2CH_3$
ethyl methyl ether

$CH_3CH_2OCH_2CH_3$
diethyl ether
often called ethyl ether

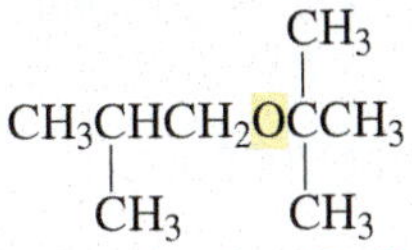
tert-butyl isobutyl ether

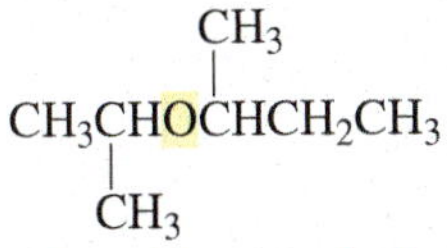
sec-butyl isopropyl ether

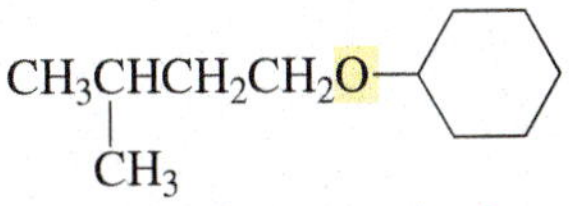
cyclohexyl isopentyl ether

Systematic Names

The IUPAC system names an ether as an alkane with an RO substituent. The substituents are named by replacing the "yl" ending in the name of the alkyl substituent with "oxy."

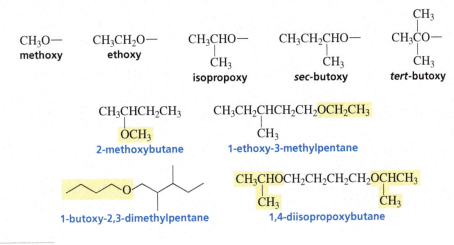

$CH_3O—$
methoxy

$CH_3CH_2O—$
ethoxy

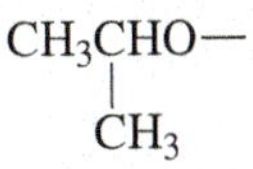
isopropoxy

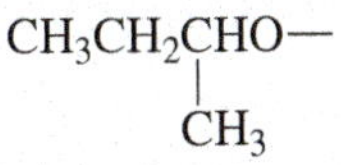
sec-butoxy

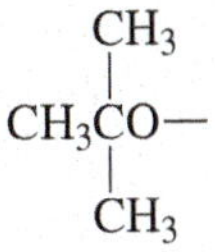
tert-butoxy

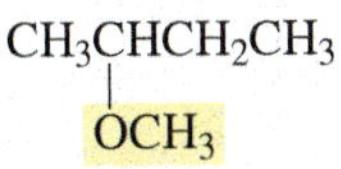
2-methoxybutane

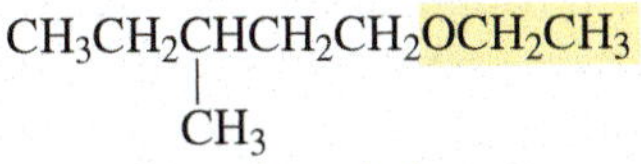
1-ethoxy-3-methylpentane

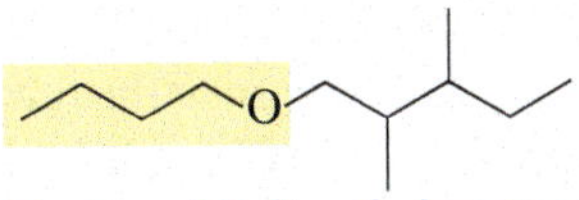
1-butoxy-2,3-dimethylpentane

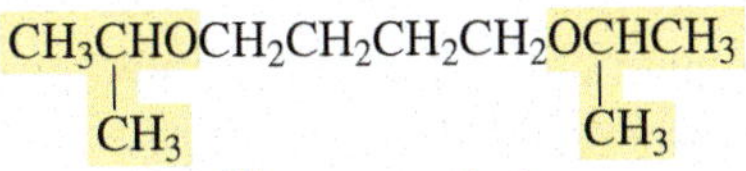
1,4-diisopropoxybutane

PROBLEM 22 ♦

a. What is each ether's systematic name?

1. $CH_3OCH_2CH_3$
2. $CH_3CH_2OCH_2CH_3$
3. 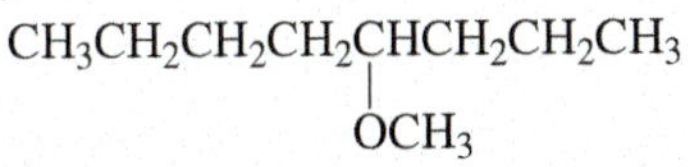
4. 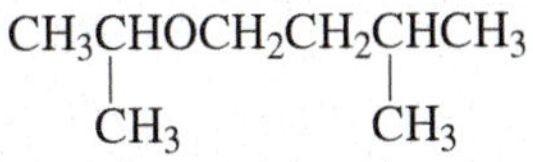
5. 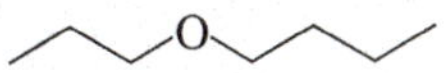
6. 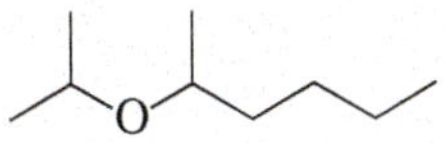

b. Do all of these ethers have common names?

c. What are their common names?

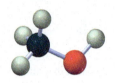

methyl alcohol

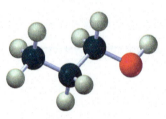

ethyl alcohol

3.6 THE NOMENCLATURE OF ALCOHOLS

An **alcohol** is a compound in which a hydrogen of an alkane has been replaced by an OH group. Alcohols are classified in the same way as alkyl halides are classified.

- A **primary alcohol** has an OH group attached to a primary carbon.
- A **secondary alcohol** has an OH group attached to a secondary carbon.
- A **tertiary alcohol** has an OH group attached to a tertiary carbon.

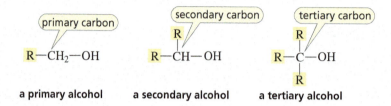

a primary alcohol   a secondary alcohol   a tertiary alcohol

Common Names

The common name of an alcohol consists of the name of the alkyl group to which the OH group is attached, followed by the word "alcohol."

$CH_3CH_2\textbf{OH}$ $CH_3CH_2CH_2\textbf{OH}$ CH_3CHOH $CH_3CHCH_2\textbf{OH}$
ethyl alcohol   propyl alcohol $\overset{|}{CH_3}$ $\overset{|}{CH_3}$
                                                isopropyl alcohol   isobutyl alcohol

Systematic Names

The **functional group** is the center of reactivity in an organic molecule. The IUPAC system uses *suffixes* to denote certain functional groups. The functional group of an alcohol is the OH group, which is denoted by the suffix "ol." Thus, the systematic name of an alcohol is obtained by replacing the "e" at the end of the name of the parent hydrocarbon with "ol."

CH_3OH CH_3CH_2OH
methanol   ethanol

propyl alcohol

The carbon to which the OH group is attached determines whether an alcohol is primary, secondary, or tertiary.

When necessary, the position of the functional group is indicated by a number immediately preceding the name of the parent hydrocarbon or immediately preceding the suffix. The most recently approved IUPAC names are those with the number immediately preceding the suffix. However, the chemical community has been slow to adopt this change, so the names most likely to appear in the literature, on reagent bottles, and on standardized tests are those with the number preceding the name of the parent hydrocarbon. They are also the ones that appear most often in this book. Because they are not the most recently approved IUPAC names, they will be referred to as systematic names.

$CH_3CH_2CHCH_2CH_3$
$\overset{|}{OH}$
3-pentanol
or
pentan-3-ol

3-methyl-1-hexanol
or
3-methylhexan-1-ol

the number precedes the name of the parent hydrocarbon

the number precedes the suffix

The following rules are used to name a compound that has a functional group suffix:

LEARN THE STRATEGY

1. The parent hydrocarbon is the longest continuous chain that *contains the functional group.*
2. The parent hydrocarbon is numbered in the direction that gives the *functional group suffix the lowest possible number.*

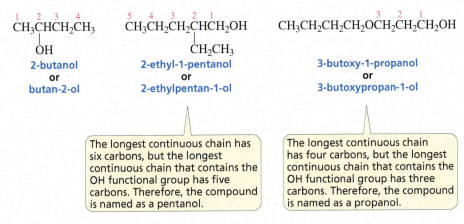

The longest continuous chain has six carbons, but the longest continuous chain that contains the OH functional group has five carbons. Therefore, the compound is named as a pentanol.

The longest continuous chain has four carbons, but the longest continuous chain that contains the OH functional group has three carbons. Therefore, the compound is named as a propanol.

3. If there are two OH groups, the suffix "diol" is added to the name of the parent hydrocarbon.

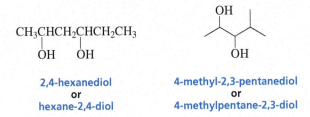

2,4-hexanediol
or
hexane-2,4-diol

4-methyl-2,3-pentanediol
or
4-methylpentane-2,3-diol

4. If there is a functional group suffix and a substituent, the functional group suffix gets the lowest possible number.

When there is only a substituent, the substituent gets the lowest possible number.

When there is only a functional group, the functional group suffix gets the lowest possible number.

When there is both a functional group and a substituent, the functional group suffix gets the lowest possible number.

$HOCH_2CH_2CH_2Br$

3-bromo-1-propanol

$ClCH_2CH_2CHCH_3$
$\quad\quad\quad\quad | $
$\quad\quad\quad\quad OH$

4-chloro-2-butanol

4,4-dimethyl-2-pentanol

5. If counting in either direction gives the same number for the functional group suffix, then the chain is numbered in the direction that gives a substituent the lowest possible number. Notice that a number is not needed to designate the position of a functional group suffix in a cyclic compound, because it is assumed to be at the 1 position.

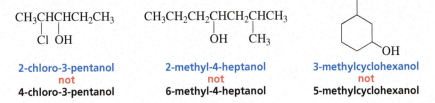

2-chloro-3-pentanol
not
4-chloro-3-pentanol

2-methyl-4-heptanol
not
6-methyl-4-heptanol

3-methylcyclohexanol
not
5-methylcyclohexanol

6. If there is more than one substituent, the substituents are listed in alphabetical order.

7-bromo-4-ethyl-2-octanol   3,4-dimethylcyclopentanol   2-ethyl-5-methylcyclohexanol

Remember that the name of a substituent is stated *before* the name of the parent hydrocarbon and a functional group suffix is stated *after* the name of the parent hydrocarbon.

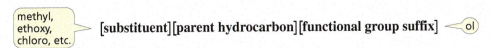

methyl, ethoxy, chloro, etc.

[substituent][parent hydrocarbon][functional group suffix] — ol

USE THE STRATEGY

PROBLEM 23 ◆

Give each of the following a systematic name and indicate whether each is a primary, secondary, or tertiary alcohol:

a. $CH_3CH_2CH_2CH_2CH_2OH$

c. $CH_3CHCH_2CHCH_2CH_3$
                $\overset{|}{CH_3}$ $\overset{|}{OH}$

b. $CH_3\overset{\overset{\textstyle CH_3}{|}}{C}CH_2CH_2CH_2Cl$
            $\underset{|}{OH}$

d. $CH_3CHCH_2CHCH_2CHCH_2CH_3$
                $\overset{|}{CH_3}$ $\overset{|}{OH}$ $\overset{|}{OH}$

PROBLEM 24

Draw the structures of a homologous series of alcohols that have from one to six carbons and give each of them a common name and a systematic name.

PROBLEM 25 ◆

Write condensed and skeletal structures for all the tertiary alcohols with molecular formula $C_6H_{14}O$ and give each a systematic name.

PROBLEM 26 ◆

Give each of the following a systematic name and indicate whether each is a primary, secondary, or tertiary alcohol:

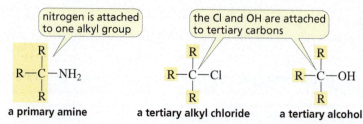

3.7 THE NOMENCLATURE OF AMINES

An **amine** is a compound in which one or more hydrogens of ammonia have been replaced by alkyl groups. Amines are classified as **primary, secondary,** and **tertiary,** depending on how many alkyl groups are attached to the nitrogen.

- A **primary amine** has one alkyl group attached to the nitrogen.
- A **secondary amine** has two alkyl groups attached to the nitrogen.
- A **tertiary amine** has three alkyl groups attached to the nitrogen.

The number of alkyl groups attached to the nitrogen determines whether an amine is primary, secondary, or tertiary.

NH_3 $R-NH_2$ $R-\overset{\overset{\textstyle R}{|}}{NH}$ $R-\overset{\overset{\textstyle R}{|}}{N}-R$

ammonia   **a primary amine**   **a secondary amine**   **a tertiary amine**

Be sure to note that the number of alkyl groups *attached to the nitrogen* determines whether an amine is primary, secondary, or tertiary. In contrast, whether the X (halogen) or OH group is *attached to a primary, secondary, or tertiary carbon* determines whether an alkyl halide or alcohol is primary, secondary, or tertiary (Sections 3.4 and 3.6).

nitrogen is attached to one alkyl group   the Cl and OH are attached to tertiary carbons

$R-\overset{\overset{\textstyle R}{|}}{\underset{\underset{\textstyle R}{|}}{C}}-NH_2$ $R-\overset{\overset{\textstyle R}{|}}{\underset{\underset{\textstyle R}{|}}{C}}-Cl$ $R-\overset{\overset{\textstyle R}{|}}{\underset{\underset{\textstyle R}{|}}{C}}-OH$

a primary amine   **a tertiary alkyl chloride**   **a tertiary alcohol**

Common Names

The common name of an amine consists of the names of the alkyl groups bonded to the nitrogen, in alphabetical order, followed by "amine." The entire name is written as one word (unlike the common names of alcohols, ethers, and alkyl halides, in which "alcohol," "ether," and "halide" are separate words).

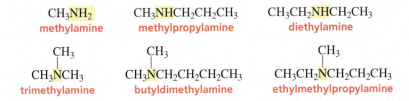

CH_3NH_2
methylamine

$CH_3NHCH_2CH_2CH_3$
methylpropylamine

$CH_3CH_2NHCH_2CH_3$
diethylamine

CH_3
|
CH_3NCH_3
trimethylamine

CH_3
|
$CH_3NCH_2CH_2CH_2CH_3$
butyldimethylamine

CH_3
|
$CH_3CH_2NCH_2CH_2CH_3$
ethylmethylpropylamine

Systematic Names

The IUPAC system uses the suffix "amine" to denote the amine functional group. The "e" at the end of the name of the parent hydrocarbon is replaced by "amine"—similar to the way alcohols are named. Also similar to the way alcohols are named, a number identifies the carbon to which the nitrogen is attached, and the number can appear before the name of the parent hydrocarbon or before "amine." The name of any alkyl group bonded to nitrogen is preceded by an "*N*" (in italics) to indicate that the group is bonded to a nitrogen rather than to a carbon.

LEARN THE STRATEGY

$\overset{4}{C}H_3\overset{3}{C}H_2\overset{2}{C}H_2\overset{1}{C}H_2NH_2$
1-butanamine
or
butan-1-amine

$\overset{1}{C}H_3\overset{2}{C}H_2\overset{3}{C}H\overset{4}{C}H_2\overset{5}{C}H_2\overset{6}{C}H_3$
|
$NHCH_2CH_3$
N-ethyl-3-hexanamine
or
N-ethylhexan-3-amine

$\overset{3}{C}H_3\overset{2}{C}H_2\overset{1}{C}H_2NCH_2CH_3$
|
CH_3
N-ethyl-*N*-methyl-1-propanamine
or
N-ethyl-*N*-methylpropan-1-amine

The substituents—regardless of whether they are attached to the nitrogen or to the parent hydrocarbon—are listed in alphabetical order, and then a number or an "*N*" is assigned to each one. The chain is numbered in the direction that gives the functional group suffix the lowest number.

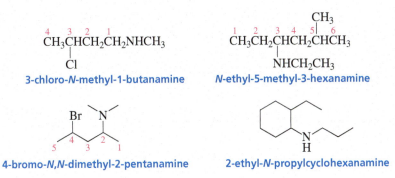

$\overset{4}{C}H_3\overset{3}{C}H\overset{2}{C}H_2\overset{1}{C}H_2NHCH_3$
|
Cl
3-chloro-*N*-methyl-1-butanamine

CH_3
|
$\overset{1}{C}H_3\overset{2}{C}H_2\overset{3}{C}H\overset{4}{C}H_2\overset{5}{C}H\overset{6}{C}H_3$
|
$NHCH_2CH_3$
N-ethyl-5-methyl-3-hexanamine

4-bromo-*N*,*N*-dimethyl-2-pentanamine

2-ethyl-*N*-propylcyclohexanamine

Nitrogen compounds with four alkyl groups attached to the nitrogen—thereby giving the nitrogen a positive formal charge—are called **quaternary ammonium salts.** Their names consist of the names of the alkyl groups in alphabetical order, followed by "ammonium" (all one word) and then the name of the accompanying anion as a separate word.

CH_3
|
$CH_3-\overset{+}{N}-CH_3$ HO^-
|
CH_3
tetramethylammonium hydroxide

CH_3
|
$CH_3CH_2CH_2-\overset{+}{N}-CH_3$ Cl^-
|
CH_2CH_3
ethyldimethylpropylammonium chloride

USE THE STRATEGY

PROBLEM 27 ◆

Give a systematic name and a common name (if it has one) for each of the following amines and indicate whether each is a primary, secondary, or tertiary amine:

a. $CH_3CH_2CH_2CH_2CH_2CH_2NH_2$

b. $CH_3CHCH_2CH_2CH_2NHCHCH_2CH_3$
      | |
      CH_3 CH_3

c. $(CH_3CH_2)_2NCH_3$

d. $CH_3CH_2CH_2NHCH_2CH_2CH_2CH_3$

e. $CH_3CH_2CH_2NCH_2CH_3$
                |
              CH_2CH_3

f.

PROBLEM 28 ◆

Are the following compounds primary, secondary, or tertiary?

a.
```
       CH3
        |
CH3 — C — Br
        |
       CH3
```

b.
```
       CH3
        |
CH3 — C — NH2
        |
       CH3
```

c. $CH_3CHCH_2CH_3$
        |
        OH

d. $CH_3NHCH_2CH_3$

PROBLEM 29 ◆

Draw condensed and skeletal structures for each of the following amines:

a. 2-methyl-*N*-propyl-1-propanamine

b. *N*-ethylethanamine

c. 5-methyl-1-hexanamine

d. methyldipropylamine

e. *N,N*-dimethyl-3-pentanamine

f. cyclohexylethylmethylamine

PROBLEM 30 ◆

For each of the following, give the systematic name and the common name (if it has one) and then indicate whether it is a primary, secondary, or tertiary amine:

a.

b.

c.

d.

Table 3.3 summarizes the ways in which alkyl halides, ethers, alcohols, and amines are named.

Table 3.3   Summary of Nomenclature

	Systematic name	Common name
Alkyl halide	substituted alkane CH_3Br bromomethane CH_3CH_2Cl chloroethane	alkyl group attached to halogen, plus *halide* CH_3Br methyl bromide CH_3CH_2Cl ethyl chloride
Ether	substituted alkane CH_3OCH_3 methoxymethane $CH_3CH_2OCH_3$ methoxyethane	alkyl groups attached to oxygen, plus *ether* CH_3OCH_3 dimethyl ether $CH_3CH_2OCH_3$ ethyl methyl ether
Alcohol	functional group suffix is *ol* CH_3OH methanol CH_3CH_2OH ethanol	alkyl group attached to OH, plus *alcohol* CH_3OH methyl alcohol CH_3CH_2OH ethyl alcohol
Amine	functional group suffix is *amine* $CH_3CH_2NH_2$ ethanamine $CH_3CH_2CH_2NHCH_3$ *N*-methyl-1-propanamine	alkyl groups attached to N, plus *amine* $CH_3CH_2NH_2$ ethylamine $CH_3CH_2CH_2NHCH_3$ methylpropylamine

Bad-Smelling Compounds

Amines are responsible for some of nature's unpleasant odors. Amines with relatively small alkyl groups have a fishy smell. For example, fermented shark, a traditional dish in Iceland, smells exactly like triethylamine. Fish is often served with lemon, because the citric acid in lemon protonates the amine, thereby converting it to its better-smelling acidic form.

The citric acid in lemon juice decreases the fishy taste.

triethylamine
bad-smelling

+ citric acid ⇌ protonated triethylamine
odorless

+ conjugate base of citric acid

The amines putrescine and cadaverine are poisonous compounds formed when amino acids are degraded in the body. Because these amines are excreted as quickly as possible, their odors may be detected in the urine and breath. Putrescine and cadaverine are also responsible for the odor of decaying flesh.

1,4-butanediamine
putrescine

1,5-pentanediamine
cadaverine

3.8 THE STRUCTURES OF ALKYL HALIDES, ALCOHOLS, ETHERS, AND AMINES

The families of compounds you just learned how to name have structural resemblances to the simpler compounds introduced in Chapter 1.

Alkyl Halides

Let's begin by looking at alkyl halides and their resemblance to alkanes. Both have the same geometry; the only difference is that a $C\!-\!X$ bond of an alkyl halide (where X denotes a halogen) has replaced a $C\!-\!H$ bond of an alkane (Section 1.7).

The $C\!-\!X$ bond of an alkyl halide is formed from the overlap of an sp^3 orbital of carbon with an sp^3 orbital of the halogen. Fluorine uses a $2sp^3$ orbital to overlap with a $2sp^3$ orbital of carbon, chlorine uses a $3sp^3$ orbital, bromine a $4sp^3$ orbital, and iodine a $5sp^3$ orbital. Thus, the $C\!-\!X$ bond becomes longer and weaker as the size of the halogen increases because the electron density of the orbital decreases with increasing volume. Notice that this is the same trend shown by the $H\!-\!X$ bond of hydrogen halides in Table 1.6 on p. 39.

1.39 Å 1.78 Å 1.93 Å 2.14 Å

Alcohols

Now let's consider the geometry of the oxygen in an alcohol; it is the same as the geometry of the oxygen in water (Section 1.12). In fact, an alcohol molecule can be thought of structurally as a water molecule with an alkyl group in place of one of the hydrogens. The oxygen in an alcohol is sp^3 hybridized, as it is in water. Of the four sp^3 orbitals of oxygen, one overlaps an sp^3 orbital of a carbon, one overlaps the s orbital of a hydrogen, and the other two each contain a lone pair.

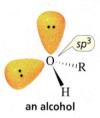

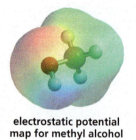

an alcohol

electrostatic potential
map for methyl alcohol

Ethers

The oxygen in an ether also has the same geometry as the oxygen in water. An ether molecule can be thought of structurally as a water molecule with alkyl groups in place of both hydrogens.

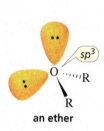

an ether

electrostatic potential
map for dimethyl ether

Amines

The nitrogen in an amine has the same geometry as the nitrogen in ammonia (Section 1.11). The nitrogen is sp^3 hybridized as in ammonia, with one, two, or three of the hydrogens replaced by alkyl groups. Remember that the number of hydrogens replaced by alkyl groups determines whether the amine is primary, secondary, or tertiary (Section 3.7).

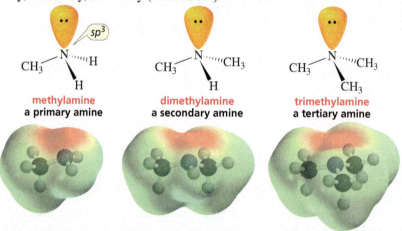

methylamine
a primary amine

dimethylamine
a secondary amine

trimethylamine
a tertiary amine

PROBLEM 31 ◆

Predict the approximate size of the following bond angles. (*Hint:* See Sections 1.11 and 1.12.)

a. the C—O—C bond angle in an ether
b. the C—N—C bond angle in a secondary amine
c. the C—O—H bond angle in an alcohol
d. the C—N—C bond angle in a quaternary ammonium salt

3.9   NONCOVALENT INTERACTIONS

Tables of physical
properties can be found
in Appendix VI.

Now we will look at noncovalent interactions—interactions that are weaker than covalent bonds—that exist between molecules and see how these interactions affect the physical properties of the five families of compounds whose names and structures we just examined. The noncovalent interactions we will look at are London dispersion forces, dipole–dipole interactions, and hydrogen bonding.

Boiling Points

The **boiling point (bp)** of a compound is the temperature at which its liquid form becomes a gas (vaporizes). For a compound to vaporize, the forces that hold the individual molecules close to each other in the liquid must be overcome. Thus, the boiling point of a compound depends on the strength of the attractive forces between the individual molecules. If the molecules are held together by strong forces, then a great deal of energy will be needed to pull the molecules away from each other and the compound will have a high boiling point. On the other hand, if the molecules are held together by weak forces, only a small amount of energy will be needed to pull the molecules away from each other and the compound will have a low boiling point.

The stronger the attractive forces holding molecules together, the higher the boiling point.

London Dispersion Forces

Alkanes contain only carbons and hydrogens. Because the electronegativities of carbon and hydrogen are similar, the bonds in alkanes are nonpolar—there are no significant partial charges on any of the atoms. Alkanes, therefore, are neutral nonpolar molecules. The nonpolar nature of alkanes gives them their oily feel.

However, it is only the average charge distribution over the alkane molecule that is neutral. Electrons move continuously and, at any instant, the electron density on one side of a molecule can be slightly greater than that on the other side, causing the molecule to have a temporary dipole. Recall that a molecule with a dipole has a negative end and a positive end (Section 1.3).

A temporary dipole in one molecule can induce a temporary dipole in a nearby molecule. As a result, the (temporarily) negative side of one molecule ends up adjacent to the (temporarily) positive side of another, as shown in Figure 3.1. Because the dipoles in the molecules are induced, the interactions between the molecules are called **induced-dipole–induced-dipole interactions.** The molecules of an alkane are held together by these induced-dipole–induced-dipole interactions, which are known as London dispersion forces (often called van der Waals forces). London dispersion forces are the weakest of all the attractive forces.

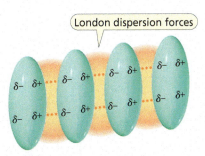

▲ **Figure 3.1**
London dispersion forces, the weakest of all the attractive forces, are induced-dipole–induced-dipole interactions.

The magnitude of the London dispersion forces that hold alkane molecules together depends on the area of contact between the molecules. The greater the area of contact, the stronger the London dispersion forces and the greater the amount of energy needed to overcome them. If you look at the boiling points of the alkanes listed in Table 3.1, you see that they increase as their molecular weight increases, because each additional methylene (CH_2) group increases the area of contact between the molecules. The four smallest alkanes have boiling points below room temperature, so they exist as gases at room temperature.

Branching lowers a compound's boiling point by reducing the area of contact. If you think of *unbranched* pentane as a cigar and *branched* 2,2-dimethylpropane as a tennis ball, you can see that branching decreases the area of contact between molecules—that is, two cigars make contact over a greater surface area than do two tennis balls. Thus, if two alkanes have the same molecular weight, the more highly branched one will have a lower boiling point.

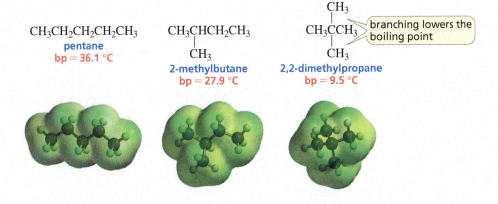

$CH_3CH_2CH_2CH_2CH_3$
pentane
bp = 36.1 °C

$CH_3CHCH_2CH_3$
$|$
CH_3
2-methylbutane
bp = 27.9 °C

$$CH_3\overset{\displaystyle CH_3}{\underset{\displaystyle CH_3}{\overset{|}{\underset{|}{C}}}}CH_3$$
2,2-dimethylpropane
bp = 9.5 °C

branching lowers the boiling point

PROBLEM 32 ◆

What is the smallest straight-chain alkane that is a liquid at room temperature (which is about 25 °C)?

Dipole–Dipole Interactions

The boiling points of a series of ethers, alkyl halides, alcohols, or amines also increase with increasing molecular weight because of the increase in the London dispersion forces. The boiling points of these compounds, however, are also affected by the polar C—Z bond. Recall that the C—Z bond is polar because nitrogen, oxygen, and the halogens are more electronegative than the carbon to which they are attached (Section 1.3).

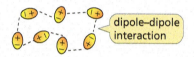

$$R-C-Z \qquad Z = N, O, F, Cl, \text{ or } Br$$
polar bond

Molecules with polar bonds are attracted to one another because they can align themselves in such a way that the positive end of one dipole is adjacent to the negative end of another dipole. These electrostatic attractive forces, called **dipole–dipole interactions,** are stronger than London dispersion forces, but not as strong as ionic or covalent bonds.

dipole–dipole interaction

Ethers generally have higher boiling points than alkanes of comparable molecular weight (Table 3.4), because *both* London dispersion forces *and* dipole–dipole interactions must be overcome for an ether to boil.

cyclopentane
bp = 49.3 °C

tetrahydrofuran
bp = 65 °C

Table 3.4   Comparative Boiling Points (°C)

Alkanes	Ethers	Alcohols	Amines
$CH_3CH_2CH_3$	CH_3OCH_3	CH_3CH_2OH	$CH_3CH_2NH_2$
−42.1	−23.7	78	16.6
$CH_3CH_2CH_2CH_3$	$CH_3OCH_2CH_3$	$CH_3CH_2CH_2OH$	$CH_3CH_2CH_2NH_2$
−0.5	10.8	97.4	47.8

For an alkyl halide to boil, both London dispersion forces and dipole–dipole interactions must be overcome. As the halogen atom increases in size, both of these interactions become stronger. A larger electron cloud means that the contact area is greater. A larger electron cloud also means that the cloud's polarizability is greater. **Polarizability** indicates how readily an electron cloud can be distorted to create a strong induced dipole. The larger the atom, the more loosely it holds the electrons in its outermost shell and the more they can be distorted. Therefore, when alkyl halides with the same alkyl group are compared—an alkyl fluoride has a lower boiling point than an alkyl chloride, which has a lower boiling point than an alkyl bromide (Table 3.5).

Table 3.5   Comparative Boiling Points of Alkanes and Alkyl Halides (°C)

—Y	H	F	Cl	Br	I
CH_3—Y	−161.7	−78.4	−24.2	3.6	42.4
CH_3CH_2—Y	−88.6	−37.7	12.3	38.4	72.3
$CH_3CH_2CH_2$—Y	−42.1	−2.5	46.6	71.0	102.5
$CH_3CH_2CH_2CH_2$—Y	−0.5	32.5	78.4	101.6	130.5
$CH_3CH_2CH_2CH_2CH_2$—Y	36.1	62.8	107.8	129.6	157.0

Hydrogen Bonds

Alcohols have much higher boiling points than ethers with similar molecular weights (Table 3.4) because, in addition to London dispersion forces and the dipole–dipole interactions of the polar

C—O bond, alcohols can form **hydrogen bonds.** A hydrogen bond is a special kind of dipole–dipole interaction that occurs between a hydrogen that is attached to an oxygen, a nitrogen, or a fluorine and a lone pair of an oxygen, a nitrogen, or a fluorine in another molecule. A hydrogen bond is not as strong as an O—H covalent bond, but it is stronger than other dipole–dipole interactions. The strongest hydrogen bonds are linear—the two electronegative atoms and the hydrogen between them lie on a straight line.

Hydrogen bonds are stronger than other dipole–dipole interactions, which are stronger than London dispersion forces.

O—H covalent bond = 0.96 Å
O—H hydrogen bond = 1.69–1.79 Å

Although each individual hydrogen bond is weak, requiring only about 5 kcal/mol to break, there are many such bonds holding alcohol molecules together. The extra energy required to break these hydrogen bonds is why alcohols have much higher boiling points than ethers with similar molecular weights.

The boiling point of water illustrates the dramatic effect that hydrogen bonding has on boiling points. Water has a molecular weight of 18 and a boiling point of 100 °C. The alkane nearest in size is methane, with a molecular weight of 16 and a boiling point of −167.7 °C.

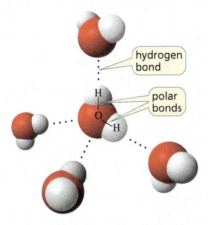

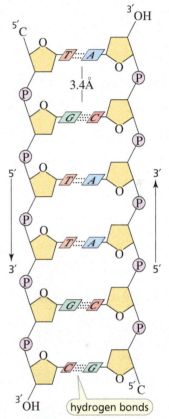

▲ Figure 3.2
DNA has two strands that run in opposite directions. The phosphates (P) and the sugars (five-membered rings) are on the outside, and the bases (A, G, T, and C) are on the inside. The two strands are held together by hydrogen bonding between the bases. A always pairs with T (using two hydrogen bonds), and G always pairs with C (using three hydrogen bonds). The structures of the bases that form the hydrogen bonds are shown in Section 26.1.

Primary and secondary amines also form hydrogen bonds, so they have higher boiling points than ethers with similar molecular weights. Nitrogen is not as electronegative as oxygen, however, so the hydrogen bonds between amine molecules are weaker than those between alcohol molecules. An amine, therefore, has a lower boiling point than an alcohol with a similar molecular weight (Table 3.4).

Because primary amines have stronger dipole–dipole interactions than secondary amines, hydrogen bonding is more significant in primary amines. Tertiary amines cannot form hydrogen bonds between their own molecules because they do not have a hydrogen attached to the nitrogen. Consequently, when we compare amines with the same molecular weight, the primary amine has a higher boiling point than the secondary amine and the secondary amine has a higher boiling point than the tertiary amine.

primary amine
bp = 97 °C

secondary amine
bp = 84 °C

tertiary amine
bp = 65 °C

Hydrogen bonds play a crucial role in biology, including making it possible for DNA to copy al its hereditary information (Figure 3.2) and holding proteins chains in the correct three-dimensiona shape (Figure 3.3). These topics are discussed in detail in Sections 26.3 and 21.14.

PROBLEM-SOLVING STRATEGY

LEARN THE STRATEGY

Predicting Hydrogen Bonding

a. Which of the following compounds forms hydrogen bonds between its molecules?
   1. $CH_3CH_2CH_2OH$ 2. $CH_3CH_2CH_2F$ 3. $CH_3OCH_2CH_3$
b. Which of these compounds forms hydrogen bonds with a solvent such as ethanol?

To solve this type of question, start by defining the kind of compound that does what is asked.

a. A hydrogen bond forms when a hydrogen attached to an O, N, or F of one molecule interacts with a lon pair on an O, N, or F of another molecule. Therefore, a compound that forms hydrogen bonds with itse must have a hydrogen attached to an O, N, or F. Only compound 1 forms hydrogen bonds with itself.
b. Ethanol has an H attached to an O, so it is able to form hydrogen bonds with a compound that has a lon pair on an O, N, or F. Therefore, all three compounds form hydrogen bonds with ethanol.

USE THE STRATEGY

PROBLEM 33 ◆

a. Which of the following compounds forms hydrogen bonds between its molecules?
   1. $CH_3CH_2OCH_2CH_2OH$ 3. $CH_3CH_2CH_2CH_2Br$ 5. $CH_3CH_2CH_2COOH$
   2. $CH_3CH_2N(CH_3)_2$ 4. $CH_3CH_2CH_2NHCH_3$ 6. $CH_3CH_2CH_2CH_2F$
b. Which of the preceding compounds forms hydrogen bonds with a solvent such as ethanol?

PROBLEM 34

Explain why

a. H_2O (100 °C) has a higher boiling point than CH_3OH (65 °C).
b. H_2O (100 °C) has a higher boiling point than NH_3 (−33 °C).
c. H_2O (100 °C) has a higher boiling point than HF (20 °C).
d. HF (20 °C) has a higher boiling point than NH_3 (−33 °C).

PROBLEM 35 ◆

Rank the following compounds from highest boiling to lowest boiling:

PROBLEM 36

Rank the compounds in each set from highest boiling to lowest boiling:

a.

b.

c.

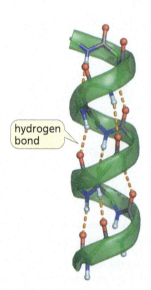

▲ Figure 3.3
Hydrogen bonds hold a segment of a protein chain in a helical structure. Notice that each hydrogen bond forms between a hydrogen (white) that is attached to a nitrogen (blue) and a lone pair on oxygen (red).

Drugs Bind to Their Receptors

Many drugs exert their physiological effects by binding to specific sites, called *receptors*, on the surface of certain cells (Section 4.17). A drug binds to a receptor using the same kinds of bonding interactions—London dispersion forces, dipole–dipole interactions, hydrogen bonding—that molecules use to bind to each other.

The most important factor in the interaction between a drug and its receptor is a snug fit. Therefore, drugs with similar shapes and polarities, which cause them to bind to the same receptor, have similar physiological effects. For example, each of the compounds shown on the next page has a nonpolar, planar, six-membered ring and substituents with similar polarities. They all have anti-inflammatory activity and are known as NSAIDs (non-steroidal anti-inflammatory drugs).

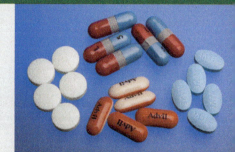

Salicylic acid has been used for the relief of fever and arthritic pain since 500 B.C. In 1897, acetylsalicylic acid (known by brand names such as Bayer Aspirin, Bufferin, Anacin, Ecotrin, and Ascriptin) was found to be a more potent anti-inflammatory agent and less irritating to the stomach; it became commercially available in 1899.

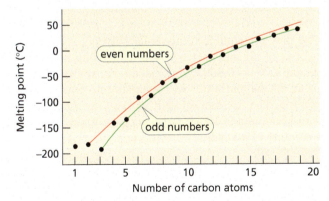

salicylic acid   acetylsalicylic acid   acetaminophen
Tylenol®

ibufenac   ibuprofen   naproxen
Advil®   Aleve®

Changing the substituents and their relative positions on the ring produced acetaminophen (Tylenol), which was introduced in 1955. It became a widely used drug because it causes no gastric irritation. However, its effective dose is not far from its toxic dose. Subsequently, ibufenac emerged; adding a methyl group to ibufenac produced ibuprofen (Advil), which is a much safer drug. Naproxen (Aleve), which has twice the potency of ibuprofen, was introduced in 1976.

Melting Points

The **melting point (mp)** of a compound is the temperature at which its solid form is converted into a liquid. The melting points of the alkanes listed in Table 3.1 show that they increase (with a few exceptions) as the molecular weight increases. The increase in melting point is less regular than the increase in boiling point because, in addition to the intermolecular attractions we just considered, the melting point is influenced by the **packing** in the crystal lattice (that is, the arrangement, including the closeness and compactness, of the molecules). The tighter the fit, the more energy required to break the lattice and melt the compound.

The melting points of straight-chain alkanes with an even number of carbons fall on a smooth curve (the red line in Figure 3.4). The melting points of straight-chain alkanes with an odd number of carbons also fall on a smooth curve (the green line). The two curves do not overlap, however, because alkanes with an odd number of carbons pack less tightly than alkanes with an even number of carbons.

◀ **Figure 3.4**
Straight-chain alkanes with an even number of carbons fall on a melting-point curve that is higher than the melting-point curve for straight-chain alkanes with an odd number of carbons.

Alkanes with an odd number of carbons pack less tightly because the molecules (each a zigzag chain with its ends tilted the same way) can lie next to each other with a methyl group on the end of one facing and repelling the methyl group on the end of the other, thus increasing the average distance between the chains. Consequently, they have weaker intermolecular attractions and correspondingly lower melting points.

odd number of carbons   **even number of carbons**

3.10 THE SOLUBILITY OF ORGANIC COMPOUNDS

The general rule that governs **solubility** is "like dissolves like." In other words:

*Polar compounds dissolve in polar solvents,
and nonpolar compounds dissolve in nonpolar solvents.*

"Polar dissolves polar" because a polar solvent such as water has partial charges that can interact with the partial charges on a polar compound. The negative poles of the solvent molecules surround the positive pole of the polar compound, and the positive poles of the solvent molecules surround the negative pole of the polar compound. The clustering of the solvent molecules around the polar molecules separates them from each other, which is what makes them dissolve. The interaction between solvent molecules and solute molecules (molecules dissolved in a solvent) is called **solvation.**

"Like dissolves like."

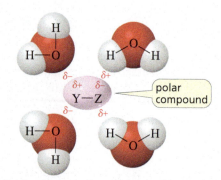

solvation of a polar compound by water

Because nonpolar compounds have no charge, polar solvents are not attracted to them. For a nonpolar molecule to dissolve in a polar solvent such as water, the nonpolar molecule would have to push the water molecules apart, disrupting their hydrogen bonding. Hydrogen bonding, however, is strong enough to exclude the nonpolar compound. On the other hand, nonpolar solutes dissolve in nonpolar solvents because the London dispersion forces between solvent molecules and solute molecules are about the same as those between solvent molecules or those between solute molecules.

Alkanes

Alkanes are nonpolar, so they are soluble in nonpolar solvents and insoluble in polar solvents such as water. The densities of alkanes increase with increasing molecular weight (Table 3.1), but even a 30-carbon alkane ($d^{20°} = 0.8097$ g/mL) is less dense than water ($d^{20°} = 1.00$ g/mL). Therefore, a mixture of an alkane and water will separate into two distinct layers, with the less dense alkane floating on top. The Alaskan oil spill in 1989, the Gulf War oil spill in 1991, and the oil spill in the Gulf of Mexico in 2010 are large-scale examples of this phenomenon because crude oil is primarily a mixture of alkanes.

Alcohols

Is an alcohol nonpolar because of its alkyl group, or is it polar because of its OH group? It depends on the size of the alkyl group. As the alkyl group increases in size, becoming a more significant fraction of the entire alcohol molecule, the compound becomes less and less soluble in water. In other words, the molecule becomes more and more like an alkane. Groups with four carbons tend to straddle the dividing line at room temperature, so alcohols with fewer than four carbons are soluble in water, but alcohols with more than four carbons are insoluble in water. Thus, an OH group can drag about three or four carbons into solution in water.

Smoke billows from a controlled burn of spilled oil off the Louisiana coast in the Gulf of Mexico.

The four-carbon dividing line is only an approximate guide because the solubility of an alcohol also depends on the structure of the alkyl group. Alcohols with branched alkyl groups are more soluble in water than alcohols with unbranched alkyl groups with the same number of carbons, because branching minimizes the contact surface of the non-polar portion of the molecule. Thus, *tert*-butyl alcohol is more soluble than *n*-butyl alcohol in water.

Ethers

The oxygen of an ether, like the oxygen of an alcohol, can drag only about three carbons into solution in water (Table 3.6). The photo on p. 73 shows that diethyl ether—an ether with four carbons—is not fully soluble in water.

Table 3.6 Solubilities of Ethers in Water		
2C s	CH_3OCH_3	soluble
3C s	$CH_3OCH_2CH_3$	soluble
4C s	$CH_3CH_2OCH_2CH_3$	slightly soluble (10 g/100 g H_2O)
5C s	$CH_3CH_2OCH_2CH_2CH_3$	minimally soluble (1.0 g/100 g H_2O)
6C s	$CH_3CH_2CH_2OCH_2CH_2CH_3$	insoluble (0.25 g/100 g H_2O)

Amines

Low-molecular-weight amines are soluble in water because primary, secondary, and tertiary amines have a lone pair they use to form a hydrogen bond. Primary amines are more soluble than secondary amines with the same number of carbons, because primary amines have two hydrogens that can engage in hydrogen bonding with water. Tertiary amines do not have hydrogens to donate for hydrogen bonds, so they are less soluble in water than are secondary amines with the same number of carbons.

Alkyl Halides

Alkyl halides have some polar character, but only alkyl fluorides have an atom that can form a hydrogen bond with water. Alkyl fluorides, therefore, are the most water soluble of the alkyl halides. The other alkyl halides are less soluble in water than ethers or alcohols with the same number of carbons (Table 3.7).

Table 3.7 Solubilities of Alkyl Halides in Water			
CH_3F very soluble	CH_3Cl soluble	CH_3Br slightly soluble	CH_3I slightly soluble
CH_3CH_2F soluble	CH_3CH_2Cl slightly soluble	CH_3CH_2Br slightly soluble	CH_3CH_2I slightly soluble
$CH_3CH_2CH_2F$ slightly soluble	$CH_3CH_2CH_2Cl$ slightly soluble	$CH_3CH_2CH_2Br$ slightly soluble	$CH_3CH_2CH_2I$ slightly soluble
$CH_3CH_2CH_2CH_2F$ insoluble	$CH_3CH_2CH_2CH_2Cl$ insoluble	$CH_3CH_2CH_2CH_2Br$ insoluble	$CH_3CH_2CH_2CH_2I$ insoluble

PROBLEM 37 ◆

In which solvent would cyclohexane have the lowest solubility: 1-pentanol, diethyl ether, ethanol, or hexane?

Cell Membranes

Cell membranes demonstrate how nonpolar molecules are attracted to other nonpolar molecules, whereas polar molecules are attracted to other polar molecules. All cells are enclosed by a membrane that prevents the aqueous (polar) contents of the cell from pouring out into the aqueous fluid that surrounds the cell. The membrane consists of two layers of phospholipid molecules—called a lipid bilayer. A phospholipid molecule has a polar head and two long nonpolar hydrocarbon tails (Section 25.5). The phospholipids are arranged so that the nonpolar tails meet in the center of the membrane. The polar heads are on both the outside surface and the inside surface, where they face the polar solutions on the outside and inside of the cell. Nonpolar cholesterol molecules are found between the tails in order to keep the nonpolar tails from moving around too much. The structure of cholesterol is shown and discussed in Section 3.16.

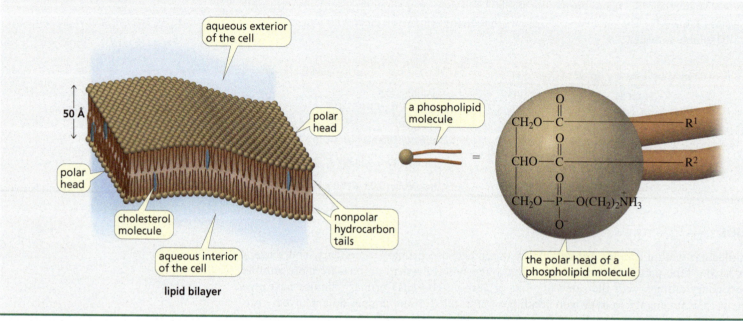

lipid bilayer

PROBLEM 38 ◆

Rank the following compounds in each set from most soluble to least soluble in water:

a. [structures: pentanoic acid, pentanol, pentane, 3-hydroxypentanoic acid]

b. $CH_3CH_2CH_2CH_2Cl$ $CH_3CH_2CH_2CH_2OH$ $HOCH_2CH_2CH_2OH$

PROBLEM 39 ◆

The effectiveness of a barbiturate as a sedative is related to its ability to penetrate the nonpolar membrane of a cell. Which of the following barbiturates would you expect to be the more effective sedative?

hexethal   barbital

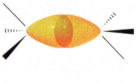

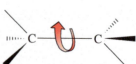

▲ Figure 3.5

A carbon–carbon single bond is formed by the overlap of cylindrically symmetrical sp^3 orbitals, so rotation about the bond can occur without changing the amount of orbital overlap.

3.11   ROTATION OCCURS ABOUT CARBON–CARBON SINGLE BONDS

We saw that a carbon–carbon single bond (a σ bond) is formed when an sp^3 orbital of one carbon overlaps an sp^3 orbital of another carbon (Section 1.7). Figure 3.5 shows that rotation about a carbon–carbon single bond can occur without any change in the amount of orbital overlap. The different spatial arrangements of the atoms that result from rotation about a single bond are called **conformational isomers** or **conformers.**

Chemists commonly use *Newman projections* to represent the three-dimensional structures that result from rotation about a σ bond. A **Newman projection** assumes that the viewer is looking along the longitudinal axis of a particular C—C bond. The carbon in front is represented by a point (where three lines are seen to intersect), and the carbon at the back is represented by a circle. The three lines emanating from each of the carbons represent its other three bonds. (Compare the three-dimensional structures shown in the margin with the two-dimensional Newman projections.)

$$H_3C—CH_3$$
ethane

Newman projections

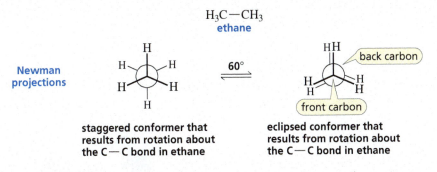

staggered conformer that results from rotation about the C—C bond in ethane

eclipsed conformer that results from rotation about the C—C bond in ethane

staggered conformer

eclipsed conformer

A staggered conformer is more stable than an eclipsed conformer.

Staggered and Eclipsed Conformers

The **staggered conformer** and the **eclipsed conformer** represent two extremes because rotation about a C—C bond can produce an infinite number (a continuum) of conformers between the two extremes.

A *staggered conformer* is more stable, and, therefore, lower in energy, than an *eclipsed conformer*. Thus, rotation about a C—C bond is not completely free because an energy barrier must be overcome when rotation occurs (Figure 3.6). However, the energy barrier is small enough (2.9 kcal/mol) to allow continuous rotation at room temperature.

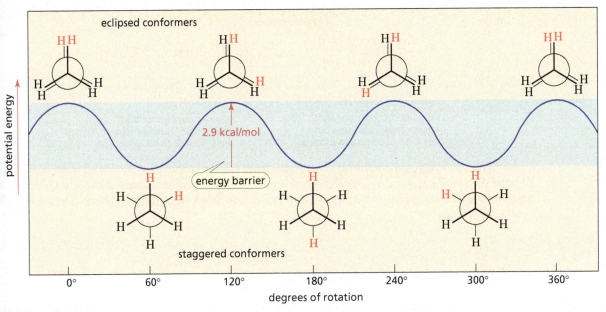

▲ **Figure 3.6**
The potential energies of the conformers of ethane obtained in one complete 360° rotation about the C—C bond. Notice that staggered conformers are at energy minima, whereas eclipsed conformers are at energy maxima.

A molecule's conformation changes from staggered to eclipsed millions of times per second at room temperature. As a result, the conformers cannot be separated from each other. At any one time, approximately 99% of ethane molecules will be in a staggered conformation because of the staggered conformer's greater stability, leaving only 1% in less stable conformations.

Why is a staggered conformer more stable than an eclipsed conformer? The major contributions to the energy difference are stabilizing interactions between the C—H σ bonding molecular orbital on one carbon and the C—H σ^* antibonding molecular orbital on the other carbon: the electrons in the filled bonding MO move partially into the unoccupied antibonding MO. Only in a staggered conformation are the two orbitals parallel, so staggered conformers maximize these stabilizing interactions. The delocalization of electrons by the overlap of a σ orbital with an empty orbital is called **hyperconjugation.**

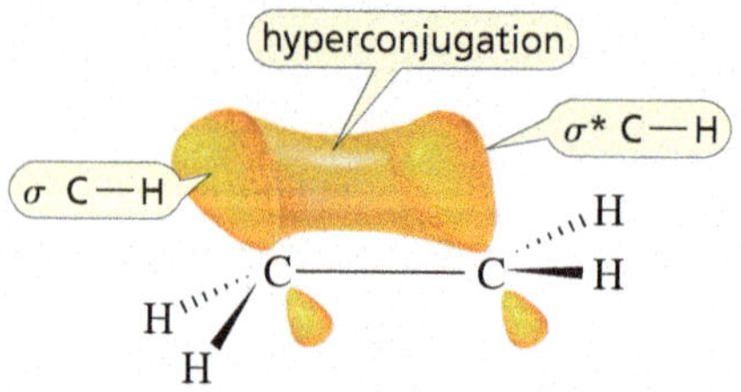

Butane has three carbon–carbon single bonds, and rotation can occur about each of them.

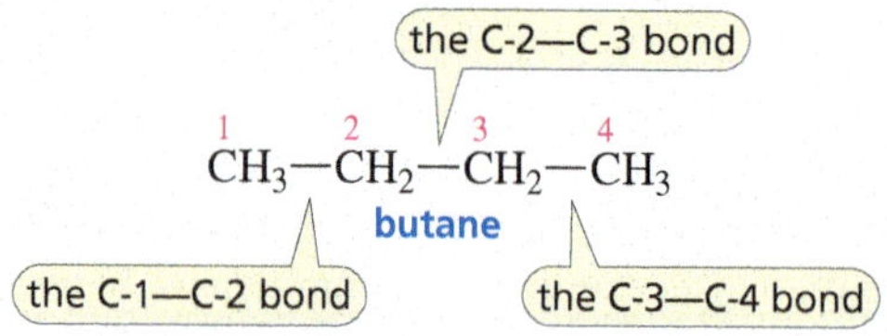

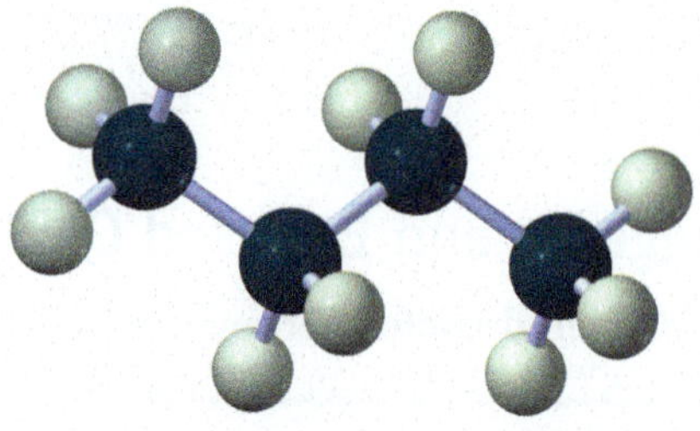

ball-and-stick model of butane

The Newman projections below show the staggered and eclipsed conformers that result from rotation about butane's C-1—C-2 bond. Notice that the carbon with the lower number is placed in the foreground in a Newman projection.

staggered conformer that
results from rotation about
the C-1—C-2 bond in butane

eclipsed conformer that
results from rotation about
the C-1—C-2 bond in butane

Although the staggered conformers that result from rotation about the C-1—C-2 bond in butane all have the same energy, the staggered conformers that result from rotation about the C-2—C-3 bond do not. The staggered and eclipsed conformers that result from rotation about the C-2—C-3 bond in butane are

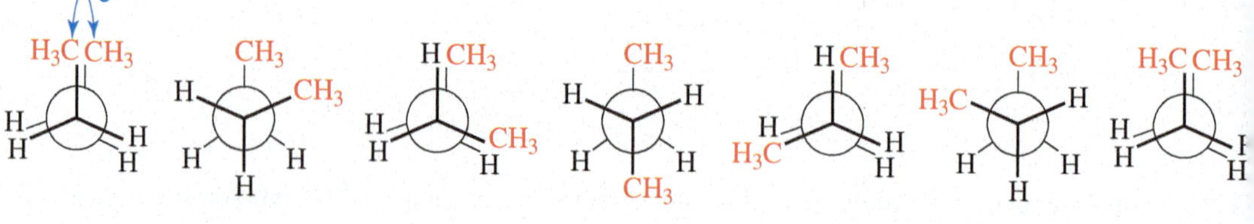

A   B   C   D   E   F   A

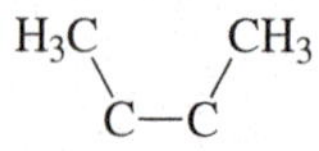

dihedral angle = 0°

dihedral angle = 180°

The relative energies of the conformers are shown in Figure 3.7. The letters in the figure correspond to the letters that identify the above structures. The degree of rotation of each conformer is identified by the dihedral angle—the angle formed in a Newman projection by a bond on the front carbon and a bond on the back carbon. For example, the conformer (A) in which one methyl group stands directly in front of the other has a dihedral angle of 0°, whereas the conformer (D) in which the methyl groups are opposite each other has a dihedral angle of 180°.

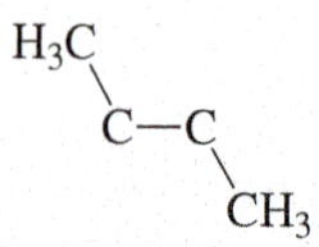

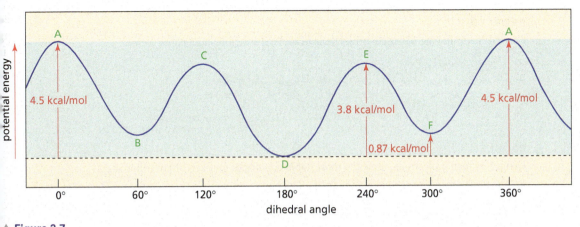

▲ **Figure 3.7**
Potential energy of butane conformers as a function of the degree of rotation about the C-2—C-3 bond. Green letters refer to the conformers (A–F) shown on p. 120.

Anti and Gauche Conformers

Of the three staggered conformers, D has the two methyl groups as far apart as possible, so D is more stable (has lower energy) than the other two staggered conformers (B and F); D is called the **anti conformer,** and B and F are called **gauche ("goesh") conformers.** (*Anti* is Greek for "opposite of"; *gauche* is French for "left.") The two gauche conformers have the same energy.

The anti and gauche conformers have different energies because of steric strain. **Steric strain** is the strain experienced by a molecule (that is, the additional energy it possesses) when atoms or groups are close enough for their electron clouds to repel each other. There is greater steric strain in a gauche conformer because the two substituents (in this case, the two methyl groups) are closer to each other. This type of steric strain is called a **gauche interaction.** In general, steric strain in molecules increases as the size of the interacting atoms or groups increases.

The eclipsed conformers also have different energies. The eclipsed conformer in which the two methyl groups are closest to each other (A) is less stable than the eclipsed conformers in which they are farther apart (C and E).

Because there is continuous rotation about all the C—C single bonds in a molecule, organic molecules are not static balls and sticks—they have many interconvertible conformers. The relative number of molecules in a particular conformation at any one time depends on its stability—the more stable it is, the greater the number of molecules that will be in that conformation. Most molecules, therefore, are in staggered conformations at any given instant, and there are more anti conformers than gauche conformers.

The preference for the staggered conformation gives carbon chains a tendency to adopt zigzag arrangements, as shown in the ball-and-stick model of decane (and in skeletal structures; see page 19).

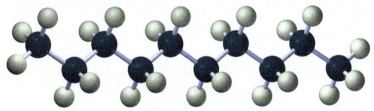

ball-and-stick model of decane

PROBLEM 40

a. Draw all the staggered and eclipsed conformers that result from rotation about the C-2—C-3 bond of pentane.

b. Draw a potential-energy diagram for rotation about the C-2—C-3 bond of pentane through 360°, starting with the least stable conformer.

PROBLEM 41

Convert the following Newman projections to skeletal structures and name them.

a. CH_3CH_2 ... CH_2CH_3 / CH_3 / H OH / H

b. H ... CH_2CH_3 / NH_2 / CH_3 CH_3 / H

PROBLEM 42 ◆

Using Newman projections, draw the most stable conformer for each of the following:

a. 3-methylpentane, viewed along the C-2—C-3 bond
b. 3-methylhexane, viewed along the C-3—C-4 bond
c. 3,3-dimethylhexane, viewed along the C-3—C-4 bond

3.12 SOME CYCLOALKANES HAVE ANGLE STRAIN

We know that, ideally, an sp^3 carbon has bond angles of 109.5° (Section 1.7). In 1885, the German chemist Adolf von Baeyer, believing that all cyclic compounds were planar, proposed that the stability of a cycloalkane could be predicted by determining the difference between this ideal bond angle and the bond angle in the planar cycloalkane. For example, the bond angles in cyclopropane are 60°, representing a 49.5° deviation from 109.5°. According to Baeyer, this deviation causes **angle strain,** which decreases cyclopropane's stability.

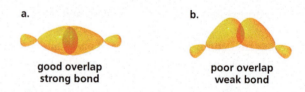

60°   90°   108°   120°

the bond angles of planar cycloalkanes

The angle strain in cyclopropane can be understood by looking at the overlap of the orbitals that form the σ bonds (Figure 3.8). Normal σ bonds are formed by the overlap of two sp^3 orbitals that point directly at each other. In cyclopropane, the overlapping orbitals cannot point directly at each other, so the amount of overlap between them is less than in a normal C—C bond. Decreasing the amount of overlap weakens the C—C bonds, and this weakness is what is known as angle strain.

a.   b.

**good overlap
strong bond**   **poor overlap
weak bond**

▲ **Figure 3.8**

(a) Overlap of sp^3 orbitals in a normal σ bond.   **(b) Overlap of sp^3 orbitals in cyclopropane.**

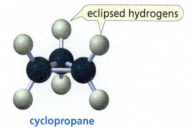

eclipsed hydrogens

cyclopropane

In addition to the angle strain of the C—C bonds, all the adjacent C—H bonds in cyclopropane are eclipsed rather than staggered, making it even more unstable.

Although planar cyclobutane would have less angle strain than cyclopropane (because the bond angles in cyclobutane would be only 19.5° [not 49.5°] less than the ideal bond angle), it would have eight pairs of eclipsed hydrogens, compared with six pairs in cyclopropane. Because of the eclipsed hydrogens, cyclobutane is not planar—one of the CH_2 groups is bent away from the plane defined by the other three carbons. Although bent cyclobutane has more angle strain than does planar cyclobutane, the increase in angle strain is more than compensated by the decrease in eclipsed hydrogens.

cyclobutane

PROBLEM 43 ◆

The bond angles in a regular polygon with *n* sides are equal to

$$180° - \frac{360°}{n}$$

a. What are the bond angles in a regular octagon?
b. What are the bond angles in a regular nonagon?

cyclopentane

If cyclopentane were planar, as Baeyer had predicted, it would have essentially no angle strain, but it would have 10 pairs of eclipsed hydrogens. Therefore, cyclopentane puckers, allowing some of the hydrogens to become nearly staggered. However, in the process, the molecule acquires some angle strain.

Contrary to what Baeyer predicted, cyclohexane is more stable than cyclopentane. Furthermore, cyclic compounds do not become less and less stable as the number of sides increases beyond six. The mistake Baeyer made was to assume that all cyclic molecules are planar.

Because three points define a plane, the carbons of cyclopropane must lie in a plane. The other cycloalkanes, however, twist and bend out of a planar arrangement in order to attain a structure that maximizes their stability by minimizing ring strain and the number of eclipsed hydrogens.

Von Baeyer, Barbituric Acid, and Blue Jeans

Johann Friedrich Wilhelm Adolf von Baeyer (1835–1917) was a professor of chemistry at the University of Strasbourg and later at the University of Munich. In 1864, he discovered barbituric acid—the first of a group of sedatives known as barbiturates—and named it after a woman named Barbara. Who Barbara was is not certain. Some say she was his girlfriend, but because Baeyer discovered barbituric acid in the same year that Prussia defeated Denmark, some believe he named it after Saint Barbara, the patron saint of artillerymen.

Baeyer was also the first to synthesize indigo, the dye used in the manufacture of blue jeans. He received the Nobel Prize in Chemistry in 1905 for his work in synthetic organic chemistry.

indigo dye

PROBLEM-SOLVING STRATEGY

Calculating the Strain Energy of a Cycloalkane

LEARN THE STRATEGY

If we assume that cyclohexane is completely free of strain, then we can use the **heat of formation**—the heat given off when a compound is formed from its elements under standard conditions—to calculate the strain energy of the other cycloalkanes. Taking the heat of formation of cyclohexane (−29.5 kcal/mol in Table 3.8) and dividing by 6 for its six CH_2 groups gives us a value of −4.92 kcal/mol for a "strainless" CH_2 group. With this value, we can calculate the heat of formation of any other "strainless" cycloalkane by multiplying the number of CH_2 groups in its ring by −4.92 kcal/mol. The strain in the compound is the difference between its actual heat of formation and its calculated "strainless" heat of formation (Table 3.8). For example, cyclopentane has an actual heat of formation of −18.4 kcal/mol and a "strainless" heat of formation of (5)(−4.92) = −24.6 kcal/mol. Therefore, cyclopentane has a strain energy of 6.2 kcal/mol, because [−18.4 − (−24.6) = 6.2]. Dividing the strain energy by the number of CH_2 groups in the cyclic compound gives the strain energy per CH_2 group for that compound.

Table 3.8	Heats of Formation and Strain Energies of Cycloalkanes			
	Heat of formation (kcal/mol)	"Strainless" heat of formation (kcal/mol)	Strain energy (kcal/mol)	Strain energy per CH_2 group (kcal/mol)
cyclopropane	+12.7	−14.6	27.3	9.1
cyclobutane	+6.8	−19.7	26.5	6.6
cyclopentane	−18.4	−24.6	6.2	1.2
cyclohexane	−29.5	−29.5	0	0
cycloheptane	−28.2	−34.4	6.2	0.9
cyclooctane	−29.7	−39.4	9.7	1.2
cyclononane	−31.7	−44.3	12.6	1.4
cyclodecane	−36.9	−49.2	12.3	1.2
cycloundecane	−42.9	−54.1	11.2	1.0

USE THE STRATEGY

PROBLEM 44

Verify the strain energy shown in Table 3.8 for cycloheptane.

3.13 CONFORMERS OF CYCLOHEXANE

Chair Conformer

The cyclic compounds most commonly found in nature contain six-membered rings because carbon rings of that size can exist in a conformation—called a *chair conformer*—that is almost completely free of strain.

A chair conformer is not planar, which would require bond angles of 120°. Instead, the carbons form a three-dimensional shape similar to a lounge chair. All the bond angles in a **chair conformer** are 111° (which is very close to the ideal tetrahedral bond angle of 109.5°), and all the adjacent bonds are staggered (Figure 3.9).

chair conformer of cyclohexane

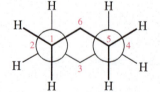

Newman projection of the chair conformer looking down the C-1—C-2 and C-5—C-4 bonds

ball-and-stick model of the chair conformer

▲ Figure 3.9
The chair conformer of cyclohexane, a Newman projection of the chair conformer showing that all the bonds are staggered, and a ball-and-stick model.

The chair conformer is so important that you should learn how to draw it:

1. Draw two parallel lines of the same length that slant slightly downward. Start the bottom line at the midpoint of the top line.

2. Connect the left ends of the parallel lines with a V that points downward; connect the right end of the parallel lines with a V that points upward. This completes the framework of the six-membered ring.

Your final chair conformer should have three sets of parallel lines.

3. Each carbon has an axial bond and an equatorial bond. The **axial bonds** (red lines with red triangles) are vertical and alternate above and below the ring. The axial bond on one of the uppermost carbons is up, the next is down, the next is up, and so on.

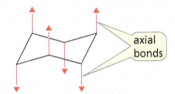

4. The **equatorial bonds** (red lines with blue balls) point outward from the ring. Because the bond angles are greater than 90°, the equatorial bonds are on a slant. If the axial bond points up, then the equatorial bond on the same carbon is on a downward slant. If the axial bond points down, then the equatorial bond on the same carbon is on an upward slant.

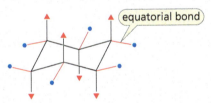

Notice that each equatorial bond is parallel to the two ring bonds one black bond away.

Remember that in this depiction, cyclohexane is viewed edge-on. The lower ring bonds are in front, and the upper bonds are in back.

PROBLEM 45

Draw 1,2,3,4,5,6-hexachlorocyclohexane with

a. all the chloro groups in axial positions.

b. all the chloro groups in equatorial positions.

Ring Flip

Cyclohexane rapidly interconverts between two stable chair conformers because of the ease of rotation about its C—C bonds. This interconversion is called **ring flip.** (See Figure 3.10 on the next page.) When the two chair conformers interconvert, bonds that are equatorial in one chair conformer become axial in the other chair conformer and bonds that are axial become equatorial.

Bonds that are equatorial in one chair conformer are axial in the other chair conformer.

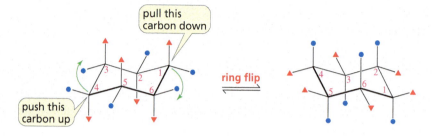

▶ **Figure 3.10**
Ring flip causes equatorial bonds to become axial bonds and axial bonds to become equatorial bonds.

Boat Conformer

Cyclohexane can also exist as a **boat conformer,** shown in Figure 3.11. Like the chair conformer, the boat conformer is free of angle strain. However, the boat conformer is not as stable because some of the C—H bonds are eclipsed. The boat conformer is further destabilized by the close proximity of the **flagpole hydrogens**—the hydrogens at the "bow" and "stern" of the boat—which cause steric strain.

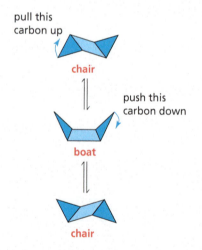

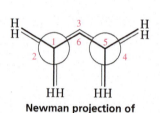

▲ **Figure 3.11**
The boat conformer of cyclohexane, a Newman projection of the boat conformer showing that some of the C—H bonds are eclipsed, and a ball-and-stick model.

boat conformer of cyclohexane

Newman projection of the boat conformer

ball-and-stick model of the boat conformer

Twist-Boat and Half-Chair Conformers

The conformers that cyclohexane assumes when interconverting from one chair conformer to the other are shown in Figure 3.12. To convert from the boat conformer to a chair conformer, one of the two topmost carbons of the boat conformer must be pulled down so that it becomes the bottommost carbon of the chair conformer. When the carbon is pulled down just a little, the **twist-boat conformer** is obtained, which is more stable than the boat conformer because the flagpole hydrogens have moved away from each other, thus relieving some steric strain. When the carbon is pulled down to the point where it is in the same plane as the sides of the boat, the very unstable **half-chair conformer** is obtained. Pulling the carbon down farther produces the *chair conformer.*

Figure 3.12 shows the relative energy of a cyclohexane molecule as it interconverts from one chair conformer to the other. The energy barrier for interconversion is 12.1 kcal/mol. Using this value, it has been calculated that cyclohexane undergoes 10^5 ring flips per second at room temperature. In other words, the two chair conformers are in rapid equilibrium.

pull this carbon up

chair

push this carbon down

boat

chair

NOTE TO THE STUDENT

• Build a model of cyclohexane. Convert it from one chair conformer to the other by pulling the bottommost carbon up and pushing the topmost carbon down.

▶ **Figure 3.12**
The conformers of cyclohexane—and their relative energies—as one chair conformer interconverts to the other chair conformer.

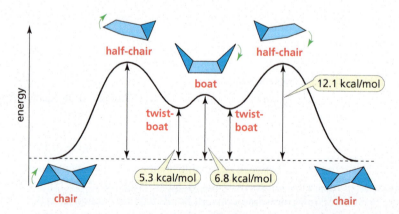

half-chair   half-chair

boat

12.1 kcal/mol

energy

twist-boat   twist-boat

5.3 kcal/mol   6.8 kcal/mol

chair   chair

Because the chair conformers are so much more stable than any of the other conformers, most molecules of cyclohexane are chair conformers at any given instant. For example, for every 10,000 chair conformers of cyclohexane, there is no more than one twist-boat conformer, which is the next most stable conformer.

3.14 CONFORMERS OF MONOSUBSTITUTED CYCLOHEXANES

Unlike cyclohexane, which has two equivalent chair conformers, the two chair conformers of a monosubstituted cyclohexane (such as methylcyclohexane) are not equivalent. The methyl substituent is in an equatorial position in one conformer and in an axial position in the other (Figure 3.13), because as we just saw, substituents that are equatorial in one chair conformer are axial in the other (Figure 3.10).

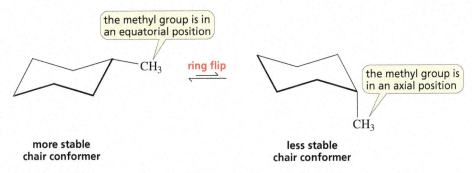

the methyl group is in an equatorial position

ring flip

the methyl group is in an axial position

more stable chair conformer

less stable chair conformer

◀ **Figure 3.13**
A substituent is in an equatorial position in one chair conformer and in an axial position in the other. The conformer with the substituent in the equatorial position is more stable.

The chair conformer with the methyl substituent in an equatorial position is the more stable of the two conformers because a substituent has more room and, therefore, fewer steric interactions when it is in an equatorial position. This can be understood by looking at Figure 3.14a, which shows that a methyl group in an equatorial position extends into space, away from the rest of the molecule.

a.

equatorial substituent points into space

CH₃

b.

axial substituent is parallel to 2 Hs

H H CH₃

◀ **Figure 3.14**
Newman projections of methylcyclohexane:
(a) the methyl substituent is equatorial
(b) the methyl substituent is axial

In contrast, any axial substituent will be relatively close to the axial substituents on the other two carbons on the same side of the ring because all three axial bonds are parallel to each other (Figure 3.14b). Because the interacting axial substituents are in 1,3-positions relative to each other, these unfavorable steric interactions are called **1,3-diaxial interactions.**

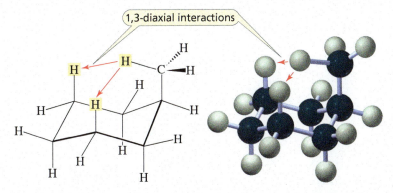

1,3-diaxial interactions

NOTE TO THE STUDENT
• Build a model of methylcyclohexane so you can see that a substituent has more room when it is in an equatorial position than when it is in an axial position.

A gauche conformer of butane and the axially substituted conformer of methylcyclohexane are compared in Figure 3.15 on the next page. Notice that the gauche interaction in butane is the same as a 1,3-diaxial interaction in methylcyclohexane.

▶ **Figure 3.15**
Butane has one gauche interaction between a methyl group and a hydrogen, whereas methylcyclohexane has two 1,3-diaxial interactions between a methyl group and a hydrogen.

gauche butane
one gauche interaction

axial methylcyclohexane
two 1,3-diaxial interactions

Table 3.9 Equilibrium Constants for Several Monosubstituted Cyclohexanes at 25 °C	
Substituent	$K_{eq} = \dfrac{[\text{equatorial}]}{[\text{axial}]}$
H	1
CH₃	18
CH₃CH₂	21
CH₃CH with CH₃	35
CH₃C with CH₃ (two)	4800
CN	1.4
F	1.5
Cl	2.4
Br	2.2
I	2.2
HO	5.4

In Section 3.11, we saw that the gauche interaction between the methyl groups of butane causes a gauche conformer to be 0.87 kcal/mol less stable than the anti conformer. Because there are two such interactions in the chair conformer of methylcyclohexane when the methyl group is in an axial position, this conformer is 1.74 kcal/mol less stable than the chair conformer with the methyl group in an equatorial position.

Because of the difference in stability of the two chair conformers, a sample of methylcyclohexane (or any other monosubstituted cyclohexane) will contain more chair conformers with the substituent in an equatorial position than with the substituent in an axial position. The relative amounts of the two chair conformers depend on the substituent (Table 3.9).

The substituent with the greater bulk in the vicinity of the 1,3-diaxial hydrogens will have a greater preference for an equatorial position because it will have stronger 1,3-diaxial interactions. For example, the experimental equilibrium constant (K_{eq}) for the conformers of methylcyclohexane (Table 3.9) indicates that 95% of methylcyclohexane molecules have the methyl group in an equatorial position at 25 °C:

LEARN THE STRATEGY

$$K_{eq} = \frac{[\text{equatorial conformer}]}{[\text{axial conformer}]} = \frac{18}{1}$$

$$\% \text{ of equatorial conformer} = \frac{[\text{equatorial conformer}]}{[\text{equatorial conformer}] + [\text{axial conformer}]} \times 100$$

$$\% \text{ of equatorial conformer} = \frac{18}{18 + 1} \times 100 = 95\%$$

In *tert*-butylcyclohexane, where the 1,3-diaxial interactions are even more destabilizing because a *tert*-butyl group is larger than a methyl group, more than 99.9% of the molecules have the *tert*-butyl group in an equatorial position.

Starch and Cellulose—Axial and Equatorial

Polysaccharides are compounds formed by linking many sugar molecules together. Two of the most common naturally occurring polysaccharides are amylose (an important component of starch) and cellulose. Both are formed by linking glucose molecules together. Starch, a water-soluble compound, is found in many of the foods we eat—potatoes, rice, flour, beans, corn, and peas. Cellulose, a water-insoluble compound, is the major structural component of plants. Cotton, for example, is composed of about 90% cellulose and wood is about 50% cellulose.

a glucose molecule
an equatorial bond
an axial bond
three glucose subunits of amylose

foods rich in starch

three glucose subunits of cellulose

cotton plant and cotton towel

How can two compounds with such different physical properties both be formed by linking together glucose molecules? If you examine their structures, you will see that the linkages in the two polysaccharides are different. In starch, an oxygen on an *axial* bond of one glucose is linked to an equatorial bond of another glucose, whereas in cellulose, an oxygen on an *equatorial* bond of one glucose is linked to an equatorial bond of another glucose. The axial bonds cause starch to form a helix that promotes hydrogen bonding with water molecules—as a result, starch is soluble in water. The equatorial bonds cause cellulose to form linear arrays that are held together by intermolecular hydrogen bonds, so it cannot form hydrogen bonds with water—as a result, cellulose is not soluble in water (Section 20.15).

Mammals have digestive enzymes that can break the axial linkages in starch but not the equatorial linkages in cellulose. Grazing animals have bacteria in their digestive tracts that possess the enzyme that can break the equatorial bonds, so cows and horses can eat hay to meet their nutritional need for glucose.

PROBLEM 46 ♦

Using the data in Table 3.9, calculate the percentage of molecules of cyclohexanol that have the OH group in an equatorial position at 25 °C.

PROBLEM 47 ♦

The chair conformer of fluorocyclohexane is 0.25 kcal/mol more stable when the fluoro substituent is in an equatorial position than when it is in an axial position. How much more stable is the anti conformer than a gauche conformer of 1-fluoropropane, considering rotation about the C-1—C-2 bond?

USE THE STRATEGY

3.15 CONFORMERS OF DISUBSTITUTED CYCLOHEXANES

If a cyclohexane ring has two substituents, we must take both substituents into account when predicting which of the two chair conformers is more stable. Let's use 1,4-dimethylcyclohexane as an example.

Geometric Isomers

First notice that there are two different dimethylcyclohexanes. One has both methyl substituents on the *same side* of the cyclohexane ring (both point downward in the example below)—it is called the **cis isomer** (*cis* is Latin for "on this side"). The other has the two methyl substituents on *opposite sides* of the ring (one points upward and one points downward in the example below)—it is called the **trans isomer** (*trans* is Latin for "across").

two methyl groups are on the same side of the ring

two methyl groups are on opposite sides of the ring

cis-1,4-dimethylcyclohexane

trans-1,4-dimethylcyclohexane

The cis isomer of a disubstituted cyclic compound has its substituents on the same side of the ring.

The trans isomer of a disubstituted cyclic compound has its substituents on opposite sides of the ring.

cis-1,4-Dimethylcyclohexane and *trans*-1,4-dimethylcyclohexane are examples of **cis–trans isomers** or **geometric isomers.** Geometric isomers have the same atoms, and the atoms are connected to each other in the same order, but they have different spatial arrangements—the cis isomer has its substituents on the same side of the ring, whereas the trans isomer has its substituent on opposite sides of the ring. The cis and trans isomers are different compounds with different melting and boiling points, so they can be separated from each other.

PROBLEM-SOLVING STRATEGY

LEARN THE STRATEGY

Differentiating Cis–Trans Isomers

Does the cis isomer or the trans isomer of 1,2-dimethylcyclohexane have one methyl group in an equatorial position and the other in an axial position?

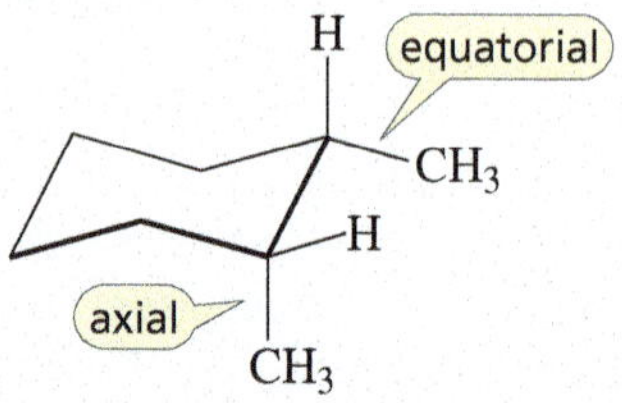

Is this the cis isomer or the trans isomer?

To solve this problem, we need to determine whether the two substituents are on the same side of the ring (cis) or on opposite sides of the ring (trans). If the bonds bearing the substituents both point upward or both point downward, then the compound is the cis isomer; if one bond points upward and the other downward, then the compound is the trans isomer. Because the conformer in question has both methyl groups attached to downward-pointing bonds, it is the cis isomer.

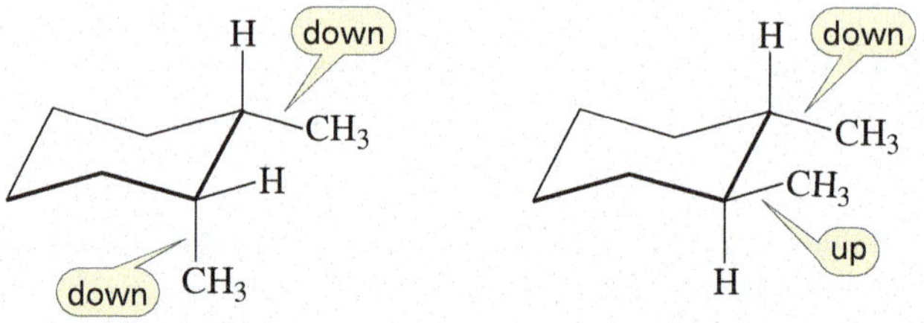

cis isomer   trans isomer

The isomer that is the most misleading when drawn in two dimensions is a *trans*-1,2-disubstituted isomer. At first glance, the methyl groups of *trans*-1,2-dimethylcyclohexane (on the right in the above image) appear to be on the same side of the ring, so you might think the compound is the cis isomer. Close inspection shows, however, that one bond points upward and the other downward, so we know that it is the trans isomer. Alternatively, if you look at the two axial hydrogens, they are clearly trans (one points straight up and the other straight down), so the methyl groups must also be trans.

USE THE STRATEGY

PROBLEM 48 ◆

Is each of the following a cis isomer or a trans isomer?

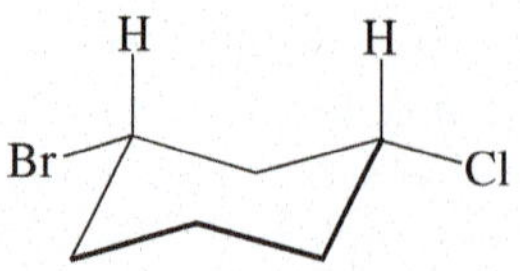

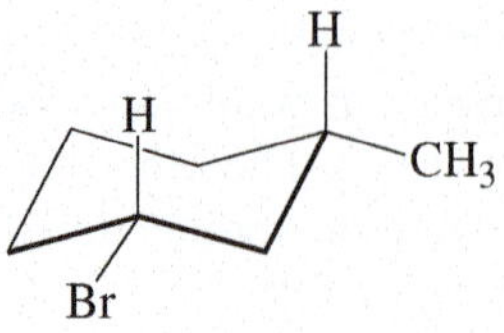

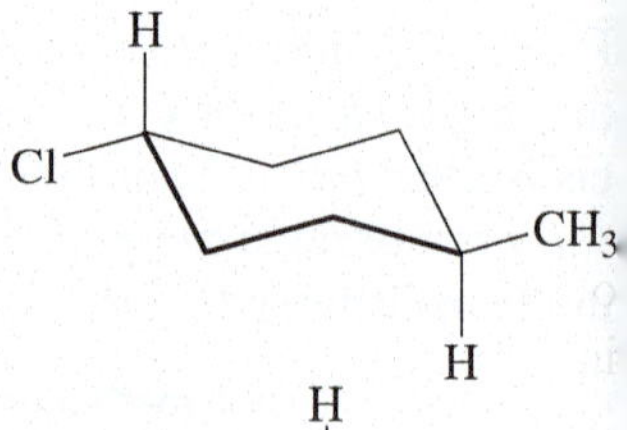

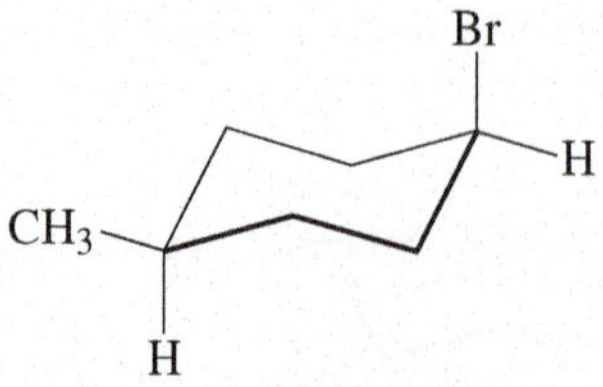

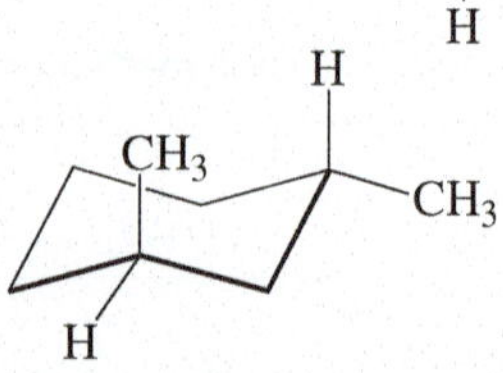

Relative Stability of Conformers

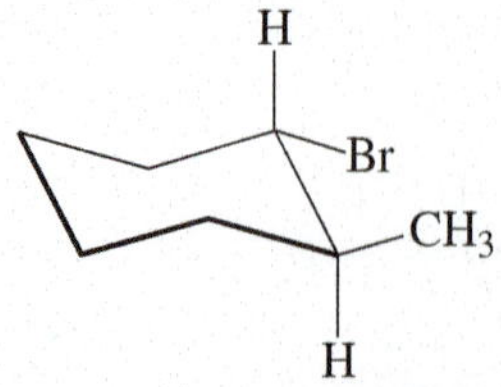

Every compound with a cyclohexane ring has two chair conformers. Thus, both the cis isomer and the trans isomer of a disubstituted cyclohexane have two chair conformers. Let's compare the structures of the two chair conformers of *cis*-1,4-dimethylcyclohexane to see if we can predict any difference in their stabilities.

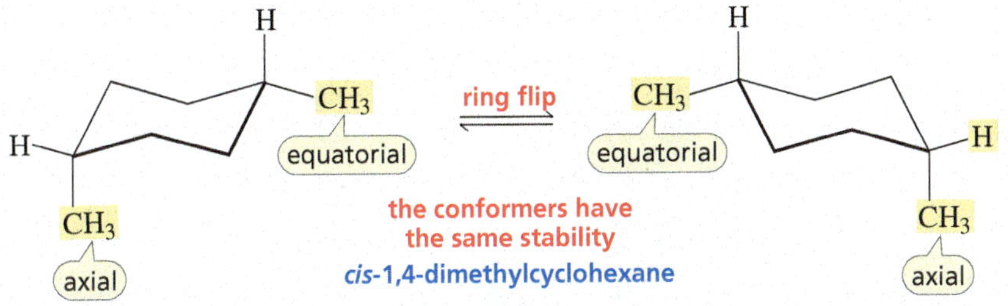

the conformers have
the same stability

cis-1,4-dimethylcyclohexane

The conformer shown on the left has one methyl group in an equatorial position and one methyl group in an axial position. The conformer on the right also has one methyl group in an equatorial position and one methyl group in an axial position. Therefore, both chair conformers are equally stable.

In contrast, the two chair conformers of *trans*-1,4-dimethylcyclohexane have different stabilities because one has both methyl substituents in equatorial positions and the other has both methyl groups in axial positions. *The conformer with both substituents in equatorial positions is more stable.*

trans-1,4-dimethylcyclohexane

If one chair conformer has both groups in equatorial positions and the other has both groups in axial positions, the more stable conformer is the one with both groups in equatorial positions.

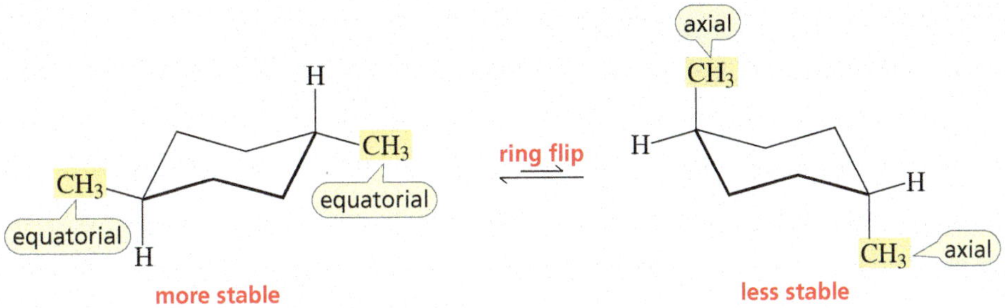

The chair conformer with both substituents in axial positions has four 1,3-diaxial interactions, causing it to be about 4×0.87 kcal/mol $= 3.5$ kcal/mol less stable than the chair conformer with both methyl groups in equatorial positions. Thus, almost all the molecules of *trans*-1,4-dimethylcyclohexane are chair conformers with both substituents in equatorial positions.

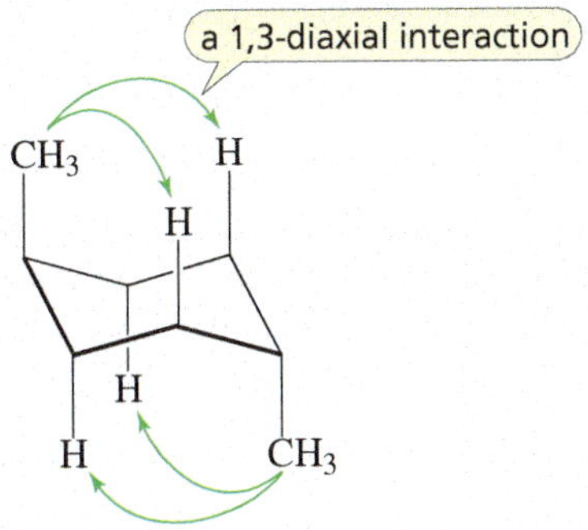

this chair conformer has four 1,3-diaxial interactions

Now let's look at the geometric isomers of 1-*tert*-butyl-3-methylcyclohexane. Both substituents of the cis isomer are in equatorial positions in one chair conformer, and both are in axial positions in the other. *The conformer with both substituents in equatorial positions is more stable.*

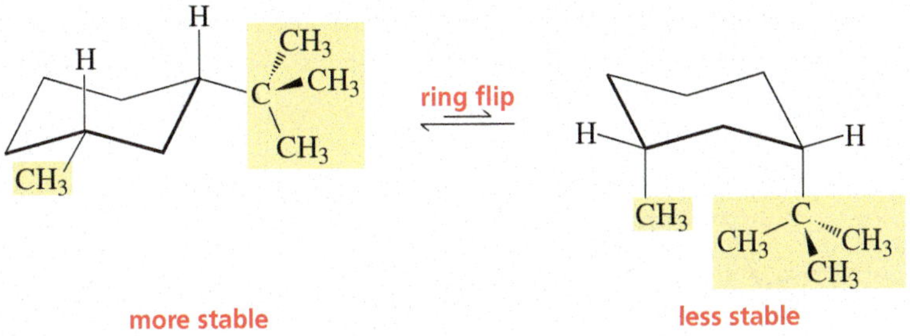

cis-1-*tert*-butyl-3-methylcyclohexane

Both chair conformers of the trans isomer have one substituent in an equatorial position and the other in an axial position. Because the *tert*-butyl group is larger than the methyl group, the 1,3-diaxial interactions will be stronger when the *tert*-butyl group is in an axial position. *Therefore, the conformer with the* tert-*butyl group in an equatorial position is more stable.*

If the chair conformers have one group in an equatorial position and one in an axial position, the more stable conformer is the one with the larger group in the equatorial position.

more stable   less stable

trans-1-*tert*-butyl-3-methylcyclohexane

PROBLEM 49 ◆

Which has a higher percentage of the diequatorial-substituted conformer compared with the diaxial-substituted conformer: *trans*-1,4-dimethylcyclohexane or *cis*-1-*tert*-butyl-3-methylcyclohexane?

PROBLEM-SOLVING STRATEGY

LEARN THE STRATEGY

Drawing the More Stable Conformer

Draw the more stable conformer of *cis*-1-methyl-4-propylcyclohexane.

First draw cyclohexane. Then add the axial bonds (because they are the easiest to draw) starting with a bond pointing upward from a back corner and then alternating down, up, down, and so on. Look at the axial bonds on the carbons where you will be placing the substituents (in this case, look at the axial bonds on the 1- and 4-positions). You can see that one axial bond points upward and the other points downward. Therefore, substituents bonded to them will be trans.

Now that you know that one conformer of a trans-1,4-disubstituted isomer has both substituents in axial positions, you also know that the other has both substituents in equatorial positions. That means that each conformer of the cis isomer (the one you want to draw) has one substituent in an equatorial position and one in an axial position. Put the smaller substituent on one of the axial positions (either axial position is fine).

Now you need to draw an equatorial bond on which to place the larger substituent. If the axial bond on the carbon where you will draw the equatorial bond points upward, then the equatorial bond needs to be on a downward-pointing slant; if the axial bond points downward, then the equatorial bond needs to be on an upward-pointing slant.

Erase any bond that does not have a substituent. (Otherwise, it indicates that a methyl group is attached to the bond.)

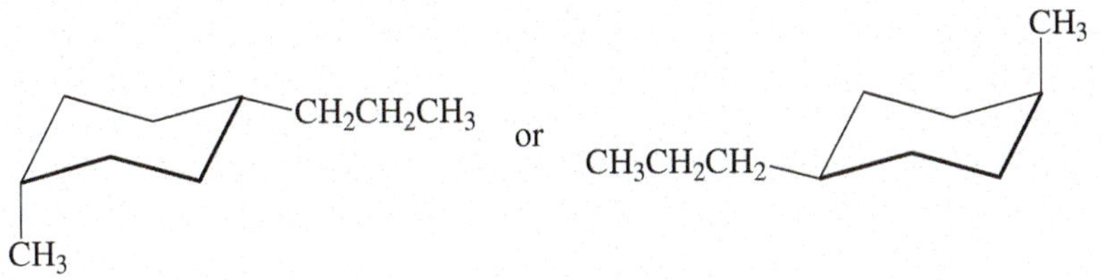

PROBLEM 50

a. Draw the more stable chair conformer of *cis*-1-ethyl-2-methylcyclohexane.
b. Draw the more stable chair conformer of *trans*-1-ethyl-2-methylcyclohexane.
c. Which compound is more stable: *cis*-1-ethyl-2-methylcyclohexane or *trans*-1-ethyl-2-methylcyclohexane?

USE THE STRATEGY

PROBLEM 51 ◆

For each of the following disubstituted cyclohexanes, indicate whether the substituents in the two chair conformers are both equatorial in one chair conformer and both axial in the other *or* one equatorial and one axial in each of the chair conformers:

a. *cis*-1,2-   **b.** *trans*-1,2-   **c.** *cis*-1,3-   **d.** *trans*-1,3-   **e.** *cis*-1,4-   **f.** *trans*-1,4-

PROBLEM 52 SOLVED

a. Draw Newman projections of the two conformers of *cis*-1,3-dimethylcyclohexane.
b. Which of the conformers predominates at equilibrium?

SOLUTION TO 52 a. Draw two staggered conformers adjacent to one another, connect them, and number the carbons so you know where to put the substituents.

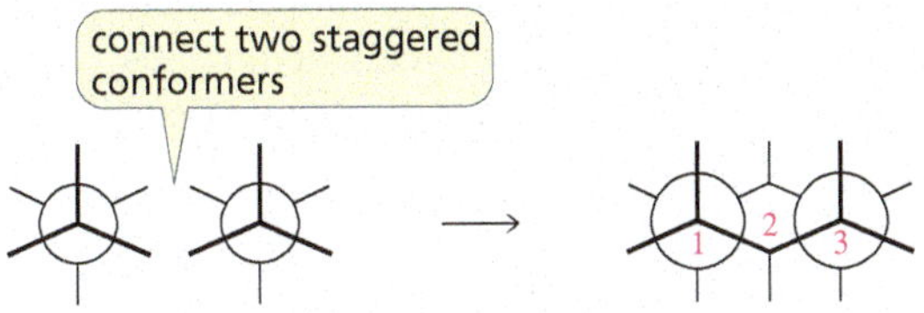

LEARN THE STRATEGY

Attach the methyl groups to the 1 and 3 carbons. Because you are drawing the cis isomer, the two substituents should both be on upward-pointing bonds in one conformer and both be on downward-pointing bonds in the other conformer.

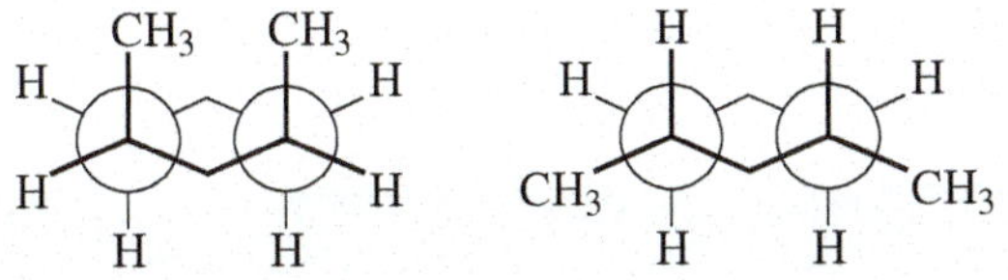

SOLUTION TO 52 b. The conformer on the right predominates at equilibrium because it is more stable—both methyl groups are on equatorial bonds.

PROBLEM 53

USE THE STRATEGY

a. Draw Newman projections of the two conformers of *trans*-1,3-dimethylcyclohexane.
b. Which of the conformers predominates at equilibrium?

PROBLEM 54 ◆

a. Calculate the energy difference between the two chair conformers of *trans*-1,4-dimethylcyclohexane.
b. What is the energy difference between the two chair conformers of *cis*-1,4-dimethylcyclohexane?

3.16 FUSED CYCLOHEXANE RINGS

When two cyclohexane rings are fused—**fused rings** share two adjacent carbons—one ring can be considered to be a pair of substituents bonded to the other ring. As with any disubstituted cyclohexane, the two substituents can be either cis or trans. The trans isomer (in which one substituent bond points upward and the other downward) has both substituents in the equatorial position. The cis isomer has one substituent in the equatorial position and one in the axial position. **Trans-fused** rings, therefore, are more stable than **cis-fused** rings.

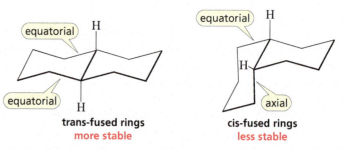

trans-fused rings
more stable

cis-fused rings
less stable

Hormones are chemical messengers—organic compounds synthesized in glands and delivered by the bloodstream to target tissues in order to stimulate or inhibit some process. Many hormones are **steroids.** Steroids have four rings designated here by A, B, C, and D. The B, C, and D rings are all trans fused and, in most naturally occurring steroids, the A and B rings are also trans fused.

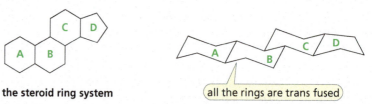

the steroid ring system

all the rings are trans fused

The most abundant member of the steroid family in animals is **cholesterol,** the precursor of all other steroids. Cholesterol is an important component of cell membranes. (See the box on p. 118.) Because its rings are locked in a specific conformation, it is more rigid than other membrane components.

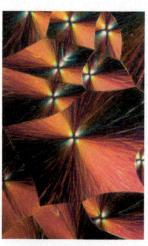

cholesterol crystals

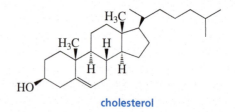

cholesterol

Cholesterol and Heart Disease

Cholesterol is probably the best-known steroid because of the widely publicized correlation between cholesterol levels in the blood and heart disease. Cholesterol is synthesized in the liver and is present in almost all body tissues. It is also found in many foods, but we do not require cholesterol in our diet because the body can synthesize all we need. A diet high in cholesterol can lead to high levels of cholesterol in the bloodstream, and the excess can accumulate on the walls of arteries, restricting the flow of blood. This disease of the circulatory system is known as *atherosclerosis* and is a primary cause of heart disease.

   Cholesterol travels through the bloodstream packaged in particles that are classified according to their density. Low-density lipoprotein (LDL) particles transport cholesterol from the liver to other tissues. Receptors on the surfaces of cells bind LDL particles, allowing them to be brought into the cell so it can

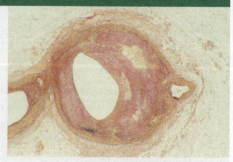

cholesterol (brown) blocking
an artery

use the cholesterol. High-density lipoprotein (HDL) is a cholesterol scavenger, removing cholesterol from the surfaces of membranes and delivering it back to the liver, where it is converted into bile acids. LDL is the so-called "bad" cholesterol, whereas HDL is the "good" cholesterol. The more cholesterol we eat, the less the body synthesizes. But this does not mean that dietary cholesterol has no effect on the total amount of cholesterol in the bloodstream, because dietary cholesterol inhibits the synthesis of the LDL receptors. So the more cholesterol we eat, the less the body synthesizes, but also the less the body can get rid of by transporting it to target cells.

How High Cholesterol Is Treated Clinically

Statins are drugs that reduce serum cholesterol levels by inhibiting the enzyme that catalyzes the formation of a compound needed for the synthesis of cholesterol. As a consequence of diminished cholesterol synthesis in the liver, the liver forms more LDL receptors—the receptors that help clear LDL (the so-called "bad" cholesterol) from the bloodstream. Studies show that for every 10% that cholesterol is reduced, deaths from coronary heart disease are reduced by 15% and total death risk is reduced by 11%.

lovastatin
Mevacor®

simvastatin
Zocor®

atorvastatin
Lipitor®

Lovastatin and simvastatin are natural statins used clinically under the trade names Mevacor and Zocor. Atorvastatin (Lipitor), a synthetic statin, is the most popular statin. It has greater potency and lasts longer in the body than natural statins because the products of its breakdown are as active as the parent drug in reducing cholesterol levels. Therefore, smaller doses of the drug may be administered. In addition, Lipitor is less polar than lovastatin and simvastatin; so it persists longer in liver cells, where it is needed. Lipitor has been one of the most widely prescribed drugs in the United States for the past several years.

ESSENTIAL CONCEPTS

Section 3.0

- **Alkanes** are **hydrocarbons** that contain only single bonds. Their general molecular formula is C_nH_{2n+2}.
- **Constitutional isomers** have the same molecular formula, but their atoms are linked differently.

Section 3.1

- Removing a hydrogen from an alkane creates an **alkyl group.**
- An **iso group** contains a carbon bonded to a hydrogen and two CH_3 groups.
- A compound can have more than one name, but a name must specify only one compound.

Section 3.2

- To name an alkane, first determine the number of carbons in the **parent hydrocarbon.**
- Substituents are listed as prefixes in alphabetical order, with a number to designate their position on the parent hydrocarbon.
- **Systematic names** can contain numbers; **common names** never do.

Section 3.3

- A single substituent on a ring does not need a number.

Sections 3.4 and 3.5

- **Alkyl halides** and **ethers** are named as substituted alkanes.

Sections 3.6 and 3.7

- **Alcohols** and **amines** are named using a functional group suffix.
- A **functional group** is a center of reactivity in a molecule.
- When there is only a substituent, the substituent prefix gets the lower of the possible numbers; when there is only a functional group, the functional group suffix gets the lower of the possible numbers; when there is both a functional group and a substituent, the functional group suffix gets the lower of the possible numbers.
- Whether alkyl halides or alcohols are **primary, secondary,** or **tertiary** depends on whether the X (halogen) or OH is attached to a primary, secondary, or tertiary carbon.
- Whether amines are **primary, secondary,** or **tertiary** depends on whether one, two, or three alkyl groups are attached to the nitrogen.
- Compounds with four alkyl groups attached to a nitrogen are called **quaternary ammonium salts.**

Section 3.8

- The oxygen of an alcohol or an ether has the same geometry as the oxygen of water; the nitrogen of an amine has the same geometry as the nitrogen of ammonia.

Section 3.9

- The boiling point of a compound increases as the attractive forces between its molecules—**London dispersion forces, dipole–dipole interactions,** and **hydrogen bonds**—increase.
- Hydrogen bonds are stronger than other dipole–dipole interactions, which are stronger than London dispersion forces.
- A **hydrogen bond** is an interaction between a hydrogen bonded to an O, N, or F and a lone pair of an O, N, or F in another molecule.
- In a series of **homologues,** the boiling point increases with increasing molecular weight. Branching lowers the boiling point.
- **Polarizability** indicates the ease with which an electron cloud can be distorted. Larger atoms are more polarizable.

Section 3.10

- Polar compounds dissolve in polar solvents; nonpolar compounds dissolve in nonpolar solvents.
- **Solvation** is the interaction between a solvent and a molecule or an ion dissolved in that solvent.
- The oxygen of an alcohol or an ether can drag three or four carbons into solution in water.

Section 3.11

- Rotation about a C—C bond results in staggered and eclipsed conformers that rapidly interconvert.
- **Conformers** are different conformations of the same compound. They cannot be separated.
- A **staggered conformer** is more stable than an **eclipsed conformer** because of **hyperconjugation.**
- The **anti conformer** is more stable than a **gauche conformer** because of steric strain.
- **Steric strain** results from repulsion between the electron clouds of atoms or groups.
- A **gauche interaction** causes steric strain in a gauche conformer.

Section 3.12

- Five- and six-membered rings are more stable than three- and four-membered rings because of the **angle strain** that results when bond angles deviate markedly from the ideal bond angle of 109.5° and the presence of eclipsed hydrogens.

Section 3.13

- Cyclohexane rapidly interconverts between two stable chair conformers—this is called **ring flip.**
- Bonds that are **axial** in one chair conformer are **equatorial** in the other and vice versa.

Section 3.14

- A chair conformer with an equatorial substituent has less steric strain and is, therefore, more stable than a chair conformer with an axial substituent.
- An axial substituent experiences unfavorable **1,3-diaxial interactions.**

Section 3.15

- Cis and trans isomers (**geometric isomers**) are different compounds and can be separated.
- A **cis isomer** has its two substituents on the same side of the ring; a **trans isomer** has its two substituents on opposite sides of the ring.
- When one conformer of a disubstituted cyclohexane has both substituents in equatorial positions and the other has both substituents in axial positions, the more stable conformer is the one with both substituents in equatorial positions.
- When each conformer of a disubstituted cyclohexane has one substituent in an equatorial position and the other in an axial position, the more stable conformer is the one that has its larger substituent in the equatorial position.

PROBLEMS

55. Draw a condensed structure and a skeletal structure for each of the following:
   a. *sec*-butyl *tert*-butyl ether
   b. isoheptyl alcohol
   c. *sec*-butylamine
   d. isopentyl bromide
   e. 5-(1-methylethyl)nonane
   f. triethylamine
   g. 4-(1,1-dimethylethyl)heptane
   h. 5,5-dibromo-2-methyloctane
   i. 3-ethoxy-2-methylhexane
   j. 5-(1,2-dimethylpropyl)nonane
   k. 3,4-dimethyloctane
   l. 5-isopropyldecane

56. List the following compounds from highest boiling to lowest boiling:

57. a. What is each compound's systematic name?

b. Draw a skeletal structure for each condensed structure given and draw a condensed structure for each skeletal structure.

1. $(CH_3)_3CCH_2CH_2CH_2CH(CH_3)_2$

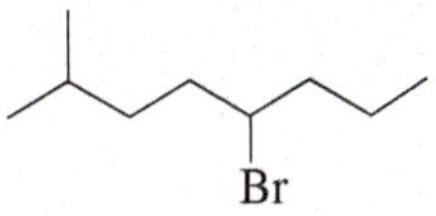

5. $BrCH_2CH_2CH_2CH_2CH_2NHCH_2CH_3$

9.

2.
Br

6.

10.
OH

3. $CH_3CHCH_2CHCH_2CH_3$
      | |
    CH_3 OH

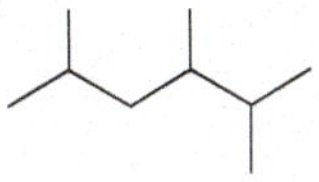

7. $CH_3CH_2CHOCH_2CH_3$
           |
        $CH_2CH_2CH_2CH_3$

11.
Br

4. $(CH_3CH_2)_4C$

8. $CH_3OCH_2CH_2CH_2OCH_3$

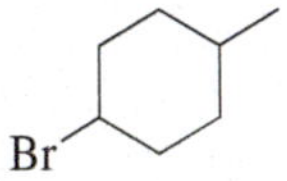

58. Which of the following represents a cis isomer?

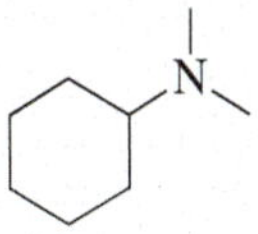

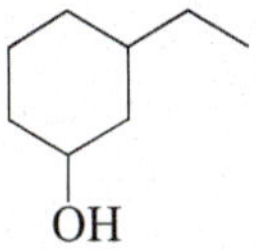

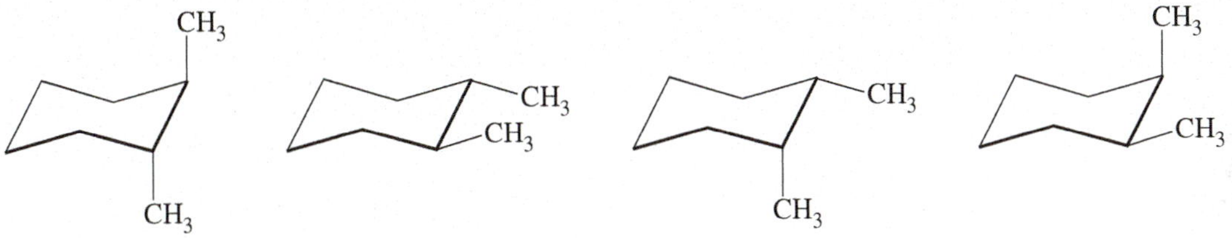

    A   B   C   D

59. a. How many primary carbons does each of the following compounds have?

b. How many secondary carbons does each one have?

c. How many tertiary carbons does each one have?

1.

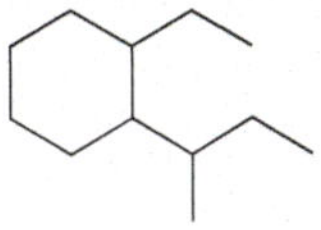

2.

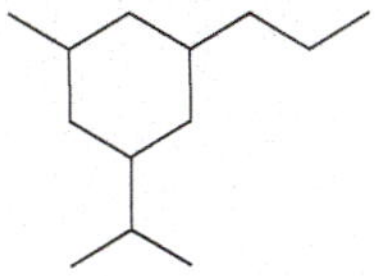

60. Which of the following conformers of isobutyl chloride is the most stable?

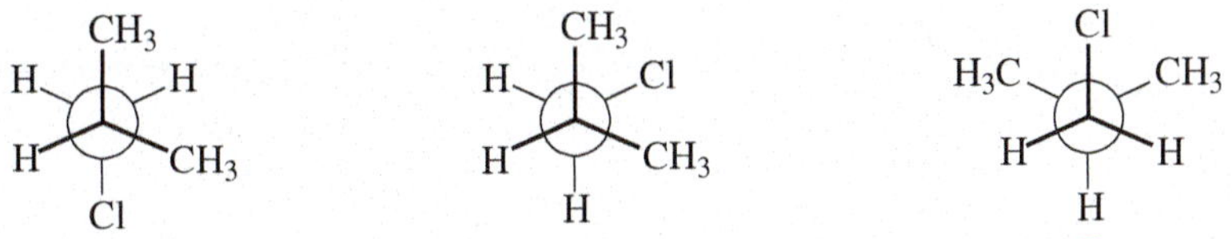

    A   B   C

61. Draw a skeletal structure for an alkane that has

a. six carbons, all secondary.

b. eight carbons and only primary hydrogens.

c. seven carbons with two isopropyl groups.

62. What is each compound's systematic name?

a. $CH_3CH_2CHCH_3$
           |
       NH_2

e. $CH_3CHCH_2CH_2CH_3$
     |
    CH_3

h. $CH_3CHCH_2CH_2CH_2OH$
     |
    CH_3

b. $CH_3CH_2CHCH_3$
          |
      Cl

f. CH_3CHNH_2
     |
   CH_3

i. 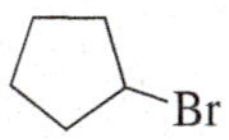
Br

c. $CH_3CH_2CHNHCH_2CH_3$
           |
       CH_3

            CH_3
            |
g. CH_3CBr
       |
    CH_2CH_3

j. 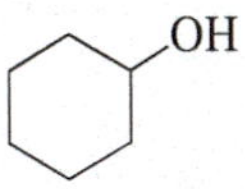
OH

d. $CH_3CH_2CH_2OCH_2CH_3$

137

63. Which has
  a. the higher boiling point: 1-bromopentane or 1-bromohexane?
  b. the higher boiling point: pentyl chloride or isopentyl chloride?
  c. the greater solubility in water: 1-butanol or 1-pentanol?
  d. the higher boiling point: 1-hexanol or 1-methoxypentane?
  e. the higher melting point: hexane or isohexane?
  f. the higher boiling point: 1-chloropentane or 1-pentanol?
  g. the higher boiling point: 1-bromopentane or 1-chloropentane?
  h. the higher boiling point: diethyl ether or butyl alcohol?
  i. the greater density: heptane or octane?
  j. the higher boiling point: isopentyl alcohol or isopentylamine?
  k. the higher boiling point: hexylamine or dipropylamine?

64. a. Draw Newman projections of the two conformers of *cis*-1,3-dimethylcyclohexane.
  b. Which of the conformers predominates at equilibrium?
  c. Draw Newman projections of the two conformers of the trans isomer.
  d. Which of the conformers predominates at equilibrium?

65. Ansaid and Motrin belong to the group of drugs known as nonsteroidal anti-inflammatory drugs (NSAIDs). Both are only slightly soluble in water, but one is a little more soluble than the other. Which of the drugs has the greater solubility in water?

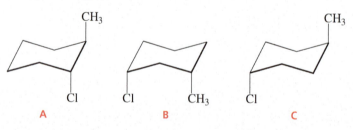

66. Draw a picture of the hydrogen bonding in methanol.

67. A student was given the structural formulas of several compounds and was asked to give them systematic names. How many did the student name correctly? Correct those that are misnamed.
  a. 4-bromo-3-pentanol
  b. 2,2-dimethyl-4-ethylheptane
  c. 5-methylcyclohexanol
  d. 1,1-dimethyl-2-cyclohexanol
  e. 5-(2,2-dimethylethyl)nonane
  f. isopentylbromide
  g. 3,3-dichlorooctane
  h. 5-ethyl-2-methylhexane
  i. 1-bromo-4-pentanol
  j. 3-isopropyloctane
  k. 2-methyl-2-isopropylheptane
  l. 2-methyl-*N*,*N*-dimethyl-4-hexanamine

68. Which of the following conformers has the highest energy (is the least stable)?

69. Give the systematic names for all alkanes with molecular formula C_7H_{16} that do not have any secondary hydrogens.

70. Draw skeletal structures for the following:
  a. 5-ethyl-2-methyloctane
  b. 1,3-dimethylcyclohexane
  c. 2,3,3,4-tetramethylheptane
  d. propylcyclopentane
  e. 2-methyl-4-(1-methylethyl)octane
  f. 2,6-dimethyl-4-(2-methylpropyl)decane

71. For rotation about the C-3 — C-4 bond of 2-methylhexane, do the following:
  a. Draw the Newman projection of the most stable conformer.
  b. Draw the Newman projection of the least stable conformer.
  c. About which other carbon–carbon bonds may rotation occur?
  d. How many of the carbon–carbon bonds in the compound have staggered conformers that are all equally stable?

72. Draw all the isomers that have molecular formula $C_5H_{11}Br$. (*Hint:* There are eight.)
  a. Give the systematic name for each of the isomers.
  b. Give a common name for each isomer that has a common name.
  c. How many of the isomers are primary alkyl halides?
  d. How many of the isomers are secondary alkyl halides?
  e. How many of the isomers are tertiary alkyl halides?

73. What is each compound's systematic name?

a.

b.   OH

c.

d.   OH

e.

f.   Cl

g.   O

h.

i.   NH₂

j.

74. Draw the two chair conformers for each of the following and indicate which conformer is more stable:
  a. *cis*-1-ethyl-3-methylcyclohexane
  b. *trans*-1-ethyl-2-methylcyclohexane
  c. *trans*-1-ethyl-2-isopropylcyclohexane
  d. *cis*-1,2-diethylcyclohexane
  e. *cis*-1-ethyl-3-isopropylcyclohexane
  f. *cis*-1-ethyl-4-isopropylcyclohexane

75. Why are lower molecular weight alcohols more soluble in water than higher molecular weight alcohols?

76. a. Draw a potential energy diagram for rotation about the C—C bond of 1,2-dichloroethane through 360°, starting with the least stable conformer. The anti conformer is 1.2 kcal/mol more stable than a gauche conformer. A gauche conformer has two energy barriers, 5.2 kcal/mol and 9.3 kcal/mol.
  b. Draw the conformer that is present in greatest concentration.
  c. How much more stable is the most stable staggered conformer than the most stable eclipsed conformer?
  d. How much more stable is the most stable staggered conformer than the least stable eclipsed conformer?

77. For each of the following compounds, determine whether the cis isomer or the trans isomer is more stable.

a.

b.

c.

78. How many ethers have molecular formula $C_5H_{12}O$? Draw their structures and give each a systematic name. What are their common names?

79. Draw the most stable conformer of the following molecule. (A solid wedge points out of the plane of the paper toward the viewer. A hatched wedge points back from the plane of the paper away from the viewer.)

$$CH_3$$

$$CH_3 \quad\quad CH_3$$

80. What is each compound's systematic name?

a. $CH_3CH_2CHCH_2CH_2CHCH_3$
              $|$ $|$
           $NHCH_3$ CH_3

b. CH_3
                         $|$
     $CH_3CH_2CHCH_2CHCH_2CH_3$
              $|$
           CH_3CHCH_3

c. CH_2CH_3
              $|$
     $CH_3CHCHCH_2Cl$
              $|$
             Cl

d. $CH_3CH_2CHCH_3$
              $|$
           CH_3CHCH_3

e.

f.

139

81. Calculate the energy difference between the two chair conformers of *trans*-1,2-dimethylcyclohexane.

82. The most stable form of glucose (blood sugar) is a six-membered ring in a chair conformation with its five substituents all in equatorial positions. Draw the most stable conformer of glucose by putting the OH groups and hydrogens on the appropriate bonds in the structure on the right.

glucose

83. What is each compound's systematic name?

a. $CH_3CHCH_2CHCH_2CH_3$
    | |
    CH_3 OH

d. $CH_3CHCH_2CH_2CHCH_2CH_2CH_3$
    | |
    CH_3 Br

g.

b.

e.

h.

c. $CH_3CHCHCH_2CH_3$
          |
          OH

f.

i.

84. Explain the following:
  a. 1-Hexanol has a higher boiling point than 3-hexanol.
  b. Diethyl ether has very limited solubility in water, but tetrahydrofuran is completely soluble.

tetrahydrofuran

85. One of the chair conformers of *cis*-1,3-dimethylcyclohexane is 5.4 kcal/mol less stable than the other. How much steric strain does a 1,3-diaxial interaction between two methyl groups introduce into the conformer?

86. Bromine is a larger atom than chlorine, but the equilibrium constants in Table 3.9 indicate that a chloro substituent has a greater preference for the equatorial position than does a bromo substituent. Suggest an explanation for this fact.

87. Name the following compounds:

a.

b.

c.

88. Using the data obtained in Problem 81, calculate the percentage of molecules of *trans*-1,2-dimethylcyclohexane that will have both methyl groups in equatorial positions.

89. Using the data obtained in Problem 85, calculate the amount of steric strain in each of the chair conformers of 1,1,3-trimethylcyclohexane. Which conformer predominates at equilibrium?

90. Draw the conformers for the following trisubstituted cyclohexane. Calculate the strain energy of each conformer. (The gauche interaction between a methyl group and an ethyl group is 0.96 kcal/mol; the 1,3-diaxial interaction between a methyl group and a H is 0.87 kcal/mol and between an ethyl group and a H is 1.00 kcal/mol.)

PART TWO

Electrophilic Addition Reactions, Stereochemistry, and Electron Delocalization

The reactions of organic compounds can be divided into three main types: **addition reactions, substitution reactions,** and **elimination reactions.** The particular type of reaction a compound undergoes depends on the functional group in the compound. **Part 2** discusses the reactions of compounds whose functional group is a carbon–carbon double bond or a carbon–carbon triple bond. We will see that these compounds undergo addition reactions or, more precisely, **electrophilic addition reactions. Part 2** also examines stereochemistry, thermodynamics and kinetics, as well as electron delocalization—topics that can be important when trying to determine the outcome of a reaction.

Chapter 4 Isomers: The Arrangement of Atoms in Space

Chapter 4 discusses the various kinds of isomers that are possible for organic compounds.

Chapter 5 Alkenes: Structure, Nomenclature, and an Introduction to Reactivity • Thermodynamics and Kinetics

Chapter 5 begins with a look at the structure, nomenclature, and stability of alkenes—*compounds that contain carbon–carbon double bonds*—and then introduces some fundamental principles that govern the reactions of organic compounds. You will revisit how to draw curved arrows to show how electrons move during the course of a reaction as new covalent bonds are formed and existing covalent bonds are broken. Chapter 5 also discusses the principles of thermodynamics and kinetics, which are central to an understanding of how and why organic reactions take place.

Chapter 6 The Reactions of Alkenes • The Stereochemistry of Addition Reactions

Organic compounds can be divided into families and, fortunately, all members of a family react in the same way. In **Chapter 6,** you will learn how the family of compounds known as alkenes reacts and what kinds of products are formed from the reactions. Although many different reactions are covered, you will see that they all take place by similar pathways.

Chapter 7 The Reactions of Alkynes • An Introduction to Multistep Synthesis

Chapter 7 covers the reactions of alkynes—*compounds that contain carbon–carbon triple bonds*. Because alkenes and alkynes both have reactive carbon–carbon π bonds, you will discover that their reactions have many similarities. This chapter also introduces you to some of the techniques chemists use to design the synthesis of organic compounds, and you will have your first opportunity to design a multistep synthesis.

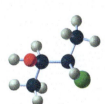

Chapter 8 Delocalized Electrons and Their Effect on Stability, pK_a, and the Products of a Reaction • Aromaticity, Electronic Effects, and an Introduction to the Reactions of Benzene

In **Chapter 8,** you will learn more about delocalized electrons, which were introduced in Chapter 2. You will see how delocalized electrons influence some of the chemical properties with which you are already familiar, such as pK_a values, the stability of carbocations, and the products obtained from the reactions of certain alkenes. Then we will turn to the reactions of dienes, com-pounds *that have two carbon–carbon double bonds*. You will see that if the two double bonds are conjugated, then electron delocalization will play a role in the products that are obtained. We also will examine the structural features that cause a compound to be aromatic and see why aromatic compounds do not undergo the reactions characteristic of alkenes and dienes.

TUTORIAL

Enhanced by
MasteringChemistry®

USING MOLECULAR MODELS

Build the models suggested as you proceed through the chapter.

1. Build a model of each of the enantiomers of 2-bromobutane (see p. 153).
   a. Try to superimpose them.
   b. Turn them so you can see that they are mirror images.
   c. Which one is (R)-2-bromobutane?

2. Build models of the stereoisomers of 3-chloro-2-butanol that are labeled **1** and **2** on pp. 164–165.
   a. Where are the Cl and OH substituents (relative to each other) in the Fischer projection? (Recall that in a Fischer projection, the horizontal lines represent bonds that point out of the plane of the paper toward the viewer, whereas the vertical lines represent bonds that point back from the plane of the paper away from the viewer.)
   b. Where are the Cl and OH substituents (relative to each other) in the most stable conformer (considering rotation about the C-2 — C-3 bond)?

3. a. Build models of the stereoisomers of 2,3-dibromobutane labeled **1** and **2** on the top of p. 169.
   b. Build models of their mirror images.
   c. Show that the stereoisomer labeled **1** is superimposable on its mirror image but the stereoisomer labeled **2** is not superimposable on its mirror image.

4. Build a model of each of the four stereoisomers of 2,3-dibromopentane. Why does 2,3-dibromopentane have four stereoisomers, whereas 2,3-dibromobutane has only three?

5. Build a model of (S)-2-pentanol.

6. Build a model of (2S,3S)-3-bromo-2-butanol. Rotate the model so its conformation is displayed as a Fischer projection. Compare this structure with that shown on p. 174.

7. Build a model of each of the compounds shown in Problem 68 on p. 182. Name the compounds.

8. a. Build a model of cis-1-bromo-4-chlorocyclohexane. Build its mirror image. Are they superimposable?
   b. Build a model of cis-1-bromo-2-chlorocyclohexane. Build its mirror image. Are they superimposable?

9. Build models of cis-1,2-dichlorocyclohexane and trans-1,2-dichlorocyclohexane. Build their mirror images. Notice that the mirror images of the cis stereoisomers are superimposable but the mirror images of the trans stereoisomers are not superimposable.

10. Build models of the molecules shown in Problems 99a and 99c on p. 185. What is the configuration of the asymmetric center in each of the molecules?

Do the *last two* problems after you study Chapter 6.

11. Build two models of trans-2-pentene. To each model, add Br_2 to opposite sides of the double bond, first adding Br^+ to the top of the double bond in one model and adding it to the bottom of the double bond in the other model, thereby forming the enantiomers shown on p. 272. Rotate the models so they represent Fischer projections. Are they erythro or threo enantiomers? Compare your answer with that given on p. 272.

12. See the box titled "Cyclic Alkenes" on p. 269. Build models of the following compounds. Can any of them not be built?
    a. cis-cyclooctene
    b. trans-cyclooctene
    c. cis-cyclohexene
    d. trans-cyclohexene

4 Isomers: The Arrangement of Atoms in Space

mirror image

In this chapter, we will see why interchanging two groups bonded to a carbon can have a profound effect on the physiological properties of a compound. For example, interchanging a hydrogen and a methyl group converts the active ingredient in Vicks vapor inhaler to methamphetamine, the street drug known as meth. The same change converts the active ingredient in Aleve, a common drug for pain, to a compound that is highly toxic to the liver.

We will now turn our attention to **isomers**—compounds with the same molecular formula but different structures. Isomers fall into two main classes: *constitutional isomers* and *stereoisomers*.

- **Constitutional isomers** differ in the way their atoms are connected. For example, ethanol and dimethyl ether are constitutional isomers because they both have molecular formula C_2H_6O, but their atoms are connected differently (the oxygen in ethanol is bonded to a carbon and to a hydrogen, whereas the oxygen in dimethyl ether is bonded to two carbons).

constitutional isomers

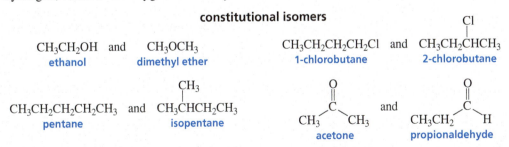

- The atoms in **stereoisomers** are connected in the same way. Stereoisomers differ in the way their atoms are arranged in space. There are two kinds of stereoisomers: conformational isomers and configurational isomers.

143

- **Conformational isomers** (also called **conformers**) are stereoisomers that rapidly interconvert at room temperature. Because they interconvert, they cannot be separated.

- **Configurational isomers** are stereoisomers that cannot interconvert unless covalent bonds are broken. Because they cannot interconvert, configurational isomers can be separated.

- There are two kinds of **conformational isomers**—those due to **rotation about carbon–carbon single bonds** and those due to **amine inversion**. We looked at conformational isomers due to rotation about carbon–carbon single bonds in Sections 3.11–3.15 (staggered, eclipsed, chair, boat, and so on). We will look at amine inversion in Section 4.16.

- There are two kinds of **configurational isomers**—**cis–trans isomers** and **isomers that contain asymmetric centers.**

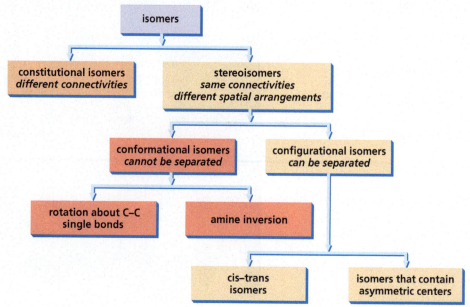

NOTE TO THE STUDENT

- By convention, chemists use the term stereoisomers to refer only to configurational isomers. So when you are asked to draw the stereoisomers for a particular compound (for example, Problems 44 and 45), draw only configurational isomers; do not draw conformational isomers.

Do not confuse conformational isomers and configurational isomers.

- Conformational isomers (or conformers) are different spatial arrangements of the same compound (for example, anti and gauche conformers; Section 3.11). They cannot be separated. Some conformers are more stable than others.

Different Conformations

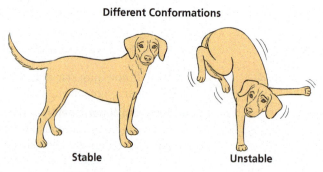

Stable   Unstable

- Configurational isomers are different compounds (for example, cis and trans isomers). They can be separated from each other. Bonds have to be broken to interconvert compounds with different configurations.

Different Configurations

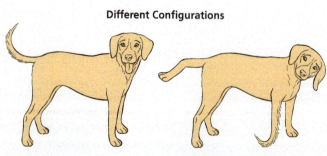

PROBLEM 1 ◆

a. Draw three constitutional isomers with molecular formula C_3H_8O.

b. How many constitutional isomers can you draw for $C_4H_{10}O$?

4.1 CIS–TRANS ISOMERS RESULT FROM RESTRICTED ROTATION

The first type of configurational isomers we will look at are **cis–trans isomers** (also called **geometric isomers**). These isomers result from restricted rotation. Restricted rotation can be caused by either a *cyclic structure* or a *double bond.*

Cyclic Compounds

You saw examples of cis–trans isomers due to restricted rotation about the bonds in a ring when we looked at disubstituted cyclohexanes in Section 3.15. *The cis isomer has its substituents on the same side of the ring; the trans isomer has its substituents on opposite sides of the ring.*

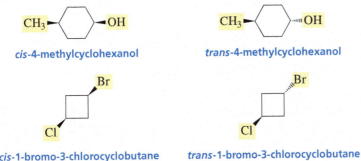

cis-4-methylcyclohexanol   *trans*-4-methylcyclohexanol

cis-1-bromo-3-chlorocyclobutane   *trans*-1-bromo-3-chlorocyclobutane

NOTE TO THE STUDENT

• A solid wedge represents a bond that points out of the plane of the paper toward the viewer, and a hatched wedge represents a bond that points into the plane of the paper away from the viewer.

PROBLEM 2

Draw the cis and trans isomers for the following:

a. 1-bromo-4-chlorocyclohexane   **b.** 1-ethyl-3-methylcyclobutane

Compounds with Double Bonds

Compounds with carbon–carbon double bonds can also have cis and trans isomers. The structure of the smallest compound with a carbon–carbon double bond (ethene) was described in Section 1.8, where we saw that the double bond was composed of a σ bond and a π bond. We saw that the π bond was formed by side-to-side overlap of two parallel p orbitals—one from each carbon. Other compounds with carbon–carbon double bonds have similar structures.

Rotation about a double bond occurs only if the π bond breaks—that is, only if the p orbitals are no longer parallel (Figure 4.1). Consequently, the energy barrier to rotation about a carbon–carbon double bond is much greater than the energy barrier to rotation about a carbon–carbon single bond: 62 kcal/mol versus 2.9 kcal/mol (Section 3.11).

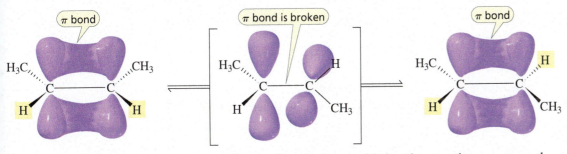

◀ **Figure 4.1**
Rotation about the carbon–carbon double bond breaks the π bond.

This high energy barrier to rotation about a carbon–carbon double bond means that a compound with a carbon–carbon double bond can have two distinct structures—the hydrogens bonded to the sp^2 carbons can be on the same side of the double bond or on opposite sides of the double bond.

- The compound with the hydrogens on the same side of the double bond is called the **cis isomer.**
- The compound with the hydrogens on opposite sides of the double bond is called the **trans isomer.**

this cis isomer has the hydrogens on the same side of the double bond

this trans isomer has the hydrogens on opposite sides of the double bond

cis isomer

trans isomer

Notice that the cis and trans isomers have the same molecular formula and the same bonds, but have different *configurations*—they differ in the way their atoms are oriented in space.

If one of the sp^2 carbons is attached to two identical substituents, then the compound cannot have cis and trans isomers.

NOTE TO THE STUDENT

- Build models to see why cis and trans isomers are not possible if one of the sp^2 carbons is attached to two identical substituents.

cis and trans isomers are not possible for these compounds because two substituents on an sp^2 carbon are the same

Cis and Trans Isomers Can Be Separated

Cis and trans isomers can be separated from each other because they are different compounds with different physical properties—for example, they have different boiling points and different dipole moments.

cis isomer
bp = 3.7 °C
$\mu = 0.33$ D

trans isomer
bp = 0.9 °C
$\mu = 0$ D

cis isomer
bp = 99.3 °C

trans isomer
bp = 91.8 °C

Notice that the trans isomers of the alkene have a dipole moment (μ) of zero because the dipole moments of their individual bonds cancel (Section 1.16).

PROBLEM 3 ◆

a. Which of the following compounds can exist as cis–trans isomers?

b. For those compounds that can exist as cis and trans isomers, draw and label the isomers.

1. $CH_3CH=CHCH_2CH_2CH_3$

3. $CH_3CH=CHCH_3$

2. $CH_3CH_2C=CHCH_3$
        |
       CH_2CH_3

4. $CH_3CH_2CH=CH_2$

PROBLEM 4

Draw skeletal structures for the compounds in Problem 3, including any cis–trans isomers.

PROBLEM 5 ◆

Draw four compounds with molecular formula C_5H_{10} that have carbon–carbon double bonds but do not have cis–trans isomers.

PROBLEM 6 ◆

Which of the following compounds have a dipole moment of zero?

A   B   C   D

Cis–Trans Interconversion in Vision

Our ability to see depends in part on an interconversion of cis and trans isomers that takes place in our eyes. A protein called opsin binds to *cis*-retinal (formed from vitamin A) in photoreceptor cells (called rod cells) in the retina to form rhodopsin. When rhodopsin absorbs light, a double bond interconverts between the cis and trans configurations, triggering a nerve impulse that plays an important role in vision. *trans*-Retinal is then released from opsin. *trans*-Retinal isomerizes back to *cis*-retinal, and another cycle begins. To trigger the nerve impulse, a group of about 500 rod cells must register five to seven rhodopsin isomerizations per cell within a few tenths of a second.

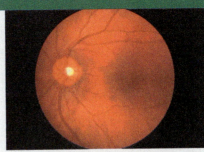

view inside the human eye

cis double bond

cis-retinal binds to opsin (a protein)

cis double bond

+ H₂N—opsin ⇌

cis-retinal

rhodopsin

N—opsin

absorption of light converts *cis*-retinal to *trans*-retinal

light

trans double bond

trans double bond

O + H₂N—opsin ⇌

trans-retinal

the protein releases *trans*-retinal

N—opsin

<table>
<tr><td>4.2</td><td>

USING THE *E,Z* SYSTEM TO DISTINGUISH ISOMERS

</td></tr>
</table>

We saw that as long as each of the *sp*² carbons of an alkene is bonded to one hydrogen, we can use the terms *cis* and *trans* to designate the geometric isomers of an alkene: *if the hydrogens are on the same side of the double bond, it is the cis isomer; if the hydrogens are on opposite sides of the double bond, it is the trans isomer* (Section 4.1).

cis isomer

trans isomer

But how do we designate the isomers of a compound such as 1-bromo-2-chloropropene?

Which isomer is cis and which is trans?

The *E,Z* system of nomenclature was devised for alkenes that do not have a hydrogen attached to each of the *sp*² carbons.*

---

*The IUPAC prefers the *E* and *Z* designations because they can be used for all alkene isomers. Many chemists, however, continue to use the "cis" and "trans" designations for simple molecules.

To name an isomer by the *E,Z* system, we first determine the relative priorities of the two groups bonded to one of the sp^2 carbons and then the relative priorities of the two groups bonded to the other sp^2 carbon. (The rules for assigning relative priorities are explained below.)

The *Z* isomer has the high-priority groups on the same side.

The *E* isomer has the high-priority groups on opposite sides.

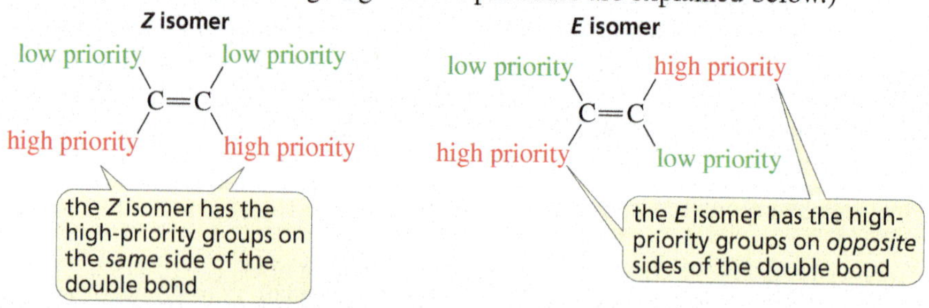
Z isomer
low priority
low priority
C=C
high priority
high priority
the Z isomer has the high-priority groups on the same side of the double bond
E isomer
low priority
high priority
C=C
high priority
low priority
the E isomer has the high-priority groups on opposite sides of the double bond

If the two high-priority groups (one from each carbon) are on the same side of the double bond, the isomer is the *Z* isomer (*Z* is for *zusammen*, German for "together"). If the high-priority groups are on opposite sides of the double bond, the isomer is the *E* isomer (*E* is for *entgegen*, German for "opposite").

Determining Relative Priorities

LEARN THE STRATEGY

1. The relative priorities depend on the atomic numbers of the atoms bonded directly to the sp^2 carbon. The greater the atomic number, the higher the priority. Now let's look at how to determine that the isomer below on the left is the *Z* isomer:
   - The sp^2 carbon on the left is bonded to a Br and to an H; Br has a greater atomic number than H, so **Br** has the higher priority.

The greater the atomic number of the atom bonded to the sp^2 carbon, the higher the priority of the substituent.

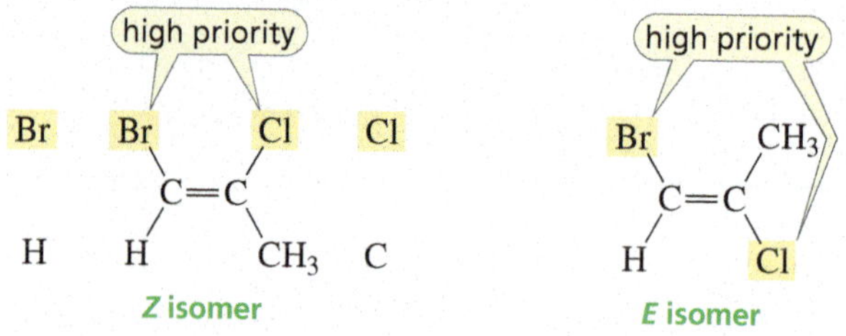
high priority
Br
Br
Cl
Cl
C=C
H
H
CH3
C
Z isomer
high priority
Br
CH3
C=C
H
Cl
E isomer

   - The sp^2 carbon on the right is bonded to a Cl and to a C; Cl has the greater atomic number, so **Cl** has the higher priority. (Notice that you use the atomic number of C, not the mass of the CH_3 group, because the priorities are based on the atomic numbers of atoms, *not* on the masses of groups.)

Thus, the isomer on the left has the high-priority groups (Br and Cl) on the same side of the double bond, so it is the **Z isomer.** (Zee groups are on Zee Zame Zide.) The isomer on the right has the high-priority groups on opposite sides of the double bond, so it is the ***E* isomer.**

If the atoms attached to the sp^2 carbon are the same, the atoms attached to the tied atoms are compared; the one with the greater atomic number belongs to the group with the higher priority.

2. If the two atoms attached to an sp^2 carbon are the same (there is a tie), then consider the atomic numbers of the atoms that are attached to the "tied" atoms. Now we will see why the isomer below on the left is the *Z* isomer:
   - Both atoms bonded to the sp^2 carbon on the left are carbons (in a CH_2Cl group and in a CH_2CH_2Cl group), so there is a tie.

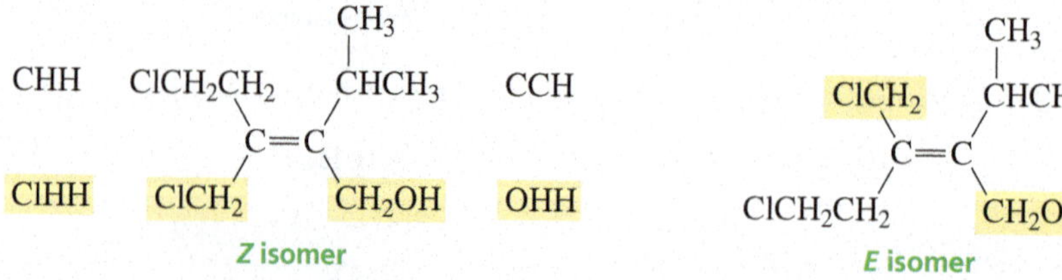
CH3
CHH
ClCH2CH2
CHCH3
CCH
C=C
ClHH
ClCH2
CH2OH
OHH
Z isomer
CH3
ClCH2
CHCH3
C=C
ClCH2CH2
CH2OH
E isomer

The C of the CH_2Cl group is bonded to **Cl, H, H,** and the C of the CH_2CH_2Cl group is bonded to **C, H, H.** Cl has a greater atomic number than C, so the CH_2Cl group has the higher priority.

   - Both atoms attached to the sp^2 carbon on the right are Cs (in a CH_2OH group and in a $CH(CH_3)_2$ group), so there is a tie on this side as well. The C of the CH_2OH group is bonded to **O, H, H,** and the C of the $CH(CH_3)_2$ group is bonded to **C, C, H.** Of these six atoms, O has the greatest atomic number, so **CH_2OH** has the higher priority. (Note that you do not add the atomic numbers—you take the single atom with the greatest atomic number.)

The *E* and *Z* isomers are as shown above.

3. If an atom is doubly bonded to another atom, the priority system treats it as if it were singly bonded to two of those atoms. If an atom is triply bonded to another atom, it is treated as if it were singly bonded to three of those atoms.

- For example, in the isomer shown next on the left, the sp^2 carbon on the left is bonded to a CH_2CH_2OH group and to a $CH_2C\equiv CH$ group:

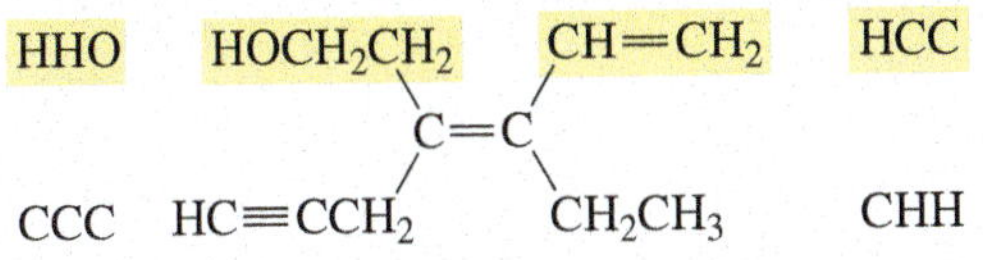

If an atom is doubly bonded to another atom, treat it as if it were singly bonded to two of those atoms.

If an atom is triply bonded to another atom, treat it as if it were singly bonded to three of those atoms.

Z isomer 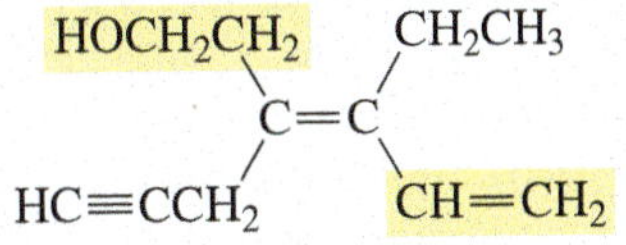   *E* isomer

Because the atoms bonded to the sp^2 carbon are both carbons, there is a tie. Each of the carbons is bonded to **C, H, H,** so there is another tie. We turn our attention to the groups attached to the CH_2 groups to break the tie. One of these groups is CH_2OH, and the other is $C\equiv CH$; the C of the CH_2OH group is bonded to **H, H, O;** the triple-bonded C is considered to be bonded to **C, C, C.** Of the six atoms, O has the greatest atomic number, so CH_2OH has the higher priority.

- Both atoms bonded to the sp^2 carbon on the right are Cs, so they are tied. The first carbon of the CH_2CH_3 group is bonded to **C, H, H;** the first carbon of the $CH=CH_2$ group is bonded to an H and doubly bonded to a C, so it is considered to be bonded to **H, C, C.** One C cancels in each of the two groups, leaving **H** and **H** in the CH_2CH_3 group and **H** and **C** in the $CH=CH_2$ group. C has a greater atomic number than H, so $CH=CH_2$ has the higher priority.

Cancel atoms that are identical in the two groups; use the remaining atoms to determine the group with the higher priority.

4. If two isotopes (atoms with the same atomic number but different mass numbers) are being compared, the mass number is used to determine the relative priorities.

- For example, in the isomer shown next on the left, the sp^2 carbon on the left is bonded to a deuterium (D) and to a hydrogen (H): D and H have the same atomic number, but D has a greater mass number, so **D** has the higher priority.

If atoms have the same atomic number but different mass numbers, the atom with the greater mass number has the higher priority.

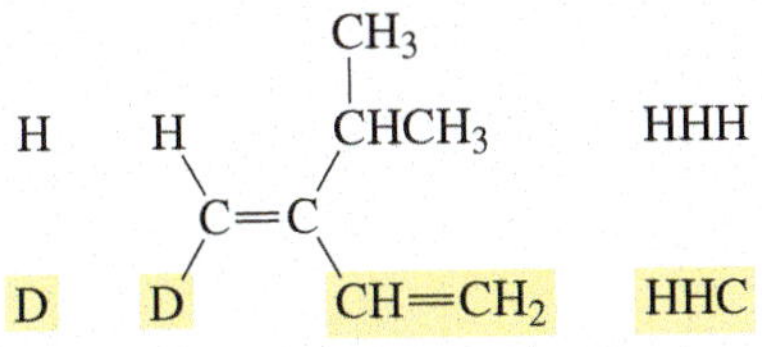

Z isomer 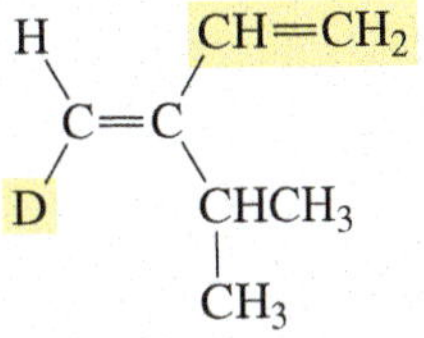   *E* isomer

- The Cs that are attached to the sp^2 carbon on the right are *both* bonded to **C, C, H,** so we must go to the next set of atoms to break the tie. The second carbon of the $CH(CH_3)_2$ group is bonded to **H, H, H,** whereas the second carbon of the $CH=CH_2$ group is bonded to **H, H, C.** (*Notice that to get the third atom, you go back along the double bond.*) Therefore, $CH=CH_2$ has the higher priority.

USE THE STRATEGY

PROBLEM 7

Draw and label the *E* and *Z* isomers for each of the following:

a. $CH_3CH_2CH=CHCH_3$

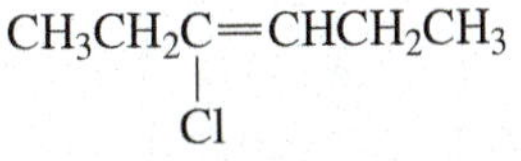

c. 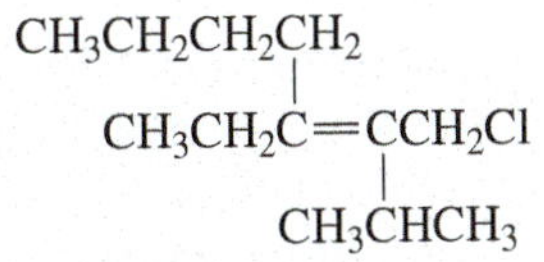
$$CH_3CH_2CH_2CH_2$$
$$|$$
$$CH_3CH_2C=CCH_2Cl$$
$$|$$
$$CH_3CHCH_3$$

b.
$$CH_3CH_2C=CHCH_2CH_3$$
$$|$$
$$Cl$$

d. 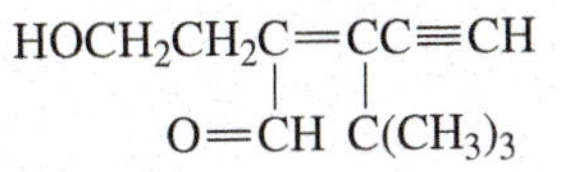
$$HOCH_2CH_2C=CC\equiv CH$$
$$| \quad |$$
$$O=CH \quad C(CH_3)_3$$

PROBLEM 8 ◆

Assign relative priorities to each set of substituents:

a. —Br   —I   —OH   —CH_3

b. —CH_2CH_2OH —OH   —CH_2Cl —$CH=CH_2$

tamoxifen

PROBLEM 9 ♦

Tamoxifen slows the growth of some breast tumors by binding to estrogen receptors. Is tamoxifen an *E* or a *Z* isomer?

PROBLEM 10

Draw skeletal structures for each pair of isomers in Problem 7.

PROBLEM 11 ♦

Name each of the following:

a.   b.   c.

PROBLEM-SOLVING STRATEGY

LEARN THE STRATEGY

Drawing *E,Z* Structures

Draw the structure of (*E*)-1-bromo-2-methyl-2-butene.

First draw the compound without specifying the isomer so you can see what substituents are bonded to the sp^2 carbons. Then determine the relative priorities of the two groups bonded to each of the sp^2 carbons.

$$\overset{\overset{\displaystyle CH_3}{|}}{BrCH_2C}=CHCH_3$$

The sp^2 carbon on the left is attached to a CH_3 and a CH_2Br; CH_2Br has the higher priority. The sp^2 carbon on the right is attached to a CH_3 and an H; CH_3 has the higher priority. To draw the *E* isomer, put the two high-priority substituents on opposite sides of the double bond.

$$\underset{CH_3}{\overset{BrCH_2}{\diagdown}}C=C\underset{CH_3}{\overset{H}{\diagup}}$$

USE THE STRATEGY

PROBLEM 12

Draw the *Z* isomer of an alkene that has a CH_3 and an H on one sp^2 carbon and isopropyl and butyl groups on the other sp^2 carbon.

4.3 A CHIRAL OBJECT HAS A NONSUPERIMPOSABLE MIRROR IMAGE

Why can't you put your right shoe on your left foot? Why can't you put your right glove on your left hand? It is because hands, feet, gloves, and shoes have right-handed and left-handed forms. An object with a right-handed and a left-handed form is said to be **chiral** (ky-ral), a word derived from the Greek word *cheir,* which means "hand."

A chiral object has a *nonsuperimposable mirror image.* In other words, its mirror image *does not look the same* as an image of the object itself. A hand is chiral because when you look at your left hand in a mirror, you see a right hand, not a left hand (Figure 4.2a).

chiral objects

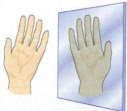 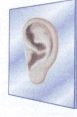 

▶ **Figure 4.2a**
A chiral object is not the same as its mirror image—they are nonsuperimposable.

left hand   right hand

In contrast, a chair is not chiral; the reflection of the chair in the mirror looks the same as the chair itself. Objects that are not chiral are said to be **achiral.** An achiral object has a *superimposable mirror image* (Figure 4.2b).

achiral objects

◀ **Figure 4.2b**
An achiral object is the same as its mirror image—they are superimposable.

PROBLEM 13 ◆

Which of the following objects are chiral?

a. a mug with DAD written to one side of the handle
b. a mug with MOM written to one side of the handle
c. a mug with DAD written opposite the handle
d. a mug with MOM written opposite the handle

e. a wheelbarrow
f. a remote control device
g. a nail
h. screw

4.4 AN ASYMMETRIC CENTER IS A CAUSE OF CHIRALITY IN A MOLECULE

Objects are not the only things that can be chiral. Molecules can be chiral too. The usual *cause of chirality in a molecule is an asymmetric center.* (Other features that cause chirality are relatively uncommon and beyond the scope of this book, but you can see one example in Problem 103.)

An **asymmetric center** (also called a chiral center) is an atom bonded to four different groups. Each of the following compounds has an asymmetric center that is indicated by a star.

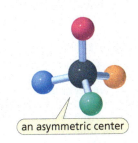

an asymmetric center

A molecule with an asymmetric center is chiral.

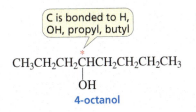

C is bonded to H, OH, propyl, butyl

CH₃CH₂CH₂CHCH₂CH₂CH₂CH₃
|
OH
4-octanol

C is bonded to H, Br, ethyl, methyl

CH₃CHCH₂CH₃
|
Br
2-bromobutane

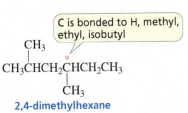

C is bonded to H, methyl, ethyl, isobutyl

$$CH_3$$
|
CH₃CHCH₂CHCH₂CH₃
|
CH₃
2,4-dimethylhexane

PROBLEM 14 ◆

Which of the following compounds has an asymmetric center?

a. CH₃CH₂CHCH₃
          |
          Cl

b. CH₃CH₂CHCH₃
          |
          CH₃

          CH₃
          |
c. CH₃CH₂CCH₂CH₂CH₃
          |
          Br

d. CH₃CH₂OH

e. CH₃CH₂CHCH₂CH₃
          |
          Br

f. CH₂=CHCHCH₃
          |
          NH₂

PROBLEM 15 SOLVED

Tetracycline is called a broad-spectrum antibiotic because it is active against a wide variety of bacteria. How many asymmetric centers does tetracycline have?

SOLUTION Because an asymmetric center must have four different groups attached to it, only sp^3 carbon can be asymmetric centers. Therefore, we start by locating all the sp^3 carbons in tetracycline. (They are numbered in red.) Tetracycline has nine sp^3 carbons. Four of them (1, 2, 5, and 8) are not asymmetric centers because they are not bonded to four different groups. Tetracycline, therefore, has five asymmetric centers (3, 4, 6, 7, and 9).

tetracycline

4.5 ISOMERS WITH ONE ASYMMETRIC CENTER

A compound with one asymmetric center, such as 2-bromobutane, can exist as two stereoisomers. The two stereoisomers are analogous to a left and a right hand. If we imagine a mirror between the two stereoisomers, we can see that they are mirror images of each other. Moreover, they are nonsuperimposable mirror images, which makes them different molecules.

$$CH_3\overset{*}{C}HCH_2CH_3$$
$$|$$
$$Br$$

2-bromobutane

mirror

two stereoisomers of 2-bromobutane
enantiomers

Molecules that are nonsuperimposable mirror images of each other are called **enantiomers** (from the Greek *enantion*, which means "opposite"). Thus, the two stereoisomers of 2-bromobutane are enantiomers.

A molecule that has a *nonsuperimposable* mirror image, like an object that has a *nonsuperimposable* mirror image, is *chiral* (Figure 4.3a). Therefore, each member of a pair of enantiomers is chiral. A molecule that has a *superimposable* mirror image, like an object that has a *superimposable* mirror image, is *achiral* (Figure 4.3b). Notice that chirality is a property of an entire object or an entire molecule.

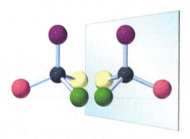

**nonsuperimposable
mirror images**

NOTE TO THE STUDENT

• Prove to yourself that the two stereoisomers of 2-bromobutane are not identical by building ball-and-stick models to represent them and then trying to superimpose one on the other. The tutorial on p. 142 tells you what other models you should build as you go through this chapter.

▶ **Figure 4.3**
(a) A chiral molecule has a nonsuperimposable mirror image.
(b) An achiral molecule has a superimposable mirror image. To see that the achiral molecule is superimposable on its mirror image, mentally rotate the molecule clockwise.

a.

a chiral molecule   nonsuperimposable mirror image

b.

an achiral molecule   superimposable mirror image

PROBLEM 16 ♦

Which of the compounds in Problem 14 can exist as enantiomers?

4.6 ASYMMETRIC CENTERS AND STEREOCENTERS

An asymmetric center is also called a **stereocenter** (or a **stereogenic center**), but they do not mean quite the same thing. A stereocenter is an atom at which the interchange of two groups produces a stereoisomer. Thus, stereocenters include both (1) *asymmetric centers,* where the interchange of two groups produces an enantiomer, and (2) the sp^2 carbons of an alkene or the sp^3 carbons of a cyclic compound, where the interchange of two groups converts a cis isomer to a trans isomer or vice versa. This means that although *all asymmetric centers are stereocenters,* not all stereocenters are asymmetric centers.

PROBLEM 17 ♦

a. How many asymmetric centers does the following compound have?
b. How many stereocenters does it have?

$$CH_3CHCH{=}CHCH_3$$
$$\underset{\displaystyle Cl}{|}$$

4.7 HOW TO DRAW ENANTIOMERS

Chemists draw enantiomers using either *perspective formulas* or *Fischer projections*.

To draw the perspective formula of an enantiomer:

- show two of the bonds to the asymmetric center in the plane of the paper. (Make sure the two bonds are adjacent to each other.)
- show one bond as a solid wedge protruding forward out of the paper.
- show the fourth bond as a hatched wedge extending behind the paper. (The solid and hatched wedges must be adjacent to each other and the solid wedge must be below the hatched wedge.)

When you draw the first enantiomer, the four groups bonded to the asymmetric center can be placed around it in any order. You can then draw the second enantiomer by drawing the mirror image of the first.

perspective formulas of the enantiomers
of 2-bromobutane

A **Fischer projection,** devised in the late 1800s by Emil Fischer when printing techniques could handle lines but not wedges, represents an asymmetric center as the point of intersection of two perpendicular lines.

LEARN THE STRATEGY

A solid wedge represents a bond that extends out of the plane of the paper toward the viewer.

A hatched wedge represents a bond that points back from the plane of the paper away from the viewer.

When you draw a perspective formula, make sure the two bonds in the plane of the paper are adjacent to each other; neither the solid wedge nor the hatched wedge should be drawn between them.

- Horizontal lines represent the bonds that project out of the plane of the paper toward the viewer.
- Vertical lines represent the bonds that extend back from the plane of the paper away from the viewer.
- The carbon chain is usually drawn vertically, with C-1 at the top.

In a Fischer projection, horizontal
lines project out of the plane of the
paper toward the viewer, and vertical
lines extend back from the plane of
the paper away from the viewer.

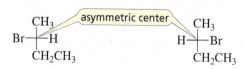

**Fischer projections of the enantiomers
of 2-bromobutane**

LEARN THE STRATEGY

When you draw an enantiomer using a Fischer projection, you can put the four atoms or groups bonded to the asymmetric center in any order around that center. You can then draw the second enantiomer by interchanging two of the atoms or groups. It does not matter which two you interchange. (Make models to convince yourself that this is true.) It is best to interchange the groups on the two horizontal bonds, because then the enantiomers look like mirror images on your paper.

Whether you are drawing perspective formulas or Fischer projections, interchanging two atoms or groups will produce the other enantiomer. Interchanging two atoms or groups a second time brings you back to the original molecule.

USE THE STRATEGY

PROBLEM 18

Draw enantiomers for each of the following using

a. perspective formulas.   **b.** Fischer projections.

 1. $CH_3\overset{\overset{\displaystyle Br}{|}}{C}HCH_2OH$ 2. $ClCH_2CH_2\overset{\overset{\displaystyle CH_3}{|}}{C}HCH_2CH_3$ 3. $CH_3\overset{\overset{\displaystyle CH_3}{|}}{C}H\underset{\underset{\displaystyle OH}{|}}{C}HCH_3$

PROBLEM 19 SOLVED

Do the following structures represent identical compounds or a pair of enantiomers?

$$CH_3CH_2{-}\overset{\overset{\displaystyle CH=CH_2}{|}}{\underset{\underset{\displaystyle H}{}}{C}}{\cdots}CH_3 \quad \text{and} \quad H{-}\overset{\overset{\displaystyle CH_2CH_3}{|}}{\underset{\underset{\displaystyle CH=CH_2}{}}{C}}{\cdots}CH_3$$

SOLUTION Interchanging two atoms or groups attached to an asymmetric center produces an enantiomer (the middle structure below). Interchanging two atoms or groups a second time brings you back to the original compound (the structure on the right below). Because groups have to be interchanged twice to get from one structure to the other, the two structures above represent identical compounds.

$$CH_3CH_2{-}\overset{\overset{\displaystyle CH=CH_2}{|}}{\underset{\underset{\displaystyle H}{}}{C}}{\cdots}CH_3 \quad \xrightarrow[\text{CH=CH}_2 \text{ and H}]{\text{interchange}} \quad CH_3CH_2{-}\overset{\overset{\displaystyle H}{|}}{\underset{\underset{\displaystyle CH=CH_2}{}}{C}}{\cdots}CH_3 \quad \xrightarrow[\text{ethyl and H}]{\text{interchange}} \quad H{-}\overset{\overset{\displaystyle CH_2CH_3}{|}}{\underset{\underset{\displaystyle CH=CH_2}{}}{C}}{\cdots}CH_3$$

In Section 4.8, you will learn another way to determine whether two structures represent identical compounds or enantiomers.

4.8 NAMING ENANTIOMERS BY THE *R,S* SYSTEM

How do we name the enantiomers of a compound such as 2-bromobutane so that we know which one we are talking about? We need a system of nomenclature that indicates the arrangement of the atoms or groups around the asymmetric center. Chemists use the letters *R* and *S* for this purpose. For any pair of enantiomers, one member will have the **R configuration** and the other will have the **S configuration.** This system of nomenclature is called the **Cahn-Ingold-Prelog system** after the three scientists who devised it.

Naming a Compound Drawn as a Perspective Formula

First, let's look at how you can determine the configuration of a compound drawn as a perspective formula. We will use the enantiomers of 2-bromobutane as an example.

Br   Br
| |
CH₃CH₂—C—''''H   H''''—C—CH₂CH₃
| |
CH₃   CH₃

enantiomers of 2-bromobutane

1. **Rank the groups (or atoms) bonded to the asymmetric center in order of priority.** The atomic numbers of the *atoms* directly attached to the asymmetric center determine the relative priorities. The higher the atomic number of the atom, the higher the priority. If there is a tie, you need to consider the atoms to which the tied atoms are attached. How to do this is explained next to the structures.

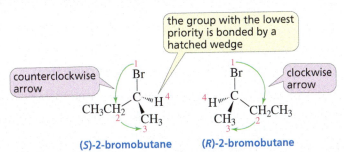

The Cs of the methyl and ethyl group tie. Because the methyl group C is attached to H, H, and H and the ethyl group C is attached to C, H, and H, the ethyl group has the higher priority. The priorities are:

(1) bromine   (2) ethyl   (3) methyl   (4) hydrogen.

2. **If the group (or atom) with the lowest priority (4) is bonded by a hatched wedge:**

   - Draw an arrow in the direction of decreasing priority—(1) to (2) to (3).
   - If the arrow points clockwise, then the compound has the *R* configuration (*R* is for *rectus*, which is Latin for "right").
   - If the arrow points counterclockwise, then the compound has the *S* configuration (*S* is for *sinister*, which is Latin for "left").
   - The letter *R* or *S* (in parentheses) precedes the systematic name of the compound.

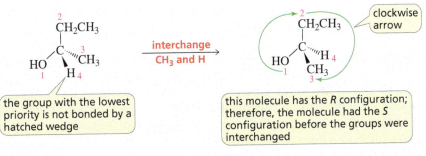

(*S*)-2-bromobutane   (*R*)-2-bromobutane

3. **If the group (or atom) with the lowest priority (4) is not bonded by a hatched wedge:**

   - Interchange group 4 with the group that is bonded by a hatched wedge.
   - Proceed as in step 2—namely, draw an arrow from (1) to (2) to (3).

Because the arrow points clockwise, the compound with the interchanged groups has the *R* configuration. Therefore, the original compound, before the groups were interchanged, has the *S* configuration.

LEARN THE STRATEGY

NOTE TO THE STUDENT

- **This should remind you of the way relative priorities are determined for *E* and *Z* isomers because the system of priorities was originally devised for the *R,S* system and was later adopted for the *E,Z* system (Section 4.2).**

Clockwise specifies *R* if the lowest-priority substituent is on a hatched wedge.

Counterclockwise specifies *S* if the lowest-priority substituent is on a hatched wedge.

right turn

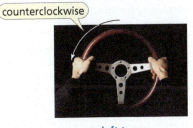

left turn

PROBLEM 20 ◆

Assign relative priorities to the groups or atoms in each of the following sets:

a. —CH_2OH —CH_3 —H   —CH_2CH_2OH
b. —CH_2Br —OH   —CH_3 —CH_2OH
c. —$CH(CH_3)_2$ —CH_2CH_2Br —CI   —$CH_2CH_2CH_2Br$

d. —$CH=CH_2$ —CH_2CH_3 ⬡   —CH_3

PROBLEM 21 ◆ **SOLVED**

Does the following compound have the *R* or the *S* configuration?

SOLUTION Start by adding the missing solid wedge and the H to which it is bonded. The solid wedge can be drawn either to the right or to the left of the hatched wedge. (Recall that the solid and hatched wedges must be adjacent.)

Because the group with the lowest priority is not on the hatched wedge, interchange the Cl and H so that H is on the hatched wedge. An arrow drawn from (1) to (2) to (3) indicates that the compound has the *S* configuration. Therefore, the compound before the pair was interchanged had the *R* configuration.

PROBLEM 22

Do the following compounds have the *R* or the *S* configuration?

Naming a Compound Drawn as a Fischer Projection

Now let's see how you can determine the configuration of a compound drawn as a Fischer projection.

1. **Rank the groups (or atoms) that are bonded to the asymmetric center in order of priority.**
2. **Draw an arrow from (1) to (2) to (3).** If the arrow points clockwise, then the compound has the *R* configuration; if it points counterclockwise, then the compound has the *S* configuration, *provided that the group with the lowest priority (4) is on a vertical bond.*

clockwise signifies *R*, because H is on a vertical bond

(*R*)-3-chlorohexane   (*S*)-3-chlorohexane

Clockwise specifies *R* if the lowest-priority substituent is on a vertical bond.

Clockwise specifies *S* if the lowest-priority substituent is on a horizontal bond.

If you assume that a clockwise arrow indicates the *R* configuration, then you get a VERy good answer if the group with the lowest priority is on a VERtical bond, but a HORribly bad answer if the group with the lowest priority is on a HORizontal bond.

If the group (or atom) with the lowest priority is on a *horizontal bond,* the answer you get from the direction of the arrow will be the opposite of the correct answer. For example, if the arrow points clockwise, suggesting the *R* configuration, then the compound actually has the *S* configuration; if the arrow points counterclockwise, suggesting the *S* configuration, then it actually has the *R* configuration. In the example on the top of the next page, the group with the lowest priority is on a horizontal bond, so clockwise signifies the *S* configuration.

clockwise signifies *S*, because H is on a horizontal bond

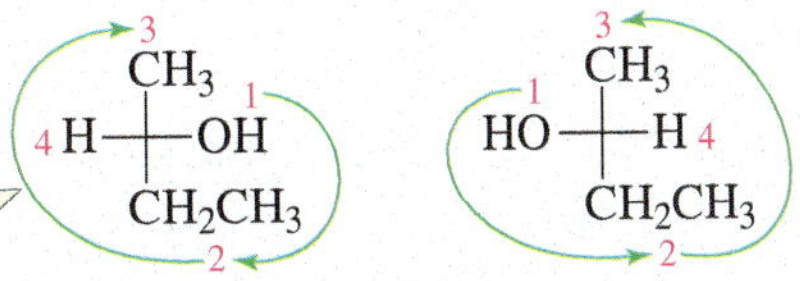

(*S*)-2-butanol   (*R*)-2-butanol

USE THE STRATEGY

PROBLEM 23 ◆

What is the configuration of each of the following?

a. 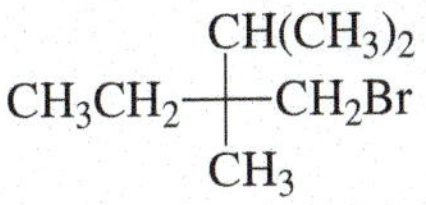
$$CH_3CH_2 \overset{CH(CH_3)_2}{\underset{CH_3}{\rule{0pt}{1em}|\!\!-\!\!|}} CH_2Br$$

b. 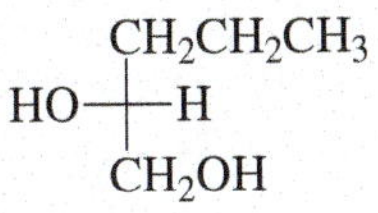
$$HO \overset{CH_2CH_2CH_3}{\underset{CH_2OH}{\rule{0pt}{1em}|\!\!-\!\!|}} H$$

c. 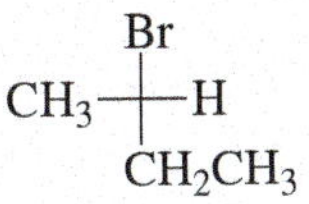
$$CH_3 \overset{Br}{\underset{CH_2CH_3}{\rule{0pt}{1em}|\!\!-\!\!|}} H$$

d. 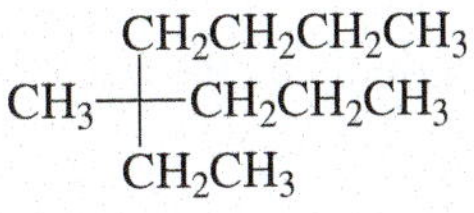
$$CH_3 \overset{CH_2CH_2CH_2CH_3}{\underset{CH_2CH_3}{\rule{0pt}{1em}|\!\!-\!\!|}} CH_2CH_2CH_3$$

PROBLEM-SOLVING STRATEGY

Recognizing Pairs of Enantiomers

Do the structures represent identical compounds or a pair of enantiomers?

LEARN THE STRATEGY

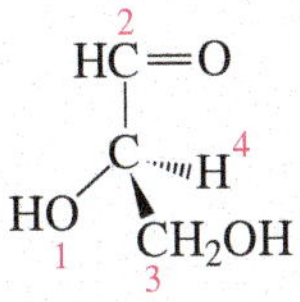

and

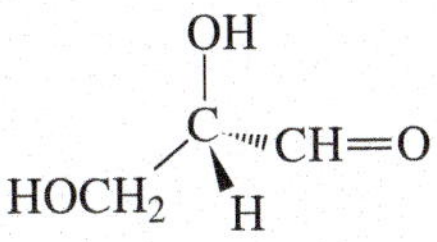

The easiest way to answer this question is to determine their configurations. If one has the *R* configuration and the other has the *S* configuration, then they are enantiomers. If both have the same configuration, then they are identical compounds.

The OH group has the highest priority, the H has the lowest priority, and the Cs of the other two groups tie. The CH=O group has a higher priority than the CH$_2$OH group, because if an atom is doubly bonded to another atom, the priority system treats it as if it were singly bonded to two of those atoms. Thus, the C of the CH=O group is considered to be bonded to O, O, H, whereas the C of the CH$_2$OH group is considered to be bonded to O, H, H. An O cancels in each group, leaving O, H in the CH=O group and H, H in the CH$_2$OH group.

Because the structure on the left has the *R* configuration and the structure on the right has the *S* configuration, these two structures represent a pair of enantiomers.

PROBLEM 24 ◆

Do the following structures represent identical compounds or a pair of enantiomers?

USE THE STRATEGY

a. 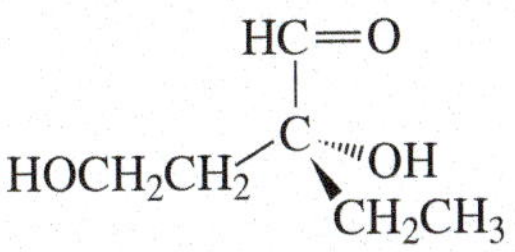
and

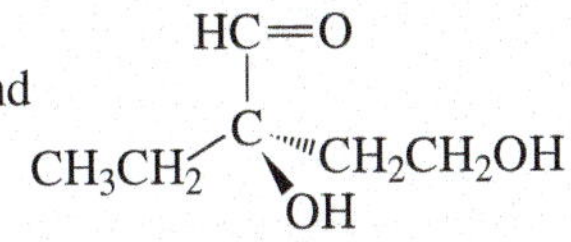

b. 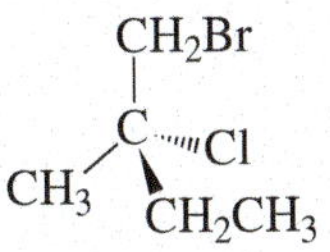
and

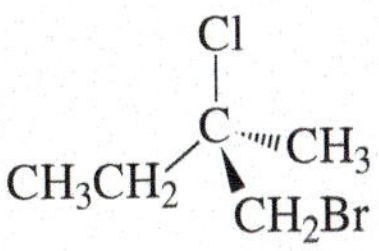

c. 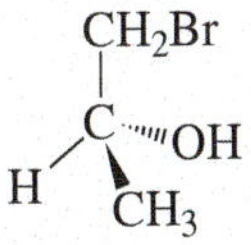
and

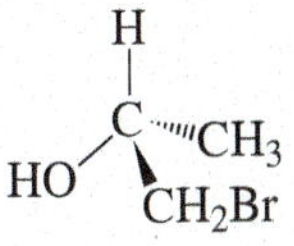

d. 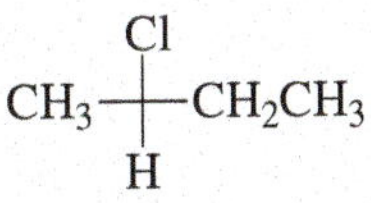
$$CH_3 \overset{Cl}{\underset{H}{\rule{0pt}{1em}|\!\!-\!\!|}} CH_2CH_3 \quad \text{and} \quad H \overset{CH_3}{\underset{CH_2CH_3}{\rule{0pt}{1em}|\!\!-\!\!|}} Cl$$

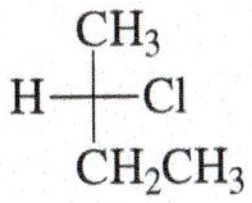

NOTE TO THE STUDENT

• When comparing two Fischer projections to see if they are the same or different, never rotate one 90° or flip it "front-to-back" — that is a quick way to get a wrong answer. A Fischer projection can be rotated 180° in the plane of the paper, but that is the only way to move it without risking an incorrect answer.

PROBLEM-SOLVING STRATEGY

Drawing an Enantiomer with a Desired Configuration

(S)-Alanine is a naturally occurring amino acid. Draw its structure using a perspective formula.

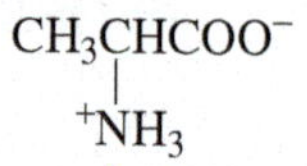

$$CH_3CHCOO^-$$
$$|$$
$$^+NH_3$$

alanine

First draw the bonds about the asymmetric center. (Remember that the solid wedge and the hatched wedge must be adjacent to each other and the solid wedge must be below the hatched wedge.)

Put the group with the lowest priority on the hatched wedge. Put the group with the highest priority on any remaining bond.

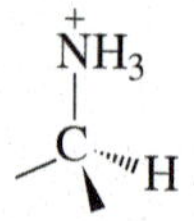

Because you have been asked to draw the S enantiomer, draw an arrow counterclockwise from the group with the highest priority to the next available bond and put the group with the second highest priority on that bond.

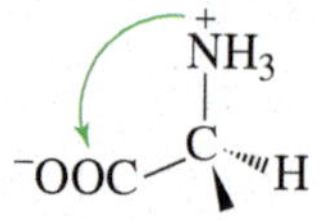

Put the remaining substituent (the one with the third highest priority) on the last available bond.

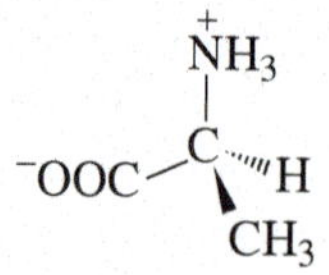

PROBLEM 25

Draw a perspective formula for each of the following:

a. (S)-2-chlorobutane   **b.** (R)-1,2-dibromobutane

PROBLEM 26   SOLVED

Convert the Fischer projection to a perspective formula.

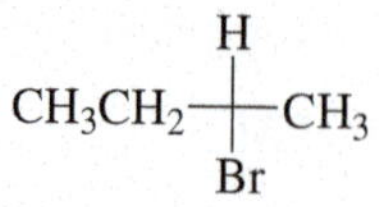

$$CH_3CH_2 \overset{H}{\underset{Br}{|}} CH_3$$

SOLUTION First determine the configuration of the Fischer projection: it has the R configuration. Then draw the perspective formula with that configuration by following the steps described in the preceding Problem-Solving Strategy.

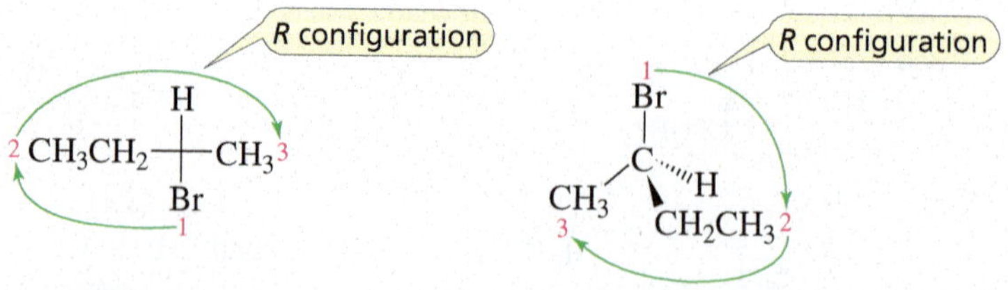

PROBLEM 27

Convert the Fischer projection to a perspective formula.

$$\begin{array}{c} COO^- \\ H\!\!-\!\!\!\!-\!\!CH_3 \\ CH_2CH_3 \end{array}$$

4.9 CHIRAL COMPOUNDS ARE OPTICALLY ACTIVE

Enantiomers share many of the same properties, including the same boiling points, the same melting points, and the same solubilities. In fact, all the physical properties of enantiomers are the same except those that stem from how groups bonded to the asymmetric center are arranged in space. For example, one property that enantiomers do not share is the way they interact with plane-polarized light.

Normal light, such as that coming from a light bulb or the Sun, consists of rays that oscillate in all directions. In contrast, all the rays in a beam of **plane-polarized light** oscillate in a single plane. Plane-polarized light is produced by passing normal light through a polarizer (Figure 4.4).

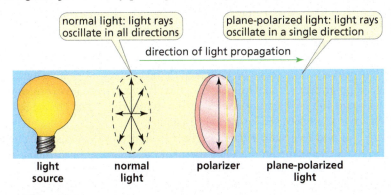

normal light: light rays oscillate in all directions

plane-polarized light: light rays oscillate in a single direction

direction of light propagation

light source · normal light · polarizer · plane-polarized light

▲ **Figure 4.4**
Only light oscillating in a single plane can pass through a polarizer.

You can experience the effect of a polarizer by wearing a pair of polarized sunglasses. Polarized sunglasses allow only light oscillating in a single plane to pass through, which is why they block reflections (glare) more effectively than nonpolarized sunglasses do.

In 1815, the physicist Jean-Baptiste Biot discovered that certain naturally occurring organic compounds are able to rotate the **plane of polarization** of plane-polarized light. He noted that some compounds rotated it clockwise and some rotated it counterclockwise. He proposed that the ability to rotate the plane of polarization of plane-polarized light was due to some asymmetry in the molecules. It was later determined that the asymmetry was associated with compounds having one or more asymmetric centers.

When plane-polarized light passes through a solution of achiral molecules, the light emerges from the solution with its plane of polarization unchanged (Figure 4.5).

When light is filtered through two polarizers (polarized lenses) at a 90° angle to each other, none of the light passes through.

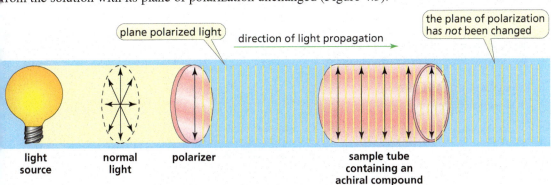

plane polarized light

direction of light propagation

the plane of polarization has *not* been changed

light source · normal light · polarizer · sample tube containing an achiral compound

▲ **Figure 4.5**
An achiral compound does not rotate the plane of polarization of plane-polarized light.

On the other hand, when plane-polarized light passes through a solution of chiral molecules, the light emerges with its plane of polarization rotated either clockwise or counterclockwise (Figure 4.6). If one enantiomer rotates it clockwise, its mirror image will rotate it exactly the same amount counterclockwise.

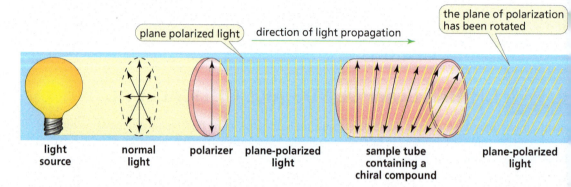

▲ **Figure 4.6**
A chiral compound rotates the plane of polarization of plane-polarized light.

An achiral compound does not rotate the plane of polarization of plane-polarized light.

A chiral compound rotates the plane of polarization of plane-polarized light.

A compound that rotates the plane of polarization of plane-polarized light is said to be **optically active.** In other words, chiral compounds are optically active, and achiral compounds are **optically inactive.**

If an optically active compound rotates the plane of polarization clockwise, then the compound is said to be **dextrorotatory,** which can be indicated in the compound's name by the prefix (+). If it rotates the plane of polarization counterclockwise, then it is said to be **levorotatory,** which can be indicated by (−). *Dextro* and *levo* are Latin prefixes for "to the right" and "to the left," respectively. Sometimes lowercase *d* and *l* are used instead of (+) and (−).

Do not confuse (+) and (−) with *R* and *S*. The (+) and (−) symbols indicate the direction in which an optically active compound rotates the plane of polarization of plane-polarized light, whereas *R* and *S* indicate the arrangement of the groups about an asymmetric center. Some compounds with the *R* configuration are (+) and some are (−). Likewise, some compounds with the *S* configuration are (+) and some are (−).

For example, (*S*)-lactic acid and (*S*)-sodium lactate both have an *S* configuration, but (*S*)-lactic acid is dextrorotatory whereas (*S*)-sodium lactate is levorotatory. When we know which direction an optically active compound rotates the plane of polarization, we can incorporate (+) or (−) into its name.

CH$_3$ CH$_3$

C⸺H   C⸺H

HO   COOH   HO   COO$^-$ Na$^+$

(S)-(+)-lactic acid   **(S)-(−)-sodium lactate**

We can tell by looking at the structure of a compound whether it has the *R* or the *S* configuration, but the only way we can tell whether a compound is dextrorotatory (+) or levorotatory (−) is to put the compound in a polarimeter (Section 4.10).

PROBLEM 28 ◆

a. Is (*R*)-lactic acid dextrorotatory or levorotatory?

b. Is (*R*)-sodium lactate dextrorotatory or levorotatory?

LEARN THE STRATEGY

PROBLEM 29 SOLVED

What is the configuration of (−)-glyceraldehyde?

HC=O

C⸺H

HO   CH$_2$OH

(+)-glyceraldehyde

SOLUTION We know that (+)-glyceraldehyde has the *R* configuration because the group with the lowest priority is on the hatched wedge and the arrow drawn from the OH group to the HC=O group is clockwise. Therefore, (−)-glyceraldehyde has the *S* configuration.

PROBLEM 30 ◆

USE THE STRATEGY

What is the configuration of the following compounds? (Use the given structures to answer the question.)

a. (−)-glyceric acid   **b.** (+)-isoserine   **c.** (+)-lactic acid

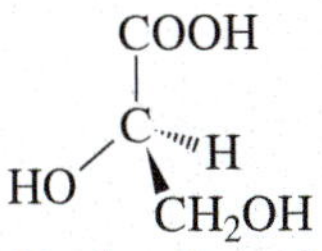

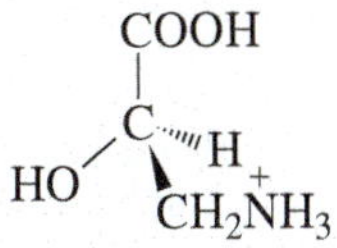

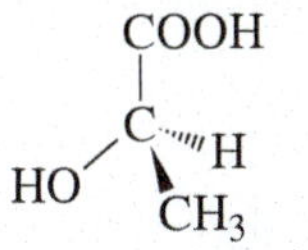

COOH   COOH   COOH

HO C'''H CH₂OH   HO C'''H⁺ CH₂NH₃   HO C'''H CH₃

(−)-glyceric acid   (+)-isoserine   (−)-lactic acid

PROBLEM 31 SOLVED

LEARN THE STRATEGY

(*S*)-(−)-2-Methyl-1-butanol can be converted to (+)-2-methylbutanoic acid without breaking any of the bonds to the asymmetric center. What is the configuration of (−)-2-methylbutanoic acid?

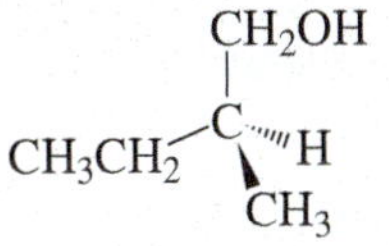

CH₂OH   COOH

CH₃CH₂ C'''H CH₃   CH₃CH₂ C'''H CH₃

(*S*)-(−)-2-methyl-1-butanol 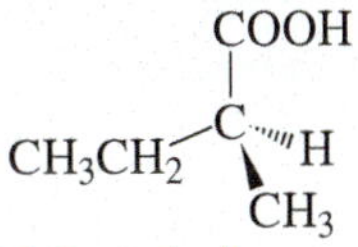   (+)-2-methylbutanoic acid

SOLUTION We know that (+)-2-methylbutanoic acid has the configuration shown here because it was formed from (*S*)-(−)-2-methyl-1-butanol without breaking any bonds to the asymmetric center. From its structure, we can determine that (+)-2-methylbutanoic acid has the *S* configuration. Therefore, (−)-2-methylbutanoic acid has the *R* configuration.

PROBLEM 32 ◆

USE THE STRATEGY

The reaction of (*R*)-1-iodo-2-methylbutane with hydroxide ion forms an alcohol without breaking any bonds to the asymmetric center. The alcohol rotates the plane of polarization of plane-polarized light counterclockwise. What is the configuration of (+)-2-methyl-1-butanol?

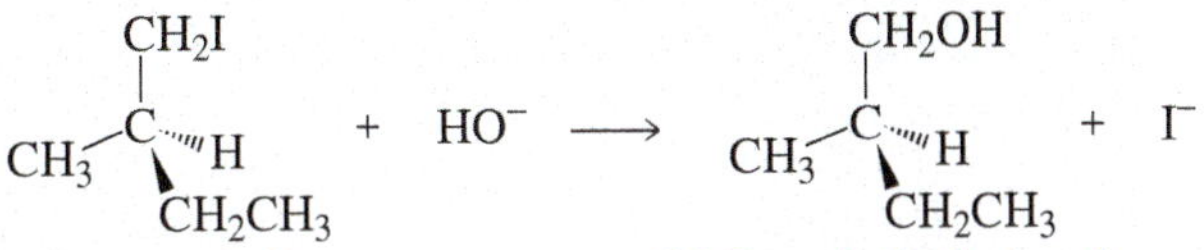

CH₂I   CH₂OH

CH₃ C'''H CH₂CH₃ + HO⁻ ⟶   CH₃ C'''H CH₂CH₃ + I⁻

(*R*)-1-iodo-2-methylbutane   (−)-2-methyl-1-butanol

4.10 HOW SPECIFIC ROTATION IS MEASURED

The direction and amount an optically active compound rotates the plane of polarization of plane-polarized light can be measured with an instrument called a **polarimeter.** Figure 4.7 on the next page provides a simplified description of how a polarimeter functions. The amount of rotation caused by an optically active compound varies with the wavelength of the light being used, so the light source must produce monochromatic (single-wavelength) light. Most polarimeters use light from a sodium arc (called the sodium D-line; wavelength = 589 nm).

The monochromatic light passes through a polarized lens and emerges as plane-polarized light, which then passes through a sample tube. If the tube is empty, the light emerges from it with its plane of polarization unchanged. The light then passes through an analyzer, which is a second polarized lens mounted on an eyepiece with a dial marked in degrees. The user looks through the eyepiece and rotates the analyzer until he or she sees total darkness. At this point, the analyzer is at a right angle to the polarizer, so no light passes through. This analyzer setting corresponds to zero rotation.

The sample to be measured is then placed in the sample tube. If the sample is optically active, it will rotate the plane of polarization. The analyzer, therefore, will no longer block all the light, so some light will reach the user's eye. The user now rotates the analyzer again until no light passes through. The amount the analyzer is rotated can be read from the dial. This value, which is measured in degrees, is called the **observed rotation** (α) (Figure 4.7). The observed rotation depends on the number of optically active molecules that the light encounters in the sample, which in turn depends on the concentration of the sample and the length of the sample tube. The observed rotation also depends on the temperature and the wavelength of the light source.

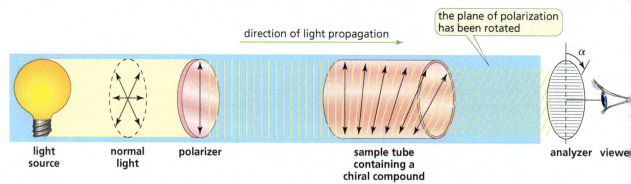

▲ **Figure 4.7**
A schematic drawing of a polarimeter.

Each optically active compound has a characteristic specific rotation. The **specific rotation** at a specified temperature and wavelength can be calculated from the observed rotation using the formula

$$[\alpha]_\lambda^T = \frac{\alpha}{l \times c}$$

where $[\alpha]$ is the specific rotation, T is temperature in degrees Celsius, λ is the wavelength of the incident light (when the sodium D-line is used, λ is indicated as D), α is the observed rotation, l is the length of the sample tube in decimeters, and c is the concentration of the sample in grams in 100 milliliters of solution (or grams/ml if it is a pure liquid).

If one enantiomer has a specific rotation of $+ 5.75^*$, the specific rotation of the other enantiomer must be -5.75, because the mirror-image molecule rotates the plane of polarization the same amount but in the opposite direction. The specific rotations of some compounds are listed in Table 4.1.

Table 4.1	Specific Rotation of Some Naturally Occurring Compounds
Cholesterol	-31.5
Cocaine	-16
Codeine	-136
Morphine	-132
Penicillin V	$+233$
Progesterone	$+172$
Sucrose (table sugar)	$+66.5$
Testosterone	$+109$

CH₂OH structures:

(R)-2-methyl-1-butanol   (S)-2-methyl-1-butanol

$[\alpha]_D^{20\,°C} = +5.75$ $[\alpha]_D^{20\,°C} = -5.75$

A mixture of equal amounts of two enantiomers—such as (R)-(−)-lactic acid and (S)-(+)-lactic acid—is called a **racemic mixture** or a **racemate**. Racemic mixtures are optically inactive because for every molecule in a racemic mixture that rotates the plane of polarization in one direction, there is a mirror-image molecule that rotates the plane in the opposite direction. As a result, the light emerges from a racemic mixture with its plane of polarization unchanged. The symbol ($\pm$) is used to specify a racemic mixture. Thus, ($\pm$)-2-bromobutane indicates a mixture of 50% (+)-2-bromobutane and 50% (−)-2-bromobutane.

---

*Unlike observed rotation, which is measured in degrees, specific rotation has units of 10^{-1} deg cm² g⁻¹. In this book, values of specific rotation will be given without units.

PROBLEM 33 ◆

The observed rotation of 2.0 g of a compound in 50 mL of solution in a polarimeter tube 20 cm long is +138°. What is the specific rotation of the compound?

PROBLEM 34 ◆

(S)-(+)-Monosodium glutamate (MSG) is a flavor enhancer used in many foods. Some people have an allergic reaction to MSG (including headache, chest pain, and an overall feeling of weakness). Fast food often contains substantial amounts of MSG, which is widely used in Chinese food as well. (S)-(+)-MSG has a specific rotation of +24.

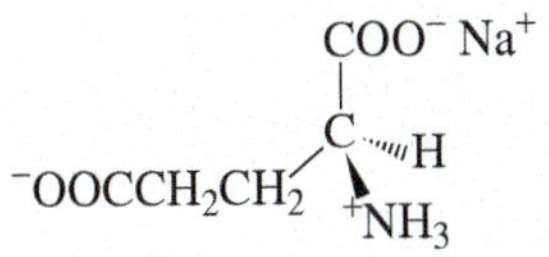

$$\text{COO}^- \text{Na}^+$$
$$\overset{|}{\underset{\text{+NH}_3}{\overset{\text{C}\text{\tiny\textbackslash\textbackslash\textbackslash}\text{H}}{}}}$$
$$^-\text{OOCCH}_2\text{CH}_2$$

(S)-(+)-monosodium glutamate

a. What is the specific rotation of (R)-(−)-monosodium glutamate?
b. What is the specific rotation of a racemic mixture of MSG?

4.11   ENANTIOMERIC EXCESS

Whether a particular sample of a compound consists of a single enantiomer, a racemic mixture, or a mixture of enantiomers in unequal amounts can be determined by its **observed specific rotation,** which is the specific rotation of the particular sample. For example:

- If a sample of (S)-(+)-2-bromobutane is **enantiomerically pure** (meaning only one enantiomer is present), it will have an *observed specific rotation* of +23.1 because its *specific rotation* is +23.1.
- If the sample is a racemic mixture, it will have an observed specific rotation of 0.
- If the observed specific rotation is positive but less than +23.1, the sample is a mixture of enantiomers but contains more S enantiomer than R enantiomer, because the S enantiomer is dextrorotatory.

The **enantiomeric excess (ee),** also called the **optical purity,** tells us how much of an excess of one enantiomer is in the mixture. It can be calculated from the observed specific rotation:

LEARN THE STRATEGY

$$\text{enantiomeric excess} = \frac{\text{observed specific rotation}}{\text{specific rotation of the pure enantiomer}} \times 100\%$$

For example, if the sample of 2-bromobutane has an observed specific rotation of +9.2, then the enantiomeric excess is 40%. In other words, the excess of one of the enantiomers comprises 40% of the mixture.

$$\text{enantiomeric excess} = \frac{+9.2}{+23.1} \times 100\% = 40\%$$

If the mixture has a 40% enantiomeric excess, 40% of the mixture is excess S enantiomer and 60% is a racemic mixture. Half of the racemic mixture plus the amount of excess S enantiomer equals the amount of the S enantiomer present in the mixture. Therefore, 70% of the mixture is the S enantiomer $[(1/2 \times 60) + 40]$ and 30% is the R enantiomer.

USE THE STRATEGY

PROBLEM 35 ♦

(+)-Mandelic acid has a specific rotation of +158. What would be the observed specific rotation of each of the following mixtures?
a. 50% (−)-mandelic acid and 50% (+)-mandelic acid
b. 25% (−)-mandelic acid and 75% (+)-mandelic acid
c. 75% (−)-mandelic acid and 25% (+)-mandelic acid

PROBLEM 36 ♦

Naproxen, a nonsteroidal anti-inflammatory drug that is the active ingredient in Aleve (p. 115), has a specific rotation of +66. One commercial preparation results in a mixture with a 97% enantiomeric excess.
a. Does naproxen have the R or the S configuration?
b. What percent of each enantiomer is obtained from the commercial preparation?

PROBLEM 37 **SOLVED**

A solution prepared by mixing 10 mL of a 0.10 M solution of the R enantiomer of a compound and 30 mL of a 0.10 M solution of the S enantiomer was found to have an observed specific rotation of +4.8. What is the specific rotation of each of the enantiomers? (*Hint:* mL × M = millimole, abbreviated as mmol)

SOLUTION One mmol (10 mL × 0.10 M) of the R enantiomer is mixed with 3 mmol (30 mL × 0.10 M) of the S enantiomer; 1 mmol of the R enantiomer plus 1 mmol of the S enantiomer will form 2 mmol of a racemic mixture, so there will be 2 mmol of S enantiomer left over. Because 2 out of 4 mmol is excess S enantiomer, the solution has a 50% enantiomeric excess. Knowing the enantiomeric excess and the observed specific rotation allows us to calculate the specific rotation.

$$\text{enantiomeric excess} = \frac{\text{observed specific rotation}}{\text{specific rotation of the pure enantiomer}} \times 100\%$$

$$50\% = \frac{+4.8}{x} \times 100\%$$

$$\frac{50}{100} = \frac{+4.8}{x}$$

$$\frac{1}{2} = \frac{+4.8}{x}$$

$$x = 2(+4.8)$$

$$x = 9.6$$

The S enantiomer has a specific rotation of +9.6, so the R enantiomer has a specific rotation of −9.6.

4.12 COMPOUNDS WITH MORE THAN ONE ASYMMETRIC CENTER

Many organic compounds have more than one asymmetric center. The more asymmetric centers a compound has, the more stereoisomers it can have. If we know the number of asymmetric centers, we can calculate the *maximum* number of stereoisomers for that compound: *a compound can have a maximum of* 2^n *stereoisomers, where* n *equals the number of asymmetric centers* (provided it does not also have stereocenters that would cause it to have cis–trans isomers; see Problem 38).

> The maximum number of stereoisomers = 2^n, where n equals the number of asymmetric centers.

For example, 3-chloro-2-butanol has two asymmetric centers. Therefore, it can have a maximum of four ($2^2 = 4$) stereoisomers. The four stereoisomers are shown below as perspective formulas and on the next page as Fischer projections.

$$\overset{*}{\text{CH}_3}\overset{*}{\text{CH}}\text{CHCH}_3$$
$$\text{Cl} \quad \text{OH}$$

3-chloro-2-butanol

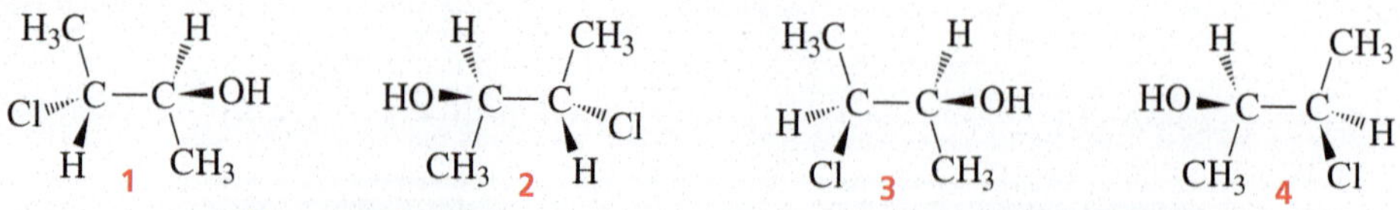

erythro enantiomers   threo enantiomers
perspective formulas of the stereoisomers of 3-chloro-2-butanol (staggered)

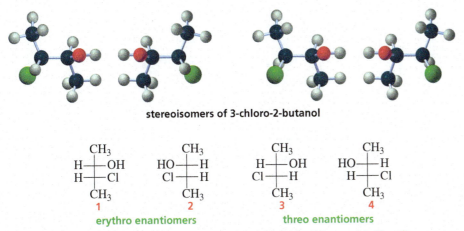

stereoisomers of 3-chloro-2-butanol

Fischer projections of the stereoisomers of 3-chloro-2-butanol (eclipsed)

Diastereomers

The four stereoisomers of 3-chloro-2-butanol consist of two pairs of enantiomers. Stereoisomers **1** and **2** are nonsuperimposable mirror images. They, therefore, are enantiomers. Stereoisomers **3** and **4** are also enantiomers. Stereoisomers **1** and **3** are not identical, and they are not mirror images. Such stereoisomers are called **diastereomers.** Stereoisomers **1** and **4**, **2** and **3**, and **2** and **4** are also pairs of diastereomers.

Notice that in a pair of diastereomers, the configuration of one of the asymmetric centers is the same in both but the configuration of the other asymmetric center is different. *Diastereomers are stereoisomers that are not enantiomers.* (Note that cis–trans isomers are also considered to be diastereomers, because they are stereoisomers but not enantiomers.)

When Fischer projections are drawn for stereoisomers with two adjacent asymmetric centers (such as those for 3-chloro-2-butanol), the enantiomers with the hydrogens on the same side of the carbon chain are called the **erythro enantiomers** (see Problem 62), whereas those with the hydrogens on opposite sides are called the **threo enantiomers.** Therefore, **1** and **2** are the erythro enantiomers of 3-chloro-2-butanol (the hydrogens are on the same side), whereas **3** and **4** are the threo enantiomers.

We saw that enantiomers have *identical physical properties.* They also have *identical chemical properties*—that is, they react with a given achiral reagent at the same rate. Diastereomers, on the other hand, have *different physical properties* (meaning different melting points, boiling points, solubilities, specific rotations, and so on) and *different chemical properties*—that is, they react with a given achiral reagent at different rates.

Diastereomers are stereoisomers that are not enantiomers.

Fischer Projections versus Perspective Formulas

In each of the Fischer projections shown above, the horizontal bonds project out of the paper toward the viewer and the vertical bonds extend behind the paper away from the viewer. Groups can rotate freely about the carbon–carbon single bonds, but Fischer projections show the stereoisomers in their eclipsed conformations.

Because a Fischer projection does not show the three-dimensional structure of the molecule and because it represents the molecule in a relatively unstable eclipsed conformation, most chemists prefer to use perspective formulas. Perspective formulas (those shown in the first group of images in this section) show the molecule's three-dimensional structure in a stable, staggered conformation, so they provide a more accurate representation of structure.

When perspective formulas are drawn (by rotating the right side of the molecule about the horizontal C—C bond) to show the stereoisomers in their less stable eclipsed conformations (those shown next), we can easily see that the erythro enantiomers have similar groups on the same side. We will use both perspective formulas and Fischer projections to depict the arrangement of groups bonded to an asymmetric center.

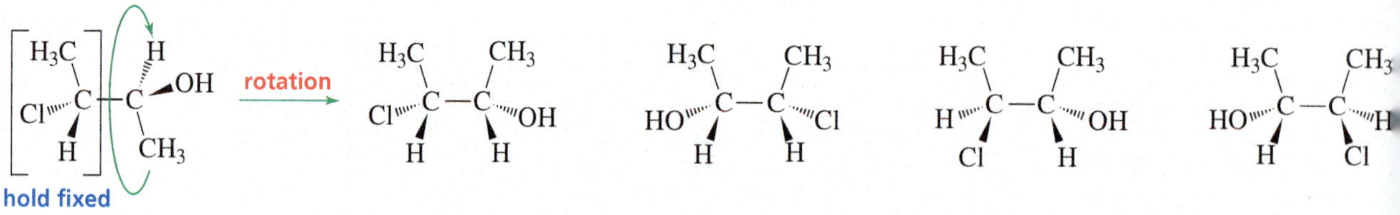

perspective formulas of the stereoisomers of 3-chloro-2-butanol (eclipsed)

PROBLEM 38

The following compound has only one asymmetric center. Why then does it have four stereoisomers?

$$CH_3CH_2\overset{*}{C}HCH_2CH\!=\!CHCH_3$$
$$\qquad\qquad\quad |$$
$$\qquad\qquad\quad Br$$

PROBLEM 39 ◆

a. Stereoisomers with two asymmetric centers are called ___ if the configuration of both asymmetric centers in one stereoisomer is the opposite of the configuration of the asymmetric centers in the other stereoisomer.

b. Stereoisomers with two asymmetric centers are called ___ if the configuration of both asymmetric centers in one stereoisomer is the same as the configuration of the asymmetric centers in the other stereoisomer.

c. Stereoisomers with two asymmetric centers are called ___ if one of the asymmetric centers has the same configuration in both stereoisomers and the other asymmetric center has the opposite configuration in the two stereoisomers.

PROBLEM 40 ◆

The stereoisomer of cholesterol found in nature is shown here.

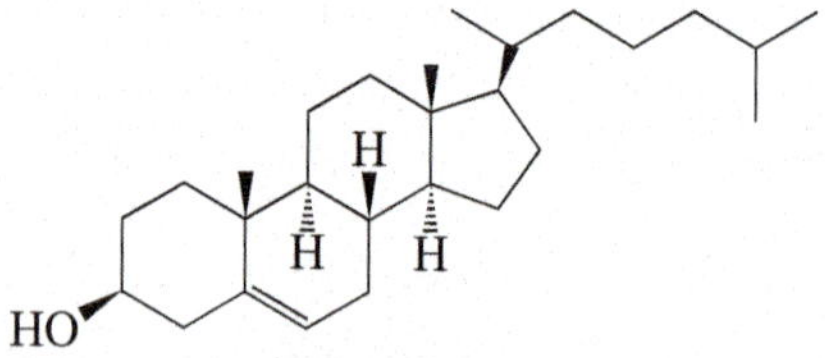

cholesterol

a. How many asymmetric centers does cholesterol have?

b. What is the maximum number of stereoisomers that cholesterol can have?

PROBLEM 41

Draw the stereoisomers of the following amino acids. Indicate pairs of enantiomers and pairs of diastereomers.

a. $CH_3CHCH_2\!-\!CHCOO^-$
   $\qquad |\qquad\qquad |$
   $\qquad CH_3\qquad\ ^+NH_3$
   leucine

b. $CH_3CH_2CH\!-\!CHCOO^-$
   $\qquad\qquad\ |\qquad |$
   $\qquad\qquad CH_3\ ^+NH_3$
   isoleucine

4.13 STEREOISOMERS OF CYCLIC COMPOUNDS

Cyclic compounds can also have multiple stereoisomers. For example, 1-bromo-2-methylcyclopentane has two asymmetric centers and four stereoisomers. Because the compound is cyclic, the substituents can be either cis or trans (Section 3.15). Enantiomers can be drawn for both the cis isomer and the trans isomer. Each of the four stereoisomers is *chiral*.

cis-1-bromo-2-methylcyclopentane
enantiomers

trans-1-bromo-2-methylcyclopentane
enantiomers

PROBLEM 42

1-Bromo-2-methylcyclopentane has four pairs of diastereomers. Draw the four pairs.

1-Bromo-3-methylcyclohexane also has two asymmetric centers. The carbon that is bonded to a Br and an H is also bonded to two different carbon-containing groups ($-CH_2CH(CH_3)CH_2CH_2CH_2-$ and $-CH_2CH_2CH_2CH(CH_3)CH_2-$), so it is an asymmetric center. The carbon that is bonded to a CH_3 and an H is also bonded to two different carbon-containing groups, so it, too, is an asymmetric center.

these two groups are different Br
asymmetric center   asymmetric center
CH_3

Because the compound has two asymmetric centers, it has four stereoisomers. Enantiomers can be drawn for both the cis isomer and the trans isomer. Each of the four stereoisomers is *chiral*.

cis-1-bromo-3-methylcyclohexane
enantiomers

trans-1-bromo-3-methylcyclohexane
enantiomers

1-Bromo-3-methylcyclobutane does not have any asymmetric centers. The C-1 carbon has a Br and an H attached to it, but its other two groups [$-CH_2CH(CH_3)CH_2-$] are identical; C-3 has a CH_3 and an H attached to it, but its other two groups [$-CH_2CH(Br)CH_2-$] are identical. Because the compound does not have a carbon with four different groups attached to it, it has only two stereoisomers, the cis isomer and the trans isomer. Both stereoisomers are *achiral*.

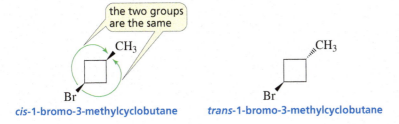

the two groups
are the same
CH_3

CH_3

Br

Br

cis-1-bromo-3-methylcyclobutane   trans-1-bromo-3-methylcyclobutane

1-Bromo-4-methylcyclohexane also has no asymmetric centers. Therefore, the compound has only two stereoisomers: the cis isomer and the trans isomer. Both stereoisomers are *achiral*.

Br —⟨ ⟩— CH_3 Br —⟨ ⟩—‴CH_3

cis-1-bromo-4-methylcyclohexane   trans-1-bromo-4-methylcyclohexane

---

PROBLEM 43 ♦

Which of the following compounds has one or more asymmetric centers?

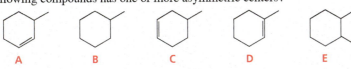

A   B   C   D   E

PROBLEM 44

Draw all possible stereoisomers for each of the following:

a. 2-chloro-3-hexanol

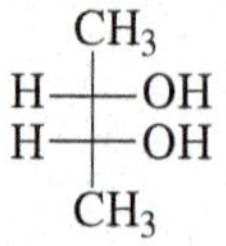

b. 2-bromo-4-chlorohexane

c. 2,3-dichloropentane

d. 1,3-dibromopentane

PROBLEM 45

Draw the stereoisomers of 2-methylcyclohexanol.

PROBLEM 46 ♦

Of all the possible cyclooctanes that have one chloro substituent and one methyl substituent, which ones do *not* have any asymmetric centers?

PROBLEM-SOLVING STRATEGY

LEARN THE STRATEGY

Drawing Enantiomers and Diastereomers

Draw an enantiomer and a diastereomer for the following compound:

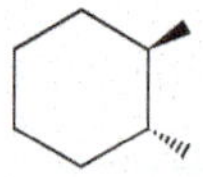

You can draw an enantiomer in one of two ways. You can change the configuration of both asymmetric centers by changing the solid wedge to a hatched wedge and the hatched wedge to a solid wedge, as in **A.** Or you can draw a mirror image of the given compound as in **B.** Notice that because **A** and **B** are each an enantiomer of the given compound, **A** and **B** are identical. (You can see that they are identical by rotating **B** 180° clockwise.)

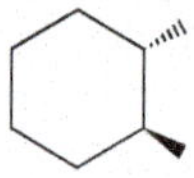

or

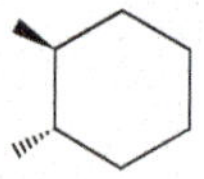

A   B

You can draw a diastereomer by changing the configuration of only one of the asymmetric centers as in **C** or **D.**

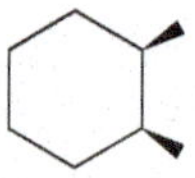

or

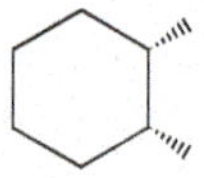

C   D

USE THE STRATEGY

PROBLEM 47

Draw a diastereomer for each of the following:

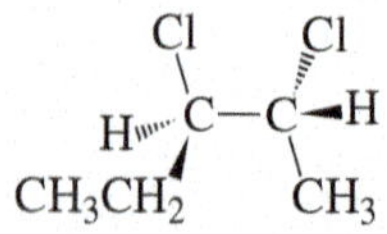

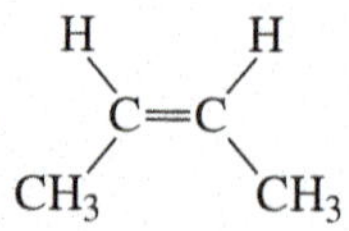

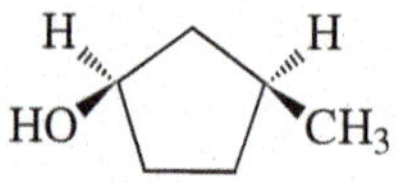

PROBLEM 48 ♦

Indicate whether each of the structures in the second row is an enantiomer of, is a diastereomer of, or is identical to the structure in the top row.

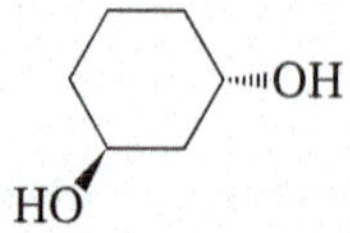

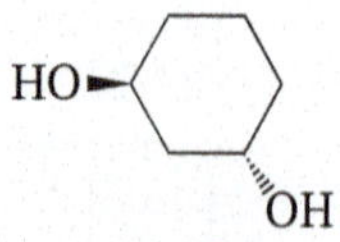

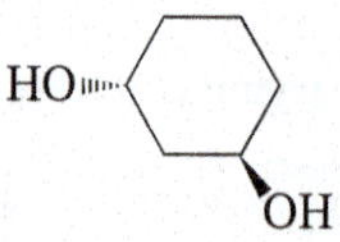

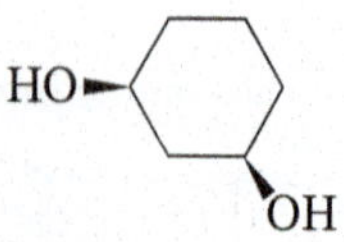

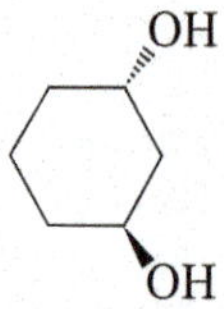

4.14 MESO COMPOUNDS HAVE ASYMMETRIC CENTERS BUT ARE OPTICALLY INACTIVE

In the examples we have just seen, all the compounds with two asymmetric centers had four stereo-isomers. Some compounds with two asymmetric centers, however, like the one shown below, have only three stereoisomers.

$$CH_3CHCHCH_3$$
$$\quad | \quad |$$
$$\quad Br \ Br$$

2,3-dibromobutane

perspective formulas of the stereoisomers of 2,3-dibromobutane (staggered)

The "missing" stereoisomer is the mirror image of **1** because **1** and its mirror image are the same molecule. This can be seen more clearly when the perspective formulas are drawn in eclipsed conformations or when Fischer projections are used.

perspective formulas of the stereoisomers of 2,3-dibromobutane (eclipsed)

Fischer projections of the stereoisomers of 2,3-dibromobutane (eclipsed)

It is easy to see from the following perspective formulas that **1** and its mirror image are identical. To convince yourself that the Fischer projection of **1** and its mirror image are identical, rotate the mirror image 180°.

NOTE TO THE STUDENT

• Remember, you can move Fischer projections only by rotating them 180° in the plane of the paper.

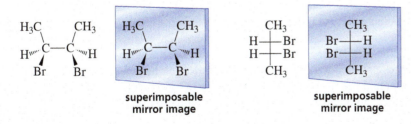

superimposable mirror image   superimposable mirror image

Stereoisomer **1** is called a meso compound. Even though a **meso** (mee-zo) **compound** has asymmetric centers, it is achiral because it is superimposable on its mirror image. *Mesos* is the Greek word for "middle."

Notice that a **plane of symmetry** cuts a meso compound in half so that one half is the mirror image of the other half. As an example, look at the first compound on the top of the next page. If the top half of the molecule rotates polarized light to the right, then the bottom half will rotate the light the same amount to the left. Thus, they will cancel each other and the compound will not be optically active.

A meso compound has two or more asymmetric centers and a plane of symmetry.

Meso compounds are not optically active.

If a compound has a plane of symmetry, it will not be optically active even though it has asymmetric centers.

meso compounds

It is easy to recognize when a compound with two asymmetric centers has a stereoisomer that is a meso compound because the four atoms or groups bonded to one asymmetric center are identical to those bonded to the other asymmetric center. For example, both of the asymmetric centers in the following compound are bonded to an H, an OH, a CH_3, and a $CH(OH)CH_3$.

meso compound   enantiomers

meso compound   enantiomers

A compound with the same four atoms or groups bonded to two different asymmetric centers will have three stereoisomers: one will be a meso compound, and the other two will be enantiomers.

In the case of cyclic compounds that have two asymmetric centers with identical substituents, the cis isomer is a meso compound and the trans isomer is a pair of enantiomers.

cis-1,3-dimethylcyclopentane
meso compound

trans-1,3-dimethylcyclopentane
enantiomers

cis-1,2-dibromocyclohexane
meso compound

trans-1,2-dibromocyclohexane
enantiomers

In the perspective formula above, *cis*-1,2-dibromocyclohexane appears to have a plane of symmetry. Remember, however, that cyclohexane is not a planar hexagon but exists preferentially in the chair conformation. The chair conformation does not have a plane of symmetry, but the much less stable boat conformation does. So, is *cis*-1,2-dibromocyclohexane a meso compound? The answer is yes. If a compound has a conformer with a plane of symmetry, the compound is achiral, even if the conformer with the plane of symmetry is not the most stable conformer.

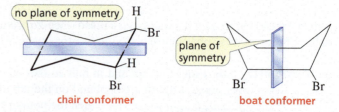

chair conformer   boat conformer

This rule holds for acyclic compounds as well. Earlier we saw that 2,3-dibromobutane is an achiral meso compound because it has a plane of symmetry. To see its plane of symmetry, however, we had to look at a relatively unstable eclipsed conformer. The more stable staggered conformer does not have a plane of symmetry. 2,3-Dibromobutane is still a meso compound, however, because it has a conformer with a plane of symmetry.

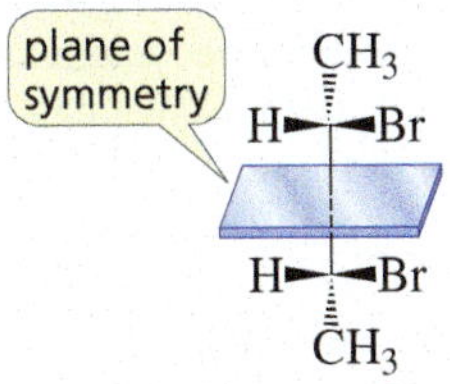

eclipsed conformer 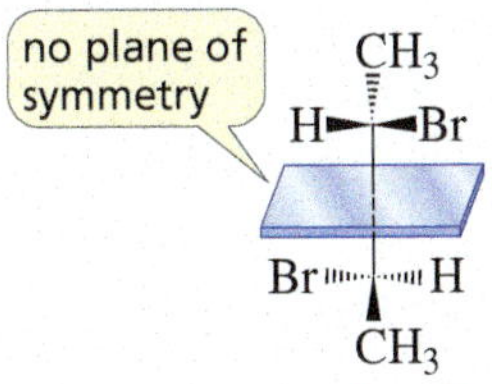   staggered conformer

PROBLEM-SOLVING STRATEGY

Recognizing Whether a Compound Has a Stereoisomer That Is a Meso Compound

LEARN THE STRATEGY

Which of the following compounds has a stereoisomer that is a meso compound?

A 2,3-dimethylbutane
B 3,4-dimethylhexane
C 2-bromo-3-methylpentane
D 1,3-dimethylcyclohexane

E 1,4-dimethylcyclohexane
F 1,2-dimethylcyclohexane
G 3,4-diethylhexane
H 1-bromo-2-methylcyclohexane

Check each compound to see if it has the necessary requirements for having a stereoisomer that is a meso compound. That is, does it have two asymmetric centers, and if so, do they each have the same four substituents attached to them?

Compounds **A**, **E**, and **G** do *not* have a stereoisomer that is a meso compound because they do not have any asymmetric centers.

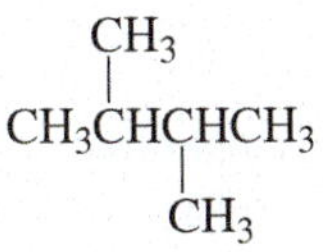

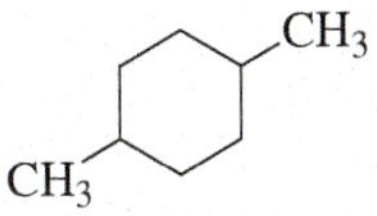

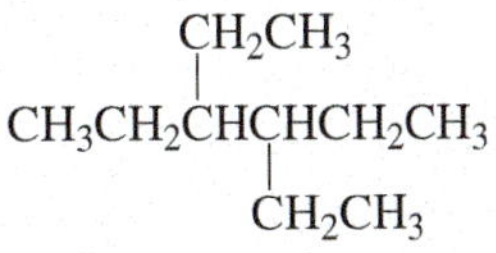

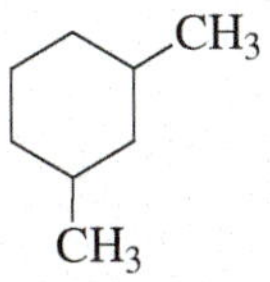

Compounds **C** and **H** have two asymmetric centers. They do *not* have a stereoisomer that is a meso compound, however, because the two asymmetric centers in each compound are *not* bonded to the same four substituents.

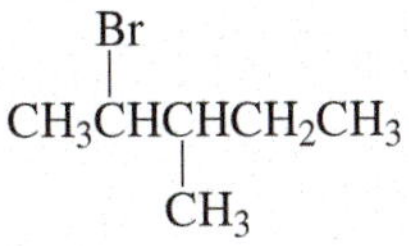

Compounds **B**, **D**, and **F** have two asymmetric centers, and the two asymmetric centers in each compound are bonded to the same four substituents. Therefore, these compounds have a stereoisomer that is a meso compound.

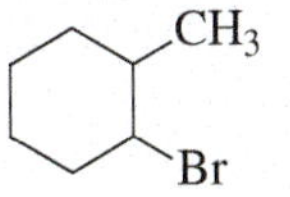

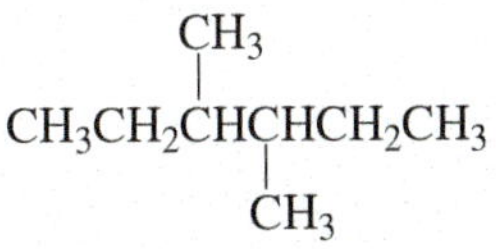

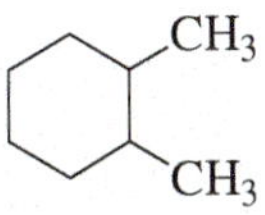

In the case of the acyclic compound, the meso compound is the stereoisomer with a plane of symmetry when drawn as an eclipsed conformer (**B**). For the cyclic compounds, the meso compound is the cis isomer (**D** and **F**).

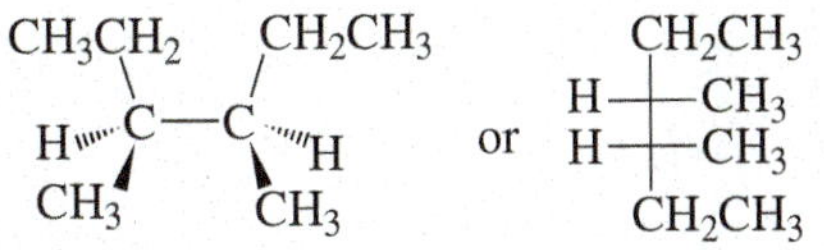

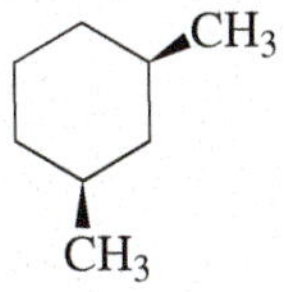

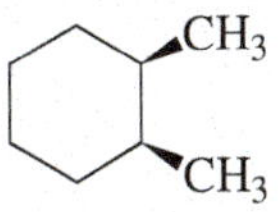

USE THE STRATEGY

PROBLEM 49 ◆

Which of the following compounds has a stereoisomer that is a meso compound?

A 2,4-dibromohexane

B 2,4-dibromopentane

C 2,4-dimethylpentane

D 1,3-dichlorocyclohexane

E 1,4-dichlorocyclohexane

F 1,2-dichlorocyclobutane

PROBLEM 50 SOLVED

Which of the following are optically active?

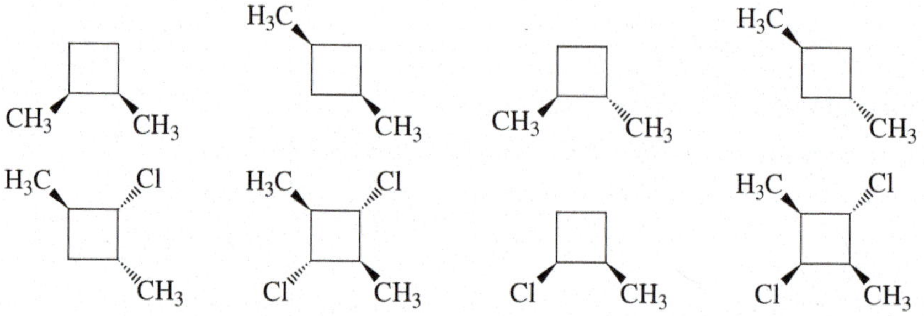

SOLUTION In the *top row*, only the *third* compound is optically active. The first compound has a plane of symmetry, and an optically active compound cannot have a plane of symmetry; the second and fourth compounds do not have any asymmetric centers. In the *bottom row*, the *first* and *third* compounds are optically active. The second and fourth compounds do not have any asymmetric centers.

PROBLEM 51

Draw all the stereoisomers for each of the following:

a. 1-chloro-3-methylpentane

b. 2-methyl-1-propanol

c. 2-bromo-1-butanol

d. 3-bromo-2-butanol

e. 3,4-dichlorohexane

f. 1,2-dichlorocyclobutane

g. 1,3-dichlorocyclohexane

h. 1,4-dichlorocyclohexane

i. 1-bromo-2-chlorocyclobutane

j. 1-bromo-3-chlorocyclobutane

4.15 HOW TO NAME ISOMERS WITH MORE THAN ONE ASYMMETRIC CENTER

If a compound has more than one asymmetric center, the steps used to determine whether a given asymmetric center has the *R* or the *S* configuration must be applied to each of the asymmetric centers individually. As an example, let's name one of the stereoisomers of 3-bromo-2-butanol.

Naming a Compound Drawn as a Perspective Formula

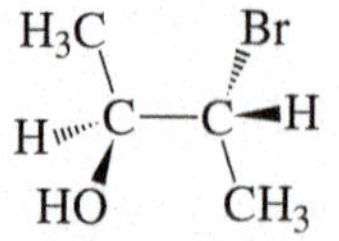

a stereoisomer of 3-bromo-2-butanol

LEARN THE STRATEGY

First, we will determine the configuration at C-2. The OH has priority (1), the C-3 carbon (the C attached to Br, C, H) is (2), CH_3 is (3), and H has priority (4). Because the group with the lowest priority is bonded by a hatched wedge, we can immediately draw an arrow from (1) to (2) to (3). The arrow points counterclockwise, so the configuration at C-2 is *S*.

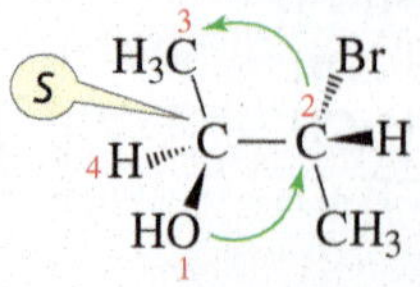

Now we need to determine the configuration at C-3. Because the group with the lowest priority (H) is not bonded by a hatched wedge, we must put it there by interchanging it with the group that is bonded by the hatched wedge.

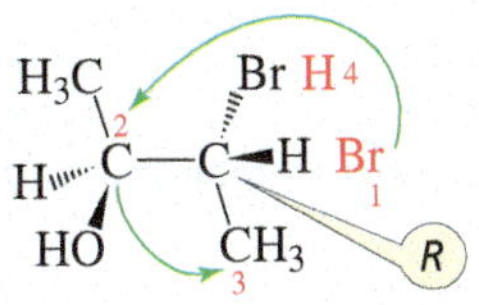

The arrow going from (1) (Br) to (2) the C-2 carbon (the C attached to O, C, H) to (3) (the methyl group) points counterclockwise, indicating that C-3 has the *S* configuration. However, because we interchanged two groups before we drew the arrow, C-3 in the compound before the groups were interchanged had the *R* configuration. Thus, the isomer is named (2*S*,3*R*)-3-bromo-2-butanol. Notice that the configurations are placed in parentheses in numerical order.)

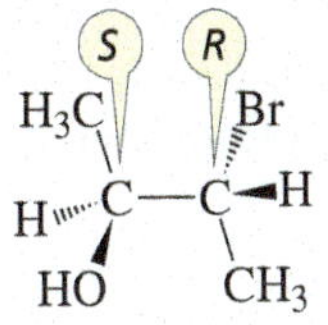

(2*S*,3*R*)-3-bromo-2-butanol

Naming a Compound Drawn as a Fischer Projection

When Fischer projections are used, the procedure is similar. Just apply the steps that you learned for a Fischer projection with one asymmetric center to each asymmetric center individually. At C-2, the arrow from (1) to (2) to (3) points clockwise, suggesting an *R* configuration. However, the group with the lowest priority is on a horizontal bond, so C-2 has the *S* configuration instead (Section 4.8).

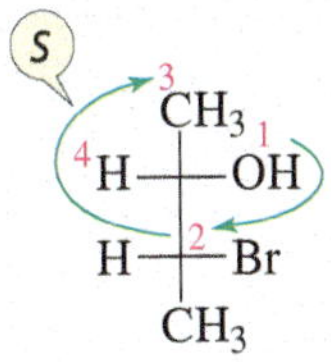

Repeating these steps for C-3 identifies that asymmetric center as having the *R* configuration. Thus, the isomer is named (2*S*,3*R*)-3-bromo-2-butanol.

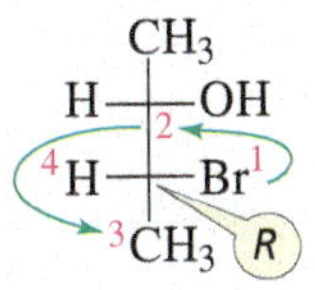

(2*S*,3*R*)-3-bromo-2-butanol

Examples

The four stereoisomers of 3-bromo-2-butanol are named as shown on the next page. Take a few minutes to verify their names.

- Notice that enantiomers have the opposite configuration at both asymmetric centers.
- Notice that diastereomers have the same configuration at one asymmetric center and the opposite configuration at the other asymmetric center.

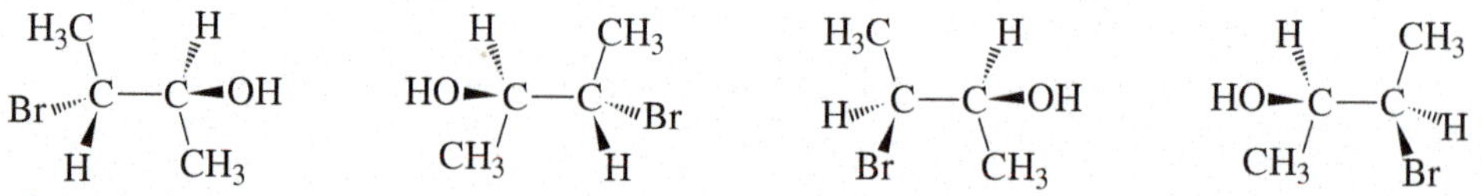

perspective formulas of the stereoisomers of 3-bromo-2-butanol

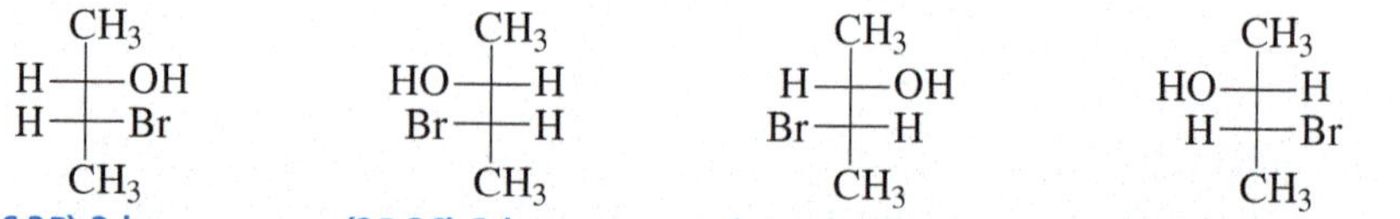

(2S,3R)-3-bromo-2-butanol   (2R,3S)-3-bromo-2-butanol   (2S,3S)-3-bromo-2-butanol   (2R,3R)-3-bromo-2-butanol

Fischer projections of the stereoisomers of 3-bromo-2-butanol

(2S,3R)-3-bromo-2-butanol   (2R,3S)-3-bromo-2-butanol   (2S,3S)-3-bromo-2-butanol   (2R,3R)-3-bromo-2-butanol

Tartaric acid has three stereoisomers because each of its two asymmetric centers has the same set of four substituents. The meso compound and the pair of enantiomers are named as shown.

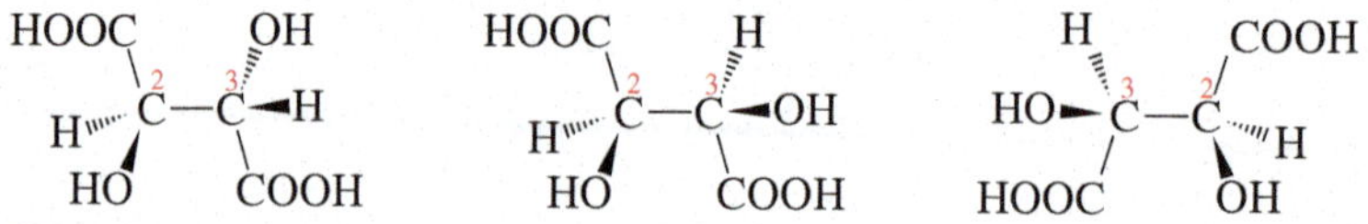

(2R,3S)-tartaric acid   (2R,3R)-tartaric acid   (2S,3S)-tartaric acid

meso compound   enantiomers

perspective formulas of the stereoisomers of tartaric acid

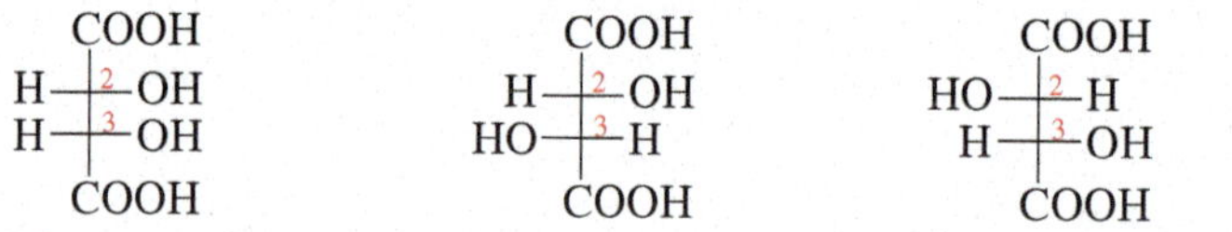

(2R,3S)-tartaric acid   (2R,3R)-tartaric acid   (2S,3S)-tartaric acid

meso compound   enantiomers

Fischer projections of the stereoisomers of tartaric acid

The physical properties of the three stereoisomers of tartaric acid are listed in Table 4.2. The meso compound and either of the enantiomers are diastereomers. Notice that the physical properties of enantiomers are the same, whereas the physical properties of diastereomers are different.

Table 4.2   Physical Properties of the Stereoisomers of Tartaric Acid

	Melting point, °C	Specific rotation	Solubility, g/100 g H_2O at 15 °C
(2R,3R)-(+)-Tartaric acid	171	+11.98	139
(2S,3S)-(−)-Tartaric acid	171	−11.98	139
(2R,3S)-Tartaric acid (meso)	146	0	125
(±)-Tartaric acid	206	0	139

PROBLEM 52

Draw the four stereoisomers of 1,3-dichloro-2-pentanol using

a. Fischer projections.   **b.** perspective formulas.

USE THE STRATEGY

PROBLEM 53

Name the isomers you drew in Problem 52.

PROBLEM 54 ◆

Chloramphenicol is a broad-spectrum antibiotic that is particularly useful against typhoid fever. What is the configuration of each of its asymmetric centers?

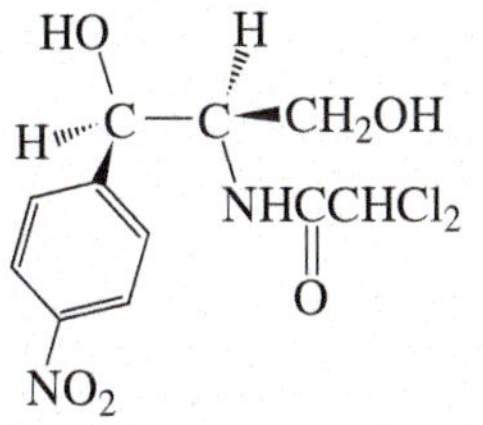

chloramphenicol

PROBLEM-SOLVING STRATEGY

Drawing a Perspective Formula for a Compound with Two Asymmetric Centers

LEARN THE STRATEGY

Draw a perspective formula for (2S,3R)-3-chloro-2-pentanol.

First write a condensed structure for the compound so you know what groups are attached to the asymmetric centers.

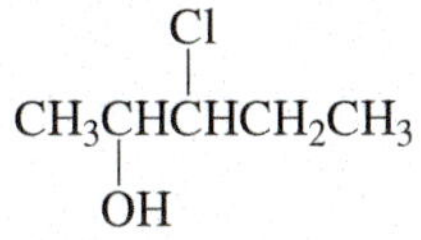

3-chloro-2-pentanol

Now draw the bonds that are in the plane of the paper and then add the wedges. Remember that the solid and hatched wedges must be adjacent and that the hatched wedge must be above the solid wedge.

At each asymmetric center, put the group with the lowest priority on the hatched wedge.

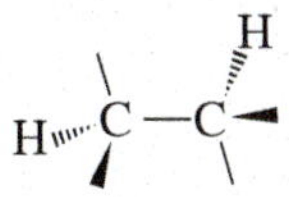

At each asymmetric center, put the group with the highest priority on a bond so that an arrow will point clockwise to the group with the next highest priority (if you want the *R* configuration) or counterclockwise (if you want the *S* configuration).

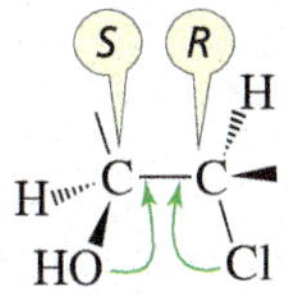

Put the remaining substituents on the last available bonds.

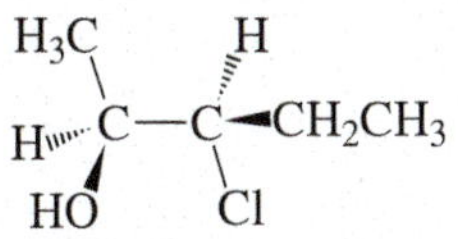

(2S,3R)-3-chloro-2-pentanol

PROBLEM 55

USE THE STRATEGY

Draw a perspective formula for each of the following:

a. (S)-3-chloro-1-pentanol

b. (2R,3R)-2,3-dibromopentane

c. (2S,3R)-3-methyl-2-pentanol

d. (R)-1,2-dibromobutane

PROBLEM 56 ♦

Name the following:

a.

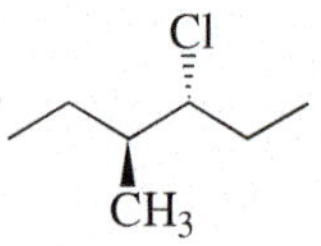

c.

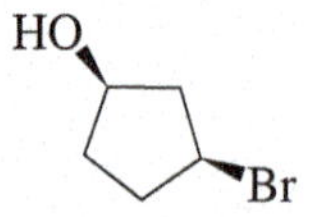

b. 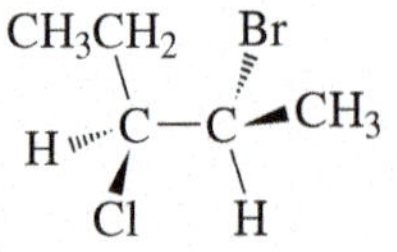 CH_3CH_2 Br
$H_{\text{\tiny{''''}}}C-C_{\text{\tiny{'''}}}CH_3$
Cl H

d. 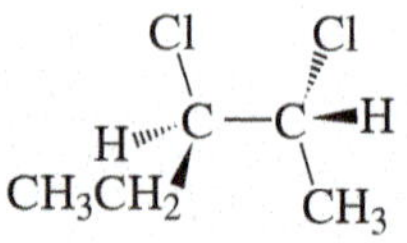 Cl Cl
$H_{\text{\tiny{''''}}}C-C_{\text{\tiny{'''}}}H$
CH_3CH_2 CH_3

PROBLEM 57 ♦

Threonine, an amino acid, has four stereoisomers. The stereoisomer found in nature is (2*S*,3*R*)-threonine. Which of the following structures represents the naturally occurring amino acid?

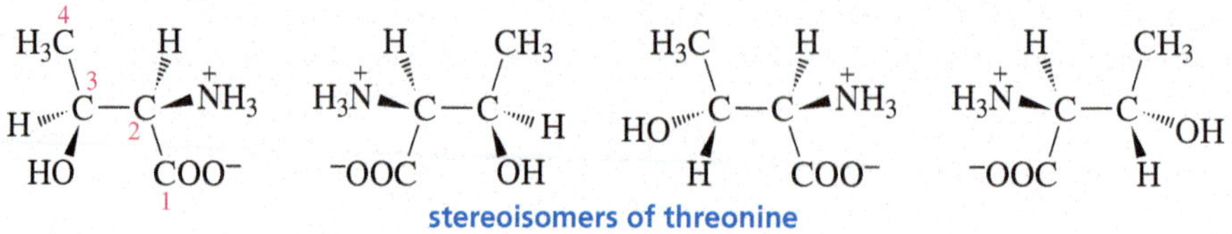

stereoisomers of threonine

LEARN THE STRATEGY

PROBLEM 58 SOLVED

Convert the Fischer projection to a perspective formula.

CH_3
H——OH
H——Cl
CH_3

NOTE TO THE STUDENT

• For practice interconverting between Newman projections, Fischer projections, perspective formulas, and skeletal structures, see the tutorial on p. 187.

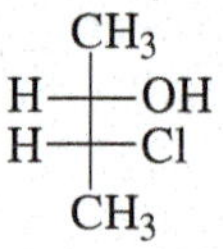

SOLUTION Determine the configuration of the two asymmetric centers and then draw the perspective formula with the same configurations, following the steps in the Problem-Solving Strategy on p. 175.

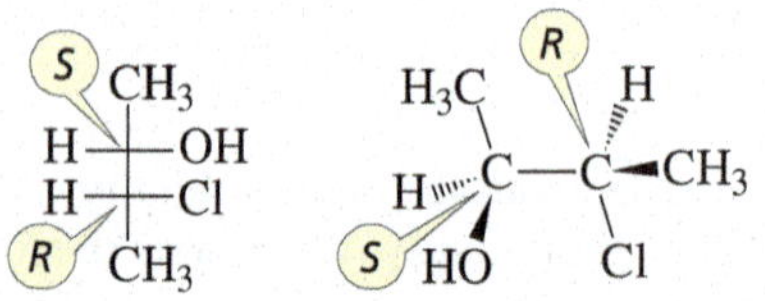

USE THE STRATEGY

PROBLEM 59

Convert the Fischer projection to a perspective formula.

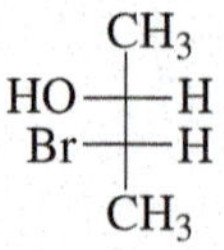

CH_3
HO——H
Br——H
CH_3

LEARN THE STRATEGY

PROBLEM 60 SOLVED

Convert the perspective formula to a skeletal structure.

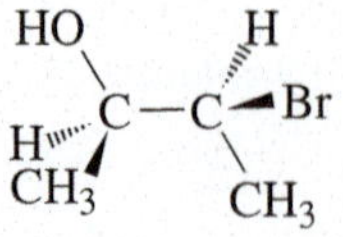

HO H
$H_{\text{\tiny{''''}}}C-C_{\text{\tiny{'''}}}Br$
CH_3 CH_3

SOLUTION Determine the configuration of the two asymmetric centers in the perspective formula. Now draw the skeletal structure, making sure the asymmetric centers have the same configurations they had in the perspective formulas.

PROBLEM 61

USE THE STRATEGY

Convert the perspective formula to a skeletal structure.

PROBLEM 62 ◆

The following compound has two asymmetric centers and four stereoisomers. Two of these are D-erythrose and D-threose, which are naturally occurring sugars. The configuration of D-erythrose is (2R,3R), and the configuration of D-threose is (2S,3R).

a. Which structure represents D-erythrose? **b.** Which represents D-threose?

4.16 NITROGEN AND PHOSPHORUS ATOMS CAN BE ASYMMETRIC CENTERS

Atoms other than carbon can be asymmetric centers. Any atom that has four different groups or atoms attached to it is an asymmetric center. For example, the following pairs of compounds, with nitrogen and phosphorus asymmetric centers, are enantiomers.

If one of the four "groups" attached to nitrogen is a lone pair, the enantiomers cannot be separated because they interconvert rapidly at room temperature. This rapid interconversion is called **amine inversion** (Figure 4.8). The lone pair is necessary for inversion: quaternary ammonium ions—ions with four bonds to nitrogen and hence no lone pair—do not invert. One way to picture amine inversion is to think of an umbrella that turns inside out in a windstorm.

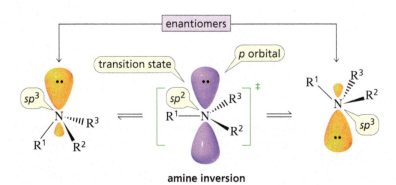

amine inversion

◀ **Figure 4.8**
Amine inversion takes place through a transition state in which the sp^3 nitrogen becomes an sp^2 nitrogen. The three groups bonded to the sp^2 nitrogen lie in a plane, and the lone pair is in a p orbital perpendicular to the plane. The "inverted" and "noninverted" amine molecules are enantiomers.

The energy required for amine inversion is approximately 6 kcal/mol, about twice the energy required for rotation about a carbon–carbon single bond, but still small enough to allow the enantiomers to interconvert rapidly at room temperature. As a result, the enantiomers cannot be separated.

PROBLEM 63

Explain why compound **A** has two stereoisomers but compounds **B** and **C** exist as single compounds.

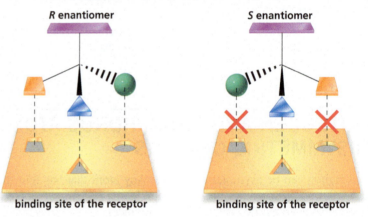

4.17 RECEPTORS

A **receptor** is a protein that binds a particular molecule. Because proteins are chiral, a receptor binds one enantiomer better than the other. In Figure 4.9, the receptor binds the *R* enantiomer, but it does not bind the *S* enantiomer, just as a right hand prefers a right glove.

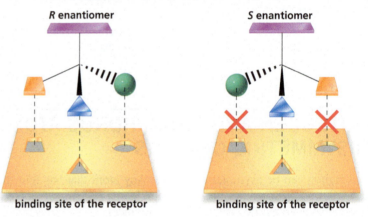

▲ **Figure 4.9**
A schematic diagram showing why only one enantiomer is bound by a receptor. One enantiomer fits into the binding site, and one does not.

The fact that a receptor typically recognizes only one enantiomer causes enantiomers to have different physiological properties. For example, receptors located on the exteriors of nerve cells in the nose are able to perceive and differentiate the estimated 10,000 smells to which they are exposed. The reason (*R*)-(−)-carvone (found in spearmint oil) and (*S*)-(+)-carvone (the main constituent of caraway seed oil) have such different odors is that each enantiomer fits into a different receptor.

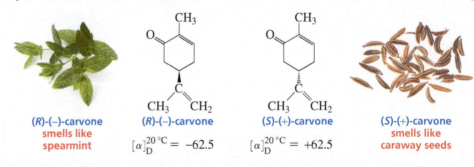

(*R*)-(–)-carvone
smells like
spearmint

(*R*)-(–)-carvone
$[\alpha]_D^{20\,°C} = -62.5$

(*S*)-(+)-carvone
$[\alpha]_D^{20\,°C} = +62.5$

(*S*)-(+)-carvone
smells like
caraway seeds

Many drugs exert their physiological activity by binding to cell-surface receptors. If the drug has an asymmetric center, the receptor can bind one of the enantiomers preferentially. Thus, enantiomers of a drug can have the same physiological activities, different degrees of the same activity, or very different activities, depending on the drug. For example, the enantiomers shown on the top of the next page have very different physiological activities.

Dr. Frances O. Kelsey receives the President's medal for Distinguished Federal Civilian Service from President John F. Kennedy in 1962 for preventing the sale of thalidomide. Kelsey was born in British Columbia in 1914. She received a B.Sc. in 1934 and a M.Sc. in pharmacology in 1936 from McGill University. In 1938, she received a Ph.D. and an M.D. from the University of Chicago, where she became a member of the faculty. She married a fellow faculty member, and they had two daughters. She joined the FDA in 1960 and worked there until 2005, when she retired at the age of 90. Each year the FDA selects a staff member to receive the Dr. Frances O. Kelsey Award for Excellence and Courage in Protecting Public Health.

the active ingredient
in Vicks Vapor Inhaler®

methamphetamine
"meth"

PROBLEM 64 ◆

Limonene exists as two different stereoisomers. The *R* enantiomer is found in oranges and lemons, and the *S* enantiomer is found in spruce trees. Which of the following is found in oranges and lemons?

(+)-limonene

(–)-limonene

The Enantiomers of Thalidomide

Thalidomide was developed in West Germany and was first marketed (as Contergan) in 1957 for insomnia, tension, and morning sickness during pregnancy. At that time, it was available in more than 40 countries but had not been approved for use in the United States because Frances O. Kelsey, a physician-scientist for the Food and Drug Administration (FDA), had insisted upon additional tests to explain a British study that had found nervous system side effects.

The (+)-isomer of thalidomide has stronger sedative properties, but the commercial drug was a racemic mixture. No one knew that the (–)-isomer is a teratogen—a compound that causes congenital deformations—until women who had been given the drug during the first three months of pregnancy gave birth to babies with a wide variety of defects, with deformed limbs being most common. By the time the danger was recognized and the drug withdrawn from the market on November 27, 1961, about 10,000 children had been damaged. It was eventually determined that the (+)-isomer also has mild teratogenic activity and that each of the enantiomers can racemize (interconvert) in the body. Thus, it is not clear whether the birth defects would have been less severe if the women had been given only the (+)-isomer.

Because thalidomide damaged fast-growing cells in the developing fetus, it has recently been approved—with restrictions and with tight controls—for the eradication of certain kinds of cancer cells.

asymmetric center

thalidomide

4.18 HOW ENANTIOMERS CAN BE SEPARATED

Enantiomers cannot be separated by the usual separation techniques such as fractional distillation or crystallization because their identical boiling points and solubilities cause them to distill or crystallize simultaneously.

Louis Pasteur was the first to succeed in separating a pair of enantiomers. While working with crystals of sodium ammonium tartrate, he noted that the crystals were not identical—some were "right-handed" and some were "left-handed." After painstakingly separating the two kinds of crystals with a pair of tweezers, he found that a solution of the right-handed crystals rotated the plane of polarization of plane-polarized light clockwise, whereas a solution of the left-handed crystals rotated it counterclockwise.

Crystals of potassium hydrogen tartrate, a naturally occurring salt found in wines. Sometime the crystals can be found on a wine cork. Most fruits produce citric acid, but grapes produce large quantities of tartaric acid instead. Potassium hydrogen tartrate, also called cream of tartar, is used in place of vinegar or lemon juice in some recipes.

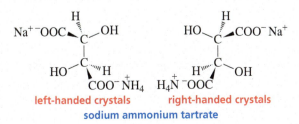

left-handed crystals

right-handed crystals

sodium ammonium tartrate

Pasteur, only 26 years old at the time and unknown in scientific circles, was concerned about the accuracy of his observations because a few years earlier, Eilhardt Mitscherlich, a well-known German organic chemist, had reported that crystals of sodium ammonium tartrate were all identical. Pasteur immediately reported his findings to Jean-Baptiste Biot and repeated the experiment with Biot present. Biot was convinced that Pasteur had successfully separated the enantiomers.

Later, chemists recognized how lucky Pasteur had been. Sodium ammonium tartrate forms asymmetric crystals only under the precise conditions that Pasteur happened to employ. Under other conditions, the symmetric crystals that Mitscherlich had obtained are formed. But to quote Pasteur, "Chance favors the prepared mind."

Pasteur's experiment gave rise to a new chemical term. Tartaric acid is obtained from grapes, so it was also called racemic acid (*racemus* is Latin for "a bunch of grapes"). This is how a mixture of equal amounts of enantiomers came to be known as a **racemic mixture** (Section 4.10). Separation of enantiomers is called the **resolution of a racemic mixture.**

Separating Enantiomers by Chromatography

Separating enantiomers by hand, as Pasteur did, is not a universally useful method because few compounds form asymmetric crystals. Until relatively recently, separating enantiomers was a very tedious process. Fortunately, enantiomers can now be separated relatively easily by a technique called **chromatography.**

In this method, the mixture to be separated is dissolved in a solvent and the solution is passed through a column packed with a chiral material that adsorbs organic compounds. The two enantiomers move through the column at different rates because they have different affinities for the chiral material—just as a right hand prefers a right-hand glove to a left-hand glove—so one enantiomer emerges from the column before the other.

Because it is now so much easier to separate enantiomers, many drugs are being sold as single enantiomers rather than as racemic mixtures (see the box "Chiral Drugs").

The chiral material used in chromatography is one example of a **chiral probe,** something capable of distinguishing between enantiomers. A polarimeter is another example of a chiral probe (Section 4.10). Two kinds of biological molecules—receptors (Section 4.17) and enzymes (Section 6.15)—are also chiral probes. You will see that certain enzymes also can be used to separate enantiomers (Section 21.7).

The French chemist and microbiologist Louis Pasteur (1822–1895) was the first person to demonstrate that microbes cause specific diseases. Asked by the French wine industry to find out why wine often went sour while aging, he showed that the microorganisms that cause grape juice to ferment, producing wine, also cause wine to become sour. Gently heating the wine after fermentation, a process called pasteurization, kills the organisms before they can sour the wine.

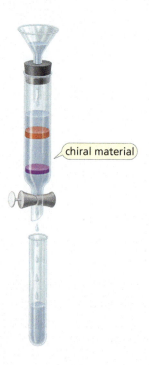

chiral material

Chiral Drugs

Until relatively recently, most drugs with one or more asymmetric centers were marketed as racemic mixtures because of the difficulty of synthesizing single enantiomers and the high cost of separating enantiomers. In 1992, however, the Food and Drug Administration (FDA) issued a policy statement encouraging drug companies to use recent advances in synthesis (see the box on Chiral Catalysts on p. 278) and separation techniques to develop single-enantiomer drugs. Now most new drugs sold are single enantiomers. Drug companies have been able to extend their patents by marketing a single enantiomer of a drug that was previously available only as a racemate (see p. 290).

If a drug is sold as a racemate, the FDA requires both enantiomers to be tested because drugs bind to receptors and, because receptors are chiral, the enantiomers of a drug can bind to different receptors (Section 4.17). Therefore, enantiomers can have similar or very different physiological properties. Examples are numerous. Testing has shown that (S)-(+)-ketamine is four times more potent an anesthetic than (R)-(−)-ketamine, and the disturbing side effects are apparently associated only with the (R)-(−)-enantiomer. Only the S isomer of the beta-blocker propranolol shows activity; the R isomer is inactive. The S isomer of Prozac, an antidepressant, is better at blocking serotonin but is used up faster than the R isomer. The activity of ibuprofen, the popular analgesic marketed as Advil, Nuprin, and Motrin, resides primarily in the (S)-(+)-enantiomer. Heroin addicts can be maintained with (−)-α-acetylmethadol for a 72-hour period compared with 24 hours with racemic methadone. This means less frequent visits to an outpatient clinic, because a single dose can keep an addict stable through an entire weekend.

Prescribing a single enantiomer spares the patient from having to metabolize the less potent enantiomer and decreases the chance of unwanted drug interactions. Drugs that could not be given as racemates because of the toxicity of one of the enantiomers can now be used. For example, (S)-penicillamine can be used to treat Wilson's disease even though (R)-penicillamine causes blindness.

ESSENTIAL CONCEPTS

Section 4.0

- **Stereochemistry** is the field of chemistry that deals with the structures of molecules in three dimensions.
- **Isomers** are compounds with the same molecular formula but different structures.
- **Constitutional isomers** differ in the way their atoms are connected.
- **Stereoisomers** differ in the way their atoms are arranged in space.
- There are two kinds of stereoisomers, **conformational isomers** that cannot be separated from one another, and **configurational isomers** that can be separated from one another.
- There are two kinds of configurational isomers: **cis–trans isomers** and isomers that contain **asymmetric centers.**

Section 4.1

- Because rotation about the bonds in a cyclic compound is restricted, disubstituted cyclic compounds can exist as **cis–trans isomers.** The **cis isomer** has the substituents on the same side of the ring; the **trans isomer** has the substituents on opposite sides of the ring.
- Because rotation about a double bond is restricted, an alkene can exist as **cis–trans isomers.** The **cis isomer** has its hydrogens on the same side of the double bond; the **trans isomer** has its hydrogens on opposite sides of the double bond.

Section 4.2

- The **Z isomer** has the high-priority substituents on the same side of the double bond; the **E isomer** has the high-priority substituents on opposite sides of the double bond.

Section 4.3

- A **chiral** molecule has a nonsuperimposable mirror image; an **achiral** molecule has a superimposable mirror image.

Section 4.4

- An **asymmetric center** is an atom bonded to four different atoms or groups.

Section 4.5

- **Enantiomers** are nonsuperimposable mirror images.

Section 4.8

- The letters **R** and **S** indicate the **configuration** about an asymmetric center.

- If one member of a pair of stereoisomers has the *R* configuration and the other has the *S* configuration, they are enantiomers; if both have the same configuration, they are identical.

Section 4.9

- Chiral compounds are **optically active;** achiral compounds are **optically inactive.**
- If one enantiomer rotates the plane of polarization clockwise (+), its mirror image rotates it the same amount counterclockwise (−).

Section 4.10

- A **racemic mixture** is a mixture of equal amounts of two enantiomers.
- Each optically active compound has a characteristic **specific rotation.**
- A **racemic mixture,** indicated by (±), is optically inactive.

Sections 4.12 and 4.13

- **Diastereomers** are stereoisomers that are not enantiomers.
- Enantiomers have identical physical and chemical properties; diastereomers have different physical and chemical properties.
- In the case of compounds with two asymmetric centers, enantiomers have the opposite configuration at both asymmetric centers; diastereomers have the same configuration at one asymmetric center and the opposite configuration at the other asymmetric center.

Section 4.14

- A **meso compound** has two or more asymmetric centers and a plane of symmetry; it is optically inactive.
- A compound with the same four groups bonded to two different asymmetric centers has three stereoisomers—namely, a meso compound and a pair of enantiomers.

Section 4.16

- Atoms other than carbons (such as N and P) can be asymmetric centers if they are bonded to four different atoms or groups.

Section 4.17

- A receptor binds one enantiomer better than another.

Section 4.18

- Enantiomers can be separated by chromatography.

PROBLEMS

65. Disregarding stereoisomers, draw the structures of all hydrocarbons with molecular formula C_5H_{10}. Which ones can exist as cis–trans isomers?

66. Draw all possible stereoisomers for each of the following. Indicate those compounds for which no stereoisomers are possible.
- **a.** 1-bromo-2-chlorocyclohexane
- **b.** 2-bromo-4-methylpentane
- **c.** 1,2-dichlorocyclohexane
- **d.** 2-bromo-4-chloropentane
- **e.** 1-bromo-4-chlorocyclohexane
- **f.** 1,2-dimethylcyclopropane
- **g.** 4-bromo-2-pentene
- **h.** 3,3-dimethylpentane
- **i.** 1-bromo-2-chlorocyclobutane
- **j.** 1-bromo-3-chlorocyclobutane

67. Which of the following has an asymmetric center?

CHBr$_2$Cl   BHFCl   CH$_3$CHCl$_2$ CHFBrCl   BeHCl

68. Name the following compounds using *R,S* designations:

a.
$$\begin{array}{c} CH_3 \\ HO{-}\!\!\!-\!\!\!-H \\ H{-}\!\!\!-\!\!\!-Cl \\ CH_2CH_3 \end{array}$$

b.
$$\begin{array}{c} CH_2OH \\ HO{-}\!\!\!-\!\!\!-CH_3 \\ CH_2CH_2CH_2OH \end{array}$$

c.
$$\begin{array}{c} CH_3CH_2 \quad Br \\ H\!\!\cdots\!C{-}C\!\!\cdots\!CH_3 \\ HO \quad\quad CH_2Br \end{array}$$

69. Mevacor is used clinically to lower serum cholesterol levels. How many asymmetric centers does Mevacor have?

Mevacor®

70. Do the following compounds have the *E* or the *Z* configuration?

a.
$$\begin{array}{c} CH_3CH_2 \quad\; CH_2CH_2Cl \\ C{=}C \\ CH_3 \quad\quad CH_2CH_3 \end{array}$$

b.
$$\begin{array}{c} CH_3CH_2CH_2 \quad\; CH(CH_3)_2 \\ C{=}C \\ HC{\equiv}CCH_2 \quad\; CH_2CH{=}CH_2 \end{array}$$

c.
$$\begin{array}{c} Br \quad\quad CH_2CH_2CH_2CH_3 \\ C{=}C \\ CH_3 \quad\quad CH_2Br \end{array}$$

d.
$$\begin{array}{c} O \\ \| \\ C \\ CH_3 \quad C{=}C \quad CH_2Br \\ HOCH_2 \quad\; CH_2CH_2Cl \end{array}$$

e.
$$\begin{array}{c} \quad\quad\quad CH_3 \\ BrCH_2CH_2 \quad CH_2CH_2CHCH_3 \\ C{=}C \\ Br \quad\quad CH_3 \end{array}$$

f.
$$\begin{array}{c} CH_3 \\ CH_3CH \quad\; CH_2CH_2CH_2CH_3 \\ C{=}C \\ CH_3 \quad\quad CH_2CH_2Cl \end{array}$$

71. Are the following pairs identical, enantiomers, diastereomers, or constitutional isomers?

a. [cyclopropane with Br, Br] and [cyclopropane with Br, Br]

b. [cyclopropane with Br, Br] and [cyclopropane with Br, Br]

c. [cyclopropane with CH$_3$, CH$_3$] and [cyclopropane with CH$_3$, CH$_3$]

d. [cyclohexane with CH$_3$, CH$_3$, CH$_3$] and [cyclohexane with CH$_3$, CH$_3$, CH$_3$]

e.
$$\begin{array}{c} CH_3CH_2 \quad Br \\ H\!\!\cdots\!C{-}C\!\!\cdots\!CH_3 \\ HO \quad\; CH(CH_3)_2 \end{array} \text{ and } \begin{array}{c} CH_3CH_2 \quad CH_3 \\ HO\!\!\cdots\!C{-}C\!\!\cdots\!CH(CH_3)_2 \\ H \quad\; Br \end{array}$$

f. [cyclohexane with Cl, Cl] and [cyclohexane with Cl, Cl]

g.
$$\begin{array}{c} H \quad\; CH_3 \\ C{=}C \\ CH_3 \quad Br \end{array} \text{ and } \begin{array}{c} H \quad\; Br \\ C{=}C \\ CH_3 \quad CH_3 \end{array}$$

h.
$$\begin{array}{c} Br \quad\; CH_3 \\ C{=}C \\ H \quad\; CH_3 \end{array} \text{ and } \begin{array}{c} H \quad\; CH_3 \\ C{=}C \\ Br \quad CH_3 \end{array}$$

72. Assign relative priorities to each set of substituents:

a. —CH$_2$CH$_2$CH$_3$ —CH(CH$_3$)$_2$ —CH=CH$_2$ —CH$_3$
b. —CH$_2$NH$_2$ —NH$_2$ —OH   —CH$_2$OH
c. —C(=O)CH$_3$ —CH=CH$_2$ —Cl   —C≡N

73. Draw the structure for a compound with molecular formula C$_2$H$_2$I$_2$F$_2$
   a. that is optically inactive because it does not have an asymmetric center.
   b. that is optically inactive because it is a meso compound.
   c. that is optically active.

74. Which of the following are optically active?

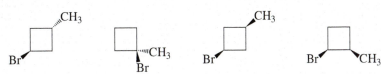

75. For many centuries, the Chinese have used extracts from a group of herbs known as ephedra to treat asthma. A compound named ephedrine has been isolated from these herbs and found to be a potent dilator of air passages in the lungs.

a. How many stereoisomers does ephedrine have?
b. The stereoisomer shown here is the one that is pharmacologically active. What is the configuration of each of the asymmetric centers?

$$CH_3$$
$$-CHCHNHCH_3$$
$$OH$$
ephedrine

$$HO^{\text{'''}}C-C\text{-}NHCH_3$$

76. Name the following:

a.

b.

c.

Br, OH, Cl, OH

77. Which of the following has an achiral stereoisomer?

A 2,3-dichlorobutane
B 2,3-dichloropentane
C 2,3-dichloro-2,3-dimethylbutane
D 1,3-dichlorocyclopentane

E 1,3-dibromocyclobutane
F 2,4-dibromopentane
G 2,3-dibromopentane
H 1,4-dimethylcyclohexane

I 1,2-dimethylcyclopentane
J 1,2-dimethylcyclobutane

78. Using skeletal structures, draw the stereoisomers of

a. 2,4-dichloroheptane.

b. 2,4-dichloropentane.

79. Are the following pairs identical, enantiomers, diastereomers, or constitutional isomers?

a. Cl ... and Cl ..., CH₃, CH₃

c. Cl ... and CH₃ ..., CH₃, Cl

b. Cl ... and Cl ..., CH₃, CH₃

d. Cl ... and Cl ..., CH₃, CH₃

80. Citrate synthase, one of the enzymes in the series of enzyme-catalyzed reactions known as the citric acid cycle (Section 24.9), catalyzes the synthesis of citric acid from oxaloacetic acid and acetyl-CoA. If the synthesis is carried out with acetyl-CoA that contains radioactive carbon (^{14}C) in the indicated position (Section 1.1), the isomer shown here is obtained.
a. Which stereoisomer of citric acid is synthesized: *R* or *S*?
b. If the acetyl-CoA used in the synthesis contains ^{12}C instead of ^{14}C, will the product of the reaction be chiral or achiral?

$$HOOCCH_2CCOOH + {}^{14}CH_3CSCoA \xrightarrow{\text{citrate synthase}}$$
oxaloacetic acid   **acetyl-CoA**

$${}^{14}CH_2COOH$$
$$HO-C\text{'''}COOH$$
$$CH_2COOH$$
citric acid

81. A solution of an unknown compound (3.0 g of the compound in 200 mL of solution), when placed in a polarimeter tube 2.0 dm long, was found to rotate the plane of polarized light 18° in a counterclockwise direction. What is the specific rotation of the compound?

82. Are the following pairs identical, enantiomers, diastereomers, or constitutional isomers?

a. 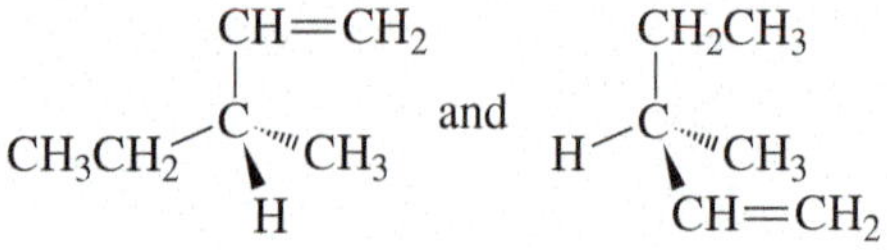 CH₃CH₂—C(CH=CH₂)("""CH₃)(H) and H—C(CH₂CH₃)("""CH₃)(CH=CH₂)

b. H—C(CH₂OH)(CH₃)(CH₂CH₃) and CH₃—C(CH₂CH₃)(H)(CH₂OH)

c. HO—H, H—Cl (CH₃ top, CH₃ bottom) and H—OH, Cl—H (CH₃ top, CH₃ bottom)

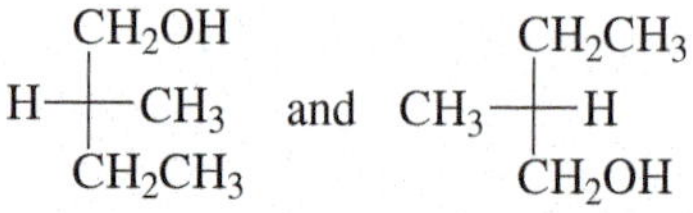

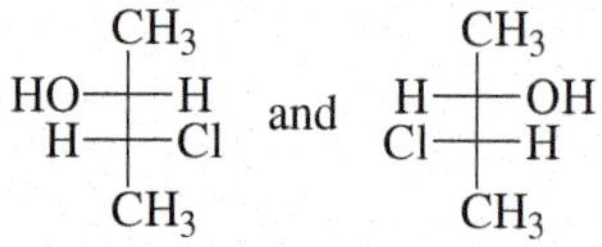

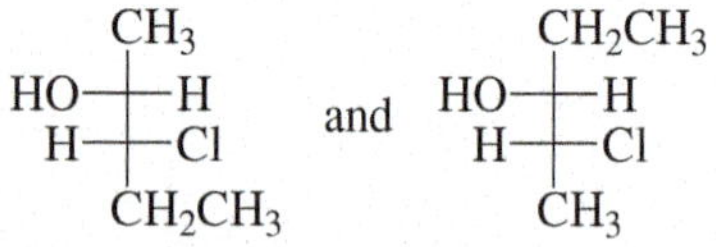

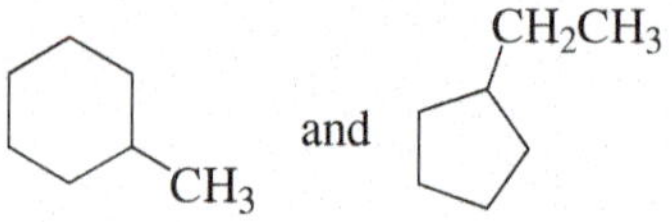

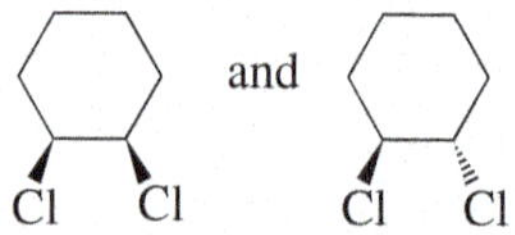

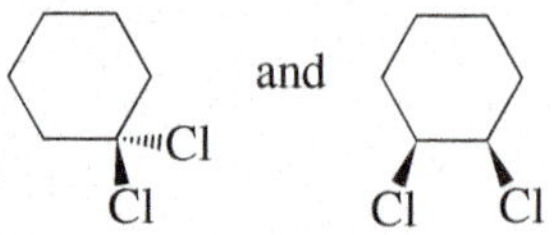

d. HO—H, H—Cl (CH₃ top, CH₂CH₃ bottom) and HO—H, H—Cl (CH₂CH₃ top, CH₃ bottom)

e. (methylcyclohexane with CH₃) and (ethylcyclopentane with CH₂CH₃)

f. (cyclohexane with Cl, Cl) and (cyclohexane with Cl, Cl)

g. (cyclohexane with ""Cl and Cl) and (cyclohexane with Cl, Cl)

h. (cyclohexane with Cl, Cl) and (cyclohexane with Cl, Cl)

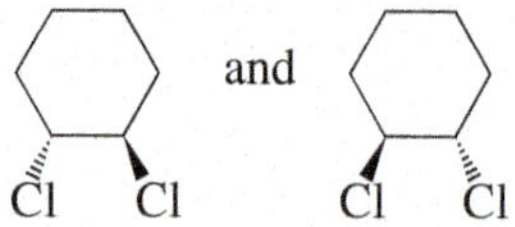

83. The specific rotation of (R)-(+)-glyceraldehyde is +8.7. If the observed specific rotation of a mixture of (R)-glyceraldehyde and (S)-glyceraldehyde is +1.4, what percent of glyceraldehyde is present as the R enantiomer?

84. Indicate whether each of the following structures is (R)-2-chlorobutane or (S)-2-chlorobutane:

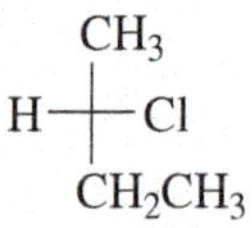

a. 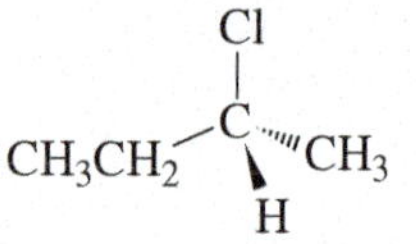 CH₃CH₂—C(Cl)("""CH₃)(H)

b. H—Cl (CH₃ top, CH₂CH₃ bottom)

c. (zig-zag with Cl)

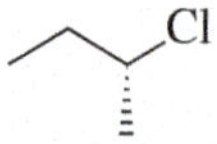

d. 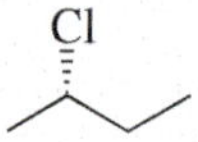 (zig-zag with Cl)

e. 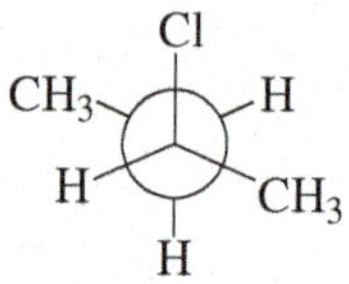 (Newman projection: Cl, CH₃, H, H, CH₃)

f. 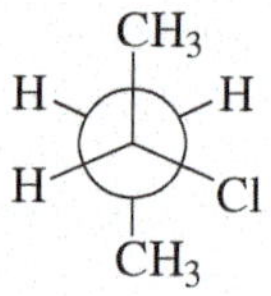 (Newman projection: CH₃, H, H, H, Cl, CH₃)

85. Butaclamol is a potent antipsychotic that has been used clinically in the treatment of schizophrenia. How many asymmetric centers does it have?

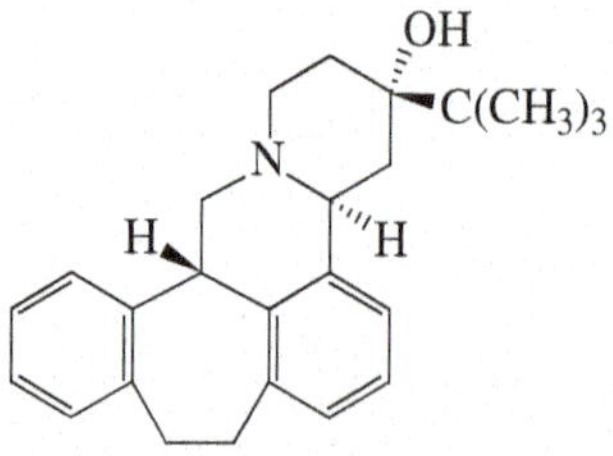

(structure of butaclamol with OH, C(CH₃)₃, N, H, H)

butaclamol

86. Explain how R and S are related to (+) and (−).

87. Are the following pairs identical, enantiomers, diastereomers, or constitutional isomers?

a. 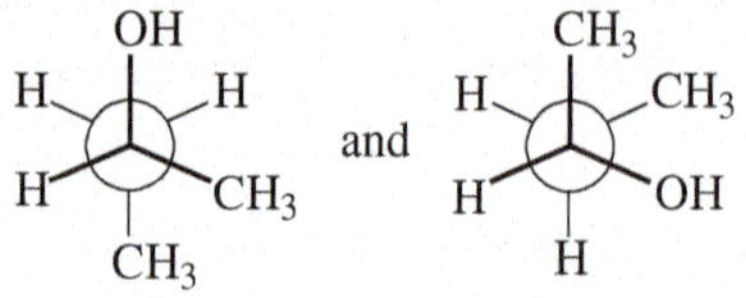 (Newman projection: OH, H, H, H, CH₃, CH₃) and (Newman projection: CH₃, H, CH₃, H, OH, H)

b. 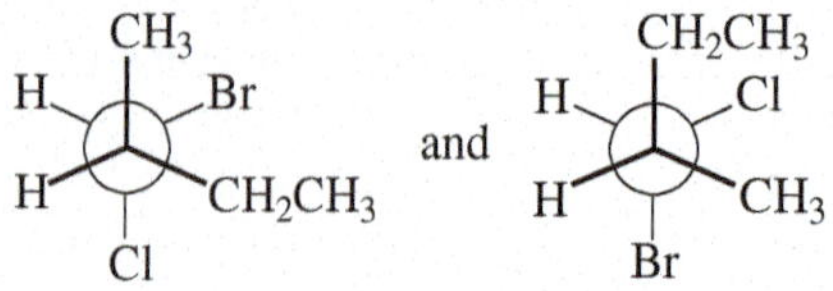 (Newman projection: CH₃, H, Br, H, CH₂CH₃, Cl) and (Newman projection: CH₂CH₃, H, Cl, H, CH₃, Br)

88. a. Draw all possible stereoisomers of the following compound:
b. Which stereoisomers are optically inactive?

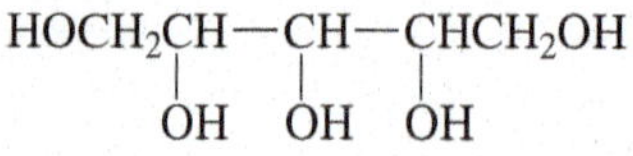

HOCH₂CH—CH—CHCH₂OH with OH, OH, OH below

89. What is the configuration of the asymmetric centers in the following compounds?

a. 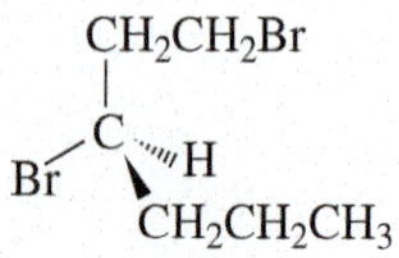 Br—C("""H)(CH₂CH₂Br)(CH₂CH₂CH₃)

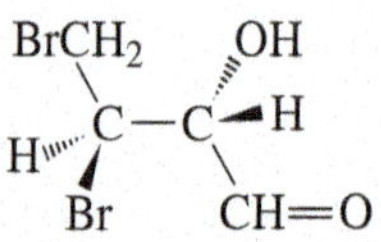

b. H""—C(BrCH₂)(Br)—C(OH)(H)(CH=O)

c. H""—C(Br)(CH₃)—C(CH₂CH₃)(H)(Br)

184

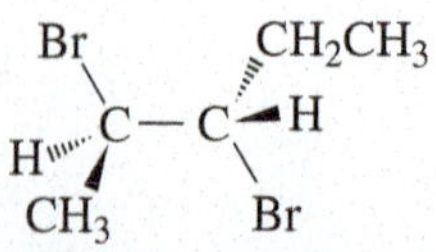

90. a. Draw all the isomers with molecular formula C_6H_{12} that contain a cyclobutane ring. (*Hint:* There are seven.)
  b. Name the compounds without specifying the configuration of any asymmetric centers.
  c. Identify
    1. constitutional isomers   **3.** cis–trans isomers   **5.** achiral compounds   **7.** enantiomers
    2. stereoisomers   **4.** chiral compounds   **6.** meso compounds   **8.** diastereomers

91. A compound has a specific rotation of -39.0. A solution of the compound (0.187 g/100 mL) has an observed rotation of $-6.52°$ when placed in a polarimeter tube 10 cm long. What is the percent of each enantiomer in the solution?

92. Are the following pairs identical, enantiomers, diastereomers, or constitutional isomers?

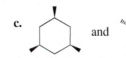

a. and   **b.** and   **c.** and   **d.** and

93. Draw structures for the following:
  a. (*S*)-1-bromo-1-chlorobutane
  b. (2*R*,3*R*)-2,3-dichloropentane
  c. an achiral stereoisomer of 1,2-dimethylcyclohexane
  d. a chiral stereoisomer of 1,2-dibromocyclobutane
  e. two achiral stereoisomers of 3,4,5-trimethylheptane

94. For each of the following structures, draw the most stable chair conformer.

a.
$CH_3CH_2CH_2$ ⎯ ⎯ $CH(CH_3)_2$
        $^{\prime\prime\prime\prime}CH_2CH_2CH_3$
CH_2CH_2CH
    |
    CH_3

b.
    H CH_3
HO ⎯
HOOC ⎯

95. Explain why the enantiomers of 1,2-dimethylaziridine can be separated even though one of the "groups" attached to nitrogen is a lone pair.

enantiomers of 1,2-dimethylaziridine

96. A sample of (*S*)-(+)-lactic acid was found to have an enantiomeric excess of 72%. How much *R* isomer is present in the sample?

97. Indicate whether each of the structures in the second row is an enantiomer of, is a diastereomer of, or is identical to the structure in the top row.

        CH_3
    H ⎯⎯ Br
    H ⎯⎯ CH_3
        CH_2CH_3

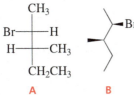
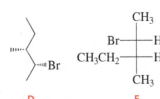

    CH_3
Br ⎯⎯ H
H ⎯⎯ CH_3
    CH_2CH_3
A

B

    CH_3
Br ⎯⎯ H
CH_3 ⎯⎯ CH_2CH_3
    H
C

D

    CH_3
Br ⎯⎯ H
CH_3CH_2 ⎯⎯ H
    CH_3
E

98. a. Using the wedge-and-dash notation, draw the nine stereoisomers of 1,2,3,4,5,6-hexachlorocyclohexane.
  b. From the nine stereoisomers, identify one pair of enantiomers.
  c. Draw the most stable conformer of the most stable stereoisomer.

99. What is the configuration of each of the asymmetric centers in the following compounds?

a.   **b.**   **c.**

185

100. Tamiflu is used for the prevention and treatment of flu. What is the configuration of each of its asymmetric centers? (How Tamiflu Works is explained in Section 22.11.)

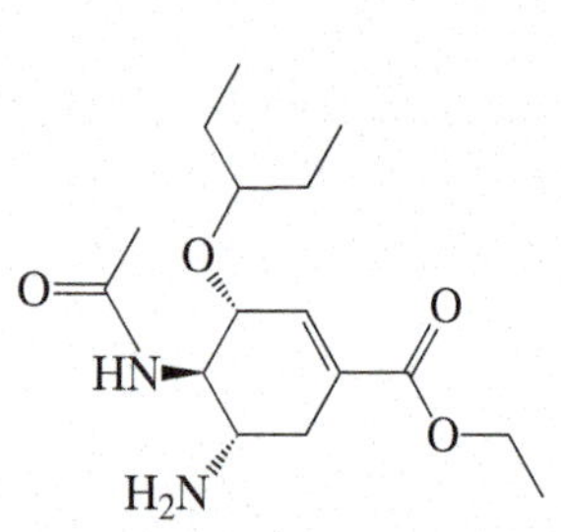

Tamiflu®

101. A student decided that the configuration of the asymmetric centers in a sugar such as D-glucose could be determined rapidly by simply assigning the *R* configuration to an asymmetric center with an OH group on the right and the *S* configuration to an asymmetric center with an OH group on the left. Is he correct? (We will see in Chapter 20 that the "D" in D-glucose means that the OH group on the bottommost asymmetric center is on the right.)

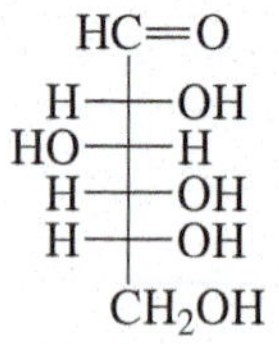

D-glucose

102. a. Draw the two chair conformers for each of the stereoisomers of *trans*-1-*tert*-butyl-3-methylcyclohexane.
b. For each pair, indicate which conformer is more stable.

103. a. Do the following compounds have any asymmetric centers?
b. Are the compounds chiral? (*Hint*: Make models.)

    1. CH_2=C=CH_2 **2.** CH_3CH=C=$CHCH_3$

104. Is the following compound optically active?

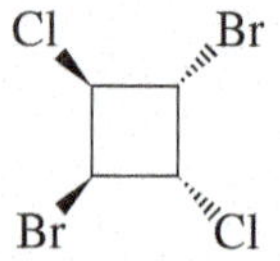

105. What is the configuration of each of the asymmetric centers in the following compounds?

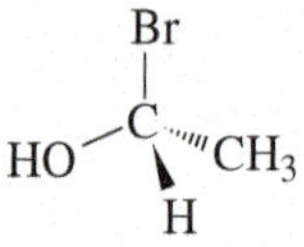

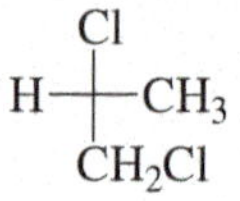

186

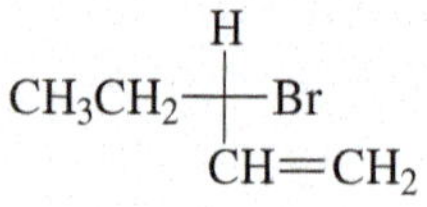

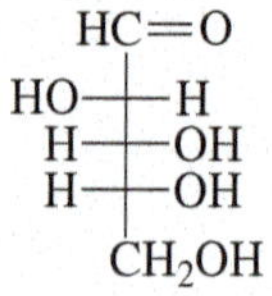

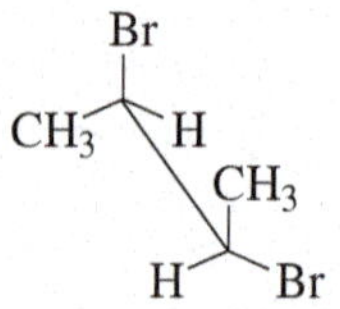

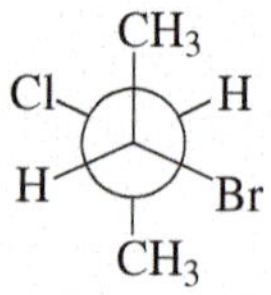

INTERCONVERTING STRUCTURAL REPRESENTATIONS

ESSENTIAL SKILL

TUTORIAL

Enhanced by
MasteringChemistry®

If you have access to a set of molecular models, converting between *perspective formulas*, *Fischer projections*, and *skeletal structures* is rather straightforward. If, however, you are interconverting these three-dimensional structures on a two-dimensional piece of paper, it is easy to make a mistake, particularly if you are not good at visualizing structures in three dimensions. Fortunately, there is a relatively foolproof method for these interconversions. All you need to know is how to determine whether an asymmetric center has the *R* or the *S* configuration (Sections 4.8 and 4.15). Look at the following examples to learn how easy it is to interconvert the various structural representations.

1. **Converting a Fischer projection to a perspective formula.** First determine the configuration of the asymmetric center. After you find that it has the *R* configuration, draw the perspective formula with that configuration. When you draw the perspective formula, start by putting the group with the lowest priority (in this case, H) on the hatched wedge.

2. **Converting a perspective formula to a skeletal structure.** After finding that the asymmetric center in the perspective formula has the *R* configuration, draw the skeletal structure with the *R* configuration.

3. **Converting a skeletal structure to a Fischer projection.** After finding that C-2 has the *S* configuration and C-3 has the *R* configuration, draw the Fischer projection with the same configurations at C-2 and C-3.

4. **Converting a perspective formula to a Fischer projection.** Determine the configuration of the asymmetric centers in the perspective formula and then draw the Fischer projection with the same configuration at those asymmetric centers.

5. **Converting a Fischer projection to a skeletal structure.** Determine the configuration of the asymmetric centers in the Fischer projection and then draw the skeletal structures with the same configuration at those asymmetric centers.

6. Converting a Fischer projection to a perspective formula. Determine the configuration of each of the two asymmetric centers in the Fischer projection. Draw the three solid lines of the perspective formula. Add a solid and hatched wedge at each carbon (making sure the solid wedge is below the hatched wedge) and then add the lowest-priority group on each carbon to the hatched wedge. Now add the other two groups to each carbon in a way that will give the desired configuration at each asymmetric carbon.

7. Converting a skeletal structure to a perspective formula. Determine the configurations of the two asymmetric centers in the skeletal structure. Continue as described in Example 6 to arrive at the desired perspective formula.

Converting from a *Newman projection* to (or from) a Fischer projection, a perspective formula, or a skeletal structure is relatively straightforward if you remember that a Fischer projection represents an eclipsed conformer. (Recall that in a Fischer projection, the horizontal lines represent bonds that point out of the plane of the paper toward the viewer and the vertical lines represent bonds that point back from the plane of the paper away from the viewer.)

8. Converting a Newman projection to a Fischer projection. First convert the staggered conformer of the Newman projection to an eclipsed conformer by rotating either the front or back carbon. (Here the carbon in front is not moved while the back carbon is rotated counterclockwise.) Now you can draw the Fischer projection by pulling down the bond attached to the CH_3 group on the back carbon. As a result, the front carbon in the Newman projection becomes the top cross in the Fischer projection and the back carbon becomes the bottom cross. Notice that the bond to the CH_3 group on the back carbon, which points to the top of the page in the eclipsed Newman projection, now points to the bottom of the page in the Fischer projection. Next, add the other atoms to their bonds, keeping the atoms on the right side of the Newman projection on the right side of the Fischer projection and the atoms on the left side of the Newman projection on the left side.

9. Converting a Newman projection to a perspective formula. First convert the staggered conformer to an eclipsed conformer and then convert the eclipsed conformer to a Fischer projection as you did in Example 8. Once you have the Fischer projection, you can convert it to a perspective formula as described in Example 6.

10. **Converting a perspective formula to a Newman projection.** Determine the configurations of the asymmetric centers in the perspective formula and then draw the Fischer projection with its asymmetric centers in the same configuration. Because the Fischer projection is in an eclipsed conformation, you can draw the Newman projection in the eclipsed conformation by making the downward-pointing bond in the Fischer projection the upward-pointing bond on the back carbon of the Newman projection. Now add the other atoms to their bonds, keeping the atoms on the right side of the Fischer projection on the right side of the Newman projection and the atoms on the left side on the Fischer projection on the left side. Next, rotate the Newman projection so it is in a staggered conformation. (Here the back carbon is rotated clockwise, but it could have been rotated counterclockwise.)

5 Alkenes

Structure, Nomenclature, and an Introduction to Reactivity

Thermodynamics and Kinetics

the gothic tower of Glasgow University overlooking the River Kelvin (see p. 203)

Some of the things you will learn about in this chapter are how the Kelvin temperature scale got its name, how trans fats get into our food, why carrots and flamingos are orange, how insect populations can be controlled, and how compounds in biological systems recognize each other.

In Chapter 3, we saw that alkanes are hydrocarbons that contain only carbon–carbon *single* bonds. Now we will take a look at **alkenes,** hydrocarbons that contain a carbon–carbon *double* bond.

Alkenes play many important roles in biology. For example, ethene ($H_2C = CH_2$), the smallest alkene, is a plant hormone—a compound that controls growth and other changes in the plant's tissues. Among other things, ethene affects seed germination, flower maturation, and fruit ripening. Many of the flavors and fragrances produced by plants also belong to the alkene family.

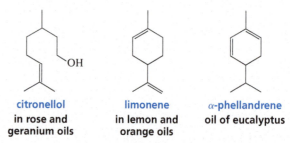

citronellol
in rose and geranium oils

limonene
in lemon and orange oils

α-phellandrene
oil of eucalyptus

We will begin our study of alkenes by looking at their structures and how they are named. Then we will examine a reaction of an alkene, paying close attention to the steps by which the reaction occurs and the energy changes that accompany them. You will see that some of the discussion in this chapter revolves around concepts with which you are already familiar, whereas some of the information is new and will broaden the foundation of knowledge that you will be building on in subsequent chapters.

Tomatoes are shipped green so they arrive unspoiled. Ripening starts when they are exposed to ethene.

Pheromones

Insects communicate by releasing pheromones—chemical substances that other insects of the same species detect with their antennae. Many of the sex, alarm, and trail pheromones are alkenes or are synthesized from alkenes. Interfering with an insect's ability to send or receive chemical signals is an environmentally safe way to control insect populations. For example, traps containing synthetic sex attractants have been used to capture crop-destroying insects such as the gypsy moth and the boll weevil.

muscalure
sex attractant of the housefly

disparlure
sex attractant of the gypsy moth

5.1 MOLECULAR FORMULAS AND THE DEGREE OF UNSATURATION

We saw that the general molecular formula for a noncyclic alkane is C_nH_{2n+2} (p. 89) and that the general molecular formula for a cyclic alkane is C_nH_{2n} because the cyclic structure reduces the number of hydrogens by two (Section 3.3).

The general molecular formula for an *acyclic alkene* is also C_nH_{2n} because the alkene's double bond causes it to have two fewer hydrogens than an alkane with the same number of carbons. Thus, the general molecular formula for a *cyclic alkene* must be C_nH_{2n-2}.

NOTE TO THE STUDENT
• Noncyclic compounds are also called acyclic compounds because the prefix *a* is Greek for "non" or "not."

CH₃CH₂CH₂CH₂CH₃
alkane
C_5H_{12}
C_nH_{2n+2}

CH₃CH₂CH₂CH=CH₂
alkene
C_5H_{10}
C_nH_{2n}

cyclic alkane
C_5H_{10}
C_nH_{2n}

cyclic alkene
C_5H_8
C_nH_{2n-2}

The general molecular formula for a hydrocarbon is C_nH_{2n+2} minus two hydrogens for every π bond or ring present in the molecule.

Therefore, *the general molecular formula for a hydrocarbon is C_nH_{2n+2} minus two hydrogens for every π bond or ring in the molecule.*

The total number of π bonds and rings is called the compound's **degree of unsaturation.** Thus, C_8H_{14}, which has four fewer hydrogens than an acyclic alkane with eight carbons ($C_nH_{2n+2} = C_8H_{18}$), has two degrees of unsaturation. Therefore, we know that the sum of the compound's π bonds and rings is two.

several compounds with molecular formula C₈H₁₄

CH₃CH=CH(CH₂)₃CH=CH₂

CH₂CH₃

CH₃(CH₂)₅C≡CH

Because *alkanes* contain the maximum number of C—H bonds possible—that is, they are saturated with hydrogen—they are called **saturated hydrocarbons.** In contrast, *alkenes* are called **unsaturated hydrocarbons** because they have fewer than the maximum number of hydrogens.

CH₃CH₂CH₂CH₃
alkane
a saturated hydrocarbon

CH₃CH=CHCH₃
alkene
an unsaturated hydrocarbon

PROBLEM 1 SOLVED

LEARN THE STRATEGY

What is the molecular formula for a 5-carbon hydrocarbon with one π bond and one ring?

SOLUTION For a 5-carbon hydrocarbon with no π bonds and no rings, $C_nH_{2n+2} = C_5H_{12}$. A 5-carbon hydrocarbon with one π bond and one ring (that is, two degrees of unsaturation) has four fewer hydrogens, because two hydrogens are subtracted for every π bond or ring in the molecule. Therefore, the molecular formula is C_5H_8.

USE THE STRATEGY

PROBLEM 2 ◆

What is the molecular formula for each of the following?

a. a 4-carbon hydrocarbon with two π bonds and no rings
b. a 10-carbon hydrocarbon with one π bond and 2 rings

LEARN THE STRATEGY

PROBLEM 3 SOLVED

Determine the degree of unsaturation for a hydrocarbon with a molecular formula of $C_{10}H_{16}$.

SOLUTION   For a 10-carbon hydrocarbon with no π bonds and no rings, $C_nH_{2n+2} = C_{10}H_{22}$. A 10-carbon compound with molecular formula $C_{10}H_{16}$ has six fewer hydrogens, so the degree of unsaturation is $6/2 = 3$.

USE THE STRATEGY

PROBLEM 4 ◆

Determine the degree of unsaturation for hydrocarbons with the following molecular formulas:

a. $C_{20}H_{34}$ **b.** C_8H_{16} **c.** $C_{12}H_{20}$ **d.** $C_{40}H_{56}$

PROBLEM 5

Determine the degree of unsaturation and then draw possible structures for noncyclic compounds with the following molecular formulas:

a. C_3H_6 **b.** C_3H_4 **c.** C_4H_6

PROBLEM 6 ◆

Several studies have shown that β-carotene, a precursor of vitamin A, may play a role in preventing cancer. β-Carotene has a molecular formula of $C_{40}H_{56}$, and it contains two rings and no triple bonds. How many double bonds does it have?

β-Carotene is an orange colored compound found in carrots, apricots, and flamingo feathers (Sections 13.21 and 16.13).

5.2   THE NOMENCLATURE OF ALKENES

We saw that the IUPAC system uses a suffix to denote certain functional groups (Sections 3.6 and 3.7). The double bond is the functional group of an alkene; its presence is denoted by the suffix "ene." Therefore, the systematic (IUPAC) name of an alkene is obtained by replacing the "ane" ending of the corresponding alkane with "ene." For example, a two-carbon alkene is called ethene, and a three-carbon alkene is called propene. Ethene also is frequently called by its common name: ethylene.

$$H_2C=CH_2 \qquad CH_3CH=CH_2$$

systematic name: **Ethene**   **Propene**   **Cyclopentene**   **Cyclohexene**
common name: **Ethylene**   **Propylene**

Most alkene names need a number to indicate the position of the double bond. (The four names above do not because there is no ambiguity.) The IUPAC rules you learned in Sections 3.3 and 3.6 apply to alkenes as well:

LEARN THE STRATEGY

1. The longest continuous chain that contains the functional group (in this case, the carbon–carbon double bond) is numbered in the direction that gives the functional group suffix the lowest possible number. For example, 1-butene signifies that the double bond is between the first and second carbons of butene; 2-hexene signifies that the double bond is between the second and third carbons of hexene.

Number the longest continuous chain containing the functional group in the direction that gives the functional group suffix the lowest possible number.

$$\overset{4}{C}H_3\overset{3}{C}H_2\overset{2}{C}H=\overset{1}{C}H_2 \qquad \overset{1}{C}H_3\overset{2}{C}H=\overset{3}{C}H\overset{4}{C}H_3 \qquad \overset{1}{C}H_3\overset{2}{C}H=\overset{3}{C}H\overset{4}{C}H_2\overset{5}{C}H_2\overset{6}{C}H_3$$

1-butene   **2-butene**   **2-hexene**

$$\overset{6}{C}H_3\overset{5}{C}H_2\overset{4}{C}H_2\overset{3}{C}H_2\overset{2}{C}CH_2CH_2CH_3$$
$$\underset{\overset{|}{1}}{\overset{||}{}}{CH_2}$$

2-propyl-1-hexene

the longest continuous chain has 8 carbons, but the longest continuous chain containing the functional group has 6 carbons, so the parent name of the compound is hexene

Notice that 1-butene does not have a common name. You might be tempted to call it "butylene," which is analogous to "propylene" for propene. Butylene, however, is not an appropriate name because it could signify either 1-butene or 2-butene, and a name must be unambiguous.

Recall that the stereoisomers of an alkene are named using a *cis* or *trans* (or *E* or *Z*) prefix (Sections 4.1 and 4.2).

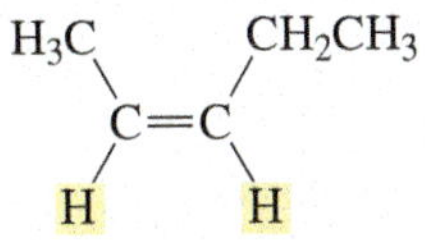

cis-2-pentene
or
(*Z*)-2-pentene

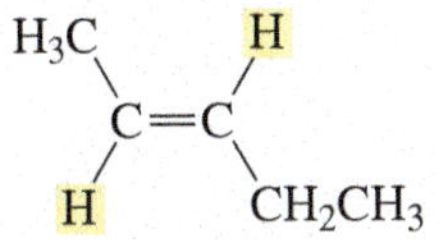

trans-2-pentene
or
(*E*)-2-pentene

2. For a compound with two or more double bonds, the "ne" ending of the corresponding alkane is replaced with "diene," "triene," "tetraene," and so on, depending on the number of double bonds in the parent hydrocarbon.

2,4-heptadiene

1,3-pentadiene

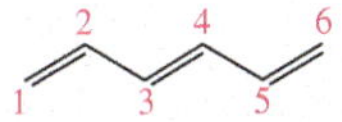

(3*E*)-1,3,5-hexatriene

3. The name of a substituent is stated before the name of the longest continuous chain that contains the functional group, together with a number to designate the carbon to which the substituent is attached. Notice that *if a compound's name contains both a functional group suffix and a substituent, the chain is numbered so that the functional group suffix gets the lowest possible number.*

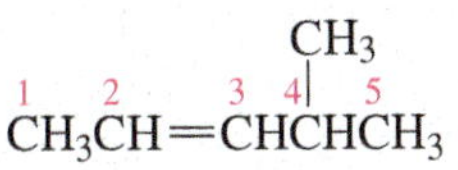

4-methyl-2-pentene

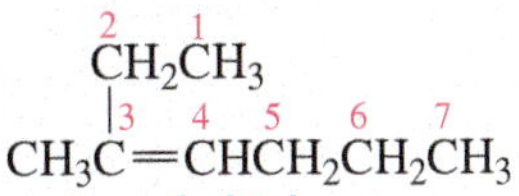

3-methyl-3-heptene

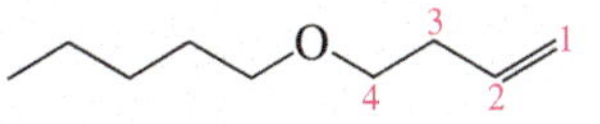

4-pentoxy-1-butene

4-methyl-1,3-pentadiene

When there are both a functional group and a substituent, the functional group suffix gets the lowest possible number.

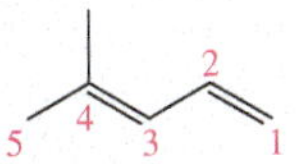

4. If a chain has more than one substituent, the substituents are stated in alphabetical order, using the same rules for alphabetizing discussed in Section 3.2. Then the appropriate number is assigned to each substituent.

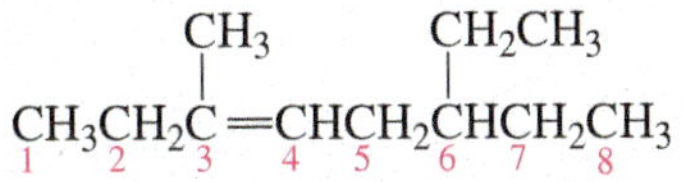

6-ethyl-3-methyl-3-octene

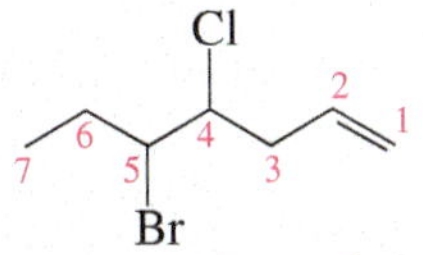

5-bromo-4-chloro-1-heptene

Substituents are stated in alphabetical order.

5. If counting in either direction results in the same number for the functional group suffix, the correct name is the one containing the lowest substituent number. For example, the compound shown below on the left is a 4-octene whether the longest continuous chain is numbered from left to right or from right to left. If you number from left to right, then the substituents are at positions 4 and 7, but if you number from right to left, they are at positions 2 and 5. Of those four substituent numbers, 2 is the lowest, so the compound is named 2,5-dimethyl-4-octene.

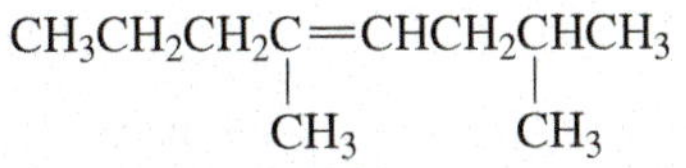

2,5-dimethyl-4-octene
not
4,7-dimethyl-4-octene
because 2 < 4

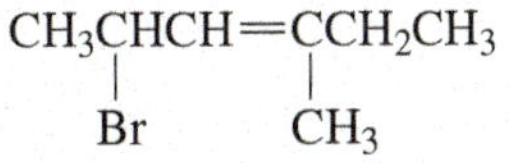

2-bromo-4-methyl-3-hexene
not
5-bromo-3-methyl-3-hexene
because 2 < 3

A substituent receives the lowest possible number only when there is no functional group suffix or the same number for the functional group suffix is obtained in both directions.

6. A number is not needed to denote the position of the double bond in a cyclic alkene because the ring is always numbered so that the double bond is between carbons 1 and 2. To assign numbers to any substituents, count around the ring in the direction (clockwise or counter-clockwise) that puts the lowest number into the name.

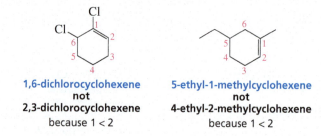

3-ethylcyclopentene   4,5-dimethylcyclohexene   4-ethyl-3-methylcyclohexene

Notice that 1,6-dichlorocyclohexene is *not* called 2,3-dichlorocyclohexene because the former has the lowest substituent number (1), even though it does not have the lowest sum of substituent numbers (1 + 6 = 7 versus 2 + 3 = 5).

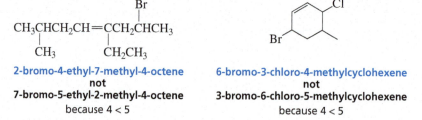

1,6-dichlorocyclohexene   5-ethyl-1-methylcyclohexene
not   not
2,3-dichlorocyclohexene   4-ethyl-2-methylcyclohexene
because 1 < 2   because 1 < 2

7. If counting in either direction leads to the same number for the alkene functional group suffix and the same lowest number or numbers for one or more of the substituents, then ignore those substituents and choose the direction that gives the lowest number to one of the remaining substituents.

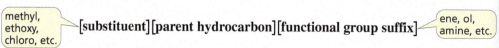

2-bromo-4-ethyl-7-methyl-4-octene   6-bromo-3-chloro-4-methylcyclohexene
not   not
7-bromo-5-ethyl-2-methyl-4-octene   3-bromo-6-chloro-5-methylcyclohexene
because 4 < 5   because 4 < 5

Remember that the name of a substituent is stated *before* the name of the parent hydrocarbon and that a functional group suffix is stated *after* the name of the parent hydrocarbon.

> methyl, ethoxy, chloro, etc.

[substituent][parent hydrocarbon][functional group suffix]

> ene, ol, amine, etc.

USE THE STRATEGY

PROBLEM 7 ◆

What is each compound's systematic name?

a. CH₃CHCH=CHCH₃
   |
   CH₃

b. CH₃CH₂C=CCHCH₃
        | |
        CH₃ Cl

c. [structure with Br on cyclopentene]

d. BrCH₂CH₂CH=CCH₃
                |
                CH₂CH₃

e. [cyclohexene structure with two CH₃ groups]

f. CH₃CH=CHCH₂OCH₂CH₂CH₂CH₃

g. [structure with Br]

h. [branched alkene structure]

The *sp²* carbons of an alkene are called **vinylic carbons.** An *sp³* carbon that is adjacent to a vinylic carbon is called an **allylic carbon.** A hydrogen bonded to a vinylic carbon is called a **vinylic hydrogen,** and a hydrogen bonded to an allylic carbon is called an **allylic hydrogen.**

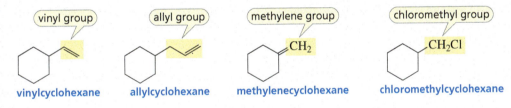

$$RCH_2-CH=CH-CH_2R$$

allylic carbons

PROBLEM 8 ◆

a. How many vinylic hydrogens does cyclopentene have?
b. How many allylic hydrogens does it have?

Two groups containing a carbon–carbon double bond are used in common names—the **vinyl group** and the **allyl group.** The vinyl group is the smallest possible group that contains a vinylic carbon, and the allyl group is the smallest possible group that contains an allylic carbon. When "vinyl" or "allyl" is used in a name, the substituent must be attached to the vinylic or allylic carbon, respectively.

$$CH_2=CH- \qquad CH_2=CHCH_2-$$
vinyl group $\qquad$ **allyl group**

$$CH_2=CHCl \qquad CH_2=CHCH_2Br$$

common name: vinyl chloride   allyl bromide
systematic name: chloroethene   3-bromopropene

Notice how these groups and some others can be used as substituent names in systematic nomenclature.

vinyl group   allyl group   methylene group   chloromethyl group

vinylcyclohexane   allylcyclohexane   methylenecyclohexane   chloromethylcyclohexane

---

PROBLEM 9 ◆

Draw the structure for each of the following:

a. 3,3-dimethylcyclopentene
b. 6-bromo-2,3-dimethyl-2-hexene

c. ethyl vinyl ether
d. allyl alcohol

5.3   THE STRUCTURE OF ALKENES

Alkenes have structures similar to that of ethene, the smallest alkene, whose structure was described in Section 1.8. Each double bonded carbon of an alkene has three *sp²* orbitals. Each of these orbitals overlaps an orbital of another atom to form a σ bond, one of which is one of the bonds in the double bond. Thus, the σ bond of the double bond is formed by the overlap of an *sp²* orbital of one carbon with an *sp²* orbital of the other carbon. The other bond of the double bond is a π bond formed from side-to-side overlap of the remaining *p* orbital on each of the *sp²* carbons.

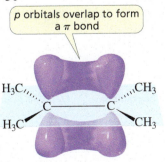

p orbitals overlap to form a π bond

Because three points determine a plane, each sp^2 carbon and the two atoms singly bonded to it lie in a plane. In order to achieve maximum orbital–orbital overlap, the two p orbitals must be parallel to each other. For the two p orbitals to be parallel, all six atoms of the double-bond system must be in the same plane.

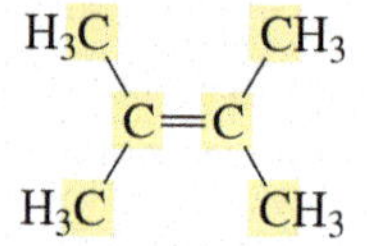

**the six carbon atoms
are in the same plane**

LEARN THE STRATEGY

PROBLEM 10 SOLVED

How many carbons are in the planar double-bond system in the following compound?

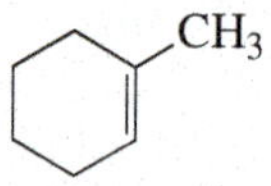

SOLUTION   Five carbons are in its planar double-bond system: the two sp^2 carbons (indicated by blue dots) and the three carbons bonded to the sp^2 carbons (indicated by red dots). The sixth atom in the plane is a hydrogen.

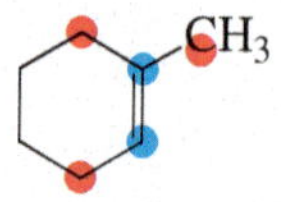

USE THE STRATEGY

PROBLEM 11 ◆

How many carbons are in the planar double-bond system in each of the following compounds?

a. 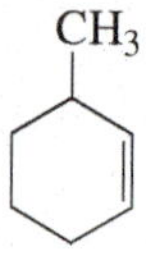   b. 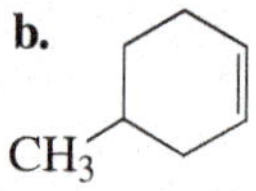   c.

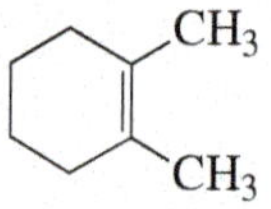

LEARN THE STRATEGY

PROBLEM-SOLVING STRATEGY

Drawing Isomers for Compounds with Two Double Bonds
How many isomers does the following compound have?

$$ClCH_2CH=CHCH=CHCH_2CH_3$$

It has four isomers because each of its double bonds can have either the E or the Z configuration. Thus, there are E-E, Z-Z, E-Z, and Z-E isomers.

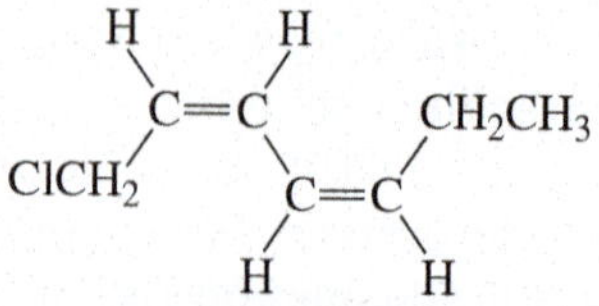

(2Z,4Z)-1-chloro-2,4-heptadiene   **(2Z,4E)-1-chloro-2,4-heptadiene**

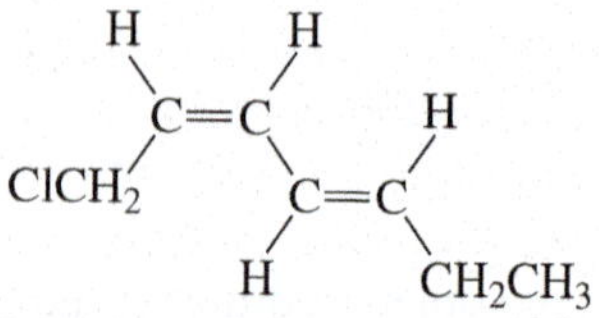

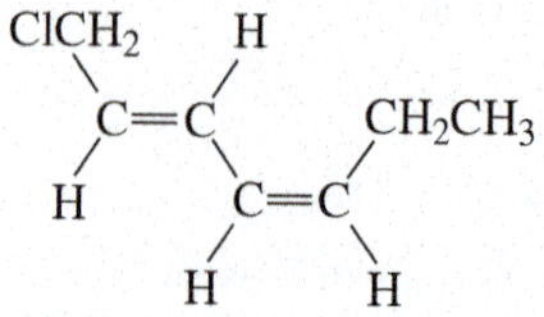

(2E,4Z)-1-chloro-2,4-heptadiene   **(2E,4E)-1-chloro-2,4-heptadiene**

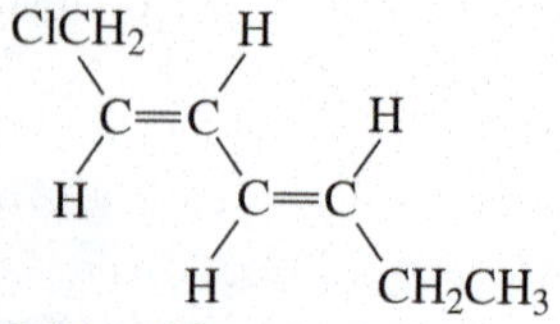

PROBLEM 12

Draw the isomers for the following compounds and then name each one:

a. 2-methyl-2,4-hexadiene   **b.** 2,4-heptadiene   **c.** 1,3-pentadiene

5.4 HOW AN ORGANIC COMPOUND REACTS DEPENDS ON ITS FUNCTIONAL GROUP

There are millions of organic compounds (and more being made each year). If you had to memorize how each of them reacts, studying organic chemistry would not be a very pleasant experience. Fortunately, organic compounds can be divided into families, and all the members of a family react in the same way.

The family to which an organic compound belongs is determined by its functional group. The **functional group** determines the kinds of reactions a compound undergoes. You are already familiar with the functional group of an alkene: the carbon–carbon double bond. All compounds with a carbon–carbon double bond react in the same way, whether the compound is a small molecule such as ethene or a large molecule such as cholesterol. (A table of common functional groups is located inside the back cover of this book.)

What makes learning organic chemistry even easier is that *all the families can be placed in one of four groups*, and *all the families in a group react in similar ways*.

We will start our study of reactions by looking at the reactions of alkenes, a family that belongs to the first of the four groups. We will revisit these groups each time we finish studying the reactions of all the families in a particular group.

5.5   HOW ALKENES REACT • CURVED ARROWS SHOW THE FLOW OF ELECTRONS

When you study the reactions of a particular functional group, you need to understand *why* the functional group reacts the way it does. It is not enough to look at the two reactions shown in Section 5.4 and see that the carbon–carbon double bond reacts with HBr to form a product in which the H and Br atoms have taken the place of the π bond. You need to understand *why* the reaction occurs. If you understand the reason for each functional group's reactivity, you will be able to look at an organic compound and predict the kind of reactions it can undergo.

In essence, organic chemistry is all about the interaction between electron-rich species and electron-deficient species. These are the forces that make chemical reactions happen. So each time you encounter a new functional group, remember that the reactions it undergoes can be explained by a very simple rule:

> *Electron-deficient species are attracted to electron-rich species.*

Therefore, to understand how a functional group reacts, you must first learn to recognize electron-deficient and electron-rich species.

Electrophiles

An electron-deficient species is called an **electrophile.** Literally, "electrophile" means "electron loving" (*phile* is the Greek suffix for "loving"). An electrophile looks for a pair of electrons. It is easy to recognize an electrophile—it has either a positive charge, a partial positive charge, or an incomplete octet that can accept electrons.

Electron-deficient species are attracted to electron-rich species.

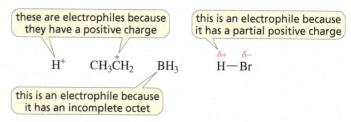

Nucleophiles

An electron-rich species is called a **nucleophile.** A nucleophile has a pair of electrons it can share. Nucleophiles and electrophiles attract each other (like negative and positive charges) because nucleophiles have electrons to share and electrophiles are seeking electrons. Thus, the preceding rule can be restated as *nucleophiles react with electrophiles.*

A nucleophile reacts with an electrophile.

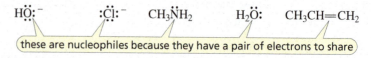

In Section 2.12, you saw that Lewis acids are compounds that accept a share in a pair of electrons and Lewis bases are compounds that donate a share in a pair of electrons. Thus, electrophiles are Lewis acids and nucleophiles are Lewis bases. So saying that an electrophile reacts with a nucleophile is the same as saying that a Lewis acid reacts with a Lewis base.

Because the Lewis definitions are so broad, we will use the term *base* when a Lewis base reacts with a proton and *nucleophile* when it reacts with something other than a proton. (This is an important distinction because base strength is a thermodynamic property and nucleophile strength is a kinetic property.)

PROBLEM 13 ♦

Which of the following are electrophiles, and which are nucleophiles?

$$H^-   \qquad CH_3O^-   \qquad CH_3C\equiv CH   \qquad CH_3\overset{+}{C}HCH_3   \qquad NH_3$$

PROBLEM 14

Identify the nucleophile and the electrophile in the following acid–base reactions:

a. $AlCl_3 + NH_3 \rightleftharpoons Cl_3\bar{Al} - \overset{+}{N}H_3$

b. $H - Br + HO^- \rightleftharpoons Br^- + H_2O$

The Mechanism of a Reaction

Let's now see how the rule "nucleophiles react with electrophiles" allows us to predict the characteristic reaction of an alkene. We saw that the π bond of an alkene consists of a cloud of electrons above and below the σ bond. As a result of this cloud of electrons, an alkene is an electron-rich molecule—it is a nucleophile. (Notice the relatively electron-rich pale orange area in the electrostatic potential maps for *cis*- and *trans*-2-butene.) We have also seen that a π bond is weaker than a σ bond (Section 1.15). The π bond, therefore, is the bond that is most easily broken when an alkene undergoes a reaction. For these reasons, we can predict that an alkene will react with an electrophile and, in the process, the π bond will break.

Thus, if a reagent such as hydrogen bromide is added to an alkene, the alkene (a nucleophile) will react with the partially positively charged hydrogen (an electrophile) of hydrogen bromide; the product of the reaction will be a carbocation. In the second step of the reaction, the positively charged carbocation (an electrophile) will react with the negatively charged bromide ion (a nucleophile) to form an alkyl halide.

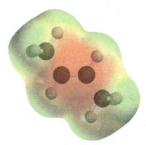

trans-2-butene

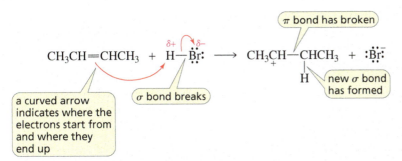

carbocation   alkyl halide

The step-by-step description of the process by which reactants (in this case, an alkene + HBr) are changed into products (an alkyl halide) is called the **mechanism of the reaction.**

We use curved arrows to help us understand a mechanism.

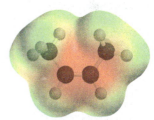

cis-2-butene

LEARN THE STRATEGY

- Curved arrows are drawn to show how the electrons move as new covalent bonds are formed and existing covalent bonds are broken (Section 2.3).

- Each arrow represents the simultaneous movement of two electrons (an electron pair) from a nucleophile (at the tail of the arrow) toward an electrophile (at the point of the arrow).

- The tail of the arrow is positioned where the electrons are in the reactant; the tail always starts at a lone pair or at a bond.

- The head of the arrow points to where these same electrons end up in the product; the arrow always points at an atom or a bond.

For the reaction of 2-butene with HBr, an arrow is drawn to show that the two electrons of the π bond of the alkene are attracted to the partially positively charged hydrogen of HBr.

$$CH_3CH{=}CHCH_3 + H-\overset{..}{\underset{..}{Br}}: \longrightarrow CH_3\overset{+}{CH}-CHCH_3 + :\overset{..}{\underset{..}{Br}}:^-$$

π bond has broken

σ bond breaks

a curved arrow indicates where the electrons start from and where they end up

new σ bond has formed

The mechanism of a reaction describes the step-by-step process by which reactants are changed to products.

The hydrogen is not immediately free to accept this pair of electrons because it is already bonded to a bromine, and hydrogen can be bonded to only one atom at a time (Section 1.4). However, as the π electrons of the alkene move toward the hydrogen, the H — Br bond breaks, with bromine keeping the bonding electrons.

A curved arrow indicates where the electrons start from and where they end up.

Notice that the π electrons are pulled away from one sp^2 carbon but remain attached to the other. Thus, the two electrons that originally formed the π bond now form a new σ bond between carbon and the hydrogen from HBr. The product is positively charged, because the sp^2 carbon that did not form the new bond with hydrogen has lost a share in an electron pair (the electrons of the π bond).

An arrowhead with two barbs signifies the movement of two electrons.

In the second step of the reaction, a lone pair on the negatively charged bromide ion forms a bond with the positively charged carbon of the carbocation. Notice that in both steps of the reaction *a nucleophile reacts with an electrophile.*

$$CH_3CH\!-\!CHCH_3 \;+\; :\!\ddot{Br}\!:^- \;\longrightarrow\; CH_3CH\!-\!CHCH_3$$

with H below the left and $:\ddot{Br}:$ H below the right, labeled "new σ bond"

The tail of the arrow is positioned where the electrons are in the reactant; the tail always starts at a lone pair or a bond.

Solely from the knowledge that a nucleophile reacts with an electrophile and a π bond is the weakest bond in an alkene, we have been able to predict that the product of the reaction is 2-bromobutane. The overall reaction involves the addition of 1 mole of HBr to 1 mole of the alkene. The reaction, therefore, is called an **addition reaction.** Because the first step of the reaction is the addition of an electrophile (H^+) to the alkene, the reaction is more precisely called an **electrophilic addition reaction.**

The head of the arrow points to where these same electrons end up in the product; the arrow always points at an atom or a bond.

Electrophilic addition reactions are the characteristic reactions of alkenes.

At this point, you may think it would be easier just to memorize the fact that 2-bromobutane is the product of the reaction without trying to understand the mechanism that explains why 2-bromobutane is the product. However, you will soon be encountering a great many reactions, and you will not be able to memorize them all. *It will be much easier to learn a few mechanisms that are based on similar rules than to try to memorize thousands of reactions.* And if you understand the mechanism of each reaction, the unifying principles of organic chemistry will soon be clear to you, making mastery of the material much easier and a lot more fun.

Alkenes undergo electrophilic addition reactions.

A Few Words About Curved Arrows

1. An arrow is used to show both the bond that forms and the bond that breaks. An arrow always starts at a lone or at a bond and points at an atom or a bond. Draw the arrows so that they point in the direction of the electron flow; the arrows should never go against the flow. This means that *an arrow flows away from a negatively charged atom and/or points toward a positively charged atom.*

2. Curved arrows are meant to indicate the movement of electrons. *Never use a curved arrow to indicate the movement of an atom.* For example, do not use an arrow as a lasso to remove a proton, as shown in the equation on the right:

3. *The head of a curved arrow always points at an atom or a bond.* Never draw the head of the arrow pointing out into space.

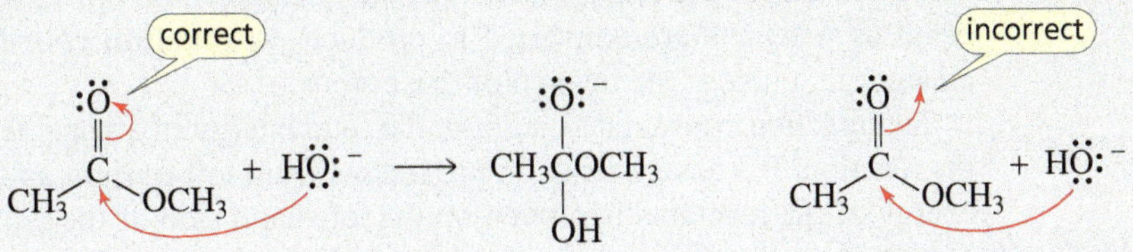

correct

incorrect

$$CH_3COCH_3$$
$$|$$
$$OH$$

$$+ \; H\ddot{O}:^-$$

4. *A curved arrow starts at an electron source; it does not start at an atom.* In the following example, the arrow starts at the electron-rich π bond, not at a carbon atom:

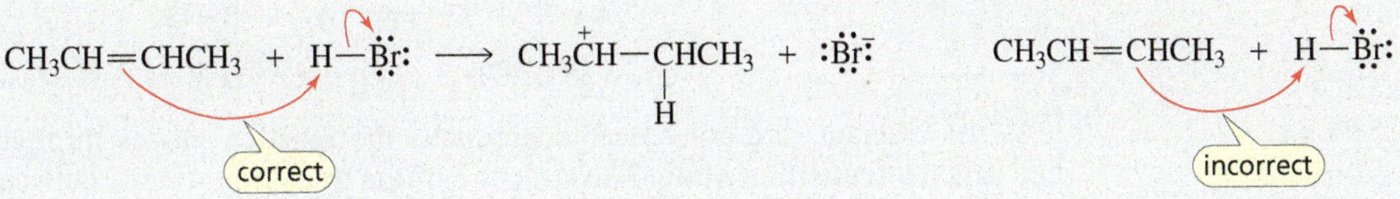

$$CH_3CH=CHCH_3 \; + \; H—\ddot{B}r: \longrightarrow CH_3\overset{+}{C}H—CHCH_3 \; + \; :\ddot{B}r:^-$$
$$|$$
$$H$$

$$CH_3CH=CHCH_3 \; + \; H—\ddot{B}r:$$

correct

incorrect

PROBLEM 15 SOLVED

Use curved arrows to show the movement of electrons in the following reaction steps. (*Hint:* Look at the reactants and look at the products and then draw the arrows to convert the reactants to products.)

a.

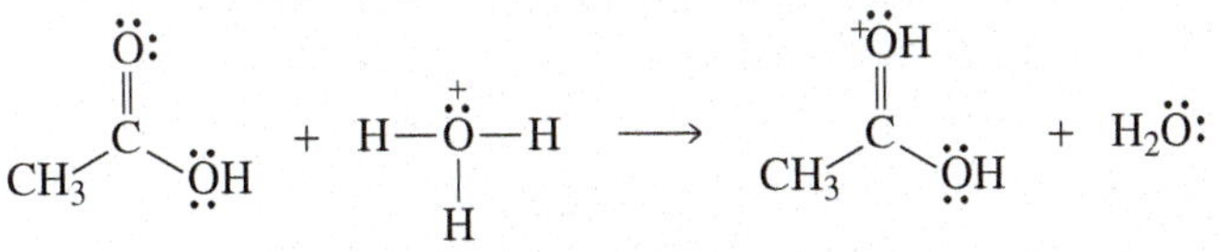

$$+ \; H—\overset{+}{\underset{H}{\ddot{O}}}—H \longrightarrow \qquad + \; H_2\ddot{O}:$$

b.

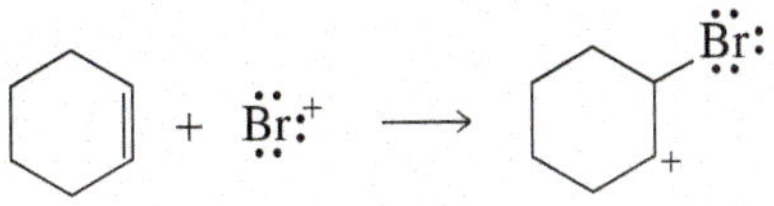

$$+ \; \ddot{B}r:^+ \longrightarrow$$

c.

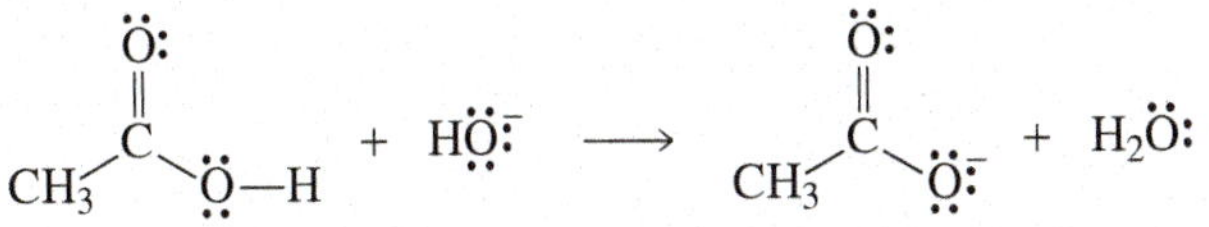

$$+ \; H\ddot{O}:^- \longrightarrow \qquad + \; H_2\ddot{O}:$$

d. 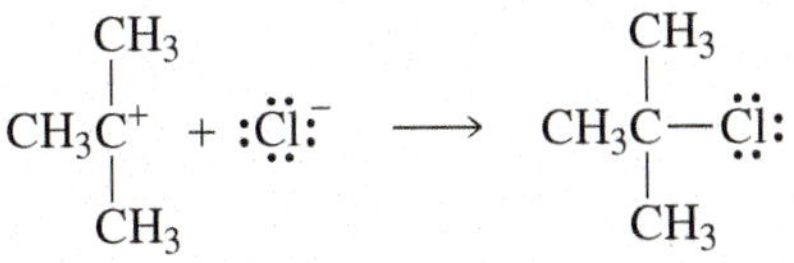
$$CH_3\overset{CH_3}{\underset{CH_3}{\overset{|}{\underset{|}{C}}}}{}^+ \; + \; :\ddot{C}l:^- \longrightarrow CH_3\overset{CH_3}{\underset{CH_3}{\overset{|}{\underset{|}{C}}}}—\ddot{C}l:$$

SOLUTION TO 15 a. From the products we see that the doubly bonded oxygen gains a proton and H_3O^+ loses a proton with oxygen retaining the electrons it shared with the proton. Now we need to draw the arrows that result in those products. Notice that the oxygen that gained a proton became positively charged and that the oxygen that lost a proton is no longer positively charged.

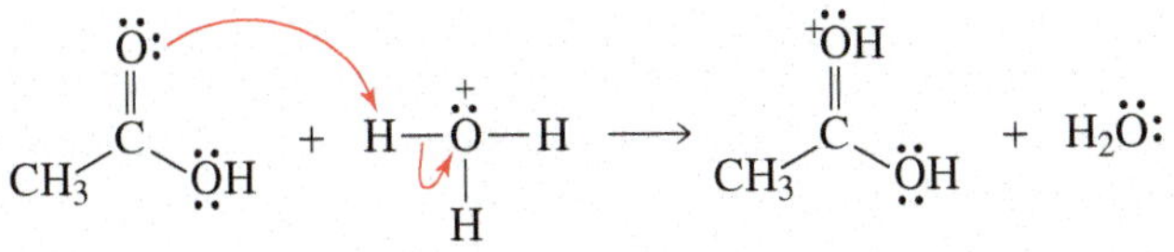

$$+ \; H—\overset{+}{\underset{H}{\ddot{O}}}—H \longrightarrow \qquad + \; H_2\ddot{O}:$$

PROBLEM 16

For each of the reactions in Problem 15, indicate which reactant is the nucleophile and which is the electrophile.

PROBLEM 17

Draw the products that would be obtained by following the incorrect arrows in the box entitled "A Few Words about Curved Arrows" and explain what is wrong with the structures you obtain.

USE THE STRATEGY

NOTE TO THE STUDENT

• It is critically important that you learn how to draw curved arrows. Be sure to do the tutorial on p. 225. It should take no more than 15 minutes, yet it can make an enormous difference to your success in this course.

A Reaction Coordinate Diagram Describes the Reaction Pathway

The mechanism of a reaction, such as the one shown on pp. 199–200, describes the steps known to occur as reactants are converted to products. A **reaction coordinate diagram** shows the energy changes that take place in each of these steps.

In a reaction coordinate diagram, the total energy of all species is plotted against the progress of the reaction. Because a reaction progresses from left to right as written in a chemical equation, the energy of the reactants is plotted on the left-hand side of the *x*-axis and the energy of the products is plotted on the right-hand side. A typical reaction coordinate diagram is shown in Figure 5.1. It describes the reaction of A—B with C to form A and B—C. Remember that *the more stable the species, the lower its energy.*

$$A-B \;+\; C \;\rightleftharpoons\; A \;+\; B-C$$

reactants   products

As the reactants are converted to products, the reaction passes through a *maximum* energy state called a **transition state.** The structure of the transition state is between the structure of the reactants and the structure of the products. As reactants are converted to products, bonds that break and bonds that form are partially broken and partially formed in the transition state. (Dashed lines are used to show partially broken or partially formed bonds.) The height of the transition state (the difference between the energy of the reactants and the energy of the transition state) tells us how likely it is that the reaction will occur. If the height is too great, the reactants will not be able to be converted to products, so no reaction will take place.

The more stable the species, the lower its energy.

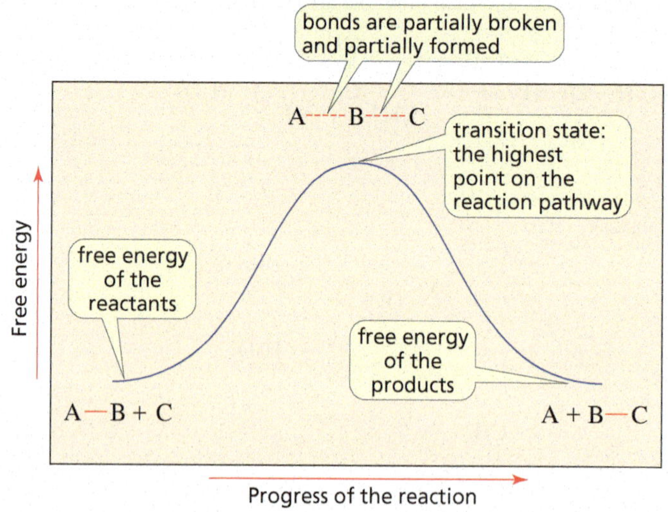

▶ **Figure 5.1**
A reaction coordinate diagram, which shows the energy changes that take place as reactants are converted to products. The dashed lines in the transition state indicate bonds that are partially formed or partially broken.

5.6   THERMODYNAMICS: HOW MUCH PRODUCT IS FORMED?

To understand the energy changes that take place in a reaction such as the addition of HBr to an alkene, you need to understand some of the basic concepts of *thermodynamics*, which describes a reaction at equilibrium, and *kinetics*, which explains the rates of chemical reactions.

Consider a reaction in which Y is converted to Z: the *thermodynamics* of the reaction tells us the relative amounts of reactants (Y) and products (Z) present when the reaction has reached equilibrium, whereas the *kinetics* of the reaction tells us how fast Y is converted to Z.

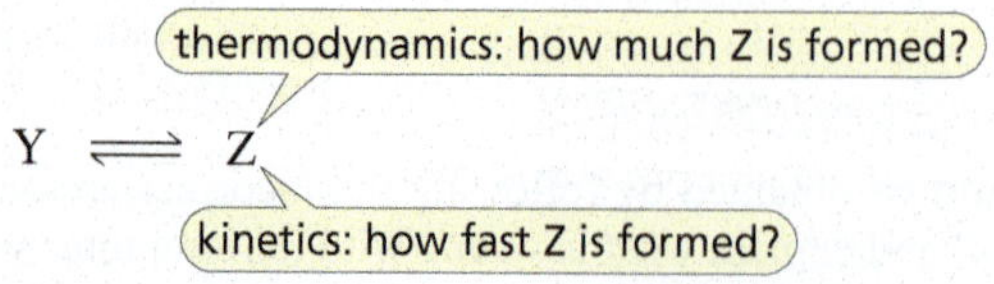

thermodynamics: how much Z is formed?

$$Y \;\rightleftharpoons\; Z$$

kinetics: how fast Z is formed?

Thermodynamics is the field of chemistry that describes the properties of a system at equilibrium. The relative concentrations of reactants and products at equilibrium can be expressed by an equilibrium constant, K_{eq} (Section 2.2).

$$A + B \rightleftharpoons C + D$$

$$K_{eq} = \frac{[\text{products}]}{[\text{reactants}]} = \frac{[C][D]}{[A][B]}$$

The relative concentrations of products and reactants at equilibrium depend on their relative stabilities: *the more stable the compound, the greater its concentration at equilibrium.*

- If the products are more stable (have a lower free energy) than the reactants (Figure 5.2a), there will be a higher concentration of products than reactants at equilibrium, so K_{eq} will be greater than 1.
- If the reactants are more stable than the products (Figure 5.2b), there will be a higher concentration of reactants than products at equilibrium, so K_{eq} will be less than 1.

Now you can understand why the strength of an acid is determined by the stability of its conjugate base (Section 2.6): as the base becomes more stable, the equilibrium constant (K_a) for its formation becomes larger—and the larger the K_a, the stronger the acid.

Gibbs Free-Energy Change

The difference between the free energy of the products and the free energy of the reactants under standard conditions is called the **Gibbs free-energy change**, or ΔG°. The symbol $^\circ$ indicates standard conditions, which means that all species are at a concentration of 1 M, a temperature of 25 °C, and a pressure of 1 atm.

$$\Delta G^\circ = \textbf{free energy of the products} - \textbf{free energy of the reactants}$$

From this equation, we can see that ΔG° will be negative if the products have a lower free energy (are more stable) than the reactants. In other words, the reaction will release more energy than it consumes; such a reaction is called an **exergonic reaction** (Figure 5.2a).

William Thomson (1824–1907) was born in Belfast, Northern Ireland. He was a professor of natural philosophy at the University of Glasgow, Scotland. For developing the Kelvin scale of absolute temperature and other important work in mathematical physics, he was given the title Baron Kelvin, which allowed him to be called **Lord Kelvin**. The name comes from the River Kelvin that flows by the University of Glasgow (see p. 190). His statue is in the botanic gardens that are adjacent to The Queen's University of Belfast.

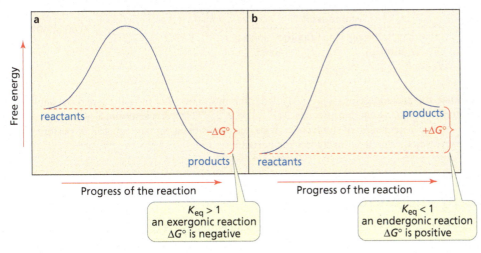

$K_{eq} > 1$
an exergonic reaction
ΔG° is negative

$K_{eq} < 1$
an endergonic reaction
ΔG° is positive

◀ **Figure 5.2**
Reaction coordinate diagrams for
(a) **a reaction in which the products are more stable than the reactants (an exergonic reaction)**
(b) **a reaction in which the products are less stable than the reactants (an endergonic reaction)**

If the products have a higher free energy (are less stable) than the reactants, ΔG° will be positive, and the reaction will consume more energy than it releases; such a reaction is called an **endergonic reaction** (Figure 5.2b).

(Notice that the terms *exergonic* and *endergonic* refer to whether the reaction has a negative ΔG° or a positive ΔG°. Do not confuse these terms with *exothermic* and *endothermic*, which we will define later in this section.)

A successful reaction is one in which the products are favored at equilibrium (that is, the products are more stable than the reactants). We saw that whether reactants or products are favored at equilibrium can be indicated by the equilibrium constant (K_{eq}) or by the change in free energy (ΔG°). These two quantities are related by the equation

$$\Delta G^\circ = -RT \ln K_{eq}$$

where R is the gas constant (1.986×10^{-3} kcal mol^{-1} K^{-1}) and T is the temperature in kelvins. (The Kelvin scale avoids negative temperatures by assigning 0 K to -273 °C, the lowest temperature known. Thus, because K = °C + 273, 25 °C = 298 K.)

The more stable the compound, the greater its concentration at equilibrium.

When products are favored at equilibrium, ΔG° is negative and K_{eq} is greater than 1.

When reactants are favored at equilibrium, ΔG° is positive and K_{eq} is less than 1.

PROBLEM 18 ◆

a. Which of the monosubstituted cyclohexanes in Table 3.9 on p. 128 has a negative $\Delta G°$ for the conversion of an axial-substituted chair conformer to an equatorial-substituted chair conformer?
b. Which monosubstituted cyclohexane has the most negative $\Delta G°$ for this conversion?
c. Which monosubstituted cyclohexane has the greatest preference for an equatorial position?
d. Calculate $\Delta G°$ for the conversion of "axial" methylcyclohexane to "equatorial" methylcyclohexane at 25 °C.

LEARN THE STRATEGY

PROBLEM 19 SOLVED

The $\Delta G°$ for conversion of "axial" fluorocyclohexane to "equatorial" fluorocyclohexane at 25 °C is −0.25 kcal/mol. Calculate the percentage of fluorocyclohexane molecules that have the fluoro substituent in an equatorial position at equilibrium.

SOLUTION   First, we must calculate the equilibrium constant for the reaction:

$$\text{fluorocyclohexane} \rightleftharpoons \text{fluorocyclohexane}$$
$$\text{axial} \qquad\qquad \text{equatorial}$$

$$\Delta G° = -0.25 \text{ kcal/mol at 25 °C}$$

$$\Delta G° = -RT \ln K_{eq}$$

$$-0.25 \, \frac{\text{kcal}}{\text{mol}} = -1.986 \times 10^{-3} \frac{\text{kcal}}{\text{mol K}} \times 298 \text{ K} \times \ln K_{eq}$$

$$\ln K_{eq} = 0.422$$

$$K_{eq} = 1.53 = \frac{[\text{fluorocyclohexane}]_{\text{equatorial}}}{[\text{fluorocyclohexane}]_{\text{axial}}} = \frac{1.53}{1}$$

Now we must determine the percentage of the total that is equatorial, as follows:

$$\frac{[\text{fluorocyclohexane}]_{\text{equatorial}}}{[\text{fluorocyclohexane}]_{\text{equatorial}} + [\text{fluorocyclohexane}]_{\text{axial}}} = \frac{1.53}{1.53 + 1} = \frac{1.53}{2.53} = 0.60 \text{ or } 60\%$$

USE THE STRATEGY

PROBLEM 20

a. Calculate the percentage of isopropylcyclohexane molecules that have the isopropyl substituent in an equatorial position at equilibrium. (Its $\Delta G°$ value at 25 °C is −2.1 kcal/mol.)
b. Why is the percentage of molecules with the substituent in an equatorial position greater for isopropylcyclohexane than for fluorocyclohexane?

Enthalpy and Entropy

The Gibbs standard free-energy change ($G°$) has an enthalpy ($H°$) component and an entropy ($\Delta S°$) component (T is the temperature in kelvins):

$$\Delta G° = \Delta H° - T\Delta S°$$

Enthalpy is the heat given off or the heat consumed during the course of a reaction. Heat is given off when bonds are formed, and heat is consumed when bonds are broken. Thus, $\Delta H°$ is a measure of the energy of the bond-making and bond-breaking processes that occur as reactants are converted to products.

$$\Delta H° = \textbf{heat required to break bonds} - \textbf{heat released from forming bonds}$$

If the bonds that are formed in a reaction are stronger than the bonds that are broken, more energy will be released in the bond-forming process than will be consumed in the bond-breaking process, and $\Delta H°$ will be negative. A reaction with a negative $\Delta H°$ is called an **exothermic reaction**. If the bonds that are formed are weaker than those that are broken, $\Delta H°$ will be positive. A reaction with a positive $\Delta H°$ is called an **endothermic reaction**.

Entropy is a measure of the freedom of motion in a system. Restricting the freedom of motion of molecule decreases its entropy. For example, in a reaction in which two molecules come together to form a single molecule, the entropy of the product will be less than the entropy of the reactants because two separate molecules can move in ways that are not possible when they are bound together in a single molecule. In such a reaction, $\Delta S°$ will be negative. In a reaction in which a single molecule is cleaved into two separate molecules, the products will have greater freedom of motion than the reactant and $\Delta S°$ will be positive.

$$\Delta S° = \text{freedom of motion of the products} - \text{freedom of motion of the reactants}$$

> Entropy is a measure of the freedom of motion in a system.

PROBLEM 21

a. For which reaction in each set will $\Delta S°$ be more significant?
b. For which reaction will $\Delta S°$ be positive?

    1. $A \rightleftharpoons B$ or $A + B \rightleftharpoons C$ **2.** $A \rightleftharpoons B + C$ or $A + B \rightleftharpoons C + D$

We saw that a favorable reaction has a negative $\Delta G°$ (and a $K_{eq} > 1$). The expression for the Gibbs standard free-energy change shows that negative values of $\Delta H°$ and positive values of $\Delta S°$ contribute to make $\Delta G°$ negative. In other words, *the formation of products with stronger bonds and greater freedom of motion causes $\Delta G°$ to be negative* and, therefore, the reaction to be a favorable one.

Notice that the entropy term is temperature dependent and, therefore, becomes more important as the temperature increases. As a result, a reaction with a positive $\Delta S°$ may be endergonic at low temperatures, but exergonic at high temperatures.

> The formation of products with stronger bonds and greater freedom of motion causes $\Delta G°$ to be negative.

PROBLEM 22 ◆

a. For a reaction with $\Delta H° = -12$ kcal/mol and $\Delta S° = 0.01$ kcal mol^{-1} K^{-1}, calculate the $\Delta G°$ and the equilibrium constant at: **1.** 30 °C and **2.** 150 °C.
b. How does $\Delta G°$ change as T increases?
c. How does K_{eq} change as T increases?

5.7 INCREASING THE AMOUNT OF PRODUCT FORMED IN A REACTION

Fortunately, there are ways to increase the amount of product formed in a reaction.

Le Châtelier's principle states that *if an equilibrium is disturbed, then the system will adjust to offset the disturbance.* In other words, if the concentration of C or D is decreased, then A and B will react to form more C and D in order to maintain the value of the equilibrium constant. (The value of a constant must be maintained—that is why it is called a *constant*.)

$$A + B \rightleftharpoons C + D$$

$$K_{eq} = \frac{[C][D]}{[A][B]}$$

> If an equilibrium is disturbed, the system will adjust to offset the disturbance.

Thus, if a product crystallizes out of solution as it is formed or if it can be distilled off as a liquid or driven off as a gas, the reactants will continue to react to replace the departing product in order to maintain the relative concentrations of products and reactants (that is, to maintain the value of the equilibrium constant). Additional product can also be formed if the equilibrium is disturbed by increasing the concentration of one or more of the reactants.

Living organisms obtain energy by carrying out a series of sequential reactions that convert complex nutrient molecules (such as glucose) into simple molecules (Section 24.6). Such a series of reactions is called a **metabolic pathway.** Some of the reactions in a metabolic pathway are endergonic and, therefore, produce very little product. However, the amount of product produced is increased if the endergonic reaction is followed by a highly exergonic reaction—another application of Le Châtelier's principle.

For example, very little B is produced in the first of the two sequential reactions shown on the top of the next page, because the conversion of A to B is endergonic. However, as the highly exergonic second reaction converts B to C, the first reaction will replenish the equilibrium concentration of B. Thus, the exergonic reaction drives the endergonic reaction that precedes it.

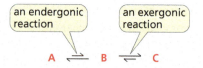

The two sequential reactions (an endergonic reaction followed by an exergonic reaction) are called **coupled reactions.** Coupled reactions are the thermodynamic basis for how metabolic pathways are regulated (Section 24.6).

5.8 CALCULATING $\Delta H°$ VALUES

Values of $\Delta H°$ can be calculated from bond dissociation enthalpies (Table 5.1), as shown in the following example. The bond dissociation enthalpy is indicated by the term *DH*. (Recall that the bond dissociation enthalpy of the π bond of ethene is estimated to be 62 kcal/mol; Section 1.15).

$$
\begin{array}{cccc}
\underset{\substack{H \\ }}{\overset{\substack{H \\ }}{C}}{=}\underset{\substack{H \\ }}{\overset{\substack{H \\ }}{C}} & + & H{-}Br & \longrightarrow & H{-}\underset{\substack{| \\ H}}{\overset{\substack{H \\ |}}{C}}{-}\underset{\substack{| \\ Br}}{\overset{\substack{H \\ |}}{C}}{-}H
\end{array}
$$

<u>bonds being broken</u>   <u>bonds being formed</u>

π bond of ethene	*DH* = 62 kcal/mol		C—H	*DH* = 101 kcal/mol
H—Br	*DH* = 88 kcal/mol		C—Br	*DH* = 71 kcal/mol
	DH_{total} = 150 kcal/mol			DH_{total} = 172 kcal/mol

$\Delta H°$ for the reaction = *DH* for bonds being broken − *DH* for bonds being formed

= 150 kcal/mol − 172 kcal/mol

= −22 kcal/mol

Because values of $\Delta H°$ are relatively easy to calculate, organic chemists frequently evaluate reactions in terms of that quantity alone, since $\Delta S°$ values cannot be calculated but must be experimentally determined. However, the entropy term can be ignored only if the reaction involves a small change in entropy and the temperature is low, because then the $T\Delta S°$ term will be small and the value of $\Delta H°$ will be very close to the value of $\Delta G°$. Remember, though, that many organic reactions occur with a significant change in entropy or occur at high temperatures and so have significant $T\Delta S°$ terms; ignoring the entropy term in such cases can lead to a wrong conclusion. It is permissible to use $\Delta H°$ values to *approximate* whether a reaction has a favorable equilibrium constant, but if a precise answer is needed, $\Delta G°$ values must be used.

Table 5.1	Experimental Bond Dissociation Enthalpies Y—Z $\rightarrow$ Y + Z			
Bond	***DH* (kcal/mol)**		**Bond**	***DH* (kcal/mol)**
CH_3—H	105.0		H—H	104.2
CH_3CH_2—H	101.1		F—F	37.7
			Cl—Cl	58.0
$(CH_3)_2CH$—H	98.6		Br—Br	53.5
$(CH_3)_3C$—H	96.5		I—I	51.0
			H—F	136.3
CH_3—CH_3	90.2		H—Cl	103.2
CH_3CH_2—CH_3	89.1		H—Br	87.5
$(CH_3)_2CH$—CH_3	88.6		H—I	71.3
$(CH_3)_3C$—CH_3	87.5			

(Continued)

Bond	DH (kcal/mol)		Bond	DH (kcal/mol)
			CH_3—F	114.8
H_2C=CH_2	174.5		CH_3—Cl	83.6
HC≡CH	230.4		CH_3CH_2—Cl	84.7
			$(CH_3)_3C$—Br	70.7
H_2C=CH—H	110.7		CH_3—I	57.1
HC≡C—H	133.3		CH_3CH_2—I	56.3
			$(CH_3)_2CH$—Cl	85.1
HO—H	118.8		$(CH_3)_3C$—Cl	84.8
CH_3O—H	104.2		CH_3—Br	70.3
CH_3—OH	92.1		CH_3CH_2—Br	70.5
			$(CH_3)_2CH$—Br	72.0

S. J. Blanksby and G. B. Ellison, *Acc. Chem. Res.*, **2003**, *36*, 255.

For example, because $\Delta H°$ for the addition of HBr to ethene is significantly negative (−22 kcal/mol), we can assume that $\Delta G°$ is also negative. However, if the value of $\Delta H°$ were close to zero, we could no longer assume that $\Delta H°$ and $\Delta G°$ have the same sign. (Keep in mind that two conditions must be met to justify using $\Delta H°$ values to predict $\Delta G°$ values. First, the entropy change must be small; second, the reaction must take place in the gas phase.)

When reactions occur in solution, which is the case for the vast majority of organic reactions, the solvent molecules can solvate the reactants and the products (Section 3.9). Solvation can have a large effect on both $\Delta H°$ and $\Delta S°$.

For example, in a reaction of a solvated polar reagent, the $\Delta H°$ for breaking the dipole–dipole interactions between the solvent and the reactant has to be taken into account, and in a reaction with a solvated polar product, the $\Delta H°$ for forming the dipole–dipole interactions between the solvent and the product has to be taken into account. In addition, solvation can greatly reduce the freedom of motion of the molecules, thereby affecting $\Delta S°$.

PROBLEM 23 ◆

a. Use the bond dissociation enthalpies in Table 5.1 to calculate the $\Delta H°$ value for the addition of HCl to ethene.
b. Calculate the $\Delta H°$ value for the addition of H_2 to ethene.
c. Are the reactions exothermic or endothermic?
d. Do you expect the reactions to be exergonic or endergonic?

5.9 USING $\Delta H°$ VALUES TO DETERMINE THE RELATIVE STABILITIES OF ALKENES

Catalytic Hydrogenation

Hydrogen (H_2) adds to the double bond of an alkene, in the presence of a metal catalyst, to form an alkane. The most common metal catalyst is palladium, which is used as a powder adsorbed on charcoal to maximize its surface area; it is referred to as "palladium on carbon" and is abbreviated as Pd/C. The metal catalyst is required to weaken the very strong H—H bond. (See Figure 1.1 on p. 22.)

$$CH_3CH=CHCH_3 + H_2 \xrightarrow{\text{Pd/C}} CH_3CH_2CH_2CH_3$$
alkene   alkane

A reduction reaction increases the number of C—H bonds.

- The addition of hydrogen to a compound is a reduction reaction. A **reduction reaction** increases the number of C—H bonds and/or decreases the number of C—O, C—N, or C—X bonds (X is a halogen).
- The addition of hydrogen is called **hydrogenation.** Because hydrogenation reactions require a catalyst, they are called **catalytic hydrogenations.**

The mechanism for catalytic hydrogenation is too complex to be easily described. We know that hydrogen is adsorbed on the surface of the metal and that the alkene interacts with the metal by overlapping its *p* orbitals with the vacant orbitals of the metal. All the bond-breaking and bond-forming events occur on the surface of the metal. As the alkane product forms, it diffuses away from the metal surface (Figure 5.3).

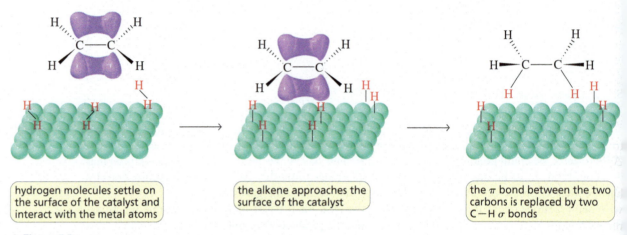

hydrogen molecules settle on the surface of the catalyst and interact with the metal atoms

the alkene approaches the surface of the catalyst

the π bond between the two carbons is replaced by two C—H σ bonds

▲ **Figure 5.3**
Catalytic hydrogenenation of an alkene to form an alkane.

We can think of catalytic hydrogenation as occurring in the following way: both the H—H bond of H_2 and the π bond of the alkene break, and then the resulting hydrogen radicals add to the resulting carbon radicals.

$$CH_3CH\!=\!CHCH_3 \longrightarrow CH_3\overset{.}{C}H\!-\!\overset{.}{C}HCH_3 \longrightarrow CH_3CH\!-\!CHCH_3$$
$$H\!-\!H \qquad\qquad H\!\cdot\ \ \cdot H \qquad\qquad H \ \ \ H$$

PROBLEM-SOLVING STRATEGY

LEARN THE STRATEGY

Choosing the Reactant for a Synthesis

What alkene would you start with if you wanted to synthesize methylcyclohexane?

You need to choose an alkene with the same number of carbons as the desired product and attached in the same way as those in the desired product. Several alkenes could be used for this synthesis because the double bond can be located anywhere in the molecule.

USE THE STRATEGY

PROBLEM 24

What alkene would you start with if you wanted to synthesize

a. pentane?   **b.** ethylcyclopentane?

PROBLEM 25

How many different alkenes can be hydrogenated to form

a. butane?   **b.** 3-methylpentane?   **c.** hexane?

Relative Stabilities of Alkenes

To determine the relative stabilities of alkenes, the three hydrogenation reactions shown here were carried out and their $\Delta H°$ values were experimentally determined.

	$\Delta H°$ (kcal/mol)	heat of hydrogenation

$$\underset{\underset{CH_3}{|}}{CH_3C}=CHCH_3 \; + \; H_2 \xrightarrow{\text{Pd/C}} \underset{\underset{CH_3}{|}}{CH_3CH}CH_2CH_3 \qquad -26.9 \qquad 26.9 \text{ kcal/mol}$$

$$\underset{\underset{CH_3}{|}}{CH_2}=CCH_2CH_3 \; + \; H_2 \xrightarrow{\text{Pd/C}} \underset{\underset{CH_3}{|}}{CH_3CH}CH_2CH_3 \qquad -28.5 \qquad 28.5 \text{ kcal/mol}$$

$$\underset{\underset{CH_3}{|}}{CH_3CH}CH=CH_2 \; + \; H_2 \xrightarrow{\text{Pd/C}} \underset{\underset{CH_3}{|}}{CH_3CH}CH_2CH_3 \qquad -30.3 \qquad 30.3 \text{ kcal/mol}$$

> the product of each of the three reactions is 2-methylbutane

The heat released in a hydrogenation reaction is called the **heat of hydrogenation.** It is customary to give it a positive value. Hydrogenation reactions, however, are exothermic (they have negative $\Delta H°$ values), so *the heat of hydrogenation is the value of $\Delta H°$ without the negative sign.*

The $\Delta H°$ values tell us the relative energies of the reactants and products for the three catalytic hydrogenation reactions. However, because we do not know the precise mechanism of the reaction, we cannot draw reaction coordinate diagrams for the reactions. So we will connect the energy of the reactants and products with dotted lines to indicate the uncertainty of the energy change that occurs between the reactants and products (Figure 5.4).

NOTE TO THE STUDENT

• Notice that when a reaction coordinate diagram shows $\Delta H°$ values, the *y*-axis is potential energy (Figure 5.4); when it shows $\Delta G°$ values, the *y*-axis is free energy (Figure 5.2).

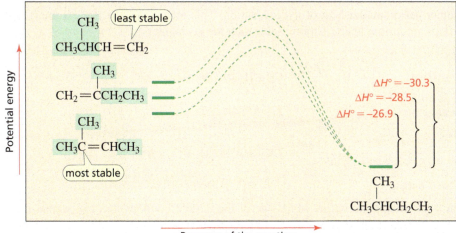

◀ **Figure 5.4**
The relative energies (stabilities) of three alkenes that can be catalytically hydrogenated to 2-methylbutane. The most stable alkene has the smallest heat of hydrogenation.

The three reactions all form the same alkane product, so the energy of the *product* in Figure 5.4 is the same for each reaction. The three reactions, however, have different heats of hydrogenation, so the three *reactants* must have different energies. For example, the alkene that releases the most heat must have the most energy to begin with (it must be the *least* stable of the three alkenes). In contrast, the alkene that releases the least heat must have the least energy to begin with (it must be the *most* stable of the three alkenes). Notice that *the most stable alkene has the smallest heat of hydrogenation.*

The most stable alkene has the smallest heat of hydrogenation.

If you look at the structures of the three alkene reactants in Figure 5.4, you will see that the stability of an alkene increases as the number of alkyl groups bonded to the sp^2 carbons increases.

For example, the most stable alkene in Figure 5.4 has two alkyl groups bonded to one sp^2 carbon and one alkyl group bonded to the other sp^2 carbon, for a total of three alkyl groups (three methyl groups) bonded to its two sp^2 carbons. The alkene of intermediate stability has two alkyl groups (a methyl group and an ethyl group) bonded to its sp^2 carbons, and the least stable of the three alkenes has only one alkyl group (an isopropyl group) bonded to an sp^2 carbon.

We can, therefore, make the following statement:

> *the stability of an alkene increases*
> *as the number of alkyl substituents bonded to its sp² carbons increases.*

the more alkyl groups bonded to the *sp²* carbons, the more stable the alkene

Alkyl groups stabilize alkenes.

most stable $\underset{R}{\overset{R}{>}}C=C\underset{R}{\overset{R}{}}$ > $\underset{R}{\overset{R}{}}C=C\underset{H}{\overset{R}{}}$ > $\underset{R}{\overset{R}{}}C=C\underset{H}{\overset{H}{}}$ > $\underset{H}{\overset{R}{}}C=C\underset{H}{\overset{H}{}}$ least stable

(Some students find it easier to understand this concept from the point of view of the hydrogen bonded to the *sp²* carbons—namely, *the stability of an alkene increases as the number of hydrogen bonded to its sp² carbons decreases.*)

PROBLEM 26 ◆

The same alkane is obtained from the catalytic hydrogenation of both alkene **A** and alkene **B**. The heat of hydrogenation of alkene **A** is 29.8 kcal/mol, and the heat of hydrogenation of alkene **B** is 31.4 kcal/mol. Which alkene is more stable?

PROBLEM 27 ◆

a. Which of the following compounds is the most stable?
b. Which is the least stable?
c. Which has the smallest heat of hydrogenation?

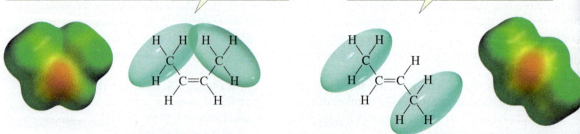

Both *trans*-2-butene and *cis*-2-butene have two alkyl groups bonded to their *sp²* carbons, but the trans isomer has a smaller heat of hydrogenation. This means that the trans isomer, in which the large groups are farther apart, is more stable than the cis isomer.

		Δ*H*° (kcal/mol)	heat of hydrogenation
$\underset{H}{\overset{H_3C}{}}C=C\underset{CH_3}{\overset{H}{}}$ + H₂ $\xrightarrow{\text{Pd/C}}$ CH₃CH₂CH₂CH₃		−27.6	27.6
trans-2-butene			
$\underset{H}{\overset{H_3C}{}}C=C\underset{H}{\overset{CH_3}{}}$ + H₂ $\xrightarrow{\text{Pd/C}}$ CH₃CH₂CH₂CH₃		−28.6	28.6
cis-2-butene			

When large groups are on the same side of the double bond, as in a cis isomer, their electron clouds can interfere with each other, causing **steric strain** in the molecule. Steric strain makes a compound less stable (Section 3.11). When the large groups are on opposite sides of the double bond, as in a trans isomer, their electron clouds cannot interact, so there is no destabilizing steric strain.

the cis isomer has steric strain due to electron cloud overlap

the trans isomer does not have steric strain

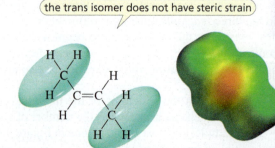

The heat of hydrogenation of *cis*-2-butene, in which the two alkyl groups are on the *same side* of the double bond, is similar to that of 2-methylpropene, in which the two alkyl groups are on the *same carbon*. The three alkenes with two alkyl groups are all *less* stable than an alkene with three alkyl groups, and they are all *more* stable than an alkene with one alkyl group.

the relative stabilities of alkenes that have two alkyl groups bonded to the *sp^2* carbons

alkyl groups are
on *opposite sides* of
the double bond

alkyl groups are
on the *same side* of
the double bond

alkyl groups
are on the same *sp^2* carbon

PROBLEM 28 ◆

Rank the following compounds from most stable to least stable:
trans-3-hexene, *cis*-3-hexene, *cis*-2,5-dimethyl-3-hexene, (*Z*)-3,4-dimethyl-3-hexene

---

Trans Fats

Oils are liquids at room temperature because the hydrocarbon chains of their fatty acid components contain several carbon–carbon double bonds, which makes it difficult for them to pack closely together. In contrast, the hydrocarbon chains of the fatty acid components of fats have fewer double bonds, so they can pack together more closely (Section 25.1). Because of their many double bonds, oils are said to be polyunsaturated.

COOH

linoleic acid
18-carbon fatty acid with two cis double bonds

Some or all of the double bonds in oils can be reduced by catalytic hydrogenation. For example, margarine and shortening are prepared by hydrogenating vegetable oils, such as soybean oil and safflower oil, until they have the desired creamy, solid consistency of butter.

All the double bonds in naturally-occurring fats and oils have the cis configuration. During catalytic hydrogenation cis–trans isomerization can occur, forming what is known as a trans fat.

COOH

oleic acid
18-carbon fatty acid with one cis double bond
before being heated

COOH

elaidic acid
18-carbon fatty acid with one trans double bond
after being heated

Trans fats are a health concern because they increase LDL, the so-called "bad" cholesterol (Section 3.16). Epidemiological studies have shown that an increase in the daily intake of trans fats significantly increases the incidence of cardiovascular disease.

5.10 KINETICS: HOW FAST IS THE PRODUCT FORMED?

Knowing that a reaction is exergonic does not tell you how fast the reaction occurs, because $\Delta G°$ describes only the difference between the stability of the reactants and the stability of the products. It does not indicate anything about the energy barrier of the reaction, which is the energy "hill" that has to be climbed for the reactants to be converted into products. **Kinetics** is the field of chemistry that studies the rates of chemical reactions and the factors that affect those rates.

The energy barrier of a reaction (indicated in Figure 5.5 by $\Delta G^{\ddagger}$) is called the **free energy of activation.** It is the difference between the free energy of the transition state and the free energy of the reactants:

$$\Delta G^{\ddagger} = \text{free energy of the transition state} - \text{free energy of the reactants}$$

As $\Delta G^{\ddagger}$ decreases, the rate of the reaction increases. Thus, *anything that makes the reactants less stable or makes the transition state more stable will make the reaction go faster.*

The rate of the reaction *increases* as the height of the energy barrier *decreases*.

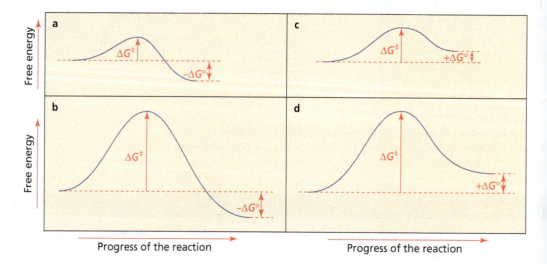

▶ **Figure 5.5**
Reaction coordinate diagrams (drawn on the same scale) for (a) a fast exergonic reaction (b) a slow exergonic reaction (c) a fast endergonic reaction (d) a slow endergonic reaction

Like ΔG°, $\Delta G^{\ddagger}$ has both an enthalpy component and an entropy component. Notice that any quantity that refers to the transition state is represented by a double-dagger superscript (‡).

$$\Delta G^{\ddagger} = \Delta H^{\ddagger} - T\Delta S^{\ddagger}$$

$\Delta H^{\ddagger}$ = **enthalpy of the transition state − enthalpy of the reactants**

$\Delta S^{\ddagger}$ = **entropy of the transition state − entropy of the reactants**

Some exergonic reactions have small free energies of activation and, therefore, can take place at room temperature (Figure 5.5a). In contrast, some exergonic reactions have free energies of activation that are so large that the reaction cannot take place unless energy is supplied in addition to that provided by the existing thermal conditions (Figure 5.5b). Endergonic reactions can also have either small free energies of activation, as in Figure 5.5c, or large free energies of activation, as in Figure 5.5d.

Let's now look at the difference between thermodynamic stability and kinetic stability.

Thermodynamic stability is indicated by ΔG°. If ΔG° is negative, then the product is *thermodynamically stable* compared with the reactant; if ΔG° is positive, then the product is *thermodynamically unstable* compared with the reactant.

Kinetic stability is indicated by $\Delta G^{\ddagger}$. If $\Delta G^{\ddagger}$ is large, then the reactant is *kinetically stable* because it reacts slowly. If $\Delta G^{\ddagger}$ is small, then the reactant is *kinetically unstable*—it reacts rapidly. Similarly, if $\Delta G^{\ddagger}$ for the reverse reaction is large, then the product is kinetically stable, but if it is small, then the product is kinetically unstable.

Generally, when chemists use the term *stability*, they are referring to thermodynamic stability.

PROBLEM 29 ◆

a. Which of the reactions in Figure 5.5 has a product that is thermodynamically stable compared with the reactant?

b. Which of the reactions in Figure 5.5 has the most kinetically stable product?

c. Which of the reactions in Figure 5.5 has the least kinetically stable product?

PROBLEM 30

Draw a reaction coordinate diagram for a reaction in which

a. the product is thermodynamically unstable and kinetically unstable.

b. the product is thermodynamically unstable and kinetically stable.

5.11 THE RATE OF A CHEMICAL REACTION

The rate of a chemical reaction is the speed at which the reactants are converted to products. The rate of a reaction depends on the following factors:

1. **The number of collisions that take place between the reacting molecules in a given period of time.** The rate of the reaction increases as the number of collisions increases.

2. **The fraction of collisions that occur with sufficient energy to get the reacting molecules over the energy barrier.** If the free energy of activation is small, then more collisions will lead to reaction than if the free energy of activation is large.

3. **The fraction of collisions that occur with the proper orientation.** 2-Butene and HBr will react only if the molecules collide with the hydrogen of HBr approaching the π bond of 2-butene. If a collision occurs with the hydrogen approaching a methyl group of 2-butene, no reaction will take place, regardless of the energy of the collision.

$$\text{rate of a reaction} = \left(\begin{array}{c}\text{number of collisions} \\ \text{per unit of time}\end{array}\right) \times \left(\begin{array}{c}\text{fraction with} \\ \text{sufficient energy}\end{array}\right) \times \left(\begin{array}{c}\text{fraction with} \\ \text{proper orientation}\end{array}\right)$$

- Increasing the concentration of the reactants increases the rate of a reaction because it increases the number of collisions that occur in a given period of time.

- Increasing the temperature at which the reaction is carried out also increases the rate of a reaction because it increases the kinetic energy of the molecules, which increases both the frequency of collisions (in a confined space, molecules that are moving faster collide more frequently) and the number of collisions that have sufficient energy to get the reacting molecules over the energy barrier (Figure 5.6).

- The rate of a reaction can also be increased by a catalyst (Section 5.13).

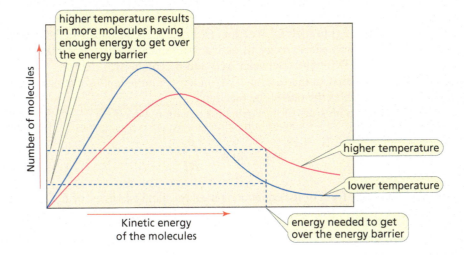

higher temperature results in more molecules having enough energy to get over the energy barrier

Number of molecules

Kinetic energy of the molecules

higher temperature

lower temperature

energy needed to get over the energy barrier

◀ **Figure 5.6**
Boltzmann distribution curves *(the number of molecules as a function of energy)* **at two different temperatures. The curve shows the distribution of molecules with a particular kinetic energy. The energy of most molecules cluster about an average, but there are some with much lower and some with much higher energy. At a higher temperature (the red line), there will be more molecules with sufficient energy to get over the energy barrier.**

The Difference Between the Rate of a Reaction and the Rate Constant for a Reaction

For a reaction in which reactant A is converted into product B, *the rate of the reaction* is proportional to the concentration of A. If the concentration of A is doubled, the rate of the reaction will double; if the concentration of A is tripled, the rate of the reaction will triple; and so on.

$$A \longrightarrow B$$

When you know the relationship between the rate of a reaction and the concentration of the reactants, you can write a **rate law** for the reaction.

$$\text{rate} \propto [A]$$

We can replace the proportionality symbol ($\propto$) with an equal sign if we use a proportionality constant k, which is called a **rate constant.**

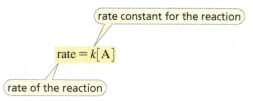

$$\text{rate} = k[\text{A}]$$

The rate of the reaction decreases as the rate constant decreases.

Do not confuse the *rate constant* for a reaction (k) with the *rate* of a reaction.

The *rate constant* tells us how easy it is to reach the transition state (how easy it is to get over the energy barrier).

- Low-energy barriers are associated with large rate constants (Figure 5.5a and c).
- High-energy barriers are associated with small rate constants (Figure 5.5b and d).

The *rate* of a reaction is a measure of the amount of product that is formed per unit of time. The preceding equations show that the *rate* is the product of the *rate constant and the reactant concentration, so the rate of the reaction depends on concentration, whereas the rate constant is independent of concentration.* Therefore, when we compare two reactions to see which one occurs more readily, we must compare their rate constants and not their concentration-dependent rates of reaction. (How rate constants are determined is explained in Appendix II.)

The Arrhenius Equation

Although rate constants are independent of concentration, they are dependent on temperature. The **Arrhenius equation** relates the rate constant of a reaction to the experimental energy of activation and to the temperature at which the reaction is carried out:

$$k = A e^{-E_a/RT}$$

In this expression, k is the rate constant, A is the frequency factor (which represents the fraction of collisions that occurs with the proper orientation for reaction), and $e^{-E_a/RT}$ is the fraction of collisions with the minimum energy (E_a) needed for reaction. (R is the gas constant, T is the temperature in kelvins, and E_a is the experimental energy of activation, which is an approximate value of the activation energy; see "The Difference Between $\Delta G^{\ddagger}$ and E_a," on p. 215.

A good rule of thumb is that an increase of 10 °C in temperature doubles the rate constant for a reaction and, therefore, doubles the rate of the reaction.

Taking the logarithm of both sides of the Arrhenius equation, we obtain

$$\ln k = \ln A - \frac{E_a}{RT}$$

Problem 65 shows how this form of the equation is used to calculate values of E_a, $\Delta G^{\ddagger}$, $\Delta H^{\ddagger}$, and $\Delta S^{\ddagger}$ for a reaction.

Ludwig Eduard Boltzmann (1844–1906) was born in Vienna, Austria. He received a Ph.D. from the University of Vienna in 1866; his dissertation was on the kinetic theory of gases. Boltzmann started his career as a professor of mathematical physics at the University of Graz. In 1872, he met Henriette van Aigentier, an aspiring teacher of mathematics and physics, who had not been allowed to audit lectures at the university because they were not open to women. Boltzmann encouraged her to appeal, which she did with success. Boltzmann returned to the University of Vienna in 1873 as a professor of mathematics. He and Henriette were married in 1876 and had five children. He went back to the University of Graz in 1887 to become its president. Boltzmann was subject to alternating moods of high elation and severe depression, which he attributed to having been born during the night between Mardi Gras and Ash Wednesday.

PROBLEM 31 SOLVED

The rate of the reaction of methyl chloride with hydroxide ion is linearly dependent on both the concentration of methyl chloride and the concentration hydroxide ion. At 30 °C, the constant (k) for the reaction is $1.0 \times 10^{-5}\,\text{M}^{-1}\text{s}^{-1}$.

a. What is the rate of the reaction when $[\text{CH}_3\text{Cl}] = 0.10$ M and $[\text{HO}^-] = 0.10$ M?
b. If the concentration of methyl chloride is decreased to 0.010 M, what will be the effect on
    1. the *rate* of the reaction?   **2.** the *rate constant* for the reaction?

SOLUTION TO 31 a.

The rate of the reaction is given by

$$\text{rate} = k[\text{methyl chloride}][\text{HO}^-]$$

Substituting the given rate constant and reactant concentrations yields

$$\text{rate} = 1.0 \times 10^{-5}\,\text{M}^{-1}\text{s}^{-1}[0.10\,\text{M}][0.10\,\text{M}] = 1.0 \times 10^{-7}\,\text{M}\,\text{s}^{-1}$$

The Difference between $\Delta G^{\ddagger}$ and E_a

The difference between the **free energy of activation** ($\Delta G^{\ddagger}$) and the **experimental energy of activation** (E_a) in the Arrhenius equation is the entropy component. The free energy of activation has both an enthalpy component and an entropy component ($\Delta G^{\ddagger} = \Delta H^{\ddagger} - T\Delta S^{\ddagger}$), whereas the experimental energy of activation has only an enthalpy component ($E_a = \Delta H^{\ddagger} + RT$), because the entropy component is implicit in the A term of the Arrhenius equation. Therefore, E_a gives only an approximate energy barrier to a reaction. The true energy barrier is given by $\Delta G^{\ddagger}$, because the energy barriers to most reactions depend on changes in both enthalpy and entropy.

PROBLEM 32 ◆

The rate constant for a reaction can be increased by _____ the stability of the reactant or by _____ the stability of the transition state.

PROBLEM 33 ◆

From the Arrhenius equation, predict how

a. increasing the experimental activation energy affects the rate constant for a reaction.
b. increasing the temperature affects the rate constant for a reaction.

How Are Rate Constants Related to the Equilibrium Constant?

The next question to consider is, how are the rate constants for a reaction related to the equilibrium constant? At equilibrium, the rate of the forward reaction must be equal to the rate of the reverse reaction because the amounts of reactants and products are no longer changing:

$$A \underset{k_{-1}}{\overset{k_1}{\rightleftharpoons}} B$$

forward rate = reverse rate

$$k_1[\mathbf{A}] = k_{-1}[\mathbf{B}]$$

Therefore,

$$K_{eq} = \frac{k_1}{k_{-1}} = \frac{[\mathbf{B}]}{[\mathbf{A}]}$$

From this equation, we see that the equilibrium constant for a reaction can be determined from the relative rate constants for the forward and reverse reactions *or* from the relative concentrations of the products and reactants at equilibrium. For example, we can say that the reaction shown in Figure 5.2a has an equilibrium constant significantly greater than one because the rate constant for the forward reaction is greater than that for the reverse reaction *or* because the products are more stable than the reactants.

PROBLEM 34 ◆

a. Which reaction has a greater equilibrium constant: one with a rate constant of $1 \times 10^{-3}\,\mathrm{sec}^{-1}$ for the forward reaction and a rate constant of $1 \times 10^{-5}\,\mathrm{sec}^{-1}$ for the reverse reaction, or one with a rate constant of $1 \times 10^{-2}\,\mathrm{sec}^{-1}$ for the forward reaction and a rate constant of $1 \times 10^{-3}\,\mathrm{sec}^{-1}$ for the reverse reaction?

b. If both reactions start with a reactant concentration of 1.0 M, which reaction will form the most product when the reactions have reached equilibrium?

NOTE TO THE STUDENT

• Additional information about kinetics and a set of problems on kinetics (and their solutions) can be found in Appendix II.

5.12 A REACTION COORDINATE DIAGRAM DESCRIBES THE ENERGY CHANGES THAT TAKE PLACE DURING A REACTION

We saw that the addition of HBr to 2-butene is a two-step process (Section 5.5). In each step, the reactants pass through a transition state as they are converted into products. The structure of the transition state for each of the steps is shown on the top of the next page in brackets.

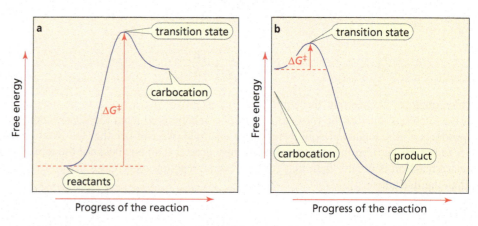

$$CH_3CH=CHCH_3 + HBr \longrightarrow \left[\overset{\delta+}{CH_3CH} \cdots CHCH_3 \atop \underset{\delta-}{\overset{|}{H}} \atop \overset{|}{Br} \right]^{\ddagger} \longrightarrow CH_3\overset{+}{C}HCH_2CH_3 + Br^-$$

‡ symbolizes the transition state

partially formed bond

partially broken bond

transition state

$$CH_3\overset{+}{C}HCH_2CH_3 + Br^- \longrightarrow \left[\overset{\delta+}{CH_3CH}CH_2CH_3 \atop \underset{\delta-}{\overset{|}{Br}} \right]^{\ddagger} \longrightarrow CH_3CHCH_2CH_3 \atop \overset{|}{Br}$$

transition state

The bonds that break and those that form during the course of the reaction are partially broken and partially formed in the transition state, as indicated by dashed lines. Atoms that either become charged or lose their charge during the course of the reaction are partially charged in the transition state. (Transition states are always shown in brackets with a double-dagger superscript.)

Reaction Coordinate Diagram

A reaction coordinate diagram can be drawn for each step of a reaction (Figure 5.7). In the first step of the electrophilic addition reaction, the alkene is converted into a carbocation that is higher in energy (less stable) than the reactants. The first step, therefore, is endergonic ($\Delta G°$ is > 0). In the second step, the carbocation reacts with a nucleophile to form a product that is lower in energy (more stable) than the carbocation reactant. This step, therefore, is exergonic ($\Delta G°$ is < 0).

▶ **Figure 5.7**

Reaction coordinate diagrams for the two steps in the addition of HBr to 2-butene:

(a) the first step (formation of the carbocation)

(b) the second step (formation of the alkyl halide)

Because the products of the first step are the reactants for the second step, we can hook the two reaction coordinate diagrams together to obtain the reaction coordinate diagram that describes the pathway for the overall reaction (Figure 5.8). The $\Delta G°$ for the overall reaction is the difference between the free energy of the final products and the free energy of the initial reactants. Figure 5.8 shows that the overall reaction is exergonic ($\Delta G°$ is negative).

Transition States and Intermediates

A chemical species that is a product of one step of a reaction and a reactant for the next step is called an **intermediate**. The carbocation intermediate formed in this reaction is too unstable to be isolated, but some reactions have more stable intermediates that can be isolated.

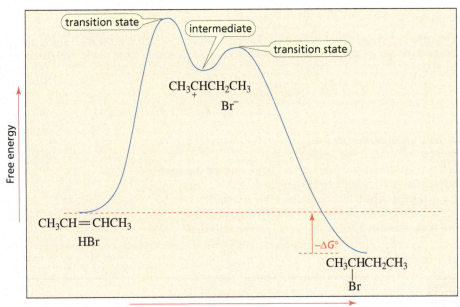

◀ **Figure 5.8**
Reaction coordinate diagram for the addition of HBr to 2-butene to form 2-bromobutane.

Transition states, in contrast, represent the highest-energy structures that are involved in the reaction. They exist only fleetingly, and they can never be isolated (Figure 5.8). Do not confuse transition states with intermediates:

- *Transition states have partially formed bonds.*
- *Intermediates have fully formed bonds.*

Figure 5.8 shows that the free energy of activation for the first step of the reaction is greater than the free energy of activation for the second step. In other words, the rate constant for the first step is smaller than the rate constant for the second step. This is what we would expect, because covalent bonds must be broken in the first step, whereas no bonds are broken in the second step.

Transition states have partially formed bonds. Intermediates have fully formed bonds.

The Rate-Determining Step

If a reaction has two or more steps, the step that has its transition state *at the highest point on the reaction coordinate* is the **rate-determining step** or **rate-limiting step.** The rate-determining step controls the overall rate of the reaction, because the overall rate cannot exceed the rate of the rate-determining step. The rate-determining step for the reaction of 2-butene with HBr is the first step—the addition of the proton (the electrophile) to the alkene (the nucleophile) to form the carbocation.

PROBLEM 35

Draw a reaction coordinate diagram for a two-step reaction in which the first step is endergonic, the second step is exergonic, and the overall reaction is endergonic. Label the reactants, products, intermediates, and transition states.

PROBLEM 36 ◆

a. Which step in the reaction coordinate diagram shown here has the greatest free energy of activation in the forward direction?
b. Is the first-formed intermediate more apt to revert to reactants or go on to form products?
c. Which step is the rate-determining step of the reaction?

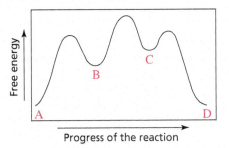

PROBLEM 37 ♦

Draw a reaction coordinate diagram for the following reaction in which C is the most stable and B the least stable of the three species and the transition state going from A to B is more stable than the transition state going from B to C:

$$A \; \underset{k_{-1}}{\overset{k_1}{\rightleftharpoons}} \; B \; \underset{k_{-2}}{\overset{k_2}{\rightleftharpoons}} \; C$$

a. How many intermediates are there?
b. How many transition states are there?
c. Which step has the greater rate constant in the forward direction?
d. Which step has the greater rate constant in the reverse direction?
e. Of the four steps, which has the greatest rate constant?
f. Which is the rate-determining step in the forward direction?
g. Which is the rate-determining step in the reverse direction?

5.13 CATALYSIS

A **catalyst** increases the rate of a reaction by giving the reactants a new pathway to follow—one with a smaller $\Delta G^{\ddagger}$. In other words, a catalyst decreases the energy barrier that must be overcome in the process of converting the reactants to products (Figure 5.9).

▲ **Figure 5.9**
A catalyst provides a pathway with a lower energy barrier, but it does not change the energy of the starting point (the reactants) or the energy of the end point (the products).

If a catalyst is going to make a reaction go faster, it must participate in the reaction, but it is not consumed or changed during the reaction. Because the catalyst is not used up or changed, only a small amount of it is needed to catalyze the reaction (typically, 1 to 10% of the number of moles of reactant). Notice in Figure 5.9 that the stability of the reactants and products is the same in both the catalyzed and uncatalyzed reactions. In other words, a catalyst does not change the relative concentrations of products and reactants when the system reaches equilibrium. Therefore, it does not change the *amount* of product formed; it changes only the *rate* at which it is formed.

A catalyst gives the reagents a new pathway with a lower "energy hill."

A catalyst does not change the *amount* of product formed; it changes only the *rate* at which it is formed.

The most common catalysts are acids, bases, and nucleophiles. Acids catalyze a reaction by giving a proton to a reactant; bases catalyze a reaction by removing a proton from a reactant, and nucleophiles catalyze reactions by forming a new covalent bond with the reactant. We will see many examples of catalyzed reactions in the following chapters. Catalysis and the kinds of species that can be used to catalyze a reaction are discussed in detail in Chapter 22.

PROBLEM 38 ♦

Which of the following parameters would be different for a reaction carried out in the presence of a catalyst compared with the same reaction carried out in the absence of a catalyst?

$$\Delta G^{\circ}, \; \Delta H^{\ddagger}, \; E_a, \; \Delta S^{\ddagger}, \; \Delta H^{\circ}, \; K_{eq}, \; \Delta G^{\ddagger}, \; \Delta S^{\circ}, \; k$$

5.14 CATALYSIS BY ENZYMES

Essentially all reactions that occur in biological systems are reactions of organic compounds. These reactions almost always require a catalyst. Most biological catalysts are proteins called **enzymes.** Enzymes are chains of amino acids linked together by covalent bonds. Each biological reaction is catalyzed by a different enzyme.

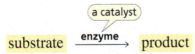

The reactant of an enzyme-catalyzed reaction is called a **substrate.** The enzyme binds the substrate in a pocket of the enzyme called the **active site.** All the bond-making and bond-breaking steps of the reaction occur while the substrate is bound to the active site.

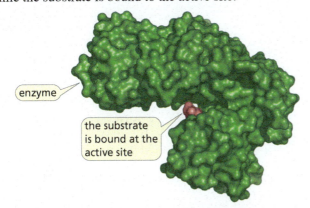

Unlike nonbiological catalysts, enzymes are specific for the substrate whose reaction they catalyze (Section 6.14). All enzymes, however, do not have the same degree of specificity. Some are specific for a single compound and will not tolerate even the slightest variation in structure, whereas some catalyze the reaction of a family of compounds with related structures. The specificity of an enzyme for its substrate is an example of **molecular recognition**—the ability of one molecule to recognize another molecule as a result of intermolecular interactions (see the introduction to Chapter 20).

The specificity of an enzyme for its substrate results from the particular amino acids that reside at the active site (Section 21.1). The amino acids bind the substrate to the active site using hydrogen bonds, London dispersion forces, and dipole–dipole interactions—the same intermolecular interactions that hold molecules together (Section 3.9). A more in-depth discussion of the interaction between the enzyme and the substrate can be found in Section 22.8.

Cell walls consist of thousands of six-membered ring molecules linked by oxygen atoms. Lysozyme is an enzyme that cleaves bacterial cell walls by breaking the bond that holds the six-membered rings together. Figure 5.10 shows a portion of lysozyme's active site and some of the amino acids that bind the substrate (the cell wall) in a precise location at the active site.

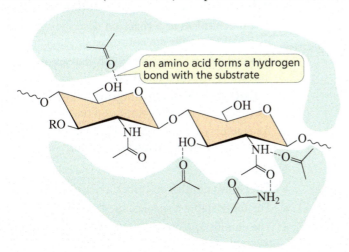

◄ **Figure 5.10**
The amino acids at the active site of the enzyme hold the substrate in the precise position necessary for reaction.

In addition to the amino acids that bind the substrate to the active site, there are also amino acids at the active site that are responsible for catalyzing the reaction. These amino acids can be acids, bases, or nucleophiles—the same kinds of species that catalyze nonbiological reactions (Section 5.13). For example, lysozyme has two catalytic groups at its active site—an acid catalyst and a nucleophilic catalyst (Figure 5.11). How these groups catalyze cleavage of the cell wall will be explained in Section 22.11, after you know more about the kind of reaction that is involved.

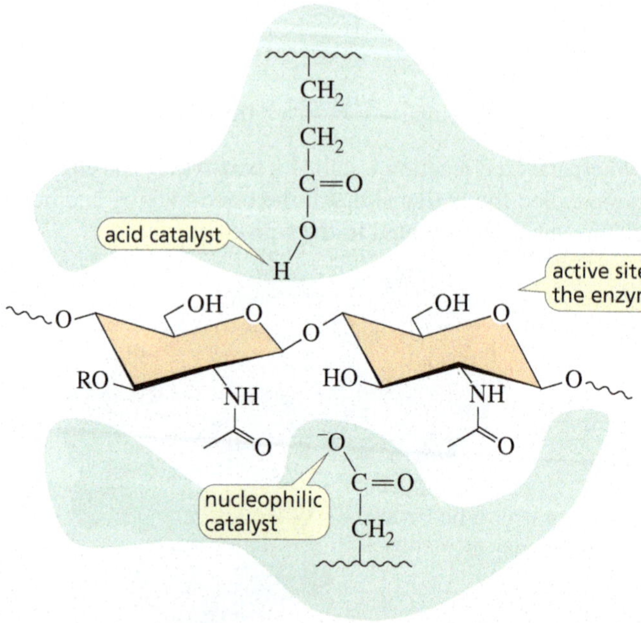

▶ **Figure 5.11**
Two amino acids at the active site of lysozyme are catalysts for the reaction that breaks the bond holding the six-membered rings together.

ESSENTIAL CONCEPTS

Section 5.1

- An **alkene** is a hydrocarbon that contains a double bond. Because alkenes contain fewer than the maximum number of hydrogens, they are called **unsaturated hydrocarbons.**
- The general molecular formula for a hydrocarbon is C_nH_{2n+2} minus two hydrogens for every π bond or ring in the molecule.
- The number of π bonds plus the number of rings is called the **degree of unsaturation.**

Section 5.2

- The **functional group suffix** of an alkene is "ene."
- When there are both a functional group and a substituent, the functional group gets the lowest possible number.

Section 5.3

- The two sp^2 carbons of an alkene and the two atoms bonded to each of them all lie in the same plane.

Section 5.4

- The double bond is the **functional group** (the center of reactivity) of an alkene.
- All compounds with a particular **functional group** react in the same way.

Section 5.5

- **Nucleophiles** are electron-rich species; **electrophiles** are electron-deficient species.
- Nucleophiles react with electrophiles.
- Due to the cloud of electrons above and below its π bond, an alkene is a nucleophile.
- Alkenes undergo **electrophilic addition reactions.**
- The **mechanism of a reaction** describes the step-by-step process by which reactants are converted to products.
- **Curved arrows** show the bonds that are formed and the bonds that are broken in a reaction.
- A **reaction coordinate diagram** shows the energy changes that take place during the course of a reaction.
- As reactants are converted to products, a reaction passes through a maximum-energy **transition state.**

Section 5.6

- **Thermodynamics** describes a reaction at equilibrium.
- The more stable a species, the lower its energy.
- The equilibrium constant, K_{eq}, gives the relative concentrations of reactants and products at equilibrium.

- The more stable the product relative to the reactant, the greater its concentration at equilibrium and the greater the K_{eq}.
- If the products are more stable than the reactants, then K_{eq} is > 1, $\Delta G°$ is negative, and the reaction is **exergonic.**
- If the reactants are more stable than the products, then K_{eq} is < 1, $\Delta G°$ is positive, and the reaction is **endergonic.**
- $\Delta G°$ and K_{eq} are related by the formula $\Delta G° = -RT \ln K_{eq}$.
- $\Delta G°$ is the **Gibbs free-energy change,** with $\Delta G° = \Delta H° - T\Delta S°$.
- $\Delta H°$ is the change in **enthalpy,** which is the heat given off or consumed as a result of bond making and bond breaking.
- An **exothermic reaction** has a negative $\Delta H°$; an **endothermic reaction** has a positive $\Delta H°$.
- $\Delta S°$ is the change in **entropy,** which is the change in the freedom of motion in a system.
- The formation of products with stronger bonds and greater freedom of motion causes $\Delta G°$ to be negative.

Section 5.7
- **Le Châtelier's principle** states that if an equilibrium is disturbed, the system will adjust to offset the disturbance.
- Two sequential reactions (an endergonic reaction followed by an exergonic reaction) are called **coupled reactions.**

Section 5.8
- $\Delta H°$ values can be calculated; $\Delta S°$ values must be experimentally determined.

Section 5.9
- **Catalytic hydrogenation** reduces alkenes to alkanes.
- The **heat of hydrogenation** is the heat released in a hydrogenation reaction. It is the $\Delta H°$ value without the negative sign.
- The **most stable alkene** has the **smallest heat of hydrogenation.**
- The stability of an alkene increases as the number of alkyl groups bonded to its sp^2 carbons increases.
- **Trans alkenes** are more stable than **cis alkenes** because of steric strain.

Section 5.10
- The free energy of activation, $\Delta G^{\ddagger}$, is the energy barrier of a reaction. It is the difference between the free energy of the transition state and the free energy of the reactants.
- The rate of the reaction increases as $\Delta G^{\ddagger}$ decreases.
- Anything that makes the reactant less stable or makes the transition state more stable increases the rate constant for the reaction (that is, decreases $\Delta G^{\ddagger}$).
- **Kinetic stability** is given by $\Delta G^{\ddagger}$; **thermodynamic stability** is given by $\Delta G°$.

Section 5.11
- The **rate** of a reaction depends on the concentration of the reactants, the temperature, and the rate constant.
- The **rate constant** for a reaction indicates how easy it is for the reactants to reach the transition state.
- The equilibrium constant can be determined from the relative rate constants for the forward and reverse reactions.

Section 5.12
- An **intermediate** is a product of one step of a reaction and a reactant of the next step.
- Transition states have partially formed bonds; intermediates have fully formed bonds.
- The **rate-determining step** has its transition state at the highest point on the reaction coordinate.

Section 5.13
- A **catalyst** decreases the energy barrier that must be overcome in the process of converting the reactants to products.
- A **catalyst** is neither consumed nor changed during the reaction.
- A catalyst does not change the *amount* of product formed; it changes only the *rate* at which the product is formed.

Section 5.14
- Most biological catalysts are proteins called **enzymes.**
- **Molecular recognition** is the ability of one molecule to recognize another molecule.

PROBLEMS

39. What is each compound's systematic name?

a. $CH_3CH_2CHCH=CHCH_2CH_2CHCH_3$
   with Br, Br substituents

c.

e.

b.

d.

f.

40. Which is more stable?

a. CH₃C=CHCH₂CH₃ or CH₃CH=CHCHCH₃ (with CH₃ substituents shown)

$$\text{a.} \quad \underset{\overset{|}{CH_3}}{CH_3C}=CHCH_2CH_3 \quad \text{or} \quad CH_3CH=CH\underset{\overset{|}{CH_3}}{CHCH_3}$$

b. 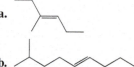 or

c. or

41. Draw the structure of a hydrocarbon that has six carbon atoms and
   a. three vinylic hydrogens and two allylic hydrogens.
   b. three vinylic hydrogens and one allylic hydrogen.
   c. three vinylic hydrogens and no allylic hydrogens.

42. Draw the condensed structure for each of the following:
   a. (Z)-1,3,5-tribromo-2-pentene
   b. (Z)-3-methyl-2-heptene
   c. (E)-1,2-dibromo-3-isopropyl-2-hexene
   d. vinyl bromide
   e. 1,2-dimethylcyclopentene
   f. diallylamine

43. Draw the skeletal structures for the compounds in Problem 42.

44. a. Draw the condensed structures and give the systematic names for all the alkenes with molecular formula C_6H_{12}, ignoring stereoisomers. (*Hint:* There are 13.)
   b. Which of the alkenes have E and Z isomers?
   c. Which of the alkenes is the most stable?
   d. Which of the alkenes is the least stable?

45. Name the following:

a.

c.

e.

b.

d.

f.

46. Of the compounds you named in Problem 45:
   a. Which is the most stable?
   b. Which is the least stable?

47. a. Which is the most stable: 3,4-dimethyl-2-hexene, 2,3-dimethyl-2-hexene, or 4,5-dimethyl-2-hexene?
   b. Which compound has the largest heat of hydrogenation?
   c. Which compound has the smallest heat of hydrogenation?

48. Draw curved arrows to show the flow of electrons responsible for the conversion of the reactants to products:

$$H-\ddot{\overset{..}{O}}{:}^- + H-\underset{\overset{|}{H}}{\overset{\overset{H}{|}}{C}}-\underset{\overset{|}{Br}}{\overset{\overset{H}{|}}{C}}-H \longrightarrow H_2O + \underset{\overset{|}{H}}{\overset{\overset{H}{|}}{C}}=\underset{\overset{|}{H}}{\overset{\overset{H}{|}}{C}} + Br^-$$

49. Draw the skeletal structure of 3,3-dimethyl-7-(1-methylethyl)-6-(1-methylpropyl)decane.

50. In a reaction in which reactant **A** is in equilibrium with product **B** at 25 °C, what relative amounts of **A** and **B** are present at equilibrium if $\Delta G°$ at 25 °C is
   a. 2.72 kcal/mol?
   b. 0.65 kcal/mol?
   c. −2.72 kcal/mol?
   d. −0.65 kcal/mol?

51. Which bond is stronger? Briefly explain why.
   a. CH₃—Cl or CH₃—Br
   b. I—Br or Br—Br

52. Squalene, a hydrocarbon with molecular formula $C_{30}H_{50}$, is obtained from shark liver. (*Squalus* is Latin for "shark.") If squalene is an acyclic compound, how many π bonds does it have?

53. For each of the following compounds, draw the possible geometric isomers and name each isomer:
   a. 2-methyl-2,4-hexadiene
   b. 1,5-heptadiene
   c. 1,4-pentadiene
   d. 3-methyl-2,4-hexadiene

54. By following the curved red arrows, draw the product(s) of each of the following reaction steps. Also indicate which species is the electrophile and which is the nucleophile.

a.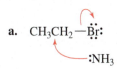

b.

c. $CH_3C \overset{CH_3 \ CH_3}{=\!\!=} CCH_3$

   $H-\overset{..}{\underset{..}{Cl}}:$

55. How many of the following names are correct? Correct the incorrect names.
- **a.** 3-pentene
- **b.** 2-octene
- **c.** 2-vinylpentane
- **d.** 1-ethyl-1-pentene
- **e.** 5-ethylcyclohexene
- **f.** 5-chloro-3-hexene
- **g.** 2-ethyl-2-butene
- **h.** (*E*)-2-methyl-1-hexene
- **i.** 2-methylcyclopentene

56. a. How many alkenes could you treat with H_2, Pd/C to prepare methylcyclopentane?
   b. Which of the alkenes is the most stable?
   c. Which of the alkenes has the smallest heat of hydrogenation?

57. Draw structures for the following:
- **a.** (2*E*,4*E*)-1-chloro-3-methyl-2,4-hexadiene
- **b.** (3*Z*,5*E*)-4-methyl-3,5-nonadiene
- **c.** (3*Z*,5*Z*)-4,5-dimethyl-3,5-nonadiene
- **d.** (3*E*,5*E*)-2,5-dibromo-3,5-octadiene

58. Given the reaction coordinate diagram for the reaction of **A** to form **G**, answer the following questions:

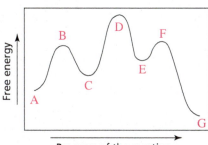

- **a.** How many intermediates are formed in the reaction?
- **b.** Which letters represent transition states?
- **c.** What is the fastest step in the reaction?
- **d.** Which is more stable: **A** or **G**?
- **e.** Does **A** or **E** form faster from **C**?
- **f.** Which is the more stable intermediate?
- **g.** What is the reactant of the rate-determining step?
- **h.** Is the first step of the reaction exergonic or endergonic?
- **i.** Is the overall reaction exergonic or endergonic?
- **j.** Which step in the forward direction has the largest rate constant?
- **k.** Which step in the reverse direction has the smallest rate constant?

59. a. Which of the following reactions has the larger $\Delta S°$ value?
   b. Is the $\Delta S°$ value positive or negative?

A ⬡—Br $+ \ HO^- \longrightarrow$ ⬡—OH $+ \ Br^-$ B ⬡—Br $+ \ HO^- \longrightarrow$ ⬡ $+ \ H_2O \ + \ Br^-$

60. Draw the structure of a compound with molecular C_8H_{14} that reacts with one equivalent of H_2 over Pd/C to form a meso compound.

61. a. What is the equilibrium constant for a reaction that is carried out at 25 °C (298 K) with $\Delta H° = 20$ kcal/mol and $\Delta S° = 5.0 \times 10^{-2}$ kcal mol^{-1}K^{-1}?
   b. What is the equilibrium constant for the same reaction carried out at 125 °C?

62. Using curved arrows, show the mechanism of the following reaction:

63. For a reaction carried out at 25 °C with an equilibrium constant of 1×10^{-3}, to increase the equilibrium constant by a factor of 10:
   a. how much must $\Delta G°$ change?
   b. how much must $\Delta H°$ change if $\Delta S° = 0$ kcal $mol^{-1} K^{-1}$?
   c. how much must $\Delta S°$ change if $\Delta H° = 0$ kcal mol^{-1}?

64. Given that the free energy of the twist-boat conformer of cyclohexane is 5.3 kcal/mol greater than that of the chair conformer, calculate the percentage of twist-boat conformers present in a sample of cyclohexane at 25 °C. Does your answer agree with the statement made in Section 3.13 about the relative number of molecules in these two conformations?

Calculating Kinetic Parameters

After obtaining rate constants at several temperatures, you can calculate E_a, $\Delta H^{\ddagger}$, $\Delta G^{\ddagger}$, and $\Delta S^{\ddagger}$ for a reaction as follows:
- The Arrhenius equation allows E_a to be obtained from the slope of a plot of $\ln k$ versus $1/T$ because

$$\ln k_2 - \ln k_1 = -E_a / R\left(\frac{1}{T_2} - \frac{1}{T_1}\right)$$

- you can determine $\Delta H^{\ddagger}$ at a given temperature from E_a because $\Delta H^{\ddagger} = E_a - RT$.
- you can determine $\Delta G^{\ddagger}$, in kJ/mol, from the following equation, which relates $\Delta G^{\ddagger}$ to the rate constant at a given temperature:

$$-\Delta G^{\ddagger} = RT \ln \frac{kh}{Tk_B}$$

In this equation, h is Planck's constant (1.58×10^{-31} kcal s) and k_B is the Boltzmann constant (3.30×10^{-19} kcal K^{-1}).
- You can determine the entropy of activation from the other two kinetic parameters via the formula $\Delta S^{\ddagger} = (\Delta H^{\ddagger} - \Delta G^{\ddagger})/T$.

Use this information to answer Problem 65.

65. From the following rate constants, determined at five temperatures, calculate the experimental energy of activation and $\Delta G^{\ddagger}$, $\Delta H^{\ddagger}$, and $\Delta S^{\ddagger}$ for the reaction at 30 °C:

Temperature	Observed rate constant
31.0 °C	2.11×10^{-5} s^{-1}
40.0 °C	4.44×10^{-5} s^{-1}
51.5 °C	1.16×10^{-4} s^{-1}
59.8 °C	2.10×10^{-4} s^{-1}
69.2 °C	4.34×10^{-4} s^{-1}

DRAWING CURVED ARROWS

ESSENTIAL SKILL

TUTORIAL

Enhanced by MasteringChemistry®

This is an extension of what you learned about drawing curved arrows on pp. 199–201. Working through these problems will take only a little of your time. It will be time well spent, however, because curved arrows are used throughout the book and it is important that you are comfortable with them. (You will not encounter some of the reaction steps shown in this exercise for weeks or even months, so don't worry about why the chemical changes take place.)

Chemists use curved arrows to show how electrons move as covalent bonds break and/or new covalent bonds form.

- Each arrow represents the simultaneous movement of two electrons (an electron pair) from a nucleophile (at the tail of the arrow) toward an electrophile (at the point of the arrow).
- The tail of the arrow is positioned where the electrons are in the reactant; the tail always starts at a lone pair or at a bond.
- The head of the arrow points to where these same electrons end up in the product; the arrow always points at an atom or at a bond.

In the following reaction step, the bond between bromine and a carbon of the cyclohexane ring breaks and both electrons in the bond end up with bromine. Thus, **the arrow starts at the electrons that carbon and bromine share in the reactant,** and **the head of the arrow points at bromine** because this is where the two electrons end up in the product.

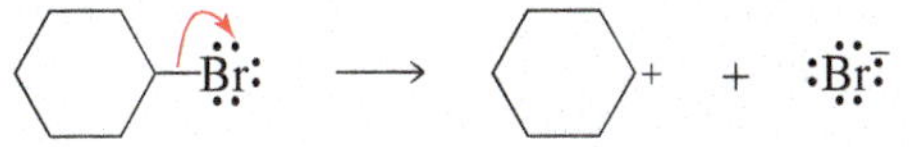

Notice that the carbon of the cyclohexane ring is positively charged in the product. This is because it has lost the two electrons it was sharing with bromine. The bromine is negatively charged in the product because it has gained the electrons that it shared with carbon in the reactant. The fact that two electrons move in this example is indicated by the two barbs on the arrowhead.

Notice that the arrow *always* starts at a bond or at a lone pair. It does *not* start at a negative charge.

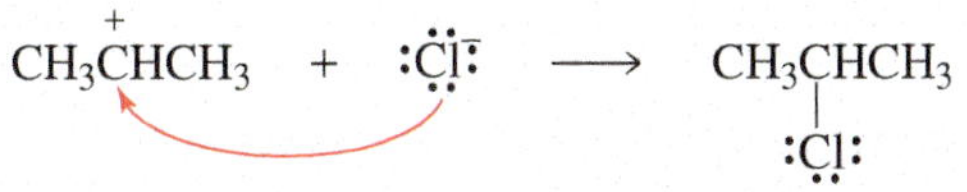

In the following reaction step, a bond is being formed between the oxygen of water and a carbon of the other reactant. The arrow starts at one of the lone pairs of the oxygen and points at the atom (the carbon) that will share the electrons in the product. The oxygen in the product is positively charged, because the electrons that oxygen had to itself in the reactant are now being shared with carbon. The carbon that was positively charged in the reactant is not charged in the product, because it has gained a share in a pair of electrons.

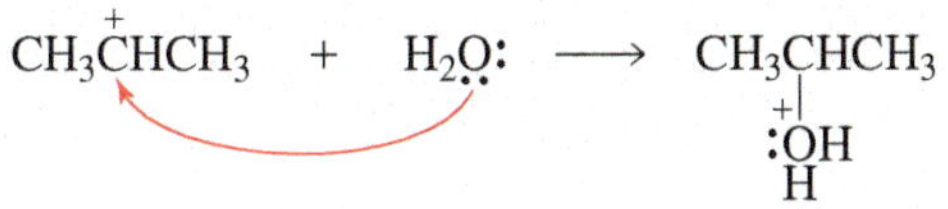

PROBLEM 1 Draw curved arrows to show the movement of the electrons in the following reaction steps. (The answers to all problems appear immediately after Problem 10.)

a.

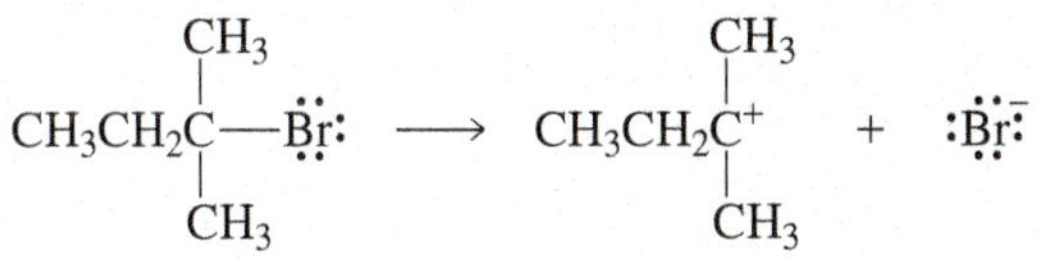

b.

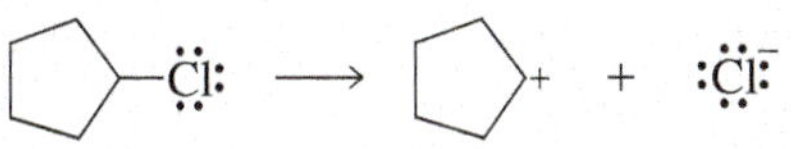

c. 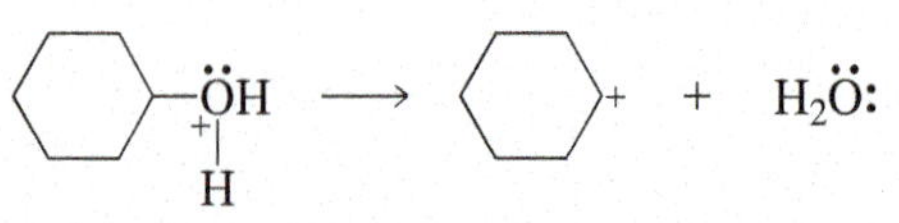 (cyclohexyl)—$\overset{\cdot\cdot}{\underset{\overset{|}{H}}{\overset{+}{O}}}$H $\longrightarrow$ (cyclohexyl)+ + + $H_2\overset{\cdot\cdot}{O}$:

d. $CH_3CH_2{-}MgBr \longrightarrow CH_3\overset{\cdot\cdot}{\overset{-}{C}}H_2$ + ^+MgBr

e. 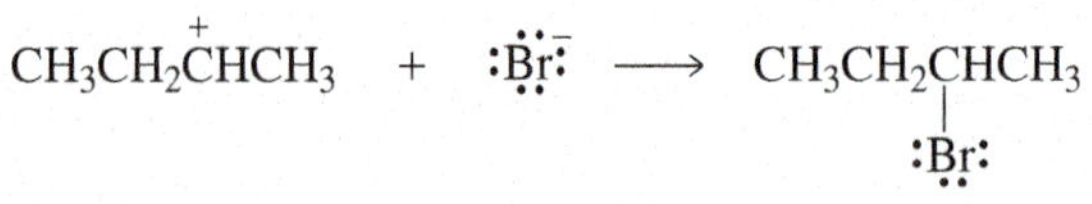 $CH_3CH_2\overset{+}{C}HCH_3$ + $:\overset{\cdot\cdot}{\underset{\cdot\cdot}{Br}}:^-$ $\longrightarrow$ $CH_3CH_2CHCH_3$ with $\underset{\overset{|}{:\overset{\cdot\cdot}{Br}:}}{}$

Frequently, chemists do not show the lone-pair electrons when they write reactions. Problem 2 shows the same reaction steps you just saw in Problem 1, except that the lone pairs are not shown.

PROBLEM 2 Draw curved arrows to show the movement of the electrons in the following reaction steps:

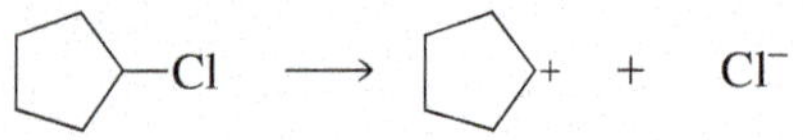

a. 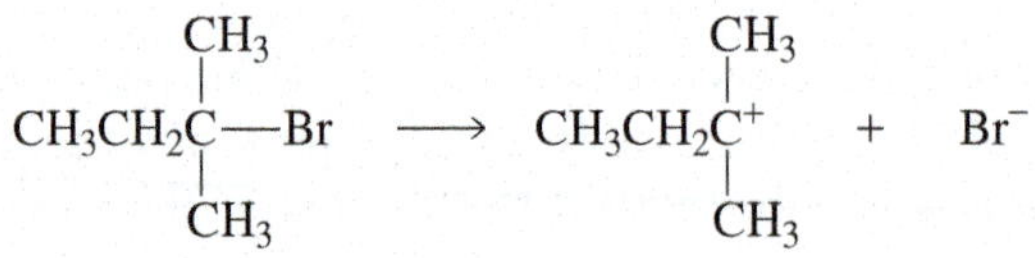 $CH_3CH_2\overset{\overset{\displaystyle CH_3}{|}}{\underset{\underset{\displaystyle CH_3}{|}}{C}}{-}Br \longrightarrow CH_3CH_2\overset{\overset{\displaystyle CH_3}{|}}{\underset{\underset{\displaystyle CH_3}{|}}{\overset{+}{C}}}$ + Br^-

b. (cyclopentyl)—Cl $\longrightarrow$ (cyclopentyl)+ + + Cl^-

c. (cyclohexyl)—$\overset{\overset{+}{O}}{\underset{\overset{|}{H}}{}}$H $\longrightarrow$ (cyclohexyl)+ + + H_2O

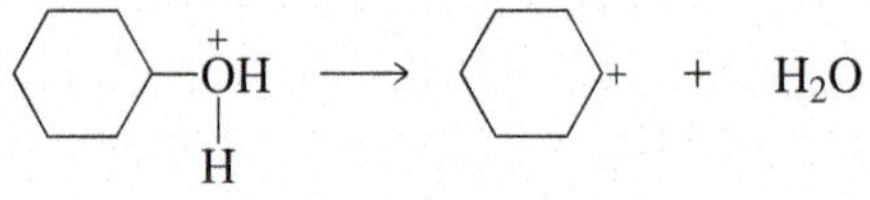

d. $CH_3CH_2{-}MgBr \longrightarrow CH_3\overset{-}{C}H_2$ + ^+MgBr

The lone-pair electrons on Br^- in part **e** must be shown in the reactant, because an arrow can start only at a bond or at a lone pair. The lone pair electrons on Br in the product do not have to be shown. It is never wrong to show lone pairs, but the only time they must be shown is when an arrow is going to start from the lone pair.

e. 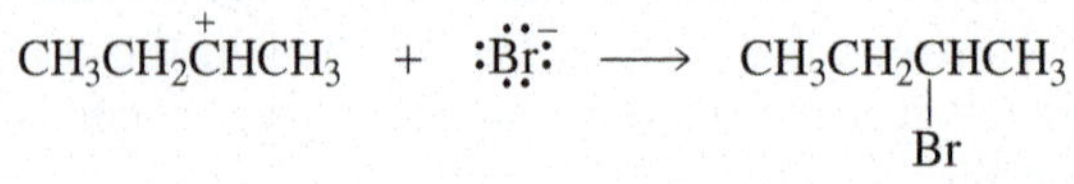 $CH_3CH_2\overset{+}{C}HCH_3$ + $:\overset{\cdot\cdot}{\underset{\cdot\cdot}{Br}}:^-$ $\longrightarrow$ $CH_3CH_2CHCH_3$ with $\underset{\overset{|}{Br}}{}$

Many reaction steps involve both breaking bonds and forming bonds. In the following examples, one bond breaks and one bond forms; the electrons in the bond that breaks are the same as the electrons in the bond that forms. Accordingly, only one arrow is needed to show how the electrons move. As in the previous examples, the arrow starts at the point where the electrons are in the reactant, and the head of the arrow points to where these same electrons end up in the product (between the CH_3 carbon and the carbon that was previously positively charged in the first example and between the two carbons in the next example). Notice that the atom that loses a share in a pair of electrons (C in the first example, H in the second) ends up with a positive charge.

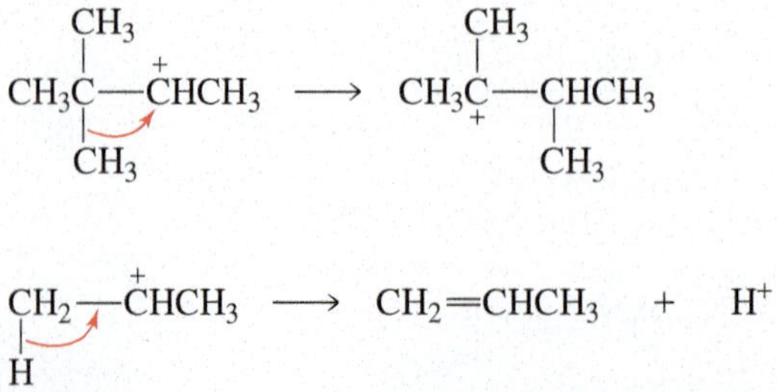

$CH_3\overset{\overset{\displaystyle CH_3}{|}}{\underset{\underset{\displaystyle CH_3}{|}}{C}}{-}\overset{+}{C}HCH_3 \longrightarrow CH_3\overset{\overset{\displaystyle CH_3}{|}}{\underset{\underset{\displaystyle CH_3}{|}}{\overset{+}{C}}}{-}CHCH_3$

$\underset{\overset{|}{H}}{CH_2}{-}\overset{+}{C}HCH_3 \longrightarrow CH_2{=}CHCH_3$ + H^+

226

Frequently, the electrons in the bond that breaks are not the same as the electrons in the bond that forms. In such cases, two arrows are needed to show the movement of the electrons—one to show the bond that forms and one to show the bond that breaks. In each of the following examples, look at the arrows that illustrate how the electrons move. Notice how the movement of the electrons allows you to determine both the structure of the products and the charges on the reactants and products.

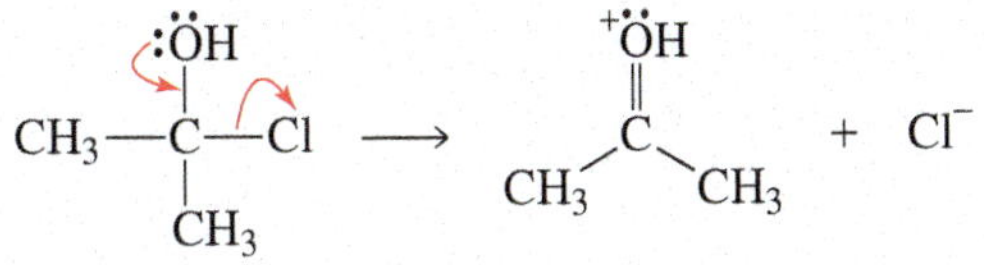

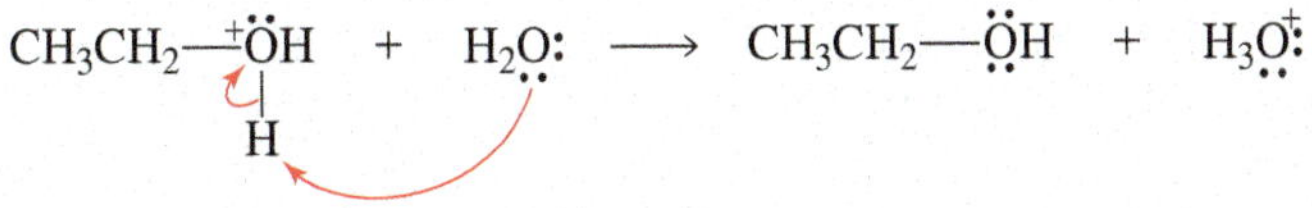

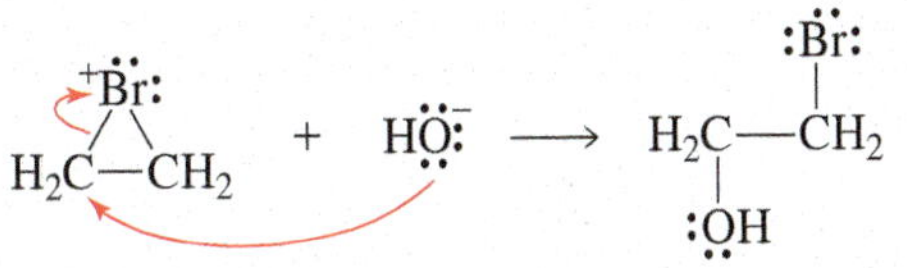

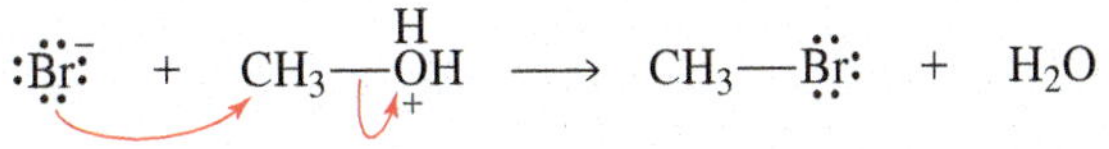

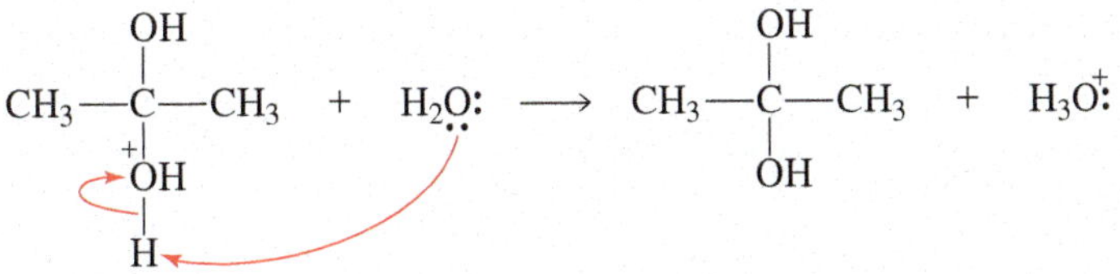

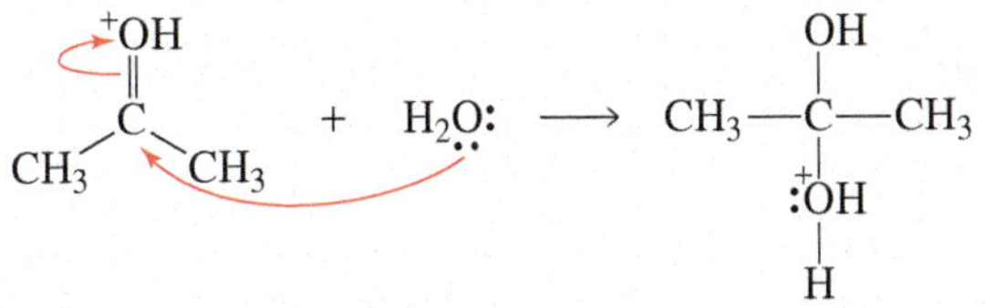

$$CH_3-\overset{\overset{\displaystyle :\ddot{O}H}{|}}{\underset{\underset{\displaystyle CH_3}{|}}{C}}-Cl \longrightarrow \underset{CH_3 \quad CH_3}{\overset{\overset{+}{\ddot{O}H}}{C}} + Cl^-$$

$$CH_3CH_2-\overset{\pm}{\ddot{O}}\underset{\underset{H}{|}}{H} + H_2\ddot{O}: \longrightarrow CH_3CH_2-\ddot{O}H + H_3\overset{+}{\ddot{O}}:$$

$$\underset{H_2C-CH_2}{\overset{+\ddot{B}r:}{\triangle}} + H\ddot{O}:^- \longrightarrow \underset{\underset{:\underset{..}{O}H}{|}}{H_2C-\overset{\overset{:\ddot{B}r:}{|}}{CH_2}}$$

$$:\ddot{\underset{..}{B}r:^- + CH_3-\overset{\overset{H}{|}}{\underset{\underset{+}{|}}{O}H} \longrightarrow CH_3-\ddot{B}r: + H_2O$$

$$CH_3-\overset{\overset{OH}{|}}{\underset{\underset{\underset{H}{|}}{\overset{+}{O}H}}{C}}-CH_3 + H_2\ddot{O}: \longrightarrow CH_3-\overset{\overset{OH}{|}}{\underset{\underset{OH}{|}}{C}}-CH_3 + H_3\overset{+}{O}:$$

$$H\ddot{O}:^- + CH_3CH_2-Br \longrightarrow CH_3CH_2-\ddot{O}H + Br^-$$

$$\underset{CH_3 \quad CH_3}{\overset{\overset{+}{OH}}{C}} + H_2\ddot{O}: \longrightarrow CH_3-\overset{\overset{OH}{|}}{\underset{\underset{\underset{H}{|}}{\overset{+}{:}{O}H}}{C}}-CH_3$$

In the next reaction, two bonds break and one bond forms; two arrows are needed to show the movement of the electrons.

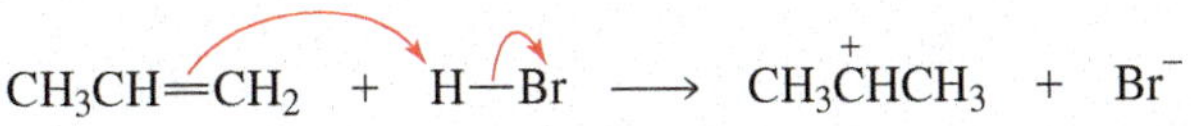

$$CH_3CH{=}CH_2 + H-Br \longrightarrow CH_3\overset{+}{C}HCH_3 + Br^-$$

In the next reaction, two bonds break and two bonds form; three arrows are needed to show the movement of the electrons.

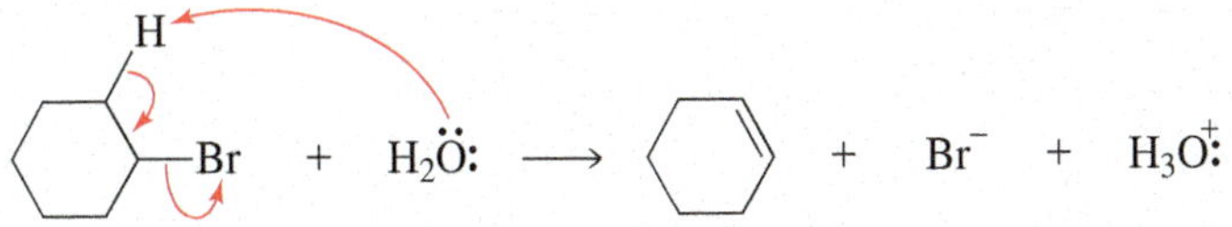

PROBLEM 3 Draw curved arrows to show the movement of the electrons that result in the formation of the given product(s). (*Hint:* Look at the structure of the product to see what bonds need to be formed and broken in order to arrive at the structure of the desired product.)

a. $CH_3\!-\!\overset{:\ddot{O}H}{\underset{CH_3}{\overset{|}{C}}}\!-\!\overset{+}{O}H \longrightarrow \overset{\overset{+}{\ddot{O}H}}{\underset{CH_3}{\overset{||}{C}}}\underset{CH_3}{} \quad + \quad H_2O$

b. $CH_3CH_2CH\!=\!CH_2 \quad + \quad H\!-\!Cl \longrightarrow CH_3CH_2\overset{+}{C}H\!-\!CH_3 \quad + \quad Cl^-$

c. $CH_3CH_2\!-\!Br \quad + \quad \overset{..}{N}H_3 \longrightarrow CH_3CH_2\!-\!\overset{+}{N}H_3 \quad + \quad Br^-$

d. $CH_3\overset{CH_3}{\underset{H}{\overset{|}{\underset{|}{C}}}}\!-\!\overset{+}{C}HCH_3 \longrightarrow CH_3\overset{CH_3}{\underset{+}{\overset{|}{C}}}\!-\!CH_2CH_3$

PROBLEM 4 Draw curved arrows to show the movement of the electrons that result in formation of the given product(s).

a. $CH_3CH\!=\!CHCH_3 \quad + \quad H\!-\!\overset{H}{\underset{|}{\overset{+}{O}}}\!-\!H \longrightarrow CH_3\overset{+}{C}H\!-\!CH_2CH_3 \quad + \quad H_2O$

b. $CH_3CH_2CH_2CH_2\!-\!Cl \quad + \quad \overset{-..}{C}\!\equiv\!N \longrightarrow CH_3CH_2CH_2CH_2\!-\!C\!\equiv\!N \quad + \quad Cl^-$

c. $CH_3\!-\!\overset{:\ddot{O}H}{\underset{OH}{\overset{|}{\underset{|}{C}}}}\!-\!\overset{+}{O}CH_3 \longrightarrow \overset{\overset{+}{\ddot{O}H}}{\underset{CH_3}{\overset{||}{C}}}\underset{OH}{} \quad + \quad CH_3OH$
   (with H on the central carbon)

d. $\underset{CH_3}{\overset{\overset{\ddot{O}:}{||}}{C}}\underset{H}{} \quad + \quad CH_3\!-\!MgBr \longrightarrow CH_3\!-\!\overset{:\ddot{O}:^-}{\underset{CH_3}{\overset{|}{\underset{|}{C}}}}\!-\!H \quad + \quad {}^+MgBr$

PROBLEM 5 Draw curved arrows to show the movement of the electrons that result in formation of the given product(s).

a. $\underset{CH_3}{\overset{\overset{O}{||}}{C}}\underset{CH_3}{} \quad + \quad CH_3CH_2\!-\!MgBr \longrightarrow CH_3\!-\!\overset{O^-}{\underset{CH_2CH_3}{\overset{|}{\underset{|}{C}}}}\!-\!CH_3 \quad + \quad {}^+MgBr$

b. $CH_3CH_2CH_2\!-\!Br \quad + \quad CH_3\overset{..}{\underset{..}{O}}{}^- \longrightarrow CH_3CH_2CH_2\!-\!OCH_3 \quad + \quad Br^-$

c. [cyclohexane ring with $\overset{+}{C}$ bearing CH_3 and CH_3] $\longrightarrow$ [cyclohexane ring with CH_3, CH_3 substituents and $\overset{+}{}$]

d. $CH_3\!-\!\overset{:\ddot{O}:^-}{\underset{CH_3}{\overset{|}{\underset{|}{C}}}}\!-\!OCH_2CH_3 \longrightarrow \underset{CH_3}{\overset{\overset{:\ddot{O}}{||}}{C}}\underset{CH_3}{} \quad + \quad CH_3CH_2O^-$

228

PROBLEM 6 Draw curved arrows to show the movement of the electrons that result in formation of the given product(s).

a. 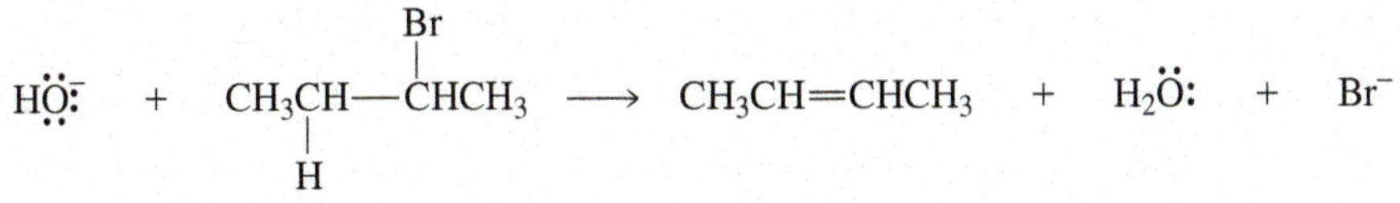 $HO\colonminus$ + CH_3CH—$\overset{\underset{\textstyle H}{|}}{\overset{\textstyle Br}{|}}CHCH_3$ $\longrightarrow$ CH_3CH=$CHCH_3$ + $H_2O\ddot{\colon}$ + Br^-

b. CH_3CH_2C≡C—H + $\colon\ddot{N}H_2$ $\longrightarrow$ CH_3CH_2C≡$C\colonminus$ + $\dot{\ddot{N}}H_3$

c. $CH_3\overset{\underset{\textstyle CH_3}{|}}{\overset{\textstyle CH_3}{|}}\overset{+}{C}$—$CHCH_2CH_3$ $\longrightarrow$ $CH_3\overset{\underset{\textstyle \overset{+}{}}{|}}{\overset{\textstyle CH_3}{|}}C$—$\overset{\underset{\textstyle CH_3}{|}}{}CHCH_2CH_3$

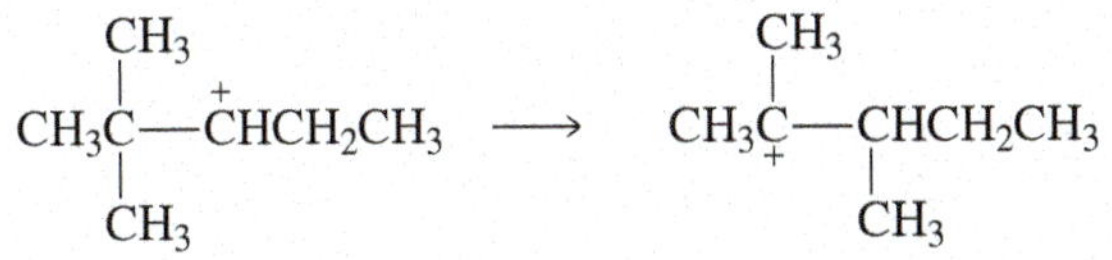

d. 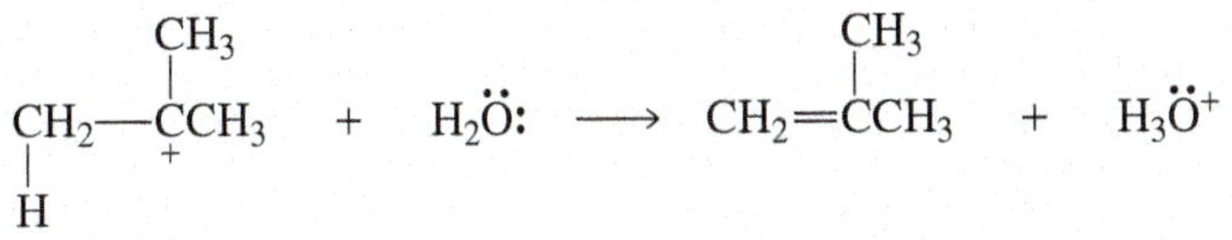 CH_2—$\overset{\underset{\textstyle H}{|}}{\overset{\textstyle CH_3}{|}}\overset{+}{C}CH_3$ + $H_2\ddot{O}\ddot{\colon}$ $\longrightarrow$ CH_2=$\overset{\underset{\textstyle CH_3}{|}}{}CCH_3$ + $H_3\ddot{O}^+$

PROBLEM 7 Draw curved arrows to show the movement of the electrons that result in formation of the given product(s).

a. 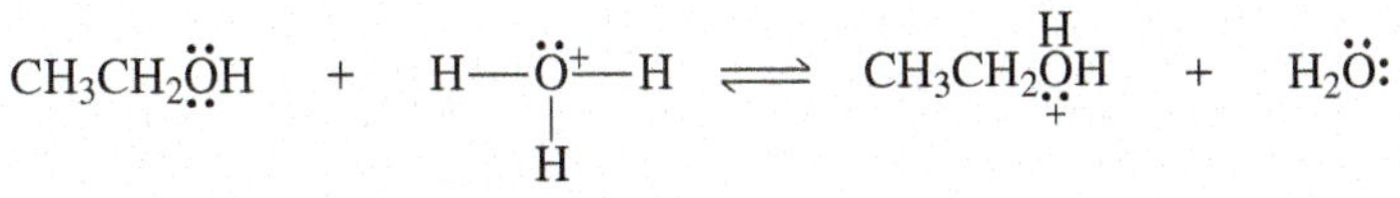 $CH_3CH_2\ddot{O}H$ + H—$\overset{\underset{\textstyle H}{|}}{\overset{+}{\ddot{O}}}$—$H$ $\rightleftharpoons$ $CH_3CH_2\overset{\underset{\textstyle +}{}}{\overset{\textstyle H}{\ddot{O}}}H$ + $H_2\ddot{O}\ddot{\colon}$

b.  $CH_3\overset{\underset{\textstyle H}{|}}{\overset{+}{N}}H_2$ + $H_2\ddot{O}\ddot{\colon}$ $\rightleftharpoons$ CH_3NH_2 + $H_3\overset{+}{\ddot{O}}\ddot{\colon}$

PROBLEM 8 Draw curved arrows to show the movement of the electrons in each step of the following reaction sequences. (*Hint:* You can tell how to draw the arrows for each step by looking at the products that are formed in that step as a result of the movement of electrons.)

a. 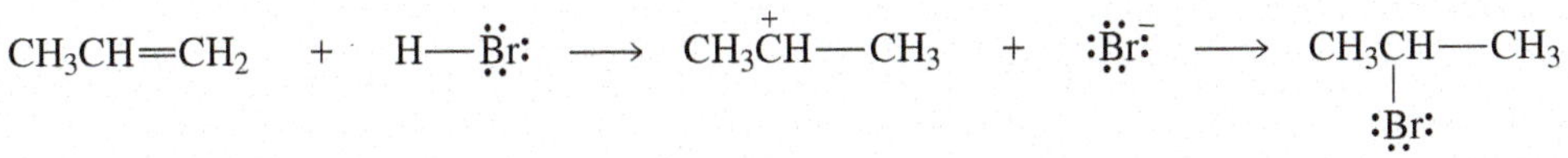 CH_3CH=CH_2 + H—$\ddot{B}r\ddot{\colon}$ $\longrightarrow$ $CH_3\overset{+}{C}H$—CH_3 + $\colon\ddot{B}r\colonminus$ $\longrightarrow$ $CH_3\overset{\underset{\textstyle \colon\ddot{B}r\ddot{\colon}}{|}}{}CH$—$CH_3$

b. $CH_3\overset{\underset{\textstyle CH_3}{|}}{\overset{\textstyle CH_3}{|}}C$—$Cl$ $\rightleftharpoons$ $CH_3\overset{\underset{\textstyle CH_3}{|}}{\overset{\textstyle CH_3}{|}}\overset{+}{C}$ + Cl^- $\xrightarrow{\dot{\ddot{N}}H_3}$ $CH_3\overset{\underset{\textstyle CH_3}{|}}{\overset{\textstyle CH_3}{|}}C$—$\overset{+}{N}H_3$

c. $\overset{\underset{\textstyle CH_3}{}}{\underset{\textstyle Cl}{}}\overset{\ddot{O}\colon}{\overset{||}{C}}$ + $HO\colonminus$ $\longrightarrow$ CH_3—$\overset{\underset{\textstyle \colon\ddot{O}H}{|}}{\overset{\textstyle \colon\ddot{O}\colonminus}{|}}C$—$Cl$ $\longrightarrow$ $\overset{\underset{\textstyle CH_3 \quad OH}{}}{\overset{\ddot{O}\colon}{\overset{||}{C}}}$ + Cl^-

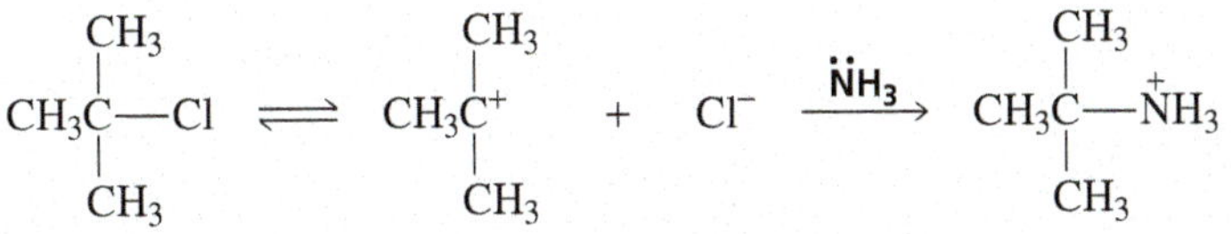

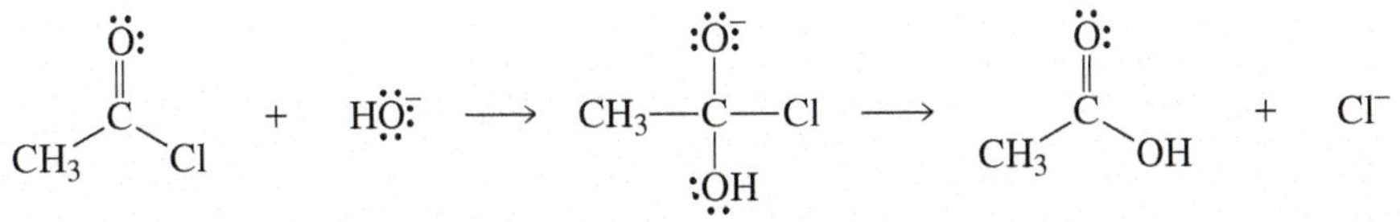

PROBLEM 9 Draw curved arrows to show the movement of the electrons in each step of the following reaction sequences.

a.

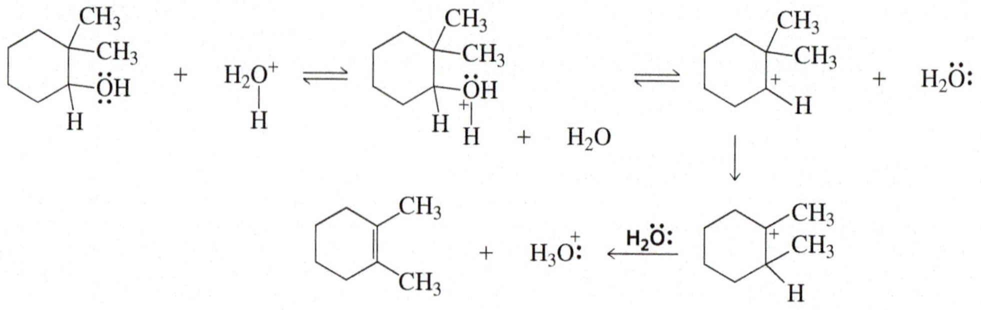

b. 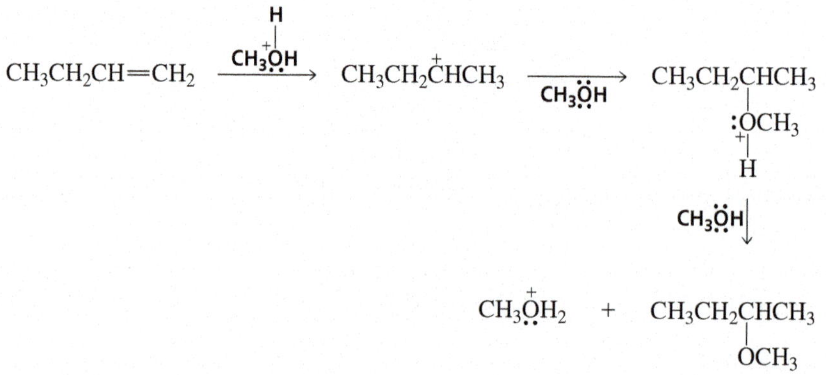 $CH_3CH_2CH=CH_2 \xrightarrow{CH_3\overset{+}{\underset{H}{O}}H} CH_3CH_2\overset{+}{C}HCH_3 \xrightarrow{CH_3\overset{..}{O}H} CH_3CH_2CHCH_3$

with $:\overset{+}{\underset{H}{OCH_3}}$

$\xrightarrow{CH_3\overset{..}{O}H}$

$CH_3\overset{..}{O}H_2 \;+\; CH_3CH_2CHCH_3$ with OCH_3

c.

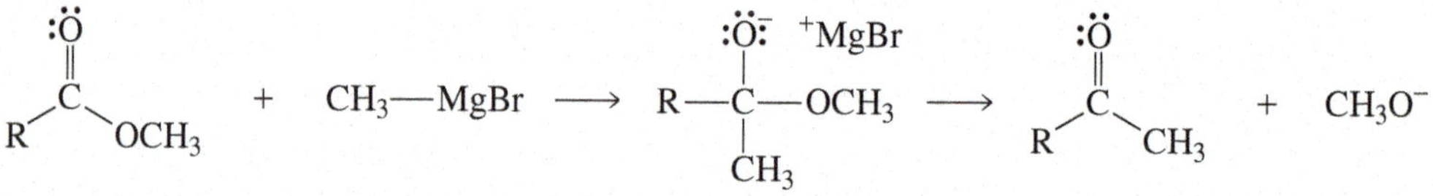

PROBLEM 10 Use what the curved arrows tell you about electron movement to determine the product(s) of each reaction step.

a. 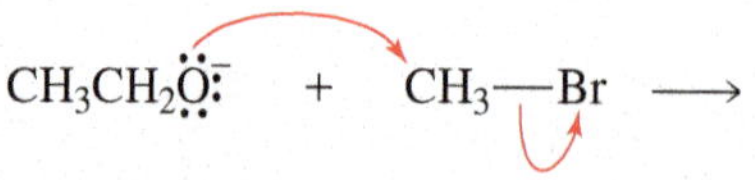 $CH_3CH_2\overset{..}{\underset{..}{O}}:^- \;+\; CH_3{-}Br \longrightarrow$

b.

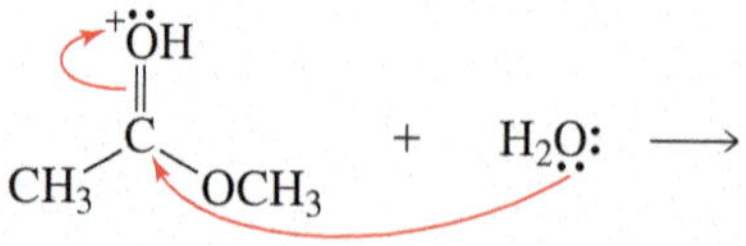

$+\; H_2\overset{..}{O}: \longrightarrow$

c. 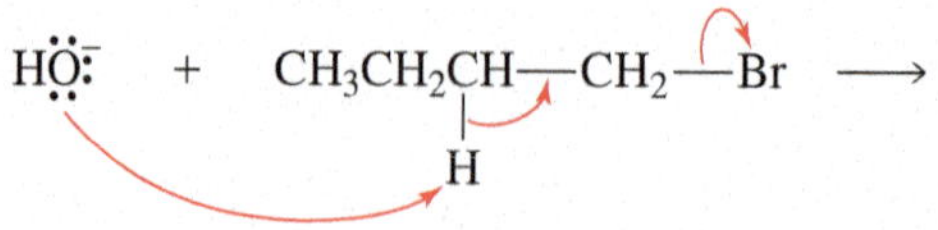 $H\overset{..}{\underset{..}{O}}:^- \;+\; CH_3CH_2CH{-}CH_2{-}Br \longrightarrow$ with H

d. 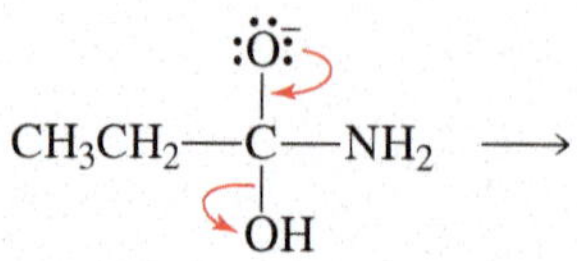 $CH_3CH_2{-}\overset{\overset{\displaystyle :\overset{..}{\underset{..}{O}}:^-}{|}}{C}{-}NH_2 \longrightarrow$ with OH

e.

$+\; CH_3{-}MgBr \longrightarrow$

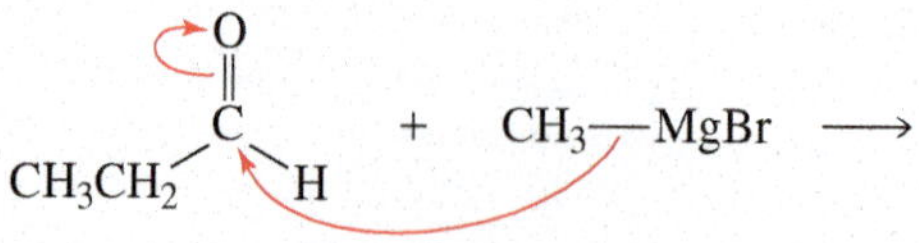

f. $CH_3 - \overset{\overset{\displaystyle CH_3}{|}}{\underset{\underset{\displaystyle CH_3}{|}}{C}} - \overset{+}{O}H \longrightarrow$

g. $CH_3 - \overset{\overset{\displaystyle :\ddot{O}H}{|}}{\underset{\underset{\displaystyle OH}{|}}{C}} - \overset{+}{O}CH_3 \longrightarrow$

ANSWERS TO PROBLEMS ON DRAWING CURVED ARROWS

PROBLEM 1 SOLVED

a. $CH_3CH_2\overset{\overset{\displaystyle CH_3}{|}}{\underset{\underset{\displaystyle CH_3}{|}}{C}} - \ddot{B}r: \longrightarrow CH_3CH_2\overset{\overset{\displaystyle CH_3}{|}}{\underset{\underset{\displaystyle CH_3}{|}}{C}}{}^+ \quad + \quad :\ddot{B}\ddot{r}:^-$

b. ⬠$-\ddot{C}l: \longrightarrow$ ⬠$^+ \quad + \quad :\ddot{C}\ddot{l}:^-$

c. ⬡$-\overset{+}{\underset{\underset{\displaystyle H}{|}}{\ddot{O}}}H \longrightarrow$ ⬡$^+ \quad + \quad H_2\ddot{O}:$

d. $CH_3CH_2 - MgBr \longrightarrow CH_3\ddot{C}H_2{}^- \quad + \quad {}^+MgBr$

e. $CH_3CH_2\overset{+}{C}HCH_3 \quad + \quad :\ddot{B}\ddot{r}:^- \longrightarrow CH_3CH_2\underset{\underset{\displaystyle :\ddot{B}\ddot{r}:}{|}}{C}HCH_3$

PROBLEM 2 SOLVED

a. $CH_3CH_2\overset{\overset{\displaystyle CH_3}{|}}{\underset{\underset{\displaystyle CH_3}{|}}{C}} - Br \longrightarrow CH_3CH_2\overset{\overset{\displaystyle CH_3}{|}}{\underset{\underset{\displaystyle CH_3}{|}}{C}}{}^+ \quad + \quad Br^-$

b. ⬠$-Cl \longrightarrow$ ⬠$^+ \quad + \quad Cl^-$

c. ⬡$-\overset{+}{\underset{\underset{\displaystyle H}{|}}{O}}H \longrightarrow$ ⬡$^+ \quad + \quad H_2O$

d. $CH_3CH_2 - MgBr \longrightarrow CH_3\ddot{C}H_2{}^- \quad + \quad {}^+MgBr$

e. $CH_3CH_2\overset{+}{C}HCH_3 \quad + \quad :\ddot{B}\ddot{r}:^- \longrightarrow CH_3CH_2\underset{\underset{\displaystyle Br}{|}}{C}HCH_3$

PROBLEM 3 SOLVED

a.

$$CH_3\!-\!\overset{\displaystyle :\ddot{O}H}{\underset{\displaystyle CH_3}{\overset{\displaystyle |}{\underset{\displaystyle |}{C}}}}\!-\!\overset{+}{O}H \longrightarrow \underset{CH_3\quad\ CH_3}{C}\overset{\overset{+}{O}H}{} \quad + \quad H_2O$$

b. $CH_3CH_2CH\!=\!CH_2 \quad + \quad H\!-\!Cl \longrightarrow CH_3CH_2\overset{+}{C}H\!-\!CH_3 \quad + \quad Cl^-$

c. $CH_3CH_2\!-\!Br \quad + \quad \overset{..}{N}H_3 \longrightarrow CH_3CH_2\!-\!\overset{+}{N}H_3 \quad + \quad Br^-$

d.

$$CH_3\overset{\overset{\displaystyle CH_3}{|}}{\underset{\underset{\displaystyle H}{|}}{C}}\!-\!\overset{+}{C}HCH_3 \longrightarrow CH_3\overset{\overset{\displaystyle CH_3}{|}}{\underset{+}{C}}\!-\!CH_2CH_3$$

PROBLEM 4 SOLVED

a.

$$CH_3CH\!=\!CHCH_3 \quad + \quad H\!-\!\overset{\overset{\displaystyle |}{\underset{\displaystyle H}{\overset{+}{O}}}}{}\!-\!H \longrightarrow CH_3\overset{+}{C}H\!-\!CH_2CH_3 \quad + \quad H_2O$$

b. $CH_3CH_2CH_2CH_2\!-\!Cl \quad + \quad \overset{..}{:}C\!\equiv\!N \longrightarrow CH_3CH_2CH_2CH_2\!-\!C\!\equiv\!N \quad + \quad Cl^-$

c.

$$CH_3\!-\!\overset{\overset{\displaystyle :\ddot{O}H}{|}}{\underset{\underset{\displaystyle OH}{|}}{\overset{|}{C}}}\!-\!\overset{+}{\underset{\displaystyle H}{\overset{..}{O}}}CH_3 \longrightarrow \underset{CH_3\quad\ OH}{C}\overset{\overset{+}{O}H}{} \quad + \quad CH_3OH$$

d.

$$\underset{CH_3\quad\ H}{\overset{\overset{\displaystyle \ddot{O}:}{\|}}{C}} \quad + \quad CH_3\!-\!MgBr \longrightarrow CH_3\!-\!\overset{\overset{\displaystyle :\ddot{O}^-}{|}}{\underset{\underset{\displaystyle CH_3}{|}}{C}}\!-\!H \quad + \quad {}^+MgBr$$

PROBLEM 5 SOLVED

a.

$$\underset{CH_3\quad\ CH_3}{\overset{\overset{\displaystyle O}{\|}}{C}} \quad + \quad CH_3CH_2\!-\!MgBr \longrightarrow CH_3\!-\!\overset{\overset{\displaystyle O^-}{|}}{\underset{\underset{\displaystyle CH_2CH_3}{|}}{C}}\!-\!CH_3 \quad + \quad {}^+MgBr$$

b. $CH_3CH_2CH_2\!-\!Br \quad + \quad CH_3\overset{..}{\underset{..}{O}}{}^- \longrightarrow CH_3CH_2CH_2\!-\!OCH_3 \quad + \quad Br^-$

c.

d.

$$CH_3\!-\!\overset{\overset{\displaystyle :\ddot{O}^-}{|}}{\underset{\underset{\displaystyle CH_3}{|}}{C}}\!-\!OCH_2CH_3 \longrightarrow \underset{CH_3\quad\ CH_3}{\overset{\overset{\displaystyle O}{\|}}{C}} \quad + \quad CH_3CH_2O^-$$

PROBLEM 6 SOLVED

a. 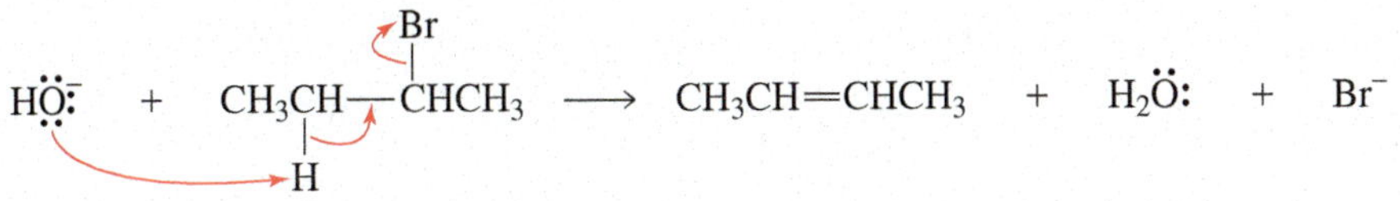 $HO:^-$ + $CH_3CH-CHCH_3$ (with Br and H) $\longrightarrow$ $CH_3CH=CHCH_3$ + $H_2O:$ + Br^-

b. 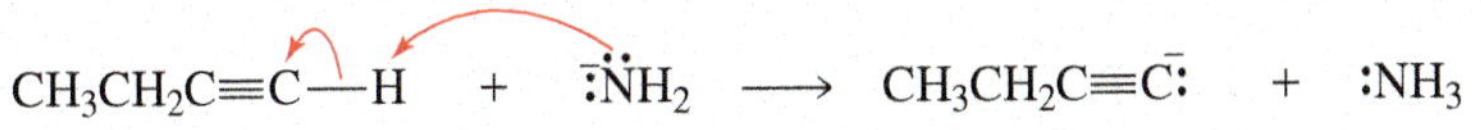 $CH_3CH_2C\equiv C-H$ + $:NH_2$ $\longrightarrow$ $CH_3CH_2C\equiv C:^-$ + $:NH_3$

c. $CH_3\overset{CH_3}{\underset{CH_3}{C}}-\overset{+}{C}HCH_2CH_3$ $\longrightarrow$ $CH_3\overset{CH_3}{\underset{+}{C}}-\overset{CH_3}{CH}CH_2CH_3$

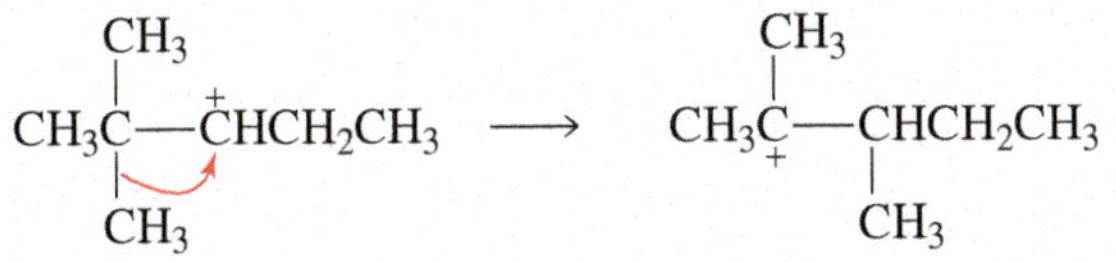

d. 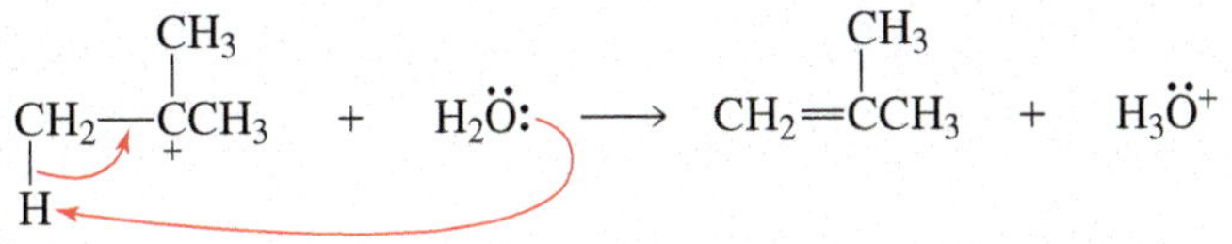 $CH_2-\overset{CH_3}{\underset{+}{C}}CH_3$ (with H) + $H_2O:$ $\longrightarrow$ $CH_2=\overset{CH_3}{C}CH_3$ + H_3O^+

PROBLEM 7 SOLVED

a. 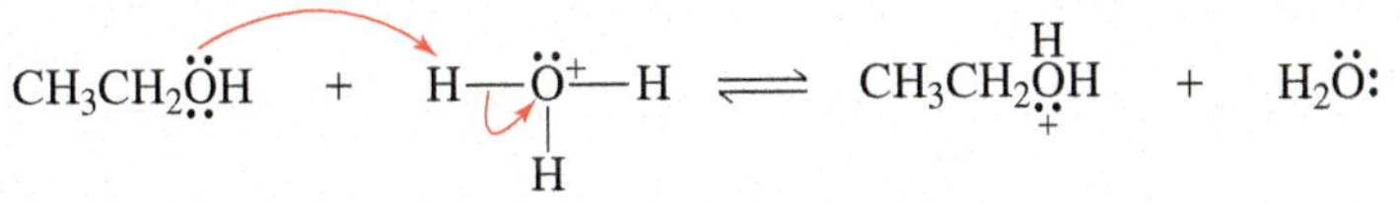 $CH_3CH_2\overset{..}{O}H$ + $H-\overset{+}{\underset{H}{O}}-H$ $\rightleftharpoons$ $CH_3CH_2\overset{H}{\underset{+}{O}}H$ + $H_2\overset{..}{O}:$

b. 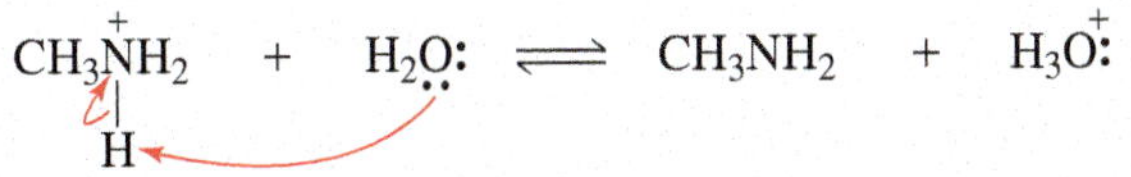 $CH_3\overset{+}{N}H_2$ (with H) + $H_2\overset{..}{O}:$ $\rightleftharpoons$ CH_3NH_2 + $H_3\overset{+}{O}:$

PROBLEM 8 SOLVED

a. 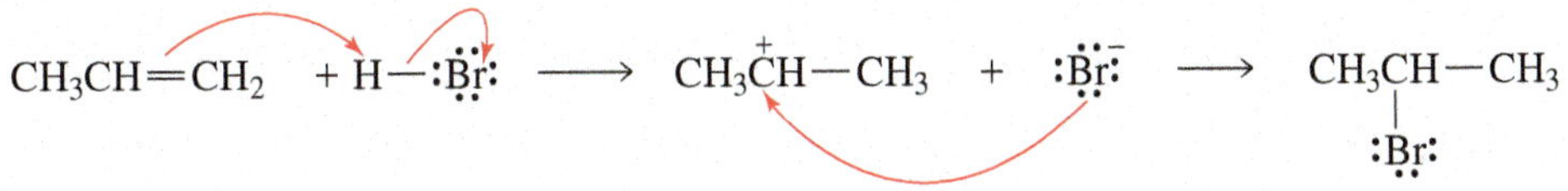 $CH_3CH=CH_2$ + $H-:\overset{..}{\underset{..}{Br}}:$ $\longrightarrow$ $CH_3\overset{+}{C}H-CH_3$ + $:\overset{..}{\underset{..}{Br}}:^-$ $\longrightarrow$ $CH_3\overset{}{\underset{:\overset{..}{\underset{..}{Br}}:}{CH}}-CH_3$

b. $CH_3\overset{CH_3}{\underset{CH_3}{C}}-Cl$ $\rightleftharpoons$ $CH_3\overset{CH_3}{\underset{CH_3}{\overset{+}{C}}}$ + Cl^- $\xrightarrow{:NH_3}$ $CH_3\overset{CH_3}{\underset{CH_3}{C}}-\overset{+}{N}H_3$

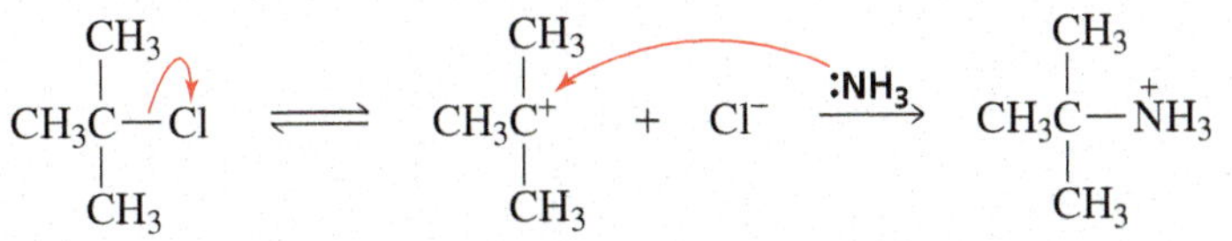

c. $\overset{\overset{..}{O}:}{\underset{CH_3}{\underset{\quad}{C}}}Cl$ + $H\overset{..}{O}:^-$ $\longrightarrow$ $CH_3-\overset{:\overset{..}{O}:^-}{\underset{:\overset{..}{O}H}{C}}-Cl$ $\longrightarrow$ $\overset{\overset{..}{O}:}{\underset{CH_3}{\underset{\quad}{C}}}OH$ + Cl^-

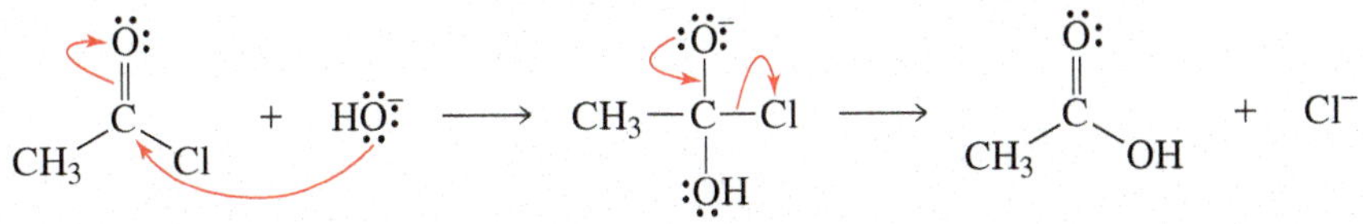

PROBLEM 9 SOLVED

a.

233

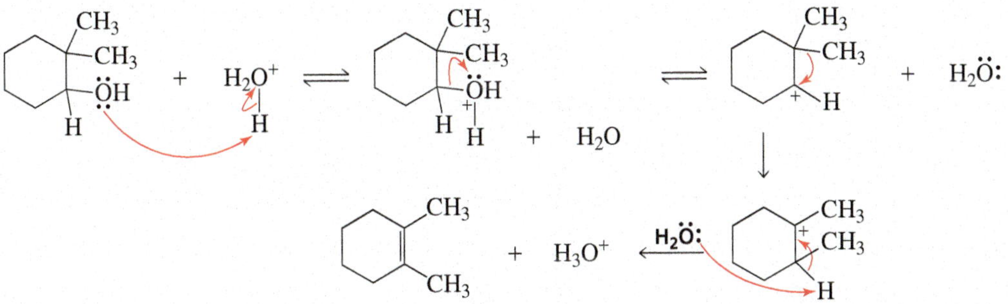

b. 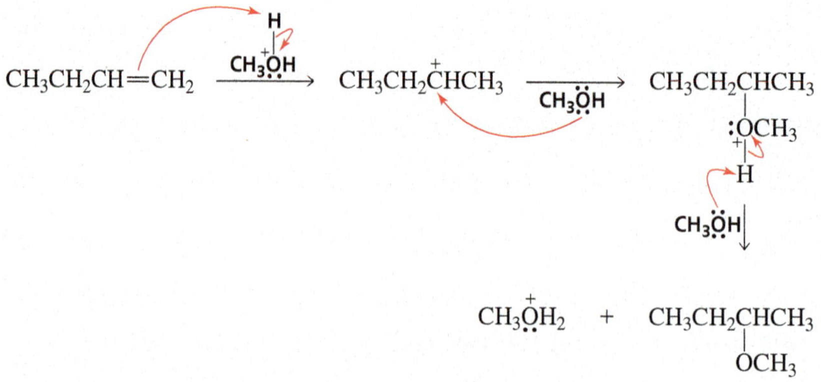

c. 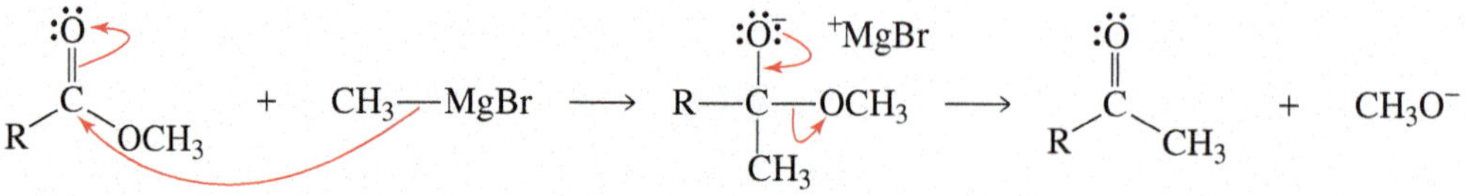

PROBLEM 10 **SOLVED**

a. 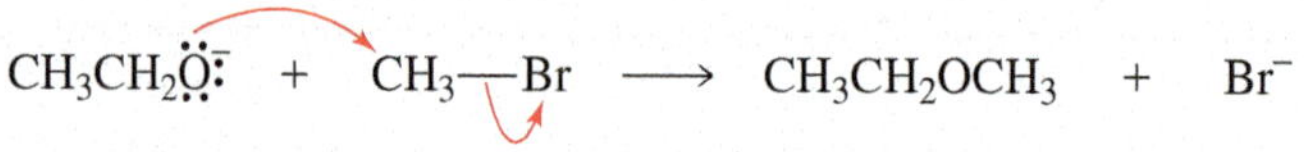

b. 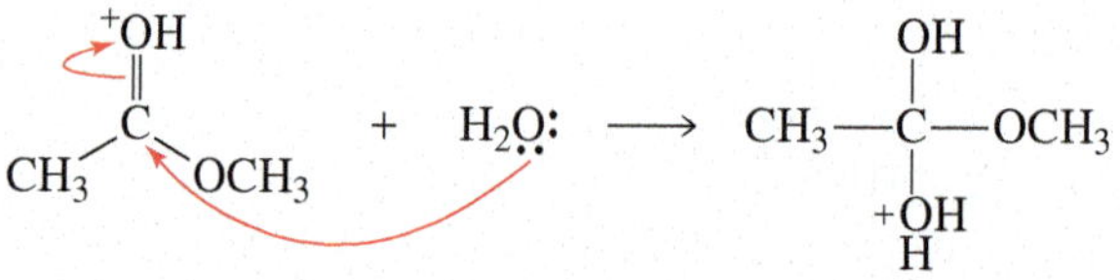

c. 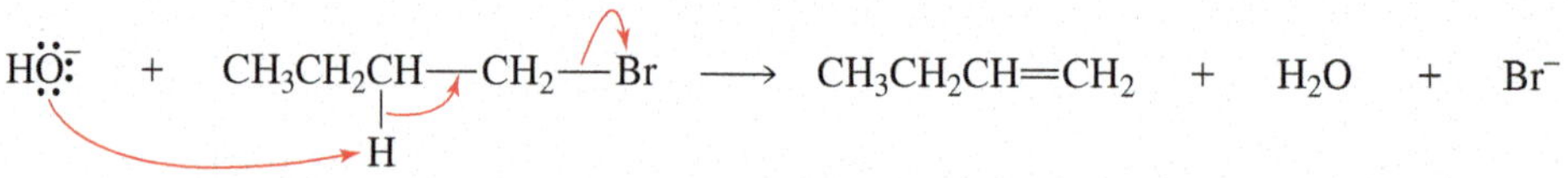

d. 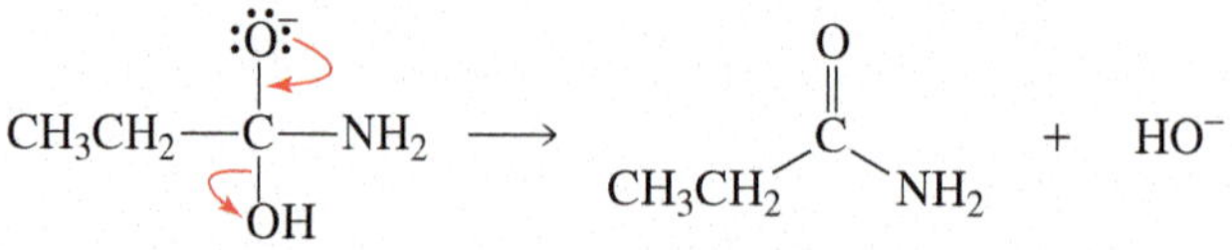

e. 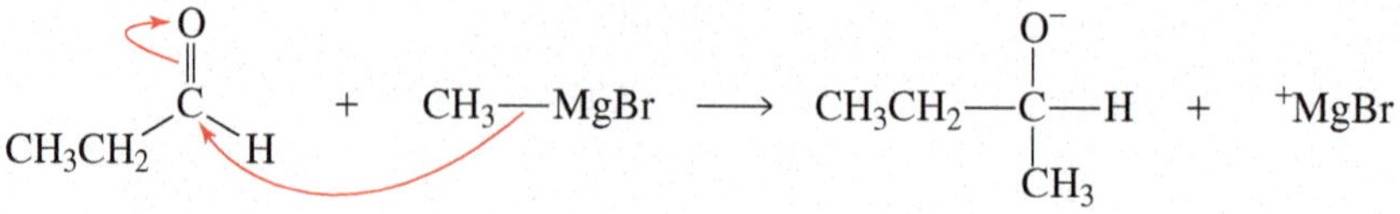

f. 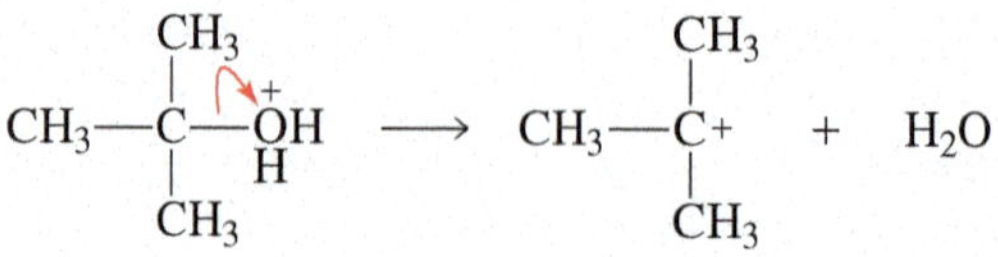

g. 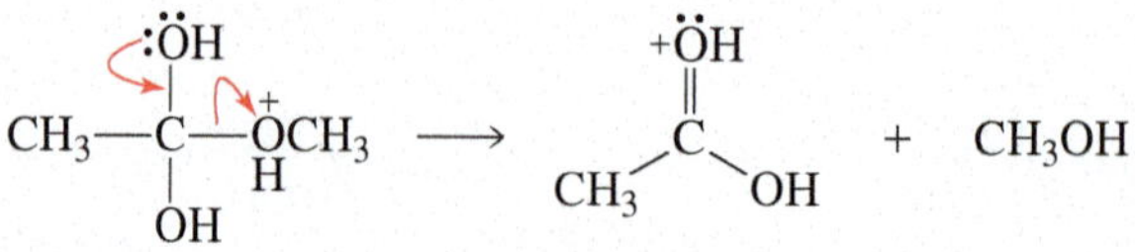

6

The Reactions of Alkenes • The Stereochemistry of Addition Reactions

A question we will consider in this chapter: Which are more harmful: the pesticides farmers spray on our food or the pesticides plants make to ward off predators?

We saw that organic compounds can be divided into families and that all the members of a family react in the same way (Section 5.4). One of these families consists of compounds with a carbon–carbon double bond—compounds known as **alkenes.**

The π bond of a double bond is weak, so it is easily broken. This allows alkenes to undergo addition reactions. Because an alkene is a nucleophile, the first species it reacts with is an electrophile. Therefore, we can more precisely say that alkenes undergo *electrophilic* addition reactions (Section 5.5).

When an alkene undergoes an electrophilic addition reaction with HBr, the first step is a relatively slow addition of a proton (an electrophile) to the alkene (a nucleophile). A **carbocation intermediate** (an electrophile) is formed, which then reacts rapidly with a bromide ion (a nucleophile) to form an alkyl halide. Notice that each step involves the reaction of an electrophile with a nucleophile. *The overall reaction is the addition of an electrophile to one of the sp^2 carbons of the alkene and the addition of a nucleophile to the other sp^2 carbon (Section 5.5).*

A curved arrow with a two-barbed arrowhead signifies the movement of two electrons. The arrow always points *from* the electron donor *to* the electron acceptor.

the electrophile adds to an sp^2 carbon of the alkene

carbocation intermediate

the nucleophile adds to the carbocation

In this chapter, you will see that alkenes react with a wide variety of electrophiles. You will also see that some of the reactions form carbocation intermediates (like the one formed when an alkene reacts with HBr), some form other kinds of intermediates, and some form no intermediate at all. At first glance, the reactions covered in this chapter may appear to be quite different, but you will see that they all occur by similar mechanisms.

As you study each reaction, look for the features they all have in common:

- The relatively loosely held π electrons of the carbon–carbon double bond are attracted to an electrophile.
- Each reaction starts with the addition of an electrophile to one of the sp^2 carbons of the alkene.
- Each reaction concludes with the addition of a nucleophile to the other sp^2 carbon.

The end result is that the π bond breaks and the electrophile and nucleophile form new σ bonds with the sp^2 carbons. Notice that the sp^2 carbons in the reactant become sp^3 carbons in the product.

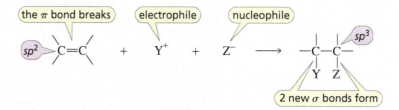

This reactivity makes it possible to synthesize a wide variety of compounds from alkenes. For example, we will see that alkyl halides, alcohols, ethers, epoxides, alkanes, aldehydes, and ketones can all be synthesized from alkenes by electrophilic addition reactions. The particular product obtained depends only on the *electrophile* and the *nucleophile* used in the addition reaction.

6.1 THE ADDITION OF A HYDROGEN HALIDE TO AN ALKENE

If the electrophilic reagent that adds to an alkene is a hydrogen halide (HF, HCl, HBr, or HI), the product of the reaction will be an alkyl halide:

Because the alkenes in the preceding reactions have the same substituents on both sp^2 carbons, it is easy to predict the product of the reaction: the electrophile (H^+) adds to either one of the sp^2 carbons, and the nucleophile (X^-; X = a halogen) adds to the other sp^2 carbon. It does not matter which sp^2 carbon the electrophile adds to because the same product is obtained in either case.

But what happens if the alkene does *not* have the same substituents on both sp^2 carbons? Which sp^2 carbon gets the hydrogen? For example, does the following reaction form *tert*-butyl chloride or isobutyl chloride?

tert-butyl chloride   isobutyl chloride

To answer this question, we need to carry out this reaction, isolate the products, and identify them. When we do, we find that the only product is *tert*-butyl chloride. If we can find out *why* it is the only product, then we can use this knowledge to predict the products of other alkene reactions. To do this, we need to look again at the **mechanism of the reaction** (Section 5.5).

Recall that the first step of the reaction—the addition of H^+ to an sp^2 carbon to form either the *tert*-butyl cation or the isobutyl cation—is the slow rate-determining step (Section 5.12). If there is any difference in the rate of formation of these two carbocations, then the one that is formed faster will be the predominant product of the first step. And the carbocation formed in the first step determines the final product of the reaction. That is, if the *tert*-butyl cation is formed, it will react rapidly with Cl^- to form *tert*-butyl chloride. On the other hand, if the isobutyl cation is formed, it will react rapidly with Cl^- to form isobutyl chloride. Because we know that the only product of the reaction is *tert*-butyl chloride, the *tert*-butyl cation must be formed faster than the isobutyl cation.

The carbocation's positive charge is on the sp^2 carbon that does *not* become attached to the proton.

Now we need to find out why the *tert*-butyl cation is formed faster. To answer this question, we need to look at two things: (1) what factors affect the stability of a carbocation and (2) how the stability of a carbocation affects the rate at which it is formed.

PROBLEM 1

Draw the mechanism for the reaction of cyclohexene with HCl.

6.2 CARBOCATION STABILITY DEPENDS ON THE NUMBER OF ALKYL GROUPS ATTACHED TO THE POSITIVELY CHARGED CARBON

Carbocations are classified based on the carbon that carries the positive charge:

- A **primary carbocation** has a positive charge on a primary carbon.
- A **secondary carbocation** has a positive charge on a secondary carbon.
- A **tertiary carbocation** has a positive charge on a tertiary carbon.

Carbocation Stability

Tertiary carbocations are more stable than secondary carbocations, and secondary carbocations are more stable than primary carbocations. Thus, we see that the stability of a carbocation increases as the number of alkyl substituents attached to the positively charged carbon increases. These are relative stabilities, however, because carbocations are rarely stable enough to isolate.

relative stabilities of carbocations

The more alkyl groups attached to the positively charged carbon, the more stable the carbocation.

Alkyl groups stabilize carbocations because they decrease the concentration of positive charge on the carbon. Notice that the blue area in the following electrostatic potential maps (representing positive charge) is least intense for the most stable *tert*-butyl cation (a tertiary carbocation) and most intense for the least stable methyl cation.

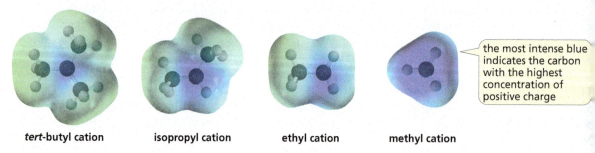

the most intense blue indicates the carbon with the highest concentration of positive charge

tert-butyl cation   isopropyl cation   ethyl cation   methyl cation

Hyperconjugation

Carbocation stability: 3° > 2° > 1°

How do alkyl groups decrease the concentration of positive charge on the carbon? Recall that the positive charge on a carbon signifies an empty *p* orbital (Section 1.10). Figure 6.1 shows that in the ethyl cation, the orbital of an adjacent C—H σ bond (the orange orbital) can overlap the empty *p* orbital (the purple orbital). This movement of electrons from a σ bond orbital toward the vacant *p* orbital decreases the charge on the sp^2 carbon and causes a partial positive charge to develop on the two atoms bonded by the overlapping σ bond orbital (the H and the C). With three atoms sharing the positive charge, the carbocation is stabilized because a charged species is more stable if its charge is dispersed over more than one atom (Section 2.8). In contrast, the positive charge in the methyl cation is concentrated solely on one atom.

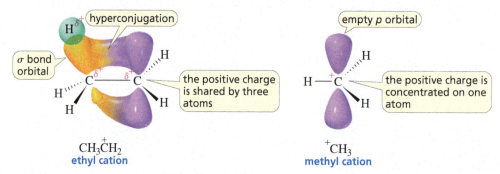

hyperconjugation

σ bond orbital

the positive charge is shared by three atoms

empty *p* orbital

the positive charge is concentrated on one atom

$CH_3\overset{+}{C}H_2$
ethyl cation

$\overset{+}{C}H_3$
methyl cation

▶ **Figure 6.1**
Stabilization of a carbocation by hyperconjugation. In the ethyl cation, the electrons of an adjacent C—H σ bond orbital are delocalized into the empty *p* orbital. Hyperconjugation cannot occur in the methyl cation.

Recall that delocalization of electrons by the overlap of a σ bond orbital with an empty orbital on an adjacent carbon is called **hyperconjugation** (Section 3.11). The molecular orbital diagram in Figure 6.2 is another way of depicting the stabilization achieved by the overlap of a filled C—H σ bond orbital with an empty *p* orbital.

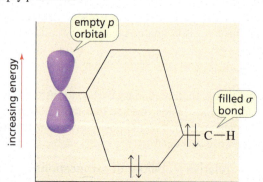

empty *p* orbital

filled σ bond

C—H

increasing energy

▶ **Figure 6.2**
A molecular orbital diagram showing the stabilization achieved by overlapping the electrons of a C—H σ bond orbital with an empty *p* orbital.

Hyperconjugation occurs only if the orbital of the σ bond and the empty *p* orbital have the proper orientation. The proper orientation is easily achieved, though, because there is free rotation about the carbon–carbon σ bond (Section 3.11). Notice that the σ bond orbitals that can overlap the empty *p* orbital are those *attached to an atom that is attached to the positively charged carbon*. In the *tert*-butyl cation, nine σ bond orbitals can potentially overlap the empty *p* orbital of the positively charged carbon. (The nine σ bonds are indicated by red dots.)

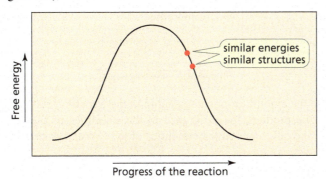

each red dot indicates a σ bond that can engage in hyperconjugation

tert-butyl cation
tertiary
carbocation

isopropyl cation
secondary
carbocation

ethyl cation
primary
carbocation

propyl cation
primary
carbocation

The isopropyl cation has six such σ bond orbitals, whereas the ethyl and propyl cations each have three. Therefore, hyperconjugation stabilizes the tertiary carbocation more than the secondary carbocation, and it stabilizes the secondary carbocation more than either of the primary carbocations. Notice that both C—H and C—C σ bond orbitals can overlap the empty p orbital.

PROBLEM 2 ◆

a. How many σ bond orbitals are available for overlap with the vacant p orbital in the methyl cation?
b. Which is more stable: a methyl cation or an ethyl cation? Why?

PROBLEM 3 ◆

a. How many σ bond orbitals are available for overlap with the vacant p orbital in

   1. the isobutyl cation?   **2.** the n-butyl cation?   **3.** the sec-butyl cation?

b. Which of the carbocations in part **a** is most stable?

PROBLEM 4 ◆

Rank the following carbocations in each set from most stable to least stable:

$$\text{a. } \overset{\displaystyle CH_3}{\underset{\displaystyle +}{CH_3CH_2\overset{|}{C}CH_3}} \qquad CH_3CH_2\overset{+}{C}HCH_3 \qquad CH_3CH_2CH_2\overset{+}{C}H_2$$

$$\text{b. } \underset{\displaystyle Cl}{CH_3\overset{|}{C}HCH_2\overset{+}{C}H_2} \qquad \underset{\displaystyle CH_3}{CH_3\overset{|}{C}HCH_2\overset{+}{C}H_2} \qquad \underset{\displaystyle F}{CH_3\overset{|}{C}HCH_2\overset{+}{C}H_2}$$

6.3   WHAT DOES THE STRUCTURE OF THE TRANSITION STATE LOOK LIKE?

Now that we know that a tertiary carbocation is more stable than a primary carbocation, the next question we need to answer is why the more stable carbocation is formed more rapidly.

We know that the rate of a reaction is determined by the free energy of activation—the difference between the free energy of the transition state and the free energy of the reactant (Section 5.11). Therefore, to understand how the stability of a carbocation affects the rate at which it is formed, we need to understand how its stability affects the stability of the transition state for its formation.

If two points on the uphill leg or the downhill leg of a reaction coordinate diagram have similar energies, they will also have similar structures—and the closer their energies, the more similar their structures will be (Figure 6.1).

similar energies
similar structures

Free energy

Progress of the reaction

We can use this information to understand what the transition state of a reaction looks like (Figure 6.2).

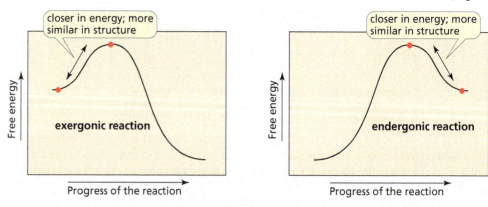

- In an exergonic reaction, the energy of the transition state is closer to the energy of the reactant than to the energy of the product, so the structure of the transition state more closely resembles the structure of the reactant.
- In an endergonic reaction, the energy of the transition state is closer to the energy of the product, so the structure of the transition state more closely resembles the structure of the product.

The above can be summarized by the Hammond postulate, which states the following:

The transition state is more similar in structure
to the species to which it is more similar in energy.

Because formation of a carbocation is an endergonic reaction (Figure 6.3), the structure of the transition state resembles the structure of the carbocation product. This means that the transition state has a significant amount of positive charge on a carbon. The same factors that stabilize the positively charged carbocation stabilize the partially positively charged transition state. Therefore, the transition state leading to the *tert*-butyl cation (a tertiary carbon with a partial positive charge) is more stable (lower in energy) than the transition state leading to the isobutyl cation (a primary carbon with a partial positive charge). Thus, the *tert*-butyl cation, with a smaller energy of activation, is formed faster than the isobutyl cation.

In an electrophilic addition reaction,
the more stable carbocation is formed more rapidly.

Now we know *why* the *tert*-butyl cation is formed faster than the isobutyl cation when 2-methylpropene reacts with HCl (Section 6.1).

The more stable carbocation is formed more rapidly.

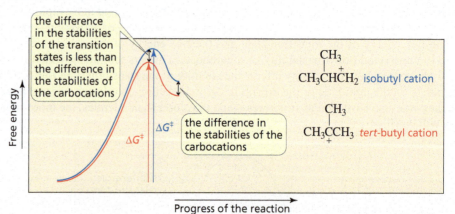

▶ **Figure 6.3**
Reaction coordinate diagram for the addition of H⁺ to 2-methylpropene to form the primary isobutyl cation and the tertiary *tert*-butyl cation.

Notice that because the amount of positive charge in the transition state is less than the amount of positive charge in the product, the difference in the stabilities of the two transition states in Figure 6.3 is less than the difference in the stabilities of the two carbocation products.

Formation of the carbocation is the rate-limiting step, so the relative rates of formation of the two carbocations determine the relative amounts of products formed. If the difference in the rates is small, both products will be formed; the major product will be the one formed from reaction of the nucleophile with the more rapidly formed carbocation. If the difference in the rates is sufficiently large, however, the product formed from reaction of the nucleophile with the more rapidly formed carbocation will be the only product. For example, in the reaction we first looked at on page 237, the rates of formation of the two possible carbocation intermediates—one primary and the other tertiary—are sufficiently different to cause *tert*-butyl chloride to be the only product of the reaction.

$$CH_3\overset{\overset{\displaystyle CH_3}{|}}{C}{=}CH_2 \ + \ HCl \ \longrightarrow \ CH_3\overset{\overset{\displaystyle CH_3}{|}}{\underset{\underset{\displaystyle Cl}{|}}{C}}CH_3 \qquad CH_3\overset{\overset{\displaystyle CH_3}{|}}{C}HCH_2Cl$$

not formed

only product formed

PROBLEM 5 ◆

Is the structure of the transition state in the following reaction coordinate diagrams more similar to the structure of the reactant or to the structure of the product?

a.   b.   c.   d.

6.4   ELECTROPHILIC ADDITION REACTIONS ARE REGIOSELECTIVE

Now that we know that the major product of an electrophilic addition reaction is the one obtained by adding the electrophile to the sp^2 carbon that results in formation of the more stable carbocation, we can predict the major product of the reaction of any unsymmetrical alkene.

For example, in the following reaction, the proton can add to C-1 to form a secondary carbocation or it can add to C-2 to form a primary carbocation. Because the secondary carbocation is more stable, it is formed more rapidly. (Primary carbocations are so unstable that they form only with great difficulty.) As a result, the only product is 2-chloropropane.

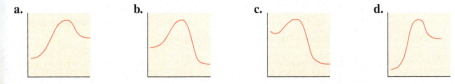

H⁺ adds to C-1

HCl

$$\overset{2}{C}H_3\overset{1}{C}H{=}CH_2$$

$$CH_3\overset{+}{C}HCH_3 \xrightarrow{\ Cl^-\ } CH_3\overset{\overset{\displaystyle Cl}{|}}{C}HCH_3$$

secondary carbocation

2-chloropropane

HCl

H⁺ adds to C-2

$$CH_3CH_2\overset{+}{C}H_2$$

primary carbocation

The sp^2 carbon that does not get the proton is the one that is positively charged in the carbocation intermediate.

Regioselective Reactions

Both of the following reactions form two products. In each case, the major product is the one obtained from reaction of the nucleophile with the more stable carbocation—that is, the carbocation that is formed more rapidly.

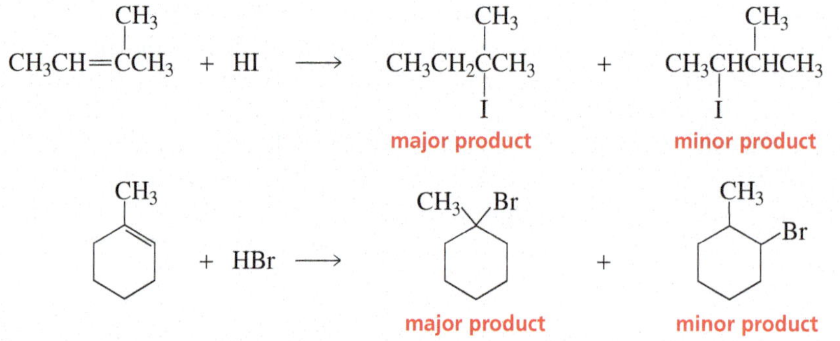

$$CH_3CH=\overset{\underset{\displaystyle CH_3}{|}}{C}CH_3 \;+\; HI \;\longrightarrow\; CH_3CH_2\overset{\underset{\displaystyle CH_3}{|}}{\underset{\displaystyle I}{C}}CH_3 \;+\; CH_3\overset{\underset{\displaystyle CH_3}{|}}{\underset{\displaystyle I}{C}H}CHCH_3$$

major product   minor product

$$\text{(methylcyclohexene)} \;+\; HBr \;\longrightarrow\; \text{(1-bromo-1-methylcyclohexane)} \;+\; \text{(1-bromo-2-methylcyclohexane)}$$

major product   minor product

The two products of each of the preceding reactions are *constitutional isomers.* That is, they have the same molecular formula but differ in how their atoms are connected. A reaction in which two or more constitutional isomers could be obtained as products but one of them predominates is called a **regioselective reaction.**

Degrees of Regioselectivity

There are degrees of **regioselectivity:** a reaction can be *moderately regioselective, highly regioselective,* or *completely regioselective.* For example, the addition of HCl to 2-methylpropene (where the two possible carbocations are tertiary and primary) is more highly regioselective than the addition of HCl to 2-methyl-2-butene (where the two possible carbocations are tertiary and secondary) because the two carbocations formed in the latter reaction are closer in stability. In a completely regioselective reaction, only one of the possible products is formed.

> more highly regioselective

$$CH_3\overset{\underset{\displaystyle CH_3}{|}}{C}=CH_2 \;+\; HCl \;\longrightarrow\; CH_3\overset{\underset{\displaystyle CH_3}{|}}{\underset{\displaystyle Cl}{C}}-CH_3 \;+\; CH_3\overset{\underset{\displaystyle CH_3}{|}}{C}H-\overset{\underset{\displaystyle Cl}{|}}{C}H_2$$

only product   not formed

Regioselectivity is the preferential formation of one constitutional isomer over another.

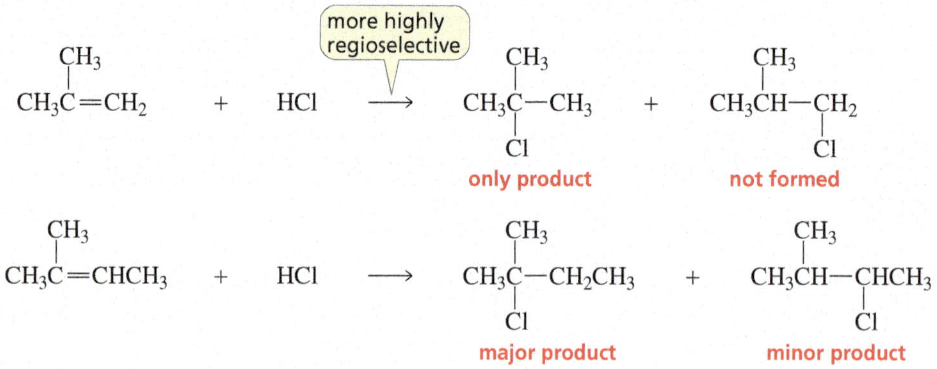

$$CH_3\overset{\underset{\displaystyle CH_3}{|}}{C}=CHCH_3 \;+\; HCl \;\longrightarrow\; CH_3\overset{\underset{\displaystyle CH_3}{|}}{\underset{\displaystyle Cl}{C}}-CH_2CH_3 \;+\; CH_3\overset{\underset{\displaystyle CH_3}{|}}{C}H-\overset{\underset{\displaystyle Cl}{|}}{C}HCH_3$$

major product   minor product

The following reaction is not regioselective. Because the addition of H^+ to either of the sp^2 carbons produces a secondary carbocation, both carbocations are formed at about the same rate. Therefore approximately equal amounts of the two alkyl halides are obtained.

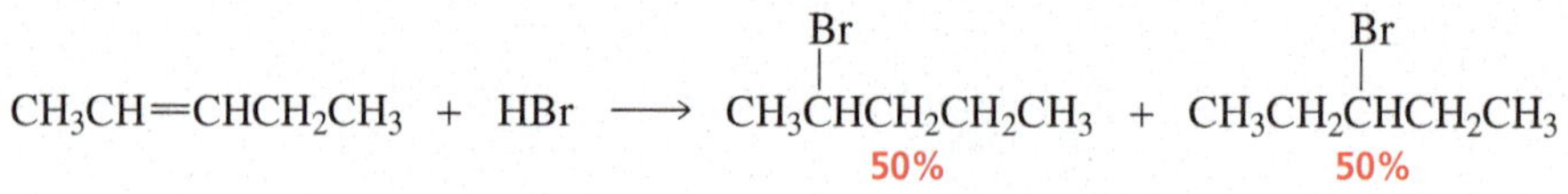

$$CH_3CH=CHCH_2CH_3 \;+\; HBr \;\longrightarrow\; CH_3\overset{\underset{\displaystyle Br}{|}}{C}HCH_2CH_2CH_3 \;+\; CH_3CH_2\overset{\underset{\displaystyle Br}{|}}{C}HCH_2CH_3$$

50%   50%

PROBLEM 6 ◆

To which compound is the addition of HBr more highly regioselective?

a. $CH_3CH_2\overset{\underset{\displaystyle CH_3}{|}}{C}=CH_2$ or $CH_3\overset{\underset{\displaystyle CH_3}{|}}{C}=CHCH_3$ **b.** (methylenecyclohexane) or (methylcyclohexene)

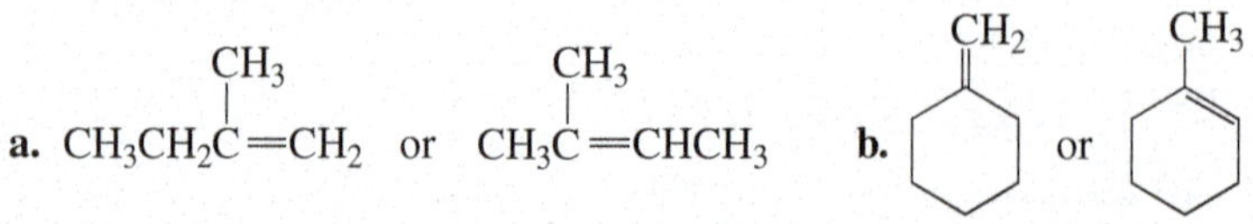

The Rule That Determines the Product of an Electrophilic Addition Reaction

Vladimir Markovnikov was the first to recognize that the major product obtained when a hydrogen halide adds to an alkene results from adding the H^+ to the sp^2 carbon bonded to the most hydrogens. Consequently, this is often referred to as **Markovnikov's rule.** However, Markovnikov's rule is alid only for addition reactions in which the electrophile is H$^+$. A better rule—and the one we ill use in this book—applies to all electrophilic addition reactions. Be sure to remember this rule ecause you will see that it applies to every reaction in this chapter.

The electrophile adds preferentially to the sp^2 carbon bonded to the most hydrogens.

LEARN THE STRATEGY

This rule provides a quick way to determine to which *sp^2* carbon the electrophile adds in the first tep of the addition reaction. The answer you get using this rule is the same as the answer you get y determining relative carbocation stabilities. In the following reaction, for example,

$$\overset{2}{CH_3CH_2}\overset{1}{CH}=CH_2 + HCl \longrightarrow CH_3CH_2\underset{\underset{Cl}{|}}{CH}CH_3$$

The electrophile adds preferentially to the *sp^2* carbon bonded to the most hydrogens.

ve can say that the electrophile (in this case, H$^+$) adds preferentially to C-1 because it is the *sp^2* car-•on bonded to the most hydrogens. Or we can say that H$^+$ adds to C-1 to form a secondary carboca-ion, which is more stable than the primary carbocation that would be formed if H$^+$ added to C-2.
The previous examples illustrate the way organic reactions are typically written. The reactants ⸜re written to the left of the reaction arrow, and the products are written to the right of the arrow. Any conditions that need to be stipulated, such as the solvent, the temperature, or a catalyst, are written ⸜bove or below the arrow. Sometimes, as shown below, only the organic (carbon-containing) reagent s written to the left of the arrow, and any other reagents are written above or below the arrow.

$$CH_3CH_2CH=CH_2 \xrightarrow{\text{HCl}} CH_3CH_2\underset{\underset{Cl}{|}}{CH}CH_3$$

PROBLEM 7 ◆

USE THE STRATEGY

What is the major product obtained from the addition of HBr to each of the following compounds?

a. $CH_3CH_2CH=CH_2$

b. $CH_3CH=\underset{\underset{CH_3}{|}}{C}CH_3$

c. 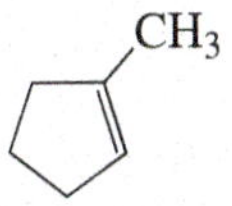 (cyclopentene with CH$_3$ substituent)

d. $CH_2=\underset{\underset{CH_3}{|}}{C}CH_2CH_2CH_3$

e. 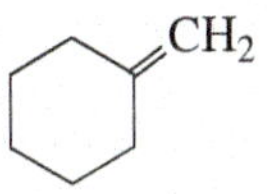 (cyclohexane with =CH$_2$ substituent)

f. $CH_3CH=CHCH_3$

PROBLEM-SOLVING STRATEGY

Planning the Synthesis of an Alkyl Halide

LEARN THE STRATEGY

a. What alkene should be used to synthesize 3-bromohexane?

$$? + HBr \longrightarrow CH_3CH_2\underset{\underset{Br}{|}}{CH}CH_2CH_2CH_3$$
3-bromohexane

The best way to answer this kind of question is to begin by listing all the alkenes that could be used. Because you want to synthesize an alkyl halide that has a bromo substituent at the 3-position, the alkene should have an *sp^2* carbon at that position. Two alkenes fit the description: 2-hexene and 3-hexene.

$CH_3CH=CHCH_2CH_2CH_3$ $CH_3CH_2CH=CHCH_2CH_3$
2-hexene   3-hexene

Because there are two possibilities, we next need to decide whether there is any advantage to using one over the other. The addition of H$^+$ to 2-hexene forms two different secondary carbocations. Because the carbocations have the same stability, approximately equal amounts of each are formed. Therefore, half the product is the desired 3-bromohexane and half is 2-bromohexane.

$$CH_3CH=CHCH_2CH_2CH_3 \xrightarrow{\;HBr\;}$$
2-hexene

$$\xrightarrow{HBr} CH_3CH_2\overset{+}{C}HCH_2CH_2CH_3 \xrightarrow{Br^-} CH_3CH_2CHCH_2CH_2CH_3$$

secondary carbocation

$$\underset{Br}{\overset{|}{\;}}$$
3-bromohexane

$$\xrightarrow{HBr} CH_3\overset{+}{C}HCH_2CH_2CH_2CH_3 \xrightarrow{Br^-} CH_3CHCH_2CH_2CH_2CH_3$$

secondary carbocation

$$\underset{Br}{\overset{|}{\;}}$$
2-bromohexane

The addition of H⁺ to either of the sp^2 carbons of 3-hexene, on the other hand, forms the same carbocation because the alkene is symmetrical. Therefore, all the product (not just half) is the desired 3-bromohexane. Thus, 3-hexene should be used for the synthesis of the desired compound.

$$CH_3CH_2CH=CHCH_2CH_3 \xrightarrow{\;HBr\;} CH_3CH_2\overset{+}{C}HCH_2CH_2CH_3 \xrightarrow{\;Br^-\;} CH_3CH_2CHCH_2CH_2CH_3$$
3-hexene

only one carbocation is formed

$$\underset{Br}{\overset{|}{\;}}$$
3-bromohexane

b. What alkene should be used to synthesize 2-bromopentane?

$$? \quad + \quad HBr \quad \longrightarrow \quad CH_3CHCH_2CH_2CH_3$$

$$\underset{Br}{\overset{|}{\;}}$$
2-bromopentane

Either 1-pentene or 2-pentene could be used because both have an sp^2 carbon at the 2-position.

$$CH_2=CHCH_2CH_2CH_3 \qquad CH_3CH=CHCH_2CH_3$$
1-pentene **2-pentene**

When H⁺ adds to 1-pentene, a secondary and a primary carbocation could be formed. The primary carbocation is so unstable, however, that none is formed. Thus, 2-bromopentane is the only product of the reaction.

$$CH_2=CHCH_2CH_2CH_3 \xrightarrow{\;HBr\;}$$
1-pentene

$$\xrightarrow{HBr} CH_3\overset{+}{C}HCH_2CH_2CH_3 \xrightarrow{Br^-} CH_3CHCH_2CH_2CH_3$$

$$\underset{Br}{\overset{|}{\;}}$$
2-bromopentane

$$\xrightarrow{HBr} \;\; \overset{+}{C}H_2CH_2CH_2CH_2CH_3$$

When H⁺ adds to 2-pentene, on the other hand, two different secondary carbocations can be formed. Because they have the same stability, they are formed in approximately equal amounts. Thus, only about half of the product is 2-bromopentane. The other half is 3-bromopentane.

$$CH_3CH=CHCH_2CH_3 \xrightarrow{\;HBr\;}$$
2-pentene

$$\xrightarrow{HBr} CH_3\overset{+}{C}HCH_2CH_2CH_3 \xrightarrow{Br^-} CH_3CHCH_2CH_2CH_3$$

$$\underset{Br}{\overset{|}{\;}}$$
2-bromopentane

$$\xrightarrow{HBr} CH_3CH_2\overset{+}{C}HCH_2CH_3 \xrightarrow{Br^-} CH_3CH_2CHCH_2CH_3$$

$$\underset{Br}{\overset{|}{\;}}$$
3-bromopentane

Because all the alkyl halide formed from 1-pentene is the desired product but only half the alkyl halide formed from 2-pentene is the desired product, 1-pentene is the best alkene to use for the synthesis.

PROBLEM 8 ◆

What alkene should be used to synthesize each of the following alkyl bromides?

a. $CH_3\underset{\underset{Br}{|}}{\overset{\overset{CH_3}{|}}{C}}CH_3$
b. ⬡—CH_2CHCH_3 with Br
c. ⬡—$\underset{\underset{Br}{|}}{\overset{\overset{CH_3}{|}}{C}}CH_3$
d. ⬡ with CH_2CH_3 and Br

6.5 THE ADDITION OF WATER TO AN ALKENE

An alkene does not react with water, because there is no electrophile present to start the reaction by adding to the alkene. The O—H bonds of water are too strong—water is too weakly acidic—to allow the hydrogen to act as an electrophile.

$$CH_3CH=CH_2 \;+\; H_2O \;\longrightarrow\; \text{no reaction}$$

If an acid (the acid used most often is H_2SO_4) is added to the solution, then a reaction will occur because the acid provides the electrophile. The product of the reaction is an *alcohol*. The addition of water to a molecule is called **hydration,** so we can say that an alkene is *hydrated* in the presence of water and acid.

$$R-CH=CH_2 \;+\; H_2O \;\overset{H_2SO_4}{\rightleftharpoons}\; R-\underset{\underset{OH}{|}}{CH}-\underset{\underset{H}{|}}{CH_2}$$

(π bond breaks) (new σ bond) **an alcohol** (new σ bond)

When you look at the mechanism for this reaction, notice that the first two steps are the same (except for the nucleophile employed) as the two steps of the mechanism for the addition of a hydrogen halide to an alkene (Section 6.1).

MECHANISM FOR THE ACID-CATALYZED ADDITION OF WATER TO AN ALKENE

$$R-CH=CH_2 + H-OSO_3H \overset{slow}{\rightleftharpoons} R-\underset{+}{CH}CH_3 \overset{H_2\ddot{O}}{\underset{fast}{\rightleftharpoons}} R-\underset{\underset{\overset{+}{:}\ddot{O}H}{|}}{CH}CH_3 \overset{H_2\ddot{O}}{\underset{fast}{\rightleftharpoons}} R-\underset{\underset{OH}{|}}{CH}CH_3 + H_3\overset{+}{O}$$

addition of the electrophile HSO_4^- addition of the nucleophile   **protonated alcohol** H_2O removes a proton, regenerating the acid catalyst

- H^+ (an electrophile) adds to the sp^2 carbon of the alkene (a nucleophile) that is bonded to the most hydrogens.
- H_2O (a nucleophile) adds to the carbocation (an electrophile), forming a protonated alcohol.
- The protonated alcohol loses a proton because the pH of the solution is greater than the pK_a of the protonated alcohol (Section 2.10). (We saw that protonated alcohols are very strong acids; Section 2.6.)

When H_2O adds to an electron-deficient carbon, it is called a nucleophile; when it shares its electrons with a proton, it is called a base.

Thus, the overall reaction is the addition of an electrophile to the sp^2 carbon bonded to the most hydrogens and addition of a nucleophile to the other sp^2 carbon.

The addition of the electrophile to the alkene is relatively slow because the π bond must be broken (Section 5.12). The subsequent addition of the nucleophile to the carbocation occurs rapidly because no bonds need to be broken in this step. In fact, the reaction of the carbocation with a nucleophile is so fast that the carbocation combines with whatever nucleophile it collides with first. There are two nucleophiles in solution: water and HSO_4^- (the conjugate base of the acid used to start the reaction).* Because the concentration of water is much greater than the concentration of HSO_4^-, the carbocation is much more likely to collide with water. The final product of the addition reaction, therefore, is an alcohol.

H_2SO_4 catalyzes the hydration reaction. We saw that a catalyst increases the rate of a reaction but is not consumed during the course of the reaction (Section 5.13). Thus, the proton adds to the alkene in the first step and is returned to the reaction mixture in the final step. Overall, then, a proton is not consumed. Because the catalyst employed in the hydration of an alkene is an acid, hydration is an **acid-catalyzed reaction.**

Remember that catalysts increase the reaction rate by decreasing the free energy of activation, but they do *not* affect the equilibrium constant of the reaction (Section 5.13). In other words, a catalyst increases the *rate* at which a product is formed but does not affect the *amount* of product formed when the reaction has reached equilibrium.

PROBLEM 9 ♦

The pK_a of a protonated alcohol is about -2.5, and the pK_a of an alcohol is about 15. Therefore, as long as the pH of the solution is greater than _____ and less than _____, more than 50% of 2-propanol (the product of the reaction on p. 244) will be in its neutral, nonprotonated form.

PROBLEM 10 ♦

Answer the following questions about the mechanism for the acid-catalyzed hydration of an alkene:
a. How many transition states are there?
b. How many intermediates are there?
c. Which step in the forward direction has the smallest rate constant?

PROBLEM 11 ♦

What is the major product obtained from the acid-catalyzed hydration of each of the following alkenes?
a. $CH_3CH_2CH_2CH{=}CH_2$ **c.** $CH_3CH_2CH_2CH{=}CHCH_3$

b.   **d.**

6.6 THE ADDITION OF AN ALCOHOL TO AN ALKENE

Alcohols react with alkenes in the same way that water does, so this reaction, too, requires an acid catalyst. The product of the reaction is an *ether.*

MECHANISM FOR THE ACID-CATALYZED ADDITION OF AN ALCOHOL TO AN ALKENE

---

* HO^- cannot be a nucleophile in this reaction because there is no appreciable concentration of HO^- in an acidic solution.

- H^+ (the electrophile) adds to the sp^2 carbon bonded to the most hydrogens.
- CH_3OH (the nucleophile) adds to the carbocation, forming a protonated ether.
- The protonated ether loses a proton, because the pH of the solution is greater than the pK_a of the protonated ether ($pK_a \sim -3.6$).

The *mechanism for the acid-catalyzed addition of an alcohol* is the same as the *mechanism for the acid-catalyzed addition of water.* The only difference in the two reactions is that ROH is the nucleophile instead of H_2O.

PROBLEM 12

a. What is the major product of each of the following reactions?
b. What do all the reactions have in common?
c. How do all the reactions differ?

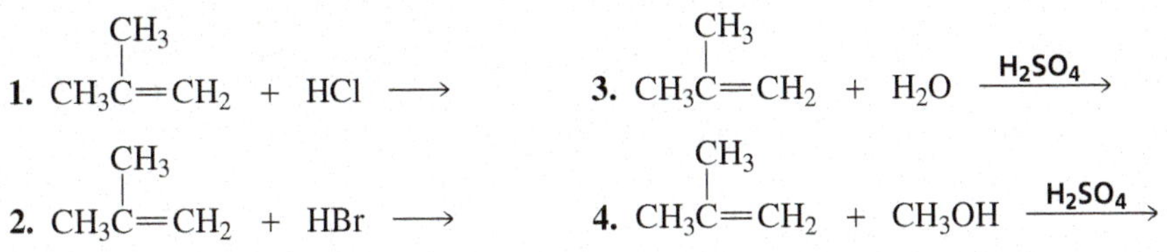

$$\text{1. } CH_3\overset{\overset{\displaystyle CH_3}{|}}{C}=CH_2 \ + \ HCl \longrightarrow$$

$$\text{2. } CH_3\overset{\overset{\displaystyle CH_3}{|}}{C}=CH_2 \ + \ HBr \longrightarrow$$

$$\text{3. } CH_3\overset{\overset{\displaystyle CH_3}{|}}{C}=CH_2 \ + \ H_2O \xrightarrow{H_2SO_4}$$

$$\text{4. } CH_3\overset{\overset{\displaystyle CH_3}{|}}{C}=CH_2 \ + \ CH_3OH \xrightarrow{H_2SO_4}$$

PROBLEM 13 SOLVED

LEARN THE STRATEGY

How could the following compound be prepared using an alkene as one of the starting materials?

$$CH_3\overset{\overset{\displaystyle CH_3}{|}}{C}HO\underset{\underset{\displaystyle CH_3}{|}}{C}HCH_2CH_3$$

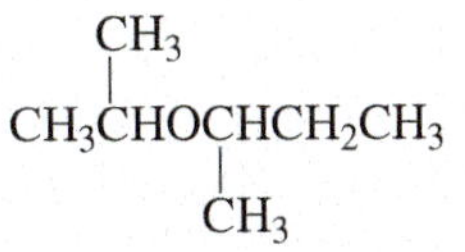

SOLUTION The desired compound—called a **target molecule**—can be prepared by the acid-catalyzed addition of an alcohol to an alkene. There are three different combinations of an alkene and an alcohol that could be used for the synthesis of this particular ether. One requires a 3-carbon alkene and a 4-carbon alcohol; the other two require a 4-carbon alkene and a 3-carbon alcohol.

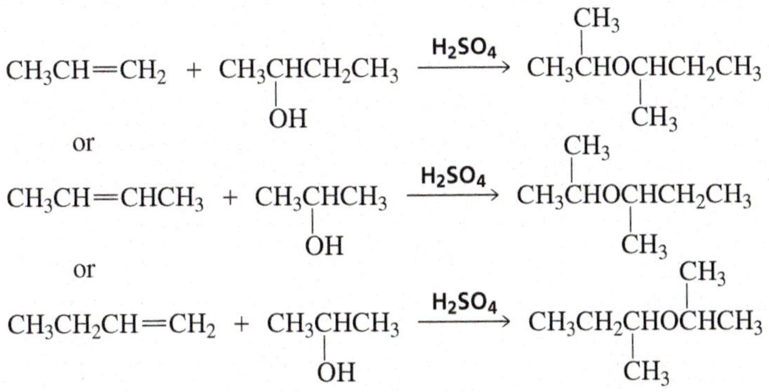

$$CH_3CH=CH_2 \ + \ CH_3\underset{\underset{\displaystyle OH}{|}}{C}HCH_2CH_3 \xrightarrow{H_2SO_4} CH_3\overset{\overset{\displaystyle CH_3}{|}}{C}HO\underset{\underset{\displaystyle CH_3}{|}}{C}HCH_2CH_3$$

or

$$CH_3CH=CHCH_3 \ + \ CH_3\underset{\underset{\displaystyle OH}{|}}{C}HCH_3 \xrightarrow{H_2SO_4} CH_3\overset{\overset{\displaystyle CH_3}{|}}{C}HO\underset{\underset{\displaystyle CH_3}{|}}{C}HCH_2CH_3$$

or

$$CH_3CH_2CH=CH_2 \ + \ CH_3\underset{\underset{\displaystyle OH}{|}}{C}HCH_3 \xrightarrow{H_2SO_4} CH_3CH_2\overset{\overset{\displaystyle CH_3}{|}}{C}HO\underset{\underset{\displaystyle CH_3}{|}}{C}HCH_3$$

PROBLEM 14

USE THE STRATEGY

How could the following compounds be prepared using an alkene as one of the starting materials?

a. (cyclopentane)—OH

b. $CH_3O\overset{\overset{\displaystyle CH_3}{|}}{\underset{\underset{\displaystyle CH_3}{|}}{C}}CH_3$

c. $CH_3\overset{\overset{\displaystyle CH_3}{|}}{C}HCH_2CH_3$, OH

d. (cyclohexane)—$\overset{\overset{\displaystyle CH_3}{}}{}$ OCH$_3$

e. $CH_3\overset{\overset{\displaystyle CH_3}{|}}{\underset{\underset{\displaystyle OH}{|}}{C}}CH_2CH_3$

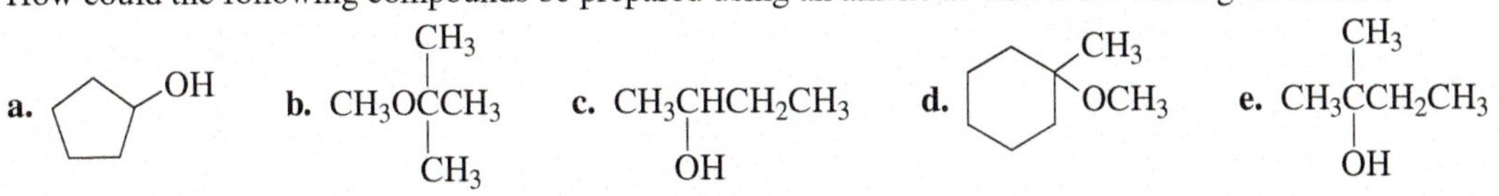

PROBLEM 15

Propose a mechanism for the following reaction (remember to use curved arrows to show the movement of electrons from the nucleophile to the electrophile):

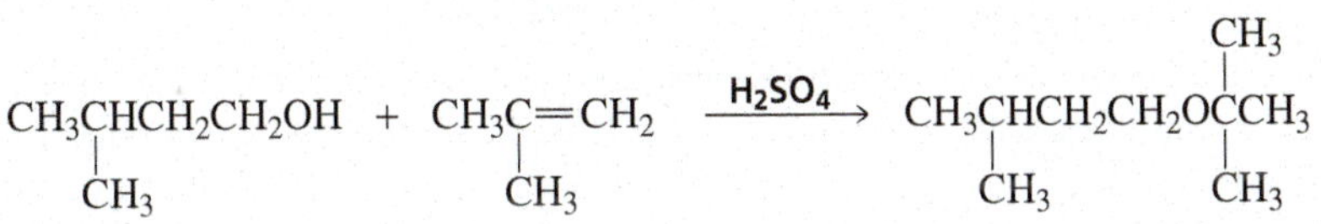

$$CH_3\underset{\underset{\displaystyle CH_3}{|}}{C}HCH_2CH_2OH \ + \ CH_3\underset{\underset{\displaystyle CH_3}{|}}{C}=CH_2 \xrightarrow{H_2SO_4} CH_3\underset{\underset{\displaystyle CH_3}{|}}{C}HCH_2CH_2O\overset{\overset{\displaystyle CH_3}{|}}{\underset{\underset{\displaystyle CH_3}{|}}{C}}CH_3$$

6.7 A CARBOCATION WILL REARRANGE IF IT CAN FORM A MORE STABLE CARBOCATION

Some electrophilic addition reactions form unexpected products—products that are not what you would get by adding an electrophile to the sp^2 carbon bonded to the most hydrogens and a nucleophile to the other sp^2 carbon.

For example, in the following reaction, 2-bromo-3-methylbutane is the expected product—the product obtained from adding H^+ to the sp^2 carbon bonded to the most hydrogens and adding Br^- to the other sp^2 carbon. This, however, is a minor product. 2-Bromo-2-methylbutane is the major product of the reaction.

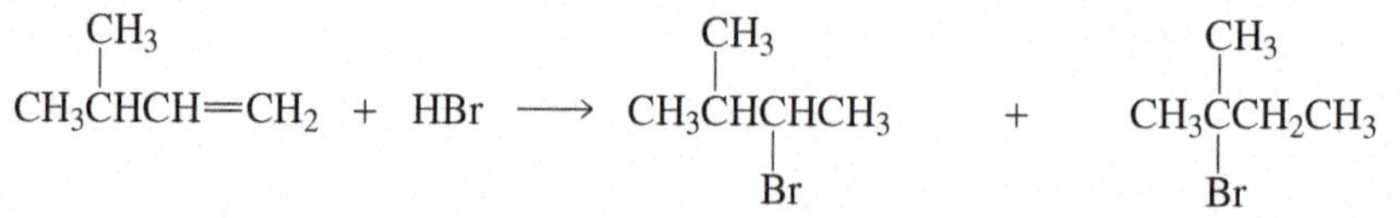

2-bromo-3-methylbutane
minor product

2-bromo-2-methylbutane
major product

In another example, the following reaction forms both 3-chloro-2,2-dimethylbutane (the expected product) and 2-chloro-2,3-dimethylbutane (the unexpected product). Again, the unexpected product is the major product of the reaction.

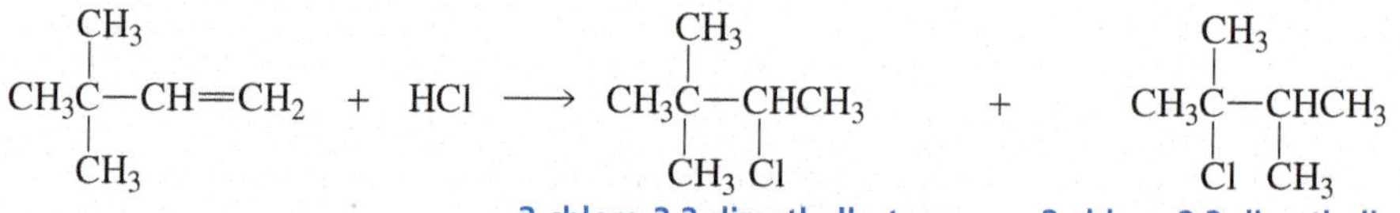

3-chloro-2,2-dimethylbutane
minor product

2-chloro-2,3-dimethylbutane
major product

In each reaction, the unexpected product results from a *rearrangement* of the carbocation intermediate. Not all carbocations rearrange.

Carbocations rearrange only if they become more stable as a result of the rearrangement.

1,2-Hydride Shift

Now let's look at the preceding reactions to see why the carbocations rearrange. In the first reaction, a *secondary* carbocation is formed initially. However, the secondary carbocation has a hydrogen that can shift with its pair of electrons to the adjacent positively charged carbon, creating a more stable *tertiary* carbocation.

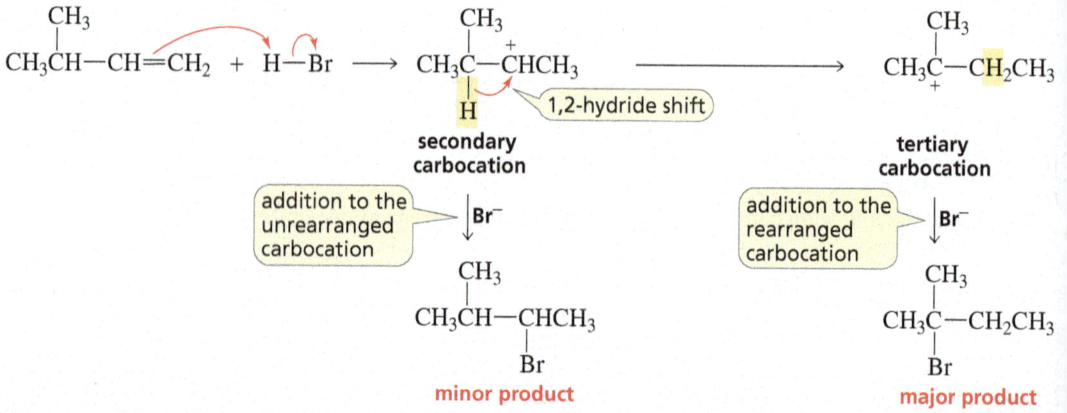

Carbocations rearrange if they become more stable as a result of the rearrangement.

secondary carbocation

1,2-hydride shift

tertiary carbocation

addition to the unrearranged carbocation

addition to the rearranged carbocation

minor product

major product

Because a hydrogen shifts with its pair of electrons, the rearrangement is called a hydride shift. (Recall that $H:^-$ is a hydride ion.) More specifically, it is called a **1,2-hydride shift** because the hydride ion moves from one carbon to an *adjacent* carbon.

As a result of the **carbocation rearrangement,** two alkyl halides are formed: one from adding the nucleophile to the unrearranged carbocation and one from adding the nucleophile to the rearranged carbocation. Rearrangement is so fast that little of the unrearranged carbocation exists in solution long enough to react with a nucleophile. Therefore, major product results from adding the nucleophile to the rearranged carbocation.

1,2-Methyl Shift

In the second reaction, again a *secondary* carbocation is formed initially. Then one of the methyl groups, with its pair of electrons, shifts to the adjacent positively charged carbon to form a more stable *tertiary* carbocation. This rearrangement is called a **1,2-methyl shift**—the methyl group moves with its electrons from one carbon to an *adjacent* carbon. Again, the major product is the one formed by adding the nucleophile to the rearranged carbocation.

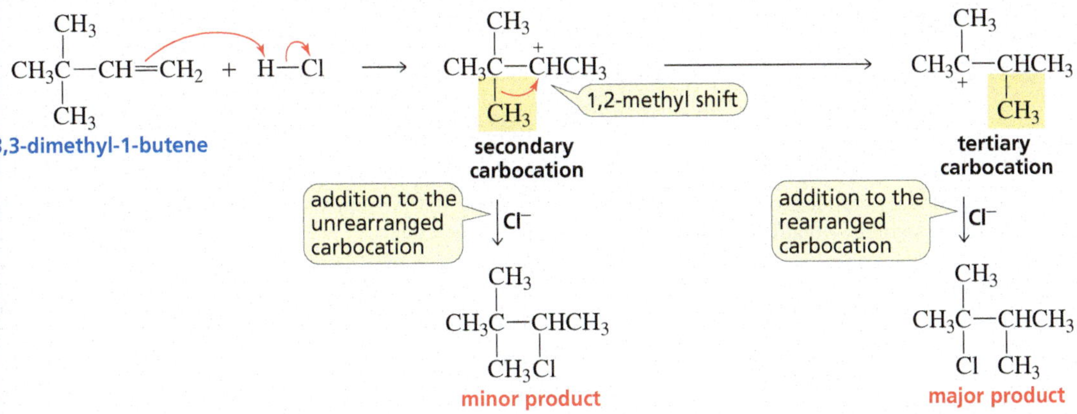

If a rearrangement does not lead to a more stable carbocation, then it typically does not occur. For example, the following reaction forms a secondary carbocation. A 1,2-hydride shift would form a different secondary carbocation. Because both carbocations are equally stable, there is no energetic advantage to the rearrangement. Consequently, the rearrangement does not occur, and only one alkyl halide is formed.

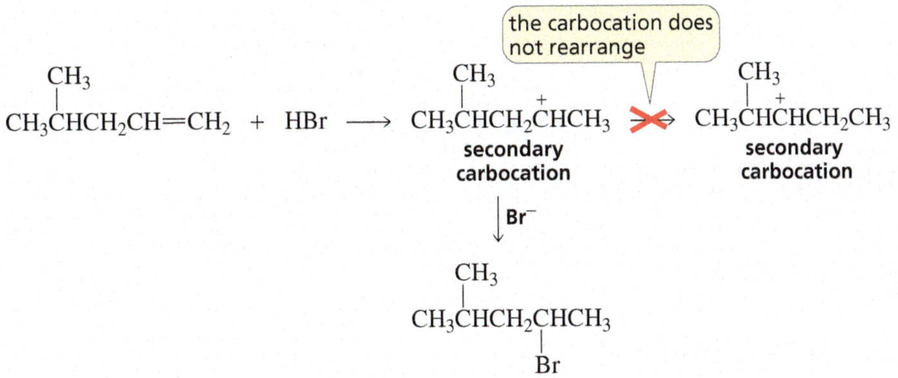

NOTE TO THE STUDENT
• Whenever a reaction forms a carbocation intermediate, check to see if it will rearrange to a more stable carbocation.

In subsequent chapters, you will study other reactions that form carbocation intermediates. Keep in mind that *whenever a reaction leads to the formation of a carbocation, you must check its structure to see if it will rearrange.*

PROBLEM 16 SOLVED

LEARN THE STRATEGY

Which of the following carbocations would you expect to rearrange?

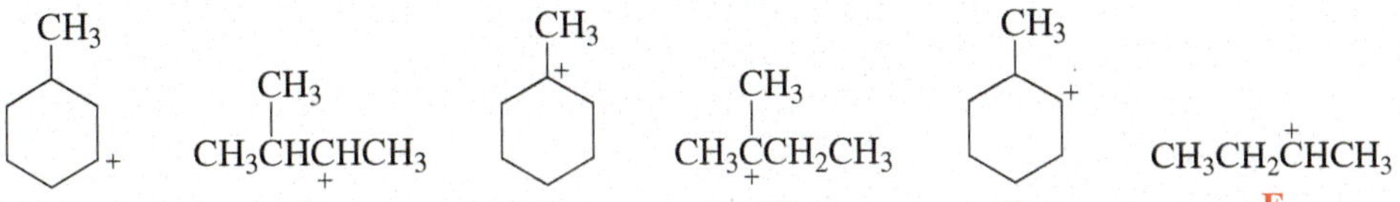

SOLUTION

A is a secondary carbocation. It does not rearrange because a 1,2-hydride shift would convert it to a different secondary carbocation, so there is no energetic advantage to the rearrangement.

B is a secondary carbocation. It rearranges because a 1,2-hydride shift converts it to a more stable tertiary carbocation.

$$CH_3\overset{\underset{|}{CH_3}}{C}-\overset{+}{C}HCH_3 \longrightarrow CH_3\overset{\underset{|}{CH_3}}{C}CH_2CH_3$$
$$\underset{H}{|}$$

C is a tertiary carbocation. It does not rearrange because its stability is not improved by rearrangement.
D is a tertiary carbocation. It does not rearrange because its stability is not improved by rearrangement.
E is a secondary carbocation. It does rearrange because a 1,2-hydride shift converts it to a more stable tertiary carbocation.

F is a secondary carbocation. It does not rearrange because rearrangement would form a carbocation with the same stability—that is, another secondary carbocation.

USE THE STRATEGY

PROBLEM 17 ◆

What is the major product obtained from the reaction of HBr with each of the following?

a. $CH_3\overset{\underset{|}{CH_3}}{C}HCH{=}CH_2$

b. $CH_3\overset{\underset{|}{CH_3}}{C}HCH_2CH{=}CH_2$

c. [cyclohexane with =CH₂ group]

d. [cyclohexene with CH₃ group]

e. $CH_2{=}CHC\overset{\underset{|}{CH_3}}{\underset{|}{CH_3}}CH_3$

f. [cyclohexene with CH₃ group]

6.8 THE ADDITION OF BORANE TO AN ALKENE: HYDROBORATION–OXIDATION

Another way to convert an alkene to an alcohol is by two successive reactions known as **hydroboration–oxidation.**

Hydroboration–Oxidation

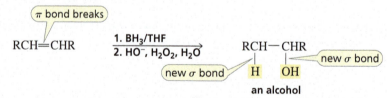

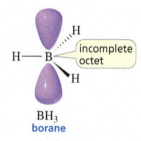

BH₃
borane

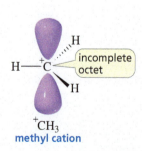

⁺CH₃
methyl cation

An atom or a molecule does not need a positive charge to be an electrophile. Borane (BH_3), for example, is an electrophile because boron has an incomplete octet. Boron uses its three sp^2 orbitals to form three bonds with hydrogen, so it has an empty p orbital that can accept a share in a pair of electrons (see p. 40). Thus, BH_3 is the electrophile that adds to the alkene and we will see that the nucleophile is an H^- bonded to boron.

When the addition reaction is over, an aqueous solution of sodium hydroxide and hydrogen peroxide is added to the reaction mixture and OH takes boron's place. The product of the reaction, therefore, is an alcohol. The numbers 1 and 2 in front of the reagents above and below the reaction arrow indicate two sequential reactions; the second set of reagents is not added until reaction with the first set is over.

Borane and Diborane

Borane exists primarily as a colorless gas called diborane. Diborane is a **dimer**—a molecule formed by joining two identical molecules. Because boron does not have a complete octet—it is surrounded by only six electrons—it has a strong tendency to acquire an additional electron pair. In the dimer, therefore, two boron atoms share the two electrons in a hydrogen–boron bond by means of unusual half-bonds. These hydrogen–boron bonds are shown as dotted lines to indicate that they consist of fewer than the normal two electrons.

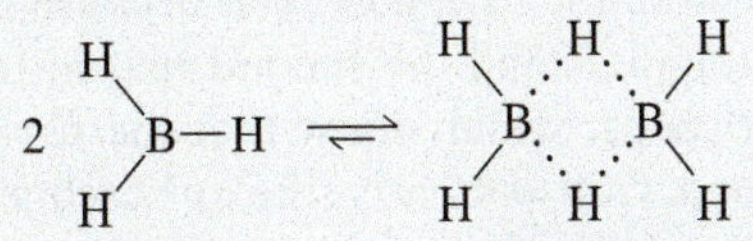

Diborane (B_2H_6) is a flammable, toxic, and explosive gas. A complex—prepared by dissolving diborane in THF—is a more convenient and less dangerous reagent. One of the lone pairs of the O atom in THF provides the two electrons that boron needs to complete its octet. The borane–THF complex is used as the source of BH_3 for hydroboration.

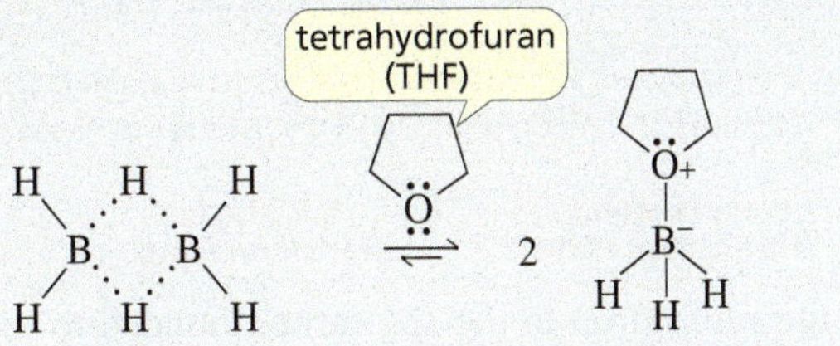

As the following reactions show, the alcohol formed from hydroboration–oxidation of an alkene has the H and OH groups switched compared with the alcohol formed from the acid-catalyzed addition of water.

acid-catalyzed addition of water

$$CH_3CH{=}CH_2 \xrightarrow[\text{H}_2\text{O}]{\text{H}_2\text{SO}_4} CH_3CHCH_3$$

OH ⟵ OH is the nucleophile

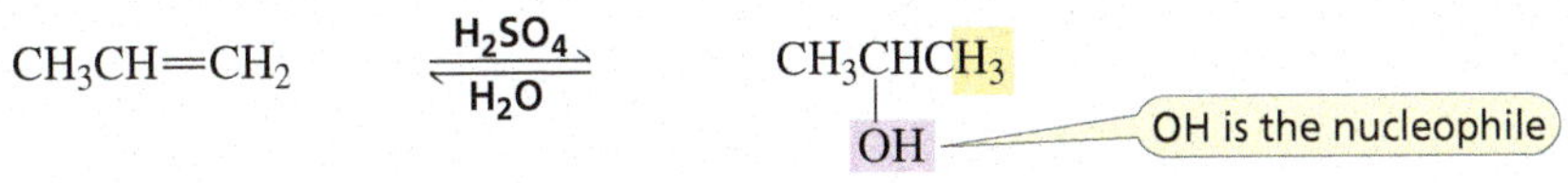

2-propanol

hydroboration-oxidation

$$CH_3CH{=}CH_2 \xrightarrow[\text{2. HO}^-,\ \text{H}_2\text{O}_2,\ \text{H}_2\text{O}]{\text{1. BH}_3/\text{THF}} CH_3CH_2CH_2OH$$ ⟵ OH replaces the electrophile

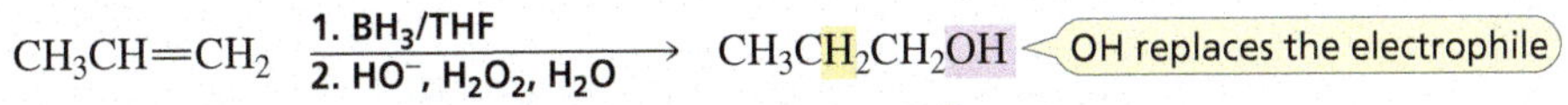

1-propanol

In the acid-catalyzed addition of water, H^+ is the electrophile; in hydroboration–oxidation, H^- is the nucleophile.

In both of the preceding reactions, *the electrophile adds to the sp^2 carbon bonded to the most hydrogens.*

- In the *acid-catalyzed addition of water*, H^+ is the electrophile and H_2O is the nucleophile.
- In *hydroboration–oxidation*, we will see that BH_3 is the electrophile (with HO subsequently taking its place) and H^- is the nucleophile.

Hydroboration

To understand why the hydroboration–oxidation of propene forms 1-propanol and not 2-propanol, we must look at the mechanism for hydroboration, the first of the two successive reactions known as hydroboration–oxidation.

MECHANISM FOR HYDROBORATION WITH BH₃

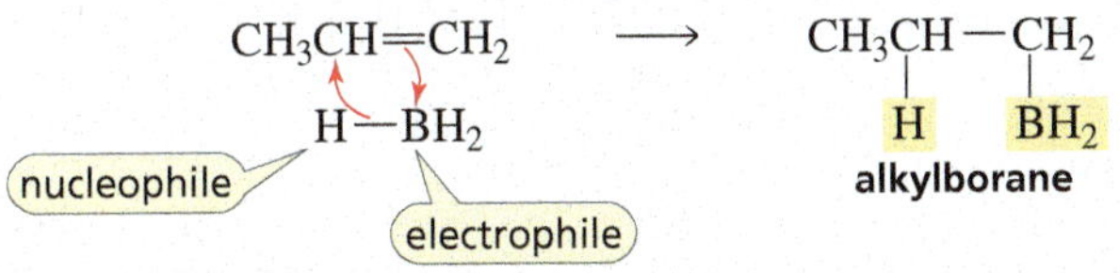

In a concerted reaction, all the bond-making and bond-breaking processes occur in the same step.

- As boron (an electrophile) accepts the π electrons from the alkene and forms a bond with one sp^2 carbon, it gives a hydride ion (a nucleophile) to the other sp^2 carbon.

Hydroboration is an example of a concerted reaction. In a **concerted reaction,** all the bond-making and bond-breaking processes occur in the same step (all the events occur "in concert"). Because both the boron and the hydride ion are added to the alkene in a single step, no intermediate is formed.

Boron, like the other electrophiles we have looked at, adds to the sp^2 carbon bonded to the most hydrogens. There are two reasons for the regioselectivity. First, there is more room at this sp^2 carbon for the electrophile to attach itself because it is the less substituted sp^2 carbon.

We can understand the second reason for the regioselectivity if we examine the two possible transition states, which show that the C—B bond has formed to a greater extent than has the C—H bond. Consequently, the sp^2 carbon that does *not* become attached to boron has a partial positive charge.

addition of BH_3		addition of HBr	

$$CH_3-\overset{\overset{H}{|}}{\underset{\overset{|}{\underset{\delta-}{H}---BH_2}}{C}}\!\!\!=\!\!\!\overset{\overset{H}{|}}{\underset{\delta+}{C}}-H$$

$$CH_3-\overset{\overset{H}{|}}{\underset{\delta+}{C}}\!\!\!=\!\!\!\overset{\overset{H}{|}}{\underset{\overset{|}{\underset{\delta-}{H_2B----H}}}{C}}-H$$

$$CH_3-\overset{\overset{H}{|}}{\underset{\overset{|}{\underset{\delta-}{Br----H}}}{C}}\!\!\!=\!\!\!\overset{\overset{H}{|}}{\underset{\delta+}{C}}-H$$

$$CH_3-\overset{\overset{H}{|}}{\underset{\delta+}{C}}\!\!\!=\!\!\!\overset{\overset{H}{|}}{\underset{\overset{|}{\underset{\delta-}{H----Br}}}{C}}-H$$

| more stable transition state | less stable transition state | more stable transition state | less stable transition state |

Addition of boron to the sp^2 carbon bonded to the most hydrogens forms a more stable transition state because the partial positive charge is on a secondary carbon. In contrast, if boron had added to the other sp^2 carbon, the partial positive charge would have been on a primary carbon. Thus, BH_3 and an electrophile such as H^+ add to the sp^2 carbon bonded to the most hydrogens for the same reason—in order to form the more stable transition state.

The alkylborane (RBH_2) formed in the first step of the reaction still has two hydrogens attached to the boron. Therefore, it reacts with another molecule of alkene to form a dialkylborane (R_2BH), which then reacts with yet another molecule of alkene to form a trialkylborane (R_3B). In each of these reactions, boron adds to the sp^2 carbon bonded to the most hydrogens and the hydride ion adds to the other sp^2 carbon.

$$CH_3CH=CH_2 \longrightarrow CH_3CH-CH_2-BHR$$
$$\underset{\text{alkylborane}}{H-BHR} \qquad\qquad \underset{\text{dialkylborane}}{\overset{|}{H}}$$

$$CH_3CH=CH_2 \longrightarrow CH_3CH-CH_2-BR_2$$
$$\underset{\text{dialkylborane}}{H-BR_2} \qquad\qquad \underset{\text{trialkylborane}}{\overset{|}{H}}$$

The alkylborane (RBH_2) is bulkier than BH_3 because R is larger than H. The dialkylborane (R_2BH) is even bulkier than the alkylborane. Therefore, the alkylborane and dialkylborane have an even stronger preference for addition to the less substituted sp^2 carbon (that is, the one bonded to the most hydrogens).

Only one hydride ion is needed for hydroboration. As a result, a reagent with only one hydrogen attached to boron, such as 9-BBN, is often used instead of BH_3. Because 9-BBN has two relatively bulky R groups, it has a stronger preference for the less substituted sp^2 carbon than BH_3 has. In mechanisms, we will write this compound as R_2BH. The mechanisms for the addition of R_2BH and BH_3 to an alkene are the same.

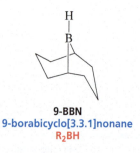

9-BBN
9-borabicyclo[3.3.1]nonane
R_2BH

MECHANISM FOR HYDROBORATION WITH R_2BH

$$CH_3CH=CH_2 \longrightarrow CH_3CH-CH_2$$
$$\underset{\text{nucleophile}}{H-BR_2} \qquad\qquad \underset{\text{trialkylborane}}{\overset{|}{H}\quad\overset{|}{BR_2}}$$

electrophile

Oxidation

When the hydroboration reaction is over, an aqueous solution of sodium hydroxide and hydrogen peroxide (HOOH) is added to the reaction mixture in order to replace BR_2 with an OH group.

An oxidation reaction increases the number of C—O, C—N, or C—X bonds and/or decreases the number of C—H bonds.

$$CH_3CH_2CH_2-BR_2 \xrightarrow{HO^-,\ H_2O_2,\ H_2O} CH_3CH_2CH_2-OH\ +\ HOBR_2$$

This is an oxidation reaction. An **oxidation reaction** decreases the number of C—H bonds or increases the number of C—O, C—N, or C—X bonds in a compound (where X is a halogen). Therefore, the overall reaction is called hydroboration–oxidation.

MECHANISM FOR THE OXIDATION REACTION

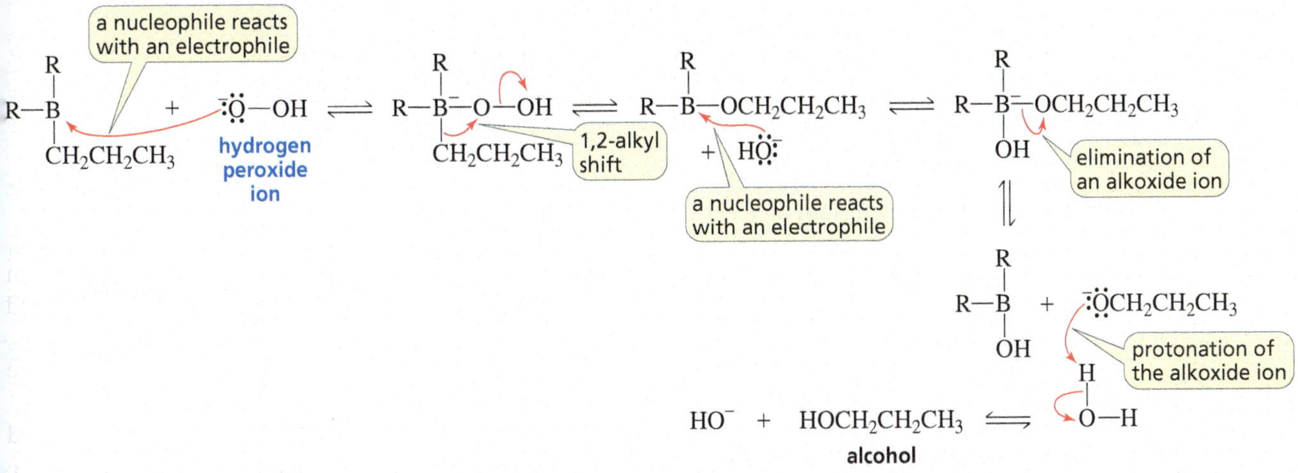

- A hydrogen peroxide ion (a nucleophile) shares a pair of electrons with the boron of $R_2BCH_2CH_2CH_3$ (an electrophile).
- A 1,2-alkyl shift displaces a hydroxide ion, allowing boron to no longer be negatively charged. (Recall that boron is to the left of carbon in the periodic table, so it is highly unstable with a negative charge.)
- A hydroxide ion shares a pair of electrons with the boron of $R_2BOCH_2CH_2CH_3$.
- An alkoxide ion is eliminated, allowing boron to no longer be negatively charged.
- Protonating the alkoxide ion forms the alcohol.

Notice that the OH group ends up attached to the sp^2 carbon bonded to the most hydrogens because it replaces boron, which was the electrophile in the hydroboration reaction.

Because carbocation intermediates are not formed in the hydroboration reaction, carbocation rearrangements do not occur.

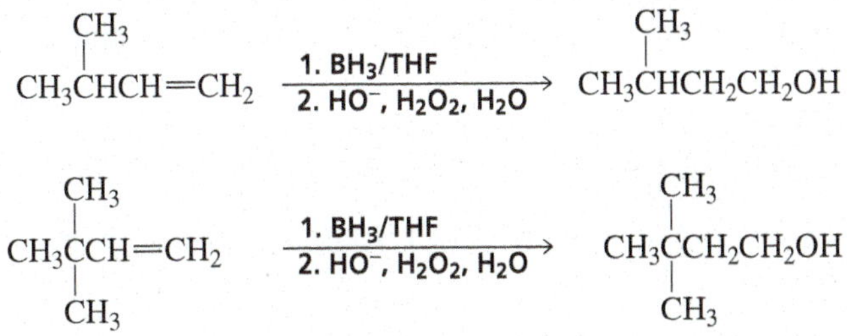

PROBLEM 18 ◆

Which is more highly regioselective: reaction of an alkene with BH_3 or with 9-BBN?

PROBLEM 19 ◆

What is the major product obtained from hydroboration–oxidation of the following alkenes?
a. 2-methyl-2-butene **b.** 1-methylcyclohexene

6.9 THE ADDITION OF A HALOGEN TO AN ALKENE

The halogens Br_2 and Cl_2 add to alkenes. The product of the reaction is a *vicinal dibromide* (or dichloride). **Vicinal** means that the two halogens are on adjacent carbons (*vicinus* is the Latin word for "near"). Reactions of alkenes with Br_2 or Cl_2 are generally carried out by mixing the alkene and the halogen in an inert solvent such as dichloromethane (CH_2Cl_2), which readily dissolves both reactants but does not participate in the reaction.

$$CH_3CH{=}CH_2 + \boxed{Br_2} \xrightarrow{\ \mathbf{CH_2Cl_2}\ } \underset{\underset{\displaystyle \boxed{Br}\quad \boxed{Br}}{|\qquad |}}{CH_3CH{-}CH_2}$$

vicinal dibromide

$$CH_3CH{=}CH_2 + \boxed{Cl_2} \xrightarrow{\ \mathbf{CH_2Cl_2}\ } \underset{\underset{\displaystyle \boxed{Cl}\quad \boxed{Cl}}{|\qquad |}}{CH_3CH{-}CH_2}$$

vicinal dichloride

These reactions might surprise you because it is not immediately apparent that an electrophile—which is necessary to start an electrophilic addition reaction—is present.

The reaction is possible because the bond joining the two halogen atoms is easily broken because Br_2 and Cl_2 are polarizable (Section 3.9). As the nucleophilic alkene approaches Br_2 (or Cl_2), it induces a dipole, which causes one of the halogen atoms to have a partial positive charge. This is the electrophile that adds to the sp^2 carbon of the alkene. The mechanism of the reaction is shown below.

MECHANISM FOR THE ADDITION OF BROMINE TO AN ALKENE

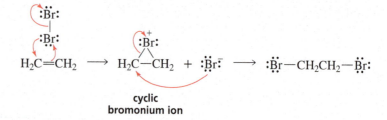

cyclic bromonium ion

cyclic bromonium ion

▲ **Figure 6.4**
The cyclic bromonium ion intermediate formed from the reaction of Br_2 with *cis*-2-butene shows that the electron-deficient region (the blue area) encompasses the carbons, even though the formal positive charge is on the bromine.

- As the π electrons of the alkene approach a molecule of Br_2, one of the bromines accepts those electrons and releases the shared electrons to the other bromine, which leaves as a bromide ion. Because bromine's electron cloud is close enough to the other sp^2 carbon to form a bond, a cyclic bromonium ion intermediate is formed rather than a carbocation intermediate.

- The cyclic bromonium ion intermediate is unstable because of the strain in the three-membered ring and the positively charged bromine, which withdraws electrons strongly from the ring carbons (Figure 6.4). Therefore, the cyclic bromonium ion reacts rapidly with a nucleophile (Br^-).

- The mechanism for the addition of Cl_2 is the same as the mechanism for the addition of Br_2.

Because a carbocation is not formed when Br_2 or Cl_2 adds to an alkene, carbocation rearrangements do not occur in these reactions.

$$\underset{\underset{\displaystyle |}{CH_3}}{CH_3CHCH{=}CH_2} + \boxed{Br_2} \xrightarrow{\ \mathbf{CH_2Cl_2}\ } \underset{\underset{\displaystyle Br}{\underset{\displaystyle |}{}}}{CH_3\overset{\overset{\displaystyle CH_3}{\overset{\displaystyle |}{}}}{C}HCHCH_2Br}$$

the carbon skeleton does not rearrange

PROBLEM 20 ♦

What will be the product of the preceding reaction if HBr is used in place of Br_2?

PROBLEM 21

a. How does the first step in the reaction of propene with Br_2 differ from the first step in the reaction of propene with HBr?

b. To understand why Br^- adds to a carbon of the bromonium ion rather than to the positively charged bromine, draw the product that would be obtained if Br^- *did* add to bromine.

Although F_2 and I_2 are halogens, they are not used as reagents in electrophilic addition reactions. Fluorine reacts explosively with alkenes, so the reaction with F_2 is not useful for synthesizing new compounds. The addition of I_2 to an alkene, on the other hand, is thermodynamically unfavorable (Section 5.10). A vicinal diiodide is unstable at room temperature, so it decomposes back to the alkene and I_2.

$$CH_3CH=CHCH_3 \ + \ I_2 \ \underset{CH_2Cl_2}{\rightleftharpoons} \ CH_3\overset{|}{C}H\overset{|}{C}HCH_3$$
$$\phantom{CH_3CH=CHCH_3 \ + \ I_2 \ \underset{CH_2Cl_2}{\rightleftharpoons} \ }\underset{\text{unstable}}{I \ \ I}$$

Competing Nucleophiles

If H_2O is present in the reaction mixture, the major product of the reaction will be a vicinal halohydrin (or, more specifically, a bromohydrin or a chlorohydrin). A **halohydrin** is an organic molecule that contains both a halogen and an OH group. In a vicinal halohydrin, the halogen and the OH group are bonded to adjacent carbons.

$$CH_3CH=CH_2 \ + \ Br_2 \ \xrightarrow{H_2O} \ CH_3\underset{OH}{\overset{|}{C}}HCH_2Br \ + \ CH_3\underset{Br}{\overset{|}{C}}HCH_2Br \ + \ HBr$$

a bromohydrin
major product | **minor product**

$$CH_3CH=\overset{\overset{\displaystyle CH_3}{|}}{C}CH_3 \ + \ Cl_2 \ \xrightarrow{H_2O} \ CH_3\overset{\overset{\displaystyle CH_3}{|}}{\underset{Cl \ \ OH}{C}}CH_3 \ + \ CH_3\overset{\overset{\displaystyle CH_3}{|}}{\underset{Cl \ \ Cl}{C}}CH_3 \ + \ HCl$$

a chlorohydrin
major product | **minor product**

MECHANISM FOR HALOHYDRIN FORMATION

- A cyclic bromonium ion (or chloronium ion) intermediate is formed in the first step because Br^+ (or Cl^+) is the only electrophile in the reaction mixture.

- The relatively unstable cyclic bromonium ion intermediate rapidly reacts with any nucleophile it collides with. Two nucleophiles—H_2O and Br^-—are present in the solution, but the concentration of H_2O far exceeds that of Br^-. Consequently, the bromonium ion is much more likely to collide with water than with Br^-.

- The protonated halohydrin is a strong acid (Section 2.3), so it readily loses a proton.

We can understand the regioselectivity—why the electrophile (Br^+) ends up attached to the sp^2 carbon bonded to the most hydrogens—if we examine the two possible transition states. They show that the C—Br bond has broken to a greater extent than the C—O bond has formed. As a result, there is a partial positive charge on the carbon that is forming a bond with the nucleophile.

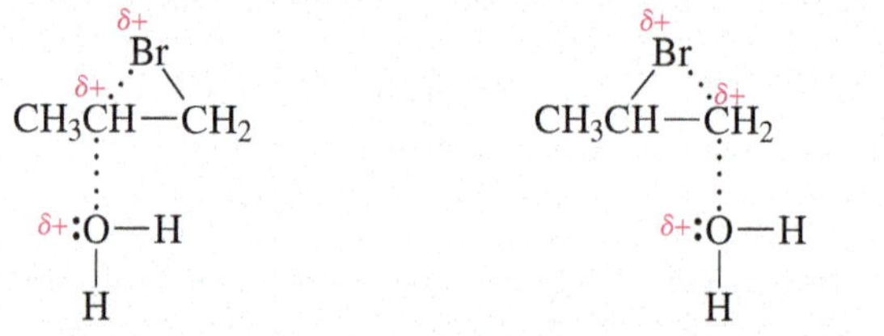

$$CH_3\overset{\delta+}{CH}-CH_2 \qquad CH_3CH-\overset{\delta+}{CH_2}$$

(more stable transition state)   (less stable transition state)

Water adds to the more substituted ring carbon because it leads to the more stable transition state because the partial positive charge is on a secondary rather than a primary carbon. Therefore, this reaction, too, follows the general rule for electrophilic addition reactions: *the electrophile* (here, Br^+) *adds to the* sp^2 *carbon that is bonded to the most hydrogens, and the nucleophile* (H_2O) *adds to the other* sp^2 *carbon.*

When nucleophiles other than H_2O are added to the reaction mixture (for example, CH_3OH or Cl^-), they, too, change the product of the reaction. Because the concentration of the added nucleophile is greater than the concentration of the halide ion generated from Br_2 or Cl_2, the added nucleophile is the one more likely to participate in the second step of the reaction.

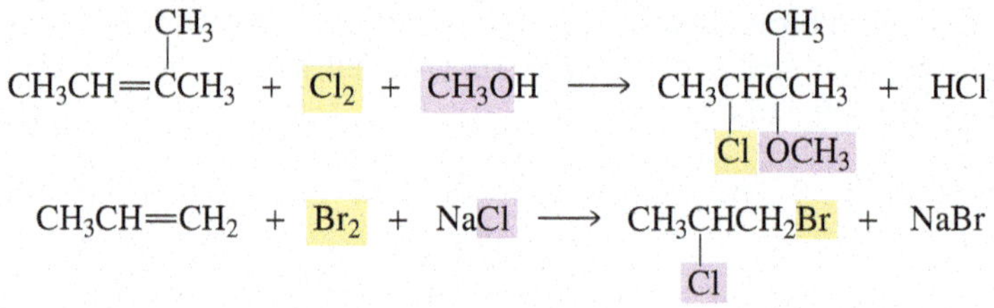

$$CH_3CH=\overset{CH_3}{\underset{}{C}}CH_3 + Cl_2 + CH_3OH \longrightarrow CH_3\overset{CH_3}{\underset{Cl\ OCH_3}{C}}CH_3 + HCl$$

$$CH_3CH=CH_2 + Br_2 + NaCl \longrightarrow CH_3\underset{Cl}{CH}CH_2Br + NaBr$$

Remember that ions such as Na^+ and K^+ cannot form covalent bonds (Section 1.3), so they do not react with organic compounds. They serve only as counterions to negatively charged species, so their presence generally is ignored when writing the reactions of organic compounds.

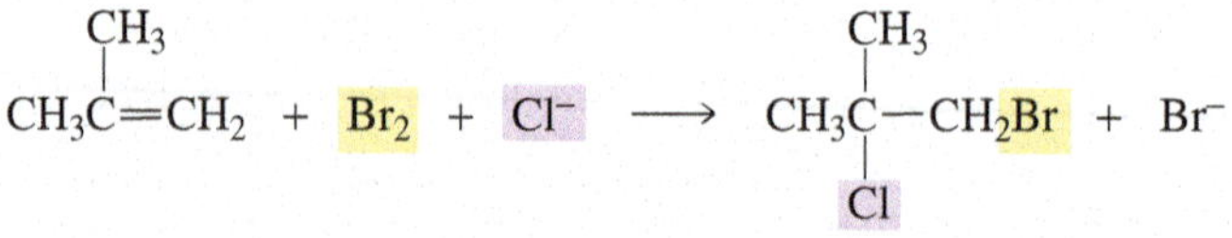

$$CH_3\overset{CH_3}{\underset{}{C}}=CH_2 + Br_2 + Cl^- \longrightarrow CH_3\overset{CH_3}{\underset{Cl}{C}}-CH_2Br + Br^-$$

PROBLEM 22

Why are Na^+ and K^+ unable to form covalent bonds?

PROBLEM 23

Each of the following reactions has two nucleophiles that could add to the intermediate formed by the reaction of the alkene with an electrophile. What is the major product of each reaction?

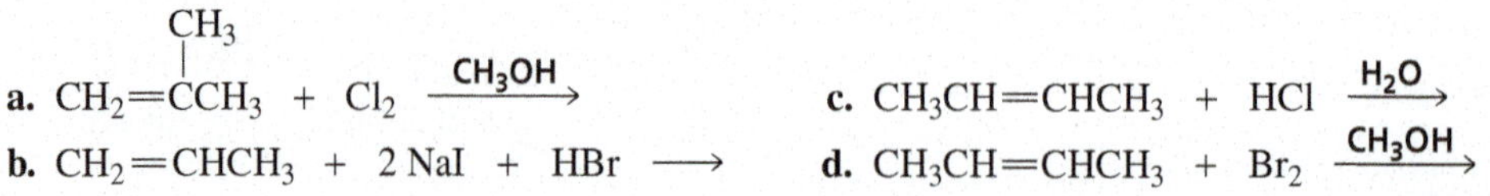

a. $CH_2=\overset{CH_3}{\underset{}{C}}CH_3 + Cl_2 \xrightarrow{CH_3OH}$ c. $CH_3CH=CHCH_3 + HCl \xrightarrow{H_2O}$

b. $CH_2=CHCH_3 + 2\,NaI + HBr \longrightarrow$ d. $CH_3CH=CHCH_3 + Br_2 \xrightarrow{CH_3OH}$

PROBLEM 24

What is the product of the addition of I—Cl to 1-butene? (*Hint:* Chlorine is more electronegative than iodine [Table 1.3].)

PROBLEM 25 ◆

What will be the major product obtained from the reaction of Br_2 with 1-butene if the reaction is carried out in

a. dichloromethane?   b. water?   c. ethyl alcohol?   d. methyl alcohol?

PROBLEM-SOLVING STRATEGY

Proposing a Mechanism

LEARN THE STRATEGY

Propose a mechanism for the following reaction:

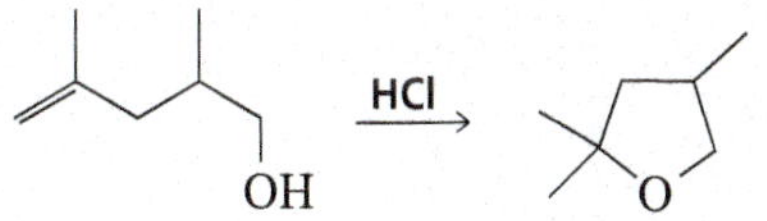

When forming a cyclic compound, start by numbering the atoms in the reactant and the product. This allows you to see the atoms that become attached to each other in the cyclic product.

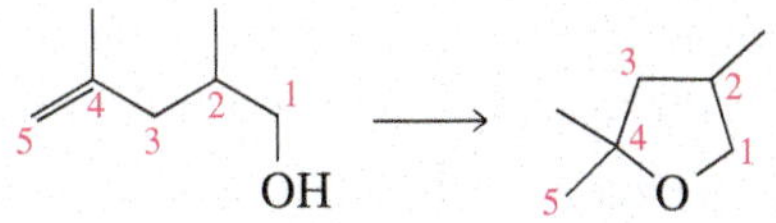

We see that the oxygen forms a bond with C-4. Adding the electrophile (H^+ from HCl) to the sp^2 carbon of the alkene that is bonded to the most hydrogens forms a carbocation (with the positive charge on C-4) that will react with a nucleophile.

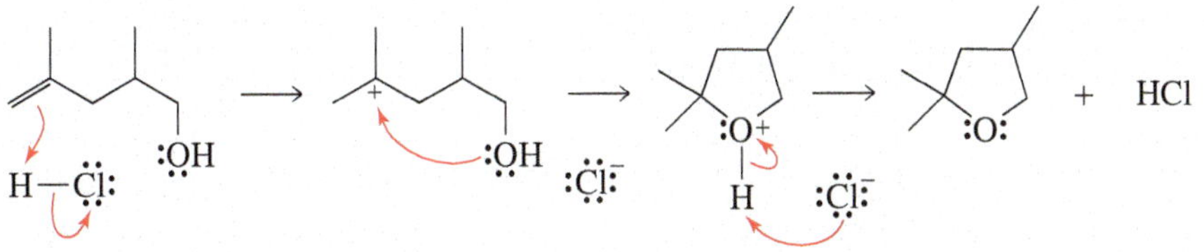

There are two nucleophiles that could add to the carbocation, OH and Cl^-. The OH group is tethered to the carbocation, so unlike Cl^-, it does not have to diffuse through the solvent to find the carbocation. Therefore, the OH group is more apt to react with the carbocation.

PROBLEM 26

USE THE STRATEGY

Propose a mechanism for the following reaction:

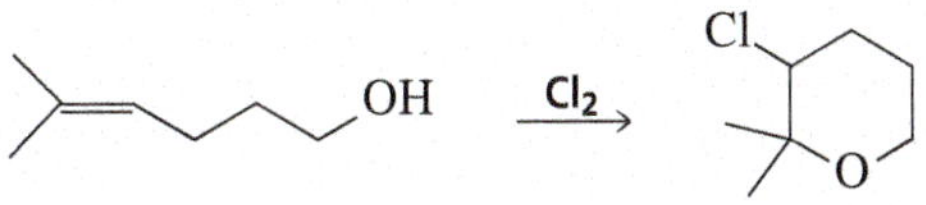

6.10 THE ADDITION OF A PEROXYACID TO AN ALKENE

An alkene can be converted to an *epoxide* by a peroxyacid.

- An **epoxide** is an ether in which the oxygen is incorporated into a three-membered ring.
- A **peroxyacid** is a carboxylic acid with an extra oxygen.

The overall reaction amounts to the transfer of an oxygen from the peroxyacid to the alkene. It is an oxidation reaction because it increases the number of C—O bonds.

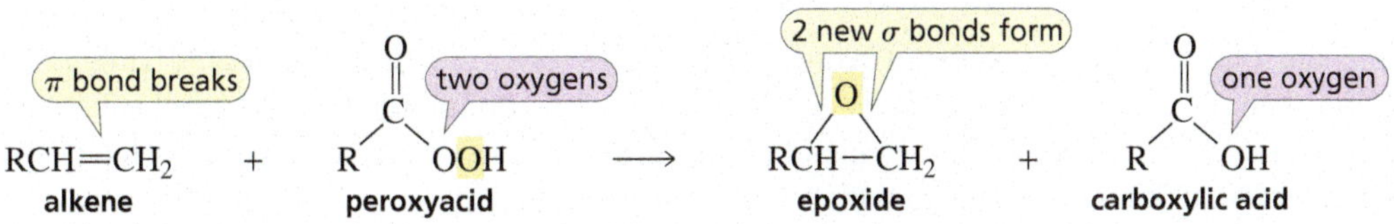

Remember that an O—O bond is weak and is, therefore, easily broken (Section 1.4).

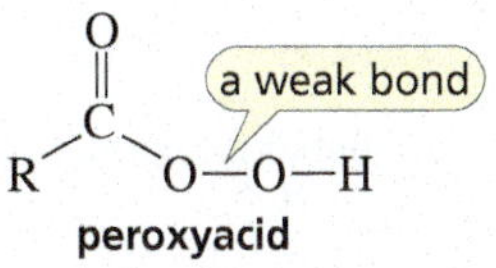

The peroxyacid commonly used for epoxidation is MCPBA (*meta*-chloroperoxybenzoic acid).

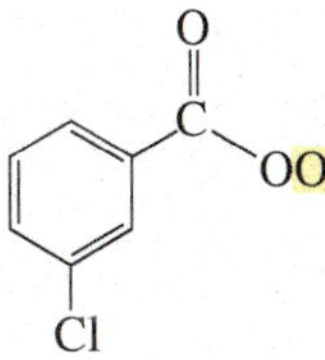

MCPBA

MECHANISM FOR THE EPOXIDATION OF AN ALKENE

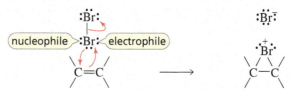

- An oxygen atom of the OOH group (the O that is attached to the H) of the peroxyacid is electron-deficient and is, therefore, an electrophile. It accepts the electrons from the π bond of the alkene, which causes the weak O—O bond of the peroxyacid to break.
- The electrons from the O—O bond are delocalized, causing the π bond of the C=O group to break and pick up a proton (Section 2.8).
- As the O—H bond breaks, the bonding electrons (the nucleophile) add to the other sp^2 carbon of the alkene.

Thus, the oxygen atom is both the electrophile and the nucleophile. This is another example of a concerted reaction—that is, all bond making and bond breaking occur in the same step.

The mechanism for the addition of oxygen to a double bond to form an epoxide is analogous to the mechanism we just saw for the addition of bromine to a double bond to form a cyclic intermediate (Section 6.9).

The unstable cyclic bromonium ion intermediate is an electrophile that subsequently reacts with a nucleophile. The epoxide, however, is stable enough to isolate because none of the ring atoms has a positive charge.

Nomenclature of Epoxides

The common name of an epoxide is obtained by adding "oxide" to the common name of the corresponding alkene; the oxygen is where the π bond of an alkene would be. The simplest epoxide is ethylene oxide.

H_2C=CH_2
ethylene

H_2C—CH_2 (O)
ethylene oxide

H_2C=$CHCH_3$
propylene

H_2C—$CHCH_3$ (O)
propylene oxide

There are two systematic ways to name epoxides. In one, the three-membered ring is called "oxirane," and the oxygen is given the 1-position. Thus, 2-ethyloxirane has an ethyl substituent at the 2-position of the oxirane ring. In the other, an epoxide is named as an alkane, with an "epoxy" prefix that identifies the carbons to which the oxygen is attached.

H_2C—$CHCH_2CH_3$ (O)
2-ethyloxirane
1,2-epoxybutane

CH_3CH—$CHCH_3$ (O)
2,3-dimethyloxirane
2,3-epoxybutane

2,2-dimethyloxirane
1,2-epoxy-2-methylpropane

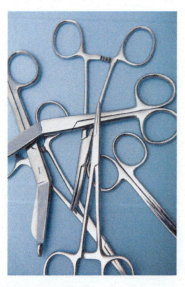

Ethylene oxide is used for the gaseous sterilization of medical instruments. When the microorganism's DNA reacts with ethylene oxide, it's alkylated (see p. 496), and the microorganism can no longer reproduce.

PROBLEM 27 ◆

Draw structures for the following:

a. 2-propyloxirane   **c.** 2,2,3,3-tetramethyloxirane
b. cyclohexene oxide   **d.** 2,3-epoxy-2-methylpentane

PROBLEM 28 ◆

What alkene would you treat with a peroxyacid in order to obtain each of the epoxides in Problem 27?

PROBLEM 29 SOLVED

Identify each of the following reactions as an oxidation reaction, a reduction reaction, or neither.

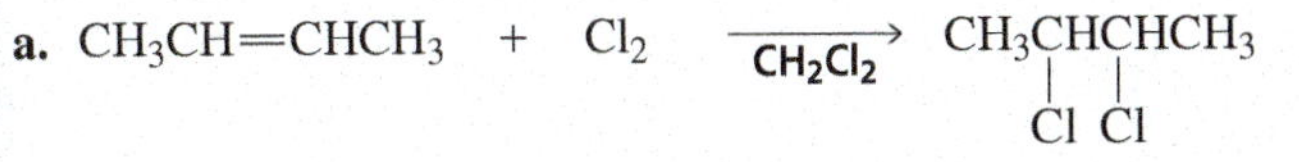

a. $CH_3CH{=}CHCH_3$ + Cl_2 $\xrightarrow{CH_2Cl_2}$ $CH_3CHCHCH_3$
                                                                  $\underset{Cl}{|}\ \underset{Cl}{|}$

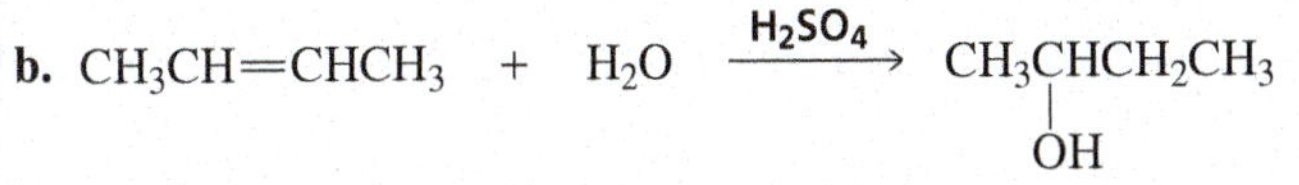

b. $CH_3CH{=}CHCH_3$ + H_2O $\xrightarrow{H_2SO_4}$ $CH_3CHCH_2CH_3$
                                                                  $\underset{OH}{|}$

SOLUTION TO 29 a. Recall that an oxidation reaction decreases the number of C—H bonds and/or increases the number of C—O, C—N, or C—X bonds (where X = a halogen) (Section 6.8), whereas a reduction reaction increases the number of C—H bonds and/or decreases the number of C—O, C—N, or C—X bonds (Section 5.9). Therefore, this is an oxidation reaction because the number of C—Cl bonds increases.

SOLUTION TO 29 b. The product has both a new C—O bond (signifying an oxidation) and a new C—H bond (signifying a reduction). Thus, the two cancel each other; the reaction is neither an oxidation nor a reduction.

6.11 THE ADDITION OF OZONE TO AN ALKENE: OZONOLYSIS

When an alkene is treated with ozone (O_3) at a low temperature, both the σ and π bonds of the double bond break and the carbons that were doubly bonded to each other are now doubly bonded to oxygens instead. This is an oxidation reaction—called **ozonolysis**—because the number of C—O bonds increases.

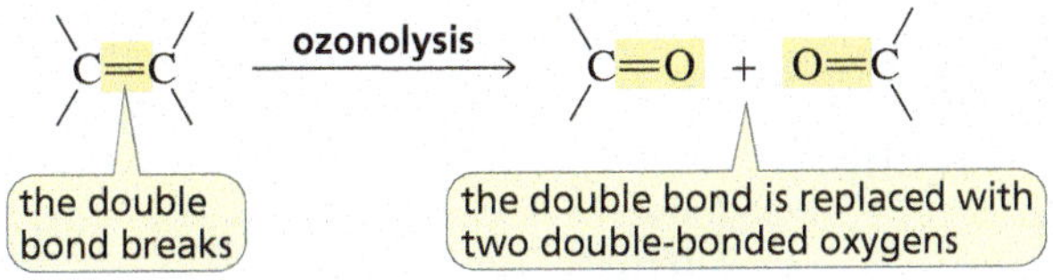

> the double bond breaks

> the double bond is replaced with two double-bonded oxygens

To determine the product of ozonolysis, replace C=C with C=O O=C.

Ozonolysis is an example of **oxidative cleavage**—an oxidation reaction that cleaves the reactant into pieces (*lysis* is Greek for "breaking down").

MECHANISM FOR OZONIDE FORMATION

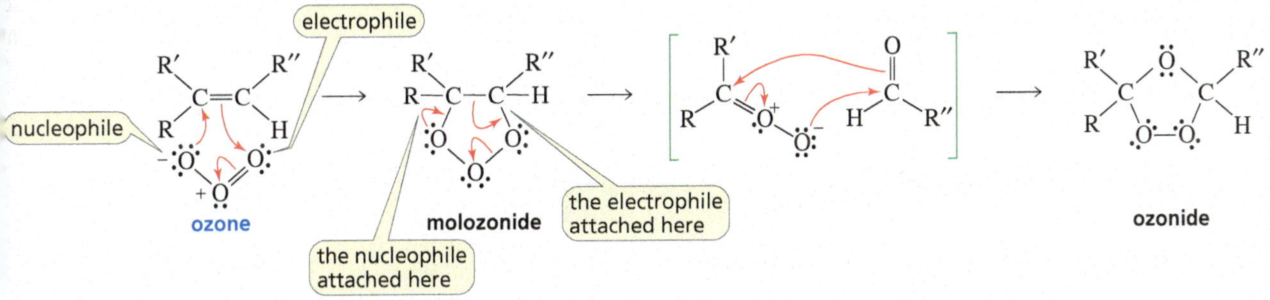

ozone

Ozone, a major constituent of smog, is a health hazard at ground level, increasing the risk of death from lung or heart disease. In the strato-sphere, however, a layer of ozone shields the Earth from harmful solar radiation.

■ The electrophile (the doubly bonded oxygen at one end of the ozone molecule) adds to one of the sp^2 carbons, and a nucleophile (the negatively charged oxygen at the other end) adds to the other sp^2 carbon. The product is a **molozonide.**

■ The molozonide is unstable because it has two O—O bonds; it immediately rearranges to a more stable **ozonide.**

Because ozonides are explosive, they are not isolated. Instead, they are immediately con-verted to ketones and/or aldehydes by dimethyl sulfide (CH_3SCH_3) or zinc in acetic acid (CH_3CO_2H).

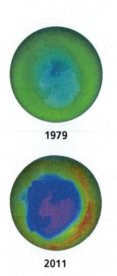

1979

2011

Maps showing the loss of stratospheric ozone over Antarctica—called the "ozone hole." The lowest ozone density is represented by dark blue (Section 12.12).

The product is a ketone if the sp^2 carbon of the alkene is bonded to two carbon-containing sub-stituents; the product is an aldehyde if one or both of the substituents bonded to the sp^2 carbon is a hydrogen.

■ A **ketone** (key-tone) has two alkyl groups bonded to a **carbonyl** (car-bo-neel) **group.**

■ An **aldehyde** has an alkyl group and a hydrogen (or has two hydrogens) bonded to a carbonyl group.

The following are additional examples of the oxidative cleavage of alkenes by ozonolysis. [Many organic reactions that need to be carried out in the cold are done conveniently at −78 °C, because that is the temperature of a mixture of dry ice (solid CO_2) in acetone.]

Ozonolysis oxidizes alkenes to aldehydes and ketones.

PROBLEM-SOLVING STRATEGY

Determining the Products of Oxidative Cleavage

LEARN THE STRATEGY

What products would you would expect to obtain when the following compounds react with ozone and then with dimethylsulfide?

a.

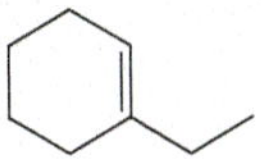

b. $CH_3CH_2CH{=}CHCH_2CH_3$

c.

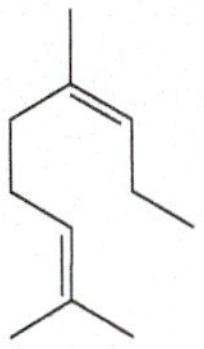

SOLUTION TO a. Break the double bond and replace it with two double-bonded oxygens.

break the double bond

$$\xrightarrow[\text{2. (CH}_3)_2\text{S}]{\text{1. O}_3, -78\ °C}$$

replace with two double-bonded oxygens

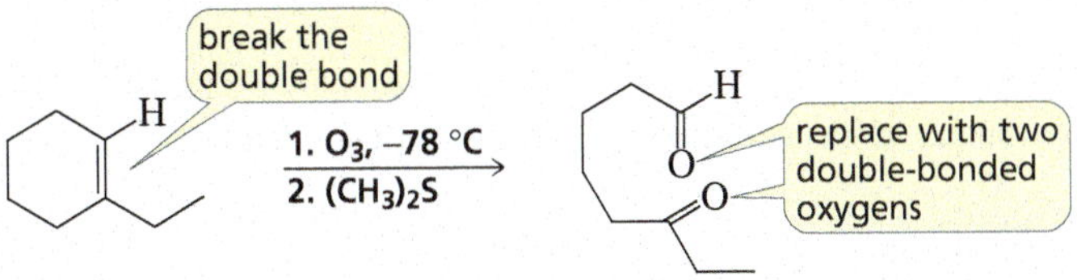

SOLUTION TO b. Break the double bond and replace it with two double-bonded oxygens. Because the alkene is symmetrical, only one product is formed.

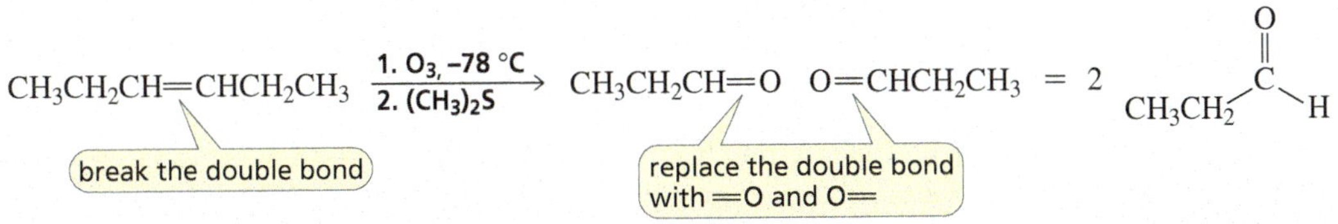

$$CH_3CH_2CH{=}CHCH_2CH_3 \xrightarrow[\text{2. (CH}_3)_2\text{S}]{\text{1. O}_3, -78\ °C} CH_3CH_2CH{=}O \quad O{=}CHCH_2CH_3 \quad = \quad 2 \quad \underset{CH_3CH_2}{\overset{\overset{\displaystyle O}{\|}}{C}}{\diagdown}_H$$

break the double bond

replace the double bond with =O and O=

SOLUTION TO c. Because the reactant has two double bonds, each one must be replaced with two double-bonded oxygens.

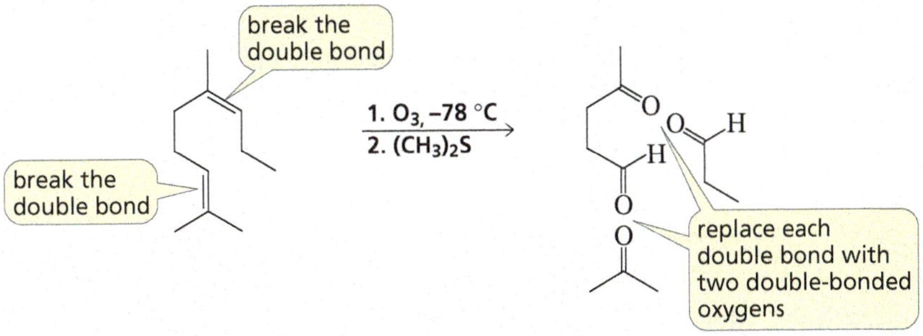

break the double bond

break the double bond

$$\xrightarrow[\text{2. (CH}_3)_2\text{S}]{\text{1. O}_3, -78\ °C}$$

replace each double bond with two double-bonded oxygens

PROBLEM 30

USE THE STRATEGY

What products are formed when the following compounds react with ozone and then with dimethyl sulfide?

a.

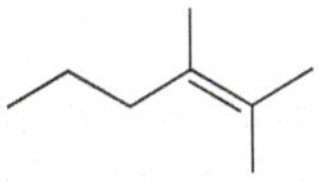

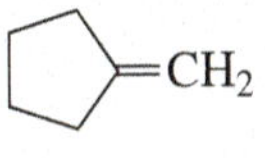

c.

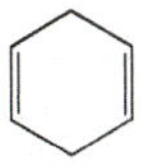

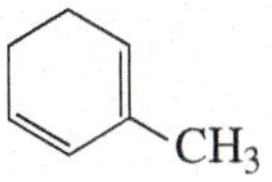

e.

b. $=CH_2$

d. CH_3

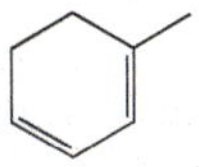

f.

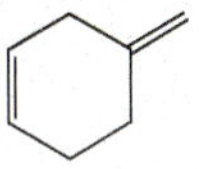

Ozonolysis can be used to determine the structure of an unknown alkene. If we know what aldehydes and/or ketones are formed by ozonolysis, we can mentally work backward to deduce the structure of the alkene. In other words, delete the =O and O= and join the carbons with a double bond. (Recall that working backward is indicated by an open arrow.)

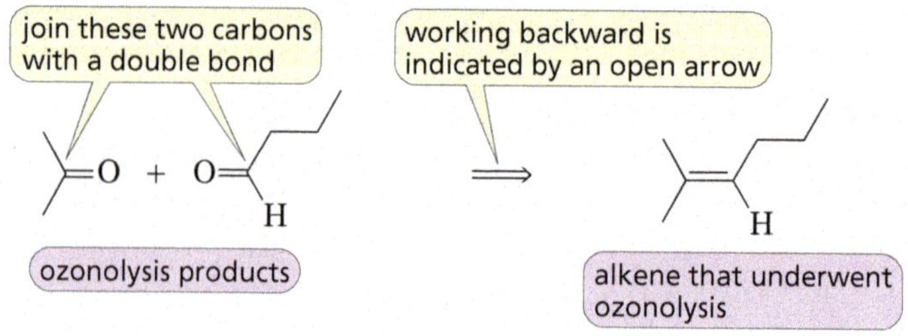

join these two carbons with a double bond

working backward is indicated by an open arrow

ozonolysis products

alkene that underwent ozonolysis

PROBLEM 31 ◆

a. What alkene would give only a ketone with three carbons as a product of oxidative cleavage?
b. What alkenes would give only an aldehyde with four carbons as a product of oxidative cleavage?

PROBLEM 32 ◆

What aspect of the structure of the alkene does ozonolysis not tell you?

LEARN THE STRATEGY

PROBLEM 33 **SOLVED**

a. The following product was obtained from the ozonolysis of an alkene followed by treatment with dimethyl sulfide. What is the structure of the alkene?

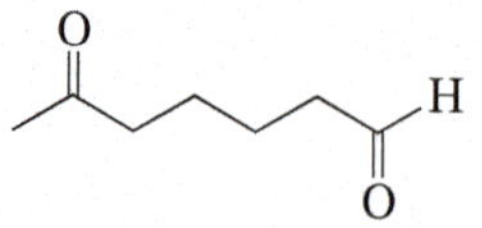

b. The following products were obtained from the oxidative cleavage of a diene. What is the structure of the diene?

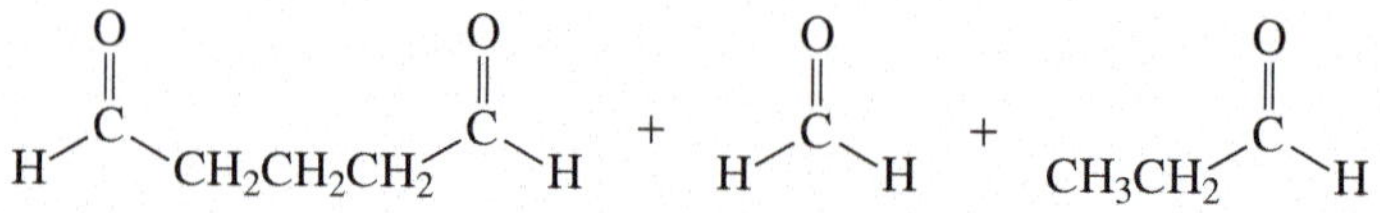

SOLUTION TO 33 a. Because only one product is obtained, the reactant must be a cyclic alkene. Numbering the product shows that the carbonyl groups are at C-1 and C-6, so the double bond in the alkene must be between C-1 and C-6.

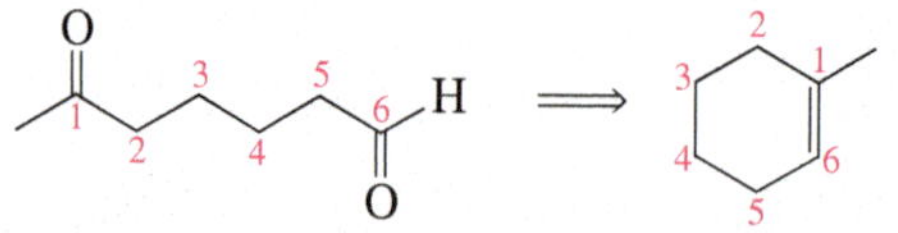

SOLUTION TO 33 b. The five-carbon product with two carbonyl groups indicates that the diene must contain five carbons flanked by two double bonds.

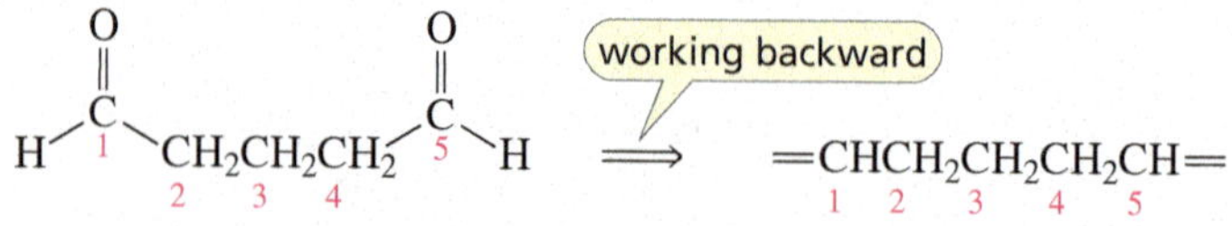

working backward

One of the other two products obtained from ozonolysis has one carbon, and the other has three carbons. Therefore, one carbon must be added to one end of the diene, and three carbons must be added to the other end.

$$CH_2=CHCH_2CH_2CH_2CH=CHCH_2CH_3$$

USE THE STRATEGY

PROBLEM 34

The following product was obtained from the ozonolysis of an alkene followed by treatment with dimethyl sulfide. What is the structure of the alkene?

a. 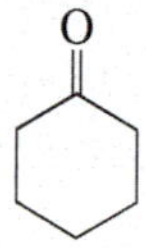 **b.** 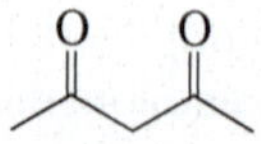 **c.**

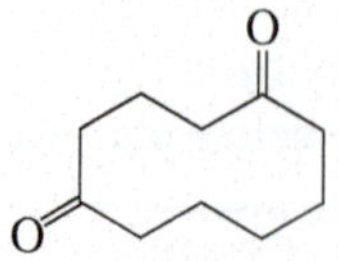

6.12 REGIOSELECTIVE, STEREOSELECTIVE, AND STEREOSPECIFIC REACTIONS

When we looked at the electrophilic addition reactions that alkenes undergo, we examined the step-by-step process by which each reaction occurs (the mechanism of the reaction), and we determined what products are formed. However, we did not consider the stereochemistry of the reactions.

Stereochemistry is the field of chemistry that deals with the structures of molecules in three dimensions. When we study the stereochemistry of a reaction, we are concerned with the following questions:

1. If the *product* of a reaction can exist as two or more stereoisomers, does the reaction produce a single stereoisomer, a set of particular stereoisomers, or all possible stereoisomers?

2. If the *reactant* can exist as two or more stereoisomers, do all stereoisomers of the reactant form the same stereoisomers of the product or does each stereoisomer of the reactant form a different stereoisomer or set of stereoisomers of the product?

Before we examine the stereochemistry of electrophilic addition reactions, we need to become familiar with some terms used in describing the stereochemistry of a reaction.

We saw that a **regioselective** reaction is one in which two *constitutional isomers* can be obtained as products but more of one is obtained than the other (Section 6.4). In other words, a regioselective reaction selects for a particular constitutional isomer. Recall that a reaction can be *moderately regioselective*, *highly regioselective*, or *completely regioselective* depending on the relative amounts of the constitutional isomers formed in the reaction.

a regioselective reaction

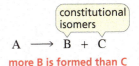

more B is formed than C

> A regioselective reaction forms more of one constitutional isomer than of another.

Stereoselective is a similar term, but it refers to the preferential formation of a *stereoisomer* rather than a *constitutional isomer*. A stereoselective reaction forms one stereoisomer preferentially over another. In other words, it selects for a particular stereoisomer. Depending on the degree of preference for a particular stereoisomer, a reaction can be described as being *moderately stereoselective*, *highly stereoselective*, or *completely stereoselective*.

a stereoselective reaction

$$A \longrightarrow B + C$$

more B is formed than C

> A stereoselective reaction forms more of one stereoisomer than of another.

A reaction is **stereospecific** if the *reactant* can exist as stereoisomers and each stereoisomer of the reactant forms a different stereoisomer or a different set of stereoisomers of the product.

stereospecific reactions

$$A \longrightarrow B$$
$$C \longrightarrow D$$

In the preceding reaction, stereoisomer A forms stereoisomer B but does not form D, so the reaction is stereospecific. Because a stereospecific reaction does not form all possible stereoisomers of the product, *all stereospecific reactions are also stereoselective. However, not all stereoselective reactions are stereospecific*, because there are stereoselective reactions with reactants that do not have stereoisomers.

> In a stereospecific reaction, each stereoisomer forms a different stereoisomeric product or a different set of stereoisomeric products.

> A stereospecific reaction is also stereoselective, but a stereoselective reaction is not necessarily stereospecific.

PROBLEM 35 ♦

What characteristics must the reactant of a stereospecific reaction have?

6.13 THE STEREOCHEMISTRY OF ELECTROPHILIC ADDITION REACTIONS

Now that you are familiar with stereoisomers (Chapter 4) *and* with the electrophilic addition reactions of alkenes, we can combine the two topics and look at the stereochemistry of these reactions. In other words, we will now look at the stereoisomers that are formed in the electrophilic addition reactions you learned about in this chapter.

We know that when an alkene reacts with an electrophilic reagent such as HBr, the major product of the addition reaction is the one obtained by adding the electrophile (H^+) to the sp^2 carbon bonded to the most hydrogens and adding the nucleophile (Br^-) to the other sp^2 carbon (Section 6.4). For example, the major product obtained from the following reaction is 2-bromopropane. Because 2-bromobutane does not have an asymmetric center, it does not have stereoisomers. Therefore, we do not need to be concerned with the stereochemistry of the reaction.

$$CH_3CH{=}CH_2 \xrightarrow{\text{HBr}} CH_3\underset{\underset{Br}{|}}{C}HCH_3$$

no stereoisomers

The following reaction, however, forms a product with an asymmetric center, so now we have to be concerned with the stereochemistry of the reaction. What is the configuration of the product? In other words, do we get the *R* enantiomer, the *S* enantiomer, or both enantiomers?

$$CH_3CH_2CH{=}CH_2 \xrightarrow{\text{HBr}} CH_3CH_2\underset{\underset{Br}{|}}{C}HCH_3$$

asymmetric center

We will begin our discussion of the stereochemistry of electrophilic addition reactions by looking at reactions that form a product with one asymmetric center. Then we will look at reactions that form a product with two asymmetric centers.

The Stereochemistry of Addition Reactions That Form a Product with One Asymmetric Center

When a reactant that does *not* have an asymmetric center undergoes a reaction that forms a product with *one* asymmetric center, the product is always a racemic mixture. For example, the reaction of 1-butene with HBr that we just looked at forms identical amounts of (*R*)-2-bromobutane and (*S*)-2-bromobutane. Thus, the reaction is not stereoselective because it does not select for a particular stereoisomer. Why is this so?

A racemic mixture contains equal amounts of a pair of enantiomers.

The addition reaction forms a carbocation intermediate. The three groups bonded to the sp^2 carbon of the carbocation intermediate lie in a plane (Section 1.10). When the bromide ion approaches the intermediate from above the plane, one enantiomer is formed; when it approaches from below the plane, the other enantiomer is formed. Because the bromide ion has equal access to both sides of the plane, identical amounts of the *R* and *S* enantiomers are formed (Figure 6.5).

▶ **Figure 6.5**
Because the products of the reaction are enantiomers, the transition states that lead to the products are also enantiomers. Therefore, the two transition states have the same stability, so the two products are formed at the same rate. The product, therefore, is a racemic mixture.

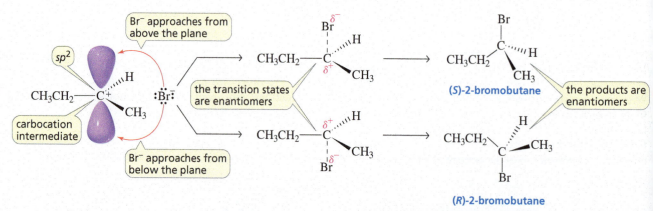

A racemic mixture is formed by any reaction that forms a product with an asymmetric center from a reactant that does not have an asymmetric center.

asymmetric center

$$CH_3CH_2CH_2CH=CH_2 \xrightarrow[H_2O]{H_2SO_4} CH_3CH_2CH_2\overset{*}{C}HCH_3$$
$$\underset{OH}{|}$$

When a reactant that does not have an asymmetric center forms a product with *one* asymmetric center, the product is always a racemic mixture.

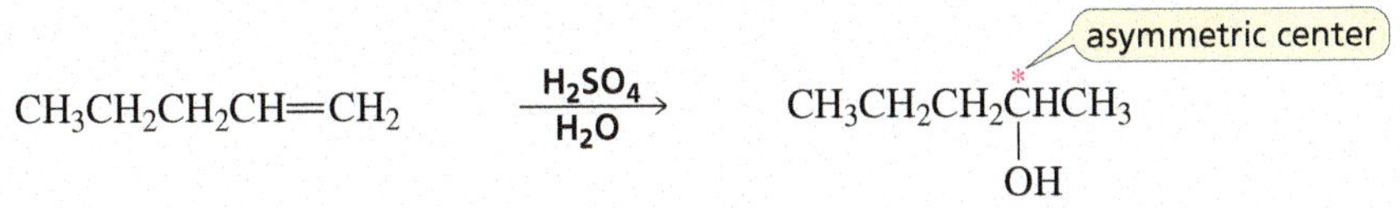

The products of the above reaction, therefore, have the configurations shown here.

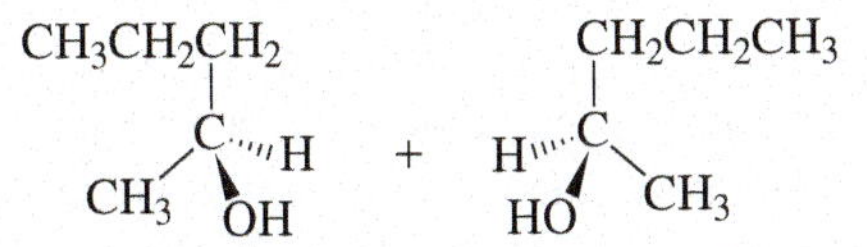

(S)-2-pentanol   (R)-2-pentanol

Notice that the same product is obtained from the E and Z stereoisomers of an alkene because they both form the same carbocation. Because this reaction forms a new asymmetric center, the product is a racemic mixture. The reaction, therefore, is neither stereoselective nor stereospecific.

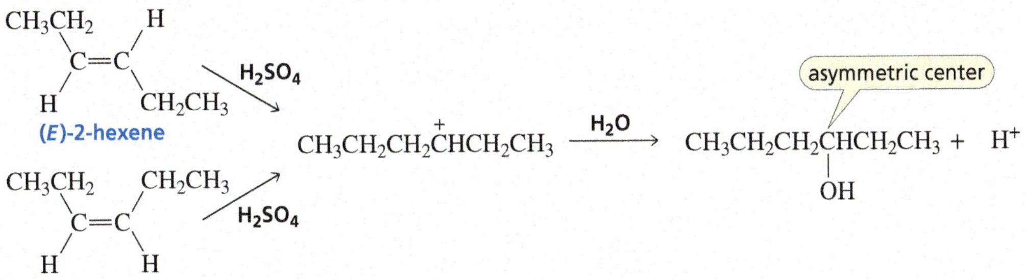

(E)-2-hexene

asymmetric center

$$CH_3CH_2CH_2\overset{+}{C}HCH_2CH_3 \xrightarrow{H_2O} CH_3CH_2CH_2\overset{}{C}HCH_2CH_3 + H^+$$
$$\underset{OH}{|}$$

(Z)-2-hexene

PROBLEM 36 ◆

a. Is the reaction of 2-butene with HBr regioselective?
b. Is it stereoselective?
c. Is it stereospecific?
d. Is the reaction of 1-butene with HBr regioselective?
e. Is it stereoselective?
f. Is it stereospecific?

PROBLEM 37

What stereoisomers are obtained from each of the following reactions?

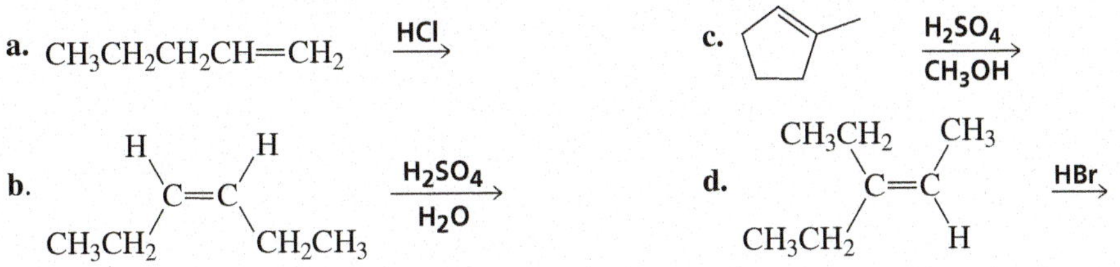

a. $CH_3CH_2CH_2CH=CH_2 \xrightarrow{HCl}$

c. [cyclopentene] $\xrightarrow[CH_3OH]{H_2SO_4}$

b. [alkene] $\xrightarrow[H_2O]{H_2SO_4}$

d. [alkene] $\xrightarrow{HBr}$

If an addition reaction creates an asymmetric center in a compound that already has an asymmetric center (and the reactant is a single enantiomer of that compound), a pair of diastereomers will be formed.

For example, let's look at the reaction on the top of the next page. Because none of the bonds to the asymmetric center in the reactant is broken during the reaction, the configuration of this asymmetric center does not change. The chloride ion can approach the planar carbocation intermediate formed in the reaction from above or from below, so two stereoisomers are formed. The stereoisomers are diastereomers because one of the asymmetric centers has the same configuration in both stereoisomers and the other has opposite configurations.

When a reactant that has an asymmetric center forms a product with *a second* asymmetric center, the products are diastereomers in unequal amounts.

Because the products of the preceding reaction are diastereomers, the transition states that lead to them are also diastereomers. The two transition states, therefore, do not have the same stability, so they are formed at different rates. Thus, the reaction is stereoselective—it forms more of one stereoisomer than of the other. The reaction is also stereospecific; the alkene with an asymmetric center in the *S* configuration forms a different pair of diastereomers than the alkene with an asymmetric center in the *R* configuration.

PROBLEM 38 SOLVED

Which product is obtained in greater yield from the preceding reaction?

SOLUTION The methyl group attached to the sp^3 carbon in the reactant is sticking up, so it provides some steric hindrance to the approach of the chloride ion from above the planar carbocation. As a result, the product obtained in greater yield is the one formed by the approach of the chloride ion from below the plane.

The Stereochemistry of Addition Reactions That Form Products with Two Asymmetric Centers

When a reactant that does not have an asymmetric center undergoes a reaction that forms a product with *two* asymmetric centers, the stereoisomers that are formed depend on the mechanism of the reaction.

Addition Reactions That Form a Carbocation Intermediate

When an addition reaction that results in a product with *two new asymmetric centers* forms a carbocation intermediate, *four stereoisomers* are formed.

Now let's see why four stereoisomers are formed. In the first step of the reaction, the proton can approach the plane containing the doubly bonded carbons of the alkene from either above or below to form the carbocation. Once the carbocation is formed, the chloride ion can approach the positively charged carbon from above or below. Thus, the addition of the proton and the chloride ion can be described as above-above (both adding from above), above-below, below-above, and below-below. As a result, four stereoisomers are obtained as products.

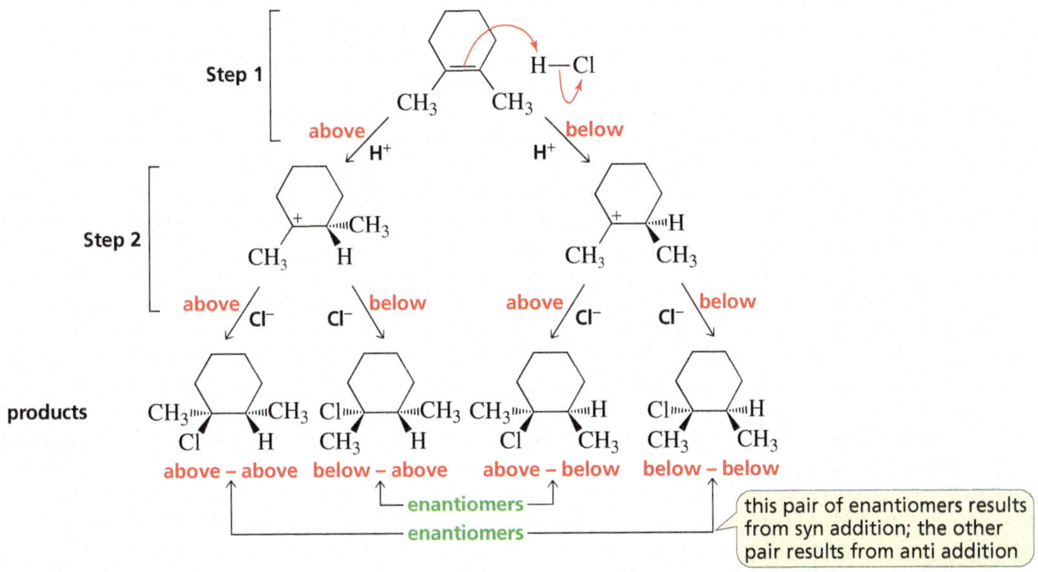

When an addition reaction that results in a product with two new asymmetric centers forms a carbocation intermediate, four stereoisomers are obtained.

When two substituents add to the same side of a double bond, the addition is called **syn addition.** When two substituents add to opposite sides of a double bond, the addition is called **anti addition.** Both syn and anti addition occur in electrophilic addition reactions that form a carbocation intermediate. The reaction does not select for a particular set of stereoisomers, so it is not stereoselective.

PROBLEM 39 SOLVED

Using perspective formulas, draw the products obtained from the reaction of *(Z)*-3,4-dimethyl-3-hexene with HCl.

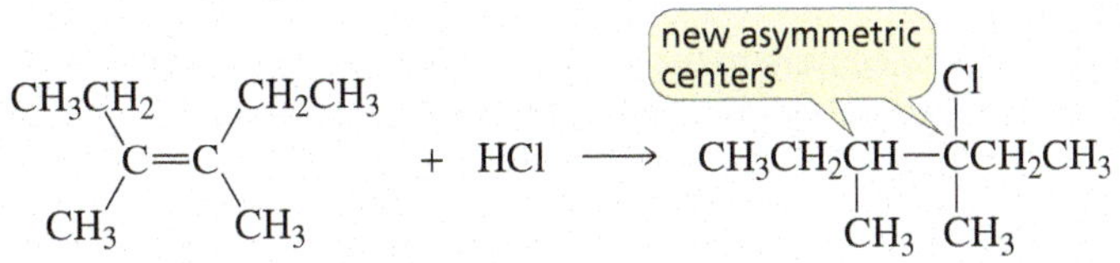

(Z)-3,4-dimethyl-3-hexene

Because the reaction forms a carbocation intermediate on the way to forming a product with two new asymmetric centers, four stereoisomers are formed. First draw the bonds about the asymmetric centers. (Remember that the lines in the plane must be adjacent and that the hatched wedge must be above the solid wedge.) Attach the groups to the bonds in any order; then draw the mirror image of the structure you just drew.

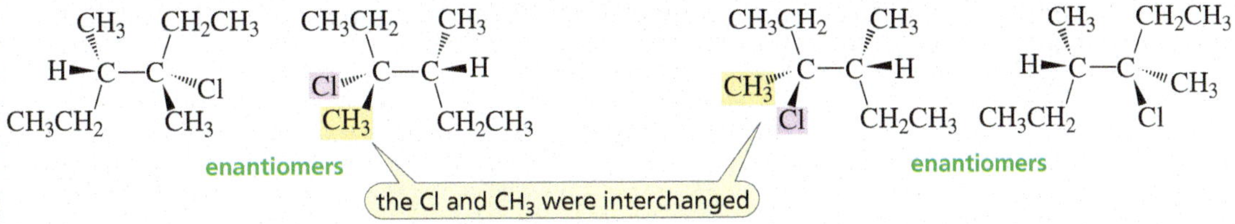

Next, draw a third structure by interchanging any two groups bonded to one of the asymmetric enters in one of the first two structures. Then draw the mirror image of the third structure.

Because the four stereoisomers formed by the cis alkene are identical to the four stereoisomers formed by the trans alkene, the reaction is also not stereospecific.

The Stereochemistry of Hydrogen Addition

We saw that in a catalytic hydrogenation reaction, the alkene sits on the surface of a metal catalyst onto which H_2 has been absorbed (Section 5.9). As a result, both hydrogen atoms add to the same side of the double bond. Thus, the addition of H_2 is a syn addition.

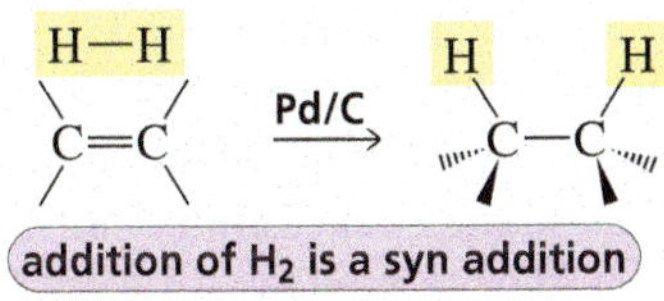

If addition of H_2 to an alkene forms a product with two asymmetric centers, then only two of the four possible stereoisomers are obtained because only syn addition can occur. (The other two stereoisomers would have to come from anti addition.)

Because the alkene can sit on the metal catalyst with either side facing up, one of the two stereoisomers results from addition of both hydrogens from below the plane of the double bond and the other results from addition of both hydrogens from above the plane.

The particular pair of stereoisomers that is formed depends on whether the reactant is a *cis* alkene or a *trans* alkene. Syn addition of H_2 to a *cis* alkene forms only the erythro enantiomers. (In Section 4.12, we saw that the erythro enantiomers are the ones with the hydrogens on the same side of the carbon chain in the eclipsed conformers.)

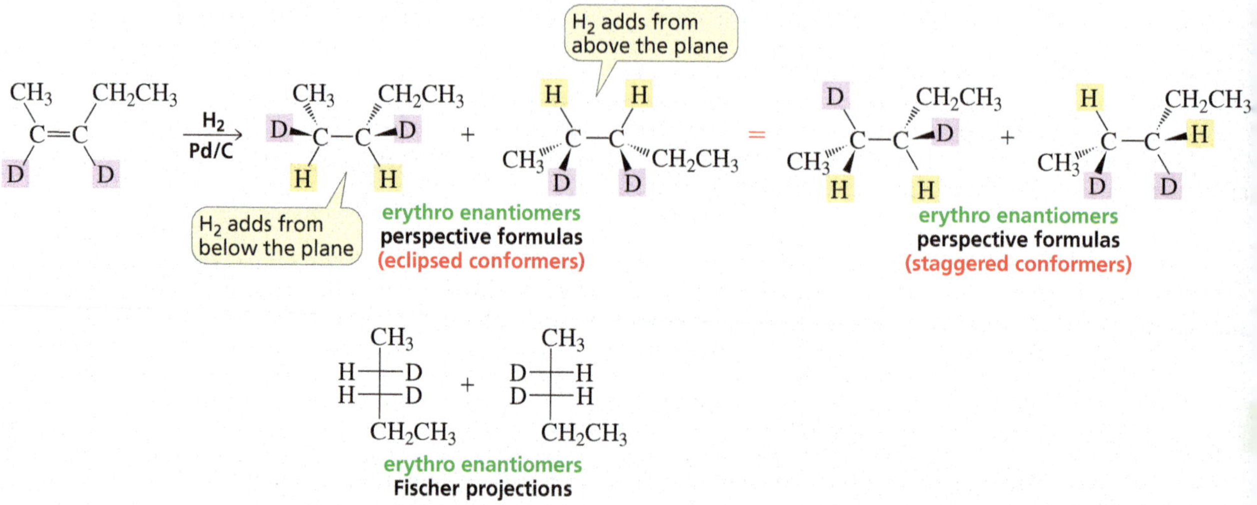

erythro enantiomers
perspective formulas
(eclipsed conformers)

erythro enantiomers
perspective formulas
(staggered conformers)

erythro enantiomers
Fischer projections

If each of the two asymmetric centers in the product is bonded to the same four substituents, a meso compound will be formed instead of the erythro enantiomers.

meso compound

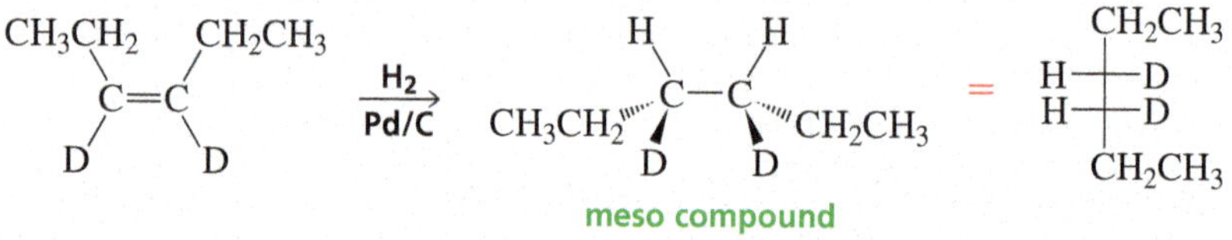

In contrast, syn addition of H_2 to a *trans* alkene forms only the threo enantiomers. Thus, the addition of hydrogen is a stereospecific reaction—the products obtained from addition to the cis isomer are different from the products obtained from addition to the trans isomer. It is also a stereoselective reaction because only two of the four stereoisomers are formed.

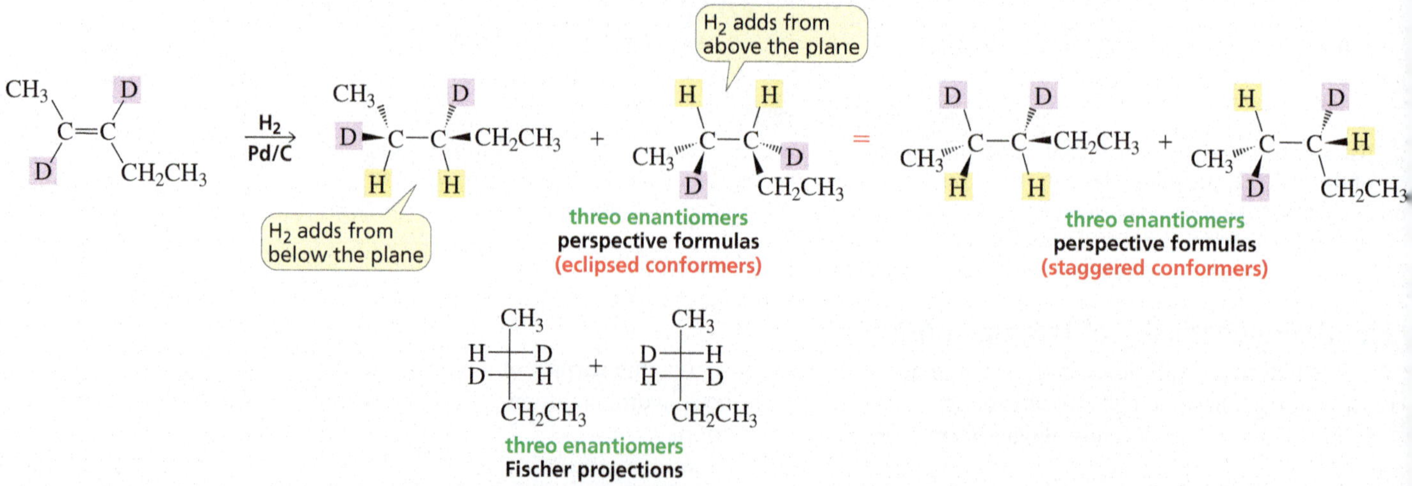

threo enantiomers
perspective formulas
(eclipsed conformers)

threo enantiomers
perspective formulas
(staggered conformers)

threo enantiomers
Fischer projections

If the reactant is cyclic, the addition of H_2 will form the cis enantiomers because the two hydrogens add to the same side of the double bond. You can draw the products by showing one with H_2 having been delivered from below the plane of the double bond and the other with H_2 having been delivered from above the plane.

H₂ adds from below the plane of the double bond

H₂ adds from above the plane of the double bond

Or you can draw one product showing H₂ having been delivered from *either* below or above the plane of the double bond and then draw the mirror image of that product.

Each of the two asymmetric centers in the product of the following reaction is bonded to the same four substituents. Therefore, syn addition forms a meso compound.

Cyclic Alkenes

Cyclic alkenes with fewer than seven carbons in the ring, such as cyclopentene and cyclohexene, exist only in the cis configuration because they do not have enough carbons to form a trans double bond. Therefore, it is not necessary to use the cis designation with their names. If the ring has seven or more carbons, however, then both cis and trans isomers are possible, so the configuration of the compound must be specified in its name.

cyclopentene   cyclohexene   *cis*-cyclooctene   *trans*-cyclooctene

---

PROBLEM 40 SOLVED

a. What stereoisomers are formed in the following reaction? Are they enantiomers or diastereomers?
b. Which stereoisomer is formed in greater yield?

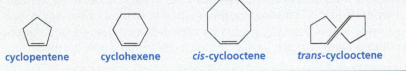

SOLUTION 40 a. The reaction forms two stereoisomers because H₂ can approach the plane of the double bond from above and below. The stereoisomers are diastereomers because the reactant has an asymmetric center and the product has a new second asymmetric center.

SOLUTION 40 b. Because the methyl group is pointing upward, it provides steric hindrance to H₂ approaching the double bond from above. Therefore, the major product is A, the compound formed by H₂ approaching the double bond from below.

The Stereochemistry of Peroxyacid Addition

The addition of a peroxyacid to an alkene to form an epoxide is a concerted reaction: the oxygen atom adds to the two sp^2 carbons at the same time (Section 6.10). Therefore, it must be a syn addition.

addition of peroxyacid is a syn addition

The oxygen can add from above or from below the plane containing the double bond. Therefore, addition of a peroxyacid to an alkene can form two stereoisomers. Syn addition to a cis alkene forms the cis enantiomers. Because only syn addition occurs, the reaction is stereoselective. You can draw the products by showing one with the oxygen having been delivered from above the plane of the double bond and the other with the oxygen having been delivered from below the plane.

Or, you can draw one product showing the oxygen having been delivered from *either* above or below the plane of the double bond and then draw the mirror image of that product.

Syn addition to a trans alkene forms the trans enantiomers. Because the cis and trans isomers each forms a different pair of enantiomers, the reaction is stereospecific. Again, you can draw the products by showing one with the oxygen having been delivered from above the plane of the double bond and the other with the oxygen having been delivered from below the plane.

Or, you can draw one product showing the oxygen having been delivered from either above or below the plane of the double bond and then draw the mirror image of that product.

Addition of a peroxyacid to *cis*-2-butene forms a meso compound because both asymmetric centers are attached to the same four groups (Section 4.14).

PROBLEM 41 ◆

a. What alkene is required to synthesize each of the following compounds?
b. What other epoxide is formed in each synthesis?
c. Assign an R or S configuration to each asymmetric center.

1. 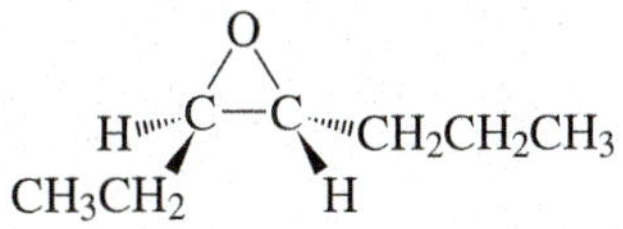   H⬝⬝⬝C—C⬝⬝⬝CH₂CH₂CH₃
              CH₃CH₂   H

2. 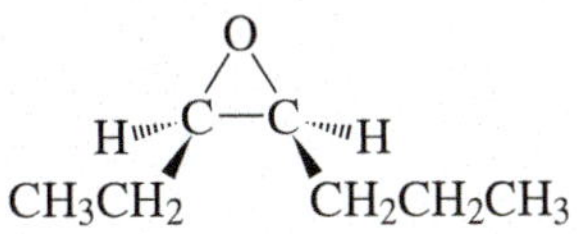   H⬝⬝⬝C—C⬝⬝⬝H
              CH₃CH₂   CH₂CH₂CH₃

The Stereochemistry of Hydroboration–Oxidation

The addition of borane (or R_2BH) to an alkene is also a concerted reaction (Section 6.8). The boron and the hydride ion add to the two sp^2 carbons of the double bond at the same time. Because the two species add simultaneously, they must add to the same side of the double bond—that is, it is a syn addition.

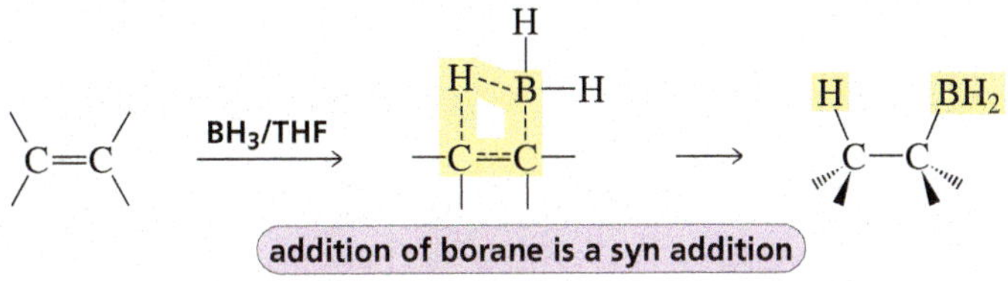

addition of borane is a syn addition

When the resulting alkylborane is oxidized by reaction with hydrogen peroxide and hydroxide ion, the OH group ends up in the same position as the boron group it replaces. Consequently, the overall reaction, called hydroboration–oxidation, amounts to a syn addition of water to a carbon–carbon double bond.

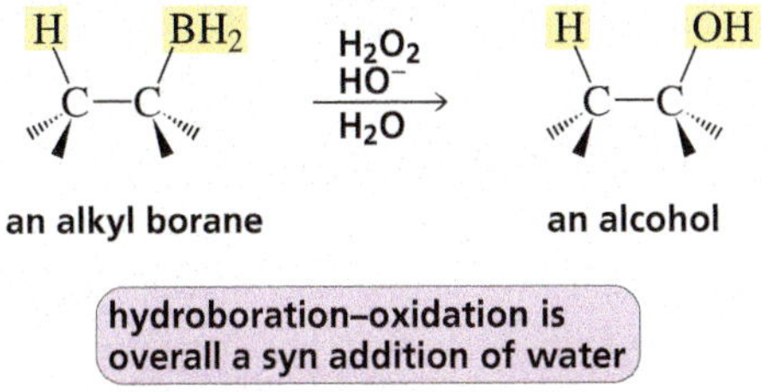

an alkyl borane   an alcohol

hydroboration–oxidation is overall a syn addition of water

Because only syn addition occurs, hydroboration–oxidation is stereoselective—only two of the four possible stereoisomers are formed. If the reactant is cyclic, syn addition forms only the enantiomers that have the added groups on the same side of the ring. You can draw the products by showing H and OH having been delivered from below the plane of the double bond and the other with the H and OH having been delivered from above the plane.

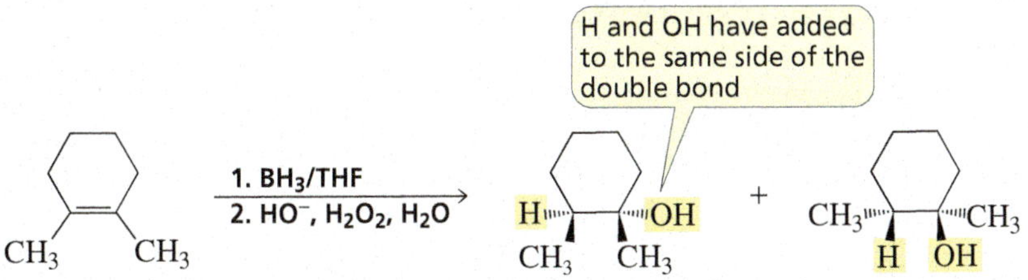

H and OH have added to the same side of the double bond

Or, you can draw one product showing the H and OH having been delivered from either above or below the plane of the double bond and then draw the mirror image of that product.

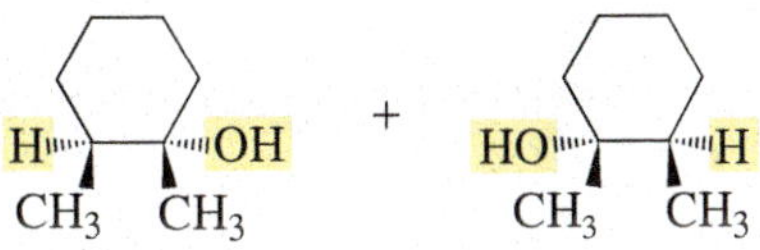

PROBLEM 42 ◆

What stereoisomers are obtained from hydroboration–oxidation of the following compounds? Assign an *R* or *S* configuration to each asymmetric center.

a. cyclohexene

b. 1-ethylcyclohexene

c. *cis*-2-butene

d. (Z)-3,4-dimethyl-3-hexene

The Stereochemistry of Addition Reactions That Form a Cyclic Bromonium or Chloronium Ion Intermediate

If two asymmetric centers are created from an addition reaction that forms a bromonium (or chloronium) ion intermediate, only one pair of enantiomers will be formed. For example, the addition of Br_2 to the cis alkene forms only the threo enantiomers.

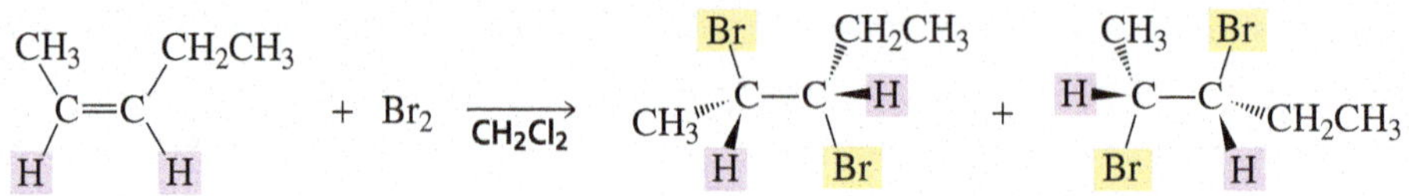

cis-2-pentene

threo enantiomers
perspective formulas

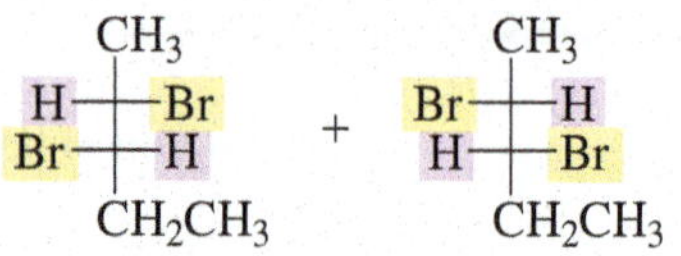

threo enantiomers
Fischer projections

Similarly, the addition of Br_2 to the trans alkene forms only the erythro enantiomers. Because the cis and trans isomers form different products, the reaction is stereospecific as well as stereoselective.

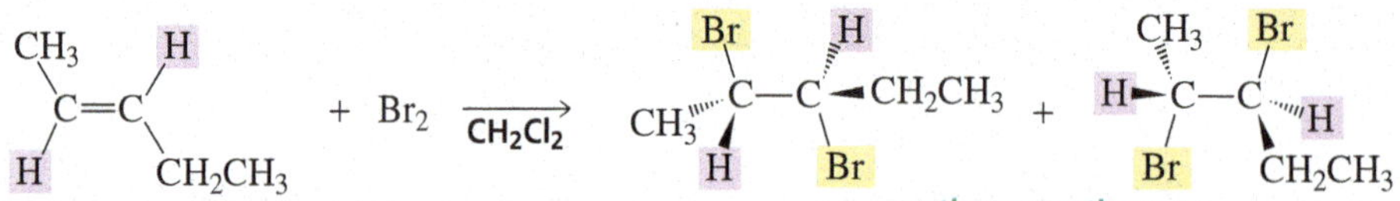

trans-2-pentene

erythro enantiomers
perspective formulas

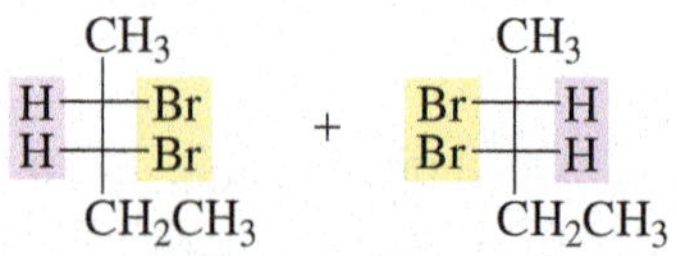

erythro enantiomers
Fischer projections

Because the addition of Br_2 to the cis alkene forms the threo enantiomers, we know that anti addition must have occurred. (Recall that we just saw that syn addition forms the erythro enantiomers.) The addition of Br_2 is anti because the two bromine atoms add to opposite sides of the double bond (Figure 6.6).

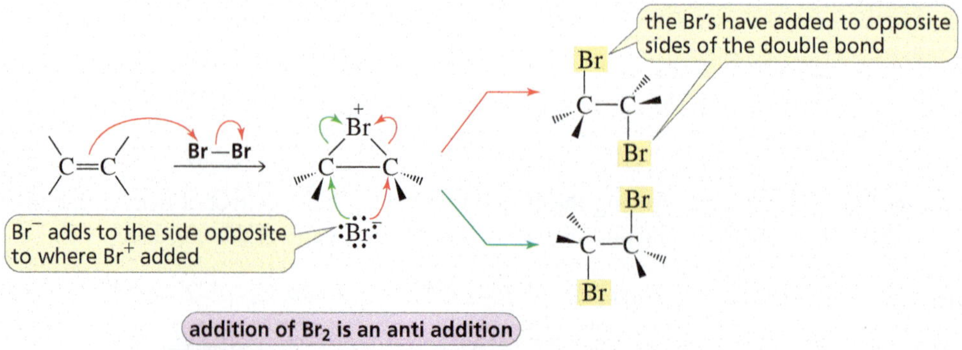

▶ **Figure 6.6**
A cyclic bromonium ion is formed in the first step of the reaction (Section 6.9). Because the bromine atom blocks one side of the intermediate, Br^- must approach from the opposite side (following either the green arrows *or* the red arrows). Thus, only anti addition of Br_2 can occur (the two bromine atoms add to opposite sides of the double bond), so only two of the four possible stereoisomers are formed.

the Br's have added to opposite sides of the double bond

Br^- adds to the side opposite to where Br^+ added

addition of Br_2 is an anti addition

If the two asymmetric centers in the product each have the same four substituents, the erythro isomers are identical and constitute a meso compound. Therefore, addition of Br_2 to *trans*-2-butene forms a meso compound.

trans-2-butene

a meso compound
perspective formula

plane of symmetry

a meso compound
Fischer projection

Because only anti addition occurs, addition of Br_2 to cyclohexene forms only the enantiomers that have the bromines on opposite sides of the ring (that is, only the trans stereoisomers are formed).

The stereochemistry of the products obtained from electrophilic addition reactions to alkenes is summarized in Table 6.1.

Table 6.1   Stereochemistry of Alkene Addition Reactions

Reaction	Type of addition	Stereoisomers formed
Addition reactions that create one asymmetric center in the product		If the reactant does not have an asymmetric center, then a racemic mixture is formed.
		If the reactant has an asymmetric center, then unequal amounts of a pair of diastereomers are formed.
Addition reactions that create two asymmetric centers in the product		
Addition of reagents that form a carbocation intermediate	syn and anti	Four stereoisomers are formed; the cis and trans isomers each form the same products.
Addition of H_2	syn	cis $\rightarrow$ erythro or cis enantiomers*
Addition of a peroxyacid	syn	
Addition of BH_3 or BHR_2	syn	trans $\rightarrow$ threo or trans enantiomers
Addition of Br_2, $Br_2 + H_2O$, $Br_2 + ROH$ (any reaction that forms a cyclic bromonium or chloronium ion intermediate)	anti	cis $\rightarrow$ threo or trans enantiomers
		trans $\rightarrow$ erythro or cis enantiomers*

*An acyclic compound forms the erythro enantiomers; a cyclic compound forms the cis enantiomers. If the two asymmetric centers have the same substituents, a meso compound will be formed instead of the pair of erythro or cis enantiomers.

PROBLEM 43

The reaction of 2-ethyl-1-pentene with Br_2, with $H_2 + Pd/C$, or with R_2BH/THF followed by aqueous $HO^- + H_2O_2$ leads to a racemic mixture. Explain why a racemic mixture is obtained in each case.

PROBLEM 44 **SOLVED**

Using a sample of *trans*-2-pentene, how could you prove that the addition of Br_2 forms a cyclic bromonium ion intermediate rather than a carbocation intermediate?

SOLUTION You could distinguish between the two intermediates by determining the number of products obtained from the reaction. If a cyclic intermediate is formed, then two products will be obtained, because only anti addition can occur. If a carbocation intermediate is formed, then four products will be obtained because both syn and anti addition can occur.

A Mnemonic to the Rescue

One way to remember what stereoisomers are obtained from a reaction that creates a product with two asymmetric centers is the mnemonic **CIS-SYN-(ERYTHRO or CIS).** (The third term in parentheses is "erythro" if the product is acyclic and "cis" if it is cyclic.) The three terms are easy to remember because they all mean "on the same side." Thus, if you have a cis reactant that undergoes the addition of H_2 (which is syn), the erythro products are obtained if the products are acyclic and the cis products are obtained if the product is cyclic.

You can change any two of the three terms, but you cannot change just one. (Note that the term in parentheses is considered to be one term.) For example, **TRANS-ANTI-(ERYTHRO or CIS)** and **CIS-ANTI-(THREO or TRANS)** are allowed because in each case, two terms were changed. In other words, anti addition to a trans alkene forms the erythro (or cis) enantiomers, and anti addition to a cis alkene forms the threo (or trans) enantiomers. **TRANS-SYN-(ERYTHRO or CIS)** is *not* allowed, because only one term was changed. Thus, syn addition to a trans alkene does *not* form the erythro (or cis) enantiomers.

This mnemonic works for all reactions that have products with structures that can be described by erythro and threo or by cis and trans.

PROBLEM-SOLVING STRATEGY

LEARN THE STRATEGY

Predicting the Stereoisomers Obtained from the Addition Reactions of Alkenes

What stereoisomers are obtained from the following reactions?

a. 1-butene + H_2O + H_2SO_4 **c.** *cis*-3-heptene + Br_2
b. cyclohexene + HBr   **d.** *trans*-3-hexene + Br_2

Start by drawing the product without regard to its configuration to check whether the reaction has created any asymmetric centers. Then determine the stereoisomers of the products, paying attention to the configuration (if any) of the reactant, how many asymmetric centers are formed, and the mechanism of the reaction. Let's start with part **a**.

a. CH₃CH₂CHCH₃
           |
          OH

The product has one asymmetric center, so equal amounts of the R and S enantiomers are obtained.

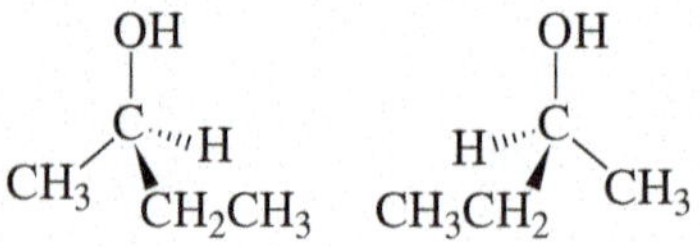

b.

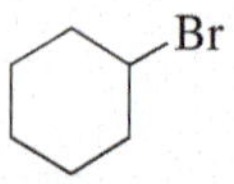

The product does not have an asymmetric center, so it has no stereoisomers.

c. CH₃CH₂CHCHCH₂CH₂CH₃
           | |
           Br Br

Two asymmetric centers have been created in the product. Because the reactant is cis and the addition of Br_2 is anti, the threo enantiomers are formed.

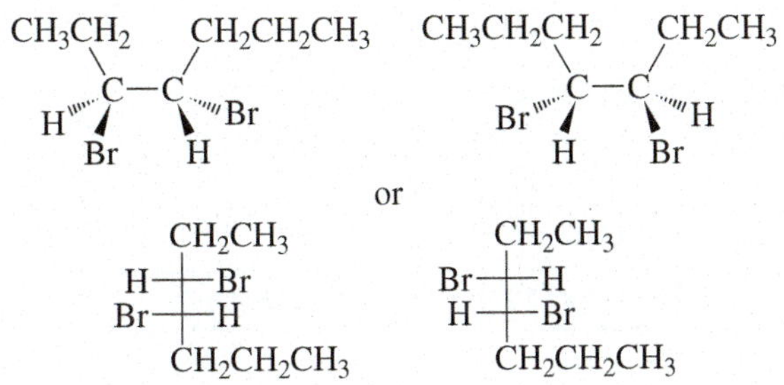

$$\begin{array}{cc}
\underset{H\overset{\displaystyle |}{\underset{\displaystyle Br}{}}}{CH_3CH_2}\!\!-\!\!C\!-\!C\!\!\underset{\displaystyle H}{\overset{\displaystyle CH_2CH_2CH_3}{}}\!\!Br
&
\underset{Br\overset{\displaystyle |}{\underset{\displaystyle H}{}}}{CH_3CH_2CH_2}\!\!-\!\!C\!-\!C\!\!\underset{\displaystyle Br}{\overset{\displaystyle CH_2CH_3}{}}\!\!H
\end{array}$$

or

$$\begin{array}{cc}
CH_2CH_3 & CH_2CH_3 \\
H\!\!-\!\!Br & Br\!\!-\!\!H \\
Br\!\!-\!\!H & H\!\!-\!\!Br \\
CH_2CH_2CH_3 & CH_2CH_2CH_3
\end{array}$$

d. $CH_3CH_2\underset{\overset{\displaystyle |}{Br}}{C}H\underset{\overset{\displaystyle |}{Br}}{C}HCH_2CH_3$

Two asymmetric centers have been created in the product. Because the reactant is trans and the addition of Br_2 is anti, one would expect the erythro enantiomers. However, the two asymmetric centers are bonded to the same four groups, so the erythro product is a meso compound. Thus, only one stereoisomer is formed.

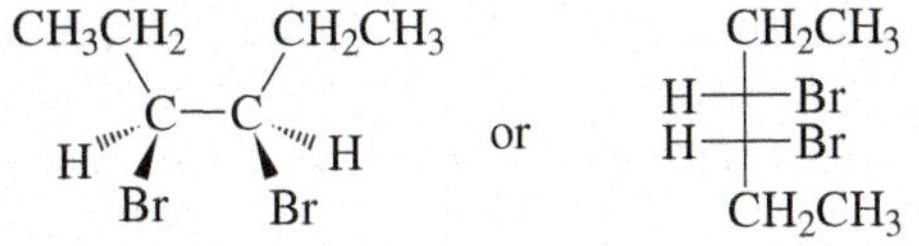

$$\begin{array}{ccc}
\underset{H\overset{\displaystyle |}{\underset{\displaystyle Br}{}}}{CH_3CH_2}\!\!-\!\!C\!-\!C\!\!\underset{\displaystyle Br}{\overset{\displaystyle CH_2CH_3}{}}\!\!H
& \text{or} &
\begin{array}{c}
CH_2CH_3 \\
H\!\!-\!\!Br \\
H\!\!-\!\!Br \\
CH_2CH_3
\end{array}
\end{array}$$

PROBLEM 45

What stereoisomers are obtained from the following reactions?

a. *trans*-2-butene + HBr
b. (*Z*)-3-methyl-2-pentene + HBr
c. (*E*)-3-methyl-2-pentene + HBr
d. *cis*-3-hexene + HBr
e. *cis*-2-pentene + Br_2
f. 1-hexene + Br_2

PROBLEM 46

When Br_2 adds to a cis alkene that has different substituents attached to each of the two sp^2 carbons, such as *cis*-2-heptene, identical amounts of the two threo enantiomers are obtained even though Br^- is more likely to add to the less sterically hindered carbon of the bromonium ion. Explain why identical amounts of the two enantiomers are obtained.

PROBLEM 47

a. What products will be obtained from the addition of Br_2 to cyclohexene if H_2O is added to the reaction mixture?
b. Propose a mechanism for the reaction.

PROBLEM 48

What stereoisomers would you expect to obtain from each of the following reactions?

a.
$$\underset{CH_3}{\overset{CH_3CH_2}{}}C\!\!=\!\!C\underset{CH_2CH_3}{\overset{CH_3}{}} \quad \xrightarrow[CH_2Cl_2]{Br_2}$$

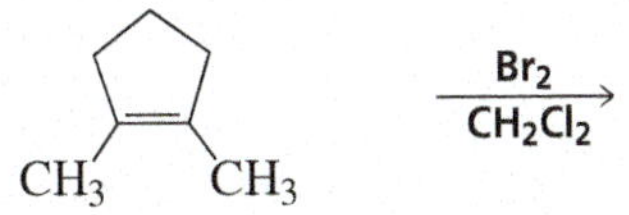

b.
$$\underset{CH_3CH_2}{\overset{CH_3CH_2}{}}C\!\!=\!\!C\underset{CH_2CH_3}{\overset{CH_3}{}} \quad \xrightarrow[CH_2Cl_2]{Br_2}$$

c.
$$\underset{CH_3\quad CH_3}{\text{cyclopentene}} \quad \xrightarrow[CH_2Cl_2]{Br_2}$$

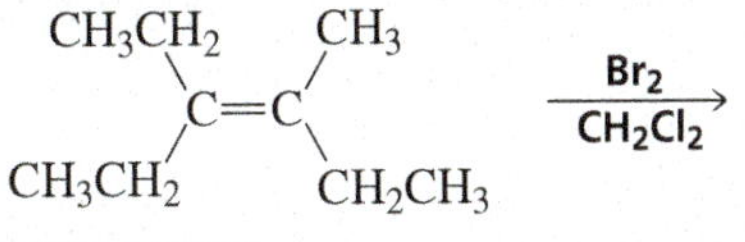

d.
$$\underset{CH_3\quad CH_2CH_3}{\text{cyclopentene}} \quad \xrightarrow[CH_2Cl_2]{Br_2}$$

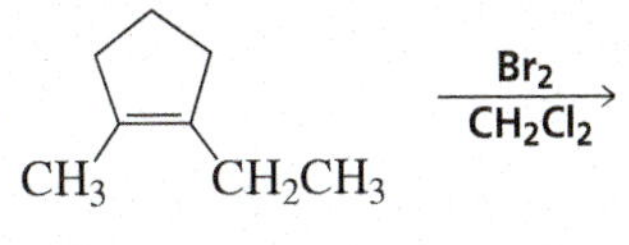

PROBLEM 49

What stereoisomers would you expect to obtain from each of the following reactions?

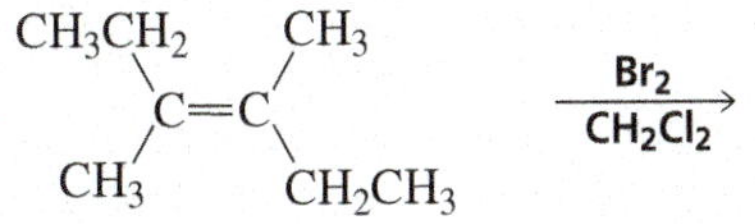

a.
$$\underset{CH_3}{\overset{CH_3CH_2}{}}C\!\!=\!\!C\underset{CH_2CH_3}{\overset{CH_3}{}} \quad \xrightarrow[Pd/C]{H_2}$$

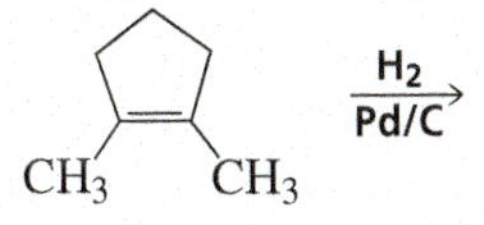

b.
$$\underset{CH_3}{\overset{CH_3CH_2}{}}C\!\!=\!\!C\underset{CH_3}{\overset{H}{}} \quad \xrightarrow[Pd/C]{H_2}$$

c.
$$\underset{CH_3\quad CH_3}{\text{cyclopentene}} \quad \xrightarrow[Pd/C]{H_2}$$

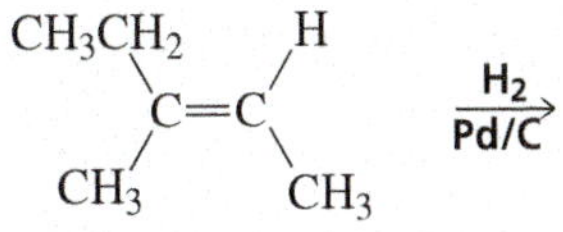

d.
$$\underset{CH_3\quad CH_2CH_3}{\text{cyclopentene}} \quad \xrightarrow[Pd/C]{H_2}$$

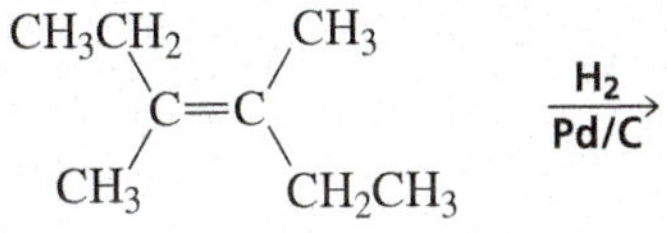

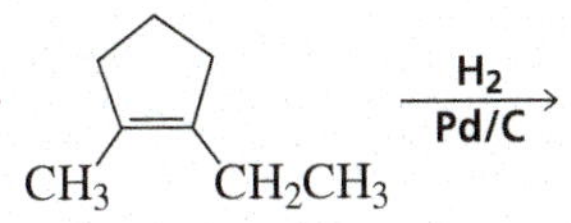

6.14 THE STEREOCHEMISTRY OF ENZYME-CATALYZED REACTIONS

The chemistry associated with living organisms is called **biochemistry.** When you study biochemistry, you study the structures and functions of the molecules found in the biological world and the reactions involved in the synthesis and degradation of these molecules. Because the compounds in living organisms are organic compounds, it is not surprising that many of the reactions encountered in an organic chemistry laboratory also occur in cells.

Cells do not contain molecules such as Cl_2, HBr, or BH_3, so you would not expect to find the addition of such reagents to alkenes in living organisms. However, cells do contain water, so some alkenes found in cells undergo the addition of water.

Most reactions that occur in cells are catalyzed by proteins called **enzymes** (Section 5.14). When an enzyme catalyzes a reaction that forms a product with an asymmetric center, only one stereoisomer is formed because

enzyme-catalyzed reactions are completely stereoselective.

For example, the enzyme fumarase catalyzes the addition of water to fumarate to form malate, a compound with one asymmetric center.

> When an enzyme catalyzes a reaction that forms a product with an asymmetric center, only one stereoisomer is formed.

However, the reaction forms only (*S*)-malate; the *R* enantiomer is not formed.

An enzyme-catalyzed reaction forms only one stereoisomer, unlike a nonenzyme-catalyzed reaction that forms a racemic mixture, because the reactant can undergo a reaction only when it is bound to the active site of the enzyme (Section 5.14). The active site can restrict delivery of the reagents to only one side of the reactant.

Enzyme-catalyzed reactions are also stereospecific; an enzyme typically catalyzes the reaction of only one stereoisomer. For example, fumarase catalyzes the addition of water to fumarate (the trans isomer just shown) but not to maleate (the cis isomer).

$$^-OOC \quad COO^- $$
$$C=C$$
$$H \qquad H$$

maleate

$$+ \ H_2O \xrightarrow{\textbf{fumarase}} \text{no reaction}$$

An enzyme is able to differentiate between the two stereoisomers because only one of them has the structure that allows it to fit into the enzyme's active site where the reaction takes place (Section 5.14).

PROBLEM 51 ◆

a. What is the product of the reaction of fumarate and H_2O when H_2SO_4 is used as a catalyst instead of fumarase?

b. What is the product of the reaction of maleate and H_2O when H_2SO_4 is used as a catalyst instead of fumarase?

6.15 ENANTIOMERS CAN BE DISTINGUISHED BY BIOLOGICAL MOLECULES

Enzymes can tell the difference between enantiomers because enzymes are proteins and proteins are chiral molecules.

Enzymes

An achiral reagent such as hydroxide ion cannot distinguish between enantiomers. Thus, it reacts with (R)-2-bromobutane at the same rate that it reacts with (S)-2-bromobutane.

Because an enzyme is *chiral*, it can distinguish not only between cis–trans isomers such as maleate and fumarate (Section 6.14) but also between enantiomers and catalyze the reaction of only one of them.

Chemists can use an enzyme's ability to distinguish between enantiomers to separate them. For example, the enzyme D-amino acid oxidase catalyzes the oxidation of the *R* enantiomer but leaves the *S* enantiomer unchanged. The oxidized product of the enzyme-catalyzed reaction can be easily separated from the unreacted enantiomer because they are different compounds.

R enantiomer   *S* enantiomer $\xrightarrow{\textbf{D-amino acid oxidase}}$ oxidized *R* enantiomer   +   unreacted *S* enantiomer

An achiral reagent reacts identically with both enantiomers. A sock, which is achiral, fits on either foot.

A chiral reagent reacts differently with each enantiomer. A shoe, which is chiral, fits on only one foot.

An enzyme is able to differentiate between enantiomers because its binding site is chiral. Therefore, the enzyme binds only the stereoisomer whose substituents are in the correct positions to interact with the substituents in the chiral binding site. If you look at Figure 4.9 on p. 178 and replace "binding site of the receptor" with "binding site of the enzyme," you can see that the enzyme binds the *R* enantiomer but not the *S* enantiomer. Why? Because the *S* enantiomer does not have its substituents in the proper positions to bind efficiently to the enzyme.

Like a right-handed glove, which fits only the right hand,
an enzyme forms only one stereoisomer and reacts with only one stereoisomer.

Chiral Catalysts

The problem of having to separate enantiomers can be avoided if a synthesis is carried out that forms one of the enantiomers preferentially. **Chiral catalysts** are being developed for the synthesis of one enantiomer in great excess over the other. For example, the catalytic hydrogenation of 2-ethyl-1-pentene forms a racemic mixture because H_2 can be delivered equally easily to both faces of the double bond (p. 268).

(R)-3-methylhexane   (S)-3-methylhexane
50%   50%

If, however, a metal catalyst is complexed to a chiral organic molecule, then H_2 will be delivered preferentially to only one face of the double bond. One such chiral catalyst—using Ru(II) as the metal and BINAP (2,2'-bis[diphenylphosphino]-1,1'-binaphthyl) as the chiral molecule—has been used to synthesize (S)-naproxen, the active ingredient in Aleve and in several other over-the-counter nonsteroidal anti-inflammatory drugs (p. 114), in greater than 98% enantiomeric excess. This is an example of an **enantioselective reaction**—a reaction that forms more of one enantiomer than another. This is a medically important enantioselective reaction, because a racemic mixture of naproxen cannot be given to a patient because (R)-naproxen is highly toxic to the liver.

(S)-naproxen
>98% ee

The Sharpless epoxidation is another example of an enantioselective reaction achieved by using a chiral catalyst. In this reaction, a peroxide (*tert*-butylhydroperoxide) delivers its oxygen to only one side of the double bond of an allylic alcohol, in the presence of a metal catalyst [titanium(IV) isopropoxide] and a chiral organic molecule (diethyl tartarate; DET).

The structure of the epoxide depends on which isomer of diethyl tartarate is used.

6.16 REACTIONS AND SYNTHESIS

This chapter has focused on the reactions of alkenes. You have seen why alkenes react, the kinds of reagents with which they react, the mechanisms by which the reactions occur, and the products that are formed. Keep in mind, however, that when you are studying reactions, you are simultaneously studying synthesis. When you learn that compound **A** reacts with a certain reagent to form compound **B**, you are learning not only about the reactivity of **A** but also about one way that compound **B** can be synthesized.

$$A \longrightarrow B$$

A reacts;
B is synthesized

For example, you saw that many different reagents can add to alkenes and that compounds such as alkyl halides, vicinal dihalides, halohydrins, alcohols, ethers, epoxides, alkanes, aldehydes, and ketones are synthesized as a result.

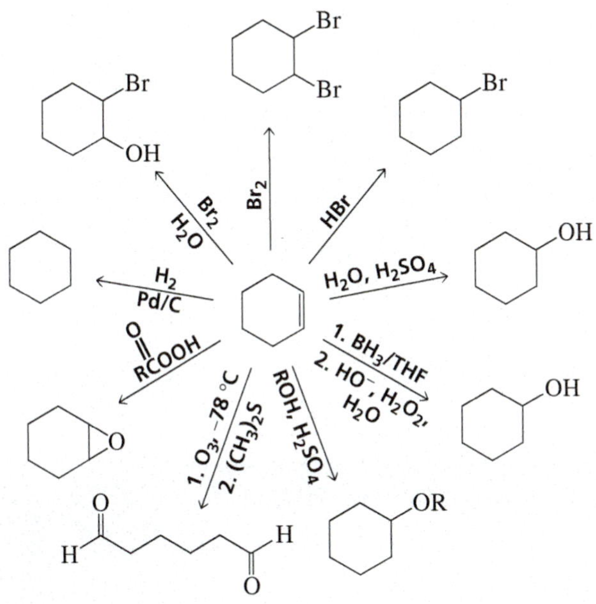

Although you have seen how alkenes react and have learned about the kinds of compounds that are synthesized when alkenes undergo reactions, you have not yet seen how alkenes are synthesized. The reactions of alkenes involve the *addition* of atoms (or groups of atoms) to the two sp^2 carbons of the double bond. Reactions that synthesize alkenes are exactly the opposite; they involve the *elimination* of atoms (or groups of atoms) from two adjacent sp^3 carbons.

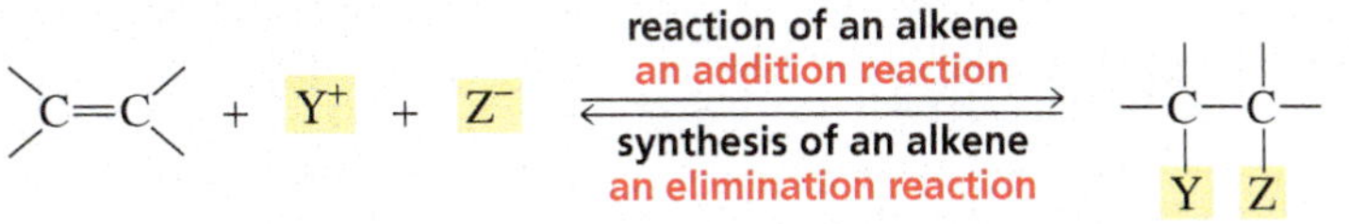

You will learn how alkenes are synthesized when you study compounds that undergo elimination reactions. The various reactions that can be used to synthesize alkenes are listed in Appendix III "Summary of Methods Used to Synthesize a Particular Functional Group."

PROBLEM 52 SOLVED

Show how each of the following compounds can be synthesized from an alkene:

LEARN THE STRATEGY

a. **b.**

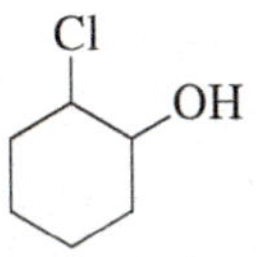

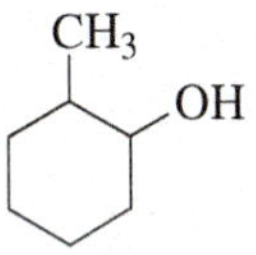

SOLUTION TO 52 a. The only alkene that can be used for this synthesis is cyclohexene. To get the desired substituents on the ring, cyclohexene must react with Cl_2 in an aqueous solution so that water will be the nucleophile.

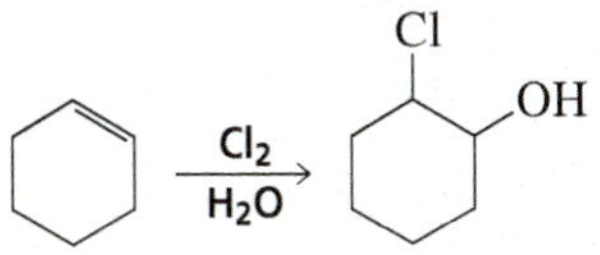

SOLUTION TO 52 b. The alkene that should be used here is 1-methylcyclohexene. To get the substituents in the desired locations, the electrophile must be R_2BH or BH_3, with HO taking boron's place in the subsequent oxidation reaction.

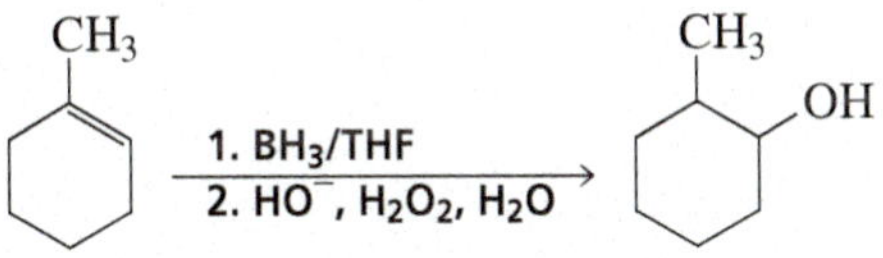

USE THE STRATEGY

PROBLEM 53

Show how each of the following compounds can be synthesized from an alkene:

a. CH₃CHOCH₃
   |
   CH₃

b.

c.

d.

e.

f.

PROBLEM 54 ♦

Explain why 3-methylcyclohexene should *not* be used as the starting material in Problem 52b.

Which Are More Harmful: Natural Pesticides or Synthetic Pesticides?

Learning to synthesize new compounds is an important part of organic chemistry. Long before chemists learned to synthesize compounds that would protect plants from predators, plants were doing the job themselves. Plants have every incentive to synthesize pesticides. When you cannot run, you need to find another way to protect yourself. But which pesticides are more harmful: those synthesized by chemists or those synthesized by plants? Unfortunately, we do not know because although federal laws require all human-made pesticides to be tested for any adverse effects, they do not require plant-made pesticides to be tested. Besides, risk evaluations of chemicals are usually done on rats, and something that is harmful to a rat may or may not be harmful to a human. Furthermore, when rats are tested, they are exposed to much higher concentrations of the chemical than would be experienced by a human, and some chemicals are harmful only at high doses. For example, we all need sodium chloride for survival, but high concentrations are poisonous, and although we associate alfalfa sprouts with healthy eating, monkeys fed very large amounts of alfalfa sprouts have been found to develop an immune system disorder.

pesticides being sprayed on a crop

ESSENTIAL CONCEPTS

Section 6.0

- **Alkenes** undergo **electrophilic addition reactions.**
- A curved arrow always points from the electron donor to the electron acceptor.

Section 6.2

- **Tertiary carbocations** are more stable than **secondary carbocations,** which are more stable than **primary carbocations.**

Section 6.3

- The **Hammond postulate** states that a transition state is more similar in structure to the species to which it is more similar in energy.
- The more stable carbocation is formed more rapidly.

Section 6.4

- Electrophilic addition reactions start with the addition of an *electrophile* to the sp^2 carbon bonded to the most hydrogens and end with the addition of a nucleophile to the other sp^2 carbon.
- **Regioselectivity** is the preferential formation of one **constitutional isomer** over another.

Sections 6.4–6.6

- The addition of hydrogen halides and the acid-catalyzed addition of water and alcohols form **carbocation intermediates.**

Section 6.7

- A carbocation will rearrange if it becomes more stable as a result of the rearrangement.
- **Carbocation rearrangements** occur by **1,2-hydride shifts** and **1,2-methyl shifts.**

Section 6.8

- **Hydroboration** does not form an intermediate.
- An **oxidation** reaction decreases the number of C—H bonds and/or increases the number of C—O, C—N, or C—X bonds (where X = a halogen).

Section 6.9

- The addition of Br₂ or Cl₂ forms an intermediate with a three-membered ring that reacts with nucleophiles.
- If a reaction does not form a carbocation intermediate, a carbocation rearrangement cannot occur.

Section 6.10

- **Epoxidation** does not form an intermediate.

Section 6.11

- **Ozonolysis** forms an intermediate with a five-membered ring.
- To determine the product of oxidative cleavage, replace $C=C$ with $C=O\ O=C$.
- A **ketone** has two alkyl groups bonded to a carbonyl group ($C=O$); an **aldehyde** has an alkyl group and one hydrogen (or has two hydrogens) bonded to a carbonyl group.

Section 6.12

- A **regioselective** reaction selects for a particular constitutional isomer.
- A **stereoselective** reaction selects for a particular stereoisomer.
- A reaction is **stereospecific** if the reactant can exist as stereoisomers and each stereoisomer forms a different stereoisomer or a different set of stereoisomers.

Section 6.13

- When a reactant that does not have an asymmetric center forms a product with one asymmetric center, the product is always a racemic mixture.

- When a reactant that has an asymmetric center forms a product with *a second* asymmetric center, the products are diastereomers in unequal amounts.

- In **syn addition,** the substituents add to the same side of a double bond; in **anti addition,** they add to opposite sides.

- Both syn and anti addition occur in electrophilic addition reactions that form a carbocation intermediate.

- The addition of H_2 or a peroxyacid to an alkene is a syn addition.

- Hydroboration–oxidation is overall a syn addition of water.

- The addition of Br_2 or Cl_2 is an anti addition.

Section 6.14

- An enzyme-catalyzed reaction forms only one stereoisomer, and an enzyme typically catalyzes the reaction of only one stereoisomer.

Section 6.15

- An achiral reagent reacts identically with both geometric isomers or with both enantiomers.

- A chiral reagent reacts differently with each geometric isomer or with each enantiomer.

- An **enantioselective** reaction selects for a particular enantiomer.

SUMMARY OF REACTIONS

As you review the electrophilic addition reactions of alkenes, keep in mind that the first step in each of them is the addition of an electrophile to the sp^2 carbon bonded to the most hydrogens.

1. Addition of hydrogen halides: H^+ is the electrophile; the halide ion is the nucleophile (Sections 6.1 and 6.4). Addition is syn and anti. The mechanism is on p. 237.

$$RCH=CH_2\ +\ HX\ \longrightarrow\ RCHCH_3$$
$$\underset{X}{|}$$

HX = HF, HCl, HBr, HI

2. Acid-catalyzed addition of water and alcohols: H^+ is the electrophile; water or an alcohol is the nucleophile (Sections 6.5 and 6.6). Addition is syn and anti. The mechanisms are on pp. 245 and 246.

$$RCH=CH_2\ +\ H_2O\ \underset{}{\overset{H_2SO_4}{\rightleftharpoons}}\ RCHCH_3$$
$$\underset{OH}{|}$$

$$RCH=CH_2\ +\ CH_3OH\ \underset{}{\overset{H_2SO_4}{\rightleftharpoons}}\ RCHCH_3$$
$$\underset{OCH_3}{|}$$

3. Hydroboration–oxidation: BH_3 is the electrophile, and H^- is the nucleophile; boron is subsequently replaced with OH (Section 6.8). Addition is syn. The mechanism is on p. 251.

$$RCH=CH_2\ \xrightarrow[\text{2. HO}^-,\ H_2O_2,\ H_2O]{\text{1. BH}_3/\text{THF}}\ RCH_2CH_2OH$$

4. Addition of halogen: Br^+ or Cl^+ is the electrophile; Br^- or Cl^- (or water) is the nucleophile (Section 6.9). Addition is anti. The mechanisms are on pp. 254 and 255.

$$RCH{=}CH_2 \ + \ Cl_2 \ \xrightarrow{CH_2Cl_2} \ RCHCH_2Cl \ \ (Cl)$$

$$RCH{=}CH_2 \ + \ Br_2 \ \xrightarrow{CH_2Cl_2} \ RCHCH_2Br \ \ (Br)$$

$$RCH{=}CH_2 \ + \ Br_2 \ \xrightarrow{H_2O} \ RCHCH_2Br \ \ (OH)$$

5. Addition of a peroxyacid: O is both the electrophile and the nucleophile (Section 6.10). Addition is syn. The mechanism is on p. 258.

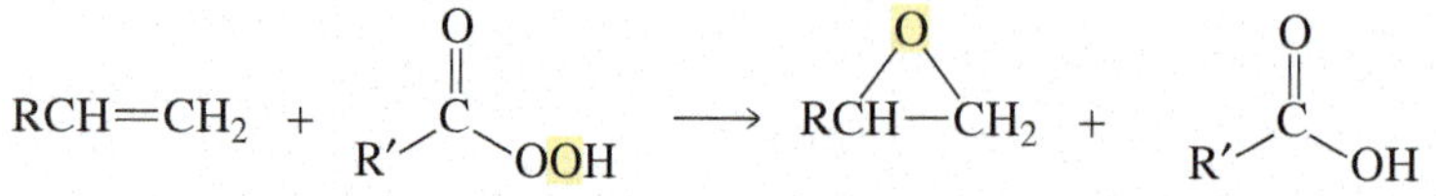

6. Ozonolysis of alkenes: one O is the electrophile and another O is the nucleophile (Section 6.11). The mechanism is on p. 259.

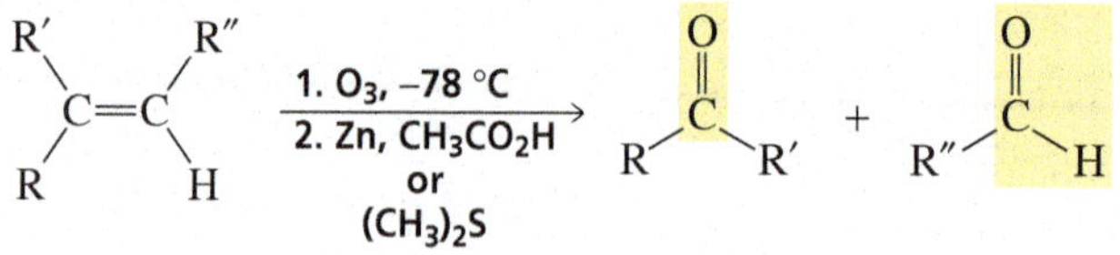

7. Addition of hydrogen is syn (Section 6.13).

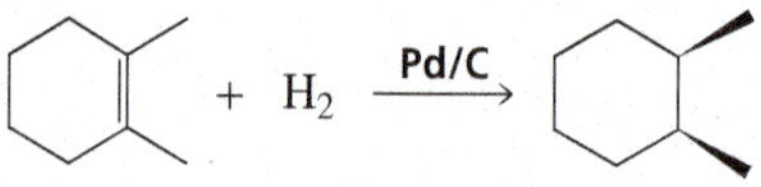

PROBLEMS

55. What is the major product of each of the following reactions?

a.

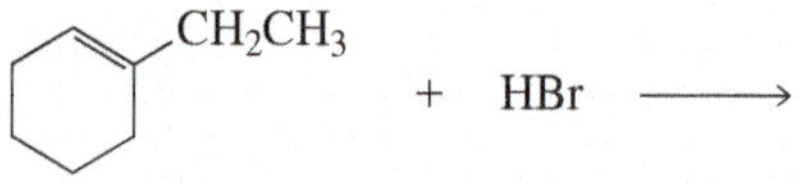

$\quad + \quad$ HBr $\longrightarrow$

c.

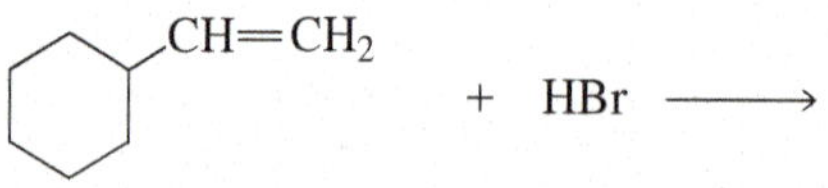

$\quad + \quad$ HBr $\longrightarrow$

b. $CH_2{=}\overset{\overset{\displaystyle CH_3}{|}}{C}CH_2CH_3 \ + \ HBr \longrightarrow$

d. $CH_3CH_2\overset{\overset{\displaystyle CH_3}{|}}{\underset{\underset{\displaystyle CH_3}{|}}{C}}CH{=}CH_2 \ + \ HBr \longrightarrow$

56. Which electrophilic addition reactions
   a. form a carbocation intermediate?
   b. form no intermediate?
   c. form a three-membered ring intermediate?
   d. form a five-membered ring intermediate?

57. Identify the electrophile and the nucleophile in each of the following reaction steps and then draw curved arrows to illustrate the bond-making and bond-breaking processes.

a. $CH_3\overset{+}{C}HCH_3 \ + \ :\!\ddot{\underset{\cdot\cdot}{Cl}}\!:^- \longrightarrow CH_3CHCH_3 \ \ (:\!\ddot{\underset{\cdot\cdot}{Cl}}\!:)$

b. $CH_3CH{=}CH_2 \ + \ H{-}Br \longrightarrow CH_3\overset{+}{C}H{-}CH_3 \ + \ Br^-$

c. $CH_3CH{=}CH_2 \ + \ BH_3 \longrightarrow CH_3CH_2{-}CH_2BH_2$

58. What is the major product of the reaction of 2-methyl-2-butene with each of the following reagents?

a. HBr
b. HI
c. Cl_2/CH_2Cl_2
d. O_3, −78 °C, followed by $(CH_3)_2S$

e. H_2/Pd
f. MCPBA (a peroxyacid)
g. $H_2O + H_2SO_4$
h. Br_2/CH_2Cl_2

i. Br_2/H_2O
j. Br_2/CH_3OH
k. BH_3/THF, followed by H_2O_2, HO^-, H_2O

59. Give two names for each of the following:

a. 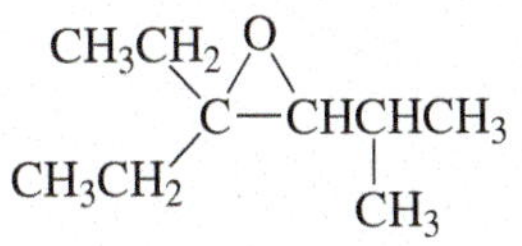
b.

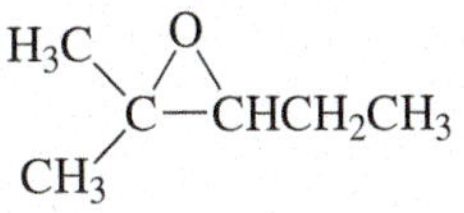

60. What reagents are needed to synthesize the following alcohols?

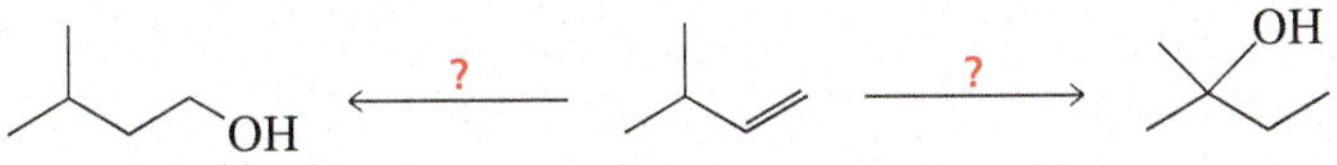

61. What are the products of the following reactions? Indicate whether each reaction is an oxidation or a reduction.

a. $CH_3CH_2\overset{\underset{\displaystyle CH_3}{|}}{C}=CHCH_2CH_3$ $\xrightarrow[\text{2. Zn, CH}_3\text{CO}_2\text{H}]{\text{1. O}_3,\ -78\ °C}$

b. $CH_3\overset{\underset{\displaystyle CH_3}{|}}{C}HCH=CHCH_3$ $\xrightarrow{\overset{\displaystyle H_2}{Pd/C}}$

c. $\xrightarrow[\text{2. (CH}_3)_2\text{S}]{\text{1. O}_3,\ -78\ °C}$

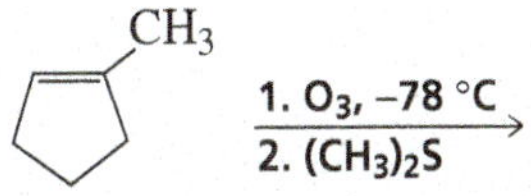

62. When 3-methyl-1-butene reacts with HBr, two alkyl halides are formed: 2-bromo-3-methylbutane and 2-bromo-2-methylbutane. Propose a mechanism that explains the formation of these two products.

63. Draw curved arrows to show the flow of electrons responsible for the conversion of the following reactants into products:

a. 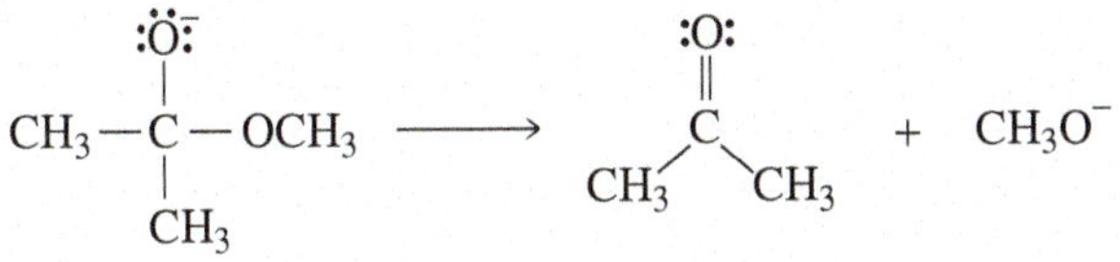 $CH_3-\overset{\underset{\displaystyle CH_3}{|}}{\overset{\overset{\displaystyle :\ddot{O}^-}{|}}{C}}-OCH_3$ $\longrightarrow$ $\underset{CH_3\ \ \ \ CH_3}{\overset{:O:}{\overset{\|}{C}}}$ $+\ CH_3O^-$

b. $CH_3C{\equiv}C-H\ +\ :\ddot{N}H_2\ \longrightarrow\ CH_3C{\equiv}C^-\ +\ \dot{N}H_3$

c. $CH_3CH_2-Br\ +\ CH_3\ddot{\ddot{O}}:^-\ \longrightarrow\ CH_3CH_2-\ddot{O}CH_3\ +\ Br^-$

64. What reagents are needed to carry out the following syntheses?

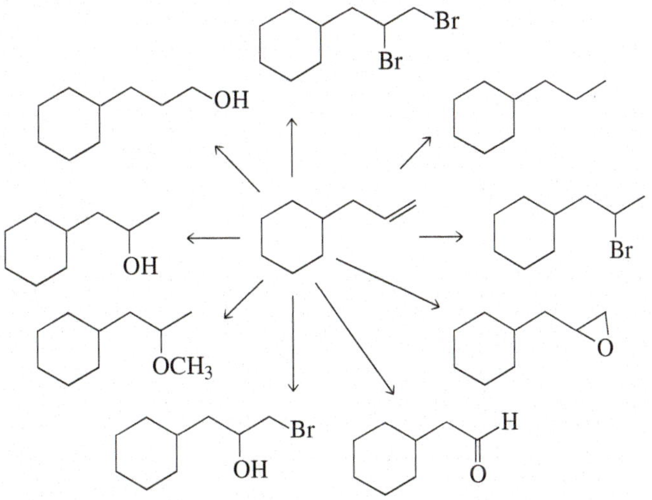

65. a. Identify two alkenes that react with HBr to form 1-bromo-1-methylcyclohexane without undergoing a carbocation rearrangement.
b. Would both alkenes form the same alkyl halide if DBr were used instead of HBr? (D is an isotope of H, so D^+ reacts like H^+.)

66. What is the major product of each of the following reactions?

a. 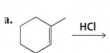 $\xrightarrow{\text{HCl}}$

c. cyclohexene $\xrightarrow{\text{RCOOH}}$ (with O double bonded structure above)

e. cyclohexene $\xrightarrow[\text{H}_2\text{O}]{\text{Cl}_2}$

g. cyclohexene $\xrightarrow[\text{CH}_3\text{OH}]{\text{H}_2\text{SO}_4}$

b. $\xrightarrow[\text{CH}_3\text{OH}]{\text{Br}_2}$

d. cyclohexene $\xrightarrow[\text{H}_2\text{O}]{\text{H}_2\text{SO}_4}$

f. cyclohexene $\xrightarrow[\text{2. (CH}_3)_2\text{S}]{\text{1. O}_3, -78°\text{C}}$

h. cyclohexene $\xrightarrow[\text{CH}_2\text{Cl}_2]{\text{Cl}_2}$

67. Using any alkene and any other reagents, how would you prepare the following compounds?

a. (cyclohexane)

c. (cyclohexane with CH_2OH)

e. (cyclohexane with CH_2CHCH_3 and OH)

b. $CH_3CH_2CH_2CHCH_3$ with Cl

d. $CH_3CH_2CHCHCH_2CH_3$ with Br OH

f. $CH_3CH_2CHCHCH_2CH_3$ with Br Cl

68. Which is more stable?

a. $CH_3\overset{+}{C}CH_3$ (with CH_3 above) or $CH_3\overset{+}{C}HCH_2CH_3$

b. $CH_3\overset{+}{C}HCH_3$ or $CH_3\overset{+}{C}HCH_2Cl$

c. 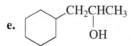 or (branched structure)

69. a. Use curved arrows to show the flow of electrons that occurs in each step of the following mechanism.
b. Draw a reaction coordinate diagram for the reaction. (*Hint*: An alkyl halide is more stable than an alkene.)

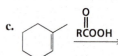

70. a. Draw the product or products that will be obtained from the reaction of *cis*-2-butene and *trans*-2-butene with each of the following reagents. If a product can exist as stereoisomers, show which stereoisomers are formed.

1. HCl
2. BH_3/THF, followed by HO^-, H_2O_2, H_2O
3. a peroxyacid
4. Br_2 in CH_2Cl_2
5. $Br_2 + H_2O$
6. $H_2 + Pd/C$
7. $H_2O + H_2SO_4$
8. $CH_3OH + H_2SO_4$

b. With which reagents do the two alkenes react to form different products?

71. 1-Methylcyclohexene forms two products when it reacts with bromine in methanol.
a. Draw the mechanism for the formation of the products.
b. Describe their stereochemical relationship—that is, are they enantiomers or diastereomers?

72. The second-order rate constant (in units of $M^{-1}s^{-1}$) for acid-catalyzed hydration at 25 °C is given for each of the following alkenes:

$$CH_3\text{-}C(H)=CH_2 \quad 4.95 \times 10^{-8}$$
$$CH_3(H)C=C(CH_3)(H) \quad 8.32 \times 10^{-8}$$
$$CH_3(H)C=C(H)(CH_3) \quad 3.51 \times 10^{-8}$$
$$CH_3(H)C=C(CH_3)(CH_3) \quad 2.15 \times 10^{-4}$$
$$CH_3(H_3C)C=C(CH_3)(CH_3) \quad 3.42 \times 10^{-4}$$

a. Calculate the relative rates of hydration of the alkenes. (*Hint*: Divide each rate constant by the smallest rate constant of the series: 3.51×10^{-8}.)
b. Why does (Z)-2-butene react faster than (E)-2-butene?
c. Why does 2-methyl-2-butene react faster than (Z)-2-butene?
d. Why does 2,3-dimethyl-2-butene react faster than 2-methyl-2-butene?

73. Which compound has the greater dipole moment?

a. $Cl(H)C=C(H)(Cl)$ or $H(H)C=C(Cl)(Cl)$

b. $Cl(H)C=C(H)(CH_3)$ or $Cl(CH_3)C=C(H)(H)$

4. Draw the products of the following reactions. If the products can exist as stereoisomers, show which stereoisomers are formed.

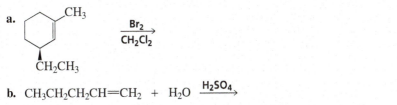

a. [structure] $\xrightarrow[\text{CH}_2\text{Cl}_2]{\text{Br}_2}$

c. [structure] $\xrightarrow[\text{CH}_2\text{Cl}_2]{\text{Br}_2}$

b. $CH_3CH_2CH_2CH{=}CH_2 + H_2O \xrightarrow{H_2SO_4}$

d. [structure] $\xrightarrow{\text{HBr}}$

5. A student was about to turn in the products he had obtained from the reaction of HI with 3,3,3-trifluoropropene when he realized that the labels had fallen off his flasks and he did not know which label belonged to which flask. His friend reminded him of the rule that says the electrophile adds to the sp^2 carbon bonded to the most hydrogens. In other words, he should label the flask containing the most product 1,1,1-trifluoro-2-iodopropane and label the flask containing the least amount of product 1,1,1-trifluoro-3-iodopropane. Should he follow his friend's advice?

6. a. Propose a mechanism for the following reaction (show all curved arrows):

$$CH_3CH_2CH{=}CH_2 + CH_3OH \xrightarrow{H_2SO_4} CH_3CH_2\underset{\underset{OCH_3}{|}}{C}HCH_3$$

b. Which step is the rate-determining step?
c. What is the electrophile in the first step?
d. What is the nucleophile in the first step?

e. What is the electrophile in the second step?
f. What is the nucleophile in the second step?

7. Draw the products, including their configurations, obtained from the reaction of 1-ethylcyclohexene with the following reagents:

a. HBr **b.** H$_2$, Pd/C **c.** R$_2$BH/THF, followed by HO$^-$, H$_2$O$_2$, H$_2$O **d.** Br$_2$/CH$_2$Cl$_2$

8. Which stereoisomer of 3-hexene forms a meso compound when it reacts with Br$_2$?

9. Which stereoisomer of 3-hexene forms (3S,4S)-4-bromo-3-hexanol and (3R,4R)-4-bromo-3-hexanol when it reacts with Br$_2$ and H$_2$O?

30. Propose a mechanism for each of the following reactions:

a. [structure] + H$_2$O $\xrightarrow{H_2SO_4}$ [structure with OH]

b. [structure] + HCl $\longrightarrow$ [structure with Cl]

31. What is the major product of each of the following reactions?

a. $HOCH_2CH_2CH_2CH{=}CH_2 + Br_2 \xrightarrow{CH_2Cl_2}$

b. $HOCH_2CH_2CH_2CH_2CH{=}CH_2 + Br_2 \xrightarrow{CH_2Cl_2}$

32. Draw the products of the following reactions. If the products can exist as stereoisomers, show which stereoisomers are formed.

a. *cis*-2-pentene + HCl
b. *trans*-2-pentene + HCl
c. 1-ethylcyclohexene + H$_2$O + H$_2$SO$_4$
d. 2,3-dimethyl-3-hexene + H$_2$, Pd/C
e. 1,2-dimethylcyclohexene + HCl

f. 1,2-dideuteriocyclohexene + H$_2$, Pd/C
g. 3,3-dimethyl-1-pentene + Br$_2$/CH$_2$Cl$_2$
h. (*E*)-3,4-dimethyl-3-heptene + H$_2$, Pd/C
i. (*Z*)-3,4-dimethyl-3-heptene + H$_2$, Pd/C
j. 1-chloro-2-ethylcyclohexene + H$_2$, Pd/C

83. a. What product is obtained from the reaction of HCl with 1-butene? With 2-butene?
b. Which of the two reactions has the greater free energy of activation?
c. Which compound reacts more rapidly with HCl: (*Z*)-2-butene or (*E*)-2-butene?

84. What is the major product of the reaction of each of the following with HBr?

a. [structure] **b.** [structure] **c.** [structure] **d.** [structure]

85. For each compound, show the products obtained from ozonolysis, followed by treatment with dimethyl sulfide.

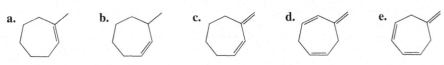

a. [structure] **b.** [structure] **c.** [structure] **d.** [structure] **e.** [structure]

86. Which stereoisomer of 3,4-dimethyl-3-hexene forms (3S,4S)-3,4-dimethylhexane and (3R,4R)-3,4-dimethylhexane when it reacts with H_2, Pd/C?

87. Draw the products of the following reactions. If the products can exist as stereoisomers show what stereoisomers are formed.

a. *cis*-2-pentene + Br_2/CH_2Cl_2
b. *trans*-2-pentene + Br_2/CH_2Cl_2
c. 1-butene + HCl
d. methylcyclohexene + HBr
e. *trans*-3-hexene + Br_2/CH_2Cl_2
f. *cis*-3-hexene + Br_2/CH_2Cl_2

g. 3,3-dimethyl-1-pentene + HBr
h. *cis*-2-butene + HBr
i. (Z)-2,3-dichloro-2-butene + H_2, Pd/C
j. (E)-2,3-dichloro-2-butene + H_2, Pd/C
k. (Z)-3,4-dimethyl-3-hexene + H_2, Pd/C
l. (E)-3,4-dimethyl-3-hexene + H_2, Pd/C

88. Of the possible products shown for the following reaction, are there any that will not be formed?

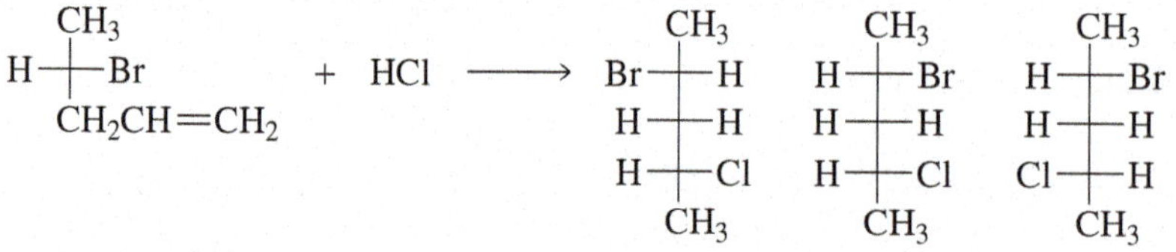

89. **A**, a compound with molecular formula C_6H_{10}, contains three adjacent methylene units. **A** reacts with one equivalent of H_2 over Pd/C to yield **B**. **A** reacts with aqueous acid to form a single product, **C**, and undergoes hydroboration/oxidation to form a pair of enantiomers, **D** and **E**. Ozonolysis of **A** followed by reaction with dimethyl sulfide forms **F** with molecular formula $C_6H_{10}O_2$. Provide structures for **A–F**.

90. The reaction of an alkene with diazomethane forms a cyclopropane ring. Propose a mechanism for the reaction. (*Hint:* It is a concerted reaction.)

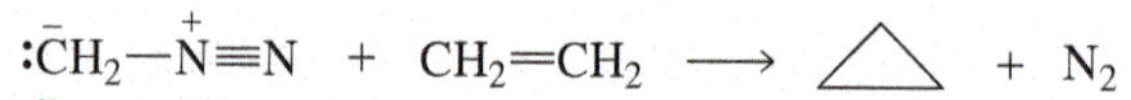

$$:\!\overset{-}{C}H_2\!-\!\overset{+}{N}\!\equiv\!N \;+\; CH_2\!=\!CH_2 \;\longrightarrow\; \triangle \;+\; N_2$$
diazomethane

Note: Diazomethane is a gas that must be handled with great care because it is both explosive and toxic.

91. Two chemists at Dupont found that ICH_2ZnI is better than diazomethane at converting a C=C bond to a cyclopropane ring. Propose a mechanism for the reaction, now known as the **Simmons–Smith reaction** in their honor.

92. a. Dichlorocarbene can be generated by heating chloroform with HO^-. Propose a mechanism for the reaction.

$$CHCl_3 \;+\; HO^- \;\xrightarrow{\Delta}\; Cl_2C\!: \;+\; H_2O \;+\; Cl^-$$
chloroform dichlorocarbene

b. Dichlorocarbene can also be generated by heating sodium trichloroacetate. Propose a mechanism for the reaction.

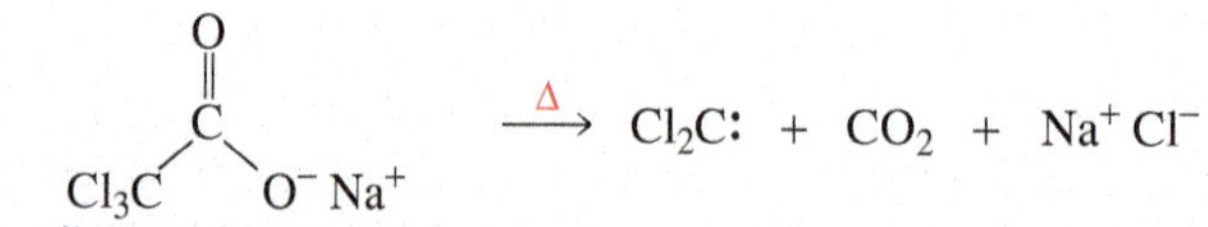

$$\xrightarrow{\Delta}\; Cl_2C\!: \;+\; CO_2 \;+\; Na^+\,Cl^-$$

93. What product is obtained from the reaction of dichlorocarbene with cyclopentene?

94. What alkene gives the product shown after reaction first with ozone and then with dimethyl sulfide?

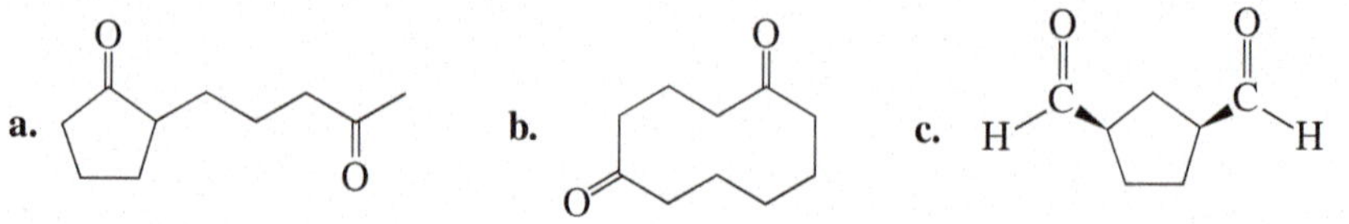

95. Draw the products of the following reactions, including their configurations:

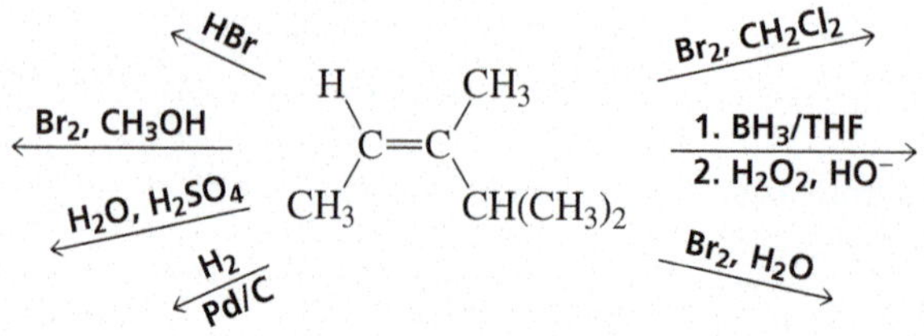

96. a. Propose a mechanism for the following reaction:

+ HBr ⟶

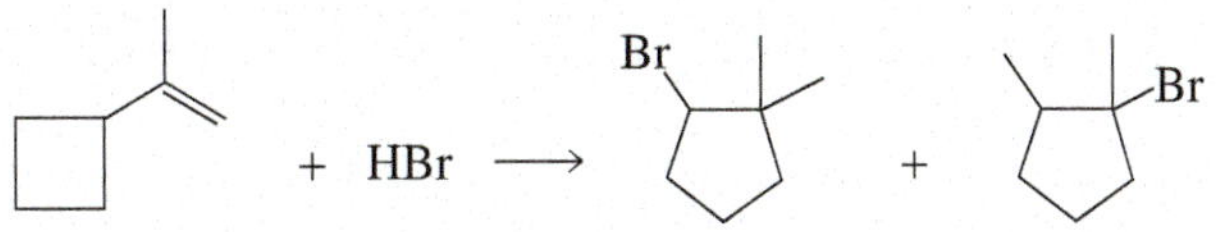

b. Is the initially formed carbocation primary, secondary, or tertiary?
c. Is the rearranged carbocation primary, secondary, or tertiary?
d. Why does the rearrangement occur?

97. Which compound is hydrated more rapidly?

$$CH_3\underset{|}{\overset{CH_3}{C}}=CH_2 \quad \text{or} \quad ClCH_2\underset{|}{\overset{CH_3}{C}}=CH_2$$

98. When the following compound is hydrated in the presence of acid, the unreacted alkene is found to have retained the deuterium atoms. What does this tell you about the mechanism for hydration?

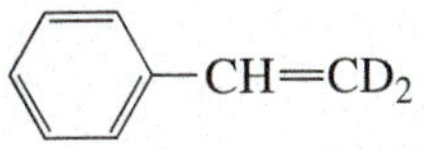

—CH=CD$_2$

99. When fumarate reacts with D$_2$O in the presence of the enzyme fumarase, only one isomer of the product is formed, as shown here. Is the enzyme catalyzing a syn or an anti addition of D$_2$O?

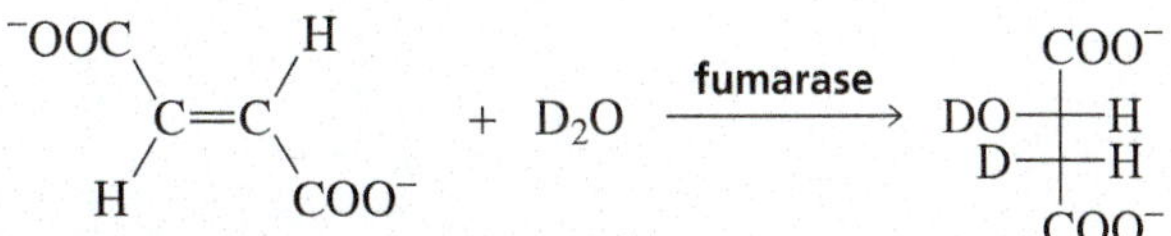

100. What stereoisomers are obtained when (*S*)-3-methyl-1-pentene reacts with Cl$_2$?

101. Propose a mechanism for the following reaction:

+ H$_2$O $\xrightarrow{\text{H}_2\text{SO}_4}$

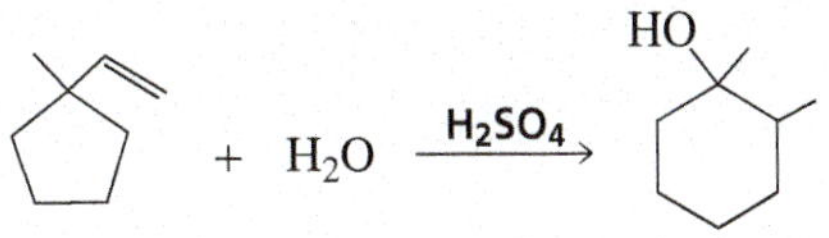

102. What hydrocarbon forms the following products after reaction first with ozone and then with dimethyl sulfide?

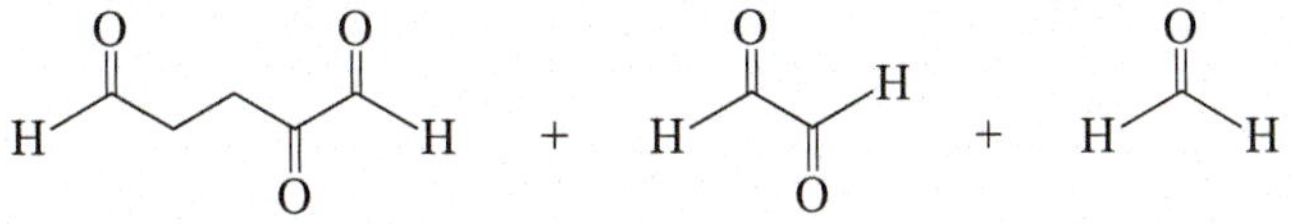

103. Ozonolysis of an alkene, followed by treatment with dimethyl sulfide, forms the following product(s). Identify the alkene in each case.

a.

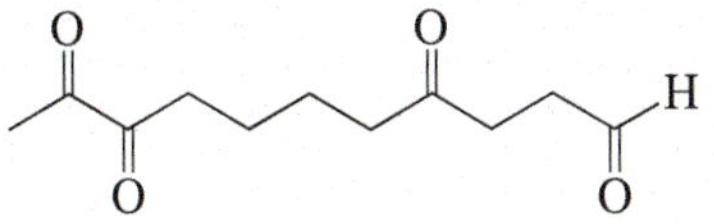

b.

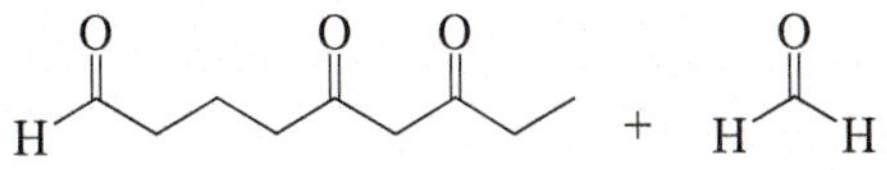

7

The Reactions of Alkynes • An Introduction to Multistep Synthesis

Currently, in order to protect our environment, the challenge facing organic chemists is to design syntheses that use reactants and generate products that cause little or no toxicity to the environment. Preventing pollution at the molecular level is known as green chemistry (see page 312). You will have your first opportunity to design multistep syntheses in this chapter.

An **alkyne** is a hydrocarbon that contains a carbon–carbon triple bond. Relatively few alkynes are found in nature. Examples include capillin, which has fungicidal activity, and ichthyothereol, a convulsant obtained from poison dart frogs that is used by the indigenous people of the Amazon for poisoned arrowheads.

poison dart frog

capillin
fungicide   **ichthyothereol**
convulsant   **an enediyne**
class of anticancer drug

A class of naturally occurring compounds called enediynes has been found to have powerful anticancer properties because they are able to cleave DNA. (You will see how they do this in Chapter 12, Problem 51.) All enediynes have a nine- or ten-membered ring that contains two triple bonds separated by a double bond. One of the first enediynes approved for clinical use is used to treat acute myeloid leukemia. Several others are currently in clinical trials. (See the application box on page 290.)

Other drugs on the market that contain an alkyne functional group are not naturally occurring compounds; they exist only because chemists have been able to synthesize them. Their trade names are shown in green. Trade names are always capitalized; only the company that holds the patent for a product can use the product's trade name for commercial purposes.

Sinovial®

Supirdyl®

parsalmide
analgesic

pargyline
antihypertensive

Synthetic Alkynes Are Used to Treat Parkinson's Disease

Parkinson's disease is a degenerative condition characterized by tremors. It is caused by the destruction of cells in the *substantia nigra*, a crescent-shaped region in the midbrain. These are the cells that release dopamine, the neurotransmitter that plays an important role in movement, muscle control, and balance. A neurotransmitter is a compound used to communicate between brain cells.

Dopamine is synthesized from tyrosine (one of the 20 common amino acids; Section 20.1). Ideally, Parkinson's disease could be treated by giving the patient dopamine. Unfortunately, dopamine is not polar enough to cross the blood–brain barrier. Therefore, L-DOPA, its immediate precursor, is the drug of choice, but it ceases to control the disease's symptoms after it has been used for a while.

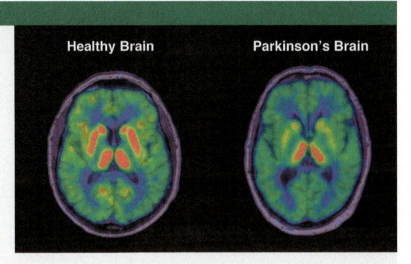

Healthy Brain   **Parkinson's Brain**

tyrosine → (tyrosine hydroxylase) → L-DOPA → (amino acid decarboxylase) → dopamine → (monoamine oxidase) →

tyrosine   **L-DOPA**   **dopamine**

Dopamine is oxidized in the body by an enzyme called monoamine oxidase. Two drugs, each containing a C≡CH group, have been developed that inhibit this enzyme, thus preventing the oxidation of dopamine and thereby increasing its availability in the brain. Both drugs have structures similar to that of dopamine, so they are able to bind to the enzyme's active site. (Recall that enzymes recognize their substrates by their shape; Section 6.15.) Because these drugs form covalent bonds with groups at the enzyme's active site, they become permanently attached to the active site, thus preventing the enzyme from binding dopamine. Patients on these drugs continue to take L-DOPA, but now this drug can be taken at longer intervals and can control the disease's symptoms for a longer period of time.

selegiline
Eldepryl®

rasagiline
Azilect®

Selegiline was approved by the FDA first, but one of the compounds to which it is metabolized has a structure similar to that of methamphetamine (the street drug "meth"; see page 179). So some patients taking the drug experience psychiatric and cardiac effects. These side effects have not been found in patients taking rasagiline.

Notice that the name of most enzymes ends in "ase," preceded by an indication of what reaction the enzyme catalyzes. Thus, tyrosine hydroxylase puts an OH group on tyrosine, amino acid decarboxylase removes a carboxyl (COO⁻) group from an amino acid (or, in this case, from a compound similar to an amino acid), and monoamine oxidase oxidizes an amine.

Why Are Drugs So Expensive?

The average cost of launching a new drug is $1.2 billion. The manufacturer has to recover this cost quickly because the patent must be filed as soon as the drug is first discovered. Although a patent is good for 20 years, it takes an average of 12 years to bring a drug to market after its initial discovery, so the patent protects the discoverer of the drug for an average of eight years. It is only during the eight years of patent protection that drug sales can provide the income needed to cover the initial costs as well as to pay for research on new drugs.

Why does it cost so much to develop a new drug? The Food and Drug Administration (FDA) has high standards that must be met before a drug is approved for a particular use. An important factor leading to the high price of many drugs is the low rate of success in progressing from the initial concept to an approved product. In fact, only 1 or 2 of every 100 compounds tested become lead compounds. A lead compound is a compound that shows promise of becoming a drug. Chemists modify the structure of a lead compound to see if doing so improves its likelihood of becoming a drug. For every 100 structural modifications of a lead compound, only one is worthy of further study. For every 10,000 compounds that are worth further study (that is, to be evaluated in animal studies), only 10 will get to clinical trials.

Clinical trials consist of four phases:

- Phase I is the first use of the drug in humans, usually involving a small number of healthy volunteers. The primary goal of a phase I study is to explore the drug's safety at different dosages.
- Phase II studies are larger in size, and their volunteers have the condition the drug is meant to treat. Phase II studies investigate effectiveness, safety, and side effects.
- Large phase III studies, involving hundreds to thousands of patients, verify effectiveness and dosage while continuing to monitor adverse reactions.
- If the drug is approved for use, phase IV studies might be conducted. This phase looks for side effects that weren't detected in phases I, II, or III.

Approximately 1 of every 10 compounds entering clinical trials satisfies the stringent requirements needed to become a marketable drug.

7.1 THE NOMENCLATURE OF ALKYNES

Because of its triple bond, an alkyne has four fewer hydrogens than an alkane with the same number of carbons. Therefore, while the general molecular formula for an acyclic alkane is C_nH_{2n+2},

- the general molecular formula for an acyclic alkyne is C_nH_{2n-2}.
- the general molecular formula for a cyclic alkyne is C_nH_{2n-4}.

The systematic name of an alkyne is obtained by replacing the "ane" ending of the alkane name with "yne." Analogous to the way compounds with other functional groups are named, the longest continuous chain containing the carbon–carbon triple bond is numbered in the direction that gives the functional group suffix as low a number as possible (Sections 3.6, 3.7, and 5.2). If the triple bond is at the end of the chain, the alkyne is classified as a **terminal alkyne.** Alkynes with triple bonds located elsewhere along the chain are **internal alkynes.**

1-hexyne
terminal alkyne

3-hexyne
internal alkyne

LEARN THE STRATEGY

	a terminal alkyne	an internal alkyne	
	$HC{\equiv}CH$	$\overset{4}{C}H_3\overset{3}{C}H_2\overset{2}{C}{\equiv}\overset{1}{C}H$	$\overset{1}{C}H_3\overset{2}{C}{\equiv}\overset{3}{C}\overset{4}{C}H_2\overset{5}{C}H_3$	$\overset{4}{C}H_3\overset{3}{C}H\overset{2}{C}{\equiv}\overset{1}{C}CH_3$

Systematic: ethyne / 1-butyne / 2-pentyne / 4-methyl-2-hexyne
Common: acetylene / ethylacetylene / ethylmethylacetylene / sec-butylmethyl-acetylene

In common nomenclature, alkynes are named as *substituted acetylenes.* The common name is obtained by stating the names of the alkyl groups (in alphabetical order) that have replaced the hydrogens of acetylene. Acetylene is an unfortunate common name for an alkyne because its "ene" ending is characteristic of a double bond rather than a triple bond.

If counting from either direction leads to the same number for the functional group suffix, the correct systematic name is the one that contains the lowest substituent number. If the compound contains more than one substituent, the substituents are listed in alphabetical order.

$$\underset{1}{C}H_3\overset{Cl}{\underset{2}{C}H}\overset{Br}{\underset{3}{C}H}\underset{4}{C}{\equiv}\underset{5}{C}\underset{6}{C}H_2\underset{7}{C}H_2\underset{8}{C}H_3$$

3-bromo-2-chloro-4-octyne
not **6-bromo-7-chloro-4-octyne**
because 2 < 6

$$\underset{6}{C}H_3\overset{CH_3}{\underset{5}{C}H}\underset{4}{C}{\equiv}\underset{3}{C}\underset{2}{C}H_2\underset{1}{C}H_2Br$$

1-bromo-5-methyl-3-hexyne
not **6-bromo-2-methyl-3-hexyne**
because 1 < 2

PROBLEM 1 ♦

Name the following:

a. $BrCH_2CH_2C{\equiv}CCH_3$

b. $CH_3CH_2CHC{\equiv}CCH_2CHCH_3$
         | |
         Br   Cl

c. $CH_3OCH_2C{\equiv}CCH_2CH_3$

d. $CH_3CH_2CHC{\equiv}CH$
         |
         $CH_2CH_2CH_3$

USE THE STRATEGY

A substituent receives the lowest possible number only if there is no functional group suffix or if counting from either direction leads to the same number for the functional group suffix.

PROBLEM 2 ♦

Name the following:

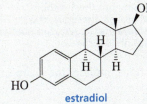

a.   b.   c.   Br

PROBLEM 3 ♦

What is the molecular formula for a monocyclic hydrocarbon with 14 carbons and 2 triple bonds?

PROBLEM 4 ♦

Draw the condensed and skeletal structures for each of the following:

a. 1-chloro-3-hexyne
b. cyclooctyne
c. isopropylacetylene

d. *sec*-butylisobutylacetylene
e. 4,4-dimethyl-1-pentyne
f. dimethylacetylene

PROBLEM 5

Draw the structures and give the common and systematic names for the seven alkynes with molecular formula C_6H_{10}.

Synthetic Alkynes Are Used for Birth Control

Estradiol and progesterone are naturally occurring female hormones. Because of their ring structures, they are classified as steroids (Section 3.16). Estradiol is responsible for the development of secondary sex characteristics in women—it affects body shape, fat deposition, bones, and joints. Progesterone is critical for the continuation of pregnancy.

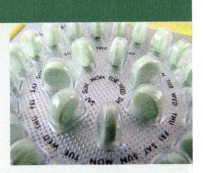

estradiol   progesterone

The four compounds shown next are synthetic steroids that are used for birth control; each contains an alkyne functional group. Most birth control pills contain ethinyl estradiol (a compound structurally similar to estradiol) and a compound structurally similar to progesterone (such as norethindrone). Ethinyl estradiol prevents ovulation, whereas norethindrone makes it difficult for a fertilized egg to attach to the wall of the uterus.

ethinyl estradiol   norethindrone   mifepristone   levonorgestrel
                           Aygestin®   RU-486   Norplant®
                                                  Mifegyne®

Mifepristone and levonorgestrel are also synthetic steroids that contain an alkyne functional group. Mifepristone, also known as RU-486, induces an abortion if taken early in pregnancy. Its name comes from Roussel-Uclaf, the French pharmaceutical company where it was first synthesized, and from an arbitrary lab serial number. Levonorgestrel is an emergency contraceptive pill. It prevents pregnancy if taken within a few days of conception.

7.2 HOW TO NAME A COMPOUND THAT HAS MORE THAN ONE FUNCTIONAL GROUP

We saw how compounds with two double bonds are named (Section 5.2). Similar rules are followed for naming compounds with two triple bonds, using the ending "diyne."

$$CH_2\!=\!C\!=\!CH_2$$

$$\overset{6}{C}H_3\overset{5}{C}H\!=\!\overset{4}{C}H\overset{3}{C}H_2\overset{2}{\underset{|}{C}}\!=\!\overset{1}{C}H_2$$
$$\underset{CH_3}{\overset{|}{}}$$

$$\overset{6}{C}H_3\overset{5}{\underset{|}{C}}H\overset{4}{C}\!\equiv\!\overset{3}{C}\overset{2}{C}H_2\overset{1}{C}\!\equiv\!\overset{}{C}H$$
$$\underset{CH_3}{\overset{|}{}}$$

systematic: propadiene   2-methyl-1,4-hexadiene   6-methyl-1,4-heptadiyne
common: allene   or   or
                2-methylhexa-1,4-diene   6-methylhepta-1,4-diyne

LEARN THE STRATEGY

1. To name an alkene in which the second functional group is not another double bond but has a functional group suffix, find the longest continuous chain containing both functional groups and put both suffixes at the end of the name. Put the "ene" ending first, with the terminal "e" omitted to avoid two adjacent vowels.

2. The number indicating the location of the first-stated functional group is usually placed before the name of the parent chain. The number indicating the location of the second-stated functional group is placed immediately before the suffix for that functional group.

3. If the two functional groups are a *double bond* and a *triple bond*, number the chain in the direction that produces a name containing the lower number. Thus, in the following examples, the lower number is given to the alkyne suffix in the compound on the left and to the alkene suffix in the compound on the right.

> When the functional groups are a double bond and a triple bond, the chain containing both groups is numbered in the direction that produces the name containing the lowest possible number, regardless of which functional group gets the lower number.

$$\overset{7}{C}H_3\overset{6}{C}H\!=\!\overset{5}{C}H\overset{4}{C}H_2\overset{3}{C}H_2\overset{2}{C}\!\equiv\!\overset{1}{C}H$$
5-hepten-1-yne
not 2-hepten-6-yne
because 1 < 2

$$\overset{1}{C}H_2\!=\!\overset{2}{C}H\overset{3}{C}H_2\overset{4}{C}H_2\overset{5}{C}\!\equiv\!\overset{6}{C}\overset{7}{C}H_3$$
1-hepten-5-yne
not 6-hepten-2-yne
because 1 < 2

$$\underset{\overset{1}{C}H_2=\overset{2}{C}H\overset{3}{C}H\overset{4}{C}\equiv\overset{5}{C}\overset{6}{C}H_3}{\overset{CH_2CH_2CH_2CH_3}{\overset{|}{}}}$$

3-butyl-1-hexen-4-yne

> the longest continuous chain has 8 carbons, but the 8-carbon chain does not contain both functional groups; therefore, the compound is named as a hexenyne because the longest continuous chain containing both functional groups has 6 carbons

4. If the same low number is obtained in both directions, number the chain in the direction that gives the double bond the lower number.

> If there is a tie between a double bond and a triple bond, the double bond gets the lower number.

$$\overset{1}{C}H_3\overset{2}{C}H\!=\!\overset{3}{C}H\overset{4}{C}\!\equiv\!\overset{5}{C}\overset{6}{C}H_3$$
2-hexen-4-yne
not 4-hexen-2-yne

$$\overset{6}{H}C\!\equiv\!\overset{5}{C}\overset{4}{C}H_2\overset{3}{C}H_2\overset{2}{C}H\!=\!\overset{1}{C}H_2$$
1-hexen-5-yne
not 5-hexen-1-yne

5. If the second functional group suffix has a higher priority than the alkene suffix, number the chain in the direction that assigns the lower number to the functional group with the higher-priority suffix. (The relative priorities of functional group suffixes are shown in Table 7.1.) The higher-priority functional group is assumed to be at the 1-position in cyclic compounds.

Table 7.1 Priorities of Functional Group Suffixes

highest priority $C=O$ > OH > NH_2 > $C=C$ = $C\equiv C$ lowest priority

the double bond is given priority over a triple bond only when there is a tie

$CH_2{=}CHCH_2OH$
2-propen-1-ol
***not* 1-propen-3-ol**

$CH_3\overset{\displaystyle CH_3}{\underset{\displaystyle |}{C}}{=}CHCH_2CH_2OH$
4-methyl-3-penten-1-ol

$CH_2{=}CHCH_2CH_2CH_2\overset{\displaystyle NH_2}{\underset{\displaystyle |}{C}}HCH_3$
6-hepten-2-amine

Number the chain so that the lowest possible number is given to the functional group with the higher priority.

$CH_3CH_2CH{=}CH\overset{\displaystyle }{\underset{\displaystyle |}{C}}HCH_3$
$\quad\quad\quad\quad\quad OH$
3-hexen-2-ol

6-methyl-2-cyclohexenol

3-cyclohexenamine

PROBLEM 6 ♦

Name the following:

a. $CH_2{=}CHCH_2C{\equiv}CCH_2CH_3$

b. $CH_3CH{=}\overset{\displaystyle CH_3}{\underset{\displaystyle |}{C}}CH_2CH{=}CH_2$

c. $CH_3CH_2CH{=}\overset{\displaystyle CH{=}CH_2}{\underset{\displaystyle |}{C}}CH_2CH_2C{\equiv}CH$

d. $HOCH_2CH_2C{\equiv}CH$

e. $CH_3CH{=}CHCH{=}CHCH{=}CH_2$

f. $CH_3CH{=}\overset{\displaystyle CH_3\;\;\;CH_3}{\underset{\displaystyle |\quad\;\;|}{C}}CH_2\overset{\displaystyle }{C}HCH_2OH$

USE THE STRATEGY

PROBLEM 7 ♦

Name the following:

a.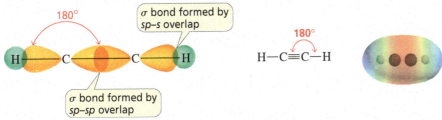

b.

c.

How a Banana Slug Knows What to Eat

Many species of mushrooms synthesize 1-octen-3-ol, a repellent that drives off predatory slugs. Such mushrooms can be recognized by small bite marks on their caps, where the slug started to nibble before the volatile compound was released. People are not put off by the release of this compound because to them it just smells like a mushroom. 1-Octen-3-ol also has antibacterial properties that may protect the mushroom from organisms that would otherwise invade the wound made by the slug. Not surprisingly, the species of mushroom that banana slugs commonly eat cannot synthesize 1-octen-3-ol.

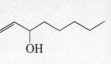

OH

1-octen-3-ol

7.3 THE STRUCTURE OF ALKYNES

The structure of ethyne was discussed in Section 1.9, where we saw that each carbon is *sp* hybridized. As a result, each carbon has two *sp* orbitals and two *p* orbitals. One *sp* orbital overlaps the *s* orbital of a hydrogen, and the other overlaps an *sp* orbital of the other carbon, to form σ bonds. (The small lobes of the *sp* orbitals are not shown.) Because the *sp* orbitals are oriented as far from each other as possible to minimize electron repulsion, ethyne is a linear molecule with bond angles of 180°.

180°

σ bond formed by *sp–s* overlap

H—C—C—H

σ bond formed by *sp–sp* overlap

180°

$H{-}C{\equiv}C{-}H$

Recall that the two π bonds of a triple bond are formed by each of the two *p* orbitals on one *sp* carbon overlapping the parallel *p* orbital on the other *sp* carbon (see Figure 7.1 on the next page). The end result can be thought of as a cylinder of electrons wrapped around the σ bond.

▶ **Figure 7.1**
(a) Each of the two π bonds of a triple bond is formed by side-to-side overlap of a p orbital of one carbon with a parallel p orbital of the adjacent carbon.

(b) The electrostatic potential map for 2-butyne shows the cylinder of electrons wrapped around the σ bond.

a.

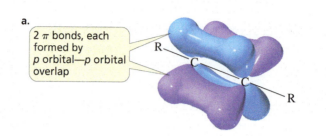

2 π bonds, each formed by p orbital—p orbital overlap

b.

cylinder of electrons

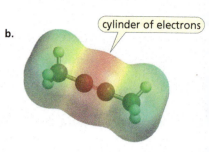

A triple bond is composed of a σ bond and two π bonds.

Also recall that a carbon–carbon triple bond is shorter and stronger than a carbon–carbon double bond, which, in turn, is shorter and stronger than a carbon–carbon single bond, and that a π bond is weaker than a σ bond (Section 1.15).

Alkyl groups stabilize alkynes, just as they stabilize alkenes and carbocations (Sections 5.9 and 6.2, respectively). Internal alkynes, therefore, are more stable than terminal alkynes.

PROBLEM 8 ◆

What orbitals are used to form the carbon–carbon σ bond between the highlighted carbons?

a. $CH_3CH{=}CHCH_3$ **d.** $CH_3C{\equiv}CCH_3$ **g.** $CH_3CH{=}CHCH_2CH_3$

b. $CH_3CH{=}CHCH_3$ **e.** $CH_3C{\equiv}CCH_3$ **h.** $CH_3C{\equiv}CCH_2CH_3$

c. $CH_3CH{=}C{=}CH_2$ **f.** $CH_2{=}CHCH{=}CH_2$ **i.** $CH_2{=}CHC{\equiv}CH$

NOTE TO THE STUDENT
- Tables of physical properties can be found in Appendix VI.

7.4 THE PHYSICAL PROPERTIES OF UNSATURATED HYDROCARBONS

All hydrocarbons—alkanes, alkenes, and alkynes—have similar physical properties.

- Their boiling points increase with increasing molecular weight (Table 7.2).
- They are insoluble in water but are soluble in nonpolar solvents (Section 3.9).
- They are less dense than water.

Alkynes are more linear than alkenes, and a triple bond is more polarizable than a double bond (Section 3.9). These two features cause an alkyne to have stronger London dispersion forces and, therefore, a higher boiling point than an alkene with the same number of carbons.

Table 7.2 Boiling Points of the Smallest Hydrocarbons					
	bp (°C)		bp (°C)		bp (°C)
CH_3CH_3 ethane	−88.6	$H_2C{=}CH_2$ ethene	−104	$HC{\equiv}CH$ ethyne	−84
$CH_3CH_2CH_3$ propane	−42.1	$CH_3CH{=}CH_2$ propene	−47	$CH_3C{\equiv}CH$ propyne	−23
$CH_3CH_2CH_2CH_3$ butane	−0.5	$CH_3CH_2CH{=}CH_2$ 1-butene	−6.5	$CH_3CH_2C{\equiv}CH$ 1-butyne	8
$CH_3(CH_2)_3CH_3$ pentane	36.1	$CH_3CH_2CH_2CH{=}CH_2$ 1-pentene	30	$CH_3CH_2CH_2C{\equiv}CH$ 1-pentyne	39
$CH_3(CH_2)_4CH_3$ hexane	68.7	$CH_3CH_2CH_2CH_2CH{=}CH_2$ 1-hexene	63.5	$CH_3CH_2CH_2CH_2C{\equiv}CH$ 1-hexyne	71
		$CH_3CH{=}CHCH_3$ cis-2-butene	3.7	$CH_3C{\equiv}CCH_3$ 2-butyne	27
		$CH_3CH{=}CHCH_3$ trans-2-butene	0.9	$CH_3CH_2C{\equiv}CCH_3$ 2-pentyne	55

PROBLEM 9 ♦

In Table 7.2, what is the smallest alkane, the smallest terminal alkene, and the smallest terminal alkyne that are liquids at room temperature, which is generally taken to be 20 °C to 25 °C?

PROBLEM 10 ♦

Why does *cis*-2-butene have a higher boiling point than *trans*-2-butene?

7.5 THE REACTIVITY OF ALKYNES

The cloud of electrons completely surrounding the σ bond makes an alkyne an electron-rich molecule. Alkynes therefore are nucleophiles, so they react with electrophiles. Thus alkynes, like alkenes, undergo *electrophilic addition reactions* because of their relatively weak π bonds. The same electrophilic reagents that add to alkenes also add to alkynes. For example, the addition of hydrogen chloride to an alkyne forms a chlorosubstituted alkene.

$$CH_3C{\equiv}CCH_3 \xrightarrow{\text{HCl}} CH_3C{=}CHCH_3$$
$$\qquad\qquad\qquad\qquad\qquad |$$
$$\qquad\qquad\qquad\qquad\quad Cl$$

π-Complex Formation

Recall that the rate-limiting step for electrophilic addition of HCl to an alkene is addition of the proton to the alkene to form an alkyl cation (Section 5.5). Similarly, addition of a proton is the rate-limiting step for electrophilic addition of HCl to an alkyne. Addition of a proton to an alkyne forms a vinylic cation. A **vinylic cation** has a positive charge on a vinylic carbon.

$$RCH{=}CH_2 + H{-}Cl \longrightarrow \overset{+}{R}CH{-}CH_3 + Cl^- \qquad\qquad RC{\equiv}CH + H{-}Cl \longrightarrow R\overset{+}{C}{=}CH_2 + Cl^-$$
$$\qquad\qquad\qquad\qquad\qquad\qquad \textbf{alkyl cation} \qquad\qquad\qquad\qquad\qquad\qquad\qquad\qquad\qquad \textbf{vinylic cation}$$

A vinylic cation is less stable than a *similarly substituted* alkyl cation because a vinylic cation has a positive charge on an *sp* carbon. By "similarly substituted," we mean that a primary vinylic cation is less stable than a primary alkyl cation and a secondary vinylic cation is less stable than a secondary alkyl cation.

An *sp* carbon is more electronegative than the sp^2 carbon of an alkyl cation and is, therefore, less able to bear a positive charge (Section 2.6).

relative stabilities of carbocations

We saw that a primary carbocation is too unstable to form. Because a secondary vinylic cation has about the same stability, it, too, is not expected to form. Some chemists think that the intermediate formed when a proton adds to an alkyne is a π-**complex** rather than a vinylic cation.

π-complex

Support for the intermediate being a π-complex comes from the observation that many (but not all) alkyne addition reactions are stereoselective. For example, the following reaction forms only (Z)-2-chloro-2-butene, which means that only anti addition of H and Cl occurs.

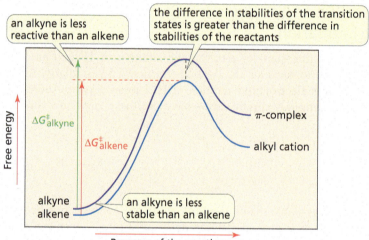

Clearly, the mechanism of the addition reaction is not completely understood. For now, we will assume that the reaction forms a π-complex.

Alkynes Are Less Reactive Than Alkenes

Figure 7.2 shows that an alkyne is less stable than an alkene and that a π-complex, although more stable than a vinylic cation, is still less stable than an alkyl cation. Because the difference in stabilities of the transition states is greater than the difference in stabilities of the reactants, the $\Delta G^\ddagger$ for formation of a π-complex is greater than the $\Delta G^\ddagger$ for formation of an alkyl cation. Therefore, an alkyne is less reactive than an alkene in an electrophilic addition reaction.

▶ **Figure 7.2**
Comparison of the free energies of activation for the addition of a proton to an alkyne and to an alkene. The $\Delta G^\ddagger$ for the reaction of an alkyne is greater than that for the reaction of an alkene, indicating that an alkyne is less reactive than an alkene.

Alkynes are less reactive than alkenes in electrophilic addition reactions.

PROBLEM 11 SOLVED

Under what circumstances can you assume that the less stable of two compounds is the more reactive compound?

SOLUTION For the less stable compound to be the more reactive compound, the less stable compound must have the more stable transition state, or the difference in the stabilities of the reactants must be greater than the difference in the stabilities of the transition states.

7.6 THE ADDITION OF HYDROGEN HALIDES AND THE ADDITION OF HALOGENS TO AN ALKYNE

Addition of a Hydrogen Halide to an Alkyne

The product of the electrophilic addition reaction of an alkyne with HCl is an alkene. Therefore, a second addition reaction can occur if excess hydrogen halide is present. The second addition—like other alkene addition reactions—is regioselective: the H^+ adds to the less substituted sp^2 carbon (that is, the one bonded to the hydrogen).

the electrophile adds here

a second electrophilic addition reaction occurs

$$CH_3C{\equiv}CCH_3 \xrightarrow{HCl} CH_3C{=}CHCH_3 \xrightarrow{HCl} CH_3\underset{\underset{Cl}{|}}{\overset{\overset{Cl}{|}}{C}}CH_2CH_3$$

halo-substituted alkene   **geminal dihalide**

The product of the second addition reaction is a **geminal dihalide,** a molecule with two halogens on the same carbon. "Geminal" comes from *geminus,* which is Latin for "twin."

The addition of a hydrogen halide to an alkyne can be stopped after the addition of one equivalent of hydrogen halide because an alkyne is more reactive than the halo-substituted alkene that is the reactant for the second addition reaction.

relative reactivity

$$\underset{\textbf{alkene}}{RCH{=}CH_2} \quad > \quad \underset{\textbf{alkyne}}{RC{\equiv}CH} \quad > \quad \underset{\underset{Cl}{|}}{RC{=}CH_2}$$

withdraws electrons inductively

halo-substituted alkene

The halo-substituted alkene is less reactive than an alkyne because the halo-substituent withdraws electrons inductively (through the σ bond), thereby decreasing the nucleophilic character of the double bond.

Addition to a Terminal Alkyne

If the alkyne is a *terminal* alkyne, the first electrophilic addition reaction is also regioselective: the H^+ adds to the less substituted *sp* carbon.

the electrophile adds here

$$CH_3CH_2C{\equiv}CH \xrightarrow{HCl} CH_3CH_2\underset{\underset{Cl}{|}}{C}{=}CH_2$$

The electrophile adds to the *sp* carbon that is bonded to the hydrogen.

The mechanism for the addition of a hydrogen halide to an alkyne is similar to the mechanism for the addition of a hydrogen halide to an alkene. The only difference is the intermediate: an alkyne forms a π-complex, whereas an alkene forms a carbocation.

MECHANISM FOR ELECTROPHILIC ADDITION OF A HYDROGEN HALIDE TO AN ALKYNE

$$RC{\equiv}CH \;+\; H{-}Cl \longrightarrow RC{\equiv}CH \;(\pi\text{-complex}) \longrightarrow RC{=}CH_2$$
$$:\ddot{Cl}:^-$$
$$\underset{Cl}{|}$$

- The alkyne (a nucleophile) reacts with an electrophile to form a π-complex.
- Chloride ion adds to the π-complex, forming a halo-substituted alkene.

The regioselectivity of electrophilic addition to an alkyne can be explained just like the regioselectivity of alkene addition reactions was explained in Section 6.4. Of the two possible transition states for the reaction, the one with a partial positive charge on the more substituted (secondary) carbon is more stable.

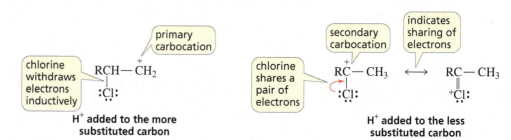

more stable transition state   less stable transition state

We saw that the second electrophilic addition reaction (the addition to the halo-substituted alkene) is also regioselective.

$$RC=CH_2 + HCl \longrightarrow RC-CH_3$$

We can see why this reaction is regioselective by comparing the stabilities of the two carbocations that could be formed. If H^+ adds to the more substituted sp^2 carbon, a *primary carbocation* would be formed; the chlorine atom further *decreases the stability* of the carbocation by withdrawing electrons inductively through the σ bond, which increases the concentration of positive charge on the carbon.

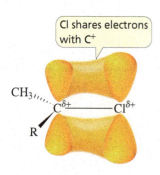

▲ **Figure 7.3**
Chlorine shares the positive charge with carbon by overlapping one of its orbitals that contains a lone pair with the empty *p* orbital of the positively charged carbon.

H^+ added to the more substituted carbon H^+ added to the less substituted carbon

In contrast, adding H^+ to the less substituted sp^2 carbon (the one bonded to the most hydrogens) forms a more stable *secondary carbon*. In addition, the chlorine atom *stabilizes* the carbocation by donating a share of its lone pair to the positively charged carbon. In this way, the positive charge is shared by carbon and chlorine (Figure 7.3).

Addition to an Internal Alkyne

Addition of excess hydrogen halide to an *unsymmetrical internal* alkyne forms two geminal dihalides because the initial addition of H^+ can occur with equal ease to either of the *sp* carbons.

$$CH_3CH_2C\equiv CCH_3 + HBr \longrightarrow CH_3CH_2CH_2\overset{Br}{\underset{Br}{C}}CH_3 + CH_3CH_2\overset{Br}{\underset{Br}{C}}CH_2CH_3$$

unsymmetrical internal alkyne   excess

Notice, however, that if the internal alkyne is symmetrical, only one geminal dihalide is formed.

$$CH_3CH_2C\equiv CCH_2CH_3 + HBr \longrightarrow CH_3CH_2CH_2\overset{Br}{\underset{Br}{C}}CH_2CH_3$$

symmetrical internal alkyne   excess

Addition of a Halogen to an Alkyne

The halogens Cl_2 and Br_2 also add to alkynes. In the presence of excess halogen, a second addition reaction occurs. The mechanism of the reaction is exactly the same as the mechanism for the addition of Cl_2 or Br_2 to an alkene (Section 6.9).

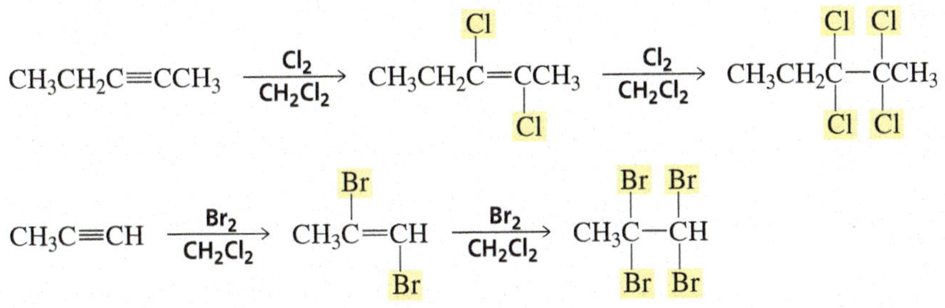

$$CH_3CH_2C\equiv CCH_3 \xrightarrow[CH_2Cl_2]{Cl_2} CH_3CH_2C = CCH_3 \xrightarrow[CH_2Cl_2]{Cl_2} CH_3CH_2C - CCH_3$$

with Cl substituents as shown

$$CH_3C\equiv CH \xrightarrow[CH_2Cl_2]{Br_2} CH_3C = CH \xrightarrow[CH_2Cl_2]{Br_2} CH_3C - CH$$

with Br substituents as shown

PROBLEM 12 ◆

What is the major product of each of the following reactions?

a. $HC\equiv CCH_3 \xrightarrow{HBr}$

b. $HC\equiv CCH_3 \xrightarrow{\text{excess HBr}}$

c. $CH_3C\equiv CCH_3 \xrightarrow[CH_2Cl_2]{Br_2}$

d. $HC\equiv CCH_3 \xrightarrow[CH_2Cl_2]{\text{excess } Br_2}$

e. $CH_3C\equiv CCH_3 \xrightarrow{\text{excess HBr}}$

f. $CH_3C\equiv CCH_2CH_3 \xrightarrow{\text{excess HBr}}$

PROBLEM 13

Drawing on what you know about the stereochemistry of alkene addition reactions,

a. write the mechanism for the reaction of 2-butyne with one equivalent of Br_2.

b. predict the configuration of the product of the reaction.

7.7 THE ADDITION OF WATER TO AN ALKYNE

In Section 6.5, we saw that alkenes undergo the acid-catalyzed addition of water. The product of the electrophilic addition reaction is an alcohol.

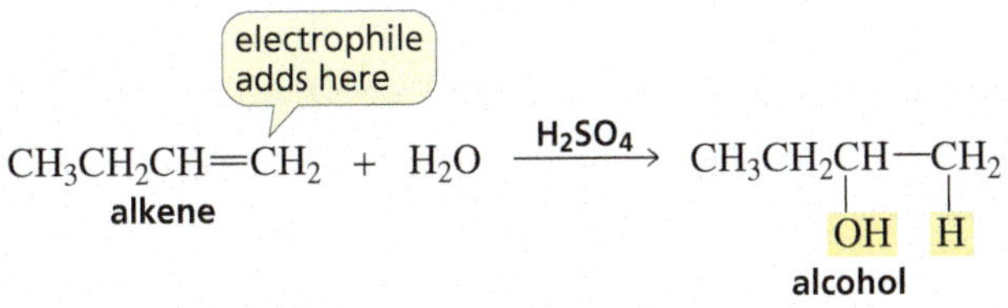

electrophile adds here

$$CH_3CH_2CH = CH_2 + H_2O \xrightarrow{H_2SO_4} CH_3CH_2CH - CH_2$$

alkene → alcohol (with OH and H)

Alkynes also undergo the acid-catalyzed addition of water. As expected, the electrophile (H^+) adds to the less substituted sp carbon.

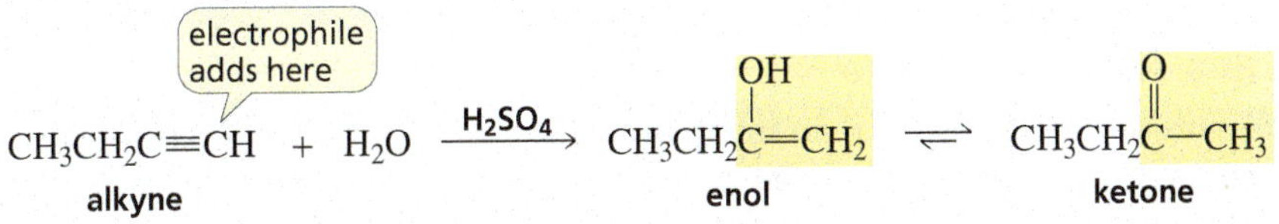

electrophile adds here

$$CH_3CH_2C\equiv CH + H_2O \xrightarrow{H_2SO_4} CH_3CH_2C = CH_2 \rightleftharpoons CH_3CH_2C - CH_3$$

alkyne → enol (with OH) ⇌ ketone (with O)

The initial product of the reaction is an *enol*. An **enol** has a carbon–carbon double bond with an OH group bonded to one of the sp^2 carbons. (The suffix "ene" signifies the double bond, and "ol" signifies the OH group. When the two suffixes are joined, the second e of "ene" is dropped to avoid two consecutive vowels, but the word is pronounced as if the second e were still there: "ene-ol.") The enol immediately rearranges to a *ketone*. A ketone and its corresponding enol are called **keto–enol tautomers.**

- **Tautomers** (taw-toe-mers) are constitutional isomers that are in rapid equilibrium.
- A ketone and an enol differ only in the location of a double bond and a hydrogen.

- The keto tautomer predominates in solution because it is usually much more stable than the enol tautomer.
- Interconversion of the tautomers is called **keto–enol interconversion** or **tautomerization.**

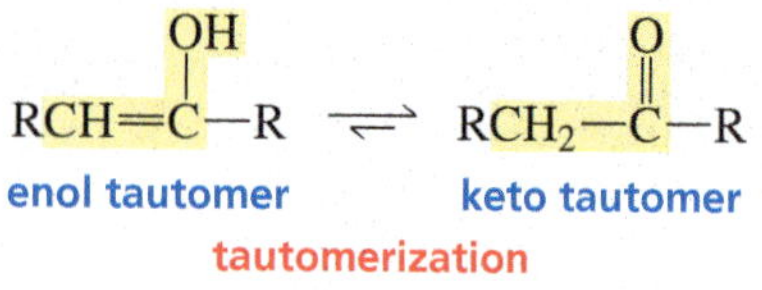

$$\underset{\substack{\text{enol tautomer}}}{RCH=\overset{\overset{\displaystyle OH}{|}}{C}-R} \;\rightleftharpoons\; \underset{\substack{\text{keto tautomer}}}{RCH_2-\overset{\overset{\displaystyle O}{\|}}{C}-R}$$

tautomerization

The mechanism for the conversion of an enol to a ketone under acidic conditions is shown below.

MECHANISM FOR ACID-CATALYZED KETO–ENOL INTERCONVERSION

$$RCH=\overset{\overset{\displaystyle :\ddot{O}-H}{|}}{C}-R \;\rightleftharpoons\; RCH_2-\overset{\overset{\displaystyle :\overset{+}{O}}{\|}}{C}-R \;\rightleftharpoons\; RCH_2-\overset{\overset{\displaystyle O}{\|}}{C}-R \;+\; H_3O^+$$

enol   ketone

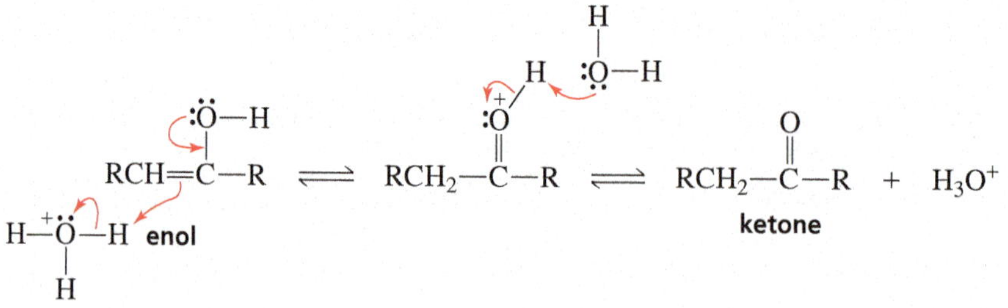

- A π bond forms between carbon and oxygen and, as the π bond between the two carbons breaks, carbon picks up a proton.
- Water removes a proton from the protonated carbonyl group.

Addition to an Internal Alkyne

The addition of water to a symmetrical internal alkyne forms a single ketone as a product. But if the alkyne is not symmetrical, then two ketones are formed because the initial addition of the proton can occur to either of the *sp* carbons.

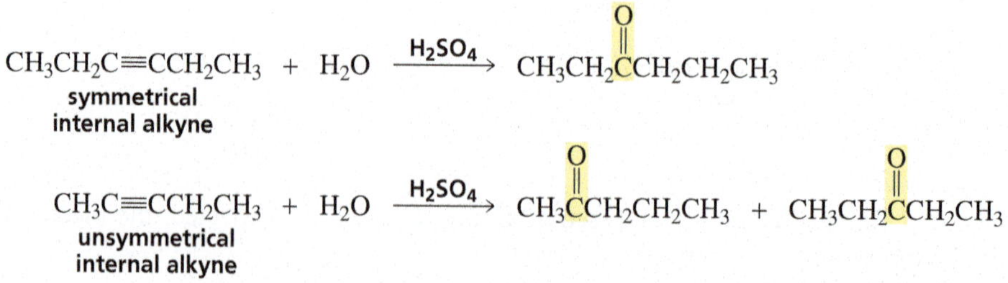

$$\underset{\substack{\text{symmetrical} \\ \text{internal alkyne}}}{CH_3CH_2C{\equiv}CCH_2CH_3} \;+\; H_2O \;\xrightarrow{\;H_2SO_4\;}\; CH_3CH_2\overset{\overset{\displaystyle O}{\|}}{C}CH_2CH_2CH_3$$

$$\underset{\substack{\text{unsymmetrical} \\ \text{internal alkyne}}}{CH_3C{\equiv}CCH_2CH_3} \;+\; H_2O \;\xrightarrow{\;H_2SO_4\;}\; CH_3\overset{\overset{\displaystyle O}{\|}}{C}CH_2CH_2CH_3 \;+\; CH_3CH_2\overset{\overset{\displaystyle O}{\|}}{C}CH_2CH_3$$

Addition to a Terminal Alkyne

Terminal alkynes are less reactive than internal alkynes toward the addition of water. Therefore, the addition of water to a terminal alkyne is generally carried out in the presence of mercuric ion (Hg^{2+}). The mercuric ion is a catalyst—it increases the rate of the addition reaction.

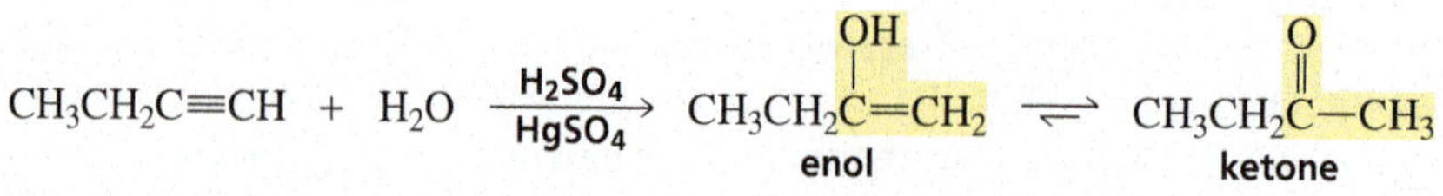

$$CH_3CH_2C{\equiv}CH \;+\; H_2O \;\xrightarrow[\text{HgSO}_4]{\;H_2SO_4\;}\; \underset{\substack{\text{enol}}}{CH_3CH_2\overset{\overset{\displaystyle OH}{|}}{C}=CH_2} \;\rightleftharpoons\; \underset{\substack{\text{ketone}}}{CH_3CH_2\overset{\overset{\displaystyle O}{\|}}{C}-CH_3}$$

The mechanism for the *mercuric-ion-catalyzed hydration* of an alkyne is shown on the top of the next page. The intermediate formed in the first step should remind you of the cyclic bromonium ion intermediate formed when Br_2 adds to an alkene (Section 6.9).

MECHANISM FOR THE MERCURIC-ION-CATALYZED HYDRATION OF AN ALKYNE

- Reaction of the alkyne with mercuric ion forms a cyclic mercurinium ion. (Two of the electrons in mercury's filled $5d$ atomic orbitals are shown.)
- Water (the nucleophile) adds to the more substituted carbon of the cyclic intermediate (Section 6.9).
- The protonated OH group loses a proton to form a mercuric enol, which immediately tautomerizes to a mercuric ketone.
- Loss of the mercuric ion forms an enol, which tautomerizes to a ketone in an acid-catalyzed reaction (see page 300).

PROBLEM 14 ♦

What ketones are formed from the acid-catalyzed hydration of 3-heptyne?

PROBLEM 15 ♦

Which alkyne should be used for the synthesis of each of the following ketones?

a. $CH_3\overset{O}{\overset{\|}{C}}CH_3$ b. $CH_3CH_2\overset{O}{\overset{\|}{C}}CH_2CH_2CH_3$ c. $CH_3\overset{O}{\overset{\|}{C}}$—⬡

PROBLEM 16 ♦

Draw all the enol tautomers for each of the ketones in Problem 15.

<table>
<tr><td>**7.8**</td><td>**THE ADDITION OF BORANE TO AN ALKYNE: HYDROBORATION–OXIDATION**</td></tr>
</table>

BH_3 or R_2BH (in THF) adds to alkynes in the same way it adds to alkenes. That is, boron is the electrophile and H^- is the nucleophile (Section 6.8). When the addition reaction is over, aqueous sodium hydroxide and hydrogen peroxide are added to the reaction mixture. The end result, as in the case of alkenes, is replacement of the boron by an OH group. The resulting enol immediately tautomerizes to a ketone.

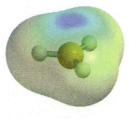

BH_3

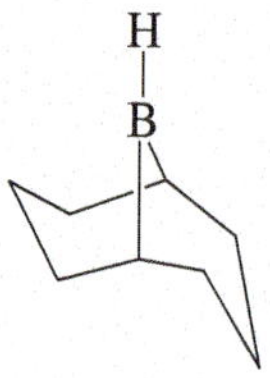

9-BBN
9-borabicyclo[3.3.1]nonane
R₂BH

Hydroboration–oxidation of a terminal alkyne forms an aldehyde.

Acid catalyzed addition of water (in the presence of mercuric ion) to a terminal alkyne forms a ketone.

A carbonyl compound will be the product of hydroboration–oxidation only if a second molecule of BH₃ or R₂BH does not add to the π bond of the boron-substituted alkene. In the case of internal alkynes, the substituents on the boron-substituted alkene prevent the approach of the second boron-containing molecule. In the case of terminal alkynes, however, there is an H instead of a bulky alkyl group on the carbon that the second boron-containing molecule would add to, so there is less steric hindrance toward the second addition reaction. Therefore, either BH₃ or R₂BH can be used with internal alkenes, but the more sterically hindered R₂BH should be used with terminal alkynes.

Boron is an electrophile because of its incomplete octet. When it reacts with a terminal alkyne it, like other electrophiles, adds preferentially to the less substituted *sp* carbon (the one bonded to the hydrogen). Because the boron-containing group is subsequently replaced by an OH group, hydroboration–oxidation of a terminal alkyne forms an *aldehyde* (the carbonyl group is on the terminal carbon), whereas the mercuric-ion-catalyzed addition of water to a terminal alkyne forms a *ketone* (the carbonyl group is *not* on the terminal carbon).

$$CH_3C\equiv CH$$

1. R₂BH/THF
2. HO⁻, H₂O₂, H₂O
→ $CH_3CH=CH$ (OH) $\rightleftharpoons$ CH_3CH_2CH (O)
aldehyde

H₂O, H₂SO₄
HgSO₄
→ $CH_3C=CH_2$ (OH) $\rightleftharpoons$ CH_3CCH_3 (O)
ketone

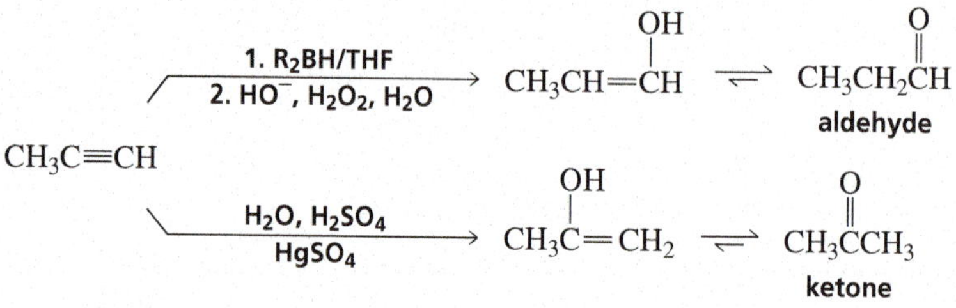

PROBLEM 17

For each of the following alkynes, draw the products obtained from (1) the acid-catalyzed addition of water (mercuric ion is added for part **a**) and from (2) hydroboration–oxidation:

a. 1-butyne   **b.** 2-butyne   **c.** 2-pentyne

PROBLEM 18 ◆

Only one alkyne forms an aldehyde when it undergoes the mercuric-ion-catalyzed addition of water. Identify the alkyne.

7.9   THE ADDITION OF HYDROGEN TO AN ALKYNE

Alkynes are reduced by catalytic hydrogenation in the same way alkenes are (Section 5.9). The initial product of hydrogenation is an alkene, but it is difficult to stop the reaction at this stage because alkenes are more reactive than alkynes. The final product of the hydrogenation reaction, therefore, is an alkane.

> an alkyne is converted to an alkane

$$CH_3CH_2C\equiv CH \xrightarrow[\text{Pd/C}]{H_2} CH_3CH_2CH=CH_2 \xrightarrow[\text{Pd/C}]{H_2} CH_3CH_2CH_2CH_3$$
alkyne   **alkene**   **alkane**

Forming a cis Alkene

The reaction can be stopped at the alkene stage if a "poisoned" (partially deactivated) palladium catalyst is used. The most common partially deactivated palladium catalyst is known as Lindlar catalyst (Figure 7.4).

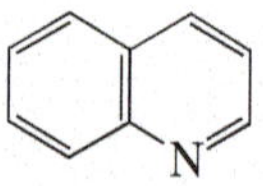

quinoline

$$CH_3CH_2C\equiv CH + H_2 \xrightarrow[\text{catalyst}]{\text{Lindlar}} CH_3CH_2CH=CH_2$$
alkyne   **alkene**

▲ **Figure 7.4**
Lindlar catalyst is prepared by precipitating palladium on barium sulfate, deactivated by quinoline. This treatment modifies the surface of palladium, making it much more effective at catalyzing the addition of hydrogen to a triple bond than to a double bond.

Because the alkyne sits on the surface of the metal catalyst and the hydrogens are delivered to the triple bond from the surface of the catalyst, both hydrogens are delivered to the same side of the double bond. In other words, syn addition of hydrogen occurs (Section 6.13). Syn addition of H₂ to an internal alkyne forms a *cis alkene.*

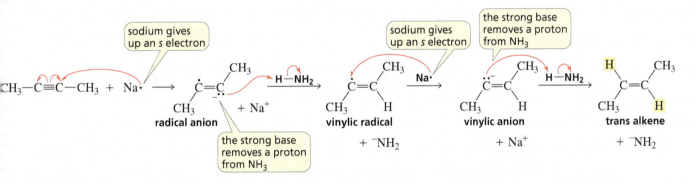

cis-2-pentene

Forming a trans Alkene

Internal alkynes can be converted to *trans alkenes* using sodium (or lithium) in liquid ammonia. The reaction stops at the alkene stage because sodium (or lithium) reacts more rapidly with triple bonds than with double bonds. This reaction is called a **dissolving metal reduction.** Ammonia is a gas at room temperature ($bp = -33\,°C$), so it is kept in the liquid state by cooling the reaction flask in a dry ice/acetone mixture, which has a temperature of $-78\,°C$.

trans-2-butene

As sodium dissolves in liquid ammonia, it forms a deep blue solution of dissolved Na^+ and electrons.

MECHANISM FOR THE CONVERSION OF AN ALKYNE TO A TRANS ALKENE

radical anion · vinylic radical · vinylic anion · trans alkene

The steps in the mechanism for the conversion of an internal alkyne to a trans alkene are:

- The single electron from the *s* orbital of sodium is transferred to an *sp* carbon of the alkyne. This forms a **radical anion**—a species with a negative charge and an unpaired electron.

  Notice that the movement of a single electron is represented by an arrowhead with a single barb. (Recall that sodium has a strong tendency to lose the single electron in its outer-shell *s* orbital; Section 1.3.)

- The radical anion is such a strong base that it can remove a proton from ammonia. This results in the formation of a **vinylic radical**—the unpaired electron is on a vinylic carbon.
- Another single-electron transfer from sodium to the vinylic radical forms a **vinylic anion.**
- The vinylic anion is also a strong base; it removes a proton from another molecule of ammonia, forming the trans alkene.

Both the radical anion and the vinylic anion can have either the cis or trans configuration. The cis and trans radical anions are in equilibrium, and so are the cis and trans vinylic anions. In each case, the equilibrium favors the trans isomer. The greater stability of the trans isomers is what causes the product to be a trans alkene (Figure 7.5).

A dissolving metal reduction cannot be used for the reduction of a terminal alkyne because sodium will cause a proton to be removed from the terminal alkyne.

$$2\ RC\equiv CH + 2\ Na \longrightarrow 2\ RC\equiv C^- + 2\ Na^+ + H_2$$

An arrowhead with a double barb signifies the movement of two electrons.

An arrowhead with a single barb signifies the movement of one electron.

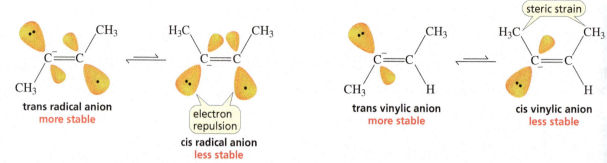

trans radical anion
more stable

cis radical anion
less stable

electron repulsion

trans vinylic anion
more stable

cis vinylic anion
less stable

steric strain

▲ **Figure 7.5**
The *trans radical anion* is more stable than the *cis radical anion*, because the nonbonded electrons are farther apart in the trans isomer. The *trans vinylic anion* is more stable than the *cis vinylic anion*, because the relatively bulky alkyl groups are farther apart in the trans isomer (Section 5.9).

PROBLEM 19 ◆
Describe the alkyne you should start with and the reagents you should use if you want to synthesize

a. pentane. **b.** *cis*-2-butene. **c.** *trans*-2-pentene. **d.** 1-hexene.

PROBLEM 20
What are products of the following reactions?

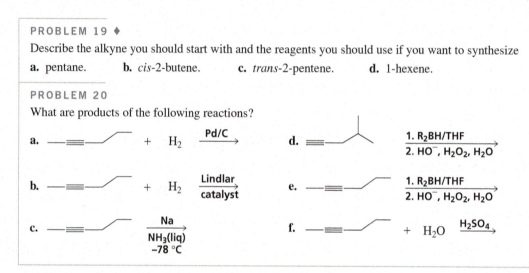

7.10	**A HYDROGEN BONDED TO AN *sp* CARBON IS "ACIDIC"**

We saw that an *sp* carbon is more electronegative than an sp^2 carbon, which is more electronegative than an sp^3 carbon (Section 2.6).

relative electronegativities of carbon atoms

most electronegative — $sp > sp^2 > sp^3$ — least electronegative

An *sp* carbon is more electronegative than an sp^2 carbon, which is more electronegative than an sp^3 carbon.

Because the most acidic compound is the one with its hydrogen attached to the most electronegative atom (when the atoms are the same size), ethyne is a stronger acid than ethene and ethene is a stronger acid than ethane.

$$HC\equiv CH \qquad H_2C=CH_2 \qquad CH_3CH_3$$

ethyne   ethene   ethane

$pK_a = 25$ $pK_a = 44$ $pK_a > 60$

We also saw that if we want to remove a proton from an acid (in a reaction that strongly favors products), we must use a base that is stronger than the base that is formed when the proton is removed because the equilibrium favors formation of the weaker base (which is equivalent to saying that the equilibrium favors formation of the weaker acid; Section 2.5).

The stronger the acid, the weaker its conjugate base.

For example, $^-NH_2$ is a stronger base than the acetylide ion that is formed because NH_3 $(pK_a = 36)$ is a weaker acid than a terminal alkyne $(pK_a = 25)$. Recall that the weaker acid has the stronger conjugate base. Therefore, an amide ion ($^-NH_2$) can be used to remove a proton from a terminal alkyne to prepare an **acetylide ion**.

$$RC\equiv CH \quad + \quad {}^-NH_2 \quad \rightleftharpoons \quad RC\equiv C^- \quad + \quad NH_3$$

   (stronger acid)   amide ion   acetylide ion

   (stronger acid) (stronger base)   **weaker base**   **weaker acid**

In contrast, if hydroxide ion is used as the base, the reaction will strongly favor reactants because hydroxide ion is a much weaker base than the acetylide ion.

$$RC\equiv CH \quad + \quad HO^- \quad \rightleftharpoons \quad RC\equiv C^- \quad + \quad H_2O$$

          hydroxide anion   acetylide anion

  weaker acid   **weaker base**   (stronger base) (stronger acid)

> **To remove a proton from an acid in a reaction that favors products, the base that removes the proton must be stronger than the base that is formed.**

Only a hydrogen bonded to an *sp* carbon is sufficiently acidic to be removed by an amide ion. An amide ion ($^-NH_2$) cannot remove a hydrogen bonded to an sp^2 or sp^3 carbon. Consequently, a hydrogen bonded to an *sp* carbon sometimes is referred to as an "acidic" hydrogen. Be careful not to misinterpret what is meant when we say that a hydrogen bonded to an *sp* carbon is "acidic." It is more acidic than most other carbon-bound hydrogens, but it is much less acidic ($pK_a = 25$) than a hydrogen of a water molecule, and we know that water is only a very weakly acidic compound ($pK_a = 15.7$).

relative acid strengths

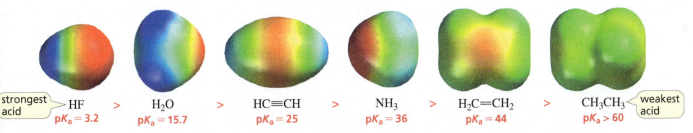

strongest acid						weakest acid
HF	>	H_2O	>	HC≡CH	>	NH_3	>	$H_2C=CH_2$	>	CH_3CH_3
$pK_a = 3.2$		$pK_a = 15.7$		$pK_a = 25$		$pK_a = 36$		$pK_a = 44$		$pK_a > 60$

Sodium Amide and Sodium in Ammonia

Take care not to confuse the compound sodium amide ($Na^+\ {}^-NH_2$), also called sodamide, with a mixture of sodium (Na) in liquid ammonia. The amide ion is the strong base used to remove a proton from a terminal alkyne. Sodium in liquid ammonia is the source of electrons and protons, respectively, used to convert an internal alkyne to a trans alkene (Section 7.9).

PROBLEM 21

Which of the following bases can remove a proton from a terminal alkyne in a reaction that favors products?

    CH_3O^- NH_3 $CH_3\bar{C}H_2$ $H_2C=\bar{C}H$ F^-

PROBLEM 22 ◆

Explain why an amide ion cannot be used to form a carbanion from an alkane in a reaction that favors products.

PROBLEM 23 ◆

Any base whose conjugate acid has a pK_a greater than _____ can remove a proton from a terminal alkyne to form an acetylide ion (in a reaction that favors products).

PROBLEM-SOLVING STRATEGY

Comparing the Acidities of Compounds

LEARN THE STRATEGY

a. Rank the following from strongest acid to weakest acid:

    $CH_3CH_2\overset{+}{N}H_3$ $CH_3CH=\overset{+}{N}H_2$ $CH_3C\equiv\overset{+}{N}H$

To compare the acidities of a group of compounds, first look at how the compounds differ. The three compounds differ in the hybridization of the nitrogen to which the acidic hydrogen is attached. Now recall what you know about hybridization and electronegativity. You know that an *sp* hybridized atom is more electronegative than an sp^2 hybridized atom, which is more electronegative than an sp^3 hybridized atom. You also know that the more electronegative the atom to which a hydrogen is attached, the more acidic the hydrogen. Now you can answer the question.

relative acidities

$$CH_3C\overset{+}{\equiv}NH \; > \; CH_3CH\overset{+}{=}NH_2 \; > \; CH_3CH_2\overset{+}{N}H_3$$

b. Draw the conjugate bases of these compounds and rank them from strongest base to weakest base.

Removing a proton from each of the compounds provides their conjugate bases. The stronger the acid, the weaker its conjugate base, so we can use the relative acid strengths obtained in part **a** to determine that the order of decreasing basicity is as follows:

relative basicities

$$CH_3CH_2NH_2 \; > \; CH_3CH=NH \; > \; CH_3C\equiv N$$

USE THE STRATEGY

PROBLEM 24 ♦

Rank the following from strongest base to weakest base:

a. $CH_3CH_2CH=\bar{C}H$ $CH_3CH_2C\equiv\bar{C}$ $CH_3CH_2CH_2\bar{C}H_2$

b. $CH_3CH_2O^-$ F^- $CH_3C\equiv C^-$ $^-NH_2$

LEARN THE STRATEGY

PROBLEM 25 SOLVED

Which carbocation is more stable?

$CH_3\overset{+}{C}H_2$ or $H_2C\overset{+}{=}CH$

SOLUTION Because an sp^2 carbon is more electronegative than an sp^3 carbon, an sp^2 carbon with a positive charge is less stable than an sp^3 carbon with a positive charge. Thus, the ethyl carbocation is more stable.

USE THE STRATEGY

PROBLEM 26 ♦

Which carbocation is more stable?

a. $H_2C\overset{+}{=}CH$ or $HC\equiv C^+$ **b.** $HC\equiv C^+$ or $CH_3\overset{+}{C}H_2$

7.11 SYNTHESIS USING ACETYLIDE IONS

Reactions that form carbon–carbon bonds are important in the synthesis of organic compounds because, without such reactions, we could not convert compounds with small carbon skeletons into compounds with larger carbon skeletons.

One reaction that forms a carbon–carbon bond is the reaction of an acetylide ion with an alkyl halide. Only primary alkyl halides or methyl halides should be used in this reaction.

> an alkylation reaction puts an alkyl group on a reactant

$$CH_3CH_2C\equiv C^- \; + \; CH_3CH_2CH_2Br \; \longrightarrow \; CH_3CH_2C\equiv CCH_2CH_2CH_3 \; + \; Br^-$$
3-heptyne

The mechanism for this reaction is well understood. Bromine is more electronegative than carbon, and as a result, the electrons in the C—Br bond are not shared equally by the two atoms: there is a partial positive charge on carbon and a partial negative charge on bromine.

$$CH_3CH_2C\equiv\ddot{C}^- \; + \; CH_3CH_2CH_2\overset{\delta+}{-}\overset{\delta-}{Br} \; \longrightarrow \; CH_3CH_2C\equiv CCH_2CH_2CH_3 \; + \; Br^-$$

The negatively charged acetylide ion (a nucleophile) is attracted to the partially positively charged carbon (an electrophile) of the alkyl halide. As the electrons of the acetylide ion approach the carbon to form the new C—C bond, they push out the bromine and its bonding electrons because carbon can bond to no more than four atoms at a time.

The previous reaction is an example of an *alkylation reaction*. An **alkylation reaction** attaches an alkyl group to a species. The mechanism for this and similar reactions is discussed in greater detail in Chapter 9. There you will learn why the reaction works best with primary alkyl halides and methyl halides.

We can convert terminal alkynes into internal alkynes of any desired chain length, simply by choosing an alkyl halide with the appropriate structure. Just count the number of carbons in the reactant (the terminal alkyne) and the number of carbons in the product to see how many carbons are needed in the alkyl halide.

$$CH_3CH_2CH_2C\equiv CH \xrightarrow{\ NaNH_2\ } CH_3CH_2CH_2C\equiv C^- \xrightarrow{\ CH_3CH_2Br\ } CH_3CH_2CH_2C\equiv CCH_2CH_3$$

1-pentyne   3-heptyne

PROBLEM 27 SOLVED

A chemist wants to synthesize 3-heptyne but cannot find any 1-pentyne, the starting material used in the synthesis just described. How else can 3-heptyne be synthesized?

SOLUTION The *sp* carbons of 3-heptyne are bonded to an *ethyl* group and to a *propyl* group. Therefore, to synthesize 3-heptyne, the acetylide ion of 1-pentyne can react with an *ethyl halide* or the acetylide ion of 1-butyne can react with a *propyl halide*. Because 1-pentyne is not available, the chemist should use 1-butyne and a propyl halide.

$$CH_3CH_2C\equiv CH \xrightarrow[\ 2.\ CH_3CH_2CH_2Cl\ ]{1.\ NaNH_2} CH_3CH_2C\equiv CCH_2CH_2CH_3$$

1-butyne   3-heptyne

(Remember that the numbers 1 and 2 in front of the reagents above and below the reaction arrow indicate two sequential reactions; the second reagent is not added until reaction with the first reagent is completely over.)

DESIGNING A SYNTHESIS I

7.12 AN INTRODUCTION TO MULTISTEP SYNTHESIS

For each reaction that has been discussed so far, we saw *why* the reaction occurs, *how* it occurs, and the *products* that are formed. A good way for you to review these reactions is to design syntheses, because when you design a synthesis, you must recall many of the reactions you have learned. And learning how to design syntheses is a very important part of organic chemistry.

Factors That Affect the Design of a Synthesis

Synthetic chemists consider *time, cost,* and *yield* in designing syntheses. In the interest of time, a well-designed synthesis consists of as few steps (sequential reactions) as possible, and each of those steps is a reaction that is easy to carry out. If two chemists in a pharmaceutical company are each asked to prepare a new drug, and one synthesizes the drug in three simple steps while the other uses six difficult steps, which chemist do you think will get a promotion?

The costs of the starting materials must also be taken into consideration. Moreover, each step in the synthesis should provide the greatest possible yield of the desired product. The more reactant needed to synthesize one gram of product, the more expensive the product is to produce. Sometimes a synthesis involving several steps is preferred because the starting materials are inexpensive, the reactions are easy to carry out, and the yield of each step is high. Such a synthesis is better than one with fewer steps if those steps require expensive starting materials and consist of reactions that are more difficult to run or give lower yields.

At this point in your chemical education, however, you are not yet familiar with the costs of different chemicals or the difficulties encountered in carrying out specific reactions. So for the time being, when you design a synthesis, just focus on finding the route with the fewest steps.

Designing a Synthesis

The following examples will give you an idea of the type of thinking required to design a successful synthesis. Problems of this kind will appear repeatedly throughout the book, because solving them is fun and is a good way to learn organic chemistry.

Example 1

Starting with 1-butyne, how could you make the ketone shown here? You can use any reagents you need.

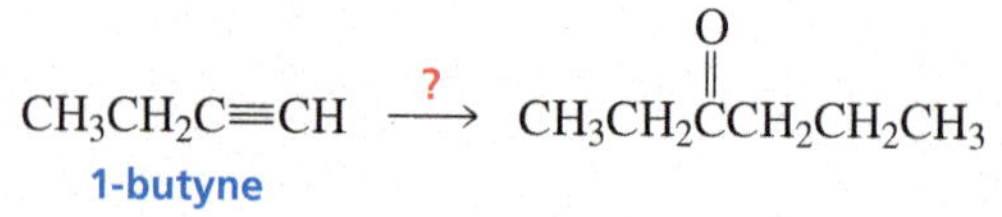

$$CH_3CH_2C{\equiv}CH \xrightarrow{?} CH_3CH_2\overset{\overset{\displaystyle O}{\|}}{C}CH_2CH_2CH_3$$

1-butyne

Many chemists find that the easiest way to design a synthesis is to work backward. Instead of looking at the reactant and deciding how to do the first step of the synthesis, look at the product and decide how to do the last step.

   The product of the synthesis is a ketone. Now you need to remember all the reactions you have learned that form a ketone. We will use the acid-catalyzed addition of water to an alkyne. (You also could use hydroboration–oxidation.) If the alkyne used in the reaction has identical substituents on both *sp* carbons, only one ketone will be obtained. Thus, 3-hexyne is the alkyne that should be used for the synthesis of the desired ketone.

$$CH_3CH_2C{\equiv}CCH_2CH_3 \xrightarrow[\text{H}_2\text{SO}_4]{\text{H}_2\text{O}} CH_3CH_2\overset{\overset{\displaystyle OH}{|}}{C}{=}CHCH_2CH_3 \rightleftharpoons CH_3CH_2\overset{\overset{\displaystyle O}{\|}}{C}CH_2CH_2CH_3$$

3-hexyne

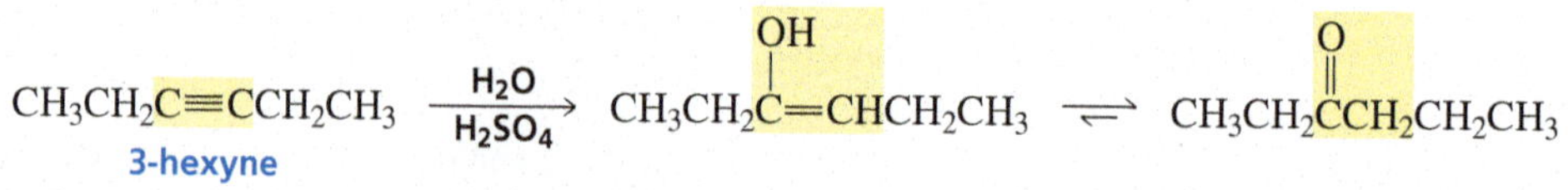

3-Hexyne can be obtained from the starting material (1-butyne) by removing the proton from its *sp* carbon, followed by alkylation. To produce the desired six-carbon product, a two-carbon alkyl halide must be used in the alkylation reaction.

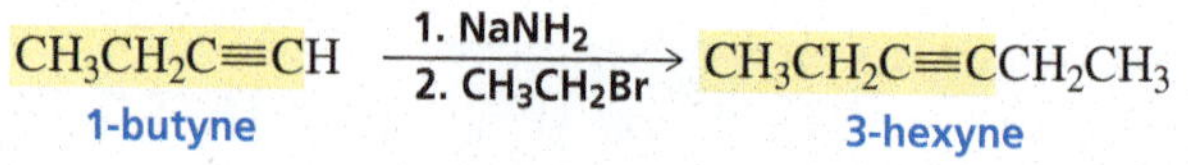

$$CH_3CH_2C{\equiv}CH \xrightarrow[\text{2. CH}_3\text{CH}_2\text{Br}]{\text{1. NaNH}_2} CH_3CH_2C{\equiv}CCH_2CH_3$$

1-butyne   **3-hexyne**

Designing a synthesis by working backward from product to reactant is not just a technique taught to organic chemistry students. It is used so frequently by experienced synthetic chemists that it has been given a name: **retrosynthetic analysis.** Chemists use open arrows when they write retrosynthetic analyses to indicate they are working backward. Typically, the reagents needed to carry out each step are not specified until the reaction is written in the forward direction. For example, the ketone synthesis just discussed is arrived at by the following retrosynthetic analysis.

retrosynthetic analysis

$$CH_3CH_2\overset{\overset{\displaystyle O}{\|}}{C}CH_2CH_2CH_3 \Longrightarrow CH_3CH_2C{\equiv}CCH_2CH_3 \Longrightarrow CH_3CH_2C{\equiv}CH$$

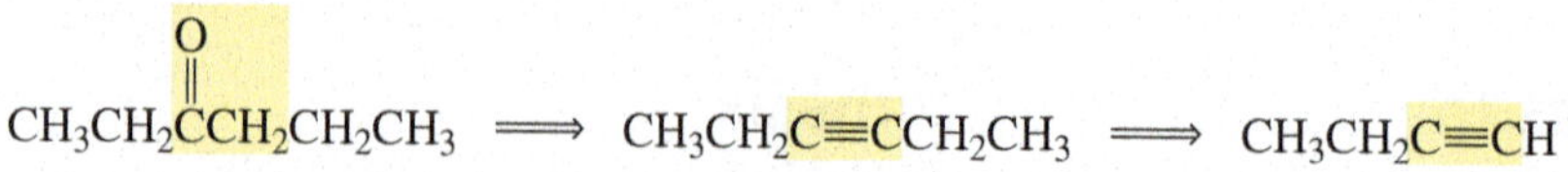

Once the sequence of reactions is worked out by retrosynthetic analysis, the synthetic scheme can be written by reversing the steps and including the reagents required for each step.

synthesis

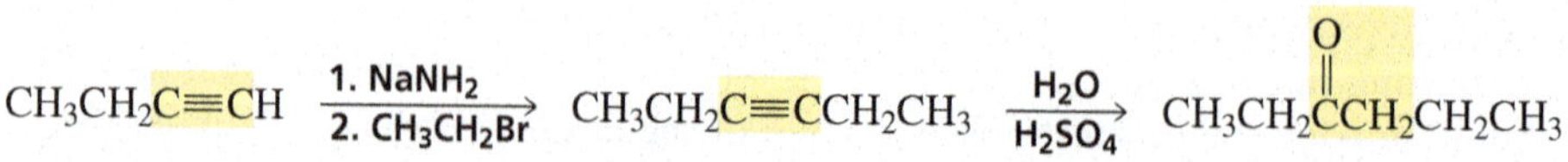

$$CH_3CH_2C{\equiv}CH \xrightarrow[\text{2. CH}_3\text{CH}_2\text{Br}]{\text{1. NaNH}_2} CH_3CH_2C{\equiv}CCH_2CH_3 \xrightarrow[\text{H}_2\text{SO}_4]{\text{H}_2\text{O}} CH_3CH_2\overset{\overset{\displaystyle O}{\|}}{C}CH_2CH_2CH_3$$

Example 2

Starting with ethyne, how could you make 2-bromopentane?

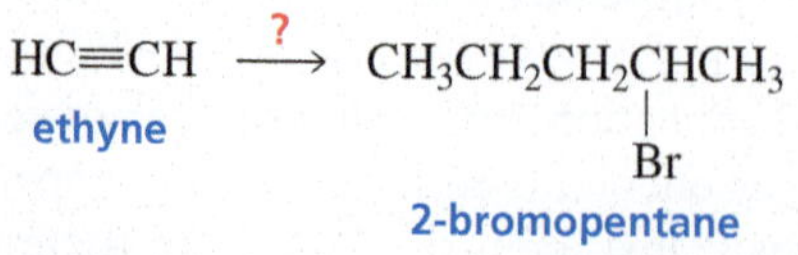

$$HC{\equiv}CH \xrightarrow{?} CH_3CH_2CH_2\overset{\overset{\displaystyle }{}}{\underset{\underset{\displaystyle Br}{|}}{C}}HCH_3$$

ethyne   **2-bromopentane**

2-Bromopentane can be prepared from 1-pentene, which can be prepared from 1-pentyne. 1-Pentyne can be prepared from ethyne and an alkyl halide with three carbons.

(retrosynthetic analysis)

$$CH_3CH_2CH_2CHCH_3 \implies CH_3CH_2CH_2CH=CH_2 \implies CH_3CH_2CH_2C\equiv CH \implies HC\equiv CH$$
    |
    Br

Now we can write the synthetic scheme:

(synthesis)

$$HC\equiv CH \xrightarrow[\text{2. } CH_3CH_2CH_2Br]{\text{1. NaNH}_2} CH_3CH_2CH_2C\equiv CH \xrightarrow[\substack{\text{Lindlar}\\\text{catalyst}}]{\text{H}_2} CH_3CH_2CH_2CH=CH_2 \xrightarrow{\text{HBr}} CH_3CH_2CH_2CHCH_3$$
                                                                                              |
                                                                                              Br

Example 3

How could 2,6-dimethylheptane be prepared from an alkyne and an alkyl halide? (The prime in R′ signifies that R and R′ can be different alkyl groups.)

$$RC\equiv CH \ + \ R'Br \xrightarrow{?} CH_3\overset{\overset{\displaystyle CH_3}{|}}{C}HCH_2CH_2CH_2\overset{\overset{\displaystyle CH_3}{|}}{C}HCH_3$$
2,6-dimethylheptane

2,6-Dimethyl-3-heptyne is the only alkyne that forms the desired alkane upon hydrogenation. (If you don't believe this, try to draw another one.) This alkyne can be prepared in two different ways: from the reaction of an acetylide ion with a primary alkyl halide (isobutyl bromide) or from the reaction of an acetylide ion with a secondary alkyl halide (isopropyl bromide).

(retrosynthetic analysis)

$$CH_3\overset{\overset{\displaystyle CH_3}{|}}{C}HCH_2CH_2CH_2\overset{\overset{\displaystyle CH_3}{|}}{C}HCH_3 \implies CH_3\overset{\overset{\displaystyle CH_3}{|}}{C}HCH_2C\equiv C\overset{\overset{\displaystyle CH_3}{|}}{C}HCH_3$$
2,6-dimethyl-3-heptyne

$$CH_3\overset{\overset{\displaystyle CH_3}{|}}{C}HCH_2Br \ + \ HC\equiv C\overset{\overset{\displaystyle CH_3}{|}}{C}HCH_3 \quad \text{or} \quad CH_3\overset{\overset{\displaystyle CH_3}{|}}{C}HBr \ + \ HC\equiv CCH_2\overset{\overset{\displaystyle CH_3}{|}}{C}HCH_3$$
isobutyl bromide   **isopropyl bromide**

Because we know that the reaction of an acetylide ion with an alkyl halide works best with primary alkyl halides and methyl halides, we should choose the synthesis that requires a primary alkyl halide.

(synthesis)

$$CH_3\overset{\overset{\displaystyle CH_3}{|}}{C}HC\equiv CH \xrightarrow[\substack{\text{2. } CH_3CHCH_2Br\\ \quad\quad | \\ \quad\quad CH_3}]{\text{1. NaNH}_2} CH_3\overset{\overset{\displaystyle CH_3}{|}}{C}HC\equiv CCH_2\overset{\overset{\displaystyle CH_3}{|}}{C}HCH_3 \xrightarrow[\text{Pd/C}]{\text{H}_2} CH_3\overset{\overset{\displaystyle CH_3}{|}}{C}HCH_2CH_2CH_2\overset{\overset{\displaystyle CH_3}{|}}{C}HCH_3$$

Example 4

How could you carry out the following synthesis using the given starting material?

$$\langle\rangle\text{--}C\equiv CH \xrightarrow{?} \langle\rangle\text{--}CH_2CH_2OH$$

An alcohol can be prepared from an alkene, and an alkene can be prepared from an alkyne.

(retrosynthetic analysis)

$$\langle\rangle\text{--}CH_2CH_2OH \implies \langle\rangle\text{--}CH=CH_2 \implies \langle\rangle\text{--}C\equiv CH$$

You can use either of the two methods you know that convert an alkyne to an alkene, because the desired alkene does not have cis–trans isomers. Hydroboration–oxidation must be used to convert the alkene to the desired alcohol because the acid-catalyzed addition of water would form a different alcohol.

synthesis

$$\text{Cyclohexyl-C}\equiv\text{CH} \xrightarrow[\substack{\text{or Na/NH}_3\text{(liq)}\\-78\,°\text{C}}]{\text{H}_2/\substack{\text{Lindlar}\\\text{catalyst}}} \text{Cyclohexyl-CH}=\text{CH}_2 \xrightarrow[\text{2. HO}^-,\text{ H}_2\text{O}_2,\text{ H}_2\text{O}]{\text{1. BH}_3/\text{THF}} \text{Cyclohexyl-CH}_2\text{CH}_2\text{OH}$$

Example 5

How could you prepare (*E*)-2-pentene from ethyne?

$$\text{HC}\equiv\text{CH} \xrightarrow{?} \begin{array}{c} \text{CH}_3\text{CH}_2 \quad\quad \text{H} \\ \diagdown C=C \diagup \\ \text{H} \quad\quad \text{CH}_3 \end{array}$$

(*E*)-2-pentene

A trans alkene can be prepared from the reaction of an internal alkyne with sodium and liquid ammonia. The alkyne needed to synthesize the desired alkene can be prepared from 1-butyne and a methyl halide. 1-Butyne can be prepared from ethyne and an ethyl halide.

retrosynthetic analysis

$$\begin{array}{c} \text{CH}_3\text{CH}_2 \quad\quad \text{H} \\ \diagdown C=C \diagup \\ \text{H} \quad\quad \text{CH}_3 \end{array} \Longrightarrow \text{CH}_3\text{CH}_2\text{C}\equiv\text{CCH}_3 \Longrightarrow \text{CH}_3\text{CH}_2\text{C}\equiv\text{CH} \Longrightarrow \text{HC}\equiv\text{CH}$$

synthesis

$$\text{HC}\equiv\text{CH} \xrightarrow[\text{2. CH}_3\text{CH}_2\text{Br}]{\text{1. NaNH}_2} \text{CH}_3\text{CH}_2\text{C}\equiv\text{CH} \xrightarrow[\text{2. CH}_3\text{Br}]{\text{1. NaNH}_2} \text{CH}_3\text{CH}_2\text{C}\equiv\text{CCH}_3 \xrightarrow[-78\,°\text{C}]{\text{Na/NH}_3\text{(liq)}} \begin{array}{c} \text{CH}_3\text{CH}_2 \quad\quad \text{H} \\ \diagdown C=C \diagup \\ \text{H} \quad\quad \text{CH}_3 \end{array}$$

Example 6

How could you prepare *cis*-2,3-diethyloxirane from ethyne?

$$\text{HC}\equiv\text{CH} \xrightarrow{?} \begin{array}{c} \text{O} \\ \diagup \diagdown \\ \text{H}\cdots\text{C}-\text{C}\cdots\text{H} \\ \text{CH}_3\text{CH}_2 \quad\quad \text{CH}_2\text{CH}_3 \end{array}$$

cis-2,3-diethyloxirane

A cis epoxide can be prepared from a cis alkene and a peroxyacid. The cis alkene that should be used is *cis*-3-hexene, which can be obtained from 3-hexyne. 3-Hexyne can be prepared from 1-butyne and an ethyl halide, and 1-butyne can be prepared from ethyne and an ethyl halide.

retrosynthetic analysis

$$\begin{array}{c} \text{O} \\ \diagup \diagdown \\ \text{H}\cdots\text{C}-\text{C}\cdots\text{H} \\ \text{CH}_3\text{CH}_2 \quad\quad \text{CH}_2\text{CH}_3 \end{array} \Longrightarrow \begin{array}{c} \text{H} \quad\quad \text{H} \\ \diagdown C=C \diagup \\ \text{CH}_3\text{CH}_2 \quad\quad \text{CH}_2\text{CH}_3 \end{array} \Longrightarrow \text{CH}_3\text{CH}_2\text{C}\equiv\text{CCH}_2\text{CH}_3 \Longrightarrow \text{CH}_3\text{CH}_2\text{C}\equiv\text{CH} \Longrightarrow \text{HC}\equiv\text{CH}$$

The alkyne must be converted to the alkene using H_2 and Lindlar catalyst so that the desired cis alkene is obtained.

synthesis

$$\text{HC}\equiv\text{CH} \xrightarrow[\text{2. CH}_3\text{CH}_2\text{Br}]{\text{1. NaNH}_2} \text{CH}_3\text{CH}_2\text{C}\equiv\text{CH} \xrightarrow[\text{2. CH}_3\text{CH}_2\text{Br}]{\text{1. NaNH}_2} \text{CH}_3\text{CH}_2\text{C}\equiv\text{CCH}_2\text{CH}_3 \xrightarrow[\substack{\text{Lindlar}\\\text{catalyst}}]{\text{H}_2} \begin{array}{c} \text{H} \quad\quad \text{H} \\ \diagdown C=C \diagup \\ \text{CH}_3\text{CH}_2 \quad\quad \text{CH}_2\text{CH}_3 \end{array}$$

$$\xrightarrow[\text{RCOOH}]{\substack{\text{O}\\\|}}$$

$$\begin{array}{c} \text{O} \\ \diagup \diagdown \\ \text{H}\cdots\text{C}-\text{C}\cdots\text{H} \\ \text{CH}_3\text{CH}_2 \quad\quad \text{CH}_2\text{CH}_3 \end{array}$$

Example 7

How could you prepare 3,3-dibromohexane from reagents that contain no more than two carbons?

$$\text{reagents with no more than 2 carbons} \xrightarrow{?} \underset{\overset{|}{Br}}{\overset{\overset{Br}{|}}{CH_3CH_2CCH_2CH_2CH_3}}$$

3,3-dibromohexane

A geminal dibromide can be prepared from an alkyne. 3-Hexyne is the alkyne of choice because it forms one geminal dibromide, whereas 2-hexyne forms two different geminal dibromides. 3-Hexyne can be prepared from 1-butyne and ethyl bromide, and 1-butyne can be prepared from ethyne and ethyl bromide.

(retrosynthetic analysis)

$$\underset{\overset{|}{Br}}{\overset{\overset{Br}{|}}{CH_3CH_2CCH_2CH_3}} \implies CH_3CH_2C\equiv CCH_2CH_3 \implies CH_3CH_2C\equiv CH \implies HC\equiv CH$$

(synthesis)

$$HC\equiv CH \xrightarrow[\text{2. } CH_3CH_2Br]{\text{1. NaNH}_2} CH_3CH_2C\equiv CH \xrightarrow[\text{2. } CH_3CH_2Br]{\text{1. NaNH}_2} CH_3CH_2C\equiv CCH_2CH_3 \xrightarrow{\text{excess HBr}} \underset{\overset{|}{Br}}{\overset{\overset{Br}{|}}{CH_3CH_2CCH_2CH_2CH_3}}$$

Example 8

How could you prepare a three-carbon aldehyde from 1-butyne?

$$CH_3CH_2C\equiv CH \xrightarrow{?} CH_3CH_2\overset{\overset{O}{\|}}{C}-H$$

Because the desired compound has fewer carbons than the starting material, we know that a cleavage reaction must occur. The desired aldehyde can be prepared by ozonolysis (an oxidative cleavage) of 3-hexene. 3-Hexene can be prepared from 3-hexyne, which can be prepared from 1-butyne and ethyl bromide.

(retrosynthetic analysis)

$$CH_3CH_2\overset{\overset{O}{\|}}{CH} \implies CH_3CH_2CH=CHCH_2CH_3 \implies CH_3CH_2C\equiv CCH_2CH_3 \implies CH_3CH_2C\equiv CH$$

(synthesis)

$$CH_3CH_2C\equiv CH \xrightarrow[\text{2. } CH_3CH_2Br]{\text{1. NaNH}_2} CH_3CH_2C\equiv CCH_2CH_3 \xrightarrow[\substack{\text{Lindlar} \\ \text{catalyst}}]{\text{H}_2} CH_3CH_2CH=CHCH_2CH_3 \xrightarrow[\text{2. } (CH_3)_2S]{\text{1. O}_3, -78\,°C} CH_3CH_2\overset{\overset{O}{\|}}{CH}$$

PROBLEM 28

How could the following compounds be synthesized from acetylene?

a. $CH_3CH_2CH_2C\equiv CH$ c. $CH_3CH=CH_2$ e. $CH_3CH_2CH_2CH_2\overset{\overset{O}{\|}}{CH}$

b. $\underset{\overset{|}{H}}{\overset{\overset{CH_3CH_2}{}}{}}C=C\underset{\overset{|}{H}}{\overset{\overset{CH_2CH_3}{}}{}}$ d. $CH_3\underset{\overset{|}{Cl}}{\overset{\overset{Cl}{|}}{C}}CH_3$ f. $CH_3\underset{\overset{|}{Br}}{CH}CH_3$

Green Chemistry: Aiming for Sustainability

Chemical innovations have improved the quality of virtually every aspect of life: food, shelter, medicine, transportation, communication, and the availability of new materials. These improvements, however, have come with a price—namely, the damage that the development and disposal of chemicals has inflicted on the environment.

Chemists now are focused on sustainability, which is defined as "meeting the needs of the current generation without sacrificing the ability to meet the needs of future generations." One way to achieve sustainability is through the use of green chemistry.

Green chemistry is pollution prevention at the molecular level. It involves the design of chemical products and processes so that the generation of polluting substances is reduced or eliminated. For example, chemists are now creating products not only for function, but also for biodegradability. They are designing syntheses that use and generate substances that cause little or no toxicity to health or to the environment. Green chemical syntheses can be cost effective because they reduce the expense for such things as waste disposal, regulatory compliance, and liability. Applying the principles of green chemistry can help us achieve a sustainable future.

ESSENTIAL CONCEPTS

Section 7.1

- An **alkyne** is a hydrocarbon that contains a carbon–carbon triple bond. The functional group suffix of an alkyne is "yne."
- A **terminal alkyne** has the triple bond at the end of the chain; an **internal alkyne** has the triple bond located elsewhere along the chain.

Section 7.2

- If a compound has a double bond and a triple bond, the "ene" ending is stated first (en-yne) and the chain is numbered to produce a name with the lowest possible number. If the same low number is produced in both directions, the double bond gets the lower number.
- If the second functional group has a higher priority than an alkene, the higher priority functional group gets the lower number.

Section 7.3

- A triple bond is composed of one σ bond and two π bonds.

Section 7.4

- All hydrocarbons—alkanes, alkenes, alkynes—have similar physical properties.

Section 7.5

- Alkynes undergo electrophilic addition reactions. The same reagents that add to alkenes also add to alkynes.
- Alkynes are less reactive than alkenes.

Section 7.6

- If excess reagent is available, alkynes undergo a second addition reaction with hydrogen halides and halogens because the product of the first reaction has a double bond.

Section 7.7

- The product of the reaction of an alkyne with water under acidic conditions is an **enol,** which immediately rearranges to a ketone. Terminal alkynes require a mercuric ion catalyst.

- The ketone and enol are called **keto–enol tautomers;** they differ in the location of a double bond and a hydrogen. The keto tautomer usually predominates at equilibrium.
- Interconversion of the tautomers is called **tautomerization** or **keto–enol interconversion.**

Section 7.8

- Hydroboration–oxidation of an internal alkyne forms a ketone; hydroboration–oxidation of a terminal alkyne forms an aldehyde. Either BH_3 or R_2BH can be used for internal alkynes; R_2BH should be used for terminal alkynes.

Section 7.9

- Catalytic hydrogenation of an alkyne forms an alkane.
- Catalytic hydrogenation with Lindlar catalyst converts an internal alkyne to a *cis alkene.*
- Sodium in liquid ammonia converts an internal alkyne to a *trans alkene.*

Section 7.10

- Electronegativity decreases in the order $sp > sp^2 > sp^3$, so ethyne is a stronger acid than ethene and ethene is a stronger acid than ethane.
- An amide ion removes a proton from an sp carbon of a terminal alkyne to form an **acetylide ion.**

Section 7.11

- An acetylide ion can undergo an alkylation reaction with a methyl halide or a primary alkyl halide to form an alkyne.
- An **alkylation reaction** attaches an alkyl group to a nucleophile.

Section 7.12

- Designing a synthesis by working backward is called **retrosynthetic analysis.**

SUMMARY OF REACTIONS

1. Electrophilic addition reactions

   a. Addition of hydrogen halides (H^+ is the electrophile; Section 7.6). The mechanism is on page 297.

$$RC\equiv CH \xrightarrow{\text{HX}} \underset{\underset{\text{X}}{|}}{RC}=CH_2 \xrightarrow{\text{excess HX}} \underset{\underset{\text{X}}{|}}{\overset{\overset{\text{X}}{|}}{RC}}-CH_3$$

   HX = HF, HCl, HBr, HI

   b. Addition of halogens (Section 7.6). The mechanism is the same as that for the reaction of alkenes with halogens (page 254).

$$RC\equiv CH \xrightarrow[\text{CH}_2\text{Cl}_2]{\text{Cl}_2} \underset{\underset{\text{Cl}}{|}}{\overset{\overset{\text{Cl}}{|}}{RC}}=CH \xrightarrow[\text{CH}_2\text{Cl}_2]{\text{Cl}_2} \underset{\underset{\text{Cl Cl}}{|\ |}}{\overset{\overset{\text{Cl Cl}}{|\ |}}{RC}}-CH$$

$$RC\equiv CR' \xrightarrow[\text{CH}_2\text{Cl}_2]{\text{Br}_2} \underset{\underset{\text{Br}}{|}}{\overset{\overset{\text{Br}}{|}}{RC}}=CR' \xrightarrow[\text{CH}_2\text{Cl}_2]{\text{Br}_2} \underset{\underset{\text{Br Br}}{|\ |}}{\overset{\overset{\text{Br Br}}{|\ |}}{RC}}-CR'$$

   c. Acid-catalyzed addition of water and hydroboration–oxidation (Sections 7.7 and 7.8). Except for the formation of a π-complex in the acid-catalyzed reaction, the mechanisms are the same as those for the reaction of alkenes with these same reagents (pages 245, 252, and 253). The mechanism for acid-catalyzed keto–enol interconversion is on page 300. The mechanism for mercuric-ion catalyzed hydration is on page 301.

$$\underset{\substack{\text{internal}\\\text{alkyne}}}{RC\equiv CR'} \xrightarrow[\substack{\text{1. R}_2\text{BH/THF or BH}_3\text{/THF}\\\text{2. HO}^-, \text{H}_2\text{O}_2, \text{H}_2\text{O}}]{\substack{\text{H}_2\text{O, H}_2\text{SO}_4\\\text{or}}} \underset{\text{ketones}}{RCCH_2R' + RCH_2CR'}$$

$$\underset{\substack{\text{terminal}\\\text{alkyne}}}{RC\equiv CH}$$

$$\xrightarrow[\text{HgSO}_4]{\text{H}_2\text{O, H}_2\text{SO}_4} \overset{\overset{\text{OH}}{|}}{RC}=CH_2 \rightleftharpoons \underset{\text{ketone}}{RCCH_3}$$

$$\xrightarrow[\text{2. HO}^-, \text{H}_2\text{O}_2, \text{H}_2\text{O}]{\text{1. R}_2\text{BH/THF}} RCH=\overset{\overset{\text{OH}}{|}}{CH} \rightleftharpoons \underset{\text{aldehyde}}{RCH_2CH}$$

2. Addition of hydrogen (Section 7.9). For H_2 addition, see page 208. The mechanism for formation of an alkane and a cis alkene is on page 208. The mechanism for formation of a trans alkene is on page 303.

$$RC\equiv CR' + 2 H_2 \xrightarrow{\text{Pd/C}} \underset{\text{alkane}}{RCH_2CH_2R'}$$

$$R-C\equiv C-R' + H_2 \xrightarrow[\text{catalyst}]{\text{Lindlar}} \underset{\text{cis alkene}}{\overset{H \diagdown \quad \diagup H}{\underset{R \diagup \quad \diagdown R'}{C=C}}}$$

$$R-C\equiv C-R' \xrightarrow[\substack{\text{NH}_3\text{(liq)}\\-78\ °\text{C}}]{\text{Na or Li}} \underset{\text{trans alkene}}{\overset{R \diagdown \quad \diagup H}{\underset{H \diagup \quad \diagdown R'}{C=C}}}$$

3. Removal of a proton from a terminal alkyne, followed by alkylation (Sections 7.10 and 7.11). The mechanism is on page 306.

$$RC\equiv CH \xrightarrow{\text{NaNH}_2} RC\equiv C^- \xrightarrow{\text{R'CH}_2\text{Br}} RC\equiv CCH_2R'$$

PROBLEMS

29. What is the major product obtained from the reaction of each of the following compounds with excess HCl?

   a. $CH_3CH_2C{\equiv}CH$ **b.** $CH_3CH_2C{\equiv}CCH_2CH_3$ **c.** $CH_3CH_2C{\equiv}CCH_2CH_2CH_3$

30. Draw a condensed structure for each of the following:

a. 2-hexyne	**e.** methoxyethyne	**i.** diethylacetylene
b. 5-ethyl-3-octyne	**f.** *sec*-butyl-*tert*-butylacetylene	**j.** di-*tert*-butylacetylene
c. methylacetylene	**g.** 1-bromo-1-pentyne	**k.** cyclopentylacetylene
d. vinylacetylene	**h.** 5-methyl-2-cyclohexenol	**l.** 5,6-dimethyl-2-heptyne

31. A student was given the structural formulas of several compounds and was asked to give them systematic names. How many did she name correctly? Correct those that are misnamed.

   a. 4-ethyl-2-pentyne   **b.** 1-bromo-4-heptyne   **c.** 2-methyl-3-hexyne   **d.** 3-pentyne

32. Identify the electrophile and the nucleophile in each of the following reaction steps. Then draw curved arrows to illustrate the bond-making and bond-breaking processes.

$$CH_3CH_2\overset{+\ddot{B}r:}{\overset{\diagup\diagdown}{C{=}CH}} \quad + \quad :\ddot{B}\overline{r}: \quad \longrightarrow \quad CH_3CH_2\underset{\underset{:\ddot{B}r:}{|}}{\overset{:\ddot{B}r:}{\overset{|}{C}}}{=}CH$$

$$CH_3C{\equiv}C{-}H \quad + \quad :\ddot{N}H_2 \quad \longrightarrow \quad CH_3C{\equiv}\ddot{C}\overline{:} \quad + \quad \ddot{N}H_3$$

$$CH_3C{\equiv}\ddot{C}\overline{:} \quad + \quad CH_3{-}\ddot{B}r: \quad \longrightarrow \quad CH_3C{\equiv}CCH_3 \quad + \quad :\ddot{B}\overline{r}:$$

33. What is each compound's systematic name?

   a. $CH_3C{\equiv}CCH_2\underset{\underset{Br}{|}}{CH}CH_3$

   b. $CH_3C{\equiv}CCH_2\underset{\underset{CH_2CH_2CH_3}{|}}{CH}CH_3$

   c. $CH_3C{\equiv}CCH_2\underset{\underset{CH_3}{|}}{\overset{\overset{CH_3}{|}}{C}}CH_3$

   d. $CH_3\underset{\underset{Cl}{|}}{CH}CH_2C{\equiv}C\underset{\underset{CH_3}{|}}{CH}CH_3$

   e.

   f.

34. What reagents should be used to carry out the following syntheses?

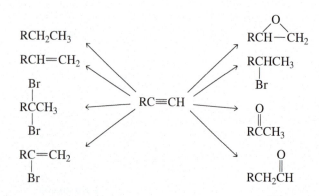

35. a. Draw the structures and give the common and systematic names for alkynes with molecular formula C_7H_{12}. Ignore stereosiomers. (*Hint:* There are 14.)
   b. How many would there be if stereoisomers are included?

36. Draw the mechanism for the following reaction:

$$CH_3CH_2CH_2C\equiv CH \xrightarrow[\text{H}_2\text{O}]{\text{Br}_2} CH_3CH_2CH_2\overset{\overset{\displaystyle O}{\|}}{C}CH_2Br$$

37. How can the following compounds be synthesized, starting with a hydrocarbon that has the same number of carbons as the desired product?

a. $CH_3CH_2CH_2CH_2\overset{\overset{\displaystyle O}{\|}}{C}H$

b. $CH_3CH_2CH_2CH_2OH$

c. $CH_3CH_2CH_2\overset{\overset{\displaystyle O}{\|}}{C}CH_2CH_2CH_2CH_3$

38. What reagents would you use for the following syntheses?
a. (Z)-3-hexene from 3-hexyne
b. (E)-3-hexene from 3-hexyne
c. hexane from 3-hexyne

39. What is the major product of the reaction of 1 mol of propyne with each of the following reagents?
a. HBr (1 mol)
b. HBr (2 mol)
c. Br_2 (1 mol)/CH_2Cl_2
d. Br_2 (2 mol)/CH_2Cl_2

e. aqueous H_2SO_4, $HgSO_4$
f. R_2BH in THF followed by H_2O_2/HO^-/H_2O
g. excess H_2, Pd/C

h. H_2/Lindlar catalyst
i. sodium amide
j. the product of part **i** followed by 1-chloropropane

40. Answer Problem 39, parts **a–h**, using 2-butyne as the starting material instead of propyne.

41. What is each compound's systematic name?

a. $CH_3C\equiv CCH_2CH_2CH_2CH=CH_2$

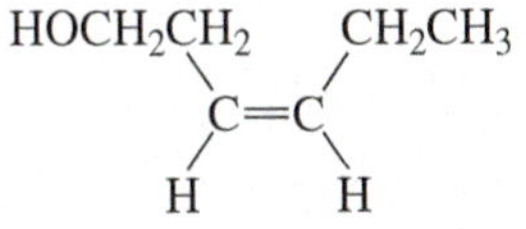

c. $CH_3CH_2C\equiv CCH_2CH_2C\equiv CH$

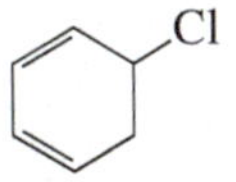

e. 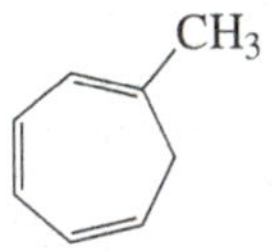 (structure: seven-membered ring with CH_3 substituent and double bonds)

b. (structure)
$$\underset{H}{\overset{HOCH_2CH_2}{\diagdown}}C=C\underset{H}{\overset{CH_2CH_3}{\diagup}}$$

d. (six-membered ring with Cl and double bonds)

f. 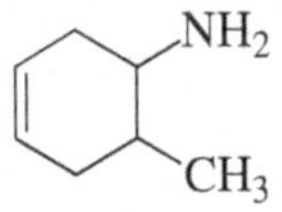 (six-membered ring with NH_2 and CH_3 substituents)

42. What is the molecular formula of a hydrocarbon that has 1 triple bond, 2 double bonds, 1 ring, and 32 carbons?

43. **a.** Starting with 3-methyl-1-butyne, how can you prepare the following alcohols?
    1. 2-methyl-2-butanol   **2.** 3-methyl-1-butanol
  b. In each case, a second alcohol would also be obtained. What alcohol would it be?

44. Which of the following names are correct? Correct those that are not correct.
a. 4-heptyne
b. 2-ethyl-3-hexyne
c. 4-chloro-2-pentyne
d. 2,3-dimethyl-5-octyne
e. 4,4-dimethyl-2-pentyne
f. 2,5-dimethyl-3-hexyne

45. Which of the following pairs are keto–enol tautomers?

a. $CH_3CH_2CH=CHCH_2OH$ and $CH_3CH_2CH_2CH_2\overset{\overset{\displaystyle O}{\|}}{C}H$

d. $CH_3CH_2CH_2CH=CHOH$ and $CH_3CH_2CH_2\overset{\overset{\displaystyle O}{\|}}{C}CH_3$

b. $CH_3\overset{\overset{\displaystyle OH}{|}}{C}HCH_3$ and $CH_3\overset{\overset{\displaystyle O}{\|}}{C}CH_3$

e. $CH_3CH_2CH_2\overset{\overset{\displaystyle OH}{|}}{C}=CH_2$ and $CH_3CH_2CH_2\overset{\overset{\displaystyle O}{\|}}{C}CH_3$

c. $CH_3CH_2CH=CHOH$ and $CH_3CH_2CH_2\overset{\overset{\displaystyle O}{\|}}{C}H$

46. How can the following compounds be prepared using ethyne as the starting material?

a. $CH_3\overset{\overset{\displaystyle O}{\|}}{C}H$

b. $CH_3CH_2\underset{\underset{\displaystyle Br}{|}}{C}HCH_2Br$

c. $CH_3\overset{\overset{\displaystyle O}{\|}}{C}CH_3$

d.  (structure: trans-alkene)

e. 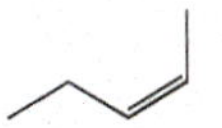 (structure: cis-alkene)

f.  (structure: alkane)

47. Do the equilibria of the following acid–base reactions lie to the right or the left? (The pK_a of H_2O_2 is 11.6.)

$$HOOH \;+\; HO^- \;\rightleftharpoons\; HOO^- \;+\; H_2O$$

$$RC\equiv CH \;+\; HOO^- \;\rightleftharpoons\; RC\equiv C^- \;+\; HOOH$$

48. What is each compound's systematic name?

a.    b. 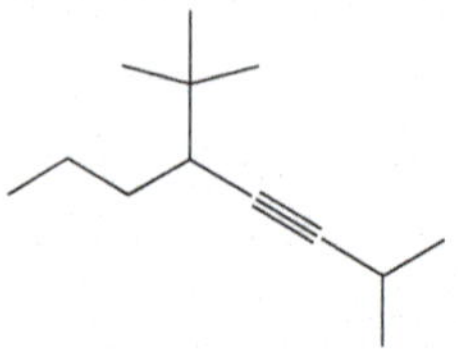   c. 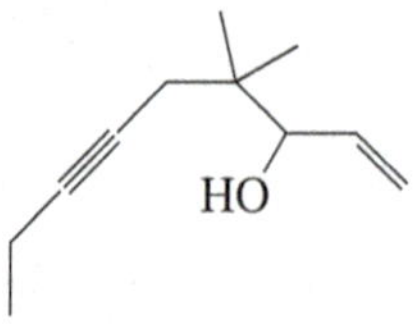 HO   d. 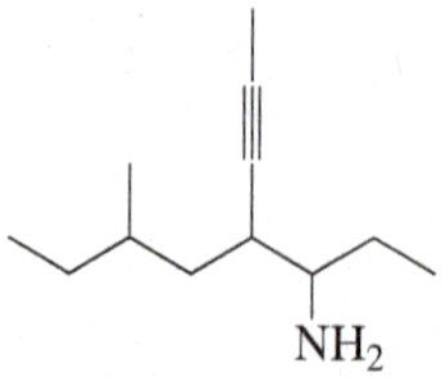 NH_2

49. What stereoisomers are obtained when 2-butyne undergoes each of the following reaction sequences?
a. 1. H_2/Lindlar catalyst 2. Br_2/CH_2Cl_2 b. 1. Na/NH_3(liq), $-78\,°C$ 2. Br_2/CH_2Cl_2 c. 1. Cl_2/CH_2Cl_2 2. Br_2/CH_2Cl_2

50. Show how the following compounds can be synthesized starting with ethyne:
a. *cis*-2-octene   b. *trans*-3-heptene

51. Draw the keto tautomer for each of the following:

a. $CH_3CH{=}\overset{\displaystyle OH}{C}CH_3$ b. $CH_3CH_2CH_2\overset{\displaystyle OH}{C}{=}CH_2$ c. 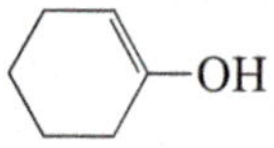 —OH   d. 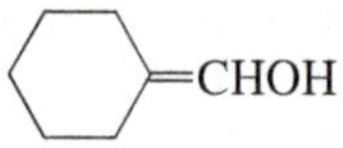 =CHOH

52. Show how each of the following compounds can be prepared using the given starting material, any needed inorganic reagents, and any organic compound that has no more than four carbons:

a. $HC\equiv CH \;\longrightarrow\; CH_3CH_2CH_2CH_2\overset{\displaystyle O}{\overset{\displaystyle \|}{C}}CH_3$

b. $HC\equiv CH \;\longrightarrow\; CH_3CH_2\underset{\displaystyle Br}{CH}CH_3$

c. $HC\equiv CH \;\longrightarrow\; CH_3CH_2CH_2\underset{\displaystyle OH}{CH}CH_3$

d.

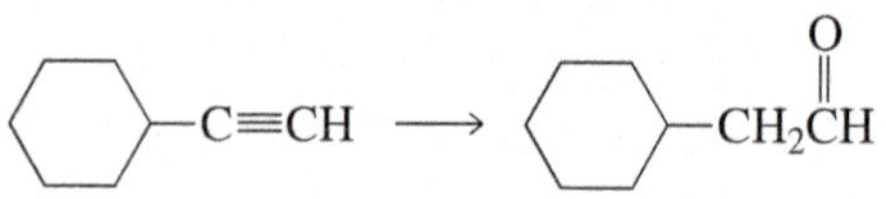

e.

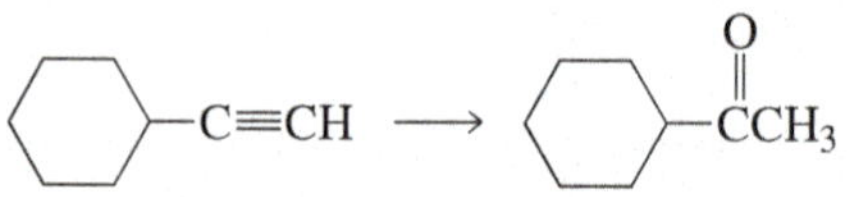

f.

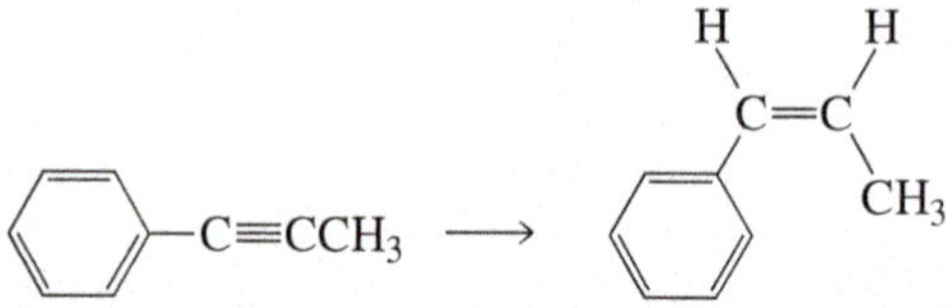

53. A chemist is planning to synthesize 3-octyne by adding 1-bromobutane to the product obtained from the reaction of 1-butyne with sodium amide. Unfortunately, however, he forgot to order 1-butyne. How else can he prepare 3-octyne?

54. a. Explain why a single pure product is obtained from hydroboration–oxidation of 2-butyne, whereas two products are obtained from hydroboration–oxidation of 2-pentyne.
b. Name two other internal alkynes that yield only one product upon hydroboration–oxidation.

55. What stereoisomers are obtained from the following reactions?

a. $CH_3CH_2C\equiv CCH_2CH_3$ $\dfrac{1.\ \mathbf{Na,\ NH_3(liq),\ -78\ °C}}{2.\ \mathbf{D_2,\ Pd/C}}$ b. $CH_3CH_2C\equiv CCH_2CH_3$ $\dfrac{1.\ \mathbf{H_2/Lindlar\ catalyst}}{2.\ \mathbf{D_2,\ Pd/C}}$

56. Explain why, in hydroboration–oxidation, HO^- and HOOH cannot be added until after the hydroboration reaction is over.

57. Starting with ethyne, describe how the following compounds can be synthesized:
a. (3S,4R)- 4-bromo-3-hexanol and (3R,4S)- 4-bromo-3-hexanol
b. (3R,4R)- 4-bromo-3-hexanol and (3S,4S)- 4-bromo-3-hexanol

8. α-Farnesene is a dodecatetraene found in the waxy coating of apple skins. What is its systematic name? Include *E* and *Z* where necessary to indicate the configuration of the double bonds.

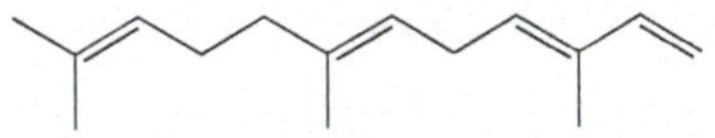

α-**farnesene**

9. Show how the following compound can be prepared from the given starting material. Draw the structure of the compound that is formed in each step of the synthesis.

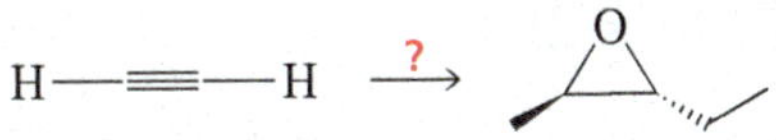

0. What are the products of the following reactions?

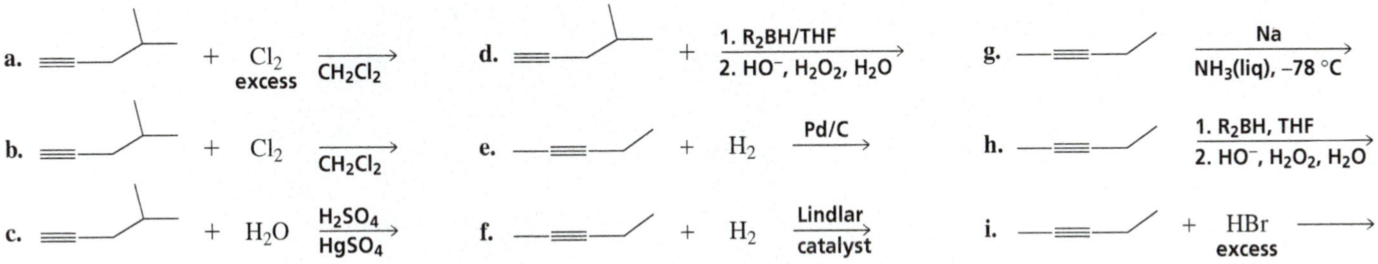

a. ≡ + Cl_2 excess $\xrightarrow{CH_2Cl_2}$

b. ≡ + Cl_2 $\xrightarrow{CH_2Cl_2}$

c. ≡ + H_2O $\xrightarrow{\substack{H_2SO_4 \\ HgSO_4}}$

d. ≡ + $\xrightarrow{\substack{1.\ R_2BH/THF \\ 2.\ HO^-,\ H_2O_2,\ H_2O}}$

e. ≡ + H_2 $\xrightarrow{Pd/C}$

f. ≡ + H_2 $\xrightarrow{\substack{Lindlar \\ catalyst}}$

g. ≡ $\xrightarrow{\substack{Na \\ NH_3(liq),\ -78\ °C}}$

h. ≡ $\xrightarrow{\substack{1.\ R_2BH,\ THF \\ 2.\ HO^-,\ H_2O_2,\ H_2O}}$

i. ≡ + HBr excess $\longrightarrow$

8 Delocalized Electrons

Their Effect on Stability, pK_a, and the Products of a Reaction

Aromaticity and Electronic Effects

An Introduction to the Reactions of Benzene

Kekulé's Dream (see p. 321)

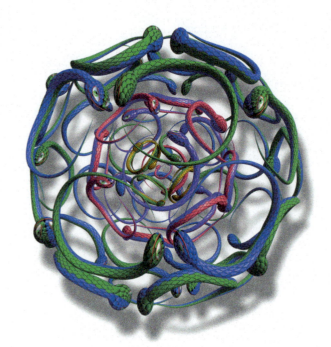

Delocalized electrons play such an important role in organic chemistry that they are part of all the remaining chapters in this book. This chapter starts by showing you how delocalized electrons are depicted. Then you will see how they affect things that are now familiar to you, such as pK_a values, the stability of carbocations, and the products formed from electrophilic addition reactions. You will also find out what it means for a compound to be aromatic and how aromaticity causes a compound with carbon–carbon double bonds to undergo substitution reactions rather than addition reactions.

Electrons that are restricted to a particular region are called **localized electrons.** Localized electrons either belong to a single atom or are shared by two atoms.

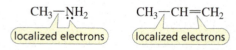

Many organic compounds have *delocalized* electrons. **Delocalized electrons** are shared by three or more atoms. You were first introduced to delocalized electrons in Section 2.8, where you saw that

the two electrons represented by the π bond of the COO⁻ group are shared by three atoms—the carbon and both oxygens. The dashed lines in the chemical structure shown here indicate that the two π electrons are delocalized over three atoms.

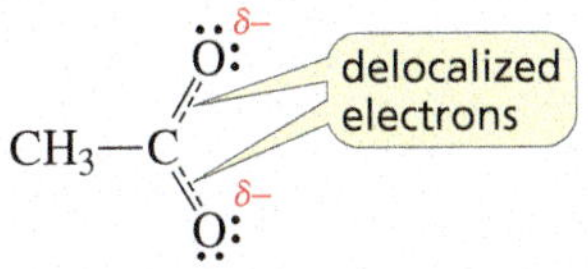

$$CH_3-C \begin{array}{c} \ddot{O}:^{\delta-} \\ \\ \ddot{O}:_{\delta-} \end{array}$$

delocalized electrons

In this chapter, you will learn how to recognize compounds that have delocalized electrons and how to draw structures that represent the electron distribution in molecules with delocalized electrons. You will also be introduced to some of the special characteristics of compounds that have delocalized electrons. Then you will be able to understand some of the wide-ranging effects that delocalized electrons have on the reactions and properties of organic compounds. We will begin by looking at benzene, a compound whose structure is ideal to illustrate the concept of delocalized electrons.

8.1 DELOCALIZED ELECTRONS EXPLAIN BENZENE'S STRUCTURE

The Puzzle of Benzene's Structure

Because early organic chemists did not know about delocalized electrons, they were puzzled by benzene's structure. They knew that benzene had a molecular formula of C_6H_6, that it was an unusually stable compound, and that it did not undergo the electrophilic addition reactions characteristic of alkenes (Section 6.0). They also knew that, when a different atom was substituted for any one of benzene's hydrogens, only *one* product was obtained and that, when the substituted product underwent a second substitution, *three* products were obtained.

$$C_6H_6 \xrightarrow[\text{with an X}]{\text{replace a hydrogen}} C_6H_5X \xrightarrow[\text{with an X}]{\text{replace a hydrogen}} C_6H_4X_2 + C_6H_4X_2 + C_6H_4X_2$$

one monosubstituted compound   three disubstituted compounds

What kind of structure would you predict for benzene if you knew only what the early chemists knew? The molecular formula (C_6H_6) tells us that benzene has eight fewer hydrogens than an acyclic alkane with six carbons ($C_nH_{2n+2} = C_6H_{14}$). Benzene, therefore, has a degree of unsaturation of four. In other words, the total number of rings and π bonds in benzene is four (Section 5.1).

Because only one product is obtained regardless of which of the six hydrogens of benzene is replaced with another atom, we know that all the hydrogens must be identical. Two structures with a degree of unsaturation of four and six identical hydrogens are shown here:

$$CH_3C\equiv C-C\equiv CCH_3$$

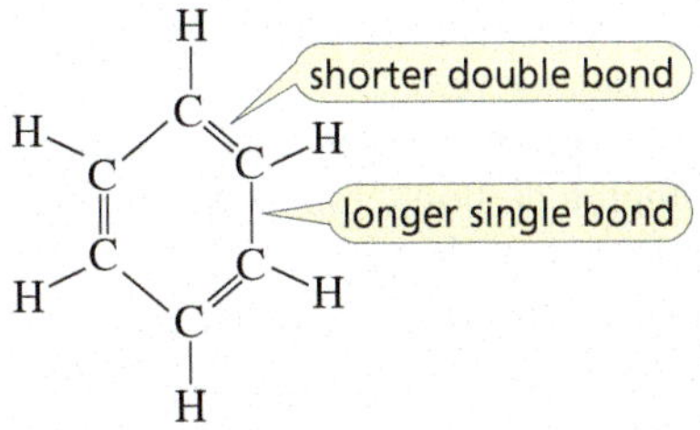

shorter double bond

longer single bond

> For every two hydrogens that are missing from the general molecular formula, C_nH_{2n+2}, a hydrocarbon has either a π bond or a ring.

Neither of these structures, however, is consistent with the observation that three compounds are obtained if a second hydrogen is replaced with another atom. The acyclic structure yields only two disubstituted products.

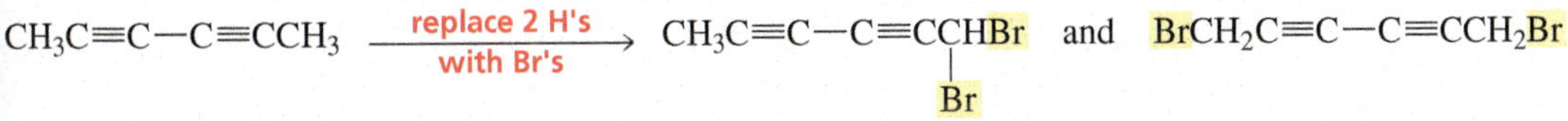

$$CH_3C\equiv C-C\equiv CCH_3 \xrightarrow[\text{with Br's}]{\text{replace 2 H's}} CH_3C\equiv C-C\equiv CCHBr \quad \text{and} \quad BrCH_2C\equiv C-C\equiv CCH_2Br$$
$$\qquad\qquad\qquad\qquad\qquad\qquad\qquad\qquad\qquad\qquad | $$
$$\qquad\qquad\qquad\qquad\qquad\qquad\qquad\qquad\qquad Br$$

The cyclic structure, with alternating single and slightly shorter double bonds, yields four disubstituted products—a 1,3-disubstituted product, a 1,4-disubstituted product, and two 1,2-disubstituted products—because the two substituents can be placed either on two adjacent carbons joined by a single bond or on two adjacent carbons joined by a double bond.

| 1,3-disubstituted product | 1,4-disubstituted product | 1,2-disubstituted product | 1,2-disubstituted product |

In 1865, the German chemist Friedrich Kekulé suggested a way of resolving this dilemma. He proposed that benzene was not a single compound, but a mixture of two compounds in rapid equilibrium.

Kekulé, Sabatier, and X-ray Diffraction Solve the Puzzle

Kekulé's proposal explained why only three disubstituted products are obtained. According to Kekulé, there actually *are* four disubstituted products, but the two 1,2-disubstituted products interconvert too rapidly to be distinguished and separated from each other.

In 1901, it was confirmed that benzene has a six-membered ring when Paul Sabatier found that catalytic hydrogenation (under extreme conditions) of benzene produced cyclohexane.

Controversy over the structure of benzene continued until the 1930s, when the new techniques of X-ray and electron diffraction (Section 14.24) produced a surprising result: they showed that *benzene is a planar molecule and that the six carbon–carbon bonds all have the same length*. The length of each carbon–carbon bond is 1.39 Å, which is shorter than a carbon–carbon single bond (1.54 Å) but longer than a carbon–carbon double bond (1.33 Å). In other words, benzene does not have alternating single and double bonds.

If the carbon–carbon bonds in benzene all have the same length, they must also have the same number of electrons between the carbons. This can be true, however, only if the π electrons are delocalized around the ring, rather than each pair of π electrons being localized between two carbons. To better understand the concept of delocalized electrons, we need to take a close look at the bonding in benzene.

Kekulé's Dream

Friedrich August Kekulé von Stradonitz (1829–1896) was born in Germany. He entered the University of Giessen to study architecture but switched to chemistry after taking a course in the subject. He was a professor of chemistry at the University of Heidelberg, at the University of Ghent in Belgium, and then at the University of Bonn. In 1890, he gave an extemporaneous speech at the twenty-fifth-anniversary celebration of his first paper on the cyclic structure of benzene. In this speech, he claimed that he had arrived at the structures as a result of dozing off in front of a fire while working on a textbook. He dreamed of chains of carbon atoms twisting and turning in a snakelike motion, when suddenly the head of one snake seized hold of its own tail and formed a spinning ring (see p. 318).

Recently, the veracity of Kekulé's snake story has been questioned by those who point out that there is no written record of the dream from the time he experienced it in 1861 until the time he related it in 1890. Others counter that dreams are not the kind of evidence one publishes in scientific papers, and it is not uncommon for scientists to experience creative ideas emerging from their subconscious at moments when they were not thinking about science. Also, Kekulé warned against publishing dreams when he said, "Let us learn to dream, and perhaps then we shall learn the truth. But let us also beware not to publish our dreams until they have been examined by the wakened mind." In 1895, he was made a nobleman by Emperor William II of Germany. This allowed him to add "von Stradonitz" to his name. Kekulé's students received three of the first five Nobel Prizes in Chemistry: van't Hoff in 1901, Fischer in 1902, and Baeyer in 1905.

Friedrich August Kekulé von Stradonitz

PROBLEM 1 ◆

a. The following compounds have the same molecular formula as benzene. How many monobrominated products could each form?

1. $HC{\equiv}CC{\equiv}CCH_2CH_3$ **2.** $CH_2{=}CHC{\equiv}CCH{=}CH_2$

b. How many dibrominated products could each of the preceding compounds form? (Do not include stereoisomers.)

c. How many dibrominated products could each of the compounds form if stereoisomers are included?

PROBLEM 2

Between 1865 and 1890, other possible structures were proposed for benzene, such as those shown here. Considering what nineteenth-century chemists knew about benzene, which is a better proposal for benzene's structure: Dewar benzene or Ladenburg benzene? Why?

Dewar benzene   **Ladenburg benzene**

8.2 THE BONDING IN BENZENE

Each of benzene's six carbons is sp^2 hybridized. An sp^2 carbon has bond angles of 120°, which is identical to the size of the angles in a planar hexagon. Thus, benzene is a planar molecule (Figure 8.1a on the next page). Because benzene is planar, the six p orbitals are parallel (Figure 8.1b) and are close enough for each p orbital to overlap the p orbital on either side of it (Figure 8.1c).

Each of the six π electrons of benzene, therefore, is localized neither on a single carbon nor in a bond between two carbons (as in an alkene). Instead, each π electron is shared by all six carbons. In other words, the six π electrons are delocalized—they roam freely within the doughnut-shaped

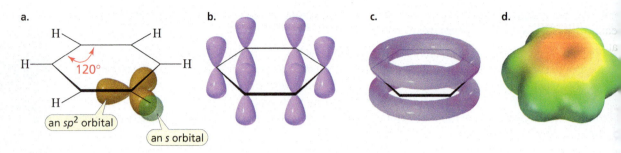

▲ **Figure 8.1**

a. Each of the carbons in benzene uses two *sp*² orbitals to bond to two other carbons; its third *sp*² orbital overlaps the *s* orbital of a hydrogen.

b. Each carbon has a *p* orbital at right angles to the *sp*² orbitals. The parallel *p* orbitals are close enough for side-to-side overlap, so each *p* orbital overlaps the *p* orbitals on *both* adjacent carbons.

c. The overlapping *p* orbitals form two continuous doughnut-shaped clouds of electrons—one above the plane of the benzene ring and one below it.

d. The electrostatic potential map shows that all the carbon–carbon bonds have the same electron density.

clouds that lie above and below the planar ring of carbons (Figure 8.1c and d). Benzene is often drawn as a hexagon containing either dashed lines or a circle to symbolize the six delocalized π electrons.

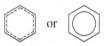

This type of representation makes it clear that there are no double bonds in benzene. Kekulé's structure was very nearly correct. The actual structure of benzene is Kekulé's structure with delocalized electrons.

8.3 RESONANCE CONTRIBUTORS AND THE RESONANCE HYBRID

A disadvantage to using dashed lines (or a circle) to represent delocalized electrons is that they do not tell us how many π electrons they represent. For example, the dashed lines inside the hexagon shown above indicate that the π electrons are shared equally by all six carbons and that all the carbon–carbon bonds have the same length, but they do not show how many π electrons are in the ring. Consequently, chemists prefer to use structures that portray the electrons as localized (and therefore, show the number of π electrons), even though the electrons in the compound's actual structure are delocalized.

The *approximate* structure with localized electrons is called a **resonance contributor,** a **resonance structure,** or a **contributing resonance structure.** The *actual* structure with delocalized electrons is called a **resonance hybrid.** We can easily see that there are six π electrons in each of benzene's resonance contributors.

NOTE TO THE STUDENT
• Electron delocalization is shown by double-headed arrows (⟷), whereas equilibrium is shown by two arrows pointing in opposite directions (⇌).

resonance contributor ⟷ resonance contributor

resonance hybrid

Resonance contributors are shown with a double-headed arrow between them. The double-headed arrow does *not* mean that the structures are in equilibrium with one another. Rather, it indicates that the actual structure lies somewhere *between* the structures of the resonance contributors. Resonance contributors are merely a convenient way to show the π electrons; they do not represent any real distribution of electrons.

The Difference Between a Resonance Contributor and a Resonance Hybrid

The following analogy illustrates the difference between resonance contributors and the resonance hybrid. Imagine that you are trying to describe to a friend what a rhinoceros looks like. You might tell your friend that a rhinoceros looks like a cross between a unicorn and a dragon. Like resonance contributors, the unicorn and the dragon do not really exist. Furthermore, like resonance contributors, they are not in equilibrium: a rhinoceros does not change back and forth between the two forms, looking like a unicorn one minute and a dragon the next. The unicorn and dragon are simply ways to describe what the actual animal—the rhinoceros—looks like. *Resonance contributors, like unicorns and dragons, are imaginary. Only the resonance hybrid, like the rhinoceros, is real.*

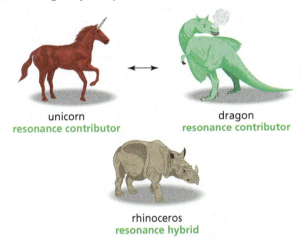

unicorn
resonance contributor

dragon
resonance contributor

rhinoceros
resonance hybrid

Electron delocalization is most effective if all the atoms sharing the delocalized electrons lie in the same plane so that their *p* orbitals can maximally overlap. For example, the electrostatic potential map shows that cyclooctatetraene is tub-shaped, not planar—its sp^2 carbons have bond angles of 120°, whereas a planar eight-membered ring would have bond angles of 135°. Because the ring is not planar, a *p* orbital can overlap with one adjacent *p* orbital, but it cannot overlap with the other adjacent *p* orbital. As a result, the eight π electrons are localized in four double bonds and not delocalized over the entire eight-membered ring. Thus, the carbon–carbon bonds do not all have the same length.

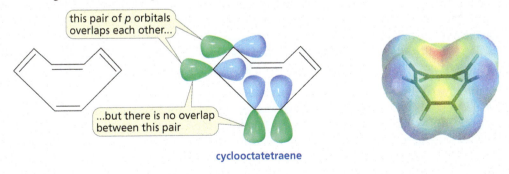

this pair of *p* orbitals overlaps each other...

...but there is no overlap between this pair

cyclooctatetraene

8.4 HOW TO DRAW RESONANCE CONTRIBUTORS

You just learned that an organic compound with delocalized electrons is generally represented as a structure with localized electrons to let us know how many π electrons it has. For example, nitroethane is usually represented with a nitrogen–oxygen double bond and a nitrogen–oxygen single bond.

$$CH_3CH_2—\overset{+}{N}\diagdown\begin{smallmatrix}:\ddot{O}\\ \\ :\ddot{O}:^-\end{smallmatrix}$$

nitroethane

However, the two nitrogen–oxygen bonds in nitroethane actually have the same length. Thus, a more accurate description of the molecule's structure is obtained by drawing the two resonance contributors. Both resonance contributors show the compound with a nitrogen–oxygen double bond and a nitrogen–oxygen single bond; they indicate that the electrons are delocalized by depicting the double bond in one contributor as a single bond in the other.

$$CH_3CH_2—\overset{+}{N}\qquad\longleftrightarrow\qquad CH_3CH_2—\overset{+}{N}\qquad\qquad CH_3CH_2—\overset{+}{N}$$

resonance contributor   **resonance contributor**   **resonance hybrid**

The resonance hybrid shows that the two π electrons are shared by three atoms. The resonance hybrid also shows that the two nitrogen–oxygen bonds are identical and that the negative charge is shared equally by both oxygens. Thus, we need to visualize and mentally average both resonance contributors to appreciate what the actual molecule—the resonance hybrid—looks like.

Notice that

> *delocalized electrons result from a* p *orbital overlapping*
> *the* p *orbitals of two adjacent atoms.*

For example, in nitroethane, the *p* orbital of nitrogen overlaps the *p* orbital of each of two adjacent oxygens; in the carboxylate ion on p. 66, the *p* orbital of carbon overlaps the *p* orbital of each of two adjacent oxygens; and in benzene, the *p* orbital of carbon overlaps the *p* orbital of each of two adjacent carbons.

Delocalized electrons result from a *p* **orbital overlapping the** *p* **orbitals of two adjacent atoms.**

Rules for Drawing Resonance Contributors

LEARN THE STRATEGY To draw a set of resonance contributors for a molecule, first draw a Lewis structure. This is the first resonance contributor. Then, following the rules listed below, move electrons to generate the next resonance contributor.

- Only electrons move. Atoms never move.
- Only π electrons (electrons in π bonds) and lone-pair electrons can move. (Never move σ electrons.)
- The total number of electrons in the molecule does not change. Therefore, each of the resonance contributors for a particular compound must have the same net charge. If one has a net charge of 0, all the others must also have net charges of 0. (A net charge of 0 does not necessarily mean that there is no charge on any of the atoms, because a molecule with a positive charge on one atom and a negative charge on another atom has a net charge of 0.)

As you study the following resonance contributors and practice drawing them, notice that electrons (π electrons or lone pairs) always move to an sp^2 or sp atom. (Remember that an sp^2 carbon is either a positively charged carbon or a doubly bonded carbon and that an sp carbon is generally a triply bonded carbon; Sections 1.8, 1.9, and 1.10.) Electrons cannot move to an sp^3 carbon because an sp^3 carbon has a complete octet and does not have a π bond that can break. Therefore, it cannot accommodate any more electrons.

Example 1

The carbocation shown at the top of the next page has delocalized electrons. To draw its resonance contributor, *move the* π *electrons to an* sp^2 *carbon*. The curved arrow shows you how to draw the second resonance contributor. Remember that the tail of the curved arrow shows where the electrons start from and the head shows where the electrons end up. The resonance hybrid shows that the π electrons are shared by three carbons and that the positive charge is shared by two carbons.

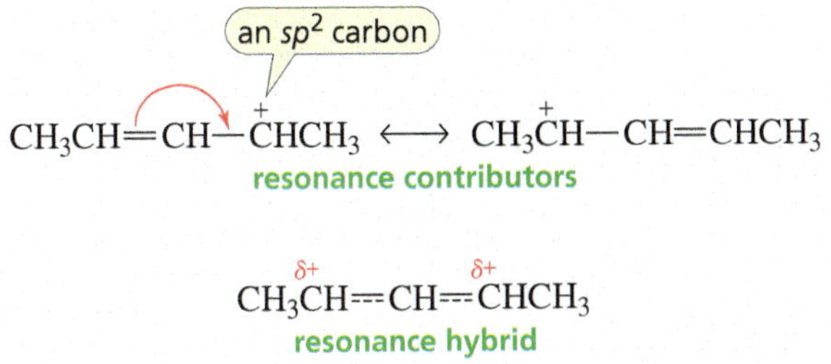

$$CH_3CH=CH-\overset{+}{C}HCH_3 \longleftrightarrow CH_3\overset{+}{C}H-CH=CHCH_3$$

an sp^2 carbon

resonance contributors

To draw resonance contributors, move only π electrons or lone pairs toward an sp^2 (or sp) carbon.

$$CH_3\overset{\delta+}{CH}=\!=\!=CH=\!=\!=\overset{\delta+}{C}HCH_3$$

resonance hybrid

Let's compare this carbocation with a similar compound in which all the electrons are localized. The π electrons in the carbocation shown below cannot move, because the carbon they would move to is an sp^3 carbon (sp^3 carbons cannot accept any more electrons).

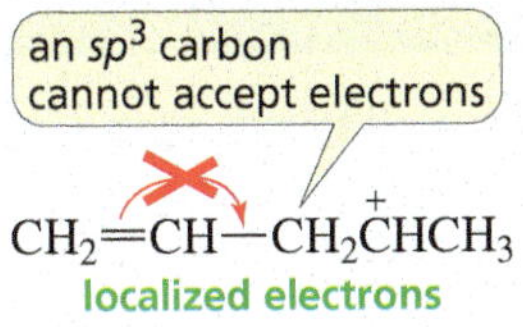

an sp^3 carbon cannot accept electrons

$$CH_2=CH-CH_2\overset{+}{C}HCH_3$$

localized electrons

Example 2

In the next example, *π electrons again move to an* sp^2 *carbon*. The resonance hybrid shows that the π electrons are shared by five carbons and that positive charge is shared by three carbons. The resonance hybrid is obtained by mentally combining the three resonance contributors.

an sp^2 carbon

$$CH_3CH=CH-CH=CH-\overset{+}{C}H_2 \longleftrightarrow CH_3CH=CH-\overset{+}{C}H-CH=CH_2 \longleftrightarrow CH_3\overset{+}{C}H-CH=CH-CH=CH_2$$

resonance contributors

$$CH_3\overset{\delta+}{CH}=\!=\!=CH=\!=\!=\overset{\delta+}{CH}=\!=\!=CH=\!=\!=\overset{\delta+}{C}H_2$$

resonance hybrid

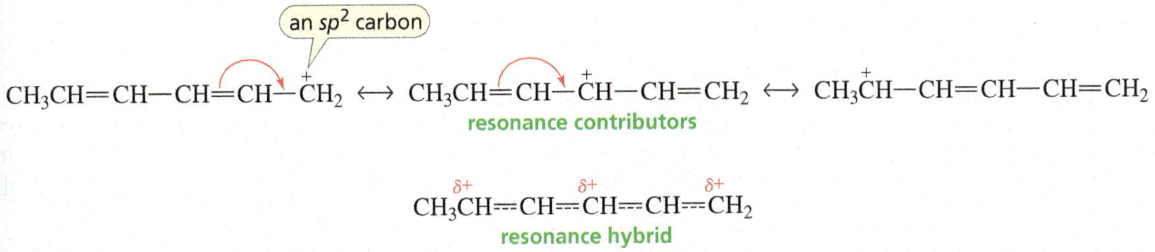

Example 3

The resonance contributor for the next compound is obtained by *moving lone-pair electrons to an* sp^2 *carbon*. The sp^2 carbon can accommodate the new electrons by breaking a π bond. The lone-pair electrons in the compound on the far right are not delocalized because they would have to move toward an sp^3 carbon.

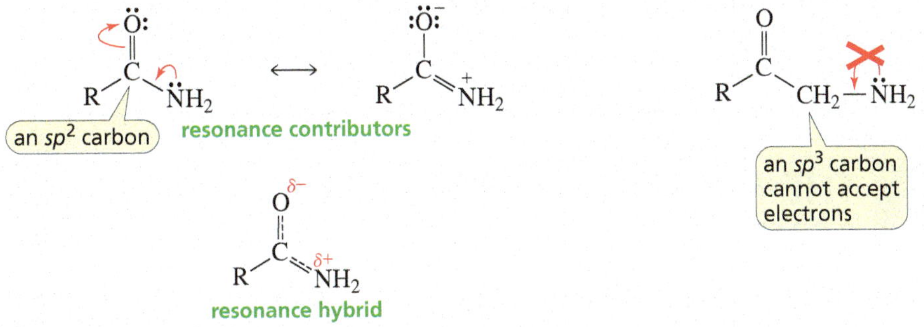

$$\underset{\underset{\text{an } sp^2 \text{ carbon}}{}}{\overset{\overset{..}{O}:}{\underset{R}{\overset{|}{C}}\overset{}{\underset{..}{N}H_2}}} \longleftrightarrow \underset{R}{\overset{:\overset{..}{O}:^-}{\underset{}{\overset{|}{C}}\overset{+}{\underset{}{N}H_2}}}$$

resonance contributors

$$\underset{R}{\overset{O}{\underset{}{\overset{\|}{C}}\underset{CH_2-\overset{..}{N}H_2}{}}}$$

an sp^3 carbon cannot accept electrons

$$\underset{R}{\overset{\overset{\delta-}{O}}{\underset{}{\overset{\|}{C}}\overset{\delta+}{\underset{}{N}H_2}}}$$

resonance hybrid

Example 4

The resonance contributor for the next compound is obtained by *moving electrons to an* sp *carbon*.

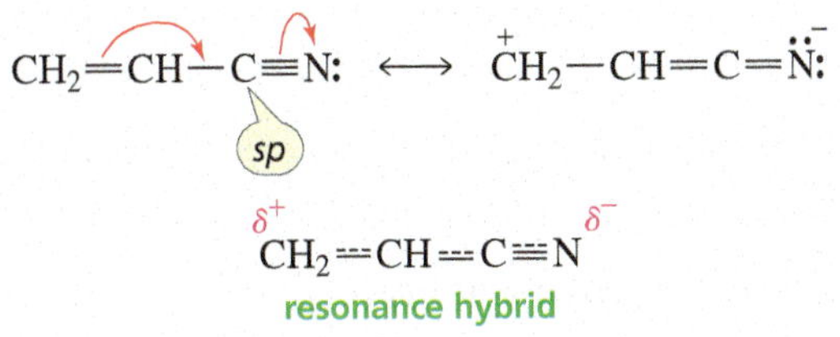

$$CH_2=CH-C\equiv N: \longleftrightarrow \overset{+}{C}H_2-CH=C=\overset{..}{N}:^-$$

sp

$$\overset{\delta+}{C}H_2=\!=\!=CH=\!=\!=C\equiv N^{\delta-}$$

resonance hybrid

PROBLEM 3

a. Which of the following compounds have delocalized electrons?
b. Draw resonance contributors for the compounds that have delocalized electrons.

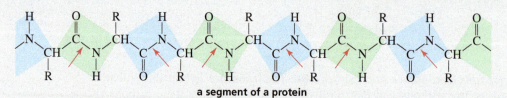

1. $CH_2{=}CHCH_2CH{=}CH_2$
2. $CH_3CH{=}CHCH{=}\overset{+}{CH_2}$
3. $CH_3CH_2\overset{\cdot\cdot}{N}HCH_2CH{=}CH_2$

4.
5.
6.
7.

Electron Delocalization Affects the Three-Dimensional Shape of Proteins

A protein consists of amino acids joined together by peptide bonds. Every third bond in a protein is a peptide bond, as indicated by the red arrows.

a segment of a protein

A resonance contributor can be drawn for a peptide bond by moving the lone pair on nitrogen toward the sp^2 carbon. The second resonance contributor shows that a peptide bond has partial double bond character.

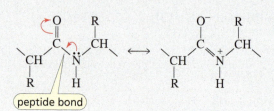

Because of the partial double-bond character of the peptide bond, the carbon and nitrogen atoms and the two atoms bonded to each of them are held rigidly in a plane, as represented in the protein segment by the blue and green boxes. Despite the rigid orientation of the peptide bond, the single bonds in the protein chain are free to rotate. Because of this, the chain is free to fold into a myriad of complex and highly intricate shapes. (Two conceptual representations of proteins are shown here; see Figure 21.10 on p. 1021).

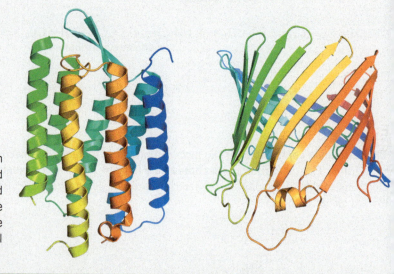

8.5 THE PREDICTED STABILITIES OF RESONANCE CONTRIBUTORS

All resonance contributors do not necessarily contribute equally to the resonance hybrid. The degree to which each one contributes depends on its predicted stability. Because resonance contributors are not real, their stabilities cannot be measured. The stabilities of resonance contributors must be predicted based on molecular features found in real molecules.

The greater the predicted stability of a resonance contributor, the more it contributes to the structure of the resonance hybrid.

The more the resonance contributor contributes to the structure of the resonance hybrid, the more similar the contributor is to the real molecule.

The examples that follow illustrate these points.

Example 1

The two resonance contributors for a carboxylic acid are shown in Figure 8.2. **B** has two features that cause it to be predicted less stable than **A**:

- One of the oxygens has a positive charge—not a stable situation for an electronegative atom.
- It has separated charges.

A molecule with **separated charges** has a positive charge and a negative charge that can be neutralized by the movement of electrons. Resonance contributors with separated charges are relatively unstable (relatively high in energy) because energy is required to keep the opposite charges separated. **A**, therefore, is predicted to be more stable than **B.** Consequently, **A** makes a greater contribution to the resonance hybrid, so the resonance hybrid looks more like **A** than like **B.**

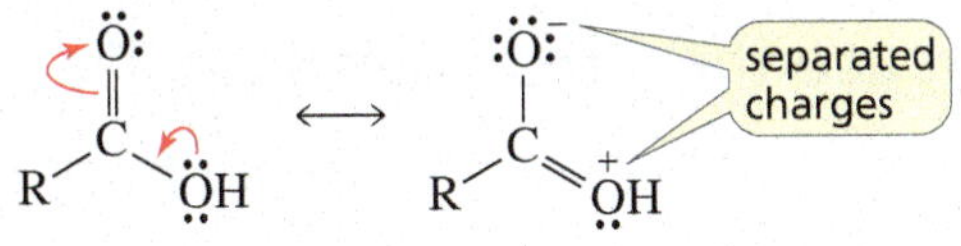

a carboxylic acid

◀ **Figure 8.2**
B is predicted to be less stable than A because B has separated charges and one of its oxygens has a positive charge.

Example 2

The two resonance contributors for a carboxylate ion are shown in Figure 8.3. **C** and **D** are predicted to be equally stable, so they contribute equally to the resonance hybrid.

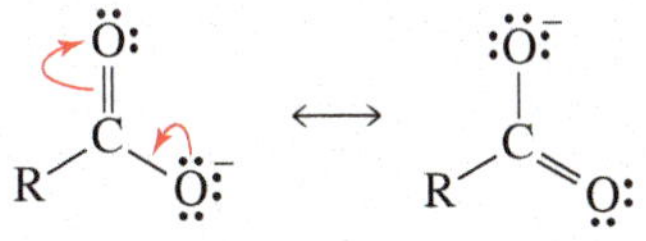

a carboxylate ion

◀ **Figure 8.3**
C and D are predicted to be equally stable. Therefore, they contribute equally to the resonance hybrid.

Example 3

When electrons can be moved in more than one direction, the most stable resonance contributor is obtained by moving the electrons *toward* the more electronegative atom. For example, **F** in Figure 8.4 results from moving the π electrons toward oxygen—the most electronegative atom in the molecule. **F** is predicted to be less stable than **E**:

- **F** has separated charges and
- one of its carbons has an incomplete octet.

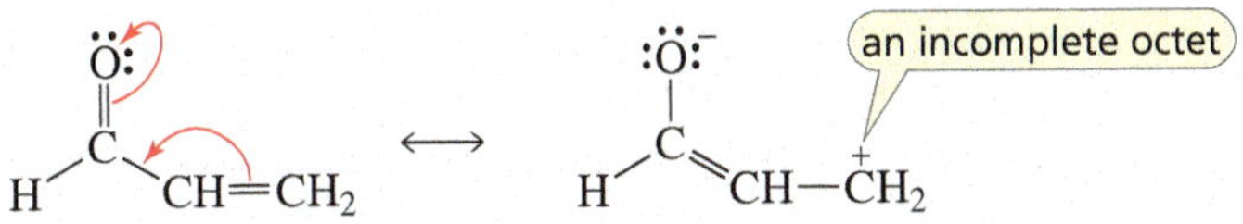

◀ **Figure 8.4**
F is obtained by moving the electrons toward oxygen, the most electronegative atom in the molecule. F is predicted to be less stable than E because F has separated charges and an incomplete octet. Therefore, E makes the greater contribution to the hybrid.

Example 4

The only time you need to show a resonance contributor that is obtained by moving electrons *away* from the most electronegative atom is when that is the only way the electrons can move. In other words, movement of electrons away from the most electronegative atom is better than no movement at all because electron delocalization makes a molecule more stable (as we will see in Section 8.6). For example, the only resonance contributor that can be drawn for **G** in Figure 8.5 results from movement of the electrons away from oxygen. **H** is predicted to be less stable than **G**.

◀ **Figure 8.5**
H is predicted to be relatively unstable because it has separated charges and its oxygen has a positive charge. Therefore, the structure of the resonance hybrid is similar to G, with only a small contribution from H.

Example 5

Let's now see which of the resonance contributors shown in Figure 8.6 has a greater predicted stability. **I** has a negative charge on carbon, whereas **J** has a negative charge on oxygen; oxygen can better accommodate the negative charge (because it is more electronegative than carbon), so **J** is predicted to be more stable than **I**.

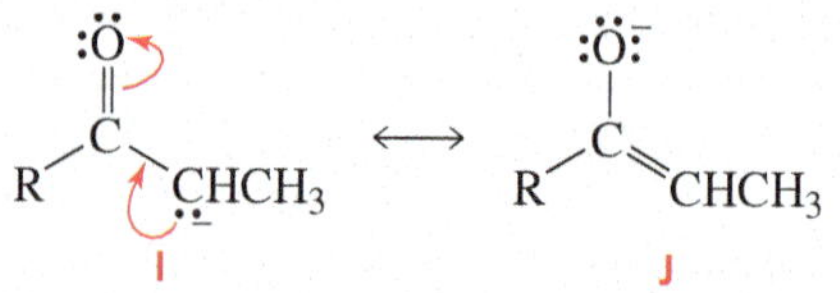

▶ **Figure 8.6**
The resonance hybrid more closely resembles J; that is, the resonance hybrid has a greater concentration of negative charge on the oxygen than on the carbon.

Features That Decrease the Predicted Stability

We can summarize the features that decrease the predicted stability of a resonance contributor as follows:

1. an atom with an incomplete octet
2. a negative charge that is not on the most electronegative atom
3. a positive charge that is on an electronegative atom
4. charge separation

When we compare the relative stabilities of resonance contributors, an atom with an incomplete octet (feature 1 in the preceding list) generally makes a structure more unstable than the other three features do.

LEARN THE STRATEGY

PROBLEM-SOLVING STRATEGY

Determining Relative Stabilities

Which carbocation is more stable?

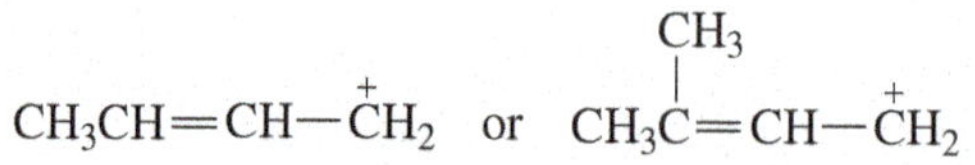

$$CH_3CH=CH-\overset{+}{C}H_2 \quad \text{or} \quad CH_3\overset{\underset{\displaystyle CH_3}{|}}{C}=CH-\overset{+}{C}H_2$$

Start by drawing the resonance contributors for each carbocation.

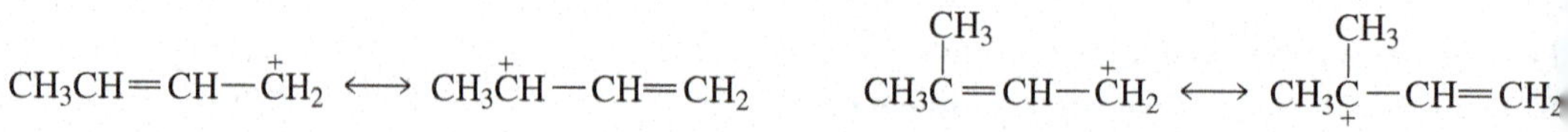

$$CH_3CH=CH-\overset{+}{C}H_2 \longleftrightarrow CH_3\overset{+}{C}H-CH=CH_2 \qquad CH_3\overset{\underset{\displaystyle CH_3}{|}}{C}=CH-\overset{+}{C}H_2 \longleftrightarrow CH_3\overset{\underset{\displaystyle CH_3}{|}}{\underset{+}{C}}-CH=CH_2$$

Now look at how the two sets of resonance contributors differ and think about how those differences affect the relative stabilities of the two resonance hybrids.

Each carbocation has two resonance contributors. The positive charge of the carbocation on the left is shared by a primary carbon and a secondary carbon. The positive charge of the carbocation on the right is shared by a primary carbon and a tertiary carbon. Because a tertiary carbon is more stable than a secondary carbon (Section 6.2), the carbocation on the right is more stable.

USE THE STRATEGY

PROBLEM 4 ◆

Which species in each pair is more stable?

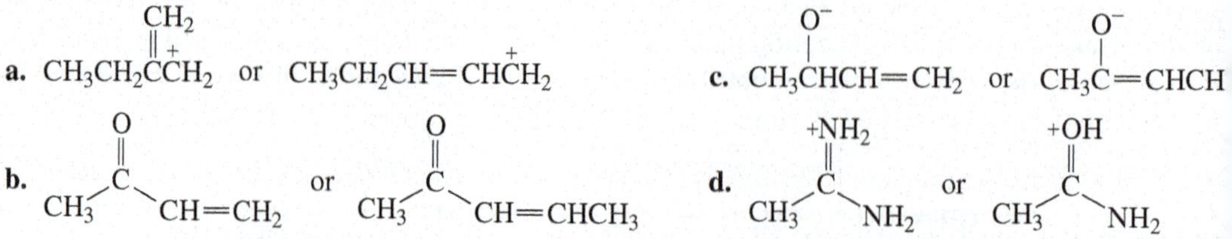

a. $CH_3CH_2\overset{\underset{\displaystyle \overset{\displaystyle CH_2}{||}}{}}{\overset{+}{C}}CH_2$ or $CH_3CH_2CH=\overset{+}{C}HCH_2$

b. $CH_3-\overset{\underset{\displaystyle CH=CH_2}{}}{\overset{\displaystyle O}{||}}{C}$ or $CH_3-\overset{\underset{\displaystyle CH=CHCH_3}{}}{\overset{\displaystyle O}{||}}{C}$

c. $CH_3\overset{\underset{\displaystyle }{}}{C}HCH=CH_2$ or $CH_3\overset{\underset{\displaystyle }{}}{C}=CHCH$ (with O^- groups)

d. $CH_3-\overset{\underset{\displaystyle NH_2}{}}{\overset{\displaystyle +NH_2}{||}}{C}$ or $CH_3-\overset{\underset{\displaystyle NH_2}{}}{\overset{\displaystyle +OH}{||}}{C}$

PROBLEM 5 SOLVED

Draw resonance contributors for the following species and rank them in order of decreasing contribution to the resonance hybrid. Then draw the resonance hybrid.

$$CH_3\overset{+}{C}-CH=CHCH_3$$
$$\underset{CH_3}{|}$$

SOLUTION   **A** is more stable than **B** because the positive charge is on a tertiary carbon in **A**, whereas it is on a secondary carbon in **B**, and a tertiary carbocation is more stable than a secondary carbocation (Section 6.2).

$$CH_3\overset{+}{C}-CH=CHCH_3 \longleftrightarrow CH_3C=CH-\overset{+}{C}HCH_3$$
$$\underset{CH_3}{|} \qquad\qquad \underset{CH_3}{|}$$
$$\quad\text{A} \qquad\qquad\qquad\qquad \text{B}$$

Draw the resonance hybrid by representing each bond that is a double bond in one contributor and a single bond in the other as a single bond plus a dashed bond. The positive charge in each contributor is represented by a partial positive charge in the hybrid.

$$CH_3\overset{\delta+}{C}==CH==\overset{\delta+}{C}HCH_3$$
$$\underset{CH_3}{|}$$

PROBLEM 6

Draw resonance contributors for each of the following species and rank them in order of decreasing contribution to the resonance hybrid. Then draw the resonance hybrid.

a.
$$\underset{CH_3}{}\overset{:\overset{..}{O}:}{\underset{}{\overset{||}{C}}}\overset{}{\underset{\ddot{O}CH_3}{}}$$

b. $CH_3\overset{..}{N}H-CH=CHCH_3$

c.
cyclohexadienone ring $=\overset{..}{\underset{..}{O}}:$ with negative charge

d.
cyclohexadienone ring $=\overset{..}{\underset{..}{O}}:$

e.
$$\underset{CH_3}{}\overset{+\overset{..}{O}H}{\underset{}{\overset{||}{C}}}\overset{}{\underset{\overset{..}{N}HCH_3}{}}$$

f. $CH_3\overset{+}{C}H-CH=CHCH_3$

LEARN THE STRATEGY

The greater the predicted stability of a resonance contributor, the more it contributes to the structure of the resonance hybrid.

The more a resonance contributor contributes to the structure of the resonance hybrid, the more similar that contributor is to the real molecule.

USE THE STRATEGY

NOTE TO THE STUDENT

• The tutorial on p. 382 will give you additional practice drawing resonance contributors and predicting their relative stabilities.

8.6 DELOCALIZATION ENERGY IS THE ADDITIONAL STABILITY DELOCALIZED ELECTRONS GIVE TO A COMPOUND

Delocalized electrons stabilize a compound. The extra stability a compound gains from having delocalized electrons is called the **delocalization energy. Electron delocalization** is also called **resonance,** so delocalization energy is also called **resonance energy.** Because delocalized electrons increase the stability of a compound, we can conclude that:

a resonance hybrid is more stable than any of its resonance contributors is predicted to be.

The delocalization energy associated with a compound that has delocalized electrons depends on the number *and* the predicted stability of the resonance contributors.

The greater the number of relatively stable resonance contributors, the greater the delocalization energy.

For example, the delocalization energy of a carboxylate ion with two relatively stable resonance contributors is significantly greater than the delocalization energy of a carboxylic acid with only one relatively stable resonance contributor.

relatively stable   relatively unstable   relatively stable   relatively stable
resonance contributors of   resonance contributors of
a carboxylic acid   a carboxylate ion

The delocalization energy is a measure of how much more stable a compound with delocalized electrons is than it would be if its electrons were localized.

The greater the number of relatively stable resonance contributors, the greater the delocalization energy.

The fact that the two resonance contributors of the carboxylate ion are equivalent also contribute to its delocalization energy.

> *The more nearly equivalent the structures of the resonance contributors, the greater the delocalization energy.*

Notice that it is the number of *relatively stable* resonance contributors—not the total number of resonance contributors—that is important in determining the delocalization energy.

For example, the delocalization energy of a carboxylate ion with two relatively stable resonance contributors is greater than the delocalization energy of the following compound with three resonance contributors because only one of its resonance contributors is relatively stable:

$$\overset{-}{C}H_2-CH\!=\!CH-\overset{+}{C}H_2 \quad\longleftrightarrow\quad CH_2\!=\!CH-CH\!=\!CH_2 \quad\longleftrightarrow\quad \overset{+}{C}H_2-CH\!=\!CH-\overset{-}{C}H_2$$

relatively unstable   relatively stable   relatively unstable

Let's now summarize what we know about resonance contributors:

- The greater the predicted stability of a resonance contributor, the more it contributes to the resonance hybrid.
- The greater the number of relatively stable resonance contributors and the more nearly equivalent their structures, the greater the delocalization energy.

PROBLEM 7 ◆

a. Predict the relative bond lengths of the three carbon–oxygen bonds in the carbonate ion.

b. What is the charge on each oxygen?

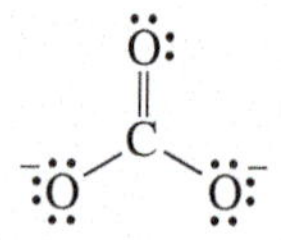

carbonate ion

PROBLEM 8

List the following in order of decreasing delocalization energy:

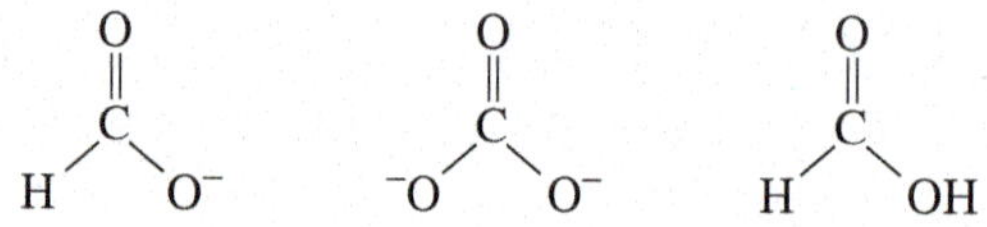

PROBLEM 9

Which has the greater delocalization energy?

$$CH_2\!=\!CH-CH\!=\!CH_2 \quad \text{or}$$

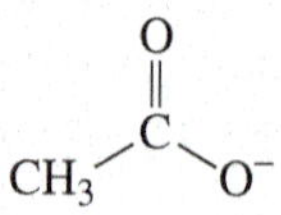

PROBLEM 10 ◆

Which has the greater delocalization energy?

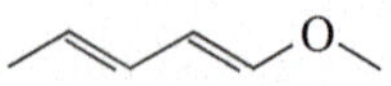

or

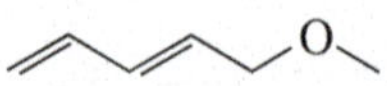

The more nearly equivalent the structures of the resonance contributors, the greater the delocalization energy.

8.7   DELOCALIZED ELECTRONS INCREASE STABILITY

We will now look at two examples that illustrate the extra stability a molecule acquires as a result of having delocalized electrons.

Stability of Dienes

Dienes are hydrocarbons with two double bonds.

- **Isolated dienes** have isolated double bonds; **isolated double bonds** are separated by more than one single bond.
- **Conjugated dienes** have conjugated double bonds; **conjugated double bonds** are separated by one single bond.

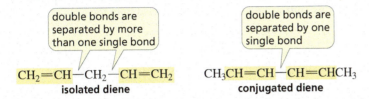

We saw in Section 5.9 that the relative stabilities of alkenes can be determined by their heats of hydrogenation. Recall that the most stable alkene has the smallest heat of hydrogenation; it gives off the least heat when it is hydrogenated because it has less energy to begin with.

Conjugated Dienes

The heat of hydrogenation of 1,3-pentadiene (a conjugated diene) is smaller than that of 1,4-pentadiene (an isolated diene). *A conjugated diene, therefore, is more stable than an isolated diene.*

	$\Delta H°$ (kcal/mol)	Heat of hydrogenation

$$CH_2{=}CH{-}CH_2{-}CH{=}CH_2 \ + \ 2\,H_2 \ \xrightarrow{\text{Pd/C}} \ CH_3CH_2CH_2CH_2CH_3 \qquad -60.2 \qquad 60.2 \text{ kcal/mol}$$

1,4-pentadiene
isolated diene

The most stable alkene has the smallest heat of hydrogenation.

$$CH_2{=}CH{-}CH{=}CH{-}CH_3 \ + \ 2\,H_2 \ \xrightarrow{\text{Pd/C}} \ CH_3CH_2CH_2CH_2CH_3 \qquad -54.1 \qquad 54.1 \text{ kcal/mol}$$

1,3-pentadiene
conjugated diene

Why is a conjugated diene more stable than an isolated diene? Two factors contribute to the difference. The first is *electron delocalization*. The π electrons in each of the double bonds of an isolated diene are *localized* between two carbons. In contrast, the π electrons in a conjugated diene are *delocalized* (Figure 8.7) and electron delocalization stabilizes a compound.

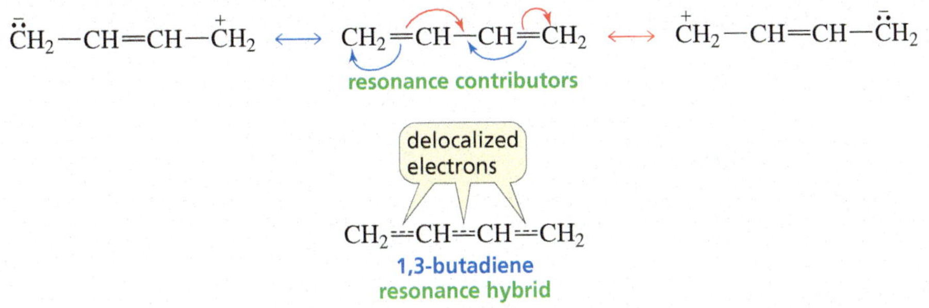

$$\overset{..}{\overset{-}{C}}H_2{-}CH{=}CH{-}\overset{+}{C}H_2 \longleftrightarrow CH_2{=}CH{-}CH{=}CH_2 \longleftrightarrow \overset{+}{C}H_2{-}CH{=}CH{-}\overset{..}{\overset{-}{C}}H_2$$

resonance contributors

delocalized electrons

$$CH_2{\cdots}CH{\cdots}CH{\cdots}CH_2$$

1,3-butadiene
resonance hybrid

An increase in delocalization energy means an increase in stability.

▲ **Figure 8.7**
The resonance hybrid shows that the single bond in a conjugated diene such as 1,3-butadiene is not a pure single bond, but has partial double-bond character due to electron delocalization. Notice that because the compound does *not* have an electronegative atom to determine the direction in which the electrons move, they can move both to the left (indicated by the blue arrows) and to the right (indicated by the red arrows).

The second factor is the *hybridization of the orbitals* that form the carbon–carbon single bonds. Because a $2s$ electron is closer to the nucleus, on average, than a $2p$ electron, a bond formed by sp^2–sp^2 overlap is *shorter* and *stronger* than one formed by sp^3–sp^2 overlap. This is because an sp^2 orbital has more s character than an sp^3 orbital (Section 1.15 and Table 8.1). Thus, one of the single bonds in a conjugated diene is a stronger single bond than those in an isolated diene, and stronger bonds cause a compound to be more stable (Figure 8.8).

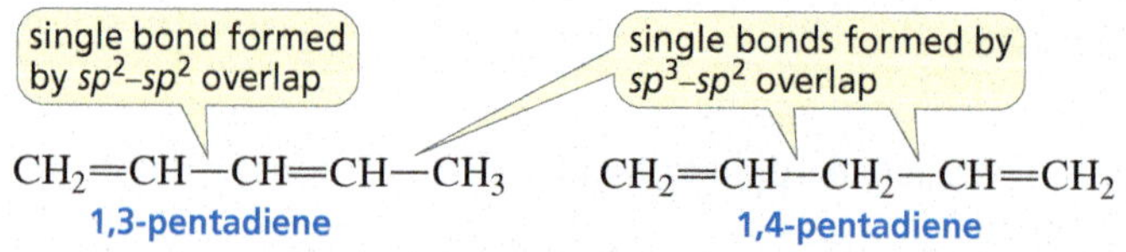

single bond formed
by sp^2–sp^2 overlap

single bonds formed by
sp^3–sp^2 overlap

$$CH_2=CH-CH=CH-CH_3 \qquad CH_2=CH-CH_2-CH=CH_2$$
1,3-pentadiene   **1,4-pentadiene**

▲ **Figure 8.8**
One carbon–carbon single bond in 1,3-pentadiene is formed from the overlap of an sp^2 orbital with another
sp^2 orbital, and the other is formed from the overlap of an sp^3 orbital with an sp^2 orbital. In contrast,
both carbon–carbon single bonds in 1,4-pentadiene are formed from the overlap of an sp^3 orbital with an
sp^2 orbital.

Table 8.1	Dependence of the Length of a Carbon–Carbon Single Bond on the Hybridization of the Orbitals Used in Its Formation	
Compound	**Hybridization**	**Bond length (Å)**
H_3C-CH_3	sp^3–sp^3	1.54
$H_3C-\overset{\overset{\displaystyle H}{\mid}}{C}=CH_2$	sp^3–sp^2	1.50
$H_2C=\overset{\overset{\displaystyle H}{\mid}}{C}-\overset{\overset{\displaystyle H}{\mid}}{C}=CH_2$	sp^2–sp^2	1.47
$H_3C-C\equiv CH$	sp^3–sp	1.46
$H_2C=\overset{\overset{\displaystyle H}{\mid}}{C}-C\equiv CH$	sp^2–sp	1.43
$HC\equiv C-C\equiv CH$	sp–sp	1.37

Allenes

Allenes are compounds that have **cumulated double bonds.** These are double bonds that are adjacent to one another. The cumulated double bonds give allenes an unusual geometry because the central carbon is sp hybridized, just like the central carbon in CO_2. The sp carbon forms two π bonds by overlapping one of its p orbitals with a p orbital of an adjacent sp^2 carbon and overlapping its second p orbital with a p orbital of the other adjacent sp^2 carbon (Figure 8.9a).

a.   **b.**

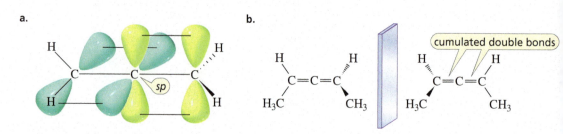

▲ **Figure 8.9**
a. The two p orbitals on the central carbon are perpendicular, so the plane containing one H—C—H group is perpendicular to the plane containing the other H—C—H group. As a result, an allene is a nonplanar molecule.
b. 2,3-Pentadiene has a nonsuperimposable mirror image, which causes it to be a chiral molecule even though it does not have an asymmetric center (see Problem 103 on p. 186).

Organic Compounds That Conduct Electricity

For an organic compound to conduct electricity, its electrons must be delocalized so they can move through the compound just like electrons can move along a copper wire. The first organic compound able to conduct electricity was prepared by hooking together a large number of acetylene molecules to form a compound called polyacetylene—a process called **polymerization.** Polyacetylene is an example of a **polymer**—a large molecule made by linking together many small molecules. Polymer chemistry is such an important field of organic chemistry that an entire chapter of this book (Chapter 27) is devoted to it.

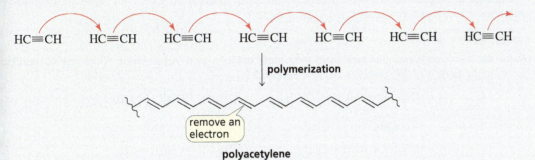

polyacetylene

The electrons in polyacetylene do not move along the length of the chain easily enough to conduct electricity well. However, if electrons are removed from or added to the chain (a process called "doping"), then the electrons can move easily down the chain and the polymer (with some refinements) can conduct electricity as well as copper can.

Polyacetylene is very sensitive to air and moisture, which limits its technological applications. However, many other conducting polymers have been developed (three of which are shown here) that have many practical uses. Notice that all these conducting polymers have a chain of conjugated double bonds.

poly(*p*-phenylene vinylene)

polythiophene

polypyrrole

An important property of conducting polymers is that they are very light. As a result, they are used to coat airplanes to prevent lightning from damaging the interior of the aircraft. The buildup of static electricity can be prevented by coating an insulator with a thin coating of conducting polymer. Conducting polymers are also used in LED (light emitting diode) displays. LEDs emit light in response to an electric current—a process known as

(Continued)

electroluminescence. LEDs are used for full-color displays in flat-screen TVs, cell phones, and the instrument panels in cars and airplanes. Continued research should lead to many more applications for conducting polymers. One such area is the development of "smart structures," such as golf clubs that adapt to the golfer's swing. Smart skis (that do not vibrate while the person is skiing) have already been created.

Alan Heeger (University of California, Santa Barbara), Alan MacDiarmid (University of Pennsylvania), and Hideki Shirakawa (University of Tsukuba, Japan) shared the Nobel Prize in Chemistry in 2000 for their work on conducting polymers.

PROBLEM 11 ♦

The heat of hydrogenation of 2,3-pentadiene, a cumulated diene, is 70.5 kcal/mol. What are the relative stabilities of cumulated, conjugated, and isolated dienes?

PROBLEM 12 ♦

Name the following dienes and rank them from most stable to least stable. (*Hint:* Alkyl groups stabilize dienes in the same way they stabilize alkenes; Section 5.9.)

$$CH_3CH\!=\!CHCH\!=\!CHCH_3 \qquad\qquad CH_2\!=\!CHCH_2CH\!=\!CH_2$$

$$\overset{\displaystyle CH_3}{\underset{}{|}}\qquad\quad \overset{\displaystyle CH_3}{\underset{}{|}}$$
$$CH_3C\!=\!CHCH\!=\!CCH_3 \qquad\qquad CH_3CH\!=\!CHCH\!=\!CH_2$$

Stability of Allylic and Benzylic Cations

Now we will look at allylic and benzylic cations. These carbocations have delocalized electrons and are, therefore, more stable than similar carbocations with localized electrons.

- An **allylic cation** is a carbocation with a positive charge on an allylic carbon; an **allylic carbon** is a carbon adjacent to an sp^2 carbon of an alkene (Section 5.2).

- A **benzylic cation** is a carbocation with a positive charge on a benzylic carbon; a **benzylic carbon** is a carbon adjacent to an sp^2 carbon of a benzene ring.

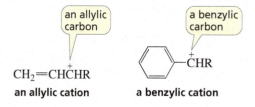

an allylic carbon

a benzylic carbon

$$CH_2\!=\!CH\overset{+}{C}HR$$

an allylic cation

a benzylic cation

allyl cation

The *allyl cation* is an unsubstituted allylic cation, and the *benzyl cation* is an unsubstituted benzylic cation.

$$CH_2\!=\!CH\overset{+}{C}H_2$$

the allyl cation

the benzyl cation

An allylic cation has two resonance contributors. The positive charge is not localized on a single carbon but is shared by two carbons.

$$RCH\!=\!CH\!-\!\overset{+}{C}H_2 \longleftrightarrow R\overset{+}{C}H\!-\!CH\!=\!CH_2$$

an allylic cation

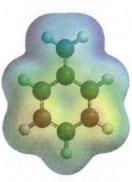

benzyl cation

A benzylic cation has five resonance contributors. Notice that the positive charge is shared by four carbons.

a benzylic cation

Because the allyl and benzyl cations have delocalized electrons, they are more stable than other primary carbocations in solution. We can add them to the list of carbocations whose relative stabilities were shown in Sections 5.9 and 7.5.

relative stabilities of carbocations

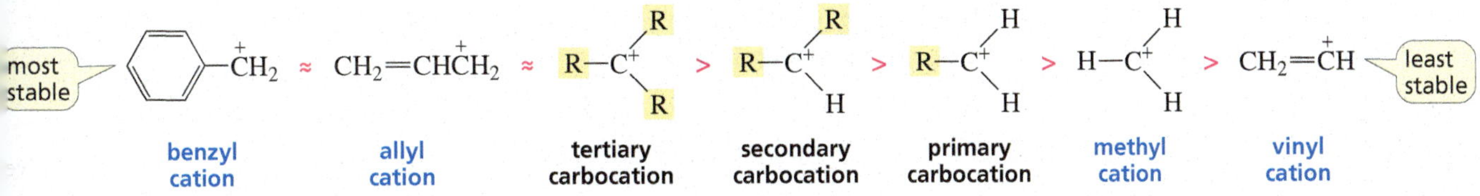

| benzyl cation | allyl cation | tertiary carbocation | secondary carbocation | primary carbocation | methyl cation | vinyl cation |

Not all allylic and benzylic cations have the same stability. Just as a tertiary alkyl carbocation is more stable than a secondary alkyl carbocation, a tertiary allylic cation is more stable than a secondary allylic cation, which in turn is more stable than the (primary) allyl cation. Similarly, a tertiary benzylic cation is more stable than a secondary benzylic cation, which is more stable than the (primary) benzyl cation.

relative stabilities

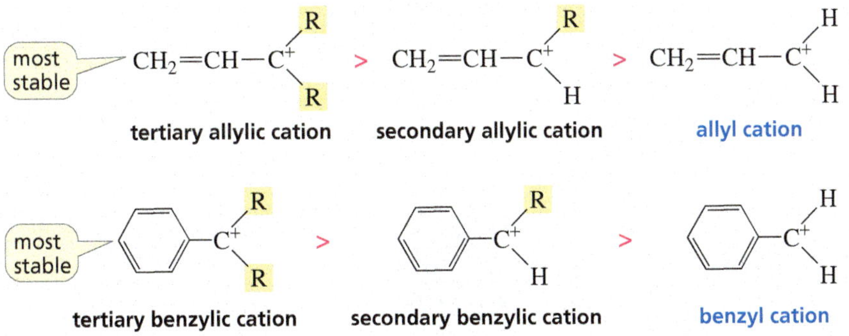

tertiary allylic cation   secondary allylic cation   allyl cation

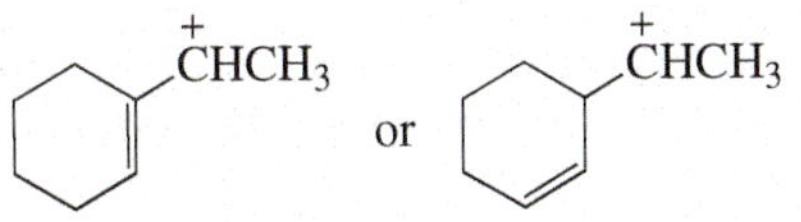

tertiary benzylic cation   secondary benzylic cation   benzyl cation

---

PROBLEM 13 ◆

Which carbocation in each pair is more stable?

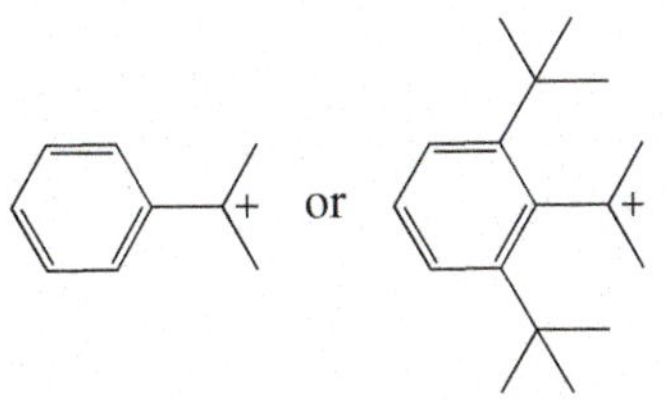

8.8   A MOLECULAR ORBITAL DESCRIPTION OF STABILITY

So far in our discussion, we have used resonance contributors to explain why compounds are stabilized by electron delocalization. Molecular orbital (MO) theory can also explain why electron delocalization stabilizes compounds.

We saw in Section 1.6 that the two lobes of a *p* orbital have opposite phases. We also saw that when two in-phase *p* orbitals overlap, a covalent bond is formed, and when two out-of-phase *p* orbitals interact, they cancel each other and produce a node between the two nuclei.

NOTE TO THE STUDENT
- Take a few minutes to review Section 1.6.
- See Special Topic II in the *Study Guide and Solutions Manual* for additional information on molecular orbital theory.

Ethene

Let's start by reviewing how the MOs of ethene are constructed. The two *p* atomic orbitals can be either in-phase or out-of-phase. (The different phases are indicated by different colors.) Notice that the number of orbitals is conserved—the number of molecular orbitals equals the number of atomic orbitals that produced the molecular orbitals. Thus, the two p atomic orbitals of ethene produce two MOs (Figure 8.10). Side-to-side overlap of in-phase p orbitals (lobes of the same color) produces a π **bonding molecular orbital,** designated ψ_1 (the Greek letter *psi*). The side-to-side interaction between out-of-phase p orbitals produces a $\pi*$ **antibonding molecular orbital,** ψ_2.

The overlap of in-phase orbitals holds atoms together;
it is a bonding interaction.

The interaction between out-of-phase orbitals pulls atoms apart;
it is an antibonding interaction.

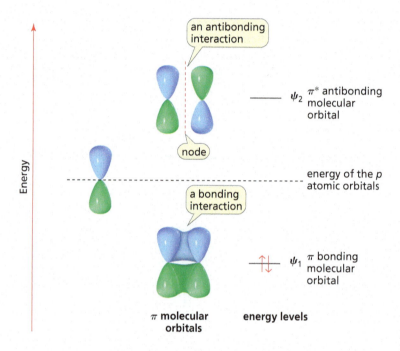

▲ **Figure 8.10**
An MO diagram of ethene. The π bonding MO is lower in energy than the p atomic orbitals, and it encompasses both carbons. In other words, each electron in the bonding MO spreads over both carbons. The $\pi*$ antibonding MO is higher in energy than the p atomic orbitals; it has a node between the lobes of opposite phases.

The π electrons are placed in molecular orbitals according to the same rules that govern the placement of electrons in atomic orbitals—namely, the aufbau principle (orbitals are filled in order of increasing energy), the Pauli exclusion principle (each orbital can hold two electrons of opposite spin), and Hund's rule (an electron will occupy an empty degenerate orbital before it will pair up with an electron already present in an orbital); see Section 1.2. Thus, ethene's two π electrons are both in ψ_1.

1,3-Butadiene

The four π electrons in 1,3-butadiene are delocalized over four carbons.

$$\overset{-}{C}H_2 - CH = CH - \overset{+}{C}H_2 \quad \longleftrightarrow \quad CH_2 = CH - CH = CH_2 \quad \longleftrightarrow \quad \overset{+}{C}H_2 - CH = CH - \overset{-}{C}H_2$$

1,3-butadiene
resonance contributors

$$CH_2 \text{---} CH \text{---} CH \text{---} CH_2$$

resonance hybrid

Each of the four carbons contributes one p atomic orbital, and the four p atomic orbitals combine to produce four MOs: ψ_1, ψ_2, ψ_3, and ψ_4 (Figure 8.11). Thus, we see that a molecular orbital results from the **linear combination of atomic orbitals (LCAO).** Half of the MOs are π-bonding MOs (ψ_1 and ψ_2), and the other half are π*-antibonding MOs (ψ_3 and ψ_4). The energies of the bonding and antibonding MOs are symmetrically distributed above and below the energy of the p atomic orbitals.

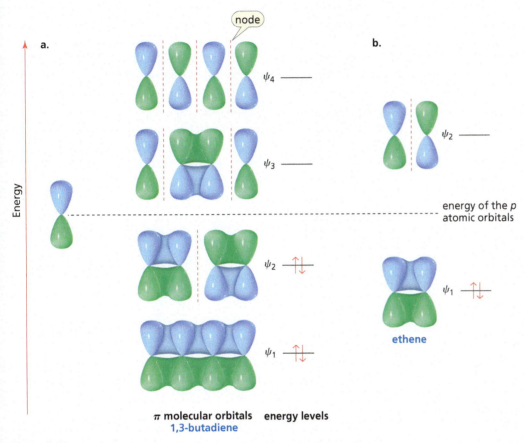

▲ **Figure 8.11**

(a) Four p atomic orbitals of 1,3-butadiene overlap to produce four MOs; two are π bonding MOs, and two are π* MOs.

(b) Two p atomic orbitals of ethene overlap to produce two MOs; one is a π MO, and the other is a π* antibonding MO.

Notice that the average energy of the four electrons in 1,3-butadiene is lower than the energy of the two electrons in ethene, because ψ_1 of ethene is closer to ψ_2 of 1,3-butadiene than to ψ_1 of 1,3-butadiene. This difference is the delocalization energy. In other words, 1,3-butadiene is stabilized by electron delocalization. Also notice that both ethene and 1,3-butadiene have filled bonding MOs. Figure 8.11 shows that *as the MOs increase in energy, the number of nodes within them increases and the number of bonding interactions decreases.*

- ψ_1 (1,3-butadiene's lowest-energy MO) has no nodes between the nuclei (it has only the node that bisects the orbitals), and it has **three bonding** interactions.

- ψ_2 has one node between the nuclei and two bonding interactions (for a net of **one bonding** interaction).

- ψ_3 has two nodes between the nuclei and one bonding interaction (for a net of **one antibonding** interaction).

- ψ_4 has three nodes between the nuclei and no bonding interactions (for a net of **three antibonding** interactions).

The four π electrons of 1,3-butadiene reside in the bonding MOs (ψ_1 and ψ_2).

ψ_1 is the most stable MO because it has three bonding interactions; its two electrons are delocalize[d] over all four nuclei, thus encompassing all four carbons. ψ_2 is also a bonding MO because it ha[s] one more bonding interaction than antibonding interaction; it is not as strongly bonding as ψ_1. Overall, ψ_3 is an antibonding MO, because it has one more antibonding interaction than bondin[g] interaction. It is not as strongly antibonding as ψ_4 that has no bonding interactions and three ant[i]bonding interactions.

The two bonding MOs in Figure 8.11a show that the greatest π electron density in a com[-]pound with two double bonds joined by one single bond is between C-1 and C-2 and betwee[n] C-3 and C-4, but there is some π electron density between C-2 and C-3—just as the resonanc[e] contributors show. The MOs also show why 1,3-butadiene is most stable in a planar conforma[-]tion: if 1,3-butadiene were not planar, there would be little or no overlap between the p orbital[s] on C-2 and C-3.

Both ψ_1 and ψ_3 are **symmetric molecular orbitals;** they have a plane of symmetry, so on[e] half is the mirror image of the other half. In contrast, ψ_2 and ψ_4 are **antisymmetric;** they do n[o]t have a plane of symmetry (but would have one, if one half of the MO were turned upside down[).] Notice in Figure 8.11 that as the MOs increase in energy, they alternate between symmetric an[d] antisymmetric.

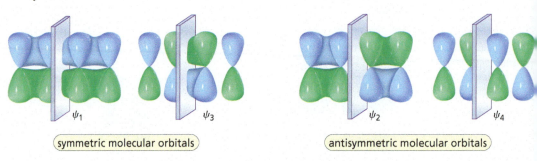

symmetric molecular orbitals   antisymmetric molecular orbitals

The highest-energy molecular orbital of 1,3-butadiene that contains electrons is ψ_2. Therefor[e] ψ_2 is called the **highest occupied molecular orbital (HOMO).** The lowest-energy molecula[r] orbital of 1,3-butadiene that does not contain electrons is ψ_3; ψ_3 is called the **lowest unoccupie[d] molecular orbital (LUMO).**

HOMO = the highest occupied molecular orbital

LUMO = the lowest unoccupied molecular orbital

The MO description of 1,3-butadiene shown in Figure 8.11 represents the electronic con[-]figuration of the molecule in its *ground state*. If the molecule absorbs light of an appropriat[e] wavelength, the light will promote an electron from its HOMO to its LUMO (from ψ_2 to ψ_3[).] The molecule will then be in an *excited* state (Section 1.2). We will see that the excitation o[f] an electron from the HOMO to the LUMO is the basis of ultraviolet and visible spectroscop[y] (Section 13.19).

PROBLEM 14 ◆

What is the total number of nodes in the ψ_3 and ψ_4 MOs of 1,3-butadiene?

PROBLEM 15 ◆

Answer the following questions for the MOs of 1,3-butadiene:

a. Which are π bonding MOs, and which are π^* antibonding MOs?
b. Which MOs are symmetric, and which are antisymmetric?
c. Which MO is the HOMO and which is the LUMO in the ground state?
d. Which MO is the HOMO and which is the LUMO in the excited state?
e. What is the relationship between the HOMO and the LUMO and symmetric and antisymmetric orbitals[?]

1,4-Pentadiene

Now let's look at look at 1,4-pentadiene. 1,4-Pentadiene, like 1,3-butadiene, has four π electron[s]. However, unlike the two pairs of π electrons in 1,3-butadiene that are *delocalized*, the two pairs o[f] π electrons in 1,4-pentadiene are *localized*—completely separate from one another.

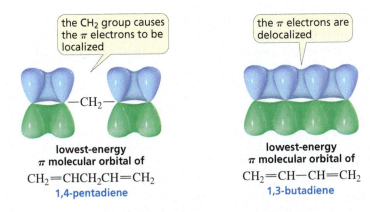

lowest-energy
π molecular orbital of
CH$_2$=CHCH$_2$CH=CH$_2$
1,4-pentadiene

lowest-energy
π molecular orbital of
CH$_2$=CH—CH=CH$_2$
1,3-butadiene

Because its electrons are localized, the lowest-energy MO of 1,4-pentadiene has the same energy as the lowest MO of ethene, a compound with one pair of localized π electrons. We have now seen that molecular orbital theory and resonance contributors are two different ways to show that the π electrons in 1,3-butadiene are delocalized and that electron delocalization stabilizes a compound.

PROBLEM 16 ◆

The most stable MO of 1,3,5-hexatriene and the most stable MO of benzene are shown here. Which compound is more stable? Why?

most stable MO of 1,3,5-hexatriene

most stable MO of benzene

8.9 DELOCALIZED ELECTRONS AFFECT pK_a VALUES

Acetic Acid versus Ethanol

We saw in Section 2.7 that a carboxylic acid is a stronger acid than an alcohol, because the carboxylate ion (the conjugate base of the carboxylic acid) is a more stable (weaker) base than the alkoxide ion (the conjugate base of an alcohol). Recall that the more stable the base, the stronger its conjugate acid.

> **The more stable the base, the stronger its conjugate acid.**

$$\underset{\substack{\text{acetic acid}\\ \text{p}K_a = 4.76}}{CH_3\overset{\overset{\displaystyle O}{\|}}{C}OH} \qquad \underset{\substack{\text{ethanol}\\ \text{p}K_a = 15.9}}{CH_3CH_2OH}$$

> **A nearby electronegative atom stabilizes an anion by inductive electron withdrawal.**

We also saw that the greater stability of the carboxylate ion is attributable to two factors:

- inductive electron withdrawal and
- electron delocalization.

That is, the doubly bonded oxygen stabilizes the carboxylate ion by decreasing the electron density of the negatively charged oxygen by inductive *electron withdrawal* and by an increase in *delocalization energy*.

Although both the carboxylic acid and the carboxylate ion have delocalized electrons, the delocalization energy of the carboxylate ion is greater than that of the carboxylic acid. This is because

acetic acid

the ion has two equivalent resonance contributors that are predicted to be relatively stable, whereas the carboxylic acid has only one (Section 8.6). Therefore, loss of a proton from a carboxylic acid is accompanied by an increase in delocalization energy—in other words, an increase in stability.

relatively stable   relatively unstable   relatively stable   relatively stable

In contrast, all the electrons in the conjugate base of an alcohol are localized, so loss of a proton from an alcohol is not accompanied by an increase in delocalization energy.

$$CH_3CH_2\ddot{O}H \rightleftharpoons CH_3CH_2\ddot{O}:^- + H^+$$

ethanol   **ethoxide ion**

acetate ion

Phenol versus Cyclohexanol

The same two factors that cause a carboxylic acid to be a stronger acid than an alcohol also cause phenol to be a stronger acid than an alcohol such as cyclohexanol—namely, stabilization of phenol's conjugate base both by inductive *electron withdrawal* and by an increase in *delocalization energy*.

phenol   **cyclohexanol**
pK_a = 10   **pK_a = 16**

The OH group of phenol is attached to an sp^2 carbon, which is more electronegative than the sp^3 carbon to which the OH group of cyclohexanol is attached (Section 2.6). The greater *inductive electron withdrawal* by the more electronegative sp^2 carbon stabilizes the conjugate base by decreasing the electron density of its negatively charged oxygen.

While both phenol and the phenolate ion have delocalized electrons, the delocalization energy of the phenolate ion is greater than that of phenol because three of phenol's resonance contributors have separated charges as well as a positive charge on an oxygen. The loss of a proton from phenol, therefore, is accompanied by an increase in delocalization energy (that is, an increase in stability).

An increase in delocalization energy means an increase in stability.

phenol

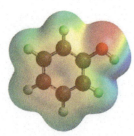

phenol

phenolate ion

In contrast, the conjugate base of cyclohexanol does *not* have any delocalized electrons to stabilize it.

cyclohexanol   **conjugate base of cyclohexanol**

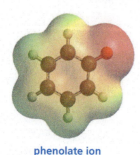

phenolate ion

Phenol is a *weaker* acid than a carboxylic acid, again because of inductive electron withdrawal and electron delocalization:

- Inductive electron withdrawal by the carbon in the phenolate ion is not as great as inductive electron withdrawal by the oxygen in the carboxylate ion.
- The increased delocalization energy when a proton is lost is not as great in the phenolate ion as in the carboxylate ion, where the negative charge is shared equally by two oxygens.

Aniline versus Cyclohexylamine

Inductive electron withdrawal and electron delocalization also explain why protonated aniline is a stronger acid than protonated cyclohexylamine.

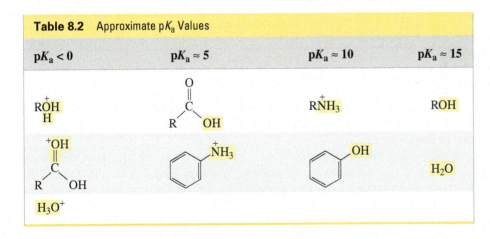

protonated aniline
pK_a = 4.60

protonated cyclohexylamine
pK_a = 11.2

First, aniline's nitrogen is attached to an sp^2 carbon, whereas cyclohexylamine's nitrogen is attached to a less electronegative sp^3 carbon. Second, the nitrogen atom of protonated aniline lacks a lone pair that can be delocalized. However, when the nitrogen loses a proton, the lone pair that formerly held the proton can be delocalized. Loss of a proton, therefore, is accompanied by an increase in delocalization energy.

protonated aniline

aniline

In contrast, cyclohexylamine does not have any delocalized electrons to stabilize it.

protonated
cyclohexylamine

cyclohexylamine

+ H$^+$

protonated aniline

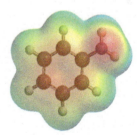

aniline

We can now add phenol and protonated aniline to the list of organic compounds whose approximate pK_a values you should know (Table 8.2). They are also listed inside the back cover for easy reference.

Table 8.2 Approximate pK_a Values

pK_a < 0	pK_a ≈ 5	pK_a ≈ 10	pK_a ≈ 15
ROH (protonated)	(carboxylic acid)	R$\overset{+}{N}H_3$	ROH
(protonated carboxylic acid)	(anilinium)	(phenol) OH	H$_2$O
H$_3$O$^+$			

LEARN THE STRATEGY

PROBLEM-SOLVING STRATEGY

Determining Relative Acidities

Which is a stronger acid?

$$CH_3CH_2OH \quad \text{or} \quad CH_2{=}CHOH$$

ethyl alcohol   vinyl alcohol

The strength of an acid depends on the stability of its conjugate base. So you can answer the question by comparing the stabilities of the two conjugate bases and remembering that the more stable base has the stronger conjugate acid.

$$\boxed{\text{localized electrons}} \qquad \boxed{\text{delocalized electrons}}$$

$$CH_3CH_2{-}\ddot{\text{O}}\colon^- \qquad CH_2{=}CH{-}\ddot{\text{O}}\colon^- \longleftrightarrow {}^-\ddot{C}H_2{-}CH{=}\text{O}\colon$$

All the electrons in ethanol's conjugate base are localized. However, vinyl alcohol's conjugate base is stabilized by electron delocalization. In addition, the oxygen in vinyl alcohol is attached to an sp^2 carbon, which is more electronegative than the sp^3 carbon to which the oxygen in ethanol is attached. As a result, vinyl alcohol is a stronger acid than ethanol.

USE THE STRATEGY

PROBLEM 17 ♦

Which member of each pair is the stronger acid?

a.

$$\underset{H}{\overset{\overset{\text{O}}{\|}}{C}}{-}CH_2OH \quad \text{or} \quad \underset{CH_3}{\overset{\overset{\text{O}}{\|}}{C}}{-}OH$$

b. $CH_3CH_2CH_2\overset{+}{N}H_3$ or $CH_3CH{=}CH\overset{+}{N}H_3$

c. $CH_3CH{=}CHCH_2OH$ or $CH_3CH{=}CHOH$

PROBLEM 18 ♦

Which member of each pair is the stronger base?

a. ethylamine or aniline
b. ethylamine or ethoxide ion
c. phenolate ion or ethoxide ion
d. phenolate ion or acetate ion

PROBLEM 19 ♦

Rank the following compounds from strongest acid to weakest acid:

$$\text{C}_6\text{H}_5{-}OH \qquad \text{C}_6\text{H}_5{-}CH_2OH \qquad \text{C}_6\text{H}_5{-}COOH$$

8.10 ELECTRONIC EFFECTS

If a substituent can donate electrons to or withdraw electrons from a benzene ring, then the pK_a values of substituted phenols, protonated anilines, and benzoic acids will change to reflect this withdrawal or donation.

Electron-donating substituents destabilize a base and, therefore, decrease the strength of its conjugate acid; electron-withdrawing substituents stabilize a base, which increases the strength of its conjugate acid (Section 2.7). Remember: the stronger the acid, the more stable (weaker) its conjugate base.

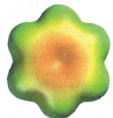

Electron donation decreases acidity. Electron withdrawal increases acidity.

electron donation destabilizes the base by increasing the electron density on the oxygen

electron withdrawal stabilizes the base by decreasing the electron density on the oxygen

Inductive Electron Withdrawal

If a substituent that is bonded to a benzene ring is *more electron withdrawing than a hydrogen*, then it will draw the σ electrons away from the benzene ring more strongly than a hydrogen will. Withdrawal of electrons through a σ bond is called **inductive electron withdrawal** (Section 2.7). The $^+NH_3$ group is an example of a substituent that withdraws electrons inductively because it is more electronegative than a hydrogen.

substituent withdraws electrons inductively (compared with a hydrogen)

Electron Donation by Hyperconjugation

We saw that an alkyl substituent (such as CH_3) stabilizes alkenes and carbocations by hyperconjugation—that is, by donating electrons to a p orbital (Section 6.2).

Electron Donation by Resonance

If a substituent has a lone pair on the atom directly attached to a benzene ring, then the lone pair can be delocalized into the ring. These substituents are said to **donate electrons by resonance.** Substituents such as NH_2, OH, OR, and Cl donate electrons by resonance. These substituents also withdraw electrons inductively because the atom attached to the benzene ring is more electronegative than a hydrogen.

donation of electrons by resonance into a benzene ring

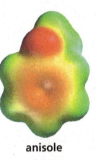

anisole

Electron Withdrawal by Resonance

If a substituent is attached to a benzene ring by an atom that is doubly or triply bonded to a more electronegative atom, then the electrons of the ring can be delocalized onto the substituent; these substituents are said to **withdraw electrons by resonance.** Substituents such as $C=O$, $C\equiv N$, SO_3H, and NO_2 withdraw electrons by resonance. These substituents also withdraw electrons inductively because the atom attached to the benzene ring has a full or partial positive charge and is, therefore, more electronegative than a hydrogen.

withdrawal of electrons by resonance from a benzene ring

benzene

nitrobenzene

Take a minute to compare the electrostatic potential maps for anisole, benzene, and nitrobenzene. Notice that an electron-donating substituent (OCH_3) makes the ring more red (more negative), whereas an electron-withdrawing substituent (NO_2) makes the ring less red (less negative).

Effect of Electron Donation and Electron Withdrawal on pK_a Values

Now let's look at how substituents affect the pK_a of phenol.

- We saw that a methyl group donates electrons by *hyperconjugation* (Section 6.2). This causes methyl-substituted phenol to be a weaker acid than unsubstituted phenol.
- Because the methoxy group (CH_3O) has a lone pair on the atom attached to the ring, it can *donate electrons by resonance*. Because oxygen is more electronegative than hydrogen, the methoxy group *withdraws electrons inductively*. The fact that methoxy-substituted phenol is a weaker acid than phenol indicates that the substituent's resonance *electron donation* into the ring *is more significant* than its inductive electron withdrawal from the ring.

OH	OH	OH	OH	OH	OH
OCH_3	CH_3		Cl	$HC{=}O$	NO_2
$pK_a = 10.20$	$pK_a = 10.09$	$pK_a = 9.95$	$pK_a = 9.38$	$pK_a = 7.66$	$pK_a = 7.14$
		phenol			

- A Cl also has a lone pair that can *donate electrons by resonance* and, because it is more electronegative than hydrogen, Cl *withdraws electrons inductively*. The fact that chloro-substituted phenol is a stronger acid than phenol indicates that Cl's inductive *electron withdrawal* from the ring *is more significant* than its resonance electron donation into the ring.
- The $HC{=}O$ and NO_2 groups *withdraw electrons both by resonance* and *inductively*. Therefore, these substituents increase the acidity of phenol. Recall that an NO_2 group has a positive charge on the nitrogen (see p. 343), which causes it to strongly withdraw electrons. This is reflected in the pK_a value of nitro-substituted phenol.

A similar substituent effect on pK_a is observed for substituted protonated anilines and substituted benzoic acids—that is, electron-withdrawing substituents increase acidity, whereas electron-donating substituents decrease acidity.

The more electron donating the substituent, the more it decreases the acidity of a COOH, an OH, or an $^+NH_3$ group attached to a benzene ring.

$\overset{+}{N}H_3$	$\overset{+}{N}H_3$	$\overset{+}{N}H_3$	$\overset{+}{N}H_3$	$\overset{+}{N}H_3$	$\overset{+}{N}H_3$
OCH_3	CH_3		Br	$HC{=}O$	NO_2
$pK_a = 5.29$	$pK_a = 5.07$	$pK_a = 4.58$	$pK_a = 3.91$	$pK_a = 1.76$	$pK_a = 0.98$
		protonated aniline			

The more electron withdrawing the substituent, the more it increases the acidity of a COOH, an OH, or an $^+NH_3$ group attached to a benzene ring.

COOH	COOH	COOH	COOH	COOH	COOH
OCH_3	CH_3		Br	$CH_3C{=}O$	NO_2
$pK_a = 4.47$	$pK_a = 4.34$	$pK_a = 4.20$	$pK_a = 4.00$	$pK_a = 3.70$	$pK_a = 3.44$
		benzoic acid			

PROBLEM 20 ♦

For each of the following substituents, indicate whether it withdraws electrons inductively, donates electrons by hyperconjugation, withdraws electrons by resonance, or donates electrons by resonance. (Effects should be compared with that of a hydrogen; remember that many substituents can be characterized in more than one way.)

a. Br   **b** CH_2CH_3 **c.** $\overset{O}{\overset{\|}{C}}CH_3$ **d.** $NHCH_3$ **e.** OCH_3 **f.** $^+N(CH_3)_3$

PROBLEM 21 ♦

Which acid in each of the following pairs is stronger?

a. CH_3COOH or $ClCH_2COOH$

b. O_2NCH_2COOH or $O_2NCH_2CH_2COOH$

c.

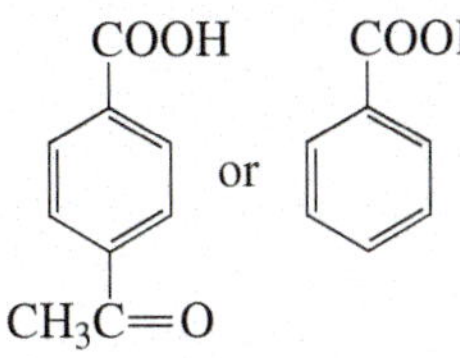

d. CH_3CH_2COOH or $H_3\overset{+}{N}CH_2COOH$

e. $HCOOH$ or CH_3COOH

f.

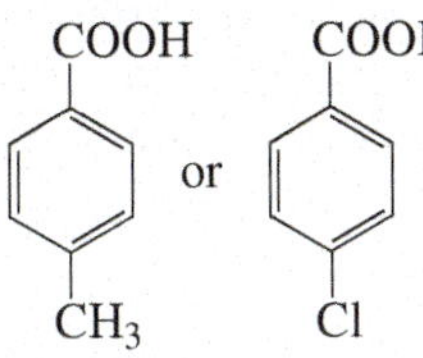

PROBLEM 22 SOLVED

Which is a stronger acid?

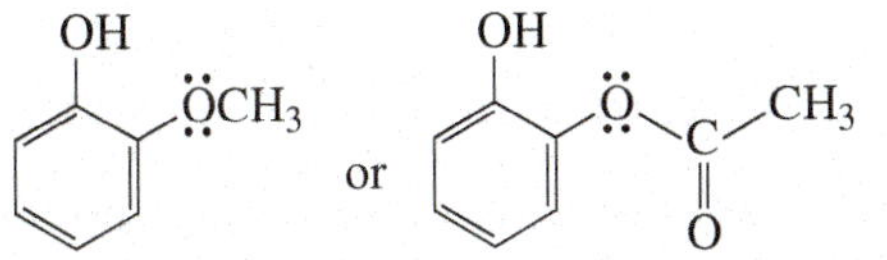

Both phenols have a substituent that can donate electrons by resonance and withdraw electrons inductively. The substituent of the compound on the right can donate electrons by resonance in two competing directions: into the ring and away from the ring.

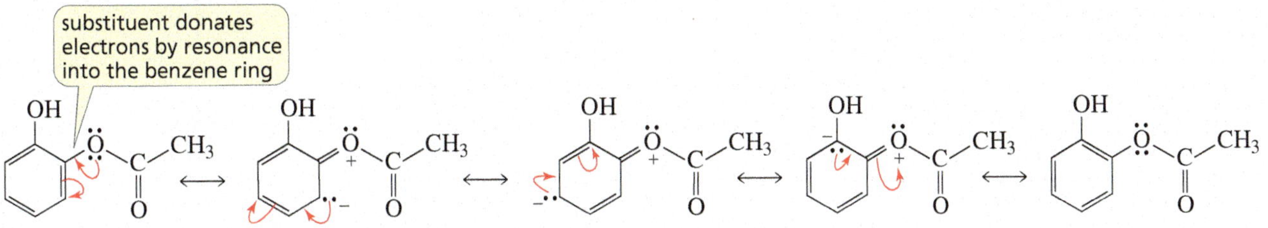

substituent donates electrons by resonance into the benzene ring

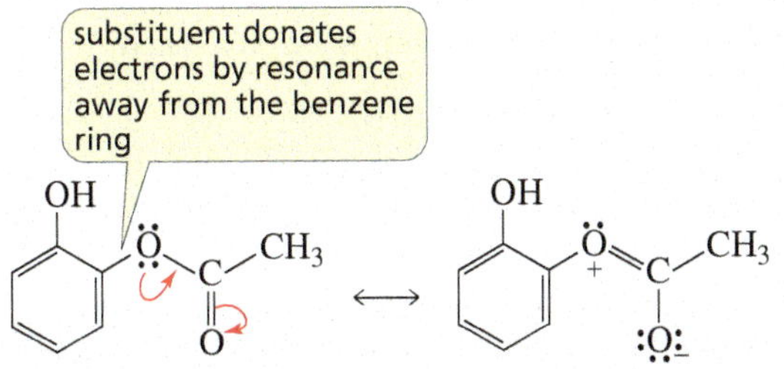

substituent donates electrons by resonance away from the benzene ring

In contrast, the substituent (CH_3O) of the compound on the left can donate electrons by resonance only into the ring (see p. 343). Therefore, overall it is better at donating electrons into the ring than is the substituent on the compound on the right. Therefore, the compound on the right is the stronger acid.

PROBLEM 23

Explain why the pK_a of *p*-nitrophenol is 7.14, whereas the pK_a of *m*-nitrophenol is 8.39. (*Hint*: Draw the resonance contributors.)

8.11 DELOCALIZED ELECTRONS CAN AFFECT THE PRODUCT OF A REACTION

Our ability to predict the product of an organic reaction correctly often depends on recognizing when compounds have delocalized electrons. For example, the alkene in the following reaction has the same number of hydrogens on both of its sp^2 carbons:

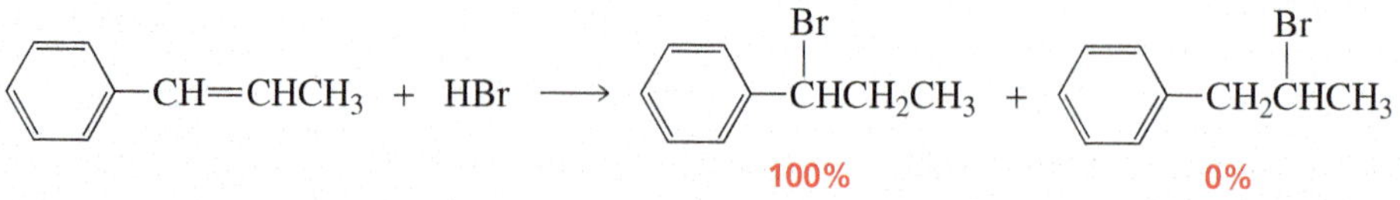

100%   0%

Therefore, the rule that tells us to add the electrophile to the sp^2 carbon bonded to the most hydrogens predicts that approximately equal amounts of the two products will be formed. When the reaction is carried out, however, only one of the products is obtained. (Notice that the stability of the benzene ring prevents its double bonds from undergoing electrophilic addition reactions; see Section 8.21.)

The rule leads us to an incorrect prediction of the reaction product because it does not take electron delocalization into consideration. It presumes that both carbocation intermediates are equally stable because they are both secondary carbocations. It does not take into account that one intermediate is a secondary alkyl carbocation, whereas the other is a secondary benzylic cation. Because the secondary benzylic cation is stabilized by electron delocalization, it is formed more readily. The difference in the rates of formation of the two carbocations is sufficient to cause only one product to be obtained.

secondary alkyl carbocation   **secondary benzylic cation**

Let this example serve as a warning. The rule that states that the electrophile adds to the sp^2 carbon bonded to the most hydrogens cannot be applied to reactions that form carbocations that can be stabilized by electron delocalization. In such cases, you must look at the relative stabilities of the individual carbocations to predict the major product of the reaction.

PROBLEM 24 ◆

What is the major product obtained from the addition of HBr to the following compound?

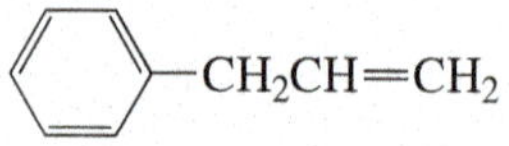

LEARN THE STRATEGY

PROBLEM 25 SOLVED

Which atom in the following compound is most likely to be protonated?

$$CH_3CH = CHOCH_3 + H^+$$

SOLUTION The resonance contributors reveal that one of the carbons has a partial negative charge. Therefore, it is the atom most likely to be protonated.

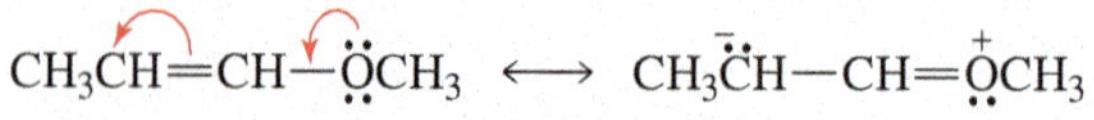

resonance contributors

USE THE STRATEGY

PROBLEM 26

Which atom in the following compound is most likely to be protonated?

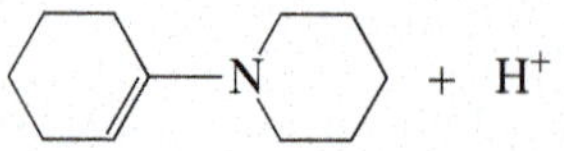

8.12 REACTIONS OF DIENES

For another example of how delocalized electrons can affect the product of a reaction, we will compare the products formed when *isolated dienes* (dienes that have only localized electrons) undergo electrophilic addition reactions to the products formed when *conjugated dienes* (dienes that have delocalized electrons) undergo the same reactions.

$$CH_2{=}CHCH_2CH_2CH{=}CH_2 \qquad CH_3CH{=}CH{-}CH{=}CHCH_3$$
$$\textbf{isolated diene} \qquad\qquad \textbf{conjugated diene}$$

Reactions of Isolated Dienes

The reactions of *dienes with isolated double bonds* are just like the reactions of alkenes. If an excess of the electrophilic reagent is present, two independent electrophilic addition reactions will occur. In each reaction, *the electrophile adds to the sp² carbon bonded to the most hydrogens* because that results in formation of the most stable carbocation.

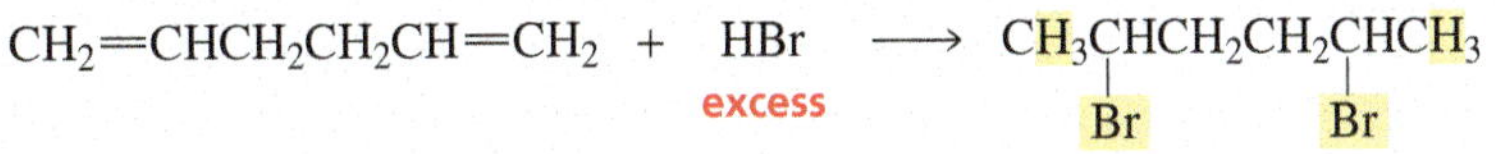

$$CH_2{=}CHCH_2CH_2CH{=}CH_2 \; + \; HBr \; \longrightarrow \; CH_3\underset{Br}{C}HCH_2CH_2\underset{Br}{C}HCH_3$$
$$\qquad\qquad\qquad\qquad\qquad \textbf{excess}$$

The reaction proceeds exactly as we would predict from our knowledge of the mechanism for the reaction of alkenes with electrophilic reagents.

MECHANISM FOR THE REACTION OF AN ISOLATED DIENE WITH EXCESS HBr

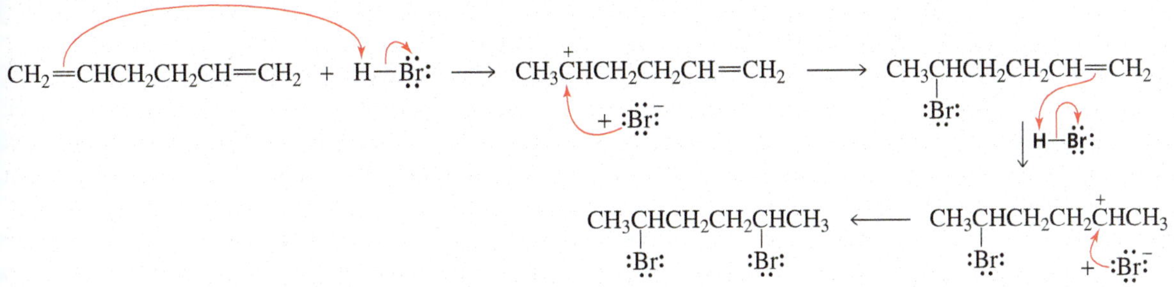

- The electrophile (H⁺) adds to the *sp²* carbon bonded to the most hydrogens in order to form the more stable carbocation (Section 6.4).
- The bromide ion adds to the carbocation.
- Because there is an excess of the electrophilic reagent, there is enough reagent to add to the other double bond; again the H⁺ adds to the *sp²* carbon bonded to the most hydrogens.
- The bromide ion adds to the carbocation.

If there is only enough electrophilic reagent to add to one of the double bonds, it will add preferentially to the more reactive one. For example, in the following reaction, addition of HCl to the double bond on the left forms a secondary carbocation, whereas addition to the double bond on the right forms a tertiary carbocation. Because the tertiary carbocation is more stable and is, therefore, formed faster, the major product of the reaction in the presence of a limited amount of HCl is 5-chloro-5-methyl-1-hexene (Section 6.4).

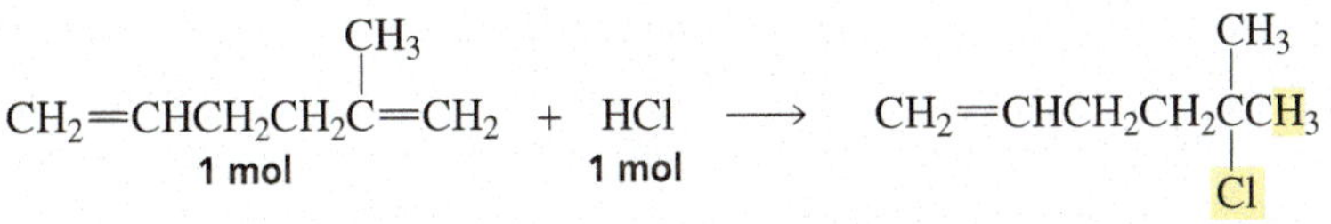

$$CH_2{=}CHCH_2CH_2\overset{\overset{\displaystyle CH_3}{|}}{C}{=}CH_2 \; + \; HCl \; \longrightarrow \; CH_2{=}CHCH_2CH_2\overset{\overset{\displaystyle CH_3}{|}}{\underset{\underset{\displaystyle Cl}{|}}{C}}CH_3$$
$$\qquad\textbf{1 mol}\qquad\qquad\quad\textbf{1 mol}$$
$$\qquad\qquad\qquad\qquad\qquad\textbf{5-chloro-5-methyl-1-hexene}$$
$$\qquad\qquad\qquad\qquad\qquad\textbf{major product}$$

PROBLEM 27 ◆

What is the major product of each of the following reactions, assuming that one equivalent of each reagent is used in each reaction?

a. CH_2=$CHCH_2CH_2CH$=$\overset{\overset{\displaystyle CH_3}{|}}{C}CH_3$ $\xrightarrow{HBr}$

b. HC≡CCH_2CH_2CH=CH_2 $\xrightarrow{Cl_2}$

c. 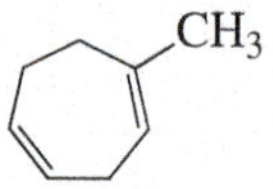 [structure: methyl cycloheptadiene] $\xrightarrow{HCl}$

Reactions of Conjugated Dienes

When a diene with *conjugated double bonds*, such as 1,3-butadiene, reacts with a limited amount of electrophilic reagent so that addition can occur at only one of the double bonds, two addition products are formed. One is a **1,2-addition product**—the result of addition at the 1- and 2-positions. The other is a **1,4-addition product**—the result of addition at the 1- and 4-positions. **1,2-Addition** is called **direct addition,** and 1,4-addition is called **conjugate addition.**

$$\overset{1}{CH_2}=\overset{2}{CH}-\overset{3}{CH}=\overset{4}{CH_2} + Cl_2 \longrightarrow CH_2-CH-CH=CH_2 + CH_2-CH=CH-CH_2$$

1,3-butadiene   1 mol

1 mol

1,2-addition product   1,4-addition product

> An isolated diene undergoes only 1,2-addition.

> A conjugated diene undergoes both 1,2- and 1,4-addition.

$$CH_2=CH-CH=CH_2 + HBr \longrightarrow CH_3CH-CH=CH_2 + CH_3-CH=CH-CH_2$$

1,3-butadiene   1 mol

1 mol

1,2-addition product   1,4-addition product

Based on your knowledge of how electrophilic reagents add to double bonds, you would expect the 1,2-addition product. The 1,4-addition product, however, is surprising, because the reagent did not add to adjacent carbons and a double bond changed its position.

When we talk about addition at the 1- and 2-positions or at the 1- and 4-positions, the numbers refer to the four carbons of the conjugated system. Thus, the carbon in the 1-position is the sp^2 carbon at one of the ends of the conjugated system—it is not necessarily the first carbon in the molecule.

$$R-\overset{1}{CH}=\overset{2}{CH}-\overset{3}{CH}=\overset{4}{CH}-R$$

the conjugated system

For example, the 1- and 4-positions in the conjugated system of 2,4-hexadiene are actually C-2 and C-5.

$$CH_3\overset{1}{CH}=CH-CH=\overset{4}{CH}CH_3 \xrightarrow{Br_2} CH_3\overset{1}{CH}-\overset{2}{CH}-CH=CHCH_3 + CH_3\overset{1}{CH}-CH=CH-\overset{4}{CH}CH_3$$

2,4-hexadiene

1,2-addition product   1,4-addition product

To understand why an electrophilic addition reaction to a conjugated diene forms both 1,2-addition and 1,4-addition products, we need to look at the mechanism of the reaction.

MECHANISM FOR THE REACTION OF A CONJUGATED DIENE WITH HBr

allylic cation

$$CH_2=CH-CH=CH_2 \ + \ H-\overset{..}{B}r\colon \ \longrightarrow \ CH_3-\overset{+}{C}H-CH=CH_2 \ \longleftrightarrow \ CH_3-CH=CH-\overset{+}{C}H_2$$

$$+ \ \colon\!\overset{..}{B}r\colon^-$$

$$CH_3-CH-CH=CH_2 \ + \ CH_3-CH=CH-CH_2$$
$$\overset{|}{\colon\!\overset{..}{B}r\colon} \qquad\qquad\qquad\qquad \overset{|}{\colon\!\overset{..}{B}r\colon}$$

1,2-addition product   **1,4-addition product**

$$CH_3\overset{\delta+}{-}CH\!=\!\!=\!\!CH\overset{\delta+}{=\!\!=}CH_2$$

- The proton adds to C-1, forming an allylic cation. The allylic cation has delocalized electrons.
- The resonance contributors of the allylic cation show that the positive charge is shared by C-2 and C-4. Consequently, the bromide ion can add to either C-2 or C-4 to form the 1,2-addition product or the 1,4-addition product, respectively.

Notice in the first step of the reaction that adding H^+ to C-1 is the same as adding it to C-4 because 1,3-butadiene is symmetrical.

As we look at more examples, notice that the first step in all electrophilic additions to conjugated dienes is the addition of the electrophile to one of the sp^2 carbons *at the end of the conjugated system*. This is the only way to form a carbocation that is stabilized by electron delocalization. If the electrophile were to add to one of the internal sp^2 carbons, the resulting carbocation would not have delocalized electrons.

PROBLEM 28 ◆

What are the products of the following reactions, assuming that one equivalent of each reagent is used in each reaction?

a. $CH_3CH=CH-CH=CHCH_3 \xrightarrow{\ Cl_2\ }$

b. $CH_3CH=\overset{\overset{\displaystyle CH_3}{|}}{C}-\underset{\underset{\displaystyle CH_3}{|}}{C}=CHCH_3 \xrightarrow{\ HBr\ }$

c. ⬠ $\xrightarrow{\ Br_2\ }$

d. ⬡ $\xrightarrow{\ HCl\ }$

PROBLEM 29 ◆

Which of the double bonds in zingiberene, the compound responsible for the aroma of ginger, is most reactive in an electrophilic addition reaction with HBr?

zingiberene   **ginger**

PROBLEM 30

What stereoisomers do the first two reactions on the top of p. 348 form? (*Hint:* Review Section 6.13.)

If the conjugated diene is not symmetrical, the major products of the reaction are those obtained by adding the electrophile to whichever sp^2 carbon at the end of the conjugated system results in formation of the more stable carbocation.

For example, in the reaction shown below, the proton adds preferentially to C-1 because the positive charge on the resulting carbocation is shared by a tertiary allylic and a primary allylic carbon. Adding the proton to C-4 would form a carbocation in which the positive charge is shared by a secondary allylic and a primary allylic carbon.

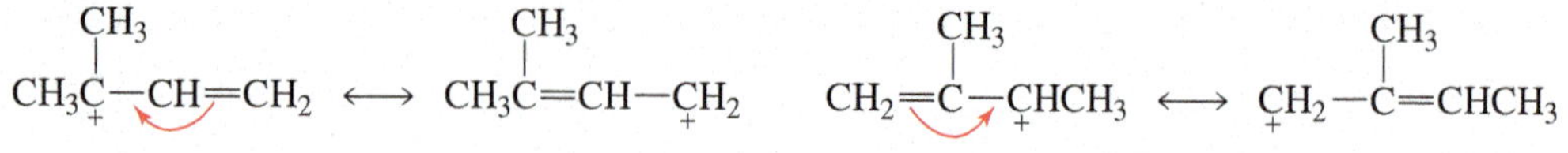

$$CH_3\overset{+}{C}-CH=CH_2 \longleftrightarrow CH_3\overset{CH_3}{C}=CH-\overset{+}{CH_2} \qquad CH_2=\overset{CH_3}{\overset{|}{C}}-\overset{+}{CH}CH_3 \longleftrightarrow \overset{+}{CH_2}-\overset{CH_3}{\overset{|}{C}}=CHCH_3$$

carbocation formed by adding H⁺ to C-1   **carbocation formed by adding H⁺ to C-4**

Because addition to C-1 forms the more stable carbocation, the major products of the reaction are the ones shown.

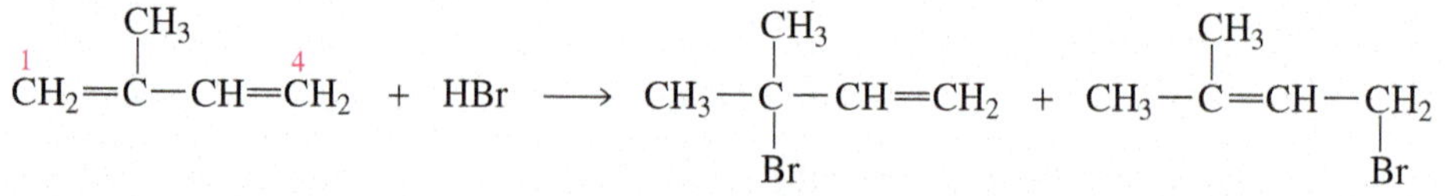

$$\underset{1}{CH_2}=\overset{CH_3}{\overset{|}{C}}-CH=\underset{4}{CH_2} + HBr \longrightarrow CH_3-\overset{CH_3}{\underset{Br}{\overset{|}{\underset{|}{C}}}}-CH=CH_2 + CH_3-\overset{CH_3}{\overset{|}{C}}=CH-\underset{Br}{\overset{|}{CH_2}}$$

1,2-addition product   **1,4-addition product**

PROBLEM 31

What products would be obtained from the reaction of 1,3,5-hexatriene with one equivalent of HBr? Disregard stereoisomers.

PROBLEM 32 ♦

What are the products of the following reactions, assuming that one equivalent of each reagent is used in each reaction? Disregard stereoisomers.

a. $CH_3CH=CH-\overset{|}{\underset{CH_3}{C}}=CH_2 \overset{Cl_2}{\longrightarrow}$ **b.** $CH_3CH=CH-\overset{|}{\underset{CH_3}{C}}=CHCH_3 \overset{HBr}{\longrightarrow}$

8.13 THERMODYNAMIC VERSUS KINETIC CONTROL

When a conjugated diene undergoes an electrophilic addition reaction, two factors—the *structure of the reactant* and the *temperature* at which the reaction is carried out—determine whether the 1,2-addition product or the 1,4-addition product will be the major product of the reaction.

Kinetic and Thermodynamic Products

When a reaction produces more than one product,

- *the more rapidly formed product* is called the **kinetic product.**
- *the more stable product* is called the **thermodynamic product.**
- reactions that produce the kinetic product as the major product are said to be **kinetically controlled.**
- reactions that produce the thermodynamic product as the major product are said to be **thermodynamically controlled.**

In many organic reactions, the more rapidly formed product is also the more stable product. Electrophilic addition to 1,3-butadiene is an example of a reaction in which the kinetic product and the thermodynamic product are *not* the same: the 1,2-addition product is the kinetic product, and the 1,4-addition product is the thermodynamic product.

The kinetic product is the more rapidly formed product.

The thermodynamic product is the more stable product.

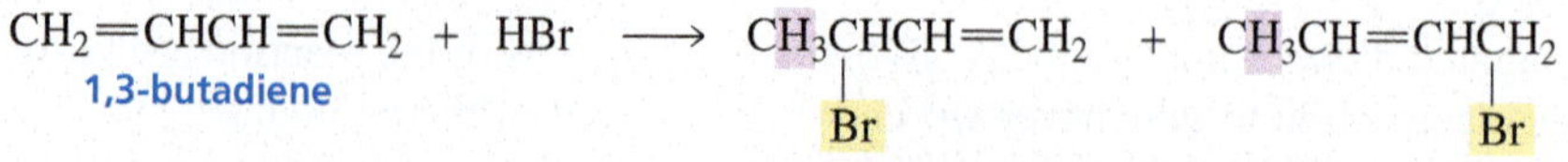

$$\underset{\textbf{1,3-butadiene}}{CH_2=CHCH=CH_2} + HBr \longrightarrow CH_3\overset{|}{\underset{Br}{CHCH}}=CH_2 + CH_3CH=CH\overset{|}{\underset{Br}{CH_2}}$$

1,2-addition product   **1,4-addition product**
kinetic product   **thermodynamic product**

Mild Conditions Favor the Kinetic Product

For a reaction such as the one just shown, in which the kinetic and thermodynamic products are not the same, the product that predominates depends on the conditions under which the reaction is carried out. If the reaction is carried out under conditions that are sufficiently mild (at a low temperature) to cause the reaction to be *irreversible*, the major product will be the *kinetic product*—that is, the faster-formed product.

$$CH_2=CHCH=CH_2 + HBr \xrightarrow{-80\ °C} \underset{\substack{\text{kinetic product} \\ 80\%}}{CH_3\underset{\underset{Br}{|}}{C}HCH=CH_2} + \underset{\substack{\text{thermodynamic product} \\ 20\%}}{CH_3CH=CHCH_2\underset{\underset{}{}}{}}$$

The kinetic product predominates when the reaction is irreversible.

Vigorous Conditions Favor the Thermodynamic Product

If, on the other hand, the reaction is carried out under conditions that are sufficiently vigorous (at a higher temperature) to cause the reaction to be *reversible*, the major product will be the *thermodynamic product*—that is, the more stable product.

$$CH_2=CHCH=CH_2 + HBr \underset{}{\overset{45\ °C}{\rightleftharpoons}} \underset{\substack{\text{kinetic product} \\ 15\%}}{CH_3\underset{\underset{Br}{|}}{C}HCH=CH_2} + \underset{\substack{\text{thermodynamic product} \\ 85\%}}{CH_3CH=CHCH_2}$$

The thermodynamic product predominates when the reaction is reversible.

Reaction Coordinate Diagrams Explain the Temperature Dependence of the Products

A reaction coordinate diagram helps explain why different products predominate under different reaction conditions (Figure 8.12). The first step of the addition reaction is the same whether the 1,2-addition product or the 1,4-addition product is ultimately formed: a proton adds to C-1. The second step of the reaction is the one that determines whether the nucleophile (Br⁻) adds to C-2 or to C-4.

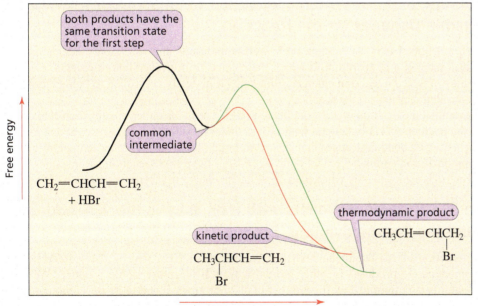

◀ **Figure 8.12**
A reaction coordinate diagram for the addition of HBr to 1,3-butadiene.

At low temperatures (−80 °C), there is enough energy for the reactants to overcome the energy barrier for the first step of the reaction and there is enough energy for the intermediate formed in the first step to form the two addition products. However, there is not enough energy for the reverse reaction to occur: the products cannot overcome the large energy barriers separating them from the intermediate. Consequently, the relative amounts of the two products obtained at −80 °C reflect the relative energy barriers for the second step of the reaction. So the major product is the more rapidly formed (kinetic) product.

In contrast, at 45 °C, there is enough energy for one or more of the products to revert back to the intermediate. The intermediate is called a **common intermediate** because it is an intermediate that both products have in common. Even if both products can revert back to the common intermediate, it is easier for the 1,2-addition product to do so. Each time the kinetic product reverts back to the common intermediate, the common intermediate can reform the kinetic product or form the thermodynamic product, so more and more thermodynamic product is formed.

The ability to revert to a common intermediate allows the products to interconvert. When two products can interconvert, their relative amounts at the end of the reaction depend on their relative stabilities. So the major product at equilibrium is the thermodynamic product.

In summary:

- A reaction that is *irreversible* under the conditions employed in the experiment will be *kinetically controlled.*

- When a reaction is under *kinetic control*, the relative amounts of the products *depend on the rates* at which they are formed.

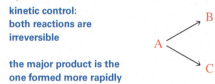

kinetic control:
both reactions are
irreversible

the major product is the
one formed more rapidly

- When sufficient energy is available to make one or more of the *reactions reversible, the reaction will be thermodynamically controlled.*

- When a reaction is under *thermodynamic control*, the relative amounts of the products *depend on their stabilities.*

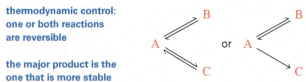

thermodynamic control:
one or both reactions
are reversible

the major product is the
one that is more stable

The Temperature at Which a Reaction Changes from Irreversible to Reversible Depends on the Reaction

For each reaction that is irreversible under mild conditions and reversible under more vigorous conditions, there is a temperature at which the change from irreversible to reversible occurs. The temperature at which a reaction changes from being kinetically controlled to being thermodynamically controlled depends on the reaction.

For example, the reaction of 1,3-butadiene with HCl remains under kinetic control at 45 °C, even though the reaction of 1,3-butadiene with HBr is under thermodynamic control at that temperature. Because a C—Cl bond is stronger than a C—Br bond (Table 5.1), a higher temperature is required for the products containing a C—Cl bond to undergo the reverse reaction. (Remember, thermodynamic control is achieved only when there is sufficient energy to reverse one or both of the reactions.)

When 1,3-Butadiene Reacts with HBr, Why Is the 1,4-Addition Product the Thermodynamic Product?

For the reaction of 1,3-butadiene with one equivalent of HBr, why is the 1,4-addition product the thermodynamic product? In other words, why is it the more stable product? We know that alkene stability is determined by the number of alkyl groups bonded to its sp^2 carbons—the greater the number of alkyl groups, the greater its stability (Section 5.9). The two products formed from the reaction of 1,3-butadiene with HBr have different stabilities because the 1,2-addition product has one alkyl group bonded to its sp^2 carbons, whereas the 1,4-product has two alkyl groups bonded to its sp^2 carbons. Thus, the 1,4-addition product is the more stable (the thermodynamic) product.

CH₃CHCH=CH₂ ... Br — 1,2-addition product — kinetic product

CH₃CH=CHCH₂ ... Br — 1,4-addition product — thermodynamic product

Why Is the 1,2-Addition Product Always the Kinetic Product?

The next question we need to answer is: why is the 1,2-addition product formed faster? For many years, chemists thought it was because the transition state for formation of the 1,2-addition product resembles the resonance contributor in which the positive charge is on a secondary allylic carbon, whereas the transition state for formation of the 1,4-addition product resembles the resonance contributor in which the positive charge is on a less stable primary allylic carbon.

$$CH_2=CHCH=CH_2 \xrightarrow{\text{HBr}} \overset{\text{secondary allylic carbon}}{CH_3\overset{+}{C}HCH=CH_2} \longleftrightarrow \overset{\text{primary allylic carbon}}{CH_3CH=CH\overset{+}{C}H_2}$$

$$:\!\overset{..}{\underset{..}{Br}}\!:\ \downarrow$$

$$\left[\overset{\delta^+}{CH_3CHCH=CH_2} \atop \underset{\delta^-}{}:\!\overset{..}{\underset{..}{Br}}\!: \right]^{\ddagger} \qquad \left[CH_3CH=CH\overset{\delta^+}{CH_2} \atop \underset{\delta^-}{}:\!\overset{..}{\underset{..}{Br}}\!: \right]^{\ddagger}$$

transition state for formation of the 1,2-addition product   **transition state for formation of the 1,4-addition product**

If this explanation were correct, the kinetically controlled reaction of 1,3-pentadiene with DCl would form equal amounts of the 1,2- and 1,4-addition products, because their transition states would both have the same stability (both would have a partial positive charge on a secondary allylic carbon). However, the kinetically controlled reaction forms mainly the 1,2-addition product. Why is the 1,2-addition product formed faster?

$$\underset{\text{1,3 pentadiene}}{CH_2=CHCH=CHCH_3} + DCl \xrightarrow{-78\ °C} \underset{\substack{D\ \ Cl \\ \text{1,2-addition product} \\ 78\%}}{CH_2CHCH=CHCH_3} + \underset{\substack{D\ \ \ \ \ \ \ Cl \\ \text{1,4-addition product} \\ 22\%}}{CH_2CH=CHCHCH_3}$$

The answer is that after D^+ adds to the double bond, the chloride ion can form a bond with C-2 faster than it can form a bond with C-4 simply because it is closer to C-2 than to C-4. So it is a *proximity effect* that causes the 1,2-addition product to be formed faster. A **proximity effect** is an effect caused by one species being close to another.

$$\overset{\text{secondary allylic cation}}{\underset{\substack{D\ \ \ Cl^- \\ }}{CH_2-\overset{2}{\underset{+}{C}H}-CH=CHCH_3}} \longleftrightarrow \underset{\substack{D\ \ \ Cl^- }}{CH_2-\overset{2}{C}H=CH-\overset{4}{\underset{+}{C}H}CH_3}$$

Cl⁻ is closer to C-2 than to C-4

PROBLEM 33 ◆

a. Why does deuterium add to C-1 rather than to C-4 in the preceding reaction?

b. Why was DCl rather than HCl used in the reaction?

PROBLEM 34 ◆

A student wanted to know whether the greater proximity of the nucleophile to the C-2 carbon in the transition state is what causes the 1,2-addition product to be formed faster when 1,3-butadiene reacts with HCl. Therefore, she decided to investigate the reaction of 2-methyl-1,3-cyclohexadiene with HCl. Her friend told her that she should use 1-methyl-1,3-cyclohexadiene instead. Should she follow her friend's advice?

Identifying the Kinetic and Thermodynamic Products

Because a proximity effect is what causes the 1,2-addition product to be formed faster, we ca
assume that

> *the 1,2-addition product is always the kinetic product*
> *for the reactions of conjugated dienes.*

Do *not* assume, however, that the 1,4-addition product is *always* the thermodynamic product.

> *The structure of the conjugated diene is what ultimately determines*
> *the thermodynamic product.*

For example, in the following reaction, the 1,2-addition product is both the kinetic product and th
thermodynamic product because it is formed faster and it is the more stable product. (It has 3 alk
groups bonded to the sp^2 carbons.)

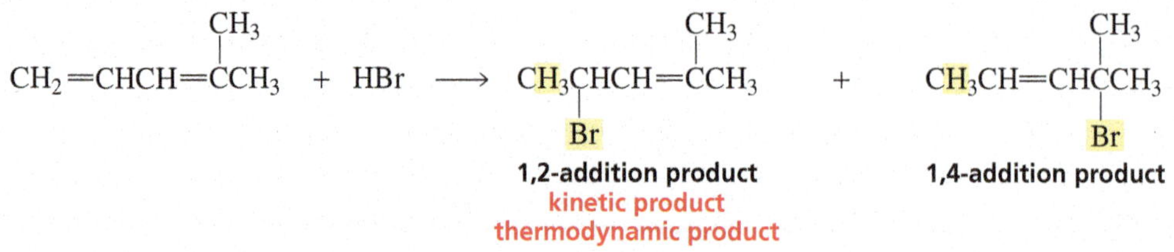

$$CH_2{=}CHCH{=}CCH_3 \ + \ HBr \ \longrightarrow \ CH_3CHCH{=}CCH_3 \ + \ CH_3CH{=}CHCCH_3$$

with CH_3 substituents and Br groups as shown

1,2-addition product
kinetic product
thermodynamic product

1,4-addition product

The 1,2- and 1,4-addition products obtained from the next reaction have the same stabilit
because both have the same number of alkyl groups bonded to their sp^2 carbons. Therefor
neither is the thermodynamic product. Approximately equal amounts of both products wi
be formed under conditions that cause the addition reaction to be reversible. The 1,2-additio
product is the kinetic product.

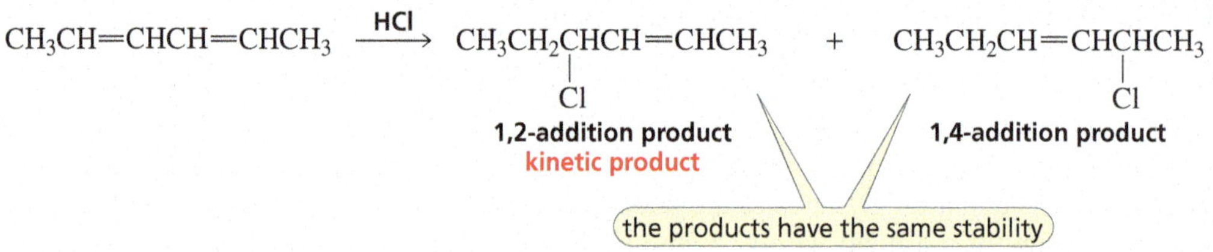

$$CH_3CH{=}CHCH{=}CHCH_3 \ \xrightarrow{\text{HCl}} \ CH_3CH_2CHCH{=}CHCH_3 \ + \ CH_3CH_2CH{=}CHCHCH_3$$

with Cl substituents as shown

1,2-addition product
kinetic product

1,4-addition product

the products have the same stability

PROBLEM 35 ◆

a. When HBr adds to a conjugated diene, what is the rate-determining step?
b. When HBr adds to a conjugated diene, what is the product-determining step?

LEARN THE STRATEGY

PROBLEM 36 SOLVED

What are the major 1,2- and 1,4-addition products of the following reaction? Indicate the kinetic and th
thermodynamic products.

$$\text{(cyclohexene ring with } CH_2 \text{ substituent)} \ + \ HCl \ \longrightarrow$$

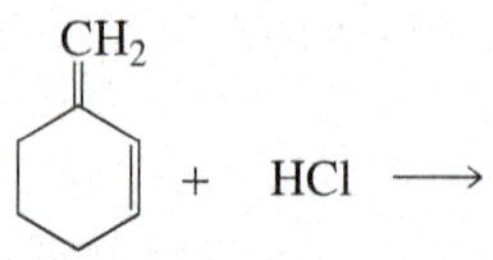

SOLUTION First, we need to determine which of the sp^2 carbons at the ends of the conjugated syste
gets the proton. The proton will preferentially add to the sp^2 carbon indicated on the top of the next pag
because the resulting carbocation shares its positive charge with a tertiary allylic and a secondary allyl
carbon. If the proton were to add to the sp^2 carbon at the other end of the conjugated system, the carboca
tion that formed would be less stable because its positive charge would be shared by a primary allylic and
secondary allylic carbon. Therefore, the major products are the ones shown. The 1,2-addition product, is th
kinetic product because of chloride ion's proximity to C-2. The 1,4-addition product, is the thermodynami
product because its more highly substituted double bond makes it more stable.

H⁺ adds here

$$CH_2 \xrightarrow{\text{HCl}} CH_3 \longleftrightarrow CH_3 \longrightarrow Cl\ CH_3 + CH_3$$

+ Cl⁻

1,2-addition product
kinetic
product

1,4-addition product
thermodynamic
product

PROBLEM 37

USE THE STRATEGY

What are the major 1,2- and 1,4-addition products of the following reactions? For each reaction, indicate the kinetic and the thermodynamic product.

a. (structure) CH₃ + HCl ⟶

b. (structure) CH=CHCH₃ + HCl ⟶

PROBLEM 38

Identify the kinetic and thermodynamic products of the following reaction:

(structure) HCl⟶

8.14 THE DIELS–ALDER REACTION IS A 1,4-ADDITION REACTION

Reactions that create new carbon–carbon bonds are very important to synthetic organic chemists because it is only through such reactions that small carbon skeletons can be converted into larger ones (Section 7.11). The Diels–Alder reaction is a particularly important reaction because it creates *two* new carbon–carbon bonds and in the process forms a cyclic compound. In recognition of the importance of this reaction to synthetic organic chemistry, Otto Diels and Kurt Alder shared the Nobel Prize in Chemistry in 1950.

In a **Diels–Alder reaction,** a conjugated diene reacts with a compound containing a carbon–carbon double bond. The latter compound is called a **dienophile** because it "loves a diene." (Recall that Δ signifies heat.)

$$CH_2{=}CH{-}CH{=}CH_2 \ + \ CH_2{=}CH{-}R \ \xrightarrow{\Delta} \ \text{(cyclohexene with R)}$$

conjugated diene   dienophile

The Diels–Alder reaction is a pericyclic reaction. A **pericyclic reaction** takes place by a cyclic shift of electrons. It is also a cycloaddition reaction. In a **cycloaddition reaction,** two reactants form a cyclic product. More precisely, the Diels–Alder reaction is a **[4 + 2] cycloaddition reaction** because *four* of the six π electrons that participate in the cyclic transition state come from the conjugated diene and *two* come from the dienophile.

The Diels–Alder reaction and other cycloaddition reactions are discussed in greater detail in Section 28.4.

MECHANISM FOR THE DIELS–ALDER REACTION

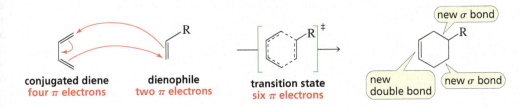

conjugated diene
four π electrons

dienophile
two π electrons

transition state
six π electrons

new σ bond

new double bond

new σ bond

- Although this reaction may not look like any reaction you have seen before, it is simply the 1,4-addition of an electrophile and a nucleophile to a conjugated diene. However, unlike the other 1,4-addition reactions you have seen—where the electrophile adds to the diene in the first step and the nucleophile adds to the carbocation in the second step—the Diels–Alder reaction is a **concerted reaction:** the addition of the electrophile and the addition of the nucleophile occur in a single step.

The Diels–Alder reaction looks odd at first glance because the electrophile and the nucleophile that add to the conjugated diene are the adjacent sp^2 carbons of a double bond. As with other 1,4-addition reactions, the double bond in the product is between the 2- and 3-positions of the diene's conjugated system.

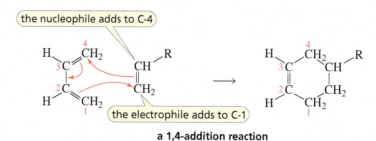

a 1,4-addition reaction

An Electron Withdrawing Group Increases the Reactivity of the Dienophile

The reactivity of the dienophile is increased when an electron-withdrawing group is attached to one of its sp^2 carbons. An electron-withdrawing group, such as a carbonyl group (C=O) or a cyano group (C≡N), withdraws electrons from the dienophile's double bond. This puts a partial positive charge on the sp^2 carbon that the π electrons of the conjugated diene add to. Thus, the electron-withdrawing group makes the dienophile a better electrophile. Consequently, the reaction can be carried out at room temperature (Figure 8.13).

resonance contributors of the dienophile   resonance hybrid

Examples of Diels–Alder Reactions

A wide variety of cyclic compounds can be obtained from Diels–Alder reactions by varying the structures of the conjugated diene and the dienophile.

Compounds containing carbon–carbon triple bonds can also be used as dienophiles in Diels–Alder reactions to prepare compounds with two isolated double bonds.

▲ Figure 8.13
The bluish color at the bottom of the electrostatic potential maps for compounds with an electron withdrawing substituent shows that an electron-withdrawing substituent makes the bottom sp^2 carbon a better electrophile.

PROBLEM 39 ◆

What are the products of the following reactions?

a. $CH_2{=}CH{-}CH{=}CH_2$ + $CH_3\overset{\overset{O}{\|}}{C}{-}C{\equiv}C{-}\overset{\overset{O}{\|}}{C}CH_3$ $\longrightarrow$

b. $CH_2{=}CH{-}CH{=}CH_2$ + $HC{\equiv}C{-}C{\equiv}N$ $\longrightarrow$

c. $CH_2{=}\overset{\overset{CH_3}{|}}{C}{-}\overset{\overset{CH_3}{|}}{C}{=}CH_2$ + (maleic anhydride) $\longrightarrow$

d. $CH_3\overset{\overset{CH_3}{|}}{C}{=}CH{-}CH{=}\overset{\overset{CH_3}{|}}{C}CH_3$ + (cyclohexene-1,4-dione) $\longrightarrow$

A Molecular Orbital Description of the Diels–Alder Reaction

The two new σ bonds formed in a Diels–Alder reaction result from electrons being transferred from one reactant to another. Molecular orbital theory provides insight into this process.

In a cycloaddition reaction, the orbitals of one reactant must overlap the orbitals of the other. Because the new σ bonds in the product are formed as a result of transfer of electrons from one reactant to another, we must consider the HOMO of one reactant and the LUMO of the other because only an empty orbital can accept electrons. It does not matter whether we envision the overlap as being between the HOMO of the dienophile and the LUMO of the diene or vice versa. We just need to consider the HOMO of one and the LUMO of the other.

To construct the HOMO and LUMO needed to illustrate the transfer of electrons, we need to look back at Figure 8.11 on p. 337. It shows that the HOMO of the diene and the LUMO of the dienophile are both antisymmetric (Figure 8.11a) and that the LUMO of the diene and the HOMO of the dienophile are both symmetric (Figure 8.11b).

Pericyclic reactions such as the Diels–Alder reaction can be described by a theory called the *conservation of orbital symmetry* (Section 28.2). This simple theory says that pericyclic reactions occur as a result of the overlap of in-phase orbitals. The phase of an orbital in Figure 8.14 is indicated by its color. Thus, each new bond formed in a Diels–Alder reaction must be created by the overlap of orbitals of the same color.

Figure 8.14 shows that regardless of which pair of HOMO and LUMO we choose, the overlapping orbitals that form the two new σ bonds have the same color. Thus, a Diels–Alder reaction occurs with relative ease. The Diels–Alder reaction and other cycloaddition reactions are discussed in greater detail in Section 28.4.

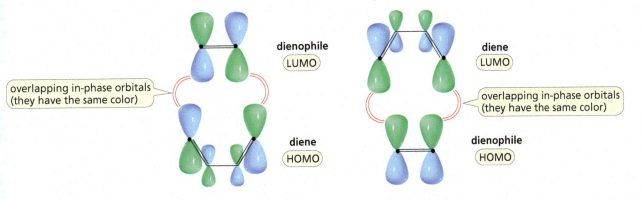

▲ **Figure 8.14**

The new σ bonds formed in a Diels–Alder reaction result from overlap of in-phase orbitals.

(a) Overlap of the HOMO of the diene and the LUMO of the dienophile.

(b) Overlap of the HOMO of the dienophile and the LUMO of the diene.

Predicting the Product When Both Reagents Are Unsymmetrically Substituted

In each of the preceding Diels–Alder reactions, only one product is formed (disregarding stereo-isomers) because at least one of the reacting molecules is symmetrically substituted. If both the diene and the dienophile are unsymmetrically substituted, however, two products are possible. The products are constitutional isomers.

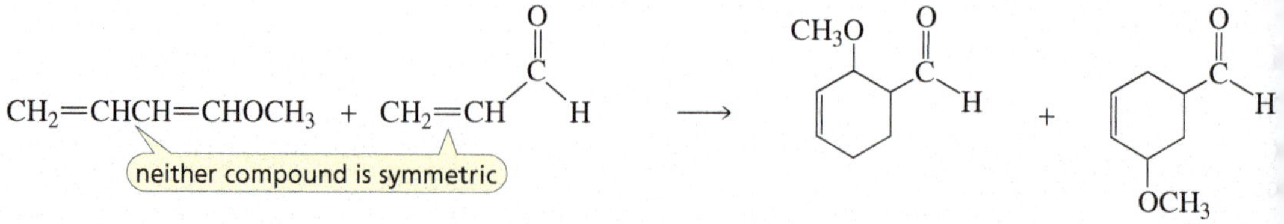

Two products are possible because the reactants can align in two different ways. (To determine the structure of the second product, don't change the position of one of the reactants and turn the other upside down.)

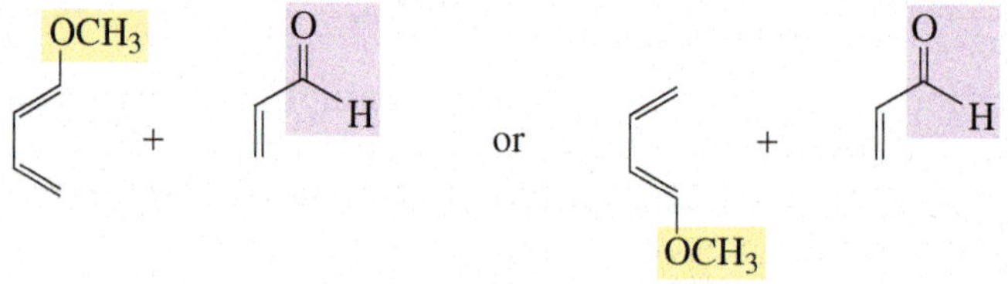

The product formed in greater yield depends on the charge distribution in each of the reactants. To determine the charge distribution, we need to draw the resonance contributors of the reactants. The methoxy group of the diene *donates electrons by resonance.* As a result, its terminal carbon has a partial negative charge. The carbonyl group of the dienophile, on the other hand, *withdraws electrons by resonance,* so its terminal carbon has a partial positive charge.

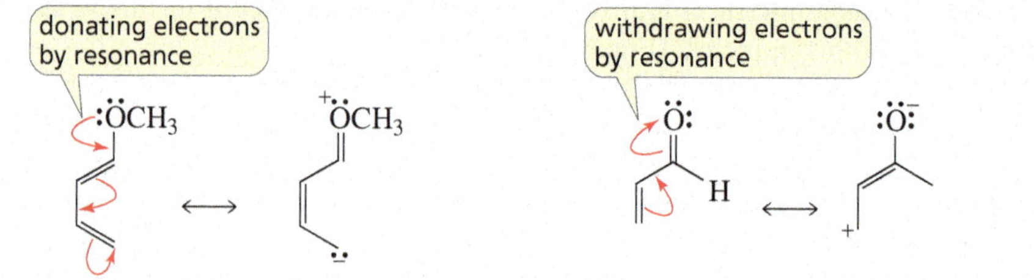

resonance contributors of the diene   resonance contributors of the dienophile

The partially positively charged carbon of the dienophile bonds preferentially to the partially negatively charged carbon of the diene. Therefore, the compound shown here is the major product.

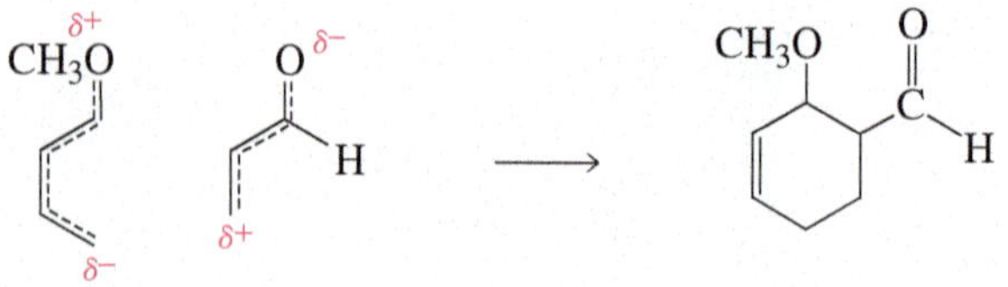

PROBLEM 40 ◆

What is the major product when the methoxy substituent in the preceding reaction is bonded to C-2 of the diene rather than to C-1?

PROBLEM 41

Write a general rule that can be used to predict the major product of a Diels–Alder reaction between an alkene with an electron-withdrawing substituent and a diene with a substituent that can donate electrons by resonance depending on the location of the substituent on the diene.

What two products are formed from each of the following reactions?

a. $CH_2{=}CH{-}CH{=}CH{-}CH_3 \;+\; HC{\equiv}C{-}C{\equiv}N \;\longrightarrow$

b. $CH_2{=}CH{-}\underset{\underset{\displaystyle CH_3}{|}}{C}{=}CH_2 \;+\; HC{\equiv}C{-}C{\equiv}N \;\longrightarrow$

Conformation of the Diene

We saw that a conjugated diene, such as 1,3-butadiene, is most stable in a *planar* conformation (Section 8.8). However, there are two different planar conformers: an *s*-cis conformer and an *s*-trans conformer. (Recall that a conformer results from rotation about single bonds; Section 3.11.)

In an **s-cis conformer**, the double bonds are cis about the single bond (*s* = single), whereas they are trans about the single bond in an *s*-trans conformer. An **s-trans conformer** is a little more stable (by 2.3 kcal/mol) because the close proximity of the hydrogens in the *s*-cis conformer causes some steric strain (Section 3.11). The rotational barrier between the *s*-cis and *s*-trans conformers is low enough (4.9 kcal/mol) to allow the conformers to interconvert rapidly at room temperature.

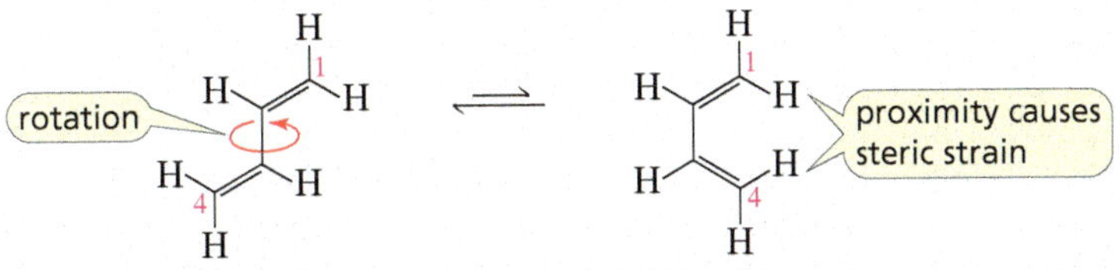

s-trans conformer   *s*-cis conformer

In order to participate in a Diels–Alder reaction, the conjugated diene must be in an *s*-cis conformation because when it is in an *s*-trans conformation, C-1 and C-4 are too far apart to react with the dienophile in a concerted reaction. Therefore, a conjugated diene that is locked in an *s*-trans conformation cannot undergo a Diels–Alder reaction because it cannot achieve the required *s*-cis conformation.

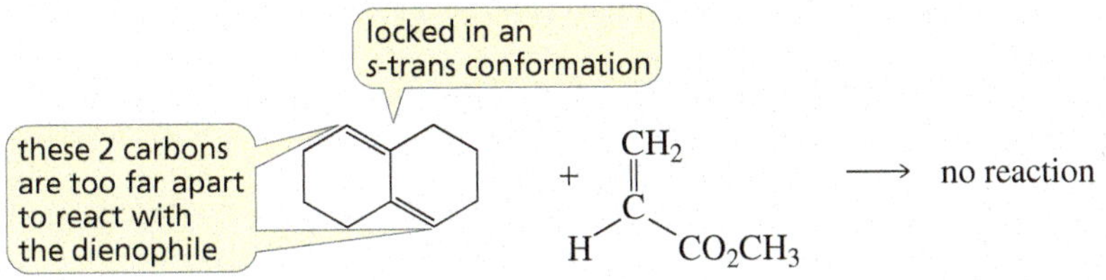

A conjugated diene that is locked in an *s*-cis conformation, such as 1,3-cyclopentadiene, is highly reactive in a Diels–Alder reaction. When the diene is a cyclic compound, the product of a Diels–Alder reaction is a **bridged bicyclic compound** (a compound that contains two rings that share two nonadjacent carbons).

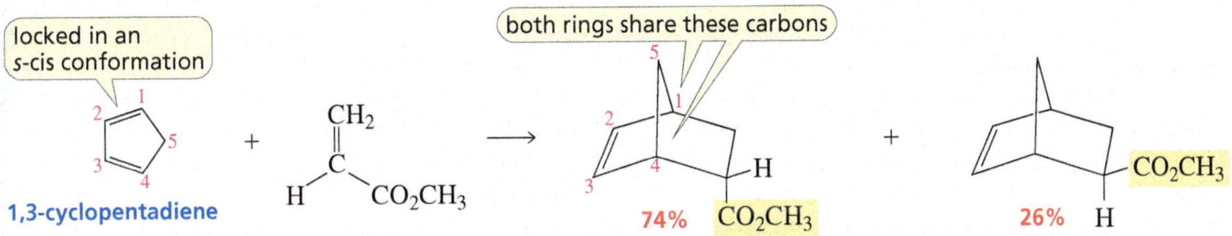

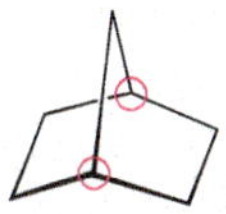

bridged bicyclic rings
2 rings share 2 nonadjacent carbons

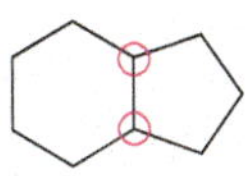

fused bicyclic rings
2 rings share 2 adjacent carbons

1,3-cyclopentadiene

bridged bicyclic compounds

Exo and Endo Products

There are two possible configurations for substituted bridged bicyclic compounds, because the substituent (R) can either point away from the double bond (the **exo** configuration) or not point away from the double bond (the **endo** configuration).

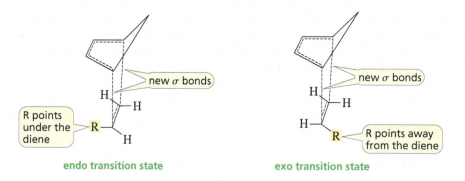

does not point away from the double bond

points away from the double bond

endo   exo

Figure 8.15 shows why both endo and exo products are formed.

▶ **Figure 8.15**
The transition states for formation of the endo and exo products show that two products are formed because the dienophile can line up in two different ways. The substituent (R) can point either under the diene (endo) or away from the diene (exo).

new σ bonds

new σ bonds

R points under the diene

R points away from the diene

endo transition state   exo transition state

When the dienophile has a substituent with π electrons, more of the endo product is formed.

The endo product is formed faster when the dienophile has a substituent with π electrons. Recent studies suggest that the increased rate of endo product formation is due to interaction between the π electrons of the substituent and the π electrons of the ring, which stabilizes the transition state. A substituent in the exo position cannot engage in such stabilizing interactions.

PROBLEM 43 ◆

Which of the following conjugated dienes will not react with a dienophile in a Diels–Alder reaction?

A   B   C   D   E   F

PROBLEM 44

What are the products of the following reactions?

a. ⬠ + CH₂=CH—C(=O)—CH₃ ⟶   b. ⬡ + CH₂=CH—C(=O)—CH₃ ⟶

PROBLEM 45 SOLVED

Rank the following dienes from most reactive to least reactive in a Diels–Alder reaction:

SOLUTION The most reactive diene has its double bonds locked in an *s*-cis conformation, whereas the least reactive diene has its double bonds locked in an *s*-trans conformation. The other two compounds are of intermediate reactivity because they can exist in both *s*-cis and *s*-trans conformations. 1,3-Pentadiene is less apt to be in the required *s*-cis conformation because of steric interference between the hydrogen and the methyl group, so it is less reactive than 2-methyl-1,3-butadiene.

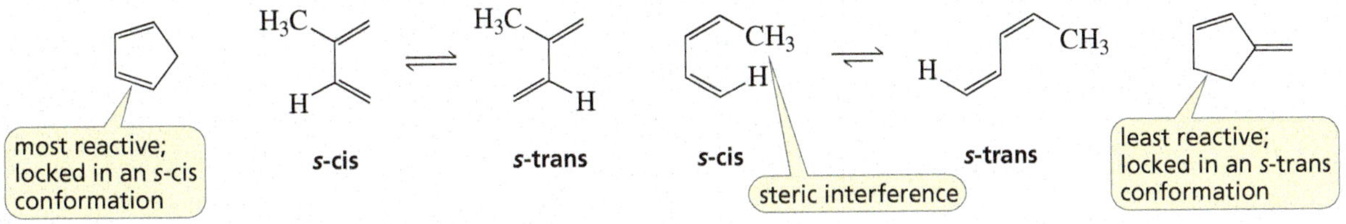

most reactive; locked in an s-cis conformation

s-cis

s-trans

s-cis

steric interference

s-trans

least reactive; locked in an s-trans conformation

Thus, the four dienes have the following order of reactivity:

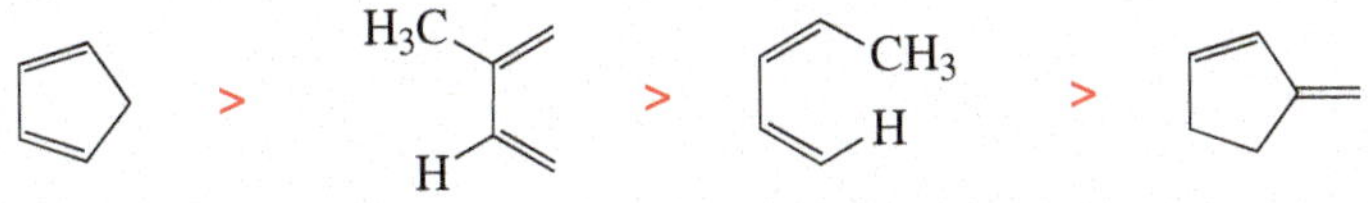

The Stereochemistry of the Diels–Alder Reaction

As with all the other reactions we have seen, if a Diels–Alder reaction creates a product with an asymmetric center, the product will be a racemic mixture (Section 6.13).

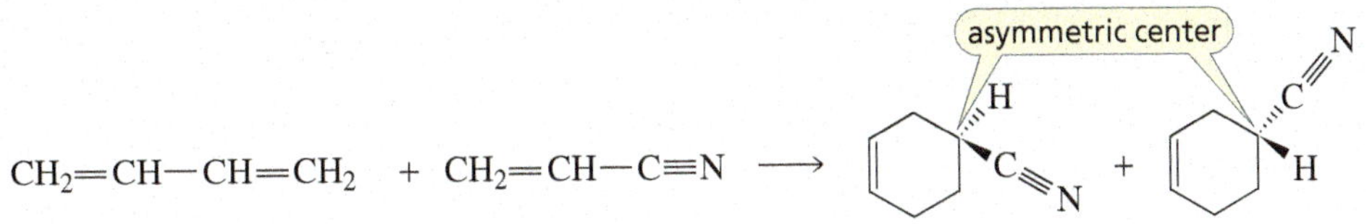

asymmetric center

$$CH_2{=}CH{-}CH{=}CH_2 \ + \ CH_2{=}CH{-}C{\equiv}N \longrightarrow$$

The Diels–Alder reaction is a syn addition reaction. One face of the diene adds to one face of the dienophile. Therefore, if the substituents in the *dienophile* are cis, then they will be cis in the product; if the substituents in the *dienophile* are trans, then they will be trans in the product. Because each of the following syn addition reactions forms a product with two new asymmetric centers, each reaction forms a pair of enantiomers (Section 6.13). The stereochemistry of the reaction will be discussed in greater detail in Section 28.4.

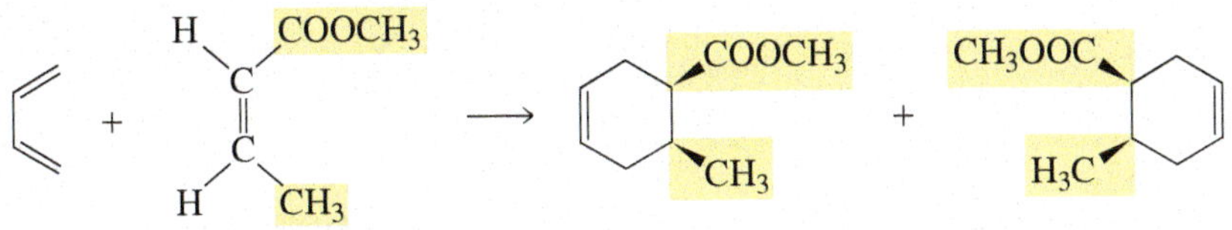

cis dienophile

cis products

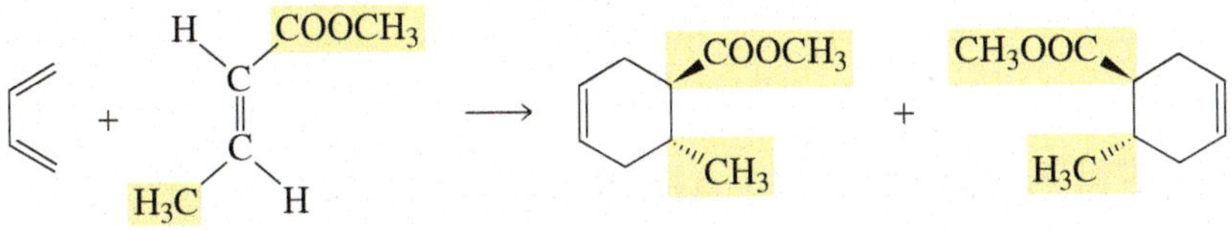

trans dienophile

trans products

PROBLEM 46 ◆

Explain why the following compounds are not optically active:

a. the product obtained from the reaction of 1,3-butadiene with *cis*-1,2-dichloroethene
b. the product obtained from the reaction of 1,3-butadiene with *trans*-1,2-dichloroethene

8.15 RETROSYNTHETIC ANALYSIS OF THE DIELS–ALDER REACTION

To determine the reactants needed to synthesize a Diels-Alder product:

LEARN THE STRATEGY

1. locate the double bond in the product. The diene that was used to form the cyclic product had double bonds on either side of this double bond, so draw in those double bonds and remove the original double bond.

2. the new σ bonds are now on the other side of these double bonds. Deleting these σ bonds and putting a π bond between the two carbons whose σ bonds were deleted gives the needed reactants—that is, the diene and the dienophile.

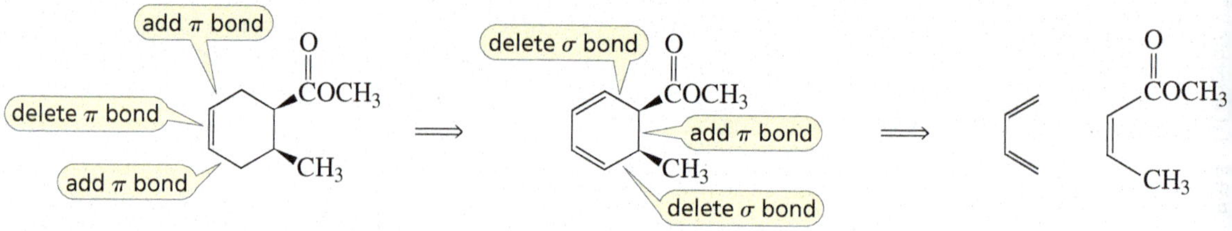

Now let's use these two rules to determine the reactants needed for the synthesis of a bridged bicyclic compound.

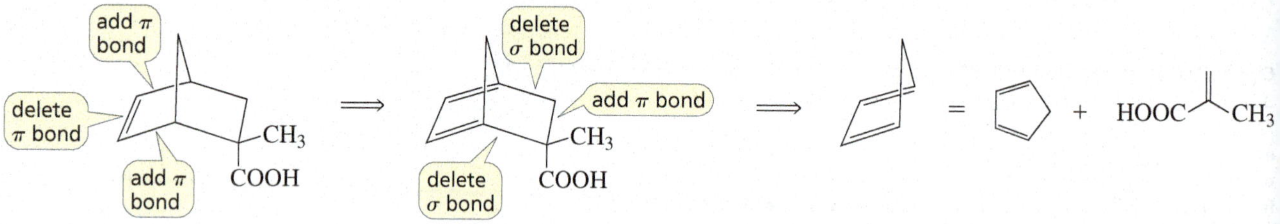

USE THE STRATEGY

PROBLEM 47 ◆

What diene and what dienophile should be used to synthesize the following?

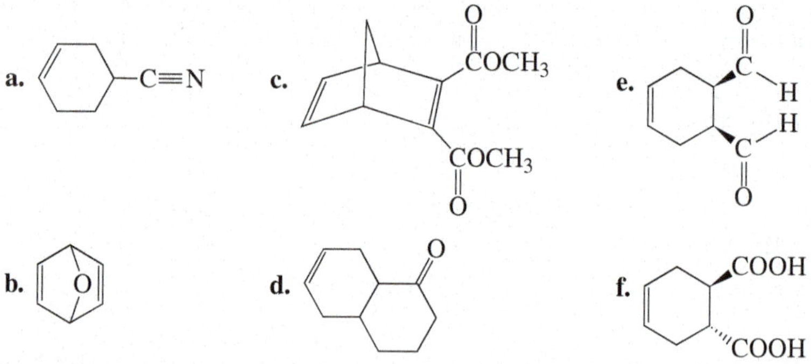

8.16 BENZENE IS AN AROMATIC COMPOUND

The two resonance contributors of benzene are identical, so we expect benzene to have a relatively large delocalization energy (Section 8.6).

The heat of hydrogenation data shown in Figure 8.16 indicate that benzene's delocalization energy is even larger (36 kcal/mol) than that expected for a compound with two equivalent resonance contributors. Most often, compounds with large delocalization energies, such as benzene, are **aromatic compounds** (Section 8.17).

Because of its large delocalization energy, benzene is an extremely stable compound. Therefore, it does not undergo the electrophilic addition reactions that are characteristic of alkenes except under extreme conditions. (Notice the conditions that Sabatier had to use in order to reduce benzene's double bonds on page 320.) Now we can understand why benzene's unusual stability puzzled nineteenth-century chemists, who did not know about delocalized electrons (Section 8.1).

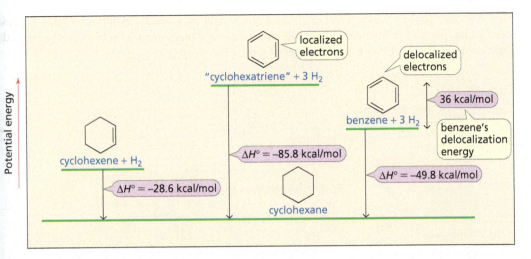

▲ **Figure 8.16**
Cyclohexene, a compound with one double bond with *localized* π electrons, has an experimental Δ*H°* = –28.6 kcal/mol for its reaction with H₂ to form cyclohexane. Therefore, the Δ*H°* of "cyclohexatriene," an unknown hypothetical compound with three double bonds with *localized* π electrons, would be three times that value (Δ*H°* = 3 × –28.6 = –85.8) for the same reaction. Benzene, which has three double bonds with *delocalized* π electrons, has an experimental Δ*H°* = –49.8 kcal/mol for its reaction with H₂ to form cyclohexane. The difference in the energies of "cyclohexatriene" and benzene (36 kcal/mol) is the delocalization energy of benzene—the extra stability benzene has as a result of having delocalized electrons.

Aromatic compounds are particularly stable.

8.17 THE TWO CRITERIA FOR AROMATICITY

How can we tell whether a compound is aromatic by looking at its structure? In other words, what structural features do aromatic compounds have in common? To be classified as aromatic, a compound must meet both of the following criteria:

LEARN THE STRATEGY

1. *It must have an uninterrupted cyclic cloud of π electrons* (called a π cloud) *above and below the plane of the molecule.* Let's look a little more closely at what this means.

   - For the π cloud to be cyclic, *the molecule must be cyclic.*
   - For the π cloud to be uninterrupted, *every atom in the ring must have a* p *orbital.*
   - For the π cloud to form, each *p* orbital must overlap the *p* orbitals on either side of it. As a result, *the molecule must be planar.*

2. *The π cloud must contain an odd number of pairs of π electrons.*

Thus, benzene is an aromatic compound because it is cyclic and planar, every carbon in the ring has a *p* orbital, and the π cloud contains *three* pairs of π electrons (Figure 8.1).

For a compound to be aromatic, it must be cyclic and planar, and it must have an uninterrupted cloud of π electrons. The π cloud must contain an odd number of pairs of π electrons.

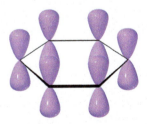

benzene's *p* orbitals

benzene's π cloud

benzene has 3 pairs of π electrons

The German physicist Erich Hückel was the first to recognize that an aromatic compound must have an *odd number of pairs of π electrons.* In 1931, he described this requirement in what has come to be known as **Hückel's rule,** or the **4*n* + 2 rule.** The rule states that for a planar, cyclic compound to be aromatic, its uninterrupted π cloud must contain (4*n* + 2) π electrons, where *n* is any whole number. According to Hückel's rule, then, an aromatic compound must have 2 (*n* = 0), 6 (*n* = 1), 10 (*n* = 2), 14 (*n* = 3), 18 (*n* = 4), and so on, π electrons. Because there are two electrons in a pair, Hückel's rule requires that an aromatic compound have 1, 3, 5, 7, 9, and so on, pairs of π electrons. Thus, Hückel's rule is a mathematical way of saying that an aromatic compound must have an *odd* number of pairs of π electrons.

USE THE STRATEGY

PROBLEM 48 ♦

Which of the following are aromatic?

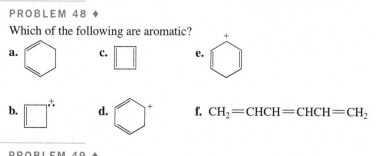

a.   c.   e.

b.   d.   f. $CH_2\!=\!CHCH\!=\!CHCH\!=\!CH_2$

PROBLEM 49 ♦

a. What is the value of *n* in Hückel's rule when a compound has nine pairs of π electrons?
b. Is such a compound aromatic?

8.18 APPLYING THE CRITERIA FOR AROMATICITY

Cyclobutadiene has two pairs of π electrons, and cyclooctatetraene has four pairs of π electrons. These compounds, therefore, are *not* aromatic because they have an *even* number of pairs of π electrons. There is an additional reason why cyclooctatetraene is not aromatic—it is not planar, it is tub-shaped (see p. 323). Because cyclobutadiene and cyclooctatetraene are not aromatic, they do not have the unusual stability of aromatic compounds.

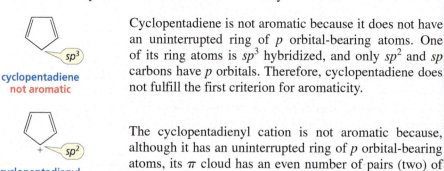

cyclobutadiene   cyclooctatetraene

Now let's look at some other compounds and determine whether they are aromatic.

cyclopentadiene
not aromatic

Cyclopentadiene is not aromatic because it does not have an uninterrupted ring of *p* orbital-bearing atoms. One of its ring atoms is sp^3 hybridized, and only sp^2 and sp carbons have *p* orbitals. Therefore, cyclopentadiene does not fulfill the first criterion for aromaticity.

cyclopentadienyl
cation
not aromatic

The cyclopentadienyl cation is not aromatic because, although it has an uninterrupted ring of *p* orbital-bearing atoms, its π cloud has an even number of pairs (two) of π electrons.

cyclopentadienyl
anion
aromatic

The cyclopentadienyl anion is aromatic; it has an uninterrupted ring of *p* orbital-bearing atoms, and the π cloud contains an odd number of pairs (three) of π electrons.

How do we know that the cyclopentadienyl anion's lone-pair electrons are π electrons? There is an easy way to determine this: if a lone pair can be used to form a π bond in the ring of a resonance contributor of the compound, then the lone-pair electrons are π electrons.

lone pair forms a π bond

resonance contributors of the cyclopentadienyl anion

resonance hybrid

When drawing resonance contributors, remember that only electrons move; atoms never move.

The resonance hybrid shows that all the carbons in the cyclopentadienyl anion are equivalent. Each carbon has exactly one-fifth of the negative charge associated with the anion.

The criteria that determine whether a monocyclic hydrocarbon is aromatic can also be used to determine whether a polycyclic hydrocarbon is aromatic. Naphthalene (five pairs of π electrons), phenanthrene (seven pairs of π electrons), and chrysene (nine pairs of π electrons) are aromatic.

naphthalene   **phenanthrene**   **chrysene**

Buckyballs

We saw that diamond, graphite, and graphene are forms of pure carbon (Section 1.8). Another form of pure carbon was discovered unexpectedly in 1985, while scientists were conducting experiments designed to understand how long-chain molecules are formed in outer space. R. E. Smalley, R. F. Curl, Jr., and H. W. Kroto shared the 1996 Nobel Prize in Chemistry for discovering this new form of carbon. They named the substance *buckminsterfullerene* (often shortened to *fullerene*) because its structure reminded them of the geodesic domes popularized by R. Buckminster Fuller, an American architect and philosopher. Buckminsterfullerene's nickname is "buckyball."

Consisting of a hollow cluster of 60 carbons, fullerene is the most symmetrical large molecule known. Like graphite and graphene, fullerene has only sp^2 carbons, but instead of being arranged in layers, the carbons are arranged in rings that fit together like the seams of a soccer ball. Each molecule has 32 interlocking rings (20 hexagons and 12 pentagons). At first glance, fullerene appears to be aromatic because of its benzene-like rings. However, the curvature of the ball prevents the molecule from fulfilling the first criterion for aromaticity—it must be planar. Therefore, fullerene is not aromatic.

Buckyballs have extraordinary chemical and physical properties. For example, they are exceedingly rugged, as shown by their ability to survive the extreme temperatures of outer space. Because they are essentially hollow cages, they can be manipulated to make new materials. For example, when a buckyball is "doped" by inserting potassium or cesium into its cavity, it becomes an excellent organic superconductor. These molecules are now being studied for use in many other applications, including the development of new polymers, catalysts, and drug-delivery systems. The discovery of buckyballs is a strong reminder of the technological advances that can be achieved as a result of basic research.

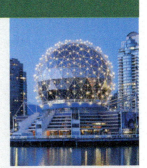

a geodesic dome.

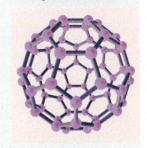

C_{60}
buckminsterfullerene
"buckyball"

PROBLEM 50 ◆

Which compound in each set is aromatic? Explain your choice.

a.

cyclopropene   **cyclopropenyl cation**   **cyclopropenyl anion**

b.

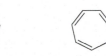

cycloheptatriene   **cycloheptatrienyl cation**   **cycloheptatrienyl anion**

LEARN THE STRATEGY

PROBLEM 51 SOLVED

The pK_a of cyclopentane is > 60, which is about what is expected for a hydrogen that is bonded to an sp^3 carbon. Explain why cyclopentadiene is a much stronger acid (pK_a of 15) even though it too involves the loss of a proton from an sp^3 carbon.

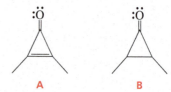

cyclopentane
$pK_a > 60$

cyclopentyl anion

cyclopentadiene
$pK_a = 15$

cyclopentadienyl anion

SOLUTION   To answer this question, we must look at the stabilities of the anions that are formed when the compounds lose a proton. (Recall that the strength of an acid is determined by the stability of its conjugate base: the more stable the base, the stronger its conjugate acid; Section 2.6). All the electrons in the cyclopentyl anion are localized. In contrast, the cyclopentadienyl anion is aromatic. As a result of its aromaticity, the cyclopentadienyl anion is an unusually stable carbanion, causing its conjugate acid to be an unusually strong acid compared to other compounds with hydrogens attached to sp^3 carbons.

USE THE STRATEGY

PROBLEM 52 ◆

a. Predict the relative pK_a values of cyclopentadiene and cycloheptatriene.
b. Predict the relative pK_a values of cyclopropene and cyclopropane.

PROBLEM 53 ◆

Which is more soluble in water: 3-bromocyclopropene or bromocyclopropane?

PROBLEM-SOLVING STRATEGY

LEARN THE STRATEGY

Analyzing Electron Distribution in Compounds

Which of the following compounds has the greater dipole moment?

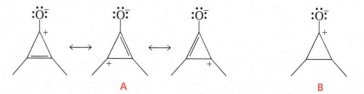

A   B

The dipole moment of these compounds results from the unequal sharing of electrons by carbon and oxygen. The dipole moment increases as the electron sharing becomes more unequal. So now the question becomes, which compound has a greater negative charge on its oxygen? To find out, we need to draw the structures with a negative charge on oxygen and determine their relative stabilities.

A   B

When the π bond is broken in order to put a negative charge on oxygen, we see that **A** has three resonance contributors, all of which are aromatic; **B** has no resonance contributors. Therefore, **A** has a greater concentration of negative charge on its oxygen, which gives **A** a greater dipole moment.

PROBLEM 54

a. In what direction is the dipole moment in fulvene? Explain.
b. In what direction is the dipole moment in calicene? Explain.

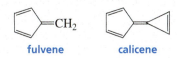

fulvene   calicene

8.19 A MOLECULAR ORBITAL DESCRIPTION OF AROMATICITY

The relative energies of the molecular orbitals of a compound that has a π cloud can be determined—without having to use any math—by drawing the cyclic compound with one of its vertices pointed down. The relative levels of the vertices correspond to the relative energies of the molecular orbitals (Figure 8.17). Notice that the number of molecular orbitals is the same as the number of atoms in the ring because each ring atom contributes a p orbital.

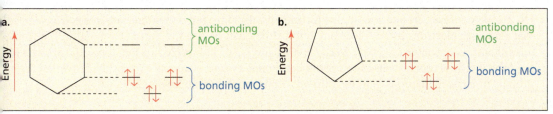

Figure 8.17
The distribution of electrons in the molecular orbitals of (a) benzene and (b) the cyclopentadienyl anion.

Aromatic compounds are stable because they have filled bonding molecular orbitals.

Molecular orbitals below the midpoint of the cyclic structure are bonding molecular orbitals, and those above the midpoint are antibonding molecular orbitals.

The six π electrons of benzene occupy its three bonding molecular orbitals, and the six π electrons of the cyclopentadienyl anion occupy its three bonding molecular orbitals. Notice that there is always an odd number of bonding molecular orbitals because one corresponds to the lowest vertex and the others come in degenerate pairs. Consequently, aromatic compounds—such as benzene and the cyclopentadienyl anion, with their odd number of pairs of π electrons—have completely filled bonding molecular orbitals. This is what gives aromatic molecules their stability.

Antiaromatic Compounds

A compound is classified as **antiaromatic** if it fulfills the first criterion for aromaticity but does not fulfill the second. In other words, it must be a planar, cyclic compound with an uninterrupted ring of p orbital-bearing atoms, and the π cloud must contain an *even* number of pairs of π electrons. Because antiaromatic compounds cannot fill their bonding molecular orbitals, they are very unstable and highly reactive compounds. Thus, an aromatic compound is *more stable* than a cyclic compound with localized electrons, whereas an antiaromatic compound is *less stable* than a cyclic compound with localized electrons.

Antiaromatic compounds are very unstable.

relative stabilities

aromatic compound > cyclic compound with localized electrons > antiaromatic compound

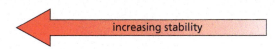

PROBLEM 55

Following the instructions for drawing the energy levels of the molecular orbitals for the compounds shown in Figure 8.17, draw the energy levels of the molecular orbitals for the cycloheptatrienyl cation, the cycloheptatrienyl anion, and the cyclopropenyl cation. For each compound, show the distribution of the π electrons. Which of the compounds are aromatic?

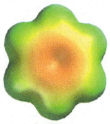

benzene

pyridine

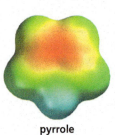

pyrrole

8.20 AROMATIC HETEROCYCLIC COMPOUNDS

A compound does not have to be a hydrocarbon to be aromatic. Many *heterocyclic compounds* are aromatic. A **heterocyclic compound** is a cyclic compound in which one or more of the ring atoms is an atom other than carbon. The atom that is not carbon is called a **heteroatom.** The name comes from the Greek word *heteros*, which means "different." The most common heteroatoms encountered in organic compounds are N, O, and S.

heterocyclic compounds

pyridine   pyrrole   furan   thiophene

Pyridine

Pyridine is an aromatic heterocyclic compound. Each of the six ring atoms of pyridine is sp^2 hybridized, which means that each has a p orbital, and the molecule contains three pairs of π electrons. Do not be confused by the lone-pair electrons on the nitrogen—they are not π electrons. Recall that lone-pair electrons are π electrons only if they can be used to form a π bond in the ring of a resonance contributor (p. 365). Because nitrogen is sp^2 hybridized, it has three sp^2 orbitals and a p orbital. The p orbital is used to form the π bond. Two of nitrogen's sp^2 orbitals overlap the sp^2 orbitals of adjacent carbons, and its third sp^2 orbital contains the lone pair.

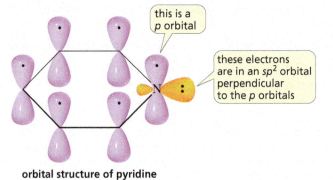

this is a
p orbital

these electrons
are in an sp^2 orbital
perpendicular
to the p orbitals

orbital structure of pyridine

Pyrrole, Furan, and Thiophene

The nitrogen atom of pyrrole is sp^2 hybridized. Thus, it has three sp^2 orbitals and a p orbital. It uses its three sp^2 orbitals to bond to two carbons and one hydrogen. The lone-pair electrons are in the p orbital that overlaps the p orbitals of adjacent carbons. Pyrrole, therefore, has three pairs of π electrons and is aromatic.

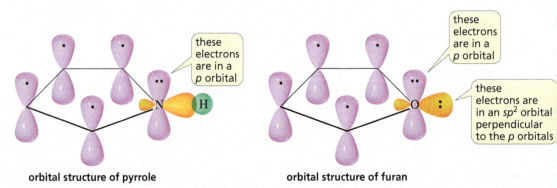

these
electrons
are in a
p orbital

these
electrons
are in a
p orbital

these
electrons are
in an sp^2 orbital
perpendicular
to the p orbitals

orbital structure of pyrrole   **orbital structure of furan**

The resonance contributors show that lone-pair electrons of pyrrole form a π bond in the ring of a resonance contributor; thus, they are π electrons.

resonance contributors of pyrrole

lone pair forms a π bond

Similarly, furan and thiophene are aromatic compounds. Both the oxygen in furan and the sulfur in thiophene are sp^2 hybridized and have one lone pair in an sp^2 orbital. The orbital picture of furan on the previous page shows that the second lone pair is in a p orbital that overlaps the p orbitals of adjacent carbons, forming a π bond. Thus, they are π electrons.

resonance contributors of furan

lone pair forms a π bond

Quinoline, indole, imidazole, purine, and pyrimidine are other examples of heterocyclic aromatic compounds. The heterocyclic compounds discussed in this section will be examined in greater detail in Chapter 19.

quinoline   indole   imidazole   purine   pyrimidine

PROBLEM 56 ◆

What orbital do the lone-pair electrons occupy in each of the following compounds?

a. $CH_3CH_2\ddot{N}H_2$ **b.**    **c.** $CH_3CH_2C\equiv N\colon$

PROBLEM 57 ◆

What orbitals contain the electrons represented as lone pairs in the structures of quinoline, indole, imidazole, purine, and pyrimidine?

PROBLEM 58 SOLVED

Which of the following compounds could be protonated without destroying its aromaticity?

LEARN THE STRATEGY

SOLUTION Pyridine's lone-pair electrons are sp^2 electrons. They, therefore, are not part of the π cloud (only π electrons are in the π cloud), so they can be protonated without destroying the aromaticity of the pyridine ring.

We know that pyrrole's lone-pair electrons are π electrons because they can be used to form a π bond in a resonance contributor.

Because pyrrole's lone-pair electrons are one of the three pairs of π electrons in the π cloud, pyrrole's aromaticity is destroyed when it is protonated and, therefore, is no longer part of the π cloud.

USE THE STRATEGY

PROBLEM 59

Which of the following compounds could be protonated without destroying its aromaticity?

PROBLEM 60

Refer to the electrostatic potential maps on p. 368 to answer the following questions:

a. Why is the bottom part of the electrostatic potential map of pyrrole blue?
b. Why is the bottom part of the electrostatic potential map of pyridine red?
c. Why is the center of the electrostatic potential map of benzene more red than the center of the electro static potential map of pyridine?

8.21 HOW BENZENE REACTS

Aromatic compounds such as benzene undergo **electrophilic aromatic substitution reactions**: a electrophile substitutes for one of the hydrogens attached to the benzene ring.

Now let's look at why this substitution reaction occurs. The cloud of π electrons above an below the plane of its ring makes benzene a nucleophile, so it reacts with an electrophile (Y^+). Whe an electrophile attaches itself to a benzene ring, a carbocation intermediate is formed.

carbocation intermediate

This description should remind you of the first step in an *electrophilic addition reaction* of a alkene: the nucleophilic alkene reacts with an electrophile and forms a carbocation intermediate (Section 6.0). In the second step of the reaction, the carbocation reacts with a nucleophile (Z^-) t form an addition product.

$$RCH=CHR + Y^+ \rightleftharpoons RCH-CHR \xrightarrow{Z^-} RCH-CHR$$

carbocation intermediate

product of electrophilic addition

If the carbocation intermediate that is formed when benzene reacts with an electrophile were t react similarly (depicted as path *a* in Figure 8.18), then the *addition product* would not be aromatic But, if the carbocation instead were to lose a proton from the site of electrophilic addition and form a *substitution product* (depicted as path *b* in Figure 8.18), then the aromaticity of the benzene ring would be restored.

Because the aromatic substitution product is much more stable than the nonaromatic addi tion product (Figure 8.19), benzene undergoes *electrophilic substitution reactions* that pre serve aromaticity, rather than *electrophilic addition reactions*—the reactions characteristic o kenes—that would destroy aromaticity. The substitution reaction is more accurately called an **electrophilic aromatic substitution reaction,** because the electrophile substitutes for a hydrogen of an aromatic compound.

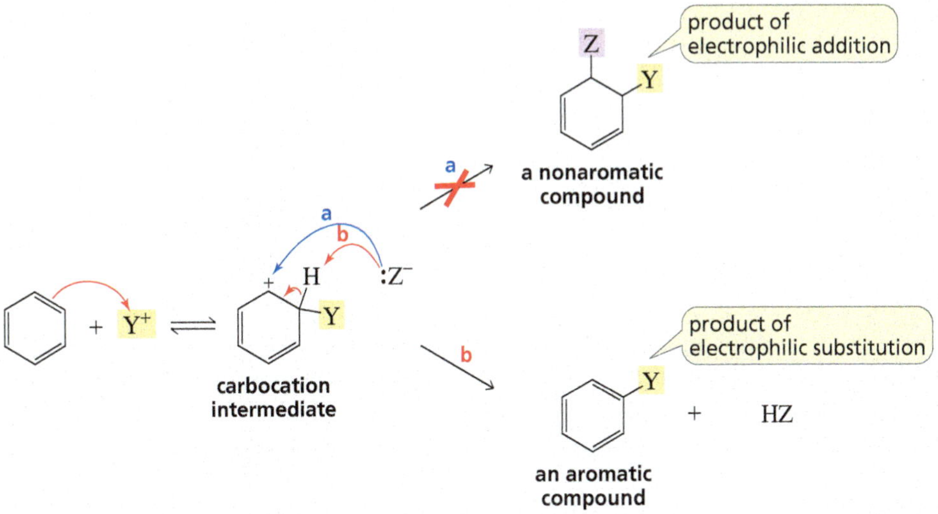

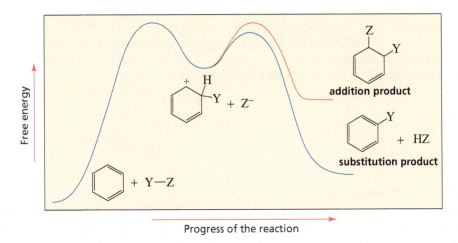

▲ Figure 8.18
Reaction of benzene with an electrophile. Because of the greater stability of the aromatic product, electrophilic substitution (path *b*) occurs rather than electrophilic addition (path *a*).

▲ Figure 8.19
Reaction coordinate diagrams for electrophilic aromatic substitution and for electrophilic addition.

Thus, when an alkene reacts with Br_2, an electrophilic addition reaction occurs.

$$CH_3CH{=}CHCH_3 \;+\; Br_2 \longrightarrow CH_3CH{-}CHCH_3$$
$$\qquad\qquad\qquad\qquad\qquad\quad\; |\quad\;\; |$$
$$\qquad\qquad\qquad\qquad\qquad Br \;\; Br$$

But when benzene reacts with Br_2, an electrophilic aromatic substitution reaction occurs. Notice that the reaction of benzene with Br_2, unlike the reaction of an alkene, requires a catalyst ($FeBr_3$) because benzene's aromaticity causes it to be unusually stable and, therefore, relatively unreactive.

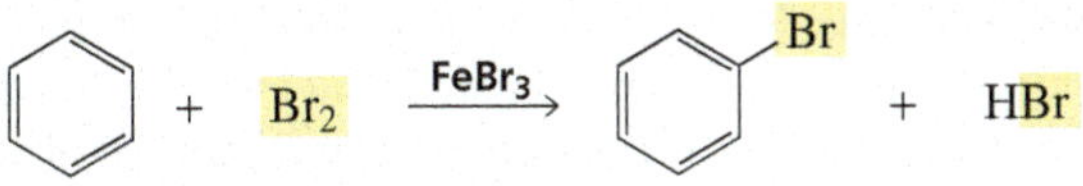

Now we can understand why early chemists found that benzene did not undergo the characteristic electrophilic addition reactions of alkenes (Section 8.1).

The mechanism for an electrophilic aromatic substitution reaction is shown on the next page.

THE MECHANISM FOR AN ELECTROPHILIC AROMATIC SUBSTITUTION REACTION

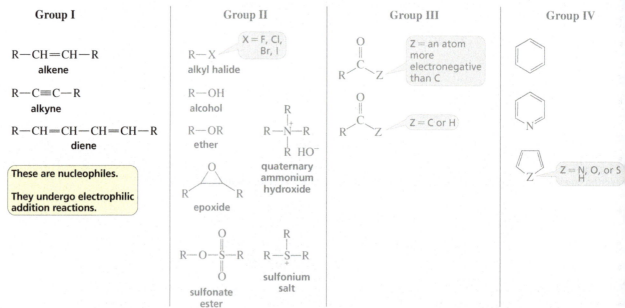

the proton is removed from the carbon that formed the bond with the electrophile

a base in the reaction mixture

- The electrophile (Y^+) adds to the nucleophilic benzene ring, thereby forming a carbocation intermediate.

- A base in the reaction mixture (:B) removes a proton from the carbon that formed the bond with the electrophile, and the electrons that held the proton move into the ring to reestablish its aromaticity.

The electrophilic aromatic substitution reactions of benzene are discussed in detail in Chapter 18.

8.22 ORGANIZING WHAT WE KNOW ABOUT THE REACTIONS OF ORGANIC COMPOUNDS

When you were first introduced to the reactions of organic compounds in Section 5.4, you saw that organic compounds can be classified into families determined by their functional group and that all the members of a family react in the same way. You also saw that each family can be placed into one of four groups and that all the families in a group react in similar ways. Now that you have finished studying the reactions of the first group, let's revisit that group.

Group I	Group II	Group III	Group IV

Group I

R—CH=CH—R
alkene

R—C≡C—R
alkyne

R—CH=CH—CH=CH—R
diene

These are nucleophiles.

They undergo electrophilic addition reactions.

Group II

R—X (X = F, Cl, Br, I)
alkyl halide

R—OH
alcohol

R—OR
ether

epoxide

R—O—S—R (sulfonate ester)

R—N⁺—R (R) HO⁻
quaternary ammonium hydroxide

R—S⁺—R
sulfonium salt

Group III

R—C(=O)—Z (Z = an atom more electronegative than C)

R—C(=O)—Z (Z = C or H)

Group IV

(benzene ring)

(pyridine ring, N)

(five-membered ring, Z = N, O, or S)

All the families in the first group are nucleophiles because of their electron-rich carbon–carbon double or triple bonds. And because double and triple bonds have relatively weak π bonds, the families in this group undergo addition reactions. Because the first species that reacts with a nucleophile is an electrophile, the reactions that the families in this group undergo are more precisely called *electrophilic addition reactions*.

- Alkenes have one π bond, so they undergo one electrophilic addition reaction.

$$\text{R—CH=CH—R} + Y^+ + Z^- \longrightarrow \text{R—CH—CH—R} \quad (\text{with } Y \text{ and } Z)$$

- Alkynes have two π bonds, so they can undergo two electrophilic addition reactions.

$$R-C\equiv C-R \ + \ 2Y^+ \ + \ 2Z^- \ \longrightarrow \ R-\overset{\overset{\displaystyle Y}{|}}{\underset{\underset{\displaystyle Y}{|}}{C}}-\overset{\overset{\displaystyle Z}{|}}{\underset{\underset{\displaystyle Z}{|}}{C}}-R$$

- However, if the first addition reaction forms an enol, the enol immediately rearranges to a ketone (or to an aldehyde), so a second addition reaction cannot occur.

$$R-C\equiv C-R \ \xrightarrow{H_3O^+} \ R-CH=\overset{\overset{\displaystyle OH}{|}}{C}-R \ \rightleftharpoons \ R-CH_2-\overset{\overset{\displaystyle O}{\|}}{C}\diagdown R$$

- If the double bonds of a diene are isolated, they react just like alkenes. If, however, the double bonds are conjugated, they undergo both 1,2- and 1,4-addition reactions because the carbocation intermediate has delocalized electrons.

$$R-CH=CH-CH=CH-R \ + \ Y^+ \ + \ Z^- \ \longrightarrow \ R-\overset{\overset{\displaystyle }{|}}{\underset{\underset{\displaystyle Y}{|}}{CH}}-\overset{\overset{\displaystyle }{|}}{\underset{\underset{\displaystyle Z}{|}}{CH}}-CH=CH-R \ + \ R-\overset{\overset{\displaystyle }{|}}{\underset{\underset{\displaystyle Y}{|}}{CH}}-CH=CH-\overset{\overset{\displaystyle }{|}}{\underset{\underset{\displaystyle Z}{|}}{CH}}-R$$

ESSENTIAL CONCEPTS

Section 8.0
- **Localized electrons** belong to a single atom or are shared by two atoms. **Delocalized electrons** are shared by more than two atoms.

Section 8.2
- The six π electrons of benzene are shared by all six carbons. Thus, benzene is a planar molecule with six delocalized π electrons.

Section 8.3
- **Resonance contributors**—structures with localized electrons—approximate the structure of a compound that has delocalized electrons—the **resonance hybrid.**
- Electron delocalization occurs only if all the atoms sharing the delocalized electrons lie in or close to the same plane.

Section 8.4
- To draw resonance contributors, move π electrons or lone-pair electrons to an sp^2 or sp hybridized atom.
- Electron delocalization (resonance) results when a p orbital overlaps the p orbitals of two adjacent atoms.

Section 8.5
- The greater the predicted stability of a resonance contributor, the more it contributes to the structure of the hybrid and the more similar its structure is to the real molecule.
- Predicted stability is decreased by (1) an atom with an incomplete octet, (2) a negative charge that is not on the most electronegative atom, (3) a positive charge that is on an electronegative atom, and (4) charge separation.

- A **resonance hybrid** is more stable than the predicted stability of any of its resonance contributors.

Section 8.6
- **Delocalization energy** (or **resonance energy**) is the extra stability a compound gains from having delocalized electrons. It tells us how much more stable a compound with delocalized electrons is than it would be if all its electrons were localized.
- The greater the number of relatively stable resonance contributors and the more nearly equivalent they are, the greater the delocalization energy.

Section 8.7
- Allylic and benzylic cations have delocalized electrons, so they are more stable than similarly substituted carbocations with localized electrons.
- **Conjugated double bonds** are separated by one single bond. **Isolated double bonds** are separated by more than one single bond.
- Because dienes with conjugated double bonds have delocalized electrons, conjugated dienes are more stable than isolated dienes.

Section 8.8
- The number of molecular orbitals equals the number of atomic orbitals that produced them.
- Side-to-side overlap of in-phase p orbitals produces a **bonding molecular orbital,** which is more stable than the p atomic orbitals. The side-to-side interaction between out-of-phase p orbitals produces an **antibonding molecular orbital,** which is less stable than the p atomic orbitals.

- A **molecular orbital** results from the **linear combination of atomic orbitals (LCAO).**
- The **highest occupied molecular orbital (HOMO)** is the highest-energy MO that contains electrons. The **lowest unoccupied molecular orbital (LUMO)** is the lowest-energy MO that does not contain electrons.
- As MOs increase in energy, the number of nodes increases, the number of bonding interactions decreases, and the MOs alternate between **symmetric** and **antisymmetric.**
- Molecular orbital theory and resonance contributors both show that electrons are delocalized and that electron delocalization makes a molecule more stable.

Section 8.9

- A carboxylic acid and a phenol are more acidic than an alcohol, and a protonated aniline is more acidic than a protonated amine because inductive electron withdrawal and an increase in delocalization energy stabilize the conjugate bases of carboxylic acids, phenols, and protonated anilines.

Section 8.10

- Donation of electrons through π bonds is called **resonance electron donation;** withdrawal of electrons through π bonds is called **resonance electron withdrawal.**
- Substituents on a benzene ring can donate electrons into the ring by hyperconjugation and by resonance electron donation and can withdraw electrons from the ring inductively or by resonance electron withdrawal.

Section 8.12

- An isolated diene, like an alkene, undergoes only 1,2-addition. If there is only enough electrophilic reagent to add to one of the double bonds, it will add preferentially to the double bond that forms the more stable carbocation.
- A conjugated diene reacts with one equivalent of an electrophilic reagent to form a **1,2-addition product** and a **1,4-addition product.** The first step is addition of the electrophile to an sp^2 carbon at one end of the conjugated system.

Section 8.13

- The more rapidly formed product is the **kinetic product;** the more stable product is the **thermodynamic product.**
- If both reactions are irreversible, the major product will be the kinetic product; if one or both of the reactions are reversible, the major product will be the thermodynamic product.
- When a reaction is **kinetically controlled,** the relative amounts of the products depend on the rates at which they are formed; when a reaction is **thermodynamically controlled,** the relative amounts of the products depend on their stabilities.
- A **common intermediate** is an intermediate that both products have in common.
- In electrophilic addition to a conjugated diene, the 1,2-product is always the kinetic product; either the 1,2- or the 1,4-product can be the thermodynamic product, depending on their relative stability.

Section 8.14

- In a **Diels–Alder reaction,** a **conjugated diene** reacts with a **dienophile** to form a cyclic compound; in this concerted **[4 + 2] cycloaddition reaction,** two new σ bonds and a π bond are formed at the expense of two π bonds.
- The reactivity of the **dienophile** is increased by electron-withdrawing groups attached to an sp^2 carbon.
- If both the diene and the dienophile are unsymmetrically substituted, two products are possible because the reactants can be aligned in two different ways.
- To participate in a Diels–Alder reaction, the conjugated diene must be in an **s-cis conformation.**
- The Diels–Alder reaction is a syn addition.
- In **bridged bicyclic compounds,** a substituent can be **endo** or **exo;** endo is favored if the substituent has π electrons.

Section 8.15

- An **aromatic compound** has an uninterrupted cyclic cloud of π electrons that contains an *odd number of pairs* of π electrons.

Section 8.19

- An **antiaromatic compound** has an uninterrupted cyclic cloud of π electrons that contains an *even number of pairs* of π electrons.
- Aromatic compounds are very stable; antiaromatic compounds are very unstable.

Section 8.20

- A **heterocyclic compound** is a cyclic compound in which one or more of the ring atoms is a **heteroatom**—that is, an atom other than carbon.
- Pyridine, pyrrole, furan, and thiophene are aromatic heterocyclic compounds.

Section 8.21

- In order to preserve its aromaticity, benzene undergoes electrophilic aromatic substitution reactions rather than electrophilic addition reactions.

SUMMARY OF REACTIONS

1. In the presence of excess electrophilic reagent, both double bonds of an *isolated diene* undergo electrophilic addition (Section 8.12). The mechanism is on p. 347.

$$CH_2{=}CHCH_2CH_2\overset{\overset{\displaystyle CH_3}{|}}{C}{=}CH_2 \;+\; \underset{\text{excess}}{HBr} \;\longrightarrow\; CH_3\underset{\underset{\displaystyle Br}{|}}{C}HCH_2CH_2\overset{\overset{\displaystyle CH_3}{|}}{\underset{\underset{\displaystyle Br}{|}}{C}}CH_3$$

In the presence of one equivalent of electrophilic reagent, only the most reactive double bond of an *isolated diene* will undergo electrophilic addition.

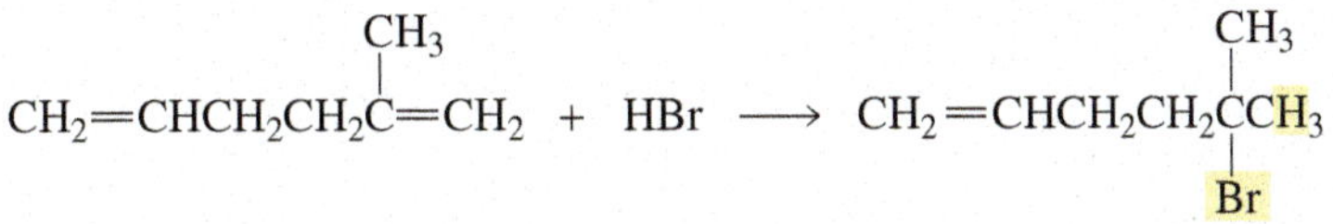

$$CH_2=CHCH_2CH_2\overset{\overset{\displaystyle CH_3}{|}}{C}=CH_2 \ + \ HBr \ \longrightarrow \ CH_2=CHCH_2CH_2\overset{\overset{\displaystyle CH_3}{|}}{\underset{\underset{\displaystyle Br}{|}}{C}}CH_3$$

2. Conjugated dienes undergo 1,2- and 1,4-addition in the presence of one equivalent of an electrophilic reagent (Section 8.12). The mechanism is on p. 349.

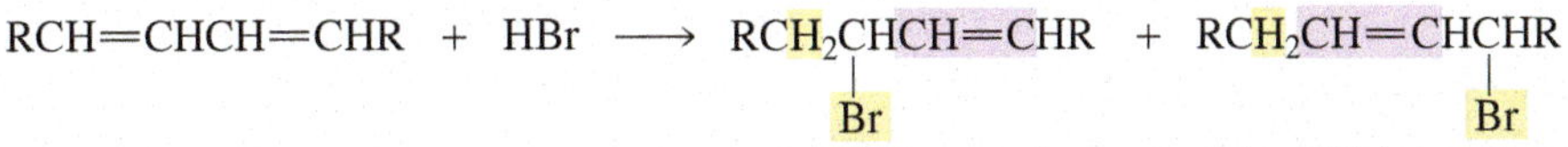

$$RCH=CHCH=CHR \ + \ HBr \ \longrightarrow \ RCH_2\underset{\underset{\displaystyle Br}{|}}{CH}CH=CHR \ + \ RCH_2CH=CHCHR$$
$$\underset{\underset{\displaystyle Br}{|}}{}$$

$$\text{1,2-addition product} \qquad \text{1,4-addition product}$$

3. Conjugated dienes undergo 1,4-addition with a dienophile (a Diels–Alder reaction; see Section 8.14). The mechanism is on p. 355.

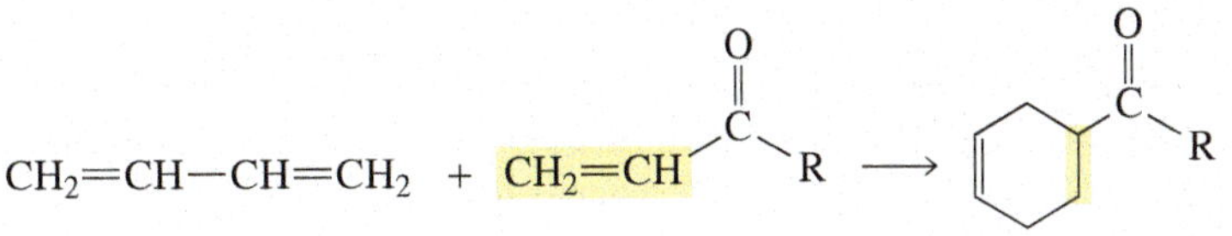

$$CH_2=CH-CH=CH_2 \ + \ CH_2=CH\overset{\overset{\displaystyle O}{\|}}{C}{-}R \ \longrightarrow$$

PROBLEMS

61. Which of the following has delocalized electrons?

A $CH_2=CH\overset{\overset{\displaystyle O}{\|}}{C}CH_3$

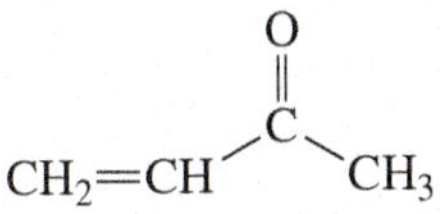

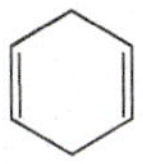

B $CH_3CH=CHOCH_2CH_3$

C $CH_3CH=CHCH=CH\overset{+}{C}H_2$

D $CH_3\overset{+}{C}HCH_2CH=CH_2$

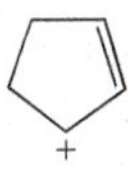

E (fused bicyclic structure)

F (cyclopentenyl cation)

G $CH_3\overset{+}{\underset{\underset{\displaystyle }{}}{C}}CH_2CH=CH_2$ with CH_3 group

H $CH_2=CHCH_2CH=CH_2$

I $CH_3CH_2NHCH_2CH=CHCH_3$

J (cyclopentenyl cation)

K (cyclohexene ring)

L (cyclohexadiene ring)

M $CH_3CH_2\overset{+}{C}HCH=CH_2$

N $CH_3CH_2NHCH=CHCH_3$

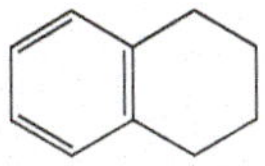

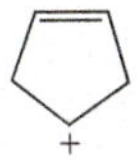

O (fused bicyclic structure)

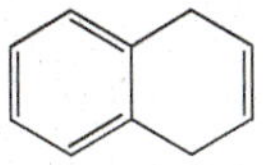

62. a. Draw resonance contributors for the following species, showing all the lone pairs:

   1. CH_2N_2 **2.** N_2O **3.** NO_2^-

   b. For each species, indicate the most stable resonance contributor.

63. What is the major product of each of the following reactions? Assume that there is an equivalent amount of each reagent.

   a. 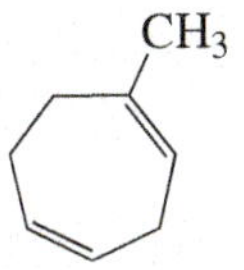 (cycloheptadiene with CH_3) $+ \ HBr \ \longrightarrow$

   b. 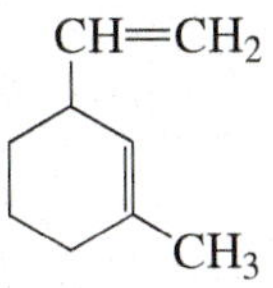 (cyclohexene with $CH=CH_2$ and CH_3) $+ \ HBr \ \longrightarrow$

64. Draw resonance contributors for the following ions:

   a. 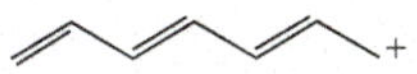 (structure with $+$)
   b. 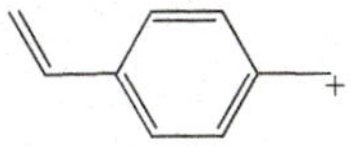 (structure with $+$)
   c. 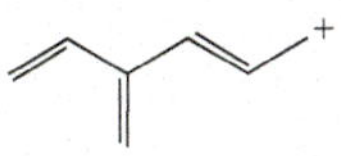 (structure with $+$)
   d. 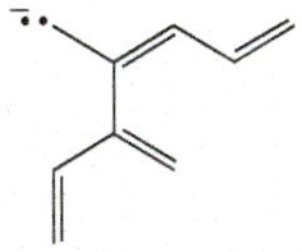 (structure with lone pair)

65. Draw all the products of the following reaction:

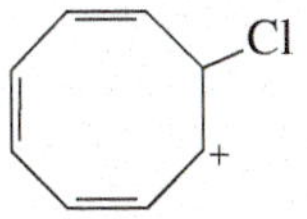

   (cyclooctadiene with Cl) $+ \ HO^- \ \longrightarrow$

66. Are the following pairs of structures resonance contributors or different compounds?

a. 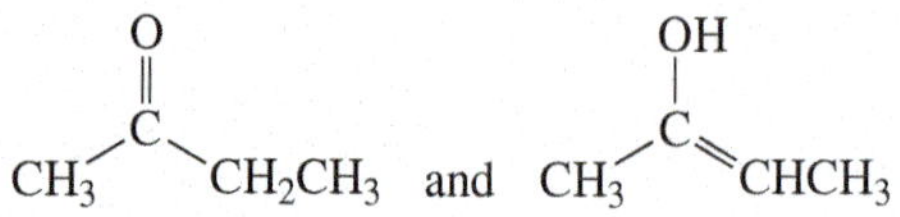

b. $CH_3\overset{+}{C}HCH{=}CHCH_3$ and $CH_3CH{=}CHCH_2\overset{+}{C}H_2$

c. $CH_3CH{=}CH\overset{+}{C}HCH{=}CH_2$ and $CH_3\overset{+}{C}HCH{=}CHCH{=}CH_2$

d. 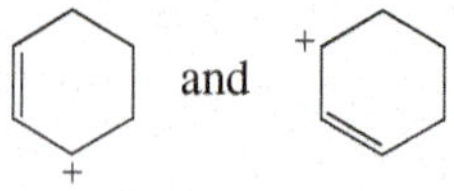

e. 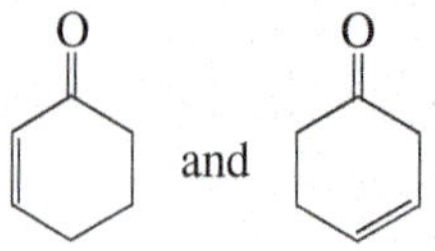

67. a. How many linear dienes have molecular formula C_6H_{10}? (Disregard cis–trans isomers.)
b. How many of the linear dienes in part **a** are conjugated dienes?
c. How many are isolated dienes?
d. How many are cumulated dienes?

68. a. Draw resonance contributors for the following species. Do not include structures that are so unstable that their contributions to the resonance hybrid would be negligible. Indicate which are major contributors and which are minor contributors to the resonance hybrid.
b. Do any of the species have resonance contributors that all contribute equally to the resonance hybrid?

1. $CH_3CH{=}CHOCH_3$
2. 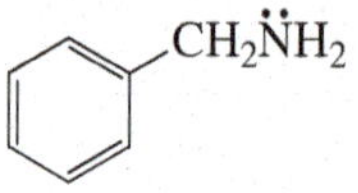
3. 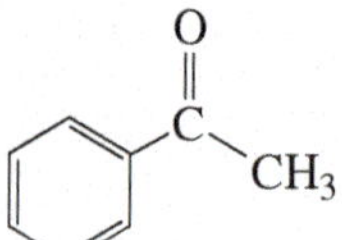
4. 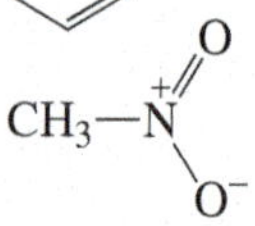
5. 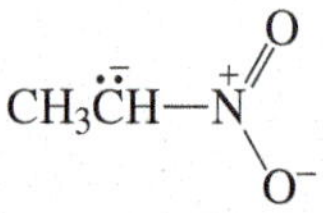
6. $CH_3CH{=}CH\overset{+}{C}H_2$
7. 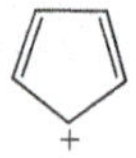
8. 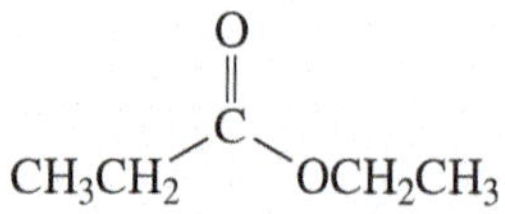
9. $CH_3CH{=}CHCH{=}CH\overset{+}{C}H_2$
10. 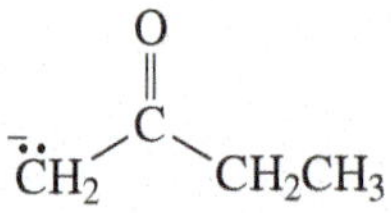
11. $CH_3\ddot{C}^{-}HCN$
12. 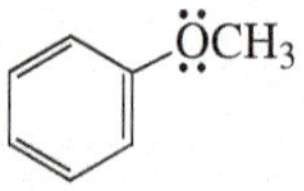
13. 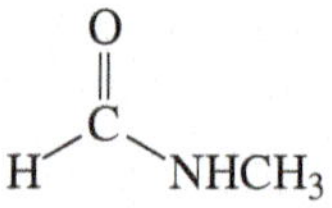
14. 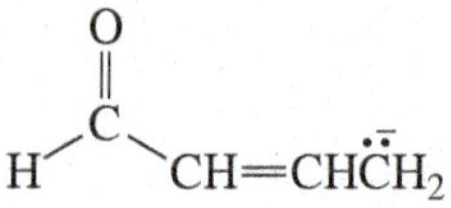
15. 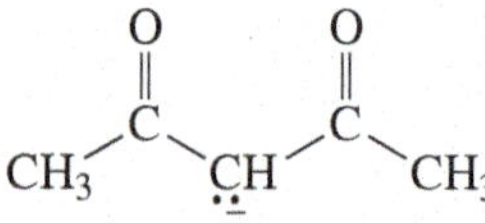

69. Which ion in each of the following pairs is more stable?

a. 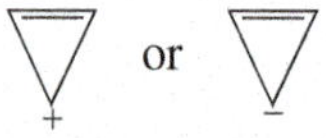

b. 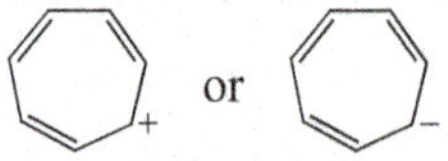

c. 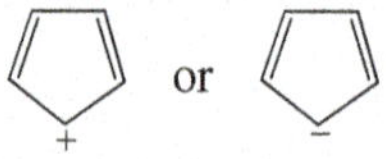

d. 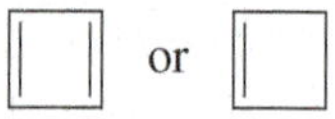

70. Which compound would you expect to have the greater heat of hydrogenation: 1,2-pentadiene or 1,4-pentadiene?

71. Which resonance contributor in each pair makes the greater contribution to the resonance hybrid?

a. $CH_3\overset{+}{C}HCH{=}CH_2$ or $CH_3CH{=}\overset{+}{C}HCH_2$

b. 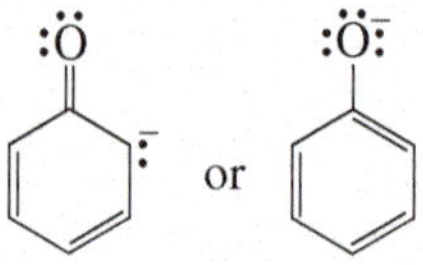

c. 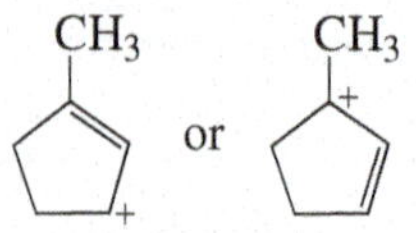

d. 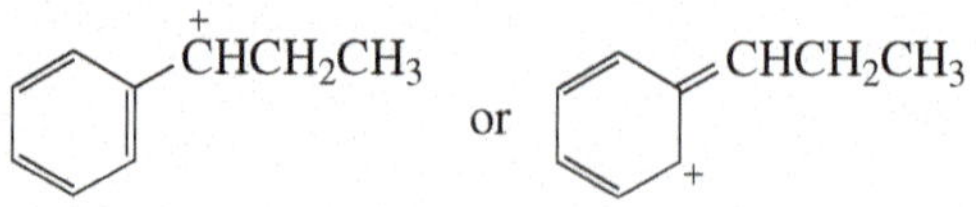

72. Which of the following are aromatic?

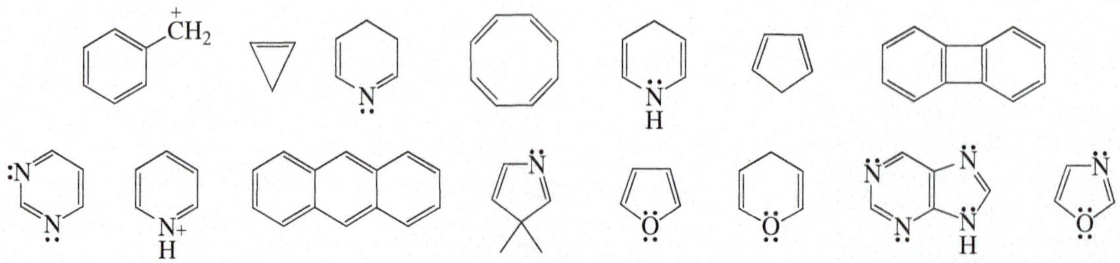

73. a. Which oxygen atom has the greater electron density?

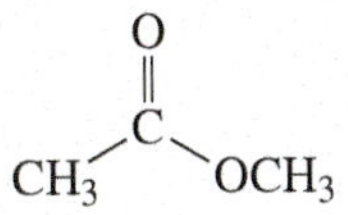

b. Which compound has the greater electron density on its nitrogen atom?

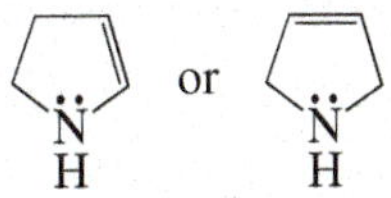

c. Which compound has the greater electron density on its oxygen atom?

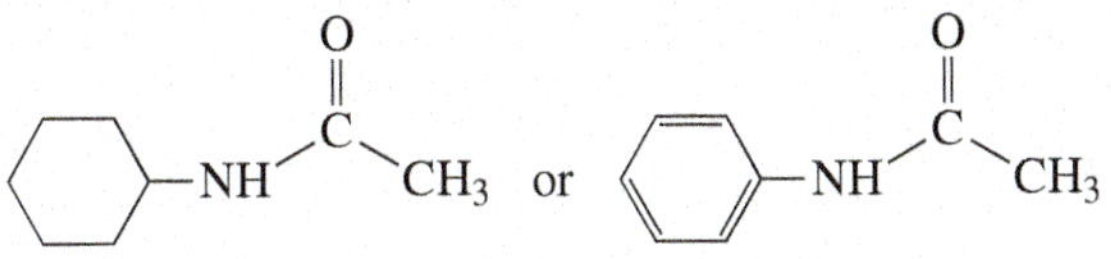

74. Which compound is the strongest base?

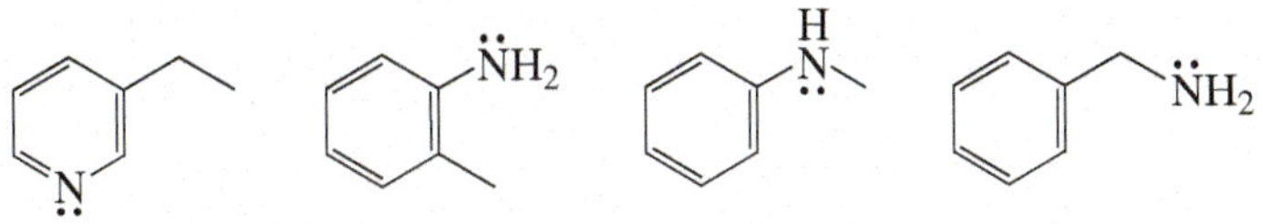

75. Which loses a proton more readily: a methyl group bonded to cyclohexane or a methyl group bonded to benzene?

76. The triphenylmethyl cation is so stable that a salt such as triphenylmethyl chloride can be isolated and stored. Why is this carbocation so stable?

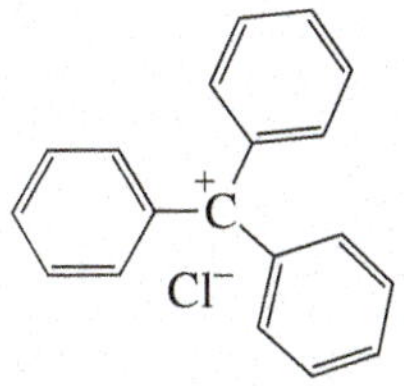

triphenylmethyl chloride

77. a. The A ring (Section 3.16) of cortisone (a steroid) is formed by a Diels–Alder reaction using the two reactants shown here. What is the product of this reaction?

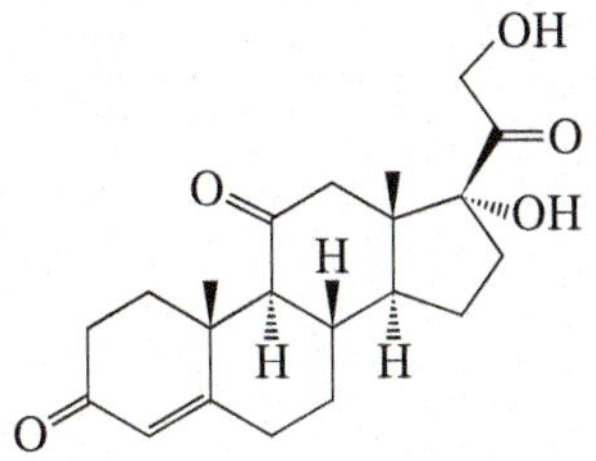

cortisone

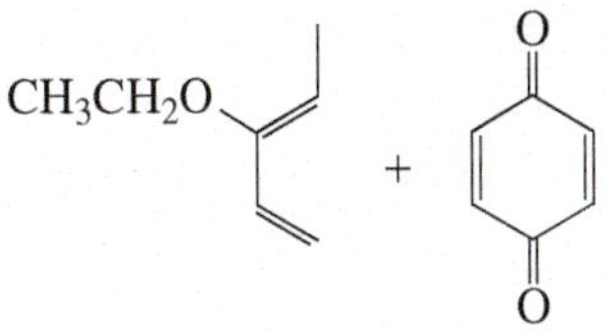

b. The C ring of estrone (a steroid) is formed by a Diels–Alder reaction using the two reactants shown here. What is the product of this reaction?

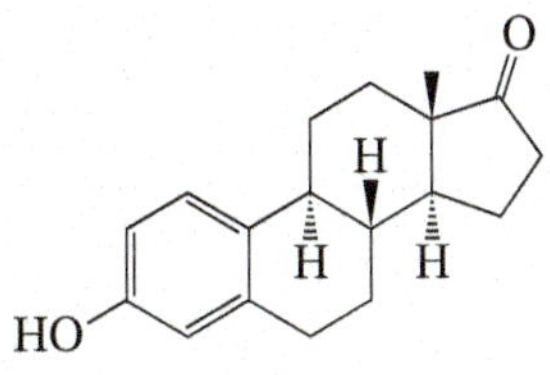

estrone

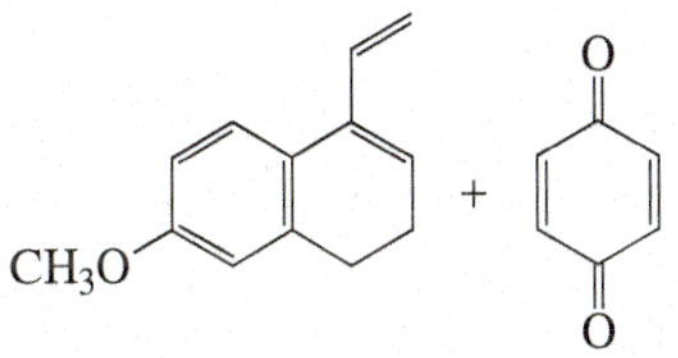

377

78. a. Which is a stronger acid?

or

b. Which is a stronger base?

1.

2. or

79. Rank the following carbocations from most stable to least stable:

80. Which species in each pair is more stable?

a.

$$\underset{H}{\overset{O}{\underset{\|}{C}}}\text{—}CH_2O^-$$ or $$CH_3\overset{O}{\overset{\|}{C}}O^-$$

c.

$$CH_3\overset{O}{\overset{\|}{C}}\ddot{C}HCH_2\overset{O}{\overset{\|}{C}}H$$ or $$CH_3\overset{O}{\overset{\|}{C}}\ddot{C}H\overset{O}{\overset{\|}{C}}CH_3$$

b.

$$CH_3\ddot{C}HCH_2\overset{O}{\overset{\|}{C}}CH_3$$ or $$CH_3CH_2\ddot{C}H\overset{O}{\overset{\|}{C}}CH_3$$

d. N:⁻ or N:⁻

81. Which species in each of the pairs in Problem 80 is the stronger base?

82. Purine is a heterocyclic compound with four nitrogen atoms.
    a. Which nitrogen is most apt to be protonated?   **b.** Which nitrogen is least apt to be protonated?

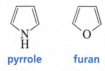

purine

83. Which of the following is the strongest acid?

84. Why is the delocalization energy of pyrrole (21 kcal/mol) greater than that of furan (16 kcal/mol)?

pyrrole   furan

85. Rank the indicated hydrogen in the following compounds from most acidic to least acidic:

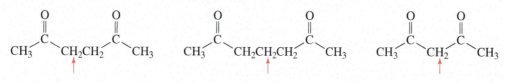

86. Answer the following questions for the molecular orbitals (MOs) of 1,3,5,7-octatetraene:
   a. How many MOs does the compound have?
   b. Which are the bonding MOs, and which are the antibonding MOs?
   c. Which MOs are symmetric, and which are antisymmetric?
   d. Which MO is the HOMO and which is the LUMO in the ground state?
   e. Which MO is the HOMO and which is the LUMO in the excited state?
   f. What is the relationship between HOMO and LUMO and symmetric and antisymmetric orbitals?
   g. How many nodes does the highest-energy MO of 1,3,5,7-octatetraene have between the nuclei?

87. How could you synthesize the following compound from starting materials containing no more than six carbons? (*Hint:* A 1,6-diketone can be synthesized by oxidative cleavage of a 1,2-disubstituted cyclohexene.)

88. A student obtained two products from the reaction of 1,3-cyclohexadiene with Br_2 (disregarding stereoisomers). His lab partner was surprised when he obtained only one product from the reaction of 1,3-cyclohexadiene with HBr (disregarding stereoisomers). Account for these results.

89. How could the following compounds be synthesized using a Diels–Alder reaction?

90. a. How could each of the following compounds be prepared from a hydrocarbon in a single step?
    b. What other organic compound would be obtained from each synthesis?

91. Draw the products obtained from the reaction of one equivalent of HBr with one equivalent of 1,3,5-hexatriene.
   a. Which product(s) will predominate if the reaction is under kinetic control?
   b. Which product(s) will predominate if the reaction is under thermodynamic control?

92. How would the following substituents affect the rate of a Diels–Alder reaction?
   a. an electron-donating substituent in the diene
   b. an electron-donating substituent in the dienophile
   c. an electron-withdrawing substituent in the diene

93. Draw the major products obtained from the reaction of one equivalent of HCl with the following compounds. For each reaction, indicate the kinetic and thermodynamic products.
   a. 2,3-dimethyl-1,3-pentadiene   b. 2,4-dimethyl-1,3-pentadiene

94. The acid dissociation constant (K_a) for loss of a proton from cyclohexanol is 1×10^{-16}.
   a. Draw a reaction coordinate diagram for loss of a proton from cyclohexanol.

   b. Draw the resonance contributors for phenol.
   c. Draw the resonance contributors for the phenolate ion.
   d. On the same plot with the energy diagram for loss of a proton from cyclohexanol, draw an energy diagram for loss of a proton from phenol.

   e. Which has a greater K_a: cyclohexanol or phenol?
   f. Which is a stronger acid: cyclohexanol or phenol?

95. Protonated cyclohexylamine has a $K_a = 1 \times 10^{-11}$. Using the same sequence of steps as in Problem 94, determine which is a stronger base: cyclohexylamine or aniline.

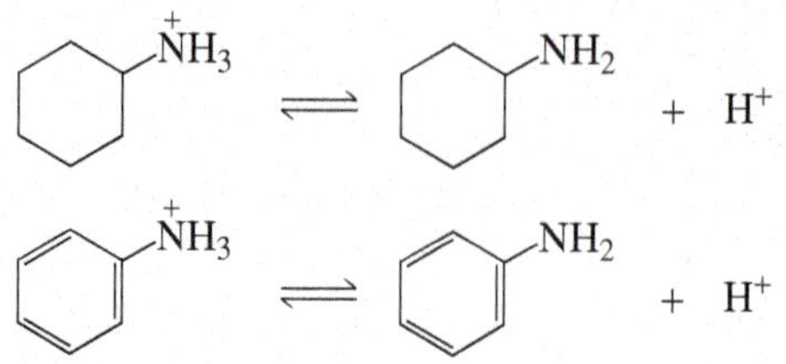

$$\text{cyclohexyl-}\overset{+}{N}H_3 \rightleftharpoons \text{cyclohexyl-}NH_2 + H^+$$

$$\text{phenyl-}\overset{+}{N}H_3 \rightleftharpoons \text{phenyl-}NH_2 + H^+$$

96. Draw the product or products that would be obtained from each of the following reactions:

a. 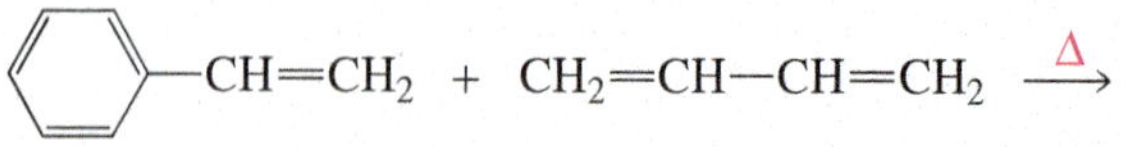 phenyl—CH=CH$_2$ + CH$_2$=CH—CH=CH$_2$ $\xrightarrow{\Delta}$

b. 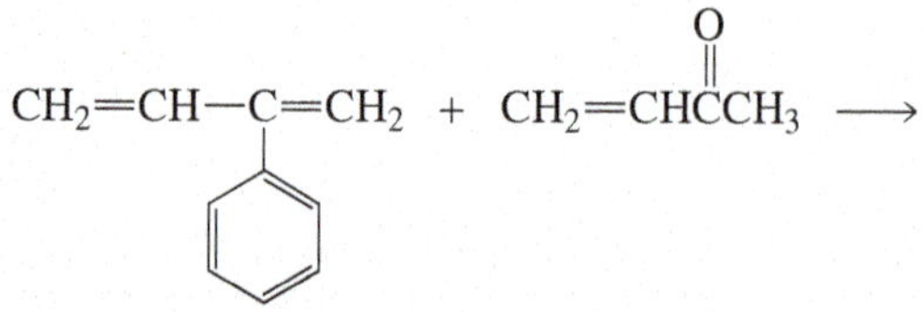 CH$_2$=CH—C(phenyl)=CH$_2$ + CH$_2$=CHCCH$_3$ (O) $\longrightarrow$

97. What two sets of a conjugated diene and a dienophile could be used to prepare the following compound?

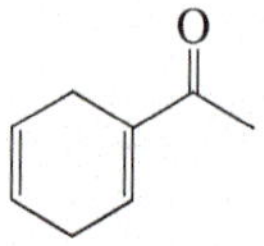

98. a. Which dienophile in each pair is more reactive in a Diels–Alder reaction?

1. CH$_2$=CHCH (O) or CH$_2$=CHCH$_2$CH (O)   **2.** CH$_2$=CHCH (O) or CH$_2$=CHCH$_3$

b. Which diene is more reactive in a Diels–Alder reaction?

CH$_2$=CHCH=CHOCH$_3$ or   CH$_2$=CHCH=CHCH$_2$OCH$_3$

99. Draw the major products obtained from the reaction of one equivalent of HBr with the following compounds. For each reaction, indicate the kinetic product and the thermodynamic product.

a. 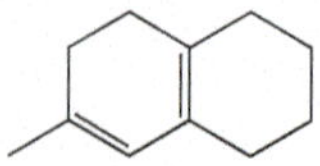   **b.**

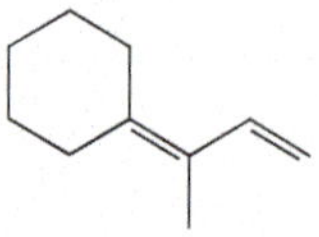

100. Cyclopentadiene can react with itself in a Diels–Alder reaction. Draw the endo and exo products.

101. Which diene and which dienophile could be used to prepare each of the following?

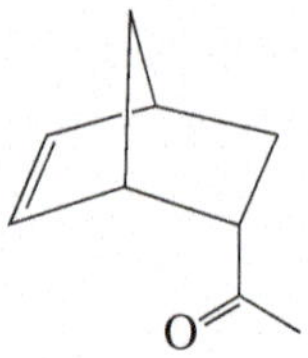

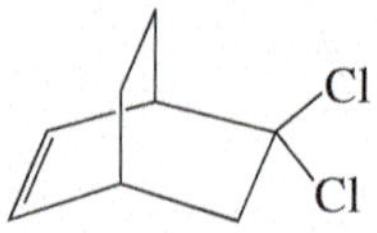

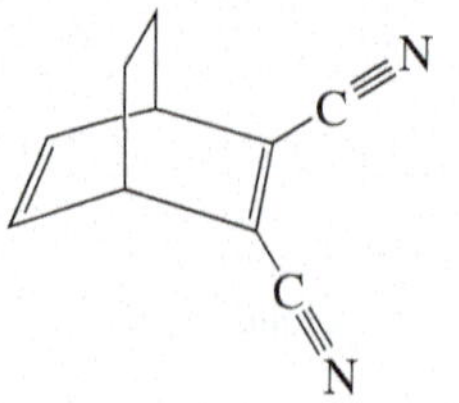

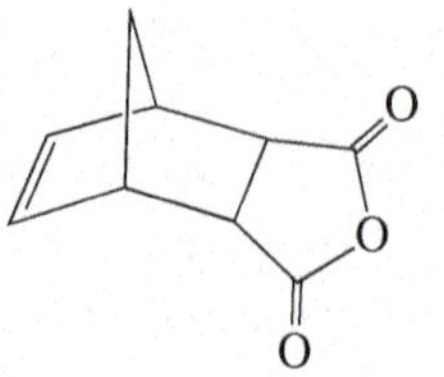

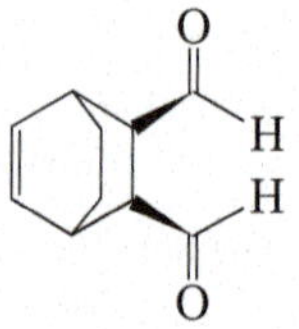

102. a. Propose a mechanism for the following reaction:

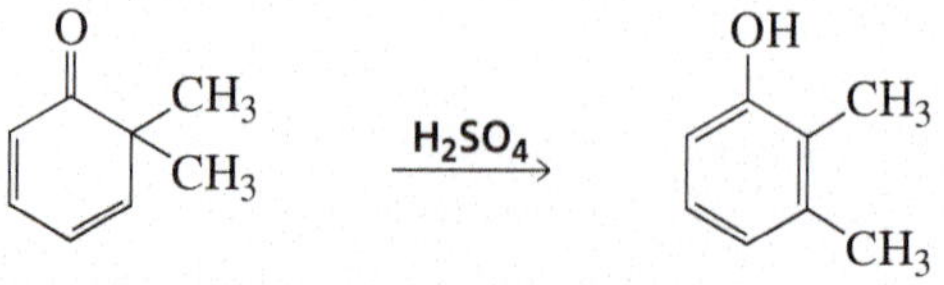

b. What is the product of the following reaction?

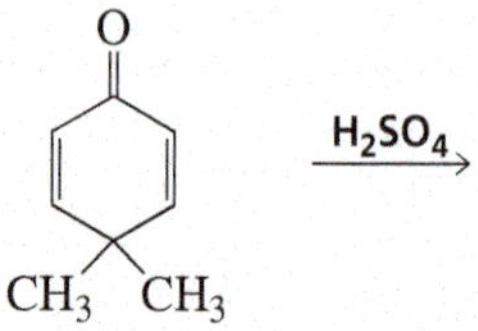

03. **a.** What are the products of the following reaction?
b. How many stereoisomers of each product could be obtained?

$$\text{(cyclohexenyl ethylene)} + Br_2 \longrightarrow$$

04. As many as 18 different Diels–Alder products can be obtained by heating a mixture of 1,3-butadiene and 2-methyl-1,3-butadiene. Identify the products.

05. On a single graph, draw the reaction coordinate diagram for the addition of one equivalent of HBr to 2-methyl-1,3-pentadiene and for the addition of one equivalent of HBr to 2-methyl-1,4-pentadiene. Which reaction is faster?

06. While attempting to recrystallize maleic anhydride, a student dissolved it in freshly distilled cyclopentadiene rather than in freshly distilled cyclopentane. Was her recrystallization successful?

maleic anhydride

07. The following equilibrium is driven to the right if the reaction is carried out in the presence of maleic anhydride (see Problem 106). What is the function of maleic anhydride?

08. In 1935, J. Bredt, a German chemist, proposed that a bicycloalkene could not have a double bond at a bridgehead carbon unless one of the rings contains at least eight carbons. This is known as Bredt's rule. Explain why there cannot be a double bond at this position.

bridgehead carbon

09. The experiment shown next and discussed in Section 8.13 shows that the proximity of the chloride ion to C-2 in the transition state causes the 1,2-addition product to form more rapidly than the 1,4-addition product.

$$CH_2{=}CHCH{=}CHCH_3 + DCl \xrightarrow{-78\ ^\circ C} \underset{\underset{D\ \ \ Cl}{|\ \ \ |}}{CH_2CHCH{=}CHCH_3} + \underset{\underset{D\ \ \ \ \ \ \ \ \ Cl}{|\ \ \ \ \ \ \ \ \ |}}{CH_2CH{=}CHCHCH_3}$$

a. Why was it important for the investigators to know that the preceding reaction was being carried out under kinetic control?
b. How could the investigators know that the reaction was being carried out under kinetic control?

10. Identify the product of the following reaction. (*Hint:* It has a fused bicyclic ring.)

11. Draw the resonance contributors of the cyclooctatrienyl dianion.
a. Which of the resonance contributors is the least stable?
b. Which of the resonance contributors makes the smallest contribution to the hybrid?

12. Investigation has shown that cyclobutadiene is actually a rectangular molecule rather than a square molecule. Explain the reason for this observation.

cyclobutadiene

DRAWING RESONANCE CONTRIBUTORS

We saw that chemists use curved arrows to show how electrons move when reactants are converted into products (see the Tutorial on p. 225). Chemists also use curved arrows when they draw resonance contributors.

We also saw that delocalized electrons are electrons that are shared by more than two atoms. When electrons are shared by more than two atoms, we cannot use solid lines to represent the location of the electrons accurately. For example, in the carboxylate ion, a pair of electrons is shared by a carbon and two oxygens. We show the pair of delocalized electrons by a dotted line spread over the three atoms. We saw that this structure is called a **resonance hybrid.** The resonance hybrid shows that the negative charge is shared by the two oxygens.

**carboxylate ion
resonance hybrid**

resonance contributors

Chemists do not like to use dotted lines when drawing structures because, unlike a solid line that represents two electrons, the dotted lines do not specify the number of electrons they represent. Therefore, chemists use structures with localized electrons (indicated by solid lines) to approximate the resonance hybrid that has delocalized electrons (indicated by dotted lines). These approximate structures are called **resonance contributors.** Curved arrows are used to show the movement of electrons in going from one resonance contributor to the next.

RULES FOR DRAWING RESONANCE CONTRIBUTORS

Now we will look at three simple rules for drawing resonance contributors:
1. Only electrons move; atoms *never* move.
2. The only electrons that can move are π electrons (electrons in π bonds) and lone-pair electrons.
3. Electrons always move toward an sp^2 or sp hybridized atom. An sp^2 carbon is a positively charged carbon or a doubly bonded carbon; an sp carbon is a triply bonded carbon.

(In Chapter 12, we will see that electrons can also move toward a carbon with an unpaired electron, which is an sp^2 carbon as well.)

π Electrons Move Toward an sp^2 Carbon That Is a Positively Charged Carbon

In the following example, π electrons move toward a positively charged carbon. Because the atom does not have a complete octet of electrons, it can accept the electrons. The carbon that is positively charged in the first resonance contributor is neutral in the second resonance contributor because it has received electrons. The carbon in the first resonance contributor that loses its share of the π electrons is positively charged in the second resonance contributor.

$$CH_3CH=CH-\overset{+}{C}HCH_3 \longleftrightarrow CH_3\overset{+}{C}H-CH=CHCH_3$$

We see that the following carbocation has three resonance contributors.

$$CH_2=CH-CH=CH-\overset{+}{C}HCH_3 \longleftrightarrow CH_2=CH-\overset{+}{C}H-CH=CHCH_3 \longleftrightarrow \overset{+}{C}H_2-CH=CH-CH=CHCH_3$$

Notice that in going from one resonance contributor to the next, the total number of electrons in the structure does not change. Therefore, each of the resonance contributors must have the same net charge.

PROBLEM 1 Draw the resonance contributors for the following carbocation (the answers can be found immediately after Problem 12):

$$CH_3CH=CH-CH=CH-CH=CH-\overset{+}{C}H_2 \longleftrightarrow$$

$$\longleftrightarrow \qquad\qquad \updownarrow$$

PROBLEM 2 Draw the resonance contributors for the following carbocation:

$$\longleftrightarrow \qquad \longleftrightarrow \qquad \longleftrightarrow \qquad \longleftrightarrow$$

π Electrons Move Toward an sp^2 Carbon That Is a Doubly Bonded Carbon

In the following example, π electrons move toward a doubly bonded carbon. The atom to which the electrons move can accept them because a π bond can break.

PROBLEM 3 Draw the resonance contributor for the following compound:

$$\longleftrightarrow$$

In the next example, the electrons can move equally easily to the left (indicated by the red arrows) or to the right (indicated by the blue arrows). When comparing the charges on the resonance contributors, we see that the charges on each of the end carbons cancel, so there is no charge on any of the carbons in the resonance hybrid.

$$\overset{-}{\ddot{C}}H_2-CH=CH-\overset{+}{C}H_2 \longleftrightarrow CH_2=CH-CH=CH_2 \longleftrightarrow \overset{+}{C}H_2-CH=CH-\overset{-}{\ddot{C}}H_2$$

$$CH_2\text{---}CH\text{---}CH\text{---}CH_2$$
resonance hybrid

When electrons can move in two directions and there is a difference in the electronegativity of the atoms to which they can move, they always move toward the more electronegative atom. For instance, in the following example, the electrons move toward oxygen, not toward carbon.

Notice that the first resonance contributor has a charge of 0. Because the number of electrons in the molecule does not change, the other resonance contributor must have a net charge of 0. (A net charge of 0 does not mean that there is no charge on any of the atoms; a resonance contributor with a positive charge on one atom and a negative charge on another has a net charge of 0.)

PROBLEM 4 Draw the resonance contributors for the following compound:

$$\longleftrightarrow \qquad \longleftrightarrow$$

A Lone Pair Moves Toward an sp^2 Carbon That Is a Doubly Bonded Carbon

In the following examples, lone-pair electrons move toward a doubly bonded carbon. Notice that the arrow starts at a pair of electrons, not at a negative charge. In the first example, each of the resonance contributors has a charge of −1; in the second example, each of the resonance contributors has no charge or a net charge of 0.

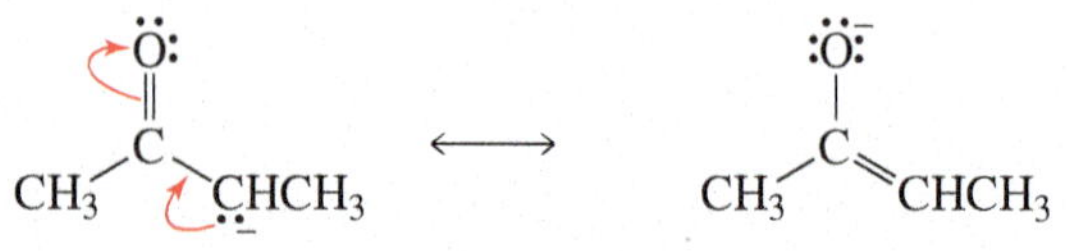

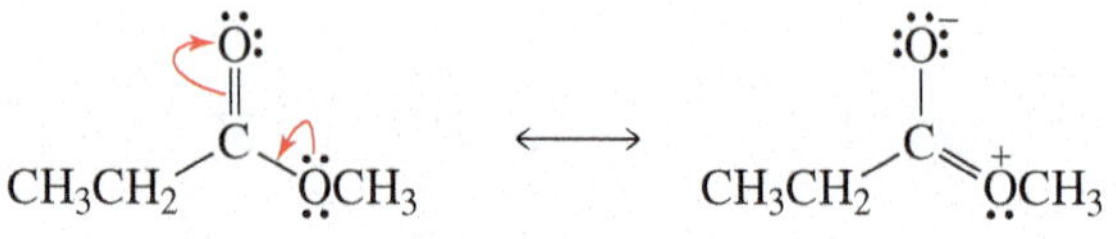

The following species has three resonance contributors. Notice again that the arrow starts at a lone pair, not at a negative charge. The three oxygen atoms share the two negative charges. Therefore, each oxygen atom in the hybrid has two-thirds of a negative charge.

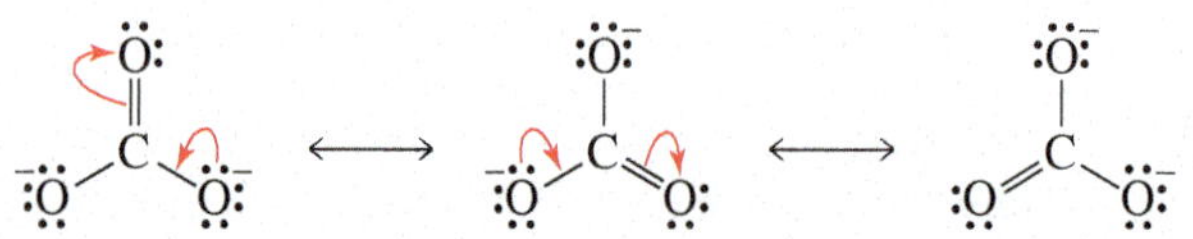

PROBLEM 5 Draw the resonance contributor for the following compound:

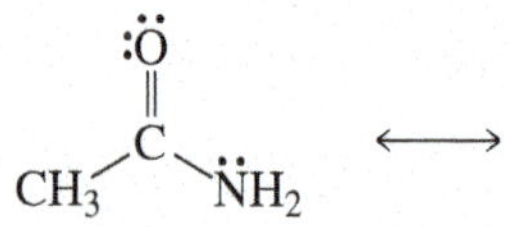

Notice in the next example that the lone pair moves away from the most electronegative atom in the molecule. This is the only way electron delocalization can occur (and any electron delocalization is better than none). The π electrons cannot move toward the oxygen because the oxygen atom has a complete octet (it is sp^3 hybridized). Recall that electrons can move only toward an sp^2 or sp hybridized atom.

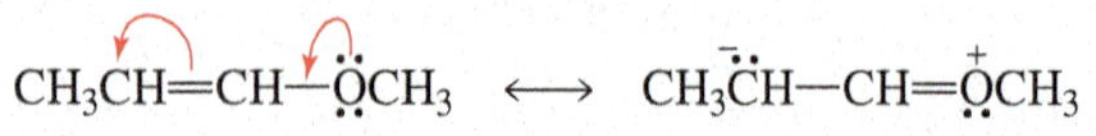

$$CH_3CH{=}CH{-}\overset{..}{\underset{..}{O}}CH_3 \longleftrightarrow CH_3\overset{-}{\overset{..}{C}}H{-}CH{=}\overset{+}{\underset{..}{O}}CH_3$$

The compound in the next example has five resonance contributors. To get to the second resonance contributor, a lone pair on nitrogen moves toward an sp^2 carbon. Notice that the first and fifth resonance contributors are not the same; they are similar to the two resonance contributors of benzene. (See p. 383.)

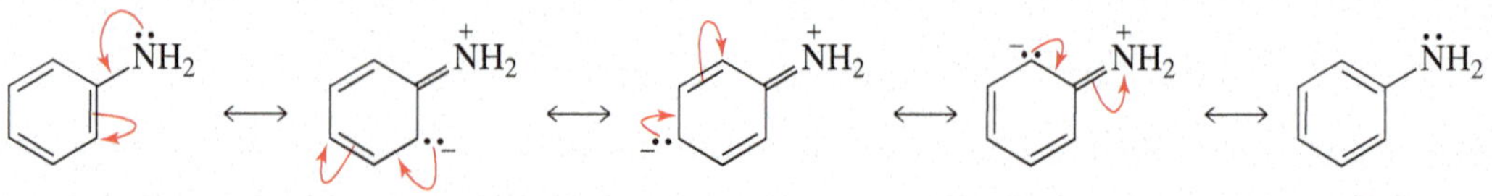

PROBLEM 6 Draw the resonance contributors for the following compound:

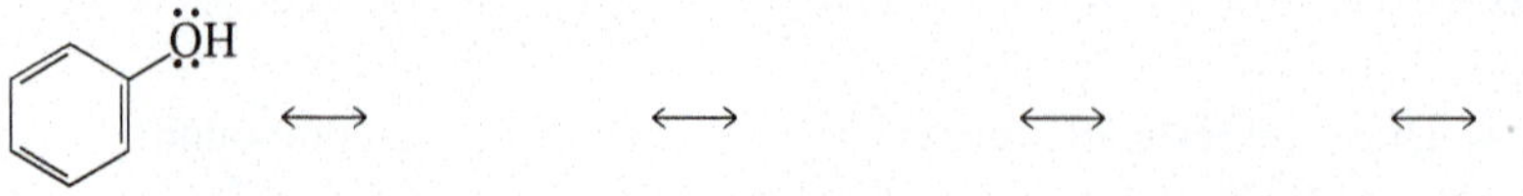

The following species do not have delocalized electrons. Electrons cannot move toward an sp^3 hybridized atom because an sp^3 hybridized atom has a complete octet and it does not have a π bond that can break, so it cannot accept any more electrons.

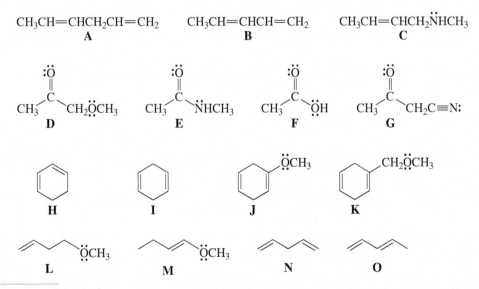

an *sp³* hybridized carbon cannot accept electrons

$CH_2{=}CH{-}CH_2{-}\overset{+}{C}H_2CH_3$

an *sp³* hybridized carbon cannot accept electrons

$CH_3CH{=}CH{-}CH_2{-}\overset{+}{N}H_2$

an *sp³* hybridized carbon cannot accept electrons

$CH_3\overset{\overset{\ddot{O}:}{\|}}{C}{-}CH_2{-}CH{=}CHCH_3$

Notice the difference in the resonance contributors for the next two examples. In the first example, electrons *move into* the benzene ring. That is, a lone pair on the atom attached to the ring moves toward an *sp²* carbon.

In the next example, electrons *move out* of the benzene ring. First, a π bond moves toward an *sp²* carbon. The electron movement is toward the oxygen because oxygen is more electronegative than carbon. Then, the other resonance contributors result from a π bond moving toward a positive charge.

In the next two examples, the atom attached to the ring has neither a lone pair nor a π bond. Therefore, the substituent can neither donate electrons into the ring nor accept electrons from the ring. Thus, these compounds have only two resonance contributors—the ones that are similar to the two resonance contributors of benzene.

PROBLEM 7 Which of the following species have delocalized electrons?

$CH_3CH{=}CHCH_2CH{=}CH_2$ **A**

$CH_3CH{=}CHCH{=}CH_2$ **B**

$CH_3CH{=}CHCH_2\overset{\cdot\cdot}{N}HCH_3$ **C**

$CH_3\overset{\overset{:\ddot{O}}{\|}}{C}CH_2\overset{\cdot\cdot}{\underset{\cdot\cdot}{O}}CH_3$ **D**

$CH_3\overset{\overset{:\ddot{O}}{\|}}{C}\overset{\cdot\cdot}{N}HCH_3$ **E**

$CH_3\overset{\overset{:\ddot{O}}{\|}}{C}\overset{\cdot\cdot}{\underset{\cdot\cdot}{O}}H$ **F**

$CH_3\overset{\overset{:\ddot{O}}{\|}}{C}CH_2C{\equiv}N:$ **G**

H   **I**   **J** ($\overset{\cdot\cdot}{\underset{\cdot\cdot}{O}}CH_3$)   **K** ($CH_2\overset{\cdot\cdot}{\underset{\cdot\cdot}{O}}CH_3$)

L ($\overset{\cdot\cdot}{\underset{\cdot\cdot}{O}}CH_3$)   **M** ($\overset{\cdot\cdot}{\underset{\cdot\cdot}{O}}CH_3$)   **N**   **O**

PROBLEM 8 Draw the resonance contributors for those compounds in Problem 7 that have delocalized electrons.

PROBLEM 9 Draw curved arrows to show how one resonance contributor leads to the next one

a.

b.

PROBLEM 10 Draw the resonance contributors for each of the following:

a.

b.

c.

d. $CH_2=CH-CH=CH-\ddot{N}H_2$

e.

f.

PROBLEM 11 Draw the resonance contributors for each of the following:

a.

b.

c.

d.

e.

f.

PROBLEM 12 Draw the resonance contributors for each of the following:

a.

b.

c.

d.

e.

f.

ANSWERS TO PROBLEMS ON DRAWING RESONANCE CONTRIBUTORS

PROBLEM 1

$$CH_3CH=CH-CH=CH-CH=CH-\overset{+}{C}H_2 \longleftrightarrow CH_3CH=CH-CH=CH-\overset{+}{C}H-CH=CH_2$$

$$CH_3\overset{+}{C}H-CH=CH-CH=CH-CH=CH_2 \longleftrightarrow CH_3CH=CH-\overset{+}{C}H-CH=CH-CH=CH_2$$

PROBLEM 2

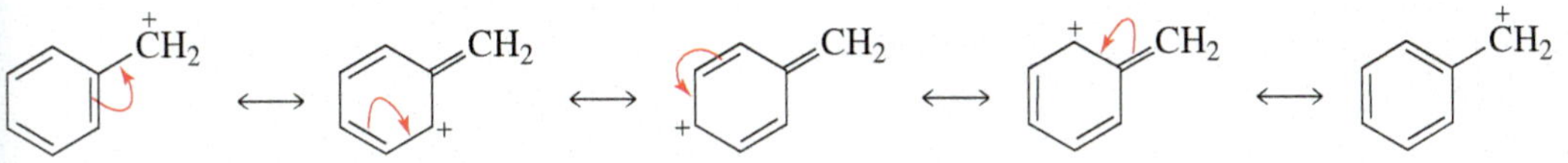

PROBLEM 3

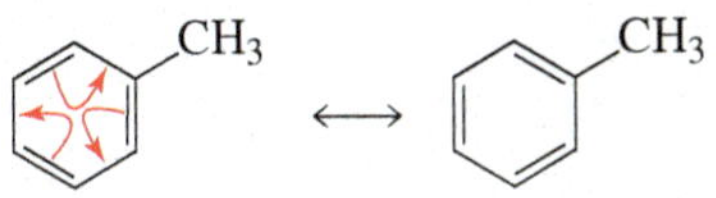

PROBLEM 4

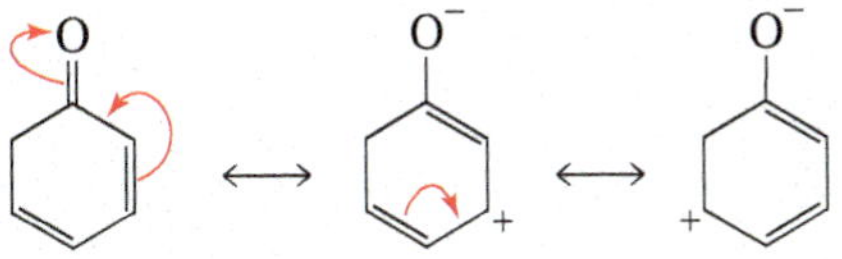

PROBLEM 5

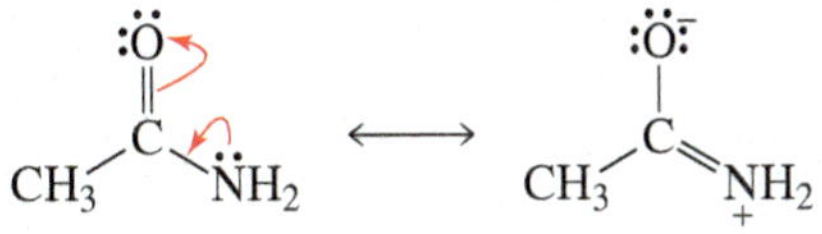

PROBLEM 6

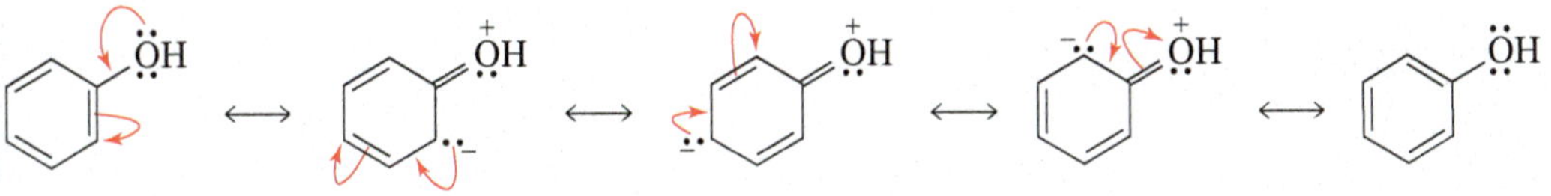

PROBLEM 7

B, E, F, H, J, M, and O

PROBLEM 8

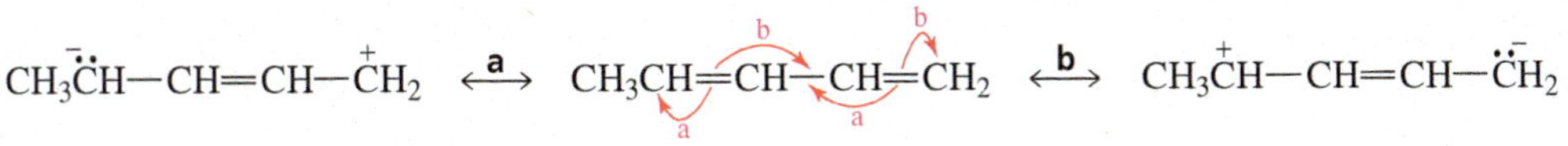

$$CH_3\overset{-}{C}H-CH=CH-\overset{+}{C}H_2 \quad \overset{a}{\longleftrightarrow} \quad CH_3CH=CH-CH=CH_2 \quad \overset{b}{\longleftrightarrow} \quad CH_3\overset{+}{C}H-CH=CH-\overset{-}{C}H_2$$

B

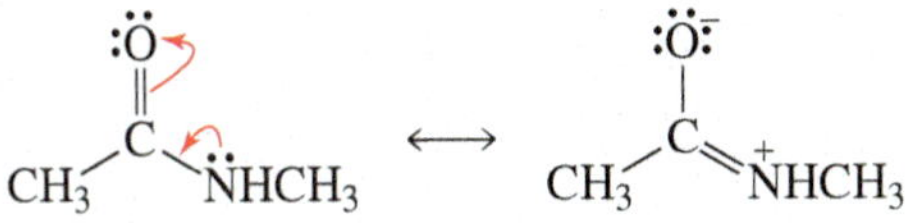

E 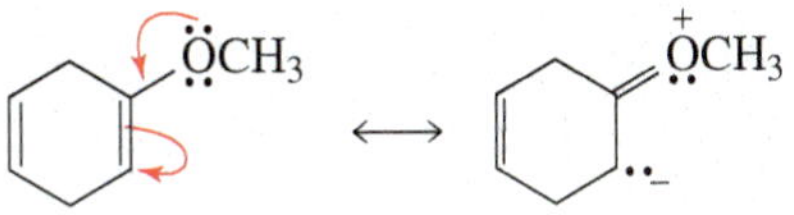   **J**

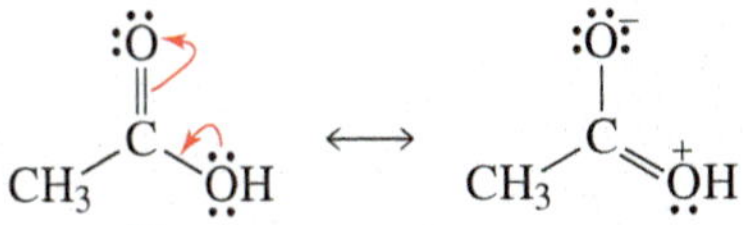

F 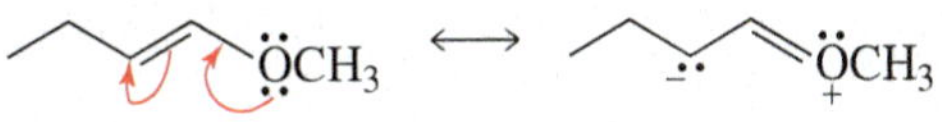   **M**

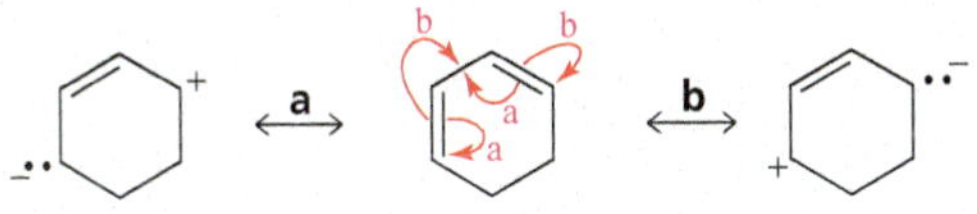

H 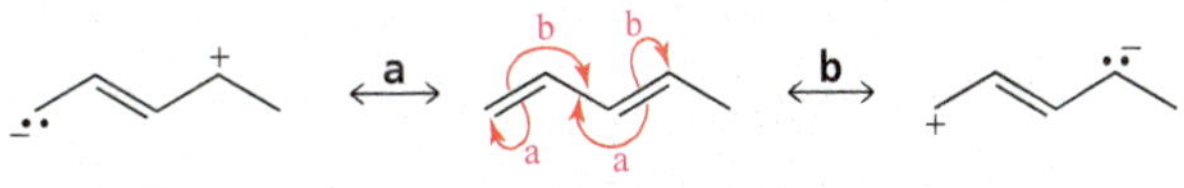   **O**

PROBLEM 9

a. 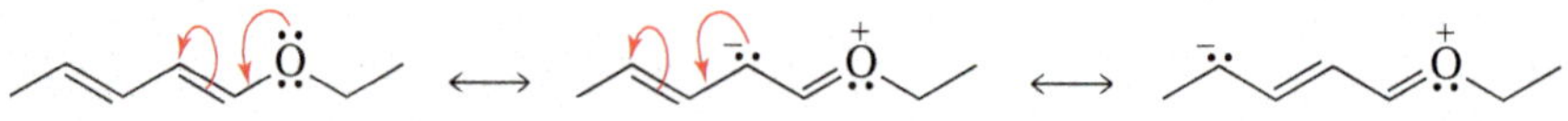 (resonance structures)

b. 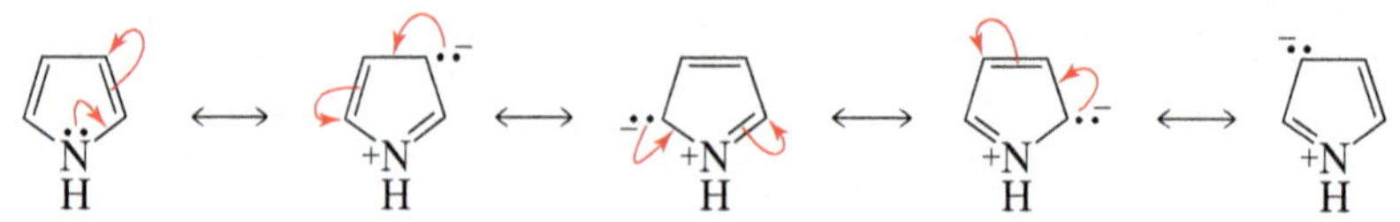 (resonance structures of pyrrole)

PROBLEM 10

a. 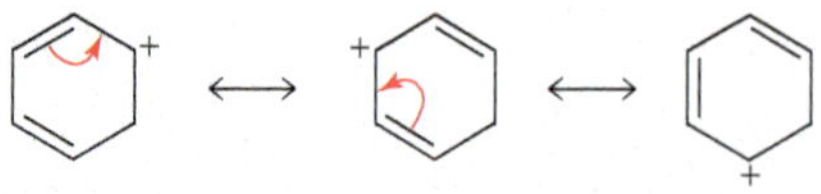 (resonance structures)

b. Notice in the following example that the electrons can move either clockwise or counterclockwise:

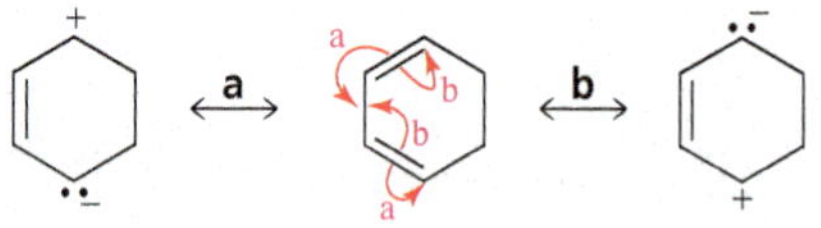

c. 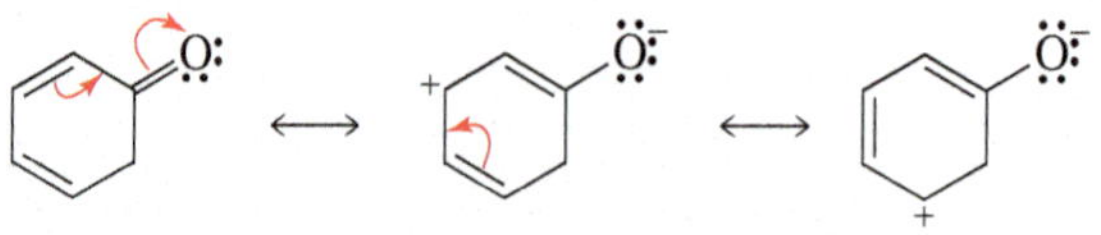 (resonance structures)

d.  $CH_2{=}CH{-}CH{=}CH{-}\ddot{N}H_2 \longleftrightarrow CH_2{=}CH{-}\overset{..}{C}H{-}CH{=}\overset{+}{N}H_2 \longleftrightarrow \overset{-}{C}H_2{-}CH{=}CH{-}CH{=}$

e. Notice in the following example that the lone-pair electrons can move toward either of the two sp carbons:

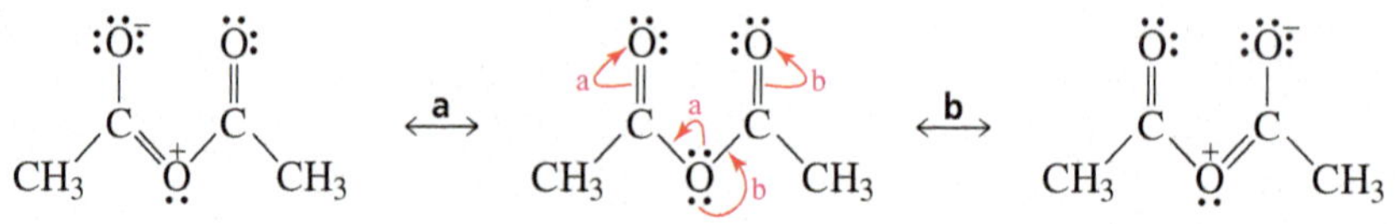

f. 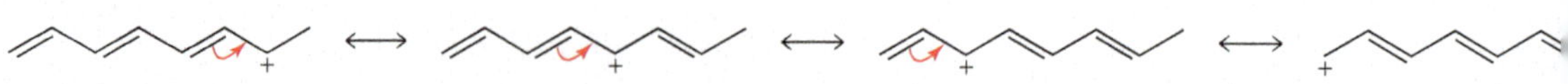 (resonance structures)

PROBLEM 11

a. 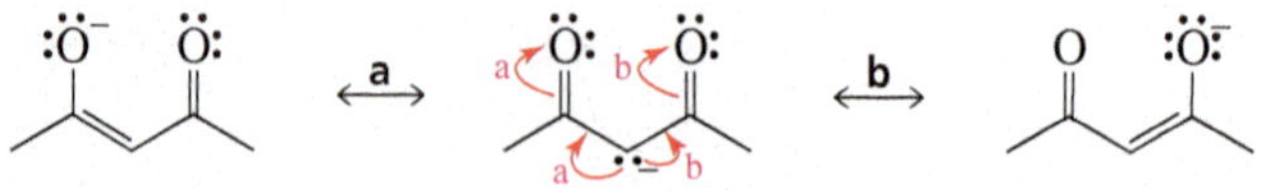 (resonance structures)

b. 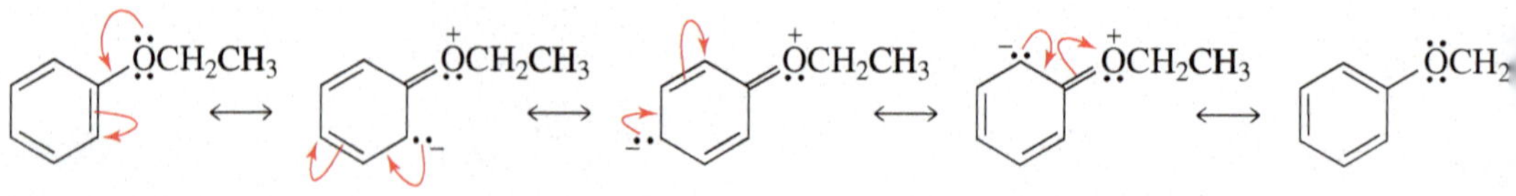 (resonance structures with $\ddot{O}CH_2CH_3$)

c. 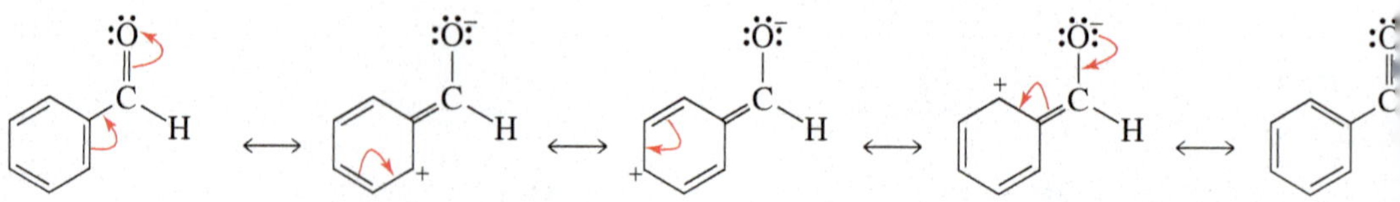 (resonance structures with $C{-}H$)

d.

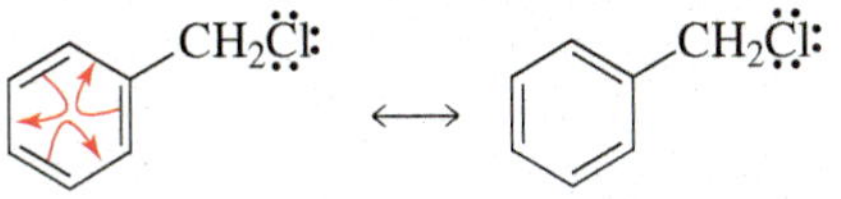

e.

f.

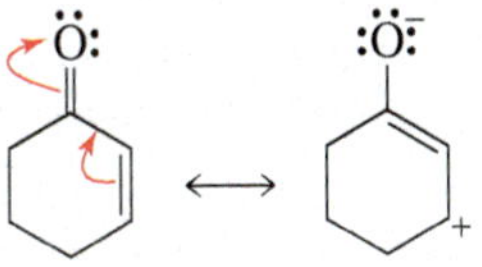

PROBLEM 12

a.

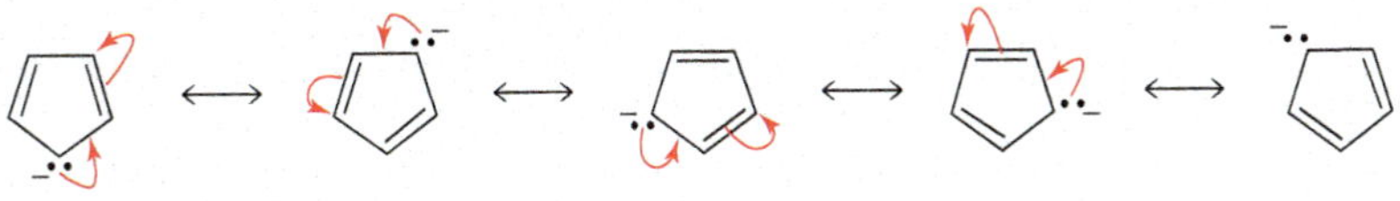

b.

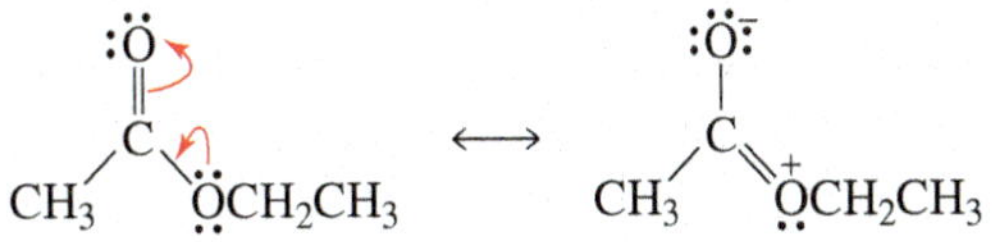

d.

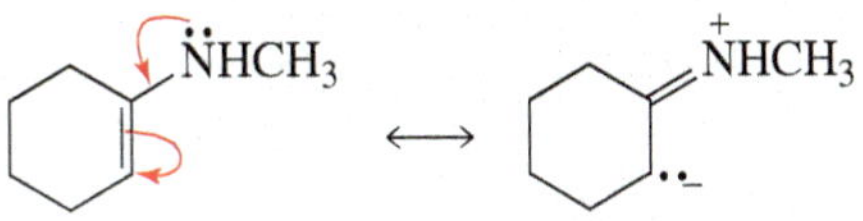

c.

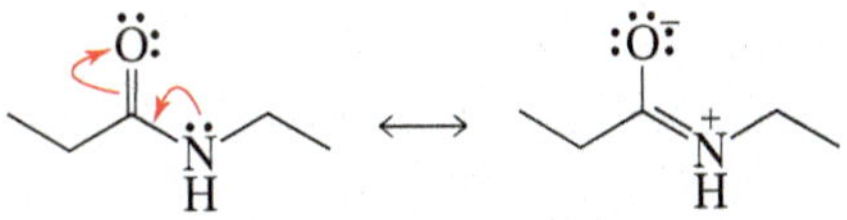

e.

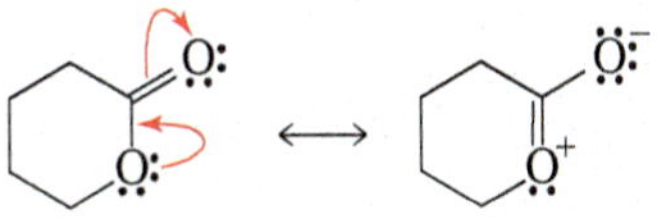

f.

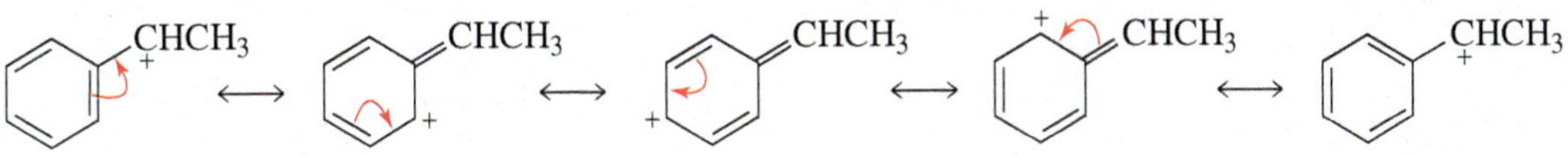

MasteringChemistry®
for Organic Chemistry

MasteringChemistry tutorials guide you through the toughest topics in chemistry with self-paced tutorials that provide individualized coaching. These assignable, in-depth tutorials are designed to coach you with hints and feedback specific to your individual misconceptions. For additional practice on Drawing Resonance Contributors, go to MasteringChemistry where the following tutorials are available:

- Drawing Resonance Contributors: Moving π Electrons
- Drawing Resonance Contributors: Predicting Aromaticity
- Drawing Resonance Contributors: Substituted Benzene Rings

PART THREE

Substitution and Elimination Reactions

The first two chapters in Part 3 (Chapters 9 and 10) discuss the reactions of compounds that have an electron withdrawing atom or group—a potential leaving group—bonded to an sp^3 carbon. These compounds ca undergo substitution reactions, elimination reactions, or both substitution and elimination reactions. The thir chapter in Part 3 (Chapter 11) introduces you to organometallic compounds, a class of nucleophiles that ca participate in substitution reactions. The last chapter in Part 3 (Chapter 12) discusses the reactions of alkanes which are compounds that do not have a leaving group but can undergo a substitution reaction under specia conditions.

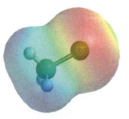

CH_3F

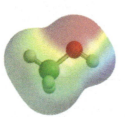

CH_3OH

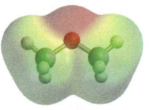

CH_3OCH_3

CH_3SH

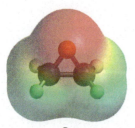

$H_2C—CH_2$ (with O)

Chapter 9 Substitution and Elimination Reactions of Alkyl Halides

Chapter 9 discusses the substitution and elimination reactions of alkyl halides and the factor that determine whether a given alkyl halide will undergo a substitution reaction, an eliminatior reaction, or both substitution and elimination reactions. Of the different compounds that underg substitution and elimination reactions, alkyl halides are examined first because they have relatively good leaving groups.

Chapter 10 Reactions of Alcohols, Ethers, Epoxides, Amines, and Sulfur-Containing Compounds

Chapter 10 discusses compounds other than alkyl halides that undergo substitution and eliminatior reactions. Here you will see that, compared with the leaving groups of alkyl halides, the leaving groups of alcohols and ethers are relatively poor. As a result, the leaving groups must be activatec before alcohols and ethers can undergo substitution or elimination reactions. Several methods commonly used to activate leaving groups will be examined. Chapter 10 also looks at the reactions of epoxides; these cyclic ethers show how ring strain affects leaving group ability. Amines have such poor leaving groups that they cannot undergo substitution or elimination reactions. However, quaternary ammonium ions, which have much better leaving groups than amines, do undergo elimination reactions. Chapter 10 also compares the reactions of thiols and sulfides with those of alcohols and ethers and discusses a sulfonium salt that biological systems use for substitution reactions, because alkyl halides are not readily available in nature.

Chapter 11 Organometallic Compounds

Organometallic compounds are very important to synthetic organic chemists. These compounds contain a carbon–metal bond. While Chapter 9 shows how an alkyl halide has an *electrophilic alkyl group* (because carbon is less electronegative than the halogen to which it is bonded), **Chapter 11** shows how an organometallic compound has a *nucleophilic alkyl group* (because carbon is more electronegative than the metal to which it is bonded). Organometallic compounds can be used to form new carbon–carbon single bonds as well as new carbon–carbon double and triple bonds.

Chapter 12 Radicals • Reactions of Alkanes

Chapter 12 discusses the substitution reactions of alkanes—hydrocarbons that contain only single bonds. In previous chapters, we saw that when a compound reacts, the weakest bond in the molecule breaks first. Alkanes, however, have only strong bonds. Therefore, special conditions that generate radicals are required for alkanes to react. Chapter 12 also looks at radical substitution reactions and radical addition reactions of alkenes. The chapter concludes with a discussion of some radical reactions that occur in the biological world.

9

Substitution and Elimination Reactions of Alkyl Halides

This cartoon was published in *Time Magazine* on June 30, 1947.

In this chapter you will see how the widespread use of DDT gave birth to the environmental movement. You will also learn why life is based on carbon rather than on silicon, even though silicon is just below carbon in the periodic table and is far more abundant than carbon in Earth's crust.

We saw that the families of organic compounds can be placed in one of four groups, and that all the families in a group react in similar ways (Section 5.5). This chapter begins our discussion of the families of compounds in Group II.

Notice that all the families in Group II have an electronegative atom or an electron-withdrawing group attached to an sp^3 carbon. This atom or group creates a polar bond that allows the compound to undergo substitution and/or elimination reactions.

Group II

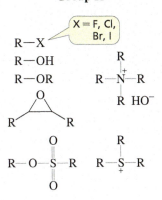

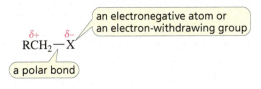

*In a **substitution reaction**, the electronegative atom or electron-withdrawing group is replaced by another atom or group.*

*In an **elimination reaction**, the electronegative atom or electron-withdrawing group is eliminated, along with a hydrogen from an adjacent carbon.*

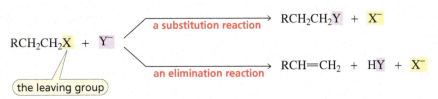

The atom or group that is *substituted* or *eliminated* is called a **leaving group.**

This chapter focuses on the *substitution and elimination reactions* of alkyl halides—compound in which the leaving group is a halide ion (F^-, Cl^-, Br^-, or I^-).

alkyl halides

R—F	R—Cl	R—Br	R—I
alkyl fluoride	alkyl chloride	alkyl bromide	alkyl iodide

Alkyl halides are a good family of compounds with which to start the study of substitution an elimination reactions because they have relatively good leaving groups; that is, the halide ions ar easily displaced. After learning about the reactions of alkyl halides, you will be prepared to move o to Chapter 10, which discusses the substitution and elimination reactions of compounds with poore leaving groups (those that are more difficult to displace) as well as a few with better leaving groups

Substitution and elimination reactions are important in organic chemistry because they make i possible to convert readily available alkyl halides into a wide variety of other compounds. These reactions are also important in the cells of plants and animals. We will see, however, that because cells exist in predominantly aqueous environments and alkyl halides are insoluble in water, biologi cal systems use compounds in which the group that is replaced is more polar than a halogen and therefore, more water soluble (Section 10.12).

The Birth of the Environmental Movement

Alkyl halides have been used as insecticides since 1939, when it was discovered that DDT (first synthesized in 1874) has a high toxicity to insects and a relatively low toxicity to mammals. DDT was used widely in World War II to control typhus and malaria in both the military and civilian populations. It saved millions of lives, but no one realized at that time that, because it is a very stable compound, it is resistant to biodegradation. In addition, DDT and DDE, a compound formed as a result of elimination of HCl from DDT, are not water soluble. Therefore, they accumulate in the fatty tissues of birds and fish and can be passed up the food chain. Most older adults have a low concentration of DDT or DDE in their bodies.

In 1962, Rachel Carson, a marine biologist, published *Silent Spring*, where she pointed out the environmental impacts of the widespread use of DDT. The book was widely read, so it brought the problem of environmental pollution to the attention of the general public for the first time. Consequently, its publication was an important event in the birth of the environmental movement. Because of the concern it raised, DDT was banned in the United States in 1972. In 2004, the Stockholm Convention banned the worldwide use of DDT except for the control of malaria in countries where the disease is a major health problem.

In Section 12.12, we will look at the environmental effects caused by synthetic alkyl halides known as chlorofluorohydrocarbons (CFCs).

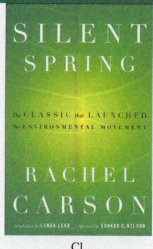

DDT

PROBLEM 1

Draw the structure of DDE.

PROBLEM 2

Methoxychlor is an insecticide that was intended to take DDT's place because it is not as soluble in fatty tissues and is more readily biodegradable. It, too, can accumulate in the environment, however, so its use was also banned—in 2002 in the European Union and in 2003 in the United States. Why is methoxychlor less soluble in fatty tissues than DDT?

methoxychlor

9.1   THE S$_N$2 REACTION

You will see that there are two different mechanisms by which a substitution reaction of an alkyl halide can take place. As you would expect, each involves the *reaction of a nucleophile with an electrophile*. In both mechanisms, the nucleophile replaces the leaving group, so the substitution reaction is more precisely called a **nucleophilic substitution reaction.**

Now that you have seen the mechanisms for many different reactions, you might be wondering how mechanisms are determined. Remember that a mechanism describes the step-by-step process by which reactants are converted into products. It is a theory that fits the accumulated experimental evidence pertaining to the reaction. Thus, *mechanisms are determined experimentally*. They are not something that chemists make up in an attempt to explain how a reaction occurs.

Experimental Evidence for the Mechanism for an S$_N$2 Reaction

We can learn a great deal about a reaction's mechanism by studying its **kinetics**—the factors that affect the rate of the reaction.

For example, the rate of the following nucleophilic substitution reaction depends on the concentrations of both reactants.

$$CH_3Br \ + \ HO^- \ \longrightarrow \ CH_3OH \ + \ Br^-$$

- Doubling the concentration of the alkyl halide (CH_3Br) doubles the rate of the reaction.
- Doubling the concentration of the nucleophile (HO^-) doubles the rate of the reaction.
- Doubling the concentration of both reactants quadruples the rate of the reaction.

Because we know the relationship between the rate of the reaction and the concentration of the reactants, we can write a **rate law** for the reaction:

$$\text{rate} \ \propto \ [\,\textbf{alkyl halide}\,][\,\textbf{nucleophile}\,]$$

The proportionality sign ($\propto$) can be replaced by an equal sign and a proportionality constant (k). This is a **second-order reaction** because its rate depends linearly on the concentration of each of the two reactants.

$$\text{rate} = k\,[\,\textbf{alkyl halide}\,][\,\textbf{nucleophile}\,]$$

the rate constant

The **rate law** tells us which molecules are involved in the transition state of the rate-determining step of the reaction. Thus, the rate law for this substitution reaction tells us that *both* the alkyl halide and the nucleophile are involved in the rate-determining transition state.

The proportionality constant is called a **rate constant.** The magnitude of the rate constant for a particular reaction indicates how difficult it is for the reactants to overcome the energy barrier of the reaction—that is, how hard it is to reach the transition state (Section 5.11).

the lower the energy barrier, the larger the rate constant

A lower energy barrier means it is easier for the reactants to reach the transition state (see Figure 9.3 on page 397).

PROBLEM 3 ◆

How will the rate of the reaction between bromomethane and hydroxide ion be affected if the following changes in concentration are made?

a. The concentration of the alkyl halide is not changed and the concentration of the nucleophile is tripled.
b. The concentration of the alkyl halide is cut in half and the concentration of the nucleophile is not changed.
c. The concentration of the alkyl halide is cut in half and the concentration of the nucleophile is doubled.

The reaction of bromomethane with hydroxide ion is an example of an **S$_N$2 reaction**, where "S" stands for *substitution*, "N" for *nucleophilic*, and "2" for *bimolecular*. **Bimolecular** means that two molecules are involved in the transition state of the rate-determining step. In 1937, Edward Hughes and Christopher Ingold proposed a mechanism for an S$_N$2 reaction. They based their mechanism on three pieces of *experimental evidence*:

1. The rate of the substitution reaction depends on the concentration of the alkyl halide *and* on the concentration of the nucleophile, indicating that both reactants are involved in the transition state of the rate-determining step.

2. As the alkyl group becomes larger (for example, as the hydrogens of bromomethane are successively replaced with methyl groups), the rate of the substitution reaction with a given nucleophile decreases.

relative rates of an S$_N$2 reaction

$$CH_3-Br \ > \ CH_3CH_2-Br \ > \ CH_3CH_2CH_2-Br \ > \ \underset{\underset{CH_3}{|}}{CH_3CH-Br} \ > \ \underset{\underset{CH_3}{|}}{\overset{\overset{CH_3}{|}}{CH_3C-Br}}$$

3. The substitution reaction of an alkyl halide in which the halogen is bonded to an asymmetric center leads to the formation of only one stereoisomer, and the configuration of the asymmetric center in the product is inverted relative to its configuration in the reacting alkyl halide.

the configuration is inverted relative to that of the reactant

(S)-2-bromobutane   (R)-2-butanol

The Mechanism for an S$_N$2 Reaction

Using the preceding evidence, Hughes and Ingold proposed that an S$_N$2 reaction is a *concerted* reaction (that is, it takes place in a single step), so no intermediates are formed.

MECHANISM FOR THE S$_N$2 REACTION OF AN ALKYL HALIDE

An S$_N$2 reaction is a one-step (concerted) reaction.

electrophile

$$HO:^- \ + \ CH_3-\ddot{B}r: \ \longrightarrow \ CH_3-\ddot{O}H \ + \ :\ddot{B}r:^-$$

nucleophile   leaving group

- The nucleophile attacks the back side of the carbon (the electrophile) that bears the leaving group and displaces it.

A productive collision is a collision that leads to the formation of the product. A productive collision in an S$_N$2 reaction requires the nucleophile to hit the carbon on the side opposite the side that is bonded to the leaving group. Therefore, the carbon is said to undergo **back-side attack.**

Why must the nucleophile attack from the back side? The simplest explanation is that the leaving group blocks the approach of the nucleophile to the front side of the molecule.

Molecular orbital theory explains why back-side attack is required. In order to form a bond, the HOMO (the highest occupied molecular orbital) of one species must interact with the LUMO (the lowest unoccupied molecular orbital) of the other (Section 8.8). Therefore, when the nucleophile (Nuc) approaches the alkyl halide to form a new bond, the nonbonding molecular orbital of the nucleophile (its HOMO) must interact with the empty σ^* antibonding molecular orbital associated with the C—Br bond (its LUMO).

Figure 9.1a shows that in a back-side attack, a bonding interaction (the interacting lobes are both green) occurs between the nucleophile and the larger lobe of the σ^* antibonding MO. But when the nucleophile approaches the front side of the carbon (Figure 9.1b), both a bonding and an antibonding interaction occur, so the two cancel each other and no bond forms. Therefore, an S_N2 reaction is successful only if the nucleophile approaches the sp^3 carbon from its back side.

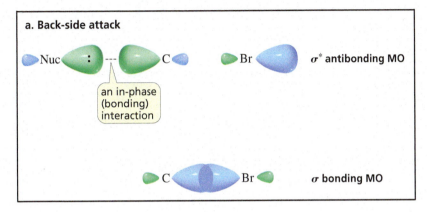

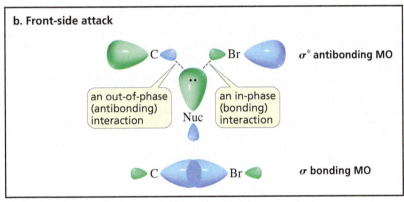

A nucleophile attacks the back side of the carbon that is bonded to the leaving group.

◀ **Figure 9.1**
(a) Back-side attack results in a bonding interaction between the HOMO (the nonbonding orbital) of the nucleophile and the LUMO (the σ^* antibonding orbital) of C—Br.

(b) Front-side attack results in both a bonding and an antibonding interaction that cancel each other.

How the Mechanism Accounts for the Experimental Evidence

How does Hughes and Ingold's mechanism account for the three pieces of experimental evidence? The mechanism shows that the alkyl halide and the nucleophile are both in the transition state of the one-step reaction. Therefore, increasing the concentration of either of them makes their collision more probable, so the rate of the reaction depends on the concentration of both, exactly as observed.

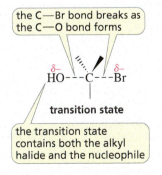

Bulky substituents attached to the carbon that undergoes back-side attack decrease the nucleophile's access to the back side of the carbon and, therefore, decrease the rate of the reaction (Figure 9.2). This explains why, as the size of the alkyl group increases, the rate of the substitution reaction decreases.

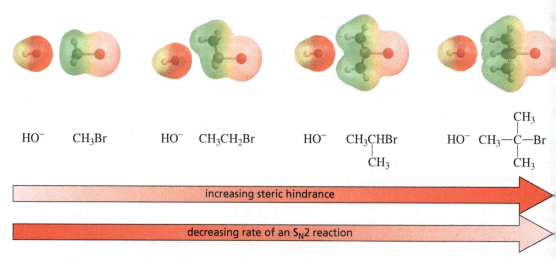

$$HO^- \quad CH_3Br \qquad HO^- \quad CH_3CH_2Br \qquad HO^- \quad CH_3CHBr \qquad HO^- \quad CH_3{-}\overset{\displaystyle CH_3}{\underset{\displaystyle CH_3}{C}}{-}Br$$

increasing steric hindrance

decreasing rate of an S$_N$2 reaction

▲ **Figure 9.2**
The approach of HO⁻ (shown by the yellow and red electrostatic potential map) to the back sides of the carbon of methyl bromide, a primary alkyl bromide, a secondary alkyl bromide, and a tertiary alkyl bromide. Increasing the bulk of the substituents bonded to the carbon that is undergoing nucleophilic attack decreases access to the back side of the carbon, thereby decreasing the rate of the S$_N$2 reaction.

Steric effects are caused by the fact that groups occupy a certain volume of space (Section 3.11). A steric effect that decreases reactivity is called **steric hindrance.** This occurs when groups block the reaction site.

It is steric hindrance that causes alkyl halides to have the following relative reactivities in an S$_N$2 reaction because primary alkyl halides are usually less sterically hindered than secondary alkyl halides, and secondary alkyl halides are less sterically hindered than tertiary alkyl halides (Figure 9.3).

> The relative lack of steric hindrance causes methyl halides and primary alkyl halides to be the most reactive alkyl halides in S$_N$2 reactions.

> Tertiary alkyl halides cannot undergo S$_N$2 reactions.

relative reactivities of alkyl halides in an S$_N$2 reaction

most reactive → methyl halide > 1° alkyl halide > 2° alkyl halide > 3° alkyl halide → too unreactive to undergo an S$_N$2 reaction

In fact, the three alkyl groups of a tertiary alkyl halide make it impossible for the nucleophile to come within bonding distance of the tertiary carbon, so tertiary alkyl halides are unable to undergo S$_N$2 reactions.

The rate of an S$_N$2 reaction depends not only on the *number* of alkyl groups attached to the carbon that is undergoing nucleophilic attack but also on their size. For example, bromoethane and 1-bromopropane are both primary alkyl halides, but bromoethane is more than twice as reactive in an S$_N$2 reaction, because the bulkier alkyl group on the carbon undergoing nucleophilic attack in 1-bromopropane provides greater steric hindrance to back-side attack.

$$CH_3\overset{\displaystyle CH_3}{\underset{\displaystyle CH_3}{C}}CH_2Br$$

Although this is a primary alkyl halide, it undergoes S$_N$2 reactions very slowly because its single alkyl group is unusually bulky.

more reactive in an S$_N$2 reaction

less reactive in an S$_N$2 reaction

$$CH_3CH_2Br \qquad CH_3CH_2CH_2Br$$
bromoethane   **1-bromopropane**

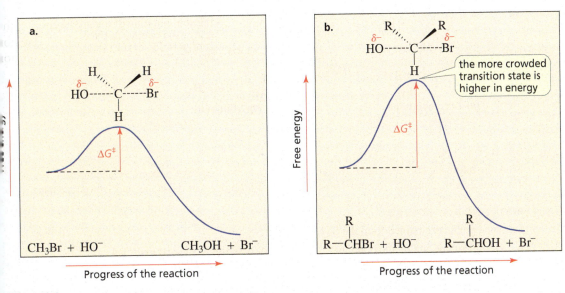

Figure 9.3

The reaction coordinate diagrams show that steric hindrance decreases the rate of the reaction by increasing the energy of the transition state:

a) the S$_N$2 reaction of *unhindered* bromomethane with hydroxide ion

b) an S$_N$2 reaction of a *sterically hindered* secondary alkyl bromide with hydroxide ion

---

PROBLEM 4 ♦

Does increasing the energy barrier for an S$_N$2 reaction increase or decrease the magnitude of the rate constant for the reaction?

Figure 9.4 illustrates the third piece of experimental evidence used by Hughes and Ingold to arrive at their proposed mechanism—namely, the inversion of configuration at the carbon undergoing substitution. This **inversion of configuration** is called a *Walden inversion*. It was named for Paul Walden, who first discovered that the configuration of a compound becomes inverted in an S$_N$2 reaction.

> An S$_N$2 reaction causes the carbon at which substitution occurs to invert its configuration.

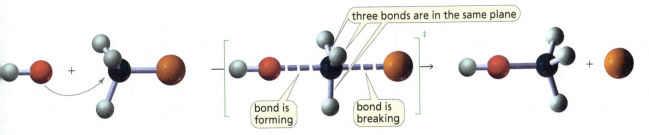

As the nucleophile approaches the back side of the tetrahedral carbon, the C—H bonds begin to move away from the nucleophile and its attacking electrons.

The C—H bonds in the transition state are all in the same plane, and the carbon is pentacoordinate (fully bonded to three atoms and partially bonded to two) rather than tetrahedral.

The C—H bonds have continued to move in the same direction. When the bond between carbon and the nucleophile is fully formed, and the bond between carbon and bromine is completely broken, carbon is once again tetrahedral.

▲ **Figure 9.4**
The reaction between hydroxide ion and bromomethane, showing that the carbon at which substitution occurs in an S$_N$2 reaction inverts its configuration, just like an umbrella can invert in a windstorm.

Because an S$_N$2 reaction takes place with inversion of configuration, only one substitution product is formed when an alkyl halide whose halogen atom is bonded to an asymmetric center undergoes an S$_N$2 reaction. The configuration of that product is inverted relative to the configuration of the alkyl halide. For example, the substitution product obtained from the reaction of hydroxide ion with (*R*)-2-bromopentane is (*S*)-2-pentanol. Thus, the mechanism proposed by Hughes and Ingold also accounts for the third piece of experimental evidence—the observed configuration of the product.

If the leaving group is attached to an asymmetric center, an S_N2 reaction forms only the stereoisomer with the inverted configuration.

NOTE TO THE STUDENT

• To draw the inverted product of an S_N2 reaction, draw the mirror image of the reactant and replace the halogen with the nucleophile.

the configuration of the product is inverted relative to the configuration of the reactant

(R)-2-bromobutane   (S)-2-butanol

PROBLEM 5 ◆

Rank the following alkyl bromides from most reactive to least reactive in an S_N2 reaction: 1-bromo-2-methylbutane, 1-bromo-3-methylbutane, 2-bromo-2-methylbutane, and 1-bromopentane.

PROBLEM 6 ◆ **SOLVED**

Draw the products obtained from the S_N2 reaction of

a. 2-bromobutane and methoxide ion.   **c.** (S)-3-chlorohexane and hydroxide ion.
b. (R)-2-bromobutane and methoxide ion.   **d.** 3-iodopentane and hydroxide ion.

SOLUTION TO 6 a. The product is 2-methoxybutane. Because the reaction is an S_N2 reaction, we know that the configuration of the product is inverted relative to the configuration of the reactant. The configuration of the reactant is not specified, however, so we cannot specify the configuration of the product. In other words, because we also do not know if the reactant is R or S or a mixture of the two, we also do not know if the product is R or S or a mixture of the two.

the configuration is not specified

$$CH_3CHCH_2CH_3 \ + \ CH_3O^- \longrightarrow CH_3CHCH_2CH_3 \ + \ Br^-$$
$$| \qquad\qquad\qquad\qquad\qquad\qquad\qquad\qquad | $$
$$Br \qquad\qquad\qquad\qquad\qquad\qquad\qquad OCH_3$$

LEARN THE STRATEGY

PROBLEM 7 SOLVED

Draw the substitution product formed when *cis*-1-bromo-4-methylcyclohexane and hydroxide ion undergo an S_N2 reaction.

SOLUTION Only the trans product is formed in this S_N2 reaction because the carbon bonded to the leaving group is attacked by the nucleophile on its back side.

the configuration of this carbon has been inverted

cis-1-bromo-4-methylcyclohexane   *trans*-4-methylcyclohexanol

---

USE THE STRATEGY

PROBLEM 8

Draw the substitution product formed by each of the following S_N2 reactions:

a. *trans*-1-iodo-4-ethylcyclohexane and methoxide ion
b. *cis*-1-chloro-3-methylcyclobutane and ethoxide ion

9.2 FACTORS THAT AFFECT S_N2 REACTIONS

We will now look at how the nature of the leaving group and the nature of the nucleophile affect an S_N2 reaction.

The Leaving Group in an S_N2 Reaction

If an alkyl iodide, an alkyl bromide, an alkyl chloride, and an alkyl fluoride with the same alkyl group were allowed to react with the same nucleophile under the same conditions, we would find that the alkyl iodide is the most reactive and the alkyl fluoride is the least reactive.

			relative rates of reaction	pK$_a$ values of HX
HO⁻ + RCH$_2$I	⟶	RCH$_2$OH + I⁻	30,000	−10
HO⁻ + RCH$_2$Br	⟶	RCH$_2$OH + Br⁻	10,000	−9
HO⁻ + RCH$_2$Cl	⟶	RCH$_2$OH + Cl⁻	200	−7
HO⁻ + RCH$_2$F	⟶	RCH$_2$OH + F⁻	1	3.2

The only difference between these four reactions is the leaving group. From the relative reaction rates, we see that iodide ion is the best leaving group and fluoride ion is the worst. This brings us to an important rule in organic chemistry that you will encounter frequently: when comparing bases of the same type,

the weaker the basicity of a group, the better is its leaving propensity

Leaving propensity depends on basicity because weak bases are stable bases; they readily bear the electrons they formerly shared with a proton. Therefore, they do not share their electrons well. Thus, a weak base is not bonded as strongly to the carbon as a strong base would be, and a weaker bond is more easily broken.

We saw that iodide ion is the weakest base of the halide ions (it has the strongest conjugate acid; Section 2.6) and fluoride ion is the strongest base (it has the weakest conjugate acid). Therefore, when comparing alkyl halides with the same alkyl group, we find that the alkyl iodide is the most reactive and the alkyl fluoride is the least reactive. In fact, the fluoride ion is such a strong base that alkyl fluorides essentially do not undergo S_N2 reactions.

relative reactivities of alkyl halides in an S_N2 reaction

most reactive ⟶ RI > RBr > RCl > RF ⟵ too unreactive to undergo an S_N2 reaction

The weaker the base, the better it is as a leaving group.

Stable bases are weak bases.

At the beginning of this chapter, we saw that the polar carbon–halogen bond causes alkyl halides to undergo substitution reactions. Carbon and iodine, however, have the same electronegativity. (See Table 1.3 on page 10.) Why, then, does an alkyl iodide undergo a substitution reaction?

We know that larger atoms are more polarizable than smaller atoms. (Recall from Section 3.9 that polarizability is a measure of how easily an atom's electron cloud can be distorted.) The high polarizability of the large iodine atom allows the C—I bond to become polarized by the approaching electron-rich nucleophile, causing a partial charge to develop on the carbon. Therefore, an alkyl iodide reacts as if it were polar, even though carbon and iodine have the same electronegativities.

PROBLEM 9 ◆

Which alkyl halide is more reactive in an S_N2 reaction with a given nucleophile?

The Nucleophile in an S_N2 Reaction

When we talk about atoms or molecules that have lone-pair electrons, sometimes we call them bases and sometimes we call them nucleophiles (Table 9.1). What is the difference between a base and a nucleophile?

Table 9.1	Common Nucleophiles/Bases			
HO^-	RO^-	H_2O	ROH	$RCOO^-$
HS^-	RS^-	H_2S	RSH	
$^-NH_2$	RNH^-	NH_3	RNH_2	
$^-C\equiv N$	$RC\equiv C^-$			
Cl^-	Br^-	I^-		

Basicity is a measure of how well a compound (a **base**) shares its lone pair with a proton. The stronger the base, the better it shares its electrons. Basicity is measured by an *equilibrium constant* (the acid dissociation constant, K_a) that indicates the tendency of the conjugate acid of the base to lose a proton (Section 2.2).

Nucleophilicity is a measure of how readily a compound (a **nucleophile**) is able to attack an electron-deficient atom. It is measured by a *rate constant* (k). In the case of an S_N2 reaction, nucleophilicity is a measure of how readily the nucleophile attacks an sp^3 carbon bonded to a leaving group.

Because the nucleophile attacks an sp^3 carbon in the rate-determining step of an S_N2 reaction, the rate of the reaction depends on the strength of the nucleophile: the better the nucleophile, the faster the rate of the S_N2 reaction.

Note that a species with a negative charge is a stronger base *and* a better nucleophile than a species that has the same attacking atom but is neutral. Thus, HO^- is a stronger base and a better nucleophile than H_2O. Also note that bases are described as being strong or weak, whereas nucleophiles are described as being good or poor.

stronger base, better nucleophile		weaker base, poorer nucleophile
HO^-	>	H_2O
CH_3O^-	>	CH_3OH
$^-NH_2$	>	NH_3
$CH_3CH_2NH^-$	>	$CH_3CH_2NH_2$

Effect of Basicity on Nucleophilicity

If the attacking atoms are the same size, *stronger bases are better nucleophiles*. For example, comparing attacking atoms in the second row of the periodic table (so they are the same size), the amide ion is both the strongest base and the best nucleophile.

relative base strengths and relative nucleophilicities

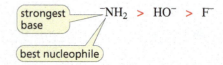

If, however, the attacking atoms of the nucleophiles are *very different in size*, another factor comes into play: the polarizability of the atom. Because the electrons are farther away in the larger tom, they are not held as tightly and can, therefore, move more freely toward a positive charge. As a result, the electrons are able to overlap the orbital of carbon from farther away, as shown in Figure 9.5. This results in a greater degree of bonding in the transition state, which makes the transition state more stable.

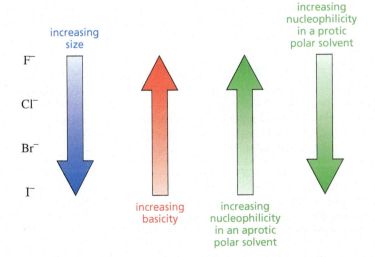

transition state

transition state

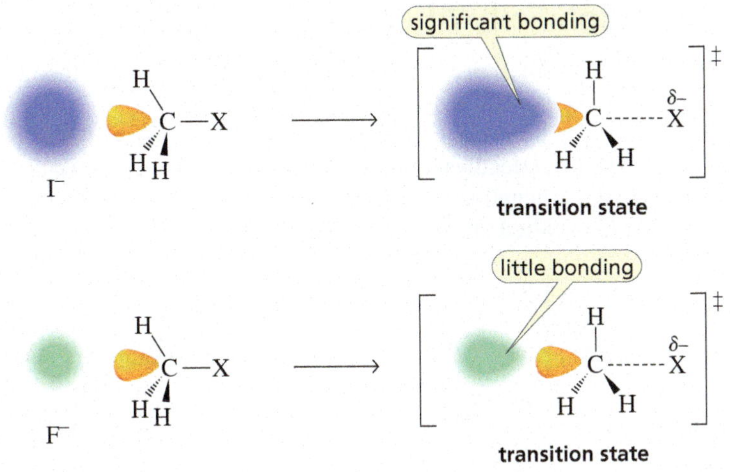

◀ Figure 9.5
An iodide ion is larger and more polarizable than a fluoride ion. Therefore, when an iodide ion approaches a carbon, the relatively loosely held electrons of the ion can overlap the orbital of carbon from farther away. The tightly bound electrons of the fluoride ion cannot start to overlap the orbital of carbon until the reactants are closer together.

Now the question becomes, does the greater polarizability that helps the larger atoms to be better nucleophiles make up for the decreased basicity that causes them to be poorer nucleophiles? The answer depends on the solvent.

Effect of Solvent on Nucleophilicity

If the reaction is carried out in an **aprotic polar solvent**—meaning the polar solvent molecules *do not have* a hydrogen bonded to an oxygen or to a nitrogen—the direct relationship between basicity and nucleophilicity is maintained: the strongest bases are still the best nucleophiles. In other words, the greater polarizability of the larger atoms does not make up for their decreased basicity. *Therefore, iodide ion, the weakest base, is the poorest nucleophile of the halide ions in an aprotic polar solvent.*

If, however, the reaction is carried out in a **protic polar solvent**—meaning the polar solvent molecules *have* a hydrogen bonded to an oxygen or to a nitrogen—the relationship between basicity and nucleophilicity becomes inverted (Figure 9.6). The largest atom (the most polarizable one) is the best nucleophile even though it is the weakest base. *Therefore, iodide ion, the weakest base, is the best nucleophile of the halide ions in a protic polar solvent.*

An aprotic solvent does not contain a hydrogen bonded to either an oxygen or a nitrogen.

A protic solvent contains a hydrogen bonded to an oxygen or a nitrogen.

When the attacking bases are different in size, the strongest bases are the best nucleophiles in aprotic polar solvents and the poorest nucleophiles in protic polar solvents.

F$^-$

Cl$^-$

Br$^-$

I$^-$

increasing size

increasing basicity

increasing nucleophilicity in an aprotic polar solvent

increasing nucleophilicity in a protic polar solvent

◀ Figure 9.6
The strongest bases are the best nucleophiles except when
(1) the bases differ in size, and
(2) the bases are used in a protic polar solvent.
Only when both conditions are met is the strongest base *not* the best nucleophile.

PROBLEM 10 ♦

Indicate whether each of the following solvents is protic or aprotic:

a. chloroform (CHCl₃)   **b.** diethyl ether   **c.** acetic acid   **d.** hexane

Why Is the Nucleophilicity Affected by the Solvent?

Why, in a protic solvent, is the smallest atom the poorest nucleophile even though it is the stronge
base? *How does a protic solvent make strong bases less nucleophilic?*

Protic solvents are hydrogen bond donors. Therefore, when a negatively charged species is place
in a protic solvent, the solvent molecules arrange themselves with their partially positively charge
hydrogens pointing toward the negatively charged species. The interaction between the ion and th
dipole of the protic solvent is called an **ion–dipole interaction.**

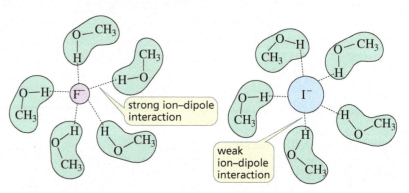

Because the solvent shields the nucleophile, at least one of the ion–dipole interactions mu
be broken before the nucleophile can participate in an S_N2 reaction. Weak bases interact weak
with protic solvents, whereas strong bases interact strongly because they are better at sharing the
electrons. It is easier, therefore, to break the ion–dipole interactions between an iodide ion (a wea
base) and the solvent than between a fluoride ion (a stronger base) and the solvent. In a prot
solvent, therefore, an iodide ion, even though it is a weaker base, is a better nucleophile than
fluoride ion (Table 9.2).

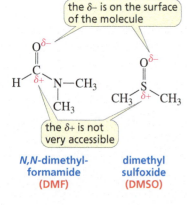

N,N-dimethyl-formamide (DMF)   dimethyl sulfoxide (DMSO)

Table 9.2   Relative Nucleophilicity toward CH₃I in Methanol
CH_3S^- > I^- > $^-C{\equiv}N$ > CH_3O^- > Br^- > NH_3 > Cl^- > F^- > CH_3OH

increasing nucleophilicity

An **aprotic polar solvent** does not have any hydrogens with partial positive charges to for
ion–dipole interactions. The molecules of an aprotic polar solvent (such as DMF or DMSO) have
partial negative charge on their surface that can solvate cations, but the partial positive charge is o
the *inside* of the molecule and, therefore, less accessible to solvate anions. Thus, fluoride ion is
good nucleophile in DMSO and a poor nucleophile in water.

Fluoride ion would be an even better nucleophile in a *nonpolar solvent* (such as hexane) becau
there would not be any ion–dipole interactions between the ion and the nonpolar solvent. Ion
compounds, however, are insoluble in nonpolar solvents, but they dissolve in aprotic polar solvent
Fluoride ion is also a good nucleophile in the gas phase, where there are no solvent molecules.

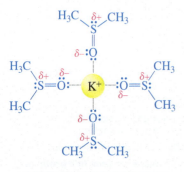

DMSO solvates a cation better than it solvates an anion

PROBLEM 11 ♦

a. Which is a stronger base: RO⁻ or RS⁻?
b. Which is a better nucleophile in an aqueous solution?
c. Which is a better nucleophile in DMSO?

PROBLEM 12 ◆

Which is a better nucleophile?

a. Br⁻ or Cl⁻ in H$_2$O

b. Br⁻ or Cl⁻ in DMSO

c. CH$_3$O⁻ or CH$_3$OH in H$_2$O

d. CH$_3$O⁻ or CH$_3$OH in DMSO

e. HO⁻ or ⁻NH$_2$ in H$_2$O

f. HO⁻ or ⁻NH$_2$ in DMSO

g. I⁻ or Br⁻ in H$_2$O

h. I⁻ or Br⁻ in DMSO

Nucleophilicity Is Affected by Steric Effects

Nucleophilicity is *affected* by steric effects because a bulky nucleophile cannot approach the back side of a carbon as easily as a less sterically hindered nucleophile can. Basicity, on the other hand, is relatively *unaffected* by steric effects because a base removes an unhindered proton.

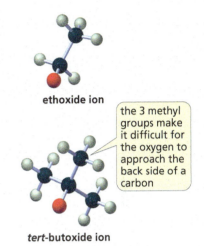

ethoxide ion

the 3 methyl groups make it difficult for the oxygen to approach the back side of a carbon

tert-butoxide ion

$$CH_3CH_2{-}O^-$$

ethoxide ion
better nucleophile

$$CH_3{-}\overset{\displaystyle CH_3}{\underset{\displaystyle CH_3}{\overset{|}{\underset{|}{C}}}}{-}O^-$$

tert-butoxide ion
stronger base

Therefore, *tert*-butoxide ion, with its three methyl groups, is a poorer nucleophile than ethoxide ion, even though *tert*-butoxide ion is a stronger base (pK_a of *tert*-butanol = 18) than ethoxide ion (pK_a of ethanol = 16).

PROBLEM 13 SOLVED

Rank the following species from best nucleophile to poorest nucleophile in an aqueous solution:

SOLUTION Let's first divide the nucleophiles into groups. There is one nucleophile with a negatively charged sulfur, three with negatively charged oxygens, and one with a neutral oxygen. We know that in the polar aqueous solvent, the compound with the negatively charged sulfur is the best nucleophile because sulfur is larger than oxygen. We also know that the poorest nucleophile is the one with the neutral oxygen. To complete the problem, we need to rank the three nucleophiles with negatively charged oxygens, which we can do by looking at the pK_a values of their conjugate acids. A carboxylic acid is a stronger acid than phenol, which is a stronger acid than water (Section 8.9). Because water is the weakest acid, its conjugate base is the strongest base and the best nucleophile. Thus, the relative nucleophilicities are:

$$CH_3S^- \; > \; HO^- \; > \; \text{C}_6\text{H}_5{-}O^- \; > \; CH_3\overset{\displaystyle O}{\overset{\|}{C}}O^- \; > \; CH_3OH$$

PROBLEM 14 ◆

Which substitution reaction takes place more rapidly?

a. CH$_3$CH$_2$Br + H$_2$O or CH$_3$CH$_2$Br + HO⁻

b. CH$_3$CHCH$_2$Br + HO⁻ or CH$_3$CH$_2$CHBr + HO⁻
   | |
   CH$_3$ CH$_3$

c. CH$_3$CH$_2$Cl + CH$_3$O⁻ or CH$_3$CH$_2$Cl + CH$_3$S⁻
   (in ethanol)

d. CH$_3$CH$_2$Cl + I⁻ or CH$_3$CH$_2$Br + I⁻

LEARN THE STRATEGY

USE THE STRATEGY

A Wide Variety of Compounds Can Be Synthesized by S$_N$2 Reaction

Many different kinds of nucleophiles can react with alkyl halides. Therefore, a wide variety of organic compounds can be synthesized by means of S$_N$2 reactions. Notice that the last reaction is the reaction of an alkyl halide with an acetylide ion. This is the reaction that we used in Section 7.1 to create longer carbon chains. Now that you know this is an S$_N$2 reaction, you can understand why you were told that it is best to use methyl halides and primary alkyl halides in this reaction.

$$CH_3CH_2Cl + HO^- \longrightarrow CH_3CH_2OH + Cl^-$$
alcohol

$$CH_3CH_2I + RO^- \longrightarrow CH_3CH_2OR + I^-$$
ether

$$CH_3CH_2Br + HS^- \longrightarrow CH_3CH_2SH + Br^-$$
thiol

$$CH_3CH_2Br + RS^- \longrightarrow CH_3CH_2SR + Br^-$$
thioether

$$CH_3CH_2Cl + NH_3 \longrightarrow CH_3CH_2\overset{+}{N}H_3 + Cl^-$$
primary ammonium ion

$$CH_3CH_2Br + CH_3\overset{\overset{\displaystyle CH_3}{|}}{N}CH_3 \longrightarrow CH_3CH_2\overset{\overset{\displaystyle CH_3}{|+}}{\underset{\underset{\displaystyle CH_3}{|}}{N}}CH_3 + Br^-$$
tertiary amine   quaternary ammonium ion

$$CH_3CH_2I + {}^-C{\equiv}N \longrightarrow CH_3CH_2C{\equiv}N + I^-$$
nitrile

$$CH_3CH_2Br + {}^-C{\equiv}CR \longrightarrow CH_3CH_2C{\equiv}CR + Br^-$$
alkyne

At first glance, it may seem that the reverse of each of these reactions satisfies the requirement for a nucleophilic substitution reaction. For example, the reverse of the first reaction would be the reaction of chloride ion (a nucleophile) with ethanol (that has an HO$^-$ leaving group). But ethanol and chloride ion do *not* react.

Why doesn't the reverse reaction take place? We can answer this question by comparing the leaving propensity of Cl$^-$ and HO$^-$, the leaving groups in the forward and reverse directions, respectively. Comparing leaving propensities means comparing basicities. Because HCl is a much stronger acid than H$_2$O, Cl$^-$ is a much weaker base than HO$^-$; because it is a weaker base, Cl$^-$ is a better leaving group. Consequently, HO$^-$ can displace Cl$^-$ (a good leaving group) in the forward reaction, but Cl$^-$ cannot displace HO$^-$ (a poor leaving group) in the reverse reaction.

PROBLEM 15 ♦

What is the product of the reaction of bromoethane with each of the following nucleophiles?

a. CH$_3$CH$_2$CH$_2$O$^-$ **b.** CH$_3$C$\equiv$C$^-$ **c.** (CH$_3$)$_3$N   **d.** CH$_3$CH$_2$S$^-$

LEARN THE STRATEGY

PROBLEM 16 SOLVED

What product is obtained when ethylamine reacts with excess methyl iodide in a basic solution of potassium carbonate?

SOLUTION Ethylamine and methyl iodide undergo an S$_N$2 reaction. The product of the reaction is a secondary amine that is predominantly in its basic (neutral) form because the pH of the basic solution is greater than the pK_a of the protonated amine (Section 2.9). The secondary amine can undergo an

S_N2 reaction with another equivalent of methyl iodide, forming a tertiary amine. The tertiary amine can react with methyl iodide in yet another S_N2 reaction. The final product of the reaction is a quaternary ammonium iodide.

$$CH_3CH_2\ddot{N}H_2 + CH_3{-}I \longrightarrow CH_3CH_2\overset{+}{N}H_2CH_3 \quad I^- \xrightarrow{K_2CO_3} CH_3CH_2\underset{\cdot\cdot}{N}HCH_3$$

$$\downarrow CH_3{-}I$$

$$\underset{\underset{CH_3}{|}}{\overset{\overset{CH_3}{|}}{CH_3CH_2\overset{+}{N}CH_3}} \quad I^- \xleftarrow{CH_3{-}I} \underset{\underset{CH_3}{|}}{CH_3CH_2\ddot{N}CH_3} \xleftarrow{K_2CO_3} \underset{\underset{CH_3}{|}}{CH_3CH_2\overset{+}{N}HCH_3} \quad I^-$$

a quaternary ammonium iodide

The reaction of an amine with sufficient methyl iodide to convert the amine into a quaternary ammonium iodide is called **exhaustive methylation.**

PROBLEM 17   USE THE STRATEGY

a. Explain why the reaction of an alkyl halide with ammonia gives a low yield of primary amine.
b. Explain why a much better yield of primary amine is obtained from the reaction of an alkyl halide with azide ion ($^-N_3$), followed by catalytic hydrogenation. (*Hint*: An alkyl azide is not nucleophilic.)

$$CH_3CH_2CH_2Br \xrightarrow{^-N_3} CH_3CH_2CH_2N{=}\overset{+}{N}{=}N^- \xrightarrow[Pd/C]{H_2} CH_3CH_2CH_2NH_2 + N_2$$

an alkyl azide

Why Are Living Organisms Composed of Carbon Instead of Silicon?

There are two reasons living organisms are composed primarily of carbon, oxygen, nitrogen, and hydrogen: the *fitness* of these elements for specific roles in life processes and their *availability* in the environment. Fitness apparently was more important than availability because carbon rather than silicon became the fundamental building block of living organisms despite the fact that silicon, which is just below carbon in the periodic table, is more than 140 times more abundant than carbon in Earth's crust.

Why are carbon, oxygen, nitrogen, and hydrogen so well suited for the roles they play in living organisms? First and foremost, they are among the smallest atoms that form covalent bonds and they can form multiple bonds. Because of these factors, they form strong bonds (which means the molecules containing them are stable). The compounds that make up living organisms must be stable and, therefore, slow to react if the organisms are to survive.

Silicon has almost twice the diameter of carbon, so silicon forms longer and weaker bonds. Consequently, an S_N2 reaction at silicon would occur much more rapidly than an S_N2 reaction at carbon. Moreover, silicon has another problem. The end product of carbon metabolism is CO_2. The analogous product of silicon metabolism would be SiO_2. But unlike carbon, which is doubly bonded to oxygen in CO_2, silicon is only singly bonded to oxygen in SiO_2. Therefore, silicon dioxide molecules polymerize to form quartz (sand). It is hard to imagine that life could exist, much less proliferate, if animals exhaled sand instead of CO_2!

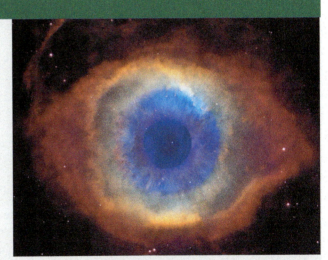

| | Abundance (atoms/100 atoms) | |
Element	In living organisms	In Earth's crust
H	49	0.22
C	25	0.19
O	25	47
N	0.3	0.1
Si	0.03	28

9.3 THE S$_N$1 REACTION

Given our understanding of S$_N$2 reactions, we would expect the rate of the following reaction to be very slow because water is a poor nucleophile and the alkyl halide is sterically hindered t back-side attack.

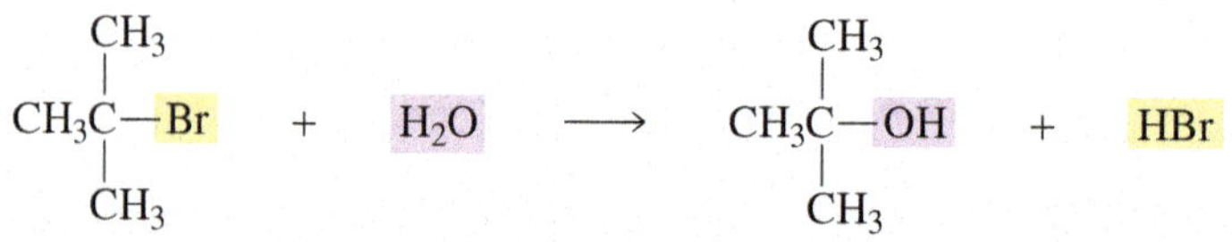

$$CH_3C—Br \quad + \quad H_2O \quad \longrightarrow \quad CH_3C—OH \quad + \quad HBr$$

It turns out, however, that the reaction is surprisingly fast. In fact, it is much faster than the reaction of bromomethane (a much less sterically hindered alkyl halide) with water. The reaction, therefore must be taking place by a mechanism different from that of an S$_N$2 reaction.

Experimental Evidence for the Mechanism for an S$_N$1 Reaction

We saw that in order to determine the mechanism of a reaction, we need to find out what factor affect the rate of the reaction, and we need to know the configuration of the products of the reaction.

Doubling the concentration of the alkyl halide doubles the rate of the reaction, but changing the concentration of the nucleophile has no effect on its rate. This knowledge allows us to write a rat law for the reaction:

$$\text{rate} = k[\text{alkyl halide}]$$

The rate of the reaction depends linearly on the concentration of only one reactant, so the reaction is a **first-order reaction.**

Because the rate law for the reaction of 2-bromo-2-methylpropane with water differs from the rate law for the reaction of bromomethane with hydroxide ion (Section 9.1), the two reactions mus have different mechanisms.

The reaction between 2-bromo-2-methylpropane and water is an **S$_N$1 reaction,** where "S" stand for *substitution,* "N" stands for *nucleophilic,* and "1" stands for *unimolecular.* **Unimolecular** mean that only one molecule is involved in the transition state of the rate-determining step.

In the S$_N$1 reactions of alkyl halides, the solvent is the nucleophile. For example, in the following reaction, methanol is both the nucleophile and the solvent. Reaction with a solvent is called **solvolysis**

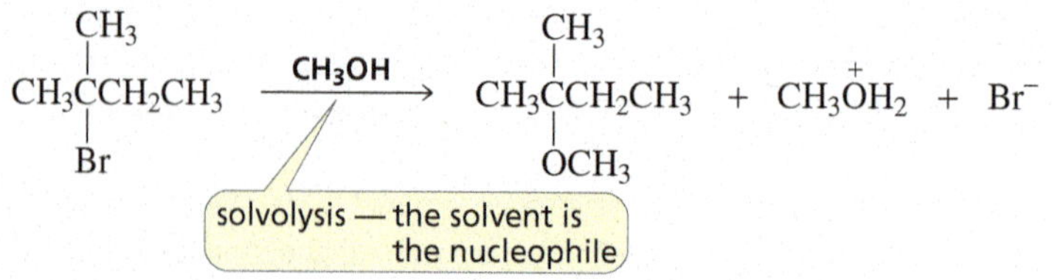

$$CH_3CCH_2CH_3 \xrightarrow{CH_3OH} CH_3CCH_2CH_3 \quad + \quad CH_3\overset{+}{O}H_2 \quad + \quad Br^-$$

solvolysis — the solvent is the nucleophile

The mechanism for an S$_N$1 reaction is based on the following experimental evidence:

1. The rate law shows that the rate of the reaction depends only on the concentration of the alky halide, so only the alkyl halide is involved in the transition state of the rate-determining step.

2. Tertiary alkyl halides undergo S$_N$1 solvolysis reactions with poor nucleophiles such as water and alcohols, but methyl halides and primary alkyl halides do not. A recent investigation of Hughes and Ingold's data showed that secondary alkyl halides also do not undergo S$_N$1 solvolysis reactions.* Thus, methyl halides, primary alkyl halides, and secondary alkyl halides undergo only S$_N$2 reactions.

3. The substitution reaction of an alkyl halide in which the halogen is bonded to an asymmetric center forms two stereoisomers: one with the same relative configuration as that of the reacting alkyl halide and the other with the inverted configuration.

Only tertiary alkyl halides undergo S$_N$1 solvolysis reactions.

---

*Murphy, T.J. *J. Chem. Ed.* **2009,** *86,* 519–524.

The Mechanism for an S$_N$1 Reaction

Unlike an S$_N$2 reaction, where the leaving group departs and the nucleophile approaches *at the same time*, the leaving group in an S$_N$1 reaction departs *before* the nucleophile approaches.

MECHANISM FOR THE S$_N$1 REACTION OF AN ALKYL HALIDE

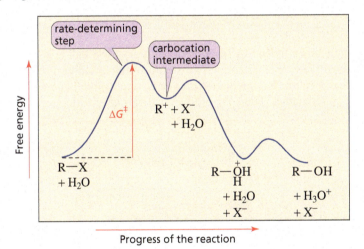

- In the first step, the carbon–halogen bond breaks and the previously shared pair of electrons leaves with the halogen. As a result, a carbocation intermediate is formed.
- In the second step, the nucleophile reacts rapidly with the carbocation (an electrophile) to form a protonated alcohol.
- Whether the alcohol product exists in its protonated (acidic) form or neutral (basic) form depends on the pH of the solution. At pH = 7, the alcohol exists only in its neutral form (Section 2.10).

Because the rate of an S$_N$1 reaction depends only on the concentration of the alkyl halide, the first step must be the slow (rate-determining) step (Figure 9.7). The nucleophile is not involved in the rate-determining step, so its concentration has no effect on the rate of the reaction.

An S$_N$1 reaction is a two-step reaction.

◀ **Figure 9.7**
The reaction coordinate diagram for an S$_N$1 reaction shows why increasing the rate of the second step will not make an S$_N$1 reaction go any faster.

We will see that the solvent plays an important role in an S$_N$1 reaction, other than reacting as a nucleophile in the second step of the reaction. The ion dipole interactions between the charged species in the solution and the solvent provide the energy necessary for dissociation of the carbon-halogen bond (Section 9.14).

How the Mechanism Accounts for the Experimental Evidence

How does the mechanism for an S$_N$1 reaction account for the three pieces of experimental evidence?

First, because the alkyl halide is the only species that participates in the rate-determining step, the mechanism agrees with the observation that the rate of the reaction depends only on the concentration of the alkyl halide; it does not depend on the concentration of the nucleophile.

Second, the mechanism shows that a carbocation is formed in the rate-determining step. Th explains why tertiary alkyl halides undergo S_N1 reactions, but primary and secondary alk halides do not. Tertiary carbocations are more stable than primary and secondary carbocation and, therefore, are the most easily formed. (In Section 9.5 we will see that allylic and benzyl halides undergo S_N1 reactions, because they too form relatively stable carbocations.)

Third, the positively charged carbon of the carbocation intermediate is sp^2 hybridized, whic means that the three bonds connected to it are in the same plane (Figure 9.8). In the second ste of the S_N1 reaction, the nucleophile can approach the carbocation from either side of the plane, s some of the product has the same configuration as the reacting alkyl halide and some has an inverte configuration.

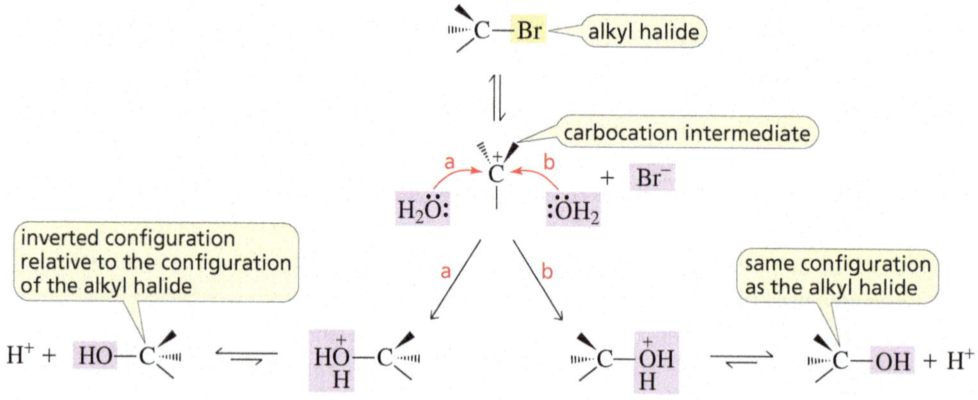

▲ Figure 9.8
If the nucleophile adds to the *opposite side* of the carbon from which the leaving group departed (labeled a) then the product will have the *inverted* configuration relative to the configuration of the alkyl halide.

If the nucleophile adds to the side of the carbon from which the leaving group departed (labeled b), then the product will have the *same* relative configuration as that of the reacting alkyl halide.

We can now understand why an S_N1 reaction of a tertiary alkyl halide in which the leaving grou is attached to an asymmetric center forms two stereoisomers: addition of the nucleophile to on side of the planar carbocation intermediate forms one stereoisomer, and addition to the other sid produces the other stereoisomer. Thus, the product is a pair of enantiomers.

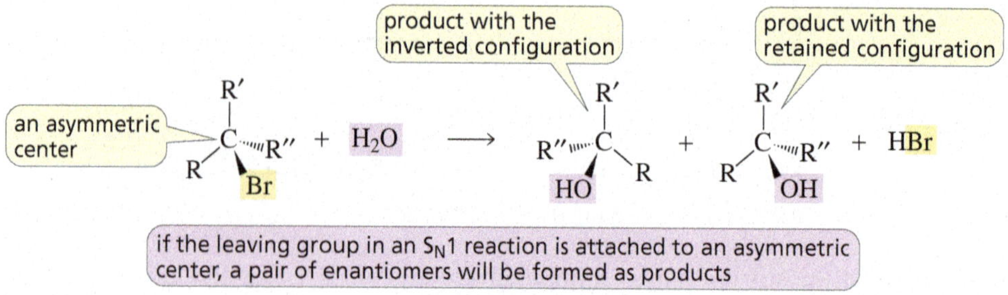

Most S_N1 Reactions Lead to Partial Racemization

Although you probably expect that equal amounts of both products are formed in an S_N1 reaction a greater amount of the product with the inverted configuration is obtained in most cases. Typically 50 to 70% of the product of an S_N1 reaction is the inverted product. If the reaction does lead to equal amounts of the two stereoisomers, the reaction is said to take place with **complete racemization.** When more of the inverted product is formed, the reaction is said to take place with **partial racemization.**

Carbocation stability: 3° > 2° > 1°

Tertiary alkyl halides undergo S_N1 reactions. Primary and secondary alkyl halides undergo S_N2 reactions.

An S_N1 reaction takes place with inversion and retention of configuration.

If the leaving group is attached to an asymmetric center, an S_N1 reaction forms a pair of enantiomers.

Why does an S$_N$1 reaction generally form more inverted product? Dissociation of the alkyl halide initially results in formation of an **intimate ion pair.** In an intimate ion pair, the bond between the carbon and the leaving group has broken, but the cation and anion remain next to each other. When they move slightly farther apart, they become a *solvent-separated ion pair,* meaning an ion pair with one or more solvent molecules between the cation and the anion. As the ions separate further, they become dissociated ions.

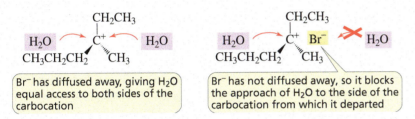

The nucleophile can attack any of these four species. If the nucleophile attacks only the fully dissociated carbocation, the product will be completely racemized. If the nucleophile attacks the carbocation of either the intimate ion pair or the solvent-separated ion pair, the leaving group will be in position to partially block the approach of the nucleophile to that side of the carbocation. As a result, more of the product with the inverted configuration will be formed.

An S$_N$1 reaction takes place with racemization.

Notice that the reaction of a nucleophile with the undissociated alkyl halide is an S$_N$2 reaction.)

PROBLEM 18 ♦

Draw the stereoisomers that are formed from the following S$_N$1 reactions:

a. 3-bromo-3-methylpentane and methanol **b.** 3-chloro-3-methylhexane and methanol

---

9.4 FACTORS THAT AFFECT S$_N$1 REACTIONS

We will now look at how the leaving group and the nucleophile affect S$_N$1 reactions.

The Leaving Group in an S$_N$1 Reaction

Because the rate-determining step of an S$_N$1 reaction is the formation of a carbocation, two factors affect the rate of the reaction:

1. the ease with which the leaving group dissociates, and
2. the stability of the carbocation that is formed.

As in an S$_N$2 reaction, there is a direct relationship between basicity and leaving propensity in an S$_N$1 reaction: the weaker the base, the less tightly it is bonded to the carbon and the more easily the carbon–halogen bond can be broken. As a result, comparing alkyl halides with the same alkyl group, an alkyl iodide is the most reactive and an alkyl fluoride is the least reactive in both S$_N$1 and S$_N$2 reactions.

relative reactivities of alkyl halides in an S$_N$1 reaction

most reactive ▷ RI > RBr > RCl > RF ◁ least reactive

The Nucleophile in an S_N1 Reaction

Because the nucleophile does not participate in an S_N1 reaction until *after* the rate-determining step, the reactivity of the nucleophile has no effect on the rate of an S_N1 reaction.

PROBLEM 19 ♦

Rank the following alkyl halides from most reactive to least reactive in an S_N1 reaction:
2-bromo-2-methylpentane, 2-chloro-2-methylpentane, 3-chloropentane, and 2-iodo-2-methylpentane.

9.5 COMPETITION BETWEEN S_N2 AND S_N1 REACTIONS

The characteristics of S_N2 and S_N1 reactions are compared in Table 9.3. Remember that the "2" in "S_N2" and the "1" in "S_N1" refer to the molecularity of the reaction (the number of molecules involved in the transition state of the rate-determining step), and *not* to the number of steps in the mechanism. In fact, the opposite is true: an S_N2 reaction proceeds by a *one*-step concerted mechanism, whereas an S_N1 reaction proceeds by a *two*-step mechanism with a carbocation intermediate.

Table 9.3 Comparing S_N2 and S_N1 Reactions

S_N2	S_N1
one-step mechanism	two-step mechanism with a carbocation intermediate
bimolecular rate-determining step	unimolecular rate-determining step
rate decreases with increasing steric hindrance	rate decreases with decreasing stability of the carbocation
product has the inverted configuration relative to that of the reactant	products have both the retained and inverted configurations relative to that of the reactant
leaving group: $I^- > Br^- > Cl^- > F^-$	leaving group: $I^- > Br^- > Cl^- > F^-$
alkyl halide reactants: methyl, primary, secondary	alkyl halide reactants: tertiary
the better the nucleophile, the faster the rate of the reaction	the strength of the nucleophile does not affect the rate of the reaction

Because an alkyl halide can potentially undergo both S_N2 and S_N1 reactions simultaneously, the rate law for the reaction of an alkyl halide with a nucleophile is given by the sum of the rate laws for the S_N2 and S_N1 reactions (Sections 9.1 and 9.3). The rate constants have been given subscripts that indicate the reaction order.

$$\text{rate} = k_2[\text{alkyl halide}][\text{nucleophile}] + k_1[\text{alkyl halide}]$$

contribution to the rate by an S_N2 reaction

contribution to the rate by an S_N1 reaction

The rate law allows us to determine whether an S_N2 reaction or an S_N1 reaction is more likely for a given alkyl halide:

Primary and secondary alkyl halides undergo S_N2 reactions.

- Primary alkyl halides and methyl halides cannot form carbocations because primary carbocations and methyl cations are too unstable to be formed. This causes k_1 to be approximately zero. Therefore, the only substitution reactions **primary alkyl halides** and **methyl halides** undergo are **S_N2 reactions.**

- Secondary carbocations are somewhat more stable than primary carbocations but not sufficiently more stable to make up for the large concentration of the nucleophile in a solvolysis reaction. Therefore, the only substitution reactions **secondary alkyl halides** undergo are **S_N2 reactions.**

Tertiary alkyl halides undergo S_N1 reactions.

- Steric hindrance prevents tertiary halides from undergoing S_N2 reactions (Section 9.2). This causes k_2 to be close to zero. Therefore, the only substitution reactions **tertiary alkyl halides** undergo are **S_N1 reactions.**

herefore, it is easy to tell whether an alkyl halide will undergo an S_N2 reaction or an S_N1 reaction. ust look at its structure. If the alkyl halide is a methyl halide or a primary or secondary alkyl halide, t will undergo S_N2 reactions. If it is a tertiary alkyl halide, it will undergo S_N1 reactions.

PROBLEM-SOLVING STRATEGY

Predicting Whether a Nucleophilic Substitution Reaction Will Be an S_N1 Reaction or an S_N2 Reaction and Determining the Product(s) of the Substitution Reaction

LEARN THE STRATEGY

Draw the configuration(s) of the substitution product(s) that are formed from the reactions of the following compounds with the indicated nucleophile:

a. Because the reactant is a secondary alkyl halide, we know it is an S_N2 reaction. Therefore, the product has the inverted configuration relative to the configuration of the reactant.

$$CH_3-\overset{\overset{\displaystyle CH_2CH_3}{|}}{\underset{\underset{\displaystyle Br}{|}}{C}}\text{—}H \quad + \quad CH_3O^- \quad \longrightarrow \quad H\text{—}\overset{\overset{\displaystyle CH_2CH_3}{|}}{\underset{\underset{\displaystyle CH_3O}{|}}{C}}\text{—}CH_3 \quad + \quad Br^-$$

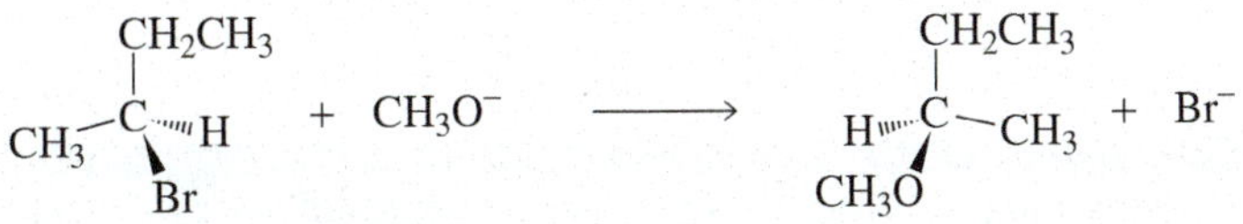

b. Because the reactant is a tertiary alkyl halide, we know it is an S_N1 reaction. Therefore, there are two substitution products, one with the retained configuration and one with the inverted configuration relative to the configuration of the reactant.

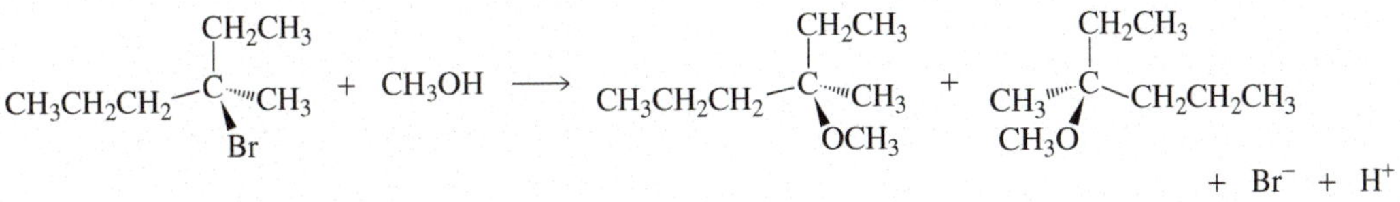

NOTE TO THE STUDENT
- One way to draw the inverted product is to draw the mirror image of the reacting alkyl halide and then put the nucleophile in the same location as the leaving group.

$$CH_3CH_2CH_2-\overset{\overset{\displaystyle CH_2CH_3}{|}}{\underset{\underset{\displaystyle Br}{|}}{C}}\text{—}CH_3 \quad + \quad CH_3OH \quad \longrightarrow \quad CH_3CH_2CH_2-\overset{\overset{\displaystyle CH_2CH_3}{|}}{\underset{\underset{\displaystyle OCH_3}{|}}{C}}\text{—}CH_3 \quad + \quad CH_3-\overset{\overset{\displaystyle CH_2CH_3}{|}}{\underset{\underset{\displaystyle CH_3O}{|}}{C}}\text{—}CH_2CH_2CH_3$$

$$+ \quad Br^- \quad + \quad H^+$$

c. Because the reactant is a tertiary alkyl halide, we know it is an S_N1 reaction. The product does not have an asymmetric center, so it does not have stereoisomers. Therefore, only one product is formed.

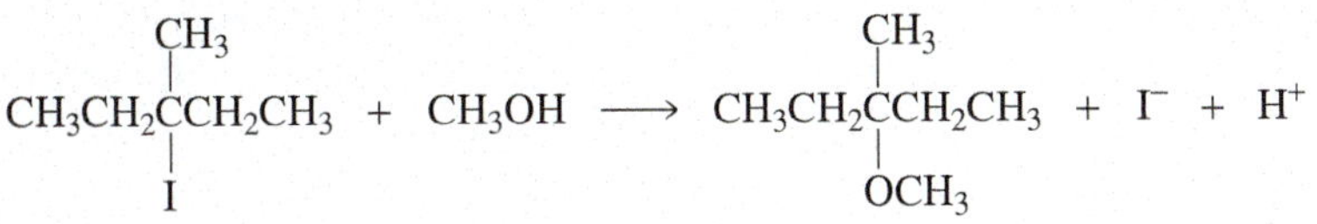

$$CH_3CH_2\overset{\overset{\displaystyle CH_3}{|}}{\underset{\underset{\displaystyle I}{|}}{C}}CH_2CH_3 \quad + \quad CH_3OH \quad \longrightarrow \quad CH_3CH_2\overset{\overset{\displaystyle CH_3}{|}}{\underset{\underset{\displaystyle OCH_3}{|}}{C}}CH_2CH_3 \quad + \quad I^- \quad + \quad H^+$$

d. Because the reactant is a secondary alkyl halide, we know it is an S_N2 reaction. Therefore, the configuration of the product is inverted relative to the configuration of the reactant. But because the configuration of the reactant is not indicated, we do not know the configuration of the product.

$$CH_3CH_2\overset{\overset{\displaystyle }{}}{\underset{\underset{\displaystyle Cl}{|}}{C}HCH_3} \quad + \quad CH_3O^- \quad \longrightarrow \quad CH_3CH_2\overset{\overset{\displaystyle }{}}{\underset{\underset{\displaystyle OCH_3}{|}}{C}HCH_3} \quad + \quad Cl^-$$

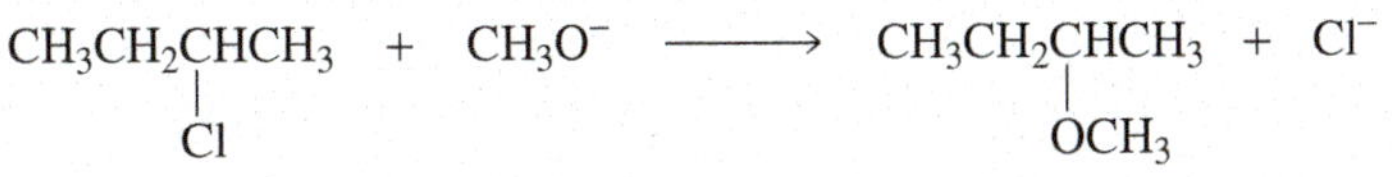

PROBLEM 20

USE THE STRATEGY

Draw the configuration(s) of the substitution product(s) formed from the reaction of the following compounds with the indicated nucleophile:

a. 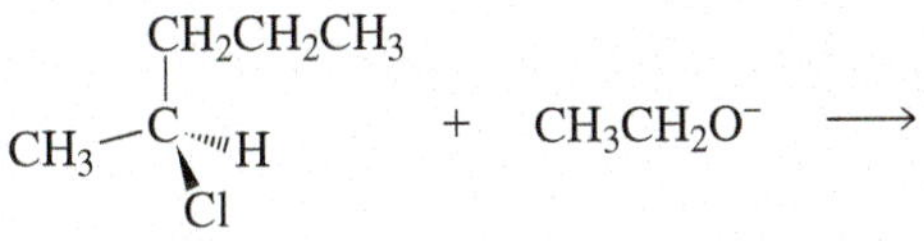 $CH_3-\overset{\overset{\displaystyle CH_2CH_2CH_3}{|}}{\underset{\underset{\displaystyle Cl}{|}}{C}}\text{—}H \quad + \quad CH_3CH_2O^- \quad \longrightarrow$

b.

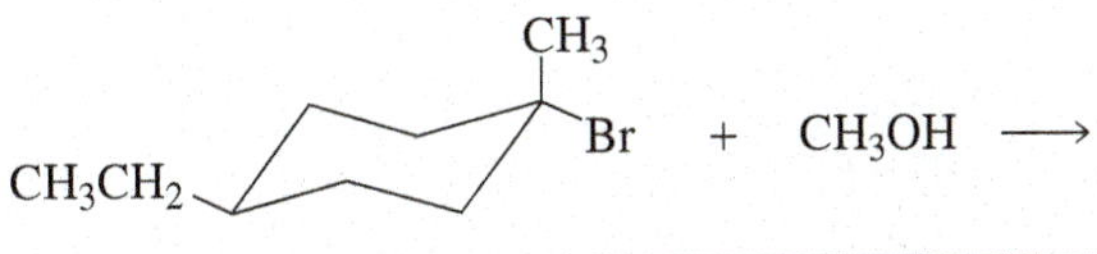

CH_3CH_2 ⬡ with CH_3 and $Br \quad + \quad CH_3OH \quad \longrightarrow$

PROBLEM 21 ◆

Which of the following reactions take place more rapidly when the concentration of the nucleophile is increased?

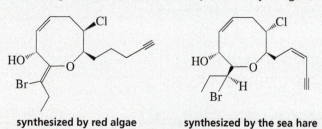

Naturally Occurring Alkyl Halides That Defend Against Predators

For a long time, chemists thought that only a few alkyl halides were found in nature. Now, however, over 5000 naturally occurring alkyl halides are known. Several marine organisms, including sponges, corals, and algae, synthesize alkyl halides that they use to deter predators. For example, red algae synthesize a toxic, foul-tasting alkyl halide that keeps predators from eating them.

One predator that is not deterred, however, is a mollusk called a sea hare. After consuming red algae, a sea hare converts the algae's alkyl halide into a structurally similar compound that the sea hare uses for its own defense. Unlike other mollusks, a sea hare does not have a shell. Its method of defense is to surround itself with a slimy substance that contains the alkyl halide, thereby protecting itself from carnivorous fish.

Humans also synthesize alkyl halides to defend against infection. The human immune system has an enzyme that kills invading bacteria—another kind of predator—by halogenating them.

synthesized by red algae   synthesized by the sea hare   a sea hare

9.6 ELIMINATION REACTIONS OF ALKYL HALIDES

In addition to undergoing nucleophilic substitution reactions, alkyl halides also undergo elimination reactions. In an **elimination reaction,** atoms or groups are removed from a reactant.

The product of an elimination reaction is an alkene.

$$CH_3CH_2CH_2X + Y^- \xrightarrow{\text{substitution}} CH_3CH_2CH_2Y + X^-$$

$$\xrightarrow{\text{elimination}} CH_3CH=CH_2 + HY + X^-$$

new double bond

Notice that when an alkyl halide undergoes an elimination reaction, the halogen (X) is removed from one carbon and a hydrogen is removed from an adjacent carbon. A double bond is formed between the two carbons from which the atoms are eliminated. Therefore, *the product of an elimination reaction is an alkene.*

9.7 THE E2 REACTION

st as there are two nucleophilic substitution reactions, S_N1 and S_N2, there are two important imination reactions, E1 and E2. The following reaction is an example of an **E2 reaction,** where ∴" stands for *elimination* and "2" stands for *bimolecular* (Section 9.1).

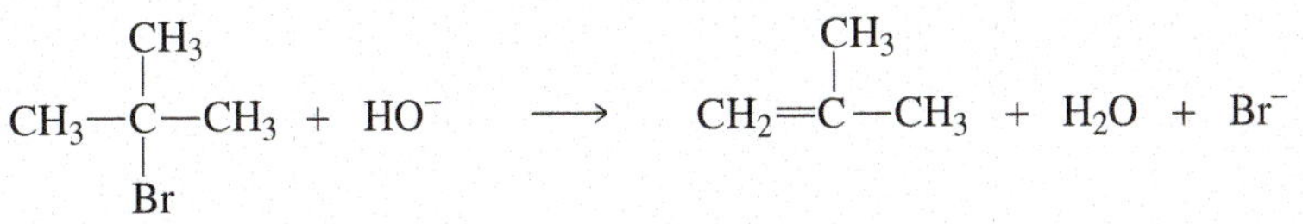

$$CH_3-\underset{\underset{Br}{|}}{\overset{\overset{CH_3}{|}}{C}}-CH_3 + HO^- \longrightarrow CH_2=\underset{}{\overset{\overset{CH_3}{|}}{C}}-CH_3 + H_2O + Br^-$$

The rate of an E2 reaction depends linearly on the concentrations of both the alkyl halide and the ∴se (in this case, hydroxide ion).

$$\text{rate} = k[\text{alkyl halide}][\text{base}]$$

he rate law tells us that both the alkyl halide and the base are involved in the transition state of the te-determining step, indicating a one-step reaction. The following mechanism—which portrays an ∴2 reaction as a concerted one-step reaction—agrees with the observed second-order kinetics:

MECHANISM FOR THE E2 REACTION OF AN ALKYL HALIDE

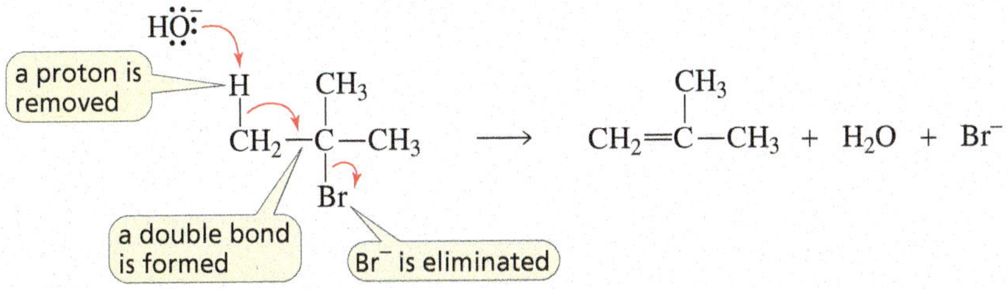

- The base removes a proton from a carbon that is adjacent to the carbon bonded to the halogen. As the proton is removed, the electrons that it shared with carbon move toward the carbon that is bonded to the halogen. As these electrons move toward the carbon, the halogen leaves (because carbon can form no more than four bonds), taking its bonding electrons with it.

√hen the reaction is over, the electrons that were originally bonded to the hydrogen in the ∴actant have formed a π bond in the product. The removal of a proton and a halide ion is called **∴ehydrohalogenation.**

The carbon to which the halogen is attached is called the **α-carbon.** A carbon adjacent to an ∴-carbon is called a **β-carbon.** Because the elimination reaction is initiated by removing a proton ∴om a β-carbon, an E2 reaction is sometimes called a **β-elimination reaction.** It is also called a **∴,2-elimination reaction** because the atoms being removed are on adjacent carbons.

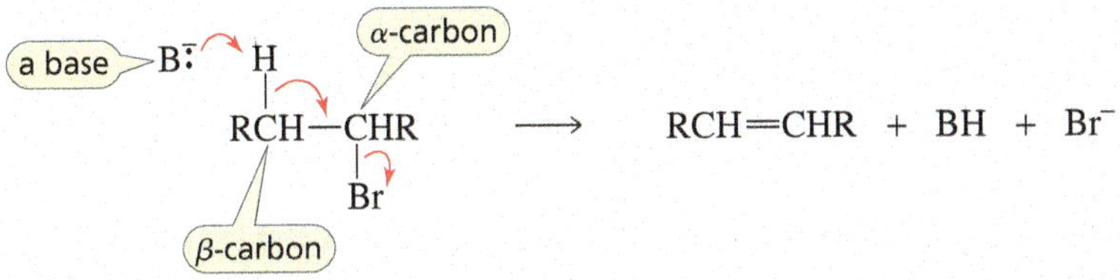

An E2 Reaction is Regioselective

∴n alkyl halide such as 2-bromopropane has two β-carbons from which a proton can be removed in ∴n E2 reaction. Because the two β-carbons are identical, the proton can be removed equally easily ∴rom either one.

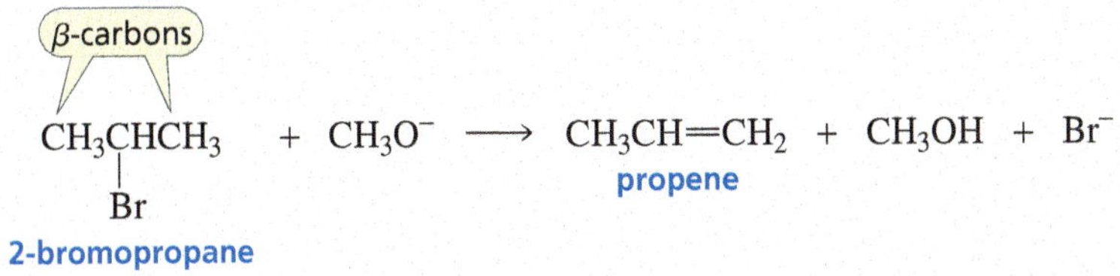

In contrast, 2-bromobutane has two structurally different β-carbons from which a proton can b$^{}$ removed. Therefore, when this alkyl halide reacts with a base, two elimination products are forme$^{}$ 2-butene (80%) and 1-butene (20%). Thus, this E2 reaction is *regioselective* because more of or$^{}$ constitutional isomer is formed than of the other (Section 6.4).

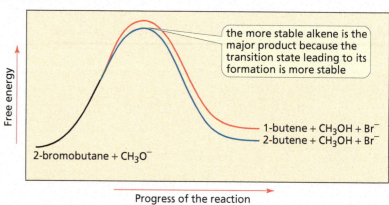

Alkyl Chlorides, Alkyl Bromides, and Alkyl Iodides Preferentially Form the More Stable Product

What factors dictate which of the two alkenes is formed in greater yield? In other words, wh$^{}$ causes the regioselectivity of an E2 reaction? We can answer this question by looking at the reac$^{}$ tion coordinate diagram in Figure 9.9.

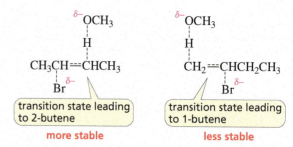

▶ **Figure 9.9**
The major product of the E2 reaction of 2-bromobutane and methoxide ion is 2-butene (indicated by the blue line), because the transition state leading to its formation is more stable than the transition state leading to formation of 1-butene (indicated by the red line).

In the transition state leading to the alkene, the C—H and C—Br bonds are partially broken an$^{}$ the double bond is partially formed (these partial bonds are indicated by dashed lines), giving th$^{}$ transition state an alkene-like structure. Therefore, factors that stabilize the alkene also stabilize th$^{}$ alkene-like transition state, causing the more stable alkene to be formed faster.

Figure 9.9 shows that the difference in the rate of formation of the two alkenes is not very grea$^{}$ Consequently, both are formed, but the *more stable* alkene is the major product. We saw that the sta$^{}$ bility of an alkene depends on the number of alkyl substituents bonded to its sp^2 carbons: the greate$^{}$ the number of alkyl substituents, the more stable the alkene (Section 5.9). Therefore, 2-butene$^{}$ with two methyl substituents bonded to its sp^2 carbons, is more stable than 1-butene, with one ethy$^{}$ substituent. Thus, 2-butene is the major product.

The following reaction also forms two elimination products. Because 2-methyl-2-butene is th$^{}$ more substituted alkene (it has a greater number of alkyl substituents bonded to its sp^2 carbons), i$^{}$ is the more stable of the two alkenes and, therefore, is the major product of the elimination reaction

Zaitsev's Rule

Alexander M. Zaitsev, a nineteenth-century Russian chemist, devised a shortcut to predict the more substituted alkene product. He pointed out *that the more substituted alkene is obtained when a hydrogen is removed from the β-carbon that is bonded to the fewest hydrogens.* This is called **Zaitsev's rule.**

$$\underbrace{CH_3CH_2CH_2\overset{\underset{\displaystyle Cl}{|}}{C}HCH_3}_{} + CH_3O^- \longrightarrow \underset{\substack{\text{2-pentene}\\67\%\\(\text{mixture of }E\text{ and }Z)}}{CH_3CH_2CH\!=\!CHCH_3} + \underset{\substack{\text{1-pentene}\\33\%}}{CH_3CH_2CH_2CH\!=\!CH_2}$$

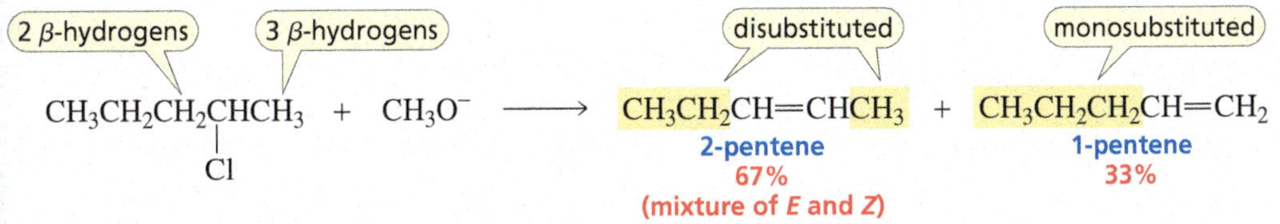

(2 β-hydrogens) (3 β-hydrogens) (disubstituted) (monosubstituted)

The major product of an E2 reaction is generally the more stable alkene.

For example, in the preceding reaction, one β-carbon is bonded to three hydrogens and the other β-carbon is bonded to two hydrogens. According to **Zaitsev's rule,** the more substituted alkene is the one formed by removing a proton from the β-carbon bonded to two hydrogens. Therefore, 2-pentene (a disubstituted alkene) is the major product and 1-pentene (a monosubstituted alkene) is the minor product.

PROBLEM 22 ♦

What is the major elimination product obtained from the reaction of each of the following alkyl halides with hydroxide ion?

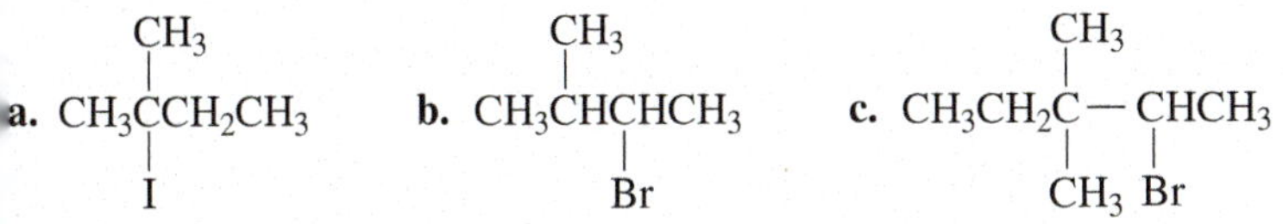

a. $CH_3\overset{\underset{\displaystyle I}{|}}{\overset{\displaystyle |}{\underset{\displaystyle CH_3}{C}}}CH_2CH_3$
b. $CH_3\overset{\underset{\displaystyle Br}{|}}{\overset{\displaystyle |}{\underset{\displaystyle CH_3}{C}}HCHCH_3}$
c. $CH_3CH_2\overset{\underset{\displaystyle CH_3}{|}}{\overset{\displaystyle CH_3}{\underset{}{C}}}-\overset{\underset{\displaystyle Br}{|}}{C}HCH_3$

Limitations of Zaitsev's Rule

You must be careful when using Zaitsev's rule. Keep in mind that the major product of an E2 reaction is generally the *more stable alkene,* and Zaitsev's rule is just a shortcut to determine which of the possible alkene products is the *more substituted alkene.* The more substituted alkene, however, is not always the more stable alkene, and in such cases, Zaitsev's rule cannot be used to predict the major product.

For example, in each of the following reactions, the major product is the alkene with conjugated double bonds because it is the more stable alkene, even though it is not the more substituted alkene.

Zaitsev's rule leads to the more substituted alkene.

$$CH_2\!=\!CHCH_2\overset{\underset{\displaystyle Cl}{|}}{\overset{\displaystyle |}{\underset{\displaystyle CH_3}{C}}}HCHCH_3 \xrightarrow{CH_3O^-} \underset{\substack{\text{conjugated diene}\\ \text{major product}}}{CH_2\!=\!CHCH\!=\!CHCH_3} + \underset{\substack{\text{isolated diene}\\ \text{minor product}}}{CH_2\!=\!CHCH_2CH\!=\!\overset{\underset{\displaystyle CH_3}{|}}{C}CH_3} + CH_3OH + Cl^-$$

conjugated double bonds

the double bond is conjugated with the benzene ring
major product

the double bond is not conjugated with the benzene ring
minor product

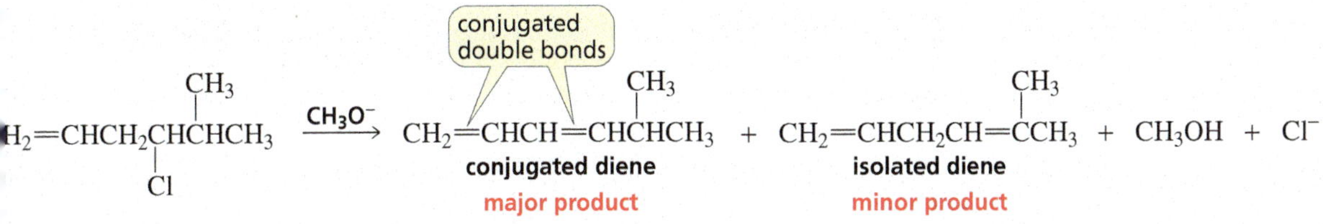

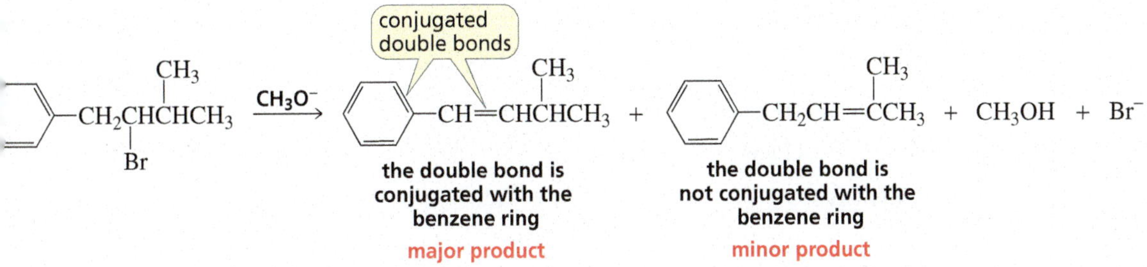

Compounds with conjugated double bonds are more stable than those with isolated double bonds.

Zaitsev's rule cannot be used to predict the major products of the foregoing reactions because does not take account of the fact that conjugated double bonds are more stable than isolated double bonds (Section 8.7). Therefore, if the alkyl halide has a double bond or a benzene ring, do not u Zaitsev's rule to predict the major product of an elimination reaction.

If the base in an E2 reaction is bulky *and* its approach to the hydrogen that leads to the mo stable alkene is sterically hindered, it will preferentially remove the most accessible hydroge Therefore, the major product is the less stable alkene.

$$
\underset{\underset{Br}{\overset{CH_3}{|}}}{CH_3CCH_2CH_3} \;+\; \underset{\underset{CH_3}{\overset{CH_3}{|}}}{CH_3CO^-} \longrightarrow \underset{28\%}{CH_3C{=}CHCH_3} \;+\; \underset{72\%}{CH_2{=}CCH_2CH_3} \;+\; \underset{\underset{CH_3}{\overset{CH_3}{|}}}{CH_3COH} \;+\; B
$$

(more accessible hydrogens / less accessible hydrogens / bulky base)

For example, in the preceding reaction, it is easier for the bulky *tert*-butoxide ion to remove on of the more exposed hydrogens from one of the two methyl groups. This leads to formation of th less stable alkene. Because the less stable alkene is more easily formed, it is the major product the reaction.

If the alkyl halide is not sterically hindered and the base is only moderately hindered, the mo stable alkene will be the major product, as expected. In other words, it takes a lot of steric hindran for the less stable product to be the major product. Thus, the major product of the following reacti is 2-butene.

$$
\underset{\underset{I}{\overset{}{|}}}{CH_3CHCH_2CH_3} \;+\; \underset{\underset{CH_3}{\overset{CH_3}{|}}}{CH_3CO^-} \longrightarrow \underset{79\%}{CH_3CH{=}CHCH_3} \;+\; \underset{21\%}{CH_2{=}CHCH_2CH_3} \;+\; \underset{\underset{CH_3}{\overset{CH_3}{|}}}{CH_3COH} \;+
$$

Table 9.4 shows the amount of each product obtained from the reaction of a sterically hindere alkyl halide with different bases. Notice that the percentage of the less stable (less substituted alkene increases as the size of the base increases.

Table 9.4 Effect of the Steric Properties of the Base on the Distribution of Products in an E2 Reaction

$$
\underset{\underset{CH_3\;Br}{\overset{CH_3}{|}}}{CH_3CH{-}CCH_3} \;+\; RO^- \longrightarrow \underset{\underset{CH_3}{\overset{CH_3}{|}}}{CH_3C{=}CCH_3} \;+\; \underset{\underset{CH_3}{\overset{CH_3}{|}}}{CH_3CHC{=}CH_2}
$$

Base	More stable alkene	Less stable alkene
$CH_3CH_2O^-$	79%	21%
$CH_3\overset{\overset{CH_3}{|}}{\underset{\underset{CH_3}{|}}{C}}O^-$	27%	73%
$CH_3\overset{\overset{CH_3}{|}}{\underset{\underset{CH_2CH_3}{|}}{C}}O^-$	19%	81%
$CH_3CH_2\overset{\overset{CH_2CH_3}{|}}{\underset{\underset{CH_2CH_3}{|}}{C}}O^-$	8%	92%

lkyl Fluorides Preferentially Form the Less Stable Alkene

though the major product of the E2 reaction of alkyl chlorides, alkyl bromides, and alkyl iodides normally the more stable alkene, the major product of the E2 reaction of alkyl fluorides is the ss stable alkene.

$$CH_3CHCH_2CH_2CH_3 + CH_3O^- \longrightarrow CH_3CH=CHCH_2CH_3 + CH_2=CHCH_2CH_2CH_3 + CH_3OH + F^-$$

| 2-fluoropentane | 2-pentene 30% (mixture of *E* and *Z*) | 1-pentene 70% |

Why do alkyl fluorides form the less stable alkene? To answer this question, we must look at the ansition states of the reactions.

When a hydrogen and a chlorine, bromine, or iodine are eliminated from an alkyl halide, the logen starts to leave as soon as the base begins to remove the proton. Consequently, the transition te resembles an alkene (see page 414).

The fluoride ion, however, is the strongest base of the halide ions and, therefore, the poorest aving group. So when a base begins to remove a proton from an alkyl fluoride, fluorine does not ve as strong a propensity to leave as another halogen would have. As a result, a negative charge velops on the carbon that is losing the proton, causing the transition state to resemble a carbanion ther than an alkene. To determine which of the carbanion-like transition states is more stable, we ed to look at the factors that affect carbanion stability.

transition state leading to 1-pentene — more stable

transition state leading to 2-pentene — less stable

arbocation and Carbanion Stability

e saw that carbocations, because they are positively charged, are *stabilized* by electron-donating kyl groups. Thus, tertiary carbocations are the most stable, and methyl cations are the least stable ection 6.2).

relative stabilities of carbocations

tertiary carbocation > secondary carbocation > primary carbocation > methyl cation

arbanions, on the other hand, are negatively charged, so they are *destabilized* by electron-donating kyl groups. Therefore, carbanions have the opposite relative stabilities. (Differences in solvation ergies also contribute to this trend because the smaller ions are the most solvated.)

relative stabilities of carbanions

tertiary carbanion < secondary carbanion < primary carbanion < methyl anion

Carbocation stability: 3° > 2° > 1°

Carbanion stability: 1° > 2° > 3°

Looking back at the transition states for the E2 reaction of an alkyl fluoride, we see that the veloping negative charge in the transition state leading to 1-pentene is on a primary carbon. hus, this transition state is more stable than the transition state leading to 2-pentene, in which the developing negative charge is on a secondary carbon. Because the transition state lead to 1-pentene is more stable, 1-pentene is formed more rapidly and is the major product of E2 reaction.

Summary of the Regioselectivity of E2 Reactions

The data in Table 9.5 show that as the halide ion increases in basicity (and so decreases in leav propensity), the yield of the more stable alkene product decreases. However, the more stable alk remains the major elimination product in all cases, except when the halogen is fluorine.

Table 9.5 Products Obtained From the E2 Reaction of CH_3O^- and 2-Halohexanes

$$CH_3CHCH_2CH_2CH_2CH_3 \ + \ CH_3O^- \longrightarrow CH_3CH{=}CHCH_2CH_2CH_3 \ + \ CH_2{=}CHCH_2CH_2CH_2CH_3$$
(with X above first carbon)

(mixture of *E* and *Z*)

Leaving group	Conjugate acid	pK_a	More stable product	Less stable product
X = I	HI	−10	81%	19%
X = Br	HBr	−9	72%	28%
X = Cl	HCl	−7	67%	33%
X = F	HF	3.2	30%	70%

We can summarize as follows: *the major product of an E2 reaction is the more stable alke except when the reactants are sterically hindered or the leaving group is poor, in which case major product is the less stable alkene.* In Section 9.11, you will see that the more stable alkene not always the major product in the case of certain cyclic alkyl halides.

PROBLEM 23 ♦

What is the major elimination product obtained from an E2 reaction of each of the following alkyl hali with hydroxide ion?

a. $CH_3CHCH_2CH_3$
    $|$
    Cl

b. $CH_3CHCHCH_2CH_3$
    with CH_3 and Cl substituents

c. $CH_3CHCH_2CH{=}CH_2$
    $|$
    Cl

d. $CH_3CHCH_2CH_3$
    $|$
    F

e. 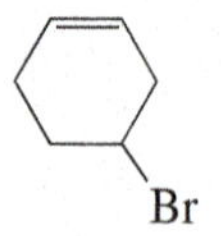 (cyclohexene ring with Br)

f. $CH_3CHCHCH_2CH_3$
    with CH_3 and F substituents

Relative Reactivities in an E2 Reaction

Because elimination from a tertiary alkyl halide typically leads to a more substituted alkene than do elimination from a secondary alkyl halide, and because elimination from a secondary alkyl hali generally leads to a more substituted alkene than does elimination from a primary alkyl halide, relative reactivities of alkyl halides in an E2 reaction are:

relative reactivities of alkyl halides in an E2 reaction

tertiary alkyl halide > secondary alkyl halide > primary alkyl halide

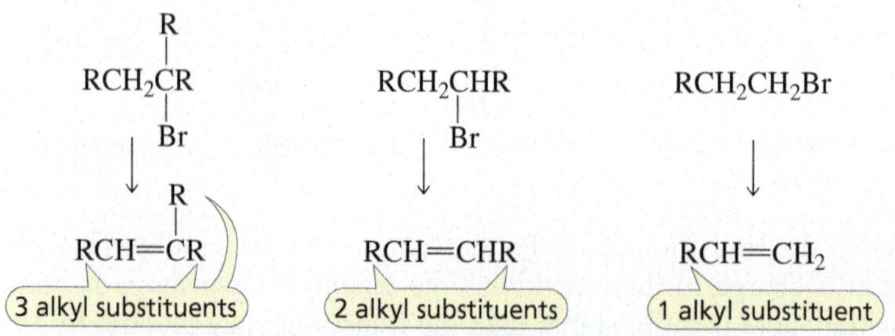

$$\underset{\text{Br}}{\overset{R}{RCH_2CR}} \qquad \underset{\text{Br}}{RCH_2CHR} \qquad RCH_2CH_2Br$$

$$\underset{R}{RCH{=}CR} \qquad RCH{=}CHR \qquad RCH{=}CH_2$$

3 alkyl substituents | 2 alkyl substituents | 1 alkyl substituent

PROBLEM 24 ◆

Which alkyl halide in each pair is more reactive in an E2 reaction with hydroxide ion?

a. 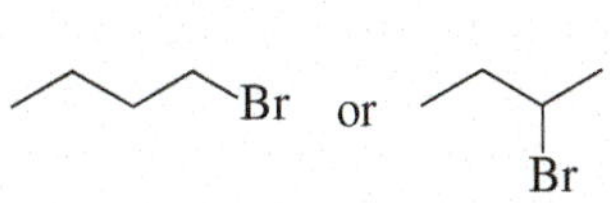 [structure: butyl bromide] or [structure: sec-butyl bromide]

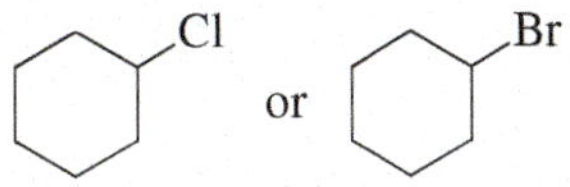

c. 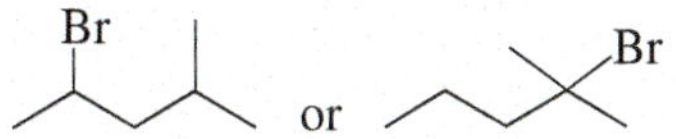 [structure with Br] or [structure with Br]

b. [cyclohexyl Cl] or [cyclohexyl Br]

d. 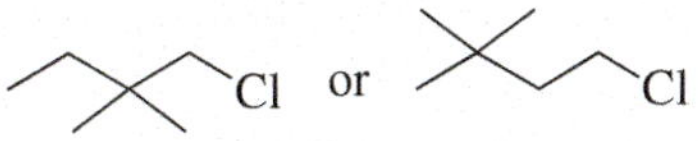 [structure with Cl] or [structure with Cl]

PROBLEM 25 ◆

Which alkyl halide in each pair is more reactive in an E2 reaction with hydroxide ion?

a. 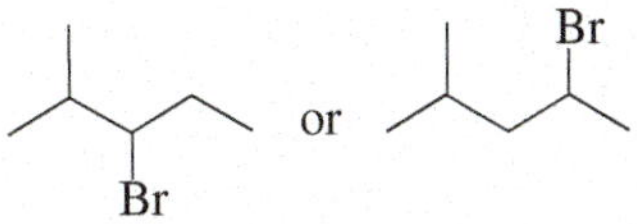 [structure with Br] or [structure with Br]

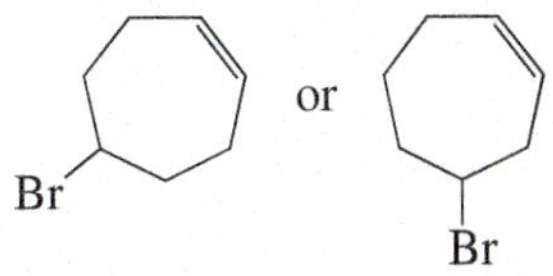

c. 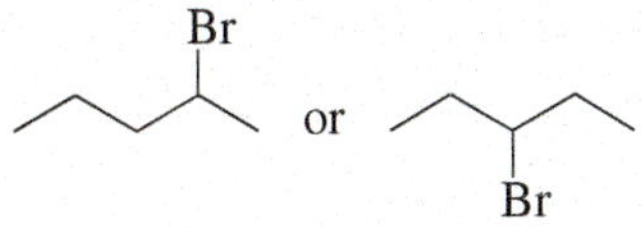 [structure with Br] or [structure with Br]

b. [cycloheptene with Br] or [cycloheptene with Br]

d. 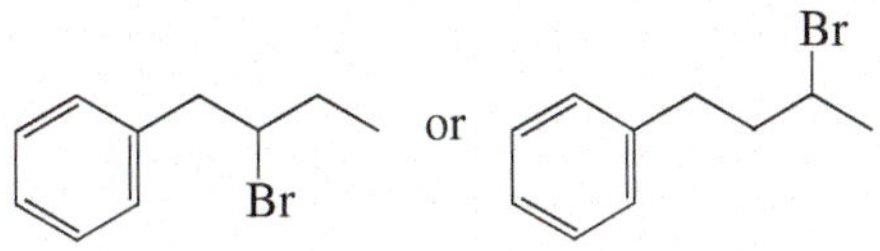 [benzyl structure with Br] or [benzyl structure with Br]

9.8 THE E1 REACTION

The second kind of elimination reaction that alkyl halides can undergo is an **E1 reaction,** where "E" stands for *elimination* and "1" stands for *unimolecular.*

$$CH_3-\underset{\underset{Br}{|}}{\overset{\overset{CH_3}{|}}{C}}-CH_3 + H_2O \longrightarrow CH_2=\underset{}{\overset{\overset{CH_3}{|}}{C}}-CH_3 + H_3O^+ + Br^-$$

The rate of an E1 reaction depends only on the concentration of the alkyl halide.

$$\text{rate} = k[\text{alkyl halide}]$$

Therefore, we know that only the alkyl halide takes part in the rate-determining step of the reaction, so an E1 reaction must have at least two steps. The following mechanism agrees with the observed first-order kinetics. Because the first step is the rate-determining step, an increase in the concentration of the base—which participates only in the second step of the reaction—has no effect on the rate of the reaction. Generally, the solvent is the base in an E1 reaction.

MECHANISM FOR THE E1 REACTION OF AN ALKYL HALIDE

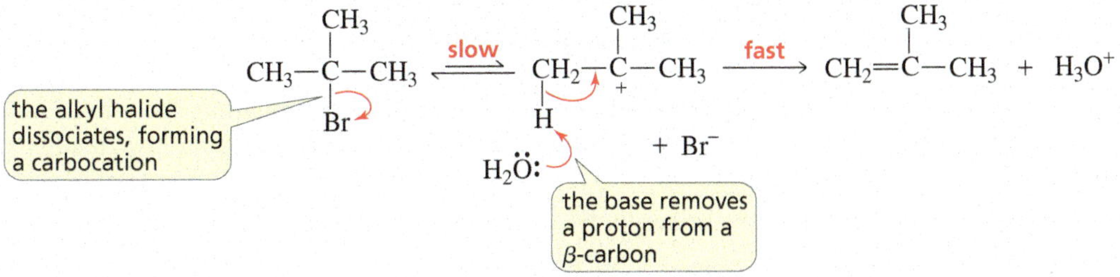

the alkyl halide dissociates, forming a carbocation

the base removes a proton from a β-carbon

- The alkyl halide dissociates, forming a carbocation.
- The base forms the elimination product by removing a proton from a β-carbon.

We saw that the pK_a of a compound such as ethane, which has hydrogens attached only to sp^3 carbons, is > 60 (Section 2.6). How, then, can a weak base such as water remove a proton from an sp^3 carbon in the second step of the preceding reaction?

First of all, the pK_a is greatly reduced by the adjacent positively charged carbon that can accept the electrons left behind when the proton is removed. Second, the β-carbon shares the positive charge as a result of hyperconjugation. Hyperconjugation drains electron density from the C—H bond, thereby weakening it. Recall that hyperconjugation (where the σ electrons of a bond attached to a carbon adjacent to a positively charged carbon spread into the empty p orbital) is also responsible for the greater stability of a tertiary carbocation compared with a secondary carbocation (Section 6.2).

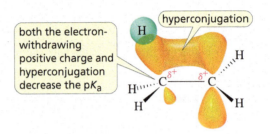

The E1 Reaction Is Regioselective

When more than one alkene can be formed, the E1 reaction, like the E2 reaction, is regioselective. And, like an E2 reaction, the major product is the *more stable alkene.*

$$CH_3CH_2\overset{\overset{\displaystyle CH_3}{|}}{\underset{\underset{\displaystyle Cl}{|}}{C}}CH_3 \ + \ H_2O \ \longrightarrow \ CH_3CH{=}\overset{\overset{\displaystyle CH_3}{|}}{C}CH_3 \ + \ CH_3CH_2\overset{\overset{\displaystyle CH_3}{|}}{C}{=}CH_2 \ + \ H_3O^+ \ + \ Cl^-$$

major product   **minor product**

The more stable alkene is the major product because its greater stability causes the transition state leading to its formation to be more stable (Figure 9.10). Therefore, it is formed more rapidly. Notice that the more stable alkene is formed by removing the hydrogen from the β-carbon bonded to the fewest hydrogens, in accordance with Zaitsev's rule.

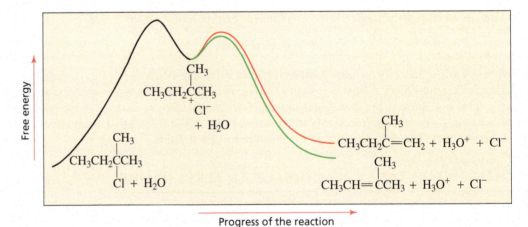

▶ **Figure 9.10**
The major product of the E1 reaction is the more stable alkene (green line) because its greater stability causes the transition state leading to its formation to be more stable.

Tertiary alkyl halides undergo E1 and E2 reactions.
Primary and secondary alkyl halides undergo only E2 reactions.

Because the rate-determining step of an E1 reaction is carbocation formation, the rate of an E1 reaction depends both on the ease with which the carbocation is formed *and* on how readily the leaving group leaves. Therefore:

- Tertiary alkyl halides readily undergo E1 reactions because they form relatively stable carbocations.

- Primary and secondary alkyl halides *do not* undergo E1 reactions because their carbocations are less stable and because high concentrations of base (the solvent) are used in these reactions, both of which increase the fraction of the reaction that takes place by an E2 pathway (see the rate law on page 422). Therefore, primary and secondary alkyl halides undergo only E2 reactions.

The Leaving Group in E2 and E1 Reactions

In a series of alkyl halides that have the same alkyl group, alkyl iodides are the most reactive and alkyl fluorides the least reactive in both E2 and E1 reactions because, as we saw in Section 9.2, weaker bases are better leaving groups.

relative reactivities of alkyl halides in E2 and E1 reactions

most reactive — RI > RBr > RCl > RF — least reactive

The weaker the base, the better it is as a leaving group.

PROBLEM 26 ◆

Four alkenes are formed from the E1 reaction of 3-bromo-2,3-dimethylpentane and methanol. Draw the structures of the alkenes and rank them according to the amount that would be formed.

PROBLEM 27 ◆

If 2-fluoropentane could undergo an E1 reaction, would you expect the major product to be the more stable alkene or the less stable alkene? Explain your answer.

PROBLEM 28 ◆

Which of the following compounds would react faster in an

a. E1 reaction? **b.** E2 reaction? **c.** S_N1 reaction? **d.** S_N2 reaction?

A

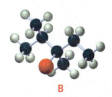
B

PROBLEM-SOLVING STRATEGY

Proposing a Mechanism

LEARN THE STRATEGY

Propose a mechanism for the following reaction:

When we see that one of the reactants is an acid, we know we should start by protonating the other reactant. We need to protonate it at the position that allows the most stable carbocation to be formed. Therefore, we protonate the CH_2 group because that forms a tertiary doubly-allylic carbocation with a positive charge that is shared by three carbons as a result of electron delocalization. A 1,2-methyl shift results in a carbocation with the desired carbon skeleton. Loss of a proton forms the final product.

PROBLEM 29

USE THE STRATEGY

Propose a mechanism for the following reaction:

9.9 COMPETITION BETWEEN E2 AND E1 REACTIONS

Primary and secondary alkyl halides undergo only E2 reactions.

Tertiary alkyl halides undergo both E1 and E2 reactions.

Notice that the same bonds are broken and formed in E2 and E1 reactions. The only difference is the timing—that is, the proton is lost in the first (and only) step of an E2 reaction, and it is lost in the second step of an E1 reaction.

We saw that:

- The only elimination reactions **primary** and **secondary alkyl halides** undergo are **E2 reactions** because of their relatively unstable carbocations and the large concentration of the base in solvolysis reactions (Section 9.8).
- **Tertiary alkyl halides** undergo both **E2** and **E1 reactions.**

Because a tertiary alkyl halide can undergo both E2 and E1 reactions, the rate law for its reaction is given by the sum of the rate laws for the E2 and E1 reactions (Sections 9.7 and 9.8):

$$\text{rate} = k_2[\text{alkyl halide}]\,[\text{base}] + k_1[\text{alkyl halide}]$$

The rate law tells us that:

An E2 reaction is favored by a high concentration of a strong base.

An E1 reaction is favored by a low concentration of a weak base.

Notice that a tertiary alkyl halide and a strong base were chosen to illustrate the E2 reaction in Section 9.1, whereas a tertiary alkyl halide and a weak base were used to illustrate the E1 reaction in Section 9.3.

PROBLEM 30

For each of the following reactions, (1) decide whether an E2 or an E1 occurs, and (2) draw the major elimination product:

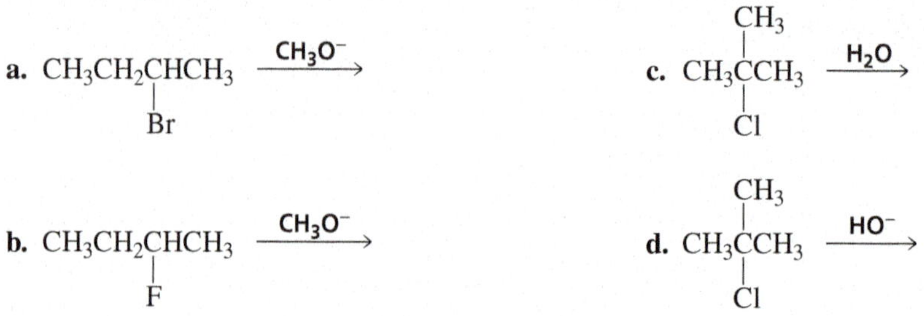

a. $CH_3CH_2CHCH_3$ (with Br) $\xrightarrow{CH_3O^-}$

b. $CH_3CH_2CHCH_3$ (with F) $\xrightarrow{CH_3O^-}$

c. CH_3CCH_3 (with CH_3 and Cl) $\xrightarrow{H_2O}$

d. CH_3CCH_3 (with CH_3 and Cl) $\xrightarrow{HO^-}$

LEARN THE STRATEGY

PROBLEM 31 SOLVED

The rate law for the reaction of HO^- with *tert*-butyl bromide to form an elimination product in 75% ethanol/ 25% water at 30 °C is the sum of the rate laws for the E2 and E1 reactions. What percentage of the reaction takes place by an E2 pathway when $[HO^-] = 5.0$ M? ($k_2 = 7.1 \times 10^{-5}$, and $k_1 = 1.5 \times 10^{-5}$)

$$\text{rate} = \text{rate law for the E2 reaction} + \text{rate law for the E1 reation}$$

$$\text{rate} = k_2\left[\textit{tert}\text{-butyl bromide}\right]\left[HO^-\right] + k_1\left[\textit{tert}\text{-butyl bromide}\right]$$

SOLUTION

$$\frac{E2}{E2 + E1} = \frac{k_2[\textit{tert}\text{-butyl bromide}][HO^-]}{k_2[\textit{tert}\text{-butyl bromide}][HO^-] + k_1[\textit{tert}\text{-butyl bromide}]}$$

$$\frac{E2}{E2 + E1} = \frac{7.1 \times 10^{-5} \times 5.0}{7.1 \times 10^{-5} \times 5.0 + 1.5 \times 10^{-5}} = \frac{35.5 \times 10^{-5}}{35.5 \times 10^{-5} + 1.5 \times 10^{-5}} = \frac{35.5}{37} = 0.96 = 96\%$$

USE THE STRATEGY

PROBLEM 32

What percentage of the reaction described in Problem 31 takes place by the E2 pathway when $[HO^-] = 0.0025$ M?

9.10 E2 AND E1 REACTIONS ARE STEREOSELECTIVE

In addition to being *regioselective*, meaning that more of one constitutional isomer is formed than the other, E2 and E1 reactions are also *stereoselective*, meaning that more than one stereosiomer is formed than another.

The Stereoisomers Formed in an E2 Reaction

Because an E2 reaction is concerted, the bonds to the groups to be eliminated must be parallel because the sp^3 orbital of the carbon bonded to H and the sp^3 orbital of the carbon bonded to X become overlapping *p* orbitals in the alkene product. For maximum overlap in the transition state, the orbitals must be parallel.

Syn and Anti Elimination

There are two ways in which the C—H and C—X bonds can be parallel: they can be either on the same side of the molecule or on opposite sides of the molecule.

    If an elimination reaction removes two substituents from opposite sides of the molecule, it is called an **anti elimination.** If the substituents are removed from the same side of the molecule, the reaction is called a **syn elimination.** Both types of elimination can occur, but *anti elimination is highly favored in an E2 reaction.* **Sawhorse projections,** which show the C—C bond from an oblique angle, reveal why this is true.

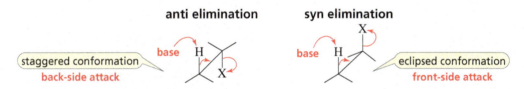

anti elimination   **syn elimination**

staggered conformation   back-side attack   base H X

base H eclipsed conformation   front-side attack

Anti elimination predominates in an E2 reaction.

- Anti elimination requires the molecule to be in a staggered conformation, whereas syn elimination requires it to be in a less stable, eclipsed conformation.

- In anti elimination, the electrons of the departing hydrogen move to the *back* side of the carbon bonded to X, whereas in syn elimination, the electrons move to the *front* side of the carbon bonded to X. We saw that displacement reactions occur through back-side attack because that is how the interacting orbitals achieve the best overlap (see Figure 9.1 on page 395).

- In anti elimination, the electron-rich base is spared the repulsion that it experiences when it is on the same side of the molecule as the electron-rich departing halide ion.

A Reactant with Two Hydrogens on the β-Carbon

An E2 reaction is *stereoselective.* In other words, more of one stereoisomer is formed than the other. For example, the 2-pentene obtained as the major constitutional isomer from the E2 reaction of 2-chloropentane on page 416 is actually a pair of stereoisomers, and more (*E*)-2-pentene is formed than (*Z*)-2-pentene.

$$\underset{\substack{(E)\text{-2-pentene}\\ \text{major product}}}{\overset{H \quad\quad CH_2CH_3}{\underset{CH_3 \quad\quad H}{C=C}}} \qquad \underset{\substack{(Z)\text{-2-pentene}\\ \text{minor product}}}{\overset{H \quad\quad H}{\underset{CH_3 \quad\quad CH_2CH_3}{C=C}}}$$

    We can make the following general statement about the stereoselectivity of an E2 reaction: if the reactant has *two hydrogens* bonded to the carbon from which a hydrogen is removed, both the *E* and *Z* products will be formed, because the reactant has two conformers in which the groups to be eliminated are anti.

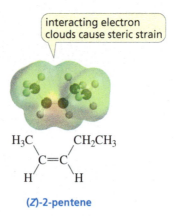

interacting electron clouds cause steric strain

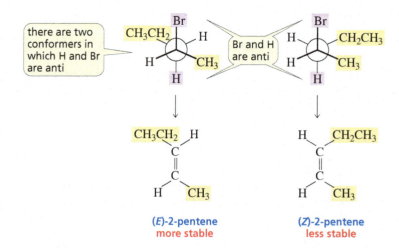

there are two conformers in which H and Br are anti

Br and H are anti

(E)-2-pentene
more stable

(Z)-2-pentene
less stable

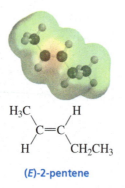

(Z)-2-pentene

(E)-2-pentene

Of the two stereoisomers, the one with the *largest groups on opposite sides of the double bond* is formed in greater yield because it is more stable, so it has the more stable transition state leading to its formation (Figure 9.11). (Recall that the alkene with the largest groups on the *same* side of the double bond is *less stable* because the electron clouds of the large substituents can interfere with each other, causing steric strain; Section 5.9).

▶ **Figure 9.11**
The major stereoisomer formed in an E2 reaction is the one with the largest groups on opposite sides of the double bond (here, the *E* isomer). This is because the more stable alkene (indicated by the blue line) has the more stable transition state and, therefore, is formed more rapidly.

Free energy

2-bromopentane
+
$CH_3CH_2O^-$

(Z)-2-pentene
(E)-2-pentene
+ CH_3CH_2OH + Br^-

Progress of the reaction

Thus, the following elimination reaction leads predominantly to the *E* isomer because this stereoisomer has the methyl group (the larger group on one sp^2 carbon) opposite the *tert*-butyl group (the larger group on the other sp^2 carbon).

If the β-carbon in an E2 reaction is bonded to two hydrogens, then two alkenes are formed and the major product is the one with the largest substituents on opposite sides of the double bond.

If the β-carbon in an E2 reaction is bonded to only one hydrogen, then one alkene is formed. Its structure depends on the structure of the alkyl halide.

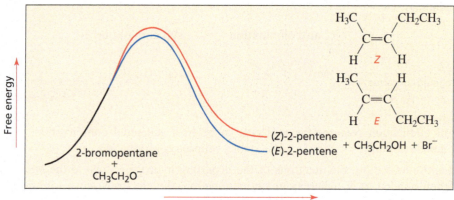

major product

minor product

A Reactant with One Hydrogen on the β-Carbon

If, however, the β-carbon from which a hydrogen is to be removed is bonded to only *one hydrogen*, then there is only one conformer in which the groups to be eliminated are anti. Therefore, only one alkene product is formed. The particular stereoisomer formed depends on the configuration of the reactant. For example, anti elimination of HBr from (2S,3S)-2-bromo-3-phenylbutane forms the *E* isomer.

CH₃Ö⁻

H ⟍ ⫽CH₃
CH₃—C—C⟍H
C₆H₅ Br
(2S,3S)-2-bromo-3-phenylbutane

⟶

CH₃ÖH

CH₃ ⟍ ⟋CH₃
C=C
C₆H₅ ⟋ ⟍H Br⁻
(E)-2-phenyl-2-butene

But anti elimination of HBr from (2S,3R)-2-bromo-3-phenylbutane forms the Z isomer.

CH₃Ö⁻

H ⟍ ⫽CH₃
C₆H₅—C—C⟍H
CH₃ Br
(2S,3R)-2-bromo-3-phenylbutane

⟶

CH₃ÖH

C₆H₅ ⟍ ⟋CH₃
C=C
CH₃ ⟋ ⟍H Br⁻
(Z)-2-phenyl-2-butene

Notice that the two groups bonded by solid wedges are on the same side of the alkene in the product.

PROBLEM-SOLVING STRATEGY

Determining the Major Product of an E2 Reaction

What is the major product formed when the following compounds undergo an E2 reaction?

a. **b.**

The alkyl halide in part **a** has two hydrogens on the β-carbon from which a hydrogen is removed in the elimination reaction. Therefore, both the E and Z isomers are formed, but the E isomer is the major product.

E isomer
major product

Z isomer

The alkyl halide in part **b** has only one hydrogen on the β-carbon from which a hydrogen is removed, so there is only one elimination product. First, we need to determine the configuration of the two asymmetric centers. Next, we need to draw the perspective formula showing only the H and Br that are eliminated and putting them on bonds drawn in the plane of the paper. Because we know the configurations of the asymmetric centers, we can add the other groups. When H and Br are eliminated, the two groups bonded to solid wedges are on the same side of the double bond in the product.

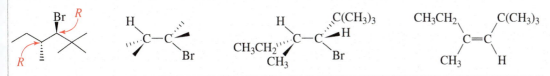

PROBLEM 33 ◆

a. What is the major product obtained when each of the following compounds undergoes an E2 reaction with methoxide ion? Show the configuration of the product.
b. Does the product obtained depend on whether you start with the R or S enantiomer of the reactant?

1. CH₃CH₂CHCHCH₃
　　　　|　　|
　　　　Br　CH₃

2. CH₃CH₂CHCH₂CH=CH₂
　　　　　　|
　　　　　　Cl

3. CH₃CH₂CHCH₂—⬡
　　　　　　|
　　　　　　Cl

The Stereoisomers Formed in an E1 Reaction

An E1 reaction, like an E2 reaction, is stereoselective; and again, like an E2 reaction, both the *E* and *Z* products are formed and the major product is the one with the *largest groups on opposite sides of the double bond*. Let's see why this is so.

We know that an E1 reaction takes place in two steps. The leaving group leaves in the first step, and a proton is lost from an adjacent carbon in the second step, following Zaitsev's rule for forming the more stable alkene. The carbocation created in the first step is planar, so the electrons from a departing proton can move toward the positively charged carbon from *either side*. Therefore, both syn and anti elimination occur.

> β-carbon has one hydrogen

(*E*)-3,4-dimethyl-3-hexene
major product

(*Z*)-3,4-dimethyl-3-hexene
minor product

The major stereoisomer obtained in an E1 reaction is the one with the largest groups on opposite sides of the double bond.

Because both syn and anti elimination occur in an E1 reaction, both the *E* and *Z* products are formed, *regardless* of whether the β-carbon from which the proton is removed is bonded to one or to two hydrogens. The major product is the one with the largest groups on opposite sides of the double bond, because that is the more stable alkene and, therefore, it is formed more rapidly.

Table 9.6 summarizes the stereochemical outcomes of substitution and elimination reactions.

Table 9.6 Stereochemistry of Substitution and Elimination Reactions

Reaction	Products
S_N2	Only the inverted product is formed.
E2	Both *E* and *Z* stereoisomers are formed (with more of the stereoisomer with the largest groups on opposite sides of the double bond) unless the β-carbon from which the hydrogen is removed is bonded to only one hydrogen, in which case only one stereoisomer is formed. The stereoisomer's configuration depends on the configuration of the reactant.
S_N1	Both stereoisomers (*R* and *S*) are formed (generally with more inverted product than retained).
E1	Both *E* and *Z* stereoisomers are formed (with more of the stereoisomer with the largest groups on opposite sides of the double bond).

LEARN THE STRATEGY

PROBLEM 34 SOLVED

What is the major product formed when the following compound undergoes an E1 reaction?

SOLUTION First, we need to consider the *regiochemistry* of the reaction: the major product is 3,4-dimethyl-3-hexene because it is the most stable of the three possible alkene products.

Next, we need to consider the *stereochemistry* of the reaction: the major product has two stereoisomers, and more (*E*)-3,4-dimethyl-3-hexene is formed because it is more stable than (*Z*)-3,4-dimethyl-3-hexene.

(*E*)-3,4-dimethyl-3-hexene   (*Z*)-3,4-dimethyl-3-hexene

PROBLEM 35

What is the major product formed when the following compounds undergo an E1 reaction?

a. $CH_3CH_2CH_2\overset{\overset{\displaystyle CH_3}{|}}{\underset{\underset{\displaystyle Cl}{|}}{C}}CH_3$

b. (benzene ring)$-\overset{\overset{\displaystyle CH_3}{|}}{\underset{\underset{\displaystyle I}{|}}{C}}-CH_2CH_3$

c. (cyclohexane with Cl and CH_3)

9.11 ELIMINATION FROM SUBSTITUTED CYCLOHEXANES

Elimination from substituted cyclohexanes follows the same stereochemical rules as elimination from open-chain compounds.

E2 Reactions of Substituted Cyclohexanes

We saw that the anti elimination favored in an E2 reaction requires that the two groups being eliminated are parallel (Section 9.9). For the two groups on a cyclohexane ring to be parallel, they both must be in *axial positions*.

(structure of cyclohexane with H and X labeled) groups to be eliminated must both be in axial positions

> In an E2 reaction of a substituted cyclohexane, the groups being eliminated must both be in axial positions.

The more stable conformer of chlorocyclohexane does not undergo an E2 reaction, because the chloro substituent is in an equatorial position. (Recall from Section 3.14 that the more stable conformer of a monosubstituted cyclohexane is the one in which the substituent is in an equatorial position because there is more room for a substituent in that position.) The less stable conformer, with the chloro substituent in the axial position, readily undergoes an E2 reaction.

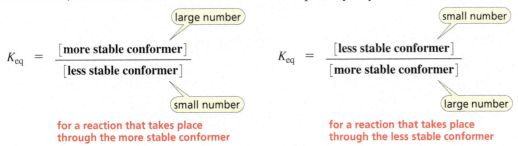

more stable $\xrightarrow{K_{eq}}$ less stable

$k' \downarrow HO^-$

(cyclohexene) $+ H_2O + Cl^-$

Because only one of the two conformers undergoes an E2 reaction, the rate constant for the elimination reaction is given by $k'K_{eq}$, where k' is a rate constant and K_{eq} is an equilibrium constant. Therefore, the reaction is faster when K_{eq} is large. Most molecules are in the more stable conformer at any given time. Therefore, K_{eq} will be large when elimination takes place by way of the more stable conformer, and it will be small when elimination takes place by way of the less stable conformer.

$$K_{eq} = \frac{[\text{more stable conformer}]}{[\text{less stable conformer}]} \qquad K_{eq} = \frac{[\text{less stable conformer}]}{[\text{more stable conformer}]}$$

large number (numerator) / small number (denominator)

small number (numerator) / large number (denominator)

for a reaction that takes place through the more stable conformer

for a reaction that takes place through the less stable conformer

For example, neomenthyl chloride undergoes an E2 reaction with ethoxide ion about 200 times faster than menthyl chloride does. The conformer of neomenthyl chloride that undergoes elimination is the *more* stable conformer because when the Cl and H are in the required axial positions, the methyl and isopropyl groups are in equatorial positions.

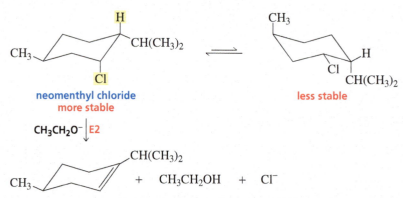

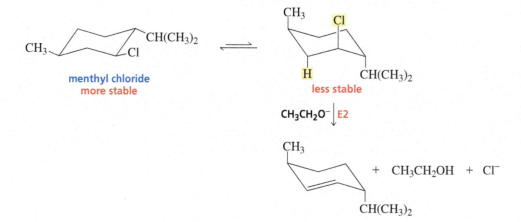

In contrast, the conformer of menthyl chloride that undergoes elimination is the *less* stable conformer because when the Cl and H are in the required axial positions, the methyl and isopropyl groups are also in axial positions.

Notice that when menthyl chloride undergoes an E2 reaction, the hydrogen that is eliminated is *not* removed from the β-carbon bonded to the fewest hydrogens. This may seem like a violation of Zaitsev's rule, but this compound has only one axial hydrogen on a β-carbon. Therefore, that hydrogen is the one that is removed, even though it is not bonded to the β-carbon with the fewest hydrogens.

PROBLEM 36 ♦

Why do *cis*-1-bromo-2-ethylcyclohexane and *trans*-1-bromo-2-ethylcyclohexane form different major products when they undergo an E2 reaction?

PROBLEM 37 ♦

Which isomer reacts more rapidly in an E2 reaction: *cis*-1-bromo-4-*tert*-butylcyclohexane or *trans*-1-bromo-4-*tert*-butylcyclohexane? Explain your answer.

E1 Reactions of Substituted Cyclohexanes

When a substituted cyclohexane undergoes an E1 reaction, the two groups that are eliminated do not both have to be in axial positions because the elimination reaction is not concerted. In the following reaction, a carbocation is formed in the first step. It then loses a proton from the adjacent carbon that is bonded to the fewest hydrogens—in other words, Zaitsev's rule is followed.

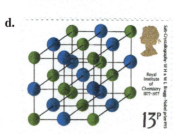

Stamps issued in honor of English Nobel Laureates:

a. Sir Derek Barton for conformational analysis, 1969

b. Sir Walter Haworth for the synthesis of vitamin C, 1937

c. A. J. P. Martin and Richard L. M. Synge for chromatography, 1952

d. William H. Bragg and William L. Bragg for crystallography, 1915 (The Braggs are one of six father–son pairs who received Nobel Prizes.)

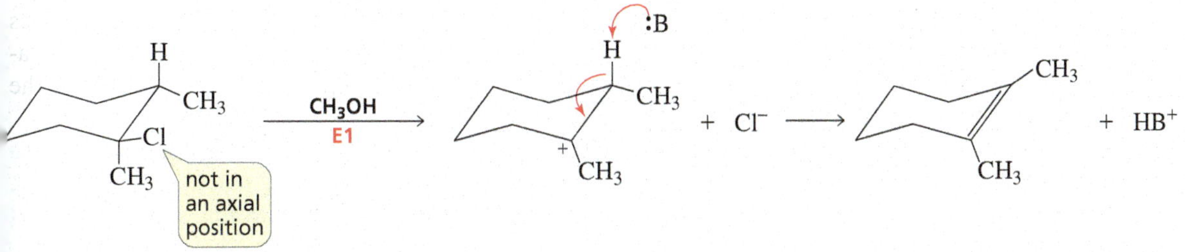

PROBLEM 38

Draw the substitution and elimination products for the following reactions, showing the configuration of each product:

a. *trans*-1-chloro-2-methylcyclohexane + CH_3O^-
b. *cis*-1-chloro-2-methylcyclohexane + CH_3O^-
c. 1-chloro-1-methylcyclohexane + CH_3O^-
d. 1-chloro-1-methylcyclohexane + CH_3OH

9.12 PREDICTING THE PRODUCTS OF THE REACTION OF AN ALKYL HALIDE WITH A NUCLEOPHILE/BASE

We saw that alkyl halides can undergo four types of reactions: S_N2, S_N1, E2, and E1. Before we can predict the products of the reaction of an alkyl halide with a nucleophile/base, we have to decide whether it undergoes S_N2 and E2 reactions or S_N1 and E1 reactions. (Conditions that favor S_N2 reactions also favor E2 reactions, and conditions that favor S_N1 reactions also favor E1 reactions, so S_N1/E2 and S_N2/E1 combinations are not possible.) To decide whether it undergoes S_N2 and E2 reactions or S_N1 and E1 reactions, we have to first look at the structure of the alkyl halide.

- Primary and secondary alkyl halides undergo S_N2 and E2 reactions.

- Tertiary alkyl halides undergo either an E2 reaction *or* S_N1 and E1 reactions. (Recall that because of steric hindrance, tertiary alkyl halides do not undergo S_N2 reactions.)

Thus, only if the alkyl halide is tertiary do we have to make another decision, which we can make by looking at the nucleophile/base.

- If it is a good nucleophile/strong base, the tertiary alkyl halide will undergo an E2 reaction.

- If the weakly basic solvent is the only nucleophile/base, the tertiary alkyl halide will undergo S_N1 and E1 reactions.

Now we have to decide whether the substitution product or the elimination product predominates when an alkyl halide undergoes both substitution *and* elimination reactions.

NOTE TO THE STUDENT
- Notice that HO^- is called a *nucleophile* in a substitution reaction (because it attacks a carbon) and it is called a *base* in an elimination reaction (because it removes a proton).

S_N2/E2 Reactions of Primary Alkyl Halides

The following reactions show that hydroxide ion can act as a nucleophile and attack the back side of the α-carbon of the primary alkyl halide to form a substitution product, or it can act as a base and remove a hydrogen from a β-carbon to form an elimination product.

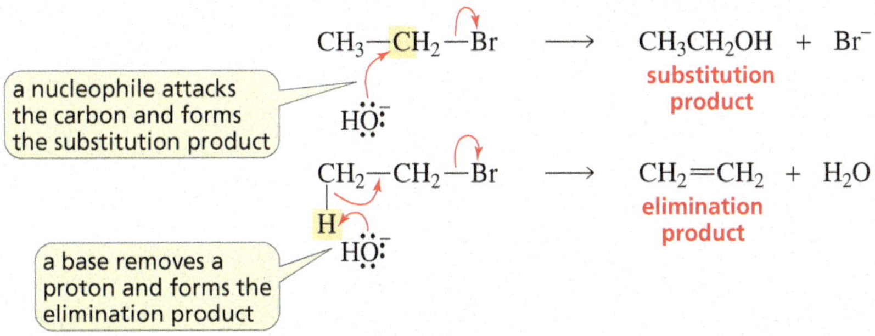

Thus, the two reactions compete with each other. Both reactions take place for the same reason: the electron-withdrawing halogen gives the carbon to which it is bonded a partial positive charge.

The relative reactivities of alkyl halides in S_N2 and E2 reactions are shown here.

In an S_N2 reaction: primary > secondary > tertiary

In an E2 reaction: tertiary > secondary > primary

Primary alkyl halides undergo primarily substitution.

Because a *primary* alkyl halide is the most reactive in an S_N2 reaction (the back side of the α-carbon is relatively unhindered; Section 9.1) and the least reactive in an E2 reaction (Section 9.7), a primary alkyl halide forms principally the substitution product. In other words, substitution will win the competition.

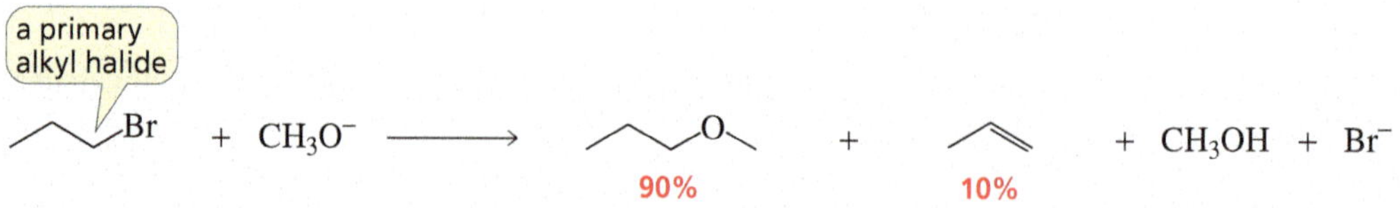

Steric Hindrance Favors the Elimination Product

However, if either the primary alkyl halide or the nucleophile/base is sterically hindered, then the nucleophile will have difficulty getting to the back side of the α-carbon but will be able to remove the more accessible proton. As a result, elimination will win the competition.

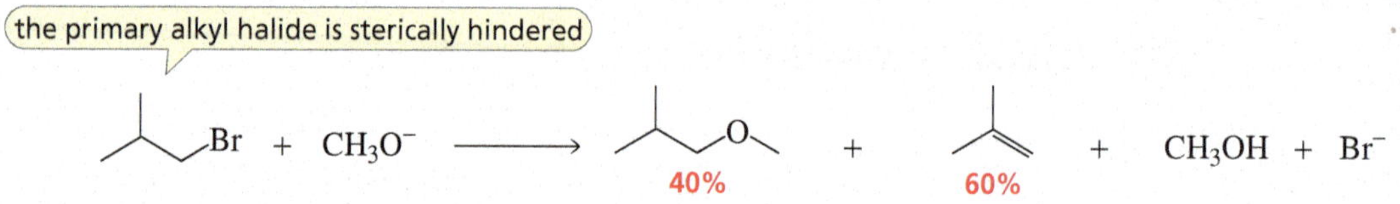

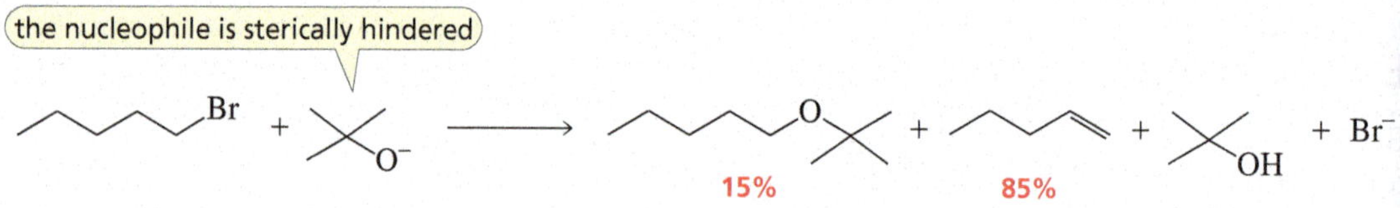

S_N2/E2 Reactions of Secondary Alkyl Halides

A *secondary* alkyl halide, compared with a primary alkyl halide, reacts slower in an S_N2 reaction and faster in an E2 reaction, so we expect both substitution and elimination reactions to be formed. The relative amounts of the two products depend on the strength and bulk of the nucleophile/base.

The stronger and bulkier the nucleophile/base, the greater the percentage of the elimination product.

A Strong Base Favors the Elimination Product

In the reactions that follow, acetate ion is a weaker base than ethoxide ion because acetic acid is a stronger acid ($pK_a = 4.76$) than ethanol ($pK_a = 15.9$). No elimination product is formed from the reaction of 2-chloropropane with the weakly basic acetate ion, whereas the elimination product is the major product formed with the strongly basic ethoxide ion.

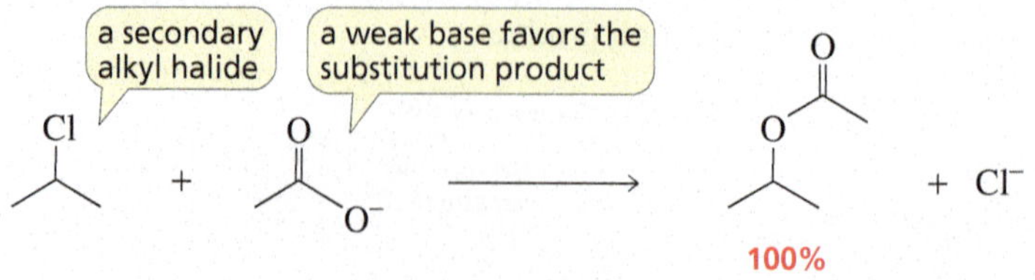

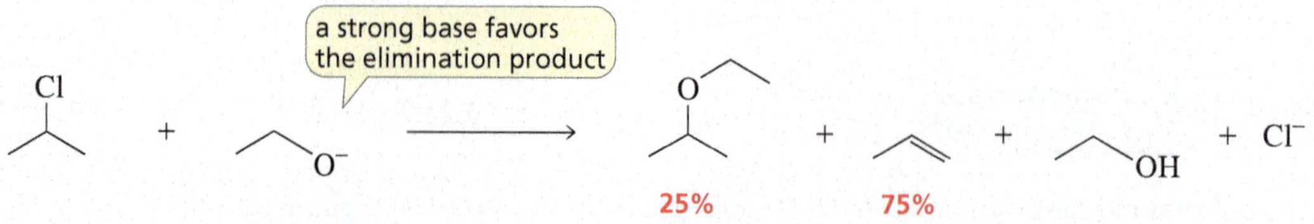

A Bulky Base Favors the Elimination Product

DBN and DBU are bulky bases commonly used to encourage elimination over substitution. Like other amines, they are relatively strong bases even though they are neutral compounds. These compounds are so bulky that only the elimination reaction occurs with a secondary alkyl halide.

DBN
1,5-diazabicyclo[4.3.0]non-5-ene

A High Temperature Favors the Elimination Product

Higher temperatures favor elimination because the $\Delta S°$ is greater for the elimination reaction since an elimination reaction forms more product molecules than a substitution reaction (Section 5.6): an elimination reaction forms three products (the alkene, the leaving group, and the conjugate acid of the base), whereas a substitution reaction forms two products (the substitution product and the leaving group).

DBU
1,8-diazabicyclo[5.4.0]undec-7-ene

a higher temperature favors elimination

Secondary alkyl halides undergo S_N2 and E2 reactions forming substitution and elimination products.

Strong and bulky bases and high temperatures favor elimination over substitution.

E2 Reaction of a Tertiary Alkyl Halide

Because a tertiary alkyl halide cannot undergo an S_N2 reaction, only an elimination product is formed when a tertiary alkyl halide reacts with a strong base.

a tertiary alkyl halide

Tertiary alkyl halides undergo E2 reactions but not S_N2 reactions, so form only elimination products.

S_N1/E1 Reactions of Tertiary Alkyl Halides

Both S_N1 and E1 reactions occur when a tertiary alkyl halide reacts with a poor nucleophile/weak base. Recall that in S_N1/E1 reactions, the alkyl halide dissociates to form a carbocation, which can then either react with the nucleophile to form a substitution product or lose a proton to form an elimination product.

Tertiary alkyl halides undergo S_N1 and E1 reactions, with substitution favored.

combines with the carbocation

substitution

rate-determining step

removes a proton

elimination

Primary and secondary alkyl halides do not undergo S_N1 and E1 reactions.

The substitution product is favored because bonds do not need to be broken in order for it to be formed. This is fortunate, because we saw that when a tertiary alkyl halide reacts with a strong base, only the elimination product is formed.

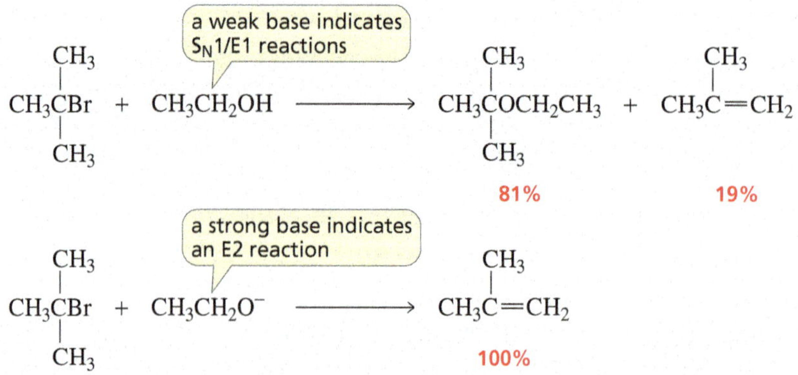

a weak base indicates
$S_N1/E1$ reactions

$$CH_3CBr(CH_3)(CH_3) + CH_3CH_2OH \longrightarrow CH_3COCH_2CH_3(CH_3)(CH_3) + CH_3C=CH_2(CH_3)$$

81%   19%

a strong base indicates
an E2 reaction

$$CH_3CBr(CH_3)(CH_3) + CH_3CH_2O^- \longrightarrow CH_3C=CH_2(CH_3)$$

100%

Table 9.7 summarizes the products obtained when alkyl halides react with nucleophiles/bases.

Table 9.7   Summary of the Products Expected in Substitution and Elimination Reactions

Class of alkyl halide	S_N2 and E2	S_N1 and E1
primary	primarily substitution, unless there is steric hindrance in the alkyl halide or nucleophile, in which case elimination is favored	cannot undergo $S_N1/E1$ solvolysis reactions
secondary	both substitution and elimination; the stronger and bulkier the base and the higher the temperature, the greater the percentage of elimination	cannot undergo $S_N1/E1$ solvolysis reactions
tertiary	only elimination	both substitution and elimination with substitution favored

---

PROBLEM 39 ♦

a. Which reacts faster in an S_N2 reaction?

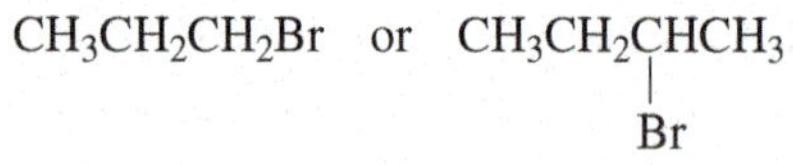

$$CH_3CH_2CH_2Br \quad or \quad CH_3CH_2CHCH_3(Br)$$

b. Which reacts faster in an E1 reaction?

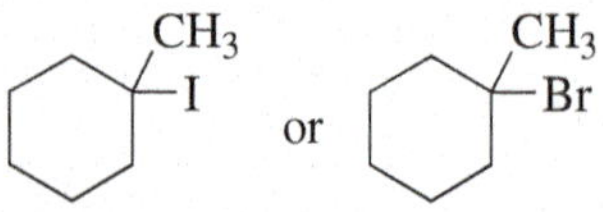

c. Which reacts faster in an S_N1 reaction?

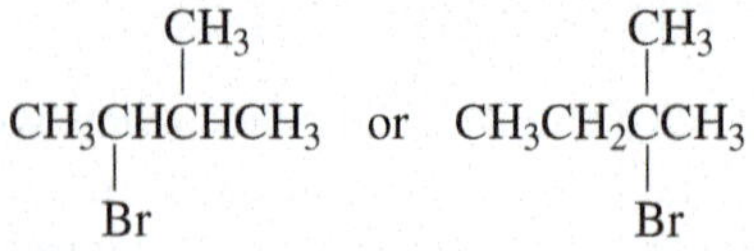

$$CH_3CHCHCH_3(CH_3)(Br) \quad or \quad CH_3CH_2CCH_3(CH_3)(Br)$$

PROBLEM 40

You were told in Section 7.11 that is best to use a methyl halide or a primary alkyl halide for the reaction of an acetylide ion with an alkyl halide. Explain why this is so.

PROBLEM 41 ♦

How does the ratio of substitution product to elimination product formed from the reaction of propyl bromide with CH_3O^- in methanol change if the nucleophile is changed to CH_3S^-?

PROBLEM 42 ♦

Explain why only a substitution product and no elimination product is obtained when the following compound reacts with sodium methoxide:

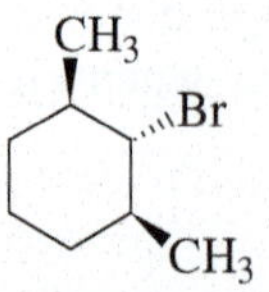

PROBLEM 43 ♦

a. Explain why 1-bromo-2,2-dimethylpropane has difficulty undergoing both S_N2 and S_N1 reactions.
b. Can it undergo E2 and E1 reactions?

9.13 BENZYLIC HALIDES, ALLYLIC HALIDES, VINYLIC HALIDES, AND ARYL HALIDES

Up to this point, our discussion of substitution reactions has been limited to methyl halides and primary, secondary, and tertiary alkyl halides. But what about benzylic, allylic, vinylic, and aryl halides?

Benzylic and Allylic Halides

Substitution Reactions

Unless they are tertiary, benzylic and allylic halides readily undergo S_N2 reactions. Tertiary benzylic and tertiary allylic halides, like tertiary alkyl halides, do *not* undergo S_N2 reactions because of steric hindrance. (Recall that a benzylic halide has a halogen on a benzylic carbon and an allylic halide has a halogen on an allylic carbon; Section 8.7.)

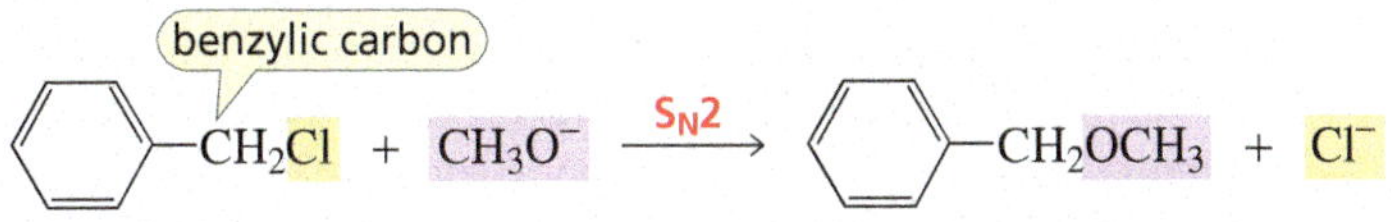

a primary
benzylic halide

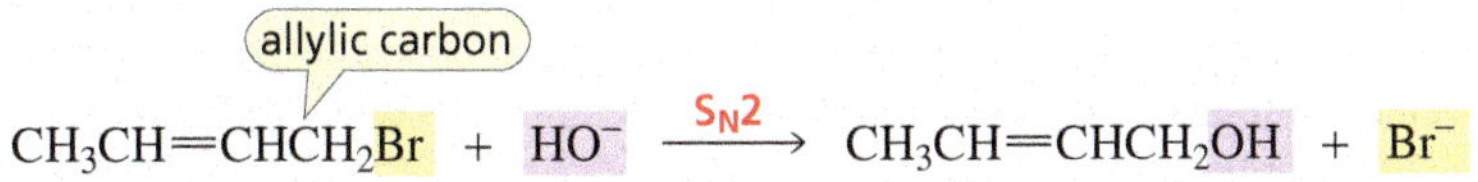

a primary
allylic halide

Recall that a benzene ring attached to a methylene group is called a benzyl group. A substitutent that consists of just a benzene ring is called a phenyl group.

a benzyl group   a phenyl group

benzyl bromide   1-bromo-2-phenylethane

Benzylic and allylic halides readily undergo S_N1 reactions as well, because they form carbocations that are stabilized by electron delocalization (Section 8.7).

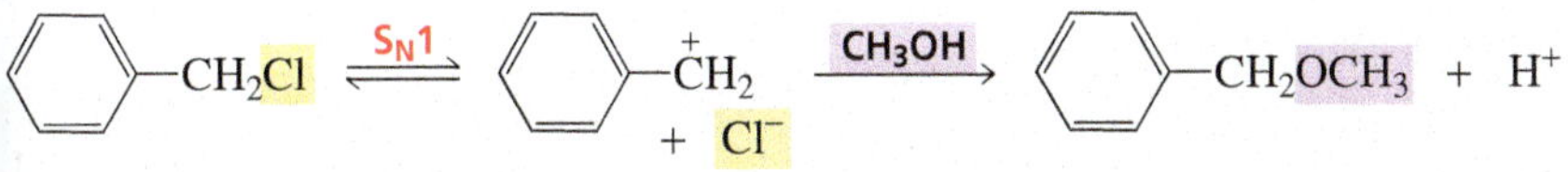

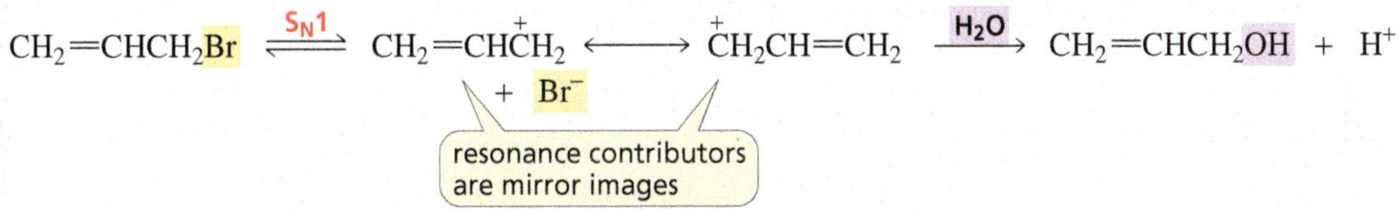

Benzylic and allylic halides undergo S_N1 and S_N2 reactions.

resonance contributors
are mirror images

If the two resonance contributors of the allylic carbocation formed in an S_N1 reaction are not mirror images (as they are in the preceding example), two substitution products will be formed. This is another example of how electron delocalization can affect the nature of the products formed in a reaction (Section 8.11).

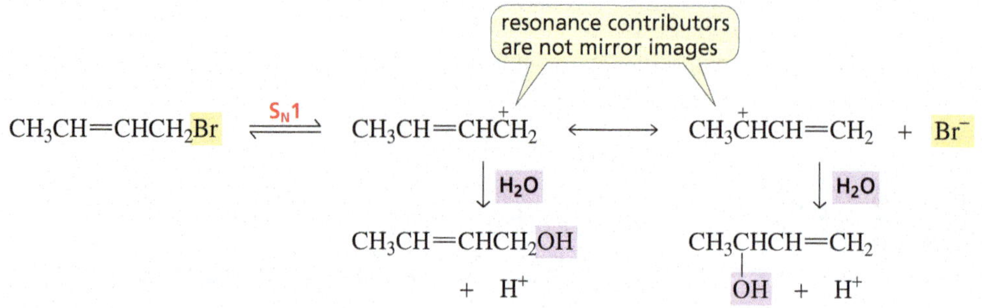

> resonance contributors
> are not mirror images

$$CH_3CH=CHCH_2Br \quad \underset{\longleftrightarrow}{\overset{S_N1}{\longrightarrow}} \quad CH_3CH=CH\overset{+}{C}H_2 \quad \longleftrightarrow \quad CH_3\overset{+}{C}HCH=CH_2 \quad + \quad Br^-$$

$\downarrow H_2O$ $\quad\quad\quad$ $\downarrow H_2O$

$$CH_3CH=CHCH_2OH \quad\quad\quad CH_3CHCH=CH_2$$
$$+ \quad H^+ \quad\quad\quad\quad\quad \overset{|}{OH} \quad + \quad H^+$$

Notice that all the reactions in this section use a good nucleophile (CH_3O^-, HO^-) to encourage S_N reactions and a poor nucleophile (CH_3OH, H_2O) to encourage S_N1 reactions.

LEARN THE STRATEGY

PROBLEM-SOLVING STRATEGY

Predicting Relative Reactivities

Which compound would you expect to be more reactive in an S_N1 reaction?

or

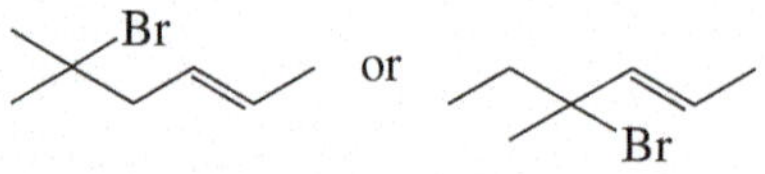

When asked to determine the relative reactivities of two compounds, we need to compare the $\Delta G^\ddagger$ value of their rate-determining steps. The faster-reacting compound will have the *smaller* $\Delta G^\ddagger$ value—that is, th smaller difference between its free energy and the free energy of its rate-determining transition state.

Both alkyl halides have approximately the same stability, so the difference in their reaction rates is du primarily to the difference in the stabilities of the transition states of their rate-determining steps. Th rate-determining step is carbocation formation. This is an endothermic reaction, so the transition stat resembles the carbocation more than it resembles the reactant (Section 6.3). Therefore, the compound tha forms the more stable carbocation is the one that reacts more rapidly.

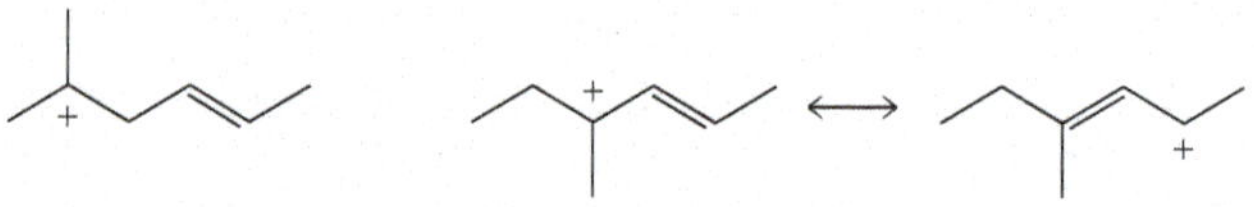

Unlike the carbocation formed by the alkyl halide on the left (which does not have delocalized electrons) the carbocation formed by the alkyl halide on the right is stabilized by electron delocalization. Thus, th alkyl halide on the right undergoes as S_N1 reaction more rapidly.

USE THE STRATEGY

PROBLEM 44

a. Assuming that the two compounds shown below have the same stability, which one would you expect t be more reactive in an S_N1 reaction?

or

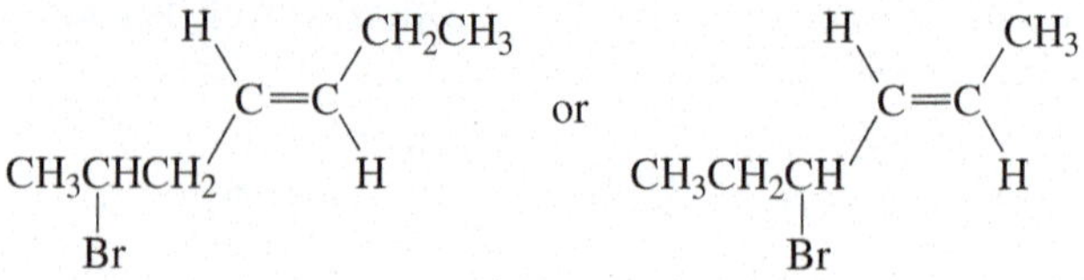

b. Draw the products that each would form when the solvent is ethanol.

Elimination Reactions

Benzylic and allylic halides that have β-hydrogens readily undergo E2 reactions because the new double bond in the product is relatively stable and, therefore, easily formed since it is conjugated with a benzene ring or with a double bond.

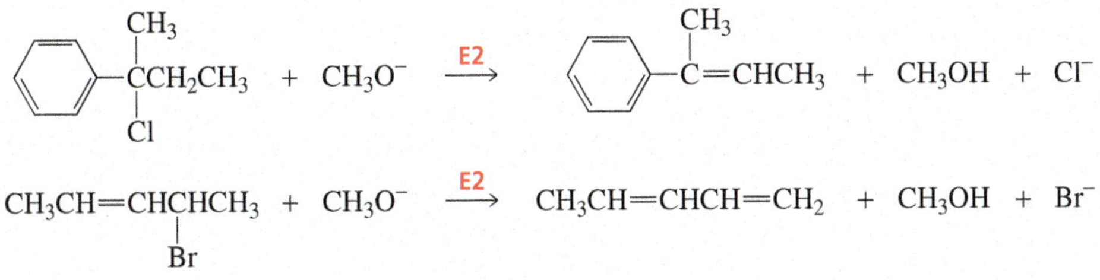

$$\text{C}_6\text{H}_5\text{C(CH}_3)\text{(Cl)CH}_2\text{CH}_3 + \text{CH}_3\text{O}^- \xrightarrow{\text{E2}} \text{C}_6\text{H}_5\text{C(CH}_3)=\text{CHCH}_3 + \text{CH}_3\text{OH} + \text{Cl}^-$$

$$\text{CH}_3\text{CH}=\text{CHCHCH}_3\text{(Br)} + \text{CH}_3\text{O}^- \xrightarrow{\text{E2}} \text{CH}_3\text{CH}=\text{CHCH}=\text{CH}_2 + \text{CH}_3\text{OH} + \text{Br}^-$$

Benzylic halides and allylic halides also undergo E1 reactions, because they form relatively stable carbocations.

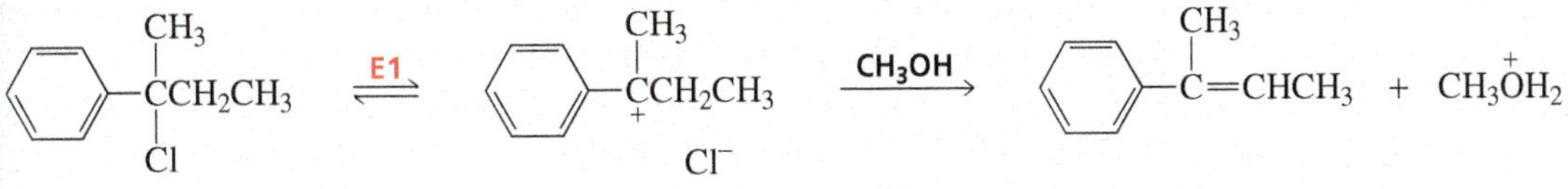

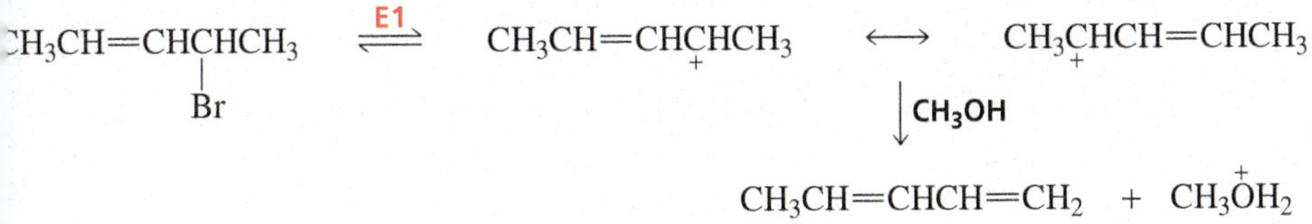

$$\text{CH}_3\text{CH}=\text{CHCHCH}_3\text{(Br)} \xrightleftharpoons{\text{E1}} \text{CH}_3\text{CH}=\text{CHCHCH}_3 \longleftrightarrow \text{CH}_3\overset{+}{\text{C}}\text{HCH}=\text{CHCH}_3 + \text{Br}^-$$

$$\downarrow \text{CH}_3\text{OH}$$

$$\text{CH}_3\text{CH}=\text{CHCH}=\text{CH}_2 + \text{CH}_3\overset{+}{\text{OH}}_2$$

If the two resonance contributors are not mirror images, as they are in the preceding example, two dienes will be formed.

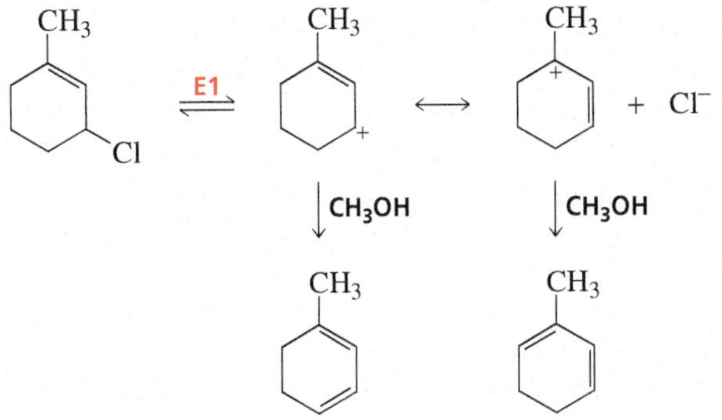

Notice that all the reactions in this section use a strong base to encourage E2 reactions and a weak base to encourage E1 reactions.

PROBLEM 45 ◆

What products will be obtained from the E2 reaction of the following alkyl halides?

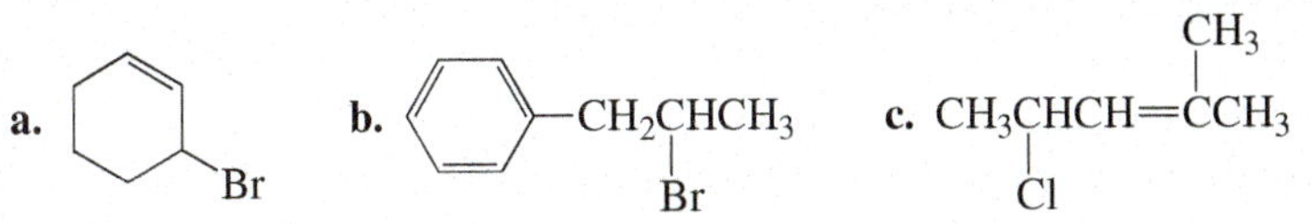

a.   b. $\text{C}_6\text{H}_5-\text{CH}_2\text{CHCH}_3\text{(Br)}$ c. $\text{CH}_3\text{CHCH}=\text{CCH}_3$ (with CH₃ and Cl substituents)

PROBLEM 46 ◆

What products will be obtained from the E1 reaction of the alkyl halides in Problem 45?

PROBLEM 47

Draw the elimination products that are formed when 3-bromo-3-methyl-1-butene reacts with

a. CH_3O^-.   b. CH_3OH.

Vinylic and Aryl Halides

Substitution Reactions

Vinylic halides and aryl halides (compounds in which the halogen is attached to a benzene ring) do not undergo S_N2 or S_N1 reactions. They do not undergo S_N2 reactions because, as the nucleophile approaches the back side of the sp^2 carbon, it is repelled by the π electrons of the double bond or the benzene ring.

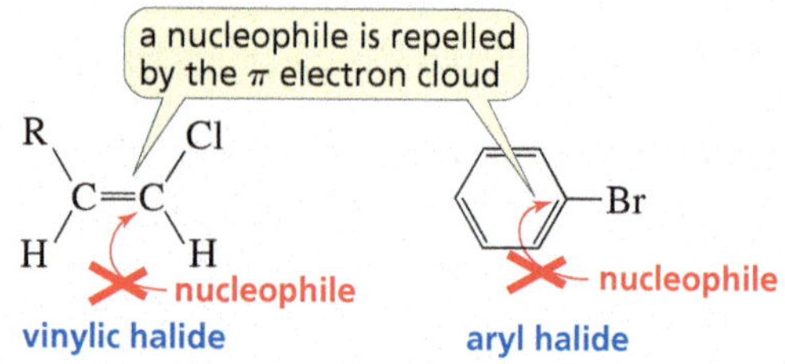

Vinylic halides and aryl halides do not undergo S_N1 reactions because vinylic and aryl cations are even more unstable than primary carbocations. The positive charge on a vinylic or aryl cation would be on an sp carbon—sp carbons are more electronegative than the sp^2 carbons that carry the positive charge of alkyl carbocations, so sp carbons are more resistant to becoming positively charged (Section 7.5). In addition, a ring carbon cannot form the 180° bond angles required for sp hybridization.

> Vinylic and aryl halides do not undergo S_N1 or S_N2 reactions.

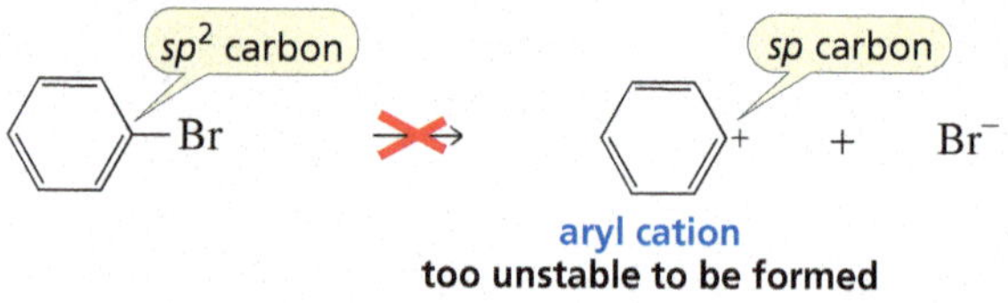

PROBLEM 48 ♦

Which compound is more reactive in an S_N1 reaction? In each case, you can assume that both alkyl halides have the same stability.

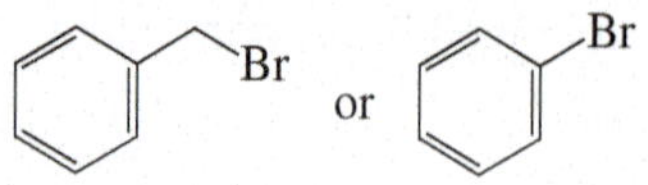

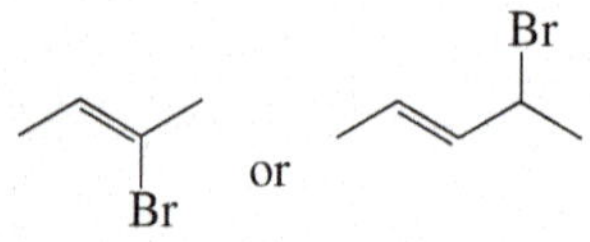

PROBLEM 49 ♦

For the pairs of compounds in Problem 48, which is more reactive in an S_N2 reaction?

Elimination Reactions

Because vinylic halides cannot undergo S_N1 reactions, they also cannot undergo E1 reactions. However, they can undergo E2 reactions. But because they are relatively unreactive, a very strong base ($^-NH_2$) is needed for the elimination reaction.

$$CH_3CH\!=\!CCH_3 \xrightarrow{\ ^-NH_2\ } CH_3C\!\equiv\!CCH_3 + 2\,NH_3 + 2\,Br^-$$
$$\underset{Br}{|}$$

PROBLEM 50

Why is a cumulated diene not formed in the reaction shown above?

PROBLEM 51 SOLVED

Describe how a double bond can be converted to a triple bond. For example, how can 2-butene be converted to 2-butyne?

$$CH_3CH{=}CHCH_3 \xrightarrow{?} CH_3C{\equiv}CCH_3$$
2-butene   **2-butyne**

SOLUTION Because we just saw how a vinylic halide can be converted to an alkyne, the question becomes, how can an alkene be converted to a vinylic halide? The vinylic halide can be obtained by adding Br_2 to the alkene followed by an E2 reaction. $^-NH_2$ is used for the E2 reaction so the two successive E2 reactions can be carried out simply by adding excess $^-NH_2$ to the vicinal dibromide.

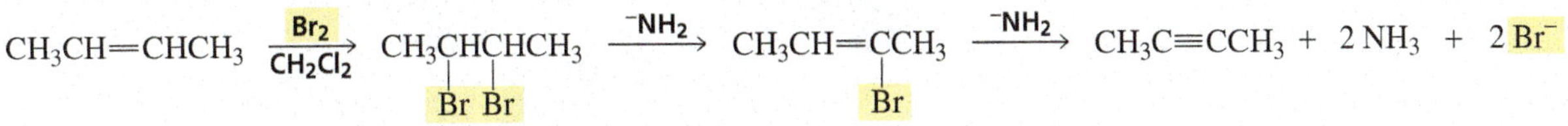

$$CH_3CH{=}CHCH_3 \xrightarrow[CH_2Cl_2]{Br_2} \underset{\underset{Br\ Br}{|\ \ |}}{CH_3CHCHCH_3} \xrightarrow{^-NH_2} \underset{\underset{Br}{|}}{CH_3CH{=}CCH_3} \xrightarrow{^-NH_2} CH_3C{\equiv}CCH_3 \ + \ 2\,NH_3 \ + \ 2\,Br^-$$

PROBLEM 52

What product is obtained when the following compound undergoes two successive elimination reactions?

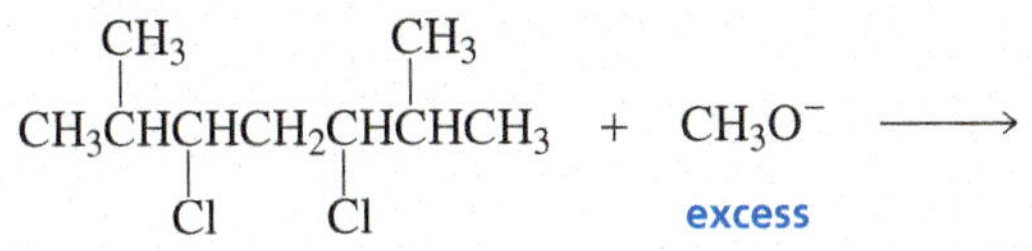

$$\underset{\underset{Cl}{|} \qquad \underset{Cl}{|}}{\overset{\overset{CH_3}{|} \qquad \overset{CH_3}{|}}{CH_3CHCHCH_2CHCHCH_3}} \ + \ CH_3O^- \longrightarrow$$
excess

PROBLEM-SOLVING STRATEGY

Predicting Whether $S_N2/E2$ or $S_N1/E1$ Reactions Occur

LEARN THE STRATEGY

Draw the substitution and elimination products that are formed from the reactions of the following compounds with the indicated nucleophile:

a. 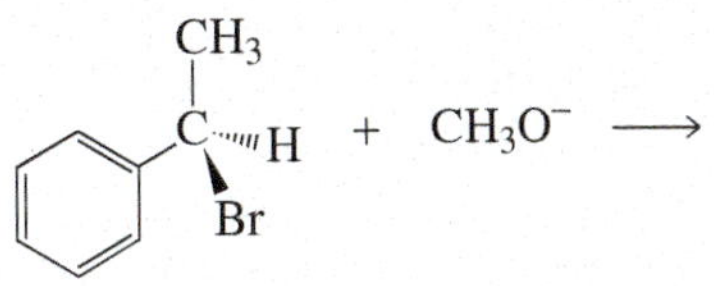 [structure: phenyl-CH(CH$_3$)(Br) with H] $+ \ CH_3O^- \longrightarrow$

b. 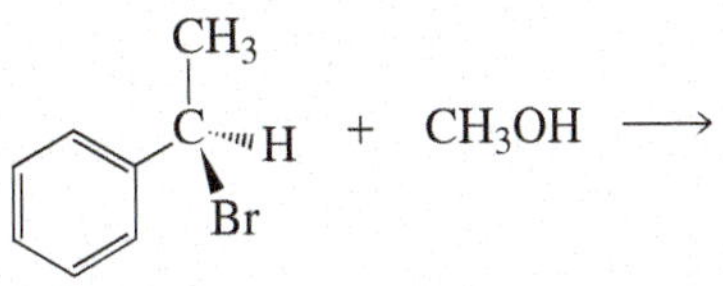 [structure: phenyl-CH(CH$_3$)(Br) with H] $+ \ CH_3OH \longrightarrow$

SOLUTION TO a. The reactant is a secondary benzylic halide, so it can undergo $S_N2/E2$ and $S_N1/E1$ reactions. Because a good nucleophile is employed, we know that S_N2 and E2 reactions occur. The substitution product has a configuration that is inverted relative to that of the reactant. Stereoisomers are not possible for the elimination product.

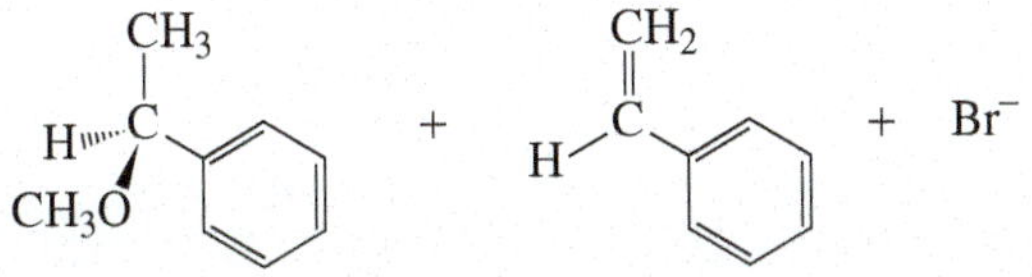

[structures: CH$_3$ / H–C–OCH$_3$ phenyl + CH$_2$=C(H)–phenyl + Br$^-$]

SOLUTION TO b. The reactant is a secondary benzylic halide so it can undergo $S_N2/E2$ and $S_N1/E1$ reactions. Because a poor nucleophile is employed, we know that S_N1 and E1 reactions occur. Therefore, two substitution products are formed, one with the retained configuration and one with the inverted configuration, relative to the configuration of the reactant.

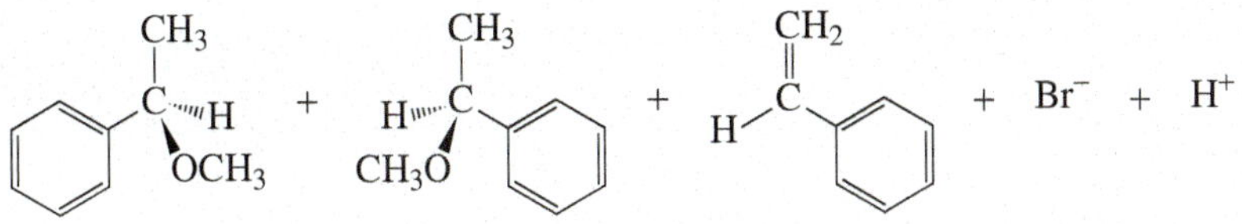

[structures: CH$_3$ / C–H / OCH$_3$ phenyl + H–C / CH$_3$ / CH$_3$O phenyl + CH$_2$=C(H)–phenyl + Br$^-$ + H$^+$]

PROBLEM 53

USE THE STRATEGY

What products are formed from the following reactions?

a. 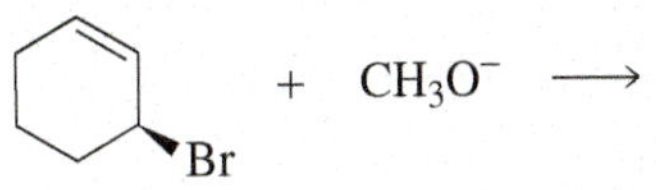 [cyclohexene with Br] $+ \ CH_3O^- \longrightarrow$

b. 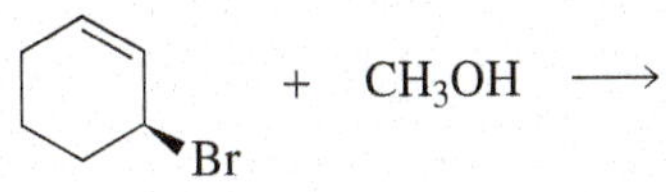 [cyclohexene with Br] $+ \ CH_3OH \longrightarrow$

9.14   SOLVENT EFFECTS

Polar solvents such as water and alcohols cluster around ions with the positive poles of the solvent molecules surrounding negative charges and the negative poles of the solvent molecules surrounding positive charges. Recall that the interaction between a solvent and an ion or a molecule dissolved in that solvent is called *solvation* (Section 3.10).

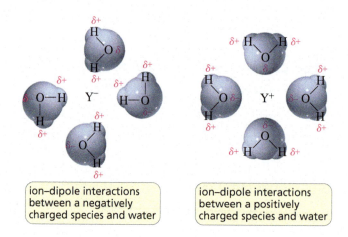

ion–dipole interactions between a negatively charged species and water

ion–dipole interactions between a positively charged species and water

When an ion interacts with a polar solvent, the charge is no longer localized solely on the ion, but is spread out to the surrounding solvent molecules. Spreading out the charge stabilizes the charged species.

The stabilization of charges by solvent interaction plays an important role in organic reactions. For example, when an alkyl halide undergoes an S_N1 reaction, the first step is dissociation of the carbon–halogen bond to form a carbocation and a halide ion. Energy is required to break the bond, but with no bonds being formed, where does the energy come from?

If the reaction is carried out in a polar solvent, the ions that are produced are solvated. The energy associated with a single ion–dipole interaction is small, but the additive effect of all the ion–dipole interactions that take place when a solvent stabilizes a charged species represents a great deal of energy. These ion–dipole interactions provide much of the energy necessary for dissociation of the carbon–halogen bond. So the alkyl halide does not fall apart spontaneously in an S_N1 reaction—polar solvent molecules pull it apart. An S_N1 reaction, therefore, cannot take place in a nonpolar solvent.

Solvation Effects

The tremendous amount of energy provided by solvation can be appreciated by considering the energy required to break the crystal lattice of sodium chloride (table salt). In the absence of a polar solvent, sodium chloride must be heated to more than 800 °C to overcome the forces that hold the oppositely charged ions together. However, sodium chloride readily dissolves in water at room temperature because solvation of the Na^+ and Cl^- ions by water molecules provides the energy necessary to separate the ions.

How a Solvent Affects Reaction Rates in General

How increasing the polarity of the solvent affects the rate of most chemical reactions depends *only* on whether or not a reactant that participates in the rate-limiting step is charged:

If a reactant that participates in the rate-determining step is charged, increasing the polarity of the solvent will decrease the rate of the reaction.

If none of the reactants that participates in the rate-determining step is charged, increasing the polarity of the solvent will increase the rate of the reaction.

Now let's see why this is true. The rate of a reaction depends on the difference between the free energy of the reactants and the free energy of the transition state of the rate-determining step. We can predict, therefore, how increasing the polarity of the solvent will affect the rate of a reaction simply by looking at the reactants and the transition state of the rate-determining step to see which will be more stabilized by a more polar solvent.

The greater or the more concentrated the charge on a molecule, the stronger its interactions with a polar solvent and the more the charge is stabilized. Therefore, if the size or concentration of the charge on the reactants is greater than that on the transition state, then a polar solvent will stabilize the reactants more than it will stabilize the transition state. Therefore, *increasing the polarity of the solvent* increases the difference in energy ($\Delta G^{\ddagger}$) between the transition state and the reactants, which *decreases the rate of the reaction*, as shown in Figure 9.12.

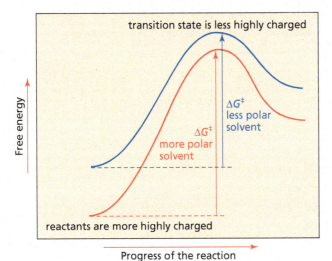

Progress of the reaction

Increasing the polarity of the solvent will decrease the rate of the reaction if a reactant in the rate-determining step is charged.

◀ **Figure 9.12**
The charge on the reactants is greater than the charge on the transition state.
As a result, increasing the polarity of the solvent increases the stability of the reactants more than the stability of the transition state, so the reaction will be slower.

On the other hand, if the size of the charge on the transition state is greater than the size of the charge on the reactants, then a polar solvent will stabilize the transition state more than it will stabilize the reactants. Therefore, *increasing the polarity of the solvent* decreases the difference in energy ($\Delta G^{\ddagger}$) between the transition state and the reactants, which *increases the rate of the reaction*, as shown in Figure 9.13.

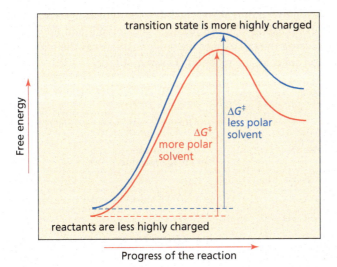

Progress of the reaction

Increasing the polarity of the solvent will increase the rate of the reaction if none of the reactants in the rate-determining step is charged.

◀ **Figure 9.13**
The charge on the transition state is greater than the charge on the reactants.
As a result, increasing the polarity of the solvent increases the stability of the transition state more than the stability of the reactants, so the reaction will be faster.

How a Solvent Affects the Rate of an S_N1 or E1 Reaction of an Alkyl Halide

Now let's look at specific reactions, beginning with an S_N1 or E1 reaction of an alkyl halide. The alkyl halide, which is the only reactant in the rate-determining step of an S_N1 or E1 reaction, is a neutral molecule with a small dipole moment. The partial charges in the transition state are greater than those in the alkyl halide because as the carbon–halogen bond breaks, the carbon becomes more positive and the halogen becomes more negative. Because the partial charges in the transition state are greater than those in the reactant, increasing the polarity of the solvent stabilizes the transition state more than the reactant, which increases the rate of the S_N1 or E1 reaction (Figure 9.13 and Table 9.8).

rate-determining step of an S_N1 or E1 reaction

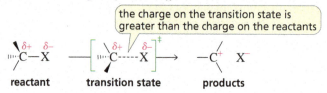

the charge on the transition state is greater than the charge on the reactants

reactant   transition state   products

Table 9.8 The Effect of the Polarity of the Solvent on the Rate of Reaction of 2-Bromo-2-Methylpropane (a Tertiary Alkyl Halide) in an S_N1 Reaction

Solvent	Relative rate
100% water	1200
80% water/20% ethanol	400
50% water/50% ethanol	60
20% water/80% ethanol	10
100% ethanol	1

Increasing polarity ↑

How a Solvent Affects the Rate of an S_N2 or E2 Reaction of an Alkyl Halide

How increasing the polarity of the solvent affects the rate of an S_N2 or E2 reaction of an alkyl halide depends on whether the nucleophile/base is charged or neutral.

The nucleophile/base in most S_N2 or E2 reactions of alkyl halides is negatively charged. Increasing the polarity of the solvent has a strong stabilizing effect on the negatively charged nucleophile/base. The transition state of an S_N2 or E2 reaction also has a negative charge, but that charge is dispersed over two atoms. Consequently, the interactions between the solvent and the transition state are not as strong as those between the solvent and the fully charged nucleophile/base. Therefore, increasing the polarity of the solvent stabilizes the nucleophile/base more than it stabilizes the transition state, so the reaction is slower (Figure 9.12).

rate-determining step of an S_N2 or E2 reaction with a negatively charged nucleophile/base

reactants   transition state of an S_N2 reaction   products

reactants   transition state of an E2 reaction   products

If, however, the S_N2 or E2 reaction of an alkyl halide is with a neutral nucleophile/base (such as an amine), then increasing the polarity of the solvent will increase the rate of the reaction (Figure 9.13).

Because a polar solvent decreases the rate of an S_N2 or E2 reaction when the nucleophile is negatively charged, we would like to carry out the reaction in a nonpolar solvent. However, negatively charged nucleophiles do not dissolve in nonpolar solvents such as hexane. Instead, an aprotic polar solvent such as DMF or DMSO is used. Because these solvents are not hydrogen bond donors, they are less effective than solvents such as water and alcohols at solvating negative charges; indeed, DMSO and DMF solvate negative charges very poorly because their partial positive charge is on the inside of the molecule (p. 402).

In summary, the way a change in the polarity of the solvent affects the rate of a reaction does not depend on the mechanism of the reaction. It depends only on whether the reactants that participate in the rate-determining step are charged or neutral. Therefore,

- an S_N2 or E2 reaction of an alkyl halide with a charged nucleophile/base should be carried out in an aprotic polar solvent.

- an S_N2 or E2 reaction of an alkyl halide with a neutral nucleophile/base should be carried out in a polar solvent.

- an S_N1 or E1 reaction of an alkyl halide should be carried out in a polar solvent.

LEARN THE STRATEGY

PROBLEM 54 ♦

Amines are good nucleophiles, even though they are neutral. How would the rate of an S_N2 reaction between an amine and an alkyl halide be affected if the polarity of the solvent is increased?

USE THE STRATEGY

PROBLEM 55 ♦

How will the rate of each of the following S_N2 reactions change if it is carried out in a more polar solvent?

a. $CH_3CH_2CH_2CH_2Br + HO^- \longrightarrow CH_3CH_2CH_2CH_2OH + Br^-$

b. $CH_3\overset{+}{S}CH_3 + NH_3 \longrightarrow CH_3\overset{+}{N}H_3 + CH_3SCH_3$
    $\quad\;\; |$
    $\quad\; CH_3$

c. $CH_3CH_2I + NH_3 \longrightarrow CH_3CH_2\overset{+}{N}H_3\, I^-$

An S_N2 reaction of an alkyl halide is favored by a high concentration of a good (negatively charged) nucleophile in an aprotic polar solvent or by a high concentration of a good (neutral) nucleophile in a protic polar solvent.

An S_N1 reaction of an alkyl halide is favored by a low concentration of a poor nucleophile in a protic polar solvent.

Environmental Adaptation

The microorganism *Xanthobacter* has learned to use alkyl halides that reach the ground as industrial pollutants as a source of carbon. The microorganism synthesizes an enzyme that uses the alkyl halide as a starting material to produce other carbon-containing compounds that the microorganism needs.

The first step of the enzyme-catalyzed reaction is an S_N2 reaction with a charged nucleophile. The enzyme has several nonpolar groups at its active site (the pocket in the enzyme where the reaction it catalyzes takes place; Section 5.14). The nonpolar groups on the surface of the enzyme provide the nonpolar environment needed to maximize the rate of the reaction.

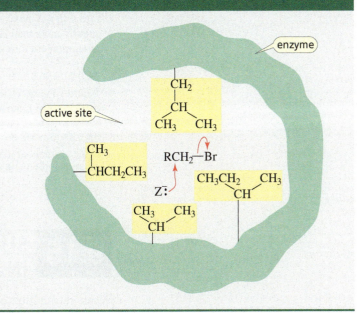

PROBLEM 56 ◆

Which reaction in each of the following pairs takes place more rapidly? (EtOH is ethyl alcohol; Et$_2$O is diethyl ether.)

a. $CH_3Br + HO^- \longrightarrow CH_3OH + Br^-$
$CH_3Br + H_2O \longrightarrow CH_3OH + HBr$

b. $CH_3I + HO^- \longrightarrow CH_3OH + I^-$
$CH_3Cl + HO^- \longrightarrow CH_3OH + Cl^-$

c. $CH_3Br + NH_3 \longrightarrow CH_3\overset{+}{N}H_3 + Br^-$
$CH_3Br + H_2O \longrightarrow CH_3OH + HBr$

d. $CH_3Br + HO^- \xrightarrow{\text{DMSO}} CH_3OH + Br^-$
$CH_3Br + HO^- \xrightarrow{\text{EtOH}} CH_3OH + Br^-$

e. $CH_3Br + NH_3 \xrightarrow{\text{Et}_2\text{O}} CH_3\overset{+}{N}H_3 + Br^-$
$CH_3Br + NH_3 \xrightarrow{\text{EtOH}} CH_3\overset{+}{N}H_3 + Br^-$

LEARN THE STRATEGY

PROBLEM 57 SOLVED

Most of the pK_a values given in this text were determined in water. How would the pK_a values of carboxylic acids, alcohols, ammonium ions $(R\overset{+}{N}H_3)$, phenol, and an anilinium ion $(C_6H_5\overset{+}{N}H_3)$ change if they were determined in a solvent less polar than water?

SOLUTION A pK_a is the negative logarithm of an equilibrium constant, K_a (Section 2.2). Because we are determining how decreasing the polarity of a solvent affects an equilibrium constant, we must look at how decreasing the polarity of the solvent affects the stability of the reactants and products (Section 5.6).

A neutral acid is a weaker acid in a solvent that is less polar than water.

$$K_a = \frac{[B^-][H^+]}{[HB]} \qquad K_a = \frac{[B][H^+]}{[HB^+]}$$

a neutral acid $\qquad$ a positively charged acid

Carboxylic acids, alcohols, and phenol are neutral in their acidic forms (HB) and charged in their basic forms (B$^-$). A polar solvent stabilizes B$^-$ and H$^+$ more than it stabilizes HB, thereby increasing K_a. Therefore, K_a is larger (a stronger acid) in water than in a less polar solvent, so the K_a values of carboxylic acids, alcohols, and phenol are smaller (they are weaker acids) and, therefore, their pK_a values are larger in a less polar solvent.

Ammonium ions and an anilinium ion are charged in their acidic forms (HB$^+$) and neutral in their basic forms (B). A polar solvent stabilizes HB$^+$ and H$^+$ more than it stabilizes B. Because HB$^+$ is stabilized slightly more than H$^+$, K_a is smaller (a weaker acid) in water than in a less polar solvent, so the pK_a values of ammonium ions and an anilinium ion are smaller (they are stronger acids) in a less polar solvent.

PROBLEM 58 ◆

Would you expect acetate ion $(CH_3CO_2^-)$ to be a better nucleophile in an S$_N$2 reaction with an alkyl halide carried out in methanol or in dimethyl sulfoxide?

PROBLEM 59 ◆

Under which of the following reaction conditions will (R)-1-chloro-1-phenylethane form the most (R)-1-phenyl-1-ethanol: in water or in 1.0 M HO$^-$?

9.15 SUBSTITUTION AND ELIMINATION REACTIONS IN SYNTHESIS

When substitution or elimination reactions are used in synthesis, care must be taken to choose reactants that maximize the yield of the desired product.

Using Substitution Reactions to Synthesize Compounds

In Section 9.2, you saw that nucleophilic substitution reactions of alkyl halides can lead to a wide variety of organic compounds. For example, ethers are synthesized by the reaction of an alkyl halide with an alkoxide ion. This reaction, called the **Williamson ether synthesis** (after Alexander Williamson, who discovered it in 1850) is still considered one of the best ways to synthesize an ether. Because it uses a good nucleophile, it is an S_N2 reaction.

Williamson ether synthesis

$$R-Br + R-O^- \longrightarrow R-O-R + Br^-$$
alkyl halide   alkoxide ion   ether

The alkoxide ion (RO^-) for a Williamson ether synthesis can be prepared by using sodium hydride (NaH) to remove a proton from an alcohol.

$$ROH + NaH \longrightarrow RO^- + Na^+ + H_2$$

If you want to synthesize an ether such as the one shown next, you have a choice of starting materials: you can use either a propyl halide and butoxide ion or a butyl halide and propoxide ion.

$$CH_3CH_2CH_2Br + CH_3CH_2CH_2CH_2O^- \longrightarrow CH_3CH_2CH_2OCH_2CH_2CH_2CH_3 + Br^-$$
propyl bromide   butoxide ion   butyl propyl ether

$$CH_3CH_2CH_2CH_2Br + CH_3CH_2CH_2O^- \longrightarrow CH_3CH_2CH_2CH_2OCH_2CH_2CH_3 + Br^-$$
butyl bromide   propoxide ion   butyl propyl ether

However, if you want to synthesize *tert*-butyl ethyl ether, the starting materials must be an ethyl halide and *tert*-butoxide ion.

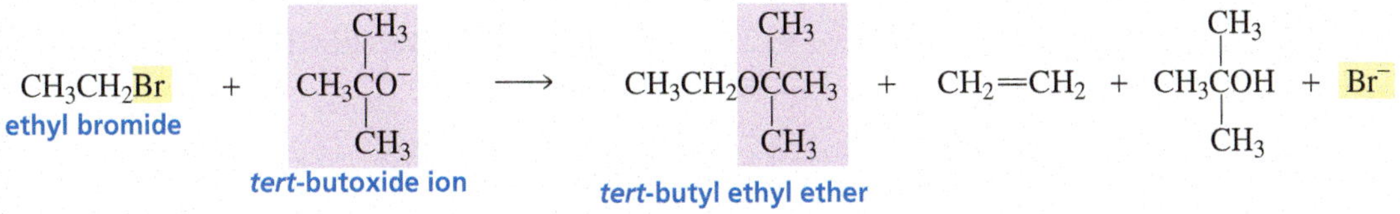

$$CH_3CH_2Br + \underset{\underset{CH_3}{|}}{\overset{\overset{CH_3}{|}}{CH_3CO^-}} \longrightarrow CH_3CH_2O\underset{\underset{CH_3}{|}}{\overset{\overset{CH_3}{|}}{CCH_3}} + CH_2=CH_2 + CH_3\underset{\underset{CH_3}{|}}{\overset{\overset{CH_3}{|}}{COH}} + Br^-$$
ethyl bromide   *tert*-butoxide ion   *tert*-butyl ethyl ether

If, instead, you used a *tert*-butyl halide and ethoxide ion, you would not obtain any ether because the reaction of a tertiary alkyl halide with a strong base forms only the elimination product.

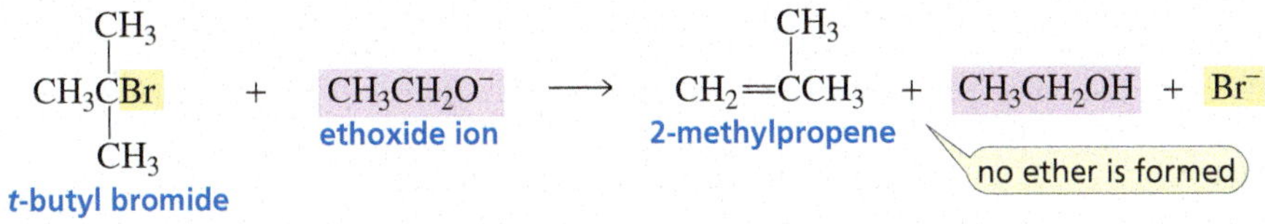

$$\underset{\underset{CH_3}{|}}{\overset{\overset{CH_3}{|}}{CH_3CBr}} + CH_3CH_2O^- \longrightarrow \underset{\underset{CH_3}{|}}{CH_2=CCH_3} + CH_3CH_2OH + Br^-$$
t-butyl bromide   ethoxide ion   2-methylpropene   (no ether is formed)

In ether synthesis, the less hindered group should be provided by the alkyl halide.

Consequently, a Williamson ether synthesis should be designed so that the *less hindered alkyl group* is provided by the *alkyl halide* and the *more hindered alkyl group* comes from the *alkoxide ion*.

PROBLEM 60 ◆

A small amount of another organic product is formed in a Williamson ether synthesis. What is this product when the alkyl halide used in the synthesis of butyl propyl ether is

a. propyl bromide?   **b.** butyl bromide?

PROBLEM 61

What is the best way to prepare the following ethers using an alkyl halide and an alkoxide ion?

a. 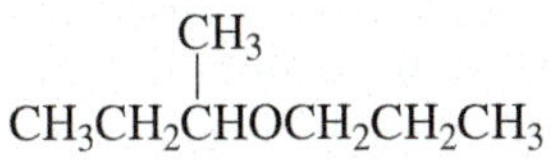 $CH_3CH_2\underset{\overset{|}{CH_3}}{\overset{\overset{CH_3}{|}}{CH}}OCH_2CH_2CH_3$

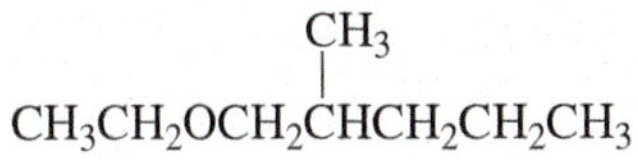

c. 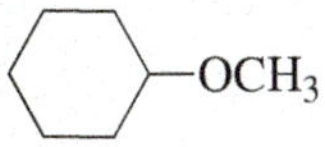 cyclohexyl—OCH₃

b. $CH_3CH_2OCH_2\underset{\overset{|}{CH_3}}{\overset{\overset{CH_3}{|}}{CH}}CH_2CH_2CH_3$

d. 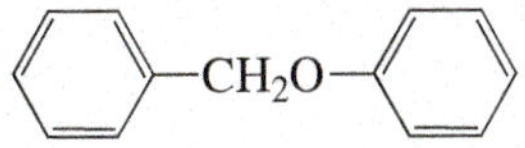 phenyl—CH₂O—phenyl

Using Elimination Reactions to Synthesize Alkenes

If you want to synthesize an alkene, you should choose the most hindered alkyl halide possible in order to maximize the elimination product and minimize the substitution product. For example, 2-bromopropane is a better starting material than 1-bromopropane for the synthesis of propene because the secondary alkyl halide gives a higher yield of the desired elimination product and a lower yield of the competing substitution product. The percentage of alkene could be further increased by using a sterically hindered base such as *tert*-butoxide ion or DBN instead of hydroxide ion, and by carrying out the reaction at a high temperature. (Section 9.11).

$$
\underset{\text{2-bromopropane}}{\overset{\overset{\displaystyle Br}{|}}{CH_3CHCH_3}} + HO^- \longrightarrow \underset{\text{major product}}{CH_3CH{=}CH_2} + \underset{\text{minor product}}{\overset{\overset{\displaystyle OH}{|}}{CH_3CHCH_3}} + H_2O + \boxed{Br^-}
$$

$$
\underset{\text{1-bromopropane}}{CH_3CH_2CH_2\boxed{Br}} + HO^- \longrightarrow \underset{\text{minor product}}{CH_3CH{=}CH_2} + \underset{\text{major product}}{CH_3CH_2CH_2OH} + H_2O + \boxed{Br^-}
$$

To synthesize 2-methyl-2-butene from 2-bromo-2-methylbutane, you would use a high concentration of a strong base to promote an E2 reaction so that the tertiary alkyl halide would form *only* the elimination product (Section 9.9). If a weak base were used instead, then both elimination and substitution products would be formed.

$$
\underset{\text{2-bromo-2-methylbutane}}{\overset{\overset{\displaystyle CH_3}{|}}{\underset{\overset{\displaystyle |}{Br}}{CH_3CCH_2CH_3}}}
$$

$$
\xrightarrow[\text{E2}]{CH_3O^-} \underset{}{\overset{\overset{\displaystyle CH_3}{|}}{CH_3C{=}CHCH_3}} + CH_3OH + \boxed{Br^-}
$$

$$
\xrightarrow[\text{E1}]{CH_3OH} \underset{\text{2-methyl-2-butene}}{\overset{\overset{\displaystyle CH_3}{|}}{CH_3C{=}CHCH_3}} + \underset{\overset{\overset{\displaystyle |}{OCH_3}}{}}{\overset{\overset{\displaystyle CH_3}{|}}{CH_3CCH_2CH_3}} + H_2O + \boxed{Br^-}
$$

PROBLEM 62 ◆

Identify the three products formed when 2-bromo-2-methylpropane is dissolved in a mixture of 80% ethanol and 20% water.

PROBLEM 63

What products (including stereoisomers, if applicable) are formed from the reaction of 3-bromo-3-methylpentane:

a. with HO⁻? **b.** with H₂O?

9.16 INTERMOLECULAR VERSUS INTRAMOLECULAR REACTIONS

A molecule with two functional groups is called a **bifunctional molecule.** If the two functional groups are able to react with each other, then two kinds of reactions can occur—an *intermolecular* reaction and an *intramolecular* reaction. To understand the difference, let's look at a molecule with two functional groups that can react in an S_N2 reaction—namely, a good nucleophile (such as an alkoxide ion) and an alkyl bromide.

If the alkoxide ion of one molecule displaces the bromide ion of a second molecule, then the reaction is an intermolecular reaction. *Inter* is Latin for "between," so an **intermolecular reaction** takes place between two molecules. If the product of this reaction subsequently reacts with a third bifunctional molecule (and then a fourth, and so on), a polymer will be formed. A **polymer** is a large molecule formed by linking together repeating units of small molecules.

an intermolecular reaction

BrCH$_2$(CH$_2$)$_n$CH$_2$Ö:⁻   Br—CH$_2$(CH$_2$)$_n$CH$_2$Ö:⁻ ⟶ BrCH$_2$(CH$_2$)$_n$CH$_2$ÖCH$_2$(CH$_2$)$_n$CH$_2$Ö:⁻ + Br⁻

nucleophile   electrophile

Alternatively, if the alkoxide ion of a molecule displaces the bromide ion of the *same* molecule (thereby forming a cyclic compound), then the reaction is an intramolecular reaction. *Intra* is Latin for "within," so an **intramolecular reaction** takes place within a single molecule.

an intramolecular reaction

Br—CH$_2$(CH$_2$)$_n$CH$_2$Ö:⁻ ⟶

(CH$_2$)$_n$ H$_2$C   CH$_2$ Ö   +   Br⁻

electrophile   nucleophile

How to Determine Whether the Intermolecular or Intramolecular Reaction Predominates

Which reaction is more likely to occur: an intermolecular reaction or an intramolecular reaction? The answer depends on the *concentration* of the bifunctional molecule and the *size of the ring* that would be formed in the intramolecular reaction.

The intramolecular reaction has an advantage: the reacting groups are tethered together, so they do not have to diffuse through the solvent to find a group with which to react. Therefore, a low concentration of reactant favors an intramolecular reaction because the two functional groups have a better chance of finding each other if they are in the same molecule. A high concentration of reactant helps compensate for some of the advantage gained by tethering, thereby increasing the likelihood of an intermolecular reaction.

How much of an advantage an intramolecular reaction has over an intermolecular reaction also depends on the size of the ring that is formed—that is, on the length of the tether. If the intramolecular reaction forms a five- or six-membered ring, then it will be favored over the intermolecular reaction because five- and six-membered rings are stable and, therefore, are easily formed. (Numbering the atoms in the reactant can help you determine the size of the ring in the product.)

Three- and four-membered rings are strained (Section 3.12), which makes them less stable than five- and six-membered rings and, therefore, less easily formed. Thus, the higher activation energy for formation of three- and four-membered rings cancels some of the advantage gained by tethering.

The likelihood of the reacting groups finding each other decreases sharply when the groups are in compounds that would form seven-membered and larger rings. Therefore, the intramolecular reaction becomes less favored as the ring size increases beyond six.

PROBLEM 64 ♦

After a proton is removed from the OH group, which compound in each pair forms a cyclic ether more rapidly?

a. 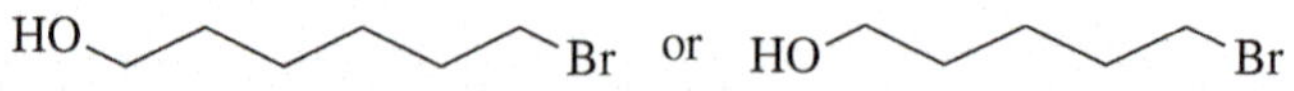 HO⁓⁓⁓⁓Br   or   HO⁓⁓⁓Br

b. 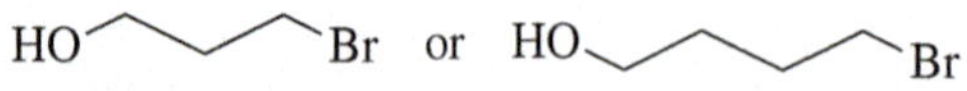 HO⁓⁓Br   or   HO⁓⁓⁓Br

c. 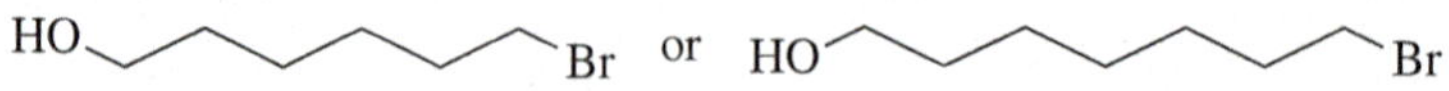 HO⁓⁓⁓⁓Br   or   HO⁓⁓⁓⁓⁓Br

---

PROBLEM-SOLVING STRATEGY

LEARN THE STRATEGY

Investigating How Stereochemistry Affects Reactivity

Which of the following compounds forms an epoxide as a result of reacting with sodium hydride (NaH)?

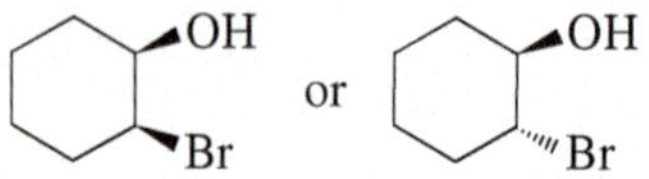

Hydride ion removes a proton from the OH group, forming a good nucleophile that can react with the secondary alkyl halide in an intramolecular S_N2 reaction to form an epoxide. An S_N2 reaction requires back-side attack. Only when the alkoxide ion and Br are on opposite sides of the cyclohexane ring is the alkoxide ion able to attack the back side of the carbon that is attached to Br. Therefore, only the trans isomer is able to form an epoxide.

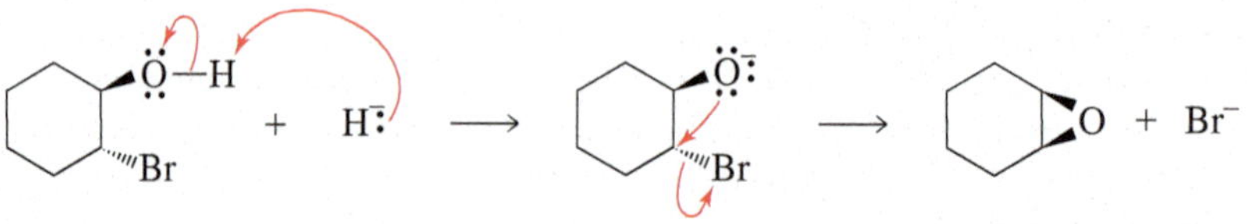

USE THE STRATEGY

PROBLEM 65

Draw the products of the following intramolecular reactions:

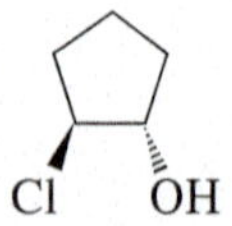

a. 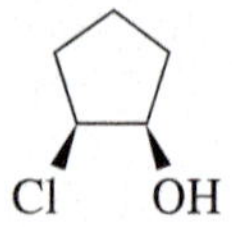 $\xrightarrow{\text{NaH}}$

Cl   OH

c. $BrCH_2CH_2CH_2CH_2CH_2OH \xrightarrow{\text{NaH}}$

d. $CH_3CH_2\overset{\overset{\displaystyle CH_3}{|}}{\underset{\underset{\displaystyle OH}{|}}{C}}CH_2Cl \xrightarrow{\text{NaH}}$

b. $\xrightarrow{\text{NaH}}$

Cl   OH

e. $CH_3CH_2CH_2CH{=}CH_2 \xrightarrow[\text{2. NaH}]{\text{1. Cl}_2,\ \text{H}_2\text{O}}$

DESIGNING A SYNTHESIS II

9.17 APPROACHING THE PROBLEM

When you are asked to design a synthesis, one way to approach the task is to think about the starting material you have been given and ask yourself if there is an obvious series of reactions beginning with the starting material that can get you on the road to the **target molecule** (the desired product). Sometimes this is the best way to approach a *simple* synthesis. The following examples give you practice employing this strategy.

Example 1

How is the target molecule prepared from the given starting material?

Adding HBr to the alkene forms a compound with a leaving group that can be replaced by a nucleophile. Because $^-C\equiv N$ is a relatively weak base (the pK_a of $HC\equiv N$ is 9.1), the desired substitution reaction is favored over the competing elimination reaction.

synthesis

target molecule

Example 2

How is trans-2-methylcyclohexanol prepared from 1-bromo-1-methylcyclohexane?

Elimination of HBr from the reactant forms an alkene that can add water *via* an electrophilic addition reaction. The elimination reaction should be carried out under S_N2 conditions (that is, with a strong base), so there will be no competing substitution reaction. (Recall that tertiary alkyl halides do not undergo S_N2 reactions.) Hydroboration-oxidation puts the OH on the desired carbon. Because R_2BH adds preferentially to the less sterically hindered side of the double bond and the overall hydroboration–oxidation reaction results in the syn addition of water, the target molecule (as well as its enantiomer) is obtained.

synthesis

target molecule

Example 3

How is ethyl methyl ketone prepared from 1-bromobutane?

$$CH_3CH_2CH_2CH_2Br \xrightarrow{?} CH_3CH_2\overset{\overset{\displaystyle O}{\displaystyle \|}}{C}CH_3$$

As you learned in Section 7.12, working backward can be a useful way to design a synthesis, particularly when the starting material does not clearly indicate how to proceed as in this example.

Look at the target molecule and ask yourself how it could be prepared. Once you have an answer, look at the precursor you have identified for the target molecule and ask yourself how the precursor could be prepared. Keep working backward one step at a time, until you get to the given starting material. Recall that this technique is called *retrosynthetic analysis*.

At this point in your study of organic chemistry, you know only three ways to synthesize a ketone: (1) the addition of water to an *alkyne* (Section 7.7), (2) hydroboration–oxidation of an *alkyne* (Section 7.8), and (3) ozonolysis of an *alkene* (Section 6.11). Because the target molecule has the same number of carbons as the starting material, we can rule out ozonolysis. Now we know that the precursor molecule must be an alkyne. The alkyne needed to prepare the ketone can be prepared from two successive E2 reactions of a vicinal dihalide, which in turn can be synthesized from an alkene. The desired alkene can be prepared from the given starting material by an elimination reaction, using a bulky base to maximize the elimination product.

retrosynthetic analysis

$$\underset{\text{target molecule}}{CH_3CH_2\overset{\displaystyle O}{\overset{\|}{C}}CH_3} \Longrightarrow CH_3CH_2C\equiv CH \Longrightarrow \underset{\underset{Br}{|}}{CH_3CH_2CHCH_2Br} \Longrightarrow CH_3CH_2CH=CH_2 \Longrightarrow CH_3CH_2CH_2CH_2Br$$

open arrow indicates you are working backward

Now you can write the reaction sequence in the forward direction, along with the reagents needed to carry out each reaction. Notice that a bulky base (DBN) is used in the elimination reaction in order to maximize the amount of elimination product. (Three equivalents of $-NH_2$ are used because $-NH_2$ can deprotonate the terminal alkyne, which is then reprotonated before the alkyne is converted to the ketone.)

synthesis

$$CH_3CH_2CH_2CH_2Br \xrightarrow{\text{DBN}} CH_3CH_2CH=CH_2 \xrightarrow[\text{CH}_2\text{Cl}_2]{\text{Br}_2} \underset{\underset{Br}{|}}{CH_3CH_2CHCH_2Br} \xrightarrow{3\ ^-NH_2} CH_3CH_2C\equiv C^- \xrightarrow[\substack{\text{H}_2\text{SO}_4 \\ \text{HgSO}_4}]{\text{H}_2\text{O}} \underset{\text{target molecule}}{CH_3CH_2\overset{\displaystyle O}{\overset{\|}{C}}CH_3}$$

Example 4

How is the following cyclic ether prepared from the given starting material?

$$BrCH_2CH_2CH_2CH=CH_2 \xrightarrow{?} \text{[cyclic ether with CH}_3\text{]}$$

To prepare a cyclic ether, the alkyl halide and alcohol must both be part of the same molecule. To determine the precursor to the target molecule, find the new bond that was formed in the target molecule and number the atoms in the ring formed by the new bond.

retrosynthetic analysis

new bond
target molecule
new bond

$$\Longrightarrow \underset{\underset{5\,\text{OH}}{|}}{\overset{1\ \ \ 2\ \ \ 3\ \ \ 4}{BrCH_2CH_2CH_2CHCH_3}} \Longrightarrow BrCH_2CH_2CH_2CH=CH_2$$

Addition of water to the given starting material creates the required bifunctional compound, which after deprotonation forms the cyclic ether *via* an intramolecular reaction.

synthesis

$$BrCH_2CH_2CH_2CH=CH_2 \xrightarrow[\text{H}_2\text{O}]{\text{H}_2\text{SO}_4} \underset{\underset{OH}{|}}{BrCH_2CH_2CH_2CHCH_3} \xrightarrow{\text{NaH}} \text{[cyclic ether with CH}_3\text{]}$$

target molecule

PROBLEM 66

For each of the following target molecules, design a multistep synthesis to show how it can be prepared from the given starting material:

a. [cyclohexyl–Br] ⟶ [cyclohexene oxide (epoxide)]

b. [cyclohexyl–Br] ⟶ [trans-1,2-dibromocyclohexane]

c. [cyclohexyl–CH=CH₂] ⟶ [cyclohexyl–CH₂CH with O (aldehyde)]

d. [cyclohexyl–CH=CH₂] ⟶ [cyclohexyl–CH₂CH₂CH₂CH₃]

ESSENTIAL CONCEPTS

Section 9.1

- An S_N2 reaction is bimolecular: both the alkyl halide and the nucleophile are involved in the transition state of the rate-limiting step, so the rate of the reaction depends on the concentration of both of them.

- An S_N2 reaction has a one-step mechanism: the nucleophile attacks the back side of the carbon that is attached to the halogen. Therefore, it takes place with inversion of configuration.

- Because of steric hindrance, the relative reactivities of alkyl halides in an S_N2 reaction are primary > secondary > tertiary. Tertiary alkyl halides cannot undergo S_N2 reactions.

Section 9.2

- The relative reactivities of alkyl halides that differ only in the halogen atom are RI > RBr > RCl > RF in S_N2, S_N1, E2, and E1 reactions.

- Basicity is a measure of how well a compound shares its lone pair with a proton; nucleophilicity is a measure of how readily a species with a lone pair is able to attack an electron-deficient atom.

- Protic solvents (H_2O, ROH) have a hydrogen attached to an O or an N; aprotic solvents (DMF, DMSO) do not have a hydrogen attached to an O or an N.

- In general, a stronger base is a better nucleophile. However, if the attacking atoms are very different in size *and* the reaction is carried out in a protic solvent, the stronger bases are poorer nucleophiles because of ion–dipole interactions between the ion and the solvent.

Section 9.3

- An S_N1 reaction is unimolecular; only the alkyl halide is involved in the transition state of the rate-limiting step, so the rate of the reaction is dependent only on the concentration of the alkyl halide.

- An S_N1 reaction has a two-step mechanism: the halogen departs in the first step, forming a carbocation intermediate that is attacked by a nucleophile in the second step. S_N1 reactions of alkyl halides are solvolysis reactions, meaning that the solvent is the nucleophile.

- The rate of an S_N1 reaction depends on the ease of carbocation formation.

- An S_N1 reaction takes place with racemization.

Section 9.5

- The only substitution reactions that primary and secondary alkyl halides and methyl halides undergo are S_N2 reactions.

- The only substitution reactions that tertiary alkyl halides undergo are S_N1 reactions.

Section 9.7

- Removal of a proton and a halide ion is called dehydrohalogenation. The product of an elimination reaction is an alkene. An E2 reaction is a concerted, one-step reaction in which the proton and the halide ion are removed in the same step.

- An E2 reaction is regioselective; the major product is the more stable alkene, unless the reactants are sterically hindered or the leaving group is poor.

- The more stable alkene is generally the more substituted alkene.

- Alkyl substitution increases the stability of a carbocation and decreases the stability of a carbanion.

Section 9.8

- An E1 reaction is a two-step reaction in which the alkyl halide dissociates, forming a carbocation intermediate. Then, a base removes a proton from a carbon adjacent to the positively charged carbon.

- An E1 reaction is regioselective; the major product is the more stable alkene.

Section 9.9

- The only elimination reactions that primary and secondary alkyl halides undergo are E2 reactions.

- Tertiary alkyl halides undergo both E2 and E1 reactions.

- For alkyl halides that can undergo both E2 and E1 reactions, the E2 reaction is favored by a strong base) and the E1 reaction is favored by a weak base (that is, a weakly basic solvent).

Section 9.10

- An E2 reaction is stereoselective: anti elimination is favored. If the β-carbon has two hydrogens, then both *E* and *Z* stereoisomers are formed; the one with the largest groups on opposite sides of the double bond is formed in greater yield because it is more stable.

- If the β-carbon is bonded to only one hydrogen, then only one alkene is formed in an E2 reaction; its structure depends on the structure of the alkyl halide.

- An E1 reaction is stereoselective. Both *E* and *Z* stereoisomers are formed regardless of the number of hydrogens bonded to the β-carbon.

Section 9.11

- The two groups eliminated from a six-membered ring must both be in axial positions in an E2 reaction (but not in an E1 reaction).

- Elimination is more rapid when H and X are diaxial in the more stable conformer.

Section 9.12

- Primary alkyl halides undergo S_N2/E2 reactions: the substitution reaction is favored unless the nucleophile/base is sterically hindered.

- Secondary alkyl halides undergo S_N2/E2 reactions: both substitution and elimination products are formed; strong bases, bulky bases, and high temperatures favor the elimination product.

- Tertiary alkyl halides undergo E2 reactions with strong bases and S_N1/E1 reactions with weak bases.

Section 9.13

- Benzylic and allylic halides undergo S_N2, E2, S_N1, and E1 reactions. (If they are tertiary, they cannot undergo S_N2 reactions.) The S_N2/E2 reactions are favored by a high concentration of a good nucleophile. The S_N1/E1 reactions are favored by a low concentration of a poor nucleophile.

- Vinylic and aryl halides cannot undergo S_N2, S_N1, or E1 reactions. Vinylic halides can undergo E2 reactions with a strong base.

Section 9.14

- Polar solvents insulate opposite charges from one another.
- Increasing the polarity of the solvent decreases the rate of the reaction if one or more reactants that participate in the rate-determining step are charged, and increases the rate of the reaction if none of the reactants that participate in the rate-determining step is charged.

Section 9.15

- The Williamson ether synthesis prepares ethers from the reaction of an alkyl halide with an alkoxide ion. The less hindered alkyl group should be provided by the alkyl halide.

- The most hindered alkyl halide should be used for the synthesis of an alkene.

Section 9.16

- If the two functional groups of a bifunctional molecule can react with each other, both intermolecular (between two molecules) and intramolecular (within one molecule) reactions can occur. The reaction more likely to occur depends on the concentration of the bifunctional molecule and the size of the ring that would be formed in the intramolecular reaction.

SUMMARY OF REACTIONS

1. S_N2 reaction: a one-step reaction

$$\overset{..}{Nuc}^- \ + \ -\overset{|}{\underset{|}{C}}-X \ \longrightarrow \ -\overset{|}{\underset{|}{C}}-Nuc \ + \ X^-$$

- Relative reactivities of alkyl halides: CH_3X > primary > secondary > tertiary. Tertiary alkyl halides cannot undergo S_N2 reactions.
- Only the inverted product is formed.

2. S_N1 reaction: a two-step reaction with a carbocation intermediate

$$-\overset{|}{\underset{|}{C}}-X \ \longrightarrow \ -\overset{|}{\underset{|}{C}}{}^+ \ \xrightarrow{\ \overset{..}{Nuc}^-\ } \ -\overset{|}{\underset{|}{C}}-Nuc$$
$$+ \ X^-$$

- Reactivity: only tertiary alkyl halides, allylic halides, and benzylic halides undergo S_N1 reactions.
- Products with both inverted and retained configurations are formed.

3. E2 reaction: a one-step reaction

$$\overset{..}{B}^- \ + \ -\overset{|}{\underset{|}{C}}-\overset{H}{\underset{|}{C}}-X \ \longrightarrow \ \overset{}{\underset{}{C}}=\overset{}{\underset{}{C} }\ + \ ^+BH \ + \ X^-$$

- Relative reactivities of alkyl halides: tertiary > secondary > primary
- Anti elimination only: if the β-carbon from which the hydrogen is removed *is bonded to two hydrogens,* then both *E* and *Z* stereoisomers are formed. The isomer with the largest groups on opposite sides of the double bond is the major product. If the β-carbon from which the hydrogen is removed *is bonded to only one hydrogen,* then only one elimination product is formed. Its configuration depends on the configuration of the reactant.

4. E1 reaction: a two-step reaction with a carbocation intermediate

$$-\overset{|}{\underset{H}{C}}-\overset{|}{\underset{|}{C}}-X \ \longrightarrow \ -\overset{|}{\underset{H}{C}}-\overset{|}{\underset{|}{C}}{}^+ \ \longrightarrow \ \overset{}{\underset{}{C}}=\overset{}{\underset{}{C}} \ + \ ^+BH$$
$$\overset{..}{B} \ + \ X^-$$

- Reactivity: only tertiary alkyl halides, allylic halides, and benzylic halides undergo E1 reactions.
- Anti and syn elimination: both *E* and *Z* stereoisomers are formed. The isomer with the largest groups on opposite sides of the double bond is the major product.

For alkyl halides (tertiary, benzylic, and allylic) that can undergo both S_N2/E2 and S_N1/E1 reactions
- a good nucleophile/strong base favors S_N2/E2.
- a poor nucleophile/weak base favors S_N1/E1.

Competing S$_N$2 and E2 Reactions
- Primary alkyl halides: primarily substitution.
- Secondary alkyl halides: substitution and elimination. (Strong, bulky bases and high temperatures favor elimination over substitution.)
- Tertiary alkyl halides: only elimination.
- Allylic and benzylic halides: substitution and elimination (unless they are tertiary).

Competing S$_N$1 and E1 Reactions
Tertiary alkyl halides, allylic halides, and benzylic halides: substitution and elimination with substitution favored.

PROBLEMS

67. What product is formed when 1-bromopropane reacts with each of the following nucleophiles?
a. HO$^-$ c. CH$_3$S$^-$ e. CH$_3$O$^-$
b. $^-$NH$_2$ d. HS$^-$ f. CH$_3$NH$_2$

68. Which member of each pair is a better nucleophile in methanol?
a. H$_2$O or HO$^-$ c. H$_2$O or H$_2$S   e. I$^-$ or Br$^-$
b. NH$_3$ or H$_2$O   d. HO$^-$ or HS$^-$ f. Cl$^-$ or Br$^-$

69. Which member in each pair in Problem 68 is a better leaving group?

70. What nucleophiles would form the following compounds as a result of reacting with 1-iodobutane?

a. 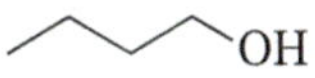 ~~~~OH   d. ~~~~SH

b. ~~~~O~   e. ~~~~S~   g. 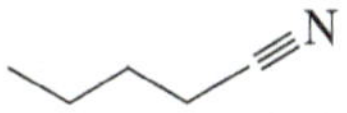 ~~~~≡N

c. ~~~~N~   f. ~~~~O—C(=O)~   h. ~~~~≡~
        H

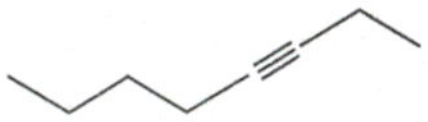

71. Explain how each of the following changes affect the rate of the reaction of 1-bromobutane with ethoxide ion in DMF.
a. The concentration of both the alkyl halide and the nucleophile are tripled.   c. The alkyl halide is changed to 1-chlorobutane.
b. The solvent is changed to ethanol.   d. The alkyl halide is changed to 2-bromobutane.

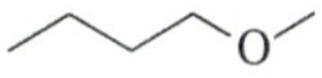

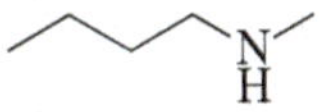

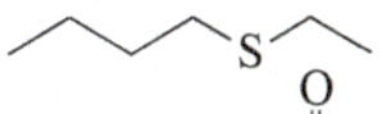

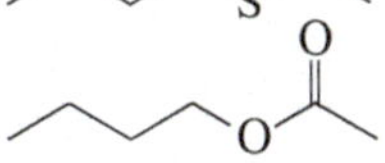

72. Explain how the following changes affect the rate of the reaction of 2-bromo-2-methylbutane with methanol:
a. The alkyl halide is changed to 2-chloro-2-methylbutane.   b. The alkyl halide is changed to 2-chloro-3-methylbutane.

73. Starting with cyclohexene, how can the following compounds be prepared?
a. methoxycyclohexane   b. cyclohexylmethylamine   c. dicyclohexyl ether

74. Rank the following species in each set from best nucleophile to poorest nucleophile.

a. CH$_3$CO$^-$(C=O), CH$_3$CH$_2$S$^-$, CH$_3$CH$_2$O$^-$ in methanol   c. H$_2$O and NH$_3$ in methanol

b. [phenyl]—O$^-$ and [cyclohexyl]—O$^-$ in DMSO   d. Br$^-$, Cl$^-$, I$^-$ in methanol

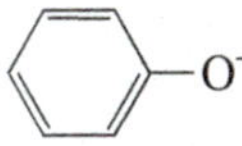

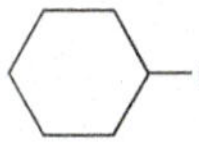

75. The pK$_a$ of acetic acid in water is 4.76. What effect will a decrease in the polarity of the solvent have on the pK$_a$? Why?

76. a. Identify the substitution products that form when 2-bromo-2-methylpropane is dissolved in a mixture of 80% ethanol and 20% water.
b. Explain why the same products are obtained when 2-chloro-2-methylpropane is dissolved in a mixture of 80% ethanol and 20% water.

77. Draw the substitution products for each of the following reactions; if the products can exist as stereoisomers, show what stereoisomers are obtained:
a. (R)-2-bromopentane + CH$_3$O$^-$ d. allyl chloride + CH$_3$OH
b. (R)-3-bromo-3-methylheptane + CH$_3$OH   e. 1-bromo-2-butene + CH$_3$O$^-$
c. benzyl chloride + CH$_3$CH$_2$OH   f. 1-bromo-2-butene + CH$_3$OH

78. Draw the products obtained from the solvolysis of each of the following compounds in ethanol:

a. 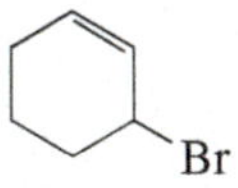 [cyclohexene with Br]   b. 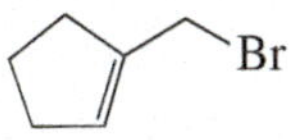 [cyclopentene—Br]   c. 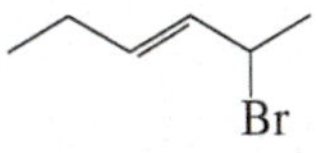 ~~~~=~Br

79. Would you expect methoxide ion to be a better nucleophile if it is dissolved in CH_3OH or if it is dissolved in DMSO? Why?

80. Which reaction in each of the following pairs takes place more rapidly?

a.

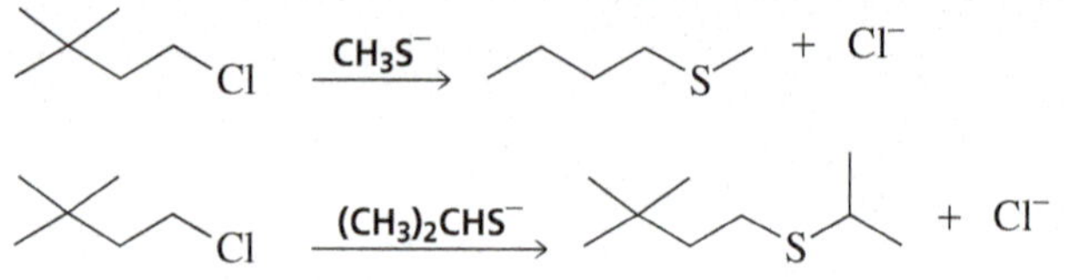

$\xrightarrow{CH_3S^-}$ $+\ Cl^-$

$\xrightarrow{(CH_3)_2CHS^-}$ $+\ Cl^-$

c.

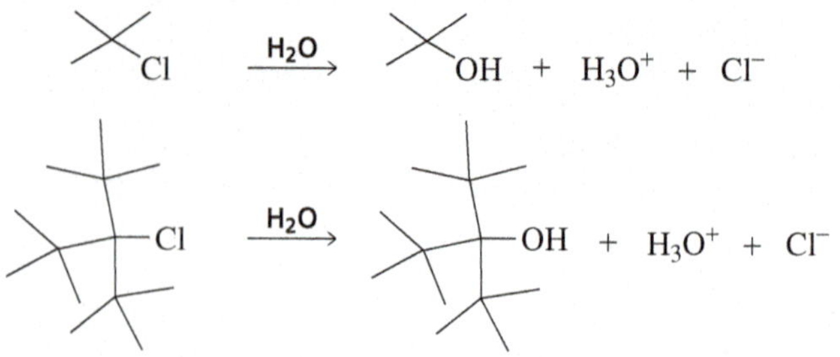

$\xrightarrow{H_2O}$ $OH\ +\ H_3O^+\ +\ Cl^-$

$-Cl\ \xrightarrow{H_2O}$ $-OH\ +\ H_3O^+\ +\ Cl^-$

b.

$Cl\ \xrightarrow{HO^-}$ $OH\ +\ Cl^-$

$O\ Cl\ \xrightarrow{HO^-}$ $O\ OH\ +\ Cl^-$

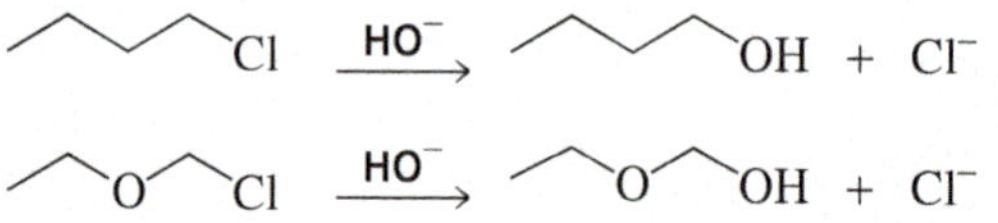

d. $(CH_3)_3CBr\ \xrightarrow{H_2O}\ (CH_3)_3COH\ +\ H_3O^+\ +\ Br^-$

$(CH_3)_3CBr\ \xrightarrow{CH_3CH_2OH}\ (CH_3)_3COCH_2CH_3\ +\ CH_3CH_2\overset{+}{O}H_2\ +\ Br^-$

81. Alkylbenzyldimethyl ammonium chloride is a leave-on skin antiseptic used to treat such things as cuts and cold sores. It is also the antiseptic in many hand sanitizers. It is actually a mixture of compounds that differ in the number of carbons (any even number between 8 and 18) in the alkyl group. Show three different sets of reagents (each set composed of an alkyl chloride and an amine) that can be used to synthesize the alkylbenzyldimethyl ammonium chloride shown here.

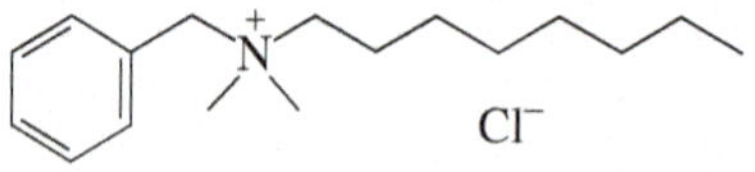

82. Fill in the squares in the following chemical equations:

a. $Br\ +\ \square\ \longrightarrow$ OCH_3

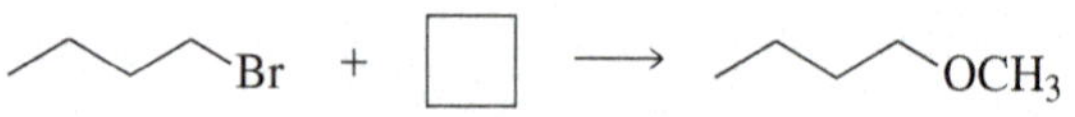

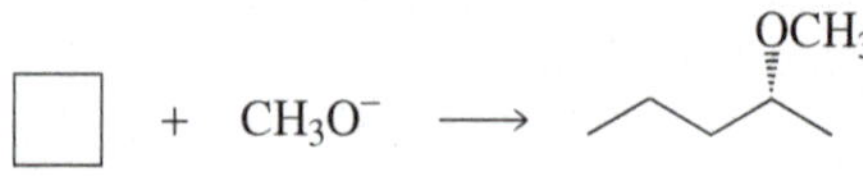

c. $+\ \square\ \longrightarrow$

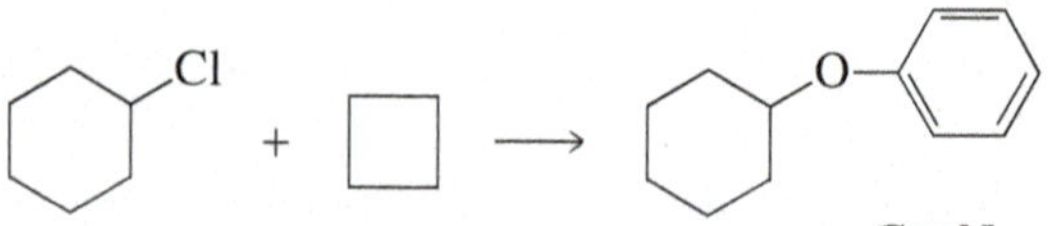

b. $\square\ +\ CH_3O^-\ \longrightarrow$

d. $\square\ +\ ^-C\equiv N\ \longrightarrow$

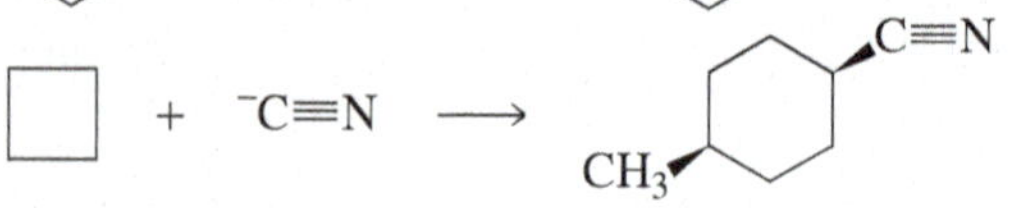

83. Draw the major product obtained when each of the following alkyl halides undergoes an E2 reaction:

a. $CH_3CHCH_2CH_3$ with Br

c. $CH_3CHCH_2CH_3$ with Cl

e. $CH_3CHCH_2CH_2CH_3$ with Cl

b. Cl

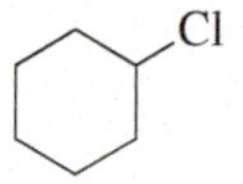

d. CH_2Cl

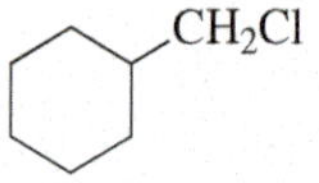

f. $CH_3\ Cl$

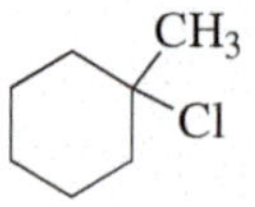

84. Draw the major product obtained when an alkyl halide in Problem 83 undergoes an E1 reaction.

85. a. Indicate how each of the following factors affects an E1 reaction:
   1. the strength of the base   **2.** the concentration of the base   **3.** the solvent
   b. Indicate how each of the same factors affects an E2 reaction.

86. Which species in each pair is more stable?

a. $CH_3\overset{-}{C}HCH_2CH_3$ or $CH_3CH_2CH_2\overset{-}{C}H_2$

e. $\overset{+}{C}H_2CH_2CH=CH_2$ or $CH_3\overset{+}{C}HCH=CH_2$

b. $CH_3\overset{+}{C}HCH_2CH_3$ or $CH_3CH_2CH_2\overset{+}{C}H_2$

f. $CH_3CH\overset{-}{C}HCH_3$ with CH_3 or $CH_3\overset{-}{C}CH_2CH_3$ with CH_3

c. $\overset{-}{C}H_2CH_2CH=CH_2$ or $CH_3\overset{-}{C}HCH=CH_2$

d. $CH_3CHCH=CH_2$ with CH_3 or $CH_3CH_2C=CH_2$ with CH_3

g. $CH_3CH\overset{+}{C}HCH_3$ with CH_3 or $CH_3\overset{+}{C}CH_2CH_3$ with CH_3

87. A chemist wanted to synthesize the anesthetic 2-ethoxy-2-methylpropane. He used ethoxide ion and 2-chloro-2-methylpropane for his synthesis and ended up with no ether. What was the product of his synthesis? What reagents should he have used?

88. Which reactant in each of the following pairs undergoes an elimination reaction more rapidly? Explain your choice.

a. $(CH_3)_3CCl \xrightarrow{CH_3O^-}$ or $(CH_3)_3CBr \xrightarrow{CH_3O^-}$

b. 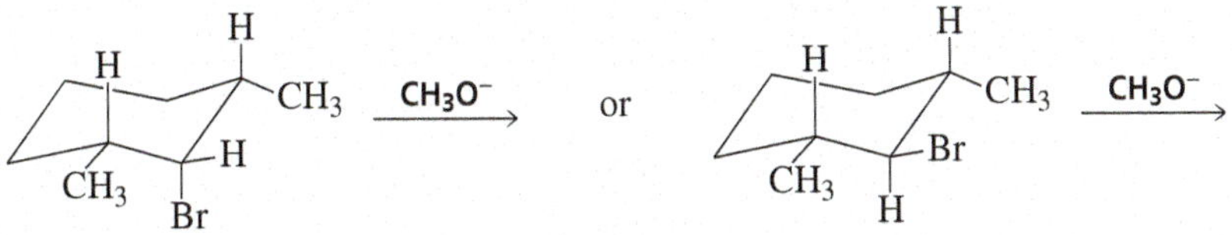 $\xrightarrow{CH_3O^-}$ or $\xrightarrow{CH_3O^-}$

89. For each of the following reactions, draw the major elimination product; if the product can exist as stereoisomers, indicate which stereoisomer is obtained in greater yield.
   a. (R)-2-bromohexane + high concentration of CH_3O^-
   b. (R)-3-bromo-3-methylhexane + CH_3OH
   c. trans-1-chloro-2-methylcyclohexane + high concentration of CH_3O^-
   d. trans-1-chloro-3-methylcyclohexane + high concentration of CH_3O^-
   e. 3-bromo-3-methylpentane + high concentration of $CH_3CH_2O^-$
   f. 3-bromo-3-methylpentane + CH_3CH_2OH

90. a. Which reacts faster in an E2 reaction: 3-bromocyclohexene or bromocyclohexane?
   b. Which reacts faster in an E1 reaction?

91. When the following reactions are carried out under the same conditions, the rate constant for the first reaction (k_H) is found to be 7 times greater than the rate constant for the second reaction (k_D). What does that tell you about the mechanism of the reaction? (*Hint*: a C—D bond is 1.2 kcal/mol stronger than a C—H bond.)

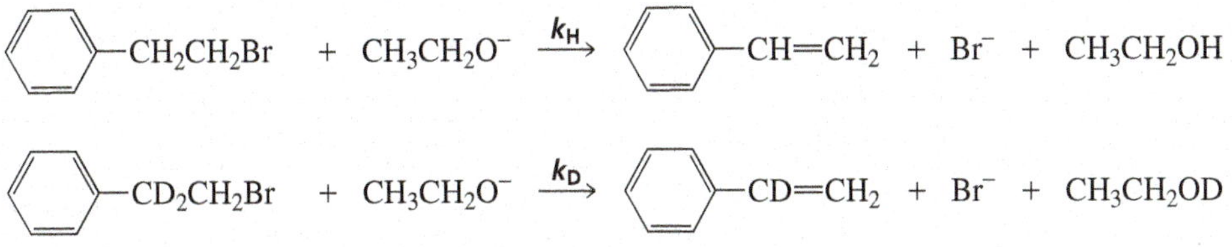

92. Starting with an alkyl halide, how could the following compounds be prepared?
   a. 2-methoxybutane   **b.** 1-methoxybutane   **c.** butylmethylamine

93. Indicate which species in each pair gives a higher substitution-product-to-elimination-product ratio when it reacts with isopropyl bromide:
   a. ethoxide ion or *tert*-butoxide ion   **b.** ^-OCN or ^-SCN **c.** Cl^- or Br^- **d.** CH_3S^- or CH_3O^-

94. Rank the following from most reactive to least reactive in an E2 reaction:

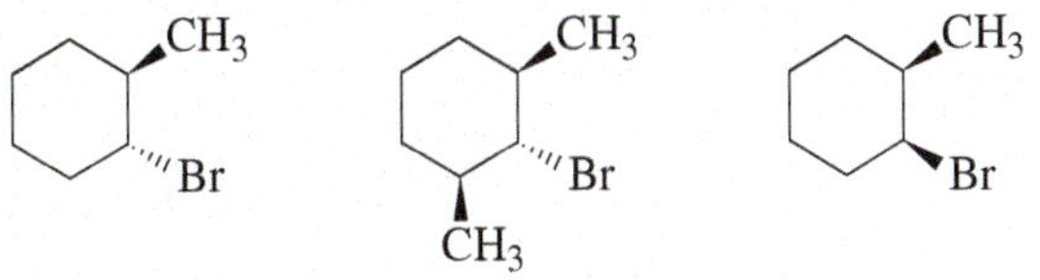

95. Rank the following from most reactive to least reactive in an S_N1 reaction.

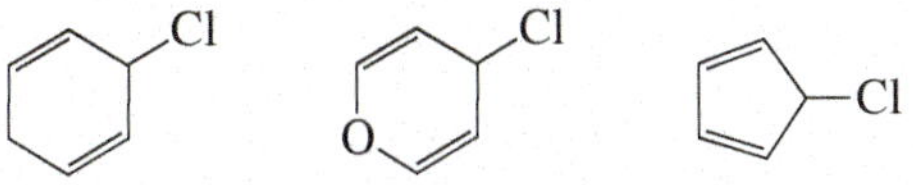

96. For each of the following alkyl halides, indicate the stereoisomer that would be obtained in greatest yield in an E2 reaction.
   a. 3-bromo-2,2,3-trimethylpentane   **c.** 3-bromo-2,3-dimethylpentane
   b. 4-bromo-2,2,3,3-tetramethylpentane   **d.** 3-bromo-3,4-dimethylhexane

97. Which of following ethers cannot be made by a Williamson ether synthesis?

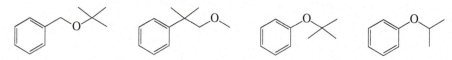

98. When 2-bromo-2,3-dimethylbutane reacts with a strong base, two alkenes (2,3-dimethyl-1-butene and 2,3-dimethyl-2-butene) are formed.
   a. Which of the bases (A, B, C, or D) would form the highest percentage of the 1-alkene?
   b. Which would give the highest percentage of the 2-alkene?

$$CH_3CO^- \quad CH_3CH_2CO^- \quad CH_3CH_2O^- \quad CH_3CH_2CO^-$$

A   B   C   D

99. What are the products of the following reactions?

a. ~~~~I $\xrightarrow{(CH_3CH_2)_3N}$

b. (Cl) $\xrightarrow{CH_3S^-}$

c. (Br, cyclohexane) $\xrightarrow{HO^-}$

d. (Br, cyclohexane) $\xrightarrow{HO^-}$

e. ~~~~Cl $\xrightarrow{CH_3O^-}$

f. (Br) $\xrightarrow{CH_3O^-}$

g. (cyclohexyl I) $\xrightarrow{}$ (−O−)

h. (phenyl-CH₂CH₂Cl) $\xrightarrow{CH_3O^-}$

100. **a.** Draw the structures of the products obtained from the reaction of each enantiomer of *cis*-1-chloro-2-isopropylcyclopentane with sodium methoxide in methanol.
   b. Are all the products optically active?
   c. How would the products differ if the starting material were the trans isomer? Are these products optically active?
   d. Will the cis enantiomers or the trans enantiomers form substitution products more rapidly?
   e. Will the cis enantiomers or the trans enantiomers form elimination products more rapidly?

101. When the following compound undergoes solvolysis in ethanol, three products are obtained. Propose a mechanism to account for the formation of these products.

(cyclohexene-CH₂Br) $\xrightarrow{CH_3CH_2OH}$ (=CH₂) + (=CH₂, OCH₂CH₃) + (CH₂OCH₂CH₃)

102. *cis*-1-Bromo-4-*tert*-butylcyclohexane and *trans*-1-bromo-4-*tert*-butylcyclohexane both react with sodium ethoxide in ethanol to form 4-*tert*-butylcyclohexene. Explain why the cis isomer reacts much more rapidly than the trans isomer.

103. Draw the substitution and elimination products.

a. ~~~~Br $\xrightarrow{CH_3O^-}$

b. ~~~(Br)~~~ $\xrightarrow{CH_3O^-}$

c. ~~~(Br)~~~ $\xrightarrow{CH_3O^-}$

d. ~~~(Br)~~~ $\xrightarrow{CH_3OH}$

e. ~~~(Cl)~~~ $\xrightarrow{CH_3O^-}$

f. ~~~(Cl)~~~ $\xrightarrow{CH_3OH}$

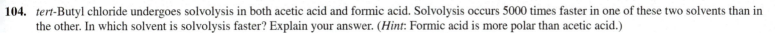

104. *tert*-Butyl chloride undergoes solvolysis in both acetic acid and formic acid. Solvolysis occurs 5000 times faster in one of these two solvents than in the other. In which solvent is solvolysis faster? Explain your answer. (*Hint*: Formic acid is more polar than acetic acid.)

$$\text{(CH}_3)_3\text{C—Cl} + \text{CH}_3\text{—C(=O)—OH} \longrightarrow \text{(CH}_3)_3\text{C—O—C(=O)—CH}_3 + \text{Cl}^-$$

$$\text{(CH}_3)_3\text{C—Cl} + \text{H—C(=O)—OH} \longrightarrow \text{(CH}_3)_3\text{C—O—C(=O)—H} + \text{Cl}^-$$

105. What substitution products are obtained when each of the following compounds is added to a solution of sodium acetate in acetic acid?
  a. 2-chloro-2-methyl-3-hexene   **b.** 3-bromo-1-methylcyclohexene

106. Show how each of the following compounds can be synthesized from the given starting materials:

  a. $CH_3CH_2CH_2Br \longrightarrow CH_3CH_2CH_2CH_2CH_3$ **c.** $CH_3CH_2CH_2Br \longrightarrow CH_3CH_2CH_2CH_2\overset{O}{\overset{\|}{C}}H$

  b. $CH_3CH_2CH_2Br \longrightarrow CH_3CH_2CH_2\overset{O}{\overset{\|}{C}}CH_3$ **d.** $CH_3CH_2CH_2Br \longrightarrow CH_3CH_2CH_2\overset{O}{\overset{/\backslash}{CH—CH_2}}$

107. In which solvent—ethanol or diethyl ether—would the equilibrium for the following S_N2 reaction lie farther to the right?

$$CH_3SCH_3 + CH_3Br \rightleftharpoons CH_3\overset{CH_3}{\underset{+}{S}}CH_3 + Br^-$$

108. The rate of the reaction of methyl iodide with quinuclidine was measured in nitrobenzene, and then the rate of the reaction of methyl iodide with triethylamine was measured in the same solvent. The concentration of the reagents was the same in both experiments.
  a. Which reaction had the larger rate constant?
  b. The same experiment was done using isopropyl iodide instead of methyl iodide. Which reaction had the larger rate constant?
  c. Which alkyl halide has the larger $k_{quinuclidine}/k_{triethylamine}$ ratio?

quinuclidine $CH_3CH_2\overset{CH_2CH_3}{\overset{|}{N}}CH_2CH_3$
                                triethylamine

109. Two bromoethers are obtained from the reaction of the following alkyl dihalide with methanol. Draw the structures of the ethers.

110. Draw the substitution products for each of the following S_N2 reactions. If the products can exist as stereoisomers, show which stereoisomers are formed:
  a. (3S,4S)-3-bromo-4-methylhexane + CH_3O^- **c.** (3R,4R)-3-bromo-4-methylhexane + CH_3O^-
  b. (3S,4R)-3-bromo-4-methylhexane + CH_3O^- **d.** (3R,4S)-3-bromo-4-methylhexane + CH_3O^-

111. Draw the elimination products for each of the following E2 reactions; if the products can exist as stereoisomers, indicate which stereoisomers are obtained.
  a. (2S,3S)-2-chloro-3-methylpentane + high concentration of CH_3O^-
  b. (2S,3R)-2-chloro-3-methylpentane + high concentration of CH_3O^-
  c. (2R,3S)-2-chloro-3-methylpentane + high concentration of CH_3O^-
  d. (2R,3R)-2-chloro-3-methylpentane + high concentration of CH_3O^-
  e. 3-chloro-3-ethyl-2,2-dimethylpentane + high concentration of $CH_3CH_2O^-$

112. Draw the major elimination product that would be obtained from each of the following reactants with a strong base and with a weak base:

  a.   **b.**

113. Which of the following hexachlorocyclohexanes is the least reactive in an E2 reaction?

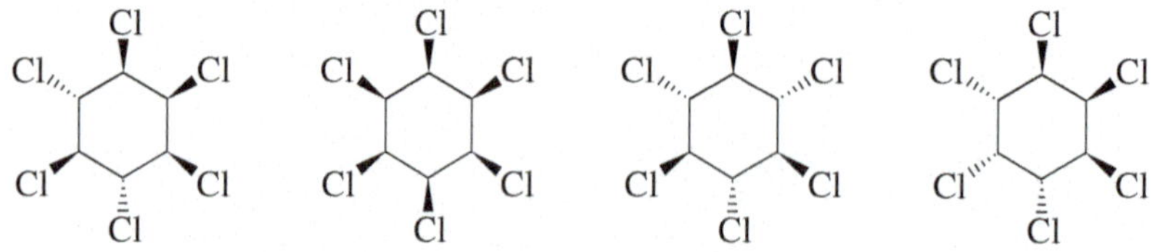

114. Explain why the rate of the reaction of 1-bromo-2-butene with ethanol is increased if silver nitrate is added to the reaction mixture.

115. Draw the products of each of the following S_N2/E2 reactions. If the products can exist as stereoisomers, show which stereoisomers are formed.
   a. (3S,4S)-3-bromo-4-methylhexane + CH_3O^-
   b. (3R,4R)-3-bromo-4-methylhexane + CH_3O^-
   c. (3S,4R)-3-bromo-4-methylhexane + CH_3O^-
   d. (3R,4S)-3-bromo-4-methylhexane + CH_3O^-

116. Two elimination products are obtained from the following E2 reaction:

$$CH_3CH_2CHDCH_2Br \xrightarrow{CH_3O^-}$$

   a. What are the elimination products?
   b. Which is formed in greater yield? (*Hint:* See Problem 91.)

117. Draw the structures of the products obtained from the following reaction:

$+ \ CH_3OH \longrightarrow$

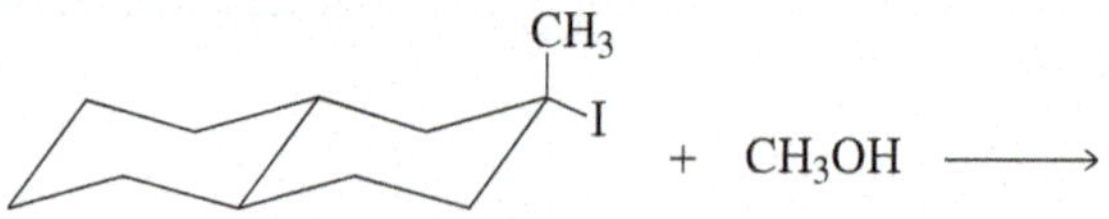

118. How could you prepare the following compounds from the given starting materials?

   a. $CH_3CH_2CH_2CH_2Br \longrightarrow CH_3CH_2\overset{\overset{\displaystyle O}{\|}}{C}CH_2CH_2CH_3$

   b. $BrCH_2CH_2CH_2CH_2Br \longrightarrow$

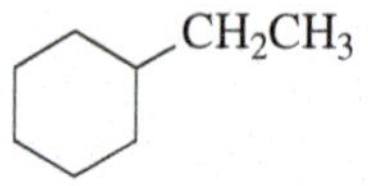

119. *cis*-4-Bromocyclohexanol and *trans*-4-bromocyclohexanol form the same elimination product but a different substitution product when they react with HO^-.

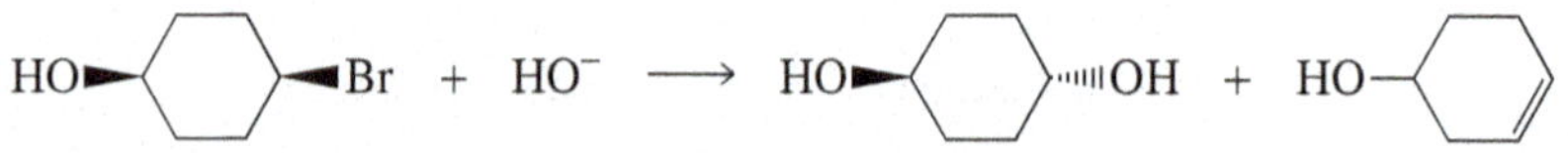

cis-4-bromocyclohexanol

trans-4-bromocyclohexanol

   a. Why do they form the same elimination product?
   b. Explain, by showing the mechanisms, why different substitution products are obtained.
   c. How many stereoisomers does each of the elimination and substitution reactions form?

120. a. Draw the product of each of the following reactions:

   1. $CH_3CH_2C{\equiv}CH \xrightarrow[\text{2. } CH_3CH_2Br]{\text{1. NaNH}_2}$

   2. $CH_3CH_2CH_2OH \xrightarrow[\text{2. } CH_3Br]{\text{1. NaH}}$

   b. Give two sets of reactants (each set including an alkyl halide and a nucleophile) that could be used to synthesize the following alkyne:

$$CH_3CH_2C{\equiv}CCH_2CH_2CH_2CH_3$$

   c. Give two sets of reactants (each set including an alkyl halide and a nucleophile) that could be used to synthesize the following ether:

$$CH_3CH_2OCH_2CH_2\underset{\underset{\displaystyle CH_3}{|}}{C}HCH_3$$

456

121. Propose a mechanism for the following reaction:

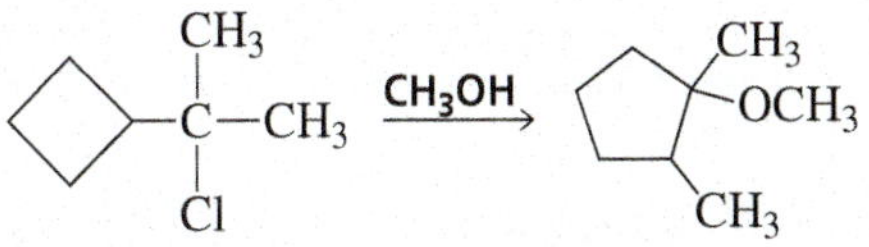

122. Explain why tetrahydrofuran can solvate a positively charged species better than diethyl ether can.

tetrahydrofuran   diethyl ether

123. The reaction of an alkyl chloride with potassium iodide is generally carried out in acetone to maximize the amount of alkyl iodide that is formed. Why does the solvent increase the yield of alkyl iodide? (*Hint:* Potassium iodide is soluble in acetone, but potassium chloride is not.)

124 **a.** Propose a mechanism for the following reaction.
    b. Explain why two products are formed.
    c. Explain why methanol substitutes for only one of the bromines.

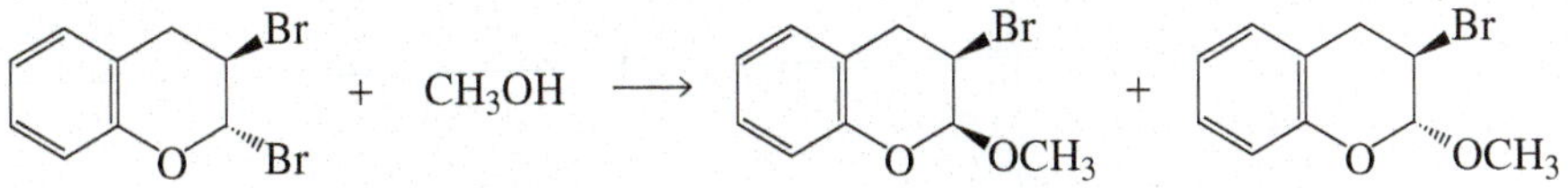

125. What products are formed when the following stereoisomer of 2-chloro-1,3-dimethylcyclohexane reacts with methoxide ion?

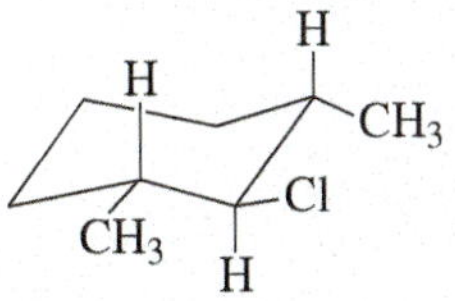

126. Predict the product for the following reaction and write a mechanism to explain how it is formed.

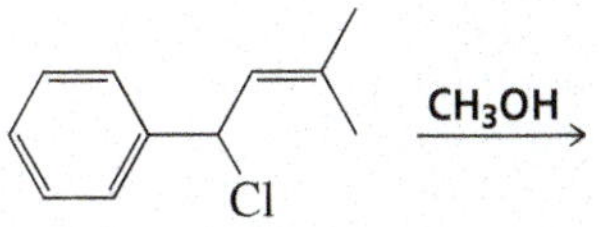

127. For each of the following compounds, draw the product that forms in an E2 reaction and indicate its configuration:
    a. (1*S*,2*S*)-1-bromo-1,2-diphenylpropane   **b.** (1*S*,2*R*)-1-bromo-1,2-diphenylpropane

128. Chlordane, like DDT, is an alkyl halide that was used as an insecticide for crops such as corn and citrus and for lawns. In 1983, it was banned for all uses except against termites, and in 1988, it was banned for use against termites as well. Chlordane can be synthesized from two reactants in one step. One of the reactants is hexachlorocyclopentadiene. What is the other reactant? (*Hint:* See Section 8.14.)

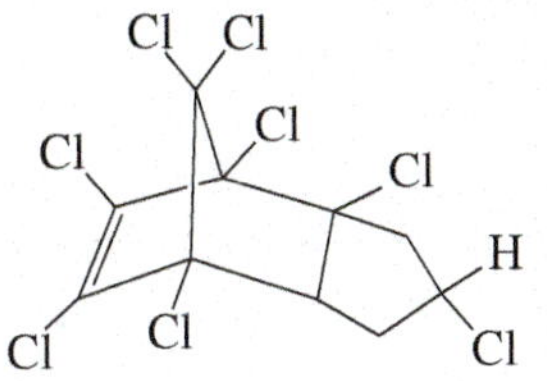

chlordane

129. When equivalent amounts of methyl bromide and sodium iodide are dissolved in methanol, the concentration of iodide ion quickly decreases and then slowly returns to its original concentration. Account for this observation.

130. Explain why the following alkyl halide does not undergo a substitution reaction, regardless of the base that is used.

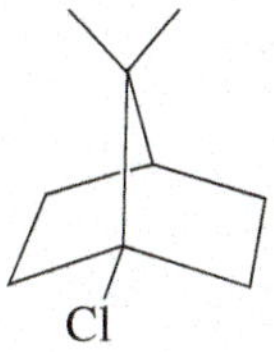

131. The reaction of chloromethane with hydroxide ion at 30 °C has a $\Delta G°$ value of -21.7 kcal/mol. What is the equilibrium constant for the reaction?

10 Reactions of Alcohols, Ethers, Epoxides, Amines, and Sulfur-Containing Compounds

dried coca leaves

Group II

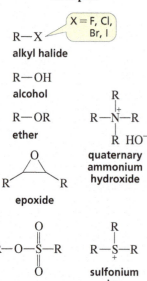

R—X   X = F, Cl, Br, I
alkyl halide

R—OH
alcohol

R—OR
ether

R—N⁺—R with R groups and HO⁻
quaternary ammonium hydroxide

epoxide (O bridging R and R)
epoxide

sulfonate ester (R—O—S(=O)(=O)—R)
sulfonate ester

R—S⁺—R
sulfonium ion

Chemists search the world for plants and berries and the ocean for flora and fauna that might be used as the source of a lead compound for the development of a new drug. In this chapter, we will see how cocaine, which is obtained from the leaves of Erythroxylon coca—a bush native to the highlands of the South American Andes, was used as the source of a lead compound for the development of some common anesthetics (see page 491).

We saw that alkyl halides, a family of compounds in Group II, undergo substitution and/or elimination reactions because of their electron-withdrawing halogen atoms (Chapter 9). Other families of compounds in Group II also have electron-withdrawing groups, and they too undergo substitution and/or elimination reactions. The relative reactivity of these compounds depends on the electron-withdrawing group—that is, on the leaving group.

Reactivity Depends on the Basicity of the Leaving Group

- The leaving groups of alcohols and ethers (HO^-, RO^-) are much stronger bases than the leaving group of an alkyl halide. Because they are stronger bases, they are poorer leaving groups and, therefore, are harder to displace. Consequently, alcohols and ethers are less reactive than alkyl halides in substitution and elimination reactions. We will see that alcohols and ethers must be "activated" before they can undergo a substitution or elimination reaction.

- The leaving group of an amine ($^-NH_2$) is so basic that amines cannot undergo substitution and elimination reactions but they are important bases and nucleophiles.

- Tertiary amines, the leaving groups of **quaternary ammonium ions,** not only are less basic than the leaving groups of alcohols and ethers but also have a positive charge that enhances their leaving ability. Therefore, quaternary ammonium ions undergo elimination reactions as long as a strong base is present and the reaction is heated.

- **Sulfonate esters** and **sulfonium ions** have very good leaving groups, the first because of electron delocalization and the second because of its positive charge. Thus, they undergo substitution and/or elimination reactions with ease.

> When bases with similar features are compared:
> the weaker the base,
> the more easily it is displaced.
> Recall that the stronger the acid,
> the weaker its conjugate base.

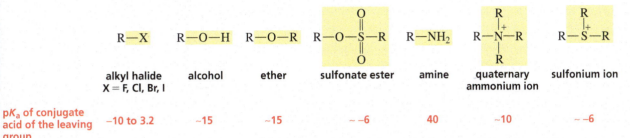

	R—X	R—O—H	R—O—R	R—O—S—R	R—NH₂	R—N⁺—R	R—S⁺—R
	alkyl halide X = F, Cl, Br, I	alcohol	ether	sulfonate ester	amine	quaternary ammonium ion	sulfonium ion
pK_a of conjugate acid of the leaving group	−10 to 3.2	~15	~15	~ −6	40	~10	~ −6

10.1 NUCLEOPHILIC SUBSTITUTION REACTIONS OF ALCOHOLS: FORMING ALKYL HALIDES

An **alcohol** has a strongly basic leaving group (HO⁻) that cannot be displaced by a nucleophile. Therefore, an alcohol cannot undergo a nucleophilic substitution reaction.

$$CH_3-\ddot{O}H + Br^- \;\;\cancel{\longrightarrow}\;\; CH_3-Br + HO^-$$

(a strongly basic leaving group)

strong base

However, if the alcohol's OH group is converted into a group that is a weaker base (and, therefore, a better leaving group), a nucleophilic substitution reaction can occur.

Converting an OH Group into a Better Leaving Group

One way to convert an OH group into a weaker base is to protonate it by adding acid to the reaction mixture. Protonation changes the leaving group from HO⁻ to H₂O, which is a weak enough base to be displaced by a nucleophile. The substitution reaction is slow and requires heat (except in the case of tertiary alcohols) if it is to take place at a reasonable rate.

$$CH_3-\ddot{O}H + HBr \rightleftharpoons CH_3-\overset{+}{\underset{Br^-}{O}}H \overset{\Delta}{\longrightarrow} CH_3-Br + H_2O$$

(poor leaving group)   (good leaving group)   (a weakly basic leaving group)   **weak base**

Because the OH group of the alcohol must be protonated before it can be displaced by a nucleophile, only weakly basic nucleophiles (I⁻, Br⁻, Cl⁻) can be used in the substitution reaction. Moderately and strongly basic nucleophiles (NH₃, RNH₂, and CH₃O⁻) cannot be used because they too would be protonated in the acidic solution and, once protonated, would no longer be nucleophiles (⁺NH₄, RN⁺H₃) or would be poor nucleophiles (CH₃OH).

PROBLEM 1 ◆

Why are NH₃ and CH₃NH₂ no longer nucleophiles when they are protonated?

Primary, secondary, and tertiary alcohols all undergo nucleophilic substitution reactions with HI, HBr, and HCl to form alkyl halides.

$$CH_3CH_2CH_2OH + HI \xrightarrow{\Delta} CH_3CH_2CH_2I + H_2O$$
primary alcohol

secondary alcohol

$$\underset{\overset{|}{OH}}{\overset{\overset{CH_3}{|}}{CH_3CCH_2CH_3}} + HBr \longrightarrow \underset{\overset{|}{Br}}{\overset{\overset{CH_3}{|}}{CH_3CCH_2CH_3}} + H_2O$$
tertiary alcohol

The S$_N$1 Reaction of Secondary and Tertiary Alcohols

The mechanism of the substitution reaction depends on the structure of the alcohol. Secondary and tertiary alcohols undergo S$_N$1 reactions.

MECHANISM FOR THE S$_N$1 REACTION OF AN ALCOHOL

An acid protonates the most basic atom in a molecule.

- An acid always reacts with an organic molecule in the same way: it protonates the most basic atom in the molecule.

- Weakly basic water is the leaving group that is expelled, forming a carbocation.

- The carbocation, like the carbocation formed when an alkyl halide dissociates in an S$_N$1 reaction, has two possible fates: it can combine with a nucleophile and form a substitution product, or it can lose a proton and form an elimination product (Section 9.12).

Although the reaction can form both a substitution product and an elimination product, little elimination product is actually obtained because the alkene formed in an elimination reaction can undergo a subsequent electrophilic addition reaction with HBr to form more of the substitution product (Section 6.1).

Tertiary alcohols undergo substitution reactions with hydrogen halides faster than secondary alcohols do because tertiary carbocations are more stable and, therefore, are formed more rapidly than secondary carbocations. (Recall that alkyl groups stabilize carbocations by hyperconjugation; Section 6.2) As a result, the reaction of a tertiary alcohol with a hydrogen halide proceeds readily at room temperature, whereas the reaction of a secondary alcohol with a hydrogen halide must be heated to have the reaction occur at a reasonable rate.

The S_N2 Reaction of Primary Alcohols

Primary alcohols cannot undergo S_N1 reactions because primary carbocations are too unstable to be formed, even when the reaction is heated (Section 9.3). Therefore, when a primary alcohol reacts with a hydrogen halide, it must do so in an S_N2 reaction.

MECHANISM FOR THE S_N2 REACTION OF AN ALCOHOL

- The acid protonates the most basic atom in the reactant.
- The nucleophile attacks the back side of the carbon and displaces the leaving group.

Only a substitution product is obtained. No elimination product is formed because the halide ion, although a good nucleophile, is a weak base in a reaction mixture that contains alcohol and water (that is, in a polar protic solvent). Recall that a strong base is required to remove a hydrogen from a β-carbon in an E2 reaction (Section 9.9).

When HCl is used instead of HBr or HI, the S_N2 reaction is slower because Cl^- is a poorer nucleophile than Br^- or I^- (Section 9.2). However, the rate of the reaction can be increased if $ZnCl_2$ is used as a catalyst.

$$CH_3CH_2CH_2OH + HCl \xrightarrow[\Delta]{ZnCl_2} CH_3CH_2CH_2Cl + H_2O$$

$ZnCl_2$ is a Lewis acid that complexes strongly with oxygen's lone-pair electrons. This interaction weakens the C—O bond, thereby creating a better leaving group than water.

Carbocation stability: $3° > 2° > 1°$

Secondary and tertiary alcohols undergo S_N1 reactions with hydrogen halides.

Primary alcohols undergo S_N2 reactions with hydrogen halides.

The Lucas Test

Before spectroscopy became available for structural analysis, chemists had to identify compounds by tests that gave visible results. The Lucas test is one such test. It determines whether an alcohol is primary, secondary, or tertiary by taking advantage of the relative rates at which the three classes of alcohols react with $HCl/ZnCl_2$.

To carry out the test, the alcohol is added to a mixture of HCl and $ZnCl_2$ (known as Lucas reagent). Low-molecular-weight alcohols are soluble in Lucas reagent, but the alkyl halide products are not, so they cause the solution to turn cloudy. If the alcohol is tertiary, the solution turns cloudy immediately. If the alcohol is secondary, the solution turns cloudy in approximately one to five minutes. If the alcohol is primary, the solution turns cloudy only if it is heated. Because the test relies on the complete solubility of the alcohol in Lucas reagent, it is limited to alcohols with fewer than six carbons.

PROBLEM 2 SOLVED   **LEARN THE STRATEGY**

Using the pK_a values of the conjugate acids of the leaving groups (the pK_a of HBr is −9, and the pK_a of H_2O is 15.7), explain the difference in reactivity between CH_3Br and CH_3OH in a nucleophilic substitution reaction.

SOLUTION The conjugate acid of the leaving group of CH_3Br is HBr; the conjugate acid of the leaving group of CH_3OH is H_2O. Because HBr is a much stronger acid ($pK_a = -9$) than H_2O ($pK_a = 15.7$), Br^- is a much weaker base than HO^-. (Recall that the stronger the acid, the weaker its conjugate base.) Therefore, Br^- is a much better leaving group than HO^-, causing CH_3Br to be much more reactive than CH_3OH in a nucleophilic substitution reaction.

USE THE STRATEGY

PROBLEM 3

Explain the difference in reactivity between $CH_3\overset{+}{O}H_2$ and CH_3OH in a nucleophilic substitution reaction. (The pK_a of H_3O^+ is −1.7.)

LEARN THE STRATEGY

PROBLEM 4 SOLVED

Show how 1-butanol can be converted into the following compound:

SOLUTION Because the OH group of 1-butanol is too basic to allow the alcohol to undergo a nucleophilic substitution reaction with CH_3O^-, the alcohol must first be converted into an alkyl halide. The alkyl halide has a leaving group that can be substituted by CH_3O^-, the nucleophile required to obtain the desired product.

$$\text{OH} \xrightarrow[\Delta]{\textbf{HBr}} \text{Br} \xrightarrow{\textbf{CH}_3\text{O}^-} \text{O}$$

USE THE STRATEGY

PROBLEM 5

Explain how 1-butanol can be converted into the following compounds:

a. b. c.

Because the reaction of a secondary alcohol with a hydrogen halide is an S_N1 reaction, a carbocation is formed as an intermediate. Therefore, we must check for the possibility of a carbocation rearrangement when determining the product of the substitution reaction. Remember that a carbocation rearrangement will occur if it leads to formation of a more stable carbocation (Section 6.7).

For example, the major product of the following reaction is 2-bromo-2-methylbutane, because a 1,2-hydride shift converts the initially formed secondary carbocation into a more stable tertiary carbocation.

$$\begin{array}{c} CH_3 \\ | \\ CH_3CHCHCH_3 \\ | \\ OH \end{array} + \text{HBr} \rightleftharpoons \begin{array}{c} CH_3 \\ | \\ CH_3CHCHCH_3 \\ | \\ \overset{+}{O}H \\ H \end{array} \rightleftharpoons \begin{array}{c} CH_3 \\ | \\ CH_3CH\overset{+}{C}HCH_3 \\ \end{array} + H_2O \quad \xrightarrow[\text{shift}]{\textbf{1,2-hydride}} \begin{array}{c} CH_3 \\ | \\ CH_3\overset{+}{C}CH_2CH_3 \\ \end{array}$$

secondary carbocation   tertiary carbocation

$\downarrow Br^-$ $\downarrow Br^-$

$$\begin{array}{c} CH_3 \\ | \\ CH_3CHCHCH_3 \\ | \\ Br \end{array} \qquad\qquad \begin{array}{c} CH_3 \\ | \\ CH_3CCH_2CH_3 \\ | \\ Br \end{array}$$

2-bromo-3-methyl-
butane
minor product

2-bromo-2-methyl-
butane
major product

Grain Alcohol and Wood Alcohol

When ethanol is ingested, it acts on the central nervous system. Moderate amounts affect judgment and lower inhibitions. Higher amounts interfere with motor coordination and cause slurred speech and amnesia. Still higher amounts cause nausea and loss of consciousness. Ingesting very large amounts of ethanol interferes with spontaneous respiration and can be fatal.

The ethanol in alcoholic beverages is produced by the fermentation of glucose, generally obtained from grapes or from grains such as corn, rye, and wheat (which is why ethanol is also known as grain alcohol). Grains are cooked in the presence of malt (sprouted barley) to convert much of their starch into glucose. Yeast enzymes are added to convert the glucose into ethanol and carbon dioxide (Section 24.7).

$$\underset{\text{glucose}}{C_6H_{12}O_6} \xrightarrow{\textbf{yeast enzymes}} 2\ \underset{\text{ethanol}}{CH_3CH_2OH} + 2\ CO_2$$

The kind of beverage produced (white or red wine, beer, scotch, bourbon, champagne) depends on what plant species provides the glucose, whether the CO_2 formed in the fermentation is allowed to escape, whether other substances are added, and how the beverage is purified (by sedimentation, for wines; by distillation, for scotch and bourbon).

The tax imposed on liquor would make ethanol a prohibitively expensive laboratory reagent. Laboratory alcohol, therefore, is not taxed because ethanol is needed in a wide variety of commercial processes. Although not taxed, it is carefully regulated by the federal government to make certain that it is not used for the preparation of alcoholic beverages. Denatured alcohol—ethanol that has been made undrinkable by the addition of a denaturant such as benzene or methanol—is not taxed, but the added impurities make it unfit for many laboratory uses.

Methanol, also known as wood alcohol (because at one time it was obtained by heating wood in the absence of oxygen), is highly toxic. Ingesting even very small amounts can cause blindness, and ingesting as little as an ounce has been fatal. (See Methanol Poisoning on p. 477.)

PROBLEM 6 ◆

What is the major product of each of the following reactions?

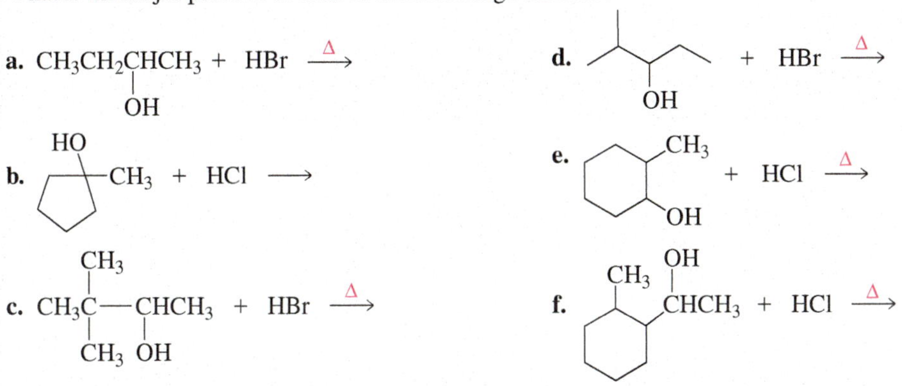

a. $CH_3CH_2\underset{\underset{OH}{|}}{C}HCH_3$ + HBr $\xrightarrow{\Delta}$

b. —CH_3 + HCl $\longrightarrow$

c. $CH_3\underset{\underset{CH_3}{|}}{\overset{\overset{CH_3}{|}}{C}}-\underset{\underset{OH}{|}}{C}HCH_3$ + HBr $\xrightarrow{\Delta}$

d. + HBr $\xrightarrow{\Delta}$

e. + HCl $\xrightarrow{\Delta}$

f. + HCl $\xrightarrow{\Delta}$

PROBLEM 7 SOLVED

LEARN THE STRATEGY

What stereoisomers does the following reaction form?

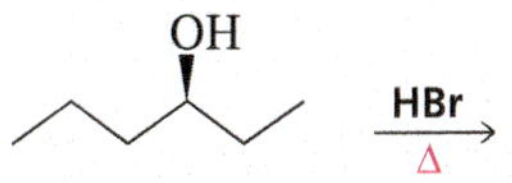

SOLUTION This must be an S_N1 reaction because the reactant is a secondary alcohol. Therefore, the reaction forms a carbocation intermediate. The bromide ion can attach to the carbocation from either the side from which water left or from the opposite side, so both the R and S stereoisomers are formed.

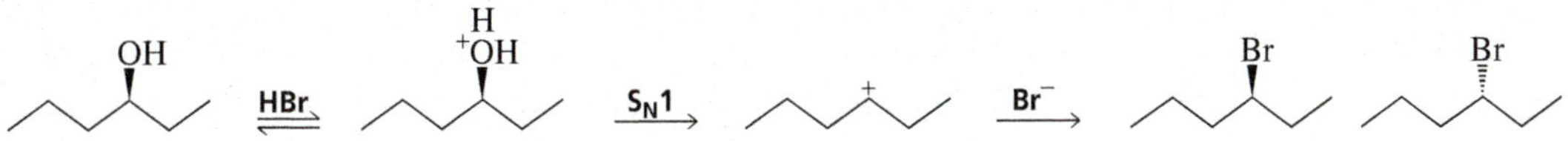

PROBLEM 8 ◆

USE THE STRATEGY

What stereoisomers does the following reaction form?

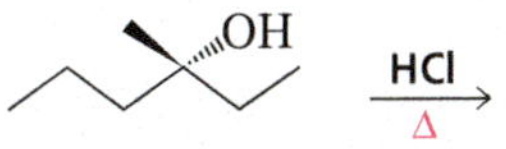

10.2 OTHER METHODS USED TO CONVERT ALCOHOLS INTO ALKYL HALIDES

Alcohols are inexpensive and readily available compounds, but they do not undergo nucleophilic substitution because the HO^- group is too basic to be displaced by a nucleophile (Section 10.1). Chemists, therefore, need ways to convert readily available but unreactive alcohols into reactive alkyl halides that can be used as starting materials for the preparation of a wide variety of organic compounds (Section 9.2).

$$R-OH \xrightarrow[\Delta]{HX} R-X$$

alcohol → alkyl halide (X = Cl, Br, I), which reacts with ⁻OCH₃ to give R—OCH₃, with ⁻C≡CCH₃ to give R—C≡CCH₃, and with ⁻C≡N to give R—C≡N

We just saw that an alcohol can be converted to an alkyl halide by treating it with a hydrogen halide. Better yields of the alkyl halide are obtained and carbocation rearrangements can be avoided if a phosphorus trihalide (PCl₃ or PBr₃) or thionyl chloride (SOCl₂) is used instead.

$$CH_3CH_2OH + PBr_3 \xrightarrow{\text{pyridine}} CH_3CH_2Br$$

$$CH_3CH_2OH + PCl_3 \xrightarrow{\text{pyridine}} CH_3CH_2Cl$$

$$CH_3CH_2OH + SOCl_2 \xrightarrow{\text{pyridine}} CH_3CH_2Cl$$

These reagents all react in the same way: they convert the alcohol into an intermediate that has a better leaving group than a halide ion, so the halide ion can displace it.

MECHANISM FOR THE CONVERSION OF AN ALCOHOL TO AN ALKYL BROMIDE (OR ALKYL CHLORIDE) USING PBr₃ (OR PCl₃)

$$CH_3CH_2-\ddot{O}H + P-Br \rightleftharpoons CH_3CH_2-\overset{+}{\ddot{O}}-PBr_2 \rightleftharpoons CH_3CH_2-\ddot{O}-PBr_2 \longrightarrow CH_3CH_2Br$$

phosphorus tribromide → a bromophosphite group + ⁻O—PBr₂

pyridine

- The first step is an S_N2 reaction on phosphorus.
- Pyridine is generally used as a solvent in these reactions because it is a poor nucleophile. However, it is sufficiently basic to remove a proton from the intermediate, which prevents the intermediate from reverting to starting materials.
- The bromophosphite group is a weaker base than a halide ion, so it is easily displaced by a halide ion.

MECHANISM FOR THE CONVERSION OF AN ALCOHOL TO AN ALKYL CHLORIDE USING SOCl₂

$$CH_3-\ddot{O}H + Cl-\overset{O}{\underset{\parallel}{S}}-Cl \rightleftharpoons CH_3-\overset{+}{\ddot{O}}-\overset{O}{\underset{\parallel}{S}}-Cl \rightleftharpoons CH_3-\ddot{O}-\overset{O}{\underset{\parallel}{S}}-Cl \longrightarrow CH_3Cl + SO_2$$

thionyl chloride → a chlorosulfite group + Cl⁻

- The first step is an S_N2 reaction on sulfur.
- Pyridine removes a proton from the intermediate, which prevents the intermediate from reverting to starting materials.
- The chlorosulfite group is a weaker base than a chloride ion, so it is easily displaced by a chloride ion.

The foregoing reactions work well for primary and secondary alcohols, but tertiary alcohols give poor yields because the intermediate formed by a tertiary alcohol is sterically hindered to back-side attack by the halide ion.

Table 10.1 summarizes some of the methods commonly used to convert alcohols into alkyl halides.

Table 10.1   Commonly Used Methods for Converting Alcohols into Alkyl Halides

$$\text{ROH} + \text{HBr} \xrightarrow{\Delta} \text{RBr}$$

$$\text{ROH} + \text{HI} \xrightarrow{\Delta} \text{RI}$$

$$\text{ROH} + \text{HCl} \xrightarrow{\text{ZnCl}_2 / \Delta} \text{RCl}$$

$$\text{ROH} + \text{PBr}_3 \xrightarrow{\text{pyridine}} \text{RBr}$$

$$\text{ROH} + \text{PCl}_3 \xrightarrow{\text{pyridine}} \text{RCl}$$

$$\text{ROH} + \text{SOCl}_2 \xrightarrow{\text{pyridine}} \text{RCl}$$

PROBLEM 9

What stereoisomers do the following reactions form?

a. 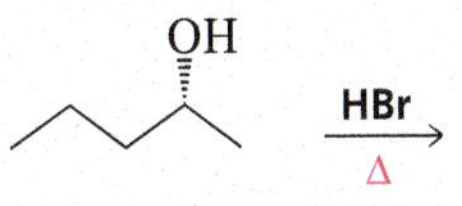 $\xrightarrow[\Delta]{\text{HBr}}$

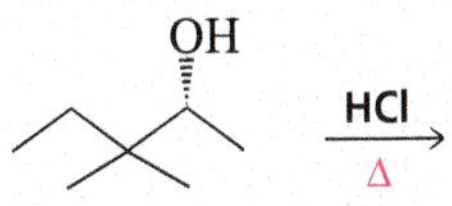

b. 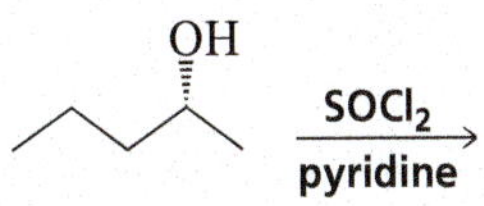 $\xrightarrow[\text{pyridine}]{\text{SOCl}_2}$

c. $\xrightarrow[\Delta]{\text{HCl}}$

d. 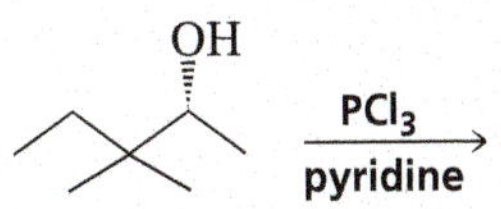 $\xrightarrow[\text{pyridine}]{\text{PCl}_3}$

10.3 CONVERTING AN ALCOHOL INTO A SULFONATE ESTER

Another way a primary or secondary alcohol can be activated for a subsequent reaction with a nucleophile—instead of converting it into an alkyl halide—is to convert it into a sulfonate ester.

Forming a Sulfonate Ester

A **sulfonate ester** is formed when an alcohol reacts with a sulfonyl chloride. (Notice that sulfur, which is in the third row of the periodic table, has an expanded valence shell—that is, it is surrounded by 12 electrons.)

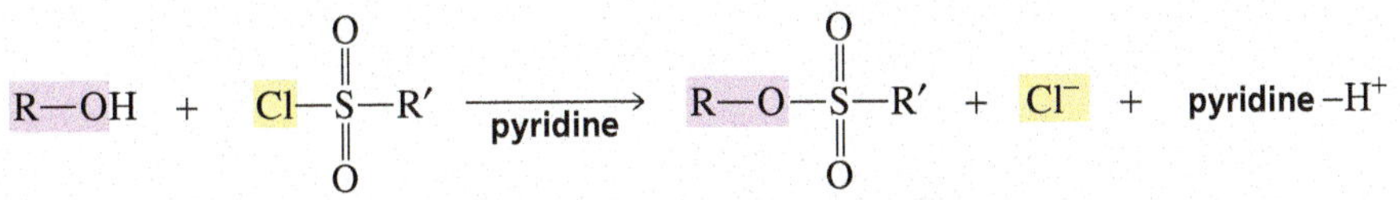

$$\text{R—OH} + \underset{\text{a sulfonyl chloride}}{\text{Cl—S(=O)}_2\text{—R}'} \xrightarrow{\text{pyridine}} \underset{\text{a sulfonate ester}}{\text{R—O—S(=O)}_2\text{—R}'} + \text{Cl}^- + \text{pyridine—H}^+$$

The reaction is a nucleophilic substitution reaction in which the alcohol displaces the chloride ion. Pyridine is the solvent and it is the base that removes a proton from the intermediate.

MECHANISM FOR THE CONVERSION OF AN ALCOHOL TO A SULFONATE ESTER

$$R-\ddot{O}H \quad + \quad Cl-\overset{\overset{O}{\|}}{\underset{\underset{O}{\|}}{S}}-R' \quad \rightleftharpoons \quad R-\overset{+}{\ddot{O}}-\overset{\overset{O}{\|}}{\underset{\underset{O}{\|}}{S}}-R' \quad \rightleftharpoons \quad R-O-\overset{\overset{O}{\|}}{\underset{\underset{O}{\|}}{S}}-R' \quad + \quad Cl^-$$

- The first step is an S_N2 reaction on sulfur.
- Pyridine removes a proton from the intermediate, which prevents the intermediate from reverting to starting materials.

Several sulfonyl chlorides are available to activate OH groups. The most common one is *para*-toluenesulfonyl chloride (abbreviated as TsCl). The sulfonate ester formed from the reaction of TsCl and an alcohol is called an **alkyl tosylate** (abbreviated as ROTs).

para-toluenesulfonyl
chloride
TsCl

methanesulfonyl
chloride

trifluoromethanesulfonyl
chloride

Sulfonate Esters in Substitution Reactions

Once the alcohol has been activated by being converted into a sulfonate ester, the appropriate nucleophile is added. (Notice that in all cases, the added nucleophile is a *much* better nucleophile than the chloride ion that also is present in the solution, because it was the leaving group in the synthesis of the sulfonate ester.) The reactions are S_N2 reactions and take place readily at room temperature because the sulfonate ester has an excellent leaving group. Sulfonate esters react with a wide variety of nucleophiles, so they can be used to synthesize a wide variety of compounds.

This substitution reaction will not take place if the alcohol that forms the sulfonate ester is tertiary because it would be too sterically hindered to undergo the subsequent S_N2 reaction. (Recall that tertiary alkyl halides cannot undergo S_N2 reactions; Section 9.2.)

A sulfonate ester has an excellent leaving group. A sulfonic acid is a very strong acid ($pK_a = -6.5$) because its conjugate base is particularly stable (weak) due to delocalization of its negative charge over three oxygens. (Recall from Section 8.6 that electron delocalization stabilizes a species.) As a result, the leaving group of a sulfonate ester is about 100 times better as a leaving group than is chloride ion.

resonance contributors

The Inability to Perform an S$_N$2 Reaction Causes a Severe Clinical Disorder

In the human body, an enzyme called HGPRT catalyzes the nucleophilic substitution reaction shown here. The pyrophosphate group is an excellent leaving group because the electrons released when the group departs, like the electrons released when a sulfonate group departs, are stabilized by electron delocalization.

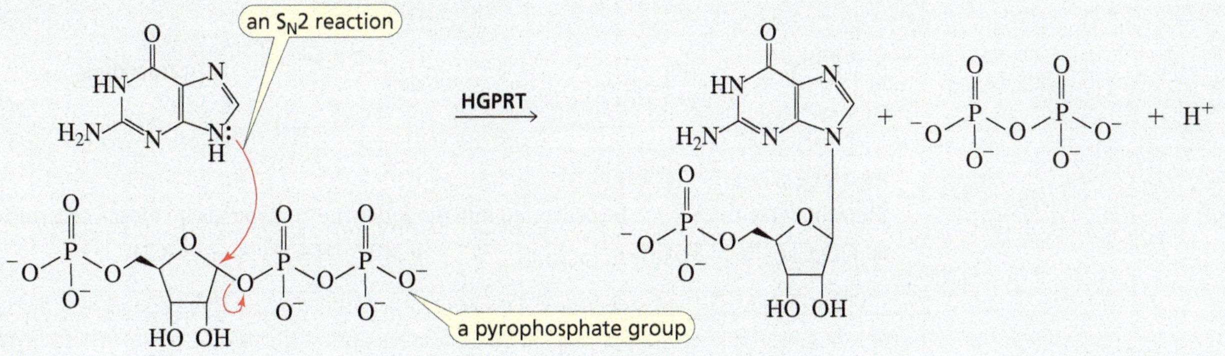

A severe deficiency in HGPRT causes Lesch-Nyhan syndrome. This congenital defect occurs mostly in males and has tragic symptoms—namely, crippling arthritis and severe malfunctions in the nervous system such as mental retardation, highly aggressive and destructive behavior, and self-mutilation. Children with Lesch-Nyhan syndrome have such a compulsive urge to bite their fingers and lips that they must be restrained. Fortunately, HGPRT deficiencies in fetal cells can be detected by amniocentesis. The condition occurs in 1 in 380,000 live births.

PROBLEM 10 SOLVED   **LEARN THE STRATEGY**

Explain why the ether obtained by treating an optically active alcohol with PBr$_3$ in pyridine followed by sodium methoxide has the *same* configuration as the alcohol, whereas the ether obtained by treating the alcohol with tosyl chloride followed by sodium methoxide has a configuration *opposite* that of the alcohol.

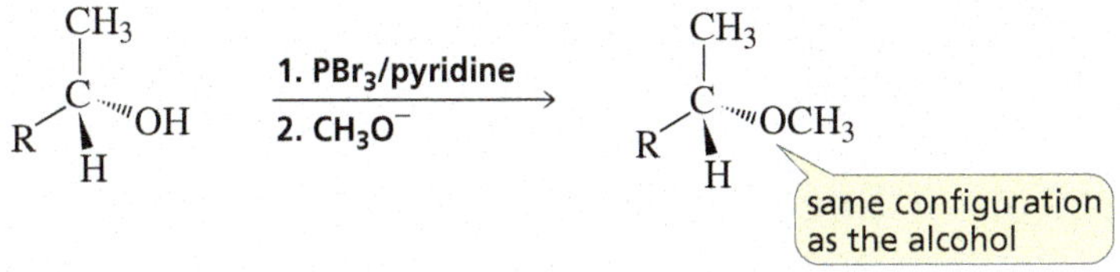

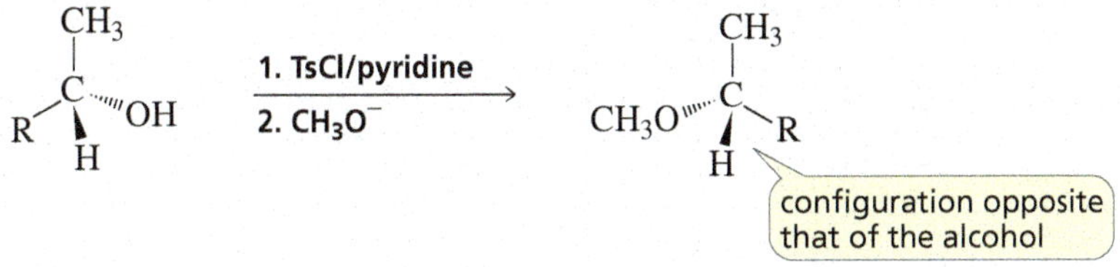

SOLUTION Conversion of the alcohol to the ether by way of an alkyl halide requires two successive S$_N$2 reactions: (1) attack of Br$^-$ on the bromophosphite intermediate and (2) attack of CH$_3$O$^-$ on the alkyl halide. Each S$_N$2 reaction takes place with inversion of configuration, so the final product has the same configuration as the starting material.

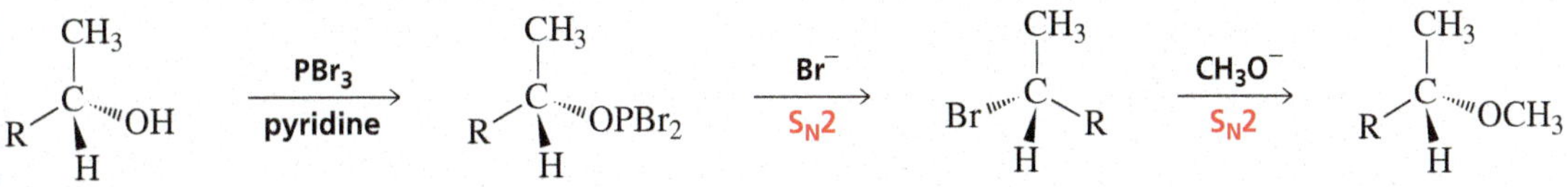

In contrast, conversion of the alcohol to the ether by way of an alkyl tosylate requires only one S$_N$2 reaction (attack of CH$_3$O$^-$ on the alkyl tosylate), so the final product and the starting material have opposite configurations.

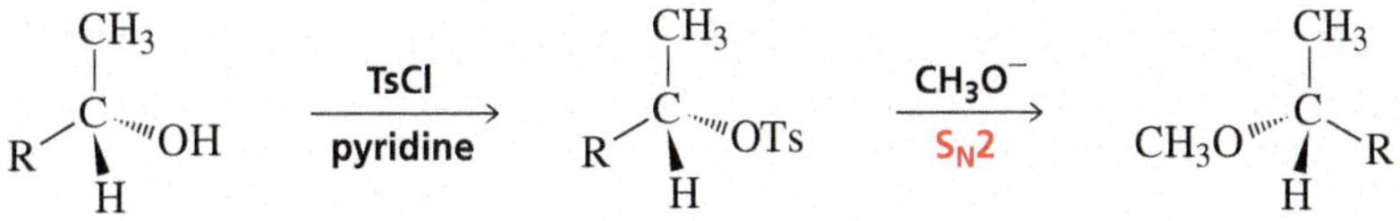

USE THE STRATEGY

PROBLEM 11

What stereoisomers do the following reactions form?

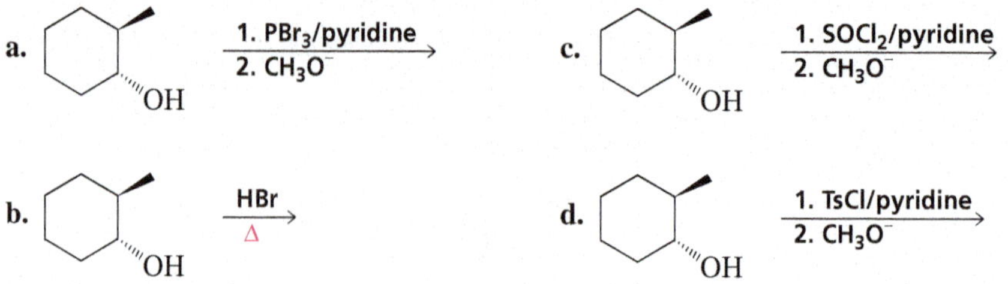

a. [structure] $\xrightarrow{\begin{array}{c}\text{1. PBr}_3\text{/pyridine}\\ \text{2. CH}_3\text{O}^-\end{array}}$

c. [structure] $\xrightarrow{\begin{array}{c}\text{1. SOCl}_2\text{/pyridine}\\ \text{2. CH}_3\text{O}^-\end{array}}$

b. [structure] $\xrightarrow[\Delta]{\text{HBr}}$

d. [structure] $\xrightarrow{\begin{array}{c}\text{1. TsCl/pyridine}\\ \text{2. CH}_3\text{O}^-\end{array}}$

PROBLEM 12

Show how 1-propanol can be converted into the following compounds by means of a sulfonate ester:

a. $CH_3CH_2CH_2SCH_2CH_3$

b. $CH_3CH_2CH_2OCH_2CHCH_3$
  $\qquad\qquad\qquad\qquad\qquad\quad |$
  $\qquad\qquad\qquad\qquad\qquad\ \ CH_3$

10.4 ELIMINATION REACTIONS OF ALCOHOLS: DEHYDRATION

An alcohol can undergo an elimination reaction by losing an OH from one carbon and an H from an adjacent carbon. The product of the reaction is an alkene. Overall, this amounts to the elimination of a molecule of water. Loss of water from a molecule is called **dehydration.**

Dehydration of an alcohol requires an acid catalyst and heat. Sulfuric acid (H_2SO_4) is the most commonly used acid catalyst. Recall that a catalyst increases the rate of a reaction but is not consumed during the course of a reaction (Section 5.13).

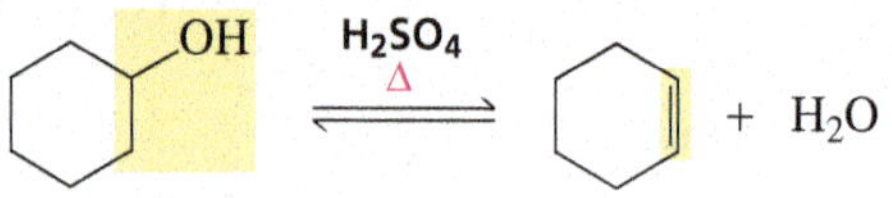

[structure: cyclohexanol] $\overset{H_2SO_4}{\underset{\Delta}{\rightleftharpoons}}$ [structure: cyclohexene] $+ H_2O$

The E1 Dehydration of Secondary and Tertiary Alcohols

The mechanism for acid-catalyzed dehydration depends on the structure of the alcohol; dehydrations of secondary and tertiary alcohols are E1 reactions.

MECHANISM FOR THE E1 DEHYDRATION OF AN ALCOHOL

Dehydration of secondary and tertiary alcohols are E1 reactions.

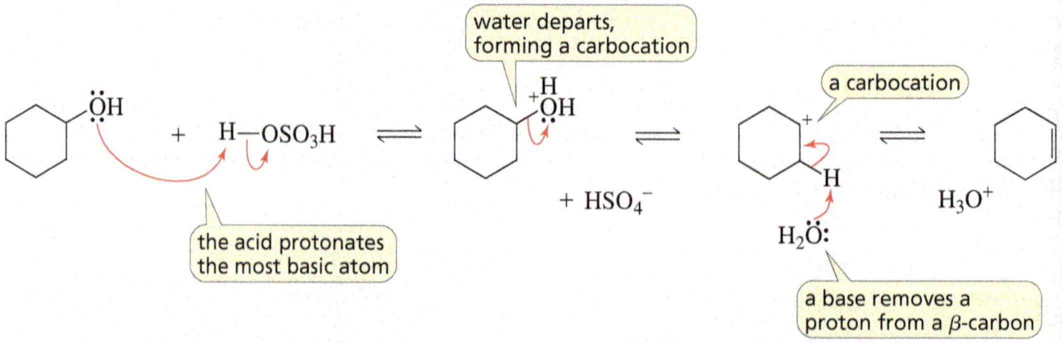

water departs, forming a carbocation

a carbocation

the acid protonates the most basic atom

a base removes a proton from a β-carbon

- The acid protonates the most basic atom in the reactant. As we saw earlier, protonation converts the very poor leaving group (HO⁻) into a good leaving group (H_2O).
- Water departs, leaving behind a carbocation.
- A base in the reaction mixture (water is the base that is present in the highest concentration) removes a proton from a β-carbon (a carbon adjacent to the positively charged carbon), forming an alkene and regenerating the acid catalyst. Notice that the dehydration reaction is an E1 reaction of a protonated alcohol.

When acid-catalyzed dehydration leads to more than one elimination product, the major product is the more stable alkene—that is, the one obtained by removing a proton from the β-carbon bonded to the fewest hydrogens (Section 9.7).

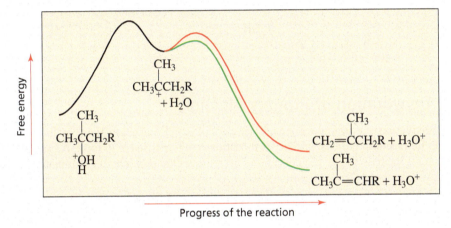

$$CH_3CCH_2CH_3 \overset{H_2SO_4}{\underset{\Delta}{\rightleftharpoons}} CH_3C=CHCH_3 + CH_2=CCH_2CH_3 + H_2O$$
$$\quad\quad 84\% \quad\quad\quad 16\%$$

The more stable alkene is the major product because it has the more stable transition state leading to its formation (Figure 10.1).

◀ **Figure 10.1**
The more stable alkene is the major product obtained from the dehydration of an alcohol because the transition state leading to its formation is more stable (indicated by the green line), allowing it to be formed more rapidly.

Notice that the acid-catalyzed dehydration of an alcohol is the reverse of the acid-catalyzed addition of water to an alkene (Section 6.5).

$$RCH_2CHR + H^+ \overset{dehydration}{\underset{hydration}{\rightleftharpoons}} RCH=CHR + H_2O + H^+$$

To prevent the alkene formed in the dehydration reaction from adding water and reforming the alcohol, the alkene is removed by distillation as it is formed because it has a lower boiling point than the alcohol (Section 3.9). Removing a product displaces the reaction to the right according to Le Châtelier's principle (Section 5.7).

Because the rate-determining step in the dehydration of a secondary or tertiary alcohol is formation of a carbocation intermediate, the rate of dehydration reflects the ease with which the carbocation is formed: tertiary alcohols are the easiest to dehydrate because tertiary carbocations are more stable and are, therefore, more easily formed than secondary and primary carbocations (Section 6.2).

relative ease of dehydration

Be sure to check the structure of the carbocation formed in a dehydration reaction for the possibility of rearrangement. Remember that a carbocation will rearrange if rearrangement produces a more stable carbocation (Section 6.7). For example, the secondary carbocation formed initially in the reaction on the top of the next page rearranges to a more stable tertiary carbocation:

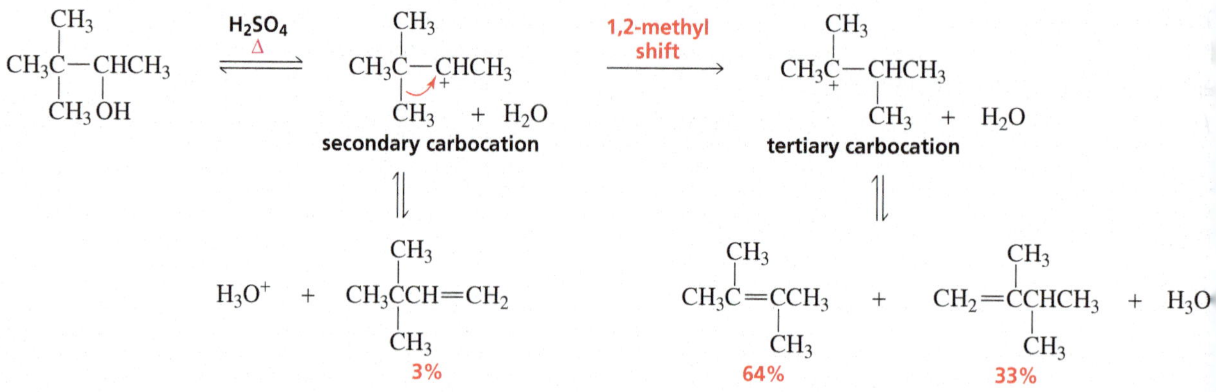

secondary carbocation   tertiary carbocation

3%   64%   33%

PROBLEM 13 ◆

Which of the following alcohols dehydrates the fastest when heated with acid?

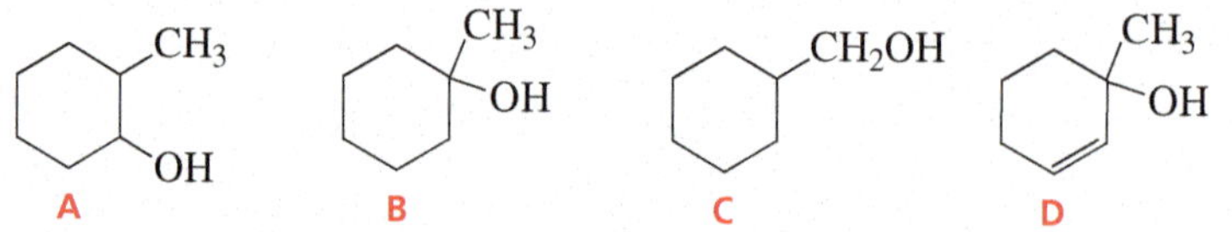

A   B   C   D

The E2 Dehydration of Primary Alcohols

While the dehydration of a secondary or a tertiary alcohol is an E1 reaction, the dehydration of a primary alcohol is an E2 reaction, because primary carbocations are too unstable to form. A base in the reaction mixture (water is the base in greatest concentration) removes the proton in the elimination reaction.

MECHANISM FOR THE E2 DEHYDRATION OF A PRIMARY ALCOHOL AND FOR THE COMPETING S$_N$2 REACTION

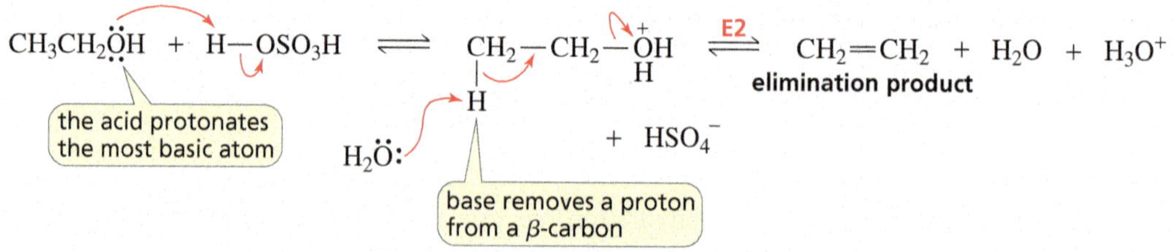

the acid protonates the most basic atom

base removes a proton from a β-carbon

Dehydration of a primary alcohol is an E2 reaction.

back-side attack by the nucleophile

proton dissociation

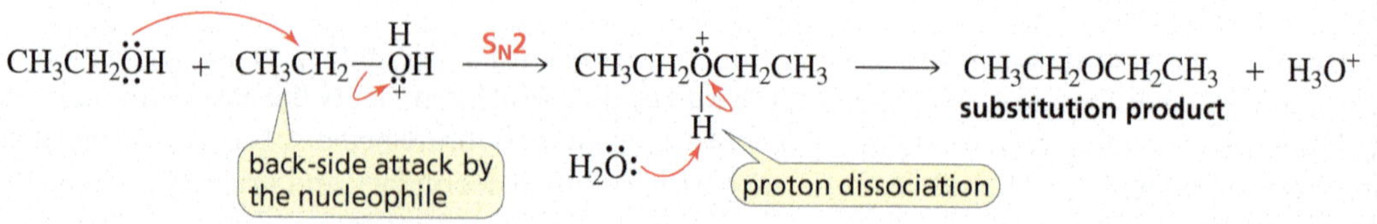

The reaction also forms an ether in a competing S$_N$2 reaction, because primary alcohols are the ones most likely to form substitution products under S$_N$2/E2 conditions. However, the elimination reaction is favored because of the high temperature required for the dehydration reaction (Section 9.12).

Because the dehydration of a primary alcohol is an E2 reaction, we expect 1-butene to be the product of the E2 dehydration of 1-butanol. The product, however, is actually 2-butene.

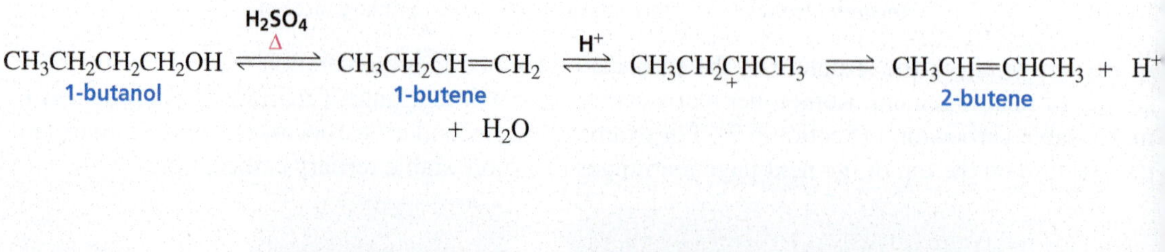

CH$_3$CH$_2$CH$_2$CH$_2$OH   1-butanol   CH$_3$CH$_2$CH=CH$_2$ 1-butene   + H$_2$O   CH$_3$CH$_2$$\overset{+}{\text{C}}HCH_3$ CH$_3$CH=CHCH$_3$ + H$^+$ 2-butene

1-Butene is the initial product but after it forms, a proton from the acidic solution adds to the double bond (adding to the sp^2 carbon bonded to the most hydrogens in accordance with the rule that governs electrophilic addition reactions), thereby forming a carbocation (Section 6.4). Loss of a proton from the β-carbon bonded to the fewest hydrogens (Zaitsev's rule) forms 2-butene (Section 9.8).

PROBLEM 14

What is the major product obtained when each of the following alcohols is heated in the presence of H_2SO_4?

a. 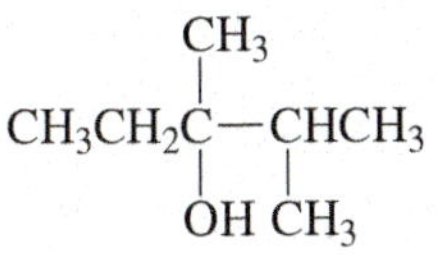 CH₃CH₂C—CHCH₃ (with CH₃ on top, OH and CH₃ below)

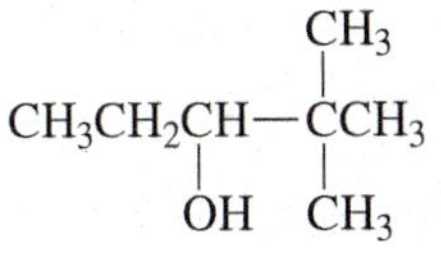

b. 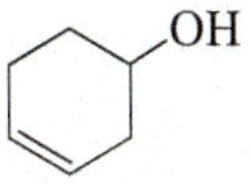 (cyclohexene ring with OH substituent)

c. CH₃CH₂CH—CCH₃ (with CH₃ on top, OH and CH₃ below)

d. $CH_3CH_2CH_2CH_2CH_2OH$

e. $CH_2{=}CHCH_2CH_2OH$

f. 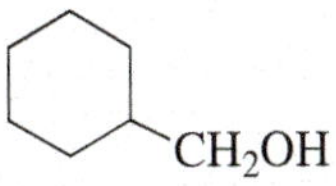 (cyclohexane ring with CH₂OH substituent)

Alcohols undergo $S_N1/E1$ reactions unless they must form a primary carbocation, in which case they undergo $S_N2/E2$ reactions.

PROBLEM 15

Heating an alcohol with sulfuric acid is a good way to prepare a symmetrical ether such as diethyl ether.

a. Explain why it is not a good way to prepare an unsymmetrical ether such as ethyl propyl ether.

b. How would you synthesize ethyl propyl ether?

PROBLEM-SOLVING STRATEGY

Proposing a Mechanism

Propose a mechanism for the following reaction:

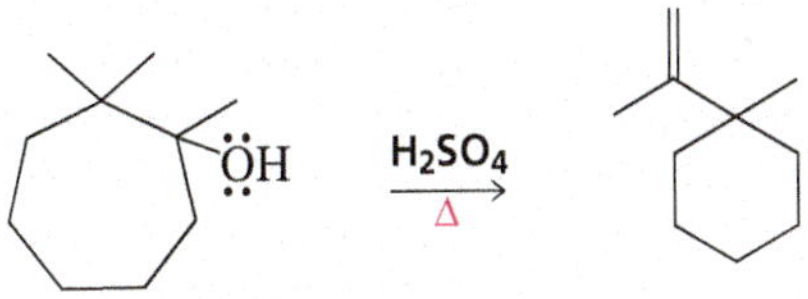

LEARN THE STRATEGY

(structure of seven-membered ring with two methyl groups and OH) $\xrightarrow[\Delta]{H_2SO_4}$ (structure of six-membered ring with isopropenyl and methyl groups)

Even the most complicated-looking mechanism can be reasoned out if you proceed one step at a time, always keeping in mind the structure of the final product. Recall that when an acid is added to a reactant, it protonates the most basic atom in the reactant. Oxygen is the only basic atom, so that is where protonation occurs. Loss of water forms a tertiary carbocation.

(mechanism structures) $\rightleftharpoons$ (carbocation structure) + $H_2\ddot{O}{:}$

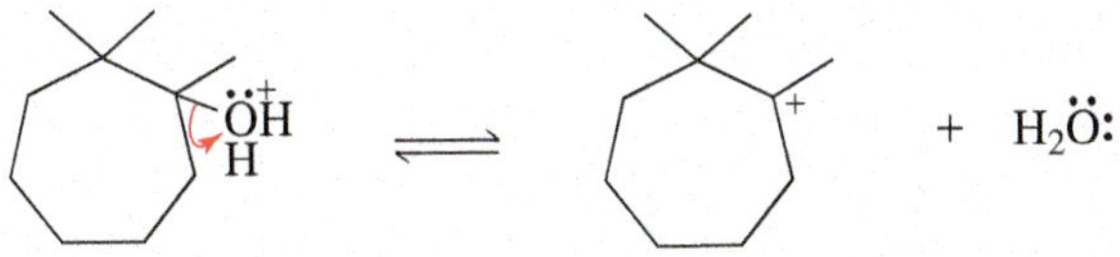

Because the reactant contains a seven-membered ring and the final product has a six-membered ring, a ring-contraction rearrangement must occur to relieve the strain in the seven-membered ring. (Notice that a ring-expansion rearrangement is a 1,2-alkyl shift; see Chapter 6, Problem 96.) When doing a *ring-contraction* or a *ring-expansion rearrangement* (Problem 70), you may find it helpful to label the equivalent carbons in the reactant and product, as shown here. Of the two possible pathways for ring contraction, one leads to a tertiary carbocation and the other leads to a primary carbocation. The correct pathway must be the one that leads to the tertiary carbocation, because only that carbocation has the same arrangement of atoms as the product and the primary carbocation would be too unstable to form.

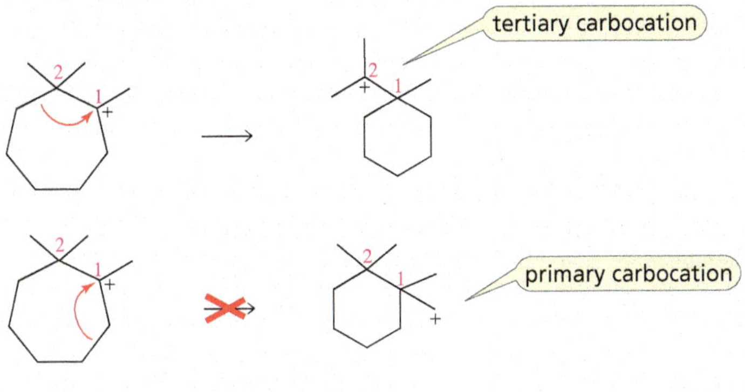

(structure) $\rightarrow$ (structure) **tertiary carbocation**

(structure) ✗ (structure) **primary carbocation**

The final product is obtained by removing a proton from a β-carbon of the rearranged carbocation.

USE THE STRATEGY

PROBLEM 16

Propose a mechanism for each of the following reactions:

a.

$$\xrightarrow[\Delta]{H_2SO_4}$$

b.

$$\xrightarrow[\Delta]{H_2SO_4}$$

PROBLEM 17

Draw the product of each of the following reactions:

a. + HBr ⟶

b. $$\xrightarrow[\Delta]{H_2SO_4}$$

PROBLEM 18

Explain why the following alcohols, when heated with acid, form the same alkene.

and

The Stereochemistry of the Dehydration Reaction

The stereochemical outcome of the E1 dehydration of an alcohol is identical to the stereochemical outcome of the E1 dehydrohalogenation of an alkyl halide. That is, both the *E* and *Z* stereoisomers are obtained as products, and the major product is the stereoisomer in which the larger group on each of the sp^2 carbons are on opposite sides of the double bond. Because that stereoisomer is more stable, it is formed more rapidly (Section 9.10).

$$CH_3CH_2CHCH_3 \ \underset{\Delta}{\overset{H_2SO_4}{\rightleftharpoons}} \ CH_3CH_2\overset{+}{C}HCH_3 \ \rightleftharpoons$$
$$\underset{OH}{|} \qquad\qquad\qquad + \ H_2O$$

trans-2-butene 74%   +   cis-2-butene 23%   + $CH_3CH_2CH{=}CH_2$ 1-butene 3%   + H^+

PROBLEM 19 ◆

What stereoisomers are formed in the following reactions? Which stereoisomer is the major product?

a. the acid-catalyzed dehydration of 1-pentanol to 2-pentene
b. the acid-catalyzed dehydration of 3,4-dimethyl-3-hexanol to 3,4-dimethyl-3-hexene

PROBLEM 20 ◆

If the compound shown in the margin is heated in the presence of H_2SO_4,

a. what constitutional isomer would be formed in greatest yield?
b. what stereoisomer would be formed in greater yield?

Biological Dehydrations

Dehydration reactions occur in cells. Instead of being catalyzed by strong acids, which would not be available in a cell, they are catalyzed by enzymes. Enolase, for example, catalyzes the dehydration of 2-phosphoglycerate in glycolysis. Glycolysis is a series of reactions that prepare glucose for entry into the citric acid cycle (Section 24.6).

2-phosphoglycerate   **phosphoenolpyruvate**

Fumarase is the enzyme that catalyzes the dehydration of malate in the citric acid cycle. The citric acid cycle is a series of reactions that oxidize compounds derived from carbohydrates, fatty acids, and amino acids (Section 24.9).

malate   **fumarate**

Changing an E1 Dehydration into an E2 Dehydration

The relatively harsh conditions (acid and heat) required for alcohol dehydration and the structural changes resulting from carbocation rearrangements in the E1 reaction may result in low yields of the desired alkene. Dehydration, however, can be carried out under milder conditions that favor E2 reactions by replacing the OH group with a good leaving group.

$$CH_3CH_2CHCH_3 \xrightarrow[\text{pyridine, 0 °C}]{\text{POCl}_3} CH_3CH=CHCH_3$$
$$\qquad\quad |$$
$$\qquad\quad OH$$

For example, reaction with phosphorus oxychloride ($POCl_3$) converts the OH group of the alcohol into $OPOCl_2$, which is a good leaving group. The reaction conditions (pyridine is the solvent so it is present at a high concentration) favor an E2 reaction and carbocations are not formed in E2 reactions, so carbocation rearrangements do not occur. Pyridine removes a proton in order to prevent the intermediate from reverting to starting materials and is the base employed in the E2 reaction. Pyridine also prevents the buildup of HCl, which would add to the alkene.

phosphorus oxychloride $+ \, Cl^-$ 2 $+ \, {}^-OPOCl_2$

Sulfonate esters also have good leaving groups. So dehydration *via* an E2 reaction can be achieved by converting the alcohol to a sulfonate ester and then adding a strong base to carry out the E2 reaction. This reaction works well with secondary and tertiary alcohols but not with primary alcohols because of the competing substitution reaction.

$$CH_3CH_2-\overset{\overset{\displaystyle CH_3}{|}}{\underset{\underset{\displaystyle CH_3}{|}}{C}}-OH \xrightarrow[\text{pyridine}]{\text{TsCl}} CH_3CH_2-\overset{\overset{\displaystyle CH_3}{|}}{\underset{\underset{\displaystyle CH_3}{|}}{C}}-OTs \xrightarrow{CH_3O^-} CH_3CH=CCH_3 + {}^-OTs + CH_3OH$$
$$\qquad\qquad\qquad\qquad\qquad + \, Cl^- \; \text{(an E2 reaction)}$$

PROBLEM 21 ◆

What alcohol would you treat with phosphorus oxychloride and pyridine to form each of the following alkenes?

a. $CH_3CH_2\overset{\overset{\displaystyle CH_3}{|}}{C}=CH_2$ b. [cyclohexene with CH₃]   c. $CH_3CH=CHCH_2CH_3$ d. [cyclohexane with =CH₂]

10.5 OXIDATION OF ALCOHOLS

Chromium-Based Oxidizing Agents

A variety of reagents are available that oxidize alcohols. For many years, a commonly used reagent was chromic acid (H_2CrO_4), which is formed when sodium dichromate ($Na_2Cr_2O_7$) is dissolved in aqueous acid. Notice that *secondary alcohols* are oxidized to *ketones*.

$$CH_3CH_2\overset{\overset{\displaystyle OH}{|}}{C}HCH_3 \xrightarrow{H_2CrO_4} CH_3CH_2\overset{\overset{\displaystyle O}{\|}}{C}CH_3$$

[cyclohexanol] $\xrightarrow[H_2SO_4]{Na_2Cr_2O_7}$ [cyclohexanone]

secondary alcohols   **ketones**

Primary alcohols are initially oxidized to *aldehydes* by chromic acid. The reaction, however, does not stop at the aldehyde. Instead, the aldehyde is further oxidized to a *carboxylic acid*. These reactions are easily recognized as oxidations because the number of C—H bonds in the reactant decreases and the number of C—O bonds increases (Section 6.8).

$$R-CH_2OH \xrightarrow{H_2CrO_4} \left[\underset{R}{\overset{O}{\underset{}{\overset{\|}{C}}}} \diagdown H\right] \xrightarrow[\text{oxidation}]{\text{further}} \underset{R}{\overset{O}{\underset{}{\overset{\|}{C}}}} \diagdown OH$$

primary alcohol   **aldehyde**   **carboxylic acid**

Pyridinium chlorochromate (PCC) is a gentler oxidizing agent. It also oxidizes secondary alcohols to ketones, but it stops the reaction at the aldehyde when it oxidizes primary alcohols. PCC must be used in an anhydrous solvent such as CH_2Cl_2 because if water is present, the aldehyde will be further oxidized to a carboxylic acid.

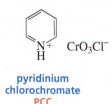

pyridinium chlorochromate
PCC

$$R-CH_2OH \xrightarrow[CH_2Cl_2]{PCC} \underset{R}{\overset{O}{\underset{}{\overset{\|}{C}}}} \diagdown H$$

primary alcohol   **aldehyde**

Notice that in the oxidation of both primary and secondary alcohols, a hydrogen is removed from the carbon to which the OH is attached. The carbon bearing the OH group in a tertiary alcohol is not bonded to a hydrogen, so its OH group cannot be oxidized to a carbonyl (C=O) group.

this C is not bonded to an H, so the alcohol cannot be oxidized to a carbonyl compound

$$CH_3-\overset{\overset{\displaystyle CH_3}{|}}{\underset{\underset{\displaystyle CH_3}{|}}{C}}-OH$$

tertiary alcohol

Hypochlorous Acid as the Oxidizing Reagent

Because of the toxicity of chromium-based reagents, other reagents for the oxidation of alcohols have been developed. One of the more common is hypochlorous acid (HOCl). Because HOCl is unstable, it is generated in the reaction mixture by an acid–base reaction between H^+ and $^-$OCl (using CH_3COOH and NaOCl). Secondary alcohols are oxidized to ketones, and primary alcohols are oxidized to aldehydes.

Primary alcohols are oxidized to aldehydes by HOCl.

Secondary alcohols are oxidized to ketones.

MECHANISM FOR THE OXIDATION OF AN ALCOHOL BY HOCl

the acid protonates the most basic atom

- The acid protonates the oxygen, the most basic atom in the alcohol.
- Because the reaction is not heated, water does not leave spontaneously but must be kicked out by hypochlorite ion in an S_N2 reaction.
- A base in the reaction mixture (water is the base in highest concentration) removes a proton from the carbon bonded to the O—Cl group, and the very weak O—Cl bond breaks.

The Swern Oxidation

The Swern oxidation also uses nontoxic reagents—dimethyl sulfoxide [$(CH_3)_2SO$], oxalyl chloride [$(COCl)_2$], and triethylamine. As with HOCl, primary alcohols are oxidized to aldehydes, and secondary alcohols are oxidized to ketones.

The actual oxidizing agent is the dimethylchlorosulfonium ion. (See Chapter 16, Problem 80.)

- The alcohol displaces the chloride ion in an S_N2 reaction.
- The intermediate loses a proton.
- Triethylamine removes a proton in an E2 reaction to form the aldehyde (or ketone).

Blood Alcohol Concentration

As blood passes through the arteries in our lungs, an equilibrium is established between the alcohol in our blood and the alcohol in our breath. Therefore, if the concentration of one is known, then the concentration of the other can be determined.

The test that law enforcement agencies use to determine a person's blood alcohol level is based on the oxidation of breath ethanol. In this test, a person blows into a device called a breathalyzer and a measured volume of breath passes through a solution of chromic acid, an oxidizing agent. When ethanol is oxidized, the oxidizing agent is reduced to green chromic ion. The breathalyzer contains a spectrophotometer that quantitatively measures the amount of visible light absorbed by the green chromic ion—the greater the absorbance, the greater the concentration of chromic ion and, therefore, the greater the concentration of breath ethanol (Section 13.21).

$$CH_3CH_2OH \ + \ H_2CrO_4 \ \longrightarrow \ \underset{\text{green}}{\overset{\displaystyle CH_3 \overset{\displaystyle O}{\overset{\|}{C}} OH}{}} \ + \ Cr^{3+}$$

Treating Alcoholism with Antabuse

Disulfiram, most commonly known as Antabuse, is used to treat alcoholism. It causes violently unpleasant effects if ethanol is consumed within two days after taking the drug.

Antabuse®

Antabuse works by inhibiting aldehyde dehydrogenase, the enzyme responsible for oxidizing acetaldehyde (a product of ethanol metabolism) to acetic acid. This causes a buildup of acetaldehyde. It is acetaldehyde that causes the unpleasant physiological effects of intoxication: intense flushing, nausea, dizziness, sweating, throbbing headaches, decreased blood pressure, and, ultimately, shock. Consequently, Antabuse should be, taken only under strict medical supervision. In Chapter 23, we will see what can be done to prevent a hangover.

Antabuse inhibits this enzyme, so acetaldehyde builds up.

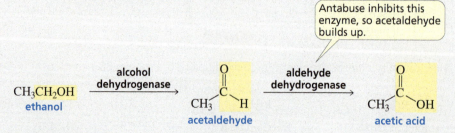

In some people, aldehyde dehydrogenase does not function properly even under normal circumstances. Their symptoms in response to ingesting alcohol are nearly the same as those of individuals who are medicated with Antabuse.

Methanol Poisoning

In addition to oxidizing ethanol to acetaldehyde, alcohol dehydrogenase can oxidize methanol to formaldehyde. Formaldehyde is damaging to many tissues, and because eye tissue is particularly sensitive, methanol ingestion can cause blindness.

$$CH_3OH \xrightarrow{\text{alcohol dehydrogenase}} \underset{\text{formaldehyde}}{\underset{H \quad\quad H}{\overset{O}{\underset{\|}{C}}}}$$

methanol

If methanol is ingested, the patient is given ethanol intravenously for several hours. Ethanol competes with methanol for binding at the active site of the enzyme. Binding ethanol minimizes the amount of methanol that can be bound, which minimizes the amount of formaldehyde that can be formed. So ethanol is given to the patient until all the ingested methanol has been excreted in the urine.

PROBLEM 22 ◆

What product is obtained from the reaction of each of the following alcohols with

a. H_2CrO_4?   **b.** HOCl?   **c.** the regents required for a Swern oxidation?

   1. 3-pentanol   **3.** 2-methyl-2-pentanol   **5.** cyclohexanol

   2. 1-pentanol   **4.** 2,4-hexanediol   **6.** 1,4-butanediol

---

10.6 NUCLEOPHILIC SUBSTITUTION REACTIONS OF ETHERS

The OR group of an **ether** and the OH group of an alcohol have nearly the same basicity because the conjugate acids of these two groups have similar pK_a values. (The pK_a of CH_3OH is 15.5, and the pK_a of H_2O is 15.7.) Both groups are strong bases, so both are very poor leaving groups. Consequently, ethers, like alcohols, need to be activated before they undergo a nucleophilic substitution reaction.

$$R—\overset{..}{\underset{..}{O}}—H \qquad R—\overset{..}{\underset{..}{O}}—R$$

    an alcohol   **an ether**

Like alcohols, ethers are activated by protonation. Ethers, therefore, undergo nucleophilic substitution reactions with HBr or HI. (HCl cannot be used because Cl^- is too poor a nucleophile.) The reaction of ethers with hydrogen halides, like the reactions of alcohols with hydrogen halides, is slow. The reaction mixture must be heated to cause the reaction to occur at a reasonable rate.

$$R—\overset{..}{\underset{..}{O}}—R' + HI \rightleftharpoons R—\overset{\overset{H}{|}}{\underset{..}{\overset{+}{O}}}—R' \xrightarrow{\Delta} R—I + R'—\overset{..}{\underset{..}{O}}H$$

                             I^-

    (poor leaving group)   (good leaving group)

What happens *after* the ether is protonated depends on the structure of the ether. If departure of ROH creates a relatively stable carbocation (such as a tertiary carbocation), an S_N1 reaction will occur—that is, the ROH group will leave.

MECHANISM FOR ETHER CLEAVAGE: AN S_N1 REACTION

the acid protonates the most basic atom

methanol departs, forming a carbocation

addition of a nucleophile

$+ CH_3\overset{..}{\underset{..}{O}}H$

- The acid protonates the oxygen, thereby converting the very basic RO⁻ leaving group into the less basic ROH leaving group.
- The leaving group departs, forming a carbocation.
- The halide ion combines with the carbocation.

However, if departure of ROH would create an unstable carbocation (such as a methyl, vinyl, aryl, or primary carbocation), the leaving group cannot depart. Therefore, the reaction will be an S_N2 reaction—that is, the leaving group will be displaced by a nucleophile.

MECHANISM FOR ETHER CLEAVAGE: AN S_N2 REACTION

$$CH_3-\overset{..}{\underset{..}{O}}-CH_2CH_2CH_3 \; + \; H-\overset{..}{\underset{..}{I}}: \; \rightleftharpoons \; CH_3-\overset{\overset{H}{|}}{\underset{+}{\overset{..}{O}}}-CH_2CH_2CH_3 \; \xrightarrow{S_N2} \; CH_3-\overset{..}{\underset{..}{I}}: \; + \; CH_3CH_2CH_2-\overset{..}{\underset{..}{O}}H$$

the acid protonates the most basic atom

:Ï:⁻ nucleophile attacks the less sterically hindered carbon

- Protonation converts the very basic RO⁻ leaving group into the less basic ROH leaving group.
- The halide ion preferentially attacks the less sterically hindered of the two alkyl groups.

Ethers are cleaved by an S_N1 reaction unless the instability of the carbocation requires the cleavage to be an S_N2 reaction.

Ether cleavage forms only a substitution product because any alkene that would be formed in an elimination reaction would undergo electrophilic addition with HBr or HI to form the same alkyl halide that is obtained from the substitution reaction.

The reagents (such as $SOCl_2$, PCl_3, or $TsCl$) used to activate alcohols so that they can undergo nucleophilic substitution reactions cannot be used to activate ethers. When an alcohol reacts with one of these activating agents, a proton dissociates from the intermediate in the second step of the reaction and a stable product results.

However, when an ether reacts with one of these activating agents, the oxygen atom does not have a proton that can dissociate, so a stable product cannot be formed. Instead, the stable starting materials are reformed.

Because hydrogen halides are the only reagents that react with ethers, ethers are frequently used as solvents. Some common ether solvents are shown in Table 10.2.

Table 10.2   Some Ethers Are Used as Solvents

diethyl ether "ether"	tetrahydrofuran THF	tetrahydropyran THP	1,4-dioxane	1,2-dimethoxyethane DME	tert-butyl methyl ether TBME

Anesthetics

Because diethyl ether (commonly known as ether) is a short-lived muscle relaxant, it was at one time widely used as an *inhalation anesthetic*. However, it takes effect slowly and has a slow and unpleasant recovery period so, over time, other anesthetics, such as isoflurane, enflurane, and halothane, replaced it. Even so, diethyl ether is still used where trained anesthesiologists are scarce because it is the safest anesthetic for an untrained person to administer. Anesthetics interact with the nonpolar molecules of cell membranes, causing the membranes to swell, which interferes with their permeability.

Sodium pentothal (also called thiopental sodium) is an *intravenous anesthetic*. The onset of anesthesia and the loss of consciousness occur within seconds of its administration. Care must be taken when administering sodium pentothal because the dose for effective anesthesia is 75% of the lethal dose. Because of this high level of toxicity, it cannot be used as the sole anesthetic but, instead, is generally used to induce anesthesia before an inhalation anesthetic is administered.

Propofol, in contrast, has all the properties of the "perfect anesthetic": it can be administered as the sole anesthetic by intravenous drip, it has a rapid and pleasant induction period, and it has a wide margin of safety in trained hands. Recovery from the drug is also rapid and pleasant.

(Continued)

"ether" isoflurane enflurane halothane thiopental sodium / sodium pentothal diprivan / propofol

amputation of a leg without anesthetic in 1528

a painting showing the first use of anesthesia (ether) during surgery in 1846 at Massachusetts General Hospital by surgeon John Collins Warren

---

PROBLEM 23 SOLVED

LEARN THE STRATEGY

What are the major products obtained when the following ether is heated with one equivalent of HI?

SOLUTION The reaction takes place by an S_N2 pathway because neither alkyl group will form a relatively stable carbocation (one would be vinylic and the other primary). Iodide ion attacks the carbon of the ethyl group because otherwise it would have to attack a vinylic carbon, and vinylic carbons are not attacked by nucleophiles (Section 9.13). Thus, the major products are ethyl iodide and an enol that immediately rearranges to an aldehyde (Section 7.7).

PROBLEM 24

USE THE STRATEGY

What are the major products obtained when each of the following ethers is heated with one equivalent of HI?

a.

b.

c.

d.

e. OCH₃

f. CH₃ CH₃

PROBLEM 25 SOLVED

Explain why methyl propyl ether forms both methyl iodide and propyl iodide when it is heated with excess HI.

SOLUTION On page 478, we saw that the S_N2 reaction of methyl propyl ether with an equivalent amount of HI forms methyl iodide and propyl alcohol because the methyl group is less sterically hindered than the propyl group to attack by the iodide ion. When there is excess HI, the alcohol product of this first reaction reacts with HI in another S_N2 reaction. Thus, the products are two alkyl iodides.

$$\diagup\!\!\diagdown\!\!\diagup OCH_3 \xrightarrow[\Delta]{HI} \diagup\!\!\diagdown\!\!\diagup OH \xrightarrow[\Delta]{HI} \diagup\!\!\diagdown\!\!\diagup I \;+\; H_2O$$
$$+\;\; CH_3I$$

PROBLEM 26 ◆

Explain why HF and HCl cannot be used to cleave ethers in an S_N2 reaction.

10.7 NUCLEOPHILIC SUBSTITUTION REACTIONS OF EPOXIDES

We saw that an alkene can be converted into an **epoxide,** a three-membered ring ether, by a peroxy-acid (Section 6.10).

$$
\underset{\text{alkene}}{RCH=CH_2} + \underset{\text{peroxyacid}}{R\overset{O}{\overset{\|}{C}}OOH} \longrightarrow \underset{\text{epoxide}}{RCH-CH_2} + \underset{\text{carboxylic acid}}{R\overset{O}{\overset{\|}{C}}OH}
$$

(π bond breaks) (2 oxygens) (2 new σ bonds form) (1 oxygen)

An alkene can also be converted into an epoxide using Cl_2 and H_2O, followed by reaction with NaH (Sections 6.9 and 9.15).

an intramolecular S_N2 reaction

Although an epoxide and an ether have the same leaving group, epoxides are much more reactive than ethers in nucleophilic substitution reactions because the strain in their three-membered ring is relieved when the ring opens (Figure 10.2). Epoxides, therefore, undergo nucleophilic substitution reactions with a wide variety of nucleophiles.

H$_2$C—CH$_2$ (with O bridging)
ethylene oxide

CH$_3$CH$_2$OCH$_2$CH$_3$
diethyl ether

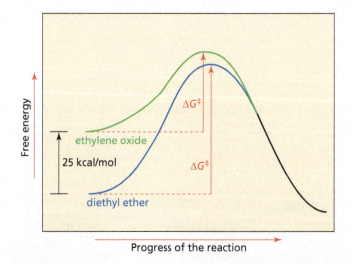

▶ **Figure 10.2**
The reaction coordinate diagrams for nucleophilic attack of hydroxide ion on ethylene oxide and on diethyl ether. The greater reactivity of the epoxide is a result of the strain in the three-membered ring, which increases the epoxide's free energy.

Nucleophilic Substitution: Acidic Conditions

Epoxides, like other ethers, undergo substitution reactions with hydrogen halides. The mechanism of the reaction depends on whether it is carried out under acidic conditions or neutral/basic conditions. Under acidic conditions, the mechanism shown next is followed.

MECHANISM FOR NUCLEOPHILIC SUBSTITUTION: ACIDIC CONDITIONS

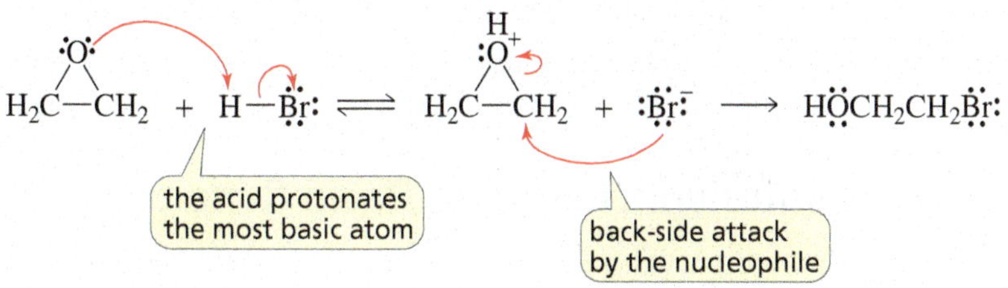

$$H_2C\!-\!CH_2 + H\!-\!\ddot{B}r\!: \rightleftharpoons H_2C\!-\!CH_2 + :\ddot{B}r\!:^- \longrightarrow H\ddot{O}CH_2CH_2\ddot{B}r\!:$$

the acid protonates the most basic atom

back-side attack by the nucleophile

- The acid protonates the oxygen of the epoxide.
- The protonated epoxide undergoes back-side attack by the halide ion.

Because epoxides are so much more reactive than ethers, the reaction takes place readily at room temperature, unlike the reaction of an ether with a hydrogen halide that requires heat.

Protonated epoxides are so reactive that they can be opened by poor nucleophiles such as H_2O and alcohols. (HB^+ is any acid in the solution, and :B is any base.)

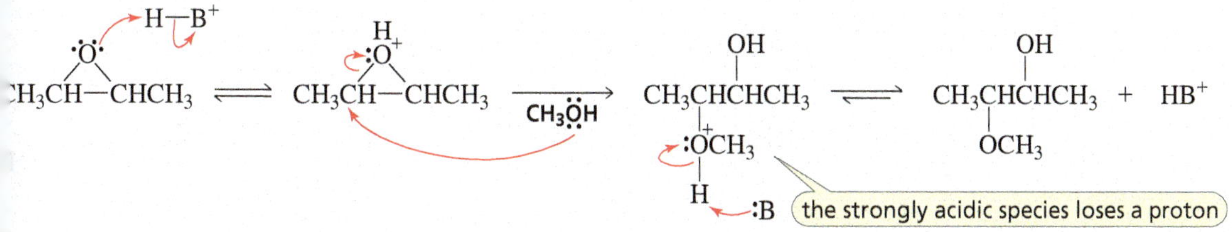

$$CH_3CH\!-\!CHCH_3 \rightleftharpoons CH_3CH\!-\!CHCH_3 \xrightarrow{CH_3\ddot{O}H} CH_3CHCHCH_3 \rightleftharpoons CH_3CHCHCH_3 + HB^+$$

the strongly acidic species loses a proton

If different substituents are attached to the two ring carbons of the protonated epoxide (and the nucleophile is something other than H_2O), the product obtained from nucleophilic attack on the 2-position of the oxirane ring will be different from that obtained from nucleophilic attack on the 3-position. The major product is the one resulting from nucleophilic attack on the *more substituted* ring carbon.

obtained by nucleophilic attack on the more substituted ring carbon

$$CH_3CH\!-\!CH_2 \xrightarrow{HCl} CH_3CH\!-\!CH_2 \xrightarrow{CH_3OH} CH_3CHCH_2OH + CH_3CHCH_2OCH_3 + HCl$$

major product   minor product

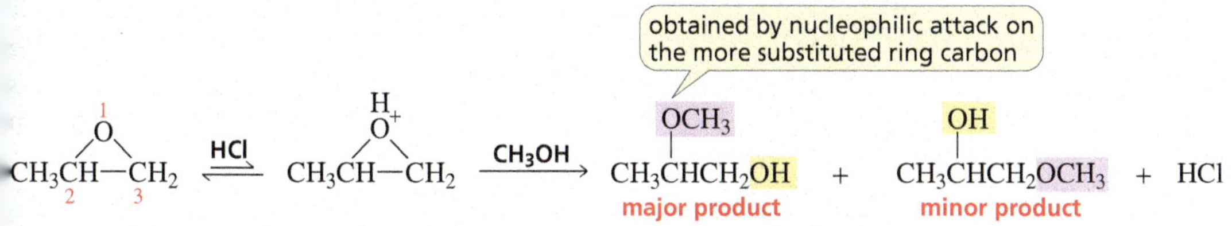

The more substituted carbon is more likely to be attacked because after the epoxide is protonated, it is so reactive that one of the C—O bonds begins to break even before the nucleophile has an opportunity to attack.

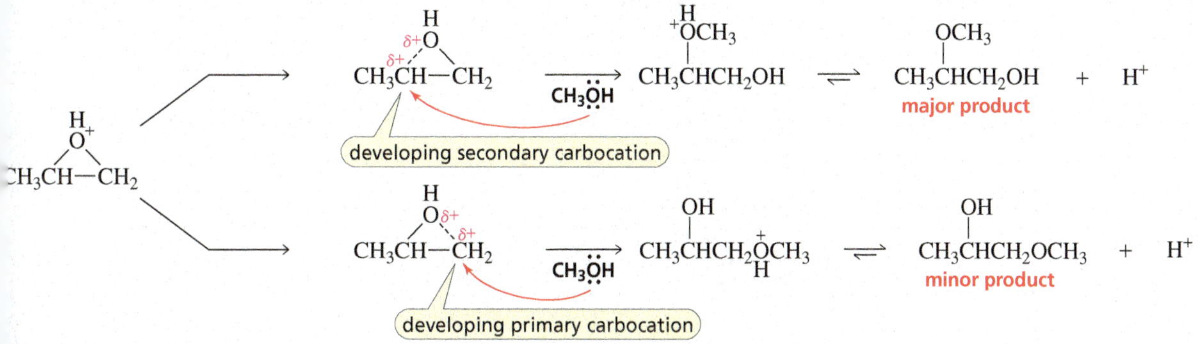

$$CH_3CH\!-\!CH_2 \xrightarrow{CH_3\ddot{O}H} CH_3CHCH_2OH \rightleftharpoons CH_3CHCH_2OH + H^+$$

developing secondary carbocation

major product

$$CH_3CH\!-\!CH_2 \xrightarrow{CH_3\ddot{O}H} CH_3CHCH_2\overset{+}{O}CH_3 \rightleftharpoons CH_3CHCH_2OCH_3 + H^+$$

developing primary carbocation

minor product

As the bond starts to break, a partial positive charge develops on the carbon that is losing its share of oxygen's electrons. Therefore, the protonated epoxide breaks preferentially in the direction that puts the partial positive charge on the more substituted carbon, because a more substituted carbocation is more stable. (Recall that tertiary carbocations are more stable than secondary carbocations, which are more stable than primary carbocations.)

> **Under acidic conditions, the nucleophile preferentially attacks the more substituted ring carbon.**

The best way to describe the reaction is to say that it occurs by a pathway that is partially S_N1 and partially S_N2.

- It is not a pure S_N1 reaction because a carbocation intermediate is not fully formed.
- It is not a pure S_N2 reaction because the leaving group begins to depart before the compound is attacked by the nucleophile.

Nucleophilic Substitution: Neutral or Basic Conditions

Although an ether must be protonated before it can undergo a nucleophilic substitution reaction (Section 10.6), the strain in the three-membered ring allows an epoxide to undergo nucleophilic substitution reactions without first being protonated (Figure 10.2). When a nucleophile attacks an unprotonated epoxide, the reaction is a pure S_N2 reaction.

MECHANISM FOR NUCLEOPHILIC SUBSTITUTION: NEUTRAL OR BASIC CONDITIONS

- The C—O bond does not begin to break until the carbon is attacked by the nucleophile. The nucleophile is more likely to attack the *less substituted* carbon because it is less sterically hindered.
- The alkoxide ion picks up a proton from the solvent or from an acid added after the reaction is over.

> **Under neutral or basic conditions, the nucleophile preferentially attacks the less sterically hindered ring carbon.**

Thus, the site of nucleophilic attack on an unsymmetrical epoxide under neutral or basic conditions (when the epoxide *is not* protonated) is different from the site of nucleophilic attack under acidic conditions (when the epoxide *is* protonated).

LEARN THE STRATEGY

Epoxides are useful reagents because they can react with a wide variety of nucleophiles, leading to the formation of a wide variety of products.

USE THE STRATEGY

PROBLEM 27 ◆

Draw the major product of each of the following reactions:

a. 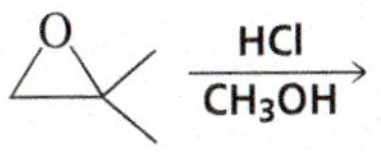 $\xrightarrow[\text{CH}_3\text{OH}]{\text{HCl}}$

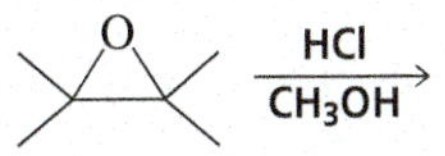

b. 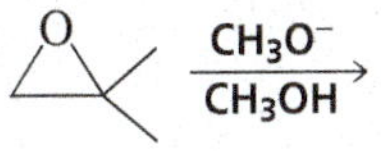 $\xrightarrow[\text{CH}_3\text{OH}]{\text{CH}_3\text{O}^-}$

c. $\xrightarrow[\text{CH}_3\text{OH}]{\text{HCl}}$

d. 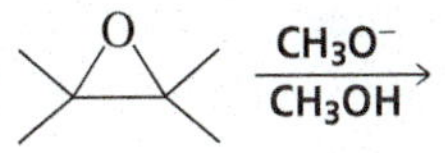 $\xrightarrow[\text{CH}_3\text{OH}]{\text{CH}_3\text{O}^-}$

PROBLEM 28 ◆

Would you expect the reactivity of a five-membered ring ether such as tetrahydrofuran (Table 10.2) to be more similar to the reactivity of an epoxide or to the reactivity of a noncyclic ether? Why?

Converting an Alkene to an Alcohol Without a Carbocation Rearrangement

In Section 6.6, we saw that an *alcohol* can be prepared from the acid-catalyzed addition of water to an *alkene*. The reaction forms a carbocation intermediate, which will rearrange if the rearrangement leads to a more stable carbocation.

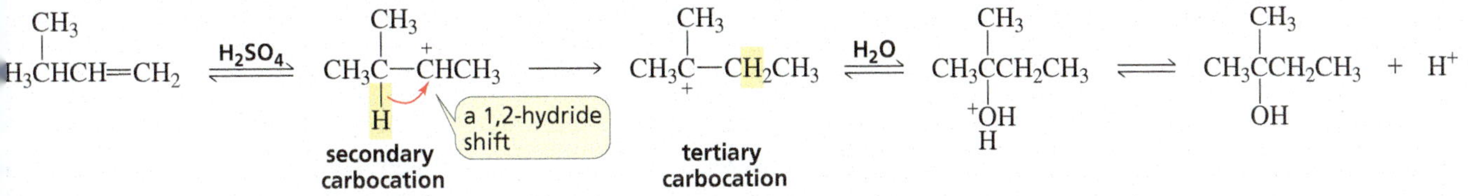

The alkene can be converted to the alcohol without a carbocation rearrangement, if the alkene is first converted to an epoxide (Section 6.10), which is then treated with a hydride ion donor such as lithium aluminum hydride (LAH). Addition of acid protonates the alkoxide ion.

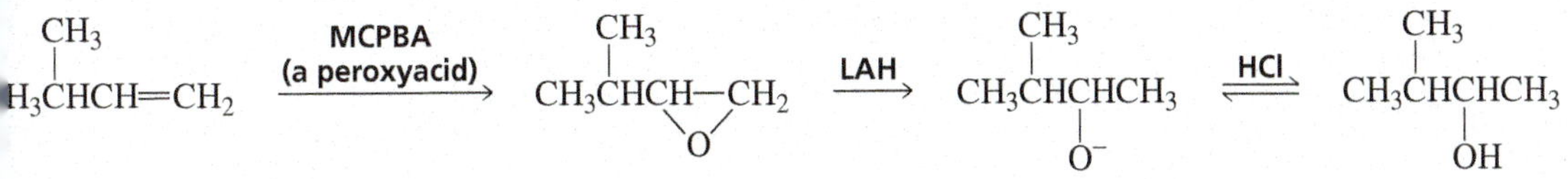

The hydride ion is a nucleophile that attacks the less substituted carbon of the epoxide.

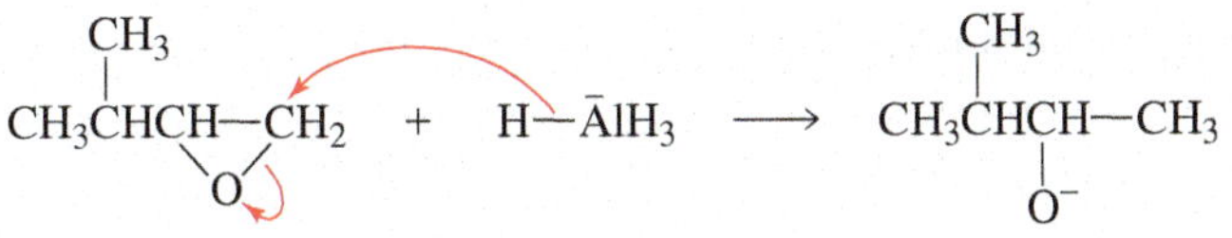

PROBLEM 29

How can the following compounds be prepared from 3,3-dimethyl-1-butene?

a. 2,3-dimethyl-2-butanol   **b.** 3,3-dimethyl-2-butanol

Trans and Cis Diols

The reaction of cyclohexene oxide with hydroxide ion leads to a **trans 1,2-diol,** because the S_N2 reaction involves back-side attack. A diol is also called a **glycol.** Because the OH groups are on adjacent carbons, 1,2-diols are also known as **vicinal diols** or **vicinal glycols.** (Recall that *vicinal* means that two substituents are on adjacent carbons; see Section 6.9.)

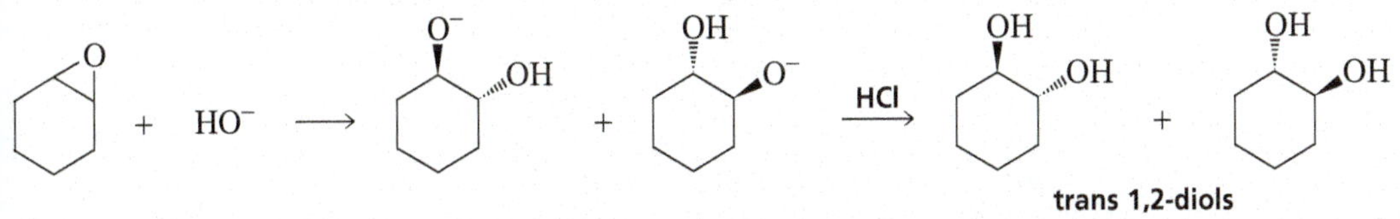

trans 1,2-diols

Notice that two stereoisomers are formed because the reaction forms two new asymmetric centers and only anti addition occurs (Section 6.13).

PROBLEM 30

What products are obtained from the reaction of cyclohexene oxide with

a. methoxide ion? **b.** methylamine?

A **cis 1,2-diol** can be obtained by oxidizing an alkene with osmium tetroxide (OsO_4) followed b hydrolysis with aqueous hydrogen peroxide.

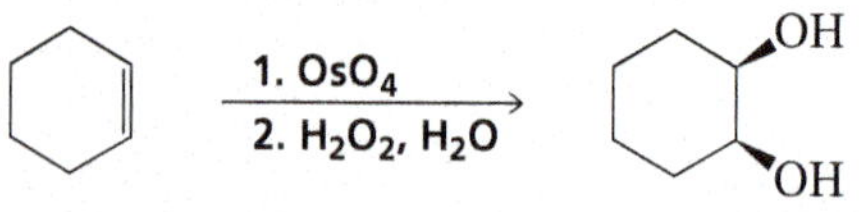

a cis 1,2-diol

The 1,2-diol is cis because addition of osmium tetroxide to the alkene is a syn addition—that is both oxygens are delivered to the same side of the double bond.

MECHANISM FOR CIS-GLYCOL FORMATION

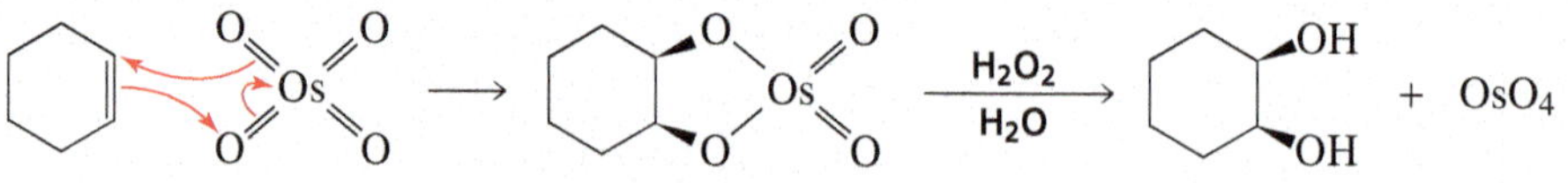

a cyclic osmate
intermediate

- Osmium tetroxide forms a cyclic intermediate when it reacts with an alkene.
- The intermediate is hydrolyzed with aqueous hydrogen peroxide. Hydrogen peroxide re-oxidizes the osmium reagent back to osmium tetroxide. (Because osmium tetroxide i recycled, only a catalytic amount of this expensive and toxic oxidizing agent is needed.)

PROBLEM 31 ◆

What products are obtained from the reaction of each of the following alkenes with OsO_4 followed by aqueous H_2O_2?

a. $CH_3C{=}CHCH_2CH_3$
      |
      CH_3

b.

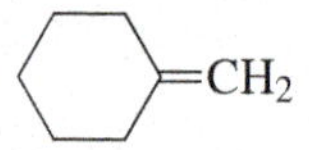

PROBLEM 32

What stereoisomers are obtained from the reaction of each of the following alkenes with OsO_4 followed by aqueous H_2O_2?

a. *trans*-2-butene   **b.** *cis*-2-butene   **c.** *cis*-2-pentene   **d.** *trans*-2-pentene

PROBLEM 33

What stereoisomers are obtained from the reaction of the alkenes in Problem 32 with a peroxyacid followed by reaction with hydroxide ion?

Crown Ethers—Another Example of Molecular Recognition

Crown ethers are cyclic compounds that contain several ether linkages around a central cavity. A crown ether specifically binds certain metal ions or organic molecules, depending on the cavity's size. The crown ether is called the **host**, and the species it binds is called the **guest**. Because the ether linkages are chemically inert, the crown ether can bind the guest without reacting with it. The *host–guest complex* is called an **inclusion compound**.

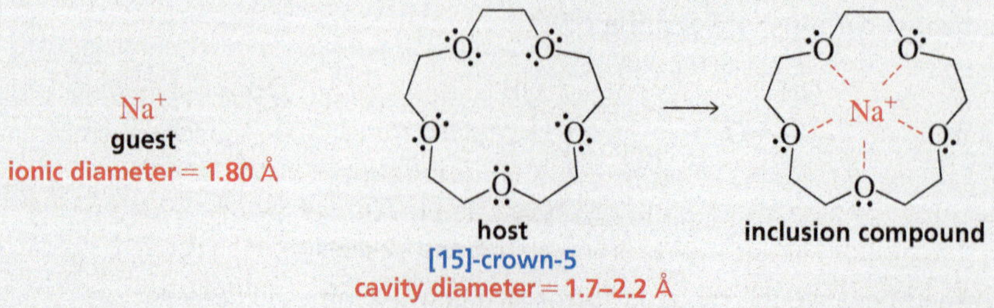

Na$^+$
guest
ionic diameter = 1.80 Å

host
[15]-crown-5
cavity diameter = 1.7–2.2 Å

inclusion compound

Crown ethers are named [X]-crown-Y, where X is the total number of atoms in the ring and Y is the number of oxygen atoms in the ring. Thus, [15]-crown-5 has 15 atoms in the ring, five of which are oxygens. [15]-Crown-5 selectively binds Na^+ because the ether's cavity diameter is 1.7 to 2.2 Å and Na^+ has an ionic diameter of 1.80 Å. Binding occurs through the interaction of the positively charged ion with the lone-pair electrons of the oxygen atoms that point into the cavity. The ability of a host to bind only certain guests is another example of molecular recognition (Section 5.14).

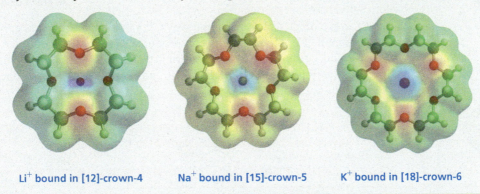

Li^+ bound in [12]-crown-4   **Na^+ bound in [15]-crown-5**   **K^+ bound in [18]-crown-6**

Crown Ethers Can Be Used to Catalyze S_N2 Reactions

A problem that often arises in the laboratory is finding a solvent that will dissolve all the reactants needed for a given reaction. For example, if we want cyanide ion to react with 1-bromohexane, we must find a way of mixing sodium cyanide (an ionic compound soluble only in water) with the alkyl halide (an organic compound that is insoluble in water). If we mix an aqueous solution of sodium cyanide with a solution of the alkyl halide in a nonpolar solvent, there will be two distinct phases—an aqueous phase and a nonpolar phase—because the two solutions are immiscible. How, then, can a reaction between sodium cyanide and 1-bromohexane take place?

$$\text{1-bromohexane} \quad \overset{\wedge\wedge\wedge}{\text{Br}} + Na^{+} \, ^{-}C{\equiv}N \xrightarrow{?} \quad \overset{\wedge\wedge\wedge}{\text{C}}{\equiv}N + Na^{+} Br^{-}$$

1-bromohexane

The two compounds can react with each other if [15]-crown-5 is added to the reaction mixture.

Na^+ binds in the cavity of [15]-crown-5 and the inclusion compound is soluble in the nonpolar solvent because the outside of the crown is composed primarily of nonpolar C—H bonds. The inclusion compound must carry a counterion to balance its positive charge. Thus, cyanide ion will also be in the nonpolar solvent, where it is a powerful nucleophile because it is not solvated. In this way, nucleophilic substitution reactions with alkyl halides that are soluble only in nonpolar solvents can readily take place.

$$\text{[15]-crown-5} \quad + \; NaC{\equiv}N \longrightarrow \quad Na^{+} \qquad ^{-}C{\equiv}N$$

[15]-crown-5   (powerful nucleophile)

In the next reaction, the counter ion is K^+, which has a larger ionic diameter than Na^+, so a larger crown ether ([18]-crown-6) must be used.

$$\overset{\wedge}{\text{Br}} + \overset{O}{\underset{O^{-}K^{+}}{\|}} \xrightarrow{\text{[18]-crown-6}} \overset{\wedge}{\text{O}}\overset{O}{\|} + K^{+} + Br^{-}$$

10.8 ARENE OXIDES

An **arene oxide** is a compound in which one of the "double bonds" of an aromatic hydrocarbon (also called an **arene**) has been converted to an epoxide. Formation of an arene oxide is the first step in changing an aromatic compound that enters the body as a foreign substance (for example, a drug, cigarette smoke, automobile exhaust) into a more water-soluble compound that can eventually be eliminated. The enzyme that converts arenes into arene oxides is called cytochrome P_{450}.

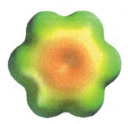

benzene

benzene oxide

benzene $\xrightarrow[\text{O}_2]{\text{cytochrome P}_{450}}$ benzene oxide
an arene oxide

An arene oxide can react in two ways. It can react as a typical epoxide, undergoing attack by a nucleophile (Y^-) to form addition products (Section 10.7). Notice that two addition products are formed because the reaction forms two new asymmetric centers and only anti addition occurs (Section 6.13). Alternatively, it can rearrange to form a phenol, which other epoxides cannot do.

addition products

rearranged product

PROBLEM 34

Draw the mechanism for formation of the two addition products.

The mechanism for the rearrangement is shown next.

MECHANISM FOR ARENE OXIDE REARRANGEMENT

the acid protonates
the most basic atom

rate-determining
step

a carbocation   NIH shift   a protonated dienone   phenol   + HB⁺

- An acid protonates the arene oxide.
- The three-membered ring opens, forming a resonance-stabilized carbocation.
- The carbocation forms a protonated *dienone* as a result of a 1,2-hydride shift. This is called an *NIH shift* because it was first observed in a laboratory at the National Institutes of Health.
- Removal of a proton from the protonated dienone forms phenol.

Because formation of the carbocation is the rate-determining step, the rate of phenol formation depends on the stability of the carbocation. The more stable the carbocation, the more easily the ring opens to form the rearranged product.

Only one arene oxide can be formed from naphthalene because the "double bond" shared by the two rings cannot be epoxidized. Remember that benzene rings are particularly stable, so naphthalene is epoxidized only at a position that leaves one of the benzene rings intact.

Naphthalene oxide can rearrange to form either 1-naphthol or 2-naphthol. The carbocation leading to 1-naphthol is more stable because its positive charge can be stabilized by electron delocalization without destroying the aromaticity of the benzene ring on the left of the structure. In contrast, the positive charge on the carbocation leading to 2-naphthol can be stabilized by electron delocalization only if the aromaticity of the benzene ring is destroyed. (This can be seen by comparing the predicted stabilities of the resonance contributors of the two carbocations; see Problem 35.) Consequently, rearrangement leads predominantly to 1-naphthol.

PROBLEM 35

Draw all possible resonance contributors for the two carbocations in the preceding reaction. Use the resonance contributors to explain why 1-naphthol is the major product of the reaction.

PROBLEM 36

The existence of the NIH shift was established by determining the major product obtained from rearrangement of the following arene oxide, in which a hydrogen has been replaced by a deuterium.

a. What would be the major product if the NIH shift occurs? (*Hint:* A C—H bond is easier to break than a C—D bond.)
b. What would be the major product if the carbocation forms phenol by losing H^+ or D^+, rather than by going through the NIH shift?

PROBLEM 37

How do the major products obtained from rearrangement of the following arene oxides differ?

Arene Oxides Can Be Carcinogens

Some aromatic hydrocarbons are carcinogens—that is, compounds that cause cancer. Investigation has revealed, however, that the hydrocarbons themselves are not carcinogenic; the actual carcinogens are the arene oxides into which the hydrocarbons are converted in the body.

How do arene oxides cause cancer? We saw that nucleophiles react with epoxides to form addition products. 2′-Deoxyguanosine, a component of DNA (Section 26.1), has a nucleophilic NH_2 group that is known to react with certain arene oxides. Once a molecule of 2′-deoxyguanosine becomes covalently attached to an arene oxide, the 2′-deoxyguanosine can no longer fit into the DNA double helix. As a result, the genetic code will not be properly transcribed (Section 26.7), which can lead

segment of DNA

The more stable the carbocation formed when the arene oxide opens, the less likely it is that the arene oxide is carcinogenic.

to mutations that cause cancer. Cancer results when cells lose their ability to control their growth and reproduction.

an arene oxide

2'-deoxyguanosine

covalently attached to the arene oxide

Carcinogenicity is Determined by Carbocation Stability

Not all arene oxides are carcinogenic. Whether a particular arene oxide is carcinogenic depends on the relative rates of its two reaction pathways: rearrangement and reaction with a nucleophile.

Arene oxide rearrangement leads to phenols that are not carcinogenic, whereas formation of addition products from nucleophilic attack by DNA can lead to cancer-causing products. Thus, if the rate of arene oxide rearrangement is faster than the rate of nucleophilic attack by DNA, then the arene oxide will be harmless. However, if the rate of nucleophilic attack is faster than the rate of rearrangement, the arene oxide will likely be a carcinogen.

We saw that the rate-limiting step of arene oxide rearrangement is formation of the carbocation. Thus, the rate of the rearrangement reaction and, therefore, an arene oxide's cancer-causing potential depend on the stability of the carbocation. If the carbocation is relatively stable, then it will be formed relatively easily, so rearrangement will be fast and the arene oxide will most likely not be carcinogenic. On the other hand, if the carbocation is relatively unstable, then rearrangement will be slow and the arene oxide will more likely exist long enough to be attacked by nucleophiles and thus be carcinogenic. This means that *the more stable the carbocation formed when the epoxide ring of an arene oxide opens, the less likely it is that the arene oxide is carcinogenic.*

Benzo[*a*]pyrene and Cancer

Benzo[*a*]pyrene is one of the most carcinogenic arenes. It is formed whenever an organic compound is not completely burned. For example, benzo[*a*]pyrene is found in cigarette smoke, automobile exhaust, and charcoal-broiled meat. Several arene oxides can be formed from benzo[*a*]pyrene. The two most harmful are the 4,5-oxide and the 7,8-oxide.

$$\text{benzo[}a\text{]pyrene} \xrightarrow[\text{O}_2]{\text{cytochrome P}_{450}} \text{4,5-benzo[}a\text{]pyrene oxide} + \text{7,8-benzo[}a\text{]pyrene oxide}$$

benzo[*a*]pyrene   **4,5-benzo[*a*]pyrene oxide**   **7,8-benzo[*a*]pyrene oxide**

The 4,5-oxide is harmful because it forms a carbocation that cannot be stabilized by electron delocalization without destroying the aromaticity of an adjacent benzene ring. Thus, the carbocation is relatively unstable, so the epoxide tends not to open until it is attacked by a nucleophile (the carcinogenic pathway).

The 7,8-oxide is harmful because it reacts with water (a nucleophile) to form a diol, which then forms a diol epoxide. The diol epoxide does not readily undergo rearrangement (the harmless pathway), because it opens to a carbocation that is destabilized by the electron-withdrawing OH groups. Because carbocation formation is slow, the diol epoxide can exist long enough to be attacked by nucleophiles (the carcinogenic pathway).

$$\xrightarrow[\text{epoxide hydrolase}]{\text{H}_2\text{O}} \qquad \xrightarrow[\text{cytochrome P}_{450}]{\text{O}_2}$$

a diol epoxide

PROBLEM 38 SOLVED

Which compound is more likely to be carcinogenic?

or

SOLUTION The nitro-substituted compound is more likely to be carcinogenic. The nitro group destabilizes the carbocation formed when the ring opens by withdrawing electrons from the ring by resonance (Section 8.10). In contrast, the methoxy group stabilizes the carbocation by donating electrons to the ring by resonance. Because carbocation formation leads to the harmless product, the nitro-substituted compound with a less stable (less easily formed) carbocation is less likely to undergo rearrangement to a harmless product. In addition, the electron-withdrawing nitro group increases the arene oxide's susceptibility to nucleophilic attack, which is the cancer-causing pathway.

PROBLEM 39

Explain why the two arene oxides in Problem 38 open in opposite directions.

PROBLEM 40 ♦

Which compound is more likely to be carcinogenic?

or

Chimney Sweeps and Cancer

In 1775, British physician Percival Pott became the first to recognize that environmental factors can cause cancer when he observed that chimney sweeps had a higher incidence of scrotum cancer than the male population as a whole. He theorized that something in the chimney soot was causing cancer. We now know that it was benzo[*a*]pyrene.

Percival Pott

A Victorian chimney sweep and his assistant—a boy small enough to fit inside narrow passages.

PROBLEM 41

Three arene oxides can be obtained from phenanthrene.

phenanthrene

a. Draw the structures of the three phenanthrene oxides.

b. Draw the structures of the phenols that can be obtained from each phenanthrene oxide.

c. If a phenanthrene oxide can lead to the formation of more than one phenol, which phenol will be obtained in greater yield?

d. Which of the three phenanthrene oxides is most likely to be carcinogenic?

10.9 AMINES DO NOT UNDERGO SUBSTITUTION OR ELIMINATION REACTIONS

Although **amines,** like alkyl halides, alcohols, and ethers, have an electron-withdrawing group bonded to an sp^3 carbon, they are not included in Group II because they do not undergo substitution and elimination reactions.

An amine's lack of reactivity in substitution and elimination reactions can be understood by comparing the leaving propensity of its electron-withdrawing group with the leaving propensity of the electron-withdrawing groups of the compounds that do undergo substitution and/or elimination reactions.

The relative leaving propensities of the groups can be determined by comparing the pK_a values of their conjugate acids, recalling that the weaker the acid, the stronger its conjugate base and the poorer the base is as a leaving group. The pK_a values of the conjugate acids show that the leaving group of an amine ($^-NH_2$) is such a strong base that amines cannot undergo substitution or elimination reactions. (HF has been used for the comparison because F is in the same row of the periodic chart as O and N, but recall that an alkyl fluoride has the poorest leaving group of the alkyl halides.)

The stronger the base, the poorer it is as a leaving group.

relative reactivities

most reactive $\quad RCH_2F \quad > \quad RCH_2OH \quad > \quad RCH_2OR \quad > \quad RCH_2NH_2 \quad$ least reactive

$$
\begin{array}{cccc}
\text{HF} & H_2O & \text{ROH} & NH_3 \\
pK_a = 3.2 & pK_a = 15.7 & pK_a \sim 16 & pK_a = 36
\end{array}
$$

Protonating the amino group makes it a better leaving group, but not nearly as good as a protonated alcohol, which is almost 14 pK_a units more acidic than a protonated amine.

$$
\begin{array}{cc}
CH_3CH_2\overset{+}{O}H_2 & > & CH_3CH_2\overset{+}{N}H_3 \\
pK_a = -2.4 & & pK_a = 11.2
\end{array}
$$

Therefore, unlike protonated alcohols, protonated amines cannot undergo substitution and elimination reactions because even when protonated, the group is too basic to be displaced.

Amines React as Bases and Nucleophiles

Although they cannot undergo substitution or elimination reactions, amines are extremely important organic compounds. The lone pair on its nitrogen allows an amine to react as both a base and a nucleophile.

Amines are the most common organic bases. We saw that protonated amines have pK_a values of about 11 (Section 2.3) and that protonated anilines have pK_a values of about 5 (Section 8.9). Neutral amines have very high pK_a values. For example, the pK_a of methylamine is 40.

$$CH_3CH_2CH_2\overset{+}{N}H_3 \qquad CH_3\overset{+}{N}H_2 \qquad CH_3CH_2\overset{+}{N}H \qquad$$

$$
\begin{array}{cccccc}
pK_a = 10.8 & pK_a = 10.9 & pK_a = 11.1 & pK_a = 4.58 & pK_a = 5.07 & pK_a = 40
\end{array}
$$

Amines react as nucleophiles in a wide variety of reactions. For example, we saw that they react s nucleophiles with alkyl halides and epoxides in S_N2 reactions.

$$\text{CH}_3\text{CH}_2\text{Br} + \text{CH}_3\text{NH}_2 \longrightarrow \text{CH}_3\text{CH}_2\overset{+}{\text{NH}}_2\text{CH}_3 + \text{Br}^-$$

an S_N2 reaction

$$\text{CH}_3\text{CH}\overset{\text{O}}{-}\text{CH}_2 + \text{CH}_3\text{NH}_2 \longrightarrow \text{CH}_3\overset{\text{O}^-}{\text{CH}}\text{CH}_2\overset{+}{\text{NH}}_2\text{CH}_3 \longrightarrow \text{CH}_3\overset{\text{OH}}{\text{CH}}\text{CH}_2\text{NHCH}_3$$

We will also see that they react as nucleophiles with a wide variety of carbonyl compounds Sections 16.8, 16.9, 17.10, 17.18, and 17.19).

Alkaloids

Alkaloids are amines found in the leaves, bark, roots, or seeds of many plants. Examples include caffeine (found in tea leaves, coffee beans, and cola nuts) and nicotine (found in tobacco leaves). Nicotine causes brain cells to release dopamine and endorphins, compounds that makes us feel good, thereby making nicotine addictive. Ephedrine, a bronchodilator, is an alkaloid obtained from *Ephedra sinica*, a plant found in China. Morphine, an analgesic, is an alkaloid obtained from opium, a milky fluid exuded by a species of poppy (p. 3).

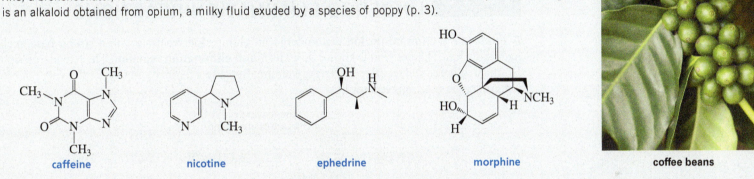

caffeine   nicotine   ephedrine   morphine   coffee beans

Lead Compounds for the Development of Drugs

Medicinal agents used by humans since ancient times provided the starting point for the development of our current arsenal of drugs. The active ingredients were isolated from the herbs, berries, roots, and bark used by medicine men and women, shamans, and witch doctors. Scientists still search the world for plants, berries, flora, and fauna that might yield new medicinal compounds.

Once a naturally occurring drug is isolated and its structure determined, it can serve as a prototype in a search for other biologically active compounds. The prototype is called a **lead compound** (that is, it plays a leading role in the search). Analogues of the lead compound are synthesized and tested to see if they are more effective or have fewer side effects than the lead compound. An analogue may have a different substituent than the lead compound, a branched chain instead of a straight chain, a different functional group, or some other structural difference. Producing analogues by changing the structure of a lead compound is called **molecular modification.**

In a classic example of molecular modification, a number of synthetic local anesthetics were developed from cocaine, an alkaloid obtained from the leaves of *Erythroxylon coca,* a bush native to the highlands of the South American Andes (see p. 458). Cocaine is a highly effective local anesthetic, but it produces undesirable effects on the central nervous system (CNS), ranging from initial euphoria to severe depression. By dissecting the cocaine molecule step by step—removing the methoxycarbonyl group and cleaving the seven-membered-ring system—scientists identified the portion of the molecule that carries the local anesthetic activity but does not induce the damaging CNS effects. This knowledge provided an improved lead compound.

cocaine
lead compound

improved lead compound

(Continued)

Hundreds of analogues were then synthesized. Successful anesthetics obtained by molecular modification were Benzocaine (a topical anesthetic), Novocain (used by dentists), and Xylocaine (one of the most widely used injectable anesthetics).

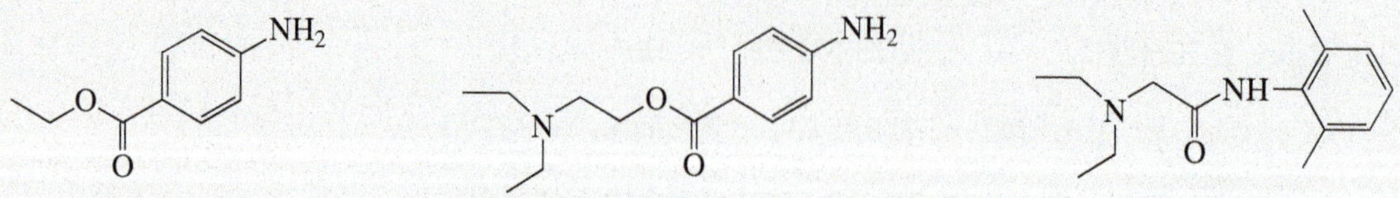

Benzocaine®

procaine
Novocain®

lidocaine
Xylocaine®

PROBLEM 42

Explain why the half-life (the time it takes for one-half of the compound to be metabolized) of Xylocaine is longer than that of Novocaine.

10.10 QUATERNARY AMMONIUM HYDROXIDES UNDERGO ELIMINATION REACTIONS

A **quaternary ammonium ion** can undergo an elimination reaction with a strong base such as hydroxide ion. The reaction is known as a **Hofmann elimination reaction.** The leaving group in a Hofmann elimination reaction is a tertiary amine. Because a tertiary amine is a relatively poor leaving group, the reaction requires heat.

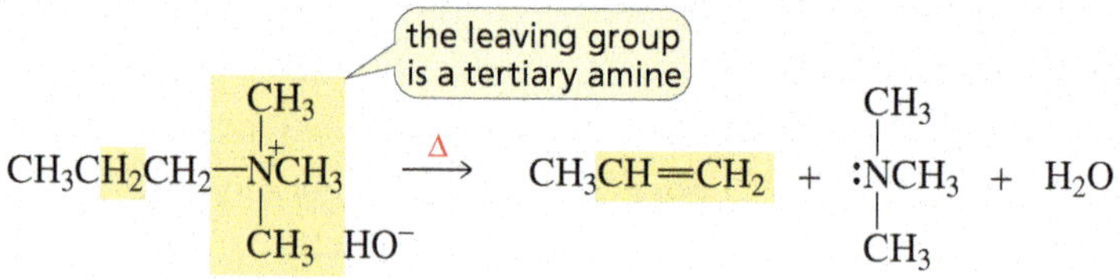

the leaving group is a tertiary amine

$$CH_3CH_2CH_2 - \overset{\overset{\displaystyle CH_3}{|}}{\underset{\underset{\displaystyle CH_3}{|}}{N^+}}CH_3 \quad HO^- \quad \overset{\Delta}{\longrightarrow} \quad CH_3CH=CH_2 \; + \; :N\overset{\overset{\displaystyle CH_3}{|}}{\underset{\underset{\displaystyle CH_3}{|}}{}}CH_3 \; + \; H_2O$$

quaternary ammonium ion

A Hofmann elimination reaction is an E2 reaction, which means the proton and the tertiary amine are removed in the same step (Section 9.7). Very little substitution product is formed.

MECHANISM FOR THE HOFMANN ELIMINATION

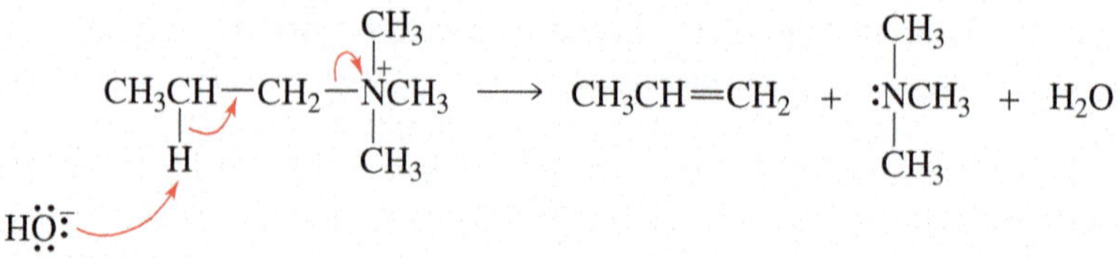

$$CH_3CH - CH_2 - \overset{\overset{\displaystyle CH_3}{|}}{\underset{\underset{\displaystyle CH_3}{|}}{N^+}}CH_3 \quad \longrightarrow \quad CH_3CH=CH_2 \; + \; :N\overset{\overset{\displaystyle CH_3}{|}}{\underset{\underset{\displaystyle CH_3}{|}}{}}CH_3 \; + \; H_2O$$

$$HO^-$$

- The tertiary amine is attached to the α-carbon, and the proton is removed from the adjacent carbon (the β-carbon).

If the quaternary ammonium ion has more than one β-carbon, the major alkene product is the one obtained by removing a proton from the β-carbon bonded to the *most* hydrogens. For example, in the following reaction, the major alkene product is obtained by removing a proton from the β-carbon bonded to three hydrogens, and the minor alkene product results from removing a proton from the β-carbon bonded to two hydrogens.

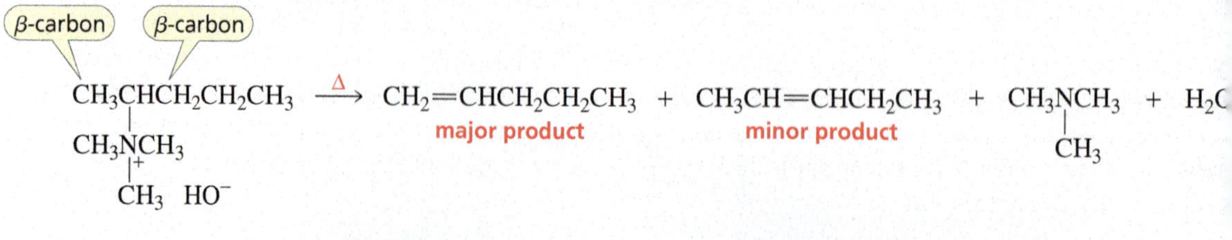

β-carbon β-carbon

$$CH_3CHCH_2CH_2CH_3 \quad \overset{\Delta}{\longrightarrow} \quad CH_2=CHCH_2CH_2CH_3 \; + \; CH_3CH=CHCH_2CH_3 \; + \; CH_3N\overset{}{\underset{\underset{\displaystyle CH_3}{|}}{}}CH_3 \; + \; H_2O$$
$$\underset{\underset{\displaystyle CH_3 \; HO^-}{|}}{CH_3NCH_3}$$

major product   minor product

Bitrex®

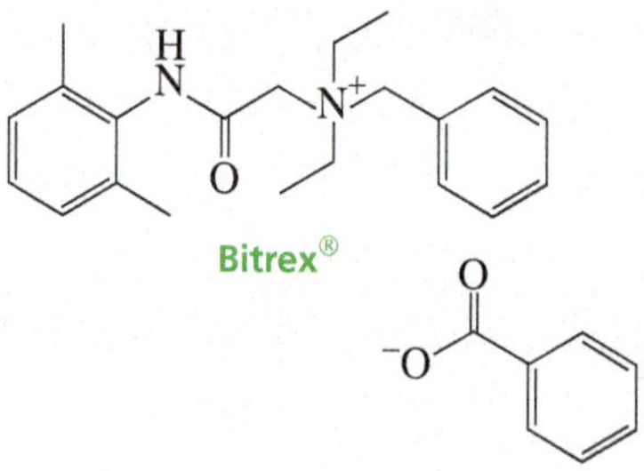

Bitrex, a quaternary ammonium salt, is nontoxic and one of the most bitter-tasting substances known. It is used to encourage deer to look elsewhere for food, it is put on the backs of animals to keep them from biting one another and on children's fingers to persuade them to stop sucking their thumbs or biting their fingernails, and it is added to toxic substances to keep them from being ingested accidentally.

We saw that in an E2 reaction of an alkyl chloride, alkyl bromide, or alkyl iodide, the proton removed from the β-carbon *bonded to the fewest hydrogens* (*Zaitsev's rule*; Section 9.7). Now, however, we see that in an E2 reaction of a quaternary ammonium ion, the proton is removed from the β-carbon *bonded to the most hydrogens* (anti-Zaitsev elimination).

PROBLEM 43

If a quaternary ammonium ion can undergo an elimination reaction with a strong base, why can't a protonated tertiary amine undergo the same reaction?

PROBLEM 44 ◆

What are the major products of the following reaction?

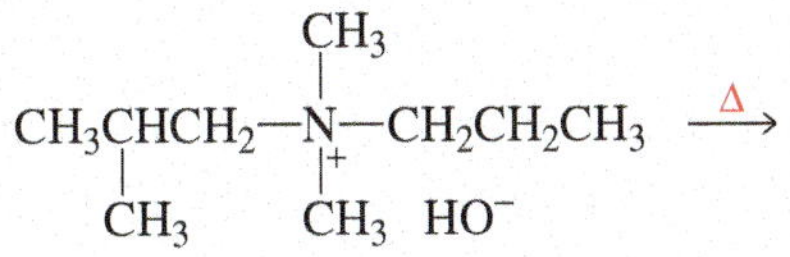

$$
\begin{array}{c}
\quad\quad CH_3 \\
\quad\quad | \\
CH_3CHCH_2-N-CH_2CH_2CH_3 \xrightarrow{\Delta} \\
\quad\quad | \quad\quad | \\
\quad\quad CH_3 \quad CH_3 \;\; HO^-
\end{array}
$$

PROBLEM 45 ◆

What are the minor products of the preceding Hofmann elimination reaction?

PROBLEM 46 ◆

What is the difference between the reaction that occurs when isopropyltrimethylammonium hydroxide is heated and the reaction that occurs when 2-bromopropane is treated with hydroxide ion?

> In a Hofmann elimination reaction, the proton is removed from the β-carbon bonded to the most hydrogens.

The Reason for anti-Zaitsev Elimination

Quaternary amines violate Zaitsev's rule for the same reason that alkyl fluorides violate it (Section 9.7). Alkyl halides, other than alkyl fluorides, have relatively good leaving groups that immediately start to depart when hydroxide ion starts to remove the proton, forming a transition state with an *alkene-like* structure. The proton is removed *from the β-carbon bonded to the fewest hydrogens* in order to achieve the most stable *alkene-like* transition state.

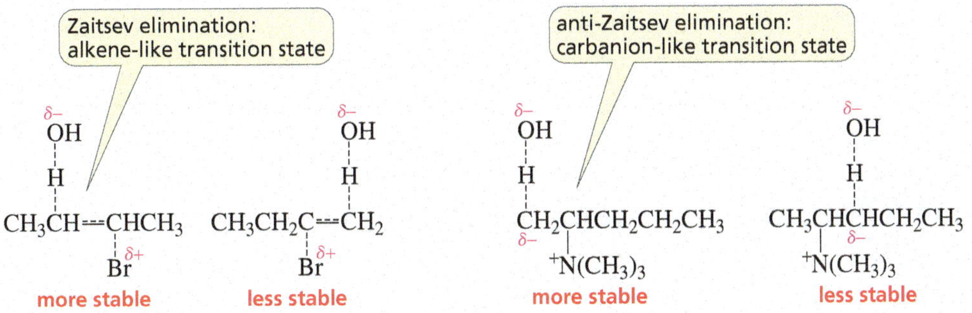

Zaitsev elimination: alkene-like transition state

anti-Zaitsev elimination: carbanion-like transition state

Quaternary ammonium ions and alkyl fluorides have poorer leaving groups that do not start to leave when hydroxide ion starts to remove a proton. Therefore, a partial negative charge builds up on the carbon from which the proton is being removed, giving the transition state a *carbanion-like* structure (Section 9.7). The proton is removed *from the β-carbon bonded to the most hydrogens* in order to achieve the more stable *carbanion-like* transition state. (Recall that primary carbanions are more stable than secondary carbanions, which are more stable than tertiary carbanions.) Steric factors also favor anti-Zaitsev elimination.

Because anti-Zaitsev elimination occurs in the Hofmann elimination reaction, *anti-Zaitsev elimination* is also known as *Hofmann elimination*.

For a quaternary ammonium ion to undergo an elimination reaction, the counterion must be hydroxide ion, because a strong base is needed to remove the proton from the β-carbon. Halide ions are weak bases, so quaternary ammonium *halides* cannot undergo Hofmann elimination reactions.

However, a quaternary ammonium *halide* can be converted into a quaternary ammonium *hydroxide* by treatment with silver oxide and water. The silver halide precipitates, and the halide ion is replaced by hydroxide ion.

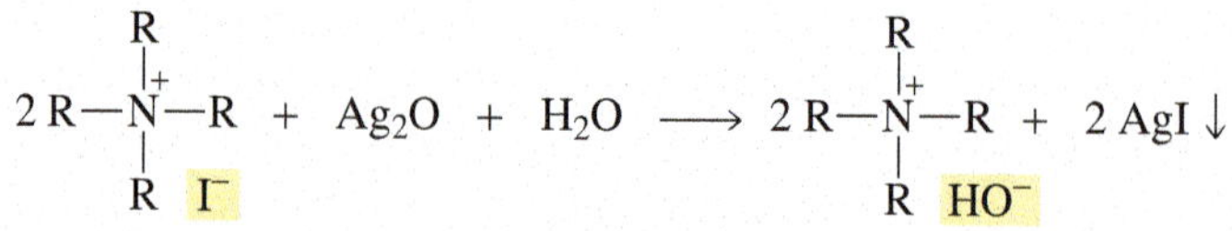

$$2\ R-\overset{R}{\underset{R}{\overset{|}{\underset{|}{N}}}}-R\ +\ Ag_2O\ +\ H_2O\ \longrightarrow\ 2\ R-\overset{R}{\underset{R}{\overset{|}{\underset{|}{N}}}}-R\ +\ 2\ AgI\downarrow$$

PROBLEM 47 ♦

What is the major product of each of the following reactions?

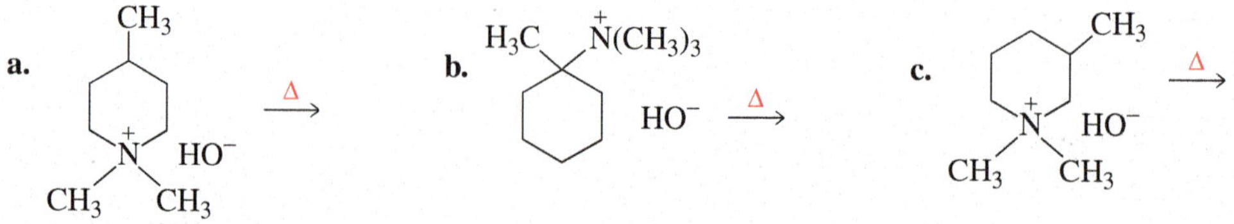

a. (structure) $\overset{\Delta}{\longrightarrow}$ b. (structure) $\overset{\Delta}{\longrightarrow}$ c. (structure) $\overset{\Delta}{\longrightarrow}$

LEARN THE STRATEGY

PROBLEM 48 SOLVED

Describe a synthesis for the following compound, using the given starting material and any necessar reagents:

$$CH_3CH_2CH_2CH_2NH_2\ \longrightarrow\ CH_3CH_2CH=CH_2$$

SOLUTION Although an amine cannot undergo an elimination reaction, a quaternary ammonium hydroxid can. The amine, therefore, must first be converted into a quaternary ammonium iodide by allowing it to reac with excess methyl iodide in a basic solution of potassium carbonate (see Problem 16 on pp. 404–405).

Treatment with aqueous silver oxide forms the quaternary ammonium hydroxide. Heat is required for th elimination reaction.

$$CH_3CH_2CH_2CH_2NH_2\ \xrightarrow[\text{K}_2\text{CO}_3]{\overset{\text{CH}_3\text{I}}{\text{excess}}}\ CH_3CH_2CH_2CH_2\overset{+}{N}(CH_3)_3\ \underset{\text{I}^-}{}\ \xrightarrow[\text{H}_2\text{O}]{\text{Ag}_2\text{O}}\ CH_3CH_2CH_2CH_2\overset{+}{N}(CH_3)_3\ \underset{\text{HO}^-}{}\ \overset{\Delta}{\longrightarrow}\ CH_3CH_2CH=CH_2\ +\ H$$

USE THE STRATEGY

PROBLEM 49

Describe a synthesis for each of the following compounds, using the given starting material and any neces sary reagents:

a. $CH_3CH_2CH_2\underset{\underset{Br}{|}}{C}HCH_3\ \longrightarrow\ CH_3CH_2CH_2CH=CH_2$ b.  (structure) $\longrightarrow\ CH_2=CH-CH=CH_2$

10.11 THIOLS, SULFIDES, AND SULFONIUM IONS

Thiols

Thiols are sulfur analogues of alcohols. They used to be called mercaptans because they form strong complexes with heavy metal cations such as arsenic and mercury—that is, they capture mercury.

$$\underset{\text{thiol}}{2\ CH_3CH_2SH}\ +\ \underset{\text{mercuric ion}}{Hg^{2+}}\ \longrightarrow\ CH_3CH_2S-Hg-SCH_2CH_3\ +\ 2\ H^+$$

Thiols are named by adding the suffix *thiol* to the name of the parent hydrocarbon. If there is a second functional group in the molecule with a higher priority that is identified by a suffix, the SI group can be indicated by its substituent name, *mercapto*. Like other substituent names, it is placec before the name of the parent hydrocarbon.

CH_3CH_2SH	$CH_3CH_2CH_2SH$	$\overset{\overset{CH_3}{|}}{CH_3CHCH_2CH_2SH}$	$HSCH_2CH_2OH$
ethanethiol	**1-propanethiol**	**3-methyl-1-butanethiol**	**2-mercaptoethanol**

Because sulfur is not as electronegative as oxygen, thiols are not good at hydrogen bonding. Consequently, they have weaker intermolecular attractions and, therefore, considerably lower boiling points than alcohols (Section 3.9). For example, the boiling point of CH_3CH_2SH is 37 °C, whereas the boiling point of CH_3CH_2OH is 78 °C.

Sulfur atoms are larger than oxygen atoms, so the negative charge of the thiolate ion is spread over a larger volume of space than the negative charge of an alkoxide ion, causing the thiolate ion to be more stable (Section 2.6). Thiols, therefore, are stronger acids ($pK_a \sim 10$) than alcohols ($pK_a \sim 15$). The less basic thiolate ions are less well solvated than alkoxide ions, so in protic solvents, the larger thiolate ions are better nucleophiles than alkoxide ions (Section 9.2).

$$CH_3-\overset{..}{\underset{..}{S}}: + CH_3CH_2-Br \xrightarrow{CH_3OH} CH_3-\overset{..}{\underset{..}{S}}-CH_2CH_3 + Br^-$$

Sulfides

The sulfur analogues of ethers are called **sulfides** or **thioethers**. Sulfur is an excellent nucleophile because its electron cloud is polarizable (Section 9.2). As a result, a thioether reacts readily with an alkyl halide to form a **sulfonium ion**, whereas an ether does not do the equivalent reaction because oxygen is not as nucleophilic as sulfur and cannot accommodate a positive charge as well as sulfur can.

$$CH_3-\overset{..}{\underset{..}{S}}-CH_3 + CH_3-I \longrightarrow \underset{\text{iodide}}{CH_3-\overset{+}{\underset{CH_3}{S}}-CH_3} \; I^-$$

dimethyl sulfide
thioether

trimethylsulfonium iodide
sulfonium ion

Low-molecular-weight thiols are noted for their strong and pungent odors, such as the odors associated with onions, garlic, and skunks. Natural gas is completely odorless and can cause deadly explosions if a leak goes undetected. As a result, a small amount of a thiol is added to natural gas to give it an odor so that gas leaks can be detected.

Sulfonium Ions

The positively charged group of a sulfonium ion is an excellent leaving group, so a sulfonium ion readily undergoes nucleophilic substitution reactions. Like other S_N2 reactions, the reaction works best if the group undergoing nucleophilic attack is a methyl group or a primary alkyl group.

$$HO:^- + CH_3-\overset{+}{\underset{CH_3}{S}}-CH_3 \longrightarrow CH_3-\overset{..}{\underset{..}{O}}H + CH_3-\overset{..}{\underset{..}{S}}-CH_3$$

PROBLEM 50

Using an alkyl halide and a thiol as starting materials, how would you prepare the following thioethers?

a. [structure]   **b.** [structure]   **c.** $CH_2=CHCHSCH_2CH_3$ with CH_3 branch   **d.** [structure]

Mustard Gas—A Chemical Warfare Agent

Chemical warfare occurred for the first time in 1915, when Germany released chlorine gas against French and British forces in the Battle of Ypres. For the remainder of World War I, both sides used a variety of chemical agents as weapons. One of the more common was mustard gas, a reagent that produces large blisters on exposed skin. Mustard gas is extremely reactive because its highly nucleophilic sulfur atom easily displaces a chloride ion by an intramolecular S_N2 reaction, forming a cyclic sulfonium ion that reacts rapidly with a nucleophile. The sulfonium ion is particularly reactive because of its strained three-membered ring and its excellent (positively charged) leaving group.

[reaction scheme]

mustard gas

sulfonium ion
+ Cl⁻

H₂Ö:

+ H⁺

H⁺ + [structure with HO and OH]   H₂Ö: [structure]   + Cl⁻

(Continued)

The blistering caused by mustard gas results from the high local concentrations of HCl produced when mustard comes into contact with water—or any other nucleophile—on skin or in lung tissue. Autopsies of soldiers killed by mustard gas in World War I revealed that they had extremely low white blood cell counts and defects in bone marrow development, indicating profound effects on rapidly dividing cells.

Alkylating Agents as Cancer Drugs

Because cancer is characterized by the uncontrolled growth and proliferation of cells, the discovery that mustard gas affected rapidly dividing cells suggested that it might be an effective anticancer agent. Therefore, chemists started looking for less reactive analogues of mustard gas that might be used in chemotherapy—that is, the use of chemicals in the treatment of cancer.

Because mustard gas forms a three-membered ring that can react rapidly with nucleophiles, its clinical reactivity is thought to be due to its ability to alkylate groups on the surface of DNA. Alkylating DNA can destroy it, which means that the rapidly growing cells of cancerous tumors are killed. Unfortunately, compounds used for chemotherapy can also kill normal cells. That is why many side effects, such as nausea and hair loss, are associated with cancer chemotherapy. The challenge for chemists now is to find drugs that target only cancer cells (Section 24.7).

The cancer drugs shown here are all alkylating agents—they attach an alkyl group to a nucleophile on DNA under physiological conditions.

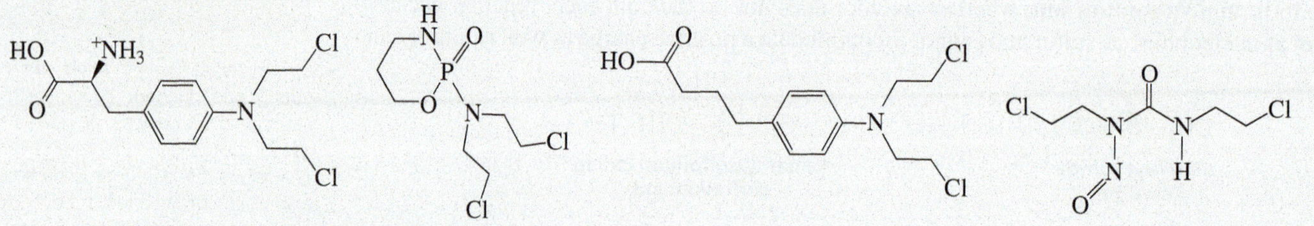

melphalan   cyclophosphamide   chloroambucil   carmustine

PROBLEM 51 ◆

The following three nitrogen mustards were studied for possible clinical use. One is now used clinically, one was found to be too unreactive, and one was found to be too insoluble in water to be injected intravenously. Which is which? (*Hint:* Draw resonance contributors.)

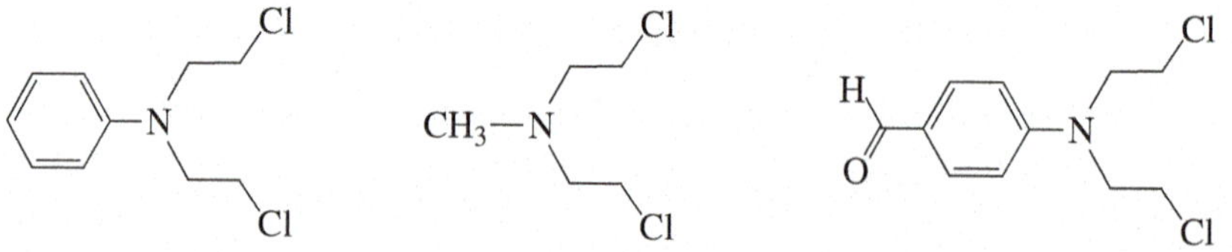

PROBLEM 52 ◆

Why is melphalan a good cancer drug?

PROBLEM 53

Mechlorethamine, the drug in Problem 51 that is in clinical use, is so highly reactive that it can be administered only by physicians who are experienced in its use. Explain why the four cancer drugs in the box at the top of this page are less reactive alkylating agents.

10.12 METHYLATING AGENTS USED BY CHEMISTS VERSUS THOSE USED BY CELLS

Methyl Halides Are the Methylating Agents Used by Chemists

If an organic chemist wanted to put a methyl group on a nucleophile, methyl iodide would most likely be used as the methylating agent. Of the methyl halides, methyl iodide has the most easily displaced leaving group because I⁻ is the weakest base of the halide ions. In addition, methyl iodide is a liquid at room temperature, so it is easier to handle than methyl bromide or methyl chloride, which are gases at room temperature. The reaction would be a simple S_N2 reaction.

$$\text{Nuc} + \text{CH}_3\text{—I} \longrightarrow \text{CH}_3\text{—Nuc} + \text{I}^-$$

ethyl halides, however, are not available in a cell. Because they are only slightly soluble in water,
xyl halides are not found in the predominantly aqueous environments of biological systems.

Eradicating Termites

Alkyl halides can be very toxic to biological organisms. For example, methyl bromide has been used to kill termites and other pests. Methyl bromide works by methylating the NH_2 and SH groups of enzymes, thereby destroying the enzymes' ability to catalyze important biological reactions. Unfortunately, methyl bromide has been found to deplete the ozone layer (Section 12.12), so its use has been banned.

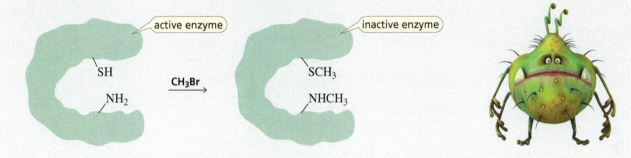

-Adenosylmethionine Is a Methylating Agent Used by Cells

lls use S-adenosylmethionine (SAM; also called AdoMet), a water-soluble sulfonium ion, as a
ethylating agent. (A less common biological methylating agent is discussed in Section 23.7.)
Although SAM is a much larger and more complicated looking molecule than methyl iodide, it
rforms the same function—namely, it transfers a methyl group to a nucleophile. Remember that
ological molecules are typically more complex than the molecules chemists use because of the
ed for molecular recognition (Section 5.14).

$$\longrightarrow \text{CH}_3\text{—Nuc} +$$

S-adenosylmethionine
SAM
AdoMet

this was the leaving group

S-adenosylhomocysteine
SAH

otice that SAM is sulfonium ion (Section 10.11). The positively charged sulfur readily accepts
e electrons left behind when the methyl group is transferred. In other words, the methyl group
attached to a very good leaving group, allowing biological methylation to readily take place.

xamples of Biological Methylation Reactions

specific example of a biological methylation reaction that uses SAM is the conversion of noradren-
ne (norepinephrine) to adrenaline (epinephrine). The reaction uses SAM to provide the methyl
oup. Noradrenaline and adrenaline are hormones that stimulate the breakdown of glycogen—the
dy's primary fuel source (see p. 973). You may have felt this "adrenaline rush" when preparing for
hallenging activity. Adrenaline is about six times more potent than noradrenaline. This methyla-
n reaction, therefore, is very important physiologically.

experiencing an adrenaline rush

The conversion of phosphatidylethanolamine, a component of cell membranes, into phosphtidylcholine, another cell membrane component, requires three methylations by three equivale of SAM (Section 25.5).

phosphatidylethanolamine + 3 SAM ⟶ phosphatidylcholine + 3 SA

S-Adenosylmethionine: A Natural Antidepressant

Marketed under the name SAMe (pronounced Sammy), *S*-adenosylmethionine is sold in many health food and drug stores as a treatment for depression and arthritis. Although SAMe has been used clinically in Europe for more than three decades, it has not been rigorously evaluated in the United States and, therefore, has not been approved by the FDA. It can be sold, however, because the FDA does not prohibit the sale of most naturally occurring substances as long as the marketer does not make therapeutic claims.

SAMe has also been found to be effective in the treatment of liver diseases, such as those caused by alcohol and the hepatitis C virus. The attenuation of injury to the liver is accompanied by an increase in the concentration of glutathione in the liver. Glutathione is an important biological antioxidant (Section 22.1). SAM is required for the biosynthesis of cysteine, one of the 20 most common naturally occurring amino acids (Section 22.9), which is required for the biosynthesis of glutathione.

PROBLEM 54

Propose a mechanism for the following reaction:

10.13 ORGANIZING WHAT WE KNOW ABOUT THE REACTIONS OF ORGANIC COMPOUNDS

We saw that organic compounds can be put into families and all the members of a family react in the same way. We also saw that the families can be put into one of four groups and all the families in a group react in similar ways. Now that we have finished studying the families in Group II, let's revisit it.

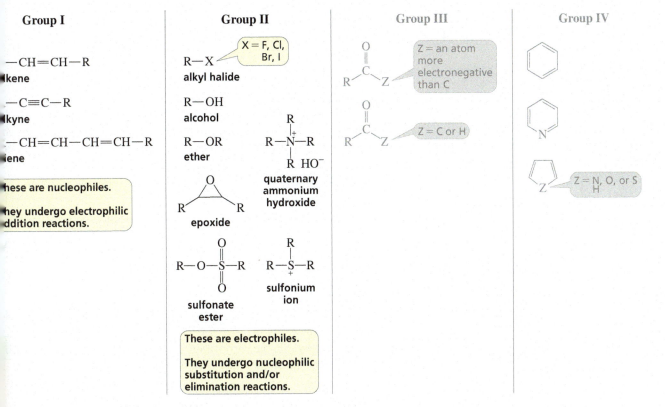

All the families in Group II are *electrophiles*, due to the partial positive charge on the carbon attached to the electron-withdrawing leaving group. As a result, the families in this group react with *nucleophiles*. The nucleophile can either attack the carbon to which the electron-withdrawing group is attached and substitute for it, or it can remove a hydrogen from an adjacent carbon and eliminate the electron-withdrawing group by forming an alkene. Thus, the families in Group II undergo nucleophilic substitution reactions and/or elimination reactions.

- Alkyl halides have excellent leaving groups, so they undergo substitution and/or elimination reactions with ease.

- Alcohols have much poorer leaving groups, so they need to be activated before they can undergo nucleophilic substitution and/or elimination reactions.

- Ethers, like alcohols, have poor leaving groups, but unlike alcohols, they can be activated only by protonation. They undergo only substitution reactions.

- Epoxides are more reactive than acyclic ethers because of the angle strain in the three-membered ring. Thus, they readily undergo substitution reactions whether or not they are activated by protonation.

- Quaternary ammonium hydroxide ions undergo elimination reactions. Because the leaving group is poor, the reaction requires heat and forms the less substituted elimination product.

- Sulfonate esters and sulfonium salts have good leaving groups, so they readily undergo substitution reactions. If the sulfonate ester is made from a tertiary alcohol, it undergoes elimination instead of substitution.

ESSENTIAL CONCEPTS

Section 10.1

- The leaving groups of alcohols and ethers are stronger bases than halide ions, so alcohols and ethers must be "activated" before they can undergo a substitution or elimination reaction.
- An alcohol can be activated by protonation. Therefore, it undergoes nucleophilic substitution reactions with HI, HBr, and HCl to form alkyl halides. These are S_N1 reactions in the case of secondary and tertiary alcohols and S_N2 reactions in the case of primary alcohols.
- S_N1 reactions form carbocation intermediates, so carbocation rearrangements can occur.

Section 10.2

- An alcohol can also be converted into an alkyl halide by PBr_3, PCl_3, or $SOCl_2$. These reagents convert the alcohol into an intermediate that has a leaving group that is easily displaced by a halide ion.

Section 10.3

- Converting an alcohol into a sulfonate ester is another way to activate an alcohol for subsequent reaction with a nucleophile that is a better nucleophile than a halide ion.
- Tertiary alcohols can form sulfonate esters but they are too sterically hindered to undergo a subsequent reaction with a nucleophile.
- Converting an alcohol to an alkyl halide with PBr_3, PCl_3, or $SOCl_2$ followed by reaction with a nucleophile forms a substitution product with the same configuration as the alcohol, whereas converting an alcohol to a sulfonate ester followed by reaction with a nucleophile forms a substitution product with a configuration opposite that of the alcohol.

Section 10.4

- An alcohol undergoes dehydration (elimination of a water molecule) when it is heated with an acid.
- Dehydration is an E1 reaction in the case of secondary and tertiary alcohols and an E2 reaction in the case of primary alcohols.
- Tertiary alcohols are the easiest to dehydrate, and primary alcohols are the hardest.
- The major product of alcohol dehydration is the more stable alkene.
- If the alkene has stereoisomers, the stereoisomer in which the largest groups are on opposite sides of the double bond will be the major product.
- E1 reactions form carbocation intermediates, so carbocation rearrangements can occur.
- Dehydration of secondary and tertiary alcohols can take place by an E2 reaction if a good leaving group is put on the alcohol (by allowing it to react with $POCl_3$, or in the case of tertiary alcohol, by converting it to a sulfonate ester) before the elimination reaction.

Section 10.5

- Chromic acid oxidizes primary alcohols to carboxylic acids and secondary alcohols to ketones.

- PCC, hypochlorous acid, and the Swern oxidation oxidizes primary alcohols to aldehydes and secondary alcohols to ketones.

Section 10.6

- Ethers undergo nucleophilic substitution reactions with HBr or HI and heat; if departure of the leaving group creates a relatively stable carbocation, an S_N1 reaction occurs; otherwise, an S_N2 reaction occurs with the nucleophile attacking the less sterically hindered of the two alkyl groups.

Section 10.7

- Ring strain causes an epoxide to be more reactive than an ether.
- Epoxides undergo nucleophilic substitution reactions. Under acidic conditions, the nucleophile adds to the more substituted ring carbon; under neutral or basic conditions, it adds to the less sterically hindered ring carbon.
- To convert an alkene to an alcohol without a carbocation rearrangement, first convert the alkene to an epoxide and then react the epoxide with LAH.
- Epoxidation of cyclohexene followed by reaction with hydroxide ion forms a trans 1,2-diol.
- Reaction of cyclohexene with osmium tetroxide followed by hydrolysis with hydrogen peroxide forms a cis 1,2-diol.

Section 10.8

- Aromatic hydrocarbons (arenes) are oxidized to arene oxides that undergo nucleophilic addition or rearrange to form phenols.
- The more stable the carbocation formed during rearrangement, the less likely the arene oxide is carcinogenic.

Section 10.9

- Amines cannot undergo substitution or elimination reactions because their leaving groups are very strong bases. Amines react as bases and nucleophiles.

Section 10.10

- A quaternary ammonium hydroxide undergoes an E2 reaction (a Hofmann elimination) if it is heated. Hydroxide ion removes a proton from the β-carbon bonded to the most hydrogens.

Section 10.11

- Thiols are stronger acids and have lower boiling points than alcohols.
- Thiolate ions are weaker bases and better nucleophiles than alkoxide ions in protic solvents.
- Thioethers react with alkyl halides to form sulfonium ions, which have excellent leaving groups, so they undergo substitution reactions with ease.

Section 10.12

- S-Adenosylmethionine, the most common biological methylating agent, is a sulfonium ion.

SUMMARY OF REACTIONS

1. Converting an *alcohol* to an *alkyl halide* (Sections 10.1 and 10.2). The mechanisms are shown on pages 460, 461, 464, and 465.

$$R-OH + HBr \xrightarrow{\Delta} R-Br$$

$$R-OH + HI \xrightarrow{\Delta} R-I$$

$$R-OH + HCl \xrightarrow{\Delta} R-Cl$$

relative rate: tertiary > secondary > primary

$$R-OH + PBr_3 \xrightarrow{pyridine} R-Br$$

$$R-OH + PCl_3 \xrightarrow{pyridine} R-Cl$$

$$R-OH + SOCl_2 \xrightarrow{pyridine} R-Cl$$

only for primary and secondary alcohols

2. Converting an alcohol to a *sulfonate ester* (Section 10.3). The mechanism is shown on page 466.

$$R-OH + Cl-\overset{O}{\underset{O}{\overset{\|}{\underset{\|}{S}}}}-R' \xrightarrow{pyridine} R-O-\overset{O}{\underset{O}{\overset{\|}{\underset{\|}{S}}}}-R' + HCl$$

3. Using an S_N2 reaction to convert an *activated alcohol* (an alkyl halide or a sulfonate ester) to a *compound with a new group bonded to the sp³ carbon* (Section 10.3). The mechanism is shown on p. 466.

$$R-Br + Y^- \longrightarrow R-Y + Br^-$$

$$R-O-\overset{O}{\underset{O}{\overset{\|}{\underset{\|}{S}}}}-R' + Y^- \longrightarrow R-Y + {}^-O-\overset{O}{\underset{O}{\overset{\|}{\underset{\|}{S}}}}-R'$$

4. Elimination reactions of alcohols: dehydration (Section 10.4). The mechanisms are shown on pages 468 and 470.

$$-\overset{|}{\underset{H}{C}}-\overset{|}{\underset{OH}{C}}- \underset{\Delta}{\overset{H_2SO_4}{\rightleftharpoons}} C=C + H_2O$$

relative rate: tertiary > secondary > primary

$$-\overset{|}{\underset{H}{C}}-\overset{|}{\underset{OH}{C}}- \xrightarrow[\text{pyridine, 0 °C}]{POCl_3} C=C + {}^-OPOCl_2$$

dehydration without carbocation rearrangement

5. Oxidation of alcohols (Section 10.5). The mechanism is shown on page 475.

primary alcohols $\quad RCH_2OH \xrightarrow{H_2CrO_4} \left[\overset{O}{\underset{}{\overset{\|}{R}}\overset{}{C}_{H}} \right] \xrightarrow[\text{oxidation}]{\text{further}} \overset{O}{\underset{OH}{\overset{\|}{R}\overset{}{C}}}$

$$RCH_2OH \xrightarrow[\text{CH}_2\text{Cl}_2]{PCC} \overset{O}{\underset{H}{\overset{\|}{R}\overset{}{C}}}$$

$$RCH_2OH \xrightarrow[\text{0 °C}]{NaOCl, CH_3COOH} \overset{O}{\underset{H}{\overset{\|}{R}\overset{}{C}}}$$

RCH$_2$OH $\xrightarrow[\text{2. triethylamine}]{\text{1. CH}_3\text{SCH}_3,\ \text{Cl}-\overset{\text{O}}{\underset{}{\text{C}}}-\overset{\text{O}}{\underset{}{\text{C}}}-\text{Cl},\ -60\ ^\circ\text{C}}$ $\overset{\text{O}}{\underset{\text{R}\ \ \ \ \text{H}}{\text{C}}}$

secondary alcohols

$\overset{\text{OH}}{\underset{\text{RCHR}}{|}}$ $\xrightarrow{\text{H}_2\text{CrO}_4}$ $\overset{\text{O}}{\underset{\text{R}\ \ \ \text{R}}{\overset{||}{\text{C}}}}$

$\overset{\text{OH}}{\underset{\text{RCHR}}{|}}$ $\xrightarrow[\text{CH}_2\text{Cl}_2]{\text{PCC}}$ $\overset{\text{O}}{\underset{\text{R}\ \ \ \text{R}}{\overset{||}{\text{C}}}}$

$\overset{\text{OH}}{\underset{\text{RCHR}}{|}}$ $\xrightarrow[0\ ^\circ\text{C}]{\text{NaOCl, CH}_3\text{COOH}}$ $\overset{\text{O}}{\underset{\text{R}\ \ \ \text{R}}{\overset{||}{\text{C}}}}$

$\overset{\text{OH}}{\underset{\text{RCHR}}{|}}$ $\xrightarrow[\text{2. triethylamine}]{\text{1. CH}_3\text{SCH}_3,\ \text{Cl}-\overset{\text{O}}{\text{C}}-\overset{\text{O}}{\text{C}}-\text{Cl},\ -60\ ^\circ\text{C}}$ $\overset{\text{O}}{\underset{\text{R}\ \ \ \text{R}}{\overset{||}{\text{C}}}}$

6. Nucleophilic substitution reactions of ethers (Section 10.6). The mechanisms are shown on pages 477 and 478.

$$\text{R}-\text{O}-\text{R}' \ +\ \text{HX} \ \xrightarrow{\Delta}\ \text{R}-\text{OH} \ +\ \text{R}'-\text{X}$$

HX = HBr or HI

7. Nucleophilic substitution reactions of *epoxides* (Section 10.7). The mechanisms are shown on pages 481 and 482.

$\underset{\text{CH}_3}{\overset{\text{CH}_3}{\text{C}}}\overset{\text{O}}{\diagup\!\diagdown}\text{CH}_2$ $\xrightarrow[\text{CH}_3\text{OH}]{\text{HCl}}$ $\text{CH}_3\overset{\text{OCH}_3}{\underset{\text{CH}_3}{\text{CCH}_2\text{OH}}}$ under acidic conditions, the nucleophile attacks the more substituted ring carbon

$\underset{\text{CH}_3}{\overset{\text{CH}_3}{\text{C}}}\overset{\text{O}}{\diagup\!\diagdown}\text{CH}_2$ $\xrightarrow[\text{CH}_3\text{OH}]{\text{CH}_3\text{O}^-}$ $\text{CH}_3\overset{\text{OH}}{\underset{\text{CH}_3}{\text{CCH}_2\text{OCH}_3}}$ under neutral or basic conditions, the nucleophile attacks the less sterically hindered ring carbon

8. Formation of cis and trans 1,2-diols (Section 10.7). The mechanism is shown on page 484.

cyclohexene $\xrightarrow[\text{2. H}_2\text{O}_2,\ \text{H}_2\text{O}]{\text{1. OsO}_4}$ cyclohexane-1,2-diol (OH, OH)

cyclohexene oxide $\xrightarrow[\text{2. HCl}]{\text{1. HO}^-}$ (OH, OH) + (OH, OH)

9. Reactions of *arene oxides:* ring opening and rearrangement (Section 10.8). The mechanism is shown on page 486.

benzene oxide $\xrightarrow[\text{2. HB}^+]{\text{1. Y}^-}$ (Y, OH) + (OH, Y) + (OH)

10. Elimination reactions of *quaternary ammonium hydroxides;* the proton is removed from the β-carbon bonded to the most hydrogens (Section 10.10). The mechanism is shown on page 492.

$\text{RCH}_2\text{CH}_2\overset{\text{CH}_3}{\underset{\text{CH}_3}{\overset{|+}{\text{N}}}}\text{CH}_3$ HO^- $\xrightarrow[\substack{\text{Hofmann}\\\text{elimination}}]{\Delta}$ $\text{RCH}=\text{CH}_2\ +\ \overset{\text{CH}_3}{\underset{\text{CH}_3}{\overset{|}{\text{N}}}}\text{CH}_3\ +\ \text{H}_2\text{O}$

11. Reactions of thiols, sulfides, and sulfonium ions (Section 10.11). The mechanisms are shown on page 495.

$$2\,RSH \; + \; Hg^{2+} \; \longrightarrow \; RS-Hg-SR \; + \; 2\,H^{+}$$

$$RS^{-} \; + \; R'-Br \; \longrightarrow \; RS-R' \; + \; Br^{-}$$

$$R-S-R \; + \; R'-I \; \longrightarrow \; R-\overset{\overset{R'}{|}}{\underset{+}{S}}-R \; + \; I^{-}$$

$$R-\overset{\overset{R}{|}}{\underset{+}{S}}-R \; + \; Y^{-} \; \longrightarrow \; R-Y \; + \; R-S-R$$

PROBLEMS

55. What is the product of each of the following reactions?

a. $CH_3CH_2CH_2OH$ $\xrightarrow[\substack{2.\ CH_3\overset{O}{\overset{\|}{C}}O^-}]{1.\ \text{methanesulfonyl chloride}}$

b. $CH_3CH_2CH_2CH_2OH + PBr_3 \xrightarrow{\text{pyridine}}$

c. $CH_3\underset{\underset{CH_3}{|}}{C}HCH_2CH_2OH$ $\xrightarrow[\substack{2.\ \langle\text{Ar}\rangle-O^-}]{1.\ p\text{-toluenesulfonyl chloride}}$

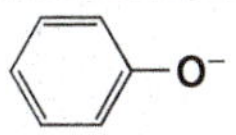

d. $CH_3CH_2CH_2CH_2OH \xrightarrow[\Delta]{H_2SO_4}$

e. 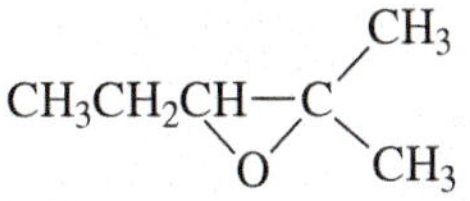 $CH_3CH_2CH{-}\overset{\overset{CH_3}{|}}{\underset{\underset{O}{\diagup\diagdown CH_3}}{C}}$ $\xrightarrow[CH_3OH]{CH_3O^-}$

f. 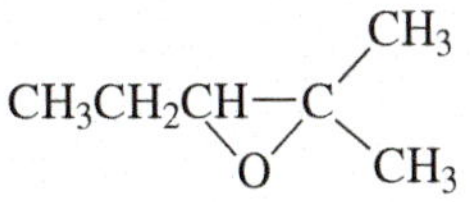 $CH_3CH_2CH{-}\overset{\overset{CH_3}{|}}{\underset{\underset{O}{\diagup\diagdown CH_3}}{C}}$ $\xrightarrow[CH_3OH]{HCl}$

g. $CH_3\underset{\underset{CH_3}{|}}{C}HCH_2CH_2OH$ $\xrightarrow[\text{pyridine}]{SOCl_2}$

56. Indicate which alcohol in each pair undergoes an elimination reaction more rapidly when heated with H_2SO_4.

a. 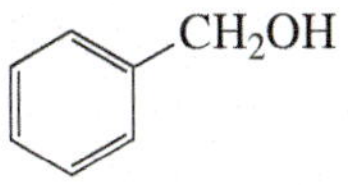 $\langle\text{Ar}\rangle$–CH$_2$OH or $\langle\text{Ar}\rangle$–CH$_2$CH$_2$OH

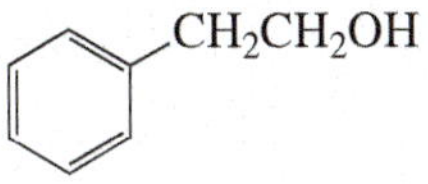

b. 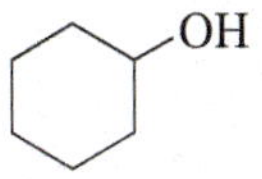 $\langle\text{cyclohexane}\rangle$–OH or $\langle\text{cyclohexene}\rangle$–OH

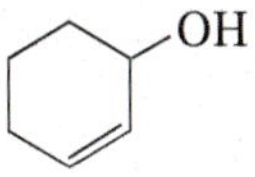

c. 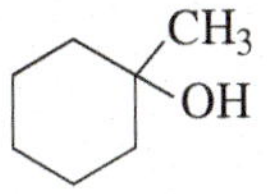 $\langle\text{cyclohexane}\rangle$(CH$_3$)(OH) or $\langle\text{cyclohexane}\rangle$(CH$_3$)(OH)

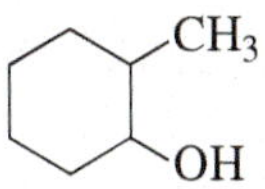

d. 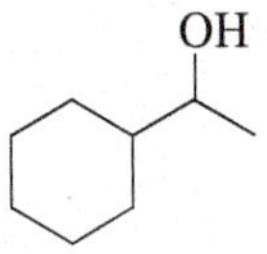 $\langle\text{cyclohexyl}\rangle$–CH(OH)CH$_3$ or $\langle\text{phenyl}\rangle$–CH(OH)CH$_3$

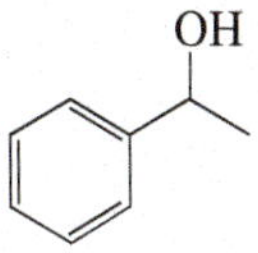

e. 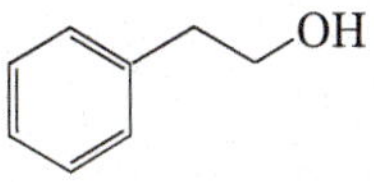 $\langle\text{phenyl}\rangle$–CH$_2CH_2$OH or $\langle\text{phenyl}\rangle$–CH(OH)CH$_3$

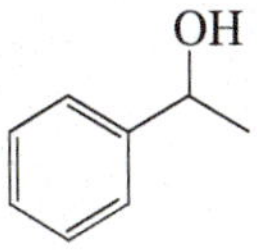

57. Identify A–E.

$\langle\text{cyclohexene}\rangle$ $\xrightarrow[H_2O]{Br_2}$ **A** + **B** $\xrightarrow{HO^-}$ **C** $\xrightarrow[H_2O]{HO^-}$ **D** + **E**

58. Starting with (R)-1-deuterio-1-propanol, how could you prepare

a. (S)-1-deuterio-1-propanol?   **b.** (S)-1-deuterio-1-methoxypropane?   **c.** (R)-1-deuterio-1-methoxypropane?

59. When heated with H_2SO_4, both 3,3-dimethyl-2-butanol and 2,3-dimethyl-2-butanol are dehydrated to form 2,3-dimethyl-2-butene. Which alcohol dehydrates more rapidly?

60. What is the major product obtained from the reaction of 2-ethyloxirane with each of the following reagents?

a. 0.1 M HCl   **b.** CH_3OH/HCl **c.** 0.1 M NaOH   **d.** CH_3OH/CH_3O^-

61. Write the appropriate reagent over each arrow.

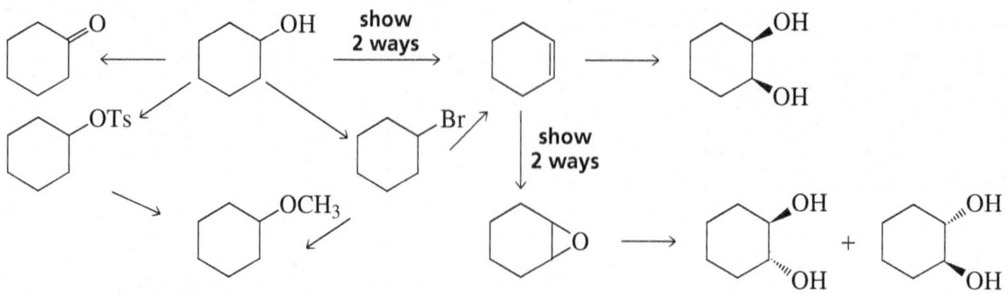

62. What alkenes would you expect to be obtained from the acid-catalyzed dehydration of 1-hexanol?

63. What is the major product(s) of each of the following reactions?

a. $CH_3\overset{\displaystyle CH_3}{\underset{\displaystyle CH_3}{C}}OCH_2CH_3$ + HBr $\overset{\Delta}{\longrightarrow}$

b. $CH_3\overset{\displaystyle CH_3}{CH}CH_2OCH_3$ + HI $\overset{\Delta}{\longrightarrow}$

c. cyclohexane–CH_2CH_2OH $\overset{H_2CrO_4}{\longrightarrow}$

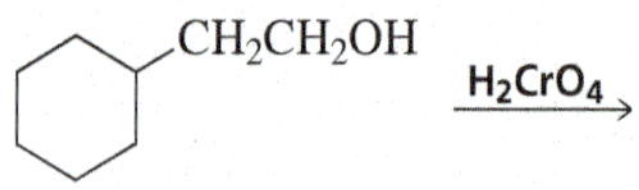

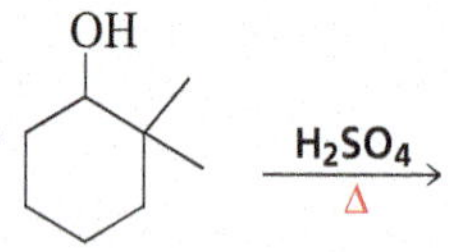

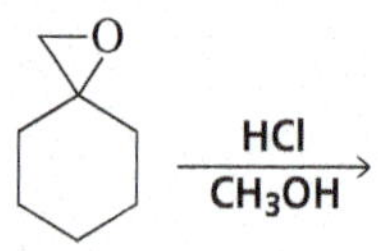

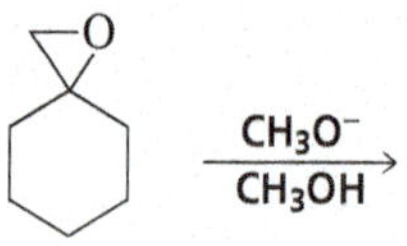

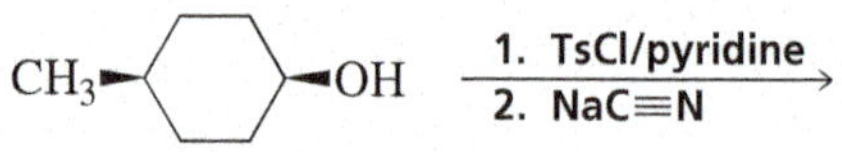

d. $\overset{H_2SO_4}{\underset{\Delta}{\longrightarrow}}$

e. $\overset{HCl}{\underset{CH_3OH}{\longrightarrow}}$

f. $\overset{CH_3O^-}{\underset{CH_3OH}{\longrightarrow}}$

g. CH_3–cyclohexane–OH $\overset{\text{1. TsCl/pyridine}}{\underset{\text{2. NaC}\equiv\text{N}}{\longrightarrow}}$

h. $\overset{NaOCl}{\underset{\underset{0\,°C}{CH_3COOH}}{\longrightarrow}}$

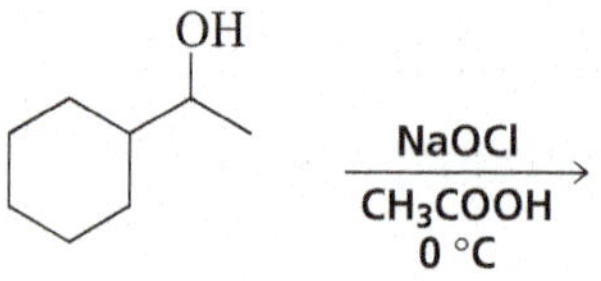

i. $CH_3\overset{}{CH}CH_2\overset{\overset{\displaystyle CH_3}{|+}}{N}CH_2CH_3$ $\overset{\Delta}{\longrightarrow}$
  with CH_3 CH_3 HO^-

j. + HBr $\overset{\Delta}{\longrightarrow}$

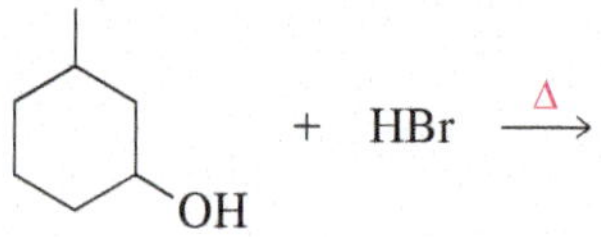

64. In Section 10.12, we saw that *S*-adenosylmethionine (SAM) methylates the nitrogen atom of noradrenaline to form adrenaline, a more potent hormone. If SAM methylates an OH group attached to the benzene ring instead, it completely destroys noradrenaline's activity.

a. Show the mechanism for the methylation of the OH group by SAM.
b. Which reaction is more apt to occur, methylation on nitrogen or methylation on oxygen?

noradrenaline + *S*-adenosymethionine → a biologically inactive compound + SAH

65. When deuterated phenanthrene oxide undergoes a rearrangement in water to form a phenol, 81% of the deuterium is retained in the product.

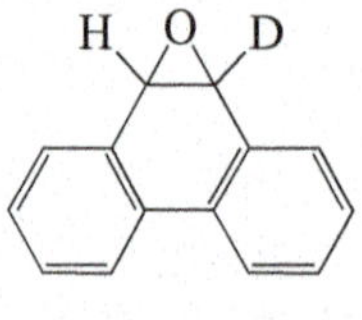

a. What percentage of the deuterium is retained if an NIH shift occurs?
b. What percentage of the deuterium is retained if an NIH shift does not occur?

66. An unknown alcohol with a molecular formula of $C_7H_{14}O$ was oxidized to an aldehyde with HOCl. When an acidic solution of the alcohol was distilled, two alkenes were obtained. The alkene formed in greater yield was determined to be 1-methylcyclohexene. The other alkene formed the original unknown alcohol when treated with BH_3/THF followed by H_2O_2, HO^-, and H_2O. Identify the unknown alcohol.

67. Explain why the acid-catalyzed dehydration of an alcohol is a reversible reaction, whereas the base-promoted dehydrohalogenation of an alkyl halide is an irreversible reaction.

68. Explain why (*S*)-2-butanol forms a racemic mixture when it is heated in sulfuric acid.

504

69. Fill in each box with the appropriate reagent:

a. $CH_3CH_2CH=CH_2$ $\xrightarrow[\text{2.}]{\text{1.}\ \square}$ $CH_3CH_2CH_2CH_2OH$ $\xrightarrow{\square}$ $CH_3CH=CHCH_3$ $\xrightarrow{\square}$ $CH_3\overset{O}{\underset{}{\overset{\|}{C}}}H$

b. $CH_3CH_2CH=CH_2$ $\xrightarrow[\text{2.}]{\text{1.}\ \square}$ $CH_3CH_2CH_2CH_2OH$ $\xrightarrow{\square}$ $CH_3CH_2CH_2\overset{O}{\underset{}{\overset{\|}{C}}}H$

c. 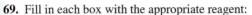 ⬡ $\xrightarrow{\square}$ ⬡(Br) $\xrightarrow{\square}$ ⬡(OH) $\xrightarrow{\square}$ ⬡(=O)

70. Propose a mechanism for the following reaction:

$$\square\text{-}\overset{CH_3}{\underset{}{\overset{|}{CHOH}}} \xrightarrow[\Delta]{H_2SO_4} \text{(cyclopentene-}CH_3\text{)} + HB^+$$

71. What product would be formed if the four-membered ring alcohol in Problem 70 were heated with an equivalent amount of HBr rather than with a catalytic amount of H_2SO_4?

72. Which of the following ethers would be obtained in greatest yield directly from alcohols?

73. Using the given starting material, any necessary inorganic reagents, and any carbon-containing compounds with no more than two carbons, indicate how the following syntheses could be carried out:

a. (cyclohexane-OH) $\longrightarrow$ (cyclohexane)

b. $CH_3CH_2CH_2CH_2Br \longrightarrow CH_3CH_2CH_2\overset{O}{\underset{}{\overset{\|}{C}}}OH$

74. When piperidine undergoes the series of reactions shown here, 1,4-pentadiene is obtained as the product. When the four different methyl-substituted piperidines undergo the same series of reactions, each forms a different diene: 1,5-hexadiene; 1,4-pentadiene; 2-methyl-1,4-pentadiene; and 3-methyl-1,4-pentadiene. Which methyl-substituted piperidine forms which diene?

$$\text{piperidine} \xrightarrow[\substack{\text{2. Ag}_2\text{O, H}_2\text{O} \\ \text{3. }\Delta}]{\text{1. excess CH}_3\text{I/K}_2\text{CO}_3} CH_3\overset{}{\underset{CH_3}{\overset{|}{N}}}CH_2CH_2CH_2CH=CH_2 \xrightarrow[\substack{\text{2. Ag}_2\text{O, H}_2\text{O} \\ \text{3. }\Delta}]{\text{1. excess CH}_3\text{I/K}_2\text{CO}_3} CH_2=CHCH_2CH=CH_2$$

1,4-pentadiene

75. When 3-methyl-2-butanol is heated with concentrated HBr, a rearranged product is obtained. When 2-methyl-1-propanol reacts under the same conditions, a rearranged product is not obtained. Explain.

76. Draw structures for compounds **A–F.**

$$\text{(alcohol)} \xrightarrow[\text{pyridine}]{PCl_3} \mathbf{A} \xrightarrow{tert\text{-BuO}^-} \mathbf{B} \xrightarrow[\text{H}_2\text{O}]{H_2SO_4} \mathbf{C} \xrightarrow[\Delta]{H_2SO_4} \mathbf{D} \xrightarrow[\text{2. (CH}_3)_2\text{S}]{\text{1. O}_3,\ -78\ °C} \mathbf{E} + \mathbf{F}$$

77. Propose a mechanism for each of the following reactions:

a. (epoxide with CH$_2$Cl) $\xrightarrow{CH_3O^-}$ CH_3OCH_2—(product)—CH_3 $+ Cl^-$

b. (benzene ring with OH groups) $\xrightarrow[\Delta]{H_2SO_4}$ (cyclic product) $+ H_2O$

78. How could you synthesize isopropyl propyl ether, using isopropyl alcohol as the only carbon-containing reagent?

79. When ethyl ether is heated with excess HI for several hours, the only organic product obtained is ethyl iodide. Explain why ethyl alcohol is not obtained as a product.

80. When the following seven-membered ring alcohol is dehydrated, three alkenes are formed. Propose a mechanism for their formation.

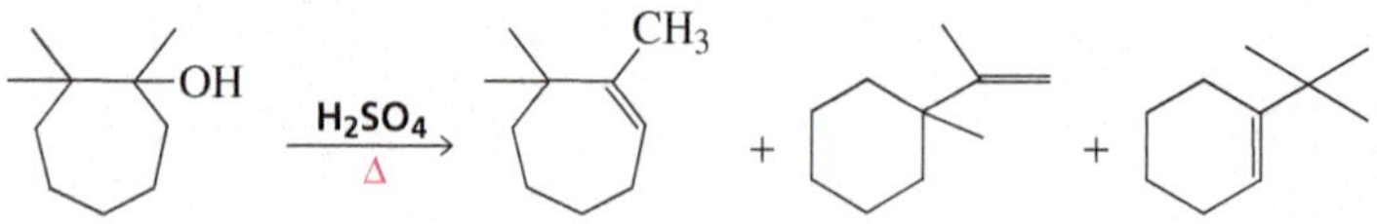

81. Ethylene oxide reacts readily with HO⁻ because of the strain in the three-membered ring. Explain why cyclopropane, a compound with approximately the same amount of strain, does not react with HO⁻.

82. Describe how each of the following compounds could be synthesized from the given starting material.

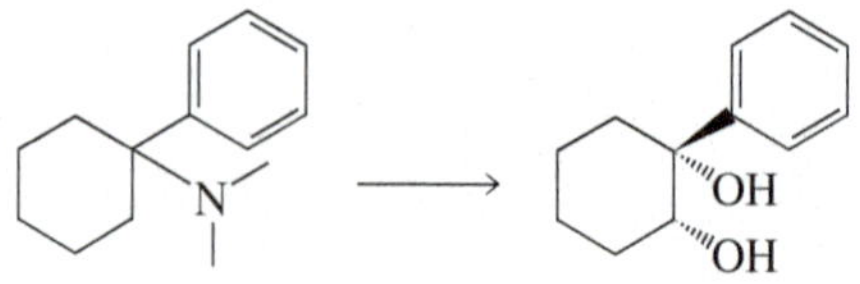

83. Propose a mechanism for each of the following reactions:

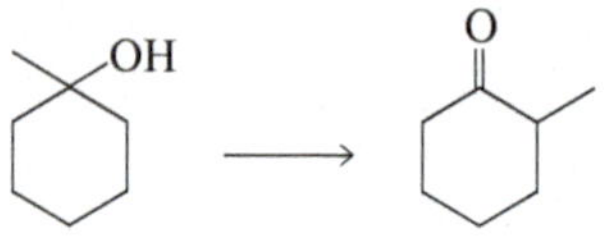

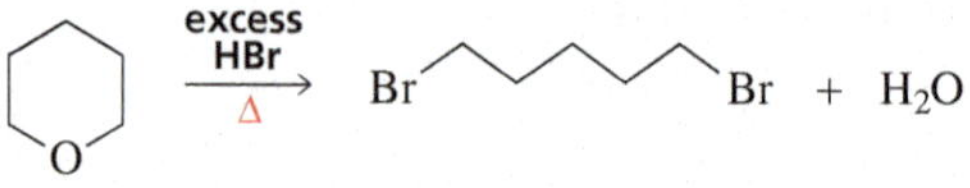

84. Triethylene glycol is one of the products obtained from the reaction of excess ethylene oxide and hydroxide ion. Propose a mechanism for its formation.

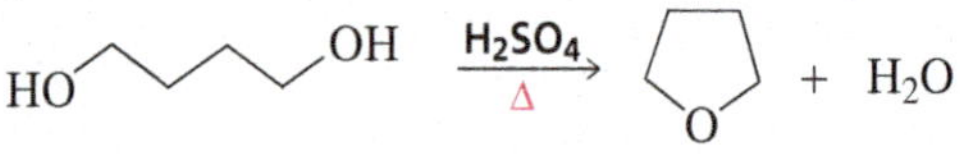

$$H_2C\!\!-\!\!CH_2 + HO^- \longrightarrow HOCH_2CH_2OCH_2CH_2OCH_2CH_2OH$$
triethylene glycol

85. a. Propose a mechanism for the following reaction:

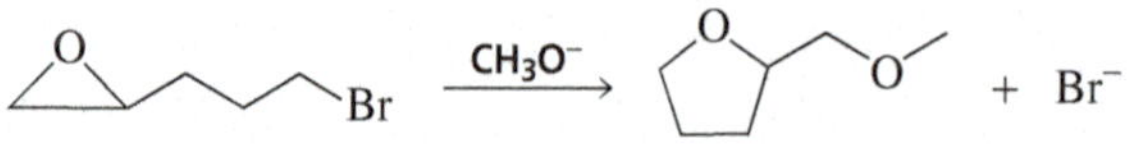

b. A small amount of a product containing a six-membered ring is also formed. Draw the structure of that product.
c. Why is so little six-membered ring product formed?

86. Propose a mechanism for the following reaction:

87. Early organic chemists used the Hofmann elimination reaction as the last step of a process known as a **Hofmann degradation**—a method used to identify amines. In a *Hofmann degradation,* an amine is methylated with excess methyl iodide in a basic solution, treated with silver oxide to convert the quaternary ammonium iodide to a quaternary ammonium hydroxide, and then heated to allow it to undergo a Hofmann elimination. Once the alkene product is identified, working backward gives the structure of the amine. Identify the amine in each of the following cases:

a. 4-Methyl-2-pentene is obtained from the Hofmann degradation of a primary amine.
b. 3-Methyl-1-butene is obtained from the Hofmann degradation of a primary amine.
c. 2-Methyl-1-3-butadiene is obtained from two successive Hofmann degradations of a secondary amine.

88. An ion with a positively charged nitrogen atom in a three-membered ring is called an aziridinium ion. The following aziridinium ion reacts with sodium methoxide to form compounds **A** and **B**:

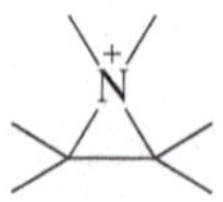

aziridinium ion

If a small amount of aqueous Br₂ is added to **A**, the reddish color of Br₂ persists, but the color disappears when Br₂ is added to **B**. When the aziridinium ion reacts with methanol, only **A** is formed. Identify **A** and **B**.

89. The following reaction takes place several times faster than the reaction of 2-chlorobutane with HO⁻:

$$(CH_3CH_2)_2\overset{..}{N}\!\!-\!\!CH_2CHCH_2CH_3 \xrightarrow{\ HO^-\ } (CH_3CH_2)_2\overset{..}{N}\!\!-\!\!CHCH_2CH_3$$
$$\underset{Cl}{|} \qquad\qquad\qquad \underset{CH_2OH}{|}$$

a. Explain the enhanced reaction rate.
b. Explain why the OH group in the product is not bonded to the carbon that was bonded to the Cl group in the reactant.

90. Which of the following reactions occurs more rapidly?

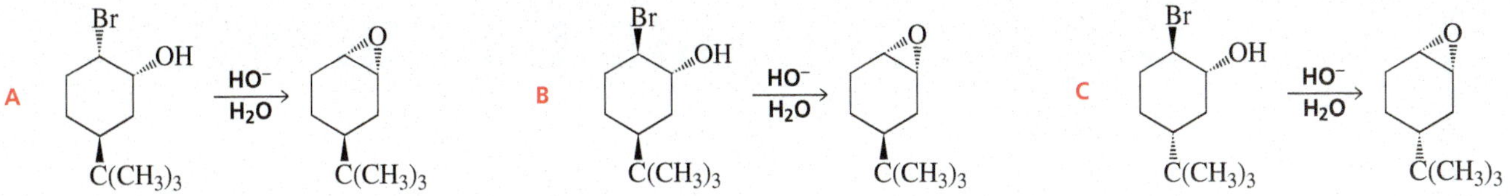

A   [structure: cyclohexane with Br, OH, C(CH₃)₃] $\xrightarrow[\text{H}_2\text{O}]{\text{HO}^-}$ [epoxide product with C(CH₃)₃]

B   [structure: cyclohexane with Br, OH, C(CH₃)₃] $\xrightarrow[\text{H}_2\text{O}]{\text{HO}^-}$ [epoxide product with C(CH₃)₃]

C   [structure: cyclohexane with Br, OH, C(CH₃)₃] $\xrightarrow[\text{H}_2\text{O}]{\text{HO}^-}$ [epoxide product with C(CH₃)₃]

91. Propose a mechanism for each of the following reactions:

a. [cyclohexene with alcohol side chain] $\xrightarrow[\Delta]{\text{H}_2\text{SO}_4}$ [bicyclic product]

c. [spiro alcohol] $\xrightarrow[\text{H}_2\text{O}]{\text{H}_2\text{SO}_4}$ [octahydronaphthalene]

b. [epoxide with diene] $\xrightarrow[\text{H}_2\text{O}]{\text{H}_2\text{SO}_4}$ [decalin diol product]

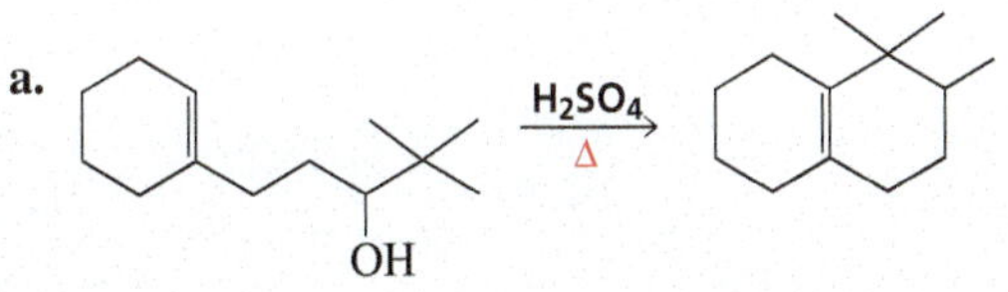

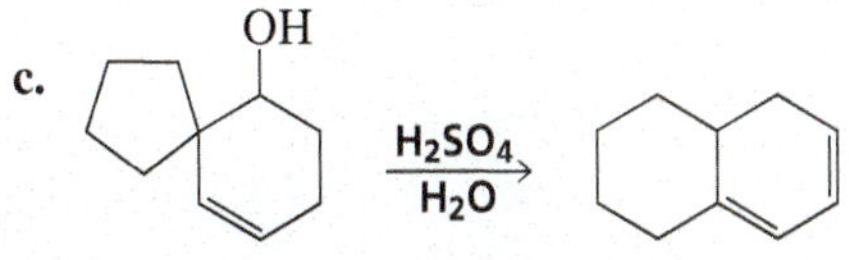

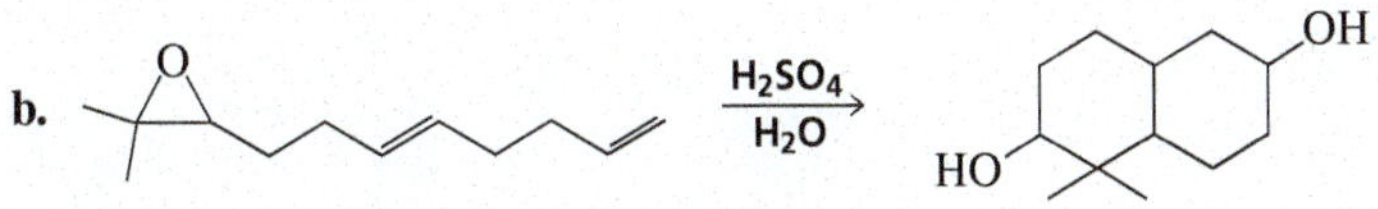

92. A vicinal diol has OH groups on adjacent carbons. The dehydration of a vicinal diol is accompanied by a rearrangement called the **pinacol rearrangement.** Propose a mechanism for this reaction.

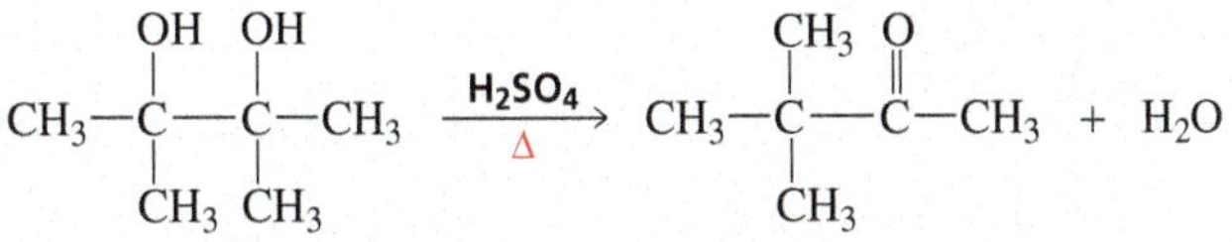

$$\underset{\underset{\text{CH}_3 \ \text{CH}_3}{|\quad\quad|}}{\overset{\overset{\text{OH} \ \text{OH}}{|\quad\quad|}}{\text{CH}_3-\text{C}-\text{C}-\text{CH}_3}} \xrightarrow[\Delta]{\text{H}_2\text{SO}_4} \underset{\underset{\text{CH}_3}{|}}{\overset{\overset{\text{CH}_3 \ \ \text{O}}{|\quad\ \ ||}}{\text{CH}_3-\text{C}-\text{C}-\text{CH}_3}} + \text{H}_2\text{O}$$

93. Although 2-methyl-1,2-propanediol is an unsymmetrical vicinal diol, only one product is obtained when it is dehydrated in an acidic solution.

  a. What is this product?   **b.** Why is only one product obtained?

94. What product is obtained when the following vicinal diol is heated in an acidic solution?

[structure: two cyclopentane rings joined with HO and OH groups]

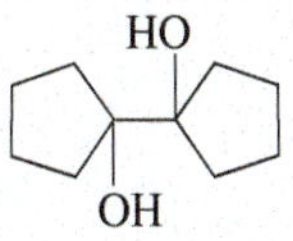

95. Two stereoisomers are obtained from the reaction of cyclopentene oxide and dimethylamine. The *R,R*-isomer is used in the manufacture of eclanamine, an antidepressant. What other isomer is obtained?

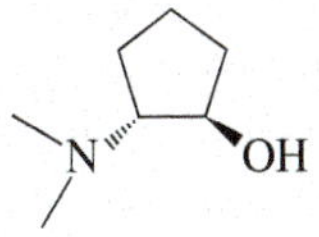

[structure: *R,R*-isomer — cyclopentane with N(CH₃)₂ and OH]   [structure: eclanamine]

R,R-isomer 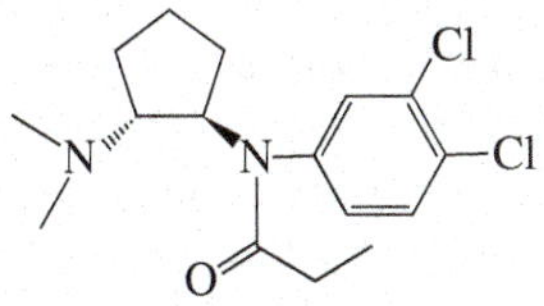   eclanamine

96. Propose a mechanism for each of the following reactions:

a. [cyclohexane diol] $\xrightarrow[\Delta]{\text{H}_2\text{SO}_4}$ [cyclohexanone product] + [cyclopentyl ketone product]

b. [cyclopropyl alcohol] $\xrightarrow[\Delta]{\text{H}_2\text{SO}_4}$ [cyclobutanone product]

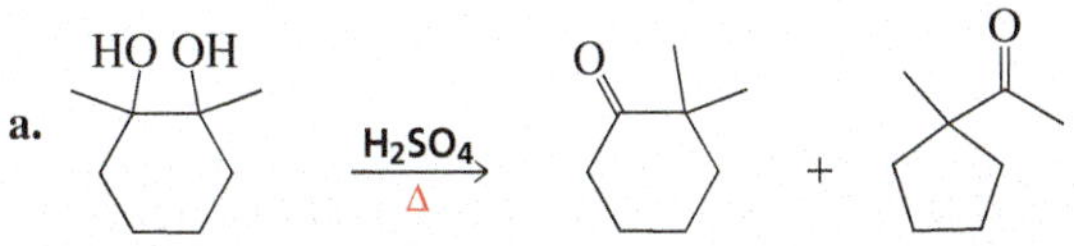

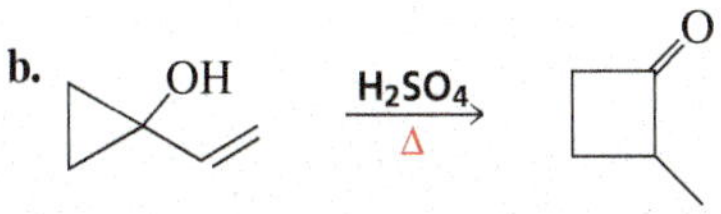

97. Triethylenemelamine (TEM) is an antitumor agent. Its activity is due to its ability to cross-link DNA.

  a. Explain why it can be used only under slightly acidic conditions.
  b. Explain why it can cross-link DNA.

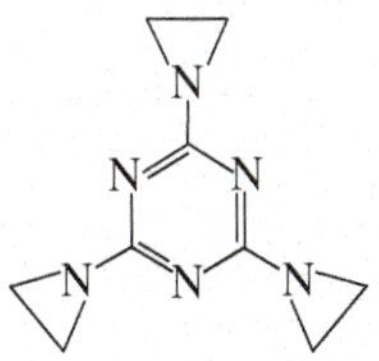

[structure: triethylenemelamine (TEM) — triazine ring with three aziridine groups]

triethylenemelamine (TEM)

11 Organometallic Compounds

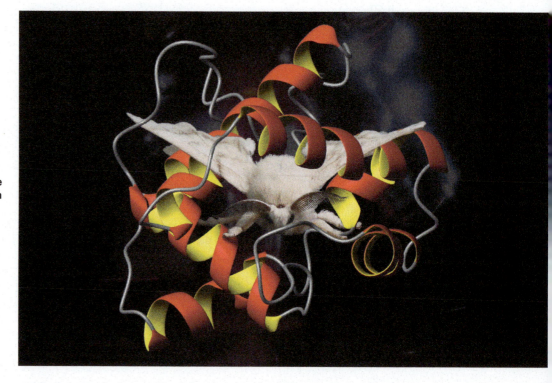

Bombyx mori superimposed on the pheromone binding protein

In this chapter, you will be asked to design a synthesis for bombykol, the sex pheromone of the silk moth (bombyx mori); see page 191. Molecules of bombykol diffuse through open pores in the male moth's antennae. When bombykol binds to its receptor, an electrical charge is produced that causes a nerve impulse to be sent to the brain. Bombykol, however, is a nonpolar molecule (page 531) and has to cross an aqueous solution to get to its receptor. This problem is solved by the pheromone binding protein. The protein binds bombykol in a hydrophobic pocket and then carries it to the receptor. The area around the receptor is relatively acidic. The acidic environment causes the pheromone binding protein to unfold and release bombykol to the receptor.

We saw that compounds in Group II—alkyl halides, alcohols, ethers, epoxides and sulfonate esters—contain a carbon that is bonded to a *more* electronegative atom. The carbon, therefore, is *electrophilic* and reacts with a nucleophile.

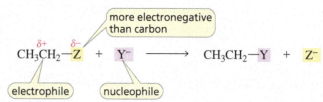

But what if you wanted a carbon to react with an electrophile? For that, you need a compound with a nucleophilic carbon. To be *nucleophilic,* carbon has to be bonded to a *less* electronegative atom.

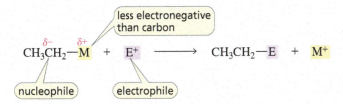

Because metals are less electronegative than carbon (Table 11.1), one way to create a nucleophilic carbon is to bond the carbon to a metal (M). A compound that contains a carbon–metal bond is called **an organometallic compound.**

The electrostatic potential maps show that the carbon attached to the more electronegative chlorine in the alkyl chloride is an electrophile (it is blue-green), whereas the carbon attached to the less electronegative metal (Li) in the organometallic compound is a nucleophile (it is red).

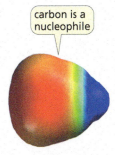

carbon is an electrophile

CH₃Cl
alkyl halide

A carbon is an electrophile
if it is attached to a
more electronegative atom.

carbon is a nucleophile

CH₃Li
organometallic compound

A carbon is a nucleophile
if it is attached to a
less electronegative atom.

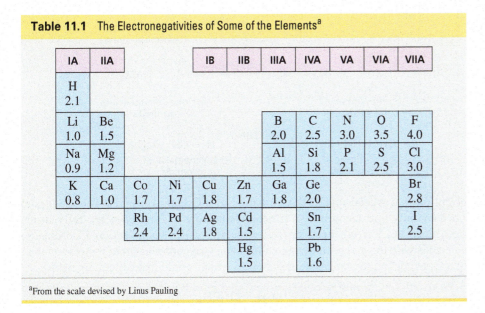

Table 11.1 The Electronegativities of Some of the Elements[a]

IA	IIA			IB	IIB	IIIA	IVA	VA	VIA	VIIA
H 2.1										
Li 1.0	Be 1.5					B 2.0	C 2.5	N 3.0	O 3.5	F 4.0
Na 0.9	Mg 1.2					Al 1.5	Si 1.8	P 2.1	S 2.5	Cl 3.0
K 0.8	Ca 1.0	Co 1.7	Ni 1.7	Cu 1.8	Zn 1.7	Ga 1.8	Ge 2.0			Br 2.8
		Rh 2.4	Pd 2.4	Ag 1.8	Cd 1.5		Sn 1.7			I 2.5
					Hg 1.5		Pb 1.6			

[a]From the scale devised by Linus Pauling

A wide variety of metals can be attached to carbon to form organometallic compounds.

$$\overset{\delta^-}{C}-\overset{\delta^+}{Mg} \qquad \overset{\delta^-}{C}-\overset{\delta^+}{Li} \qquad \overset{\delta^-}{C}-\overset{\delta^+}{Cu} \qquad \overset{\delta^-}{C}-\overset{\delta^+}{Cd} \qquad \overset{\delta^-}{C}-\overset{\delta^+}{Pd}$$

$$\overset{\delta^-}{C}-\overset{\delta^+}{Zn} \qquad \overset{\delta^-}{C}-\overset{\delta^+}{Al} \qquad \overset{\delta^-}{C}-\overset{\delta^+}{Pb} \qquad \overset{\delta^-}{C}-\overset{\delta^+}{Hg} \qquad \overset{\delta^-}{C}-\overset{\delta^+}{Ru}$$

The name of an organometallic compound usually begins with the name of the alkyl group, followed by the name of the metal.

$CH_3CH_2CH_2CH_2Li$	CH_3CH_2MgBr	$(CH_3CH_2CH_2)_2Cd$	$(CH_3CH_2)_4Pb$
butyllithium	**ethylmagnesium bromide**	**dipropylcadmium**	**tetraethyllead**

11.1 ORGANOLITHIUM AND ORGANOMAGNESIUM COMPOUNDS

Two of the most common organometallic compounds are *organolithium compounds* and *organomagnesium compounds.*

Preparing Organolithium and Organomagnesium Compounds

Organolithium compounds are prepared by adding lithium metal to an alkyl halide in a nonpolar solvent such as hexane.

$$CH_3CH_2CH_2CH_2\text{—Br} + 2\,Li \xrightarrow{\text{hexane}} CH_3CH_2CH_2CH_2\text{—Li} + LiBr$$

1-bromobutane   butyllithium

carbon–metal bond

chlorobenzene + 2 Li $\xrightarrow{\text{hexane}}$ phenyllithium + LiCl

Organomagnesium compounds (commonly called **Grignard reagents** after their discoverer, Victor Grignard) are prepared by adding an alkyl halide to magnesium metal shavings being stirred in an ether—usually diethyl ether or tetrahydrofuran (THF)—under anhydrous conditions. The reaction inserts magnesium between the carbon and the halogen.

carbon–metal bond

cyclohexyl bromide + Mg $\xrightarrow{\text{diethyl ether}}$ cyclohexylmagnesium bromide

$$CH_2{=}CH\text{—Br} + Mg \xrightarrow{\text{THF}} CH_2{=}CH\text{—MgBr}$$

vinyl bromide   vinylmagnesium bromide

The solvent plays a crucial role in the formation of a Grignard reagent. The magnesium atom is surrounded by only four electrons, so it needs four more to form an octet. Solvent molecules provide these electrons. Sharing electrons with a metal is called *coordination*. Coordination of the solvent molecules with the magnesium atom allows the Grignard reagent to dissolve in the solvent, preventing it from coating the magnesium shavings, which would make them unreactive.

$$CH_3CH_2 \quad CH_2CH_3$$
$$\diagdown \ddot{O} \diagup$$
$$CH_3\text{—}\overset{..}{Mg}\text{—Br}$$
$$\diagup \ddot{O} \diagdown$$
$$CH_3CH_2 \quad CH_2CH_3$$

Alkyl halides, vinylic halides, and aryl halides can all be used to form organolithium and organomagnesium compounds. Alkyl bromides are used most often, because they react more readily than alkyl chlorides and are less expensive than alkyl iodides.

Organolithium and Organomagnesium Compounds Are Nucleophiles

Because carbon is more electronegative than the metal to which it is bonded, organolithium and organomagnesium compounds react as if they were carbanions. Thus, they are nucleophiles and bases.

$$CH_3CH_2\text{—MgBr} \qquad \text{reacts as if it were} \qquad CH_3\overset{..}{C}H_2 \quad \overset{+}{MgBr}$$

ethylmagnesium bromide

phenyllithium —Li   reacts as if it were $\overset{-}{:}$ Li$^+$

Organomagnesium and organolithium compounds are very strong bases. Therefore, they react immediately with any acid present in the reaction mixture, even with very weak acids such as water and alcohols. When this happens, the organometallic compound is converted to an alkane. Although this is not how one would normally synthesize an alkane, it is a useful way to prepare deuterated hydrocarbons. Just use D_2O instead of H_2O.

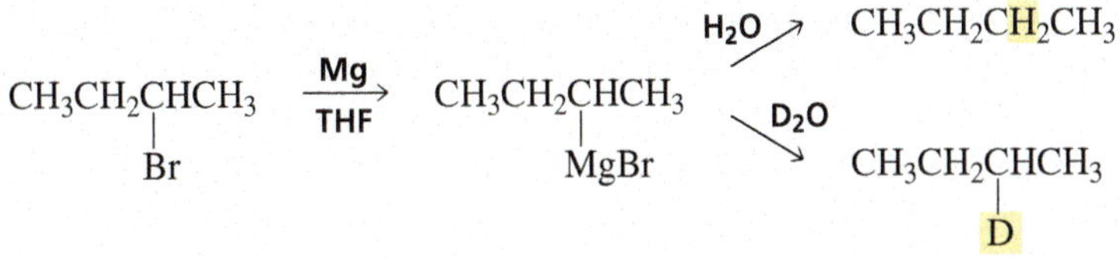

$$CH_3CH_2CHCH_3 \quad \xrightarrow[\text{THF}]{\text{Mg}} \quad CH_3CH_2CHCH_3 \quad \begin{array}{c} \xrightarrow{H_2O} \quad CH_3CH_2CH_2CH_3 \\ \\ \xrightarrow{D_2O} \quad CH_3CH_2CHCH_3 \end{array}$$

This means that Grignard reagents and organolithium compounds cannot be prepared from compounds that contain acidic groups (such as OH, NH_2, NHR, SH, $C{\equiv}CH$, or COOH). Because even trace amounts of moisture can convert an organometallic compound into an alkane, it is important that all reagents are dry when organometallic compounds are synthesized and when they react with other reagents.

PROBLEM 1 ♦

Which of the following reactions favor formation of the products? (For the pK_a values necessary to solve this problem, see Appendix I. Recall that the equilibrium favors formation of the weaker acid; see Section 2.5.)

$$CH_3MgBr + H_2O \rightleftharpoons CH_4 + HOMgBr$$
$$CH_3MgBr + CH_3OH \rightleftharpoons CH_4 + CH_3OMgBr$$
$$CH_3MgBr + NH_3 \rightleftharpoons CH_4 + H_2NMgBr$$
$$CH_3MgBr + CH_3NH_2 \rightleftharpoons CH_4 + CH_3NHMgBr$$
$$CH_3MgBr + HC{\equiv}CH \rightleftharpoons CH_4 + HC{\equiv}CMgBr$$

11.2 TRANSMETALLATION

The reactivity of an organometallic compound depends on the polarity of the carbon–metal bond: *the greater the polarity of the bond, the more reactive the compound is as a nucleophile.* The polarity of the bond depends on the difference in electronegativity between the metal and carbon (Table 11.1).

For example, magnesium has an electronegativity of 1.2, compared with 2.5 for carbon. This large difference in electronegativity makes the carbon–magnesium bond highly polar. Lithium (1.0) is even less electronegative than magnesium. Thus, the carbon–lithium bond is more polar than the carbon–magnesium bond, so an organolithium reagent is a more reactive nucleophile than a Grignard reagent.

An organometallic compound will undergo **transmetallation** (metal exchange) if it is added to a metal halide whose metal is more electronegative than the metal in the organometallic compound. In other words, transmetallation occurs if the alkyl group can be transferred to a metal with an electronegativity closer to that of carbon, thereby forming a less polar carbon–metal bond and, therefore, a less reactive nucleophile.

For example, cadmium is more electronegative (1.5) than magnesium (1.2). Consequently, a carbon–cadmium bond is less polar than a carbon–magnesium bond, so metal exchange occurs.

> The more polar the carbon–metal bond, the more reactive the organometallic compound is as a nucleophile.

$$2\ CH_3CH_2MgCl + CdCl_2 \longrightarrow (CH_3CH_2)_2Cd + 2\ MgCl_2$$
ethylmagnesium chloride — diethylcadmium

> Transmetallation occurs if the alkyl group can be transferred to a more electronegative metal.

PROBLEM 2 ♦

Which is more reactive, an organolithium compound or an organosodium compound? Explain your answer.

PROBLEM 3 ♦

What organometallic compound is formed from the reaction of excess methylmagnesium chloride and $GaCl_3$? (*Hint:* See Table 11.1.)

11.3 ORGANOCUPRATES

New carbon–carbon bonds can be made using an organometallic reagent that has a *transition metal* as its metal atom. Transition metals are indicated by purple in the periodic table at the back of this book. The reactions are called **coupling reactions** because two CH-containing groups are joined (coupled) together.

Coupling reactions join two CH-containing groups.

The first organometallic compounds used in coupling reactions were copper-containing **organocuprates (R_2CuLi)**, also called **Gilman reagents** after their discoverer, Henry Gilman. Organocuprates are less reactive than organolithium reagents or Grignard reagents because a carbon–copper bond is less polar than a carbon–lithium or carbon–magnesium bond—that is, Cu is closer in electronegativity to C than is Li or Mg. Only one of the two alkyl groups in an organocuprate is used as a nucleophile in reactions.

Forming Organocuprates

An organocuprate is prepared by the reaction of an organolithium compound with cuprous iodide in diethyl ether or in THF. Notice that because Cu is more electronegative (1.8) than Li (1.0), trans-metallation occurs.

$$2\ CH_3Li\ +\ CuI\ \xrightarrow{\text{THF}}\ (CH_3)_2CuLi\ +\ LiI$$

an organolithium
compound

an organocuprate
a Gilman reagent

An Organocuprate Replaces Cl, Br, or I with an Alkyl Group

The organocuprate reacts by coupling one of its alkyl groups to the alkyl group of an alkyl halide (with the exception of alkyl fluorides, which do not undergo this reaction) and displacing the halogen. This means that an alkane can be formed from two alkyl halides—one alkyl halide is used to form the organocuprate, which then reacts with the second alkyl halide in a coupling reaction. The precise mechanism of the substitution reaction is unknown but is thought to involve radicals.

$$CH_3CH_2CH_2CH_2{-}Br\ +\ (CH_3CH_2CH_2)_2CuLi\ \longrightarrow\ CH_3CH_2CH_2CH_2{-}CH_2CH_2CH_3\ +\ CH_3CH_2CH_2Cu\ +\ LiBr$$

new C–C bond

an alkyl group of an organocuprate replaces a Cl, Br, or I

heptane

The R groups of the organocuprate *and* the alkyl halide can be primary alkyl, methyl, aryl, vinylic, or allylic. In other words, any R group except secondary or tertiary alkyl.

$RCH_2{-}$	$CH_3{-}$	(aryl)	$RCH{=}CH{-}$	$RCH{=}CHCH_2{-}$
primary alkyl	**methyl**	**aryl**	**vinylic**	**allylic**

Because organocuprates can react with vinylic halides and aryl halides, they can be used to prepare compounds that cannot be prepared by S_N2 reactions with Grignard reagents or organolithium compounds. (Remember that vinylic and aryl halides cannot undergo nucleophilic attack; Section 9.13).

a vinylic halide

an aryl halide

The substitution reaction is stereospecific. In other words, the configuration of the double bond is retained in the product—that is, the groups bonded to the sp^2 carbons maintain their positions.

$$\underset{\textbf{cis}}{\underset{H}{\overset{R}{>}}C=C\underset{H}{\overset{Br}{<}}} + (CH_3)_2CuLi \longrightarrow \underset{\textbf{cis}}{\underset{H}{\overset{R}{>}}C=C\underset{H}{\overset{CH_3}{<}}}$$

$$\underset{\textbf{trans}}{\underset{H}{\overset{R}{>}}C=C\underset{Br}{\overset{H}{<}}} + (CH_3)_2CuLi \longrightarrow \underset{\textbf{trans}}{\underset{H}{\overset{R}{>}}C=C\underset{CH_3}{\overset{H}{<}}}$$

Organocuprates can even replace halogens in compounds that contain other functional groups such as compounds with a bromo or chloro substituent attached to a carbon adjacent to a carbonyl group.

$$\underset{\overset{|}{CH_3}}{BrCH_2CH}\overset{\overset{O}{\|}}{C}CH_3 + (CH_3)_2CuLi \longrightarrow \underset{\overset{|}{CH_3}}{CH_3CH_2CH}\overset{\overset{O}{\|}}{C}CH_3 + CH_3Cu + LiBr$$

PROBLEM 4 SOLVED   LEARN THE STRATEGY

Describe two ways to synthesize ethylcyclopentane from alkyl bromides.

SOLUTION   First we need to choose the two alkyl bromides whose alkyl groups will be coupled. Recalling that secondary alkyl halides cannot be used, the two alkyl bromides must be bromomethane and bromomethylcyclopentane. Either of the alkyl bromides can be used to form the organocuprate that then reacts with the other alkyl bromide.

$$CH_3Br \xrightarrow[\text{2. CuI}]{\text{1. Li}} (CH_3)_2CuLi \quad + \quad \text{[cyclopentane–}CH_2Br\text{]} \longrightarrow \text{[cyclopentane–}CH_2CH_3\text{]}$$

or

$$\text{[cyclopentane–}CH_2Br\text{]} \xrightarrow[\text{2. CuI}]{\text{1. Li}} \left(\text{[cyclopentane–}CH_2\text{]}\right)_2\!CuLi \xrightarrow{CH_3Br} \text{[cyclopentane–}CH_2CH_3\text{]}$$

PROBLEM 5   USE THE STRATEGY

Muscalure is the sex attractant of the common housefly. Flies are lured to traps filled with bait that contain muscalure and an insecticide. Eating the bait is fatal. How could you synthesize muscalure using 1-bromopentane as one of the starting materials?

$$\underset{H}{\overset{CH_3(CH_2)_7}{>}}C=C\underset{H}{\overset{(CH_2)_{12}CH_3}{<}}$$
muscalure

PROBLEM 6 SOLVED   LEARN THE STRATEGY

What bromo-substituted compound is required to react with $(CH_2{=}CH)_2CuLi$ in order to form the following compound?

SOLUTION Delete the CH_2=CH group and replace it with a bromine.

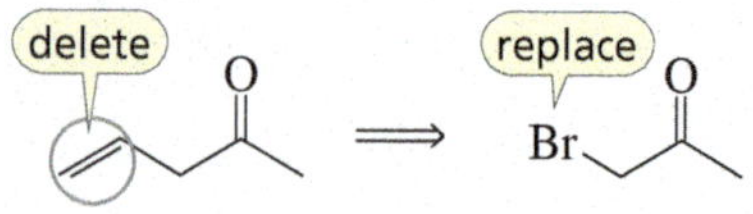

USE THE STRATEGY

PROBLEM 7

What bromo-substituted compound would be required to react with $(CH_2$=$CH)_2CuLi$ in order to form each of the following compounds?

a. 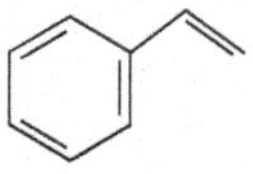   b. 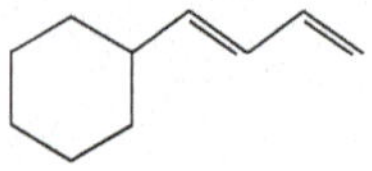   c.

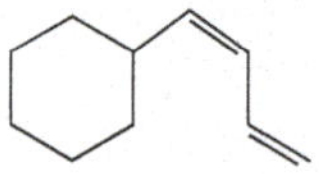

An Organocuprate Is a Nucleophile

Because they are nucleophiles, organocuprates react with electrophiles. For example, in the following reactions, the organocuprate reacts with an epoxide in a nucleophilic substitution reaction.

LEARN THE STRATEGY

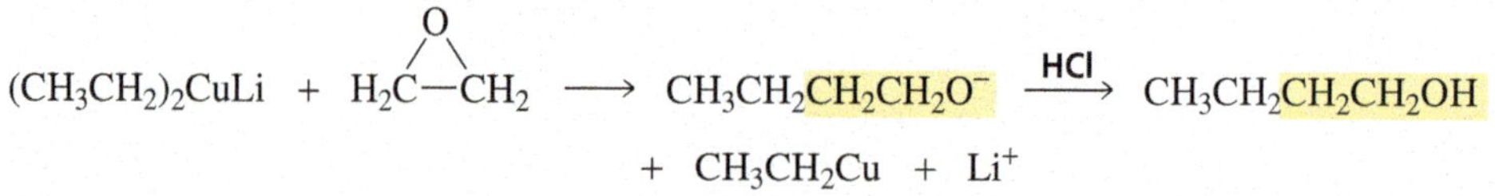

$$(CH_3CH_2)_2CuLi \ + \ H_2C\overset{O}{-}CH_2 \ \longrightarrow \ CH_3CH_2CH_2CH_2O^- \ \xrightarrow{HCl} \ CH_3CH_2CH_2CH_2OH$$

$$+ \ CH_3CH_2Cu \ + \ Li^+$$

Notice that when an organocuprate reacts with ethylene oxide, a primary alcohol is formed that contains two more carbons than the organometallic compound.

USE THE STRATEGY

PROBLEM 8 ♦

What alcohols are formed from the reaction of ethylene oxide with the following organocuprates followed by the addition of acid?

a. $(CH_3CH_2CH_2)_2CuLi$ b. $(CH_3CH$=$CH)_2CuLi$ c.

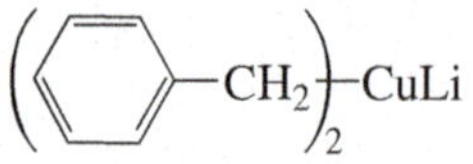

PROBLEM 9

What are the products of the following reactions?

a. 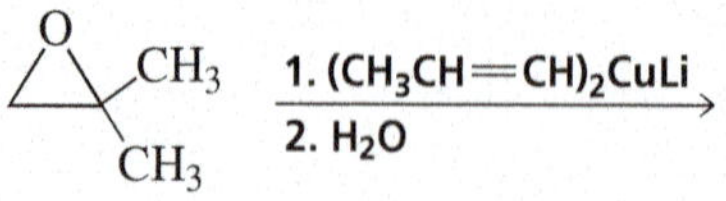 $\xrightarrow[\text{2. H}_2\text{O}]{\text{1. (CH}_3\text{CH}=\text{CH)}_2\text{CuLi}}$ b. 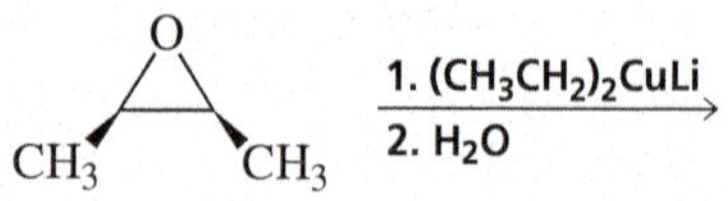 $\xrightarrow[\text{2. H}_2\text{O}]{\text{1. (CH}_3\text{CH}_2\text{)}_2\text{CuLi}}$

PROBLEM 10

How could the following compounds be prepared, using cyclohexene as a starting material?

a. 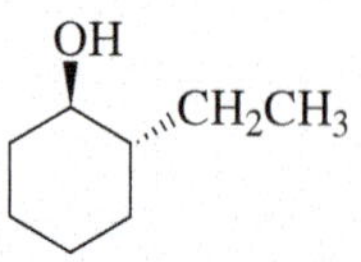   b. 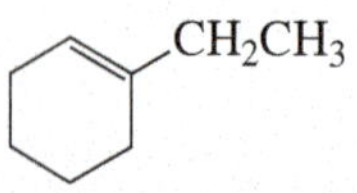   c. 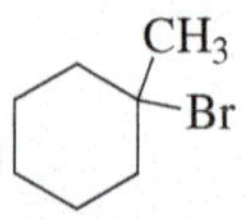   d.

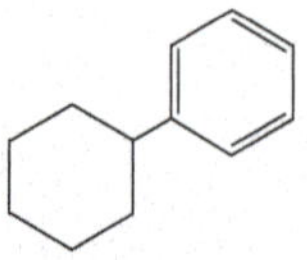

PROBLEM 11 SOLVED

Using any necessary reagents, show how the following compounds could be prepared using ethylene oxide as one of the reactants:

a. $CH_3CH_2CH_2CH_2OH$
b. $CH_3CH_2CH_2CH_2Br$

c. $CH_3CH_2CH_2CH_2D$
d. $CH_3CH_2CH_2CH_2CH_2CH_2OH$

SOLUTION

a. CH_3CH_2Br $\xrightarrow[\text{2. CuI}]{\text{1. Li}}$ $(CH_3CH_2)_2CuLi$ $\xrightarrow[\text{2.HCl}]{\text{1. } \triangle O}$ $CH_3CH_2CH_2CH_2OH$

b. product of **a** $\xrightarrow{\text{PBr}_3}$ $CH_3CH_2CH_2CH_2Br$

c. product of **b** $\xrightarrow[\text{2. CuI}]{\text{1. Li}}$ $(CH_3CH_2CH_2CH_2)_2CuLi$ $\xrightarrow{\text{D}_2\text{O}}$ $CH_3CH_2CH_2CH_2D$

d. the same reaction sequence as in part **a**, but with butyl bromide in the first step

---

11.4 PALLADIUM-CATALYZED COUPLING REACTIONS

Many new methods that use transition metals to carry out coupling reactions have been developed in recent years. These methods have greatly expanded the synthetic chemist's arsenal of reactions that form new carbon–carbon bonds. We will look at two reactions that use a palladium catalyst (L_2Pd): the *Suzuki reaction* and the *Heck reaction*. The two reactions have several features in common:

- Both reactions form a new C—C bond by replacing a halogen of a *vinylic halide* or an *aryl halide* (bromides and iodides work best) with a carbon-containing group (R). Thus, they are substitution reactions.

a vinylic halide

an aryl halide

- The palladium atom of the catalyst is coordinated with ligands (L_2Pd). Several different ligands (L) can be used; a common ligand is triphenylphospine [$P(C_6H_5)_3$].
- Both reactions occur via a catalytic cycle in which the palladium atom participates in breaking bonds in the reactants and forming the new C—C bond in the product.
- The first step of each cycle is insertion of palladium between the carbon and halogen to form an **organopalladium compound.**

$$R—X \ + \ L_2Pd \ \longrightarrow \ L_2Pd\begin{smallmatrix} R \\ \\ X \end{smallmatrix}$$

an organopalladium compound

- If the organopalladium compound has a β-hydrogen on an sp^3 carbon, the organopalladium compound will rapidly undergo an elimination reaction before the coupling reaction has a chance to occur. This explains why vinylic and aryl halides are the reactants in these reactions—they cannot undergo an elimination reaction under the conditions used to carry out the coupling reactions.

$$L_2Pd\begin{smallmatrix} CH_2CH_2R \\ \\ X \end{smallmatrix} \ \xrightarrow{\beta\text{-elimination}} \ R—CH=CH_2$$

β-hydrogens

- The reactions can be carried out if the reactants have other functional groups.
- The reactions are stereospecific: the configuration of the double bond in a vinylic halide is retained in the product.
- The reactions are very efficient, often requiring less than 0.1% of the palladium catalyst (compared to the reactants) and giving high yields of products (80–98%).

The Suzuki Reaction

The **Suzuki reaction** couples the R group of a *vinylic or aryl halide* with the R′ group of an **organoboron compound** in a basic solution in the presence of a palladium catalyst (PdL_2). The general reaction is shown below.

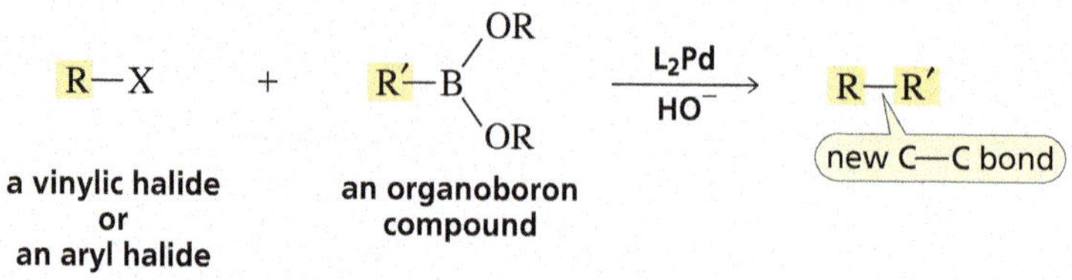

a vinylic halide
or
an aryl halide

an organoboron compound

Examples of Suzuki Reactions

The R′ group of the organoboron compound can be either an *alkyl* group, an *alkenyl* group, or an *aryl* group. Notice that the carbons that were bonded to the halogen and to the boron are joined by a new C—C bond. (If the catalysts are modified, halides other than vinylic or aryl can be used, but that is beyond the scope of this book.)

alkyl-organoboron compound

In a Suzuki reaction, the R′ group of an organoboron compound replaces a halogen.

In a Suzuki reaction, the new C—C bond joins the carbon that was bonded to the halogen with the carbon that was bonded to boron.

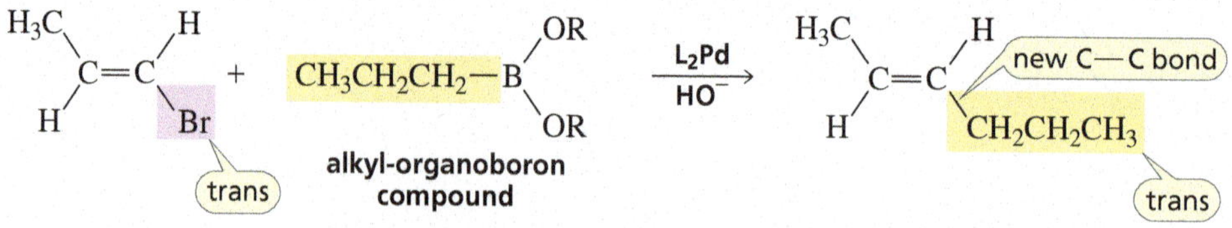

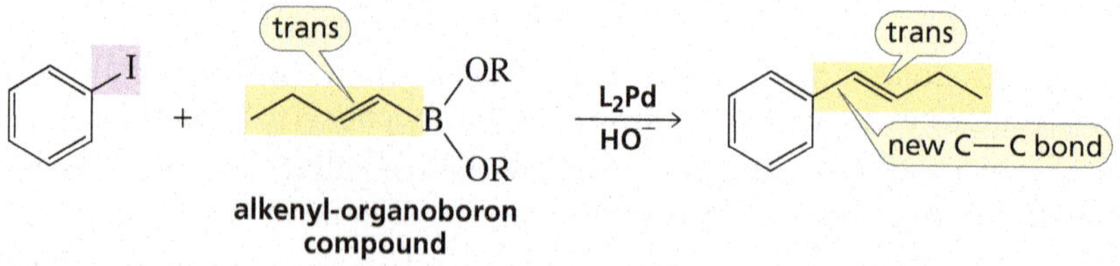

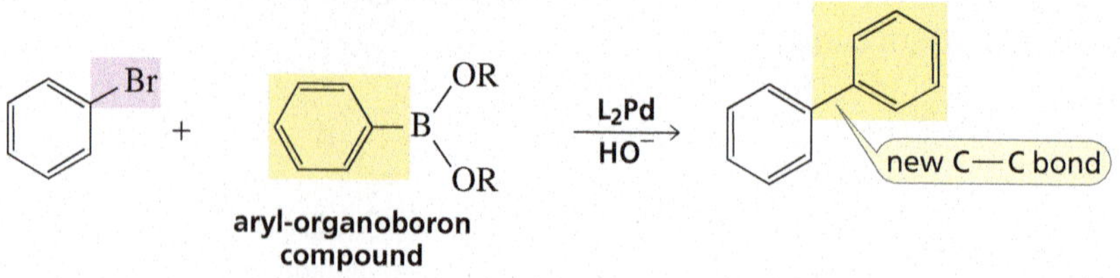

alkenyl-organoboron compound

aryl-organoboron compound

When the reactant is a vinylic halide, the configuration of its double bond is retained in the product. When an alkenyl-organoboron compound is used, the new double bond in the product is always trans because an alkenyl-organoboron compound always has a *trans configuration* (see pages 517–518).

Mechanism for the Suzuki Reaction

The mechanism for the Suzuki reaction is under active investigation; it has been found to depend on the ligand (L) and the base. The mechanism for the reaction when the ligand is triphenylphosphine, $[P(C_6H_5)_3]$ and the base is hydroxide ion is shown on the next page.

MECHANISM FOR THE SUZUKI REACTION

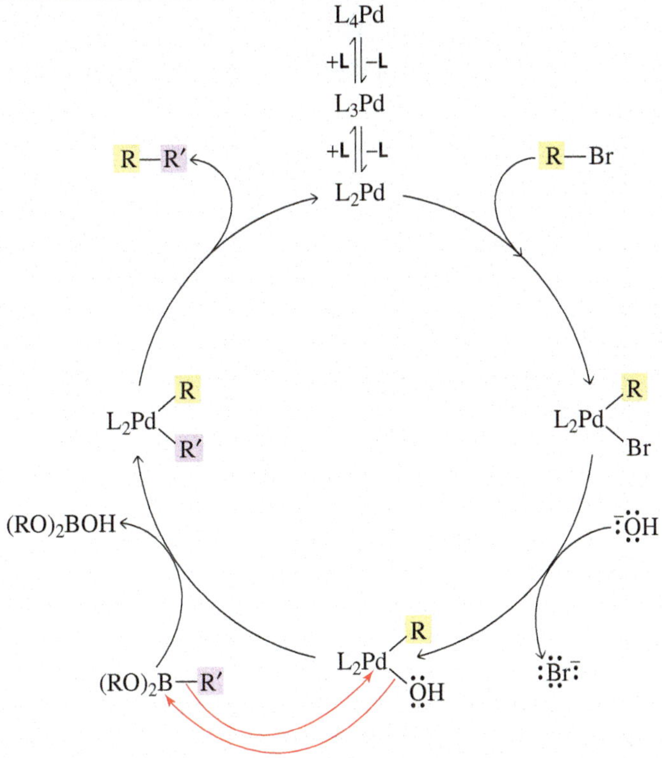

L$_4$Pd
+L ‖ −L
L$_3$Pd
+L ‖ −L
L$_2$Pd

R—R′

R—Br

L$_2$Pd〈R / R′

L$_2$Pd〈R / Br

(RO)$_2$BOH

:ÖH

(RO)$_2$B—R′

L$_2$Pd〈R / ÖH

:Br:⁻

- The catalyst (L$_2$Pd) is very reactive, so it must be generated in situ (in the reaction mixture) by two successive dissociations of a ligand from the more stable precatalyst (L$_4$Pd).
- Palladium is inserted between the carbon and the halogen.
- Hydroxide ion displaces the halide ion.
- The R′ group is transferred from boron to palladium as hydroxide ion is transferred from palladium to boron.
- The last step is effectively the reverse of the first step: palladium is eliminated and a new C—C bond is formed. The catalyst can now react with another molecule of alkyl halide.

The insertion of a metal between two atoms is called **oxidative addition**—two new groups are added to the metal; it is an oxidation because the oxidation state of palladium is 0 in L$_2$Pd and +2 in L$_2$Pd(X)R. (Recall that oxidation is loss of electrons; reduction is gain of electrons.) **Reductive elimination** eliminates two groups from the metal. Thus, the first step in a Suzuki reaction is an *oxidative addition*, and the last step is a *reductive elimination*.

Oxidative addition: two groups are added to the metal

Reductive elimination: two groups are eliminated from the metal

Preparing the Organoboron Compound for a Suzuki Reaction

The alkyl- or alkenyl-organoboron compound used in a Suzuki reaction is prepared by hydroboration of a terminal alkene or a terminal alkyne, respectively. Often the boron-containing compound is catecholborane.

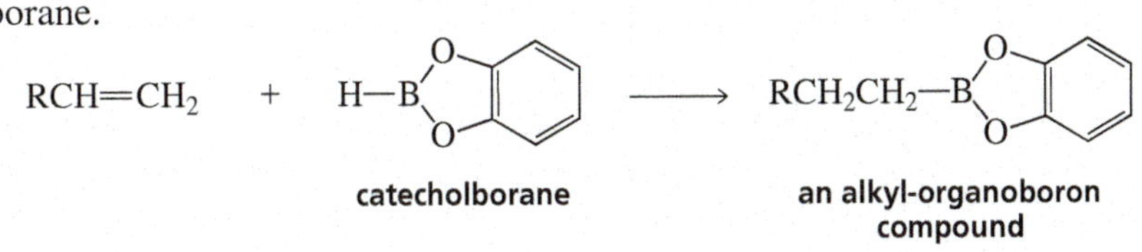

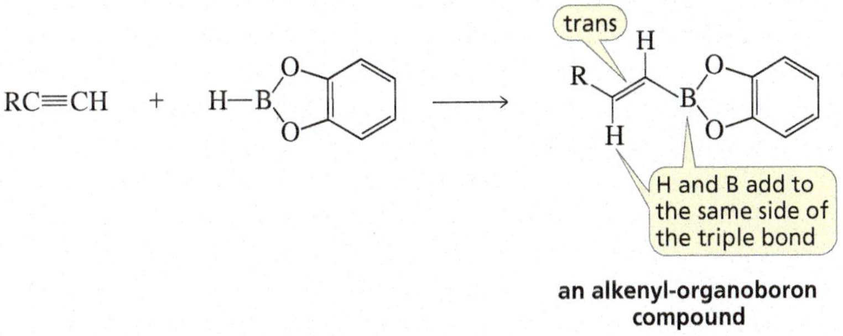

RCH=CH$_2$ + H—B(O—O) ⟶ RCH$_2$CH$_2$—B(O—O)

catecholborane

an alkyl-organoboron compound

RC≡CH + H—B(O—O) ⟶

trans

R / H   H / B(O—O)

H and B add to the same side of the triple bond

an alkenyl-organoboron compound

Recall that boron adds to the triply bonded carbon that is bonded to the hydrogen and that B and the H add to the same side of the triple bond (Section 7.8). Therefore, the boron in an alkenyl-organoboron compound is always trans to the substituent on the adjacent sp^2 carbon.

An aryl-organoboron compound is prepared from an organolithium compound and trimethylborate.

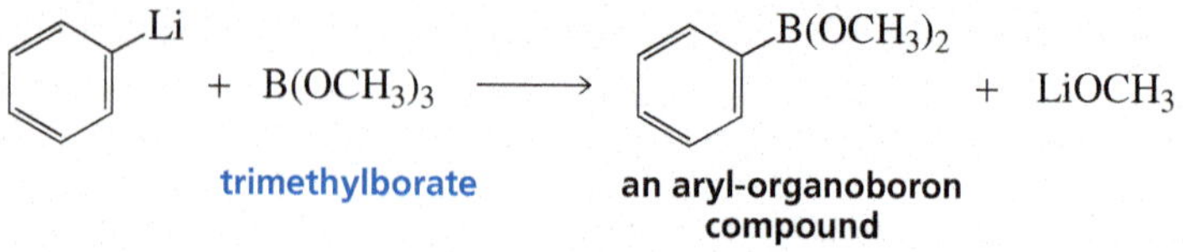

trimethylborate   an aryl-organoboron compound

LEARN THE STRATEGY

PROBLEM 12 SOLVED

What is the product of each of the following reactions?

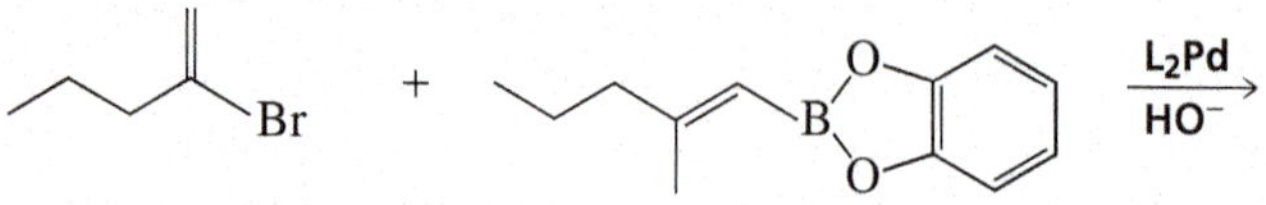

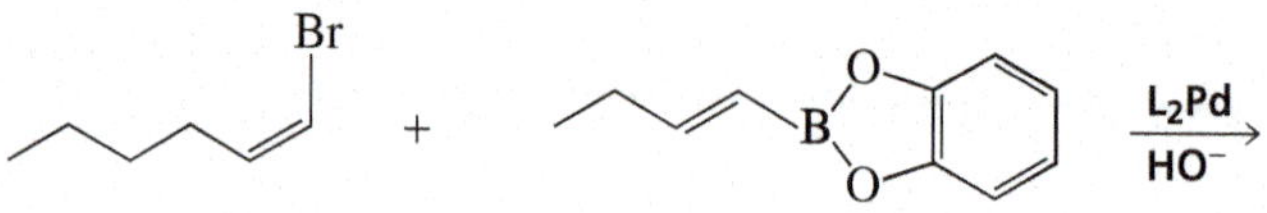

SOLUTION TO 12 a. Attach the carbon that is bonded to boron to the carbon that is bonded to the halogen, maintaining the configuration of the double bonds.

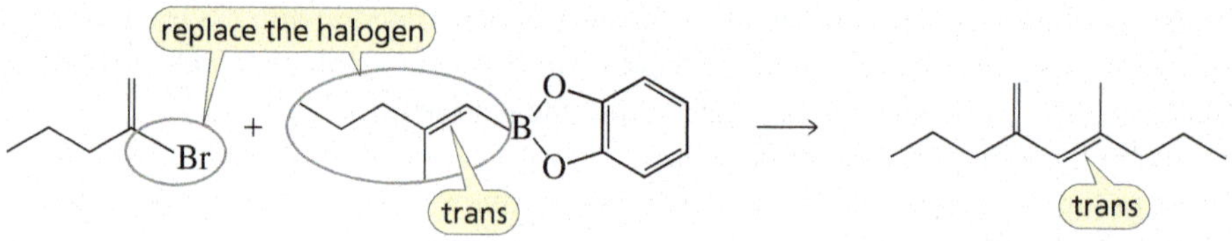

SOLUTION TO 12 b. Attach the carbon that is bonded to boron to the carbon that is bonded to the halogen, maintaining the configuration of both double bonds.

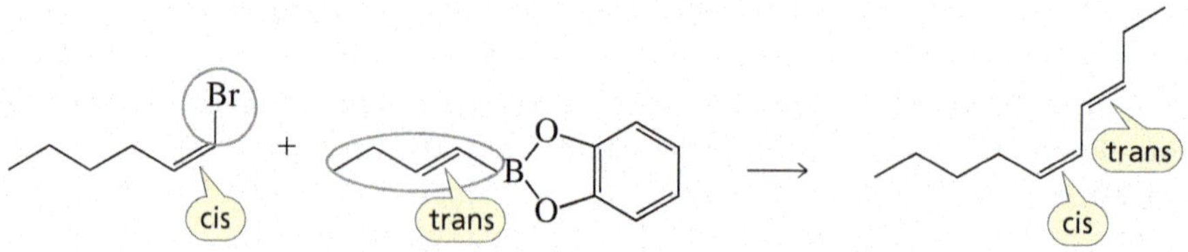

USE THE STRATEGY

PROBLEM 13

What is the product of each of the following reactions?

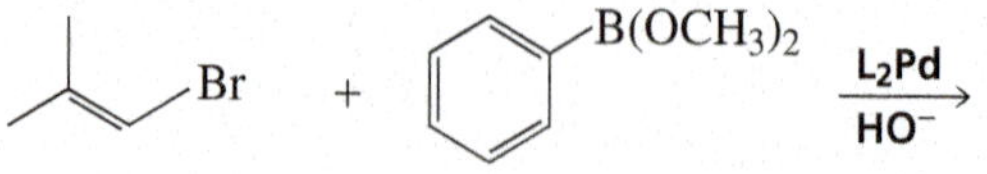

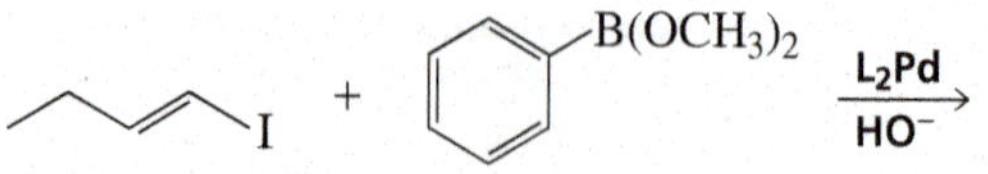

PROBLEM 14 ◆

What aryl or vinylic halides would you use to synthesize the following compounds, using the alkenyl-organoboron compound shown here?

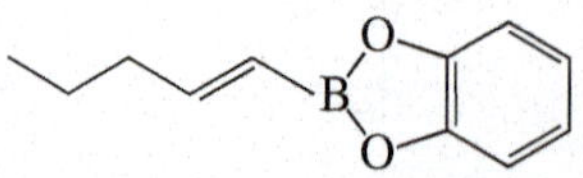

a. 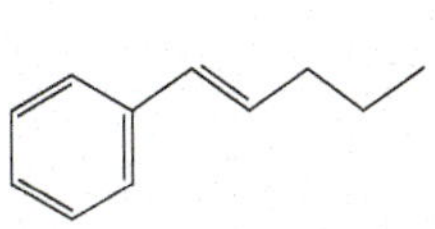 **b.** 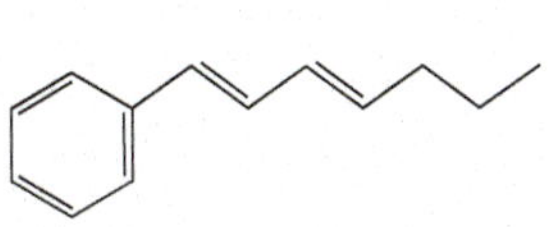 **c.**

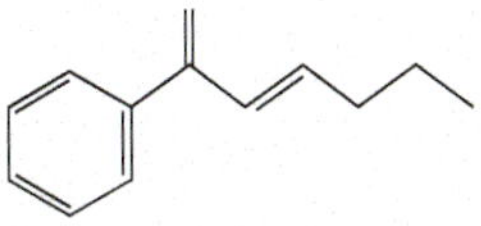

PROBLEM 15 ◆

What hydrocarbon would you use to prepare the organoboron compound in Problem 14?

PROBLEM 16

The Stille reaction is similar to the Suzuki reaction. It replaces the alkenyl-organoboron compound of the Suzuki reaction with an alkenyl-organotin compound. (R is an alkyl group such as methyl or butyl.) Unlike the alkenyl-organoboron compound that always has a trans configuration, the alkenyl-organotin compound can have a cis configuration. What is the product of the Stille reaction shown here?

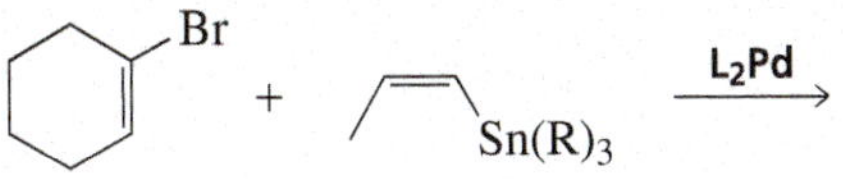

$$\text{(cyclohexenyl)–Br} \quad + \quad \text{(cis-alkenyl)–Sn(R)}_3 \quad \xrightarrow{\textbf{L}_2\textbf{Pd}}$$

The Heck Reaction

The **Heck reaction** couples a *vinylic* or an *aryl halide* with an *alkene* in the presence of a base (such as triethylamine) and a palladium catalyst (L_2Pd). Like the Suzuki reaction, the Heck reaction is a substitution reaction: the R group of the halide *replaces a vinylic hydrogen* of an alkene. If there is a substituent attached to the alkene (Z), the R group will be trans to that substituent in the product. Notice in the reactions shown below that the new C—C bond joins two sp^2 carbons.

H is substituted by R   R is trans to Z

$$\text{R—X} \quad + \quad \text{H} \diagdown \diagdown_Z \quad \xrightarrow[\textbf{(CH}_3\textbf{CH}_2)_3\textbf{N}]{\textbf{L}_2\textbf{Pd}} \quad \text{R} \diagdown \diagdown_Z$$

a vinylic halide
or
an aryl halide

new C—C bond

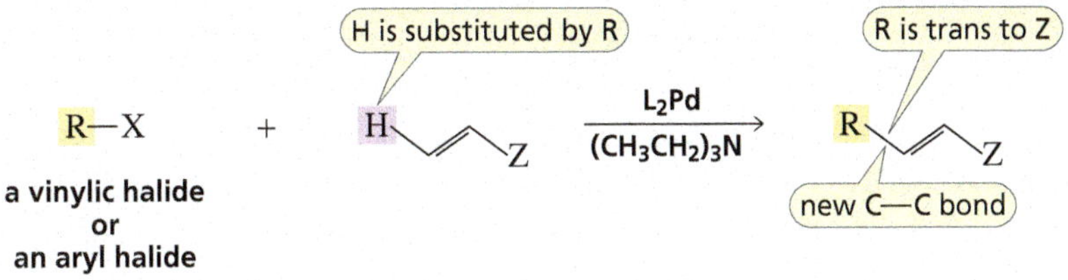

Examples of Heck Reactions

$$\text{(phenyl)–Br} \quad + \quad \text{CH}_2\text{=CH}_2 \quad \xrightarrow[\textbf{(CH}_3\textbf{CH}_2)_3\textbf{N}]{\textbf{L}_2\textbf{Pd}} \quad \text{(phenyl)–CH=CH}_2$$

the new C—C bond joins two sp^2 carbons

In a Heck reaction, the R group of the halide replaces a vinylic hydrogen of an alkene and the R is trans to the alkene's substituent.

In a Heck reaction, the new C—C bond joins two sp^2 carbons.

trans   trans   the new C—C bond joins two sp^2 carbons

$$\text{(trans-alkenyl)–I} \quad + \quad \text{CH}_2\text{=CH–C(=O)OCH}_3 \quad \xrightarrow[\textbf{(CH}_3\textbf{CH}_2)_3\textbf{N}]{\textbf{L}_2\textbf{Pd}} \quad \text{product}$$

the R group of the halide is trans to the substituent

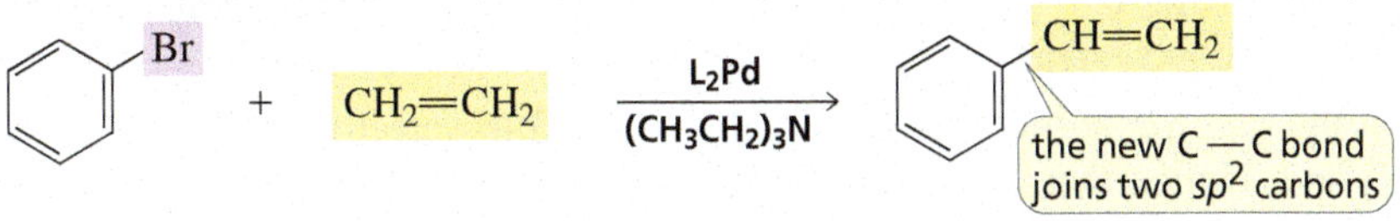

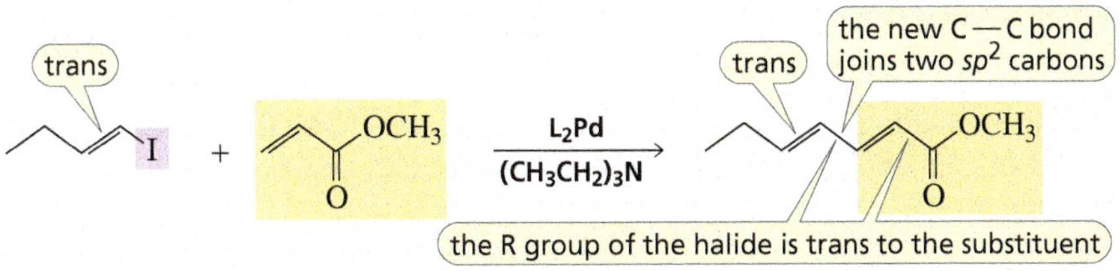

When the reactant is a vinylic halide, notice that the configuration of its double bond is retained in the product.

Mechanism for the Heck Reaction

The currently accepted mechanism for the Heck reaction is shown on the next page.

MECHANISM FOR THE HECK REACTION

$$L_4Pd$$
$$+L \updownarrow -L$$
$$L_3Pd$$
$$+L \updownarrow -L$$
$$L_2Pd$$

$(CH_3CH_2)_3\overset{+}{N}H \;+\; Br^- \quad\xleftarrow{(CH_3CH_2)_3N}\quad HBr$

$R\!-\!Br$

(reaction cycle with intermediates: L_2Pd with Br and R; ligand substitution with alkene Z–CH$_2$; migratory insertion; rotation between the two LPd conformers; syn elimination; L_2Pd with Br and H)

- The catalyst (L_2Pd) is generated as a result of two successive dissociations of a ligand from the precatalyst (L_4Pd).
- Next is an oxidative addition: palladium is inserted between the carbon and the halogen.
- The next step is a ligand substitution: the alkene takes the place of one of the ligands (L).
- Because carbon is more electronegative than palladium, R is a nucleophile. It adds to the sp^2 carbon of the alkene and the electrons of the π bond add to palladium. This is a concerted syn addition.
- Rotation about the C—C bond, as indicated by the green arrow, provides the conformer required for the next step.
- Next, a hydride ion is transferred to palladium and the electrons of the Pd—C bond form the π bond. Notice that this intramolecular reaction is a concerted syn elimination with a hydride ion being eliminated from a β-carbon. (The concerted elimination reactions we have seen previously are anti eliminations where a proton, (not a hydride ion), is eliminated from a β-carbon; Section 9.7.)
- Another ligand substitution occurs: a ligand (L) takes the place of the alkene.
- Reductive elimination regenerates the catalyst and forms HBr. Triethylamine reacts with HBr to prevent it from adding to the alkene. The catalyst can now react with another molecule of alkyl halide.

The Preferred Alkene for a Heck Reaction

The nucleophilic R group can add to either sp^2 carbon of the alkene. Therefore, the reaction leads to a high yield of a single product only in the following situations:

- if the alkene is symmetrical.
- if one of the sp^2 carbons is sterically hindered to the addition of the nucleophile. Therefore, internal alkenes are much less reactive than less sterically hindered terminal alkenes, so internal alkenes are generally not employed in Heck reactions.

■ if one of the sp^2 carbons is bonded to a group that can withdraw electrons by resonance (such as C=O or C≡N), causing the other sp^2 carbon to have a partial positive charge that makes it more susceptible to nucleophilic addition.

PROBLEM-SOLVING STRATEGY

Determining the Product of a Heck Reaction

LEARN THE STRATEGY

What is the product of each of the following reactions?

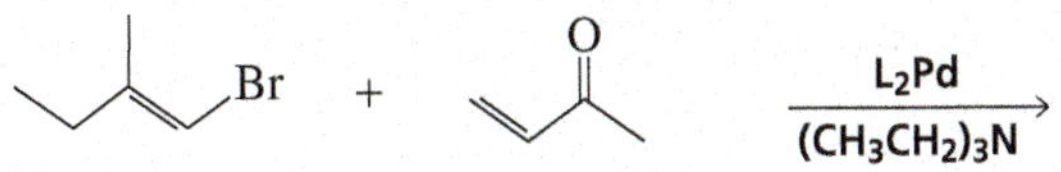

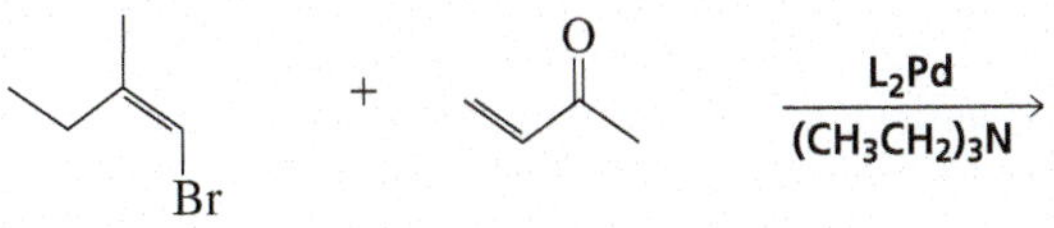

Delete the halogen and then attach its bond to the unsubstituted sp^2 carbon of the vinyl ketone, making sure that the new C—C bond is trans to the substituent on the alkene.

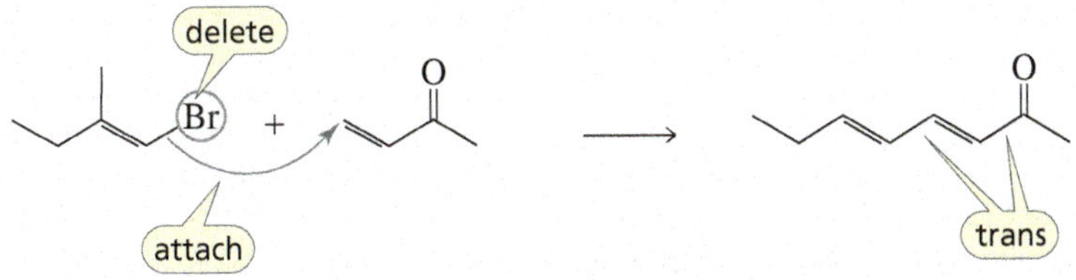

Again, delete the halogen and attach its bond to the unsubstituted sp^2 carbon of the vinyl ketone. Notice that in order to have the new C—C bond trans to the substituent, the vinyl ketone is redrawn so that the substituent is pointing down.

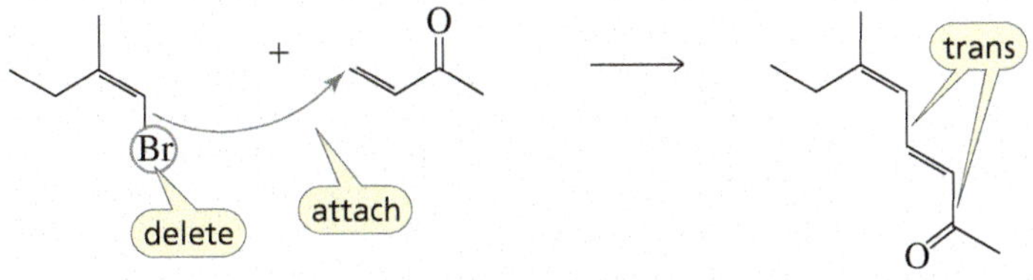

PROBLEM 17 ◆

USE THE STRATEGY

What is the product of each of the following reactions?

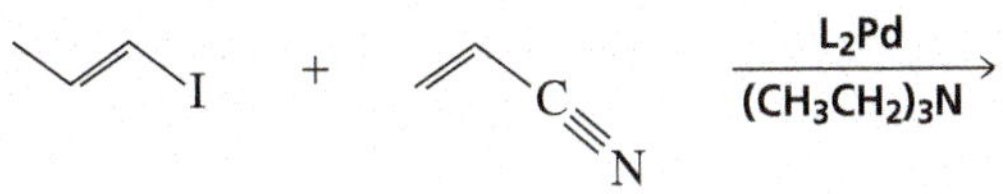

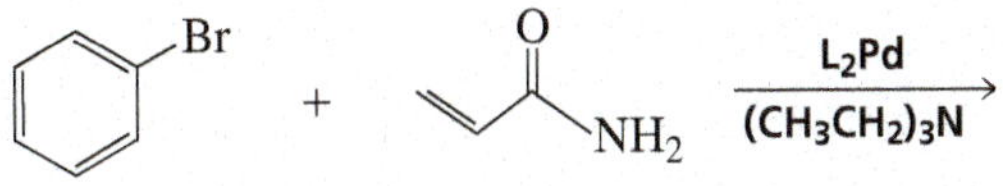

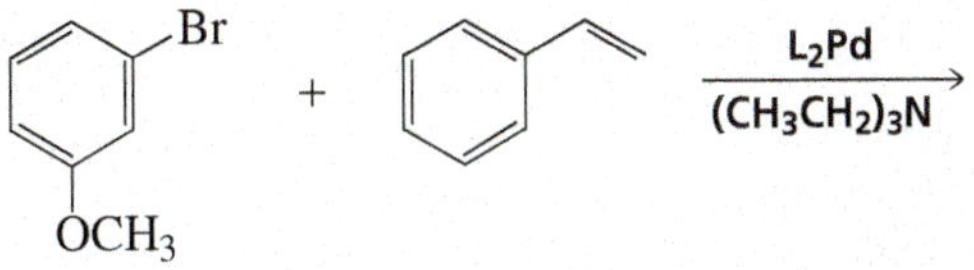

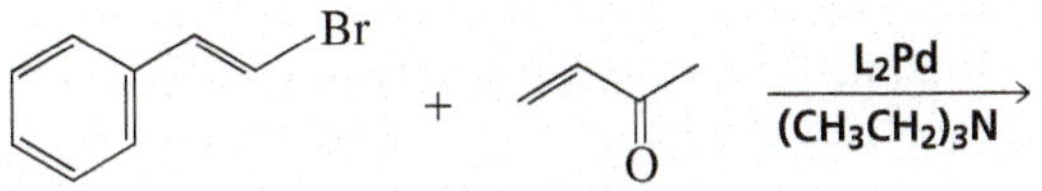

PROBLEM 18 SOLVED

What reactants are needed to synthesize each of the following compounds using a Heck reaction?

a.

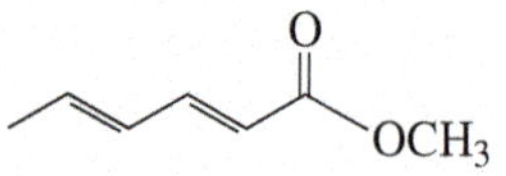

b.

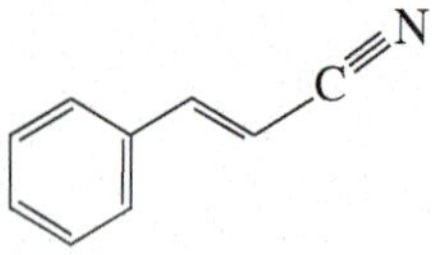

SOLUTION TO 18 a. Cleave the molecule at the end of the C—C bond that joins two sp^2 carbons and add a halogen to the C—C bond.

cleave here

add Br

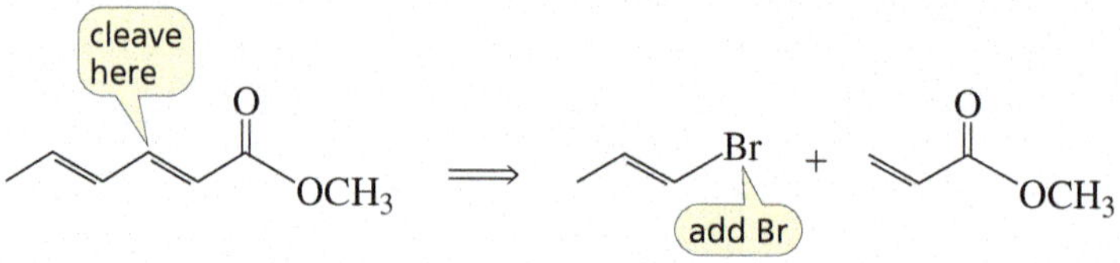

SOLUTION TO 18 b. As you just did in solving the previous problem, cleave the molecule at the end of the C—C bond that joins two sp^2 carbons and add a halogen to the C—C bond.

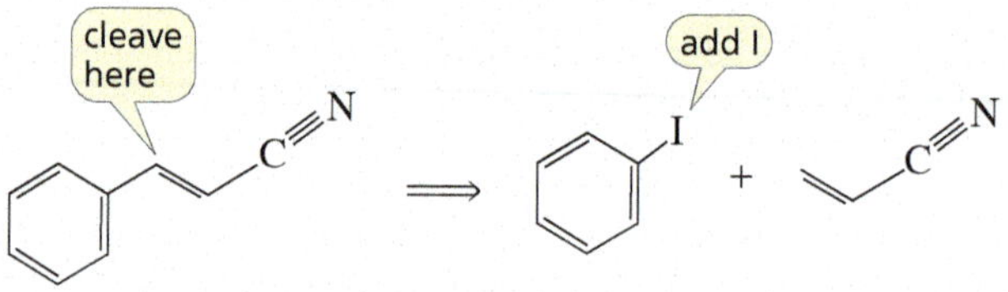

cleave here

add I

PROBLEM 19

What reactants are needed to synthesize each of the following compounds using a Heck reaction?

a.

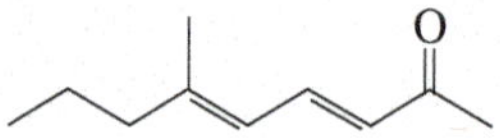

b.

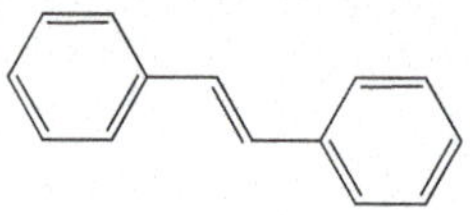

PROBLEM 20

Show how the Suzuki and/or Heck reactions can be used to prepare the following compounds:

a.

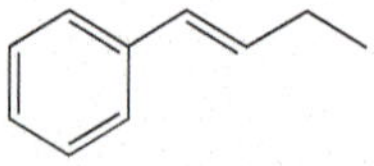

b.

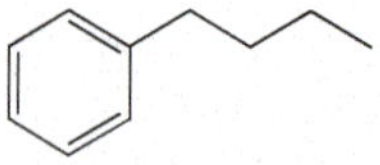

PROBLEM 21 ◆

Identify two pairs of an alkyl bromide and an alkene that could be used in a Heck reaction to prepare the following compound:

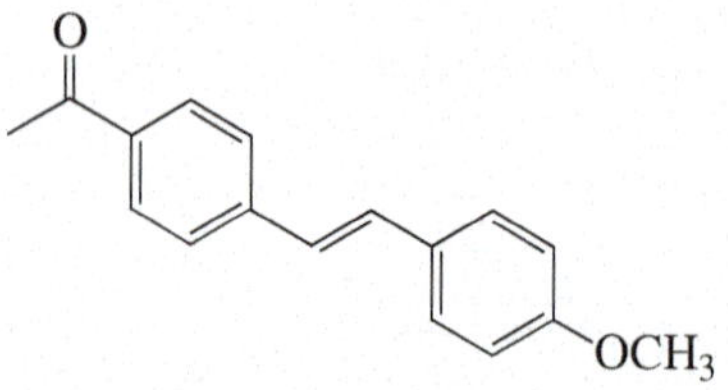

11.5 ALKENE METATHESIS

Alkene metathesis, also called **olefin metathesis,** is a reaction that breaks the strongest bond in an alkene (the double bond) and then rejoins the fragments. *Metathesis* is a Greek word that means "transposition." When the fragments are joined, each new double bond is formed between two sp^2 carbons that were not previously bonded.

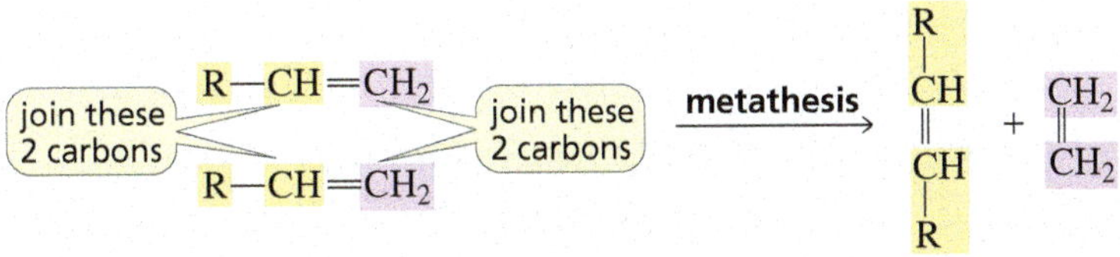

There are two ways the fragments can be joined. If they are joined as shown above, then two new alkene products are obtained. If they are joined as shown below, then the starting material is re-formed. All is not lost, however, because the re-formed starting material can undergo another round of metathesis. Note that both reactions form a double bond between two sp^2 carbons that were not previously bonded.

Terminal alkenes give the best yield of a single alkene product because one of the new alkene products is ethene. Ethene is a gas, so it is removed from the reaction mixture as it is formed. This shifts the equilibrium that exists between the two pathways for joining the fragments toward the pathway that forms ethene and the other new alkene.

A Catalyst Used for Alkene Metathesis

Alkene metathesis requires a transition metal catalyst. If the catalyst does not affect other functional groups that are in the starting alkene, then a large variety of alkenes can be used for metathesis. Grubbs catalysts—catalysts that contain ruthenium—have been found to be the ones most tolerant of other functional groups. The ligands (L) in the Grubbs catalyst shown here are not identified because there are several generations of Grubbs catalysts, each with different ligands.

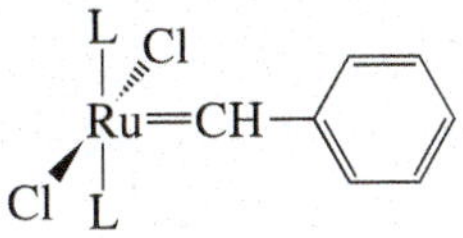

Grubbs catalyst

Examples of Alkene Metathesis

If *E* and *Z* isomers are possible for the product of metathesis, both will be formed.

$$2\ CH_3CH_2CH{=}CH_2 \xrightarrow{\text{Grubbs catalyst}} CH_3CH_2CH{=}CHCH_2CH_3 + CH_2{=}CH_2$$

both *E* and *Z*

Metathesis can also be done using two different alkenes as the starting materials.

$$CH_3CH{=}CHCH_3 + CH_3CH_2CH{=}CHCH_2CH_3 \xrightarrow{\text{Grubbs catalyst}} 2\ CH_3CH{=}CHCH_2CH_3$$

If the reactant is a diene, **ring-closing metathesis** can occur.

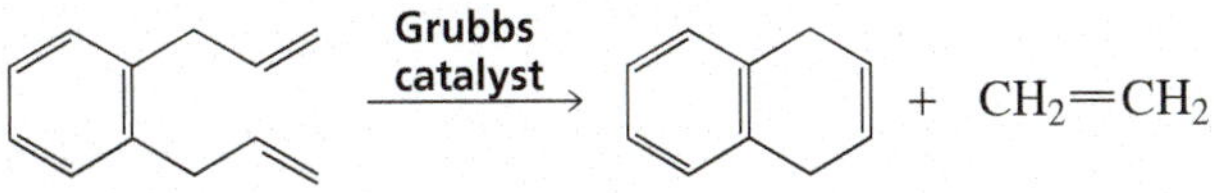

Mechanism for Alkene Metathesis

Metathesis occurs in two phases. The first phase creates two intermediates (I and II). Each intermediate has the metal atom of the catalyst in place of a group that was attached by the double bond in the starting alkene.

$$R{-}CH{=}CH_2 \xrightarrow{\text{first phase}} R{-}CH{=}M + M{=}CH_2$$

I　　　**II**

The two-step mechanism for this first phase is shown here.

MECHANISM FOR ALKENE METATHESIS (Phase I)

Grubbs catalyst is written as M=CH—⟨phenyl⟩

the metal bonds to the more substituted sp^2 carbon of the alkene

starting alkene

I

or

the metal bonds to the less substituted sp^2 carbon of the alkene

starting alkene

II

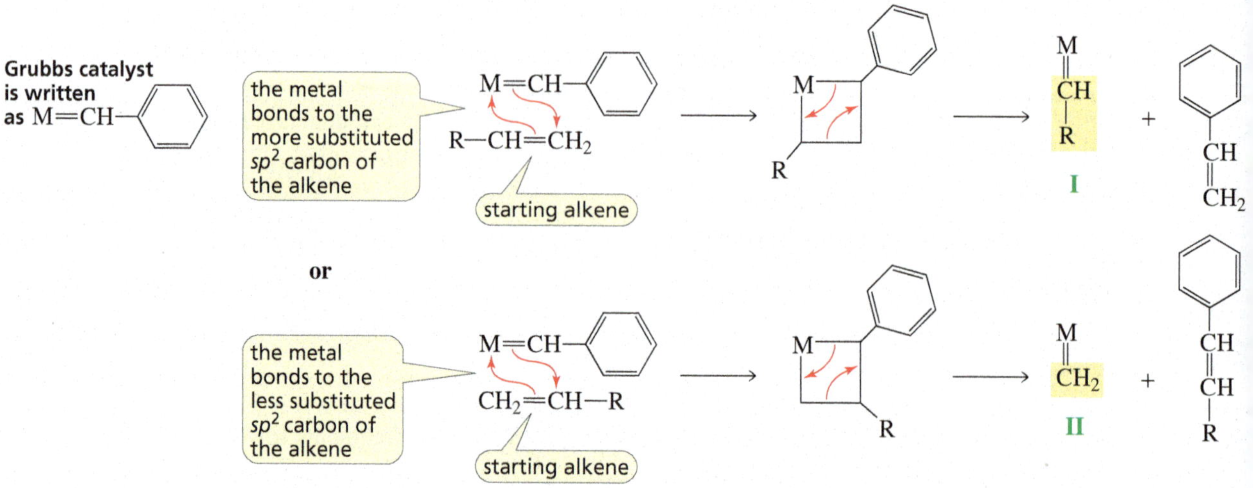

- The Grubbs catalyst and the starting alkene undergo a [2 + 2] cycloaddition reaction (Section 8.14). This reaction forms two different metallocyclobutane intermediates because the metal can bond to either sp^2 carbon of the alkene.

- Each of the metallocyclobutane intermediates undergoes a ring-opening reaction. Two metal-containing intermediates are formed (I and II).

In the second phase of metathesis, each of the metal-containing intermediates formed in the first phase reacts with the starting alkene to form a new alkene and the other metal-containing intermediate. Notice that the mechanism for each of the two phases is the same.

MECHANISM FOR ALKENE METATHESIS (Phase II)

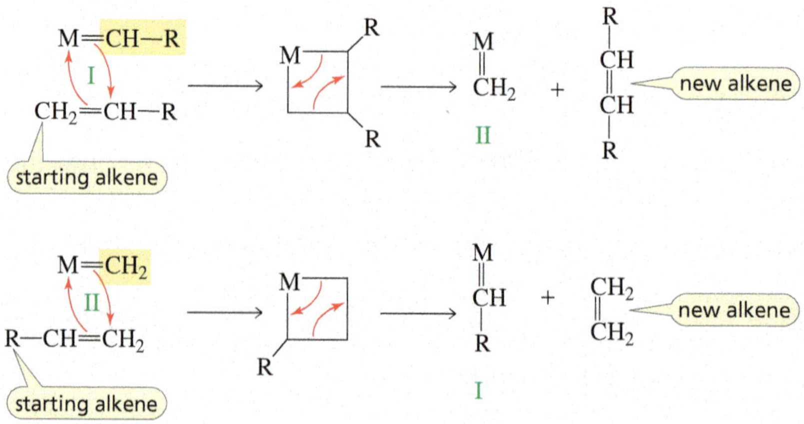

starting alkene

new alkene

II

starting alkene

new alkene

I

- Each metal-containing intermediate undergoes a [2 + 2] cycloaddition reaction to form a metallocyclobutane that undergoes a ring-opening reaction to form the other metal-containing intermediate and a new alkene. Thus, metal-containing intermediate I reacts to form metal-containing intermediate II, and metal-containing intermediate II reacts to form metal-containing intermediate I.

Because the metal atom of a metal-containing intermediate can bond to either sp^2 carbon of the alkene in the first step of the second phase, two different metallocyclobutanes can form. Only if the metal in intermediate I bonds to the *less substituted* sp^2 carbon of the alkene is a new alkene formed. If the metal in intermediate I bonds to the more substituted sp^2 carbon, then the starting alkene is re-formed. Similarly, a new alkene is formed only if the metal in intermediate II bonds to the *more substituted* sp^2 carbon. If the metal in intermediate II bonds to the less substituted sp^2 carbon, then the starting alkene is re-formed.

Each of the metal-containing intermediates formed in the second phase can now react with another molecule of the alkene starting material.

PROBLEM 22

What products are obtained from metathesis of each of the following alkenes?

a. $CH_3CH_2CH{=}CH_2$ **b.** 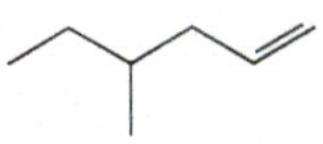 [structure] **c.** 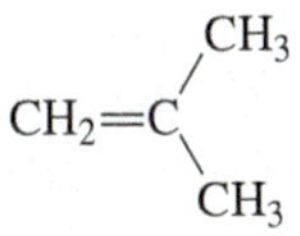 $CH_2{=}C$ with CH_3 and CH_3 groups **d.** 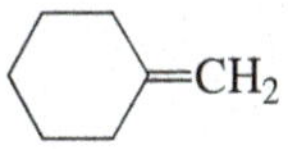 [cyclohexane ring]=CH_2

PROBLEM 23

Draw the product of ring-closing metathesis for each of the following compounds:

a. 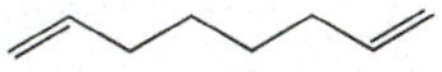 [structure] **b.** 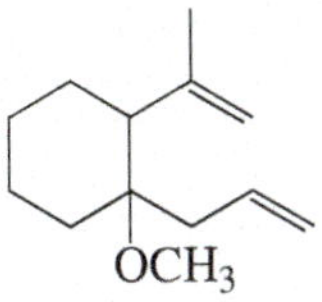 [structure with OCH$_3$] **c.** 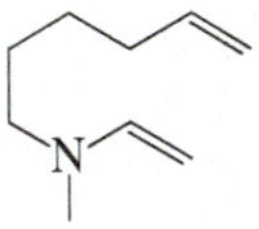 [structure with N]

PROBLEM 24 SOLVED

What compound undergoes metathesis to form each of the following compounds?

a.  [cyclopentane] **b.**  [structure]

LEARN THE STRATEGY

SOLUTION TO 24 a. Break the double bond that formed during metathesis, and then add $=CH_2$ to each end.

a. [cyclopentane] $\Longrightarrow$ [structure with CH$_2$] with labels "break the double bond" and "add $=CH_2$ to each end"

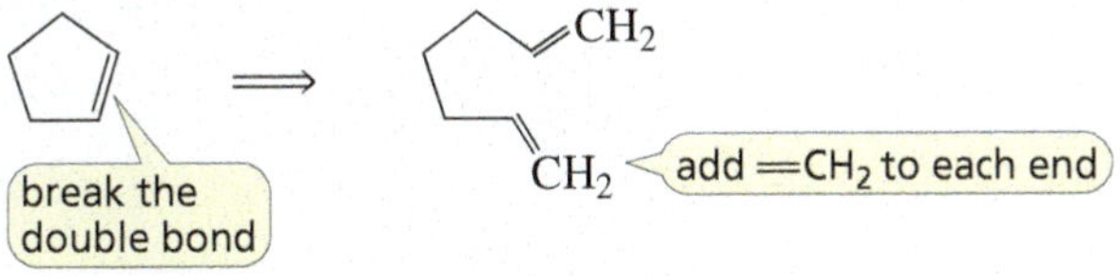

SOLUTION TO 24 b. As you just did in solving the previous problem, break the double bond that formed during metathesis, and then add $=CH_2$ to each end.

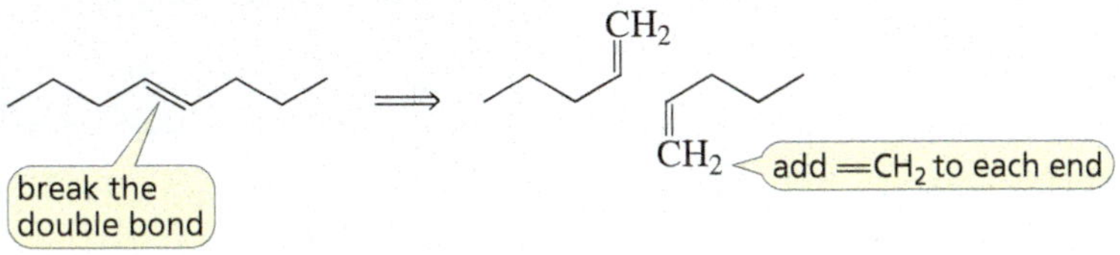

b. [structure] $\Longrightarrow$ [structure with CH$_2$] with labels "break the double bond" and "add $=CH_2$ to each end"

PROBLEM 25

What compound undergoes metathesis to form each of the following compounds?

a. 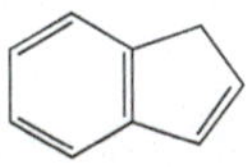 [bicyclic structure] **b.** 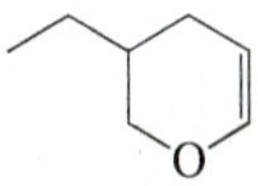 [structure with O ring]

USE THE STRATEGY

Alkyne Metathesis

Alkynes can also undergo metathesis. The preferred catalysts for alkyne metathesis are Schrock catalysts—catalysts that contain molybdenum or tungsten as the transition metal.

$$CH_3CH_2C{\equiv}CH \xrightarrow{\text{Schrock catalyst}} CH_3CH_2C{\equiv}CCH_2CH_3 \; + \; HC{\equiv}CH$$

PROBLEM 26 ♦

What new products are obtained from metathesis of the following alkyne?

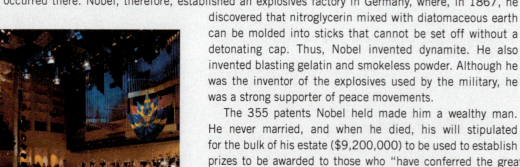

Grubbs, Schrock, Suzuki, and Heck Receive the Nobel Prize

Robert H. Grubbs, a professor at the California Institute of Technology, Richard R. Schrock, a professor at the Massachusetts Institute of Technology, and Yves Chauvin of the Institut Francais du Pétrol each received a share in the 2005 Nobel Prize in Chemistry for their work on olefin metathesis. Akira Suzuki, a professor at Hokkaido University in Sapporo, Japan, Richard F. Heck, a professor at the University of Delaware, and Ei-ichi Negishi, a professor at Purdue University each received a share in the 2010 Nobel Prize in Chemistry for their work on palladium-catalyzed coupling reactions. It is widely believed that John K. Stille of Colorado State University would have received a share in that prize if he had not been killed in the 1989 crash of United Airlines Flight 232 in Iowa.

Robert Grubbs

The Nobel Prize

The Nobel Prize is generally considered the highest honor a scientist can receive. These awards were established by **Alfred Bernhard Nobel (1833–1896)** and were first conferred in 1901.

Nobel was born in Stockholm, Sweden. When he was nine, he moved with his parents to St. Petersburg, Russia, where his father worked for the Russian government, manufacturing torpedoes and land and water mines that he had invented. As a young man, Alfred did research on explosives in a factory his father owned near Stockholm. In 1864, an explosion in the factory killed five people, including his younger brother, causing Alfred to look for ways to make explosives easier to handle and transport. After the explosion, the Swedish government would not allow the factory to be rebuilt because so many accidents had occurred there. Nobel, therefore, established an explosives factory in Germany, where, in 1867, he discovered that nitroglycerin mixed with diatomaceous earth can be molded into sticks that cannot be set off without a detonating cap. Thus, Nobel invented dynamite. He also invented blasting gelatin and smokeless powder. Although he was the inventor of the explosives used by the military, he was a strong supporter of peace movements.

The 355 patents Nobel held made him a wealthy man. He never married, and when he died, his will stipulated for the bulk of his estate ($9,200,000) to be used to establish prizes to be awarded to those who "have conferred the greatest benefit on mankind." He instructed that the money be invested and the interest earned each year be divided into five equal portions "to be awarded to the persons having made the most important contributions in the fields of chemistry, physics, physiology or medicine, literature, and to the one who had done the most toward fostering fraternity among nations, the abolition of standing armies, and the holding and promotion of peace congresses." Nobel also directed that no consideration be given to the nationality of the prize candidates, that each prize be shared by no more than three persons, and that no prize be awarded posthumously.

Alfred Bernhard Nobel

Nobel's instructions stipulated that the prizes for chemistry and physics were to be awarded by the Royal Swedish Academy of Sciences, the prizes for physiology or medicine by the Karolinska Institute in Stockholm, the prize for literature by the Swedish Academy, and the prize for peace by a five-person committee appointed by the Norwegian Parliament. The deliberations are secret, and the decisions cannot be appealed. In 1969, the Swedish Central Bank established a prize in economics in Nobel's honor. The recipient of this prize is selected by the Royal Swedish Academy of Sciences. On December 10—the anniversary of Nobel's death—the prizes are awarded in Stockholm, except for the peace prize, which is awarded in Oslo.

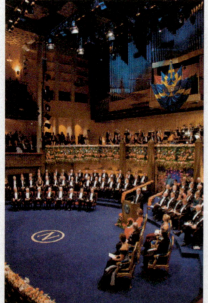

The Golden Hall inside the City Hall in Stockholm, where the Nobel prize winners have a celebratory dinner.

ESSENTIAL CONCEPTS

Section 11.0

- An **organometallic compound** contains a carbon–metal bond.
- The carbon attached to the halogen in an alkyl halide is an electrophile, whereas the carbon attached to the metal in an organometallic compound is a nucleophile.

Section 11.1

- **Organomagnesium compounds (Grignard reagents)** and **organolithium compounds** are the most common organometallic compounds. They cannot be prepared from or used to react with compounds that contain acidic groups.
- The greater the polarity of the carbon–metal bond, the more reactive the organometallic compound is as a nucleophile.

Section 11.2

- **Transmetallation** occurs when an organometallic compound is added to a metal halide whose metal is more electronegative (closer to the electronegativity of carbon) than the metal in the organometallic compound.

Section 11.3

- **Coupling reactions** join two carbon-containing groups together.
- An **organocuprate (a Gilman reagent)** substitutes an R group for a Cl, Br, or I of an alkyl halide. The R group of the organocuprate and the R group of the alkyl halide cannot be secondary or tertiary.
- An organocuprate reacts with an epoxide in a nucleophilic substitution reaction.

Section 11.4

- The **Suzuki** and **Heck reactions** are coupling reactions that occur via a catalytic cycle and require a palladium catalyst.
- The first step in both Suzuki and Heck reactions is **oxidative addition**—two groups are added to a metal. The last step in both reactions is a **reductive elimination**—two groups are eliminated from a metal.
- A **Suzuki reaction** couples the R group of a vinylic or aryl halide with the R′ group of an **organoboron** compound. The new C—C bond joins the carbon that was bonded to the halogen with the carbon that was bonded to boron. When an alkenyl-organoboron compound is used, the new double bond in the product will be trans.
- A **Heck reaction** couples a vinylic or an aryl halide with an alkene. The R group of the vinylic or aryl halide replaces a vinylic hydrogen of an alkene. The new C—C bond joins two sp^2 carbons, and the R group of the aryl or vinyl halide is trans to any substituent on the alkene. If the reactant is a vinylic halide, the configuration of the vinyl group will be retained in the product.

Section 11.5

- **Alkene metathesis** (or **olefin metathesis**) breaks the double bond of an alkene and then rejoins the fragments. When the fragments are joined, the new double bond is formed between two sp^2 carbons that were not previously bonded. Alkynes also undergo metathesis.
- Terminal alkenes give the best yield of a single alkene product in metathesis because one of the products is ethene, which is easily removed from the reaction mixture, thus shifting the equilibrium in favor of the other new alkene product.

SUMMARY OF REACTIONS

1. Formation of an *organolithium* or *organomagnesium* compound (Section 11.1)

$$\text{RBr} \xrightarrow[\text{diethyl ether}]{\text{Mg}} \underset{\substack{\text{Grignard} \\ \text{reagent}}}{\text{RMgBr}} \qquad \text{RBr} \xrightarrow[\text{hexane}]{\text{2 Li}} \underset{\substack{\text{organolithium} \\ \text{compound}}}{\text{RLi}} + \text{LiBr}$$

2. Organocuprates (Section 11.3)
 a. Forming an organocuprate

$$2 \text{ R′Li} + \text{CuI} \xrightarrow{\text{THF}} \underset{\text{organocuprate}}{\text{R′}_2\text{CuLi}} + \text{LiI}$$

 b. Coupling of an alkyl halide and an organocuprate

$$\underset{\text{R and R′ cannot be secondary or tertiary alkyl}}{\text{R—X} + \text{R′}_2\text{CuLi}} \longrightarrow \underset{\text{X = Cl, Br, or I}}{\text{R—R′} + \text{R′Cu} + \text{LiX}}$$

 c. Reaction of an organocuprate with an epoxide

$$\text{R}_2\text{CuLi} + \overset{\text{O}}{\overbrace{\text{H}_2\text{C—CH}_2}} \longrightarrow \text{RCH}_2\text{CH}_2\text{O}^- \xrightarrow{\text{HCl}} \text{RCH}_2\text{CH}_2\text{OH}$$

product alcohol contains two more carbons than the organocuprate

527

3. Palladium-Catalyzed Coupling Reactions (Section 11.4)

    a. Coupling of a *vinylic* or an *aryl halide* with an *organoboron compound*: the Suzuki reaction. The mechanism is shown on page 517.

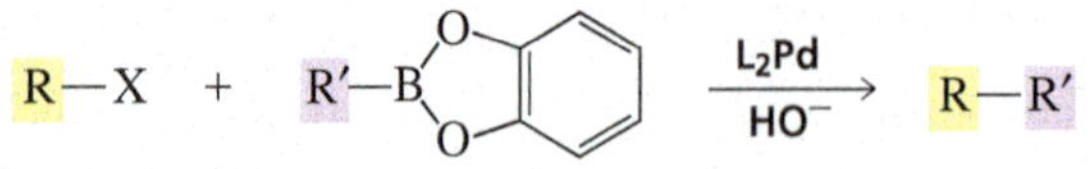

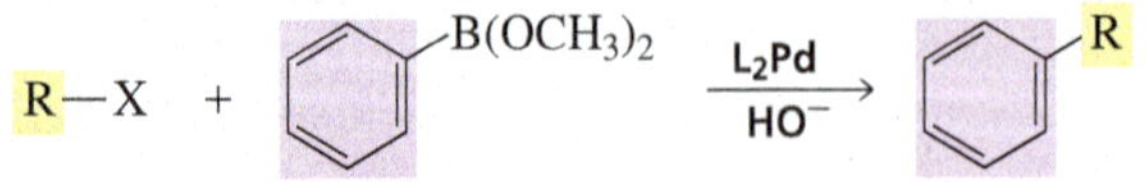

$$R-X \ + \ R'-B\underset{O}{\overset{O}{\Big\langle}} \quad \xrightarrow[\text{HO}^-]{\text{L}_2\text{Pd}} \quad R-R'$$

$$R-X \ + \ \underset{}{\overset{}{\bigcirc}}-B(OCH_3)_2 \quad \xrightarrow[\text{HO}^-]{\text{L}_2\text{Pd}} \quad \overset{R}{\bigcirc}$$

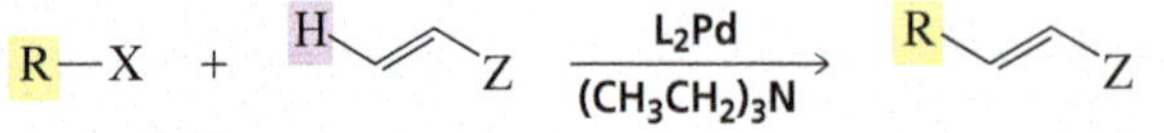

R = vinylic or aryl
X = Br or I

    b. Coupling of a *vinylic* or an *aryl halide* with an *alkene*: the Heck reaction. The mechanism is shown on page 520.

$$R-X \ + \ \overset{H}{\diagdown}\diagup\diagdown_Z \quad \xrightarrow[\text{(CH}_3\text{CH}_2)_3\text{N}]{\text{L}_2\text{Pd}} \quad \overset{R}{\diagdown}\diagup\diagdown_Z$$

R = vinylic or aryl   the product is
X = Br or I   a trans alkene

4. Alkene metathesis (Section 11.5). The mechanism is shown on page 524.

    a. The starting material is an alkene.

$$2 \ R-CH{=}CH_2 \quad \xrightarrow{\substack{\text{Grubbs}\\ \text{catalyst}}} \quad R-CH{=}CH-R \ + \ CH_2{=}CH_2$$

the reaction works best with a terminal alkene

    b. The starting material is a diene.

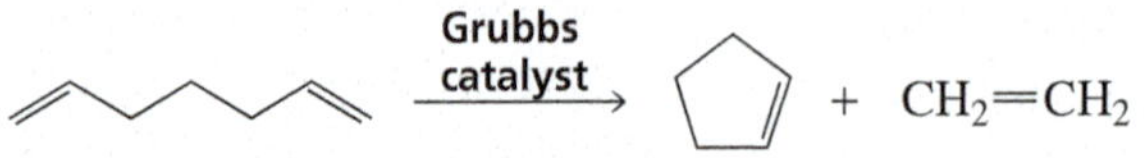

$$\diagup\diagdown\diagup\diagdown\diagup\diagdown \quad \xrightarrow{\substack{\text{Grubbs}\\ \text{catalyst}}} \quad \bigcirc \ + \ CH_2{=}CH_2$$

    c. The starting material is an alkyne.

$$RC{\equiv}CH \quad \xrightarrow{\substack{\text{Schrock}\\ \text{catalyst}}} \quad RC{\equiv}CR \ + \ HC{\equiv}CH$$

PROBLEMS

27. What are the products of the following reactions?

a.  $\bigcirc$—I $+$ $\diagup\diagdown$ $\xrightarrow[\text{(CH}_3\text{CH}_2)_3\text{N}]{\text{L}_2\text{Pd}}$

b. 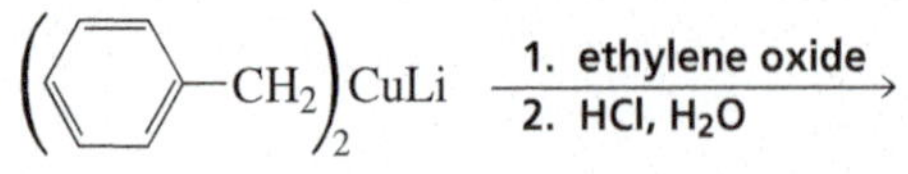 $\left(\bigcirc{-}CH_2 \right)_2 CuLi$ $\xrightarrow[\text{2. HCl, H}_2\text{O}]{\text{1. ethylene oxide}}$

c. 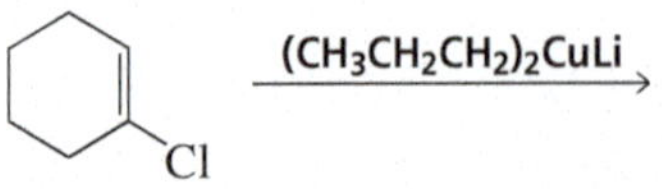 $\bigcirc$ (with Cl) $\xrightarrow{\text{(CH}_3\text{CH}_2\text{CH}_2)_2\text{CuLi}}$

d. 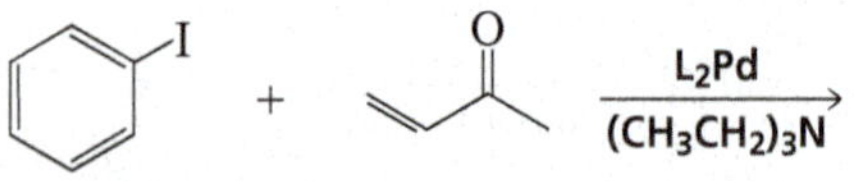 $\bigcirc$—I $+$ $\diagup\diagdown\diagup\overset{O}{\diagdown}$ $\xrightarrow[\text{(CH}_3\text{CH}_2)_3\text{N}]{\text{L}_2\text{Pd}}$

e. 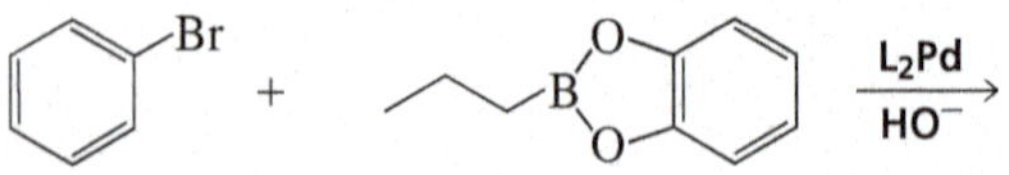 $\bigcirc$—Br $+$ $\diagup\diagdown\diagup{-}B\underset{O}{\overset{O}{\Big\langle}}$ $\xrightarrow[\text{HO}^-]{\text{L}_2\text{Pd}}$

f. 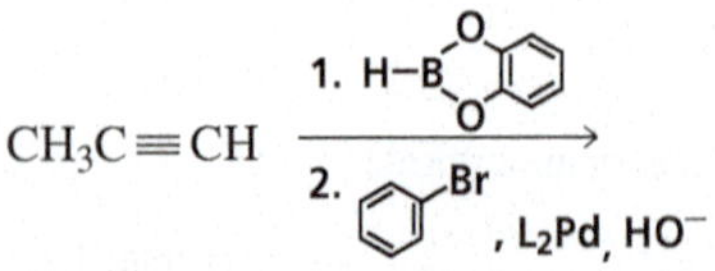 $CH_3C{\equiv}CH$ $\xrightarrow[\text{2. } \bigcirc{-}\text{Br} , \text{ L}_2\text{Pd, HO}^-]{\text{1. H}-B\underset{O}{\overset{O}{\Big\langle}}}$

28. Which of the following alkyl halides could be successfully used to form a Grignard reagent?

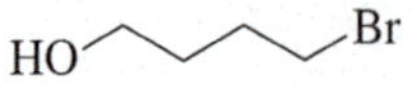

HO$\diagup\diagdown\diagup$Br   Br$\diagup\diagdown\underset{O}{\overset{}{\diagup}}$OH $\underset{}{\overset{}{\text{N}}}\diagup\diagdown\diagup$Br   Br$\diagup\diagdown\diagup$NH$_2$

         A 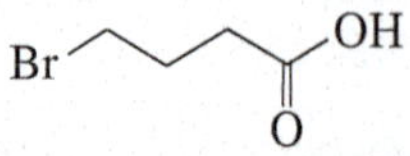   **B** 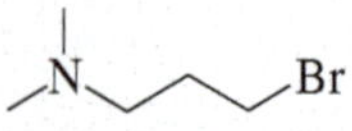   **C**    **D**

29. The coupling of an alkyne with an aryl halide in the presence of a palladium catalyst and triethylamine is called a **Sonogashira reaction.** What reactants couple to form this product?

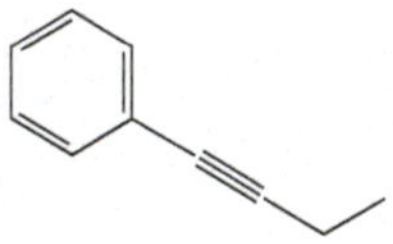

30. Identify **A** through **H**.

$$2 \text{ CH}_3\text{Br} \xrightarrow[\text{2. B}]{\text{1. A}} \text{C} \xrightarrow[\text{2. E}]{\text{1. D}} \text{CH}_3\text{CH}_2\text{CH}_2\text{OH} \xrightarrow[\substack{\text{2. G} \\ \text{3. H}}]{\text{1. F}} \text{CH}_3\text{CH}_2\text{CH}_2\text{OCH}_2\text{CH}_2\text{OH}$$

31. Using the given starting material, any necessary inorganic reagents and catalysts, and any carbon-containing compounds with no more than three carbons, indicate how each of the following compounds can be prepared:

a.

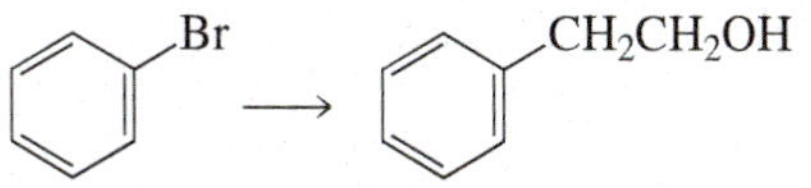

d.

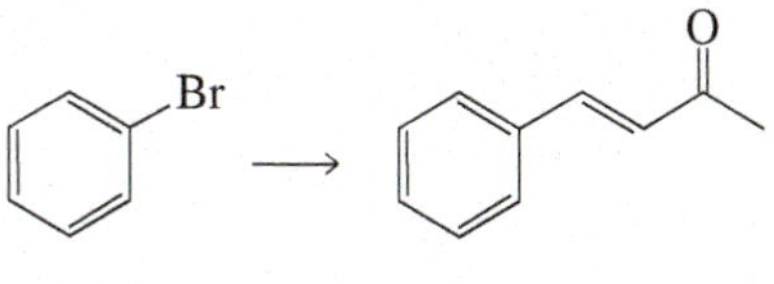

b. $\text{CH}_3\text{CHCH}_2\text{OH} \longrightarrow \text{CH}_3\text{CHCH}_2\text{CH}_2\text{CH}_2\text{OH}$
     │ CH₃ │ CH₃

e.

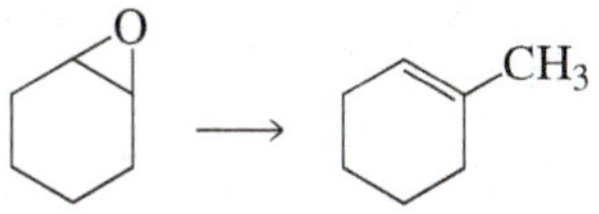

c. $\text{CH}_3\text{CH}_2\text{C}\equiv\text{CH} \longrightarrow \text{CH}_3\text{CH}_2\text{C}\equiv\text{CCH}_2\text{CH}_3$

f.

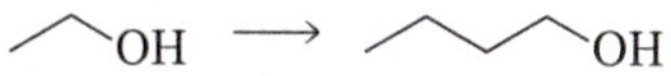

32. What alkyl halide reacts with lithium divinylcuprate $[(\text{CH}_2\text{=CH})_2\text{CuLi}]$ for the synthesis of each of the following compounds?

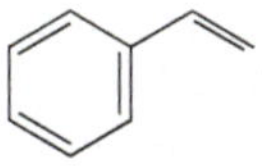

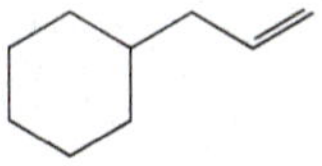

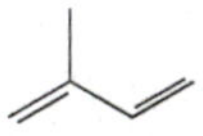

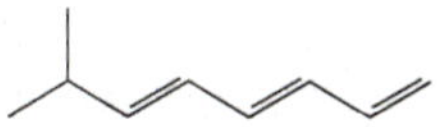

33. What are the products of the following reactions?

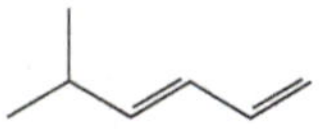

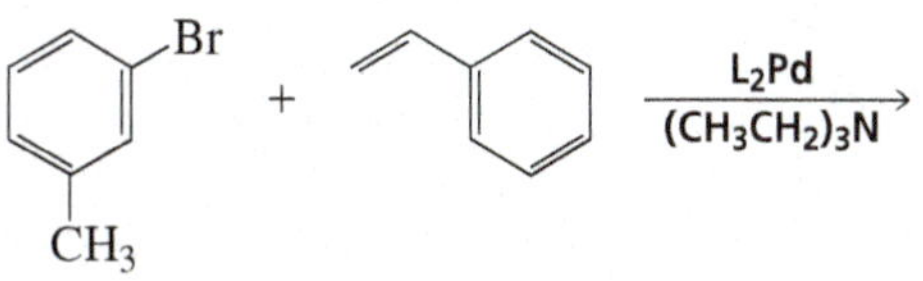

34. What vinylic halide couples with styrene $(\text{C}_6\text{H}_5\text{-CH}\text{=CH}_2)$ in order to synthesize each of the following compounds using a Heck reaction?

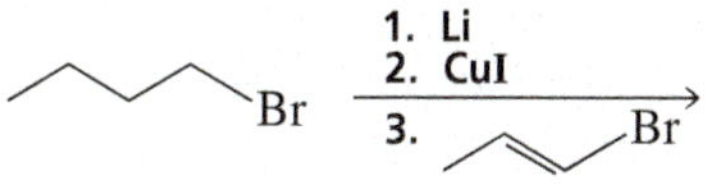

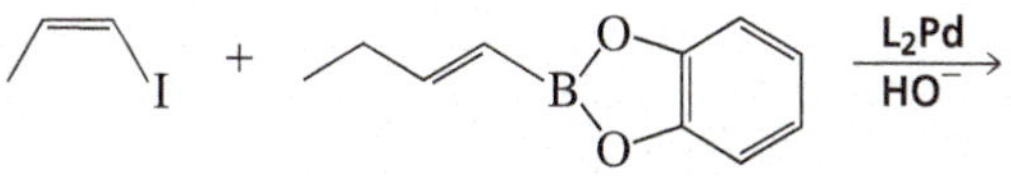

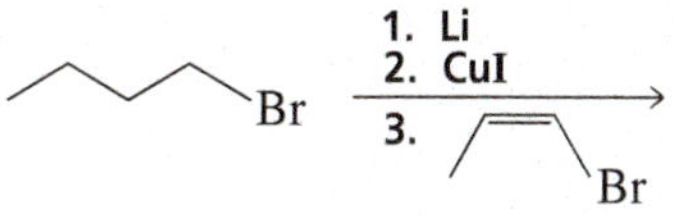

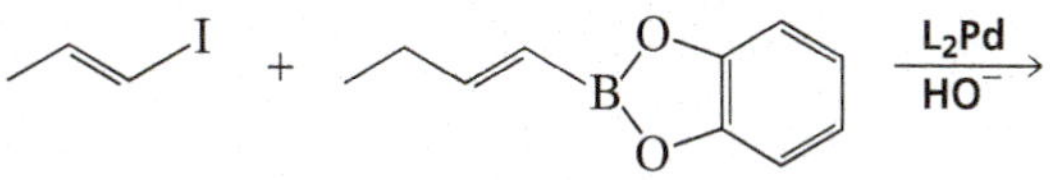

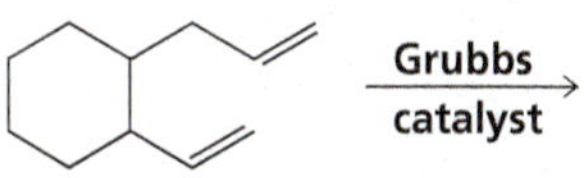

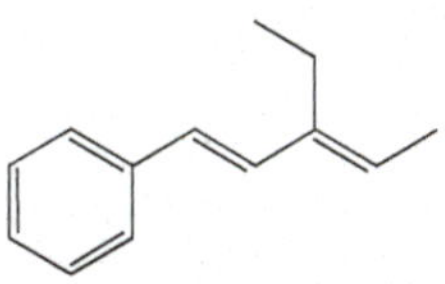

35. The following compound undergoes an intramolecular reaction to form ethene and a product with a five-membered ring. Identify the product and the catalyst used to carry out the reaction.

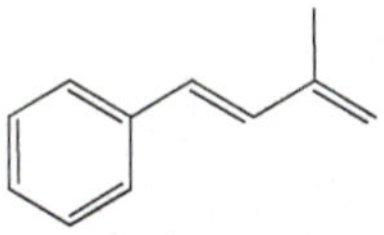

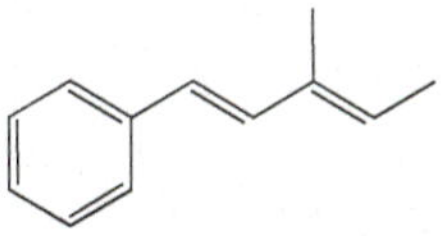

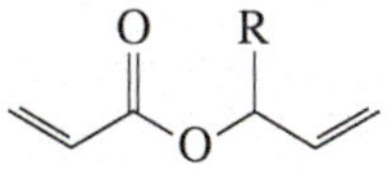

36. Using ethynylcyclohexane as a starting material and any other needed reagents, how can the following compounds be synthesized?

a. **b.** **c.**

37. What are the products of the following reactions?

a.

$$\text{(ArBr)} + \text{(styrene)} \xrightarrow[\text{(CH}_3\text{CH}_2)_3\text{N}]{\text{L}_2\text{Pd}}$$

c.

$$\text{(ArBr)} + \text{(methyl vinyl ketone)} \xrightarrow[\text{(CH}_3\text{CH}_2)_3\text{N}]{\text{L}_2\text{Pd}}$$

b.

$$\text{(ArI)} + \text{B(OR)}_2 \xrightarrow[\text{HO}^-]{\text{L}_2\text{Pd}}$$

d.

$$\text{(ArI)} \xrightarrow{\substack{\text{1. Li} \\ \text{2. B(OR)}_3 \\ \text{3. CH}_2\text{=CHBr, L}_2\text{Pd, HO}^-}}$$

38. Using the given starting material, any necessary inorganic reagents and catalysts, and any carbon-containing compounds with no more than two carbons, indicate how each of the following compounds can be prepared:

a. $CH_3CH_2CH_2CH_2Br \longrightarrow CH_3CH_2CH_2CH_2CHCH$ (with O)

c. $CH_3CH{=}CHCH_3 \longrightarrow CH_3CHCCH_3$ (with O, CH₃)

b. $CH_3CH{=}CHCH_3 \longrightarrow CH_3CH{=}CCH_2CH_3$ (with CH₃)

39. Dimerization is a side reaction that occurs during the preparation of a Grignard reagent. Propose a mechanism that accounts for the formation of the dimer.

$$\text{(cyclohexyl)}{-}\text{Br} \xrightarrow[\text{THF}]{\text{Mg}} \text{(cyclohexyl)}{-}\text{MgBr} + \text{(dicyclohexyl)}$$

a dimer

40. A student added an equivalent of 3,4-epoxy-4-methylcyclohexanol to a solution of methylmagnesium bromide in diethyl ether, and then added dilute hydrochloric acid. He expected that the product would be 1,2-dimethyl-1,4-cyclohexanediol. He did not get any of the expected product. What product did he get?

3,4-epoxy-4-methyl-cyclohexanol   1,2-dimethyl-1,4-cyclohexanediol

41. Using the given starting material, any necessary inorganic reagents, and any carbon-containing compounds with no more than two carbons, indicate how each of the following compounds can be prepared:

a. (cyclohexanol, OH) $\longrightarrow$ (cyclohexyl-CH₂CH₂OH)

b. (isobutanol, OH) $\longrightarrow$ D

42. a. Which of the following compounds cannot be prepared by a Heck reaction?
b. For those compounds that can be prepared by a Heck reaction, what starting materials are required?

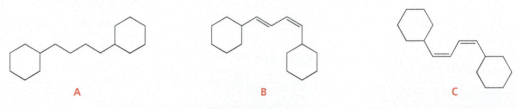

A   B   C

43. What is the maximum yield of 2-pentene that can be formed in the reaction shown on page 523?

44. Bombykol is the sex pheromone of the silk moth.

bombykol

Show how bombykol can be synthesized from the following compounds.

HO ⟶≡ + ⟶ Br

45. a. Metathesis of which of the following sets of alkenes leads to the highest yield of a single alkene?

    1. 1-butene and 1-pentene   **2.** 2-butene and 3-hexene   **3.** 2-butene and 1-pentene

b. What kinds of alkenes should be used in metathesis reactions that use two different alkenes as starting materials?

46. A key step in the synthesis of naproxen, an NSAID more commonly known by its brand name, Aleve (Section 3.9), is a coupling reaction of 2-bromo-6-methoxynaphthalene to form 2-methoxy-6-vinylnaphthalene. Show three different coupling reactions, and the required reagents, that could be used to carry out this step.

2-bromo-6-methoxynaphthalene   2-methoxy-6-vinylnaphthalene

47. A dibromide loses only one bromine when it reacts with sodium hydroxide. The dibromide forms toluene $(C_6H_5\text{-}CH_3)$ when it reacts with magnesium shavings in ether followed by treatment with dilute acid. Give possible structures for the dibromide.

48. What starting material is required in order to synthesize each of the following compounds by ring-closing metathesis?

a.   **b.**   CH₃   **c.**

49. What product is obtained from ring-opening metathesis polymerization of each of the following compounds? (*Hint:* In each case, the product is an unsaturated hydrocarbon with a high molecular weight.)

a.

b.

12 Radicals

The world needs a renewable, nonpolluting, and economically affordable source of energy (see page 533).

Alkanes are widespread both on Earth and on other planets. The atmospheres of Jupiter, Saturn, Uranus, and Neptune contain large quantities of methane (CH_4), the smallest alkane, which is an odorless and flammable gas. The blue colors of Uranus and Neptune are the result of methane in their atmospheres (see page 26). Alkanes on Earth are found in natural gas and petroleum, which are formed by the decomposition of plant and animal material that have been buried for long periods of time in Earth's crust, where oxygen is scarce. As a result, natural gas and petroleum are known as fossil fuels.

We have seen that there are three classes of hydrocarbons:

- *alkanes*, which contain only carbon–carbon single bonds
- *alkenes*, which contain carbon–carbon double bonds
- *alkynes*, which contain carbon–carbon triple bonds

Because **alkanes** do not contain any double or triple bonds, they are called **saturated hydrocarbons,** meaning they are saturated with hydrogen. A few examples of alkanes are

$CH_3CH_2CH_2CH_3$
butane

ethylcyclopentane

4-ethyl-3,3-dimethyldecane

(*R,R*)-1,3-dimethyl-cyclohexane

12.1 ALKANES ARE UNREACTIVE COMPOUNDS

We have seen that the carbon–carbon double and triple bonds of *alkenes* and *alkynes* are composed of strong σ bonds and weaker π bonds and that, because of the relative weakness of the π bond, alkenes and alkynes undergo electrophilic addition reactions (Sections 6.0 and 7.5).

Alkanes, on the other hand, have only strong σ bonds. In addition, the electrons in the C—C and C—H σ bonds are shared equally or almost equally by the bonding atoms, so none of the atoms in an alkane has a significant charge. This means that alkanes are neither nucleophiles nor electrophiles, so neither electrophiles nor nucleophiles are attracted to them. Alkanes, therefore, are relatively unreactive compounds. The failure of alkanes to undergo reactions prompted early organic chemists to call them *paraffins,* from the Latin *parum affinis,* which means "little affinity" (for other compounds).

Natural Gas and Petroleum

Natural gas is approximately 75% methane. The remaining 25% is composed of other small alkanes such as ethane, propane, and butane. In the 1950s, natural gas replaced coal as the main energy source for domestic and industrial heating in many parts of the United States.

Petroleum is a complex mixture of alkanes and cycloalkanes that can be separated into fractions by distillation. Natural gas (hydrocarbons containing fewer than 5 carbons) is the fraction that boils off at the lowest temperature. The fraction that boils at somewhat higher temperatures (hydrocarbons containing 5–11 carbons) is gasoline; the next fraction (9–16 carbons) includes kerosene and jet fuel. The fraction with 15–25 carbons is used for heating oil and diesel oil, and the highest-boiling fraction is used for lubricants and greases. After distillation, a nonvolatile residue called asphalt or tar is left behind.

The 5–11 carbon fraction that is used for gasoline is actually a poor fuel for internal combustion engines. To become a high-performance gasoline, it requires a process known as catalytic cracking. Catalytic cracking converts straight-chain hydrocarbons that are poor fuels into branched-chain compounds that are high-performance fuels (Section 3.2). Originally, cracking (also called pyrolysis) required heating the gasoline to very high temperatures to obtain hydrocarbons with three to five carbons. Modern cracking methods use catalysts to accomplish the same thing at much lower temperatures.

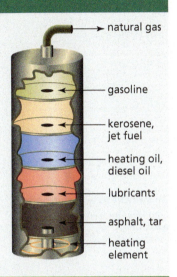

natural gas
gasoline
kerosene, jet fuel
heating oil, diesel oil
lubricants
asphalt, tar
heating element

Fossil Fuels: A Problematic Energy Source

Modern society faces three major problems as a consequence of our dependence on fossil fuels for energy. First, these fuels are a nonrenewable resource and the world's supply is continually decreasing. Second, a group of Middle Eastern and South American countries controls a large portion of the world's supply of petroleum. These countries have formed a cartel called the *Organization of Petroleum Exporting Countries (OPEC)* that controls both the supply and the price of crude oil. Political instability in any OPEC country can seriously affect the world's oil supply.

Third, burning fossil fuels—particularly coal—increases the concentration of CO_2 in the atmosphere; burning coal also increases the concentration of atmospheric SO_2. Scientists have established experimentally that SO_2 causes "acid rain," a threat to plants and, therefore, to our food and oxygen supplies (see page 50 and Section 2.2).

The concentration of atmospheric CO_2 at Mauna Loa, Hawaii, has been periodically measured since 1958. The concentration has increased 28% since the first measurements were taken, causing scientists to predict an increase in Earth's temperature as a result of the absorption of infrared radiation by CO_2 (the *greenhouse effect*). A steady increase in the temperature of Earth would have devastating consequences, including the formation of new deserts, massive crop failure, and the melting of glaciers with a concomitant rise in sea level. Clearly, what we need is a renewable, nonpolitical, nonpolluting, and economically affordable source of energy.

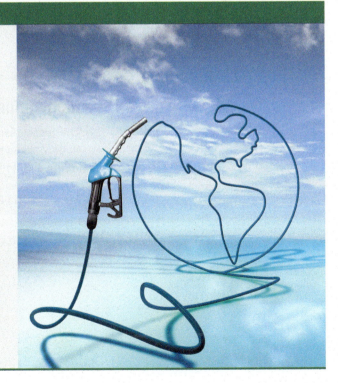

12.2 THE CHLORINATION AND BROMINATION OF ALKANES

Alkanes react with chlorine (Cl_2) or bromine (Br_2) to form alkyl chlorides and alkyl bromides. These **halogenation reactions** take place only at high temperatures or in the presence of light with an appropriate wavelength. (Irradiation with light is symbolized by $h\nu$.)

$$CH_4 \ + \ Cl_2 \ \xrightarrow[\substack{or \\ h\nu}]{\Delta} \ CH_3Cl \ + \ HCl$$
chloromethane

$$CH_3CH_3 \ + \ Br_2 \ \xrightarrow[\substack{or \\ h\nu}]{\Delta} \ CH_3CH_2Br \ + \ HBr$$
bromoethane

Halogenation is the only reaction (other than combustion) that alkanes undergo (without the assistance of a metal catalyst). In a **combustion** reaction, alkanes react with oxygen at high temperatures to form carbon dioxide and water.

Heterolysis and Homolysis

When a bond breaks so that both of its electrons stay with one of the atoms, the process is called **heterolytic bond cleavage** or **heterolysis.**

in heterolytic bond cleavage, both electrons remain with one atom...

...which is why the arrowhead is drawn with 2 barbs

$$A \!-\! B \ \longrightarrow \ A^+ \ + \ :B^-$$

When a bond breaks so that each of the atoms retains one of the bonding electrons, the process is called **homolytic bond cleavage** or **homolysis.** Homolysis results in the formation of radicals. A **radical** (often called a **free radical**) is a species containing an atom with an unpaired electron. A radical is highly reactive because acquiring an electron will complete its octet.

in homolytic bond cleavage, each atom gets one electron...

...which is why each arrowhead is drawn with 1 barb

$$A \!-\! B \ \longrightarrow \ A\cdot \ + \ \cdot B$$

radicals

Monochlorination of an Alkane

The mechanism for the halogenation of an alkane is well understood. As an example, let's look at the mechanism for the monochlorination (substituting one Cl for one H) of methane. The monochlorination of alkanes other than methane has the same mechanism.

MECHANISM FOR THE MONOCHLORINATION OF METHANE

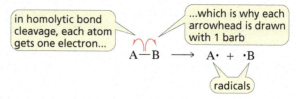

$$:\!\ddot{C}l\!-\!\ddot{C}l\!: \ \xrightarrow[\substack{or \\ h\nu}]{\Delta} \ 2\ :\!\ddot{C}l\cdot \quad \text{(initiation step)}$$
homolytic cleavage

$$:\!\ddot{C}l\cdot \ + \ H\!-\!CH_3 \ \longrightarrow \ H\ddot{C}l\!: \ + \ \cdot CH_3$$
methyl radical

$$\cdot CH_3 \ + \ :\!\ddot{C}l\!-\!\ddot{C}l\!: \ \longrightarrow \ CH_3Cl \ + \ :\!\ddot{C}l\cdot$$

propagation steps

$$:\!\ddot{C}l\cdot \ + \ :\!\ddot{C}l\cdot \ \longrightarrow \ Cl_2$$

$$\cdot CH_3 \ + \ \cdot CH_3 \ \longrightarrow \ CH_3CH_3$$

$$:\!\ddot{C}l\cdot \ + \ \cdot CH_3 \ \longrightarrow \ CH_3Cl$$

termination steps

NOTE TO THE STUDENT
- See the tutorial on page 563 for additional information on drawing curved arrows in radical systems.

- Heat or light supplies the energy required to break the $Cl-Cl$ bond homolytically. This is the **initiation step** of the reaction because it creates radicals from a molecule in which all the electrons are paired.

- The chlorine radical formed in the initiation step removes a hydrogen atom from the alkane (in this case methane), forming HCl and a methyl radical.

- The methyl radical removes a chlorine atom from Cl_2, forming the monohalogenated alkane (chloromethane) and another chlorine radical, which can then remove a hydrogen atom from another molecule of methane.

  Steps 2 and 3 are **propagation steps** because the radical created in the first propagation step reacts in the second propagation step to produce the radical that participates in the first propagation step. A propagation step propagates the chain. Thus, the two propagation steps are repeated over and over. The first propagation step determines the rate of the overall reaction.

- Any two radicals in the reaction mixture can combine to form a molecule in which all the electrons are paired. The combination of two radicals is called a **termination step** because it helps bring the reaction to an end by decreasing the number of radicals available to propagate the reaction. Any two radicals can combine, so a radical reaction produces a mixture of products.

> Radical chain reactions have initiation, propagation, and termination steps.

Because the reaction has radical intermediates and repeating propagation steps, it is called a **radical chain reaction.** This particular radical chain reaction is called a **radical substitution reaction** because it substitutes a chlorine for one of the hydrogens of the alkane.

Maximizing the Monohalogneated Product

To maximize the amount of monohalogenated product formed, a radical substitution reaction should be carried out in the presence of excess alkane. Excess alkane in the reaction mixture increases the probability that the halogen radical will collide with a molecule of alkane rather than with a molecule of alkyl halide—even toward the end of the reaction, by which time a considerable amount of alkyl halide will have been formed.

If the halogen radical removes a hydrogen from a molecule of alkyl halide rather than from a molecule of alkane, a dihalogenated product will be obtained.

$$Cl\cdot \ + \ CH_3Cl \ \longrightarrow \ \cdot CH_2Cl \ + \ HCl$$
$$\cdot CH_2Cl \ + \ Cl_2 \ \longrightarrow \ CH_2Cl_2 \ + \ Cl\cdot$$

(alkyl halide) — refers to CH_3Cl
(a dihalogenated compound) — refers to CH_2Cl_2

Monobromination of an Alkane

The bromination of alkanes has the same mechanism as the chlorination of alkanes.

MECHANISM FOR THE MONOBROMINATION OF ETHANE

$$:\ddot{Br}-\ddot{Br}: \ \xrightarrow[\text{or } h\nu]{\Delta} \ 2:\ddot{Br}\cdot \qquad \text{(initiation step)}$$

$$:\ddot{Br}\cdot \ + \ H-CH_2CH_3 \ \longrightarrow \ H\ddot{Br}: \ + \ CH_3\dot{C}H_2$$
$$\qquad\qquad\qquad\qquad\qquad \text{ethyl radical}$$
$$CH_3\dot{C}H_2 \ + \ :\ddot{Br}-\ddot{Br}: \ \longrightarrow \ CH_3CH_2Br \ + \ :\ddot{Br}\cdot$$

(propagation steps)

$$:\ddot{Br}\cdot \ + \ :\ddot{Br}\cdot \ \longrightarrow \ Br_2$$
$$CH_3\dot{C}H_2 \ + \ CH_3\dot{C}H_2 \ \longrightarrow \ CH_3CH_2CH_2CH_3$$
$$CH_3\dot{C}H_2 \ + \ :\ddot{Br}\cdot \ \longrightarrow \ CH_3CH_2\ddot{Br}:$$

(termination steps)

Why Radicals No Longer Have to Be Called Free Radicals

At one time an "R" group was called a radical. For example, the OH substituent in CH_3CH_2OH was said to be attached to an ethyl radical. To distinguish this kind of ethyl radical from $CH_3\dot{C}H_2$, which has an unpaired electron and is not attached to a substituent, $CH_3\dot{C}H_2$ was called a "free radical"—it was free from attachment to a substituent. Now that we call "R" a *substituent* or a *group* instead of a *radical*, we no longer need to call a compound with an unpaired electron a "free radical"; the word *radical* is now unambiguous.

PROBLEM 1

Write the initiation, propagation, and termination steps for the monochlorination of cyclohexane.

PROBLEM 2

Write the steps for formation of tetrachloromethane (CCl_4) from the reaction of methane with $Cl_2 + h\nu$.

12.3 RADICAL STABILITY DEPENDS ON THE NUMBER OF ALKYL GROUPS ATTACHED TO THE CARBON WITH THE UNPAIRED ELECTRON

Radicals are classified according to the carbon that bears the unpaired electron.

- Primary radicals have the unshared electron on a primary carbon.
- Secondary radicals have the unshared electron on a secondary carbon.
- Tertiary radicals have the unshared electron on a tertiary carbon.

Alkyl groups stabilize radicals the same way they stabilize carbocations—that is, by hyperconjugation (Section 6.2). Therefore, the relative stabilities of **primary, secondary, and tertiary alkyl radicals** follow the same order as the relative stabilities of primary, secondary, and tertiary carbocations.

Stability of alkyl radicals: 3° > 2° > 1°

relative stabilities of alkyl radicals

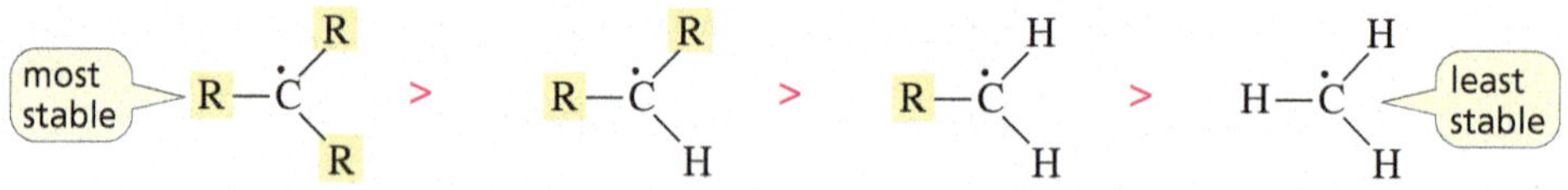

tertiary radical > secondary radical > primary radical > methyl radical

The differences in the relative stabilities of the radicals are much smaller than the differences in the relative stabilities of the carbocations because alkyl groups do not stabilize radicals as well as they stabilize carbocations.

The MO diagrams in Figure 12.1 explain why alkyl groups stabilize carbocations better than they stabilize radicals. Stabilization of a carbocation results from overlap between a filled orbital of a C—H or C—C σ bond and an empty *p* orbital—a two-electron system (Figure 12.1a). In contrast, stabilization of a radical results from overlap between a filled orbital of a C—H or C—C σ bond and a *p* orbital that contains one electron—a three-electron system (Figure 12.1b).

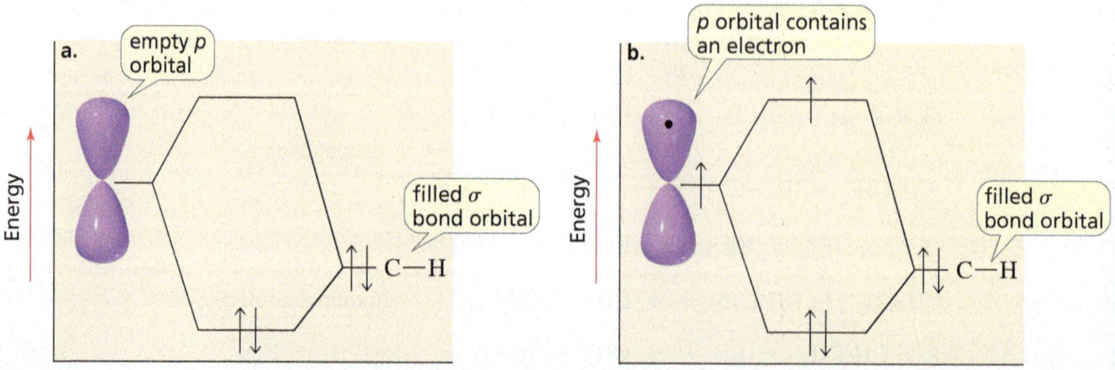

▶ **Figure 12.1**
MO diagrams showing the stabilization achieved when the electrons of an orbital of a C—H or C—C σ bond overlap
(a) an empty *p* orbital.
(b) a *p* orbital that contains one electron.

Both electrons are in a bonding MO in the two-electron system, whereas one of the electrons has to go into an antibonding MO in the three-electron system. Overall, the three-electron system is stabilizing because there are more electrons in the bonding MO than in the antibonding MO, but it is not as stabilizing as the two-electron system, which does not have an electron in the antibonding MO. Consequently, an alkyl group stabilizes a carbocation about 5–10 times better than it stabilizes a radical.

PROBLEM 3 ◆

a. Which of the hydrogens in the structure in the margin is the easiest for a chlorine radical to remove?

b. How many secondary hydrogens does the structure have?

12.4 THE DISTRIBUTION OF PRODUCTS DEPENDS ON PROBABILITY AND REACTIVITY

Two different alkyl halides are obtained from the monochlorination of butane. Substitution of a hydrogen bonded to one of the primary carbons produces 1-chlorobutane, whereas substitution of a hydrogen bonded to one of the secondary carbons forms 2-chlorobutane.

$$CH_3CH_2CH_2CH_3 \; + \; Cl_2 \; \xrightarrow{h\nu} \; CH_3CH_2CH_2CH_2Cl \; + \; CH_3CH_2\overset{\displaystyle Cl}{\underset{|}{C}}HCH_3 \; + \; HCl$$

butane   1-chlorobutane   2-chlorobutane
                   expected = 60%   expected = 40%
                  experimental = 29%   experimental = 71%

The expected (statistical) distribution of products is 60% 1-chlorobutane and 40% 2-chlorobutane because 6 of butane's 10 hydrogens can be substituted to form 1-chlorobutane, whereas only 4 can be substituted to form 2-chlorobutane.

When the reaction is carried out in the laboratory, however, the products are found to be 29% 1-chlorobutane and 71% 2-chlorobutane. In other words, the product distribution does not depend solely on the *probability* of a chlorine radical colliding with a primary or a secondary hydrogen.

Because more 2-chlorobutane is obtained than expected and *the rate-determining step of the overall reaction is removal of the hydrogen atom*, we can conclude that it is easier to remove a hydrogen atom from a secondary carbon to form a secondary radical than it is to remove a hydrogen atom from a primary carbon to form a primary radical.

We should not be surprised that it is easier to form a secondary radical, because a secondary radical is more stable than a primary radical. The more stable the radical, the more easily it is formed, because the stability of the radical is reflected in the stability of the transition state leading to its formation (Section 6.3).

After experimentally determining the amount of each chlorination product obtained from various hydrocarbons, chemists were able to assign values for the relative rates of alkyl radical formation. The experimental data showed that *at room temperature* it is 5.0 times easier for a chlorine radical to form a tertiary radical than a primary radical, and it is 3.8 times easier to form a secondary radical than a primary radical. The precise ratios differ at different temperatures. (How relative rates are determined experimentally is described in Problem 44.)

relative rates of alkyl radical formation by a chlorine radical at room temperature

tertiary > secondary > primary
5.0   3.8   1.0

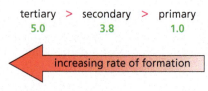

increasing rate of formation

LEARN THE STRATEGY

Now we see that both *probability* (the number of hydrogens that can be removed that lead to the formation of a particular product) and *reactivity* (the relative rate at which a particular radical is formed) must be taken into account when determining the relative amounts of the products obtained from the radical chlorination of an alkane.

Both probability and reactivity must be taken into account when calculating the relative amounts of products.

relative amount of 1-chlorobutane

number of hydrogens × relative reactivity
$6 \times 1.0 = 6.0$

relative amount of 2-chlorobutane

number of hydrogens × relative reactivity
$4 \times 3.8 = 15$

The anticipated percent yield of each alkyl chloride (as a percentage of all the monochlorinated products) is calculated by dividing the relative amount of the particular product by the sum of the relative amounts of all the alkyl chloride products $(6 + 15 = 21)$.

percent yield of 1-chlorobutane

$$\frac{6.0}{21} = 29\%$$

percent yield of 2-chlorobutane

$$\frac{15}{21} = 71\%$$

USE THE STRATEGY

PROBLEM 4 ◆

How many alkyl chlorides are obtained from monochlorination of the following alkanes? Disregard stereoisomers.

a.

b.

c.

d.

e.

f.

g.

h.

i.

Using Radical Halogenation in Synthesis

Because the radical chlorination of an alkane can yield several different monosubstitution products as well as products that contain more than one chlorine atom, it is not the best method to use to synthesize an alkyl halide. The addition of a hydrogen halide to an alkene or the conversion of an alcohol to an alkyl halide are both much better ways to make an alkyl halide (Sections 6.1, 10.1, and 10.2).

Radical halogenation of an alkane is still a useful reaction because it is the only way to convert an inert alkane into a reactive compound. And once the halogen is introduced into the alkane, it can be replaced by a variety of other substituents (Section 11.2).

LEARN THE STRATEGY

PROBLEM 5 SOLVED

If cyclopentane reacts with more than one equivalent of Cl_2 at a high temperature, how many dichlorocyclopentanes would you expect to obtain as products?

SOLUTION Seven dichlorocyclopentanes could be obtained as products. Only one isomer is possible for the 1,1-dichloro compound. The 1,2- and 1,3-dichloro compounds have two asymmetric centers. Each has three stereoisomers because the cis isomer is a meso compound and the trans isomer is a pair of enantiomers.

   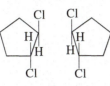

1,1-dichloro compound **meso compound** **enantiomers** **meso compound** **enantiomers**
1,2-dichloro compounds   **1,3-dichloro compounds**

USE THE STRATEGY

PROBLEM 6

How many alkyl halides are obtained from monochlorination of the alkanes in Problem 4 if stereoisomers are included?

12.5 THE REACTIVITY–SELECTIVITY PRINCIPLE

The relative rates of radical formation by a bromine radical are different from the relative rates of radical formation by a chlorine radical. For example, at 125 °C, a bromine radical forms a tertiary radical 1600 times faster than a primary radical and it forms a secondary radical 82 times faster than a primary radical.

relative rates of alkyl radical formation by a bromine radical at 125 °C

tertiary > secondary > primary
1600   82   1

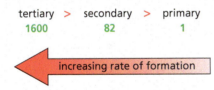

The differences in these relative rates are so great that the *reactivity factor* is vastly more important than the *probability factor* in determining the relative amounts of products obtained in a radical substitution reaction.

For example, the radical bromination of butane gives a 98% yield of 2-bromobutane, compared with a 71% yield of 2-chlorobutane obtained when butane is chlorinated (Section 12.4).

$$CH_3CH_2CH_2CH_3 + Br_2 \xrightarrow{h\nu} CH_3CH_2CH_2CH_2Br + CH_3CH_2\overset{\displaystyle Br}{\underset{\displaystyle |}{C}}HCH_3 + HBr$$

1-bromobutane   **2-bromobutane**
2%   **98%**

LEARN THE STRATEGY

PROBLEM 7 SOLVED

a. What is the major monobromination product of the following reaction? Disregard stereoisomers.
b. What is the anticipated percent yield of the major product (as a percentage of all the monobrominated products)?

$$\underset{\displaystyle CH_3CHCH_2CH_3}{\overset{\displaystyle \overset{\textstyle CH_3}{|}}{}} \xrightarrow[h\nu]{Br_2}$$

SOLUTION

a. The relative amount of each product is determined by multiplying the number of hydrogens that can be removed that lead to the formation of that product by the relative rate for formation of the radical (see the top of the next page). Therefore, the major product is 2-bromo-2-methylbutane.

$$\underset{\substack{\text{1-bromo-2-methylbutane}}}{\overset{\overset{\displaystyle CH_3}{|}}{CH_2CHCH_2CH_3}}\quad \underset{\substack{\text{2-bromo-2-methylbutane}}}{\overset{\overset{\displaystyle CH_3}{|}}{CH_3CCH_2CH_3}}\quad \underset{\substack{\text{2-bromo-3-methylbutane}}}{\overset{\overset{\displaystyle CH_3}{|}}{CH_3CHCHCH_3}}\quad \underset{\substack{\text{1-bromo-3-methylbutane}}}{\overset{\overset{\displaystyle CH_3}{|}}{CH_3CHCH_2CH_2}}$$
Br · · · · · Br · · · · · Br · · · · · Br

6 × 1.0 = 6 · · · · 1 × 1600 = 1600 · · · · 2 × 82 = 164 · · · · 3 × 1.0 = 3

b. The anticipated percent yield of the major product is obtained by dividing its relative amount by the relative amounts of all the alkyl halides produced in the reaction. The anticipated yield is 90%.

$$6.0 + 1600 + 164 + 3.0 = 1773 \qquad \frac{1600}{1773} = 0.90 = 90\%$$

USE THE STRATEGY

PROBLEM 8 ◆

a. What is the major product of the reaction in Problem 7 when the alkane reacts with Cl_2 instead of with Br_2? Disregard stereoisomers.

b. What is the anticipated percent yield of the major product?

Explaining the Difference in Relative Rates

Why are the relative rates of radical formation so different when a bromine radical rather than a chlorine radical is used as the hydrogen-removing reagent?

To answer this question, we must compare the $\Delta H°$ values for forming primary, secondary, and tertiary radicals by a chlorine radical and by a bromine radical. These $\Delta H°$ values can be calculated using the bond dissociation enthalpies in Table 5.1 on pages 206–207.

	$\Delta H°$ (kcal/mol)
Cl· + ⌃ ⟶ ⌃· + HCl	101 − 103 = −2
Cl· + ⌃ ⟶ ⌃· + HCl	99 − 103 = −4
Cl· + ⋏ ⟶ ⋏· + HCl	97 − 103 = −6

	$\Delta H°$ (kcal/mol)
Br· + ⌃ ⟶ ⌃· + HBr	101 − 88 = 13
Br· + ⌃ ⟶ ⌃· + HBr	99 − 88 = 11
Br· + ⋏ ⟶ ⋏· + HBr	97 − 88 = 9

$\Delta H°$ is equal to the energy of the bond being broken minus the energy of the bond being formed; Section 5.6.

We must also be aware that bromination is a much slower reaction than chlorination. The activation energy for removing a hydrogen atom by a bromine radical is about 4.5 times greater than that for removing a hydrogen atom by a chlorine radical.

Using the calculated $\Delta H°$ values and the experimental activation energies, we can draw reaction coordinate diagrams for formation of primary, secondary, and tertiary radicals by a chlorine radical (Figure 12.2a) and by a bromine radical (Figure 12.2b).

▶ **Figure 12.2**

(a) Reaction coordinate diagrams for formation of primary, secondary, and tertiary alkyl radicals by a chlorine radical. The transition states have relatively little radical character because they resemble the reactants.

(b) Reaction coordinate diagrams for formation of primary, secondary, and tertiary alkyl radicals by a bromine radical. The transition states have a relatively high degree of radical character because they resemble the products.

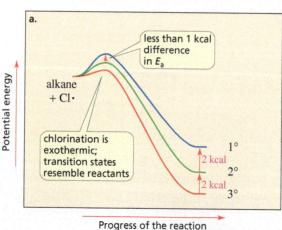

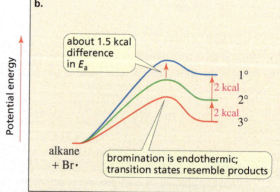

Because the reaction of a chlorine radical with an alkane to form a primary, secondary, or tertiary radical is exothermic, the transition states resemble the reactants (see the Hammond postulate, Section 6.3). The reactants all have approximately the same energy, so there is only small difference in the activation energies for the formation of a primary, secondary, or tertiary radical.

In contrast, the reaction of a bromine radical with an alkane is endothermic, so the transition states resemble the products. Because there are significant differences in the energies of the product radicals—depending on whether they are primary, secondary, or tertiary—there are significant differences in the activation energies.

The Reactivity-Selectivity Principle

Because there is little difference in the activation energies, a chlorine radical makes primary, secondary, and tertiary radicals with almost equal ease (Figure 12.2a). A bromine radical, on the other hand, has a clear preference for forming the easiest-to-form tertiary radical (Figure 12.2b). In other words, because a bromine radical is relatively unreactive, it is highly selective about which radical it forms. In contrast, the more reactive chlorine radical is considerably less selective.

These observations illustrate the **reactivity–selectivity principle:**

The greater the reactivity of a species, the less selective it will be.

Because chlorination is relatively nonselective, it is a useful reaction only when there is only one kind of hydrogen in the alkane.

A bromine radical is less reactive and more selective than a chlorine radical.

The more reactive a species is, the less selective it will be.

PROBLEM-SOLVING STRATEGY

LEARN THE STRATEGY

Planning the Synthesis of an Alkyl Halide

Would chlorination or bromination of methylcyclohexane produce a greater yield of 1-halo-1-methylcyclohexane?

To solve this kind of problem, first draw the structures of the compounds being discussed.

1-Halo-1-methylcyclohexane is a tertiary alkyl halide, so the question becomes, "Would bromination or chlorination produce a greater yield of a tertiary alkyl halide?" Because bromination is more selective, it will produce a greater yield of the desired compound. Chlorination will form some of the tertiary alkyl halide, but it will also form significant amounts of primary and secondary alkyl halides.

PROBLEM 9 ♦

USE THE STRATEGY

a. Would chlorination or bromination produce a greater yield of 1-halo-2,3-dimethylbutane?
b. Would chlorination or bromination produce a greater yield of 2-halo-2,3-dimethylbutane?
c. Would chlorination or bromination be a better way to make 1-halo-2,2-dimethylpropane?

Why Alkanes Undergo Only Chlorination and Bromination

By comparing the $\Delta H°$ values for the first propagation step in the monohalogenation of methane, we can understand why alkanes undergo chlorination and bromination but not iodination, and why fluorination is too violent a reaction to be useful.

F_2

Cl_2

Br_2

I_2
Halogens

first propagation step for monohalogenation

$$F\cdot + CH_4 \longrightarrow \cdot CH_3 + HF \quad \Delta H° = 105 - 136 = -31$$

$$Cl\cdot + CH_4 \longrightarrow \cdot CH_3 + HCl \quad \Delta H° = 105 - 103 = 2$$

$$Br\cdot + CH_4 \longrightarrow \cdot CH_3 + HBr \quad \Delta H° = 105 - 88 = 17$$

$$I\cdot + CH_4 \longrightarrow \cdot CH_3 + HI \quad \Delta H° = 105 - 71 = 34$$

The fluorine radical is the most reactive of the halogen radicals—it reacts violently with alkanes ($\Delta H° = -31$ kcal/mol). In contrast, the iodine radical is so unreactive ($\Delta H° = 34$ kcal/mol) that it is unable to remove a hydrogen atom from an alkane. Consequently, it reacts with another iodine radical and reforms I_2.

LEARN THE STRATEGY

PROBLEM 10 SOLVED

How could butanone be prepared from butane?

butane $\xrightarrow{?}$ butanone

SOLUTION We know that the first reaction has to be a radical halogenation because that is the only reaction that an alkane undergoes. Bromination will lead to a greater yield of the desired 2-halo-substituted compound than will chlorination because a bromine radical is more selective than a chlorine radical. A nucleophilic substitution reaction forms the alcohol, which forms the target molecule when it is oxidized.

$\xrightarrow[h\nu]{Br_2}$ $\xrightarrow{HO^-}$ $\xrightarrow[\substack{CH_3COOH \\ 0\,°C}]{NaOCl}$

USE THE STRATEGY

PROBLEM 11

Show how the following compounds could be prepared from 2-methylpropane:

a. 2-bromo-2-methylpropane   **b.** 2-methyl-1-propene   **c.** 2-iodo-2-methylpropane

---

12.6   FORMATION OF EXPLOSIVE PEROXIDES

Ethers are a laboratory hazard because they form explosive peroxides by reacting with O_2 when exposed to air. We will see that this reaction is similar to the reaction that causes fats to become rancid (Section 12.11).

MECHANISM FOR PEROXIDE FORMATION

an α-carbon

$$R-O-\underset{\underset{H}{|}}{C}H-R + Y\cdot \longrightarrow R-O-\dot{C}H-R + HY$$

$$R-O-\dot{C}H-R + \cdot\ddot{O}-\ddot{O}\cdot \longrightarrow R-O-CH-R$$
$$\underset{:\ddot{O}-\ddot{O}\cdot}{|}$$
a peroxide radical

$$R-O-\underset{:\ddot{O}-\ddot{O}\cdot}{|}{CH}-R + R-O-\underset{H}{|}{C}H-R \longrightarrow R-O-\underset{:\ddot{O}-\ddot{O}H}{|}{CH}-R + R-O-\dot{C}H-R$$
a peroxide

- A chain-initiating radical removes a hydrogen atom from an α-carbon of the ether. (The α-carbon is the carbon attached to the oxygen.) This is an initiation step because it creates the radical that is used in the first propagation step.
- The radical formed in the initiation step reacts with oxygen in a propagation step, forming a peroxide radical.
- In the second propagation step, the peroxide radical removes a hydrogen atom from an α-carbon of another molecule of ether to form a peroxide and regenerate the radical used in the first propagation step.

A **peroxide** is a compound with an O—O bond. Because an O—O bond is easily cleaved homolytically, a peroxide forms radicals that can create new radicals—it is a **radical initiator.** Thus, the peroxide product of the preceding radical chain reaction can initiate another radical chain reaction—an explosive situation. To prevent the formation of explosive peroxides, ethers contain a stabilizer that traps the chain-initiating radical. Once an ether is purified (in which case it no longer contains the stabilizer), it has to be used or discarded within 24 hours.

PROBLEM 12 ◆

a. Which ether is most apt to form a peroxide?   **b.** Which ether is least apt to form a peroxide?

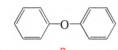
A

B

C

D

12.7   THE ADDITION OF RADICALS TO AN ALKENE

The addition of HBr to 1-butene forms 2-bromobutane, because the electrophile (H^+) adds to the sp^2 carbon bonded to the most hydrogens (Section 6.4). If, however, you want to synthesize 1-bromobutane, then you need to find a way to make bromine an electrophile so it, instead of H^+, adds to the sp^2 carbon bonded to the most hydrogens.

$$CH_3CH_2CH=CH_2 + HBr \longrightarrow CH_3CH_2\overset{\overset{\displaystyle Br}{|}}{C}HCH_3$$
1-butene   2-bromobutane

> The electrophile adds to the sp^2 carbon that is bonded to the most hydrogens.

$$CH_3CH_2CH=CH_2 + HBr \xrightarrow{\text{peroxide}} CH_3CH_2CH_2CH_2Br$$
1-butene   1-bromobutane

If a peroxide (ROOR) is added to the reaction mixture, the product of the addition reaction will be the desired 1-bromobutane. Thus, the peroxide changes the mechanism of the reaction in a way that causes a bromine radical to be the electrophile.

The following mechanism for the addition of HBr to an alkene in the presence of a peroxide shows that it is a radical chain reaction with characteristic initiation, propagation, and termination steps:

MECHANISM FOR THE ADDITION OF HBr TO AN ALKENE IN THE PRESENCE OF A PEROXIDE

$$R\ddot{O}-\ddot{O}R \xrightarrow[\Delta]{\substack{\text{light} \\ \text{or}}} 2 \, R\ddot{O}\cdot$$
a peroxide   alkoxy radicals

⎫
⎬ ← initiation steps
⎭

$$R-\ddot{O}\cdot + H-\ddot{B}\ddot{r}: \longrightarrow R-\ddot{O}-H + \cdot\ddot{B}\ddot{r}:$$
                                                            bromine
                                                            radical

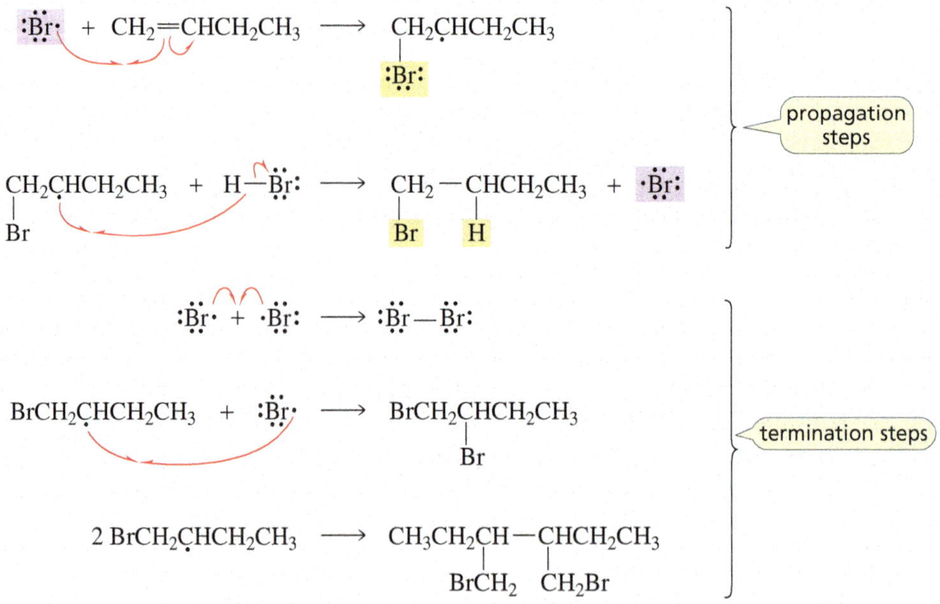

$$:\ddot{Br}\cdot \ + \ CH_2=CHCH_2CH_3 \ \longrightarrow \ CH_2\overset{|}{C}HCH_2CH_3$$
$$\overset{|}{:\ddot{Br}:}$$

$$CH_2\overset{\cdot}{C}HCH_2CH_3 \ + \ H-\ddot{Br}: \ \longrightarrow \ CH_2-CHCH_2CH_3 \ + \ \cdot\ddot{Br}:$$
$$\overset{|}{Br} \qquad\qquad\qquad\qquad \overset{|}{Br}\ \ \overset{|}{H}$$

propagation steps

$$:\ddot{Br}\cdot \ + \ \cdot\ddot{Br}: \ \longrightarrow \ :\ddot{Br}-\ddot{Br}:$$

$$BrCH_2\overset{\cdot}{C}HCH_2CH_3 \ + \ :\ddot{Br}\cdot \ \longrightarrow \ BrCH_2CHCH_2CH_3$$
$$\overset{|}{Br}$$

termination steps

$$2\ BrCH_2\overset{\cdot}{C}HCH_2CH_3 \ \longrightarrow \ CH_3CH_2CH-CHCH_2CH_3$$
$$\overset{|}{BrCH_2}\ \ \overset{|}{CH_2Br}$$

- The weak O—O bond of the peroxide readily breaks homolytically in the presence of light or heat to form alkoxy radicals. This is an initiation step because it creates radicals.

- The alkoxy radical completes its octet by removing a hydrogen atom from a molecule of HBr, thus forming a bromine radical. This too is an initiation step because it creates the radical that is used in the first propagation step.

- The bromine radical now seeks an electron to complete its octet. Because the π bond of an alkene is the weakest bond in the molecule, the bromine radical completes its octet by combining with one of the electrons of the π bond to form a C—Br bond. The second electron of the π bond is the unpaired electron in the product.

  If the bromine radical adds to the sp^2 carbon of 1-butene that is bonded to the most hydrogens, a secondary alkyl radical is formed. If the bromine radical adds to the other sp^2 carbon, a primary alkyl radical is formed. *The bromine radical, therefore, adds to the sp^2 carbon that is bonded to the most hydrogens in order to form the more stable radical.*

- The product of the first propagation step removes a hydrogen atom from another molecule of HBr to produce a molecule of the alkyl halide product and another bromine radical.

  The preceding two steps are propagation steps. As we saw with other pairs of propagation steps, a radical (Br·) reacts in the first propagation step to form a radical that reacts in the second propagation step to regenerate the radical (Br·) that is the reactant in the first propagation step.

- The last three steps are termination steps.

Because the first species that adds to the alkene is a radical (Br·), the addition of HBr in the presence of a peroxide is called a **radical addition reaction.**

When HBr reacts with an alkene in the absence of a peroxide, the electrophile—the first species to add to the alkene—is H$^+$. In the presence of a peroxide, the electrophile is Br·. In both reactions, the electrophile adds to the sp^2 carbon that is bonded to the most hydrogens in order to form the more stable intermediate, thereby following the general rule for electrophilic addition reactions.

The radical intermediate formed in the following reaction does not rearrange, because radicals do not rearrange as readily as carbocations do.

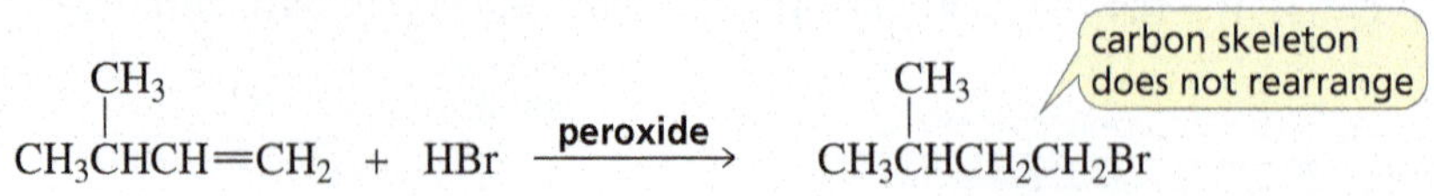

$$\overset{\displaystyle CH_3}{\underset{\displaystyle CH_3\overset{|}{C}HCH=CH_2}{}} \ + \ HBr \ \xrightarrow{\text{peroxide}} \ \overset{\displaystyle CH_3}{\underset{\displaystyle CH_3\overset{|}{C}HCH_2CH_2Br}{}}$$

carbon skeleton does not rearrange

Without a *radical initiator* (in this case, peroxide), the radical addition reaction just described would not occur. Any compound that can readily undergo homolysis (dissociate to form radicals) can act as a radical initiator. We will encounter radical initiators again when we discuss polymers in Chapter 27.

While radical initiators cause radical reactions to occur, *radical inhibitors* have the opposite effect; they trap radicals as they are formed, thereby preventing reactions that depend on the presence of radicals. We will see how radical inhibitors trap radicals in Section 12.11.

PROBLEM 13

Write the propagation steps for the addition of HBr to 1-methylcyclohexene in the presence of a peroxide.

Peroxide Affects Only the Addition of HBr

A peroxide has no effect on the addition of HCl or HI to an alkene; the product that forms in the presence of a peroxide is the same as the product that forms in the absence of a peroxide.

$$CH_3CH{=}CH_2 \ + \ HCl \ \xrightarrow{\text{peroxide}} \ CH_3CHCH_3$$
$$\text{Cl}$$

$$\overset{\displaystyle CH_3}{CH_3C}{=}CH_2 \ + \ HI \ \xrightarrow{\text{peroxide}} \ \overset{\displaystyle CH_3}{CH_3CCH_3}$$
$$\text{I}$$

Why is the **peroxide effect** observed for the addition of HBr, but not for the addition of HCl or HI? This question can be answered by looking at the $\Delta H°$ values for the two propagation steps in the radical chain reaction (Table 5.1 on pages 206–207).

propagation steps for radical addition

$Cl\cdot \ + \ CH_2{=}CH_2 \ \longrightarrow \ ClCH_2\dot{C}H_2$ $\Delta H° = 63 - 85 = -22 \text{ kcal/mol}$ — exothermic

$ClCH_2\dot{C}H_2 \ + \ HCl \ \longrightarrow \ ClCH_2CH_3 \ + \ Cl\cdot$ $\Delta H° = 103 - 101 = +2 \text{ kcal/mol}$ — endothermic

$Br\cdot \ + \ CH_2{=}CH_2 \ \longrightarrow \ BrCH_2\dot{C}H_2$ $\Delta H° = 63 - 72 = -9 \text{ kcal/mol}$ — exothermic

$BrCH_2\dot{C}H_2 \ + \ HBr \ \longrightarrow \ BrCH_2CH_3 \ + \ Br\cdot$ $\Delta H° = 87 - 101 = -14 \text{ kcal/mol}$

$I\cdot \ + \ CH_2{=}CH_2 \ \longrightarrow \ ICH_2\dot{C}H_2$ $\Delta H° = 63 - 57 = +6 \text{ kcal/mol}$ — endothermic

$ICH_2\dot{C}H_2 \ + \ HI \ \longrightarrow \ ICH_2CH_3 \ + \ I\cdot$ $\Delta H° = 71 - 101 = -30 \text{ kcal/mol}$ — exothermic

For the radical addition of HCl, the first propagation step is exothermic and the second is endothermic. For the radical addition of HI, the first propagation step is endothermic and the second is exothermic. Only for the radical addition of HBr are both propagation steps exothermic.

In a radical reaction, the steps that propagate the chain reaction compete with the steps that terminate it. Termination steps are always exothermic, because only bond making (and no bond breaking) occurs. Therefore, only when both propagation steps are exothermic can propagation compete successfully with termination. When HCl or HI adds to an alkene in the presence of a peroxide, any chain reaction that is initiated is then terminated rather than propagated because propagation cannot compete successfully with termination. Consequently, the radical chain reaction does not take place, and the only reaction that occurs is ionic addition (H$^+$ followed by Cl$^-$ or I$^-$).

PROBLEM 14 ♦

What is the major product of the reaction of 2-methyl-2-butene with each of the following reagents?

a. HBr   **b.** HCl   **c.** HBr + peroxide   **d.** HCl + peroxide

12.8 THE STEREOCHEMISTRY OF RADICAL SUBSTITUTION AND RADICAL ADDITION REACTIONS

We have seen that when a reactant that does not have an asymmetric center undergoes a reaction that forms a product with one asymmetric center, the product is a racemic mixture (Section 6.13). Thus, the following *radical substitution reaction* forms a racemic mixture (that is, an equal amount of each enantiomer).

When a reactant that does not have an asymmetric center undergoes a reaction that forms a product with one asymmetric center, the product is a racemic mixture.

$$CH_3CH_2CH_2CH_3 \ + \ Br_2 \ \xrightarrow{h\nu} \ CH_3CH_2\overset{\displaystyle |}{\underset{\displaystyle Br}{C}}HCH_3 \ + \ HBr$$

an asymmetric center

configuration of the products

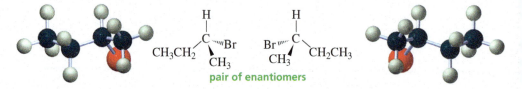

pair of enantiomers

Similarly, the product of the following *radical addition reaction* is a racemic mixture:

$$CH_3CH_2\overset{\displaystyle CH_3}{\overset{\displaystyle |}{C}}=CH_2 \ + \ HBr \ \xrightarrow{peroxide} \ CH_3CH_2\overset{\displaystyle CH_3}{\overset{\displaystyle |}{C}}HCH_2Br$$

an asymmetric center

configuration of the products

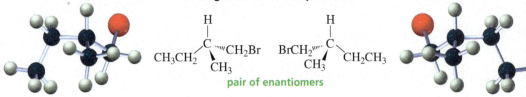

pair of enantiomers

Why Radical Substitution and Radical Addition Form a Racemic Mixture

Both the radical substitution and radical addition reactions form a racemic mixture because both reactions form a radical intermediate, and the reaction of the intermediate determines the configuration of the products. The radical intermediate in the substitution reaction is formed when the bromine radical removes a hydrogen atom from the reactant; the radical intermediate in the addition reaction is formed when the bromine radical adds to one of the sp^2 carbons of the double bond.

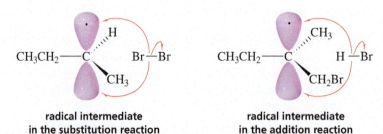

radical intermediate
in the substitution reaction

radical intermediate
in the addition reaction

The carbon that bears the unpaired electron in the radical intermediate is sp^2 hybridized, so the three atoms to which it is bonded lie in a plane (Section 1.10). Therefore, the incoming atom has equal ccess to both sides of the plane. Consequently, identical amounts of the *R* and *S* enantiomers are ormed in both the substitution and addition reactions.

Identical amounts of the *R* and *S* enantiomers are also obtained if a hydrogen bonded to an symmetric center is substituted by a halogen. Breaking the bond to the asymmetric center destroys he configuration at the asymmetric center and forms a planar radical intermediate. The incoming ralogen has equal access to both sides of the plane, so a racemic mixture is formed.

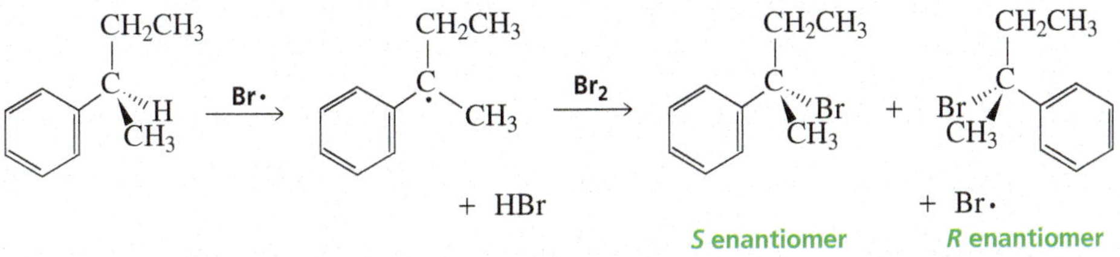

+ HBr   *S* enantiomer   *R* enantiomer
                                  + Br·

PROBLEM 15 ◆

a. What hydrocarbon with molecular formula C_4H_{10} forms only two monochlorinated products? Both products are achiral.

b. What hydrocarbon with the same molecular formula as in part **a** forms three monochlorinated products? One is achiral and two are chiral.

PROBLEM 16

Draw the stereoisomers of the major monobromination products obtained from the following reaction.

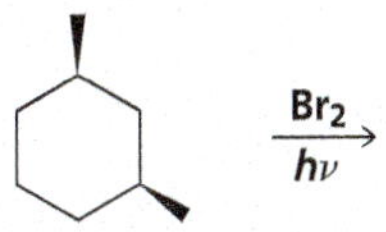

12.9 RADICAL SUBSTITUTION OF ALLYLIC AND BENZYLIC HYDROGENS

The Stability of Allylic and Benzylic Radicals

An **allylic radical** has an unpaired electron on an allylic carbon and, like an allylic cation, has two esonance contributors (Section 8.7).

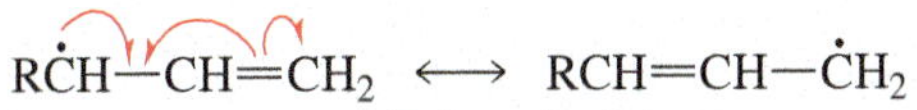

$$R\dot{C}H-CH=CH_2 \longleftrightarrow RCH=CH-\dot{C}H_2$$
an allylic radical

Electron delocalization increases the stability of a molecule.

A **benzylic radical** has an unpaired electron on a benzylic carbon and, like a benzylic cation, has ive resonance contributors (Section 8.7).

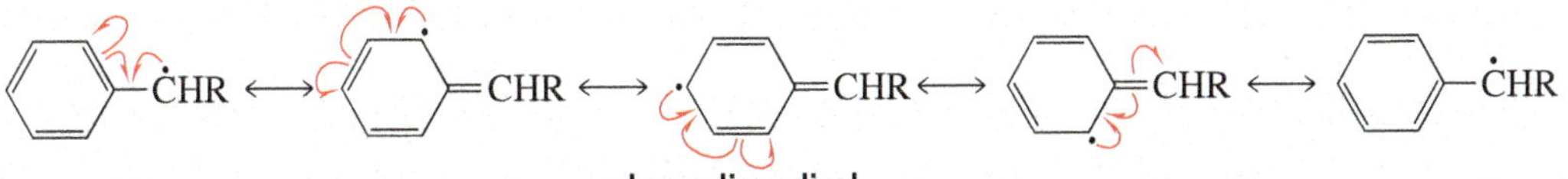

a benzylic radical

Because electron delocalization stabilizes a molecule (Section 8.6), allyl and benzyl radicals are more stable than other primary radicals. They are even more stable than tertiary radicals.

relative stabilities of radicals

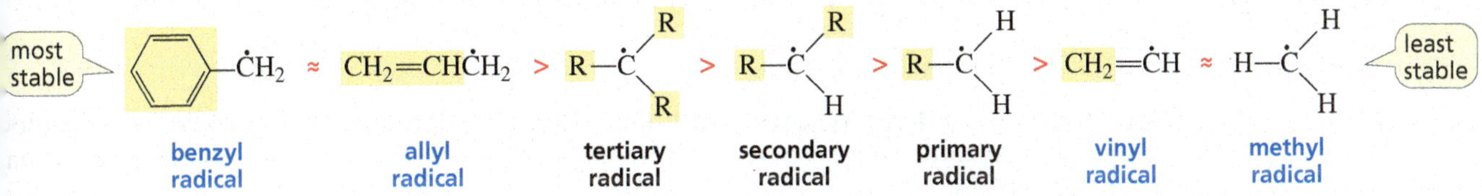

| benzyl radical | allyl radical | tertiary radical | secondary radical | primary radical | vinyl radical | methyl radical |

We know that the more stable the radical, the faster it can be formed. This means that a hydrogen bonded to either a benzylic carbon or an allylic carbon is preferentially substituted in a halogenation reaction. Because bromination is more selective than chlorination, the percent of substitution at the benzylic or allylic carbon is greater for bromination.

$$CH_3CH{=}CH_2 \ + \ X_2 \ \xrightarrow{\Delta} \ \overset{\overset{\displaystyle X}{|}}{CH_2}CH{=}CH_2 \ + \ HX$$

allylic substituted product

NBS Is Used to Brominate Allylic Carbons

N-Bromosuccinimide (NBS) is frequently used to brominate allylic carbons because it allows a radical substitution reaction to be carried out in the presence of a low concentration of Br_2 and a low concentration of HBr. If a high concentration is present, addition of Br_2 or HBr to the double bond competes with allylic substitution.

cyclohexene   *N*-bromosuccinimide
NBS   3-bromocyclohexene   succinimide

The bromination reaction begins with homolytic cleavage of the N—Br bond of NBS. This generates the bromine radical needed to initiate the radical reaction. Light or heat and a radical initiator such as a peroxide are used to promote the homolytic cleavage.

In the first propagation step, the bromine radical removes an allylic hydrogen to form HBr and an allylic radical. Notice that the allylic radical is stabilized by electron delocalization. The allylic radical reacts with Br_2 in the second propagation step, thus forming the allylic bromide and the chain-propagating bromine radical.

The Br_2 used in the second propagation step is produced in low concentration from a fast ionic reaction between NBS and the HBr that is produced in the first propagation step.

Even though there are two resonance contributors in the preceding reaction, only one substitution product (disregarding stereoisomers) is formed because the resonance contributors are mirror images. However, if the resonance contributors are not mirror images, then two substitution products (disregarding stereoisomers) are formed:

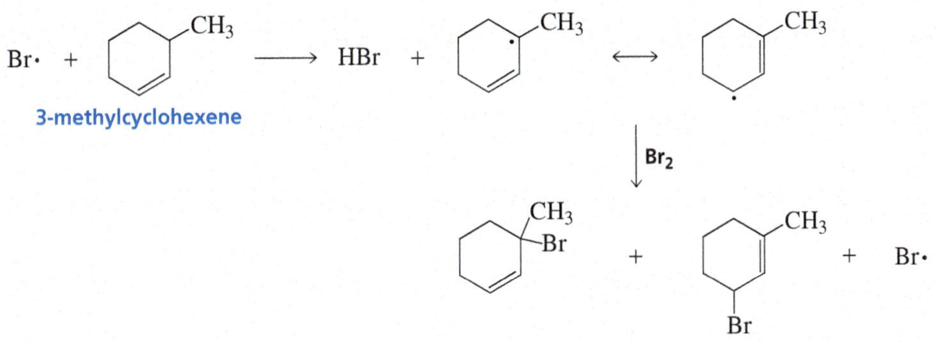

3-methylcyclohexene

PROBLEM 17 SOLVED

LEARN THE STRATEGY

How many allylic substituted bromoalkenes are formed from the reaction of 2-pentene with NBS? Disregard stereoisomers.

SOLUTION Because of the high selectivity of the bromine radical, it will remove a secondary allylic hydrogen from C-4 of 2-pentene more easily than it will remove a primary allylic hydrogen from C-1. The resonance contributors of the resulting radical intermediate are mirror images, so only one bromoalkene is formed (disregarding stereoisomers).

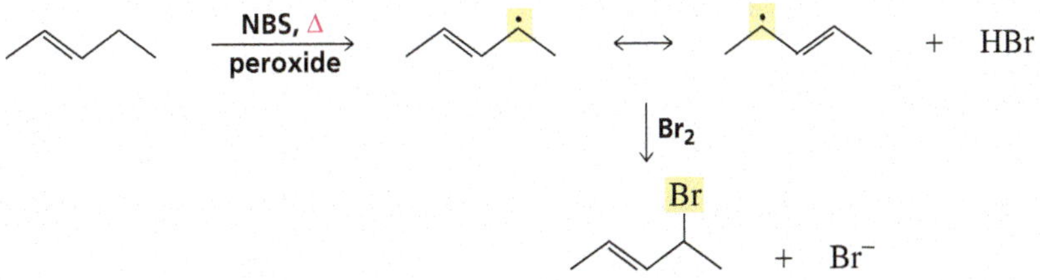

PROBLEM 18

USE THE STRATEGY

Two products are formed when methylenecyclohexane reacts with NBS? Show how each is formed. Disregard stereoisomers.

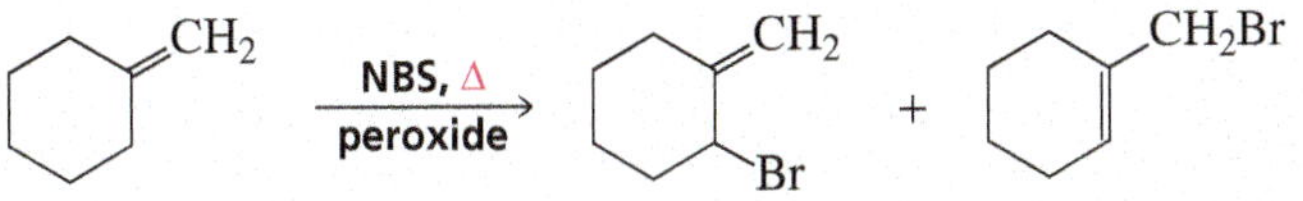

PROBLEM 19 ◆

How many allylic substituted bromoalkenes are formed from the reactions in Problems 17 and 18 if stereoisomers are included?

PROBLEM 20

a. How many stereoisomers are formed from the reaction of cyclohexene with NBS?
b. How many stereoisomers are formed from the reaction of 3-methylcyclohexene with NBS?

PROBLEM 21

Draw the resonance contributors for the following radicals:

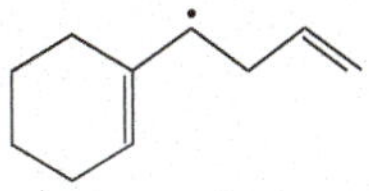

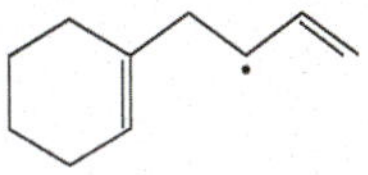

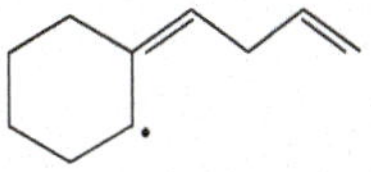

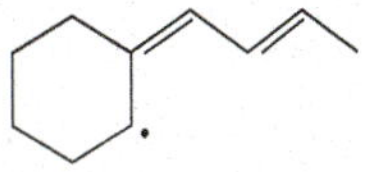

PROBLEM 22

a. Draw the major product(s) of the reaction of 1-methylcyclohexene with the following reagents, disregarding stereoisomers:

    1. NBS/Δ/peroxide   **2.** Br$_2$/CH$_2$Cl$_2$ **3.** HBr   **4.** HBr/peroxide

b. For each reaction, show which stereoisomers are obtained.

A Bromine Radical Can Remove an Allylic Hydrogen and It Can Add to a Double Bond

Why does the bromine radical generated from NBS remove an allylic hydrogen, whereas the bromine radical generated from HBr + peroxide adds to the double bond? The bromine radical can in fact do both.

When NBS is used there is little HBr present to complete the addition reaction after the bromine radical has added to the double bond. Because addition of a bromine radical to a double bond is reversible, the reactant is reformed and allylic substitution becomes the major reaction pathway.

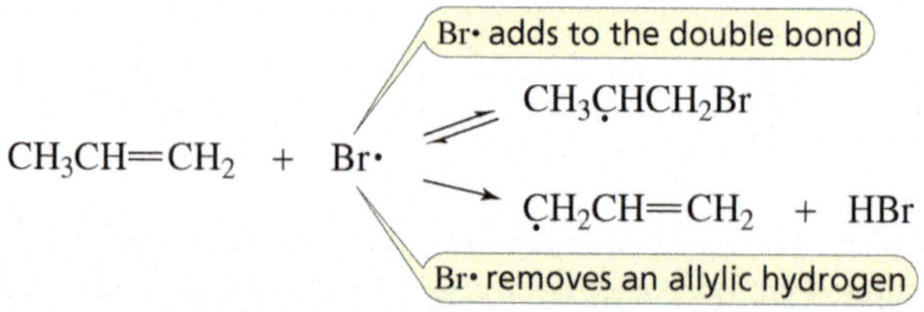

$$CH_3CH=CH_2 \ + \ Br\cdot$$

> Br· adds to the double bond
> $CH_3\overset{\cdot}{C}HCH_2Br$
> $\overset{\cdot}{C}H_2CH=CH_2 \ + \ HBr$
> Br· removes an allylic hydrogen

Cyclopropane

Although it is an alkane, cyclopropane undergoes electrophilic addition reactions as if it were an alkene. Cyclopropane is more reactive than propene toward the addition of acids such as HBr and HCl, but is less reactive toward the addition of Cl_2 and Br_2, so a Lewis acid ($FeCl_3$ or $FeBr_3$) is needed to catalyze halogen addition (Section 8.17).

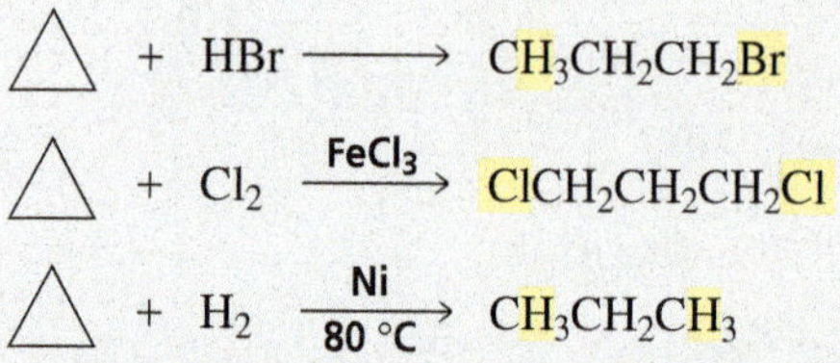

$$\triangle \ + \ HBr \ \longrightarrow \ CH_3CH_2CH_2Br$$

$$\triangle \ + \ Cl_2 \ \xrightarrow{FeCl_3} \ ClCH_2CH_2CH_2Cl$$

$$\triangle \ + \ H_2 \ \xrightarrow[80\,°C]{Ni} \ CH_3CH_2CH_3$$

It is the strain in the small ring that makes it possible for cyclopropane to undergo electrophilic addition reactions. Because of the 60° bond angles in its three-membered ring, the compound's sp^3 orbitals cannot overlap head-on. Thus, the C—C bonds in cyclopropane are considerably weaker than normal C—C bonds (see Figure 3.8 on page 122). Consequently, the three-membered ring undergoes a ring-opening reaction with electrophilic reagents.

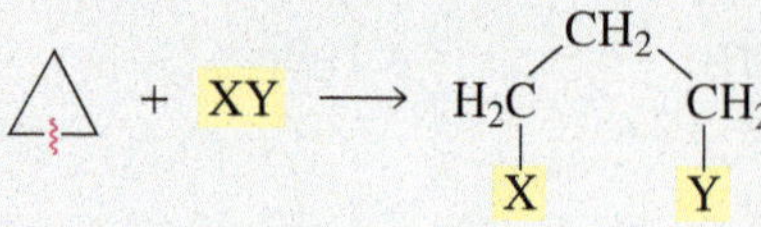

$$\triangle \ + \ XY \ \longrightarrow \ H_2C\overset{CH_2}{\underset{X \quad\ Y}{\diagdown\!\diagup}}CH_2$$

DESIGNING A SYNTHESIS III

12.10 MORE PRACTICE WITH MULTISTEP SYNTHESIS

Now that the number of reactions with which you are familiar has increased, you can design the synthesis of a wide variety of compounds.

Example 1

Starting with the ether shown here, how could you prepare the aldehyde?

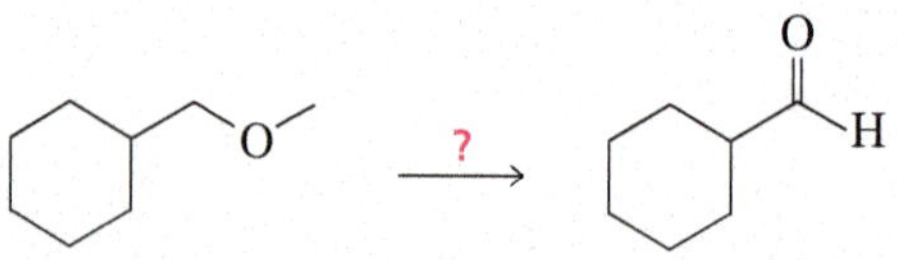

Heating the ether with one equivalent of HI forms a primary alcohol that, when oxidized, forms the desired aldehyde.

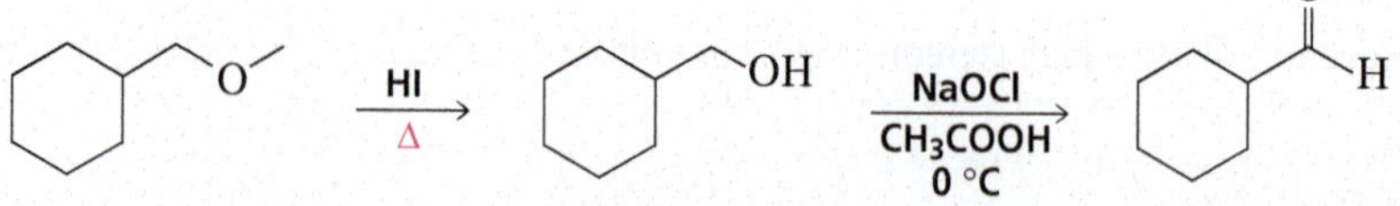

Example 2

Suggest a way to prepare 1,3-cyclohexadiene from cyclohexane.

Deciding what the first reaction should be is easy, because the only reaction that an alkane can undergo is a radical substitution reaction with Cl_2 or Br_2. Next, an E2 reaction, using a strong and bulky base and carried out at a relatively high temperature to encourage elimination over substitution, forms cyclohexene. Radical bromination of cyclohexene forms an allylic bromide, which then forms the desired target molecule by undergoing another E2 reaction.

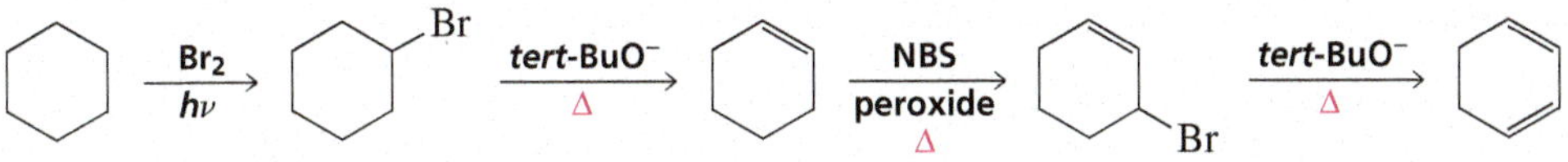

Example 3

Starting with methylcyclohexane, how could the following vicinal *trans*-dihalide be prepared?

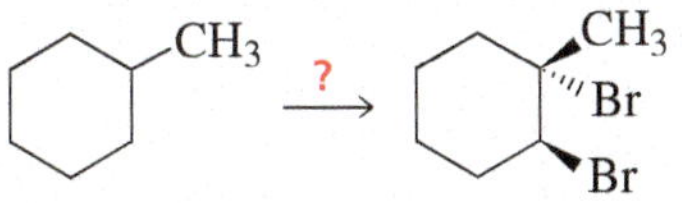

Again, because the starting material is an alkane, the first reaction must be a radical substitution. Bromination leads to selective substitution of the tertiary hydrogen. Under E2 conditions, tertiary alkyl halides undergo only elimination, so there is no competing substitution product formed in the next reaction. A relatively unhindered base should be used to favor removal of a proton from the secondary carbon over removal of a proton from the methyl group. The final step is addition of Br_2; only anti addition occurs in this reaction, so the target molecule (along with its enantiomer) is obtained.

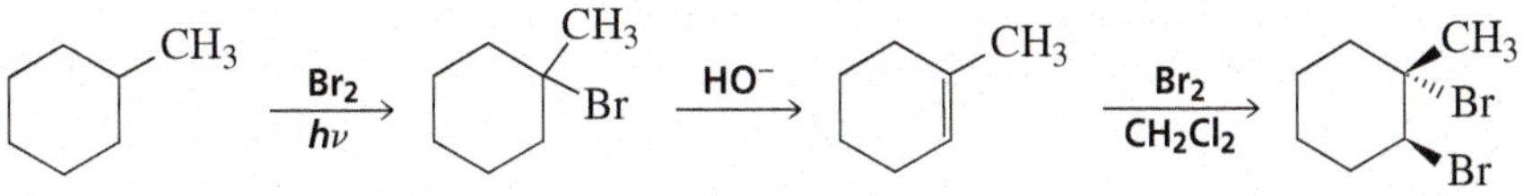

Example 4

Design a synthesis for the target molecule from the indicated starting material.

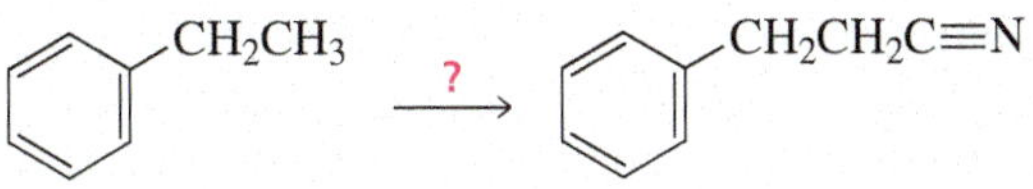

It is not immediately obvious how to carry out this synthesis, so let's use retrosynthetic analysis to find a way (see the top of the next page). The only method you know for introducing a $C\equiv N$ group into a molecule is nucleophilic substitution. The alkyl halide for that substitution reaction can be obtained from the addition of HBr to an alkene in the presence of a peroxide. The alkene for that addition reaction can be obtained from an elimination reaction using an alkyl halide obtained by benzylic substitution.

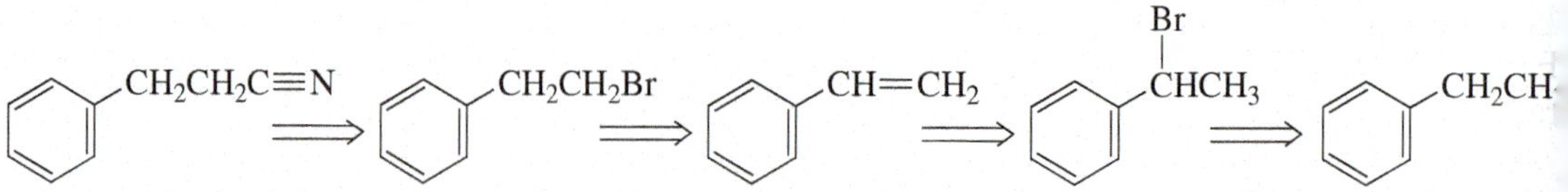

The reaction sequence can now be written in the forward direction along with the reagents required to carry out each step. Notice that a strong and bulky base at a relatively high temperature is used to encourage elimination over substitution.

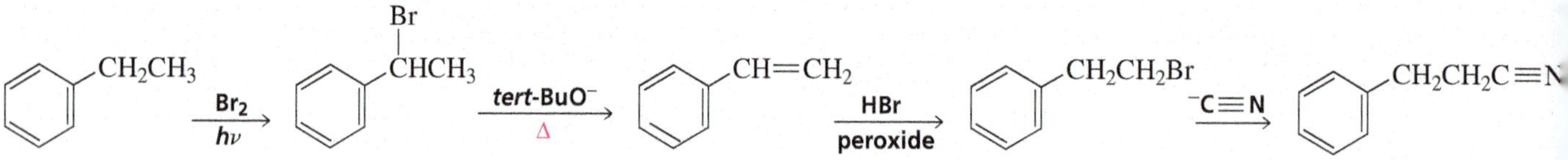

PROBLEM 23

Design a multistep synthesis to show how the following compounds can be prepared from the given starting material:

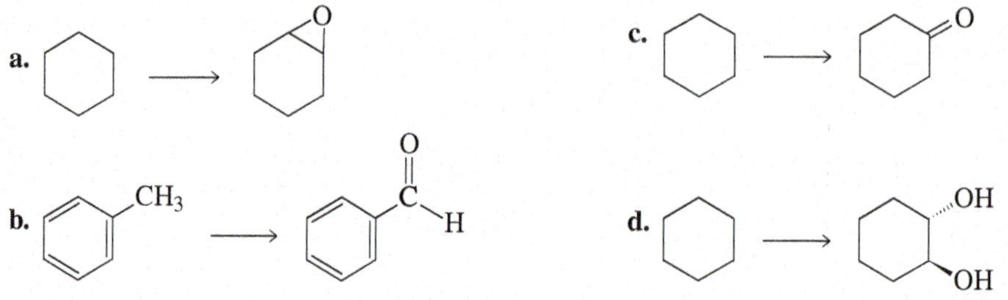

a.

b.

c.

d.

12.11 RADICAL REACTIONS IN BIOLOGICAL SYSTEMS

Because of the large amount of heat or light energy required to initiate a radical reaction and the difficulty in controlling a chain reaction once it is initiated, scientists assumed for a long time that radical reactions were not important in biological systems. It is now widely recognized, however, that many biological reactions involve radicals. The radicals in these reactions, instead of being generated by heat or light, are formed by the interaction of organic molecules with metal ions. The radical reactions take place at the active sites of enzymes (Section 5.14). Containing the chain reaction at a specific site allows the reaction to be controlled.

Converting Nonpolar Compounds to Polar Compounds

Water-soluble (polar) compounds are readily eliminated by the body. In contrast, water-insoluble (nonpolar) compounds are not eliminated but, instead, accumulate in the nonpolar components of cells. For cells to avoid becoming "toxic dumps," nonpolar compounds that are ingested (such as drugs, foods, and environmental pollutants) must be converted into polar compounds that can be excreted.

A radical reaction carried out in the liver converts nonpolar hydrocarbons into less toxic polar alcohols by substituting an H in the hydrocarbon with an OH. The reaction is catalyzed by an iron-containing enzyme called cytochrome P_{450}.

A radical intermediate is created when Fe^V=O removes a hydrogen atom from the alkane. Then Fe^{IV}—OH dissociates homolytically into Fe^{III} and ȮH, and the ȮH immediately combines with the radical intermediate to form the alcohol.

This reaction can also have the opposite toxicological effect. For example, studies found that when animals inhale dichloromethane (CH_2Cl_2), it becomes a carcinogen as a result of an H being substituted by an OH.

Decaffeinated Coffee and the Cancer Scare

Animal studies revealing that dichloromethane becomes a carcinogen when inhaled immediately led to a study of thousands of workers who inhaled dichloromethane daily. However, no increased risk of cancer was found in this group. (This shows that the results of studies done on humans do not always agree with the results of those done on laboratory animals.)

Because dichloromethane was the solvent used to extract caffeine from coffee beans in the manufacture of decaffeinated coffee, a study was done to see what happened to animals that drank dichloromethane. When dichloromethane was added to the drinking water given to laboratory rats and mice, researchers found no toxic effects, even in rats that had consumed an amount of dichloromethane equivalent to the amount that would be ingested by drinking 120,000 cups of decaffeinated coffee per day and in mice that had consumed an amount equivalent to drinking 4.4 million cups of decaffeinated coffee per day.

However, because of the initial concern, researchers sought alternative methods for extracting caffeine from coffee beans. Extraction by CO_2 at supercritical temperatures and pressures was found to be a better method because it extracts caffeine without simultaneously extracting some of the flavor compounds, as dichloromethane does. This was one of the first green (environmentally benign) commercial chemical processes to be developed. After the caffeine has been removed, the CO_2 can be recycled, whereas dichloromethane is not a substance that should be released into the environment (p. 312).

Oxidation of Fats and Oils

Fats and oils are easily oxidized by O_2 by means of a radical chain reaction to form compounds that have strong odors. These compounds are responsible for the unpleasant taste and smell associated with sour milk and rancid butter.

Fats and oils have double bonds (—CH=CH—CH$_2$—CH=CH—) that are separated by two single bonds (Section 16.13). The mechanism for their oxidation by O_2 is shown on the top of the next page. Notice the similarity of this mechanism to that shown for peroxide formation in Section 12.6.

MECHANISM FOR THE OXIDATION OF FATS AND OILS BY OXYGEN

- In the initiation step, a radical (· R) removes a hydrogen atom from a methylene group that is flanked by two double bonds. This hydrogen is the one most easily removed because the resulting radical is relatively stable, since the unpaired electron is shared by three carbons.
- In the first propagation step, the radical created in the initiation step reacts with O_2, forming a peroxy radical.
- In the second propagation step, the peroxy radical removes a hydrogen atom from another molecule of fat or oil. The two propagation steps are repeated over and over.
- The alkyl hydroperoxide undergoes further oxidation to form butyric acid and other short-chain carboxylic acids.

Radical Inhibitors

The molecules that form cell membranes (Section 25.5) can undergo the same radical oxidation reaction that fats and oils undergo, which leads to their degradation. Radical reactions in cells have been implicated in the aging process.

Clearly, unwanted radicals in cells must be destroyed before they damage cell membranes or other cell components. Radical reactions can be prevented by **radical inhibitors,** compounds that destroy reactive radicals by converting them to relatively stable radicals or to compounds with only paired electrons. Radical inhibitors are *antioxidants*—that is, they prevent oxidation reactions by radicals.

Hydroquinone is an example of a radical inhibitor. When hydroquinone traps a radical, it forms semiquinone. Semiquinone can trap another radical and form quinone, a compound whose electrons are all paired. Hydroquinones are found in the cells of all aerobic organisms.

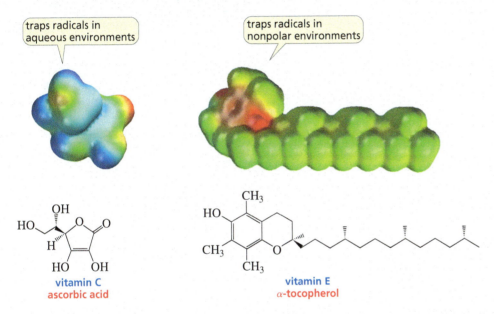

Two other examples of radical inhibitors in living systems are vitamins C and E. Vitamin C (also called ascorbic acid) is a water-soluble compound that traps radicals formed in the interior of cells and in blood plasma (both of which have aqueous environments).

traps radicals in aqueous environments

traps radicals in nonpolar environments

vitamin C
ascorbic acid

vitamin E
α-tocopherol

Vitamin E (also called α-tocopherol) is a fat-soluble compound that traps radicals formed in cell membranes, which are nonpolar. Vitamin E is the primary antioxidant for fat tissue in humans and is, therefore, important in preventing the development of atherosclerosis.

Why one vitamin functions in aqueous environments and the other in nonaqueous environments is apparent from their structures and their electrostatic potential maps; both show that vitamin C is a relatively polar compound and vitamin E is a nonpolar compound.

Because radicals are implicated in the aging process, many products are available that contain antioxidants.

Nuts are a natural source of vitamin E.

Food Preservatives

Radical inhibitors found in food are known as *preservatives*. They preserve food by preventing radical chain reactions. Vitamin E is a naturally occurring preservative found in such things as vegetable oil, sunflower seeds, and spinach. BHA and BHT are synthetic preservatives that are added to many packaged foods. Notice that, like hydroquinone, vitamin E and all the synthetic preservatives are phenols.

butylated hydroxyanisole
BHA

butylated hydroxytoluene
BHT

food preservatives

Is Chocolate a Health Food?

We have long been told that our diets should include lots of fruits and vegetables because they are good sources of antioxidants. Antioxidants protect against cardiovascular disease, cancer, and cataracts, and they are thought to slow the effects of aging. Chocolate is made up of hundreds of organic compounds, including high levels of antioxidants called catechins. (Catechins are also phenols.)

catechins

On a weight basis, the concentration of antioxidants in chocolate is higher than in red wine or green tea, and 20 times higher than in tomatoes. Another piece of good news for chocolate lovers is that stearic acid, the main fatty acid in chocolate, does not appear to raise blood cholesterol levels the way other saturated fatty acids do. Dark chocolate contains more than twice the level of antioxidants found in milk chocolate. Unfortunately, white chocolate contains no antioxidants.

PROBLEM 24 ♦

How many atoms share the unpaired electron in semiquinone?

PROBLEM 25

Using resonance structures, explain why a catechin is an antioxidant.

12.12 RADICALS AND STRATOSPHERIC OZONE

Ozone (O_3), a major constituent of smog, is a health hazard at ground level—it inflames the airways, worsens lung ailments, and increases the risk of death from heart or lung disease. In the stratosphere, however, a layer of ozone shields Earth from harmful solar radiation, with the greatest concentrations of ozone lying between 12 and 15 miles above Earth's surface.

In the stratosphere, ozone acts as a filter for biologically harmful ultraviolet (UV) light that otherwise would reach Earth's surface. Among other effects, short-wavelength UV light can damage DNA in skin cells, causing mutations that trigger skin cancer (Section 28.6). We owe our very existence to this protective ozone layer. According to current theories of evolution, life could not have developed on land without it. Instead most, if not all, living things would have had to remain in the ocean, where water screens out the harmful UV light.

The ozone layer is thinnest at the equator and densest toward the poles. Since about 1985, scientists have noted a precipitous drop in stratospheric ozone over Antarctica. This area of ozone depletion, dubbed the "ozone hole," is unprecedented in the history of ozone observations. Scientists subsequently noted a similar decrease in ozone over Arctic regions; then, in 1988, they detected a depletion of ozone over the United States for the first time. Three years later, scientists determined that the rate of ozone depletion was two to three times faster than originally anticipated.

Polar stratospheric clouds increase the rate of ozone destruction. These clouds form over Antarctica during the cold winter months. Ozone depletion in the Arctic is less severe because the temperature generally does not get low enough for stratospheric clouds to form there.

Strong circumstantial evidence implicated synthetic chlorofluorocarbons (CFCs)—alkanes in which all the hydrogens have been replaced by fluorine and chlorine—as a major cause of ozone depletion. These gases, known commercially as Freon, had been used extensively as cooling fluids in refrigerators and air conditioners. They were also once widely used as propellants in aerosol spray cans (deodorant, hair spray, and so on) because of their odorless, nontoxic, and nonflammable properties and because, being chemically inert, they do not react with the contents of the can. Now such use is banned and propane and butane are used as propellants instead.

The global agreement to phase out CFCs and other ozone-depleting agents seems to be working. The ozone hole is stabilizing and it is hoped that it will regain its density by 2070.

Ozone is formed from the interaction of molecular oxygen with very short wavelength ultraviolet light.

$$O_2 \xrightarrow{h\nu} O + O$$

$$O + O_2 \longrightarrow O_3$$
$$\text{ozone}$$

Chlorofluorocarbons are stable until they reach the stratosphere. There they encounter wavelengths of ultraviolet light that cause the C—Cl bond to break homolytically, generating chlorine radicals.

$$F-\overset{\overset{\displaystyle Cl}{|}}{\underset{\underset{\displaystyle F}{|}}{C}}-Cl \xrightarrow{h\nu} F-\overset{\overset{\displaystyle Cl}{|}}{\underset{\underset{\displaystyle F}{|}}{C}}\cdot + Cl\cdot$$

These chlorine radicals are the ozone-removing agents. They react with ozone to form chlorine monoxide radicals and molecular oxygen. The chlorine monoxide radical then reacts with more ozone to form a chlorine dioxide radical, which dissociates to regenerate a chlorine radical. These three steps—two of which each destroy an ozone molecule—are the propagating steps that are repeated over and over. It has been calculated that a single chlorine radical destroys 100,000 ozone molecules!

$$Cl\cdot + O_3 \longrightarrow ClO\cdot + O_2$$

$$ClO\cdot + O_3 \longrightarrow \cdot ClO_2 + O_2$$

$$\cdot ClO_2 \longrightarrow Cl\cdot + O_2$$

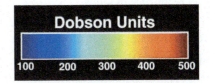

Dobson Units

100   200   300   400   500

The growth of the Antarctic ozone hole, located mostly over the continent of Antarctica, since 1979. The images were made from data supplied by total ozone-mapping spectrometers (TOMSs). The color scale depicts the total ozone values in Dobson units, with the lowest ozone densities represented by dark blue. (The ozone over a given area is compressed to 0 °C and 1 atm pressure, and the thickness of the slab is measured. 1 Dobson unit = 0.01 mm thickness.)

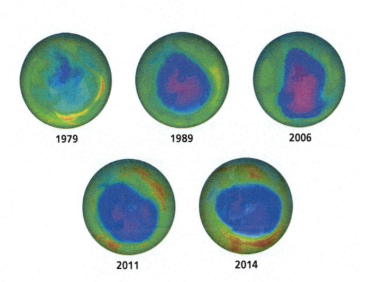

1979   1989   2006

2011   2014

Artificial Blood

Clinical trials are underway to test the use of perfluorocarbons—alkanes in which all the hydrogens have been replaced by fluorines—as compounds to replace blood volume and mimic hemoglobin's ability to carry oxygen to cells and transport carbon dioxide to the lungs.

These compounds are not a true blood substitute, because blood performs many functions that artificial blood cannot. For example, white blood cells fight against infection and platelets are involved in blood clotting. However, artificial blood has several advantages in trauma situations until an actual transfusion can be done: it is safe from disease, it can be administered to any blood type, its availability does not depend on blood donors, and it can be stored longer than whole blood, which is good for only about 40 days.

ESSENTIAL CONCEPTS

Section 12.1

- **Alkanes** are **saturated hydrocarbons.** They do not contain any carbon–carbon double or triple bonds. Thus, they are saturated with hydrogen.
- Alkanes are unreactive compounds because they have only strong σ bonds and atoms with no partial charges.

Section 12.2

- In **heterolytic bond cleavage,** a bond breaks so that one of the atoms retains both of the bonding electrons; in **homolytic bond cleavage,** a bond breaks so that each of the atoms retains one of the bonding electrons.
- Alkanes undergo **radical substitution reactions** with chlorine (Cl_2) or bromine (Br_2) at high temperatures or in the presence of light to form alkyl chlorides or alkyl bromides. This substitution reaction is a **radical chain reaction** with **initiation, propagation,** and **termination steps.**
- The rate-determining step of a radical substitution reaction is removal of a hydrogen atom to form an alkyl radical.

Section 12.3

- The relative stabilities of radicals and, therefore, their relative rates of formation: $3° > 2° > 1° >$ methyl.

Section 12.4

- Calculation of the relative amounts of products obtained from the radical halogenation of an alkane must take into account both probability and the relative rate at which a particular radical is formed.

Section 12.5

- The **reactivity–selectivity principle** states that the more reactive a species is, the less selective it will be.
- A bromine radical is less reactive than a chlorine radical, so a bromine radical is more selective about which hydrogen atom it removes.

Section 12.6

- Ethers form explosive peroxides when they are exposed to air.
- A peroxide is a **radical initiator** because it creates radicals.

Section 12.7

- **Radical addition reactions** are chain reactions with **initiation, propagation,** and **termination steps.**
- A peroxide reverses the order of addition of H and Br to an alkene because it causes $Br\cdot$, instead of H^+, to be the electrophile. **The peroxide effect** is observed only for the addition of HBr.

Section 12.8

- If a reactant that does not have an asymmetric center undergoes a radical substitution or a radical addition reaction that forms a product with an asymmetric center, then a racemic mixture will be obtained.
- A racemic mixture is also obtained if a hydrogen bonded to an asymmetric center is substituted by a halogen.

Section 12.9

- Allylic and benzylic radicals are more stable than tertiary radicals.
- *N*-Bromosuccinimide (NBS) is used to brominate allylic carbons.

Section 12.11

- Some biological reactions involve radicals formed by the interaction of organic molecules with metal ions. The reactions take place at the active sites of enzymes.
- Fats, oils, and membranes are oxidized by O_2 in a radical chain reaction.
- A radical inhibitor (an antioxidant) destroys radicals by converting them to relatively stable radicals or to compounds that have only paired electrons.

Section 12.12

- The interaction of CFCs with UV light generates chlorine radicals, which are ozone-removing agents.

SUMMARY OF REACTIONS

. Alkanes undergo radical substitution reactions with Cl_2 or Br_2 in the presence of heat or light (Sections 12.2–12.5). The mechanisms of the reaction are shown on pages 534 and 535.

$$CH_3CH_3 + Cl_2 \xrightarrow{\Delta \text{ or } h\nu} CH_3CH_2Cl + HCl$$
excess

$$CH_3CH_3 + Br_2 \xrightarrow{\Delta \text{ or } h\nu} CH_3CH_2Br + HBr$$
excess
bromination is more selective than chlorination

. A radical initiator removes a hydrogen atom from an α-carbon of an ether to form a peroxide (Section 12.6). The mechanism of the reaction is shown on page 542.

$$R\!-\!O\!-\!\underset{\underset{H}{|}}{C}H\!-\!R + O_2 \xrightarrow{\cdot R} R\!-\!O\!-\!\underset{\underset{O-O-H}{|}}{C}H\!-\!R$$

. Alkenes undergo radical addition of hydrogen bromide in the presence of a peroxide ($Br\cdot$ is the electrophile; Section 12.7). The mechanism of the reaction is shown on page 543.

$$RCH\!=\!CH_2 + HBr \xrightarrow{\text{peroxide}} RCH_2CH_2Br$$

. Alkyl-substituted benzenes undergo radical substitution at the benzylic position (Section 12.9).

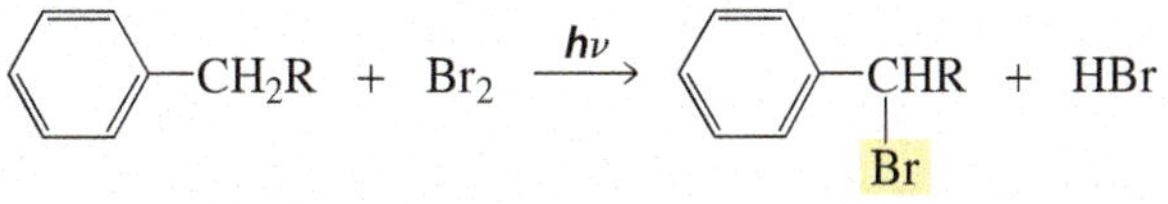

. Alkenes undergo radical substitution at allylic carbons. NBS is used for bromination at allylic carbons (Section 12.9). The mechanism of the reaction is shown on page 548.

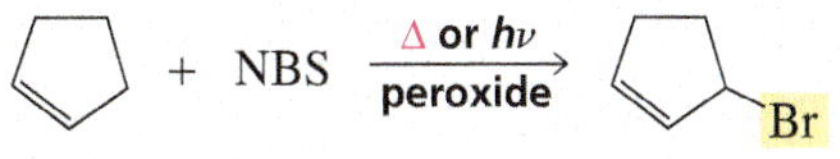

$$RCH_2CH\!=\!CH_2 + NBS \xrightarrow[\text{peroxide}]{\Delta \text{ or } h\nu} R\underset{\underset{Br}{|}}{C}HCH\!=\!CH_2 + RCH\!=\!CH\underset{\underset{Br}{|}}{C}H_2 + HBr$$

PROBLEMS

26. What are the product(s) of each of the following reactions? Disregard stereoisomers.

a. $CH_2\!=\!CHCH_2CH_2CH_3 + NBS \xrightarrow[\text{peroxide}]{\Delta}$

b. $CH_3\underset{\underset{CH_3}{|}}{C}\!=\!CHCH_3 + NBS \xrightarrow[\text{peroxide}]{\Delta}$

c. 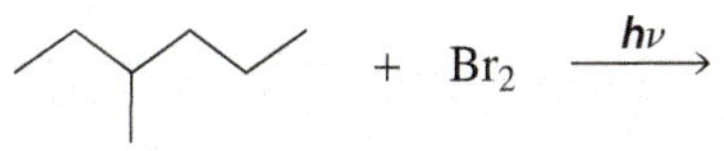 $+ Br_2 \xrightarrow{h\nu}$

d. 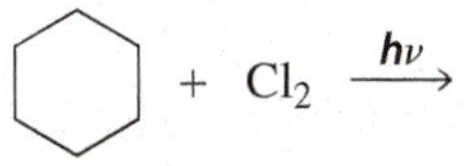 $+ Cl_2 \xrightarrow{h\nu}$

e. 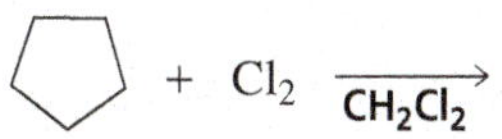 $+ Cl_2 \xrightarrow{CH_2Cl_2}$

f. 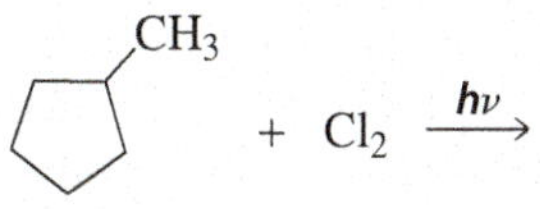 $+ Cl_2 \xrightarrow{h\nu}$

27. a. What alkane, with molecular formula C_5H_{12}, forms only one monochlorinated product when it is heated with Cl_2?
b. What alkane, with molecular formula C_7H_{16}, forms seven monochlorinated products (disregarding stereoisomers) when heated with Cl_2?

28. Explain why iodine (I_2) does not react with ethane, even though I_2 is more easily cleaved homolytically than the other halogens.

29. What is the major product obtained from treating an excess of each of the following compounds with Cl_2 in the presence of ultraviolet light at room temperature? Disregard stereoisomers.

a.

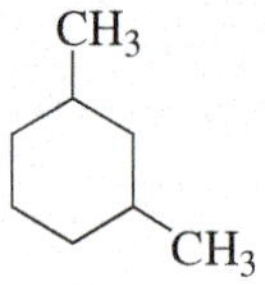

b.

c.

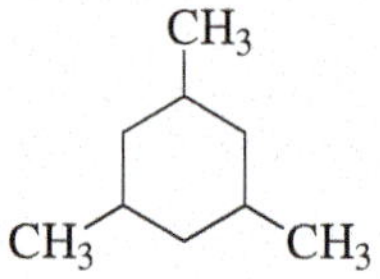

30. What are the answers to Problem 29 when the same compounds are treated with Br_2 at 125 °C?

31. What is the major product of the following reactions? Disregard stereoisomers:

a. 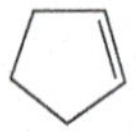 $\xrightarrow[\text{peroxide}]{\Delta}$ + NBS

b.  + NBS $\xrightarrow[\text{peroxide}]{\Delta}$

c. 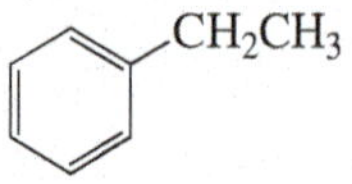 + NBS $\xrightarrow[\text{peroxide}]{\Delta}$

d.  CH_3 CH_3 + NBS $\xrightarrow[\text{peroxide}]{\Delta}$

e. 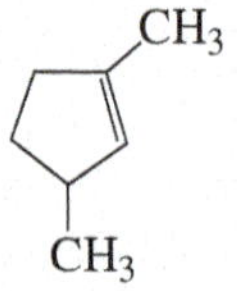 + NBS $\xrightarrow[\text{peroxide}]{\Delta}$

f. 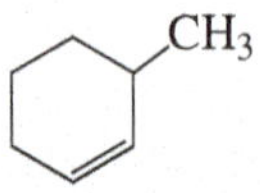 + NBS $\xrightarrow[\text{peroxide}]{\Delta}$

32. When 2-methylpropane is monochlorinated in the presence of light at room temperature, 36% of the product is 2-chloro-2-methylpropane and 64% is 1-chloro-2-methylpropane. From these data, calculate how much easier it is to remove a hydrogen atom from a tertiary carbon than from a primary carbon under these conditions.

33. Draw all the monochlorination products, including stereoisomers, that are formed from radical chlorination of the following compound:

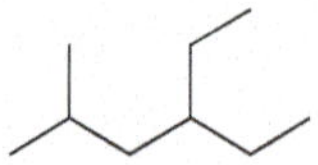

34. Propose a mechanism to account for the products formed in the following reaction:

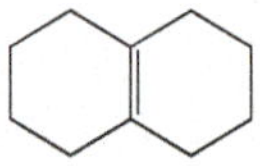

+ NBS $\xrightarrow[\text{peroxide}]{\Delta}$ +

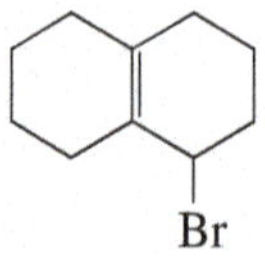

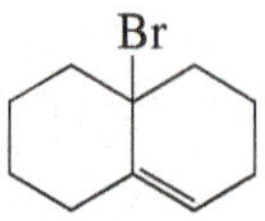

35. The deuterium kinetic isotope effect for the halogenation of an alkane is defined in the following equation, where $X\cdot = Cl\cdot$ or $Br\cdot$

$$\text{deuterium kinetic isotope effect} = \frac{\text{rate of homolytic cleavage of a C—H bond by X·}}{\text{rate of homolytic cleavage of a C—D bond by X·}}$$

Predict whether chlorination or bromination would have a greater deuterium kinetic isotope effect.

36. What are the major products of the following reaction?

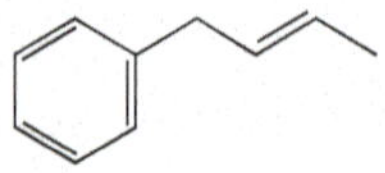

$\xrightarrow[\text{peroxide}]{\text{NBS, }\Delta}$

37. a. How many monochlorination products can be obtained from the radical chlorination of methylcyclohexane? Disregard stereoisomers.
b. Which product would be obtained in greatest yield? Explain.
c. How many monochlorination products would be obtained if all stereoisomers are included?

38. a. What five-carbon alkene forms the same product whether it reacts with HBr in the *presence* of a peroxide or with HBr in the *absence* of a peroxide?
b. Draw the structures of four six-carbon alkenes that form the same product, whether they react with HBr in the *presence* of a peroxide or with HBr in the *absence* of a peroxide.

39. What alkyl halide will be obtained in greatest yield? Ignore stereoisomers.

a. 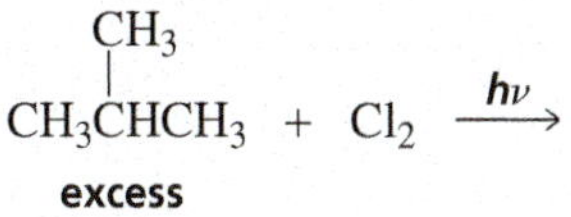 $\overset{\overset{\displaystyle CH_3}{|}}{CH_3CHCH_3}$ + Cl_2 $\xrightarrow{h\nu}$

excess

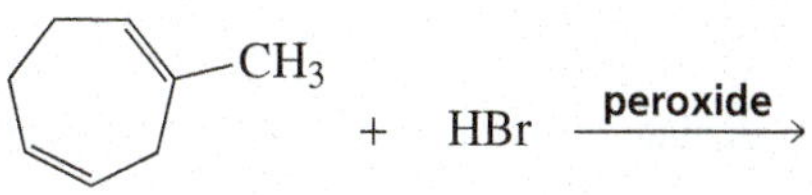

b. 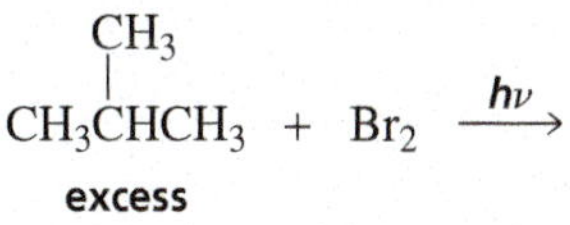 $\overset{\overset{\displaystyle CH_3}{|}}{CH_3CHCH_3}$ + Br_2 $\xrightarrow{h\nu}$

excess

c. 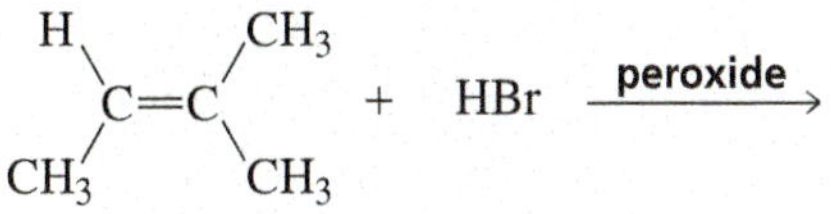 $\underset{\underset{\displaystyle CH_3}{}}{\overset{\overset{\displaystyle H}{}}{C}}=\underset{\underset{\displaystyle CH_3}{}}{\overset{\overset{\displaystyle CH_3}{}}{C}}$ + HBr $\xrightarrow{peroxide}$

d. 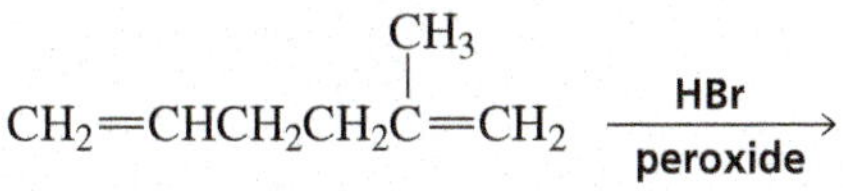 $CH_2{=}CHCH_2CH_2\overset{\overset{\displaystyle CH_3}{|}}{C}{=}CH_2$ $\xrightarrow[peroxide]{HBr}$

e. (cycloheptene with CH_3) + HBr $\xrightarrow{peroxide}$

f. 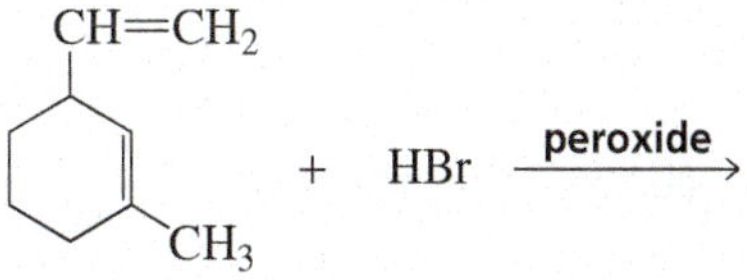 (cyclohexene with $CH{=}CH_2$ and CH_3) + HBr $\xrightarrow{peroxide}$

40. Starting with cyclohexane, how could the following compounds be prepared?

a.  (cyclohexene)

b. 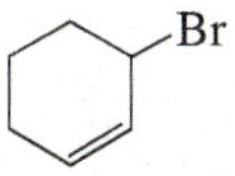 (cyclohexene with Br)

c. 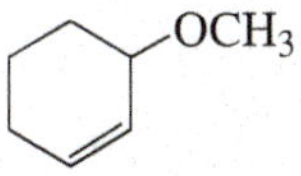 (cyclohexene with OCH_3)

d. 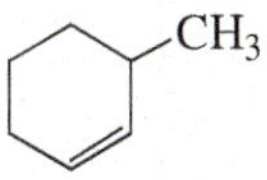 (cyclohexene with CH_3)

e. 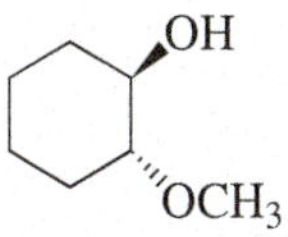 (cyclohexane with OH and OCH_3)

41. a. Propose a mechanism for the following reaction:

$$CH_3CH_3 + CH_3{-}\overset{\overset{\displaystyle CH_3}{|}}{\underset{\underset{\displaystyle CH_3}{|}}{C}}{-}OCl \xrightarrow{\Delta} CH_3CH_2Cl + CH_3{-}\overset{\overset{\displaystyle CH_3}{|}}{\underset{\underset{\displaystyle CH_3}{|}}{C}}{-}OH$$

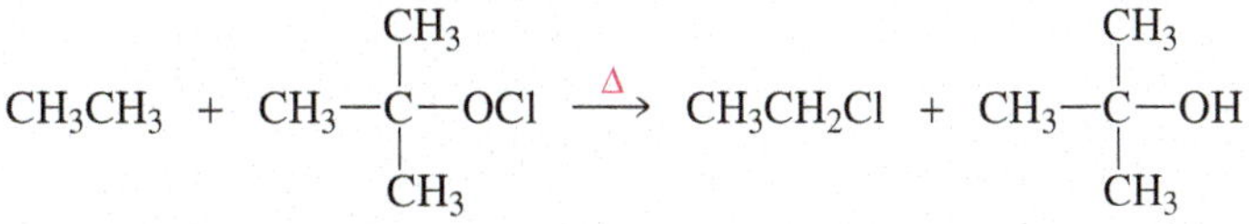

b. Given that $\Delta H°$ for the reaction is -42 kcal/mol and the bond dissociation enthalpies for the C—H, C—Cl, and O—H bonds are 101, 85, and 105 kcal/mol respectively, calculate the bond dissociation enthalpy of the O—Cl bond.

42. What stereoisomers are obtained from the following reaction?

$$\underset{\underset{\displaystyle CH_3}{}}{\overset{\overset{\displaystyle CH_3CH_2}{}}{C}}=\underset{\underset{\displaystyle CH_3}{}}{\overset{\overset{\displaystyle CH_2CH_3}{}}{C}} + HBr \xrightarrow{peroxide}$$

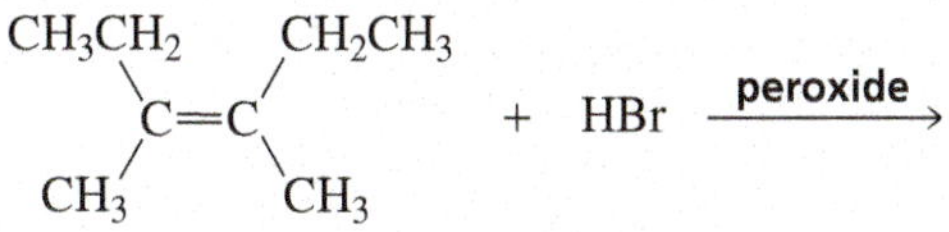

43. Using the given starting material and any necessary organic or inorganic reagents, indicate how the desired compounds could be synthesized:

a. 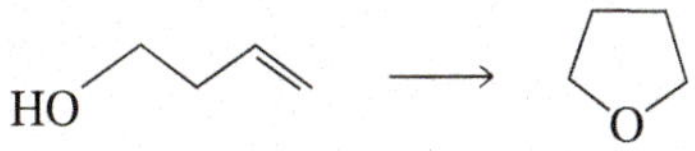 $\underset{HO}{\diagup\diagup\diagdown\diagup\diagup}$ $\longrightarrow$ (tetrahydrofuran)

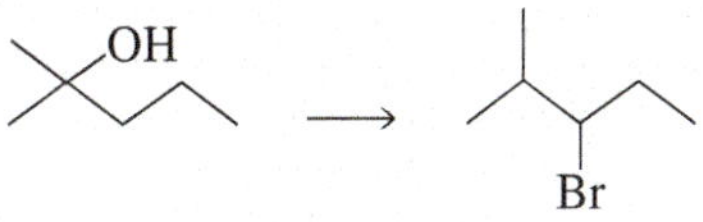

b. 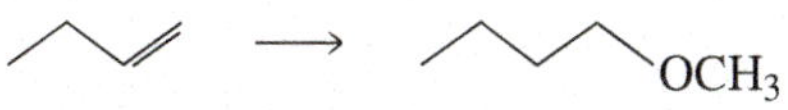 $\diagup\diagdown\diagup$ $\longrightarrow$ $\diagup\diagdown\diagup\diagdown_{OCH_3}$

c. $\overset{OH}{\diagup\diagdown\diagup\diagdown}$ $\longrightarrow$ $\underset{Br}{\diagup\diagdown\diagup\diagdown}$

d.  $\diagup\diagdown\diagup\diagup$ $\longrightarrow$ $\diagup\diagdown\diagup\diagdown\diagup\diagdown$

44. A chemist wanted to determine experimentally the relative ease of removing a hydrogen atom from a tertiary, a secondary, and a primary carbon by a chlorine radical. He allowed 2-methylbutane to undergo chlorination at 300 °C and obtained as products 36% 1-chloro-2-methylbutane, 18% 2-chloro-2-methylbutane, 28% 2-chloro-3-methylbutane, and 18% 1-chloro-3-methylbutane. What values did he obtain for the relative ease of removing a hydrogen atom from tertiary, secondary, and primary hydrogen carbons by a chlorine radical under the conditions of his experiment?

45. At 600 °C, the ratio of the relative rates of formation of a tertiary, a secondary, and a primary radical by a chlorine radical is 2.6 : 2.1 : 1. Explain the change in the degree of regioselectivity compared to what was found in Problem 44.

46. Draw the products of the following reactions, including all stereoisomers:

a. 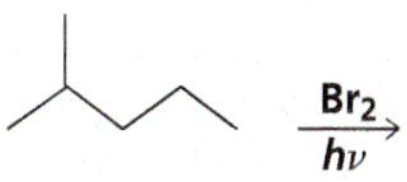 (2-methylpentane) $\xrightarrow[h\nu]{Br_2}$

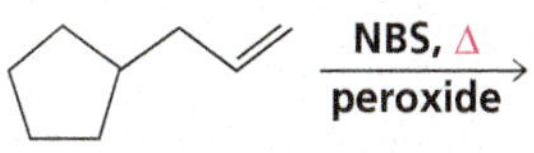

b.  $\diagup\diagdown\diagup$ $\xrightarrow[h\nu]{Br_2}$

c. $\diagup\diagdown\diagup\diagup$ $\xrightarrow[peroxide]{NBS, \Delta}$

d. 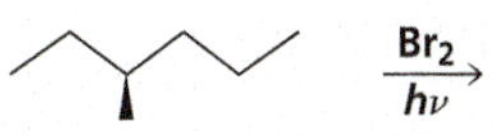 $\diagup\diagdown\diagup\diagup$ $\xrightarrow[h\nu]{Br_2}$

e. (cyclopentane with allyl group) $\xrightarrow[peroxide]{NBS, \Delta}$

f. 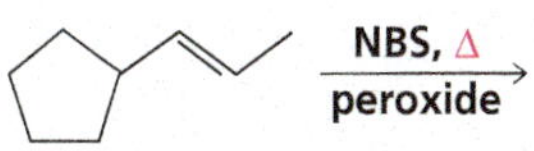 (cyclopentane with propenyl group) $\xrightarrow[peroxide]{NBS, \Delta}$

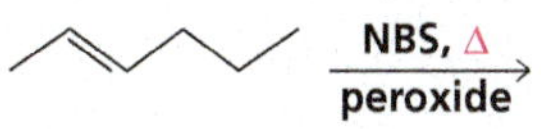

47. a. Calculate the $\Delta H°$ value for the following reaction:

$$CH_4 + Cl_2 \xrightarrow{h\nu} CH_3Cl + HCl$$

b. Calculate the sum of the $\Delta H°$ values for the following two propagation steps:

$$CH_3-H + \cdot Cl \longrightarrow \cdot CH_3 + H-Cl$$
$$\cdot CH_3 + Cl-Cl \longrightarrow CH_3-Cl + \cdot Cl$$

c. Why do both calculations give you the same value of $\Delta H°$?

48. A possible alternative mechanism to that shown in Problem 47 for the monochlorination of methane involves the following propagation steps:

$$CH_3-H + \cdot Cl \longrightarrow CH_3-Cl + \cdot H$$
$$\cdot H + Cl-Cl \longrightarrow H-Cl + \cdot Cl$$

How do you know that the reaction does not take place by this mechanism?

49. Propose a mechanism for the following reaction:

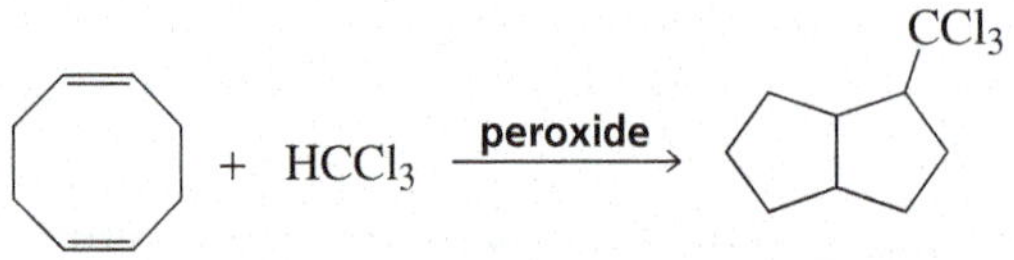

50. Explain why the rate of bromination of methane decreases if HBr is added to the reaction mixture.

51. Enediynes are natural products with potent antitumor properties because they are able to cleave DNA (page 288). Their cytotoxic properties are due to the enediyne undergoing a cyclization to form a highly reactive diradical intermediate. The intermediate abstracts hydrogen atoms from the backbone of DNA, which triggers its damage. Draw the structure of the diradical intermediate.

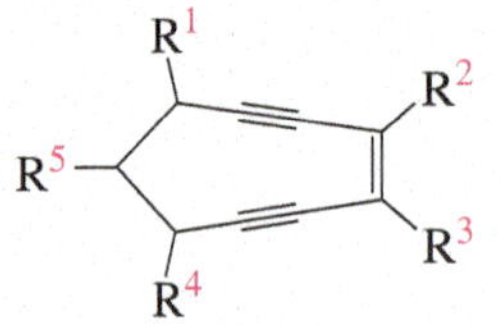

an enediyne

DRAWING CURVED ARROWS IN RADICAL SYSTEMS

We have seen that chemists use curved arrows

- to show how electrons move when going from a reactant to a product.
- to show how electrons move when going from one resonance contributor to the next.

The two previous tutorials on drawing curved arrows showed the simultaneous movement of two electrons. The movement of two electrons is indicated by an arrowhead with two barbs. Electron movement in radical systems involves the movement of only one electron. The movement of a single electron is indicated by an arrowhead with one barb.

DRAWING CURVED ARROWS IN RADICAL REACTIONS

When a bond breaks in a way that allows each of the bonded atoms to retain one of the bonding electrons, an arrowhead with one barb is used to represent the movement of each of the single electrons.

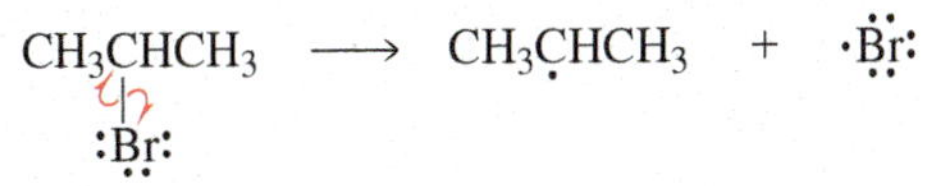

$$CH_3CHCH_3 \longrightarrow CH_3\dot{C}HCH_3 + \cdot\ddot{B}r:$$
$$\overset{|}{:}\ddot{B}r:$$

Sometimes, the lone pairs are not shown.

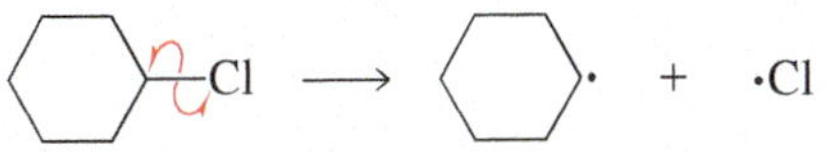

When a bond is formed using one electron from one atom and one electron from another, an arrowhead with one barb is used to represent the movement of each of the single electrons.

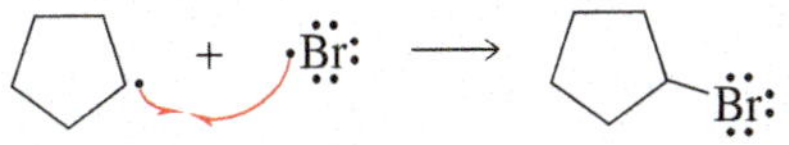

Sometimes, the lone pairs are not shown.

$$CH_3CH_2\dot{C}HCH_3 + \cdot Cl \longrightarrow CH_3CH_2\overset{|}{C}HCH_3$$
$$\overset{|}{Cl}$$

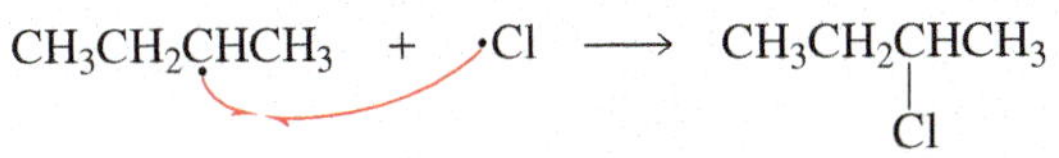

PROBLEM 1 Draw curved arrows to show the movement of the electrons as the bond breaks.

a. $CH_3CH_2—\ddot{C}l: \longrightarrow CH_3\dot{C}H_2 + \cdot\ddot{C}l:$ **c.** $CH_3\ddot{O}—\ddot{O}CH_3 \longrightarrow CH_3\ddot{O}\cdot + \cdot\ddot{O}CH_3$

b.

PROBLEM 2 Draw curved arrows to show the movement of the electrons as the bond forms.

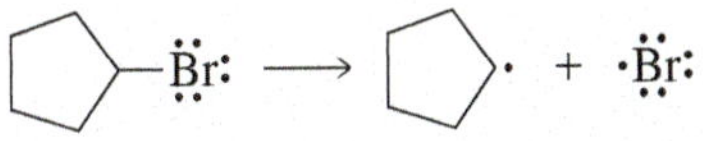

a. 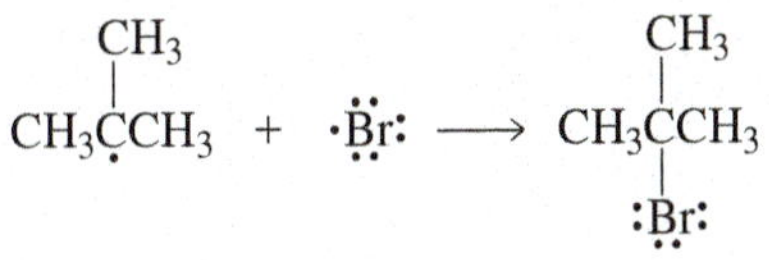 $CH_3\overset{\overset{\displaystyle CH_3}{|}}{\underset{|}{\dot{C}}}CH_3 + \cdot\ddot{B}r: \longrightarrow CH_3\overset{\overset{\displaystyle CH_3}{|}}{\underset{\underset{\displaystyle :\ddot{B}r:}{|}}{C}}CH_3$ **c.**

b. $\cdot Br + \cdot Br \longrightarrow Br_2$ **d.** $CH_3\dot{C}H_2 + CH_3\dot{C}H_2 \longrightarrow CH_3CH_2CH_2CH_3$

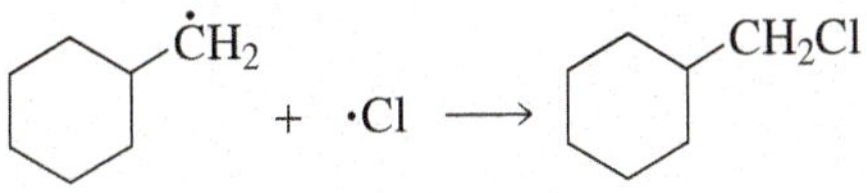

Often one bond breaks and another bond forms in the same step.

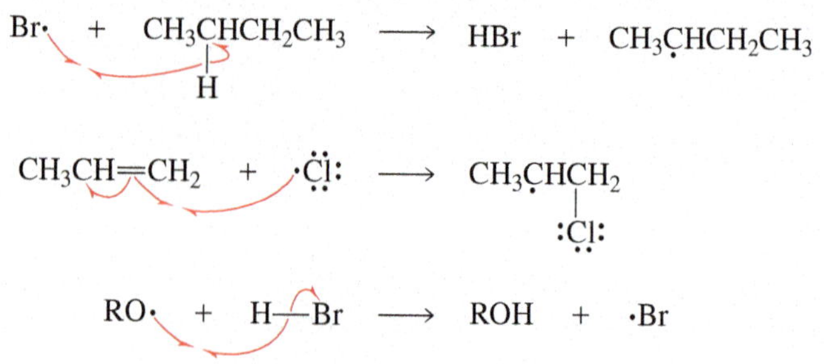

$$Br\cdot \ + \ CH_3CHCH_2CH_3 \ \longrightarrow \ HBr \ + \ CH_3\overset{\centerdot}{C}HCH_2CH_3$$
$$\underset{H}{|}$$

$$CH_3CH{=}CH_2 \ + \ \cdot\ddot{C}l\colon \ \longrightarrow \ CH_3\overset{\centerdot}{C}HCH_2$$
$$\underset{:\ddot{C}l:}{|}$$

$$RO\cdot \ + \ H{-}Br \ \longrightarrow \ ROH \ + \ \cdot Br$$

PROBLEM 3 Draw curved arrows to show the movement of the electrons as one (or more) bond breaks and another bond forms.

a. 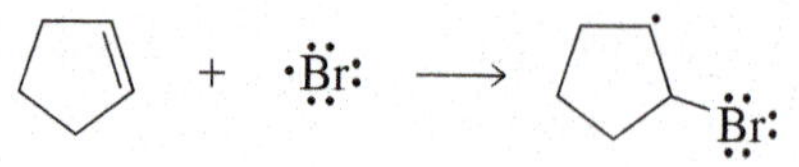 (cyclopentene) $+ \ \cdot\ddot{B}r\colon \ \longrightarrow$ (cyclopentyl radical) $\overset{\centerdot}{\ddot{B}r}\colon$

b. $CH_3\overset{\centerdot}{C}HCH_2CH_3 \ + \ H{-}Cl \ \longrightarrow \ CH_3CH_2CH_2CH_3 \ + \ \cdot Cl$

c. $CH_3\overset{\centerdot}{C}H_2 \ + \ Br{-}Br \ \longrightarrow \ CH_3CH_2Br \ + \ \cdot Br$

d. 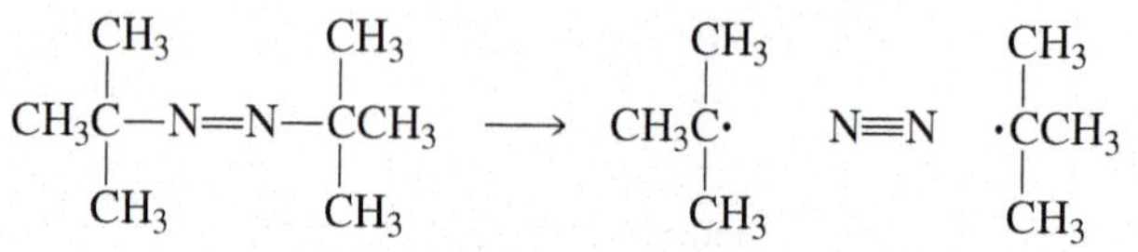 $\underset{CH_3}{\overset{CH_3}{|}}CH_3\overset{|}{C}{-}N{=}N{-}\overset{|}{C}CH_3\underset{CH_3}{\overset{CH_3}{|}} \ \longrightarrow \ CH_3\overset{\centerdot}{C}\cdot \quad N{\equiv}N \quad \cdot\overset{\centerdot}{C}CH_3$

(with CH₃ substituents)

DRAWING CURVED ARROWS IN CONTRIBUTING RESONANCE STRUCTURES THAT ARE RADICALS

The arrows that represent electron movement in resonance contributors of radicals have only one barb on the arrowhead because the arrow represents the movement of only one electron.

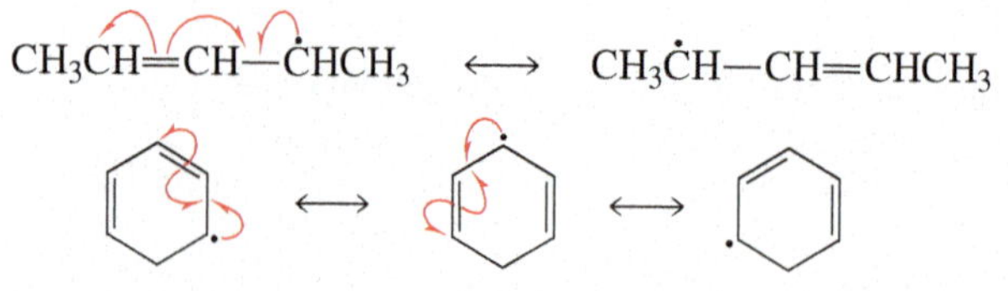

$$CH_3CH{=}CH{-}\overset{\centerdot}{C}HCH_3 \ \longleftrightarrow \ CH_3\overset{\centerdot}{C}H{-}CH{=}CHCH_3$$

(benzene radical resonance structures)

PROBLEM 4 Draw curved arrows to show the movement of the electrons as one resonance contributor is converted to the next.

a.  (pentadienyl radical) $\longleftrightarrow$

b. 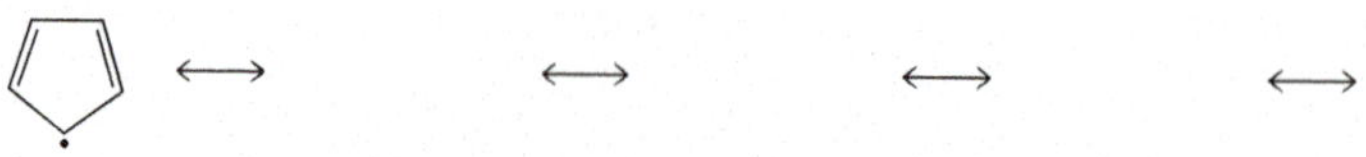 (cyclopentadienyl radical) $\longleftrightarrow \quad \quad \longleftrightarrow \quad \quad \longleftrightarrow \quad \quad \longleftrightarrow$

c. (hexatrienyl radical) $\longleftrightarrow \quad \quad \longleftrightarrow \quad \quad \longleftrightarrow$

d. 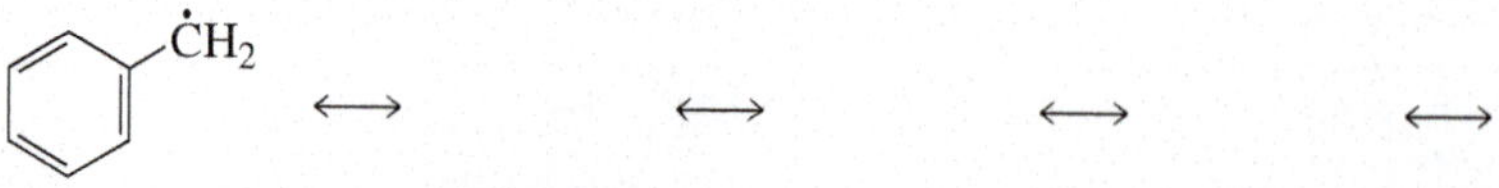 (benzyl radical) $\overset{\centerdot}{C}H_2 \quad \longleftrightarrow \quad \quad \longleftrightarrow \quad \quad \longleftrightarrow \quad \quad \longleftrightarrow$

ANSWERS TO PROBLEMS ON DRAWING CURVED ARROWS IN RADICAL SYSTEMS

PROBLEM 1

a. CH_3CH_2—$\ddot{C}l:$ $\longrightarrow$ $CH_3\dot{C}H_2$ + $\cdot\ddot{C}l:$ c. $CH_3\ddot{O}$—$\ddot{O}CH_3$ $\longrightarrow$ $CH_3\ddot{O}\cdot$ + $\cdot\ddot{O}CH_3$

b. [cyclopentyl]—$\ddot{B}r:$ $\longrightarrow$ [cyclopentyl]$\cdot$ + $\cdot\ddot{B}r:$

PROBLEM 2

a. $CH_3\overset{\underset{\displaystyle |}{CH_3}}{C}CH_3$ + $\cdot\ddot{B}r:$ $\longrightarrow$ $CH_3\overset{\underset{\displaystyle :\ddot{B}r:}{\overset{\displaystyle CH_3}{|}}}{C}CH_3$ c. [cyclohexyl]$\dot{C}H_2$ + $\cdot Cl$ $\longrightarrow$ [cyclohexyl]CH_2Cl

b. $\cdot Br$ + $\cdot Br$ $\longrightarrow$ Br_2 d. $CH_3\dot{C}H_2$ + $CH_3\dot{C}H_2$ $\longrightarrow$ $CH_3CH_2CH_2CH_3$

PROBLEM 3

a. [cyclopentene] + $\cdot\ddot{B}r:$ $\longrightarrow$ [cyclopentyl]$\ddot{B}r:$

b. $CH_3\dot{C}HCH_2CH_3$ + H—Cl $\longrightarrow$ $CH_3CH_2CH_2CH_3$ + $\cdot Cl$

c. $CH_3\dot{C}H_2$ + Br—Br $\longrightarrow$ CH_3CH_2Br + $\cdot Br$

d. $CH_3\overset{\underset{\displaystyle CH_3}{\overset{\displaystyle CH_3}{|}}}{C}$—$N$=$N$—$\overset{\underset{\displaystyle CH_3}{\overset{\displaystyle CH_3}{|}}}{C}CH_3$ $\longrightarrow$ $CH_3\overset{\underset{\displaystyle CH_3}{\overset{\displaystyle CH_3}{|}}}{C}\cdot$ N≡N $\cdot\overset{\underset{\displaystyle CH_3}{\overset{\displaystyle CH_3}{|}}}{C}CH_3$

PROBLEM 4

a.

b.

c.

d.

PART FOUR

Identification of Organic Compounds

Unless your professor chose to teach the chapters in this section at the very beginning of the course, at this point you have had the opportunity to solve problems that asked you to design the synthesis of an organic compound. But if you were actually to go into the laboratory to carry out a synthesis you designed, how would you know that the compound you obtained was the one you had set out to prepare?

When a scientist discovers a new compound with physiological activity, its structure must be ascertained. Only after its structure is known can the scientist design methods to synthesize the compound so that more of it can be made available than nature can provide.

These are just two reasons why chemists need to determine the structures of organic compounds. **Part 4** discusses some of the techniques they use to do this.

Chapter 13 Mass Spectrometry; Infrared Spectroscopy; Ultraviolet/Visible Spectroscopy

In **Chapter 13** you will learn about mass spectrometry, infrared spectroscopy, and ultraviolet/visible spectroscopy, three instrumental techniques that chemists use to analyze compounds. **Mass spectrometry** is used to find the molecular mass and the molecular formula of an organic compound; it is also used to identify certain structural features of the compound by identifying the fragments produced when the molecule breaks apart. **Infrared (IR) spectroscopy** allows us to identify the kinds of bonds and, therefore, the kinds of functional groups in an organic compound. **Ultraviolet and visible (UV/Vis) spectroscopy** provides information about compounds that have conjugated double bonds.

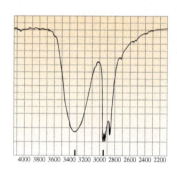

Chapter 14 NMR Spectroscopy

Chapter 14 discusses **nuclear magnetic resonance (NMR) spectroscopy,** which provides information about the carbon–carbon and carbon–hydrogen framework of an organic compound. The chapter also includes descriptions of **2-D NMR** and **X-ray crystallography,** techniques that are used to determine the structures of large molecules.

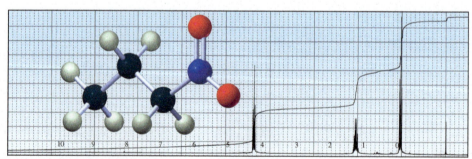

1-Nitropropane

13

Mass Spectrometry; Infrared Spectroscopy; UV/Vis Spectroscopy

The red, purple, and blue colors of many flowers, fruits, and vegetables are due to a class of compounds called anthocyanins (see pages 607–608).

Determining the structures of organic compounds is a fundamental part of organic chemistry. Scientists search the world for compounds with physiological activity. When such a compound is formed, its structure must be determined. Without knowing its structure, a chemist cannot synthesize the compound or make analogues with improved properties. Likewise, the structures of newly synthesized compounds must be confirmed. For example, you know that a ketone should be formed when an alkyne undergoes the acid-catalyzed addition of water (Section 7.7), but how can you verify that the product of the reaction is actually a ketone?

NOTE TO THE STUDENT

• There are additional spectroscopy problems in the *Study Guide and Solutions Manual.*

Before the structure of a compound can be determined, the compound must be isolated. For example, the product of a reaction carried out in the laboratory must first be isolated from the solvent, from any unreacted starting materials, and from any side products that might have formed. A compound found in nature must be isolated from the organism that synthesized it.

Isolating products and figuring out their structures used to be daunting tasks. The only tools chemists had for isolating products were distillation (for liquids) and sublimation or fractional recrystallization (for solids). Today, a variety of chromatographic techniques allows compounds to be isolated with relative ease. You will learn about these techniques in your laboratory course.

At one time, determining the structure of an organic compound required finding out its molecular formula by elemental analysis, determining the compound's physical properties (melting point, boiling point, and so on), and conducting simple chemical tests that indicate the presence (or absence) of certain functional groups (see page 958).

Unfortunately, these simple procedures were inadequate for characterizing molecules with complex structures, and because a relatively large sample of the compound was needed to perform all the tests, they were impractical for the analysis of compounds that were difficult to obtain in large amounts.

Today, a number of different instrumental techniques are used to identify organic compounds. These techniques can be performed quickly on small amounts of a compound and can provide much more information about the compound's structure than can simple chemical tests.

- **Mass spectrometry** allows us to determine the *molecular mass* and the *molecular formula* of a compound, as well as some of its *structural features*.

- **Infrared (IR) spectroscopy** tells us the *kinds of functional groups* a compound has.

- **Ultraviolet/Visible (UV/Vis) spectroscopy** provides information about organic compounds with conjugated double bonds.

- **Nuclear magnetic resonance (NMR) spectroscopy** provides information about the carbon–carbon and carbon–hydrogen framework of an organic compound. This technique is discussed in Chapter 14.

In this chapter, we will look at mass spectrometry, infrared spectroscopy, and ultraviolet/visible spectroscopy. Sometimes more than one technique is required to deduce the structure of a compound. Several problems in this book will require you to use two or three techniques at the same time.

We will be referring to different classes of organic compounds as we discuss various analytical techniques; they are listed in Table 13.1. (They are also listed inside the back cover of the book for easy reference.)

Table 13.1 Classes of Organic Compounds

Class	Structure		Class	Structure
Alkane	$-\overset{\mid}{\underset{\mid}{C}}-$	contains only C—C and C—H bonds	Aldehyde	$\underset{R}{\overset{O}{\parallel}}{}^{\nwarrow}H$
Alkene	$\overset{/}{\underset{/}{C}}=\overset{\backslash}{\underset{\backslash}{C}}$		Ketone	$R\overset{O}{\overset{\parallel}{C}}R$
Alkyne	$-C\equiv C-$		Carboxylic acid	$R\overset{O}{\overset{\parallel}{C}}OH$
Nitrile	$-C\equiv N$		Ester	$R\overset{O}{\overset{\parallel}{C}}OR$
Alkyl halide	R—X X = F, Cl, Br, or I		Amides	$R\overset{O}{\overset{\parallel}{C}}NH_2$ $R\overset{O}{\overset{\parallel}{C}}NHR$
Ether	R—O—R			
Alcohol	R—OH			$R\overset{O}{\overset{\parallel}{C}}NR_2$
Benzene	(benzene ring)		Amine (primary)	R—NH$_2$
Phenol	(benzene ring)—OH		Amine (secondary)	$R-\overset{R}{\underset{\mid}{N}}H$
Aniline	(benzene ring)—NH$_2$		Amine (tertiary)	$R-\overset{R}{\underset{\mid}{N}}-R$

13.1 | MASS SPECTROMETRY

At one time, the molecular mass of a compound was determined by its vapor density or freezing-point depression, and molecular formulas were determined by elemental analysis, a technique for measuring the relative proportions of the elements in the compound. These were long and tedious procedures that required a relatively large amount of a very pure sample of the compound. Today, molecular masses and molecular formulas can be rapidly determined from a very small sample by mass spectrometry.

Ionization of the Sample

In mass spectrometry, a small amount of a compound is introduced into an instrument called a mass spectrometer, where it is vaporized and then ionized (an electron is removed from each molecule). The sample can be ionized in several ways. Electron ionization (EI), the most common method, bombards the vaporized molecules with a beam of high-energy electrons. The energy of the beam can be varied but is typically about 70 electron volts. When the electron beam hits a molecule, it knocks out an electron, producing a **molecular ion.** A molecular ion is a **radical cation,** a species with an unpaired electron *and* a positive charge.

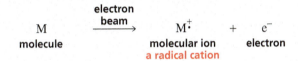

$$\underset{\substack{\text{M} \\ \textbf{molecule}}}{\text{M}} \xrightarrow{\substack{\textbf{electron} \\ \textbf{beam}}} \underset{\substack{\textbf{molecular ion} \\ \textcolor{red}{\textbf{a radical cation}}}}{\text{M}^{+\bullet}} + \underset{\textbf{electron}}{\text{e}^-}$$

Electron bombardment injects so much kinetic energy into the molecular ions that most of them break apart (fragment) into cations, radicals, neutral molecules, and other radical cations. Not surprisingly, the bonds most likely to break are the weakest ones and those that result in formation of the most stable products.

Detecting the Positively-Charged Fragments

All the *positively charged fragments* of the molecule are drawn between two negatively charged plates, which accelerate the fragments into an analyzer tube (Figure 13.1). Neutral fragments are not attracted to the negatively charged plates and, therefore, are not accelerated. They are eventually pumped out of the spectrometer.

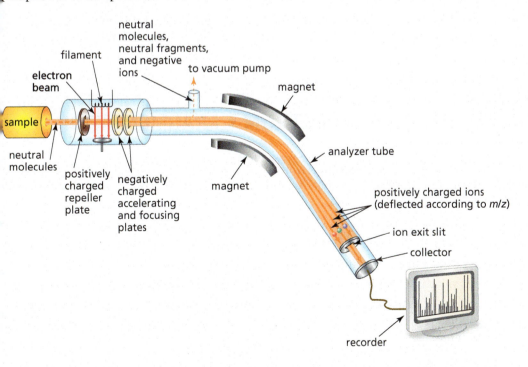

◀ **Figure 13.1**
Schematic diagram of an electron ionization mass spectrometer. A beam of high-energy electrons causes molecules to ionize and fragment. Positively charged fragments pass through the analyzer tube. Changing the magnetic field strength makes it possible to separate fragments of varying mass-to-charge ratios.

The analyzer tube is surrounded by a magnet whose magnetic field deflects the positively charged fragments in a curved path. At a given magnetic field strength, the degree to which the path is curved depends on the mass-to-charge ratio (m/z) of the fragment: the path of a fragment with a smaller m/z value bends more than that of a heavier fragment. In this way, the particles with the same m/z values are separated from all the others. If a fragment's path matches the curvature of the analyzer tube, the fragment will pass through the tube and out the ion exit slit.

A collector records the relative number of fragments with a particular m/z value passing through the slit. The more stable the fragment, the more likely it is to arrive at the collector without breaking down further. The strength of the magnetic field is gradually increased, so fragments with progressively larger m/z values are guided through the tube and out the exit slit.

Output from a Mass Spectrometer

The mass spectrometer records a **mass spectrum**—a graph of the relative abundance of each fragment plotted against its m/z value (Figure 13.2). Because the charge (z) on essentially all the fragments that reach the collector plate is +1, m/z is the mass (m) of the fragment. *Remember that only positively charged species reach the collector.*

A mass spectrum records only positively charged fragments.

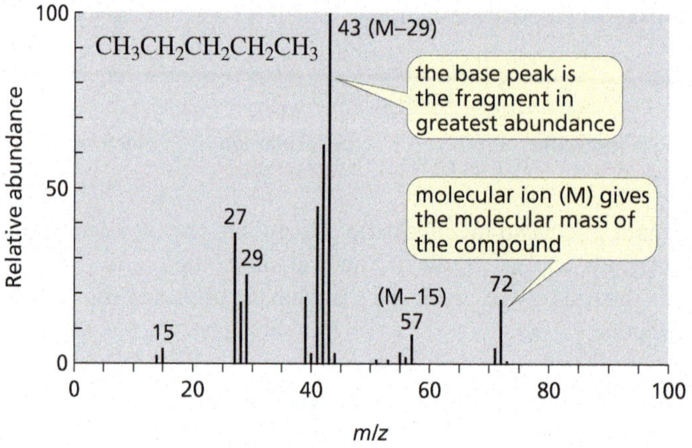

▶ **Figure 13.2**
The mass spectrum of pentane. The base peak represents the fragment that appears in greatest abundance. The m/z value of the molecular ion (M) gives the molecular mass of the compound.

PROBLEM 1 ◆

Which of the following fragments produced in a mass spectrometer will be accelerated through the analyzer tube?

$$CH_3\dot{C}H_2 \qquad CH_3CH_2\overset{+}{C}H_2 \qquad [CH_3CH_2CH_3]^{\overset{+}{\cdot}} \qquad \dot{C}H_2CH{=}CH_2 \qquad \overset{+}{C}H_2CH{=}CH_2$$

13.2 THE MASS SPECTRUM • FRAGMENTATION

The molecular ion and fragment ions produced in a mass spectrometer and recorded by it are unique for each compound. A mass spectrum, therefore, is like a fingerprint of the compound. Therefore, a compound can be identified by comparing its mass spectrum with those of known compounds determined under the same conditions.

Interpreting a Mass Spectrum

The mass spectrum of pentane is shown in Figure 13.2.

- Each m/z value is the m/z value of one of the fragments to the nearest whole number.
- The peak with the highest m/z value (in this case, $m/z = 72$) is the molecular ion (M), the fragment that results when the molecule loses an electron as a result of being bombarded by the electron beam.

- The m/z value of the molecular ion gives the molecular mass of the compound.

- Peaks with smaller m/z values—called fragment ion peaks—represent positively charged fragments of the molecular ion.

- The extremely tiny peak at $m/z = 73$ will be explained later.

Because it is not known which bond loses the electron, the molecular ion is written in brackets and the positive charge and unpaired electron are assigned to the entire structure.

$$CH_3CH_2CH_2CH_2CH_3 \xrightarrow[\text{beam}]{\textit{electron}} [CH_3CH_2CH_2CH_2CH_3]^{\stackrel{+}{\bullet}} + e^-$$

$$\text{molecular ion}$$
$$m/z = 72$$

> The m/z value of the molecular ion gives the molecular mass of the compound.

The **base peak** is the tallest peak, because it has the greatest relative abundance. It has the greatest relative abundance because it is generally the most stable positively charged fragment. The base peak is assigned a relative abundance of 100%, and the relative abundance of each of the other peaks is shown as a percentage of the base peak.

A mass spectrum gives us structural information about the compound because the m/z values and relative abundances of the fragments depend on the strength of the molecular ion's bonds and the stability of the fragments.

> *Weak bonds break in preference to strong bonds, and bonds that break to form more stable fragments break in preference to those that form less stable fragments.*

> The way a molecular ion fragments depends on the strength of its bonds and the stability of the fragments.

For example, all the C—C bonds in the molecular ion formed from pentane have about the same strength. However, the C-2–C-3 bond is the one most likely to break because it leads to formation of a *primary* carbocation and a *primary* radical (or a *primary* radical and a *primary* carbocation), which together are more stable than the *primary* carbocation and *methyl* radical (or *primary* radical and *methyl* cation) obtained from C-1–C-2 fragmentation.

Ions formed by C-2–C-3 fragmentation have m/z values of 43 and 29, whereas ions formed by C-1–C-2 fragmentation have m/z values 57 and 15. The base peak of 43 in the mass spectrum of pentane indicates the greater likelihood of C-2–C-3 fragmentation.

> Relative stabilities of carbocations:
> tertiary > secondary > primary > methyl

> Relative stabilities of radicals:
> tertiary > secondary > primary > methyl

Identifying Fragments

One method used to identify fragment ions makes use of the difference between the m/z value of a given fragment ion and that of the molecular ion. For example, the fragment ion with $m/z = 43$ in the mass spectrum of pentane is 29 units smaller than the molecular ion ($72 - 43 = 29$). An ethyl radical (CH_3CH_2) has a m/z value of 29 (because the mass numbers of C and H are 12 and 1, respectively). Thus, the peak at 43 can be attributed to loss of an ethyl radical from the molecular ion. Similarly, the fragment ion with $m/z = 57$ can be attributed to loss of a methyl radical from the molecular ion ($72 - 57 = 15$). Peaks at $m/z = 15$ and $m/z = 29$ are readily recognizable as being due to methyl and ethyl cations, respectively.

Peaks are commonly observed at m/z values two units below the m/z values of the carbocations because a carbocation can lose two hydrogen atoms.

> **NOTE TO THE STUDENT**
> • Appendix V contains a table of common fragment ions and a table of common fragments lost.

$$CH_3CH_2\overset{+}{C}H_2 \longrightarrow \overset{+}{C}H_2CH=CH_2 + 2\,H\cdot$$
$$m/z = 43 \qquad\qquad m/z = 41$$

Comparing the Mass Spectra of Pentane and Isopentane

Isopentane has the same molecular formula as pentane. Thus, it too has a molecular ion with an m/z value of 72 (Figure 13.3). Its mass spectrum is similar to that of pentane, with one notable exception: the peak at $m/z = 57$, indicating loss of a methyl radical, is much more intense than the same peak in pentane.

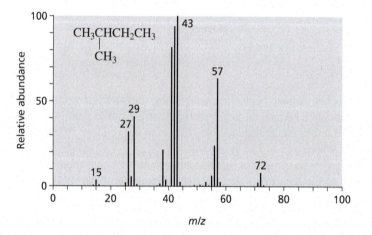

▶ **Figure 13.3**
The mass spectrum of isopentane. The peak at $m/z = 57$ **corresponds to the loss of a methyl group and the formation of a relatively stable secondary carbocation.**

Isopentane is more likely than pentane to lose a methyl radical because when it does, a *secondary* carbocation is formed. In contrast, when pentane loses a methyl radical, a less stable *primary* carbocation is formed.

$$\left[\begin{array}{c} CH_3 \\ | \\ CH_3CHCH_2CH_3 \end{array} \right]^{+\cdot} \longrightarrow CH_3\overset{+}{C}HCH_2CH_3 + \overset{\cdot}{C}H_3$$

molecular ion
$m/z = 72$ **$m/z = 57$**

PROBLEM 2

What distinguishes the mass spectrum of 2,2-dimethylpropane from the mass spectra of pentane and isopentane?

PROBLEM 3 ♦

What is the most likely m/z value for the base peak in the mass spectrum of 3-methylpentane?

13.3 USING THE m/z VALUE OF THE MOLECULAR ION TO CALCULATE THE MOLECULAR FORMULA

The **rule of 13** allows possible molecular formulas to be determined from the m/z value of the molecular ion. Remember that the m/z value of the molecular ion gives the molecular mass of the compound.

First the **base value** must be determined. To do this, divide the m/z value of the molecular ion by 13. The answer gives the number of carbons in the compound. For example, if the m/z value is 142, dividing 142 by 13 gives 10 as the number of carbons (with 12 left over). The number of hydrogens is determined by adding the number left over to the number of carbons ($10 + 12 = 22$). Thus, the base value is $C_{10}H_{22}$.

If the compound has one oxygen, then one O (16 amu) must be added to the base value and one C and four Hs (16 amu) must be subtracted from it. Thus, the molecular formula is $C_9H_{18}O$. If the compound has two oxygens, the process must be repeated, in which case the molecular formula is $C_8H_{14}O_2$. (Notice that to maintain the m/z value, you need to subtract the same number of atomic mass units that you add.)

LEARN THE STRATEGY

PROBLEM 4 SOLVED

Draw possible structures for an ester that has a molecular ion with an *m/z* value of 74.

SOLUTION Dividing 74 by 13 gives 5 with 9 left over. Thus, the base value is C_5H_{14}. We know that an ester has two oxygens. For each oxygen, add one O and subtract one C and four Hs. This gives a molecular formula of $C_3H_6O_2$. Possible structures are

USE THE STRATEGY

PROBLEM 5 ♦

Determine the molecular formula for each of the following:

a. a compound that contains only C and H and has a molecular ion with an *m/z* value of 72
b. a compound that contains C, H, and one O and has a molecular ion with an *m/z* value of 100
c. a compound that contains C, H, and two Os and has a molecular ion with an *m/z* value of 102
d. an amide that has a molecular ion with an *m/z* value of 115

PROBLEM 6 ♦

a. Suggest possible molecular formulas for a compound that has a molecular ion with an *m/z* value of 86.
b. Can one of the possible molecular formulas contain a nitrogen atom?

PROBLEM 7

If a compound has a molecular ion with an odd-numbered mass, then the compound contains an odd number of nitrogen atoms. This is known as the **nitrogen rule.**

a. Calculate the *m/z* value for the molecular ion of each of the following compounds:

1.   2. H_2N NH_2

b. Explain why the nitrogen rule holds.
c. State the rule in terms of a molecular ion with an even-numbered mass.

PROBLEM-SOLVING STRATEGY

Using Mass Spectra to Determine Structures

LEARN THE STRATEGY

The mass spectra of two very stable cycloalkanes both show a molecular ion peak at *m/z* = 98. One spectrum shows a base peak at *m/z* = 69, whereas the other shows a base peak at *m/z* = 83. Identify the cycloalkanes.

First, let's determine the molecular formula of the compounds from the *m/z* value of their molecular ions. Dividing 98 by 13 results in 7 with 7 left over. Thus, each has a molecular formula of C_7H_{14}.

Now let's see what fragment is lost to give the base peak. A base peak of 69 means the loss of an ethyl radical (98 − 69 = 29), whereas a base peak of 83 means the loss of a methyl radical (98 − 83 = 15).

Because the two cycloalkanes are known to be very stable, we can assume they do not have three- or four-membered rings. A seven-carbon cycloalkane with a base peak signifying the loss of an ethyl radical must be ethylcyclopentane. A seven-carbon cycloalkane with a base peak signifying the loss of a methyl radical must be methylcyclohexane.

USE THE STRATEGY

PROBLEM 8 ♦

Identify the hydrocarbon that has a molecular ion with an *m/z* value of 128, a base peak with an *m/z* value of 43, and significant peaks with *m/z* values of 57, 71, and 85.

13.4   ISOTOPES IN MASS SPECTROMETRY

M+1 Peak

The molecular ions of pentane and isopentane both have *m/z* values of 72, but each spectrum shows a very small peak at $m/z = 73$ (Figures 13.2 and 13.3). This peak is called the M+1 peak because the ion responsible for it is one unit heavier than the molecular ion. The M+1 peak owes its presence to the fact that there are two naturally occurring isotopes of carbon: ^{12}C and ^{13}C (98.89% and 1.11% of naturally occurring carbon, respectively; see Section 1.1). Because mass spectrometry records individual molecules, any molecule containing a ^{13}C will appear at M+1.

The isotopic distributions of several elements commonly found in organic compounds are listed in Table 13.2.

Table 13.2 The Natural Abundance of Isotopes Commonly Found in Organic Compounds			
Element	**Natural abundance**		
Carbon	^{12}C 98.89%	^{13}C 1.11%	
Hydrogen	^{1}H 99.99%	^{2}H 0.01%	
Nitrogen	^{14}N 99.64%	^{15}N 0.36%	
Oxygen	^{16}O 99.76%	^{17}O 0.04%	^{18}O 0.20%
Sulfur	^{32}S 95.0%	^{33}S 0.76%	^{34}S 4.22% ^{36}S 0.02%
Fluorine	^{19}F 100%		
Chlorine	^{35}Cl 75.77%		^{37}Cl 24.23%
Bromine	^{79}Br 50.69%		^{81}Br 49.31%
Iodine	^{127}I 100%		

M+2 Peak

A mass spectrum can also show a M+2 peak due to ^{18}O or from having two heavy isotopes in the same molecule (such as ^{13}C and ^{2}H, or two ^{13}Cs). These situations are unusual, so M+2 peaks are generally very small.

The presence of a large M+2 peak is evidence of a compound containing either chlorine or bromine, because each of these elements has a high percentage of a naturally occurring isotope that is two units heavier than the most abundant isotope. From the natural abundance of the isotopes of chlorine and bromine shown in Table 13.2, we can conclude that,

- if the M+2 peak is one-third the height of the M peak, then the compound contains a chlorine atom (the natural abundance of ^{37}Cl is one-third that of ^{35}Cl).

- if the M and M+2 peaks are about the same height, then the compound contains a bromine atom (the natural abundances of ^{79}Br and ^{81}Br are about the same).

In calculating the *m/z* values of molecular ions and fragments, the *atomic mass* of a single isotope of the atom must be used (for example, Cl = 35 or 37) because mass spectrometry records the *m/z* value of an *individual* fragment. The *atomic masses* in the periodic table (Cl = 35.453) cannot be used, because they are the *weighted averages* of all the naturally occurring isotopes for that element.

PROBLEM 9 ◆

Predict the relative intensities of the molecular ion peak, the M+2 peak, and the M+4 peak for a compound that contains two bromine atoms.

13.5 HIGH-RESOLUTION MASS SPECTROMETRY CAN REVEAL MOLECULAR FORMULAS

All the mass spectra shown in this book were produced with a low-resolution mass spectrometer. These spectrometers give the *m/z* value of a fragment to the nearest whole number. High-resolution mass spectrometers can determine the *exact molecular mass* of a fragment to a precision of 0.0001 amu, making it possible to distinguish between compounds that have the same molecular mass to the nearest whole number. For example, the following listing shows six compounds that have a molecular mass of 122 amu, but each of them has a different exact molecular mass.

Exact Molecular Masses and Molecular Formulas for Several Compounds with a Molecular Mass of 122 amu

Exact molecular mass (amu)	122.1096	122.0845	122.0732	122.0368	122.0579	122.0225
Molecular formula	C_9H_{14}	$C_7H_{10}N_2$	$C_8H_{10}O$	$C_7H_6O_2$	$C_4H_{10}O_4$	$C_4H_{10}S_2$

Table 13.3	The Exact Masses of Some Common Isotopes
Isotope	**Mass**
^1H	1.007825 amu
^{12}C	12.00000 amu
^{14}N	14.0031 amu
^{16}O	15.9949 amu
^{32}S	31.9721 amu
^{35}Cl	34.9689 amu
^{79}Br	78.9183 amu

The exact masses of some common isotopes are listed in Table 13.3. There are computer programs that can determine the molecular formula of a compound from the compound's exact molecular mass.

PROBLEM 10 ◆

Which molecular formula has an exact molecular mass of 86.1096 amu: C_6H_{14}, $C_4H_{10}N_2$, or $C_4H_6O_2$?

PROBLEM 11 ◆

a. Can a low-resolution mass spectrometer distinguish between $C_2H_5^+$ and CHO^+?
b. Can a high-resolution mass spectrometer distinguish between them?

13.6 THE FRAGMENTATION PATTERNS OF FUNCTIONAL GROUPS

Each functional group has characteristic fragmentation patterns that can help identify a compound. The patterns began to be recognized after the mass spectra of many compounds containing a particular functional group had been studied. We will look at the fragmentation patterns of alkyl halides, ethers, alcohols, and ketones as examples.

Alkyl Halides

Mass Spectrum of 1-Bromopropane

Let's look first at the mass spectrum of 1-bromopropane (Figure 13.4). The relative heights of the M and M+2 peaks are about equal, which we have seen is characteristic of a compound containing a bromine atom.

- Electron bombardment is most likely to dislodge a lone-pair electron if the molecule has any, because a molecule does not hold onto its lone-pair electrons as tightly as it holds onto its bonding electrons. Thus, the molecular ion is created when electron bombardment dislodges one of bromine's lone-pair electrons.

$$CH_3CH_2CH_2-{}^{79}\ddot{\ddot{Br}}: \ + \ CH_3CH_2CH_2-{}^{81}\ddot{\ddot{Br}}: \ \xrightarrow{-e^-} \ CH_3CH_2CH_2-{}^{79}\overset{+}{\ddot{Br}}: \ + \ CH_3CH_2CH_2-{}^{81}\overset{+}{\ddot{Br}}: \ \longrightarrow \ CH_3CH_2\overset{+}{C}H_2 \ + \ {}^{79}\ddot{\ddot{Br}}: \ + \ {}^{81}\ddot{\ddot{Br}}$$

1-bromopropane   *m/z* = 122   *m/z* = 124   *m/z* = 43

Bond Dissociation Energy
C–Br = 71 kcal/mol
C–Cl = 85 kcal/mol
C–C = 89 kcal/mol
C–H = 99 kcal/mol

- The weakest bond in the molecular ion is the one most apt to break. In this case, the C—Br bond is the weakest. When the C—Br bond breaks, it breaks heterolytically, with both electrons going to the more electronegative of the atoms that were joined by the bond, forming a propyl cation and a bromine atom. As a result, the base peak in the mass spectrum of 1-bromopropane is at *m/z* = 43 (M − 79 = 43 or [M+2] − 81 = 43).

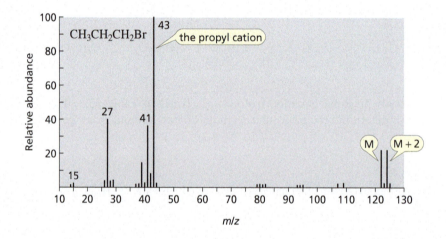

▶ **Figure 13.4**
The mass spectrum of 1-bromopropane. The M and M + 2 peaks are about equal because ^{79}Br and ^{81}Br have almost equal abundances.

Mass Spectrum of 2-Chloropropane

The mass spectrum of 2-chloropropane is shown in Figure 13.5. The M+2 peak is one-third the height of the molecular ion peak, which is consistent with a compound that contains a chlorine atom (Section 13.4).

The way a molecular ion fragments depends on the strength of its bonds and the stability of the fragments.

- The C—Cl (85 kcal/mol) and C—C (89 kcal/mol) bonds have similar strengths, so both bonds can break.

- The C—Cl bond breaks heterolytically (because the atoms joined by the bond have different electronegativities), giving a base peak at *m/z* = 43.

- The C—C bond breaks homolytically (because the atoms joined by the bond have the same electronegativity); the peaks at *m/z* = 63 and *m/z* = 65 have a 3:1 ratio, indicating that these fragments contain a chlorine atom. They result from *homolytic cleavage* (called **α-cleavage**) of a group bonded to the α-carbon. (The α-carbon is the carbon bonded to the chlorine).

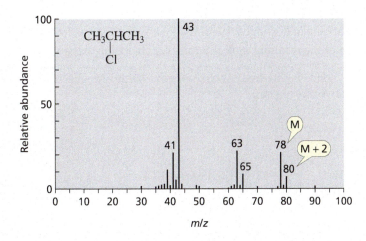

$$CH_3CH\overset{..}{\underset{..}{\overset{35}{\text{Cl}}}}: \quad + \quad CH_3CH\overset{..}{\underset{..}{\overset{37}{\text{Cl}}}}:$$

2-chloropropane

A bond between a carbon and a more electronegative atom breaks heterolytically.

A bond between a carbon and an atom of similar electronegativity breaks homolytically.

◀ Figure 13.5
The mass spectrum of 2-chloropropane. The M+2 peak is one-third the size of the M peak because ^{37}Cl has one-third the natural abundance of ^{35}Cl.

α-Cleavage occurs because it leads to a cation that is relatively stable because all its atoms have complete octets:

product of
α-cleavage

all the atoms have
complete octets

α-cleavage

α-Cleavage occurs in alkyl chlorides because the C—C and C—Cl bonds have similar strengths.

α-Cleavage is less likely to occur in alkyl bromides because the C—C bond is much stronger than the C—Br bond.

α-Cleavage occurs in *alkyl chlorides* because the C—C (89 kcal/mol) and C—Cl (85 kcal/mol) bonds have similar energies. α-Cleavage is much less likely to occur in *alkyl bromides* because the C—C (89 kcal/mol) bond is much stronger than the C—Br bond (71 kcal/mol).

PROBLEM 12

Sketch the mass spectrum expected for 1-chloropropane.

Ethers

The mass spectrum of 2-isopropoxybutane is shown in Figure 13.6.

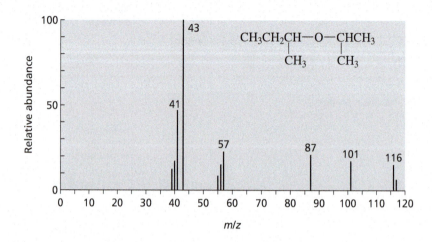

▶ **Figure 13.6**
The mass spectrum of 2-isopropoxybutane.

The fragmentation pattern of an ether is similar to that of an alkyl halide.

1. Electron bombardment dislodges one of the lone-pair electrons from oxygen.
2. Fragmentation of the resulting molecular ion occurs mainly in two ways:

- A C—O bond is cleaved *heterolytically,* with the electrons going to the mor electronegative oxygen atom.

$$CH_3CH_2\overset{\underset{|}{CH_3}}{CH}-\overset{..}{\underset{..}{O}}-\overset{\underset{|}{CH_3}}{CHCH_3} \xrightarrow{-e^-} CH_3CH_2\overset{\underset{|}{CH_3}}{CH}-\overset{+}{\underset{..}{O}}-\overset{\underset{|}{CH_3}}{CHCH_3}$$

2-isopropoxybutane   **m/z = 116**

$$\longrightarrow CH_3CH_2\overset{\underset{|}{CH_3}}{\overset{+}{CH}} \;+\; \overset{..}{\underset{..}{O}}-\overset{\underset{|}{CH_3}}{CHCH} \quad m/z = 57$$

$$\longrightarrow CH_3CH_2\overset{\underset{|}{CH_3}}{CH}-\overset{..}{\underset{..}{O}}: \;+\; \overset{\underset{|}{CH_3}}{\overset{+}{CHCH_3}} \quad m/z = 43$$

- A C—C bond is cleaved *homolytically* at an α-carbon because this leads to a relativel stable cation since all its atoms have complete octets. (An α-carbon is the carbon bonde to the oxygen.)

$$CH_3CH_2\overset{\underset{|}{CH_3}}{CH}\overset{+}{\underset{..}{O}}-\overset{\underset{|}{CH_3}}{CHCH_3} \xrightarrow{\alpha\text{-cleavage}} CH=\overset{+}{\underset{..}{O}}-\overset{\underset{|}{CH_3}}{CHCH_3} \;+\; CH_3\dot{C}H_2$$

                                        m/z = 87

$$CH_3CH_2\overset{\underset{|}{CH_3}}{CH}\overset{+}{\underset{..}{O}}-\overset{\underset{|}{CH_3}}{CHCH_3} \xrightarrow{\alpha\text{-cleavage}} CH_3CH_2CH=\overset{+}{\underset{..}{O}}-\overset{\underset{|}{CH_3}}{CHCH_3} \;+\; \dot{C}H_3$$

                                        m/z = 101

$$CH_3CH_2\overset{\underset{|}{CH_3}}{CH}-\overset{+}{\underset{..}{O}}\overset{\underset{|}{CH_3}}{CHCH_3} \xrightarrow{\alpha\text{-cleavage}} CH_3CH_2\overset{\underset{|}{CH_3}}{CH}-\overset{+}{\underset{..}{O}}=CHCH_3 \;+\; \dot{C}H_3$$

                                        m/z = 101

> An arrowhead with a double barb signifies the movement of two electrons.

> An arrowhead with a single barb signifies the movement of one electron.

α-carbon

The alkyl group most easily cleaved is the one that forms the most stable radical. Thus, th peak at *m/z* = 87 is more abundant than the one at *m/z* = 101 (even though the compound ha three methyl groups bonded to α-carbons that can be cleaved to produce the peak at *m/z* = 101), because a primary radical is more stable than a methyl radical.

PROBLEM 13 ♦

The mass spectra of 1-methoxybutane, 2-methoxybutane, and 2-methoxy-2-methylpropane are shown below. Match each compound with its spectrum.

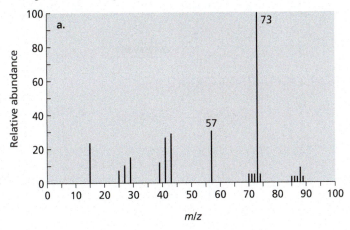

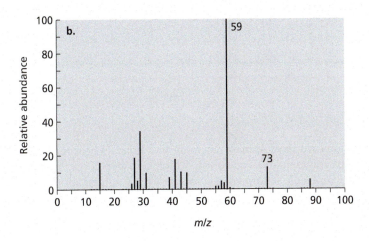

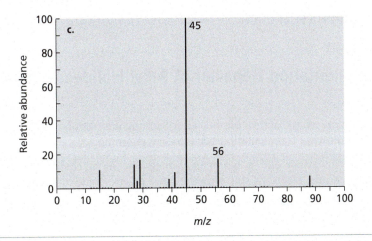

Alcohols

As with alkyl halides and ethers, the molecular ion of an alcohol is created by knocking out a lone-pair electron. The molecular ions obtained from alcohols fragment so readily that few of them survive to reach the collector. As a result, the mass spectrum of an alcohol shows a small molecular ion peak, if any. Notice the tiny molecular ion peak at $m/z = 102$ in the mass spectrum of 2-hexanol (Figure 13.7).

α-Cleavage

The molecular ion of an alcohol, like those of alkyl halides and ethers, undergoes α-cleavage. Consequently, the mass spectrum of 2-hexanol shows a base peak at $m/z = 45$ (α-cleavage leading to a more stable butyl radical) and a smaller peak at $m/z = 87$ (α-cleavage leading to a less stable methyl radical).

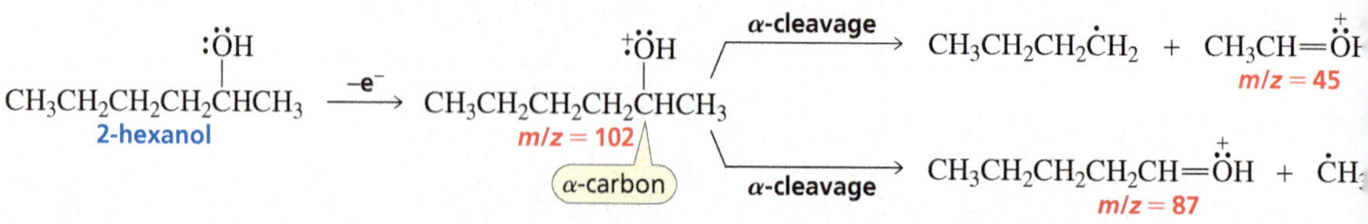

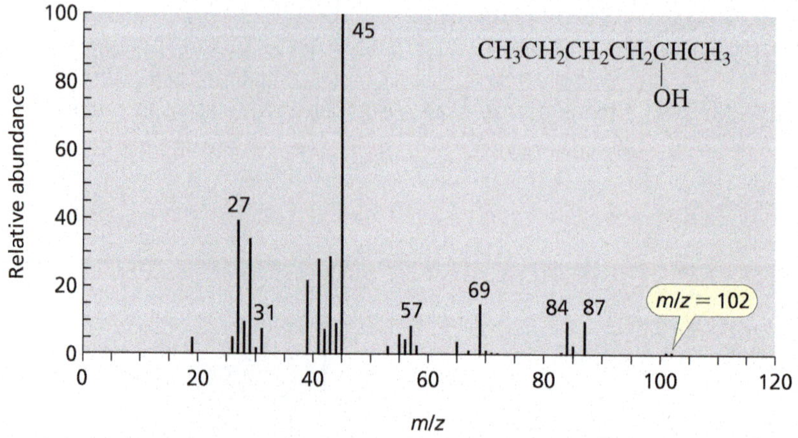

$$CH_3CH_2CH_2CH_2\overset{:\ddot{O}H}{\underset{|}{C}}HCH_3 \xrightarrow{-e^-} CH_3CH_2CH_2CH_2\overset{:\ddot{O}H}{\underset{|}{C}}HCH_3$$

2-hexanol $m/z = 102$

α-carbon α-cleavage

$\xrightarrow{\alpha\text{-cleavage}} CH_3CH_2CH_2\dot{C}H_2 + CH_3CH=\overset{+}{\ddot{O}}H$ $m/z = 45$

$\xrightarrow{\alpha\text{-cleavage}} CH_3CH_2CH_2CH_2CH=\overset{+}{\ddot{O}}H + \dot{C}H_3$ $m/z = 87$

▶ **Figure 13.7**
The mass spectrum of 2-hexanol.
The molecular ion peak at $m/z = 102$
is small because alcohols fragment easily.

Loss of Water

In all the fragmentations we have seen so far, only one bond is broken. In the case of alcohols, however, an important fragmentation occurs in which two bonds are broken. Two bonds break because a stable water molecule is formed as a result of the fragmentation. The water is composed of the OH group and a γ-hydrogen. Loss of water results in a fragmentation peak at $m/z = M-18$.

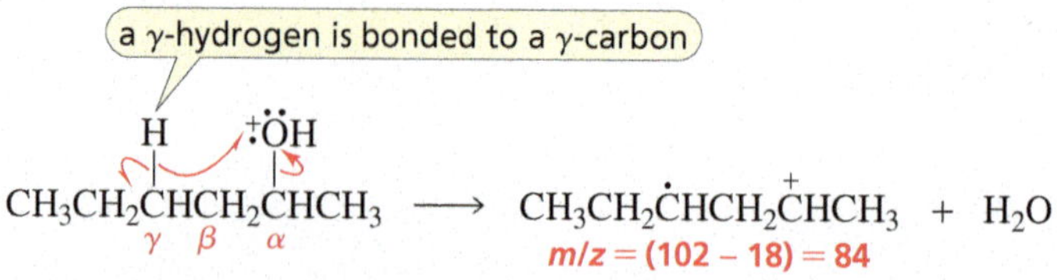

a γ-hydrogen is bonded to a γ-carbon

$$CH_3CH_2\underset{\gamma}{C}H\underset{\beta}{C}H_2\underset{\alpha}{C}HCH_3 \longrightarrow CH_3CH_2\dot{C}HCH_2\overset{+}{C}HCH_3 + H_2O$$

$m/z = (102 - 18) = 84$

Common Fragmentation Behavior of Alkyl Halides, Ethers, and Alcohols

Notice that alkyl halides, ethers, and alcohols—after forming a molecular ion by losing a lone-pair electron—have the following fragmentation behaviors in common:

- A bond between carbon and a *more electronegative* atom (halogen or oxygen) breaks heterolytically and the electrons stay with the more electronegative atom.

- A bond between carbon and an atom of *similar electronegativity* (carbon or hydrogen) breaks homolytically.

- The bonds most likely to break are the weakest bonds and those that lead to formation of the most stable products. (Look for a fragmentation that results in a cation with atoms that all have complete octets.)

PROBLEM 14 ◆

Primary alcohols have a strong peak at $m/z = 31$.

a. What fragment is responsible for this peak?   **b.** Draw the mechanism for its formation.

Ketones

A ketone generally has an intense molecular ion peak that is formed by knocking out a lone-pair electron. The molecular ion fragments homolytically at the C—C bond adjacent to the C=O bond because this cleavage (also called α-cleavage) results in formation of a cation with atoms that all have complete octets. The alkyl group leading to the more stable radical is the one more easily cleaved. Thus, the peak at $m/z = 43$ is more abundant than the one at $m/z = 71$.

$$CH_3CH_2CH_2CCH_3 \xrightarrow{-e^-} CH_3CH_2CH_2CCH_3$$
2-pentanone $m/z = 86$

$$\longrightarrow CH_3CH_2\dot{C}H_2 + CH_3C\equiv\overset{+}{O}:$$
$m/z = 43$

$$\longrightarrow CH_3CH_2CH_2C\equiv\overset{+}{O}: + \dot{C}H_3$$
$m/z = 71$

If one of the alkyl groups attached to a carbonyl carbon has a γ-hydrogen, then a cleavage that goes through a favorable six-membered-ring transition state may occur. This fragmentation—known as a **McLafferty rearrangement**—breaks two bonds. The bond between the α-carbon and the β-carbon breaks homolytically and a hydrogen atom from the γ-carbon migrates to the oxygen. Notice that fragmentation occurs in a way that produces a stable alkene and a cation with an oxygen atom that has a complete octet.

$$\text{McLafferty rearrangement}$$
$m/z = 86$ $H_2C=CH_2 + \dot{C}H_2-\overset{O}{\underset{}{C}}-CH_3$
$m/z = 58$

$CH_3-CH_2-\overset{+}{\overset{..}{O}}CH_3$
α-cleavage of an ether

$CH_3-CH_2-\overset{+}{\overset{..}{O}}H$
α-cleavage of an alcohol

$CH_3-\overset{\overset{+\overset{..}{O}}{\|}}{C}R$
α-cleavage of a ketone

PROBLEM 15 ◆

Identify the ketones responsible for the mass spectra below.

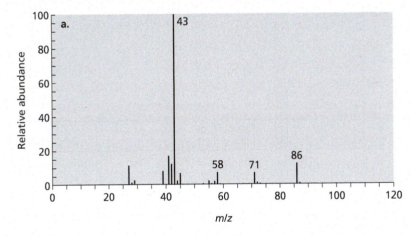

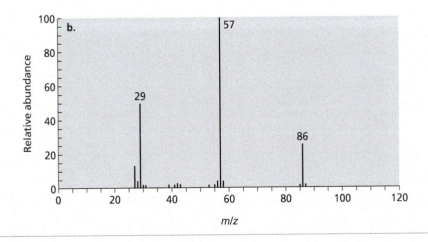

PROBLEM 16

Draw the mechanism for formation of the fragments that would allow these ketones to be distinguished by their mass spectra.

PROBLEM 17

Using curved arrows, show the principal fragments you would expect to see in the mass spectrum of each of the following compounds:

a. $CH_3CH_2CH_2CH_2CH_2OH$

c. $CH_3CH_2OCCH_2CH_2CH_3$ with CH_2CH_3 and CH_3 groups

e. CH_3CH_2CHCl with CH_3 group

b. $CH_3CH_2CHCH_2CH_2CH_2CH_3$ with OH group

d. $CH_3CCH_2CH_2CH_2CH_3$ with O double bond

f. CH_3-C-Br with CH_3 and CH_3 groups

PROBLEM 18 ♦

The reaction of (Z)-2-pentene with water and a trace of H_2SO_4 forms two products. Identify the products from their mass spectra.

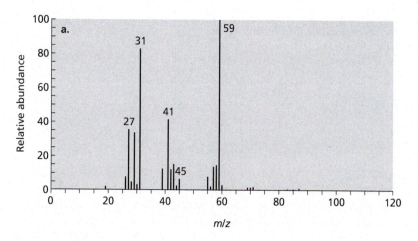

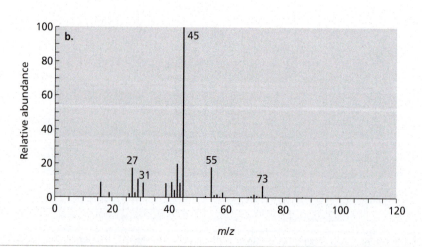

13.7 OTHER IONIZATION METHODS

Methods other than electron ionization (EI–MS) can be used to obtain mass spectral data. In chemical ionization–mass spectrometry (CI–MS), the sample is sprayed with a pre-ionized gas such as methane or ammonia that causes the sample to ionize by electron transfer or proton transfer from the gas to the sample. Because the molecular ions produced by this technique are less apt to undergo fragmentation, the ability to obtain the molecular mass (and, therefore, the molecular formula) of the sample is enhanced.

Both EI–MS and CI–MS require the sample to be vaporized before it is ionized. Therefore, these methods are used only for samples with relatively low molecular masses. Mass spectra of large molecules, even large biological molecules such as enzymes or oligonucleotides, can be obtained through desorption ionization (DI). In this technique, the sample is dissolved on a matrix compound, which is then ionized by one of several methods. The matrix compound transfers energy to the sample, causing some of the sample molecules to ionize and be ejected from the matrix. The ejected ions are accelerated in the mass spectrometer and recorded to give a mass spectrum as in EI–MS (Section 13.1). Ionization methods include bombarding the matrix compound with a high-energy beam of ions (called secondary ion MS [SIMS]), photons (called matrix-assisted laser desorption ionization [MALDI]), or neutral atoms (called fast-atom bombardment [FAB]).

13.8 GAS CHROMATOGRAPHY–MASS SPECTROMETRY

Mixtures of compounds are often analyzed using gas chromatography and mass spectrometry (GC–MS) at the same time. The sample is injected into a gas chromatograph and the various components of the mixture travel through the column at different rates, based on their boiling points. The lowest boiling component of the mixture exits first. As each compound exits, it enters the mass spectrometer where it is ionized, forming a molecular ion and fragments of the molecular ion. The mass spectrometer records a mass spectrum for each of the components of the mixture. GC–MS is widely used to analyze forensic samples.

Mass Spectrometry in Forensics

Forensic science is the application of science for the purpose of answering questions related to a civil or criminal case. Mass spectrometry is an important tool of the forensic scientist. It is used to analyze body fluids for the presence and levels of drugs and other toxic substances. It can also identify the presence of drugs in hair, which increases the window of detection from hours and days (after which body fluids are no longer useful) to months and even years. It was employed for the first time at an athletic event in 1955 to detect drugs in athletes at a cycling competition in France. (Twenty percent of those tests were positive.) Mass spectrometry is also used to identify residues of arson fires and explosives from post-explosion residues, and to analyze such things as paints, adhesives, and fibers.

13.9 SPECTROSCOPY AND THE ELECTROMAGNETIC SPECTRUM

Spectroscopy is the study of the interaction of matter and electromagnetic radiation. A continuum of different types of **electromagnetic radiation**—each associated with a particular energy range—makes up the electromagnetic spectrum (Figure 13.8). Visible light is the electromagnetic radiation we are most familiar with, but it represents only a fraction of the full electromagnetic spectrum. X-rays, microwaves, and radio waves are other familiar types of electromagnetic radiation.

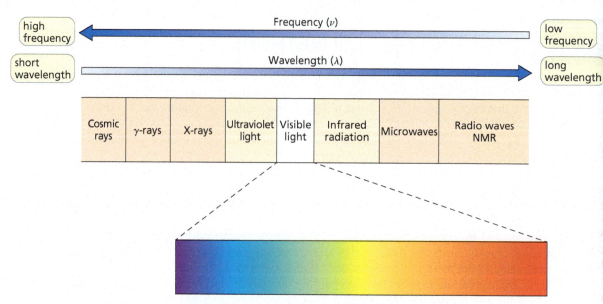

▲ Figure 13.8
The electromagnetic spectrum. Electromagnetic radiation with the highest energy (the highest frequency and the shortest wavelength) is located on the left. Electromagnetic radiation with the lowest energy (the lowest frequency and the longest wavelength) is located on the right.

The various kinds of electromagnetic radiation are characterized briefly as follows:

- *Cosmic rays* are discharged by the sun; they have the highest energy of the various kinds of electromagnetic radiation.

- *γ-rays* (gamma rays) are emitted by the nuclei of certain radioactive elements. Because of their high energy, they can severely damage biological organisms.

- *X-rays,* somewhat lower in energy than γ-rays, are less harmful, except in high doses. Low-dose X-rays are used to examine the internal structure of organisms. The denser the tissue, the more it blocks X-rays.

- *Ultraviolet (UV) light,* a component of sunlight, causes sunburns, and repeated exposure to it can cause skin cancer by damaging DNA molecules in skin cells (Section 28.6).

- *Visible light* is the electromagnetic radiation you see.

- You feel *infrared radiation* as heat.

- You cook with *microwaves* and use them in radar.

- *Radio waves* have the lowest energy of the various kinds of electromagnetic radiation. They are used for radio and television communication, digital imaging, remote control devices, and wireless linkages for computers. Radio waves are also used in NMR spectroscopy and in magnetic resonance imaging (MRI).

Each spectroscopic technique discussed in this book employs a different kind of electromagnetic radiation. In this chapter, we will look at infrared (IR) spectroscopy and ultraviolet/visible (UV/Vis) spectroscopy. In Chapter 14, we will see how compounds can be identified using nuclear magnetic resonance (NMR) spectroscopy.

Characterizing Electromagnetic Radiation

Because electromagnetic radiation has wave-like properties, it is characterized, as a wave is, by either its frequency (ν) or its wavelength (λ).

- **Frequency** is the number of wave crests that pass by a given point in one second; frequency has units of hertz (Hz).

- **Wavelength** is the distance from any point on one wave to the corresponding point on the next wave; wavelength is generally measured in nanometers; 1 nm $= 10^{-9}$ m.

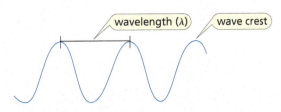

The relationship between the energy (E) and the frequency (ν) or wavelength (λ) of the electromagnetic radiation is described by the equation

LEARN THE STRATEGY

$$E = h\nu = \frac{hc}{\lambda}$$

where h is *Planck's constant* (a proportionality constant named after the German physicist who discovered the relationship,) and c is the speed of light. The equation shows that short wavelengths have high energies and high frequencies, and long wavelengths have low energies and low frequencies.

One way to describe the *frequency* of the electromagnetic radiation—and the one most often used in infrared spectroscopy—is **wavenumber** ($\tilde{\nu}$), which is the number of waves in 1 cm. Wavenumbers, therefore, have units of reciprocal centimeters (cm^{-1}).

> *High frequencies, large wavenumbers*, and *short wavelengths*
> are associated with *high energies*.

USE THE STRATEGY

PROBLEM 19 ◆

a. Which is higher in energy: electromagnetic radiation with wavenumber 100 cm^{-1} or with wavenumber 2000 cm^{-1}?

b. Which is higher in energy: electromagnetic radiation with wavelength 950 nm or with wavelength 850 nm?

PROBLEM 20 ◆

One of the following depicts the waves associated with infrared radiation, and one depicts the waves associated with visible light. Which is which?

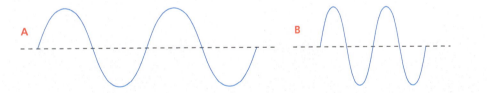

13.10 INFRARED SPECTROSCOPY

The length reported for a bond between two atoms is an average length, because in reality a bond behaves as if it were a vibrating spring.

Stretching and Bending Vibrations

A bond vibrates with both stretching and bending motions.

- A *stretch* is a vibration occurring along the line of the bond; a stretching vibration changes the bond length.

- A *bend* is a vibration that does *not* occur along the line of the bond; a bending vibration changes the bond angle.

A diatomic molecule such as H—Cl can undergo only a **stretching vibration** because it has no bond angles.

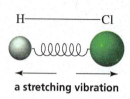

a stretching vibration

The vibrations of a molecule containing three or more atoms are more complex because they include stretches and bends (Figure 13.9). The stretching and bending vibrations, moreover, can be symmetric or asymmetric, and the bending vibrations can be either in-plane or out-of-plane. **Bending vibrations** are often referred to by the terms *rock, scissor, wag*, and *twist*.

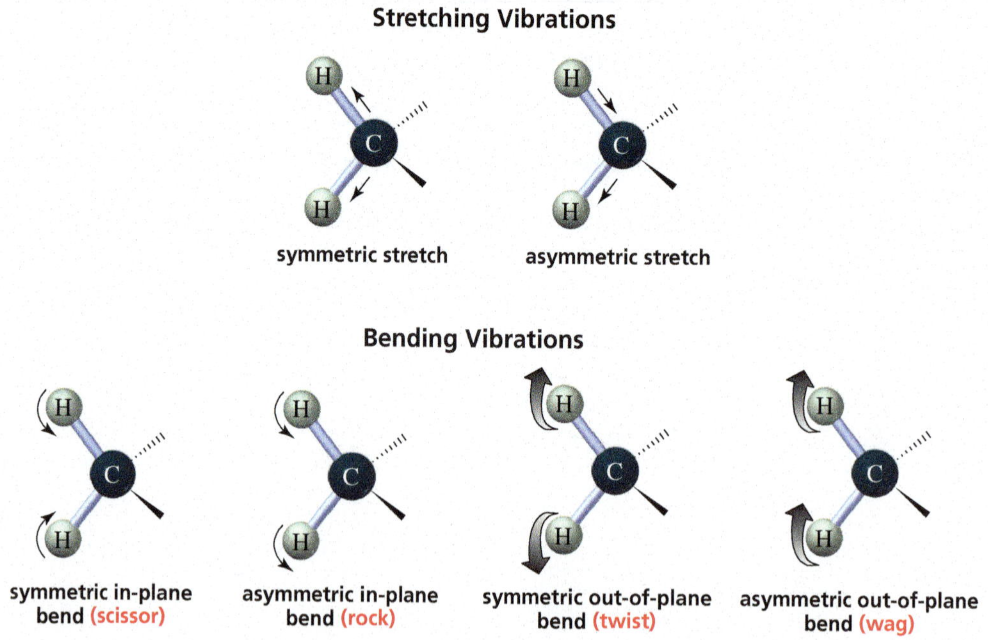

Stretching Vibrations

symmetric stretch   asymmetric stretch

Bending Vibrations

symmetric in-plane bend **(scissor)**   asymmetric in-plane bend **(rock)**   symmetric out-of-plane bend **(twist)**   asymmetric out-of-plane bend **(wag)**

▶ **Figure 13.9**
Stretching and bending vibrations of bonds in organic molecules containing three or more atoms.

Infrared radiation has just the right range of frequencies (4000 to 600 cm^{-1}) to correspond to the frequencies of the stretching and bending vibrations of the bonds in organic molecules. The range is just below (to the right of) the "red region" of visible light (Figure 13.11), meaning that infrared light has lower energy than visible light. (*Infra* is Latin for "below.")

Each stretching and bending vibration of a given bond occurs with a characteristic frequency. When a molecule is bombarded with radiation of a frequency that exactly matches the frequency of the vibration of one of its bonds, the molecule absorbs energy. This allows the bond to stretch and bend a bit more. By experimentally determining the wavenumbers of the energy absorbed by a particular compound, we can ascertain what kinds of bonds it has. For example, the stretching vibration of a C=O bond absorbs energy with wavenumber ~1700 cm^{-1}, whereas the stretching vibration of an O—H bond absorbs energy with wavenumber ~3400 cm^{-1} (Figure 13.10).

C=O   O—H

$\tilde{\nu} = {\sim}1700 \text{ cm}^{-1}$ $\tilde{\nu} = {\sim}3400 \text{ cm}^{-1}$

The Infrared Spectrum

An **infrared spectrum**, obtained by passing infrared radiation through a sample of a compound, is a plot of the percent transmission of radiation versus the wavenumber (or wavelength) of the radiation transmitted (Figure 13.10). The instrument used to obtain an infrared spectrum is called an IR spectrophotometer. At 100% transmission all the energy of the radiation (of a particular wavenumber) passes through the molecule. Lower values of percent transmission mean that some of the energy is being absorbed by the compound. Each downward spike in an IR spectrum represents absorption of energy. The spikes are called **absorption bands.** Most chemists report the location of absorption bands using wavenumbers.

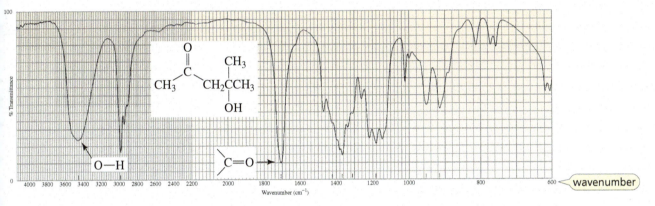

▲ **Figure 13.10**
An infrared spectrum shows the percent transmission of radiation versus the wavenumber of the radiation. The (C═O) stretch absorbs at 1705 cm⁻¹ and the (O─H) stretch absorbs at 3450 cm⁻¹.

A Fourier transform IR (FT-IR) spectrophotometer measures all frequencies simultaneously. This allows time for multiple measurements, which are averaged. The information is then digitized and Fourier transformed by a computer to produce an FT-IR spectrum. The infrared spectra shown in this text are FT-IR spectra.

The Functional Group and Fingerprint Regions

An infrared spectrum can be divided into two areas. The area on the left (4000−1400 cm⁻¹) is where most of the functional groups show absorption bands. This is called the **functional group region.**

The area on the right (1400−600 cm⁻¹) is called the **fingerprint region** because it is characteristic of the compound as a whole, just as a fingerprint is characteristic of an individual. Even if two different molecules have the same functional groups, their IR spectra will not be identical because the functional groups are not in exactly the same environment in both compounds. This difference is reflected in the patterns of the absorption bands in the fingerprint region.

For example, 2-pentanol and 3-pentanol have the same functional groups, so they show similar absorption bands in the functional group region. Their fingerprint regions are different, however, because the compounds are different (Figure 13.11). Thus, a compound can be positively identified by comparing its fingerprint region with the fingerprint region of the spectrum of a known sample of the compound.

a.

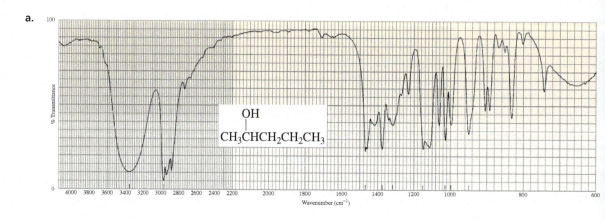

b.

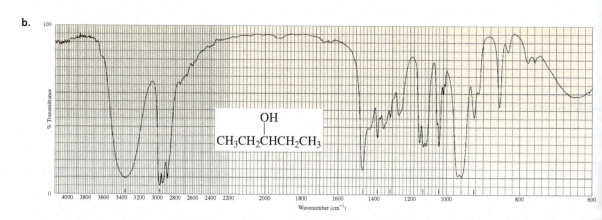

▲ **Figure 13.11**

The IR spectra of (a) 2-pentanol and (b) 3-pentanol. The functional group regions (4000–1400 cm⁻¹) are very similar because the two compounds have the same functional groups. The fingerprint region (1400–600 cm⁻¹) is unique for each compound.

IR spectra can be quite complex because the stretching and bending vibrations of each bond in a molecule can produce an absorption band. Organic chemists, however, do not try to identify all the absorption bands in an IR spectrum. They tend to focus on the functional groups. In this chapter, we will look at several characteristic absorption bands so you will be able to tell something about the structure of a compound that gives a particular IR spectrum.

When identifying an unknown compound, chemists often use IR spectroscopy in conjunction with information obtained from other spectroscopic techniques. Some of the problems in this chapter and many of those in Chapter 14 provide practice in using information from two or more analytical methods to identify compounds.

13.11 CHARACTERISTIC INFRARED ABSORPTION BANDS

It takes more energy to stretch a bond than to bend it, so stretching vibrations are found at higher wavenumbers than bending vibrations.

More energy is required to stretch a bond than to bend it, so absorption bands for stretching vibrations are found in the functional group region (4000−1400 cm⁻¹), whereas those for bending vibrations are typically found in the fingerprint region (1400−600 cm⁻¹). As a result, stretching vibrations are the ones most often used to determine what kinds of bonds a molecule has. The *frequencies of the stretching vibrations* associated with different types of bonds are listed in Table 13.4 and are discussed in Sections 13.12 and 13.13.

Table 13.4   Frequencies of Important IR Stretching Vibrations

Type of bond	Wavenumber (cm^{-1})	Intensity
C≡N	2260–2220	medium
C≡C	2260–2100	medium to weak
C=C	1680–1600	medium
C=N	1650–1550	medium
⬡	~1600 and ~1500–1430	strong to weak
C=O	1780–1650	strong
C—O	1250–1050	strong
C—N	1230–1020	medium
O—H (alcohol)	3650–3200	strong, broad
O—H (carboxylic acid)	3300–2500	strong, very broad
N—H	3500–3300	medium, broad
C—H	3300–2700	medium

NOTE TO THE STUDENT
• You can find an extensive table of characteristic functional group frequencies in Appendix V.

13.12   THE INTENSITY OF ABSORPTION BANDS

When a bond stretches, the increasing distance between the atoms increases its dipole moment. The intensity of the absorption band depends on the size of this change in dipole moment:

the greater the change in dipole moment, the more intense the absorption.

For example, absorption bands for the stretching vibrations of C=O and C=C bonds appear at similar frequencies, but they are easily distinguished: the one for C=O is much more intense because it is associated with a much greater change in dipole moment since the bond is more polar. (Compare the C=O absorption bands in Figures 13.12–13.15 with the C=C absorption band in Figure 13.19.)

The stretching vibration of an O—H bond is associated with a greater change in dipole moment than that of an N—H bond, because the O—H bond is more polar. Consequently, an O—H bond shows more intense absorption than an N—H bond. Similarly, an N—H bond shows more intense absorption than a C—H bond, because the N—H bond is more polar.

relative bond polarities
relative intensities of IR absorption

(most polar; most intense)   (least polar; least intense)

O—H   >   N—H   >   C—H

The intensity of an absorption band also depends on the number of bonds responsible for the absorption. For example, the absorption band for a C—H stretch is more intense for a compound such as octyl iodide, which has 17 C—H bonds, than for methyl iodide, which has only three C—H bonds.

The concentration of the sample used to obtain an IR spectrum also affects the intensity of the absorption bands. Concentrated samples have greater numbers of absorbing molecules and, therefore, more intense absorption bands. In the chemical literature, you will find intensities referred to as strong (s), medium (m), weak (w), broad, and sharp.

Recall that the dipole moment of a bond is equal to the *magnitude of the charge* on one of the bonded atoms multiplied by the *distance* between the two bonded atoms (Section 1.3).

The greater the change in the dipole moment, the more intense the absorption.

The more polar the bond, the more intense the absorption.

13.13 THE POSITION OF ABSORPTION BANDS

The frequency of a stretching vibration—the amount of energy required to stretch a bond—depends on the *strength* of the bond and the *masses* of the bonded atoms. The stronger the bond, the greater the energy required to stretch it. The frequency of the stretching vibration is also inversely related to the mass of the atoms joined by the bond; thus, heavier atoms vibrate at lower frequencies.

Hooke's Law

The approximate wavenumber of an absorption band can be calculated from the following equation derived from **Hooke's law,** which describes the motion of a vibrating spring:

$$\widetilde{\nu} = \frac{1}{2\pi c}\left[\frac{f(m_1 + m_2)}{m_1 m_2}\right]^{1/2}$$

Lighter atoms show absorption bands at larger wavenumbers.
C—H ~3000 cm^{-1}
C—D ~2200 cm^{-1}
C—O ~1100 cm^{-1}
C—Cl ~ 700 cm^{-1}

where $\widetilde{\nu}$ is the wavenumber of an absorption band, c is the speed of light, f is the force constant of the bond (a measure of the strength of the bond), and m_1 and m_2 are the masses of the atoms (in grams) joined by the bond. This equation shows that *stronger bonds* and *lighter atoms* give absorption bands with higher frequencies.

The Originator of Hooke's Law

Robert Hooke (1635–1703) was born on the Isle of Wight off the southern coast of England. A brilliant scientist, he contributed to almost every scientific field. He was the first to suggest that light had wave-like properties. He discovered that Gamma Arietis is a double star, and he discovered Jupiter's Great Red Spot. In a lecture published posthumously, he suggested that earthquakes are caused by the cooling and contracting of Earth. He examined cork under a microscope and coined the term *cell* to describe what he saw. He wrote about evolutionary development based on his studies of microscopic fossils, and he produced some highly regarded studies of insects. Hooke also invented the balance spring for watches and the universal joint currently used in automobiles.

Robert Hooke's drawing of a "blue fly" appeared in *Micrographia*, the first book on microscopy, published by Hooke in 1665.

The Effect of Bond Order

Bond order—whether a bond is a single bond, a double bond, or a triple bond—affects bond strength. Therefore, bond order affects the position of absorption bands (Table 13.4).

Stronger bonds show absorption bands at larger wavenumbers.
C≡N ~2200 cm^{-1}
C=N ~1600 cm^{-1}
C—N ~1100 cm^{-1}

A C≡C bond is stronger than a C=C bond, so a C≡C bond stretches at a higher frequency (~2100 cm^{-1}) than does a C=C bond (~1650 cm^{-1}); C—C bonds show stretching vibrations in the region from 1300 to 800 cm^{-1}, but because they are weak and very common, these vibrations are of little value in identifying organic compounds.

Similarly, a C=O bond stretches at a higher frequency (~1700 cm^{-1}) than does a C—O bond (~1100 cm^{-1}), and a C≡N bond stretches at a higher frequency (~2200 cm^{-1}) than does a C=N bond (~1600 cm^{-1}), which in turn stretches at a higher frequency than does a C—N bond (~1100 cm^{-1}).

PROBLEM 21 ◆

a. Which occurs at a larger wavenumber:
   1. a C≡C stretch or a C=C stretch?   **3.** a C—N stretch or a C=N stretch?
   2. a C—H stretch or a C—H bend?   **4.** a C=O stretch or a C—O stretch?

b. Assuming that the force constants are the same, which occurs at a larger wavenumber:
   1. a C—O stretch or a C—Cl stretch?   **2.** a C—O stretch or a C—C stretch?

13.14 THE POSITION AND SHAPE OF AN ABSORPTION BAND IS AFFECTED BY ELECTRON DELOCALIZATION AND HYDROGEN BONDING

Table 13.4 shows a range of wavenumbers for the frequency of the stretching vibration for each functional group because the exact position and shape of a group's absorption band depends on other structural features of the molecule, such as electron delocalization, the electronic effect of neighboring substituents, and hydrogen bonding. In fact, the exact position and shape of a compound's absorption bands reveals important details about its structure.

Electron Delocalization

The IR spectrum in Figure 13.12 shows that the carbonyl group (C=O) of 2-pentanone absorbs at 1720 cm^{-1}, whereas the IR spectrum in Figure 13.13 shows that the carbonyl group of 2-cyclohexenone absorbs at a lower frequency (1680 cm^{-1}). 2-Cyclohexenone's carbonyl group absorbs at a lower frequency because it has more single-bond character due to electron delocalization. A single bond is weaker than a double bond, so a carbonyl group with significant single-bond character stretches at a lower frequency than one with little or no single-bond character.

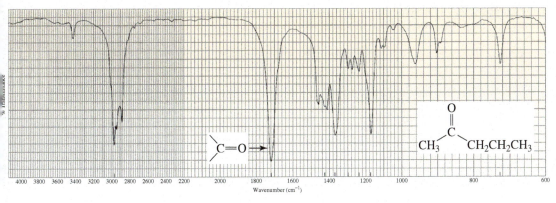

Figure 13.12
The intense absorption band at ~1720 cm^{-1} indicates a C=O bond.

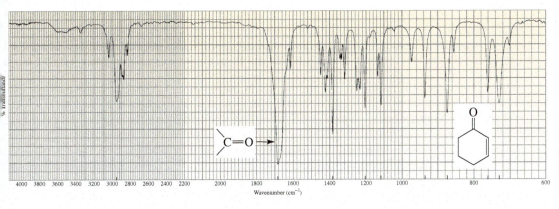

Figure 13.13
Electron delocalization gives the carbonyl group less double-bond character, so it absorbs at a lower frequency (~1680 cm^{-1}) than does a carbonyl group with localized electrons (~1720 cm^{-1}).

Putting an atom other than carbon next to the carbonyl group also causes the position of the carbonyl absorption band to shift. Whether it shifts to a lower or to a higher frequency depends on whether the predominant effect of the atom is to donate electrons by resonance or to withdraw electrons inductively.

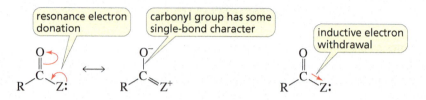

The predominant effect of the nitrogen of an amide is electron donation by resonance. In contrast, oxygen is less able than nitrogen to accommodate a positive charge because of oxygen's greater electronegativity, so the predominant effect of the oxygen of the OR group of an ester is inductive electron withdrawal (Sections 2.7 and 8.10). As a result, the carbonyl group of an ester has less single-bond character, so it requires more energy to stretch (1740 cm^{-1} in Figure 13.14) than the carbonyl group of an amide (1660 cm^{-1} in Figure 13.15).

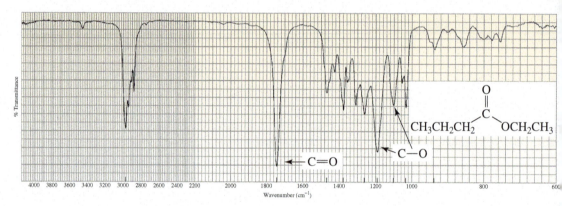

▲ **Figure 13.14**
The electron-withdrawing oxygen atom of the OR group makes the carbonyl group of an ester harder to stretch (~1740 cm^{-1}) than the carbonyl group of a ketone (~1720 cm^{-1}).

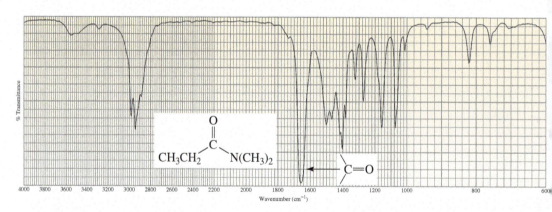

▲ **Figure 13.15**
The carbonyl group of an amide has less double-bond character than the carbonyl group of a ketone, so the carbonyl group of an amide stretches more easily (~1660 cm^{-1}) than the carbonyl group of a ketone (~1720 cm^{-1}).

When we compare the frequency for the stretching vibration of the carbonyl group of an ester (1740 cm^{-1} in Figure 13.14) with that of the carbonyl group of a ketone (1720 cm^{-1} in Figure 13.12), we see how important inductive electron withdrawal is to the position of the stretching vibration of the carbonyl group.

A C—O bond shows a stretching vibration between 1250 and 1050 cm^{-1}. If the C—O bond is in an alcohol or an ether, the stretch will lie toward the lower end of the range (Figure 13.16). If, however, the C—O bond is in a carboxylic acid, the stretch will lie at the higher end of the range (Figure 13.17).

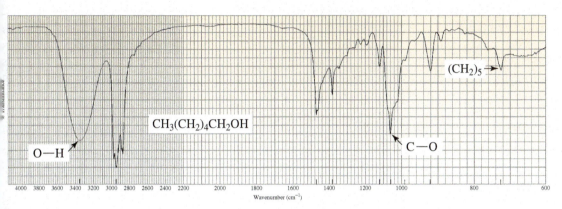

The position of the C—O absorption varies because the C—O bond in an *alcohol* or in an *ether* is a pure single bond, whereas the C—O bond in a *carboxylic acid* has partial double-bond character due to electron delocalization. *Esters* show C—O stretches at both ends of the range because esters have two C—O single bonds: one that is a pure single bond and one that has partial double-bond character (Figure 13.14).

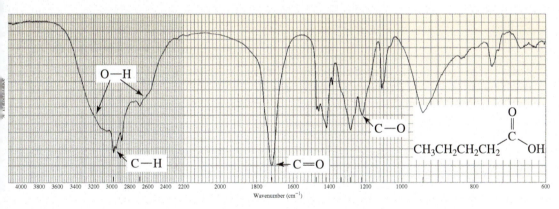

▲ Figure 13.16
The IR spectrum of 1-hexanol. The C—O absorption lies at the lower end of the range of C—O stretches (1250–1050 cm⁻¹) because the C—O bond in an alcohol is a pure single bond.

▲ Figure 13.17
The IR spectrum of pentanoic acid. The C—O absorption lies at the higher end of the range of C—O stretches (1250–1050 cm⁻¹) because the C—O bond in a carboxylic acid has some double-bond character.

PROBLEM-SOLVING STRATEGY

Differences in IR Spectra

LEARN THE STRATEGY

Which occurs at a larger wavenumber: the C—N stretch of an amine or the C—N stretch of an amide?

To answer this question, we need to determine what effect electron delocalization has on the C—N bond in amines and amides. When we do that we see that the C—N bond of the amine is a pure single bond, whereas electron delocalization causes the C—N bond of the amide to have partial double-bond character. The C—N stretch of an amide, therefore, occurs at a larger wavenumber.

USE THE STRATEGY

PROBLEM 22 ◆

Which occurs at a larger wavenumber:

a. the C—O stretch of phenol or the C—O stretch of cyclohexanol?
b. the C=O stretch of a ketone or the C=O stretch of an amide?
c. the C—N stretch of cyclohexylamine or the C—N stretch of aniline?

PROBLEM 23 ◆

Which shows an absorption band at a larger wavenumber: a carbonyl group bonded to an sp^3 carbon or carbonyl group bonded to an sp^2 carbon of an alkene?

PROBLEM 24 ◆

Why is the C—O absorption band of 1-hexanol at a smaller wavenumber (1060 cm^{-1}) than the C—absorption band of pentanoic acid (1220 cm^{-1})?

PROBLEM 25 ◆

Rank the following compounds from highest wavenumber to lowest wavenumber for their C=O absorptio bands:

a.

b.

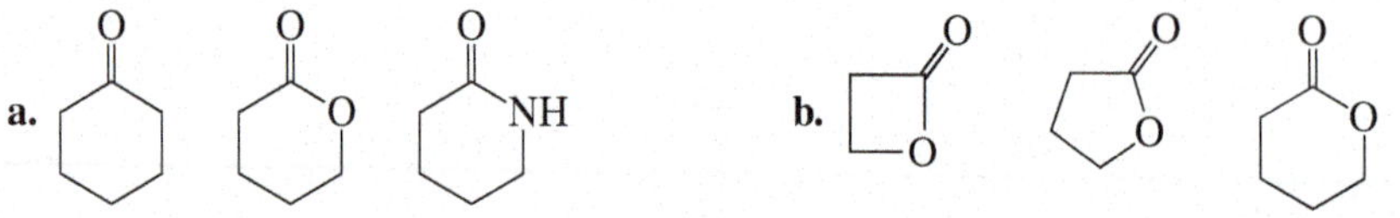

Hydrogen Bonding

Because O—H bonds are polar, they show intense absorption bands (Figures 13.16 and 13.17 Both the position and the shape of an O—H absorption band depend on hydrogen bonding. It i easier for an O—H bond to stretch if it is hydrogen bonded, because the hydrogen is attracted t the oxygen of a neighboring molecule. Hydrogen-bonded OH groups have broad absorption band because hydrogen bonds vary in strength, and bonds with different strengths absorb at differer frequencies.

Carboxylic acids exist as hydrogen-bonded dimers. The additional hydrogen bonding of carboxyli acids compared with the hydrogen bonding of alcohols causes the O—H stretch of a carboxyli acid to occur at a lower frequency and to be broader ($3300–2500 \text{ cm}^{-1}$) than the O—H stretch of a alcohol ($3550–3200 \text{ cm}^{-1}$).

hydrogen-bonded dimer

hydrogen bond

$3300–2500 \text{ cm}^{-1}$

$3550–3200 \text{ cm}^{-1}$

The position, intensity, and shape of an absorption band are helpful in identifying functional groups.

Because the position and breadth of an O—H stretch depend on hydrogen bonding, they als depend on the concentration of the solution. The more concentrated the solution, the more likely i is for the OH-containing molecules to form intermolecular hydrogen bonds. Therefore, the O—H stretch in a concentrated (hydrogen-bonded) solution of an alcohol occurs at $3550–3200 \text{ cm}^{-1}$ wherea the O—H stretch in a dilute solution (with little or no hydrogen bonding) occurs at $3650–3590 \text{ cm}^{-}$

R—O—H------O—R
concentrated solution
$3550–3200 \text{ cm}^{-1}$

R—O—H
dilute solution
$3650–3590 \text{ cm}^{-1}$

N—H bonds are less polar and form weaker hydrogen bonds than O—H bonds, so the absorptio band for an N—H stretch is less intense and narrower than that for an O—H stretch (Figure 13.22)

PROBLEM 26 ◆

Which shows an O—H stretch at a larger wavenumber: ethanol dissolved in carbon disulfide or an undilute sample of ethanol?

13.15 | C—H ABSORPTION BANDS

Important information about the identity of a compound is provided by the stretching and bending vibrations of its C—H bonds.

Stretching Vibrations

The strength of a C—H bond depends on the hybridization of the carbon—namely, the greater the *s* character of the carbon, the stronger is the bond that it forms (see Table 1.7 on page 42).

A C—H bond, therefore, is stronger when the carbon is *sp* hybridized than when it is sp^2 hybridized, which in turn is stronger than when the carbon is sp^3 hybridized. Because more energy is needed to stretch a stronger bond, the absorption band for a C—H stretch is at ~3300 cm^{-1} for an *sp* carbon, at ~3100 cm^{-1} for an sp^2 carbon, and at ~2900 cm^{-1} for an sp^3 carbon (Table 13.5).

Table 13.5   Carbon–Hydrogen Stretching Vibrations

	Wavenumber (cm^{-1})
C≡C—H	~3300
C=C—H	3100–3020
C—C—H	2960–2850
R—$\overset{\overset{O}{\|\|}}{C}$—H	~2820 and ~2720

A useful step in the analysis of an IR spectrum is to look at the absorption bands in the vicinity of 3000 cm^{-1}. The only absorption band in the vicinity of 3000 cm^{-1} in Figure 13.18 is slightly to the right of that value. This tells us that the compound has hydrogens bonded to sp^3 carbons, but none bonded to sp^2 or to *sp* carbons. Both Figures 13.19 and 13.20 show absorption bands slightly to the left and slightly to the right of 3000 cm^{-1}, indicating that the compounds that produced those spectra contain hydrogens bonded to both sp^2 and sp^3 carbons.

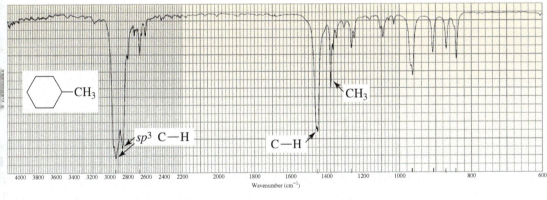

◀ **Figure 13.18**
The IR spectrum of methylcyclohexane. The absorptions at 2940 and 2860 cm^{-1} indicate that methylcyclohexane has hydrogens bonded to sp^3 carbons.

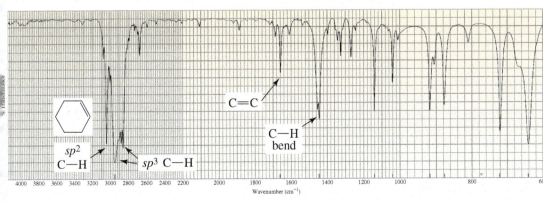

◀ **Figure 13.19**
The IR spectrum of cyclohexene. The absorptions at 3040, 2950, and 2860 cm^{-1} indicate that cyclohexene has hydrogens bonded to both sp^2 and sp^3 carbons.

Once we know that a compound has hydrogens bonded to sp^2 carbons, we need to determine whether those carbons are the sp^2 carbons of an alkene or of a benzene ring. A benzene ring is indicated by two sharp absorption bands, one at ~1600 cm^{-1} and one at 1500−1430 cm^{-1}, whereas an alkene is indicated by a band only at ~1600 cm^{-1} (Table 13.4). The compound whose spectrum is shown in Figure 13.19 is, therefore, an alkene, whereas the one whose spectrum is shown in Figure 13.20 has a benzene ring. (If you also have an NMR spectrum of the compound, the presence of a benzene ring is very easy to detect; see Section 14.12.)

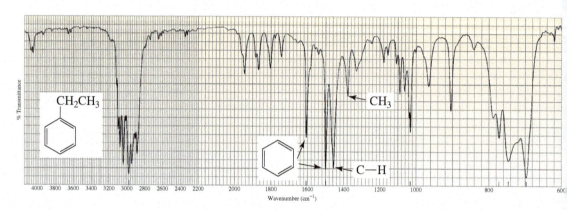

▲ **Figure 13.20**
The IR spectrum of ethylbenzene. The absorptions in the 3100–2880 cm^{-1} region indicate that ethylbenzene has hydrogens bonded to both sp^2 and sp^3 carbons. The two sharp absorptions at 1610 and 1500 cm^{-1} indicate that the sp^2 carbons are due to a benzene ring.

Be aware that N—H bending vibrations also occur at 1600 cm^{-1}, so absorption at that wavelength does not always indicate a C=C bond. However, absorption bands resulting from N—H bends tend to be broader (due to hydrogen bonding) and more intense (due to being more polar) than those caused by C=C stretches, and they are accompanied by N—H stretches at 3500−3300 cm^{-1} (Figure 13.22).

The stretch of the C—H bond of an aldehyde group shows two absorption bands—one at ~2820 cm^{-1} and the other at ~2720 cm^{-1} (Figure 13.21). This makes aldehydes relatively easy to identify because essentially no other absorption occurs at these wavenumbers.

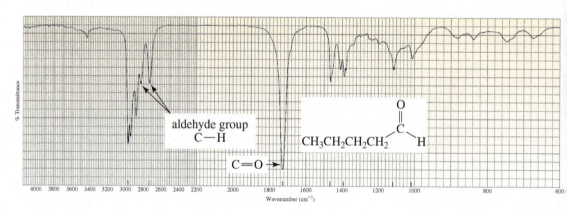

▲ **Figure 13.21**
The absorptions at ~2820 cm^{-1} and ~2720 cm^{-1} readily identify an aldehyde group. Note also the intense absorption band at ~1730 cm^{-1} indicating a C=O bond.

Bending Vibrations

If a compound has sp^3 carbons, a look at 1400 cm^{-1} will tell you whether the compound has a methyl group. All hydrogens bonded to sp^3 carbons show a C—H bending vibration slightly to the *left* of 1400 cm^{-1}. Methyl groups show an additional C—H bending vibration slightly to the *right* of 1400 cm^{-1}. So if a compound has a methyl group, absorption bands will appear *both* to the left and to the right of 1400 cm^{-1}; otherwise, only the band to the left of 1400 cm^{-1} will be present.

Table 13.6 Carbon–Hydrogen Bending Vibrations

	Wavenumber (cm^{-1})
CH$_3$— —CH$_2$— —CH—	1450–1420
CH$_3$—	1385–1365
terminal alkene (monosubstituted)	990 and 910
terminal alkene (disubstituted)	890
trans	980–960
cis	730–675
trisubstituted	840–800

You can see evidence for a methyl group in Figure 13.18 (methylcyclohexane) and in Figure 13.20 (ethylbenzene), but not in Figure 13.19 (cyclohexene). Two methyl groups attached to the same carbon can sometimes be detected by a split in the methyl peak at ~1380 cm^{-1} (Figure 13.22).

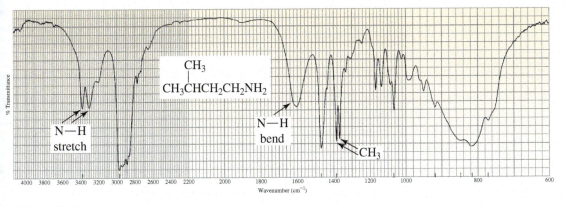

CH$_3$
|
CH$_3$CHCH$_2$CH$_2$NH$_2$

▲ **Figure 13.22**
The IR spectrum of isopentylamine. The two-peak absorption band at ~1380 cm^{-1} indicates the presence of an isopropyl group. The N—H bend around 1600 cm^{-1} is broad due to intermolecular hydrogen bonding.

The C—H bending vibrations for hydrogens bonded to sp^2 carbons show absorption bands in the 1000−600 cm^{-1} region. As Table 13.5 shows, the frequency of the C—H bending vibration of an alkene depends on the number of alkyl groups attached to the sp^2 carbons and on the configuration of the alkene. It is important to realize that these absorption bands can be shifted out of the characteristic regions if strongly electron-withdrawing or electron-donating substituents are close to the double bond. Acyclic compounds with more than four adjacent methylene (CH$_2$) groups show a characteristic absorption band at 720 cm^{-1} that results from in-phase rocking of the methylene groups (Figure 13.16).

13.16 THE ABSENCE OF ABSORPTION BANDS

The absence of an absorption band can be as useful as the presence of one in identifying a compound by IR spectroscopy.

For example, the spectrum in Figure 13.23 shows a strong absorption at ~1100 cm^{-1}, indicating the presence of a C—O bond. Clearly, the compound is not an alcohol because there is no absorption above 3100 cm^{-1}. Nor is it a carbonyl compound because there is no absorption at ~1700 cm^{-1}. The compound has no C≡C, C=C, C≡N, C=N, or C—N bonds. We may deduce, then, that the compound is an ether. Its C—H absorption bands show that it has hydrogens bonded only to sp^3 carbons (2950 cm^{-1}) and that it has a methyl group (1385 cm^{-1}). We also know that the compound has fewer than four adjacent methylene groups, because there is no absorption at ~720 cm^{-1}. The compound is diethyl ether.

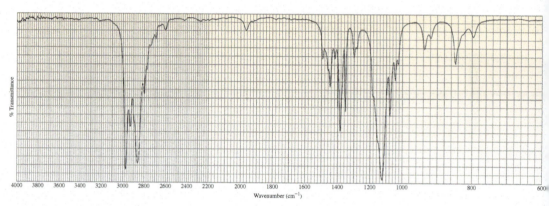

▲ **Figure 13.23**
The IR spectrum of diethyl ether.

PROBLEM 27 ◆

How do you know that the absorption band at ~1100 cm^{-1} in Figure 13.23 is due to a C—O bond and not to a C—N bond?

PROBLEM 28 ◆

a. An oxygen-containing compound shows an absorption band at ~1700 cm^{-1} and no absorption bands at ~3300 cm^{-1}, ~2700 cm^{-1}, or ~1100 cm^{-1}. What class of compound is it?

b. A nitrogen-containing compound shows no absorption band at ~3400 cm^{-1} and no absorption bands between ~1700 cm^{-1} and ~1600 cm^{-1} or between 2260 cm^{-1} and 2220 cm^{-1}. What class of compound is it?

PROBLEM 29

How can IR spectroscopy be used to distinguish between the following compounds?

a. a ketone and an aldehyde
b. a cyclic ketone and an open-chain ketone
c. benzene and cyclohexene

d. *cis*-2-hexene and *trans*-2-hexene
e. cyclohexene and cyclohexane
f. a primary amine and a tertiary amine

PROBLEM 30

For each of the following pairs of compounds, name one absorption band that can be used to distinguish between them.

a. CH₃CH₂CH₂CH₃ and CH₃CH₂OCH₃

b. (structure) and (structure)

c. (cyclohexane) and (methylcyclohexane)

d. CH₃CH₂C≡CCH₃ and CH₃CH₂C≡CH

e. (structure) and (structure)

f. (benzene) and (toluene)

13.17 SOME VIBRATIONS ARE INFRARED INACTIVE

In order for the vibration of a bond to absorb IR radiation, the dipole moment of the bond must change when it vibrates. Therefore, not all bond vibrations give rise to an absorption band.

For example, the C=C bond in 1-butene has a dipole moment because the molecule is not symmetrical. So when the bond stretches, the dipole moment changes and an absorption band is observed.

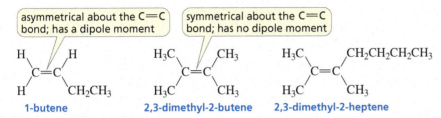

| asymmetrical about the C=C bond; has a dipole moment | symmetrical about the C=C bond; has no dipole moment |

1-butene 2,3-dimethyl-2-butene 2,3-dimethyl-2-heptene

In contrast, 2,3-dimethyl-2-butene is symmetrical, so its C=C bond has no dipole moment. When the bond stretches, it still has no dipole moment. Because stretching is not accompanied by a change in dipole moment, no absorption band is observed. The vibration is said to be *infrared inactive.*

2,3-Dimethyl-2-heptene experiences a very small change in dipole moment when its C=C bond stretches, so only an extremely weak absorption band (if any) is detected for the stretching vibration of the bond.

PROBLEM 31 ◆

Which of the following compounds has a vibration that is infrared inactive?
1-butyne, 2-butyne, H_2, H_2O, Cl_2, and ethene

PROBLEM 32 ◆

Identify the compound that gives the mass spectrum and infrared spectrum shown here.

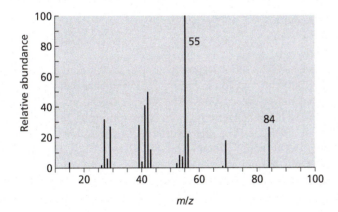

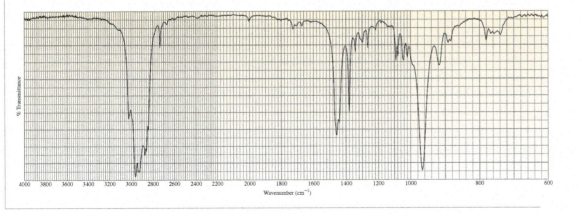

13.18   HOW TO INTERPRET AN INFRARED SPECTRUM

We will now look at some IR spectra to see what we can deduce about the structures of the compounds that give the spectra. We might not know how to identify the compound precisely, but when we are told what it is, its structure should fit our observations.

Compound 1

The absorptions in the 3000 cm^{-1} region in Figure 13.24 indicate that hydrogens are attached both to sp^2 carbons (3075 cm^{-1}) and to sp^3 carbons (2950 cm^{-1}). Now we need to determine whether the sp^2 carbons belong to an alkene or to a benzene ring. The absorption at 1650 cm^{-1} and the absorption at ~890 cm^{-1} (Table 13.5) suggest that the compound is a terminal alkene with two alkyl substituents at the 2-position. The absence of absorption at ~720 cm^{-1} indicates that the compound has fewer than four adjacent methylene groups. We are not surprised to find that the compound is 2-methyl-1-pentene.

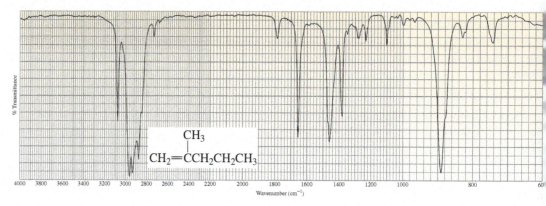

▲ **Figure 13.24**
The IR spectrum of Compound 1.

Compound 2

The absorptions in the 3000 cm^{-1} region in Figure 13.25 indicate that hydrogens are attached to sp^2 carbons (3050 cm^{-1}) but not to sp^3 carbons. The sharp absorptions at 1600 cm^{-1} and 1460 cm^{-1} indicate that the compound has a benzene ring. The absorptions at 2810 cm^{-1} and 2730 cm^{-1} show that the compound is an aldehyde. The characteristically strong absorption band for the carbonyl group (C=O) is lower (~1700 cm^{-1}) than normal (1720 cm^{-1}), so the carbonyl group has partial single-bond character. Thus, it must be attached directly to the benzene ring, so electron delocalization from the ring can reduce its double bond character. The compound is benzaldehyde.

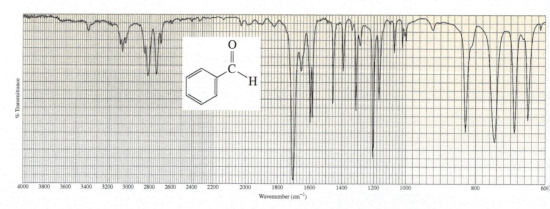

▲ **Figure 13.25**
The IR spectrum of Compound 2.

Compound 3

The absorptions in the 3000 cm^{-1} region in Figure 13.26 indicate that hydrogens are attached to sp^3 carbons (2950 cm^{-1}) but not to sp^2 carbons. The shape of the strong absorption band at 3300 cm^{-1} is characteristic of an O—H group of an alcohol. The absorption at 2100 cm^{-1} indicates that the compound has a triple bond. The sharp absorption band at 3300 cm^{-1} indicates that the compound has a hydrogen attached to an sp carbon, so we know it is a terminal alkyne. The structure of the compound is shown on the spectrum.

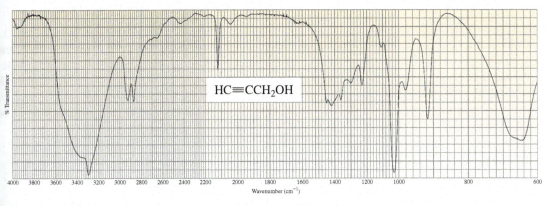

▲ **Figure 13.26**
The IR spectrum of Compound 3.

Compound 4

The absorptions in the 3000 cm^{-1} region in Figure 13.27 indicate that hydrogens are attached to sp^3 carbons (2950 cm^{-1}). The relatively strong absorption band at 3300 cm^{-1} suggests that the compound has an N—H bond. The presence of the N—H bond is confirmed by the absorption band at 1560 cm^{-1}. The C=O absorption at 1660 cm^{-1} indicates that the compound is an amide. The structure of the compound is shown on the spectrum.

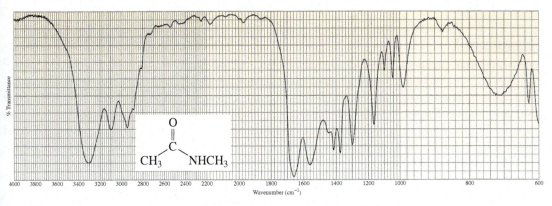

▲ **Figure 13.27**
The IR spectrum of Compound 4.

Compound 5

The absorptions in the 3000 cm^{-1} region in Figure 13.28 indicate that the compound has hydrogens attached to sp^2 carbons (>3000 cm^{-1}) and to sp^3 carbons (<3000 cm^{-1}). The absorptions at 1605 cm^{-1} and 1500 cm^{-1} indicate that the compound contains a benzene ring. The absorption at 1720 cm^{-1} for the carbonyl group indicates that the compound is a ketone and that the carbonyl group is not directly attached to the benzene ring. The absorption at ~1380 cm^{-1} indicates a methyl group. The structure of the compound is shown on the spectrum.

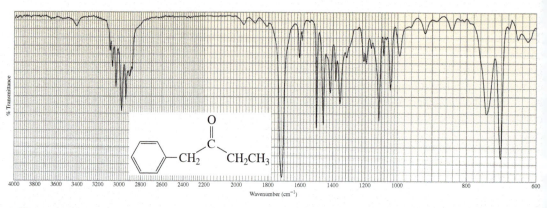

▲ **Figure 13.28**
The IR spectrum of Compound 5.

PROBLEM 33 ◆

A compound with molecular formula C_4H_6O gives the infrared spectrum shown here. Identify the compound.

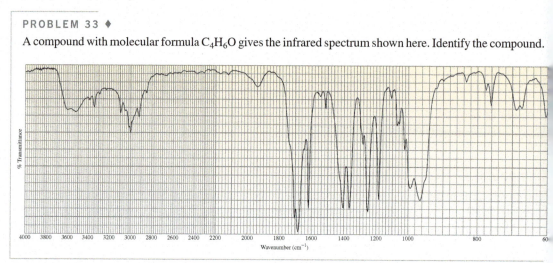

13.19 ULTRAVIOLET AND VISIBLE SPECTROSCOPY

Ultraviolet and visible (UV/Vis) spectroscopy provides information about compounds that have conjugated double bonds. Ultraviolet light and visible light have just the right energy to cause an electronic transition in a molecule—that is, to promote an electron from one molecular orbital to another of higher energy.

Depending on the energy needed for the electronic transition, a molecule absorbs either ultraviolet or visible light. If it absorbs **ultraviolet light,** a UV spectrum is obtained; if it absorbs **visible light,** a visible spectrum is obtained.

- Ultraviolet light has wavelengths ranging from 180 to 400 nm (nanometers).
- Visible light has wavelengths ranging from 400 to 780 nm.

Wavelength (λ) is inversely related to the energy of the radiation, so the shorter the wavelength, the greater the energy of the radiation. Ultraviolet light, therefore, has greater energy than visible light.

UV/Vis Light Causes an Electronic Transition

$$E = \frac{hc}{\lambda}$$

The shorter the wavelength, the greater the energy of the radiation.

In the ground-state electronic configuration of a molecule, all the electrons are in the lowest-energy molecular orbitals (Section 1.2). When a molecule absorbs light with sufficient energy (ΔE in Figure 13.29) to promote an electron to a higher-energy molecular orbital—that is, when it undergoes an **electronic transition**—the molecule is then in an excited state.

Ultraviolet and visible light has just enough energy to promote an electron from a π bonding MO to a π^* antibonding MO. (Figure 13.29). Therefore, *only compounds with π bonds can produce UV/Vis spectra.*

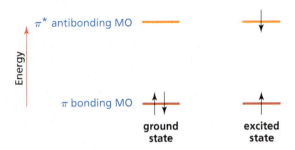

◀ **Figure 13.29**
Ultraviolet and visible light promote an electron from a π bonding MO to a π^* antibonding MO. The electron's spin does not change.

The UV spectrum of methyl vinyl ketone is shown in Figure 13.30. (The absorption below 205 nm is due to the absorption of UV light by oxygen.) The λ_{max} (stated as "lambda max") is the wavelength at which the absorption band for the $\pi \rightarrow \pi^*$ transition (stated as "π to π star") has its maximum absorbance. For methyl vinyl ketone, λ_{max} = 219 nm.

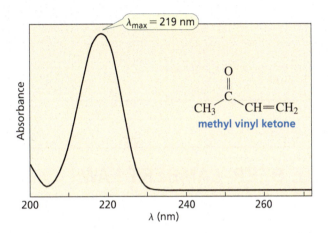

Only compounds with π electrons can produce ultraviolet and visible spectra.

◀ **Figure 13.30**
UV spectrum of methyl vinyl ketone

The absorption band is broad because each electronic state has vibrational sublevels (Figure 13.31), and electronic transitions can occur from and to these different sublevels. As a result, an electronic transition spans a range of wavelengths.

A **chromophore** is the part of a molecule that absorbs UV or visible light. The following four compounds all have the same chromophore, so they all have approximately the same λ_{max}.

▲ **Figure 13.31**
UV/Vis absorption bands are broad because each electronic state has vibrational sublevels.

  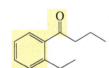 

Ultraviolet Light and Sunscreens

Exposure to ultraviolet (UV) light stimulates specialized cells in the skin to produce a black pigment known as melanin, which causes the skin to look tan (Section 24.8). Melanin absorbs UV light, so it protects our bodies from the harmful effects of the sun. If more UV light reaches the skin than melanin can absorb, the light will "burn" the skin and cause photochemical reactions that can result in skin cancer (Section 28.7).

UV-A is the lowest-energy UV light (315 to 400 nm). It is the light that causes skin to wrinkle. Much of the more dangerous, higher-energy light—namely, UV-B (290 to 315 nm) and UV-C (180 to 290 nm)—is filtered out by the ozone layer in the stratosphere, which is why the thinning of the ozone layer became such an important issue (Section 12.12).

Applying a sunscreen can protect skin from UV light. The amount of protection from UV-B light (the light that causes skin to burn) is indicated by the sunscreen's SPF (sun protection factor); the higher the SPF, the greater the protection. Some sunscreens contain an inorganic component, such as zinc oxide, which reflects the light as it reaches the skin. Others contain a compound that absorbs UV light.

(Continued)

para-Aminobenzoic acid (PABA) was the first commercially available UV-absorbing sunscreen. It absorbs UV-B light, but is not very soluble in oily skin lotions. Thus, the next generation of sunscreens contained Padimate O, a less polar compound. Subsequent research showed that sunscreens need to absorb both UV-B and UV-A light in order to provide adequate protection against skin cancer. Now the FDA requires that sunscreens, such as Give-Tan F, protect against both UV-A and UV-B light.

para-aminobenzoic acid
PABA

2-ethylhexyl 4-(dimethylamino)benzoate
Padimate O

2-ethoxyethyl (*E*)-3-(4-methoxyphenyl)-2-propenoate
Give-Tan F

13.20 THE BEER–LAMBERT LAW

Wilhelm Beer and Johann Lambert independently proposed that the absorbance of a sample at a given wavelength depends on the amount of absorbing species that the light encounters as it passes through a solution of the sample. In other words, absorbance depends on both the concentration of the sample and the length of the light path through the sample. The relationship between absorbance, concentration, length of the light path, and molar absorptivity, known as the **Beer–Lambert law,** is given by

$$A = c\,l\,\varepsilon, \text{ where}$$

A = absorbance of the sample

c = concentration of the sample, in moles/liter

l = length of the light path through the sample, in centimeters

ε = molar absorptivity ($M^{-1}cm^{-1}$)

The **molar absorptivity** (ε) is a constant that is characteristic of the compound at a particular wavelength. It is the absorbance that would be observed for a 1.00 M solution in a cell with a 1.00 cm path length. (The abbreviation ε comes from the fact that molar absorptivity was formerly called the extinction coefficient.)

For example, the molar absorptivity of methyl vinyl ketone dissolved in hexane is $14{,}000\,M^{-1}\,cm^{-1}$ at 219 nm. The solvent in which the sample is dissolved is reported because molar absorptivity is not exactly the same in all solvents. Therefore, the UV spectrum of methyl vinyl ketone in hexane would be reported as λ_{max} = 219 nm (ε_{max} = 14,000, hexane).

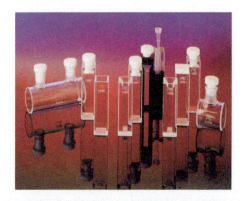

The solution whose UV or visible spectrum is to be taken is put into a cell, such as one of those shown here. Most cells have 1 cm path lengths. Either glass or quartz cells can be used for visible spectra, but quartz cells (made of high-purity fused silica) must be used for UV spectra because glass absorbs UV light.

PROBLEM 34 ◆

A solution of a compound in ethanol shows an absorbance of 0.52 at 236 nm in a cell with a 1 cm light path. Its molar absorptivity in ethanol at that wavelength is $12{,}600\,M^{-1}\,cm^{-1}$. What is the concentration of the compound?

PROBLEM 35 ◆

A 4.0×10^{-5} M solution of a compound in hexane shows an absorbance of 0.40 at 252 nm in a cell with a 1 cm light path. What is the molar absorptivity of the compound in hexane at 252 nm?

13.21 THE EFFECT OF CONJUGATION ON λ_{max}

The wavelength at which the $\pi \rightarrow \pi*$ transition occurs increases as the number of conjugated double bonds in the compound increases. (Recall that conjugated double bonds are separated by one single bond; Section 8.7.) For example, the λ_{max} of 3,5-hexadien-2-one is at a longer wavelength (249 nm) than the λ_{max} of methyl vinyl ketone (219 nm), because 3,5-hexadien-2-one has three conjugated double bonds, whereas methyl vinyl ketone has two conjugated double bonds.

methyl vinyl ketone
$\lambda_{max} = 219$ nm

3,5-hexadien-2-one
$\lambda_{max} = 249$ nm

The λ_{max} values of the $\pi \rightarrow \pi*$ transition for several conjugated polyenes are listed in Table 13.7. Thus, the λ_{max} of a compound can be used to estimate the number of conjugated double bonds in a compound. Notice that both the λ_{max} and the molar absorptivity increase as the number of conjugated double bonds increases.

Table 13.7 Values of λ_{max} and ε for Conjugated Polyenes

Compound	λ_{max} (nm)	ε (M^{-1} cm^{-1})
	217	21,000
	256	50,000
	290	85,000
	334	125,000
	364	138,000

The λ_{max} increases as the number of conjugated double bonds increases.

In a $\pi \rightarrow \pi*$ transition, an electron is promoted from the **HOMO (highest occupied molecular orbital)** to the **LUMO (lowest unoccupied molecular orbital)**. Conjugation raises the energy of the HOMO and lowers the energy of the LUMO (Figure 8.11 on page 337). Therefore, the more conjugated double bonds in a compound, the less energy required for the electronic transition, so the longer the wavelength at which it occurs (Figure 13.33).

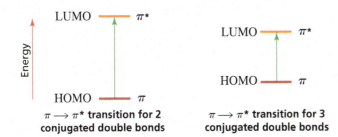

$\pi \rightarrow \pi*$ transition for 2
conjugated double bonds

$\pi \rightarrow \pi*$ transition for 3
conjugated double bonds

◀ **Figure 13.32**
Conjugation raises the energy of the HOMO and lowers the energy of the LUMO.

If a compound has enough conjugated double bonds, it will absorb visible light (light with wavelengths > 400 nm) and the compound will be colored. For example, β-carotene, a precursor of vitamin A with a $\lambda_{max} = 455$ nm, is an orange substance found in carrots, apricots, and the feathers of flamingos. Lycopene with a $\lambda_{max} = 474$ nm—found in tomatoes, watermelon, and pink grapefruit—is red.

Watermelon contains lycopene.

β-**carotene**
λ_{max} = **455 nm**

lycopene
λ_{max} = **474 nm**

An **auxochrome** is a substituent that, when attached to a chromophore, alters both the λ_{max} and the intensity of the absorption, usually increasing both. For example, OH and NH_2 groups are auxochromes. The lone-pair electrons on oxygen and nitrogen in the compounds shown below are available to interact with the π electron cloud of the benzene ring by electron delocalization; such an interaction increases the λ_{max}.

R is CH_3 in chlorophyll *a*

R is CH in chlorophyll *b*

benzene	phenol	phenolate ion	aniline	anilinium ion
λ_{max} = **255 nm**	**270 nm**	**287 nm**	**280 nm**	**254 nm**

Removing a proton from phenol increases the λ_{max} because the phenolate ion has an additional lone pair. Protonating aniline decreases the λ_{max} because the lone pair is no longer available to interact with the π cloud of the benzene ring. Because the anilinium ion does not have an auxochrome, its λ_{max} is similar to that of benzene.

PROBLEM 36 ◆

Predict the λ_{max} of the following compound:

PROBLEM 37 ◆

Rank each set of compounds in order of decreasing λ_{max}:

a.

b.

Chlorophyll *a* and *b* are highly conjugated compounds that absorb visible light, causing green light to be reflected from the surface tissues of plants.

13.22 THE VISIBLE SPECTRUM AND COLOR

White light is a mixture of all visible wavelengths. If any of these wavelengths are removed from white light, the eye registers the remaining light as colored. Therefore, any compound that absorbs visible light appears colored. The wavelengths that the compound does *not* absorb are reflected back to the viewer, producing the color the viewer sees.

The relationship between the wavelengths of the light that a substance absorbs and the substance's observed color is shown in the color wheel on the next page, where complementary colors are connected by a line.

When a molecule absorbs light, all the non-absorbed (reflected) light combines to produce the complement of the absorbed color. Therefore,

- when a substance absorbs red light, it appears to be green (red's complementary color); when a substance absorbs violet light, it appears to be yellow.

- when a substance absorbs two different colors of light, then your eyes perceive the combination of all the reflected colors. For example, when it absorbs red and violet light, it appears to be yellow-green.

- when a substance absorbs two complementary colors (for example, red and green), then the colors cancel and the substance appears to be gray or white.

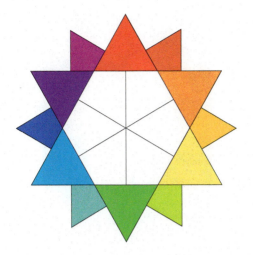

The human eye is able to distinguish more than a million different shades of color!

Azobenzenes (benzene rings connected by an N=N bond) have an extended conjugated system that causes them to absorb visible light. The two shown here are used commercially as dyes. Changing the number of conjugated double bonds and the substituents attached to them creates a large number of different colors. Notice that the only difference between butter yellow and methyl orange is an SO_3^- group.

butter yellow
an azobenzene

methyl orange
an azobenzene

When margarine was first produced, it was colored with butter yellow to make it look more like butter. (White margarine would not be very appetizing.) This dye was abandoned after it was found to be carcinogenic. β-Carotene (pages 605–606) is now used to color margarine.

What Makes Blueberries Blue and Strawberries Red?

A class of highly conjugated compounds called *anthocyanins* is responsible for the red, purple, and blue colors of many flowers (poppies, peonies, cornflowers), fruits (cranberries, rhubarb, strawberries, blueberries, the red skin of apples, the purple skin of grapes), and vegetables (beets, radishes, red cabbage).

In a neutral or basic solution, the monocyclic fragment (on the right-hand side of the anthocyanin) is not conjugated with the rest of the molecule, so the anthocyanin does not absorb visible light and is, therefore, a colorless compound. In an acidic environment, however, the OH group becomes protonated and water is eliminated. (Recall that water, being a weak base, is a good leaving group; see Section 10.1.) Loss of water results in the third ring becoming conjugated with the rest of the molecule.

monocyclic fragment

(conjugation is disrupted)
colorless

(conjugation is disrupted)
colorless

anthocyanin
(three rings are conjugated)
red, blue, or purple

R = H, OH, or OCH$_3$
R' = H, OH, or OCH$_3$

As a result of the increase in conjugation, the anthocyanin absorbs visible light with wavelengths between 480 and 550 nm. The exact wavelength of light absorbed depends on the substituents (R and R') on the anthocyanin. Thus, the flower, fruit, or vegetable appears red, purple, or blue, depending on what R and R' are. You can see this color change if you alter the pH of cranberry juice so that it is no longer acidic.

Lycopene, β-carotene, and anthocyanins are found in the leaves of trees, but their characteristic colors are usually obscured by the green color of chlorophyll. Chlorophyll is an unstable molecule, so plants must continually synthesize it. Its synthesis requires sunlight and warm temperatures. As the weather becomes colder in the fall, plants can no longer replace chlorophyll as it degrades, so the other colors become apparent.

13.23 SOME USES OF UV/VIS SPECTROSCOPY

UV/Vis spectroscopy is not nearly as useful as other instrumental techniques for determining the structures of organic compounds. However, UV/Vis spectroscopy has many other important uses.

UV/Vis spectroscopy is often used to measure reaction rates. The rate of any reaction can be measured, as long as one of the reactants or one of the products absorbs UV or visible light at a wavelength at which the other reactants and products have little or no absorbance.

For example, the anion of nitroethane has a λ_{max} at 240 nm, but neither the other product (H_2O) nor the reactants show any significant absorbance at this wavelength. To measure the rate at which hydroxide ion removes a proton from nitroethane to form the nitroethane anion, the UV spectrophotometer is adjusted to measure absorbance at 240 nm as a function of time (Figure 13.33) instead of absorbance as a function of wavelength (Figure 13.30).

$$CH_3CH_2NO_2 \;+\; HO^- \;\rightleftharpoons\; CH_3\overset{-}{C}HNO_2 \;+\; H_2O$$

nitroethane

nitroethane anion
$\lambda_{max} = 240$ nm

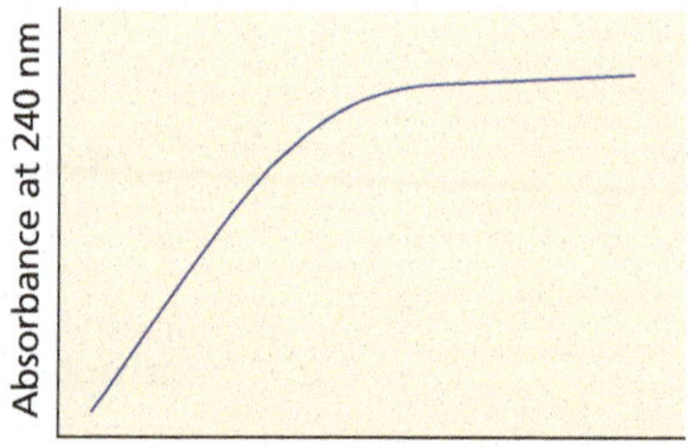

▲ **Figure 13.33**
The rate at which a proton is removed from nitroethane is determined by monitoring the increase in absorbance at 240 nm.

The enzyme lactate dehydrogenase catalyzes the reduction of pyruvate to lactate by NADH (Section 25.8). NADH is the only species in the reaction mixture that absorbs light at 340 nm, so the rate of the reaction can be determined by monitoring the decrease in absorbance at 340 nm (Figure 13.34).

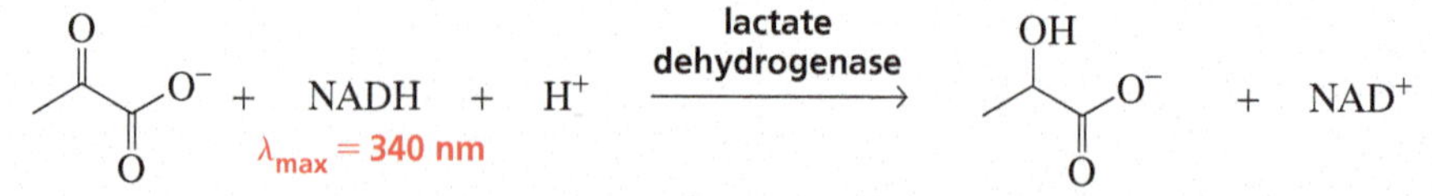

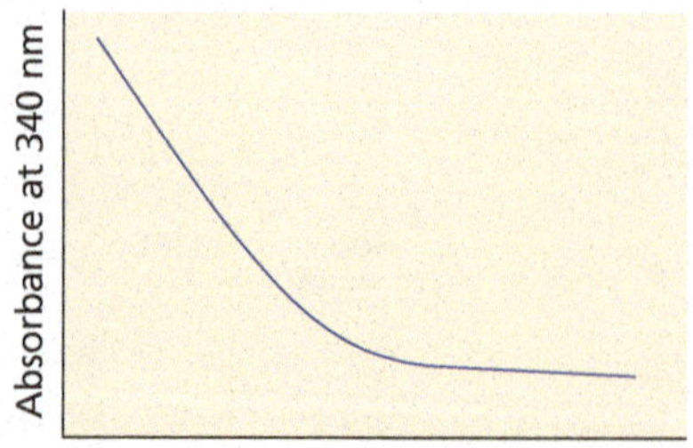

$$\text{pyruvate} + NADH + H^+ \xrightarrow{\text{lactate dehydrogenase}} \text{lactate} + NAD^+$$

$\lambda_{max} = 340$ nm

pyruvate

lactate

▲ **Figure 13.34**
The rate of reduction of pyruvate by NADH is measured by monitoring the decrease in absorbance at 340 nm.

Recall that the Henderson–Hasselbalch equation states that the pK_a of a compound is the pH at which half the compound exists in its acidic form and half exists in its basic form (Section 2.10).

The pK_a of a compound can be determined by UV/Vis spectroscopy if either the acidic form or the basic form of the compound absorbs UV or visible light. For example, the phenolate ion has a λ_{max} at 287 nm. If the absorbance at 287 nm is monitored as a function of pH, the pK_a of phenol can be ascertained by determining the pH at which exactly one-half the increase in absorbance has occurred (Figure 13.35). At this pH, half of the phenol has been converted into phenolate ion, so this pH is equal to the pK_a of the compound.

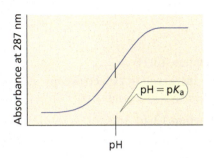

◀ **Figure 13.35**
**The absorbance of an aqueous solution
of phenol as a function of pH.**

UV spectroscopy can also be used to estimate the nucleotide composition of DNA. The two strands of DNA are held together by hydrogen bonds between the bases in one strand and the bases in the other (see Figure 26.3 on page 1161): each guanine forms three hydrogen bonds with a cytosine (a G–C pair); each adenine forms two hydrogen bonds with a thymine (an A–T pair). When DNA is heated, the strands break apart and the absorbance increases because single-stranded DNA has a greater molar absorptivity at 260 nm than does double-stranded DNA. The melting temperature (T_m) of DNA is the midpoint of an absorbance-versus-temperature curve (Figure 13.36).

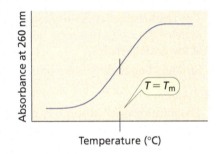

◀ **Figure 13.36**
**The absorbance of a solution of DNA
as a function of temperature.**

The T_m increases with increasing numbers of G–C pairs, because they are held together by three hydrogen bonds, whereas A–T pairs are held together by only two hydrogen bonds. Therefore, T_m can be used to estimate the fraction of G–C pairs. These are just a few examples of the many uses of UV/Vis spectroscopy.

PROBLEM 38 ◆

a. At pH = 7, one of the ions shown here is purple and the other is blue. Which is which? (*Hint:* refer to the color spectrum in Figure 13.8 on page 584.)
b. What would be the difference in the colors of the compounds at pH = 3?

$(CH_3)_2N$ — ... — $\overset{+}{N}(CH_3)_2$ $(CH_3)_2N$ — ... — $\overset{+}{N}(CH_3)_2$

$N(CH_3)_2$

PROBLEM 39 ◆

Describe a way to determine the rate of the alcohol dehydrogenase catalyzed oxidation of ethanol by NAD^+.

PROBLEM 40 ◆

The absorbance of a solution of a weak acid was measured under the same conditions at a series of pH values. Its conjugate base is the only species in the solution that absorbs UV light at the wavelength used. Estimate the pK_a of the acid from the data obtained.

pH	1.0	2.0	3.0	4.0	5.0	6.0	7.0	8.0	9.0	10.0
Absorbance	0	0	0.10	0.50	0.80	1.10	1.50	1.60	1.60	1.60

ESSENTIAL CONCEPTS

Section 13.1

- **Mass spectrometry** allows us to determine the molecular mass and the molecular formula of a compound and some of its structural features.

- The **molecular ion** (a **radical cation**), which is formed by removing an electron from a molecule, can break apart. The bonds most likely to break are the weakest ones and those that result in formation of the most stable products.

- A **mass spectrum** is a graph of the relative abundance of each positively charged fragment plotted against its m/z value. The m/z value of the molecular ion (M) gives the molecular mass of the compound.

Section 13.2

- Peaks with smaller m/z values—**fragment ion peaks**—represent positively charged fragments of the molecular ion. The **base peak** is the peak with the greatest abundance. It the most stable fragment.

Section 13.3

- The **rule of 13** allows possible molecular formulas to be determined from the m/z value of the molecular ion.

- The **nitrogen rule** states that if a compound has a molecular ion with an odd-numbered mass, the compound contains an odd number of nitrogen atoms.

Section 13.4

- The **M + 1 peak** occurs because of the naturally occurring ^{13}C isotope.

- If the **M + 2 peak** is one-third the height of the M peak, the compound contains a chlorine atom; if the M and M + 2 peaks are about the same height, the compound contains a bromine atom.

Section 13.5

- High-resolution mass spectrometers determine the exact molecular mass, which allows a compound's molecular formula to be determined.

Section 13.6

- Electron bombardment is most likely to dislodge a lone-pair electron.

- A bond between carbon and a more electronegative atom breaks heterolytically, with the electrons going to the more electronegative atom, whereas a bond between carbon and an atom of similar electronegativity breaks homolytically.

- α-**Cleavage** occurs because it forms a fragment with atoms that all have complete octets.

Section 13.9

- **Spectroscopy** is the study of the interaction of matter and **electromagnetic radiation.**

- High-energy radiation is associated with high frequencies, large wavenumbers, and short wavelengths.

Section 13.10

- **Infrared (IR) spectroscopy** identifies the kinds of functional groups in a compound.

Section 13.11

- It takes more energy to stretch a bond than to bend it.

Section 13.12

- To absorb IR radiation, the dipole moment of the bond must change when the vibration occurs.

- The intensity of an **absorption band** depends on the size of the change in dipole moment (more polar bonds show more intense absorptions), the number of bonds giving rise to the absorption, and the concentration.

Section 13.13

- The frequency of an absorption band is inversely related to the mass of the atoms that form the bond, so heavier atoms vibrate at lower frequencies.

- Stronger bonds show absorption bands at larger wavenumbers.

Section 13.14

- The position of an absorption band depends on bond order, hybridization, inductive electron donation and withdrawal, electron delocalization, and hydrogen bonding.

- The shape of an absorption band depends on hydrogen bonding. Hydrogen bonds vary in strength so hydrogen-bonded groups show broad absorption bands.

Section 13.18

- **Ultraviolet** and **visible (UV/Vis) spectroscopy** provide information about compounds with conjugated double bonds.

- UV light (180–400 nm) has greater energy than visible light (400–780 nm)—the shorter the wavelength, the greater the energy.

- UV and visible light cause a $\pi \rightarrow \pi^*$ **electronic transition.**

- A **chromophore** is the part of a molecule that absorbs UV or visible light.

Section 13.19

- The **Beer–Lambert law** is the relationship between absorbance, concentration, length of the light path, and molar absorptivity: $A = c\,l\,\varepsilon$.

Section 13.20

- The more conjugated double bonds in a compound, the longer the λ_{max} at which absorption occurs and the greater the molar absorptivity.

- An **auxochrome** alters the λ_{max} and the intensity of the absorption.

Section 13.21

- When a molecule absorbs light, all the non-absorbed (reflected) light combines to produce the complement of the absorbed color.

PROBLEMS

1. In the mass spectrum of the following compounds, which is the tallest—the peak at $m/z = 57$ or the peak at $m/z = 71$?
   a. 3-methylpentane
   b. 2-methylpentane

2. List three factors that influence the intensity of an IR absorption band.

3. Draw structures for a saturated hydrocarbon that has a molecular ion with an m/z value of 128.

4. Rank the following compounds in order of increasing λ_{max}:

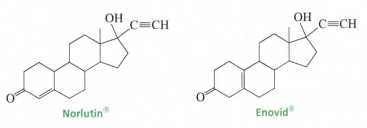

5. For each of the following pairs of compounds, identify one IR absorption band that could be used to distinguish between them:

a. $CH_3CH_2\overset{\displaystyle O}{\overset{\|}{C}}OCH_3$ and $CH_3CH_2\overset{\displaystyle O}{\overset{\|}{C}}CH_3$

g. $CH_3\overset{\displaystyle O}{\overset{\|}{C}}OCH_2CH_3$ and $CH_3\overset{\displaystyle O}{\overset{\|}{C}}CH_2OCH_3$

b. [cyclohexane-CH₃] and $CH_3CH_2CH_2CH_2CH_2CH_2CH_3$

h. [cyclohexanone] and [cyclohexenone]

c. $CH_3CH_2CH_2OH$ and $CH_3CH_2OCH_3$

i. $CH_3CH_2CH=CHCH_3$ and $CH_3CH_2C\equiv CCH_3$

d. $CH_3CH_2\overset{\displaystyle O}{\overset{\|}{C}}NH_2$ and $CH_3CH_2\overset{\displaystyle O}{\overset{\|}{C}}OCH_3$

j. $CH_3CH_2\overset{\displaystyle O}{\overset{\|}{C}}H$ and $CH_3CH_2\overset{\displaystyle O}{\overset{\|}{C}}CH_3$

e. [cyclohexyl]–CH_2CH_2OH and [cyclohexyl]–$CHCH_3$ with OH

k. [cyclohexyl]–$\overset{\displaystyle O}{\overset{\|}{C}}H$ and [phenyl]–$\overset{\displaystyle O}{\overset{\|}{C}}H$

f. *cis*-2-butene and *trans*-2-butene

l. $CH_3CH_2CH=CH_2$ and $CH_3CH_2CH=\overset{\displaystyle CH_3}{\overset{|}{C}}CH_3$

NOTE TO THE STUDENT
• There are additional spectroscopy problems in the *Study Guide and Solutions Manual.*

6. a. How could you use IR spectroscopy to determine whether the following reaction had occurred?
   b. After purifying the product, how could you determine whether all the NH_2NH_2 had been removed?

[phenyl]–$\overset{\displaystyle O}{\overset{\|}{C}}H$ $\xrightarrow[\text{HO}^-,\ \Delta]{\text{NH}_2\text{NH}_2}$ [phenyl]–CH_3

7. Assuming that the force constant is approximately the same for C—C, C—N, and C—O bonds, predict the relative positions of their stretching vibrations in an IR spectrum.

8. Norlutin and Enovid are ketones that suppress ovulation, so they have been used clinically as contraceptives. For which of these compounds would you expect the infrared carbonyl absorption (C=O stretch) to be at a higher frequency? Explain.

Norlutin®   Enovid®

49. In the following boxes, list the types of bonds and the approximate wavenumber at which each type of bond is expected to show an IR absorption:

3600   3000   1800   1400   1000

Wavenumber (cm⁻¹)

50. A mass spectrum shows significant peaks at $m/z = 87, 115, 140,$ and 143. Which of the following compounds is responsible for that mass spectrum?

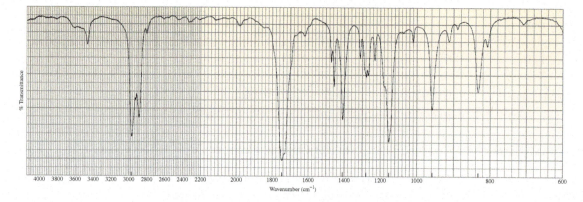

51. How could IR spectroscopy distinguish between 1,5-hexadiene and 2,4-hexadiene?

52. A compound gives a mass spectrum with essentially only three peaks at $m/z = 77$ (40%), 112 (100%), and 114 (33%). Identify the compound.

53. What hydrocarbons that contain a six-membered ring will have a molecular ion peak at $m/z = 112$?

54. The IR spectrum of a compound with molecular formula C_5H_8O is shown below. Identify the compound.

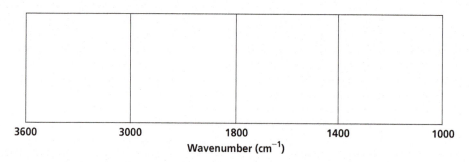

55. Rank the following compounds from highest wavenumber to lowest wavenumber for their C=O absorption band:

56. Rank the following compounds from highest wavenumber to lowest wavenumber for their C—O absorption band:

57. What peaks in their mass spectra can be used to distinguish between the following compounds?

58. The mass spectrum for a compound with molecular weight of 102 is shown below. Its IR spectrum has a broad, strong absorption at 3600 cm^{-1} and a medium absorption at 1360 cm^{-1}.
  a. Identify the compound.
  b. Show the mechanism for formation of the peak at $m/z = 84$.

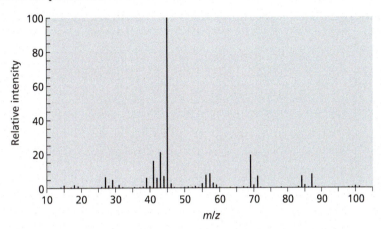

m/z

59. Which one of the following five compounds produced the IR spectrum shown below?

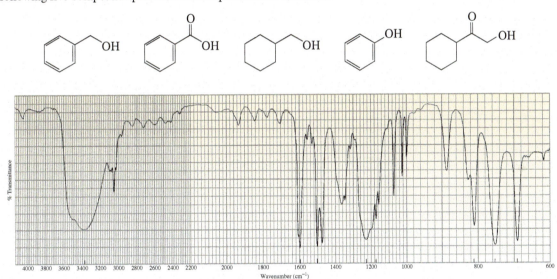

Wavenumber (cm^{-1})

60. The IR spectrum of compound **A** with a molecular formula of $C_5H_{12}O$ is shown below. Compound **A** is oxidized to give compound **B**, a ketone with a molecular formula of $C_5H_{10}O$. When compound **A** is heated with H_2SO_4, compounds **C** and **D** are obtained. Considerably more **D** is obtained than **C**. Reaction of compound **C** with O_3, followed by treatment with dimethyl sulfide, gives two products: formaldehyde and compound **E**, with a molecular formula of C_4H_8O. Reaction of compound **D** with O_3, followed by treatment with dimethyl sulfide, gives two products: compound **F**, with a molecular formula of C_3H_6O, and compound **G**, with a molecular formula of C_2H_4O. What are the structures of compounds **A** through **G**?

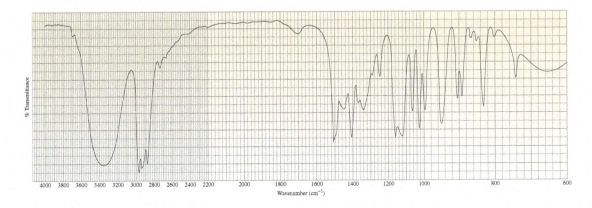

Wavenumber (cm^{-1})

61. Each of the IR spectra shown below is accompanied by a set of four compounds. In each case, indicate which of the four compounds is responsible for the spectrum.

a. $CH_3CH_2CH_2C\equiv CCH_3$ $CH_3CH_2CH_2CH_2OH$ $CH_3CH_2CH_2CH_2C\equiv CH$ $CH_3CH_2CH_2\overset{\displaystyle O}{\overset{\|}{C}}OH$

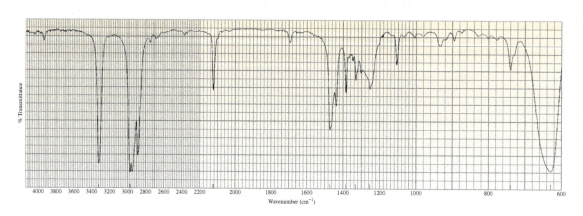

b.

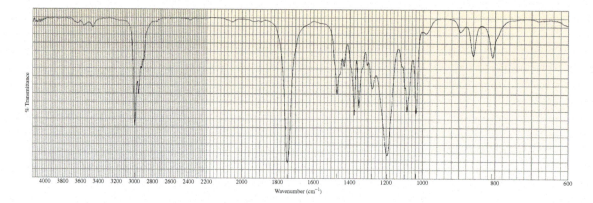

c.

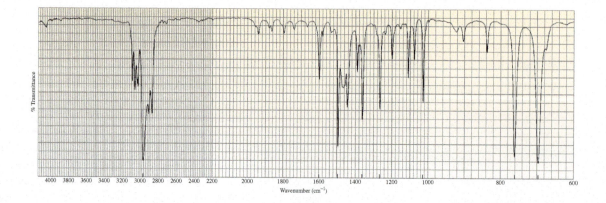

62. Five compounds are shown for each of the IR spectra below. Indicate which of the five compounds is responsible for each spectrum.

a.

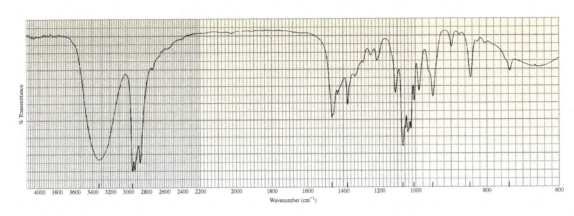

b.

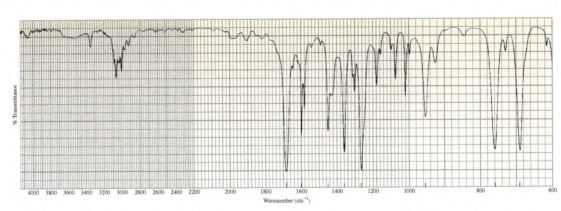

c.

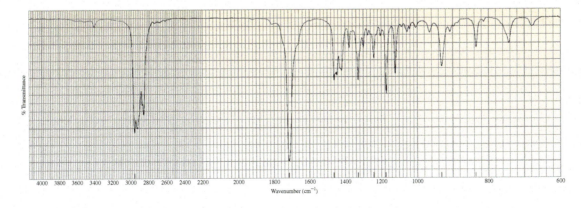

63. A compound is known to be one of those shown here. What absorption bands in its IR spectrum allow you to identify the compound?

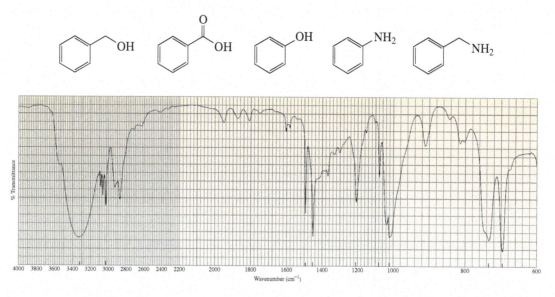

A   B   C

64. How can IR spectroscopy distinguish between 1-hexyne, 2-hexyne, and 3-hexyne?

65. Draw a structure for a carboxylic acid that has a molecular ion with an m/z value of 116.

66. A solution of ethanol has been contaminated with benzene—a technique employed to make ethanol unfit to drink. Benzene has a molar absorptivity of 230 M^{-1} cm^{-1} at 260 nm in ethanol, and ethanol shows no absorbance at 260 nm. How can the concentration of benzene in the solution be determined?

67. Give approximate wavenumbers for the major characteristic IR absorption bands that would be given by each of the following compounds:

a.

c.   OH

e.   CH$_2$C≡CH

b.

d.   NH$_2$

f.   OH

68. Given that the force constants are similar for C—H and C—C bonds, explain why the stretching vibration of a C—H bond occurs at a larger wavenumber.

69. Which one of the following five compounds produced the IR spectrum shown below?

70. Phenolphthalein is an acid–base indicator. In solutions of pH < 8.5, it is colorless; in solutions of pH > 8.5, it is deep red-purple. Account for the change in color.

phenolphthalein

71. Each of the IR spectra shown below is the spectrum of one of the following compounds. Identify the compound that produced each spectrum.

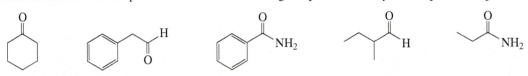

a.

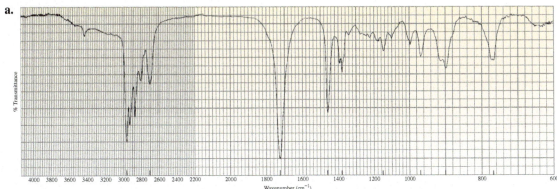

b.

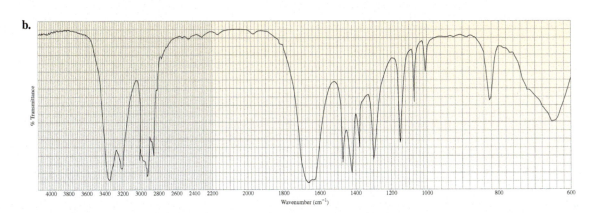

c.

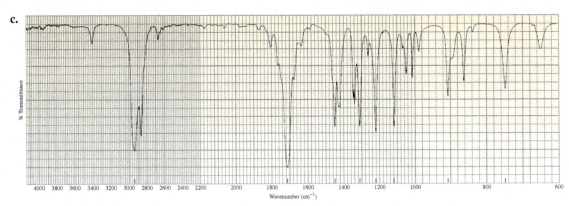

72. How can you use UV spectroscopy to distinguish between the compounds in each of the following pairs?

73. Calculate the approximate wavenumber at which a C=C stretch occurs, given that the force constant for the C=C bond is 1×10^6 g s^{-2}.

74. The IR and mass spectra for three different compounds are shown below. Identify each compound.

a.

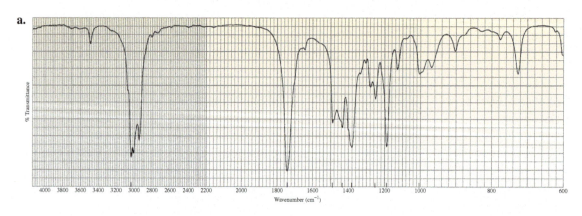

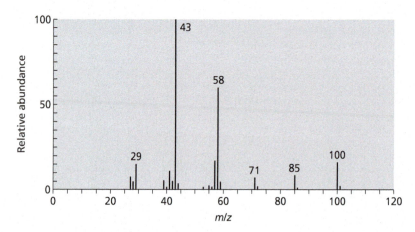

b.

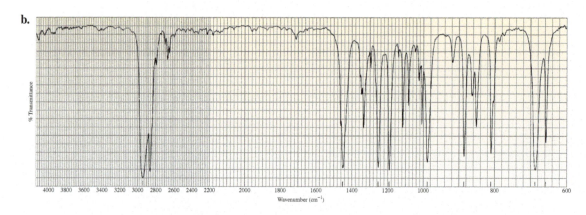

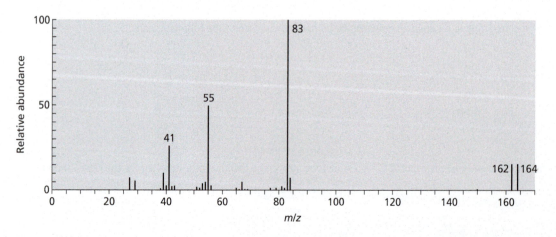

c.

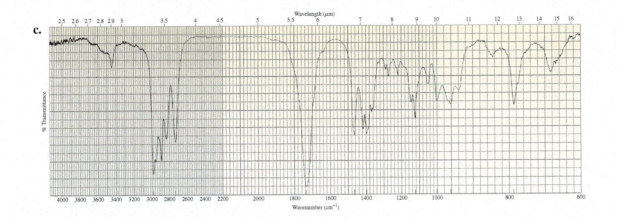

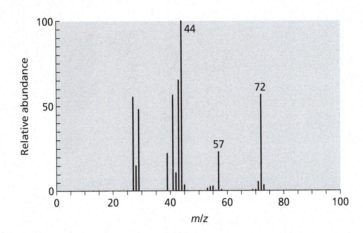

14 NMR Spectroscopy

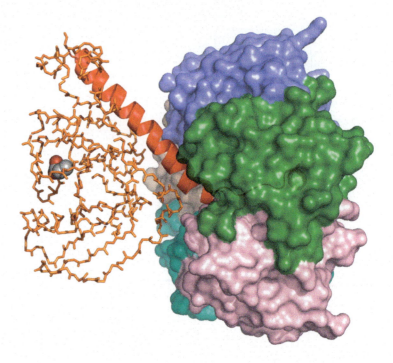

The structure of *E.coli* heat-labile enterotoxin determined by X-ray crystallography and downloaded from the Protein Data Bank (see page 667). The protein has seven subunits–that is, seven protein chains (Section 21.16). Five of them (blue, green purple, and so on) attach to a cell membrane. The toxin uses the red helical spike to deliver the catalytic subunit (orange) into the cell. The red and gray space-filling model is a protein side chain that acts as a base catalyst (Section 22.8). This toxin is responsible for what is known as traveler's diarrhea.

Chapter 13 introduced three analytical techniques that are used to determine the structures of organic compounds: mass spectrometry, IR spectroscopy, and ultraviolet/visible spectroscopy. A fourth technique, *nuclear magnetic resonance (NMR) spectroscopy*, helps to identify the carbon–hydrogen and carbon–carbon framework of an organic compound.

NMR spectroscopy has an advantage over the other analytical techniques you have looked at. It not only allows us to identify the functionality at a specific carbon, it also enables us to connect neighboring carbons. In many cases, NMR spectroscopy can be used to determine a molecule's entire structure.

14.1   AN INTRODUCTION TO NMR SPECTROSCOPY

NMR spectroscopy was developed by physicists in the late 1940s to study the properties of atomic nuclei. In 1951, chemists realized that NMR spectroscopy could also be used to study the structures of organic compounds.

Nuclei that have an odd number of protons or an odd number of neutrons (or both) have a property called spin that allows them (^{1}H, ^{13}C, ^{15}N, ^{19}F, and ^{31}P) to be studied by NMR. Nuclei such as ^{12}C and ^{16}O do not have spin and, therefore, cannot be studied by NMR. Because hydrogen nuclei (protons) were the first nuclei studied by NMR, the acronym *NMR* is generally assumed to mean **^{1}H NMR (proton magnetic resonance).**

α- and β-Spin States

As a result of its charge, a nucleus with spin has a magnetic moment and generates a magnetic field similar to that generated by a small bar magnet. In the absence of an applied magnetic field, the magnetic moments of the nuclei are randomly oriented. However, when placed between the poles of a strong magnet, the magnetic moments of the nuclei align either *with* or *against* the applied magnetic field (Figure 14.1).

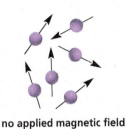
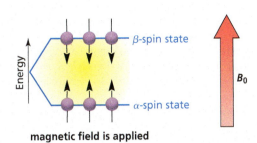

no applied magnetic field   magnetic field is applied

β-spin state
α-spin state
B_0
Energy

◀ **Figure 14.1**
In the absence of an applied magnetic field, the magnetic moments of the nuclei are randomly oriented. In the presence of an applied magnetic field, the magnetic moments of the nuclei line up with (the α-spin state) or against (the β-spin state) the applied magnetic field.

Nuclei with magnetic moments that align with the field are in the lower-energy **α-spin state,** whereas those with magnetic moments that align against the field are in the higher-energy **β-spin state.** The β-spin state is higher in energy because more energy is needed to align the magnetic moments against the field than with it. As a result, more nuclei are in the α-spin state. The difference in the populations is very small (about 20 out of 1 million protons), but it is sufficient to form the basis of NMR spectroscopy.

The energy difference (ΔE) between the α- and β-spin states depends on the strength of the **applied magnetic field** (B_0): the greater the strength of the applied magnetic field, the greater the ΔE (Figure 14.2).

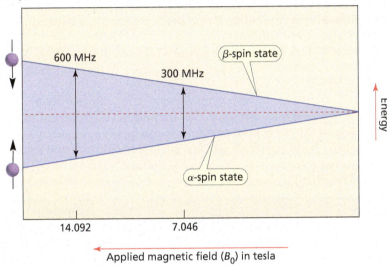

600 MHz

300 MHz

β-spin state

α-spin state

Energy

14.092   7.046

Applied magnetic field (B_0) in tesla

◀ **Figure 14.2**
The difference in energy between the α- and β-spin states increases as the strength of the applied magnetic field increases.

Flipping the Spin

When a sample is subjected to a pulse of radiation whose energy corresponds to the difference in energy (ΔE) between the α- and β-spin states, nuclei in the α-spin state are promoted to the β-spin state. This transition is called "flipping" the spin.

With currently available magnets, the energy difference between the α- and β-spin states is small, so only a small amount of energy is needed to flip the spin. The radiation used to supply this energy is in the radio frequency (rf) region of the electromagnetic spectrum and is called **rf radiation** (see Figure 13.8 on page 584). When the nuclei absorb rf radiation and flip their spins, they generate signals whose frequency depends on the difference in energy (ΔE) between the α- and β-spin states. The NMR spectrometer detects these signals and plots their frequency versus their intensity; this plot is an NMR spectrum.

The nuclei are said to be *in resonance* with the rf radiation, hence the term **nuclear magnetic resonance.** In this context, "resonance" refers to the nuclei flipping back and forth between the α- and β-spin states in response to the rf radiation; it has nothing to do with the "resonance" associated with electron delocalization.

Energy Difference Between Spin States Depends on the Operating Frequency

The following equation shows that the energy difference between the spin states (ΔE) depends on the operating frequency of the spectrometer (ν), which depends, in turn, on the strength of the magnetic field (B_0), measured in tesla (T), and the *gyromagnetic ratio* (γ); h is Planck's constant (Section 13.9).

$$\Delta E = h\nu = h\frac{\gamma}{2\pi}B_0$$

The **gyromagnetic ratio** is a constant that depends on the particular kind of nucleus. In the case of a proton, $\gamma = 2.675 \times 10^8\ \mathrm{T^{-1}s^{-1}}$; in the case of a $^{13}\mathrm{C}$ nucleus, it is $6.688 \times 10^7\ \mathrm{T^{-1}s^{-1}}$. Canceling Planck's constant on both sides of the equation gives

$$\nu = \frac{\gamma}{2\pi}B_0$$

Earth's magnetic field is 5×10^{-5} T, measured at the equator. Its maximum surface magnetic field is 7×10^{-5} T, measured at the south magnetic pole.

The magnetic field is proportional to the operating frequency of the spectrometer.

The following calculation shows that if an $^1\mathrm{H}$ NMR spectrometer is equipped with a magnet that generates a magnetic field of 7.046 T, then the spectrometer will require an operating frequency of 300 MHz (megahertz):

$$\nu = \frac{\gamma}{2\pi}B_0$$

$$= \frac{2.675 \times 10^8}{2(3.1416)}\mathrm{T^{-1}s^{-1}} \times 7.046\ \mathrm{T}$$

$$= 300 \times 10^6\ \mathrm{Hz} = 300\ \mathrm{MHz}$$

The equation shows that the strength of the *magnetic field* (B_0) *is proportional to the operating frequency* (ν), in MHz, of the spectrometer. Therefore, if the spectrometer has a more powerful magnet, then it must have a higher **operating frequency.** For example, a magnetic field of 14.092 T requires an operating frequency of 600 MHz.

Today's **NMR spectrometers** operate at frequencies between 300 and 1000 MHz. The resolution of the NMR spectrum increases as the operating frequency of the instrument—and, therefore, the strength of the magnet—increases.

Because each kind of nucleus has its own gyromagnetic ratio, different frequencies are required to bring different kinds of nuclei into resonance. For example, an NMR spectrometer that requires a frequency of 300 MHz to flip the spin of an $^1\mathrm{H}$ nucleus requires a frequency of 75 MHz to flip the spin of a $^{13}\mathrm{C}$ nucleus. NMR spectrometers are equipped with radiation sources that can be tuned to different frequencies so that they can be used to obtain NMR spectra of different kinds of nuclei ($^1\mathrm{H}$, $^{13}\mathrm{C}$, $^{15}\mathrm{N}$, $^{19}\mathrm{F}$, $^{31}\mathrm{P}$).

Nikola Tesla (1856–1943)

The tesla, used to measure the strength of a magnetic field, was named in honor of Nikola Tesla. Tesla was born in Croatia, emigrated to the United States in 1884, and became a citizen in 1891. He was a proponent of using alternating current to distribute electricity and bitterly fought Thomas Edison, who promoted direct current. Tesla was granted a patent for developing the radio in 1900, but Guglielmo Marconi was also given a patent for its development in 1904. Not until 1943—a few months after his death—was Tesla's patent upheld by the U.S. Supreme Court.

Tesla held over 800 patents and is credited with developing neon and fluorescent lighting, the electron microscope, the refrigerator motor, and the Tesla coil (a type of transformer for changing the voltage of alternating current). Perhaps his most important contribution was polyphase electric power, which became the prototype for all large power systems. He made most of his equipment himself, including insulators, a technology that was kept classified until recently because the same technology was being used for part of the U.S. Strategic Defense Initiative. Tesla frequently staged flamboyant high-voltage demonstrations, which may explain why he did not receive proper recognition for his work.

Nikola Tesla in his laboratory

PROBLEM 1 ◆

What frequency (in MHz) is required to cause a proton to flip its spin when it is exposed to a magnetic field of 1 T?

PROBLEM 2 ◆

a. Calculate the magnetic field (in tesla) required to flip an $^1\mathrm{H}$ nucleus in an NMR spectrometer that operates at 360 MHz.

b. What strength magnetic field is required when a 500-MHz instrument is used for $^1\mathrm{H}$ NMR?

14.2 FOURIER TRANSFORM NMR

The most common way to obtain an **NMR spectrum** is to dissolve a small amount of a compound in about 0.5 mL of solvent. This solution is put into a long, thin glass tube, which is then placed within a powerful magnetic field (Figure 14.3). Spinning the sample tube about its long axis averages the position of the molecules in the magnetic field, which increases the resolution of the spectrum.

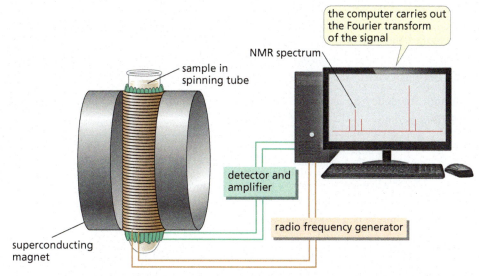

the computer carries out the Fourier transform of the signal

NMR spectrum

sample in spinning tube

detector and amplifier

radio frequency generator

superconducting magnet

◀ **Figure 14.3**
Schematic diagram of an NMR spectrometer

In modern instruments called *pulsed Fourier transform (FT) spectrometers,* the magnetic field is held constant and an rf pulse of short duration excites all the protons simultaneously. The rf pulse covers a range of frequencies, so each nucleus can absorb the frequency it requires to come into resonance (flip its spin) and produce a signal—called a free induction decay (FID)—at a frequency corresponding to ΔE. The intensity of the FID signal decays as the nuclei lose the energy they gained from the rf pulse.

A computer measures the change in intensity over time and converts it into intensity-versus-frequency data, in a mathematical operation known as a *Fourier transform,* to produce a spectrum called a **Fourier transform NMR (FT–NMR) spectrum.** An FT–NMR spectrum can be recorded in about 2 seconds—and large numbers of FIDs can be averaged in a few minutes—using less than 5 mg of compound. The NMR spectra in this book are FT–NMR spectra that were taken on a spectrometer with an operating frequency of 300 MHz.

14.3 SHIELDING CAUSES DIFFERENT NUCLEI TO SHOW SIGNALS AT DIFFERENT FREQUENCIES

Because the frequency of an NMR signal depends on the strength of the magnetic field experienced by the nucleus (Figure 14.2), if all the hydrogens in a compound were to experience the same magnetic field, they would all give signals of the same frequency. If this were the case, all ¹H NMR spectra would consist of one signal, which would tell us nothing about the structure of the compound, except that it contains hydrogens.

A nucleus, however, is embedded in a cloud of electrons that partly *shields* it from the **applied magnetic field.** Fortunately for chemists, the **shielding** varies for different hydrogens in a molecule. In other words, all the hydrogens do not experience the same magnetic field.

What causes shielding? In a magnetic field, the electrons circulate about the nuclei and induce a **local magnetic field** that acts in opposition to the applied magnetic field and, therefore, subtracts from it. As a result, the **effective magnetic field**—the amount of magnetic field filtered through the surrounding electrons that the nuclei actually "sense"—is somewhat smaller than the applied magnetic field:

The electron density of the environment in which the proton is located shields the proton from the applied magnetic field.

$$B_{\text{effective}} = B_{\text{applied}} - B_{\text{local}}$$

This means that the greater the electron density of the environment in which the proton* is located, the greater is B_{local}, and the smaller is $B_{effective}$. This type of shielding is called **diamagnetic shielding.**

The larger the magnetic field sensed by the proton, the higher the frequency of the signal.

- Protons in electron-rich environments sense a *smaller effective magnetic field*. Therefore, they require a *lower frequency* to come into resonance—that is, flip their spin—because ΔE is smaller (Figure 14.2).

- Protons in electron-poor environments sense a *larger effective magnetic field* and so require a *higher frequency* to come into resonance because ΔE is larger.

An NMR spectrum exhibits a signal for each proton in a different environment.

deshielded = less shielded

- Protons in electron-rich environments are more shielded and appear at lower frequencies (on the right-hand side of the spectrum; Figure 14.4).

- Protons in electron-poor environments are less shielded and appear at higher frequencies (on the left-hand side of the spectrum).

Notice that high frequency in an NMR spectrum is on the left-hand side, just as it is in IR and UV/Vis spectra.

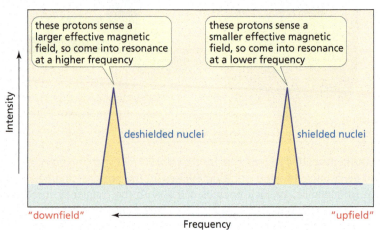

▶ **Figure 14.4**
Shielded protons come into resonance at lower frequencies than deshielded nuclei.

14.4 THE NUMBER OF SIGNALS IN AN ¹H NMR SPECTRUM

Protons in the same environment are called **chemically equivalent protons.** For example, 1-bromopropane has three different sets of chemically equivalent protons: the three methyl protons are chemically equivalent because of rotation about the C—C bond; the two methylene (CH₂) protons on the middle carbon are chemically equivalent; and the two methylene protons on the carbon bonded to the bromine make up the third set of chemically equivalent protons.

Each set of chemically equivalent protons produces an NMR signal.

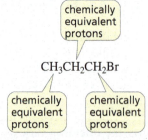

Each set of chemically equivalent protons in a compound produces a separate signal in its ¹H NMR spectrum. Thus, 1-bromopropane has three signals in its ¹H NMR spectrum because it has three sets of chemically equivalent protons. (Sometimes the signals are not sufficiently separated and overlap each other. When this happens, one sees fewer signals than anticipated.)

---

* In discussions of NMR spectroscopy, the terms *proton* and *hydrogen* are both used to describe a covalently bonded hydrogen.

2-Bromopropane has two sets of chemically equivalent protons, so it has two signals in its ¹H NMR spectrum. The six methyl protons are equivalent so they produce only one signal, and the hydrogen bonded to the middle carbon gives the second signal. The chemically equivalent protons in the compounds shown below are designated by the same letter.

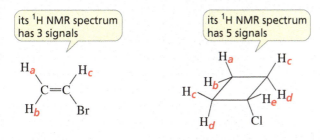

> You can tell how many sets of chemically equivalent protons a compound has from the number of signals in its ¹H NMR spectrum.

Ethyl methyl ether has three sets of chemically equivalent protons: the methyl protons on the carbon adjacent to the oxygen, the methylene protons on the carbon adjacent to the oxygen, and the methyl protons on the carbon that is one carbon removed from the oxygen.

If bonds are prevented from freely rotating, as in a compound with a double bond or in a cyclic compound, two protons on the same carbon may not be equivalent. For example, the H_a and H_b protons of bromoethene are not equivalent because they are not in the same environment: H_a is trans to Br, whereas H_b is cis to Br. Thus, the ¹H NMR spectrum of bromoethene has three signals.

its ¹H NMR spectrum has 3 signals

its ¹H NMR spectrum has 5 signals

H_a and H_b are not equivalent

H_a and H_b are not equivalent
H_c and H_d are not equivalent

The ¹H NMR spectrum of chlorocyclobutane has five signals. The H_a and H_b protons are not equivalent: H_a is trans to Cl, whereas H_b is cis to Cl. Similarly, the H_c and H_d protons are not equivalent.

PROBLEM-SOLVING STRATEGY

Determining the Number of Signals in an ¹H NMR Spectrum

How many signals would you expect to see in the ¹H NMR spectrum of ethylbenzene?

CH_3CH_2—⬡

LEARN THE STRATEGY

To determine the number of signals in the spectrum, replace each hydrogen in turn by another atom (here we use Br) and name the resulting compound. The number of different names corresponds to the number of signals in the ¹H NMR spectrum. We get five different names for the bromosubstituted ethylbenzenes, so we expect to see five signals in the ¹H NMR spectrum of ethylbenzene.

$BrCH_2CH_2$—⬡

1-bromo-2-phenylethane

CH_3CH—⬡
|
Br

1-bromo-1-phenylethane

CH_3CH_2—⬡(Br)

1-bromo-2-ethylbenzene

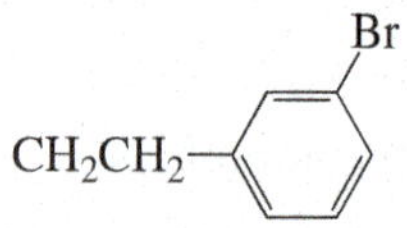

Br
CH$_2$CH$_2$—⬡   CH$_3$CH$_2$—⬡—Br   CH$_3$CH$_2$—⬡   CH$_3$CH$_2$—⬡
                                                                        | |
                                                                        Br   Br

1-bromo-3-ethylbenzene 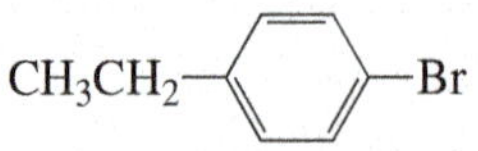   **1-bromo-4-ethylbenzene** 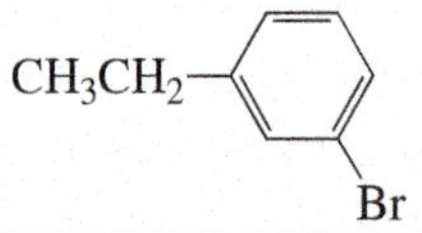   **1-bromo-3-ethylbenzene** 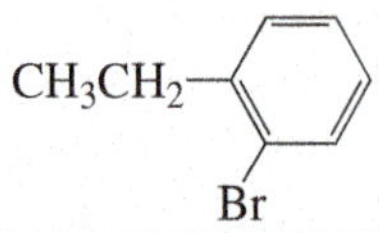   **1-bromo-2-ethylbenzene**

USE THE STRATEGY

PROBLEM 3

How many signals would you expect to see in the ^{1}H NMR spectrum of each of the five compounds with molecular formula C_6H_{14}?

PROBLEM 4 ◆

How many signals would you expect to see in the ^{1}H NMR spectrum of each of the following compounds

a. CH$_3$CH$_2$CH$_2$CH$_3$

f.
$$\begin{array}{c} O \\ \| \\ CH_3CH_2CH_2 \quad C \quad CH_3 \end{array}$$

k.
$$\begin{array}{c} O \\ \| \\ CH_2{=}CH \quad C \quad H \end{array}$$

b. BrCH$_2$CH$_2$Br

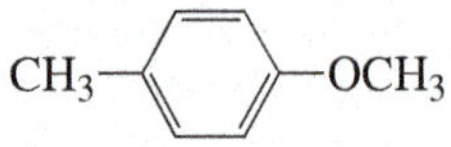

g. CH$_3$CH$_2$CHCH$_2$CH$_3$
                   |
                   Cl

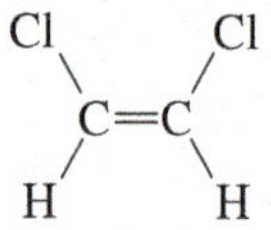

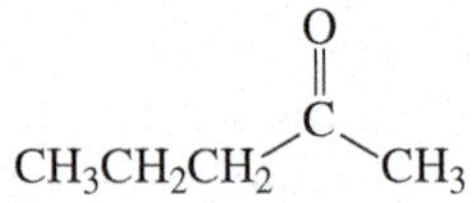

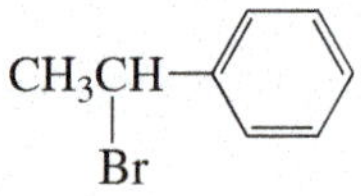

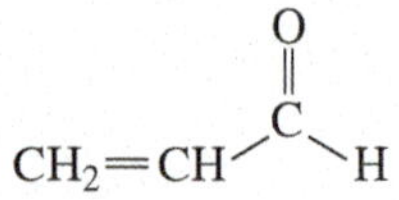

l. ⬡ with Br, Br

c. CH$_2$=CCl$_2$

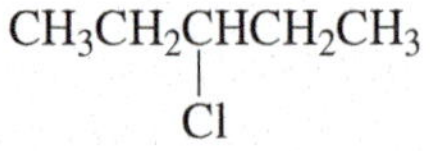

h. 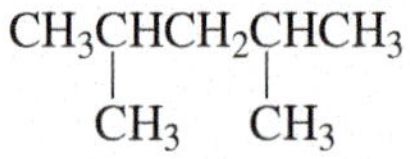 CH$_3$CHCH$_2$CHCH$_3$
           | |
           CH$_3$ CH$_3$

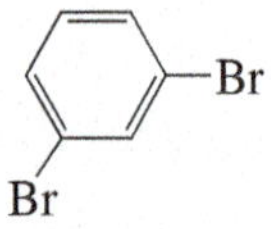

m. 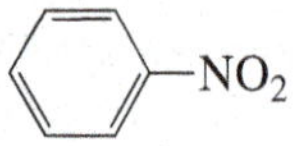 ⬡—NO$_2$

d. CH$_3$—⬡—OCH$_3$

i. CH$_3$CH—⬡
           |
           Br

n.  CH$_3$—⬡—CH$_3$

e.
$$\begin{array}{c} Cl \quad\quad Cl \\ \diagdown \quad\quad \diagup \\ C{=}C \\ \diagup \quad\quad \diagdown \\ H \quad\quad\quad H \end{array}$$

j. ⬡ (cyclohexene)

o. 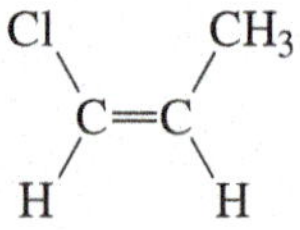
$$\begin{array}{c} Cl \quad\quad CH_3 \\ \diagdown \quad\quad \diagup \\ C{=}C \\ \diagup \quad\quad \diagdown \\ H \quad\quad\quad H \end{array}$$

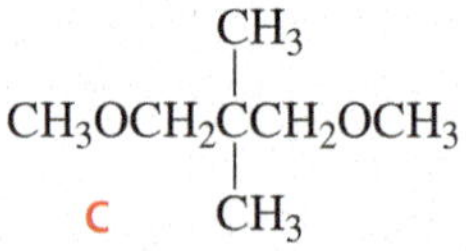

PROBLEM 5

How could you distinguish the ^{1}H NMR spectra of the following compounds?

CH$_3$OCH$_2$OCH$_3$ CH$_3$OCH$_3$ CH$_3$OCH$_2$CCH$_2$OCH$_3$
                                                              |
A   **B**   **C**   CH$_3$

with the CH$_3$ on top of the central carbon.

PROBLEM 6

Draw an isomer of dichlorocyclopropane that gives an ^{1}H NMR spectrum

a. with one signal.   **b.** with two signals.   **c.** with three signals.

14.5 THE CHEMICAL SHIFT TELLS HOW FAR THE SIGNAL IS FROM THE REFERENCE SIGNAL

The Reference Compound

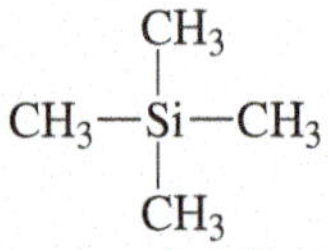

$$\begin{array}{c} CH_3 \\ | \\ CH_3{-}Si{-}CH_3 \\ | \\ CH_3 \end{array}$$

tetramethylsilane
TMS

A small amount of an inert **reference compound** is added to the sample tube containing the com pound whose NMR spectrum is to be taken. The most commonly used reference compound i tetramethylsilane (TMS). Because it is highly volatile (bp = 26.5 °C), it can easily be remove from the sample by evaporation after the NMR spectrum is taken.

The methyl protons of TMS are in a more electron-rich environment than are most protons i organic molecules because silicon is less electronegative than carbon (their electronegativities ar 1.8 and 2.5, respectively). Consequently, the signal for the methyl protons of TMS is at a lowe frequency than most other signals (that is, the TMS signal appears to the right of the other signals)

The Chemical Shift

The position at which a signal occurs in an NMR spectrum is called the *chemical shift*.

> The **chemical shift** is a measure of how far the signal is from the signal for the reference compound.

Most proton chemical shifts are between 0 and 12 ppm.

The most common scale for chemical shifts is the δ (delta) scale. The TMS signal defines the zero position on the δ scale (Figure 14.5).

The chemical shift is determined by measuring the distance from the TMS signal in hertz and dividing by the operating frequency of the instrument in megahertz. Because the units are Hz/MHz, a chemical shift has units of parts per million (ppm) of the operating frequency:

$$\delta = \text{chemical shift (ppm)} = \frac{\text{distance from the TMS signal (Hz)}}{\text{operating frequency of the spectrometer (MHz)}}$$

The ^{1}H NMR spectrum in Figure 14.5 shows that the chemical shift (δ) of the methyl protons is at 1.05 ppm and the chemical shift of the methylene protons, which are deshielded by the electron-withdrawing bromine, is at 3.28 ppm. *Notice that low-frequency (shielded) signals have small δ (ppm) values, whereas high-frequency (deshielded) signals have large δ values.*

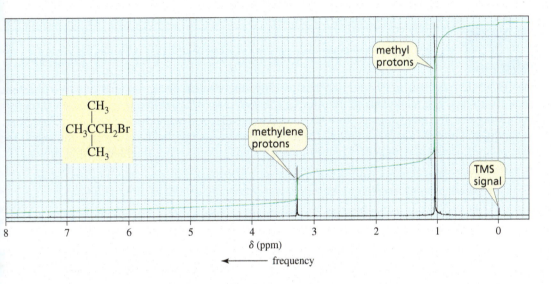

The greater the value of the chemical shift (δ), the higher the frequency.

◀ **Figure 14.5**
The ^{1}H NMR spectrum of 1-bromo-2,2-dimethylpropane. The TMS signal is a reference signal from which chemical shifts are measured; it defines the zero position on the scale.

δ Is Independent of the Operating Frequency

The advantage of the δ scale is that the chemical shift is *independent of the operating frequency of the NMR spectrometer*. Thus, the chemical shift of the methyl protons of 1-bromo-2,2-dimethylpropane is at 1.05 ppm in both a 300 MHz and a 500 MHz instrument. If the chemical shift were reported in hertz instead, the chemical shift would be at 315 Hz in a 300 MHz instrument and at 525 Hz in a 500 MHz instrument (315/300 = 1.05, 525/500 = 1.05).

The following diagram will help you keep track of the terms associated with NMR spectroscopy:

NOTE TO THE STUDENT

• The terms *upfield* and *downfield*, which came into use when continuous-wave (CW) spectrometers were used (before the advent of Fourier transform spectrometers), are so entrenched in the vocabulary of NMR that you should know what they mean. *Upfield* means farther to the right-hand side of the spectrum, and *downfield* means farther to the left-hand side of the spectrum.

protons in electron-poor environments	protons in electron-dense environments
deshielded protons	shielded protons
downfield	upfield
high frequency	low frequency
large δ values	small δ values

◀———— increasing δ ppm
◀———— increasing frequency

PROBLEM 7 ◆

How many hertz from the TMS signal is the signal occurring at 2.0 ppm

a. in a 300 MHz spectrometer? **b.** in a 500 MHz spectrometer?

PROBLEM 8 ◆

A signal is seen at 600 Hz from the TMS signal in an NMR spectrometer with a 300-MHz operating frequency.

a. What is the chemical shift of the signal?
b. What is its chemical shift in an instrument operating at 500 MHz?
c. How many hertz from the TMS signal is the signal in a 500-MHz spectrometer?

PROBLEM 9 ◆

a. If two signals differ by 1.5 ppm in a 300 MHz spectrometer, by how much do they differ in a 500 MHz spectrometer?
b. If two signals differ by 90 Hz in a 300 MHz spectrometer, by how much do they differ in a 500 MHz spectrometer?

PROBLEM 10 ◆

Where would you expect to find the ^{1}H NMR signal of $(CH_3)_2Mg$ relative to the TMS signal? (*Hint:* Magnesium is less electronegative than silicon.)

14.6 THE RELATIVE POSITIONS OF ^{1}H NMR SIGNALS

The ^{1}H NMR spectrum in Figure 14.5 has two signals because the compound has two different kinds of protons. The methylene protons are in a *less electron-rich environment* than are the methyl protons because the methylene protons are closer to the electron-withdrawing bromine. Therefore, the methylene protons are *less shielded* from the applied magnetic field. As a result, the signal for these protons occurs at a higher frequency than the signal for the more shielded methyl protons.

> *Remember that the right-hand side of an NMR spectrum is the low-frequency side, where protons in electron-rich environments (more shielded) show a signal.*

> *The left-hand side is the high-frequency side, where protons in electron-poor environments (less shielded) show a signal (Figure 14.4).*

We expect the ^{1}H NMR spectrum of 1-nitropropane to have three signals because the compound has three different kinds of protons. The closer the protons are to the electron-withdrawing nitro group, the less they are shielded from the applied magnetic field, so the higher the frequency at which their signals appear. Thus, the protons closest to the nitro group show a signal at the highest frequency (4.37 ppm), and the ones farthest from the nitro group show a signal at the lowest frequency (1.04 ppm).

Electron withdrawal causes NMR signals to appear at higher frequencies (at larger δ values).

1.04 ppm   2.07 ppm
4.37 ppm

$$CH_3CH_2CH_2NO_2$$

Compare the chemical shifts of the methylene protons immediately adjacent to the halogen in the following alkyl halides. The position of the signal depends on the electronegativity of the halogen—as the electronegativity of the halogen increases, the shielding of the protons decreases, so the frequency of the signal increases. Thus, the signal for the methylene protons adjacent to fluorine (the most electronegative of the halogens) occurs at the highest frequency, whereas the signal for the methylene protons adjacent to iodine (the least electronegative of the halogens) occurs at the lowest frequency.

$$CH_3CH_2CH_2CH_2CH_2F \qquad CH_3CH_2CH_2CH_2CH_2Cl \qquad CH_3CH_2CH_2CH_2CH_2Br \qquad CH_3CH_2CH_2CH_2CH_2I$$

4.50 ppm   3.50 ppm   3.40 ppm   3.20 ppm

PROBLEM 11 ◆

a. Which proton or set of protons in each of the following compounds is the least shielded?

b. Which proton or set of protons in each compound is the most shielded?

1. $CH_3CH_2CH_2Cl$ **2.** $CH_3CH_2 \overset{\displaystyle O}{\overset{\displaystyle \|}{C}} OCH_3$ **3.** $CH_3CHCHBr$ with Br Br

14.7 THE CHARACTERISTIC VALUES OF CHEMICAL SHIFTS

Approximate values of chemical shifts for different kinds of protons are listed in Table 14.1. (A more extensive compilation can be found in Appendix V.)

Table 14.1 Approximate Values of Chemical Shifts (ppm) for ${}^1H$ NMR*

Type of proton	ppm	Type of proton	ppm	Type of proton	ppm	Type of proton	ppm
$-CH_3$	0.85	⬡$-CH_3$	2.3	$I-\overset{\|}{\underset{\|}{C}}-H$	2.5–4	$R-OH$	Variable, 2–5
$-CH_2-$	1.20	$-C\equiv C-H$	2.4	$Br-\overset{\|}{\underset{\|}{C}}-H$	2.5–4	⬡$-OH$	Variable, 4–7
$-\overset{\|}{\underset{\|}{CH}}-$	1.55	$R-O-CH_3$	3.3	$Cl-\overset{\|}{\underset{\|}{C}}-H$	3–4	⬡$-H$	6.5–8
$-\overset{\|}{\underset{\|}{C}}=C-CH_3$	1.7	$R-\overset{\|}{\underset{R}{C}}=CH_2$	4.7	$F-\overset{\|}{\underset{\|}{C}}-H$	4–4.5	$-\overset{\displaystyle O}{\overset{\displaystyle \|}{C}}-H$	9.0–10
$-\overset{\displaystyle O}{\overset{\displaystyle \|}{C}}-CH_3$	2.1	$R-\overset{\|}{\underset{R}{C}}=\overset{\|}{\underset{R}{C}}-H$	5.3	$R-NH_2$	Variable, 1.5–4	$-\overset{\displaystyle O}{\overset{\displaystyle \|}{C}}-OH$	Variable, 10–12
						$-\overset{\displaystyle O}{\overset{\displaystyle \|}{C}}-NH_2$	Variable, 5–8

*The values are approximate because they are affected by neighboring substituents.

An ${}^1H$ NMR spectrum can be divided into seven regions, one of which is empty. If you can remember the kinds of protons that appear in each region, you will know how to tell what kinds of protons a molecule has from a quick look at its NMR spectrum.

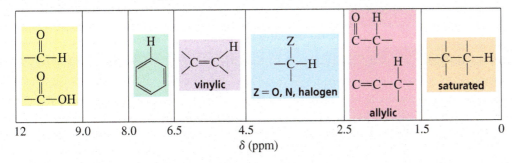

Methine, Methylene, and Methyl Protons

Carbon is more electronegative than hydrogen (Table 1.3 on page 10). Therefore, the chemical shift of a **methine proton** (a hydrogen bonded to an sp^3 carbon that is attached to *three* carbons) is more deshielded and so shows a chemical shift at a higher frequency than the chemical shift

$-\overset{\|}{\underset{\|}{CH}}-$
methine

$-CH_2-$
methylene

$-CH_3$
methyl

of **methylene protons** (hydrogens bonded to an sp^3 carbon that is attached to *two* carbons) in a similar environment. Likewise, the chemical shift of methylene protons is at a higher frequency than the chemical shift of **methyl protons** (hydrogens bonded to an sp^3 carbon that is attached to *one* carbon) in a similar environment (Table 14.1).

In a similar environment, the signal for a methine proton occurs at a higher frequency than the signal for methylene protons, which occurs at a higher frequency than the signal for methyl protons.

methine proton — methylene proton — methyl proton

$$\underset{\text{1.55 ppm}}{\overset{\displaystyle C}{\underset{\displaystyle C}{C-C-H}}} \qquad \underset{\text{1.20 ppm}}{\overset{\displaystyle C}{\underset{\displaystyle C}{H-C-H}}} \qquad \underset{\text{0.85 ppm}}{\overset{\displaystyle H}{\underset{\displaystyle C}{H-C-H}}}$$

LEARN THE STRATEGY

For example, the ^{1}H NMR spectrum of butanone shows three signals. The signal at the lowest frequency is the signal for the *a* protons; these protons are farthest from the electron-withdrawing carbonyl group. The *b* and *c* protons are the same distance from the carbonyl group, but the signal for the *c* protons is at a higher frequency because methylene protons appear at a higher frequency than do methyl protons in a similar environment.

$$\overset{\displaystyle O}{\underset{\underset{a \qquad c}{CH_3CH_2}}{\overset{\|}{C}}\underset{b}{CH_3}}$$
butanone

$$\overset{b \quad\; c \;\; a}{CH_3OCHCH_3} \atop \underset{a}{CH_3}$$
2-methoxypropane

(In correlating an NMR spectrum with a structure, this text labels the set of protons responsible for the signal at the lowest frequency *a,* the next set *b,* the next set *c,* and so on.)

The signal for the *a* protons of 2-methoxypropane is the one at the lowest frequency because these protons are farthest from the electron-withdrawing oxygen. The *b* and *c* protons are the same distance from the oxygen, but the signal for the *c* protons appears at a higher frequency because, in a similar environment, a methine proton appears at a higher frequency than do methyl protons.

USE THE STRATEGY

PROBLEM 12 ♦

Which underlined proton (or sets of protons) has the greater chemical shift (that is, the higher frequency signal)?

a. $CH_3\underline{CH}\underline{CH}Br$ with Br Br

c. $CH_3\underline{CH_2}\underline{CH}CH_3$ with Cl

e. $CH_3\underline{CH_2}CH{=}CH_2$

b. $CH_3\underline{CH}OCH_3$ with CH_3

d. $CH_3\underline{CH}\overset{O}{\underset{\|}{C}}\underline{CH_2}CH_3$ with CH_3

f. $\underline{CH_3}O\underline{CH_2}CH_2CH_3$

PROBLEM 13 ♦

Which underlined proton (or sets of protons) has the greater chemical shift (that is, the higher frequency signal)?

a. $CH_3CH_2\underline{CH_2}Cl$ or $CH_3CH_2\underline{CH_2}Br$

b. $CH_3CH_2\underline{CH_2}Cl$ or $CH_3CH_2\underline{CH}CH_3$ with Cl

c. $CH_3CH_2\overset{O}{\underset{\|}{C}}\underline{H}$ or $CH_3CH_2\overset{O}{\underset{\|}{C}}O\underline{CH_3}$

PROBLEM 14

Without referring to Table 14.1, label the proton or set of protons in each compound that gives the signal at the lowest frequency *a,* at the next lowest *b,* and so on.

a. $CH_3CH_2\overset{O}{\underset{\|}{C}}H$

b. $CH_3CH_2CHCH_3$ with OCH_3

c. $ClCH_2CH_2CH_2Cl$

d. $CH_3CH_2CHCH_2CH_3$
  |
  OCH_3

f. $CH_3CH_2CH_2OCHCH_3$
  |
  CH_3

h. $CH_3CHCH_2OCH_3$
  |
  CH_3

e. $CH_3CH_2CH_2 \overset{\displaystyle O}{\overset{\|}{C}} OCH_3$

g. $CH_3CH_2CH_2 \overset{\displaystyle O}{\overset{\|}{C}} CH_3$

i. $CH_3CHCHCH_3$
  | |
  CH_3 Cl

14.8 DIAMAGNETIC ANISOTROPY

The chemical shifts of hydrogens bonded to sp^2 carbons are at higher frequencies than one would predict from the electronegativities of the sp^2 carbons. For example, a hydrogen bonded to a benzene ring appears at 6.5 to 8.0 ppm, a hydrogen bonded to the terminal sp^2 carbon of an alkene appears at 4.7 to 5.3 ppm, and a hydrogen bonded to a carbonyl carbon appears at 9.0 to 10.0 ppm (Table 14.1).

$$\text{(7.3 ppm)} \quad C_6H_5\text{-H} \qquad \overset{\text{(5.3 ppm)}}{CH_3CH_2CH{=}CH_2} \text{(4.7 ppm)} \qquad CH_3CH_2 \overset{\displaystyle O}{\overset{\|}{C}} \text{H (9.0 ppm)}$$

The unusual chemical shifts associated with hydrogens bonded to carbons that form π bonds are due to **diamagnetic anisotropy.** This term describes an environment in which different magnetic fields are found at different points in space. (*Anisotropic* is Greek for "different in different directions.")

Because π electrons are less tightly held by nuclei than are σ electrons, π electrons are freer to move in response to a magnetic field. When a magnetic field is applied to a compound with π electrons, the π electrons move in a circular path that induces a small local magnetic field. How this induced magnetic field affects the chemical shift of a proton depends on the direction of the induced magnetic field in the region where the proton is located, relative to the direction of the applied magnetic field.

Benzene Ring Protons

The magnetic field induced by the π electrons of a benzene ring in the region where benzene's protons are located is oriented in the same direction as the applied magnetic field (Figure 14.6). As a result, the protons sense a larger effective magnetic field—namely, the sum of the strengths of the applied field and the induced field. Because frequency is proportional to the strength of the magnetic field experienced by the protons (Figure 14.2), the protons show signals at *higher frequencies* than they would if the π electrons did not induce a magnetic field.

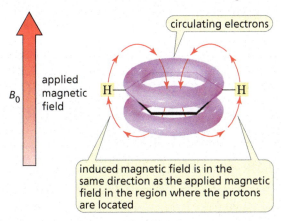

circulating electrons

B_0 applied magnetic field

induced magnetic field is in the same direction as the applied magnetic field in the region where the protons are located

◀ **Figure 14.6**
The magnetic field induced by the π electrons of a benzene ring in the vicinity of the protons attached to the sp^2 carbons has the same direction as the applied magnetic field. As a result, these protons sense a larger effective magnetic field, so their signals appear at higher frequencies.

Alkene and Aldehyde Protons

The magnetic field induced by the π electrons of an alkene or by the π electrons of an aldehyde (in the region where protons bonded to sp^2 carbons are located) is also oriented in the *same direction* as the applied magnetic field (Figure 14.7). These protons, too, show signals at higher than expected frequencies.

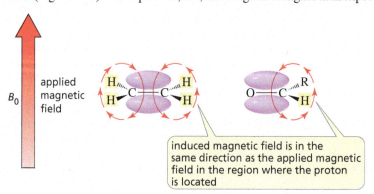

▶ **Figure 14.7**
▶ **Figure 14.7**
The magnetic fields induced by the π electrons of an alkene and by the π electrons of a carbonyl group in the vicinity of the vinylic and aldehydic protons have the same direction as the applied magnetic field. Because a larger effective magnetic field is sensed by the protons, their signals appear at higher frequencies.

induced magnetic field is in the same direction as the applied magnetic field in the region where the proton is located

Alkyne Proton

In contrast, the chemical shift of a hydrogen bonded to an sp carbon is at a lower frequency (~1.9 ppm) than one would predict from the electronegativity of the sp carbon. This is because the direction of the magnetic field induced by the alkyne's cylinder of π electrons, in the region where the proton is located, is *opposite* to the direction of the applied magnetic field (Figure 14.8). Thus, the proton senses a smaller effective magnetic field and, therefore, shows a signal at a *lower frequency* than it would if the π electrons did not induce a magnetic field.

induced magnetic field is in the direction opposite to that of the applied magnetic field in the region where the proton is located

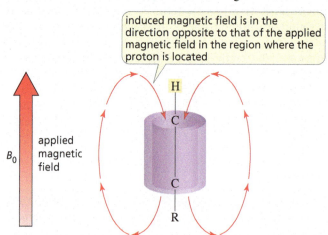

▶ **Figure 14.8**
The magnetic field induced by the π electrons of an alkyne in the vicinity of the proton bonded to the sp carbon is in a direction opposite to that of the applied magnetic field. Because a smaller effective magnetic field is sensed by the proton, its signal appears at a lower frequency.

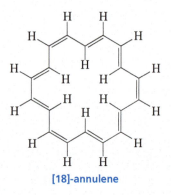

[18]-annulene

PROBLEM 15 ◆

[18]-Annulene shows two signals in its ^{1}H NMR spectrum: one at 9.25 ppm and the other to the right of the TMS signal at −2.88 ppm. What hydrogens are responsible for each of the signals? (*Hint:* Look at the direction of the induced magnetic field outside and inside the benzene ring in Figure 14.6.)

14.9 THE INTEGRATION OF NMR SIGNALS REVEALS THE RELATIVE NUMBER OF PROTONS CAUSING EACH SIGNAL

The two signals in the ^{1}H NMR spectrum in Figure 14.9 on the top of the next page are not the same size because *the area under each signal is proportional to the number of protons that produce the signal.*

- The lower frequency signal has a larger area because it is produced by 9 protons.
- The higher frequency signal has a smaller area because it is produced by 2 protons.

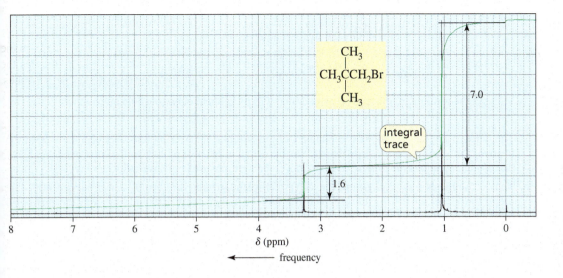

Analysis of the integration line in the ^{1}H NMR spectrum of 1-bromo-2,2-dimethylpropane. The peak at 3.3 ppm has a smaller integral trace than the peak at 1.0 ppm because the peak at 3.3 ppm is produced by two methylene protons, whereas the peak at 1.0 ppm is produced by nine methyl protons.

Integration of the Signals

You might remember from a calculus course that the area under a curve can be determined by an integral. An NMR spectrometer is equipped with a computer that calculates the integrals electronically and then displays them as an integral trace superimposed on the original spectrum (green line in Figure 14.9). The height of each step in the integral trace is proportional to the area under the corresponding signal, which, in turn, is proportional to the number of protons producing the signal.

For example, the heights of the integral traces in Figure 14.9 tell you that the ratio of the integrals is approximately 1.6 : 7.0. Dividing by the smallest number gives a new ratio (1 : 4.4). You then need to multiply this ratio by a number that makes all the numbers in the ratio close to whole numbers—in this case that number is 2. The ratio of protons in the compound is 2 : 8.8, which is rounded to 2 : 9, because there can be only whole numbers of protons. (The heights of the integral traces are approximate because of experimental error.) Modern spectrometers print the integrals as numbers on the spectrum; see Figure 14.10 on page 635.

Integration indicates the *relative* number of protons that produce each signal, not the *absolute* number. In other words, integration could not distinguish between the following two compounds because both would show an integral ratio of 1 : 3.

$$CH_3-\underset{\underset{Cl}{|}}{C}H-Cl \qquad\qquad CH_3-\underset{\underset{Cl}{|}}{\overset{\overset{CH_3}{|}}{C}}-CH_2Cl$$

1,1-dichloroethane   **1,2-dichloro-2-methylpropane**
ratio of protons = 1 : 3   **ratio of protons = 2 : 6 = 1 : 3**

PROBLEM 16 ◆

How would integration distinguish the ^{1}H NMR spectra of the following compounds?

$$CH_3-\underset{\underset{CH_3}{|}}{\overset{\overset{CH_3}{|}}{C}}-CH_2Br \qquad CH_3-\underset{\underset{Br}{|}}{\overset{\overset{CH_3}{|}}{C}}-CH_2Br \qquad CH_3-\underset{\underset{CH_2Br}{|}}{\overset{\overset{CH_2Br}{|}}{C}}-CH_2Br$$

PROBLEM 17 SOLVED

a. Calculate the ratios of the different kinds of protons in a compound with an integral ratio of 6 : 4 : 18.4 (going from left to right across the spectrum).

b. Determine the structure of a compound with molecular formula $C_7H_{14}O_2$ that gives these relative integrals in the observed order.

LEARN THE STRATEGY

SOLUTION

a. Divide each number in the ratio by the smallest number:

$$\frac{6}{4} = 1.5 \qquad \frac{4}{4} = 1 \qquad \frac{18.4}{4} = 4.6$$

Multiply the resulting numbers by a number that will make them close to whole numbers:

$$1.5 \times 2 = 3 \qquad 1 \times 2 = 2 \qquad 4.6 \times 2 = 9$$

The ratio 3 : 2 : 9 gives the relative numbers of the different kinds of protons. Because the sum of the relative number of protons (14) is the same as the actual number of protons in the compound, we know that the actual ratio is the relative ratio and not some multiple of the relative ratio.

b. The "3" suggests a methyl group, the "2" a methylene group, and the "9" a *tert*-butyl group. The methyl group is closest to the group in the molecule that causes deshielding, and the *tert*-butyl group is farthest away from the group that causes deshielding. The following compound meets these requirements:

USE THE STRATEGY

PROBLEM 18 ◆

Which of the following compounds is responsible for the ¹H NMR spectrum shown below?

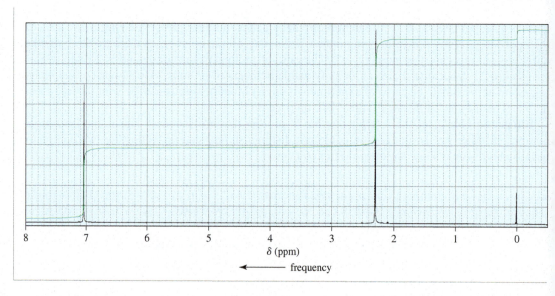

14.10 THE SPLITTING OF SIGNALS IS DESCRIBED BY THE N + 1 RULE

Notice that the shapes of the signals in the ¹H NMR spectrum in Figure 14.10 are different from the shapes of the signals in the ¹H NMR spectrum in Figure 14.9. Both signals in Figure 14.9 are **singlets,** meaning each is composed of a single peak. In contrast, the signal for the methyl protons in Figure 14.10 (the lower-frequency signal) is split into two peaks (a **doublet**), and the signal for the methine proton is split into four peaks (a **quartet**). (Magnifications of the doublet and quartet are shown as insets in Figure 14.10; integration numbers are shown in green.)

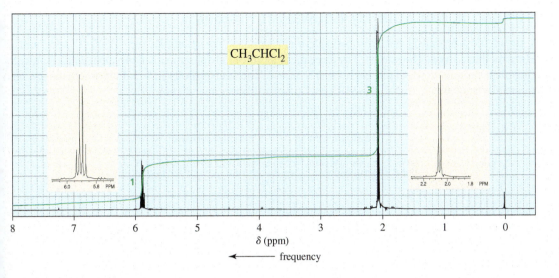

CH₃CHCl₂

◀ **Figure 14.10**
The ¹H NMR spectrum of 1,1-dichloroethane. The higher-frequency signal (due to C*H*Cl₂) is an example of a quartet; the lower-frequency signal (due to C*H₃*) is a doublet.

The *N* + 1 Rule

Splitting is caused by protons bonded to *adjacent* carbons. The splitting of a signal is described by the *N* **+ 1 rule,** where *N* is the number of equivalent protons bonded to *adjacent* carbons that are not equivalent to the proton producing the signal. Both signals in Figure 14.9 are singlets; the three methyl groups give an unsplit signal because they are attached to a carbon that is not bonded to a hydrogen; the methylene group also gives an unsplit signal because it too is attached to a carbon that is not bonded to a hydrogen ($N = 0$, so $N + 1 = 1$).

In contrast, the carbon adjacent to the methyl group in Figure 14.10 is bonded to one proton (C*H*Cl₂), so the signal for the methyl protons is split into a doublet ($N = 1$, so $N + 1 = 2$). The carbon adjacent to the carbon bonded to the methine proton is bonded to three equivalent protons (C*H₃*), so the signal for the methine proton is split into a quartet ($N = 3$, so $N + 1 = 4$).

The number of peaks in a signal is called the **multiplicity** of the signal. Splitting is always mutual: if the *a* protons split the *b* protons, then the *b* protons must split the *a* protons. The *a* and *b* protons, in this case, are *coupled protons*. **Coupled protons** split each other's signal. Notice that coupled protons are bonded to adjacent carbons.

Keep in mind that it is not the number of protons producing a signal that determines the multiplicity of the signal; rather, it is the number of protons bonded to the immediately adjacent carbons that determines the multiplicity. For example, the signal for the *a* protons in the following compound is split into three peaks (a **triplet**) because the adjacent carbon is bonded to two protons. The signal for the *b* protons is a quartet because the adjacent carbon is bonded to three protons, and the signal for the *c* protons is a singlet.

An ¹H NMR signal is split into *N* + 1 peaks, where *N* is the number of equivalent protons bonded to adjacent carbons.

Coupled protons split each other's signal.

Coupled protons are bonded to adjacent carbons.

$$
\underset{a \qquad b \qquad\qquad c}{CH_3CH_2\overset{\overset{\overset{\textstyle O}{\|}}{C}}{}OCH_3}
$$

The signal for the *a* protons in the following compound is a triplet, and the signal for the *d* proton is also a triplet. Because the *a* and *c* protons are not equivalent, the *N* + 1 rule must be applied separately to each set to determine the splitting of the *b* protons. Thus, the signal for the *b* protons is split into a quartet by the *a* protons, and each of the four peaks of the quartet is split into a triplet by the *c* protons: $(N_a + 1)(N_c + 1) = (4)(3) = 12$. As a result, the signal for the *b* protons is a **multiplet** (a signal that is more complex than a triplet, quartet, and so on). The signal for the *c* protons is also a multiplet: $(N_b + 1)(N_d + 1) = (3)(2) = 6$. The actual number of peaks seen in a multiplet depends on how many of them overlap (Section 14.14).

$$
\underset{a \quad\ b \quad c \qquad d}{CH_3CH_2CH_2\overset{\overset{\overset{\textstyle O}{\|}}{C}}{}H}
$$

A Proton Is Not Split By Equivalent Protons

A signal for a proton is not split by *equivalent* protons. For example, the ^{1}H NMR spectrum of bromomethane shows one singlet. The three methyl protons are chemically equivalent, and chemically equivalent protons do not split each other's signal. The four protons in 1,2-dichloroethane are also chemically equivalent, so its ^{1}H NMR spectrum also shows one singlet.

Equivalent protons do not split each other's signal.

CH_3Br $ClCH_2CH_2Cl$

bromomethane   **1,2-dichloroethane**

each compound shows one singlet in its ^{1}H NMR spectrum because equivalent protons do not split each other's signals

PROBLEM 19 ◆

One of the spectra below is produced by 1-chloropropane and the other by 1-iodopropane. Which is which?

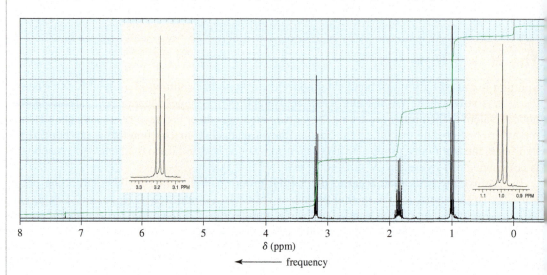

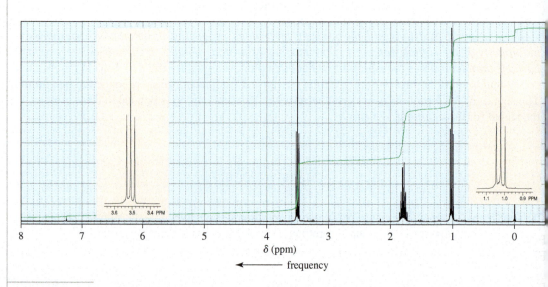

PROBLEM 20

Explain how the following compounds, each with the same molecular formula, could be distinguished by their ^{1}H NMR spectra.

$ClCH_2-CH_2-CHCl_2$ $CH_3-\underset{\underset{Cl}{|}}{CH}-CHCl_2$ $CH_3-CH_2-CCl_3$

A   **B**   **C**

PROBLEM 21 ◆

The ^1H NMR spectra of two carboxylic acids with molecular formula $C_3H_5O_2Cl$ are shown below. Identify the carboxylic acids. (The "offset" notation means that the farthest-left signal has been moved to the right by the indicated amount to fit on the spectrum; thus, the signal at 9.8 ppm offset by 2.4 ppm has an actual chemical shift of $9.8 + 2.4 = 12.2$ ppm.)

a.

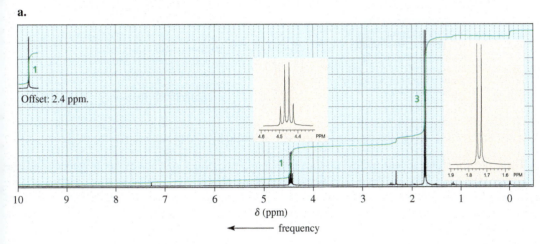

b.

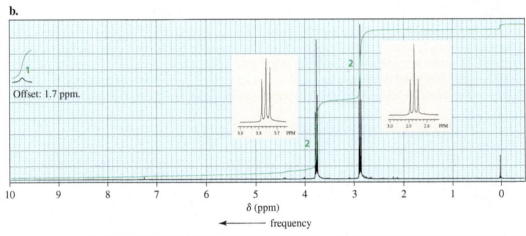

14.11 WHAT CAUSES SPLITTING?

Splitting occurs when different kinds of protons are close enough for their magnetic fields to influence one another—a situation called **spin–spin coupling.**

Forming a Doublet

For example, the frequency at which the methyl protons of 1,1-dichloroethane show a signal in Figure 14.10 is influenced by the magnetic field of the methine proton (Figure 14.11). If the magnetic field of the methine proton aligns *with* that of the applied magnetic field (indicated in Figure 14.11 by the up arrow), then it will add to the applied magnetic field, which will cause the methyl protons to show a signal at a slightly higher frequency. (Recall that the frequency is proportional to the strength of the magnetic field.) On the other hand, if the magnetic field of the methine proton aligns *against* the applied magnetic field (indicated in Figure 14.11 by the down arrow), then it will subtract from the applied magnetic field and the methyl protons will show a signal at a lower frequency.

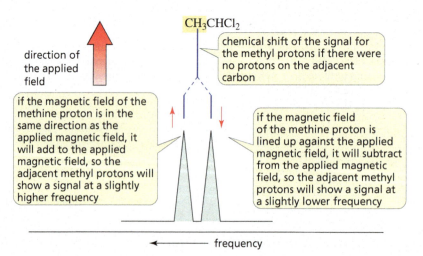

The signal for the methyl protons of 1,1-dichloroethane is split into a doublet by the methine proton.

Therefore, the signal for the methyl protons is split into two peaks, one at a higher frequency and one at a lower frequency. Because the α- and β-spin states have almost the same population (Figure 14.1 on page 621), about half the methine protons are lined up with the applied magnetic field and about half are lined up against it. As a result, the two peaks of the *doublet* have approximately the same height and area.

Forming a Quartet

Similarly, the frequency at which the methine proton shows a signal is influenced by the magnetic fields of the three protons bonded to the adjacent carbon. The magnetic fields of all three methyl protons can align with the applied magnetic field, two can align with the field and one against it, one can align with it and two against it, or all three can align against it (Figure 14.12). Because the magnetic field that the methine proton senses is affected in four different ways, its signal is a *quartet* (Figure 14.13).

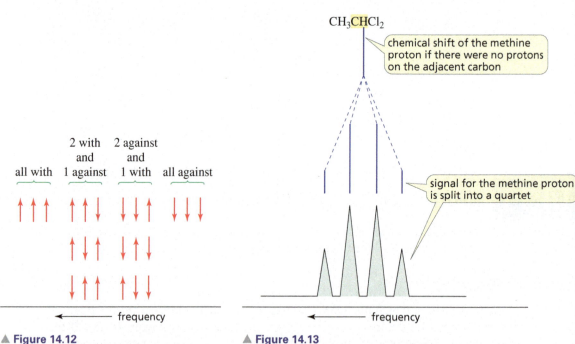

▲ **Figure 14.12**
The different ways in which the magnetic fields of three protons can be aligned.

▲ **Figure 14.13**
The signal for the methine proton of 1,1-dichloroethane is split into a quartet by the methyl protons.

Why does the signal for the methine proton in Figure 14.13 have peaks of different intensities—that is, why are the inner peaks more intense than the outer peaks?

The relative intensities of the peaks in a signal reflect the number of ways the neighboring protons can be aligned relative to the applied magnetic field (Figure 14.12). There is only one way to align the magnetic fields of three protons so that they are all lined up with the applied magnetic field and only one way to align them so that they are all lined up against the applied field. However, there are three ways to align the magnetic fields of three protons so that two are lined up with the applied field and one is lined up against it, and there are three ways to align them so that one is lined up with the applied field and two are lined up against it. Therefore, a quartet has relative peak intensities of 1 : 3 : 3 : 1.

Long-Range Coupling

Normally, *nonequivalent* protons split each other's signal only if they are on the same or *adjacent* carbons. Splitting is a "through-bond" effect, not a "through-space" effect, and it is rarely observed if the protons are separated by more than three σ bonds. However, if they are separated by four bonds and one of the bonds is a double or a triple bond, then splitting is sometimes observed. This phenomenon is called **long-range coupling.**

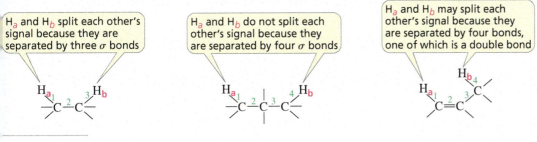

H$_a$ and H$_b$ split each other's signal because they are separated by three σ bonds

H$_a$ and H$_b$ do not split each other's signal because they are separated by four σ bonds

H$_a$ and H$_b$ may split each other's signal because they are separated by four bonds, one of which is a double bond

PROBLEM 22

Draw a diagram like the one shown in Figure 14.12 to predict

a. the relative intensities of the peaks in a triplet.
b. the relative intensities of the peaks in a quintet.

14.12   MORE EXAMPLES OF ^{1}H NMR SPECTRA

Let's now look at a few more spectra to give you additional practice in analyzing ^{1}H NMR spectra.

Example 1

There are two signals in the ^{1}H NMR spectrum of 1,3-dibromopropane (Figure 14.14). The signal for the **b** protons is split into a triplet by the **a** protons. The protons on the two carbons adjacent to the one bonded to the **a** protons are equivalent. Because the two sets of protons are equivalent, the $N + 1$ rule is applied to both sets at the same time when determining the splitting of the signal for the **a** protons. In other words, N is equal to the sum of the equivalent protons on both carbons. Thus, the signal for the **a** protons is split into a quintet ($4 + 1 = 5$). Integration confirms that two methylene groups contribute to the higher frequency signal because it shows that twice as many protons produce that signal than the lower frequency signal.

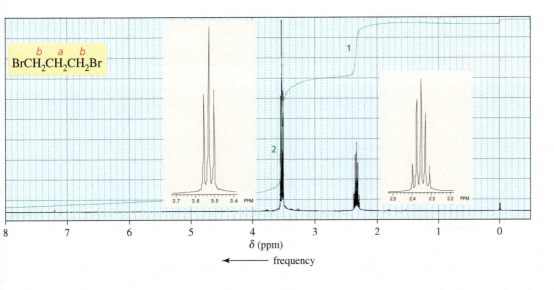

BrCH$_2$CH$_2$CH$_2$Br

δ (ppm)

frequency

◀ **Figure 14.14**
The ^{1}H NMR spectrum of
1,3-dibromopropane.
The quintet corresponds to H$_a$ and
the triplet corresponds to H$_b$.

Example 2

The ^{1}H NMR spectrum in Figure 14.15 shows five signals. The signal for the **a** protons is split into a triplet by the **c** protons and the signal for the **b** protons is split into a doublet by the **e** proton. The signal for the **c** protons is split into a multiplet by the **a** and **d** protons (the $N + 1$ rule is applied separately to the **a** and **d** protons). The signal for the **d** protons is split into a triplet by the **c** protons; and the signal for the **e** proton is split into a septet by the **b** protons (the $N + 1$ rule is applied to both sets of **b** protons at the same time).

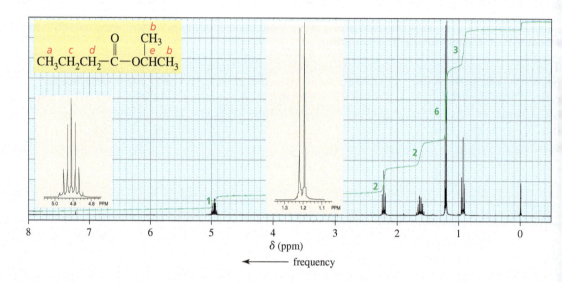

▶ **Figure 14.15**
The ^{1}H NMR spectrum of isopropyl butanoate.

PROBLEM 23

Indicate the number of signals and the multiplicity of each signal in the ^{1}H NMR spectrum of each of the following compounds:

a. $CH_3CH_2CH_2CH_2CH_2CH_3$ **b.** $ICH_2CH_2CH_2Br$ **c.** $ClCH_2CH_2CH_2Cl$ **d.** $ICH_2CH_2CHBr_2$

Example 3

The ^{1}H NMR spectrum of 3-bromo-1-propene shows four signals (Figure 14.16). The signal for the **a** protons is split into a doublet by the **d** proton. Although the **b** and **c** protons are bonded to the same carbon, they are not equivalent (one is cis to the bromomethyl group, and the other is trans to the bromomethyl group), so each produces a separate signal. The signal for the **d** proton is a multiplet because it is split separately by the **a**, **b**, and **c** protons. Notice that the signals for the three vinylic protons are at relatively high frequencies because of diamagnetic anisotropy (Section 14.8).

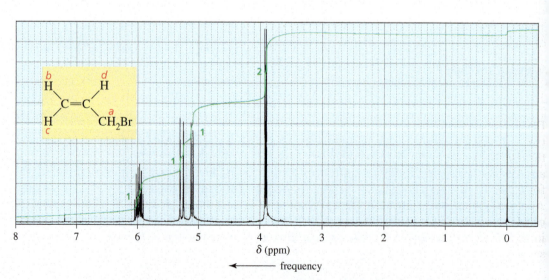

▶ **Figure 14.16**
The ^{1}H NMR spectrum of 3-bromo-1-propene.

Because the **b** and **c** protons in Figure 14.16 are not equivalent, they split one another's signal. This means that the signal for the **b** proton is split into a doublet by the **d** proton and that each of the peaks in the doublet is split into a doublet by the **c** proton. The signal for the **b** proton should, therefore, be what is called a **doublet of doublets,** and so should the signal for the **c** proton. However, the mutual splitting of the signals of two nonidentical protons bonded to the same car-bon—called **geminal coupling**—is often too small to be observed if they are bonded to an sp^2 carbon (see Table 14.2 on page 644). Therefore, the signals for the **b** and **c** protons in Figure 14.16 appear as doublets rather than as doublets of doublets.

A Quartet versus a Doublet of Doublets

There is a clear difference between a quartet and a doublet of doublets, even though both have four peaks. A quartet results from splitting by *three equivalent* adjacent protons; it therefore has relative peak intensities of 1 : 3 : 3 : 1, and the individual peaks are equally spaced (Figures 14.13). A dou-blet of doublets, on the other hand, results from splitting by *two nonequivalent* adjacent protons; it has relative peak intensities of 1 : 1 : 1 : 1, and the individual peaks are not necessarily equally spaced (see Figure 14.21 on page 648).

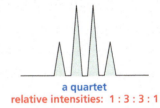

a quartet
relative intensities: 1 : 3 : 3 : 1

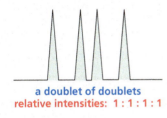

a doublet of doublets
relative intensities: 1 : 1 : 1 : 1

Example 4

Ethylbenzene has five sets of chemically equivalent protons (Figure 14.17). Notice the expected triplet for the **a** protons and the quartet for the **b** protons. (This is the characteristic pattern for an ethyl group.) It is reasonable to expect the signal for the **c** protons to be a doublet and the signal for the **e** proton to be a triplet.

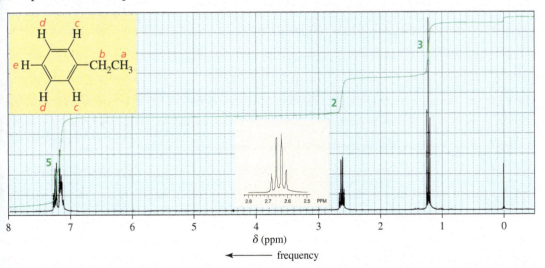

◀ **Figure 14.17**
The ^{1}H NMR spectrum of ethylbenzene. The signals for the *c, d,* and *e* protons overlap.

Because the **c** and **e** protons are not equivalent, they must be considered separately in determining the splitting of the signal for the **d** protons $(N_c + 1)(N_e + 1)$. Therefore, you expect the signal for the **d** protons to be split into a doublet by the **c** protons and each peak of the doublet to be split into another doublet by the **e** proton, forming a doublet of doublets. However, three distinct signals for the **c, d,** and **e** protons are not apparent in Figure 14.17. Instead, you see overlapping signals. Apparently, the electronic effect (that is, the electron-donating or electron-withdrawing ability) of an ethyl substituent is not sufficiently different from that of a hydrogen to cause a difference in the environments of the **c, d,** and **e** protons that is large enough to allow them to appear as separate signals.

Example 5

Unlike the benzene protons of ethylbenzene (*c*, *d*, and *e*), the benzene protons of nitrobenzen
(*a*, *b*, and *c*) show three distinct signals (Figure 14.18), and the multiplicity of each signal is wha
you predicted for the signals for the benzene ring protons in ethylbenzene (*c* is a doublet, *b* i
a triplet, and *a* is a doublet of doublets with some overlapping peaks). In contrast to the ethy
substituent, the nitro substituent is sufficiently electron withdrawing to cause the *a*, *b*, and *c* proton
to be in sufficiently different environments for their signals not to overlap.

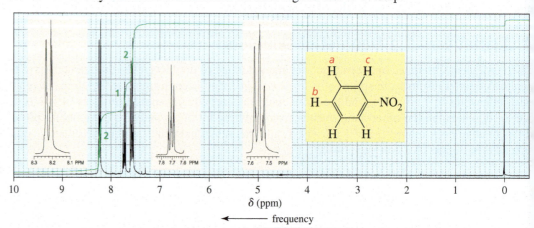

▶ **Figure 14.18**
**The ^{1}H NMR spectrum of nitrobenzene.
The signals for the *a*, *b*, and *c* protons
do not overlap and their multiplicities are
what the *N* + 1 rule predicts.**

Notice that the signals for the benzene ring protons in Figures 14.17 and 14.18 occur in the 7.0 t
8.5 ppm region. Other kinds of protons usually do not resonate in this region, so signals in thi
region indicate that the compound probably contains a benzene ring.

**Benzene ring protons occur
in the 7.0-8.5 ppm region.**

PROBLEM 24

Explain the relative chemical shifts of the benzene ring protons in Figure 14.18.

PROBLEM 25

How can their ^{1}H NMR spectra distinguish the following compounds?

A   B   C

PROBLEM 26 ◆

Identify each compound from its molecular formula and its ^{1}H NMR spectrum:

a. C_9H_{12}

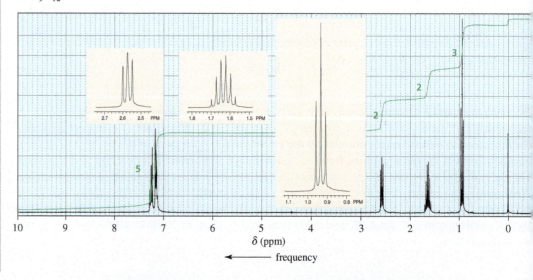

b. $C_5H_{10}O$

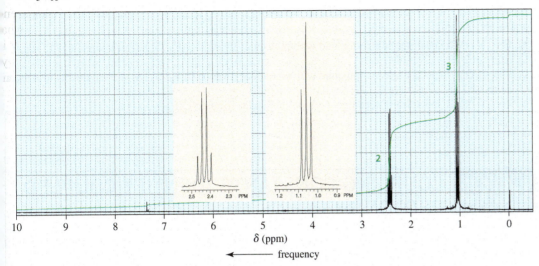

c. $C_9H_{10}O_2$

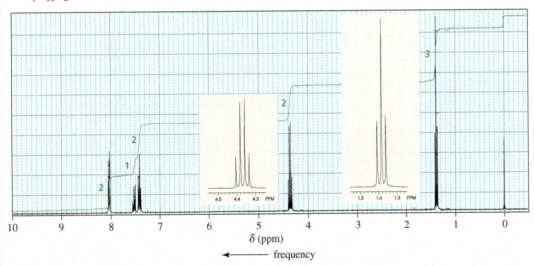

PROBLEM 27
Predict the splitting patterns for the signals given by the compounds in Problem 4.

PROBLEM 28
Describe the ^{1}H NMR spectrum you would expect for each of the following compounds, indicating the relative positions of the signals:

a. $BrCH_2CH_2CH_2CH_2Br$

b. $CH_3OCH_2CH_2CH_2Br$

c. $O={\Large\bigcirc}=O$

d. $CH_3\overset{\displaystyle CH_3}{\underset{\displaystyle Br}{C}}CH_2CH_3$

e. $CH_3\overset{O}{\overset{\|}{C}}CH_2\overset{O}{\overset{\|}{C}}OCH_3$

f. $\overset{H}{\underset{H}{}}C=C\overset{H}{\underset{Cl}{}}$

g. $CH_3CH_2OCH_2CH_3$

h. $CH_3CH_2OCH_2Cl$

i. $CH_3CHCHCl_2$
        $\quad\ \ |$
        $\quad\ \ Cl$

j. $\square{-}O$

k. $CH_3CH\overset{O}{\overset{\|}{C}}H$
      $\quad\ |$
      $\quad\ CH_3$

l. $CH_3OCH_2CH_2CH_2OCH_3$

m. $\overset{Cl}{\underset{H}{}}C=C\overset{Cl}{\underset{Cl}{}}$

n. $\overset{Cl}{\underset{H}{}}C=C\overset{H}{\underset{Cl}{}}$

o. ⬠

p. ⬠—Cl

PROBLEM 29

Propose structures that are consistent with the following spectra. (Integral ratios are given from left to right across the spectrum.)

a. The ^{1}H NMR spectrum of a compound with molecular formula $C_4H_{10}O_2$ has two singlets with integral ratios of $2:3$.

b. The ^{1}H NMR spectrum of a compound with molecular formula $C_6H_{10}O_2$ has two singlets with integral ratios of $2:3$.

c. The ^{1}H NMR spectrum of a compound with molecular formula $C_8H_6O_2$ has two singlets with integral ratios of $1:2$.

14.13 COUPLING CONSTANTS IDENTIFY COUPLED PROTONS

The distance, in hertz, between two adjacent peaks of a split NMR signal is called the **coupling constant** (denoted by **J**). The coupling constant for the **a** protons being split by the **b** proton is denoted by J_{ab}. The signals of coupled protons (protons that split each other's signal) have the same coupling constant; in other words, $J_{ab} = J_{ba}$ (Figure 14.19). Coupling constants are useful in analyzing complex NMR spectra because protons on adjacent carbons can be identified by their identical coupling constants.

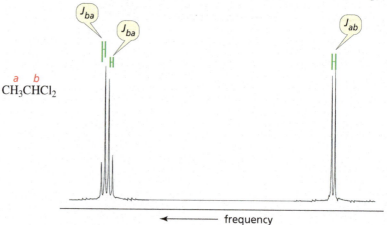

▶ **Figure 14.19**
The a and b protons of 1,1-dichloroethane are coupled protons—that is, they split each other's signals, so their signals have the same coupling constant, $J_{ab} = J_{ba}$.

J has the same value regardless of the operating frequency of the spectrometer. The magnitude of a coupling constant is a measure of how strongly the nuclear spins of the coupled protons influence each other. It depends, therefore, on the number of bonds and the type of bonds that connect the coupled protons, as well as on the geometric relationship of the protons. Characteristic coupling constants are shown in Table 14.2; they range from 0 to 15 Hz.

Table 14.2 Approximate Values of Coupling Constants

Types of protons	Approximate J_{ab} (Hz)	Types of protons	Approximate J_{ab} (Hz)	Types of protons	Approximate J_{ab} (Hz)
H_a $-C-$ H_b	12	H_a $C=C$ H_b	15 (trans)	$C=C$ H_a H_b	2
H_a H_b $-C-C-$	7	H_a $C=C$ H_b	10 (cis)	H_a $C=C$ H_b	1 (long-range coupling)
H_a H_b $-C-C-C-$	0				

The coupling constant for two nonequivalent hydrogens bonded to the *same sp³* carbon is large (12 Hz). In contrast, the coupling constant for two nonequivalent hydrogens bonded to the *same sp²* carbon is often too small to see (2 Hz) (see Figure 14.16), but it is large if the nonequivalent hydrogens are bonded to *adjacent sp²* carbons (10–15 Hz). We saw that π electrons allow long-range coupling—that is, coupling through four bonds (Section 14.11).

Coupling Constants Distinguish Cis and Trans Isomers

Coupling constants can be used to distinguish the ^{1}H NMR spectra of cis and trans alkenes. The coupling constant of trans vinylic protons is significantly greater than that of cis vinylic protons, because the coupling constant depends on the dihedral angle between the two C—H bonds in the H—C=C—H unit (Figure 14.20). The coupling constant is greatest when the angle between the two C—H bonds is 180° (trans) and smallest when the angle is 0° (cis).

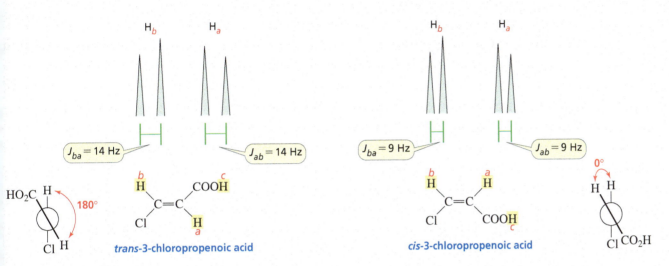

▲ **Figure 14.20**
The doublets observed for the *a* and *b* protons in the ^{1}H NMR spectra of a trans alkene and a cis alkene. The coupling constant for trans protons (14 Hz) is greater than the coupling constant for cis protons (9 Hz) because it depends on the dihedral angle (180° for trans protons; 0° for cis protons).

The peaks in the doublets in Figure 14.20 are not the same height because of **leaning**—an arrow drawn from the top of the smaller peak of the doublet to the top of the larger peak points at the signal of the proton responsible for the splitting. Thus, the figure shows that H_a is split by H_b and H_b is split by H_a.

The coupling constant for trans vinylic protons is greater than that for cis vinylic protons.

Summary

Let's now summarize the information that can be obtained from an ^{1}H NMR spectrum:

1. The number of signals indicates the minimum number of different kinds of protons in the compound. (There could be more if there is overlap.)
2. The position of a signal indicates the kind of proton(s) that produce the signal (methyl, methylene, methine, allylic, vinylic, benzene, and so on) and the kinds of neighboring substituents.
3. The integration of the signal tells the relative number of protons that produce the signal.
4. The multiplicity of the signal ($N + 1$) tells the number of protons (N) bonded to adjacent carbons.
5. The coupling constants identify coupled protons.

PROBLEM 30 ◆

Why is there no coupling between the *a* and *c* protons or between the *b* and *c* protons in the cis and trans alkenes shown in Figure 14.20?

PROBLEM-SOLVING STRATEGY

LEARN THE STRATEGY

Using IR and ^{1}H NMR Spectra to Deduce a Chemical Structure

Identify the compound with molecular formula $C_9H_{10}O$ that gives the IR and ^{1}H NMR spectra shown here.

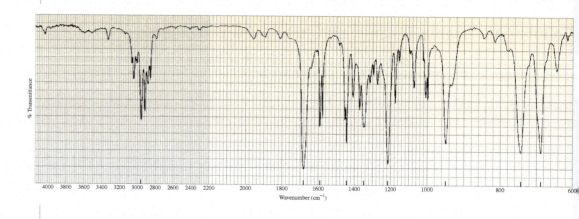

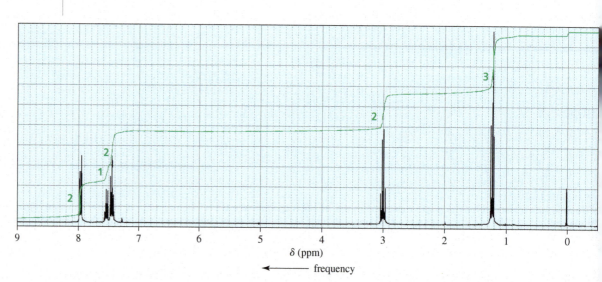

One way to approach this kind of problem is to identify whatever structural features you can from the ^{1}H NMR spectrum and then use the information from the molecular formula and the IR spectrum to expand on that knowledge.

The signals in the 7.4 to 8.0 ppm region of the NMR spectrum indicate a benzene ring; because the signals integrate to 5H, it is a monosubstituted benzene ring. The triplet at ~1.2 ppm and the quartet at ~3.0 ppm indicate an ethyl group that is attached to an electron-withdrawing group.

Based on the molecular formula and the IR spectrum, it is clear that the compound is a ketone: it has a carbonyl group at ~1680 cm^{-1}, only one oxygen, and no absorption bands at ~2820 and ~2720 cm^{-1} that would indicate an aldehyde. The carbonyl group absorption band is at a lower frequency than is typical, which suggests that it has partial single-bond character as a result of electron delocalization, indicating that it is attached to an sp^2 carbon. Now you can conclude that the compound is the ketone shown here. The integration ratio (5 : 2 : 3) confirms this answer.

PROBLEM 31 ♦

Identify the compound with molecular formula $C_8H_{10}O$ that gives the IR and ^1H NMR spectra shown here.

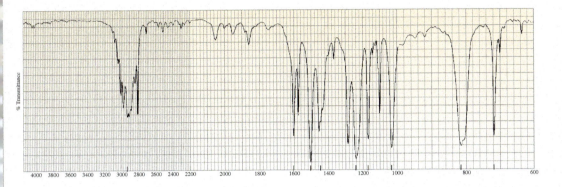

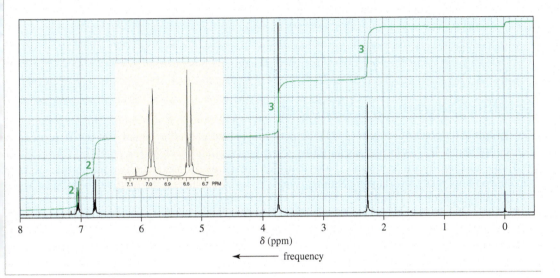

14.14   SPLITTING DIAGRAMS EXPLAIN THE MULTIPLICITY OF A SIGNAL

The splitting pattern obtained when a signal is split by more than one set of protons can best be understood by using a splitting diagram. In a **splitting diagram** (also called a **splitting tree**), the NMR peaks are shown as vertical lines, and the effect of each of the splittings is shown one at a time.

Splitting Diagram for a Doublet of Doublets

The splitting diagram in Figure 14.21 depicts the splitting of the signal for the **c** proton of the compound shown on the top of the next page into a doublet of doublets by the **b** and **d** protons. Notice that the diagram begins with the largest (J_{cb}) of the two J values.

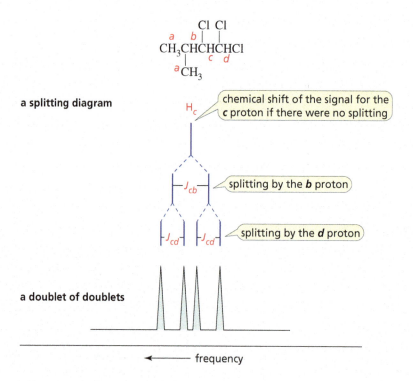

► Figure 14.21
A splitting diagram for a doublet of doublets. The diagram begins with J_{cb} because $J_{cb} > J_{cd}$.

Splitting Diagram for a Multiplet

The splitting diagram in Figure 14.22 depicts the splitting of the signal for the **b** protons of 1-bromopropane. It is split into a quartet by the **a** protons, and each of the resulting four peaks is split into a triplet by the **c** protons. How many of the 12 peaks are actually seen in the spectrum depends on how many overlap one another, which in turn depends on the relative magnitudes of the two coupling constants, J_{ba} and J_{bc}.

For example, Figure 14.22 shows that there are 12 peaks when J_{ba} is much greater than J_{bc}, 9 peaks when $J_{ba} = 2J_{bc}$, and only 6 peaks when $J_{ba} = J_{bc}$. When peaks overlap, their intensities add together.

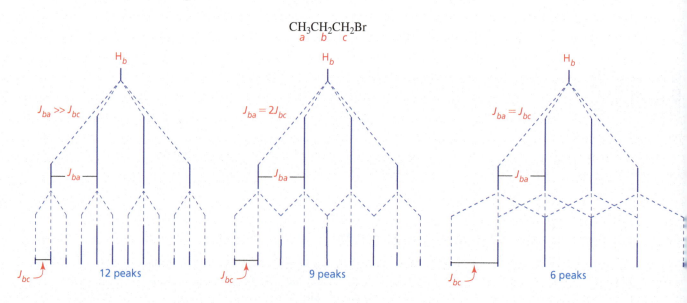

▲ Figure 14.22
A splitting diagram for a quartet of triplets. The number of peaks actually observed when a signal is split by two sets of protons depends on the relative magnitudes of the two coupling constants.

When $J_{ab} = J_{ac}$

You expect the signal for the *a* protons of 1-chloro-3-iodopropane to be a triplet of triplets (nine peaks) because the signal would be split into a triplet by the *b* protons and each of the three resulting peaks would be split into a triplet by the *c* protons. The signal, however, in Figure 14.23 is actually a quintet.

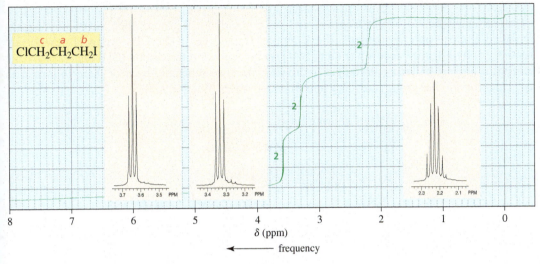

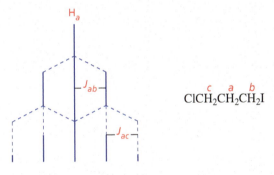

◀ **Figure 14.23**
The ^1H NMR spectrum of 1-chloro-3-iodopropane. The signal at 2.2 ppm is a quintet instead of a triplet of triplets because $J_{ab} = J_{ac}$.

Finding that the signal for the *a* protons is a quintet indicates that J_{ab} and J_{ac} have about the same value. The splitting diagram shows that a quintet results if $J_{ab} = J_{ac}$. Therefore, we can conclude that if the coupling constants are the same, the $N + 1$ rule can be applied to the two sets of nonequivalent protons at the same time. (Notice that the *b* protons in Figure 14.22 are a sextet when $J_{ba} = J_{bc}$.)

Summary

We can conclude that *when two different sets of protons split a signal, the multiplicity of the signal can be determined by using*

- *the $N + 1$ rule simultaneously for both sets, if the two sets are equivalent and, therefore, have the same coupling constants.*
- *the $N + 1$ rule simultaneously for both sets, if the two sets are nonequivalent but have similar coupling constants.*
- *the $N + 1$ rule separately for each set of protons—that is, $(N_a + 1)(N_b + 1)$, if the two sets are nonequivalent and have different coupling constants.*

PROBLEM 32

Draw a splitting diagram for H_b, where

a. $J_{ba} = 12$ Hz and $J_{bc} = 6$ Hz.   **b.** $J_{ba} = 12$ Hz and $J_{bc} = 12$ Hz.

14.15 ENANTIOTOPIC AND DIASTEREOTOPIC HYDROGENS

Enantiotopic Hydrogens

If a carbon is bonded to two hydrogens *and* to two nonidentical groups, then the two hydrogens are called **enantiotopic hydrogens.** For example, H_a and H_b in the CH_2 group of ethanol are enantiotopic hydrogens because the other two groups bonded to the carbon (CH_3 and OH) are not identical. They are called enantiotopic hydrogens, because replacing each of them in turn with deuterium (or another group) creates a pair of enantiomers (Section 4.5).

Enantiotopic hydrogens show one NMR signal because they are chemically equivalent.

Enantiotopic hydrogens are chemically equivalent,
so they show only one NMR signal.

Prochiral Carbons

The carbon to which the enantiotopic hydrogens are attached is called a **prochiral carbon.** If H_a is replaced by a deuterium, the asymmetric center will have the *R* configuration. Thus, H_a is called the **pro-*R*-hydrogen.** Similarly, H_b is called the **pro-*S*-hydrogen** because replacing it with a deuterium will generate an asymmetric center with the *S* configuration.

Diastereotopic Hydrogens

LEARN THE STRATEGY

Diastereotopic hydrogens are two hydrogens bonded to a carbon that result in a pair of diastereomers when each is replaced by a deuterium (or another group).

We know that diastereotopic hydrogens are not chemically equivalent because they do not react at the same rate with achiral reagents. For example, removal of H_b and Br to form (*E*)-2-butene occurs more rapidly than removal of H_a and Br to form (*Z*)-2-butene because (*E*)-2-butene is more stable than (*Z*)-2-butene (Section 9.10). Because diastereotopic hydrogens are not equivalent, the $N + 1$ rule must be applied to them separately.

Diastereotopic hydrogens show two NMR signals because they are not chemically equivalent.

Diastereotopic hydrogens are not chemically equivalent,
so they show two NMR signals.

Typically, the chemical shifts of diastereotopic hydrogens are similar and, like other nonequivalent hydrogens, can even be the same by chance. The farther the diastereotopic hydrogens are from the asymmetric center, the more similar their chemical shifts are expected to be.

PROBLEM 33 ♦   USE THE STRATEGY

a. For the following compounds, which pairs of hydrogens (H_a and H_b) are enantiotopic hydrogens?
b. Which pairs are diastereotopic hydrogens?

$$\overset{H_a}{\underset{H_b}{CH_3CH_2\overset{|}{\underset{|}{C}}CH_3}}$$

$$\underset{H}{\overset{CH_3}{\diagdown}}C=C\underset{H_b}{\overset{H_a}{\diagup}}$$

A   B   C   D

PROBLEM 34 **SOLVED**

How many signals would you expect to see for the indicated hydrogens in the following compound's ¹H NMR spectrum?

a.   c.   e.

b.   d.   f.

SOLUTION You know that unless the two hydrogens of a CH_2 group are diastereotopic, they are chemically equivalent and will produce one signal. If the two hydrogens are diastereotopic, then they are not chemically equivalent and will produce two signals.

a. These hydrogens are enantiotopic, so they produce one signal.
b. These hydrogens are diastereotopic, so they produce two signals.
c. These hydrogens are enantiotopic, so they produce one signal.
d. These hydrogens are neither enantiotopic nor diastereotopic, so they produce one signal.
e. These hydrogens are diastereotopic, so they produce two signals.
f. These hydrogens are enantiotopic, so they produce one signal.

PROBLEM 35

How would the ¹H NMR spectra for the four compounds with molecular formula $C_3H_6Br_2$ differ?

PROBLEM 36 **SOLVED**

We expect the signal for the methyl protons adjacent to the diastereotopic hydrogens of 2-bromobutane to be a doublet of doublets as a result of splitting by the nonequivalent diastereotopic hydrogens. The signal, however, is a triplet. Use a splitting diagram to explain why it is a triplet rather than a doublet of doublets.

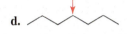

$$CH_3\overset{*}{C}HCH_2CH_3$$
$$\underset{Br}{|}$$

diastereotopic hydrogens

SOLUTION The observation of a triplet means that the $N + 1$ rule did not have to be applied to the diastereotopic hydrogens separately, but could have been applied to the two protons as a set ($N = 2$, so $N + 1 = 3$). This indicates that the coupling constant for splitting of the methyl signal by one of the diastereotopic hydrogens is similar to the coupling constant for splitting by the other diastereotopic hydrogen—that is, $J_{ab} \approx J_{ac}$.

H_a

J_{ab}

J_{ac}

14.16 THE TIME DEPENDENCE OF NMR SPECTROSCOPY

We saw that the three methyl hydrogens of ethyl bromide produce just one signal in the ^{1}H NMR spectrum because they are chemically equivalent due to rotation about the C—C bond. At any given instant, however, the three hydrogens can be in quite different environments. For example, one can be anti to the bromine, one can be gauche to the bromine, and one can be eclipsed with the bromine.

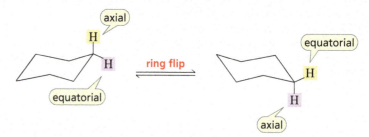

anti   gauche   eclipsed

The average time to take an NMR spectrum is about one second. Any process (in this case rotation about a C—C bond) that happens faster than once every second causes the signals to average. Therefore, there is only one signal in the ^{1}H NMR spectrum for the three methyl hydrogens of ethyl bromide. The signal represents an average of their environments.

Similarly, the ^{1}H NMR spectrum of cyclohexane shows only one signal, even though cyclohexane has both axial and equatorial hydrogens. This is because the chair conformers of cyclohexane undergo ring flip too rapidly at room temperature for the NMR spectrometer to detect them individually. Because axial hydrogens in one chair conformer are equatorial hydrogens in the other chair conformer (Section 3.14), all the hydrogens in cyclohexane have the same average environment on the NMR time scale, so the NMR spectrum shows one signal.

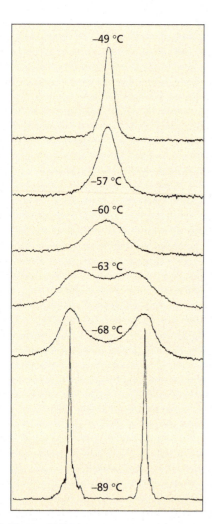

▲ Figure 14.24
A series of ^{1}H NMR spectra of cyclohexane-d_{11} obtained as the temperature is lowered from −49 °C to −89 °C.

Cyclohexane-d_{11} has 11 deuterium atoms, so it has only one hydrogen. Several ^{1}H NMR spectra of cyclohexane-d_{11} taken at various temperatures are shown in Figure 14.24. Cyclohexane with only one hydrogen was used for this experiment to prevent splitting, which would have complicated the spectrum. Deuterium signals are not detectable in ^{1}H NMR and do not split other signals (Section 14.18).

At 30 °C, the ^{1}H NMR spectrum of cyclohexane-d_{11} shows one sharp signal, which is an average for the axial hydrogen of one chair and the equatorial hydrogen of the other chair. As the temperature decreases, the signal becomes broader and eventually separates into two signals, which are equidistant from the original signal. At −89 °C, two sharp singlets are observed. At that temperature the rate of ring flip, which is temperature dependent, has decreased sufficiently to allow the two kinds of hydrogens (axial and equatorial) to be individually detected on the NMR time scale.

14.17 PROTONS BONDED TO OXYGEN AND NITROGEN

Table 14.1 on p. 629 shows that the chemical shift of a proton bonded to an oxygen or a nitrogen occurs within a range of chemical shifts. For example, the chemical shift of the OH proton of an alcohol ranges from 2 to 5 ppm. The chemical shift is variable because it depends on the extent of hydrogen bonding that the proton experiences. The chemical shift increases as the extent of hydrogen bonding increases, because hydrogen bonding decreases the electron density around the proton.

PROBLEM 37

Explain why the chemical shift of the OH proton of a carboxylic acid is at a higher frequency than the chemical shift of an OH proton of an alcohol.

The ^{1}H NMR spectrum of pure dry ethanol is shown in Figure 14.25a, and the ^{1}H NMR spectrum of ethanol with a trace amount of acid is shown in Figure 14.25b.

The spectrum shown in Figure 14.25a is what you would predict based on what you have learned so far. The signal for the proton bonded to oxygen is the signal at the highest frequency and is split into a triplet by the neighboring methylene protons; the signal for the methylene protons is split into a multiplet by the combined effects of the methyl protons and the OH proton.

a.

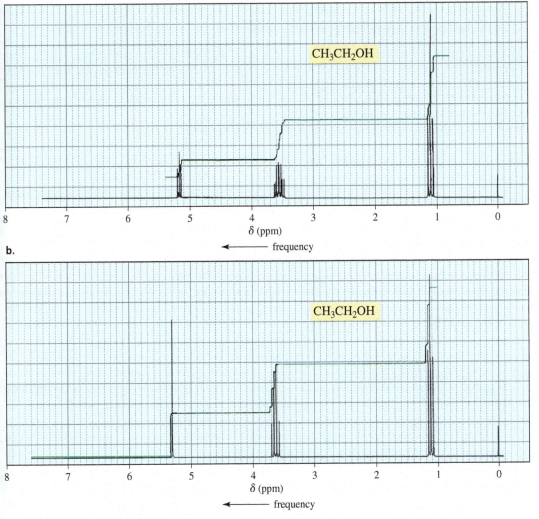

b.

◀ **Figure 14.25**

(a) The ^{1}H NMR spectrum of pure ethanol.

(b) The ^{1}H NMR spectrum of ethanol containing a trace amount of acid.

However, the spectrum shown in Figure 14.25b is the type of spectrum most often obtained for alcohols. The signal for the proton bonded to oxygen is not split, and this proton does not split the signal of the adjacent methylene protons. Therefore, the signal for the OH proton is a singlet, and the signal for the methylene protons is a quartet because it is split only by the three methyl protons.

Proton Exchange

The two spectra in Figure 14.25 are different because protons bonded to oxygen undergo **proton exchange,** which means that they are transferred from one molecule to another. Whether the OH proton and the methylene protons split each other's signals depends on how long a particular proton stays attached to the oxygen.

In a sample of pure alcohol, the rate of proton exchange is very slow. This causes the spectrum to look no different from one obtained if proton exchange did not occur.

Acids and bases catalyze proton exchange—the alcohol and the acid (or base) exchange protons. Therefore, if the alcohol is contaminated with just a trace amount of acid or base, proton exchange becomes rapid. When proton exchange is rapid, the spectrum records only an average of all possible environments. Therefore, a rapidly exchanging proton is recorded as a singlet. The effect of a rapidly exchanging proton on adjacent protons is also averaged. Thus, not only is its signal not split by adjacent protons, a rapidly exchanging proton does not cause splitting.

MECHANISM FOR ACID-CATALYZED PROTON EXCHANGE

RÖ—H + HÖH ⇌ RO—H + HÖH ⇌ RÖ: + HÖH

- A proton is transferred to the alcohol.
- A different proton is removed from the alcohol.

The signal for an OH proton is often easy to spot in an ^{1}H NMR spectrum because it is frequently broader than other signals (see the signal at δ 4.9 in Figure 14.26b on the next page). The broadening occurs because the rate of proton exchange is not slow enough to result in a cleanly split signal, as in Figure 14.25a, or fast enough for a cleanly averaged signal, as in Figure 14.25b. NH protons also show broad signals, not because of proton exchange, which is generally quite slow for NH protons, but for reasons that are beyond the scope of this book.

> **Unless the sample is pure, the hydrogen of an OH group is not split by its neighbors and does not split its neighbors.**

PROBLEM 38 ♦

Which has a greater chemical shift for the OH proton, the ^{1}H NMR spectrum of pure ethanol or the ^{1}H NMR spectrum of ethanol dissolved in CH_2Cl_2?

PROBLEM 39

Propose a mechanism for proton exchange of an alcohol in aqueous base.

PROBLEM 40 ♦

Identify the compound with molecular formula C_3H_7NO responsible for the ^{1}H NMR spectrum shown here.

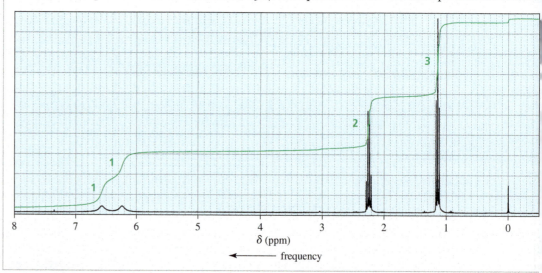

14.18 THE USE OF DEUTERIUM IN ^{1}H NMR SPECTROSCOPY

Deuterium signals are not seen in an ^{1}H NMR spectrum. Therefore, substituting a deuterium for a hydrogen is a technique used to identify signals and to simplify ^{1}H NMR spectra. For example, if the ^{1}H NMR spectrum of $CH_3CH_2OCH_3$ were compared with the ^{1}H NMR spectrum of $CH_3CD_2OCH_3$, the signal at the highest frequency in the first spectrum would be absent in the second spectrum, indicating that this signal corresponds to the methylene group.

The OH signal of an alcohol can be identified by taking an ^{1}H NMR spectrum of the alcohol and then taking another spectrum after a few drops of D_2O have been added to the sample. The OH signal is the one that becomes less intense (or disappears) in the second spectrum because of the proton exchange process just discussed. This technique can be used with any proton that undergoes exchange.

$$R-O-H \ + \ D-O-D \ \longrightarrow \ R-O-D \ + \ D-O-H$$

$$\text{(seen in } ^1\text{H NMR)} \qquad\qquad\qquad \text{(not seen in } ^1\text{H NMR)}$$

The sample used to obtain an ^{1}H NMR spectrum is made by dissolving the compound in an appropriate solvent. Solvents with protons cannot be used because the signals for solvent protons would be very intense, as there is much more solvent than compound in a solution. Instead, deuterated solvents such as $CDCl_3$ (rather than $CHCl_3$) and D_2O (rather than H_2O) are commonly used.

14.19 THE RESOLUTION OF ^{1}H NMR SPECTRA

An ^{1}H NMR spectrum taken on a 60 MHz NMR spectrometer is shown in Figure 14.26a (60 MHz spectrophotometers are no longer being manufactured.); an ^{1}H NMR spectrum of the same compound taken on a 300 MHz instrument is shown in Figure 14.26b. Why is the resolution of the second spectrum so much better?

To produce separate signals with "clean" splitting patterns, the difference in the chemical shifts (Δv in Hz) of two coupled protons must be *at least 10 times* the value of their coupling constant (J).

a. (from a 60-MHz spectrophotometer)

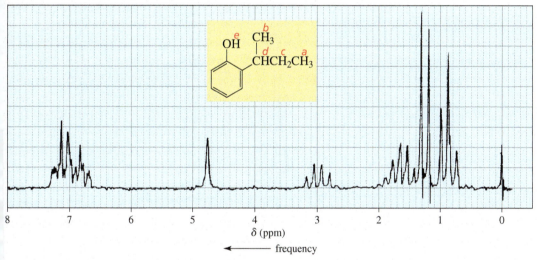

b. (from a 300-MHz spectrophotometer)

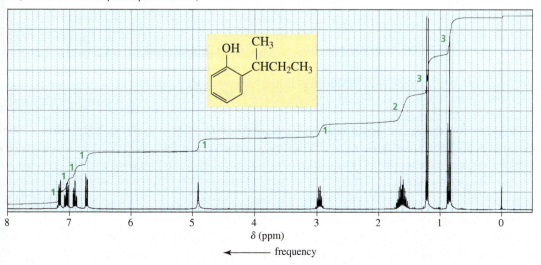

◀ **Figure 14.26**
(a) A 60 MHz ^{1}H NMR spectrum of 2-*sec*-butylphenol.
(b) A 300 MHz ^{1}H NMR spectrum of 2-*sec*-butylphenol.

Figure 14.27 shows that as $\Delta v/J$ decreases, the two signals produced by the H_a and H_b protons in an ethyl group appear closer to each other and the outer peaks of the signals become less intense while the inner peaks become more intense. The quartet and triplet of the ethyl group are clearly observed only when $\Delta v/J$ is greater than 10.

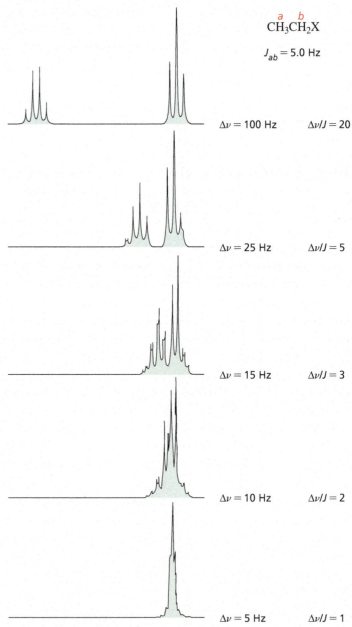

▶ Figure 14.27
The splitting pattern of an ethyl group as a function of $\Delta v/J$.

Now you can understand why the resolution of an ^{1}H NMR spectrum increases as the operating frequency of the spectrometer increases. *Δv values depend on the operating frequency of the spectrometer.* For example, the difference in the chemical shifts of the H_a and H_c protons of 2-*sec*-butylphenol is 0.8 ppm, which corresponds to 240 Hz in a 300 MHz spectrometer but only 48 Hz in a 60 MHz spectrometer.

In contrast, *J values are independent of the operating frequency,* so J_{ac} is 7 Hz, whether the spectrum is taken on a 300 MHz or a 60 MHz spectrometer. Only with the 300 MHz spectrometer is the difference in chemical shift more than 10 times the value of the coupling constant, so only in the 300 MHz spectrum do the signals show clean splitting patterns (Figure 14.26b).

in a 300 MHz spectrometer

$$\frac{\Delta v}{J} = \frac{240}{7} = 34$$

in a 60 MHz spectrometer

$$\frac{\Delta v}{J} = \frac{48}{7} = 6.9$$

14.20 ^{13}C NMR SPECTROSCOPY

The number of signals in a ^{13}**C NMR** spectrum tells how many different kinds of carbons a compound has—just as the number of signals in an ^{1}H NMR spectrum tells how many different kinds of hydrogens a compound has. The principles behind ^{1}H NMR and ^{13}C NMR spectroscopy are essentially the same.

The use of ^{13}C NMR spectroscopy as a routine analytical procedure was not possible until computers were available that could carry out a Fourier transform. ^{13}C NMR requires Fourier transform techniques because the signals obtained from a single scan are too weak to be distinguished from background electronic noise. However, ^{13}C FT–NMR scans can be repeated rapidly, so a large number of scans can be recorded and added together. When hundreds of scans are combined, ^{13}C signals stand out because electronic noise is random, so its sum is close to zero.

The individual ^{13}C signals are weak for two reasons. First, the isotope of carbon (^{13}C) that produces ^{13}C NMR signals constitutes only 1.11% of the carbon in nature (Section 1.1). (The most abundant isotope of carbon, ^{12}C, has no nuclear spin and, therefore, cannot produce an NMR signal.) The low abundance of ^{13}C causes the intensities of the signals in ^{13}C NMR to be weaker than those in ^{1}H NMR by a factor of approximately 100. Second, the gyromagnetic ratio (γ) of ^{1}H is about four times that of ^{13}C. The intensity of a signal is proportional to γ^3, so the overall intensity of a ^{1}H signal is about 6400 times ($100 \times 4 \times 4 \times 4$) stronger than the intensity of a ^{13}C signal.

One advantage to ^{13}C NMR spectroscopy is that the chemical shifts of carbon atoms range over about 220 ppm (Table 14.3), compared with a range of about 12 ppm for hydrogens (Table 14.1). This means that signals for carbons in different environments are more easily distinguished. For example, the data in Table 14.3 show that aldehyde (190 to 200 ppm) and ketone (205 to 220 ppm) carbonyl groups can be distinguished from each other and from other carbonyl groups.

Table 14.3 Approximate Values of Chemical Shifts (ppm) for ^{13}C NMR

Type of carbon	ppm	Type of carbon	ppm	Type of carbon	ppm	Type of carbon	ppm
$(CH_3)_4Si$	0	C≡C	70–90	C—Cl	25–50	R—C(HO)=O	175–185
R—CH$_3$	0–35	C≡N	110–120	C—N	40–60		
R—CH$_2$—R	15–55	C=C	100–150	C—O	50–90	R—C(H)=O	190–200
R—CH(R)—R	25–55	C=N	150–170	R(R)N—C=O	165–175		
		(benzene ring) C	110–170			R—C(R)=O	205–220
R—C(R)(R)—R	30–40	C—I	−20–10	RO—C=O	165–175		
		C—Br	10–40				

The reference compound used in ^{13}C NMR is TMS, the same reference compound used in ^{1}H NMR. You will find it helpful when analyzing a ^{13}C NMR spectrum to divide it into five regions and remember the kind of carbons that show signals in each.

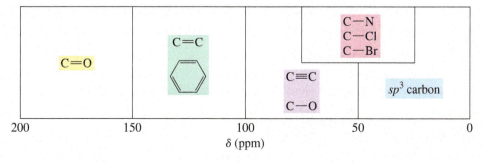

A disadvantage of ^{13}C NMR spectroscopy is that, unless special techniques are used, the area under a ^{13}C NMR signal is *not* proportional to the number of carbons that produce the signal. Thus, the number of carbons that produce a particular ^{13}C NMR signal cannot routinely be determined by integration.

A ^{13}C NMR Spectrum

The ^{13}C NMR spectrum of 2-butanol shows four signals (Figure 14.28), so you know that it has carbons in four different environments. The relative positions of the signals depend on the same factors that determine the relative positions of the proton signals in an ^1H NMR spectrum—namely, carbons in electron-rich environments produce low-frequency signals, whereas carbons close to electron-withdrawing groups produce high-frequency signals. This means that the signals for the carbons of 2-butanol are in the same relative order as the signals of the protons bonded to those carbons in its ^1H NMR spectrum.

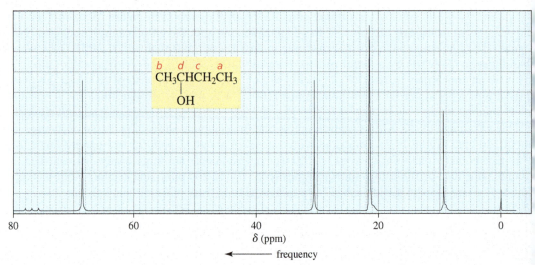

▶ **Figure 14.28**
The ^{13}C NMR spectrum of 2-butanol.

Thus, the carbon of the methyl group farthest away from the electron-withdrawing OH group gives the lowest-frequency signal. The other methyl carbon comes next in order of increasing frequency, followed by the methylene carbon; the carbon attached to the OH group gives the highest-frequency signal.

The signals in ^{13}C NMR are not normally split by neighboring carbons because there is little likelihood of an adjacent carbon being a ^{13}C; the probability of two ^{13}C carbons being next to each other is 1.11% × 1.11% (or about 1 in 10,000). ^{12}C does not have a magnetic moment, so it cannot split the signal of an adjacent ^{13}C.

The signals in a ^{13}C NMR spectrum can be split by hydrogens, but this is not usually observed because the spectra are recorded using spin-decoupling, which obliterates the carbon–proton interactions. Thus, all the signals are singlets in an ordinary ^{13}C NMR spectrum (Figure 14.28).

A Proton-Coupled ^{13}C NMR Spectrum

If the spectrometer is run in a *proton-coupled* mode, then each signal will be split by the *hydrogens* bonded to the carbon that produces the signal. The multiplicity of the signal is determined by the $N + 1$ rule.

The coupling between ^{13}C and its attached hydrogens is called **heteronuclear coupling** because the coupling is between different types of nuclei. (The coupling between nuclei of the same type, such as the coupling between adjacent protons in ^1H NMR, is called **homonuclear coupling**.)

If the spectrometer is run in a proton-coupled mode, splitting by the directly attached protons is observed in a ^{13}C NMR spectrum.

The **proton-coupled ^{13}C NMR spectrum** of 2-butanol is shown in Figure 14.29. The signals for the two methyl carbons are each split into a quartet because each methyl carbon is bonded to three hydrogens (3 + 1 = 4). The signal for the methylene carbon is split into a triplet because the carbon is bonded to two hydrogens (2 + 1 = 3), and the signal for the carbon bonded to the OH group is split into a doublet because the carbon is bonded to one hydrogen (1 + 1 = 2). (The signal at 77 ppm is produced by the solvent, $CDCl_3$.)

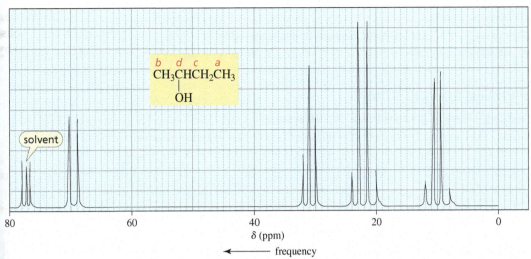

The proton-coupled ^{13}C NMR spectrum of 2-butanol. Each signal is split by the hydrogens bonded to the carbon that produces the signal, according to the **N + 1** rule.

In the following ^{13}C NMR spectrum, the three methyl groups at one end of the molecule are equivalent, so they give one signal (Figure 14.30). Because the intensity of a signal is somewhat related to the number of carbons producing it (and somewhat related to the number of hydrogens on the carbons), the signal for these three methyl groups is the most intense signal in the spectrum. The tiny signal at ~31 ppm is for the quaternary carbon; carbons that are not attached to hydrogens give very small signals.

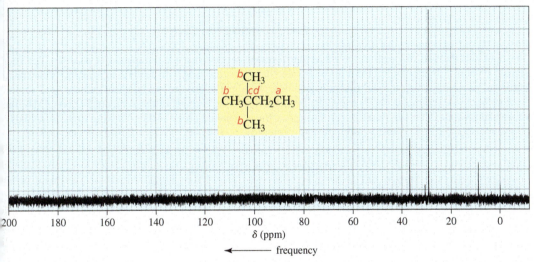

◀ **Figure 14.30**
The ^{13}C NMR spectrum of 2,2-dimethylbutane.

PROBLEM 41

Answer the following questions for each compound:

a. How many signals are in its ^{13}C NMR spectrum?
b. Which signal is at the lowest frequency?

1. $CH_3CH_2CH_2Br$

2. $(CH_3)_2C=CH_2$

3. CH_3CHCH_3
       |
       Br

4.

5.

6.

7.

8.

9. $CH_2=CHBr$

PROBLEM 42

Describe the proton-coupled ^{13}C NMR spectra for compounds 1, 3, and 5 in Problem 41, indicating the relative positions of the signals.

PROBLEM 43

How can 1,2-, 1,3-, and 1,4-dinitrobenzene be distinguished by

a. ^{1}H NMR spectroscopy? **b.** ^{13}C NMR spectroscopy?

LEARN THE STRATEGY

PROBLEM-SOLVING STRATEGY

Deducing a Chemical Structure from a ^{13}C NMR Spectrum

Identify the compound with molecular formula $C_9H_{10}O_2$ that gives the following ^{13}C NMR spectrum:

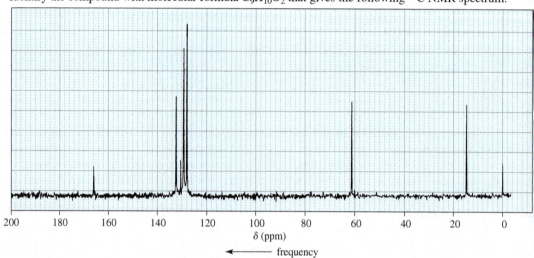

δ (ppm)

← frequency

First, pick out the signals that can be identified easily. For example, the signal for the carbonyl carbon at 166 ppm and the two oxygens in the molecular formula suggest that the compound is an ester. The four signals at about 130 ppm indicate that the compound has a benzene ring with a single substituent. (One signal is for the carbon to which the substituent is attached, one signal is for the two adjacent carbons, and so on.) Subtracting those fragments (C_6H_5 and CO_2) from the molecular formula of the compound leaves C_2H_5, the molecular formula of an ethyl substituent. Therefore, the compound is one of the following two compounds.

Since the signal for the methylene group is at ~60 ppm, it must be adjacent to an oxygen. Thus, the compound is the one on the left.

USE THE STRATEGY

PROBLEM 44 ◆

Identify each compound below from its molecular formula and its ^{13}C NMR spectrum.

a. $C_{11}H_{22}O$

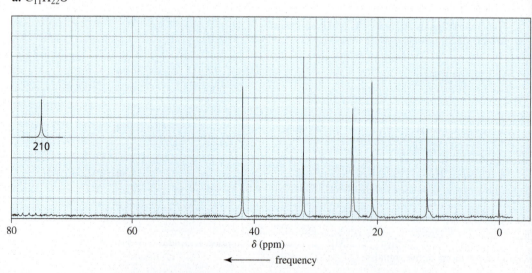

210

δ (ppm)

← frequency

b. C_8H_9Br

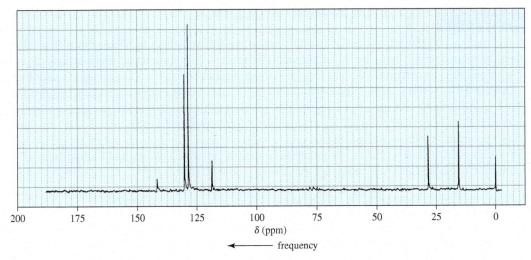

c. $C_6H_{10}O$

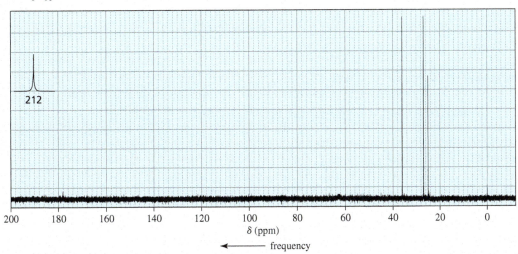

d. C_6H_{12}

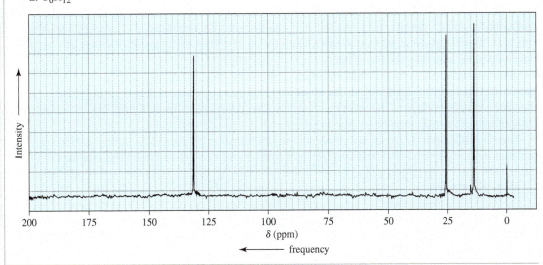

14.21 DEPT ^{13}C NMR SPECTRA

A technique called DEPT ^{13}C NMR (for Distortionless Enhancement by Polarization Transfer) has been developed to distinguish CH_3, CH_2, and CH groups. It is now much more widely used than proton coupling to determine the number of hydrogens attached to each carbon in a compound.

A **DEPT ^{13}C NMR** recording shows four spectra produced by the same compound (Figure 14.31). The top spectrum is run under conditions that allow signals produced only by CH_3 carbons. The next spectrum is run under conditions that allow signals produced only by CH_2 carbons, and the third spectrum shows signals produced only by CH carbons. The bottom spectrum shows signals for all carbons, thereby making it possible to detect carbons *not* bonded to hydrogens. Thus, a DEPT ^{13}C NMR spectrum indicates whether a signal is produced by a CH_3, CH_2, CH, or C.

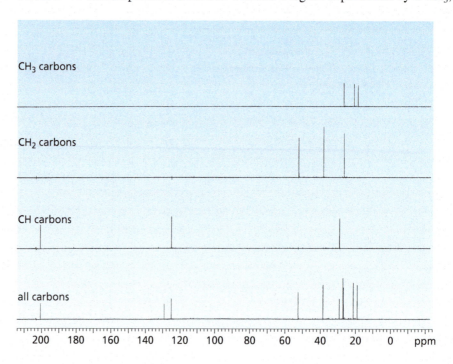

citronellal

▶ **Figure 14.31**
The DEPT ^{13}C NMR spectrum of citronellal.

14.22 TWO-DIMENSIONAL NMR SPECTROSCOPY

Complex molecules such as proteins and nucleic acids are difficult to analyze by NMR because the signals in their spectra overlap. Such compounds can be analyzed by **two-dimensional NMR spectroscopy (2-D NMR),** a technique that determines the structures of complex molecules in solution. This is a particularly important technique for studying biological molecules whose properties depend on how they fold in water.

More recently, 3-D and 4-D NMR spectroscopy have been developed and can be used to determine the structures of highly complex molecules. A thorough discussion of 2-D NMR is beyond the scope of this book, but this chapter would not be complete without a brief introduction to this increasingly important spectroscopic technique.

The ^{1}H NMR and ^{13}C NMR spectra discussed in the preceding sections have *one frequency* axis and one intensity axis; 2-D NMR spectra have *two frequency* axes and one intensity axis.

COSY Spectra

The most common 2-D spectra involve ^{1}H–^{1}H shift correlations, which identify coupled protons (that is, protons that split each other's signal). This is called ^{1}H–^{1}H shift-COrrelated SpectroscopY, known by the acronym COSY.

A portion of the **COSY spectrum** of ethyl vinyl ether is shown in Figure 14.32a; it looks like a mountain range because intensity is the third axis. These "mountain-like" spectra, called *stack plots,* are not the spectra actually used to identify a compound. Instead, the compound is identified using a contour plot (Figure 14.32b) that represents each mountain in Figure 14.32a by a large dot (as if its top had been cut off). The two mountains shown in Figure 14.32a correspond to the dots labeled Y and Z in Figure 14.32b. The usual one-dimensional ¹H NMR spectrum is plotted on both the *x*- and *y*-axes.

Analyzing COSY Spectra

To analyze the spectrum in Figure 14.32b, a diagonal line is drawn through the dots that bisect the spectrum. Dots that are *not* on the diagonal (X, Y, Z) are called *cross peaks;* they reveal pairs of protons that are coupled.

For example, if you start at the cross peak labeled X and draw a straight line, parallel to the *y*-axis, from X to the diagonal, you hit the dot on the diagonal at ~1.1 ppm, the signal produced by the H_a protons. If you then go back to X and draw a straight line, parallel to the *x*-axis, from X to the diagonal, you hit the dot on the diagonal at ~3.8 ppm, the signal produced by the H_b protons. This means that the H_a and H_b protons are coupled.

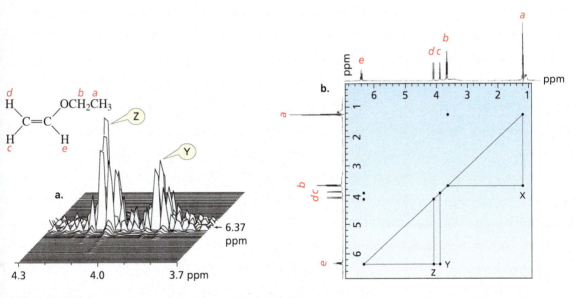

▲ **Figure 14.32**
In a COSY spectrum, an ¹H NMR spectrum is plotted on both the *x*- and *y*-axes.
(a) A portion of a COSY spectrum of ethyl vinyl ether (stack plot).
(b) A COSY spectrum of ethyl vinyl ether (contour plot). Cross peaks Y and Z represent the two mountains in (a).

If you go next to the cross peak labeled Y and draw another two perpendicular lines back to the diagonal, you see that the H_c and H_e protons are coupled. Similarly, the cross peak labeled Z shows that the H_d and H_e protons are coupled.

Although you have been using cross peaks below the diagonal as examples, the cross peaks above the diagonal give the same information because a COSY spectrum is symmetrical with respect to the diagonal. The absence of a cross peak due to coupling of H_c and H_d is consistent with expectations—the two nonidentical protons bonded to an sp^2 carbon of ethyl vinyl ether are also not coupled in the one-dimensional ¹H NMR spectrum shown in Figure 14.19. The power of a COSY experiment is that it reveals coupled protons without having to do a coupling constant analysis.

The COSY spectrum of 1-nitropropane is shown in Figure 14.33 on the next page. Cross peak X shows that the H_a and H_b protons are coupled, and cross peak Y shows that the H_b and H_c protons are coupled. Notice that the two triangles in the figure have a common vertex, because the H_b protons are coupled to both the H_a and H_c protons.

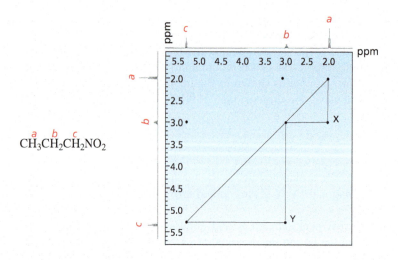

CH$_3$CH$_2$CH$_2$NO$_2$
(labeled *a b c*)

▲ **Figure 14.33**
The COSY spectrum of 1-nitropropane. A COSY spectrum identifies coupled protons, such as H$_a$/H$_b$ and H$_b$/H$_c$ in 1-nitropropane.

PROBLEM 45

Identify pairs of coupled protons in the compound whose COSY spectrum is shown below.

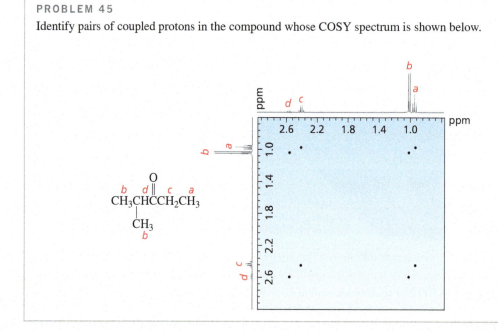

HETCOR Spectra

HETCOR spectra (from HETeronuclear CORrelation) are 2-D NMR spectra that show ^{13}C–^1H shift correlations and, therefore, reveal coupling between protons and the carbon to which they are attached. In a HETCOR spectrum, a compound's ^{13}C NMR spectrum is shown on the x-axis, and its ^1H NMR spectrum is shown on the y-axis. The cross peaks in a HETCOR spectrum identify which hydrogens are attached to which carbons.

Analyzing HETCOR Spectra

Cross peak W in Figure 14.34 indicates that the hydrogens that produce a signal at ~0.9 ppm in the ^1H NMR spectrum are bonded to the carbon that produces a signal at ~8 ppm in the ^{13}C NMR spectrum. Similarly, cross peak Y shows that the hydrogens that produce a signal at ~2.5 ppm are bonded to the carbon that produces a signal at ~34 ppm.

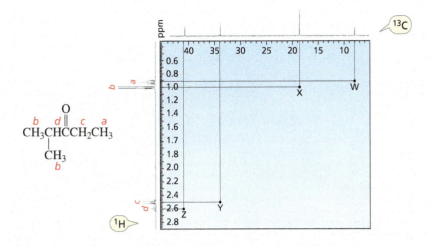

Figure 14.34
The HETCOR spectrum for the compound shown. A HETCOR spectrum reveals coupling between protons and the carbons to which they are attached.

PROBLEM 46 ◆

What does cross peak X in Figure 14.34 tell you?

Clearly, 2-D NMR techniques are not necessary for interpreting the NMR spectra of simple compounds such as those shown here. However, in the case of many complicated molecules, signals cannot be assigned without the aid of 2-D NMR. A number of different 2-D NMR techniques are now available. One involves the use of 2-D ^{13}C INADEQUATE* spectra, which show ^{13}C–^{13}C shift correlations and thus identify directly bonded carbons. Another plots chemical shifts on one frequency axis and coupling constants on the other. Finally, there are techniques that use the Nuclear Overhauser Effect (NOESY for very large molecules, ROESY for midsize molecules)† to locate protons that are close together in space.

14.23   NMR USED IN MEDICINE IS CALLED MAGNETIC RESONANCE IMAGING

NMR has become an important tool in medical diagnosis because it allows physicians to examine internal organs and structures without resorting to surgery or to the harmful ionizing radiation of X-rays. When NMR was first introduced into clinical practice in 1981, the selection of an appropriate name was a matter of some debate. Because many members of the general public associate the word *nuclear* with harmful radiation or radioactivity, the "N" was dropped from the medical application of NMR, which is known as **magnetic resonance imaging (MRI).** The spectrometer is called an **MRI scanner.**

An MRI scanner consists of a magnet large enough to surround a person, along with an apparatus for exciting the nuclei, modifying the magnetic field, and receiving signals. (By comparison, the NMR spectrometer used by chemists is only large enough to accommodate a 5-mm glass tube.) Different tissues yield different signals, which are separated into components by Fourier transform analysis (Section 14.2). Each component can be attributed to a specific location within the part of the body being scanned, so that a set of images through the scanned volume is generated. MRI can produce an image showing any cross section of the body, regardless of the person's position within the machine, which allows optimal visualization of the anatomical feature of interest.

---

* INADEQUATE is an acronym for Incredible Natural Abundance Double Quantum Transfer Experiment.

† NOESY and ROESY are the acronyms for Nuclear Overhauser Effect SpectroscopY and Rotation-frame Overhauser Effect SpectroscopY, respectively.

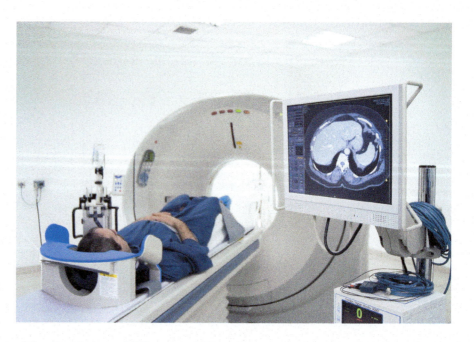

Most of the signals in an MRI scan originate from the hydrogens of water molecules because tissues contain far more of these hydrogens than they do hydrogens of organic compounds. The difference in the way water is bound in different tissues is what produces much of the variation in signal between different organs, as well as the variation between healthy and diseased tissue. MRI scans, therefore, can provide much more information than images obtained by other means.

For example, MRI can provide detailed images of blood vessels. Flowing fluids, such as blood, respond differently to excitation in an MRI scanner than do stationary tissues, and proper processing results in only the flowing fluids being displayed. The quality of these images has become high enough that it can often eliminate the need for more invasive diagnostic techniques.

The versatility of MRI has been enhanced by using gadolinium as a contrast agent. Gadolinium modifies the magnetic field in its immediate vicinity, altering the signal from nearby hydrogens. The distribution of gadolinium, which is infused into a patient's veins, may be affected by certain disease processes such as cancer and inflammation. Any abnormal patterns of distribution are revealed in the MRI images.

A brain tumor and a brain abscess may have very similar appearances in an MRI (Figure 14.35a). Suppressing the signal from water makes it possible to detect signals from specific compounds such as choline and acetate. A tumor will produce an elevated choline signal, whereas an abscess is more likely to produce an elevated acetate signal (Figure 14.35b).

 Figure 14.35

(a) The white circle indicates a brain lesion that could be caused by either a tumor (elevated choline) or an abscess (elevated acetate).

(b) The major peak in the spectrum corresponds to acetate, supporting the diagnosis of an abscess.

a.
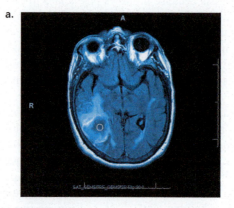

b.
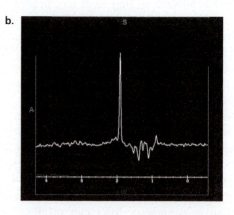

14.24 X-RAY CRYSTALLOGRAPHY

X-ray crystallography is a technique used to determine the arrangement of atoms within a crystal. With this technique, the structure of any material that can form a crystal can be determined. Dorothy Crowfoot Hodgkin was a pioneer of X-ray crystallography, using it to determine the structures of cholesterol (1937), penicillin (1945), vitamin B$_{12}$ (1954), and insulin (1969). (The structures of these compounds can be found in Sections 3.16, 15.12, 23.6, and 21.8, respectively.) In 1953, Rosalind Franklin's X-ray images allowed James Watson and Francis Crick to accurately describe the structure of DNA (Section 26.1).

A crystal is a solid composed of atoms and molecules in a regular three-dimensional array, called a unit cell, that is repeated indefinitely throughout the crystal. Often the most difficult aspect of X-ray crystallography is obtaining the crystal, which should be at least 0.5 mm long in all three dimensions to provide good structural data. Fortunately, crystallographers can get a head start by looking up over 14,000 crystallization conditions on the *Biological Macromolecule Crystallization Database* (xpdb.nist.gov:8060/BMCD4).

In order to resolve atomic distances, the wavelength of the light used and the distance between the atoms to be resolved must be about the same order of magnitude. The length of a C—C bond is 1.54 Å. Visible light, therefore, with wavelengths of 4000–8000 Å cannot be used. The X-rays produced by bombarding a Cu-anode with electrons in a vacuum tube have a wavelength of 1.542 Å, which make them ideal for resolving most atomic distances. For high-resolution structures, crystallographers use more intense sources of X-rays. The high intensity coupled with the superior dispersion of the X-rays in a synchrotron allows experiments to be carried out at less than 1 Å resolution.

Once a suitable crystal has been obtained, it is bombarded with X-rays while being gradually rotated. X-rays, however, cannot be focused by conventional lenses to form an image of a molecule. This is where X-ray diffraction comes in. While most of the X-rays pass through the crystal, some are scattered by the electron clouds of the atoms and land on a detector, creating a diffraction pattern of regularly spaced dots (Figure 14.36).

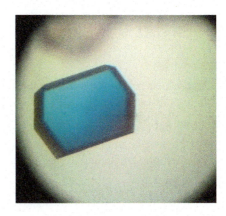

A crystal of lysozyme from hen egg white (Section 5.14). It appears blue because it was photographed under polarized light.

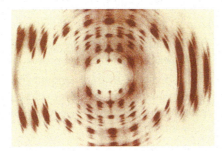

▲ **Figure 14.36**
an X-ray diffraction pattern

By means of complex mathematical methods, now facilitated by computers, a three-dimensional model of the electron density within the crystal is produced from the two-dimensional **X-ray diffraction** images taken at different angles of rotation. Electrons are found around atoms, so any two regions of electron density within bonding distance of each other can be assumed to represent atoms bonded to one another. The greater the density of the electron cloud around the atom, the more precisely the location of the atom can be assigned. Therefore, the position of electron-deficient hydrogen atoms cannot be assigned unambiguously. However, hydrogen atoms now can be located in the final structure by computational methods.

Structural Databases

There are a variety of databases that provide structural information to chemists. Two important examples are the Chemical Abstracts Services Database (CAS) and the Research Collaboratory for Structural Bioinformatics Protein Data Bank (RCSB PDB).

CAS (http://www.cas.org/) is a subscription-based service with over 100 million organic, inorganic, protein, and nucleic acid structures. It can be accessed through SciFinder, which allows searches via a structural drawing program as well as searches by subject, author, or structure name.

The RCSB Protein Data Bank (www.pdb.org) is the repository of more than 57,000 three-dimensional structures that have been determined by X-ray crystallography (one is shown on p. 620) or by multi-dimensional NMR (Section 14.22). The structures are publicly available for download in a PDB format. The structures can be visualized and manipulated with free programs such as Visual Molecular Dynamics (http://www.ks.uiuc.edu/Research/vmd/) or Chimera (http://www.cgl.ucsf.edu/chimera).

Dorothy Crowfoot Hodgkin (1910–1994) *was born in Egypt to English parents. She received an undergraduate degree from Somerville College at Oxford University and earned a Ph.D. from Cambridge University. She performed the first three-dimensional calculations in crystallography, and she was the first to use computers to determine the structures of compounds. She received the 1964 Nobel Prize in Chemistry for her work on vitamin B$_{12}$. Hodgkin was a professor of chemistry at Somerville, where one of her research students was Margaret Thatcher, the future prime minister of England. Hodgkin was a founding member of Pugwash, an organization whose purpose was to further communication between scientists on both sides of the Iron Curtain.*

ESSENTIAL CONCEPTS

Section 14.0

- **NMR spectroscopy** identifies the carbon–hydrogen framework of an organic compound.

Section 14.1

- The greater the magnetic field produced by the magnet, the greater the operating frequency of the spectrophotometer.

Section 14.3

- The larger the magnetic field sensed by the proton, the higher the frequency of its signal.
- The electron density of the environment in which the proton is located **shields** the proton from the applied magnetic field. Therefore, a proton in an electron-dense environment shows a signal at a lower frequency than a proton near electron-withdrawing groups.

Section 14.4

- Each set of chemically equivalent protons produces a signal, so the number of signals in an ^1H NMR spectrum indicates the number of different kinds of protons in a compound.

Section 14.5

- The **chemical shift** (δ), which is independent of the operating frequency of the spectrometer, is a measure of how far the signal is from the reference TMS signal.
- Low-frequency signals have small δ (ppm) values; high frequency signals have large δ values. The position of a signal indicates the kind of proton(s) responsible for the signal and the kinds of neighboring substituents.

Section 14.7

- In a similar environment, the chemical shift of a methine proton is at a higher frequency than that of methylene protons, which is at a higher frequency than that of methyl protons.

Section 14.8

- **Diamagnetic anisotropy** causes unusual chemical shifts for hydrogens bonded to carbons with π bonds.

Section 14.9

- **Integration** tells us the relative number of protons that produce each signal.

Section 14.10

- The **multiplicity** of a signal indicates the number of protons bonded to adjacent carbons. Multiplicity is described by the $N + 1$ **rule,** where N is the number of equivalent protons bonded to an adjacent carbon.
- Coupled protons split each other's signal.

Section 14.11

- A **splitting diagram** describes the splitting pattern obtained when a signal is split by more than one set of protons.

Section 14.13

- The **coupling constant** (J), which is independent of the operating frequency of the spectrometer, is the distance between two adjacent peaks of a split NMR signal. Coupled protons have the same coupling constant.
- The coupling constant for trans alkene protons is greater than that for cis alkene protons.

Section 14.14

- When two different sets of protons split a signal, the multiplicity of the signal is determined by using the $N + 1$ rule separately $(N_a + 1)(N_b + 1)$ for each set of protons when the coupling constants for the two sets are different. When the coupling constants are similar, the $N + 1$ rule is applied to both sets at the same time.

Section 14.15

- **Enantiotopic hydrogens** are chemically equivalent, so they show only one signal; **diastereotopic hydrogens** are not chemically equivalent, so they show two signals.

Section 14.17

- The chemical shift of a proton bonded to an O or to an N depends on the extent to which the proton is hydrogen bonded.
- In the presence of trace amounts of acid or base, protons bonded to oxygen undergo **proton exchange.** In that case, the signal for a proton bonded to an O is not split and does not split the signal of adjacent protons.

Section 14.20

- The number of signals in a ^{13}C NMR spectrum corresponds to the number of different kinds of carbons in the compound. Carbons in electron-rich environments produce low-frequency signals, whereas carbons close to electron-withdrawing groups produce high-frequency signals.
- ^{13}C NMR signals are not split by attached protons unless the spectrometer is run in a proton-coupled mode.

Section 14.21

- A DEPT ^{13}C NMR spectrum tells whether a signal is produced by CH_3, CH_2, CH, or C.

Section 14.22

- **2-D NMR** spectra have two frequency axes and one intensity axis.

Section 14.23

- NMR (known in medical applications as **MRI**) is an important tool in medical diagnosis because it allows internal structures to be examined without surgery or harmful X-rays.

Section 14.24

- 2-D NMR and **X-ray crystallography** are techniques that can be used to determine the structures of large molecules.

PROBLEMS

47. How many signals are produced by each of the following compounds in its

   a. ^{1}H NMR spectrum?   **b.** ^{13}C NMR spectrum?

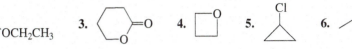

1. CH_3—⬡—$OCHCH_3$ **2.**   **3.**   **4.**   **5.**   **6.**
                      |
                    CH_3

NOTE TO THE STUDENT

• There are additional spectroscopy problems in the *Study Guide and Solutions Manual.*

48. Draw a splitting diagram for the H_b proton and give its multiplicity if

   a. $J_{ba} = J_{bc}$.   **b.** $J_{ba} = 2J_{bc}$.

$$
\begin{array}{c}
H_a\;\;X\;\;X \\
|\;\;\;\;\;|\;\;\;\;\;| \\
H_a-C-C-C-X \\
|\;\;\;\;\;|\;\;\;\;\;| \\
H_a\;\;H_b\;\;H_c
\end{array}
$$

49. Label each set of chemically equivalent protons, using **a** for the set that will be at the lowest frequency in the ^{1}H NMR spectrum, **b** for the next lowest, and so on. Indicate the multiplicity of each signal.

   a. CH_3CHNO_2
            |
            CH_3

   c. CH_3CH — $C(=O)$ — $CH_2CH_2CH_3$, with CH_3 below CH_3CH

   e. $ClCH_2CCHCl_2$, with CH_3 above and CH_3 below the central C

   b. $CH_3CH_2CH_2OCH_3$

   d. $CH_3CH_2CH_2$ — $C(=O)$ — CH_2Cl

   f. $ClCH_2CH_2CH_2CH_2CH_2Cl$

50. Determine the ratios of the chemically nonequivalent protons in a compound if the steps of the integration curves measure 40.5, 27, 13, and 118 mm, from left to right across the spectrum. Draw the structure of a compound whose ^{1}H NMR spectrum would show these integrals in the observed order.

51. How can ^{1}H NMR distinguish between the compounds in each of the following pairs?

   a. ⌒⌒O⌒ and ⌒O⌒⌒

   f. ⬡ and ⬡

   b. Br⌒⌒⌒Br and Br⌒⌒⌒NO_2

   g. CH_3CHCl and CH_3CDCl , each with CH_3 below

   c. CH_3CH—$CHCH_3$ (with CH_3 and CH_3 above) and $CH_3CCH_2CH_3$ (with CH_3 above and CH_3 below)

   h. H—Cl—CH₃ over D—Cl—H with Cl below and D—CH₃ over H—H with Cl below

   d. CH_3—C(CH_3)(OCH_3)—$C(=O)$—OCH_3 and CH_3—C(OCH_3)(OCH_3)—CH_3

   i. cyclopropane (H, H / Cl, Cl) and cyclopropane (H, Cl / Cl, H)

   e. CH_3O—⬡—CH_2CH_3 and CH_3—⬡—OCH_2CH_3

   j. CH_3—⬡—CCH_3 (with CH_3 above and CH_3 below) and ⬡—CH_2CCH_3 (with CH_3 above and CH_3 below)

52. Answer the following questions:

   a. What is the relationship between chemical shift in ppm and operating frequency?
   b. What is the relationship between chemical shift in hertz and operating frequency?
   c. What is the relationship between coupling constant in hertz and operating frequency?
   d. How does the operating frequency in NMR spectroscopy compare with the operating frequency in IR and UV/Vis spectroscopy?

53. Match each of the ^{1}H NMR spectra with one of the following compounds:

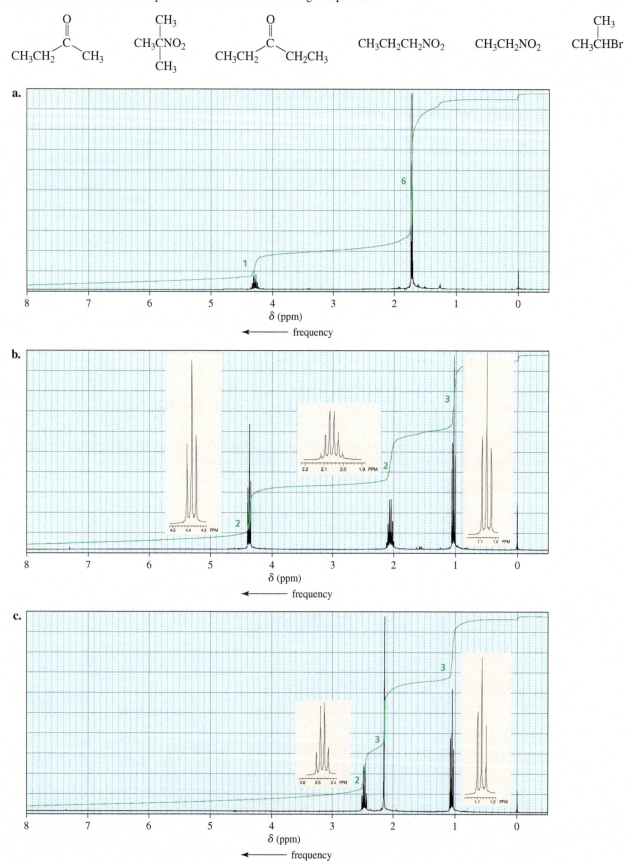

54. The ^{1}H NMR spectra of three isomers with molecular formula C_4H_9Br are shown here. Which isomer produces which spectrum?

a.

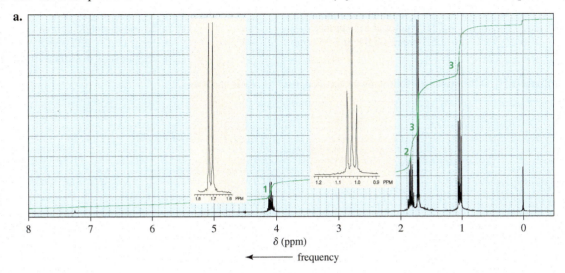

b.

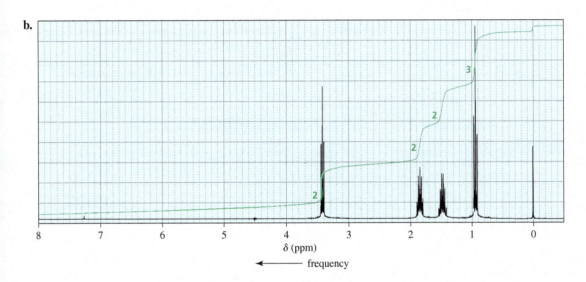

c.

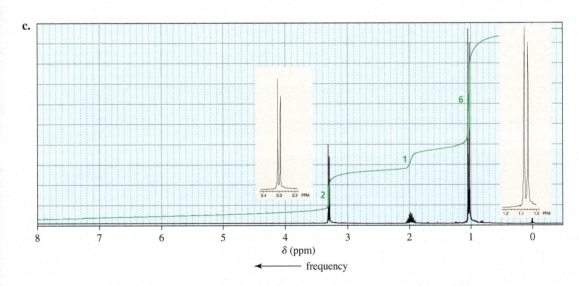

55. Identify each of the following compounds from the ^{1}H NMR data and molecular formula:

a. $C_4H_8Br_2$: a 6H singlet at 1.97 ppm
         a 2H singlet at 3.89 ppm

b. C_8H_9Br: a 3H doublet at 2.01 ppm
         a 1H quartet at 5.14 ppm
         a 5H broad singlet at 7.35 ppm

c. $C_5H_{10}O_2$: a 3H triplet at 1.15 ppm
         a 3H triplet at 1.25 ppm
         a 2H quartet at 2.33 ppm
         a 2H quartet at 4.13 ppm

56. Identify the compound with molecular formula $C_7H_{14}O$ that gives the following proton-coupled ^{13}C NMR spectrum:

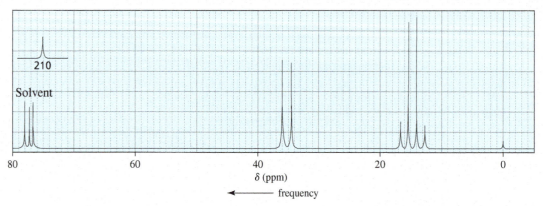

57. Compound **A**, with molecular formula C_4H_9Cl, shows two signals in its ^{13}C NMR spectrum. Compound **B**, an isomer of compound **A**, shows four signals, and in the proton-coupled mode, the signal farthest downfield is a doublet. Identify compounds **A** and **B**.

58. Would it be better to use ^{1}H NMR or ^{13}C NMR spectroscopy to distinguish 1-butene, *cis*-2-butene, and 2-methylpropene? Explain your answer.

59. There are four esters with molecular formula $C_4H_8O_2$. How can they be distinguished by ^{1}H NMR?

60. Identify the compound with molecular formula C_6H_{14} that is responsible for the following ^{1}H NMR spectrum:

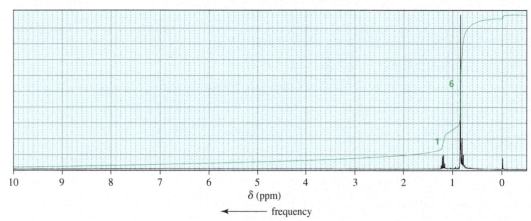

61. An alkyl halide reacts with an alkoxide ion to form a compound whose ^{1}H NMR spectrum is shown here. Identify the alkyl halide and the alkoxide ion. (*Hint:* See Section 9.15.)

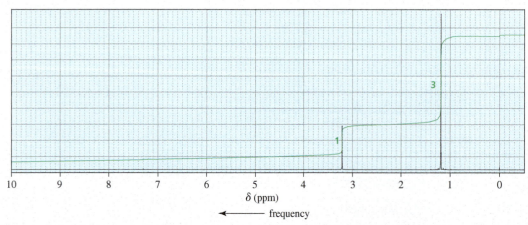

62. The ^{1}H NMR spectra of three isomers with molecular formula $C_7H_{14}O$ are shown here. Which isomer produces which spectrum?

a.

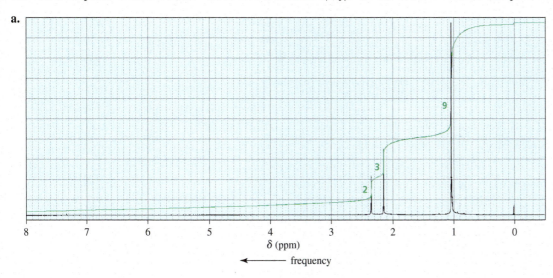

b.

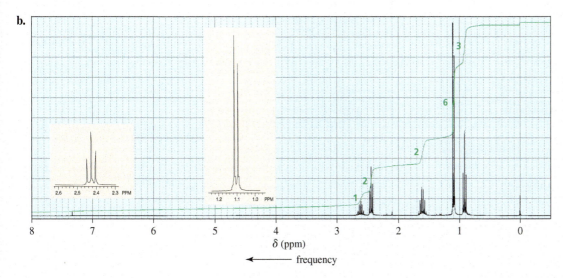

c.

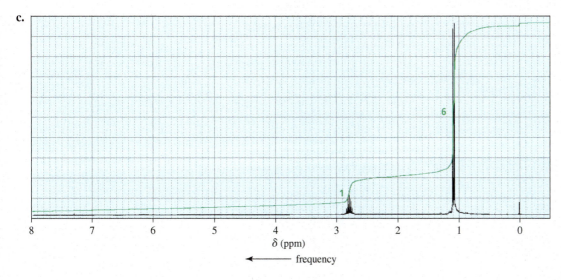

63. Identify each of the following compounds from its molecular formula and its IR and ^{1}H NMR spectra:

a. $C_5H_{12}O$

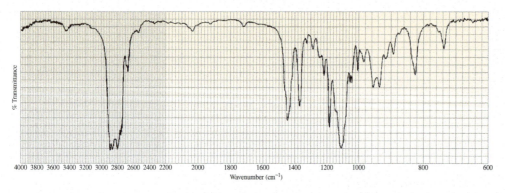

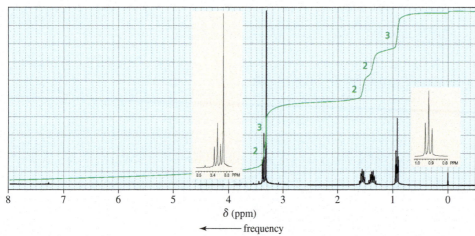

b. $C_6H_{12}O_2$

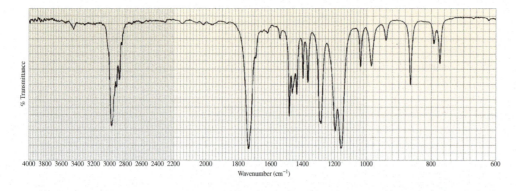

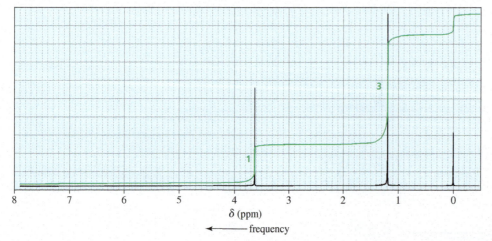

c. $C_4H_7ClO_2$

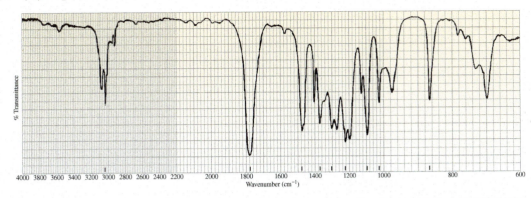

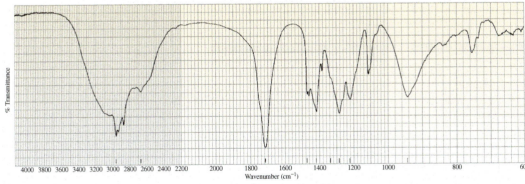

d. $C_4H_8O_2$

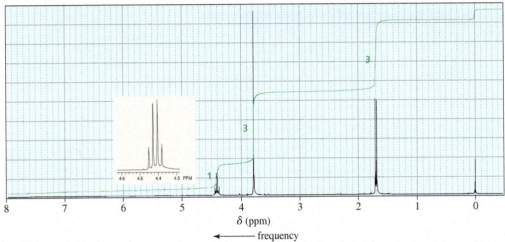

Offset: 2.0 ppm

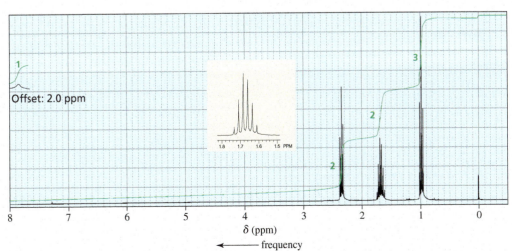

64. Identify each of the following compounds from its molecular formula and its ^{13}C NMR spectrum:

a. $C_4H_{10}O$

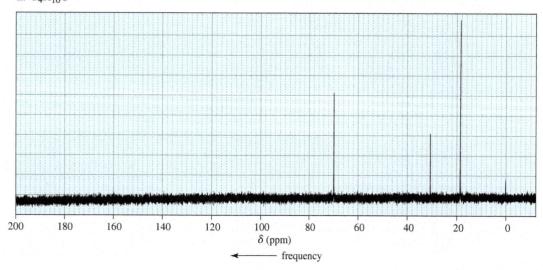

b. $C_6H_{12}O$

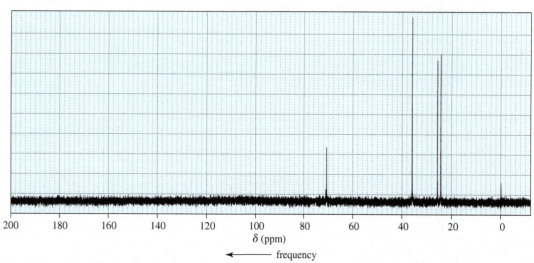

65. The ^1H NMR spectrum of 2-propen-1-ol is shown here. Indicate the protons in the molecule that are responsible for each of the signals in the spectrum.

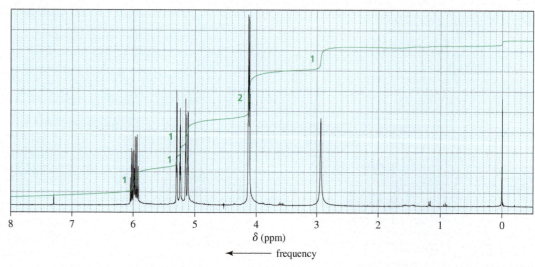

66. How can the signals in the 6.5 to 8.1 ppm region of their ^{1}H NMR spectra distinguish the following compounds?

OCH$_3$ ... NO$_2$ OCH$_3$ ... NO$_2$ OCH$_3$ ... NO$_2$

67. The ^{1}H NMR spectra of two compounds, each with molecular formula $C_{11}H_{16}$, are shown here. Identify the compounds.

a.

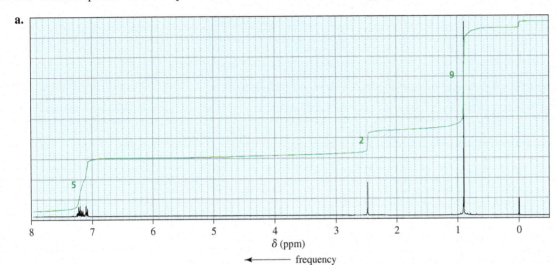

δ (ppm)

frequency

b.

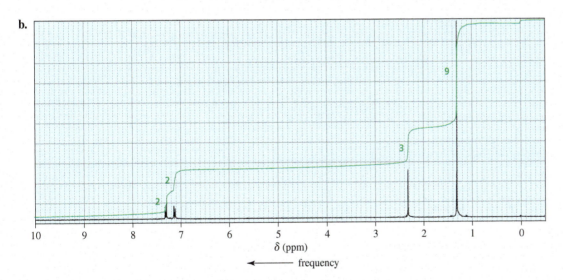

δ (ppm)

frequency

68. Draw a splitting diagram for the H_b proton if $J_{bc} = 10$ and $J_{ba} = 5$.

Cl   CH$_2$Cl
   C=C *a*
H   H
c   *b*

69. Sketch the following spectra that would be obtained for 2-chloroethanol:

  a. The ^{1}H NMR spectrum for an anhydrous sample of the alcohol.
  b. The ^{1}H NMR spectrum for a sample of the alcohol that contains a trace amount of acid.
  c. The ^{13}C NMR spectrum.
  d. The proton-coupled ^{13}C NMR spectrum.
  e. The four parts of a DEPT ^{13}C NMR spectrum.

70. How can ^1H NMR be used to prove that the addition of HBr to propene follows the rule that says that the electrophile adds to the sp^2 carbon bonded to the most hydrogens?

71. Identify each of the following compounds from its molecular formula and its ^1H NMR spectrum:

a. C_8H_8

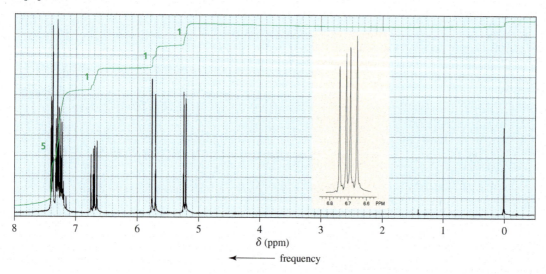

b. $C_6H_{12}O$

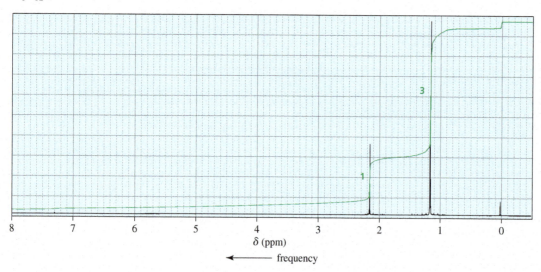

c. $C_9H_{18}O$

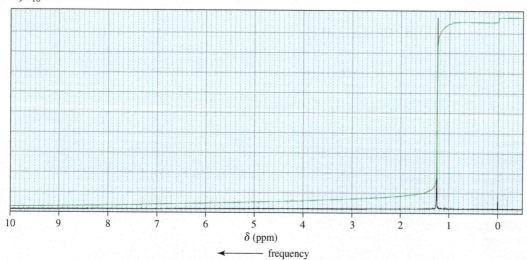

d. C_4H_8O

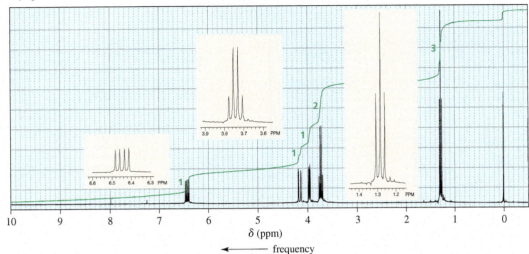

72. Dr. N. M. Arr was called in to help analyze the ${}^1H$ NMR spectrum of a mixture of compounds known to contain only C, H, and Br. The mixture showed two singlets—one at 1.8 ppm and the other at 2.7 ppm—with relative integrals of 1 : 6, respectively. Dr. Arr determined that the spectrum was that of a mixture of bromomethane and 2-bromo-2-methylpropane. What was the ratio of bromomethane to 2-bromo-2-methylpropane in the mixture?

73. Calculate the amount of energy (in calories) required to flip an ${}^1H$ nucleus in an NMR spectrometer that operates at 300 MHz.

74. The following ${}^1H$ NMR spectra are for four compounds, each with molecular formula of $C_6H_{12}O_2$. Identify the compounds.

a.

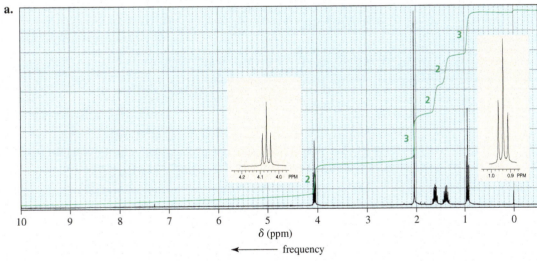

b.

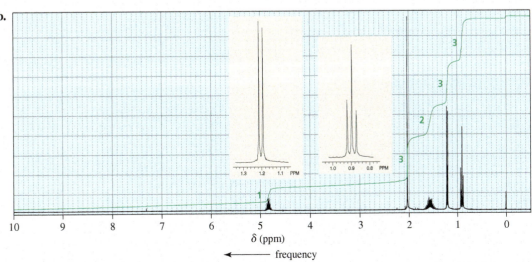

c.

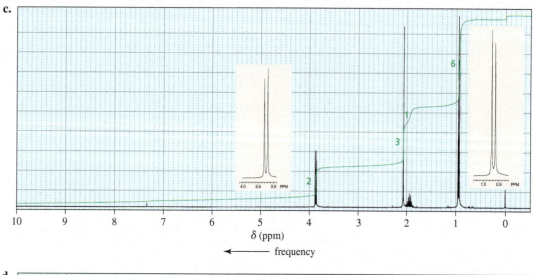

δ (ppm)

← frequency

d.

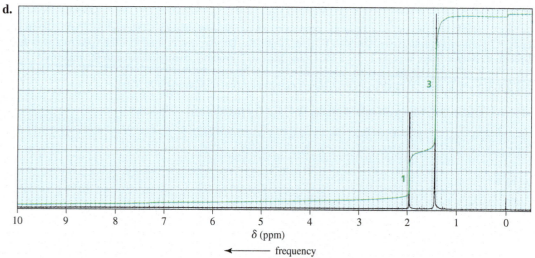

δ (ppm)

← frequency

75. When compound **A** ($C_5H_{12}O$) is treated with HBr, it forms compound **B** ($C_5H_{11}Br$). The ^1H NMR spectrum of compound **A** has a 1H singlet, a 3H doublet, a 6H doublet, and two 1H multiplets. The ^1H NMR spectrum of compound **B** has a 6H singlet, a 3H triplet, and a 2H quartet. Identify compounds **A** and **B**.

76. Identify the compound with molecular formula $C_6H_{10}O$ that gives the following DEPT ^{13}C NMR spectrum:

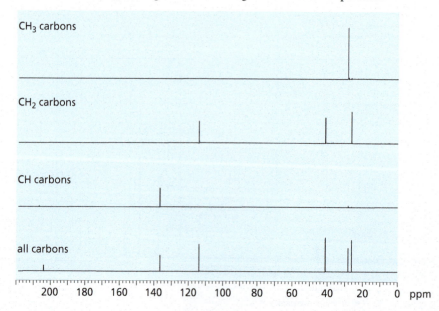

77. Identify each of the following compounds from its molecular formula and its IR and ^{1}H NMR spectra:

a. $C_6H_{12}O$

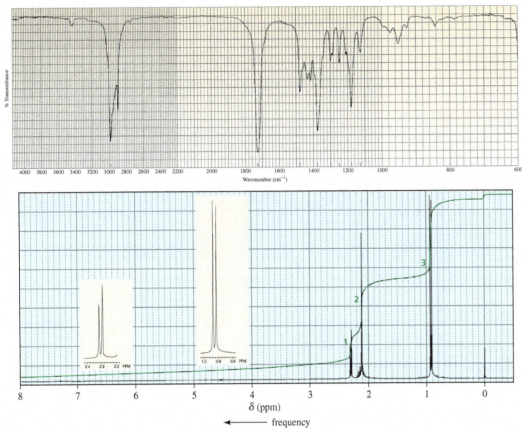

b. $C_6H_{14}O$

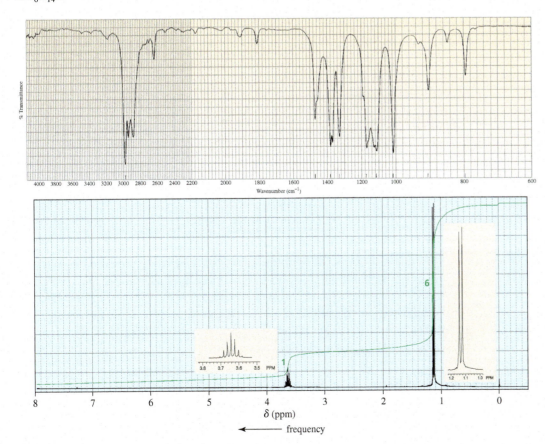

c. $C_{10}H_{13}NO_3$

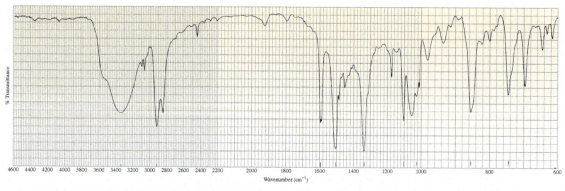

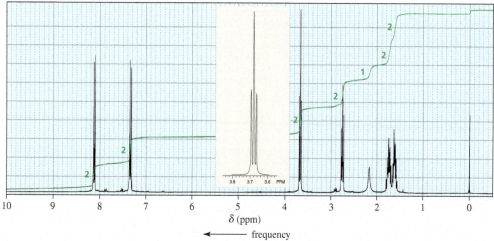

δ (ppm)

← frequency

d. $C_{11}H_{14}O_2$

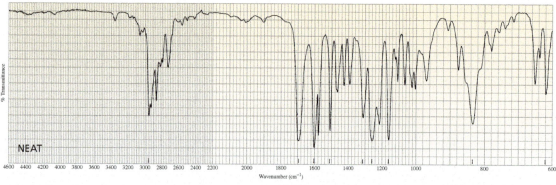

NEAT

Wavenumber (cm⁻¹)

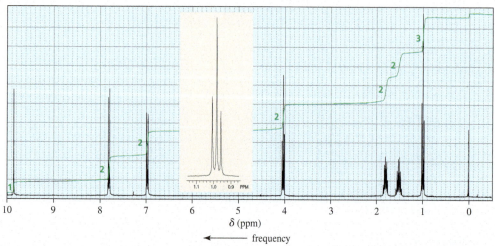

δ (ppm)

← frequency

78. How many signals are produced by each of the following compounds in its

a. ¹H NMR spectrum? b. ¹³C NMR spectrum?

 1.

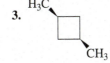 2.

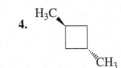 3.

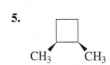 4.

5.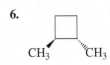

6.

79. Identify each of the following compounds from its mass spectrum, IR spectrum, and ¹H NMR spectrum:

a.

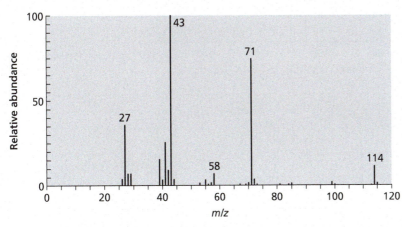

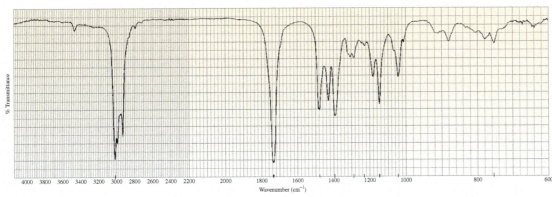

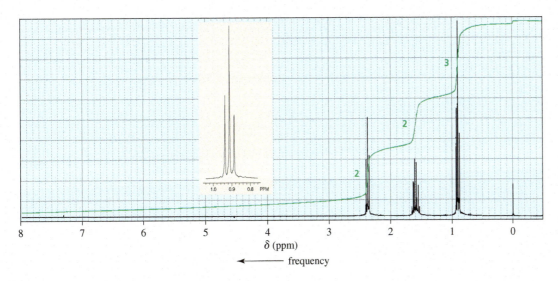

b.

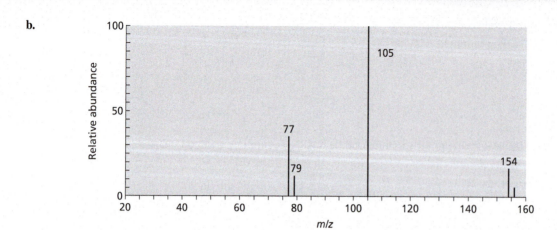

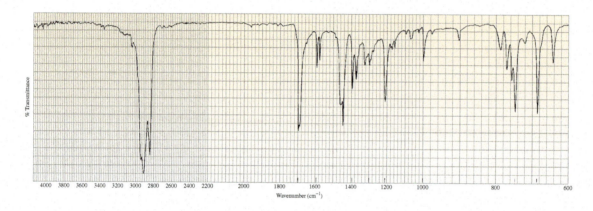

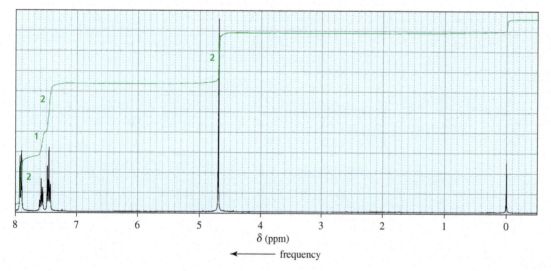

Carbonyl Compounds

The three chapters in Part 5 focus on the reactions of compounds that contain a carbonyl group. Carbonyl compounds can be classified as either those that contain a group that can be replaced by another group (carboxylic acids and carboxylic acid derivatives) or those that contain a group that cannot be replaced by another group (aldehydes and ketones).

Chapter 15 Reactions of Carboxylic Acids and Carboxylic Acid Derivatives

The reactions of carboxylic acids and carboxylic acid derivatives are discussed in **Chapter 15,** where you will see that they all react with nucleophiles in the same way—they undergo nucleophilic acyl substitution reactions. In a nucleophilic acyl substitution reaction, the nucleophile adds to the carbonyl carbon, forming an unstable tetrahedral intermediate that collapses by eliminating the weaker of two bases. As a result, all you need to know to determine the product of one of these reactions—or even whether a reaction will occur—is the relative basicity of the two potential leaving groups in the tetrahedral intermediate.

Chapter 16 Reactions of Aldehydes and Ketones •
More Reactions of Carboxylic Acid Derivatives

Chapter 16 starts by comparing the reactions of carboxylic acids and carboxylic acid derivatives with the reactions of aldehydes and ketones. This comparison is made by discussing their reactions with carbon nucleophiles and hydride ion. You will see that carboxylic acids and carboxylic acid derivatives undergo *nucleophilic acyl substitution* reactions with carbon nucleophiles and hydride ion, just as they did with nitrogen and oxygen nucleophiles in Chapter 15. Aldehydes and ketones, on the other hand, undergo *nucleophilic addition* reactions with carbon nucleophiles and hydride ion and *nucleophilic addition-elimination* reactions with oxygen and nitrogen nucleophiles (and the species eliminated is always water). What you learned in Chapter 15 about the partitioning of tetrahedral intermediates is revisited in this chapter. The reactions of α,β-unsaturated carbonyl compounds are also discussed.

Chapter 17 Reactions at the α-Carbon

Many carbonyl compounds have two sites of reactivity: the carbonyl group *and* a hydrogen bonded to an α-carbon. Chapters 15 and 16 discuss the reactions of carbonyl compounds that take place at the carbonyl group, whereas **Chapter 17** examines the reactions of carbonyl compounds that involve the α-carbon.

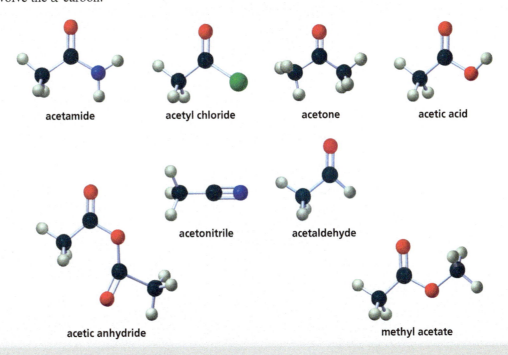

acetamide   acetyl chloride   acetone   acetic acid

acetonitrile   acetaldehyde

acetic anhydride   methyl acetate

15

Reactions of Carboxylic Acids and Carboxylic Acid Derivatives

Some of the things you will learn in this chapter are how aspirin decreases inflammation and fever, why Dalmatians are the only dogs that excrete uric acid, how bacteria become resistant to penicillin, and why young people sleep better than adults.

Group III

We saw that the families of organic compounds can be placed into one of four groups, and that all the families in a group react in similar ways (Section 5.4). This chapter begins our discussion of the families of compounds in Group III—compounds that contain a carbonyl group.

The **carbonyl group** (a carbon doubly bonded to an oxygen) is probably the most important functional group. Compounds containing carbonyl groups—called **carbonyl** ("car-bo-neel") **compounds**—are abundant in nature, and many play important roles in biological processes.

An **acyl** ("a-sil") **group** consists of a carbonyl group attached to an alkyl group (R) or to an aromatic group (Ar), such as benzene.

The group (or atom) attached to the acyl group strongly affects the reactivity of the carbonyl compound. In fact, carbonyl compounds can be divided into two classes determined by that group.

Carboxylic Acids and Carboxylic Acid Derivatives

One class of carbonyl compounds is made up of those in which the acyl group is attached to a group (or atom) that *can* be replaced by another group. Carboxylic acids, esters, acyl chlorides, and amides belong to this class. All of these compounds contain a group (OH, OR, Cl, NH_2, NHR, NR_2) that can be replaced by a nucleophile.

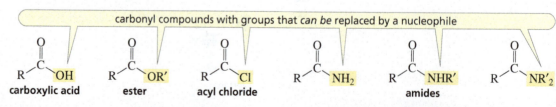

Esters, acyl chlorides, and amides are called **carboxylic acid derivatives** because they differ from a carboxylic acid only in the nature of the group or atom that has replaced the OH group of the carboxylic acid.

Aldehydes and Ketones

The second class of carbonyl compounds are those in which the acyl group is attached to a group that *cannot* readily be replaced by another group. Aldehydes and ketones belong to this class. The H bonded to the acyl group of an aldehyde and the R group bonded to the acyl group of a ketone cannot be readily replaced by a nucleophile.

carbonyl compounds with groups that *cannot be* replaced by a nucleophile

$$\underset{\text{aldehyde}}{\overset{\displaystyle O}{\underset{R}{\overset{\|}{\underset{}{C}}}\diagdown H}} \qquad \underset{\text{ketone}}{\overset{\displaystyle O}{\underset{R'}{\overset{\|}{\underset{}{C}}}\diagdown R}}$$

The Weaker the Base, the Easier It Can Be Replaced

We saw that, when comparing bases of the same type, weak bases are good leaving groups and strong bases are poor leaving groups (Section 9.2). The pK_a values of the conjugate acids of the leaving groups of various carbonyl compounds are listed in Table 15.1. Notice that the acyl groups of

Table 15.1 The pK_a Values of the Conjugate Acids of the Leaving Groups of Carbonyl Compounds

Carbonyl compound	Leaving group	Conjugate acid of the leaving group	pK_a
Carboxylic Acids and Carboxylic Acid Derivatives			
R–C(=O)–Cl	Cl⁻	HCl	-7
R–C(=O)–OR′	⁻OR′	R′OH	~15–16
R–C(=O)–OH	⁻OH	H_2O	15.7
R–C(=O)–NH₂	⁻NH₂	NH_3	36*
Aldehydes and Ketones			
R–C(=O)–H	H⁻	H_2	35
R–C(=O)–R	R⁻	RH	> 60

*An amide can undergo substitution reactions only when its leaving group is converted to NH_3, giving its conjugate acid ($^+NH_4$) a pK_a value of 9.4.

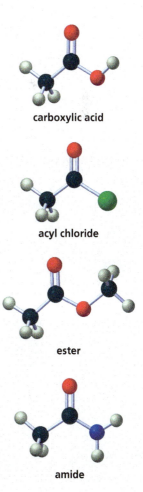

carboxylic acid

acyl chloride

ester

amide

carboxylic acids and carboxylic acid derivatives are attached to weaker bases than are the acyl groups of aldehydes and ketones. (Remember that the lower the pK_a, the stronger the acid and the weaker its conjugate base.) The hydrogen of an aldehyde and the alkyl group of a ketone are too basic to be replaced by another group.

This chapter discusses the reactions of carboxylic acids and carboxylic acid derivatives. We will see that these compounds undergo substitution reactions because they have an acyl group attached to a group that can be replaced by a nucleophile. The reactions of aldehydes and ketones are discussed in Chapter 16, where we will see that these compounds *do not* undergo substitution reactions because their acyl group is attached to a group that *cannot* be replaced by a nucleophile.

15.1 THE NOMENCLATURE OF CARBOXYLIC ACIDS AND CARBOXYLIC ACID DERIVATIVES

Let's start by looking at how carboxylic acids are named, because their names form the basis of the names of the other carbonyl compounds.

Naming Carboxylic Acids

The functional group of a carboxylic acid is called a **carboxyl group.**

carboxyl group

$-COOH$ $-CO_2H$

carboxyl groups are frequently shown in these abbreviated forms

LEARN THE STRATEGY

In systematic (IUPAC) nomenclature, a **carboxylic acid** is named by replacing the terminal "e" of the alkane name with "oic acid." For example, the one-carbon alkane is methan*e*, so the one-carbon carboxylic acid is methan*oic acid*.

systematic name:	methanoic acid	ethanoic acid	propanoic acid	butanoic acid
common name:	formic acid	acetic acid	propionic acid	butyric acid

pentanoic acid
valeric acid

hexanoic acid
caproic acid

propenoic acid
acrylic acid

valerian flowers

Carboxylic acids containing six or fewer carbons are frequently called by their common names. These names were chosen by early chemists to describe some feature of the compound, usually its origin.

- Formic acid is found in ants, bees, and other stinging insects; its name comes from *formica*, which is Latin for "ant."
- Acetic acid—contained in vinegar—got its name from *acetum*, the Latin word for "vinegar."
- Propionic acid is the smallest acid that shows some of the characteristics of the larger fatty acids (Section 25.1); its name comes from the Greek words *pro* ("the first") and *pion* ("fat").
- Butyric acid is found in rancid butter; the Latin word for "butter" is *butyrum*.
- Valeric acid got its name from valerian, an herb that has been used as a sedative since Greco/Roman times.
- Caproic acid is found in goat's milk. If you have ever smelled a goat, then you know what caproic acid smells like. *Caper* is the Latin word for "goat."

happy goat

In systematic nomenclature, the position of a substituent is designated by a number. The carbonyl carbon is always the C-1 carbon. In common nomenclature, the position of a substituent is designated by a lowercase Greek letter, and the carbonyl carbon is not given a designation. Thus, the carbon adjacent to the carbonyl carbon is the α-carbon, the carbon adjacent to the α-carbon is the β-carbon, and so on.

$$\underset{6\quad5\quad4\quad3\quad2}{CH_3CH_2CH_2CH_2CH_2}\overset{O}{\underset{1}{\overset{\parallel}{C}}}OH \qquad \underset{\varepsilon\quad\delta\quad\gamma\quad\beta\quad\alpha}{CH_3CH_2CH_2CH_2CH_2}\overset{O}{\overset{\parallel}{C}}OH$$

systematic nomenclature   common nomenclature

α = **alpha**
β = **beta**
γ = **gamma**
δ = **delta**
ε = **epsilon**

Take a careful look at the following examples to make sure that you understand the difference between systematic (IUPAC) and common nomenclature:

systematic name:	2-methoxybutanoic acid	3-bromopentanoic acid	4-chlorohexanoic acid
common name:	α-methoxybutyric acid	β-bromovaleric acid	γ-chlorocaproic acid

Carboxylic acids in which a carboxyl group is attached to a ring are named by adding "carboxylic acid" to the name of the cyclic compound.

cyclohexanecarboxylic acid   benzenecarboxylic acid
benzoic acid

α-**Hydroxycarboxylic acids are found in skin products that claim to reduce wrinkles by penetrating the top layer of the skin, causing it to flake off.**

Naming Acyl Chlorides

Acyl chlorides have a Cl in place of the OH group of a carboxylic acid. Acyl chlorides are named by replacing "ic acid" of the acid name with "yl chloride." For cyclic acids that end with "carboxylic acid," "carboxylic acid" is replaced with "carbonyl chloride." (Acyl bromides exist too, but are less common than acyl chlorides.)

systematic name:	ethanoyl chloride	3-methylpentanoyl bromide	cyclopentanecarbonyl
common name:	acetyl chloride	β-methylvaleryl bromide	chloride

Naming Esters

LEARN THE STRATEGY

An **ester** has an OR′ group in place of the OH group of a carboxylic acid. In naming an ester, the name of the group (R′) attached to the **carboxyl oxygen** is stated first, followed by the name of the acid, with "ic acid" replaced by "ate." (The prime on R′ indicates that the alkyl group it designates does not have to be the same as the alkyl group designated by R.) Recall that a benzene ring is called a phenyl group and a benzene ring attached to a methylene group is called a benzyl group (see the top of the next page).

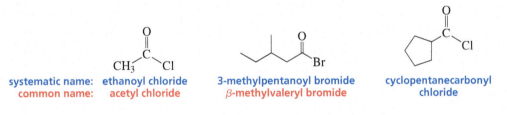

systematic name:	ethyl ethanoate	phenyl propanoate	methyl 3-bromobutanoate
common name:	ethyl acetate	phenyl propionate	methyl β-bromobutyrate

$$\overset{O}{\underset{R}{\overset{\parallel}{\underset{}{C}}}}\underset{OR'}{}$$

carbonyl oxygen
carboxyl oxygen

The double-bonded oxygen is the carbonyl oxygen.

The single-bonded oxygen is the carboxyl oxygen.

USE THE STRATEGY

a phenyl group

a benzyl group

—CH$_2$—

PROBLEM 1 ◆

The aromas of many flowers and fruits are due to esters such as those shown in this problem. What are the common names of these esters? (Also see Problem 57.)

a.

jasmine

b.

banana

c.

apple

Salts of carboxylic acids are named similarly to the way esters are named. That is, the cation is named first, followed by the name of the acid, again with "ic acid" replaced by "ate."

systematic name: **sodium methanoate**   **potassium ethanoate**   **sodium benzenecarboxylate**
common name: **sodium formate**   **potassium acetate**   **sodium benzoate**

Frequently, the name of the cation is omitted.

acetate   **pyruvate**   **(S)-(+)-lactate**

Cyclic esters are called **lactones.** In systematic nomenclature, they are named as "2-oxacycloalkan ones" ("oxa" designates the oxygen atom; "one" designates the doubly bonded oxygen). For their common names, the length of the carbon chain is indicated by the common name of the carboxylic acid, and a Greek letter specifies the carbon to which the oxygen is attached. Thus, six-membered ring lactones are δ-lactones (the carboxyl oxygen is on the δ-carbon), five-membered ring lactones are γ-lactones, and four-membered ring lactones are β-lactones.

2-oxacyclohexanone   **3-methyl-2-oxacyclohexanone**   **2-oxacyclopentanone**   **3-ethyl-2-oxacyclopentanone**
δ-valerolactone   **δ-caprolactone**   **γ-butyrolactone**   **γ-caprolactone**
a δ-lactone   **a δ-lactone**   **a γ-lactone**   **a γ-lactone**

Naming Amides

LEARN THE STRATEGY

An **amide** has an NH$_2$, NHR, or NR$_2$ group in place of the OH group of a carboxylic acid. Amides are named by replacing "oic acid," "ic acid," or "ylic acid" of the acid name with "amide."

systematic name: **ethanamide**   **4-chlorobutanamide**   **benzenecarboxamide**
common name: **acetamide**   **γ-chlorobutyramide**   **benzamide**

If a substituent is bonded to the nitrogen, the name of the substituent is stated first (if there is more than one substituent bonded to the nitrogen, they are stated alphabetically), followed by the name of the amide. The name of each substituent is preceded by an *N* to indicate that the substituent is bonded to a nitrogen.

N-cyclohexylpropanamide   **N-ethyl-N-methylpentanamide**   **N,N-diethylbutanamide**

Cyclic amides are called **lactams.** Their nomenclature is similar to that of lactones. In systematic nomenclature, they are named as "2-azacycloalkanones" ("aza" designates the nitrogen atom). For their common names, the length of the carbon chain is indicated by the common name of the carboxylic acid, and a Greek letter specifies the carbon to which the nitrogen is attached.

2-azacyclohexanone   **2-azacyclopentanone**   **2-azacyclobutanone**
δ-valerolactam   **γ-butyrolactam**   **β-propiolactam**
a δ-lactam   **a γ-lactam**   **a β-lactam**

USE THE STRATEGY

PROBLEM 2 ◆

Name the following:

a. $CH_3CH_2CH_2$—C(=O)—O⁻ K⁺

b.

c.

d.

e.

f. CH_3CH_2—C(=O)—NH_2

g.

h. COOH

i.

PROBLEM 3

Draw the structure for each of the following:

a. phenyl acetate
b. γ-caprolactam
c. sodium formate

d. N-benzylethanamide
e. γ-methylcaproic acid
f. β-bromobutyramide

g. ethyl 2-chloropentanoate
h. cyclohexanecarbonyl chloride
i. α-chlorovaleric acid

Nature's Sleeping Pill

Melatonin, a naturally occurring amide, is a hormone synthesized by the pineal gland from the amino acid tryptophan. (An amino acid is an α-aminocarboxylic acid.) Melatonin regulates the dark–light clock in our brains that governs such things as the sleep–wake cycle, body temperature, and hormone production.

melatonin   **tryptophan**

Melatonin levels increase from evening to night and then decrease as morning approaches. People with high levels of melatonin sleep longer and more soundly than those with low levels. The concentration of the hormone in our bodies varies with age—6-year-olds have more than five times the concentration that 80-year-olds have—which is one of the reasons young people have less trouble sleeping than older people. Melatonin supplements are used to treat insomnia, jet lag, and seasonal affective disorder.

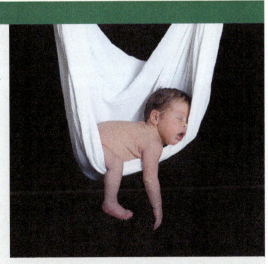

Derivatives of Carbonic Acid

Carbonic acid—a compound with two OH groups bonded to a carbonyl carbon—is unstable, readily breaking down to CO_2 and H_2O. The reaction is reversible, so carbonic acid is formed when CO_2 is bubbled into water (Section 2.11).

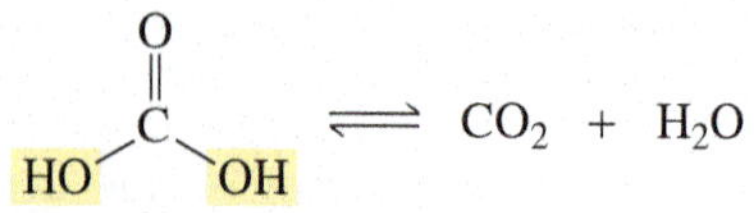

The OH groups of carbonic acid, just like the OH group of a carboxylic acid, can be substituted by other groups.

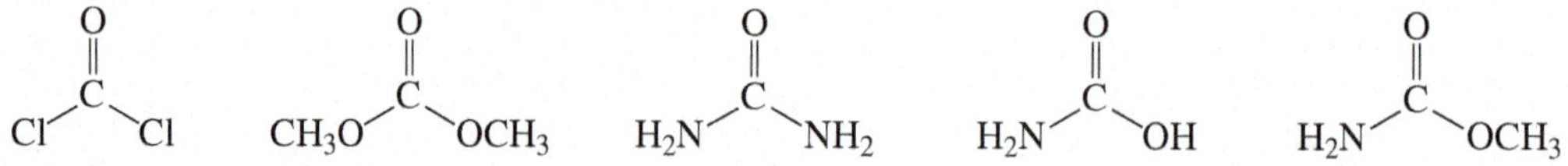

| phosgene | dimethyl carbonate | urea | carbamic acid | methyl carbamate |

15.2 THE STRUCTURES OF CARBOXYLIC ACIDS AND CARBOXYLIC ACID DERIVATIVES

The **carbonyl carbon** in carboxylic acids and carboxylic acid derivatives is sp^2 hybridized. It uses its three sp^2 orbitals to form σ bonds to the carbonyl oxygen, the α-carbon, and a substituent (Y). The three atoms attached to the carbonyl carbon are in the same plane, and the bond angles are each approximately 120°.

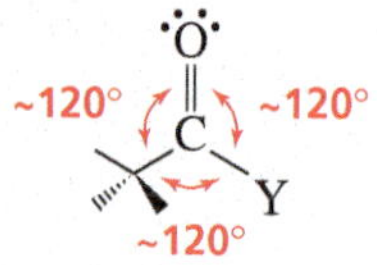

The **carbonyl oxygen** is also sp^2 hybridized. One of its sp^2 orbitals forms a σ bond with the carbonyl carbon, and each of the other two sp^2 orbitals contains a lone pair. The remaining p orbital of the carbonyl oxygen overlaps the remaining p orbital of the carbonyl carbon to form a π bond (Figure 15.1). Esters, carboxylic acids, and amides each have two resonance contributors. The resonance contributor on the right makes a greater contribution to the hybrid in the amide than in the ester or the carboxylic acid. This is due to the fact that the amide's resonance contributor is more stable because nitrogen, being less electronegative than oxygen, can better accommodate a positive charge.

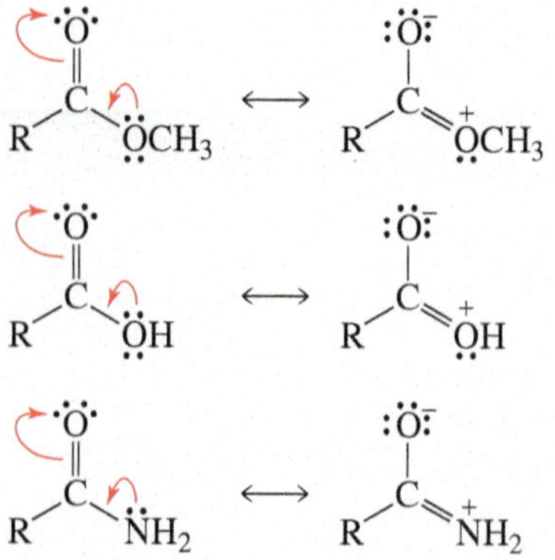

The resonance contributor on the right makes an insignificant contribution to an acyl chloride (Section 15.5), so there is no need to show it.

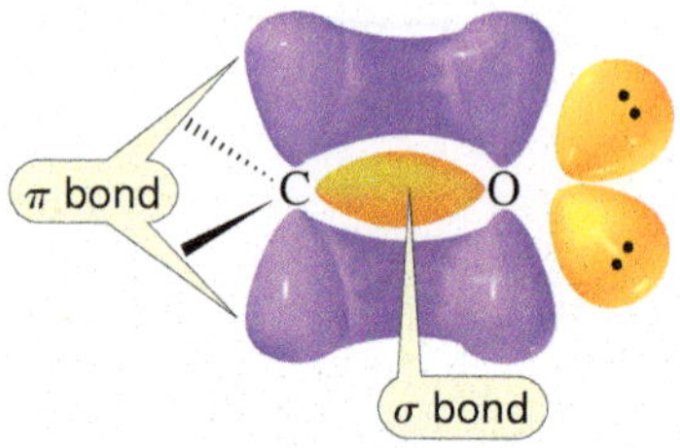

▲ **Figure 15.1**
Bonding in a carbonyl group. The π bond is formed by the side-to-side overlap of a p orbital of carbon with a p orbital of oxygen.

LEARN THE STRATEGY

USE THE STRATEGY

PROBLEM 4 ◆

Which is longer, the carbon–oxygen single bond in a carboxylic acid or the carbon–oxygen bond in an alcohol? Why?

PROBLEM 5 ♦

There are three carbon–oxygen bonds in methyl acetate.

a. What are their relative bond lengths?
b. What are the relative infrared (IR) stretching frequencies of these bonds?

PROBLEM 6 ♦

Which is a correct statement?

A. The delocalization energy of an ester is about 18 kcal/mol, and the delocalization energy of an amide is about 10 kcal/mol.
B. The delocalization energy of an ester is about 10 kcal/mol, and the delocalization energy of an amide is about 18 kcal/mol.

15.3 THE PHYSICAL PROPERTIES OF CARBONYL COMPOUNDS

The acid properties of carboxylic acids were discussed in Sections 2.3 and 8.9. Recall that carboxylic acids have pK_a values of approximately 5.

Boiling Points

Carbonyl compounds have the following relative boiling points:

relative boiling points

amide > carboxylic acid ≫ ester ~ acyl chloride ~ ketone ~ aldehyde

The boiling points of an ester, acyl chloride, ketone, and aldehyde of comparable molecular weight are similar and are *lower* than the boiling point of an alcohol of similar molecular weight because only the alcohol molecules can form hydrogen bonds with each other. The boiling points of these four carbonyl compounds are *higher* than the boiling point of the same-sized ether because of the relatively strong dipole–dipole interactions between the polar carbonyl groups.

Tables of physical properties can be found in Appendix VI.

CH₃CH₂CH₂OH
alcohol
bp = 97.4 °C

H—C(=O)—OCH₃
ester
bp = 32 °C

CH₃—C(=O)—Cl
acyl chloride
bp = 51 °C

CH₃—C(=O)—CH₃
ketone
bp = 56 °C

CH₃CH₂—C(=O)—H
aldehyde
bp = 49 °C

CH₃CH₂OCH₃
ether
bp = 10.8 °C

CH₃—C(=O)—OH
carboxylic acid
bp = 118 °C

CH₃—C(=O)—NH₂
amide
bp = 221 °C

Carboxylic acids have relatively high boiling points because each molecule has two groups that can form hydrogen bonds. Amides have the highest boiling points because they have strong dipole–dipole interactions. This is due to the fact that the resonance contributor with separated charges contributes significantly to the overall structure of the compound (Section 15.2). In addition, if the nitrogen of an amide is bonded to a hydrogen, hydrogen bonds can form between the molecules.

intermolecular hydrogen bonds

dipole–dipole interactions

Solubility

Carboxylic acid derivatives are soluble in solvents such as ethers, chloroalkanes, and aromatic hydrocarbons. Like alcohols and ethers, carbonyl compounds with fewer than four carbons are soluble in water.

Esters, *N,N*-disubstituted amides, and nitriles are often used as solvents because they are polar but do not have reactive OH or NH$_2$ groups. We saw that dimethylformamide (DMF) is a common aprotic polar solvent (Section 9.2).

15.4 HOW CARBOXYLIC ACIDS AND CARBOXYLIC ACID DERIVATIVES REACT

The reactivity of carbonyl compounds is due to the polarity of the carbonyl group, which results from oxygen being more electronegative than carbon. The carbonyl carbon is, therefore, electron deficient (it is an electrophile), so it reacts with nucleophiles.

Formation of a Tetrahedral Intermediate

When a nucleophile adds to the carbonyl carbon of a carboxylic acid derivative, the weakest bond in the molecule—the π bond—breaks, and an intermediate is formed. It is called a **tetrahedral intermediate** because the sp^2 carbon in the reactant has become an sp^3 carbon (that is, a tetrahedral carbon) in the intermediate.

A compound that has an sp^3 carbon bonded to an oxygen atom generally is unstable if the sp^3 carbon is bonded to another electronegative atom.

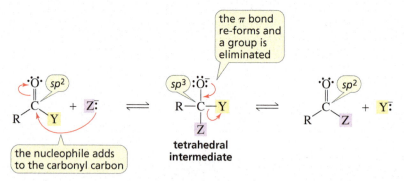

The tetrahedral compound is an intermediate rather than a final product because it is not stable. Generally, *a compound that has an* sp^3 *carbon bonded to an oxygen atom is unstable if the* sp^3 *carbon is bonded to another electronegative atom.* The tetrahedral intermediate, therefore, is unstable because Y and Z are both electronegative atoms. A lone pair on the oxygen re-forms the π bond, and either Y$^-$ or Z$^-$ is eliminated along with its bonding electrons. (Here we show Y$^-$ being eliminated.)

The Weaker Base Is Eliminated from the Tetrahedral Intermediate

Whether Y$^-$ or Z$^-$ is eliminated from the tetrahedral intermediate depends on their relative basicities. The weaker base is eliminated preferentially, making this another example of the principle we first saw in Section 9.2: when comparing bases of the same type, *the weaker base is a better leaving group*. Because a weak base does not share its electrons as well as a strong base does, a weaker base forms a weaker bond—one that is easier to break.

If Z^- is a weaker base than Y^-, then Z^- is eliminated.

tetrahedral intermediate

> Z^- is a weaker base than Y^-, so Z^- is eliminated and the reactants are re-formed

The weaker the base, the better it is as a leaving group.

In this case, no new product is formed. The nucleophile adds to the carbonyl carbon, but the tetrahedral intermediate eliminates the nucleophile and re-forms the reactants.

On the other hand, if Y^- is a weaker base than Z^-, then Y^- is eliminated and a new product is formed.

> Y^- is a weaker base than Z^-, so Y^- is eliminated and the products are formed

tetrahedral intermediate

This reaction is a **nucleophilic acyl substitution reaction** because a nucleophile (Z^-) has replaced the substituent (Y^-) that was attached to the acyl group in the reactant. It is also called an **acyl transfer reaction** because an acyl group has been transferred from one group to another. Often chemists call it a **nucleophilic addition–elimination reaction** because a nucleophile adds to the carbonyl group to form a tetrahedral intermediate, and then a group is eliminated from the tetrahedral intermediate.

If the basicities of Y^- and Z^- are similar, some molecules of the tetrahedral intermediate will eliminate Y^- and others will eliminate Z^-. When the reaction is over, both the reactants and the products are present.

> the basicities of Y^- and Z^- are similar, so a mixture of reactants and products is obtained

tetrahedral intermediate

A carboxylic acid derivative will undergo a nucleophilic acyl substitution reaction if the newly added group in the tetrahedral intermediate is not a weaker base than the group attached to the acyl group in the reactant.

We can, therefore, make the following general statement about the reactions of carboxylic acid derivatives:

A carboxylic acid derivative will undergo a nucleophilic acyl substitution reaction, provided that the newly added group in the tetrahedral intermediate is not a weaker base than the group attached to the acyl group in the reactant.

The remaining sections of this chapter show specific examples of this general principle. Therefore, you can always determine the outcome of the reactions of carboxylic acids and carboxylic acid derivatives presented in this chapter by examining the tetrahedral intermediate and remembering that the weaker base is preferentially eliminated (Section 15.4).

Comparing Nucleophilic Acyl Substitution with Nucleophilic Substitution

Let's compare this two-step nucleophilic acyl substitution reaction with a one-step S_N2 reaction. When a nucleophile attacks an electrophilic carbon, the weakest bond in the molecule breaks. The weakest bond in an S_N2 reaction is the bond to the leaving group, so this is the bond that breaks in the first and only step of the reaction (Section 9.1). In contrast, the weakest bond in a nucleophilic acyl substitution reaction is the π bond, so this bond breaks first and the leaving group is eliminated in a subsequent step.

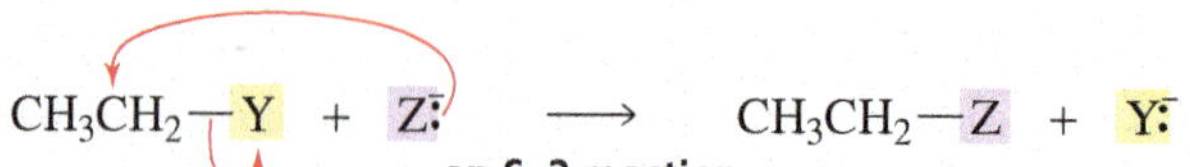

$$CH_3CH_2-Y + Z\colon^- \longrightarrow CH_3CH_2-Z + Y\colon^-$$

an S_N2 reaction

LEARN THE STRATEGY

PROBLEM-SOLVING STRATEGY

Using Basicity to Predict the Outcome of a Nucleophilic Acyl Substitution Reaction

What is the product of the reaction of acetyl chloride with CH_3O^-? The pK_a of HCl is –7; the pK_a of CH_3OH is 15.5.

To identify the product of the reaction, we need to compare the basicities of the two groups in the tetrahedral intermediate so we can determine which one will be eliminated. Because HCl is a stronger acid than CH_3OH, Cl^- is a weaker base than CH_3O^-. Therefore, Cl^- is eliminated from the tetrahedral intermediate and methyl acetate is the product of the reaction.

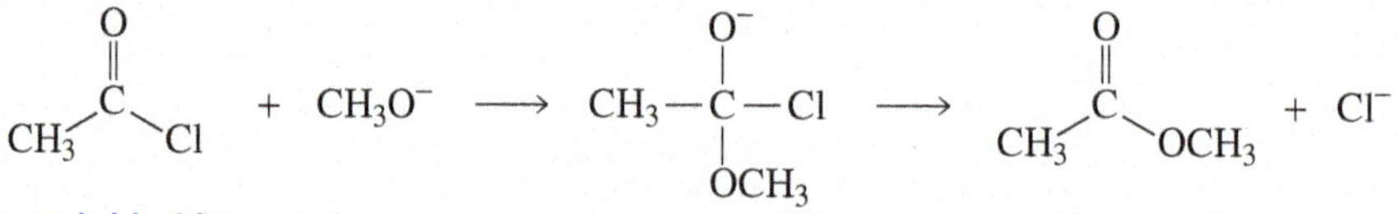

acetyl chloride   methyl acetate

USE THE STRATEGY

PROBLEM 7 ♦

a. What is the product of the reaction of acetyl chloride with HO^-? The pK_a of HCl is –7; the pK_a of H_2O is 15.7.
b. What is the product of the reaction of acetamide with HO^-? The pK_a of NH_3 is 36; the pK_a of H_2O is 15.7.

PROBLEM 8 ♦

What is the product of an acyl substitution reaction—a new carboxylic acid derivative, a mixture of two carboxylic acid derivatives, or no reaction—if the new group in the tetrahedral intermediate is the following?

a. a stronger base than the substituent that is attached to the acyl group
b. a weaker base than the substituent that is attached to the acyl group
c. similar in basicity to the substituent that is attached to the acyl group

15.5 THE RELATIVE REACTIVITIES OF CARBOXYLIC ACIDS AND CARBOXYLIC ACID DERIVATIVES

We just saw that there are two steps in a nucleophilic acyl substitution reaction: *formation* of a tetrahedral intermediate and *collapse* of the tetrahedral intermediate. The weaker the base attached to the acyl group (Table 15.1), the easier it is for *both steps* of the reaction to take place.

relative basicities of the leaving groups

$$\text{weakest base} \quad Cl^- < {}^-OR \approx {}^-OH < {}^-NH_2 \quad \text{strongest base}$$

Therefore, carboxylic acid derivatives have the following relative reactivities:

relative reactivities of carboxylic acid derivatives

most reactive — acyl chloride   ester   carboxylic acid   amide — least reactive

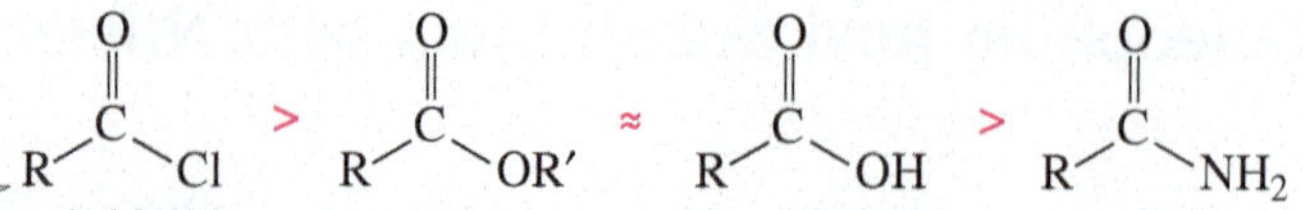

A Weak Base Makes the First Step Easier

How does having a weak base attached to the acyl group make the *first* step of the acyl substitution reaction easier? The key factor is the extent to which the lone-pair electrons on Y are delocalized onto the carbonyl oxygen.

Weak bases do not share their electrons well, so the weaker the basicity of Y, the smaller the contribution from the resonance contributor with a positive charge on Y. In addition, when Y = Cl, delocalization of chlorine's lone pair is minimal due to the poor orbital overlap between the large $3p$ orbital on chlorine and the smaller $2p$ orbital on carbon. The less the contribution from the resonance contributor with the positive charge on Y, the greater the contribution from the resonance contributor with the electrophilic carbonyl carbon. Thus, weak bases cause the carbonyl carbon to be more electrophilic and, therefore, more reactive toward nucleophiles.

relative reactivity: acyl chloride >
ester ~ carboxylic acid > amide

resonance contributors of a carboxylic acid or carboxylic acid derivative

---

PROBLEM 9 ◆

a. Which compound has the stretching vibration for its carbonyl group at the highest frequency: acetyl chloride, methyl acetate, or acetamide?

b. Which one has the stretching vibration for its carbonyl group at the lowest frequency?

A Weak Base Makes the Second Step Easier

A weak base attached to the acyl group also makes the *second* step of the acyl substitution reaction easier, because weak bases are easier to eliminate when the tetrahedral intermediate collapses.

the weaker the base, the easier it is to eliminate

In Section 15.5 we saw that in a nucleophilic acyl substitution reaction, the nucleophile that adds to the carbonyl carbon must be a stronger base than the substituent that is attached to the acyl group. This means that

a carboxylic acid derivative can be converted to a less reactive carboxylic acid derivative in a nucleophilic acyl substitution reaction, but not to one that is more reactive.

For example, an acyl chloride can be converted to the less reactive ester because an alkoxide ion is a stronger base than a chloride ion.

An ester, however, cannot be converted to the more reactive acyl chloride because a chloride ion is a weaker base than an alkoxide ion.

Reaction Coordinate Diagrams for Nucleophilic Acyl Substitution

Reaction coordinate diagrams for nucleophilic acyl substitution reactions with nucleophiles of varying basicity are shown in Figure 15.2 on the next page (where TI is the tetrahedral intermediate).

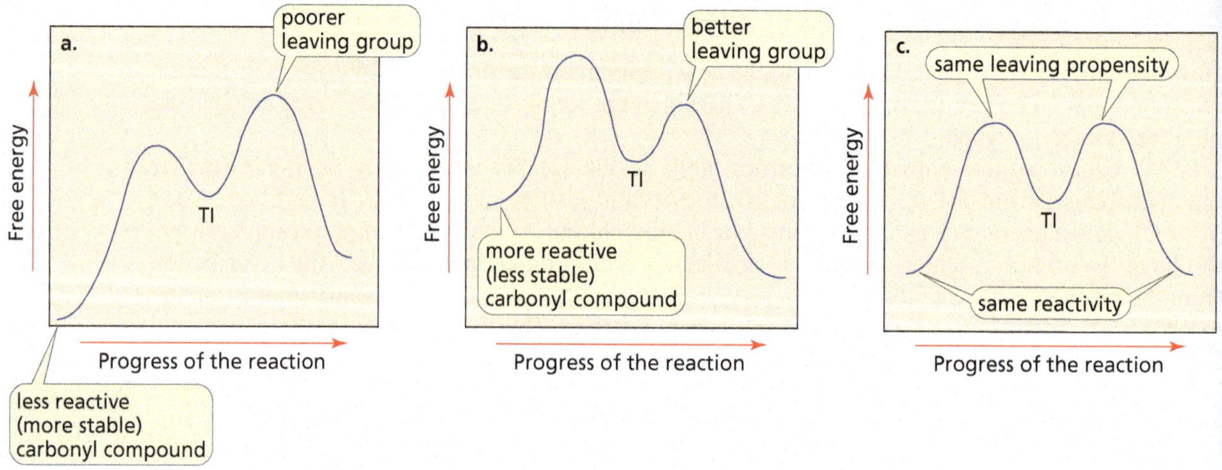

▲ **Figure 15.2**
(a) The nucleophile is a weaker base than the group attached to the acyl group in the reactant.
(b) The nucleophile is a stronger base than the group attached to the acyl group in the reactant.
(c) The nucleophile and the group attached to the acyl group in the reactant have similar basicities.

LEARN THE STRATEGY

- To form a more reactive compound from a less reactive compound, the new group in the tetrahedral intermediate has to be a weaker base than the group attached to the acyl group in the reactant. However, the lower energy pathway is for the tetrahedral intermediate (TI) to eliminate the newly added group and re-form the reactants, so no reaction takes place (Figure 15.2a).

- To form a less reactive compound from a more reactive compound, the new group in the tetrahedral intermediate has to be a stronger base than the group attached to the acyl group in the reactant. The lower energy pathway is for the tetrahedral intermediate (TI) to eliminate the group attached to the acyl group in the reactant, so a substitution product is formed (Figure 15.2b).

- If the reactant and product have similar reactivities, then both groups in the tetrahedral intermediate will have similar basicities. In this case, the tetrahedral intermediate can eliminate either group with similar ease, so a mixture of the reactant and the substitution product results (Figure 15.2c).

USE THE STRATEGY

PROBLEM 10 ◆

Using the pK_a values listed in Table 15.1, predict the products of the following reactions:

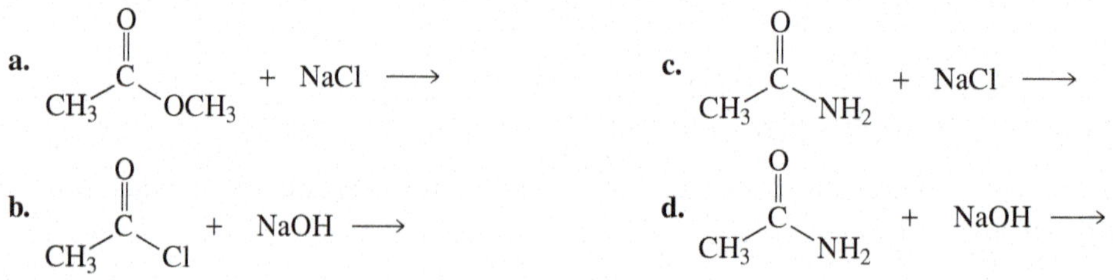

a.

$$\underset{CH_3}{\overset{O}{\underset{\|}{C}}}OCH_3 \quad + \quad NaCl \quad \longrightarrow$$

c.

$$\underset{CH_3}{\overset{O}{\underset{\|}{C}}}NH_2 \quad + \quad NaCl \quad \longrightarrow$$

b.

$$\underset{CH_3}{\overset{O}{\underset{\|}{C}}}Cl \quad + \quad NaOH \quad \longrightarrow$$

d.

$$\underset{CH_3}{\overset{O}{\underset{\|}{C}}}NH_2 \quad + \quad NaOH \quad \longrightarrow$$

PROBLEM 11 ◆

Is the following statement true or false?

If the newly added group in the tetrahedral intermediate is a stronger base than the group attached to the acyl group in the reactant, then formation of the tetrahedral intermediate is the rate-limiting step of a nucleophilic acyl substitution reaction.

15.6 REACTIONS OF ACYL CHLORIDES

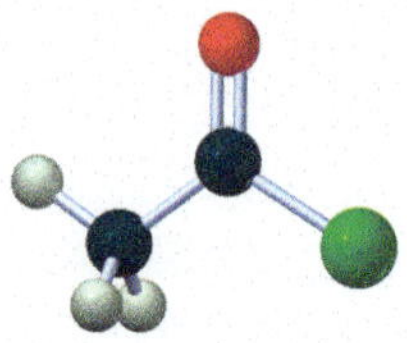

Acyl chlorides react with alcohols to form esters, with water to form carboxylic acids, and with amines to form amides because, in each case, the incoming nucleophile is a stronger base than the departing halide ion (Table 15.1). Methanol is both the nucleophile and the solvent in the first reaction, and water is both the nucleophile and the solvent in the second.

acetyl chloride

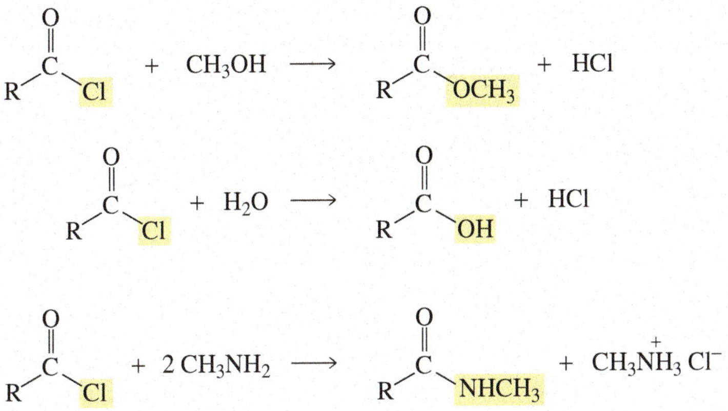

All carboxylic acid derivatives undergo nucleophilic acyl substitution reactions by one of the two following mechanisms. The mechanism followed depends only on whether the nucleophile is charged or neutral.

MECHANISM FOR THE REACTION OF AN ACYL CHLORIDE WITH A NEGATIVELY CHARGED NUCLEOPHILE

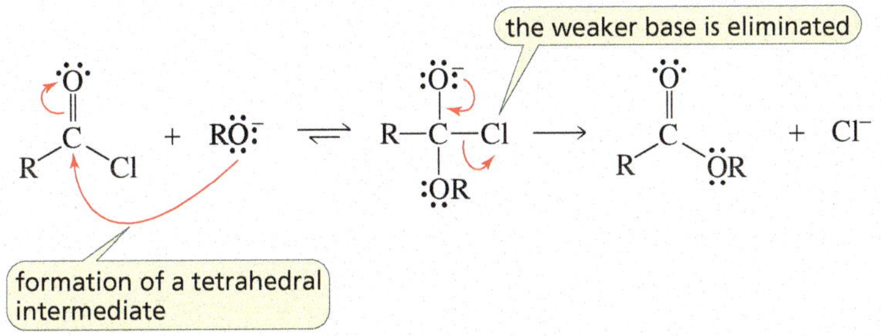

- The nucleophile adds to the carbonyl carbon, forming a tetrahedral intermediate.
- The unstable tetrahedral intermediate collapses, eliminating the chloride ion because it is a weaker base than the alkoxide ion.

If the nucleophile is neutral, the mechanism has an additional step in which a proton is lost.

MECHANISM FOR THE REACTION OF AN ACYL CHLORIDE WITH A NEUTRAL NUCLEOPHILE

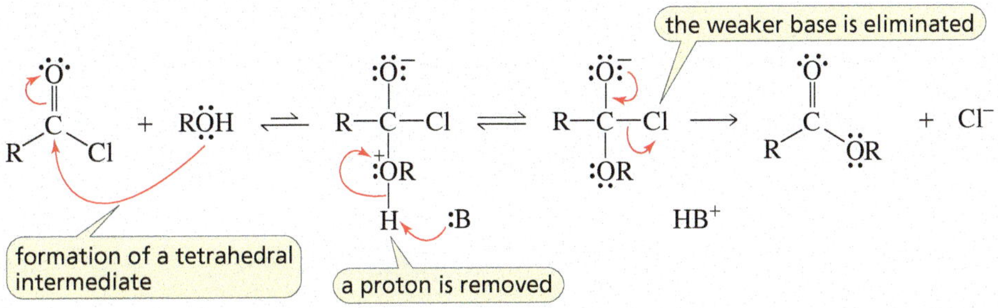

The weaker base is eliminated from the tetrahedral intermediate.

- The nucleophilic alcohol adds to the carbonyl carbon, forming a tetrahedral intermediate.
- Because the protonated ether group is a strong acid, the tetrahedral intermediate loses a proton. Proton transfers to and from oxygen are diffusion controlled, that is, they occur very rapidly. (:B represents any species in the solution that can remove a proton.)
- The unstable tetrahedral intermediate collapses, eliminating the chloride ion because it is a weaker base than the alkoxide ion.

LEARN THE STRATEGY

Notice that the reaction of an acyl chloride with an amine (on the top of this page) or with ammonia (shown next) to form an amide is carried out with twice as much amine or ammonia as acyl chloride because the proton produced in the reaction protonates any amine or ammonia that has yet to react (When the nucleophile in the mechanism on the bottom of the previous page is an amine or ammonia, :B is most likely to be the amine or ammonia.) Once protonated, the amine or ammonia is no longer a nucleophile, so it cannot react with the acyl chloride. Using twice as much amine or ammonia as acyl chloride guarantees that there is enough unprotonated amine or ammonia to react with all the acyl chloride.

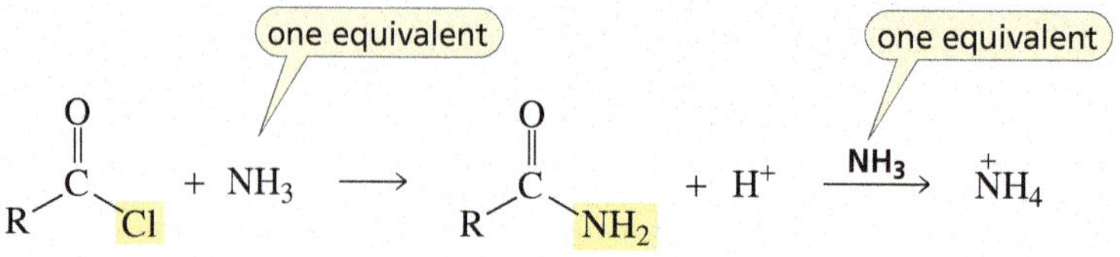

USE THE STRATEGY

PROBLEM 12

Starting with acetyl chloride, what neutral nucleophile would you use to synthesize each of the following compounds?

a. 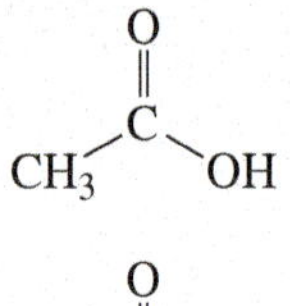

b. 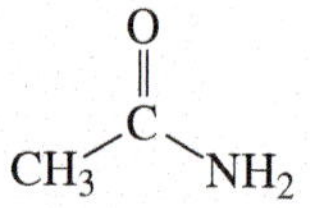

c. 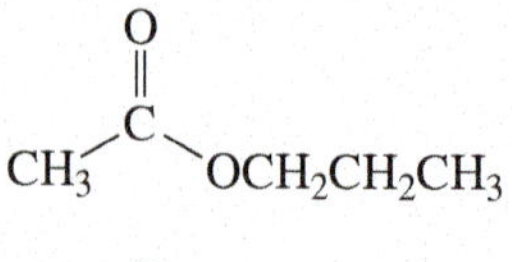

d. 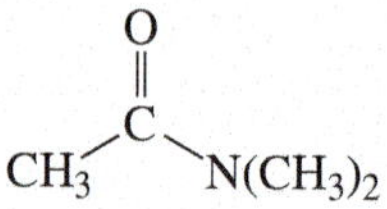

e. 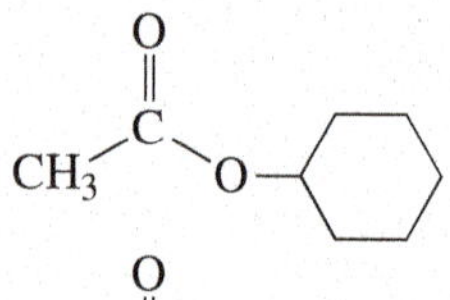

f. 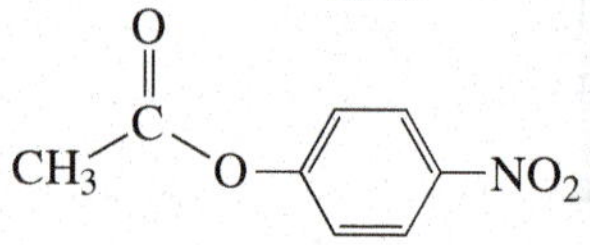

PROBLEM 13

Write the mechanism for each of the following reactions:

a. the reaction of acetyl chloride with water to form acetic acid

b. the reaction of benzoyl chloride with excess methylamine to form *N*-methylbenzamide

PROBLEM 14 SOLVED

a. What two amides are obtained from the reaction of acetyl chloride with an equivalent of ethylamine and an equivalent of propylamine?

b. Why is only one amide obtained from the reaction of acetyl chloride with an equivalent of ethylamine and an equivalent of triethylamine?

SOLUTION TO 14 a. Either of the amines can react with acetyl chloride, so both *N*-ethylacetamide and *N*-propylacetamide are formed.

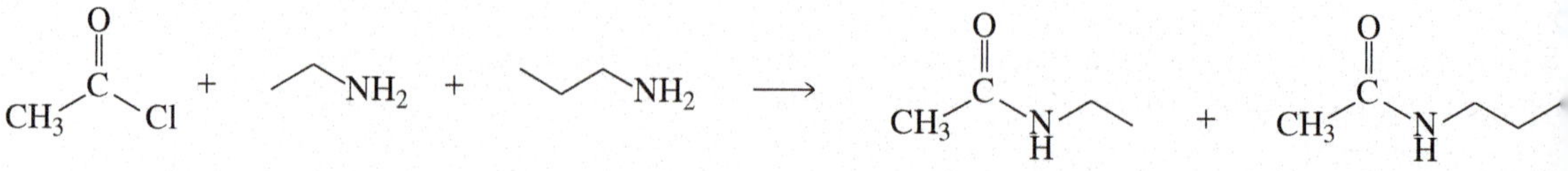

N-ethylacetamide *N*-propylacetamide

SOLUTION TO 14 b. Either of the amines can react with acetyl chloride, so two compounds are formed initially. However, the compound formed by triethylamine is very reactive because it has a positively charged nitrogen, which is an excellent leaving group. Therefore, the compound reacts immediately with any unreacted ethylamine, so *N*-ethylacetamide is the only amide product of the reaction.

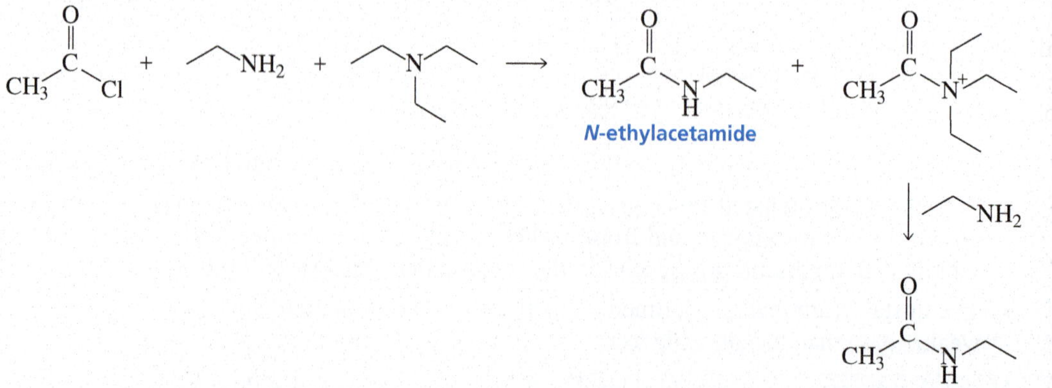

N-ethylacetamide

N-ethylacetamide

15.7 REACTIONS OF ESTERS

Esters do not react with chloride ion because it is a much weaker base than the RO^- group of the ester, so Cl^- (not RO^-) would be the base eliminated from the tetrahedral intermediate, re-forming the ester (Table 15.1).

An ester reacts with water to form a carboxylic acid and an alcohol. A reaction with water that converts one compound into two compounds is called a **hydrolysis reaction** (*lysis* is Greek for "breaking down").

methyl acetate

> a hydrolysis reaction

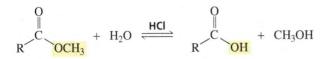

An ester reacts with an alcohol to form a new ester and a new alcohol. A reaction with an alcohol that converts one compound into two compounds is called an **alcoholysis reaction.** This particular alcoholysis reaction is also called a **transesterification reaction** because one ester is converted to another ester.

> a transesterification reaction

Both the hydrolysis and the transesterification of an ester are very slow reactions because water and alcohols are poor nucleophiles and the RO^- group of an ester is a poor leaving group. Therefore, these reactions are always catalyzed when carried out in the laboratory. Both the hydrolysis and transesterification of an ester can be catalyzed by acids (Section 15.10). The rate of hydrolysis can also be increased by hydroxide ion and the rate of alcoholysis can be increased by the conjugate base (RO^-) of the reactant alcohol (Section 15.11).

An ester reacts with an amine to form an amide. A reaction with an amine that converts one compound into two compounds is called an **aminolysis reaction.** Notice that the aminolysis of an ester requires only one equivalent of amine, unlike the aminolysis of an acyl halide, which requires two equivalents (Section 15.6). This is because the leaving group of an ester (RO^-) is more basic than the amine, so the alkoxide ion—rather than unreacted amine—picks up the proton generated in the reaction.

> an aminolysis reaction

The reaction of an ester with an amine is not as slow as the reaction of an ester with water or an alcohol because an amine is a better nucleophile. This is fortunate because the reaction cannot be catalyzed by an acid. The acid would protonate the amine, and a protonated amine is not a nucleophile. The rate of the reaction, however, can be increased by heat.

PROBLEM 15 ◆

Starting with methyl acetate, what neutral nucleophile would you use to synthesize each of the following compounds?

a. ethyl acetate   **b.** acetamide   **c.** phenyl acetate   **d.** benzyl acetate

---

PROBLEM 16 ◆

We saw that it is necessary to use excess amine in the reaction of an acyl chloride with an amine. Explain why it is not necessary to use excess alcohol in the reaction of an acyl chloride with an alcohol.

PROBLEM 17

Write a mechanism for each of the following reactions:

a. the uncatalyzed hydrolysis of methyl propionate.
b. the aminolysis of phenyl formate, using methylamine.

PROBLEM 18 SOLVED

Rank the following esters from most reactive to least reactive toward hydrolysis:

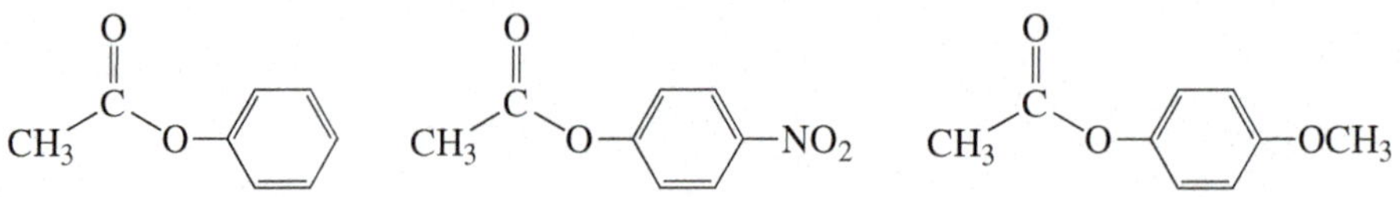

LEARN THE STRATEGY

SOLUTION We know that the reactivity of a carboxylic acid derivative depends on the basicity of the group attached to the acyl group—the weaker the base, the easier it is for *both steps* of the reaction to take place (Section 15.6). So now we need to compare the basicities of the three phenolate ions.

The nitro-substituted phenolate ion is the weakest base because the nitro group withdraws electrons inductively and by resonance (see page 343), which decreases the concentration of negative charge on the oxygen. The methoxy-substituted phenolate ion is the strongest base because the methoxy group donates electrons by resonance more than it withdraws electrons inductively (see page 343), so the concentration of negative charge on the oxygen is increased. Therefore, the three esters have the following relative reactivity toward hydrolysis.

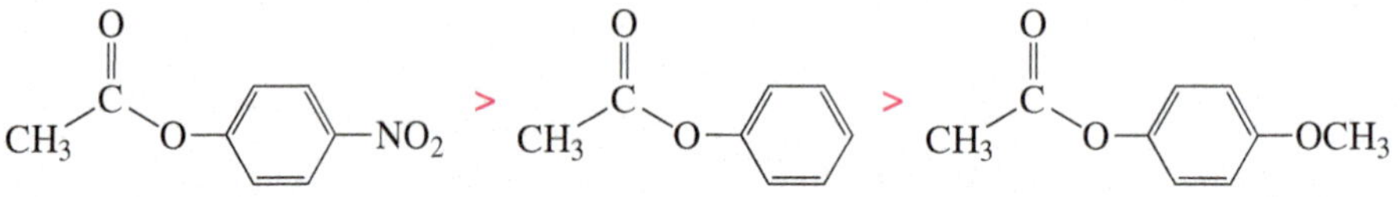

USE THE STRATEGY

PROBLEM 19 ◆

Which ester hydrolyzes more rapidly?

a. methyl acetate or phenyl acetate? **b.** phenyl acetate or benzyl acetate?

PROBLEM 20 ◆

a. State three factors that cause the uncatalyzed hydrolysis of an ester to be a slow reaction.
b. Which is faster, the hydrolysis of an ester or the aminolysis of the same ester? Explain.

15.8 ACID-CATALYZED ESTER HYDROLYSIS AND TRANSESTERIFICATION

We saw that esters hydrolyze slowly because water is a poor nucleophile and esters have relatively basic (poor) leaving groups. The rate of hydrolysis can be increased by either acid or hydroxide ion. When you examine the mechanisms for these reactions, notice the following features that hold for all organic reactions:

> *All organic reactants, intermediates, and products in acidic solutions are positively charged or neutral; negatively charged species are not formed in acidic solutions.*

> *All organic reactants, intermediates, and products in basic solutions are negatively charged or neutral; positively charged species are not formed in basic solutions.*

Hydrolysis of an Ester with a Primary or Secondary Alkyl Group

When an acid is added to a reaction, the first thing that happens is the acid protonates the atom in the reactant that has the greatest electron density. Therefore, when an acid is added to an ester, the acid protonates the carbonyl oxygen.

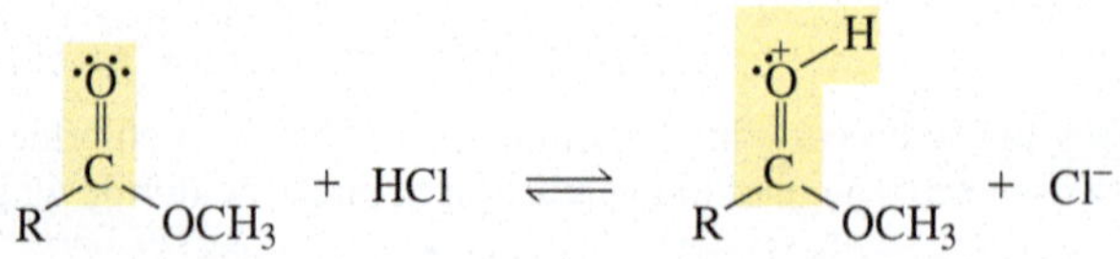

The resonance contributors of the ester show that one oxygen has a negative charge and one has a positive charge. The oxygen with the negative charge (the carbonyl oxygen) is the one that is protonated.

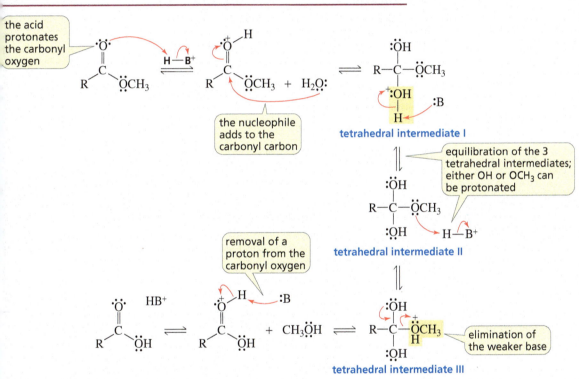

MECHANISM FOR ACID-CATALYZED ESTER HYDROLYSIS

NOTE TO THE STUDENT

- Pay attention to the three tetrahedral intermediates that occur in this mechanism:

  protonated tetrahedral intermediate I ⇌ neutral tetrahedral intermediate II ⇌ protonated tetrahedral intermediate III

  This pattern is repeated in many more acid-catalyzed reactions.

- The acid protonates the carbonyl oxygen.
- The nucleophile (H_2O) adds to the carbonyl carbon of the protonated carbonyl group, forming a protonated tetrahedral intermediate.
- The protonated tetrahedral intermediate (I) is in equilibrium with its nonprotonated form (II).
- The nonprotonated tetrahedral intermediate can be re-protonated on OH, which re-forms tetrahedral intermediate I, or protonated on OCH_3, which forms tetrahedral intermediate III. (From Section 2.10, we know that the relative amounts of the three tetrahedral intermediates depend on the pH of the solution and the pK_a values of the protonated intermediates.)
- When tetrahedral intermediate I collapses, it eliminates H_2O in preference to CH_3O^- (because H_2O is a weaker base) and re-forms the ester. When tetrahedral intermediate III collapses, it eliminates CH_3OH rather than HO^- (because CH_3OH is a weaker base) and forms the carboxylic acid. Because H_2O and CH_3OH have approximately the same basicity, it is as likely for tetrahedral intermediate I to collapse to re-form the ester as it is for tetrahedral intermediate III to collapse to form the carboxylic acid. (Tetrahedral intermediate II is much less likely to collapse because both HO^- and CH_3O^- are strong bases and, therefore, poor leaving groups.)
- Removal of a proton from the protonated carboxylic acid forms the carboxylic acid and re-forms the acid catalyst.

Because tetrahedral intermediates I and III are equally likely to collapse, both ester and carboxylic acid are present when the reaction has reached equilibrium. Excess water can be used to force the equilibrium to the right (Le Châtelier's principle; Section 5.7). Or, if the boiling point of the product alcohol is significantly lower than the boiling points of the other components of the reaction, the reaction can be driven to the right by distilling off the alcohol as it is formed.

In Section 15.11, we will see that the mechanism for the acid-catalyzed reaction of a carboxylic acid and an alcohol to form an ester and water is the exact reverse of the mechanism for the acid-catalyzed hydrolysis of an ester to form a carboxylic acid and an alcohol.

PROBLEM 21 ♦

In the mechanism for the acid-catalyzed hydrolysis of an ester,

a. what species could be represented by HB^+?
b. what species could be represented by :B?
c. what species is HB^+ most likely to be in the hydrolysis reaction?
d. what species is HB^+ most likely to be in the reverse reaction?

PROBLEM 22

Using the mechanism for the acid-catalyzed hydrolysis of an ester as your guide, write the mechanism—showing all the curved arrows—for the acid-catalyzed reaction of acetic acid and methanol to form methyl acetate. Use HB^+ and :B to represent the proton-donating and proton-removing species, respectively.

How an Acid Increases the Rate of Ester Hydrolysis

Now let's see how the acid catalyst increases the rate of ester hydrolysis. For a catalyst to increase the rate of a reaction, it must increase the rate of the slow steps of a reaction. Four of the six steps in the mechanism for acid-catalyzed ester hydrolysis are proton transfer steps. Proton transfer to or from an electronegative atom such as oxygen or nitrogen is always a fast step. The other two steps in the mechanism—namely, formation of the tetrahedral intermediate and collapse of the tetrahedral intermediate—are relatively slow. The acid increases the rates of both these steps.

The acid increases *the rate of formation of the tetrahedral intermediate* by protonating the carbonyl oxygen. Protonated carbonyl groups are more susceptible than nonprotonated carbonyl groups to nucleophilic addition, because a positively charged oxygen is more electron withdrawing than an uncharged oxygen. Increased electron withdrawal by the positively charged oxygen makes the carbonyl carbon more electron deficient, which increases its reactivity toward nucleophiles.

protonating the carbonyl oxygen increases the susceptibility of the carbonyl carbon to nucleophilic addition

The acid increases *the rate of collapse of the tetrahedral intermediate* by decreasing the basicity of the leaving group, which makes it easier to eliminate: in the acid-catalyzed hydrolysis of an ester, the leaving group is CH_3OH, which is a weaker base than CH_3O^-, the leaving group in the uncatalyzed reaction.

protonating the leaving group makes it a better leaving group

PROBLEM 23 ◆

What products are formed from the acid-catalyzed hydrolysis of the following esters?

a. 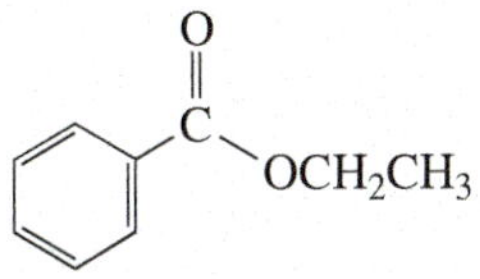   b. 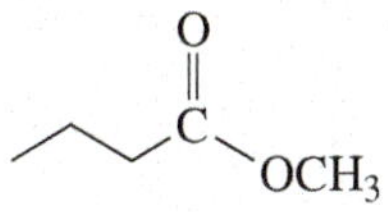   c.

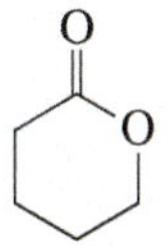

PROBLEM 24

Show the mechanism for the acid-catalyzed formation of **23c** starting with the product obtained from its hydrolysis.

Hydrolysis of an Ester with a Tertiary Alkyl Group

The hydrolysis of an ester with a *tertiary alkyl group* forms the same products as the hydrolysis of an ester with *a primary or secondary alkyl group*—namely, a carboxylic acid and an alcohol—but does so by a completely different mechanism. The hydrolysis of an ester with a tertiary alkyl group is an S_N1 reaction rather than a nucleophilic acyl substitution reaction, because the leaving group of the protonated ester is a relatively stable tertiary carbocation.

MECHANISM FOR THE HYDROLYSIS OF AN ESTER WITH A TERTIARY ALKYL GROUP

departure of the leaving group to form a tertiary carbocation

reaction of the carbocation with a nucleophile

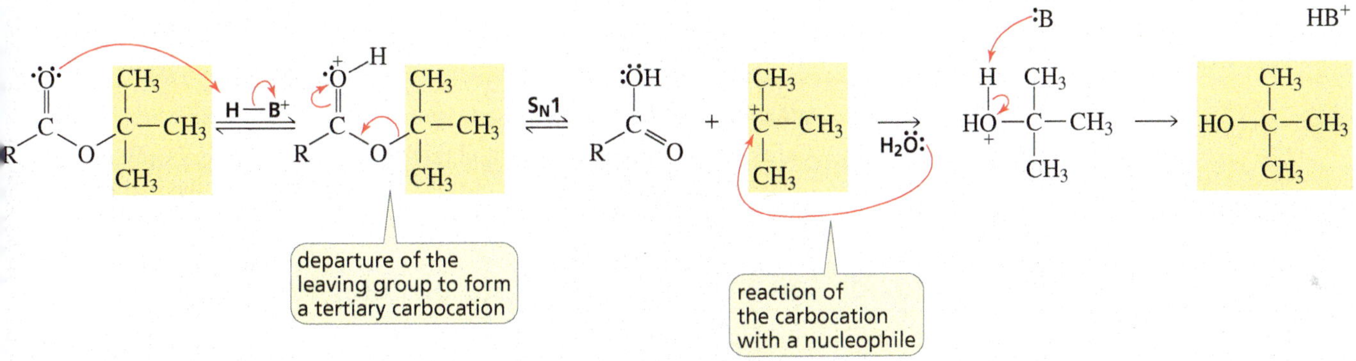

- An acid protonates the carbonyl oxygen.
- The leaving group departs, forming a tertiary carbocation.
- The nucleophile (H_2O) reacts with the carbocation.
- A base removes a proton from the strongly acidic protonated alcohol.

Transesterification

Transesterification—the reaction of an ester with an alcohol—is also catalyzed by acid. The mechanism for acid-catalyzed transesterification is identical to the mechanism for acid-catalyzed ester hydrolysis, except that the nucleophile is ROH rather than H_2O. As in ester hydrolysis, the leaving groups in the tetrahedral intermediate have approximately the same basicity. Consequently, an excess of the reactant alcohol must be used in order to produce a good yield of the desired product.

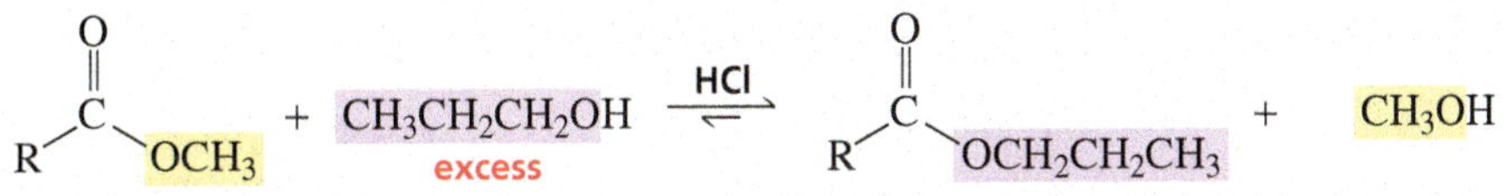

PROBLEM 25 ◆

What products are obtained from the following reactions?

a. ethyl benzoate + excess isopropanol + HCl

b. phenyl acetate + excess ethanol + HCl

PROBLEM 26

Write the mechanism for the acid-catalyzed transesterification of ethyl acetate with methanol.

PROBLEM 27

Write the mechanism for the acid-catalyzed reaction of *tert*-butyl acetate with methanol.

15.9 HYDROXIDE-ION-PROMOTED ESTER HYDROLYSIS

The rate of hydrolysis of an ester can be increased by hydroxide ion. Like an acid catalyst, hydroxide ion increases the rates of the two slow steps of the reaction—namely, formation of the tetrahedral intermediate and collapse of the tetrahedral intermediate.

MECHANISM FOR THE HYDROXIDE-ION-PROMOTED HYDROLYSIS OF AN ESTER

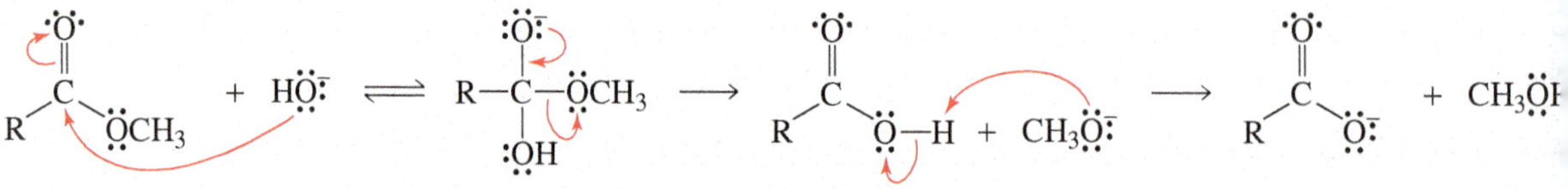

- Hydroxide ion adds to the carbonyl carbon of the ester.
- The two potential leaving groups in the tetrahedral intermediate (HO⁻ and CH₃O⁻) have the same basicity, so they have the same leaving propensity. Elimination of HO⁻ re-forms the ester, whereas elimination of CH₃O⁻ forms a carboxylic acid.
- The final products are not the carboxylic acid and methoxide ion because if only one base is protonated, it will be the stronger base. Therefore, the final products are the carboxylate ion and methanol because CH_3O^- is a stronger base than $RCOO^-$.

How Hydroxide Ion Increases the Rate of Ester Hydrolysis

Hydroxide ion is a better nucleophile than water.

Hydroxide ion increases the rate of formation of the tetrahedral intermediate because HO⁻ is a better nucleophile than H_2O. Hydroxide ion increases the rate of collapse of the tetrahedral intermediate because, in a basic solution, the tetrahedral intermediate is negatively charged. The transition state for expulsion of CH_3O^- by a negatively charged oxygen is more stable than the transition state for expulsion of CH_3O^- by a neutral oxygen because, in the former, the electronegative oxygen does not develop a partial positive charge.

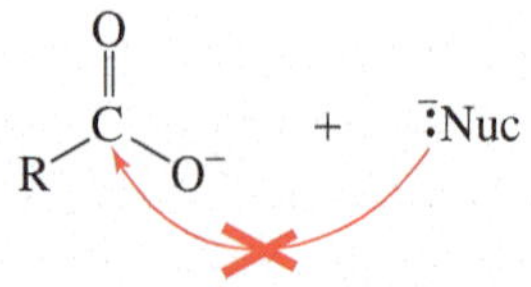

Because carboxylate ions are negatively charged, they do not react with nucleophiles. Therefore, the hydroxide-ion-promoted hydrolysis of an ester, unlike the acid-catalyzed hydrolysis of an ester, is *not* a reversible reaction.

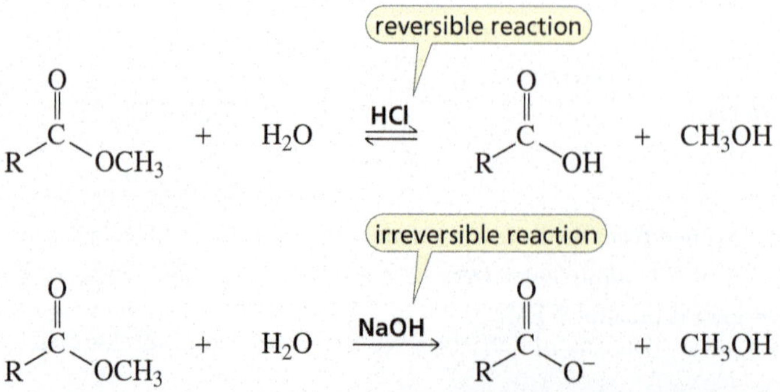

The hydrolysis of an ester in the presence of hydroxide ion is called a *hydroxide-ion-promoted reaction* rather than a base-catalyzed reaction because hydroxide ion increases the rate of the first step of the reaction by being a better nucleophile than water—not by being a stronger base than water—and because hydroxide ion is consumed in the overall reaction. To be a catalyst, a species must not be changed by or consumed in the reaction. Therefore, hydroxide ion is actually a reagent rather than a catalyst, so it is more accurate to call the reaction a hydroxide-ion-*promoted* reaction than a hydroxide-ion-*catalyzed* reaction.

Hydroxide Ion Promotes Only Hydrolysis Reactions

Hydroxide ion promotes only hydrolysis reactions. Hydroxide ion cannot promote reactions of carboxylic acid derivatives with alcohols or with amines because one function of hydroxide ion is to provide a good nucleophile for the first step of the reaction. When the nucleophile is supposed to be an alcohol or an amine, nucleophilic addition by hydroxide ion would form a different product from the one that would be formed by nucleophilic addition of an alcohol or an amine. Hydroxide can be used to promote a hydrolysis reaction because the same product is formed, whether the nucleophile that adds to the carbonyl carbon is H_2O or HO^-.

Reactions in which the nucleophile is an alcohol can be promoted by the conjugate base of the alcohol. One function of the alkoxide ion is to provide a good nucleophile for the first step of the reaction, so only reactions in which the nucleophile is an alcohol can be promoted by the conjugate base of the alcohol.

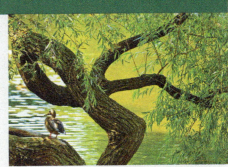

Aspirin, NSAIDs, and COX-2 Inhibitors

Salicylic acid, found in willow bark and myrtle leaves, is perhaps the oldest known drug. As early as the fifth century B.C., Hippocrates wrote about the curative powers of willow bark. In 1897, Felix Hoffmann, a scientist working at Bayer and Co. in Germany, found that acylating salicylic acid produced a more potent drug to control fever and pain (see pages 114–115). They called it *aspirin*; "a" for acetyl, "spir" for the spiraea flower that also contains salicylic acid, and "in," which was a common ending for drugs at that time. It soon became the world's best-selling drug. However, its mode of action was not discovered until 1971, when it was found that the anti-inflammatory and fever-reducing activity of aspirin were due to a transesterification reaction that blocks the synthesis of prostaglandins and the related thromboxanes.

Prostaglandins have several different physiological functions (Section 25.6). One is to stimulate inflammation and another to induce fever. The enzyme prostaglandin synthase catalyzes the conversion of arachidonic acid to PGH_2, a precursor of all prostaglandins.

willow tree

arachidonic acid

prostaglandin synthase

PGH$_2$

PGE$_2$
a prostaglandin

a thromboxane

Prostaglandin synthase is composed of two enzymes. One of them—cyclooxygenase—has a CH_2OH group at its active site that is necessary for enzymatic activity. When the CH_2OH group reacts with aspirin in a transesterification reaction, the enzyme is inactivated (see the top of the next page). This prevents prostaglandins from being synthesized, so inflammation is suppressed and fever is reduced. Notice that the carboxyl group of aspirin is a basic catalyst. It removes a proton from the CH_2OH group, which makes it a better nucleophile. This is why aspirin is maximally active in its basic form (see page 74). (The red arrows show the formation of the tetrahedral intermediate; the blue arrows show its collapse.) *(Continued)*

Because aspirin inhibits the formation of PGH_2, it also inhibits the synthesis of thromboxanes, compounds involved in blood clotting. Presumably, this is why low levels of aspirin have been reported to reduce the incidence of strokes and heart attacks that result from the formation of blood clots. Because of aspirin's activity as an anticoagulant, doctors caution patients not to take aspirin for several days before surgery.

Other NSAIDs (nonsteroidal anti-inflammatory drugs), such as ibuprofen (the active ingredient in Advil, Motrin, and Nuprin) and naproxen (the active ingredient in Aleve), also inhibit the synthesis of prostaglandins (see pages 114–115).

There are two forms of prostaglandin synthase: one carries out the normal production of prostaglandin, and the other synthesizes additional prostaglandin in response to inflammation. NSAIDs inhibit the synthesis of all prostaglandins. One prostaglandin regulates the production of acid in the stomach, so when prostaglandin synthesis stops, the acidity of the stomach can rise above normal levels. Celebrex, a relatively new drug, inhibits only the prostaglandin synthase that produces prostaglandin in response to inflammation. Thus, inflammatory conditions now can be treated without some of the harmful side effects.

Celebrex®

PROBLEM 28 ♦

a. What species other than an acid can be used to increase the rate of the transesterification reaction that converts methyl acetate to propyl acetate?

b. Explain why the rate of aminolysis of an ester cannot be increased by H^+, HO^-, or RO^-.

PROBLEM 29 ♦

D. N. Kursanov, a Russian chemist, proved that the bond that is broken in the hydroxide-ion-promoted hydrolysis of an ester is the acyl C—O bond, rather than the alkyl C—O bond, by studying the hydrolysis of the following ester under basic conditions:

a. What products contained the ^{18}O label?

b. What product would have contained the ^{18}O label if the alkyl C—O bond had broken?

PROBLEM 30 SOLVED

Early chemists could envision three possible mechanisms for hydroxide-ion-promoted ester hydrolysis. Devise an experiment that would show which of the three is the actual mechanism.

1. a nucleophilic acyl substitution reaction

2. an S_N2 reaction

$$R-\overset{\cdot\overset{\cdot\cdot}{O}\cdot}{\underset{}{C}}-\overset{}{O}-R' \; + \; H\overset{\cdot\cdot}{\underset{\cdot\cdot}{O}}{:}^{-} \longrightarrow R-\overset{\cdot\overset{\cdot\cdot}{O}\cdot}{\underset{}{C}}-O^{-} \; + \; R'OH$$

3. an S_N1 reaction

$$R-\overset{\cdot\overset{\cdot\cdot}{O}\cdot}{\underset{}{C}}-O-R' \longrightarrow R-\overset{\cdot\overset{\cdot\cdot}{O}\cdot}{\underset{}{C}}-O^{-} \; + \; R'^{+} \; \overset{H\overset{\cdot\cdot}{\underset{\cdot\cdot}{O}}{:}^{-}}{\longrightarrow} \; R'OH$$

SOLUTION Start with a single stereoisomer of an alcohol with the OH group bonded to an asymmetric center and determine its specific rotation. Then convert the alcohol to an ester using an acyl chloride such as acetyl chloride. Next, hydrolyze the ester under basic conditions, isolate the alcohol obtained as a product, and determine its specific rotation.

$$\underset{\underset{OH}{\overset{CH_2CH_3}{\underset{|}{CH_3}}}}{CH_3-\overset{}{\underset{}{C}}{\cdots H}} \; \overset{CH_3\overset{O}{\overset{\|}{C}}Cl}{\longrightarrow} \; \underset{\underset{\underset{O}{\overset{\|}{OCCH_3}}}{\overset{CH_2CH_3}{\underset{|}{CH_3}}}}{CH_3-\overset{}{\underset{}{C}}{\cdots H}} \; \overset{HO^-}{\underset{H_2O}{\longrightarrow}} \; \underset{\underset{OH}{|}}{CH_3CHCH_2CH_3} \; + \; CH_3\overset{O}{\overset{\|}{C}}O^-$$

- If the reaction is a nucleophilic acyl substitution reaction, the product alcohol will have the same specific rotation as the reactant alcohol because no bonds to the asymmetric center are broken during formation or hydrolysis of the ester.

- If the reaction is an S_N2 reaction, the product alcohol and the reactant alcohol will have opposite specific rotations because the mechanism requires back-side attack of hydroxide ion on the asymmetric center (Section 9.1).

- If the reaction is an S_N1 reaction, the product alcohol will have a small (or zero) specific rotation because the mechanism requires carbocation formation, which leads to racemization of the alcohol (Section 9.3).

15.10 REACTIONS OF CARBOXYLIC ACIDS

Carboxylic acids can undergo nucleophilic acyl substitution reactions only when they are in their acidic forms. The basic form of a carboxylic acid is not reactive because its negative charge makes it resistant to approach by a nucleophile. Therefore, carboxylate ions are even less reactive than amides in nucleophilic acyl substitution reactions.

relative reactivities toward nucleophilic acyl substitution

$$\boxed{\text{most reactive}}\;\underset{R}{\overset{O}{\overset{\|}{C}}}\text{OH} \; > \; \underset{R}{\overset{O}{\overset{\|}{C}}}\text{NH}_2 \; > \; \underset{R}{\overset{O}{\overset{\|}{C}}}\text{O}^-\;\boxed{\text{least reactive}}$$

Carboxylic acids have approximately the same reactivity as esters, because the HO^- leaving group of a carboxylic acid has about the same basicity as the RO^- leaving group of an ester.

A Carboxylic Acid and an Alcohol Undergo a Nucleophilic Acyl Substitution Reaction

Carboxylic acids, therefore, react with alcohols to form esters. The reaction must be carried out in an acidic solution, not only to catalyze the reaction but also to keep the carboxylic acid in its acidic form so that the nucleophile will react with it. Because the tetrahedral intermediate formed in this reaction has two potential leaving groups with approximately the same basicity, the reaction must be carried out with excess alcohol to drive it toward products.

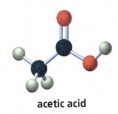

acetic acid

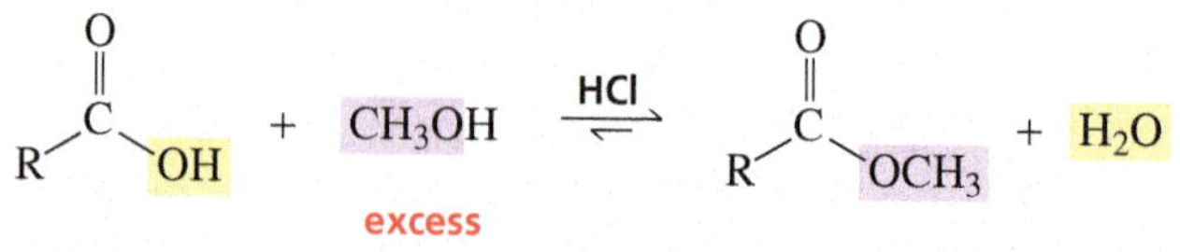

$$R-\overset{\overset{\displaystyle O}{\|}}{C}-OH \ + \ CH_3OH \ \underset{excess}{\overset{HCl}{\rightleftharpoons}} \ R-\overset{\overset{\displaystyle O}{\|}}{C}-OCH_3 \ + \ H_2O$$

Emil Fischer was the first to discover that an ester could be prepared by treating a carboxylic acid with excess alcohol in the presence of an acid catalyst, so the reaction is called a **Fischer esterification.** Its mechanism is the exact reverse of the mechanism for the acid-catalyzed hydrolysis of an ester shown on page 705. Also see Problem 22.

A Carboxylic Acid and an Amine Undergo an Acid–Base Reaction

Carboxylic acids do not undergo nucleophilic acyl substitution reactions with amines. A carboxylic acid is an acid and an amine is a base, so the carboxylic acid immediately loses a proton to the amine when the two compounds are mixed. The resulting ammonium carboxylate salt is the final product of the reaction; the carboxylate ion is not reactive and the protonated amine is not a nucleophile.

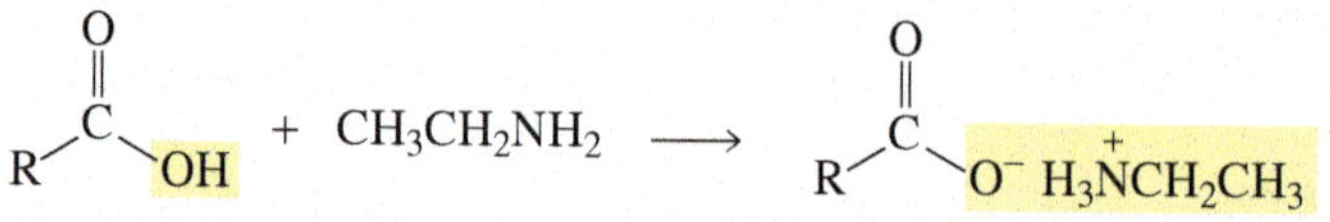

$$R-\overset{\overset{\displaystyle O}{\|}}{C}-OH \ + \ CH_3CH_2NH_2 \ \longrightarrow \ R-\overset{\overset{\displaystyle O}{\|}}{C}-O^-\ \overset{+}{H_3N}CH_2CH_3$$

an ammonium carboxylate salt

An ammonium carboxylate salt can lose water to form an amide if it is heated.

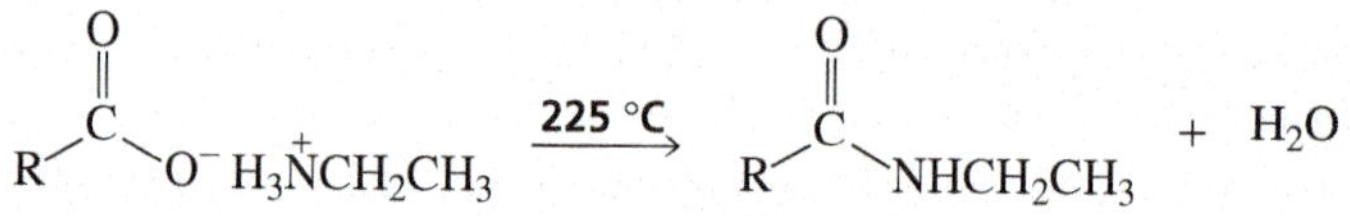

$$R-\overset{\overset{\displaystyle O}{\|}}{C}-O^-\ \overset{+}{H_3N}CH_2CH_3 \ \overset{225\ °C}{\longrightarrow} \ R-\overset{\overset{\displaystyle O}{\|}}{C}-NHCH_2CH_3 \ + \ H_2O$$

PROBLEM 31 ♦

Show how each of the following esters could be prepared using a carboxylic acid as one of the starting materials:

a. methyl butyrate (odor of apples)　　　　**b.** octyl acetate (odor of oranges)

LEARN THE STRATEGY

PROBLEM-SOLVING STRATEGY

Proposing a Mechanism

Propose a mechanism for the following reaction:

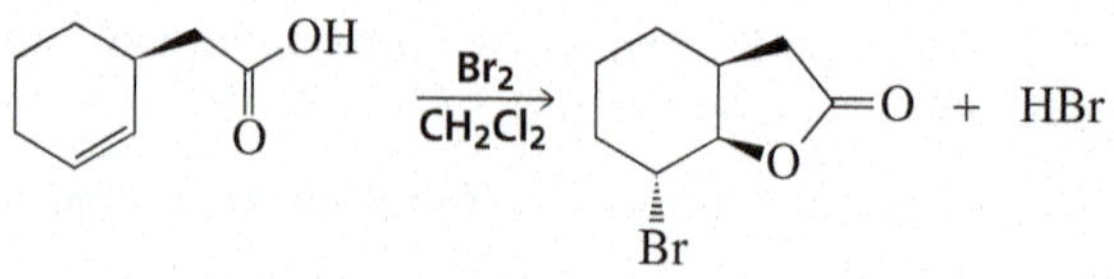

When you are asked to propose a mechanism, look carefully at the reactants to determine the first step. One of the reactants has two functional groups: a carboxyl group and a carbon–carbon double bond. The other reactant, Br_2, does not react with carboxylic acids but does react with alkenes (Section 6.9). Approach to one side of the double bond is sterically hindered by the carboxyl group, so Br_2 will add to the other side of the double bond, forming a cyclic bromonium ion.

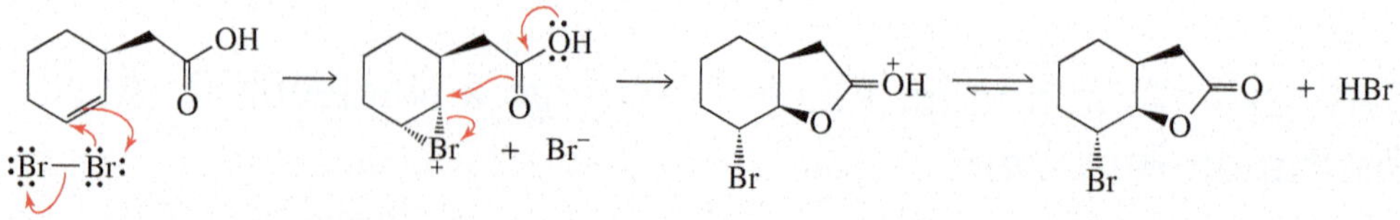

We know that in the second step of this addition reaction, a nucleophile attacks the bromonium ion. Of the two nucleophiles present, the carbonyl oxygen is more likely than the bromide ion to attack the back side of the bromonium ion because the carbonyl group and the bromonium ion are tethered together, resulting in a compound with the observed configuration. Loss of a proton forms the final product of the reaction.

PROBLEM 32

USE THE STRATEGY

Propose a mechanism for the following reaction. (*Hint:* Number the carbons to help you see where they end up in the product.)

$$CH_2{=}CHCH_2CH_2CH{=}\overset{\overset{\displaystyle CH_3}{|}}{C}CH_3 \ + \ \underset{CH_3}{\overset{\displaystyle O}{\overset{\|}{C}}}\!\!\!\diagdown OH \ \xrightarrow{\ H_2SO_4\ }$$

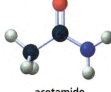

15.11 REACTIONS OF AMIDES

Amides are very unreactive compounds, which is comforting, because the proteins that impart strength to biological structures and catalyze the reactions that take place in cells are composed of amino acids linked together by amide bonds (Section 21.0). Amides do not react with halide ions, alcohols, or water because, in each case, the incoming nucleophile is a weaker base than the leaving group of the amide (Table 15.1).

$$\underset{R}{\overset{\displaystyle O}{\overset{\|}{C}}}\!\!\!\diagdown NHCH_2CH_2CH_3 \ + \ Cl^- \ \longrightarrow \ \text{no reaction}$$

$$\underset{R}{\overset{\displaystyle O}{\overset{\|}{C}}}\!\!\!\diagdown NHCH_3 \ + \ CH_3OH \ \longrightarrow \ \text{no reaction}$$

$$\underset{R}{\overset{\displaystyle O}{\overset{\|}{C}}}\!\!\!\diagdown NHCH_2CH_3 \ + \ H_2O \ \longrightarrow \ \text{no reaction}$$

We will see, however, that amides do react with water under acidic or strongly basic conditions, and with alcohols under acidic conditions (Sections 15.12 and 15.13).

acetamide

Dalmatians: Do Not Fool with Mother Nature

When amino acids are metabolized, the excess nitrogen is concentrated into uric acid, a compound with five amide bonds. A series of enzyme-catalyzed hydrolysis reactions degrade uric acid—one amide bond at a time—all the way to ammonium ion. The extent to which uric acid is degraded depends on the species. Primates, birds, reptiles, and insects excrete excess nitrogen as uric acid. Other mammals excrete excess nitrogen as allantoin. Excess nitrogen in aquatic animals is excreted as allantoic acid, urea, or as ammonium salts.

uric acid	allantoin	allantoic acid	urea	ammonium salt
excreted by: birds, reptiles, insects, primates	mammals (except primates)	marine vertebrates	cartilaginous fish, amphibia	marine invertebrates

urate oxidase → allantoinase → allantoicase → urease → $^+NH_4\ X^-$

Dalmatians, unlike other dogs, excrete high levels of uric acid. This is because breeders of Dalmatians have selected dogs that have no white hairs in their black spots, and the gene that causes the white hairs is linked to the gene that causes uric acid to be hydrolyzed to allantoin. Dalmatians, therefore, are susceptible to gout, a painful buildup of uric acid in joints.

PROBLEM 33 ♦

What acyl chloride and amine are required to synthesize the following amides?

a. *N*-ethylbutanamide

b. *N,N*-dimethylbenzamide

PROBLEM 34 ♦

Which of the following reactions lead to the formation of an amide?

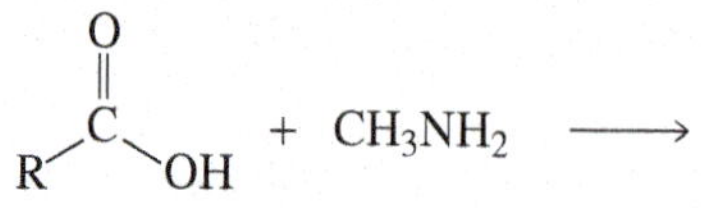

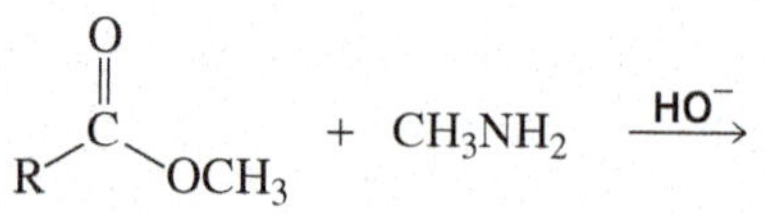

15.12 ACID-CATALYZED AMIDE HYDROLYSIS AND ALCOHOLYSIS

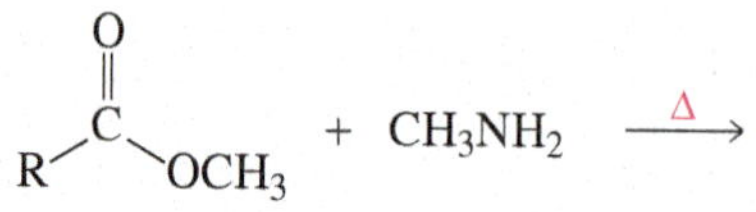

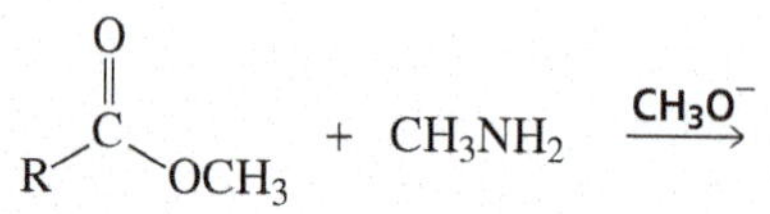

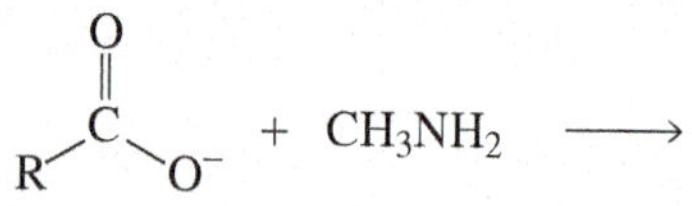

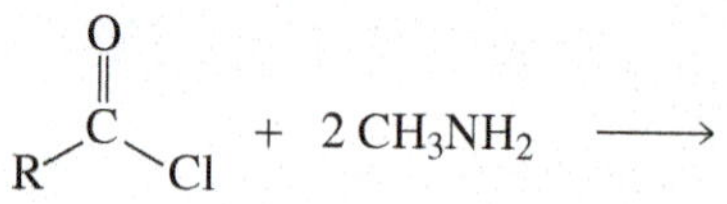

Amides react with water to form carboxylic acids and with alcohols to form esters, if the reaction mixture is heated in the presence of an acid.

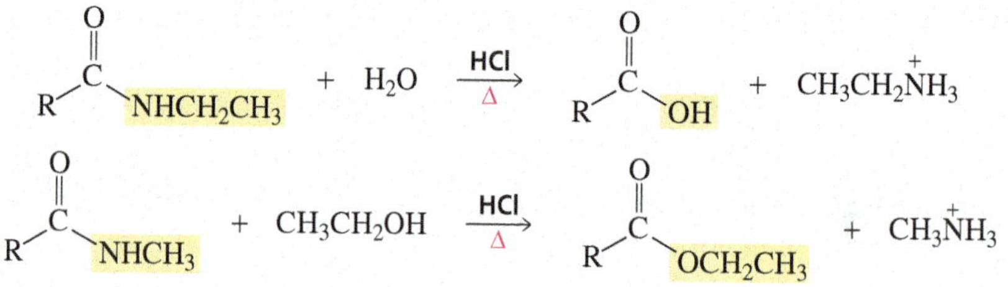

The mechanism for the acid-catalyzed hydrolysis of an amide is exactly the same as the mechanism for the acid-catalyzed hydrolysis of an ester shown on page 705.

MECHANISM FOR THE ACID-CATALYZED HYDROLYSIS OF AN AMIDE

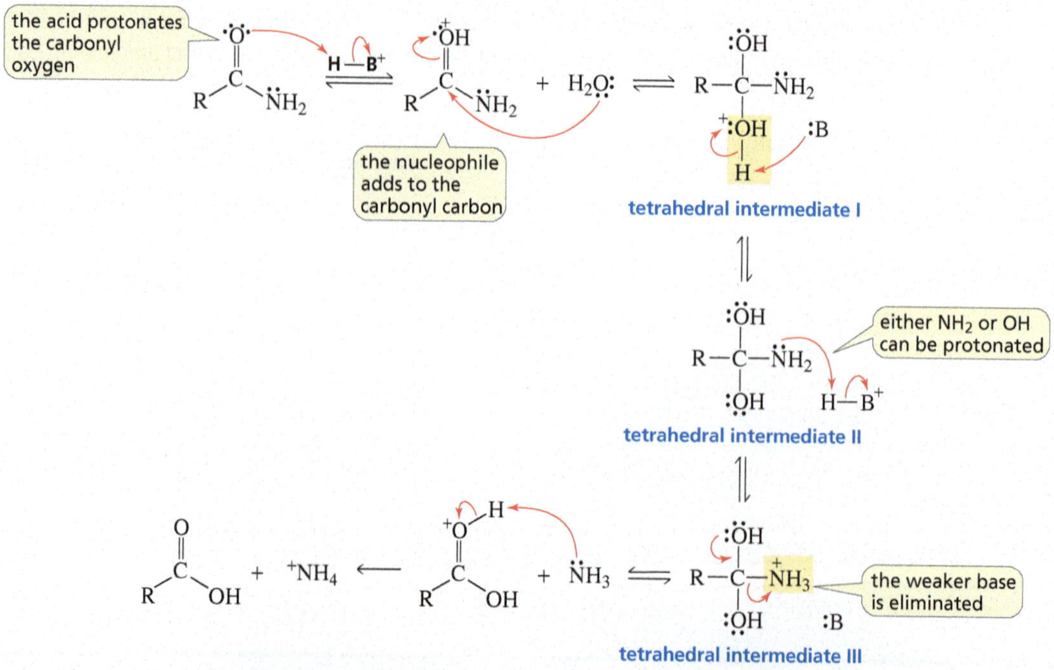

NOTE TO THE STUDENT

• Notice again the pattern of the three tetrahedral intermediates:

protonated tetrahedral intermediate I ⇌
neutral tetrahedral intermediate II ⇌
protonated tetrahedral intermediate III

- The acid protonates the carbonyl oxygen, which increases the susceptibility of the carbonyl carbon to nucleophilic addition.
- Addition of the nucleophile (H_2O) to the carbonyl carbon leads to tetrahedral intermediate I, which is in equilibrium with its nonprotonated form, tetrahedral intermediate II.
- Re-protonation can occur either on oxygen to re-form tetrahedral intermediate I or on nitrogen to form tetrahedral intermediate III. Protonation on nitrogen is favored because the NH_2 group is a stronger base than the OH group.
- Of the two possible leaving groups in tetrahedral intermediate III (HO^- and NH_3), NH_3 is the weaker base, so it is the one eliminated.
- NH_3 is protonated after it is eliminated from the tetrahedral intermediate because if only one base is protonated, it will be the stronger one. This prevents the reverse reaction from occurring because $^+NH_4$ is not a nucleophile.

NOTE TO THE STUDENT
- Notice that the acid-catalyzed hydrolysis of an ester and an amide take place by the same 6 steps:
- protonation
- addition of the nucleophile
- deprotonation of tetrahedral intermediate I
- protonation of tetrahedral intermediate II
- elimination
- deprotonation

Why a Catalyst Is Required for Hydrolysis and Alcoholysis of an Amide

Let's take a minute to see why an amide cannot be hydrolyzed without a catalyst. In an uncatalyzed reaction, the NH_2 group of the tetrahedral intermediate will not be protonated. Therefore, HO^- will be eliminated from the tetrahedral intermediate (because HO^- is a weaker base than $^-NH_2$), which will re-form the amide.

$$CH_3-\overset{\overset{\displaystyle :\ddot{O}H}{|}}{\underset{\underset{\displaystyle :\ddot{O}H}{|}}{C}}-\overset{+}{N}H_3 \qquad CH_3-\overset{\overset{\displaystyle :\ddot{O}H}{|}}{\underset{\underset{\displaystyle :\ddot{O}H}{|}}{C}}-\ddot{N}H_2$$

the leaving group in acid-catalyzed amide hydrolysis

the leaving group in (unsuccessfully) uncatalyzed amide hydrolysis

When an amide reacts with an alcohol in the presence of acid to form an ester, it follows the same mechanism as it does when it reacts with water to form a carboxylic acid.

The Discovery of Penicillin

Sir Alexander Fleming (1881–1955) was a professor of bacteriology at the University of London. The story is told that one day Fleming was about to throw away a culture of staphylococcal bacteria that had been contaminated by a rare strain of the mold *Penicillium notatum.* He noticed that the bacteria had disappeared wherever there was a particle of mold. This suggested to him that the mold must have produced an antibacterial substance. Ten years later, in 1938, Howard Florey and Ernest Chain isolated the active substance—penicillin G—but the delay allowed the sulfa drugs to be the first antibiotics (Section 18.19). After penicillin G was found to cure bacterial infections in mice, it was used successfully in 1941 on nine cases of human bacterial infections. By 1943, it was being produced for the military and was first used for war casualties in Sicily and Tunisia. The drug became available to the civilian population in 1944. The pressure of the war made the determination of penicillin G's structure a priority because once its structure was determined, large quantities of the drug could conceivably be synthesized.

Fleming, Florey, and Chain shared the 1945 Nobel Prize in Physiology or Medicine. Chain also discovered penicillinase, the enzyme that destroys penicillin (see the next page). Although Fleming is generally given credit for the discovery of penicillin, there is clear evidence that the germicidal activity of the mold was recognized in the nineteenth century by Lord Joseph Lister (1827–1912), the English physician renowned for the introduction of aseptic surgery in 1865. Unfortunately, it took several years for the surgical profession to follow his example.

Penicillin and Drug Resistance

The antibiotic activity of penicillin results from its ability to acylate (put an acyl group on) a CH_2OH group of an enzyme that has a role in the synthesis of bacterial cell walls. Acylation occurs by a nucleophilic acyl substitution reaction (see the next page): the CH_2OH group adds to the carbonyl carbon of the β-lactam of penicillin, forming a tetrahedral intermediate (red arrows). The four-membered ring amide is more reactive than a noncyclic amide because when the π bond re-forms, the strain in the four-membered ring is released when the amino group is eliminated (blue arrows).

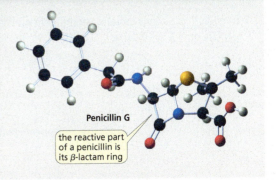

Penicillin G

the reactive part of a penicillin is its β-lactam ring

(Continued)

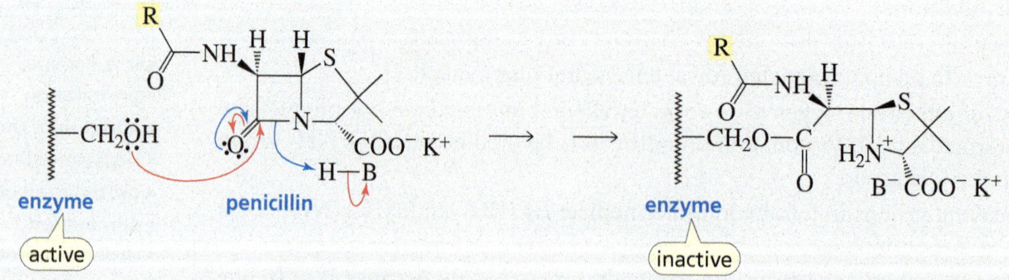

Acylation inactivates the enzyme, and actively growing bacteria die because they are unable to synthesize functional cell walls. Penicillin has no effect on mammalian cells because they are not enclosed by cell walls. Penicillins are stored at cold temperatures to minimize hydrolysis of the β-lactam.

Bacteria that are resistant to penicillin secrete penicillinase, an enzyme that catalyzes hydrolysis of the β-lactam of penicillin. The ring-opened product has no antibacterial activity.

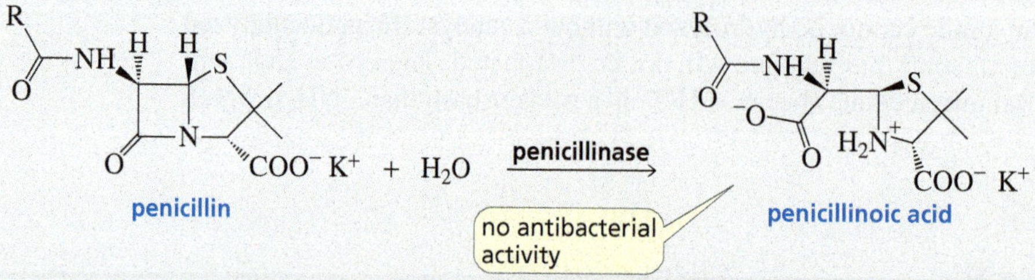

Penicillins in Clinical Use

More than 10 different penicillins are currently in clinical use. They differ only in the group (R) attached to the carbonyl group. The variable groups (R) of these penicillins are shown here. In addition to their structural differences, the penicillins differ in the organisms against which they are most effective. They also differ in their susceptibility to penicillinase. For example, methicillin, a synthetic penicillin, is effective against bacteria that are resistant to penicillin G, a naturally occurring penicillin. Almost 19% of humans are allergic to penicillin G.

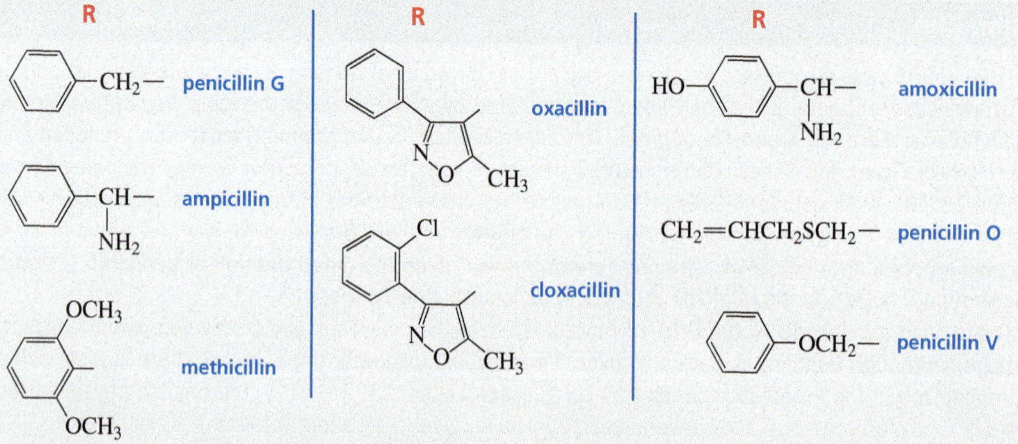

A Semisynthetic Penicillin

Penicillin V is a semisynthetic penicillin in clinical use. It is not a naturally occurring penicillin, but it is also not a true synthetic penicillin because chemists do not synthesize it. The *Penicillium* mold synthesizes it after being fed 2-phenoxyethanol, the compound needed for the R group.

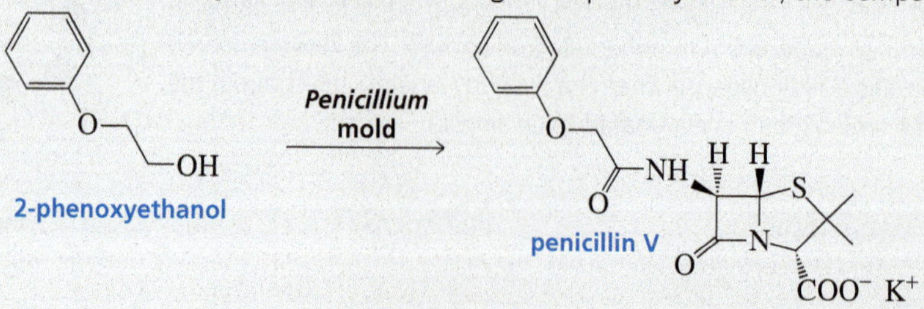

PROBLEM 35

Write the mechanism for the acid-catalyzed reaction of an amide with an alcohol to form an ester.

PROBLEM 36 ◆

Rank the following amides from greatest reactivity to least reactivity toward acid-catalyzed hydrolysis:

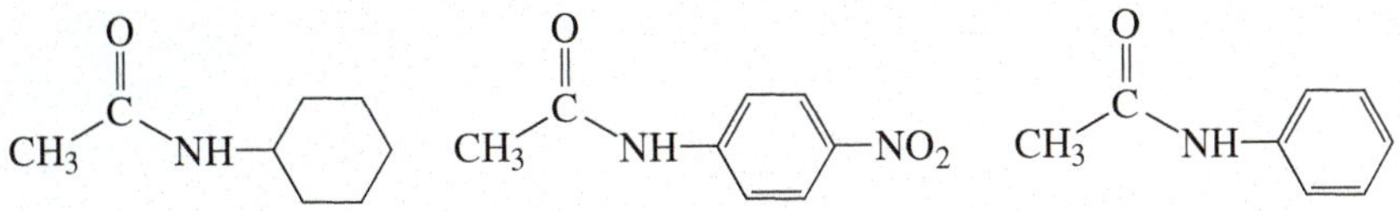

A   B   C

15.13 HYDROXIDE-ION-PROMOTED HYDROLYSIS OF AMIDES

Amides can also be hydrolyzed when heated under strongly basic conditions.

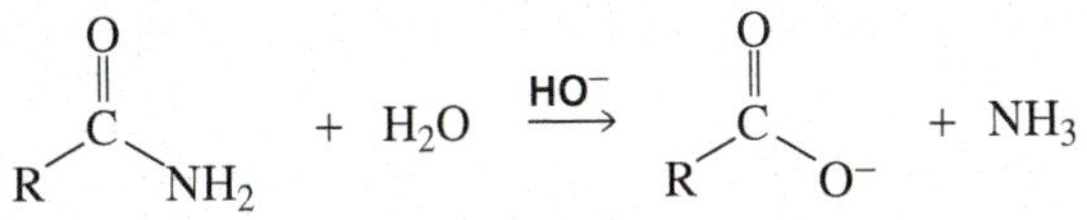

MECHANISM FOR THE HYDROXIDE-ION-PROMOTED HYDROLYSIS OF AN AMIDE

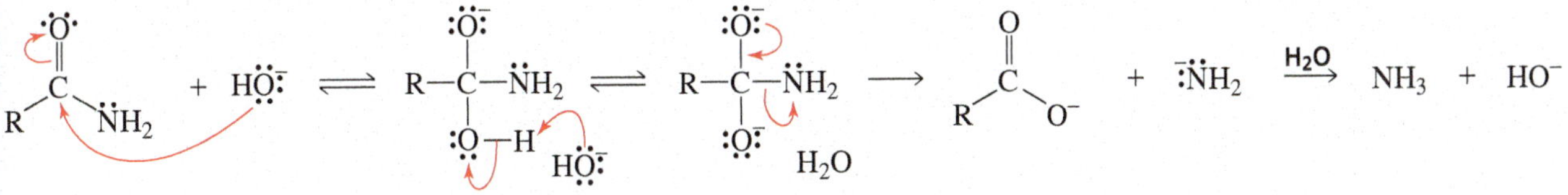

- Hydroxide ion is the nucleophile instead of water. Because it is a better nucleophile than water, it forms the tetrahedral intermediate more rapidly. Of the two potential leaving groups in the tetrahedral intermediate, $^-$OH is the weaker base and, therefore, the one more likely to be eliminated, thereby reforming the amide.
- In strongly basic solutions, the reaction is second order in hydroxide ion. That is, two equivalents of hydroxide ion participate in the reaction. The second equivalent of hydroxide ion removes a proton from the tetrahedral intermediate.
- Now the possible leaving groups are $^-NH_2$ and O^{2-}. Because $^-NH_2$ is the weaker base, it is eliminated and the carboxylate ion is formed.
- The amide ion is protonated because it is a stronger base than hydroxide ion.

Notice that one equivalent of hydroxide ion is not a catalyst (it is not regenerated) but the second equivalent is regenerated, so it is a catalyst.

Synthetic Polymers

Synthetic polymers play important roles in our daily lives. **Polymers** are compounds that are made by linking together many small molecules called **monomers**. The monomers of many synthetic polymers are held together by ester and amide bonds. For example, Dacron is a polyester and nylon is a polyamide.

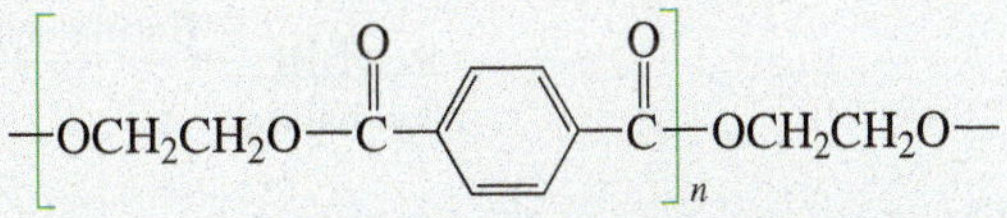

Dacron®

(Continued)

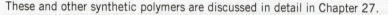

nylon 6

Synthetic polymers can take the place of fabrics, metals, glass, wood, and paper, allowing us to have a greater variety and larger quantities of materials than nature can provide. New polymers are continually being designed to fit human needs. For example, Kevlar (a polyamide) has a tensile strength greater than steel. It is used for high-performance skis and bulletproof vests. Lexan (a polyester) is a strong and transparent polymer used for such things as traffic light lenses and compact disks.

Kevlar®

Lexan®

These and other synthetic polymers are discussed in detail in Chapter 27.

Dissolving Sutures

Dissolving sutures, such as dexon and poly(dioxanone) (PDS), are synthetic polymers that are now routinely used in surgery. The many ester groups they contain are slowly hydrolyzed to small molecules that are then metabolized to compounds easily excreted by the body. Patients no longer have to undergo a second medical procedure that was required to remove the sutures when traditional suture materials were used.

Depending on their structures, these synthetic sutures lose 50% of their strength after two to three weeks and are completely absorbed within three to six months.

Dexon®   **PDS®**

> **PROBLEM 37 ◆**
>
> Why, in the last step of the mechanism for hydroxide-ion promoted hydrolysis of an amide, is the amide ion protonated?

15.14 HYDROLYSIS OF AN IMIDE: A WAY TO SYNTHESIZE A PRIMARY AMINE

Although primary amines can be prepared by S_N2 reactions with alkyl halides, the yields are poor because it is difficult to stop the reaction after one alkyl group has been placed on the nitrogen (see Problem 16 on page 405). A much better way to prepare a primary amine from an alkyl halide is by means of a **Gabriel synthesis.** The Gabriel synthesis involves the hydrolysis of an imide. An **imide** is a compound with two acyl groups bonded to a nitrogen.

$$RCH_2Br \xrightarrow{\text{Gabriel synthesis}} RCH_2NH_2$$

alkyl halide   **primary amine**

The steps involved in the synthesis are shown on the top of the next page. Notice that the alkyl group of the alkyl halide used in the second step of the reaction is identical to the alkyl group of the desired primary amine.

- A base removes a proton from the nitrogen of phthalimide.
- The resulting nucleophile reacts with an alkyl halide. Because this is an S_N2 reaction, it works best with primary alkyl halides (Section 9.1).
- Hydrolysis of the two amide bonds of the *N*-substituted imide is catalyzed by acid. Because the solution is acidic, the final products are a primary alkyl ammonium ion and phthalic acid.
- Reaction of the alkyl ammonium ion with base forms the primary amine.

Only one alkyl group can be placed on the nitrogen because there is only one hydrogen bonded to the nitrogen of phthalimide. This means that the Gabriel synthesis can be used only for the preparation of primary amines.

LEARN THE STRATEGY

PROBLEM 38 ◆

What alkyl bromide would you use in a Gabriel synthesis to prepare each of the following amines?

a. pentylamine   **b.** isohexylamine   **c.** benzylamine   **d.** cyclohexylamine

USE THE STRATEGY

PROBLEM 39

Primary amines can also be prepared by the reaction of an alkyl halide with azide ion, followed by catalytic hydrogenation. What advantage do this method and the Gabriel synthesis have over the synthesis of a primary amine using an alkyl halide and ammonia?

$$CH_3CH_2CH_2Br \xrightarrow{\ ^-N_3\ } CH_3CH_2CH_2N{=}\overset{+}{N}{=}\overset{-}{N} \xrightarrow{\underset{Pd/C}{H_2}} CH_3CH_2CH_2NH_2 + N_2$$

15.15 NITRILES

Nitriles are compounds that contain a cyano ($C{\equiv}N$) group. They are considered to be carboxylic acid derivatives because, like all the other carboxylic acid derivatives, they can be hydrolyzed to carboxylic acids, as shown on the next page.

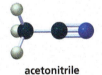

acetonitrile

Naming Nitriles

In systematic nomenclature, nitriles are named by adding "nitrile" to the name of the parent alkane. Notice in the following examples that the triply bonded carbon of the nitrile group is included in the number of carbons in the longest continuous chain.

LEARN THE STRATEGY

$$CH_3C{\equiv}N \qquad \langle\!\!\rangle{-}C{\equiv}N \qquad \overset{\displaystyle CH_3}{\overset{|}{CH_3CHCH_2CH_2CH_2C{\equiv}N}} \qquad CH_2{=}CHC{\equiv}N$$

systematic name:	ethanenitrile	benzenecarbonitrile	5-methylhexanenitrile	propenenitrile
common name:	acetonitrile	benzonitrile	δ-methylcapronitrile	acrylonitrile
	methyl cyanide	phenyl cyanide	isohexyl cyanide	

In common nomenclature, nitriles are named by replacing "ic acid" of the carboxylic acid name with "onitrile." They can also be named as alkyl cyanides—using the name of the alkyl group that is attached to the triply bonded carbon.

PROBLEM 40 ♦

Give two names for each of the following nitriles:

a. $CH_3CH_2CH_2C{\equiv}N$

b. $CH_3CHCH_2CH_2C{\equiv}N$
$\quad\quad\ \ |$
$\quad\quad CH_3$

Reactions of Nitriles

Nitriles are even harder to hydrolyze than amides, but they slowly hydrolyze to carboxylic acids when heated with water and an acid.

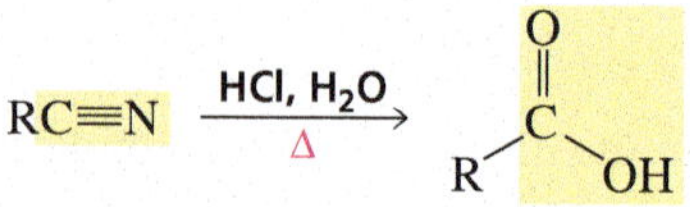

MECHANISM FOR THE ACID-CATALYZED HYDROLYSIS OF A NITRILE

Notice again the pattern of the three intermediates:
protonated intermediate ⇌
neutral intermediate ⇌
protonated intermediate

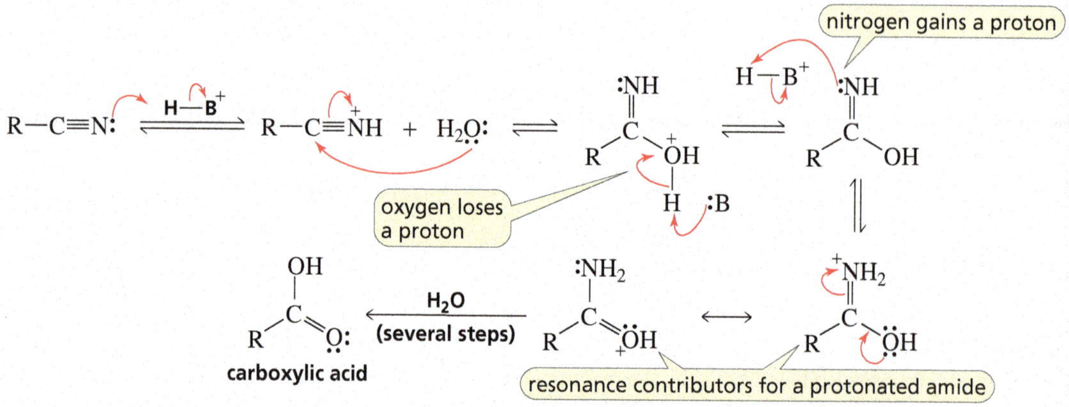

- The acid protonates the nitrogen of the cyano group, which makes the carbon of the cyano group more susceptible to the addition of water. (The addition of water to a protonated cyano group is analogous to the addition of water to a protonated carbonyl group.)
- A base removes a proton from oxygen, forming a neutral species that can be reprotonated on oxygen or protonated on nitrogen. Protonation on nitrogen forms a species, whose two resonance contributors are shown. (Notice that the resonance contributor on the left is a protonated amide.)
- The protonated amide is immediately hydrolyzed to a carboxylic acid—because an amide is easier to hydrolyze than a nitrile—by means of the acid-catalyzed mechanism shown on page 712.

Using Nitriles in Synthesis

We saw that nitriles can be prepared from an S_N2 reaction of alkyl halide with cyanide ion (see page 404). Because a nitrile can be hydrolyzed to a carboxylic acid, you now know how to convert an alkyl halide to a carboxylic acid. Notice that the carboxylic acid has one more carbon than the alkyl halide.

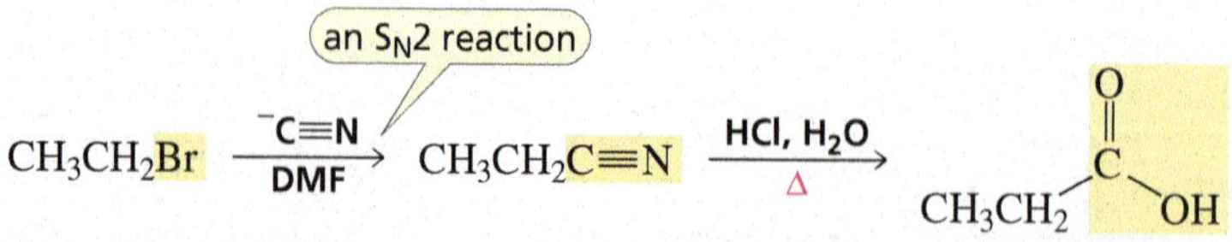

Catalytic hydrogenation of a nitrile is another way to make a primary amine. Raney nickel is the preferred metal catalyst for this reduction.

USE THE STRATEGY

PROBLEM 41 ♦

Which alkyl halides form the carboxylic acids listed here after reaction with sodium cyanide followed by heating the product in an acidic aqueous solution?

a. butyric acid **b.** isovaleric acid **c.** cyclohexanecarboxylic acid

PROBLEM 42 SOLVED

An amide with an NH_2 group can be dehydrated to a nitrile with thionyl chloride ($SOCl_2$). Propose a mechanism for this reaction.

$$\underset{R}{\overset{O}{\underset{\quad}{\overset{\|}{C}}}}\text{—}NH_2 \xrightarrow{SOCl_2} RC\equiv N$$

SOLUTION

- Thionyl chloride is an electrophile, so we need to draw the resonance contributors of the amide to see which atom is the nucleophile.
- The nucleophilic oxygen adds to the electrophilic thionyl chloride eliminating chloride ion.
- Loss of a proton forms a compound that can form the triple bond because of its very good leaving group.
- Loss of another proton forms the neutral nitrile.

15.16 ACID ANHYDRIDES

Loss of water from two molecules of a carboxylic acid results in an **acid anhydride.** "Anhydride" means "without water." An anhydride is a *carboxylic acid derivative*, because the OH of a carboxylic acid has been replaced by a carboxylate ion.

carboxylate ion

$$\underset{R}{\overset{O}{\overset{\|}{C}}}\text{—}OH \quad HO\text{—}\underset{R}{\overset{O}{\overset{\|}{C}}} \xrightarrow{\Delta} \underset{R}{\overset{O}{\overset{\|}{C}}}\text{—}O\text{—}\underset{R}{\overset{O}{\overset{\|}{C}}} + H_2O$$

an acid anhydride

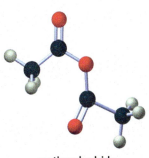

acetic anhydride

Naming Anhydrides

If the two carboxylic acid molecules that form the acid anhydride are the same, then the anhydride is a **symmetrical anhydride.** If they are different, then it is a **mixed anhydride.** Symmetrical anhydrides are named by replacing "acid" in the acid name with "anhydride." Mixed anhydrides are named by stating the names of both acids in alphabetical order, followed by "anhydride."

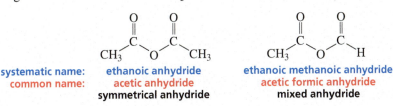

systematic name: ethanoic anhydride | ethanoic methanoic anhydride
common name: acetic anhydride | acetic formic anhydride
symmetrical anhydride | mixed anhydride

Reactions of Anhydrides

The leaving group of an anhydride is a carboxylate ion (its conjugate acid has a pK_a of ~5), which means that an anhydride is less reactive than an acyl chloride but more reactive than an ester or a carboxylic acid.

relative reactivities of carboxylic acid derivatives

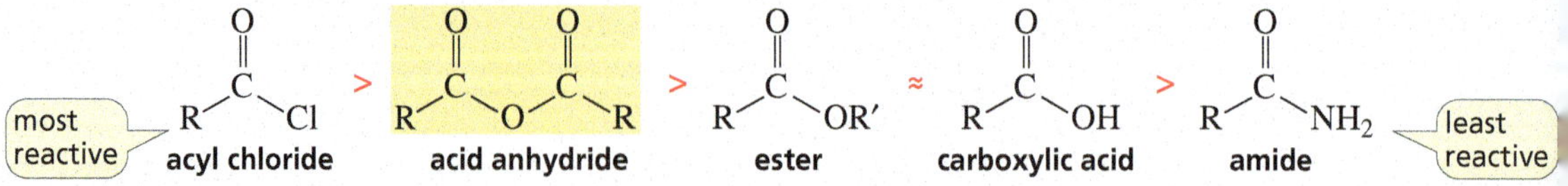

Therefore, as shown below, an acid anhydride reacts with an alcohol to form an ester and a carboxylic acid, with water to form two equivalents of a carboxylic acid, and with an amine to form an amide and a carboxylate ion. In each case, the incoming nucleophile—after it loses a proton—is a stronger base than the departing carboxylate ion. (Recall that a carboxylic acid derivative can be converted to one that is less reactive but not to one that is more reactive.)

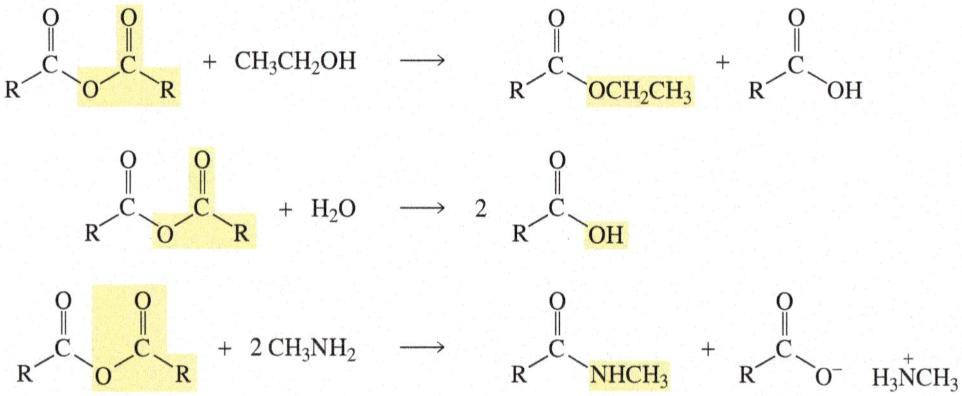

For the reaction of an amine with an anhydride, two equivalents of amine must be used so that sufficient amine is present to react with both the carbonyl compound and the proton produced in the reaction (Section 15.6). (The amine is a stronger base than the carboxylate ion product, so the amine will be protonated by the acid generated in the reaction. When an anhydride reacts with an alcohol or water, the carboxylate ion product is a stronger base than an alcohol or water, so the carboxylate ion will be protonated by the acid generated in the reaction.)

The reactions of acid anhydrides follow the general mechanisms described in Section 15.6. For example, compare the mechanism for the reaction of an acid anhydride with an alcohol to the mechanism for the reaction of an acyl chloride with an alcohol on page 699.

MECHANISM FOR THE REACTION OF AN ACID ANHYDRIDE WITH AN ALCOHOL

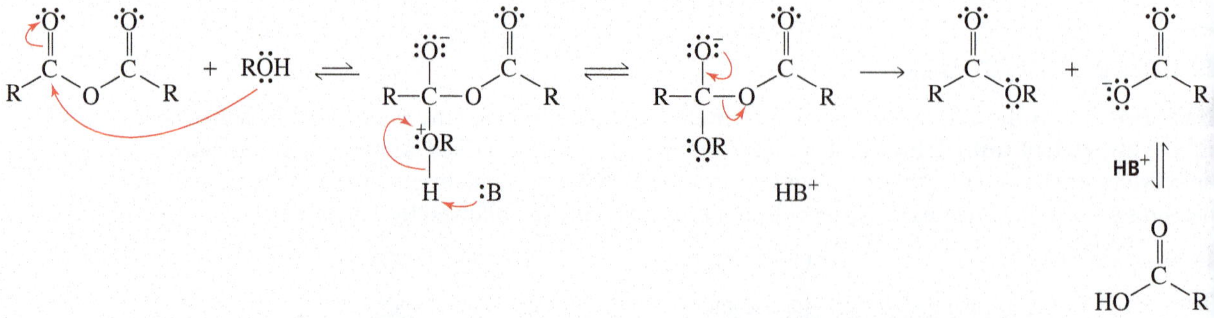

LEARN THE STRATEGY

- The nucleophile adds to the carbonyl carbon, forming a tetrahedral intermediate.
- A proton is removed from the tetrahedral intermediate.
- The carboxylate ion, the weaker of the two bases in the tetrahedral intermediate, is eliminated.

PROBLEM 43

a. Propose a mechanism for the reaction of acetic anhydride with water.

b. How does this mechanism differ from the mechanism for the reaction of acetic anhydride with an alcohol?

PROBLEM 44

Propose a mechanism for the reaction of an acyl chloride with acetate ion to form an acid anhydride.

PROBLEM 45 ♦

We saw that acid anhydrides react with alcohols, water, and amines. In which of these reactions can the tetrahedral intermediate eliminate the carboxylate ion even if it does not lose a proton before the elimination step? Explain.

What Drug-Enforcement Dogs Are Really Detecting

Morphine, the most widely used analgesic for severe pain, is the standard by which other painkilling medications are measured. Although scientists have learned how to synthesize morphine, most commercial morphine is obtained from opium, a milky fluid exuded by a species of poppy (see page 3). Morphine occurs in opium at concentrations as high as 10%. Opium was used for its analgesic properties as early as 4000 B.C. In Roman times both opium use and opium addiction were widespread. Methylating one of the OH groups of morphine produces codeine, which has one-tenth the analgesic activity of morphine. Codeine profoundly inhibits the cough reflex.

Heroin, which is much more potent (and more widely abused) than morphine, is synthesized by treating morphine with acetic anhydride. This puts an acetyl group on each of the OH groups of morphine. Therefore, acetic acid is also formed as a product. To detect heroin, drug-enforcement agencies use dogs trained to recognize the pungent odor of acetic acid.

15.17 DICARBOXYLIC ACIDS

pK_a Values

The structures of some common dicarboxylic acids and their pK_a values are listed in Table 15.2 on the next page.

Table 15.2 Structures, Names, and pK_a Values of Some Simple Dicarboxylic Acids

Dicarboxylic acid	Common name	pK_{a1}	pK_{a2}
	Oxalic acid	1.27	4.27
	Malonic acid	2.86	5.70
	Succinic acid	4.21	5.64
	Glutaric acid	4.34	5.27
	Adipic acid	4.41	5.28
	Phthalic acid	2.95	5.41

Although the two carboxyl groups of a dicarboxylic acid are identical, the two pK_a values are differen[t] because the protons are lost one at a time and, therefore, leave from different species—namely, the firs[t] proton is lost from a neutral molecule, then the second proton is lost from a negatively charged ion.

A COOH group withdraws electrons inductively (compared to an H) and, therefore, increases th[e] acidity of the first COOH group (Section 2.7). The pK_{a1} values of the dicarboxylic acids sho[w] that the acid-strengthening effect of the COOH group decreases as the separation between the tw[o] carboxyl groups increases.

Dehydration

Dicarboxylic acids readily lose water (dehydrate) when heated, if they can form a cyclic anhydrid[e] with a five- or a six-membered ring.

glutaric acid   glutaric anhydride

phthalic acid   phthalic anhydride

Cyclic anhydrides are more easily prepared if the dicarboxylic acid is heated in the presence o[f] acetyl chloride or acetic anhydride.

succinic acid + acetic anhydride → succinic anhydride + 2 CH₃COOH

PROBLEM 46

a. Propose a mechanism for the formation of succinic anhydride in the presence of acetic anhydride.

b. How does acetic anhydride make it easier to form the anhydride?

15.18 HOW CHEMISTS ACTIVATE CARBOXYLIC ACIDS

Of the various classes of carbonyl compounds discussed in this chapter—acyl halides, acid anhydrides, esters, carboxylic acids, and amides—carboxylic acids are the most commonly available, both in the laboratory and in cells. Therefore, carboxylic acids are the reagents most likely to be available when a chemist or a cell needs to synthesize a carboxylic acid derivative.

However, we saw that carboxylic acids are relatively unreactive toward nucleophilic acyl substitution reactions because the OH group of a carboxylic acid is a strong base and, therefore, a poor leaving group. And at physiological pH (pH = 7.4), a carboxylic acid is even more resistant to nucleophilic acyl substitution reactions because it exists predominantly in its unreactive, negatively charged basic form.

Therefore, both organic chemists and cells need a way to activate carboxylic acids so that they can readily undergo nucleophilic acyl substitution reactions. First we will look at how chemists activate carboxylic acids, and then we will see how cells do it.

Activating Carboxylic Acids in the Lab

One way organic chemists activate carboxylic acids is by converting them to acyl chlorides, the most reactive of the carboxylic acid derivatives. A carboxylic acid can be converted to an acyl chloride by heating it either with thionyl chloride (SOCl₂) or phosphorus trichloride (PCl₃).

These reagents convert the leaving group of a carboxylic acid to a better leaving group than the chloride ion.

good leaving groups

As a result, when the chloride ion subsequently adds to the carbonyl carbon and forms a tetrahedral intermediate, the chloride ion is *not* the group that is eliminated.

a better leaving group than a chloride ion

Notice that these reagents are the same reagents that cause the OH group of an alcohol to be replaced by a chlorine (Section 10.2).

Once the acyl halide has been prepared, a wide variety of carboxylic acid derivatives can be synthesized by adding the appropriate nucleophile.

$$\underset{R}{\overset{O}{\|}}\!\!\!\!\!\underset{Cl}{C} \; + \; \underset{R}{\overset{O}{\|}}\!\!\!\!\!\underset{O^-}{C} \; \longrightarrow \; \underset{R}{\overset{O}{\|}}\!\!\!\!\!\underset{O}{C}\!\!\!\!\!\underset{R}{\overset{O}{\|}}\!\!\!\!\!C \; + \; Cl^-$$

anhydride

$$\underset{R}{\overset{O}{\|}}\!\!\!\!\!\underset{Cl}{C} \; + \; ROH \; \longrightarrow \; \underset{R}{\overset{O}{\|}}\!\!\!\!\!\underset{OR}{C} \; + \; HCl$$

ester

$$\underset{R}{\overset{O}{\|}}\!\!\!\!\!\underset{Cl}{C} \; + \; 2\,RNH_2 \; \longrightarrow \; \underset{R}{\overset{O}{\|}}\!\!\!\!\!\underset{NHR}{C} \; + \; \overset{+}{R}NH_3\;Cl^-$$

amide

Carboxylic acids can also be activated for nucleophilic acyl substitution reactions by being converted to acid anhydrides by a dehydrating agent, such as P_2O_5.

$$2 \; \underset{R}{\overset{O}{\|}}\!\!\!\!\!\underset{OH}{C} \; \xrightarrow{P_2O_5} \; \underset{R}{\overset{O}{\|}}\!\!\!\!\!\underset{O}{C}\!\!\!\!\!\underset{R}{\overset{O}{\|}}\!\!\!\!\!C \; + \; H_2O$$

Carboxylic acids and carboxylic acid derivatives can be prepared by methods other than nucleophilic acyl substitution reactions. A summary of the methods used to synthesize these compounds is provided in Appendix III.

PROBLEM 47 ◆

How could you synthesize the following compounds starting with a carboxylic acid?

a. $CH_3CH_2\underset{}{\overset{O}{\overset{\|}{C}}}\!-\!O\!-\!\text{(phenyl)}$

b. $\text{(phenyl)}\underset{}{\overset{O}{\overset{\|}{C}}}\!-\!NHCH_2CH_3$

15.19 HOW CELLS ACTIVATE CARBOXYLIC ACIDS

Biosynthesis is the synthesis of compounds by a living organism. Acyl halides and acid anhydrides are not useful reagents in cells because cells live in an aqueous environment, and acyl halides and acid anhydrides are rapidly hydrolyzed in water. So cells cannot activate carboxylic acids the way chemists do.

Cells Use ATP to Activate Carboxylic Acids

One way cells can activate a carboxylic acid is to use adenosine triphosphate (ATP) to convert the carboxylic acid to an **acyl phosphate** or an **acyl adenylate**—carbonyl compounds with good leaving groups. ATP is an ester of triphosphoric acid. Its structure is shown here, both in its entirety and with "Ad" in place of the adenosyl group.

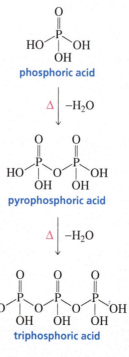

$$HO\!-\!\underset{\underset{OH}{|}}{\overset{\overset{O}{\|}}{P}}\!-\!OH$$

phosphoric acid

$\Delta \mid -H_2O$

$$HO\!-\!\underset{\underset{OH}{|}}{\overset{\overset{O}{\|}}{P}}\!-\!O\!-\!\underset{\underset{OH}{|}}{\overset{\overset{O}{\|}}{P}}\!-\!OH$$

pyrophosphoric acid

$\Delta \mid -H_2O$

$$HO\!-\!\underset{\underset{OH}{|}}{\overset{\overset{O}{\|}}{P}}\!-\!O\!-\!\underset{\underset{OH}{|}}{\overset{\overset{O}{\|}}{P}}\!-\!O\!-\!\underset{\underset{OH}{|}}{\overset{\overset{O}{\|}}{P}}\!-\!OH$$

triphosphoric acid

adenosine triphosphate
ATP

Acyl phosphates and acyl adenylates are mixed anhydrides of a carboxylic acid and phosphoric acid.

an acyl phosphate   an acyl adenylate
mixed anhydrides

Forming an Acyl Phosphate

An acyl phosphate is formed by nucleophilic attack of a carboxylate ion on the γ-phosphorus (the phosphorus farthest away from the adenosyl group) of ATP. Attack of the nucleophile breaks the **phosphoanhydride bond** (rather than the π bond), so an intermediate is not formed. Essentially, it is an S_N2 reaction with an adenosine diphosphate (ADP) leaving group.

adenosine triphosphate
ATP

acyl phosphate

adenosine diphosphate
ADP

Forming an Acyl Adenylate

An acyl adenylate is formed by nucleophilic attack of a carboxylate ion on the α-phosphorus of ATP. The enzyme that catalyzes the reaction determines which phosphorus is attacked by the nucleophile and, therefore, whether an acyl phosphate or an acyl adenylate is formed. In Section 25.2, we will see how the reactions that require acyl phosphate intermediates differ from those that require acyl adenylate intermediates.

adenosine triphosphate
ATP

acyl adenylate

pyrophosphate

Reaction with ATP Occurs at the Active Site of an Enzyme

Because both the carboxylate anion and ATP are negatively charged, they cannot react with each other unless they are at the active site of an enzyme. One of the functions of the enzymes that catalyze these reactions is to neutralize the negative charges of ATP so it can react with a nucleophile (Figure 15.3). Another function of the enzyme is to exclude water from the active site where the reaction takes place. Otherwise, hydrolysis of the mixed anhydride would compete with the desired nucleophilic acyl substitution reaction.

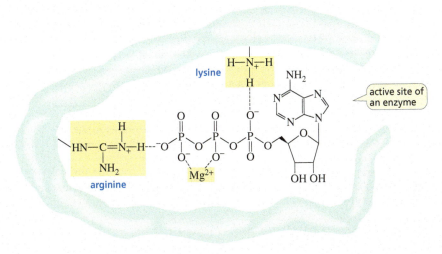

◄ **Figure 15.3**
The interactions between ATP, Mg^{2+}, and positively charged groups at the active site of an enzyme.

Enzyme-catalyzed reactions that have ATP as one of the reactants require Mg^{2+}, which helps reduce the negative charge on ATP at the active site.

Activating a Carboxylic Acid by Converting It to a Thioester

Cells can also activate a carboxylic acid by converting it to a thioester. A **thioester** is an ester with a sulfur in place of the carboxyl oxygen.

thioester

Thioesters are the most common forms of activated carboxylic acids in a cell. Although thioesters hydrolyze at about the same rate as oxygen esters, they are much more reactive than oxygen esters toward the addition of nitrogen and carbon nucleophiles. This allows a thioester to survive in the aqueous environment of the cell—without being hydrolyzed—while waiting to be used as a reactant in a nucleophilic acyl substitution reaction.

Why Thioesters Are More Reactive Than Esters

The carbonyl carbon of a thioester is more susceptible to nucleophilic addition than is the carbonyl carbon of an oxygen ester, because electron delocalization onto the carbonyl oxygen that reduces the carbonyl group's electrophilicity is weaker when Y is S than when Y is O. Electron delocalization is weaker because less overlap occurs between the $3p$ orbital of sulfur and the $2p$ orbital of carbon than between the $2p$ orbital of oxygen and the $2p$ orbital of carbon. In addition, the tetrahedral intermediate formed from a thioester undergoes elimination more rapidly than the tetrahedral intermediate formed from an oxygen ester because a thiolate ion is a weaker base and is, therefore, easier to eliminate than an alkoxide ion.

CH_3CH_2SH
$pK_a = 10.5$

CH_3CH_2OH
$pK_a = 15.9$

Coenzyme A Is the Thiol Used by Cells

The thiol used in biological systems for the formation of thioesters is coenzyme A. The compound is written "CoASH" to emphasize that the thiol group is the reactive part of the molecule. CoASH is composed of a decarboxylated cysteine (an amino acid), pantothenate (a vitamin), and phosphorylated adenosine diphosphate.

coenzyme A
CoASH

decarboxylated cysteine

pantothenate

phosphorylated ADP

When a cell converts a carboxylic acid to a thioester, it first converts the carboxylic acid to an acyl adenylate. The acyl adenylate then reacts with CoASH to form the thioester. The most common thioester in cells is acetyl-CoA.

Acetylcholine (an ester) is an example of an ester that cells synthesize using acetyl-CoA. Acetylcholine is a *neurotransmitter*—that is, it transmits nerve impulses across the synapses (spaces) between nerve cells.

Nerve Impulses, Paralysis, and Insecticides

After an impulse is transmitted between two nerve cells, acetylcholine must be hydrolyzed immediately to enable the recipient cell to receive another impulse. Acetylcholinesterase, the enzyme that catalyzes this hydrolysis, has a CH_2OH group that is necessary for its catalytic activity. The CH_2OH group participates in a transesterification reaction with acetylcholine, which releases choline. Hydrolysis of the ester group attached to the enzyme restores its active form.

Diisopropyl fluorophosphate (DFP), a military nerve gas used during World War II, inactivates acetylcholinesterase by reacting with its CH_2OH group. When the enzyme is inactivated, nerve impulses cannot be transmitted properly and paralysis occurs. DFP is extremely toxic. Its LD_{50} (the lethal dose for 50% of the test animals) is only 0.5 mg/kg of body weight.

Malathion and parathion, widely used as insecticides, are compounds related to DFP. The LD_{50} of malathion is 2800 mg/kg. Parathion is much more toxic, with an LD_{50} of 2 mg/kg.

ESSENTIAL CONCEPTS

Section 15.0

- A **carbonyl group** is a carbon doubly bonded to an oxygen; an **acyl group** is a carbonyl group attached to an alkyl (R) or an aromatic (Ar) group.
- **Acyl chlorides, acid anhydrides, esters,** and **amides** are called **carboxylic acid derivatives** because they differ from a carboxylic acid only in the nature of the group that has replaced the OH group of the carboxylic acid.

Section 15.1

- The functional group ending of a carboxylic acid is "oic acid," so the name of a two carbon carboxylic acid is ethanoic acid. Common names are: formic, acetic, propionic, butyric, valeric, and caproic acid.
- Cyclic esters are called **lactones;** cyclic amides are called **lactams.**

Section 15.4

- The reactivity of carbonyl compounds resides in the polarity of the carbonyl group; the carbonyl carbon has a partial positive charge (it is an electrophile) that is attractive to nucleophiles.
- Carboxylic acids and carboxylic acid derivatives undergo nucleophilic acyl substitution reactions in which a nucleophile replaces the substituent attached to the acyl group in the reactant.
- A carboxylic acid or carboxylic acid derivative will undergo a nucleophilic acyl substitution reaction provided that the newly added group in the tetrahedral intermediate is not a weaker base than the group attached to the acyl group in the reactant.
- Generally, a compound with an sp^3 carbon bonded to an oxygen is unstable if the sp^3 carbon is bonded to another electronegative atom.

Section 15.5

- The weaker the base attached to the acyl group, the more easily both slow steps of the nucleophilic acyl substitution reaction take place.
- The relative reactivities toward nucleophilic acyl substitution are acyl chlorides > acid anhydrides > esters ~ carboxylic acids > amides > carboxylate ions.

Section 15.6

- Nucleophilic acyl substitution reactions occur by one of two mechanisms, depending on whether the nucleophile is charged or neutral.

Section 15.7

- **Hydrolysis, alcoholysis,** and **aminolysis** are reactions where water, alcohols, and amines, respectively, convert one compound into two compounds.
- A **transesterification reaction** converts one ester to another ester.

Section 15.8

- All organic reactants, intermediates, and products in acidic solutions are positively charged or neutral; in basic solutions they are negatively charged or neutral.
- An acid increases the rate of formation of the tetrahedral intermediate by protonating the carbonyl oxygen, which increases the electrophilicity of the carbonyl carbon.
- An acid increases the rate of collapse of the tetrahedral intermediate by protonating the leaving group, which decreases its basicity and makes it easier to eliminate.
- The hydrolysis of an ester with a tertiary alkyl group occurs via an S_N1 reaction.

Section 15.9

- Hydroxide (or alkoxide) ion increases the rate of formation of the tetrahedral intermediate—it is a better nucleophile than water (or an alcohol)—and increases the rate of collapse of the tetrahedral intermediate by making the transition state more stable.
- Hydroxide ion promotes only hydrolysis reactions; alkoxide ion promotes only alcoholysis reactions.
- The rate of hydrolysis can be increased by acid or by HO^-; the rate of alcoholysis can be increased by acid or by RO^-.

Section 15.10

- The reaction of a carboxylic acid with excess alcohol and an acid catalyst is called a **Fischer esterification.**

Sections 15.11–15.13

- Amides are unreactive compounds but do react with water and alcohols if the reaction mixture is heated in an acidic solution. Amides are also hydrolyzed in strongly basic solutions.

Section 15.14

- The **Gabriel synthesis,** which converts an alkyl halide to a primary amine, involves the hydrolysis of an **imide.**

Section 15.15

- Nitriles are harder to hydrolyze than amides.

Section 15.16

- Acid anhydrides are less reactive than acyl chlorides but more reactive than esters. Therefore, acid anhydrides react with alcohols, water, and amines to form esters, carboxylic acids, and amides, respectively.

Section 15.17

- Dicarboxylic acids lose water when they are heated if they can form a cyclic anhydride with a five- or a six-membered ring.

Section 15.18

- Organic chemists activate carboxylic acids by converting them to acyl chlorides or acid anhydrides.

Section 15.19

- Cells activate carboxylic acids by converting them to **acyl phosphates, acyl adenylates,** or **thioesters.**

SUMMARY OF REACTIONS

The general mechanisms for nucleophilic acyl substitution reactions are shown on pages 694–695.

1. **Reactions of acyl chlorides (Section 15.6). The mechanisms are shown on page 699.**

$$\underset{\text{R}}{\overset{\text{O}}{\overset{\|}{\text{C}}}}\text{Cl} + \text{CH}_3\text{OH} \longrightarrow \underset{\text{R}}{\overset{\text{O}}{\overset{\|}{\text{C}}}}\text{OCH}_3 + \text{HCl}$$

$$\underset{\text{R}}{\overset{\text{O}}{\overset{\|}{\text{C}}}}\text{Cl} + \text{H}_2\text{O} \longrightarrow \underset{\text{R}}{\overset{\text{O}}{\overset{\|}{\text{C}}}}\text{OH} + \text{HCl}$$

$$\underset{\text{R}}{\overset{\text{O}}{\overset{\|}{\text{C}}}}\text{Cl} + 2\ \text{CH}_3\text{NH}_2 \longrightarrow \underset{\text{R}}{\overset{\text{O}}{\overset{\|}{\text{C}}}}\text{NHCH}_3 + \text{CH}_3\overset{+}{\text{N}}\text{H}_3\ \text{Cl}^-$$

2. **Reactions of esters (Sections 15.8–15.9). The mechanisms are shown on pages 703, 705, and 706.**

$$\underset{\text{R}}{\overset{\text{O}}{\overset{\|}{\text{C}}}}\text{OR} + \text{CH}_3\text{OH} \underset{}{\overset{\text{HCl}}{\rightleftharpoons}} \underset{\text{R}}{\overset{\text{O}}{\overset{\|}{\text{C}}}}\text{OCH}_3 + \text{ROH}$$

$$\underset{\text{R}}{\overset{\text{O}}{\overset{\|}{\text{C}}}}\text{OR} + \text{CH}_3\text{OH} \overset{\text{CH}_3\text{O}^-}{\longrightarrow} \underset{\text{R}}{\overset{\text{O}}{\overset{\|}{\text{C}}}}\text{OCH}_3 + \text{ROH}$$

$$\underset{\text{R}}{\overset{\text{O}}{\overset{\|}{\text{C}}}}\text{OR} + \text{H}_2\text{O} \underset{}{\overset{\text{HCl}}{\rightleftharpoons}} \underset{\text{R}}{\overset{\text{O}}{\overset{\|}{\text{C}}}}\text{OH} + \text{ROH}$$

$$\underset{\text{R}}{\overset{\text{O}}{\overset{\|}{\text{C}}}}\text{OR} + \text{H}_2\text{O} \overset{\text{HO}^-}{\longrightarrow} \underset{\text{R}}{\overset{\text{O}}{\overset{\|}{\text{C}}}}\text{O}^- + \text{ROH}$$

$$\underset{\text{R}}{\overset{\text{O}}{\overset{\|}{\text{C}}}}\text{OR} + \text{CH}_3\text{NH}_2 \longrightarrow \underset{\text{R}}{\overset{\text{O}}{\overset{\|}{\text{C}}}}\text{NHCH}_3 + \text{ROH}$$

3. **Reactions of carboxylic acids (Section 15.10)**

$$\underset{\text{R}}{\overset{\text{O}}{\overset{\|}{\text{C}}}}\text{OH} + \text{CH}_3\text{OH} \underset{}{\overset{\text{HCl}}{\rightleftharpoons}} \underset{\text{R}}{\overset{\text{O}}{\overset{\|}{\text{C}}}}\text{OCH}_3 + \text{H}_2\text{O}$$

$$\underset{\text{R}}{\overset{\text{O}}{\overset{\|}{\text{C}}}}\text{OH} + \text{CH}_3\text{NH}_2 \longrightarrow \underset{\text{R}}{\overset{\text{O}}{\overset{\|}{\text{C}}}}\text{O}^-\,\text{H}_3\overset{+}{\text{N}}\text{CH}_3 \overset{225°\text{C}}{\longrightarrow} \underset{\text{R}}{\overset{\text{O}}{\overset{\|}{\text{C}}}}\text{NH}_2 + \text{H}_2\text{O}$$

4. **Reactions of amides (Sections 15.11–15.13). The mechanisms are shown on pages 712, 715, and 719.**

$$\underset{\text{R}}{\overset{\text{O}}{\overset{\|}{\text{C}}}}\text{NH}_2 + \text{H}_2\text{O} \overset{\text{HCl}}{\underset{\Delta}{\longrightarrow}} \underset{\text{R}}{\overset{\text{O}}{\overset{\|}{\text{C}}}}\text{OH} + \overset{+}{\text{N}}\text{H}_4\text{Cl}^-$$

$$\underset{\text{R}}{\overset{\text{O}}{\overset{\|}{\text{C}}}}\text{NH}_2 + \text{CH}_3\text{OH} \overset{\text{HCl}}{\underset{\Delta}{\longrightarrow}} \underset{\text{R}}{\overset{\text{O}}{\overset{\|}{\text{C}}}}\text{OCH}_3 + \overset{+}{\text{N}}\text{H}_4\text{Cl}^-$$

$$\underset{\text{R}}{\overset{\text{O}}{\overset{\|}{\text{C}}}}\text{NH}_2 + \text{H}_2\text{O} \overset{\text{HO}^-}{\underset{\Delta}{\longrightarrow}} \underset{\text{R}}{\overset{\text{O}}{\overset{\|}{\text{C}}}}\text{O}^- + \text{NH}_3$$

$$\underset{\text{R}}{\overset{\text{O}}{\overset{\|}{\text{C}}}}\text{NH}_2 \overset{\text{SOCl}_2}{\longrightarrow} \text{RC}\equiv\text{N}$$

5. **Gabriel synthesis of a primary amine (Section 15.14)**

$$\text{phthalimide}\ \overset{\text{NH}}{} \quad \overset{\begin{array}{l}1.\ \text{HO}^-\\2.\ \text{RCH}_2\text{Br}\\3.\ \text{HCl, H}_2\text{O, }\Delta\\4.\ \text{HO}^-\end{array}}{\longrightarrow} \text{RCH}_2\text{NH}_2$$

phthalimide

729

6. Hydrolysis of nitriles (Section 15.15). The mechanism is shown on page 718.

$$RC \equiv N \ + \ H_2O \ \xrightarrow[\Delta]{HCl} \ \underset{R}{\overset{O}{\|}}C-OH \ + \ \overset{+}{N}H_4Cl^-$$

7. Reactions of acid anhydrides (Section 15.16). The mechanism is shown on page 720.

8. Reactions of dicarboxylic acids (Section 15.17)

9. Activation of carboxylic acids by chemists (Section 15.18). The mechanism is shown on page 723.

10. Activation of carboxylic acids by cells (Section 15.19). The mechanisms are shown on page 725.

acetyl-CoA

PROBLEMS

48. Draw a structure for each of the following:

   a. *N,N*-dimethylhexanamide
   b. 3,3-dimethylhexanamide
   c. cyclohexanecarbonyl chloride
   d. propanenitrile

   e. propionamide
   f. sodium acetate
   g. benzoic anhydride
   h. β-valerolactone

   i. 3-methylbutanenitrile
   j. cycloheptanecarboxylic acid
   k. benzoyl chloride

49. Name the following:

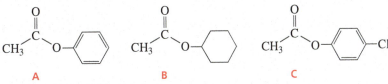

50. What compounds are formed from the reaction of benzoyl chloride with the following reagents?

   a. sodium acetate
   b. water
   c. excess dimethylamine
   d. aqueous HCl

   e. aqueous NaOH
   f. cyclohexanol
   g. excess benzylamine
   h. 4-chlorophenol

   i. isopropyl alcohol
   j. excess aniline
   k. potassium formate

51. What compounds are obtained from the following hydrolysis reactions?

52. a. Rank the following esters in order of decreasing reactivity in the first slow step of a nucleophilic acyl substitution reaction (formation of the tetrahedral intermediate):

           A   B   C

   b. Rank the same esters in order of decreasing reactivity in the second slow step of a nucleophilic acyl substitution reaction (collapse of the tetrahedral intermediate).

53. Because bromocyclohexane is a secondary alkyl halide, both cyclohexanol and cyclohexene are formed when the alkyl halide reacts with hydroxide ion. Suggest a method to synthesize cyclohexanol from bromocyclohexane that forms little or no cyclohexene.

54. a. Which compound would you expect to have a greater dipole moment, methyl acetate or butanone?
   b. Which would you expect to have a higher boiling point?

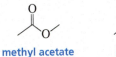

         methyl acetate   butanone

55. How could you use ^{1}H NMR spectroscopy to distinguish the following esters?

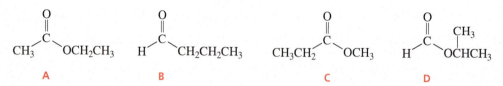

A   B   C   D

56. Rank the following compounds in order of decreasing frequency of the carbon–oxygen double-bond stretch:

57. Using an alcohol for one method and an alkyl halide for the other, show two ways to make each of the following esters:

a. propyl acetate (odor of pears)
b. isopentyl acetate (odor of bananas)

c. ethyl butyrate (odor of pineapple)
d. methyl phenylethanoate (odor of honey)

58. What reagents would you use to convert methyl propanoate to the following compounds?

a. isopropyl propanoate
b. sodium propanoate

c. *N*-ethylpropanamide
d. propanoic acid

59. What products would you expect to obtain from the following reactions?

a. malonic acid + 2 acetyl chloride
b. methyl carbamate + methylamine

c. urea + water
d. β-ethylglutaric acid + acetyl chloride + Δ

60. A compound with molecular formula $C_5H_{10}O_2$ gives the following IR spectrum. When it undergoes acid-catalyzed hydrolysis, the compound with the ^{1}H NMR spectrum shown below is formed. Identify the compounds.

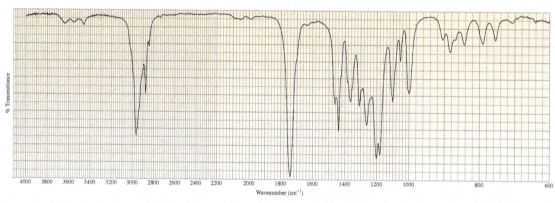

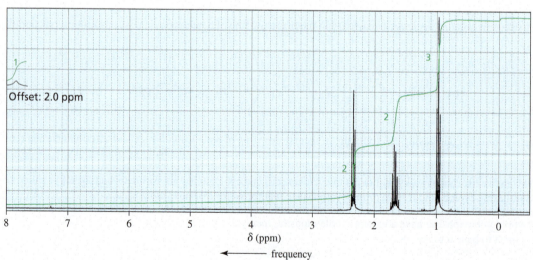

1. Aspartame, the sweetener used in the commercial products NutraSweet and Equal, is 200 times sweeter than sucrose. What products will be obtained if aspartame is hydrolyzed completely in an aqueous solution of HCl?

aspartame

2. To prove that the reaction of an ester with hydroxide ion is not an S_N2 reaction but, instead, forms a tetrahedral intermediate that subsequently collapses, the products obtained after allowing hydroxide ion to react with an ester containing an isotopically labeled carbonyl oxygen (^{18}O) were isolated. One of the products was isotopically labeled hydroxide ion. Explain how obtaining this product proves that a tetrahedral intermediate was formed.

$$\overset{^{18}O}{\underset{}{}}\text{C}_6\text{H}_5\text{—C(=}^{18}\text{O)—OCH}_2\text{CH}_3 \ + \ ^-\text{:ÖH} \ \rightleftharpoons \ \text{C}_6\text{H}_5\text{—C(=O)—OCH}_2\text{CH}_3 \ + \ ^{18}\text{:ÖH}^-$$

3. a. Which of the following reactions does not give the carbonyl product shown?

   b. Which of the reactions that do not occur can be made to occur if an acid catalyst is added to the reaction mixture?

   1. $CH_3COOH + CH_3COO^- \longrightarrow$ (anhydride)

   2. $CH_3COCl + CH_3COO^- \longrightarrow$ (anhydride)

   3. $CH_3CONH_2 + Cl^- \longrightarrow CH_3COCl$

   4. $CH_3COOH + CH_3NH_2 \longrightarrow CH_3CONHCH_3$

   5. $CH_3COOCH_3 + CH_3NH_2 \overset{\Delta}{\longrightarrow} CH_3CONHCH_3$

   6. $CH_3COOCH_3 + Cl^- \longrightarrow CH_3COCl$

   7. $CH_3CONHCH_3 + CH_3COO^- \longrightarrow$ (anhydride)

   8. $CH_3COCl + H_2O \longrightarrow CH_3COOH$

   9. $CH_3CONHCH_3 + H_2O \longrightarrow CH_3COOH$

   10. (anhydride) $+ CH_3OH \longrightarrow CH_3COOCH_3$

54. Describe how the target molecule (butanone) can be synthesized in a high yield from butane.

butane $\overset{?}{\longrightarrow}$ butanone

55. 1,4-Diazabicyclo[2.2.2]octane (abbreviated DABCO) is a tertiary amine that catalyzes transesterification reactions. Explain how it does this.

1,4-diazabicyclo[2.2.2]octane
DABCO

66. a. If the equilibrium constant for the reaction of acetic acid and ethanol to form ethyl acetate is 4.02, what is the concentration of ethyl acetate at equilibrium if the reaction is carried out with equal amounts of acetic acid and ethanol?

   b. What is the concentration of ethyl acetate at equilibrium if the reaction is carried out with 10 times more ethanol than acetic acid? *Hint:* Recall the quadratic equation: for $ax^2 + bx + c = 0$,

$$x = \frac{-b \ \pm \ (b^2 - 4ac)^{1/2}}{2a}$$

   c. What is the concentration of ethyl acetate at equilibrium if the reaction is carried out with 100 times more ethanol than acetic acid?

67. Two products, **A** and **B**, are obtained from the reaction of 1-bromobutane with NH_3. Compound **A** reacts with acetyl chloride to form **C**, and compound **B** reacts with acetyl chloride to form **D**. The IR spectra of **C** and **D** are shown. Identify **A**, **B**, **C**, and **D**.

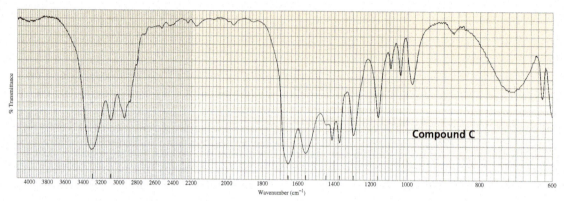

68. What are the products of the following reactions?

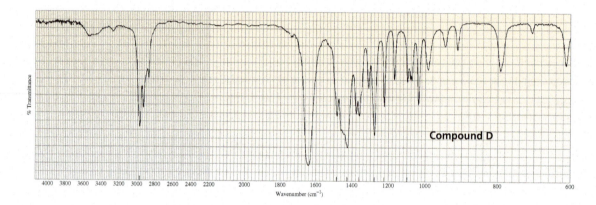

a.
$$CH_3-\overset{O}{\underset{}{C}}-Cl \;+\; KF \longrightarrow$$

b.
(pyrrolidinone) $+\; H_2O \xrightarrow{\text{HCl}\; \Delta}$

c.
(benzoic acid) $\text{OH} \xrightarrow[\text{2. 2 CH}_3\text{NH}_2]{\text{1. SOCl}_2}$

d.
(succinic anhydride) $+\; H_2O \longrightarrow$

e.
$$Cl-\overset{O}{\underset{}{C}}-Cl \;+\; \text{(catechol)} \longrightarrow$$

f.
(γ-butyrolactone) $+\; H_2O \text{ excess} \xrightarrow{\text{HCl}}$

g.
$$CH_3-\overset{O}{\underset{}{C}}-O-CH_2-O-\overset{O}{\underset{}{C}}-CH_3 \;+\; CH_3OH \text{ excess} \xrightarrow{CH_3O^-}$$

h.
(2-(carboxymethyl)benzoic acid) CH_2COOH, COH $\xrightarrow[\Delta]{(CH_3C)_2O}$

i.
(phthalic anhydride) $+\; NH_3 \text{ excess} \longrightarrow$

j.
(isochroman-1-one) $+\; CH_3OH \text{ excess} \xrightarrow{\text{HCl}}$

69. Phosgene ($COCl_2$) was used as a poison gas in World War I. What product would be formed from the reaction of phosgene with each of the following reagents?

    a. one equivalent of methanol   **b.** excess methanol   **c.** excess propylamine   **d.** excess water

70. What reagent should be used to carry out the following reaction?

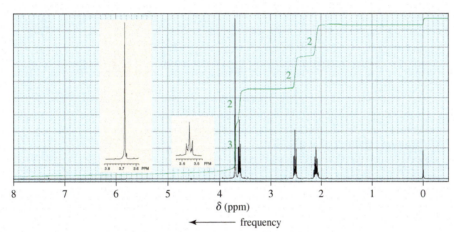

71. When a student treated butanedioic acid with thionyl chloride, she was surprised to find that the product she obtained was an anhydride rather than an acyl chloride. Propose a mechanism to explain why she obtained an anhydride.

72. When butanoic acid and ^{18}O-labeled methanol react under acidic conditions, what compounds are labeled when the reaction has reached equilibrium?

73. When treated with an equivalent of methanol, compound **A**, with molecular formula $C_4H_6Cl_2O$, forms the compound whose ^1H NMR spectrum is shown here. Identify compound **A**.

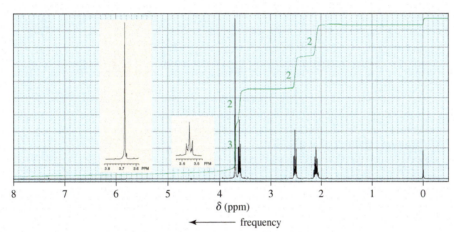

74. a. Identify the two products obtained from the following reaction:

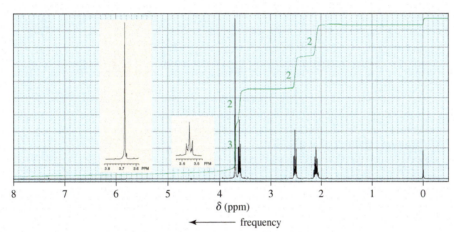

    b. A student carried out the preceding reaction, but stopped it before it was half over, whereupon he isolated the major product. He was surprised to find that the product he isolated was neither of the products obtained when the reaction was allowed to go to completion. What product did he isolate?

75. An aqueous solution of a primary or secondary amine reacts with an acyl chloride to form an amide as the major product. However, if the amine is tertiary, an amide is not formed. What product *is* formed? Explain.

76. If propionyl chloride is added to one equivalent of methylamine, only a 50% yield of *N*-methylpropanamide is obtained. If, however, the acyl chloride is added to two equivalents of methylamine, the yield of *N*-methylpropanamide is almost 100%. Explain these observations.

77. a. When a carboxylic acid is dissolved in isotopically labeled water ($H_2^{18}O$) and an acid catalyst is added, the label is incorporated into both oxygens of the acid. Propose a mechanism to account for this.

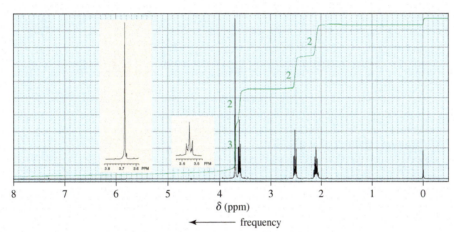

    b. If a carboxylic acid is dissolved in isotopically labeled methanol ($CH_3^{18}OH$) and an acid catalyst is added, where will the label reside in the product?

    c. If an ester is dissolved in isotopically labeled water ($H_2^{18}O$) and an acid catalyst is added, where will the label reside in the product?

78. a. A student did not obtain any ester when he added 2,4,6-trimethylbenzoic acid to an acidic solution of ethanol. Why? (*Hint:* Build models.)

    b. Would he have encountered the same problem if he had tried to synthesize the methyl ester of 4-methylbenzoic acid in the same way?

79. Identify the major and minor products of the following reaction:

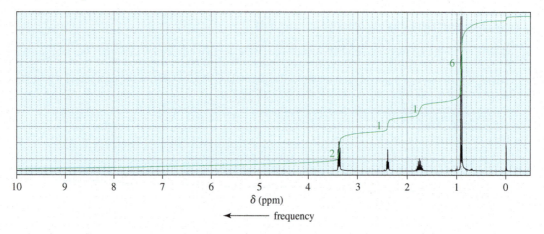

80. When a compound with molecular formula $C_{11}H_{14}O_2$ undergoes acid-catalyzed hydrolysis, one of the products that is isolated gives the following ^1H NMR spectrum. Identify the compound.

81. Cardura, a drug used to treat hypertension, is synthesized as shown here.

a. Identify the intermediate (**A**) and show the mechanism for its formation.
b. Show the mechanism for conversion of **A** to **B**. Which is formed more rapidly, **A** or **B**?

82. The reaction of a nitrile with an alcohol in the presence of a strong acid forms an *N*-substituted amide. This reaction, known as the **Ritter reaction,** does not work with primary alcohols.

$$RC{\equiv}N + R'OH \xrightarrow{HCl} \text{the Ritter reaction}$$

a. Propose a mechanism for the Ritter reaction.
b. Why does the Ritter reaction not work with primary alcohols?
c. How does the Ritter reaction differ from the acid-catalyzed hydrolysis of a nitrile to form an amide?

83. Some credit card sales slips have a top sheet of "carbonless paper" that transfers an imprint of a signature to a sheet lying underneath (the customer receipt). The paper contains tiny capsules filled with the following colorless compound. When you press on the paper, the capsules burst, and the colorless compound comes into contact with the acid-treated bottom sheet, forming a highly colored compound. What is the structure of the colored compound?

84. The ^{1}H NMR spectra for two esters with molecular formula $C_8H_8O_2$ are shown next. Which of the esters is hydrolyzed more rapidly in an aqueous solution with a pH of 10?

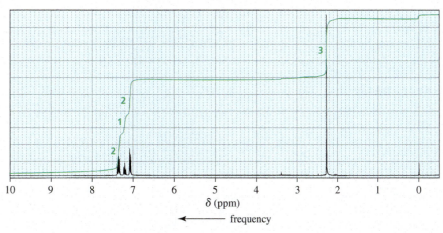

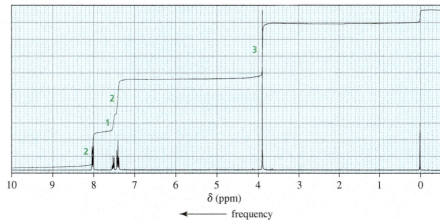

85. Show how the following compounds could be prepared from the given starting materials. You can use any necessary organic or inorganic reagents.

a.

c.

b.

d.

86. Is the acid-catalyzed hydrolysis of acetamide a reversible or an irreversible reaction? Explain.

87. What product do you expect to obtain from each of the following reactions?

a.

b.

88. The intermediate shown here is formed during the hydroxide-ion-promoted hydrolysis of the ester group. Propose a mechanism for the reaction.

89. The following compound has been found to be an inhibitor of penicillinase. The enzyme can be reactivated by hydroxylamine (NH_2OH). Propose a mechanism to account for the inhibition and for the reactivation.

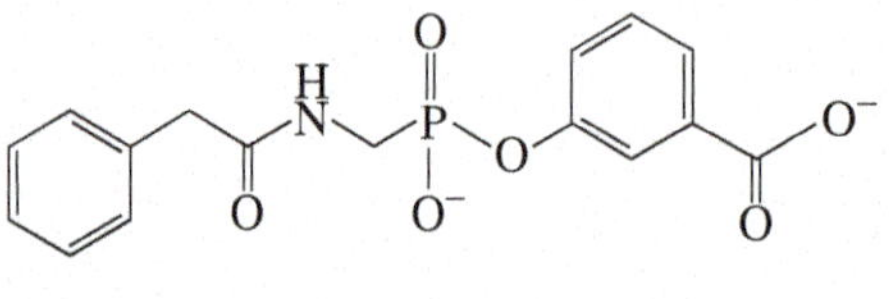

90. Propose a mechanism that accounts for the formation of the product.

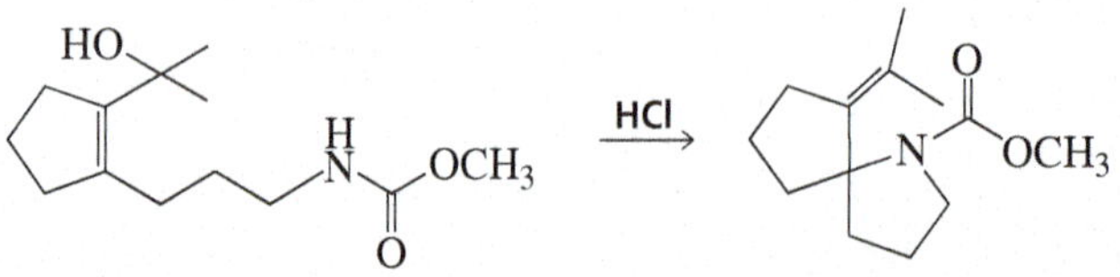

91. Catalytic antibodies catalyze a reaction by forcing the conformation of the substrate in the direction of the transition state. The synthesis of the antibody is carried out in the presence of a transition state analog—a stable molecule that structurally resembles the transition state. This causes an antibody to be generated that recognizes and binds to the transition state, thereby stabilizing it. For example, the following transition state analog has been used to generate a catalytic antibody that catalyzes the hydrolysis of the structurally similar ester:

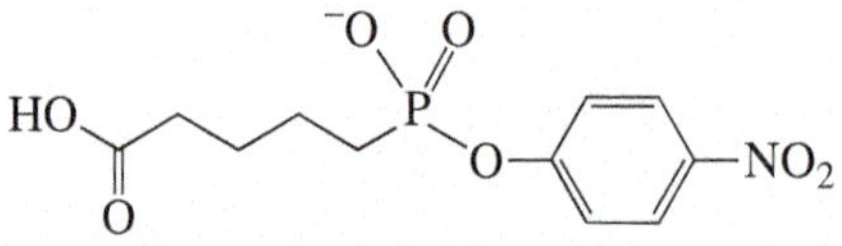

transition state analog

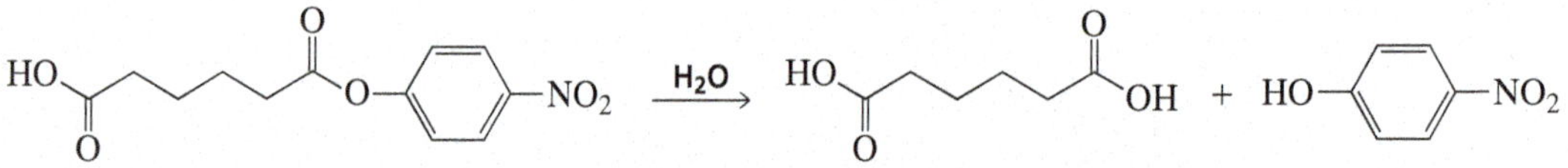

a. Draw a possible transition state for the hydrolysis reaction.

b. The following transition state analog is used to generate a catalytic antibody for the catalysis of ester hydrolysis. Draw the structure of an ester whose rate of hydrolysis would be increased by this catalytic antibody.

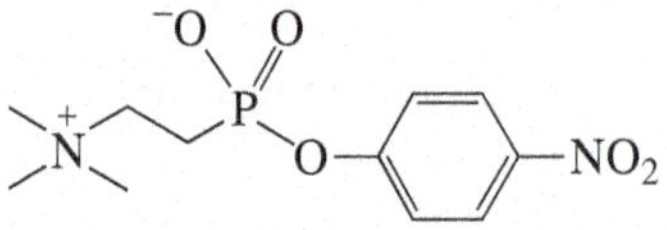

c. Design a transition state analog that catalyzes amide hydrolysis at the amide group indicated.

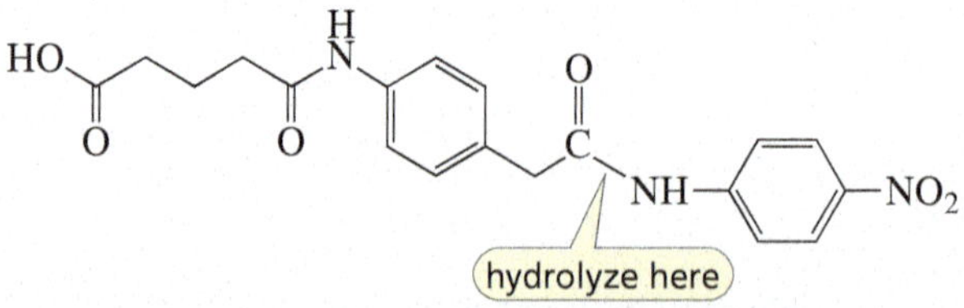

hydrolyze here

92. Information about the mechanism of the reaction undergone by a series of substituted benzenes can be obtained by plotting the logarithm of the observed rate constant determined at a particular pH against the Hammett substituent constant (σ) for the particular substituent. The σ value for hydrogen is 0. Electron-donating substituents have negative σ values; the more strongly electron donating the substituent, the more negative its σ value. Electron-withdrawing substituents have positive σ values; the more strongly electron withdrawing the substituent, the more positive its σ value. The slope of a plot of the logarithm of the rate constant versus σ is called the ρ (rho) value. The ρ value for the hydroxide-ion-promoted hydrolysis of a series of meta- and para-substituted ethyl benzoates is +2.46; the ρ value for amide formation for the reaction of a series of meta- and para-substituted anilines with benzoyl chloride is −2.78.

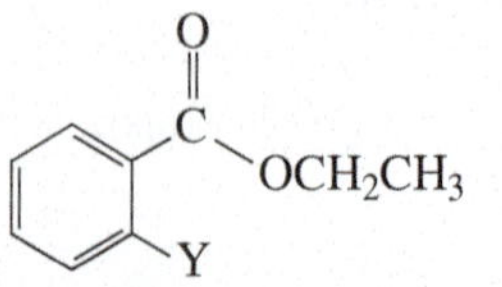

ortho-substituted 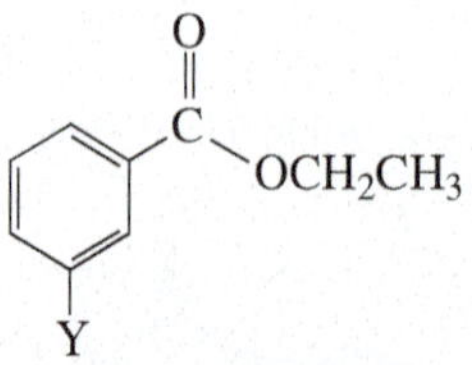   **meta-substituted** 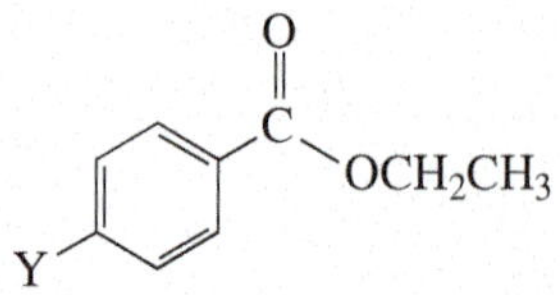   **para-substituted**

a. Why does one set of experiments give a positive ρ value, whereas the other set of experiments gives a negative ρ value?

b. Why were ortho-substituted compounds not included in the experiment?

c. What do you predict the sign of the ρ value to be for the ionization of a series of meta- and para-substituted benzoic acids?

16

Reactions of Aldehydes and Ketones •
More Reactions of
Carboxylic Acid Derivatives

a yew tree forest

Taxol, a compound extracted from the bark of yew trees, was found to be an effective drug against several kinds of cancer. However, the trees grow very slowly, the bark of one tree provides only a small amount of the drug, and removing the bark kills the tree. Yew tree forests, moreover, are the home of the spotted owl—an endangered species. When chemists determined the structure of the drug they were disappointed to find that it would be a very difficult compound to synthesize. Nevertheless, the current supply of taxol is sufficient to meet medical demands. In this chapter, you will see how this was accomplished.

Group III

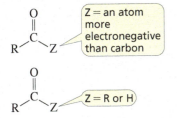

This chapter continues the discussion of the families of compounds in Group III. Here we will look at the reactions of aldehydes and ketones—that is, carbonyl compounds that do not have a group that can be substituted by another group, and we will compare their reactions with those of the carboxylic acid derivatives that you studied in Chapter 15.

The carbonyl carbon of the simplest aldehyde, formaldehyde, is bonded to two hydrogens. The carbonyl carbon of all other **aldehydes** is bonded to a hydrogen and to an alkyl (or aryl) group (R). The carbonyl carbon of a **ketone** is bonded to two alkyl groups.

formaldehyde · an aldehyde · a ketone

Many compounds found in nature have aldehyde or ketone functional groups. Aldehydes have pungent odors, whereas ketones tend to smell sweet. Vanillin and cinnamaldehyde are examples of naturally occurring aldehydes. A whiff of vanilla extract allows you to appreciate the pungent odor of vanillin. The ketones camphor and carvone are responsible for the characteristic sweet odors of the leaves of camphor trees, spearmint leaves, and caraway seeds.

vanillin
vanilla flavoring

cinnamaldehyde
cinnamon flavoring

camphor

(R)-(−)-carvone
spearmint oil

(S)-(+)-carvone
caraway seed oil

Progesterone and testosterone are two biologically important ketones that illustrate how a small difference in structure can be responsible for a large difference in biological activity. Both are sex hormones, but progesterone is synthesized primarily in the ovaries, whereas testosterone is synthesized primarily in the testes.

progesterone

testosterone

The physical properties of aldehydes and ketones were discussed in Section 15.3 (see also Appendix VI), and the methods used to prepare aldehydes and ketones are summarized in Appendix III.

16.1 THE NOMENCLATURE OF ALDEHYDES AND KETONES

Naming Aldehydes

LEARN THE STRATEGY

The systematic (IUPAC) name of an aldehyde is obtained by replacing the final "e" on the name of the parent hydrocarbon with "al." For example, a one-carbon aldehyde is called methan*al*, and a two-carbon aldehyde is called ethan*al*. The position of the carbonyl carbon does not have to be designated because it is always at the end of the parent hydrocarbon (or else the compound would not be an aldehyde), so it always has the 1-position.

The common name of an aldehyde is the same as the common name of the corresponding carboxylic acid, except that "aldehyde" is substituted for "oic acid" (or "ic acid"). Recall that the position of a substituent is designated by a lowercase Greek letter when common names are used. The carbonyl carbon is not given a designation, so the carbon adjacent to the carbonyl carbon is the α-carbon (Section 15.1).

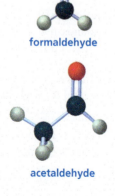

formaldehyde

acetaldehyde

systematic name: **methanal**   **ethanal**
common name: **formaldehyde**   **acetaldehyde**

2-bromopropanal
α-bromopropionaldehyde

3-hydroxybutanal   methyl 3-cyanopropanoate   4-amino-3-methyl-2-butanol

If a ketone or an aldehyde has a second functional group of higher naming priority, the carbonyl oxygen is indicated by the prefix "oxo."

systematic name: 4-oxopentanal   methyl 3-oxobutanoate   5-oxopentanamide

We saw that if one of the functional groups is an alkene (or an alkyne), suffix endings are used for both functional groups; the alkene (or alkyne) functional group is stated first, with its "e" ending omitted to avoid two successive vowels (Section 7.2).

$$CH_3CH=CHCH_2\overset{O}{\underset{}{\overset{\parallel}{C}}}H$$

3-pentenal

PROBLEM 3

Name the following:

a.   b.   c.

USE THE STRATEGY

16.2 THE RELATIVE REACTIVITIES OF CARBONYL COMPOUNDS

We saw that the carbonyl group is polar because oxygen is more electronegative than carbon, so oxygen has a greater share of the double bond's electrons (Section 15.5). As a result, the carbonyl carbon is electron deficient (it is an electrophile). Therefore, it reacts with nucleophiles. The electron deficiency of the carbonyl carbon in formaldehyde, acetaldehyde, and acetone is indicated by the faint blue region in each compound's electrostatic potential map.

formaldehyde

An aldehyde has a greater partial positive charge on its carbonyl carbon than a ketone does because an alkyl group is more electron donating than a hydrogen (Section 6.2). An aldehyde, therefore, is more reactive than a ketone toward nucleophilic addition. Steric factors also contribute to the greater reactivity of an aldehyde. The carbonyl carbon of an aldehyde is more accessible to a nucleophile because the hydrogen attached to the carbonyl carbon of an aldehyde is smaller than the second alkyl group attached to the carbonyl carbon of a ketone.

acetaldehyde

relative reactivities

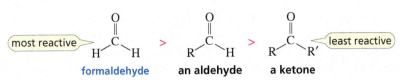

formaldehyde   an aldehyde   a ketone

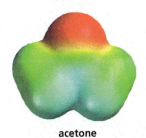

acetone

For the same reason, ketones with small alkyl groups bonded to the carbonyl carbon are more reactive than those with large alkyl groups.

relative reactivities

Steric factors are also important in the transition state, which is tetrahedral (so it has bond angles of 109.5°). This causes the alkyl groups to be closer to one another than they are in the carbonyl compound, where the bond angles are 120°. As a result of the greater steric crowding in their transition states, ketones have less stable transition states than aldehydes have.

Aldehydes are more reactive than ketones.

In summary, alkyl groups stabilize the carbonyl compound and destabilize the transition state. Both factors increase $\Delta G^{\ddagger}$, causing ketones to be less reactive than aldehydes.

PROBLEM 4 ♦

Which ketone in each pair is more reactive?

a. 2-heptanone or 4-heptanone
b. bromomethyl phenyl ketone or chloromethyl phenyl ketone

Aldehydes and ketones are less reactive than acyl chlorides and acid anhydrides but they are more reactive than esters, carboxylic acids, and amides.

How does the reactivity of an aldehyde or a ketone toward nucleophiles compare with the reactivity of the carbonyl compounds whose reactions you studied in Chapter 15? Aldehydes and ketones are in the middle—they are *less* reactive than acyl halides and anhydrides, but they are *more* reactive than esters, carboxylic acids, and amides.

relative reactivities of carbonyl compounds

acyl halide > acid anhydride > aldehyde > ketone > ester ~ carboxylic acid > amide > carboxylate ion

most reactive   least reactive

16.3 HOW ALDEHYDES AND KETONES REACT

Nucleophilic Acyl Substitution

In Section 15.5, we saw that the carbonyl group of a carboxylic acid or a carboxylic acid derivative is attached to a group that can be replaced with another group. Therefore, these compounds undergo **nucleophilic acyl substitution reactions**—they react with nucleophiles to form substitution products.

Carboxylic acid derivatives undergo nucleophilic acyl substitution reactions with nucleophiles.

In contrast, the carbonyl group of an aldehyde or a ketone is attached to a group that is too strong a base (H^- or R^-) to be eliminated under normal conditions, so it cannot be replaced with another group. Consequently, aldehydes and ketones react with nucleophiles to form addition products, not substitution products.

Nucleophilic Addition

The addition of a nucleophile to the carbonyl carbon of an aldehyde or a ketone forms a tetrahedral compound. If the nucleophile is a strong base, such as R⁻ or H⁻, then the tetrahedral compound will be stable because it will not have a group that can be eliminated. Thus, the reaction will be an *irreversible* **nucleophilic addition reaction.** Recall that generally a tetrahedral compound is unstable if the sp^3 carbon is attached to two oxygens or an oxygen and another electronegative atom (a nitrogen or a halogen); see Section 15.5.

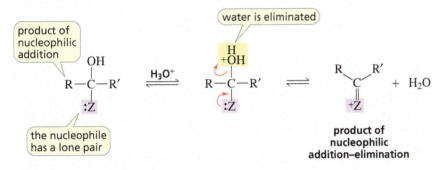

> the reaction is irreversible because Z⁻ is too basic to be eliminated

product of nucleophilic addition

Aldehydes and ketones undergo irreversible nucleophilic addition reactions with nucleophiles that are strong bases.

Nucleophilic Addition-Elimination

If the attacking atom of the nucleophile is an oxygen or a nitrogen and there is enough acid in the solution to protonate the OH group of the tetrahedral compound, then water can be eliminated from the addition product by a lone pair on the oxygen or nitrogen.

Aldehydes and ketones undergo reversible nucleophilic addition–elimination reactions with nucleophiles that have a lone pair on the attacking atom.

Thus, the reaction is a nucleophilic addition-elimination reaction. The reaction is reversible because either of the two atoms with a lone pair that are attached to the tetrahedral carbon can be protonated and, therefore, eliminated. We will see that the fate of the dehydrated product depends on the identity of Z.

16.4 REACTIONS OF CARBONYL COMPOUNDS WITH CARBON NUCLEOPHILES

Reactions with Grignard Reagents

Addition of a Grignard reagent to a carbonyl compound is a versatile reaction that forms a new C—C bond. This reaction can produce compounds with a variety of structures because both the structure of the carbonyl compound and the structure of the Grignard reagent can be varied.

We saw that a Grignard reagent is prepared by adding an alkyl halide to magnesium shavings in diethyl ether under anhydrous conditions (Section 11.1). We also saw that a Grignard reagent reacts as if it were a carbanion; therefore, it is a strong base and a good nucleophile.

$$CH_3CH_2Br \xrightarrow[\text{Et}_2\text{O}]{\text{Mg}} CH_3CH_2MgBr$$

$$CH_3CH_2MgBr \quad \textbf{\textcolor{red}{reacts as if it were}} \quad CH_3\ddot{C}H_2 \ \overset{+}{M}gBr$$

Consequently, aldehydes and ketones undergo nucleophilic addition reactions with Grignard reagents (Section 16.3).

Reactions of Aldehydes and Ketones with Grignard Reagents

The reaction of an aldehyde or a ketone with a Grignard reagent is a nucleophilic addition reaction—the nucleophilic Grignard reagent adds to the carbonyl carbon. The tetrahedral alkoxide ion is stable because it does not have a group that can be eliminated. (Recall that a tetrahedral compound is unstable only if the sp^3 carbon is attached to an oxygen *and* to another electronegative atom.)

MECHANISM FOR THE REACTION OF AN ALDEHYDE OR A KETONE WITH A GRIGNARD REAGENT

- Nucleophilic addition of the alkyl fragment of the Grignard reagent to the carbonyl carbon forms an alkoxide ion that is complexed with the magnesium ion.
- Addition of dilute acid breaks up the complex.

When a Grignard reagent reacts with formaldehyde, the product of the nucleophilic addition reaction is a *primary alcohol* that has one more carbon than the Grignard reagent.

When a Grignard reagent reacts with an aldehyde other than formaldehyde, the product of the nucleophilic addition reaction is a *secondary alcohol*.

When a Grignard reagent reacts with a ketone, the product of the nucleophilic addition reaction is a *tertiary alcohol*.

A Grignard reagent can also react with carbon dioxide. The product of the reaction is a *carboxylic acid* that has one more carbon than the Grignard reagent.

In the following reactions, the reagents above and below the reaction arrows are numbered in order of use, indicating that the acid is not added until after the Grignard reagent has reacted with the carbonyl compound.

O
‖
CH₃CH₂ — C — CH₂CH₃   →[1. **CH₃MgBr**][2. H₃O⁺]   CH₃CH₂C̈CH₂CH₃ (OH above, CH₃ below)

O
‖
CH₃CH₂CH₂ — C — H   →[1. ⬡—MgBr][2. H₃O⁺]   CH₃CH₂CH₂CH— (OH above) ⬡

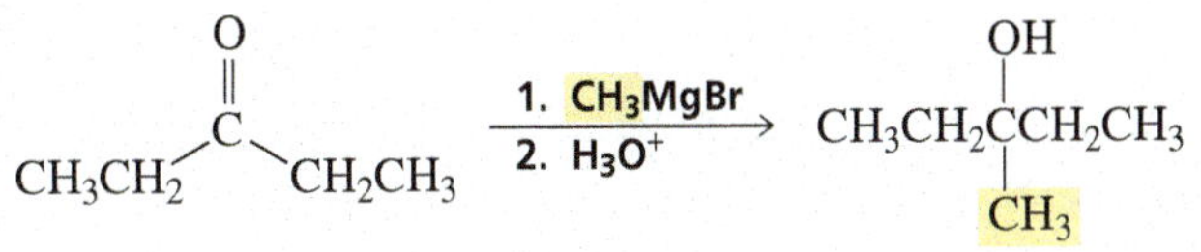

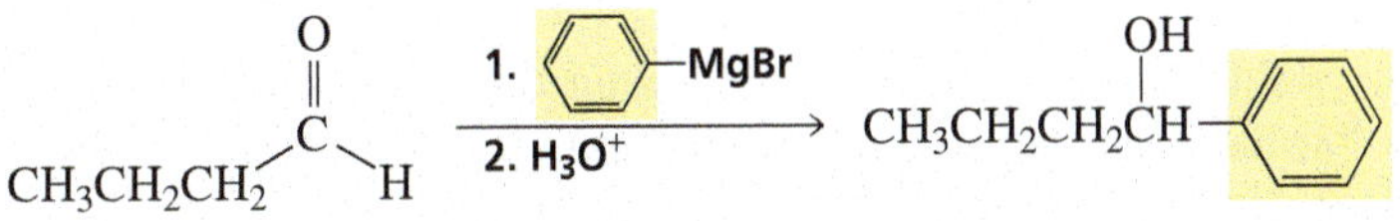

If the reaction with the carbonyl compound forms a product with an asymmetric center, such as the preceding reaction, the product will be a racemic mixture. (Recall that when a reactant without an asymmetric center undergoes a reaction that forms a product with an asymmetric center, the product is a racemic mixture; Section 6.13.)

Enzyme-Catalyzed Carbonyl Additions

If the reaction with a carbonyl compound that forms a product with an asymmetric center is catalyzed by an enzyme, then a racemic mixture is not formed. Instead, only one of the enantiomers is formed because the enzyme can block one face of the carbonyl compound so that it cannot be attacked, or the enzyme can position the nucleophile so it can attack the carbonyl group from only one side. Recall that enzyme-catalyzed reactions are completely stereoselective (Section 6.14).

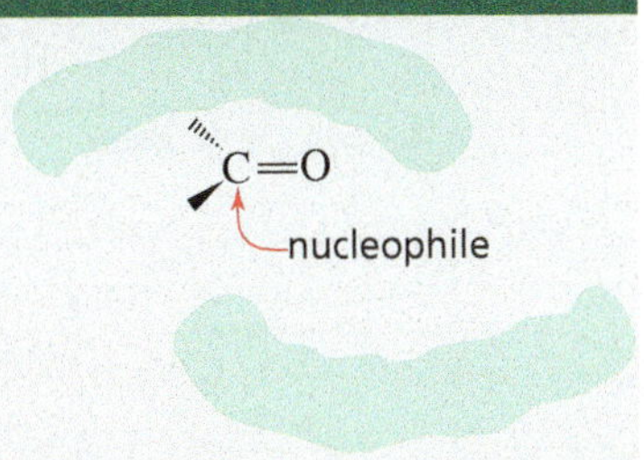

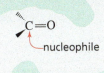

PROBLEM 5 ◆

What products are formed when the following compounds react with CH₃MgBr, followed by the addition of dilute acid? Disregard stereoisomers.

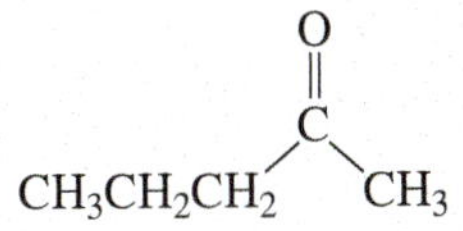

a. 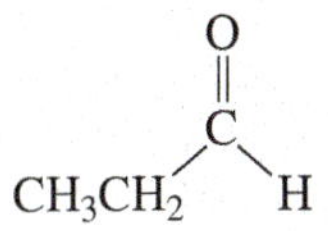
O
‖
CH₃CH₂ — C — H

b.
O
‖
CH₃CH₂CH₂ — C — CH₃

c.
(cyclohexanone structure)

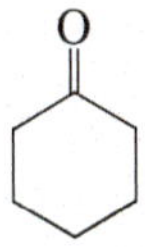

PROBLEM 6 ◆

We saw on the previous page that 3-methyl-3-hexanol can be synthesized from the reaction of 2-pentanone with ethylmagnesium bromide. What other combinations of ketone and Grignard reagent could be used to prepare the same tertiary alcohol?

PROBLEM 7 ◆

a. How many stereoisomers are obtained from the reaction of 2-pentanone with ethylmagnesium bromide followed by the addition of dilute acid?

b. How many stereoisomers are obtained from the reaction of 2-pentanone with methylmagnesium bromide followed by the addition of dilute acid?

Reactions of Esters and Acyl Chlorides with Grignard Reagents

Grignard reagents also react with esters and acyl chlorides, compounds you studied in Chapter 15. Esters and acyl chlorides undergo two successive reactions with the Grignard reagent. The first reaction is a *nucleophilic acyl substitution reaction* because an ester or an acyl chloride, unlike an aldehyde or a ketone, has a group that can be replaced by the alkyl fragment of the Grignard reagent (Section 15.5). The second reaction is a *nucleophilic addition reaction*.

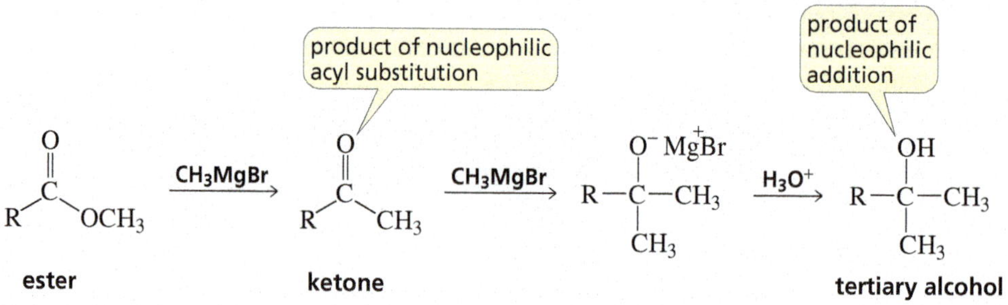

product of nucleophilic acyl substitution

product of nucleophilic addition

The product of the reaction of an ester with a Grignard reagent is a tertiary alcohol. Because the tertiary alcohol is formed from two successive reactions with the Grignard reagent, the alcohol has at least two identical alkyl groups bonded to the tertiary carbon.

MECHANISM FOR THE REACTION OF AN ESTER WITH A GRIGNARD REAGENT

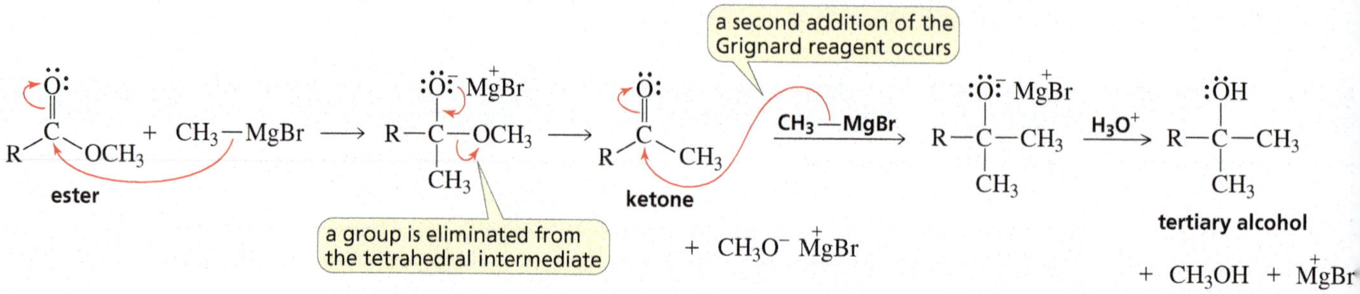

a second addition of the Grignard reagent occurs

a group is eliminated from the tetrahedral intermediate

- Nucleophilic addition of the Grignard reagent to the carbonyl carbon forms a tetrahedral intermediate that is unstable because it has a group that can be eliminated.
- The tetrahedral intermediate eliminates methoxide ion, forming a ketone.
- The ketone reacts with a second molecule of the Grignard reagent, forming an alkoxide ion that forms a tertiary alcohol when it is protonated.

Tertiary alcohols are also formed from the reaction of two equivalents of a Grignard reagent with an acyl chloride (or with an acid anhydride). The mechanism for the reaction is the same as the mechanism for the reaction of a Grignard reagent with an ester.

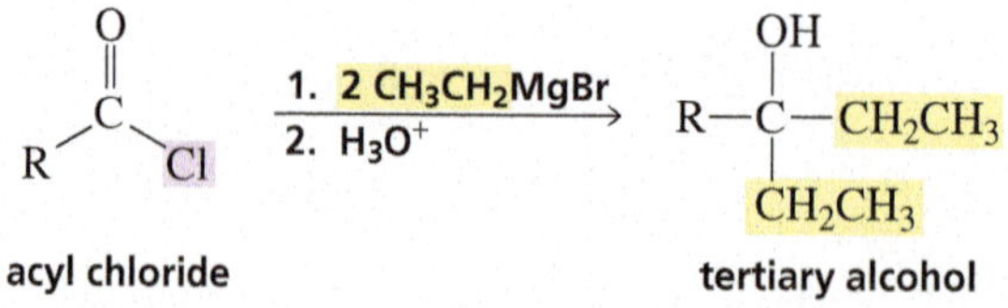

An acyl chloride forms a ketone if an organocuprate is used instead of a Grignard reagent. (Recall that an organocuprate replaces a halogen with an alkyl group.)

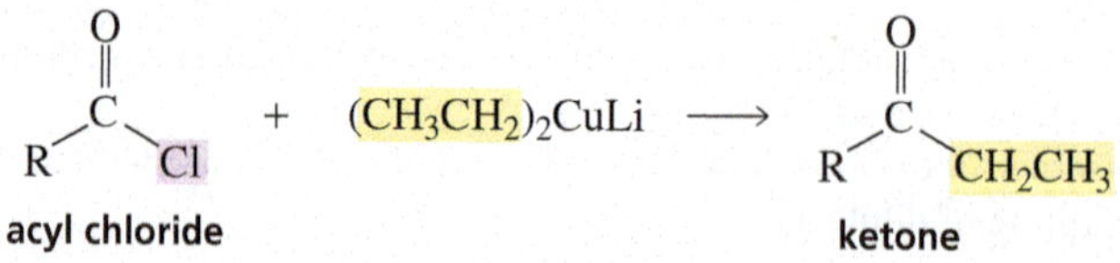

LEARN THE STRATEGY

PROBLEM 8 SOLVED

a. Which of the following tertiary alcohols cannot be prepared by the reaction of an ester with excess Grignard reagent?

b. For those alcohols that can be prepared by the reaction of an ester with excess Grignard reagent, what ester and what Grignard reagent should be used?

OH
|
A CH₃CCH₂CH₃
|
CH₃

OH
|
C CH₃CH₂CCH₂CH₂CH₃
|
CH₃

OH
|
E CH₃CCH₂CH₂CH₂CH₃
|
CH₂CH₃

OH
|
B CH₃CCH₃
|
CH₃

OH
|
D CH₃CH₂CCH₂CH₃
|
CH₃

OH
|
F ⬡—C—⬡
|
CH₃

SOLUTION TO 8 a. A tertiary alcohol prepared by the reaction of an ester with two equivalents of a Grignard reagent must have at least two identical substituents bonded to the carbon to which the OH is attached because two of the three substituents come from the Grignard reagent. Alcohols C and E do not have two identical substituents, so they cannot be prepared in this way.

SOLUTION TO 8 b. (A) An ester of propanoic acid and excess methylmagnesium bromide.

PROBLEM 9 ◆

USE THE STRATEGY

Which of the following secondary alcohols can be prepared by the reaction of methyl formate with excess Grignard reagent?

CH₃CH₂CHCH₃
|
OH

A

CH₃CHCH₃
|
OH

B

CH₃CHCH₂CH₂CH₃
|
OH

C

CH₃CH₂CHCH₂CH₃
|
OH

D

PROBLEM 10

Write the mechanism for the reaction of acetyl chloride with two equivalents of ethylmagnesium bromide.

PROBLEM-SOLVING STRATEGY

Predicting the Products of a Reaction with a Grignard Reagent

LEARN THE STRATEGY

Why doesn't a Grignard reagent add to the carbonyl carbon of a carboxylic acid?

We know that Grignard reagents add to carbonyl carbons, so if we find that a Grignard reagent does not add to the carbonyl carbon, we can conclude that it must react more rapidly with another part of the molecule. A carboxylic acid has an acidic proton that reacts rapidly with the Grignard reagent, converting the Grignard reagent to an alkane.

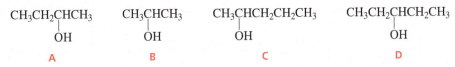

PROBLEM 11 ◆

USE THE STRATEGY

Which of the following compounds does not form an alcohol when it reacts with excess Grignard reagent?

O
‖
CH₃CH₂—C—NHCH₃

A

O
‖
CH₃CH₂—C—OCH₃

B

O
‖
HOCH₂CH₂—C—OCH₃

C

Retrosynthetic Analysis

LEARN THE STRATEGY

We saw that a Grignard reagent reacts with a carbonyl compound to form an alcohol with more carbons than those in either of the reactants. This allows us to use retrosynthetic analysis to determine how to synthesize an alcohol or any compound, such as a ketone, that can be formed from an alcohol. For example, let's see how 3-hexanone can be synthesized from 1-propanol and no other carbon-containing reagents.

3-Hexanone can be synthesized from 3-hexanol, which can be synthesized from a three-carbon aldehyde and a three-carbon Grignard reagent. Oxidation of 1-propanol forms the three-carbon aldehyde, and conversion of 1-propanol to a propyl halide allows the three-carbon Grignard reagent to be synthesized.

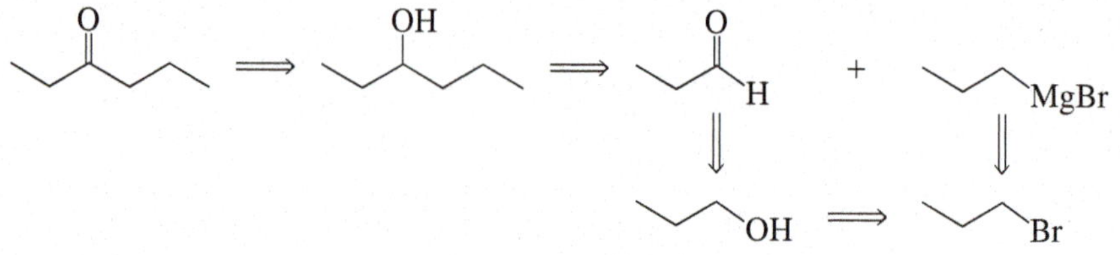

Now we can write the reaction in the forward direction, showing the reagents needed for each step of the synthesis. Note that acetic acid protonates the alkoxide ion and the resulting alcohol is oxidized to a ketone.

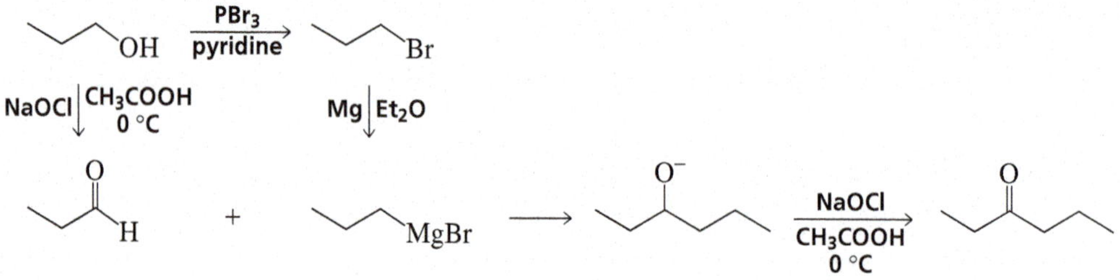

USE THE STRATEGY

PROBLEM 12

Show how the following compounds can be synthesized from cyclohexanol.

a.

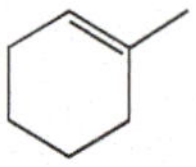

b.

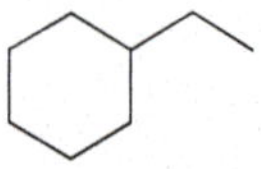

c.

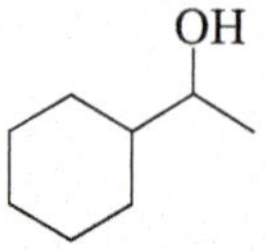

Reaction with Acetylide Ions

We saw that a terminal alkyne can be converted to an acetylide ion by a strong base (Section 7.11).

$$CH_3C\equiv CH \xrightarrow{\text{NaNH}_2} CH_3C\equiv C:^-$$

An acetylide ion is another example of a carbon nucleophile that reacts with an aldehyde or a ketone to form a nucleophilic addition product. When the reaction is over, a weak acid (one that does not react with the triple bond, such as the pyridinium ion shown here) is added to the reaction mixture to protonate the alkoxide ion forming an α-hydroxy alkyne.

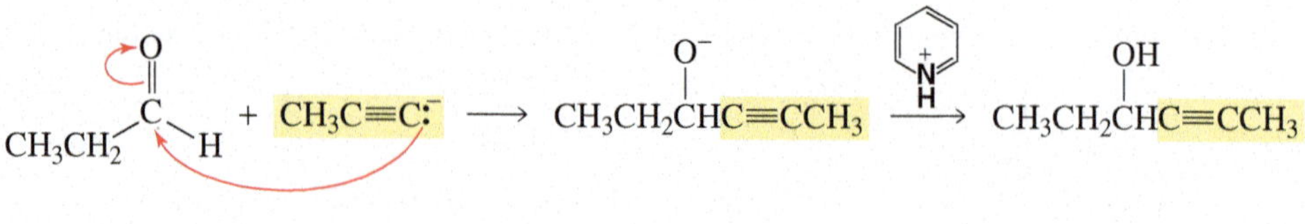

PROBLEM 13

a. Show how the following compounds can be prepared, using ethyne as one of the starting materials:

1. 1-pentyn-3-ol   **2.** 1-phenyl-2-butyn-1-ol   **3.** 2-methyl-3-hexyn-2-ol

b. Explain why ethyne should be alkylated before, rather than after, nucleophilic addition.

PROBLEM 14 ♦

What is the product of the reaction of an ester with excess acetylide ion followed by the addition of pyridinium chloride?

Reaction with Cyanide Ion

Cyanide ion is another carbon nucleophile that can add to an aldehyde or a ketone. The product of the reaction is a **cyanohydrin.** Unlike the addition reactions of other carbon nucleophiles, the addition of cyanide ion has to be carried out under acidic conditions. (The reason is explained below.) Excess cyanide ion is used to ensure that HCl does not convert all the cyanide ion to HCN, so some cyanide is available to act as a nucleophile.

a cyanohydrin

MECHANISM FOR THE REACTION OF AN ALDEHYDE OR KETONE WITH HYDROGEN CYANIDE

- Cyanide ion adds to the carbonyl carbon.
- The alkoxide ion is protonated by hydrogen cyanide.

Compared with Grignard reagents and acetylide ions, cyanide ion is a relatively weak base. The pK_a of CH_3CH_3 is > 60 and the pK_a of $RC \equiv CH$ is 25, but the pK_a of $HC \equiv N$ is 9.14. (Recall that the stronger the acid, the weaker the conjugate base.) Therefore, the cyano group, unlike the R^- or $RC \equiv C^-$ groups, can be eliminated from the addition product.

Cyanohydrins, however, are stable. The OH group will not eliminate the cyano group because the transition state for the elimination reaction would have a partial *positive* charge on the oxygen, making the transition state relatively unstable.

If, however, the OH group loses its proton, then the cyano group will be eliminated because the oxygen atom would then have a partial *negative* charge in the transition state. Therefore, in basic solutions, a cyanohydrin is converted back to the carbonyl compound.

Cyanide ion does not react with esters because the cyanide ion is a weaker base than an alkoxide ion, so the cyanide ion would be eliminated from the tetrahedral intermediate. (Recall that the incoming nucleophile cannot be a weaker base than the substituent attached to the acyl group; Section 15.5).

PROBLEM 15 ♦

In the mechanism for cyanohydrin formation, why is HCN the acid that protonates the alkoxide ion instead of HCl?

PROBLEM 16 ♦

Can a cyanohydrin be prepared by treating a ketone with sodium cyanide?

PROBLEM 17 ♦

Explain why aldehydes and ketones react with a weak acid such as hydrogen cyanide but do not react with strong acids such as HCl or H_2SO_4 (other than being protonated by them).

Using Cyanohydrins in Synthesis

The addition of hydrogen cyanide to aldehydes and ketones is a synthetically useful reaction because of the subsequent reactions that can be carried out on the cyanohydrin. For example, the acid-catalyzed hydrolysis of a cyanohydrin forms an α-hydroxycarboxylic acid (Section 15.15).

$$
\underset{\substack{\text{a cyanohydrin}}}{
R-\overset{\overset{\displaystyle OH}{|}}{\underset{\underset{\displaystyle R}{|}}{C}}-C\equiv N}
\quad\xrightarrow[\Delta]{HCl,\ H_2O}\quad
\underset{\substack{\text{an } \alpha\text{-hydroxycarboxylic acid}}}{
R-\overset{\overset{\displaystyle OH}{|}}{\underset{\underset{\displaystyle R}{|}}{C}}-\overset{\overset{\displaystyle O}{\|}}{C}OH}
$$

The catalytic addition of two moles of hydrogen to the triple bond of a cyanohydrin produces a primary amine with an OH group on the β-carbon. (Raney nickel, rather than palladium, is the catalyst used to reduce a nitrile group.)

$$
R-\overset{\overset{\displaystyle OH}{|}}{CH}-C\equiv N
\quad\xrightarrow[\text{Raney nickel}]{H_2}\quad
R-\overset{\overset{\displaystyle OH}{|}}{CH}-CH_2NH_2
$$

LEARN THE STRATEGY

PROBLEM 18 SOLVED

How can the following compounds be prepared from a carbonyl compound that has one less carbon than the desired product?

a. $HOCH_2CH_2NH_2$

b. $CH_3\overset{\overset{\displaystyle O}{\|}}{\underset{\underset{\displaystyle OH}{|}}{CH}}\overset{C}{}OH$

SOLUTION TO 18 a. The starting material for the synthesis of this two-carbon compound must have one carbon. Therefore, it must be formaldehyde. Addition of hydrogen cyanide followed by addition of H_2 to the triple bond of the cyanohydrin forms the target molecule.

$$
\underset{H}{\overset{O}{\underset{\|}{\overset{\|}{C}}}}\underset{H}{}
\quad\xrightarrow[HCl]{NaC\equiv N}\quad
HOCH_2C\equiv N
\quad\xrightarrow[\text{Raney nickel}]{H_2}\quad
HOCH_2CH_2NH_2
$$

SOLUTION TO 18 b. The starting material for the synthesis of this three-carbon compound must have two carbons. Therefore, it must be acetaldehyde. Addition of hydrogen cyanide, followed by hydrolysis of the resulting cyanohydrin, forms the target molecule.

$$
CH_3\overset{O}{\overset{\|}{C}}H
\quad\xrightarrow[HCl]{NaC\equiv N}\quad
CH_3\overset{}{\underset{\underset{\displaystyle OH}{|}}{CH}}C\equiv N
\quad\xrightarrow[\Delta]{HCl,\ H_2O}\quad
CH_3\overset{}{\underset{\underset{\displaystyle OH}{|}}{CH}}\overset{\overset{\displaystyle O}{\|}}{C}OH
$$

USE THE STRATEGY

PROBLEM 19

Show two ways to convert an alkyl halide into a carboxylic acid that has one more carbon than the alkyl halide.

16.5 REACTIONS OF CARBONYL COMPOUNDS WITH HYDRIDE ION

Reactions of Aldehydes and Ketones with Hydride Ion

A hydride ion is another strong base and good nucleophile that reacts with an aldehyde or a ketone to form a nucleophilic addition product. Usually, sodium borohydride ($NaBH_4$) is used for the source of the hydride ion.

$$CH_3CH_2CH_2 \overset{O}{\underset{}{\overset{\|}{C}}} H \xrightarrow[\text{2. } H_3O^+]{\text{1. } NaBH_4} CH_3CH_2CH_2CH_2OH$$

aldehyde   **primary alcohol**

$$CH_3CH_2CH_2 \overset{O}{\underset{}{\overset{\|}{C}}} CH_3 \xrightarrow[\text{2. } H_3O^+]{\text{1. } NaBH_4} CH_3CH_2CH_2\overset{OH}{\underset{|}{C}}HCH_3$$

ketone   **secondary alcohol**

Recall that the addition of hydrogen to a compound is a **reduction reaction** (Section 5.9). Aldehydes are reduced to primary alcohols, and ketones are reduced to secondary alcohols. Notice that the acid is not added to the reaction mixture until after the hydride ion has reacted with the carbonyl compound.

MECHANISM FOR THE REACTION OF AN ALDEHYDE OR A KETONE WITH HYDRIDE ION

- Addition of a hydride ion to the carbonyl carbon of an aldehyde or a ketone forms an alkoxide ion.
- Protonation with a dilute acid forms an alcohol.

PROBLEM 20 ♦

What alcohols are obtained from the reduction of the following compounds with sodium borohydride?

a. 2-methylpropanal   **b.** cyclohexanone   **c.** 4-*tert*-butylcyclohexanone   **d.** acetophenone

Reaction of an Acyl Chloride with Hydride Ion

Because an acyl chloride has a group that can be replaced by another group, it undergoes two successive reactions with hydride ion, just as it undergoes two successive reactions with a Grignard reagent (Section 16.4). Therefore, the reaction of an acyl chloride with sodium borohydride forms a primary alcohol with the same number of carbons as the acyl chloride.

$$CH_3CH_2CH_2 \overset{O}{\underset{}{\overset{\|}{C}}} Cl \xrightarrow[\text{2. } H_3O^+]{\text{1. } 2\ NaBH_4} CH_3CH_2CH_2CH_2OH$$

acyl chloride   **primary alcohol**

MECHANISM FOR THE REACTION OF AN ACYL CHLORIDE WITH HYDRIDE ION

- The acyl chloride undergoes a nucleophilic acyl substitution reaction because it has a group (Cl^-) that can be replaced by hydride ion. The product of this reaction is an aldehyde.
- The aldehyde undergoes a nucleophilic addition reaction with a second equivalent of hydride ion, forming an alkoxide ion.
- Protonation of the alkoxide ion forms a primary alcohol.

$LiAlH_4$ is a stronger reducing agent than $NaBH_4$. However, replacing three of the hydrogens on $LiAlH_4$ with alkoxy (OR) groups makes it a weaker reducing agent than $NaBH_4$. For example, lithium tri-*tert*-butoxyaluminum hydride reduces an acyl chloride only as far as the aldehyde (recall that an acyl chloride is more reactive than an aldehyde), whereas $NaBH_4$ reduces the acyl chloride all the way to an alcohol.

$$\underset{\text{acyl chloride}}{\overset{\displaystyle O}{\underset{R}{\overset{\|}{C}}{-}Cl}} \xrightarrow[\text{2. } H_2O]{\text{1. } LiAl[OC(CH_3)_3]_3H,\ -78\ ^\circ C} \underset{\text{aldehyde}}{\overset{\displaystyle O}{\underset{R}{\overset{\|}{C}}{-}H}}$$

Reaction of an Ester with Hydride Ion

Esters, carboxylic acids, and amides are less reactive than aldehydes and ketones. Therefore, esters, carboxylic acids, and amides must be reduced with lithium aluminum hydride ($LiAlH_4$), a stronger reducing agent than $NaBH_4$. Lithium aluminum hydride (also abbreviated as LAH) is not as safe or as easy to use as sodium borohydride. It reacts violently with protic solvents, so it must be used in a dry, aprotic solvent, and it is never used if $NaBH_4$ can be used instead.

The reaction of an ester with $LiAlH_4$ produces two alcohols, one corresponding to the acyl portion of the ester and one corresponding to the alkyl portion.

> **Acyl chlorides and esters undergo two successive reactions with hydride ion and with Grignard reagents.**

$$\underset{\text{ester}}{\overset{\displaystyle O}{\underset{CH_3CH_2}{\overset{\|}{C}}{-}OCH_3}} \xrightarrow[\text{2. } H_3O^+]{\text{1. } 2\ LiAlH_4} \underset{\text{primary alcohol}}{CH_3CH_2CH_2OH} + CH_3OH$$

MECHANISM FOR THE REACTION OF AN ESTER WITH HYDRIDE ION

$$\underset{\text{ester}}{\overset{\displaystyle \overset{..}{\underset{..}{O}}:}{\underset{R}{\overset{\|}{C}}{-}OCH_3}} + H{-}\bar{A}lH_3 \longrightarrow R{-}\underset{\underset{\text{a group is eliminated}}{H}}{\overset{\overset{\displaystyle :\overset{..}{\underset{..}{O}}:{-}}{|}}{C}}{-}OCH_3 \longrightarrow \underset{\underset{+\ CH_3O^-}{\text{aldehyde}}}{\overset{\displaystyle \overset{..}{\underset{..}{O}}:}{\underset{R}{\overset{\|}{C}}{-}H}} \xrightarrow{H{-}\bar{A}lH_3} RCH_2\overset{..}{\underset{..}{O}}:^- \xrightarrow{H_3O^+} \underset{\underset{\text{alcohol}}{\text{primary}}}{RCH_2OH} + CH_3OH$$

product of nucleophilic acyl substitution / **product of nucleophilic addition**

- The ester undergoes a nucleophilic acyl substitution reaction because an ester has a group (CH_3O^-) that can be replaced by hydride ion. The product of this reaction is an aldehyde.
- The aldehyde undergoes a nucleophilic addition reaction with a second equivalent of hydride ion, forming an alkoxide ion.
- Protonation of the two alkoxide ions forms two alcohols.

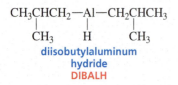

CH_3CHCH_2—Al—CH_2CHCH_3
| | |
CH_3   H   CH_3
diisobutylaluminum hydride
DIBALH

The reaction of an ester with hydride ion cannot be stopped at the aldehyde because an aldehyde is more reactive than an ester toward nucleophilic addition (Section 16.2). However, if diisobutylaluminum hydride (DIBALH) is used as the hydride donor at a low temperature, then the reaction can be stopped at the aldehyde. DIBALH, therefore, makes it possible to convert esters into aldehydes.

$$CH_3CH_2CH_2-\overset{\overset{\displaystyle O}{\|}}{C}-OCH_3 \quad \xrightarrow[\text{2. H}_2\text{O}]{\text{1. [(CH}_3\text{)}_2\text{CHCH}_2]_2\text{AlH, }-78\text{ °C}} \quad CH_3CH_2CH_2-\overset{\overset{\displaystyle O}{\|}}{C}-H \;+\; CH_3OH$$

ester   **aldehyde**

The reaction is carried out at −78 °C (the temperature of a dry ice/acetone bath). At this temperature, the initially formed tetrahedral intermediate is stable, so it does not eliminate the alkoxide ion. If all of the unreacted hydride donor is removed before the solution warms up, then there will be no hydride ion available to react with the aldehyde that is formed when the tetrahedral intermediate eliminates the alkoxide ion.

Reaction of a Carboxylic Acid with Hydride Ion

The reaction of a carboxylic acid with hydride ion forms a primary alcohol with the same number of carbons as the carboxylic acid.

$$CH_3CH_2-\overset{\overset{\displaystyle O}{\|}}{C}-OH \quad \xrightarrow[\text{2. H}_3\text{O}^+]{\text{1. LiAlH}_4} \quad CH_3CH_2CH_2OH$$

propanoic acid   **1-propanol**
carboxylic acid   **primary alcohol**

The reaction of a carboxylic acid with $LiAlH_4$ forms a primary alcohol.

MECHANISM FOR THE REACTION OF A CARBOXYLIC ACID WITH HYDRIDE ION

- A hydride ion reacts with the acidic hydrogen of the carboxylic acid, because this reaction is faster than the addition of hydride ion to the carbonyl carbon. The products of the reaction are H_2, a carboxylate ion and AlH_3.
- The nucleophilic carboxylate ion reacts with the electrophilic AlH_3, which creates a new hydride donor.
- Addition of hydride ion to the carbonyl carbon, followed by elimination from the unstable tetrahedral intermediate forms an aldehyde.
- Addition of hydride ion to the aldehyde forms the primary alcohol.

PROBLEM 21 ◆

What products are obtained from the reaction of the following compounds with $LiAlH_4$ followed by treatment with dilute acid?

a. ethyl butanoate   **c.** methyl benzoate
b. benzoic acid   **d.** pentanoic acid

Reaction of an Amide with Hydride Ion

Amides also undergo two successive additions of hydride ion when they react with $LiAlH_4$. Overall, the reaction converts a carbonyl group into a methylene (CH_2) group, so the product of the reaction is an amine. Primary, secondary, or tertiary amines can be formed, depending on the number of substituents bonded to the nitrogen of the amide. (Notice that H_2O rather than H_3O^+ is used in the second step of the reaction. If H_3O^+ is used, the product will be an ammonium ion rather than an amine.)

The mechanism for the reaction shows why the product of the reaction is an amine. Take a minute to note the similarities between this mechanism and the mechanism for the reaction of hydride ion with a carboxylic acid.

MECHANISM FOR THE REACTION OF AN *N*-SUBSTITUTED AMIDE WITH HYDRIDE ION

- A hydride ion removes the acidic hydrogen from the nitrogen of the amide, and the electrons left behind are delocalized onto oxygen.
- An electrophile (AlH_3) accepts a pair of electrons from the anion, forming a new hydride donor.
- Addition of hydride ion, elimination from the unstable tetrahedral intermediate, and a second addition of hydride ion followed by protonation forms the amine.

The mechanisms for the reaction of $LiAlH_4$ with unsubstituted and *N,N*-disubstituted amides are somewhat different, but have the same result—the conversion of a carbonyl group to a methylene group.

Biological reactions also need reagents that can deliver hydride ions to carbonyl groups. The most common hydride donor in cells is NADPH. (Sodium borohydride and lithium aluminum hydride are too reactive to be used in cells.) This hydride donor is discussed in Section 24.1.

PROBLEM 22 ◆

What amides would you react with LiAlH₄ to form the following amines?

a. benzylmethylamine

c. diethylamine

b. ethylamine

d. triethylamine

PROBLEM 23

How would you make the following compounds from *N*-benzylbenzamide?

a. dibenzylamine   **b.** benzoic acid   **c.** benzyl alcohol

---

16.6   MORE ABOUT REDUCTION REACTIONS

An organic compound is reduced when hydrogen (H₂) is added to it. A molecule of H₂ can be thought of as being composed of (1) a hydride ion and a proton, (2) two hydrogen atoms, or (3) two electrons and two protons. Remember that the product of a reduction reaction has more C—H bonds than the reactant.

components of H:H

H:⁻ H⁺   H· ·H   ·⁻ H⁺ ·⁻ H⁺

(a hydride ion and a proton)   (two hydrogen atoms)   (two electrons and two protons)

Reduction by Addition of a Hydride Ion and a Proton

When aldehydes and ketones are reduced to alcohols by NaBH₄, we saw that the reduction occurs as a result of the addition of a hydride ion followed by the addition of a proton.

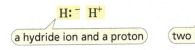 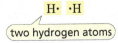 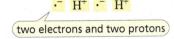

Imines and nitriles can also be reduced by NaBH₄, because the carbon attached to the nitrogen has a partial positive charge.

$$RCH{=}NCH_3 \xrightarrow[\text{2. H}_2\text{O}]{\text{1. NaBH}_4} RCH_2NHCH_3$$

$$RC{\equiv}N \xrightarrow[\text{2. H}_2\text{O}]{\text{1. NaBH}_4} RCH_2NH_2$$

However, NaBH₄ cannot reduce the double and triple bonds of alkenes and alkynes. These compounds are nucleophiles (they do not possess a partial charge) and, therefore, will not react with the nucleophilic hydride ion.

$$RCH{=}CH_2 \xrightarrow{\text{NaBH}_4} \text{no reaction}$$

$$RC{\equiv}CR \xrightarrow{\text{NaBH}_4} \text{no reaction}$$

Reduction by Addition of Two Hydrogen Atoms

Recall that hydrogen can be added to carbon–carbon double and triple bonds in the presence of a metal catalyst (Sections 5.9 and 7.9). In these reactions, called **catalytic hydrogenations,** the H—H bond breaks homolytically, so reduction results from the addition of two hydrogen atoms to the reactant.

$$CH_3CH_2CH{=}CH_2 + H_2 \xrightarrow{\text{Pd/C}} CH_3CH_2CH_2CH_3$$

$$CH_3CH_2CH_2C{\equiv}CH + 2 H_2 \xrightarrow{\text{Pd/C}} CH_3CH_2CH_2CH_2CH_3$$

Catalytic hydrogenation can also be used to reduce carbon–nitrogen double and triple bonds. The reaction products are amines (Sections 15.15 and 16.6).

$$CH_3CH_2CH{=}NCH_3 + H_2 \xrightarrow{\text{Pd/C}} CH_3CH_2CH_2NHCH_3$$

$$CH_3CH_2CH_2C{\equiv}N + 2 H_2 \xrightarrow{\text{Raney nickel}} CH_3CH_2CH_2CH_2NH_2$$

The C=O group of ketones and aldehydes can also be reduced by catalytic hydrogenation. A palladium catalyst is not very effective at reducing these carbonyl groups (or at reducing nitrile groups; see the previous reaction). However, they are reduced easily over a nickel catalyst. (Raney nickel is finely dispersed nickel with adsorbed hydrogen, so an external source of H_2 is not needed. However, H_2 is shown in the reactions to remind you that it is the reducing agent.) Aldehydes are reduced to primary alcohols, and ketones are reduced to secondary alcohols. (Raney nickel also reduces carbon-carbon double and triple bonds.)

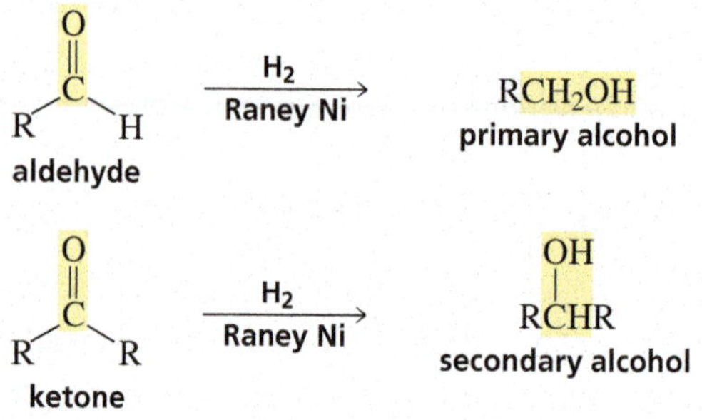

The C=O group of carboxylic acids, esters, and amides are less reactive than the C=O group of aldehydes and ketones and are, therefore, harder to reduce (Section 16.5). They cannot be reduced by catalytic hydrogenation (except under extreme conditions).

PROBLEM 24 ◆

What are the products of the following reactions?

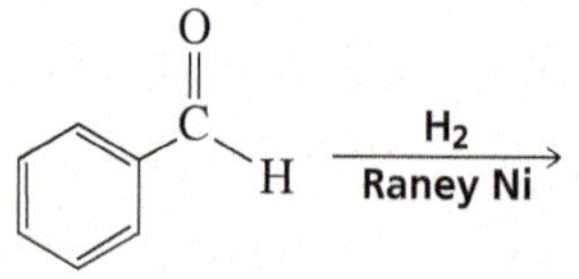

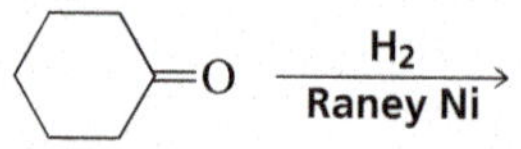

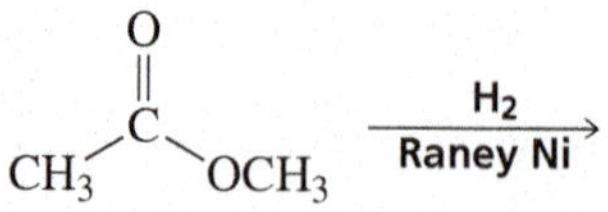

Reduction by Addition of an Electron, a Proton, an Electron, a Proton

We saw that an alkyne can be reduced to a trans alkene using sodium in liquid ammonia (Section 7.9). In this reaction, called a **dissolving metal reduction,** sodium donates an electron to the alkyne and ammonia donates a proton. This sequence is then repeated, so the overall reaction adds an electron, a proton, an electron, and a proton to the alkyne. The mechanism is on page 303.

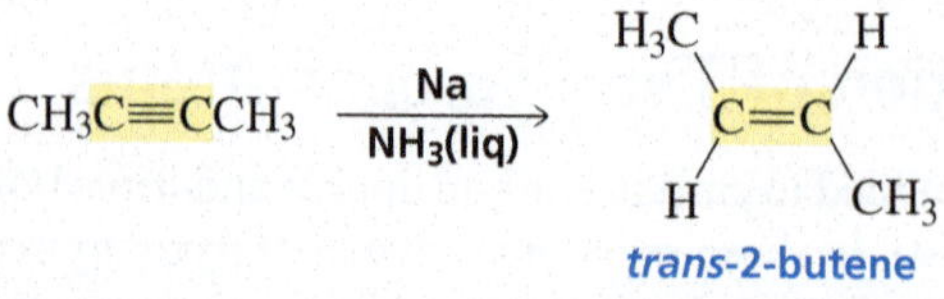

trans-2-butene

16.7 CHEMOSELECTIVE REACTIONS

A **chemoselective reaction** is a reaction in which a reagent reacts with one functional group in preference to another. For example, because sodium borohydride cannot reduce an ester, an amide, or a carboxylic acid, it can be used to selectively reduce an aldehyde or a ketone in a compound that also contains a less reactive carbonyl group. Water, not aqueous acid, is used in the second step of the reaction to avoid hydrolyzing the ester.

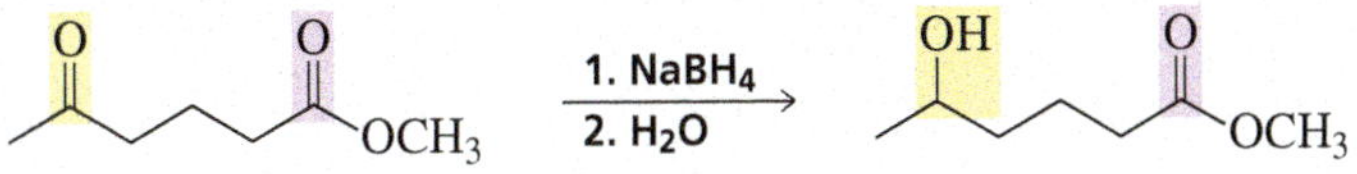
O
O
OCH3
1. NaBH4
2. H2O
OH
O
OCH3

Sodium in liquid ammonia can reduce a carbon–carbon triple bond, but *not* a carbon–carbon double bond. This reagent is, therefore, useful for reducing a triple bond in a compound that also contains a double bond.

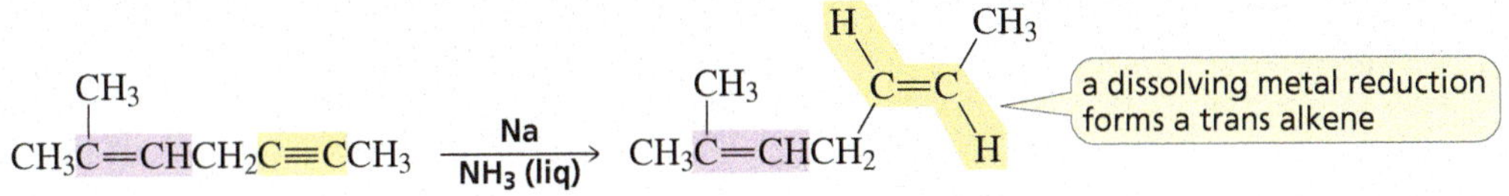
CH3
CH3C=CHCH2C≡CCH3
Na
NH3 (liq)
CH3
CH3C=CHCH2
H
CH3
C=C
H
a dissolving metal reduction forms a trans alkene

Reducing reagents that are hydride ion donors, such as sodium borohydride, cannot reduce carbon–carbon double bonds or carbon–carbon triple bonds, because both the hydride ion and the double or triple bond are nucleophiles. Therefore, a carbonyl group in a compound that also has an alkene functional group can be selectively reduced. Water, not aqueous acid, is used in the second step of the reaction to avoid addition of the acid to the double bond.

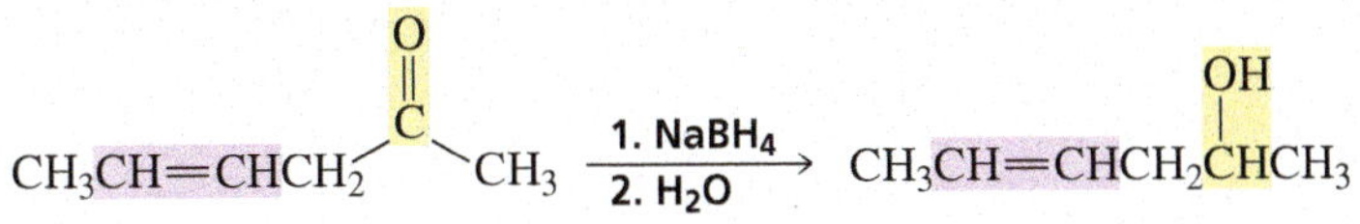
O
C
CH3CH=CHCH2
CH3
1. NaBH4
2. H2O
OH
CH3CH=CHCH2CHCH3

PROBLEM 25

What reducing agents should be used to obtain the desired target molecules?

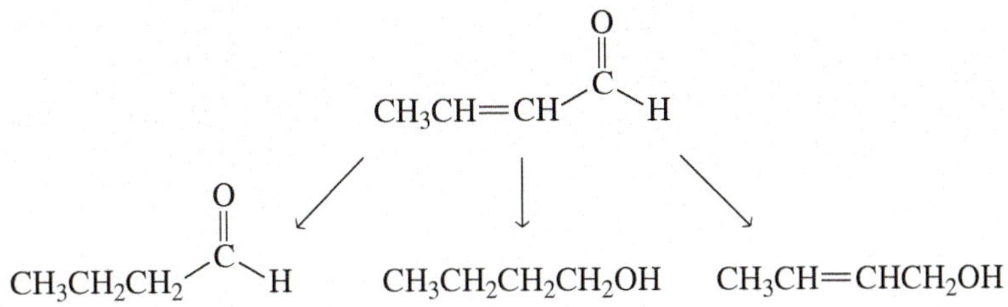
O
C
CH3CH=CH
H
O
C
CH3CH2CH2
H
CH3CH2CH2CH2OH
CH3CH=CHCH2OH

PROBLEM 26

What are the products of the following reactions?

a. 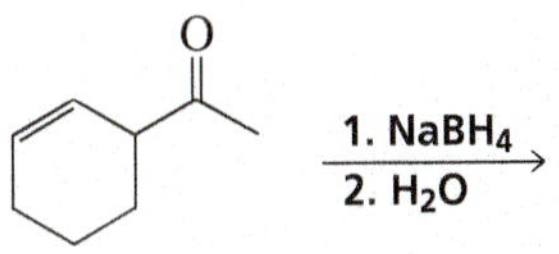
O
1. NaBH4
2. H2O

c. 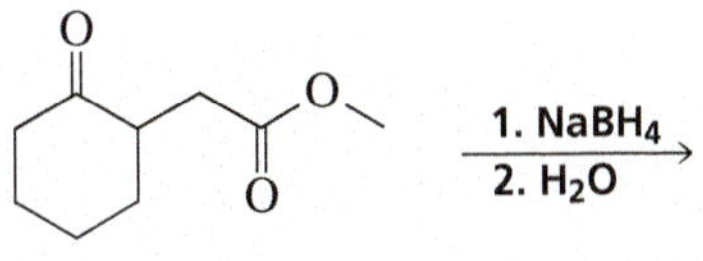
O
O
O
1. NaBH4
2. H2O

b. 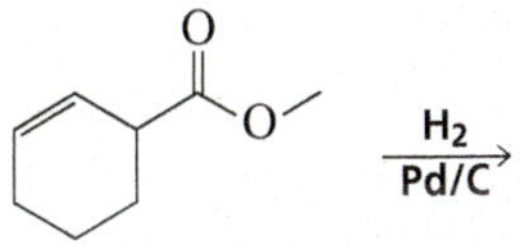
O
O
H2
Pd/C

d. 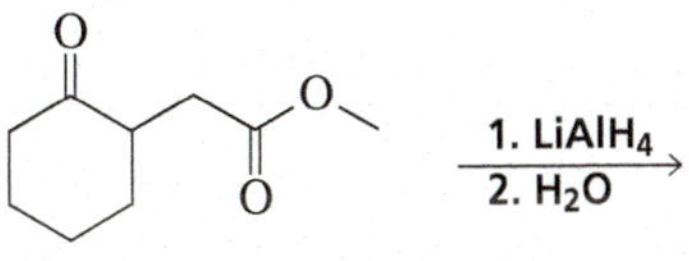
O
O
O
1. LiAlH4
2. H2O

16.8 REACTIONS OF ALDEHYDES AND KETONES WITH NITROGEN NUCLEOPHILES

Reactions with Primary Amines

An aldehyde or a ketone reacts with a *primary* amine to form an imine (sometimes called a **Schiff base**). An **imine** is a compound with a carbon–nitrogen double bond. The reaction requires a trace amount of acid. Notice that imine formation replaces a C=O with a C=NR.

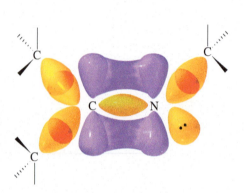

▲ **Figure 16.1**
Bonding in an imine. The π bond is formed by side-to-side overlap of a p orbital of carbon with a p orbital of nitrogen; it is perpendicular to the orange orbitals.

$$
\underset{\text{an aldehyde}}{\overset{\text{R}}{\underset{\text{H}}{\text{C}}}=\text{O}} \;+\; \underset{\text{a primary amine}}{\text{CH}_3\text{CH}_2\text{NH}_2} \underset{\text{acid}}{\overset{\text{trace}}{\rightleftharpoons}} \underset{\text{an imine}}{\overset{\text{R}}{\underset{\text{H}}{\text{C}}}=\text{NCH}_2\text{CH}_3} \;+\; \text{H}_2\text{O}
$$

$$
\underset{\text{a ketone}}{\overset{\text{R}}{\underset{\text{R}}{\text{C}}}=\text{O}} \;+\; \underset{\text{a primary amine}}{\bigcirc\!\!-\text{CH}_2\text{NH}_2} \underset{\text{acid}}{\overset{\text{trace}}{\rightleftharpoons}} \underset{\text{an imine}}{\overset{\text{R}}{\underset{\text{R}}{\text{C}}}=\text{NCH}_2\!-\!\bigcirc} \;+\; \text{H}_2\text{O}
$$

A C=N group (Figure 16.1) is similar to a C=O group (Figure 15.1 on page 692). The imine nitrogen is sp^2 hybridized. One of its sp^2 orbitals forms a σ bond with the imine carbon, one forms a σ bond with a substituent, and the third sp^2 orbital contains a lone pair. The p orbital of nitrogen and the p orbital of carbon overlap to form a π bond.

Forming Imine Derivatives

Compounds such as hydroxylamine and hydrazine, like primary amines, have an NH_2 group. Thus, like primary amines, they react with aldehydes and ketones to form imines—called *imine derivatives* because the substituent attached to the imine nitrogen is not an R group. The imine derivative obtained from the reaction with hydroxylamine is called an **oxime,** and the imine derivative obtained from the reaction with hydrazine is called a **hydrazone.**

Imine formation replaces C=O with C=NR.

$$
\underset{\text{R}}{\overset{\text{R}}{\text{C}}}=\text{O} \;+\; \underset{\text{hydroxylamine}}{\text{H}_2\text{NOH}} \underset{\text{acid}}{\overset{\text{trace}}{\rightleftharpoons}} \underset{\text{an oxime}}{\overset{\text{R}}{\underset{\text{R}}{\text{C}}}=\text{NOH}} \;+\; \text{H}_2\text{O}
$$

$$
\underset{\text{R}}{\overset{\text{R}}{\text{C}}}=\text{O} \;+\; \underset{\text{hydrazine}}{\text{H}_2\text{NNH}_2} \underset{\text{acid}}{\overset{\text{trace}}{\rightleftharpoons}} \underset{\text{a hydrazone}}{\overset{\text{R}}{\underset{\text{R}}{\text{C}}}=\text{NNH}_2} \;+\; \text{H}_2\text{O}
$$

The Mechanism

The mechanism for imine formation is shown on the top of the next page. Because the addition of the amine is followed by elimination of water, the reaction is a **nucleophilic addition-elimination** reaction. We will see that the pH of the reaction mixture must be carefully controlled. (Recall that HB^+ represents any species in the solution that is capable of donating a proton, and :B represents any species in the solution that is capable of removing a proton.)

MECHANISM FOR IMINE FORMATION

nucleophile adds to the carbonyl carbon

N–protonated carbinolamine

neutral tetrahedral intermediate
a carbinolamine

O–protonated carbinolamine

elimination of water

imine

HB+

protonated imine

+ H₂O

removal of a proton

NOTE TO THE STUDENT

- Notice that the pattern of three tetrahedral intermediates that we saw in the acid-catalyzed mechanisms in Chapter 15 also occurs in this mechanism:

protonated tetrahedral intermediate ⇌ neutral tetrahedral intermediate ⇌ protonated tetrahedral intermediate.

- The amine adds to the carbonyl carbon.
- Protonation of the alkoxide ion and deprotonation of the ammonium ion form a neutral tetrahedral intermediate, called a *carbinolamine*.
- The carbinolamine is in equilibrium with two protonated forms because either its oxygen (forward step) or its nitrogen (reverse step) can be protonated.
- Because the nucleophile has a lone pair, water is eliminated from the oxygen-protonated intermediate, thereby forming a protonated imine.
- A base removes a proton from the nitrogen to form the imine.

Recall that a stable tetrahedral compound is formed when a Grignard reagent or a hydride ion adds to an aldehyde or a ketone. This is not true for the tetrahedral compound formed when an amine adds to an aldehyde or a ketone. This tetrahedral intermediate is *unstable* because it contains a group with a lone pair that can eliminate water when the OH group is protonated.

Aldehydes and ketones react with primary amines to form imines.

$$R-\underset{\underset{H}{|}}{\overset{\overset{OH}{|}}{C}}-R \qquad R-\underset{\underset{CH_3}{|}}{\overset{\overset{OH}{|}}{C}}-R$$

stable tetrahedral compounds

$$R-\underset{\underset{:NHCH_3}{|}}{\overset{\overset{\overset{H}{|}}{+OH}}{C}}-R$$

unstable tetrahedral compound

Imine formation is reversible because there are two protonated tetrahedral intermediates that can eliminate a group. The nitrogen of the carbinolamine is more basic than the oxygen, so the equilibrium favors the nitrogen-protonated tetrahedral intermediate. However, the equilibrium can be forced toward the oxygen-protonated tetrahedral intermediate and, therefore, toward the imine by removing water as it is formed.

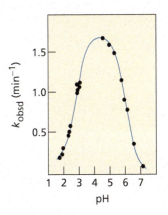

▲ **Figure 16.2**
A pH-rate profile for the reaction of acetone with hydroxylamine. It shows the dependence of the reaction rate on the pH of the reaction mixture.

Controlling the pH

A plot of the observed rate constant for the reaction of acetone with hydroxylamine to form an imine as a function of the pH of the reaction mixture is shown in Figure 16.2.

This type of plot is called a **pH–rate profile.** For this reaction, the maximum rate occurs at about pH 4.5, which is 1.5 pH units below the pK_a of protonated hydroxylamine ($pK_a = 6.0$). As the acidity increases below pH 4.5, the rate of the reaction decreases because more and more of the amine becomes protonated. As a result, less and less of the amine is present in the nucleophilic nonprotonated form that is necessary for the first step of the reaction. As the acidity decreases above pH 4.5, the rate decreases because less and less of the tetrahedral intermediate is present in the reactive oxygen-protonated form.

Therefore, the pH at which imine formation is carried out must be carefully controlled. There must be sufficient acid present to protonate the oxygen atom of the carbinolamine so that H_2O rather than the much more basic HO^- can be the leaving group. This is denoted in the reactions by "trace acid." If no acid is added to the solution, the pH of the solution will be greater than the pK_a value of the protonated amine. Thus, some acid must be added to allow the reaction to be carried out at about 1.5 pH units below the pK_a value of the protonated amine—the maximum rate indicated by the pH-rate profile. Most protonated amines have pK_a values of about 10.5, in which case trace acid must be added so the reaction can be carried out at pH = 9.

If too much acid is present, all of the amine will be protonated. Protonated amines are not nucleophiles, so they cannot react with carbonyl groups. Therefore, unlike the acid-catalyzed reactions you saw in Chapter 15, there is not enough acid present to protonate the carbonyl group in the first step of the reaction (see Problem 30).

Imine Hydrolysis

An imine can be hydrolyzed back to the carbonyl compound and the amine in an acidic solution. This reaction is irreversible because the amine is protonated in the acidic solution, so it is unable to react with the carbonyl compound to re-form the imine.

An imine undergoes acid-catalyzed hydrolysis to form a carbonyl compound and a primary amine.

$$\overset{R}{\underset{R}{\diagdown}}C=NCH_2CH_3 + H_2O \xrightarrow{HCl} \overset{R}{\underset{R}{\diagdown}}C=O + CH_3CH_2\overset{+}{N}H_3$$

Imine formation and hydrolysis are important reactions in biological systems. For example, we will see that all the reactions that require vitamin B_6 involve imine formation (Section 24.5), and imine hydrolysis is the reason that DNA contains T nucleotides instead of U nucleotides (Section 26.10).

PROBLEM 27

A ketone can be prepared from the reaction of a nitrile with a Grignard reagent. Describe the intermediate formed in this reaction, and show how it can be converted to a ketone.

PROBLEM 28 ◆

Why is the pK_a value of protonated hydroxylamine (6.0) so much lower than the pK_a value of a protonated primary amine such as protonated methylamine (10.7)?

PROBLEM 29 ◆

At what pH should imine formation be carried out if the amine's protonated form has a pK_a value of 9.0?

PROBLEM 30 ◆

The pK_a of protonated acetone is about -7.5, and the pK_a of protonated hydroxylamine is 6.0.

a. In a reaction with hydroxylamine at pH 4.5 (Figure 16.2), what fraction of acetone is present in its acidic, protonated form? (*Hint:* See Section 2.10.)

b. In a reaction with hydroxylamine at pH 1.5, what fraction of acetone is present in its acidic, protonated form?

c. In a reaction with acetone at pH 1.5 (Figure 16.2), what fraction of hydroxylamine is present in its reactive basic form?

PROBLEM 31

Imines can exist as stereoisomers. The isomers are named using the E,Z system of nomenclature (Section 4.2). The lone pair has the lowest priority.

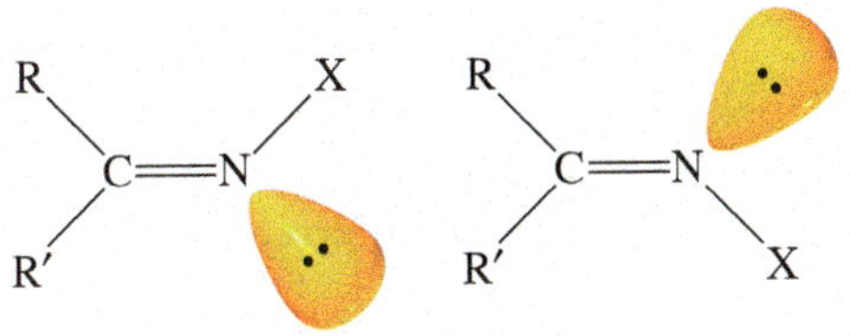

Draw the structure of each of the following compounds:

a. the (*E*)-hydrazone of benzaldehyde **b.** the (*Z*)-oxime of propiophenone

Reactions with Secondary Amines

An aldehyde or a ketone reacts with a *secondary amine* to form an enamine (pronounced "ENE-amine"). The name comes from joining "ene" and "amine," with the second "e" in "ene" omitted in order to avoid two successive vowels.

An **enamine** is an α,β-unsaturated tertiary amine—that is, a tertiary amine with a double bond in the α,β-position relative to the nitrogen. Notice that the double bond is in the part of the molecule that is provided by the aldehyde or ketone, not in the part that is provided by the secondary amine. Like imine formation, the reaction requires a trace amount of an acid catalyst.

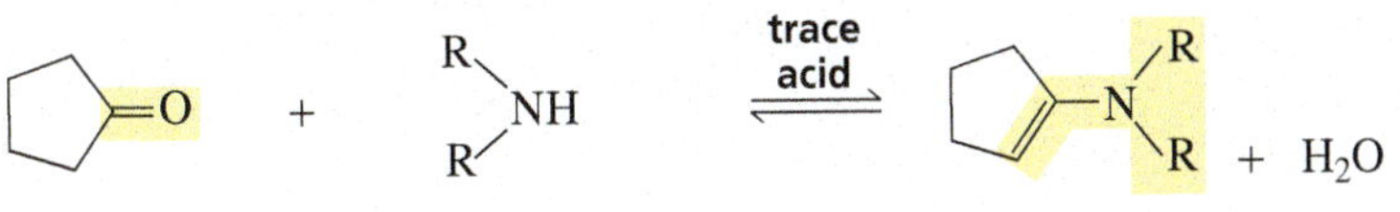

a secondary amine an enamine

Aldehydes and ketones react with secondary amines to form enamines.

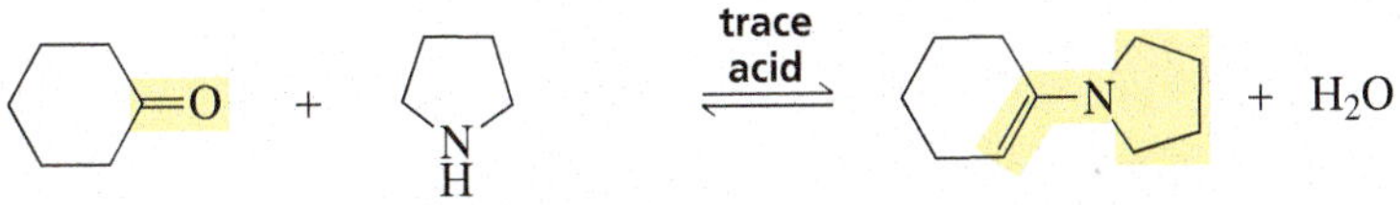

a secondary amine an enamine

The Mechanism

Notice that the mechanism for enamine formation is exactly the same as that for imine formation, except for the site from which a proton is lost in the last step.

MECHANISM FOR ENAMINE FORMATION

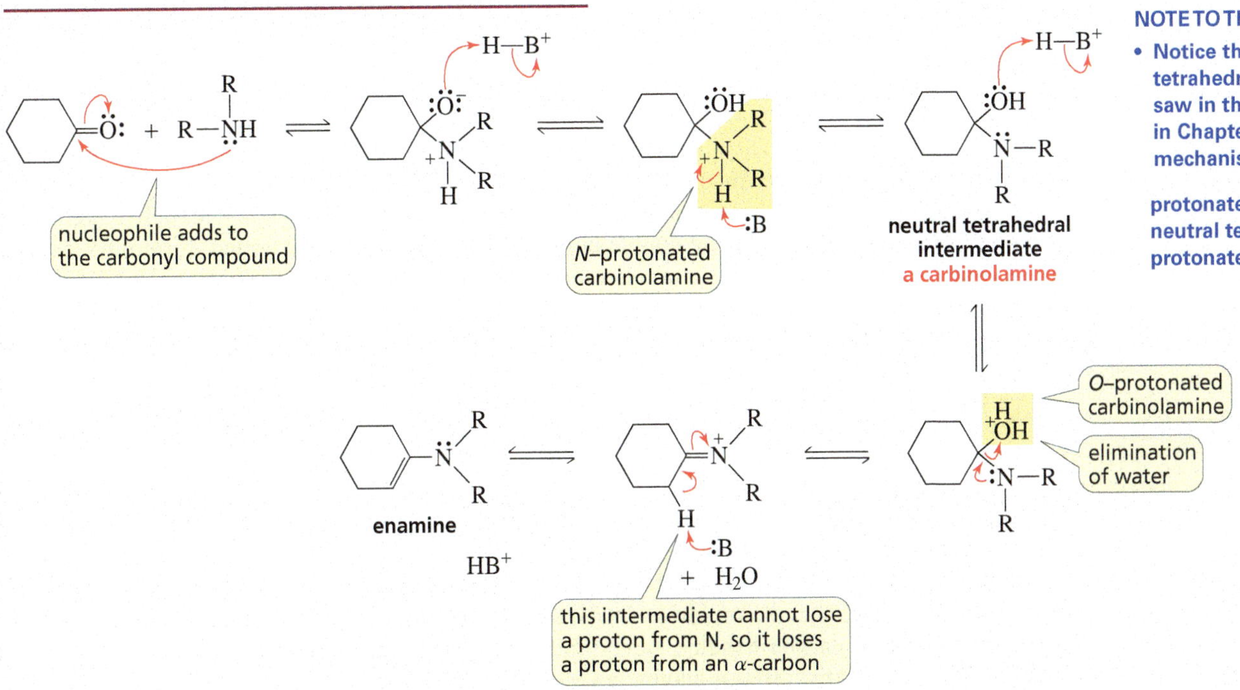

NOTE TO THE STUDENT

- **Notice that the pattern of three tetrahedral intermediates that we saw in the acid-catalyzed mechanisms in Chapter 15 also occurs in this mechanism:**

protonated tetrahedral intermediate ⇌ neutral tetrahedral intermediate ⇌ protonated tetrahedral intermediate.

- The amine adds to the carbonyl carbon.
- Protonation of the alkoxide ion and deprotonation of the ammonium ion forms a neutral tetrahedral intermediate (a carbinolamine).
- The carbinolamine is in equilibrium with two protonated forms because either its oxygen or its nitrogen can be protonated.
- Because the nucleophile has a lone pair, water is eliminated from the oxygen-protonated carbinolamine, thereby forming a compound with a positively charged nitrogen.
- When a primary amine reacts with an aldehyde or a ketone, a proton is removed from the positively charged nitrogen in the last step of the mechanism, forming a neutral imine. When the amine is secondary, however, the positively charged nitrogen is not bonded to a hydrogen. In this case, a stable neutral molecule can be obtained only by removing a proton from the α-carbon of the compound provided by the carbonyl compound. An enamine is the result.

As with imine formation, because the nitrogen of the carbinolamine is more basic than the oxygen atom, water must be removed as it is formed in order to force the equilibrium toward the enamine.

Enamine Hydrolysis

In an aqueous acidic solution, an enamine is hydrolyzed back to the carbonyl compound and secondary amine. This reaction is similar to the acid-catalyzed hydrolysis of an imine back to the carbonyl compound and primary amine (page 762).

An enamine undergoes acid-catalyzed hydrolysis to form a carbonyl compound and a secondary amine.

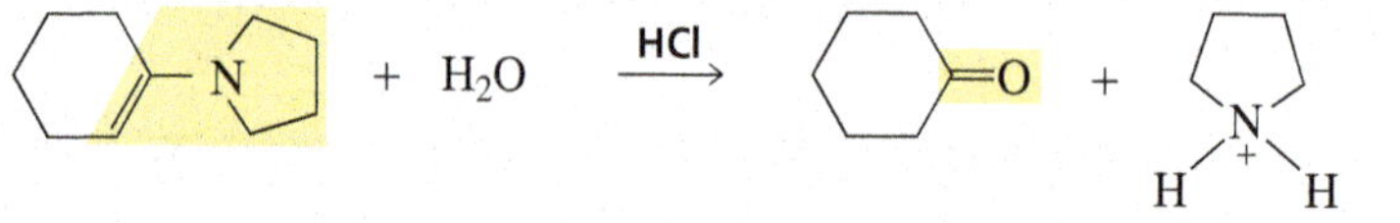

PROBLEM 32

a. Write the mechanism for the following reactions:

  1. the acid-catalyzed hydrolysis of an imine to a carbonyl compound and a primary amine
  2. the acid-catalyzed hydrolysis of an enamine to a carbonyl compound and a secondary amine

b. How do the two mechanisms differ?

PROBLEM 33

What are the products of the following reactions? (A trace amount of acid is present in each case.)

a. cyclopentanone + ethylamine   c. acetophenone + hexylamine
b. cyclopentanone + diethylamine   d. acetophenone + cyclohexylamine

---

Reductive Amination

The imine formed from the reaction of an aldehyde or a ketone with ammonia is relatively unstable because it does not have a substituent other than a hydrogen attached to the nitrogen. Nevertheless, such an imine is a useful intermediate.

For example, if the reaction with ammonia is carried out in the presence of a reducing agent such as H_2 and a metal catalyst, then the double bond will be reduced as it is formed, forming a primary amine. The reaction of an aldehyde or a ketone with excess ammonia in the presence of a reducing agent is called **reductive amination.**

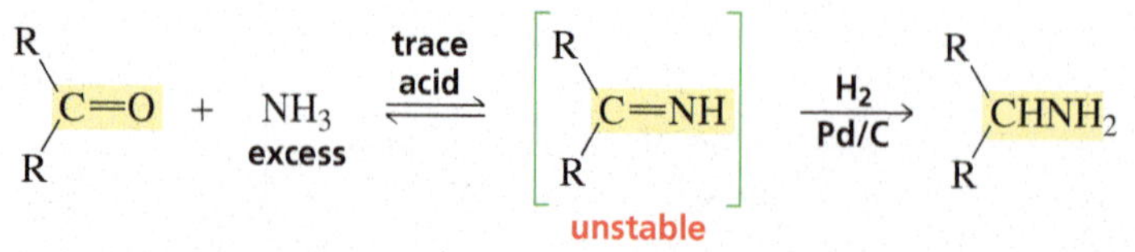

The double bond of an imine or enamine is reduced more rapidly than a $C=O$ bond, so reduction of the carbonyl group does not compete with reduction of the imine in these reactions.

Secondary and tertiary amines can be prepared from imines and enamines by reducing the imine or enamine. Sodium cyanoborohydride ($NaBH_3CN$) is a commonly used reducing agent for these reactions because it can be handled easily and it is stable even in acidic solutions. (Notice that $NaBH_3CN$ differs from $NaBH_4$ in having a $C\equiv N$ group in place of one of the hydrogens.)

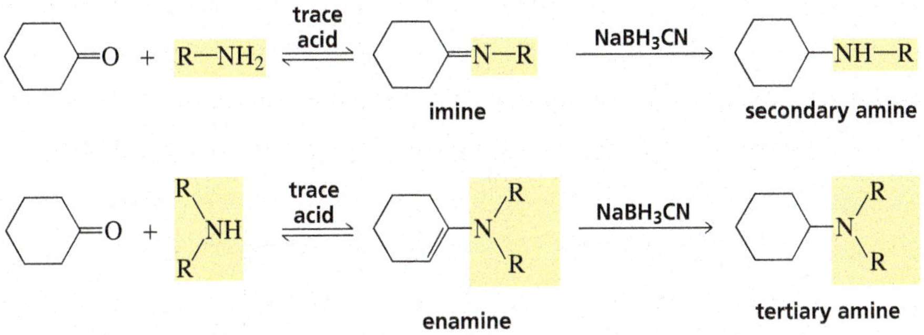

PROBLEM 34 ◆

Excess ammonia must be used when a primary amine is synthesized by reductive amination. What product will be obtained if the reaction is carried out with excess carbonyl compound?

PROBLEM 35

The compounds commonly known as "amino acids" are actually α-aminocarboxylic acids (Section 21.0). What carbonyl compounds should be used to synthesize the two amino acids shown here?

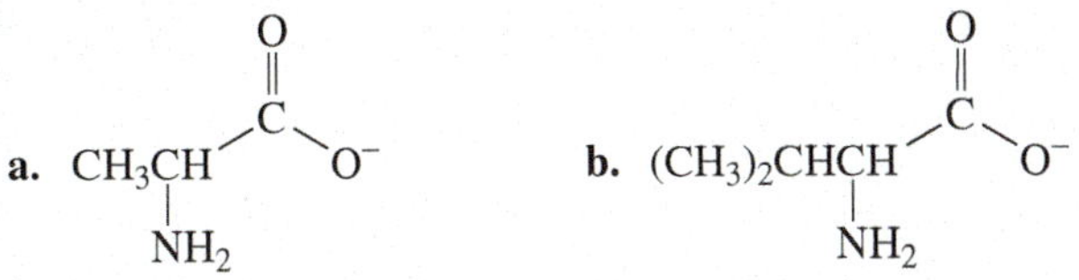

Serendipity in Drug Development

Many drugs have been discovered accidentally. Librium, a tranquilizer, is one example of such a drug. Leo Sternbach, a research chemist at Hoffmann-LaRoche, synthesized a series of quinazoline 3-oxides, but none of them showed any pharmacological activity. One of the compounds was not submitted for testing because it was not the quinazoline 3-oxide Sternbach had set out to synthesize. Two years after the project was abandoned, a laboratory worker came across this compound while cleaning up the lab, and Sternbach decided that he might as well submit it for testing before it was thrown away. The compound was found to have tranquilizing properties and, when its structure was investigated, was discovered to be a benzodiazepine 4-oxide.

Methylamine, instead of displacing the chloro substituent in an S_N2 reaction to form a quinazoline 3-oxide, had added to the imine group of the six-membered ring. This caused the ring to open and then reclose to form a benzodiazepine. The compound was given the brand name Librium when it was put into clinical use in 1960.

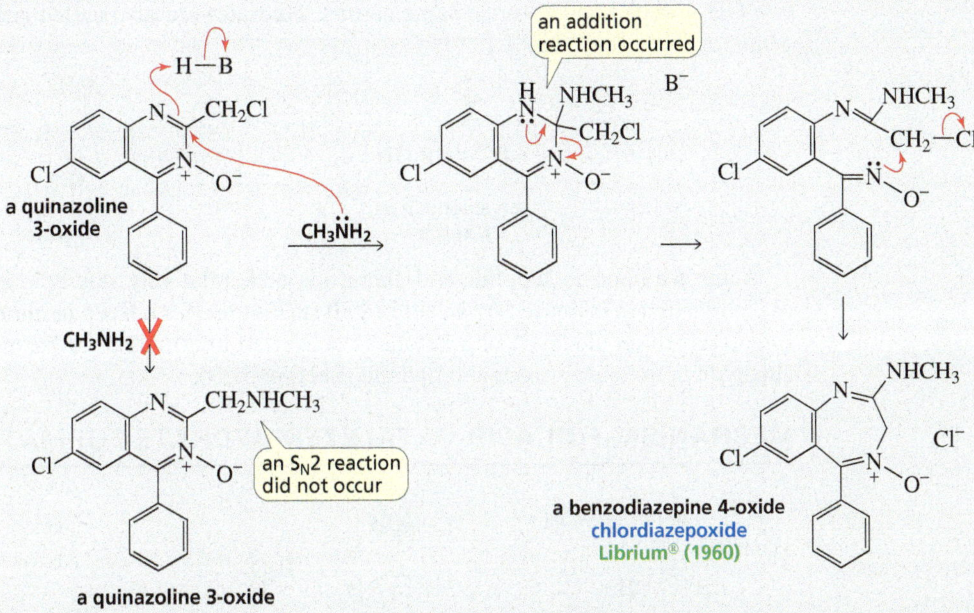

Librium was structurally modified in an attempt to find other tranquilizers (Section 10.9). One successful modification produced Valium, a tranquilizer almost 10 times more potent than Librium. Currently, there are 8 benzodiazepines in clinical use as tranquilizers in the United States and some 15 others abroad. Rohypnol is one of the so-called date-rape drugs.

(Continued)

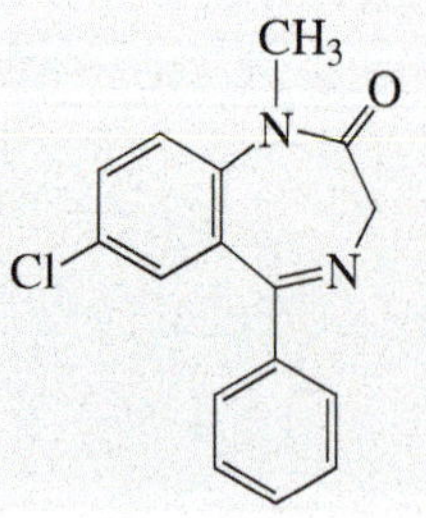

diazepam
Valium® (1963)

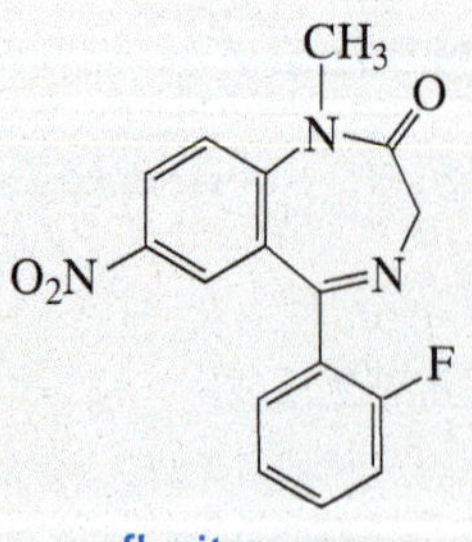

flunitrazepam
Rohypnol® (1963)

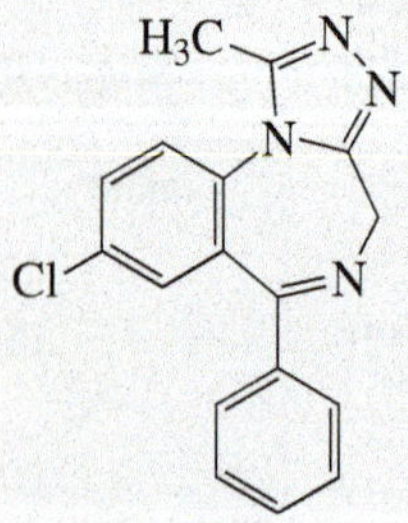

alprazolam
Xanax® (1970)

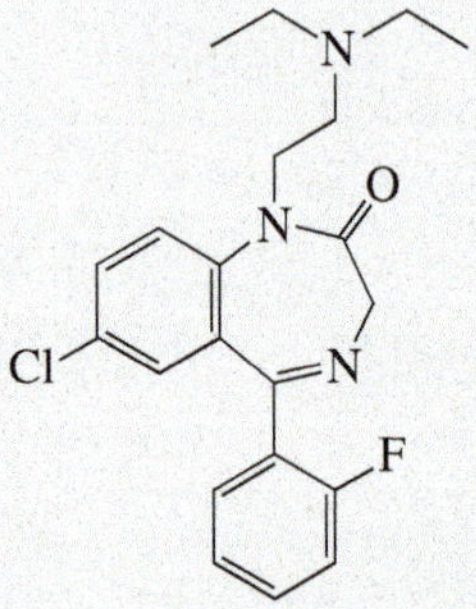

flurazepam
Dalmane® (1970)

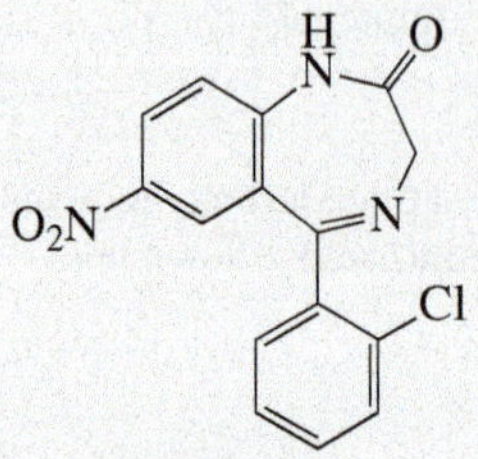

clonazepam
Klonopin® (1975)

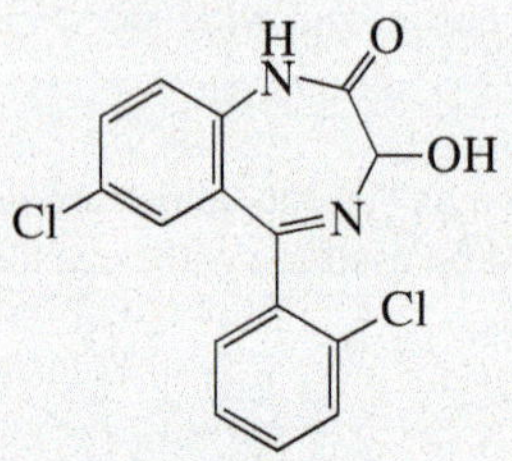

lorazepam
Ativan® (1977)

Viagra is a recent example of a drug that was discovered accidentally. Viagra was in clinical trials as a drug for heart ailments. The clinical trials were canceled when Viagra was found to be ineffective as a heart drug. However, those enrolled in the trials refused to return the remaining tablets. The pharmaceutical company then realized that the drug had other marketable effects.

16.9 REACTIONS OF ALDEHYDES AND KETONES WITH OXYGEN NUCLEOPHILES

Reaction with Water

The addition of water to an aldehyde or a ketone forms a *hydrate*. A **hydrate** is a molecule with two OH groups bonded to the same carbon. Hydrates are also called **gem-diols** (*gem* comes from *geminus*, Latin for "twin").

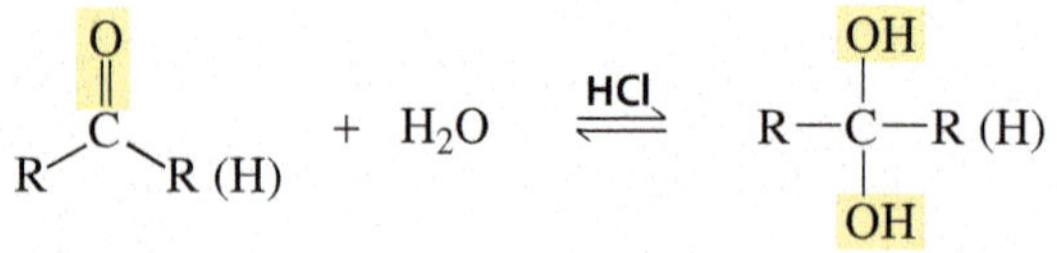

**an aldehyde or
a ketone**

**a *gem*-diol
a hydrate**

Water is a poor nucleophile and, therefore, adds relatively slowly to a carbonyl group. The rate of the reaction is increased by an acid catalyst (Figure 16.3). Keep in mind that a catalyst affects the *rate* at which an aldehyde or a ketone is converted to a hydrate; it has no effect on the *amount* of aldehyde or ketone converted to hydrate (Section 5.13).

MECHANISM FOR ACID-CATALYZED HYDRATE FORMATION

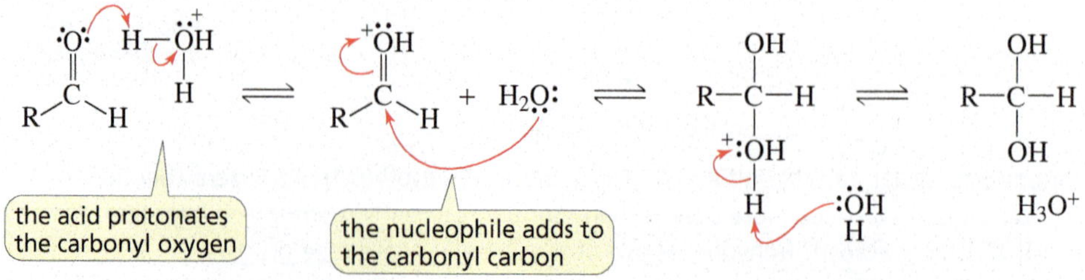

the acid protonates
the carbonyl oxygen

the nucleophile adds to
the carbonyl carbon

- The acid protonates the carbonyl oxygen, which makes the carbonyl carbon more susceptible to nucleophilic addition (Figure 16.3).
- Water adds to the carbonyl carbon.
- Removal of a proton from the protonated tetrahedral intermediate forms the hydrate.

PROBLEM 36

Hydration of an aldehyde is also catalyzed by hydroxide ion. Propose a mechanism for the reaction.

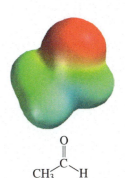

How Much Aldehyde or Ketone Is Hydrated?

The extent to which an aldehyde or a ketone is hydrated in an aqueous solution depends on the substituents attached to the carbonyl group. For example, only 0.2% of acetone is hydrated at equilibrium, but 99.9% of formaldehyde is hydrated.

▲ **Figure 16.3**
The electrostatic potential maps show that the carbonyl carbon of the protonated aldehyde is more electrophilic (the blue is more intense) than the carbonyl carbon of the unprotonated aldehyde.

$$K_{eq}$$

acetone
99.8%   + H_2O ⇌ $CH_3-\overset{OH}{\underset{OH}{\overset{|}{\underset{|}{C}}}}-CH_3$ 2×10^{-3}
0.2%

acetaldehyde
42%   + H_2O ⇌ $CH_3-\overset{OH}{\underset{OH}{\overset{|}{\underset{|}{C}}}}-H$ 1.4
58%

formaldehyde
0.1%   + H_2O ⇌ $H-\overset{OH}{\underset{OH}{\overset{|}{\underset{|}{C}}}}-H$ 2.3×10^3
99.9%

Why is there such a difference in the extent of hydration? We know that the equilibrium constant for a reaction depends on the relative stabilities of the reactants and products (Section 5.6). The equilibrium constant for hydrate formation, which is a measure of the extent of hydration, depends, therefore, on the relative stabilities of the carbonyl compound and the hydrate.

$$K_{eq} = \frac{[\textbf{products}]}{[\textbf{reactants}]} = \frac{[\textbf{hydrate}]}{[\textbf{carbonyl compound}][\textbf{H}_2\textbf{O}]}$$

We saw that electron-donating alkyl groups make a carbonyl compound *more stable* (less reactive) (Section 16.2).

most stable → $CH_3-\overset{O}{\overset{\|}{C}}-CH_3$ > $CH_3-\overset{O}{\overset{\|}{C}}-H$ > $H-\overset{O}{\overset{\|}{C}}-H$ (120°)

In contrast, alkyl groups make the hydrate *less stable* because of steric interactions between the alkyl groups when the bond angles change from 120° to 109.5° (Section 16.2).

least stable → $CH_3-\overset{OH}{\underset{OH}{\overset{|}{\underset{|}{C}}}}-CH_3$ < $CH_3-\overset{OH}{\underset{OH}{\overset{|}{\underset{|}{C}}}}-H$ < $H-\overset{OH}{\underset{OH}{\overset{|}{\underset{|}{C}}}}-H$ (109.5°)

Alkyl groups, therefore, shift the equilibrium to the left (toward reactants) because they stabilize the carbonyl compound (making the denominator larger) and destabilize the hydrate (making the numerator smaller), which makes K_{eq} smaller. As a result, less acetone than formaldehyde is hydrated at equilibrium. The relative stability of hydrates of aldehydes is why "generally" was used in the statement on pages 694 and 745: "*generally, a compound that has an sp^3 carbon bonded to two oxygens or to an oxygen and another electronegative atom is unstable.*"

In summary, the percentage of hydrate present in solution at equilibrium depends on both electronic and steric effects:

LEARN THE STRATEGY

- electron-donating substituents and bulky substituents (such as the methyl groups of acetone) decrease the percentage of hydrate present at equilibrium.
- electron-withdrawing substituents and small substituents (the hydrogens of formaldehyde) increase the percent of hydration present at equilibrium.

USE THE STRATEGY

PROBLEM 37 ◆

Which ketone forms the most hydrate in an aqueous solution?

A

B

PROBLEM 38

When trichloroacetaldehyde is dissolved in water, almost all of it is converted to the hydrate. Chloral hydrate, the product of the reaction, is a sedative that can be lethal. A cocktail laced with it is known—in detective novels, at least—as a "Mickey Finn." Explain why an aqueous solution of trichloroacetaldehyde is almost all hydrate.

$$Cl_3C-\overset{\displaystyle O}{\overset{\|}{C}}-H \;+\; H_2O \;\rightleftharpoons\; Cl_3C-\overset{\displaystyle OH}{\underset{\displaystyle OH}{\overset{|}{\underset{|}{C}}}}-H$$

trichloroacetaldehyde   chloral hydrate

Preserving Biological Specimens

A 37% solution of formaldehyde in water, known as *formalin*, was commonly used to preserve biological specimens. Formaldehyde is an eye and skin irritant, however, so formalin has been replaced in most biology laboratories with other preservatives. One frequently used preparation is a solution of 2–5% phenol in ethanol with added antimicrobial agents.

Proving that the Hydrate Is Formed

If the amount of hydrate formed from the reaction of water with a ketone is too small to detect, how do we know that the reaction has even occurred? We can prove that it occurs by adding the ketone to ^{18}O-labeled water and isolating the ketone after equilibrium has been established. Finding that ^{18}O has been incorporated into the ketone indicates that hydration has occurred.

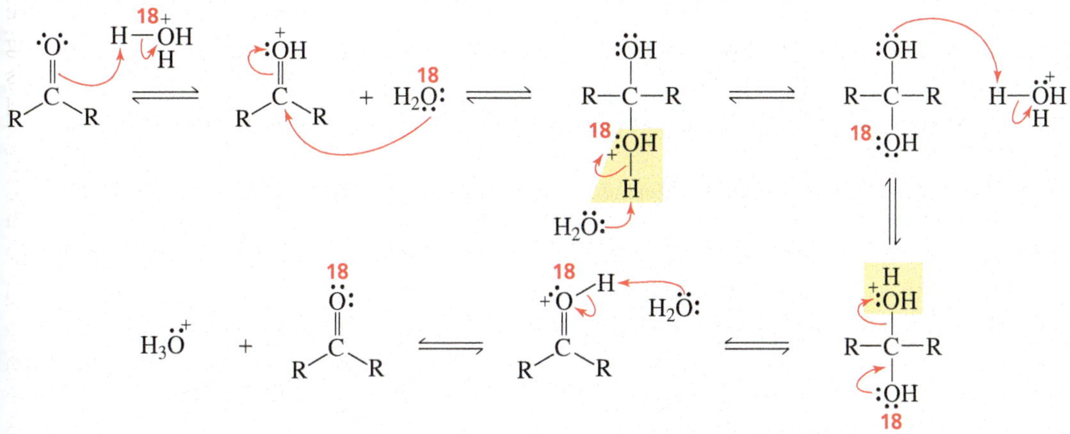

NOTE TO THE STUDENT

- Notice that the pattern of three tetrahedral intermediates occurs again in this mechanism—namely,

  protonated tetrahedral intermediate ⇌ neutral tetrahedral intermediate ⇌ protonated tetrahedral intermediate.

Reaction with Alcohols

The product formed when one equivalent of an alcohol adds to an *aldehyde* or a *ketone* is called a **hemiacetal.** The product formed when a second equivalent of alcohol is added is called an **acetal** (ass-ett-AL). Like water, an alcohol is a poor nucleophile, so an acid catalyst is required for the reaction to take place at a reasonable rate. (Occasionally you see the terms *hemiketal* and *ketal* used instead of *hemiacetal* and *acetal* for the products of the reaction of an alcohol with a ketone.)

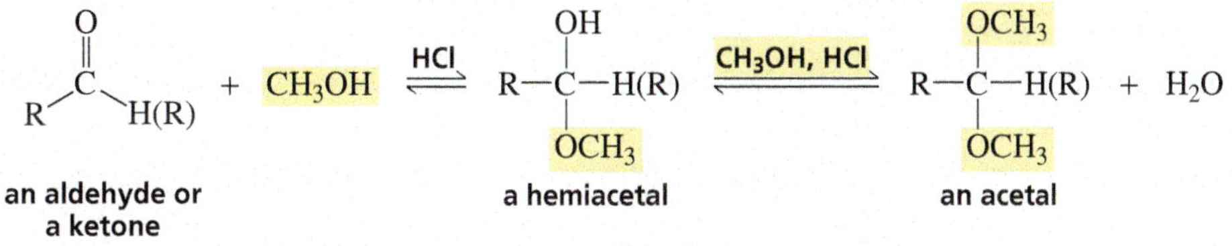

Hemi is the Greek word for "half." When one equivalent of alcohol has added to an aldehyde or a ketone, the compound is halfway to the final acetal, which contains groups from two equivalents of alcohol.

MECHANISM FOR ACID-CATALYZED ACETAL FORMATION

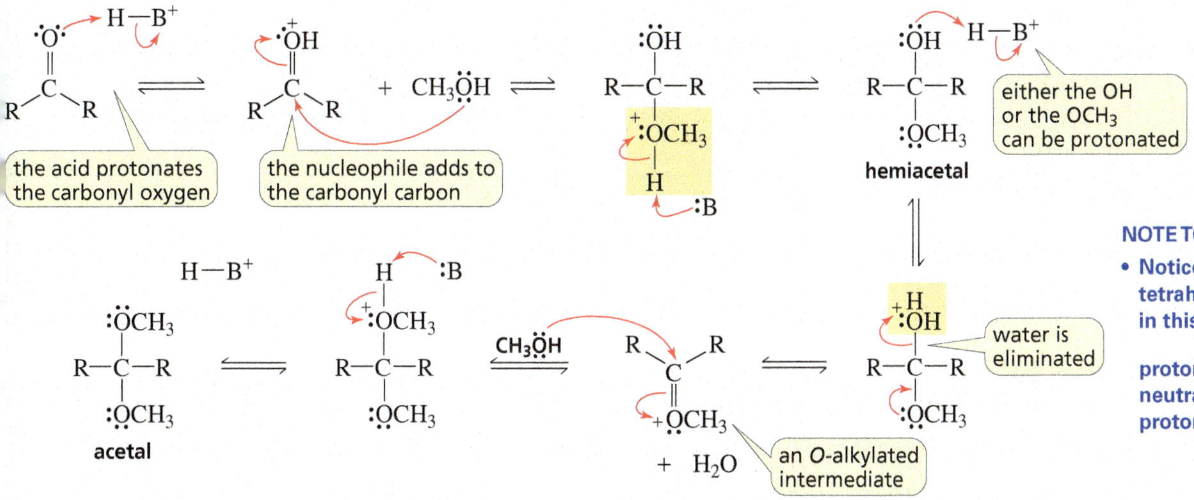

NOTE TO THE STUDENT

- Notice that the pattern of three tetrahedral intermediates occurs again in this mechanism—namely,

  protonated tetrahedral intermediate ⇌ neutral tetrahedral intermediate ⇌ protonated tetrahedral intermediate.

- The acid protonates the carbonyl oxygen, making the carbonyl carbon more susceptible to nucleophilic addition (Figure 16.3).

- The alcohol adds to the carbonyl carbon.

- Loss of a proton from the protonated tetrahedral intermediate forms the neutral tetrahedral intermediate (the hemiacetal).

- The hemiacetal is in equilibrium with its protonated form. The two oxygen atoms of the hemiacetal are equally basic, so either one can be protonated.
- Because the nucleophile has a lone pair, water is eliminated from the second protonated intermediate, thereby forming an *O*-alkylated intermediate that is very reactive because of its positively charged oxygen.
- Nucleophilic addition to this intermediate by a second molecule of alcohol, followed by loss of a proton, forms the acetal.

Although the sp^3 carbon of an acetal is bonded to two oxygens, which suggests that it is not stable, the acetal can be isolated if the water that is eliminated is removed from the reaction mixture. If water is not available, the only compound the acetal can form is the *O*-alkylated intermediate, which is even less stable than the acetal. However, if water is available, water can add to the *O*-alkylated intermediate and the aldehyde (or ketone) will be re-formed.

Therefore, the acetal can be hydrolyzed back to the aldehyde or ketone in an acidic aqueous solution.

$$\underset{\substack{| \\ OCH_3}}{\overset{\substack{OCH_3 \\ |}}{R-C-R}} + H_2O \underset{}{\overset{HCl}{\rightleftharpoons}} \underset{R \quad R}{\overset{O}{C}} + 2\,CH_3OH$$

Carbohydrates

When you study carbohydrates in Chapter 20, you will see that the individual sugar units in a carbohydrate are held together by acetal groups. For example, the reaction of the aldehyde group and an alcohol group of D-glucose forms a cyclic compound that is a hemiacetal. Molecules of the cyclic compound are then hooked together by the reaction of the hemiacetal group of one molecule with an OH group of another, resulting in the formation of an acetal. Hundreds of cyclic glucose molecules hooked together by acetal groups is the major component of both starch and cellulose (Section 20.16).

PROBLEM 39 ◆

Which of the following are

a. hemiacetals?   **b.** acetals?   **c.** hydrates?

1.
$$\underset{\substack{| \\ OCH_3}}{\overset{\substack{OH \\ |}}{CH_3-C-CH_3}}$$

3.
$$\underset{\substack{| \\ OCH_3}}{\overset{\substack{OCH_3 \\ |}}{CH_3-C-H}}$$

5.
$$\underset{\substack{| \\ OCH_3}}{\overset{\substack{OCH_3 \\ |}}{CH_3-C-CH_3}}$$

7.
$$\underset{\substack{| \\ OCH_3}}{\overset{\substack{OH \\ |}}{CH_3-C-H}}$$

2.
$$\underset{\substack{| \\ OCH_2CH_3}}{\overset{\substack{OCH_2CH_3 \\ |}}{CH_3-C-H}}$$

4.
$$\underset{\substack{| \\ OH}}{\overset{\substack{OH \\ |}}{CH_3-C-CH_3}}$$

6.
$$\underset{\substack{| \\ OH}}{\overset{\substack{OH \\ |}}{CH_3-C-H}}$$

8.
$$\underset{\substack{| \\ OCH_3}}{\overset{\substack{OH \\ |}}{CH_3-C-CH_2CH_3}}$$

PROBLEM-SOLVING STRATEGY

Analyzing the Behavior of Acetals

LEARN THE STRATEGY

Why are acetals stable in basic solutions but hydrolyzed back to the aldehyde or ketone in acidic aqueous solutions?

The best way to approach this question is to write out the mechanism that describes the situation to which the question is referring. When the mechanism is written, the answer may become apparent.

In an acidic solution, the acid protonates an oxygen of the acetal. This creates a weak base (CH_3OH) that can be eliminated. When the group is eliminated, water can add to the highly reactive intermediate, and you are on your way back to the ketone (or aldehyde), via the reverse of the mechanism on page 769.

In a basic solution, the CH_3O group of the acetal cannot be protonated. Therefore, a very basic CH_3O^- group has to be eliminated to re-form the ketone (or aldehyde), and the transition state of the elimination reaction would be very unstable because it would have a partial positive charge.

PROBLEM 40

USE THE STRATEGY

a. Would you expect hemiacetals to be stable in basic solutions? Explain your answer.
b. Acetal formation must be catalyzed by an acid. Explain why it cannot be catalyzed by CH_3O^-.
c. Can the rate of hydrate formation be increased by hydroxide ion as well as by acid? Explain.

PROBLEM 41

Explain why an acetal can be isolated but most hydrates cannot be isolated.

Summary of the Reactions of Aldehydes and Ketones with Nitrogen and Oxygen Nucleophiles

Notice that the mechanisms for imine, enamine, hydrate, and acetal formation are similar. The nucleophile in each reaction has a lone pair on its attacking atom. After the nucleophile has added to the carbonyl carbon, water is eliminated from a protonated tetrahedral intermediate, forming a positively charged species. Then the positively charged species undergoes a reaction to form a neutral product.

- In imine and hydrate formation, a neutral product is achieved through loss of a proton from a nitrogen and an oxygen, respectively. (In hydrate formation, the neutral product is the original aldehyde or ketone.)
- In enamine formation, a neutral product is achieved through the loss of a proton from an α-carbon.
- In acetal formation, a neutral compound is achieved through the addition of a second equivalent of alcohol followed by loss of a proton.

16.10 PROTECTING GROUPS

A ketone (or an aldehyde) reacts with a 1,2-diol to form a five-membered ring acetal, and it reacts with a 1,3-diol to form a six-membered ring acetal. Recall that five- and six-membered rings are formed relatively easily (Section 9.16). The mechanism is the same as that shown in Section 16.9 for acetal formation, except that instead of reacting with two OH groups (each in a separate molecule of alcohol), the carbonyl compound reacts with two OH groups a single molecule of the diol.

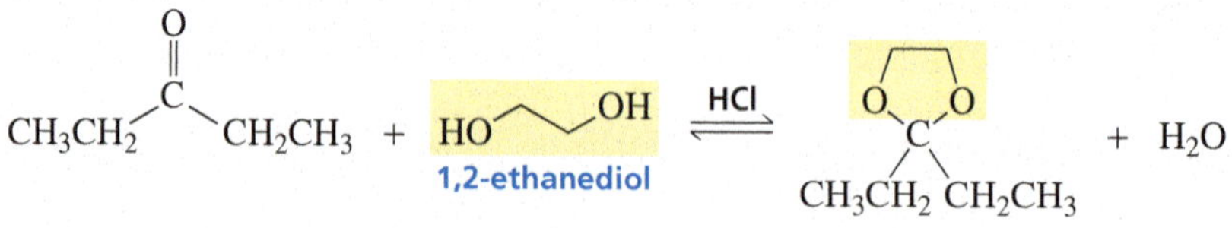

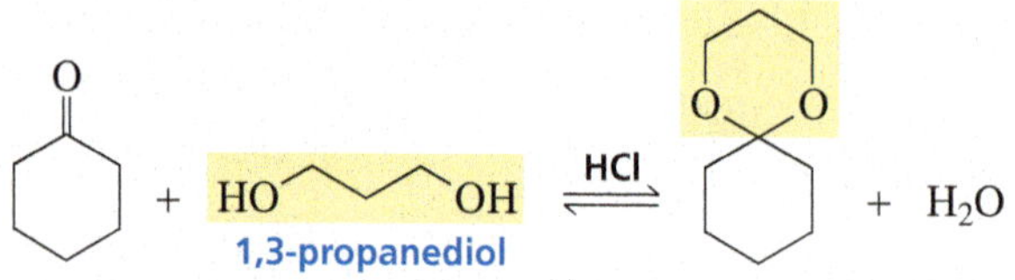

If a compound has two functional groups that will react with a given reagent, but you want only one of them to react, then you must protect the other functional group from the reagent. A group that protects a functional group from a synthetic operation that the functional group would not otherwise survive is called a **protecting group.**

If you have ever painted a room with a spray gun, you may have taped over the things you did not want to paint, such as baseboards and window frames. In a similar way, 1,2-diols and 1,3-diols are used to protect ("tape over") the carbonyl group of aldehydes and ketones.

For example, suppose you want to convert the keto ester shown below into a hydroxyketone. Both functional groups of the keto ester will be reduced by $LiAlH_4$, and the one that you don't want to react—the keto group—is the more reactive of the two groups.

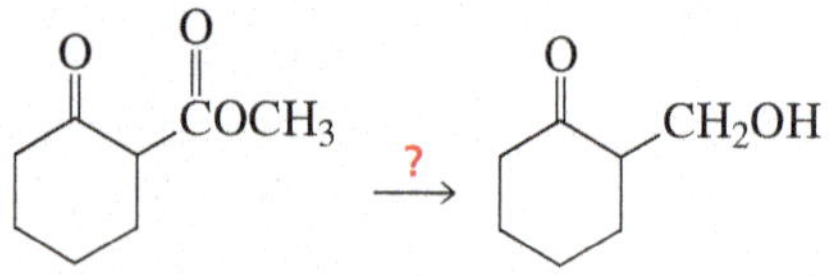

Protecting a Ketone or an Aldehyde

If, however, the keto group is first converted to an acetal, only the ester group will react with $LiAlH_4$. The protecting group can then be removed (called "deprotection") by acid-catalyzed hydrolysis after the ester has been reduced. It is critical that the conditions used to remove a protecting group do not affect other groups in the molecule. Acetals are good protecting groups because they are similar to ethers and, like ethers, they do not react with bases, reducing agents, or oxidizing agents.

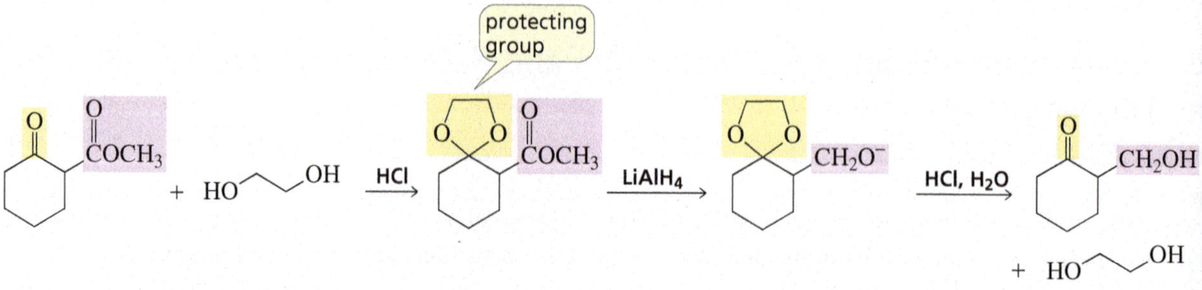

PROBLEM 42 ◆

a. What would have been the product of the preceding reaction with $LiAlH_4$ if the keto group had not been protected?

b. What reagent could you use to reduce only the keto group?

PROBLEM 43
Explain why acetals do not react with nucleophiles.

Protecting an OH Group

One of the best ways to protect an OH group of an alcohol is to convert it to a silyl ether. *tert*-Butyldimethylsilyl chloride is a commonly used reagent for this conversion. The TBDMS ether is formed by an S_N2 reaction. Although a tertiary alkyl halide does not undergo an S_N2 reaction, the tertiary silyl compound does because Si—C bonds are longer than C—C bonds, which reduces steric hindrance at the site of nucleophilic attack. An amine, generally imidazole, is included in the reaction mixture to react with the HCl generated in the reaction. After the TBDMS ether is isolated, the Grignard reagent can be synthesized because the compound no longer has an acidic OH group. The silyl ether, which is stable in neutral and basic solutions, can have its protecting group removed with tetrabutylammonium fluoride or with aqueous acid.

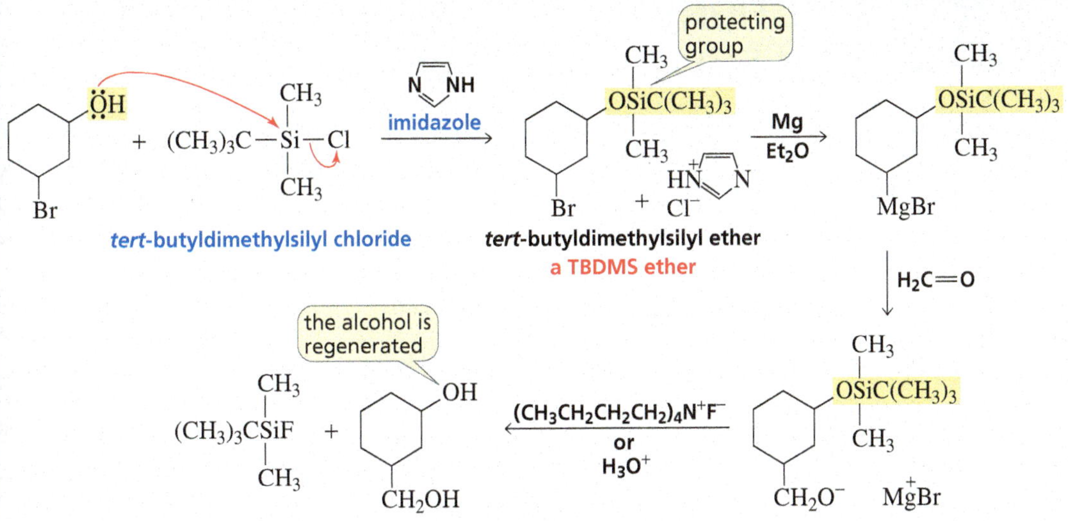

The OH group of a carboxylic acid group can be protected by converting the carboxylic acid into an ester. The ester shown below now has only one OH group that will react with thionyl chloride.

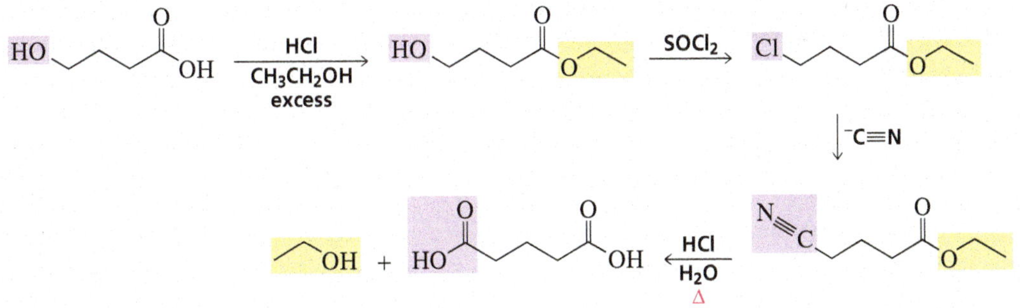

A protecting group should be used only when absolutely necessary because protection and deprotection adds two steps to the synthesis, which decreases the overall yield of the target molecule (the desired product).

PROBLEM 44
What products would be formed from the preceding reaction if the carboxylic acid group were not protected?

PROBLEM 45 ◆
a. In a six-step synthesis, what is the yield of the target molecule if each of the reactions employed gives an 80% yield?
b. What would the yield be if two more steps (each with an 80% yield) were added to the synthesis?

PROBLEM 46

Show how each of the following compounds could be prepared from the given starting material. Each requires a protecting group.

a. 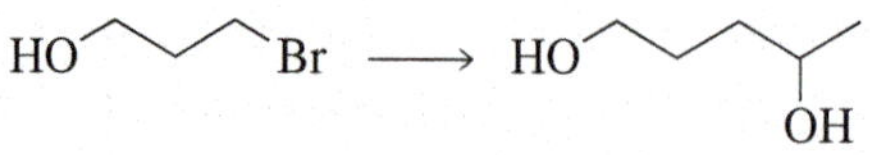 HO⌒⌒Br ⟶ HO⌒⌒⌒OH

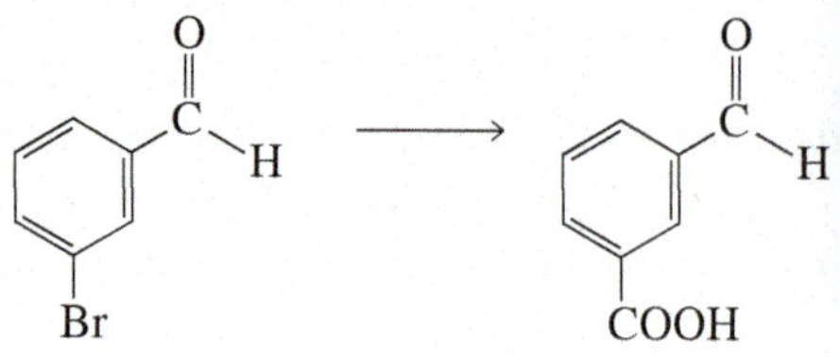

b. 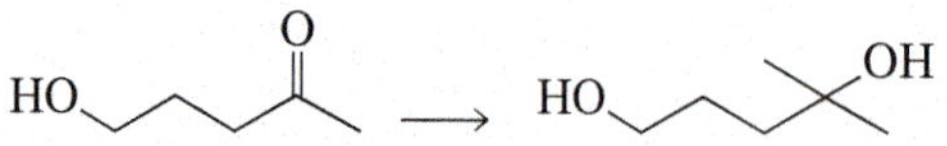 HO⌒⌒C(=O)⌒ ⟶ HO⌒⌒⌒C(CH₃)₂OH

c. (3-bromobenzaldehyde) ⟶ (3-carboxybenzaldehyde)

16.11 REACTIONS OF ALDEHYDES AND KETONES WITH SULFUR NUCLEOPHILES

Aldehydes and ketones react with thiols (the sulfur analogues of alcohols; Section 10.11) to form thioacetals. The mechanism for the addition of a thiol is the same as the mechanism for the addition of an alcohol (Section 16.9).

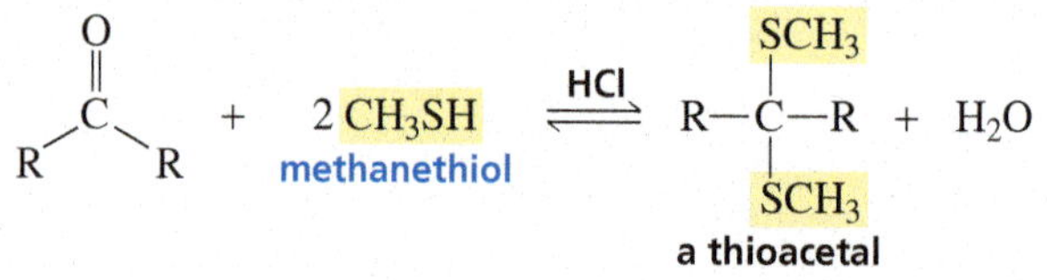

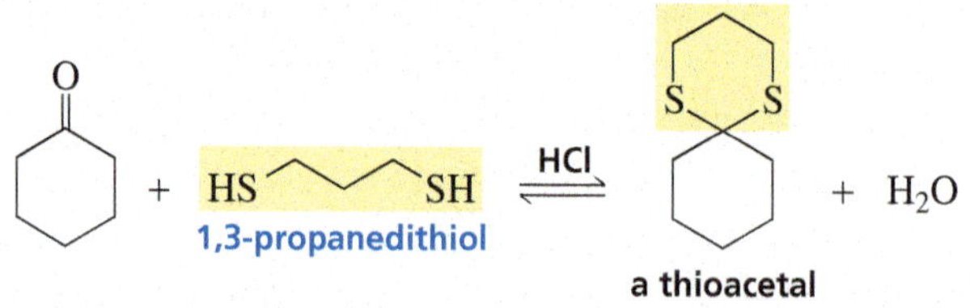

$$R-\overset{O}{\underset{}{C}}-R \;+\; 2\,CH_3SH \;\underset{}{\overset{HCl}{\rightleftharpoons}}\; R-\overset{SCH_3}{\underset{SCH_3}{C}}-R \;+\; H_2O$$

methanethiol

a thioacetal

(cyclohexanone) + HS⌒⌒SH $\overset{HCl}{\rightleftharpoons}$ (1,3-dithiane spiro) + H₂O

1,3-propanedithiol

a thioacetal

Thioacetal formation is useful in organic synthesis because a thioacetal is desulfurized when it reacts with H₂ and Raney nickel. **Desulfurization** replaces the C—S bonds with C—H bonds.

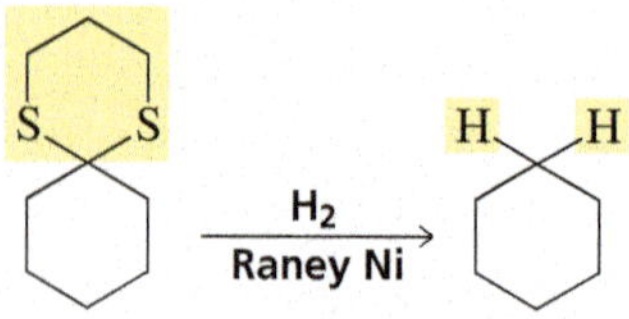

(thioacetal) $\overset{H_2}{\underset{Raney\ Ni}{\longrightarrow}}$ (CH₂ with two H's)

Thus, thioacetal formation followed by desulfurization provides a way to convert a carbonyl group into a methylene group.

16.12 REACTIONS OF ALDEHYDES AND KETONES WITH A PEROXYACID

Aldehydes *and* ketones react with the conjugate base of a peroxyacid to form carboxylic acids and esters, respectively. Recall that a **peroxyacid** contains one more oxygen than a carboxylic acid (Section 6.10), and it is this oxygen that is inserted between the carbonyl carbon and the H of an aldehyde or the R of a ketone (see the top of the next page). The reaction is called a **Baeyer–Villiger oxidation.** It is an oxidation reaction because the number of C—O bonds increases. A particularly good reagent for a Baeyer–Villiger oxidation is peroxytrifluoroacetate ion.

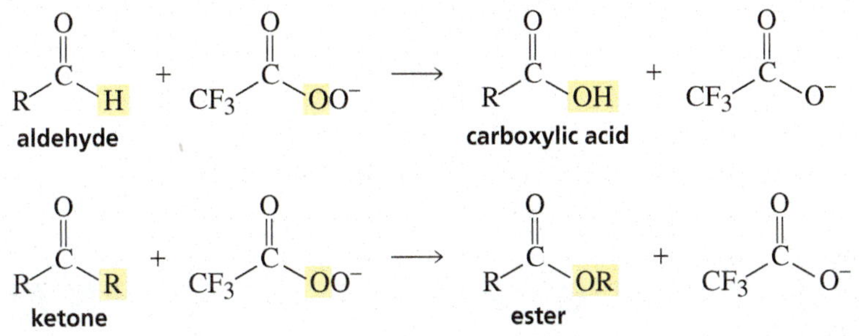

Aldehydes are oxidized to carboxylic acids by a peroxyacid.

Ketones are oxidized to esters by a peroxyacid.

If the two alkyl substituents attached to the carbonyl group of the ketone are different, then on what side of the carbonyl carbon is the oxygen inserted? For example, does the oxidation of cyclohexyl methyl ketone form methyl cyclohexanecarboxylate or cyclohexyl acetate?

cyclohexyl methyl ketone   **methyl cyclohexanecarboxylate**   or   **cyclohexyl acetate**

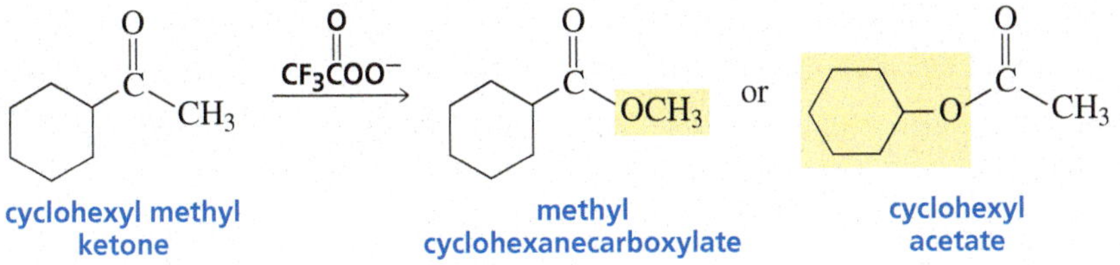

To answer this question, we need to look at the mechanism of the reaction.

MECHANISM FOR THE BAEYER–VILLIGER OXIDATION

a weak O—O bond

unstable intermediate

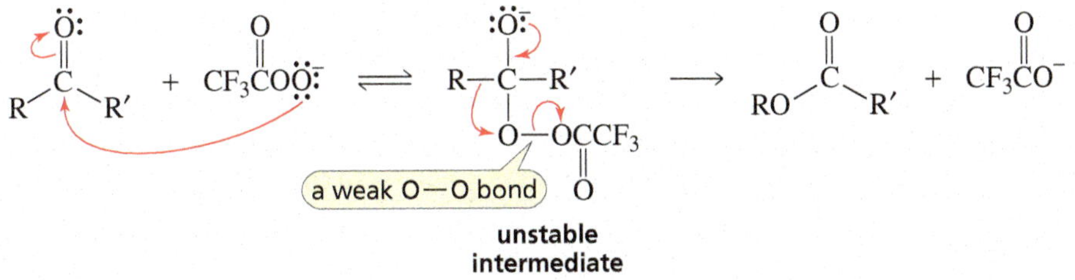

- The nucleophilic oxygen of the peroxyacid adds to the carbonyl carbon and forms an unstable tetrahedral intermediate with a weak O—O bond.
- As the π bond re-forms and the weak O—O bond breaks heterolytically, one of the alkyl groups migrates to an oxygen. This migration is similar to the 1,2-shifts that occur when carbocations rearrange (Section 6.7).

Studies of the migration tendencies of different groups have established the following order:

LEARN THE STRATEGY

relative migration tendencies

most likely to migrate — H > *tert*-alkyl > *sec*-alkyl ~ phenyl > primary alkyl > methyl — least likely to migrate

Therefore, the product of the Baeyer–Villiger oxidation of cyclohexyl methyl ketone is cyclohexyl acetate because a secondary alkyl group (the cyclohexyl group) is more likely to migrate than a methyl group. Aldehydes are always oxidized to carboxylic acids, because H has the greatest tendency to migrate.

PROBLEM 47 ◆   **USE THE STRATEGY**

What is the product of each of the following reactions?

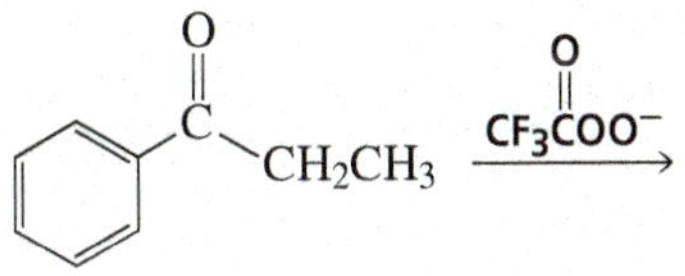

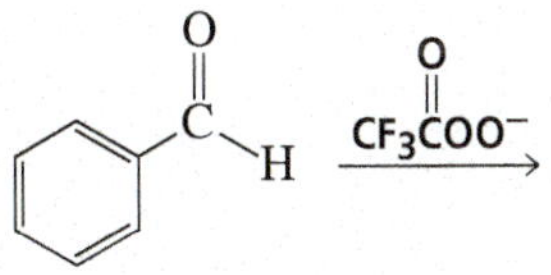

c. $\xrightarrow{\text{CF}_3\text{COO}^-}$

d. $\xrightarrow{\text{CF}_3\text{COO}^-}$

16.13 THE WITTIG REACTION FORMS AN ALKENE

An aldehyde or a ketone reacts with a *phosphonium ylide* ("ILL-id") to form an alkene. This reaction called a **Wittig reaction,** interchanges the doubly bonded oxygen of the carbonyl compound with the doubly bonded carbon group of the phosphonium ylide.

An **ylide** is a compound with opposite charges on adjacent covalently bonded atoms that have complete octets. The ylide can be written in the doubly bonded form because phosphorus can have more than eight valence electrons.

$$(C_6H_5)_3\overset{+}{P}-\overset{..}{\overset{-}{C}}HR \quad \text{or} \quad (C_6H_5)_3P=CHR$$
a phosphonium ylide

The Wittig reaction is a concerted [2 + 2] cycloaddition reaction. (Cycloaddition reactions were introduced in Section 8.14.) It is called a [2 + 2] cycloaddition reaction because, of the four electrons involved in the cyclic transition state, two come from the carbonyl group and two come from the ylide.

MECHANISM FOR THE WITTIG REACTION

- The nucleophilic carbon of the ylide adds to the carbonyl carbon while the carbonyl oxygen adds to the electrophilic phosphorus.
- Elimination of triphenylphosphine oxide forms the alkene product.

Preparing the Phosphonium Ylide

The phosphonium ylide needed for a particular synthesis is obtained by an S_N2 reaction between triphenylphosphine and an alkyl halide with the desired number of carbons. A proton on the carbon adjacent to the positively charged phosphorus atom ($pK_a = 35$) can be removed by a strong base such as butyllithium (Section 11.1).

Importance of the Wittig Reaction

The Wittig reaction is a powerful way to make an alkene because the reaction is completely regioselective—only one alkene is formed.

$$\text{⬡}=O \ + \ (C_6H_5)_3P=CH_2 \ \longrightarrow \ \text{⬡}=CH_2 \ + \ (C_6H_5)_3P=O$$

The Wittig reaction is the best way to make a terminal alkene, such as the one just shown, because other methods form a terminal alkene only as a minor product.

$$\text{⬡}\overset{CH_3}{\underset{Br}{<}} \xrightarrow{\ HO^-\ } \text{⬡}-CH_3 \ + \ \text{⬡}=CH_2$$

minor

$$\text{⬡}-CH_2Br \xrightarrow{\ HO^-\ } \text{⬡}-CH_2OH \ + \ \text{⬡}=CH_2$$

minor

A limitation of the Wittig reaction is that when it is used for the synthesis of an internal alkene, a mixture of *E* and *Z* stereoisomers is generally formed.

β-Carotene

β-Carotene is found in orange and yellow-orange fruits and vegetables such as apricots, mangoes, carrots, and sweet potatoes. It is also responsible for the characteristic color of flamingos. β-Carotene is used in the food industry to color margarine. The synthesis of β-carotene from vitamin A for use in foods is an important application of the Wittig reaction in industry.

vitamin A aldehyde   +   phosphonium ylide

β-carotene

Many people take β-carotene as a dietary supplement because there is some evidence associating high levels of β-carotene with a low incidence of cancer. More recent evidence, however, suggests that β-carotene taken in pill form does not have the cancer-preventing effects of β-carotene obtained from the diet.

Retrosynthetic Analysis

When synthesizing an alkene using a Wittig reaction, the first thing you must do is decide which part of the alkene should come from the carbonyl compound and which part should come from the ylide. If both sets of carbonyl compound and ylide are available, the better choice is the set that requires the less sterically hindered alkyl halide for the synthesis of the ylide via an S_N2 reaction. (Recall that the more sterically hindered the alkyl halide, the less reactive it is in an S_N2 reaction; see Section 9.1.)

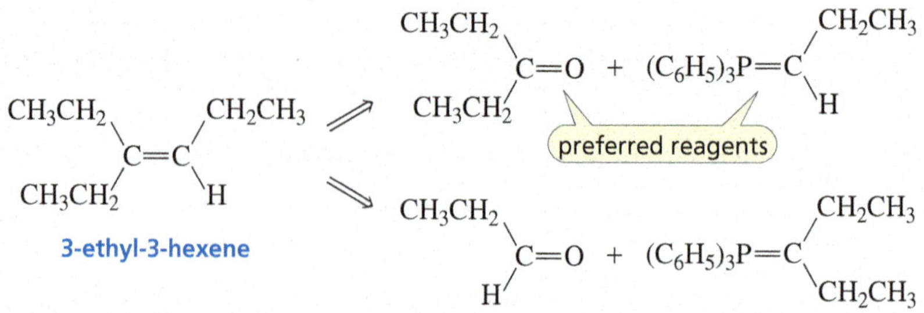

3-ethyl-3-hexene

For the synthesis of 3-ethyl-3-hexene, for example, it is better to use a three-carbon alkyl halide for the ylide and a five-carbon carbonyl compound than a five-carbon alkyl halide for the ylide and a three-carbon carbonyl compound, because it is easier to form an ylide from a primary alkyl halide (1-bromopropane) than from a secondary alkyl halide (3-bromopentane).

LEARN THE STRATEGY

PROBLEM 48 SOLVED

a. What two sets of reagents (each consisting of a carbonyl compound and phosphonium ylide) can be used for the synthesis of the following alkene?
b. What alkyl halide is required to prepare each of the phosphonium ylides?
c. What is the best set of reagents to use for the synthesis?

$$CH_3CH_2CH_2CH\!=\!\underset{\underset{CH_3}{|}}{C}CH_3$$

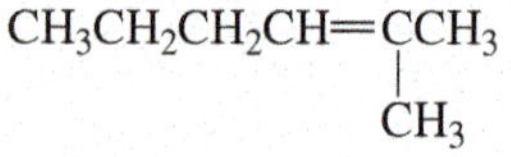

SOLUTION TO 48 a. The atoms on either side of the double bond can come from the carbonyl compound, so two pairs of compounds could be used.

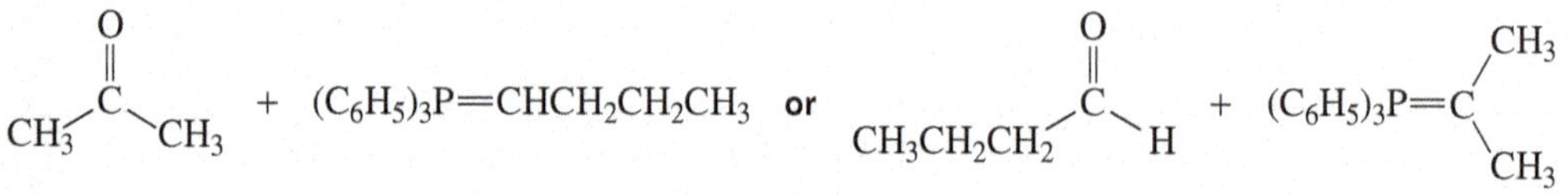

SOLUTION TO 48 b. The alkyl halide required to make the phosphonium ylide would be 1-bromobutane for the first pair of reagents or 2-bromopropane for the second pair.

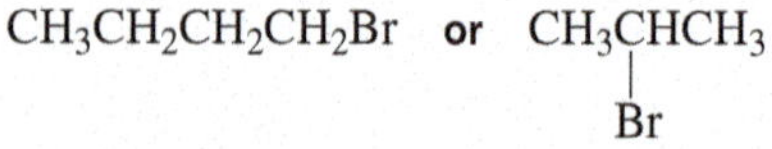

$$CH_3CH_2CH_2CH_2Br \quad \textbf{or} \quad CH_3\underset{\underset{Br}{|}}{C}HCH_3$$

SOLUTION TO 48 c. The primary alkyl halide would be more reactive in the S_N2 reaction required to make the ylide, so the best method would be to use the first set of reagents (acetone and the ylide obtained from 1-bromobutane).

USE THE STRATEGY

PROBLEM 49

a. What two sets of reagents (each consisting of a carbonyl compound and phosphonium ylide) can be used for the synthesis of each of the following alkenes?
b. What alkyl halide is required to prepare each of the phosphonium ylides?
c. What is the best set of reagents to use for the synthesis?

1. 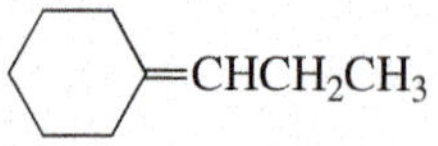 $=CHCH_2CH_3$ 2. $(C_6H_5)_2C\!=\!CHCH_3$ 3. 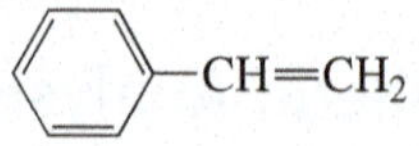 $-CH\!=\!CH_2$

DESIGNING A SYNTHESIS IV

16.14 DISCONNECTIONS, SYNTHONS, AND SYNTHETIC EQUIVALENTS

The route to the synthesis of a complicated molecule from simple starting materials is not always obvious. We saw that it is often easier to work backward from the desired product—a process called *retrosynthetic analysis* (Section 7.12). In a retrosynthetic analysis, the chemist dissects a molecule into smaller and smaller pieces to arrive at readily available starting materials.

retrosynthetic analysis

target molecule $\Longrightarrow$ Y $\Longrightarrow$ X $\Longrightarrow$ W $\Longrightarrow$ starting materials

A useful step in a retrosynthetic analysis is a **disconnection**—breaking a bond to produce two fragments. Typically, one fragment is positively charged and one is negatively charged. The fragments of a disconnection are called **synthons.** Synthons are often not real compounds. For example, if we consider the retrosynthetic analysis of cyclohexanol, a disconnection gives two synthons—namely, an α-hydroxycarbocation and a hydride ion.

retrosynthetic analysis

A **synthetic equivalent** is the reagent that is actually used as the source of a synthon. For the synthesis of cyclohexanol, cyclohexanone is the synthetic equivalent for the α-hydroxycarbocation, and sodium borohydride is the synthetic equivalent for hydride ion. Thus, cyclohexanol, the target molecule, can be prepared by treating cyclohexanone with sodium borohydride.

synthesis

When carrying out a disconnection, we must decide, after breaking the bond, which fragment gets the positive charge and which gets the negative charge. In the retrosynthetic analysis of cyclohexanol, we could have given the positive charge to the hydrogen, and many acids (HCl, HBr, and so on) could have been used for the synthetic equivalent for H^+. However, we would have been at a loss to find a synthetic equivalent for an α-hydroxycarbanion. Therefore, when we carried out the disconnection, we assigned the positive charge to the carbon and the negative charge to the hydrogen.

Cyclohexanol can also be disconnected by breaking the C—O bond instead of the C—H bond, forming a carbocation and hydroxide ion.

retrosynthetic analysis

The problem then becomes choosing a synthetic equivalent for the carbocation. A synthetic equivalent for a positively charged synthon needs an electron-withdrawing group at just the right place. Cyclohexyl bromide, with an electron-withdrawing bromine, is a synthetic equivalent for the cyclohexyl carbocation. Cyclohexanol, therefore, can be prepared by treating cyclohexyl bromide with hydroxide ion. This method, however, is not as satisfactory as the first synthesis we proposed—reduction of cyclohexanone—because some of the alkyl halide is converted to an alkene, so the overall yield of the target molecule will be lower.

synthesis

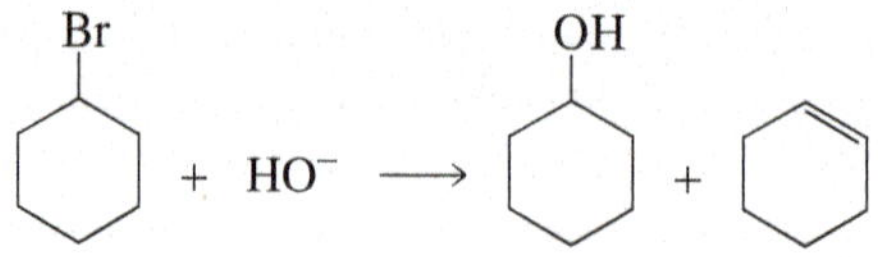

Retrosynthetic analysis shows that 1-methylcyclohexanol can be formed from the reaction of cyclohexanone, the synthetic equivalent for the α-hydroxycarbocation, with methylmagnesium bromide, the synthetic equivalent for the methyl anion.

retrosynthetic analysis

synthesis

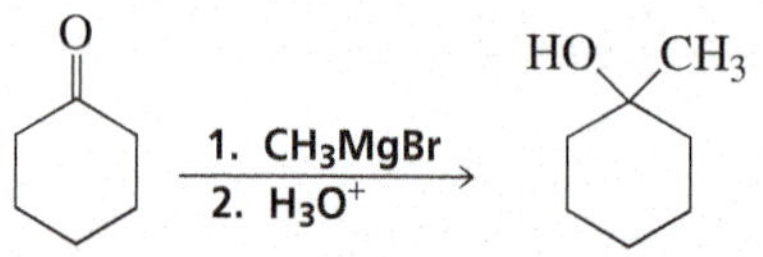

Other disconnections of 1-methylcyclohexanol are possible because any bond to carbon can serve as a disconnection site. For example, one of the ring C—C bonds could be broken. However, these are not useful disconnections because the synthetic equivalents of the synthons they produce are not easily prepared. A disconnection must lead to readily obtainable starting materials.

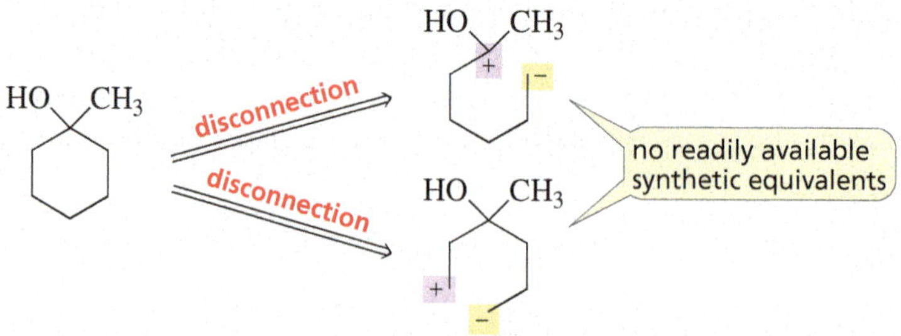

For additional practice using retrosynthetic analysis, see the tutorial on page 854.

PROBLEM 50

Using bromocyclohexane as a starting material, how could you synthesize the following compounds?

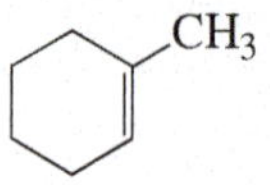

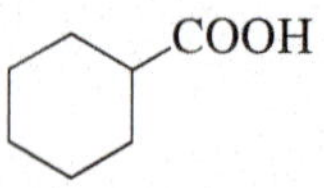

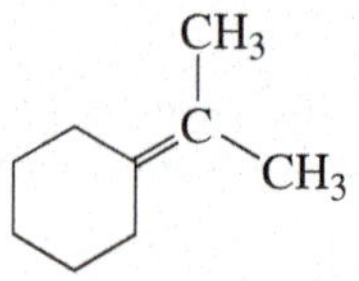

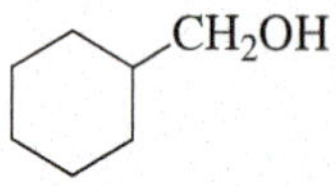

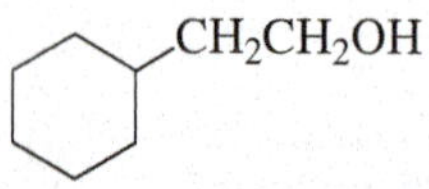

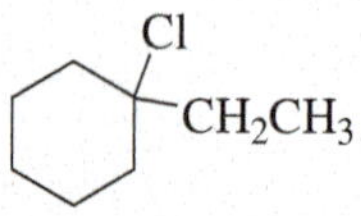

Synthesizing Organic Compounds

Organic chemists synthesize compounds for many reasons: to study their properties, to answer a variety of chemical questions, or to take advantage of one or more useful properties. One reason chemists synthesize a natural product—that is, a compound synthesized in nature—is to provide us with a larger supply of the compound than nature can produce. For example, Taxol—a compound that has successfully treated ovarian cancer, breast cancer, and certain forms of lung cancer by inhibiting mitosis—is extracted from the bark of *Taxus*, a yew tree found in the Pacific Northwest.

The supply of natural Taxol is limited because yew trees are uncommon, they grow very slowly, and stripping the bark kills the tree. Moreover, the bark of a 40-foot tree, which may have taken 200 years to grow, provides only 0.5 g of the drug. In addition, *Taxus* forests serve as habitats for the spotted owl, an endangered species, so harvesting the trees would accelerate the owl's demise. Once chemists determined the structure of Taxol, efforts were undertaken to synthesize it in order to make it more widely available as an anticancer drug. Several syntheses have been successful.

a spotted owl (*strix occidentalis*) taking off from a falconer's glove

yew tree bark

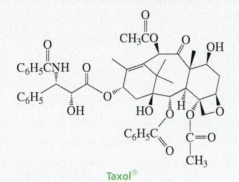

Taxol®

Once a compound has been synthesized, chemists can study its properties to learn how it works. Then they may be able to design and synthesize analogues to obtain safer or more potent drugs (Section 10.9).

Semisynthetic Drugs

Taxol is difficult to synthesize because of its many functional groups and 11 asymmetric centers. Chemists have made the synthesis a lot easier by allowing the common English yew shrub to carry out the first part of the synthesis. A precursor of the drug is extracted from the shrub's needles, and the precursor is converted to Taxol in a four-step procedure in the laboratory. Thus, the precursor is isolated from a renewable resource, whereas the drug itself could be obtained only by killing a slow-growing tree. This is an example of how chemists have learned to synthesize compounds jointly with nature.

a yew shrub

16.15 NUCLEOPHILIC ADDITION TO α,β-UNSATURATED ALDEHYDES AND KETONES

The resonance contributors for an α,β-unsaturated carbonyl compound show that the molecule has two electrophilic sites: the carbonyl carbon and the β-carbon.

$$RCH{=}CH{-}C(=\ddot{O}){-}R \quad \longleftrightarrow \quad RCH{-}CH{=}C({-}\ddot{O}^-){-}R$$

electrophilic site   an α,β-unsaturated carbonyl compound   electrophilic site

This means that a nucleophile can add either to the carbonyl carbon or to the β-carbon.

Direct Addition and Conjugate Addition

Nucleophilic addition to the carbonyl carbon is called **direct addition** or 1,2-addition.

Nucleophilic addition to the β-carbon is called **conjugate addition** or 1,4-addition because it occurs at the 1- and 4-positions. The initial product of 1,4-addition is an enol, which tautomerizes to a ketone or to an aldehyde (Section 7.7). Thus, the overall reaction is addition to the carbon–carbon double bond, with the nucleophile adding to the β-carbon and a proton adding to the α-carbon.

Whether the product obtained from nucleophilic addition to an α,β-unsaturated aldehyde or ketone is the direct addition product or the conjugate addition product depends on the nature of the nucleophile and the structure of the carbonyl compound.

When the Nucleophile Is a Weak Base

We saw that when two competing reactions are both *irreversible*, the reaction is under kinetic control, and when one or both of the reactions is *reversible*, the reaction is under thermodynamic control (Section 8.13).

> Addition to the β-carbon (conjugate addition) is generally irreversible.
>
> Addition to the carbonyl carbon (direct addition) can be reversible or irreversible.

When the nucleophile is a *weak base*, such as a halide ion, a cyanide ion, a thiol, an alcohol, or an amine, then direct addition is *reversible*, because a weak base is a good leaving group. Therefore, the reaction is under *thermodynamic control*.

The reaction that prevails when the reaction is under thermodynamic control is the one that forms the more stable product (Section 8.13). The conjugate addition product is always the more stable product because it retains the very stable carbonyl group. Therefore, weak bases form conjugate addition products.

Nucleophiles that are weak bases form conjugate addition products.

When the Nucleophile Is a Strong Base

When the nucleophile is a *strong base*, such as a Grignard reagent or a hydride ion, then direct addition is irreversible. Now, because the two competing reactions are both irreversible, the reaction is under *kinetic control*.

The reaction that prevails when the reaction is under kinetic control is the one that is faster. Therefore, the product depends on the reactivity of the carbonyl group. Compounds with reactive carbonyl groups form primarily direct addition products because for *those* compounds, direct addition is faster. Compounds with less reactive carbonyl groups form conjugate addition products because for *those* compounds, conjugate addition is faster.

For example, aldehydes have more reactive carbonyl groups than ketones do, so aldehydes form primarily direct addition products with hydride ion and Grignard reagents. Ethanol (EtOH) is used to protonate the alkoxide ion.

Compared with aldehydes, ketones form less of the direct addition product and more of the conjugate addition product, because ketones are more sterically hindered and, therefore, less reactive than aldehydes.

Nucleophiles that are strong bases form direct addition products with reactive carbonyl groups and form conjugate addition products with less reactive carbonyl groups.

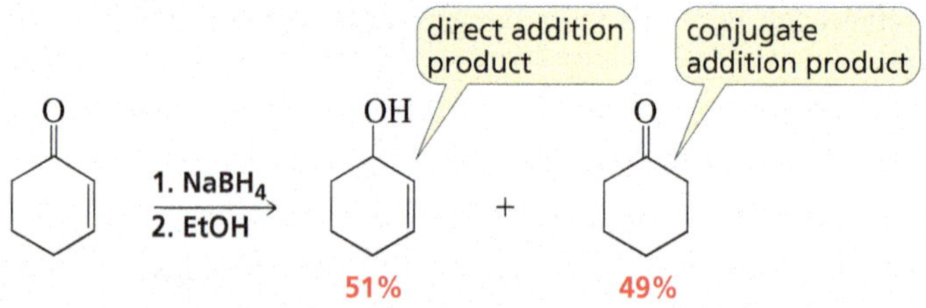
direct addition product
conjugate addition product
O
OH
O
1. NaBH4
2. EtOH
+
51%
49%

Grignard Reagents and Organocuprates

Like hydride ions, Grignard reagents are strong bases and, therefore, add irreversibly to carbonyl groups. Thus, the reaction is under kinetic control. If the carbonyl compound is reactive, reaction with the Grignard reagent will form the direct addition product.

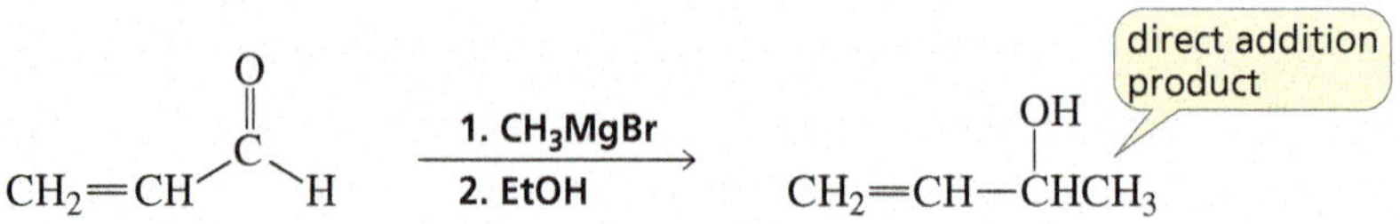
direct addition product
CH2=CH–C(=O)–H
1. CH3MgBr
2. EtOH
CH2=CH–CH(OH)CH3

If, however, the rate of direct addition is slowed down by steric hindrance, reaction with the Grignard reagent will form the conjugate addition product because conjugate addition then becomes the faster reaction.

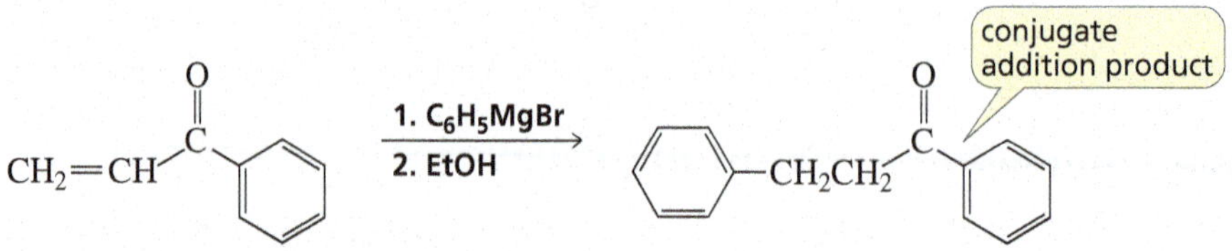
conjugate addition product
CH2=CH
1. C6H5MgBr
2. EtOH
CH2CH2

Only conjugate addition occurs when organocuprates (Section 11.3) react with α,β-unsaturated aldehydes and ketones. Therefore, Grignard reagents should be used when you want to add an alkyl group to the carbonyl carbon, whereas organocuprates should be used when you want to add an alkyl group to the β-carbon.

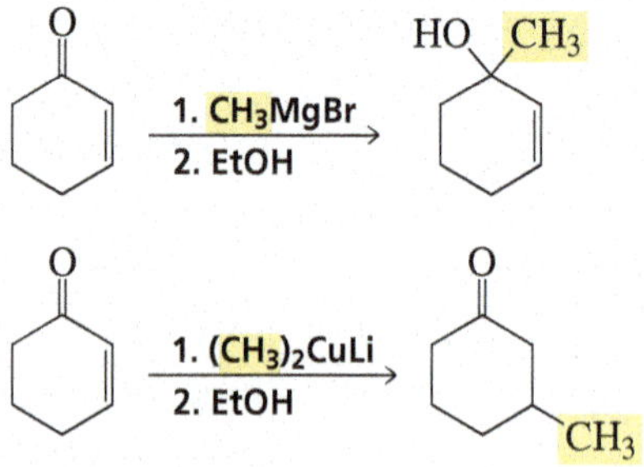
1. CH3MgBr
2. EtOH
HO CH3
1. (CH3)2CuLi
2. EtOH
CH3

PROBLEM 51

What is the major product of each of the following reactions?

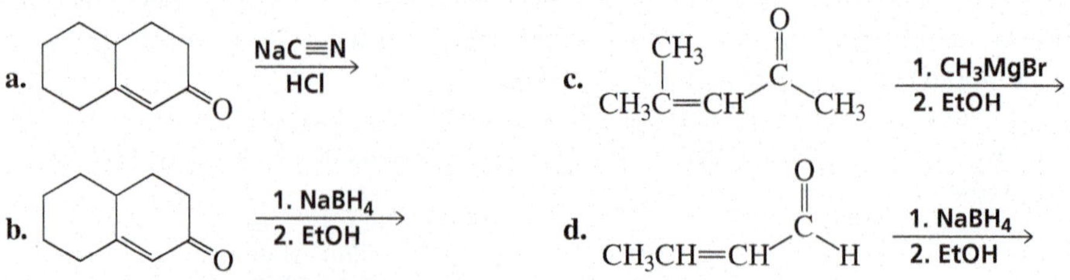
a.
NaC≡N
HCl
b.
1. NaBH4
2. EtOH
c.
CH3C(CH3)=CH–C(=O)–CH3
1. CH3MgBr
2. EtOH
d.
CH3CH=CH–C(=O)–H
1. NaBH4
2. EtOH

Hard and Soft Electrophiles and Nucleophiles

Electrophiles and nucleophiles can be classified as either *hard* or *soft*. Hard electrophiles and nucleophiles are more polarized than soft ones. Hard nucleophiles prefer to react with hard electrophiles, and soft nucleophiles prefer to react with soft electrophiles. Therefore, a Grignard reagent with a highly polarized C—Mg bond prefers to react with the harder C=O bond, whereas a Gilman reagent with a much less polarized C—Cu bond prefers to react with the softer C=C bond.

16.16 NUCLEOPHILIC ADDITION TO α,β-UNSATURATED CARBOXYLIC ACID DERIVATIVES

α,β-Unsaturated carboxylic acid derivatives, like α,β-unsaturated aldehydes and ketones, have two electrophilic sites for nucleophilic addition. They can undergo *conjugate addition* or *nucleophilic acyl substitution*. Notice that α,β-unsaturated carboxylic acid derivatives undergo *nucleophilic acyl substitution* rather than *direct nucleophilic addition* because they have a group that can be replaced by a nucleophile (Section 16.3).

Nucleophiles react with α,β-unsaturated carboxylic acid derivatives *with reactive carbonyl groups*, such as acyl chlorides, at the carbonyl group, forming nucleophilic acyl substitution products. Conjugate addition products are formed from the reaction of nucleophiles *with less reactive carbonyl groups*, such as esters and amides.

Enzyme-Catalyzed Cis–Trans Interconversion

Enzymes that catalyze the interconversion of cis and trans isomers are called cis–trans isomerases. These isomerases are all known to contain thiol (SH) groups. Thiols are weak bases and, therefore, add to the β-carbon of an α,β-unsaturated ketone (conjugate addition), forming a carbon–carbon single bond that rotates before the enol is able to tautomerize to the ketone. When tautomerization occurs, the absence of a proton at the active site of the enzyme in the vicinity of the α-carbon prevents the addition of a proton to the α-carbon. Therefore, the thiol is eliminated, leaving the compound as it was originally except for the configuration of the double bond.

PROBLEM 52

What is the major product of each of the following reactions?

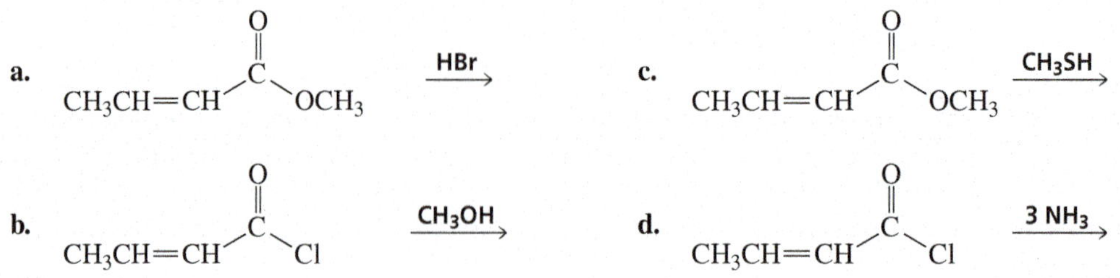

a.
$$CH_3CH=CH-\overset{\overset{\displaystyle O}{\|}}{C}-OCH_3 \quad \xrightarrow{HBr}$$

c.
$$CH_3CH=CH-\overset{\overset{\displaystyle O}{\|}}{C}-OCH_3 \quad \xrightarrow{CH_3SH}$$

b.
$$CH_3CH=CH-\overset{\overset{\displaystyle O}{\|}}{C}-Cl \quad \xrightarrow{CH_3OH}$$

d.
$$CH_3CH=CH-\overset{\overset{\displaystyle O}{\|}}{C}-Cl \quad \xrightarrow{3\ NH_3}$$

16.17 CONJUGATE ADDITION REACTIONS IN BIOLOGICAL SYSTEMS

Several reactions in biological systems involve the conjugate addition to α,β-unsaturated carbonyl compounds. Below are two examples. The first occurs in gluconeogenesis—the synthesis of glucose from pyruvate (Section 24.12). The second occurs in the oxidation of fatty acids (Section 24.5).

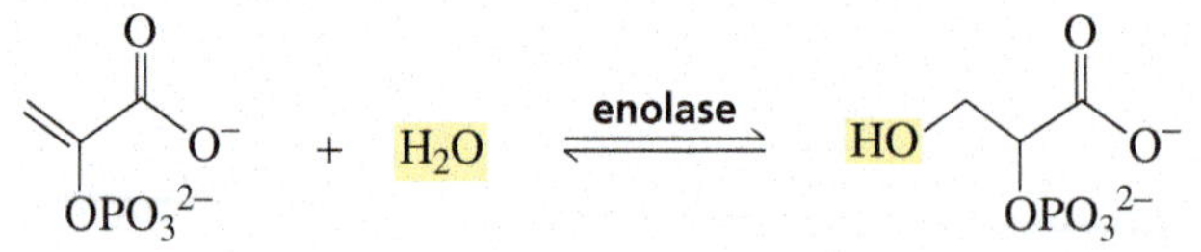

$$\text{phosphoenolpyruvate} + H_2O \underset{\text{enolase}}{\rightleftharpoons} \text{2-phosphoglycerate}$$

phosphoenolpyruvate   **2-phosphoglycerate**

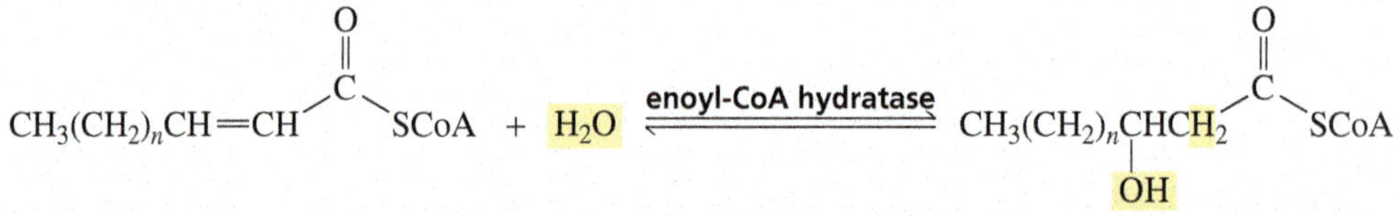

$$CH_3(CH_2)_nCH=CH-\overset{\overset{\displaystyle O}{\|}}{C}-SCoA + H_2O \underset{\text{enoyl-CoA hydratase}}{\rightleftharpoons} CH_3(CH_2)_nCHCH_2-\overset{\overset{\displaystyle O}{\|}}{C}-SCoA$$
$$\overset{|}{OH}$$

Cancer Chemotherapy

Two compounds—vernolepin and helenalin—owe their effectiveness as anticancer drugs to conjugate addition reactions.

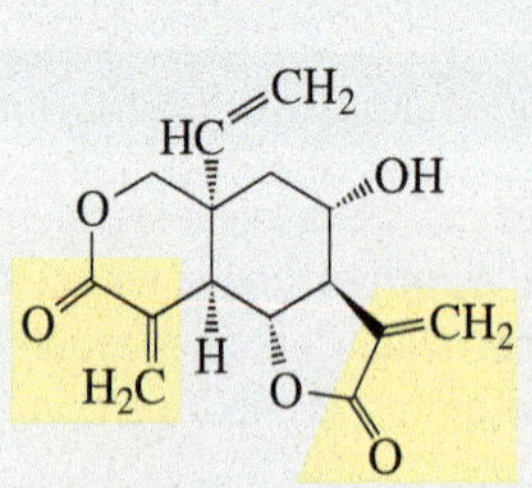

vernolepin 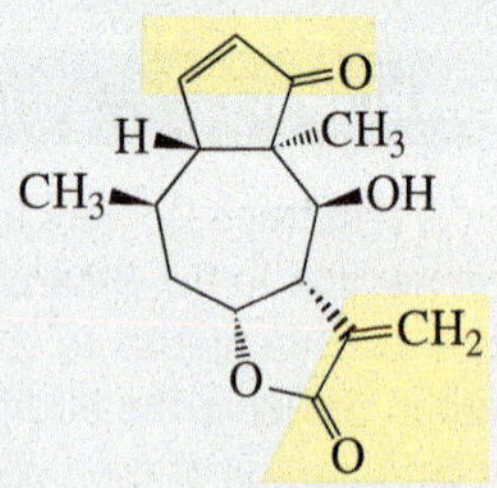   **helenalin**

Cancer cells are cells that have lost the ability to control their growth, so they proliferate rapidly. DNA polymerase is an enzyme that a cell needs to make a copy of its DNA for a new cell. DNA polymerase has an SH group at its active site, and each of these drugs has two α,β-unsaturated carbonyl groups. When an SH group of DNA polymerase adds irreversibly to the β-carbon of one of the α,β-unsaturated carbonyl groups of vernolepin or helenalin, the enzyme is inactivated because the active site of the enzyme is now blocked by the drug, so the enzyme cannot bind its substrate.

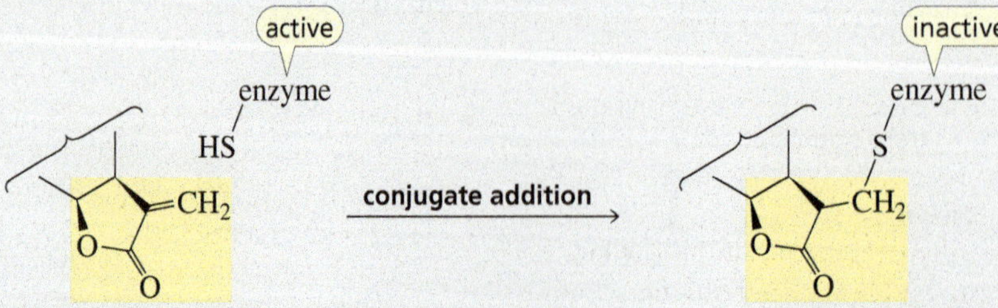

active enzyme $\xrightarrow{\text{conjugate addition}}$ inactive enzyme

ESSENTIAL CONCEPTS

Sections 16.0–16.1

- **Aldehydes** and **ketones** have an acyl group attached to an H and an R, respectively. The functional group ending of an aldehyde is "al;" the functional group ending of a ketone is "one."

Section 16.2

- Electronic and steric factors cause an aldehyde to be more reactive than a ketone toward nucleophilic addition.
- Aldehydes and ketones are less reactive than acyl halides and acid anhydrides and are more reactive than esters, carboxylic acids, and amides.

Section 16.3

- Aldehydes and ketones undergo **nucleophilic addition reactions** with strongly basic (C and H) nucleophiles, and undergo **nucleophilic addition–elimination reactions** with O and N nucleophiles.
- Acyl chlorides and esters undergo a nucleophilic acyl substitution reaction with strongly basic (C and H) nucleophiles to form a ketone or an aldehyde, which then undergoes a nucleophilic addition reaction with a second equivalent of the nucleophile to form an alcohol.

Section 16.4

- Grignard reagents react with aldehydes to form secondary alcohols; with ketones, esters, and acyl halides to form tertiary alcohols; and with carbon dioxide to form carboxylic acids. (An acyl chloride forms a ketone if an organocuprate is used instead of a Grignard reagent.) Aldehydes and ketones react with an acetylide ion to form an α-hydroxyalkyne and with cyanide ion to form a cyanohydrin.

Section 16.5

- Aldehydes, acyl chlorides, esters, and carboxylic acids are reduced by hydride ion to primary alcohols; ketones are reduced to secondary alcohols; and amides are reduced to amines.

Section 16.6

- An organic compound is reduced when H_2 is added to it. Reduction can occur by addition of a hydride ion and a proton, addition of two hydrogen atoms, or addition of an electron, a proton, an electron, a proton.

Section 16.7

- A **chemoselective reaction** is one in which a reagent reacts with one functional group in preference to another.

Section 16.8

- Aldehydes and ketones react with a primary amine to form an **imine** and with a secondary amine to form an **enamine**. The mechanisms are the same, except for the site from which a proton is lost in the last step.
- Imines and enamines are hydrolyzed under acidic conditions back to the carbonyl compound and the amine.

Section 16.9

- Aldehydes and ketones undergo acid-catalyzed addition of water to form a hydrate. Electron-donating and bulky substituents decrease the percentage of hydrate present at equilibrium. Most hydrates are too unstable to be isolated.
- Acid-catalyzed addition of an alcohol to an aldehyde or a ketone forms a **hemiacetal**; a second addition of alcohol forms an **acetal**. Acetal formation is reversible.

Section 16.10

- The carbonyl group of an aldehyde or a ketone can be protected by being converted to an acetal; the OH group of an alcohol can be protected by being converted to a TBDMS ether; and the OH group of a carboxylic acid can be protected by being converted to an ester.

Section 16.11

- Aldehydes and ketones react with a thiol to form a **thioacetal;** desulfurization replaces the C—S bonds with C—H bonds.

Section 16.12

- Aldehydes and ketones are oxidized by the conjugate base of a peroxyacid to carboxylic acids and esters, respectively.

Section 16.13

- An aldehyde or a ketone reacts with a phosphonium ylide in a **Wittig reaction** to form an alkene. A Wittig reaction is a concerted [2 + 2] cycloaddition reaction.

Section 16.14

- A useful step in a retrosynthetic analysis is a **disconnection**— breaking a bond to produce two fragments. **Synthons** are the fragments produced by a disconnection. A **synthetic equivalent** is the reagent used as the source of a synthon.

Section 16.15

- Nucleophilic addition to the carbonyl carbon of an α,β-**unsaturated aldehyde** or **ketone** is called **direct addition;** addition to the β-carbon is called **conjugate addition.**
- Whether direct or conjugate addition occurs depends on the strength of the nucleophile and the structure of the carbonyl compound.
- Nucleophiles that are weak bases—namely, halide ions, cyanide ion, water, alcohols, thiols, and amines—form conjugate addition products.
- Nucleophiles that are strong bases—namely, hydride ion and Grignard reagents—form direct addition products with reactive carbonyl groups and conjugate addition products with less reactive (sterically hindered) carbonyl groups.
- Grignard reagents form direct addition products. Organocuprates form conjugate addition products.

Section 16.16

- α,β-**Unsaturated carboxylic acid derivatives** with reactive carbonyl groups react with nucleophiles to form nucelophilic acyl substitution products; those with less reactive carbonyl groups react with nucleophiles to form conjugate addition products.

SUMMARY OF REACTIONS

1. Reactions of *carbonyl compounds* with Grignard reagents (Section 16.4)

   a. Reaction of *formaldehyde* with a Grignard reagent forms a primary alcohol. The mechanism is shown on page 746.

   b. Reaction of an *aldehyde* (other than formaldehyde) with a Grignard reagent forms a secondary alcohol. The mechanism is shown on page 746.

   c. Reaction of a *ketone* with a Grignard reagent forms a tertiary alcohol. The mechanism is shown on page 746.

   d. Reaction of CO_2 with a Grignard reagent forms a carboxylic acid. The mechanism is shown on page 746.

   e. Reaction of an *ester* with excess Grignard reagent forms a tertiary alcohol with two identical subunits. The mechanism is shown on page 748.

   f. Reaction of an *acyl chloride* with excess Grignard reagent forms a tertiary alcohol with two identical substituents.

   g. Reaction of an *acyl chloride* with an organocuprate forms a ketone.

2. Reaction of *carbonyl compounds* with an acetylide ion forms an α-hydroxy alkyne (Section 16.4). The mechanism is shown on page 750.

3. Reaction of an *aldehyde* or a *ketone* with cyanide ion under acidic conditions forms a cyanohydrin (Section 16.4). The mechanism is shown on page 751.

4. Reactions of *carbonyl compounds* with hydride ion donors (Section 16.5)

   a. Reaction of an *aldehyde* with sodium borohydride forms a primary alcohol. The mechanism is shown on page 753.

b. Reaction of a *ketone* with sodium borohydride forms a secondary alcohol. The mechanism is shown on page 753.

$$\underset{R}{\overset{O}{\underset{\|}{C}}}{}_R \quad \xrightarrow[\text{2. } H_3O^+]{\text{1. NaBH}_4} \quad R-\underset{\overset{|}{OH}}{\overset{OH}{CH}}-R$$

c. Reaction of an *acyl chloride* with sodium borohydride forms a primary alcohol. The mechanism is shown on page 753.

$$\underset{R}{\overset{O}{\underset{\|}{C}}}{}_{Cl} \quad \xrightarrow[\text{2. } H_3O^+]{\text{1. 2 NaBH}_4} \quad R-CH_2-OH$$

d. Reaction of an *acyl chloride* with lithium tri-*tert*-butoxyaluminum hydride forms an aldehyde.

$$\underset{R}{\overset{O}{\underset{\|}{C}}}{}_{Cl} \quad \xrightarrow[\text{2. } H_2O]{\text{1. LiAl[OC(CH}_3)_3]_3H, -78 °C} \quad \underset{R}{\overset{O}{\underset{\|}{C}}}{}_H$$

e. Reaction of an *ester* with lithium aluminum hydride forms two alcohols. The mechanism is shown on page 754.

$$\underset{R}{\overset{O}{\underset{\|}{C}}}{}_{OR'} \quad \xrightarrow[\text{2. } H_3O^+]{\text{1. 2 LiAlH}_4} \quad RCH_2OH \;+\; R'OH$$

f. Reaction of an *ester* with diisobutylaluminum hydride (DIBALH) forms an aldehyde.

$$\underset{R}{\overset{O}{\underset{\|}{C}}}{}_{OR'} \quad \xrightarrow[\text{2. } H_2O]{\text{1. [(CH}_3)_2CHCH_2]_2AlH, -78 °C} \quad \underset{R}{\overset{O}{\underset{\|}{C}}}{}_H$$

g. Reaction of a *carboxylic acid* with lithium aluminum hydride forms a primary alcohol. The mechanism is shown on page 755.

$$\underset{R}{\overset{O}{\underset{\|}{C}}}{}_{OH} \quad \xrightarrow[\text{2. } H_3O^+]{\text{1. LiAlH}_4} \quad R-CH_2-OH$$

h. Reaction of an *amide* with lithium aluminum hydride forms an amine. The mechanism is shown on page 756.

$$\underset{R}{\overset{O}{\underset{\|}{C}}}{}_{NH_2} \quad \xrightarrow[\text{2. } H_2O]{\text{1. LiAlH}_4} \quad R-CH_2-NH_2$$

$$\underset{R}{\overset{O}{\underset{\|}{C}}}{}_{NHR'} \quad \xrightarrow[\text{2. } H_2O]{\text{1. LiAlH}_4} \quad R-CH_2-NHR'$$

$$\underset{R}{\overset{O}{\underset{\|}{C}}}{}_{\underset{\overset{|}{R''}}{NR'}} \quad \xrightarrow[\text{2. } H_2O]{\text{1. LiAlH}_4} \quad R-CH_2-\underset{\overset{|}{R''}}{N}-R'$$

5. More reduction reactions (Section 16.6)

$$RCH{=}NCH_3 \quad \xrightarrow[\text{2. } H_2O]{\text{1. NaBH}_4} \quad RCH_2NHCH_3$$

$$RC{\equiv}N \quad \xrightarrow[\text{2. } H_2O]{\text{1. NaBH}_4} \quad RCH_2NH_2$$

$$RCH{=}NCH_3 \quad \xrightarrow[\text{Pd/C}]{H_2} \quad RCH_2NHCH_3$$

$$RC{\equiv}N \quad \xrightarrow[\text{Raney nickel}]{H_2} \quad RCH_2NH_2$$

$$\underset{R}{\overset{O}{\underset{\|}{C}}}{}_H \quad \xrightarrow[\text{Raney nickel}]{H_2} \quad RCH_2OH$$

$$\underset{R}{\overset{O}{\underset{\|}{C}}}{}_R \quad \xrightarrow[\text{Raney nickel}]{H_2} \quad \underset{\overset{|}{OH}}{RCHR}$$

6. Reactions of *aldehydes* and *ketones* with amines and amine derivatives (Section 16.8)

a. Reaction with a *primary amine* forms an imine. Hydrazine and hydroxylamine can also be used; they form a hydrazone and an oxime, respectively. The mechanism is shown on page 760.

$$\underset{R}{\overset{R'}{>}}C=O \ + \ H_2NZ \ \underset{\text{acid}}{\overset{\text{trace}}{\rightleftharpoons}} \ \underset{R}{\overset{R'}{>}}C=NZ \ + \ H_2O$$

$$(Z = R,\ NH_2,\ OH)$$

b. Reaction with a *secondary amine* forms an enamine. The mechanism is shown on page 763.

$$\underset{-CH}{\overset{R}{>}}C=O \ + \ R'NHR'' \ \underset{\text{acid}}{\overset{\text{trace}}{\rightleftharpoons}} \ \underset{-C}{\overset{R}{>}}C-N\underset{R''}{\overset{R'}{<}} \ + \ H_2O$$

7. *Reductive amination:* the imines and enamines formed from the reaction of aldehydes and ketones with ammonia and primary and secondary amines are reduced to primary, secondary, and tertiary amines, respectively (Section 16.8).

$$\underset{R}{\overset{R}{>}}C=O \ + \ NH_3 \ \underset{\text{acid}}{\overset{\text{trace}}{\rightleftharpoons}} \ \left[\underset{R}{\overset{R}{>}}C=NH\right] \ \overset{H_2}{\underset{Pd/C}{\longrightarrow}} \ \underset{R}{\overset{R}{>}}CHNH_2$$

$$\bigcirc=O \ + \ R-NH_2 \ \underset{\text{acid}}{\overset{\text{trace}}{\rightleftharpoons}} \ \bigcirc=N-R \ \overset{NaBH_3CN}{\longrightarrow} \ \bigcirc-NH-R$$

$$\bigcirc=O \ + \ \underset{R}{\overset{R}{>}}NH \ \underset{\text{acid}}{\overset{\text{trace}}{\rightleftharpoons}} \ \bigcirc-N\underset{R}{\overset{R}{<}} \ \overset{NaBH_3CN}{\longrightarrow} \ \bigcirc-N\underset{R}{\overset{R}{<}}$$

8. Reaction of an *aldehyde* or a *ketone* with water forms a hydrate (Section 16.9). The mechanism is shown on page 767.

$$\underset{R}{\overset{O}{\underset{}{\|}}}\!\!\!\!\overset{}{C}\!\!\!-R' \ + \ H_2O \ \overset{HCl}{\rightleftharpoons} \ R-\underset{OH}{\overset{OH}{\underset{|}{C}}}-R'$$

9. Reaction of an *aldehyde* or a *ketone* with excess alcohol forms first a hemiacetal and then an acetal (Section 16.9). The mechanism is shown on page 769.

$$\underset{R}{\overset{O}{\underset{}{\|}}}\!\!\!\!\overset{}{C}\!\!\!-R' \ + \ 2\,R''OH \ \overset{HCl}{\rightleftharpoons} \ R-\underset{OR''}{\overset{OH}{\underset{|}{C}}}-R' \ \rightleftharpoons \ R-\underset{OR''}{\overset{OR''}{\underset{|}{C}}}-R' \ + \ H_2O$$

10. Protecting groups (Section 16.10)

a. *Aldehydes* and *ketones* can be protected by being converted to acetals.

$$\underset{R}{\overset{O}{\underset{}{\|}}}\!\!\!\!\overset{}{C}\!\!\!-R \ + \ HO\!\!-\!\!OH \ \overset{HCl}{\rightleftharpoons} \ \underset{R}{\overset{O-C-O}{\underset{}{>}}}R \ + \ H_2O$$

b. The OH group of an *alcohol* can be protected by being converted to a silyl ether.

$$R-OH \ + \ (CH_3)_3CSi(CH_3)_2Cl \ \overset{\overset{N\ \ NH}{\diagdown\diagup}}{\longrightarrow} \ R-OSi(CH_3)_2C(CH_3)_3$$

c. The OH group of a *carboxylic acid* can be protected by being converted to an ester.

$$\underset{R}{\overset{O}{\underset{}{\|}}}\!\!\!\!\overset{}{C}\!\!\!-OH \ + \ \underset{\text{excess}}{CH_3OH} \ \overset{HCl}{\rightleftharpoons} \ \underset{R}{\overset{O}{\underset{}{\|}}}\!\!\!\!\overset{}{C}\!\!\!-OCH_3 \ + \ H_2O$$

11. Reaction of an *aldehyde* or a *ketone* with a thiol forms a thioacetal, and desulfurization of a *thioacetal* forms an alkane (Section 16.11).

$$
\underset{R}{\overset{O}{\underset{\|}{C}}}{}_{R'} + 2\,R''SH \underset{HCl}{\rightleftharpoons} R-\underset{\underset{SR''}{|}}{\overset{\overset{SR''}{|}}{C}}-R' + H_2O \xrightarrow[\text{Raney Ni}]{H_2} R-CH_2-R'
$$

12. *Aldehydes* and *ketones* are oxidized by a peroxyacid (a Baeyer–Villiger oxidation) to carboxylic acids and esters, respectively (Section 16.12). The mechanism is shown on page 775. Relative migration tendencies: H > tertiary > secondary ~ phenyl > primary > methyl

$$
R\overset{O}{\overset{\|}{C}}H + CF_3\overset{O}{\overset{\|}{C}}OO^- \longrightarrow R\overset{O}{\overset{\|}{C}}OH + CF_3\overset{O}{\overset{\|}{C}}O^-
$$

$$
R\overset{O}{\overset{\|}{C}}R + CF_3\overset{O}{\overset{\|}{C}}OO^- \longrightarrow R\overset{O}{\overset{\|}{C}}OR + CF_3\overset{O}{\overset{\|}{C}}O^-
$$

13. Reaction of an *aldehyde* or a *ketone* with a phosphonium ylide (a Wittig reaction) forms an alkene (Section 16.13). The mechanism is shown on page 776.

$$
\underset{R}{\overset{R}{}}C=O + (C_6H_5)_3P=C\underset{H}{\overset{CH_3}{}} \longrightarrow \underset{R}{\overset{R}{}}C=C\underset{H}{\overset{CH_3}{}} + (C_6H_5)_3P=O
$$

14. Reaction of an *α,β-unsaturated aldehyde* or a *ketone* with a nucleophile forms a direct addition product and/or a conjugate addition product, depending on the strength of the nucleophile and the structure of the aldehyde or ketone (Section 16.15). The mechanism is shown on page 782.

$$
RCH=CH\overset{O}{\overset{\|}{C}}R' + NuH \longrightarrow RCH=CH-\underset{\underset{Nu}{|}}{\overset{\overset{OH}{|}}{C}}-R' + RCHCH_2\overset{O}{\overset{\|}{C}}R'
$$

direct addition   conjugate addition

Nucleophiles that are weak bases ($^-C\equiv N$, RSH, RNH$_2$, Br$^-$) form conjugate addition products. Nucleophiles that are strong bases (RLi, RMgBr, H$^-$) form direct addition products with reactive carbonyl groups and conjugate addition products with sterically hindered carbonyl groups. Organocuprates (R$_2$CuLi) form conjugate addition products.

15. Reaction of an *α,β-unsaturated carboxylic acid derivative* with a nucleophile forms a nucleophilic acyl substitution product with a reactive carbonyl group and a conjugate addition product with a less reactive carbonyl group (Section 16.16).

$$
RCH=CH\overset{O}{\overset{\|}{C}}Cl + NuH \longrightarrow RCH=CH\overset{O}{\overset{\|}{C}}Nu + HCl
$$

nucleophilic
acyl substitution

$$
RCH=CH\overset{O}{\overset{\|}{C}}NHR + NuH \longrightarrow RCHCH_2\overset{O}{\overset{\|}{C}}NHR
$$

Nu

conjugate addition

PROBLEMS

53. Draw the structure for each of the following:

   a. isobutyraldehyde
   b. 4-hexenal
   c. diisopentyl ketone
   d. 3-methylcyclohexanone
   e. 2,4-pentanedione
   f. 4-bromo-3-heptanone
   g. γ-bromocaproaldehyde
   h. 2-ethylcyclopentanecarbaldehyde
   i. 4-methyl-5-oxohexanal

54. What are the products of the following reactions?

a.
$$CH_3CH_2-C(=O)-H + CH_3CH_2OH \xrightarrow[\text{excess}]{HCl}$$

b.
$$C_6H_5-C(=O)-Cl + (CH_3CH_2CH_2)_2CuLi \longrightarrow$$

c.
$$CH_3CH_2-C(=O)-CH_3 \xrightarrow[\text{2. H}_3O^+]{\text{1. NaBH}_4}$$

d.
$$CH_3CH_2-C(=O)-CH_2CH_3 + NaC\equiv N \xrightarrow[\text{excess}]{HCl}$$

e.
$$CH_3CH_2CH_2-C(=O)-OCH_2CH_3 \xrightarrow[\text{2. H}_3O^+]{\text{1. 2 LiAlH}_4}$$

f.
$$CH_3CH_2CH_2-C(=O)-CH_3 + HOCH_2CH_2OH \xrightarrow{HCl}$$

g.
(cyclohexenone with CH$_3$) $+ NaC\equiv N \xrightarrow[\text{excess}]{HCl}$

h.
(cyclohexane with $C\equiv N$) $+ H_2 \xrightarrow[\text{nickel}]{\text{Raney}}$

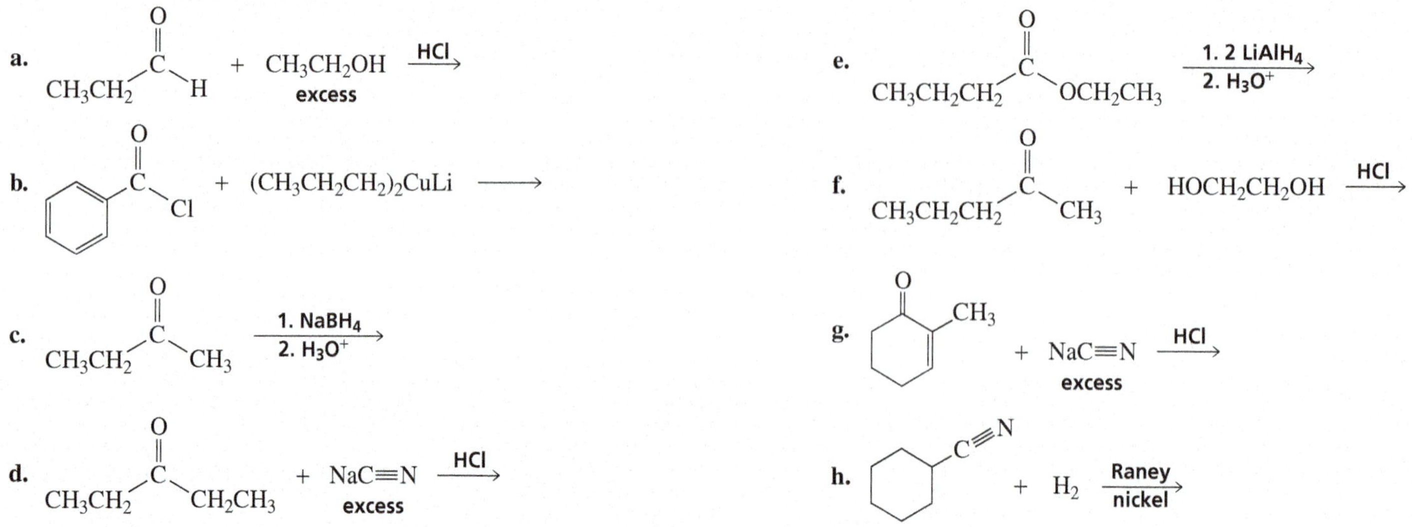

55. Rank the following compounds from most reactive to least reactive toward nucleophilic addition:

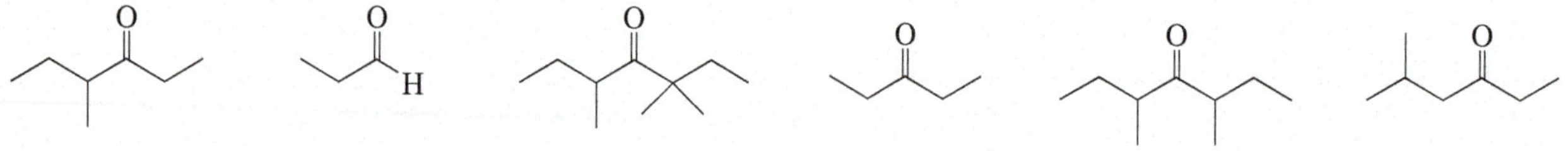

56. Draw the structure of two esters that will be reduced to propanol and butanol by LiAlH$_4$ (followed by addition of aqueous acid).

57. a. Show the reagents required to form the primary alcohol in each of the following reactions.
   b. Which of the reactions cannot be used for the synthesis of isobutyl alcohol?

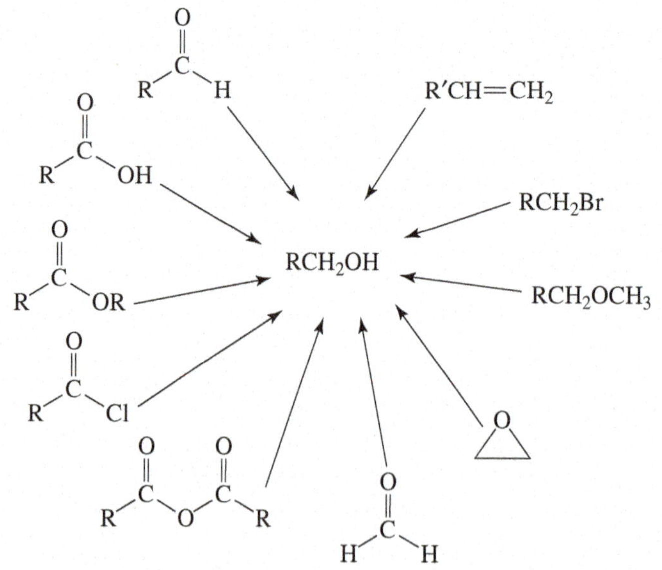

$$R-C(=O)-H$$
$$R-C(=O)-OH$$
$$R-C(=O)-OR$$
$$R-C(=O)-Cl$$
$$R-C(=O)-O-C(=O)-R$$
$$H-C(=O)-H$$
$$R'CH=CH_2$$
$$RCH_2Br$$
$$RCH_2OCH_3$$
(epoxide)

$$\longrightarrow RCH_2OH$$

58. Draw the products of the following reactions. Indicate whether each reaction is an oxidation or a reduction.

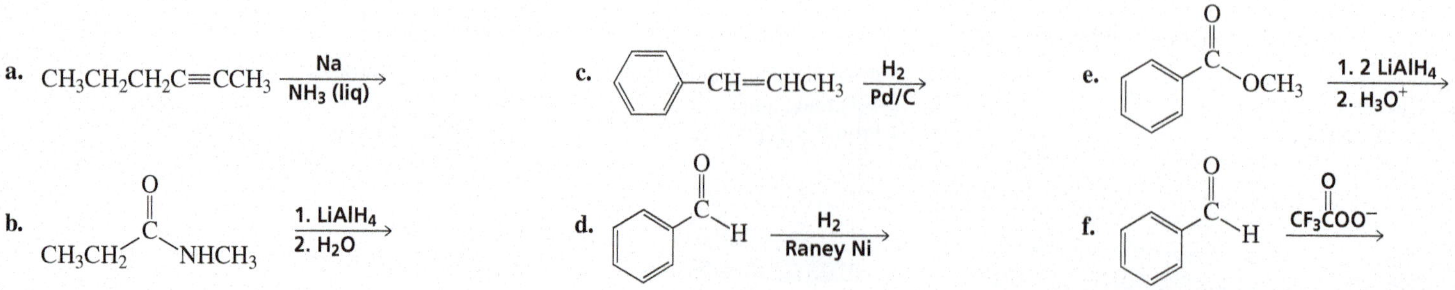

a.
$$CH_3CH_2CH_2C\equiv CCH_3 \xrightarrow[\text{NH}_3\text{ (liq)}]{\text{Na}}$$

b.
$$CH_3CH_2-C(=O)-NHCH_3 \xrightarrow[\text{2. H}_2O]{\text{1. LiAlH}_4}$$

c.
$$C_6H_5-CH=CHCH_3 \xrightarrow[\text{Pd/C}]{H_2}$$

d.
$$C_6H_5-C(=O)-H \xrightarrow[\text{Raney Ni}]{H_2}$$

e.
$$C_6H_5-C(=O)-OCH_3 \xrightarrow[\text{2. H}_3O^+]{\text{1. 2 LiAlH}_4}$$

f.
$$C_6H_5-C(=O)-H \xrightarrow{CF_3COO^-}$$

59. Name the following:

a. 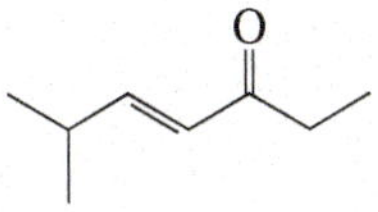 [structure] b. 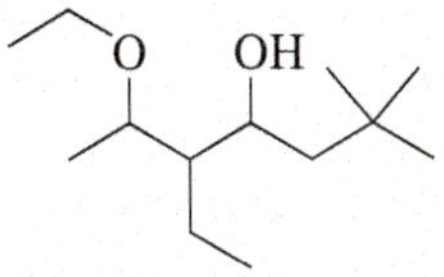 [structure] c. 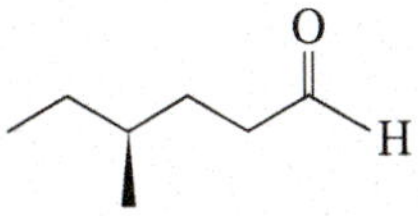 [structure] d. 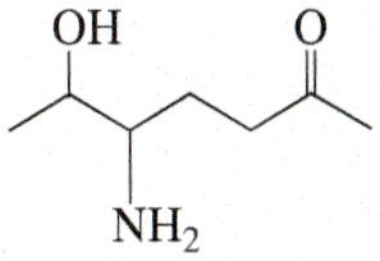 [structure]

60. Using cyclohexanone as the starting material, describe how each of the following compounds can be synthesized:

a. 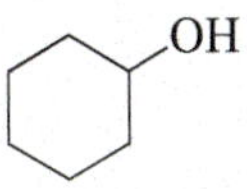 [cyclohexane-OH] d. 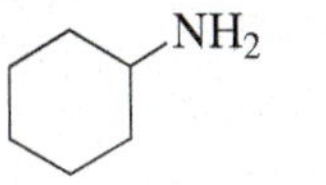 [cyclohexane-NH₂] g. 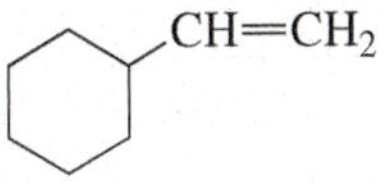 [cyclohexane-CH=CH₂]

b. [cyclohexene] e. [cyclohexane-CH₂NH₂] h. 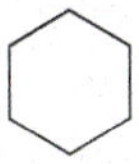 [cyclohexane] (show two methods)

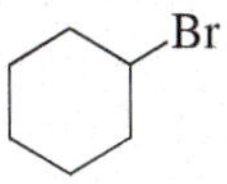

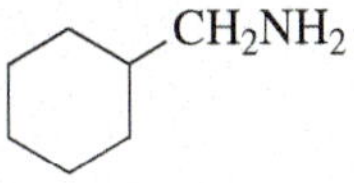

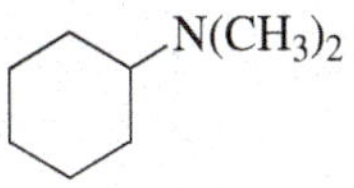

c. [cyclohexane-Br] f. [cyclohexane-N(CH₃)₂] i. 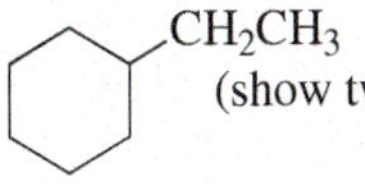 [cyclohexane-CH₂CH₃] (show two methods)

61. Propose a mechanism for each of the following reactions:

a. 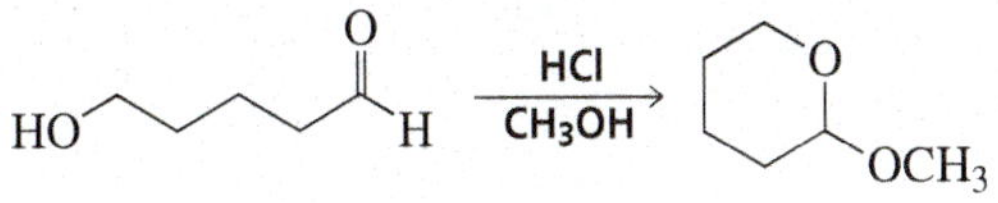 [structure] $\xrightarrow[\text{CH}_3\text{OH}]{\text{HCl}}$ [structure]

b. 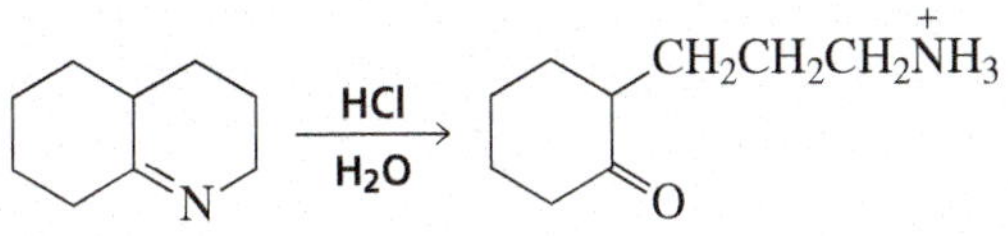 [structure] $\xrightarrow[\text{H}_2\text{O}]{\text{HCl}}$ [structure]

62. Show how each of the following compounds can be prepared, using the given starting material:

a. 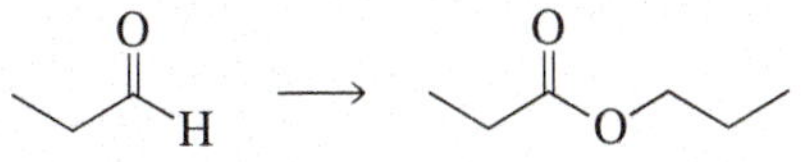 [structure] → [structure]

b. 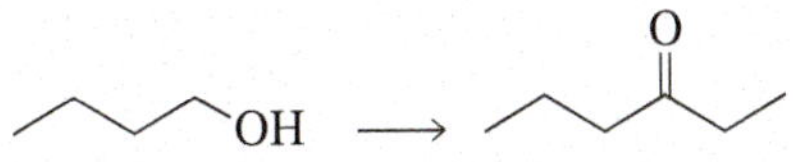 [structure] → [structure]

c. 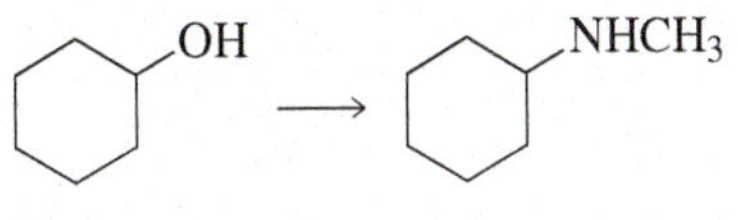 [structure] → [structure]

d. 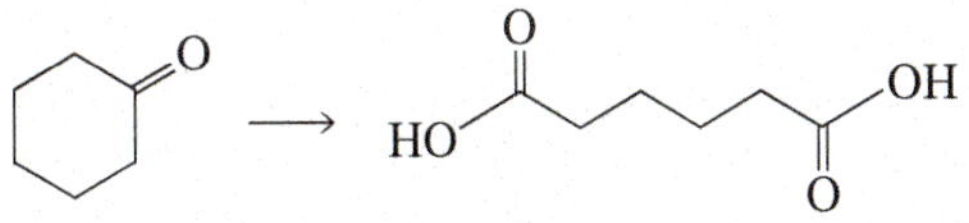 [structure] → [structure]

e. 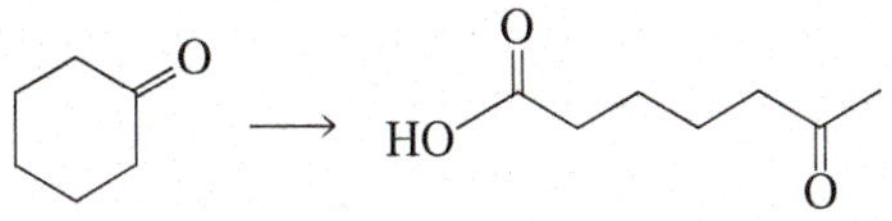 [structure] → [structure]

f. 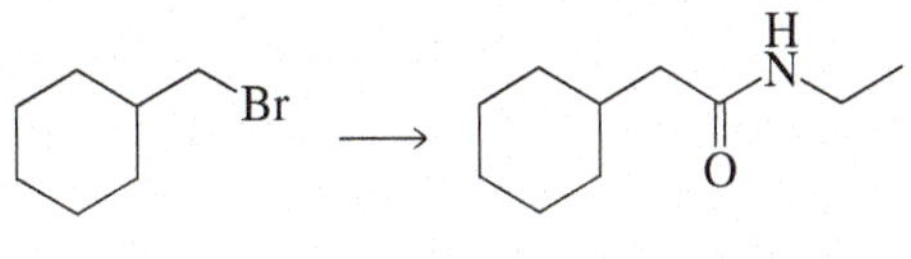 [structure] → [structure]

63. Fill in the boxes:

a. CH_3OH ⬜→ CH_3Br ⬜/⬜→ ⬜ $\xrightarrow[\text{2.}]{\text{1.}}$ $\text{CH}_3\text{CH}_2\text{OH}$

b. CH_4 ⬜→ CH_3Br ⬜/⬜→ ⬜ $\xrightarrow[\text{2.}]{\text{1.}}$ $\text{CH}_3\text{CH}_2\text{CH}_2\text{OH}$

64. Thiols can be prepared from the reaction of thiourea with an alkyl halide, followed by hydroxide-ion-promoted hydrolysis.
  a. Propose a mechanism for the reaction.
  b. What thiol will be formed if the alkyl halide employed is pentyl bromide?

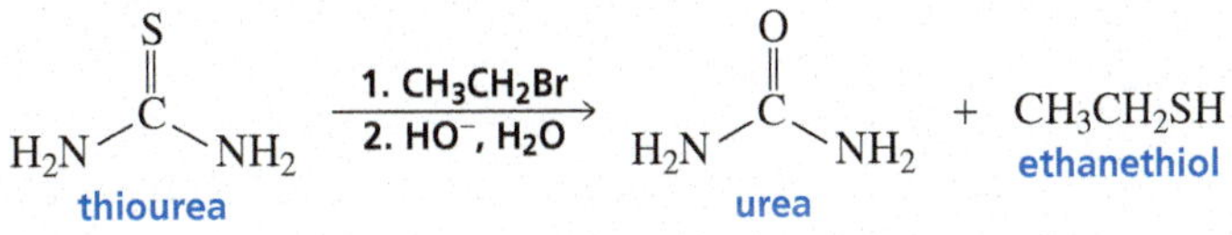

$\text{H}_2\text{N}-\overset{\overset{\displaystyle S}{\|}}{C}-\text{NH}_2$ $\xrightarrow[\text{2. HO}^-,\text{H}_2\text{O}]{\text{1. CH}_3\text{CH}_2\text{Br}}$ $\text{H}_2\text{N}-\overset{\overset{\displaystyle O}{\|}}{C}-\text{NH}_2$ $+\ \text{CH}_3\text{CH}_2\text{SH}$

thiourea urea ethanethiol

793

65. Identify **A** through **O**:

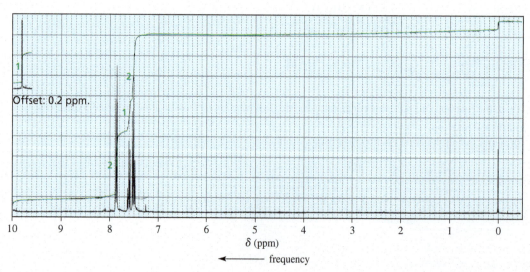

66. What are the products of the following reactions?

a.

$C_6H_5-C(CH_2CH_3)=NCH_2CH_3$ + H_2O $\xrightarrow{HCl}$

b.

$CH_3CH_2-\overset{O}{\overset{\|}{C}}-CH_3$ $\xrightarrow[\text{2. H}_3\text{O}^+]{\text{1. CH}_3\text{CH}_2\text{MgBr}}$

c.

cyclopentanone + $(C_6H_5)_3P=CHCH_3$ $\longrightarrow$

d.

$CH_3CH_2-\overset{O}{\overset{\|}{C}}-OCH_3$ $\xrightarrow[\text{2. H}_3\text{O}^+]{\substack{\text{1. CH}_3\text{CH}_2\text{MgBr}\\ \text{excess}}}$

e.

phenyl propyl ketone + CH_3OH $\xrightarrow{HCl}$

f.

2-pyrrolidinone $\xrightarrow[\text{2. H}_2\text{O}]{\text{1. LiAlH}_4}$

g.

cyclohexanone + $CH_3CH_2NH_2$ $\xrightarrow{\substack{\text{trace}\\ \text{acid}}}$

h.

cyclohexanone + $(CH_3CH_2)_2NH$ $\xrightarrow{\substack{\text{trace}\\ \text{acid}}}$

i.

$CH_2=CH-\overset{O}{\overset{\|}{C}}-OCH_3$ + CH_3NH_2 $\longrightarrow$

j. 2 $CH_2=CH-\overset{O}{\overset{\|}{C}}-OCH_3$ + CH_3NH_2 $\longrightarrow$

67. The only organic compound obtained when compound **Z** undergoes the following sequence of reactions gives the ^{1}H NMR spectrum shown. Identify compound **Z**.

Compound Z $\xrightarrow[\text{2. H}_3\text{O}^+]{\text{1. phenylmagnesium bromide}}$ $\xrightarrow[\substack{\text{CH}_3\text{COOH}\\ \text{0 °C}}]{\text{NaOCl}}$

Offset: 0.2 ppm.

δ (ppm)

⟵ frequency

68. Propose a mechanism for each of the following reactions:

a.

OCH₃ [structure with HCl / H₂O] → [cyclohexanone structure]

b. [dihydropyran structure] + CH₃CH₂OH →(HCl) [tetrahydropyran-OCH₂CH₃ structure]

69. How many signals would the product of the following reaction show in

a. its ¹H NMR spectrum? **b.** its ¹³C NMR spectrum?

[structure with OCH₃] →(1. excess CH₃MgBr / 2. H₃O⁺)

70. Fill in the boxes with the appropriate reagents:

[aldehyde structure] 1. ☐ 2. ☐ ⇄ [OH structure] ☐ → [ketone structure] ☐ → [CH₃O OCH₃ acetal structure]

71. Rank the following compounds from most reactive to least reactive toward nucleophilic addition:

[benzaldehyde] [p-methoxybenzaldehyde] [p-nitrobenzaldehyde] [p-isopropylbenzaldehyde]

72. How could you convert N-methylbenzamide to the following compounds?

a. N-methylbenzylamine **b.** benzoic acid **c.** methyl benzoate **d.** benzyl alcohol

73. What are the products of the following reactions? Show all stereoisomers that are formed.

a. [cyclohexenone] →(1. (CH₃)₂CuLi / 2. EtOH)

b. [cyclohexenone] →(1. CH₃MgBr / 2. H₃O⁺)

c. [phenyl ketone C(=O)CH₂CH₃] + [pyrrolidine N-H] →(trace acid)

d. [ketone] →(1. NaBH₄ / 2. H₃O⁺)

74. List three different sets of reagents (each set consisting of a carbonyl compound and a Grignard reagent) that could be used to prepare each of the following tertiary alcohols:

a. CH₃CH₂ĊCH₂CH₂CH₂CH₃ with OH and phenyl

b. CH₃CH₂ĊCH₂CH₂CH₃ with OH and CH₂CH₃

75. What product is formed when 3-methyl-2-cyclohexenone reacts with each of the following reagents?

a. CH₃MgBr followed by H₃O⁺ **b.** (CH₃CH₂)₂CuLi followed by H₃O⁺ **c.** HBr **d.** CH₃CH₂SH

76. What is the product of each of the following reactions?

a.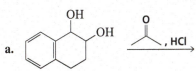

[diol structure] + [acetone O, HCl] →

b. [naphthalenone structure] + HO—OH, HCl →

77. What is the product of each of the following reactions?

a. 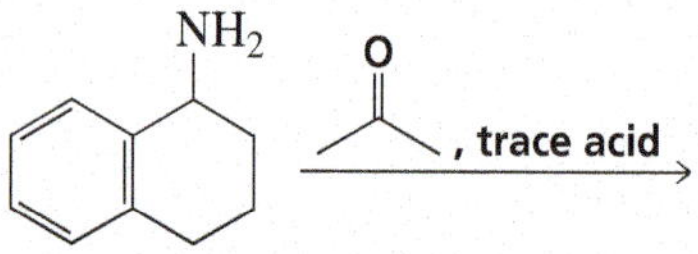 , trace acid

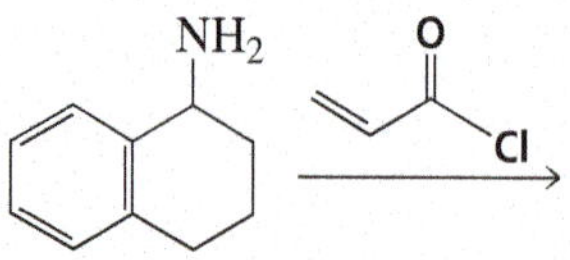

b. 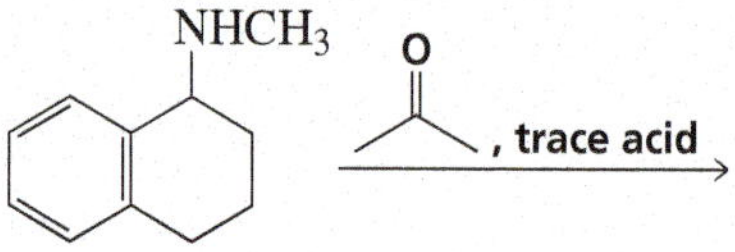 , trace acid

c.

d. 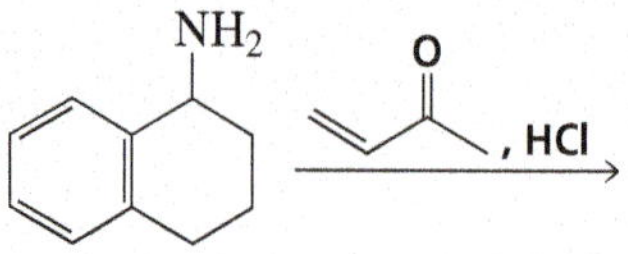 , HCl

78. Propose a mechanism for each of the following reactions:

a. 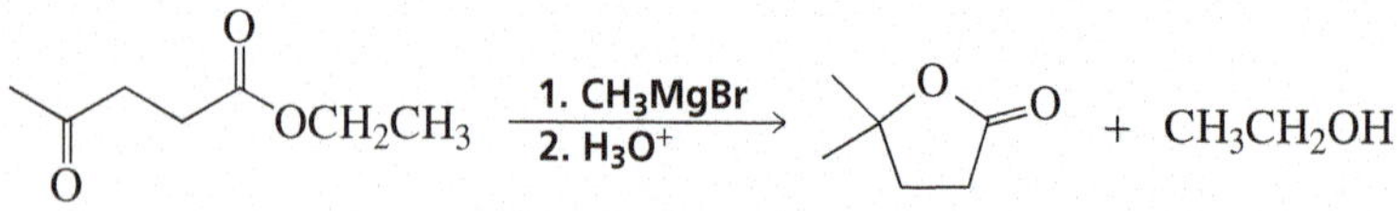 $\xrightarrow[\text{2. H}_3\text{O}^+]{\text{1. CH}_3\text{MgBr}}$ + CH₃CH₂OH

b. 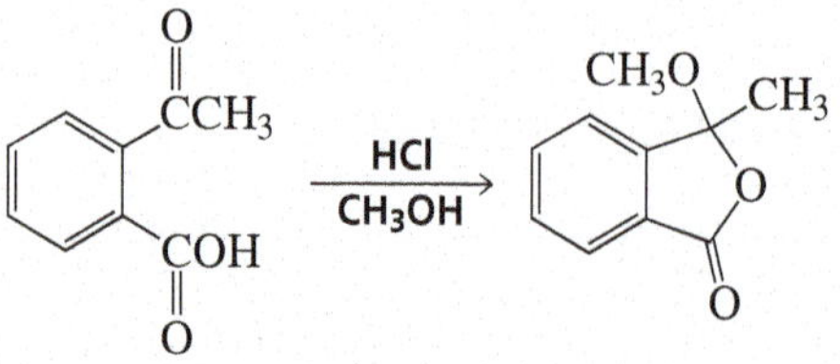 $\xrightarrow[\text{CH}_3\text{OH}]{\text{HCl}}$

79. Draw structures for **A-D** for each of the following:

a. 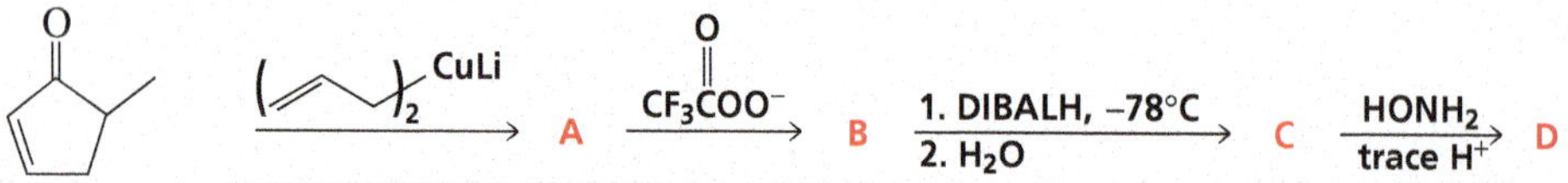 $\xrightarrow{\text{CuLi}}$ **A** $\xrightarrow{\text{CF}_3\text{COO}^-}$ **B** $\xrightarrow[\text{2. H}_2\text{O}]{\text{1. DIBALH, }-78°\text{C}}$ **C** $\xrightarrow[\text{trace H}^+]{\text{HONH}_2}$ **D**

b. $\xrightarrow[\text{2. H}_3\text{O}^+]{\text{1. NaBH}_4}$ **A** $\xrightarrow[\text{pyridine}]{\text{PBr}_3}$ **B** $\xrightarrow[\text{2. CH}_3\text{CH}_2\text{CH}_2\bar{\text{C}}\text{H}_2\text{Li}^+]{\text{1. P(C}_6\text{H}_5\text{)}_3}$ **C** $\xrightarrow{\text{H}\overset{\text{O}}{\underset{}{}}\text{H}}$ **D**

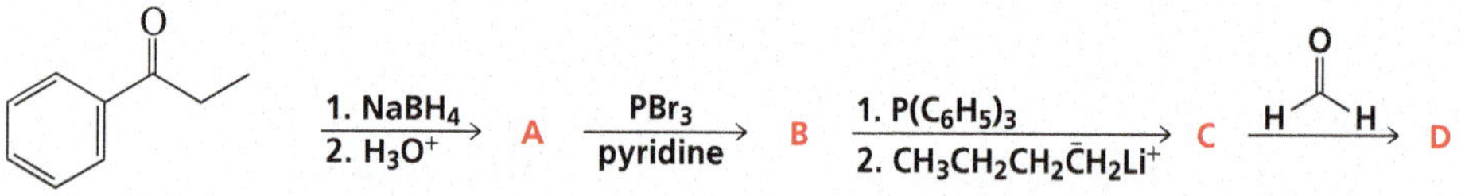

80. Propose a mechanism to explain how dimethyl sulfoxide and oxalyl chloride react to form the dimethylchlorosulfonium ion used as the oxidizing agent in the Swern oxidation (see Chapter 10, page 475).

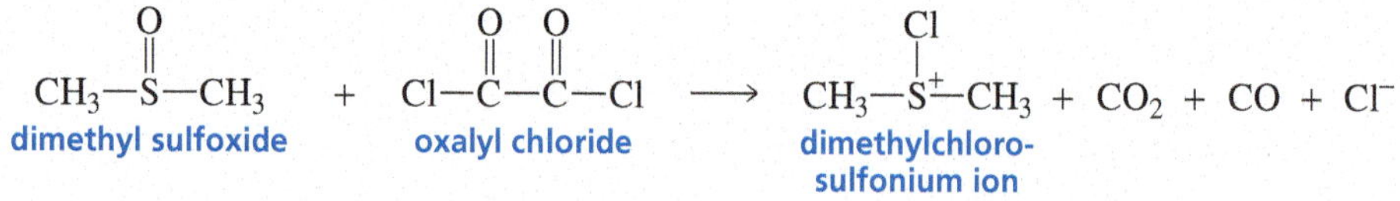

CH₃—S—CH₃ + Cl—C—C—Cl ⟶ CH₃—S⁺—CH₃ + CO₂ + CO + Cl⁻
dimethyl sulfoxide   oxalyl chloride   dimethylchlorosulfonium ion

81. a. Propose a mechanism for the following reaction:

CH₂=CHCHC≡N $\xrightarrow[\text{H}_2\text{O}]{\text{HO}^-}$ N≡CCH₂CH₂CH

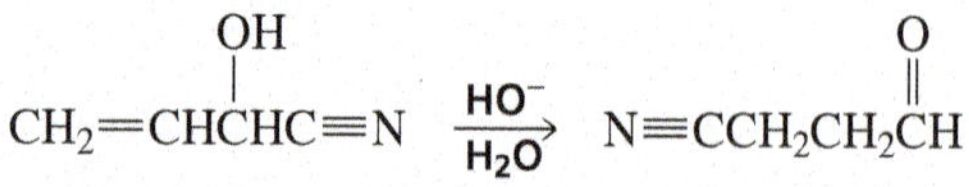

b. What is the product of the following reaction?

CH₃CCH=CH₂ $\xrightarrow[\text{H}_2\text{O}]{\text{HO}^-}$

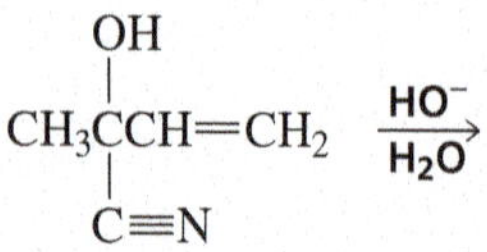

82. Unlike a phosphonium ylide that reacts with an aldehyde or a ketone to form an alkene, a sulfonium ylide reacts with an aldehyde or a ketone to form an epoxide. Explain why one ylide forms an alkene, whereas the other forms an epoxide.

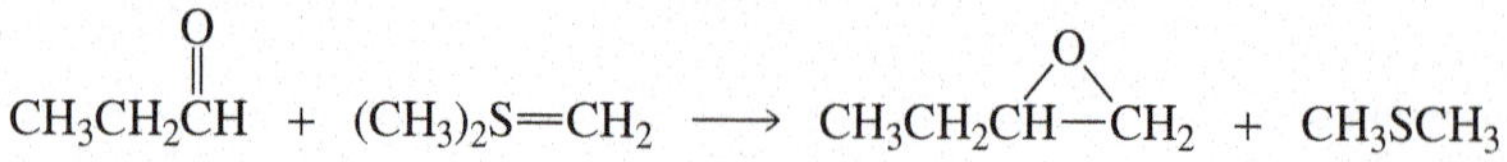

CH₃CH₂CH + (CH₃)₂S=CH₂ ⟶ CH₃CH₂CH—CH₂ + CH₃SCH₃

83. A compound gives the following IR spectrum. Upon reaction with sodium borohydride followed by acidification, it forms the product with the ^{1}H NMR spectrum shown below. Identify the starting material and the product.

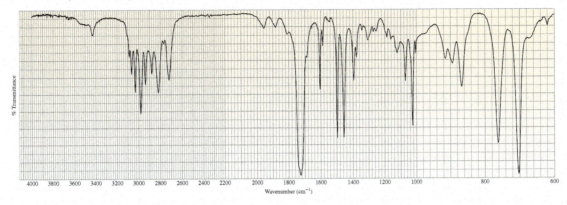

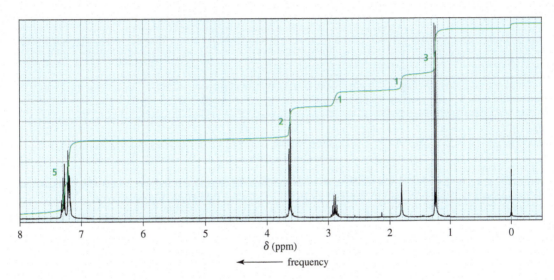

δ (ppm)

← frequency

84. How can the following compounds be prepared from the given starting materials?

a. [benzoate ester: Ph–C(=O)–OCH₃] → [Ph–C(OH)(CH₃)CH₃]

b. [benzoate ester: Ph–C(=O)–OCH₃] → [Ph–C(=O)–H]

c. [CH₃CH₂CH₂CH₂Br] → [CH₃CH₂CH₂CH₂–C(=O)–OH]

d. [δ-valerolactam: O=C–N(H)–CH(CH₃) ring] → [piperidine: N(H)–CH(CH₃) ring]

e. [cyclohexanol: OH] → [cyclohexyl–NHCH₃]

f. [keto acid: CH₃–C(=O)–(CH₂)₄–C(=O)–OH] → [lactone ring with ethyl group]

85. Propose a mechanism for each of the following reactions:

a. [cyclohexenone with gem-dimethyl and CH₃ groups] + HCl → [phenol with CH₃, CH₃, CH₃ substituents]

b. [cyclohexadienone with gem-dimethyl] + HCl → [phenol with CH₃, CH₃ substituents]

797

86. a. In an aqueous solution, D-glucose exists in equilibrium with two six-membered ring compounds. Draw the structures of these compounds.
  b. Which of the six-membered ring compounds will be the major product?

$$
\begin{array}{c}
\text{HC=O} \\
\text{H} \longrightarrow \text{OH} \\
\text{HO} \longrightarrow \text{H} \\
\text{H} \longrightarrow \text{OH} \\
\text{H} \longrightarrow \text{OH} \\
\text{CH}_2\text{OH}
\end{array}
$$

D-glucose

87. Shown below is the ^{1}H NMR spectrum of the alkyl bromide used to make the phosphonium ylide that reacts with a ketone in a Wittig reaction to form a compound with molecular formula $C_{11}H_{14}$. What product is obtained from the Wittig reaction?

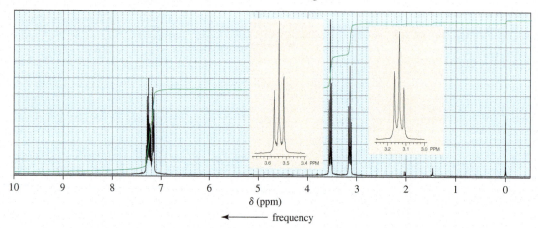

δ (ppm)

← frequency

88. In the presence of an acid catalyst, acetaldehyde forms a trimer known as paraldehyde. Because it induces sleep when it is administered to animals in large doses, paraldehyde is used as a sedative or hypnotic. Propose a mechanism for the formation of paraldehyde.

$$
\underset{\text{CH}_3}{\overset{O}{\underset{}{\parallel}}}\overset{}{\underset{\text{H}}{C}} \quad \overset{\text{HCl}}{\rightleftharpoons} \quad \text{paraldehyde}
$$

paraldehyde

89. What carbonyl compound and what phosphonium ylide are needed to synthesize the following compounds?

  a. —CH=CHCH$_2$CH$_2$CH$_3$ **b.** =CHCH$_2$CH$_3$ **c.** —CH=CH—   **d.** =CH$_2$

90. Identify compounds **A** and **B**:

$$
\textbf{A} \xrightarrow[\text{2. H}_3\text{O}^+]{\text{1. (CH}_2\text{=CH)}_2\text{CuLi}} \textbf{B} \xrightarrow[\text{2. H}_3\text{O}^+]{\text{1. CH}_3\text{Li}} \quad \text{CH}_2\text{=CHCCH}_2\text{CHCH}_3
$$

with substituents CH_3, OH, and CH_3.

91. When a cyclic ketone reacts with diazomethane, the next larger cyclic ketone is formed. This is called a ring-expansion reaction. Draw a mechanism for the following ring-expansion reaction.

cyclohexanone   diazomethane   cycloheptanone

$$
\text{cyclohexanone} \quad + \quad \bar{\ddot{\text{C}}}\text{H}_2-\overset{+}{\text{N}}\equiv\text{N} \quad \longrightarrow \quad \text{cycloheptanone} \quad + \quad \text{N}_2
$$

92. A compound reacts with methylmagnesium bromide followed by acidification to form the product with the following ¹H NMR spectrum. Identify the compound.

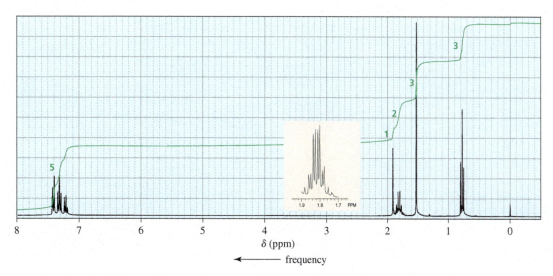

δ (ppm)

← frequency

93. Show how each of the following compounds can be prepared from the given starting material. In each case, you will need to use a protecting group.

a. CH₃CHCH₂COCH₃ $\longrightarrow$ CH₃CHCH₂CCH₃
$\quad\quad$ | $\quad$ ‖ $\quad\quad\quad\quad\quad\quad$ | $\quad\quad$ |
$\quad\quad$ OH $\;$ O $\quad\quad\quad\quad\quad\quad\quad$ OH $\;$ OH

c.

b.

94. Describe how 1-ethylcyclohexanol can be prepared from cyclohexane. You can use any inorganic reagents, any solvents, and any organic reagents as long as they contain no more than two carbons.

95. The pK_a values of the carboxylic acid groups of oxaloacetic acid are 2.22 and 3.98.

a. Which carboxyl group is the stronger acid?

b. The amount of hydrate present in an aqueous solution of oxaloacetic acid depends on the pH of the solution: 95% at pH 0, 81% at pH 1.3, 35% at pH 3.1, 13% at pH 4.7, 6% at pH 6.7, and 6% at pH 12.7. Explain this pH dependence.

oxaloacetic acid

96. The **Baylis–Hillman reaction** is a DABCO (1,4-diazabicyclo[2.2.2]octane) catalyzed reaction of an α,β-unsaturated carbonyl compound with an aldehyde to form an allylic alcohol. Propose a mechanism for the reaction. (*Hint:* DABCO serves as both a nucleophile and as a base in the reaction.)

DABCO

97. To solve this problem, you need to read the description of the Hammett σ, ρ treatment given in Chapter 15, Problem 92. When the rate constants for the hydrolysis of several morpholine enamines of para-substituted propiophenones are determined at pH 4.7, the ρ value is positive; however, when the rates of hydrolysis are determined at pH 10.4, the ρ value is negative.

    a. What is the rate-determining step of the hydrolysis reaction when it is carried out in a basic solution?

    b. What is the rate-determining step of the hydrolysis reaction when it is carried out in an acidic solution?

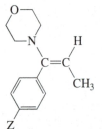

a morpholine enamine of
a para-substituted propiophenone

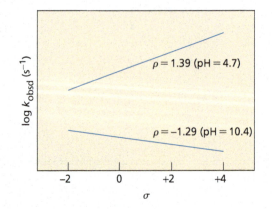

$\rho = 1.39$ (pH $= 4.7$)

$\rho = -1.29$ (pH $= 10.4$)

98. Propose a mechanism for each of the following reactions:

    a.

    b.

17

Reactions at the α-Carbon

a paper mulberry tree

Fifteen aromatase inhibitors, compounds used in the treatment of breast cancer, have been isolated from the leaves of the paper mulberry tree (see page 823).

When we looked at the reactions of carbonyl compounds in Chapters 15 and 16, we saw that their site of reactivity is the partially positively charged carbonyl carbon to which a nucleophile adds.

Many carbonyl compounds have a second site of reactivity—namely, a hydrogen bonded to *a carbon that is adjacent to the carbonyl carbon*. This hydrogen is sufficiently acidic to be removed by a strong base. The carbon adjacent to a carbonyl carbon is called an **α-carbon;** therefore, a hydrogen bonded to an α-carbon is called an **α-hydrogen.**

In Section 17.1, you will find out why a hydrogen bonded to an α-carbon is more acidic than hydrogens bonded to other sp^3 carbons, and then you will look at some reactions that result from this acidity. Later in the chapter, you will see that a hydrogen is not the only substituent that can be removed from an α-carbon: a carboxyl group bonded to an α-carbon can be removed as CO_2. At the end of the chapter, you will be introduced to some important biological reactions that rely on the ability to remove a proton or a carboxyl group from an α-carbon.

801

17.1 THE ACIDITY OF AN α-HYDROGEN

Hydrogen and carbon have similar electronegativities, which means that the two atoms share the electrons that bond them together almost equally. Consequently, a hydrogen bonded to a carbon is usually not acidic. This is particularly true for hydrogens bonded to sp^3 carbons because these carbons are the most similar to hydrogen in electronegativity (Section 7.10). For example, the pK_a of ethane is greater than 60.

$$CH_3CH_3$$

$pK_a > 60$

Hydrogens Bonded to sp^3 Carbons Adjacent to Carbonyl Carbons Are Relatively Acidic

A hydrogen bonded to an sp^3 carbon that is adjacent to a carbonyl carbon, however, is much more acidic than hydrogens bonded to other sp^3 carbons. For example:

- the pK_a value for dissociation of a proton from the α-carbon of an aldehyde or a ketone ranges from 16 to 20.
- the pK_a value for dissociation of a proton attached to the α-carbon of an ester is about 25.

The α-hydrogen of an aldehyde or a ketone is more acidic than the α-hydrogen of an ester.

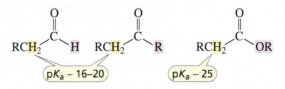

$pK_a \sim 16$–20 $pK_a \sim 25$

A compound that contains a relatively acidic hydrogen bonded to an sp^3 carbon is called a **carbon acid** (Table 17.1).

Table 17.1 The pK_a Values of Some Carbon Acids

Structure	pK_a	Structure	pK_a	Structure	pK_a
$CH_2(H)C(=O)N(CH_3)_2$	30	$CH_2(H)C{\equiv}N$	25	$CH_3CH_2O-C(=O)-CH(H)-C(=O)-OCH_2CH_3$	13.3
$CH_2(H)C(=O)OCH_2CH_3$	25	$N{\equiv}CCH(H)C{\equiv}N$	11.8	$CH_3-C(=O)-CH(H)-C(=O)-OCH_2CH_3$	10.7
$CH_2(H)C(=O)CH_3$	20	$CH_3CH(H)NO_2$	8.6	$CH_3-C(=O)-CH(H)-C(=O)-CH_3$	8.9
$CH_2(H)C(=O)H$	17	$O_2NCH(H)NO_2$	3.6	$CH_3-C(=O)-CH(H)-C(=O)-H$	5.9

A hydrogen bonded to an α-carbon is more acidic than hydrogens bonded to other sp^3 carbons because the base formed when a proton is removed from an α-carbon is relatively stable. And, as we have seen, the more stable the base, the stronger its conjugate acid (Section 2.6).

Why is the base formed by removing a proton from an α-carbon more stable than bases formed by removing a proton from other sp^3 carbons? When a proton is removed from ethane, the electrons left behind reside solely on a carbon. This carbanion with localized electrons is unstable because carbon is not very electronegative. As a result, the pK_a of its conjugate acid is very high.

localized electrons

$$CH_3CH_3 \;\rightleftharpoons\; CH_3\overset{..}{\underset{..}{C}}H_2 + H^+$$

Electron Delocalization Stabilizes the Conjugate Base

In contrast, when a proton is removed from an α-carbon, two factors combine to increase the stability of the base that is formed. First, the electrons left behind when the proton is removed are delocalized, and electron delocalization increases stability (Section 8.6). More importantly, the electrons are delocalized onto an oxygen, an atom that is better able to accommodate them because it is more electronegative than carbon.

electrons are better accommodated on O than on C

resonance contributors

delocalized electrons

Now we can understand why aldehydes and ketones (pK_a = 16–20) are more acidic than esters (pK_a = 25). The electrons left behind when a proton is removed from the α-carbon of an ester are not as readily delocalized onto the carbonyl oxygen (indicated by the red arrows) as they would be in an aldehyde or a ketone. This is because the oxygen of the OR group of the ester also has a lone pair that can be delocalized onto the carbonyl oxygen (indicated by the blue arrows). Thus, the lone pair on carbon and the lone pair on oxygen compete for delocalization onto the same oxygen.

delocalization of a lone pair on oxygen   delocalization of a lone pair on carbon

resonance contributors

Nitroalkanes, nitriles, and *N,N*-disubstituted amides also have a relatively acidic α-hydrogen (Table 17.1) because, in each case, the electrons left behind when the proton is removed can be delocalized onto an atom that is more electronegative than carbon.

$$CH_3CH_2NO_2 \qquad CH_3CH_2C\equiv N \qquad$$

nitroethane   propanenitrile   *N,N*-dimethylacetamide
pK_a = 8.6   pK_a = 26   pK_a = 30

PROBLEM 1 ◆

Explain why the pK_a of a hydrogen bonded to the sp^3 carbon of propene is greater (pK_a = 42) than that of any of the carbon acids listed in Table 17.1, but is less than the pK_a of an alkane (pK_a > 60).

If the α-carbon is *between* two carbonyl groups, the α-hydrogen is even more acidic (Table 17.1). For example, the pK_a value for dissociation of a proton from the α-carbon of 2,4-pentanedione, a compound with an α-carbon between two ketone carbonyl groups, is 8.9. And the pK_a value for dissociation of a proton from the α-carbon of ethyl 3-oxobutyrate, a compound with an α-carbon between a ketone carbonyl group and an ester carbonyl group, is 10.7. 2,4-Pentanedione is a **β-diketone** because the ketone has a second keto group at the β-position; ethyl 3-oxobutyrate is a **β-keto ester.**

$pK_a = 8.9$ $pK_a = 10.7$

2,4-pentanedione
acetylacetone
a β-diketone

ethyl 3-oxobutyrate
ethyl acetoacetate
a β-keto ester

The acidity of α-hydrogens bonded to carbons flanked by two carbonyl groups increases because the electrons left behind when the proton is removed can be delocalized onto *two* oxygens. β-Diketones have lower pK_a values than β-keto esters because, as we saw on page 803, electrons are more readily delocalized onto the carbonyl oxygen of a ketone than they are onto the carbonyl oxygen of an ester.

2,4-pentanedione

resonance contributors for the 2,4-pentanedione anion

PROBLEM 2 ◆

Give an example for each of the following:

a. a β-keto nitrile   **b.** a β-diester   **c.** a β-keto aldehyde

---

PROBLEM-SOLVING STRATEGY

LEARN THE STRATEGY

The Acid–Base Behavior of a Carbonyl Compound

Explain why a base cannot remove a proton from the α-carbon of a carboxylic acid.

If a base cannot remove a proton from the α-carbon of a carboxylic acid, then the base must react with another portion of the molecule more rapidly. Because the proton on the carboxyl group is more acidic ($pK_a \sim 5$) than the proton on the α-carbon, we can conclude that the base removes a proton from the carboxyl group rather than from the α-carbon.

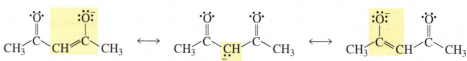

USE THE STRATEGY

PROBLEM 3 ◆

Explain why a base can remove a proton from the α-carbon of *N,N*-dimethylethanamide but not from the α-carbon of either *N*-methylethanamide or ethanamide.

N,N-dimethylethanamide   *N*-methylethanamide   ethanamide

PROBLEM 4 ◆

Explain why the α-hydrogen of an *N,N*-disubstituted amide is less acidic (pK_a = 30) than the α-hydrogen of an ester (pK_a = 25).

PROBLEM 5 ◆

Rank the compounds in each of the following groups from strongest acid to weakest acid:

a. 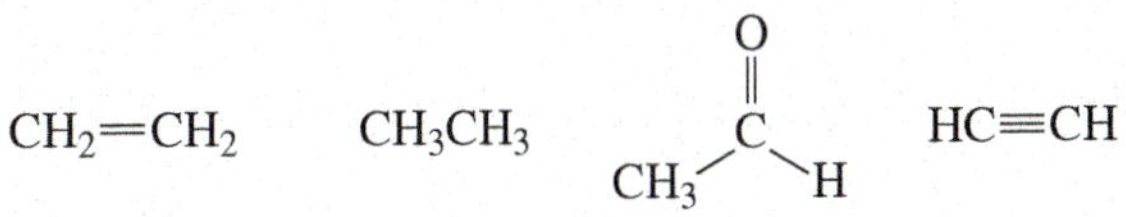 $CH_2{=}CH_2$ CH_3CH_3 (acetaldehyde) $HC{\equiv}CH$

b. 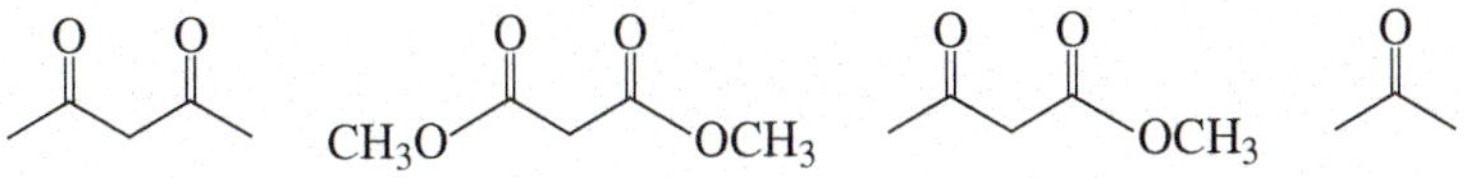 (structures)

c. 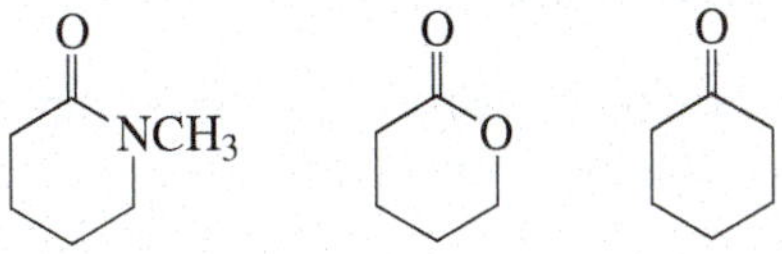 (structures)

17.2 KETO–ENOL TAUTOMERS

A ketone exists in equilibrium with its enol tautomer. Recall that **tautomers** are isomers that are in rapid equilibrium (Section 7.7). Keto–enol tautomers differ in the location of a double bond and a hydrogen.

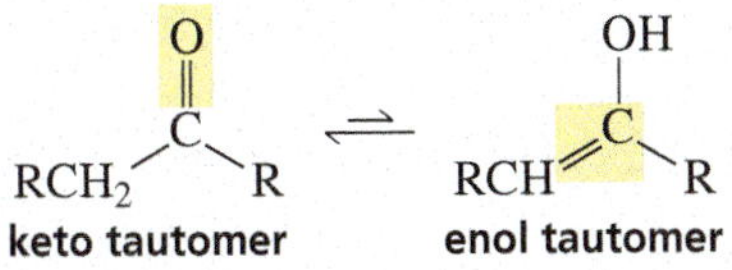

(keto tautomer ⇌ enol tautomer)

For most ketones, the **enol tautomer** is much less stable than the **keto tautomer.** For example, an aqueous solution of acetone exists as an equilibrium mixture of more than 99.9% keto tautomer and less than 0.1% enol tautomer.

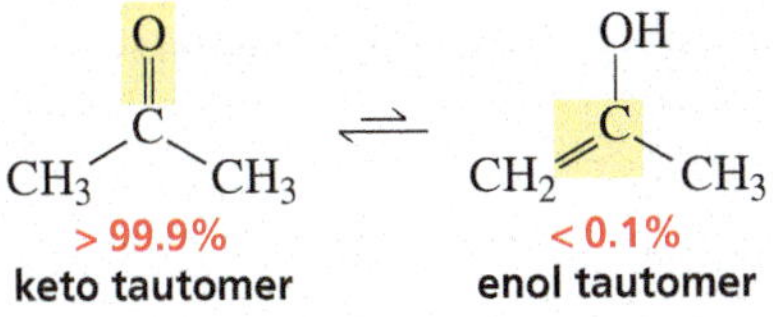

keto tautomer >99.9% ⇌ enol tautomer <0.1%

The fraction of the enol tautomer in an aqueous solution is considerably greater for a β-diketone because the enol tautomer is stabilized both by intramolecular hydrogen bonding and by conjugation of the carbon–carbon double bond with the second carbonyl group.

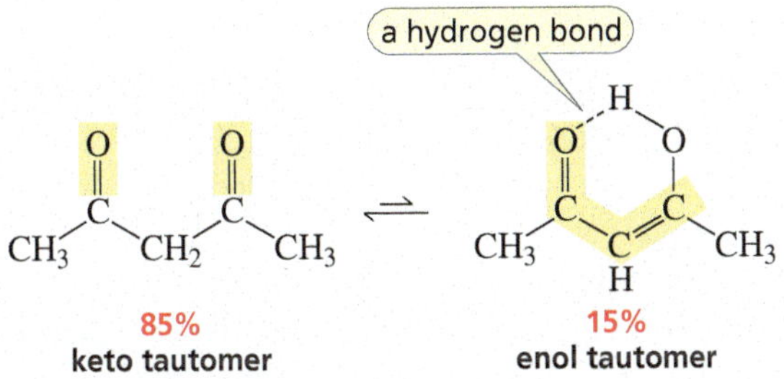

keto tautomer 85% ⇌ enol tautomer 15%

Phenol is unusual in that its enol tautomer is *more* stable than its keto tautomer because the enol tautomer is aromatic, but the keto tautomer is not (Section 8.17).

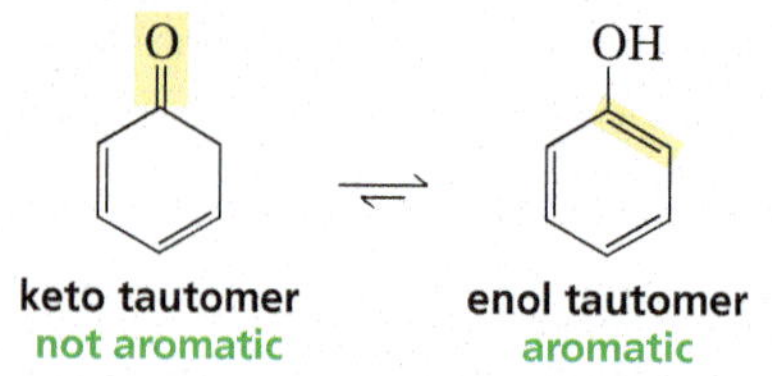

keto tautomer
not aromatic

enol tautomer
aromatic

PROBLEM 6

Explain why 92% of 2,4-pentanedione exists as the enol tautomer in hexane but only 15% of this compound exists as the enol tautomer in water.

17.3 KETO–ENOL INTERCONVERSION

Now that we know that an α-hydrogen is somewhat acidic, we can better understand why keto and enol tautomers interconvert as we first saw in Section 7.7. **Keto–enol interconversion** (also called **tautomerization**) can be catalyzed by either a base or an acid.

Base-Catalyzed Keto–Enol Interconversion

THE MECHANISM

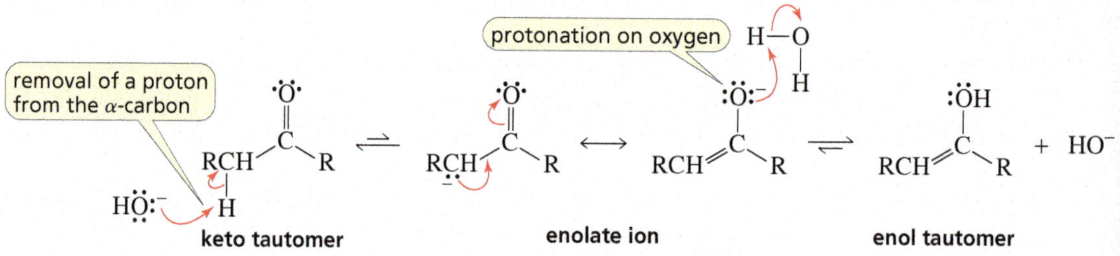

- Hydroxide ion removes a proton from the α-carbon of the keto tautomer, forming an anion called an **enolate ion.** The enolate ion has two resonance contributors.
- Protonating the oxygen forms the enol tautomer, whereas protonating the α-carbon re-forms the keto tautomer.

Acid-Catalyzed Keto–Enol Interconversion

THE MECHANISM

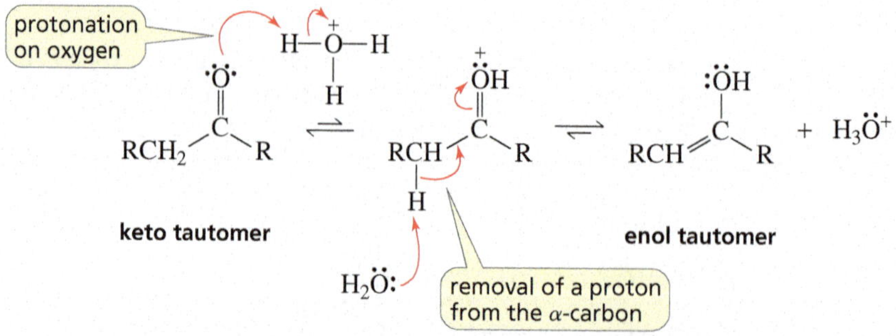

- The acid protonates the carbonyl oxygen of the keto tautomer.
- Water removes a proton from the α-carbon, forming the enol tautomer.

Notice that the steps are reversed in the base- and acid-catalyzed interconversions. In the base-catalyzed reaction, the base removes a proton from an α-carbon in the first step and the oxygen is protonated in the second step. In the acid-catalyzed reaction, the oxygen is protonated in the first step and the proton is removed from the α-carbon in the second step. Notice also that the catalyst, as expected, is regenerated in both the acid- and base-catalyzed reactions.

PROBLEM 7 ◆

Draw the enol tautomers for each of the following compounds. For compounds that have more than one enol tautomer, indicate the one that is more stable.

a.

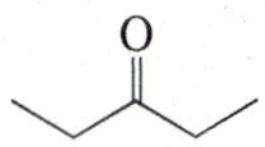

c.

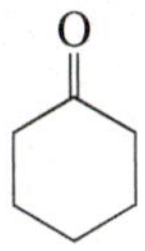

e.

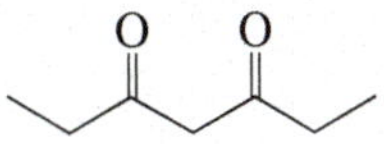

b.

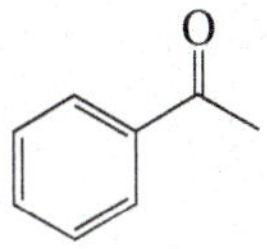

d.

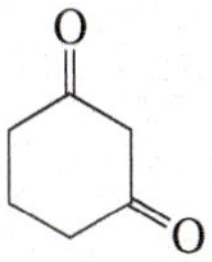

f.

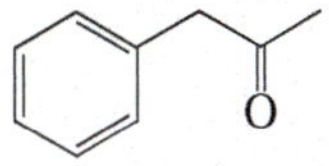

PROBLEM 8

When a dilute solution of acetaldehyde in D_2O containing NaOD is shaken, explain why the methyl hydrogens are exchanged with deuterium but the hydrogen attached to the carbonyl carbon is not.

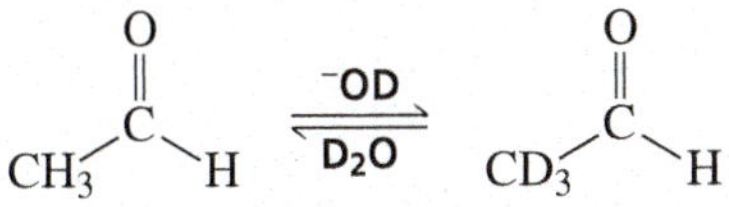

$$\underset{CH_3}{}\overset{O}{\underset{}{\|}}\overset{}{\underset{}{C}}-H \quad \underset{D_2O}{\overset{-OD}{\rightleftharpoons}} \quad \underset{CD_3}{}\overset{O}{\underset{}{\|}}\overset{}{\underset{}{C}}-H$$

17.4   HALOGENATION OF THE α-CARBON OF ALDEHYDES AND KETONES

When Br_2, Cl_2, or I_2 is added to a solution of an aldehyde or a ketone, a halogen replaces *one or more* of the α-hydrogens of the carbonyl compound. The reaction can be catalyzed by either an acid or a base. This is an **α-substitution reaction** because one electrophile (Br^+) is substituted for another (H^+) on the α-carbon.

Acid-Catalyzed Halogenation

In the acid-catalyzed reaction, the halogen replaces *one* of the α-hydrogens:

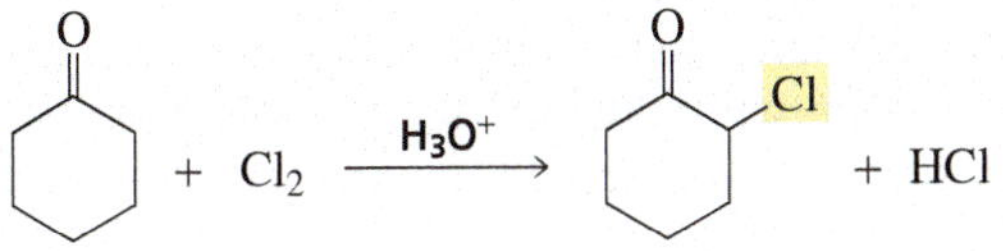

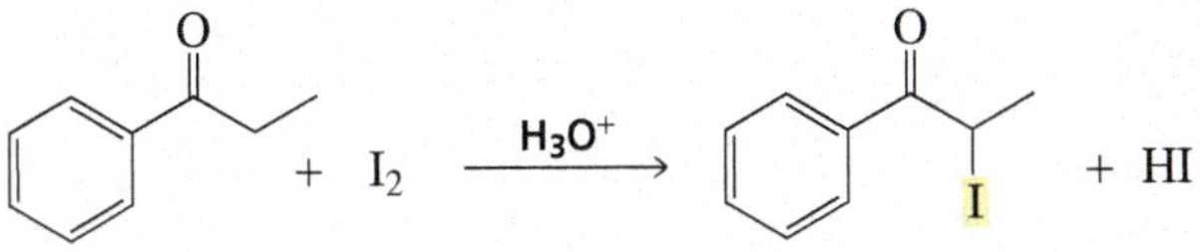

$$+ \ Cl_2 \ \xrightarrow{H_3O^+} \quad + \ HCl$$

Under acidic conditions, one α-hydrogen is substituted for a halogen.

$$+ \ I_2 \ \xrightarrow{H_3O^+} \quad + \ HI$$

THE MECHANISM

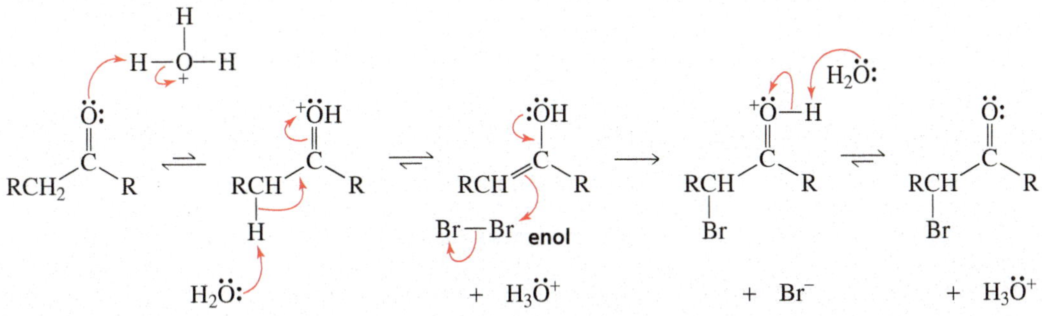

- The carbonyl oxygen is protonated.
- Water removes a proton from the α-carbon, forming an enol.
- The enol reacts with an electrophilic halogen; the other halogen atom keeps the bonding electrons.
- The very acidic protonated carbonyl group loses a proton.

Base-Promoted Halogenation

When excess Br_2, Cl_2, or I_2 is added to a *basic* solution of an aldehyde or a ketone, the halogen replaces *all* of the α-hydrogens.

Under basic conditions, *all* of the α-hydrogens are substituted for halogens.

THE MECHANISM

- Hydroxide ion removes a proton from the α-carbon, forming an enolate ion.
- The enolate ion reacts with the electrophilic halogen; the other halogen atom keeps the bonding electrons. Notice that the hydroxide ion is not regenerated; therefore, it is promoting the reaction, not catalyzing it.

These two steps are repeated until all of the α-hydrogens are replaced by a halogen. Each successive halogenation is *more rapid* than the previous one because the electron-withdrawing halogen atom increases the acidity of the remaining α-hydrogen(s). This is why *all* of the α-hydrogens are replaced by halogens.

Under acidic conditions, on the other hand, each successive halogenation is *slower* than the previous one because the electron-withdrawing halogen atom decreases the basicity of the carbonyl oxygen, thereby making protonation of the carbonyl oxygen (the first step in the acid-catalyzed reaction) less favorable.

Comparing Keto–Enol Interconversion and α-Substitution

Notice the similarity between keto–enol interconversion and α-substitution. Actually, keto–enol interconversion is an α-substitution reaction in which hydrogen serves as both the electrophile that is removed from the α-carbon and the electrophile that is added to the α-carbon when the enol or enolate ion reverts back to the keto tautomer.

PROBLEM 9

Explain why a racemic mixture is formed when (*R*)-2-methylpentanal is dissolved in an acidic or basic solution.

PROBLEM 10 ◆

A ketone undergoes acid-catalyzed bromination, acid-catalyzed chlorination, racemization, (see page 408) and acid-catalyzed deuterium exchange at the α-carbon. All of these reactions have similar rate constants. What does this tell you about the mechanisms of these reactions?

17.5 HALOGENATION OF THE α-CARBON OF CARBOXYLIC ACIDS

Carboxylic acids cannot undergo substitution reactions at the α-carbon because a base removes a proton from the OH group instead of from the α-carbon, since the OH group is more acidic. If, however, a carboxylic acid is treated with PBr_3 and Br_2, the α-carbon will be brominated. This halogenation reaction is called the **Hell–Volhard–Zelinsky reaction** or, more simply, the **HVZ reaction.**

the Hell–Volhard–Zelinsky (HVZ) reaction

You will see, when you examine the reaction, that α-substitution occurs because an acyl bromide, rather than a carboxylic acid, is the compound that undergoes α-substitution.

STEPS IN THE HELL–VOLHARD–ZELINSKY REACTION

- PBr_3 converts the carboxylic acid into an acyl bromide (Section 15.18), which is in equilibrium with its enol.
- Bromination of the enol forms a protonated α-brominated acyl bromide, which is hydrolyzed to a α-brominated carboxylic acid (Section 15.6).

Replacing the α-Halogen of Carbonyl Compounds

A halogen attached to the α-carbon of a carbonyl compound can be replaced only by a poor nucleophile (weak base). A strong base will form α,β-unsaturated carbonyl compound by removing a proton from the β-carbon and eliminating the bromide ion. Recall that a weak base favors substitution and a strong base favors elimination (Section 9.12).

PROBLEM 11

Show how the following compounds can be prepared from the given starting material:

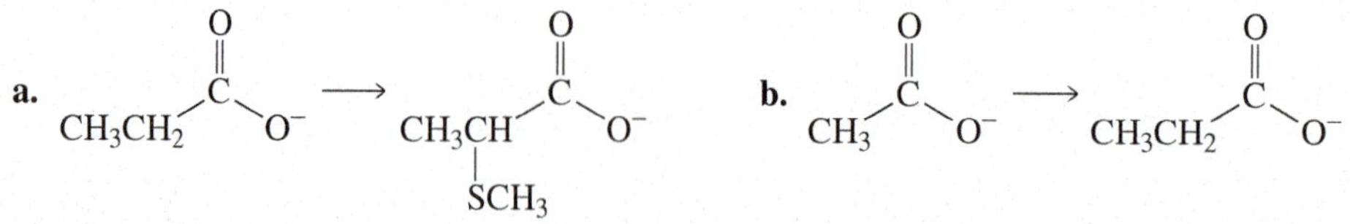

a.

b.

17.6 FORMING AN ENOLATE ION

The amount of carbonyl compound converted to an enolate ion depends on the pK_a of the carbonyl compound and the particular base used to remove the α-hydrogen.

For example, when hydroxide ion or an alkoxide ion is used to remove an α-hydrogen from cyclohexanone, only a small amount of the carbonyl compound is converted to the enolate ion because the product acid (H_2O) is a *stronger acid* than the reactant acid (the ketone). (Recall that the equilibrium of an acid–base reaction favors dissociation of the strong acid and formation of the weak acid; see Section 2.5.)

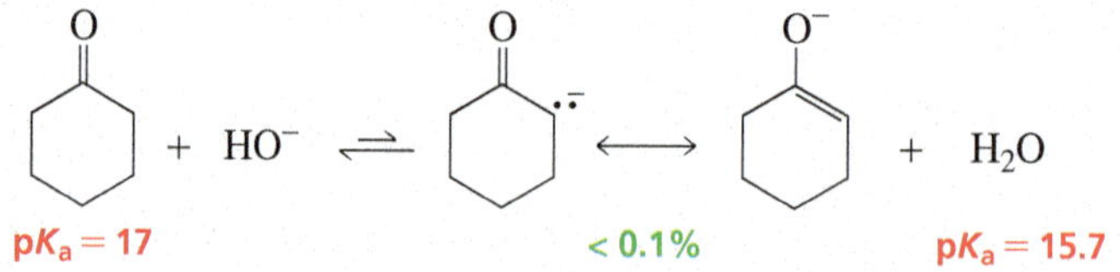

In contrast, when LDA (lithium diisopropylamide) is used to remove an α-hydrogen, essentially all the carbonyl compound is converted to the enolate ion because the product acid (diisopropylamine, or DIA) is a much *weaker acid* than the reactant acid (the ketone). This is why LDA is the base of choice for those reactions that require the carbonyl compound to be completely converted to an enolate ion before it reacts with an electrophile (Section 17.7).

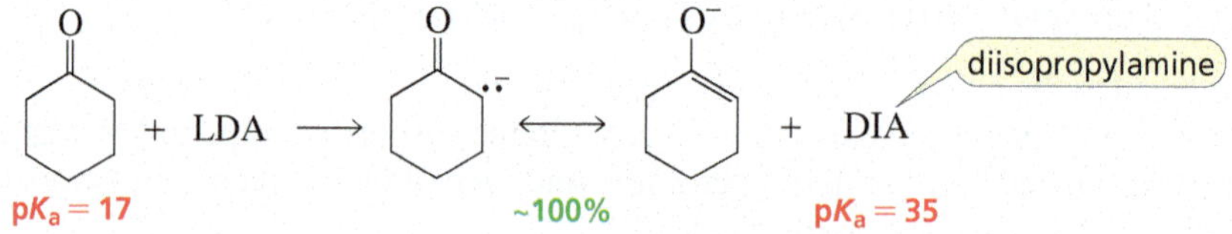

LDA is easily prepared by adding butyllithium to diisopropylamine in THF at −78 °C (that is, at the temperature of a dry ice/acetone bath.)

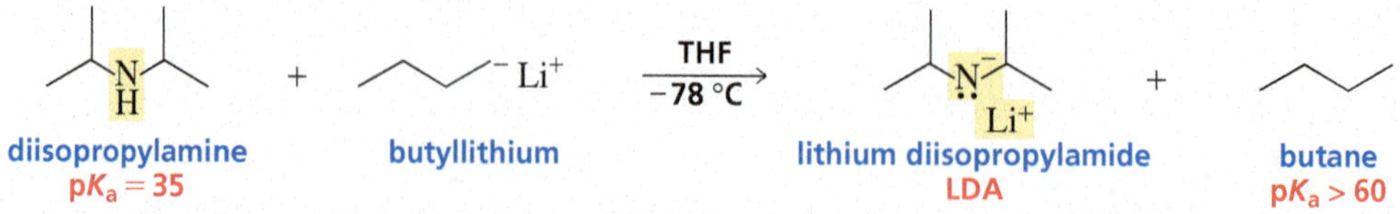

Using a nitrogen base to form an enolate ion can be a problem because a nitrogen base can also react as a nucleophile and add to the carbonyl carbon (Section 16.8). However, the two bulky isopropyl substituents attached to the nitrogen of LDA make it difficult for the nitrogen to get close enough to the carbonyl carbon to react with it. Consequently, LDA is a strong base but a poor nucleophile—that is, it removes an α-hydrogen much faster than it adds to a carbonyl carbon.

PROBLEM 12 ◆
What compound is formed when a dilute solution of cyclohexanone is shaken with NaOD in D_2O for several hours?

17.7 ALKYLATING THE α-CARBON

Putting an alkyl group on the α-carbon of a carbonyl compound is an important reaction because it gives us another way to form a carbon–carbon bond. Alkylation is carried out by first removing a proton from the α-carbon with a strong base, such as LDA, to form an enolate ion and then adding the appropriate alkyl halide. Because the alkylation is an S_N2 reaction, it works best with methyl halides and primary alkyl halides (Section 9.2).

Although an enolate ion has two resonance contributors, for the sake of simplicity, only the resonance contributor with the negative charge on carbon is shown in many of the reactions in this chapter. For example, compare the following presentation with the preceding one.

It is important that a strong base such as LDA is used to form the enolate ion. If a weaker base such as hydroxide ion or an alkoxide ion is used, very little of the desired monoalkylated product will be obtained. We saw that these weaker bases form only a small amount of enolate ion at equilibrium (Section 17.6). Therefore, most of the hydroxide will be present when the alkyl halide is added to the reaction mixture. As a result, the major product of the reaction will be an alcohol formed from an S_N2 reaction of hydroxide ion with the alkyl halide, and the alkylated ketone will be a minor product.

Esters and nitriles can also be alkylated on the α-carbon.

Ketones, aldehydes, esters, and nitriles can be alkylated on the α-carbon.

Alkylating Unsymmetrical Ketones

If the ketone is unsymmetrical and has hydrogens on both α-carbons, two monoalkylated products can be obtained because either α-carbon can be alkylated. For example, methylation of 2-methylcyclohexanone with one equivalent of methyl iodide forms both 2,6-dimethylcyclohexanone and 2,2-dimethylcyclohexanone. The relative amounts of the two products depend on the reaction conditions.

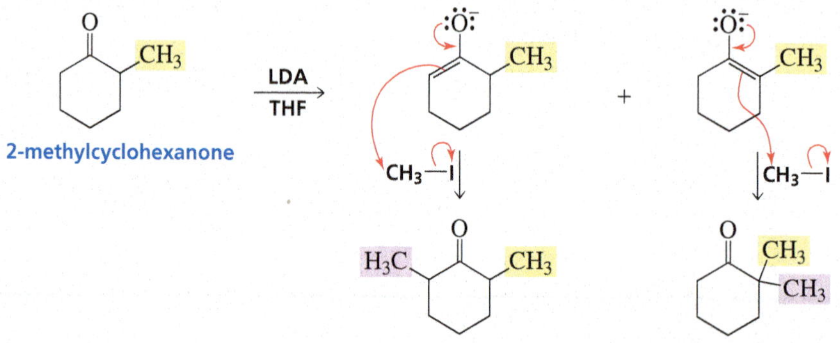

2-methylcyclohexanone

2,6-dimethylcyclohexanone   2,2-dimethylcyclohexanone

Kinetic Enolate Ion

The kinetic enolate ion is formed by removing a proton from the least substituted carbon.

LDA at −78 °C forms the kinetic enolate ion.

The enolate ion leading to 2,6-dimethylcyclohexanone is the *kinetic* enolate ion because it is formed faster. The α-hydrogen that is removed to make this enolate ion is more accessible to the base and it is slightly more acidic. Because 2,6-dimethylcyclohexanone is formed faster, it is the major product if the reaction is carried out under conditions (LDA at −78 °C) that cause the reaction to be irreversible (Section 8.13).

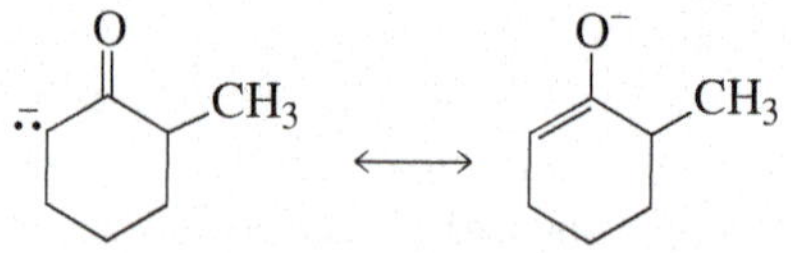

kinetic enolate ion

Thermodynamic Enolate Ion

The thermodynamic enolate ion is formed by removing a proton from the most substituted carbon.

LDA at 0 °C forms the thermodynamic enolate ion.

The enolate ion leading to 2,2-dimethylcyclohexanone is the *thermodynamic* enolate ion because it is the more stable enolate ion since it has the more substituted double bond. (Alkyl substitution increases enolate ion stability for the same reason that it increases alkene stability; Section 5.9.) Therefore, 2,2-dimethylcyclohexanone is the major product if the reaction is carried out under conditions that cause enolate ion formation to be reversible (LDA at 0 °C).

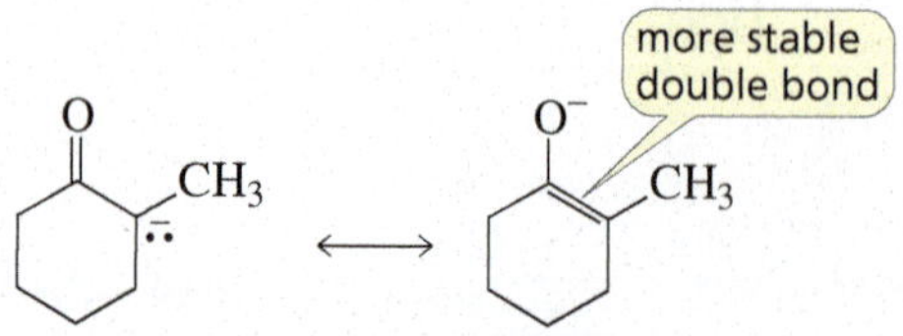

more stable double bond

thermodynamic enolate ion

The Synthesis of Aspirin

The industrial synthesis of aspirin starts with a reaction known as the **Kolbe–Schmitt carboxylation reaction.** In this reaction, a phenolate ion reacts with carbon dioxide under pressure to form salicylic acid. Reaction of salicylic acid with acetic anhydride forms acetylsalicylic acid (aspirin).

salicylic acid

acetylsalicylic acid
aspirin

During World War I, the Bayer Company in Germany bought as much phenol as it could on the international market, knowing that eventually all of it could be used to manufacture aspirin. This left little phenol available for other countries to purchase for the synthesis of 2,4,6-trinitrophenol (picric acid), a common explosive at that time.

PROBLEM-SOLVING STRATEGY

Alkylating a Carbonyl Compound

LEARN THE STRATEGY

How could 4-methyl-3-hexanone be prepared from a ketone containing no more than six carbons?

4-methyl-3-hexanone

Either of two sets of ketone and alkyl halide could be used for the synthesis: one set is 3-hexanone and a methyl halide; the other is 3-pentanone and an ethyl halide.

3-hexanone   + CH_3Br or   **3-pentanone**   + CH_3CH_2Br

3-Pentanone and an ethyl halide are the preferred starting materials because 3-pentanone is symmetrical; therefore, only one α-substituted ketone will be formed.

3-pentanone $\xrightarrow[\text{THF}]{\text{LDA}}$ $\xrightarrow{CH_3CH_2Br}$ **4-methyl-3-hexanone**

In contrast, 3-hexanone can form two different enolate ions; therefore, two α-substituted ketones will be formed, decreasing the yield of the target molecule.

3-hexanone $\xrightarrow[\text{THF}]{\text{LDA}}$ +

$\downarrow CH_3Br$ $\downarrow CH_3Br$

4-methyl-3-hexanone   +   **2-methyl-3-hexanone**

PROBLEM 14

How could each of the following compounds be prepared from a ketone and an alkyl halide?

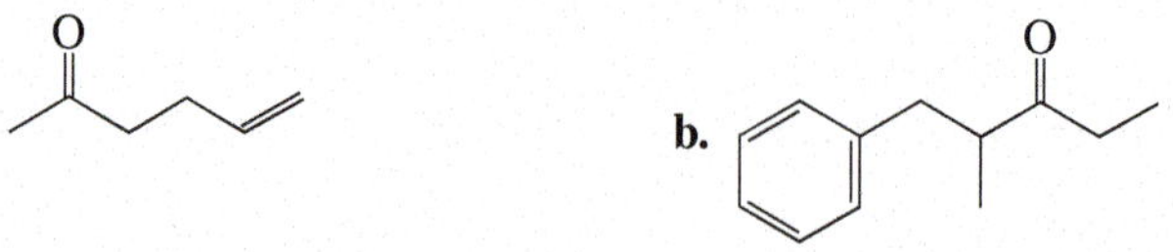

a. **b.**

PROBLEM 15 ◆

How many stereoisomers are obtained from each of the syntheses described in Problem 14?

PROBLEM 16

How could each of the following compounds be prepared from cyclohexanone?

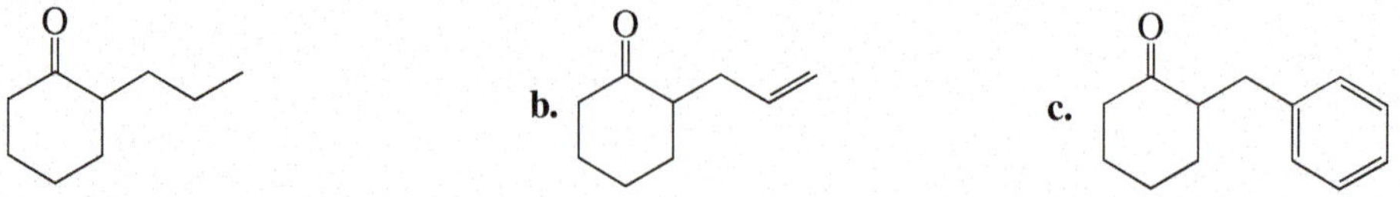

a. **b.** **c.**

17.8 ALKYLATING AND ACYLATING THE α-CARBON VIA AN ENAMINE INTERMEDIATE

We saw that an enamine is formed when an aldehyde or a ketone reacts with a secondary amine (Section 16.8).

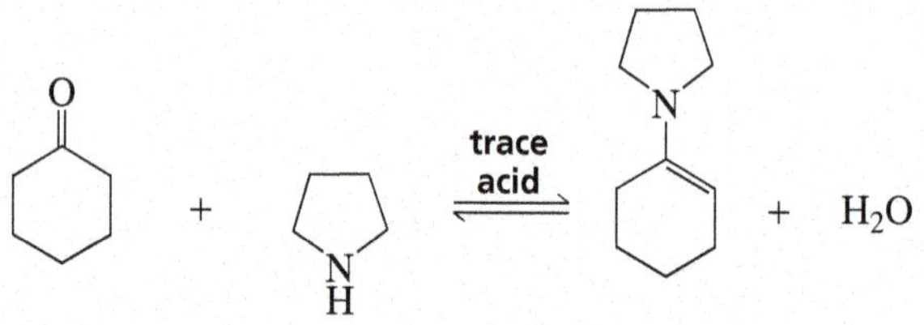

Enamines react with electrophiles in the same way that enolate ions do.

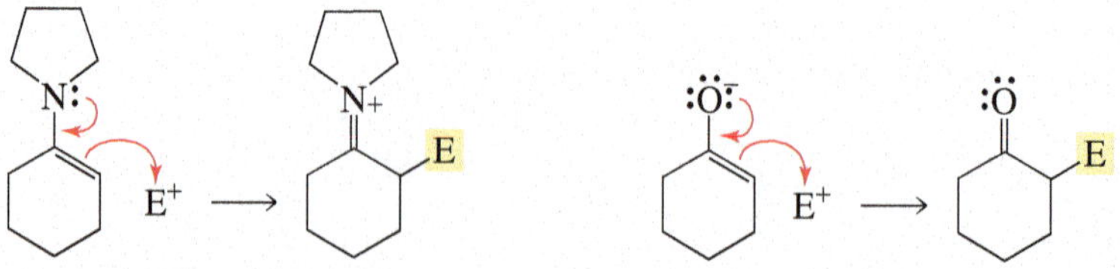

an enamine reacts with an electrophile

an enolate ion reacts with an electrophile

As a result, aldehydes and ketones can be alkylated at the α-carbon by first forming an enamine of the carbonyl compound:

STEPS IN THE ALKYLATION OF AN α-CARBON VIA AN ENAMINE

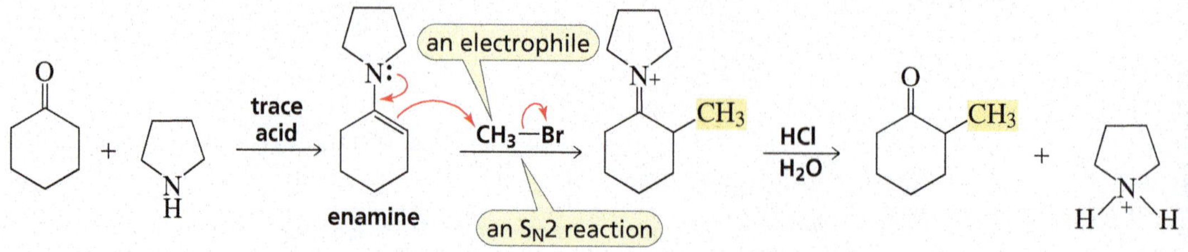

- The carbonyl compound is converted to an enamine (by treating it with a secondary amine in the presence of a trace amount of acid).
- The enamine reacts with the alkyl halide in an S_N2 reaction.
- The imine is hydrolyzed under acidic conditions to an α-alkylated ketone and the secondary amine (now protonated) that was used to form the enamine.

Because the alkylation step is an S_N2 reaction, only primary alkyl halides or methyl halides should be used (Section 9.2). The main advantage to using an enamine intermediate to alkylate an aldehyde or a ketone is that it forms a monoalkylated product without having to use a strong base (LDA).

STEPS IN THE ACYLATION OF AN α-CARBON VIA AN ENAMINE

In addition to using an enamine to alkylate the α-carbon of an aldehyde or ketone, it can also be used to acylate the α-carbon.

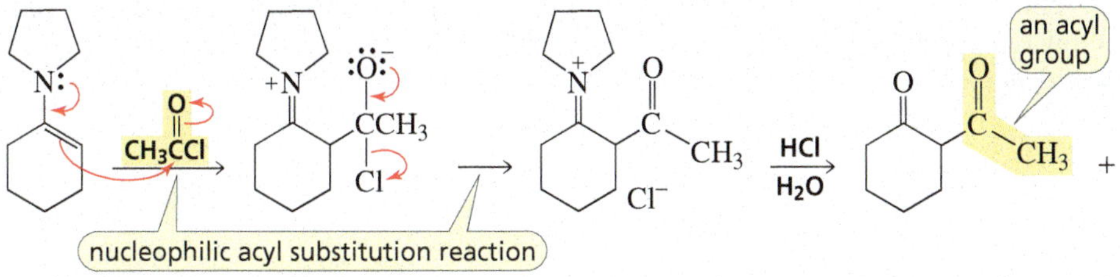

PROBLEM 17

Describe how the following compounds could be prepared from cyclohexanone using an enamine intermediate:

a.

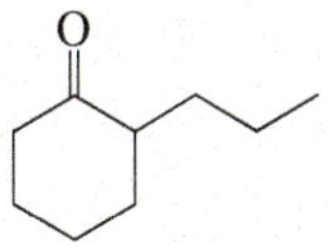

b.

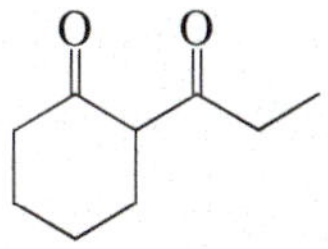

17.9 ALKYLATING THE β-CARBON

In Section 16.15, we saw that nucleophiles react with α,β-unsaturated aldehydes and ketones, forming either direct addition products and/or conjugate addition products.

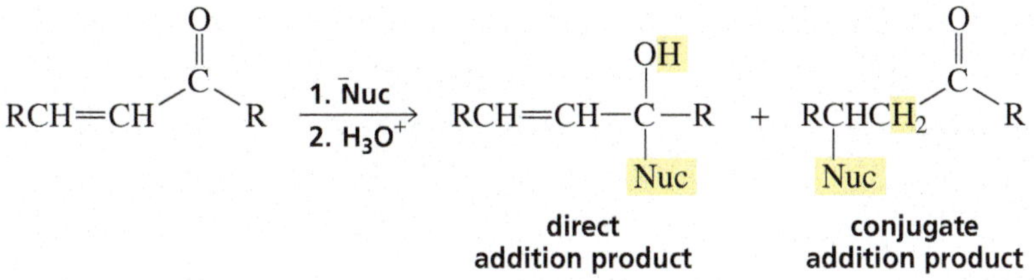

direct addition product

conjugate addition product

Recall that when the nucleophile (Nuc) is a relatively weak base, conjugate addition (adding the nucleophile to the β-carbon) predominates (Section 16.15).

Alkylating the β-Carbon via an Enamine

Enamines, which are relatively weak bases, attach the α-carbon of an aldehyde or ketone to the β-carbon of an α,β-unsaturated carbonyl compound via conjugate addition. The product is a *1,5-dicarbonyl compound*.

Alkylating the β-Carbon via a Michael Reaction

When the nucleophile in a conjugate addition reaction is an enolate ion, the reaction is called a **Michael reaction.** Enolate ions that work best in Michael reactions are formed from carbon acids that are flanked by two electron-withdrawing groups—that is, enolate ions of β-diketones, β-diesters, β-keto esters, and β-keto nitriles.

Because these enolate ions are relatively weak bases, conjugate addition occurs—that is, addition to the β-carbon of α,β-unsaturated aldehydes and ketones predominates over direct addition. These enolate ions also add to the β-carbon of α,β-unsaturated esters and amides because of the low reactivity of their carbonyl groups. Notice that *a Michael reaction also forms a 1,5-dicarbonyl compound.*

A Michael reaction forms a 1,5-dicarbonyl compound.

MECHANISM FOR A MICHAEL REACTION

- A base removes a proton from the α-carbon of the carbon acid.
- The enolate ion adds to the β-carbon of an α,β-unsaturated carbonyl compound.
- The α-carbon is protonated.

If either of the reactants in a Michael reaction has an ester group, then the base used to remove the α-proton must be the same as the leaving group of the ester (Section 17.13).

PROBLEM 18

Draw the products of the following reactions:

a. 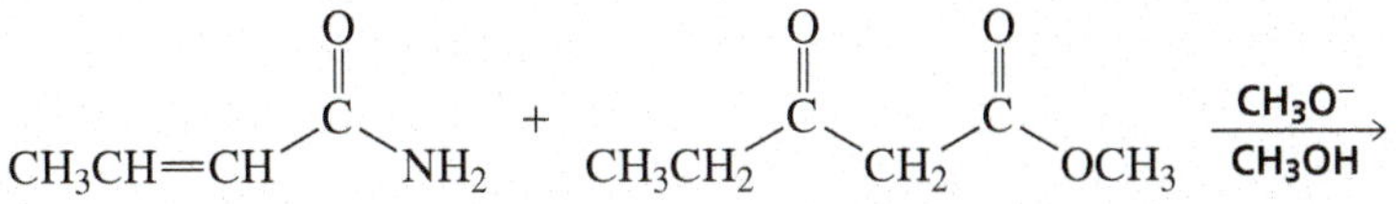 $CH_3CH=CH-\overset{\overset{\displaystyle O}{\|}}{C}-NH_2$ + $CH_3CH_2-\overset{\overset{\displaystyle O}{\|}}{C}-CH_2-\overset{\overset{\displaystyle O}{\|}}{C}-OCH_3$ $\xrightarrow[\text{CH}_3\text{OH}]{\text{CH}_3\text{O}^-}$

b. 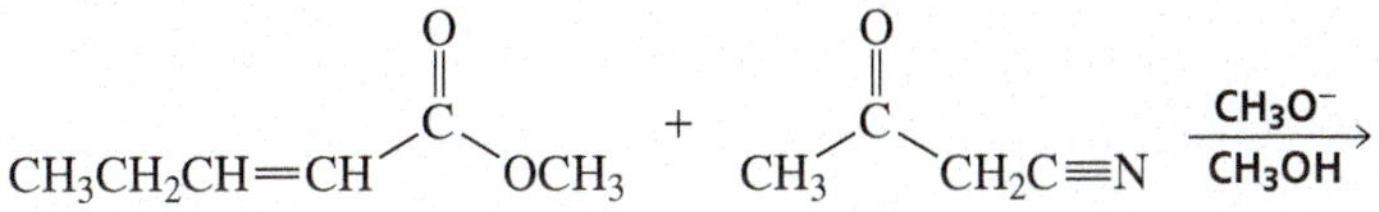 $CH_3CH_2CH=CH-\overset{\overset{\displaystyle O}{\|}}{C}-OCH_3$ + $CH_3-\overset{\overset{\displaystyle O}{\|}}{C}-CH_2C\equiv N$ $\xrightarrow[\text{CH}_3\text{OH}]{\text{CH}_3\text{O}^-}$

PROBLEM 19 ◆

What reagents should be used to prepare the following compounds?

a.

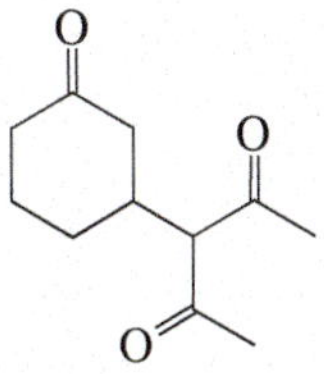

b.

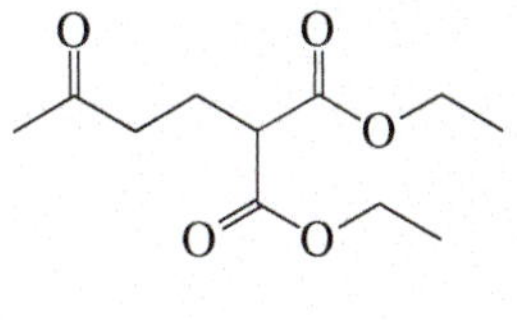

c.

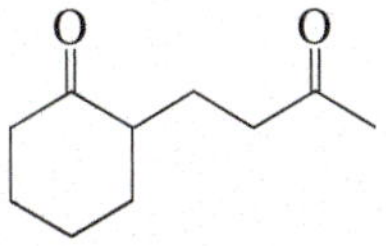

17.10 AN ALDOL ADDITION FORMS A β-HYDROXYALDEHYDE OR A β-HYDROXYKETONE

You learned in Chapter 16 that the carbonyl carbon of an aldehyde or a ketone is an electrophile. You just learned that a proton can be removed from the α-carbon of an aldehyde or a ketone, converting the α-carbon to a nucleophile.

An **aldol addition** is a reaction in which *both* of these reactivities are observed. That is, one molecule of a carbonyl compound—after a proton is removed from an α-carbon—reacts as a *nucleophile* and adds to the *electrophilic* carbonyl carbon of a second molecule of the carbonyl compound.

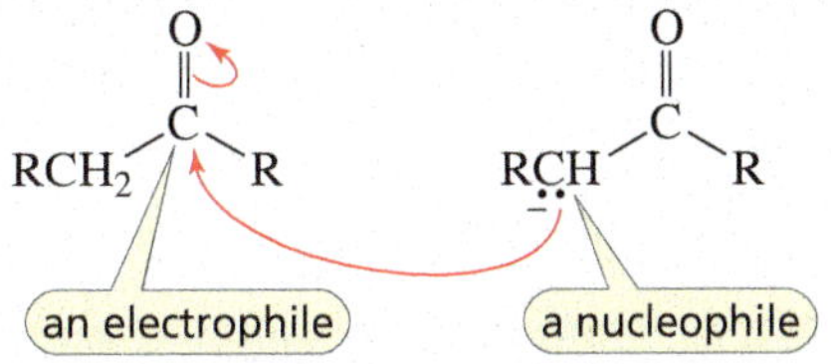

An Aldol Addition

An aldol addition is a reaction between two molecules of an *aldehyde* or two molecules of a *ketone*. Notice that the reaction forms a new C—C bond that connects the α-carbon of one molecule (the nucleophile) and the carbon that was originally the carbonyl carbon of the other molecule (the electrophile). That is, the carbon attached to the OH group is adjacent to the carbon attached to the CH=O group.

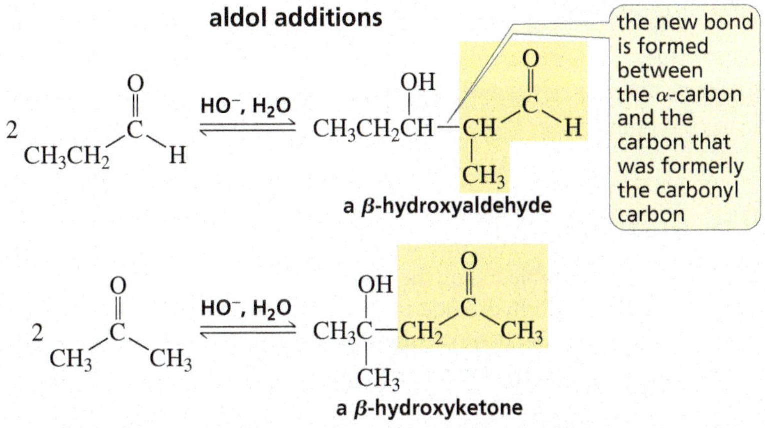

aldol additions

$2 \;\; CH_3CH_2-\overset{\overset{\displaystyle O}{\|}}{C}-H$ $\xrightarrow{\text{HO}^-,\ \text{H}_2\text{O}}$ $CH_3CH_2CH-\overset{\overset{\displaystyle OH}{|}}{CH}-\overset{\overset{\displaystyle O}{\|}}{C}-H$
                                                              $\overset{|}{CH_3}$

the new bond is formed between the α-carbon and the carbon that was formerly the carbonyl carbon

a β-hydroxyaldehyde

$2 \;\; CH_3-\overset{\overset{\displaystyle O}{\|}}{C}-CH_3$ $\xrightarrow{\text{HO}^-,\ \text{H}_2\text{O}}$ $CH_3\overset{\overset{\displaystyle OH}{|}}{C}-CH_2-\overset{\overset{\displaystyle O}{\|}}{C}-CH_3$
                                                      $\overset{|}{CH_3}$

a β-hydroxyketone

When the reactant is an aldehyde, the product is a **β-hydroxyaldehyde,** which is why the reaction is called an aldol addition ("ald" for *aldehyde*, "ol" for *alcohol*). When the reactant is a ketone, the product is a **β-hydroxyketone.** Because the addition reaction is reversible, good yields of the addition product are obtained only if it is removed from the solution as it is formed.

MECHANISM FOR THE ALDOL ADDITION

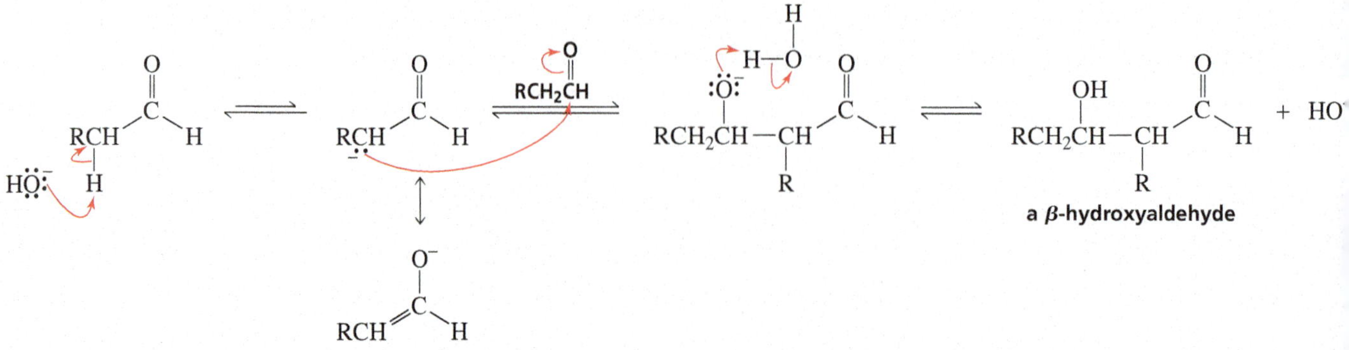

- A base removes a proton from the α-carbon, creating an enolate ion.
- The enolate ion adds to the carbonyl carbon of a second molecule of the carbonyl compound.
- The negatively charged oxygen is protonated.

A ketone forms an aldol addition product by the same mechanism.

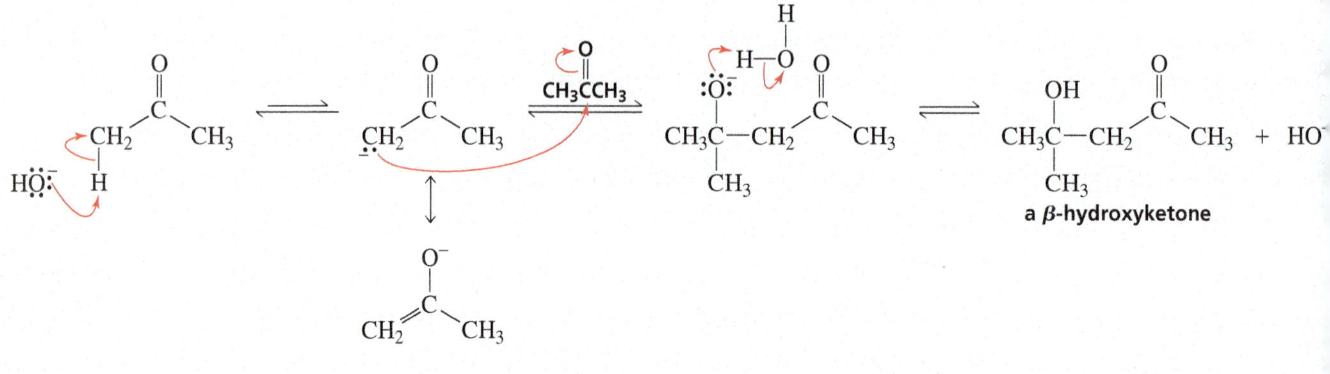

Notice that the aldol addition is a nucleophilic addition reaction. It is just like the nucleophilic addition reactions that aldehydes and ketones undergo with other carbon nucleophiles (Section 16.4). Because an aldol addition occurs between two molecules of the same carbonyl compound, the product has twice as many carbons as the reacting aldehyde or ketone.

An aldol addition forms a β-hydroxyaldehyde or a β-hydroxyketone.

PROBLEM 20

What aldol addition product is formed from each of the following compounds?

a. 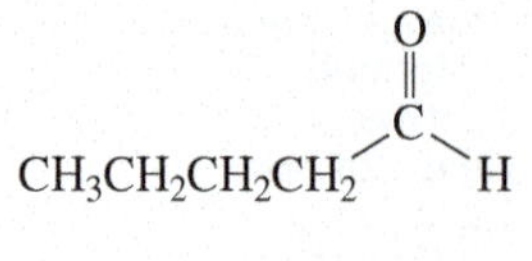 $CH_3CH_2CH_2CH_2$ — CHO

b. 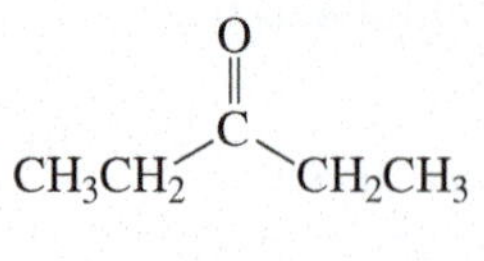 CH_3CH_2 — CO — CH_2CH_3

c. 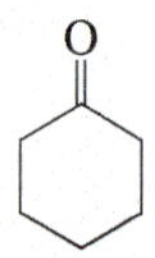 (cyclohexanone)

A Retro-Aldol Addition

Because an aldol addition is reversible, when the product of an aldol addition (the β-hydroxyaldehyde or β-hydroxyketone) is heated with hydroxide ion and water, the aldehyde or ketone that formed the aldol addition product can be regenerated (see the top of the next page). In Section 17.21 we will see that a retro-aldol addition is an important reaction in glycolysis.

PROBLEM 21 ◆

What aldehyde or ketone would be obtained when each of the following compounds is heated in a basic aqueous solution?

a. 2-ethyl-3-hydroxyhexanal

b. 4-hydroxy-4-methyl-2-pentanone

c. 2,4-dicyclohexyl-3-hydroxybutanal

d. 5-ethyl-5-hydroxy-4-methyl-3-heptanone

17.11 THE DEHYDRATION OF ALDOL ADDITION PRODUCTS FORMS α,β-UNSATURATED ALDEHYDES AND KETONES

Dehydration Under Acidic Conditions

We saw that alcohols dehydrate when they are heated with acid (Section 10.4). The β-hydroxyaldehyde and β-hydroxyketone products of aldol addition reactions are easier to dehydrate than many other alcohols because the double bond formed when the compound is dehydrated is conjugated with a carbonyl group. Conjugation increases the stability of the product and, therefore, makes it easier to form (Section 8.7).

When the product of an aldol addition is dehydrated, the overall reaction is called an **aldol condensation.** A **condensation reaction** is a reaction that combines two molecules while removing a small molecule. Water is the small molecule that is removed in an aldol condensation. Notice that an aldol condensation forms an **α,β-unsaturated aldehyde** or an **α,β-unsaturated ketone,** called enones ("ene" for the double bond and "one" for the carbonyl group).

An aldol addition product loses water to form an aldol condensation product.

Unlike alcohols that can be dehydrated only under acidic conditions, β-hydroxyaldehydes and β-hydroxyketones can also be dehydrated under basic conditions.

An aldol condensation forms an α,β-unsaturated aldehyde or an α,β-unsaturated ketone.

Dehydration Under Basic Conditions

In Chapter 9 we looked at E1 reactions (two-step elimination reactions that form a carbocation intermediate) and E2 reactions (concerted elimination reactions). The preceding base-catalyzed dehydration represents a third kind of elimination reaction—namely, an **E1cB reaction** (elimination unimolecular conjugate base), a two-step elimination reaction that forms a carbanion intermediate. E1cB reactions occur only when the carbanion can be stabilized by electron delocalization.

MECHANISM FOR THE E1cB REACTION

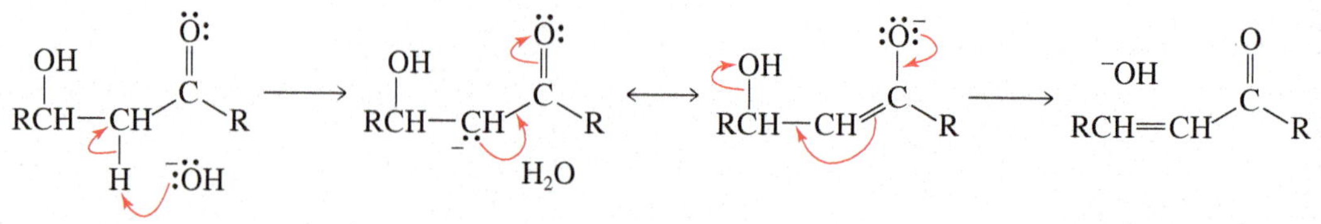

- Hydroxide ion removes a proton from the α-carbon, thereby forming an enolate ion.
- The enolate ion eliminates the OH group.

Sometimes dehydration occurs under the conditions in which the aldol addition is carried out, without requiring additional heat. For example, in the following reaction, the aldol addition product loses water as soon as it is formed because the new double bond is conjugated not only with the carbonyl group but also with the benzene ring. (Recall that the more stable the alkene, the easier it is formed.)

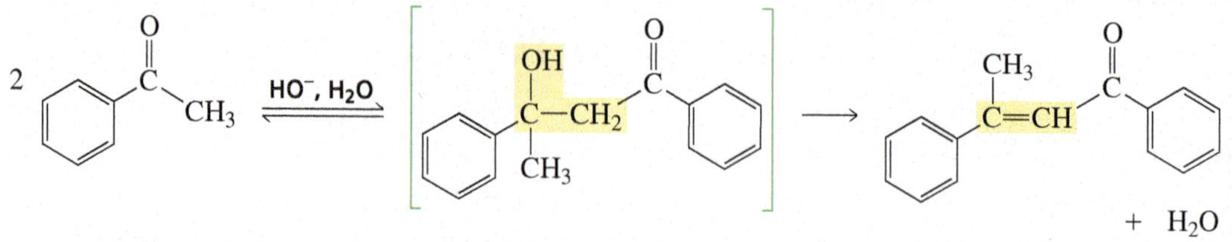

PROBLEM 22 ◆

What product is obtained from the aldol condensation of cyclohexanone?

LEARN THE STRATEGY

PROBLEM 23 SOLVED

How could you prepare the following compound using a starting material that contains no more than three carbons?

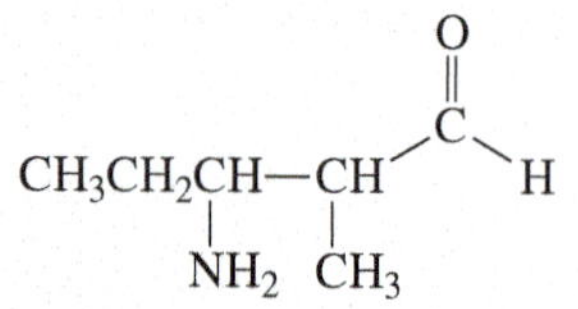

SOLUTION A compound with the correct six-carbon skeleton can be obtained if a three-carbon aldehyde undergoes an aldol addition. Dehydration of the addition product forms an α,β-unsaturated aldehyde. Conjugate addition of NH_3 (Section 16.15) forms the target molecule.

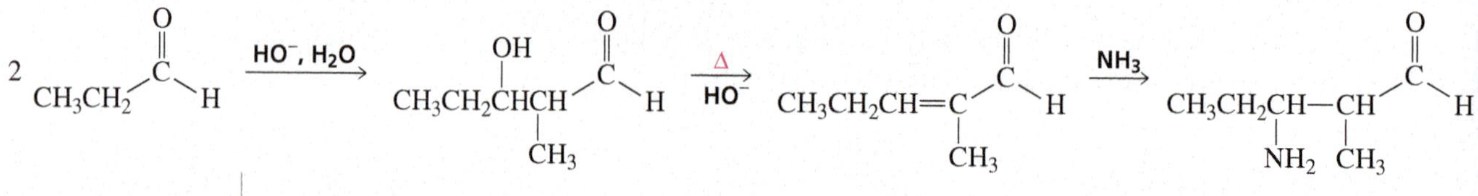

PROBLEM 24

How could you prepare the following compound using a starting material that contains no more than three carbons?

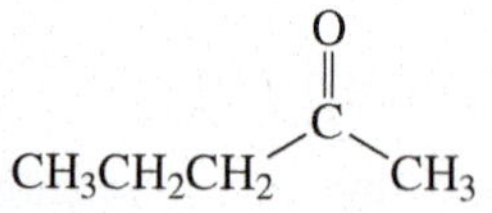

17.12 A CROSSED ALDOL ADDITION

If two different carbonyl compounds are used in an aldol addition—known as a **crossed aldol addition**—four products can be formed because the reaction with hydroxide ion can form two different enolate ions (**A⁻** and **B⁻**) and each enolate ion can react with either of the two carbonyl compounds (**A** or **B**). A reaction that forms four products clearly is not a synthetically useful reaction.

$$CH_3CH_2\overset{O}{\underset{}{\overset{\parallel}{C}}}H \;(A) \quad + \quad CH_3\overset{O}{\underset{}{\overset{\parallel}{C}}}H \;(B)$$

(reaction scheme showing formation of four products:)

$$CH_3CH_2CH-CH-\overset{O}{\overset{\parallel}{C}}H \; \underset{HO^-}{\overset{H_2O}{\rightleftharpoons}} \; CH_3CH_2CH-CH-\overset{O}{\overset{\parallel}{C}}H$$
(with O⁻ / OH and CH₃ substituents)

$$CH_3CH-CH-\overset{O}{\overset{\parallel}{C}}H \; \underset{HO^-}{\overset{H_2O}{\rightleftharpoons}} \; CH_3CH-CH-\overset{O}{\overset{\parallel}{C}}H$$

$$CH_3CH_2CH-CH_2-\overset{O}{\overset{\parallel}{C}}H \; \underset{HO^-}{\overset{H_2O}{\rightleftharpoons}} \; CH_3CH_2CH-CH_2-\overset{O}{\overset{\parallel}{C}}H$$

$$CH_3CH-CH_2-\overset{O}{\overset{\parallel}{C}}H \; \underset{HO^-}{\overset{H_2O}{\rightleftharpoons}} \; CH_3CH-CH_2-\overset{O}{\overset{\parallel}{C}}H$$

Obtaining Primarily One Product When One Carbonyl Compound Does Not Have α-Hydrogens

Primarily one product can be obtained from a crossed aldol addition if one of the carbonyl compounds does not have any α-hydrogens and, therefore, cannot form an enolate ion. That cuts the possible products from four to two.

Then, if the carbonyl compound with α-hydrogens is added slowly to a solution of the carbonyl compound without α-hydrogens and a base, the chance that the carbonyl compound with α-hydrogens, after forming an enolate ion, will react with another molecule of carbonyl compound with α-hydrogens will be minimized because there is a much greater concentration of the carbonyl compound without α-hydrogens. Thus, the possible products are cut to essentially one.

If one of the carbonyl compounds does not have any α-hydrogens, then the compound with α-hydrogens is added slowly to a solution of the carbonyl compound without α-hydrogens and a base.

(reaction scheme:)

benzaldehyde + HO⁻ → 1. CH₃—C(=O)—C(CH₃)₃ (add slowly) 2. HCl → [aldol addition product with OH, CH—CH₂—C(=O)—C(CH₃)₃] → CH=CH—C(=O)—C(CH₃)₃ + H₂O   **90%**

The aldol addition product formed in this reaction loses water as soon as it is formed because the new double bond is conjugated with the carbonyl group and the benzene ring. This crossed aldol condensation is sufficiently important to be given its own name—the **Claisen-Schmidt condensation**.

Obtaining Primarily One Product When Both Carbonyl Compounds Have α-Hydrogens

If both carbonyl compounds have α-hydrogens, primarily one aldol addition product can be formed if LDA is used to remove the α-hydrogen from the carbonyl carbon that is needed for the enolate ion. Because LDA is a strong base, all of the carbonyl compound is converted to an enolate ion, so none of that carbonyl compound is left for the enolate ion to react with in an aldol addition. Therefore, an aldol addition does not occur until the second carbonyl compound is added to the reaction mixture. If the second carbonyl compound is added slowly, the chance that it will form an enolate ion and then react with another molecule of the same carbonyl compound will be minimized.

If both reactants undergoing a condensation reaction have α-hydrogens, then LDA is used to form the enolate ion and the other carbonyl carbon is added slowly.

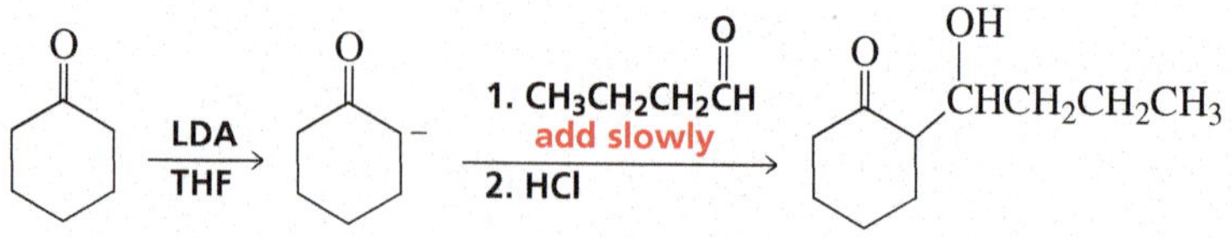

Retrosynthetic Analysis

LEARN THE STRATEGY

You can use the following strategy to determine the starting materials needed for the synthesis of a compound formed by an *aldol addition*.

retrosynthetic analysis

- Locate the new carbon–carbon bond that was formed between the α-carbon of one molecule of carbonyl compound and the carbonyl carbon of the other.
- Draw the two carbonyl compounds that were the reactants in the aldol addition.

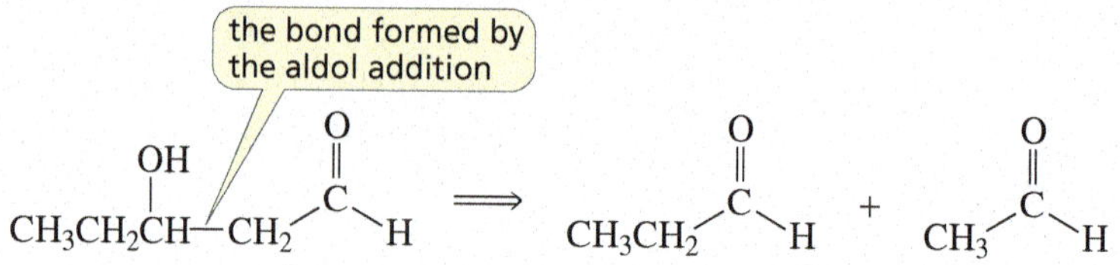

synthesis

- Add LDA to the carbonyl compound whose α-carbon will be the enolate ion.
- Add the other carbonyl compound slowly to the enolate ion.

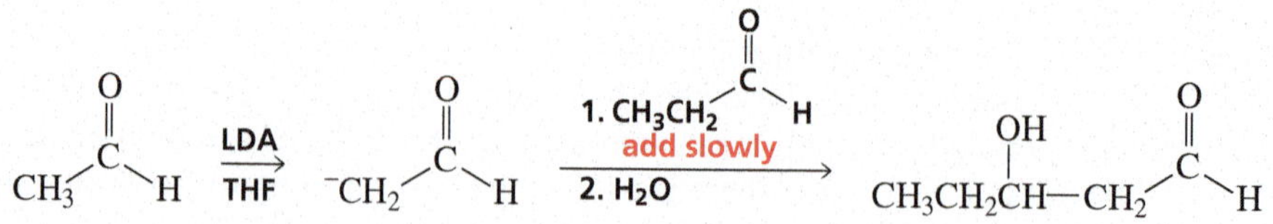

A similar strategy can be used to determine the starting materials needed for the synthesis of a compound formed by an *aldol condensation*:

retrosynthetic analysis

- convert the α,β-unsaturated carbonyl compound to a β-hydroxycarbonyl compound.
- proceed as described above.

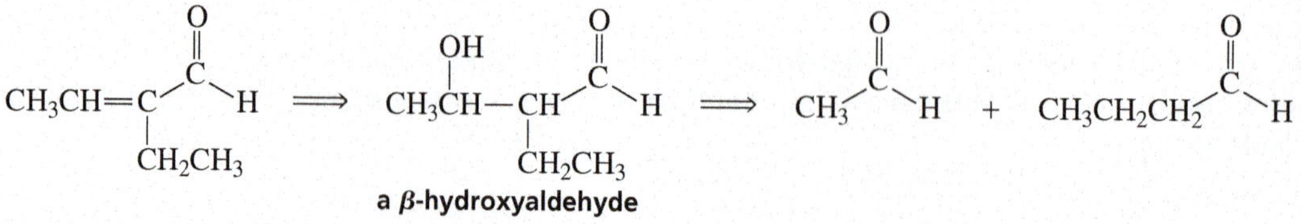

a β-hydroxyaldehyde

synthesis

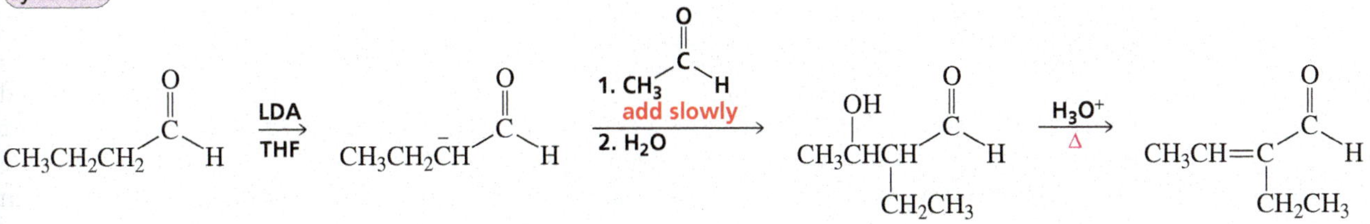

PROBLEM 25   USE THE STRATEGY

Describe how the following compounds can be prepared using an aldol addition in the first step of the synthesis:

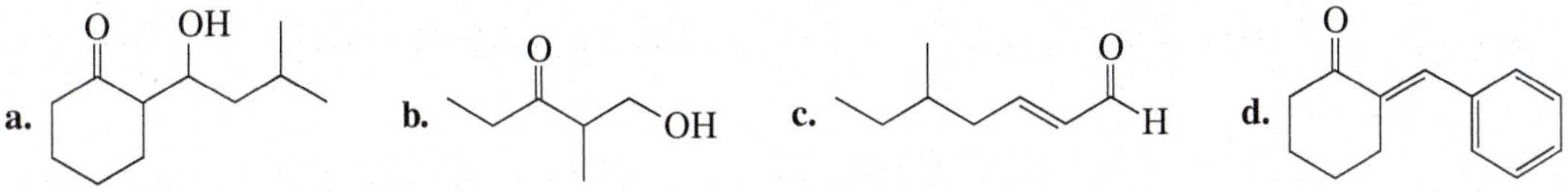

PROBLEM 26

What two carbonyl compounds are required for the synthesis of morachalcone A (the aromatase inhibitor discussed in the box below), via a Claisen–Schmidt condensation?

PROBLEM 27

Propose a mechanism for the following reaction:

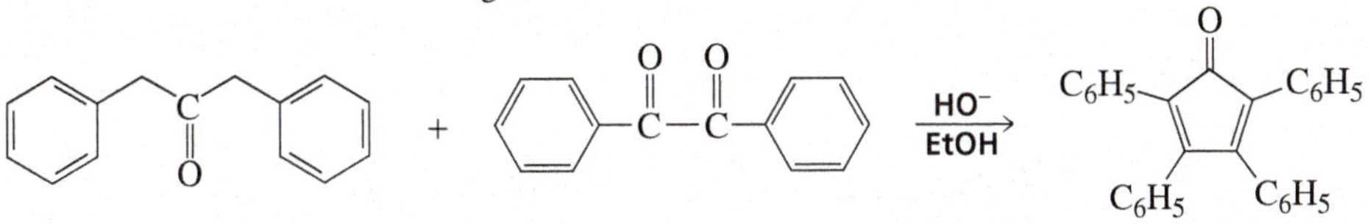

Breast Cancer and Aromatase Inhibitors

Current statistics show that one in eight women will develop breast cancer. Men also get breast cancer but are 100 times less likely to do so than women. There are several different types of tumors that cause breast cancer, some of which are estrogen dependent. An estrogen-dependent tumor has receptors that bind estrogen. Without estrogen, the tumor cannot grow. The A ring of estrogen hormones (estrone and estradiol) is an aromatic phenol (Section 3.16). One of the last steps in the biosynthesis of estrogen hormones from cholesterol is catalyzed by an enzyme called aromatase. Aromatase catalyzes the reaction that causes the A ring to become aromatic.

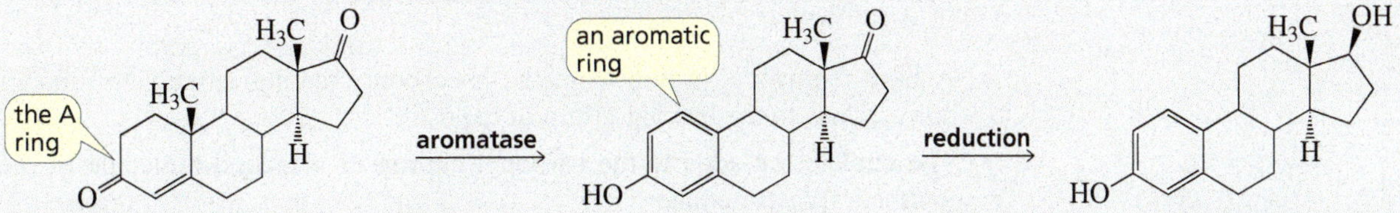

estrone   estradiol

Therefore, one approach to the treatment of breast cancer is to administer drugs that inhibit aromatase. If aromatase is inhibited, the estrogen hormones cannot be synthesized but the biosynthesis of other important hormones from cholesterol will not be affected. There are several aromatase inhibitors on the market and scientists continue to search for more potent ones. Fifteen different aromatase inhibitors have been isolated from the leaves of the paper mulberry tree (*Broussonetia papyifera*), one of which is morachalcone A (see page 801).

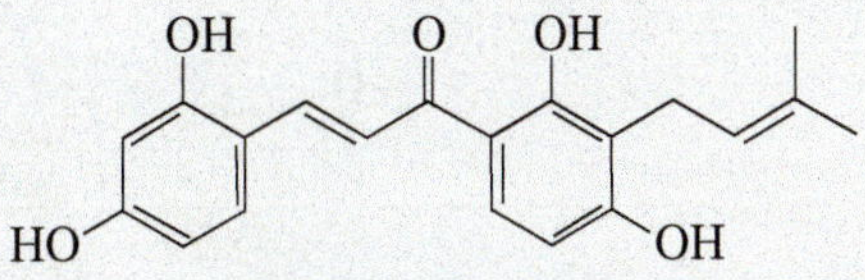

morachalcone A

17.13 A CLAISEN CONDENSATION FORMS A β-KETO ESTER

When two molecules of an *ester* undergo a condensation reaction, the reaction is called a **Claisen condensation.** The product of a Claisen condensation is a β-**keto ester.**

In a Claisen condensation, as in an aldol addition, one molecule of carbonyl compound is the nucleophile and a second molecule is the electrophile. And, as in an aldol addition, the new C—C bond connects the α-carbon of one molecule of ester to the carbonyl carbon of the other. An alcohol is the small molecule removed in a Claisen condensation.

A Claisen condensation forms a β-keto ester.

a β-keto ester

MECHANISM FOR THE CLAISEN CONDENSATION

- A base removes a proton from the α-carbon, creating an enolate ion. The base employed corresponds to the leaving group of the ester.
- The enolate ion adds to the carbonyl carbon of a second molecule of the ester, forming a tetrahedral intermediate.
- The carbon–oxygen π bond re-forms, eliminating an alkoxide ion.

Thus, like the reaction of esters with other nucleophiles, the Claisen condensation is a nucleophilic acyl substitution reaction (Section 15.9).

The base used to remove the proton from the α-carbon in a Claisen condensation should be the same as the leaving group of the ester, so the reactant will not change if the base adds to the carbonyl group.

Comparing a Claisen Condensation with an Aldol Addition

Notice that, after the nucleophilic addition, the Claisen condensation and the aldol addition reactions differ. In the Claisen condensation, the negatively charged oxygen re-forms the carbon–oxygen π bond and eliminates the ⁻OR group. In the aldol addition, the negatively charged oxygen obtains a proton from the solvent.

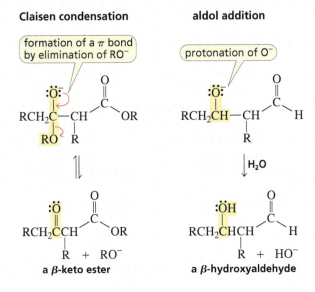

The last step of the Claisen condensation is different than the last step of the aldol addition because esters are different than aldehydes or ketones. The carbon bonded to the negatively charged oxygen in an ester is also bonded to a group that can be eliminated, whereas the carbon bonded to the negatively charged oxygen in an aldehyde or a ketone is not bonded to such a group. Thus, the Claisen condensation is a nucleophilic acyl substitution reaction, whereas the aldol addition is a nucleophilic addition reaction.

A Claisen Condensation Requires an Ester with Two α-Hydrogens

The Claisen condensation is reversible and favors the reactant because it is more stable than the β-keto ester. The condensation reaction can be driven to completion, however, if a proton is removed from the β-keto ester (Le Châtelier's principle; Section 5.7). A proton is easily removed because the central α-carbon of the β-keto ester is flanked by two carbonyl groups, making its α-hydrogen much more acidic than the α-hydrogen of the ester.

Consequently, a successful Claisen condensation requires an ester with two α-hydrogens and an equivalent amount of base rather than a catalytic amount of base. When the reaction is over, addition of acid to the reaction mixture reprotonates the β-keto ester anion and protonates the alkoxide ion, which prevents the reverse reaction from occurring.

PROBLEM 28 ◆

Draw the products of the following reactions:

a. 2 CH₃CH₂CH₂C(=O)OCH₃ → (1. CH₃O⁻ 2. HCl)

b. 2 CH₃CHCH₂C(=O)OCH₂CH₃ (with CH₃) → (1. CH₃CH₂O⁻ 2. HCl)

PROBLEM 29 ◆

Which of the following esters cannot undergo a Claisen condensation?

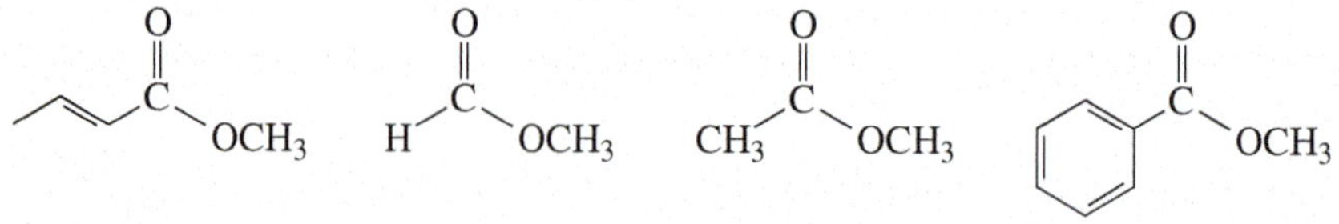

A   B   C   D

A Crossed Claisen Condensation

If one of the carbonyl compounds does not have any α-hydrogens, then the compound with α-hydrogens is added slowly to a solution of the carbonyl compound without α-hydrogens and a base.

A **crossed Claisen condensation** is a condensation reaction between two different esters. Like a crossed aldol addition, a crossed Claisen condensation is a useful reaction only if it is carried out under conditions that foster the formation of primarily one product. Otherwise, the reaction forms a mixture of products that are difficult to separate.

Primarily one product will be formed from a crossed Claisen condensation if one of the esters has no α-hydrogens (and, therefore, cannot form an enolate ion) and the ester with α-hydrogens is added slowly to a solution of the ester without α-hydrogens and the alkoxide ion.

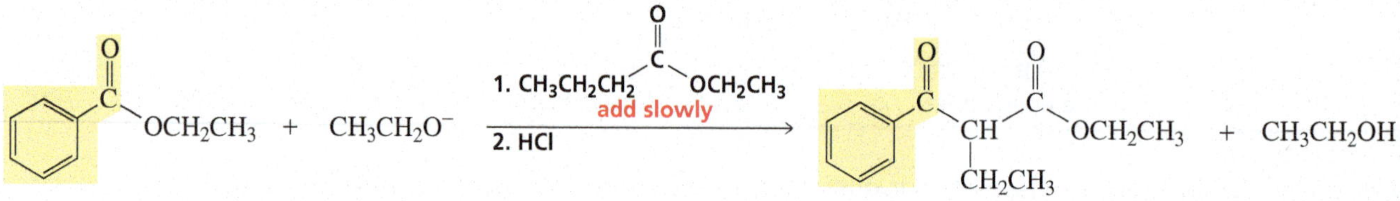

If both carbonyl compounds have α-hydrogens, then LDA is used to form the enolate ion and the other carbonyl compound is added slowly.

If both esters have α-hydrogens, primarily one product can be formed if LDA is used to remove the α-hydrogen to form the ester enolate ion. The other ester is then added slowly to maximize the chance that it will react with the enolate ion and minimize the chance that it will form an enolate ion and react with another molecule of its parent ester.

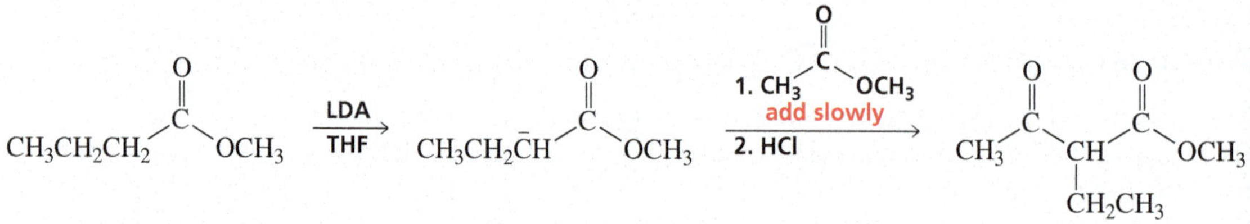

PROBLEM 30

What is the product of each of the following reactions?

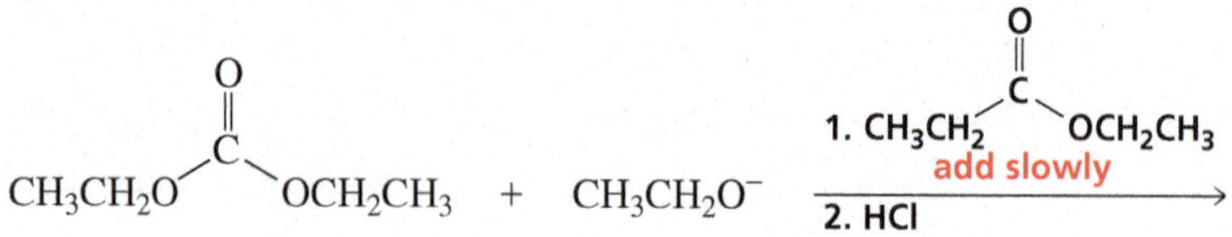

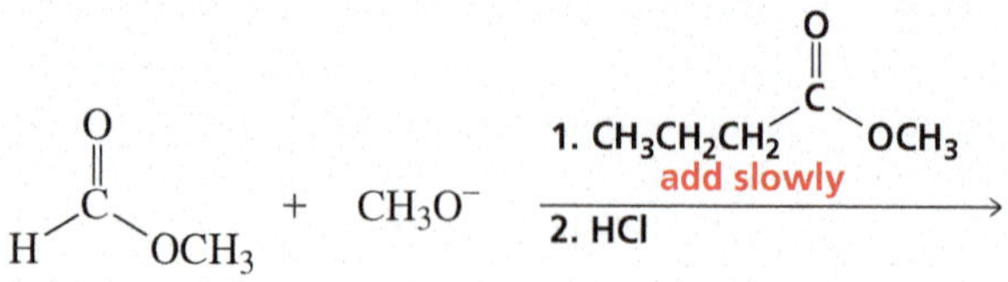

17.14 OTHER CROSSED CONDENSATIONS

In addition to crossed aldol additions and crossed Claisen condensations, a ketone can undergo a crossed condensation with an ester. The product is a *1,3-dicarbonyl compound*.

If one of the carbonyl compounds does not have any α-hydrogens, then the compound with α-hydrogens is added slowly to a solution of the compound without α-hydrogens and a base. A *β-keto aldehyde* is formed when a ketone condenses with a *formate ester*.

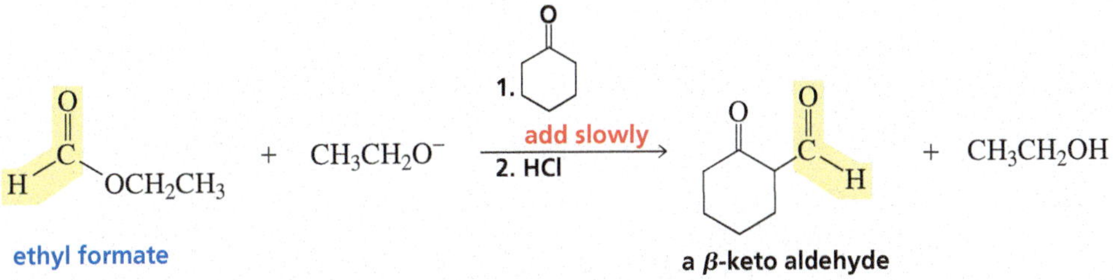

> A condensation between a ketone and an ester forms a 1,3-dicarbonyl compound.

ethyl formate   a β-keto aldehyde

> If one of the carbonyl compounds does not have any α-hydrogens, then the compound with α-hydrogens is added slowly to a solution of the compound without α-hydrogens and a base.

A *β-keto ester* is formed when a ketone condenses with *diethyl carbonate*.

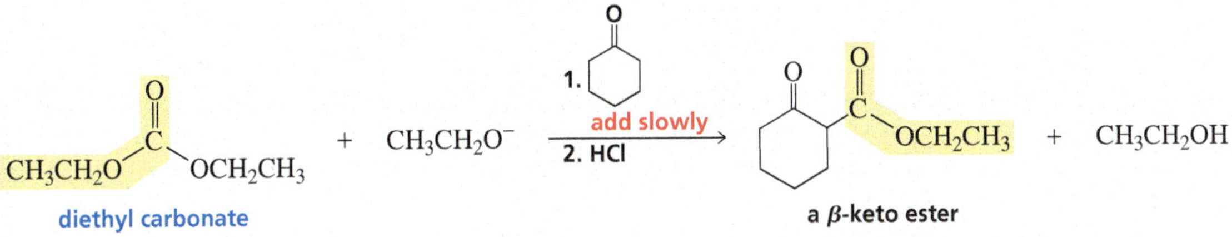

diethyl carbonate   a β-keto ester

If both the ketone and the ester have α-hydrogens, then LDA is used to form the needed enolate ion. The other carbonyl compound is then added slowly to the enolate ion to minimize the chance of it also forming an enolate ion and reacting with another molecule of its parent ester. A *β-diketone* is formed when a ketone condenses with an *ester*.

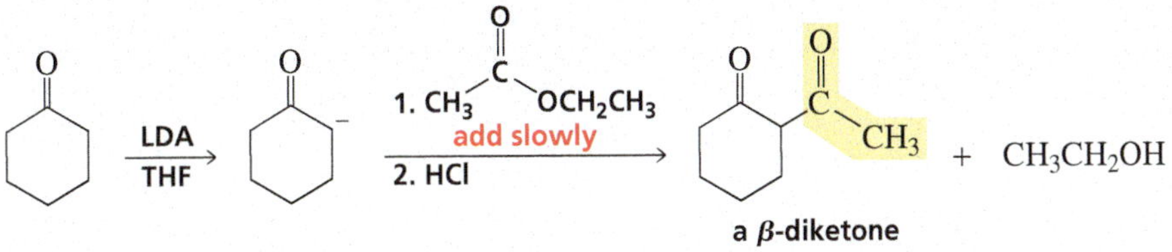

a β-diketone

> If both reactants undergoing a condensation reaction have α-hydrogens, then LDA is used to form the enolate ion and the other carbonyl carbon is added slowly.

PROBLEM 31

Show how each of the following compounds can be prepared from methyl phenyl ketone:

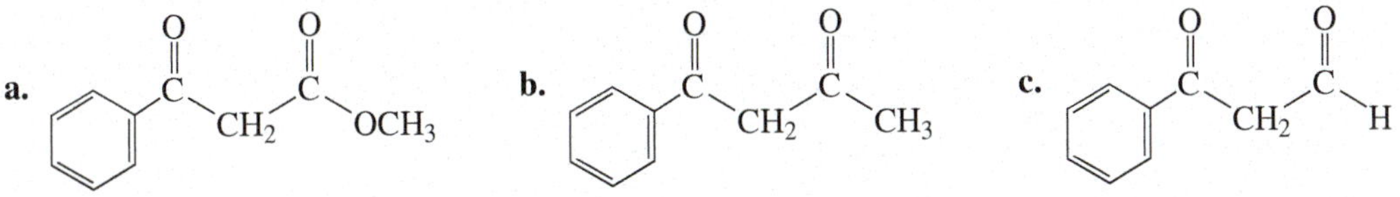

17.15 INTRAMOLECULAR CONDENSATIONS AND INTRAMOLECULAR ALDOL ADDITIONS

We saw that if a compound has two functional groups that can react with each other, an intramolecular reaction readily occurs if the reaction leads to the formation of a five- or a six-membered ring (Section 9.16). Consequently, a compound with two ester, aldehyde, or ketone groups can undergo an intramolecular reaction if a product with a five- or six-membered ring will be formed.

Intramolecular Claisen Condensations

The addition of base to a 1,6-diester causes the diester to undergo an intramolecular Claisen condensation, thereby forming a five-membered ring β-keto ester. An intramolecular Claisen condensation is called a **Dieckmann condensation.**

A six-membered ring β-keto ester is formed from the Dieckmann condensation of a 1,7-diester.

The Dieckmann condensation, like the Claisen condensation, can be driven to completion by carrying out the reaction with enough base to remove a proton from the α-carbon of the β-keto ester product. When the reaction is over, acid is added to reprotonate the condensation product and neutralize any remaining base.

The mechanism for the Dieckmann condensation is the same as the mechanism for the Claisen condensation. The only difference in the two reactions is that the enolate ion and the carbonyl group undergoing nucleophilic acyl substitution are in different molecules in the Claisen condensation and are in the same molecule in the Dieckmann condensation.

MECHANISM FOR THE DIECKMANN CONDENSATION

- A base removes a proton from an α-carbon, creating an enolate ion. The base employed corresponds to the leaving group of the ester.
- The enolate ion adds to a carbonyl carbon.
- The π bond re-forms, eliminating an alkoxide ion.

PROBLEM 32

Write the mechanism for the reaction of a 1,7-diester with an alkoxide ion to form a cyclic β-keto ester.

Intramolecular Aldol Additions

Because a 1,4-diketone has two different sets of α-hydrogens, two different β-hydroxyketones can potentially form, one with a five-membered ring and one with a three-membered ring (see the top of the next page). The greater stability of the five-membered ring draws the equilibria towards its formation (Section 3.12). In fact, the five-membered ring product is the only product formed from the intramolecular aldol addition of a 1,4-diketone.

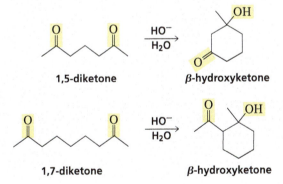

The intramolecular aldol addition of a 1,6-diketone can potentially lead to either a seven- or a five-membered ring product. Again, the more stable product—the one with the five-membered ring—is the only product of the reaction.

1,5-Diketones and 1,7-diketones undergo intramolecular aldol additions to form six-membered ring β-hydroxyketones.

1,5-diketone → **β-hydroxyketone**

1,7-diketone → **β-hydroxyketone**

PROBLEM 33 ◆

If the preference for formation of a six-membered ring were not so great, what other cyclic product would be formed from the intramolecular aldol addition of

a. 2,6-heptanedione?   **b.** 2,8-nonanedione?

PROBLEM 34

Can 2,4-pentanedione undergo an intramolecular aldol addition? If so, why? If not, why not?

PROBLEM 35 ◆

Draw the product of the reaction of each of the following compounds with a base:

a.   **b.**   **c.**   **d.**

17.16 THE ROBINSON ANNULATION

We saw that both Michael reactions and aldol additions form new carbon–carbon bonds. The **Robinson annulation** is a reaction that puts these two carbon–carbon bond-forming reactions together to form an α,β-unsaturated cyclic ketone. "Annulation" comes from *annulus*, Latin for "ring." An **annulation reaction** is a ring-forming reaction. The Robinson annulation makes it possible to synthesize many complicated organic molecules.

STEPS IN THE ROBINSON ANNULATION

- The first stage of a Robinson annulation is a Michael reaction that forms a 1,5-diketone.
- The second stage is an intramolecular aldol addition.
- Heating the basic solution dehydrates the alcohol.

Notice that a Robinson annulation results in a product that has a 2-cyclohexenone ring.

LEARN THE STRATEGY

PROBLEM-SOLVING STRATEGY

Determining the Products of a Robinson Annulation

Draw the product obtained by heating each pair of ketones in a basic solution.

a. First align the α,β-unsaturated carbonyl compound so that the carbonyl oxygen points at seven o'clock and the double bond is at the top of the structure. Draw the dicarbonyl compound to the right of the first with one ketone carbonyl oxygen also pointing at seven o'clock and the other pointing at noon. Connect the two pairs of carbons and draw the double bond that forms as a result of dehydration.

> A Robinson annulation forms an α,β-unsaturated cyclic ketone.

b. Align the carbonyl compounds as described in part **a.** The carbonyl compound on the right has two different α-carbons, so make certain that the α-carbon with the most acidic α-hydrogen is the one that points at the β-carbon of the α,β-unsaturated carbonyl compound, because that will be the α-hydrogen that is removed by hydroxide ion.

PROBLEM 36

Draw the product obtained by heating each pair of ketones in a basic solution.

a.

b.

Retrosynthetic Analysis

To determine the starting materials needed for the synthesis of a 2-cyclohexenone by a Robinson annulation, draw a line that bisects both the double bond and the bond between the β- and γ-carbons on the other side of the molecule.

- The β-carbon comes from the α,β-unsaturated carbonyl compound.
- The γ-carbon is the α-carbon of the enolate ion of the dicarbonyl compound that adds to the β-carbon in the Michael reaction.
- The sp^2 carbon of the bisected double bond closest to the γ-carbon was a carbonyl carbon and the other sp^2 carbon was an alkyl group.

α,β-unsaturated
carbonyl compound

By cutting through the following compound in the same way, we can determine the required reactants for its synthesis:

PROBLEM 37

What two carbonyl compounds are needed to synthesize each of the following compounds, using a Robinson annulation?

a.

b.

c.

CO$_2$ CAN BE REMOVED FROM A CARBOXYLIC ACID THAT HAS A CARBONYL GROUP AT THE 3-POSITION

Carboxylate ions do not lose CO$_2$ for the same reason that alkanes such as ethane do not lose a proton—namely, the leaving group would be a carbanion with localized electrons. Such carbanions are very strong bases, which makes them very poor leaving groups.

 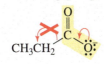

If, however, the CO_2 group is attached to a carbon that is adjacent to a carbonyl carbon, the CO_2 group can be removed because the electrons left behind can be delocalized onto the carbonyl oxygen. Consequently, carboxylate ions with a carbonyl group at the 3-position lose CO_2 when they are heated. Loss of CO_2 from a molecule is called **decarboxylation.**

A carboxylic acid with a carbonyl group at the 3-position decarboxylates when heated.

removing CO₂ from an α-carbon

Notice the similarity between removing CO_2 from an α-carbon and removing a proton from an α-carbon. In both reactions, when the substituent—CO_2 in one case, H^+ in the other—is removed, the electrons left behind are delocalized onto an oxygen.

removing a proton from an α-carbon

Decarboxylation is even easier under acidic conditions because the reaction is catalyzed by an intramolecular transfer of a proton from the carboxyl group to the carbonyl oxygen. The enol that is formed immediately tautomerizes to a ketone.

We saw in Section 17.1 that it is harder to remove a proton from an α-carbon if the electrons are delocalized onto the carbonyl group of an ester rather than onto the carbonyl group of a ketone. For the same reason, a higher temperature (~135 °C) is required to decarboxylate a β-dicarboxylic acid such as malonic acid than to decarboxylate a β-keto acid.

In summary, carboxylic acids with a carbonyl group at the 3-position lose CO_2 when they are heated.

PROBLEM 38 ◆

Which of the following compounds will decarboxylate when heated?

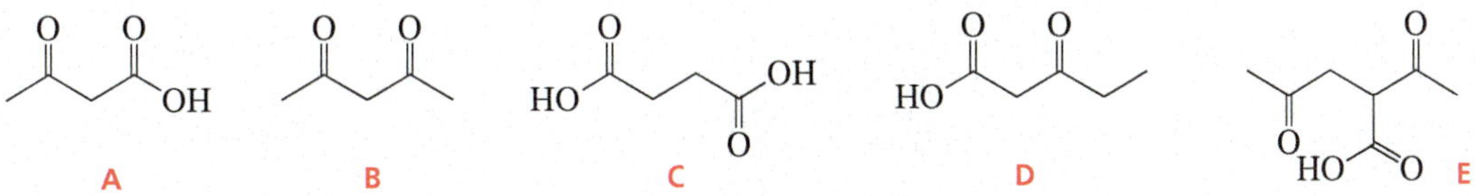

A   B   C   D   E

17.18 THE MALONIC ESTER SYNTHESIS: A WAY TO SYNTHESIZE A CARBOXYLIC ACID

A combination of two of the reactions discussed in this chapter—namely, alkylation of an α-carbon and decarboxylation of a carboxylate ion with a carbonyl group at the 3-position—can be used to prepare carboxylic acids of any desired chain length. The procedure is called the **malonic ester synthesis** because the starting material for the synthesis is the diester of malonic acid.

The carboxyl group and the α-carbon of the carboxylic acid being synthesized come from malonic ester. The rest of the carboxylic acid comes from the alkyl halide used in the second step of the synthesis. Thus, a malonic ester synthesis forms a carboxylic acid with two more carbons than the alkyl halide.

malonic ester synthesis

diethyl malonate
malonic ester

1. $CH_3CH_2O^-$
2. RBr
3. HCl, H_2O, Δ

from malonic ester
from the alkyl halide

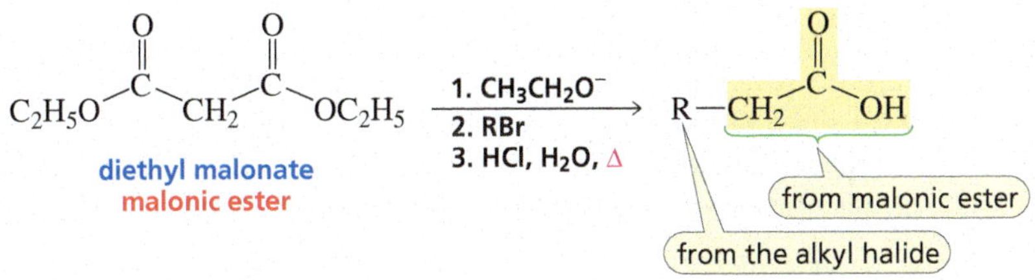

STEPS IN THE MALONIC ESTER SYNTHESIS

alkylation of the α-carbon

$CH_3CH_2O^-$

removal of a proton from the α-carbon

R—Br

α-substituted malonic ester

$+ Br^-$

HCl, H_2O Δ hydrolysis

decarboxylation

$R-CH_2$ $\overset{O}{\overset{\parallel}{C}}$ OH $+ CO_2$ $\overset{\Delta}{\longleftarrow}$ HO $\overset{O}{\overset{\parallel}{C}}$ CH $\overset{O}{\overset{\parallel}{C}}$ OH

R $+ 2\ CH_3CH_2OH$

α-substituted malonic acid

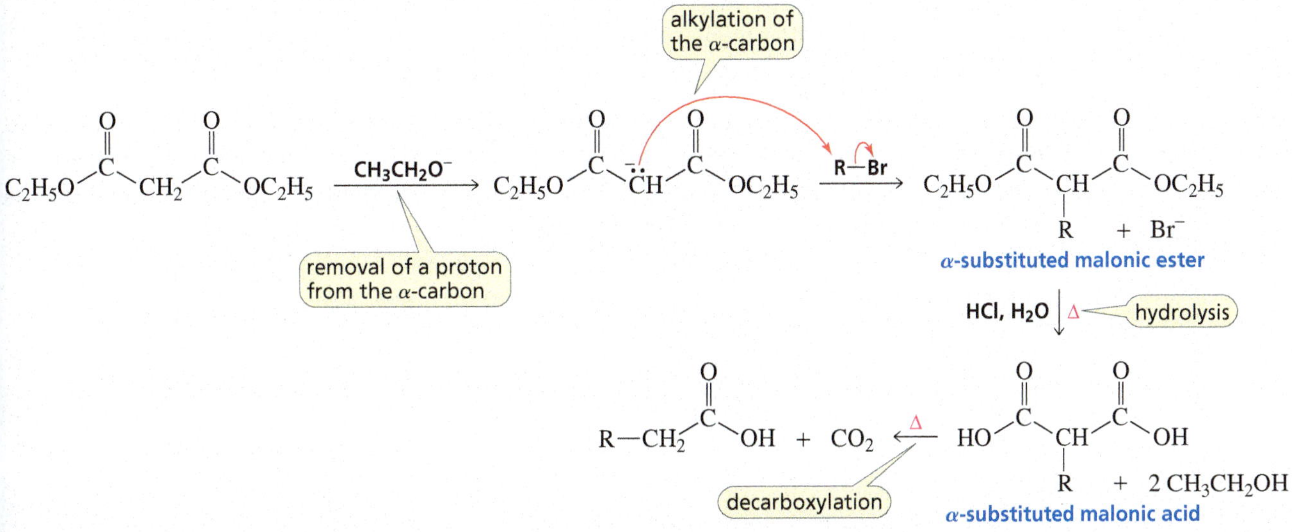

- A proton is easily removed from the α-carbon because it is flanked by two carbonyl groups.

- The resulting α-carbanion reacts with an alkyl halide, forming an α-substituted malonic ester. Because alkylation is an S_N2 reaction, it works best with primary alkyl halides and methyl halides (Section 9.2).

- Heating the α-substituted malonic ester in an acidic aqueous solution hydrolyzes both ester groups to carboxylic acid groups, forming an α-substituted malonic acid.

- Further heating decarboxylates the 3-oxocarboxylic acid.

Carboxylic acids with two substituents bonded to the α-carbon can be prepared by carrying out two successive alkylations of the α-carbon.

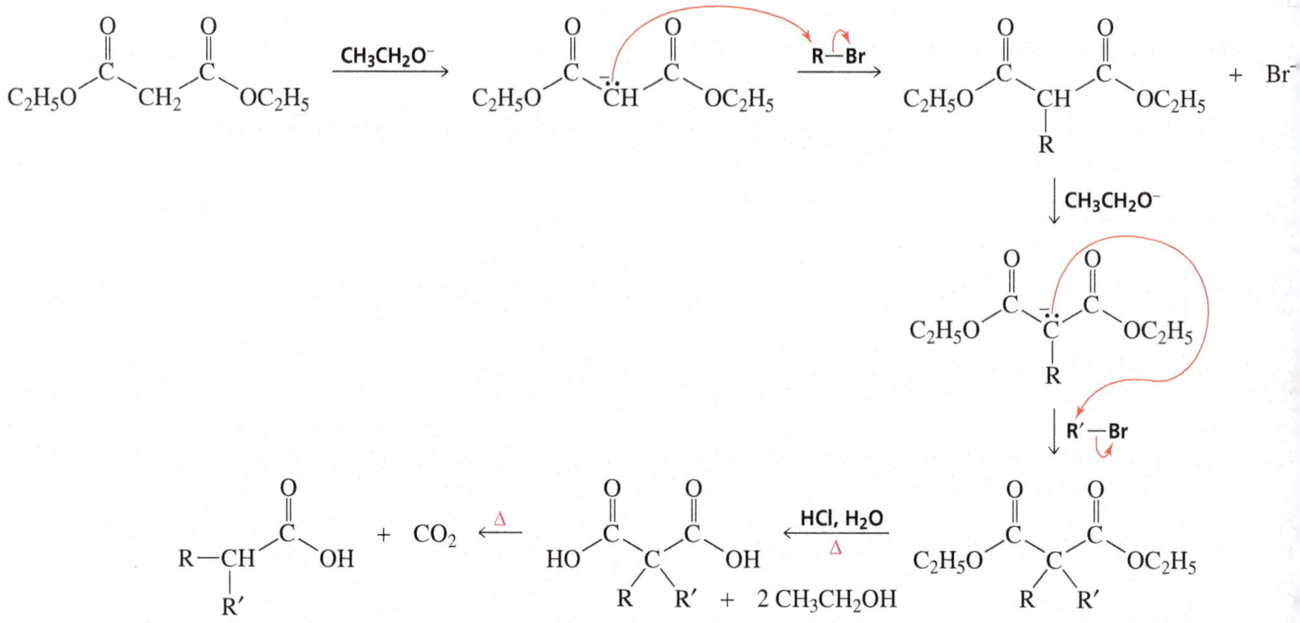

Retrosynthetic Analysis

LEARN THE STRATEGY

We saw that when a carboxylic acid is synthesized by a malonic ester synthesis, the carboxyl group and the α-carbon come from malonic ester. Any substituent attached to the α-carbon comes from the alkyl halide used in the second step of the synthesis. If the α-carbon has two substituents, then two successive alkylations of the α-carbon will form the desired carboxylic acid.

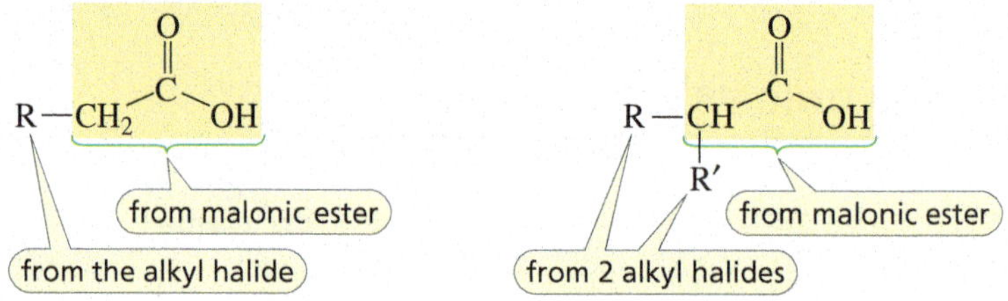

USE THE STRATEGY

> **PROBLEM 39 ◆**
>
> What alkyl bromide(s) should be used in the malonic ester synthesis of each of the following carboxylic acids?
>
> **a.** propanoic acid
>
> **b.** 2-methylpropanoic acid
>
> **c.** 3-phenylpropanoic acid
>
> **d.** 4-methylpentanoic acid
>
> ---
>
> **PROBLEM 40**
>
> Explain why the following carboxylic acids cannot be prepared by a malonic ester synthesis:
>

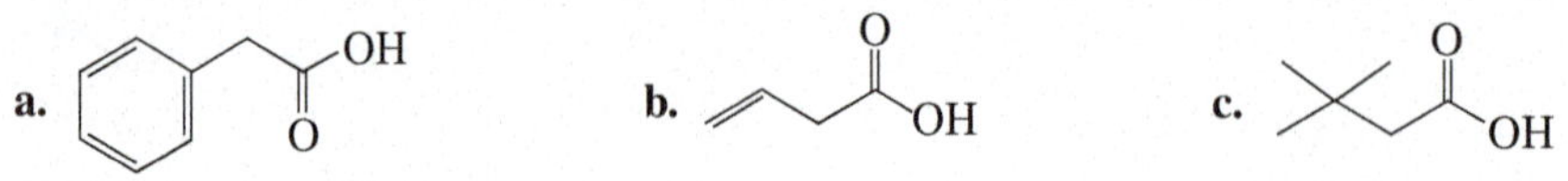

17.19 THE ACETOACETIC ESTER SYNTHESIS: A WAY TO SYNTHESIZE A METHYL KETONE

The only difference between the acetoacetic ester synthesis and the malonic ester synthesis is the use of acetoacetic ester rather than malonic ester as the starting material. The difference in starting material causes the product of the **acetoacetic ester synthesis** to be a *methyl ketone* rather than a *carboxylic acid.* The carbonyl group of the methyl ketone and the carbons on either side of it come from acetoacetic ester; the rest of the ketone comes from the alkyl halide used in the second step of the synthesis.

acetoacetic ester synthesis

An acetoacetic ester synthesis forms a methyl ketone with three more carbons than the alkyl halide.

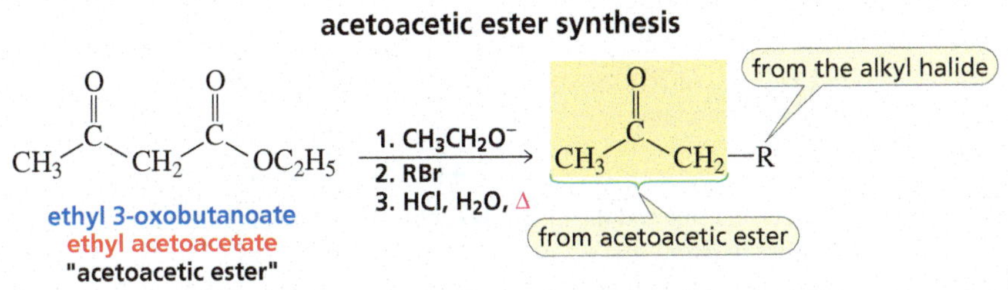

The steps in the acetoacetic ester synthesis are the same as those in the malonic ester synthesis.

STEPS IN THE ACETOACETIC ESTER SYNTHESIS

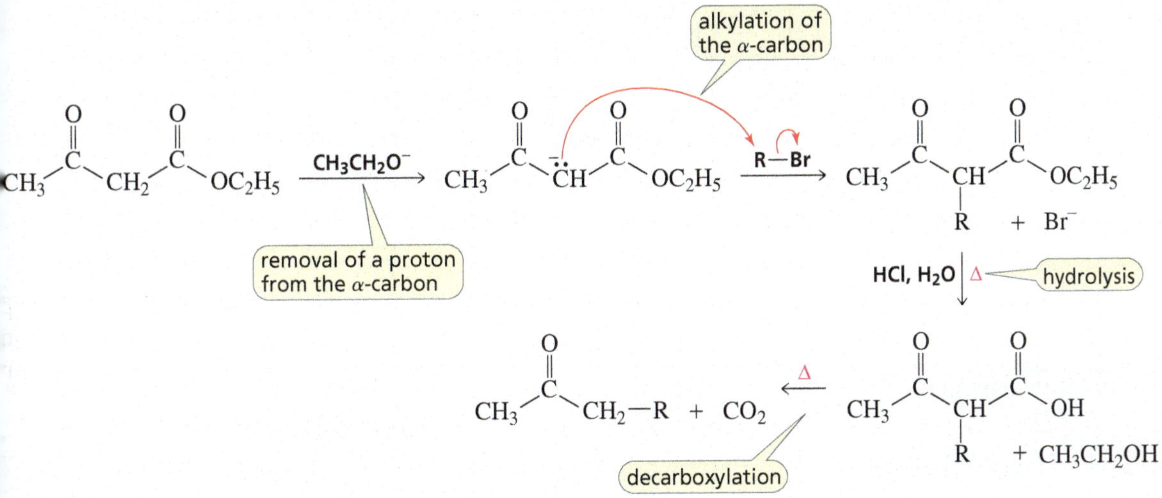

- A proton is removed from the α-carbon that is flanked by two carbonyl groups.
- The α-carbanion undergoes an S_N2 reaction with the alkyl halide.
- Hydrolysis and decarboxylation form the methyl ketone.

Retrosynthetic Analysis

When a methyl ketone is synthesized by an acetoacetic ester synthesis, the carbonyl carbon and the carbons on either side of it come from acetoacetic ester. Any substituent attached to the α-carbon comes from the alkyl halide used in the second step of the synthesis.

LEARN THE STRATEGY

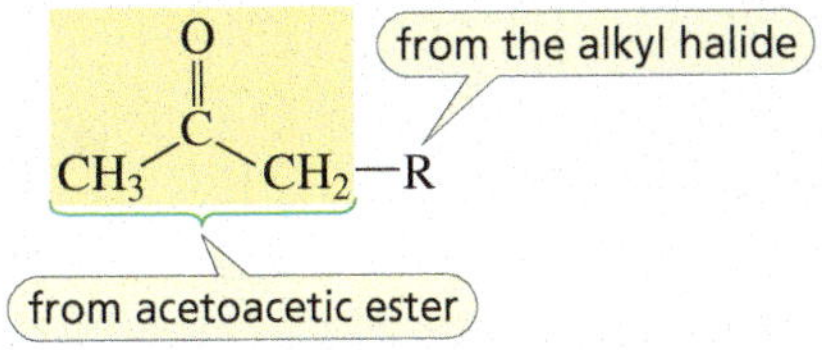

USE THE STRATEGY

PROBLEM 41 ◆

What alkyl bromide should be used in the acetoacetic ester synthesis of each of the following methyl ketones?

a. 2-pentanone   **b.** 2-octanone   **c.** 4-phenyl-2-butanone

PROBLEM 42 SOLVED

Starting with methyl propanoate, how could you prepare 4-methyl-3-heptanone?

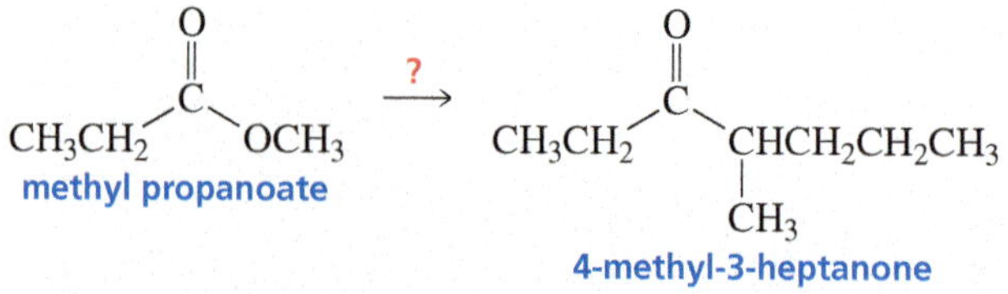

Because the starting material is an ester and the target molecule has more carbons than the starting material, a Claisen condensation appears to be a good way to start this synthesis. The Claisen condensation forms a β-keto ester that can be easily alkylated at the desired carbon because it is flanked by two carbonyl groups. Acid-catalyzed hydrolysis forms a 3-oxocarboxylic acid that decarboxylates when heated.

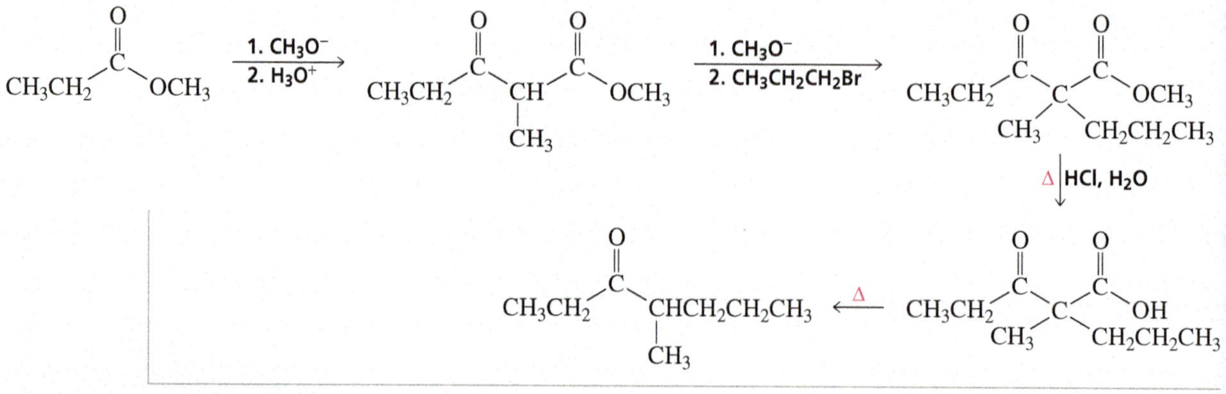

DESIGNING A SYNTHESIS V

17.20 MAKING NEW CARBON–CARBON BONDS

When planning the synthesis of a compound that requires the formation of a new carbon–carbon bond:

- locate the new bond that needs to be made and perform a disconnection—that is, break the bond to produce two fragments (Section 16.14).
- determine which of the atoms that will form the new bond should be the electrophile and which should be the nucleophile.
- choose a compound with the desired electrophilic and nucleophilic groups.

Example 1

The new bond that needs to be made in the synthesis of the following β-diketone is the one that makes the second five-membered ring:

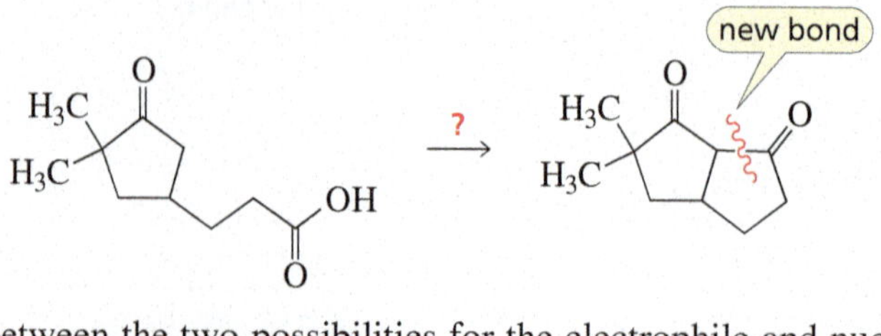

It is easy to choose between the two possibilities for the electrophile and nucleophile because we know that a carbonyl carbon is an electrophile.

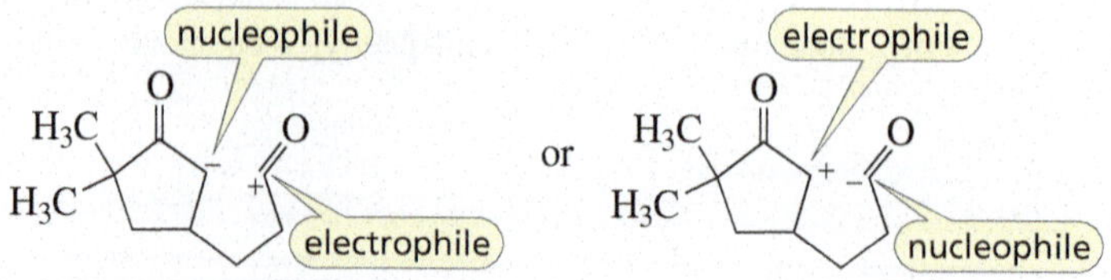

If we know what the starting material is, we can use it as a clue to arrive at the desired compound. For example, an ester carbonyl group would be a good electrophile for this synthesis because it has a group that would be eliminated. Moreover, the α-hydrogens of the ketone are more acidic than the α-hydrogens of the ester, so the desired nucleophile would be easy to obtain. Thus, converting the starting material to an ester (Section 15.18), followed by an intramolecular condensation, forms the target molecule.

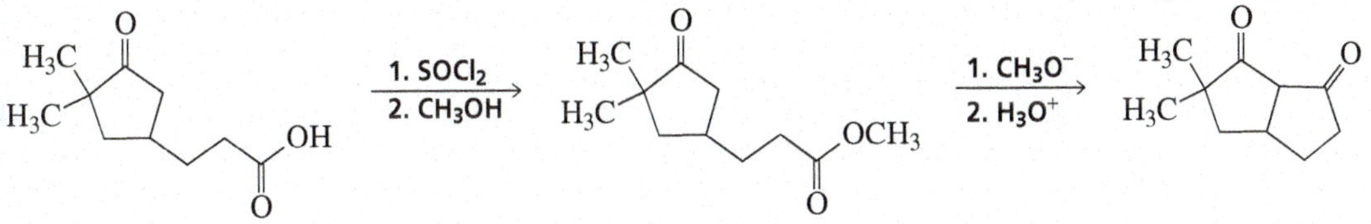

Example 2

In this example, two new carbon–carbon bonds must be formed.

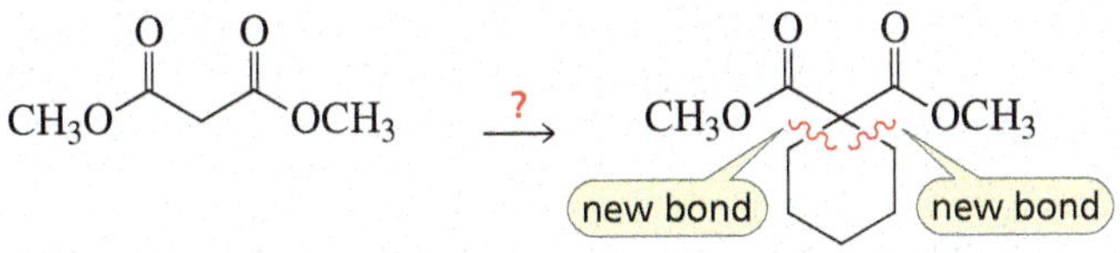

After identifying the electrophilic and nucleophilic sites, we see that two successive alkylations of a diester of malonic acid, using 1,5-dibromopentane for the alkyl halide, will produce the target molecule.

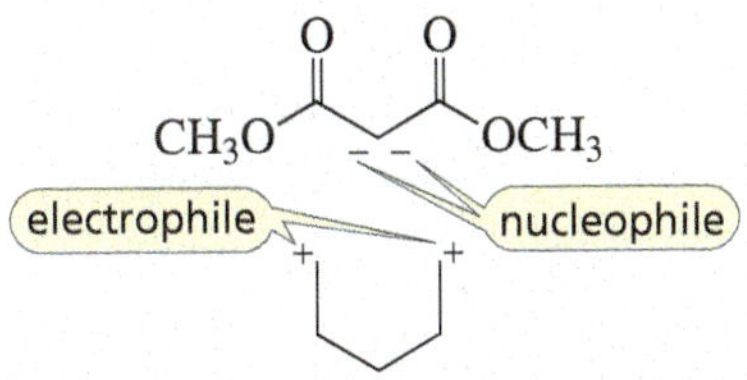

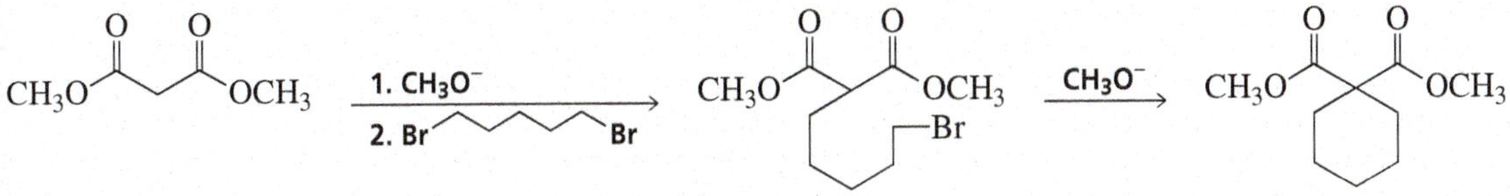

Example 3

The diester given as the starting material suggests that a Dieckmann condensation should be used to obtain the cyclic compound:

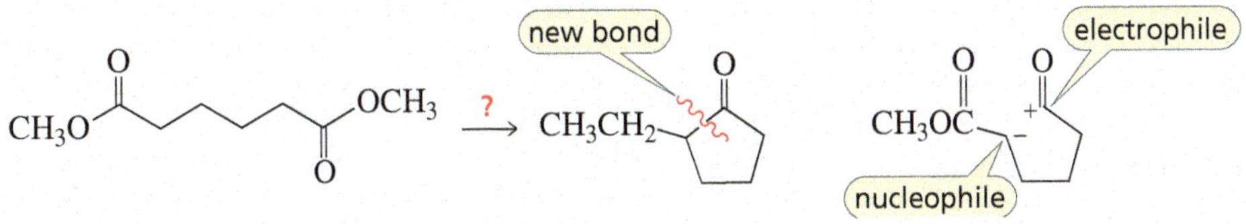

The Dieckmann condensation is followed by alkylation of the α-carbon (that is flanked by two carbonyl groups) of the cyclopentanone ring. Hydrolysis of the β-keto ester and decarboxylation forms the target molecule.

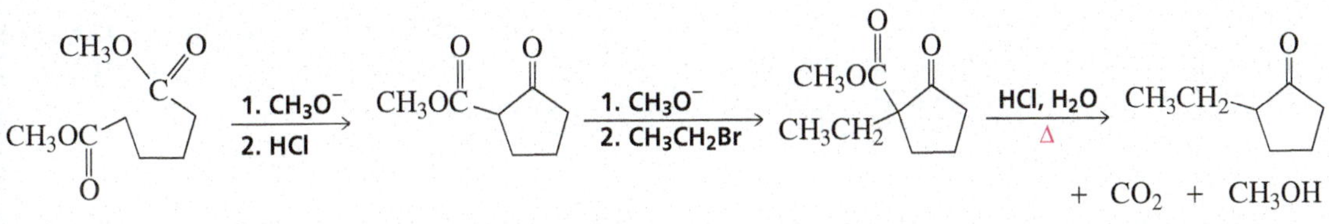

PROBLEM 43

Design a synthesis for each of the following compounds using the given starting material:

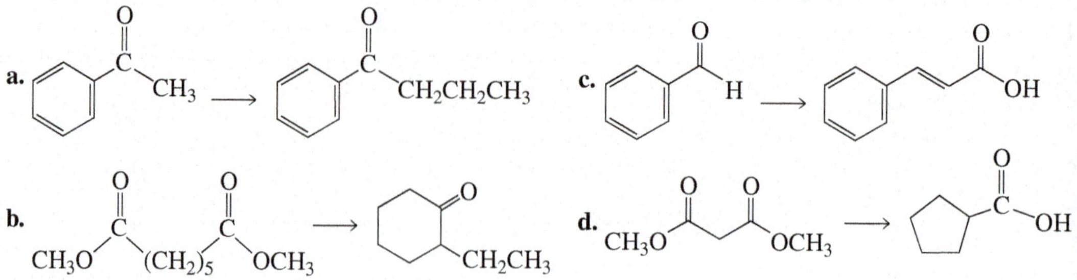

17.21 REACTIONS AT THE α-CARBON IN LIVING SYSTEMS

Many reactions that occur in cells involve reactions at the α-carbon—that is, the kinds of reactions you have been studying in this chapter. We will now look at a few examples.

A Biological Aldol Addition

Glucose, the most abundant sugar found in nature, is synthesized in cells from two molecules of pyruvate. The series of reactions that convert two molecules of pyruvate to glucose is called **gluconeogenesis** (Section 24.12). The reverse process—the breakdown of glucose to two molecules of pyruvate—is called **glycolysis** (Section 24.6).

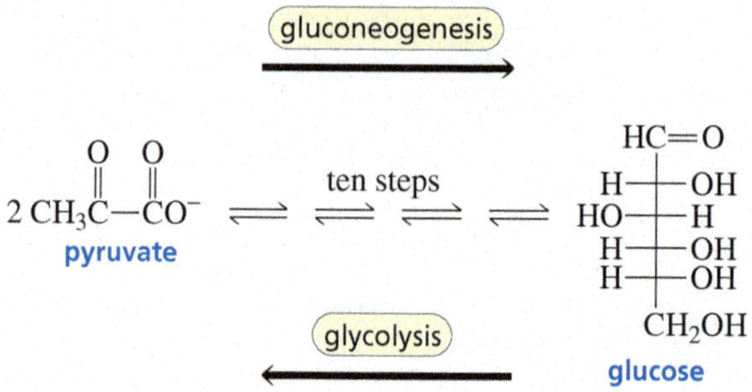

Because glucose has twice as many carbons as pyruvate, you should not be surprised to learn that one of the steps in the biosynthesis of glucose is an aldol addition. An enzyme called aldolase catalyzes a crossed aldol addition between dihydroxyacetone phosphate and glyceraldehyde-3-phosphate. The product is fructose-1,6-bisphosphate, which is subsequently converted to glucose.

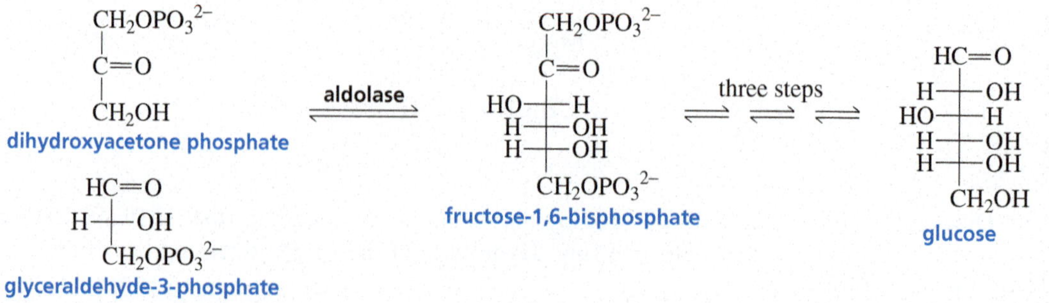

The reaction catalyzed by aldolase is reversible. The reverse reaction—the cleavage of fructose-1,6-bisphosphate to dihydroxyacetone phosphate and glyceraldehyde-3-phosphate—is a retro-aldol addition (p. 818). The mechanism for this reaction is discussed in Section 22.12.

PROBLEM 44

Propose a mechanism for the formation of fructose-1,6-bisphosphate from dihydroxyacetone phosphate and glyceraldehyde-3-phosphate, using hydroxide ion as the catalyst.

A Biological Aldol Condensation

Collagen is the most abundant protein in mammals, amounting to about one-fourth of the total protein. It is the major fibrous component of bone, teeth, skin, cartilage, and tendons. Individual collagen molecules—called tropocollagen—can be isolated only from tissues of young animals. As animals age, the individual collagen molecules become cross-linked, which is why meat from older animals is tougher than meat from younger ones. Collagen cross-linking is an example of an aldol condensation.

cross-linked collagen

Before collagen molecules can cross-link, the primary amino groups of the lysine residues of collagen must be converted to aldehyde groups. (Lysine is an amino acid.) The enzyme that catalyzes this reaction is called lysyl oxidase. An aldol condensation between two aldehyde groups results in a cross-linked protein.

A Biological Claisen Condensation

Fatty acids are long-chain carboxylic acids (Section 25.1). Naturally occurring fatty acids are unbranched and contain an even number of carbons because they are synthesized from acetate, a compound with two carbons.

In Section 15.19, you saw that carboxylic acids can be activated in cells by being converted to thioesters of coenzyme A.

One of the necessary reactants for fatty acid synthesis is malonyl-CoA, which is obtained by carboxylation of acetyl-CoA. The mechanism for this reaction will be discussed in Section 24.4.

Before fatty acid synthesis can occur, the acyl groups of acetyl-CoA and malonyl-CoA are transferred to other thiols by means of a transesterification reaction.

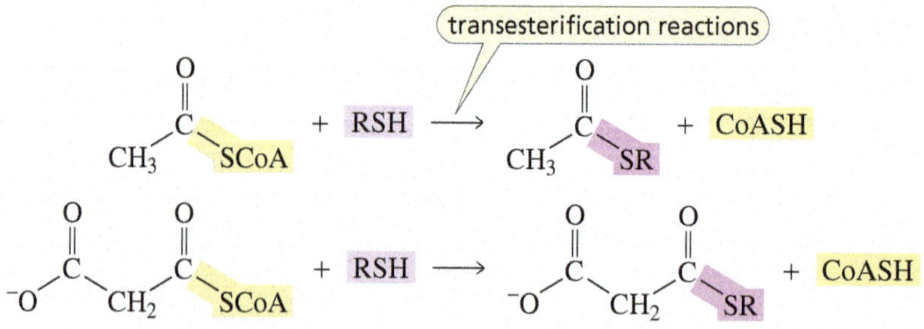

A molecule of acetyl thioester and a molecule of malonyl thioester are the reactants for the *first round* in the biosynthesis of a fatty acid.

STEPS IN FATTY ACID BIOSYNTHESIS

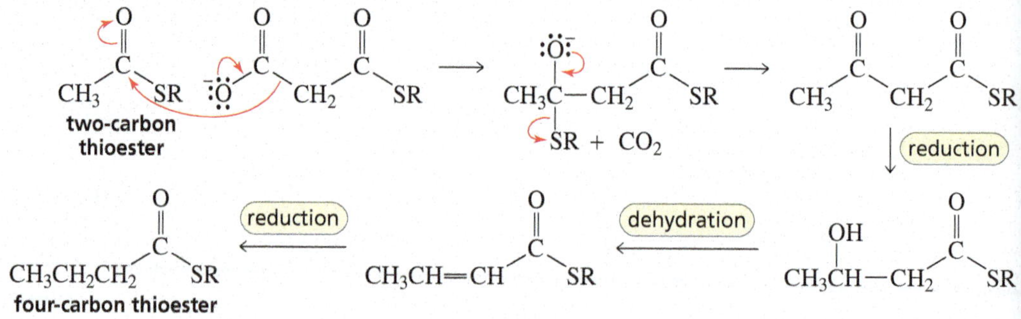

- The first step is a Claisen condensation. The nucleophile needed for a Claisen condensation is obtained by removing CO_2—rather than a proton—from the α-carbon of malonyl thioester. (Recall that a carboxylic acid with a carbonyl group at the 3-position is easily decarboxylated; Section 17.17.) Because CO_2 is a gas, loss of CO_2 drives the reaction to completion.
- The product of the condensation reaction undergoes a reduction (recall that a ketone is easier to reduce than an ester; Section 16.2), a dehydration, and a second reduction to form a four-carbon thioester. Each reaction is catalyzed by a different enzyme.

The four-carbon thioester and another molecule of malonyl thioester are the reactants for the *second round* of the biosynthesis.

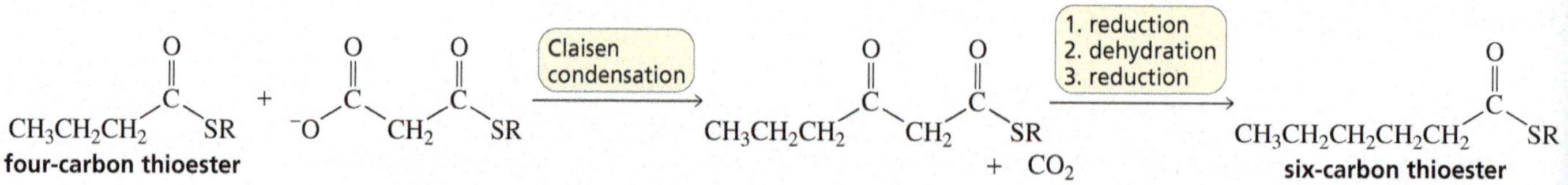

- Again, the product of the Claisen condensation undergoes a reduction, a dehydration, and a second reduction—this time to form a six-carbon thioester.
- The sequence of reactions is repeated, and each time two more carbons are added to the chain.

This sequence of reactions explains why naturally occurring fatty acids are unbranched and contain an even number of carbons. Once a thioester with the desired number of carbons is obtained, it undergoes a transesterification reaction with glycerol-3-phosphate to form fats, oils, and phospholipids (Section 24.11).

PROBLEM 45 ◆

Palmitic acid is a straight-chain saturated 16-carbon fatty acid. How many moles of malonyl-CoA are required for the synthesis of one mole of palmitic acid?

PROBLEM 46 ◆

a. If the biosynthesis of palmitic acid were carried out with CD_3COSR and nondeuterated malonyl thioester, how many deuterium atoms would be incorporated into palmitic acid?

b. If the biosynthesis of palmitic acid were carried out with $^-OOCCD_2COSR$ and nondeuterated acetyl thioester, how many deuterium atoms would be incorporated into palmitic acid?

A Biological Decarboxylation

The decarboxylation of acetoacetate is an example of a decarboxylation reaction that occurs in cells.

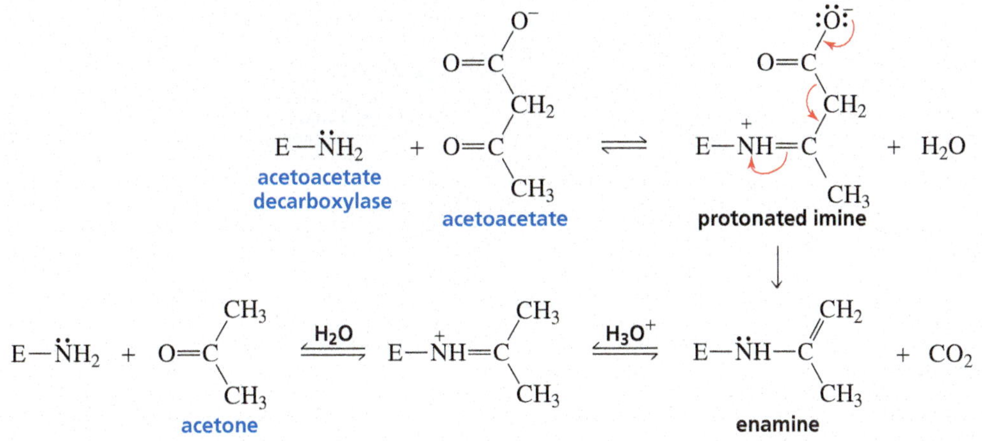

- The amino group of acetoacetate decarboxylase, the enzyme that catalyzes the reaction, forms an imine with the keto group of acetoacetate.
- The positively charged nitrogen readily accepts the pair of electrons left behind when the substrate loses CO_2, forming an enamine.
- Tautomerization and protonation of the CH_2 group forms an imine.
- Hydrolysis of the imine produces the decarboxylated product (acetone) and regenerates the enzyme (Section 16.8).

In ketosis, a pathological condition that can occur in people with diabetes, the body produces more acetoacetate than can be metabolized. The excess acetoacetate is decarboxylated, so ketosis can be recognized by the smell of acetone on a person's breath.

PROBLEM 47

When the enzymatic decarboxylation of acetoacetate is carried out in $H_2^{18}O$, all the acetone that is formed contains ^{18}O. What does this tell you about the mechanism of the reaction?

17.22 ORGANIZING WHAT WE KNOW ABOUT THE REACTIONS OF ORGANIC COMPOUNDS

We saw that the families of organic compounds can be put in one of four groups, and that all the families in a group react in similar ways. Now that we have finished studying the families in Group III, let's revisit this group.

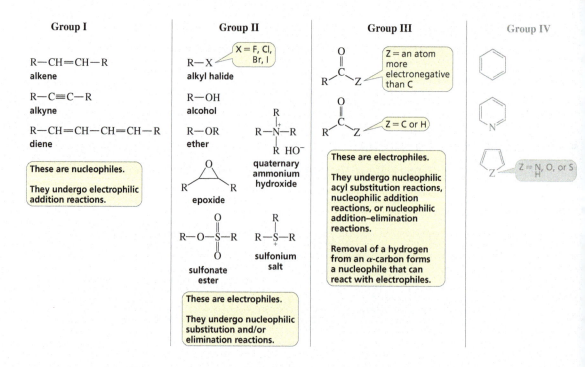

Both families in Group III have carbonyl groups, and because a carbonyl carbon is an *electrophile*, both families in this group react with *nucleophiles*.

- The first family (carboxylic acids and carboxylic acid derivatives) has a group attached to the carbonyl carbon that can be replaced by another group. Therefore, this family undergoes nucleophilic acyl substitution reactions.

product of nucleophilic acyl substitution

- The second family (aldehyde and ketones) does not have a group attached to the carbonyl carbon that can be replaced by another group. Therefore, this family undergoes nucleophilic addition reactions with strongly basic nucleophiles such as R^- or H^-.

product of nucleophilic addition

- If the attacking atom of the nucleophile is an oxygen or a nitrogen and there is enough acid in the solution to protonate the OH group of the tetrahedral compound formed by the nucleophilic addition reaction, then water is eliminated from the addition product. The fate of the dehydrated product depends on Z.

product of nucleophilic addition–elimination

- Aldehydes, ketones, esters, and *N,N*-disubstituted amides have a hydrogen on an α-carbon that can be removed by a strong base. Removal of a hydrogen from an α-carbon creates a nucleophile (an enolate ion) that can react with electrophiles.

ESSENTIAL CONCEPTS

Section 17.1

- A hydrogen bonded to an α-**carbon** of an aldehyde, ketone, ester, or *N,N*-disubstituted amide is sufficiently acidic to be removed by a strong base.

- A **carbon acid** is a compound with a relatively acidic hydrogen bonded to an sp^3 carbon.

- Aldehydes and ketones (pK_a ~16 to 20) are more acidic than esters (pK_a ~25). β-**Diketones** (pK_a ~ 9) and β-**keto esters** (pK_a ~11) are even more acidic.

Sections 17.2–17.3

- **Keto–enol interconversion** can be catalyzed by acids or by bases. Generally, the **keto tautomer** is more stable.

Section 17.4

- Aldehydes and ketones react with Br_2, Cl_2, or I_2: under acidic conditions, a halogen replaces one α-hydrogen; under basic conditions, halogens replace all the α-**hydrogens.**

Section 17.5

- The **HVZ reaction** brominates the α-carbon of a carboxylic acid.

- The α-bromo substituent of carbonyl compounds can be replaced only by poor nucleophiles (weak bases).

Section 17.6

- LDA (a strong, bulky base but a poor nucleophile) is used to form an **enolate ion** in reactions that require the carbonyl compound to be completely converted to enolate ion before it reacts with an electrophile.

Section 17.7

- LDA is the base used for monoalkylation of an α-carbon. The less substituted α-carbon is alkylated when the reaction is under kinetic control (LDA, -78 °C), whereas the more substituted α-carbon is alkylated when the reaction is under thermodynamic control (LDA, 0 °C).

Section 17.8

- Aldehydes and ketones can be alkylated or acylated on the α-carbon by first forming an enamine with a secondary amine.

Section 17.9

- α,β-Unsaturated aldehydes and ketones can be alkylated on the β-carbon by an enamine or by a **Michael reaction** using enolate ions—preferable those of β-diketones, β-diesters, β-keto esters, and β-keto nitriles. These reactions form 1,5-dicarbonyl compounds.

Section 17.10

- In an **aldol addition,** the enolate ion of an aldehyde or a ketone reacts with the carbonyl carbon of a second molecule of aldehyde or ketone, forming a β-hydroxyaldehyde or a β-hydroxyketone. The new C—C bond forms between the α-carbon of one molecule (the nucleophile) and the carbon that formerly was the carbonyl carbon of the other molecule (the electrophile). Therefore, the carbon attached to OH is adjacent to the carbon attached to the carbonyl group.

Section 17.11

- The product of an aldol addition can be dehydrated under acidic or basic conditions to form an **aldol condensation product.** Base-catalyzed dehydration is an **E1cB reaction.**

Section 17.12

- Primarily one product can be obtained from a crossed aldol addition if a carbonyl compound with α-hydrogens is added slowly to a solution of a carbonyl compounds without α-hydrogens and a base. If both carbonyl compounds have α-hydrogens, primarily one product can be obtained if LDA is used to form the enolate ion and the other carbonyl compound is added slowly.

Section 17.13

- In a **Claisen condensation,** the enolate ion of an ester reacts with a second molecule of ester, eliminating an ⁻OR group to form a β-keto ester.

Section 17.15

- An intramolecular condensation readily occurs if the reaction forms a product with a five- or six-membered ring. A **Dieckmann condensation** is an intramolecular Claisen condensation.

Section 17.16

- A **Robinson annulation** is a ring-forming reaction in which a Michael reaction and an intramolecular aldol condensation occur sequentially to form a compound with a 2-cyclohexenone ring.

Section 17.17

- Carboxylic acids with a carbonyl group at the 3-position **decarboxylate** when they are heated.

Section 17.18

- Carboxylic acids and carboxylate ions can be prepared by a **malonic ester synthesis;** the resulting carboxylic acid has two more carbons (the carboxyl group and the α-carbon) than the alkyl halide.

Section 17.19

- Methyl ketones can be prepared by an **acetoacetic ester synthesis;** the resulting methyl ketone has three more carbons (the carbonyl group and the carbon on either side of it) than the alkyl halide.

SUMMARY OF REACTIONS

1. Keto–enol interconversion (Section 17.3). The mechanisms are shown on page 806.

2. Halogenation of the α-carbon of aldehydes and ketones (Section 17.4). The mechanisms are shown on pages 807 and 808.

3. Halogenation of the α-carbon of carboxylic acids: the Hell–Volhard–Zelinsky (HVZ) reaction (Section 17.5). The steps in the reaction are shown on page 809.

4. Alkylating the α-carbon (Section 17.7). The mechanism is shown on page 811.

5. Alkylating or acylating the α-carbon of aldehydes and ketones by means of an enamine intermediate (Section 17.8). The steps in the synthesis are shown on pages 814–815.

6. Alkylating the β-carbon of an α,β-unsaturated carbonyl compound (Section 17.9)

a. via an enamine: The mechanism is shown on page 816.

b. via a Michael reaction: The mechanism is shown on page 816.

7. An aldol addition of two aldehydes, two ketones, or an aldehyde and a ketone (Section 17.10). The mechanism is shown on page 818.

8. Aldol condensation is an aldol addition followed by acid-catalyzed or base-catalyzed dehydration (Section 17.11). The mechanism for base-catalyzed dehydration is shown on page 820.

9. Claisen condensation of two esters (Section 17.13). The mechanism is shown on page 824.

10. Crossed addition and crossed condensation reactions:

a. when one of the carbonyl compounds does not have any α-hydrogens (Sections 17.12 and 17.14).

b. when both carbonyl compounds have α-hydrogens (Sections 17.12 and 17.14).

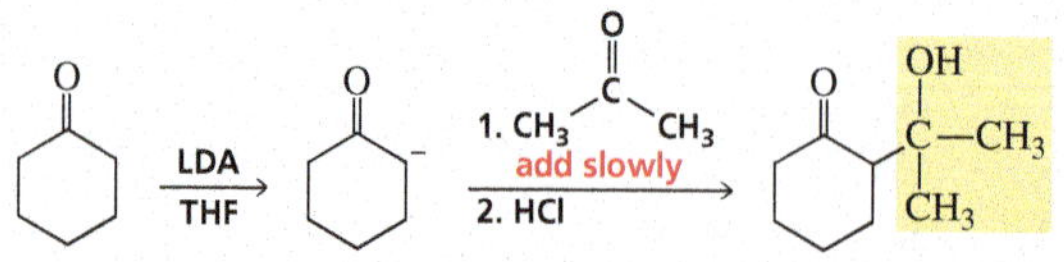

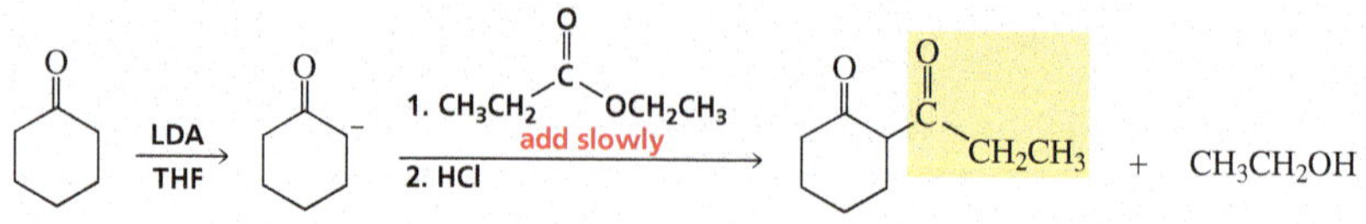

11. Robinson annulation (Section 17.16). The steps in the reaction are shown on page 830.

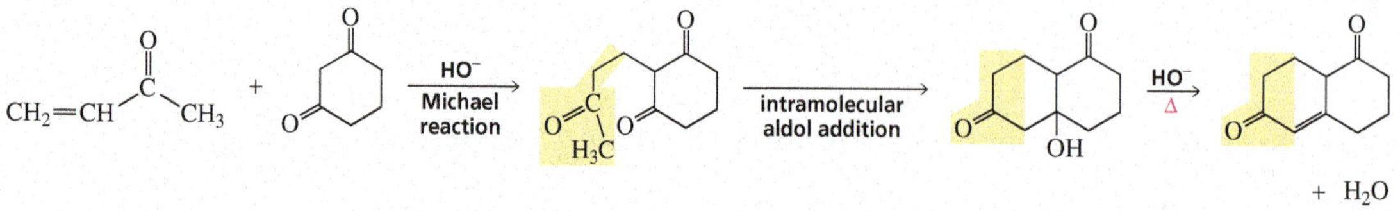

12. Decarboxylation of 3-oxocarboxylic acids and 3-oxocarboxylate ions (Section 17.17). The mechanism is shown on page 832.

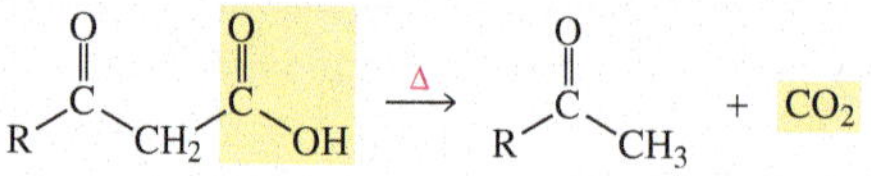

13. Malonic ester synthesis: synthesis of carboxylic acids (Section 17.18). The steps in the synthesis are shown on page 833.

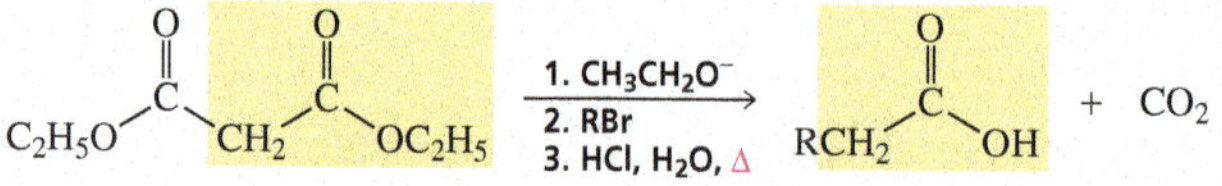

14. Acetoacetic ester synthesis: synthesis of methyl ketones (Section 17.19). The steps in the synthesis are shown on page 835.

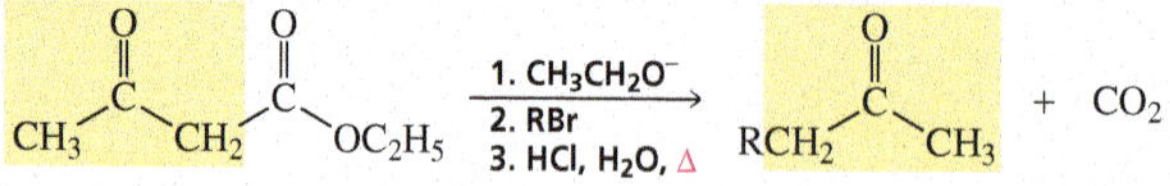

PROBLEMS

48. Draw a structure for each of the following:

    a. ethyl acetoacetate
    b. α-methylmalonic acid
    c. a β-keto ester
    d. the enol tautomer of cyclopentanone
    e. the carboxylic acid obtained from the malonic ester synthesis when the alkyl halide is propyl bromide

49. Draw the products of the following reactions:

    a. diethyl heptanedioate: (1) sodium ethoxide; (2) HCl
    b. pentanoic acid + PBr$_3$ + Br$_2$, followed by hydrolysis
    c. acetone + LDA/THF: (1) slow addition of ethyl acetate; (2) HCl
    d. diethyl 2-ethylhexanedioate: (1) sodium ethoxide; (2) HCl
    e. diethyl malonate: (1) sodium ethoxide; (2) isobutyl bromide; (3) HCl, H$_2$O + Δ
    f. acetophenone + LDA/THF: (1) slow addition of diethyl carbonate; (2) HCl

50. Number the following compounds in order of increasing pK_a value. (Number the most acidic compound 1.)

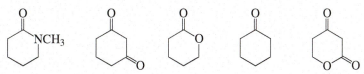

51. The ^{1}H NMR chemical shifts of nitromethane, dinitromethane, and trinitromethane are at δ 6.10, δ 4.33, and δ 7.52. Match each chemical shift with the compound. Explain how chemical shift correlates with pK_a.

52. Which of the following compounds decarboxylates when heated?

A   B   C

53. Draw the products of the following reactions:

a. 1,3-cyclohexanedione + LDA/THF, followed by allyl bromide
b. γ-butyrolactone + LDA/THF, followed by methyl iodide
c. 2,7-octanedione + sodium hydroxide
d. diethyl 1,2-benzenedicarboxylate + sodium ethoxide: (1) slow addition of ethyl acetate; (2) HCl

54. A racemic mixture of 2-methyl-1-phenyl-1-butanone is formed when (R)-2-methyl-1-phenyl-1-butanone is dissolved in an acidic or basic aqueous solution. Give an example of another ketone that undergoes acid- or base-catalyzed racemization.

55. Draw the products of the following reactions:

$$\xleftarrow{\begin{array}{c}\text{1. LDA/THF, } -78\,^\circ\text{C} \\ \hline \text{2. D}_2\text{O}\end{array}} \qquad \xrightarrow{\begin{array}{c}\text{1. LDA/THF, 0}\,^\circ\text{C} \\ \hline \text{2. D}_2\text{O}\end{array}}$$

56. What is the product of the following reaction?

$$2 \quad + \quad \text{HO}^- \quad \xrightarrow{\text{add slowly}}$$

57. An aldol addition can be catalyzed by acids as well as by bases. Propose a mechanism for the acid-catalyzed aldol addition of propanal.

58. In the presence of excess base and excess halogen, a methyl ketone is converted to a carboxylate ion. The reaction is known as the **haloform reaction** because one of the products is haloform (chloroform, bromoform, or iodoform). Before spectroscopy became a routine analytical tool, the haloform reaction served as a test for methyl ketones: the formation of iodoform, a bright yellow compound, signaled that a methyl ketone was present. Why do only methyl ketones form a haloform?

$$\underset{\substack{\text{methyl ketone}}}{R-\overset{\overset{\displaystyle O}{\|}}{C}-CH_3} \quad + \quad \underset{\substack{\text{excess}}}{I_2} \quad \xrightarrow{\text{HO}^-} \quad \underset{\substack{\text{carboxylate ion}}}{R-\overset{\overset{\displaystyle O}{\|}}{C}-O^-} \quad + \quad \underset{\substack{\text{iodoform}}}{CHI_3}$$

59. Identify **A–L**. (*Hint:* **A** shows three singlets in its ^{1}H NMR spectrum with integral ratios 3 : 2 : 3 and gives a positive iodoform test; see Problem 58.)

$$\underset{C_5H_8O_3}{A} \xrightarrow[\Delta]{\text{HCl, H}_2\text{O}} B \xrightarrow[\text{H}_2\text{O}]{\text{HO}^-} C \xrightarrow[\Delta]{\text{HCl}} D$$

$$\downarrow \begin{array}{c}\text{1. CH}_3\text{O}^- \\ \text{2. CH}_3\text{Br}\end{array}$$

$$E \xrightarrow[\Delta]{\text{HCl, H}_2\text{O}} H \xrightarrow[\substack{\text{excess} \\ \text{I}_2 \\ \text{excess}}]{\text{HO}^-} I \xrightarrow{\text{SOCl}_2} J \xrightarrow{\text{CH}_3\text{OH}} K \xrightarrow[\text{2. HCl}]{\text{1. CH}_3\text{O}^-} L$$

$$\downarrow \begin{array}{c}\text{1. CH}_3\text{O}^- \\ \text{2. CH}_3\text{Br}\end{array}$$

$$F \xrightarrow[\Delta]{\text{HCl, H}_2\text{O}} G$$

60. Using cyclopentanone as the reactant, show the product of

a. acid-catalyzed keto–enol interconversion. **b.** an aldol addition. **c.** an aldol condensation.

61. Show how 4-methyl-3-hexanol can be synthesized from 3-pentanone.

62. Show how the following compound can be synthesized from the given starting material. (*Hint:* Start with an acetoacetic ester.)

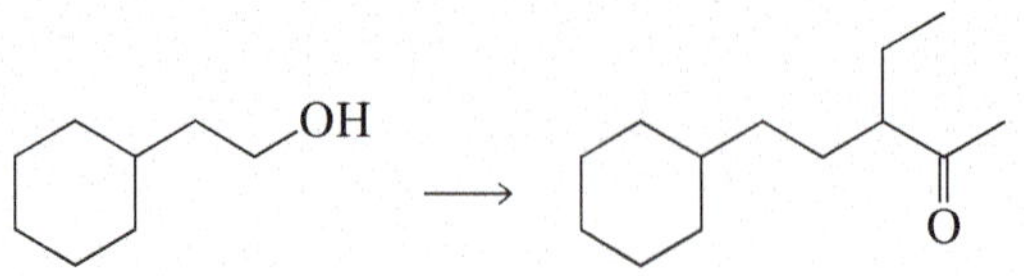

63. Show how the following compounds can be prepared from cyclohexanone:

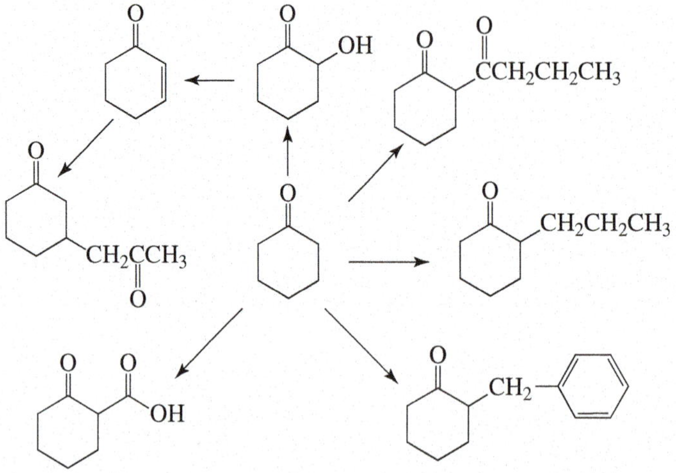

64. A β,γ-unsaturated carbonyl compound rearranges to a more stable conjugated α,β-unsaturated compound in the presence of either acid or base.

a. Propose a mechanism for the base-catalyzed rearrangement. **b.** Propose a mechanism for the acid-catalyzed rearrangement.

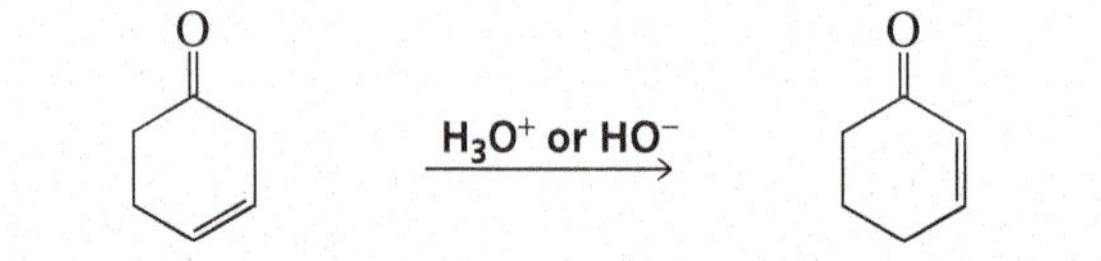

β,γ-unsaturated carbonyl compound

α,β-unsaturated carbonyl compound

65. There are other condensation reactions similar to the aldol and Claisen condensations:

a. The **Perkin condensation** is the condensation of an aromatic aldehyde and acetic anhydride. Draw the product obtained from the following Perkin condensation:

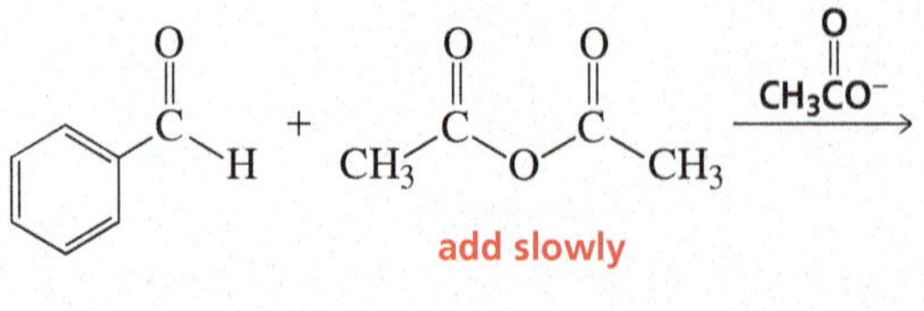

b. What compound is formed if water is added to the product of a Perkin condensation?

c. The **Knoevenagel condensation** is the condensation of an aldehyde or a ketone that has no α-hydrogens and a compound such as diethyl malonate that has an α-carbon flanked by two electron-withdrawing groups. Draw the product obtained from the following Knoevenagel condensation:

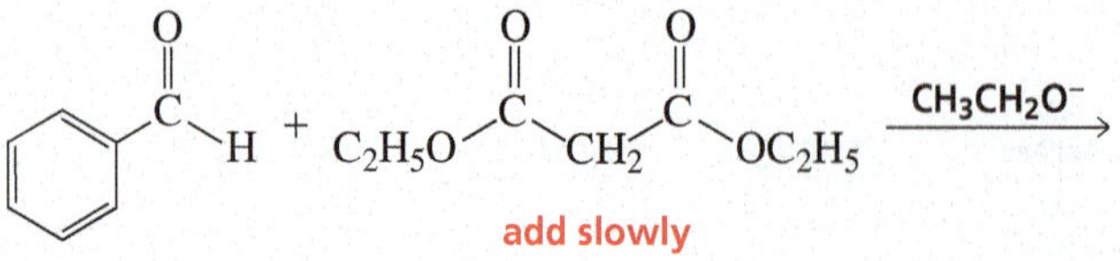

d. What product is obtained when the product of a Knoevenagel condensation is heated in an aqueous acidic solution?

66. Indicate how each of the following compounds can be synthesized from the given starting material and any other necessary reagents:

a.

CH₃–C(=O)–CH₃ ⟶ CH₃–C(=O)–CH₂–C(=O)–CH₃

c.

CH₃–C(=O)–CH₂–C(=O)–OCH₂CH₃ ⟶ CH₃–C(=O)–CH₂–⟨cyclopentane⟩

b.

CH₃–C(=O)–(CH₂)₃–C(=O)–OCH₃ ⟶ [cyclohexanedione with two CH₃ groups]

d.

CH₃CH₂O–C(=O)–(CH₂)₄–C(=O)–OCH₂CH₃ ⟶ [cyclopentanone with (CH₂)₂CH₃]

67. The **Reformatsky reaction** is an addition reaction in which an organozinc reagent is used instead of a Grignard reagent to add to the carbonyl group of an aldehyde or a ketone. Because the organozinc reagent is less reactive than a Grignard reagent, a nucleophilic addition to the ester group does not occur. The organozinc reagent is prepared by treating an α-bromo ester with zinc.

CH₃CH₂–C(=O)–H + CH₃CH(ZnBr)–C(=O)–OCH₃ ⟶ CH₃CH₂CH(O⁻ ⁺ZnBr)–CH(CH₃)–C(=O)–OCH₃ →[H₂O] CH₃CH₂CH(OH)–CH(CH₃)–C(=O)–OCH₃

organozinc reagent   β-hydroxy ester

Describe how each of the following compounds can be prepared, using a Reformatsky reaction:

a. CH₃CH₂CH₂CH(OH)CH₂–C(=O)–OCH₃

b. CH₃CH₂CH(OH)CH(CH₂CH₃)–C(=O)–OH

c. CH₃CH₂CH=C(CH₃)–C(=O)–OH

d. CH₃CH₂C(OH)(CH₂CH₃)CH₂–C(=O)–OCH₃

68. The ketone whose ¹H NMR spectrum is shown here was obtained as the product of an acetoacetic ester synthesis. What alkyl halide was used in the synthesis?

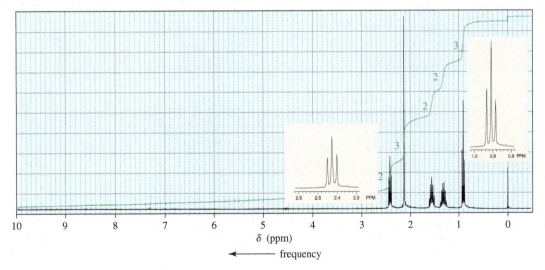

10   9   8   7   6   5   4   3   2   1   0
δ (ppm)
← frequency

69. Indicate how the following compounds can be synthesized from cyclohexanone and any other necessary reagents:

a. [cyclohexanone with CH₂CH₂CH₂CH₃]

b. [cyclohexanone with C(=O)CH₂CH₂CH₃] (two ways)

c. [cyclohexanone with C(=O)OCH₂CH₃]

d. [cyclohexanone with C(=O)H]

e. [bicyclic diketone]

70. Compound **A** with molecular formula C₆H₁₀ has two peaks in its ¹H NMR spectrum, both of which are singlets (with ratio 9 : 1). Compound **A** reacts with an acidic aqueous solution containing mercuric sulfate to form compound **B**, which gives a positive iodoform test (Problem 58) and has an ¹H NMR spectrum that shows two singlets (with ratio 3 : 1). Identify **A** and **B**.

71. a. What carboxylic acid is formed when the malonic ester synthesis is carried out with one equivalent of malonic ester, one equivalent of 1,5-dibromopentane, and two equivalents of base?

b. What carboxylic acid is formed when the malonic ester synthesis is carried out with two equivalents of malonic ester, one equivalent of 1,5-dibromopentane, and two equivalents of base?

72. Draw the products of the following reactions:

a. 2 [structure] $\xrightarrow{\text{1. CH}_3\text{CH}_2\text{O}^-}{\text{2. H}_3\text{O}^+}$

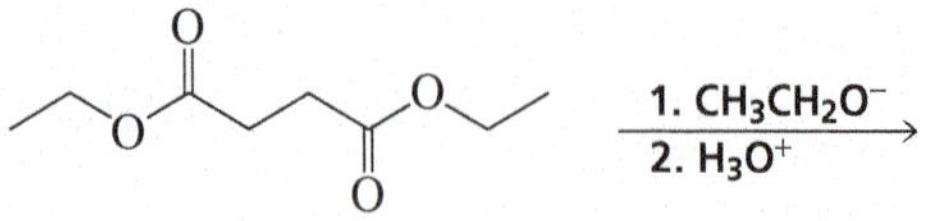

c. 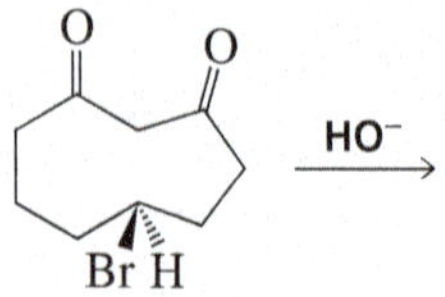 [structure] $\xrightarrow{\text{HO}^-}$

b. 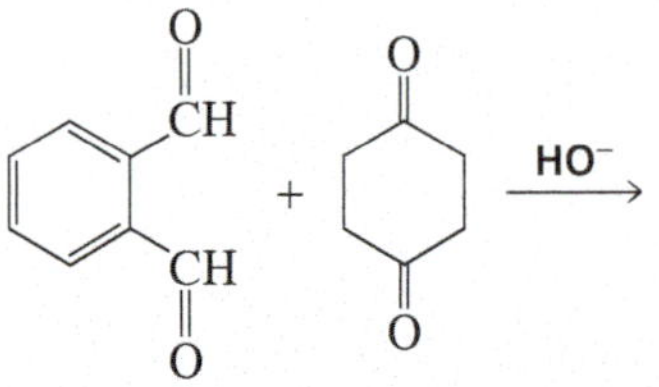 [structure] + [structure] $\xrightarrow{\text{HO}^-}$

d. 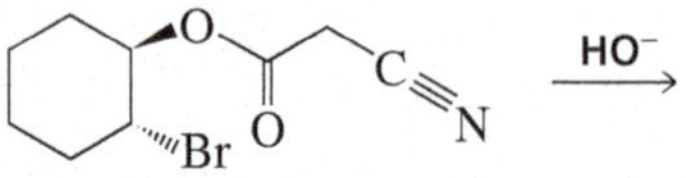 [structure] $\xrightarrow{\text{HO}^-}$

73. Show how the following compound can be prepared from starting materials that have no more than five carbons:

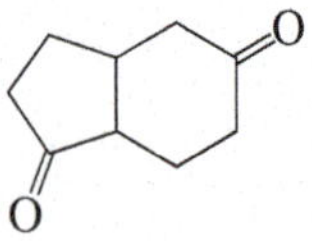

[structure]

74. a. Show how the amino acid alanine can be synthesized from propanoic acid. (The structures of the amino acids can be found on page 987.)

b. Show how the amino acid glycine can be synthesized from phthalimide and diethyl 2-bromomalonate.

75. Show how the following compounds can be synthesized. The only carbon-containing compounds available to you for each synthesis are shown.

a. 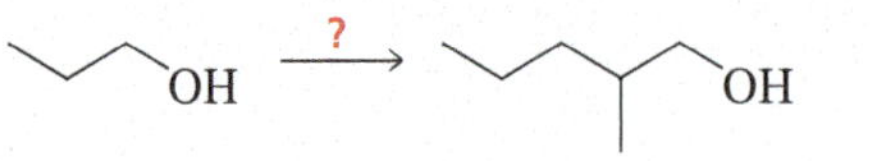 [structure] $\xrightarrow{?}$ [structure]

b. 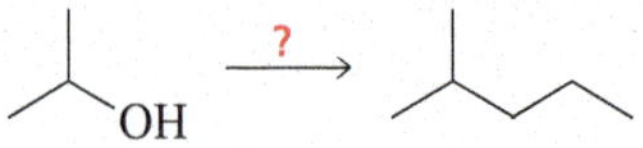 [structure] $\xrightarrow{?}$ [structure]

c. 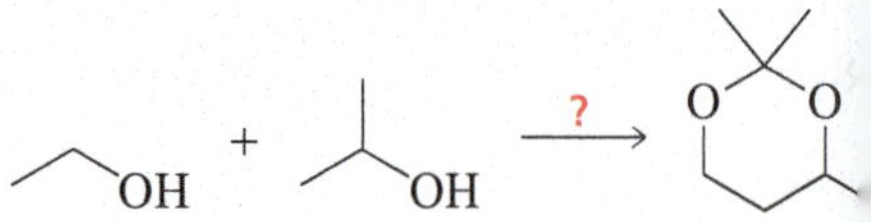 [structure] + [structure] $\xrightarrow{?}$ [structure]

76. A student tried to prepare the following compounds using aldol condensations. Which of these compounds was she successful in synthesizing? Explain why the other syntheses were not successful.

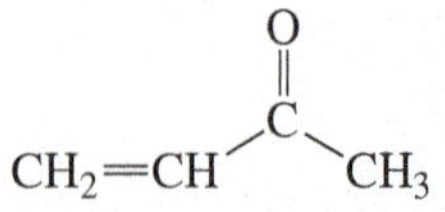

[structures A, B, C, D]

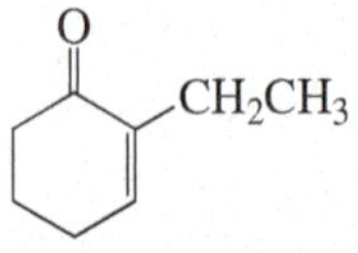

[structures E, F, G]

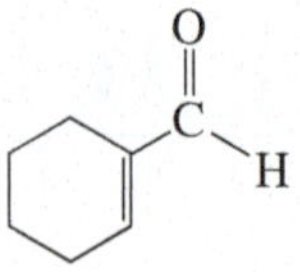

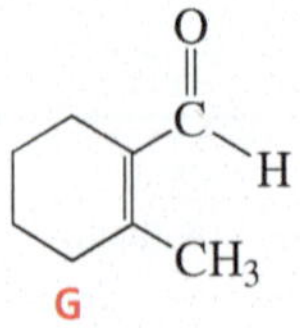

77. Explain why the following bromoketone forms different bicyclic compounds under different reaction conditions:

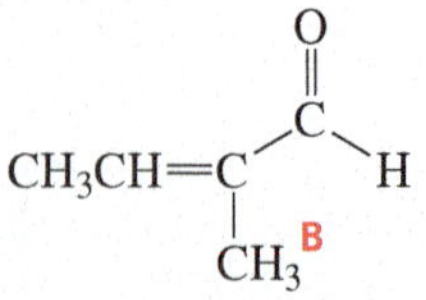

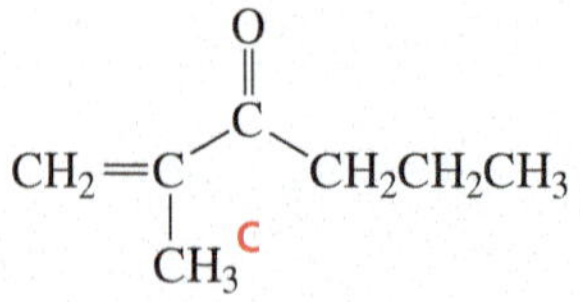

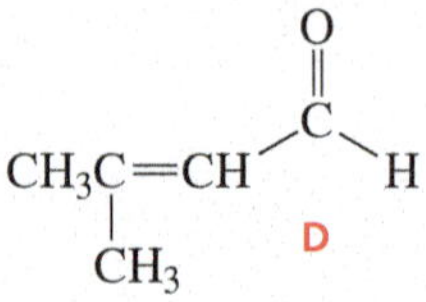

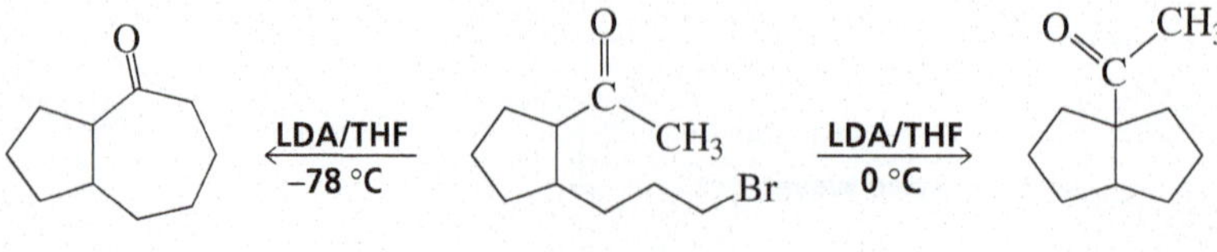

[structure] $\xleftarrow{\text{LDA/THF}\ -78\,°\text{C}}$ [structure] $\xrightarrow{\text{LDA/THF}\ 0\,°\text{C}}$ [structure]

850

78. A **Mannich reaction** puts a $\underset{\displaystyle R}{\overset{\displaystyle R}{\diagdown}}NCH_2-$ group on the α-carbon of a ketone. Propose a mechanism for the reaction.

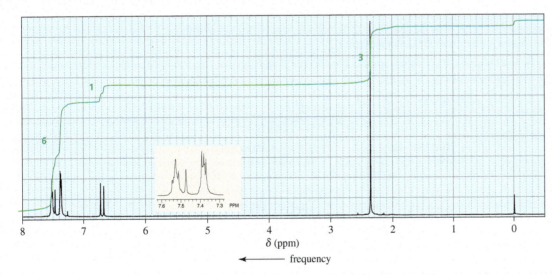

79. A carboxylic acid is formed when an α-haloketone reacts with hydroxide ion. This reaction is called a **Favorskii reaction**. Propose a mechanism for the following Favorskii reaction. (*Hint:* In the first step, HO^- removes a proton from the α-carbon that is not bonded to Br; a three-membered ring is formed in the second step; and HO^- is a nucleophile in the third step.)

80. An α,β-unsaturated carbonyl compound can be prepared by a reaction known as a **selenenylation–oxidation reaction**. A selenoxide is formed as an intermediate. Propose a mechanism for the reaction.

81. What carbonyl compounds are required to prepare a compound with molecular formula $C_{10}H_{10}O$ whose ^1H NMR spectrum is shown below?

82. Ninhydrin reacts with an amino acid to form a purple-colored compound. Propose a mechanism to account for the formation of the colored compound.

83. A **Cannizzaro reaction** is the reaction of an aldehyde that has no α-hydrogens with concentrated aqueous sodium hydroxide. In this reaction, half the aldehyde is converted to a carboxylic acid and the other half is converted to an alcohol. Propose a mechanism for the following Cannizzaro reaction:

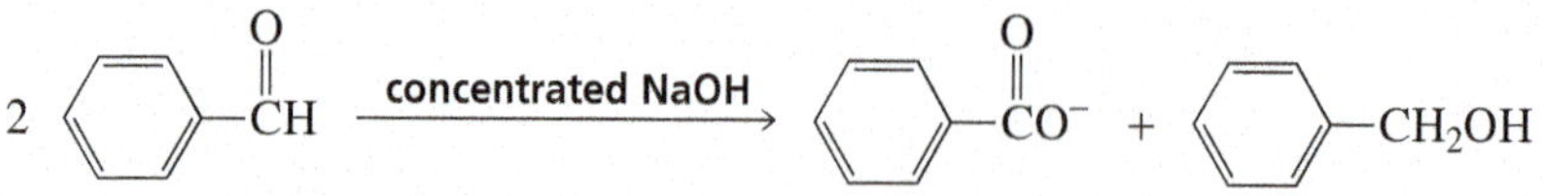

84. Propose a mechanism for each of the following reactions:

a.

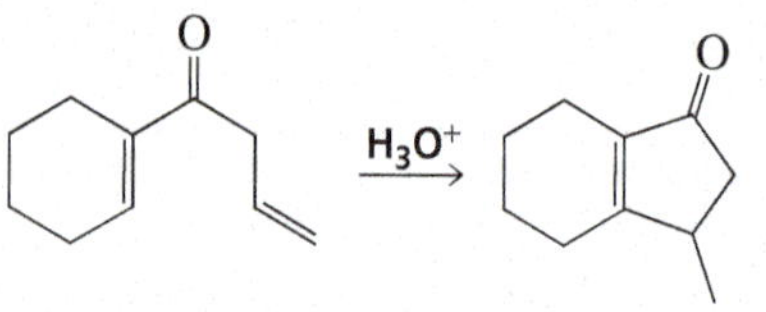

b.

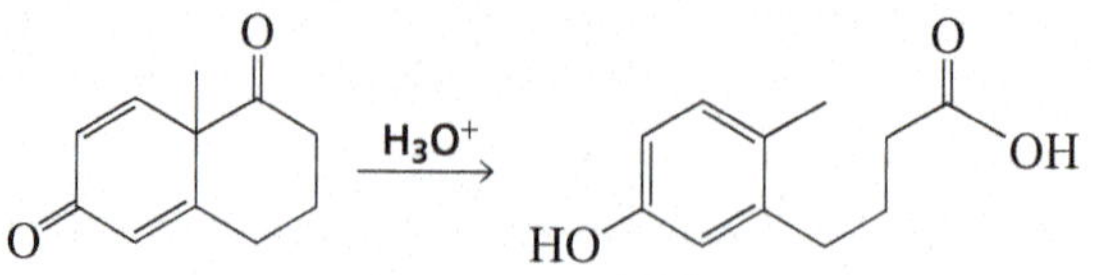

85. The following reaction is known as the **benzoin condensation.** The reaction does not take place if sodium hydroxide is used instead of sodium cyanide. Propose a mechanism for the reaction and explain why the reaction does not occur if hydroxide ion is the base.

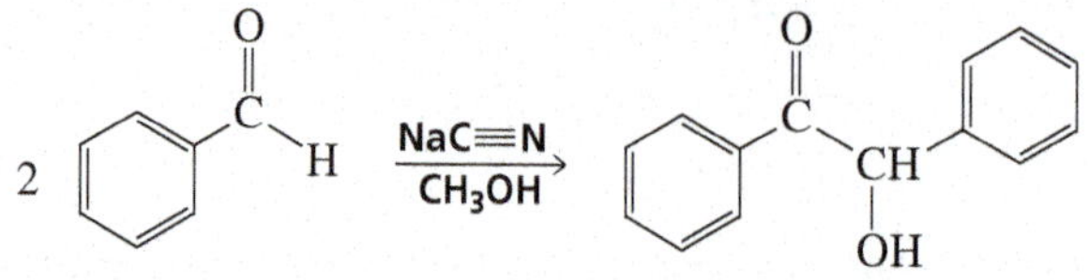

benzoin

86. Orsellinic acid, a common constituent of lichens, is synthesized from the condensation of acetyl thioester and malonyl thioester. If a lichen is grown on a medium containing acetate that was radioactively labeled with ^{14}C at the carbonyl carbon, which carbons will be labeled in orsellinic acid?

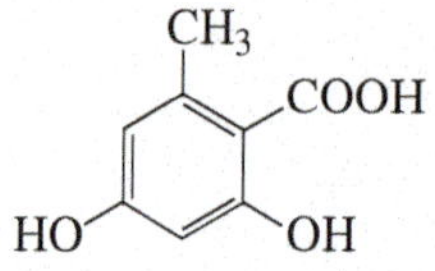

orsellinic acid

87. Alkylation of the following compound with methyl iodide under two different conditions forms two different ketoesters (**A** and **B**). Each ketoester forms a cyclic diketone (**C** and **D**) when treated with methoxide ion in methanol.

a. Draw the structures of **A** and **B,** and indicate the conditions used in the alkylation reaction that cause that ketoester to be formed.
b. Draw the mechanism for the conversion of **A** to **C.**

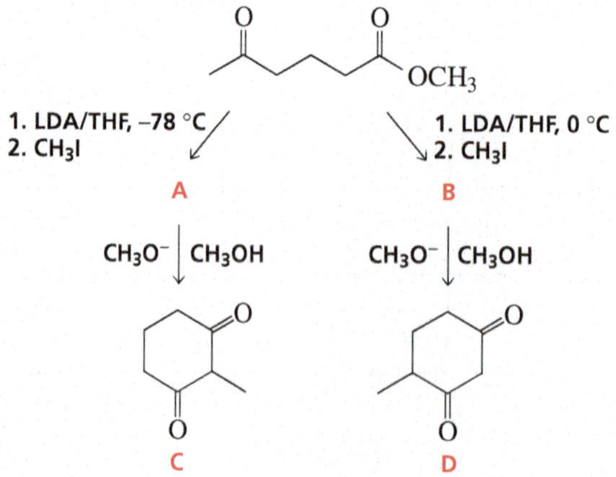

88. Propose a mechanism for the following reaction. (*Hint:* The intermediate has a cumulated double bond.)

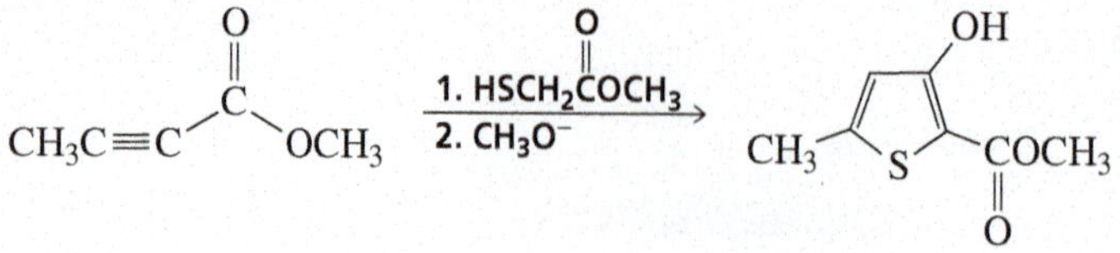

852

89. A compound known as *Hagemann's ester* can be prepared by treating a mixture of formaldehyde and ethyl acetoacetate first with base and then with acid and heat. Write the structure for the product of each of the steps.

   a. The first step is an aldol-like condensation.
   b. The second step is a Michael addition.
   c. The third step is an intramolecular aldol addition.
   d. The final transformation includes a dehydration and a hydrolysis followed by a decarboxylation.

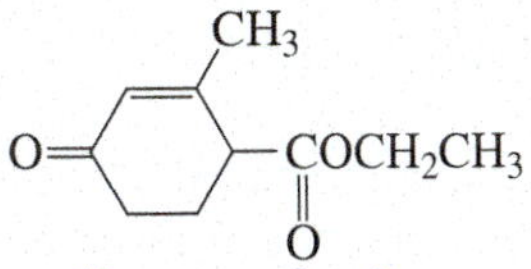

Hagemann's ester

90. Amobarbital is a sedative marketed under the trade name Amytal. Propose a synthesis of amobarbital, using diethyl malonate and urea (page 2) as two of the starting materials.

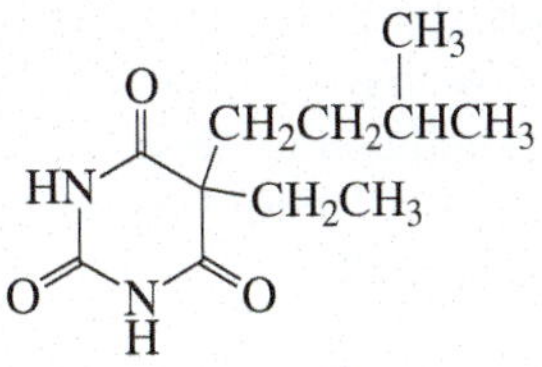

Amytal®

91. Propose a mechanism for the following reaction:

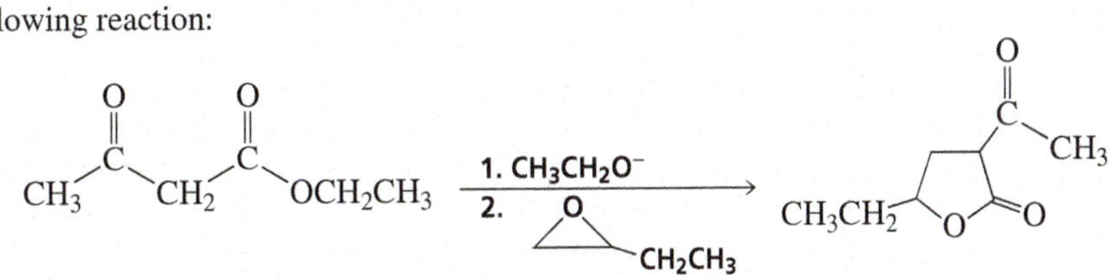

ESSENTIAL SKILL

TUTORIAL

Enhanced by MasteringChemistry®

SYNTHESIS AND RETROSYNTHETIC ANALYSIS

Organic synthesis is the preparation of organic compounds from other organic compounds. The word *synthesis* comes from the Greek word *synthesis*, which means "a putting together." You have been introduced to many aspects of organic synthesis, and have had the opportunity to design the synthesis of many organic compounds. In this tutorial, we will examine some of the strategies chemists use when designing a synthesis.

The most important factor in designing a synthesis is to have a good command of organic reactions. The more reactions you know, the better your chances of coming up with a successful synthesis. The guiding factor in planning a synthesis is to keep it as simple as possible. The simpler the synthetic plan, the greater the chance it will be successful.

CHANGING THE FUNCTIONAL GROUP

The first thing to do when designing a synthesis is to compare both the carbon skeleton and the positions of the functional group in the reactant and the product. If they both are the same, then all you need to determine is how to convert the functional group in the reactant to the functional group in the product. For examples in addition to those below, review Problem 64 in Chapter 6, Problem 34 in Chapter 7, Problems 70, 99, and 103 in Chapter 9, Problem 52 in Chapter 10, Problem 56 in Chapter 16, and Problem 63 in Chapter 17.

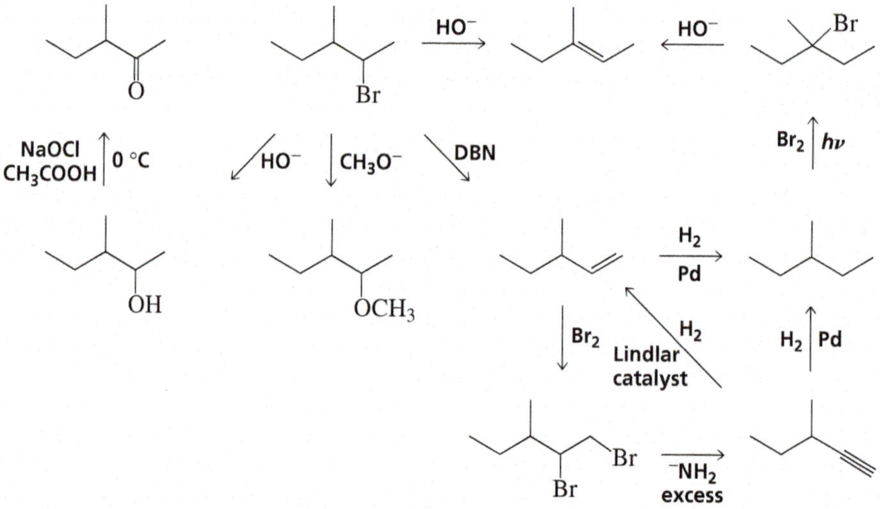

Notice that HO^- is used as the base to form a double bond between C-2 and C-3, whereas a bulky base (DBN) is needed to put the double bond between C-1 and C-2. Only two S_N2 reactions are shown here (using CH_3O^- and HO^-), but many other products can be synthesized just by changing the nucleophile. Notice, too, how single, double, and triple bonds can be interconverted.

FUNCTIONALIZING A CARBON

Recall that a carbon can be functionalized by a radical reaction.

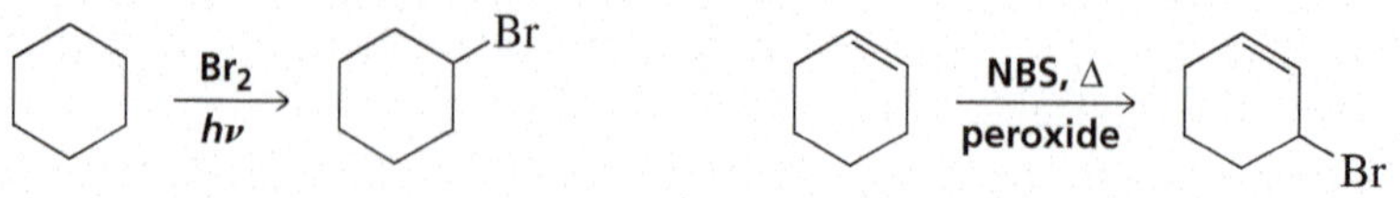

CHANGING THE POSITION OF THE FUNCTIONAL GROUP

If the carbon skeleton has not changed but the position of the functional group has, then you need to consider reactions that will change the position of the functional group. Recall that in the electrophilic addition reactions shown on the top of the next page, $\cdot$Br and R_2BH (which is replaced by OH) are electrophiles; therefore, they add to the sp^2 carbon bonded to the most hydrogens.

CHANGING THE CARBON SKELETON

If the carbon skeleton has changed but the number of carbons has not, then you need to consider reactions that will form a carbocation intermediate, because you know that a carbocation will rearrange if it can form a more stable carbocation.

PROBLEM 1 What reagents are required to convert the reactant to the product?

a.

b.

c.

d.

e.

f.

g.

h.

ADDING ONE CARBON TO THE CARBON SKELETON

There are several ways to create a product with one more carbon than the reactant. The method you choose depends on the functional group that you want to end up with in the product.

ADDING MORE THAN ONE CARBON TO THE CARBON SKELETON

There are many different ways to increase the carbon skeleton by more than one carbon. Acetylide ions, epoxides, Grignard reagents, aldol additions, Wittig reactions, and coupling reactions are just some of the methods that can be used. Common methods used to form new C—C bonds are summarized in Appendix IV.

PROBLEM 2 Starting with bromocyclohexane, how can each of the following compounds be prepared?

a. 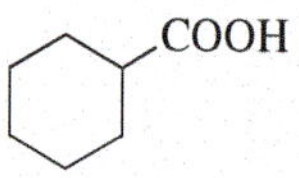 [cyclohexane]COOH   b. 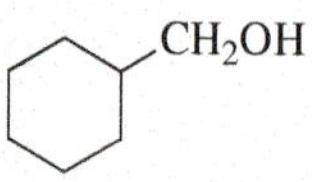 [cyclohexane]CH$_2$OH   c. 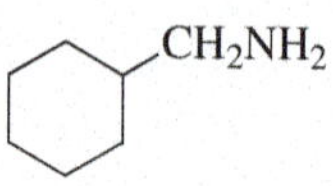 [cyclohexane]CH$_2$NH$_2$ d. 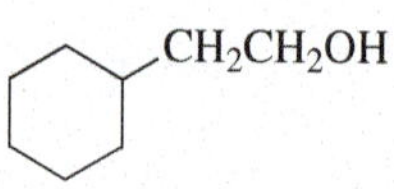 [cyclohexane]CH$_2$CH$_2$OH   e. 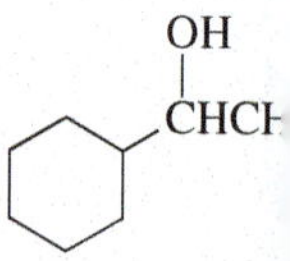 [cyclohexane]CHCH with OH

PROBLEM 3 Describe how the following compounds can be prepared from compounds containing no more than six carbons. (You can also use triphenylphosphine.)

a. 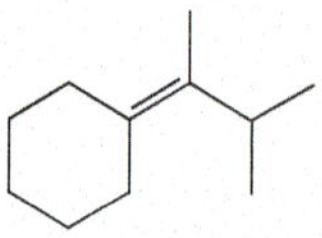 [structure]   b. 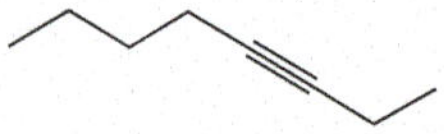 [structure]   c. 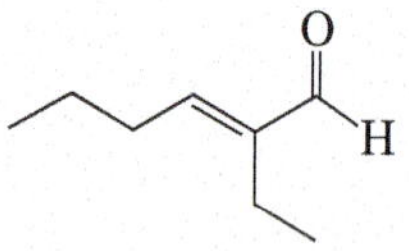 [structure with CHO]   d. 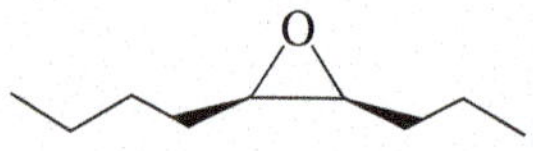 [epoxide structure]

USING RETROSYNTHETIC ANALYSIS TO CREATE A FUNCTIONAL GROUP

When you know what functional group you want to create, you can try to remember the various ways it can be synthesized. For example, a ketone can be synthesized by the acid-catalyzed addition of water to an alkyne, hydroboration–oxidation of an alkyne, oxidation of a secondary alcohol, and ozonolysis of an alkene. Notice that ozonolysis *decreases* the number of carbons in a molecule.

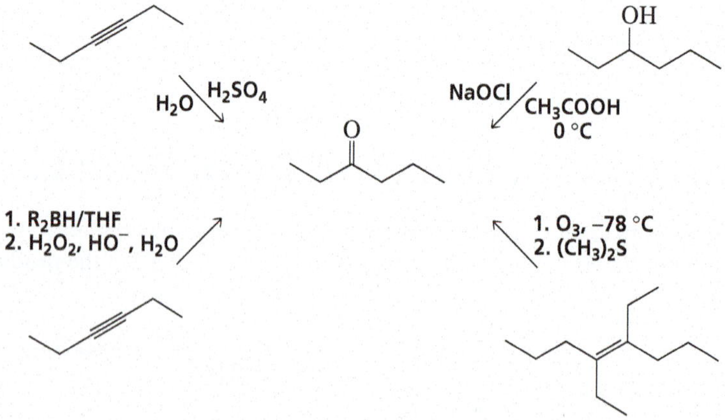

[reaction scheme with reagents: H$_2$O, H$_2$SO$_4$; 1. R$_2$BH/THF, 2. H$_2$O$_2$, HO⁻, H$_2$O; NaOCl, CH$_3$COOH, 0 °C; 1. O$_3$, −78 °C, 2. (CH$_3$)$_2$S]

In addition, methyl ketones can be synthesized by the acetoacetic ester synthesis, aromatic ketones can be synthesized by a Friedel–Crafts acylation, and a cyclic ketone, when treated with diazomethane, forms the next-size-larger cyclic ketone. Unless you have an exceptional memory, recalling all the methods you have learned to synthesize a particular functional group might be challenging. Therefore, they are listed for you in Appendix III.

PROBLEM 4 How many ways can you recall to synthesize

a. an ether?   b. an aldehyde?   c. an alkene?   d. an amine?

PROBLEM 5 Describe three ways to synthesize the following compound:

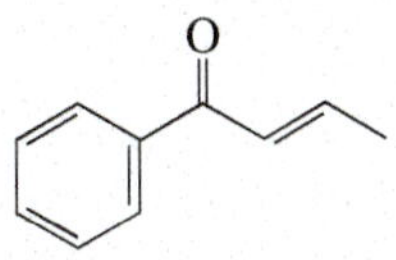

USING DISCONNECTIONS IN RETROSYNTHETIC ANALYSIS

We saw that a disconnection can be a useful step in a retrosynthetic analysis (Sections 16.14 and 17.20). Recall that a disconnection involves breaking a bond to give two fragments and then adding a positive charge to the end of one fragment and a negative charge to the end of the other. If the following compound is disconnected at bond **a**, then we see that the target molecule can be prepared from the reaction of cyclohexanone with a Grignard reagent.

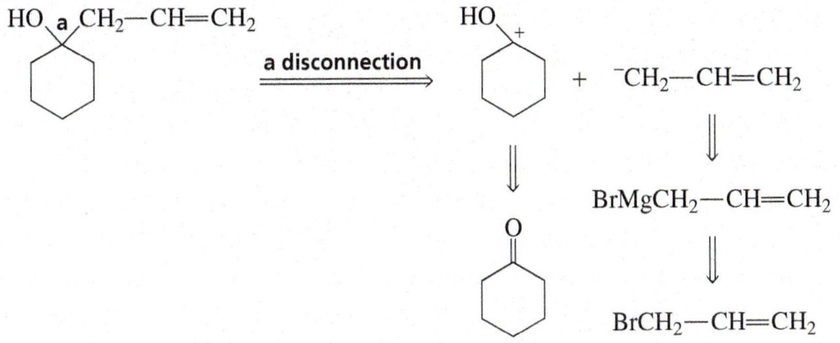

On the other hand, if the compound is disconnected at bond **b**, then we see that an epoxide is the synthetic equivalent for the electrophile and an organocuprate is the synthetic equivalent for the nucleophile.

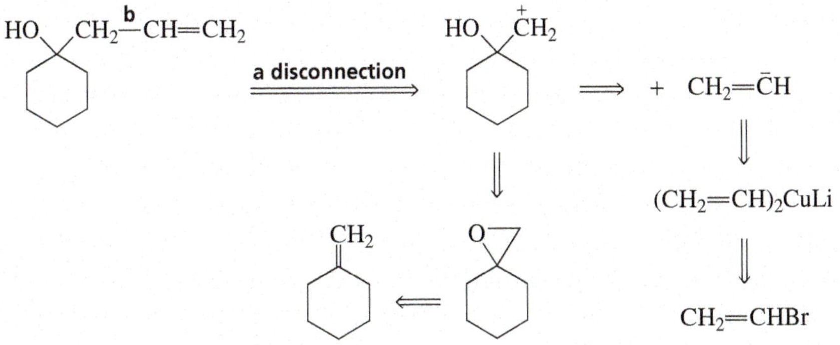

PROBLEM 6 Do a retrosynthetic analysis on each of the following compounds, ending with the given starting materials:

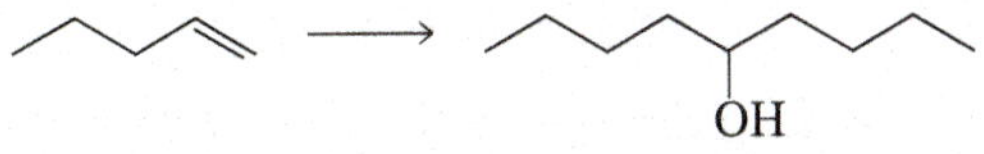

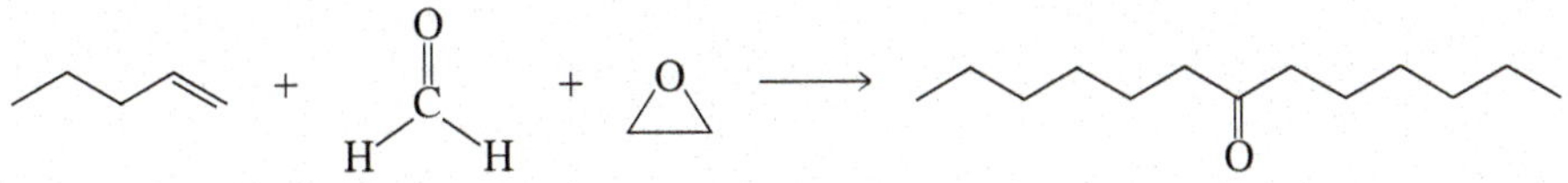

USING THE RELATIVE POSITIONS OF TWO FUNCTIONAL GROUPS TO DESIGN A SYNTHESIS

If a compound has two functional groups, then the relative positions of the two groups can provide a valuable hint as to how to approach the synthesis. For example, the following synthesis forms a **1,2-dioxygenated** compound.

a 1,2-dioxygenated
compound

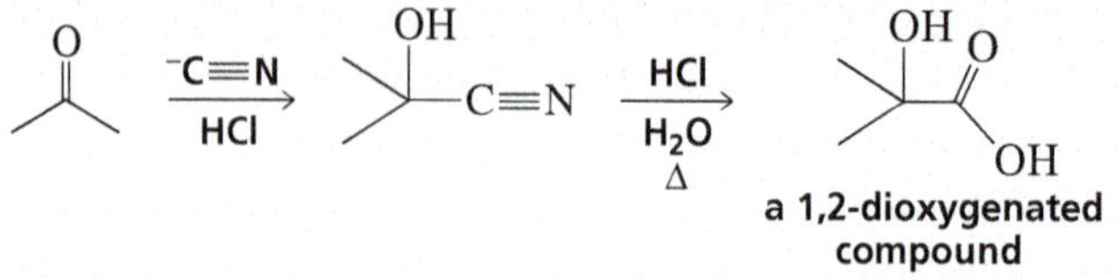

Retrosynthetic analysis shows that a **1,3-dioxygenated** compound can be formed by an aldol addition.

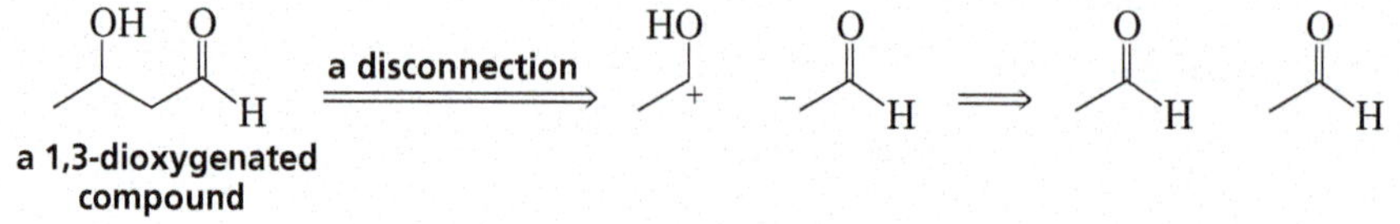

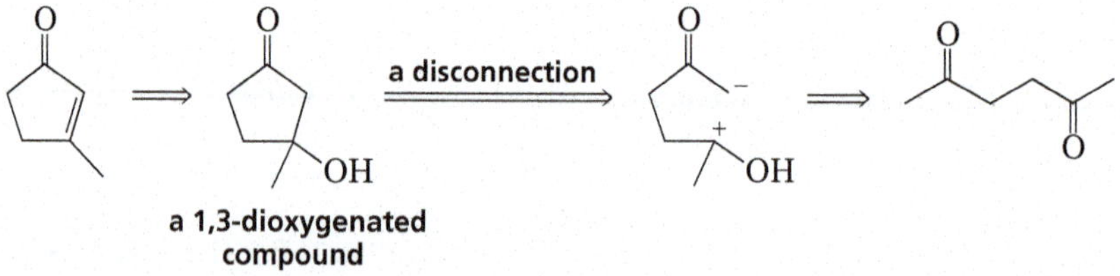

a 1,3-dioxygenated
compound

a 1,3-dioxygenated
compound

Disconnection of a **1,4-dioxgenated** compound shows it can be synthesized by nucleophilic attack of a negatively charged α-carbon (an enolate ion) on a compound with a positively charged α-carbon.

a 1,4-dioxygenated
compound

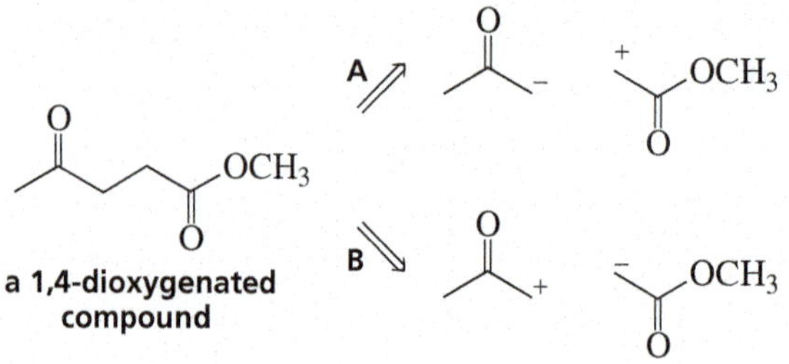

An α-brominated carbonyl compound is a synthetic equivalent for the positively charged α-carbon, and an enamine is the synthetic equivalent for the negatively charged α-carbon. Because esters cannot form enamines, path A is the preferred disconnection. Fortunately, the iminium ion hydrolyzes more readily than the ester.

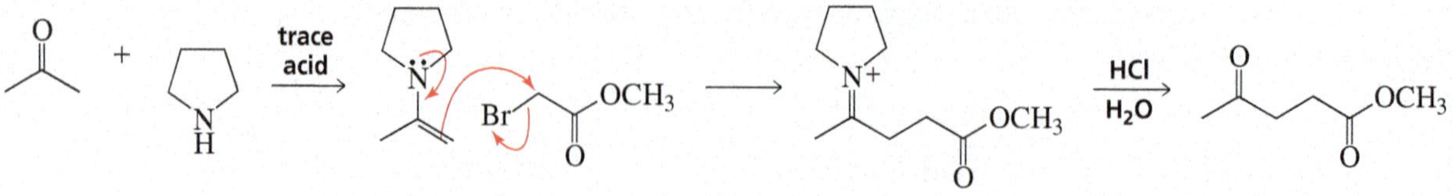

Disconnection of a **1,5-dioxgenated** compound shows it can be synthesized by nucleophilic attack of a negatively charged α-carbon on a carbonyl compound with a positively charged β-carbon.

a 1,5-dioxygenated
compound

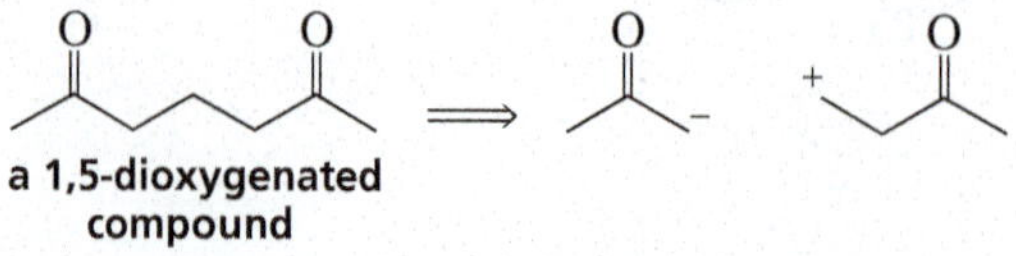

An α,β-unsaturated carbonyl compound is the synthetic equivalent of a compound with a positively charged β-carbon. And we saw that an enamine is the synthetic equivalent for the negatively
charged α-carbon.

PROBLEM 7 Describe how the following compound can be synthesized from compounds
containing no more than six carbons.

Disconnection of a **1,6-dioxgenated** compound shows it can be synthesized by nucleophilic
attack of a negatively charged α-carbon on a carbonyl compound with a positively charged γ-carbon.

**a 1,6-dioxygenated
compound**

There is no synthetic equivalent for a carbonyl compound with a positively charged γ-carbon, so
we have to consider another route. Recognizing that a 1,6-dioxygenated compound can be prepared
by oxidative cleavage of a cyclohexene provides an easy route to the target molecule, because a
compound with a six-membered ring can be readily prepared by a Diels–Alder reaction.

PROBLEM 8 Do a retrosynthetic analysis on each of the following compounds, ending with
available starting materials.

a.

b.

c.

d.

e.

PROBLEM 9 Use retrosynthetic analysis to plan a synthesis of the following target molecules.
The only carbon-containing compounds available for the syntheses are cyclohexanol, ethanol,
and carbon dioxide.

a.

b.

c.

d.

PROBLEM 10 How can you synthesize the following compounds from starting materials containing no more than four carbons?

a. 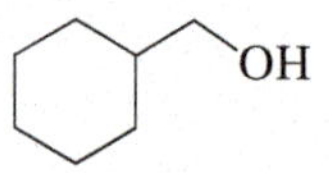 (cyclohexyl-CH2-OH)

b. 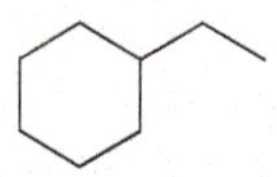 (cyclohexyl-CH2CH3)

c. 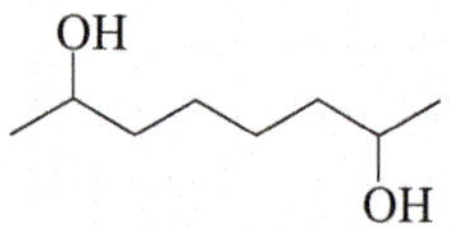 (OH-CH...CH-OH chain)

d. 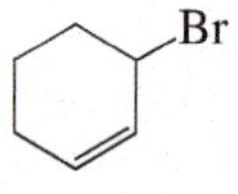 (cyclohexenyl-Br)

PROBLEM 11 Show how the following compounds can be synthesized from the given starting materials:

a. 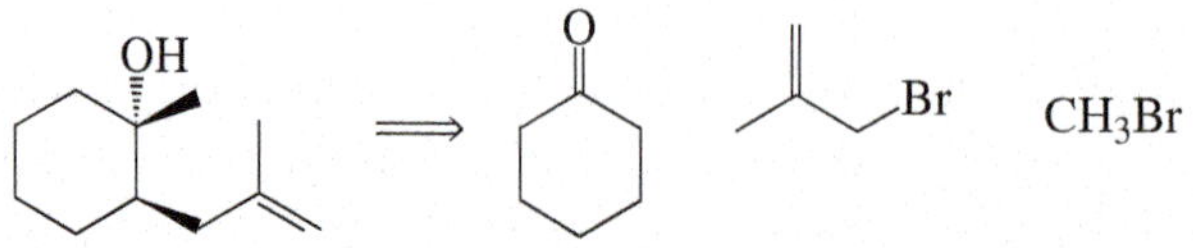 (cyclohexane with OH and isopropenyl substituents) ⟹ (cyclohexanone) (isobutenyl bromide) CH₃Br

b. 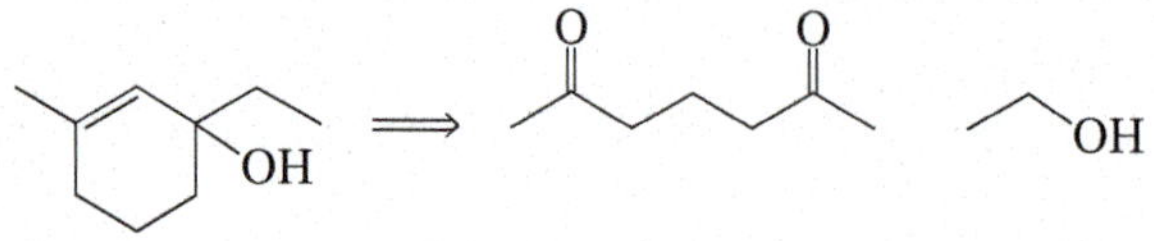 (methylcyclohexenol with ethyl) ⟹ (diketone) (CH₃CH₂OH)

EXAMPLES OF MULTISTEP ORGANIC SYNTHESIS

To give you an idea of the kind of thinking required for the synthesis of a complicated molecule, we will look at the synthesis of lysergic acid, which was done by R. B. Woodward, and the synthesis of caryophyllene, which was done by E. J. Corey. Woodward (in 1965) and Corey (in 1990) both received the Nobel Prize for their contributions to synthetic organic chemistry.

Lysergic acid was first synthesized by Woodward in 1954. The diethylamide of lysergic acid (LSD) is a better known compound because of its hallucinogenic properties. It is a tribute to Woodward's ability as a synthetic chemist that the next synthesis of lysergic acid was not accomplished until 1969. Lysergic acid has an indole ring. Because indole is unstable under acidic conditions (Section 20.5), Woodward designed a synthesis that did not form the indole ring until the final step.

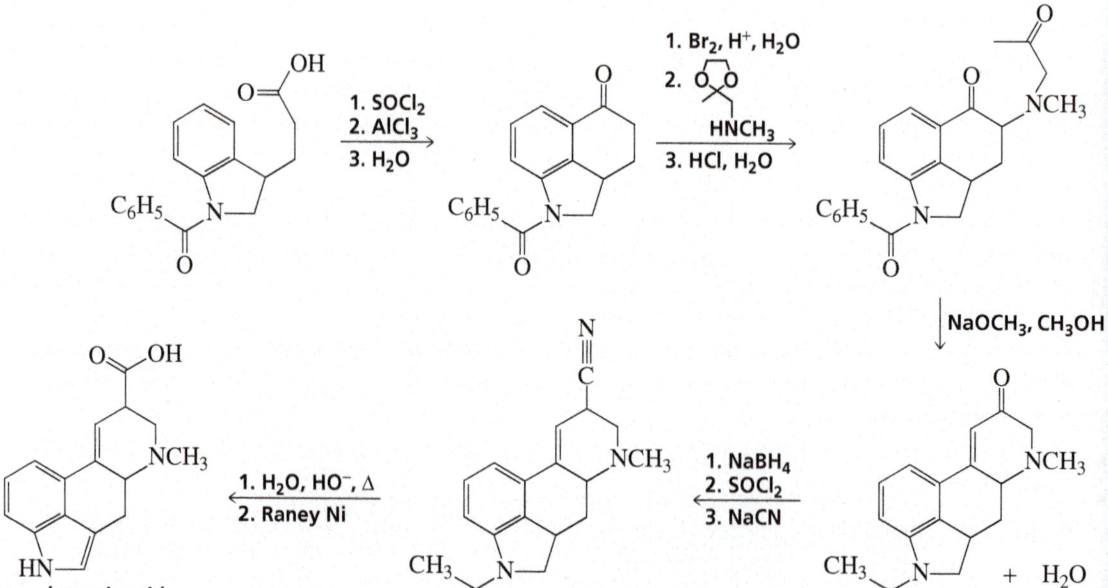

lysergic acid

- The starting material is a dihydroindole carboxylic acid with its nitrogen protected by being converted to an amide.

- An intramolecular Friedel–Crafts acylation forms the third ring (Section 18.6).

- Acid-catalyzed bromination of the α-carbon is followed by an S_N2 substitution of the bromine by an amine and removal of the ketone's protecting group.

- An aldol condensation forms the fourth ring.

- The ketone carbonyl group is reduced to an OH group, which is replaced by a Cl, and the Cl is then replaced by a cyano group.

- Both the cyano group and the protecting amide group are hydrolyzed.

- The last step is the reverse of hydrogenation of a double bond. It is carried out by using a typical hydrogenation catalyst in the absence of hydrogen.

Caryophyllene is an oil found in cloves. The most important strategic element in its synthesis was the realization that the nine-membered ring could be formed by fragmenting a six-membered ring fused to a five-membered ring. The problem then became designing the synthesis of the compound that undergoes fragmentation.

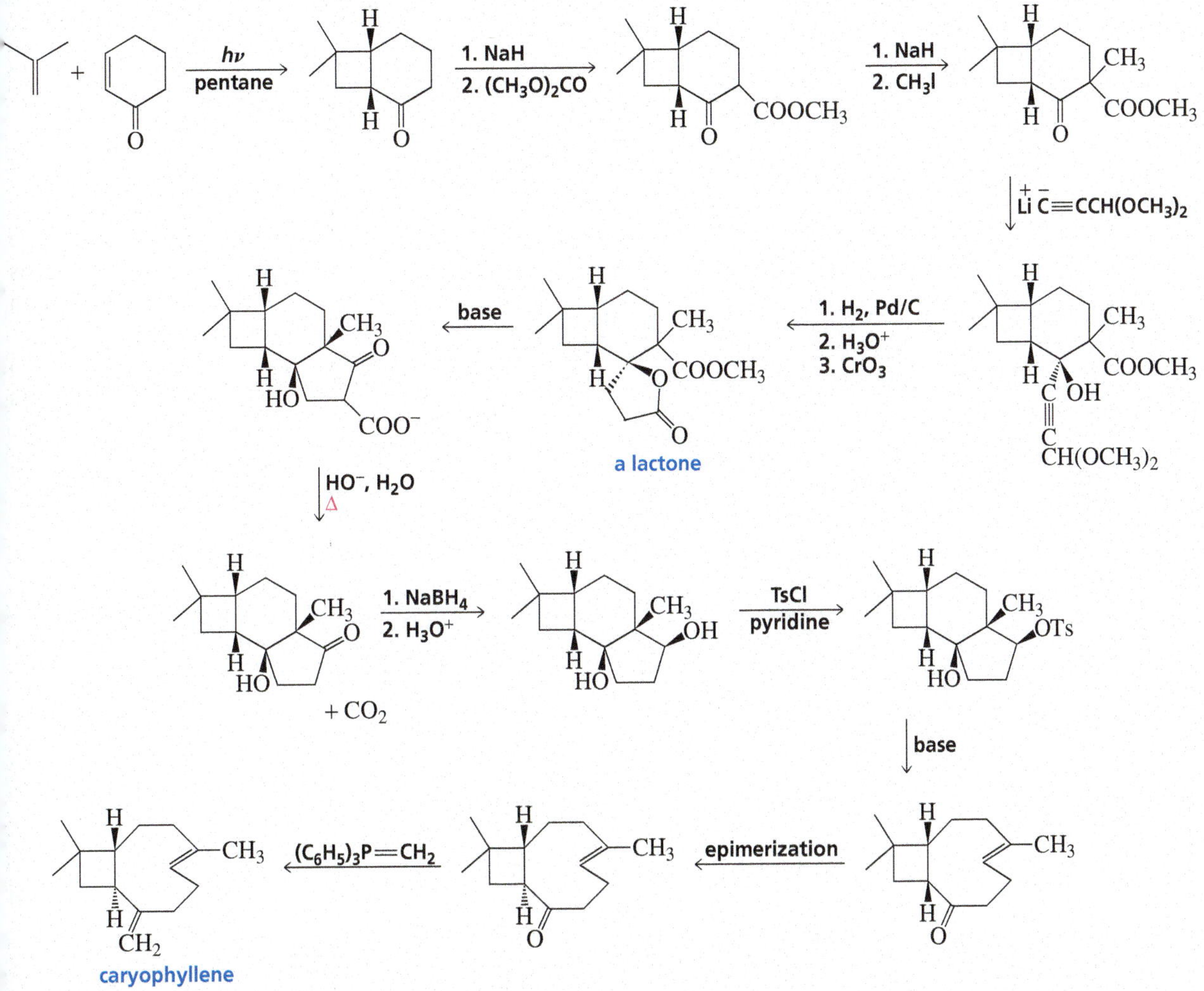

a lactone

caryophyllene

- The first step is a photochemical [2+2] cycloaddition reaction (Section 28.4).

- One α-hydrogen of the resulting ketone is substituted with a methoxycarbonyl group, and the second α-hydrogen on the same carbon is substituted with a methyl group. (Today we would use LDA, not NaH.)

- An acetylide ion is used to add a three-carbon fragment to the carbonyl group of the ketone.

- The triple bond is reduced, the acetal is hydrolyzed, and the resulting aldehyde is oxidized to a carboxylic acid that forms a lactone.

- A Claisen condensation and hydrolysis of the lactone forms the 3-oxocarboxylic acid that is decarboxylated to give the cyclic ketone.
- Reduction of the ketone forms an alcohol.
- Because the secondary alcohol is more reactive than the sterically hindered tertiary alcohol, the desired tosylate is formed.
- The stage is now set for the desired fragmentation reaction. This reaction involves removing a proton from the OH group, which allows the oxygen to form a carbonyl group and eliminate the tosyl group.
- Epimerization of the asymmetric center adjacent to the carbonyl group occurs once the ketone has been formed.
- A Wittig reaction forms the exocyclic double bond.

Answers to Problems

PROBLEM 1

PROBLEM 2

e.

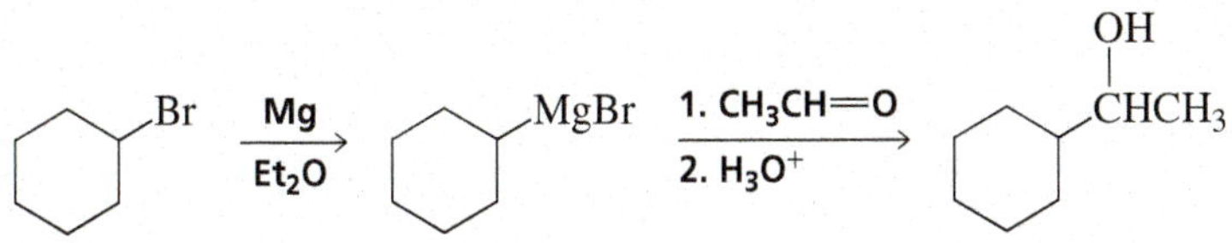

PROBLEM 3

a. An alkene can be made in one step by a Wittig reaction.

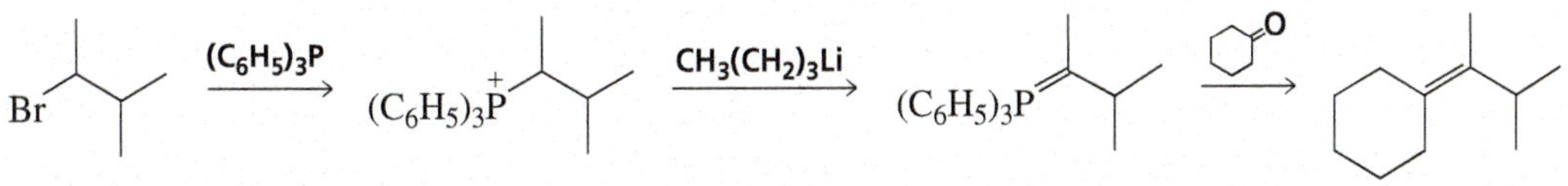

b. The alkyne can be prepared from 1-hexyne and ethyl bromide as shown here or it can be prepared from 1-butyne and butyl bromide.

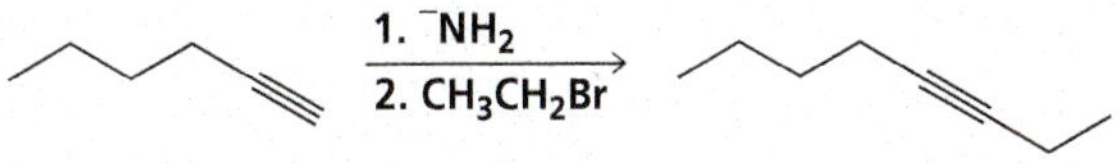

c. The desired aldehyde has eight carbons and a structure that suggests it can be prepared by an aldol condensation using an aldehyde with four carbons.

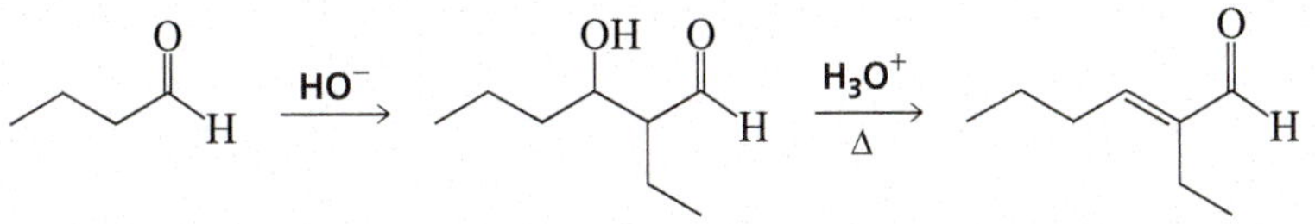

d.

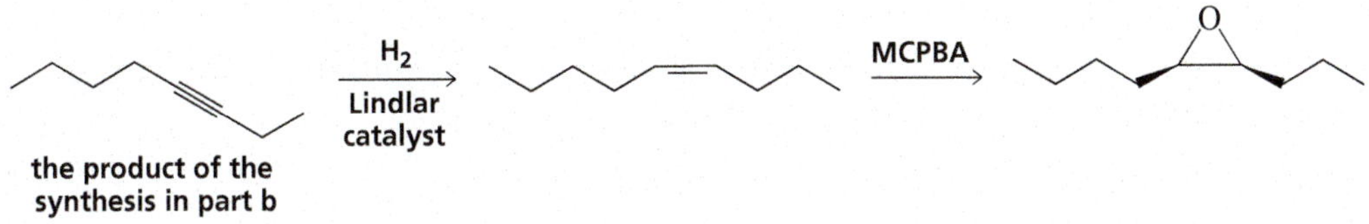

the product of the
synthesis in part b

PROBLEM 4   See Appendix III on page A-8.

PROBLEM 5

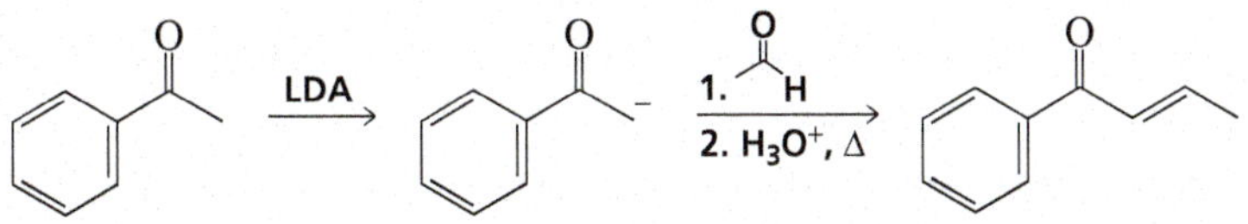

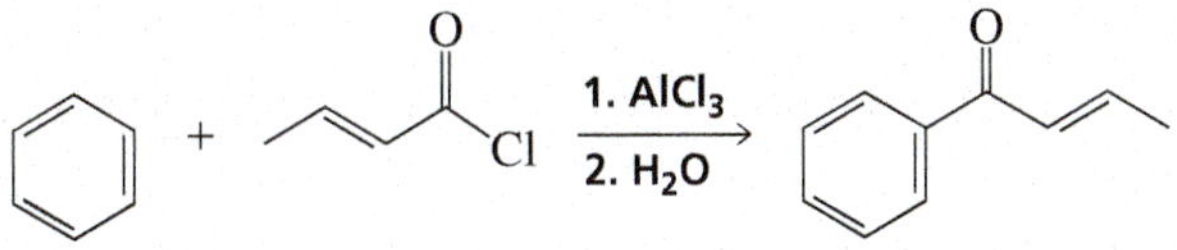

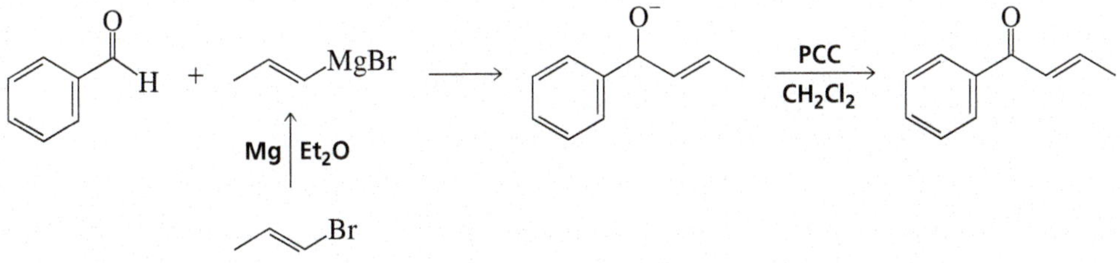

PROBLEM 6

a.

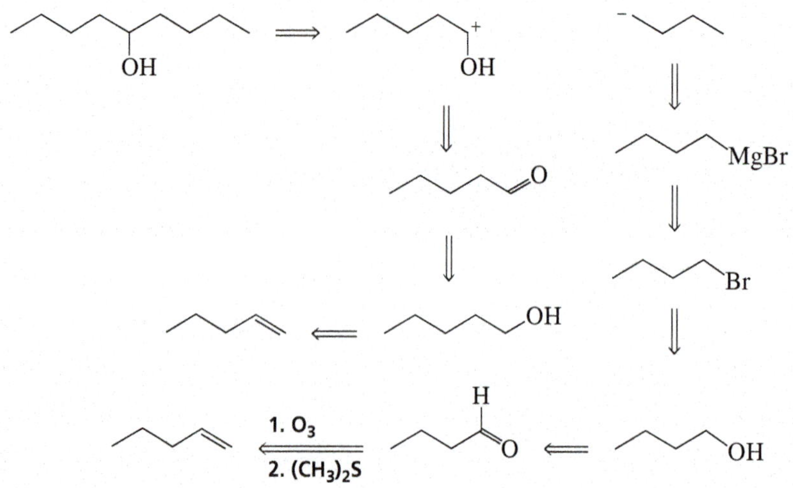

b.

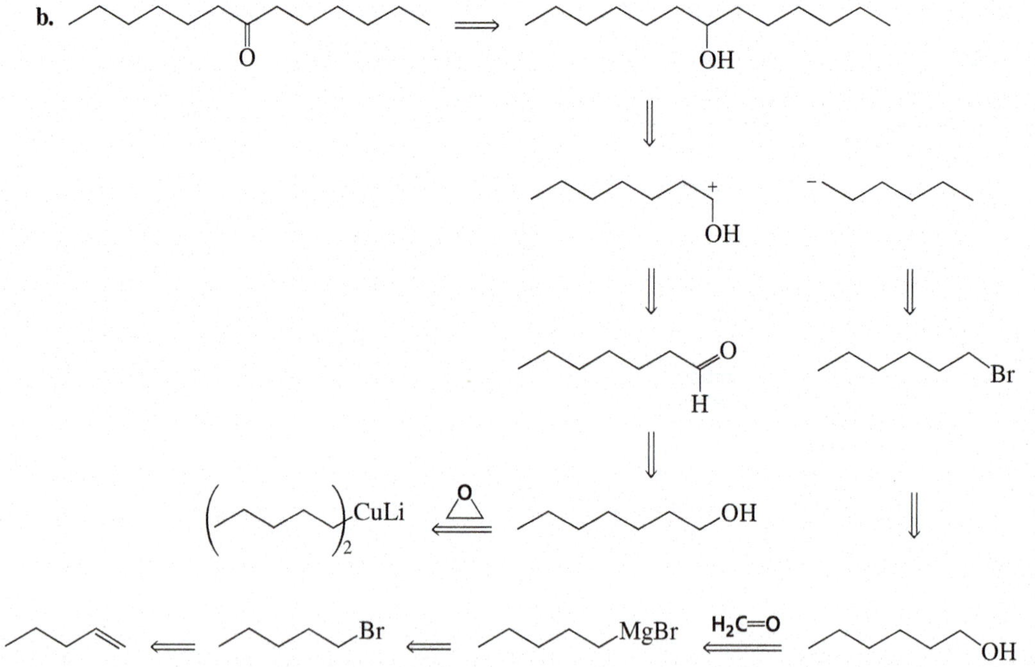

PROBLEM 7 Because the target molecule is a 1,5-dioxygenated compound, it can be synthesized from a negatively charged α-carbon (using an enamine as the synthetic equivalent) and an α,β-unsaturated ketone.

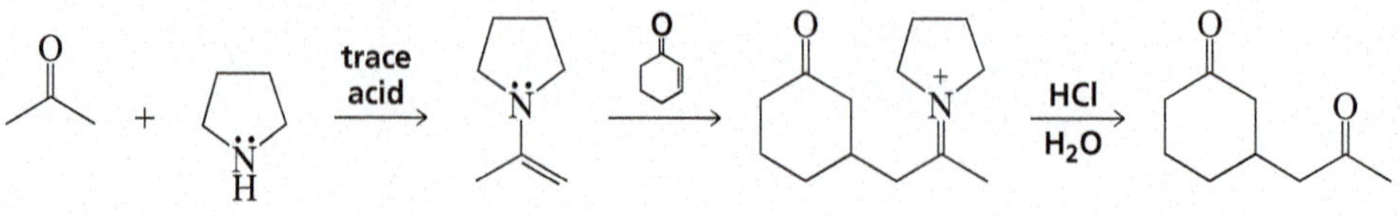

PROBLEM 8

a.

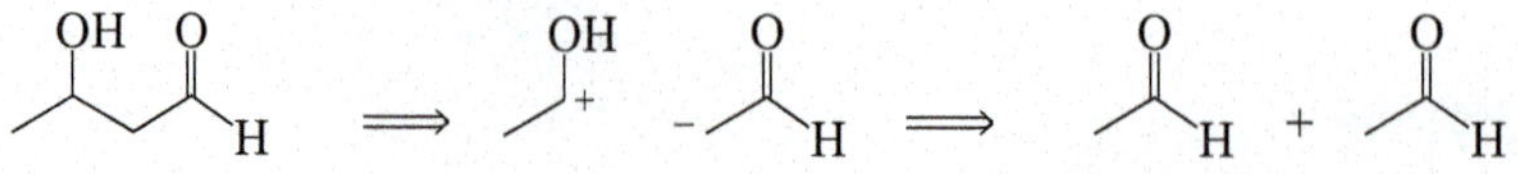

b.

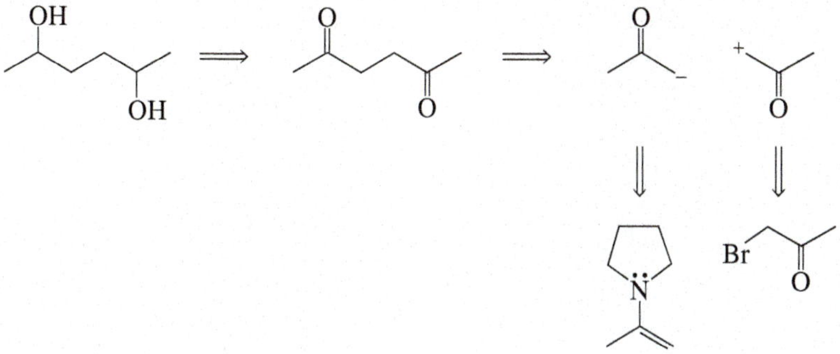

c.

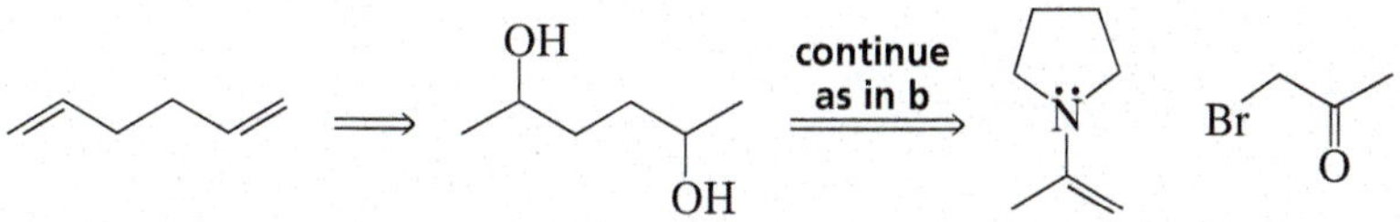

d.

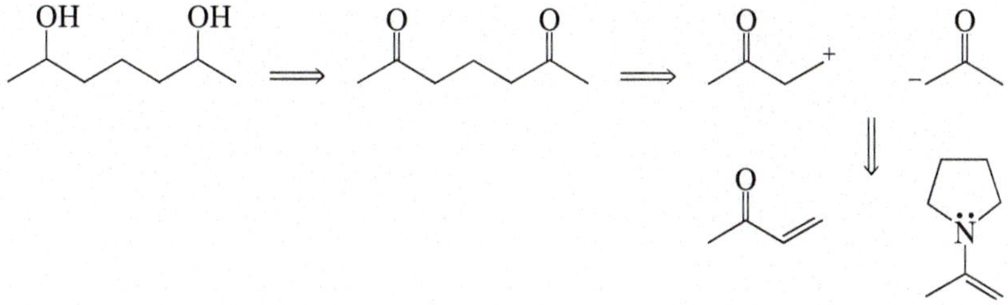

e.

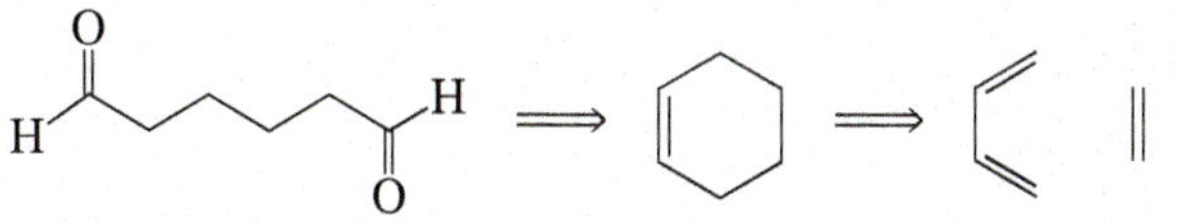

PROBLEM 9

a.

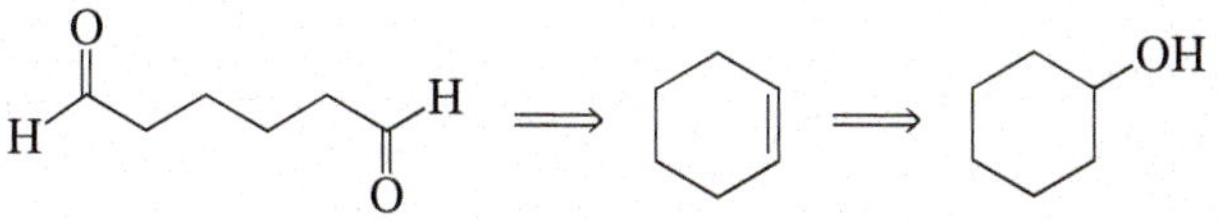

b.

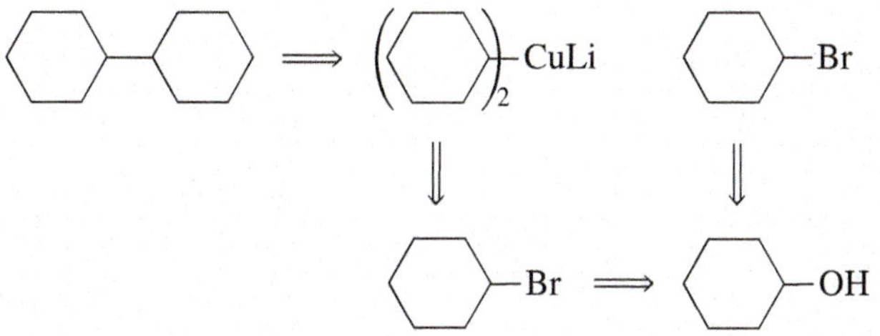

c.

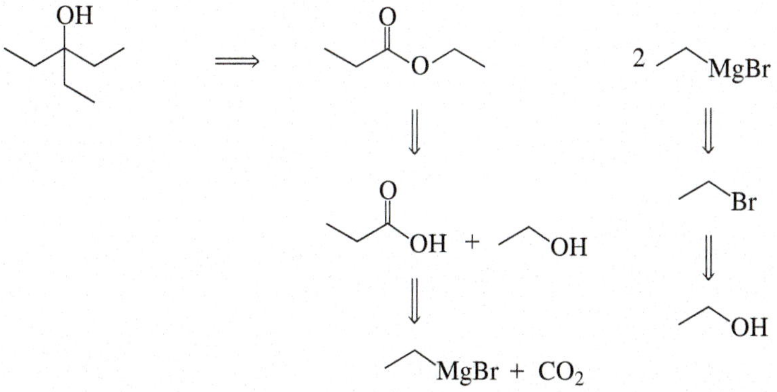

d.
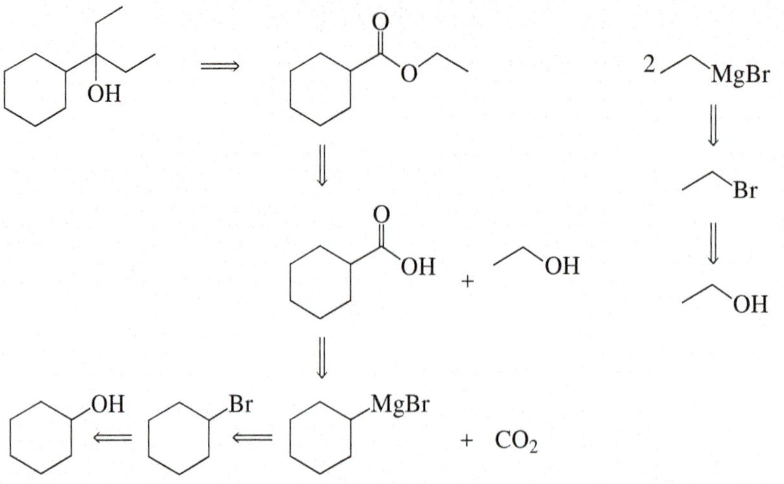

PROBLEM 10

a.
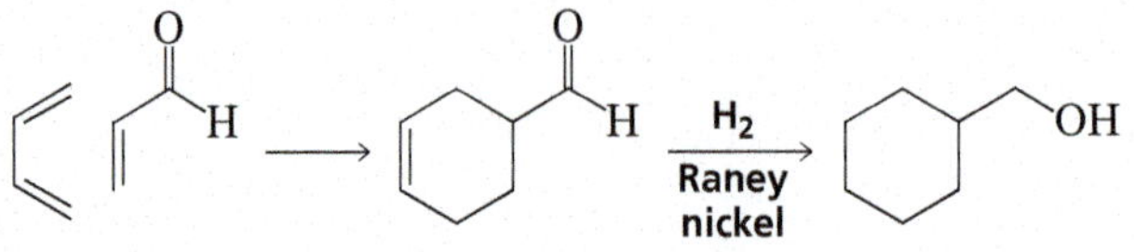

b.
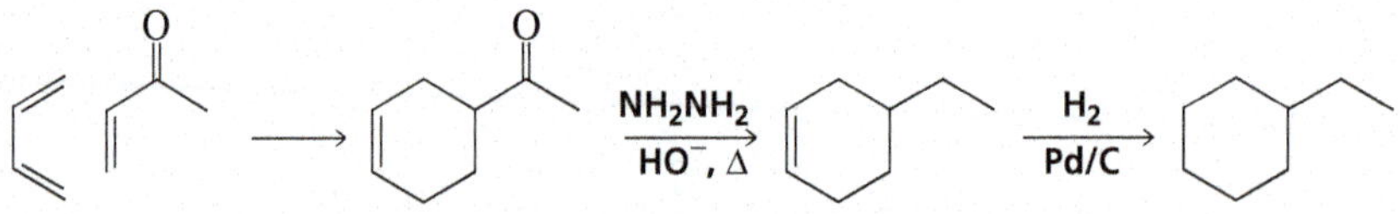

c.
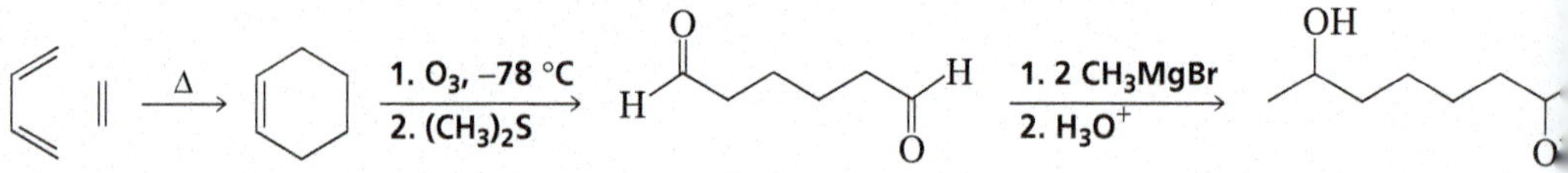

d.
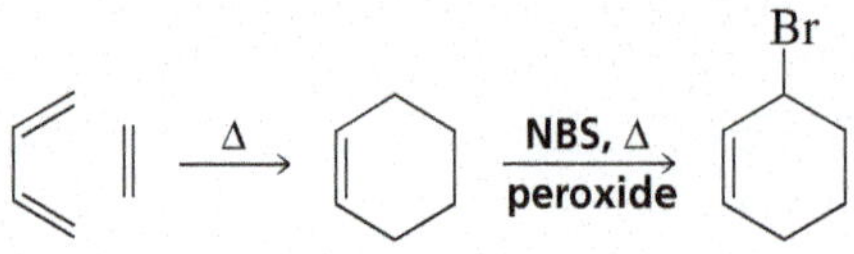

PROBLEM 11

a.
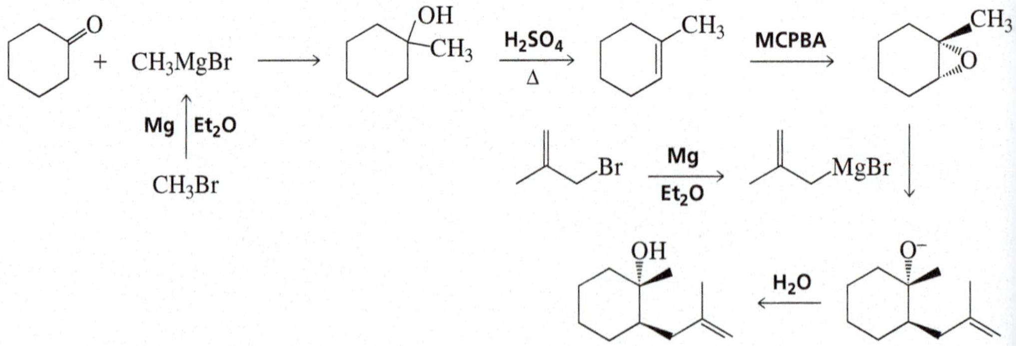

b.
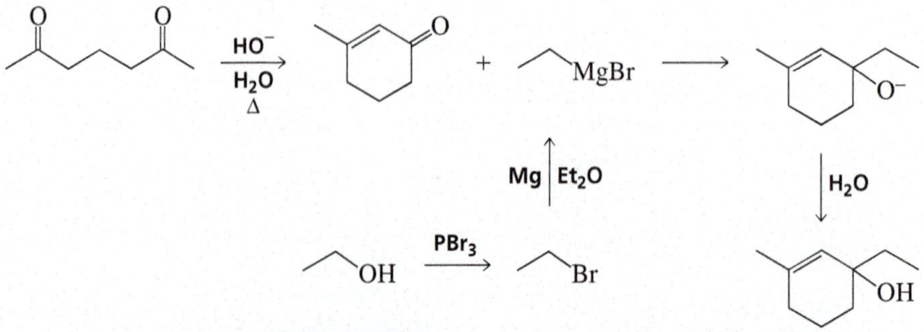

Aromatic Compounds

The two chapters in Part 6 discuss the reactions of aromatic compounds. In Chapter 8, you learned about the structure of benzene, the most common aromatic compound, why it is classified as aromatic, and the kinds of reactions it undergoes. Now you will look in greater detail at the reactions of aromatic compounds.

Chapter 18 Reactions of Benzene and Substituted Benzenes

Chapter 18 focuses on the reactions of benzene and substituted benzenes. Although benzene, alkenes, and dienes are all nucleophiles (because they all have carbon–carbon π bonds), benzene's aromaticity causes it to react in a way that is different from the way alkenes and dienes react. You will see how a substituent can be placed on a benzene ring, the kinds of substituents that can be paced on a ring, and some reactions that can change the substituent after it is on the ring. You will also learn how a substituent affects both the reactivity of a benzene ring and the placement of an incoming substituent. Chapter 18 also describes two additional types of reactions that can be used to synthesize substituted benzenes: the reactions of arene diazonium salts and nucleophilic aromatic substitution reactions.

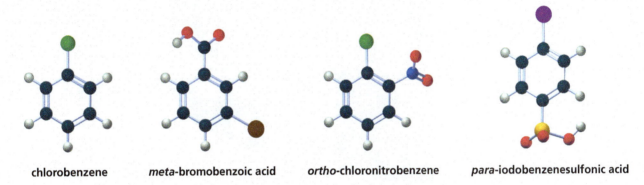

chlorobenzene   *meta*-bromobenzoic acid   *ortho*-chloronitrobenzene   *para*-iodobenzenesulfonic acid

Chapter 19 More About Amines • Reactions of Heterocyclic Compounds

You first met amines in Chapter 2, and you have continued to encounter them in almost every chapter since. **Chapter 19** starts by extending the coverage of amines. You saw that amines do not undergo addition, substitution, or elimination reactions; their importance lies in their reactions as bases and nucleophiles with other organic compounds. Chapter 19 also covers the reactions of aromatic heterocyclic compounds. You will see that they undergo the same reactions that benzene and substituted benzenes undergo and by the same mechanisms.

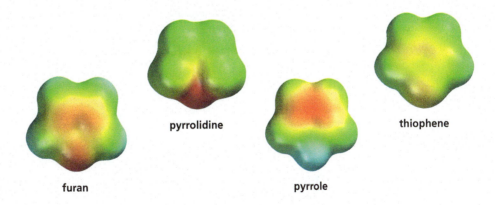

pyrrolidine

thiophene

furan

pyrrole

18 Reactions of Benzene and Substituted Benzenes

gas lamps along a street

The compound we know as benzene was first isolated in 1825 by Michael Faraday. He extracted it from the liquid residue obtained after heating whale oil under pressure to produce the gas then being used in gas lamps. In 1834, Eilhard Mitscherlich correctly determined the molecular formula (C_6H_6) of Faraday's compound and named it benzin because of its relationship to benzoic acid, a known substituted form of benzene. Later its name was changed to benzene.

Group IV

We saw that the families of organic compounds can be placed in one of four groups, and that all the families in a group react in similar ways (Section 5.4). This chapter begins our discussion of aromatic compounds—the families of compounds in Group IV.

In Chapter 8, you learned that benzene is an aromatic compound that can be represented by two resonance contributors.

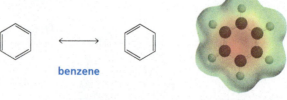

benzene

electrostatic potential
map for benzene

You also learned that aromatic compounds are unusually stable compounds (Section 8.16). You saw that to be aromatic:

- a compound must be cyclic and planar and have an uninterrupted cloud of π electrons (called a π cloud) above and below the plane of the molecule.
- the π cloud must contain an odd number of pairs of π electrons (Section 8.17).

Many substituted benzenes are found in nature. A few that have physiological activity are shown on the top of the next page.

ephedrine
a bronchodilator

chloramphenicol
an antibiotic that is particularly
effective against typhoid fever

mescaline
active agent of
the peyote cactus

adrenaline
epinephrine
a hormone released
by the body in
response to stress

Many other physiologically active substituted benzenes are not found in nature, but exist because chemists have synthesized them. In fact, more than two-thirds of the top 400 brand name and generic drugs contain a benzene ring. Three commonly prescribed drugs are shown here.

Lexapro®
antidepressant

Plavix®
inhibitor of platelet aggregation

Zyrtec®
antihistamine

We saw that when naturally occurring compounds are found to have desirable physiological activities, chemists attempt to synthesize structurally similar compounds with the hope of developing them into useful products (Section 10.9). For example, chemists have synthesized compounds with structures similar to that of adrenaline, producing amphetamine, a central nervous system stimulant, and the closely related methamphetamine (methylated amphetamine); both are used clinically as appetite suppressants. Methamphetamine, known as "meth," is also made and sold illegally. The compounds shown here represent just a few of the many substituted benzenes that have been synthesized for commercial use.

amphetamine
appetite
suppressant

methamphetamine
"meth"

acetylsalicylic acid
aspirin

saccharin
artificial
sweetener

p-dichlorobenzene
in mothballs and
air fresheners

Measuring Toxicity

Agent Orange, a defoliant widely used in the Vietnam War, is a mixture of two synthetic substituted benzenes: 2,4-D and 2,4,5-T. Dioxin (TCDD), a contaminant formed during the manufacture of Agent Orange, has been implicated as the causative agent of the various symptoms suffered by those exposed to the defoliant during the war.

2,4-dichlorophenoxyacetic acid
2,4-D

2,4,5-trichlorophenoxyacetic acid
2,4,5-T

2,3,7,8-tetrachlorodibenzo[b,e][1,4]dioxin
TCDD

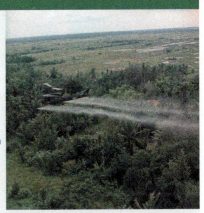

Vietnam War defoliation mission

The toxicity of a compound is indicated by its LD_{50} value—the dosage found to kill 50% of the test animals exposed to it. Dioxin, with an LD_{50} value of 0.0006 mg/kg for guinea pigs, is extremely toxic. For comparison, the LD_{50} values of some well-known but far less toxic poisons are 0.96 mg/kg for strychnine and 15 mg/kg for both arsenic trioxide and sodium cyanide. One of the most toxic agents known is the toxin that causes botulism; it has an LD_{50} value of about 1×10^{-7} mg/kg.

18.1 THE NOMENCLATURE OF MONOSUBSTITUTED BENZENES

Tables of physical properties can be found in Appendix VI.

Some monosubstituted benzenes are named simply by adding the name of the substituent to "benzene."

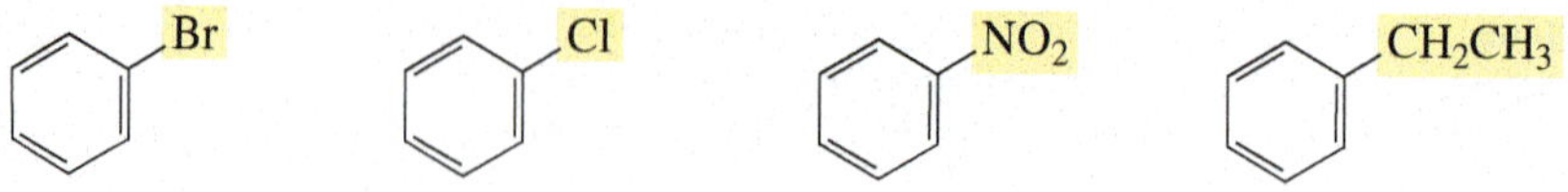

bromobenzene   chlorobenzene   nitrobenzene   ethylbenzene

Some monosubstituted benzenes have names that incorporate the substituent. Unfortunately, these names have to be memorized.

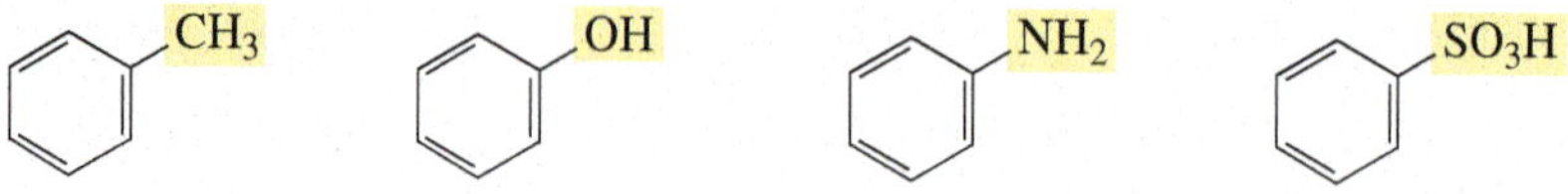

toluene   phenol   aniline   benzenesulfonic acid

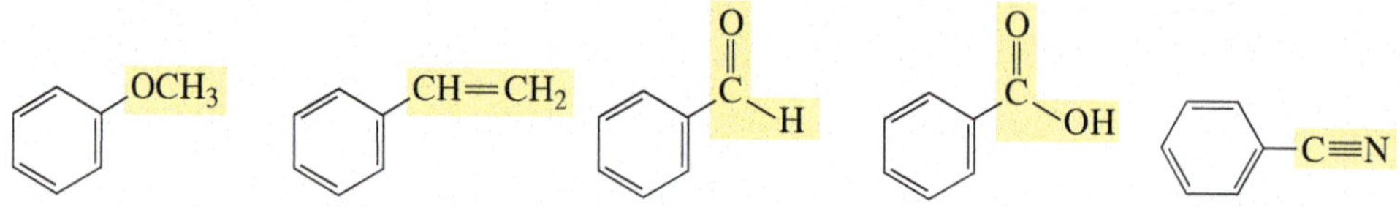

anisole   styrene   benzaldehyde   benzoic acid   benzonitrile

LEARN THE STRATEGY

Recall that when a benzene ring is a substituent, it is called a **phenyl group.** A benzene ring with a methylene group is called a **benzyl group** (Section 9.13).

phenyl group   benzyl group

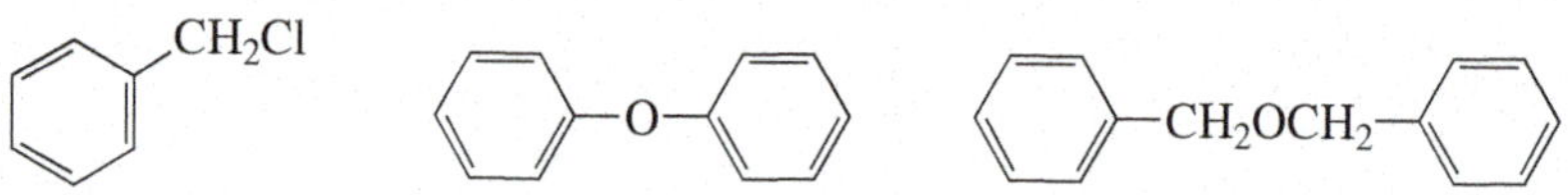

chloromethylbenzene
benzyl chloride   diphenyl ether   dibenzyl ether

With the exception of toluene, benzene rings with an alkyl substituent are named as alkyl-substituted benzenes (when the alkyl substituent has a name), or as phenyl-substituted alkanes (when the alkyl substituent does not have a name). For example, *sec*-pentyl cannot be used as a substituent name because both compounds that appear below on the right would be called *sec*-pentylbenzene, and a name must specify only one compound.

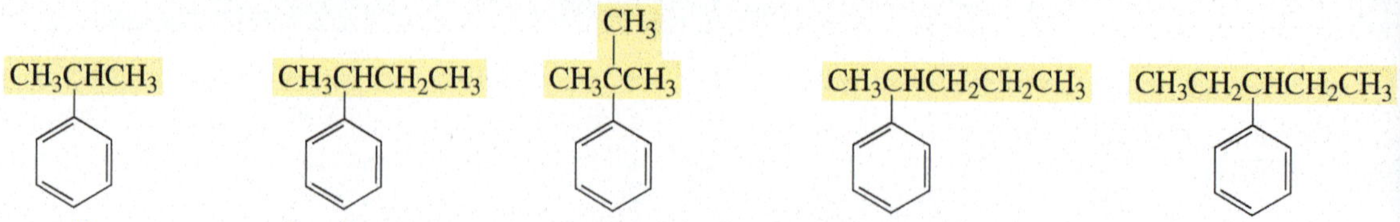

isopropylbenzene   *sec*-butylbenzene   *tert*-butylbenzene   2-phenylpentane   3-phenylpentane
cumene

Aryl (Ar) is the general term for either a phenyl group or a substituted phenyl group, just as alkyl (R) is the general term for a group derived from an alkane. Thus, ArOH can be used to designate any of the following phenols:

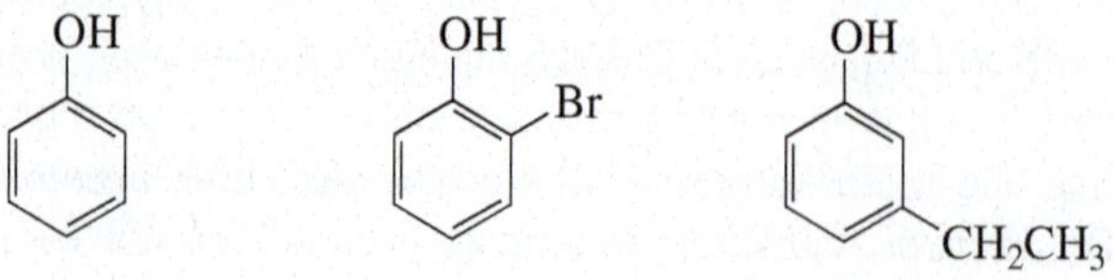

USE THE STRATEGY

PROBLEM 1 ♦

Draw the structure for each of the following:

a. 2-phenylhexane   **b.** benzyl alcohol   **c.** 3-benzylpentane   **d.** bromomethylbenzene

---

The Toxicity of Benzene

Benzene, which has been widely used in chemical synthesis and frequently used as a solvent, is a toxic substance. The major adverse effects of chronic exposure are seen in the central nervous system and bone marrow; it causes leukemia and aplastic anemia. A higher-than-average incidence of leukemia has been found in industrial workers with long-term exposure to as little as 1 ppm benzene in the atmosphere.

Toluene has replaced benzene as a solvent because, although it too is a central nervous system depressant, it does not cause leukemia or aplastic anemia. "Glue sniffing," a highly dangerous activity, produces narcotic central nervous system effects because glue contains toluene.

18.2   THE GENERAL MECHANISM FOR ELECTROPHILIC AROMATIC SUBSTITUTION REACTIONS

We saw that aromatic compounds such as benzene undergo **electrophilic aromatic substitution reactions** (Section 8.21). In an electrophilic aromatic substitution reaction, an electrophile substitutes for a hydrogen that is attached to a ring carbon of an aromatic compound.

an electrophilic aromatic substitution reaction

NOTE TO THE STUDENT
• **Review Section 8.21**

The following are the five most common electrophilic aromatic substitution reactions:

1. **Halogenation:** a bromine (Br), a chlorine (Cl), or an iodine (I) substitutes for a hydrogen.
2. **Nitration:** a nitro group (NO_2) substitutes for a hydrogen.
3. **Sulfonation:** a sulfonic acid group (SO_3H) substitutes for a hydrogen.
4. **Friedel–Crafts acylation:** an acyl group ($RC{=}O$) substitutes for a hydrogen.
5. **Friedel–Crafts alkylation:** an alkyl group (R) substitutes for a hydrogen.

All five of these reactions take place by the same two-step mechanism.

GENERAL MECHANISM FOR ELECTROPHILIC AROMATIC SUBSTITUTION

- The electrophile (Y^+) adds to the nucleophilic benzene ring, thereby forming a carbocation intermediate. The structure of the carbocation intermediate can be approximated by three resonance contributors.

■ A base in the reaction mixture (:B) removes a proton from the carbocation intermediate, and the electrons that held the proton move into the ring to reestablish its aromaticity. Notice that *the proton is removed from the carbon that has formed the bond with the electrophile.*

The first step is relatively slow and endergonic because an aromatic compound is being converted to a much less stable nonaromatic intermediate (Figure 18.1). The second step is fast and strongly exergonic because this step restores the stability-enhancing aromaticity.

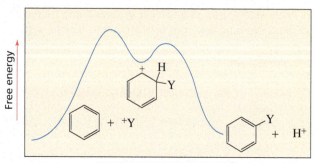

▲ **Figure 18.1**
Reaction coordinate diagram for an electrophilic aromatic substitution reaction.

We will now look at each of these five electrophilic aromatic substitution reactions individually. As you study them, notice that they differ only in how the electrophile (Y^+) that is needed to start the reaction is generated. Once the electrophile is formed, all five reactions follow the same two-step mechanism just shown.

18.3 HALOGENATION OF BENZENE

Bromination and Chlorination of Benzene

The bromination or chlorination of benzene requires a Lewis acid catalyst such as ferric bromide or ferric chloride. Recall that a *Lewis acid* is a compound that accepts a share in an electron pair (Section 2.12).

bromination

$$\text{C}_6\text{H}_6 + Br_2 \xrightarrow{FeBr_3} \text{C}_6\text{H}_5Br + HBr$$

chlorination

$$\text{C}_6\text{H}_6 + Cl_2 \xrightarrow{FeCl_3} \text{C}_6\text{H}_5Cl + HCl$$

Why does the reaction of benzene with Br_2 or Cl_2 require a catalyst when the reaction of an alkene with these reagents does not require a catalyst? Benzene's aromaticity makes it much more stable and, therefore, much less reactive than an alkene, so, benzene requires a better electrophile. Donating a lone pair to the Lewis acid weakens the Br—Br (or Cl—Cl) bond, thereby providing a better leaving group and making Br_2 (or Cl_2) a better electrophile.

generation of the electrophile

$$:\!\ddot{B}r\!-\!\ddot{B}r\!: + FeBr_3 \longrightarrow :\!\ddot{B}r\!-\!\ddot{B}r^+\!-\!FeBr_3$$

a leaving group / a better leaving group
an electrophile / a better electrophile

MECHANISM FOR BROMINATION

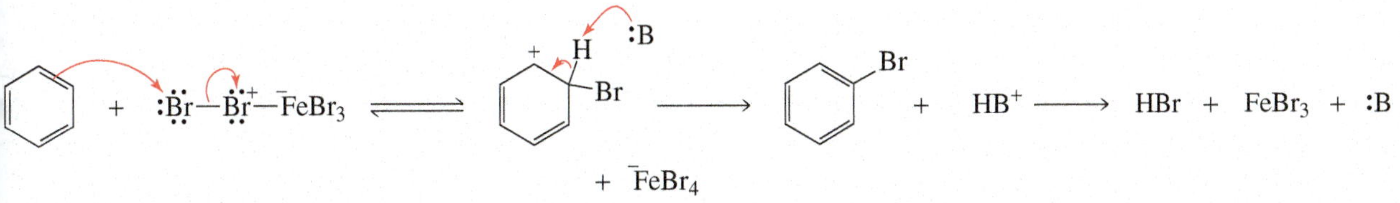

- The electrophile adds to the benzene ring.
- A base (:B) in the reaction mixture (such as $^-FeBr_4$ or the solvent) removes the proton from the carbon that formed the bond with the electrophile. Notice that the catalyst is regenerated.

For the sake of clarity, only one of the three resonance contributors of the carbocation intermediate is shown in this and subsequent mechanisms for electrophilic aromatic substitution reactions. Bear in mind, however, that each carbocation intermediate actually has the three resonance contributors shown in Section 18.3.

As you can see below, the mechanism for chlorination of benzene is the same as that for bromination.

MECHANISM FOR CHLORINATION

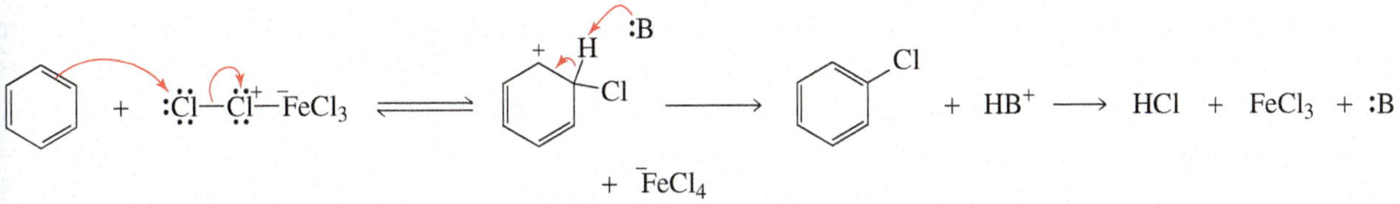

The ferric bromide and ferric chloride catalysts react readily with moisture in the air during handling, which inactivates them. To avoid this problem, they are generated in situ (in the reaction mixture) from iron filings and bromine (or chlorine). As a result, the halogen component of the Lewis acid is the same as the halogen used in the substitution reaction (Br_2 and $FeBr_3$ or Cl_2 and $FeCl_3$).

$$2\ Fe\ +\ 3\ Br_2\ \longrightarrow\ 2\ FeBr_3$$

$$2\ Fe\ +\ 3\ Cl_2\ \longrightarrow\ 2\ FeCl_3$$

PROBLEM 2 ◆

Why does hydration inactivate $FeBr_3$?

Iodination of Benzene

Iodobenzene can be prepared using I_2 and an oxidizing agent under acidic conditions. Hydrogen peroxide is commonly used as the oxidizing agent.

iodination

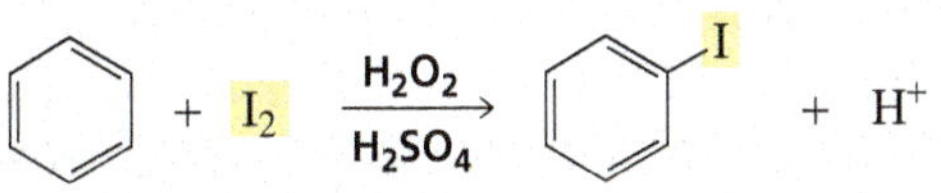

The oxidizing agent converts I_2 into the electrophilic iodonium ion (I^+).

generation of the electrophile

$$I_2\ +\ H_2O_2\ +\ 2\ H_2SO_4\ \longrightarrow\ 2\ I^+\ +\ 2\ H_2O\ +\ 2\ HSO_4^-$$

Once the electrophile is formed, iodination of benzene occurs by the same mechanism as bromination and chlorination.

MECHANISM FOR IODINATION

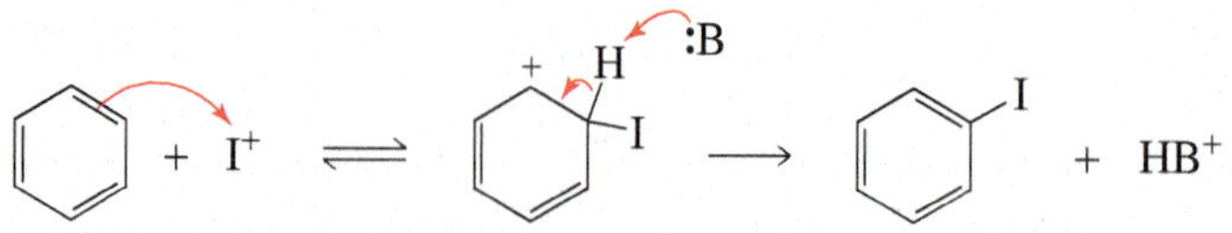

- The electrophile adds to the benzene ring.
- A base (:B) in the reaction mixture removes the proton from the carbon that formed the bond with the electrophile.

Thyroxine

Thyroxine, a hormone produced by the thyroid gland, increases the rate at which fats, carbohydrates, and proteins are metabolized. Humans obtain thyroxine from tyrosine (an amino acid) and iodine. The thyroid gland is the only part of the body that uses iodine, which we acquire primarily from seafood or iodized salt.

An enzyme called iodoperoxidase converts the I^- that we ingest to I^+, the electrophile needed to place an iodo substituent on a benzene ring. A deficiency in iodine is the number one cause of preventable intellectual disability in children.

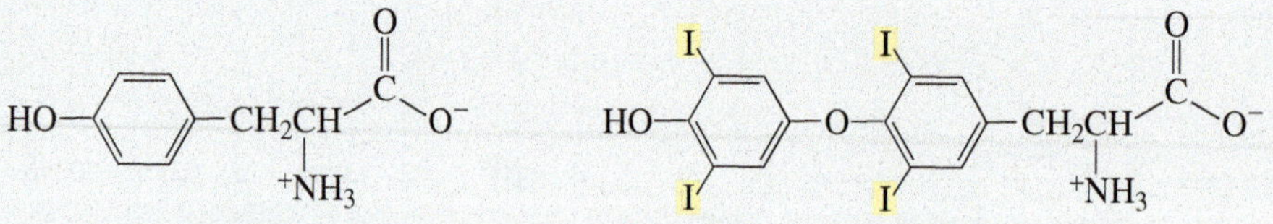

tyrosine

thyroxine

Chronically low levels of thyroxine cause enlargement of the thyroid gland as it tries in vain to make more thyroxine, a condition known as goiter. Low thyroxine levels can be corrected by taking thyroxine orally. Synthroid, the most popular brand of thyroxine, is currently one of the most-prescribed drugs in the United States.

18.4 NITRATION OF BENZENE

The nitration of benzene with nitric acid requires sulfuric acid as a catalyst.

nitration

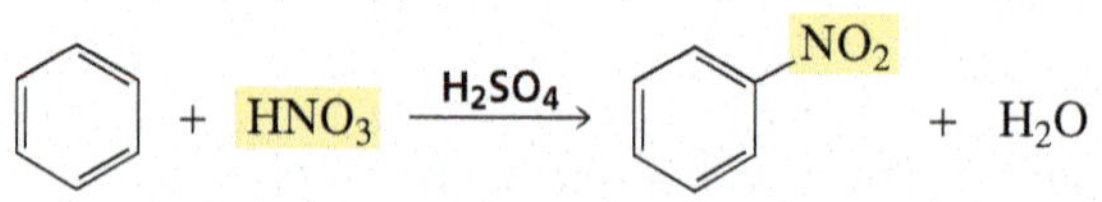

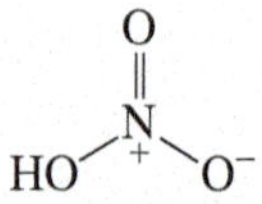

To generate the necessary electrophile, sulfuric acid protonates nitric acid. Protonated nitric acid then loses water to form a nitronium ion, the electrophile required for nitration.

generation of the electrophile

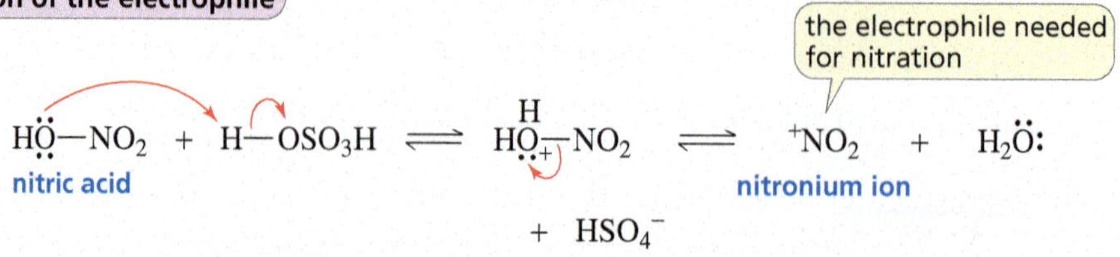

the electrophile needed for nitration

$$HO-NO_2 \ + \ H-OSO_3H \ \rightleftharpoons \ HO{+\atop}-NO_2 \ \rightleftharpoons \ {^+}NO_2 \ + \ H_2O:$$

nitric acid

nitronium ion

$$+ \ HSO_4^-$$

The mechanism for the electrophilic aromatic substitution reaction is the same as the mechanisms for the electrophilic aromatic substitution reactions we looked at in Section 18.3.

nitric acid

$$O=\overset{+}{N}=O$$

nitronium ion

MECHANISM FOR NITRATION

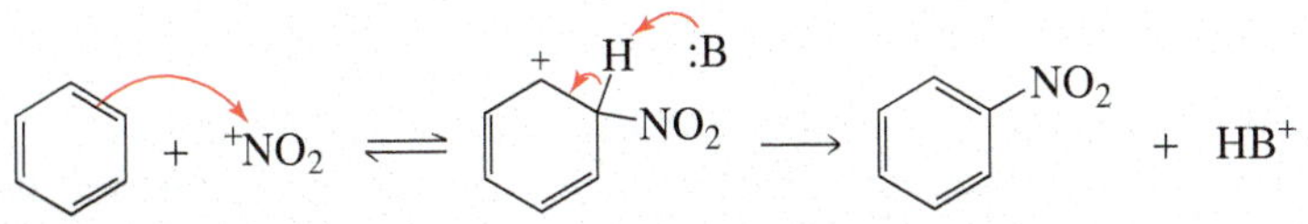

The electrophile adds to the benzene ring.

- The electrophile adds to the benzene ring.
- A base (:B) in the reaction mixture (for example, H_2O, HSO_4^-, or the solvent) removes the proton from the carbon that formed the bond with the electrophile.

PROBLEM 3 SOLVED

Propose a mechanism for the following reaction:

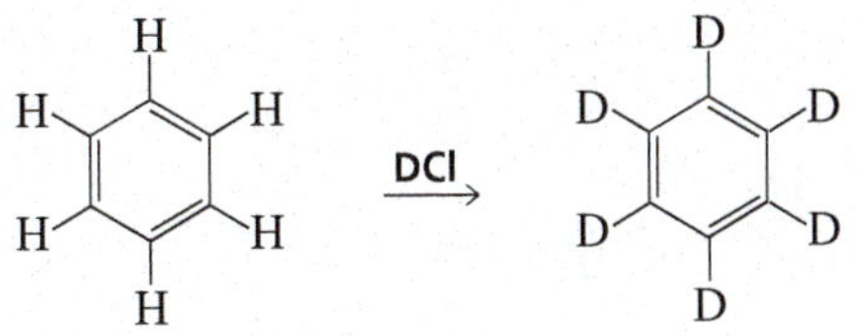

$$\xrightarrow{\text{DCl}}$$

SOLUTION The only electrophile available is D^+. Therefore, D^+ adds to a ring carbon and H^+ comes off the same ring carbon. The reaction can be repeated at each of the other five ring carbons.

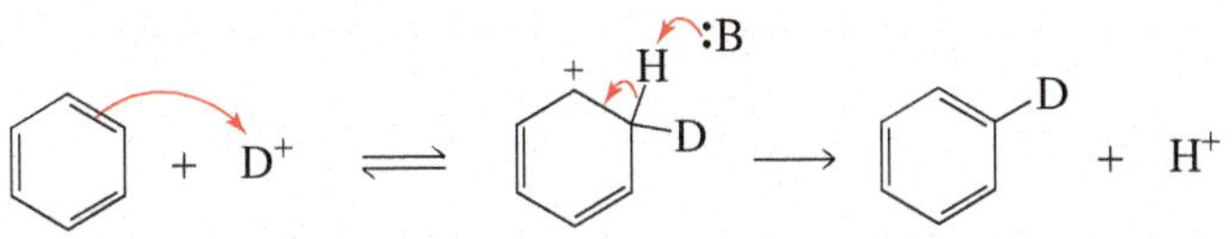

18.5 SULFONATION OF BENZENE

Concentrated sulfuric acid or fuming sulfuric acid (a solution of SO_3 in sulfuric acid) is used to sulfonate aromatic rings.

sulfonation

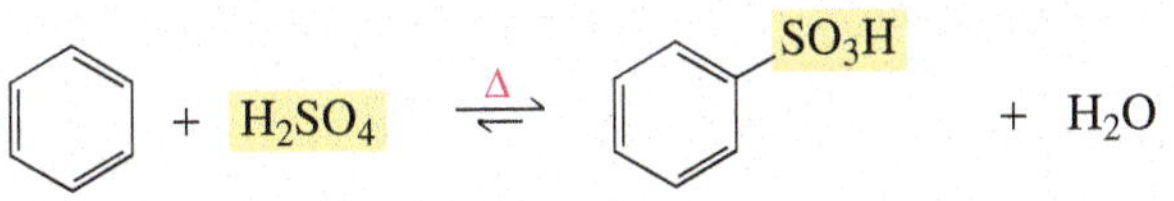

$$+ \quad H_2SO_4 \quad \underset{\Delta}{\rightleftharpoons} \quad \text{(SO}_3\text{H)} \quad + \quad H_2O$$

Take a minute to note the similarities in the mechanisms for forming the $^+SO_3H$ electrophile for sulfonation and the $^+NO_2$ electrophile for nitration.

generation of the electrophile   the electrophile needed for sulfonation

$$H\ddot{O}-SO_3H \;+\; H-OSO_3H \;\rightleftharpoons\; H\overset{+}{\underset{..}{O}}{-}SO_3H \;\rightleftharpoons\; {}^+SO_3H \;+\; H_2\ddot{O}: \;\rightleftharpoons\; SO_3 \;+\; H_3O^+$$

sulfuric acid   sulfonium ion

$$+ \quad HSO_4^-$$

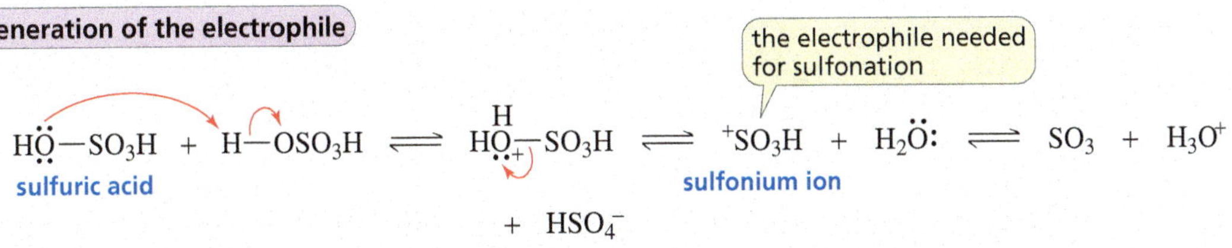

MECHANISM FOR SULFONATION

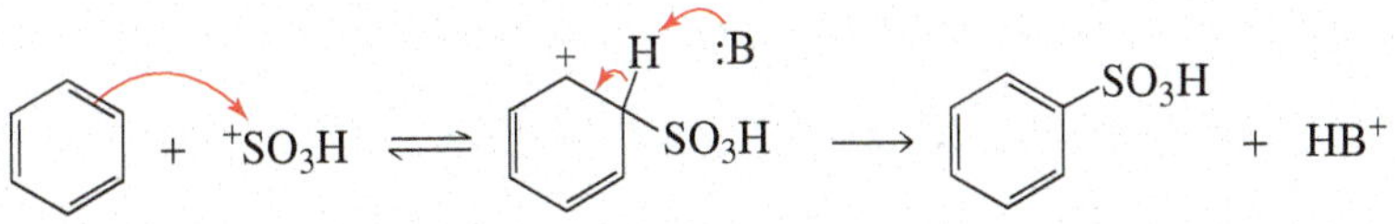

$$+ \quad {}^+SO_3H \quad \rightleftharpoons \quad \text{(SO}_3\text{H)} \quad \longrightarrow \quad \text{(SO}_3\text{H)} \quad + \quad HB^+$$

- The electrophile attaches to the ring.
- A base (:B) in the reaction mixture (for example, H_2O, HSO_4^-, or the solvent) removes the proton from the carbon that formed the bond with the electrophile.

We saw that a sulfonic acid is a strong acid because its conjugate base is particularly stable (weak) due to delocalization of its negative charge over three oxygens (Section 10.3).

benzenesulfonic acid $\xrightarrow{pK_a = -6.5}$ benzenesulfonate ion $+\ H^+$

Sulfonation is reversible. If benzenesulfonic acid is heated in dilute acid, an H^+ adds to the ring and the sulfonic acid group comes off the ring.

$\text{C}_6\text{H}_5\text{SO}_3\text{H} \xrightarrow{H_3O^+/100\ °C} \text{C}_6\text{H}_6 + SO_3 + H^+$

MECHANISM FOR DESULFONATION

(mechanism scheme showing benzenesulfonic acid reacting with $H\text{—}\overset{+}{O}H_2$ to give arenium ion intermediate, then benzene $+\ ^+SO_3H \rightleftharpoons SO_3 + H^+$, and $+\ H_2O$)

18.6 FRIEDEL–CRAFTS ACYLATION OF BENZENE

Two electrophilic substitution reactions bear the names of chemists Charles Friedel and James Crafts. *Friedel–Crafts acylation* places an acyl group on a benzene ring, and *Friedel–Crafts alkylation* places an alkyl group on a benzene ring.

An acyl chloride or an acid anhydride is used as the source of the acyl group needed for a Friedel–Crafts acylation.

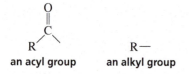

an acyl group   **an alkyl group**

Friedel–Crafts acylation

(reaction 1) benzene $+$ acyl chloride $\xrightarrow[\text{2. H}_2\text{O}]{\text{1. AlCl}_3}$ aryl ketone (an acyl group) $+\ HCl$

(reaction 2) benzene $+$ acid anhydride $\xrightarrow[\text{2. H}_2\text{O}]{\text{1. AlCl}_3}$ aryl ketone $+$ carboxylic acid

The electrophile (an acylium ion) required for the reaction is formed by the reaction of an acyl chloride (or an acid anhydride) with $AlCl_3$, a Lewis acid.

generation of the electrophile

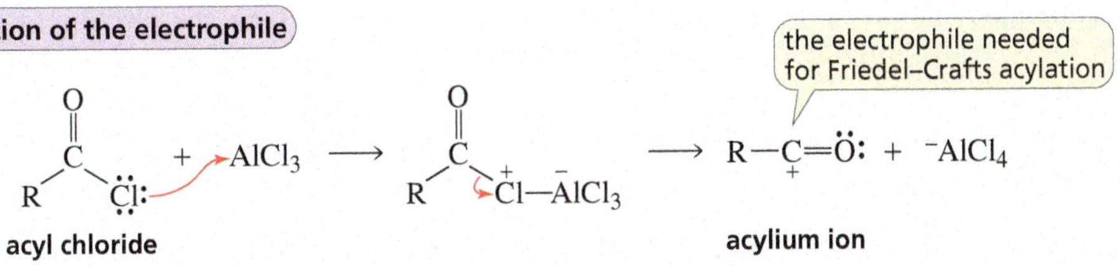

the electrophile needed
for Friedel–Crafts acylation

acyl chloride

acylium ion

MECHANISM FOR FRIEDEL–CRAFTS ACYLATION

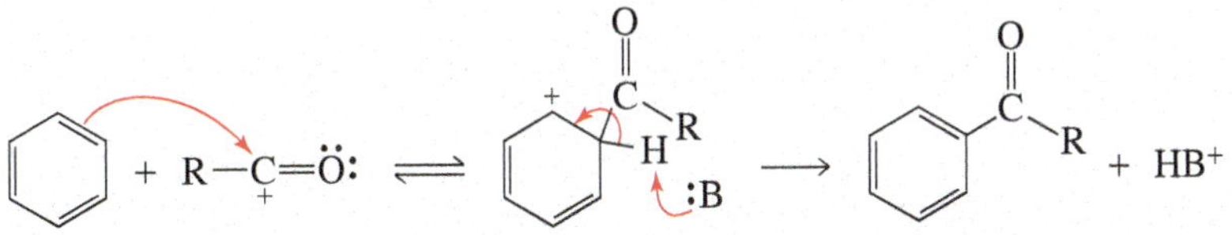

- The electrophile (in this case, an acylium ion) adds to the benzene ring.
- A base (:B) in the reaction mixture removes the proton from the carbon that formed the bond with the electrophile.

Because the product of a Friedel–Crafts acylation contains a carbonyl group that can complex with AlCl$_3$, Friedel–Crafts acylations must be carried out with more than one equivalent of AlCl$_3$. When the reaction is over, water is added to the reaction mixture to liberate the product from the complex.

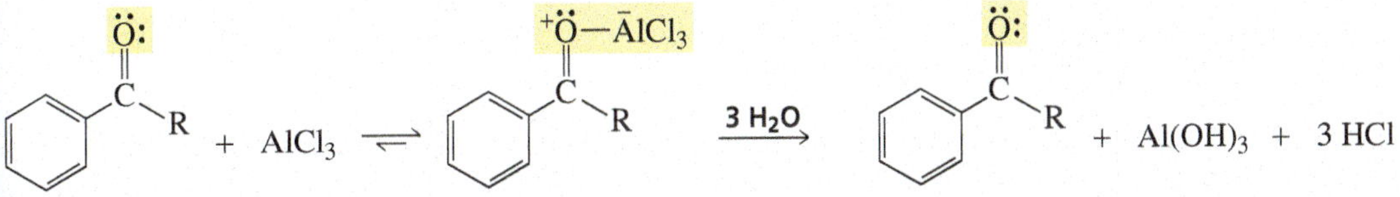

The synthesis of benzaldehyde from benzene poses a problem because formyl chloride, the acyl halide required for the reaction, is unstable and must be generated in situ. The **Gattermann–Koch reaction** uses a high-pressure mixture of carbon monoxide and HCl to generate formyl chloride and uses an aluminum chloride–cuprous chloride catalyst for the acylation.

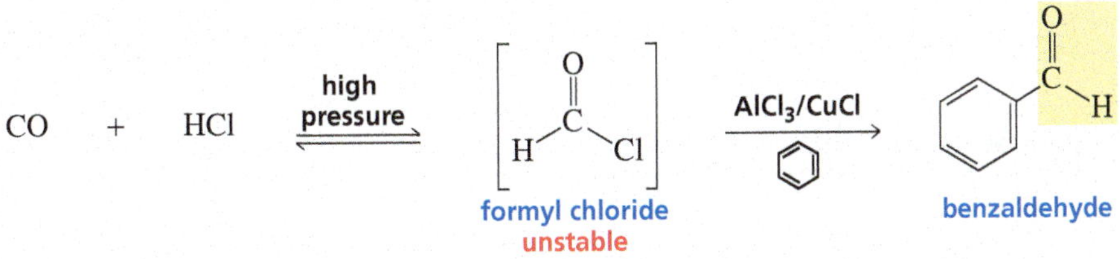

CO + HCl

high
pressure

formyl chloride
unstable

AlCl$_3$/CuCl

benzaldehyde

PROBLEM 4

Show the mechanism for the generation of the acylium ion if an acid anhydride is used instead of an acyl chloride for the source of the acylium ion.

18.7 FRIEDEL–CRAFTS ALKYLATION OF BENZENE

Friedel–Crafts alkylation substitutes an alkyl group for a hydrogen of benzene.

Friedel–Crafts alkylation

an alkyl group

+ RCl $\xrightarrow{\text{AlCl}_3}$ + HCl

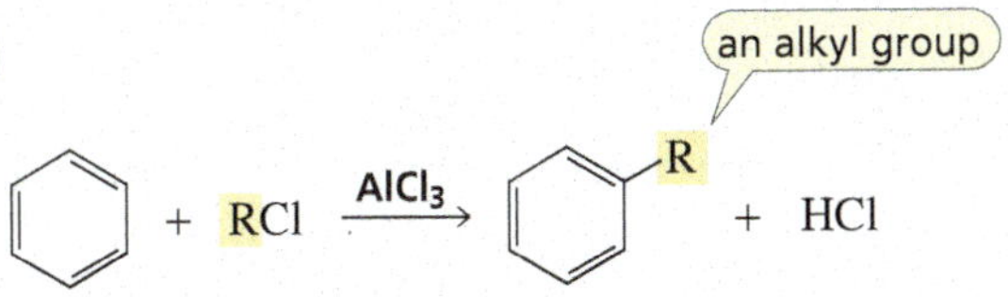

The electrophile required for the reaction (a carbocation) is formed from the reaction of an alkyl halide with $AlCl_3$. Alkyl chlorides, alkyl bromides, and alkyl iodides can all be used. Vinylic halides and aryl halides cannot be used because their carbocations are too unstable to be formed (Section 9.13).

generation of the electrophile

the electrophile needed for Friedel–Crafts alkylation

MECHANISM FOR FRIEDEL–CRAFTS ALKYLATION

- The electrophile adds to the benzene ring.
- A base (:B) in the reaction mixture removes the proton from the carbon that formed the bond with the electrophile.

In Section 18.12, we will see that an alkyl-substituted benzene is more reactive than benzene. Therefore, a large excess of benzene is used in Friedel–Crafts alkylations to prevent the alkyl-substituted product from being alkylated instead of benzene. When benzene is in excess, the electrophile is more likely to encounter a molecule of benzene than a molecule of the more reactive alkyl-substituted benzene.

Carbocation Rearrangement

We saw that a carbocation will rearrange if rearrangement leads to a more stable carbocation (Section 6.7). When the carbocation employed in a Friedel–Crafts alkylation reaction rearranges, the major product is the product with the rearranged alkyl group on the benzene ring. For example, when benzene reacts with 1-chlorobutane, 60%–80% of the product (the actual percentage depends on the reaction conditions) has the rearranged alkyl substituent. (See the box on "Incipient Primary Carbocations" on the following page).

rearrangement of the carbocation

When benzene reacts with 1-chloro-2,2-dimethylpropane, the initially formed primary carbocation rearranges to a tertiary carbocation; in this case, 100% of the product (under all reaction conditions) has the rearranged alkyl substituent.

rearrangement of the carbocation

Incipient Primary Carbocations

For simplicity, the two reactions just shown that involve carbocation rearrangements were written showing the formation of a primary carbocation. However, we know that primary carbocations are too unstable to be formed (Section 9.4). Not surprisingly, a true primary carbocation is never formed in a Friedel–Crafts alkylation. Instead, the carbocation remains complexed with the Lewis acid—a so-called "incipient" carbocation. A carbocation rearrangement occurs because the incipient carbocation has sufficient carbocation character to permit the rearrangement.

A Biological Friedel–Crafts Alkylation

A Friedel–Crafts alkylation is one of the steps in the biosynthesis of vitamin KH_2, the coenzyme required to form blood clots (Section 23.8). The pyrophosphate group, an excellent leaving group, departs in an S_N1 reaction, forming an allylic cation. A Friedel–Crafts alkylation places the long-chain alkyl group on the benzene ring. Conversion of the carboxyl group to a methyl group, which requires several steps, forms vitamin KH_2.

PROBLEM 5 ◆

What is the major product of a Friedel–Crafts alkylation using the following alkyl chlorides?

a. CH_3CH_2Cl c. $CH_3CH_2CH(Cl)CH_3$ e. $(CH_3)_2CHCH_2Cl$

b. $CH_3CH_2CH_2Cl$ d. $(CH_3)_3CCl$ f. $CH_2{=}CHCH_2Cl$

18.8 ALKYLATION OF BENZENE BY ACYLATION–REDUCTION

We just saw that a Friedel–Crafts alkylation cannot produce a good yield of an alkylbenzene containing a straight-chain alkyl group because the incipient primary carbocation rearranges to a more stable carbocation.

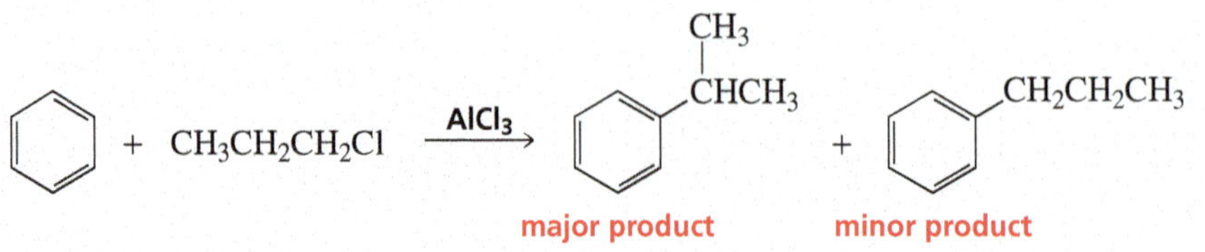

major product   **minor product**

Acylium ions, however, do not rearrange. Consequently, a straight-chain alkyl group can be placed on a benzene ring by means of a Friedel–Crafts acylation, followed by reduction of the carbonyl group to a methylene group. Conversion of a carbonyl group to a methylene group is a reduction reaction because the two C—O bonds are replaced with two C—H bonds (Section 5.9). Only a ketone carbonyl group that is adjacent to a benzene ring can be reduced to a methylene group by catalytic hydrogenation (H_2 + Pd/C).

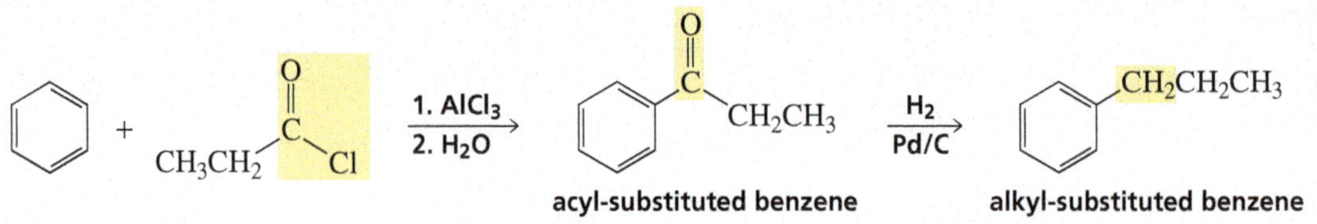

acyl-substituted benzene   **alkyl-substituted benzene**

Besides avoiding carbocation rearrangements, another advantage to preparing alkyl-substituted benzenes by acylation–reduction rather than by direct alkylation is that a large excess of benzene does not have to be used (Section 18.7). Unlike alkyl-substituted benzenes, which are more reactive than benzene, acyl-substituted benzenes are less reactive than benzene, so acyl-substituted benzenes do not undergo a second Friedel–Crafts reaction (Section 18.12).

The **Wolff–Kishner reduction** also converts the carbonyl group of a ketone to a methylene group. This method reduces all ketone carbonyl groups, not just those adjacent to benzene rings.

Wolff–Kishner reduction

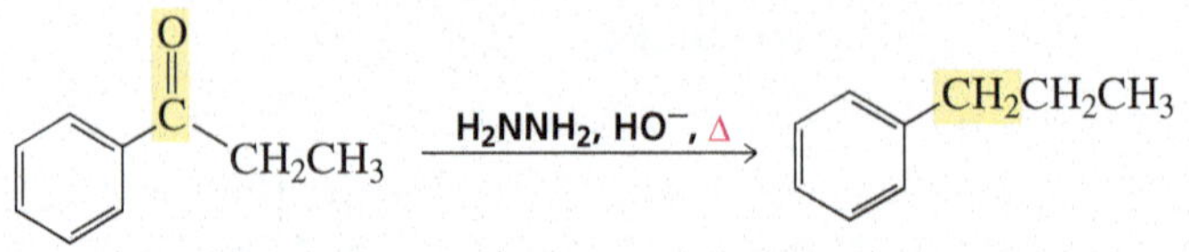

We saw that hydrazine reacts with a ketone to form a hydrazone (Section 16.8). Notice that hydroxide ion and heat are what differentiate the Wolff–Kishner reduction from ordinary hydrazone formation.

MECHANISM FOR THE WOLFF–KISHNER REDUCTION

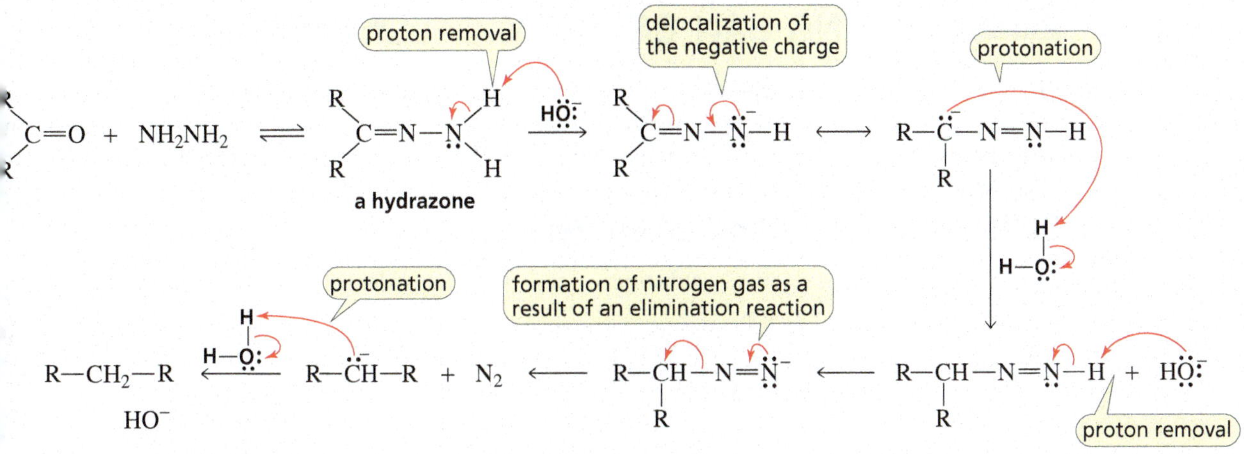

- The ketone reacts with hydrazine to form a hydrazone. The mechanism is similar to that for imine formation on page 761.
- Hydroxide ion removes a proton from the NH_2 group of the hydrazone. The reaction requires heat because this proton is only weakly acidic.
- The negative charge can be delocalized onto carbon, which removes a proton from water.
- The last two steps are repeated to form the deoxygenated product and nitrogen gas.

18.9   USING COUPLING REACTIONS TO ALKYLATE BENZENE

Alkylbenzenes with straight-chain alkyl groups can also be prepared from bromobenzene or chlorobenzene, using a Suzuki reaction or an organocuprate (Sections 11.3 and 11.4).

Suzuki reaction

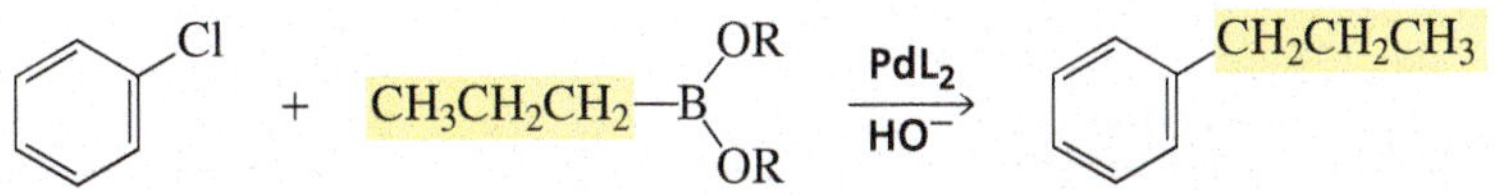

reaction with an organocuprate

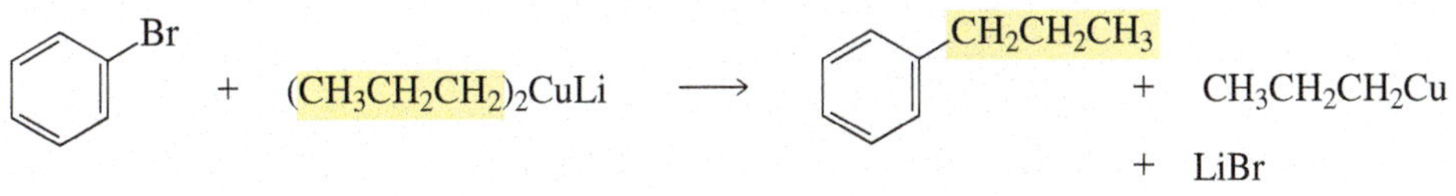

PROBLEM 6

Describe two ways to prepare each of the following compounds from benzene.

a. 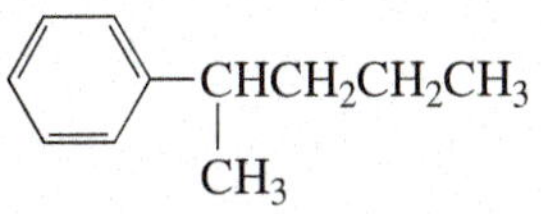 ⬡—CHCH$_2$CH$_2$CH$_3$
          |
          CH$_3$

b. 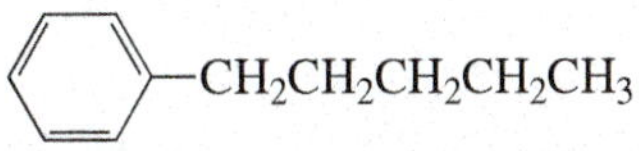 ⬡—CH$_2$CH$_2$CH$_2$CH$_2$CH$_3$

18.10 HOW SOME SUBSTITUENTS ON A BENZENE RING CAN BE CHEMICALLY CHANGED

Benzene rings with substituents other than those placed on a benzene ring by the five reactions listed in Section 18.2 can be prepared by first synthesizing one of those substituted benzenes and then chemically changing the substituent. Several of these reactions will be familiar to you.

Substitution and Elimination Reactions

We saw that bromine selectively substitutes for a benzylic hydrogen in a radical substitution reaction (Section 12.9).

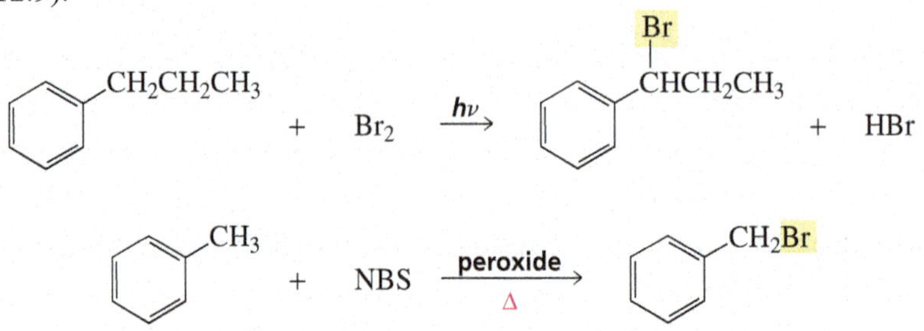

Once a halogen has been placed in the benzylic position, it can be replaced by a nucleophile via an $S_{N}2$ or $S_{N}1$ reaction (Section 9.13). A wide variety of substituted benzenes can be prepared this way.

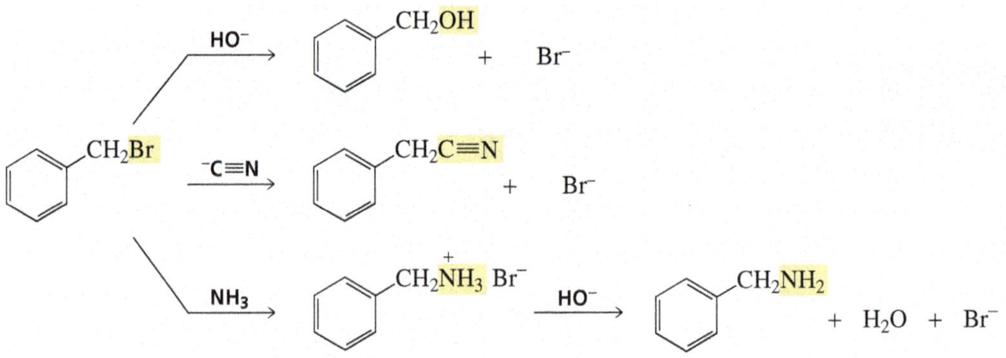

Remember that alkyl halides can also undergo elimination reactions (Sections 9.8 and 9.7). In the following reaction, the elimination product predominates because the double bond is stabilized by conjugation with the benzene ring.

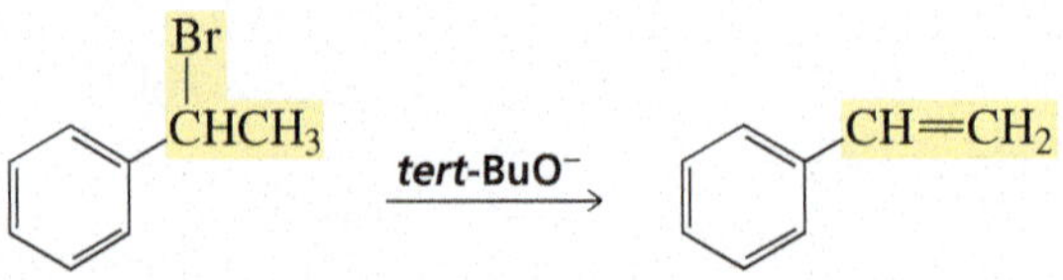

Oxidation and Reduction Reactions

Substituents with double and triple bonds can be reduced by catalytic hydrogenation (Sections 5.9 and 16.8).

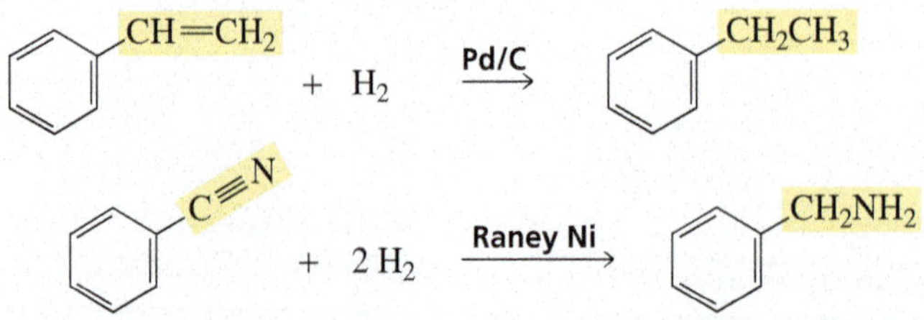

Because benzene is an unusually stable compound, it can be reduced only at high temperature and pressure (Section 8.1). Therefore, only the substituents are reduced in the preceding reactions.

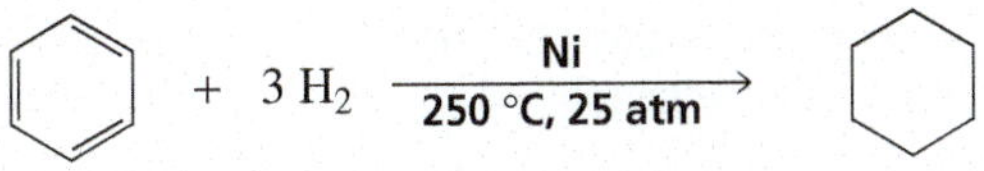

An alkyl group attached to a benzene ring can be oxidized to a carboxyl group by an oxidizing agent such as chromic acid. Notice that the benzene ring is too stable to be oxidized—only the alkyl group is oxidized.

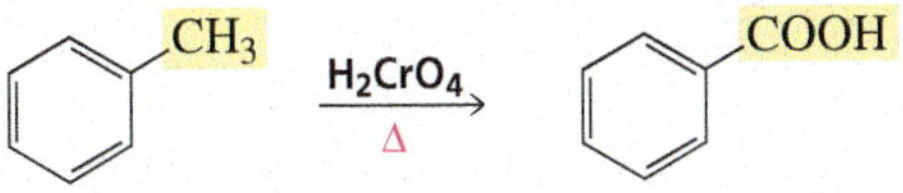

Regardless of the length of the alkyl substituent, it is oxidized to a COOH group, provided that a hydrogen is bonded to the benzylic carbon.

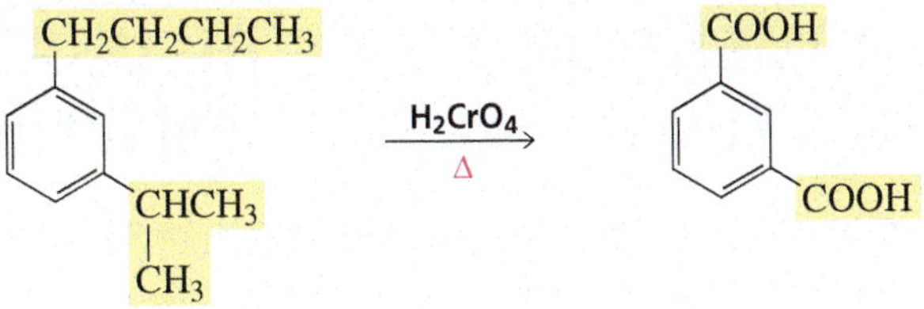

If the alkyl group lacks a benzylic hydrogen, the oxidation reaction will not occur because the first step in the oxidation reaction is removal of a hydrogen from the benzylic carbon.

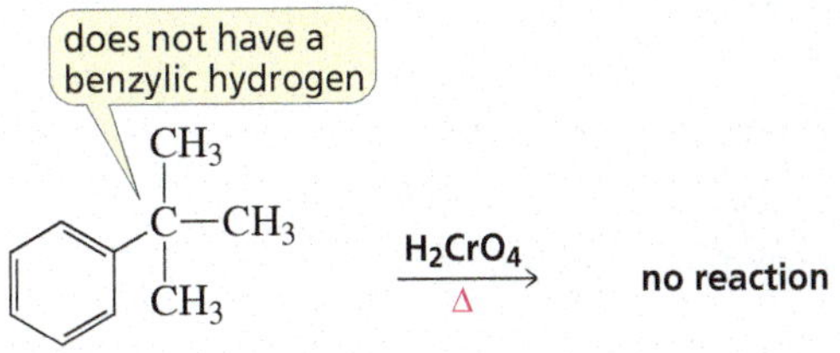

A nitro substituent can be reduced to an amino substituent. Catalytic hydrogenation is commonly used to carry out this reaction.

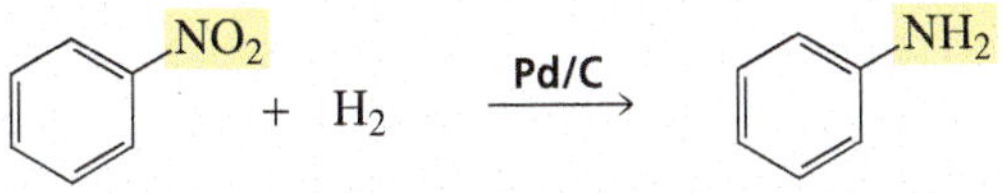

PROBLEM 7 ◆

What are the products of the following reactions?

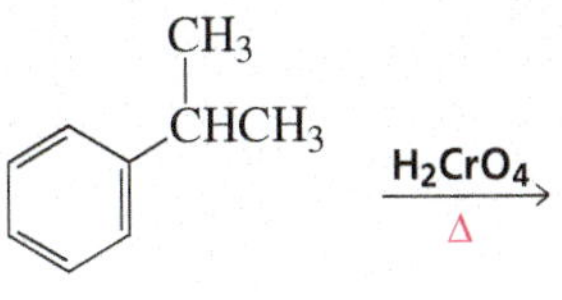

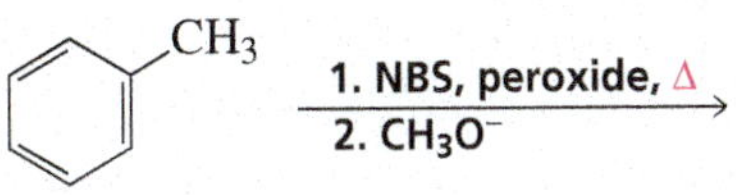

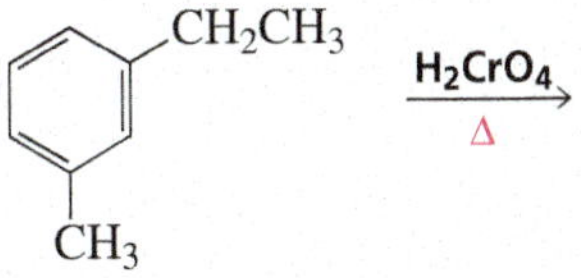

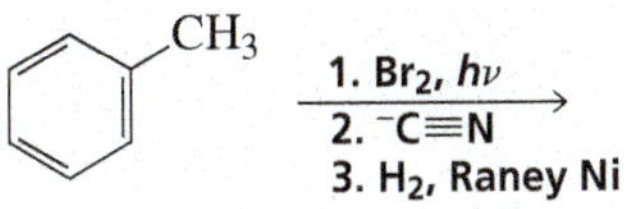

PROBLEM 8 SOLVED

Show how the following compounds can be prepared from benzene:

a. benzaldehyde   **c.** 1-bromo-2-phenylethane   **e.** aniline
b. styrene   **d.** 2-phenylethanol   **f.** benzoic acid

SOLUTION TO 8 a. Benzaldehyde can be prepared by the Gatterman–Koch reaction (page 877) or by the following sequence of reactions:

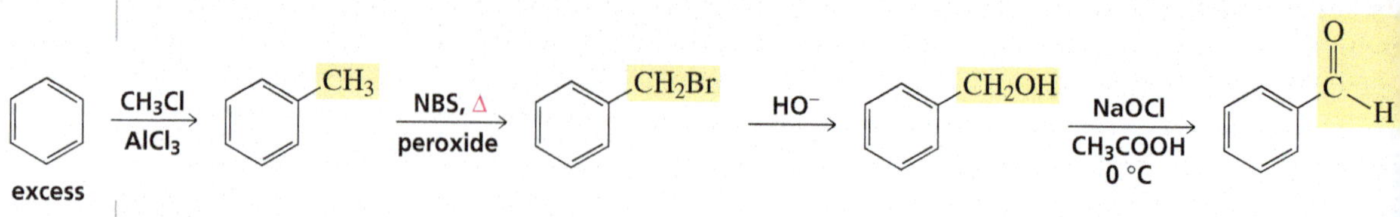

18.11 THE NOMENCLATURE OF DISUBSTITUTED AND POLYSUBSTITUTED BENZENES

Section 18.1 described how monosubstituted benzenes are named. Now we will look at how benzene rings that have more than one substituent are named.

Naming Disubstituted Benzenes

The relative positions of two substituents on a benzene ring can be indicated either by numbers or by the prefixes *ortho, meta,* and *para*:

- *ortho* substituents are adjacent to each other.
- *meta* substituents are separated by one carbon.
- *para* substituents are opposite each other.

Often, the abbreviations for these prefixes (*o, m, p*) are used in compounds' names.

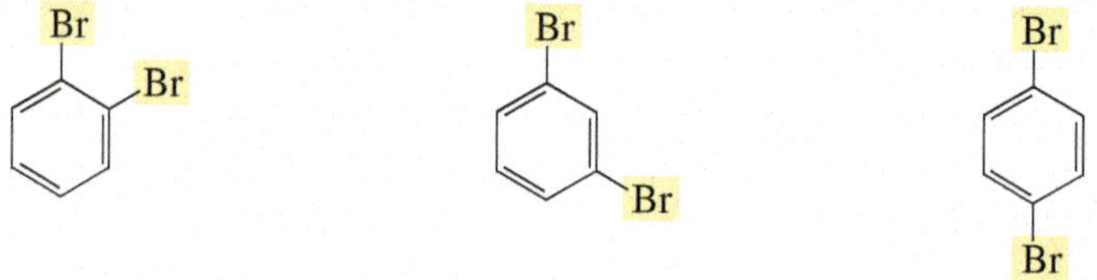

1,2-dibromobenzene
ortho-dibromobenzene
o-dibromobenzene

1,3-dibromobenzene
meta-dibromobenzene
m-dibromobenzene

1,4-dibromobenzene
para-dibromobenzene
p-dibromobenzene

LEARN THE STRATEGY   If the two substituents are different, they are listed in alphabetical order, each preceded by its assigned number. The first mentioned substituent is given the 1-position, and the ring is numbered in the direction that gives the second substituent the lowest possible number.

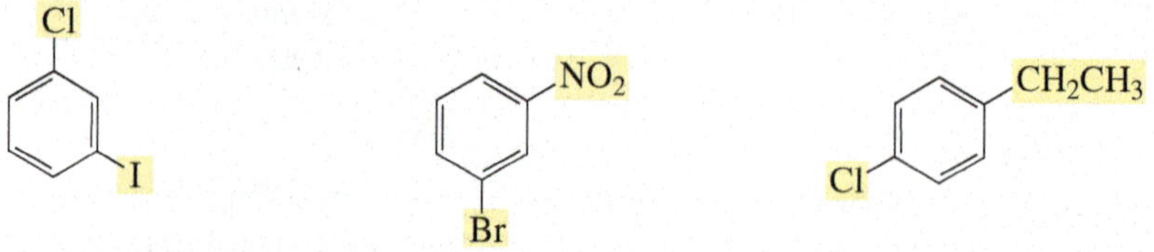

1-chloro-3-iodobenzene
meta-chloroiodobenzene

1-bromo-3-nitrobenzene
meta-bromonitrobenzene

1-chloro-4-ethylbenzene
para-chloroethylbenzene

If one of the substituents can be incorporated into a name, then that name is used and the incorporated substituent is given the 1-position. However, there is one exception. Methylbenzene, not toluene, is used to name a compound that has a methyl group and a second substituent on the ring. A benzene ring with two methyl substituents is often called xylene.

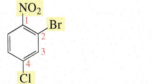

4-nitroaniline
para-nitroaniline

2-ethylphenol
ortho-ethylphenol

1-chloro-2-methylbenzene
ortho-chloromethylbenzene

1,3-dimethylbenzene
meta-xylene

USE THE STRATEGY

PROBLEM 9 ◆

Name the following:

a. **b.** **c.** **d.**

PROBLEM 10

Draw a structure for each of the following:

a. *m*-chloromethylbenzene **c.** *o*-nitroaniline **e.** *m*-dichlorobenzene
b. *p*-bromophenol **d.** *m*-chlorobenzonitrile **f.** *o*-xylene

---

Naming Polysubstituted Benzenes

If the benzene ring has more than two substituents, the substituents are numbered in the direction that results in the lowest possible numbers in the name of the compound. The substituents are listed in alphabetical order, each preceded by its assigned number.

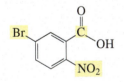

2-bromo-4-chloro-1-nitrobenzene   **4-bromo-1-chloro-2-nitrobenzene**   **1-bromo-4-chloro-2-nitrobenzene**

As with disubstituted benzenes, if one of the substituents can be incorporated into a name, that name is used and the incorporated substituent is given the 1-position. The ring is then numbered in the direction that results in the lowest possible numbers in the name of the compound.

5-bromo-2-nitrobenzoic acid   **3-bromo-4-chlorophenol**   **2-ethyl-4-iodoaniline**

PROBLEM 11

Draw the structure for each of the following:

a. 2-bromo-4-iodophenol   **b.** 2,5-dinitrobenzaldehyde   **c.** 4-bromo-3-chloroaniline

---

PROBLEM 12 ◆

Correct the following incorrect names:

a. 2,4,6-tribromobenzene   **c.** *para*-methylbromobenzene   **e.** *meta*-bromotoluene
b. 3-hydroxynitrobenzene   **d.** 1,6-dichlorobenzene   **f.** 6-ethyl-4-iodoaniline

18.12 THE EFFECT OF SUBSTITUENTS ON REACTIVITY

Like benzene, substituted benzenes undergo the five electrophilic aromatic substitution reactions listed in Section 18.2: halogenation, nitration, sulfonation, and Friedel–Crafts acylation and alkylation.

Now we need to find out whether a substituted benzene is more or less reactive than benzene itself. The answer depends on the substituent. Some substituents make the ring more reactive toward electrophilic aromatic substitution than benzene, and some make it less reactive.

Electron Donation Increases Reactivity
Electron Withdrawal Decreases Reactivity

The slow step of an electrophilic aromatic substitution reaction is the addition of an electrophile to the nucleophilic aromatic ring to form a carbocation intermediate (see Figure 18.1 on page 872). A substituent that makes benzene a better nucleophile makes it more attractive to electrophiles; a substituent that makes benzene a poorer nucleophile makes it less attractive to electrophiles.

In addition, because the transition state for the slow step is closer in energy to the carbocation intermediate than to benzene, the transition state resembles the carbocation intermediate (Section 6.3). Therefore, the transition state has a significant partial positive charge.

As a result, substituents that donate electrons to the benzene ring not only increase benzene's nucleophilicity (making it more reactive and hence less stable) but stabilize the partially positively charged transition state, thereby increasing the rate of electrophilic aromatic substitution; these are called **activating substituents.** In contrast, substituents that withdraw electrons from the benzene ring decrease benzene's nucleophilicity (making it less reactive and hence more stable) and destabilize the transition state, thereby decreasing the rate of electrophilic aromatic substitution; these are called **deactivating substituents.**

relative rates of electrophilic aromatic substitution

Relative Reactivity of Substituted Benzenes

The substituents in Table 18.1 are listed according to how they affect the reactivity of a benzene ring toward electrophilic aromatic substitution compared with benzene—where the substituent is a hydrogen.

> *The activating substituents make the benzene ring more reactive toward electrophilic aromatic substitution,*
> *and the deactivating substituents make the benzene ring less reactive.*

Remember that activating substituents donate electrons to the ring and deactivating substituents withdraw electrons from the ring.

In Section 8.10 you learned that substituents can:

- donate electrons by resonance and by hyperconjugation.
- withdraw electrons by resonance and inductively.

Strongly Activating Substituents

All the *strongly activating substituents* in Table 18.1 donate electrons to the ring by resonance because they have a lone pair on the atom attached to the ring. Additionally, they all withdraw electrons from the ring inductively, because the atom attached to the ring is more electronegative than hydrogen. The fact that these substituents have been found experimentally to increase the reactivity of the benzene ring indicates that their electron donation to the ring by resonance is more significant than their inductive electron withdrawal from the ring.

strongly activating substituents

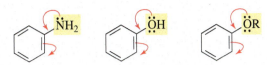

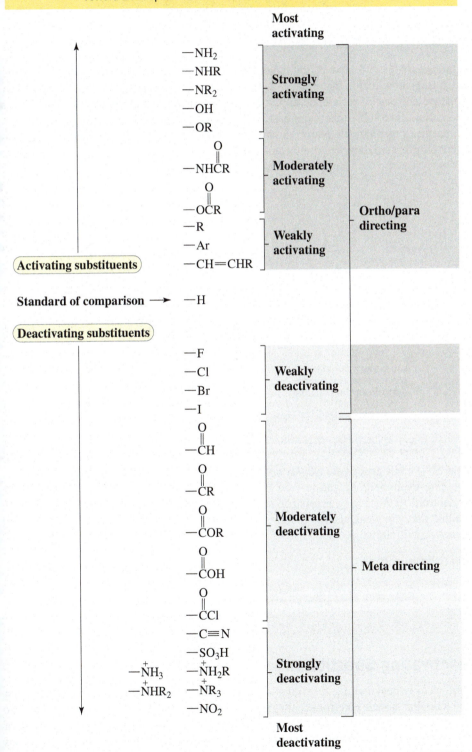

Table 18.1   The Effect of Substituents on the Reactivity of a Benzene Ring
Toward Electrophilic Aromatic Substitution

**Electron-donating substituents
increase the reactivity of the
benzene ring toward electrophilic
aromatic substitution.**

**Electron-withdrawing substituents
decrease the reactivity of the
benzene ring toward electrophilic
aromatic substitution.**

Moderately Activating Substituents

The *moderately activating substituents* also both donate electrons to the ring by resonance and withdraw electrons from the ring inductively. Because they are only moderately activating, we know that they donate electrons to the ring by resonance less effectively than do the strongly activating substituents.

moderately activating substituents

These substituents are less effective at donating electrons to the ring by resonance because, unlike the strongly activating substituents that donate electrons only to the ring, the moderately activating substituents donate electrons by resonance in two competing directions: to the ring and away from the ring. The fact that these substituents increase the reactivity of the benzene ring indicates that, despite their diminished resonance electron donation to the ring, overall they donate electrons by resonance more strongly than they withdraw electrons inductively.

substituent donates electrons by resonance to the benzene ring

substituent donates electrons by resonance away from the benzene ring

Weakly Activating Substituents

Alkyl, aryl, and CH=CHR groups are *weakly activating substituents*. An alkyl substituent donates electrons to the ring by hyperconjugation (see Figure 18.3 on page 892). Aryl and CH=CHR groups donate electrons to the ring by resonance and also withdraw electrons from the ring by resonance. The fact that they are weak activators indicates that they are slightly more electron donating than they are electron withdrawing.

weakly activating substituents

Weakly Deactivating Substituents

The halogens are weakly deactivating substituents. Like all the strongly and moderately activating substituents, the halogens donate electrons to the ring by resonance and withdraw electrons from the ring inductively. Because the halogens have been found experimentally to decrease the reactivity of the benzene ring, they must withdraw electrons inductively more strongly than they donate electrons by resonance.

weakly deactivating substituents

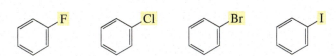

Let's look at why the halogens are *weakly* deactivating, whereas OH and OCH₃ are *strongly* activating. The electronegativities of chlorine and oxygen are similar, so they have similar inductive electron-withdrawing abilities. However, chlorine does not donate electrons by resonance as well as oxygen does because when chlorine donates electrons by resonance, it uses a 3p orbital to form a π bond with the 2p orbital of carbon. The 3p–2p orbital overlap is less effective than the 2p–2p orbital overlap that forms the π bond between oxygen and carbon.

Fluorine, like oxygen, uses a 2p orbital, so fluorine, like oxygen, donates electrons by resonance better than chlorine does. However, this is outweighed by fluorine's greater electronegativity, which causes it to strongly withdraw electrons inductively. Bromine and iodine are less effective than chlorine at withdrawing electrons inductively, but they are also less effective at donating electrons by resonance because they use 4p and 5p orbitals, respectively. Thus, all the halogens withdraw electrons inductively more strongly than they donate electrons by resonance (see Problem 13).

Moderately Deactivating Substituents

The moderately deactivating substituents all have a carbonyl group directly attached to the benzene ring. A carbonyl group withdraws electrons from a benzene ring both inductively and by resonance.

moderately deactivating substituents

Strongly Deactivating Substituents

The strongly deactivating substituents are powerful electron withdrawers. Except for the ammonium ions (⁺NH₃, ⁺NH₂R, ⁺NHR₂, and ⁺NR₃), these substituents withdraw electrons both inductively and by resonance. The ammonium ions have no resonance effect, but the positive charge on the nitrogen atom causes them to strongly withdraw electrons inductively.

strongly deactivating substituents

Take a minute to compare the electrostatic potential maps for anisole, benzene, and nitrobenzene. Notice that an electron-donating substituent (OCH₃) makes the ring more red (more negative; more nucleophilic), whereas an electron-withdrawing substituent (NO₂) makes the ring less red (less negative; less nucleophilic).

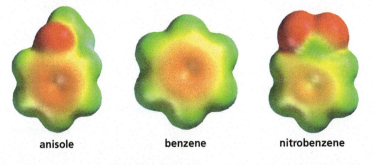

anisole   benzene   nitrobenzene

LEARN THE STRATEGY

PROBLEM 13 SOLVED

Explain why the halo-substituted benzenes have the relative reactivities shown in Table 18.1.

SOLUTION Table 18.1 shows that fluorine is the least deactivating of the halogen substituents and iodine is the most deactivating. We know that fluorine is the most electronegative of the halogens, which means that it is best at withdrawing electrons inductively. Fluorine is also best at donating electrons by resonance because its $2p$ orbital—compared with the $3p$ orbital of chlorine, the $4p$ orbital of bromine, or the $5p$ orbital of iodine—can better overlap the $2p$ orbital of carbon when it forms a π bond. Therefore, fluorine is best both at donating electrons by resonance and at withdrawing electrons inductively. Because fluorine is the weakest deactivator of the halogens, electron donation by resonance must be the more important factor in determining the relative reactivities of halo-substituted benzenes.

USE THE STRATEGY

PROBLEM 14

Explain why fluorobenzene is more reactive than chlorobenzene toward electrophilic aromatic substitution but chloromethylbenzene is more reactive than fluoromethylbenzene.

PROBLEM 15 ◆

List the compounds in each set from most reactive to least reactive toward electrophilic aromatic substitution:

a. benzene, phenol, toluene, nitrobenzene, bromobenzene
b. dichloromethylbenzene, difluoromethylbenzene, toluene, chloromethylbenzene

18.13 THE EFFECT OF SUBSTITUENTS ON ORIENTATION

When a substituted benzene undergoes an electrophilic aromatic substitution reaction, where does the new substituent attach itself? In other words, is the product of the reaction the ortho isomer, the meta isomer, or the para isomer?

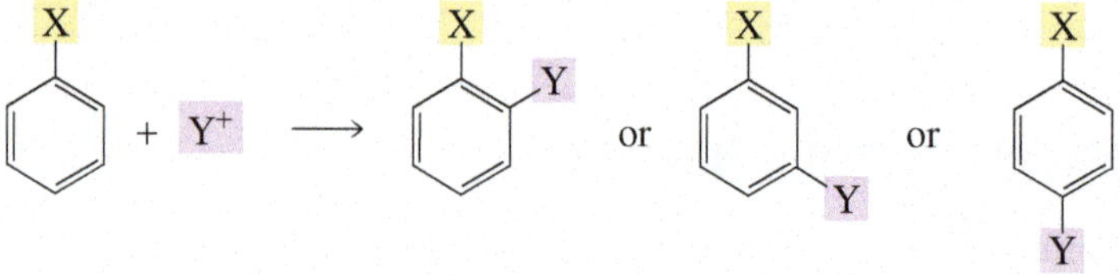

ortho isomer   meta isomer   para isomer

The substituent already attached to the benzene ring determines the location of the new substituent. The attached substituent will have one of two effects: it will direct an incoming substituent to the ortho *and* para positions, or it will direct an incoming substituent to the meta position.

All activating substituents and the weakly deactivating halogens are **ortho–para directors,** and all substituents that are more deactivating than the halogens are **meta directors.** Thus, the substituents can be divided into three groups:

- All *activating substituents* direct an incoming electrophile to the ortho and para positions.

All activating substituents are ortho–para directors.

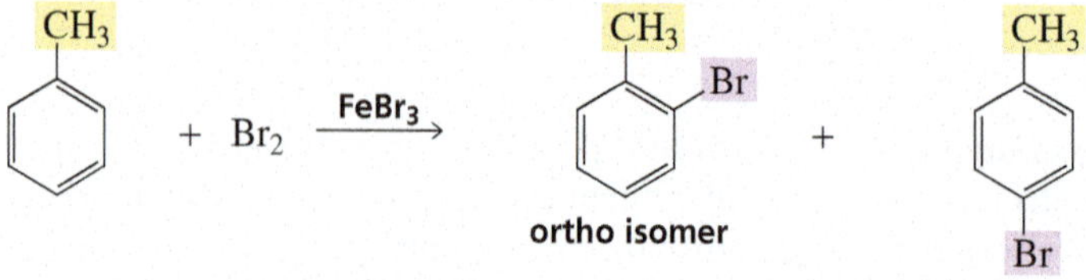

ortho isomer   para isomer

- The *weakly deactivating* halogens also direct an incoming electrophile to the ortho and para positions.

The weakly deactivating halogens are ortho–para directors.

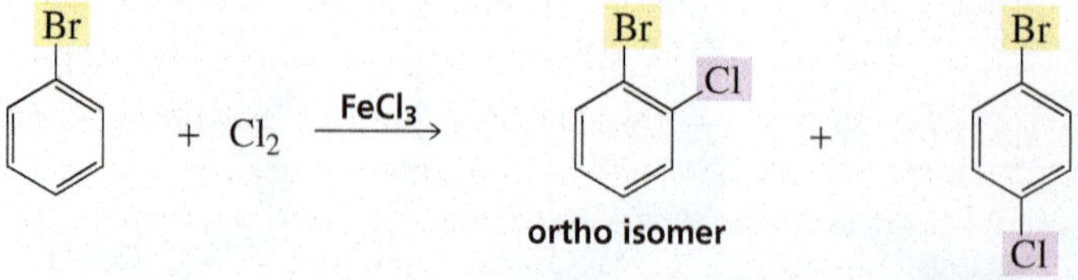

ortho isomer   para isomer

- All *moderately and strongly deactivating* substituents direct an incoming electrophile to the meta position.

meta isomer

All deactivating substituents (except the halogens) are meta directors.

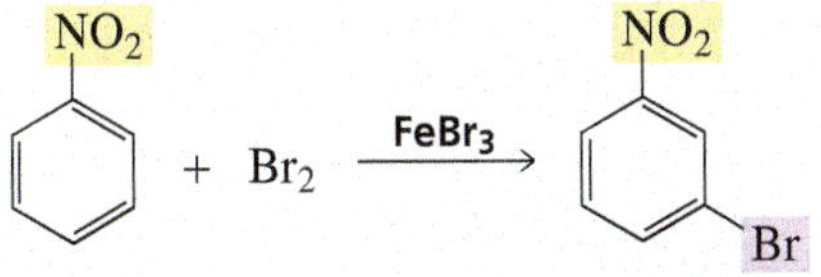

To understand why a substituent directs an incoming electrophile to a particular position, we must look at the stability of the carbocation intermediate. Because the transition state of the rate-limiting step resembles the carbocation intermediate, anything that stabilizes the carbocation intermediate also stabilizes the transition state for its formation (Figure 18.1 on page 872).

When a substituted benzene undergoes an electrophilic aromatic substitution reaction, three different carbocation intermediates potentially can be formed: an *ortho*-substituted carbocation, a *meta*-substituted carbocation, and a *para*-substituted carbocation. The relative stabilities of the three carbocations determines the preferred pathway of the reaction because the more stable the carbocation, the more stable the transition state for its formation and, therefore, the more rapidly it is formed (Section 6.3).

Substituents That Donate Electrons by Resonance are Ortho–Para Directors

When the substituent is one that can donate electrons by *resonance,* the carbocations formed by putting the incoming electrophile on the ortho and para positions have a fourth resonance contributor (Figure 18.2). This is an especially stable resonance contributor because it is the only one whose atoms (except for hydrogen) all have complete octets. It is obtained only by directing an incoming substituent to the ortho and para positions. Therefore, *all substituents that donate electrons by resonance are ortho–para directors.*

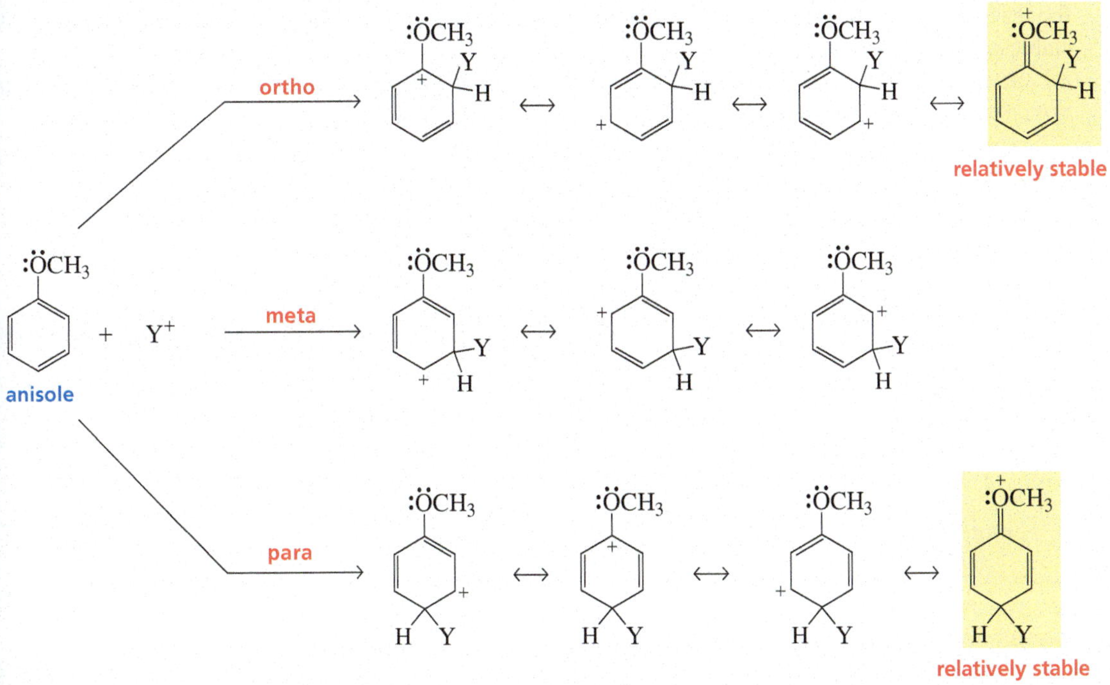

◀ **Figure 18.2**
The structures of the carbocation intermediates formed from the reaction of an electrophile with anisole at the ortho, meta, and para positions.

Substituents That Donate Electrons by Hyperconjugation are Ortho–Para Directors

When the substituent is an alkyl group, the resonance contributors that are highlighted in Figure 18.3 are the most stable. In those contributors, the alkyl group is attached directly to the positively charged carbon and can stabilize it by hyperconjugation. Therefore, *alkyl substituents are ortho–para directors* because none of the meta-substituted resonance contributors have a positively charged carbon directly attached to the alkyl group.

All substituents that donate electrons either by resonance or by hyperconjugation are ortho–para directors.

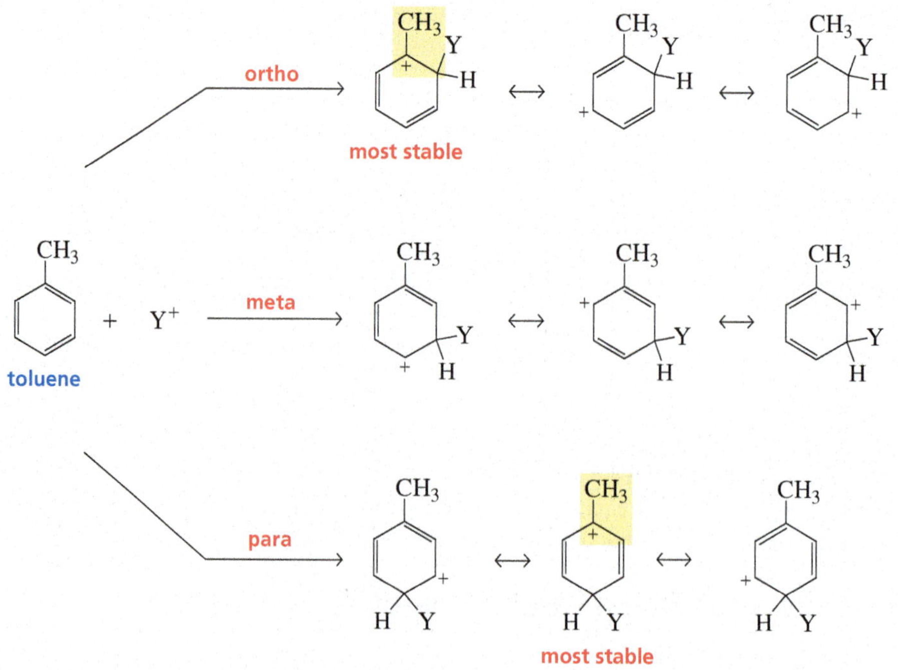

▶ **Figure 18.3**
The structures of the carbocation intermediates formed from the reaction of an electrophile with toluene at the ortho, meta, and para positions.

Substituents That Cannot Donate Electrons by Resonance or by Hyperconjugation are Meta Directors

Substituents with a positive charge or a partial positive charge on the atom attached to the benzene ring withdraw electrons inductively from the benzene ring, and most withdraw electrons by resonance as well. For all of these substituents, the resonance contributors highlighted in Figure 18.4 are the least stable because they have a positive charge on each of two adjacent atoms, so that the most stable carbocation is formed when the incoming electrophile is directed to the meta position. Thus, *all substituents that overall withdraw electrons (except for the halogens, which are ortho–para directors because they donate electrons by resonance) are meta directors.*

All substituents that cannot donate electrons are meta directors.

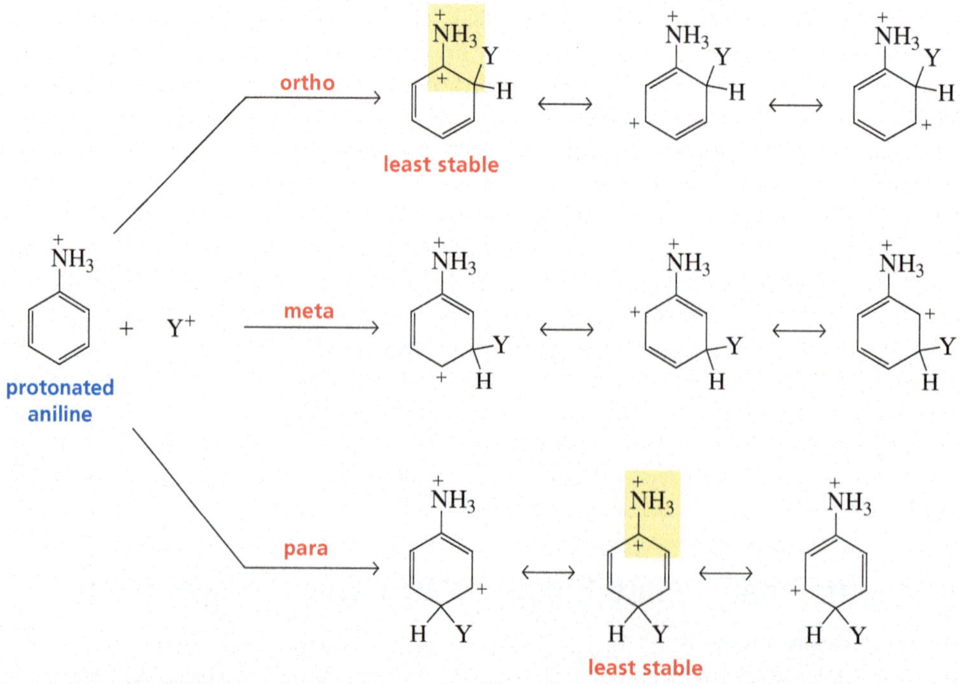

▶ **Figure 18.4**
The structures of the carbocation intermediates formed from the reaction of an electrophile with protonated aniline at the ortho, meta, and para positions.

Notice that the three possible carbocation intermediates in Figures 18.3 and 18.4 are the same, except for the substituent. The nature of that substituent determines whether the resonance contributors with the substituent directly attached to the positively charged carbon are the most stable (when the substituent is electron donating) or the least stable (when the substituent is electron withdrawing).

In summary:

all the activating substituents and the weakly deactivating halogens are ortho–para directors (Table 18.1);
substituents more deactivating than the halogens are meta directors.

In other words, all substituents that donate electrons either by resonance or by hyperconjugation are ortho–para directors, whereas all substituents that cannot donate electrons are meta directors.

You do not need to resort to memorization to distinguish an ortho–para director from a meta director. It is easy to tell them apart: all ortho–para directors, except for alkyl, aryl, and CH=CHR groups, have at least one lone pair on the atom directly attached to the ring; all meta directors have a positive charge or a partial positive charge on the atom attached to the ring. Take a few minutes to examine the substituents listed in Table 18.1 to see that this is true.

PROBLEM 16

Draw the resonance contributors for:

a. benzaldehyde   **b.** chlorobenzene

PROBLEM 17 ♦

What product(s) result from nitration of each of the following?

a. propylbenzene   **c.** benzaldehyde   **e.** benzenesulfonic acid
b. bromobenzene   **d.** benzonitrile   **f.** cyclohexylbenzene

---

PROBLEM 18 ♦

Are the following substituents ortho–para directors or meta directors?

a. CH=CHC≡N   **b.** NO_2 **c.** CH_2OH **d.** COOH   **e.** CF_3 **f.** N=O

18.14 THE ORTHO–PARA RATIO

When a benzene ring with an ortho–para-directing substituent undergoes an electrophilic aromatic substitution reaction, what percentage of the product is the ortho isomer and what percentage is the para isomer?

Solely on the basis of probability, more of the ortho product is expected because two ortho positions are available to the incoming electrophile but only one para position is available. The ortho positions, however, are sterically hindered by the substituent on the ring, whereas the para position is not. Consequently, the para isomer will be formed preferentially if either the substituent on the ring or the incoming electrophile is large. The following nitration reactions illustrate the decrease in the ortho–para ratio as the size of the substituent on the ring increases:

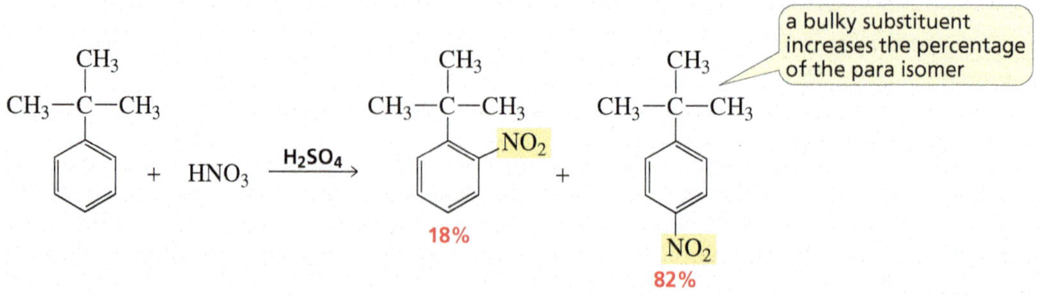

a bulky substituent increases the percentage of the para isomer

18%

82%

Fortunately, the physical properties of the ortho- and para-substituted isomers differ enough to allow them to be easily separated. Consequently, electrophilic aromatic substitution reactions that lead to both ortho and para isomers are useful in syntheses because the desired product can be easily separated from the reaction mixture.

18.15 ADDITIONAL CONSIDERATIONS REGARDING SUBSTITUENT EFFECTS

It is important to know whether a substituent is activating or deactivating to determine the conditions needed to carry out a reaction.

Halogenation

Because halogenation is the *fastest* of the electrophilic aromatic substitution reactions, it should be carried out without the Lewis acid catalyst ($FeBr_3$ or $FeCl_3$) if the ring has a strongly activating substituent.

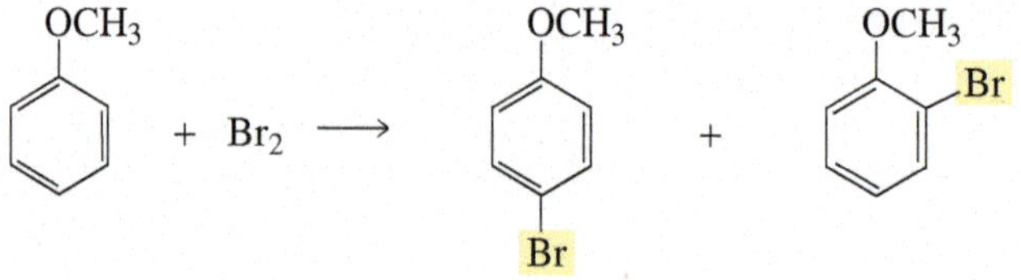

If the Lewis acid catalyst and excess bromine are used, the tribromo compound is obtained.

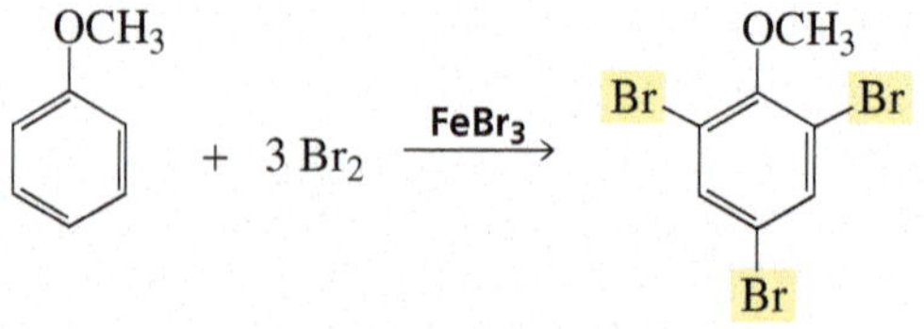

Friedel–Crafts Acylation and Alkylation

Friedel–Crafts reactions are the *slowest* of the electrophilic aromatic substitution reactions. Therefore, if a benzene ring has been moderately or strongly deactivated—that is, if it has a meta-directing substituent—it will be too unreactive to undergo either Friedel–Crafts acylation or Friedel–Crafts alkylation. In fact, nitrobenzene is so unreactive that it is often used as a solvent for Friedel–Crafts reactions.

A benzene ring with a meta director cannot undergo a Friedel–Crafts reaction.

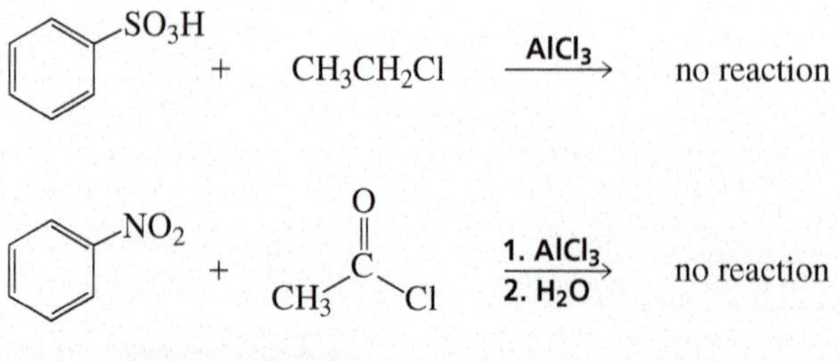

Aniline and *N*-substituted anilines also do not undergo Friedel–Crafts reactions. The lone pair on the amino group forms a complex with the Lewis acid catalyst ($AlCl_3$) needed to carry out the reaction, thereby converting the NH_2 substituent into a deactivating meta director. And, as we just saw, a benzene ring with a meta director cannot undergo a Friedel–Crafts reaction.

Phenol and anisole can undergo Friedel–Crafts reactions—at the ortho and para positions—because oxygen, being a weaker base than nitrogen, does not complex with the Lewis acid. (A Lewis acid complexes with a carbonyl oxygen because a carbonyl oxygen has a partial negative charge, unlike an oxygen directly attached to a benzene ring, which has a partial positive charge.)

Aniline also cannot be nitrated, because nitric acid is an oxidizing agent and an NH_2 group is easily oxidized. (Nitric acid and aniline can be an explosive combination.) However, if the NH_2 group is converted to an *N*-acetyl group, then the ring can be nitrated. The acetyl group can subsequently be removed by acid-catalyzed hydrolysis (Section 15.12). The base added in the last step deprotonates the anilinium ion and the carboxylic acid, which were the products of acid-catalyzed hydrolysis.

PROBLEM 19 SOLVED

LEARN THE STRATEGY

What products are obtained from the reaction of the following compound with one equivalent of Br_2, using $FeBr_3$ as a catalyst?

SOLUTION The left-hand ring is attached to a substituent that activates that ring by resonance electron donation. In contrast, the right-hand ring is attached to a substituent that deactivates that ring by resonance electron withdrawal.

substituent donates electrons into the ring by resonance

substituent withdraws electrons from the ring by resonance

Therefore, the left-hand ring is more reactive toward electrophilic aromatic substitution. The activating substituent directs the bromine to the positions that are ortho and para to it.

PROBLEM 20

USE THE STRATEGY

What products are obtained from the reaction of the following compounds with one equivalent of Br_2, using $FeBr_3$ as a catalyst?

a. CH_3—〈〉—〈〉—OCH_3

b. CH_3—〈〉—C(=O)—〈〉—$COCH_3$

c. 〈〉—CH_2O—〈〉

d. CH_3O—〈〉—〈〉—NO_2

PROBLEM 21 ◆
Give the products, if any, of each of the following reactions:

a. benzonitrile + methyl chloride + AlCl$_3$ c. benzoic acid + CH$_3$CH$_2$Cl + AlCl$_3$
b. phenol + Br$_2$ d. benzene + 2 CH$_3$Cl + AlCl$_3$

DESIGNING A SYNTHESIS VI

18.16 THE SYNTHESIS OF MONOSUBSTITUTED AND DISUBSTITUTED BENZENES

As the number of reactions you know increases, so do the possibilities you are able to choose from when you design a synthesis. For example, you can now design two very different routes for the synthesis of 2-phenylethanol from benzene.

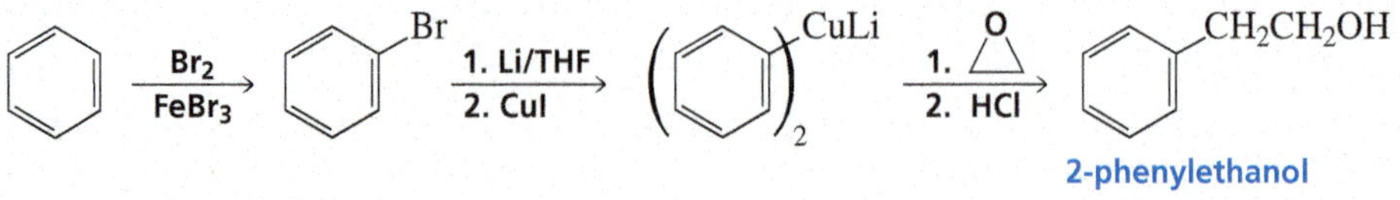

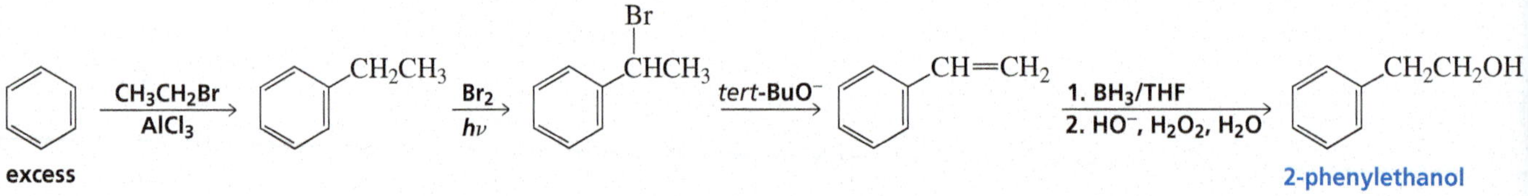

2-phenylethanol

2-phenylethanol

The preferred route depends on such factors as convenience, expense, and the expected yield of the target molecule (the desired product). For example, the first route shown for the synthesis of 2-phenylethanol is the better procedure, because the second route has more steps, requires excess benzene to prevent polyalkylation, and uses a radical reaction that can produce unwanted side products.

Example 1

Designing the synthesis of a disubstituted benzene requires careful consideration of the order in which the substituents are to be placed on the ring. For example, if you want to synthesize *m*-bromobenzenesulfonic acid, the sulfonic acid group has to be placed on the ring first, because that group will direct the bromo substituent to the desired meta position.

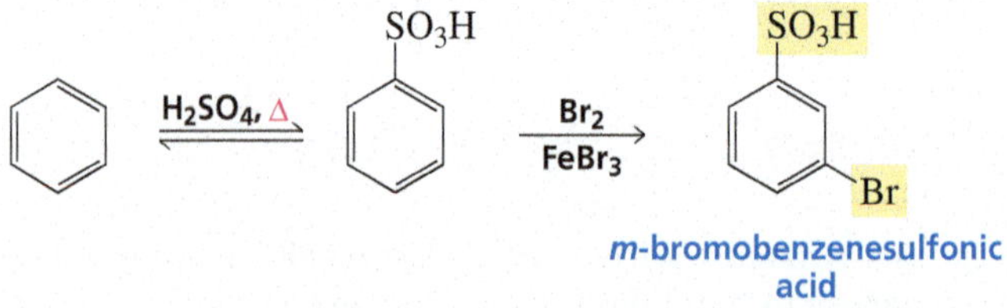

m-bromobenzenesulfonic acid

Example 2

If the desired product is *p*-bromobenzenesulfonic acid, then the order of the two reactions in Example 1 must be reversed because only the bromo substituent is an ortho–para director.

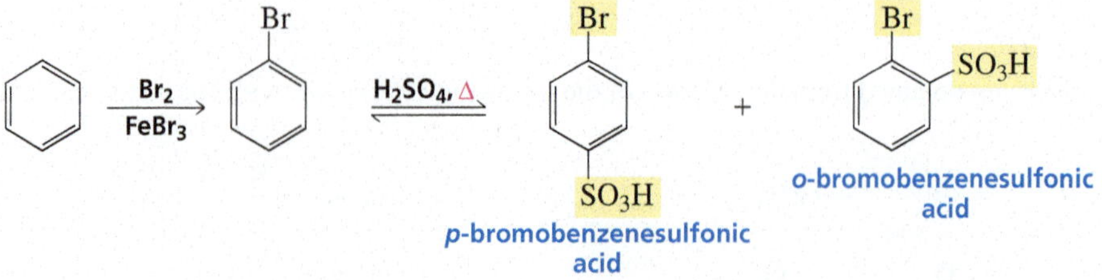

p-bromobenzenesulfonic acid

o-bromobenzenesulfonic acid

Example 3

Both substituents of *m*-nitroacetophenone are meta directors. The Friedel–Crafts acylation reaction must be carried out first because a ring with a meta director such as nitrobenzene cannot undergo a Friedel–Crafts reaction (Section 18.15).

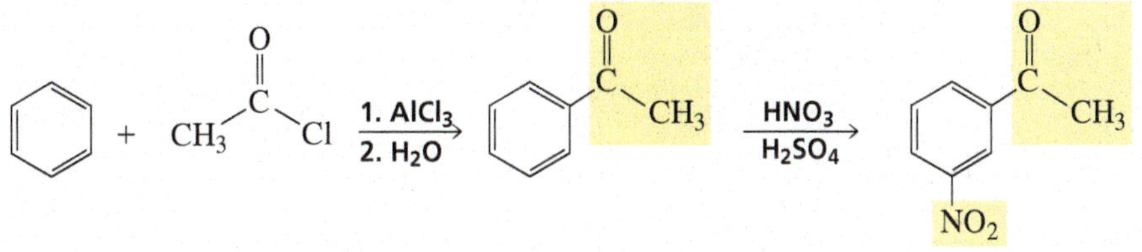

m-nitroacetophenone

Example 4

Another question that needs to be considered is, at what point in a reaction sequence should a substituent be chemically modified? For example, in the following reaction, the methyl group is oxidized *after* it directs the chloro substituent to the para position. (*o*-Chlorobenzoic acid is also formed in this reaction.)

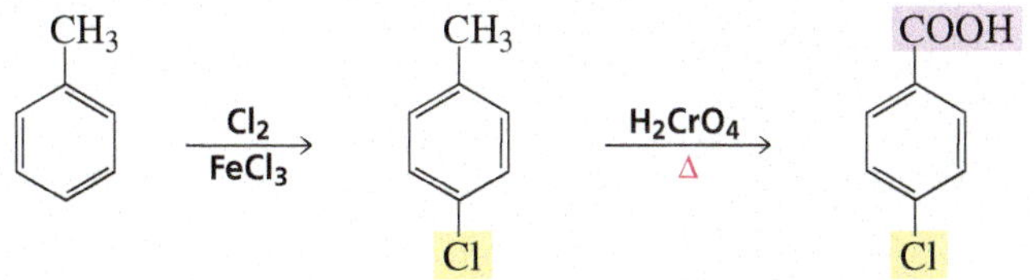

p-chlorobenzoic acid

Example 5

In the next reaction, the methyl group is oxidized *before* chlorination because a meta director is needed to obtain the desired product.

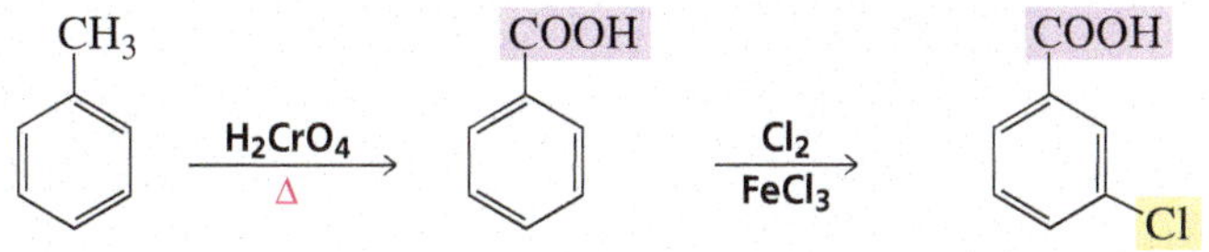

m-chlorobenzoic acid

Example 6

Although chemists often have several ways to carry out a reaction, there may be factors that require the use of one particular method. For example, we saw that there is more than one way to get a straight-chain alkyl group onto a benzene ring (Sections 18.8 and 18.9). In the next synthesis, a Friedel–Crafts alkylation/reduction can be used to put the first alkyl group on the ring but not the second one. A meta-directing group is required to get the second alkyl substituent onto the ring, and a Friedel–Crafts reaction cannot be done on a ring with a meta director. Instead, the target molecule has to be prepared using a coupling reaction.

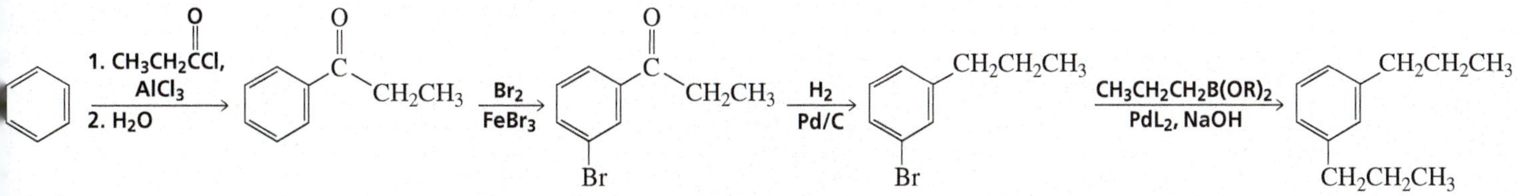

m-dipropylbenzene

PROBLEM 22 ♦

a. Does a coupling reaction have to be used to synthesize *p*-dipropylbenzene?

b. Can a coupling reaction be used to synthesize *p*-dipropylbenzene?

PROBLEM 23

Show how the following compounds can be synthesized from benzene:

a. 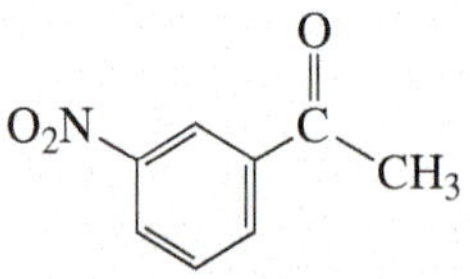 (O₂N-substituted acetophenone structure) **b.** 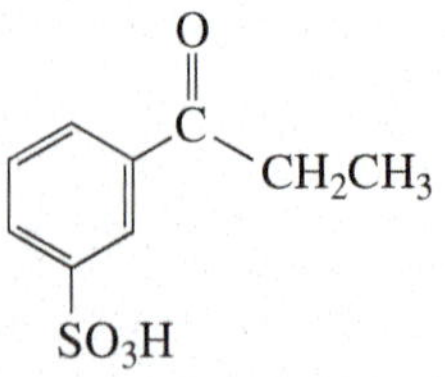 (propiophenone with SO₃H structure) **c.** 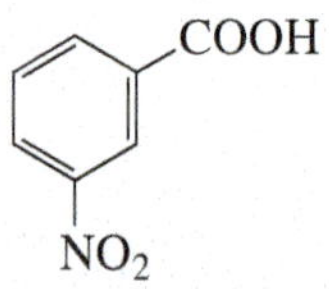 (COOH and NO₂ substituted benzene structure)

Example 7

In the next synthesis, the type of reaction employed, the order in which the substituents are put on the benzene ring, and the point at which a substituent is chemically modified must all be considered. The straight-chain propyl substituent must be put on the ring by a Friedel–Crafts acylation rather than by an alkylation reaction, to avoid the carbocation rearrangement that would occur with the alkylation reaction. The Friedel–Crafts acylation must be carried out before sulfonation because acylation cannot be carried out on a ring with a meta-directing sulfonic acid substituent. Finally, the sulfonic acid group must be put on the ring after the carbonyl group is reduced to a methylene group so that the sulfonic acid group is directed primarily to the para position by the alkyl group.

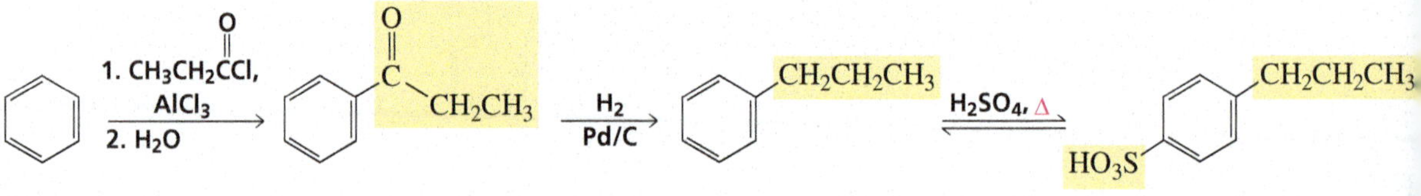

p-propylbenzenesulfonic acid

Alternatively, the propylbenzene intermediate can be prepared using bromobenzene and a coupling reaction.

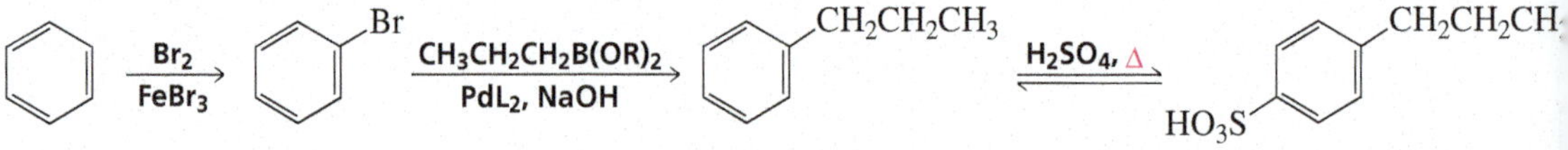

p-propylbenzenesulfonic acid

PROBLEM 24

Show how each of the following compounds can be synthesized from benzene:

a. *o*-bromopropylbenzene   **d.** *p*-chloroaniline   **g.** 2-phenylpropene
b. *m*-nitrobenzoic acid   **e.** *m*-chloroaniline   **h.** *m*-bromopropylbenzene
c. *p*-nitrobenzoic acid   **f.** *m*-xylene   **i.** 1-phenyl-2-propanol

18.17 THE SYNTHESIS OF TRISUBSTITUTED BENZENES

When a disubstituted benzene undergoes an electrophilic aromatic substitution reaction, the directing effects of both substituents have to be considered. If both substituents direct the incoming substituent to the same position, the product of the reaction is easily predicted.

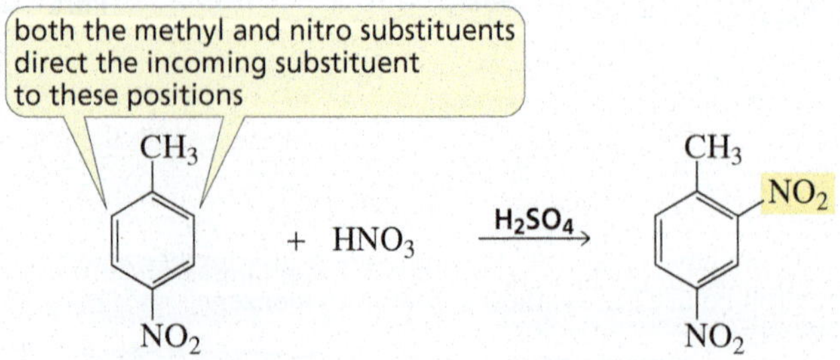

both the methyl and nitro substituents direct the incoming substituent to these positions

+ HNO₃ $\xrightarrow{H_2SO_4}$

Notice that three positions are activated in the next reaction, but the new substituent ends up primarily on only two of the three. Steric hindrance makes the position between the substituents less accessible, so only a small amount of the third product formed.

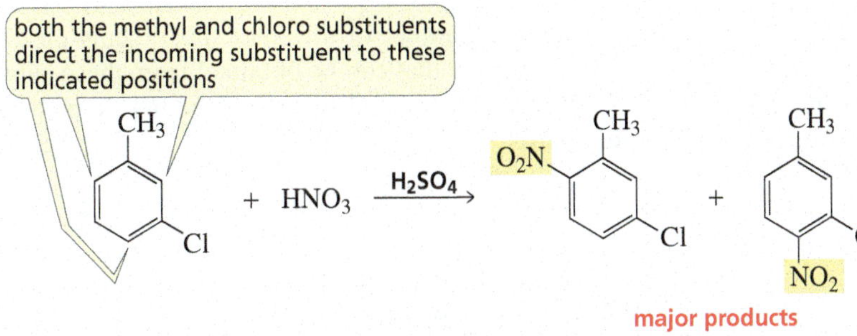

both the methyl and chloro substituents direct the incoming substituent to these indicated positions

major products

If the two substituents direct the new substituent to different positions, a strongly activating substituent will win out over a weakly activating substituent or a deactivating substituent.

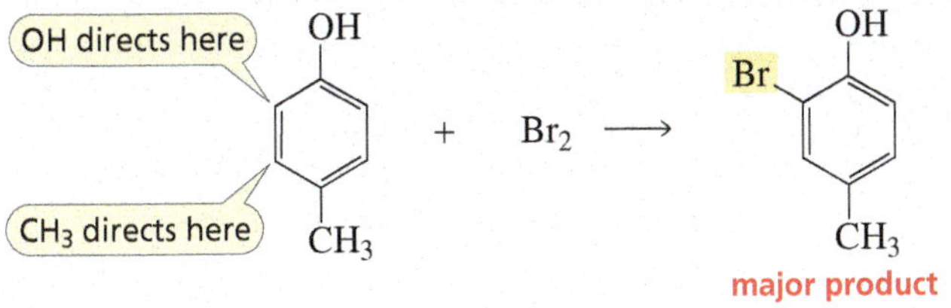

OH directs here
CH₃ directs here

major product

If the two substituents have similar activating properties, neither will dominate and a mixture of products will be obtained.

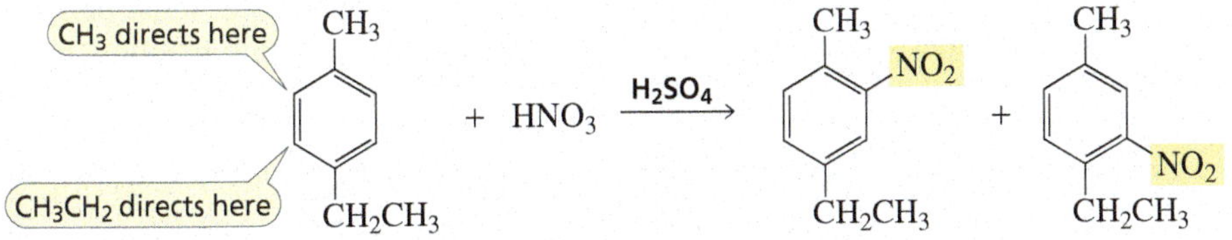

CH₃ directs here
CH₃CH₂ directs here

PROBLEM 25 ◆

What is the major product(s) of each of the following reactions?

a. bromination of *p*-methylbenzoic acid
b. chlorination of *o*-benzenedicarboxylic acid
c. bromination of *p*-chlorobenzoic acid
d. nitration of *p*-fluoroanisole
e. nitration of *p*-methoxybenzaldehyde
f. nitration of *p-tert*-butylmethylbenzene

PROBLEM 26 ◆

A student had prepared three ethyl-substituted benzaldehydes, but neglected to label them. The student at the next bench said they could be identified by brominating a sample of each and determining how many bromo-substituted products are formed. Is the student's advice sound?

PROBLEM 27 SOLVED

When phenol is treated with Br₂, a mixture of *ortho*-bromophenol and *para*-bromophenol is obtained. Design a synthesis that would convert phenol primarily to *ortho*-bromophenol.

SOLUTION In the first step, the bulky sulfonic acid group adds preferentially to the para position. Both the OH and SO₃H groups direct bromine to the position ortho to the OH group. Heating in dilute acid removes the sulfonic acid group (Section 18.5).

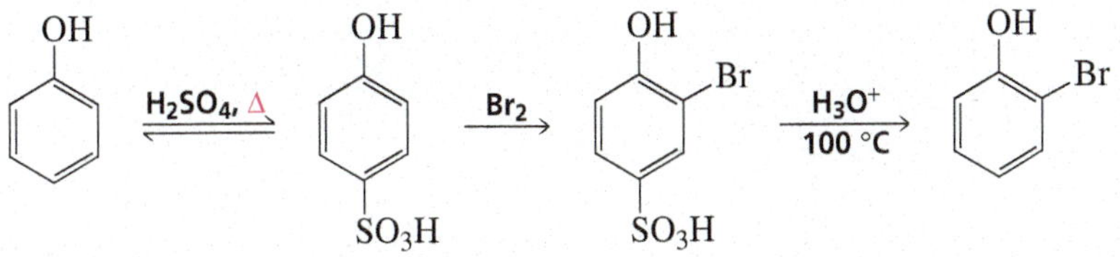

Using a sulfonic acid group to block the para position is a common strategy for synthesizing high yields of ortho-substituted compounds.

18.18 SYNTHESIZING SUBSTITUTED BENZENES USING ARENEDIAZONIUM SALTS

We have learned how to place a limited number of different substituents on a benzene ring—the substituents listed in Section 18.2 and those that can be obtained from these substituents by chemical conversion (Section 18.10). The list of substituents that can be placed on a benzene ring can be expanded using **arenediazonium salts**.

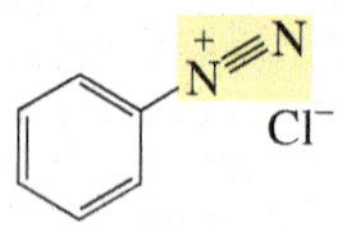

an arenediazonium salt

Displacing the leaving group of a diazonium ion by a nucleophile occurs readily because the leaving group is nitrogen gas (indicated by an upward pointing arrow), which is very stable. Some displacements involve phenyl cations, whereas others involve radicals—the actual mechanism depends on the particular nucleophile.

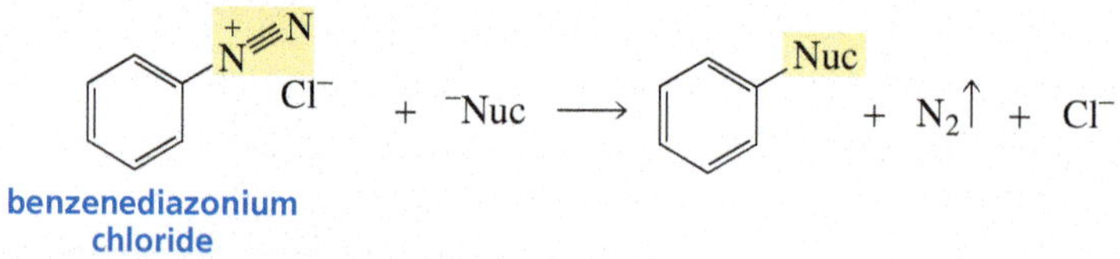

benzenediazonium chloride

Aniline can be converted to an arenediazonium salt by treatment with nitrous acid (HNO_2). Because nitrous acid is unstable, it must be formed in situ, using an aqueous solution of sodium nitrite and an acid. N_2 is such a good leaving group that the diazonium salt must be synthesized at $0\,°C$ and used immediately without isolation. The mechanism for this reaction is shown in Section 18.19.

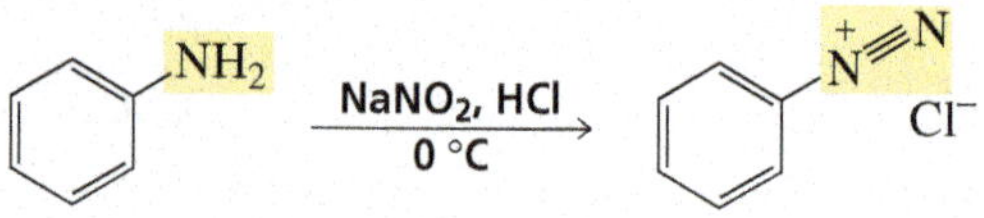

Sandmeyer Reactions

The nucleophiles $^-C{\equiv}N$, Cl^-, and Br^- will replace the diazonium group if the appropriate copper(I) salt is added to the solution containing the arenediazonium salt. The reaction of an arenediazonium salt with a copper(I) salt is known as a **Sandmeyer reaction.**

Sandmeyer reactions

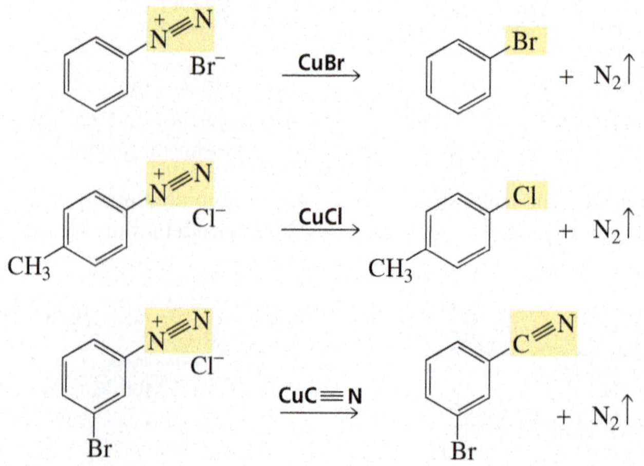

KCl and KBr cannot be used in place of CuCl and CuBr in Sandmeyer reactions. The cuprous salts are required, which indicates that the copper(I) ion has a role in the reaction. Although the precise mechanism is not known, it is thought that the copper(I) ion donates an electron to the diazonium salt, forming an aryl radical and nitrogen gas.

Although chloro and bromo substituents can be placed directly on a benzene ring by halogenation, the Sandmeyer reaction can be a useful alternative. For example, if you want to make *para*-chloroethylbenzene, the chlorination of ethylbenzene leads to a mixture of ortho and para isomers.

If, however, you start with *para*-ethylaniline, diazotize it, and use a Sandmeyer reaction for chlorination, then only the desired para product is formed.

Replacing a Diazonium Group with an Iodo Substituent

An iodo substituent will replace the diazonium group if potassium iodide is added to the solution containing the diazonium ion.

The Schiemann Reaction

Fluoro substitution occurs if the arenediazonium salt is heated with fluoroboric acid (HBF_4). This is known as the **Schiemann reaction.**

Schiemann reaction

Synthesizing a Phenol

If the acidic aqueous solution in which the diazonium salt has been synthesized is allowed to warm up, then an OH group will replace the diazonium group. (H_2O is the nucleophile.)

A better yield of phenol is obtained if aqueous copper(I) oxide and copper(II) nitrate are added to the cold solution.

Replacing a Diazonium Group with a Hydrogen

A hydrogen will replace a diazonium group if the arenediazonium salt is treated with hypophosphorous acid (H_3PO_2). This is a useful reaction if an amino group or a nitro group is needed for directing purposes and subsequently must be removed. For example, it is difficult to envision how 1,3,5-tribromobenzene could be synthesized without such a reaction.

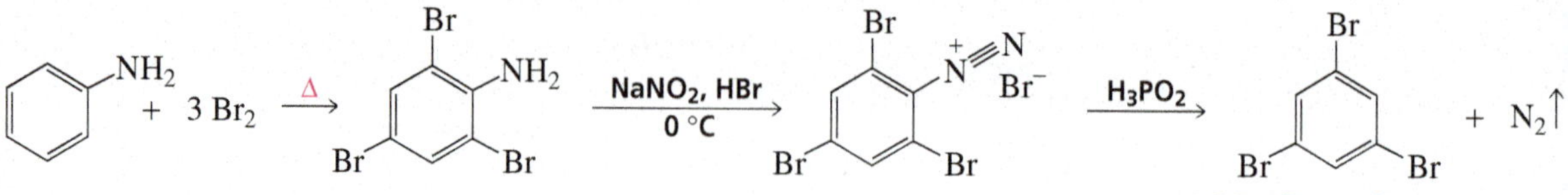

Remember that reactions involving arenediazonium ions must be carried out at 0 °C because they are unstable at higher temperatures.

Retrosynthetic Analysis

We saw that when designing a complicated synthesis, it is often easier to work backward. For example, when planning a synthesis for *meta*-dibromobenzene, we realize that, because a bromo substituent is an ortho–para director, halogenation cannot be used to put both bromo substituents on the ring. Knowing that a bromo substituent can be put on a benzene ring with a Sandmeyer reaction and that the bromo substituent in a Sandmeyer reaction replaces what originally was a meta-directing nitro substituent, we have a route to the synthesis of the target molecule.

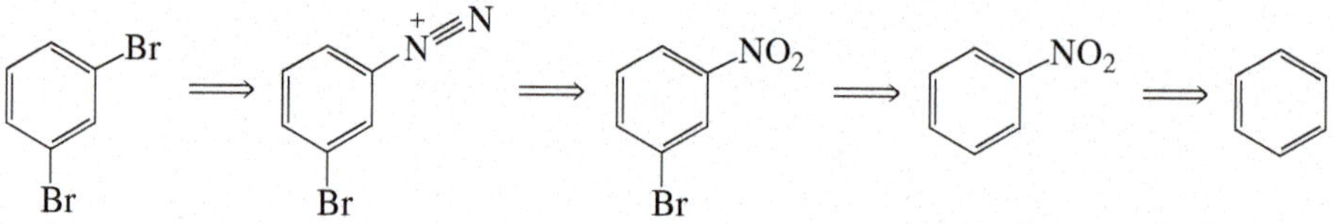

Now we can write the reaction in the forward direction, including the reagents necessary to carry out each step.

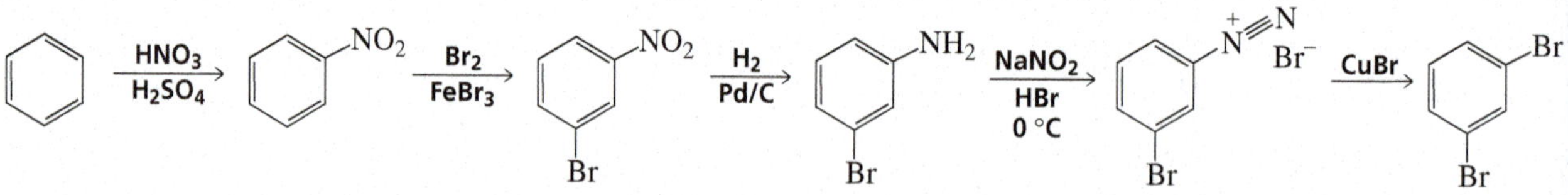

PROBLEM 28 ◆

Why isn't FeBr$_3$ used as a catalyst in the first step of the synthesis of 1,3,5-tribromobenzene?

PROBLEM 29

Explain why a diazonium group on a benzene ring cannot be used to direct an incoming substituent to the meta position.

PROBLEM 30

Write the sequence of steps required for the conversion of benzene into benzenediazonium chloride.

PROBLEM 31

Show how the following compounds can be synthesized from benzene:

a. *p*-nitrobenzoic acid   c. *o*-chlorophenol   e. *p*-methylbenzonitrile
b. *m*-bromophenol   d. *m*-methylnitrobenzene   f. *m*-chlorobenzaldehyde

18.19 AZOBENZENES

In addition to being used to synthesize substituted benzenes, arenediazonium ions can be used as electrophiles in electrophilic aromatic substitution reactions. Because an arenediazonium ion is unstable at room temperature, it can be used as an electrophile only in reactions that can be carried out well below room temperature. In other words, only highly activated benzene rings (phenols, anilines, and *N*-alkylanilines) can undergo electrophilic aromatic substitution reactions with arene-diazonium ion electrophiles. The product of the reaction is an **azobenzene.** The N=N linkage is called an **azo linkage.**

Because the electrophile is so large, substitution takes place preferentially at the less sterically hindered para position. However, if the para position already has a substituent, then substitution will occur at an ortho position.

The mechanism for electrophilic aromatic substitution with an arenediazonium ion electrophile is the same as the mechanism for electrophilic aromatic substitution with any other electrophile.

MECHANISM FOR ELECTROPHILIC AROMATIC SUBSTITUTION WITH AN ARENEDIAZONIUM ION ELECTROPHILE

- The electrophile adds to the benzene ring.
- A base in the solution removes a proton from the carbon that formed the bond with the electrophile.

Azobenzenes, like alkenes, can exist in cis and trans forms. The trans isomer is more stable because the cis isomer has steric strain (Section 5.9).

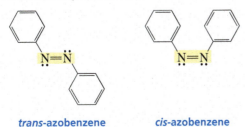

trans-azobenzene   *cis*-azobenzene

Azobenzenes are colored compounds because of their extended conjugation (Section 13.22). They are used commercially as dyes.

Discovery of the First Antibiotic

The observation that azo dyes effectively dyed wool fibers (animal protein) suggested that such dyes might bind to bacterial proteins, too, and perhaps harm the bacteria in the process. Well over 10,000 dyes were screened *in vitro* ("in glass"), but none showed any antibiotic activity. At that point, the investigators decided to screen the dyes *in vivo* ("in an organism"). *In vivo* studies were done in mice that had been infected with a bacterial culture. Several dyes turned out to counteract Gram-positive infections. In 1935, the least toxic of these, Prontosil (a bright red dye), became the first drug to treat bacterial infections.

Prontosil®

Gerhard Domagk, a scientist at German dye manufacturer I. G. Farbenindustrie, was the one whose studies showed that Protonsil was an effective antibiotic. His daughter, who was dying of a streptococcal infection as a result of cutting her finger, was the first person to receive and be cured by the drug (1935). Protonsil received wider fame when it was used in 1936 to save the life of Franklin D. Roosevelt, Jr., a son of the U.S. president.

Domagk received the Nobel Prize in Physiology or Medicine in 1939, but Adolf Hitler did not allow Germans to accept Nobel Prizes because Carl von Osdietsky, a German who was in a concentration camp, had been awarded the Nobel Prize for Peace in 1935. Domagk was eventually able to accept the prize in 1947 but, because of the time that had passed, he was not given the monetary award that normally comes with the prize.

The fact that Prontosil was inactive *in vitro* but active *in vivo* should have been recognized as a sign that the dye was converted to an active compound by the mouse, but this did not occur to the investigators, who were content just to have found a useful antibiotic.

When scientists at the Pasteur Institute later investigated Prontosil, they noted that mice given the drug did not excrete a red compound. Urine analysis showed that the mice excreted *para*-acetamidobenzenesulfonamide, a colorless compound, instead. Chemists knew that anilines are acetylated *in vivo*, so they prepared the nonacetylated compound (sulfanilamide). When sulfanilamide was tested in mice infected with streptococcus, all the mice were cured, whereas the untreated control mice died.

para-acetamidobenzenesulfonamide   *para*-aminobenzenesulfonamide
                                                                        sulfanilamide

Protonsil is an example of a **prodrug**, a compound that becomes an effective drug only after it undergoes a reaction in the body. Sulfanilamide was the first of the sulfa drugs, and the sulfa drugs were the first class of antibiotics. Sulfanilamide acts by inhibiting the bacterial enzyme that synthesizes folic acid, a compound that bacteria need for growth (Section 23.7).

Drug Safety

In October 1937, patients who had obtained sulfanilamide from a company in Tennessee were experiencing excruciating abdominal pains before slipping into fatal comas. The FDA asked Eugene Geiling, a pharmacologist at the University of Chicago, and his graduate student Frances Kelsey to investigate. They found that the drug company was dissolving sulfanilamide in diethylene glycol, a sweet-tasting liquid, to make the drug easy to swallow. However, the safety of diethylene glycol in humans had never been tested, and it turned out to be a deadly poison. Interestingly, Frances Kelsey was the person who, later in her career, prevented thalidomide from being marketed in the United States (see page 178).

At the time of the sulfanilamide investigation, there was no legislation to prevent the sale of medicines with lethal effects, so in June 1938, the Federal Food, Drug, and Cosmetic Act was enacted. This legislation required all new foods, drugs, and cosmetics to be thoroughly tested for safety before being marketed. The laws are amended from time to time to reflect changing circumstances.

PROBLEM 32

What product is formed from the reaction of *p*-methylphenol with benzenediazonium chloride?

PROBLEM 33

In the mechanism for electrophilic aromatic substitution with a diazonium ion as the electrophile, why does nucleophilic attack occur on the terminal nitrogen of the diazonium ion rather than on the nitrogen that has the formal positive charge?

PROBLEM 34

Draw the structure of the activated benzene ring and the diazonium ion used in the synthesis of each of the following compounds, whose structures can be found on page 607.

a. butter yellow **b.** methyl orange

18.20 THE MECHANISM FOR THE FORMATION OF A DIAZONIUM ION

Conversion of an NH_2 group to a diazonium group requires a *nitrosonium ion*. A nitrosonium ion is formed when water is eliminated from protonated nitrous acid, similar to the way the nitronium ion is generated from nitric acid (Section 18.5) and the sulfonium ion is generated from sulfuric acid (Section 18.6).

MECHANISM FOR FORMATION OF A DIAZONIUM ION FROM ANILINE

- Aniline shares an electron pair with the nitrosonium ion.
- Removal of a proton from nitrogen forms a **nitrosamine.**
- Delocalization of nitrogen's lone pair and protonation on oxygen form a protonated *N*-hydroxyazo compound.
- The protonated *N*-hydroxyazo compound is in equilibrium with its nonprotonated form.
- The *N*-hydroxyazo compound can be reprotonated on nitrogen (the reverse reaction) or protonated on oxygen (the forward reaction).
- Elimination of water forms the diazonium ion.

Notice the pattern of the three intermediates that you saw in several other mechanisms: protonated intermediate ⇌ neutral intermediate ⇌ protonated intermediate.

A New Cancer-Fighting Drug

Temozolomide, the drug used to treat the late Senator Ted Kennedy's brain tumor, is a relatively new member in the arsenal of cancer-fighting drugs. The drug can be taken orally. While in the circulatory system, it is converted to its active form by reacting with water and losing CO_2. Once inside the cell, the drug eliminates a methyldiazonium ion, a very reactive methylating agent. As we saw in Section 10.11, methylating DNA triggers the death of cancer cells. Temozolomide is another example of a prodrug (Section 18.19). *(Continued)*

temozolomide → activated form → methyldiazonium

$$DNA-\overset{+}{Nuc}-CH_3 + N_2 \xleftarrow[\text{S}_N\text{2 reaction}]{DNA-\ddot{N}uc} CH_3-\overset{+}{N}\equiv N +$$

methyldiazonium

PROBLEM 35 ♦

Which amide bond is hydrolyzed in the first step of the conversion of temozolomide to methyldiazonium?

PROBLEM 36

Explain why a secondary amine forms a nitrosamine rather than a diazonium ion when it reacts with a nitrosonium ion.

a secondary amine + $^+N{=}O$ ⟶ a nitrosamine + H^+

PROBLEM 37

Diazomethane can be used to convert a carboxylic acid to a methyl ester. Propose a mechanism for this reaction.

a carboxylic acid + diazomethane ⟶ a methyl ester + $N_2\uparrow$

Nitrosamines and Cancer

An outbreak of food poisoning in sheep in Norway was traced to their ingestion of nitrite-treated fish meal. This immediately raised concerns about human consumption of nitrite-treated foods, because sodium nitrite, a commonly used food preservative, can react with secondary amines present in food to produce nitrosamines, which are known to be carcinogenic. Smoked fish, cured meats, and beer all contain nitrosamines. Nitrosamines are also found in cheese because some cheeses are preserved with sodium nitrite and cheese is rich in secondary amines. When consumer groups in the United States asked the Food and Drug Administration to ban the use of sodium nitrite as a preservative, the request was vigorously opposed by the meat-packing industry.

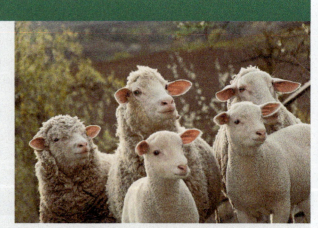

Despite extensive investigations, it has not yet been determined whether the small amounts of nitrosamines present in our food pose a hazard to our health. Until this question is answered, it will be hard to avoid sodium nitrite in our diet. Meanwhile, it is worrisome to note that Japan has both one of the highest gastric cancer rates and the highest average ingestion of sodium nitrite. Some good news, however, is that the concentration of nitrosamines in bacon has been considerably reduced in recent years by the addition of ascorbic acid—a nitrosamine inhibitor—to the curing mixture. Also, improvements in the malting process have reduced the level of nitrosamines in beer. Dietary sodium nitrite does have a redeeming feature: there is some evidence that it protects against botulism, a type of severe food poisoning.

18.21 NUCLEOPHILIC AROMATIC SUBSTITUTION

We saw that aryl halides do not react with nucleophiles because the π electron cloud repels the approach of a nucleophile (Section 9.13).

a nucleophile is repelled
by the π electron cloud

nucleophile

If, however, the aryl halide has one or more substituents that strongly withdraw electrons from the ring by resonance, a **nucleophilic aromatic substitution reaction** can take place. The electron-withdrawing groups must be positioned ortho or para to the halogen. The greater the number of electron-withdrawing substituents ortho and para to the halogen, the more easily the nucleophilic aromatic substitution reaction occurs. Notice the conditions required for each reaction.

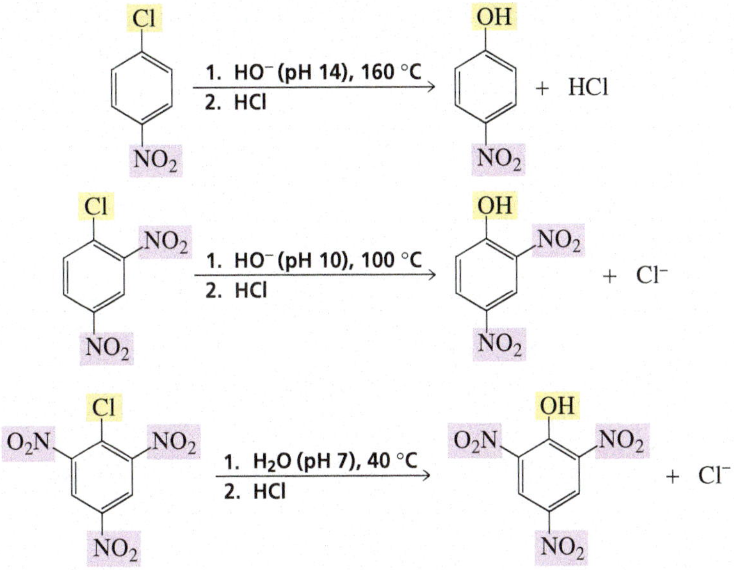

The Mechanism for Nucleophilic Aromatic Substitution

Nucleophilic aromatic substitution takes place by a two-step mechanism—nucleophilic addition, followed by elimination. The reaction is called an **S_NAr reaction** (substitution nucleophilic aromatic).

GENERAL MECHANISM FOR NUCLEOPHILIC AROMATIC SUBSTITUTION

Meisenheimer complex

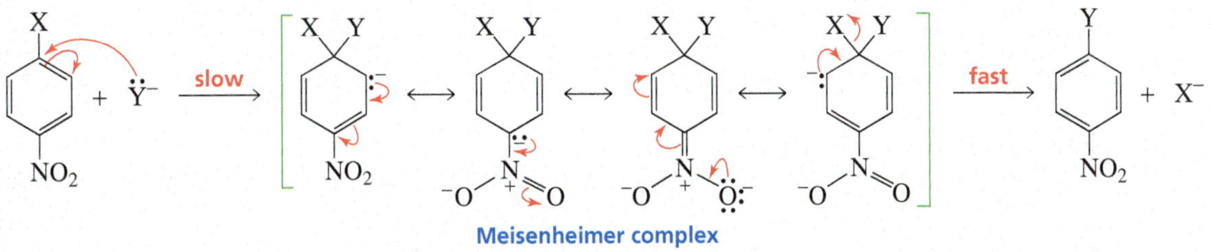

- The nucleophile attacks the carbon bearing the leaving group from a trajectory that is nearly perpendicular to the aromatic ring. (Recall from Section 9.13 that groups cannot be displaced from sp^2 carbon atoms by back-side attack.) Nucleophilic attack forms a resonance-stabilized carbanion intermediate called a *Meisenheimer complex*.
- The leaving group is eliminated, reestablishing the aromaticity of the ring.

In a nucleophilic aromatic substitution reaction, the incoming nucleophile must be a stronger base than the substituent that is being replaced, because the weaker of the two bases will be the one eliminated from the intermediate.

The electron-withdrawing substituent must be ortho or para to the site of nucleophilic attack because the electrons of the attacking nucleophile can be delocalized onto the substituent only if the substituent is in one of those positions.

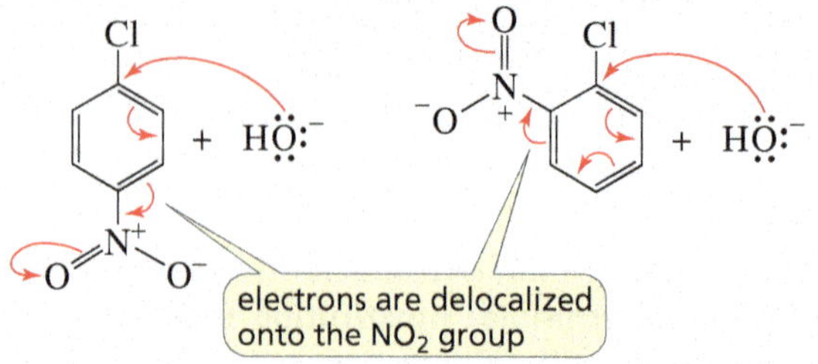

electrons are delocalized onto the NO_2 group

Examples of Nucleophilic Aromatic Substitution

A variety of substituents can be placed on a benzene ring by means of a nucleophilic aromatic substitution reaction. The only requirement is that the incoming group be a stronger base than the group that is being replaced.

An electron-withdrawing substituent increases the reactivity of the benzene ring toward nucleophilic substitution and decreases the reactivity of the benzene ring toward electrophilic substitution.

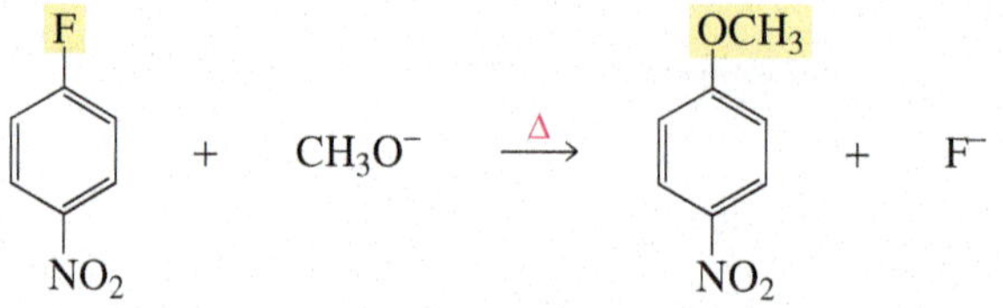

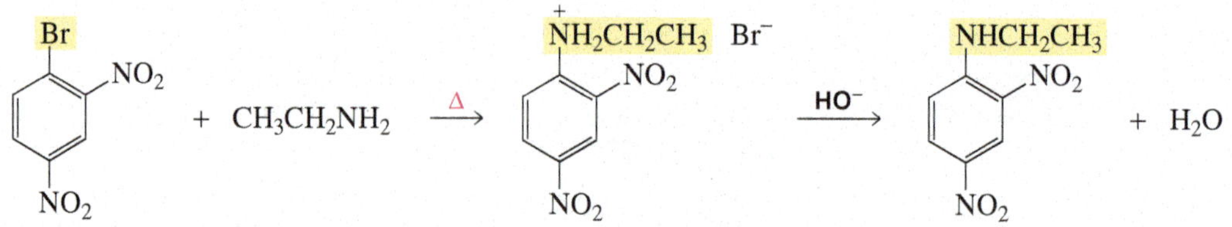

Notice that the strongly electron-withdrawing nitro substituent that *activates* the benzene ring toward *nucleophilic aromatic substitution*, *deactivates* the ring toward *electrophilic aromatic substitution* (Table 18.1). In other words, making the ring less electron rich makes it more reactive toward a nucleophile but less reactive toward an electrophile.

PROBLEM 38

Draw resonance contributors for the carbanion that would be formed if *meta*-chloronitrobenzene were to react with hydroxide ion. Why doesn't the reaction occur?

PROBLEM 39 ♦

a. Rank the following compounds from greatest tendency to least tendency to undergo nucleophilic aromatic substitution:

chlorobenzene   1-chloro-2,4-dinitrobenzene   *p*-chloronitrobenzene

b. Rank the same compounds from greatest tendency to least tendency to undergo electrophilic aromatic substitution.

PROBLEM 40

Show how each of the following compounds can be synthesized from benzene:

a. *o*-nitrophenol   **b.** *p*-nitroaniline   **c.** *p*-bromoanisole   **d.** anisole

DESIGNING A SYNTHESIS VII

18.22 THE SYNTHESIS OF CYCLIC COMPOUNDS

Most of the reactions that we have been studying are intermolecular reactions—that is, the two reacting groups are in different molecules. *Cyclic compounds are formed from intramolecular reactions*—reactions in which the two reacting groups are in the same molecule. We saw that intramolecular reactions are particularly favored if the reaction forms a compound with a five- or a six-membered ring (Section 9.16).

In designing the synthesis of a cyclic compound, we must determine the kinds of reactive groups that are necessary for a successful synthesis. For example, we know that a ketone is formed from the Friedel–Crafts acylation of a benzene ring with an acyl chloride (Section 18.7). Therefore, a cyclic ketone will result if a Lewis acid ($AlCl_3$) is added to a compound that contains both a benzene ring and an acyl chloride. The size of the ring is determined by the number of carbons between the two reacting groups.

We know that an ester is formed from the reaction of a carboxylic acid with an alcohol. Therefore, a lactone (a cyclic ester) can be prepared from a reactant that has a carboxylic acid group and an alcohol group in the same molecule separated by the appropriate number of carbons.

We saw that a cyclic ether can be prepared by an intramolecular Williamson ether synthesis (Section 9.16).

A cyclic ether can also be prepared by an intramolecular electrophilic addition reaction (Section 6.6).

$$CH_2=CHCH_2CH_2CH_2CHCH_3 \xrightarrow{\text{HCl}} CH_3CHCH_2CH_2CH_2CHCH_3 \longrightarrow$$

The product obtained from an intramolecular reaction can undergo further reactions, making it possible to synthesize many different compounds. For example, the cyclic benzylic bromide formed as the product of the next synthesis could undergo an elimination reaction, or it could undergo substitution with a wide variety of nucleophiles, or it could be converted to a Grignard reagent that could react with many different electrophiles.

PROBLEM 41

Design a synthesis for each of the following, using an intramolecular reaction:

a.

b.

c.

d.

e.

f.

ESSENTIAL CONCEPTS

Section 18.1

- Some monosubstituted benzenes are named as substituted benzenes (for example: bromobenzene, nitrobenzene); others have names that incorporate the substituent (for example: toluene, phenol, aniline).

Section 18.2

- Benzene is an aromatic compound, so it undergoes **electrophilic aromatic substitution reactions.**

- The most common electrophilic aromatic substitution reactions are **halogenation, nitration, sulfonation,** and **Friedel–Crafts acylation** and **alkylation.**

Sections 18.3–18.6

- Once the electrophile is generated, all electrophilic aromatic substitution reactions take place by the same two-step mechanism: (1) benzene forms a bond with the electrophile, forming a carbocation intermediate; and (2) a base removes a proton from the carbon that formed the bond with the electrophile.

Section 18.7

- If the carbocation formed from the alkyl halide used in a **Friedel–Crafts alkylation** reaction can rearrange, the major product will be the product with the rearranged alkyl group.

Section 18.8

- A straight-chain alkyl group can be placed on a benzene ring via a **Friedel–Crafts acylation reaction,** followed by reduction of the carbonyl group by catalytic hydrogenation or by a **Wolff–Kishner reduction.**

Section 18.9

- Alkylbenzenes with straight-chain alkyl groups also can be prepared from bromobenzene by a coupling reaction.

Section 18.11

- The relative positions of two substituents on a benzene ring are indicated in the compound's name either by numbers or by the prefixes *ortho*, *meta*, and *para*.

Section 18.12

- A substituent on a benzene ring affects the reactivity of the ring: the rate of electrophilic aromatic substitution is increased by electron-donating substituents and decreased by electron-withdrawing substituents.

Section 18.13

- A substituent affects the placement of an incoming substituent. All activating substituents and the weakly deactivating halogens are **ortho–para directors;** all substituents more deactivating than the halogens are **meta directors.**

- Ortho–para directors, with the exception of alkyl, aryl, and CH=CHR, have a lone pair on the atom attached to the ring; meta directors have a positive or partial positive charge on the atom attached to the ring.

Section 18.14

- Ortho–para-directing substituents form the para isomer preferentially if either the substituent or the incoming electrophile is large.

Section 18.15

- Benzene rings with meta-directing substituents cannot undergo Friedel–Crafts reactions.

Section 18.16

- When planning the synthesis of a disubstituted benzene, the order in which the substituents are placed on the ring and the point in a reaction sequence at which a substituent is chemically modified are important considerations.

Section 18.17

- When a disubstituted benzene undergoes an electrophilic aromatic substitution reaction, the directing effect of both substituents must be considered.

Section 18.18

- The kinds of substituents that can be placed on a benzene ring are expanded by reactions of arene diazonium salts and nucleophilic aromatic substitution reactions.

- Aniline reacts with nitrous acid to form an **arenediazonium salt;** a diazonium group can be displaced by a nucleophile.

Section 18.19

- Arenediazonium ions can be used as electrophiles with highly activated benzene rings to form **azobenzenes** that can exist in cis and trans forms.

Section 18.21

- An aryl halide with one or more strongly electron-withdrawing groups ortho or para to the halogen leaving group can undergo a **nucleophilic aromatic substitution (S_NAr) reaction.** The incoming nucleophile must be a stronger base than the halide ion that is replaced.

SUMMARY OF REACTIONS

1. Electrophilic aromatic substitution reactions

   a. Halogenation (Section 18.3). The mechanisms are shown on pages 873 and 874.

   b. Nitration, sulfonation, and desulfonation (Sections 18.4 and 18.5). The mechanisms are shown on pages 875 and 876.

   c. Friedel–Crafts acylation and alkylation (Sections 18.6 and 18.7). The mechanisms are shown on pages 877 and 878.

d. Formation of benzaldehyde by a Gatterman–Koch reaction (Section 18.6)

2. Reduction of a carbonyl group to a methylene group (Section 18.8)

3. Alkylation by a coupling reaction (Section 18.9)

a. Alkylation by a Suzuki reaction

b. Alkylation with an organocuprate

4. Reactions of a substituent on a benzene ring (Section 18.10)

Z^- = a nucleophile

5. Reaction of aniline with nitrous acid (Section 18.20). The mechanism is shown on page 905.

6. Replacing a diazonium group (Section 18.18)

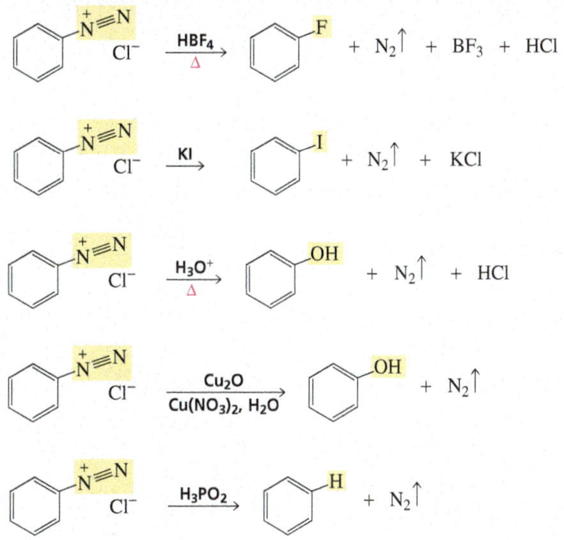

HBF₄ → $+ N_2\uparrow + BF_3 + HCl$

KI → $+ N_2\uparrow + KCl$

H₃O⁺ Δ → OH $+ N_2\uparrow + HCl$

Cu₂O / Cu(NO₃)₂, H₂O → OH $+ N_2\uparrow$

H₃PO₂ → H $+ N_2\uparrow$

7. Forming an azobenzene (Section 18.19). The mechanism is shown on page 903.

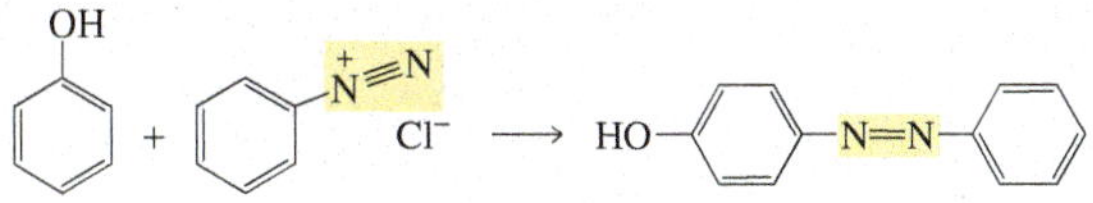

OH + $\overset{+}{N}{=}N$ Cl⁻ ⟶ HO—⟨ ⟩—N=N—⟨ ⟩

8. Nucleophilic aromatic substitution reactions of aryl halides (Section 18.21). The mechanism is shown on page 907.

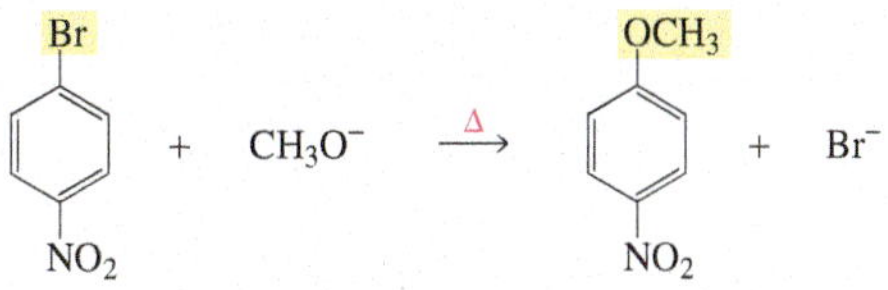

Br / NO₂ + CH₃O⁻ →Δ OCH₃ / NO₂ + Br⁻

PROBLEMS

42. Draw the structure for each of the following:

- **a.** phenol
- **b.** benzyl phenyl ether
- **c.** benzonitrile
- **d.** benzaldehyde
- **e.** anisole
- **f.** styrene
- **g.** toluene
- **h.** *tert*-butylbenzene
- **i.** benzyl chloride

43. Name the following:

a. 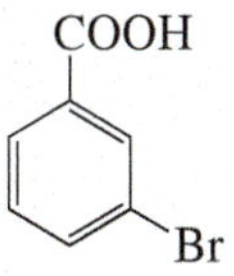 COOH / Br

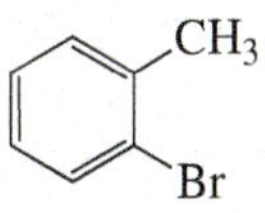

b. 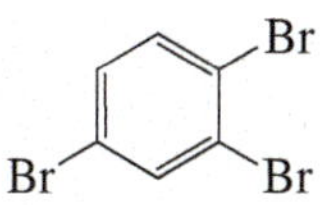 Br / Br / Br

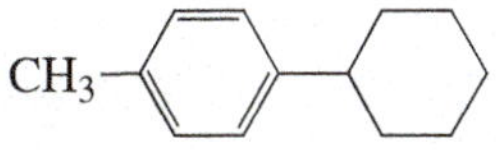

c. 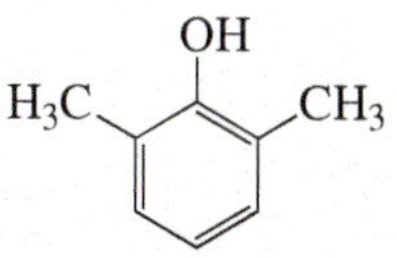 H₃C / OH / CH₃

d. CH=CH₂ / O₂N

e. 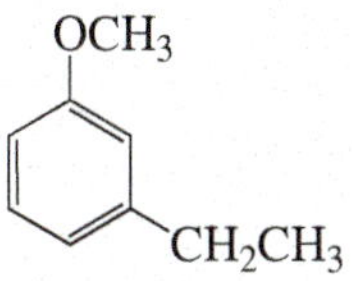 OCH₃ / CH₂CH₃

f. 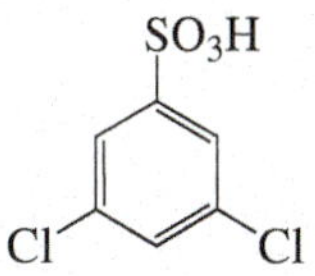 SO₃H / Cl / Cl

g. CH₃ / Br

h. CH₃—⟨ ⟩—⬡

i. 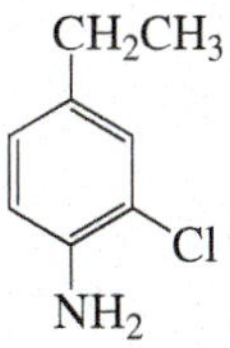 CH₂CH₃ / Cl / NH₂

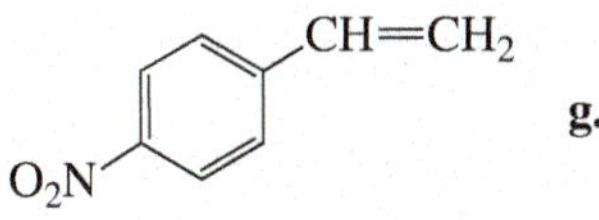

44. Provide the necessary reagents next to the arrows.

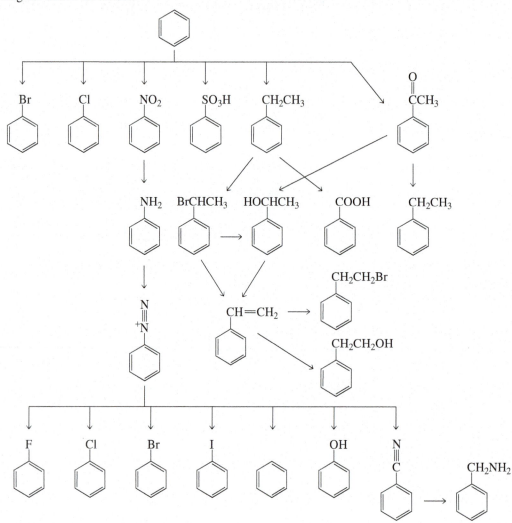

45. Draw the structure for each of the following:

a. *m*-ethylphenol
b. *p*-nitrobenzenesulfonic acid
c. (*E*)-2-phenyl-2-pentene

d. *o*-bromoaniline
e. 4-bromo-1-chloro-2-methylbenzene
f. *m*-chlorostyrene

g. *o*-nitroanisole
h. 2,4-dichloromethylbenzene
i. *m*-chlorobenzoic acid

46. For each of the statements in Column I, choose a substituent from Column II that fits the description for the compound on the right:

Column I	Column II	
a. Z donates electrons by hyperconjugation and does not donate or withdraw electrons by resonance.	OH	
b. Z withdraws electrons inductively and withdraws electrons by resonance.	Br	
c. Z deactivates the ring and directs ortho–para.	$^+NH_3$	
d. Z withdraws electrons inductively, donates electrons by resonance, and activates the ring.	CH_2CH_3	
e. Z withdraws electrons inductively and does not donate or withdraw electrons by resonance.	NO_2	

47. What product is obtained from the reaction of excess benzene with each of the following reagents?

a. isobutyl chloride + $AlCl_3$
b. 1-chloro-2,2-dimethylpropane + $AlCl_3$
c. dichloromethane + $AlCl_3$

48. Draw the product(s) of each of the following reactions:

a. benzoic acid + HNO_3/H_2SO_4
b. isopropylbenzene + Cl_2 + $FeCl_3$
c. *p*-xylene + acetyl chloride + $AlCl_3$ followed by H_2O
d. *o*-methylaniline + benzenediazonium chloride

e. cyclohexyl phenyl ether + Br_2
f. phenol + H_2SO_4 + Δ
g. ethylbenzene + $Br_2/FeBr_3$
h. *m*-xylene + $Na_2Cr_2O_7$ + HCl + Δ

49. Rank the following substituted anilines from most basic to least basic:

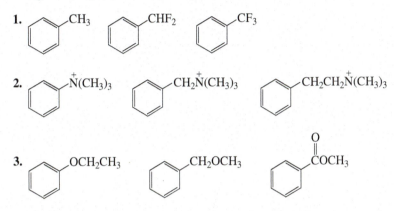

50. For each horizontal row of substituted benzenes, indicate

  a. the one that is the most reactive in an electrophilic aromatic substitution reaction.
  b. the one that is the least reactive in an electrophilic aromatic substitution reaction.
  c. the one that yields the highest percentage of meta product in an electrophilic aromatic substitution reaction.

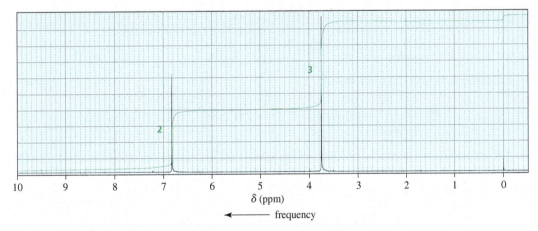

51. Draw the product of each of the following reactions:

52. Show how the following compounds can be synthesized from benzene:

  a. *m*-chlorobenzenesulfonic acid   **d.** 1-phenylpentane   **g.** *p*-cresol
  b. *m*-chloroethylbenzene   **e.** *m*-bromobenzoic acid   **h.** benzyl alcohol
  c. *m*-bromobenzonitrile   **f.** *m*-hydroxybenzoic acid   **i.** benzylamine

53. If anisole is allowed to sit in D_2O that contains a small amount of D_2SO_4, what products are formed?

54. The compound with the ^1H NMR spectrum shown below is known to be highly reactive toward electrophilic aromatic substitution. Identify the compound.

δ (ppm)

← frequency

55. Rank each group of compounds from most reactive to least reactive toward electrophilic aromatic substitution:

  a. benzene, ethylbenzene, chlorobenzene, nitrobenzene, anisole
  b. 1-chloro-2,4-dinitrobenzene, 2,4-dinitrophenol, 1-methyl-2,4-dinitrobenzene
  c. toluene, *p*-cresol, benzene, *p*-xylene
  d. benzene, benzoic acid, phenol, propylbenzene
  e. *p*-methylnitrobenzene, 2-chloro-1-methyl-4-nitrobenzene, 1-methyl-2,4-dinitrobenzene, *p*-chloromethylbenzene
  f. bromobenzene, chlorobenzene, fluorobenzene, iodobenzene

56. How many electrophilic aromatic substitution products are obtained from chlorination of

a. *o*-xylene? **b.** *p*-xylene? **c.** *m*-xylene?

57. What are the products of the following reactions?

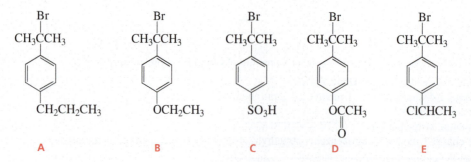

a. [phenyl acetate structure] + HNO$_3$ $\xrightarrow{\text{H}_2\text{SO}_4, \Delta}$

c. [anisole structure with OCH$_3$] + [succinic anhydride] $\xrightarrow{\text{1. AlCl}_3}{\text{2. H}_2\text{O}}$

e. [toluene structure with CH$_3$] $\xrightarrow{\begin{array}{l}\text{1. Br}_2, h\nu \\ \text{2. 2 Li/hexane} \\ \text{3. CuI/THF} \\ \text{4. ethylene oxide} \\ \text{5. HCl}\end{array}}$

b. [3-bromotoluene structure] $\xrightarrow{\begin{array}{l}\text{1. Mg/Et}_2\text{O} \\ \text{2. D}_2\text{O}\end{array}}$

d. [N-phenylpiperidine structure] + Br$_2$ $\longrightarrow$

f. [benzotrifluoride structure with CF$_3$] + Cl$_2$ $\xrightarrow{\text{FeCl}_3}$

58. For each of the following compounds, indicate the ring carbon(s) that is/are nitrated when the compound is treated with HNO$_3$/H$_2$SO$_4$:

a. [3-nitrobenzoic acid: NO$_2$ and COOH]

b. [structure with CH$_3$–O–C(=O)–phenyl–O–C(=O)–CH$_3$]

c. [4-chloroanisole: OCH$_3$ and Cl]

d. [4-hydroxybenzoic acid: OH and COOH]

e. [2-bromoacetophenone: C(=O)CH$_3$ and Br]

f. [3-methylphenol: OH and CH$_3$]

g. [structure with CH$_3$ and NO$_2$]

h. [1,3-dichloro with Cl: Cl and Cl]

59. Show two pairs of reagents that would each result in formation of the following compound:

[4-methylbenzophenone structure with C=O connecting phenyl and tolyl with CH$_3$]

60. Why is anisole nitrated more rapidly than thioanisole under the same conditions?

[anisole structure: benzene–OCH$_3$]   anisole

[thioanisole structure: benzene–SCH$_3$]   thioanisole

61. Describe two ways to prepare anisole from benzene.

62. Which of the following compounds reacts with HBr more rapidly?

CH$_3$—[benzene]—CH=CH$_2$ or   CH$_3$O—[benzene]—CH=CH$_2$

63. The following tertiary alkyl bromides undergo an S$_N$1 reaction in aqueous acetone to form the corresponding tertiary alcohols. Rank the alkyl bromides from most reactive to least reactive.

Br	Br	Br	Br	Br
CH$_3$CCH$_3$	CH$_3$CCH$_3$	CH$_3$CCH$_3$	CH$_3$CCH$_3$	CH$_3$CCH$_3$
[benzene]	[benzene]	[benzene]	[benzene]	[benzene]
CH$_2$CH$_2$CH$_3$	OCH$_2$CH$_3$	SO$_3$H	OCCH$_3$ (C=O)	ClCHCH$_3$
A	**B**	**C**	**D**	**E**

64. An aromatic hydrocarbon with a molecular formula of $C_{13}H_{20}$ has an ^1H NMR spectrum with a signal at ~7 ppm that integrates to 5H. It also has two singlets; one of the singlets has 1.5 times the area of the second. What is the structure of the aromatic hydrocarbon?

65. Show how the following compounds can be synthesized from benzene:

a. *N,N,N*-trimethylanilinium iodide
b. 2-methyl-4-nitrophenol
c. *p*-benzylchlorobenzene
d. benzyl methyl ether
e. *p*-nitroaniline
f. *m*-bromoiodobenzene
g. *p*-dideuteriobenzene
h. *p*-nitro-*N*-methylaniline
i. 1-bromo-3-nitrobenzene

66. Use the four compounds shown below to answer the following questions:

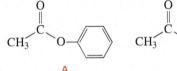 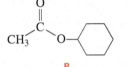 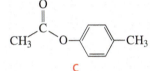 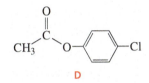

a. Why are the *ortho*-halo-substituted benzoic acids stronger acids than benzoic acid?
b. Why is *o*-fluorobenzoic acid the weakest of the *ortho*-halo-substituted benzoic acids?
c. Why do *o*-chlorobenzoic acid and *o*-bromobenzoic acid have similar pK_a values?

67. a. Rank the following esters from most reactive to least reactive in the first slow step of a nucleophilic acyl substitution reaction (formation of the tetrahedral intermediate):

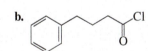

b. Rank the same esters from most reactive to least reactive in the second slow step of a nucleophilic acyl substitution reaction (collapse of the tetrahedral intermediate).

68. A mixture of 0.10 mol benzene and 0.10 mol *p*-xylene was allowed to react with 0.10 mol nitronium ion until all the nitronium ion was gone. Two products were obtained: 0.002 mol of one and 0.098 mol of the other.

a. What was the major product?
b. Why was more of one product obtained than of the other?

69. Does *m*-xylene or *p*-xylene react more rapidly with $Cl_2 + FeCl_3$? Explain your answer.

70. What are the products of the following reactions?

71. Benzene underwent a Friedel–Crafts acylation followed by a Wolff–Kishner reduction. The product gave the following ^1H NMR spectrum. What acyl chloride was used in the Friedel–Crafts acylation?

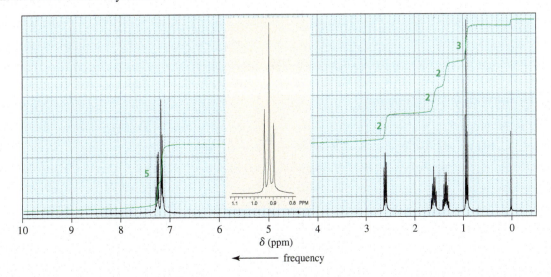

72. What products are obtained from the reaction of the following compounds with $H_2CrO_4 + \Delta$?

a. 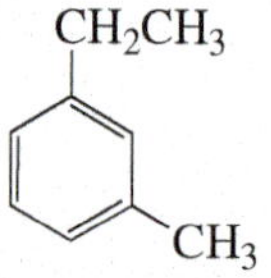
CH₂CH₃ ... CH₃

b. 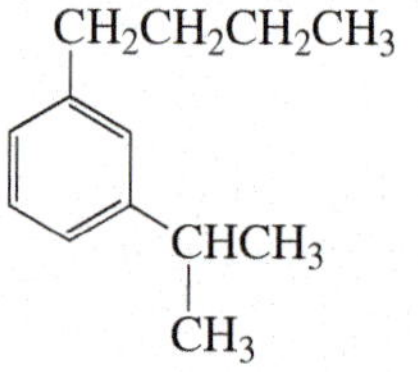
CH₂CH₂CH₂CH₃ ... CHCH₃ / CH₃

c. 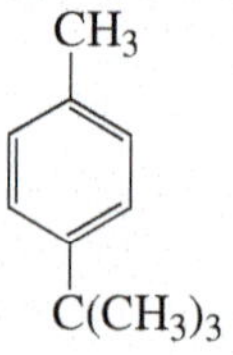
CH₃ ... C(CH₃)₃

73. Which set of underlined hydrogens has its ^1H NMR signal at a higher frequency?

a. $CH_3CH_2C\underline{H}_3$ or $CH_3OCH_2C\underline{H}_3$ b. $CH_3CH=C\underline{H}_2$ or $CH_3OCH=C\underline{H}_2$

74. Friedel–Crafts alkylations can be carried out with carbocations formed from reactions other than the reaction of an alkyl halide with $AlCl_3$. Propose a mechanism for the following reaction:

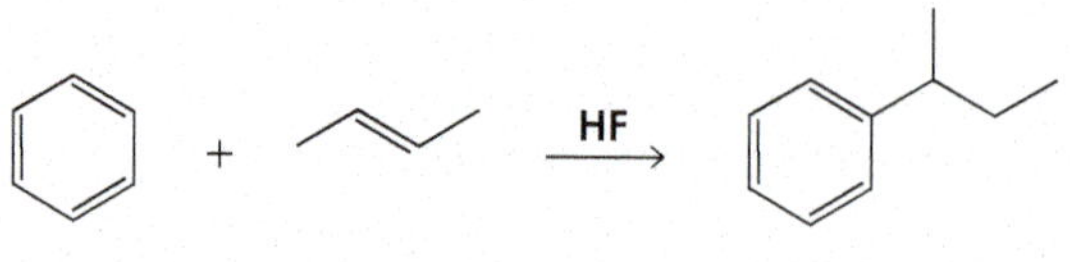

75. Show how the following compounds can be prepared from the given starting materials. You can use any necessary organic or inorganic reagents.

a.

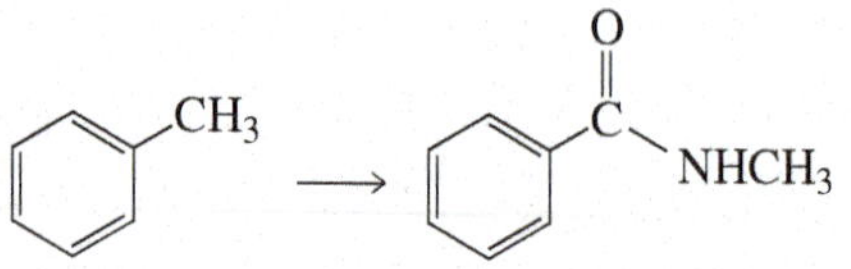

b.

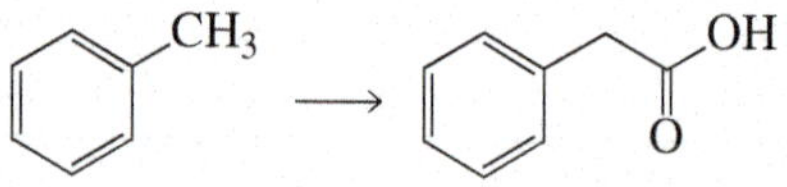

76. Rank the following compounds from largest K_{eq} to smallest K_{eq} for hydrate formation:

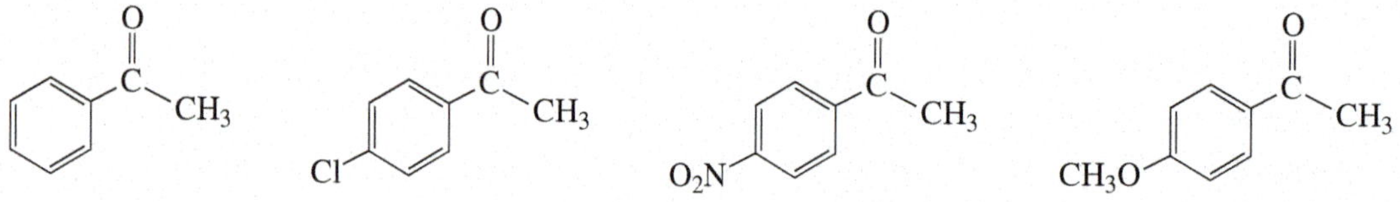

77. A chemist isolated an aromatic compound with molecular formula $C_6H_4Br_2$. He treated this compound with nitric acid and sulfuric acid and isolated three different isomers, in different amounts, with molecular formula $C_6H_3Br_2NO_2$. What was the structure of the original compound?

78. a. Describe three ways the following reaction can be carried out:

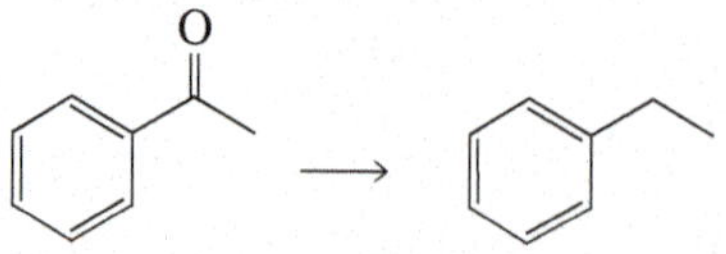

b. Describe two ways the following reaction can be carried out:

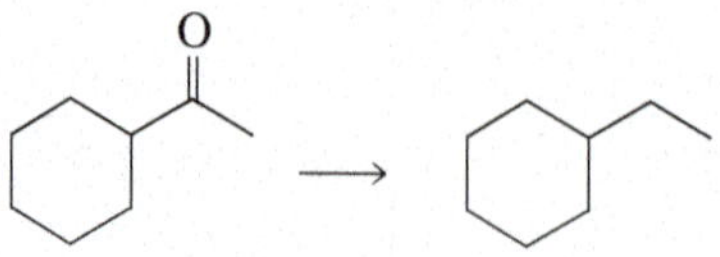

79. Propose a mechanism for each of the following reactions:

a.

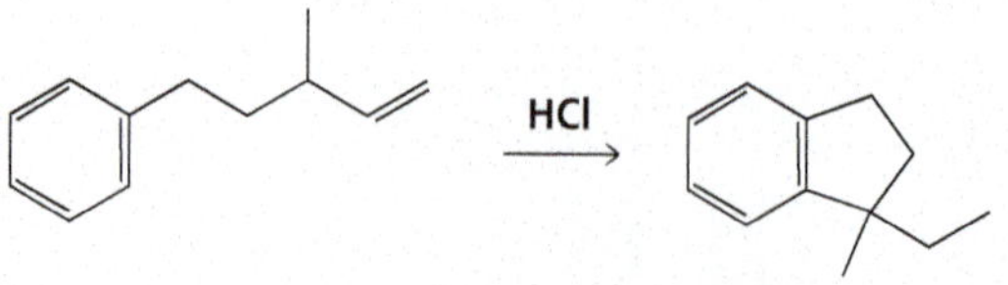

b.

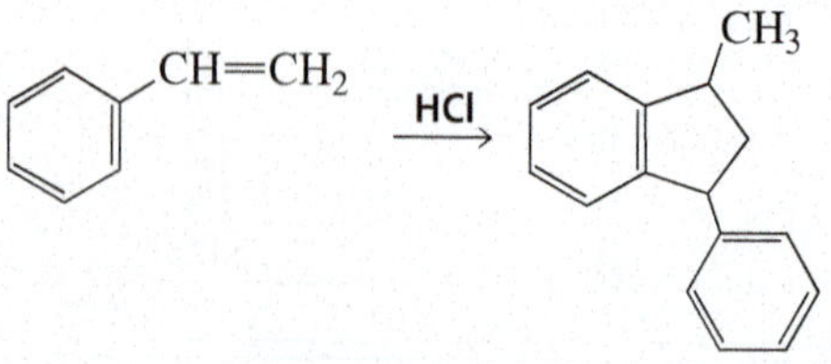

80. How can you prepare the following compounds with benzene as one of the starting materials?

a. 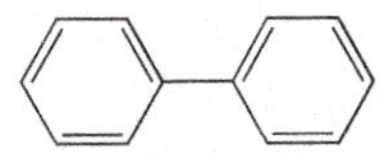 b. 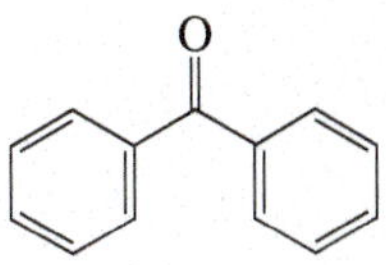 c. 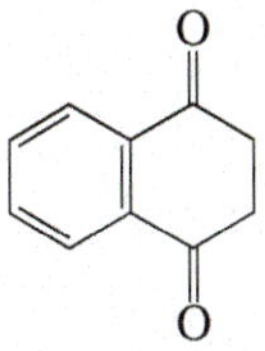 d.

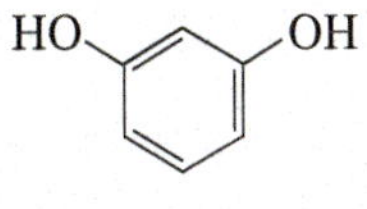

81. Describe how naphthalene can be prepared from the given starting material.

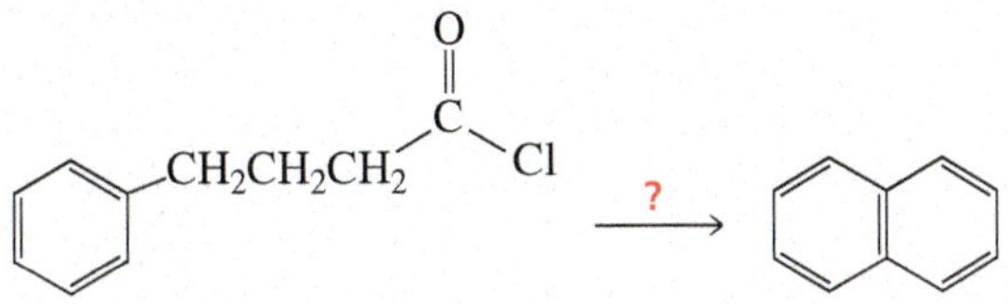

82. Using resonance contributors for the carbocation intermediate, explain why a phenyl group is an ortho–para director.

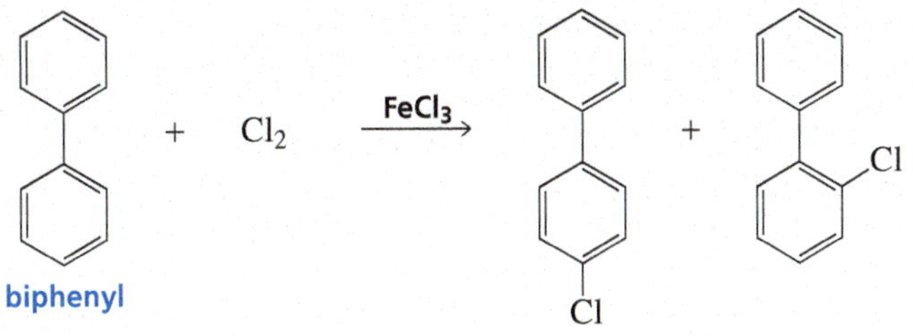

$$\text{biphenyl} + Cl_2 \xrightarrow{FeCl_3} + Cl$$

biphenyl

83. Describe two synthetic routes for the preparation of *p*-methoxyaniline from benzene.

84. What reagents are required to carry out the following synthesis?

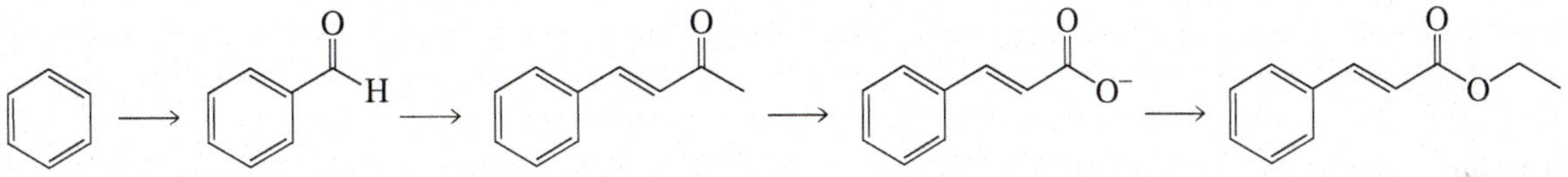

85. Identify **A–J**:

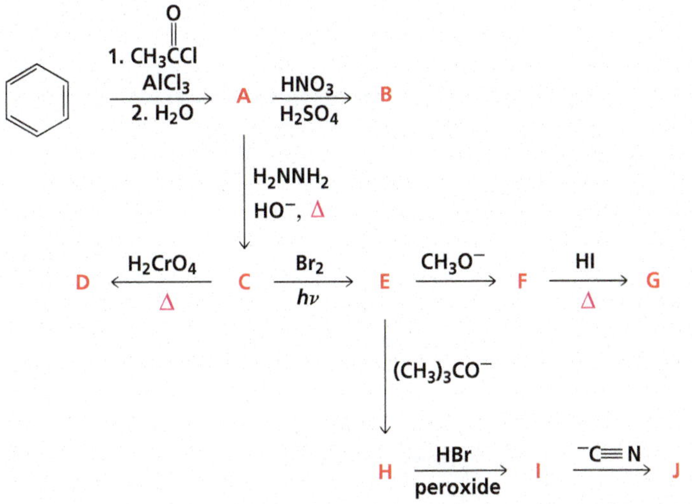

1. CH₃CCl
   AlCl₃ → **A** $\xrightarrow[H_2SO_4]{HNO_3}$ **B**
2. H₂O

H₂NNH₂
HO⁻, Δ

D $\xleftarrow[\Delta]{H_2CrO_4}$ **C** $\xrightarrow[h\nu]{Br_2}$ **E** $\xrightarrow{CH_3O^-}$ **F** $\xrightarrow[\Delta]{HI}$ **G**

(CH₃)₃CO⁻

H $\xrightarrow[\text{peroxide}]{HBr}$ **I** $\xrightarrow{^-C\equiv N}$ **J**

86. Which is a more stable intermediate in each pair?

a. 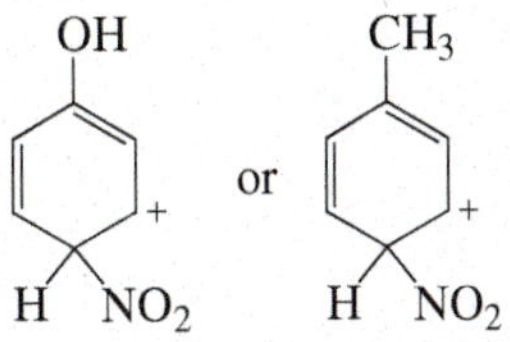 or

b. 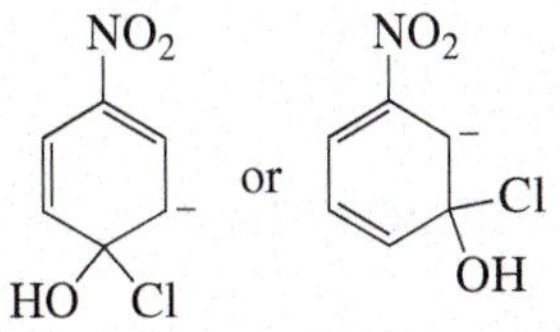 or

919

87. The pK_a values of a few ortho-, meta-, and para-substituted benzoic acids are shown below:

COOH
Cl
$pK_a = 2.94$

COOH
Cl
$pK_a = 3.83$

COOH
Cl
$pK_a = 3.99$

COOH
NO_2
$pK_a = 2.17$

COOH
NO_2
$pK_a = 3.49$

COOH
NO_2
$pK_a = 3.44$

COOH
NH_2
$pK_a = 4.95$

COOH
NH_2
$pK_a = 4.73$

COOH
NH_2
$pK_a = 4.89$

The relative pK_a values depend on the substituent. For chloro-substituted benzoic acids, the ortho isomer is the most acidic and the para isomer is the least acidic; for nitro-substituted benzoic acids, the ortho isomer is the most acidic and the meta isomer is the least acidic; and for amino-substituted benzoic acids, the meta isomer is the most acidic and the ortho isomer is the least acidic. Explain these relative acidities.

a. Cl: ortho > meta > para   **b.** NO_2: ortho > para > meta   **c.** NH_2: meta > para > ortho

88. Propose a mechanism for each of the following reactions:

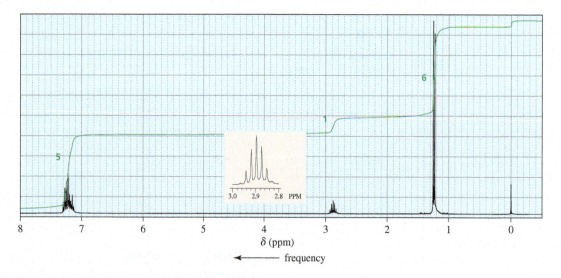

89. *p*-Fluoronitrobenzene is more reactive toward hydroxide ion than is *p*-chloronitrobenzene. What does this tell you about the rate-determining step for nucleophilic aromatic substitution?

90. When heated with chromic acid, compound **A** forms benzoic acid. Identify compound **A** from its ^{1}H NMR spectrum.

91. Show how the following compounds can be prepared from benzene:

a.
CH$_3$
CH$_3$C=CH$_2$

b.
OCH$_3$
Br
NO$_2$

c.
O
CH$_3$—C
SO$_3$H
CH$_2$CH(CH$_3$)$_2$

d.
C(CH$_3$)$_3$
CH$_3$CH$_2$
Br

92. How can you distinguish the following compounds using:

   a. their infrared spectra?   **b.** their ^{1}H NMR spectra?

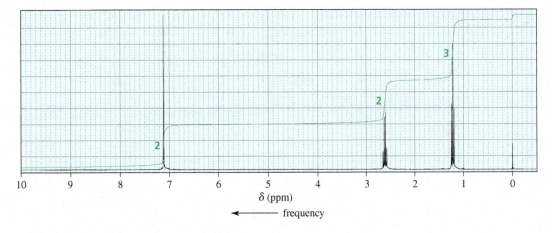

A   B   C   D   E   F   G

93. Describe how mescaline can be synthesized from benzene. The structure of mescaline is given on page 869.

94. Propose a mechanism for the following reaction that explains why the configuration of the asymmetric center in the reactant is retained in the product:

95. Propose a mechanism for each of the following reactions:

a.

b.

96. Describe how 3-methyl-1-phenyl-3-pentanol can be prepared from benzene. You can use any inorganic reagents and solvents, and any organic reagents provided they contain no more than two carbons.

97. An unknown compound reacts with ethyl chloride and aluminum trichloride to form a compound that has the following ^{1}H NMR spectrum. What is the structure of the compound?

δ (ppm)

← frequency

98. a. Explain why the following reaction leads to the products shown:

$$CH_3CHCH_2NH_2 \xrightarrow[\text{HCl}]{\text{NaNO}_2} CH_3CCH_3 + CH_3C{=}CH_2$$

b. What product is obtained from the following reaction?

$$CH_3-\overset{OH}{\underset{CH_3}{C}}-\overset{NH_2}{\underset{CH_3}{C}}-CH_3 \xrightarrow[\text{HCl}]{\text{NaNO}_2}$$

99. Explain why hydroxide ion catalyzes the reaction of piperidine with 2,4-dinitroanisole but has no effect on the reaction of piperidine with 1-chloro-2,4-dinitrobenzene.

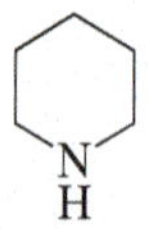

piperidine

100. What is the product of the following reaction?

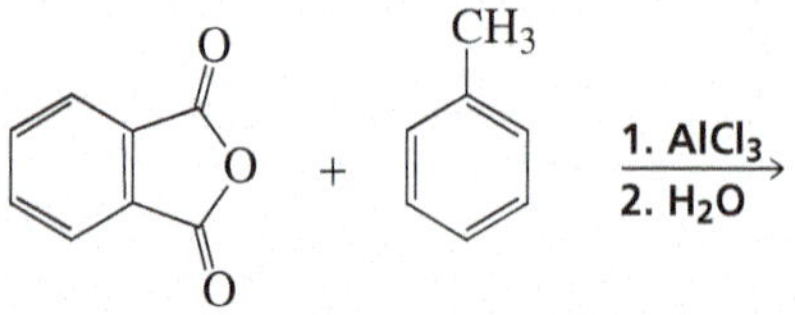

1. AlCl₃
2. H₂O

101. Tyramine is an alkaloid found in mistletoe and ripe cheese. Dopamine is a neurotransmitter involved in the regulation of the central nervous system.

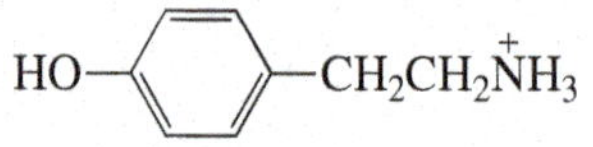

tyramine 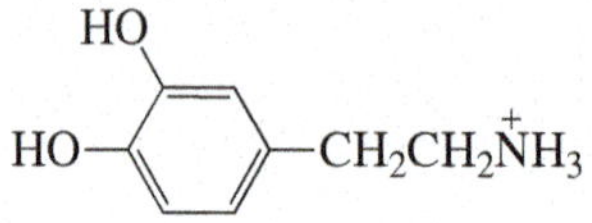   dopamine

a. How can tyramine be prepared from β-phenylethylamine?
b. How can dopamine be prepared from tyramine?
c. Give two ways to prepare β-phenylethylamine from β-phenylethyl chloride.
d. How can β-phenylethylamine be prepared from benzyl chloride?
e. How can β-phenylethylamine be prepared from benzaldehyde?

102. a. How can aspirin be synthesized from benzene?
b. Ibuprofen is the active ingredient in pain relievers such as Advil, Motrin, and Nuprin. How can ibuprofen be synthesized from benzene?
c. Acetaminophen is the active ingredient in Tylenol. How can acetominophen be synthesized from benzene?

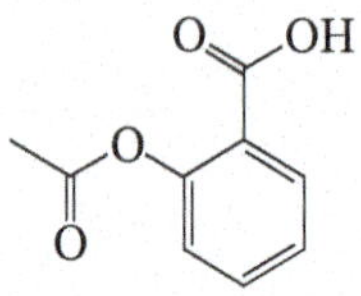

aspirin 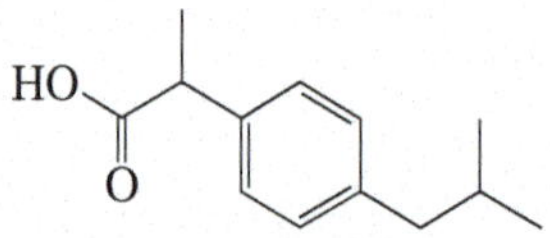   ibuprofen 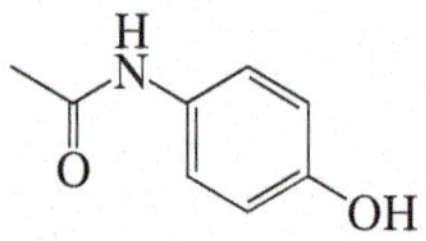   acetaminophen

103. a. Ketoprofen, like ibuprofen, is an anti-inflammatory analgesic. How can ketoprofen be synthesized from the given starting material?

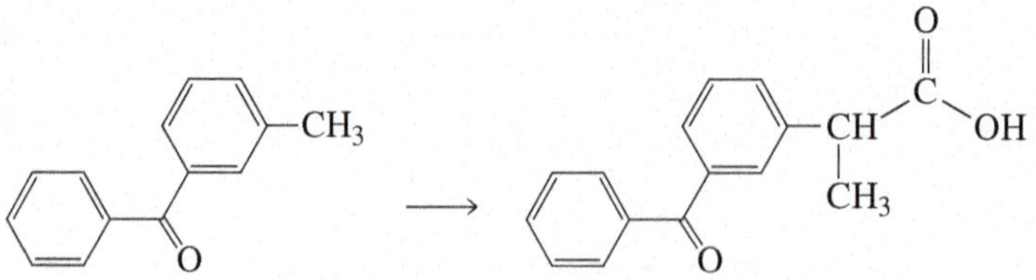

ketoprofen

b. Ketoprofen and ibuprofen both have a propanoic acid substituent (see Problem 102). Explain why the identical subunits are synthesized in different ways.

104. Show how Novocain, a painkiller used frequently by dentists, can be prepared from benzene and compounds containing no more than four carbons.

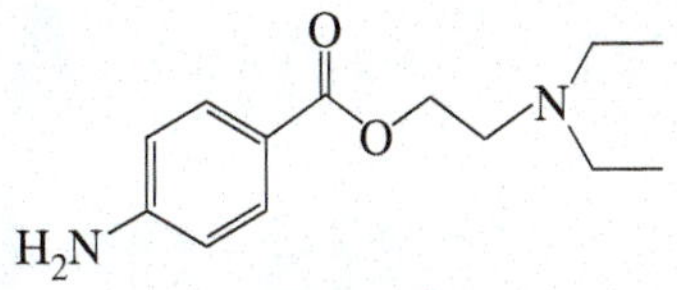

Novocain®

105. Show how lidocaine, one of the most widely used injectable anesthetics, can be prepared from benzene and compounds containing no more than four carbons.

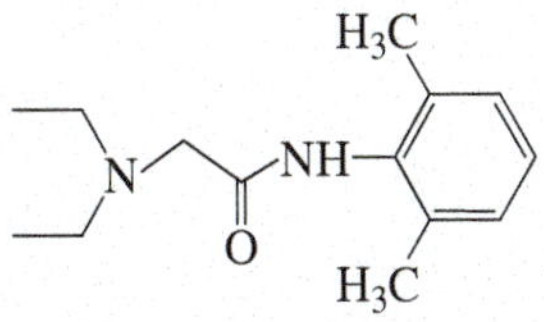

lidocaine

106. Saccharin, an artificial sweetener, is about 300 times sweeter than sucrose. Describe how saccharin can be prepared from benzene.

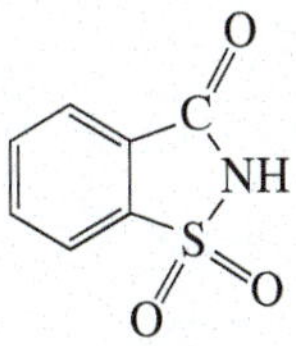

saccharin

19 More About Amines • Reactions of Heterocyclic Compounds

Auletta

In 1994, a team from UC Santa Cruz reported the structures of milnamides A and B, compounds the team had obtained from the sponge Auletta. Subsequently, other researchers in various countries isolated these compounds from three other coral reef sponges. Collaborative research between scientists at the University of British Columbia and Wyeth Pharmaceuticals led to an explanation of why breaking the saturated heterocyclic ring of milnamide A resulted in a compound (milnamide B) with greater potency against solid tumor cancer cells. Using milnamide B as a lead compound, chemists at Eisai Inc. designed and synthesized E7974 (see page 491). This compound shows broad activity against a variety of human tumors and exhibits better efficacy than the classic anticancer drug, 5-fluorouracil (see page 1091). It is now being evaluated as an anticancer drug.

milnamide A   milnamide B   E7974

Amines, compounds in which one or more of the hydrogens of ammonia (NH_3) have been replaced by an alkyl group, are among some of the most abundant compounds in the biological world. You will come to appreciate their biological importance when you

- explore the structures and properties of amino acids and proteins in Chapter 21.
- learn how enzymes catalyze reactions in Chapter 22.

- investigate the ways in which coenzymes—compounds derived from vitamins—help enzymes catalyze reactions in Chapter 23.

- learn about nucleic acids (DNA and RNA) in Chapter 26.

Amines are exceedingly important compounds to organic chemists—far too important to leave until the end of a course in organic chemistry. You have, therefore, already studied many aspects of amines and their chemistry. For example, you have looked at the physical properties of amines—their hydrogen-bonding properties, boiling points, and solubilities (Section 3.9) — and you have learned how amines are named (Section 3.7). You have also seen that:

- the nitrogen in amines is sp^3 hybridized with the lone pair residing in an sp^3 orbital (Section 3.8).

- amines invert rapidly at room temperature through a transition state in which the sp^3 nitrogen becomes an sp^2 nitrogen (Section 4.16).

- the lone-pair electrons of the nitrogen atom cause amines to react as bases (that is, to share their lone pair with a proton) and as nucleophiles (that is, to share their lone pair with an atom other than a proton).

$$\boxed{\text{an amine is a base}}$$

$$R-\overset{..}{N}H_2 \;+\; H-Br \;\longrightarrow\; R-\overset{+}{N}H_3 \;+\; Br^-$$

$$\boxed{\text{an amine is a nucleophile}}$$

$$R-\overset{..}{N}H_2 \;+\; CH_3-Br \;\longrightarrow\; R-\overset{+}{N}H_2-CH_3 \;+\; Br^-$$

In this chapter, we will revisit some of these topics and look at other aspects of amines and their chemistry that we have not considered previously.

Some amines are **heterocyclic compounds** (or **heterocycles**)—cyclic compounds in which one or more of the atoms of the ring are **heteroatoms** (Section 8.20). A variety of atoms, such as N, O, S, Se, P, Si, B, and As, can be incorporated into ring structures.

Heterocycles are an extraordinarily important class of compounds, making up more than half of all known organic compounds. Most drugs, most vitamins, and many natural products are heterocycles. In this chapter, we will consider the most prevalent heterocyclic compounds—the ones containing the heteroatom N, O, or S.

19.1   MORE ABOUT NOMENCLATURE

In Section 3.7, we saw that amines are classified as primary, secondary, or tertiary, depending on whether one, two, or three hydrogens of ammonia have been replaced by an alkyl group. We also saw that amines have both common and systematic names. Common names are obtained by citing the names of the alkyl substituents attached to the nitrogen (in alphabetical order) followed by "amine." Systematic names employ "amine" as a functional group suffix.

a primary amine	**a secondary amine**	**a tertiary amine**
systematic name:	1-pentanamine	*N*-ethyl-1-butanamine	*N*-ethyl-*N*-methyl-1-propanamine
common name:	pentylamine	butylethylamine	ethylmethylpropylamine

Naming Nitrogen-Containing Saturated Heterocycles

A saturated cyclic amine—one without any double bonds—can be named as a cycloalkane, using the prefix *aza* to denote the nitrogen atom. There are, however, other acceptable names. Some of the more commonly used names are shown on the top of the next page. Notice that heterocyclic rings are numbered to give the heteroatom the lowest possible number.

azacyclopropane	azacyclobutane	azacyclopentane	azacyclohexane	2-methylazacyclohexane	N-ethylazacyclopentane
aziridine	azetidine	pyrrolidine	piperidine	2-methylpiperidine	N-ethylpyrrolidine

Naming Oxygen- and Sulfur-Containing Saturated Heterocycles

Saturated heterocycles with oxygen and sulfur heteroatoms are named similarly. The prefix for oxygen is *oxa* and the prefix for sulfur is *thia*.

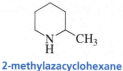

oxacyclopropane	thiacyclopropane	oxacyclobutane	oxacyclopentane	oxacyclohexane	1,4-dioxacyclohexane
oxirane	thiirane	oxetane			
ethylene oxide			tetrahydrofuran	tetrahydropyran	1,4-dioxane

PROBLEM 1 ◆

Name the following:

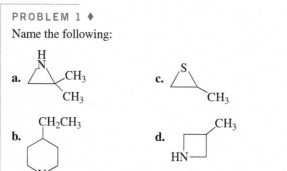

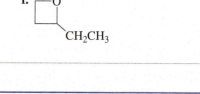

19.2 MORE ABOUT THE ACID–BASE PROPERTIES OF AMINES

Amines are the most common organic bases. We saw that ammonium ions have pK_a values of about 10 (Section 2.3) and anilinium ions have pK_a values of about 5 (Sections 8.9 and 8.10). We also saw that the greater acidity of anilinium ions compared with ammonium ions is due to the greater stability of the conjugate bases of anilinium ions as a result of electron delocalization and inductive electron withdrawal. Amines have very high pK_a values. For example, the pK_a of methylamine is 40.

$$CH_3CH_2CH_2\overset{+}{N}H_3 \qquad \underset{}{\bigcirc}-\overset{+}{N}H_3 \qquad CH_3NH_2$$

an ammonium ion	an anilinium ion	an amine
$pK_a = 10.8$	$pK_a = 4.58$	$pK_a = 40$

Saturated amine heterocycles containing five or more atoms have physical and chemical properties like those of acyclic amines. For example, pyrrolidine, piperidine, and morpholine are typical secondary amines, and N-methylpyrrolidine and quinuclidine are typical tertiary amines. The conjugate acids of these amines have the pK_a values expected for ammonium ions.

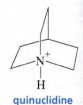

the ammonium ions of:

pyrrolidine	piperidine	morpholine	N-methylpyrrolidine	quinuclidine
$pK_a = 11.27$	$pK_a = 11.12$	$pK_a = 9.28$	$pK_a = 10.32$	$pK_a = 11.38$

Atropine

Atropine is a naturally occurring saturated heterocyclic compound found in Jimson weed and nightshade (*Atropa belladonna*). The *R* isomer is the one nature synthesizes, but it racemizes during isolation. Atropine has a variety of medicinal usages, thus landing it on the list of drugs the World Health Organization deems needed for basic health care.

atropine

Atropine blocks acetylcholine receptor sites, lowering the activity of all muscles regulated by the parasympathetic nervous system. It is used to treat low heart rates, spasms from the gastrointestinal tract, and tremors associated with Parkinson's disease. It also reduces the secretions of many organs, it is used as an antidote to organophosphate poisoning (see page 727), and it dilates pupils. The Romans used atropine in combination with opium as an anesthetic. During the Renaissance, women enhanced their appearance by using the juice from the berries of nightshade to dilate their pupils. Cleopatra was reported to do this as well. Note that *belladonna* is Italian for "beautiful woman."

Atropa belladonna

PROBLEM 2 SOLVED

LEARN THE STRATEGY

Explain the difference in the pK_a values of the piperidinium and aziridinium ions:

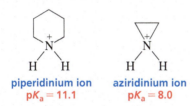

piperidinium ion
$pK_a = 11.1$

aziridinium ion
$pK_a = 8.0$

SOLUTION The piperidinium ion has a pK_a value expected for an ammonium ion, whereas the pK_a value of the aziridinium ion is considerably lower. The internal bond angle of the three-membered ring is smaller than usual, which causes the external bond angles to be larger than usual. Because of the larger bond angles, the orbital used by nitrogen to overlap the orbital of hydrogen has more *s* character than does a typical sp^3 nitrogen (Section 1.11). This makes nitrogen in the aziridinium ion more electronegative, which lowers the pK_a (Section 2.7).

PROBLEM 3 ♦

USE THE STRATEGY

Why is the conjugate acid of morpholine more acidic than the conjugate acid of piperidine?

PROBLEM 4 ♦

a. Draw the structure of 3-quinuclidinone.
b. What is the approximate pK_a of its conjugate acid?
c. Which has a lower pK_a value, the conjugate acid of 3-bromoquinuclidine or the conjugate acid of 3-chloroquinuclidine?

19.3 AMINES REACT AS BASES AND AS NUCLEOPHILES

Amines Are Bases

We saw that the leaving group of an amine ($^-NH_2$) is such a strong base that amines cannot undergo the substitution and elimination reactions that alkyl halides, alcohols, and ethers do (Section 10.9). The relative reactivities of these compounds in substitution and/or elimination reactions—each with an electron-withdrawing group bonded to an sp^3 carbon—can be appreciated by comparing the pK_a

values of the conjugate acids of their leaving groups, keeping in mind that the weaker the acid, the stronger its conjugate base. And, when comparing bases of the same type, the stronger the base, the poorer it is as a leaving group (Section 9.2).

relative reactivities

We saw, however, that amines react as bases in proton transfer reactions and in elimination reactions of alky halides (Sections 2.3 and 9.12).

Amines are Nucleophiles

The lone pair on the nitrogen of an amine makes it a nucleophile as well as a base. We saw that amines react as nucleophiles in a number of different reactions; for example, in nucleophilic substitution reactions which *alkylate* the amine (Section 9.2)

$$CH_3CH_2Br \; + \; CH_3NH_2 \; \longrightarrow \; CH_3CH_2\overset{+}{N}H_2CH_3 \; \rightleftharpoons \; CH_3CH_2NHCH_3 \; + \; HBr$$

primary amine Br^- secondary amine

and in nucleophilic acyl substitution reactions that *acylate* the amine (Sections 15.6 and 15.10).

We also saw that aldehydes and ketones react with primary amines to form imines and with secondary amines to form enamines (Section 16.8)

and amines are nucleophiles in conjugate addition reactions (Section 16.16).

We also saw that primary arylamines react with nitrous acid to form stable arenediazonium salts (Section 18.17). Arenediazonium salts are useful to synthetic chemists because the diazonium group can be replaced by a variety of nucleophiles.

PROBLEM 5

Draw the product of each of the following reactions:

a. [structure: acetophenone] $+ CH_3CH_2CH_2NH_2$ $\xrightarrow{\text{trace acid}}$

b. CH_3—C(=O)—Cl $+ 2$ [pyrrolidine N–H] $\longrightarrow$

c. [benzene ring]—NH_2 $\xrightarrow[\text{2. H}_2\text{O, Cu}_2\text{O, Cu(NO}_3)_2]{\text{1. HCl, NaNO}_2,\ 0\ °C}$

d. [structure: acetophenone] $+ CH_3CH_2NHCH_2CH_3$ $\xrightarrow{\text{trace acid}}$

19.4 SYNTHESIS OF AMINES

The many different ways to synthesize amines and the mechanisms of these reactions have been discussed in several different places in this book. These reactions are summarized and the references to where they are discussed can be found in Appendix III.

19.5 AROMATIC FIVE-MEMBERED-RING HETEROCYCLES

We will begin the discussion of aromatic heterocyclic compounds by looking at those with a five-membered ring.

Aromaticity of Pyrrole, Furan, and Thiophene

Pyrrole, furan, and **thiophene** are heterocycles with five-membered rings. We saw that these compounds are aromatic because they are cyclic and planar, every atom in the ring has a *p* orbital, and the π cloud contains *three* pairs of π electrons (Section 8.20).

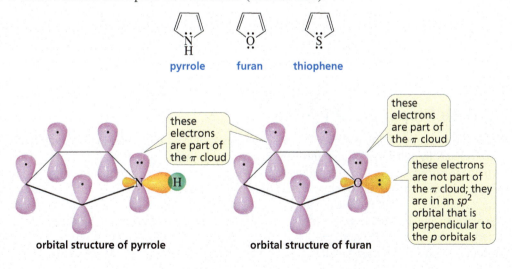

pyrrole   **furan**   **thiophene**

these electrons are part of the π cloud

these electrons are part of the π cloud

these electrons are not part of the π cloud; they are in an sp^2 orbital that is perpendicular to the *p* orbitals

orbital structure of pyrrole   **orbital structure of furan**

The resonance contributors of pyrrole and its resonance hybrid are shown below.

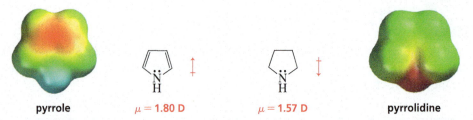

resonance contributors of pyrrole

resonance hybrid

PROBLEM 6

Draw arrows to show the movement of electrons in going from one resonance contributor of pyrrole to the next.

Pyrrole has a dipole moment of 1.80 D (Section 1.16). The saturated amine with a five-membered ring—pyrrolidine—has a slightly smaller dipole moment of 1.57 D, but as we see from the electrostatic potential maps, the two dipole moments are in opposite directions. (The red area is in the ring in pyrrole and on the nitrogen in pyrrolidine.) The dipole moment in pyrrolidine is due to inductive electron withdrawal by the nitrogen. Apparently, the ability of pyrrole's nitrogen to donate electrons into the ring by resonance more than makes up for its inductive electron withdrawal (Section 8.10).

pyrrole $\mu = 1.80\,D$ $\mu = 1.57\,D$ **pyrrolidine**

In Section 8.6, we saw that a compound's delocalization energy increases as the resonance contributors become more stable and more nearly equivalent. The delocalization energies of pyrrole, furan, and thiophene are not as great as the delocalization energies of benzene or the cyclopentadienyl anion, each a compound for which all the resonance contributors are equivalent.

relative delocalization energies of some aromatic compounds

Thiophene, with the least electronegative heteroatom, has the greatest delocalization energy of the three, and furan, with the most electronegative heteroatom, has the smallest delocalization energy. This is what we would expect, because the resonance contributors with a positive charge on the heteroatom are the most stable for the compound with the least electronegative heteroatom and the least stable for the compound with the most electronegative heteroatom. Recall that the more stable the resonance contributors, the greater the delocalization energy (Section 8.6).

Acid-Base Properties of Pyrrole

Pyrrole is an extremely weak base because the electrons shown as a lone pair in the structure are part of the π cloud. In other words, the nitrogen donates the lone-pair into the five-membered ring (as shown by the resonance contributors on the top of this page). Therefore, protonating pyrrole destroys its aromaticity. As a result, the conjugate acid of pyrrole is a very strong acid ($pK_a = -3.8$).

The resonance hybrid of pyrrole (on the top of this page) shows that there is a partial positive charge on the nitrogen and a partial negative charge on each of the carbons. As a result, pyrrole is protonated on a carbon rather than on the nitrogen. The carbon it attaches to is C-2 because a proton is an electrophile, and we will see that electrophiles attach preferentially to the C-2 position of pyrrole.

Pyrrole is unstable in strongly acidic solutions because, once protonated, it can readily polymerize.

The partial positive charge on its nitrogen atom (apparent in the structure of the resonance hybrid) causes pyrrole ($pK_a \sim 17$) to be more acidic than the analogous saturated amine ($pK_a \sim 36$). In addition, the sp^2 nitrogen in pyrrole is more electronegative than the sp^3 nitrogen in the saturated amine, contributing to pyrrole's increased acidity (Section 2.6).

Indole contains a pyrrole ring fused to a benzene ring. Indole is aromatic because it is cyclic and planar and its π cloud contains 5 pairs of π electrons. Indole is protonated at C-3, because protonation at C-2 would disturb the aromaticity of the benzene ring.

The pK_a values of several nitrogen-containing heterocycles are listed in Table 19.1.

Table 19.1   The pK_a Values of Several Nitrogen-Containing Heterocycles

$pK_a = -3.8$	$pK_a = -2.4$	$pK_a = 1.0$	$pK_a = 2.5$	$pK_a = 4.85$	$pK_a = 5.16$
$pK_a = 6.8$	$pK_a = 8.0$	$pK_a = 11.1$	$pK_a = 14.4$	$pK_a \sim 17$	$pK_a \sim 36$

PROBLEM-SOLVING STRATEGY

Determining Relative Basicity

LEARN THE STRATEGY

Rank the following compounds from least basic to most basic:

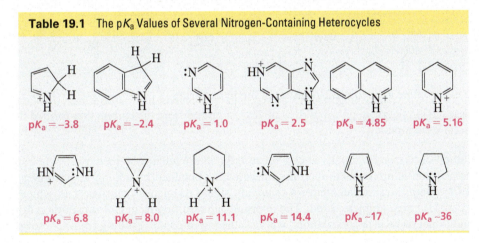

First, we need to see whether any of the compounds will lose their aromaticity if they are protonated. Such compounds will be very difficult to protonate, so will be very weak bases. Pyrrole is one such compound. Next, we need to see whether any of the compounds has a lone pair that can be delocalized.

A delocalized lone pair will be harder to protonate than a localized one. Aniline is such a compound. Finally, we need to look at the hybridization of the nitrogens. The nitrogen of the unsaturated six-membered ring is sp^2 hybridized, whereas the nitrogen of the saturated six-membered ring is sp^3 hybridized. We know that an sp^2 nitrogen is more electronegative and, therefore, harder to protonate than an sp^3 nitrogen. Thus, the order of basicity is

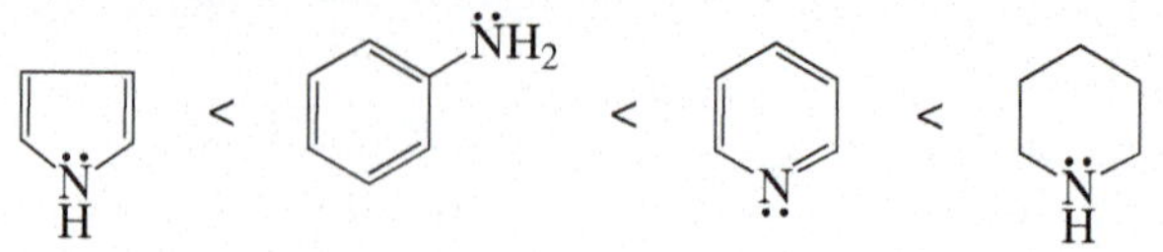

USE THE STRATEGY

PROBLEM 7 ♦

Explain why cyclopentadiene ($pK_a = 15$) is more acidic than pyrrole ($pK_a \sim 17$), even though nitrogen is more electronegative than carbon.

Electrophilic Aromatic Substitution Reactions

Because pyrrole, furan, and thiophene are aromatic, they undergo electrophilic aromatic substitution reactions.

Pyrrole, furan, and thiophene undergo electrophilic aromatic substitution, preferentially at C-2.

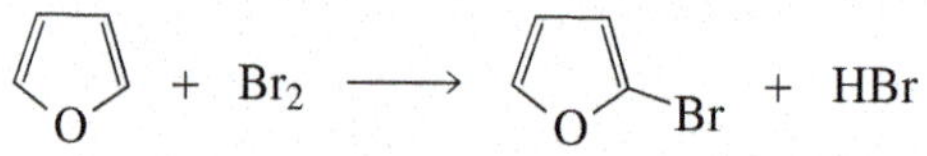

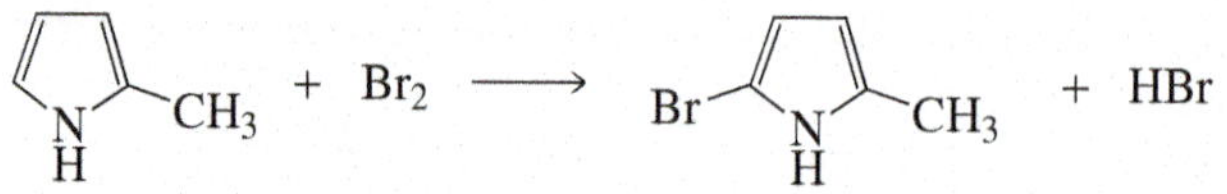

Notice that the mechanism of the reaction is the same as the mechanism for electrophilic aromatic substitution of benzene (Section 18.2).

MECHANISM FOR ELECTROPHILIC AROMATIC SUBSTITUTION

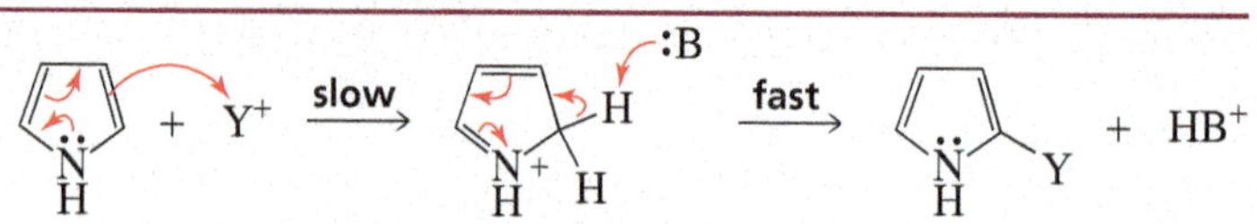

- The electrophile adds to the 2-position of the pyrrole ring.
- A base (:B) in the reaction mixture removes the proton from the carbon that formed the bond with the electrophile.

Substitution occurs preferentially at C-2 because the intermediate obtained by adding a substituent to this position is more stable than the intermediate obtained by adding a substituent to C-3 (Figure 19.1). Both intermediates have a relatively stable resonance contributor in which all the atoms (except H) have complete octets. The intermediate resulting from C-2 substitution of pyrrole has *two* additional resonance contributors, whereas the intermediate resulting from C-3 substitution has only *one* additional resonance contributor.

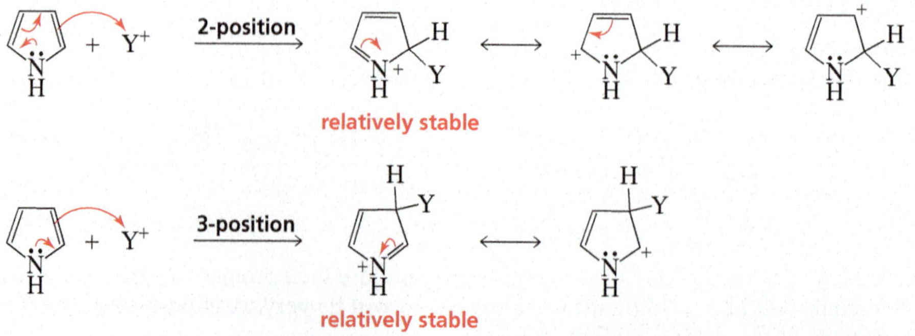

▶ **Figure 19.1**
Structures of the intermediates that would be formed from the reaction of pyrrole with an electrophile at C-2 and C-3.

If both positions adjacent to the heteroatom are occupied, electrophilic substitution will take place at C-3.

$$CH_3 \diagdown S \diagdown CH_3 + Br_2 \longrightarrow CH_3 \diagdown S \diagdown CH_3 + HBr$$

Relative Reactivity

Pyrrole, furan, and thiophene are all more reactive than benzene toward electrophilic aromatic substitution because a lone pair on the heteroatom can donate electrons into the ring by resonance. Therefore,

- they are more nucleophilic than benzene
- they are better able to stabilize the positive charge on the intermediate (Figure 19.1).

Furan is not as reactive as pyrrole in electrophilic aromatic substitution reactions. The oxygen of furan is more electronegative than the nitrogen of pyrrole, making oxygen less effective than nitrogen in increasing the nucleophicity of the ring and in stabilizing the carbocation. Thiophene is less reactive than furan because the $3p$ orbital of sulfur overlaps less effectively than the $2p$ orbital of nitrogen or oxygen with the $2p$ orbital of carbon. The electrostatic potential maps in the margin illustrate the different electron densities of the three rings.

relative reactivity toward electrophilic aromatic substitution

pyrrole > furan > thiophene > benzene

The relative reactivities of pyrrole, furan, and thiophene are reflected in the Lewis acid required to catalyze a Friedel–Crafts acylation (Section 18.6).

- Benzene requires $AlCl_3$, a relatively strong Lewis acid.

$$\text{benzene} + CH_3\overset{O}{\underset{}{C}}Cl \xrightarrow[\text{2. H}_2\text{O}]{\text{1. AlCl}_3} \text{benzene-}\overset{O}{\underset{}{C}}CH_3 + HCl$$

- Thiophene is more reactive than benzene; it can undergo a Friedel–Crafts reaction using $SnCl_4$, a weaker Lewis acid.

$$\text{thiophene} + CH_3\overset{O}{\underset{}{C}}Cl \xrightarrow[\text{2. H}_2\text{O}]{\text{1. SnCl}_4} + HCl$$

- An even weaker Lewis acid, BF_3, can be used when the reactant is furan.

$$\text{furan} + CH_3\overset{O}{\underset{}{C}}Cl \xrightarrow[\text{2. H}_2\text{O}]{\text{1. BF}_3/\text{THF}} + HCl$$

- Pyrrole is so reactive that an anhydride is used instead of a more reactive acyl chloride, and no catalyst is necessary.

$$\text{pyrrole} + CH_3\overset{O}{\underset{}{C}}O\overset{O}{\underset{}{C}}CH_3 \longrightarrow + CH_3\overset{O}{\underset{}{C}}OH$$

Pyrrole, furan, and thiophene are more reactive than benzene toward electrophilic aromatic substitution.

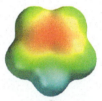

pyrrole

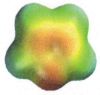

furan

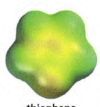

thiophene

PROBLEM 8

When pyrrole is added to a dilute solution of D_2SO_4 in D_2O, 2-deuteriopyrrole is formed. Propose a mechanism to account for the formation of this compound.

19.6 AROMATIC SIX-MEMBERED-RING HETEROCYCLES

Now we will look at heterocycles with a six-membered aromatic ring.

Aromaticity of Pyridine

As we saw, **pyridine** is an aromatic compound with a nitrogen in place of one of the carbons of benzene (Section 8.20).

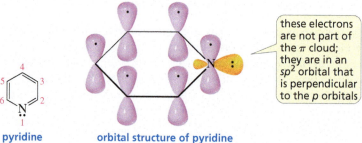

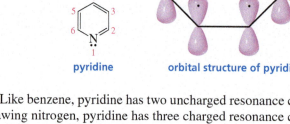

pyridine   orbital structure of pyridine

these electrons are not part of the π cloud; they are in an sp^2 orbital that is perpendicular to the p orbitals

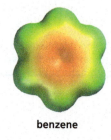

benzene

Like benzene, pyridine has two uncharged resonance contributors. Because of the electron-withdrawing nitrogen, pyridine has three charged resonance contributors that benzene does not have.

resonance contributors of pyridine

The dipole moment of pyridine is 1.57 D. As the electrostatic potential map indicates, the electron-withdrawing nitrogen is the negative end of the dipole.

pyridine

$\mu = 1.57$ D

Acid–Base Properties of Pyridine

The pyridinium ion is a stronger acid than a typical ammonium ion because its proton is attached to an sp^2 nitrogen, which is more electronegative than an sp^3 nitrogen (Section 2.6).

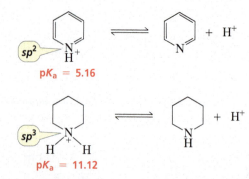

sp^2 $pK_a = 5.16$

sp^3 $pK_a = 11.12$

Pyridine Reacts as a Nucleophile

Pyridine undergoes reactions characteristic of tertiary amines. For example, pyridine undergoes an S_N2 reaction with an alkyl halide (Section 9.2).

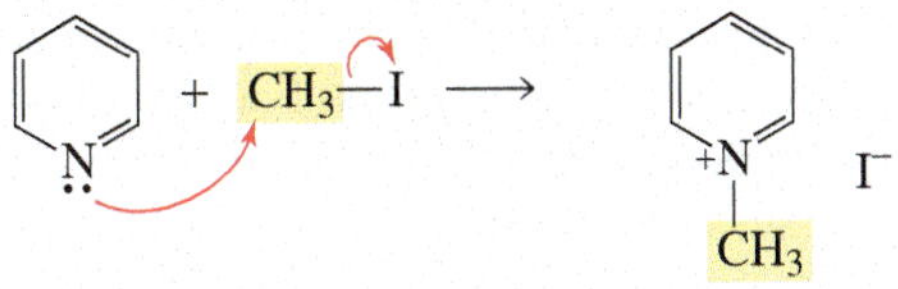

LEARN THE STRATEGY

PROBLEM 9 SOLVED

Will an amide be the final product obtained from the reaction of an acyl chloride with pyridine in an aqueous solution? Explain your answer.

SOLUTION The positively charged nitrogen in the initially formed carbonyl compound causes pyridine to be an excellent leaving group. Therefore, the compound hydrolyzes rapidly. As a result, the final product of the reaction is a carboxylic acid. (If the final pH of the solution is greater than the pK_a of the carboxylic acid and the pK_a of pyridine, both will be predominantly in their basic form.)

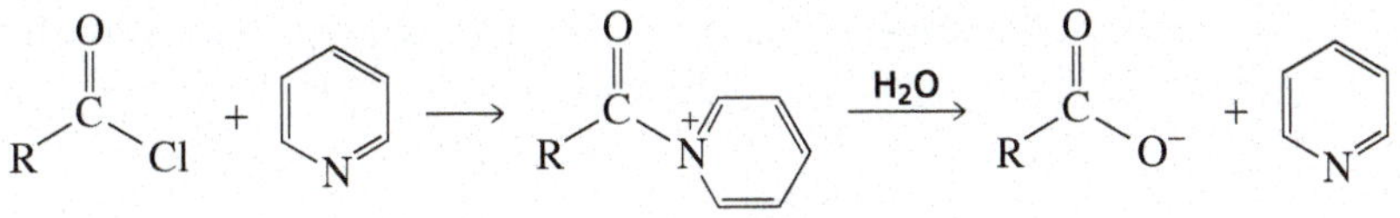

USE THE STRATEGY

PROBLEM 10 ◆

Draw the product formed when pyridine reacts with ethyl bromide.

Electrophilic Aromatic Substitution Reactions

Because it is aromatic, pyridine (like benzene) undergoes electrophilic aromatic substitution reactions and by the same mechanism.

MECHANISM FOR ELECTROPHILIC AROMATIC SUBSTITUTION

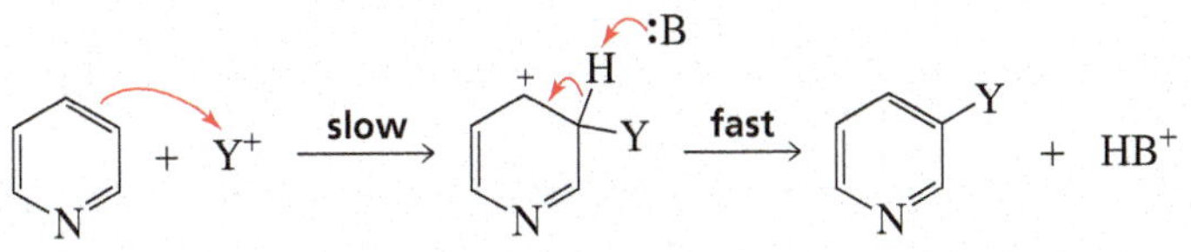

- The electrophile adds to the 3-position of the pyridine ring.
- A base (:B) in the reaction mixture removes the proton from the carbon that formed the bond with the electrophile.

Electrophilic aromatic substitution of pyridine takes place at C-3 because the most stable intermediate is obtained by placing an electrophilic substituent at that position (Figure 19.2). When the substituent is placed at C-2 or C-4, one of the resulting resonance contributors is particularly unstable because its nitrogen atom has an incomplete octet *and* a positive charge.

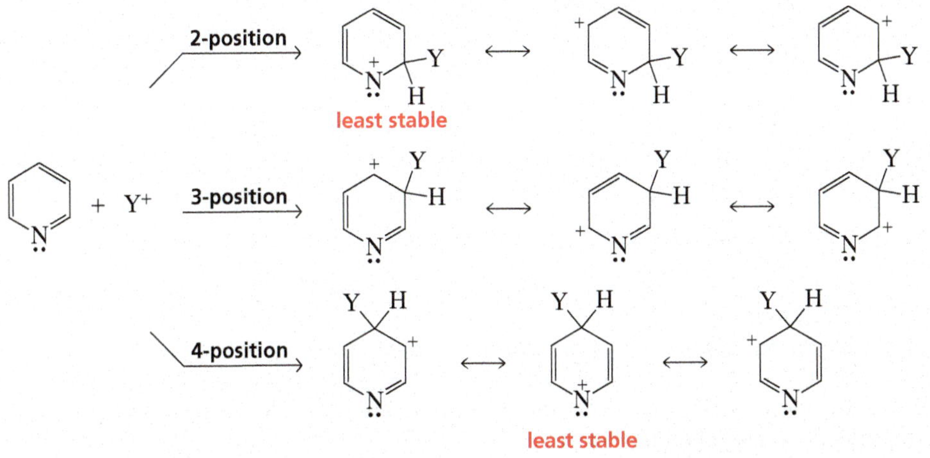

◀ **Figure 19.2**
Structures of the intermediates that would be formed from the reaction of an electrophile with pyridine.

The electron-withdrawing nitrogen atom makes pyridine less nucleophilic than benzene and the intermediate obtained from electrophilic aromatic substitution less stable than the intermediate obtained from electrophilic aromatic substitution of benzene. Pyridine, therefore, is less reactive than benzene. Indeed, it is even less reactive than nitrobenzene. (Recall from Section 18.12 that an electron-withdrawing nitro group strongly deactivates a benzene ring toward electrophilic aromatic substitution.)

relative reactivity toward electrophilic aromatic substitution

Pyridine, therefore, undergoes electrophilic aromatic substitution reactions only under vigorous conditions, and the yields of these reactions are often quite low. If the nitrogen becomes protonated under the reaction conditions, the reactivity decreases further because a positively charged nitrogen will make the pyridine ring less nucleophilic and the carbocation intermediate even less stable.

Pyridine undergoes electrophilic aromatic substitution at C-3.

We saw that highly deactivated benzene rings do not undergo Friedel–Crafts alkylation or acylation reactions (Section 18.15). Therefore, pyridine, whose reactivity is similar to a highly deactivated benzene ring, does not undergo these reactions either.

Nucleophilic Aromatic Substitution Reactions

Pyridine is *less* reactive than benzene toward electrophilic aromatic substitution and *more* reactive than benzene toward nucleophilic aromatic substitution.

Because pyridine is *less* reactive than benzene in *electrophilic* aromatic substitution reactions, it should not be surprising that pyridine is *more* reactive than benzene in *nucleophilic* aromatic substitution reactions. The electron-withdrawing nitrogen atom that destabilizes the intermediate in electrophilic aromatic substitution stabilizes the intermediate in nucleophilic aromatic substitution.

MECHANISM FOR NUCLEOPHILIC AROMATIC SUBSTITUTION

- The nucleophile adds to the ring carbon attached to the leaving group.
- The leaving group is eliminated.

Notice that, in nucleophilic aromatic substitution reactions, the ring has a leaving group that is replaced by a nucleophile.

Nucleophilic aromatic substitution of pyridine takes place at C-2 or C-4 because addition to these positions leads to the most stable intermediate. Only when addition occurs to these positions is a resonance contributor obtained that has the greatest electron density on nitrogen, the most electronegative of the ring atoms (Figure 19.3).

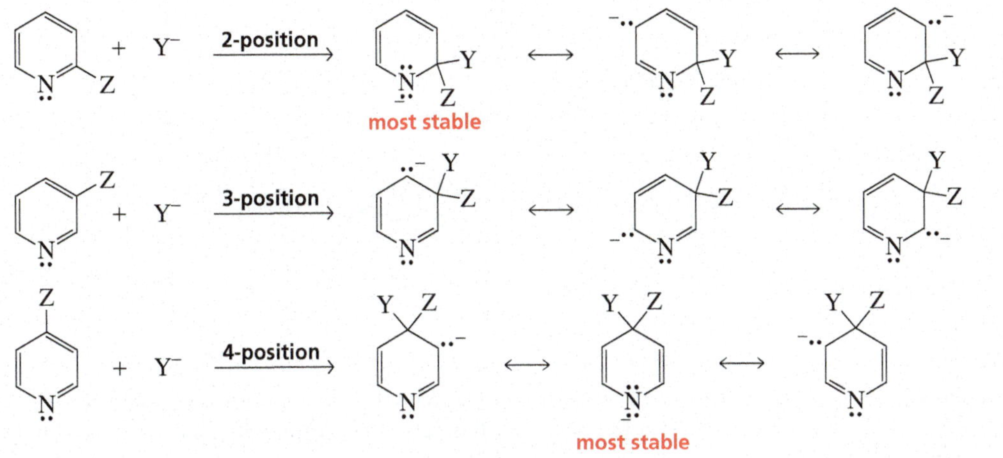

◀ **Figure 19.3**
Structures of the intermediates that would
be formed from the reaction of a nucleophile
with a substituted pyridine.

If the leaving groups at C-2 and C-4 are different, the incoming nucleophile will preferentially
substitute for the weaker base (the better leaving group).

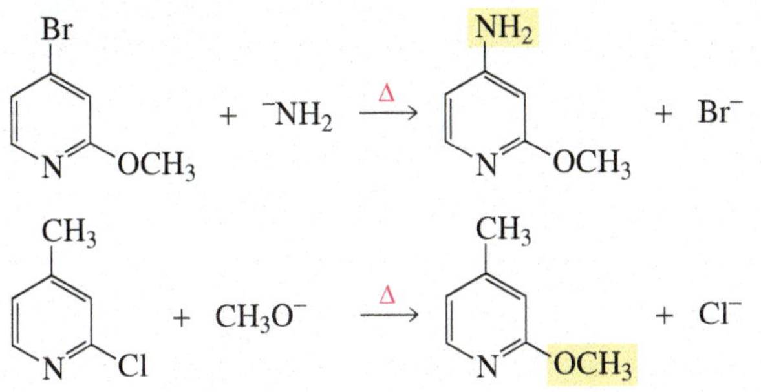

Pyridine undergoes nucleophilic
aromatic substitution at C-2 or C-4.

PROBLEM 11

How do the mechanisms of the following reactions differ?

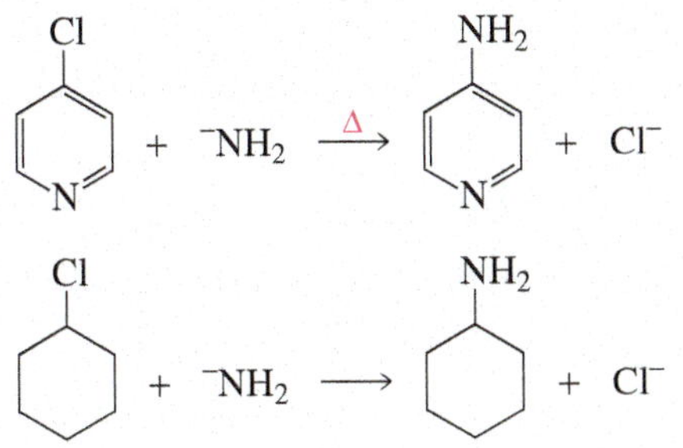

PROBLEM 12

a. Propose a mechanism for the following reaction:

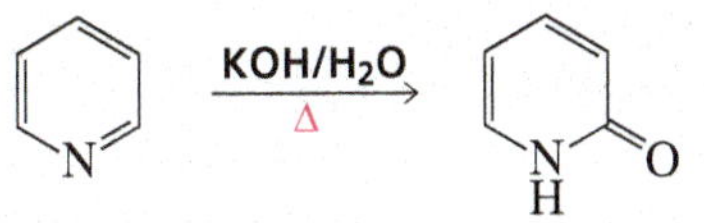

b. What other product is formed in this reaction?

Reactions of Substituted Pyridines

Substituted pyridines undergo many of the side-chain reactions that substituted benzenes undergo.
For example, alkyl-substituted pyridines can be brominated and oxidized.

When an aminopyridine is diazotized, a pyridone is formed. The diazonium salt reacts immediately with water to form a hydroxypyridine, which is in equilibrium with its keto form. (Section 18.17). (The mechanism for the conversion of a primary amino group to a diazonium group is shown in Section 18.20.)

The electron-withdrawing nitrogen and the ability to delocalize the negative charge cause the hydrogens on carbons attached to the 2- and 4-positions of the pyridine ring to have about the same acidity as the hydrogens attached to the α-carbons of ketones (Section 17.1).

Consequently, the hydrogens can be removed by base, and the resulting carbanions can react as nucleophiles.

PROBLEM 13 ♦

Rank the following compounds from easiest to hardest at removing a proton from its methyl substituent:

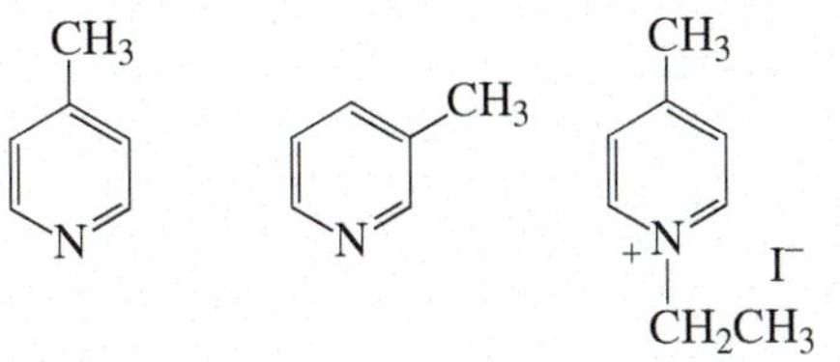

19.7 SOME HETEROCYCLIC AMINES HAVE IMPORTANT ROLES IN NATURE

Proteins are naturally occurring polymers of α-amino acids (Chapter 21). Three common amino acids found in proteins contain heterocyclic rings:

- proline has a pyrrolidine ring (Section 19.2).
- tryptophan has an indole ring (Section 19.5).
- histidine has an imidazole ring.

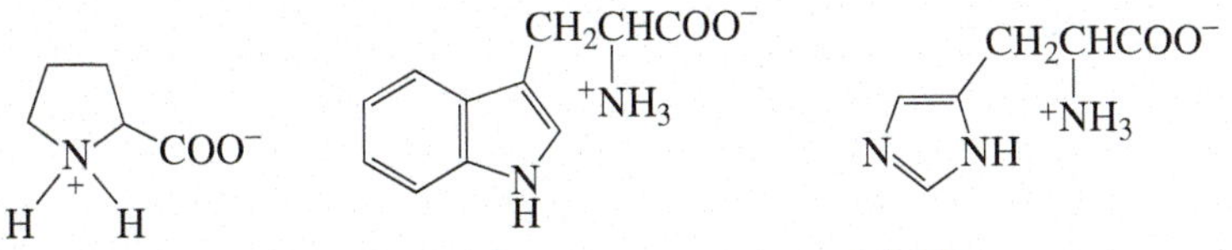

proline   tryptophan   histidine

Imidazole

Imidazole, the heterocyclic ring of histidine, is an aromatic compound because it is cyclic and planar, every atom in the ring has a p orbital, and the π cloud contains *three* pairs of π electrons (Section 8.17). The lone-pair electrons on N-1 are part of the π cloud because they are in a p orbital, whereas the lone-pair electrons on N-3 are not part of the π cloud because they are in an sp^2 orbital that is perpendicular to the p orbitals.

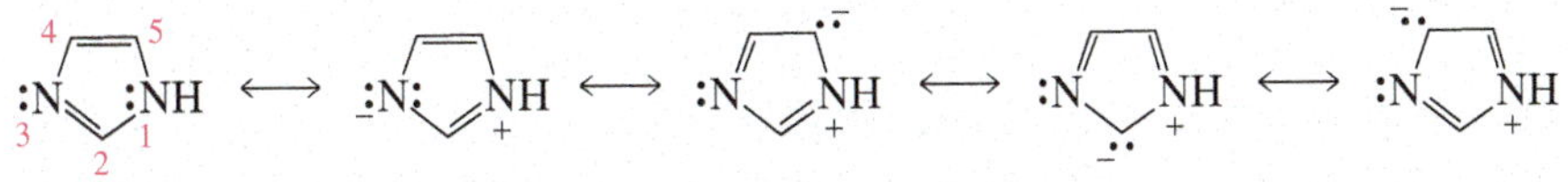

resonance contributors of imidazole

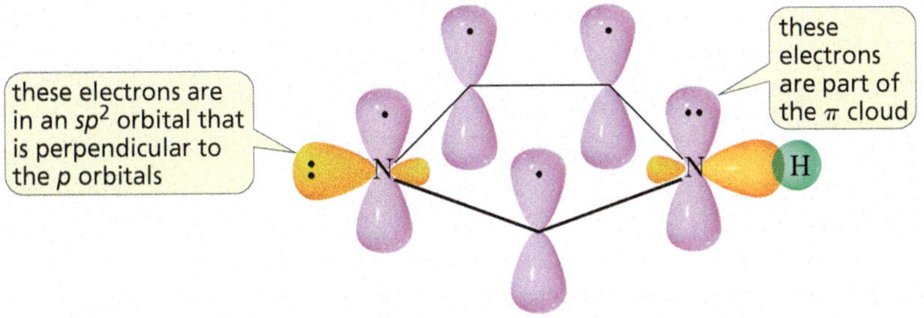

these electrons are in an sp^2 orbital that is perpendicular to the p orbitals

these electrons are part of the π cloud

orbital structure of imidazole

Because the lone-pair electrons in the sp^2 orbital are not part of the π cloud, imidazole is protonated in acidic solutions. The conjugate acid of imidazole has a $pK_a = 6.8$. Therefore, imidazole exists in both the protonated and unprotonated forms at physiological pH (7.4). This is one of the reasons that histidine, the imidazole-containing amino acid, is a catalytic component of many enzymes.

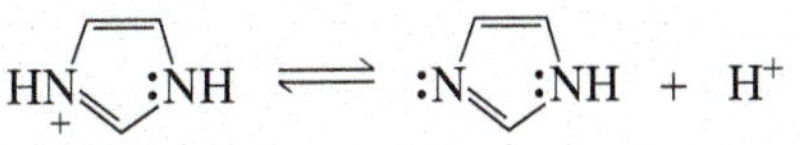

$pK_a = 6.8$

Notice that both protonated imidazole and the imidazole anion have two equivalent resonance contributors. Thus, the two nitrogens become equivalent when imidazole is either protonated or deprotonated.

protonated imidazole   **imidazole anion**

resonance hybrid   resonance hybrid

Searching for Drugs: An Antihistamine, a Nonsedating Antihistamine, and a Drug for Ulcers

When scientists know something about the molecular basis of drug action—such as how a particular drug interacts with a receptor—they can design and synthesize compounds that might have a desired physiological activity. For example, histidine can be decarboxylated in an enzyme-catalyzed reaction to form histamine (see page 939). When the body produces excess histamine, it causes the symptoms associated with the common cold and allergies. Knowing that the amine is positively charged at physiological pH gave scientists one clue as to how it might interact with its receptor.

After an extensive search, the compounds shown here (as well as several others)—called antihistamines—were found to bind to the histamine receptor but not trigger the same response as histamine. Like histamine, these drugs have a protonated amino group. They also have bulky groups that keep histamine from approaching the receptor.

antihistamines

diphenhydramine
Benadryl®

chlorpheniramine
Chlortrimetron®

These compounds, however, are able to cross the blood–brain barrier and bind to receptors in the central nervous system. Binding to these receptors causes drowsiness, the well-known side effect associated with these drugs. The search began for compounds that would bind to the histamine receptor but not cross the blood–brain barrier. Because the blood–brain barrier is nonpolar, this goal was achieved by putting polar groups on the compounds. Allegra, Claritin, and Zyrtec are nonsedating antihistamines.

nonsedating antihistamines

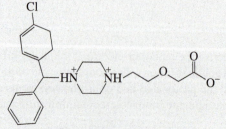

fexofenadine
Allegra®

loratadine
Claritin®

cetirizine
Zyrtec®

(Continued)

In addition to causing allergic responses, excess histamine production by the body also causes the hypersecretion of HCl by the cells of the stomach lining, which leads to the development of ulcers. The antihistamines that prevent allergic responses have no effect on HCl production. This suggested that a second kind of histamine receptor—called a histamine H_2-receptor—causes the secretion of HCl.

Because 4-methylhistamine was found to cause weak inhibition of HCl secretion, it was used as a lead compound (Section 10.9). More than 500 molecular modifications were performed over a 10-year period before a compound was found that would bind to the histamine H_2-receptor.

Tagamet, introduced in 1976, was the first drug for the treatment of peptic ulcers. Previously, the only treatment was extensive bed rest, a bland diet, and antacids. Zantac followed in 1981, and by 1988 it became the world's best-selling prescription drug. Protonix became available in 1994.

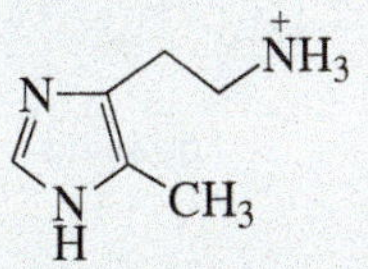

4-methylhistamine

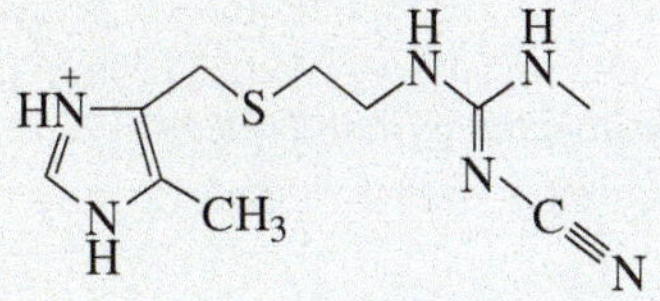

cimetidine
Tagamet®

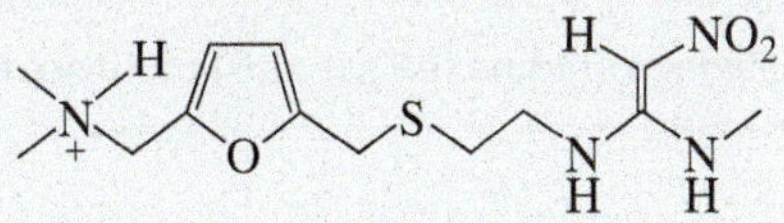

ranitidine
Zantac®

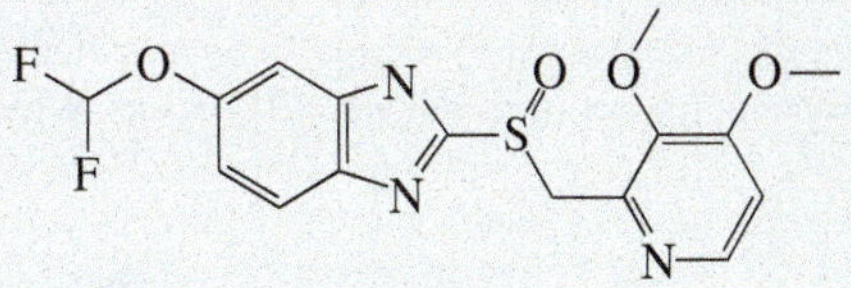

pantoprazole
Protonix®

PROBLEM 14 ◆

What is the major product of the following reaction?

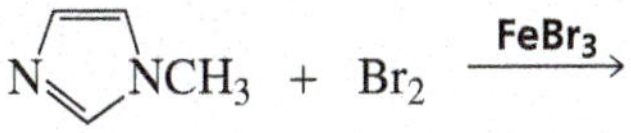

$$\text{imidazole-NCH}_3 + Br_2 \xrightarrow{\text{FeBr}_3}$$

PROBLEM 15 ◆

Rank imidazole, pyrrole, and benzene from most reactive to least reactive toward electrophilic aromatic substitution.

PROBLEM 16 ◆

Imidazole boils at 257 °C, whereas *N*-methylimidazole boils at 199 °C. Explain the difference in boiling points.

PROBLEM 17 ◆

Why is imidazole a stronger acid ($pK_a = 14.4$) than pyrrole ($pK_a \sim 17$)?

PROBLEM 18 ◆

What percent of imidazole is protonated at physiological pH (7.4)?

Purine and Pyrimidine

We know that nucleic acids (DNA and RNA) are important biological molecules. Both DNA and RNA contain substituted **purines** and substituted **pyrimidines** (Section 26.1): DNA contains adenine, guanine, cytosine, and thymine (abbreviated A, G, C, and T), whereas RNA contains adenine, guanine, cytosine, and uracil (A, G, C, and U). Why DNA contains T instead of U is explained in Section 26.10.

Guanine (a hydroxypurine) and cytosine, uracil, and thymine (hydroxypyrimidines) are more stable in the keto form than in the enol form; therefore, the keto forms are shown on the top of the next page. We will see that the preference for the keto form is crucial for proper base pairing in DNA (Section 26.3).

purine

pyrimidine

adenine   guanine   cytosine   uracil   thymine

PROBLEM 19 ◆

Draw guanine and cytosine in the enol form.

PROBLEM 20 ◆

Why is protonated pyrimidine ($pK_a = 1.0$) more acidic than protonated pyridine ($pK_a = 5.2$)?

Porphyrin

Substituted *porphyrins* are another group of important naturally occurring heterocyclic compounds. A **porphyrin ring system** consists of four pyrrole rings joined by one-carbon bridges. Heme, which is found in hemoglobin and myoglobin, contains an iron ion (Fe^{II}) ligated by the four nitrogens of a porphyrin ring system. **Ligation** is the sharing of lone-pair electrons with a metal ion.

a porphyrin ring system

heme

Hemoglobin is a protein responsible for transporting oxygen to cells, and myoglobin is a protein responsible for storing oxygen in cells. Hemoglobin has four polypeptide chains and four heme groups (page 1024); myoglobin has one polypeptide chain and one heme group.

The iron atom in the heme group, in addition to being ligated to the four nitrogens of the porphyrin ring, is also ligated to a histidine of the protein component (globin), and the sixth ligand is oxygen.

Carbon monoxide is about the same size and shape as O_2, but CO binds more tightly than O_2 to iron. In addition, once a hemoglobin molecule has bound two CO molecules, it can no longer achieve the configuration necessary to bind O_2. Consequently, breathing carbon monoxide can be fatal because it interferes with the transport of oxygen through the bloodstream.

The extensive conjugated system of heme gives blood its characteristic red color. Its high molar absorptivity (about 160,000 $M^{-1}cm^{-1}$) allows concentrations as low as 1×10^{-8} M to be detected by UV spectroscopy (Section 13.20).

The ring system in chlorophyll *a*, the substance responsible for the green color of plants (its structure is on page 606), is similar to porphyrin, but it contains a cyclopentanone ring and one of its pyrrole rings is partially reduced. The metal ion in chlorophyll *a* is magnesium (Mg^{II}).

Vitamin B_{12} also has a ring system similar to porphyrin, but in this case, the metal ion is cobalt (Co^{III}). Other aspects of vitamin B_{12}'s structure and chemistry are discussed in Section 23.6.

Porphyrin, Bilirubin, and Jaundice

The average human body breaks down about 6 g of hemoglobin each day. The protein portion (globin) and the iron are reutilized, but the porphyrin ring is cleaved between the A and B rings to form a linear tetrapyrrole called biliverdin (a green compound). Then the bridge between the C and D ring is reduced, forming bilirubin (a red-orange compound). You can witness heme degradation by observing the changing colors of a bruise.

Enzymes in the large intestine reduce bilirubin to urobilinogen (a colorless compound). Some urobilinogen is transported to the kidney, where it is oxidized to urobilin (a yellow compound). This is the compound that gives urine its characteristic color.

If more bilirubin is formed than can be metabolized and excreted by the liver, it accumulates in the blood. When its concentration there reaches a certain level, it diffuses into the tissues, giving them a yellow appearance. This condition is known as jaundice.

19.8 ORGANIZING WHAT WE KNOW ABOUT THE REACTIONS OF ORGANIC COMPOUNDS

Group I

R—CH=CH—R
alkene

R—C≡C—R
alkyne

R—CH=CH—CH=CH—R
diene

These are nucleophiles.

They undergo electrophilic addition reactions.

Group II

R—X $X = F, Cl, Br, I$
alkyl halide

R—OH
alcohol

R—OR
ether

epoxide

$R—N^+—R$, R, R HO^-
quaternary ammonium hydroxide

sulfonate ester

$R—S^+—R$, R
sulfonium salt

These are electrophiles.

They undergo nucleophilic substitution and/or elimination reactions.

Group III

$$R—\overset{O}{\overset{\|}{C}}—Z$$ $Z =$ an atom more electronegative than C

$$R—\overset{O}{\overset{\|}{C}}—Z$$ $Z = C$ or H

These are electrophiles.

They undergo nucleophilic acyl substitution reactions, nucleophilic addition reactions, or nucleophilic addition–elimination reactions.

Removal of a hydrogen from an α-carbon forms a nucleophile that can react with electrophiles.

Group IV

benzene

pyridine

pyrrole, furan, thiophene $Z = N, O,$ or S

These are nucleophiles.

They undergo electrophilic aromatic substitution reactions.

Halo-substituted benzenes and halo-substituted pyridines are electrophiles.

They undergo nucleophilic aromatic substitution reactions.

We saw that the families of organic compounds can be put in one of four groups, and that all the families in a group react in similar ways. Now that we have finished studying the families in Group IV, let's review how these compounds react.

All the compounds in Group IV are aromatic. To preserve the aromaticity of the rings, these compounds undergo electrophilic aromatic substitution reactions and/or nucleophilic aromatic substitution reactions.

- Benzene and substituted benzenes are nucleophiles; they react with electrophiles in electrophilic aromatic substitution reactions.

- If the electron density of the benzene ring is reduced by strongly electron-withdrawing groups, a halo-substituted benzene can undergo a nucleophilic aromatic substitution reaction.

- Pyrrole, furan, and thiophene are *more* nucleophilic than benzene; they are more reactive than benzene in electrophilic aromatic substitution reactions.

- Pyridine is *less* nucleophilic than benzene; it undergoes electrophilic aromatic substitution reactions only under rigorous conditions.

- 2- and 4-Halo-substituted pyridines readily undergo nucleophilic aromatic substitution reactions.

ESSENTIAL CONCEPTS

Section 19.0

- Some amines are **heterocyclic compounds**—cyclic compounds in which one or more of the atoms of the ring is an atom other than carbon.
- The lone pair on the nitrogen causes amines to be both bases and nucleophiles.

Section 19.1

- Amines are classified as primary, secondary, or tertiary, depending on whether one, two, or three hydrogens of ammonia have been replaced by alkyl groups.
- Heterocyclic rings are numbered so that the **heteroatom** has the lowest possible number.

Section 19.2

- Saturated heterocycles have physical and chemical properties like those of acyclic compounds that contain the same heteroatom.

Section 19.3

- Amines react as **bases** in proton-transfer reactions and in elimination reactions.
- Amines react as **nucleophiles** in nucleophilic substitution reactions, in nucleophilic acyl substitution reactions, in nucleophilic addition–elimination reactions, and in conjugate addition reactions.

Section 19.5

- **Pyrrole, furan,** and **thiophene** are heterocyclic aromatic compounds that undergo electrophilic aromatic substitution reactions preferentially at C-2. They are more reactive than benzene toward electrophilic aromatic substitution.
- Protonating pyrrole destroys its aromaticity.
- Pyrrole polymerizes in strongly acidic solutions.

Section 19.6

- Replacing one of benzene's carbons with a nitrogen forms **pyridine,** a heterocyclic aromatic compound that undergoes electrophilic aromatic substitution reactions at C-3 under vigorous conditions.

- 2- and 4-Halo-substituted pyridines undergo nucleophilic aromatic substitution reactions at C-2 or C-4.

- Pyridine is *less* reactive than benzene toward electrophilic aromatic substitution and *more* reactive than benzene toward nucleophilic aromatic substitution.

Section 19.7

- Three amino acids have heterocyclic rings: proline has a **pyrrolidine** ring, tryptophan has an **indole** ring, and histidine has an **imidazole** ring.

- Nucleic acids (DNA and RNA) contain substituted **purines** and substituted **pyrimidines**. Hydroxypurines and hydroxypyrimidines are more stable in the keto form than in the enol form.

- A **porphyrin ring system** consists of four pyrrole rings joined by one-carbon bridges; in hemoglobin and myoglobin, the four nitrogen atoms are ligated to iron (Fe^{II}). The metal ion in chlorophyll *a* is magnesium (Mg^{II}), and the metal ion in vitamin B_{12} is cobalt (Co^{III}).

SUMMARY OF REACTIONS

1. Reactions of amines as nucleophiles (Section 19.3)

   a. In alkylation reactions (Section 9.2)
   b. In reactions with a carbonyl group that has a leaving group (Sections 15.6, 15.7, and 15.16)
   c. In reactions with an aldehyde or ketone to form an imine or an enamine (Section 16.8)
   d. In conjugate addition reactions (Sections 16.15 and 16.16)
   e. In reactions with nitrous acid: primary arylamines form arenediazonium salts (Section 18.17)

2. Electrophilic aromatic substitution reactions

   a. Pyrrole, furan, and thiophene (Section 19.5): the mechanism is shown on page 932.

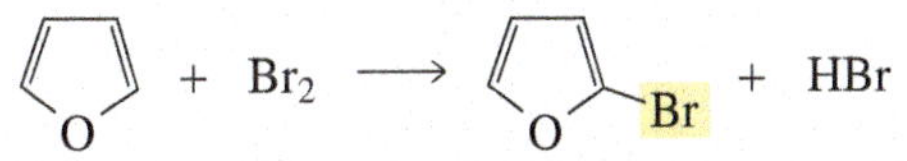

   b. Pyridine (Section 19.6): the mechanism is shown on page 935.

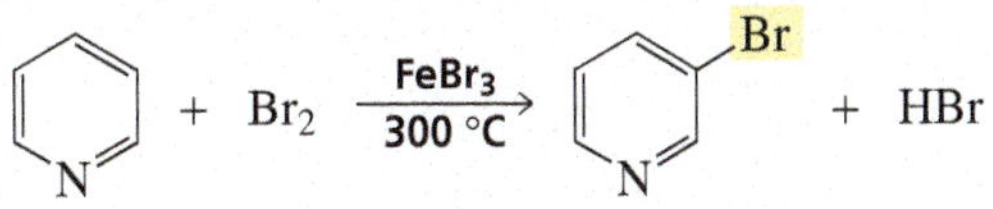

3. Nucleophilic aromatic substitution reactions of 2- and 4-halo-substituted pyridines (Section 19.6). The mechanism is shown on page 936.

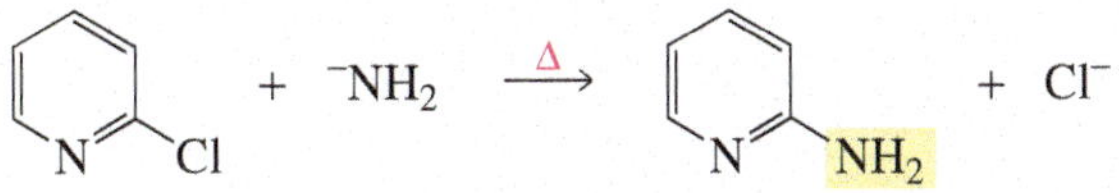

4. Reactions of substituents on pyridine (Section 19.6)

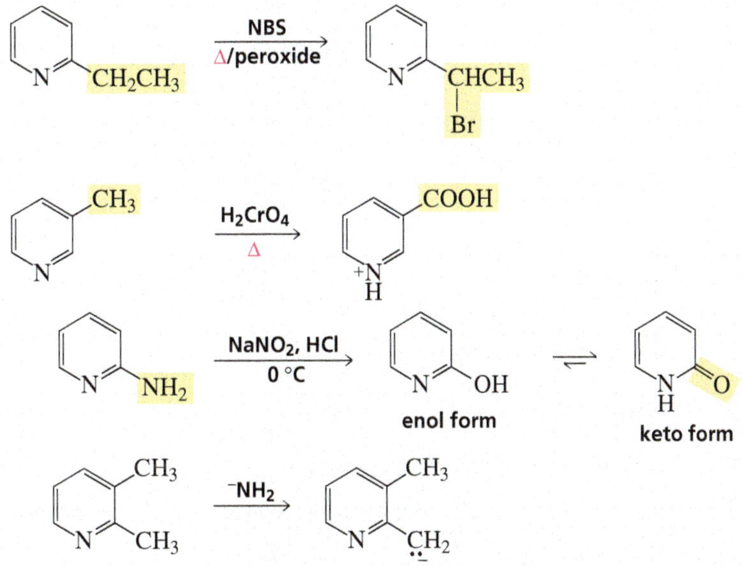

945

PROBLEMS

21. Name the following:

a. —NH with CH₃ (azetidine structure)

b. (piperidine with CH₃ groups)

c. (pyrrole with Cl)

d. (piperidine with CH₃CH₂ and CH₃)

22. What are the products of the following reactions?

a. (pyrrole) + (benzoyl chloride) ⟶

b. (2-bromopyridine) + HO⁻ $\xrightarrow{\Delta}$

c. CH₃CH₂CH₂CH₂Br $\xrightarrow[\text{2. H}_2\text{, Raney Ni}]{\text{1. }^-\text{C}\equiv\text{N}}$

d. (furan with CH₃) + (acetyl chloride) $\xrightarrow[\text{2. H}_2\text{O}]{\text{1. BF}_3\text{/THF}}$

e. (2-chloropyridine) $\xrightarrow{\text{C}_6\text{H}_5\text{Li}}$

f. (cyclohexylamine) + CH₃CH₂CH₂Br ⟶

g. (pyrrole) + C₆H₅N⁺≡N ⟶

h. (2-methylpyridine) $\xrightarrow[\text{2. CH}_3\text{CH}_2\text{CH}_2\text{Br}]{\text{1. }^-\text{NH}_2}$

i. (2-bromothiophene) $\xrightarrow[\substack{\text{2. CO}_2 \\ \text{3. HCl}}]{\text{1. Mg/Et}_2\text{O}}$

23. Rank the following compounds from strongest acid to weakest acid:

(series of nitrogen heterocyclic structures)

24. Which of the following compounds is easier to decarboxylate?

 or

25. Rank the following compounds from most reactive to least reactive in an electrophilic aromatic substitution reaction:

(C₆H₅)—OCH₃   (C₆H₅)—NHCH₃   (C₆H₅)—SCH₃

26. One of the following compounds undergoes electrophilic aromatic substitution predominantly at C-3, and one undergoes electrophilic aromatic substitution predominantly at C-4. Which is which?

(pyridine with CCH₂CH₃ group with O)   (pyridine with NHCH₂CH₃)

27. Benzene undergoes electrophilic aromatic substitution reactions with aziridines in the presence of a Lewis acid such as AlCl₃.

a. What are the major and minor products of the following reaction?

b. Would you expect epoxides to undergo similar reactions?

(benzene) + (N-methyl aziridine with CH₃) $\xrightarrow{\text{AlCl}_3}$

28. Pyrrole reacts with excess *para*-(*N,N*-dimethylamino)benzaldehyde to form a highly colored compound. Draw the structure of the colored compound.

29. The dipole moments of furan and tetrahydrofuran are in the same direction. One compound has a dipole moment of 0.70 D, and the other has a dipole moment of 1.73 D. Which is which?

30. Name the following:

a.

b.

c.

31. A tertiary amine reacts with hydrogen peroxide to form a tertiary amine oxide.

$$R-\overset{R}{\underset{\cdot\cdot}{N}}-R \;+\; HO-OH \;\longrightarrow\; R-\overset{R}{\underset{OH}{\overset{+}{N}}}-R \;+\; HO^- \;\longrightarrow\; R-\overset{R}{\underset{O^-}{\overset{+}{N}}}-R \;+\; H_2O$$

tertiary amine   **tertiary amine oxide**

Tertiary amine oxides undergo a reaction similar to the Hofmann elimination reaction (Section 10.10), called a **Cope elimination.** In this reaction, a tertiary amine oxide, rather than a quaternary ammonium ion, undergoes elimination. A strong base is not needed for a Cope elimination because the amine oxide acts as its own base.

$$CH_3CH_2CH_2\overset{CH_3}{\underset{\underset{O^-}{|+}}{N}}CH_2\overset{CH_3}{\underset{}{C}}HCH_3 \quad\overset{\Delta}{\longrightarrow}\quad CH_3CH=CH_2 \;+\; \overset{CH_3}{\underset{\underset{OH}{|}}{N}}CH_2\overset{CH_3}{\underset{}{C}}HCH_3$$

Does the Cope elimination have an alkene-like transition state or a carbanion-like transition state?

32. What products are obtained when the following tertiary amines react with hydrogen peroxide followed by heat?

a. $CH_3\overset{CH_3}{\underset{}{N}}CH_2CH_2CH_3$

b. $CH_3NCH_2CH_2CH_3$

c. $CH_3CH_2\overset{CH_3}{\underset{}{N}}CH_2\overset{CH_3}{\underset{}{C}}HCH_3$

d.

33. **a.** Draw resonance contributors to show why pyridine-*N*-oxide is more reactive than pyridine toward electrophilic aromatic substitution.
    b. At what position does pyridine-*N*-oxide undergo electrophilic aromatic substitution?

34. The chemical shifts of the C-2 hydrogen in the spectra of pyrrole, pyridine, and pyrrolidine are 2.82 ppm, 6.42 ppm, and 8.50 ppm. Match each heterocycle with its chemical shift.

35. Explain why protonating aniline has a dramatic effect on the compound's UV spectrum, whereas protonating pyridine has only a small effect on that compound's UV spectrum.

36. Explain why pyrrole ($pK_a \sim 17$) is a much stronger acid than ammonia ($pK_a = 36$).

$$\text{(pyrrole, } pK_a \sim 17) \;\rightleftharpoons\; \text{(pyrrole anion)} \;+\; H^+ \qquad NH_3 \;\rightleftharpoons\; {}^-NH_2 \;+\; H^+ \quad (pK_a = 36)$$

37. Propose a mechanism for the following reaction:

$$2\,\text{(pyrrole)} \;+\; H_2C{=}O \;\overset{HCl}{\longrightarrow}\; \text{(dipyrrolylmethane)}$$

38. Quinolines, heterocyclic compounds that contain a pyridine ring fused to a benzene ring, are commonly synthesized by a method known as the **Skraup synthesis**, in which aniline reacts with glycerol under acidic conditions. Nitrobenzene is added to the reaction mixture to serve as an oxidizing agent. The first step in the synthesis is the dehydration of glycerol to propenal.

$$\underset{OH\;\;\;\;OH\;\;\;OH}{CH_2-CH-CH_2} \;\overset{H_2SO_4}{\underset{\Delta}{\longrightarrow}}\; \underset{+\;2\,H_2O}{CH_2{=}CH-CH{=}O} \;\xrightarrow[\text{2. }\;\text{(NO}_2\text{)}]{\text{1. }\;\text{(NH}_2\text{)}}\; \text{quinoline}$$

glycerol   **quinoline**

  a. What product would be obtained if *para*-ethylaniline were used instead of aniline?
  b. What product would be obtained if 3-hexen-2-one were used instead of glycerol?
  c. What starting materials are needed for the synthesis of 2,7-diethyl-3-methylquinoline?

947

39. Propose a mechanism for the following reactions:

a. 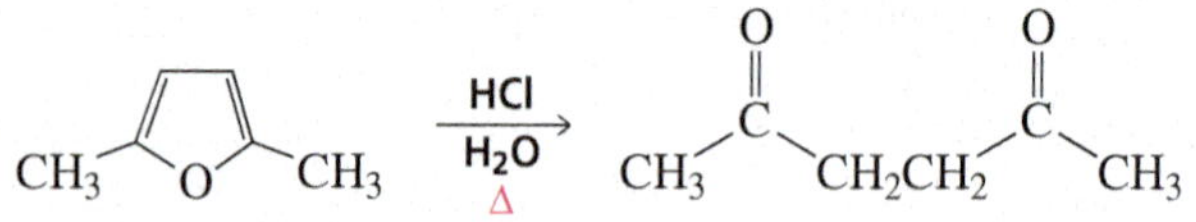 $\xrightarrow[\substack{H_2O \\ \Delta}]{HCl}$

b. 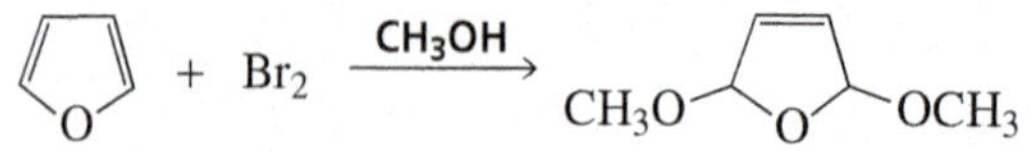 $+ \; Br_2 \xrightarrow{CH_3OH}$

40. What is the major product of the following reactions?

a. 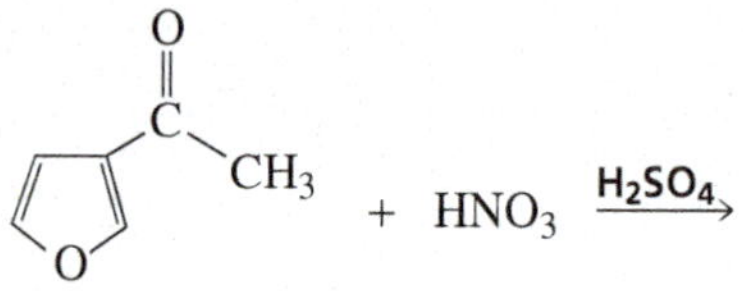 $+ \; HNO_3 \xrightarrow{H_2SO_4}$

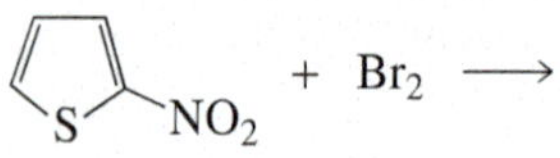

c. 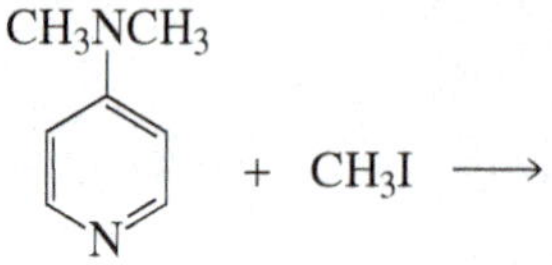 $+ \; CH_3I \longrightarrow$

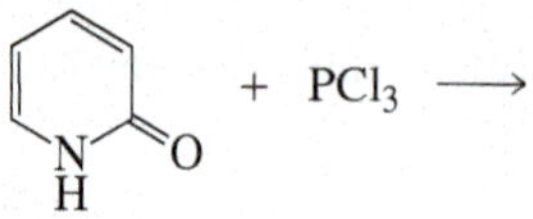

e. 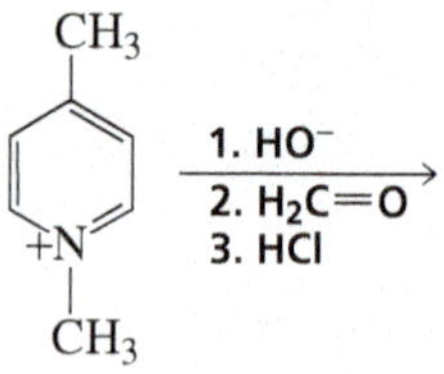 $\xrightarrow[\substack{2.\;H_2C=O \\ 3.\;HCl}]{1.\;HO^-}$

b. $+ \; Br_2 \longrightarrow$

d. $+ \; PCl_3 \longrightarrow$

f. 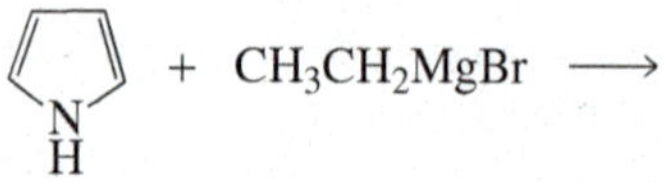 $+ \; CH_3CH_2MgBr \longrightarrow$

41. 2-Phenylindole is prepared from the reaction of acetophenone and phenylhydrazine, a method known as the **Fischer indole synthesis**. Propose a mechanism for this reaction. (*Hint:* the reactive intermediate is the enamine tautomer of the phenylhydrazone.)

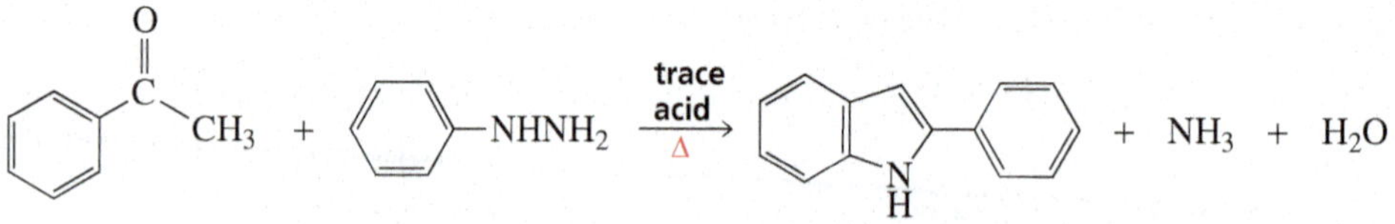

$+$ $-NHNH_2 \xrightarrow[\Delta]{\substack{trace \\ acid}}$ $+ \; NH_3 \; + \; H_2O$

2-phenylindole

42. What starting materials are required to synthesize the following compounds, using the Fischer indole synthesis? (*Hint:* see Problem 41.)

a.

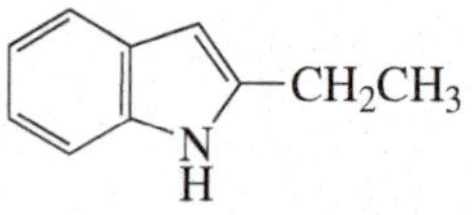

b.

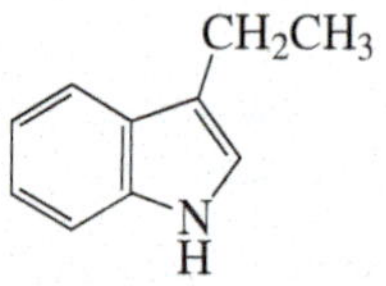

c.

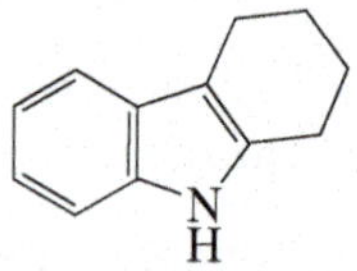

43. Organic chemists work with tetraphenylporphyrins rather than with porphyrins because tetraphenylporphyrins are much more resistant to air oxidation. Tetraphenylporphyrin can be prepared by the reaction of benzaldehyde with pyrrole. Propose a mechanism for the formation of the ring system shown here:

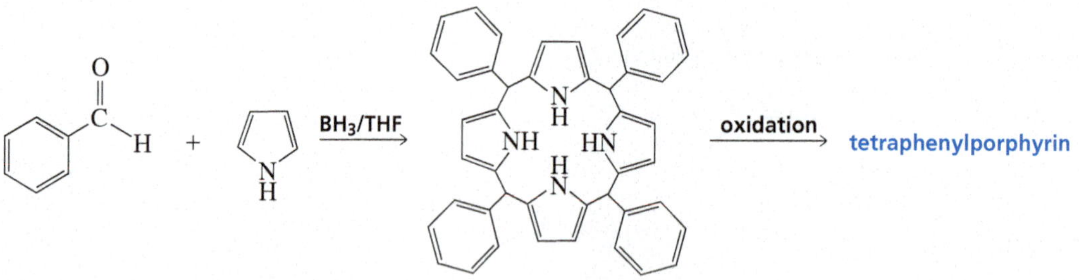

$+$ $\xrightarrow{BH_3/THF}$ $\xrightarrow{oxidation}$ **tetraphenylporphyrin**

44. Show how the following compounds can be prepared from the given starting material.

a. 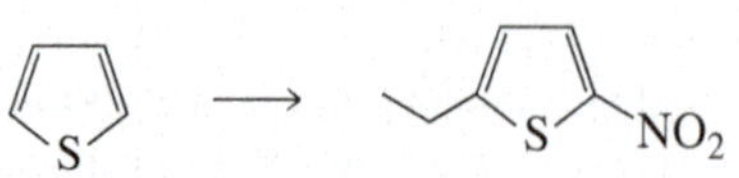 $\longrightarrow$

b. 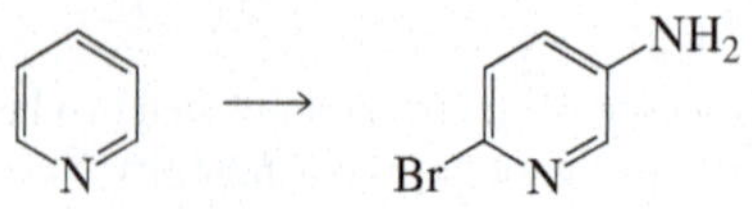 $\longrightarrow$

PART SEVEN

Bioorganic Compounds

Chapters 20 through 26 discuss the chemistry of bioorganic compounds—organic compounds found in living systems. Many of these compounds are larger than the organic compounds you have seen up to this point, and they often have more than one functional group, but the principles that govern their structure and reactivity are essentially the same as those that govern the structure and reactivity of the compounds that you have been studying. These chapters, therefore, will give you the opportunity to review much of the organic chemistry you have learned as you apply this chemistry to compounds found in the biological world.

Chapter 20 The Organic Chemistry of Carbohydrates

Chapter 20 introduces you to the organic chemistry of carbohydrates, the most abundant class of compounds in the biological world. First you will learn about the structures and reactions of monosaccharides. Then you will see how monosaccharides are linked to form disaccharides and polysaccharides. Many examples of carbohydrates that are found in nature are discussed.

Chapter 21 Amino Acids, Peptides, and Proteins

Chapter 21 starts by looking at the physical properties of amino acids. Then you will see how amino acids are linked to form peptides and proteins. You will also see how proteins are synthesized in the laboratory. Later, when you study Chapter 26, you will be able to compare the way proteins are synthesized in the laboratory with the way they are synthesized in nature. What you learn about protein structure in Chapter 21 will prepare you for understanding how enzymes catalyze chemical reactions, which is discussed in Chapter 22.

Chapter 22 Catalysis in Organic Reactions and in Enzymatic Reactions

Chapter 22 first describes the various ways that organic reactions can be catalyzed and then shows how enzymes employ these same methods to catalyze reactions in cells. It also explains why enzymes are extraodinarily good catalysts.

Chapter 23 The Organic Chemistry of the Coenzymes—Compounds Derived from Vitamins

Chapter 23 describes the chemistry of the coenzymes—organic compounds that some enzymes need to catalyze reactions. Coenzymes play a variety of chemical roles: some function as oxidizing and reducing agents, some allow electrons to be delocalized, some activate groups for further reaction, and some provide good nucleophiles or strong bases needed for reactions. Because coenzymes are derived from vitamins, you will see why vitamins are necessary for many of the reactions that occur in cells.

Chapter 24 The Organic Chemistry of the Metabolic Pathways

Chapter 24 looks at the reactions that cells carry out to obtain the energy they need and to synthesize the compounds they require. You will see that these reactions are similar to the reactions you have seen in previous chapters. You will also see why many reactions that occur in cells could not occur without ATP.

Chapter 25 The Organic Chemistry of Lipids

Chapter 25 discusses the chemistry of lipids. Lipids are water-soluble compounds found in animals and plants. First you will study the structure and function of different kinds of lipids. You will then understand such things as how aspirin prevents inflammation and how cholesterol and other terpenes are synthesized in nature.

Chapter 26 The Chemistry of the Nucleic Acids

Chapter 26 covers the structure and chemistry of nucleosides, nucleotides, and nucleic acids (RNA and DNA). You will see how nucleotides are linked to form nucleic acids, why DNA contains thymine instead of uracil, why DNA does not have the $2'$-OH group that RNA has, how the genetic messages encoded in DNA are transcribed into mRNA and then translated into proteins, and how the sequence of bases in DNA can be determined.

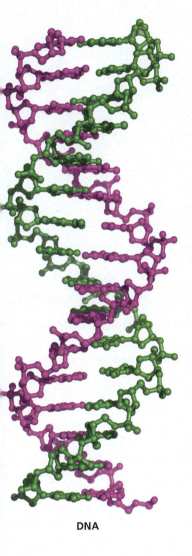

DNA

20

The Organic Chemistry of Carbohydrates

a field of sugar cane

Bioorganic compounds are organic compounds found in living systems. The first group of bioorganic compounds we will look at are carbohydrates—the most abundant class of compounds in the biological world, making up more than 50% of the dry weight of Earth's biomass. Carbohydrates are important constituents of all living organisms and have a variety of different functions. Some are important structural components of cells; others act as recognition sites on cell surfaces. For example, the first event in each and every life is when a sperm cell recognizes a carbohydrate on the outer surface of an egg. Other carbohydrates serve as a major source of metabolic energy. The leaves, fruits, seeds, stems, and roots of plants, for instance, contain carbohydrates that plants use for their own metabolic needs and these also serve the metabolic needs of the animals that eat them.

The structures of bioorganic compounds can be quite complex, yet their reactivity is governed by the same principles that govern the reactivity of the comparatively simple organic molecules we have discussed so far. In other words, the organic reactions that chemists carry out in the laboratory are in many ways just like those performed by nature inside a cell. Thus, bioorganic reactions can be thought of as organic reactions that take place in tiny flasks called cells.

Although most bioorganic compounds have more complicated structures than those of the organic compounds you are now used to seeing, do not let the structures fool you into thinking that their chemistry must be equally complicated. One reason the structures of bioorganic compounds are more complicated is that they must recognize each other. Much of their structure is for that very purpose, a function called **molecular recognition.**

Early chemists noted that carbohydrates have molecular formulas that make them appear to be hydrates of carbon, $C_n(H_2O)_n$—hence the name. Structural studies later revealed that these compounds are not hydrates because they do not contain intact water molecules. Nevertheless, the term *carbohydrate* persisted.

Carbohydrates are polyhydroxy aldehydes such as glucose, polyhydroxy ketones such as fructose, and compounds formed by linking polyhydroxy aldehydes and/or polyhydroxy ketones together (Section 20.15). The chemical structures of carbohydrates are commonly represented by Fischer projections. Notice that both glucose and fructose have the molecular formula $C_6H_{12}O_6$, consistent with the general formula $C_6(H_2O)_6$ that made early chemists think that these compounds were hydrates of carbon. Notice, too, that the structures of glucose and fructose differ only at the top two carbons.

Recall that all the horizontal bonds in Fischer projections point toward the viewer (Section 4.7).

H C=O
H—OH
HO—H
H—OH
H—OH
CH₂OH

Fischer projection
D-glucose
a polyhydroxy aldehyde

CH₂OH
C=O
HO—H
H—OH
H—OH
CH₂OH

Fischer projection
D-fructose
a polyhydroxy ketone

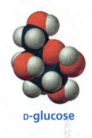

D-glucose

D-fructose

Photosynthesis

The most abundant carbohydrate in nature is glucose. Animals obtain glucose from food that contains glucose, such as plants. Plants produce glucose by *photosynthesis*. During photosynthesis, plants take up water through their roots and use carbon dioxide from the air to synthesize glucose and oxygen.

$$\underset{\text{glucose}}{C_6H_{12}O_6} + 6\,O_2 \underset{\text{photosynthesis}}{\overset{\text{oxidation}}{\rightleftharpoons}} 6\,CO_2 + 6\,H_2O + \text{energy}$$

Because photosynthesis is the reverse of the process used by organisms to obtain energy—specifically, the oxidation of glucose to carbon dioxide and water—plants require energy to carry out photosynthesis. They obtain that energy from sunlight, which is captured by chlorophyll molecules in green plants. Photosynthesis uses the CO_2 that animals exhale as waste and generates the O_2 that animals inhale to sustain life. Nearly all the oxygen in the atmosphere has been released by photosynthetic processes.

20.1 CLASSIFYING CARBOHYDRATES

The terms *carbohydrate*, *saccharide*, and *sugar* are used interchangeably. *Saccharide* comes from the word for sugar in several early languages (*sarkara* in Sanskrit, *sakcharon* in Greek, and *saccharum* in Latin).

Simple carbohydrates are **monosaccharides** (single sugars); **complex carbohydrates** contain two or more monosaccharides linked together. **Disaccharides** contain two monosaccharides linked together, **oligosaccharides** contain 3 to 10 (*oligos* is Greek for "few"), and **polysaccharides** contain 10 or more. Disaccharides, oligosaccharides, and polysaccharides can be broken down to monosaccharides by hydrolysis.

(a monosaccharide subunit)

$$\underset{\text{polysaccharide}}{-M-M-M-M-M-M-M-M-M-} \xrightarrow{\text{hydrolysis}} \underset{\text{monosaccharide}}{n\,M}$$

Monosaccharides Are Either Aldoses or Ketoses

A *monosaccharide* can be a polyhydroxy aldehyde such as glucose or a polyhydroxy ketone such as fructose. Polyhydroxy aldehydes are called **aldoses** ("ald" is for aldehyde; "ose" is the suffix for a sugar); polyhydroxy ketones are called **ketoses.**

Monosaccharides are also classified according to the number of carbons they contain: those with three carbons are **trioses,** those with four carbons are **tetroses,** those with five carbons are

pentoses, and those with six and seven carbons are **hexoses** and **heptoses.** Therefore, a six-carbon polyhydroxy aldehyde such as glucose is an aldohexose, whereas a six-carbon polyhydroxy ketone such as fructose is a ketohexose.

PROBLEM 1 ◆

Classify the following monosaccharides:

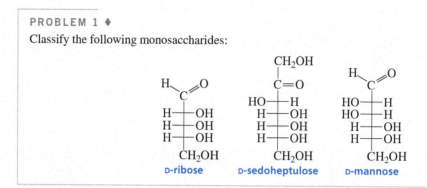

D-ribose   D-sedoheptulose   D-mannose

20.2 THE D AND L NOTATION

The smallest aldose, and the only one whose name does not end in "ose," is glyceraldehyde, an aldotriose.

asymmetric center

$$HOCH_2CH\overset{O}{\underset{OH}{\overset{||}{-}}}C\overset{}{\underset{}{-}}H$$

glyceraldehyde

A carbon to which four different groups are attached is an asymmetric center.

Because glyceraldehyde has an asymmetric center, it can exist as a pair of enantiomers. We know that the isomer shown on the far left below has the *R* configuration because an arrow drawn from the highest priority substituent (OH) to the next highest priority substituent (HC=O) is clockwise and the lowest priority group is on a hatched wedge (Section 4.8). The *R* and *S* enantiomers drawn as Fischer projections are shown on the right.

clockwise is *R*

H is on a horizontal bond so counter-clockwise is *R*

(*R*)-(+)-glyceraldehyde   (*S*)-(−)-glyceraldehyde   (*R*)-(+)-glyceraldehyde   (*S*)-(−)-glyceraldehyde

perspective formulas   **Fischer projections**

The notations D and L are used to describe the configurations of carbohydrates. In a Fischer projection of a monosaccharide, the carbonyl group is always placed on top (in the case of aldoses) or as close to the top as possible (in the case of ketoses). Examine the Fischer projection of galactose shown below and note that the compound has four asymmetric centers (C-2, C-3, C-4, and C-5).

If the OH group attached to the bottommost asymmetric center of a Fischer projection (the carbon second from the bottom) is on the right, then the compound is a D-sugar. If that same OH group is on the left, then the compound is an L-sugar.

Almost all sugars found in nature are D-sugars. The mirror image of a D-sugar is an L-sugar.

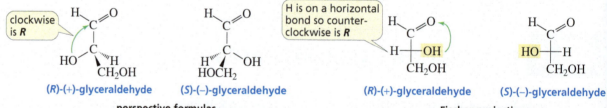

D-glyceraldehyde   L-glyceraldehyde mirror image of D-glyceraldehyde   D-galactose   L-galactose mirror image of D-galactose

the OH group is on the right   the OH group is on the right

The common name of the monosaccharide, together with the D or L designation, completely defines its structure, because the configurations of all the asymmetric centers are implicit in the common name. Thus, the structure of L-galactose is obtained by changing the configuration of *all* the asymmetric centers in D-galactose.

Emil Fischer and his colleagues studied carbohydrates in the late nineteenth century, when techniques for determining the configurations of compounds were not available. Fischer arbitrarily assigned the *R*-configuration to the dextrorotatory isomer of glyceraldehyde that we call D-glyceraldehyde. He turned out to be correct: D-glyceraldehyde is (*R*)-(+)-glyceraldehyde, and L-glyceraldehyde is (*S*)-(−)-glyceraldehyde.

Like *R* and *S*, the symbols D and L indicate the configuration about an asymmetric center, but do not indicate whether the compound rotates the plane of polarization of plane-polarized light to the right (+) or to the left (−) (Section 4.9). For example, D-glyceraldehyde is dextrorotatory, whereas D-lactic acid is levorotatory. In other words, optical rotation, like a melting point or a boiling point, is a physical property of a compound, whereas "*R, S,* D, and L" are conventions humans use to indicate the configuration about an asymmetric center.

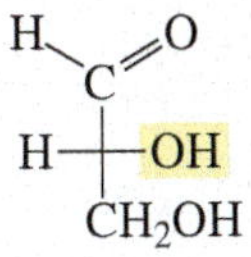

D-(+)-glyceraldehyde 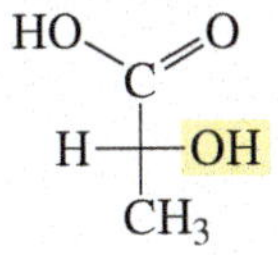   D-(−)-lactic acid

---

PROBLEM 2

Draw Fischer projections of L-glucose and L-fructose.

PROBLEM 3 ◆

Indicate whether each of the following structures is D-glyceraldehyde or L-glyceraldehyde, assuming that the horizontal bonds point toward you and the vertical bonds point away from you (Section 4.7):

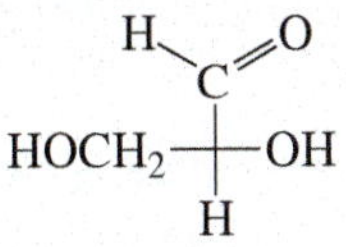

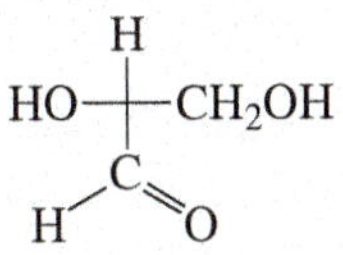

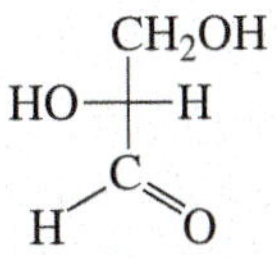

20.3 THE CONFIGURATIONS OF ALDOSES

Aldotetroses have two asymmetric centers and, therefore, four stereoisomers. Two of the stereoisomers are D-sugars and two are L-sugars. The names of the aldotetroses—erythrose and threose—were used to name the erythro and threo pairs of enantiomers described in Section 4.12.

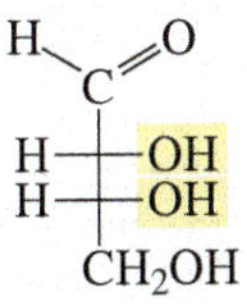

D-erythrose 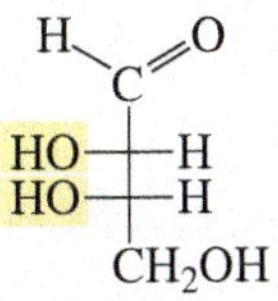   L-erythrose 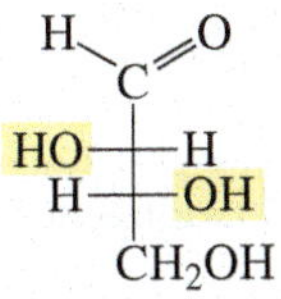   D-threose 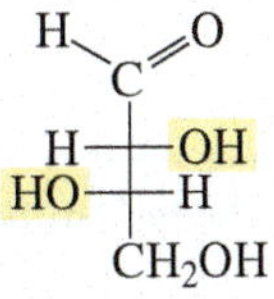   L-threose

Aldopentoses have three asymmetric centers and, therefore, 8 stereoisomers (four pairs of enantiomers); *aldohexoses* have four asymmetric centers and 16 stereoisomers (eight pairs of enantiomers). The structures of the four D-aldopentoses and the eight D-aldohexoses are shown in Table 20.1 on the next page.

Table 20.1 Configurations of the D-Aldoses

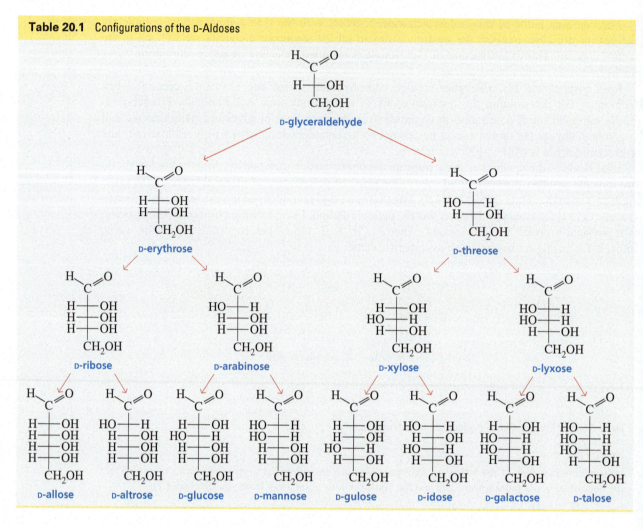

Diastereomers are stereoisomers that are not enantiomers.

Monosaccharides that differ in configuration at only one asymmetric center are called **epimers.** For example, D-ribose and D-arabinose are C-2 epimers because they differ in configuration only at C-2; D-idose and D-talose are C-3 epimers. An epimer is a particular kind of diastereomer. (Recall that diastereomers are stereoisomers that are not enantiomers; Section 4.12.)

D-Mannose is the C-2 epimer of D-glucose.

D-Galactose is the C-4 epimer of D-glucose.

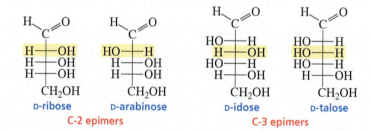

D-Glucose, D-mannose, and D-galactose are the most common aldohexoses in living systems. An easy way to learn their structures is to memorize the structure of D-glucose and then remember that D-mannose is the C-2 epimer of D-glucose and D-galactose is the C-4 epimer of D-glucose.

PROBLEM 4 ♦

a. Are D-erythrose and L-erythrose enantiomers or diastereomers?
b. Are L-erythrose and L-threose enantiomers or diastereomers?

PROBLEM 5 ♦

a. What sugar is the C-3 epimer of D-xylose?
b. What sugar is the C-5 epimer of D-allose?
c. What sugar is the C-4 epimer of L-gulose?
d. What sugar is the C-4 epimer of D-lyxose?

20.4 THE CONFIGURATIONS OF KETOSES

The structures of the naturally occurring ketoses are shown in Table 20.2—they all have a keto group in the 2-position. A ketose has one less asymmetric center than an aldose with the same number of carbons. Therefore, a ketose has only half as many stereoisomers as an aldose with the same number of carbons.

Table 20.2   Configurations of the D-Ketoses

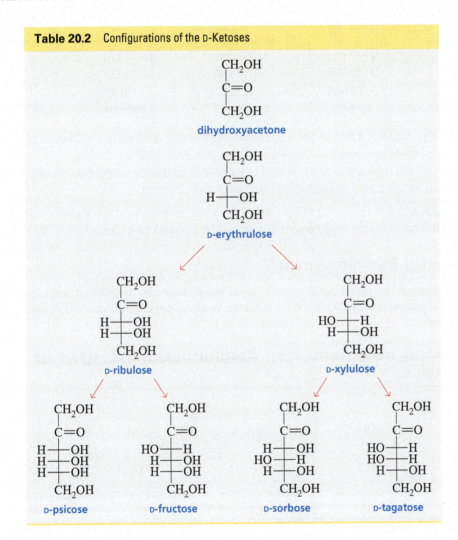

20.5 THE REACTIONS OF MONOSACCHARIDES IN BASIC SOLUTIONS

Epimerization

In a basic solution, a monosaccharide is converted to a mixture of polyhydroxy aldehydes and polyhydroxy ketones. Let's look at what happens to D-glucose in a basic solution, beginning with its conversion to its C-2 epimer.

MECHANISM FOR THE BASE-CATALYZED EPIMERIZATION OF A MONOSACCHARIDE

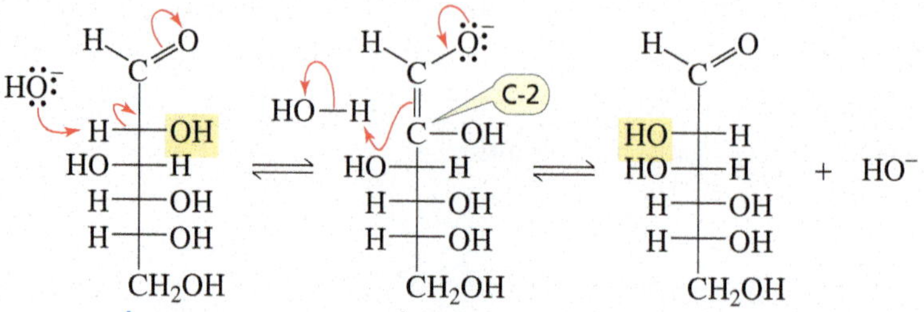

D-glucose   enolate ion   D-mannose

- The base removes a proton from an α-carbon, forming an enolate ion (Section 17.3). Notice that C-2 in the enolate ion is no longer an asymmetric center.
- When C-2 is reprotonated, the proton can come from the top or the bottom of the planar sp^2 carbon, forming both D-glucose and D-mannose (C-2 epimers).

Because the reaction forms a pair of epimers, it is called an epimerization. **Epimerization** changes the configuration of a carbon by removing a proton from it and then reprotonating it.

The Enediol Rearrangement

In addition to forming its C-2 epimer in a basic solution, D-glucose also undergoes an **enediol rearrangement,** which forms D-fructose. Subsequent enediol rearrangements form other ketohexoses.

MECHANISM FOR THE BASE-CATALYZED ENEDIOL REARRANGEMENT OF A MONOSACCHARIDE

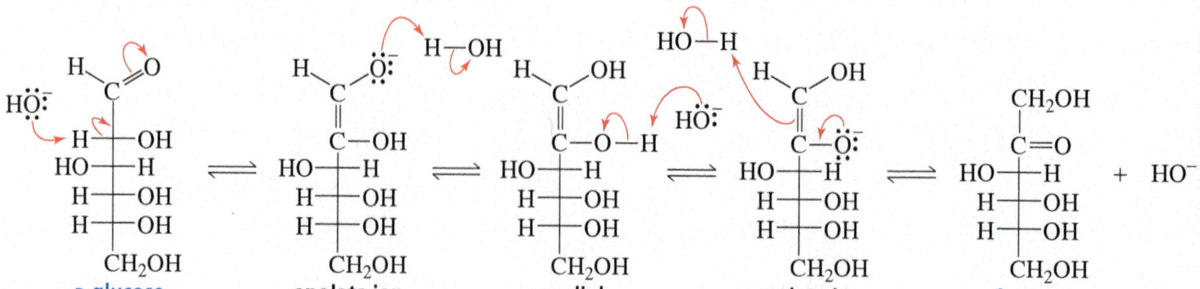

D-glucose   enolate ion   enediol   enolate ion   D-fructose

LEARN THE STRATEGY

- The base removes a proton from an α-carbon, forming an enolate ion.
- Either C-2 can be protonated (as in the epimerization mechanism shown above), or the oxygen of the enolate ion can be protonated to form an enediol (as shown here).
- The enediol has two OH groups that can form a carbonyl group. Removal of a proton from the OH at C-1 followed by tautomerization (as in base-catalyzed epimerization shown above) re-forms D-glucose or forms D-mannose; removal of a proton from the OH group at C-2 followed by tautomerization forms D-fructose.

Another enediol rearrangement, initiated by a base removing a proton from C-3 of D-fructose, forms an enediol that can tautomerize to give a ketose with the carbonyl group at C-2 or C-3. Thus, the carbonyl group can be moved up and down the chain.

In a basic solution, an aldose forms a C-2 epimer and one or more ketoses.

PROBLEM 9

Show how an enediol rearrangement can move the carbonyl carbon of fructose from C-2 to C-3.

PROBLEM 10

Write the mechanism for the base-catalyzed conversion of D-fructose to D-glucose and D-mannose.

USE THE STRATEGY

PROBLEM 11 ♦

When D-tagatose is added to a basic aqueous solution, an equilibrium mixture of monosaccharides is obtained, two of which are aldohexoses and two of which are ketohexoses. Identify the aldohexoses and ketohexoses.

20.6 # OXIDATION–REDUCTION REACTIONS OF MONOSACCHARIDES

Because they contain *alcohol* functional groups and *aldehyde* or *ketone* functional groups, the reactions of monosaccharides are an extension of what you have already learned about the reactions of alcohols, aldehydes, and ketones. For example, an aldehyde group in a monosaccharide can be oxidized or reduced and can react with nucleophiles to form imines, hemiacetals, and acetals.

Reduction

The carbonyl group in aldoses and ketoses can be reduced to an alcohol by $NaBH_4$ (Section 16.5). The product of the reduction is a polyalcohol, known as an **alditol.**

Reduction of an aldose forms one alditol. For example, the reduction of D-mannose forms D-mannitol, the alditol found in mushrooms, olives, and onions.

NOTE TO THE STUDENT

• As you read this section and those that follow, you will find cross-references to earlier discussions of simpler organic compounds that undergo the same reactions. Go back and look at these earlier discussions when you see a cross-reference; it will make learning about carbohydrates a lot easier.

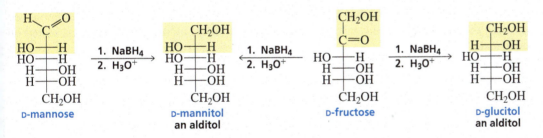

Reduction of a ketose forms two alditols because the reaction creates a new asymmetric center in the product. For example, the reduction of D-fructose forms D-mannitol and D-glucitol, the C-2 epimer of D-mannitol. D-Glucitol—also called sorbitol—is about 60% as sweet as sucrose. It is found in plums, pears, cherries, and berries. It is also used as a sugar substitute in the manufacture of candy; we will see why on page 974.

PROBLEM 12 ♦

What products are obtained from the reduction of

a. D-idose?   **b.** D-sorbose?

PROBLEM 13 ♦

a. What other monosaccharide is reduced only to the alditol obtained from the reduction of

　　1. D-talose?   **2.** D-glucose?   **3.** D-galactose?

b. What monosaccharide is reduced to two alditols, one of which is the alditol obtained from the reduction of

　　1. D-talose?   **2.** D-allose?

Oxidation

Br_2 is a mild oxidizing agent that easily oxidizes aldehydes but cannot oxidize ketones or alcohols. Therefore, adding a small amount of an aqueous solution of Br_2 to an unknown monosaccharide can tell you whether the sugar is an aldose or a ketose. The reddish-brown color of Br_2 will disappear if the monosaccharide is an aldose because when Br_2 oxidizes the aldehyde, Br_2 is reduced to Br^-, which is colorless. If the red color persists, indicating no reaction with Br_2, then the monosaccharide is a ketose. The product of the oxidation reaction is an **aldonic acid.**

D-glucose + Br_2 (red) $\xrightarrow{H_3O^+}$ D-gluconic acid an aldonic acid + $2\ Br^-$ (colorless)

Both aldoses and ketoses are oxidized to aldonic acids by Tollens' reagent (Ag^+, NH_3, HO^-). The oxidizing agent in Tollens' reagent is Ag^+. It is reduced to metallic silver that forms a shiny mirror on the inside of the test tube. Although Tollens' reagent only oxidizes aldehydes, it cannot be used to distinguish aldoses and ketoses. The oxidation reaction is carried out in a basic solution that will convert a ketose to an aldose by an enediol rearrangement (Section 20.5), and the aldose will then be oxidized by Tollens' reagent.

ketose $\underset{\xrightarrow{HO^-}}{\rightleftharpoons}$ aldose $\xrightarrow[HO^-]{Ag^+,\ NH_3}$ carboxylate ion + Ag silver

Dilute nitric acid (HNO_3) is a stronger oxidizing agent than those discussed above. It oxidizes aldehydes *and* primary alcohols, but it does not oxidize secondary alcohols. The product obtained when both the aldehyde and the primary alcohol groups of an aldose are oxidized is called an **aldaric acid.**

In an aldONic acid, ONe end is oxidized.

In an aldARic acid, both ends ARe oxidized.

D-glucose $\xrightarrow[\Delta]{HNO_3}$ D-glucaric acid an aldaric acid

---

PROBLEM 14 ◆

a. Name an aldohexose other than D-glucose that is oxidized to D-glucaric acid by nitric acid.
b. What is another name for D-glucaric acid?
c. Name another pair of aldohexoses that are oxidized to identical aldaric acids.

20.7 LENGTHENING THE CHAIN: THE KILIANI–FISCHER SYNTHESIS

The carbon chain of an aldose can be increased by one carbon by a modified **Kiliani–Fischer synthesis.** Thus, tetroses can be converted to pentoses, and pentoses can be converted to hexoses.

STEPS IN THE MODIFIED KILIANI–FISCHER SYNTHESIS

new asymmetric center

products are C-2 epimers

D-ribose

D-arabinose

The Kiliani–Fischer synthesis leads to a pair of C-2 epimers.

D-erythrose

- In the first step, hydrogen cyanide adds to the carbonyl group (Section 16.4). This reaction converts the carbonyl carbon in the starting material to an asymmetric center. Consequently, two products are formed that differ only in their configuration at C-2. The configurations of the other asymmetric centers do not change because no bond to any of the asymmetric centers is broken during the course of the reaction.

- The $C\equiv N$ bond is reduced to an imine, using a partially deactivated catalyst so that the imine is not further reduced to an amine (Section 7.9).

- The two imines are hydrolyzed to two aldoses (Section 16.8).

LEARN THE STRATEGY

Notice that the modified Kiliani–Fischer synthesis leads to a pair of C-2 epimers. The two epimers are not obtained in equal amounts because they are diastereomers, and diastereomers are generally formed in unequal amounts (Section 6.13).

PROBLEM 15 ◆

What monosaccharides are formed in a modified Kiliani–Fischer synthesis starting with each of the following monosaccharides?

a. D-xylose   **b.** L-threose

USE THE STRATEGY

20.8 SHORTENING THE CHAIN: THE WOHL DEGRADATION

The **Wohl degradation**—the opposite of the Kiliani–Fischer synthesis—shortens an aldose chain by one carbon. Thus, hexoses are converted to pentoses, and pentoses are converted to tetroses.

STEPS IN THE WOHL DEGRADATION

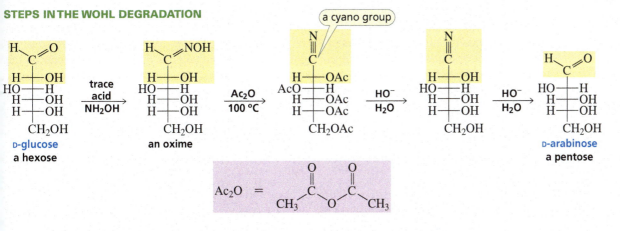

D-glucose
a hexose

an oxime

a cyano group

D-arabinose
a pentose

LEARN THE STRATEGY

- In the first step, the aldehyde reacts with hydroxylamine to form an oxime (Section 16.8).
- Heating with acetic anhydride dehydrates the oxime, forming a cyano group; in addition, all the OH groups are converted to esters as a result of reacting with acetic anhydride (Section 15.16).
- In a basic aqueous solution, all the ester groups are hydrolyzed and the cyano group is eliminated (Sections 15.9 and 16.4).

USE THE STRATEGY

PROBLEM 16 ◆

What two monosaccharides can be degraded to

a. D-ribose?　　　　**b.** D-arabinose?　　　　**c.** L-ribose?

Measuring the Blood Glucose Levels in Diabetes

Glucose in the bloodstream reacts with an NH_2 group of hemoglobin to form an imine (Section 16.8) that subsequently undergoes an irreversible rearrangement to a more stable α-aminoketone known as hemoglobin A1C.

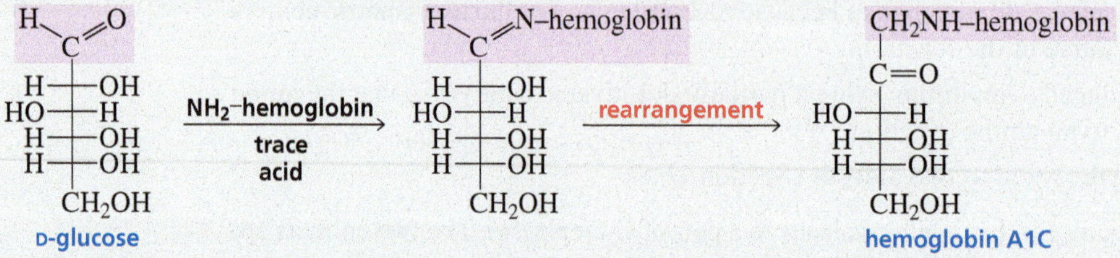

Insulin is the hormone that regulates the level of glucose—and, therefore, the amount of hemoglobin A1C—in the blood. Diabetes is a condition in which the body does not produce sufficient insulin, or in which the insulin it produces does not function properly. Because people with untreated diabetes have increased blood glucose levels, they also have a higher concentration of hemoglobin A1C than people without diabetes. Thus, measuring the hemoglobin A1C level is a way to determine whether the blood glucose level of a diabetic patient is being controlled.

Cataracts, a common complication in diabetes, are caused by the reaction of glucose with the NH_2 group of proteins in the lens of the eye. Some think the arterial rigidity common in old age may be attributable to a similar reaction of glucose with the NH_2 group of proteins.

20.9 THE STEREOCHEMISTRY OF GLUCOSE: THE FISCHER PROOF

Emil Fischer's determination of the stereochemistry of glucose, done in 1891, is an example of brilliant chemical reasoning. He chose (+)-glucose for his study because it is the most common monosaccharide found in nature.

Fischer knew that (+)-glucose is an aldohexose, but 16 different structures can be written for an aldohexose. Which of them represents the structure of (+)-glucose? The 16 stereoisomers of an aldohexose are actually eight pairs of enantiomers, so if we know the structures of one set of eight, we automatically know the structures of the other set of eight. Therefore, Fischer needed to consider only one set of eight. He considered the eight stereoisomers that have their C-5 OH group on the right in the Fischer projection (the stereoisomers shown here, which we now call the D-aldoses). One of these is (+)-glucose and its mirror image is (−)-glucose.

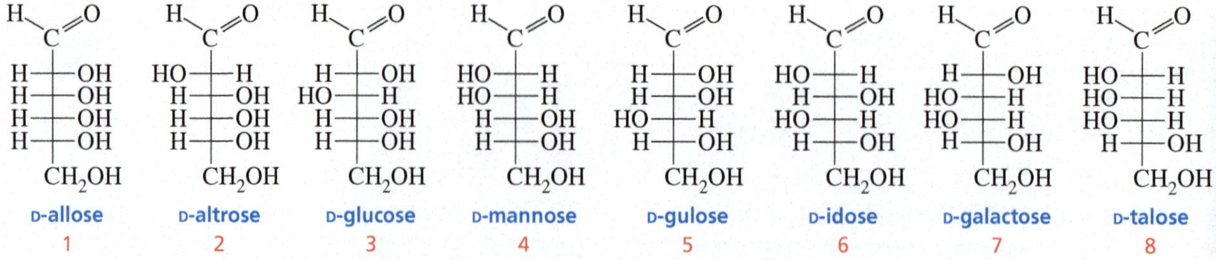

Fischer used the following information to determine glucose's stereochemistry—that is, to determine the configuration of each of its asymmetric centers.

1. When the Kiliani–Fischer synthesis is performed on the sugar known as (−)-arabinose, the two sugars known as (+)-glucose and (+)-mannose are obtained. This means that (+)-glucose and (+)-mannose are C-2 epimers. Consequently, (+)-glucose and (+)-mannose have to be one of the following pairs: sugars 1 and 2, 3 and 4, 5 and 6, or 7 and 8.

2. (+)-Glucose and (+)-mannose are both oxidized by nitric acid to optically active aldaric acids. The aldaric acids of sugars 1 and 7 would not be optically active because each has a plane of symmetry. (We saw in Section 4.14 that a compound containing a plane of symmetry is achiral.) Excluding sugars 1 and 7 means that (+)-glucose and (+)-mannose must be sugars 3 and 4 or 5 and 6.

3. Because (+)-glucose and (+)-mannose are the products obtained when the Kiliani–Fischer synthesis is carried out on (−)-arabinose, Fischer knew that if (−)-arabinose has the structure shown below on the left, then (+)-glucose and (+)-mannose are sugars 3 and 4. On the other hand, if (−)-arabinose has the structure shown on the right, then (+)-glucose and (+)-mannose are sugars 5 and 6:

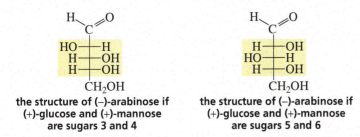

When (−)-arabinose is oxidized with nitric acid, it forms an optically active aldaric acid. This means that the aldaric acid does *not* have a plane of symmetry. Therefore, (−)-arabinose must have the structure shown above on the left because the aldaric acid of the sugar on the right would have a plane of symmetry. Thus, (+)-glucose and (+)-mannose are represented by sugars 3 and 4.

4. Now the only question remaining is whether (+)-glucose is sugar 3 or sugar 4. To answer this, Fischer had to develop a chemical method for interchanging the aldehyde and primary alcohol groups of an aldohexose. When he chemically interchanged those groups on the sugar known as (+)-glucose, he obtained an aldohexose that was different from (+)-glucose, but when he interchanged those groups on (+)-mannose, he still had (+)-mannose. Therefore, he was able to conclude that (+)-glucose is sugar 3 because interchanging its aldehyde and primary alcohol groups leads to a different sugar (L-gulose).

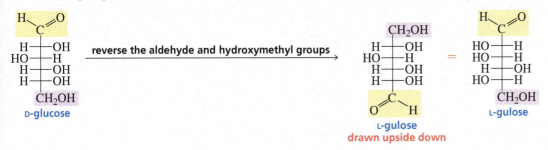

If (+)-glucose is sugar 3, then (+)-mannose must be sugar 4. As predicted, when the aldehyde and hydroxymethyl groups of sugar 4 are interchanged, the same sugar is obtained.

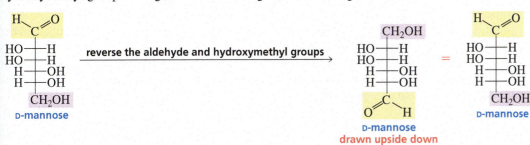

Emil Fischer (1852–1919) *was born in a village near Cologne, Germany. He became a chemist against the wishes of his father, a successful merchant, who wanted him to enter the family business. Fischer was a professor of chemistry at the Universities of Erlangen, Würzburg, and Berlin, and in 1902 he received the Nobel Prize in Chemistry for his work on sugars. Fischer organized German chemical production during World War I. He lost two of his three sons in that war.*

Using similar reasoning, Fischer went on to determine the stereochemistry of the other aldohexoses. He received the Nobel Prize in Chemistry in 1902 for this achievement. His original guess that (+)-glucose is a D-sugar was shown to be correct in 1951, using X-ray crystallography and a new technique known as anomalous dispersion, so all of his structures were correct. If he had been wrong and (+)-glucose had been an L-sugar, his contribution to the stereochemistry of aldoses would still have had the same importance, but all his stereochemical assignments would have had to be reversed.

Glucose/Dextrose

André Dumas first used the term *glucose* in 1838 to refer to the sweet compound that comes from honey and grapes. Later, Kekulé (Section 8.1) decided that it should be called dextrose because it was dextrorotatory. When Fischer studied the sugar, he called it glucose, and chemists have called it glucose ever since, although dextrose is often found on food labels.

LEARN THE STRATEGY

PROBLEM 17 SOLVED

Aldohexoses **A** and **B** are formed from aldopentose **C** via a Kiliani–Fischer synthesis. Nitric acid oxidizes **A** to an optically active aldaric acid, **B** to an optically inactive aldaric acid, and **C** to an optically active aldaric acid. Wohl degradation of **C** forms **D**, which is oxidized by nitric acid to an optically active aldaric acid. Wohl degradation of **D** forms (+)-glyceraldehyde. Identify **A, B, C,** and **D.**

SOLUTION This is the kind of problem that should be solved by working backward. The bottommost asymmetric center in **D** must have the OH group on the right because **D** is degraded to (+)-glyceraldehyde. Because **D** is oxidized to an optically active aldaric acid, **D** must be D-threose. The two bottommost asymmetric centers in **C** and **D** have the same configuration because **C** is degraded to **D**. Because **C** is oxidized to an optically active aldaric acid, **C** must be D-lyxose. Compounds **A** and **B**, therefore, must be D-galactose and D-talose. Because **A** is oxidized to an optically active aldaric acid, it must be D-talose and **B** must be D-galactose.

USE THE STRATEGY

PROBLEM 18 ◆

Identify **A, B, C,** and **D** in the preceding problem if **D** is oxidized to an optically *inactive* aldaric acid; if **A, B,** and **C** are oxidized to optically active aldaric acids; and if interchanging the aldehyde and alcohol groups of **A** leads to a different sugar.

20.10 MONOSACCHARIDES FORM CYCLIC HEMIACETALS

D-Glucose exists in three different forms: the open-chain form of D-glucose that we have been discussing, and two cyclic forms—α-D-glucose and β-D-glucose. We know that the two cyclic forms are different because they have different melting points and different specific rotations.

Structures of α-D-Glucose and β-D-Glucose

How can D-glucose exist in a cyclic form? In Section 16.9, we saw that an aldehyde reacts with an alcohol to form a hemiacetal. The reaction of the alcohol group bonded to C-5 of D-glucose with the aldehyde group forms two cyclic (six-membered ring) hemiacetals.

To see that the OH group on C-5 is in the proper position to attack the aldehyde group, we need to convert the Fischer projection of D-glucose to a flat ring structure. To do this, draw the primary alcohol group *up* from the back left-hand corner as shown on the next page. Groups on the *right* in a Fischer projection (blue) are *down* in the cyclic structure, and groups on the *left* in a Fischer projection (red) are *up* in the cyclic structure.

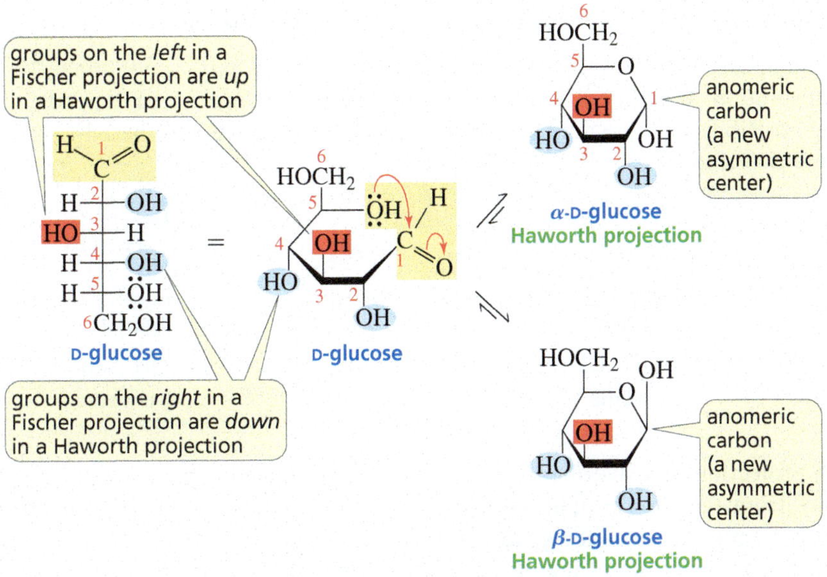

groups on the *left* in a Fischer projection are *up* in a Haworth projection

D-glucose

groups on the *right* in a Fischer projection are *down* in a Haworth projection

D-glucose

anomeric carbon (a new asymmetric center)

α-D-glucose
Haworth projection

anomeric carbon (a new asymmetric center)

β-D-glucose
Haworth projection

Groups on
the *right* in a Fischer projection
are *down* in a Haworth projection.

Groups on
the *left* in a Fischer projection
are *up* in a Haworth projection.

The cyclic hemiacetals shown here are drawn as Haworth projections. In a **Haworth projection,** the six-membered ring is represented as flat and is viewed edge-on. The ring oxygen is always placed in the back right-hand corner of the ring, with C-1 on the right-hand side, and the primary alcohol group attached to C-5 is drawn *up* from the back left-hand corner.

There are two different cyclic hemiacetals because the carbonyl carbon of the open-chain aldehyde becomes a new asymmetric center in the cyclic hemiacetal. If the OH group bonded to the new asymmetric center points down (is trans to the primary alcohol group at C-5), the hemiacetal is *α*-D-glucose; if the OH group points up (is cis to the primary alcohol group at C-5), the hemiacetal is *β*-D-glucose. The mechanism for cyclic hemiacetal formation is the same as the mechanism for hemiacetal formation between individual aldehyde and alcohol molecules (Section 16.9).

Anomers

α-D-Glucose and *β*-D-glucose are anomers. **Anomers** are two sugars that differ in configuration only at the carbon that was the carbonyl carbon in the open-chain form. This carbon is called the **anomeric carbon**. The prefixes *α*- and *β*- denote the configuration about the anomeric carbon. Because anomers, like epimers, differ in configuration at only one carbon, they too are a particular kind of diastereomer. Notice that the anomeric carbon is the only carbon in the molecule that is bonded to two oxygens.

Cyclic Compounds Are in Equilibrium with the Open-Chain Compound

In an aqueous solution, the open-chain form of D-glucose is in equilibrium with the two cyclic hemiacetals. Because formation of the cyclic hemiacetals proceeds nearly to completion (unlike formation of acyclic hemiacetals), very little glucose is in the open-chain form (about 0.02%). Even so, the sugar still undergoes the reactions discussed in previous sections (oxidation, reduction, imine formation, and so on) because the reagents react with the small amount of open-chain aldehyde that is present. As the open-chain aldehyde reacts, the equilibrium shifts to produce more open-chain aldehyde, which can then undergo reaction. Eventually, all the glucose molecules react by way of the open-chain form.

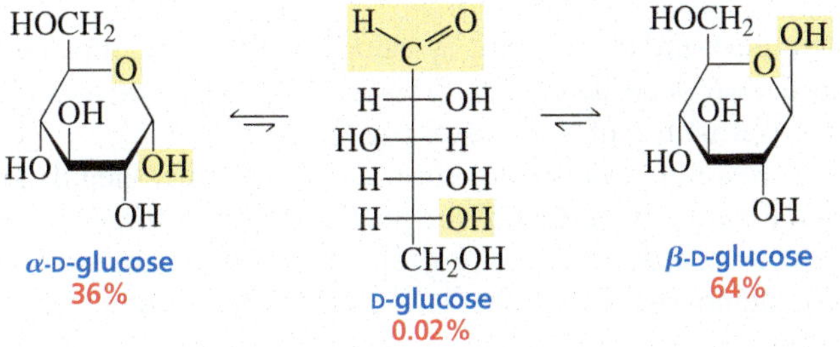

α-D-glucose
36%

D-glucose
0.02%

β-D-glucose
64%

Aldoses Exist Predominately as Cyclic Compounds

If an aldose can form a five- or a six-membered ring, it will exist predominantly as a cyclic hemiacetal in solution. Whether a five- or a six-membered ring is formed depends on their relative stabilities. D-Ribose is an example of an aldose that forms five-membered-ring hemiacetals: α-D-ribose and β-D-ribose.

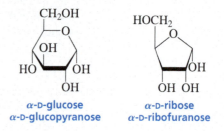

D-ribose   D-ribose α-D-ribose β-D-ribose
Haworth projections

The Haworth projection of a five-membered-ring sugar is viewed edge-on, with the ring oxygen pointing away from the viewer. The anomeric carbon is again on the right-hand side of the molecule, and the primary alcohol group is drawn up from the back left-hand corner. Again, notice that the anomeric carbon is the only carbon in the molecule that is bonded to two oxygens.

Pyranoses and Furanoses

Six-membered-ring sugars are called **pyranoses,** and five-membered-ring sugars are called **furanoses.** These names come from *pyran* and *furan,* the names of the cyclic ethers shown in the margin. Consequently, α-D-glucose is also called α-D-glucopyranose, and α-D-ribose is also called α-D-ribofuranose. The prefix "α" indicates the configuration about the anomeric carbon, and *pyranose* or *furanose* indicates the size of the ring.

pyran

furan

α-D-glucose
α-D-glucopyranose

α-D-ribose
α-D-ribofuranose

Ketoses Exist Predominately as Cyclic Compounds

Ketoses also exist in solution predominantly in cyclic forms. For example, D-fructose forms a five-membered-ring hemiacetal because its C-5 OH group reacts with its ketone carbonyl group. If the OH group bonded to the new asymmetric center is trans to the primary alcohol group, then the compound is α-D-fructofuranose; if it is cis to the primary alcohol group, the compound is β-D-fructofuranose. Notice that the anomeric carbon is C-2 in ketoses, not C-1 as in aldoses.

α-D-fructofuranose β-D-fructofuranose α-D-fructopyranose β-D-fructopyranose

D-Fructose can also form a six-membered ring by using its C-6 OH group. The pyranose form predominates in the monosaccharide, whereas the furanose form predominates when the sugar is part of a disaccharide. (See the structure of sucrose on page 972.)

Haworth projections are useful because they show clearly whether the OH groups on the ring are cis or trans to each other. Five-membered rings are nearly planar, so furanoses are represented fairly accurately by Haworth projections. Haworth projections, however, are structurally misleading for pyranoses because a six-membered ring is not flat—it exists preferentially in a chair conformation (Section 3.13).

Mutarotation

When crystals of pure α-D-glucose are dissolved in water, the specific rotation gradually changes from +112.2 to +52.7. When crystals of pure β-D-glucose are dissolved in water, the specific rotation gradually changes from +18.7 to +52.7.

This change in rotation occurs because, in water, the hemiacetal opens to form the aldehyde, and both α-D-glucose and β-D-glucose are formed when the aldehyde recyclizes. Eventually, the three forms of glucose reach equilibrium concentrations. The specific rotation of the equilibrium mixture is +52.7. This is why the same specific rotation results whether the crystals originally dissolved in water are α-D-glucose or β-D-glucose or any mixture of the two. A slow change in optical rotation to an equilibrium value is called **mutarotation.** Notice that at equilibrium, there is almost twice as much β-D-glucose (64%) as α-D-glucose (36%); see page 963.

PROBLEM 19 SOLVED

LEARN THE STRATEGY

4-Hydroxy- and 5-hydroxyaldehydes exist primarily as cyclic hemiacetals. Draw the structure of the cyclic hemiacetal formed by each of the following:

a. 4-hydroxybutanal   **b.** 4-hydroxypentanal   **c.** 5-hydroxypentanal   **d.** 4-hydroxyheptanal

SOLUTION TO 19 a. Draw the reactant with its alcohol and carbonyl groups on the same side of the molecule, and then look to see what size ring will form. Two cyclic products are obtained because the carbonyl carbon of the reactant has been converted to a new asymmetric center in the product.

PROBLEM 20

USE THE STRATEGY

Draw the following sugars using Haworth projections:

a. β-D-galactopyranose   **b.** α-D-tagatopyranose   **c.** α-L-glucopyranose

PROBLEM 21

D-Glucose most often exists as a pyranose, but it can also exist as a furanose. Draw the Haworth projection of α-D-glucofuranose.

PROBLEM 22 ◆

Draw the anomers of D-erythrofuranose.

20.11 GLUCOSE IS THE MOST STABLE ALDOHEXOSE

Drawing a Chair Conformer of D-Glucose

The chair conformer of D-glucose shows why it is the most common aldohexose in nature. To convert the Haworth projection of D-glucose to a chair conformer, start by drawing the chair so that the backrest is on the left and the footrest is on the right. Then place the ring oxygen at the back right-hand corner and the primary alcohol group in the equatorial position. Note that the primary alcohol group is the largest of all the substituents, and, therefore, it is more stable in the equatorial position where there is less steric strain (Section 3.14).

α-D-glucose
chair conformer

β-D-glucose
chair conformer

The α-position is down in a Haworth projection and axial in a chair conformation.

The β-position is up in a Haworth projection and equatorial in a chair conformation.

- Because the OH group bonded to C-4 is trans to the primary alcohol group (this is easily seen in the Haworth projection), the C-4 OH group is also in the equatorial position. (Recall from Section 3.15 that 1,2-diequatorial substituents are trans to one another.)
- The C-3 OH group is trans to the C-4 OH group; therefore, the C-3 OH group is also in the equatorial position.
- As you move around the ring, you will find that all the OH groups in β-D-glucose are in equatorial positions.
- The axial positions are all occupied by hydrogens, which require little space and, therefore, experience little steric strain. No other aldohexose exists in such a strain-free conformation. This means that D-glucose is the most stable of all the aldohexoses; therefore, we should not be surprised that it is the most prevalent aldohexose in nature.

The OH group bonded to the anomeric carbon is in the equatorial position in β-D-glucose, whereas it is in the axial position in α-D-glucose. Therefore, β-D-glucose is more stable than α-D-glucose, so β-D-glucose predominates at equilibrium in an aqueous solution.

α-D-glucose 36%   β-D-glucose 64%

Drawing Chair Conformers of Other Pyranoses

LEARN THE STRATEGY

If you remember that all the OH groups in β-D-glucose are in equatorial positions, you will find it easy to draw a chair conformer of any other pyranose. For example, if you want to draw α-D-galactose, you would put all the OH groups in equatorial positions except the OH group at C-4 (because galactose is the C-4 epimer of glucose) and the OH group at C-1 (because you want the α-anomer), which both go in axial positions.

α-D-galactose

Drawing a Chair Conformer of an L-Pyranose

To draw an L-pyranose, draw the D-pyranose first, and then draw its mirror image. For example, to draw β-L-gulose, first draw β-D-gulose, and then draw its mirror image. (Gulose differs from glucose at C-3 and C-4, so the OH groups at these positions go in axial positions.)

β-D-gulose   β-L-gulose

USE THE STRATEGY

PROBLEM 23 ◆

Which OH groups are in the axial position in β-D-mannopyranose?

Which OH groups are in the axial position in each of the following?

a. β-D-idopyranose **b.** α-D-allopyranose

20.12 FORMATION OF GLYCOSIDES

In the same way that a hemiacetal reacts with an alcohol to form an acetal (Section 16.9), the cyclic hemiacetal formed by a monosaccharide can react with an alcohol to form two acetals.

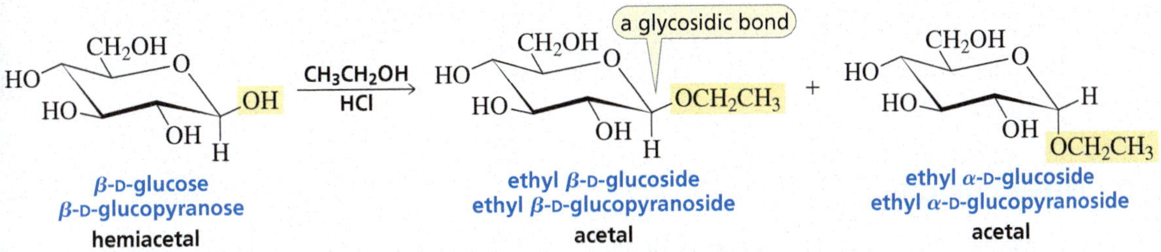

β-D-glucose
β-D-glucopyranose
hemiacetal

ethyl β-D-glucoside
ethyl β-D-glucopyranoside
acetal

ethyl α-D-glucoside
ethyl α-D-glucopyranoside
acetal

The acetal of a sugar is called a **glycoside,** and the bond between the anomeric carbon and the alkoxy oxygen is called a **glycosidic bond.** Glycosides are named by replacing the "e" ending of the sugar's name with "ide." Thus, a glycoside of glucose is a glucoside, a glycoside of galactose is a galactoside, and so on. If the pyranose or furanose name is used, the acetal is called a **pyranoside** or a **furanoside.**

Mechanism for Glycoside Formation

Notice that the reaction of a single anomer with an alcohol leads to the formation of both the α- and β-glycosides. The mechanism of the reaction shows why both glycosides are formed.

MECHANISM FOR GLYCOSIDE FORMATION

an oxocarbenium ion

alcohol approaches from the top

alcohol approaches from the bottom

a β-glycoside

an α-glycoside
major product

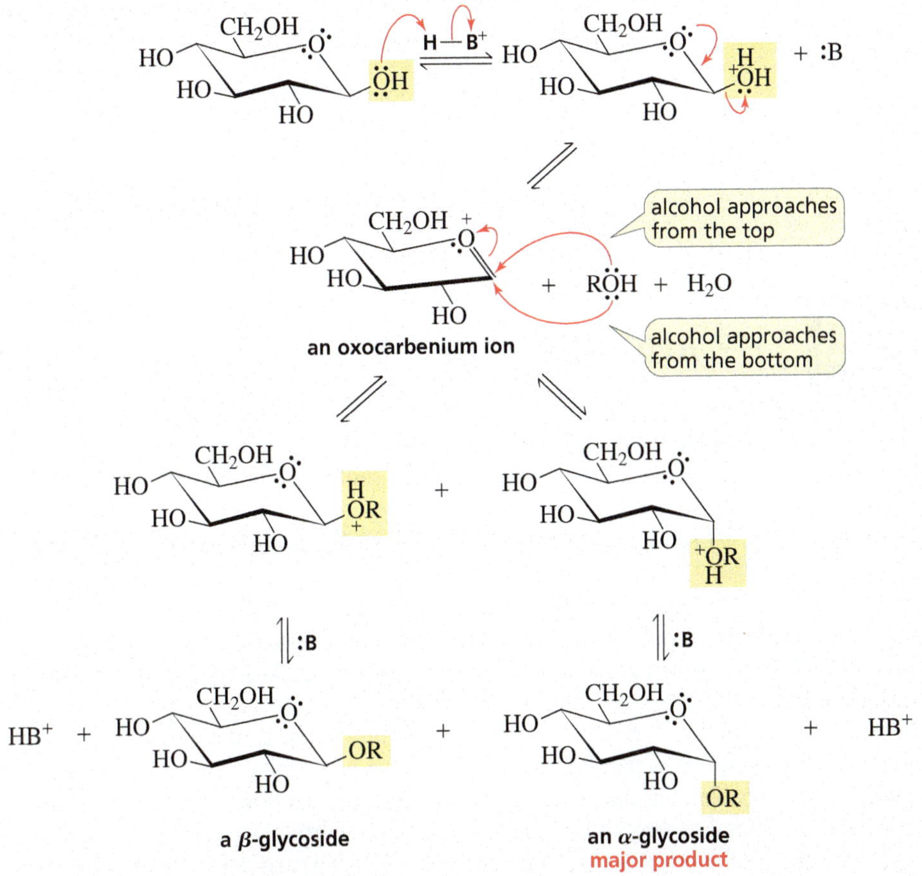

- The acid protonates the OH group bonded to the anomeric carbon.

- A lone pair on the ring oxygen helps eliminate a molecule of water. The anomeric carbon in the resulting oxocarbenium ion is sp^2 hybridized, so that part of the molecule is planar. (An **oxocarbenium** ion has a positive charge that is shared by a carbon and an oxygen.)

- When the alcohol approaches from the top of the plane, the β-glycoside is formed; when the alcohol approaches from the bottom of the plane, the α-glycoside is formed.

Notice that the mechanism is the same as that shown for acetal formation in Section 16.9. Surprisingly, D-glucose forms more α-glycoside than β-glycoside. The reason for this is explained in the next section.

PROBLEM 25 ◆

Draw the products formed when β-D-galactose reacts with ethanol and HCl. (Show all structures as chair conformers.)

N-Glycosides

The reaction of a monosaccharide with an amine is similar to the reaction of a monosaccharide with an alcohol. The product of the reaction is an *N*-glycoside. An **N-glycoside** has a nitrogen in place of the oxygen at the glycosidic linkage. The subunits of DNA and RNA are β-*N*-glycosides (Section 26.1).

N-phenyl α-D-ribosylamine
an α-*N*-glycoside

N-phenyl β-D-ribosylamine
a β-*N*-glycoside

PROBLEM 26 ◆

Why is only a trace amount of acid used in the formation of an *N*-glycoside?

20.13 THE ANOMERIC EFFECT

We saw that β-D-glucose is more stable than α-D-glucose because there is more room for a substituent in the equatorial position. However, the preference of the OH group for the equatorial position is not as large as expected. For example, the relative amounts of β-D-glucose and α-D-glucose are 2 : 1 (Section 20.10), but the relative amounts of the OH group of cyclohexanol in the equatorial and axial positions is 5.4 : 1 (Table 3.9 on page 128). Therefore, there must be a factor that stabilizes the α-position.

When glucose reacts with an alcohol to form a glucoside, the major product is the α-glucoside. Because acetal formation is reversible, the α-glucoside must be more stable than the β-glucoside. The preference of certain substituents bonded to the anomeric carbon for the axial position is called the **anomeric effect.**

What is responsible for the anomeric effect? If the substituent is axial, one of the ring oxygen's lone pairs is in an orbital that is parallel to the $\sigma*$ antibonding orbital of the C–Z bond. (See the top of page 963.) The molecule then can be stabilized by hyperconjugation—some of the electron density moves from the sp^3 orbital of oxygen into the $\sigma*$ antibonding orbital (see page 120). If the substituent is equatorial, neither of the orbitals that contain a lone pair is aligned correctly for overlap.

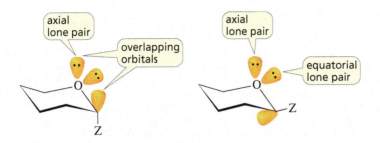

20.14 REDUCING AND NONREDUCING SUGARS

Because glycosides are acetals, they are not in equilibrium with the open-chain aldehyde (or ketone) in aqueous solutions. Without being in equilibrium with a compound that has a carbonyl group, they cannot be oxidized by Tollens' reagent (Section 20.6). Glycosides, therefore, are nonreducing sugars—they cannot reduce the oxidizing reagent.

Hemiacetals, on the other hand, are in equilibrium with the open-chain sugars in aqueous solution, so they can reduce the oxidizing reagent. In summary, as long as a sugar has an aldehyde, a ketone, or a hemiacetal group, it can reduce the oxidizing agent and is, therefore, classified as a **reducing sugar.** An acetal is a **nonreducing sugar.**

> A sugar with an aldehyde, a ketone, or a hemiacetal group is a reducing sugar. An acetal is a nonreducing sugar.

PROBLEM 27 SOLVED

LEARN THE STRATEGY

Name the following compound and indicate whether or not it is a reducing sugar:

$$\text{HO}\underset{\text{HO}}{\overset{\text{CH}_2\text{OH}}{\underset{\text{OH}}{\diagdown\!\!\diagup}}}\text{O}\;\;\text{OCH}_2\text{CH}_2\text{CH}_3$$

SOLUTION The only OH group in an axial position in part **a** is the one at C-3. Therefore, this sugar is the C-3 epimer of D-glucose, which is D-allose. The substituent at the anomeric carbon is in the β-position. Thus, the sugar's name is propyl β-D-alloside or propyl β-D-allopyranoside. Because the sugar is an acetal, it is a nonreducing sugar.

PROBLEM 28 ◆

USE THE STRATEGY

Name the following compounds and indicate whether or not each is a reducing sugar:

a.
$$\text{HO}\;\underset{\text{HO}}{\overset{\text{CH}_2\text{OH}}{\diagdown}}\underset{\text{OH}}{\overset{\text{O}}{\diagup}}\text{OH}$$

b.
$$\text{HO}\;\underset{\text{HO}}{\overset{\text{CH}_2\text{OH}}{\diagdown}}\underset{\text{OH}\quad\text{OCH}_3}{\overset{\text{O}}{\diagup}}$$

c.
$$\underset{\text{OH OH}}{\overset{\text{HOCH}_2}{\diagup}}\underset{\text{CH}_2\text{OH}}{\overset{\text{O}\quad\text{OCH}_2\text{CH}_3}{\diagdown}}$$

20.15 DISACCHARIDES

If the hemiacetal group of a monosaccharide forms an acetal by reacting with an alcohol group of another monosaccharide, the glycoside that is formed is a disaccharide. **Disaccharides** are compounds that consist of two monosaccharide subunits hooked together by a **glycosidic linkage.**

Maltose

Maltose, a disaccharide obtained from the hydrolysis of starch, contains two D-glucose subunits connected by a glycosidic linkage (see the top of the next page). This particular linkage is called an **α-1,4′-glycosidic linkage** because the linkage is between C-1 of one sugar subunit and C-4 of the other, and the oxygen bonded to the anomeric carbon in the glycosidic linkage is in the α-position. The prime superscript indicates that C-4 is not in the same ring as C-1.

Remember that the α-position is axial and the β-position is equatorial when a sugar is shown in a chair conformation.

maltose

Notice that the structure of maltose does not specify the configuration of the anomeric carbon that is not an acetal (the anomeric carbon of the subunit on the right marked with a wavy line) because maltose can exist in both the α and β forms. In α-maltose, the OH group bonded to this anomeric carbon is in the axial position. In β-maltose, this OH group is in the equatorial position.

Because maltose can exist in both α and β forms, mutarotation occurs when crystals of one form are dissolved in water. Maltose is a reducing sugar because the right-hand subunit is a hemiacetal and, therefore, is in equilibrium with the open-chain aldehyde that is easily oxidized.

Cellobiose

Cellobiose, a disaccharide obtained from the hydrolysis of cellulose, also contains two D-glucose subunits. Cellobiose is different from maltose, however, because the two glucose subunits are hooked together by a **β-1,4'-glycosidic linkage.** Thus, the only difference in the structures of maltose and cellobiose is the configuration of the glycosidic linkage. Like maltose, cellobiose exists in both α and β forms because the OH group bonded to the anomeric carbon not involved in acetal formation can be in either the axial position (in α-cellobiose) or the equatorial position (in β-cellobiose). Cellobiose is a reducing sugar and undergoes mutarotation because the subunit on the right is a hemiacetal.

cellobiose

Lactose

Lactose is a disaccharide found in milk. It constitutes 4.5% of cow's milk and 6.5% of human milk by weight. The subunits of lactose are D-galactose and D-glucose. The D-galactose subunit is an acetal, and the D-glucose subunit is a hemiacetal. The subunits are joined by a β-1,4'-glycosidic linkage. Because one of the subunits is a hemiacetal, lactose is a reducing sugar and undergoes mutarotation.

lactose

Determining the Sugar in Lactose That Has the Hemiacetal Group

A simple experiment can prove that the hemiacetal group in lactose belongs to the glucose residue and not to the galactose residue. The disaccharide is treated with excess methyl iodide in the presence of Ag_2O, reagents that methylate all the OH groups via S_N2 reactions. Because the OH group is a relatively poor nucleophile, silver oxide is used to increase the leaving propensity of the iodide ion. The product is then hydrolyzed under acidic conditions.

2,3,4,6-tetra-*O*-methylgalactose   **2,3,6-tri-*O*-methylglucose**

This treatment hydrolyzes the two acetal groups, but the ethers, formed by methylating the OH groups, are untouched. Identification of the products shows that the glucose residue contained the hemiacetal group in the disaccharide because its C-4 OH group was not able to react with methyl iodide (because it was an acetal linkage with galactose). The C-4 OH of galactose, on the other hand, was able to react with methyl iodide.

Lactose Intolerance

Lactase is an enzyme that specifically breaks the β-1,4'-glycosidic linkage of lactose. Cats and dogs lose their intestinal lactase when they become adults; they are then no longer able to digest lactose. Consequently, when they are fed milk or milk products, the undegraded lactose causes digestive problems such as bloating, abdominal pain, and diarrhea. These problems occur because only monosaccharides can pass into the bloodstream, so lactose, a disaccharide, has to pass undigested into the large intestine.

When humans have stomach flu or other intestinal disturbances, they can temporarily lose their lactase, thereby becoming lactose intolerant. About 75% of adults lose their lactase permanently as they mature—explaining why "lactose-free" products are so common. Those intolerant to lactose can take lactase in pill form before eating products that contain lactose.

Lactose intolerance is most common in people whose ancestors came from nondairy-producing countries. For example, only 3% of Danes are lactose intolerant, compared with 90% of all Chinese and Japanese and 97% of Thais. This is why you are not likely to find dairy items on menus in Chinese, Japanese, and Thai restaurants.

Galactosemia

After lactose is degraded to glucose and galactose, the galactose must be converted to glucose before it can be used by cells. Individuals who do not have the enzyme that converts galactose to glucose have the genetic disease known as galactosemia. Without this enzyme, galactose accumulates in the bloodstream, a condition that can cause mental retardation and even death in infants. Galactosemia is treated by excluding galactose from the diet.

Sucrose

The most common disaccharide, sucrose, is the substance we know as table sugar. Obtained from sugar beets and sugar cane (page 969), sucrose consists of a D-glucose subunit and a D-fructose subunit linked by a glycosidic bond between C-1 of glucose (in the α-position) and C-2 of fructose (in the β-position). About 90 million tons of sucrose are produced commercially throughout the world each year.

Unlike the other disaccharides that we have looked at, sucrose is not a reducing sugar and does not exhibit mutarotation because its glycosidic bond is between the anomeric carbon of glucose and the anomeric carbon of fructose. Sucrose, therefore, does not have a hemiacetal group, so it is not in equilibrium with the readily oxidized open-chain aldehyde or ketone form in aqueous solution.

sucrose

a mixture of sucrose, glucose, and fructose

Sucrose has a specific rotation of +66.5. When it is hydrolyzed, the resulting 1 : 1 mixture of glucose and fructose has a specific rotation of −22.0. Because the sign of the rotation changes when sucrose is hydrolyzed, a 1 : 1 mixture of glucose and fructose is called *invert sugar*. The enzyme that catalyzes the hydrolysis of sucrose is called *invertase*. Honeybees have invertase, so the honey they produce is a mixture of sucrose, glucose, and fructose.

Because fructose is sweeter than sucrose, some "lite" foods contain fructose instead of sucrose, which means that they achieve the same level of sweetness with a lower sugar (lower calorie) content.

PROBLEM 29 ◆

What is the specific rotation of an equilibrium mixture of fructose? (*Hint:* Recall that the specific rotation of an equilibrium mixture of glucose is +52.7.)

A Toxic Disaccharide

Amygdalin is a disaccharide found in apricot and peach pits. At one time, amygdalin and a modified form called laetrile were widely touted as a treatment for cancer. Subsequent studies showed them to be clinically ineffective and toxic. Amygdalin is hydrolyzed to mandelonitrile, which eliminates hydrogen cyanide (the reason amygdalin and laetrile are toxic) in an enzyme-catalyzed reaction (Section 16.4).

amygdalin   mandelonitrile   benzaldehyde   hydrogen cyanide

Apheloria corrugata

The millipede *Apheloria corrugata* stores mandelonitrile to use against predators. When it is about to be attacked, the millipede discharges mandelonitrile into a compartment that contains the enzyme that catalyzes the elimination reaction that forms benzaldehyde and hydrogen cyanide (a gas).

PROBLEM 30 ◆

a. Identify the sugars in amygdalin.
b. Identify the glycosidic linkage that connects the sugars.

20.16 POLYSACCHARIDES

Polysaccharides contain as few as 10 or as many as several thousand monosaccharide units joined together by glycosidic linkages. The most common polysaccharides are starch and cellulose.

Starch

Starch is the major component of flour, potatoes, rice, beans, corn, and peas. It is a mixture of two different polysaccharides: amylose (~20%) and amylopectin (~80%). Amylose is composed of unbranched chains of D-glucose units joined by α-1,4$'$-glycosidic linkages.

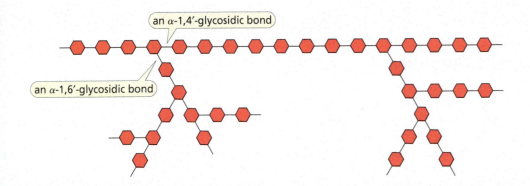

Amylopectin is a branched polysaccharide. Like amylose, it is composed of chains of D-glucose units joined by α-1,4$'$-glycosidic linkages. Unlike amylose, however, amylopectin also contains **α-1,6$'$-glycosidic linkages.** These linkages create branches in the polysaccharide (Figure 20.1). Amylopectin can contain up to 10^6 glucose units, making it one of the largest molecules found in nature.

◀ **Figure 20.1**
Branching in amylopectin. The hexagons represent glucose units. They are joined by α-1,4$'$- and α-1,6$'$-glycosidic bonds.

Cells oxidize D-glucose in the first of a series of processes that provide them with energy (Section 24.6). When animals have more D-glucose than they need for energy, they convert the excess D-glucose to a polymer called glycogen. Glycogen has a structure similar to that of amylopectin, but glycogen has more branches (Figure 20.2). The branch points in glycogen occur about every 10 residues, whereas those in amylopectin occur about every 25 residues. The high degree of branching in glycogen has important physiological consequences. When an animal needs energy, many individual glucose units can be simultaneously removed from the ends of many branches. Plants convert excess D-glucose to starch.

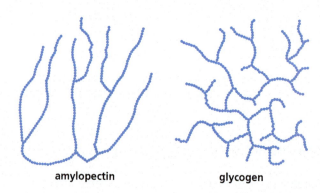

▶ **Figure 20.2**
A comparison of the branching in amylopectin and glycogen.

amylopectin   glycogen

Why the Dentist Is Right

Bacteria found in the mouth have an enzyme that converts sucrose to a polysaccharide called dextran. Dextran is made up of glucose units joined mainly through α-1,3'- and α-1,6'-glycosidic linkages. About 10% of dental plaque is composed of dextran, and bacteria hidden in the plaque attack tooth enamel. This is the chemical basis for your dentist's warning not to eat candy. This is also why sorbitol and mannitol are the saccharides added to "sugarless" gum—they cannot be converted to dextran.

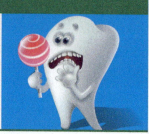

Cellulose

Cellulose is the major structural component of plants. Cotton, for example, is composed of about 90% cellulose, and wood is about 50% cellulose. Like amylose, cellulose is composed of unbranched chains of D-glucose units. Unlike amylose, however, the glucose units in cellulose are joined by β-1,4'-glycosidic linkages rather than by α-1,4'-glycosidic linkages (see page 970).

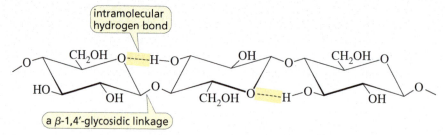

intramolecular hydrogen bond

a β-1,4'-glycosidic linkage

three subunits of cellulose

Physical Properties of Starch and Cellulose

The different glycosidic linkages in starch and cellulose give these compounds very different physical properties. The α-linkages in starch cause amylose to form a helix that promotes hydrogen bonding of its OH groups to water molecules (Figure 20.3). As a result, starch is soluble in water.

On the other hand, the β-linkages in cellulose promote the formation of intramolecular hydrogen bonds. Consequently, these molecules form linear arrays (Figure 20.4), held together by hydrogen bonds between adjacent chains (see Figure 20.4 on the next page). These large aggregates cause cellulose to be insoluble in water. The strength of these bundles of polymer chains makes cellulose an effective structural material. Processed cellulose is also used for the production of paper and cellophane.

All mammals have the enzyme (α-glucosidase) that hydrolyzes the α-1,4'-glycosidic linkages that join glucose units in amylose, amylopectin, and glycogen, but they do *not* have the enzyme (β-glucosidase) that hydrolyzes β-1,4'-glycosidic linkages. As a result, mammals *cannot* obtain the glucose they need by eating cellulose. However, bacteria that possess β-glucosidase inhabit the digestive tracts of grazing animals, so cows can eat grass and horses can eat hay to meet their nutritional requirements for glucose. Termites also harbor bacteria that break down the cellulose in the wood they eat.

▲ **Figure 20.3**
The α-1,4'-glycosidic linkages in amylose cause it to form a left-handed helix. Many of its OH groups form hydrogen bonds with water molecules.

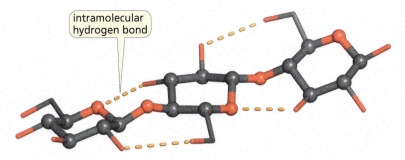

▲ Figure 20.4
The β-1,4′-glycosidic linkages in cellulose form intramolecular hydrogen bonds, which cause the molecules to assemble in linear arrays. (Notice that hydrogens are not shown in the figure.)

strands of cellulose in a plant cell wall

Chitin

Chitin (KY-tin) is a polysaccharide that is structurally similar to cellulose. It is the major structural component of the shells of crustaceans (such as lobsters, crabs, and shrimp) and the exoskeletons of insects and other arthropods. It is also the structural material of fungi. Like cellulose, chitin has β-1,4′-glycosidic linkages. Unlike cellulose, chitin has an *N*-acetylamino group instead of an OH group at the C-2 position. The β-1,4′-glycosidic linkages give chitin its structural rigidity.

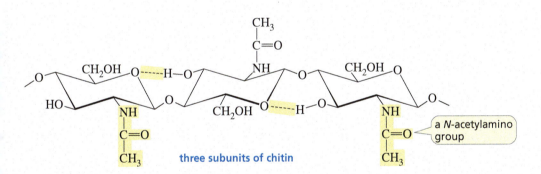

three subunits of chitin

a *N*-acetylamino group

The shell of this bright orange crab from Australia is composed largely of chitin.

Controlling Fleas

Several different drugs have been developed to help pet owners control fleas. One of these drugs is lufenuron, the active ingredient in Program. Lufenuron interferes with the flea's production of chitin. The consequences are fatal for the flea because its exoskeleton is composed primarily of chitin.

lufenuron

PROBLEM 31 ◆

What is the main structural difference between

a. amylose and cellulose?

b. amylose and amylopectin?

c. amylopectin and glycogen?

d. cellulose and chitin?

20.17 SOME NATURALLY OCCURRING COMPOUNDS DERIVED FROM CARBOHYDRATES

Deoxy Sugars

Deoxy sugars are sugars in which one of the OH groups is replaced by a hydrogen (*deoxy* means "without oxygen"). 2-Deoxyribose is a deoxy sugar that is missing the oxygen at the C-2 position. β-D-ribose is the sugar component of ribonucleic acid (RNA), whereas β-D-2-deoxyribose is the sugar component of deoxyribonucleic acid (DNA) (see Section 26.1).

β-D-ribose β-D-2-deoxyribose

Amino Sugars

In **amino sugars,** one or more of the OH groups are replaced by an amino group. *N*-Acetylglucosamine—the subunit of chitin and one of the subunits of bacterial cell walls (Section 22.10)—is an example of an amino sugar.

Some important antibiotics contain amino sugars. For example, two of the three subunits of the antibiotic gentamicin are deoxyamino sugars. Notice that the middle subunit is missing the ring oxygen, so it is not a sugar.

gentamicin
an antibiotic

Gentamicin is one of several aminoglycoside antibiotics; streptomycin and neomycin are others. The antibiotics work by binding to a bacterial ribosome where protein synthesis takes place (Section 26.8). As a result, the bacteria are not able to synthesize proteins.

Resistance to Antibiotics

A bacterial strain typically takes 15 to 20 years to become resistant to an antibiotic. For example, penicillin became widely available in 1944 and by 1952, 60% of all *Staphylococcus aureus* infections were penicillin resistant (see pages 713–714). As a result of the widespread use of the aminoglycoside antibiotics, some bacteria developed enzymes that can acetylate or phosphorylate the OH and NH_2 groups of the antibiotic. When this happens, the antibiotic can no longer bind to the bacterial ribosome, so it has no effect on the bacteria.

Until relatively recently, the last discovery of a new class of antibiotics was in the 1970s, so **drug resistance** became an increasingly important problem in medicinal chemistry. Vancomycin had been the antibiotic of last resort because there were no reported cases of vancomycin-resistant bacteria until 1989, when more and more bacteria became resistant.

The approval of Zyvox by the FDA in April 2000 was met with great relief by the medical community because it was the first in a new family of antibiotics—namely, the oxazolidinones. In clinical trials, Zyvox was found to cure 75% of the patients infected with bacteria that had become resistant to all other antibiotics. Another new class of antibiotic became available in 2005 when the FDA approved Cubicin, the first of the cyclic lipopeptide antibiotics.

Antibiotic resistance is still spreading faster than new drugs are becoming available. However, a new drug called Teixobactin, which inhibits cell wall synthesis, was discovered in 2015. The hope is that patients will be less likely to develop resistance to this new class of drug because it binds to less mutable fatty acid molecules rather than to more mutable proteins. It is scheduled to be used in human clinical trials in 2017.

Heparin—A Natural Anticoagulant

Heparin is a polysaccharide found principally in cells that line arterial walls. Some of its alcohol and amino groups are sulfonated, some of its alcohol groups are oxidized, and some of its amino groups are acetylated. Heparin is released to prevent excessive blood clot formation when an injury occurs. Heparin is widely used clinically—particularly after surgery—to prevent blood from clotting.

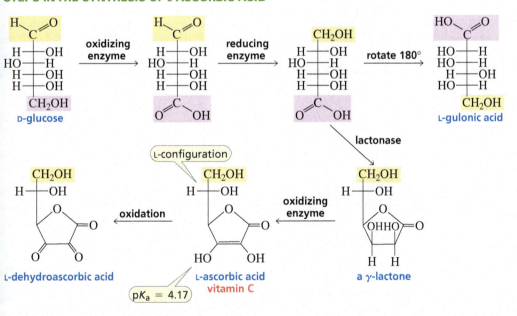

Vitamin C

L-Ascorbic acid (vitamin C) is synthesized from D-glucose in plants and in the livers of most vertebrates. Primates and guinea pigs do not have the enzymes necessary for the biosynthesis of vitamin C, so they must obtain the vitamin from their diets.

STEPS IN THE SYNTHESIS OF L-ASCORBIC ACID

- The biosynthesis of vitamin C involves the conversion of D-glucose to L-gulonic acid, reminiscent of the last step in the Fischer proof.

- L-Gulonic acid is converted to a γ-lactone by the enzyme lactonase.

- The lactone is oxidized to L-ascorbic acid. The L-designation of ascorbic acid refers to the configuration at C-5, which was C-2 in D-glucose and C-5 in L-gulonic acid.

Although L-ascorbic acid does not have a carboxylic acid group, it is an acidic compound because the pK_a of the C-3 OH group is 4.17. L-Ascorbic acid is readily oxidized to L-dehydroascorbic acid, which is also physiologically active. If the lactone ring is opened by hydrolysis, all vitamin C activity is lost. Therefore, not much intact vitamin C survives in food that has been thoroughly cooked. And if food is cooked in water and then drained, the water-soluble vitamin is thrown out with the water!

Vitamin C

Vitamin C is an antioxidant because it traps radicals formed in aqueous environments, preventing harmful oxidation reactions the radicals would cause (Section 12.11). Not all the physiological functions of vitamin C are known. However, we do know it is required for collagen fibers to form properly. Collagen is the structural protein of skin, tendons, connective tissue, and bone.

Vitamin C is abundant in citrus fruits and tomatoes. When the vitamin is not present in the diet, lesions appear on the skin, severe bleeding occurs about the gums, in the joints, and under the skin, and any wound heals slowly. The condition, known as *scurvy,* was the first disease to be treated by adjusting the diet. British sailors who shipped out to sea after the late 1700s were required to eat limes to prevent scurvy (which is how they came to be called "limeys"). Not until 200 years later did it become known that the substance preventing scurvy was vitamin C. *Scorbutus* is Latin for "scurvy"; *ascorbic,* therefore, means "no scurvy."

an English sailor circa 1829

PROBLEM 32 ♦

Explain why the C-3 OH group of vitamin C is more acidic than the C-2 OH group.

20.18 CARBOHYDRATES ON CELL SURFACES

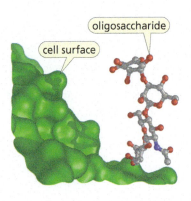

Many cells have short oligosaccharide chains on their surface that enable the cells to recognize and interact with other cells and with invading viruses and bacteria. These oligosaccharides are linked to the surface of the cell by the reaction of an OH or an NH_2 group of a cell-membrane protein with the anomeric carbon of a cyclic sugar. Proteins attached to oligosaccharides are called **glycoproteins**. The percentage of carbohydrate in glycoproteins is variable; some glycoproteins contain as little as 1% carbohydrate by weight, whereas others contain as much as 80%.

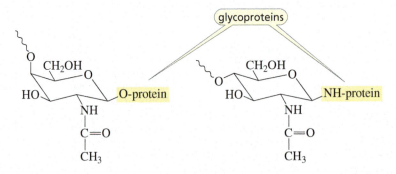

Carbohydrates on the surfaces of cells provide a way for cells to recognize one another, serving as points of attachment for other cells, viruses, and toxins. Therefore, surface carbohydrates have been found to play a role in activities as diverse as infection, prevention of infection, fertilization, inflammatory diseases such as rheumatoid arthritis and septic shock, and blood clotting. The fact that several known antibiotics contain sugars (Section 20.17) suggests that the antibiotics function by recognizing target cells. Carbohydrate interactions also are involved in the regulation of cell growth, so changes in membrane glycoproteins are thought to be correlated with malignant transformations.

Blood Types

Differences in blood type (A, B, or O) are actually differences in the sugars bound to the surfaces of red blood cells. Each type of blood is associated with a different carbohydrate structure (Figure 20.5). Type AB blood is a mixture of type A blood and type B blood.

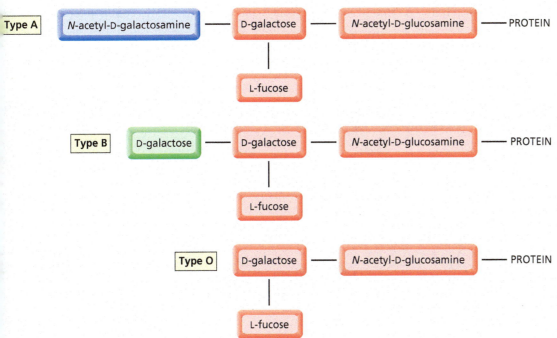

Type A — N-acetyl-D-galactosamine — D-galactose — N-acetyl-D-glucosamine — PROTEIN
                                         |
                                      L-fucose

Type B — D-galactose — D-galactose — N-acetyl-D-glucosamine — PROTEIN
                            |
                         L-fucose

Type O — D-galactose — N-acetyl-D-glucosamine — PROTEIN
              |
           L-fucose

◀ **Figure 20.5**
Blood type is determined by the sugars on the surfaces of red blood cells. Fucose is 6-deoxygalactose.

Antibodies are proteins that are synthesized by the body in response to foreign substances called *antigens*. Interaction with the antibody causes the antigen to either precipitate or be flagged for destruction by immune system cells. This is why blood cannot be transferred from one person to another unless the blood types of the donor and acceptor are compatible. Otherwise the donated blood will be considered a foreign substance and will provoke an immune response.

Looking at Figure 20.5, we can see why the immune system of people with type A blood recognizes type B blood as foreign and vice versa. The immune system of people with type A, B, or AB blood does not recognize type O blood as foreign because the carbohydrate in type O blood is also a component of types A, B, and AB. Thus, anyone can accept type O blood, so people with that blood type are called universal donors. People with type AB blood can accept types AB, A, B, and O blood, so they are called universal acceptors.

PROBLEM 33 ◆

Refer to Figure 20.5 to answer the following questions:

a. People with type O blood can donate blood to anyone, but they cannot receive blood from everyone. From whom can they *not* receive blood?

b. People with type AB blood can receive blood from anyone, but they cannot give blood to everyone. To whom can they *not* give blood?

20.19 ARTIFICIAL SWEETENERS

For a molecule to taste sweet, it must bind to a receptor on a taste bud cell on the tongue. The binding of the molecule causes a nerve impulse to pass from the taste bud cell to the brain, where the molecule is interpreted as being sweet. Artificial sweeteners differ in their degree of "sweetness" (see Table 20.3 on page 980).

Saccharin

Developers of artificial sweeteners must evaluate potential products in terms of several factors—such as toxicity, stability, and cost—in addition to taste. Saccharin (Sweet'N Low), the first artificial sweetener, was discovered accidentally by Ira Remsen in 1879. One evening he noticed that the dinner rolls initially tasted sweet and then bitter. Because his wife did not notice that the rolls had an unusual taste, Remsen tasted his fingers and found they had the same odd taste. The next day he tasted the chemicals he had been working with the day before and found one that had an

extremely sweet taste. (As strange as it may seem today, at one time it was common for chemists to taste compounds to characterize them.) He called this compound saccharin. Notice that, in spite of its name, saccharin is *not* a saccharide.

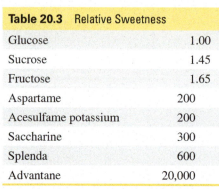

Table 20.3 Relative Sweetness

Glucose	1.00
Sucrose	1.45
Fructose	1.65
Aspartame	200
Acesulfame potassium	200
Saccharine	300
Splenda	600
Advantane	20,000

saccharin   dulcin   sodium cyclamate   aspartame

acesulfame potassium   sucralose   advantame

Because it has little caloric value, saccharin became an important substitute for sucrose when it became commercially available in 1885. The chief nutritional problem in the West was—and still is—the overconsumption of sugar and its consequences: obesity, heart disease, and dental decay. Saccharin was also a boon to people with diabetes, who must limit their consumption of sucrose and glucose. Although the toxicity of saccharin had not been studied carefully when it was first marketed (our current concern with toxicity is a fairly recent development), extensive studies since then have shown saccharin to be harmless. In 1912, saccharin was temporarily banned in the United States, not because of any concern about its toxicity, but because of a concern that people would miss out on the nutritional benefits of sugar.

Dulcin and Sodium Cyclamate

Dulcin was the second artificial sweetener to be discovered (in 1884). Even though it did not have the bitter, metallic aftertaste associated with saccharin, it never achieved much popularity. Dulcin was taken off the market in 1951 in response to concerns about its toxicity.

Sodium cyclamate became a widely used nonnutritive sweetener in the 1950s, but was banned in the United States some 20 years later in response to two studies that appeared to show that large amounts of sodium cyclamate cause liver cancer in mice.

Aspartame

Aspartame (NutraSweet, Equal) was approved by the U.S. Food and Drug Administration (FDA) in 1981. Because aspartame contains a phenylalanine subunit, it should not be used by people with the genetic disease known as phenylketonuria (PKU) (see page 1114).

Acesulfame Potassium and Sucralose

Acesulfame potassium (Sweet and Safe, Sunette, Sweet One) was approved in 1988. It has less aftertaste than saccharine and is more stable than aspartame at high temperatures.

Sucralose (Splenda) was approved in 1991. It maintains its sweetness in foods stored for long periods and at temperatures used in baking. Sucralose is made from sucrose by selectively replacing three of sucrose's OH groups with chlorines. During chlorination, the 4-position of the glucose ring becomes inverted, so sucralose is a galactopyranoside, not a glucopyranoside. Sucralose is the only artificial sweetener that has a carbohydrate-like structure. However, because of the chlorine atoms, the body does not recognize it as a carbohydrate, so it is eliminated from the body instead of being metabolized.

Advantame

Advantame is the most recently approved (2014) artificial sweetener. It has a clean sugar-like taste. Because it is 20,000 times sweeter than sucrose (Table 20.3), only small amounts of it are needed to achieve a sweet taste.

The fact that these artificial sweeteners have such different structures shows that the sensation of sweetness is not induced by a single molecular shape.

Acceptable Daily Intake

The FDA has established an acceptable daily intake (ADI) value for many of the food ingredients it clears for use. The ADI is the amount of the substance a person can consume safely, each day of his or her life. For example, the ADI for acesulfame potassium is 15 mg/kg/day. This means that each day a 132 lb person can consume the amount of acesulfame potassium that would be found in two gallons of an artificially sweetened beverage. The ADI for sucralose is 5 mg/kg/day.

ESSENTIAL CONCEPTS

Section 20.0

- **Bioorganic compounds**—organic compounds found in living systems—obey the same chemical principles as smaller organic molecules.
- Much of the structure of bioorganic compounds exists for the purpose of **molecular recognition.**
- **Carbohydrates** are polyhydroxy aldehydes (**aldoses**) and polyhydroxy ketones (**ketoses**), or compounds formed by linking aldoses and ketoses.

Section 20.2

- The notations D and L describe the configuration of the bottommost asymmetric center of a **monosaccharide** in a Fischer projection. Most naturally occurring sugars are D-sugars.

Section 20.3

- Naturally occurring ketoses have the ketone group in the 2-position.

Section 20.4

- **Epimers** differ in configuration at only one asymmetric center: D-mannose is the C-2 epimer of D-glucose, and D-galactose is the C-4 epimer of D-glucose.

Section 20.5

- In a basic solution, a monosaccharide is converted to a mixture of polyhydroxy aldehydes and polyhydroxy ketones.

Section 20.6

- Reduction of an aldose forms one **alditol**; reduction of a ketose forms two alditols.

- Br$_2$ oxidizes aldoses but not ketoses; Tollens' reagent oxidizes both.
- Aldoses are oxidized to **aldonic acids** or to **aldaric acids.**

Section 20.7

- The **Kiliani–Fischer synthesis** increases the carbon chain of an aldose by one carbon; it forms C-2 epimers.

Section 20.8

- The **Wohl degradation** decreases the carbon chain of an aldose by one carbon.

Section 20.10

- The aldehyde or keto group of a **monosaccharide** reacts with one of its OH groups to form cyclic hemiacetals: glucose forms α-D-glucose and β-D-glucose. More β-D-glucose is present than α-D-glucose in an aqueous solution at equilibrium.

- α-D-Glucose and β-D-glucose are **anomers**—they differ in configuration only at the **anomeric carbon,** which is the carbon that was the carbonyl carbon in the open-chain form.

- A slow change in optical rotation to an equilibrium value is called **mutarotation.**

- The α-position is axial when a sugar is shown in a chair conformation and down when the sugar is shown in a Haworth projection; the β-position is equatorial when a sugar is shown in a chair conformation and up when the sugar is shown in a Haworth projection.

- Six-membered-ring sugars are **pyranoses;** five-membered-ring sugars are **furanoses.**

Section 20.11

- The most abundant monosaccharide in nature is D-glucose. All the OH groups in β-D-glucose are in equatorial positions.

Section 20.12

- A cyclic hemiacetal can react with an alcohol to form an acetal, called a **glycoside.** If the name pyranose or furanose is used, the acetal is called a **pyranoside** or a **furanoside,** respectively.
- The bond between the anomeric carbon and the alkoxy oxygen is called a **glycosidic bond.**

Section 20.13

- The preference for the axial position by certain substituents bonded to the anomeric carbon is called the **anomeric effect.**

Section 20.14

- If a sugar has an aldehyde, ketone, or hemiacetal group, it is a **reducing sugar.** Acetals are nonreducing sugars.

Section 20.15

- **Disaccharides** consist of two monosaccharides hooked together by a glycosidic linkage. Maltose has an **α-1,4′-glycosidic linkage** between two glucose subunits; cellobiose has a **β-1,4′-glycosidic linkage** between two glucose subunits.
- The most common disaccharide is sucrose; it has a D-glucose subunit and a D-fructose subunit linked by their anomeric carbons.

Section 20.16

- **Polysaccharides** contain as few as ten or as many as several thousand monosaccharides joined by glycosidic linkages.
- Starch is composed of amylose and amylopectin. Amylose has unbranched chains of D-glucose units joined by **α-1,4′-glycosidic linkages.**
- Amylopectin also has chains of D-glucose units joined by α-1,4′-glycosidic linkages, but it also has **α-1,6′-glycosidic linkages** that create branches. Glycogen is similar to amylopectin but has more branches.
- Cellulose has unbranched chains of D-glucose units joined by **β-1,4′-glycosidic linkages.**
- The α-linkages cause amylose to form a helix and be water soluble; the β-linkages cause cellulose to form linear arrays and be water insoluble.

Section 20.18

- The surfaces of many cells contain short **oligosaccharides** that allow the cells to interact with each other. The oligosaccharides are attached to the cell surface by protein groups.
- Proteins bonded to oligosaccharides are called **glycoproteins.**

SUMMARY OF REACTIONS

1. Epimerization (Section 20.5). The mechanism of the reaction is shown on page 956.

2. Enediol rearrangement (Section 20.5). The mechanism of the reaction is shown on page 956.

3. Reduction (Section 20.6)

4. Oxidation (Section 20.6)

$$\text{(aldose)} \xrightarrow[\text{HO}^-]{\text{Ag}^+,\ \text{NH}_3} \text{(aldonate)} + \text{Ag}$$

$$\text{(aldose)} \xrightarrow[\text{H}_3\text{O}^+]{\text{Br}_2} \text{(aldonic acid)} + 2\ \text{Br}^-$$

$$\text{(ketose)} \xrightarrow[\text{HO}^-]{\text{Ag}^+,\ \text{NH}_3} \text{(product)} + \text{Ag}$$

$$\text{(aldose)} \xrightarrow[\Delta]{\text{HNO}_3} \text{(aldaric acid)}$$

5. Chain elongation: the Kiliani–Fischer synthesis (Section 20.7)

$$\begin{array}{c} \text{H}-\text{C}=\text{O} \\ (\text{CHOH})_n \\ \text{CH}_2\text{OH} \end{array} \xrightarrow[\substack{\textbf{1. NaC}\equiv\textbf{N/HCl} \\ \textbf{2. H}_2,\ \textbf{Pd/BaSO}_4 \\ \textbf{3. HCl/H}_2\textbf{O}}]{} \begin{array}{c} \text{H}-\text{C}=\text{O} \\ (\text{CHOH})_{n+1} \\ \text{CH}_2\text{OH} \end{array}$$

6. Chain shortening: the Wohl degradation (Section 20.8)

$$\begin{array}{c} \text{H}-\text{C}=\text{O} \\ (\text{CHOH})_n \\ \text{CH}_2\text{OH} \end{array} \xrightarrow[\substack{\textbf{1. NH}_2\textbf{OH/trace acid} \\ \textbf{2. Ac}_2\textbf{O, 100 °C} \\ \textbf{3. HO}^-,\ \textbf{H}_2\textbf{O}}]{} \begin{array}{c} \text{H}-\text{C}=\text{O} \\ (\text{CHOH})_{n-1} \\ \text{CH}_2\text{OH} \end{array}$$

7. Hemiacetal formation (Section 20.10)

8. Glycoside formation (Section 20.12). The mechanism of the reaction is shown on page 967.

$$\xrightarrow[\text{ROH}]{\text{HCl}}$$

PROBLEMS

34. What product or products are obtained when D-galactose reacts with each of the following?

   a. nitric acid + Δ

   b. Ag^+, NH_3, HO^-

   c. NaBH_4, followed by H_3O^+

   d. excess CH_3I + Ag_2O

   e. Br_2 in water

   f. ethanol + HCl

   g. 1. hydroxylamine/trace acid
       2. acetic anhydride/Δ
       3. $\text{HO}^-/\text{H}_2\text{O}$

35. Name the epimers of D-glucose.

36. Identify the sugar in each description.

   a. An aldopentose that is not D-arabinose forms D-arabinitol when it is reduced with NaBH_4.

   b. A sugar that is not D-altrose forms D-altraric acid when it is oxidized with nitric acid.

   c. A ketose that, when reduced with NaBH_4, forms D-altritol and D-allitol.

37. D-Xylose and D-lyxose are formed when D-threose undergoes a Kiliani–Fischer synthesis. D-Xylose is oxidized to an optically inactive aldaric acid, whereas D-lyxose forms an optically active aldaric acid. What are the structures of D-xylose and D-lyxose?

38. Answer the following questions about the eight aldopentoses:

a. Which are enantiomers?
b. Which are C-2 epimers?
c. Which form an optically active compound when oxidized with nitric acid?

39. What is the configuration of each of the asymmetric centers in the Fischer projection of

a. D-glucose? b. D-galactose? c. D-ribose? d. D-xylose? e. D-sorbose?

40. The reaction of D-ribose with one equivalent of methanol plus HCl forms four products. Draw the products.

41. Name the following:

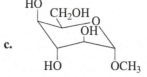

42. A student isolated a monosaccharide and determined that it had a molecular weight of 150. Much to his surprise, he found that it was not optically active. What is the structure of the monosaccharide?

43. Propose a mechanism for the formation of D-allose from D-glucose in a basic solution.

44. Treatment with sodium borohydride converts aldose **A** to an optically inactive alditol. Wohl degradation of **A** forms **B**, whose alditol is optically inactive. Wohl degradation of **B** forms D-glyceraldehyde. Identify **A** and **B**.

45. The disaccharide lactulose consists of a D-galactopyranose subunit and a D-fructofuranose subunit joined by a β-1,4′-glycosidic linkage. After treatment of lactulose with 1. excess CH_3I/Ag_2O, 2. HCl/H_2O, the D-galactopyranose subunit was found to have one nonmethylated OH group, whereas the D-fructofuranose subunit had two. Draw the structure of α-lactulose.

46. A hexose was obtained after (+)-glyceraldehyde underwent three successive Kiliani–Fischer syntheses. Identify the hexose from the following experimental information: oxidation with nitric acid forms an optically active aldaric acid; a Wohl degradation followed by oxidation with nitric acid forms an optically inactive aldaric acid; and a second Wohl degradation forms erythrose.

47. Draw the mechanism for the interconversion of α-D-glucose and β-D-glucose in dilute HCl.

48. An unknown β-D-aldohexose has only one axial substituent. A Wohl degradation forms a compound which, when treated with sodium borohydride, forms an optically active alditol. This information allows you to arrive at two possible structures for the β-D-aldohexose. What experiment can you carry out to distinguish between the two possibilities?

49. The ^1H NMR spectrum of D-glucose in D_2O exhibits two high-frequency doublets. What is responsible for these doublets?

50. D-Glucuronic acid is found widely in plants and animals. One of its functions is to detoxify poisonous HO-containing compounds by reacting with them in the liver to form glucuronides. Glucuronides are water soluble and, therefore, readily excreted. After ingestion of a poison such as turpentine or phenol, the glucuronides of these compounds are found in urine. Draw the structure of the glucuronide formed by the reaction of β-D-glucuronic acid and phenol.

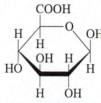

β-D-glucuronic acid

51. Monosaccharide **A** is a diastereomer of D-lyxose. Treatment of **A** with nitric acid forms an optically inactive aldaric acid. **A** undergoes a Kiliani-Fischer synthesis to form **B** and **C**. **B** is oxidized by nitric acid to an optically active aldaric acid, and **C** is oxidized to an optically inactive aldaric acid. Wohl degradation of **A** forms **D**, which is oxidized by nitric acid to an optically inactive aldaric acid. Wohl degradation of **D** forms a D-aldotriose. Identify **A**, **B**, **C**, and **D**.

52. Determine the structure of D-galactose, using arguments similar to those used by Fischer to prove the structure of D-glucose.

53. A D-aldopentose is oxidized by nitric acid to an optically active aldaric acid. A Wohl degradation of the aldopentose leads to a monosaccharide that is oxidized by nitric acid to an optically inactive aldaric acid. Identify the D-aldopentose.

54. Draw the mechanism for the formation of β-lactose from α-D-galactose and β-D-glucose in dilute HCl.

55. Hyaluronic acid, a component of connective tissue, is the fluid that lubricates joints. It is a polymer of alternating N-acetyl-D-glucosamine and D-glucuronic acid subunits joined by β-1,3'-glycosidic linkages. Draw a short segment of hyaluronic acid.

56. To synthesize D-galactose, a student went to the stockroom to get some D-lyxose to use as a starting material. She found that the labels had fallen off the bottles containing D-lyxose and D-xylose. How can she determine which bottle contains D-lyxose?

57. Draw the mechanism for the last step in the Kiliani-Fischer synthesis.

58. The aldonic acid of D-glucose forms a five-membered-ring lactone. Draw its structure.

59. The aldaric acid of D-glucose forms two five-membered-ring lactones. Draw their structures.

60. Draw the mechanism for the acid-catalyzed hydrolysis of β-maltose.

61. How many aldaric acids are obtained from the 16 aldohexoses?

62. A hexose is obtained when the residue of a shrub *Sterculia setigeria* undergoes acid-catalyzed hydrolysis. Identify the hexose from the following experimental information: it undergoes mutarotation; it does not react with Br$_2$; and D-galactonic acid and D-talonic acid are formed when it reacts with Tollens' reagent.

63. When D-fructose is dissolved in D$_2$O and the solution is made basic, the D-fructose recovered from the solution has an average of 1.7 deuterium atoms attached to the C-1 carbon per molecule. Show the mechanism that accounts for the incorporation of these deuterium atoms into D-fructose.

64. Draw each of the following:

a. β-D-talopyranose   c. α-D-tagatopyranose   e. β-L-talopyranose
b. α-D-idopyranose   d. β-D-psicofuranose   f. α-L-tagatopyranose

65. Calculate the percentages of α-D-glucose and β-D-glucose present at equilibrium from the specific rotations of α-D-glucose, β-D-glucose, and the equilibrium mixture. Compare your values with those given in Section 20.10. (*Hint:* The specific rotation of the mixture equals the specific rotation of α-D-glucose times the fraction of glucose present in the α-form plus the specific rotation of β-D-glucose times the fraction of glucose present in the β-form.)

66. The specific rotation of α-D-galactose is 150.7 and that of β-D-galactose is 52.8. When an aqueous mixture that was initially 70% α-D-galactose and 30% β-D-galactose reaches equilibrium, the specific rotation is 80.2. What is the percentage of α-D-galactose and β-D galactose at equilibrium?

67. Draw the mechanism for the elimination step in the Wohl degredation.

68. An unknown disaccharide gives a positive Tollens' test. A glycosidase hydrolyzes it to D-galactose and D-mannose. When the disaccharide is treated with methyl iodide and Ag$_2$O and then hydrolyzed with dilute HCl, the products are 2,3,4,6-tetra-*O*-methylgalactose and 2,3,4-tri-*O*-methylmannose. Propose a structure for the disaccharide.

69. Predict whether D-altrose exists preferentially as a pyranose or a furanose. (*Hint:* In the most stable arrangement for a five-membered ring, all the adjacent substituents are trans.)

70. Trehalose, C$_{12}$H$_{22}$O$_{11}$, is a nonreducing sugar that is only 45% as sweet as sugar. When hydrolyzed by aqueous acid or the enzyme maltase, it forms only D-glucose. When it is treated with excess methyl iodide in the presence of Ag$_2$O and then hydrolyzed with water under acidic conditions, only 2,3,4,6-tetra-*O*-methyl-D-glucose is formed. Draw the structure of trehalose.

71. Propose a mechanism for the rearrangement that converts an α-hydroxyimine to an α-aminoketone in the presence of a trace amount of acid (page 960).

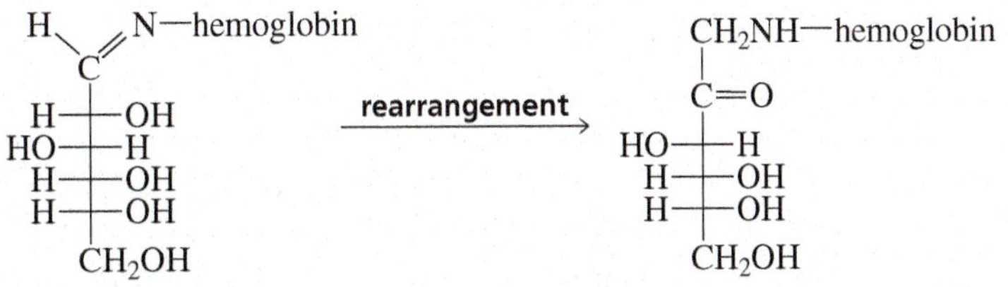

72. All the glucose units in dextran have six-membered rings. When a sample of dextran is treated with methyl iodide and Ag$_2$O and the product is hydrolyzed under acidic conditions, the final products are 2,3,4,6-tetra-*O*-methyl-D-glucose, 2,4,6-tri-*O*-methyl-D-glucose, 2,3,4-tri-*O*-methyl-D-glucose, and 2,4-di-*O*-methyl-D-glucose. Draw a short segment of dextran.

73. When a pyranose is in the chair conformation in which the CH$_2$OH group and the C-1 OH group are both in axial positions, the two groups can react to form an acetal. This is called the anhydro form of the sugar (it has "lost water"). The anhydro form of D-idose is shown here. Explain why about 80% of D-idose exists in the anhydro form in an aqueous solution at 100°C, but only about 0.1% of D-glucose exists in the anhydro form under the same conditions.

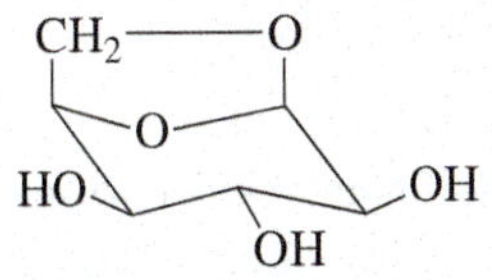

anhydro form of D-idose

21 Amino Acids, Peptides, and Proteins

Cobwebs, silk, muscles, and wool are all proteins. In this chapter you will find out why muscles and wool can be stretched but cobwebs and silk cannot. You also will see how a reduction reaction followed by an oxidation reaction can alter the structure of hair (another protein) from straight to curly or from curly to straight.

The three kinds of polymers prevalent in nature are polysaccharides, proteins, and nucleic acids. We just looked at polysaccharides (Section 20.16); now we will turn our attention to proteins and the structurally similar, but shorter, peptides. (Nucleic acids are discussed in Chapter 26.)

Peptides and **proteins** are polymers of amino acids. The amino acids are linked together by amide bonds. An **amino acid** is a carboxylic acid with a protonated amino group on the α-carbon.

a protonated
α-aminocarboxylic acid
an amino acid

amide bonds

amino acids are linked together by amide bonds

a tripeptide

Amino acid polymers can be composed of any number of amino acids. A **dipeptide** contains two amino acids linked together, a **tripeptide** contains three, an **oligopeptide** contains 4 to 10, and a

polypeptide contains many. Proteins are naturally occurring polypeptides made up of 40 to 4000 amino acids. Proteins serve many functions in living systems (Table 21.1). More than 28 million proteins are known.

Table 21.1   Examples of the Diverse Functions of Proteins in Living Systems

Structural Proteins	These proteins impart strength to biological structures or protect organisms from their environment. For example, collagen is the major component of bones, muscles, and tendons; keratin is the major component of hair, hooves, feathers, fur, and the outer layer of skin.
Protective Proteins	Snake venoms and plant toxins are proteins that protect their owners from predators. Blood-clotting proteins protect the vascular system when it is injured. Antibodies and peptide antibiotics protect us from disease.
Enzymes	Enzymes are proteins that catalyze the reactions that occur in cells.
Hormones	Some hormones, compounds that regulate the reactions that occur in living systems, are proteins.
Proteins with Physiological Functions	These proteins include those that transport and store oxygen in the body, store oxygen in the muscles, and contract muscles.

Proteins can be classified as either fibrous or globular. **Fibrous proteins** contain long chains of polypeptides arranged in threadlike bundles; these proteins are insoluble in water. **Globular proteins** tend to have roughly spherical shapes and most are soluble in water. All structural proteins are fibrous proteins; most enzymes are globular proteins.

21.1   THE NOMENCLATURE OF AMINO ACIDS

The Most Common Naturally Occurring Amino Acids

The structures of the 20 most common naturally occurring amino acids and the frequency with which each occurs in proteins are shown in Table 21.2. Other amino acids occur in nature but do so infrequently. Notice that:

- Amino acids differ only in the substituent (R) that is attached to the α-carbon. The wide variation in these substituents (called **side chains**) is what gives proteins their great structural diversity and, as a consequence, their great functional diversity.

- All amino acids—except proline—contain a primary amino group.

- Proline contains a secondary amino group incorporated into a five-membered ring.

Table 21.2   The Most Common Naturally Occurring Amino Acids Shown in the Form That Predominates at Physiological pH (7.4)

Formula		Name	Abbreviations		Average relative abundance in proteins
Aliphatic Side-Chain Amino Acids	$H-\overset{\displaystyle +NH_3}{\underset{}{\overset{\mid}{CH}}}-\overset{O}{\overset{\|}{C}}-O^-$	Glycine	Gly	G	7.5%
	$CH_3-\overset{\displaystyle +NH_3}{\underset{}{\overset{\mid}{CH}}}-\overset{O}{\overset{\|}{C}}-O^-$	Alanine	Ala	A	9.0%

Formula	Name	Abbreviations		Average relative abundance in proteins
$CH_3CH-CH-C(=O)O^-$ with CH_3 and $^+NH_3$	Valine*	Val	V	6.9%
$CH_3CHCH_2-CH-C(=O)O^-$ with CH_3 and $^+NH_3$	Leucine*	Leu	L	7.5%
$CH_3CH_2CH-CH-C(=O)O^-$ with CH_3 and $^+NH_3$	Isoleucine*	Ile	I	4.6%
Hydroxy-Containing Amino Acids $HOCH_2-CH-C(=O)O^-$ with $^+NH_3$	Serine	Ser	S	7.1%
$CH_3CH-CH-C(=O)O^-$ with OH and $^+NH_3$	Threonine*	Thr	T	6.0%
Sulfur-Containing Amino Acids $HSCH_2-CH-C(=O)O^-$ with $^+NH_3$	Cysteine	Cys	C	2.8%
$CH_3SCH_2CH_2-CH-C(=O)O^-$ with $^+NH_3$	Methionine*	Met	M	1.7%
Acidic Amino Acids $^-O-C(=O)-CH_2-CH-C(=O)O^-$ with $^+NH_3$	Aspartate (aspartic acid)	Asp	D	5.5%
$^-O-C(=O)-CH_2CH_2-CH-C(=O)O^-$ with $^+NH_3$	Glutamate (glutamic acid)	Glu	E	6.2%
Amides of Acidic Amino Acids $H_2N-C(=O)-CH_2-CH-C(=O)O^-$ with $^+NH_3$	Asparagine	Asn	N	4.4%

Formula	Name	Abbreviations		Average relative abundance in proteins
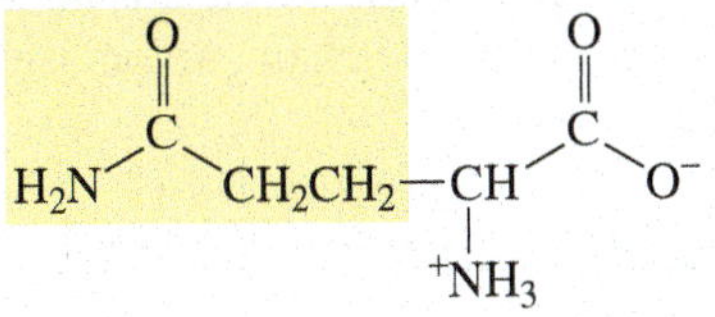	Glutamine	Gln	Q	3.9%

Basic Amino Acids

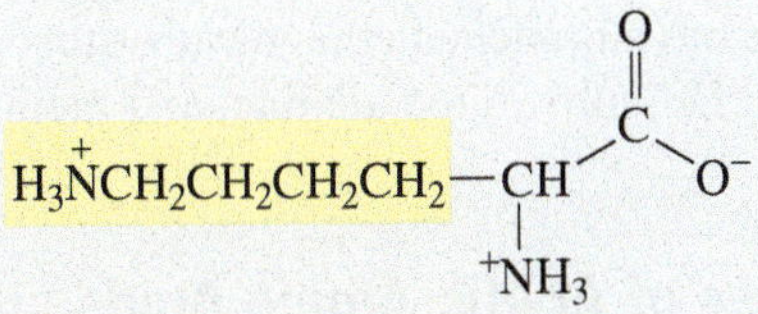

Formula	Name	Abbreviations		Average relative abundance in proteins
	Lysine*	Lys	K	7.0%
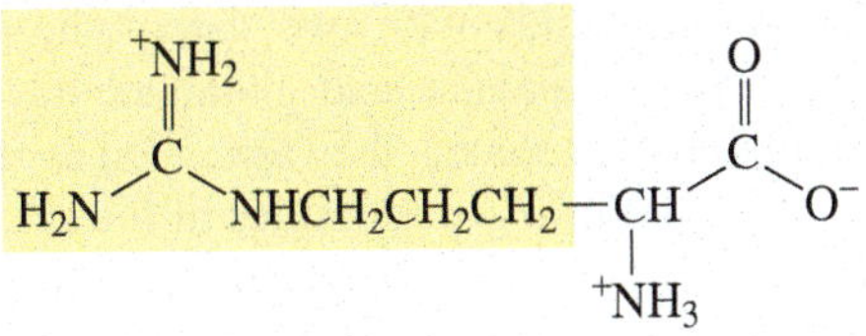	Arginine*	Arg	R	4.7%

Amino Acids With Benzene Rings

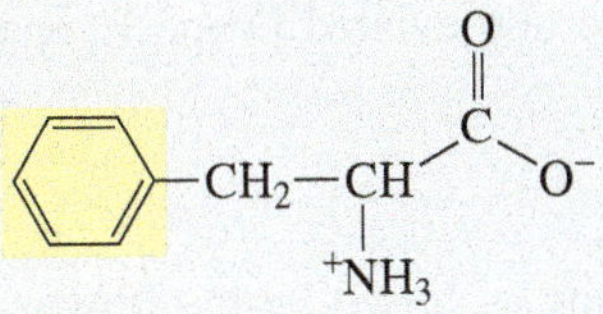

Formula	Name	Abbreviations		Average relative abundance in proteins
	Phenylalanine*	Phe	F	3.5%
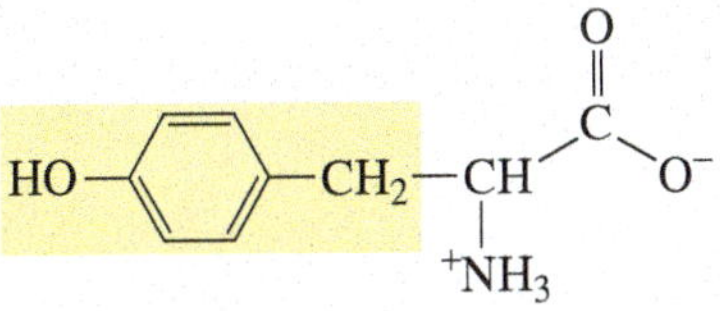	Tyrosine	Tyr	Y	3.5%

Heterocyclic Amino Acids

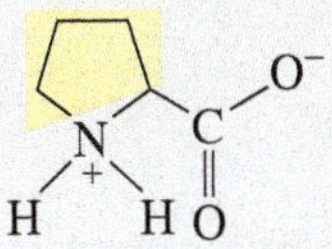

Formula	Name	Abbreviations		Average relative abundance in proteins
	Proline	Pro	P	4.6%
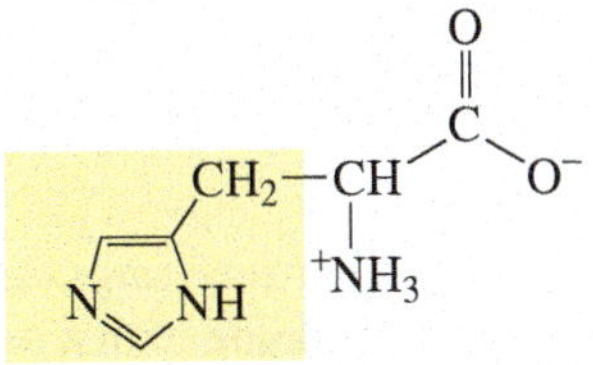	Histidine*	His	H	2.1%
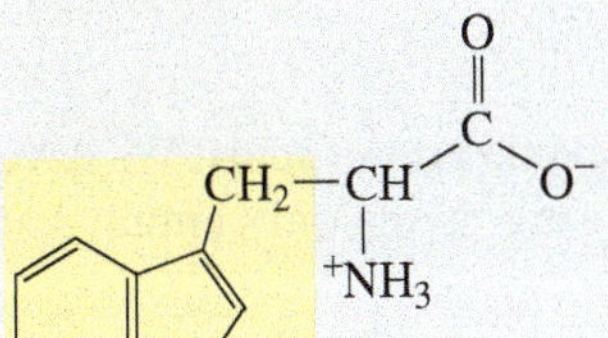	Tryptophan*	Trp	W	1.1%

*Essential amino acid

The amino acids are always called by their common names. Often, the name tells you something about the amino acid. For example, glycine got its name from its sweet taste (*glykos* is Greek for "sweet"), and valine, like valeric acid, has five carbons. Asparagine was first found in asparagus, and tyrosine was isolated from cheese (*tyros* is Greek for "cheese").

Each of the amino acids has both a three-letter abbreviation (in most cases, the first three letters of the name) and a single-letter abbreviation. Dividing the amino acids into classes, as in Table 21.2, makes them easier to learn.

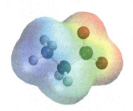

glycine

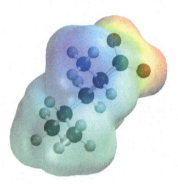

leucine

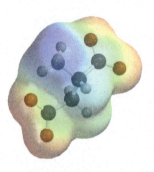

aspartate

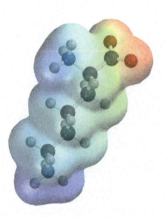

lysine

Aliphatic Side-Chain Amino Acids

These include glycine, the amino acid in which R is H, and four amino acids with alkyl side chains. Alanine is the amino acid with a methyl side chain, and valine has an isopropyl side chain. Notice that in spite of its name, isoleucine has a *sec*-butyl side chain, not an isobutyl side chain. Leucine is the amino acid that has an isobutyl side chain.

Amino Acids with Alcohol- and Sulfur-Containing Side Chains

Two amino acid side chains—serine and threonine—contain an alcohol group. Serine is an HO-substituted alanine, and threonine has a branched ethanol substituent. There are also two sulfur-containing amino acids: cysteine is an HS-substituted alanine, and methionine has a 2-(methylthio)ethyl substituent.

Acidic Amino Acids and Amides of Acidic Amino Acids

There are two acidic amino acids (amino acids with two carboxylic acid groups): aspartate and glutamate. Aspartate is a carboxy-substituted alanine, and glutamate has one more methylene group than aspartate. (If their carboxyl groups are protonated, they are called aspartic acid and glutamic acid.)

Two amino acids—asparagine and glutamine—are amides of the acidic amino acids; asparagine is the amide of aspartate, and glutamine is the amide of glutamate. Notice that the obvious one-letter abbreviations cannot be used for these four amino acids because A and G are used for alanine and glycine. Instead, aspartate and glutamate are abbreviated D and E, and asparagine and glutamine are abbreviated N and Q.

Basic Amino Acids

There are two basic amino acids (amino acids with two basic nitrogen-containing groups): lysine and arginine. Lysine has an ε-amino group, and arginine has a δ-guanidine group. At physiological pH, these groups are protonated. Use the ε and δ to remind yourself how many methylene groups each amino acid has.

$$\overset{+}{H_3}N-\underset{\varepsilon}{CH_2}\underset{\delta}{CH_2}\underset{\gamma}{CH_2}\underset{\beta}{CH_2}\underset{\alpha}{CH}\overset{O}{\underset{+NH_3}{\overset{\|}{C}}}O^-$$

a protonated ε-amino group

lysine

$$\underset{H_2N}{}\overset{\overset{+NH_2}{\|}}{\underset{\delta}{C}}\underset{NH}{-}\underset{\gamma}{CH_2}\underset{\beta}{CH_2}\underset{\alpha}{CH}\overset{O}{\underset{+NH_3}{\overset{\|}{C}}}O^-$$

a protonated δ-guanidino group

arginine

Amino Acids with Benzene Rings

Two amino acids—phenylalanine and tyrosine—contain benzene rings. As its name indicates, phenylalanine is phenyl-substituted alanine. Tyrosine is phenylalanine with a *para*-hydroxy substituent.

Heterocyclic Amino Acids

Proline, histidine, and tryptophan are heterocyclic amino acids. We have noted that proline, with its nitrogen incorporated into a five-membered ring, is the only amino acid that contains a secondary amino group.

Histidine is an imidazole-substituted alanine. Imidazole is an aromatic compound because it is cyclic and planar, each of its ring atoms has a *p* orbital, and it has three pairs of delocalized π electrons (Section 20.7).

$$\underset{\text{protonated imidazole}}{\overset{+}{HN}\diagdown\diagup:NH} \rightleftharpoons \underset{\text{imidazole}}{:N\diagdown\diagup:NH} + H^+$$

Tryptophan is an indole-substituted alanine. Like imidazole, indole is an aromatic compound (page 931). Because the lone pair on the nitrogen of indole is needed for the compound's aromaticity, indole is a very weak base. (The pK_a of protonated indole is -2.4.) Therefore, the ring nitrogen in tryptophan is never protonated under physiological conditions.

indole

Essential Amino Acids

The 10 amino acids denoted in Table 21.2 by asterisks (*) are **essential amino acids.** Humans must obtain them from their diet because they either cannot synthesize them at all or they cannot synthesize them in adequate amounts. For example, humans must have a dietary source of phenylalanine because they cannot synthesize benzene rings. However, they do not need tyrosine in their diet because they can synthesize it from phenylalanine (Section 24.8). Although humans can synthesize arginine, it is needed for growth in greater amounts than can be synthesized. Therefore, arginine is considered an essential amino acid for children but not for adults.

Proteins and Nutrition

Proteins are an important component of our diets. Dietary protein is hydrolyzed in the body to individual amino acids. Some of these amino acids are used to synthesize proteins needed by the body, some are broken down (metabolized) to supply energy to the body, and some are used as starting materials for the synthesis of nonprotein compounds that the body needs, such as thyroxine (Section 18.3), adrenaline, and melanin (Section 24.8).

Complete proteins (meat, fish, eggs, and milk) contain all 10 essential amino acids. Incomplete proteins contain too little of one or more essential amino acids to support human growth. For example, beans and peas are deficient in methionine, corn is deficient in lysine and tryptophan, and rice is deficient in lysine and threonine. Vegetarians, therefore, must have a diet that includes proteins from different sources.

PROBLEM 1

a. Explain why, when the imidazole ring of histidine is protonated, the double-bonded nitrogen is the nitrogen that accepts the proton.

b. Explain why, when the guanidino group of arginine is protonated, the double-bonded nitrogen is the nitrogen that accepts the proton.

21.2 THE CONFIGURATION OF AMINO ACIDS

The α-carbon of all the naturally occurring amino acids (except glycine) is an asymmetric center. Therefore, 19 of the 20 amino acids listed in Table 21.2 can exist as enantiomers. The D and L notation used for monosaccharides (Section 20.2) is also used for amino acids.

An amino acid drawn in a Fischer projection with the carboxyl group at the top and the R group at the bottom of the vertical axis is a **D-amino acid** if the amino group is on the right and an L-**amino acid** if the amino group is on the left (see the top of the next page). Unlike monosaccharides, where the D isomer is the one found in nature, most amino acids found in nature have the L configuration.

Naturally occurring monosaccharides have the D configuration.

Naturally occurring amino acids have the L configuration.

To date, D-amino acids have been found only in a few peptide antibiotics and in some small peptides attached to the cell walls of bacteria. (You will see how an L-amino acid can be converted to a D-amino acid in Section 23.5.)

D-glyceraldehyde   L-glyceraldehyde

D-amino acid   L-amino acid

Why D-sugars and L-amino acids? Although it made no difference which isomer nature "selected" to be synthesized, it was important that only one was selected. For example, proteins that contain both D- and L-amino acids do not fold properly, and without proper folding there can be no catalysis (Section 21.15). It was also important that the same isomer was synthesized by all organisms. For example, because mammals have L-amino acids, L-amino acids must be the isomers synthesized by the organisms that mammals depend on for food.

L-alanine
an amino acid

PROBLEM 2 ♦

a. Which isomer—(R)-alanine or (S)-alanine—is D-alanine?
b. Which isomer—(R)-aspartate or (S)-aspartate—is D-aspartate?
c. Can a general statement be made relating R and S to D and L?

Amino Acids and Disease

The Chamorro people of Guam have a high incidence of a syndrome that resembles amyotrophic lateral sclerosis (ALS, or Lou Gehrig's disease) with elements of Parkinson's disease and dementia. This syndrome developed during World War II when, as a result of food shortages, the tribe ate large quantities of *Cycas micronesica* seeds. These seeds contain β-methylamino-L-alanine, an amino acid that binds to cell receptors that bind L-glutamate. When monkeys are given β-methylamino-L-alanine, they develop some of the features of this syndrome. There is hope that, by studying the mechanism of action of β-methylamino-L-alanine, we may gain an understanding of how ALS and Parkinson's disease arise.

L-alanine   **β-methylamino-L-alanine**   **L-glutamate**

A Peptide Antibiotic

Gramicidin S, an antibiotic produced by a strain of bacteria, is a cyclic decapeptide. Notice that one of its 10 amino acids is ornithine, an amino acid not listed in Table 21.2 because it occurs rarely in nature. Ornithine resembles lysine but has one less methylene group in its side chain. Also notice that gramicidin S contains two D-amino acids.

L-Val
L-Pro   L-Orn — L-ornithine
L-Phe   L-Leu
L-Leu   D-Phe — D-phenylalanine
D-ornithine — D-Orn   L-Pro
L-Val

gramicidin S

$$H_3\overset{+}{N}CH_2CH_2CH_2CH\overset{\displaystyle O}{\underset{\overset{|}{\overset{+}{N}H_3}}{-C-O^-}}$$

ornithine

PROBLEM 3 SOLVED

Threonine has two asymmetric centers and, therefore, has four stereoisomers. Naturally occurring L-threonine is (2S,3R)-threonine. Which of the following stereoisomers is L-threonine?

SOLUTION Stereoisomer **A** has the R configuration at both C-2 and C-3 because in both cases the arrow drawn from the highest to the next-highest-priority substituent is counterclockwise. In both cases, counterclockwise signifies R because the lowest priority substituent (H) is on a horizontal bond (Section 4.7). Therefore, the configuration of (2S,3R)-threonine is the opposite of that in stereoisomer **A** at C-2 and the same as that in stereoisomer **A** at C-3. Thus, L-threonine is stereoisomer **D**. Notice that the $^+NH_3$ group is on the left, just as we would expect for the Fischer projection of a naturally occurring L-amino acid.

PROBLEM 4 ♦

Do any other amino acids in Table 21.2 have more than one asymmetric center?

21.3 ACID–BASE PROPERTIES OF AMINO ACIDS

Every amino acid has a carboxyl group and an amino group, and each group can exist in an acidic form or a basic form, depending on the pH of the solution in which the amino acid is dissolved.

We saw that compounds exist primarily in the acidic form (that is, with their protons attached) in solutions that are more acidic than their pK_a values, and primarily in the basic form (that is, without their protons) in solutions that are more basic than their pK_a values (Section 2.10).

The carboxyl groups of the amino acids have pK_a values of approximately 2, and the protonated amino groups have pK_a values near 9 (Table 21.3). Both groups, therefore, are in the acidic form in a very acidic solution (pH ~ 0). At pH = 7, the pH of the solution is greater than the pK_a of the carboxyl group, but less than the pK_a of the protonated amino group; therefore, the carboxyl group is in the basic form and the amino group is in the acidic form. In a strongly basic solution (pH ~ 12), both groups are in the basic form.

> The acidic form (with the proton) predominates if the pH of the solution is less than the pK_a of the ionizable group, and the basic form (without the proton) predominates if the pH of the solution is greater than the pK_a of the ionizable group.

> Recall from the Henderson–Hasselbalch equation (Section 2.10) that half the group is in the acidic form and half is in the basic form at pH = pK_a.

Table 21.3 The pK_a Values of Amino Acids

Amino acid	pK_a α-COOH	pK_a α-$\overset{+}{N}H_3$	pK_a Side chain
Alanine	2.34	9.69	—
Arginine	2.17	9.04	12.48
Asparagine	2.02	8.84	—
Aspartic acid	2.09	9.82	3.86
Cysteine	1.92	10.46	8.35
Glutamic acid	2.19	9.67	4.25
Glutamine	2.17	9.13	—
Glycine	2.34	9.60	—
Histidine	1.82	9.17	6.04
Isoleucine	2.36	9.68	—

(Continued)

Amino acid	pK_a α-COOH	pK_a α-$\overset{+}{N}H_3$	pK_a Side chain
Leucine	2.36	9.60	—
Lysine	2.18	8.95	10.79
Methionine	2.28	9.21	—
Phenylalanine	1.83	9.18	—
Proline	1.99	10.60	—
Serine	2.21	9.15	—
Threonine	2.63	9.10	—
Tryptophan	2.38	9.39	—
Tyrosine	2.20	9.11	10.07
Valine	2.32	9.62	—

Notice that an amino acid can never exist as an uncharged compound, regardless of the pH of the solution. To be uncharged, an amino acid would have to lose a proton from an $^+NH_3$ group with a pK_a of about 9 before it loses a proton from a COOH group with a pK_a of about 2. This is impossible because a weak acid ($pK_a = 9$) cannot lose a proton easier than a strong acid ($pK_a = 2$) can. Therefore, at physiological pH (7.4), an amino acid exists as a dipolar ion, called a zwitterion. A **zwitterion** is a compound that has a negative charge on one atom and a positive charge on a nonadjacent atom. (The name comes from *zwitter*, German for "hybrid.")

PROBLEM 5 ◆

Alanine has pK_a values of 2.34 and 9.69. Therefore, alanine exists predominately as a zwitterion in an aqueous solution with pH > ____ and pH < ____.

A few amino acids have side chains with ionizable hydrogens (Table 21.3). The protonated imidazole side chain of histidine, for example, has a pK_a of 6.04. Histidine, therefore, can exist in four different forms, and the form that predominates depends on the pH of the solution.

histidine

PROBLEM 6 ◆

Why are the carboxylic acid groups of the amino acids more acidic ($pK_a \sim 2$) than a carboxylic acid such as acetic acid ($pK_a = 4.76$)?

LEARN THE STRATEGY

PROBLEM 7 SOLVED

Draw the predominant form for each of the following amino acids at physiological pH (7.4):

a. aspartate   **c.** glutamine   **e.** arginine
b. histidine   **d.** lysine   **f.** tyrosine

SOLUTION TO 7 a. Both carboxyl groups are in the basic form because the pH of the solution is greater than their pK_a values. The protonated amino group is in the acidic form because the pH of the solution is less than its pK_a value.

USE THE STRATEGY

PROBLEM 8 ◆

Draw the predominant form for glutamate in a solution with the following pH:

a. 0   **b.** 3   **c.** 6   **d.** 11

PROBLEM 9

a. Why is the pK_a of the glutamate side chain greater than the pK_a of the aspartate side chain?
b. Why is the pK_a of the arginine side chain greater than the pK_a of the lysine side chain?

21.4 THE ISOELECTRIC POINT

The **isoelectric point** (pI) of an amino acid is the pH at which it has no net charge. In other words, it is the pH at which the amount of positive charge on an amino acid exactly balances the amount of negative charge:

$$pI = \text{pH at which there is no net charge}$$

Determining the pI of an Amino Acid without an Ionizable Side Chain

LEARN THE STRATEGY

The pI of an amino acid that does *not have an ionizable side chain*—such as alanine—is midway between its two pK_a values. This is because at pH = 2.34 half of the molecules have a negatively charged carboxyl group and half have an uncharged carboxyl group, whereas at pH = 9.69 half of the molecules have a positively charged amino group and half have an uncharged amino group. As the pH increases from 2.34, the carboxyl group of more molecules becomes negatively charged; as the pH decreases from 9.69, the amino group of more molecules becomes positively charged. Therefore, the number of negatively charged groups equals the number of positively charged groups at the intersection (average) of the two pK_a values.

alanine

$pK_a = 2.34$

$pK_a = 9.69$

$$pI = \frac{2.34 + 9.69}{2} = \frac{12.03}{2} = 6.02$$

Determining the pI of an Amino Acid with an Ionizable Side Chain

The pI of most amino acids (see Problem 12) that *have an ionizable side chain* is the average of the pK_a values of the similarly ionizing groups (either positively charged groups ionizing to uncharged groups or uncharged groups ionizing to negatively charged groups). For example, the pI of lysine is the average of the pK_a values of the two groups that are positively charged in their acidic form and uncharged in their basic form. The pI of glutamic acid is the average of the pK_a values of the two groups that are uncharged in their acidic form and negatively charged in their basic form.

lysine

$pK_a = 2.18$

$pK_a = 10.79$

$pK_a = 8.95$

glutamic acid

$pK_a = 2.19$

$pK_a = 4.25$

$pK_a = 9.67$

$$pI = \frac{8.95 + 10.79}{2} = \frac{19.74}{2} = 9.87$$

$$pI = \frac{2.19 + 4.25}{2} = \frac{6.44}{2} = 3.22$$

USE THE STRATEGY

PROBLEM 10 ◆

Calculate the pI of each of the following amino acids:

a. asparagine   **b.** arginine   **c.** serine   **d.** aspartate

PROBLEM 11 ◆

a. Which amino acid has the lowest pI value?
b. Which amino acid has the highest pI value?
c. Which amino acid has the greatest amount of negative charge at pH = 6.20?
d. Which amino acid has a greater negative charge at pH = 6.20, glycine or methionine?

PROBLEM 12

Explain why the pI values of tyrosine and cysteine cannot be determined by the method described on page 995.

PROBLEM 13 ◆

a. What percentage of the α-amino group of lysine will be protonated at its pI?

    < 25%   50%   > 75%

b. Answer the same question for the ε-amino group of lysine.

PROBLEM 14

Explain why the pI of lysine is the average of the pK_a values of its two protonated amino groups.

21.5 SEPARATING AMINO ACIDS

A mixture of amino acids can be separated by several different techniques. Electrophoresis and ion-exchange chromatography are two such techniques.

Electrophoresis

An amino acid will be overall positively charged if the pH of the solution is less than the pI of the amino acid, and it will be negatively charged if the pH of the solution is greater than the pI of the amino acid.

Electrophoresis separates amino acids *on the basis of their pI values.* A few drops of a solution of an amino acid mixture are applied to the middle of a piece of filter paper (or to a gel). When the paper (or the gel) is placed in a buffered solution between two electrodes and an electric field is applied (Figure 21.1), an amino acid with a pI greater than the pH of the solution (such as arginine) has an overall *positive charge* and migrates toward the *cathode (the negative electrode).*

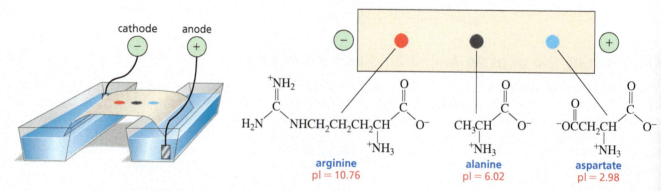

▲ **Figure 21.1**
Arginine, alanine, and aspartate are separated by electrophoresis at pH = 5.

The farther an amino acid's pI is from the pH of the solution, the more positive its overall positive charge and the farther it migrates toward the cathode in a given amount of time. An amino acid with a pI less than the pH of the solution (such as aspartate) has an overall *negative charge* and migrates toward the *anode (the positive electrode).* If two molecules have the same overall charge, the larger one will move more slowly during electrophoresis because the same charge has to move a greater mass.

Forming the Colored Product

Amino acids are colorless, so how can we detect them after they have been separated? This is accomplished by spraying the filter paper with a solution of ninhydrin and drying it in a warm oven. Amino acids form a colored product when heated with ninhydrin. The number of amino acids in the mixture is determined by the number of colored spots on the filter paper. The individual amino acids can be identified by their location on the paper compared with a standard.

The mechanism for formation of the colored product is shown below, omitting the mechanisms for dehydration, imine formation, and imine hydrolysis. (These mechanisms are shown in Sections 16.8 and 16.9.)

STEPS IN THE REACTION OF AN AMINO ACID WITH NINHYDRIN TO FORM A COLORED PRODUCT

ninhydrin

amino acid

deaminated amino acid

$+ CO_2$

purple-colored product

- Loss of water from the hydrate forms a ketone that reacts with the amino group of the amino acid to form an imine.
- Decarboxylation occurs because the electrons left behind can be delocalized onto an oxygen.
- Tautomerization followed by hydrolysis of the imine forms the deaminated amino acid and a ninhydrin-amine.
- Reaction of this amine with another molecule of dehydrated ninhydrin forms an imine. Loss of a proton from the α-carbon results in the formation of a highly conjugated (colored) product (Section 13.22).

Spraying a paper with a solution of ninhydrin allows latent fingerprints (as a consequence of amino acids left behind by the fingers) to be developed.

PROBLEM 15 ♦

What aldehyde is formed when valine is treated with ninhydrin?

Paper/Thin-Layer Chromatography

Paper chromatography once played an important role in analysis because it provided a method for separating amino acids using very simple equipment. Although more modern techniques are generally employed today, we will describe the principles behind paper chromatography because many of the same principles are employed in modern separation techniques.

Paper chromatography separates amino acids *on the basis of their polarity*. A few drops of a solution of an amino acid mixture are applied to the bottom of a strip of filter paper. The edge of the paper is then placed in a solvent. The solvent moves up the paper by capillary action, carrying the amino acids with it. Depending on their polarities, the amino acids have different affinities for the mobile (solvent) and stationary (paper) phases and, therefore, some travel up the paper farther than others.

When a solvent is used that is less polar than the paper, the more polar the amino acid, the more strongly it is adsorbed onto the relatively polar paper. The less polar amino acids travel farther up the paper because they have a greater affinity for the less polar mobile phase. Therefore, when the paper is developed with ninhydrin, the colored spot closest to the origin is the most polar amino acid and the spot farthest away from the origin is the least polar amino acid (Figure 21.2).

Less polar amino acids travel father up the paper if the solvent is less polar than the paper.

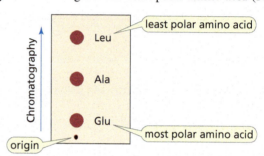

▶ **Figure 21.2**
Separation of glutamate, alanine, and leucine by paper chromatography.

- The most polar amino acids are those with charged side chains.
- The next most polar are those with side chains that can form hydrogen bonds.
- The least polar are those with hydrocarbon side chains.
- For amino acids with hydrocarbon side chains, the polarity of the amino acid decreases as the size of the alkyl group increases. In other words, leucine [R = —CH₂CH(CH₃)₂] is less polar than valine [R = —CH(CH₃)₂].

Paper chromatography has largely been replaced by **thin-layer chromatography** (TLC), which differs from paper chromatography in that TLC uses a plate with a coating of solid material instead of filter paper. How the amino acids separate depends on the solid material and the solvent chosen for the mobile phase.

We just saw that chromatography separates amino acids based on their polarity and electrophoresis separates them based on their charge. The two techniques can be applied on the same piece of filter paper—a technique called **fingerprinting**—to give a two-dimensional separation (that is, the amino acids are separated on the basis of both their polarity and their charge; see Problems 58, 72, and 79.)

PROBLEM 16 ◆

A mixture of seven amino acids (glycine, glutamate, leucine, lysine, alanine, isoleucine, and aspartate) is separated by chromatography. Explain why only six spots show up when the chromatographic plate is coated with ninhydrin and heated.

Ion-Exchange Chromatography

A technique called **ion-exchange chromatography** separates amino acids *based on their overall charge* and determines the relative amount of each amino acid in a mixture. This technique employs a column packed with an insoluble resin. A solution of a mixture of amino acids is loaded onto the top of the column, and a series of buffer solutions of increasing pH are poured through the column. The amino acids separate because they flow through the column at different rates.

The resin is a chemically inert material with charged groups. The structure of a commonly used resin is shown in Figure 21.3 on the next page.

If a mixture of lysine and glutamate in a solution of pH = 6 is loaded onto the column, glutamate will travel down the column rapidly because its negatively charged side chain would be repelled by the negatively charged sulfonic acid groups of the resin. The positively charged side chain of lysine, on the other hand, will cause that amino acid to be retained on the column longer.

This kind of resin is called a **cation-exchange resin** because it exchanges the Na^+ counterions of the SO_3^- groups for the positively charged species traveling through the column. In addition, the relatively nonpolar nature of the column causes it to retain nonpolar amino acids longer than polar amino acids.

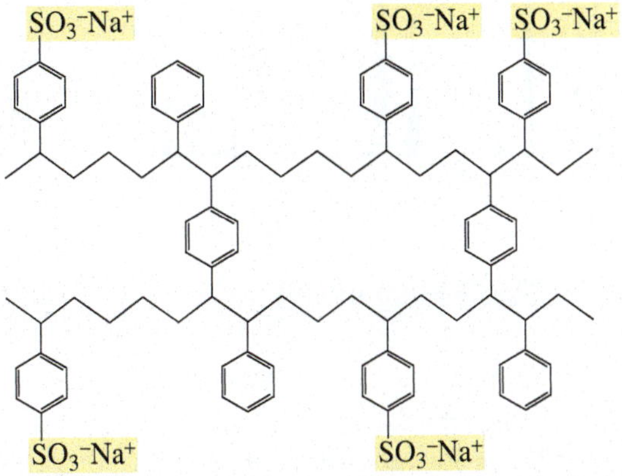

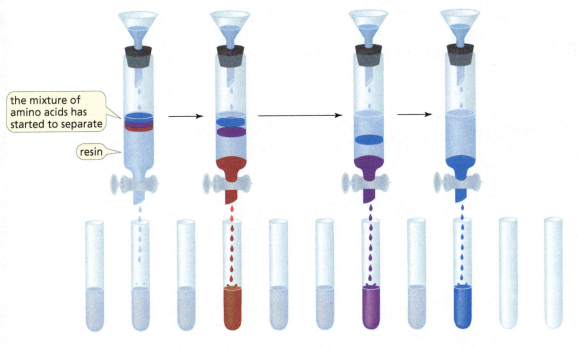

◀ **Figure 21.3**
A section of a cation-exchange resin. This particular resin is called Dowex 50.

Cations bind most strongly to cation-exchange resins.

Anions bind most strongly to anion-exchange resins.

Resins with positively charged groups are called **anion-exchange resins** because they impede the flow of anions by exchanging their negatively charged counterions for the negatively charged species traveling through the column. A common anion-exchange resin, Dowex 1, has $CH_2N^+(CH_3)_3Cl^-$ groups in place of the $SO_3^-Na^+$ groups in Figure 21.3.

An Amino Acid Analyzer

An **amino acid analyzer** is an instrument that automates ion-exchange chromatography. When a solution of amino acids passes through the column of an amino acid analyzer containing a cation-exchange resin, the amino acids move through the column at different rates, depending on their overall charge. The solution that flows out of the column (the effluent) is collected in fractions. These are collected often enough that each amino acid ends up in a different fraction (Figure 21.4).

the mixture of amino acids has started to separate

resin

▲ **Figure 21.4**
Separation of amino acids by ion-exchange chromatography.

If ninhydrin is added to each of the fractions, the concentration of amino acid in each fraction can be determined by the amount of absorbance at 570 nm, because the colored compound formed by the reaction of an amino acid with ninhydrin has a λ_{max} of 570 nm (Section 13.20). This information combined with each fraction's rate of passage through the column allows the identity and relative amount of each amino acid in the mixture to be determined (Figure 21.5).

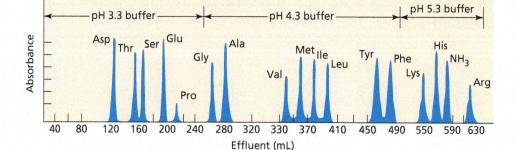

▶ **Figure 21.5**
A typical chromatogram obtained from the separation of a mixture of amino acids using an automated amino acid analyzer.

Water Softeners: Examples of Cation-Exchange Chromatography

Water-softening systems contain a column packed with a cation-exchange resin that has been flushed with concentrated sodium chloride. When "hard water" (water with high levels of Ca^{2+} and Mg^{2+}; Section 25.3) passes through the column, the resin binds Ca^{2+} and Mg^{2+} more tightly than it binds Na^+. Thus, the water-softening system removes Ca^{2+} and Mg^{2+} from the water and replaces them with Na^+. The resin must be recharged from time to time by being flushed with concentrated sodium chloride, thereby replacing the bound Ca^{2+} and Mg^{2+} with Na^+.

PROBLEM 17

Why are buffer solutions of increasingly higher pH used to elute the column that generates the chromatogram shown in Figure 21.5? (*Elute* means wash out with a solvent.)

PROBLEM 18

Explain the order of elution (with a buffer of pH 4) of the following pairs of amino acids through a column packed with Dowex 50 (Figure 21.3):

a. aspartate before serine
b. serine before alanine
c. valine before leucine
d. tyrosine before phenylalanine

PROBLEM 19 ◆

In what order would histidine, serine, aspartate, and valine be eluted with a buffer of pH 4 from a column containing an anion-exchange resin (Dowex 1)?

21.6 SYNTHESIS OF AMINO ACIDS

Chemists do not have to rely on nature to produce amino acids; they can synthesize them in the laboratory, using a variety of methods. Some of these methods are described here.

HVZ Reaction Followed by Reaction with Ammonia

One of the oldest methods used to synthesize an amino acid is to employ an HVZ reaction to replace an α-hydrogen of a carboxylic acid with a bromine (Section 17.5). The resulting α-bromocarboxylic acid can then undergo an S_N2 reaction with ammonia to form the amino acid (Section 9.2).

$$\underset{\substack{\text{carboxylic acid}}}{RCH_2\overset{O}{\underset{}{\overset{\|}{C}}}OH} \xrightarrow[\text{2. } H_2O]{\text{1. } Br_2, PBr_3} \underset{\substack{\overset{|}{Br}}}{RCH\overset{O}{\underset{}{\overset{\|}{C}}}OH} \xrightarrow[]{\overset{\text{excess}}{NH_3}} \underset{\substack{\overset{|}{\overset{+}{N}H_3}\\\text{amino acid}}}{RCH\overset{O}{\underset{}{\overset{\|}{C}}}O^-} + \overset{+}{N}H_4Br^-$$

PROBLEM 20 ◆

Why is excess ammonia used in the preceding reaction?

Reductive Amination

Amino acids can also be synthesized by reductive amination of an α-keto acid (Section 16.4).

Cells can also convert α-keto acids into amino acids, but because the reagents organic chemists use for this reaction are not available in cells, they carry out this reaction by a different mechanism (Section 24.14).

a. What amino acid is obtained from the reductive amination of each of the following metabolic intermediates in a cell by reductive amination?

pyruvic acid   oxaloacetic acid α-ketoglutaric acid

b. What amino acids are obtained from the same metabolic intermediates when the amino acids are synthesized in the laboratory?

N-Phthalimidomalonic Ester Synthesis

Amino acids can be synthesized with better yields than those obtained by the previous two methods via an *N*-phthalimidomalonic ester synthesis, a method that combines the malonic ester synthesis (Section 17.17) and the Gabriel synthesis (Section 15.14).

THE STEPS IN THE *N*-PHTHALIMIDOMALONIC ESTER SYNTHESIS

α-bromomalonic ester   potassium phthalimide   *N*-phthalimidomalonic ester

phthalic acid   amino acid

- α-Bromomalonic ester and potassium phthalimide undergo an S_N2 reaction.
- A proton is easily removed from the α-carbon of *N*-phthalimidomalonic ester because it is flanked by two carbonyl groups (Section 17.1).
- The resulting carbanion undergoes an S_N2 reaction with an alkyl halide.
- Heating in an acidic aqueous solution hydrolyzes the two esters and the two amide bonds and decarboxylates the 3-oxocarboxylic acid.

A variation of the *N*-phthalimidomalonic ester synthesis uses acetamidomalonic ester in place of *N*-phthalimidomalonic ester.

acetamidomalonic ester   acetic acid   amino acid

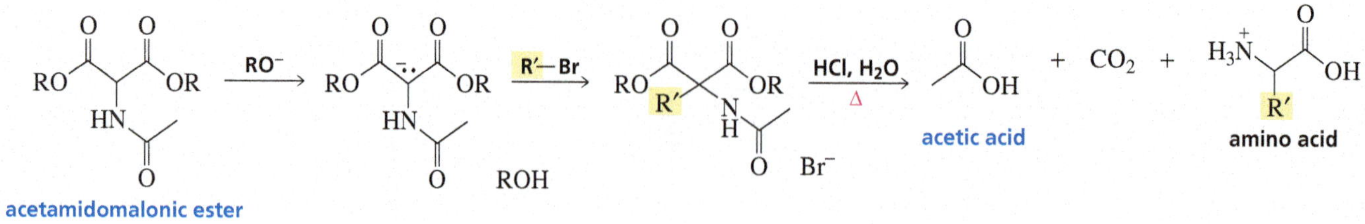

Strecker Synthesis

In the Strecker synthesis, an aldehyde reacts with ammonia to form an imine. A nucleophilic addition reaction with cyanide ion forms an intermediate, which, when hydrolyzed, forms the amino acid (Section 15.15).

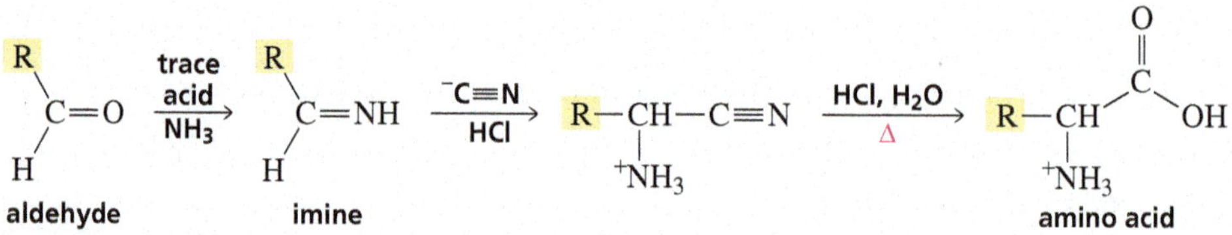

aldehyde   imine   amino acid

PROBLEM 22 ♦

What amino acid is formed using the *N*-phthalimidomalonic ester synthesis when the following alkyl halides are used in the third step?

a. CH$_3$CHCH$_2$Br
    |
    CH$_3$

b. CH$_3$SCH$_2$CH$_2$Br

PROBLEM 23 ♦

What alkyl halide is used in the acetamidomalonic ester synthesis to prepare

a. lysine?   **b.** phenylalanine?

PROBLEM 24 ♦

What amino acid is formed when the aldehyde used in the Strecker synthesis is

a. acetaldehyde?   **b.** 2-methylbutanal?   **c.** 3-methylbutanal?

21.7 RESOLUTION OF RACEMIC MIXTURES OF AMINO ACIDS

When amino acids are synthesized in nature, only the L-enantiomer is formed (Section 6.14). When amino acids are synthesized in the laboratory, however, the product is a racemic mixture of D- and L-amino acids. If only one isomer is desired, the enantiomers must be separated, which can be accomplished by an enzyme-catalyzed reaction.

Because an enzyme is chiral, it reacts at a different rate with each of the enantiomers (Section 6.14). For example, aminoacylase is an enzyme that catalyzes the hydrolysis of *N*-acetyl-L-amino acids, but not *N*-acetyl-D-amino acids. Therefore, if the racemic mixture of amino acids is converted to a pair of *N*-acetylamino acids (by a nucleophilic acyl substitution reaction) and the *N*-acetylated mixture is hydrolyzed with aminoacylase, the products will be the L-amino acid and unreacted *N*-acetyl-D-amino acid, which are easily separated.

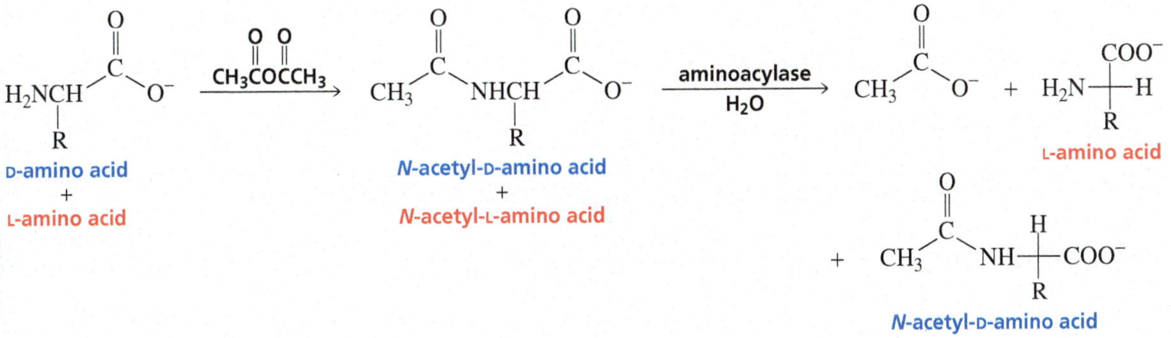

Because the resolution (separation) of the enantiomers depends on the difference in the rates of reaction of the enzyme with the two *N*-acetylated compounds, this technique is known as a **kinetic resolution.** In Section 6.14, we saw that a racemic mixture of amino acids can also be separated by the enzyme D-amino acid oxidase.

PROBLEM 25

Esterase is an enzyme that catalyzes the hydrolysis of esters. It hydrolyzes esters of L-amino acids more rapidly than esters of D-amino acids. How can this enzyme be used to separate a racemic mixture of amino acids?

21.8 PEPTIDE BONDS AND DISULFIDE BONDS

Peptide bonds and disulfide bonds are the only covalent bonds that join amino acids together in a peptide or a protein.

Peptide Bonds

The amide bonds that link amino acids are called **peptide bonds.** By convention, peptides and proteins are written with the free amino group (of the **N-terminal amino acid**) on the left and the free carboxyl group (of the **C-terminal amino acid**) on the right.

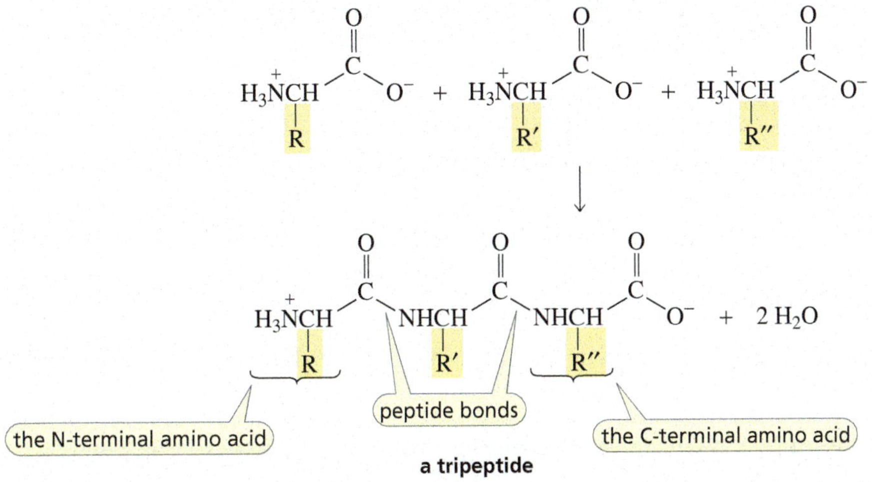

a tripeptide

- When the identities of the amino acids in a peptide are known but their sequence is not known, the amino acids are written separated by commas.
- When their sequence is known, the amino acids are written connected by hyphens.

For example, in the pentapeptide represented on the right on the top of the next page, valine is the N-terminal amino acid and histidine is the C-terminal amino acid. The amino acids are numbered starting with the N-terminal end. Alanine is, therefore, referred to as Ala 3 because it is the third amino acid from the N-terminal end.

Glu, Cys, His, Val, Ala

> the pentapeptide contains the indicated amino acids, but their sequence is not known

Val-Cys-Ala-Glu-His

> the amino acids in the pentapeptide have the indicated sequence

In naming a peptide, adjective names (ending in "yl") are used for all the amino acids except the C-terminal amino acid. Thus, the preceding pentapeptide is valylcysteinylalanylglutamylhistidine. Each amino acid has the L configuration unless otherwise specified.

A peptide bond has about 40% double-bond character because of electron delocalization (Section 15.2). Steric strain causes the configuration that has the α-carbons of adjacent amino acids on opposite side of the double bond to be more stable.

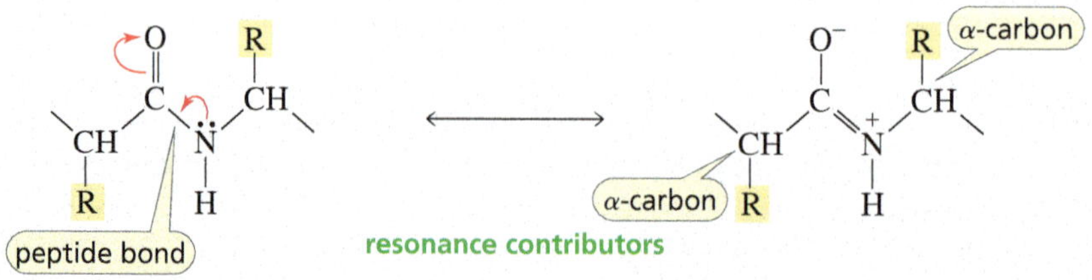

The partial double-bond character prevents free rotation about the peptide bond, so the carbon and nitrogen atoms of the peptide bond and the two atoms to which each is attached are held rigidly in a plane (Figure 21.6). This regional planarity affects the way a chain of amino acids can fold; this has important implications for the three-dimensional shapes of peptides and proteins (Section 21.14).

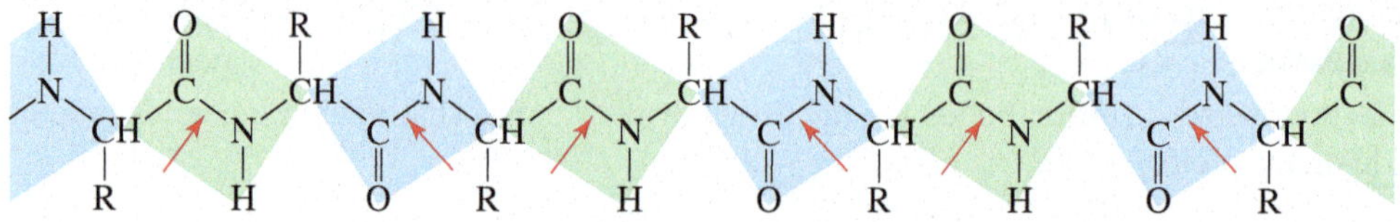

▶ **Figure 21.6**
A segment of a polypeptide chain. The red arrows indicate the peptide bonds. Colored squares indicate the plane defined by each peptide bond. Notice that the R groups bonded to the α-carbons are on alternate sides of the peptide backbone.

PROBLEM 26

Draw the tetrapeptide Ala-Thr-Asp-Asn and indicate the peptide bonds.

PROBLEM 27

Draw the resonance contributors of the peptide bond in the less stable configuration.

PROBLEM 28 ♦

Which bonds in the backbone of a peptide can rotate freely?

Disulfide Bonds

When thiols are oxidized under mild conditions, they form a **disulfide**—a compound with an S—S bond. (Like C—H bonds, the number of S—H bonds decreases in an oxidation reaction and increases in a reduction reaction.)

$$2\,R{-}SH \xrightarrow{\text{mild oxidation}} RS{-}SR$$

thiol   disulfide

The oxidizing agent commonly used for this reaction is Br_2 (or I_2) in a basic solution.

MECHANISM FOR THE OXIDATION OF A THIOL TO A DISULFIDE

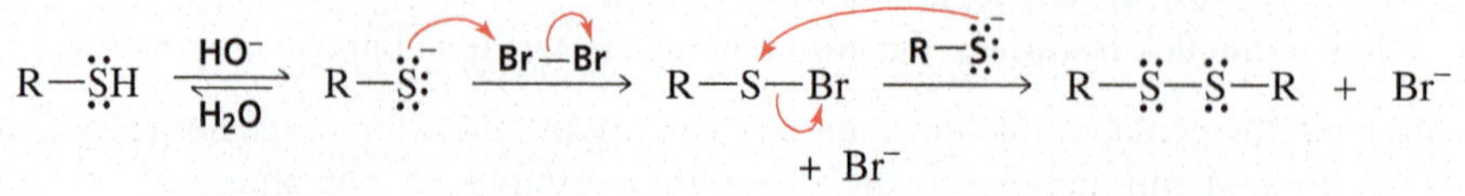

- A thiolate ion attacks the electrophilic bromine of Br_2.
- A second thiolate ion attacks the sulfur and eliminates Br^-.

Because thiols are oxidized to disulfides, disulfides are reduced to thiols.

$$RS-SR \xrightarrow{\text{reduction}} 2\,R-SH$$
disulfide   thiol

Disulfides are reduced to thiols.

The amino acid cysteine contains a thiol group, so two cysteine molecules can be oxidized to a disulfide. The disulfide is called cystine.

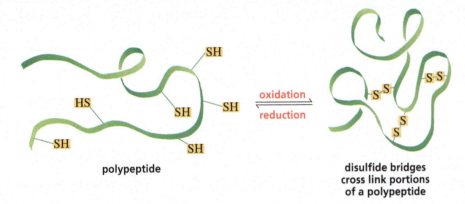

Thiols are oxidized to disulfides.

Two cysteines in a protein can be oxidized to a disulfide, creating a bond known as a **disulfide bridge**. Disulfide bridges are the only covalent bonds that are found between nonadjacent amino acids in peptides and proteins. They contribute to the overall shape of a protein by linking cysteines found in different parts of the peptide backbone, as shown in Figure 21.7; see Problem 70.

polypeptide

disulfide bridges
cross link portions
of a polypeptide

◀ **Figure 21.7**
Disulfide bridges cross link portions of a polypeptide.

The hormone insulin, synthesized in the pancreas by cells known as the islets of Langerhans, maintains the proper level of glucose in blood. Insulin is a polypeptide with two peptide chains; one contains 21 amino acids and the other 30 amino acids. The two chains are connected to each other by two **interchain disulfide bridges** (between two chains). Insulin also has an **intrachain disulfide bridge** (within a chain).

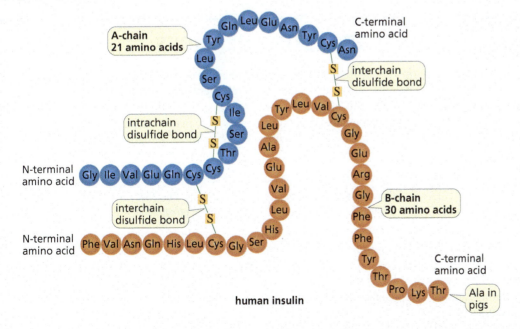

human insulin

Diabetes

Diabetes is the third leading cause of death (heart disease and cancer are the first and second) in the United States. It is caused either by insufficient secretion of insulin (type 1 diabetes) or its inability to stimulate its target cells (type 2 diabetes). Injections of insulin can control some of the symptoms associated with diabetes.

Until genetic engineering techniques became available (Section 21.11), pigs were the primary source of insulin for people with diabetes. The insulin was effective, but there were concerns about whether enough could be obtained over the long term for the growing population of people with diabetes. In addition, the C-terminal amino acid of the B-chain is alanine in pig insulin and threonine in human insulin, which caused some people to have allergic reactions. Now, however, mass quantities of synthetic insulin, chemically identical to human insulin, are produced from genetically modified host cells (Section 26.13).

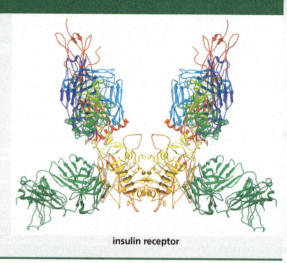

insulin receptor

Insulin binds to the insulin receptor on the surface of cells, telling the cell to transport glucose from the blood into the cell.

Hair: Straight or Curly?

Hair is made up of a protein called keratin that contains an unusually large number of cysteines (about 8% of the amino acids compared to an average of 2.8% for other proteins). These cysteines furnish keratin with many disulfide bridges that preserve its three-dimensional structure.

curly hair   straight hair

People can alter the structure of their hair (if they think it is either too straight or too curly) by changing the location of these disulfide bridges. This can be accomplished by first applying a reducing agent to the hair to reduce all the disulfide bridges in the protein. Then, after rearranging the hair into the desired shape (using curlers to curl it or combing it straight to uncurl it), an oxidizing agent is applied to form new disulfide bridges. The new disulfide bridges hold the hair in its new shape. When this treatment is applied to straight hair, it is called a "permanent." When it is applied to curly hair, it is called "hair straightening."

21.9 SOME INTERESTING PEPTIDES

Several peptide hormones are synthesized by the body to control pain. β-Endorphin has a chain of 31 amino acids, whereas leucine enkephalin and methionine enkephalin are pentapeptides. The five amino acids at the N-terminal end of β-endorphin are the same as those in methionine enkephalin (see Problem 76). These peptides control the body's sensitivity to pain by binding to receptors in certain brain cells. Part of their three-dimensional structures must be similar to that of morphine because they bind to the same receptors. The phenomenon known as "runner's high" that kicks in after vigorous exercise and the relief of pain through acupuncture are thought to be due to the release of these peptides.

Tyr-Gly-Gly-Phe-Leu   Tyr-Gly-Gly-Phe-Met
leucine enkephalin   methionine enkephalin

The nonapeptides bradykinin, vasopressin, and oxytocin are also peptide hormones. Bradykinin inhibits the inflammation of tissues. Vasopressin controls blood pressure by regulating the contraction of smooth muscle; it is also an antidiuretic. Oxytocin induces labor in pregnant women by stimulating the uterine muscle to contract; it also stimulates milk production in nursing mothers. Vasopressin and oxytocin, like β-endorphin and the enkephalins, also act on the brain. Vasopressin is a "fight-or-flight" hormone, whereas oxytocin has the opposite effect: it calms the body and promotes social bonding. In spite of their very different physiological effects, vasopressin and oxytocin differ only by two amino acids.

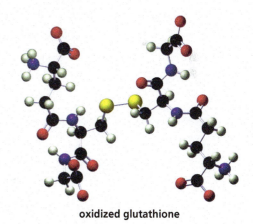

bradykinin   Arg-Pro-Pro-Gly-Phe-Ser-Pro-Phe-Arg

vasopressin   Cys-Tyr-Phe-Gln-Asn-Cys-Pro-Arg-Gly-NH$_2$
                  | |
                  S————————————————————S

oxytocin   Cys-Tyr-Ile-Gln-Asn-Cys-Pro-Leu-Gly-NH$_2$
                  | |
                  S————————————————————S

Vasopressin and oxytocin both have an intrachain disulfide bridge, and their C-terminal amino acids contain amide rather than carboxyl groups. Notice that the C-terminal amide group is indicated by writing "NH$_2$" after the name of the C-terminal amino acid.

The synthetic sweetener aspartame (or Equal or NutraSweet; see Section 20.19) is about 200 times sweeter than sucrose. Aspartame is the methyl ester of a dipeptide of L-aspartate and L-phenylalanine. The ethyl ester of the same dipeptide is not sweet. If a D-amino acid is substituted for either of the L-amino acids of aspartame, the resulting dipeptide is bitter rather than sweet.

Because aspartame contains phenylalanine, it cannot be used by people with the genetic disease known as PKU (Section 24.8).

aspartame
NutraSweet®

PROBLEM 29

An opioid pentapeptide has the following structure: Tyr-Cys-Gly-Phe-Cys

a. Draw the structure of the pentapeptide including all the side chains.

b. Write its structure following mild oxidation.

PROBLEM 30 ◆

What is the configuration about each of the asymmetric centers in aspartame?

PROBLEM 31

Glutathione is a tripeptide whose function is to destroy harmful oxidizing agents in the body. Oxidizing agents are thought to be responsible for some of the effects of aging and to play a causative role in cancer.

Glutathione removes oxidizing agents by reducing them. In the process, glutathione is oxidized, resulting in the formation of a disulfide bond between two glutathione molecules. An enzyme subsequently reduces the disulfide bond, returning glutathione to its original condition so it can react with another oxidizing agent.

glutathione   oxidizing agent ⇌ reducing agent   **oxidized glutathione**

oxidized glutathione

a. What amino acids make up glutathione?

b. What is unusual about glutathione's structure? (If you cannot answer this question, draw the structure you would expect for the tripeptide and compare your structure with the actual structure.)

21.10 THE STRATEGY OF PEPTIDE BOND SYNTHESIS: N-PROTECTION AND C-ACTIVATION

One difficulty in synthesizing a polypeptide is that the amino acids have two functional groups, enabling them to combine in different ways. Suppose, for example, that you want to make the dipeptide Gly-Ala. That dipeptide is only one of four possible dipeptides that could be formed by heating a mixture of alanine and glycine.

Gly-Ala   **Ala-Ala**   **Gly-Gly**   **Ala-Gly**

Protecting the Amino Group

If the amino group of the amino acid that is to be on the N-terminal end (in this case, Gly) is protected, then it will not be available to form a peptide bond. If its carboxyl group is activated before the second amino acid is added, then the amino group of the added amino acid (in this case, Ala) will react with the activated carboxyl group of glycine in preference to reacting with a nonactivated carboxyl group of another alanine.

The N-terminal amino acid must have its amino group protected and its carboxyl group activated.

The reagent most often used to protect the amino group of an amino acid is di-*tert*-butyl dicarbonate. Notice that the amino group rather than the carboxylate group of the amino acid reacts with di-*tert*-butyl dicarbonate because the amino group is a better nucleophile. When glycine reacts with di-*tert*-butyl dicarbonate in a nucleophilic acyl substitution reaction, the anhydride bond breaks, forming CO_2 and *tert*-butyl alcohol.

di-*tert*-butyl dicarbonate   **glycine**   **N-protected glycine**

Activating the Carboxyl Group

The reagent most often used to activate the carboxyl group is dicyclohexylcarbodiimide (DCCD). DCCD activates a carboxyl group by putting a good leaving group on the carbonyl group. Notice that in the first step, the acid protonates the most basic atom in the reactants (an N of DCCD).

N-protected Gly   **dicyclohexylcarbodiimide DCCD**

Adding the Second Amino Acid

After the amino acid's N-terminal group is protected and its C-terminal group is activated, the second amino acid is added. The unprotected amino group of the second amino acid adds to the activated carboxyl group, forming a tetrahedral intermediate (see the top of the next page). The C—O bond of the tetrahedral intermediate is easily broken because the bonding electrons are delocalized. Breaking this bond forms dicyclohexylurea, a stable diamide compound.

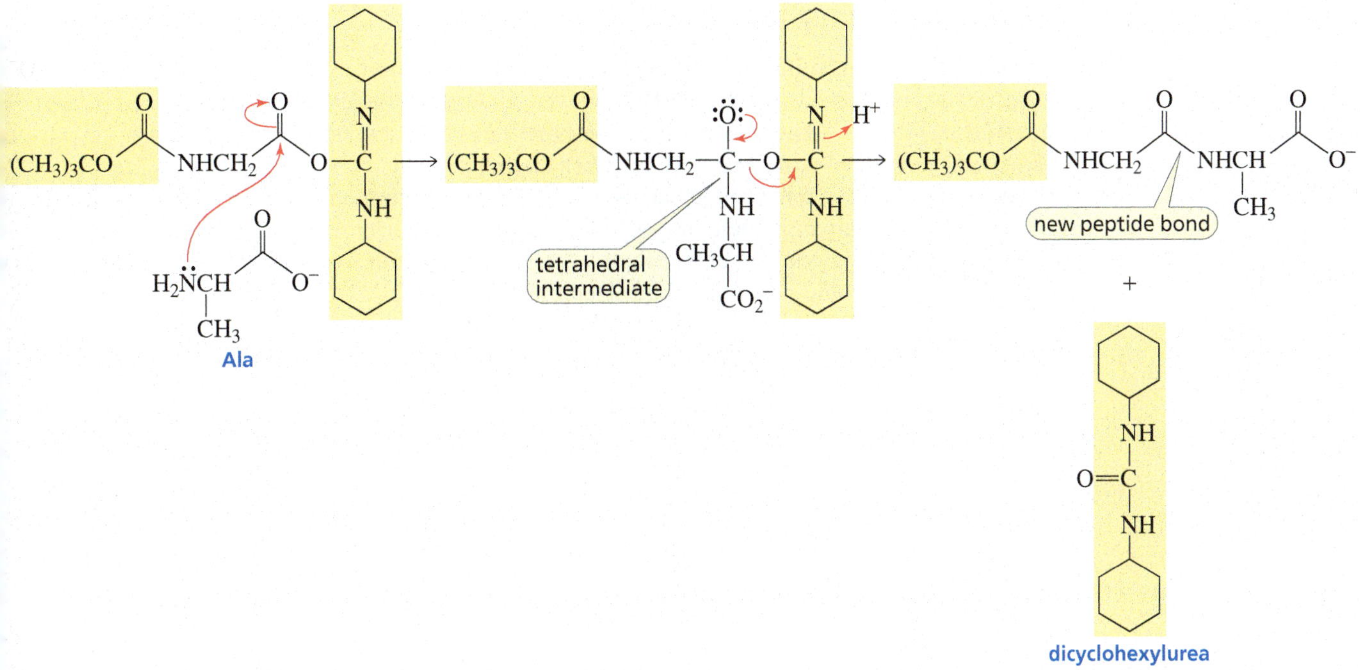

$(CH_3)_3CO$—$NHCH_2$— ... Ala (H_2NCH—CO^-, CH_3) ... tetrahedral intermediate (CH_3CH, CO_2^-) ... $(CH_3)_3CO$—$NHCH_2$—$NHCH$—O^- new peptide bond (CH_3)

+

dicyclohexylurea

Adding More Amino Acids

Amino acids can be added to the growing C-terminal end by repeating the same two steps: activating the carboxyl group of the C-terminal amino acid of the peptide by treating it with DCCD and then adding a new amino acid.

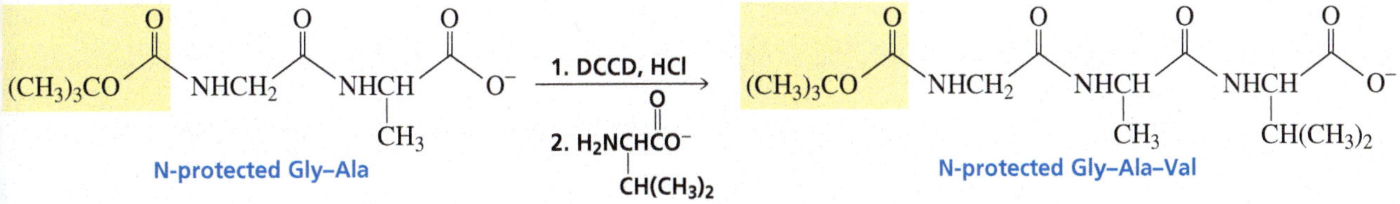

$(CH_3)_3CO$—$NHCH_2$—$NHCH$—O^- (CH_3)
N-protected Gly–Ala

1. DCCD, HCl
2. H_2NCHCO^- ($CH(CH_3)_2$)

$(CH_3)_3CO$—$NHCH_2$—$NHCH$—$NHCH$—O^- (CH_3) ($CH(CH_3)_2$)
N-protected Gly–Ala–Val

Removing the Protecting Group on the N-Terminal End

When the desired number of amino acids has been added to the chain, the protecting group, known by the acronym *t*-Boc (*tert*-butyloxycarbonyl; pronounced "tee-bok"), on the N-terminal amino acid is removed with trifluoroacetic acid in dichloromethane, a reagent that does not break any other covalent bonds. The protecting group is removed by an elimination reaction, forming isobutylene in the first step and carbon dioxide in the second step. (Follow the red arrows, then follow the blue arrows.) *t*-Boc is an ideal protecting group because it can be removed easily and, since isobutylene and carbon dioxide are gases, they escape, driving the reaction to completion.

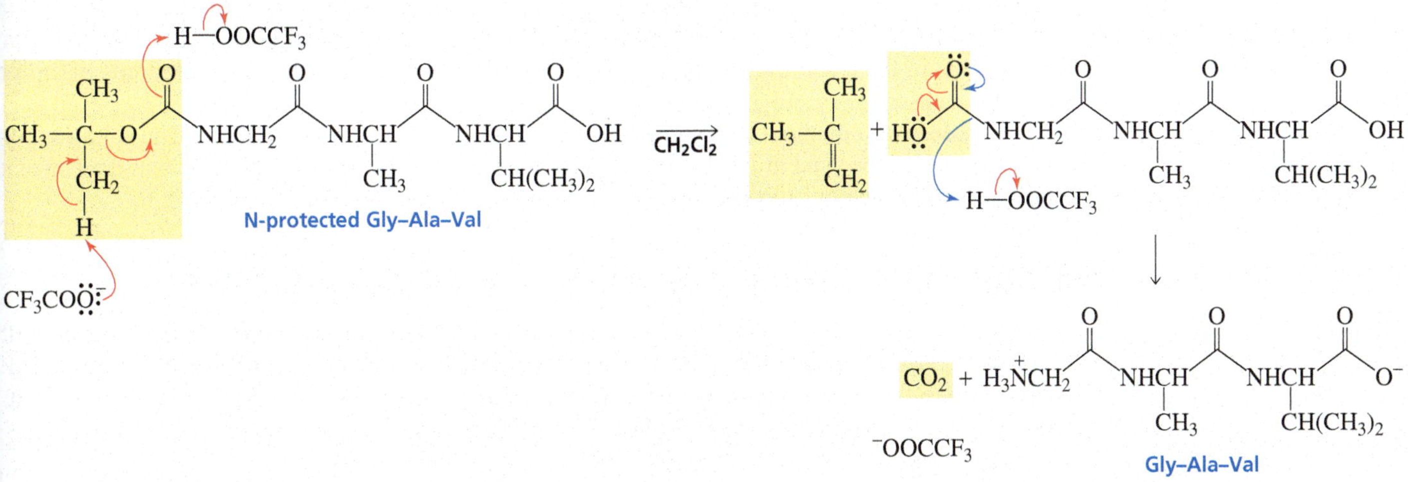

H—$OOCCF_3$

CH_3—C—O—$NHCH_2$—$NHCH$—$NHCH$—OH (CH_3)(CH_3)($CH(CH_3)_2$)
(CH_2)(H)
N-protected Gly–Ala–Val

$CF_3CO\ddot{O}^-$

$\xrightarrow{CH_2Cl_2}$

CH_3—C + $H\ddot{O}$—$NHCH_2$—$NHCH$—$NHCH$—OH (CH_2)(CH_3)($CH(CH_3)_2$)
H—$OOCCF_3$

CO_2 + $H_3\overset{+}{N}CH_2$—$NHCH$—$NHCH$—O^- (CH_3)($CH(CH_3)_2$)
$^-OOCCF_3$
Gly–Ala–Val

Yield Limitations

Theoretically, you should be able to make as long a peptide as desired with this technique. Reactions never produce 100% yields, however, and the yields are further decreased during the purification process. (The peptide must be purified after each step of the synthesis to prevent subsequent unwanted reactions with leftover reagents.)

Assuming that each amino acid can be added to the growing end of the peptide chain with an 80% yield (a relatively high yield, as you can probably appreciate from your own experience in the laboratory), the overall yield of a nonapeptide such as bradykinin would be only 17%. It is clear that large polypeptides could never be synthesized this way.

	Number of amino acids							
	2	3	4	5	6	7	8	9
Overall yield	80%	64%	51%	41%	33%	26%	21%	17%

PROBLEM 32 ♦

What dipeptides would be formed by heating a mixture of valine and N-protected leucine?

PROBLEM 33

Suppose you are trying to synthesize the dipeptide Val-Ser. Compare the product that would be obtained if thionyl chloride were used to activate the carboxyl group of N-protected valine with the product that would be obtained if it were activated with DCCD.

PROBLEM 34

Show the steps in the synthesis of the tetrapeptide Leu-Phe-Ala-Val.

PROBLEM 35 ♦

a. Calculate the overall yield of bradykinin when the yield for the addition of each amino acid to the chain is 70%.

b. What would be the overall yield of a peptide containing 15 amino acids if the yield for the incorporation of each is 80%?

21.11 AUTOMATED PEPTIDE SYNTHESIS

In addition to producing low overall yields, the method of peptide synthesis described in Section 21.10 is extremely time consuming because the product must be purified at each step of the synthesis. In 1969, Bruce Merrifield described a method that revolutionized the synthesis of peptides because it provided a much faster way to produce peptides in much higher yields. Furthermore, because it is automated, the synthesis requires fewer hours of direct attention. With this technique, bradykinin was synthesized in 27 hours with an overall yield of 85%. Subsequent refinements in the technique now allow the synthesis of a peptide containing hundreds of amino acids in a reasonable yield.

In the Merrifield method, the synthesis is done on a solid support in a column. The solid support is similar to the one used in ion-exchange chromatography (Section 21.5), except that the benzene rings have chloromethyl substituents instead of sulfonic acid substituents.

Adding the C-Terminal Amino Acid to the Solid Support

Before the C-terminal amino acid is added to the solid support, its amino group is protected with *t*-Boc to prevent the amino group from reacting with the solid support. The C-terminal amino acid is attached to the solid support by means of an S_N2 reaction—its carboxyl group attacks a benzyl carbon of the resin, displacing a chloride ion (Section 9.2).

Adding Subsequent Amino Acids

After the C-terminal amino acid is attached to the resin, the *t*-Boc protecting group is removed (Section 21.10). The next amino acid, with its amino group protected with *t*-Boc and its carboxyl group activated with DCCD, is added to the column, and then its protecting group is removed. Then the next N-terminal-protected and C-activated amino acid is added to the column. Thus, the protein is synthesized from the C-terminal end to the N-terminal end. (Proteins are synthesized in nature from the N-terminal end to the C-terminal end; Section 26.9.) Because the process uses a solid support and is automated, Merrifield's method of protein synthesis is called **automated solid-phase peptide synthesis.**

THE STEPS IN THE MERRIFIELD AUTOMATED SOLID-PHASE SYNTHESIS OF A TRIPEPTIDE

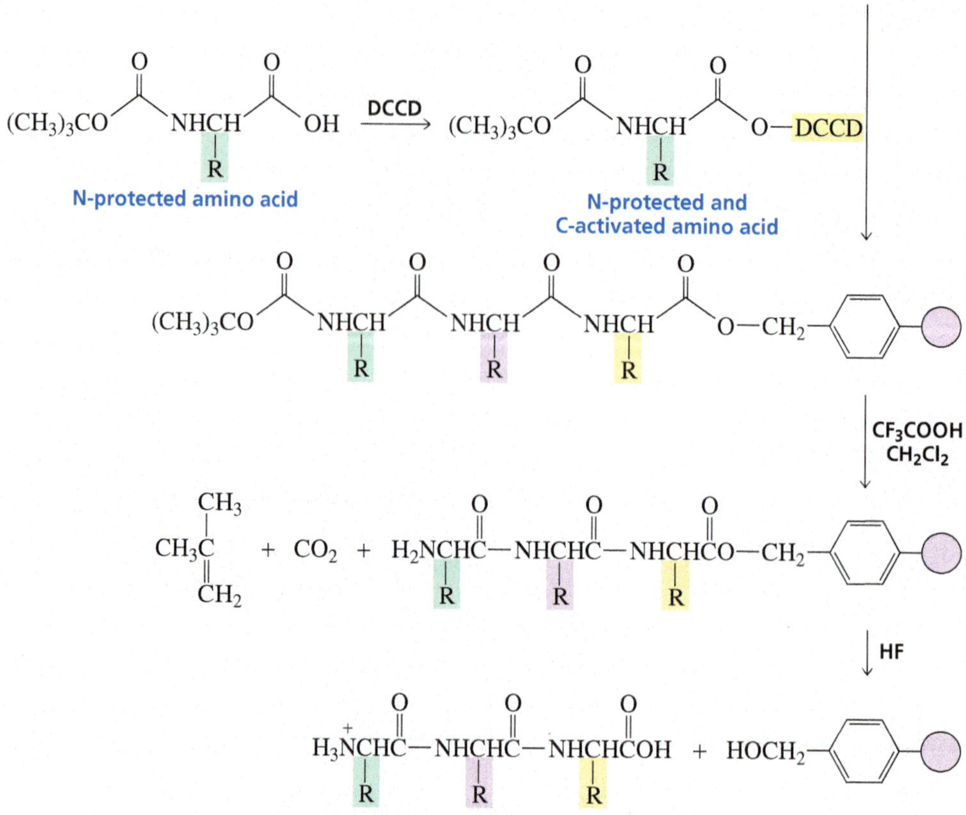

Advantage of the Merrifield Method

A huge advantage of the Merrifield method of peptide synthesis is that the growing peptide can be purified by washing the column with an appropriate solvent after each step of the procedure. The impurities are washed out of the column because they are not attached to the solid support. Because the peptide is covalently attached to the resin, none of it is lost in the purification step, leading to high yields of purified product.

After the required amino acids have been added one by one, the peptide can be removed from the resin by treatment with HF under mild conditions that do not break any peptide bonds.

Over time, Merrifield's technique has been improved so that peptides can be made more rapidly and more efficiently. However, it still cannot begin to compare with nature. A bacterial cell is able to synthesize a protein containing thousands of amino acids in seconds and can simultaneously synthesize thousands of different proteins with no mistakes.

Genetic Engineering

Since the early 1980s, it has been possible to synthesize proteins by **genetic engineering techniques** (Section 26.13). Inserting DNA into host cells causes them to produce large amounts of a desired protein. Genetic engineering techniques have also been useful in synthesizing proteins that differ in one or a few amino acids from a natural protein. Such synthetic proteins have been used, for example, to learn how a change in a single amino acid affects the properties of a protein (Section 22.9).

PROBLEM 36

Show the steps in the synthesis of the tetrapeptide in Problem 34, using Merrifield's method.

21.12 AN INTRODUCTION TO PROTEIN STRUCTURE

Proteins are described by four levels of structure.

- The **primary structure** of a protein is the sequence of the amino acids in the chain and the location of all the disulfide bridges.
- **Secondary structures** are regular conformations assumed by segments of the protein's backbone when it folds.
- The **tertiary structure** is the three-dimensional structure of the entire protein.
- If a protein has more than one polypeptide chain, it also has a quaternary structure. The **quaternary structure** is the way the individual polypeptide chains are arranged with respect to one another.

Primary Structure and Taxonomic Relationship

When scientists examine the primary structures of proteins that carry out the same function in different organisms, they can correlate the number of amino acid differences in the proteins to the closeness of the taxonomic relationship between the species. For example, cytochrome c, a protein that transfers electrons in biological oxidations, has about 100 amino acids. Yeast cytochrome c differs by 48 amino acids from horse cytochrome c, whereas duck cytochrome c differs by only two amino acids from chicken cytochrome c. Ducks and chickens, therefore, have a much closer taxonomic relationship than do horses and yeast. The cytochrome c in chickens and turkeys have identical primary structures. Human cytochrome c and chimpanzee cytochrome c are also identical and differ by one amino acid from the cytochrome c of the rhesus monkey.

21.13 HOW TO DETERMINE THE PRIMARY STRUCTURE OF A POLYPEPTIDE OR A PROTEIN

Breaking the Disulfide Bridges

The first step in determining the sequence of amino acids in a polypeptide (or a protein) is to reduce the disulfide bridges in order to obtain a single polypeptide chain. A commonly used reducing agent is 2-mercaptoethanol. Notice that when it reduces a disulfide bridge, 2-mercaptoethanol is oxidized to a disulfide (Section 21.8). Reaction of the protein thiol groups with iodoacetic acid prevents the disulfide bridges from re-forming as a result of oxidation.

reducing disulfide bridges

PROBLEM 37

Write the mechanism for the reaction of a cysteine side chain with iodoacetic acid.

Determining the Number and Kinds of Amino Acids

Now we need to determine the number and kinds of amino acids in the polypeptide (or protein) chain. To do this, a sample of the polypeptide is dissolved in 6 M HCl and heated at 100 °C for 24 hours. This treatment hydrolyzes all the amide bonds in the polypeptide, including the amide bonds in the side chains of asparagine and glutamine.

$$\text{polypeptide} \xrightarrow[\substack{100\ °\text{C} \\ 24\ \text{h}}]{6\ \text{M HCl}} \text{amino acids}$$

The mixture of amino acids is then passed through an amino acid analyzer to identify the amino acids and determine how many of each kind are in the polypeptide (Section 21.5).

Because all the asparagines and glutamines have been hydrolyzed to aspartates and glutamates, the number of aspartates or glutamates in the amino acid mixture tells us the number of aspartates plus asparagines or glutamates plus glutamines in the original polypeptide. A separate test must be conducted to distinguish between aspartate and asparagine or between glutamate and glutamine in the original polypeptide.

The strongly acidic conditions used for hydrolysis destroy all the tryptophans because the indole ring is unstable in acid (Section 19.5). However, the tryptophan content can be determined by hydroxide-ion-promoted hydrolysis of the peptide. This is not a general method for peptide bond hydrolysis because the strongly basic conditions destroy several amino acids.

Determining the N-Terminal Amino Acid

One of the most widely used methods to identify the N-terminal amino acid of a polypeptide is to treat the polypeptide with phenyl isothiocyanate (PITC), commonly known as **Edman's reagent.** Edman's reagent reacts with the N-terminal amino group, and the resulting thiazolinone derivative is cleaved from the polypeptide under mildly acidic conditions, leaving behind a polypeptide with one less amino acid.

phenyl isothiocyanate
PITC
Edman's reagent

peptide without the original N-terminal amino acid

thiazolinone

The thiazolinone rearranges in dilute acid to a more stable phenylthiohydantoin (PTH) (see Problem 78).

a thiazolinone   **a PTH-amino acid**

Because each amino acid has a different side chain (R), each amino acid forms a different PTH-amino acid. The particular PTH-amino acid can be identified by chromatography using known standards.

An automated instrument known as a *sequencer* allows about 50 successive Edman degradations of a polypeptide to be performed (100 with more advanced instruments). The entire primary structure cannot be determined in this way, however, because side products accumulate that interfere with the results.

PROBLEM 38 ◆

In determining the primary structure of insulin, what would lead you to conclude that insulin had more than one polypeptide chain?

Determining the C-Terminal Amino Acid

The C-terminal amino acid of a polypeptide can be identified using a carboxypeptidase, an enzyme that catalyzes the hydrolysis of the C-terminal peptide bond, thereby cleaving off the C-terminal amino acid.

- Carboxypeptidase A cleaves off the C-terminal amino acid, as long as it is not *arginine* or *lysine*.
- Carboxypeptidase B cleaves off the C-terminal amino acid, only if it is *arginine* or *lysine*.

Carboxypeptidases are **exopeptidases**—enzymes that catalyze the hydrolysis of a peptide bond at the end of a peptide chain.

Carboxypeptidases cannot be used to determine the amino acids at the C-terminal end of a peptide by cleaving off the C-terminal amino acids sequentially, because peptide bonds hydrolyze at different rates. For example, if the C-terminal amino acid hydrolyzed slowly and the next one hydrolyzed rapidly, then it would appear that they were being cleaved off at about the same rate.

Partial Hydrolysis

Once the N-terminal and C-terminal amino acids have been identified, a sample of the polypeptide is hydrolyzed under conditions that hydrolyze only some of the peptide bonds—a procedure known as **partial hydrolysis.** The resulting fragments are separated, and the amino acid composition of each can be determined using electrophoresis or an amino acid analyzer. The process is repeated and the sequence of the original protein can then be deduced by lining up the peptides and looking for regions of overlap. (The N-terminal and C-terminal amino acids of each fragment can also be identified, if needed.)

PROBLEM-SOLVING STRATEGY

Sequencing an Oligopeptide   LEARN THE STRATEGY

When a nonapeptide undergoes partial hydrolysis, it forms dipeptides, a tripeptide and two tetrapeptides whose amino acid compositions are shown. Reaction of the intact nonapeptide with Edman's reagent releases PTH-Leu. What is the sequence of the nonapeptide?

1. Pro, Ser	**3.** Met, Ala, Leu	**5.** Glu, Ser, Val, Pro	**7.** Met, Leu
2. Gly, Glu	**4.** Gly, Ala	**6.** Glu, Pro, Gly	**8.** His, Val

- Because we know that the N-terminal amino acid is Leu, we need to look for a fragment that contains Leu. Fragment **7** tells us that Met is next to Leu, and fragment **3** tells us that Ala is next to Met.
- Now we look for another fragment that contains Ala. Fragment **4** contains Ala and tells us that Gly is next to Ala.
- From fragment **2**, we know that Glu comes next; Glu is in both fragments **5** and **6**.
- Fragment **5** has three amino acids we have yet to place in the growing peptide (Ser, Val, Pro), but fragment **6** has only one; therefore, we know from fragment **6** that Pro is the next amino acid.

- Fragment **1** indicates that the next amino acid is Ser, so we can now use fragment **5**. Fragment **5** indicates that the next amino acid is Val, and fragment **8** tells us that His is the last (C-terminal) amino acid.
- Thus, the amino acid sequence of the nonapeptide is Leu-Met-Ala-Gly-Glu-Pro-Ser-Val-His

USE THE STRATEGY

PROBLEM 39 ◆

A decapeptide undergoes partial hydrolysis to give peptides whose amino acid compositions are shown. Reaction of the intact decapeptide with Edman's reagent releases PTH-Gly. What is the sequence of the decapeptide?

1. Ala, Trp	**3.** Pro, Val	**5.** Trp, Ala, Arg	**7.** Glu, Ala, Leu
2. Val, Pro, Asp	**4.** Ala, Glu	**6.** Arg, Gly	**8.** Met, Pro, Leu, Glu

Hydrolysis Using Endopeptidases

A polypeptide can also be partially hydrolyzed using **endopeptidases**—enzymes that catalyze the hydrolysis of a peptide bond that is *not* at the end of a peptide chain. Trypsin, chymotrypsin, and elastase are endopeptidases that catalyze the hydrolysis of only the specific peptide bonds listed in Table 21.4. Trypsin, for example, catalyzes the hydrolysis of the peptide bond on the C-side of (meaning, on the right of) positively charged side chains (arginine or lysine). These enzymes belong to the group of enzymes known as **digestive enzymes.**

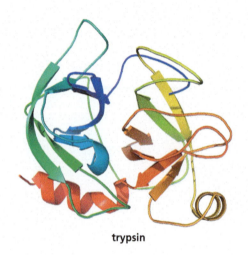

trypsin

(see the legend to Figure 21.10 on page 1021)

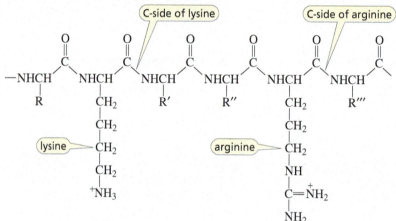

Thus, trypsin catalyzes the hydrolysis of three peptide bonds in the following polypeptide, creating a hexapeptide, a dipeptide, and two tripeptides.

Ala-Lys-Phe-Gly-Asp-Trp-Ser-Arg-Glu-Val-Arg-Tyr-Leu-His

cleavage by trypsin

Table 21.4	Specificity of Peptide or Protein Cleavage
Reagent	**Specificity**
Chemical reagents	
Edman's reagent	removes the N-terminal amino acid
Cyanogen bromide	hydrolyzes on the C-side of Met
Exopeptidases*	
Carboxypeptidase A	removes the C-terminal amino acid (not if it is Arg or Lys)
Carboxypeptidase B	removes the C-terminal amino acid (only if it is Arg or Lys)
Endopeptidases*	
Trypsin	hydrolyzes on the C-side of Arg and Lys
Chymotrypsin	hydrolyzes on the C-side of amino acids that contain aromatic six-membered rings (Phe, Tyr, Trp)
Elastase	hydrolyzes on the C-side of small amino acids (Gly, Ala, Ser, and Val)
Thermolysin	hydrolyzes on the C-side of Ile, Met, Phe, Trp, Tyr, and Val

*Cleavage will not occur if Pro is at the hydrolysis site.

Chymotrypsin catalyzes the hydrolysis of the peptide bond on the C-side of amino acids that contain aromatic six-membered rings (Phe, Tyr, Trp).

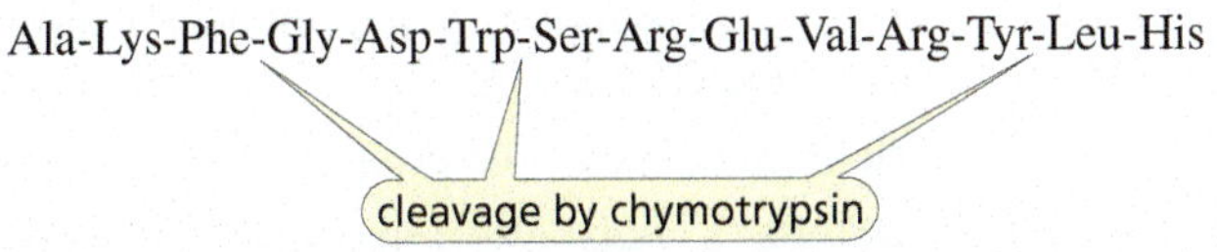
Ala-Lys-Phe-Gly-Asp-Trp-Ser-Arg-Glu-Val-Arg-Tyr-Leu-His
cleavage by chymotrypsin

Elastase catalyzes the hydrolysis of peptide bonds on the C-side of the four smallest amino acids (Gly, Ala, Ser, and Val). Chymotrypsin and elastase are much less specific than trypsin. (An explanation for the specificity of these enzymes is given in Section 23.9.)

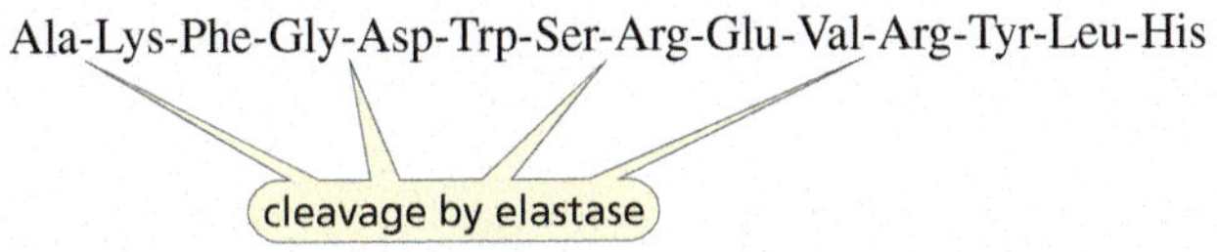
Ala-Lys-Phe-Gly-Asp-Trp-Ser-Arg-Glu-Val-Arg-Tyr-Leu-His
cleavage by elastase

None of the exopeptidases or endopeptidases that we have discussed will catalyze the hydrolysis of a peptide bond if proline is at the hydrolysis site. These enzymes recognize the appropriate hydrolysis site by its shape and charge, and the cyclic structure of proline causes the hydrolysis site to have an unrecognizable three-dimensional shape.

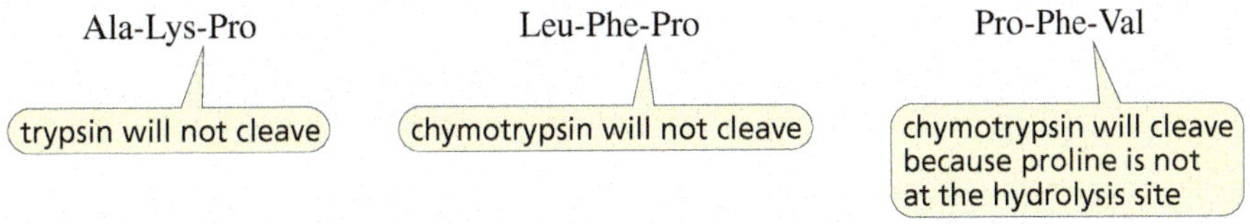
Ala-Lys-Pro
trypsin will not cleave
Leu-Phe-Pro
chymotrypsin will not cleave
Pro-Phe-Val
chymotrypsin will cleave because proline is not at the hydrolysis site

Hydrolysis Using Cyanogen Bromide

Cyanogen bromide (BrC≡N) hydrolyzes the peptide bond on the C-side of methionine. Cyanogen bromide is more specific than the endopeptidases about which peptide bonds it cleaves, so it provides more reliable information about the primary structure. Because cyanogen bromide is not a protein, it does not recognize the substrate by its shape. As a result, it will still cleave the peptide bond if proline is at the hydrolysis site.

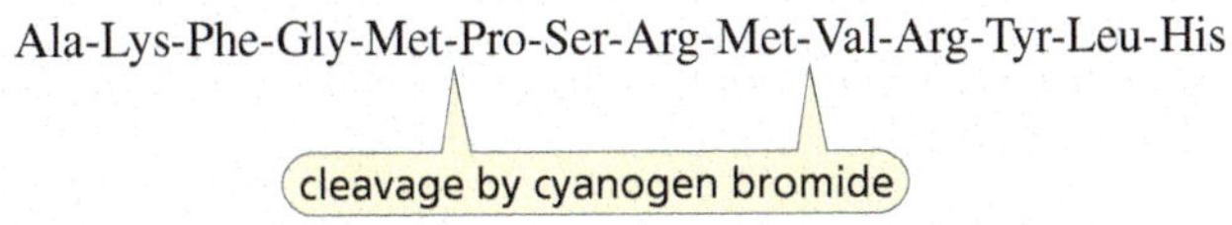
Ala-Lys-Phe-Gly-Met-Pro-Ser-Arg-Met-Val-Arg-Tyr-Leu-His
cleavage by cyanogen bromide

MECHANISM FOR THE CLEAVAGE OF A PEPTIDE BOND BY CYANOGEN BROMIDE

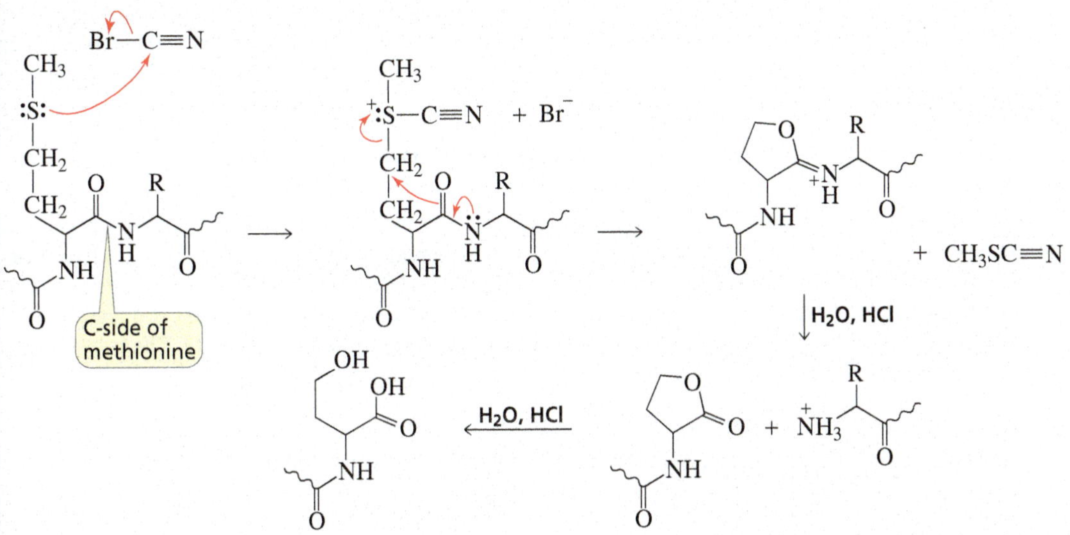
C-side of methionine
+ Br⁻
+ CH₃SC≡N
H₂O, HCl
H₂O, HCl

■ The nucleophilic sulfur of methionine attacks the carbon of cyanogen bromide and displaces a bromide ion.

■ Nucleophilic attack by oxygen on the methylene group displaces the weakly basic leaving group and forms a five-membered ring (Section 9.2).

■ Acid-catalyzed hydrolysis of the imine cleaves the protein (Section 16.8).

■ Further hydrolysis causes the lactone (a cyclic ester) to open to a carboxyl group and an alcohol group (Section 15.8).

The last step in determining the primary structure of a polypeptide is to figure out the location of any disulfide bonds. This is done by hydrolyzing a sample of the polypeptide that has intact disulfide bonds. From a determination of the amino acids in the cysteine-containing fragments, the locations of the disulfide bonds in the polypeptide can be established (see Problem 66).

PROBLEM 40

Explain why cyanogen bromide does not cleave on the C-side of cysteine.

PROBLEM 41 ◆

Indicate the peptides produced from cleavage by the indicated reagent:

a. His-Lys-Leu-Val-Glu-Pro-Arg-Ala-Gly-Ala by trypsin
b. Leu-Gly-Ser-Met-Phe-Pro-Tyr-Gly-Val by chymotrypsin

LEARN THE STRATEGY

PROBLEM 42 SOLVED

Determine the amino acid sequence of a polypeptide from the following data:

Acid-catalyzed hydrolysis gives Ala, Arg, His, Leu, 2 Lys, 2 Met, Pro, 2 Ser, Thr, and Val.
Carboxypeptidase A releases Val.
Edman's reagent releases PTH-Leu.
Treatment with cyanogen bromide gives three peptides with the following amino acid compositions:

1. His, Lys, Met, Pro, Ser   **2.** Thr, Val   **3.** Ala, Arg, Leu, Lys, Met, Ser

Trypsin-catalyzed hydrolysis gives three peptides and a single amino acid:

1. Arg, Leu, Ser   **3.** Lys
2. Met, Pro, Ser, Thr, Val   **4.** Ala, His, Lys, Met

SOLUTION Acid-catalyzed hydrolysis shows that the polypeptide has 13 amino acids. The N-terminal amino acid is Leu (revealed by Edman's reagent), and the C-terminal amino acid is Val (revealed by carboxypeptidase A).

Leu __ __ __ __ __ __ __ __ __ __ __ Val

■ Because cyanogen bromide cleaves on the C-side of Met, any peptide containing Met must have Met as its C-terminal amino acid. Thus, the peptide that does not contain Met must be the C-terminal peptide, indicating that the twelfth amino acid is Thr. We know that peptide **3** is the N-terminal peptide because it contains Leu. Because peptide **3** is a hexapeptide, we know that the sixth amino acid in the polypeptide is Met. We also know that the eleventh amino acid is Met because cyanogen bromide cleavage gave the dipeptide Thr, Val.

Ala, Arg, Lys, Ser   His, Lys, Pro, Ser

Leu __ __ __ __ Met __ __ __ __ Met Thr Val

■ Because trypsin cleaves on the C-side of Arg and Lys, any peptide containing Arg or Lys must have that amino acid as its C-terminal amino acid. Therefore, Arg is the C-terminal amino acid of peptide **1**, so we now know that the first three amino acids are Leu-Ser-Arg. We also know that the next two are Lys-Ala because if they were Ala-Lys, then trypsin cleavage would give an Ala, Lys dipeptide. The trypsin data also identify the positions of His and Lys.

Pro, Ser

Leu Ser Arg Lys Ala Met His Lys __ __ Met Thr Val

- Finally, because trypsin successfully cleaves on the C-side of Lys, Pro cannot be adjacent to Lys. Thus, the amino acid sequence of the polypeptide is

  Leu   Ser   Arg   Lys   Ala   Met   His   Lys   Ser   Pro   Met   Thr   Val

PROBLEM 43 ♦

Determine the primary structure of an octapeptide from the following data:

Acid-catalyzed hydrolysis gives 2 Arg, Leu, Lys, Met, Phe, Ser, and Tyr.

Carboxypeptidase A releases Ser.

Edman's reagent releases Leu.

Treatment with cyanogen bromide forms two peptides with the following amino acid compositions:

1. Arg, Phe, Ser   **2.** Arg, Leu, Lys, Met, Tyr

Trypsin-catalyzed hydrolysis forms the following two amino acids and two peptides:

1. Arg   **2.** Ser   **3.** Arg, Met, Phe   **4.** Leu, Lys, Tyr

USE THE STRATEGY

PROBLEM 44

Three peptides were obtained from a trypsin digestion of two different polypeptides. In each case, indicate the possible sequences from the given data and tell what further experiment should be carried out in order to determine the primary structure of the polypeptide.

a. polypeptide I: **1.** Val-Gly-Asp-Lys   **2.** Leu-Glu-Pro-Ala-Arg   **3.** Ala-Leu-Gly-Asp
b. polypeptide II: **1.** Val-Leu-Gly-Glu   **2.** Ala-Glu-Pro-Arg   **3.** Ala-Met-Gly-Lys

21.14   SECONDARY STRUCTURE

Secondary structure describes the repetitive conformations assumed by segments of the backbone chain of a polypeptide or protein. In other words, the secondary structure describes how segments of the backbone fold. Three factors determine the secondary structure of a segment of protein:

- the regional planarity about each peptide bond (as a result of the partial double-bond character of the amide bond), which limits the possible conformations of the peptide chain (Section 21.8)

- minimizing energy by maximizing the number of hydrogen bonds between peptide groups (that is, hydrogen bonds between the carbonyl oxygen of one amino acid and the amide hydrogen of another)

hydrogen bonding between peptide groups

- the need for adequate separation between neighboring R groups to avoid steric strain and repulsion of like charges

α-Helix

One type of secondary structure is an **α-helix.** In an α-helix, the backbone of the polypeptide coils around the long axis of the protein molecule. The substituents on the α-carbons of the amino acids protrude outward from the helix, thereby minimizing steric strain (Figure 21.8a). The helix is stabilized by hydrogen bonds—each hydrogen attached to an amide nitrogen is hydrogen bonded to a carbonyl oxygen of an amino acid four amino acids away (Figure 21.8b).

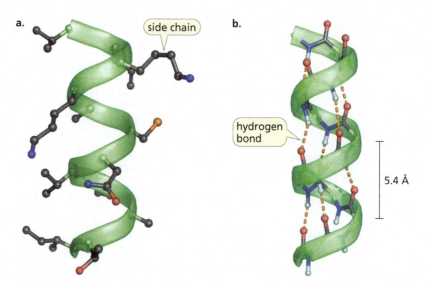

▶ Figure 21.8
(a) A segment of a protein in an α-helix. The substituents on the α-carbons point outward from the helix.
(b) The helix is stabilized by hydrogen bonding between peptide groups. Notice that each carbonyl oxygen points up and each hydrogen on the nitrogen points down.

Because the amino acids have the L-configuration, the α-helix is a right-handed helix—that is, it rotates in a clockwise direction as it spirals down. Each turn of the helix contains 3.6 amino acids, and the repeat distance of the helix is 5.4 Å.

Not all amino acids are able to fit into an α-helix. Proline, for example, causes a distortion in a helix because the bond between the proline nitrogen and the α-carbon cannot rotate to let proline fit into a helix properly. Similarly, two adjacent amino acids that have more than one substituent on a β-carbon (valine, isoleucine, or threonine) cannot fit into a helix because of steric crowding between the R groups. Finally, two adjacent amino acids with like-charged substituents cannot fit into a helix because of electrostatic repulsion between the R groups.

The percentage of amino acids coiled into an α-helix varies from protein to protein. On average about 25% of the amino acids in a globular protein are in α-helices.

Right-Handed and Left-Handed Helices

The α-helix, composed of a chain of L-amino acids, is a right-handed helix. When scientists synthesized a chain of D-amino acids, they found that it folded into a left-handed helix that was the mirror image of a right-handed α-helix. When they synthesized a peptidase that contained only D-amino acids, they found that the enzyme was just as catalytically active as a naturally occurring peptidase with L-amino acids. However, the peptidase that contained D-amino acids cleaved peptide bonds only in polypeptide chains composed of D-amino acids.

β-Pleated Sheet

The second type of secondary structure is a **β-pleated sheet.** In a β-pleated sheet, the polypeptide backbone is in a zigzag structure resembling a series of pleats. The hydrogen bonding in a β-pleated sheet occurs between neighboring peptide chains (Figure 21.9 on the next page). A β-pleated sheet is almost fully extended; the average two amino acid repeat distance is 7.0 Å.

The part of the backbone structure of a protein background that exists in a pleated sheet is indicated by a flat arrow pointing in the N → C direction (Figure 21.10). The adjacent chains of a pleated sheet can run in opposite directions (an antiparallel β-pleated sheet) or in the same direction (a parallel β-pleated sheet).

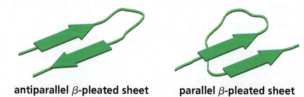

antiparallel β-pleated sheet   parallel β-pleated sheet

Because the substituents (R) on the α-carbons of the amino acids on adjacent chains of a pleated sheet are close to each other, the substituents must be small if the chains are to nestle closely enough together to maximize hydrogen-bonding interactions. Silk, for example, contains a large proportion

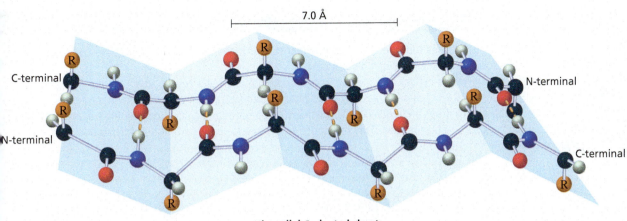

7.0 Å

C-terminal

N-terminal

N-terminal

C-terminal

antiparallel β-pleated sheet

▲ **Figure 21.9**
A segment of an antiparallel β-pleated sheet. The extended chains are held together by hydrogen bonding (orange dashes) between peptide groups. The peptide groups are in the plane of the sheet; the side chains (R) alternate between pointing above and below the plane.

of relatively small amino acids (glycine and alanine) and, therefore, has large segments of β-pleated sheet. The number of side-by-side strands in a β-pleated sheet ranges from 2 to 15 in a globular protein. The average strand in a β-pleated sheet section contains six amino acids.

Wool and the fibrous protein of muscle have secondary structures that are almost all α-helices. Consequently, these proteins can be stretched. In contrast, proteins with secondary structures that are predominantly β-pleated sheets, such as silk and spider webs, cannot be stretched because a β-pleated sheet is almost fully extended already.

Coil Conformation

Generally, less than half of the protein's backbone is arranged in a defined secondary structure—an α-helix or a β-pleated sheet. Most of the rest of the protein, though highly ordered, is nonrepetitive and, therefore, difficult to describe. Many of these ordered polypeptide fragments are said to be in **coil** or **loop conformations** (Figure 21.10).

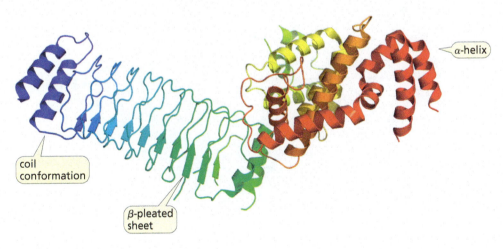

α-helix

coil conformation

β-pleated sheet

▲ **Figure 21.10**
The backbone structure of an enzyme called ligase (Section 26.5): a β-pleated sheet is indicated by a flat arrow pointing in the N → C direction, an α-helix by a helical ribbon, and a coil or loop conformation by a thin tube.

PROBLEM 45 ◆

  a. How long is an α-helix that contains 74 amino acids?
  b. How long is a fully extended peptide chain that contains the same number of amino acids?
    (The distance between consecutive amino acids in a fully extended chain is 3.5 Å; the repeat distance of an α-helix is 5.4 Å.)

β-Peptides: An Attempt to Improve on Nature

β-Peptides are polymers of β-amino acids, so they have backbones that are one carbon longer than the peptides nature synthesizes using α-amino acids. Therefore, each β-amino acid has two carbons to which side chains can be attached.

Like α-polypeptides, β-polypeptides fold into relatively stable helical and pleated sheet conformations. As a result, scientists are trying to find out whether such peptides have biological activity. Recently, a β-peptide has been synthesized that mimics the activity of the hormone somatostatin. There is hope that β-polypeptides will provide a source of new drugs and catalysts. Surprisingly, the peptide bonds in β-polypeptides are resistant to the enzymes that catalyze the hydrolysis of peptide bonds in α-polypeptides. This resistance to hydrolysis suggests that a β-polypeptide drug would have a longer duration of action in the bloodstream.

21.15 TERTIARY STRUCTURE

The *tertiary structure* of a protein is the three-dimensional arrangement of all the atoms in the protein (Figure 21.11). Proteins fold spontaneously in solution to maximize their stability. Every time there is a stabilizing interaction between atoms, free energy is released. The more free energy released (the more negative the $\Delta G°$), the more stable the protein. Consequently, a protein tends to fold in a way that maximizes the number of stabilizing interactions.

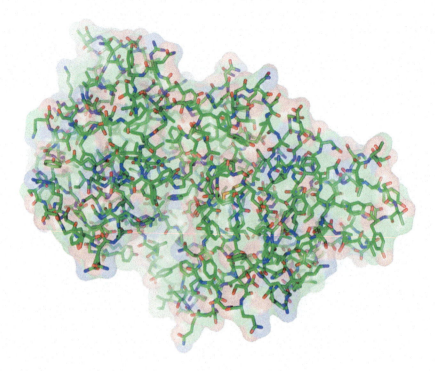

▶ **Figure 21.11**
The three-dimensional structure of thermolysin (an endopeptidase).

Stabilizing Interactions

The stabilizing interactions in a protein include disulfide bonds, hydrogen bonds, electrostatic attractions (attractions between opposite charges), and hydrophobic interactions. Stabilizing interactions can occur between peptide groups (atoms in the backbone of the protein), between side-chain groups (α-substituents), and between peptide and side-chain groups (Figure 21.12 on the next page). Because the side-chain groups help determine how a protein folds, the tertiary structure of a protein is determined by its primary structure.

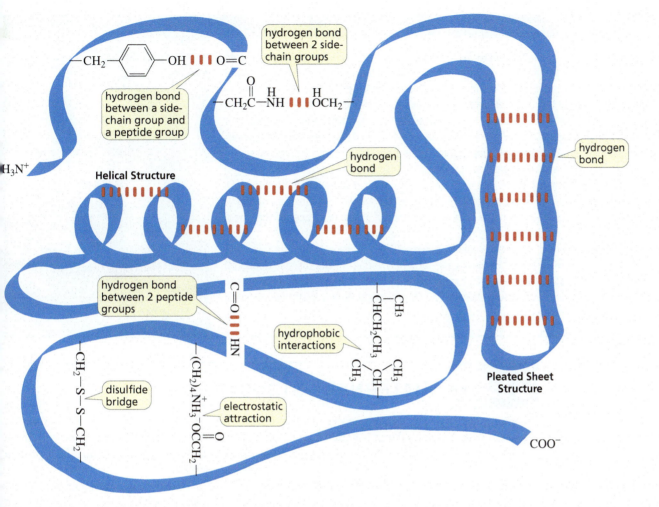

▲ **Figure 21.12**
Stabilizing interactions responsible for the tertiary structure of a protein.

Disulfide bonds are the only covalent bonds that can form when a protein folds. The other bonding interactions that occur in folding are much weaker but, because there are so many of them, they are important in determining how a protein folds.

Most proteins exist in aqueous environments. Therefore, they tend to fold in a way that exposes the maximum number of polar groups to the surrounding water and buries the nonpolar groups in the protein's interior, away from water.

Hydrophobic Interactions

The **hydrophobic interactions** between nonpolar groups in the protein increase its stability by increasing the entropy of water molecules. Water molecules that surround nonpolar groups are highly structured. When two nonpolar groups come together, the surface area in contact with water diminishes, decreasing the amount of structured water. Decreasing structure increases entropy, which in turn decreases the free energy, thereby increasing the stability of the protein. (Recall that $\Delta G° = \Delta H° - T\Delta S°$.)

The precise mechanism by which proteins fold is still an unanswered question. Protein misfolding has been linked to numerous diseases such as Alzheimer's disease and Huntington's disease.

PROBLEM 46 ◆

How would a protein that resides in the nonpolar interior of a membrane fold compared with the water-soluble protein just discussed?

Diseases Caused by a Misfolded Protein

Bovine spongiform encephalopathy (BSE), commonly known as mad cow disease, is unlike most other diseases because it is not caused by a microorganism. Instead, it is caused by a misfolded protein in the brain called prion. It is not yet known what causes the prion to become misfolded. The misfolded prion causes cells to deteriorate until the brain has a sponge-like appearance. The deterioration causes the loss of mental function, which makes cows with the disease act strangely (hence the name *mad cow disease*). It is not curable and it is fatal, but it is not contagious. It is thought that mad cow disease, first reported in the U.K. in 1985, was caused by cows eating bone meal made from scrapie-infected sheep. It takes several years from the time of first exposure until the first signs of the disease appear, but then it it progresses quickly.

There are other diseases caused by misfolded prions that have similar symptoms. Kuru—transmitted through cannibalism—has been found to occur in the Fore people of Papua New Guinea (*kuru* means "trembling"). Scrapie affects sheep and goats. This disease got its name from the tendency of sheep to scrape off their wool on fences as they lean on them in an attempt to stay upright.

The human form of the disease, called Creutzfeldt–Jakob Disease (CJD), is rare and incurable. The average age of onset of CJD is 64. In 1996, however, several cases of a new variant of the disease appeared in young adults in the U.K. To date, 229 cases have been reported, and 177 of these have been in the United Kingdom. This new variant of Creutzfeldt–Jakob Disease (vCJD) is thought to be caused by ingesting meat products of an animal infected with the disease.

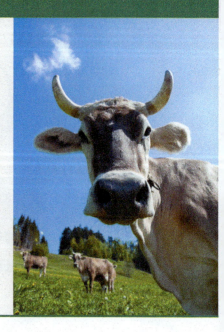

21.16 QUATERNARY STRUCTURE

Some proteins have more than one polypeptide chain. The individual chains are called **subunits.** A protein with a single subunit is called a *monomer;* one with two subunits a *dimer;* one with three subunits a *trimer;* and one with four subunits a *tetramer.* The *quaternary structure* of a protein describes the way the subunits are arranged with respect to each other.

The subunits can be the same or different. Hemoglobin, for example, is a tetramer; it has two different kinds of subunits and each hemoglobin molecule has two of each kind (Figure 21.13). Turn to page 620 to see a protein with seven subunits.

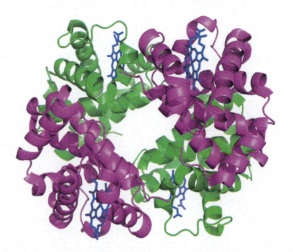

▶ **Figure 21.13**
The quaternary structure of hemoglobin. The two α-subunits are green and the two β-subunits are purple. The porphyrin rings are blue (Section 19.7).

The subunits of a protein are held together by the same kinds of interactions that hold the individual protein chains in a particular three-dimensional conformation—namely, hydrophobic interactions, hydrogen bonding, and electrostatic attractions.

> **PROBLEM 47** ◆
>
> **a.** Which would have the greatest percentage of polar amino acids, a spherical protein, a cigar-shaped protein, or a subunit of a hexamer?
>
> **b.** Which would have the smallest percentage of polar amino acids?

21.17 PROTEIN DENATURATION

Destroying the highly organized tertiary structure of a protein is called **denaturation.** Anything that breaks the interactions maintaining the three-dimensional shape of the protein will cause the protein to denature (unfold). Because these interactions are weak, proteins are easily denatured. The following are some of the ways that proteins can be denatured:

- Changing the pH denatures proteins because it changes the charges on many of the side chains. This disrupts electrostatic attractions and hydrogen bonds.

- Certain reagents such as urea and guanidine hydrochloride denature proteins by forming hydrogen bonds to protein groups that are stronger than the hydrogen bonds formed between the groups.

- Organic solvents denature proteins by associating with the nonpolar groups of the protein, thereby disrupting the normal hydrophobic interactions.

- Proteins can also be denatured by heat or by agitation. Both increase molecular motion, which can disrupt the attractive forces. A well-known example is the change that occurs to the white of an egg when it is heated or whipped.

PROBLEM 48

When apples that have been cut are exposed to oxygen, an enzyme-catalyzed reaction causes them to turn brown. Explain why coating them with lemon juice (pH 3.5) as soon as they are cut prevents the color change.

ESSENTIAL CONCEPTS

Section 21.0
- **Peptides** and **proteins** are polymers of **amino acids** linked together by **peptide (amide) bonds.**

Section 21.1
- The amino acids differ only in the substituent attached to the α-carbon.

Section 21.2
- Almost all amino acids found in nature have the L-configuration.

Section 21.3
- The carboxyl groups of the amino acids have pK_a values of ~ 2, and the protonated amino groups have pK_a values of ~ 9. At physiological pH (7.4), an amino acid exists as a **zwitterion.**

Section 21.4
- The **isoelectric point** (pI) of an amino acid is the pH at which the amino acid has no net charge.

Section 21.5
- A mixture of amino acids can be separated based on their pI values by **electrophoresis** or based on their polarities by **paper chromatography** or **thin-layer chromatography.**

- Separation based on overall charge can be accomplished by **ion-exchange chromatography,** employing an **anion-** or **cation-exchange resin.** An **amino acid analyzer** is an instrument that automates ion-exchange chromatography.

Section 21.6
- Amino acids can be synthesized by a Hell-Volhard-Zelinski reaction (followed by reaction with excess NH_3), a Strecker synthesis, reductive amination, a N-phthalimidomalonic ester synthesis, or an acetamidomalonic ester synthesis.

Section 21.7
- A racemic mixture of amino acids can be separated by a **kinetic resolution** using an enzyme-catalyzed reaction.

Section 21.8
- Rotation about a peptide bond is restricted because of its partial double-bond character.

- By convention, peptides and proteins are written with the free amino group (of the **N-terminal amino acid**) on the left and the free carboxyl group (of the **C-terminal amino acid**) on the right.

- Two cysteine side chains can be oxidized to a **disulfide bridge,** the only kind of covalent bond that is found between nonadjacent amino acids.

Section 21.10

- To synthesize a peptide bond, the amino group of the N-terminal amino acid must be protected (by *t*-Boc) and its carboxyl group activated (with DCCD). A second amino acid is added to form a dipeptide. Amino acids can be added to the growing C-terminal end by repeating the same two steps: activating the carboxyl group of the C-terminal amino acid with DCCD and then adding a new amino acid.

Section 21.11

- **Automated solid-phase peptide synthesis** allows peptides to be synthesized more rapidly and in higher yields.

Section 21.13

- The **primary structure** of a protein is the sequence of its amino acids and the location of all its disulfide bridges.

- The N-terminal amino acid can be determined with **Edman's reagent.** The C-terminal amino acid can be identified with a carboxypeptidase.

- An **exopeptidase** catalyzes the hydrolysis of a peptide bond at the end of a polypeptide or protein chain. An **endopeptidase** catalyzes the hydrolysis of a peptide bond that is not at the end of a chain.

Section 21.14

- The **secondary structure** of a protein describes how local segments of the protein's backbone fold. An **α-helix** and a **β-pleated sheet** are two types of secondary structure.

- A protein folds so as to maximize the number of stabilizing interactions—namely, **disulfide bonds, hydrogen bonds, electrostatic attractions,** and **hydrophobic interactions.**

Section 21.15

- The **tertiary structure** of a protein is the three-dimensional arrangement of all the atoms in the protein.

Section 21.16

- The **quaternary structure** of a protein describes the way the peptide chains (subunits) of a protein with more than one peptide chain are arranged with respect to each other.

Section 21.17

- Proteins can be **denatured** by a change in pH, by an organic solvent, by heat or agitation, or by certain reagents.

PROBLEMS

49. Glycine has pK_a values of 2.34 and 9.60. At what pH does glycine exist in the forms shown?

a. 50% 50% **b.** 100% **c.** 50% 50%

50. Show the peptides that would result from cleavage by the indicated reagent:
   a. Val-Arg-Gly-Met-Arg-Ala-Ser by carboxypeptidase A
   b. Ser-Phe-Lys-Met-Pro-Ser-Ala-Asp by cyanogen bromide
   c. Arg-Ser-Pro-Lys-Lys-Ser-Glu-Gly by trypsin

51. A titration curve is a plot of the pH of a solution as a function of added equivalents of hydroxide ion. As hydroxide ion is added to the aqueous solution, the pH increases because hydroxide ion removes protons from the solution. The pH flattens out when hydroxide ion can remove a proton from an ionizable group of an amino acid rather than a proton from the solution. Identify the amino acids that give the titration curves below.

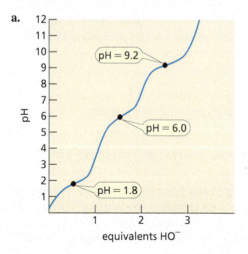

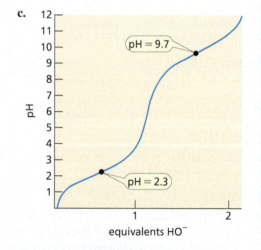

52. Which has a higher percentage of negative charge at physiological pH (7.4), leucine with pI = 5.98 or asparagine with pI = 5.43?

53. Aspartame (its structure is on page 1007) has a pI of 5.9. Draw its prevailing form at physiological pH (7.4).

54. Draw the form of aspartate that predominates at the following pH values:
  a. pH = 1.0   **b.** pH = 2.6   **c.** pH = 6.0   **d.** pH = 11.0

55. Show how phenylalanine can be prepared by reductive amination of an α-ketocarboxylic acid.

56. A professor was preparing a manuscript for publication in which she reported that the pI of the tripeptide Lys-Lys-Lys was 10.6. One of her students pointed out that there must be an error in her calculations because the pK_a of the ε-amino group of lysine is 10.8 and the pI of the tripeptide has to be greater than any of its individual pK_a values. Was the student correct?

57. What aldehydes are formed when the following amino acids are treated with ninhydrin?
  a. tyrosine   **b.** leucine   **c.** arginine

58. A mixture of amino acids that do not separate sufficiently when a single technique is used can often be separated by two-dimensional chromatography. In this technique, the mixture of amino acids is applied to a piece of filter paper and separated by chromatographic techniques. The paper is then rotated 90°, and the amino acids are further separated by electrophoresis, producing a type of chromatogram called a *fingerprint*. Identify the spots in the fingerprint obtained from a mixture of Ser, Glu, Leu, His, Met, and Thr.

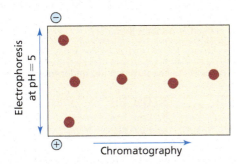

59. Determine the amino acid sequence of a polypeptide from the following data:

Complete hydrolysis of the peptide yields Arg, 2 Gly, Ile, 3 Leu, 2 Lys, 2 Met, 2 Phe, Pro, Ser, 2 Tyr, and Val.

Treatment with Edman's reagent releases PTH-Gly.

Carboxypeptidase A releases Phe.

Treatment with cyanogen bromide yields the following three peptides:
  1. Gly-Leu-Tyr-Phe-Lys-Ser-Met   **2.** Gly-Leu-Tyr-Lys-Val-Ile-Arg-Met   **3.** Leu-Pro-Phe

Treatment with trypsin yields the following four peptides:
  1. Gly-Leu-Tyr-Phe-Lys   **3.** Val-Ile-Arg
  2. Ser-Met-Gly-Leu-Tyr-Lys   **4.** Met-Leu-Pro-Phe

60. Explain why amino acids, unlike most amines and carboxylic acids, are insoluble in diethyl ether.

61. Explain the difference in the pK_a values of the carboxyl groups of alanine, serine, and cysteine.

62. Which is the more effective buffer at physiological pH, a solution of 0.1 M glycylglycylglycylglycine or a solution of 0.2 M glycine?

63. Identify the location and type of charge on the hexapeptide Lys-Ser-Asp-Cys-His-Tyr at each of the following pH values:
  a. pH = 1   **b.** pH = 5   **c.** pH = 7   **d.** pH = 12

64. Draw the product obtained when a lysine side chain in a polypeptide reacts with maleic anhydride.

65. After the polypeptide shown below was treated with maleic anhydride, it was hydrolyzed by trypsin. (After a polypeptide is treated with maleic anhydride, trypsin will cleave the polypeptide only on the C-side of arginine.)

Gly-Ala-Asp-Ala-Leu-Pro-Gly-Ile-Leu-Val-Arg-Asp-Val-Gly-Lys-Val-Glu-Val-Phe-Glu-Ala-Gly-
Arg-Ala-Glu-Phe-Lys-Glu-Pro-Arg-Leu-Val-Met-Lys-Val-Glu-Gly-Arg-Pro-Val-Gly-Ala-Gly-Leu-Trp

   a. After a polypeptide is treated with maleic anhydride, why does trypsin no longer cleave it on the C-side of lysine?
   b. How many fragments are obtained from the polypeptide?
   c. In what order will the fragments be eluted from an anion-exchange column using a buffer of pH = 5?

66. Treatment of a polypeptide with 2-mercaptoethanol yields two polypeptides with the following primary structures:

Val-Met-Tyr-Ala-Cys-Ser-Phe-Ala-Glu-Ser
Ser-Cys-Phe-Lys-Cys-Trp-Lys-Tyr-Cys-Phe-Arg-Cys-Ser

Treatment of the original intact polypeptide with chymotrypsin yields the following peptides:

1. Ala, Glu, Ser	**3.** Tyr, Val, Met	**5.** Ser, Phe, 2 Cys, Lys, Ala, Trp
2. 2 Phe, 2 Cys, Ser	**4.** Arg, Ser, Cys	**6** Tyr, Lys

Determine the positions of the disulfide bridges in the original polypeptide.

67. Show how aspartame can be synthesized using DCCD.

68. α-Amino acids can be prepared by treating an aldehyde with ammonia/trace acid, followed by hydrogen cyanide, followed by acid-catalyzed hydrolysis.
   a. Draw the structures of the two intermediates formed in this reaction.
   b. What amino acid is formed when the aldehyde that is used is 3-methylbutanal?
   c. What aldehyde is needed to prepare isoleucine?

69. Reaction of a polypeptide with carboxypeptidase A releases Met. The polypeptide undergoes partial hydrolysis to give the following peptides. What is the sequence of the polypeptide?

1. Ser, Lys, Trp	**4.** Leu, Glu, Ser	**7.** Glu, His	**10.** Glu, His, Val
2. Gly, His, Ala	**5.** Met, Ala, Gly	**8.** Leu, Lys, Trp	**11.** Trp, Leu, Glu
3. Glu, Val, Ser	**6.** Ser, Lys, Val	**9.** Lys, Ser	**12.** Ala, Met

70 a. How many different octapeptides can be made from the 20 naturally occurring amino acids?
   b. How many different proteins containing 100 amino acids can be made from the 20 naturally occurring amino acids?

71. Glycine has pK_a values of 2.3 and 9.6. Do you expect the pK_a values of glycylglycine to be higher or lower than these values?

72. A mixture of 15 amino acids gave the fingerprint shown here (see Problem 58). Identify the spots. (*Hint 1:* Pro reacts with ninhydrin to produce a yellow color; Phe and Tyr produce a green color. *Hint 2:* Count the number of spots before you start.)

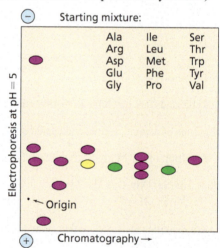

73. Write the mechanism for the reaction of an amino acid with di-*tert*-butyl dicarbonate.

74. Dithiothreitol reacts with disulfide bridges in the same way that 2-mercaptoethanol does. With dithiothreitol, however, the equilibrium lies much more to the right. Explain.

75. Show how valine can be prepared by

    a. a Hell–Volhard–Zelinski reaction.

    b. a Strecker synthesis.

    c. a reductive amination.

    d. a *N*-phthalimidomalonic ester synthesis.

    e. an acetamidomalonic ester synthesis.

76. The primary structure of β-endorphin, a peptide containing 31 amino acids synthesized by the body to control pain, is shown here:

    Tyr-Gly-Gly-Phe-Met-Thr-Ser-Glu-Lys-Ser-Gln-Thr-Pro-Leu-Val-Thr-
    Leu-Phe-Lys-Asn-Ala-Ile-Ile-Lys-Asn-Ala-Tyr-Lys-Lys-Gly-Glu

    a. What fragments are obtained as a result of treatment with each of the following?

        1. trypsin   **2.** cyanogen bromide   **3.** chymotrypsin

    b. How much of the primary structure can be determined if the amino acids contained in each fragment (but not their sequence) are known?

77. A chemist wanted to test his hypothesis that the disulfide bridges that form in many proteins do so after the minimum energy conformation of the protein has been achieved. He treated a sample of an enzyme that contained four disulfide bridges with 2-mercaptoethanol and then added urea to denature the enzyme. He slowly removed these reagents so that the enzyme could re-fold and re-form the disulfide bridges. The enzyme he recovered had 80% of its original activity. What would be the percent activity in the recovered enzyme if disulfide bridge formation were entirely random rather than determined by the tertiary structure? Does this experiment support his hypothesis?

78. Propose a mechanism for the rearrangement of the thiazoline obtained from the reaction of Edman's reagent with a peptide to a PTH-amino acid (page 1014). (*Hint:* Thioesters are very reactive toward nucleophiles.)

79. A normal polypeptide and a mutant of the polypeptide were hydrolyzed by an endopeptidase under the same conditions. The normal and mutant polypeptide differ by one amino acid. The fingerprints of the peptides obtained from the two polypeptides are shown below. What kind of amino acid substitution occurred as a result of the mutation? (That is, is the substituted amino acid more or less polar than the original amino acid? Is its pI lower or higher?) (*Hint:* Photocopy the fingerprints, cut them out, and overlay them.)

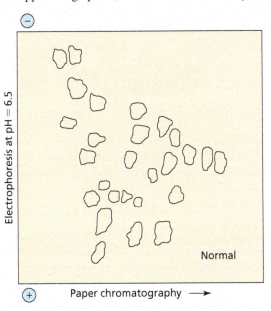

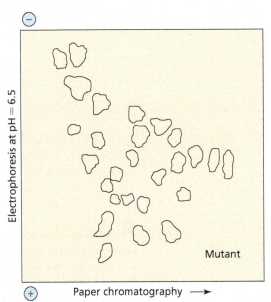

80. Determine the amino acid sequence of a polypeptide from the following data:

Complete hydrolysis of the peptide yields Ala, Arg, Gly, 2 Lys, Met, Phe, Pro, 2 Ser, Tyr, and Val.

Treatment with Edman's reagent releases PTH-Val.

Carboxypeptidase A releases Ala.

Treatment with cyanogen bromide yields the following two peptides:

1. Ala, 2 Lys, Phe, Pro, Ser, Tyr   **2.** Arg, Gly, Met, Ser, Val

Treatment with trypsin yields the following three peptides:

1. Gly, Lys, Met, Tyr   **2.** Ala, Lys, Phe, Pro, Ser   **3.** Arg, Ser, Val

Treatment with chymotrypsin yields the following three peptides:

1. 2 Lys, Phe, Pro   **2.** Arg, Gly, Met, Ser, Tyr, Val   **3.** Ala, Ser

22 Catalysis in Organic Reactions and in Enzymatic Reactions

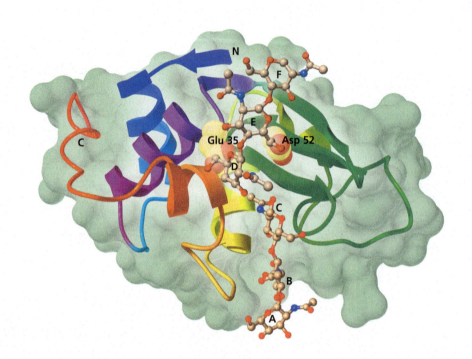

lysozyme

Lysozyme is an enzyme that catalyzes the hydrolysis of bacterial cell walls. The C on the far left indicates its C-terminal end, N indicates its N-terminal end, and the cell wall is indicated by A–F. Lysozyme's catalytic groups (Glu 35 and Asp 52) are also shown. (See Section 22.11.)

A **catalyst** is a substance that increases the rate of a chemical reaction without itself being consumed or changed in the reaction. In this chapter we will look at the types of catalysts used in organic reactions and the ways in which they provide an energetically more favorable pathway for a reaction. We will then see how the same catalysts are used in reactions that take place in cells—that is, in enzyme-catalyzed reactions. We will also see why enzymes are extraordinarily good catalysts—they can increase the rate of an intermolecular reaction by as much as 10^{16}. In contrast, nonbiological catalysts are seldom able to increase the rate of an intermolecular reaction by more than 10^4.

We saw that the rate of a chemical reaction depends on the energy barrier of the rate-determining step that must be overcome in the process of converting reactants to products. The height of the "energy hill" is indicated by the free energy of activation ($\Delta G^{\ddagger}$). A catalyst increases the rate of a chemical reaction by providing a pathway with a lower $\Delta G^{\ddagger}$ (Section 5.13). A catalyst can decrease $\Delta G^{\ddagger}$ in one of three ways:

- The catalyzed and uncatalyzed reactions can have different, but similar, mechanisms, with the catalyst providing a way to make *the reactant more reactive (less stable)* (Figure 22.1a).

- The catalyzed and uncatalyzed reactions can have different, but similar, mechanisms, with the catalyst providing a way to make *the transition state more stable* (Figure 22.1b).

- The catalyst can completely *change the mechanism* of the reaction, providing an alternative pathway with a smaller $\Delta G^{\ddagger}$ than that for the uncatalyzed reaction (Figure 22.2).

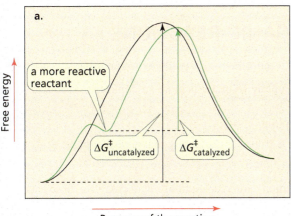

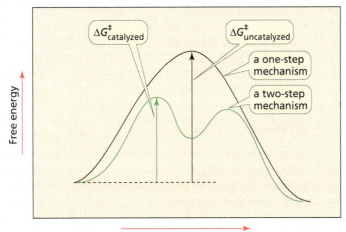

▲ **Figure 22.1**

Reaction coordinate diagrams for an uncatalyzed reaction (black) and for a catalyzed reaction (green).

(a) The catalyst converts the reactant to a more reactive species.

(b) The catalyst stabilizes the transition state.

◀ **Figure 22.2**

Reaction coordinate diagrams for an uncatalyzed reaction (black) and for a catalyzed reaction (green). The catalyzed reaction takes place by an alternative and energetically more favorable pathway.

Catalysts are Not Changed or Consumed by a Reaction

When we say that a catalyst is neither consumed nor changed by a reaction, we do not mean that it does not participate in the reaction. A catalyst *must* participate in the reaction if it is going to make it go faster. What we mean is that a catalyst has the same form after the reaction as it had before the reaction. (If a catalyst is used up in one step of the reaction, it must be regenerated in a subsequent step.) Because the catalyst is not used up during the reaction, only a small amount of the catalyst is needed. Therefore, the number of moles of catalyst added to a reaction is typically .01 to 10% of the number of moles of reactant. We call this a *catalytic* amount.

Catalysts Change the Rate Constant, Not the Equilibrium Constant

Notice in Figures 22.1 and 22.2 that the stability of the original reactants and final products is the same in both the catalyzed and corresponding uncatalyzed reactions. In other words, the catalyst does not change the equilibrium constant of the reaction. Because the catalyst does not change the equilibrium constant, it does not change the *amount* of product formed when the reaction has reached equilibrium. It changes only the *rate* at which the product is formed.

PROBLEM 1 ♦

Which of the following parameters would be different for a reaction carried out in the presence of a catalyst, compared with the same reaction carried out in the absence of a catalyst? (*Hint:* See Section 5.13.)

$$\Delta G°, \ \Delta H^{\ddagger}, \ E_a, \ \Delta S^{\ddagger}, \ \Delta H°, \ K_{eq}, \ \Delta G^{\ddagger}, \ \Delta S°, \ k_{rate}$$

22.1 CATALYSIS IN ORGANIC REACTIONS

There are several ways a catalyst can provide a more favorable pathway for an organic reaction:

- It can increase the reactivity of an electrophile so that it is more susceptible to reaction with a nucleophile.
- It can increase the reactivity of a nucleophile.
- It can increase the leaving propensity of a group by converting it to a weaker base.
- It can increase the stability of a transition state.

Now we will look at some of the most common catalysts—namely, acid catalysts, base catalysts, nucleophilic catalysts, and metal-ion catalysts—and the ways in which they catalyze the reactions of organic compounds.

22.2 ACID CATALYSIS

An **acid catalyst** increases the rate of a reaction by donating a proton to a reactant. In the preceding chapters we saw many examples of acid catalysis. For example, we saw that an acid provides the electrophile needed for the addition of water or an alcohol to an alkene (Sections 6.5 and 6.6). We also saw that an alcohol cannot undergo substitution and elimination reactions unless an acid is present to protonate the OH group, which increases its leaving propensity by making it a weaker base (Section 10.1).

Reviewing Acid-Catalyzed Hydrolysis of an Ester

To review some of the important ways an acid can catalyze a reaction, let's look again at the mechanism for the acid-catalyzed hydrolysis of an ester that we first saw in Section 15.8. The reaction has two slow steps: formation of the tetrahedral intermediate and collapse of the tetrahedral intermediate. Donation of a proton to and removal of a proton from an electronegative atom, such as oxygen or nitrogen, are always fast steps.

In an acid-catalyzed reaction, a proton is donated to the reactant.

MECHANISM FOR ACID-CATALYZED ESTER HYDROLYSIS

LEARN THE STRATEGY

The mechanism for the acid-catalyzed hydrolysis of an ester shows that the reaction can be divided into two distinct parts: formation of a tetrahedral intermediate and collapse of a tetrahedral intermediate. There are three steps in each part. Notice that in each part, the first step is a fast protonation step, the second step is a slow acid-catalyzed step that involves either breaking a π bond or forming a π bond, and the last step is a fast deprotonation step (to regenerate the catalyst).

USE THE STRATEGY

> **PROBLEM 2**
>
> Compare each of the mechanisms listed here with the mechanism for each of the two parts of the acid-catalyzed hydrolysis of an ester, indicating
>
> **a.** similarities. **b.** differences.
>
> 1. acid-catalyzed formation of a hydrate (Section 16.9)
> 2. acid-catalyzed conversion of an aldehyde to a hemiacetal (Section 16.9)
> 3. acid-catalyzed conversion of a hemiacetal to an acetal (Section 16.9)
> 4. acid-catalyzed hydrolysis of an amide (Section 15.12)

How an Acid Increases the Rate of Hydrolysis

A catalyst must increase the rate of a slow step because increasing the rate of a fast step does not increase the rate of the overall reaction. The acid increases the rates of both slow steps of the hydrolysis reaction.

- The acid increases the rate of the first slow step—formation of the tetrahedral intermediate—by protonating the carbonyl oxygen, thereby making it more susceptible to nucleophilic addition than an unprotonated carbonyl group would be. Increasing the reactivity of the carbonyl group by protonating it is an example of providing a way to convert the reactant to a more reactive species (Figure 22.1a).

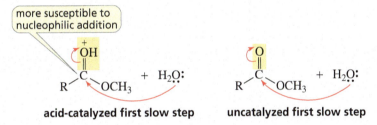

acid-catalyzed first slow step   uncatalyzed first slow step

A catalyst must increase the rate of a slow step.

Increasing the rate of a fast step will not increase the rate of the overall reaction.

- The acid increases the rate of the second slow step by decreasing the basicity—and thereby increasing the leaving propensity—of the group that is eliminated when the tetrahedral intermediate collapses. In the presence of an acid, methanol is eliminated; in the absence of an acid, methoxide ion is eliminated. Methanol is a much weaker base than methoxide ion, so methanol is much more easily eliminated.

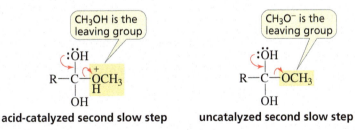

acid-catalyzed second slow step   uncatalyzed second slow step

Specific-Acid and General-Acid Catalysis

There are two types of acid catalysis:

- In **specific-acid catalysis**, the proton is fully transferred to the reactant before the slow step of the reaction begins (Figure 22.3a on the next page).
- In **general-acid catalysis**, the proton is transferred to the reactant during the slow step of the reaction (Figure 22.3b).

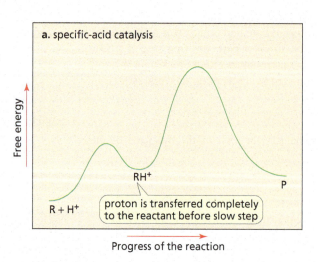

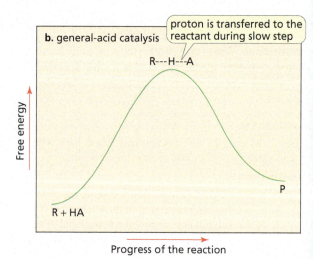

▲ **Figure 22.3**

(a) Reaction coordinate diagram for a specific-acid-catalyzed reaction. The proton is transferred completely to the reactant before the slow step of the reaction begins (R = reactant; P = product).

(b) Reaction coordinate diagram for a general-acid-catalyzed reaction. The proton is transferred to the reactant during the slow step of the reaction.

Specific-acid and general-acid catalysis increase the rate of a reaction in the same way—by donating a proton in order to make either bond making or bond breaking easier. The two types of acid catalysis differ only in the extent to which the proton is transferred in the transition state of the slow step of the reaction.

In the examples that follow, notice the difference in the extent to which the proton has been transferred when the nucleophile adds to the reactant.

- In the specific-acid-catalyzed addition of water to a carbonyl group, the nucleophile adds to a fully protonated carbonyl group. In the general-acid-catalyzed addition of water to a carbonyl group, the carbonyl group becomes protonated as the nucleophile adds to it.

specific-acid-catalyzed addition of water

general-acid-catalyzed addition of water

The proton is donated to the reactant *before* the slow step in a specific-acid-catalyzed reaction, and *during* the slow step in a general-acid-catalyzed reaction.

- In the specific-acid-catalyzed collapse of a tetrahedral intermediate, a fully protonated leaving group is eliminated, whereas in the general-acid-catalyzed collapse of a tetrahedral intermediate, the leaving group picks up a proton as the group is eliminated.

specific-acid-catalyzed elimination of the leaving group

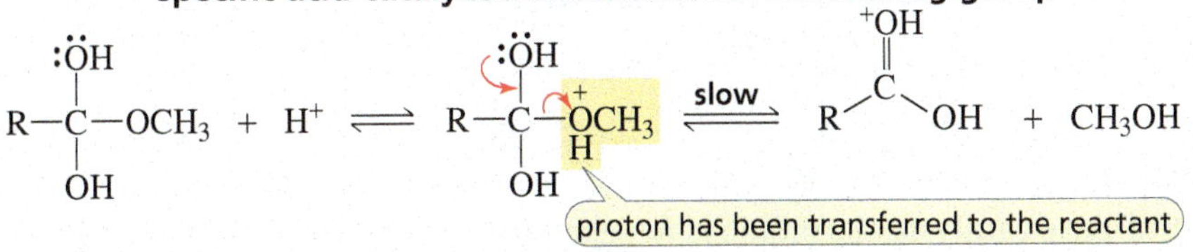

proton has been transferred to the reactant

general-acid-catalyzed elimination of the leaving group

proton is being transferred to the reactant

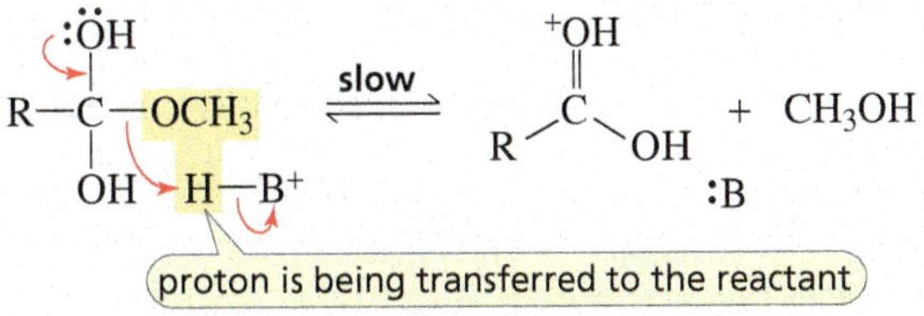

A specific-acid catalyst must be a strong enough acid (such as HCl, H_3O^+) to protonate the reactant fully before the slow step begins. A general-acid catalyst can be a weaker acid because it has only a partially transferred proton in the transition state of the slow step.

In the mechanisms shown in previous chapters, we have used strong acids as catalysts, so all the mechanisms are written as specific-acid-catalyzed reactions. Weaker acids could have been used to catalyze many of the reactions, in which case the protonation step and the subsequent slow step would have been shown as a single step. A list of acids and their pK_a values is given in Appendix I.

PROBLEM 3 ◆

Are the slow steps for the acid-catalyzed hydrolysis of an ester on page 703 general-acid catalyzed or specific-acid catalyzed?

PROBLEM 4

a. Draw the mechanism for the following reaction if it involves specific-acid catalysis.
b. Draw the mechanism if it involves general-acid catalysis.

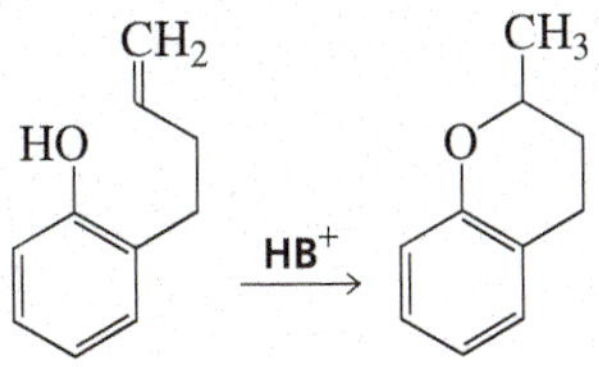

PROBLEM 5 SOLVED

An alcohol does not react with aziridine unless an acid is present. Why is the acid necessary?

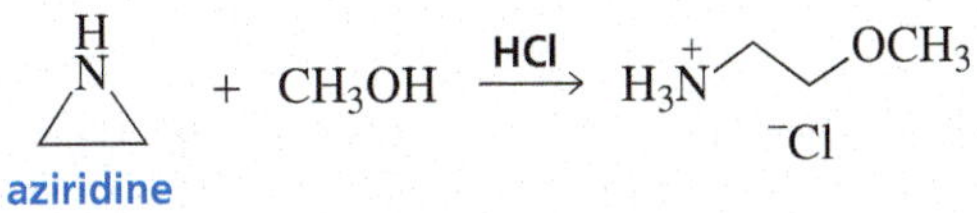

aziridine

SOLUTION Although relief of ring strain is sufficient by itself to cause an epoxide to undergo a ring-opening reaction (Section 10.7), it is not sufficient to cause an aziridine to undergo a ring-opening reaction. A negatively charged nitrogen is a stronger base and, therefore, a poorer leaving group, than a negatively charged oxygen. An acid, therefore, is needed to protonate the ring nitrogen to make it a better leaving group.

22.3 BASE CATALYSIS

We have already encountered several base-catalyzed reactions, such as the interconversion of keto and enol tautomers (Section 17.3), the Claisen condensation (Section 17.13), and the enediol rearrangement (Section 20.5). A **base catalyst** increases the rate of a reaction by removing a proton from the reactant.

Specific-Base and General-Base Catalysis

Like acid catalysis, there are two types of base catalysis:

- In **specific-base catalysis**, the proton is completely removed from the reactant before the slow step of the reaction begins.
- In **general-base catalysis**, the proton is removed from the reactant during the slow step of the reaction.

The dehydration of a hydrate in the presence of hydroxide ion is a specific-base-catalyzed reaction. Hydroxide ion (the base) increases the rate of the reaction by removing a proton from the neutral hydrate.

In a base-catalyzed reaction, a proton is removed from the reactant.

specific-base-catalyzed dehydration

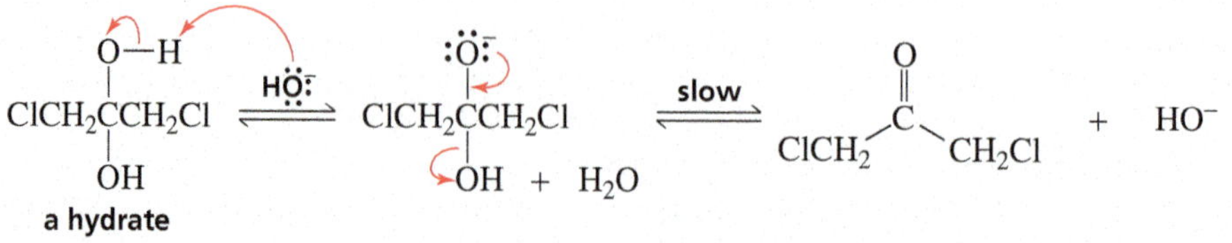

Removing a proton from the hydrate increases the rate of dehydration by providing a pathway with a more stable transition state. The transition state for the elimination of HO^- from a negatively charged tetrahedral intermediate is more stable because a positive charge does not develop on the electronegative oxygen atom, as it does in the transition state for the elimination of HO^- from a neutral tetrahedral intermediate.

transition state for elimination of HO^- from a negatively charged tetrahedral intermediate

transition state for elimination of HO^- from a neutral tetrahedral intermediate

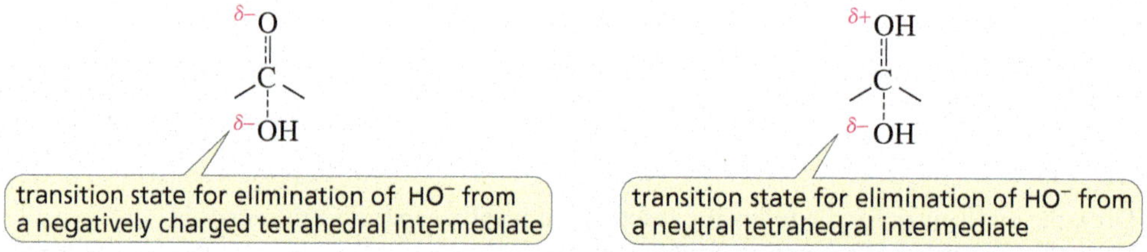

Compare the extent of proton transfer in the slow step of the preceding specific-base-catalyzed dehydration with the extent of proton transfer in the slow step of the following general-base-catalyzed dehydration:

The proton is removed from the reactant before the slow step in a specific-base-catalyzed reaction and during the slow step in a general-base-catalyzed reaction.

general-base-catalyzed dehydration

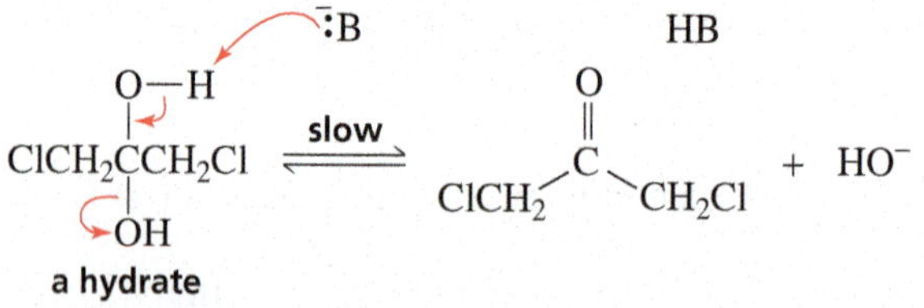

In specific-base catalysis, the base has to be strong enough to remove a proton from the reactant completely before the slow step begins. In general-base catalysis, the base can be weaker because the proton is only partially transferred to the base in the transition state of the slow step.

We will see that enzymes catalyze reactions using general-acid and general-base catalysis because at physiological pH (7.4), the concentration of H^+ (~1×10^{-7} M) is too small for effective specific-acid catalysis and the concentration of HO^- is too small for effective specific-base catalysis.

PROBLEM 6

a. Draw the mechanism for the following reaction if it involves specific-base catalysis.
b. Draw the mechanism if it involves general-base catalysis.

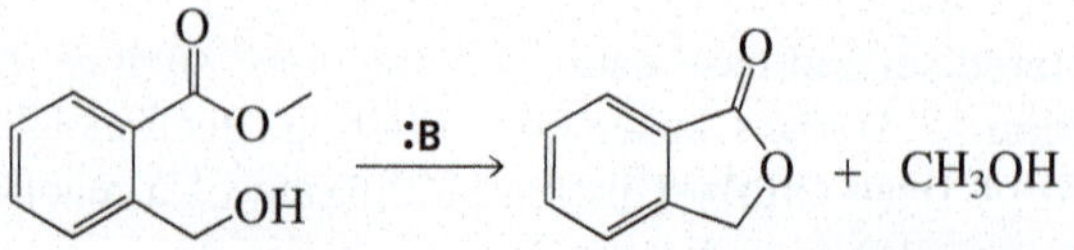

22.4 NUCLEOPHILIC CATALYSIS

A **nucleophilic catalyst** increases the rate of a reaction by reacting as a nucleophile to form a new covalent bond with the reactant. **Nucleophilic catalysis**, therefore, is also called **covalent catalysis.** A nucleophilic catalyst increases the reaction rate by completely changing the mechanism of the reaction.

Reactions Employing a Nucleophilic Catalyst

Example 1

In the following reaction, iodide ion increases the rate of conversion of ethyl chloride to ethyl alcohol by acting as a nucleophilic catalyst:

a nucleophilic catalyst

$$CH_3CH_2Cl \ + \ HO^- \ \xrightarrow[H_2O]{I^-} \ CH_3CH_2OH \ + \ Cl^-$$

A nucleophilic catalyst forms a covalent bond with the reactant.

To understand how iodide ion catalyzes this reaction, we need to compare the mechanisms for the uncatalyzed and catalyzed reactions. In the absence of iodide ion, ethyl chloride is converted to ethyl alcohol in an S_N2 reaction: the nucleophile is HO^- and the leaving group is Cl^-.

MECHANISM FOR THE UNCATALYZED REACTION

$$HO^- \ + \ CH_3CH_2-Cl \ \longrightarrow \ CH_3CH_2OH \ + \ Cl^-$$

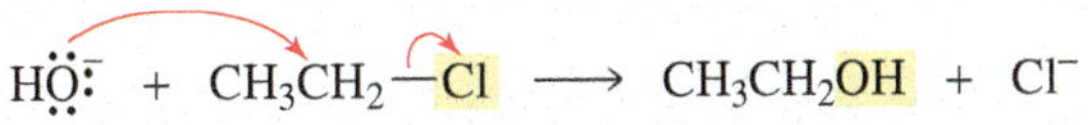

If iodide ion is present in the reaction mixture, then the reaction takes place by two successive S_N2 reactions.

MECHANISM FOR THE IODIDE-ION-CATALYZED REACTION

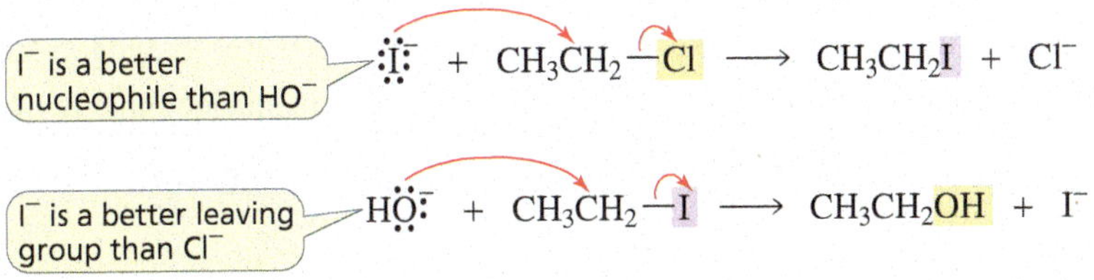

I^- is a better nucleophile than HO^-

$$I^- \ + \ CH_3CH_2-Cl \ \longrightarrow \ CH_3CH_2I \ + \ Cl^-$$

I^- is a better leaving group than Cl^-

$$HO^- \ + \ CH_3CH_2-I \ \longrightarrow \ CH_3CH_2OH \ + \ I^-$$

- The first S_N2 reaction is faster than the uncatalyzed S_N2 reaction because in a protic solvent, iodide ion is a better nucleophile than hydroxide ion, the nucleophile in the uncatalyzed reaction (Section 9.2).

- The second S_N2 reaction is also faster than the uncatalyzed S_N2 reaction because iodide ion is a weaker base and, therefore, a better leaving group than chloride ion, the leaving group in the uncatalyzed reaction.

Thus, iodide ion increases the rate of formation of ethanol by changing a relatively slow S_N2 reaction to one that involves two relatively fast S_N2 reactions (Figure 22.2).

Iodide ion is a nucleophilic catalyst because it reacts as a nucleophile, forming a covalent bond with the reactant. The iodide ion consumed in the first reaction is regenerated in the second.

Example 2

Another reaction in which a nucleophilic catalyst provides a more favorable pathway by changing the mechanism of the reaction is the imidazole-catalyzed hydrolysis of an ester.

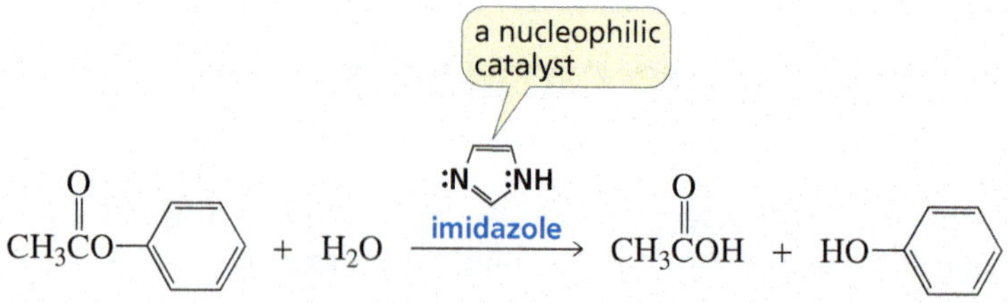

a nucleophilic catalyst

imidazole

CH_3CO—⬡ + H_2O → CH_3COH + HO—⬡

- Imidazole is a better nucleophile than water, so imidazole reacts faster with the ester than water would.
- The acyl imidazole is hydrolyzed much more rapidly than the ester would have been because the positively charged nitrogen makes imidazole a very good leaving group.

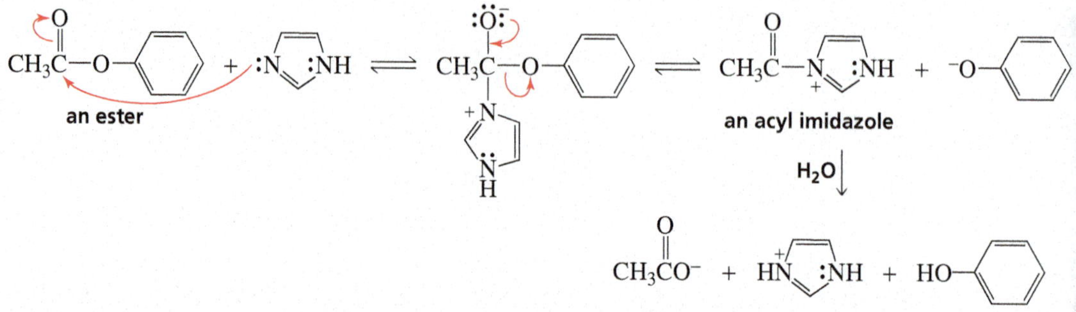

an ester **an acyl imidazole**

Because formation of the acyl imidazole and its subsequent hydrolysis are both faster than ester hydrolysis, imidazole increases the rate of hydrolysis of the ester.

22.5 METAL-ION CATALYSIS

Metal ions are Lewis acids.

Metal ions exert their catalytic effect by complexing with atoms that have lone-pair electrons. **Metal ions**, therefore, are Lewis acids (Section 2.12). A *metal ion* can increase the rate of a reaction in the following ways:

- It can make a reaction center more susceptible to receiving electrons (that is, more electrophilic) as in A.
- It can make a leaving group a weaker base and, therefore, a better leaving group, as in B.
- It can increase the rate of a hydrolysis reaction by increasing the nucleophilicity of water, as in C.

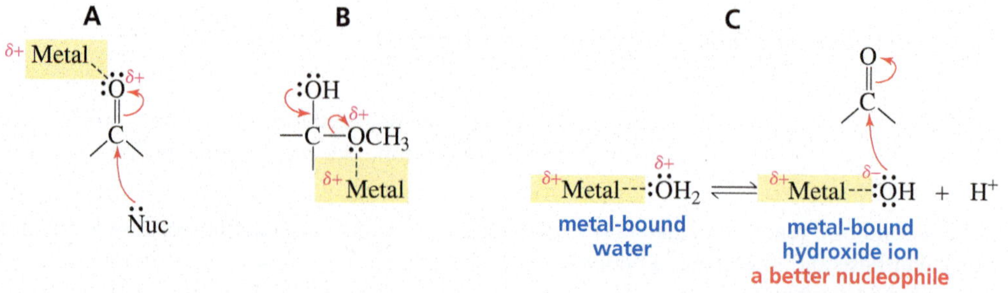

A **B** **C**

metal-bound water metal-bound hydroxide ion a better nucleophile

In **A** and **B**, the metal ion exerts the same kind of catalytic effect as a proton does. However, a metal ion can be a much more effective catalyst than a proton, because metal ions can have a charge greater than +1, and a high concentration of a metal ion can be made available at neutral pH.

In **C**, metal-ion complexation increases water's nucleophilicity by converting it to metal-bound hydroxide ion. That is, the metal-ion increases water's tendency to lose a proton as shown by the pK_a values listed in Table 22.1 on the next page. (The pK_a of water is 15.7.)

Metal-bound hydroxide ion is not as good a nucleophile as hydroxide ion but it is a better nucleophile than water. Metal ions are important catalysts in living systems because hydroxide ion itself is not available at physiological pH (7.4).

Table 22.1 The pK_a of Metal-Bound Water

M^{2+}	pK_a	M^{2+}	pK_a
Ca^{2+}	12.7	Co^{2+}	8.9
Cd^{2+}	11.6	Zn^{2+}	8.7
Mg^{2+}	11.4	Fe^{2+}	7.2
Mn^{2+}	10.6	Cu^{2+}	6.8
Ni^{2+}	9.4	Be^{2+}	5.7

Reactions Employing a Metal-Ion Catalyst

Example 1

The decarboxylation of dimethyloxaloacetate can be catalyzed by either Cu^{2+} or Al^{3+}.

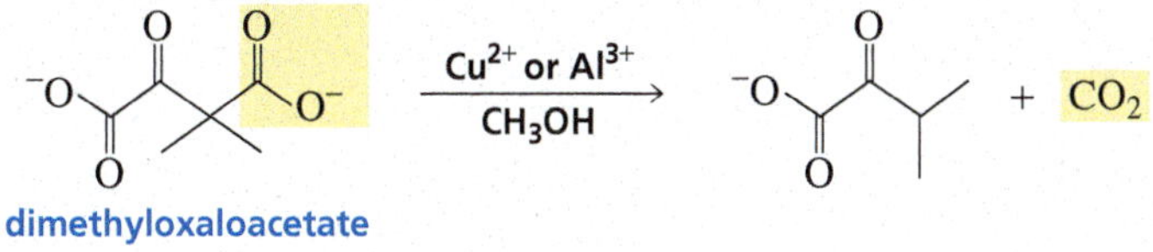

dimethyloxaloacetate

In this reaction, the metal ion complexes with two oxygen atoms of the reactant. Complexation increases the rate of decarboxylation by making the carbonyl oxygen more susceptible to receiving the electrons left behind when CO_2 is eliminated.

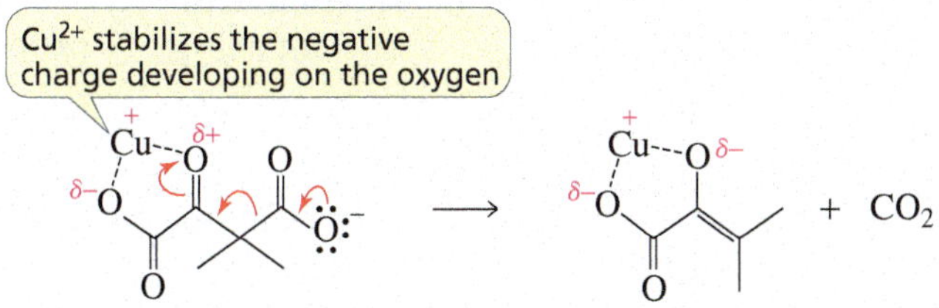

Cu^{2+} stabilizes the negative charge developing on the oxygen

Example 2

The hydrolysis of the ester shown below has two slow steps. Zn^{2+} increases the rate of the first slow step by providing metal-bound hydroxide ion, a better nucleophile than water. (It can also increase the rate of this step by complexing with the carbonyl oxygen to increase the electrophilicity of the carbonyl carbon.) Zn^{2+} increases the rate of the second slow step by decreasing the basicity of the group that is eliminated from the tetrahedral intermediate.

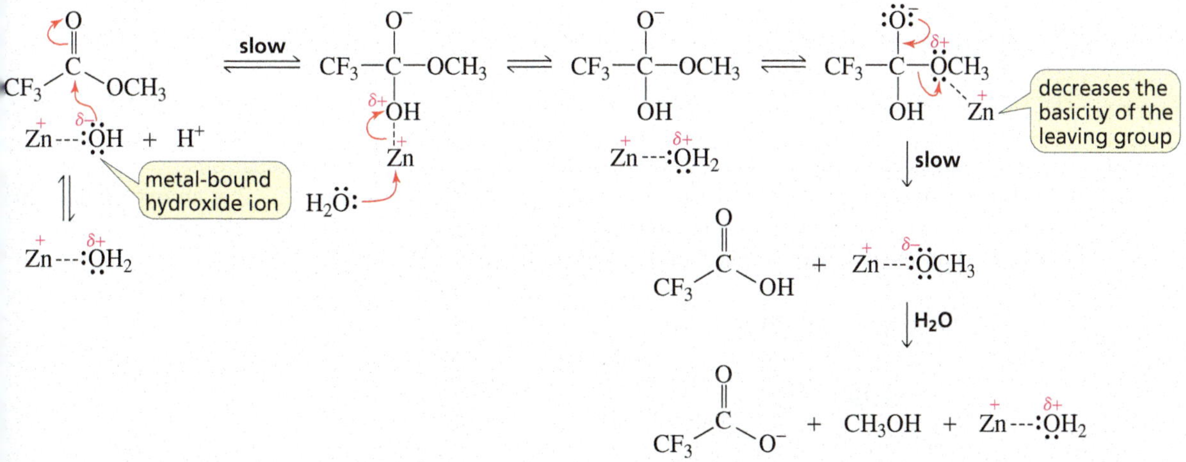

metal-bound hydroxide ion

decreases the basicity of the leaving group

PROBLEM 7 ♦

Although metal ions increase the rate of decarboxylation of dimethyloxaloacetate, they have no effect on the rate of decarboxylation of either the monoethyl ester of dimethyloxaloacetate or acetoacetate. Explain why this is so.

dimethyloxaloacetate   monoethyl ester of   acetoacetate
                            dimethyloxaloacetate

PROBLEM 8

Propose a mechanism for the Co^{2+} catalyzed hydrolysis of glycinamide.

$$H_2N \overset{O}{\diagdown} NH_2 \ + \ H_2O \ \xrightarrow{Co^{2+}} \ H_2N \overset{O}{\diagdown} O^- \ + \ ^+NH_4$$

22.6 INTRAMOLECULAR REACTIONS

The rate of a chemical reaction is determined by the number of molecular collisions with sufficient energy *and* with the proper orientation in a given period of time (Section 5.11):

$$\text{rate of reaction} = \frac{\text{number of collisions}}{\text{unit of time}} \times \frac{\text{fraction with}}{\text{sufficient energy}} \times \frac{\text{fraction with}}{\text{proper orientation}}$$

Because a catalyst decreases the energy barrier of a reaction, it increases the fraction of collisions that occur with sufficient energy to overcome the barrier.

The rate of a reaction can also be increased by increasing the frequency of the collisions, which can be achieved by increasing the concentration of the reactants. In addition, we saw that an *intramolecular reaction* that forms a five- or a six-membered ring occurs more readily than the analogous *intermolecular reaction*. This is because an intramolecular reaction has the advantage of the reacting groups being tethered together in the same molecule, which gives them a better chance of finding each other than if they were in two different molecules in a solution of the same concentration (Section 9.16). As a result, the frequency of the collisions increases.

If, in addition to being in the same molecule, the reacting groups are arranged in a way that increases the probability that they will collide with each other in the proper orientation, the rate of the reaction will be further increased. The relative rates shown in Table 22.2 on the next page demonstrate the enormous increase in the rate of a reaction when the reacting groups are properly oriented.

Relative Rates

The rate constants for a series of reactions are generally compared by looking at their relative rates because relative rates allow us to see immediately how much faster one reaction is than another. **Relative rates** are obtained by dividing the rate constant for each of the reactions by the rate constant for the slowest reaction in the series. The slowest reaction in Table 22.2 is an *intermolecular reaction*; all the others are *intramolecular reactions*.

NOTE TO THE STUDENT

• More about rate constants and reaction order can be found in Appendix II.

Because an intramolecular reaction is a first-order reaction (with units of time^{-1}) and an intermolecular reaction is a second-order reaction (with units of $M^{-1} \text{time}^{-1}$), the relative rates in Table 22.2 have units of molarity (M).

$$\text{relative rate} = \frac{\text{first-order rate constant}}{\text{second-order rate constant}} \times \frac{\text{time}^{-1}}{\text{time}^{-1}M^{-1}} = M$$

Table 22.2 Relative Rates of an Intermolecular Reaction and Five Intramolecular Reactions

Reaction	Relative rate
A	1.0
B	1×10^3 M
C	2.3×10^4 M R $= CH_3$ 1.3×10^6 M R $= (CH_3)_2CH$
D	2.2×10^5 M
E	1×10^7 M
F	5×10^7 M

Therefore, relative rates are also called *effective molarities*. **Effective molarity** is the concentration of the reactant that would be required in an *intermolecular* reaction for it to have the same rate as the *intramolecular* reaction. In other words, the effective molarity is the advantage given to a reaction by having the reacting groups in the same molecule. In some cases, having the reacting groups in the same molecule provides such an enormous increase in rate that the effective molarity is greater than the concentration of the reactant in its solid state!

- Reaction **A**, the first reaction in Table 22.2, is an intermolecular reaction between an ester and a carboxylate ion. The second reaction, **B**, has the same two reacting groups that A has, but they are in the same molecule. The rate of the intramolecular reaction is 1000 times faster than the rate of the intermolecular reaction.

- The reactant in **B** has four C—C bonds that are free to rotate, whereas the reactant in **D** has only three such bonds (see the top of the next page). Conformers in which the large groups are rotated away from each other are more stable. However, when these groups are pointed away from each other, they are in an unfavorable conformation for reaction. Because the reactant in **D** has fewer bonds that are free to rotate, the groups are more apt to be in a conformation that is favorable for a reaction. Therefore, reaction **D** is faster than reaction **B**.

- Reaction **C** is faster than reaction **B** because the alkyl substituents of the reactant in **C** decrease the available space for the reacting groups to rotate away from each other. Therefore, there is a greater probability that the molecule will be in a conformation that has the reacting groups positioned for ring closure. This is called the *gem-dialkyl effect* because the two alkyl substituents are bonded to the same (geminal) carbon. Comparing the rate when the substituents are methyl groups with the rate when the substituents are isopropyl groups, we see that the rate is further increased when the size of the alkyl groups is increased.

- The increased rate of reaction of **E** is due to the double bond that prevents the reacting groups from rotating away from each other. The bicyclic compound in **F** reacts even faster because the reacting groups are locked in the proper orientation for reaction.

PROBLEM 9 ◆

The relative rate of reaction for the cis alkene (**E**) is given in Table 22.2. What do you expect the relative rate of reaction for the trans alkene to be?

22.7 INTRAMOLECULAR CATALYSIS

Just as having two reacting groups in the same molecule increases the rate of a reaction compared with having the groups in separate molecules, having a *reacting group* and a *catalyst* in the same molecule increases the rate of a reaction compared with having them in separate molecules.

When a catalyst is part of the reacting molecule, the catalysis is called **intramolecular catalysis.** Intramolecular general-acid or general-base catalysis, intramolecular nucleophilic catalysis, and intramolecular metal-ion catalysis are all possible.

Intramolecular Nucleophilic Catalysis

When chlorocyclohexane reacts with an aqueous solution of ethanol, an alcohol and an ether are formed. Two products are formed because there are two nucleophiles (H_2O and CH_3CH_2OH) in the solution.

A 2-thiophenyl-substituted chlorocyclohexane undergoes the same reaction. However, the rate of the reaction depends on whether the thio substituent is cis or trans to the chloro substituent. If it is trans, then the 2-thiophenyl-substituted compound reacts about 70,000 times faster than the unsubstituted compound. But if it is cis, the 2-thio-substituted compound reacts at about the same rate as the unsubstituted compound.

What accounts for the much faster reaction of the trans-substituted compound? When the thio substituent is trans to the chlorine, the substituent can be an intramolecular nucleophilic catalyst—it displaces the chloro substituent by attacking the back side of the carbon to which the chloro substituent is attached (an S_N2 reaction). Back-side attack requires both substituents to be in axial positions, and only the trans isomer can have both of its substituents in axial positions (Section 3.15). Subsequent attack by water or ethanol on the sulfonium ion is rapid because breaking the three-membered ring releases strain and the positively charged sulfur is an excellent leaving group.

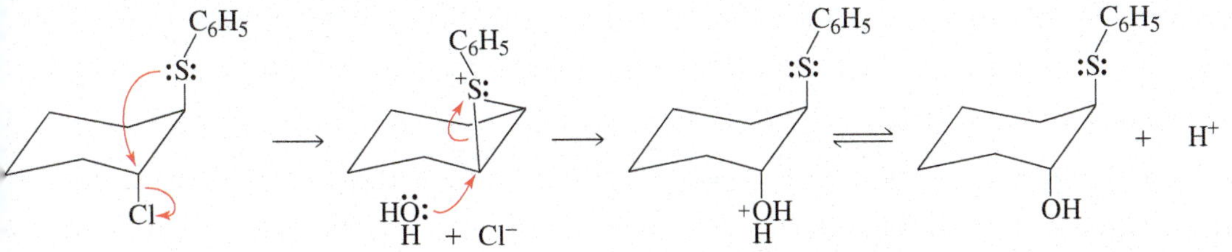

PROBLEM 10 ♦

Show all the products, including their configurations, that are obtained from the above reaction.

Intramolecular General-Base Catalysis

The rate of hydrolysis of phenyl acetate is increased about 150-fold at neutral pH by putting a carboxylate ion in the ortho position. The *ortho*-carboxy-substituted ester is commonly known as aspirin (Section 15.9). In the following reactions, each reactant and product is shown in the form that predominates at physiological pH (7.4).

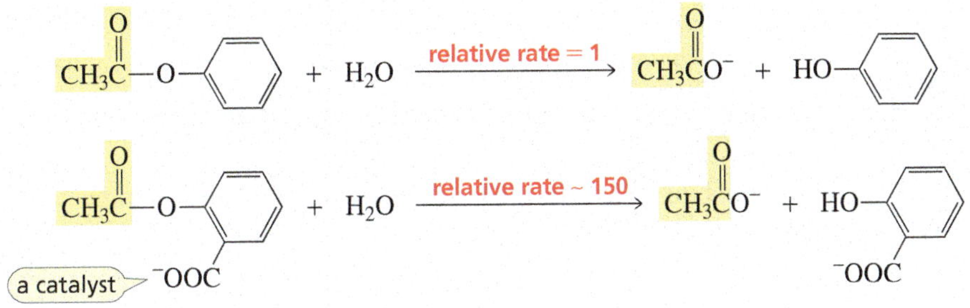

The *ortho*-carboxy group is an intramolecular general-base catalyst that increases the nucleophilicity of water, thereby increasing the rate of formation of the tetrahedral intermediate.

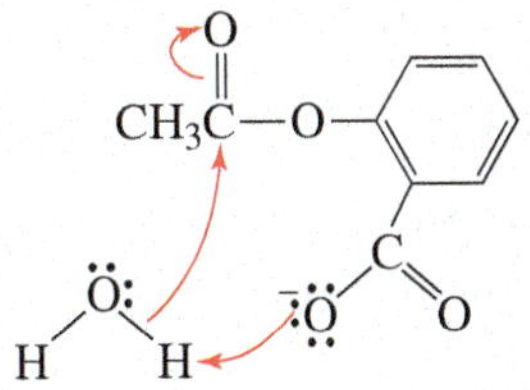

If there are nitro groups on the benzene ring, the *ortho*-carboxy substituent acts as an intramolecular *nucleophilic catalyst* instead of an intramolecular *general-base catalyst*. In this case, the carboxy group increases the rate of hydrolysis by converting the ester to an anhydride, which is more rapidly hydrolyzed than an ester (Section 15.16).

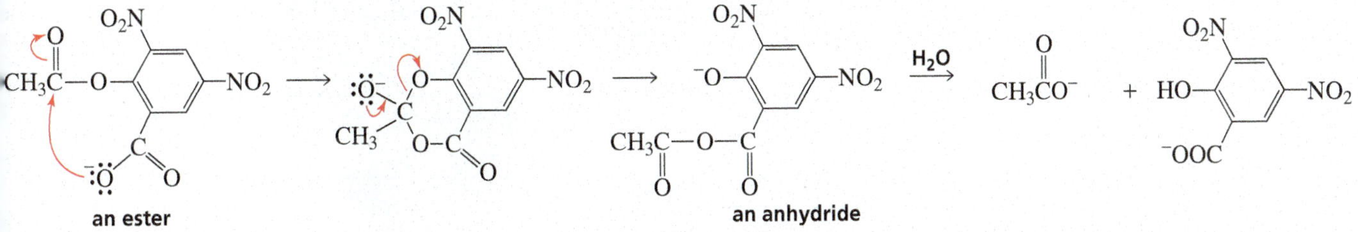

PROBLEM 11 SOLVED

Why is the *ortho*-carboxy substituent a general-base catalyst in one reaction and a nucleophilic catalyst in another?

SOLUTION Because of its location, the *ortho*-carboxy substituent forms a tetrahedral intermediate. If the tetrahedral intermediate's carboxy group is a better leaving group than its 2,4-dinitrophenoxy group, then the carboxy group will be eliminated preferentially (blue arrow). This will re-form the starting material (path **A**), which then will be hydrolyzed by a general-base-catalyzed mechanism. However, if the 2,4-dinitrophenoxy group is a better leaving group than the carboxy group, the 2,4-dinitrophenoxy group will be eliminated (green arrow), thereby forming an anhydride (path **B**), and the reaction will have occurred via nucleophilic catalysis.

tetrahedral intermediate

PROBLEM 12

Why do the nitro groups change the relative leaving tendencies of the carboxy and 2,4-dinitrophenoxy groups in the tetrahedral intermediate in Problem 11?

PROBLEM 13 ◆

Whether the *ortho*-carboxy substituent acts as an intramolecular general-base catalyst or as an intramolecular nucleophilic catalyst can be determined by carrying out the hydrolysis of aspirin with ^{18}O-labeled water and determining whether ^{18}O is incorporated into *ortho*-carboxy-substituted phenol. Explain the results that would be obtained with the two types of catalysis.

22.8 CATALYSIS IN BIOLOGICAL REACTIONS

Essentially all organic reactions that occur in cells require a catalyst. Most biological catalysts are **enzymes,** which are globular proteins (Section 21.0). Each biological reaction is catalyzed by a different enzyme.

Binding the Substrate

The reactant of an enzyme-catalyzed reaction is called a **substrate.**

$$SUBSTRATE \xrightarrow{\text{enzyme}} PRODUCT$$

The enzyme binds its substrate at its **active site,** a pocket in the cleft of the enzyme. All the bond-making and bond-breaking steps that convert the substrate to the product occur while the substrate is bound at the active site.

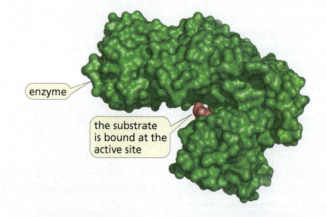

enzyme

the substrate is bound at the active site

Enzymes differ from nonbiological catalysts in that they are specific for the substrate whose reaction they catalyze (Section 6.15). Enzymes have different degrees of specificity. Some enzymes are specific for a single compound. For example, glucose-6-phosphate isomerase catalyzes only the isomerization of glucose-6-phosphate. On the other hand, some enzymes catalyze the reactions of several compounds with similar structures. For instance, hexokinase catalyzes the phosphorylation of any D-hexose. The specificity of an enzyme for its substrate is another example of **molecular recognition**—the ability of one molecule to recognize another molecule (Section 20.0).

The specificity of an enzyme results from its conformation and the particular **amino acid side chains** (α-substituents) that are at the active site. For example, an amino acid with a negatively charged side chain can at the active site associate with a positively charged group on the substrate, an amino acid side chain with a hydrogen-bond donor can associate with a hydrogen-bond acceptor on the substrate, and a hydrophobic amino acid side chain can associate with hydrophobic groups on the substrate.

In 1894, Emil Fischer proposed the **lock-and-key model** to account for the specificity of an enzyme for its substrate. This model related the specificity of an enzyme for its substrate to the specificity of a lock for a correctly shaped key.

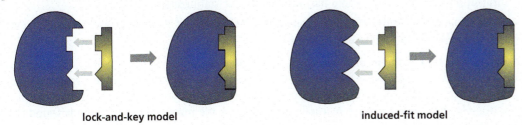

lock-and-key model   **induced-fit model**

In 1958, Daniel Koshland proposed the **induced-fit model** of substrate binding. In this model, the shape of the active site does not become completely complementary to the shape of the substrate until the enzyme has bound the substrate. The energy released as a result of binding the substrate can be used to induce a change in the conformation of the enzyme, leading to more precise binding between the substrate and the active site. An example of induced fit is shown in Figure 22.4.

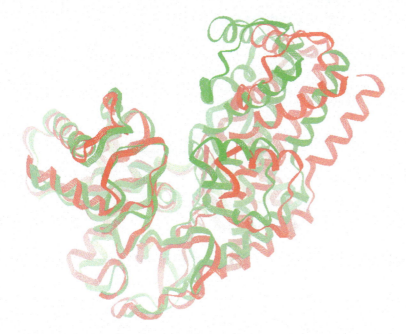

◀ **Figure 22.4**
The structure of hexokinase before binding its substrate is shown in red. The structure of hexokinase after binding its substrate is shown in green.

Catalyzing the Reaction

There is no single explanation for the remarkable catalytic ability of enzymes. Each enzyme is unique in the collection of factors it employs to catalyze a reaction. Some of the factors most enzymes have in common are:

- Reacting groups are brought together at the active site in the proper orientation for reaction. This is analogous to the way proper positioning of reacting groups increases the rate of an intramolecular reaction (Section 22.6).

- Some of the amino acid side chains of the enzyme serve as acid, base, and nucleophilic catalysts, and many enzymes also have metal ions at their active site that act as catalysts. These species are positioned relative to the substrate precisely where they are needed for catalysis. This factor is analogous to the way intramolecular catalysis by acids, bases, nucleophiles, and metal ions enhances reaction rates (Section 22.7).

    An acid catalyst increases the rate of a reaction by donating a proton to the substrate.

    A base catalyst increases the rate of a reaction by removing a proton from the substrate.

    A nucleophilic catalyst increases the rate of a reaction by forming a new covalent bond with the substrate.

- Amino acid side chains can stabilize transition states by London dispersion forces, electrostatic interactions, and hydrogen bonding. Recall that stabilizing the transition state decreases the free energy of activation ($\Delta G^{\ddagger}$) (Section 5.10). These side chains are positioned relative to the transition state precisely where they are needed for stabilization.

- The remarkable catalytic ability of enzymes stems in part from their ability to use several modes of catalysis in the same reaction.

Now we will look at the mechanisms of five enzyme-catalyzed reactions. Notice that the modes of catalysis used by enzymes are the same as the modes of catalysis used in organic reactions. Thus, if you refer back to sections referenced throughout this chapter, you will be able to see that much of the organic chemistry you have learned also applies to the reactions of compounds found in the biological world.

22.9 AN ENZYME-CATALYZED REACTION THAT IS REMINISCENT OF ACID-CATALYZED AMIDE HYDROLYSIS

The names of most enzymes end in "ase," and the enzyme's name tells you something about the reaction it catalyzes. For example, carboxypeptidase A catalyzes the hydrolysis of the C-terminal (carboxy-terminal) peptide bond in polypeptides as long as the C-terminal amino acid is not arginine or lysine (Section 21.13).

Carboxypeptidase A is a *metalloenzyme*—an enzyme that contains a tightly bound metal ion. The metal ion in carboxypeptidase A is Zn^{2+}. About one-third of all enzymes require a metal ion for catalysis; carboxypeptidase A is one of several hundred enzymes known to contain zinc.

In carboxypeptidase A, Zn^{2+} is bound to the enzyme at its active site by forming a complex with His 69, Glu 72, and His 196, as well as with a water molecule. (Glu 72 means that, starting from the N-terminal end of the enzyme, Glu is the seventy-second amino acid.)

PROPOSED MECHANISM FOR THE REACTION CATALYZED BY CARBOXYPEPTIDASE A

Overall Reaction

Binding the Substrate

Several amino acid side chains at the active site of carboxypeptidase A participate in binding the substrate in the optimal position for reaction. Arg 145 forms two hydrogen bonds and Tyr 248 forms one hydrogen bond with the C-terminal carboxyl group of the substrate. (In this example, the C-terminal amino acid is phenylalanine.) The side chain of the C-terminal amino acid is positioned in a hydrophobic pocket, which is why carboxypeptidase A is not active if the C-terminal amino acid is arginine or lysine. Apparently, the long, positively charged side chains of these amino acids (Table 21.2) cannot fit into the nonpolar pocket.

Catalyzing the Reaction

- When the substrate binds to the active site, Zn^{2+} partially complexes with the oxygen of the carbonyl group of the amide that will be hydrolyzed. Thus, Zn^{2+} polarizes the carbon–oxygen double bond, making the carbonyl carbon more susceptible to nucleophilic addition and stabilizing the negative charge that develops on the oxygen atom in the transition state that leads to the tetrahedral intermediate. Arg 127 also increases the carbonyl group's electrophilicity and stabilizes the developing negative charge on the oxygen atom. In addition, Zn^{2+} complexes with water, making it a better nucleophile. Glu 270 is a general base-catalyst, further increasing water's nucleophilicity.

- In the next step, Glu 270 is a general-acid catalyst, increasing the leaving tendency of the amino group.

- In the last step of the reaction, the carboxylic acid loses a proton and the amine gains a proton because the amine is a stronger base than the carboxylate ion.

When the reaction is over, the amino acid (phenylalanine in this example) and the peptide with one less amino acid dissociate from the enzyme, and another molecule of substrate binds to the active site.

The unfavorable electrostatic interaction between the negatively charged carboxyl group of the peptide product and the negatively charged Glu 270 side chain may facilitate the release of the products from the enzyme.

Notice that the protons are being donated and removed during (rather than before) the other bond-making and bond-breaking processes in these enzyme-catalyzed reactions, so the catalysis that occurs is general-acid and general-base catalysis (Sections 22.2 and 22.3). At physiological pH (7.4), the concentration of H^+ or HO^- ($\sim 1 \times 10^{-7}$ M) is too small for specific-acid or specific-base catalysis.

LEARN THE STRATEGY

PROBLEM 14 SOLVED

Which of the following amino acid side chains can aid the departure of a leaving group?

$$-CH_2CH_2SCH_3 \qquad -CH(CH_3)_2 \qquad -CH_2\text{—imidazole} \qquad -CH_2COOH$$

      1   **2**   **3**   **4**

SOLUTION Side chains **1** and **2** do not have an acidic proton, so they cannot aid the departure of a leaving group by protonating it. Side chains **3** and **4** each have an acidic proton, so they can aid the departure of a leaving group.

USE THE STRATEGY

PROBLEM 15 ◆

Which of the following amino acid side chains can help remove a proton from the α-carbon of an aldehyde?

$$-CH_2C(=O)NH_2 \qquad \text{—C}_6H_4\text{—}O^- \qquad -CH_2\text{—imidazole} \qquad -CH_2C(=O)O^-$$

      1   **2**   **3**   **4**

PROBLEM 16 ◆

Which of the following C-terminal peptide bonds is more readily cleaved by carboxypeptidase A? Explain your choice.

Ser-Ala-Leu   or   Ser-Ala-Asp

PROBLEM 17

Carboxypeptidase A has esterase activity as well as peptidase activity, so it can hydrolyze ester bonds as well as peptide bonds. When carboxypeptidase A hydrolyzes ester bonds, Glu 270 acts as a nucleophilic catalyst instead of a general-base catalyst. Propose a mechanism for the carboxypeptidase A–catalyzed hydrolysis of an ester bond.

22.10 ANOTHER ENZYME-CATALYZED REACTION THAT IS REMINISCENT OF ACID-CATALYZED AMIDE HYDROLYSIS

Trypsin, chymotrypsin, and elastase are members of a group of *endopeptidases* known collectively as serine proteases. (Recall that an endopeptidase cleaves a peptide bond that is not at the end of a peptide chain; see Section 21.13.) They are called *proteases* because they catalyze the hydrolysis of protein peptide bonds. They are called *serine proteases* because each one has a serine side chain at the active site that participates in the catalysis.

The Specificity of the Serine Proteases

The various serine proteases have similar primary structures, suggesting that they are evolutionarily related. Although they all have the same three catalytic side chains at the active site (that is, Asp, His, and Ser), they have one important difference—namely, the composition of the pocket at the active site that binds the side chain of the amino acid in the peptide bond that undergoes hydrolysis (Figure 22.5). This pocket is what gives the serine proteases their different specificities (Section 21.13).

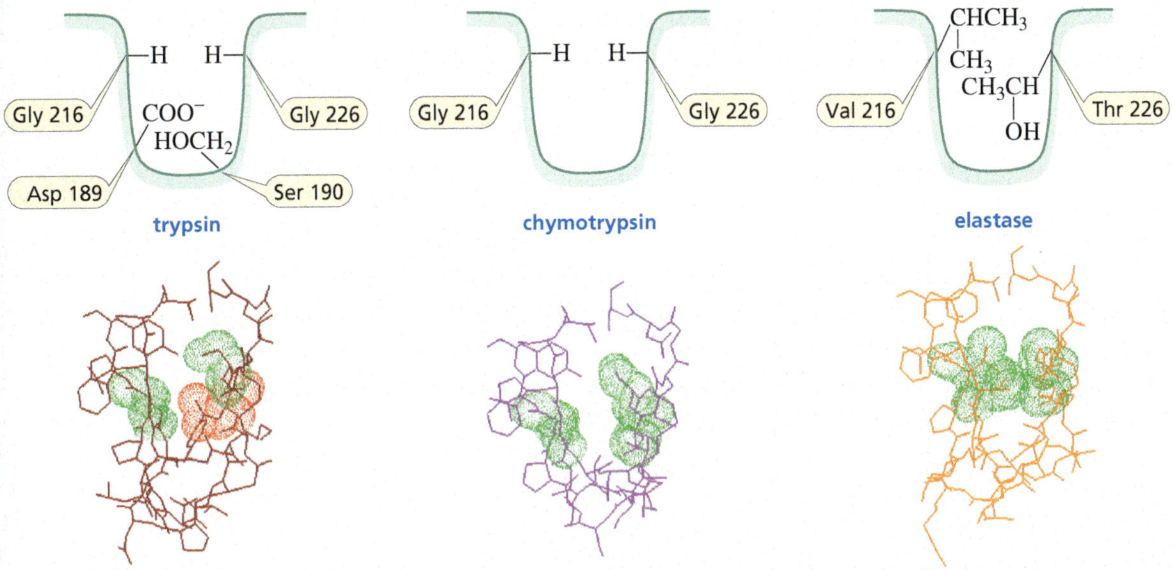

▲ **Figure 22.5**
The binding pockets in trypsin, chymotrypsin, and elastase. The negatively charged aspartate is shown in red and the relatively nonpolar amino acids are shown in green. The structures of the binding pockets explain why trypsin binds long, positively charged amino acids; chymotrypsin binds flat, nonpolar amino acids; and elastase binds only small amino acids.

- The pocket in trypsin is narrow and has a serine and a negatively charged aspartate carboxyl group at its bottom. The shape and charge of the binding pocket cause it to bind long, positively charged amino acid side chains (Lys and Arg). This is why trypsin hydrolyzes only peptide bonds on the C-side of arginine and lysine.

- The pocket in chymotrypsin is narrow and is lined with nonpolar amino acids, so chymotrypsin cleaves on the C-side of amino acids with flat, nonpolar side chains (Phe, Tyr, and Trp).

- In elastase, two glycines on the sides of the pocket in trypsin and in chymotrypsin are replaced by relatively bulky valine and threonine. Consequently, only small amino acids can fit into the pocket. Elastase, therefore, hydrolyzes peptide bonds on the C-side of small amino acids (Gly, Ala, Ser, and Val).

The Mechanism

The proposed mechanism for the chymotrypsin-catalyzed hydrolysis of a peptide bond is shown below. Notice that it involves two successive nucleophilic acyl substitution reactions: first, nucleophilic addition to an amide to form a tetrahedral intermediate, followed by elimination from the

PROPOSED MECHANISM FOR THE REACTION CATALYZED BY SERINE PROTEASES

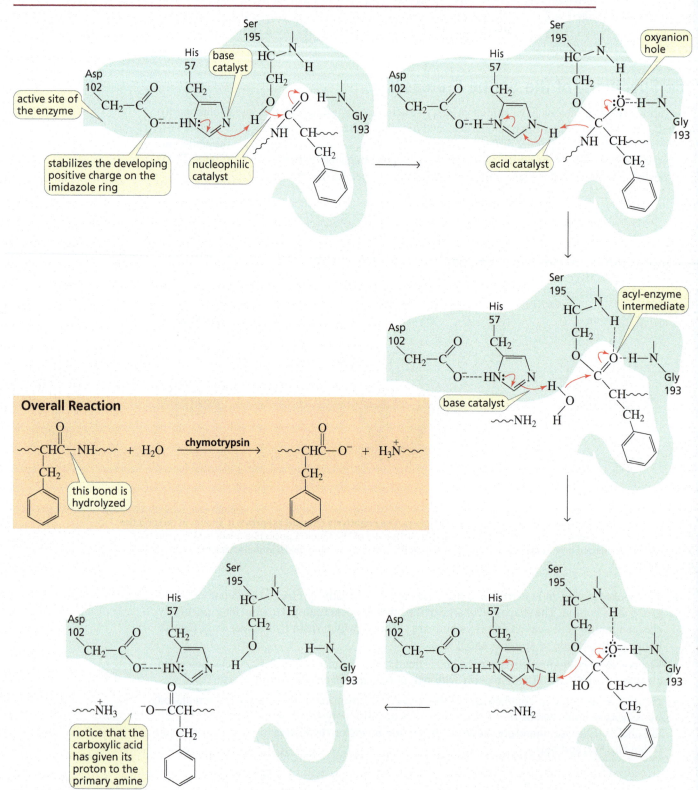

tetrahedral intermediate; and then, nucleophilic addition to an ester to form a tetrahedral intermediate, followed by elimination from the tetrahedral intermediate (Section 15.4). The other serine proteases follow the same mechanism.

- As a consequence of binding the flat, nonpolar side chain in the hydrophobic pocket, the amide linkage to be hydrolyzed is positioned very close to Ser 195. His 57 is a general-base catalyst that increases the nucleophilicity of the serine that adds to the carbonyl group. This step is helped by Asp 102, which does not remove a proton from imidazole but remains as a carboxylate ion, using its negative charge to stabilize the developing positive charge on His 57 and to position the five-membered ring so that its basic N atom is close to the OH group of serine. The stabilization of a charge by an opposite charge is called **electrostatic catalysis.** Formation of the tetrahedral intermediate causes a slight change in the conformation of the protein. This allows the negatively charged oxygen to slip into a previously unoccupied area of the active site known as the *oxyanion hole*. Once in the oxyanion hole, the negatively charged oxygen can hydrogen bond with two peptide groups (Gly 193 and Ser 195), which stabilizes the tetrahedral intermediate.

- In the next step, the tetrahedral intermediate collapses, eliminating the amino group. This is a strongly basic group that cannot be eliminated without the participation of His 57, an acid catalyst that increases the leaving propensity of the group by protonating it—thereby making it a weaker base and, therefore, a better leaving group. The product of the second step is an **acyl-enzyme intermediate** because the serine group of the enzyme has been acylated—that is, an acyl group has been put on it.

- The third step is just like the first step, except that water instead of serine is the nucleophile. Water adds to the acyl group of the acyl-enzyme intermediate, with His 57 acting as a general-base catalyst to increase water's nucleophilicity, Asp 102 again stabilizing the positively charged histidine side chain, and the negatively charged oxygen slipping into the oxyanion hole.

- In the final step, the tetrahedral intermediate collapses, eliminating serine. His 57 is a general-acid catalyst in this step, increasing serine's leaving propensity. (Notice that in this last step, the carboxylic acid loses a proton and the amine gains a proton because the amine is a stronger base than the carboxylate ion.)

Notice that the first two steps are alcoholysis of an amide and the second two steps are hydrolysis of an ester. Each of these two-step reactions requires four steps in the non-enzyme-catalyzed reactions (Sections 15.12 and 15.8). However, in the enzyme-catalyzed reaction, several steps occur simultaneously as a result of holding the reacting and catalytic groups that participate in the reaction precisely where they need to be. As a result, each of these nucleophilic acyl substitution reactions becomes a two-step reaction.

PROBLEM 18 ◆

Arginine and lysine side chains fit into trypsin's binding pocket (Figure 22.5). One of these side chains forms a direct hydrogen bond with serine and an indirect hydrogen bond (mediated through a water molecule) with aspartate. The other side chain forms direct hydrogen bonds with both serine and aspartate. Which is which?

PROBLEM 19

Explain why serine proteases do not catalyze hydrolysis if the amino acid at the hydrolysis site is a D-amino acid. Trypsin, for example, cleaves on the C-side of L-Arg and L-Lys, but not on the C-side of D-Arg and D-Lys.

Site-Specific Mutagenesis

Information about the mechanism of an enzyme-catalyzed reaction has been obtained from **site-specific mutagenesis,** a technique that replaces one amino acid of a protein with another.

For example, when Asp 102 of chymotrypsin is replaced with Asn 102, the enzyme's ability to bind the substrate is unchanged, but its ability to catalyze the reaction decreases to less than 0.05% of the value for the native enzyme. Clearly, Asp 102 must be involved in the catalytic process. We just saw that its role is to position histidine and use its negative charge to stabilize histidine's positive charge.

side chain of an aspartate (Asp) residue   side chain of an asparagine (Asn) residue

22.11 AN ENZYME-CATALYZED REACTION THAT INVOLVES TWO SEQUENTIAL S$_N$2 REACTIONS

Lysozyme is an enzyme that destroys bacterial cell walls. These cell walls are composed of alternating *N*-acetylmuramic acid (NAM) and *N*-acetylglucosamine (NAG) units linked by β-1,4'-glycosidic linkages (Section 20.15). Lysozyme destroys the cell wall by catalyzing the hydrolysis of the NAM–NAG bond.

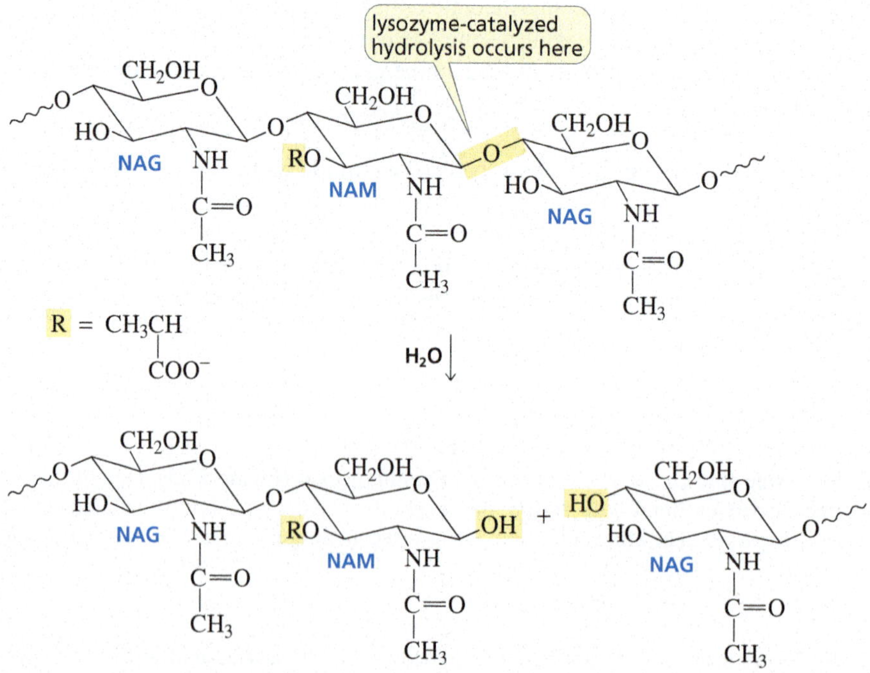

Binding the Substrate

The active site of lysozyme binds six sugar residues of the substrate. They are labeled A, B, C, D, E, and F on the next page, in Figure 22.6, and on page 1054. The many amino acid side chains involved in binding the substrate in the correct position in the active site are shown in Figure 22.6. The RO group of NAM cannot fit into the binding site for C or E. This means that NAM units must bind at the sites for B, D, and F. Hydrolysis occurs between D and E.

Catalyzing the Reaction

Lysozyme has two catalytic groups at the active site: Glu 35 and Asp 52. The discovery that the enzyme-catalyzed reaction takes place with retention of configuration at the anomeric carbon indicates that it cannot be a one-step S$_N$2 reaction. (Recall that an S$_N$2 reaction takes place with inversion of configuration; Section 9.1.) Therefore, the reaction must involve either two sequential S$_N$2 reactions or an S$_N$1 reaction with the enzyme blocking one face of an intermediate to nucleophilic addition.

Although lysozyme was the first enzyme to have its mechanism studied—and it has been studied extensively for over 40 years—only recently have data been obtained that support the mechanism involving two sequential S$_N$2 reactions.

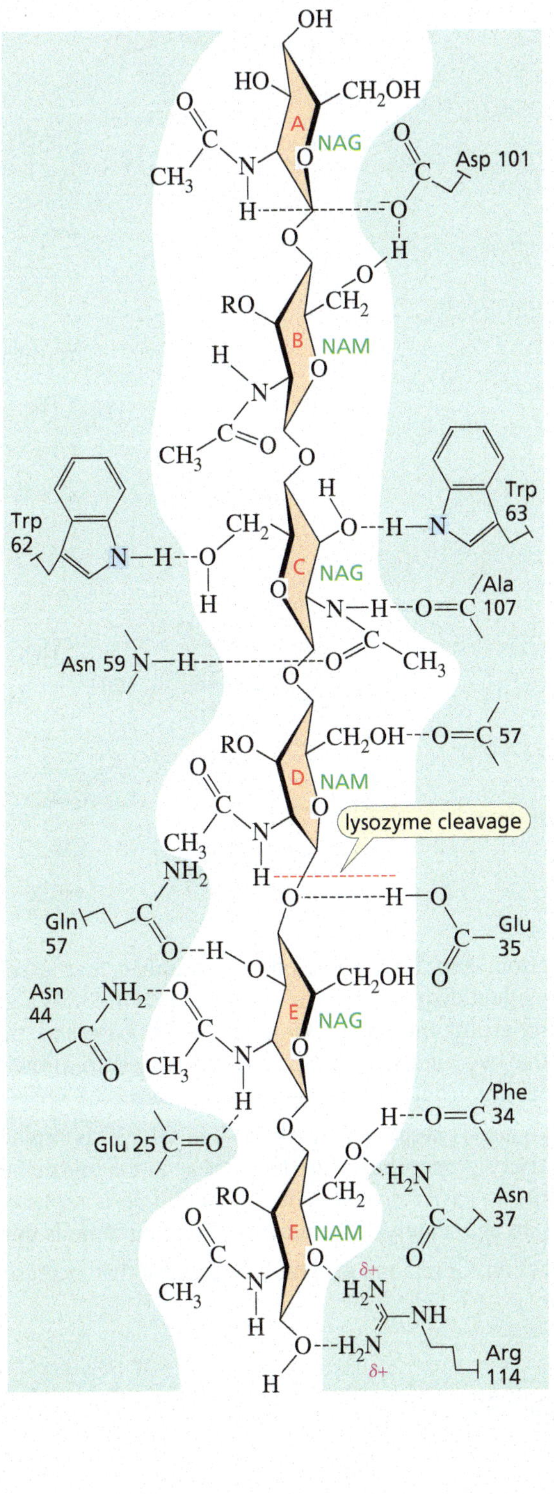

◄ **Figure 22.6**
The amino acids at the active site of lysozyme that are involved in binding the substrate.

PROPOSED MECHANISM FOR THE REACTION CATALYZED BY LYSOZYME

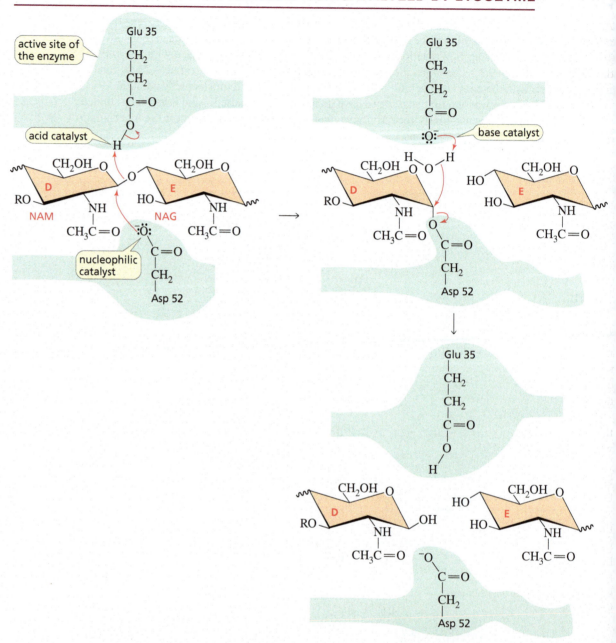

- In the first S_N2 reaction, Asp 52 is a nucleophilic catalyst that attacks the anomeric carbon (C-1) of the NAM residue, displacing the leaving group. Glu 35 is a general-acid catalyst that protonates the leaving group, thereby making it a weaker base and, therefore, a better leaving group. Notice that the two catalytic groups are positioned to allow nucleophilic attack and protonation of the leaving group to occur in the same step.

  Site-specific mutagenesis studies show that when Glu 35 is replaced by Asp, the enzyme has only weak activity. Apparently, Asp does not lie at the optimal distance from, and angle to, the oxygen atom to easily protonate it. When Glu 35 is replaced by Ala, an amino acid that cannot act as an acid catalyst, the activity of the enzyme is completely lost.

- In the second S_N2 reaction, Glu 35 is a general-base catalyst that increases water's nucleophilicity.

PROBLEM 20 ♦

If $H_2{}^{18}O$ is used in the hydrolysis reaction catalyzed by lysozyme, which ring will contain the label, NAM or NAG?

pH–Activity Profile

A plot of the activity of an enzyme as a function of the pH of the reaction mixture is called a **pH–activity profile.** The pH–activity profile for lysozyme is shown in Figure 22.7. The maximum rate occurs at about pH 5.3. The pH at which the enzyme is 50% active is 3.8 on the ascending leg of the curve and 6.7 on the descending leg. These pH values correspond to the pK_a values of the enzyme's catalytic groups.

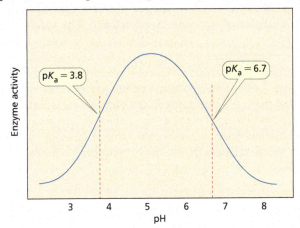

◀ **Figure 22.7**
Lysozyme's activity depends on the pH of the reaction mixture.

LEARN THE STRATEGY

The pK_a given by the ascending leg is the pK_a of a group that is catalytically active in its basic form. When that group is fully protonated (~pH = 2), the enzyme is not active. As the pH of the reaction mixture increases, a larger fraction of the group is present in its basic form, and as a result, the enzyme shows increasing activity. Similarly, the pK_a given by the descending leg is the pK_a of a group that is catalytically active in its acidic form. Maximum catalytic activity occurs when the group is fully protonated; activity decreases with increasing pH because a greater fraction of the group lacks a proton.

From the mechanism, we can conclude that Asp 52 is the group with a pK_a of 3.8 and Glu 35 is the group with a pK_a of 6.7. The pH–activity profile indicates that lysozyme is maximally active when Asp 52 is in its basic form and Glu 35 is in its acidic form.

Table 21.3 on page 993 shows that the pK_a of aspartic acid is 3.86 and the pK_a of glutamic acid is 4.25. The pK_a of Asp 52 agrees with the pK_a of aspartic acid, but the pK_a of Glu 35 is much greater than the pK_a of glutamic acid.

Why is the pK_a of the glutamic acid side chain at the active site of the enzyme so much greater than the pK_a given for glutamic acid in Table 21.3? The pK_a values in Table 21.3 were determined in water. In the enzyme, Asp 52 is surrounded by polar groups, which means that its pK_a should be close to the pK_a determined in water, a polar solvent. Glu 35, however, is situated in a predominantly nonpolar pocket, so its pK_a is greater than the pK_a determined in water. We saw that the pK_a of a carboxylic acid is greater in a nonpolar solvent because there is less tendency to form charged species in nonpolar solvents (Section 9.14).

Part of the catalytic efficiency of lysozyme results from its ability to provide different solvent environments at the active site. This allows one catalytic carboxylic acid group to exist in its acidic form and the other in its basic form. This property is unique to enzymes; chemists cannot provide different solvent environments for different parts of nonenzymatic systems.

N-acetylneuraminic acid

Tamiflu®

PROBLEM 21

Draw the pH–activity profile for an enzyme that has one catalytic group at the active site:

a. the catalytic group is a general-acid catalyst with a $pK_a = 5.6$.
b. the catalytic group is a general-base catalyst with a $pK_a = 7.2$.

USE THE STRATEGY

How Tamiflu Works

Tamiflu is one of the few antiviral drugs currently available. It is used for the prevention and treatment of influenza A and B. Before a virus particle can be released from its host cell, an enzyme called neuraminidase must cleave off a sugar residue (*N*-acetylneuraminic acid) from a glycoprotein on the surface of the cell (Section 20.18). Because *N*-acetylneuraminic acid and Tamiflu have similar shapes, the enzyme cannot distinguish between them. Therefore, it can bind either one at its active site. When the enzyme binds Tamiflu, it cannot bind *N*-acetylneuraminic acid. Thus, the virus particles are prevented from being released from their host cells, in which case they cannot infect new cells. Early treatment with Tamiflu is important because it will be less effective if a lot of cells have already been infected.

22.12 AN ENZYME-CATALYZED REACTION THAT IS REMINISCENT OF THE BASE-CATALYZED ENEDIOL REARRANGEMENT

Glycolysis is the name given to the series of ten enzyme-catalyzed reactions responsible for converting glucose to two molecules of pyruvate (Section 24.6). The second reaction in glycolysis is an isomerization reaction that converts glucose-6-phosphate to fructose-6-phosphate. Recall that glucose is an aldohexose, whereas fructose is a ketohexose (Section 20.0), so the enzyme that catalyzes this reaction—glucose-6-phosphate isomerase—converts an aldose to a ketose.

Notice that steps 2 and 3 of the mechanism are the same as the mechanism for the enediol rearrangement except that there are two steps in the enzyme-catalyzed reaction rather than four (Section 20.5). There are fewer steps because, since the reactant and the catalysts are precisely where they need to be, acid and base catalysis can occur in the same step rather than in separate steps, thereby avoiding the formation of relatively unstable, negatively charged intermediates (Section 20.5).

PROPOSED MECHANISM FOR THE REACTION CATALYZED BY GLUCOSE-6-PHOSPHATE ISOMERASE

- The first step is a ring-opening reaction of a hemiacetal (Section 20.6). A general-base catalyst (thought to be a histidine side chain) removes a proton from the OH group, and a general-acid catalyst (a protonated lysine side chain) aids the departure of the leaving group by protonating it, thereby making it a weaker base and, therefore, a better leaving group.

- In the second step, a general-base catalyst (a glutamate side chain) removes a proton from the α-carbon of the aldehyde and a general-acid catalyst protonates the oxygen, forming an enediol. Recall that the α-hydrogen of an aldehyde is relatively acidic (Section 17.1).

- In the next step, the enediol is converted to a ketone.

- In the final step, the conjugate base of the acid catalyst employed in the first step and the conjugate acid of the base catalyst employed in the first step catalyze ring closure.

PROBLEM 22

The pH–activity profile for glucose-6-phosphate isomerase indicates the participation of a group with a $pK_a = 6.7$ as a basic catalyst and a group with a $pK_a = 9.3$ as an acid catalyst. Draw the pH–activity profile and identify the amino acids that participate in the catalysis.

PROBLEM 23 ♦

When glucose undergoes base-catalyzed isomerization in the absence of the enzyme, mannose is one of the products that is formed (Section 20.5). Why is mannose not formed in the enzyme-catalyzed reaction?

22.13 AN ENZYME CATALYZED-REACTION THAT IS REMINISCENT OF A RETRO-ALDOL ADDITION

The substrate for the first enzyme-catalyzed reaction in the series of reactions known as glycolysis is D-glucose (a six-carbon compound). The final product of glycolysis is two molecules of pyruvate (a three-carbon compound). Therefore, at some point in the series of enzyme-catalyzed reactions, a six-carbon compound must be cleaved into two three-carbon compounds. The enzyme *aldolase* catalyzes this cleavage.

> *Aldolase converts fructose-1,6-bisphosphate to glyceraldehyde-3-phosphate and dihydroxyacetone phosphate.*

The enzyme is called aldolase because the reaction it catalyzes is a retro-aldol addition—that is, it is the reverse of an aldol addition (Section 17.10).

- In the first step, fructose-1,6-bisphosphate forms a protonated imine with a lysine side chain at the active site of the enzyme (Section 16.8).

LEARN THE STRATEGY

- In the next step, the bond between C-3 and C-4 is broken, with aspartate acting as a general-base catalyst. The positively charged nitrogen is more electron withdrawing than the carbonyl oxygen that would have been the electron withdrawing group if the imine had not been formed. The molecule of glyceraldehyde-3-phosphate (one of the three-carbon products) formed in this step dissociates from the enzyme.

- The enamine rearranges to a protonated imine, with the aspartic acid side chain now acting as a general-acid catalyst.

- Hydrolysis of the protonated imine releases dihydroxyacetone phosphate, the other three-carbon product.

PROPOSED MECHANISM FOR THE REACTION CATALYZED BY ALDOLASE

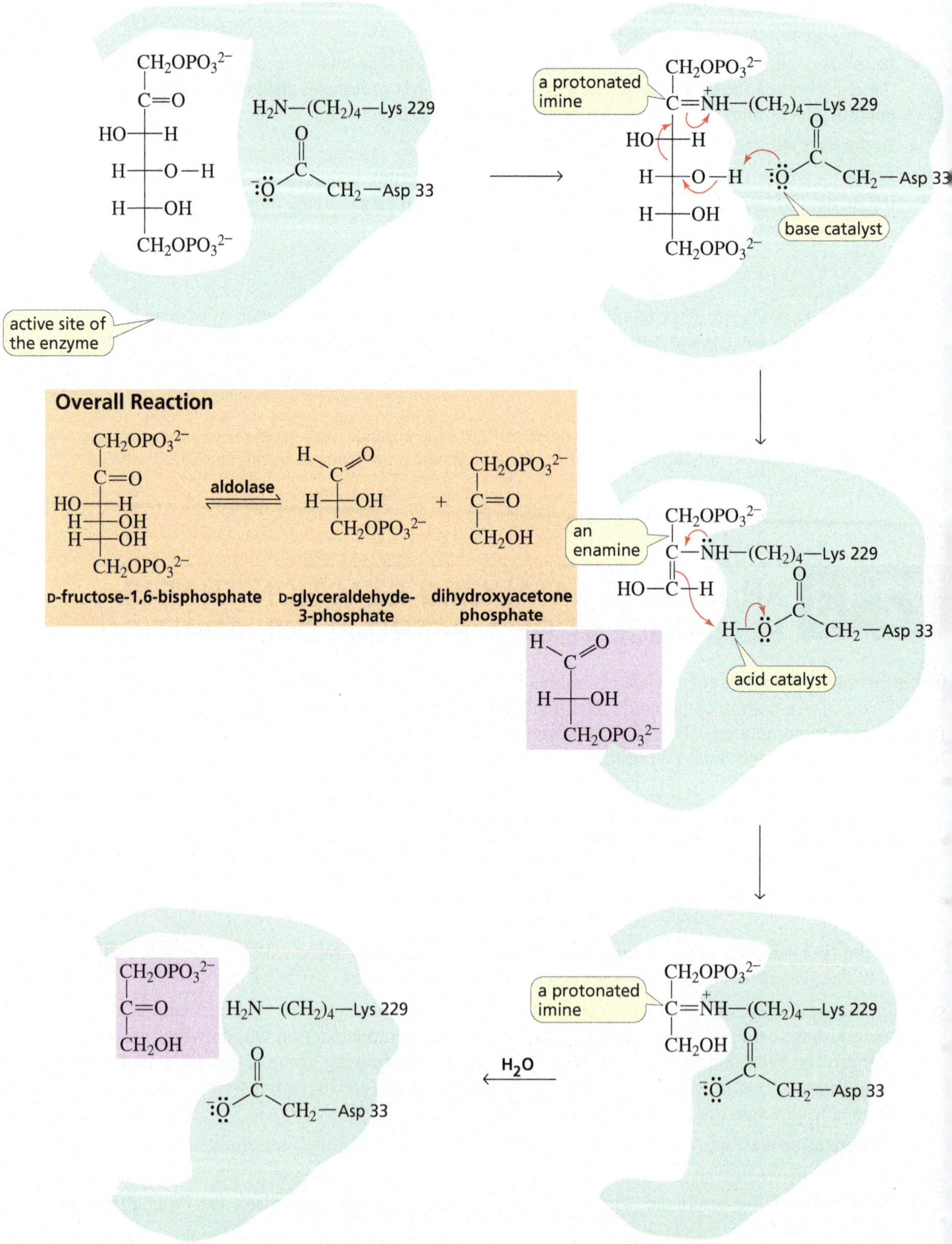

active site of the enzyme

a protonated imine

base catalyst

Overall Reaction

$$\text{D-fructose-1,6-bisphosphate} \xrightleftharpoons[]{\text{aldolase}} \text{D-glyceraldehyde-3-phosphate} + \text{dihydroxyacetone phosphate}$$

an enamine

acid catalyst

a protonated imine

H_2O

PROBLEM 24

Draw the mechanism for the hydroxide ion–catalyzed cleavage of fructose-1,6-bisphosphate.

PROBLEM 25

What advantage does the enzyme gain by forming an imine?

PROBLEM 26 ◆

Which of the following amino acid side chains can form an imine with a substrate?

$$-CH_2CNH_2 \qquad -(CH_2)_4NH_2 \qquad \qquad -CH_2OH$$

1   **2**   **3**   **4**

PROBLEM 27

In glycolysis, why must glucose-6-phosphate isomerize to fructose-6-phosphate (Section 22.12) before the cleavage reaction with aldolase occurs?

PROBLEM 28 ◆

Aldolase shows no activity if it is incubated with iodoacetic acid before fructose-1,6-bisphosphate is added to the reaction mixture. What causes this loss of activity?

ESSENTIAL CONCEPTS

Section 22.0

- A **catalyst** increases the rate of a chemical reaction but is not consumed or changed in the reaction.
- A catalyst changes the rate at which a product is formed but does not change the amount of product formed at equilibrium.
- A catalyst changes the rate of a reaction by providing a pathway with a smaller $\Delta G^{\ddagger}$, either by converting the reactant to a more reactive species, by making the transition state more stable, or by completely changing the mechanism of the reaction.

Section 22.1

- Catalysts provide more favorable pathways for a reaction by increasing the susceptibility of an electrophile to reaction with a nucleophile, by increasing the reactivity of a nucleophile, by increasing the leaving propensity of a group, or by stabilizing a transition state.

Section 22.2

- An **acid catalyst** increases the rate of a reaction by donating a proton to a reactant. In **specific-acid catalysis,** the proton is fully transferred to the reactant before the slow step of the reaction; **in general-acid catalysis,** the proton is transferred during the slow step.
- A catalyst must increase the rate of a slow step.

Section 22.3

- A **base catalyst** increases the rate of a reaction by removing a proton from a reactant. In **specific-base catalysis,** the proton is completely removed from the reactant before the slow step of the reaction; in **general-base catalysis,** the proton is removed during the slow step.

Section 22.4

- A **nucleophilic catalyst** increases the rate of a reaction by acting as a nucleophile: it forms an intermediate by forming a covalent bond with a reactant.

Section 22.5

- A **metal ion** can increase the rate of a reaction by making a reaction center more electrophilic, by making a leaving group a weaker base, or by increasing the nucleophilicity of water.

Section 22.6

- The rate of a chemical reaction is determined by the number of molecular collisions with sufficient energy and with the proper orientation in a given period of time.
- An **intramolecular reaction** that forms a five- or a six-membered ring occurs more readily than the analogous intermolecular reaction because both the frequency of the collisions and the probability that collisions will occur in the proper orientation increases.

- **Effective molarity** is the concentration of the reactant that would be required in an intermolecular reaction for it to have the same rate as the intramolecular reaction.

Section 22.7

- When a catalyst is part of the reacting molecule, the catalysis is called **intramolecular catalysis.**

Section 22.8

- Most biological catalysts are **enzymes.** The reactant of an enzyme-catalyzed reaction is called a **substrate.** The enzyme specifically binds the substrate at its active site. Bond-making and bond-breaking steps of the reaction occur while it is at the **active site.**

- The specificity of an enzyme for its substrate is an example of **molecular recognition.**

- **Induced fit** is the change in conformation of the enzyme that occurs when it binds the substrate.

- Two important factors contribute to the remarkable catalytic ability of enzymes: 1. the reacting groups are brought together at the active site in the proper orientation for reaction, and 2. the amino acid side chains (and a metal ion in the case of some enzymes) are positioned relative to the substrate, where they are needed both for catalysis and for transition state stabilization.

Section 22.9

- Stabilization of a charge by an opposite charge is called **electrostatic catalysis.**

Section 22.10

- A **pH–activity profile** is a plot of the activity of an enzyme as a function of the pH of the reaction mixture.

PROBLEMS

29. Which of the following two compounds eliminates HBr more rapidly in a basic solution?

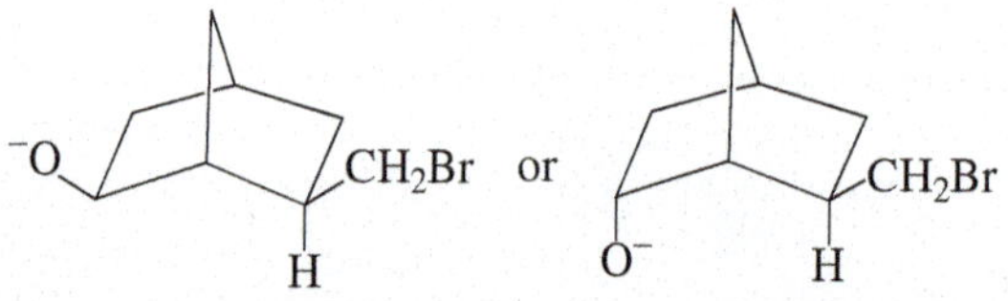

30. Which compound forms an anhydride more rapidly?

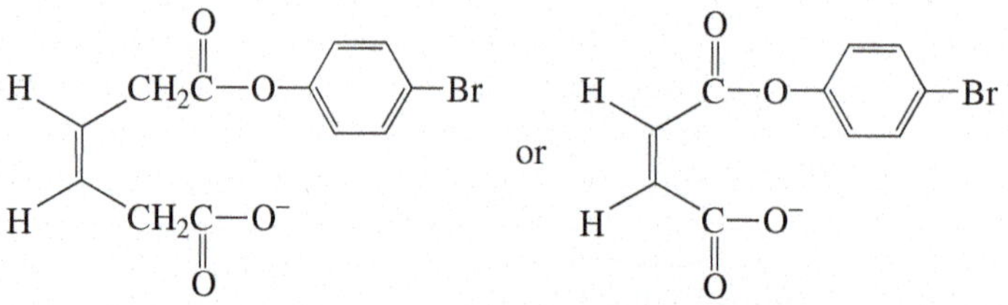

31. Which compound has the greatest rate of hydrolysis at pH = 3.5: benzamide, *o*-carboxybenzamide, *o*-formylbenzamide, or *o*-hydroxybenzamide?

32. Indicate the type of catalysis that is occurring in the slow step in each of the following reaction sequences:

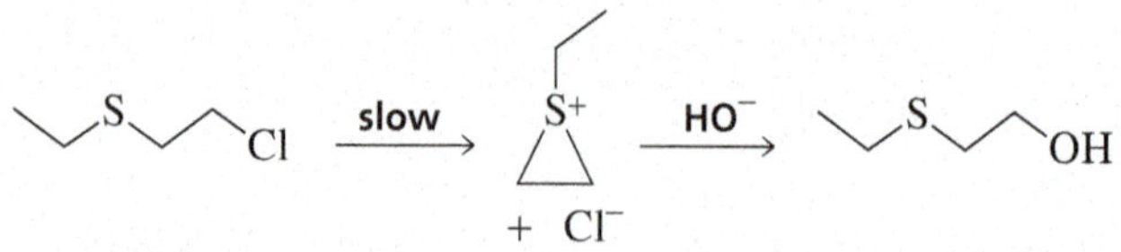

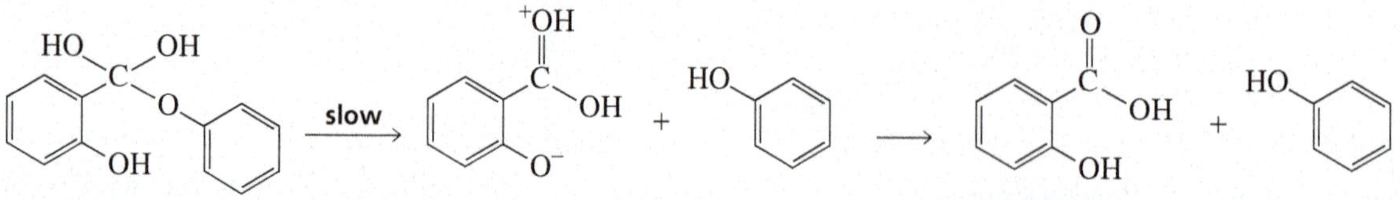

33. The deuterium kinetic isotope effect (k_{H2O}/k_{D2O}) for the hydrolysis of aspirin is 2.2. What does this tell you about the kind of catalysis exerted by the *ortho*-carboxyl substituent? (*Hint:* It is easier to break an O—H bond than an O—D bond.)

34. The rate constant for the uncatalyzed reaction of two molecules of glycine ethyl ester to form glycylglycine ethyl ester is 0.6 $M^{-1}s^{-1}$. In the presence of Co^{2+}, the rate constant is 1.5×10^6 $M^{-1}s^{-1}$. What rate enhancement does the catalyst provide?

35. Co^{2+} catalyzes the hydrolysis of the lactam shown here. Propose a mechanism for the metal-ion catalyzed reaction.

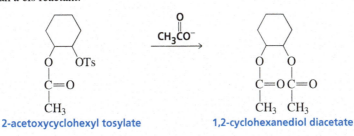

36. There are two kinds of aldolases. Class I aldolases are found in animals and plants, whereas class II aldolases are found in fungi, algae, and some bacteria. Only class I aldolases form an imine. Class II aldolases have a metal ion (Zn^{2+}) at the active site. The mechanism for catalysis by class I aldolases was described in Section 22.13. Propose a mechanism for catalysis by class II aldolases.

37. Propose a mechanism for the following reaction. (*Hint:* The rate of the reaction is much slower if the nitrogen atom is replaced by CH.)

38. The hydrolysis of the ester shown here is catalyzed by morpholine. Explain how morpholine catalyzes the reaction. (*Hint:* The pK_a of the conjugate acid of morpholine is 9.3, so morpholine is too weak a base to function as a base catalyst.)

morpholine

39. Carbonic anhydrase is an enzyme that catalyzes the conversion of carbon dioxide to bicarbonate ion (Section 2.2). It is a metalloenzyme, with Zn^{2+} coordinated at the active site by three histidine side chains. Propose a mechanism for the reaction.

$$CO_2 \ + \ H_2O \xrightarrow{\text{carbonic anhydrase}} HCO_3^- \ + \ H^+$$

40. At pH = 12, the rate of hydrolysis of ester **A** is greater than the rate of hydrolysis of ester **B**. At pH = 8, the relative rates reverse (that is, ester **B** hydrolyzes faster than ester **A**). Explain these observations.

A   B

41. 2-Acetoxycyclohexyl tosylate reacts with acetate ion to form 1,2-cyclohexanediol diacetate. The reaction is stereospecific—that is, the stereoisomers obtained as products depend on the stereoisomer used as a reactant. Recall that because 2-acetoxycyclohexyl tosylate has two asymmetric centers, it has four stereoisomers—two are cis and two are trans. Explain the following observations:

a. Both cis reactants form an optically active trans product, but each cis reactant forms a different trans product.
b. Both trans reactants form the same racemic mixture.
c. A trans reactant is more reactive than a cis reactant.

2-acetoxycyclohexyl tosylate   1,2-cyclohexanediol diacetate

42. Proof that an imine was formed between aldolase and its substrate was obtained by using D-fructose-1,6-bisphosphate labeled at the C-2 position with ^{14}C as the substrate. $NaBH_4$ was added to the reaction mixture. A radioactive product was isolated from the reaction mixture and hydrolyzed in an acidic solution. Draw the structure of the radioactive product obtained from the acidic solution. (*Hint:* $NaBH_4$ reduces an imine linkage.)

43. 3-Amino-2-oxindole catalyzes the decarboxylation of α-keto acids.

    a. Propose a mechanism for the catalyzed reaction.
    b. Would 3-aminoindole be equally effective as a catalyst?

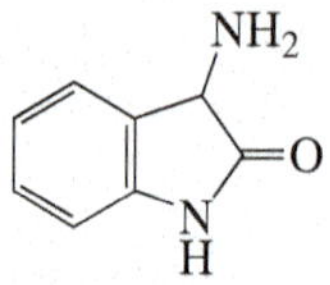

3-amino-2-oxindole

44. a. Explain why the alkyl halide shown here reacts much more rapidly with guanine than does a primary alkyl halide (such as pentyl chloride).

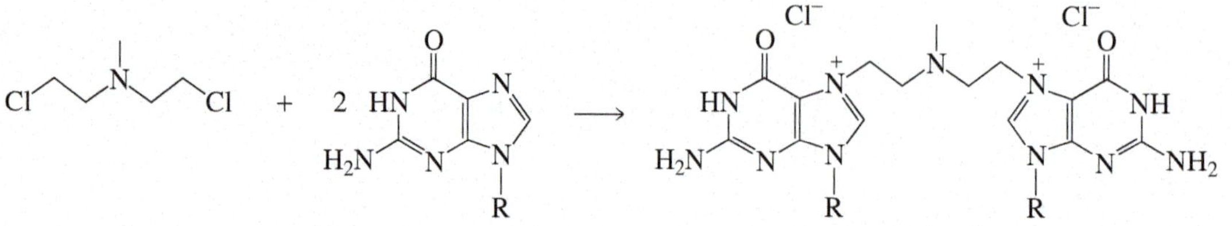

    b. The alkyl halide can react with two guanines, each in a different DNA chain, thereby cross-linking the chains. Propose a mechanism for the cross-linking reaction.

45. Triosephosphate isomerase (TIM) catalyzes the conversion of dihydroxyacetone phosphate to glyceraldehyde-3-phosphate. The enzyme's catalytic groups are Glu 165 and His 95. In the first step of the reaction, these catalytic groups function as a general-base and a general-acid catalyst, respectively. Propose a mechanism for the reaction.

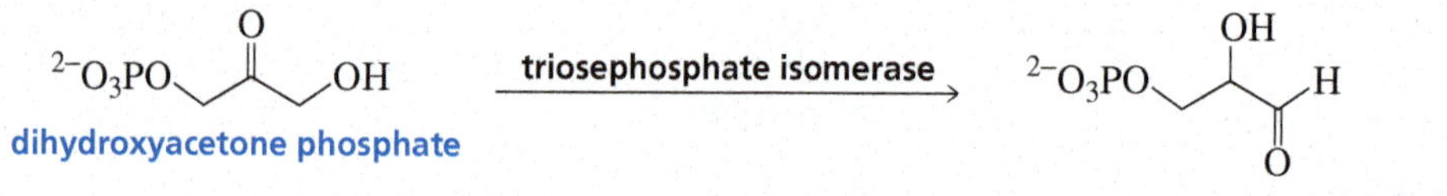

dihydroxyacetone phosphate

glyceraldehyde-3-phosphate

23

The Organic Chemistry of the Coenzymes, Compounds Derived from Vitamins

limes: vitamins A, C, K, and folate

green beans: vitamins A, B$_1$, B$_6$, C, K, folate, and riboflavin

apples: vitamins A, C, K, and folate

cucumbers: vitamins A, C, K, and pantothenate

asparagus: vitamins A, B$_1$, B$_6$, C, E, K, folate, niacin, and riboflavin

a balanced diet

In this chapter, we see why vitamins are required for normal body function, and we will look at some of the many organic reactions occurring in cells that require the participation of a vitamin.

Many enzymes cannot catalyze a reaction without the help of a **cofactor.** Some cofactors are *metal ions* and some are organic molecules.

Metal Ion Cofactors

A metal ion cofactor acts as a Lewis acid in a variety of ways to help an enzyme catalyze a reaction.

- It can coordinate with groups on the enzyme, causing them to align in a way that facilitates the reaction.
- It can help bind the substrate to the active site of the enzyme.
- It can increase the electrophilicty of the substrate.
- It can increase the nucleophilicity of water at the active site (Section 22.5).

We have seen that Zn^{2+} plays an important role in the hydrolysis reaction catalyzed by carboxypeptidase A (see page 1046).

PROBLEM 1 ♦

How does the metal ion in carboxypeptidase A increase the enzyme's catalytic activity (see pages 1046–1047)?

Coenzymes

Cofactors that are organic molecules are called **coenzymes.** Coenzymes are derived from organic compounds commonly known as *vitamins.* A **vitamin** is a substance needed in a small amount for normal body function that the body cannot synthesize at all—except for vitamins D (Section 28.6) and K (Section 18.7) that the body can synthesize—or cannot synthesize in adequate amounts. Early life forms were able to synthesize vitamins, but somewhere along the way some species—including our own—lost that ability. Now species depend on each other for vitamins.

The first such compound recognized to be essential in the diet was an amine (vitamin B_1), which led scientists to conclude incorrectly that all such compounds were amines. As a result, they were originally called vitamines ("amines required for life"). The *e* was later dropped from the name. Table 23.1 lists the vitamins and their chemically active coenzyme forms.

We have seen that the acids, bases, and nucleophiles that catalyze organic reactions in the laboratory are similar to the acidic, basic, and nucleophilic side chains that enzymes use to catalyze organic reactions that occur in cells (Sections 22.9–22.11). In this chapter, we will see that coenzymes play a variety of chemical roles that the amino acid side chains of enzymes cannot play. Some coenzymes function as oxidizing and reducing agents, some allow electrons to be delocalized, some activate groups for further reaction, and some provide good nucleophiles or strong bases needed for a particular reaction.

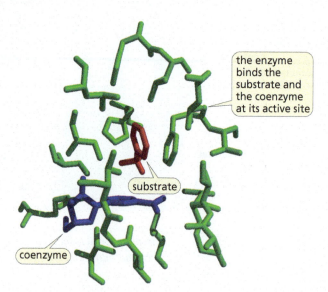

the enzyme binds the substrate and the coenzyme at its active site

substrate

coenzyme

Because it would be highly inefficient for the body to use a compound only once and then discard it, coenzymes are recycled. Therefore, we will see that any coenzyme that is changed during the course of a reaction is subsequently converted back to its original form.

Water-Soluble and Water-Insoluble Vitamins

Early nutritional studies divided vitamins into two classes: water-soluble vitamins and water-insoluble vitamins (Table 23.1). Vitamin K is the only *water-insoluble vitamin* currently known to be a precursor of a coenzyme. Vitamin A is required for proper vision, vitamin D regulates calcium and phosphate metabolism, and vitamin E is an antioxidant. Because they are not precursors of coenzymes, these vitamins are not discussed in this chapter. (Vitamins A and E are discussed in Sections 4.1 and 12.11, and vitamin D is discussed in Section 28.6.)

All the *water-soluble vitamins* except vitamin C are precursors of coenzymes. In spite of its name, vitamin C is not a vitamin because it is required in fairly high amounts and most mammals are able to synthesize it (Section 20.17). Primates and guinea pigs cannot synthesize it, however, so it must be included in their diets. We have seen that vitamins C and E are radical inhibitors and, therefore, are antioxidants: vitamin C traps radicals formed in aqueous environments, whereas vitamin E traps radicals formed in nonpolar environments (Section 12.11).

Table 23.1   The Vitamins, the Coenzymes for Which They Are Precursors, and the Chemical Functions of the Coenzymes

Vitamin	Coenzyme	Reaction catalyzed	Human deficiency disease
Water-Soluble Vitamins			
Niacin (vitamin B_3)	NAD^+, $NADP^+$ NADH, NADPH	Oxidation Reduction	Pellagra
Riboflavin (vitamin B_2)	FAD $FADH_2$	Oxidation Reduction	Skin inflammation
Thiamine (vitamin B_1)	Thiamine pyrophosphate (TPP)	Acyl group transfer	Beriberi
Lipoic acid (lipoate)	Lipoate Dihydrolipoate	Oxidation Reduction	—
Pantothenic acid (pantothenate; vitamin B_5)	Coenzyme A (CoASH)	Acyl group transfer	—
Biotin (vitamin H; vitamin B_7)	Biotin	Carboxylation	—
Pyridoxine (vitamin B_6)	Pyridoxal phosphate (PLP)	Decarboxylation Transamination Racemization C_α—C_β bond cleavage α,β-Elimination β-Substitution	Anemia
Vitamin B_{12}	Coenzyme B_{12}	Isomerization	Pernicious anemia
Folic acid (folate)	Tetrahydrofolate (THF)	One-carbon transfer	Megaloblastic anemia
Ascorbic acid (vitamin C)	—	—	Scurvy
Water-Insoluble Vitamins			
Vitamin A	—	—	
Vitamin D	—	—	Rickets
Vitamin E	—	—	—
Vitamin K	Vitamin KH_2	Carboxylation	—

It is difficult to overdose on water-soluble vitamins because the body can generally eliminate any excess. One can, however, overdose on water-insoluble vitamins because they are *not* easily eliminated by the body and can accumulate in cell membranes and other nonpolar components of the body.

Excess vitamin D, for example, causes calcification of soft tissues. The kidneys are particularly susceptible to calcification, which eventually leads to kidney failure. Vitamin D is formed in the skin as a result of a photochemical reaction caused by ultraviolet rays from the sun (Section 28.6). Because of the current, widespread use of sunscreens, a significant number of children are being found to have a vitamin D deficiency.

Vitamin B_1

Christiaan Eijkman (1858–1930) was a member of a medical team that was sent to the East Indies to study an outbreak of beriberi in 1886. At that time, all diseases were thought to be caused by microorganisms. When the microorganism that caused beriberi could not be found, the team left the East Indies but Eijkman stayed behind to become the director of a new bacteriological laboratory.

There, in 1896, Eijkman accidentally discovered the cause of the disease when he noticed that laboratory chickens had developed the symptoms characteristic of beriberi. He found that the symptoms had appeared when a cook started feeding the chickens polished (white) rice meant for hospital patients. The symptoms disappeared when a new cook resumed giving brown rice to the chickens. Later it was recognized that thiamine (vitamin B_1) is present in rice husks, which are removed when the rice is polished.

23.1 | NIACIN: THE VITAMIN NEEDED FOR MANY REDOX REACTIONS

Any enzyme that catalyzes an oxidation or a reduction reaction requires a coenzyme because none of the amino acid side chains are oxidizing or reducing agents. The coenzyme serves as the oxidizing or reducing agent. The enzyme's role is to hold the substrate and coenzyme together so that the oxidation or reduction reaction can take place (see the model on page 1064).

The Pyridine Nucleotide Coenzymes

Two coenzymes are commonly used by enzymes to catalyze oxidation reactions and reduction reactions. **Nicotinamide adenine dinucleotide (NAD$^+$)** catalyzes oxidation reactions. **Nicotinamide adenine dinucleotide phosphate (NADPH)** catalyzes reduction reactions. Because each of the coenzymes has a pyridine ring, they are known as the **pyridine nucleotide coenzymes.** We will see why **NAD$^+$** is the most common oxidizing and **NADPH** is the most common reducing agent on page 1068.

NAD$^+$ is an oxidizing agent.

NADPH is a reducing agent.

When NAD$^+$ oxidizes a substrate, the coenzyme is reduced to NADH. Because enzymes that catalyze oxidation reactions bind NAD$^+$ more tightly than they bind NADH, once the oxidation reaction is over the relatively loosely bound NADH dissociates from the enzyme.

$$\text{substrate}_{\text{reduced}} + \textbf{NAD}^+ \xrightleftharpoons{\text{enzyme}} \text{substrate}_{\text{oxidized}} + \textbf{NADH} + \text{H}^+$$

When NADPH reduces a substrate, it is oxidized to NADP$^+$. The enzymes that catalyze reduction reactions bind NADPH more tightly than they bind NADP$^+$. Once the reduction reaction is over the relatively loosely bound NADP$^+$ dissociates from the enzyme.

$$\text{substrate}_{\text{oxidized}} + \textbf{NADPH} + \text{H}^+ \xrightleftharpoons{\text{enzyme}} \text{substrate}_{\text{reduced}} + \textbf{NADP}^+$$

Many Catabolic Reactions are Oxidations
Many Anabolic Reactions are Reductions

NAD$^+$ and NADH are the coenzymes associated with **catabolic reactions**—biological reactions that break down complex biomolecules to provide the cell with energy. Many catabolic reactions are oxidation reactions. The cell maintains its [NAD$^+$]/[NADH] ratio near 1000 and its [NADP$^+$]/[NADPH] ratio at about 0.01. Therefore, an enzyme that catalyzes an oxidation reaction is much more apt to bind NAD$^+$ than NADP$^+$ because of the greater availability of NAD$^+$ in the cell.

NADP$^+$ and NADPH are the coenzymes associated with **anabolic reactions**—biological reactions involved in synthesizing complex biomolecules. Many anabolic reactions are reduction reactions. An enzyme that catalyzes a reduction reaction is more apt to bind NADPH than NADH because of the greater availability of NADPH in the cell.

The Structure of a Pyridine Nucleotide Coenzyme

NAD$^+$ is composed of two nucleotides linked together through their phosphate groups. A **nucleotide** consists of a heterocyclic compound attached, in a β-linkage, to C-1 of a phosphorylated ribose (Section 26.1). We have seen that a **heterocyclic compound** has one or more ring atoms that are not carbons (Section 8.20).

| a nucleotide | adenine | niacinamide
nicotinamide | niacin
nicotinic acid
vitamin B$_3$ |

The heterocyclic component of one of the nucleotides of NAD$^+$ is nicotinamide, and the heterocyclic component of the other is adenine. This accounts for the coenzyme's name (**n**icotinamide **a**denine **d**inucleotide). The positive charge in the NAD$^+$ abbreviation indicates the positively charged nitrogen of the pyridine ring. NADP$^+$ differs from NAD$^+$ only in having a phosphate group bonded to the 2'-OH group of the ribose of the adenine nucleotide—hence the addition of "P" to the name.

The adenine nucleotide for the coenzymes is provided by ATP. Niacin (also known as vitamin B$_3$) is the portion of the coenzyme that the body cannot synthesize and must acquire through the diet. (Humans can synthesize a small amount of vitamin B$_3$ from the amino acid tryptophan but not enough to meet the body's metabolic needs.)

adenosine triphosphate
ATP

Niacin Deficiency

A deficiency in niacin causes pellagra, a disease that begins with dermatitis and ultimately causes insanity and death. More than 120,000 cases of pellagra were reported in the United States in 1927, mainly among poor people with unvaried diets. A factor known to be present in preparations of vitamin B$_3$ prevented pellagra, but it was not until 1937 that the factor was identified as nicotinic acid. Mild deficiencies slow down metabolism, which is a potential contributing factor in obesity.

When bread companies started adding nicotinic acid to their bread, they insisted that its name be changed to niacin because nicotinic acid sounded too much like nicotine and they did not want their vitamin-enriched bread to be associated with a harmful substance.

Examples of Enzyme-Catalyzed Reactions that Require a Pyridine Nucleotide Coenzyme

Malate dehydrogenase is the enzyme that catalyzes the oxidation of the *secondary alcohol group* of malate to a *ketone*. (We will see that it is one of the reactions in the catabolic pathway known as the citric acid cycle; see Section 24.9.) The oxidizing agent in this reaction is NAD^+. Most enzymes that catalyze oxidation reactions are called *dehydrogenases*. Recall that the number of C—H bonds decreases in an oxidation reaction (Section 10.5). In other words, dehydrogenases remove hydrogen.

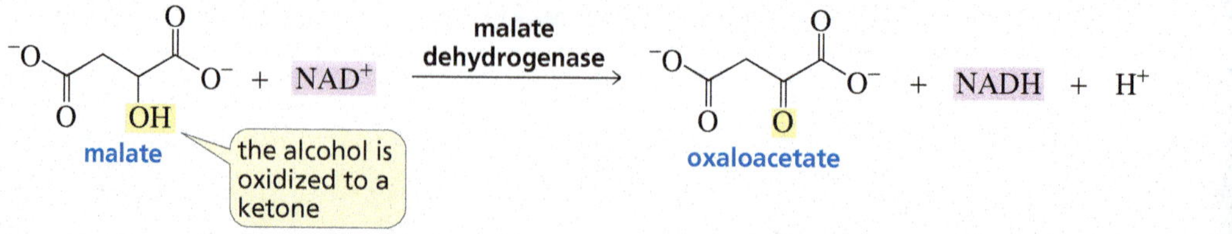

β-Aspartate-semialdehyde is reduced to homoserine in an anabolic pathway. NADPH is the reducing agent.

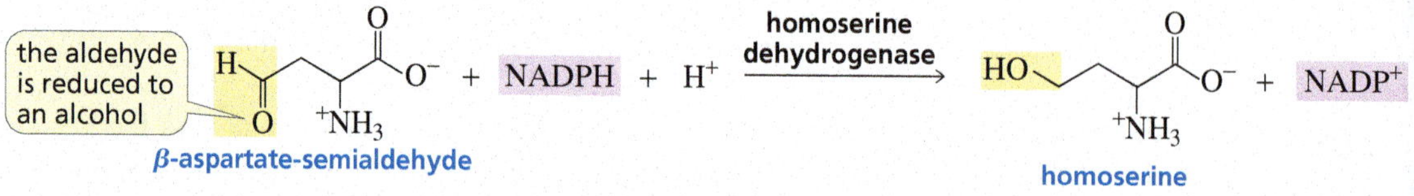

How Does NAD⁺ Oxidize a Substrate?

All the chemistry of the pyridine nucleotide coenzymes takes place at the 4-position of the pyridine ring. The purpose of the rest of the molecule is for electron delocalization and for the enzyme to be able to recognize it so it can bind it to its active site.

A substrate that is being *oxidized* donates a hydride ion (H^-) to the 4-position of the pyridine ring. In the following reaction, for example, the secondary alcohol is oxidized to a ketone. A basic amino acid side chain of the enzyme can help the reaction by removing a proton from the oxygen atom of the substrate. (In the mechanisms shown in this chapter, HB and :B⁻ represent amino acid side chains at the active site of the enzyme that can donate a proton or remove a proton, respectively.)

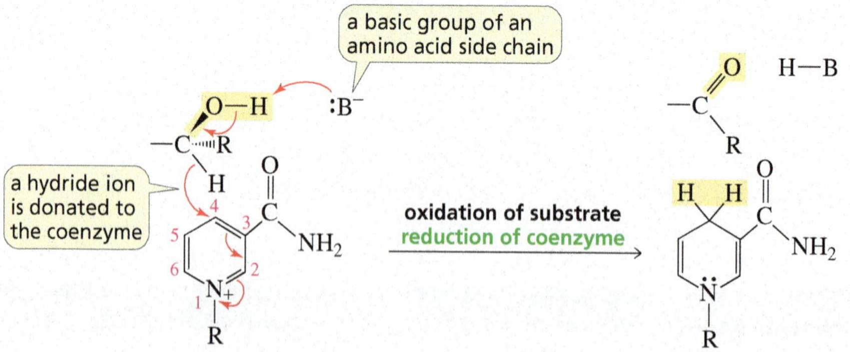

Glyceraldehyde-3-phosphate dehydrogenase (GAPDH) is another enzyme that uses NAD^+ to oxidize a substrate. The enzyme catalyzes the oxidation of the aldehyde group of glyceraldehyde-3-phosphate to a mixed anhydride of a carboxylic acid and phosphoric acid.

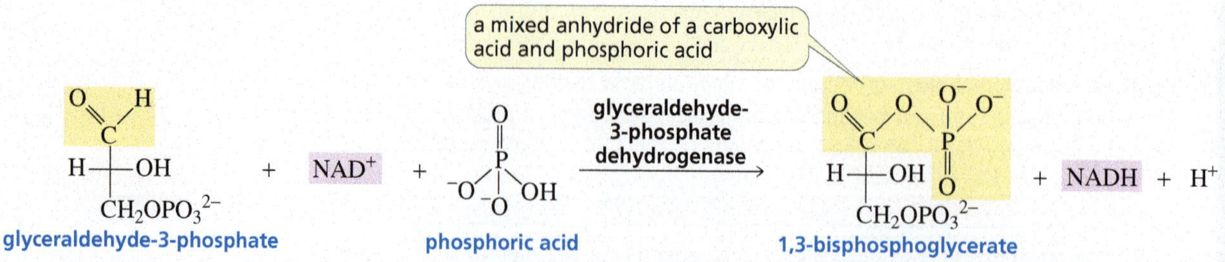

MECHANISM FOR THE REACTION CATALYZED BY GLYCERALDEHYDE-3-PHOSPHATE DEHYDROGENASE (GAPDH)

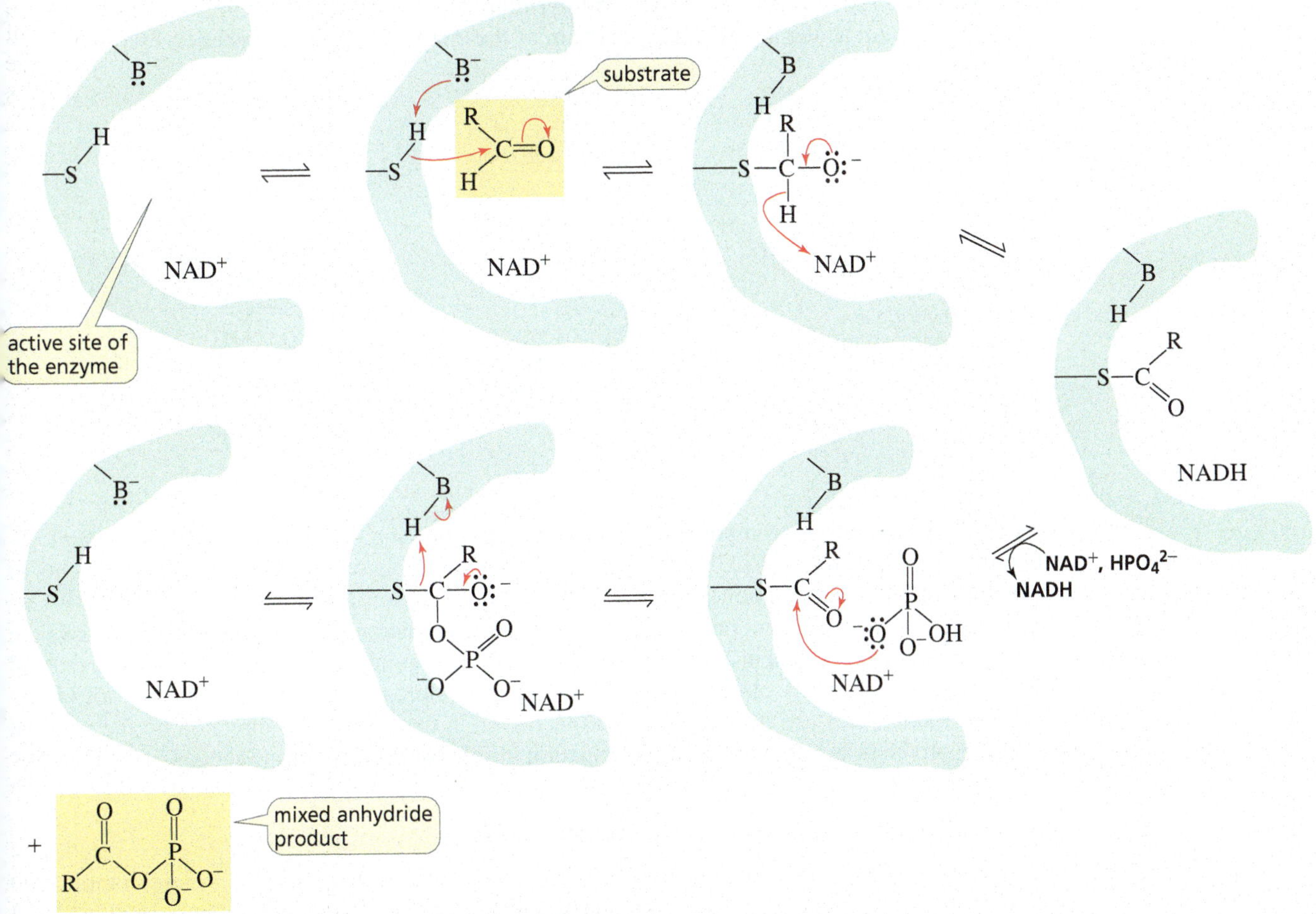

active site of the enzyme

substrate

NAD⁺

NADH

NAD⁺, HPO₄²⁻
NADH

mixed anhydride product

- The enzyme binds glyceraldehyde-3-phosphate (the substrate) at its active site.
- An SH group (a nucleophile) of a cysteine side chain adds to the carbonyl carbon of glyceraldehyde-3-phosphate to form a tetrahedral intermediate. A basic enzyme side chain increases cysteine's nucleophilicity by removing a proton (Section 22.3).
- The tetrahedral intermediate eliminates a hydride ion, transferring it to the 4-position of the pyridine ring of NAD⁺ that is also in the active site, forming NADH.
- NADH dissociates from the enzyme, and the enzyme binds a new NAD⁺.
- Phosphate adds to the thioester, forming a tetrahedral intermediate that eliminates the thiolate ion to form the mixed anhydride product (Section 15.16). The thiolate ion's leaving propensity is increased by protonation, which makes it a weaker base. (Phosphoric acid has pK_a values of 1.9, 6.7, and 12.4, so two of the groups will be primarily in their basic forms at physiological pH.)

Notice that at the end of the reaction the enzyme is exactly as it was at the beginning of the reaction, so another molecule of glyceraldehyde-3-phosphate can be converted to 1,3-bisphosphoglycerate.

PROBLEM 2 ◆

What are the products of the following reaction?

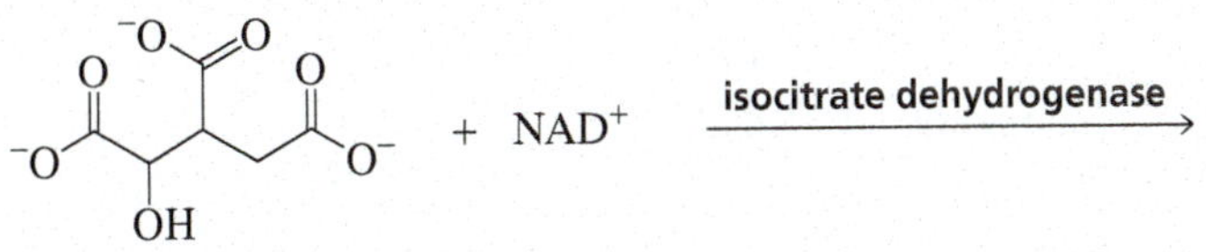

isocitrate   + NAD⁺ $\xrightarrow{\text{isocitrate dehydrogenase}}$

How Does NADPH Reduce a Substrate?

The mechanism for reduction by NADPH is the reverse of the mechanism for oxidation by NAD^+. When a substrate is being *reduced*, the dihydropyridine ring donates a hydride ion from its 4-position to the substrate. An acidic amino acid side chain of the enzyme can aid the reaction by protonating the substrate.

Because NADPH reduces compounds by donating a hydride ion, it can be considered the biological equivalent of $NaBH_4$ or $LiAlH_4$, the hydride donors we have seen used as reducing reagents in nonbiological reactions (Section 16.5).

A biological reducing agent must be recycled (rather than discarded in its oxidized form, as is the fate of a reducing agent used in a laboratory). Therefore, the equilibrium constant for its oxidized and reduced forms is generally close to unity. Consequently, biological redox reactions are not highly exergonic. Rather, they are equilibrium reactions driven in the appropriate direction by the removal of reaction products as a result of their participation in subsequent reactions (Sections 5.7 and 24.6).

Molecular Recognition and Selectivity

Why are the structures of biological oxidizing and reducing reagents so much more complicated than the structures of the reagents used to carry out the same reactions in the laboratory? NADPH is certainly more complicated than $NaBH_4$, although both reagents reduce compounds by donating a hydride ion. Much of the structural complexity of a coenzyme is for **molecular recognition**—to allow the coenzyme to be recognized and bound by the enzyme. As you study the coenzymes in this chapter, do not let the complexity of their structures intimidate you. Notice that only a small part of the coenzyme is actually involved in the chemical reaction.

Another reason for the difference in complexity is that reagents used in cells must be highly selective and, therefore, less reactive than a laboratory reagent. For example, a biological reducing agent cannot reduce just any reducible compound with which it comes into contact. Biological reactions have to be much more carefully controlled than that. Because the coenzymes are relatively unreactive compared with nonbiological reagents, the reaction between the substrate and the coenzyme does not occur at all or takes place very slowly without the enzyme. For example, NADPH does not reduce an aldehyde or a ketone unless an enzyme is present. $NaBH_4$ and $LiAlH_4$ are more reactive hydride donors—in fact, much too reactive to even exist in the aqueous environment of the cell. Similarly, NAD^+ is a much less reactive oxidizing agent than the typical oxidizing agent used in the laboratory; NAD^+ oxidizes an alcohol only in the presence of an enzyme.

Unlike a laboratory reagent, an enzyme that catalyzes an oxidation reaction can distinguish between the two hydrogens on the carbon from which it catalyzes removal of a hydride ion. For example, alcohol dehydrogenase removes only the pro-*R* hydrogen (H_a) of ethanol. It is called the pro-*R* hydrogen because if it were replaced by deuterium, the asymmetric center would have the *R* configuration; H_b is the pro-*S* hydrogen (Section 14.15).

Similarly, an enzyme that catalyzes a reduction reaction can distinguish between the two hydrogens at the 4-position of the nicotinamide ring of NADPH. An enzyme has a specific binding site for the coenzyme and, when it binds the coenzyme, it blocks one of its sides. If the enzyme blocks the B-side of NADPH, the substrate will bind to the A-side and the H_a hydride ion will be transferred to the substrate. If the enzyme blocks the A-side of the coenzyme, then the substrate will have to bind to the B-side, and the H_b hydride ion will be transferred.

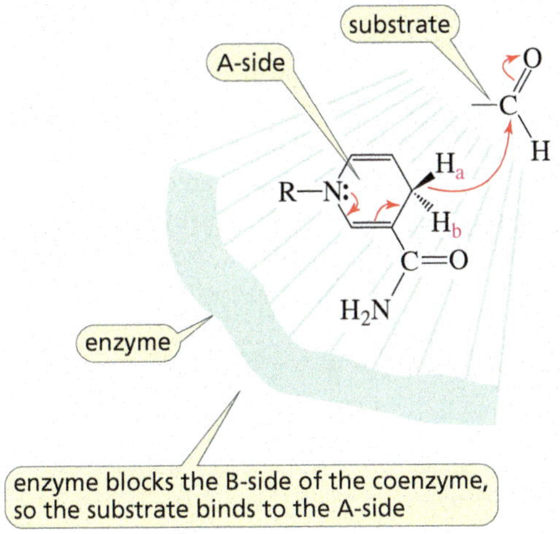

PROBLEM 3 ◆

What is the product of the following reaction?

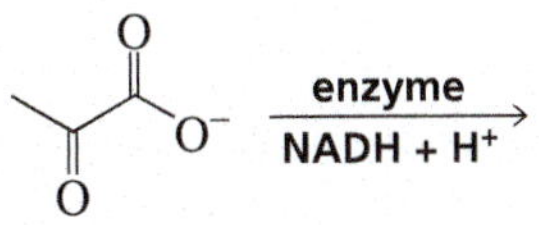

23.2 RIBOFLAVIN: ANOTHER VITAMIN USED IN REDOX REACTIONS

Flavin adenine dinucleotide (FAD), like NAD^+, is a coenzyme used to oxidize a substrate. As its name indicates, FAD is a dinucleotide in which one of the heterocyclic components is flavin and the other is adenine. Notice that instead of ribose, FAD has a reduced ribose (a ribitol group). Flavin plus ribitol is the vitamin known as *riboflavin* or vitamin B_2. Flavin is a bright yellow compound; *flavus* is Latin for "yellow." A vitamin B_2 deficiency causes inflammation of the skin.

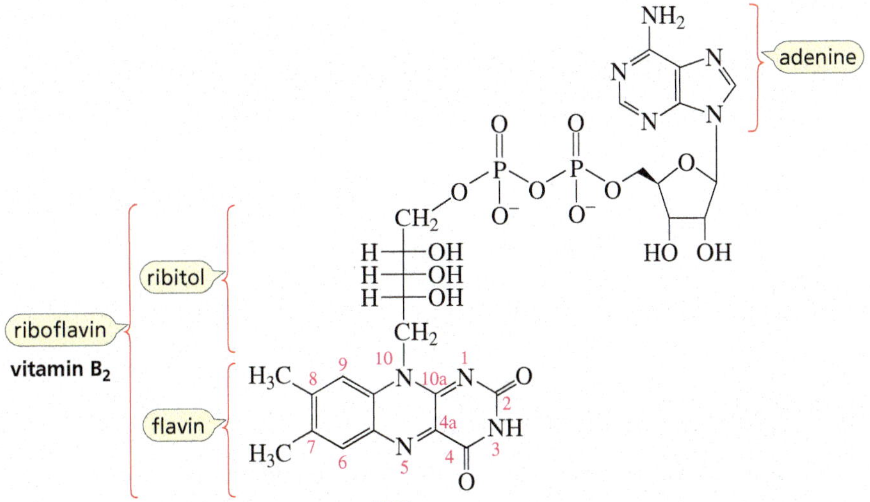

FAD

A *flavoprotein* is an enzyme that contains FAD. In most flavoproteins, FAD is bound quite tightly. Tight binding allows the enzyme to control the oxidation potential of the coenzyme. (The more positive the oxidation potential, the stronger is the oxidizing agent.) Consequently, some flavoproteins are stronger oxidizing agents than others.

Examples of Enzyme-Catalyzed Reactions that Require FAD

How can we tell which enzymes use FAD and which use NAD^+ as the oxidizing coenzyme? A rough guideline is that NAD^+ is the coenzyme used in enzyme-catalyzed oxidation reactions involving carbonyl compounds (for example, alcohols being oxidized to ketones, aldehydes, or carboxylic acids), whereas FAD is the coenzyme used in other types of oxidations. In the following reactions, FAD oxidizes a dithiol to a disulfide, a saturated alkyl group to an alkene, and an amine to an imine. (This is only an approximate guideline, however, because FAD is used in some oxidations that involve carbonyl compounds, and NAD^+ is used in some oxidations that do not involve carbonyl compounds.)

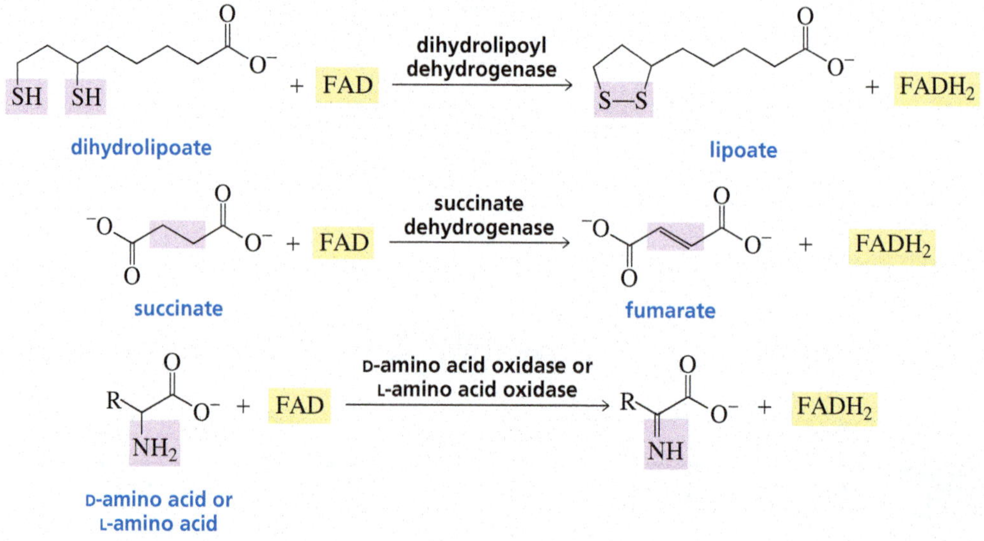

When FAD oxidizes a substrate, the coenzyme is reduced to $FADH_2$. $FADH_2$, like NADPH, is a reducing agent. All the oxidation-reduction chemistry takes place on the part flavin ring colored in yellow. Reduction of the flavin ring disrupts the conjugated system, so the reduced coenzymes are less colored than their oxidized forms (Section 13.22).

FAD is an oxidizing agent.

$FADH_2$ is a reducing agent.

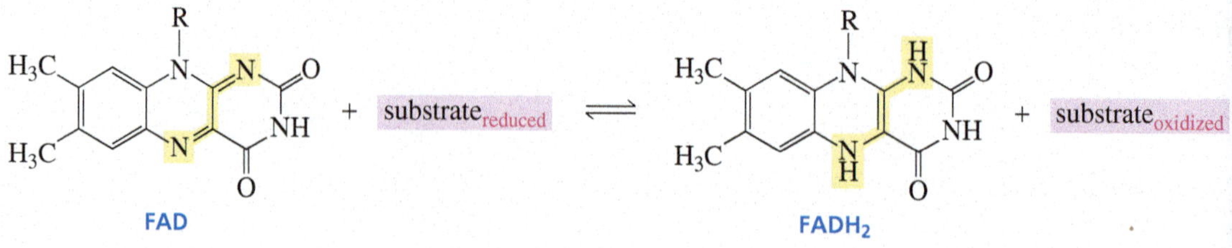

PROBLEM 4 ◆

How many conjugated double bonds are there in

a. FAD?   **b.** $FADH_2$?

Mechanisms for Oxidation by FAD

The mechanism proposed for the FAD-catalyzed oxidation of dihydrolipoate to lipoate is shown on the top of the next page.

MECHANISM PROPOSED FOR THE REACTION CATALYZED BY DIHYDROLIPOYL DEHYDROGENASE

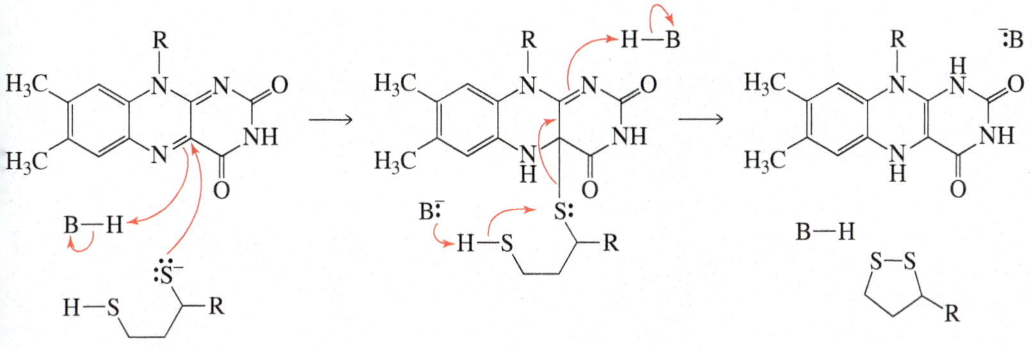

dihydrolipoate

lipoate

- The thiolate ion adds to the 4a position of the flavin ring. An acidic amino acid side chain close to the N-5 nitrogen donates a proton to it.

- A second thiolate ion attacks the sulfur that is covalently attached to the coenzyme, forming the oxidized product and $FADH_2$. A basic side chain removes a proton from the sulfur to make it a better nucleophile, and an acidic side chain protonates the nitrogen.

PROBLEM 5 ◆

Instead of adding to the 4a position and protonating N-5, the thiolate ion could have added to the 10a position and protonated N-1. (The numbering system is on page 1071.) Why is addition to the 4a position favored? (*Hint:* Which nitrogen is a stronger base?)

The mechanism proposed for the FAD-catalyzed oxidation of succinate to fumarate is similar to the mechanism you just saw for the FAD-catalyzed oxidation of dihydrolipoate to lipoate.

MECHANISM PROPOSED FOR THE REACTION CATALYZED BY SUCCINATE DEHYDROGENASE

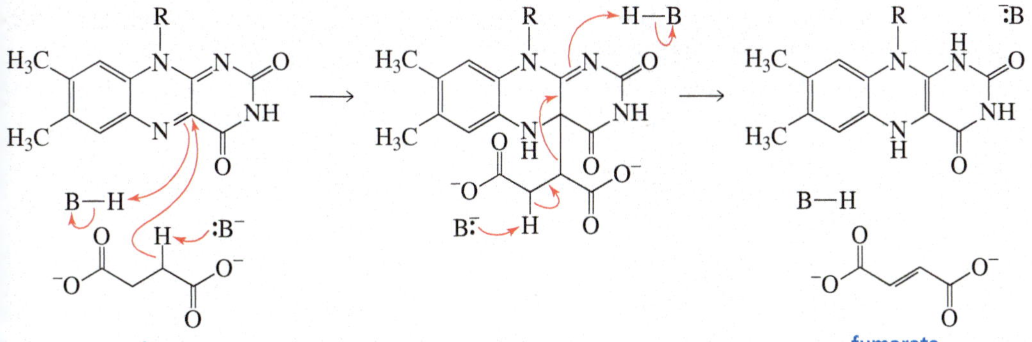

succinate

fumarate

- A base removes a proton from the α-carbon, creating a nucleophile that adds to the 4a position of the flavin ring. A proton is donated simultaneously to the N-5 nitrogen.

- A base removes a proton from the other α-carbon, forming the oxidized product and $FADH_2$.

The mechanism proposed for the FAD-catalyzed oxidation of an amino acid to an imino acid shown on the top of the next page is quite different from the two preceding FAD-catalyzed reactions. Notice that it is a concerted reaction—that is, all bond breaking and bond making occur in the same step.

MECHANISM PROPOSED FOR THE REACTION CATALYZED BY D- OR L-AMINO ACID OXIDASE

an amino acid   an imino acid

- As a basic amino acid side chain removes a proton from the nitrogen of the amino acid, the hydride ion adds to the N-5 position of the flavin ring, and an acidic amino acid side chain protonates N-1

Comparing FAD and NAD$^+$

These mechanisms show that FAD is a more versatile coenzyme than NAD$^+$. Unlike NAD$^+$ that always uses the same mechanism, flavin coenzymes use many different mechanisms to carry out oxidation reactions.

Cells contain very low concentrations of FAD and much higher concentrations of NAD$^+$. NAD$^+$ is only loosely bound to its enzyme and, after being reduced to NADH, dissociates from the enzyme. In contrast, FAD is tightly bound to its enzyme (**E**). If it remains bound after being reduced to FADH$_2$, it has to be reoxidized to FAD before the enzyme can begin another round of catalysis. The oxidizing agent used for this reaction is NAD$^+$ or O$_2$. Therefore, an enzyme that uses FAD instead of NAD$^+$ as an oxidizing agent may still require NAD$^+$ to oxidize the reduced coenzyme.

LEARN THE STRATEGY

PROBLEM 6 SOLVED

FAD is covalently bound to succinate dehydrogenase (the enzyme whose mechanism was shown on the previous page) as a result of a base removing a proton from the C-8 methyl group and an acid donating a proton to N-1. Then a histidine side chain of the enzyme adds to the methylene carbon at C-8 and a proton adds to N-5. Draw the mechanism for these two steps using B:$^-$ and HB for the base and acid, respectively.

SOLUTION

enzyme-FAD   enzyme-FADH$_2$

Notice that FAD is reduced to $FADH_2$ during the process of being attached to the enzyme. It is subsequently oxidized back to FAD. Once the coenzyme is covalently attached to the enzyme, it does not come off.

PROBLEM 7 USE THE STRATEGY

Explain why the hydrogens of the C-8 methyl group are more acidic than those of the C-7 methyl group.

23.3 VITAMIN B₁: THE VITAMIN NEEDED FOR ACYL GROUP TRANSFER

Thiamine was the first of the B vitamins to be identified, so it became known as vitamin B₁. The absence of thiamine in the diet causes a disease called beriberi, which damages the heart, impairs nerve reflexes, and in extreme cases causes paralysis (page 1065).

Vitamin B₁ is used to form the coenzyme **thiamine pyrophosphate (TPP)**. *TPP is the coenzyme required by enzymes that catalyze the transfer of an acyl group from one species to another.*

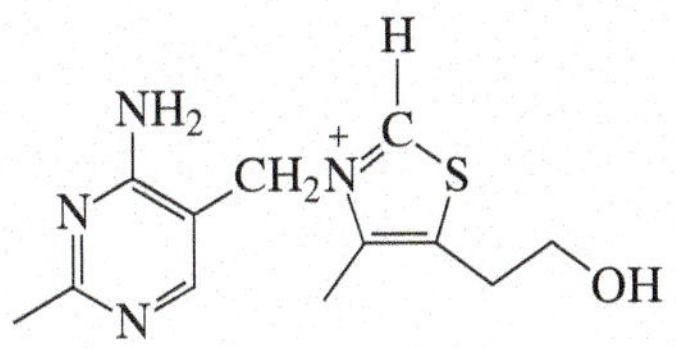

thiamine
vitamin B₁

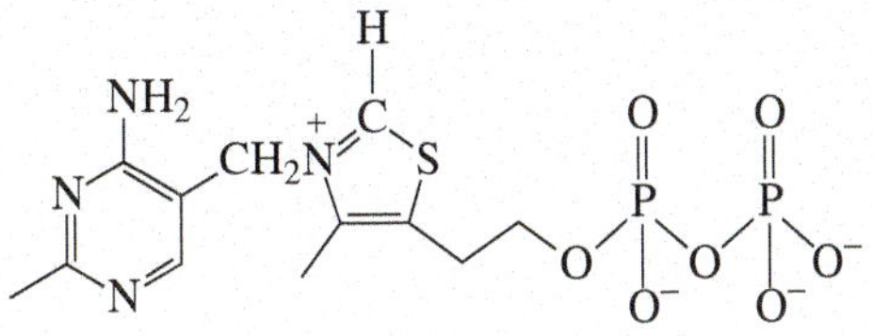

thiamine pyrophosphate
TPP

Pyruvate Decarboxylase Transfers an Acyl Group to H⁺

Pyruvate decarboxylase is an enzyme that requires thiamine pyrophosphate. This enzyme transfers the acyl group of pyruvate to a proton, resulting in the formation of acetaldehyde. A *decarboxylase* is an enzyme that catalyzes the removal of CO_2 from its substrate.

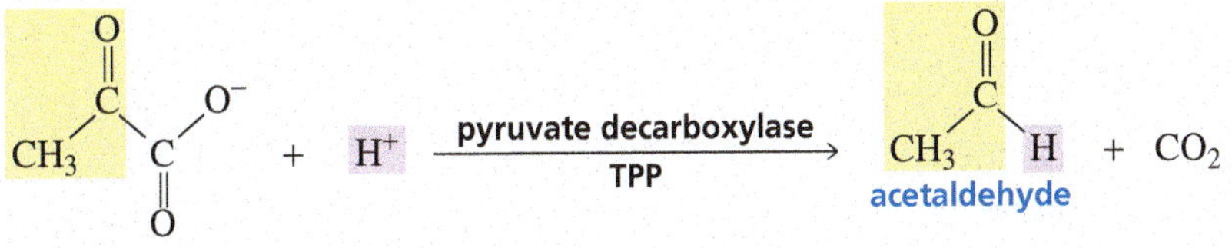

pyruvate
an α-keto acid

acetaldehyde

Thiamine pyrophosphate (TPP) is required by enzymes that catalyze the transfer of an acyl group from one group to another.

You may be wondering how an α-keto acid such as pyruvate can be decarboxylated, since the electrons left behind when CO_2 is removed cannot be delocalized onto the carbonyl oxygen. We will see that the coenzyme provides the site to which the electrons can be delocalized.

The hydrogen bonded to the imine carbon of TPP is relatively acidic ($pK_a = 12.7$) compared to a hydrogen attached to other sp^2 carbons, because the carbanion formed when the proton is removed is stabilized by the adjacent positively charged nitrogen. The TPP-ylide is a good nucleophile. (Recall that an ylide is a species with opposite charges on adjacent atoms that have complete octets.)

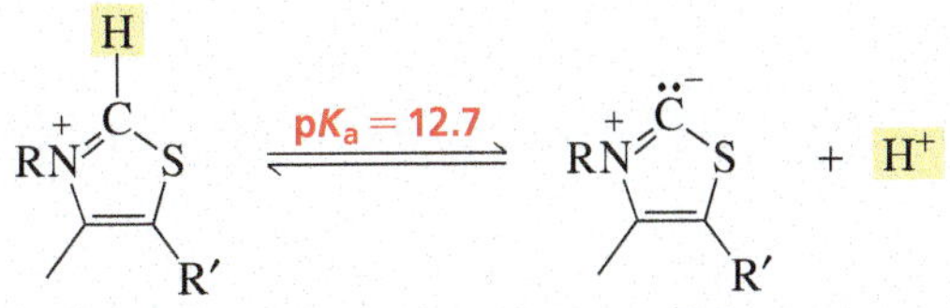

thiazolium ring

TPP-ylide

MECHANISM FOR THE REACTION CATALYZED BY PYRUVATE DECARBOXYLASE

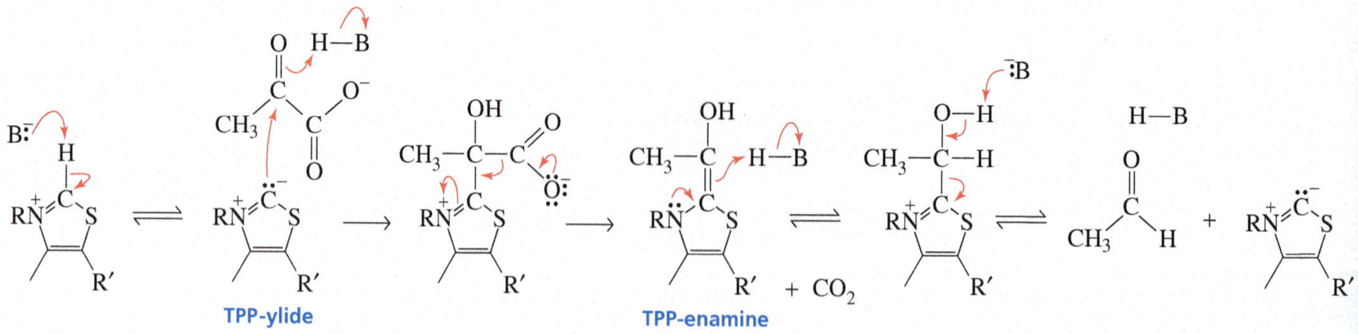

- After the proton is removed, the TPP-ylide adds to the carbonyl carbon of the α-keto acid. An acid side chain of the enzyme increases the electrophilicity of the carbonyl carbon.
- The tetrahedral intermediate thus formed can easily undergo decarboxylation because the electrons left behind when CO_2 is removed can be delocalized onto the positively charged nitrogen. The decarboxylated product is a TPP-enamine.
- Protonation of the TPP-enamine on carbon and a subsequent elimination from the tetrahedral intermediate forms acetaldehyde and regenerates the TPP-ylide.

A site to which electrons can be delocalized is called an **electron sink.** The positively charged nitrogen of TPP is a more effective electron sink than the β-keto group of a β-keto acid, a class of compounds that we have seen are fairly easily decarboxylated (Section 17.16).

PROBLEM 8

Draw structures that show the similarity between the decarboxylation of the pyruvate–TPP intermediate and the decarboxylation of a β-keto acid.

PROBLEM 9

Acetolactate synthase is another TPP-requiring enzyme. It transfers the acyl group of pyruvate to another molecule of pyruvate, forming acetolactate. This is the first step in the biosynthesis of the amino acids valine and leucine. Propose a mechanism for this reaction.

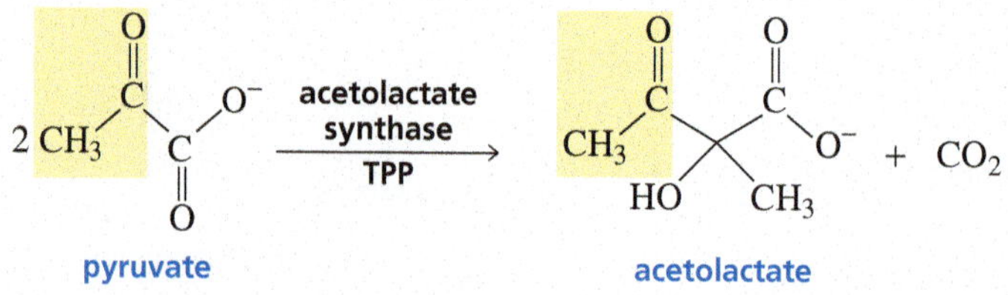

PROBLEM 10

Acetolactate synthase transfers the acyl group of pyruvate to α-ketobutyrate. This is the first step in the biosynthesis of the amino acid isoleucine. Propose a mechanism for this reaction.

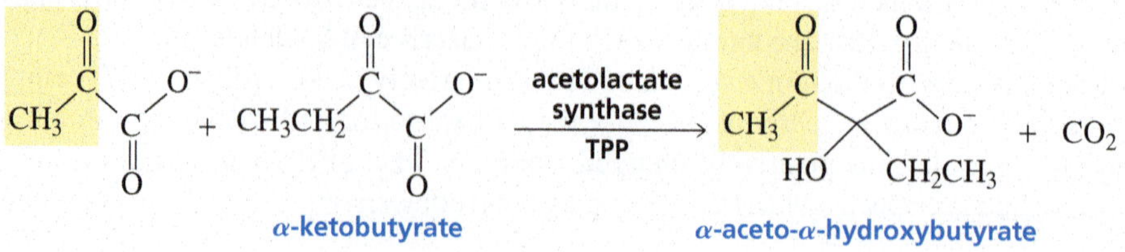

The Pyruvate Dehydrogenase Complex Transfers an Acyl Group to CoASH

We will see in Chapter 24 that the final product of carbohydrate metabolism is pyruvate. For pyruvate to be further metabolized, it must be converted to acetyl-CoA. The *pyruvate dehydrogenase complex* transfers the acyl group of pyruvate to acetyl-CoA.

The pyruvate dehydrogenase complex is a group of three enzymes and five coenzymes: TPP, lipoate, coenzyme A, FAD, and NAD$^+$.

MECHANISM FOR THE REACTION CATALYZED BY THE PYRUVATE DEHYDROGENASE COMPLEX

- The first enzyme in the pyruvate dehydrogenase complex catalyzes the reaction of the TPP-ylide with pyruvate to form the same TPP-enamine as the one formed by pyruvate decarboxylase (the reaction is shown on page 1076) and by the enzymes in Problems 9 and 10.
- The second enzyme of the complex (**E$_2$**) is attached to its coenzyme (**lipoate**) via an amide bond with a lysine side chain. The disulfide bond of lipoate is broken when it undergoes nucleophilic attack by the TPP-enamine.
- The TPP-ylide is eliminated from the tetrahedral intermediate.
- **Coenzyme A (CoASH)** reacts with the thioester in a transthioesterification reaction (one thioester is converted to another), substituting coenzyme A for dihydrolipoate (Section 15.8). At this point, the final reaction product (acetyl-CoA) has been formed.
- Before another catalytic cycle can occur, dihydrolipoate must be oxidized back to lipoate. This is done by the third enzyme (**E$_3$**), an FAD-containing enzyme. We saw the mechanism for this reaction in Section 23.2. When dihydrolipoate is oxidized by FAD, the coenzyme is reduced to FADH$_2$.
- NAD$^+$ oxidizes FADH$_2$ back to FAD.

Coenzyme A

Coenzyme A is a thiol. It is written as CoASH to emphasize that it acts as a thiol in chemical reactions. For example, we saw that CoASH is used in cells to activate carboxylic acids by converting them to thioesters (Section 15.19).

The vitamin needed to make coenzyme A is pantothenate (see Table 23.1). In coenzyme A, pantothenate is attached to a decarboxylated cysteine and a phosphorylated ADP.

decarboxylated cysteine

pantothenate

phosphate

coenzyme A
CoASH

ADP

Curing a Hangover With Vitamin B₁

An unfortunate effect of drinking too much alcohol, known as a hangover, is attributable to the acetaldehyde formed when ethanol is oxidized (Section 10.5). Vitamin B₁ can cure a hangover by getting rid of acetaldehyde. Let's see how the vitamin is able to do this.

The TPP-ylide adds to the carbonyl carbon of acetaldehyde. Removal of a proton forms the same TPP-enamine that is formed by both pyruvate decarboxylase and the pyruvate dehydrogenase complex—the only difference in the reactions is that a proton, instead of a carboxyl group, is removed from the substrate. The TPP-enamine then reacts with lipoate just as it does in the pyruvate dehydrogenase complex. The result is that the offending acetaldehyde is converted to acetyl-CoA.

TPP-enamine

There is a limit to the amount of acetaldehyde that can be converted to acetyl-CoA in a given amount of time, so the vitamin can cure only hangovers that result from moderate drinking.

LEARN THE STRATEGY

PROBLEM 11 SOLVED

TPP is a coenzyme for transketolase, the enzyme that catalyzes the conversion of a ketopentose (xylulose-5-phosphate) and an aldopentose (ribose-5-phosphate) to an aldotriose (glyceraldehyde-3-phosphate) and a ketoheptose (sedoheptulose-7-phosphate). Notice that the total number of carbons in the reactants and products is the same (5 + 5 = 3 + 7). Propose a mechanism for this reaction.

xylulose-5-P   ribose-5-P $\xrightarrow[\text{TPP}]{\text{transketolase}}$ glyceraldehyde-3-P   sedoheptulose-7-P

SOLUTION The reaction shows that an acyl group (the purple box) is transferred from xylulose-5-phosphate to ribose-5-phosphate. Because the enzyme requires TTP, we know that TPP must be the species that removes and transfers the acyl group. Thus, the reaction must start with the addition of the TPP-ylide to the carbonyl group of xylulose-5-phosphate. We can add an acid group to accept the electrons from the carbonyl group and a basic group to help form the TPP-enamine. Notice that, like in other TPP-catalyzed reactions, the electrons left behind by the group that is removed from the acyl group are delocalized onto the nitrogen of the thiazolium ring. The enamine then transfers the acyl group to the carbonyl carbon of ribose-5-phosphate. Again, an acid group accepts the electrons from the carbonyl group, and a basic group helps eliminate the TPP-ylide.

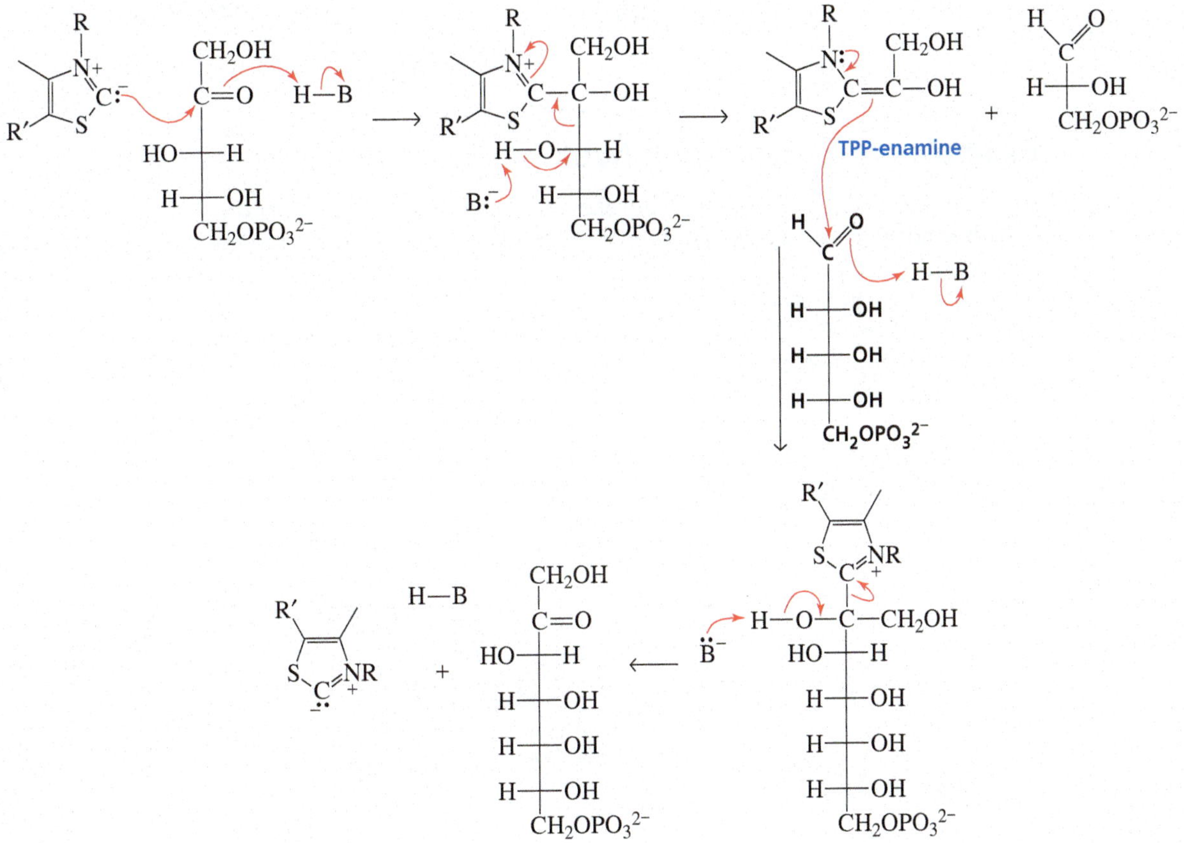

PROBLEM 12

USE THE STRATEGY

a. What acyl group does pyruvate decarboxylase transfer to a proton?
b. What acyl group does the pyruvate dehydrogenase complex transfer to coenzyme A?
c. What acyl group does transketolase transfer to ribose-5-P?

Mechanistic Similarities

Notice the similar function of TPP in all TPP-requiring enzymes. In each reaction, the TPP-ylide adds to a carbonyl carbon of the substrate and allows a bond to that carbon to be broken because the electrons left behind can be delocalized onto the nitrogen of the thiazolium ring. The acyl group is then transferred—to a proton in the case of pyruvate decarboxylase, to coenzyme A (via lipoate) in the pyruvate dehydrogenase system, and to a carbonyl group in Problems 9, 10, and 11.

23.4 BIOTIN: THE VITAMIN NEEDED FOR CARBOXYLATION OF AN α-CARBON

Biotin (also known as vitamin B_7 or vitamin H) is an unusual vitamin because it is synthesized by microorganisms that live in the intestinal tract. Consequently, biotin does not have to be included in our diet and deficiencies are rare. Biotin deficiencies, however, can be found in people who maintain a diet high in raw eggs. Egg whites contain a protein (called avidin) that binds biotin tightly and thereby prevents it from acting as a coenzyme. When eggs are cooked, avidin is denatured and the denatured protein does not bind biotin. Biotin, like lipoic acid, is attached to its enzyme (E) by forming an amide with the amino group of a lysine side chain.

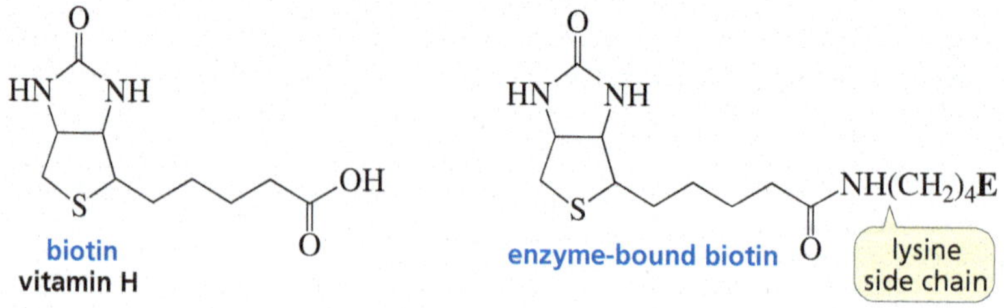

biotin
vitamin H

enzyme-bound biotin

lysine side chain

Examples of Enzyme-Catalyzed Reactions that Require Biotin

Biotin is the coenzyme required by enzymes that catalyze the carboxylation of an α-carbon (a carbon adjacent to a carbonyl group). Therefore, the enzymes that require biotin as a coenzyme are called carboxylases.

For example, pyruvate carboxylase converts pyruvate to oxaloacetate and acetyl-CoA carboxylase converts acetyl-CoA into malonyl-CoA. Biotin-requiring enzymes use bicarbonate (HCO_3^-) for the source of the carboxyl group that becomes attached to the α-carbon of the substrate.

pyruvate

oxaloacetate

a carboxyl group has been added to the α-carbon

Biotin is required by enzymes that catalyze the carboxylation of an α-carbon.

acetyl-CoA

malonyl-CoA

a carboxyl group has been added to the α-carbon

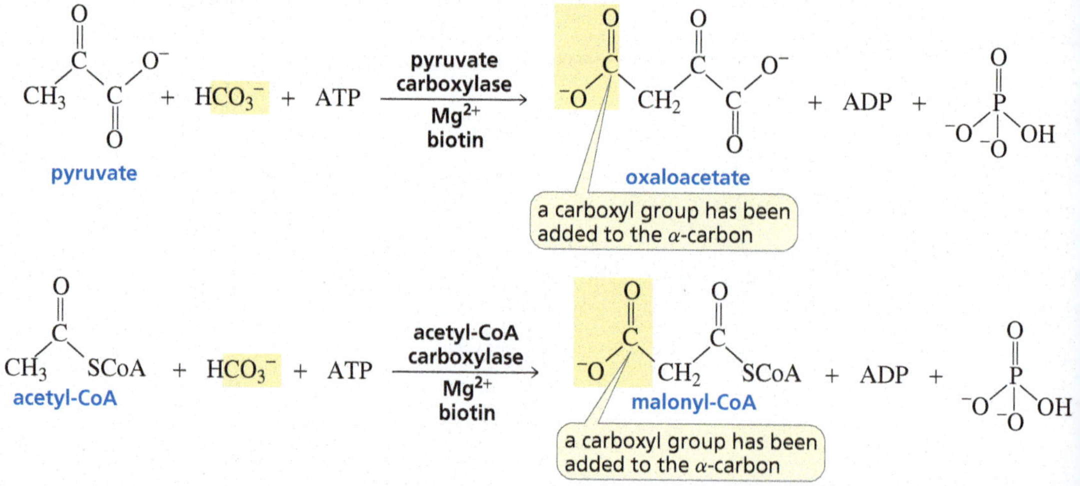

In addition to bicarbonate, biotin-requiring enzymes also require ATP and Mg^{2+}. The function of Mg^{2+} is to decrease the overall negative charge on ATP by complexing with two of its negatively charged oxygens. Unless its negative charge is reduced, ATP cannot be approached by a nucleophile (see Figure 15.3 on page 725).

Forming Activated Bicarbonate

The function of ATP is to increase the reactivity of bicarbonate by converting it to "activated bicarbonate"—a compound with a good leaving group. To form "activated bicarbonate," bicarbonate attacks the γ-phosphorous of ATP and expels ADP (Section 15.19). Notice that "activated bicarbonate" is a mixed anhydride of carbonic acid and phosphoric acid (Section 15.16).

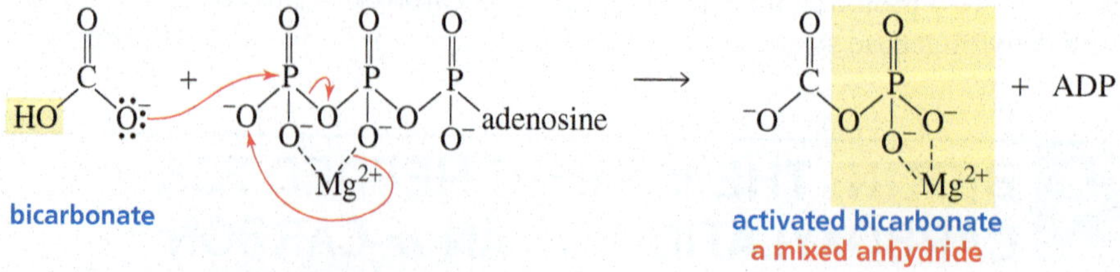

bicarbonate

activated bicarbonate
a mixed anhydride

The Mechanism for Carboxylation

Once bicarbonate has been activated, the catalytic reaction can begin. The mechanism for the carboxylation of acetyl CoA is shown on the top of the next page.

MECHANISM FOR THE REACTION CATALYZED BY ACETYL-CoA CARBOXYLASE

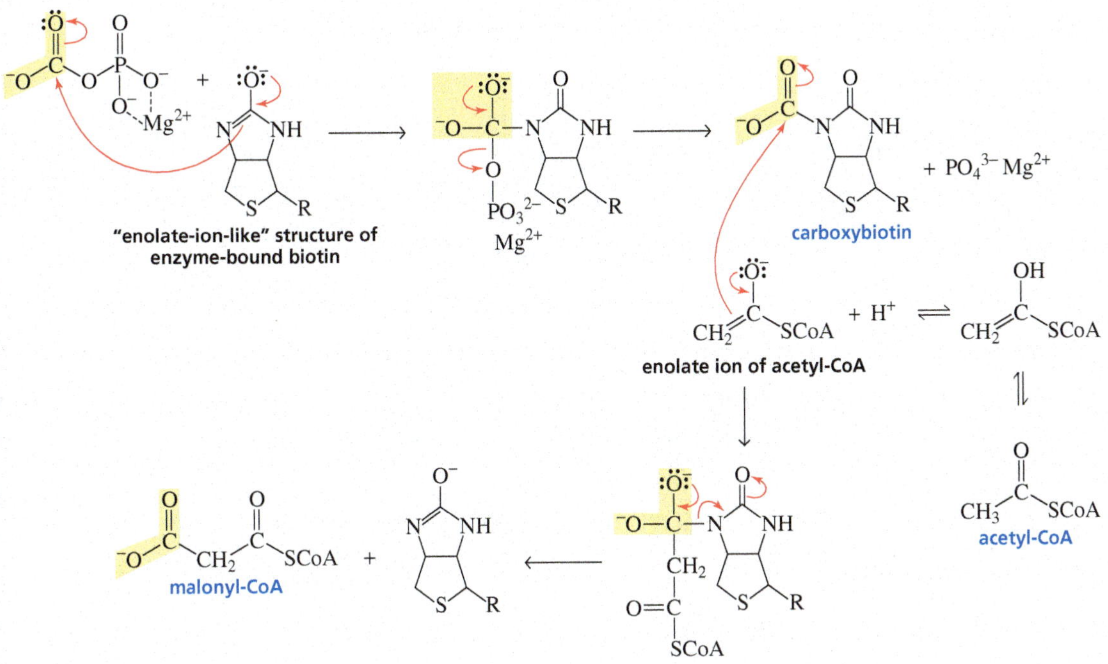

"enolate-ion-like" structure of
enzyme-bound biotin

carboxybiotin

enolate ion of acetyl-CoA

acetyl-CoA

malonyl-CoA

- Biotin reacts with activated bicarbonate in a nucleophilic acyl substitution reaction to form carboxybiotin. Because the nitrogen of an amide is not nucleophilic, the active form of biotin has an enolate-ion–like structure.

- The substrate (in this case, the enolate ion of acetyl-CoA) reacts with carboxybiotin in a second nucleophilic acyl substitution reaction that transfers the carboxyl group from carboxybiotin to the substrate and re-forms the enolate ion.

All biotin-requiring enzymes follow the same three steps: activation of bicarbonate by ATP, reaction of activated bicarbonate with biotin to form carboxybiotin, and transfer of the carboxyl group from carboxybiotin to the substrate.

23.5 VITAMIN B$_6$: THE VITAMIN NEEDED FOR AMINO ACID TRANSFORMATIONS

The coenzyme **pyridoxal phosphate (PLP)** is derived from vitamin B$_6$, which is also known as pyridoxine. Pyridoxal's "al" suffix indicates the coenzyme is an aldehyde. A deficiency in vitamin B$_6$ causes anemia; severe deficiencies can cause seizures and death.

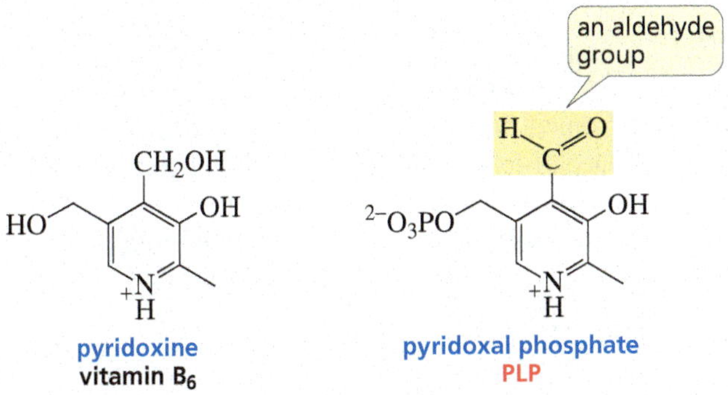

an aldehyde group

pyridoxine
vitamin B$_6$

pyridoxal phosphate
PLP

Examples of Enzyme-Catalyzed Reactions that Require PLP

PLP is required by enzymes that catalyze certain reactions of amino acids, such as decarboxylation, transamination, racemization, C$_\alpha$—C$_\beta$ bond cleavage, and α,β-elimination.

decarboxylation

$$R-CH(^{+}NH_3)-COO^- \xrightarrow[\text{PLP}]{\text{E}} R-CH_2-\overset{+}{N}H_3 \;+\; CO_2$$

transamination

$$R-CH(^{+}NH_3)-COO^- \;+\; \text{(α-ketoglutarate)} \xrightarrow[\text{PLP}]{\text{E}} R-CO-COO^- \;+\; \text{(glutamate)}$$

α-ketoglutarate   glutamate

racemization

$$\underset{\text{L-amino acid}}{R-\overset{H}{\underset{^{+}NH_3}{C}}-COO^-} \xrightarrow[\text{PLP}]{\text{E}} \underset{\text{L-amino acid}}{R-\overset{H}{\underset{^{+}NH_3}{C}}-COO^-} \;+\; \underset{\text{D-amino acid}}{R-\overset{^{+}NH_3}{\underset{H}{C}}-COO^-}$$

L-amino acid   L-amino acid   D-amino acid

C_α—C_β bond cleavage

$$\underset{^{+}NH_3}{R-\overset{OH}{C}H-CH-COO^-} \xrightarrow[\text{PLP}]{\text{E}} R-CHO \;+\; H_3\overset{+}{N}-CH_2-COO^-$$

α,β-elimination

$$\underset{^{+}NH_3}{X-CH_2-CH-COO^-} \xrightarrow[\text{PLP}]{\text{E}} CH_3-CO-COO^- \;+\; X^- \;+\; \overset{+}{N}H_4$$

Transimination

PLP becomes attached to its enzyme by forming an imine with the amino group of a lysine side chain.

The first step in all reactions catalyzed by PLP-requiring enzymes is a **transimination** reaction—a reaction that converts one imine to another imine. Thus, the amino acid substrate reacts with the imine formed by PLP and the lysine side chain, forming an imine with PLP and releasing the lysine side chain.

Pyridoxal phosphate (PLP) is required by enzymes that catalyze certain reactions of amino acids.

the coenzyme is bound to the enzyme by means of an imine linkage with a lysine side chain

amino acid   imine between PLP and the lysine side chain   imine between PLP and the amino acid   lysine side chain

Once the amino acid has formed an imine with PLP, the next step is to break a bond to the α-carbon of the amino acid. Decarboxylation breaks the bond joining the carboxyl group to the α-carbon; transamination, racemization, and α,β-elimination break the bond joining the hydrogen to the α-carbon; and C_α—C_β bond cleavage breaks the bond joining the R group to the α-carbon.

bond broken in decarboxylation

$$H_2N-\overset{COO^-}{\underset{R}{C}}-H$$

bond broken in C_α—C_β bond cleavage

bond broken in transamination, racemization, and α,β-elimination

A bond to the α-carbon can be broken because the electrons left behind when the bond breaks can be delocalized onto the positively charged protonated nitrogen of the pyridinium ring. Thus, the protonated nitrogen is an electron sink.

The coenzyme loses much of its activity if the OH group is removed from the pyridine ring. Apparently, the hydrogen bond formed by the OH group helps weaken the bond to the α-carbon.

Decarboxylation

All enzymes that catalyze the decarboxylation of an amino acid do so by the mechanism shown below.

MECHANISM FOR THE DECARBOXYLATION OF AN AMINO ACID

imine between PLP and the amino acid   + CO₂   imine between PLP and the product   imine between PLP and the lysine side chain

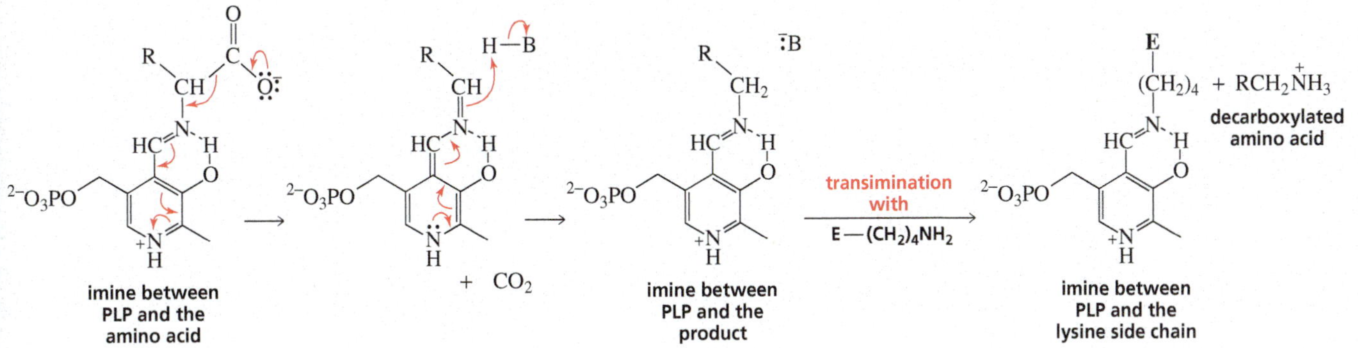

- The carboxyl group is removed from the α-carbon of the amino acid, and the electrons left behind are delocalized onto the positively charged nitrogen.

- The aromaticity of the pyridinium ring is reestablished by electron rearrangement, and what was the α-carbon of the amino acid is protonated.

- The last step in all PLP-requiring enzymes is another transimination reaction. The imine formed by PLP and the product of the reaction reacts with the lysine side chain of the enzyme, forming an imine with PLP and releasing the product.

Racemization

The mechanism for the racemization of an L-amino acid is shown next. Notice that the mechanism is the same as the mechanism for decarboxylation except for the group removed from the α-carbon of the amino acid in the first step.

MECHANISM FOR THE RACEMIZATION OF AN L-AMINO ACID

imine between PLP and the amino acid   racemized amino acid

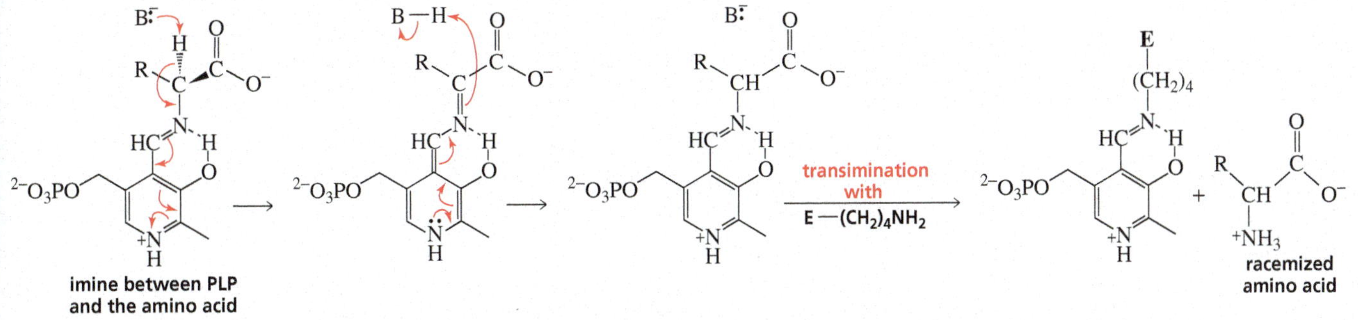

- A proton is removed from the α-carbon of the amino acid, and the electrons left behind are delocalized onto the positively charged nitrogen.

- The aromaticity of the pyridinium ring is reestablished by electron rearrangement, and what was the α-carbon of the amino acid is protonated.

- A transimination reaction with a lysine side chain releases the product of the reaction (the racemized amino acid) and regenerates the imine between the enzyme and PLP.

In the second step of the reaction, the proton can be donated to the sp^2 carbon from either side of the plane defined by the double bond. Consequently, both D- and L-amino acids are formed. In other words, the L-amino acid is racemized.

Transamination

The first reaction in the catabolism of most amino acids is replacement of the amino group of the amino acid by a ketone group. This is called a **transamination** reaction because the amino group removed from the amino acid is not lost, but is *transferred* to the ketone group of α-ketoglutarate, thereby forming glutamate.

NOTE TO THE STUDENT

• **Do not confuse** *transamination* **with** *transimination*, **discussed previously.**

The enzymes that catalyze transamination reactions are called *aminotransferases*. Each amino acid has its own aminotransferase. Transamination allows the amino groups of the various amino acids to be collected into a single amino acid (glutamate) so that excess nitrogen can be easily excreted.

MECHANISM FOR THE TRANSAMINATION OF AN AMINO ACID

• In the first step, a proton is removed from the α-carbon of the amino acid, and the electrons left behind are delocalized onto the positively charged nitrogen.

• The aromaticity of the pyridinium ring is reestablished by electron rearrangement, and the carbon attached to the pyridine ring is protonated.

• Hydrolysis of the imine forms the transaminated amino acid (an α-keto acid) and pyridoxamine (a primary amine).

At this point, the amino group has been removed from the amino acid but the coenzyme has been converted to a primary amine. The primary amine has to be converted back to a carbonyl group that can form an imine with the lysine side chain of the enzyme so another round of catalysis can occur.

- Pyridoxamine forms an imine with α-ketoglutarate, the second substrate of the reaction.
- A proton is removed from the carbon attached to the pyridine ring, and the electrons left behind are delocalized onto the positively charged nitrogen.
- The aromaticity of the pyridinium ring is reestablished by electron rearrangement, and the α-carbon is protonated.
- A transimination reaction with a lysine side chain releases the product of the reaction (glutamate) and regenerates the imine between PLP and the lysine side chain of the enzyme.

Notice that the two proton transfer steps are reversed in the two phases of the reaction. Transfer of the amino group of the amino acid to PLP requires removal of the proton from the α-carbon (of the amino acid) and donation of a proton to the carbon bonded to the pyridine ring. The steps are reversed in the transfer of the amino group of pyridoxamine to α-ketoglutarate: a proton is removed from the carbon bonded to the pyridinium ring and a proton is donated to the α-carbon (of α-ketoglutarate).

Compare the second step in a PLP-catalyzed transamination with the second step in a PLP-catalyzed decarboxylation or racemization. In an enzyme that catalyzes transamination, an acidic group at the active site is in position to donate a proton to the carbon attached to the pyridinium ring. The enzyme that catalyzes decarboxylation or racemization does not have this acidic group, so the substrate is reprotonated at the α-carbon. In other words, the *coenzyme* carries out the chemical reaction, but the *enzyme* determines the course of the reaction.

Assessing the Damage After a Heart Attack

After a heart attack, aminotransferases and other enzymes leak from the damaged cells of the heart into the bloodstream. The severity of damage done to the heart can be determined from the concentrations of alanine aminotransferase and aspartate aminotransferase in the blood.

PROBLEM 13 ◆

Which compound is more easily decarboxylated?

or

PROBLEM 14 ◆

α-Keto acids other than α-ketoglutarate can accept the amino group from pyridoxamine in enzyme-catalyzed transamination reactions. What amino acids are formed when the following α-keto acids accept the amino group?

a. pyruvate   b. oxaloacetate

PROBLEM 15 ◆

Explain why the ability of PLP to catalyze an amino acid transformation is greatly reduced if a PLP-requiring enzymatic reaction is carried out at a pH at which the pyridine nitrogen is not protonated.

PROBLEM 16 ◆

Explain why the ability of PLP to catalyze an amino acid transformation is greatly reduced if the OH substituent of pyridoxal phosphate is replaced by OCH$_3$.

PROBLEM 17

The enzyme that catalyzes the C_α—C_β bond cleavage reaction that converts serine to glycine removes the substituent (R) bonded to the α-carbon in the first step of the reaction. Starting with PLP bound to serine in an imine linkage, propose a mechanism for this reaction. (*Hint:* The first step involves removal of the proton from serine's OH group.)

PROBLEM 18

Propose a mechanism for the α,β-elimination reaction shown on page 1082.

23.6 VITAMIN B$_{12}$: THE VITAMIN NEEDED FOR CERTAIN ISOMERIZATIONS

Enzymes that catalyze certain isomerization reactions require **coenzyme B$_{12}$**, a coenzyme derived from vitamin B$_{12}$. The structure of vitamin B$_{12}$ was determined by Dorothy Crowfoot Hodgkin using X-ray crystallography (Section 14.24). The vitamin has a cyano group (or HO^- or H_2O) coordinated with cobalt. This group is replaced by a 5′-deoxyadenosyl group in the coenzyme.

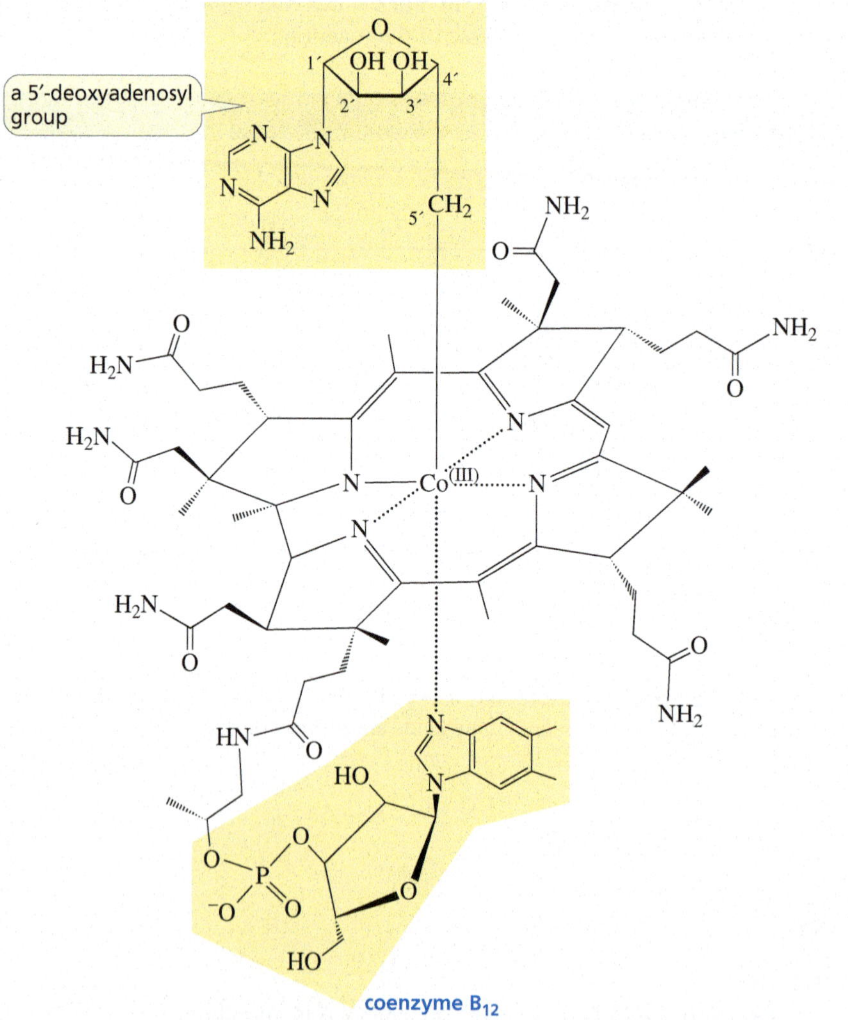

coenzyme B$_{12}$

Animals and plants cannot synthesize vitamin B$_{12}$. In fact, only a few species of bacteria found in the stomachs of cows and other animals can synthesize it. Humans must obtain all their vitamin B$_{12}$ from their diet, particularly from meat. A deficiency causes megoblastic anemia—the creation of unusually large and dysfunctional red blood cells in the bone marrow. Because humans need only a small amount of vitamin B$_{12}$, deficiencies caused by the consumption of an insufficient amount of the vitamin are rare but have been found in vegetarians who eat no animal products. Most deficiencies are caused by the inability of the intestines to absorb the vitamin.

Examples of Enzyme-Catalyzed Reactions that Require Coenzyme B₁₂

In each of these coenzyme B₁₂–requiring reactions, a hydrogen bonded to one carbon exchanges places with a group (Y) bonded to an adjacent carbon.

> Coenzyme B₁₂ is required by enzymes that catalyze the exchange of a hydrogen bonded to one carbon with a group bonded to an adjacent carbon.

For example, methylmalonyl-CoA mutase catalyzes a reaction in which the H bonded to one carbon changes places with an $C(=O)SCoA$ group bonded to an adjacent carbon. (A mutase is an enzyme that transfers a group from one position to another.) In the reaction catalyzed by dioldehydrase, an H and an OH change places.

The Mechanism for a B₁₂-Requiring Enzyme

The chemistry of coenzyme B₁₂ takes place at the bond joining the cobalt and the 5′-deoxyadenosyl group, which is an unusually weak bond (26 kcal/mol compared with 99 kcal/mol for a $C-H$ bond).

MECHANISM FOR THE REACTION CATALYZED BY DIOLDEHYDRASE

- The Co—C bond breaks homolytically, forming a 5′-deoxyadenosyl radical and reducing Co(III) to Co(II).

LEARN THE STRATEGY

- The 5′-deoxyadenosyl radical removes the hydrogen atom (in this case, from the C-1 carbon) that will change place with another group, thereby becoming 5′-deoxyadenosine.
- A hydroxyl radical (·OH) migrates from C-2 to C-1, creating a radical at C-2.
- The radical at C-2 removes a hydrogen atom from 5′-deoxyadenosine, forming the rearranged product and regenerating the 5′-deoxyadenosyl radical.
- The 5′-deoxyadenosyl radical recombines with Co(II) to regenerate the coenzyme. The unstable hydrate loses water to form an aldehyde. The enzyme–coenzyme complex is then ready for another substrate molecule.

It is likely that all coenzyme B_{12}–requiring enzymes catalyze reactions by the same general mechanism. The role of the coenzyme is to provide a way to remove a hydrogen atom from the substrate. Once the hydrogen atom has been removed, an adjacent group can migrate to take its place. The coenzyme then gives back the hydrogen atom, delivering it to the carbon that lost the migrating group.

USE THE STRATEGY

PROBLEM 19

Ethanolamine ammonia lyase, a coenzyme B_{12}–requiring enzyme, catalyzes the following reaction. Propose a mechanism for this reaction.

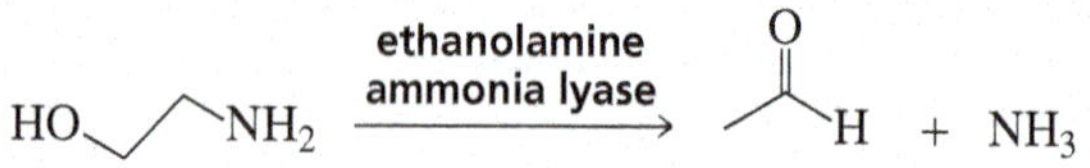

$$HO\diagdown\diagup NH_2 \xrightarrow[\text{ammonia lyase}]{\text{ethanolamine}} \overset{O}{\diagup}\diagdown H \; + \; NH_3$$

PROBLEM 20 ◆

A fatty acid (a long straight-chain carboxylic acid with an even number of carbons) is metabolized to acetyl-CoA, which can then enter the citric acid cycle to be further metabolized (Section 24.9). A fatty acid with an odd number of carbons is metabolized to acetyl-CoA and one equivalent of propionyl-CoA. Propionyl-CoA cannot enter the citric acid cycle. Two coenzyme-requiring enzymes are needed to convert it to succinyl-CoA, a compound that can enter the citric acid cycle. Write the two enzyme-catalyzed reactions and include the names of the required coenzymes.

23.7 FOLIC ACID: THE VITAMIN NEEDED FOR ONE-CARBON TRANSFER

Tetrahydrofolate (THF) is the coenzyme used by enzymes that catalyze reactions that transfer a group containing a single carbon to their substrates. The one-carbon group can be a methyl group (CH_3), a methylene group (CH_2), or a formyl group ($HC\!=\!O$). Tetrahydrofolate is produced by the reduction of two double bonds of folic acid (folate), its precursor vitamin. Bacteria can synthesize folate, but mammals cannot.

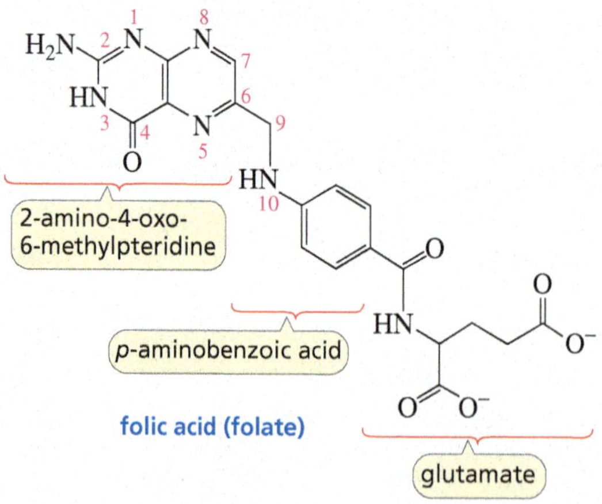

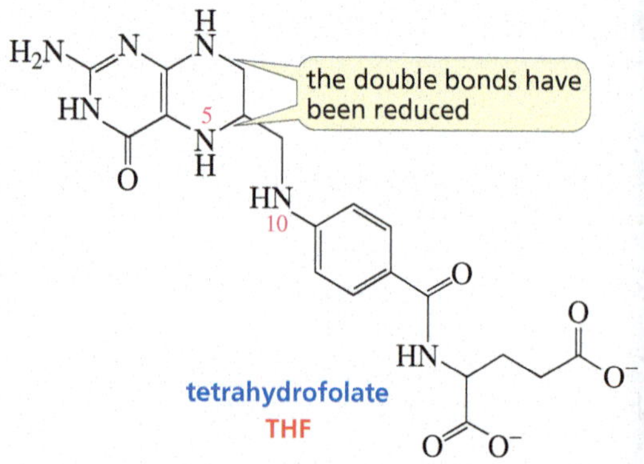

Three THF-coenzymes are shown below. R stands for the group attached to N^{10} in the above structure.

- N^{10}-formyl-THF transfers a formyl group.
- N^5-methyl-THF transfers a methyl group.
- N^5,N^{10}-methylene-THF transfers a methylene group.

A tetrahydrofolate (THF) coenzyme is required by enzymes that catalyze the transfer of a group containing one carbon to their substrates.

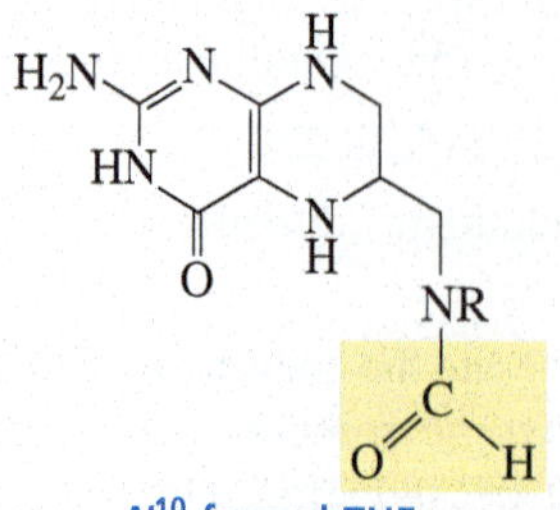

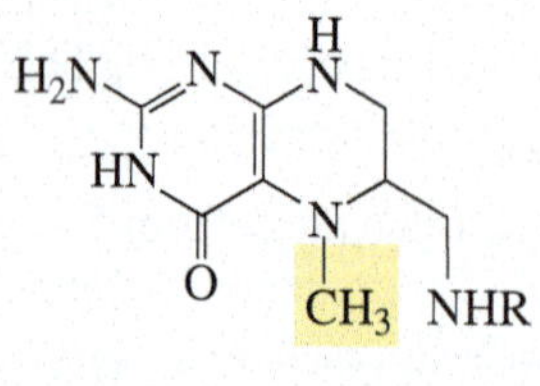

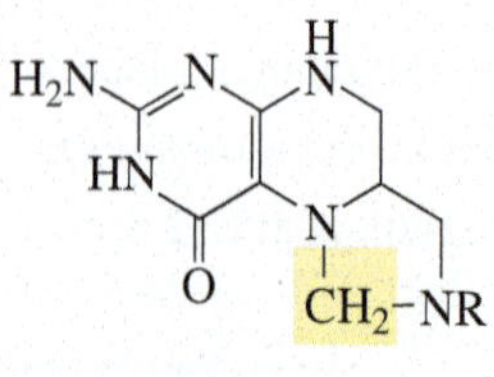

Examples of Enzyme-Catalyzed Reactions that Require a THF-Coenzyme

Glycinamide ribonucleotide (GAR) transformylase is an enzyme that requires the coenzyme N^{10}-formyl-THF. The formyl group that is given to the substrate eventually ends up being the C-8 carbon of two of the four heterocyclic bases found in DNA and RNA (Section 26.1).

ribose-5-phosphate   ribose-5-phosphate   purine

Homocysteine methyl transferase, an enzyme required for the synthesis of methionine (an amino acid), also requires N^5-methyl-THF.

homocysteine   methionine

The First Antibiotics

Sulfonamides—commonly known as sulfa drugs—were introduced clinically in 1934 as the first effective antibiotics (Section 18.19). Donald Woods, a British bacteriologist, noticed that sulfanilamide, initially the most widely used sulfonamide, was structurally similar to p-aminobenzoic acid, a compound necessary for bacterial growth.

a sulfonamide   sulfanilamide   p-aminobenzoic acid

This suggested that sulfanilamide acts by inhibiting the enzyme that incorporates p-aminobenzoic acid into folic acid. Because the enzyme cannot tell the difference between sulfanilamide and p-aminobenzoic acid, both compounds compete for the active site of the enzyme. Without sufficient folic acid, the bacteria die. Humans are not adversely affected by the drug because they do not synthesize folate; instead, they get all their folate from their diet.

Thymidylate Synthase: The Enzyme that Converts U to T

The heterocyclic bases in RNA are adenine, guanine, cytosine, and uracil (A, G, C, and U). The heterocyclic bases in DNA are adenine, guanine, cytosine, and thymine (A, G, C, and T). In other words, the heterocyclic bases in RNA and DNA are the same, with one exception: RNA contains U, whereas DNA contains T. (These bases are described in Section 26.1. Why DNA contains T instead of U is explained in Section 26.10.)

The Ts used for the biosynthesis of DNA are synthesized from Us by thymidylate synthase, an enzyme that requires the coenzyme N^5, N^{10}-methylene-THF. The actual substrate is dUMP (2'-deoxyuridine 5'-monophosphate) and the product is dTMP (2'-deoxythymidine 5'-monophosphate).

2'-deoxyuridine 5'-monophosphate N^5, N^{10}-methylene-THF   2'-deoxythymidine 5'-monophosphate   dihydrofolate DHF

dUMP   dTMP

R = 2'-deoxyribose-5'-phosphate   R = 2'-deoxyribose-5'-phosphate

Even though the only structural difference between T and U is a *methyl* group, T is synthesized by first transferring a *methylene* group to a U. The methylene group is then reduced to a methyl group. The mechanism of the reaction is shown below.

MECHANISM FOR THE REACTION CATALYZED BY THYMIDYLATE SYNTHASE

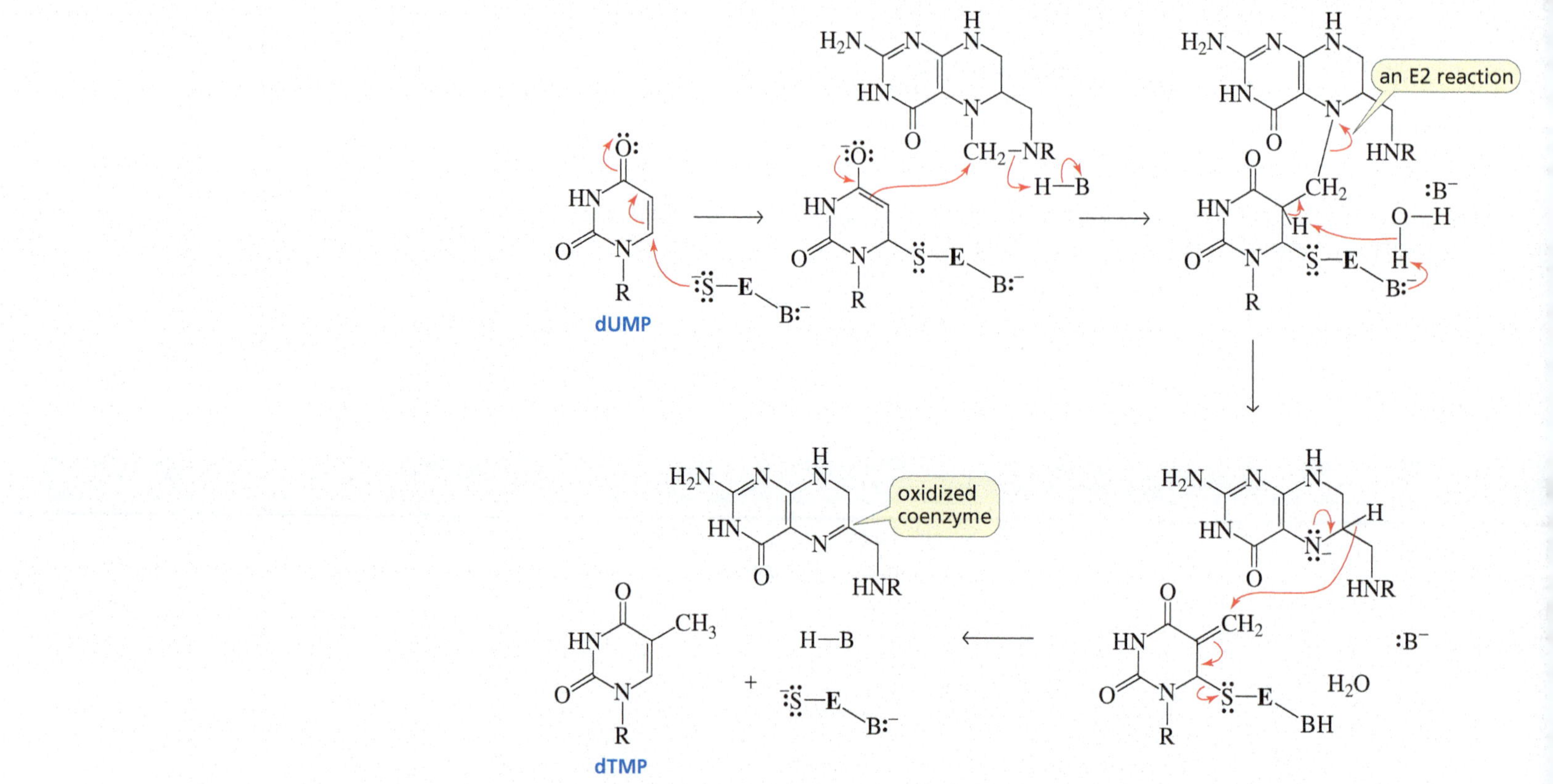

- A nucleophilic cysteine at the active site of the enzyme adds to the β-carbon of dUMP. (This is an example of conjugate addition; see Section 16.15.)
- Nucleophilic attack by the enolate ion of dUMP on the methylene group of N^5,N^{10}-methylene-THF forms a covalent bond between dUMP and the coenzyme. This is an S_N2 reaction. The leaving group has to be protonated to improve its leaving propensity.
- A base removes a proton from the α-carbon and the coenzyme is eliminated in an E2 reaction. The base is thought to be a water molecule whose basicity is increased by the O^- group of a tyrosine side chain of the enzyme (here written as :B⁻).
- Transfer of a hydride ion from the coenzyme to the methylene group and elimination of the enzyme forms dTMP and dihydrofolate (DHF).

Notice that the coenzyme that transfers the methylene group to the substrate is also the reagent that subsequently reduces the methylene group to a methyl group. Because the coenzyme is the reducing agent, it is oxidized to dihydrofolate. (Recall that oxidation decreases the number of C—H bonds.) Dihydrofolate must then be reduced back to tetrahydrofolate so that tetrahydrofolate continues to be available as a coenzyme. Dihydrofolate reductase is the enzyme that catalyzes this reaction.

$$\text{dihydrofolate} + \text{NADPH} + \text{H}^+ \xrightarrow{\substack{\text{dihydrofolate} \\ \text{reductase}}} \text{tetrahydrofolate} + \text{NADP}^+$$

Cancer Chemotherapy

Cancer is the uncontrolled growth and proliferation of cells. Because cells cannot multiply if they cannot synthesize DNA, scientists have long searched for compounds that would inhibit either thymidylate synthase or dihydrofolate reductase. If a cell cannot make Ts, it cannot synthesize DNA. Inhibiting dihydrofolate reductase also prevents the synthesis of Ts because cells have a limited amount of tetrahydrofolate. If they cannot convert dihydrofolate back to tetrahydrofolate, they cannot continue to synthesize Ts.

5-Fluorouracil is a common anticancer drug that inhibits thymidylate synthase. The enzyme reacts with 5-fluorouracil in the same way it reacts with dUMP. However, the fluorine of 5-fluorouracil causes the coenzyme to become permanently attached to the active site of the enzyme.

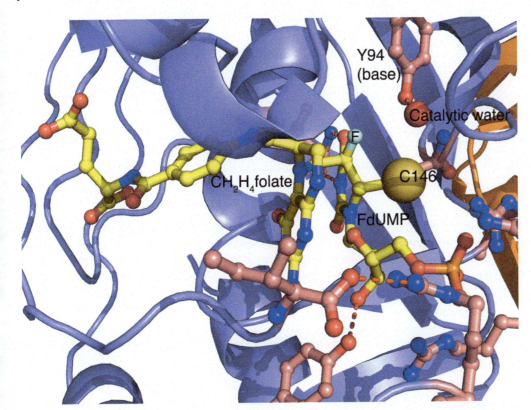

Fluorine is too electronegative to come off as F^+ in the elimination reaction (compare the illustration above with the third step in the mechanism on page 1090). Consequently, the reaction stops, leaving the enzyme permanently attached to 5-fluorouracil (Figure 23.1). Because the active site of the enzyme is now blocked with 5-fluorouracil, it cannot bind dUMP. Therefore, dTMP can no longer be synthesized and, without dTMP, DNA cannot be synthesized. Consequently, cancer cells undergo "thymineless" death.

◀ **Figure 23.1**
The tetrahydrofolate coenzyme and 5-fluoro-dUMP that is covalently bonded to cysteine 146 of the enzyme are shown in yellow. The fluorine (turquoise), tyrosine 94 tyr 94 (Y=tyrosine), and the catalytic water molecule are also visible. The side chains that hold the coenzyme in the proper position at the active site are pink.

Unfortunately, most anticancer drugs cannot discriminate between diseased and normal cells, so most cancer chemotherapy is accompanied by debilitating side effects. However, because cancer cells divide more rapidly than normal cells, they are harder hit by cancer-fighting chemotherapeutic agents than normal cells are.

Cancer Drugs and Side Effects

Scientists are searching for drugs that can discriminate between diseased and normal cells, so cancer chemotherapy will not be accompanied by debilitating side effects. A new drug now in clinical trials is able to deliver a very toxic agent only to cancer cells.

Herceptin has been used since 1998 to treat certain kinds of breast cancers. Recently, scientists have been able to attach it to another anticancer drug that is so toxic that it cannot be used directly. Once Herceptin has bound to a breast cancer cell, it releases the poisonous agent to kill the cell. The survival time for women with advanced breast cancer enrolled in the clinical trials for this combined drug has been found to be almost a year longer, with much fewer side effects, than for women treated only with Herceptin or other cancer drugs.

Inhibitors

5-Fluorouracil is a **mechanism-based inhibitor:** it inactivates the enzyme by taking part in the normal catalytic mechanism. It is also called a **suicide inhibitor** because the enzyme effectively "commits suicide" by reacting with the inhibitor. The therapeutic use of 5-fluorouracil illustrates the importance of knowing the mechanism for enzyme-catalyzed reactions. If scientists know the mechanism of a reaction, they may be able to design an inhibitor to turn the reaction off at a certain step.

Competitive Inhibitors

Aminopterin and methotrexate are anticancer drugs that are inhibitors of dihydrofolate reductase. Because their structures are similar to that of dihydrofolate, they compete with it for binding to the active site of the enzyme. Because they bind 1000 times more tightly to the enzyme than dihydrofolate does, they win the competition and, therefore, inhibit the enzyme's activity. These two compounds are examples of **competitive inhibitors.**

aminopterin R = H
methotrexate R = CH₃

trimethoprim

Because aminopterin and methotrexate inhibit the synthesis of THF, they interfere with the synthesis of any compound that requires a THF-coenzyme for its synthesis. Thus, not only do they prevent the synthesis of thymidine, they also inhibit the synthesis of adenine and guanine—other heterocyclic compounds needed for the synthesis of DNA—because their synthesis also requires a THF-coenzyme. One clinical technique used to fight cancer calls for the patient to be given a lethal dose of methotrexate. Then, after cancer cells have died, the patient is "saved" by being given N^5-formyl-THF.

Trimethoprim is used as an antibiotic because it binds to bacterial dihydrofolate reductase much more tightly than to mammalian dihydrofolate reductase.

PROBLEM 21 ◆

How do the structures of tetrahydrofolate and aminopterin differ?

PROBLEM 22

What is the source of the methyl group in thymidine?

23.8 VITAMIN K: THE VITAMIN NEEDED FOR CARBOXYLATION OF GLUTAMATE

Vitamin K is required for proper blood clotting. The letter K comes from *koagulation*, which is German for "clotting." Vitamin K is found in the leaves of green plants. Deficiencies are rare because the vitamin is also synthesized by intestinal bacteria. **Vitamin KH$_2$** (the hydroquinone of vitamin K) is the coenzyme form of the vitamin. The biosynthesis of vitamin KH$_2$ is described in Section 18.7.

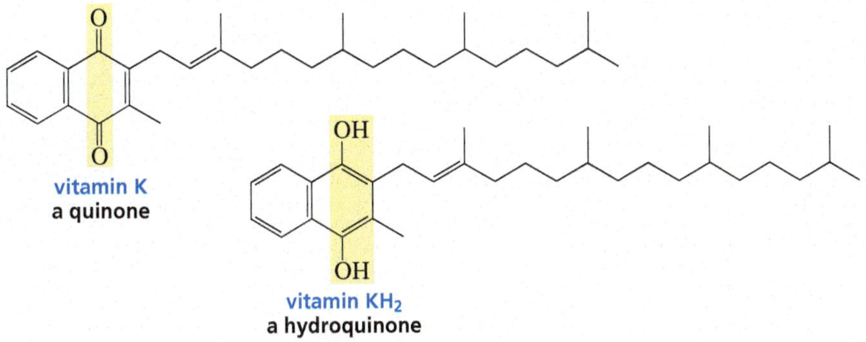

vitamin K
a quinone

vitamin KH$_2$
a hydroquinone

A series of reactions involving six proteins causes blood to clot. The process requires these six proteins to bind Ca^{2+}. γ-Carboxyglutamates bind Ca^{2+} much more effectively than glutamates do, so glutamates need to be converted to γ-glutamates in order for blood to clot.

Vitamin KH$_2$ is the coenzyme for the enzyme that converts a glutamate side chain to a γ-carboxyglutamate side chain. The enzyme uses CO_2 for the carboxyl group that it puts on the glutamate side chain.

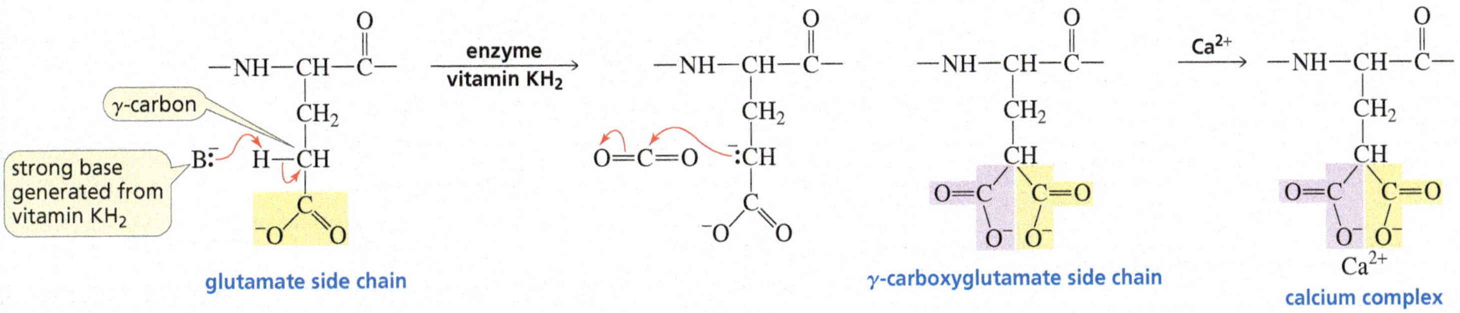

glutamate side chain

γ-carboxyglutamate side chain

calcium complex

All the proteins responsible for blood clotting have several glutamates near their N-terminal ends. For example, prothrombin, a blood-clotting protein, has glutamates at positions 7, 8, 15, 17, 20, 21, 26, 27, 30, and 33.

Vitamin KH$_2$ is required by the enzyme that catalyzes the carboxylation of the γ-carbon of a glutamate side chain.

The Mechanism

The mechanism for the vitamin KH$_2$–catalyzed carboxylation of glutamate had puzzled chemists because the γ-proton that must be removed from glutamate before it reacts with CO_2 is not very acidic. The mechanism, therefore, must involve the creation of a strong base that can remove this proton. A mechanism that has been proposed is shown on the top of the next page.

MECHANISM FOR THE CARBOXYLATION OF GLUTAMATE

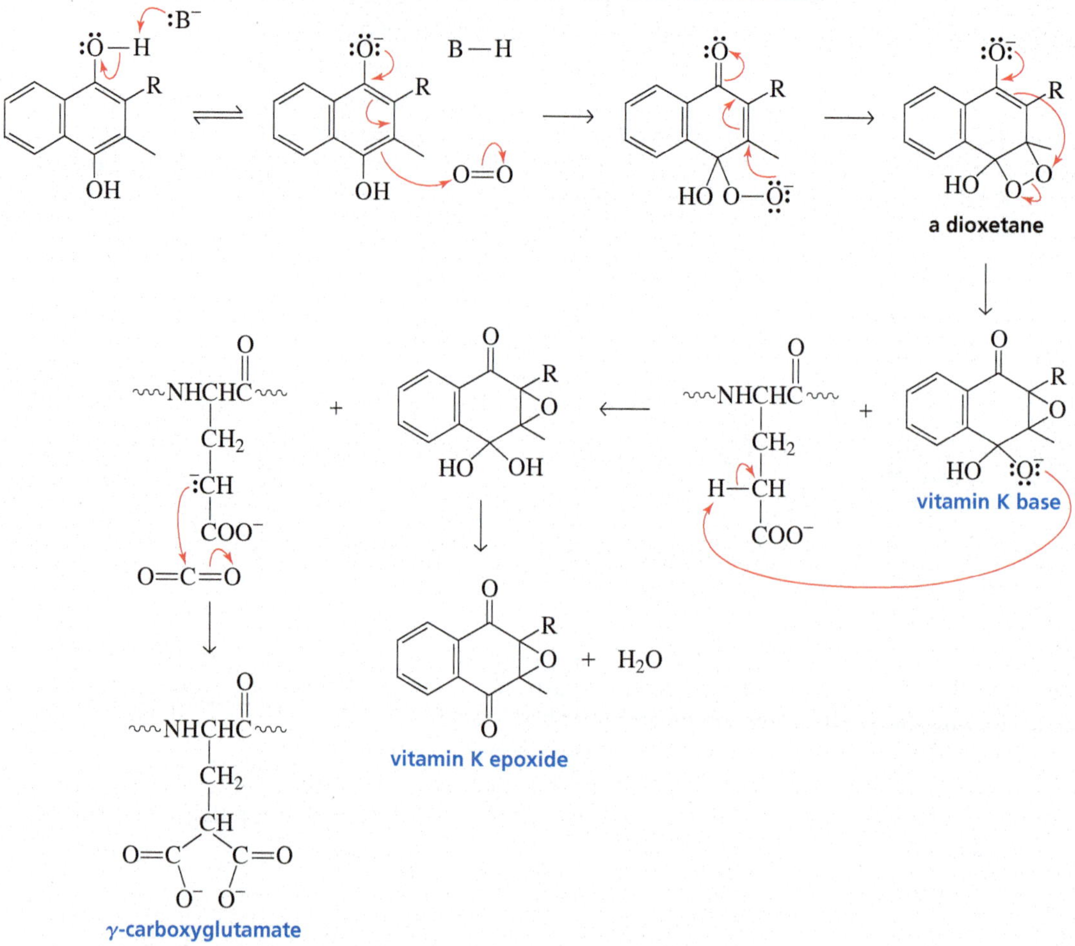

a dioxetane

vitamin K base

vitamin K epoxide

γ-carboxyglutamate

- A base removes a proton from an OH group of the coenzyme.
- The nucleophile that is formed attacks molecular oxygen.
- A conjugate addition reaction (Section 16.15) forms a dioxetane, which is a heterocyclic compound with a four-membered ring comprising two carbons and two adjacent oxygens.
- The dioxetane rearranges to an epoxide that has a strongly basic alkoxide ion. This species is called a vitamin K base.
- The vitamin K base is a strong enough base to remove a proton from the γ-carbon of glutamate.
- The glutamate carbanion adds to CO₂ to form γ-carboxyglutamate, and the protonated vitamin K base (a hydrate) loses water, forming vitamin K epoxide.

Vitamin K epoxide must now be reduced back to vitamin KH₂ so another round of catalysis can occur. The reducing agent for this enzyme-catalyzed reaction is the coenzyme, dihydrolipoate. Dihydrolipoate first reduces the epoxide to vitamin K, which is then further reduced by another dihydrolipoate to vitamin KH₂.

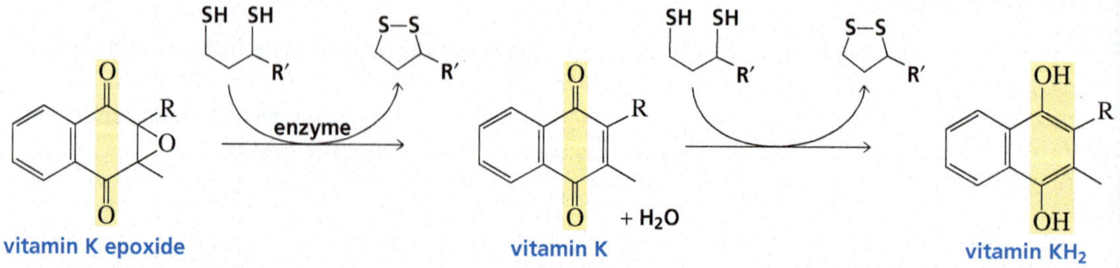

vitamin K epoxide   vitamin K   vitamin KH₂

PROBLEM 23 ◆

Thiols such as ethanethiol and propanethiol can be used to reduce vitamin K epoxide to vitamin KH₂, but they react much more slowly than dihydrolipoate. Explain why this is so.

Anticoagulants

Warfarin and dicoumarol are used clinically as anticoagulants. These compounds prevent blood clotting because they inhibit the enzyme that converts vitamin K epoxide to vitamin KH_2 by binding to the enzyme's active site. Because the enzyme cannot tell the difference between these two compounds and vitamin K epoxide, the compounds are *competitive inhibitors*. Warfarin is also commonly used as a rat poison, causing death by internal bleeding.

Vitamin E has also been found to be an anticoagulant. It inhibits the enzyme that carboxylates glutamate residues.

Too Much Broccoli

Two women with diseases characterized by abnormal blood clotting reported that they did not improve when they were given Coumadin. When questioned about their diets, one woman said that she ate at least a pound of broccoli every day, and the other ate broccoli soup and a broccoli salad every day. When broccoli was removed from their diets, Coumadin was effective in preventing the abnormal clotting of their blood. Because broccoli is high in vitamin K, these patients had been getting enough dietary vitamin K to compete successfully with the drug for the enzyme's active site, thereby making the drug ineffective.

ESSENTIAL CONCEPTS

Section 23.0

- **Cofactors** assist enzymes in catalyzing a variety of reactions that cannot be catalyzed solely by the amino acid side chains of the enzyme. Cofactors can be metal ions or organic molecules.

- Cofactors that are organic molecules are called **coenzymes;** coenzymes are derived from **vitamins,** which are substances needed in small amounts for normal body function that the body cannot synthesize or cannot synthesize in adequate amounts.

- All the water-soluble vitamins except vitamin C are precursors of coenzymes. Vitamin K is the only water-insoluble vitamin that is a precursor of a coenzyme.

- Coenzymes play a variety of chemical roles that the amino acid side chains of enzymes cannot perform. Some function as oxidizing and reducing agents; some allow electrons to be delocalized; some activate groups for further reaction; and some provide good nucleophiles or strong bases needed for a reaction.

Section 23.1

- **NAD⁺** and **NADH** are the coenzymes associated with **catabolic reactions**—biological reactions that break down complex

biomolecules in order to provide the cell with energy. Catabolic reactions are predominately oxidation reactions. NAD^+ is the most common oxidizing agent in cells.

- **NADP⁺** and **NADPH** are the coenzymes associated with **anabolic reactions**—biological reactions involved in synthesizing complex biomolecules. Anabolic reactions are predominately reduction reactions. NADPH is the most common reducing agent in cells.

Section 23.2

- **FAD** catalyzes oxidation reactions; it is reduced to **FADH₂.**

Section 23.3

- **Thiamine pyrophosphate (TPP)** is the coenzyme required by enzymes that catalyze the transfer of an acyl group.

Section 23.4

- **Biotin (vitamin H)** is the coenzyme required by enzymes that catalyze the carboxylation of a carbon adjacent to a carbonyl group.

Section 23.5

- **Pyridoxal phosphate (PLP)** is the coenzyme required by enzymes that catalyze reactions of amino acids, such as decarboxylation, transamination, racemization, C_α—C_β bond cleavage, and α,β-elimination.

- In a **transimination reaction,** one imine is converted into another imine; in a **transamination reaction,** the amino group is removed from an amino acid and transferred to another molecule, leaving a keto group in its place.

Section 23.6

- In an enzyme-catalyzed reaction that requires **coenzyme B$_{12}$,** a group bonded to one carbon changes places with a hydrogen bonded to an adjacent carbon.

Section 23.7

- **Tetrahydrofolate (THF)** is the coenzyme used by enzymes catalyzing reactions that transfer a group containing a single carbon—formyl, methyl, or methylene—to their substrates.

Section 23.8

- **Vitamin KH$_2$** is the coenzyme for the enzyme that catalyzes the carboxylation of the γ-carbon of glutamate side chains, a reaction required for blood clotting.

- A **suicide inhibitor** inactivates an enzyme by taking part in the catalytic mechanism.

- A **competitive inhibitor** inactivates an enzyme by competing with the substrate for binding at the active site of the enzyme.

PROBLEMS

24. Answer the following:
  a. What coenzyme transfers an acyl group from one substrate to another?
  b. What is the function of FAD in the pyruvate dehydrogenase complex?
  c. What is the function of NAD$^+$ in the pyruvate dehydrogenase complex?
  d. What reaction necessary for proper blood clotting is catalyzed by vitamin KH$_2$?
  e. What coenzymes are used for decarboxylation reactions?
  f. What kinds of substrates do the decarboxylating coenzymes work on?
  g. What coenzymes are used for carboxylation reactions?
  h. What kinds of substrates do the carboxylating coenzymes work on?

25. Name the coenzymes that
  a. allow electrons to be delocalized.
  b. are oxidizing agents.
  c. provide a strong base.
  d. donate one-carbon groups.

26. For each of the following reactions, name both the enzyme that catalyzes the reaction and the required coenzyme:

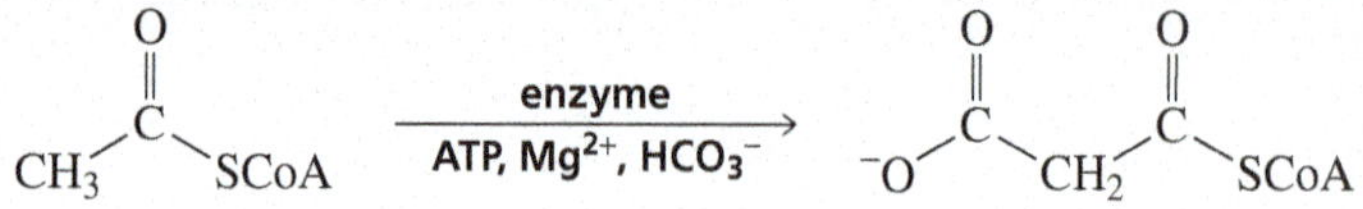

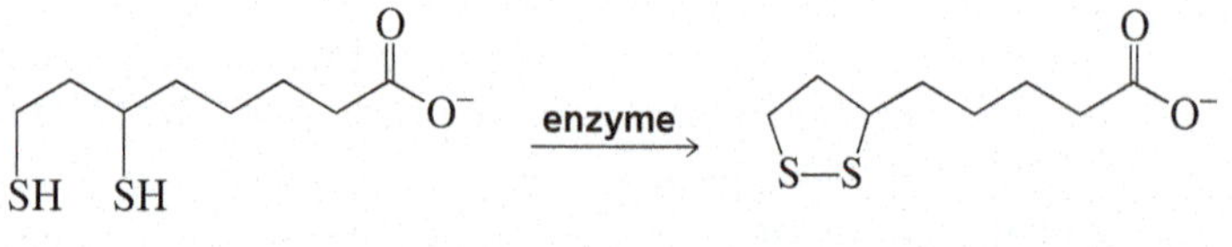

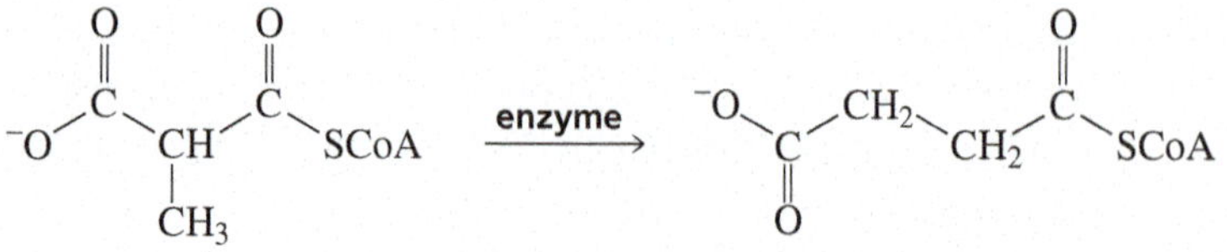

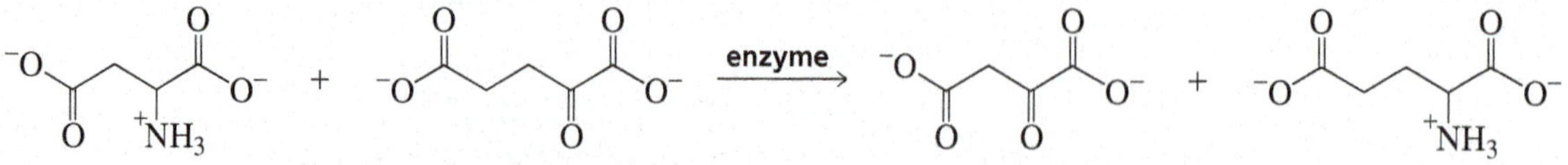

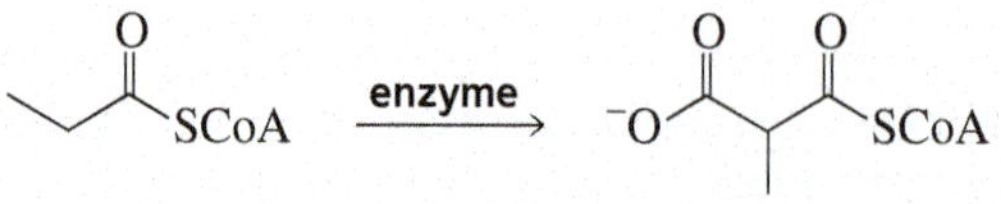

27. Five coenzymes are required by α-ketoglutarate dehydrogenase, the enzyme in the citric acid cycle that converts α-ketoglutarate to succinyl-CoA.

    a. Identify the coenzymes.

    b. Propose a mechanism for the reaction.

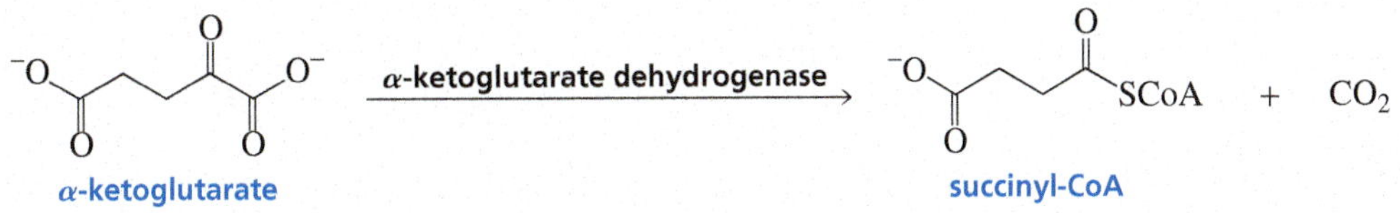

α-ketoglutarate $\xrightarrow{\alpha\text{-ketoglutarate dehydrogenase}}$ succinyl-CoA + CO_2

28. When transaminated, the three branched-chain amino acids (valine, leucine, and isoleucine) form compounds that have the characteristic odor of maple syrup. An enzyme known as branched-chain α-keto acid dehydrogenase converts these compounds into CoA esters. People who do not have this enzyme have the genetic disease known as maple syrup urine disease, so called because their urine smells like maple syrup.

    a. Draw the compounds that smell like maple syrup.

    b. Draw the CoA esters.

    c. Branched-chain α-keto acid dehydrogenase has five coenzymes. Identify them.

    d. Suggest a way to treat maple syrup urine disease.

29. What acyl groups have we seen transferred by reactions that require thiamine pyrophosphate as a coenzyme? (*Hint:* See Problems 9, 10, 11, 27, and 28.)

30. Propose a mechanism for the following reaction:

methylmalonyl-CoA $\xrightarrow[\text{coenzyme B}_{12}]{\text{methylmalonyl-CoA mutase}}$ succinyl-CoA

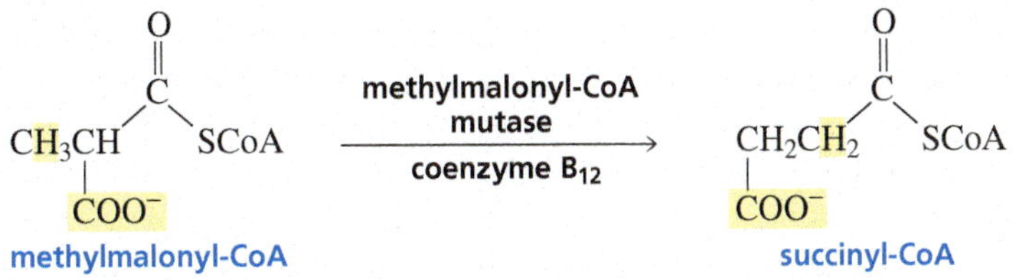

31. Draw the products of the following reaction, where T is tritium:

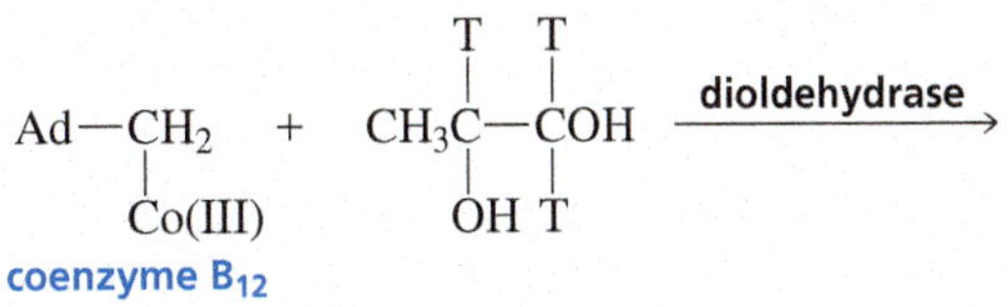

(*Hint:* Tritium is ^{3}H, a hydrogen atom with two neutrons. Although a C—T bond breaks more slowly than a C—H bond, it is still the first bond in the substrate that breaks.)

32. When UMP is dissolved in T_2O, exchange of T for H occurs at the 5-position. Propose a mechanism for this exchange. ($T = {}^3$H; see Problem 31)

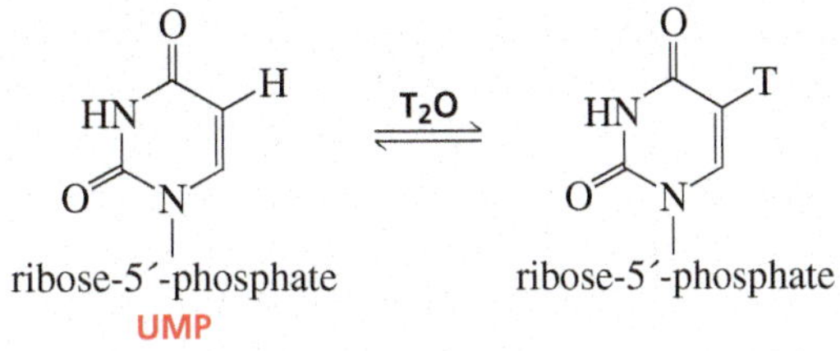

ribose-5′-phosphate $\xrightleftharpoons{T_2O}$ ribose-5′-phosphate

UMP

33. Dehydratase is a PLP-requiring enzyme that catalyzes an α,β-elimination reaction. Propose a mechanism for this reaction.

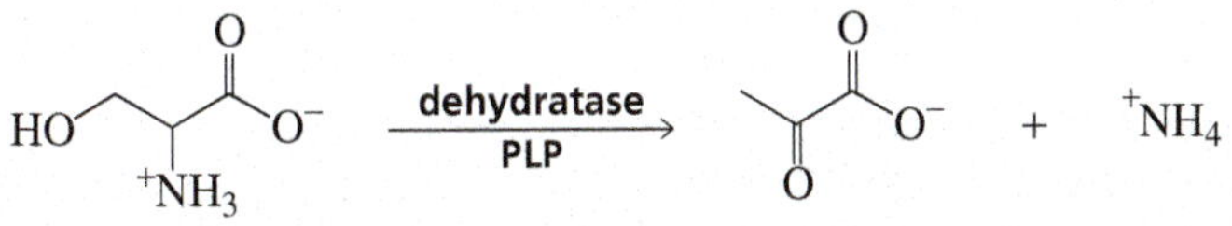

$\xrightarrow[\text{PLP}]{\text{dehydratase}}$ + $^+NH_4$

34. In addition to the reactions mentioned in Section 23.5, PLP can catalyze β-substitution reactions. Propose a mechanism for the following PLP-catalyzed β-substitution reaction:

1097

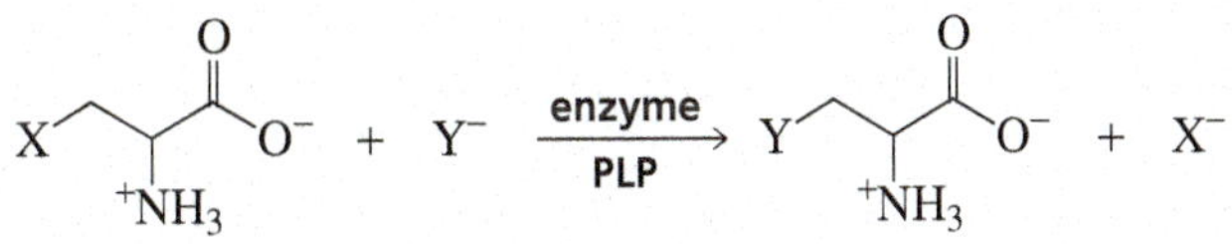

35. PLP can catalyze both α,β-elimination reactions (Problem 33) and β,γ-elimination reactions. Propose a mechanism for the following PLP-catalyzed β,γ-elimination reaction:

$$X\text{-CH}_2\text{CH}_2\text{CH(}^+\text{NH}_3\text{)CO}_2^- \xrightarrow[\text{PLP}]{\text{enzyme}} \text{CH}_3\text{COCH}_2\text{CO}_2^- + X^- + ^+\text{NH}_4$$

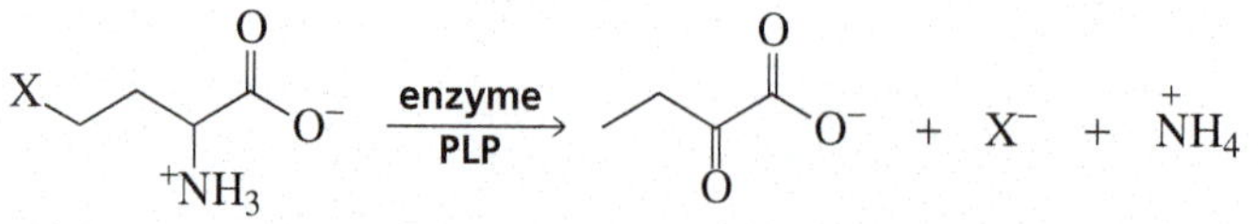

36. The glycine cleavage system is a group of four enzymes that together catalyze the following reaction:

$$\text{glycine} + \text{THF} \xrightarrow{\text{glycine cleavage system}} N^5,N^{10}\text{-methylene-THF} + \text{CO}_2$$

Use the following information to determine the sequence of reactions carried out by the glycine cleavage system:

a. The first enzyme is a PLP-requiring decarboxylase.
b. The second enzyme is aminomethyltransferase. This enzyme has a lipoate coenzyme.
c. The third enzyme synthesizes N^5,N^{10}-methylene-THF and also forms $^+\text{NH}_4$.
d. The fourth enzyme is an FAD-requiring enzyme.
e. The cleavage system also requires NAD^+.

37. Nonenzyme-bound FAD is a stronger oxidizing agent than NAD^+. How, then, can NAD^+ oxidize the reduced flavoenzyme in the pyruvate dehydrogenase complex?

38. FADH_2 reduces α,β-unsaturated thioesters to saturated thioesters. The reaction is thought to take place by a mechanism that involves radicals. Propose a mechanism for this reaction.

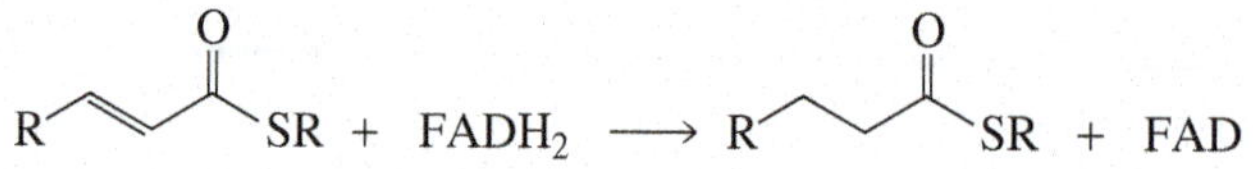

$$R\text{-CH=CH-CO-SR} + \text{FADH}_2 \longrightarrow R\text{-CH}_2\text{CH}_2\text{-CO-SR} + \text{FAD}$$

24

The Organic Chemistry of the Metabolic Pathways

You are what you eat.

The reactions that living organisms carry out to obtain the energy they need and to synthesize the compounds they require are collectively known as **metabolism.** *Metabolism can be divided into two parts: catabolism and anabolism. Catabolic reactions convert complex nutrient molecules into simple molecules that can be used for synthesis. Anabolic reactions synthesize complex biomolecules from simpler precursor molecules. Catabolism comes from the Greek word* katabol, *which means "throwing down."*

A catabolic pathway is a series of sequential reactions that converts a complex molecule into simple molecules. Catabolic pathways involve *oxidation reactions* and produce energy.

> *catabolism:*   *complex molecule*   →   *simple molecules + energy*

An anabolic pathway is a series of sequential reactions that converts simple molecules to a complex molecule. Anabolic pathways involve *reduction reactions* and require energy.

> *anabolism:*   *simple molecules + energy*   →   *complex molecule*

It is important to remember that almost every reaction that occurs in a living system is catalyzed by an enzyme. The enzyme holds the reactants and any necessary coenzymes in place, orienting the reacting functional groups and the amino acid side chain catalysts in a way that allows the enzyme-catalyzed reaction to take place (Section 22.8).

Most of the reactions described in this chapter will be familiar to you. If you take the time to refer back to the sections cited and review these reactions, you will see that many of the organic reactions done by cells are the same as the organic reactions done by chemists.

Differences in Metabolism

Humans do not necessarily metabolize compounds in the same way as other species do. This becomes a significant problem when drugs are tested on animals. For example, chocolate is metabolized to different compounds in humans and in dogs. The metabolites produced in humans are non-toxic, whereas those produced in dogs can be highly toxic. Differences in metabolism have been found even within the same species. For example, isoniazid—an antituberculosis drug—is metabolized by Eskimos much faster than by Egyptians. Current research is showing that men and women metabolize certain drugs differently. For example, kappa opioids—a class of painkillers—have been found to be about twice as effective in women as they are in men.

24.1 ATP IS USED FOR PHOSPHORYL TRANSFER REACTIONS

All cells require energy to live and reproduce. They get the energy they need from nutrients that they convert into a chemically useful form. The most important repository of chemical energy is **adenosine 5′-triphosphate (ATP)**. The importance of ATP to biological reactions is reflected in its turnover rate in humans: each day, a person uses an amount of ATP equivalent to his or her body weight.

adenosine triphosphate
ATP

ATP Reacts with Nucleophiles

A **phosphoryl transfer reaction** transfers a phosphate group to a nucleophile. We saw that phosphoryl transfer reactions can be used to activate a compound for a reaction by putting a good leaving group on it (Section 15.19). For example, a carboxylate ion does not react with a nucleophile because a carboxylate ion is negatively charged and its leaving group is a strong base (Section 15.10).

ATP, however, can activate a carboxylate ion for reaction with a nucleophile in one of three ways. Each one involves an S_N2 reaction of the carboxylate ion with ATP that breaks a phosphoanhydride bond (because the phosphoanhydride bond is weaker than the π bond) and forms a mixed anhydride of a carboxylic acid and a phosphoric acid. Thus, a good leaving group has been put on the carboxylate ion that can easily be displaced by a nucleophile.

- The carboxylate ion can attack the γ-phosphorus (the terminal phosphorus) of ATP, forming an acyl phosphate.

nucleophilic attack on the γ-phosphorus

- The carboxylate ion can attack the β-phosphorus of ATP, forming an acyl pyrophosphate.

nucleophilic attack on the β-phosphorus

ATP   an acyl pyrophosphate   AMP

- The carboxylate ion can attack the α-phosphorus of ATP, forming an acyl adenylate.

nucleophilic attack on the α-phosphorus

ATP   an acyl adenylate
a mixed anhydride
pyrophosphate

Nucleophiles other than carboxylate ions can be activated by ATP in cells.

The Enzyme Determines which Phosphorus is Attacked

Whether a nucleophile attacks the γ-phosphorus, the β-phosphorus (which rarely occurs), or the α-phosphorus depends on the enzyme catalyzing the reaction. Notice that when a nucleophile attacks the γ-phosphorus, the side product is ADP, but when it attacks the α-phosphorus, the side product is pyrophosphate. When pyrophosphate is formed, it is subsequently hydrolyzed to two equivalents of hydrogen phosphate. We saw that removing a reaction product from the reaction mixture drives the reaction to the right (see Le Châtelier's principle in Section 5.7).

pyrophosphate   phosphate

Therefore, for an enzyme-catalyzed reaction in which irreversibility is essential, the nucleophile will attack the α-phosphorus of ATP. For example, both the reaction that links nucleotide subunits to form DNA and RNA and the reaction that binds an amino acid to a tRNA (the first step in translating RNA into a protein) involve nucleophilic attack on the α-phosphorus of ATP (Sections 26.2 and 26.8, respectively). If these reactions were reversible, the genetic information in DNA would not be preserved and proteins would be synthesized that would not have the correct sequence of amino acids.

Another Way to Activate a Carboxylate Ion

We saw that a carboxylate ion can also be activated by being converted to a thioester (Section 15.19). ATP is needed for this reaction too. The carboxylate ion reacts with ATP to form an acyl adenylate, giving the carboxylate ion a good leaving group, so it can then undergo a nucleophilic acyl substitution reaction with the thiol.

ATP   + pyrophosphate   acetyl-CoA

The phosphoryl transfer reactions discussed here demonstrate the actual chemical function of ATP:

> *ATP provides a reaction pathway involving a good leaving group for a reaction that cannot occur (or would occur very slowly) because of a poor leaving group.*

Why Did Nature Choose Phosphates?

Phosphoanhydrides and esters of phosphoric acid dominate the organic chemistry of the biological world. ATP is a monoester of triphosphoric acid. (One of the OH groups of triphosphoric acid has been converted to an OR group.) Cell membranes are monoesters or diesters of phosphoric acid (Section 25.5), many of the coenzymes are monoesters or diesters of either phosphoric acid or pyrophosphoric acid (Chapter 23), and DNA and RNA are diesters of phosphoric acid (Section 26.2).

In contrast, phosphates are rarely used in organic chemistry in the laboratory. Instead, we saw that the preferred leaving groups in nonbiological reactions are such things as halide ions and sulfonate ions (Sections 9.1 and 10.3).

Why did nature choose phosphates? There are several reasons. To keep a molecule from leaking through a membrane, it should be charged; to prevent reactive nucleophiles from approaching a molecule, it should be negatively charged; and to link the bases in RNA and DNA, the linking molecule needs two functional groups (Section 26.1). Phosphoric acid, with its three OH groups, fits all these requirements; it can use two of its OH groups to link the bases, and the third OH group is negatively charged at physiological pH. In addition, the reactions of nucleophiles with phosphoanhydrides can be irreversible, which is an important attribute of many biological reactions.

24.2 WHY ATP IS KINETICALLY STABLE IN A CELL

ATP reacts readily in enzyme-catalyzed reactions, but quite slowly in the absence of an enzyme. For example, a carboxylic acid anhydride hydrolyzes in a matter of minutes, but ATP (a phosphoric acid anhydride) takes several weeks to hydrolyze. The low rate of ATP hydrolysis is important because it allows ATP to exist in a cell until it is needed for an enzyme-catalyzed reaction.

The negative charges on ATP are what make it relatively unreactive. These negative charges repel the approach of a nucleophile. When ATP is bound at an active site of an enzyme, it complexes with magnesium (Mg^{2+}), which decreases ATP's overall negative charge. (This is why ATP-requiring enzymes also require Mg^{2+}; see Section 23.4.) Interactions with positively charged groups, such as arginine or lysine side chains, at the active site of the enzyme further reduce ATP's negative charge (see Figure 15.3 on page 725). Thus, when bound at the active site of an enzyme, ATP can be readily approached by nucleophiles.

For the same reason, acyl phosphates and acyl adenylates are unreactive unless they are at the active site of an enzyme. It is important to note that the leaving groups of acyl phosphates and acyl adenylates are weaker bases than their pK_a values would indicate, because they are coordinated with magnesium ion. Metal coordination decreases their basicity, which increases their leaving propensity.

24.3 THE "HIGH-ENERGY" CHARACTER OF PHOSPHOANHYDRIDE BONDS

The reaction of a nucleophile with ATP breaks a phosphoanhydride bond. This is a highly exergonic reaction. Therefore, phosphoanhydride bonds are called **high-energy bonds.** The term *high-energy* in this context means that a great deal of energy is released when a phosphoanhydride bond is broken as a result of reacting with a nucleophile. Do not confuse it with *bond energy*, the term chemists use to describe how difficult it is to break a bond (Section 5.7).

Why is the reaction of a nucleophile with ATP so exergonic? In other words, why is the $\Delta G°'$ value large and negative?* A large negative $\Delta G°'$ means that the products of the reaction are much more stable than the reactants. Let's look at the reactants and products of the reaction on the top of the next page to see why this is so.

---

* The prime in $\Delta G°'$ indicates that two additional parameters have been added to the definition of $\Delta G°$ in Section 5.6: the reaction occurs in aqueous solution at pH = 7, and the concentration of water is assumed to be constant.

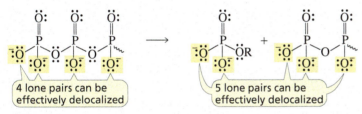

ATP   an alkyl phosphate   ADP

Three factors contribute to the greater stability of the products (ADP and the alkyl phosphate) compared to the reactants (ATP and the alcohol):

- **Greater electrostatic repulsion in ATP.** At physiological pH (7.4), ATP has 3.3 negative charges, ADP has 2.8 negative charges, and the alkyl phosphate has 1.8 negative charges (see Problems 1 and 2). Because of ATP's greater negative charge, the electrostatic repulsion is greater in ATP than in either of the products. Electrostatic repulsion destabilizes a molecule.

- **Greater solvation stabilization in the products.** Negatively charged ions are stabilized in an aqueous solution by solvation (Section 3.10). Because the reactant has 3.3 negative charges, whereas the sum of the negative charges on the products is 4.6 (2.8 + 1.8), there is greater solvation stabilization in the products than in the reactants.

- **Greater electron delocalization in the products.** A lone pair on the oxygen joining two phosphorus atoms is not effectively delocalized because delocalization would put a partial positive charge on an oxygen. When the phosphoanhydride bond breaks, one additional lone pair can be effectively delocalized. Electron delocalization stabilizes a molecule (Section 8.6).

4 lone pairs can be
effectively delocalized

5 lone pairs can be
effectively delocalized

Similar factors explain the large negative $\Delta G°'$ when a nucleophile reacts with ATP to form a substituted AMP and pyrophosphate, and when pyrophosphate is hydrolyzed to two equivalents of phosphate.

PROBLEM 1 SOLVED

LEARN THE STRATEGY

ATP has pK_a values of 0.9, 1.5, 2.3, and 7.7; ADP has pK_a values of 0.9, 2.8, and 6.8; and the alkyl phosphate has pK_a values of 1.9 and 6.7.

Do the calculation showing that at pH 7.4 the charge on ATP is −3.3.

SOLUTION Because pH 7.4 is much greater than the pK_a values of the first three ionizations of ATP, we know that these three groups are entirely in their basic forms at that pH, giving ATP three negative charges (Section 2.10). So to answer the question, we need to calculate what fraction of the group with a pK_a value of 7.7 is in its basic form at pH 7.4.

$$\frac{\text{concentration of the basic form}}{\text{total concentration}} = \frac{[A^-]}{[A^-] + [HA]}$$

$$[A^-] = \text{concentration of the basic form}$$

$$[HA] = \text{concentration of the acidic form}$$

Because this equation has two unknowns, one of the unknowns must be expressed in terms of the other unknown. Using the definition of the acid dissociation constant (K_a), we define [HA] in terms of [A⁻], K_a, and [H⁺] and now have only one unknown.

$$K_a = \frac{[A^-][H^+]}{[HA]}$$

$$[HA] = \frac{[A^-][H^+]}{K_a}$$

$$\frac{[A^-]}{[A^-] + [HA]} = \frac{[A^-]}{[A^-] + \dfrac{[A^-][H^+]}{K_a}} = \frac{K_a}{K_a + [H^+]}$$

Now we can calculate the fraction of the group with a pK_a value of 7.7 that is in the basic form. (Notice that K_a is calculated from pK_a and that $[H^+]$ is calculated from pH.)

$$\frac{K_a}{K_a + [H^+]} = \frac{2.0 \times 10^{-8}}{2.0 \times 10^{-8} + 4.0 \times 10^{-8}} = 0.3$$

Thus, the total negative charge on ATP = 3.0 + 0.3 = 3.3.

USE THE STRATEGY

PROBLEM 2

Do the calculation showing that at pH 7.4

a. the charge on ADP is −2.8.

b. the charge on the alkyl phosphate is −1.8.

24.4 THE FOUR STAGES OF CATABOLISM

The reactants required for all life processes ultimately come from our diet. In that sense, we really are what we eat. Mammalian nutrition requires fats, carbohydrates, and proteins in addition to the vitamins and minerals discussed in Chapter 23.

Catabolism can be divided into four stages (Figure 24.1).

In the first stage of catabolism, fats, carbohydrates, and proteins are hydrolyzed to fatty acids, monosaccharides, and amino acids.

In the second stage of catabolism, the products obtained from the first stage are converted to compounds that can enter the citric acid cycle.

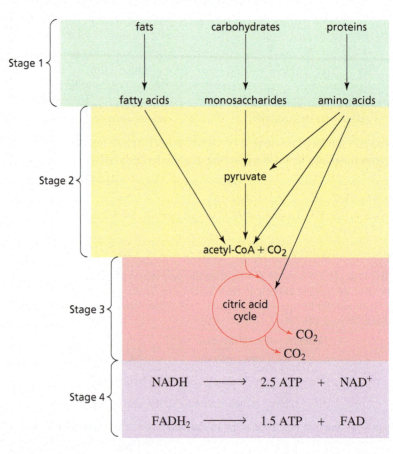

▶ **Figure 24.1**
The four stages of catabolism:
1. digestion
2. conversion of the products of the first stage to compounds that can enter the citric acid cycle
3. the citric acid cycle
4. oxidative phosphorylation.

The First Stage of Catabolism

The *first stage of catabolism* is called digestion. In this stage, the fats, carbohydrates, and proteins we consume are hydrolyzed to fatty acids (long straight-chain carboxylic acids), monosaccharides, and amino acids, respectively. These reactions occur in the mouth, stomach, and small intestine.

The Second Stage of Catabolism

In the *second stage of catabolism,* the products obtained from the first stage—fatty acids, monosaccharides, and amino acids—are converted to compounds that can enter the citric acid cycle.

The only compounds that can enter the citric acid cycle are (1) *citric acid cycle intermediates* (that is, compounds that take part in the cycle itself), (2) *acetyl-CoA*, and (3) *pyruvate* (because it can be converted to acetyl-CoA).

The Third Stage of Catabolism

The *third stage of catabolism* is the citric acid cycle. In this cycle, the acetyl group of each molecule of acetyl-CoA is converted to two molecules of CO_2.

$$CH_3-\overset{\overset{\displaystyle O}{\|}}{C}-SCoA \longrightarrow 2\ CO_2\ +\ CoASH$$

acetyl-CoA

> The citric acid cycle is the third stage of catabolism.

We saw that cells get the energy they need by using nutrient molecules to make ATP. Only a small amount of ATP is formed in the first three stages of catabolism—most is formed in the fourth stage. (You will be able to see this when you finish the chapter and can compare the answers to Problems 41, 42, 43, and 44.)

> Cells convert nutrients to adenosine triphosphate (ATP).

The Fourth Stage of Catabolism

Many catabolic reactions are oxidation reactions. In the *fourth stage of catabolism,* every molecule of NADH formed in one of the earlier stages of catabolism (formed when NAD^+ is used to carry out an oxidation reaction) is converted to 2.5 molecules of ATP in a process called *oxidative phosphorylation.* Oxidative phosphorylation also converts every molecule of $FADH_2$ formed in the earlier stages of catabolism (when FAD is used to carry out an oxidation reaction) to 1.5 molecules of ATP. Thus, most of the energy (ATP) that is provided by fats, carbohydrates, and proteins is obtained in the fourth stage of catabolism.

> Oxidative phosphorylation is the fourth stage of catabolism.

24.5   THE CATABOLISM OF FATS: STAGES 1 AND 2

Fats are Hydrolyzed to Glycerol and Fatty Acids

In the *first stage of fat catabolism*, the fat's three ester groups are hydrolyzed by an enzyme-catalyzed reaction to glycerol and three fatty acid molecules. A fatty acid is an unbranched, long-chain carboxylic acid that contains an even number of carbons (Section 25.1).

$$
\begin{array}{l}
CH_2O-\overset{\overset{\displaystyle O}{\|}}{C}-R^1 \\
CHO-\overset{\overset{\displaystyle O}{\|}}{C}-R^2 \\
CH_2O-\overset{\overset{\displaystyle O}{\|}}{C}-R^3
\end{array}
\quad + \quad 3\ H_2O \quad \xrightarrow{\text{enzyme}} \quad
\begin{array}{l}
CH_2OH \\
CHOH \\
CH_2OH
\end{array}
\quad + \quad
\begin{array}{l}
R^1-\overset{\overset{\displaystyle O}{\|}}{C}-OH \\
R^2-\overset{\overset{\displaystyle O}{\|}}{C}-OH \\
R^3-\overset{\overset{\displaystyle O}{\|}}{C}-OH
\end{array}
$$

a fat → glycerol   fatty acids

Glycerol is Converted to Dihydroxyacetone Phosphate

The following reaction sequence shows what happens to glycerol, one of the products of the preceding reaction, in the *second stage of fat catabolism.*

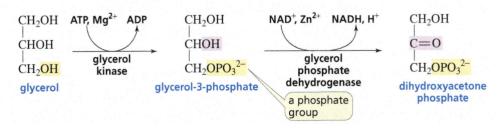

> **NOTE TO THE STUDENT**
> • Notice that when biochemical reactions are written, the only structures typically shown are those of the primary reactant and primary product. The names of other reactants and products are abbreviated and placed on a curved arrow that intersects the reaction arrow.

- In the first step, an OH group of glycerol reacts with the γ-phosphorus of ATP. The enzyme that catalyzes this reaction is called glycerol kinase. A *kinase* is an enzyme that catalyzes the transfer of a phosphoryl group to its substrate; thus, glycerol kinase puts a phosphoryl group on glycerol. Notice that this ATP-requiring enzyme also requires Mg^{2+} (Section 23.4).

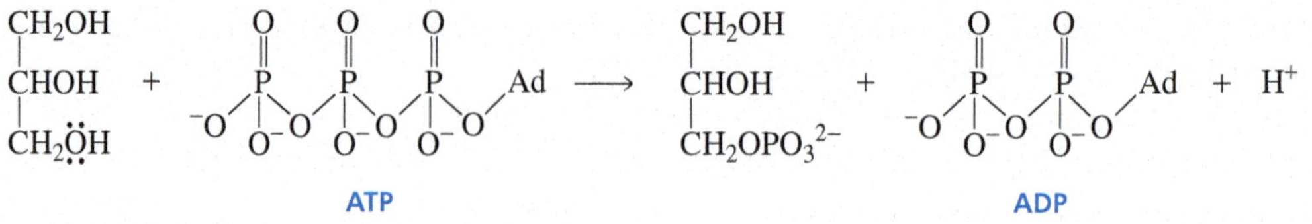

- In the second step, the secondary alcohol group of glycerol-3-phosphate is oxidized by NAD^+ to a ketone. The enzyme that catalyzes this reaction is called glycerol phosphate dehydrogenase. A *dehydrogenase* is an enzyme that oxidizes its substrate. We saw that when a substrate is oxidized by NAD^+, the substrate donates a hydride ion to the 4-position of the pyridinium ring of NAD^+ (Section 23.1). Zn^{2+} is a cofactor for the reaction; it increases the acidity of the secondary alcohol's proton by coordinating to the oxygen (Section 22.5).

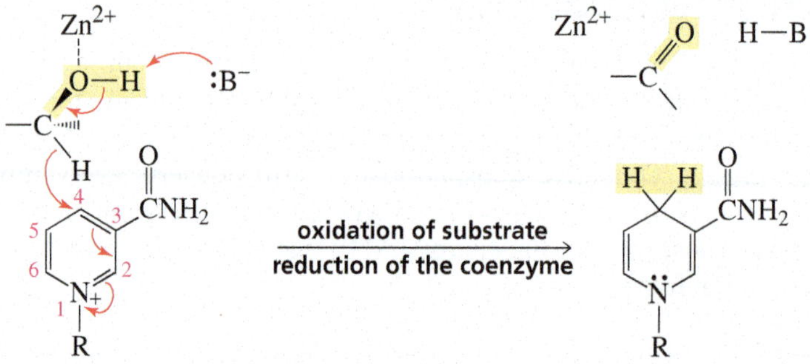

oxidation of substrate
reduction of the coenzyme

The product of the two-step reaction sequence, dihydroxyacetone phosphate, is one of the intermediates in glycolysis, so it can enter that pathway directly and be broken down further (Section 24.6).

PROBLEM 3

Show the mechanism for the reaction of glycerol with ATP to form glycerol-3-phosphate.

PROBLEM 4

The asymmetric center of glycerol-3-phosphate has the *R* configuration. Draw the structure of (*R*)-glycerol-3-phosphate.

A Fatty Acid is Activated to a Fatty Acyl-CoA

Now we will see how fatty acids, the other products formed from the hydrolysis of fats, are metabolized. Before a fatty acid can be metabolized, it must be activated. We saw that a carboxylic acid can be activated in a cell by first being converted to an acyl adenylate, which occurs when the carboxylate ion attacks the α-phosphorus of ATP. The acyl adenylate then reacts with coenzyme A in a nucleophilic acyl substitution reaction to form a thioester (Section 24.1).

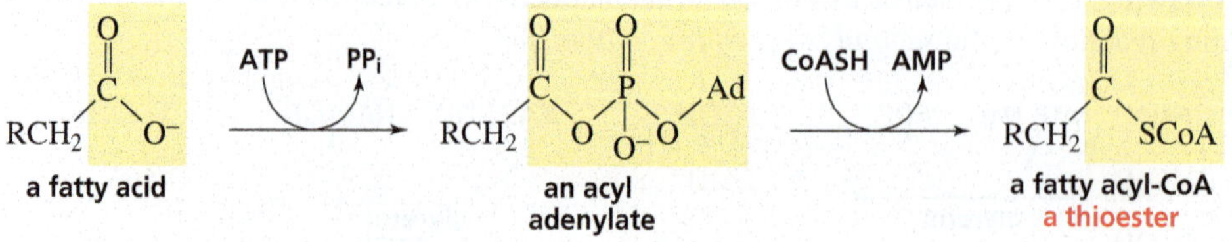

a fatty acid   an acyl adenylate   a fatty acyl-CoA
a thioester

A Fatty Acyl-CoA is Converted to Acetyl-CoA Molecules

The fatty acyl-CoA is converted to acetyl-CoA in a repeating four-step pathway called **β-oxidation.** Each passage through this series of four reactions removes two carbons from the fatty acyl-CoA by converting them to acetyl-CoA (Figure 24.2). Each of the four reactions is catalyzed by a different enzyme.

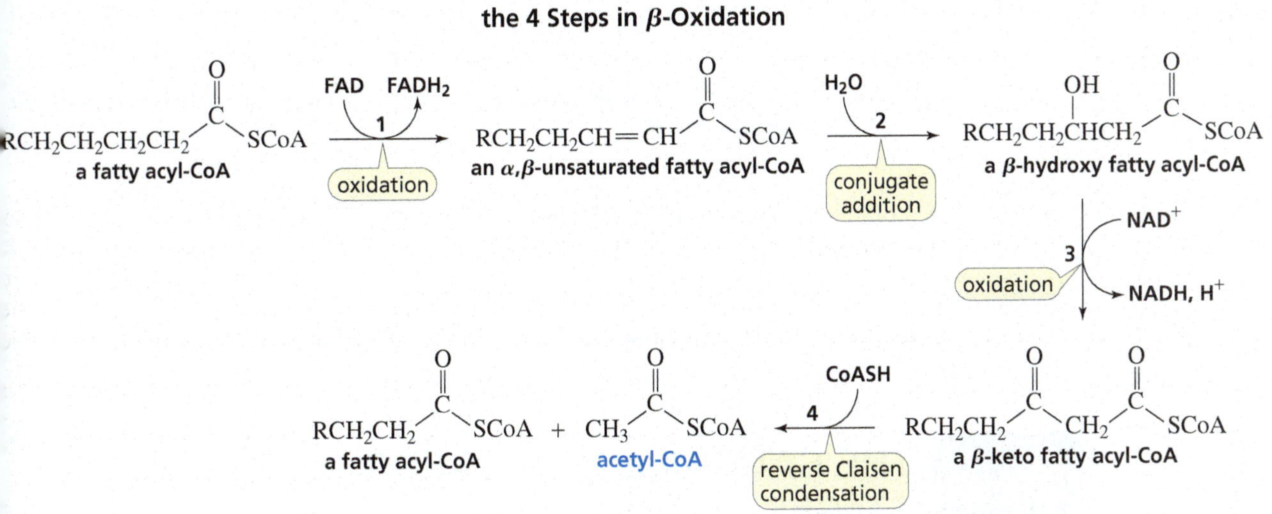

the 4 Steps in β-Oxidation

▲ **Figure 24.2**
In β-oxidation a series of four enzyme-catalyzed reactions is repeated until the entire fatty acyl-CoA molecule has been converted to acetyl-CoA molecules. The enzymes that catalyze the reactions are: 1. acyl-CoA dehydrogenase; 2. enoyl-CoA hydratase; 3. 3-L-hydroxyacyl-CoA dehydrogenase; and 4. β-ketoacyl-CoA thiolase.

1. The first reaction is an oxidation reaction that removes hydrogens from the α- and β-carbons, forming an α,β-unsaturated fatty acyl-CoA. The oxidizing agent is FAD; the mechanism of this reaction was shown in Section 23.2. The enzyme that catalyzes this reaction has been found to be deficient in 10% of babies that experience sudden infant death syndrome (SIDS), a condition where an apparently healthy baby dies, generally while sleeping. Glucose is the cell's primary fuel immediately after a meal and then the cell switches to a combination of glucose and fatty acids. The inability to oxidize fatty acids may be what causes the infant's distress.

2. The mechanism for the second reaction—the conjugate addition of water to the α,β-unsaturated fatty acyl-CoA—is shown below (Section 16.15). A glutamate side chain of the enzyme removes a proton from water, making it a better nucleophile. The enolate ion is protonated by glutamic acid.

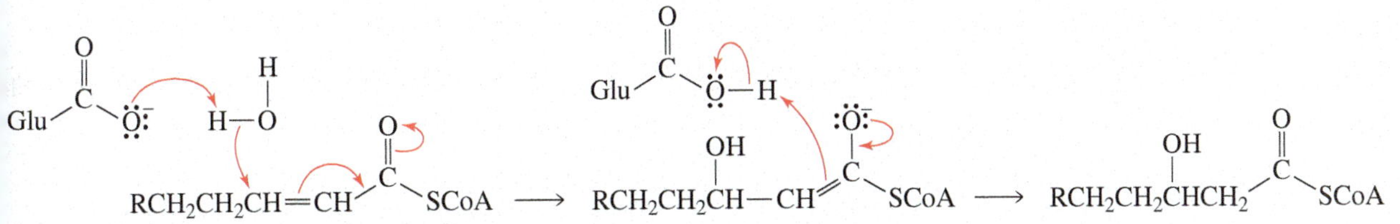

3. The third reaction is another oxidation reaction: NAD^+ oxidizes the secondary alcohol to a ketone. Recall that the mechanism for all NAD^+ oxidations involves donation of a hydride ion from the substrate to the 4-positon of the pyridinium ring of NAD^+ (see page 1106).

4. The fourth reaction is the reverse of a Claisen condensation (Section 17.13), followed by conversion of the enolate ion to the keto tautomer (Section 17.3). The mechanism for this reaction is shown here. The final product is acetyl-CoA and a fatty acyl-CoA *with two fewer carbons* than the starting fatty acyl-CoA.

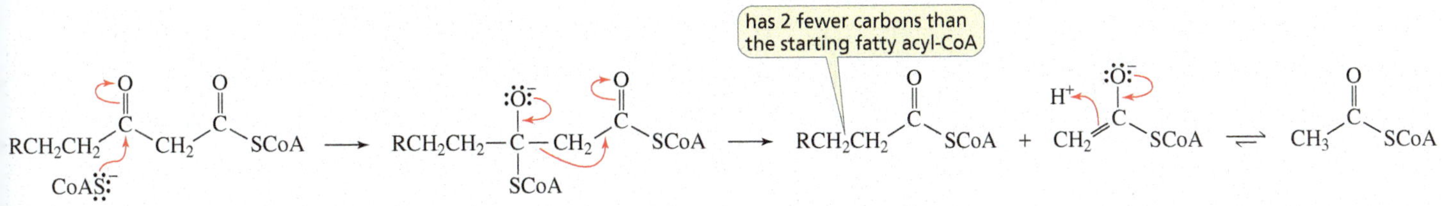

The four reactions are repeated, forming another molecule of acetyl-CoA and a fatty acyl-CoA that is now four carbons shorter than it was originally. Each time the series of four reactions is repeated, two more carbons are removed (as acetyl-CoA) from the fatty acyl-CoA. The series of reactions is repeated until the entire fatty acid has been converted to acetyl-CoA molecules. In Section 24.9 we will see how acetyl-CoA enters the citric acid cycle.

Fatty acids are converted to molecules of acetyl-CoA.

PROBLEM 5 ◆

Why does the OH group add to the β-carbon rather than to the α-carbon in the second reaction in the β-oxidation of fats? (*Hint:* See Section 16.15.)

PROBLEM 6 ◆

Palmitic acid is a 16-carbon saturated fatty acid. How many molecules of acetyl-CoA are formed from one molecule of palmitic acid?

PROBLEM 7 ◆

How many molecules of NADH are formed from the β-oxidation of one molecule of palmitic acid?

24.6 THE CATABOLISM OF CARBOHYDRATES: STAGES 1 AND 2

Carbohydrates are Hydrolyzed to Glucose Molecules

In the first stage of carbohydrate catabolism, the glycosidic bonds that hold glucose subunits together as acetals are hydrolyzed in an enzyme-catalyzed reaction, forming individual glucose molecules (Section 20.16).

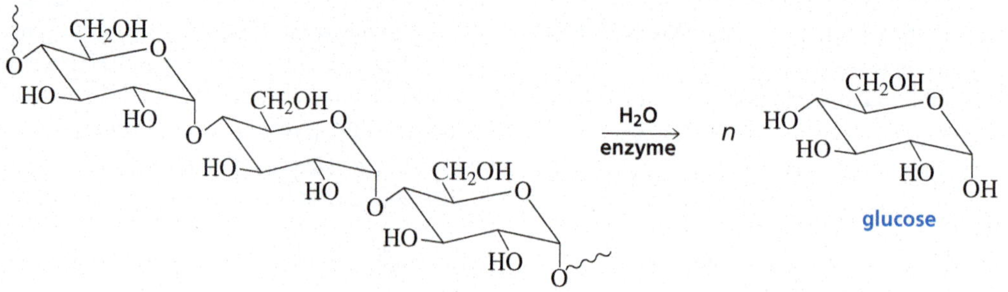

Glucose is Converted to Pyruvate

In the second stage of carbohydrate catabolism, each glucose molecule is converted to two molecules of pyruvate in a series of 10 reactions known as **glycolysis** or the *glycolytic pathway* (Figure 24.3 on the next page).

1. In the first reaction, glucose is converted to glucose-6-phosphate, by attacking the γ-phosphorus of ATP (Section 24.1)

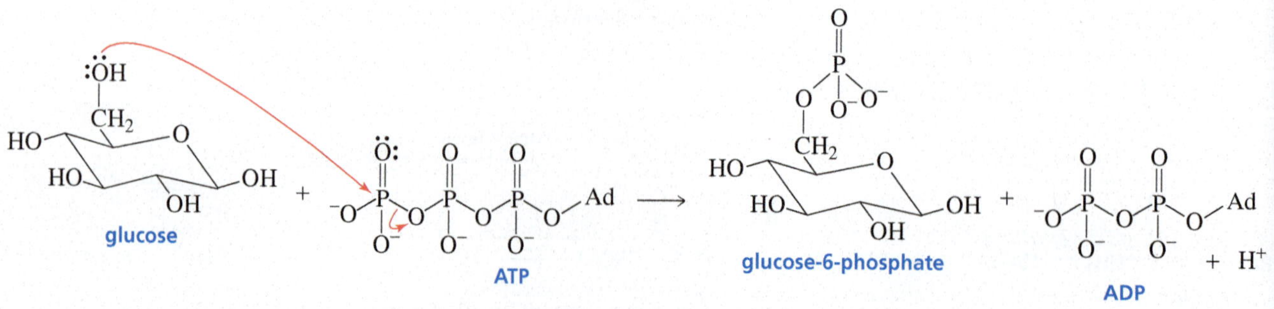

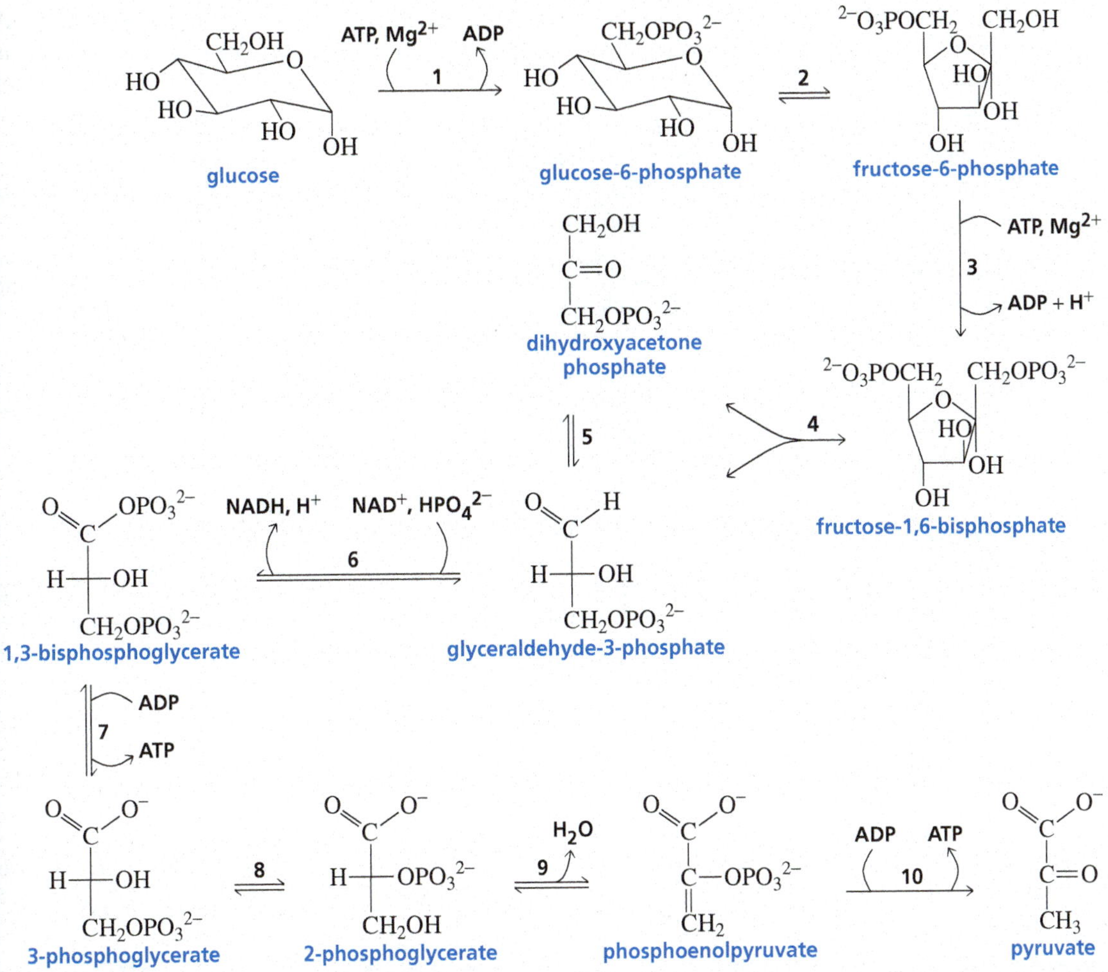

Glucose is converted to two molecules of pyruvate.

▲ **Figure 24.3**
Glycolysis, the series of enzyme-catalyzed reactions responsible for converting 1 mol of glucose to 2 mol of pyruvate. The enzymes that catalyze the reactions are: 1. hexokinase; 2. phosphoglucose isomerase; 3. phosphofructokinase; 4. aldolase; 5. triosephosphate isomerase; 6. glyceraldehyde-3-phosphate dehydrogenase; 7. phosphoglycerate kinase; 8. phosphoglycerate mutase; 9. enolase; and 10. pyruvate kinase.

2. Glucose-6-phosphate then isomerizes to fructose-6-phosphate, a reaction whose mechanism we examined in Section 22.12.
3. In the third reaction, ATP puts a second phosphate group on fructose-6-phosphate, forming fructose-1,6-bisphosphate. The mechanism of this reaction is the same as the one that converts glucose to glucose-6-phosphate.
4. The fourth reaction is the reverse of an aldol addition. We looked at the mechanism of this reaction in Section 22.13.
5. Dihydroxyacetone phosphate, produced in the fourth reaction, forms an enediol that then forms glyceraldehyde-3-phosphate (if the OH group at C-1 is the one that ketonizes) or reforms dihydroxyacetone phosphate (if the OH group at C-2 is the one that ketonizes) (Section 20.5).

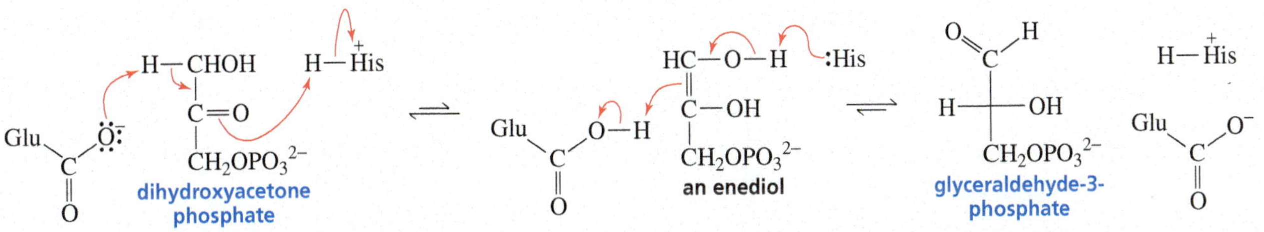

The mechanism for this reaction shows that a glutamate side chain of the enzyme removes a proton from the α-carbon, and a protonated histidine side chain donates a proton to the carbonyl oxygen. In the next step, histidine removes a proton from the C-1 OH group, and a glutamic acid protonates C-2. Compare this mechanism with the one for the enediol rearrangement shown on page 956.

> *Because each molecule of glucose is converted to a molecule of glyceraldehyde-3-phosphate and a molecule of dihydroxyacetone phosphate, and each molecule of dihydroxyacetone phosphate is converted to glyceraldehyde-3-phosphate, then overall each molecule of glucose is converted to two molecules of glyceraldehyde-3-phosphate.*

6. The aldehyde group of glyceraldehyde-3-phosphate is oxidized by NAD^+, forming 1,3-bisphosphoglycerate. In this reaction, the aldehyde is oxidized to a carboxylic acid, which then forms an ester with phosphoric acid. We looked at the mechanism of this reaction in Section 23.1.

glyceraldehyde-3-phosphate   1,3-bisphosphoglycerate

7. In the seventh reaction, 1,3-bisphosphoglycerate transfers a phosphate group to ADP, forming 3-phosphoglycerate and ATP. 1,3-Bisphosphoglycerate is called a "high-energy" intermediate, because transfer of its phosphoryl group to a nucleophile is more exergonic than transfer of a phosphoryl group from ATP to a nucleophile.

1,3-bisphosphoglycerate   ADP   3-phosphoglycerate   ATP

8. The eighth reaction is an isomerization in which 3-phosphoglycerate is converted to 2-phosphoglycerate. The enzyme that catalyzes this reaction has a phosphate group attached to a histidine side chain (see 3-phospho-His on the bottom of page 1117) that it transfers to the 2-position of 3-phosphoglycerate to form an intermediate with two phosphate groups. The intermediate transfers the phosphate group on its 3-position back to the histidine side chain.

3-phosphoglycerate   an intermediate   2-phosphoglycerate

9. The ninth reaction is a dehydration reaction that forms phosphoenolpyruvate. A lysine side chain removes a proton from the α-carbon in an E1cB reaction that forms a delocalized carbanion intermediate (Section 17.11). This proton is not very acidic because it is on the α-carbon of a carboxylate ion (Section 17.5). Two magnesium ions increase its acidity by stabilizing the conjugate base. The HO group of the intermediate is protonated by a glutamic acid side chain, which makes HO a better leaving group (Section 10.1).

2-phosphoglycerate   an intermediate   phosphoenolpyruvate

10. In the tenth and last reaction of glycolysis, phosphoenolpyruvate transfers its phosphate group to ADP, forming pyruvate and ATP. Phosphenolpyruvate is also a "high-energy" intermediate (see page 1110).

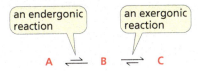

ATP is Used for Molecular Recognition

Phosphorylating glucose in the first reaction of glycolysis and phosphorylating fructose-6-phospate in the third reaction do not make glucose or fructose-6-phosphate any more reactive. The purpose of phosphorylation is to put a group on each of these compounds that allows enzymes to recognize them (and recognize the subsequent intermediates formed in glycolysis) so they can be bound at their active site. The two molecules of ATP that are used to put these "handles" on the sugar molecules are re-formed in the last step of glycolysis—namely, in the conversion of two molecules of phosphoenolpyruvate to two molecules of pyruvate and two molecules of ATP.

Coupled Reactions

Glycolysis is exergonic overall, but all the reactions in the pathway are not themselves exergonic. For example, the conversion of glyceraldehyde-3-phosphate to 1,3-bisphosphoglycerate (the sixth reaction; A to B as shown below) is an endergonic reaction. However, the subsequent reaction (the conversion of 1,3-bisphosphoglycerate to 3-phosphoglycerate; B to C as shown below) is highly exergonic. Therefore, as the second reaction converts B to C, the first reaction replenishes the equilibrium concentration of B. Thus, the exergonic reaction *drives* the endergonic reaction that precedes it. Recall that these two sequential reactions (an endergonic reaction followed by an exergonic reaction) are called **coupled reactions** (Section 5.7). Coupled reactions are the thermodynamic basis for how metabolic pathways operate.

an endergonic reaction		an exergonic reaction

A ⇌ B ⇌ C

PROBLEM 8

Draw the mechanism for the third reaction in glycolysis—the reaction of fructose-6-phosphate with ATP to form fructose-1,6-bisphosphate.

PROBLEM 9

The oxidation of glyceraldehyde-3-phosphate to 1,3-bisphosphoglycerate is an endergonic reaction, but the flow through this point in glycolysis proceeds smoothly. How is the unfavorable equilibrium constant overcome?

PROBLEM-SOLVING STRATEGY

Calculating Production of ATP

LEARN THE STRATEGY

How many molecules of ATP are obtained from each molecule of glucose that is metabolized to pyruvate?

First we need to count the number of ATPs used to convert glucose to pyruvate. We see that two are used: one to form glucose-6-phosphate and the other to form fructose-1,6-bisphosphate. Next, we need to know how many ATPs are formed. Each glyceraldehyde-3-phosphate that is metabolized to pyruvate forms two ATPs. Because each molecule of glucose forms two molecules of glyceraldehyde-3-phosphate, four molecules of ATP are formed from each molecule of glucose. Subtracting the two molecules used, we find that each molecule of glucose that is metabolized to pyruvate forms two molecules of ATP.

USE THE STRATEGY

PROBLEM 10 ◆

How many molecules of NAD$^+$ are required to convert one molecule of glucose to pyruvate?

Fats Versus Carbohydrates as a Source of Energy

Animals have a subcutaneous layer of fat cells that serves as both an energy source and an insulator. The fat content of the average man is about 21%, whereas the fat content of the average woman is about 25%. A fat provides about six times as much metabolic energy as an equal weight of hydrated glycogen because fats are less oxidized than carbohydrates and also, being nonpolar, do not bind water. In contrast, two-thirds of the weight of stored glycogen is water (Section 20.16). Humans can store sufficient fat to provide for the body's metabolic needs for two to three months, but can store only enough carbohydrate to provide for its metabolic needs for less than 24 hours. Carbohydrates, therefore, are used primarily as a quick, short-term energy source.

24.7 THE FATE OF PYRUVATE

We just saw that NAD$^+$, which cells have in limited supply, is used as an oxidizing agent in glycolysis. If glycolysis is to continue, the NADH that is produced must be oxidized back to NAD$^+$. Otherwise, no NAD$^+$ will be available as an oxidizing agent.

Under Aerobic Conditions

Under normal (aerobic) conditions (when oxygen is present), oxygen oxidizes NADH back to NAD$^+$ (this happens in *the fourth stage of catabolism*), and pyruvate is converted to acetyl-CoA, which then enters the citric acid cycle.

The conversion of pyruvate to acetyl-CoA occurs via a series of reactions catalyzed by a complex of three enzymes and five coenzymes, known collectively as the pyruvate dehydrogenase complex. The overall result of this series of reactions is to transfer the acetyl group of pyruvate to coenzyme A. We looked at the mechanisms for this series of reactions in Section 23.3.

$$\text{pyruvate} + \text{CoASH} \xrightarrow{\text{pyruvate dehydrogenase complex}} \text{acetyl-CoA} + CO_2$$

Under Anaerobic Conditions

When oxygen is in short supply (anaerobic conditions), such as when intense activity depletes all the oxygen in muscle cells, pyruvate (the product of glycolysis) oxidizes NADH back to NAD$^+$. In the process, pyruvate is reduced to lactate (lactic acid). The need to replenish oxygen is why people breathe hard during exercise.

$$\text{pyruvate} \xrightarrow[\text{lactate dehydrogenase}]{\text{NADH, H}^+ \quad \text{NAD}^+} \text{lactate}$$

The Fate of Pyruvate in Yeast

Although pyruvate is reduced to lactate under anaerobic conditions in mammals, it has a different fate in yeast—namely, it is decarboxylated to acetaldehyde by pyruvate decarboxylase (an enzyme that is not present in mammals). We looked at the mechanism of this reaction in Section 23.3.

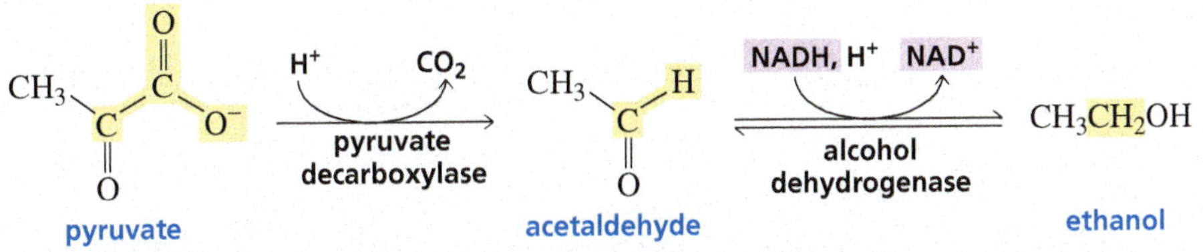

In this case, acetaldehyde is the compound that oxidizes NADH back to NAD^+ and in the process is reduced to ethanol. This reaction has been used by humankind for thousands of years to produce wine, beer, and other fermented drinks. (Notice that some enzyme names refer to the reverse reaction. For example, pyruvate decarboxylase refers to the forward reaction, whereas alcohol dehydrogenase refers to the reverse reaction.)

PROBLEM 11 ◆

Suggest a name for alcohol dehydrogenase that would refer to the forward reaction—that is, that would refer to the conversion of acetaldehyde to ethanol.

PROBLEM 12 ◆

What functional group of pyruvate is reduced when pyruvate is converted to lactate?

PROBLEM 13 ◆

What coenzyme is required to convert pyruvate to acetaldehyde?

PROBLEM 14

Propose a mechanism for the reduction of acetaldehyde by NADH to ethanol. (*Hint:* See Section 23.1.)

24.8 THE CATABOLISM OF PROTEINS: STAGES 1 AND 2

Proteins are Hydrolyzed to Amino Acids

In the first stage of protein catabolism, proteins are hydrolyzed in an enzyme-catalyzed reaction to amino acids.

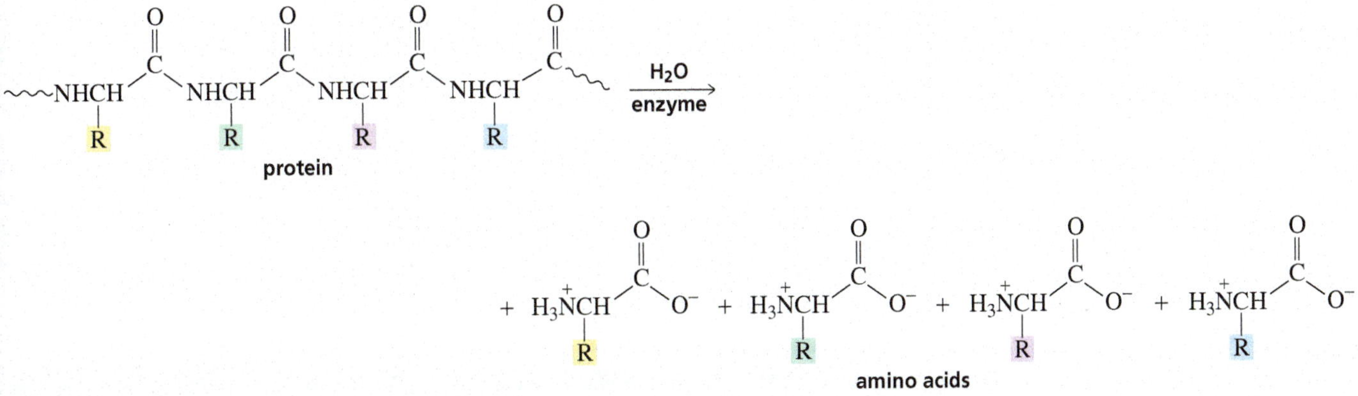

Amino Acids are Converted to Compounds that Can Enter the Citric Acid Cycle

In the second stage of protein catabolism, the amino acids are converted to acetyl-CoA, pyruvate, and/or citric acid cycle intermediates, depending on the amino acid. The products of the second stage of catabolism then enter the citric acid cycle—the third stage of catabolism—and are further metabolized.

Amino acids are converted to acetyl-CoA, pyruvate, and/or citric acid cycle intermediates.

We will use the catabolism of phenylalanine as an example of how an amino acid is metabolized (Figure 24.4). Phenylalanine is one of the essential amino acids, so it must be included in our diet (Section 21.1). The enzyme phenylalanine hydroxylase converts phenylalanine to tyrosine. Thus, tyrosine is not an essential amino acid, unless the diet lacks phenylalanine.

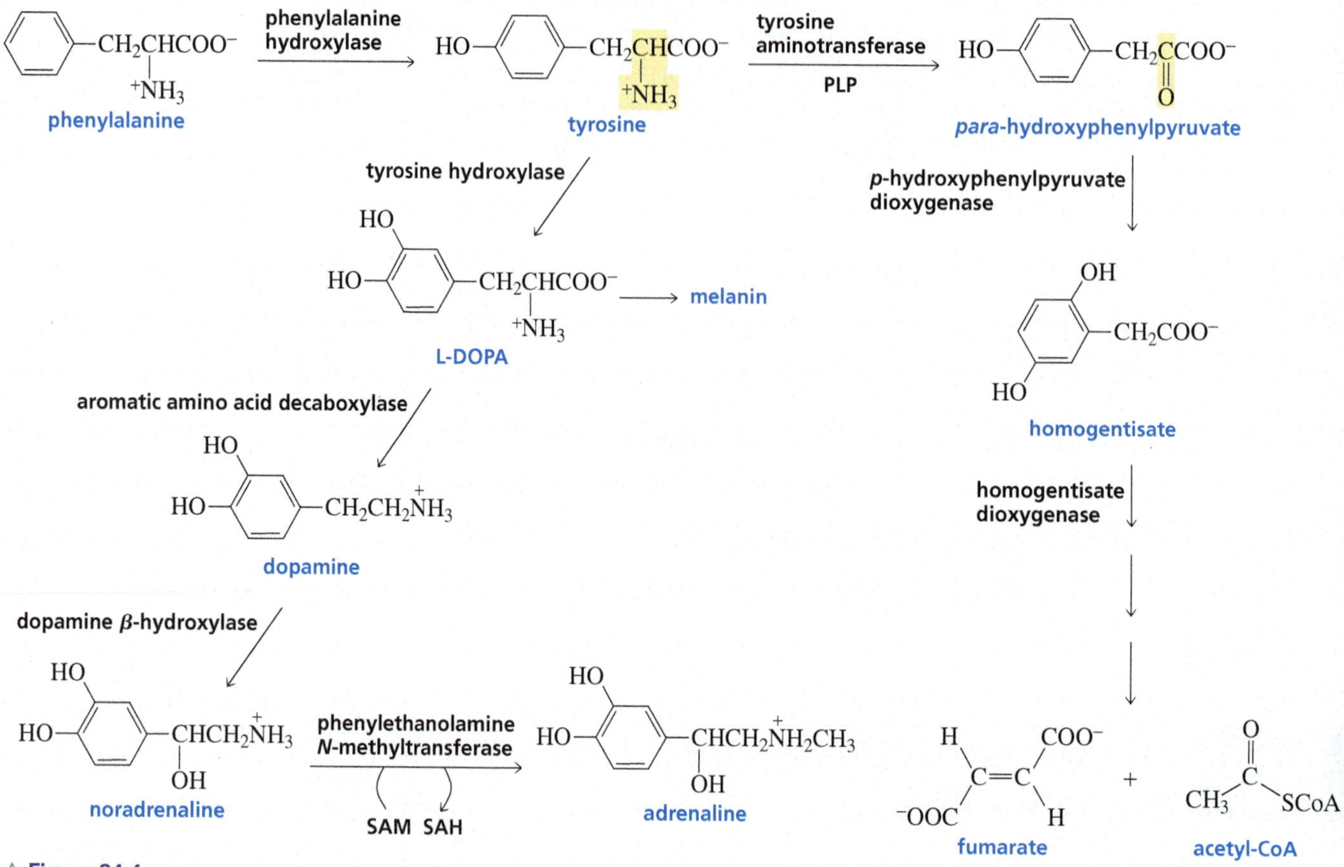

▲ Figure 24.4
The catabolism of phenylalanine.

The first reaction in the catabolism of most amino acids is transamination, a reaction that requires the coenzyme pyridoxal phosphate. We saw that transamination replaces the amino group of the amino acid with a ketone group (Section 23.5). *para*-Hydroxyphenylpyruvate, the product of the transamination of tyrosine, is converted by a series of reactions to fumarate and acetyl-CoA.

Fumarate is a citric acid cycle intermediate, so it can enter the citric acid cycle directly. We will see in Section 24.9 how acetyl-CoA enters the citric acid cycle. Remember that each of the reactions in this catabolic pathway is catalyzed by a different enzyme.

In addition to being used for energy, the amino acids that we ingest are also used for the synthesis of proteins and for the synthesis of other compounds the body needs. For example, tyrosine is used to synthesize neurotransmitters (dopamine and adrenaline) and melanin, which is the compound responsible for skin and hair pigmentation. Recall that SAM (*S*-adenosylmethionine) is the biological methylating agent that converts noradrenaline to adrenaline (Section 10.12).

Phenylketonuria (PKU): An Inborn Error of Metabolism

About one in every 20,000 babies is born without phenylalanine hydroxylase, the enzyme that converts phenylalanine to tyrosine. This genetic disease is called phenylketonuria (PKU). Without phenylalanine hydroxylase, the level of phenylalanine builds up; when it reaches a high concentration, it is transaminated to phenylpyruvate, a compound that interferes with normal brain development. The high level of phenylpyruvate found in urine gives the disease its name.

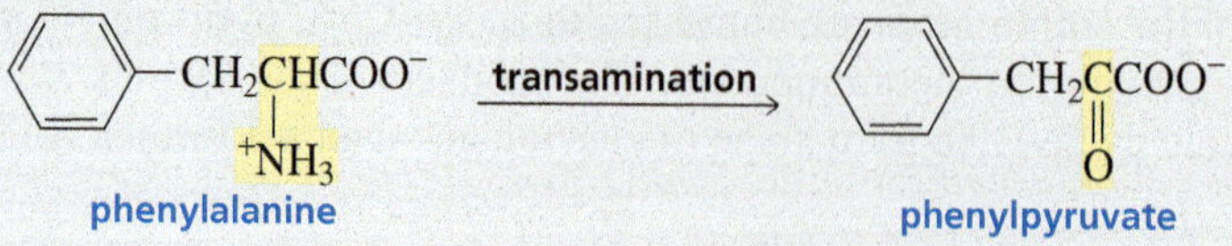

Within 24 hours after birth, all babies born in the United States are tested for high serum phenylalanine levels, which indicate a buildup of phenylalanine caused by the absence of phenylalanine hydroxylase. Babies with high levels are immediately put on a diet low in phenylalanine and high in tyrosine. As long as the phenylalanine level is kept under careful control for the first 5 to 10 years of life, the child will experience no adverse effects. You may have noticed the warning on containers of foods that contain NutraSweet, announcing that it contains phenylalanine. (Recall that this sweetener is a methyl ester of a dipeptide of L-aspartate and L-phenylalanine; see page 1007).

If phenylalanine in the diet is not controlled, then the baby will be severely mentally retarded by the time he or she is a few months old. Untreated children have paler skin and fairer hair than other members of their family because, without tyrosine, they cannot synthesize melanin, a skin and hair pigment. Half of untreated people with PKU die by age 20. When a woman with PKU becomes pregnant, she must return to the low-phenylalanine diet she had as a child because a high level of phenylalanine can cause abnormal development of the fetus.

Alcaptonuria

Another genetic disease that results from a deficiency of an enzyme in the pathway for phenylalanine degradation is alcaptonuria, which is caused by lack of homogentisate dioxygenase. The only ill effect of this enzyme deficiency is black urine. The urine of those afflicted with alcaptonuria turns black because the homogentisate they excrete immediately oxidizes in the air, forming a black compound.

PROBLEM 15 ◆

What compound is formed when alanine is transaminated?

24.9 THE CITRIC ACID CYCLE: STAGE 3

The **citric acid cycle** (the third stage of catabolism) is a series of eight reactions in which the acetyl group of each molecule of acetyl-CoA—formed by the catabolism of fats, carbohydrates, and amino acids—is converted to two molecules of CO_2 (Figure 24.5 on the next page).

> The acetyl group of each molecule of acetyl-CoA that enters the citric acid cycle is converted to two molecules of CO_2.

The series of reactions is called a *cycle* because, unlike the reactions of other metabolic pathways, they comprise a closed loop in which the product of the eighth reaction (oxaloacetate) is the reactant for the first reaction.

1. In the first reaction of the citric acid cycle, acetyl-CoA reacts with oxaloacetate to form citrate. The mechanism for the reaction shows that an aspartate side chain of the enzyme removes a proton from the α-carbon of acetyl-CoA, creating an enolate ion. This enolate ion adds to the keto carbonyl carbon of oxaloacetate and the carbonyl oxygen picks up a proton from a histidine side chain. This is similar to an aldol addition where the α-carbanion (enolate ion) of one molecule is the nucleophile and the carbonyl carbon of another is the electrophile (Section 17.10). The intermediate (a thioester) that results is hydrolyzed to citrate in a nucleophilic acyl substitution reaction (Section 15.7).

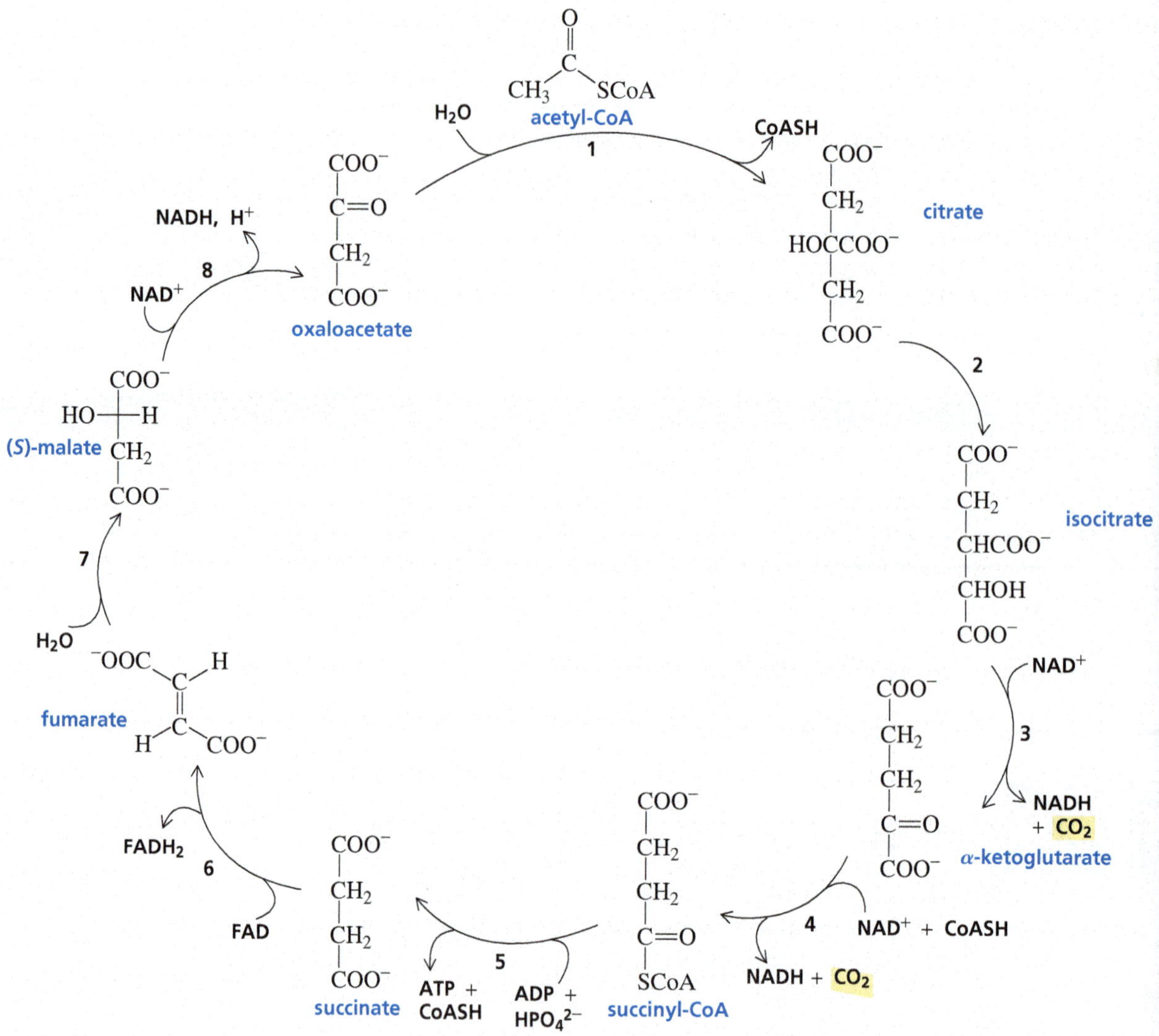

▲ **Figure 24.5**
The citric acid cycle is the series of enzyme-catalyzed reactions responsible for the oxidation of the acetyl group of acetyl-CoA to two molecules of CO_2. The enzymes that catalyze the reactions are: 1. citrate synthase; 2. aconitase; 3. isocitrate dehydrogenase; 4. α-ketoglutarate dehydrogenase; 5. succinyl-CoA synthetase; 6. succinate dehydrogenase; 7. fumarase; and 8. malate dehydrogenase.

2. In the second reaction, citrate is converted to isocitrate, its isomer. The reaction takes place in two steps: water is removed in the first step and then re-added in the second step. The first step is an *E2 dehydration* (Section 10.4) in which a serine side chain removes a proton, and the OH leaving group is protonated by a histidine side chain to make it a weaker base (H_2O) and, therefore, a better leaving group. In the second step, *conjugate addition of water* to the intermediate, with histidine removing a proton to make water a better nucleophile and serine providing the proton, forms isocitrate (Section 16.16).

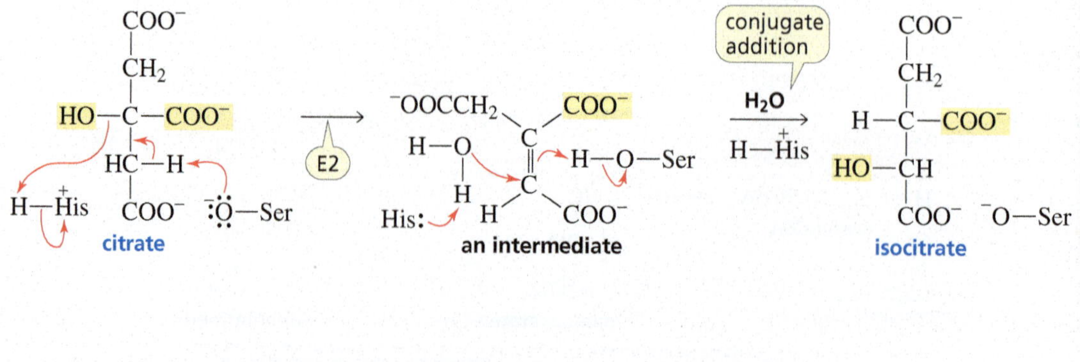

3. The third reaction is the one that releases the first molecule of CO_2. It also has two steps. In the first, the secondary alcohol group of isocitrate is *oxidized* to a ketone by NAD^+ (Section 23.1). In the second, the ketone *loses* CO_2, with Mg^{2+} acting as a catalyst (see page 1039). We saw that a CO_2 group bonded to a carbon adjacent to a carbonyl carbon can be removed because the electrons left behind can be delocalized onto the carbonyl oxygen (Section 17.17). The enolate ion tautomerizes to a ketone (Section 17.3).

isocitrate   an intermediate   α-ketoglutarate

4. The fourth reaction is the one that releases the second molecule of CO_2. The reaction requires a group of enzymes and the same five coenzymes (with the same mechanisms) required by the pyruvate dehydrogenase complex that converts pyruvate to acetyl-CoA (pages 1076–1077). Like the reaction catalyzed by the pyruvate dehydrogenase complex, the overall result of this reaction is the *transfer of an acyl group to CoASH*. Thus, the product of the reaction is succinyl-CoA.

α-ketoglutarate   succinyl-CoA

5. The fifth reaction takes place in two steps. First, hydrogen phosphate reacts with succinyl-CoA in a *nucleophilic acyl substitution reaction* to form an intermediate, which then transfers its phosphate group to ADP.

red arrows = nucleophilic addition to form a tetrahedral intermediate

blue arrows = elimination from the tetrahedral intermediate

succinyl-CoA   an intermediate   succinate

The intermediate does not transfer its phosphate group directly to ADP. Instead, it transfers the phosphate group to a histidine side chain of the enzyme, forming 3-phospho-His, which then transfers the phosphate group to ADP.

3-phospho-His   ATP

At this point, the citric acid cycle has accomplished the required transformation—that is, acetyl-CoA has been converted to CoASH and two molecules of CO_2. What remains to be done is to convert succinate to oxaloacetate, so oxaloacetate can begin the cycle again by reacting with another molecule of acetyl-CoA.

6. In the sixth reaction, FAD *oxidizes* succinate to fumarate. We looked at the mechanism of this reaction in Section 23.2.

7. *Conjugate addition of water* to the double bond of fumarate forms (S)-malate. We saw why the reaction forms only one enantiomer in Section 6.14.

8. *Oxidation of the secondary alcohol* group of (S)-malate by NAD^+ forms oxaloacetate, returning the cycle to its starting point. Oxaloacetate now begins the cycle again, reacting with another molecule of acetyl-CoA to initiate the conversion of acetyl-CoA's acetyl group to another two molecules of CO_2.

Notice that reactions 6, 7, and 8 in the citric acid cycle are similar to reactions 1, 2, and 3 in the β-oxidation of fatty acids (Section 24.5).

PROBLEM 16

Acid-catalyzed dehydration reactions are normally E1 reactions. Why is the acid-catalyzed dehydration in the second reaction of the citric acid cycle an E2 reaction?

PROBLEM 17 ◆

What functional group of isocitrate is oxidized in the third reaction of the citric acid cycle?

PROBLEM 18 ◆

The citric acid cycle is also called the tricarboxylic acid cycle (or TCA cycle). Which of the citric acid cycle intermediates are tricarboxylic acids?

PROBLEM 19 ◆

What acyl group is transferred by thiamine pyrophosphate in the fourth reaction of the citric acid cycle?

PROBLEM 20

Draw the steps for the conversion of α-ketoglutarate to succinyl-CoA.

PROBLEM 21 ◆

a. Which of the eight enzymes in the citric acid cycle has a name that refers to the reverse reaction?
b. Which of the ten enzymes in glycolysis has a name that refers to the reverse reaction?

24.10 OXIDATIVE PHOSPHORYLATION: STAGE 4

The NADH and $FADH_2$ molecules formed in the second and third stages of catabolism undergo **oxidative phosphorylation**—the fourth stage of catabolism—which oxidizes them back to NAD^+ and FAD so that they can participate in more oxidation reactions.

The electrons lost when NADH and $FADH_2$ are oxidized are transferred to a system of linked electron acceptors. One of the first electron acceptors is coenzyme Q_{10}, which is a quinone. We saw that when a quinone gains electrons (is reduced), it forms a hydroquinone (Section 12.11). When the hydroquinone passes electrons to the next electron acceptor, the hydroquinone is oxidized back to a quinone. The last electron acceptor is O_2. When O_2 accepts electrons, it is reduced to water. This chain of oxidation–reduction reactions supplies the energy that is used to convert ADP to ATP.

coenzyme Q_{10}

For each NADH that undergoes oxidative phosphorylation, 2.5 molecules of ATP are formed, and for each $FADH_2$ that undergoes oxidative phosphorylation, 1.5 molecules of ATP are formed.

$$NADH \longrightarrow NAD^+ + 2.5 \ ATP$$

$$FADH_2 \longrightarrow FAD + 1.5 \ ATP$$

Each round of the citric acid cycle forms 3 molecules of NADH, 1 molecule of $FADH_2$, and 1 molecule of ATP. Therefore, for every molecule of acetyl-CoA that enters the citric acid cycle, 7.5 molecules of ATP are formed from the 3 molecules of NADH, 1.5 molecules of ATP are formed from the 1 molecule of $FADH_2$, and 1 molecule of ATP is formed in the cycle, for a total of 10 ATP.

$$3 \ NADH + FADH_2 + ATP \longrightarrow 3 \ NAD^+ + FAD + 10 \ ATP$$

> In oxidative phosphorylation, each molecule of NADH is converted to 2.5 molecules of ATP and each molecule of $FADH_2$ is converted to 1.5 molecules of ATP.

Basal Metabolic Rate

Your basal metabolic rate (BMR) is the number of calories you would burn if you stayed in bed all day. A BMR is affected by gender, age, and genetics: it is greater for men than for women, it is greater for young people than for old people, and some people are born with a faster metabolic rate than others. The BMR is also affected by the percentage of body fat: the higher the percentage, the lower the BMR. For humans, the average BMR is about 1600 kcal/day.

In addition to consuming sufficient calories to sustain your basal metabolism, you must also consume calories for the energy needed to carry out physical activities. The more active you are, the more calories you must consume to maintain your current weight. People who consume more calories than required by their BMR plus their level of physical activity will gain weight; if they consume fewer calories, they will lose weight.

PROBLEM 22 ◆

How many molecules of ATP are obtained from the conversion of one molecule of glycerol to pyruvate

a. not including the fourth stage of catabolism?   **b.** including the fourth stage of catabolism?

24.11 ANABOLISM

Anabolism is the reverse of catabolism. In anabolism, acetyl-CoA, pyruvate, citric acid cycle intermediates, and intermediates formed in glycolysis are the starting materials for the synthesis of fatty acids, carbohydrates, and proteins.

For example, we saw how cells use acetyl-CoA to synthesize fatty acyl-CoAs (Section 17.20). Once the fatty acyl-CoAs are synthesized, they can form fats or oils by esterifying glycerol-3-phosphate, which is obtained by reducing dihydroxyacetone phosphate, an intermediate formed in glycolysis.

$$\text{dihydroxyacetone phosphate} \xrightarrow[]{\text{NADH + H}^+ \quad \text{NAD}^+} \text{glycerol-3-phosphate} \xrightarrow[\text{acyl transferase}]{R^1\text{C(O)SCoA} \quad \text{CoASH}} \xrightarrow[\text{acyl transferase}]{R^2\text{C(O)SCoA} \quad \text{CoASH}} \text{phosphatidic acid}$$

a fat or an oil ⟵ 1,2-diacylglycerol

24.12 GLUCONEOGENESIS

Gluconeogenesis—the synthesis of glucose from pyruvate—is an anabolic pathway. Glucose is the primary fuel for the body. But in times of prolonged exercise or fasting, the body runs out of glucose and must use fat for its fuel. The brain, however, cannot metabolize fat, so it has to have a continuous supply of glucose. Therefore, the body needs to have a way to synthesize glucose when a sufficient supply is not available.

As you can see by comparing gluconeogenesis and glycolysis in Figure 24.6 on the next page, many of the reactions involved in the synthesis of glucose are carried out by the same enzymes that catalyze the breakdown of glucose in glycolysis—they are just operating in reverse. However, all the reactions in gluconeogenesis cannot be just the reverse of those operating in glycolysis. Some of the enzymes in each pathway catalyze irreversible reactions, and detours must be made around these reactions when going in the other direction. By using different enzymes for the forward and reverse irreversible reactions, both pathways become thermodynamically favorable.

Reactions 1, 3, and 10 in glycolysis are irreversible (Figure 24.6). Therefore, a different enzyme is needed to catalyze the reverse of these reactions in gluconeogenesis. The reverse of the last irreversible reaction (10) in glycolysis is actually two successive enzyme-catalyzed reactions. First pyruvate is converted to oxaloacetate by pyruvate carboxylase, a biotin-dependent enzyme whose mechanism we looked at in Section 23.4. Oxaloacetate is then converted to phosphoenolpyruvate. In this reaction, oxaloacetate is decarboxylated (Section 17.17) and the oxygen of the enolate ion attacks the γ-phosphorus of GTP.

oxaloacetate + GTP ⟶ phosphoenolpyruvate + CO_2 + GDP

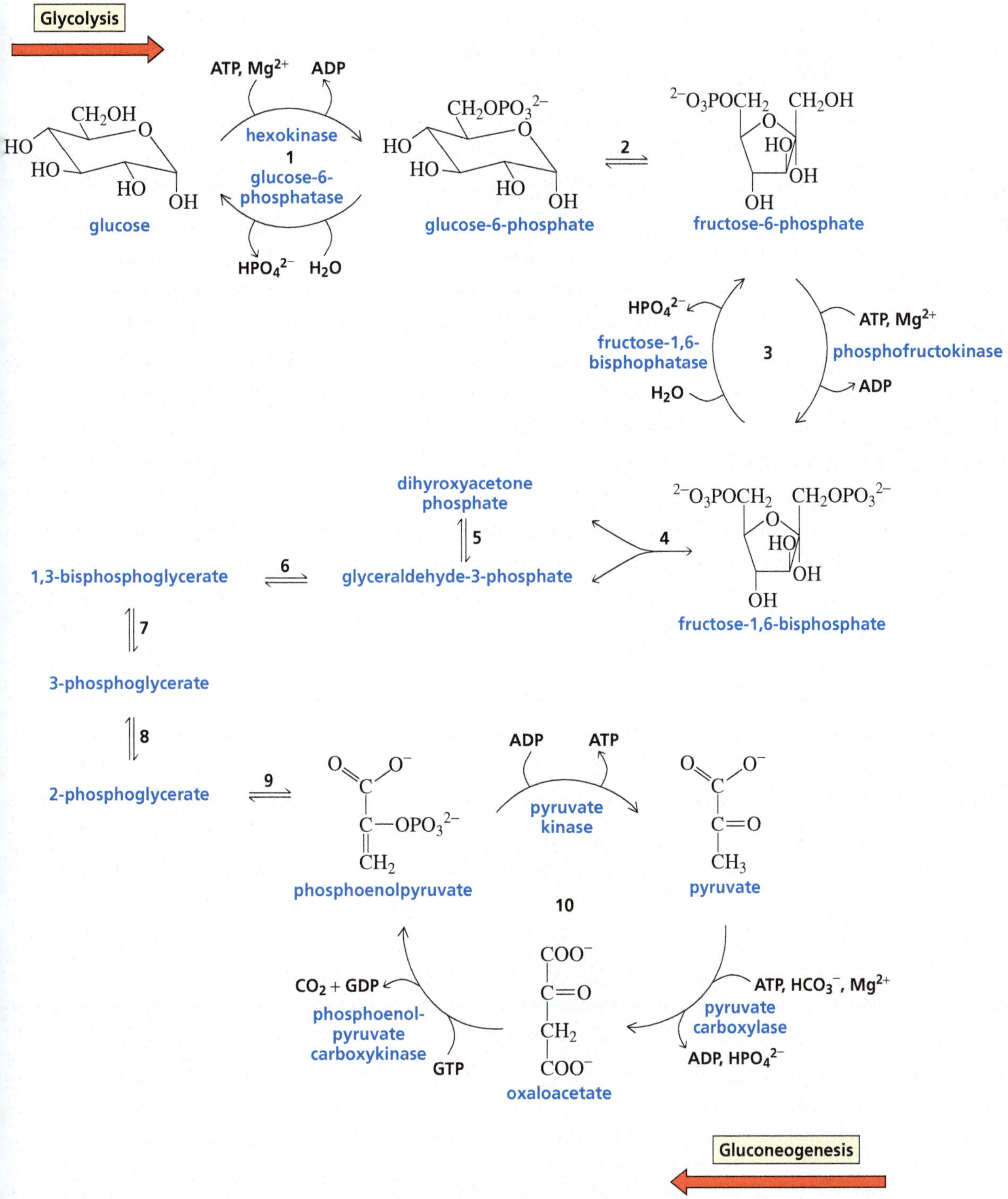

▲ Figure 24.6
Glycolysis (the conversion of glucose to pyruvate) and gluconeogenesis (the biosynthesis of glucose from pyruvate).

The hydrolysis of fructose-1,6-bisphosphate to fructose-6-phosphate, the next reaction in gluconeogenesis that needs an enzyme (3), because the reaction in the other direction is irreversible, is catalyzed by fructose-1,6-bisphosphatase. A **phosphatase** is an enzyme that catalyzes the removal of a phosphoryl group from its substrate. Finally, glucose-6-phosphatase (1) hydrolyzes glucose-6-phosphate irreversibly to glucose. Notice that both of these irreversible enzymes catalyze removal of the "handle" that was put on the molecules for the purpose of molecular recognition (page 1111).

PROBLEM 23 ◆

a. What is the name of the enzyme that converts glycerol to glycerol-3-phosphate?
b. What is the name of the enzyme that converts phosphatidic acid to 1,2-diacylglycerol?

24.13 REGULATING METABOLIC PATHWAYS

Regulatory Enzymes

The simultaneous synthesis and breakdown of glucose would be counterproductive. Therefore, the two pathways must be controlled so that glucose is synthesized and stored when the cell does not need it for energy, and glucose is broken down when it is needed for energy. To accomplish this, an enzyme, that catalyzes an irreversible reaction near the beginning of a pathway, is able to be turned on and off. This enzyme is called a **regulatory enzyme.**

Regulatory enzymes allow independent control over degradation and synthesis in response to a cell's needs. Some of the ways the three irreversible enzymes in glycolysis and the three irreversible enzymes in gluconeogenesis are controlled are quite complicated. Therefore, we will consider only a few of the control mechanisms here.

Feedback Inhibitors

Hexokinase

Hexokinase, the first enzyme that catalyzes an irreversible reaction in glycolysis (1), is a regulatory enzyme. It is inhibited by glucose-6-phosphate, its product. So if the concentration of glucose-6-phosphate rises above normal levels, then there is no reason to continue to synthesize it, so the enzyme is turned off. Glucose-6-phosphate is a **feedback inhibitor**—that is, it inhibits the enzyme that catalyzes its biosynthesis.

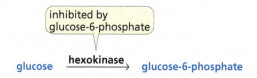

Allosteric Inhibitors and Activators

Phosphofructokinase

Phosphofructokinase, the enzyme that converts fructose-6-phosphate to fructose-1,6-bisphosphate, is the next enzyme (3) that catalyzes an irreversible reaction in glycolysis. It, too, is a regulatory enzyme. A high concentration of ATP in the cell is a signal that ATP is being produced faster than it is being consumed, so there is no reason to continue to break down glucose. Therefore, ATP is an inhibitor of phosphofructokinase. It inhibits the enzyme by binding to it and causing a conformational change that decreases its affinity for its substrate. ATP is an example of an allosteric inhibitor. An **allosteric inhibitor** inhibits an enzyme by binding to a site on the enzyme other than the active site (*allos* and *stereos* are Greek for "other" and "space"). This affects the shape of the active site, which, in turn, affects its ability to bind its substrate.

On the other hand, when the concentration of ADP and AMP in the cell are high, this is a signal that ATP is being consumed faster than it is being produced. Therefore, ADP and AMP are **allosteric activators** of phosphofructokinase. They bind to the enzyme and reverse the inhibition that was brought on by binding ATP.

Citrate is also an allosteric inhibitor of phosphofructokinase. A high concentration of citrate (a citric acid cycle intermediate) in a cell signals that the cell is currently meeting its energy needs by the oxidation of fats and proteins, so the oxidation of carbohydrates can be stopped temporarily.

Pyruvate carboxylase

Pyruvate carboxylase, the first enzyme that catalyzes an irreversible reaction in gluconeogenesis, is also a regulatory enzyme. Pyruvate has two fates. It can be converted to oxaloacetate (by pyruvate carboxylase), which then goes on to make glucose for energy storage, or it can be converted to acetyl-CoA (by the pyruvate dehydrogenase complex), which then enters the citric acid cycle to be metabolized for energy. Acetyl-CoA is an allosteric activator of pyruvate carboxylase and a feedback inhibitor of the pyruvate dehydrogenase complex. A high concentration of acetyl-CoA signals that there is currently no need for energy, so pyruvate is converted to glucose rather than prepared to enter the citric acid cycle.

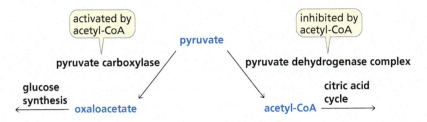

24.14 AMINO ACID BIOSYNTHESIS

The only amino acids synthesized by the body are the 10 nonessential amino acids—the others must be obtained from food. The nonessential amino acids are biosynthesized from one of four metabolic intermediates: α-ketoglutarate, pyruvate, oxaloacetate, or 3-phosphoglycerate. Each amino acid has its own pathway for its biosynthesis.

For example, we saw that glutamate is biosynthesized by a transamination reaction, using α-ketoglutarate as the nitrogen acceptor and an amino acid as the nitrogen donor (Section 23.5). Alanine and aspartate also are biosynthesized by a transamination reaction, using pyruvate and oxaloacetate, respectively, as the nitrogen acceptor and an amino acid as the nitrogen donor.

$$^-OOC-CH_2CH_2-\underset{\underset{O}{\|}}{C}-COO^-$$
α-ketoglutarate

$$CH_3-\underset{\underset{O}{\|}}{C}-COO^-$$
pyruvate

$$^-OOC-CH_2-\underset{\underset{O}{\|}}{C}-COO^-$$
oxaloacetate

amino acid → α-keto acid

amino acid → α-keto acid

amino acid → α-keto acid

$$^-OOC-CH_2CH_2-\underset{\underset{^+NH_3}{|}}{\overset{\overset{H}{|}}{C}}-COO^-$$
glutamate

$$CH_3-\underset{\underset{^+NH_3}{|}}{\overset{\overset{H}{|}}{C}}-COO^-$$
alanine

$$^-OOC-CH_2-\underset{\underset{^+NH_3}{|}}{\overset{\overset{H}{|}}{C}}-COO^-$$
aspartate

Serine is biosynthesized by oxidizing 3-phosphoglycerate (an intermediate in glycolysis), transaminating the product with glutamate and then hydrolyzing off the phosphate group.

$$\underset{\underset{CH_2OPO_3^{2-}}{|}}{\overset{\overset{COO^-}{|}}{H{-}OH}} \xrightarrow{\text{NAD}^+ \quad \text{NADH}+\text{H}^+} \underset{\underset{CH_2OPO_3^{2-}}{|}}{\overset{\overset{COO^-}{|}}{C{=}O}} \xrightarrow{\text{glutamate} \quad \alpha\text{-ketoglutarate}} \underset{\underset{CH_2OPO_3^{2-}}{|}}{\overset{\overset{COO^-}{|}}{H_3\overset{+}{N}{-}H}} \xrightarrow{\text{H}_2\text{O} \quad \text{HPO}_4^{2-}} \underset{\underset{CH_2OH}{|}}{\overset{\overset{COO^-}{|}}{H_3\overset{+}{N}{-}H}}$$

3-phosphoglycerate

serine

In Section 26.9, we will see how proteins are biosynthesized from amino acids.

PROBLEM 24

Glutamine is biosynthesized from glutamate in two steps using ATP and ammonia. Propose a mechanism for this biosynthesis.

ESSENTIAL CONCEPTS

Section 24.0

- **Metabolism** is the set of reactions living organisms carry out to obtain energy and to synthesize the compounds they need. Metabolism can be divided into **catabolism** and **anabolism.**

- A **catabolic pathway** is a series of reactions that break down a compound to provide energy and simpler compounds.

- An **anabolic pathway** is a series of reactions that leads to the synthesis of a compound from simpler compounds.

Section 24.1

- **ATP** is a cell's most important source of chemical energy; ATP provides a reaction pathway involving a good leaving group for a reaction that would not otherwise occur because of a poor leaving group. This occurs by way of a phosphoryl transfer reaction.

- A **phosphoryl transfer reaction** involves the transfer of a phosphoryl group of ATP to a nucleophile as a result of breaking a **phosphoanhydride bond.**

- A phosphoryl transfer reaction forms one of three intermediates—an **acyl (or alkyl) phosphate,** an **acyl (or alkyl) pyrophosphate,** or an **acyl (or alkyl) adenylate.**

Section 24.2

- ATP is kinetically stable in a cell because of its negative charges. ATP is reactive at the active site of an enzyme where Mg^{2+} and amino acid side chains reduce the overall negative charge.

Section 24.3

- The reaction of a nucleophile with a phosphoanhydride bond is highly exergonic because of electrostatic repulsion, solvation, and electron delocalization.

Section 24.4

- Catabolism can be divided into four stages. In the *first stage*, fats, carbohydrates, and proteins are hydrolyzed to fatty acids, monosaccharides, and amino acids.

- In the *second stage*, the products obtained from the first stage are converted to compounds that can enter the citric acid cycle. In order to enter the citric acid cycle, a compound must be either a citric acid cycle intermediate, acetyl-CoA, or pyruvate (because pyruvate can be converted to acetyl-CoA).

- The *third stage* is the **citric acid cycle,** and the *fourth stage* is **oxidative phosphorylation.**

- A **kinase** is an enzyme that transfers a phosphoryl group to its substrate.

- A **dehydrogenase** is an enzyme that oxidizes its substrate.

Section 24.5

- In the first stage of fat catabolism, a fat is hydrolyzed to fatty acids.

- In the second stage, each fatty acid is converted to a fatty acyl-CoA, which is converted to a fatty acyl-CoA with two fewer carbons and acetyl-CoA in a four-reaction pathway called **β-oxidation.** The series of four reactions is repeated until the entire fatty acyl-CoA has been converted to acetyl-CoA molecules.

Section 24.6

- In the first stage of carbohydrate catabolism, a carbohydrate is hydrolyzed to glucose molecules.

- In the second stage, glucose is converted to two molecules of pyruvate in a series of 10 reactions known as **glycolysis.**

- **Coupled reactions** consist of an endergonic reaction followed by an exergonic reaction, which drives the endergonic reaction toward products.

Section 24.7

- Under aerobic conditions, pyruvate is converted to acetyl-CoA, which then enters the citric acid cycle.

- Under anaerobic conditions, pyruvate is reduced to lactate.

- Under anaerobic conditions in yeast, pyruvate is decarboxylated to acetaldehyde, which is reduced to ethanol.

Section 24.8

- In the first stage of protein catabolism, a protein is hydrolyzed to amino acids.

- In the second stage, amino acids are metabolized to pyruvate, acetyl-CoA, and/or citric acid cycle intermediates, depending on the amino acid.

Section 24.9

- The citric acid cycle, the third stage of catabolism, is a series of 8 reactions that converts the acetyl group of each molecule of acetyl-CoA that enters the cycle to two molecules of CO_2.

Section 24.10

- In oxidative phosphorylation, the fourth stage of catabolism, each molecule of NADH and $FADH_2$ formed in oxidation reactions in the second and third stages of catabolism is converted into 2.5 molecules of ATP and 1.5 molecules of ATP, respectively.

Section 24.11

- Anabolism is the reverse of catabolism. In anabolism, acetyl-CoA, pyruvate, glycolytic intermediates, and citric acid cycle intermediates are the starting materials for the biosynthesis of fatty acids, carbohydrates, and proteins. For example, acetyl-CoA is the starting material for the biosynthesis of fats.

Section 24.12

- Many of the reactions involved in the biosynthesis of glucose from pyruvate—**gluconeogenesis**—are carried out by the same enzymes that catalyze the reactions in glycolysis—they are just operating in reverse.

Section 24.13

- Some of the enzymes near the beginning of each pathway catalyze irreversible reactions, and those reactions have to be detoured around when going in the other direction.

- The enzyme that catalyzes an irreversible reaction near the beginning of the pathway is a **regulatory enzyme**—it can be activated and inhibited.
- A **feedback inhibitor** inhibits the enzyme that catalyzes its biosynthesis.
- An **allosteric inhibitor** or **activator** inhibits or activates an enzyme by binding to a site on the enzyme other than the active site, which affects the function of the active site.

- A **phosphatase** is an enzyme that catalyzes the removal of a phosphoryl group from its substrate.

Section 24.14

- All the nonessential amino acids are biosynthesized from one of four metabolic intermediates: α-ketoglutarate, pyruvate, oxaloacetate, or 3-phosphoglycerate.

PROBLEMS

25. Indicate whether an anabolic pathway or a catabolic pathway does the following:

   a. produces energy in the form of ATP

   b. involves primarily oxidation reactions

26. Galactose can enter the glycolytic cycle but it must first react with ATP to form galactose-1-phosphate. Propose a mechanism for this reaction.

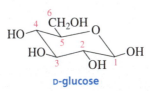

     galactose   ATP   galactose-1-phosphate   ADP

27. When pyruvate is reduced by NADH to lactate, which hydrogen in lactate comes from NADH?

28. *S*-Adenosylmethionine (SAM) is formed from the reaction of methionine with ATP (Section 10.12). The other product of the reaction is triphosphate. Propose a mechanism for this reaction.

29. Which of the ten reactions in glycolysis are

   a. phosphorylations?   **b.** isomerizations?   **c.** oxidations?   **d.** dehydrations?

30. Which reactions in the citric acid cycle form a product with a new asymmetric center?

31. a. Which reaction links glycolysis and the citric acid cycle?
    b. How is the reaction catalyzed?

32. Acyl-CoA synthase is the enzyme that activates a fatty acid by converting it to a fatty acyl-CoA (Section 24.5) in a series of two reactions. In the first reaction, the fatty acid reacts with ATP and one of the products formed is pyrophosphate. The other product reacts in a second reaction with CoASH to form the fatty acyl-CoA. Propose a mechanism for each of the reactions.

33. In some brain cancer cells, a mutated form of isocitrate dehydrogenase, instead of catalyzing the oxidation of the secondary alcohol of isocitrate, catalyzes the reduction of α-ketoglutarate. Draw the product of the reaction.

34. If the phosphorus atom in 3-phosphoglycerate is radioactively labeled, where will the label be when the reaction that forms 2-phosphoglycerate is over?

35. Which carbon atoms of glucose end up as a carboxyl group in pyruvate?

D-glucose

36. Which carbon atoms of glucose end up in ethanol under anaerobic conditions in yeast?

37. How would blood glucose levels be affected before and after a 24-hour fast if there is a deficiency of fructose-1,6-bisphosphatase?

38. Explain why the conversion of pyruvate to lactate is a reversible reaction but the conversion of pyruvate to acetaldehyde is not reversible.

39. How many molecules of acetyl-CoA are obtained from the β-oxidation of one molecule of a 16-carbon saturated fatty acyl-CoA?

40. How many molecules of CO_2 are obtained from the complete metabolism of one molecule of a 16-carbon saturated fatty acyl-CoA?

41. How many molecules of ATP are obtained from the β-oxidation of one molecule of a 16-carbon saturated fatty acyl-CoA?

42. How many molecules of NADH and $FADH_2$ are obtained from the β-oxidation of one molecule of a 16-carbon saturated fatty acyl-CoA?

43. How many molecules of ATP are obtained from the NADH and $FADH_2$ formed in the β-oxidation of one molecule of a 16-carbon saturated fatty acyl-CoA?

44. How many molecules of ATP are obtained from the complete (including the fourth stage of catabolism) metabolism of one molecule of a 16-carbon saturated fatty acyl-CoA?

45. How many molecules of ATP are obtained from complete (including the fourth stage of catabolism) metabolism of one molecule of glucose?

46. What are four possible fates of pyruvate in a mammalian cell?

47. Most fatty acids have an even number of carbons and, therefore, are completely metabolized to acetyl-CoA. A fatty acid with an odd number of carbons is metabolized to acetyl-CoA and one equivalent of propionyl-CoA. The following two reactions convert propionyl-CoA into succinyl-CoA, a citric acid cycle intermediate, so it can be further metabolized. Each of the reactions requires a coenzyme. Identify the coenzyme for each step. From what vitamins are the coenzymes derived? (*Hint:* see Chapter 23.)

propionyl-CoA   methylmalonyl-CoA   succinyl-CoA

48. If glucose is labeled with ^{14}C in the indicated position, where will the label be in pyruvate?

   a. glucose-1-^{14}C **c.** glucose-3-^{14}C **e.** glucose-5-^{14}C

   b. glucose-2-^{14}C **d.** glucose-4-^{14}C **f.** glucose-6-^{14}C

49. Write the reactions for the synthesis of citrate from two equivalents of pyruvate. What enzymes are required for the reactions?

50. Under conditions of starvation, acetyl-CoA, instead of being degraded in the citric acid cycle, is converted to acetone and 3-hydroxybutyrate, compounds called ketone bodies that the brain can use as a temporary fuel. Show how the ketone bodies are formed.

acetone   3-hydroxybutyrate

51. Shortly after adding ^{14}C-labeled glyceraldehyde-3-phosphate to a yeast extract, fructose-1,6-bisphosphate labeled at C-3 and C-4 can be isolated. Where was the ^{14}C-label in glyceraldehyde-3-phosphate?

52. UDP-galactose-4-epimerase converts UDP-galactose to UDP-glucose. The reaction requires NAD^+ as a coenzyme.

   a. Propose a mechanism for the reaction.   **b.** Why is the enzyme called an epimerase?

UDP-galactose   UDP-glucose

53. A student is trying to determine the mechanism for a reaction that uses ATP to activate a carboxylate ion, which then reacts with a thiol. If the carboxylate ion attacks the γ-phosphorus of ATP, the reaction products are the thioester, ADP, and phosphate. However, whether it attacks the α-phosphorus or the β-phosphorus of ATP cannot be determined from the reaction products because the thioester, AMP, and pyrophosphate would be the products in both reactions. The mechanisms can be distinguished by a labeling experiment in which the enzyme, the carboxylate ion, ATP, and radioactively labeled pyrophosphate are incubated, and then the ATP is isolated. If the isolated ATP is radioactive, attack occurred on the α-phosphorus. If it is not radioactive, then attack occurred on the β-phosphorus. Explain these conclusions.

54. What would be the results of the experiment in Problem 53 if radioactive AMP were added to the incubation mixture instead of radioactive pyrophosphate?

25

The Organic Chemistry of Lipids

Some of the things you will learn in this chapter are the purpose of the large deposit of fat in a whale's head, the difference between a fat and an oil, and why the venom of some snakes is poisonous. You will also see what lemon oil, geranium oil, and peppermint oil have in common and how they are biosynthesized (synthesized in nature).

Lipids are naturally occurring organic compounds that are soluble in nonpolar solvents. Because compounds are classified as lipids on the basis of a physical property—their solubility—rather than on the basis of their structures, lipids have a variety of structures and functions, as the following examples illustrate:

PGE₁
a vasodilator

cortisone
a hormone

vitamin A
a vitamin

limonene
in orange and
lemon oils

camphor
in camphor trees

tristearin
a fat

The ability of lipids to dissolve in nonpolar organic solvents results from their significant hydrocarbon component. The hydrocarbon part of a lipid molecule is responsible for its "oiliness" or "fattiness." The word *lipid* comes from the Greek *lipos*, which means "fat."

25.1 FATTY ACIDS ARE LONG-CHAIN CARBOXYLIC ACIDS

The first lipids we will look at are fatty acids. **Fatty acids** are carboxylic acids with long hydrocarbon chains that are found in nature (Table 25.1). They are unbranched and contain an even number of carbons because they are synthesized from acetate, a compound with two carbons. The mechanism for their biosynthesis is discussed in Section 17.20.

Table 25.1 Common Naturally Occurring Fatty Acids

Number of carbons	Common name	Systematic name	Structure	Melting point (°C)
Saturated				
12	lauric acid	dodecanoic acid		44
14	myristic acid	tetradecanoic acid		58
16	palmitic acid	hexadecanoic acid		63
18	stearic acid	octadecanoic acid		69
20	arachidic acid	eicosanoic acid		77
Unsaturated				
16	palmitoleic acid	(9Z)-hexadecenoic acid		0
18	oleic acid	(9Z)-octadecenoic acid		13
18	linoleic acid	(9Z,12Z)-octadecadienoic acid		−5
18	linolenic acid	(9Z,12Z,15Z)-octadecatrienoic acid		−11
20	arachidonic acid	(5Z,8Z,11Z,14Z)-eicosatetraenoic acid		−50
20	EPA	(5Z,8Z,11Z,14Z,17Z)-eicosapentaenoic acid		−50

- Fatty acids can be saturated with hydrogen (and, therefore, have no carbon–carbon double bonds).
- Fatty acids can be unsaturated (and have carbon–carbon double bonds).
- Fatty acids with more than one double bond are called **polyunsaturated fatty acids.**
- The double bonds in naturally occurring unsaturated fatty acids have the cis configuration and are always separated by one CH_2 group.

The melting points of saturated fatty acids increase with increasing molecular weight because of increased London dispersion forces between the molecules (Section 3.9). The melting points of unsaturated fatty acids with the same number of double bonds also increase with increasing molecular weight (Table 25.1).

The cis double bond in a fatty acid produces a bend in the molecule, which prevents unsaturated fatty acids from packing together as tightly as saturated fatty acids. As a result, unsaturated fatty acids have fewer intermolecular interactions and, therefore, have lower melting points than saturated fatty acids with comparable molecular weights (Table 25.1).

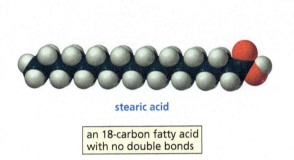

stearic acid

> an 18-carbon fatty acid with no double bonds

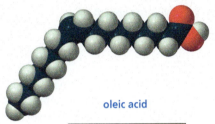

oleic acid

> an 18-carbon fatty acid with one double bond

Unsaturated fatty acids have lower melting points than saturated fatty acids.

Omega Fatty Acids

Omega indicates the position of the first double bond in an unsaturated fatty acid, counting from the methyl end. For example, linoleic acid is an omega-6 fatty acid because its first double bond is after the sixth carbon, and linolenic acid is an omega-3 fatty acid because its first double bond is after the third carbon. Mammals lack the enzyme that introduces a double bond beyond C-9, counting from the carbonyl carbon. Therefore, linoleic acid and linolenic acids are *essential fatty acids* for mammals: mammals cannot synthesize them, but because they are needed for normal body function, they must be obtained from the diet.

Omega-3 fatty acids have been found to decrease the likelihood of sudden death due to a heart attack. When under stress, the heart can develop fatal disturbances in its rhythm. Omega-3 fatty acids are incorporated into cell membranes in the heart and apparently have a stabilizing effect on heart rhythm. These fatty acids are found in fatty fish such as herring, mackerel, and salmon.

Linoleic and linolenic acids are essential fatty acids for mammals.

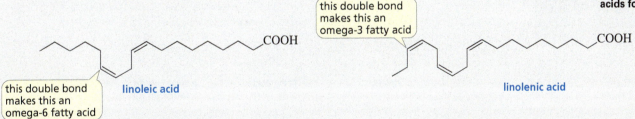

> this double bond makes this an omega-3 fatty acid

COOH

> this double bond makes this an omega-6 fatty acid

linoleic acid

linolenic acid

PROBLEM 1

Explain the difference in the melting points of the following fatty acids:

a. palmitic acid and stearic acid
b. stearic acid and oleic acid
c. palmitoleic and oleic
d. oleic and linoleic

25.2 WAXES ARE HIGH-MOLECULAR-WEIGHT ESTERS

Waxes are esters formed from long-chain carboxylic acids (fatty acids) and long-chain alcohols. For example, beeswax, the structural material of beehives, has a 26-carbon carboxylic acid component and a 30-carbon alcohol component. The word *wax* comes from the Old English *weax*, meaning "material of the honeycomb." Carnauba wax is a particularly hard wax because of its relatively high molecular weight; it has a 32-carbon carboxylic acid component and a 34-carbon alcohol component. Carnauba wax is widely used as a car wax and in floor polishes.

layers of honeycomb in a beehive

$$CH_3(CH_2)_{24}\overset{O}{\underset{}{C}}O(CH_2)_{29}CH_3$$

major component of
beeswax
structural material
of beehives

$$CH_3(CH_2)_{30}\overset{O}{\underset{}{C}}O(CH_2)_{33}CH_3$$

major component of
carnauba wax
coating on the leaves
of a Brazilian palm

$$CH_3(CH_2)_{14}\overset{O}{\underset{}{C}}O(CH_2)_{15}CH_3$$

major component of
spermaceti wax
from the heads of
sperm whales

Waxes are common in living organisms. The feathers of birds are coated with wax to make them water repellent. Some vertebrates secrete wax in order to keep their fur lubricated and water repellent. Insects secrete a waterproof, waxy layer on the outside of their exoskeletons. Wax is also found on the surfaces of certain leaves and fruits, where it serves as a protectant against parasites and minimizes the evaporation of water.

25.3 FATS AND OILS ARE TRIGLYCERIDES

Triglycerides, also called triacylglycerols, are compounds in which each of the three OH groups of glycerol has formed an ester with a fatty acid.

If the three fatty acid components of a triacylglycerol are the same, the compound is called a **simple triglyceride. Mixed triglycerides** contain two or three different fatty acid components and are more common than simple triacylglycerols.

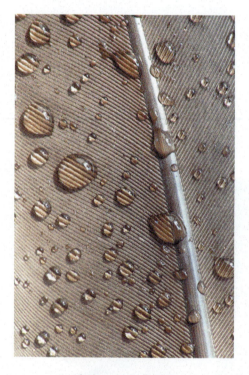

raindrops on a feather

glycerol

fatty acids

a triglyceride
a fat or an oil

Fats and Oils

Triglycerides that are solids or semisolids at room temperature are called **fats.** Most fats are obtained from animals and are composed largely of triglycerides with fatty acid components that either are saturated or have only one double bond. The saturated fatty acid tails pack closely together, giving these triglycerides relatively high melting points. They, therefore, are solids at room temperature.

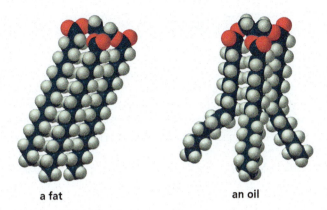

a fat   an oil

Liquid triglycerides are called **oils.** Oils typically come from plant products such as corn, soybeans, olives, and peanuts. They are composed primarily of triglycerides with unsaturated fatty acids and, therefore, cannot pack tightly together. Consequently, they have relatively low melting points and so they are liquids at room temperature. All triglyceride molecules from a single source are not necessarily identical; substances such as lard and olive oil, for example, are mixtures of several different triglycerides.

Converting Oils to Fats

Some or all of the double bonds of polyunsaturated oils can be reduced by catalytic hydrogenation (Section 5.9). Margarine and shortening are prepared by hydrogenating vegetable oils, such as soybean oil or safflower oil, until they have the desired consistency. The hydrogenation reaction must be carefully controlled, however, because reducing all the carbon–carbon double bonds would produce a hard fat with the consistency of beef tallow. We saw that trans fats can be formed during hydrogenation (Section 5.9).

This puffin's diet is high in fish oil.

Vegetable oils have become popular in food preparation because some studies have linked the consumption of saturated fats with heart disease. However, recent studies have shown that *un*saturated fats may also be implicated in heart disease. One unsaturated fatty acid—a 20-carbon fatty acid with five double bonds, known as EPA (Table 25.1) and found in high concentrations in fish oils—is thought to lower the chance of developing certain forms of heart disease.

Once consumed, dietary fat is hydrolyzed in the intestine, releasing glycerol and fatty acids (Section 24.5). We saw that fats and oils can be oxidized by O_2 to form compounds responsible for the unpleasant taste and smell associated with sour milk and rancid butter (Section 12.11).

PROBLEM 2 ◆

Which has a higher melting point, glyceryl tripalmitoleate or glyceryl tripalmitate?

PROBLEM 3

Draw the structure of an optically inactive fat that, when hydrolyzed under acidic conditions, gives glycerol, one equivalent of lauric acid, and two equivalents of stearic acid.

PROBLEM 4

Draw the structure of an optically active fat that, when hydrolyzed under acidic conditions, gives the same products as the fat in Problem 3.

Olestra: Nonfat with Flavor

Chemists have been searching for ways to reduce the caloric content of foods without decreasing their flavor. Many people who believe that "no fat" is synonymous with "no flavor" think this is a worthy endeavor. Procter & Gamble spent 30 years and more than $2 billion to develop a fat substitute they named Olestra (also called Olean). After reviewing the results of more than 150 studies, in 1996 the Federal Food and Drug Administration (page 290) approved the limited use of Olestra in snack foods.

Olestra is a semisynthetic compound. That is, Olestra itself does not exist in nature, but its components do. Developing a compound that can be made from units that are a normal part of our diet decreases the likelihood that the new compound will be toxic. Olestra is made by esterifying all the OH groups of sucrose with fatty acids obtained from cottonseed oil and soybean oil. Therefore, its component parts are table sugar and vegetable oil. Because its ester linkages are too sterically hindered to be hydrolyzed by digestive enzymes, Olestra tastes like fat but it cannot be digested and, therefore, has no caloric value.

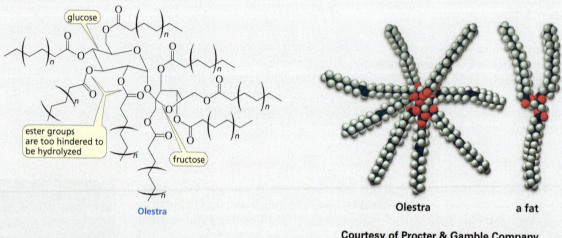

Olestra   Olestra   a fat

Courtesy of Procter & Gamble Company

Whales and Echolocation

Whales have enormous heads, accounting for 33% of their total weight. They have large deposits of fat in their heads and lower jaws. This fat is very different from both the whale's normal body fat and its dietary fat. Because major anatomical modifications were necessary to accommodate this fat, it must have some important function for the animal. It is now believed that the fat is used for echolocation—emitting sounds in pulses to gain information by analyzing the returning echoes. The fat in the whale's head focuses the emitted sound waves in a directional beam, and the echoes are received by the fat organ in the lower jaw. This organ transmits the sound to the brain for processing and interpretation, providing the whale with information about the depth of the water, changes in the sea floor, and the location of the coastline. The fat deposits in the whale's head and jaw, therefore, give the animal a unique acoustic sensory system and allow it to compete successfully for survival with the shark, which also has a well-developed sense of sound direction.

25.4 SOAPS AND MICELLES

When the ester groups of a fat or an oil are hydrolyzed in a basic solution, glycerol and fatty acids are formed. Because the solution is basic, the fatty acids are in their basic forms—namely, RCO_2^-.

Soap

The sodium or potassium salts of fatty acids are what we know as **soap.** Consequently, the hydrolysis of an ester in a basic solution is called **saponification** (the Latin word for "soap" is *sapo*). After hydrolysis, sodium chloride is added to precipitate the soap, which is dried and pressed into bars. Perfume can be added for scented soaps, dyes can be added for colored soaps, sand can be added for scouring soaps, and air can be blown into the soap to make it float in water. Three of the most common soaps are shown below:

$COO^- Na^+$

sodium stearate

$COO^- Na^+$

sodium oleate

$COO^- Na^+$

sodium linoleate

Micelles

Long-chain carboxylate ions do not exist as individual ions in aqueous solution. Instead, they arrange themselves in spherical clusters called **micelles.** Each micelle contains 50–100 long-chain carboxylate ions and resembles a large ball. The polar heads of the carboxylate ions, each accompanied by a counterion, are on the outside of the ball because of their attraction for water, whereas the nonpolar tails are buried in the interior of the ball to minimize their contact with water. The hydrophobic interactions between the nonpolar tails increase the stability of the micelle (Section 21.15).

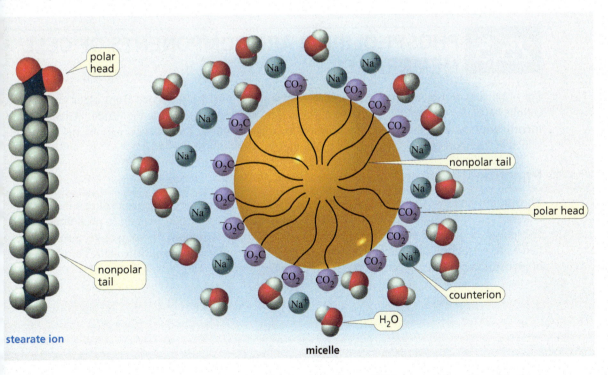

micelle

Water by itself is not a very effective cleaner because dirt is carried by nonpolar oil molecules. Soap has cleansing ability because the nonpolar oil molecules dissolve in the nonpolar interior of the micelles and are washed away with the micelle during rinsing.

Because the surface of the micelle is charged, the individual micelles repel one another instead of clustering together to form larger aggregates. However, in "hard" water—water containing high concentrations of calcium and magnesium ions—micelles do form aggregates, which we know as "bathtub ring" or "soap scum."

Detergents

The formation of soap scum in hard water led to a search for synthetic materials that would have the cleansing properties of soap but would not form scum when they encountered calcium and magnesium ions. The synthetic "soaps" that were developed, known as **detergents** (from the Latin *detergere*, which means "to wipe off"), are salts of benzenesulfonic acids. Calcium and magnesium salts of benzenesulfonic acids do not form aggregates.

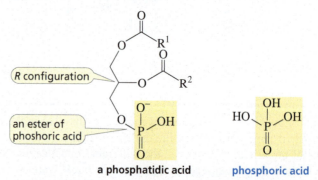

a benzenesulfonic acid   **a detergent**

After the initial introduction of detergents into the marketplace, it was discovered that those with straight-chain alkyl groups are biodegradable, whereas those with branched-chain alkyl groups are not. Therefore, to prevent detergents from polluting rivers and lakes, detergents are made only with straight-chain alkyl groups.

PROBLEM 5 SOLVED

An oil obtained from coconuts is unusual in that all three fatty acid components are identical. The molecular formula of the oil is $C_{45}H_{86}O_6$. What is the molecular formula of the carboxylate ion obtained when the oil is saponified?

SOLUTION When the oil is saponified, it forms glycerol and three equivalents of carboxylate ion. In losing glycerol, the fat loses three carbons and five hydrogens. Thus, the three equivalents of carboxylate ion have a combined molecular formula of $C_{42}H_{81}O_6$. Dividing by three gives a molecular formula of $C_{14}H_{27}O_2$ for the carboxylate ion.

25.5 PHOSPHOLIPIDS ARE COMPONENTS OF CELL MEMBRANES

For organisms to operate properly, some of their parts must be separated from other parts. On a cellular level, for example, the outside of the cell must be separated from the inside. "Greasy" lipid **membranes** serve as the barrier. In addition to isolating the cell's contents, membranes allow the selective transport of ions and organic molecules into and out of the cell.

Phosphoglycerides

Phosphoglycerides (also called **phosphoacylglycerols**), the major components of cell membranes, belong to a class of compounds called **phospholipids**—lipids that contain a phosphate group. Phosphoglycerides are similar to triglycerides except that a terminal OH group of glycerol is esterified with phosphoric acid rather than with a fatty acid, forming a **phosphatidic acid**. The C-2 carbon of glycerol in phosphoglycerides has the *R* configuration.

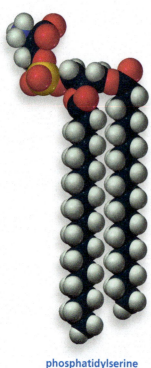

phosphatidylserine
a phosphoglyceride

a phosphatidic acid   **phosphoric acid**

Phosphatidic acids, the simplest phosphoglycerides, are present in only small amounts in membranes. The most common phosphoglycerides in membranes have a second phosphate ester linkage—they are phosphodiesters.

phosphoglycerides

The most common phosphoglycerides in cell membranes are phosphodiesters.

phosphate ester linkages

a phosphatidylethanolamine
a cephalin

a phosphatidylcholine
a lecithin

a phosphatidylserine

The alcohols most commonly used to form the second ester group are ethanolamine, choline, and serine. Phosphatidylethanolamines are also called *cephalins*, and phosphatidylcholines are called *lecithins*. Lecithins are added to foods such as mayonnaise to prevent the aqueous and fat components from separating.

PROBLEM 6 ◆

Do the identities of R^1 and R^2 in phosphatidic acid affect the configuration of the asymmetric center?

Membranes

Phosphoglycerides form membranes by arranging themselves in a **lipid bilayer.** The polar heads of the phosphoglycerides are on both surfaces of the bilayer, and the fatty acid chains form the interior of the bilayer. Cholesterol, a lipid discussed in Sections 3.16 and 25.10, is also found in the interior of the bilayer (Figure 25.1). (Compare the bilayer structure with that of the micelles formed by soap in an aqueous solution, described in Section 25.4.)

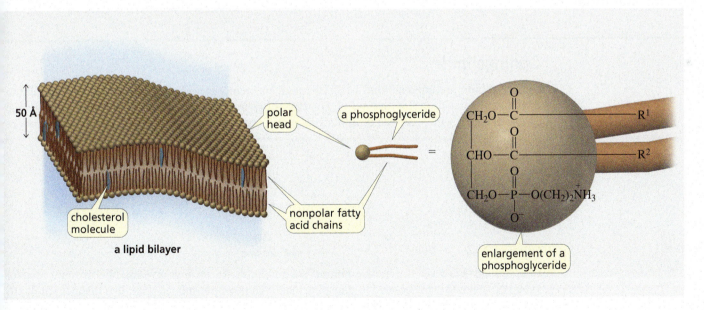

▲ **Figure 25.1**
Anatomy of a lipid bilayer.

The fluidity (viscosity) of a membrane is controlled by the fatty acid components of the phosphoglycerides. Saturated fatty acids decrease membrane fluidity because their hydrocarbon chains pack closely together. Unsaturated fatty acids increase fluidity because they pack less closely together. Cholesterol also decreases fluidity (Section 25.9). Only animal membranes contain cholesterol, so they are more rigid than plant membranes.

The unsaturated fatty acid chains of phosphoglycerides are susceptible to reaction with O_2, similar to the reaction described on page 554 for fats and oils. This oxidation reaction leads to the degradation of membranes. Vitamin E is an important antioxidant that protects fatty acid chains from degradation through oxidation. Vitamin E, also called α-tocopherol, is classified as a lipid because it is soluble in nonpolar solvents. It is, therefore, able to enter the nonpolar membranes; once there, it reacts more rapidly with oxygen than the phospholipids in the bilayer do (Section 12.11).

α-**tocopherol**
vitamin E

There are some who believe that vitamin E slows the aging process. The ability of vitamin E to react with oxygen more rapidly than fats do is the reason it is added as a preservative to many fat-containing foods.

Snake Venom

The venom of some poisonous snakes contains a phospholipase, an enzyme that hydrolyzes an ester group of a phosphoglyceride. For example, both the eastern diamondback rattlesnake and the Indian cobra contain a phospholipase that hydrolyzes an ester bond of cephalins, causing the membranes of red blood cells to rupture.

bond hydrolyzed by the phospholipase found in the Indian cobra and the eastern diamondback rattlesnake

an eastern diamondback rattlesnake

Sphingolipids

Sphingolipids are another kind of lipid found in membranes. Sphingolipids contain an amino alcohol called sphingosine instead of glycerol. In sphingolipids, the amino group of sphingosine forms an amide with a fatty acid. Both asymmetric centers in sphingosine have the *S* configuration.

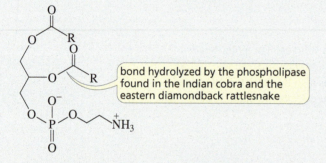

E configuration

S configuration

sphingosine

Sphingomyelins are the most common sphingolipids. In sphingomyelins, the primary alcohol group of sphingosine is bonded to phosphoethanolamine or phosphocholine, in a manner similar to the bonding in lecithins and cephalins.

sphingomyelins

Multiple Sclerosis and the Myelin Sheath

The myelin sheath is a lipid-rich covering that is wrapped around the brain's neurons—the nerve cells that control our muscles. Composed largely of sphingomyelins, the sheath increases the velocity of nerve impulses. Without the myelin sheath, nerve cells are not able to tell the muscles what to do. Multiple sclerosis is a disease characterized by loss of the myelin sheath, a consequent slowing of nerve impulses, and eventual paralysis.

ADL (adrenoleukodystrophy) is a genetic disease that destroys myelin, causing a decline in brain function as the myelin sheath that surrounds the brain's nerve cells disappears. The defect is in the X chromosome, so the disease affects males. Because women have two X chromosomes, if they inherit an abnormal one, they still have a normal X chromosome that can offset the effect of the mutation.

PROBLEM 7 ◆

Membranes contain proteins. Integral membrane proteins extend partly or completely through the membrane, whereas peripheral membrane proteins are found on the inner or outer surface of the membrane. What is the likely difference in the amino acid composition of integral and peripheral membrane proteins?

PROBLEM 8 ◆

A colony of bacteria accustomed to an environment with a temperature of 25 °C was moved to an identical environment, except that its temperature was 35 °C. The higher temperature increased the fluidity of the bacterial membranes. How can the bacteria regain their original membrane fluidity?

PROBLEM 9

The membrane phospholipids in deer have a higher degree of unsaturation in cells closer to the hoof than in cells closer to the body. Why is this trait important for survival?

25.6 PROSTAGLANDINS REGULATE PHYSIOLOGICAL RESPONSES

Prostaglandins obtained their names because the first prostaglandins were isolated from the prostate gland. Now, however, we know that prostaglandins are found in all body tissues. They are responsible for regulating a variety of physiological responses, such as inflammation, blood pressure, blood clotting, fever, pain, the induction of labor, and the sleep–wake cycle. All prostaglandins have a five-membered ring with a seven-carbon carboxylic acid substituent and an eight-carbon hydrocarbon substituent. The two substituents are trans to each other.

prostaglandin skeleton

Prostaglandins are classified using the formula PGX, where X designates the functional groups of the compound's five-membered ring.

- PGAs, PGBs, and PGCs all contain a carbonyl group and a double bond in the five-membered ring. The location of the double bond determines whether a prostaglandin is a PGA, PGB, or PGC.

PGA **PGB** **PGC**

- PGDs and PGEs are β-hydroxy ketones. A subscript indicates the total number of double bonds in the side chains.

PGD **PGE$_1$** **PGE$_2$**

- PGFs are 1,3-diols: "α" indicates a cis diol, and "β" indicates a trans diol.

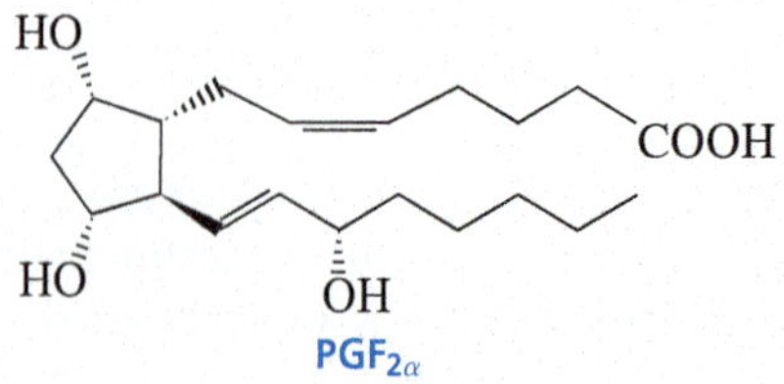

PGF$_{2\alpha}$

PROBLEM 10

Treating PGC$_2$ with a strong base such as sodium *tert*-butoxide followed by addition of acid converts it to PGA$_2$. Propose a mechanism for this reaction.

Biosynthesis of Prostaglandins

Prostaglandins are biosynthesized from arachidonic acid, a 20-carbon omega-6 fatty acid with four cis double bonds. Arachidonic acid is biosynthesized from linoleic acid, a fatty acid that must be acquired through the diet (Section 25.1).

An enzyme called prostaglandin endoperoxide synthase catalyzes the conversion of arachidonic acid to PGH$_2$, the precursor of all prostaglandins. The enzyme has two activities: a *cyclooxygenase activity* and a *hydroperoxidase activity*. It uses its cyclooxygenase activity to form the five-membered ring.

Prostaglandins are synthesized from arachidonic acid.

biosynthesis of prostaglandins, thromboxanes, and prostacyclins

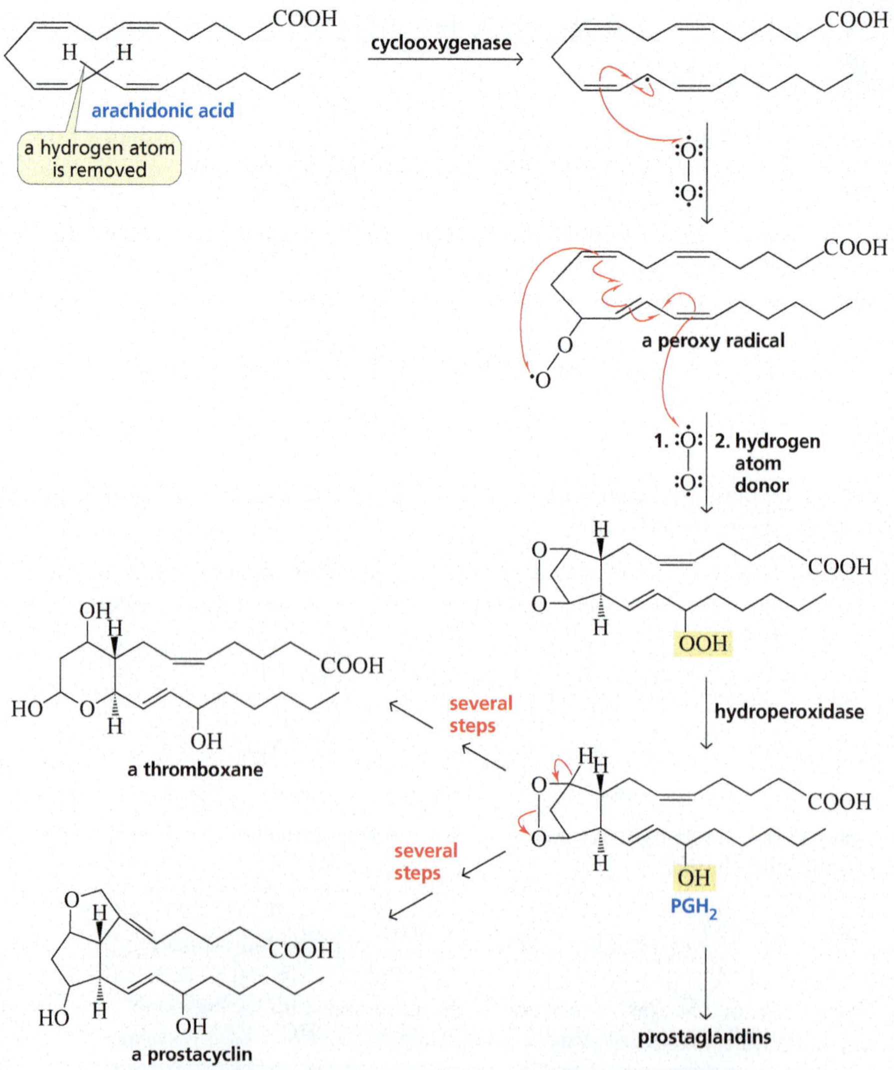

arachidonic acid

a hydrogen atom is removed

cyclooxygenase

a peroxy radical

1. :Ö: 2. hydrogen atom donor

:Ö:

OOH

hydroperoxidase

a thromboxane

several steps

several steps

OH

PGH$_2$

a prostacyclin

prostaglandins

- In the first step, a hydrogen atom is removed from a carbon flanked by two double bonds.
- The radical reacts with oxygen to form a peroxy radical. Notice that these two steps are the same as the first two steps in the reaction that causes fats to become rancid (page 554).
- The peroxy radical rearranges and reacts with a second molecule of oxygen, and the OO· group removes a H· from a hydrogen atom donor.
- The enzyme then uses its hydroperoxidase activity to convert the OOH group to an OH group, forming PGH$_2$, the compound used to synthesize prostaglandins, thromboxanes, and prostacyclins.

We saw that aspirin inhibits the synthesis of prostaglandins—thereby decreasing the fever and inflammation caused by these compounds—by acetylating cyclooxygenase. Nonsteroidal anti-inflammatory agents (NSAIDs) inhibit the synthesis of prostaglandins by competing with either arachidonic acid or the peroxy radical for the enzyme's active site (Section 15.9).

Thromboxanes and Prostacyclins

PGH$_2$ is the starting material not only for the synthesis of prostaglandins but also for the synthesis of thromboxanes and prostacyclins. Thromboxanes are vasoconstrictors and stimulate platelet aggregation. Prostacyclins are vasodilators and inhibit platelet aggregation. The opposite effects of these compounds allow them to maintain a balance in the cardiovascular system.

Leukotrienes

Arachidonic acid can also be converted to a *leukotriene*. Because they induce contraction of the muscle that lines the airways to the lungs, leukotrienes are implicated in allergic reactions, inflammatory reactions, and heart attacks. They also bring on the symptoms of asthma and contribute to the onset of anaphylactic shock, a potentially fatal allergic reaction.

arachidonic acid → a leukotriene

The compounds formed from arachidonic acid (prostaglandins, prostacyclins, thromboxanes, and leukotrienes) are known as *eicosanoids*, because they contain 20 carbons and at least one oxygen. (A 20-carbon straight-chain alkane is called eicosane; see Table 3.1 on page 89.)

25.7 TERPENES CONTAIN CARBON ATOMS IN MULTIPLES OF FIVE

Terpenes are a diverse class of compounds that contain 10, 15, 20, 25, 30, or 40 carbons. More than 20,000 terpenes are known. Many are found in oils extracted from fragrant plants. Terpenes can be hydrocarbons, or they can contain oxygen and be alcohols, ketones, or aldehydes. Oxygen-containing terpenes are sometimes called **terpenoids.** Terpenes and terpenoids have been used as spices, perfumes, and medicines for thousands of years.

menthol
peppermint oil

geraniol
geranium oil

zingiberene
oil of ginger

β-selinene
oil of celery

Terpenes are made by joining together five-carbon isoprene units, usually in a head-to-tail fashion. (As shown on the next page, the branched end of isoprene is called the head, and the unbranched end is called the tail.) **Isoprene** is the common name for 2-methyl-1,3-butadiene, a compound with five carbons.

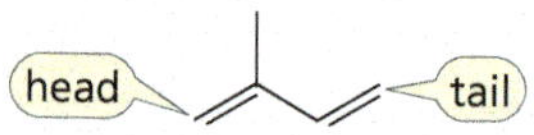

2-methyl-1,3-butadiene
isoprene

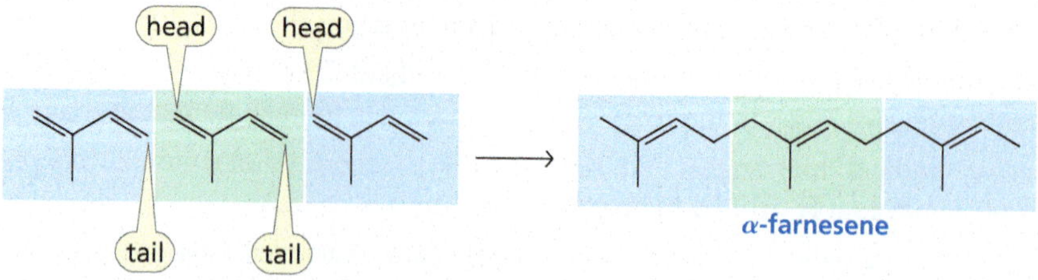

α-farnesene

In the case of cyclic compounds, linking the head of one isoprene unit to the tail of another is followed by an additional linkage to form the ring. The second linkage is not necessarily head-to-tail but is whatever linkage is necessary to form a stable five- or six-membered ring.

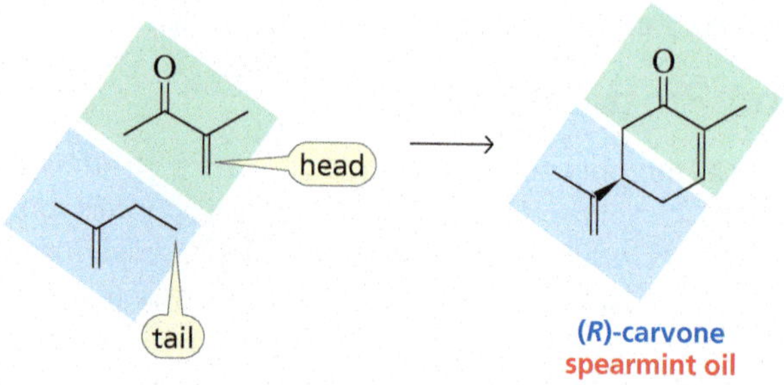

(R)-carvone
spearmint oil
a monoterpene

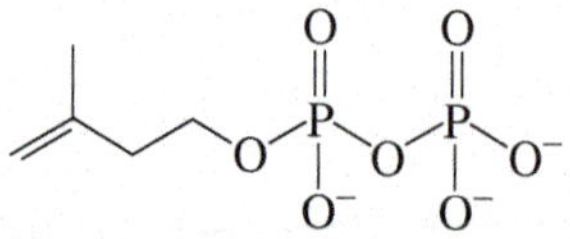

3-methyl-3-butenyl pyrophosphate
isopentenyl pyrophosphate

In Section 25.8, we will see that the compound actually used in the biosynthesis of terpenes is not isoprene, but **isopentenyl pyrophosphate**, a compound with the same carbon skeleton as isoprene, but with a leaving group that isoprene does not have. We will also look at the mechanism by which isopentenyl pyrophosphate units are joined in a head-to-tail fashion.

Classes of Terpenes

A monoterpene has 10 carbons.

Terpenes have many important biological roles. They are classified according to the number of carbons they contain.

- Monoterpenes are composed of two isoprene units, so they have 10 carbons.
- Sesquiterpenes, with 15 carbons, have three isoprene units (*sesqui* is from the Latin for "one and a half"). Many fragrances and flavorings found in plants are monoterpenes or sesquiterpenes. These compounds are known as **essential oils.**
- Diterpenes (20 carbons) are composed of four isoprene units. Cafestol and kahweol (both present in coffee beans), and vitamin A are examples of diterpenes.
- Squalene is a triterpene (30 carbons); it is the precursor of cholesterol, which is the precursor of all the other steroid hormones (Section 25.9).

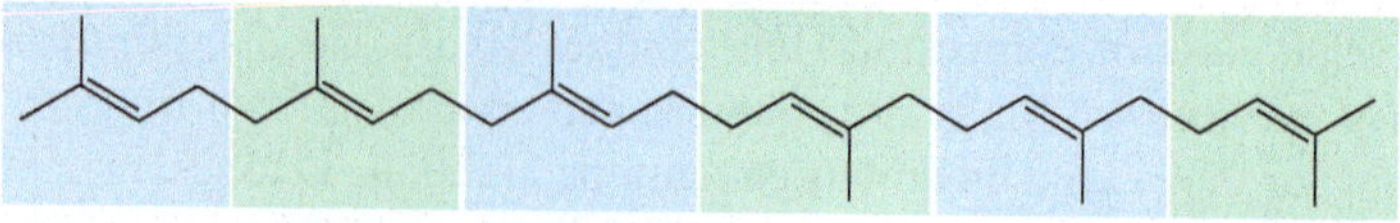

squalene

- **Carotenoids** are tetraterpenes (40 carbons). Lycopene, the compound responsible for the red coloring of tomatoes and watermelon, and β-carotene, the compound that causes carrots and apricots to be orange, are carotenoids.

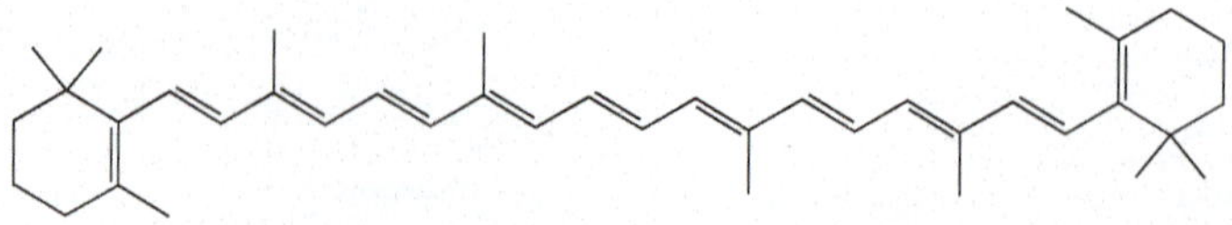

β-carotene

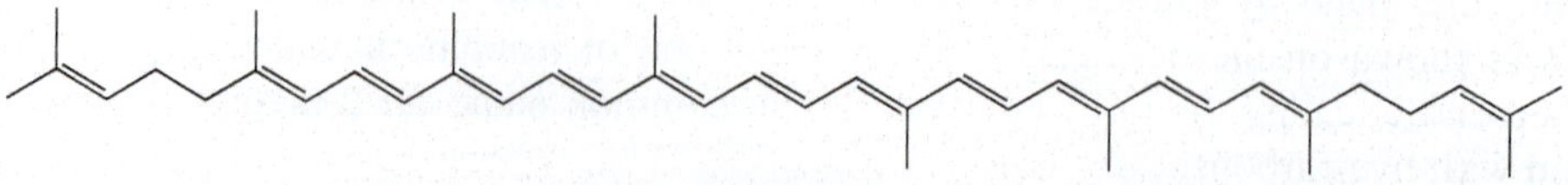

lycopene

Why carotenoids are colored compounds was discussed in Section 13.20. Cleaving β-carotene forms two molecules of vitamin A (retinol), which is oxidized to retinal—a compound that plays an important role in vision (Section 4.1).

LEARN THE STRATEGY

PROBLEM 11 SOLVED

Mark off the isoprene units in zingiberene.

SOLUTION To find an isoprene unit, start at one end and count five carbons.

USE THE STRATEGY

PROBLEM 12

Mark off the isoprene units in menthol, β-selinene, and camphor. (The structures are on pages 1139 and 1127.)

PROBLEM 13 ♦

One of the linkages in squalene is tail-to-tail, not head-to-tail. What does this suggest about how squalene is synthesized in nature? (*Hint:* Locate the position of the tail-to-tail linkage.)

PROBLEM 14

Mark off the isoprene units in lycopene and β-carotene. Can you detect a similarity in the way in which squalene, lycopene, and β-carotene are biosynthesized?

25.8 HOW TERPENES ARE BIOSYNTHESIZED

Biosynthesis of Isopentenyl Pyrophosphate

Each step in the biosynthesis of isopentenyl pyrophosphate, the five carbon compound used for the synthesis of terpenes, is catalyzed by a different enzyme.

STEPS IN THE BIOSYNTHESIS OF ISOPENTENYL PYROPHOSPHATE

2 acetyl-CoA → acetoacetyl-CoA + CoASH → (1. SCoA, 2. H_2O) → hydroxymethylglutaryl-CoA + CoASH

Claisen condensation

2 NADPH, 2 H^+ → 2 $NADP^+$

tertiary alcohol → mevalonyl pyrophosphate ← (ADP ATP) ← mevalonyl phosphate ← (ADP ATP) ← mevalonic acid + CoASH

primary alcohol

ATP → ADP

isopentenyl pyrophosphate + CO_2 + HO—P

- The first step is a Claisen condensation (Section 17.13).
- The second step is an aldol addition with a third molecule of acetyl-CoA. This is followed by hydrolysis of one of the thioester groups (see Problem 15).
- The thioester is reduced by NADPH to form mevalonic acid (see Problem 16).
- A pyrophosphate group is added by means of two successive phosphorylations of the primary alcohol with ATP (see Problem 17).
- The tertiary alcohol is phosphorylated with ATP. Subsequent decarboxylation and loss of the phosphate group form isopentenyl pyrophosphate (see Problem 18).

PROBLEM 15

Propose mechanisms for the Claisen condensation and aldol addition that comprise the first two steps of the biosynthesis of isopentenyl pyrophosphate.

PROBLEM 16 ♦

Why are two equivalents of NADPH required to reduce hydroxymethylglutaryl-CoA to mevalonic acid?

PROBLEM 17

Propose a mechanism for the conversion of mevalonic acid to mevalonyl pyrophosphate.

PROBLEM 18 **SOLVED**

Propose a mechanism for the last step in the biosynthesis of isopentenyl pyrophosphate, showing why ATP is required.

SOLUTION In order to form the double bond in isopentenyl pyrophosphate, elimination of CO_2 needs to be accompanied by elimination of hydroxide ion. But hydroxide ion, a strong base, is a poor leaving group. Therefore, ATP is used to convert the OH group into a phosphate group, which is easily eliminated because it is a weak base.

mevalonyl pyrophosphate   **ATP**

$+ \ ADP \ + \ H^+$

isopentenyl pyrophosphate

$+ \ CO_2 \ + \ HPO_4^{2-}$

How Statins Lower Cholesterol Levels

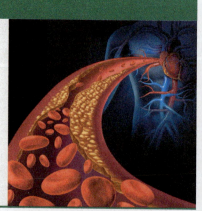

We learned in Section 3.16 that statins (Lipitor, Zocor, Mevacor) lower serum cholesterol levels. These drugs are competitive inhibitors for the enzyme that reduces hydroxymethylglutaryl-CoA to mevalonic acid (page 1141). Recall that a competitive inhibitor competes with the substrate for binding at the enzyme's active site (Section 23.7).

Decreasing the concentration of mevalonic acid decreases the concentration of isopentenyl pyrophosphate, so the synthesis of all terpenes, including cholesterol, is decreased. As a consequence of diminished cholesterol synthesis, the liver forms more LDL receptors—the receptors that help clear LDL from the bloodstream. Recall that LDL (low-density lipoprotein) is the so-called *bad* cholesterol (Section 3.16).

artery clogged by buildup of cholesterol

Biosynthesis of Dimethylallyl Pyrophosphate

Both **isopentenyl pyrophosphate** and **dimethylallyl pyrophosphate** are needed for the biosynthesis of terpenes. Therefore, some isopentenyl pyrophosphate must be converted to dimethylallyl pyrophosphate before biosynthesis can take place. This enzyme-catalyzed isomerization reaction takes place in two steps.

MECHANISM FOR THE CONVERSION OF ISOPENTYL PYROPHOSPHATE TO DIMETHYLALLYL PYROPHOSPHATE

- A cysteine side chain is in the proper position at the enzyme's active site to donate a proton to the sp^2 carbon of the alkene that is bonded to the most hydrogens (Section 6.4).
- A glutamate side chain removes a proton from a β-carbon of the carbocation intermediate that is bonded to the fewest hydrogens (in accordance with Zaitsev's rule; (Section 9.7).

Adding a proton and then removing a proton converts isopentenyl pyrophosphate to dimethylallyl pyrophosphate.

Reaction of Dimethylallyl Pyrophosphate with Isopentenyl Pyrophosphate

The enzyme-catalyzed reaction of dimethylallyl pyrophosphate with isopentenyl pyrophosphate forms geranyl pyrophosphate, a 10-carbon compound.

MECHANISM FOR THE BIOSYNTHESIS OF TERPENES

- Experimental evidence suggests that this is an S_N1 reaction (see Problem 21). Therefore, the leaving group of dimethylallyl pyrophosphate departs, forming an allylic cation.
- Isopentenyl pyrophosphate is the nucleophile that adds to the allylic cation.
- A base removes a proton, forming geranyl pyrophosphate.

Synthesizing Monoterpenes

The scheme shown here shows how some of the many monoterpenes can be synthesized from geranyl pyrophosphate:

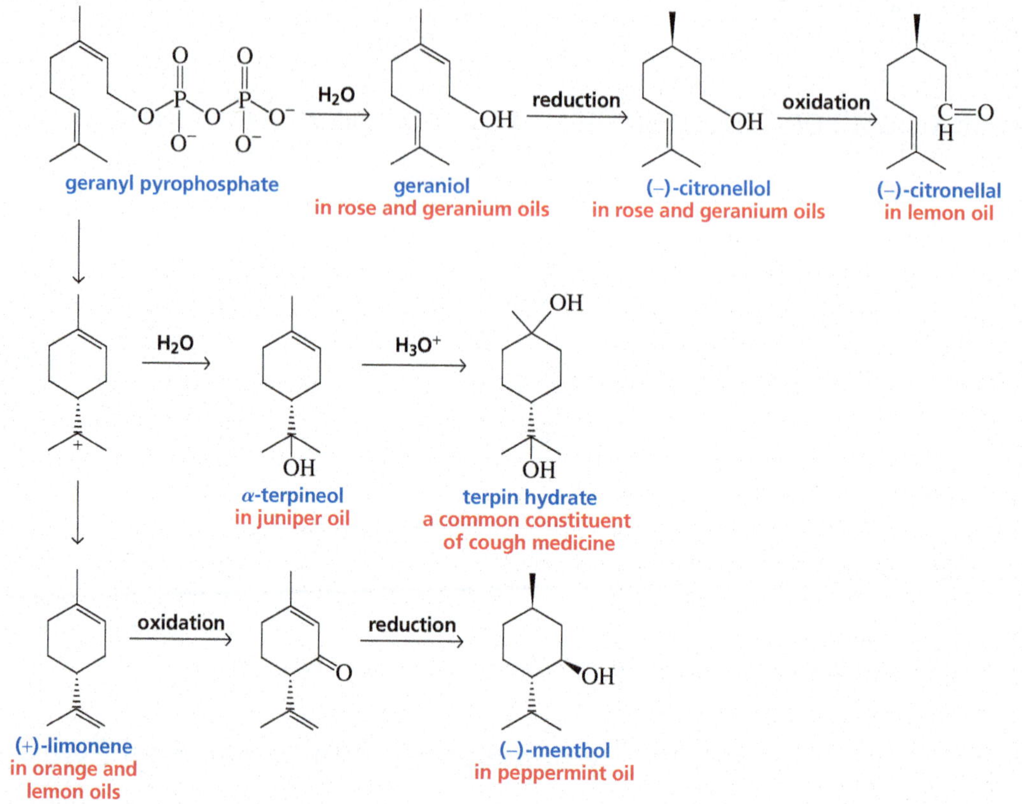

H2O
reduction
oxidation
geranyl pyrophosphate
geraniol
in rose and geranium oils
(−)-citronellol
in rose and geranium oils
(−)-citronellal
in lemon oil
H2O
H3O+
α-terpineol
in juniper oil
terpin hydrate
a common constituent
of cough medicine
oxidation
reduction
(+)-limonene
in orange and
lemon oils
(−)-menthol
in peppermint oil

LEARN THE STRATEGY

PROBLEM-SOLVING STRATEGY

Proposing a Mechanism for Biosynthesis

Propose a mechanism for the biosynthesis of limonene from geranyl pyrophosphate.

Assuming that geranyl pyrophosphate reacts like dimethylallyl pyrophosphate, the leaving group departs in an S_N1 reaction. The electrons of the π bond add to the allylic cation, forming the six-membered ring. A base removes a proton from the carbocation to form the required double bond.

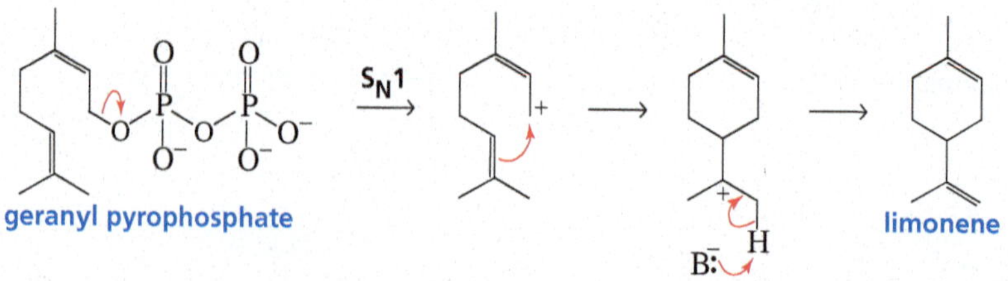

SN1
geranyl pyrophosphate
limonene

USE THE STRATEGY

PROBLEM 19

Propose a mechanism for the biosynthesis of α-terpineol from geranyl pyrophosphate.

PROBLEM 20

Propose a mechanism for the conversion of the *E* isomer of geranyl pyrophosphate to the *Z* isomer.

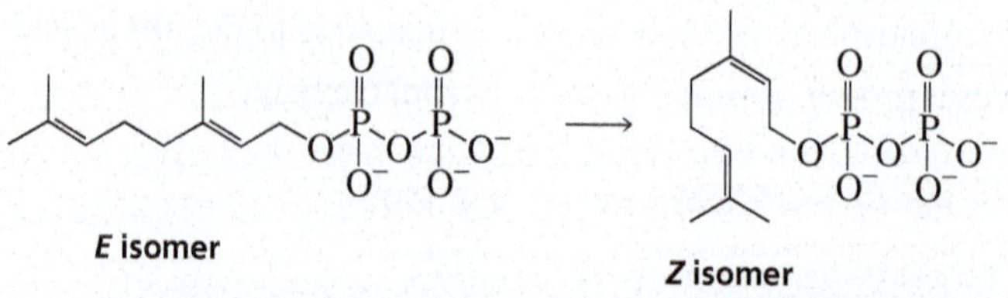

E isomer
Z isomer

Reaction of Geranyl Pyrophosphate with Isopentenyl Pyrophosphate

Geranyl pyrophosphate can react with another molecule of isopentenyl pyrophosphate to form farnesyl pyrophosphate, a 15-carbon compound.

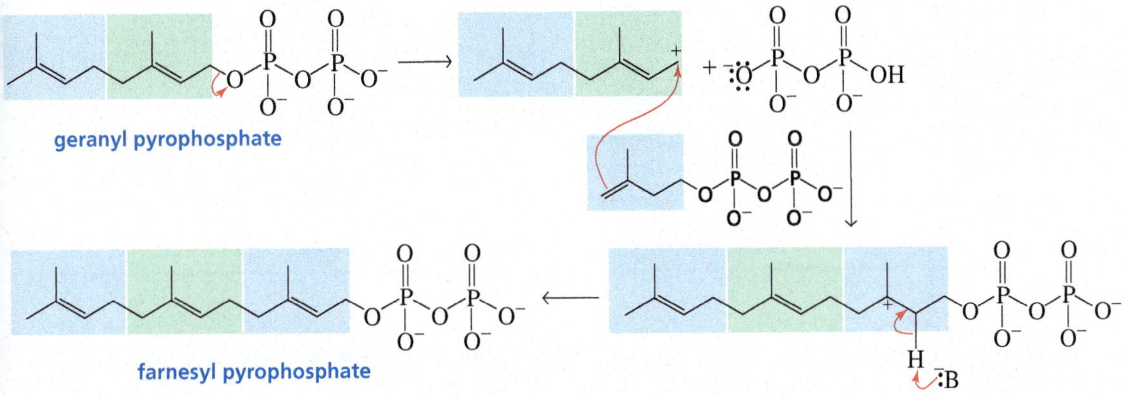

geranyl pyrophosphate

farnesyl pyrophosphate

dimethyallyl = 5 carbons

isopentenyl = 5 carbons

geranyl = 10 carbons

farnesyl = 15 carbons

geranylgeranyl = 20 carbons

PROBLEM 21 ◆

The fluoro-substituted geranyl pyrophosphate shown here reacts with isopentenyl pyrophosphate to form fluoro-substituted farnesyl pyrophosphate. The rate of the reaction is less than 1% of the rate of the reaction when unsubstituted geranyl pyrophosphate is used. What does this tell you about the mechanism of the reaction?

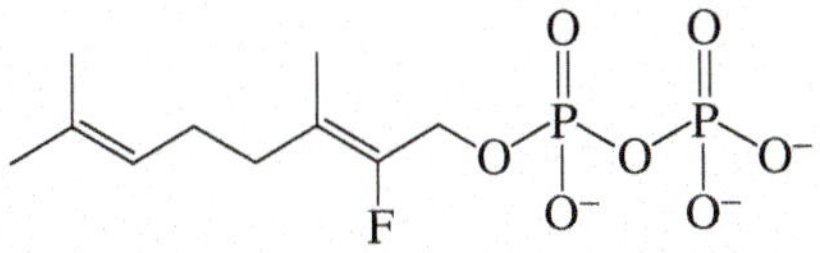

Joining Two Molecules of Farnesyl Pyrophosphate

Two molecules of farnesyl pyrophosphate form squalene, a 30-carbon compound. The reaction is catalyzed by the enzyme squalene synthase, which joins the two molecules in a tail-to-tail linkage. As we noted earlier, squalene is the precursor of cholesterol, and cholesterol is the precursor of all the steroid hormones.

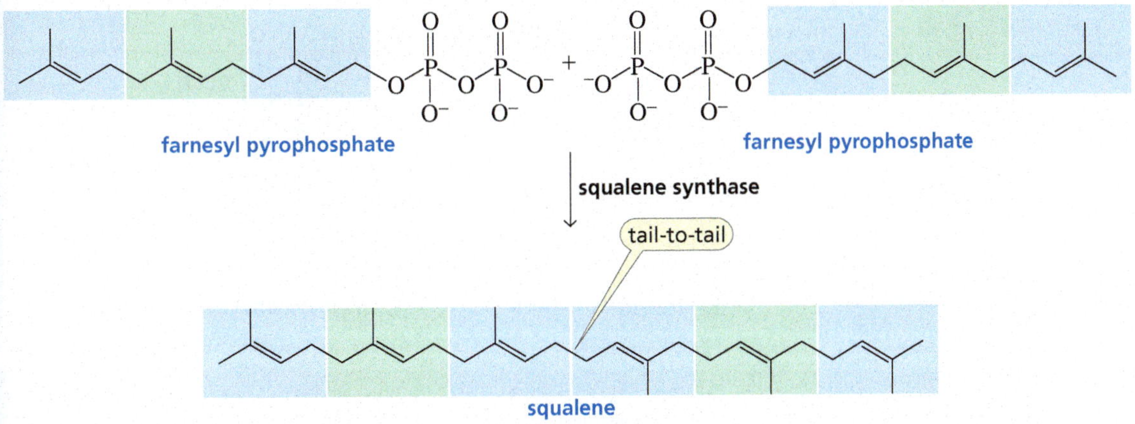

farnesyl pyrophosphate

farnesyl pyrophosphate

squalene synthase

tail-to-tail

squalene

Reaction of Farnesyl Pyrophosphate with Isopentenyl Pyrophosphate

Farnesyl pyrophosphate can react with another molecule of isopentenyl pyrophosphate, via the mechanism shown on the top of this page, to form geranylgeranyl pyrophosphate, a 20-carbon compound. Two geranylgeranyl pyrophosphates can combine tail-to-tail to form phytoene, the precursor of the carotenoinds (40-carbon compounds).

Protein Prenylation

Farnesyl and geranylgeranyl groups are put on proteins to allow them to become anchored to membranes (Section 20.18). Because "isopentenyl" is also called "prenyl," attaching these polymers of isopentenyl units to proteins is known as **protein prenylation.** The most common prenylation site on a protein is the tetrapeptide CaaX, where "C" is a cysteine, "a" is an aliphatic amino acid, and "X" can be one of several amino acids. A farnesyl group is put on the protein when X is Gln, Met, or Ser, whereas a geranylgeranyl group is put on when X is Leu.

The cysteine side chain of CaaX is the nucleophile that reacts with farnesyl pyrophosphate or geranylgeranyl pyrophosphate, forming a thioether and eliminating pyrophosphate. Once the protein has been prenylated, the C—aaX amide bond is hydrolyzed, causing cysteine to become the C-terminal amino acid. Cysteine's carboxyl group is then esterified with a methyl group.

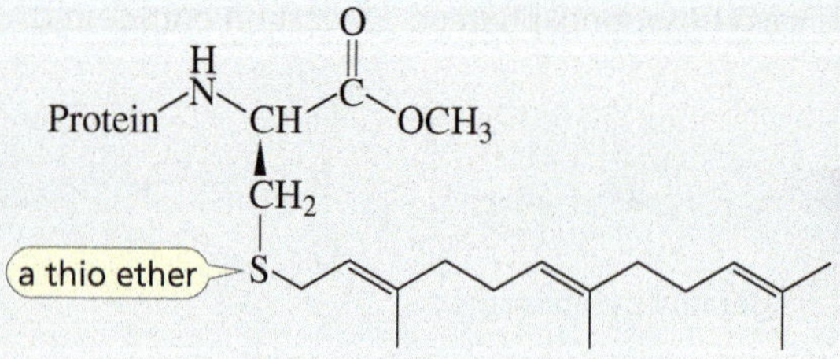

S-farnesyl cysteine methyl ester

PROBLEM 22

Farnesyl pyrophosphate forms the sesquiterpene shown here. Propose a mechanism for its formation.

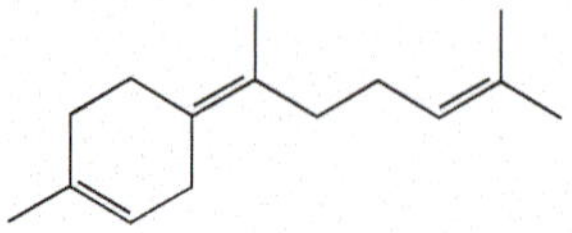

LEARN THE STRATEGY

PROBLEM 23 SOLVED

If squalene is synthesized in a medium containing acetate with a ^{14}C-labeled carbonyl carbon, which carbons in squalene will be labeled?

SOLUTION Acetate reacts with ATP to form acetyl adenylate, which then reacts with CoASH to form acetyl-CoA. Two equivalents of acetyl-CoA form acetoacetyl-CoA as a result of a Claisen condensation. Examining each step of the mechanism for the biosynthesis of isopentenyl pyrophosphate from acetyl-CoA allows you to determine the locations of the radioactively labeled carbons in isopentenyl pyrophosphate. Similarly, the locations of the radioactively labeled carbons in geranyl pyrophosphate can be determined from the mechanism for its biosynthesis from isopentenyl pyrophosphate, and the locations of the radioactively labeled carbons in farnesyl pyrophosphate can be determined from the mechanism for its biosynthesis from geranyl pyrophosphate. Knowing that squalene is obtained from a tail-to-tail linkage of two farnesyl pyrophosphates tells you which carbons in squalene will be labeled.

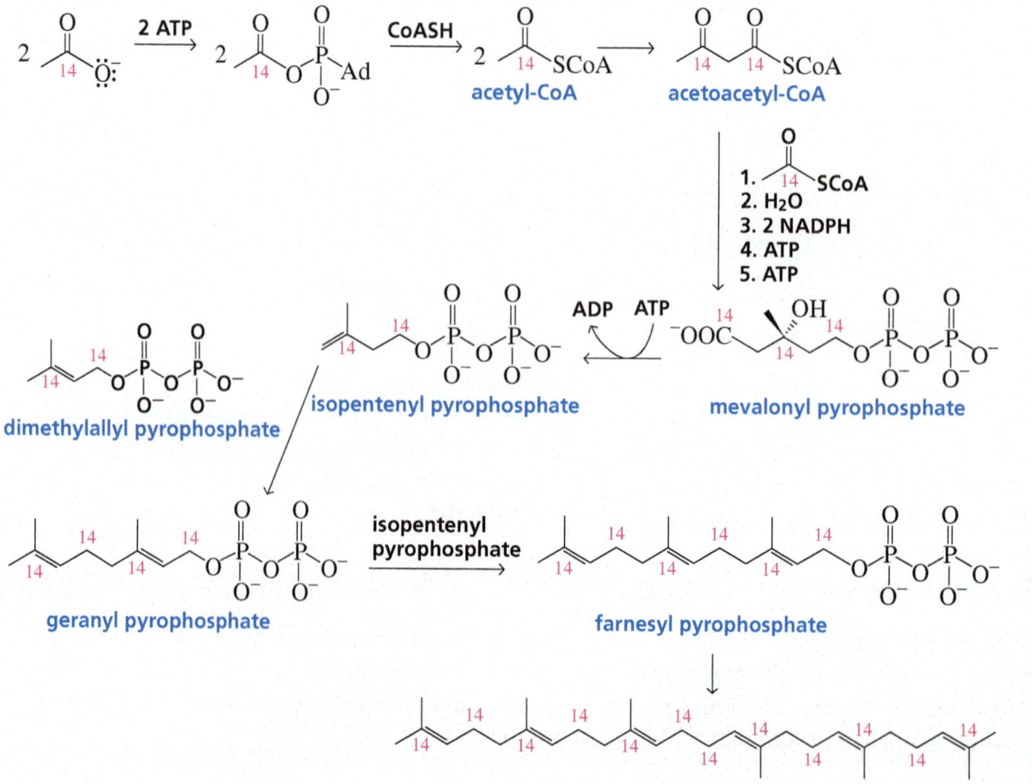

USE THE STRATEGY

PROBLEM 24

If junipers are allowed to grow in a medium containing acetate with a ^{14}C-labeled methyl carbon, which carbons in α-terpineol will be labeled?

25.9 HOW NATURE SYNTHESIZES CHOLESTEROL

You just learned that squalene is the starting material for the biosynthesis of cholesterol, an important component of cell membranes (Figure 25.1). Squalene must first be converted to lanosterol, which is converted to cholesterol in a series of 19 steps.

STEPS IN THE BIOSYNTHESIS OF LANOSTEROL AND CHOLESTEROL

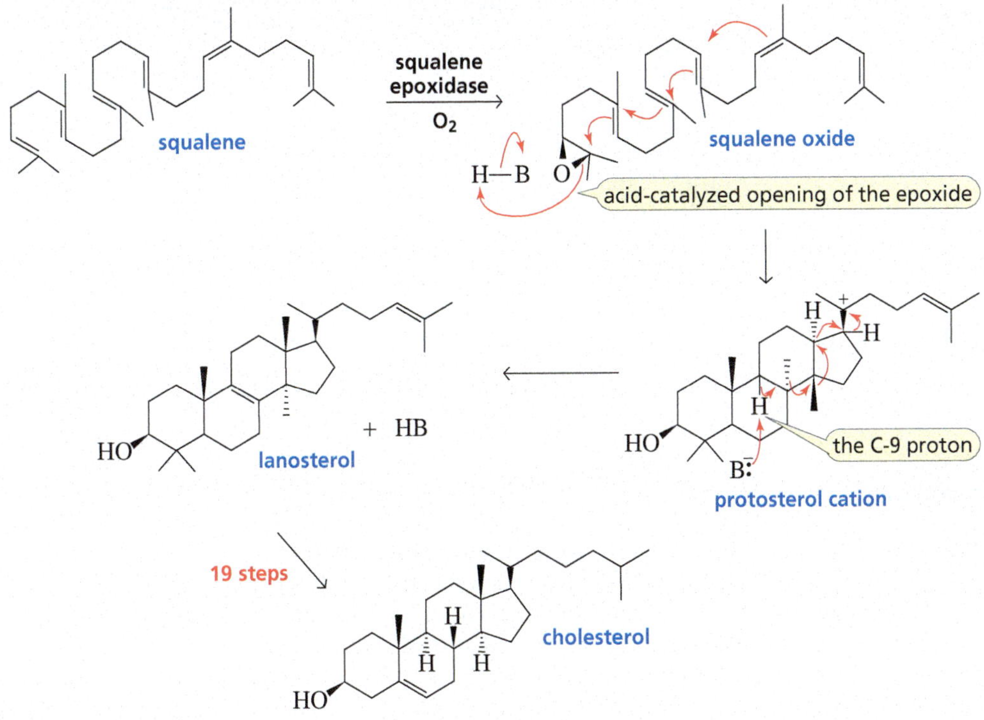

- The first step is epoxidation of the 2,3-double bond of squalene.
- Acid-catalyzed opening of the epoxide initiates a series of cyclizations resulting in the protosterol cation.
- Elimination of the C-9 proton from the cation initiates a series of 1,2-methyl and 1,2-hydride shifts, resulting in lanosterol.

Converting lanosterol to cholesterol requires removing three methyl groups from lanosterol, reducing two double bonds, and creating a new double bond. Removing methyl groups from carbon atoms is not easy, and many different enzymes are required to carry out the 19 steps. So why does nature bother? Why not just use lanosterol instead of cholesterol?

That question was answered when it was found that membranes containing lanosterol instead of cholesterol are much more permeable. Small molecules are able to pass easily through lanosterol-containing membranes. As each methyl group is removed from lanosterol, the membrane becomes less and less permeable.

PROBLEM 25 ♦

Draw the individual 1,2-hydride and 1,2-methyl shifts responsible for conversion of the protosterol cation to lanosterol. How many hydride shifts are involved? How many methyl shifts?

the steroid ring system

25.10 STEROIDS

Hormones are chemical messengers that stimulate or inhibit some process in target tissues. Many hormones are **steroids.** Because steroids are nonpolar compounds, they too are lipids. Their non-polar character allows them to cross cell membranes, so they can leave the cells in which they are synthesized and enter their target cells.

The carbons in the steroid ring system are numbered as shown in the margin. We saw that the B, C, and D rings in steroids are **trans fused** (Section 3.16). In most naturally occurring steroids, the A and B rings are also trans fused.

Many steroids have methyl groups at the 10- and 13-positions. These are called **angular methyl groups.** When steroids are drawn, both angular methyl groups are shown to be above the plane of the steroid ring system. Substituents on the same side of the steroid ring system as the angular methyl groups are designated β-substituents (indicated by a solid wedge). Those on the opposite side of the plane of the ring system are α-substituents (indicated by a hatched wedge).

PROBLEM 26 ◆

A β-hydrogen at C-5 means that the A and B rings are _____ fused; an α-hydrogen at C-5 means that they are _____ fused.

Cholesterol

The most abundant steroid in animals is **cholesterol,** which is the precursor of all the other steroids. We saw that cholesterol is biosynthesized from squalene, a triterpene (Section 25.8), and is an important component of cell membranes (Figure 25.1). Its ring structure makes it more rigid than other membrane lipids. In Section 3.16, we looked at why cholesterol is implicated in heart disease and how high cholesterol is treated clinically. Because cholesterol has eight asymmetric centers, 256 stereoisomers are possible, but only one exists in nature. (See Problem 40 in Chapter 4.)

cholesterol

PROBLEM 27 ◆

Is the OH substituent of the A ring of cholesterol an α-substituent or a β-substituent?

PROBLEM 28

The acid component of a cholesterol ester is a fatty acid such as linoleic acid. Draw the structure of a cholesterol ester.

Steroid Hormones

The steroid hormones can be divided into five classes: glucocorticoids, mineralocorticoids, andro-gens, estrogens, and progestins. Glucocorticoids and mineralocorticoids are synthesized in the adrenal cortex and are collectively known as *adrenal cortical steroids.* All adrenal cortical steroids have an oxygen at C-11.

Glucocorticoids and Mineralcorticoids

Glucocorticoids, as their name suggests, are involved in glucose metabolism; they also participate in the metabolism of proteins and fatty acids. Hydrocortisone is the most common glucocorticoid; it is used to treat itching that results from such things as insect bites and allergic reactions. Synthetic glucocorticoids are used clinically to treat arthritis and other inflammatory conditions.

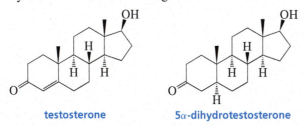

hydrocortisone   **aldosterone**

Cholesterol is the precursor of all steroids.

Mineralocorticoids cause increased reabsorption of Na^+, Cl^-, and HCO_3^- by the kidneys, leading to an increase in blood pressure. Aldosterone is an example of a mineralocorticoid.

PROBLEM 29

Aldosterone is in equilibrium with its cyclic hemiacetal. Draw the hemiacetal form of aldosterone.

Androgens

The sex hormones known as *androgens* are secreted primarily by the testes. They are responsible for the development of male secondary sex characteristics during puberty, including muscle growth. Testosterone and 5α-dihydrotestosterone are androgens.

testosterone   **5α-dihydrotestosterone**

One Drug—Two Effects

Finasteride is a drug used to treat both benign prostatic hyperplasia (BPH) and male pattern baldness (MPB).

finasteride
Proscar®
Propecia®

Finastride is a competitive inhibitor of 5α-reductase, the enzyme that reduces testosterone to dihydrotestosterone.

testosterone $\xrightarrow{\text{5}\alpha\text{-reductase}}$ dihydrotestosterone

Blocking formation of dihydrotestosterone reduces prostate volume, which reduces the risk of prostate cancer. Blocking formation of dihydrotestosterone also reduces androgen activity in the scalp, which encourages hair growth. When prescribed for BPH, the drug is called Proscar; when prescribed for MPB, it is called Propecia.

Estrogens

Estradiol and estrone are sex hormones known as *estrogens*. They are secreted primarily by the ovaries and are responsible for the development of female secondary sex characteristics. They also regulate the menstrual cycle. Progesterone is the most important member of a group of hormones called *progestins:* it is the hormone that prepares the lining of the uterus for implantation of an ovum and is essential for the maintenance of pregnancy. It also prevents ovulation during pregnancy.

estradiol   estrone   progesterone

Although the various steroid hormones have similar structures, they have remarkably different physiological effects. For example, the only difference between testosterone and progesterone is the substituent on the five-membered ring, and the only difference between 5α-dihydrotestosterone and estradiol is a methyl group and three hydrogens on the **A** ring, but these compounds make the difference between being male and being female. These examples illustrate the extreme specificity of biochemical reactions.

Bile Acids

In addition to being the precursor of all the steroid hormones, cholesterol is also the precursor of the *bile acids*. In fact, the word *cholesterol* is derived from the Greek words *chole* meaning "bile" and *stereos* meaning "solid." The bile acids—cholic acid and chenodeoxycholic acid—are synthesized in the liver, stored in the gallbladder, and secreted into the small intestine, where they act as emulsifying agents so that fats and oils can be digested by water-soluble digestive enzymes. Cholesterol is also the precursor of vitamin D (Section 28.6).

cholic acid   chenodeoxycholic acid

> **PROBLEM 30** ◆
>
> **a.** How do the structures of cholic acid and chenodeoxycholic acid differ?
> **b.** Are the three OH groups of cholic acid axial or equatorial?

25.11 SYNTHETIC STEROIDS

The potent physiological effects of steroids led scientists, in their search for new drugs, to synthesize steroids that are not available in nature and to investigate their physiological effects. Two drugs, stanozlol and Dianabol, were developed this way; they have the same muscle-building effect as testosterone.

Steroids that aid in the development of muscle are called *anabolic steroids*. These drugs, available by prescription, are used to treat people suffering from traumas accompanied by muscle deterioration. The same drugs have been administered to athletes and racehorses to increase their muscle mass. Stanozolol was the drug detected in several athletes in the 1988 Olympics. Anabolic steroids, when taken in relatively high dosages, have been found to cause liver tumors, personality disorders, and testicular atrophy.

stanozolol   Dianabol®

Many synthetic steroids have been found to be much more potent than natural steroids. Norethindrone, for example, is better than progesterone in arresting ovulation. Another synthetic steroid, RU 486, when taken along with prostaglandins, terminates pregnancy within the first nine weeks of gestation. Its name comes from Roussel-Uclaf, the French pharmaceutical company where it was first synthesized in 1980, and an arbitrary lab serial number. Notice that these compounds have structures similar to that of progesterone.

norethindrone   mefipristone
                         RU 486

ESSENTIAL CONCEPTS

Section 25.0

- **Lipids** are organic compounds, found in living organisms that are soluble in nonpolar solvents.

Section 25.1

- **Fatty acids** are carboxylic acids with long, unbranched hydrocarbon chains.
- **Saturated fatty acids** have no double bonds; **unsaturated fatty acids** have double bonds.
- The cis double bonds in naturally occurring **polyunsaturated fatty acids** are separated by one CH_2 group.

Section 25.2

- **Waxes** are esters formed from long-chain carboxylic acids and long-chain alcohols.

Section 25.3

- **Triglycerides** are compounds in which the three OH groups of **glycerol** are esterified with fatty acids.
- Triglycerides that are solids or semisolids at room temperature are **fats;** liquid triglycerides are **oils.**

Section 25.4

- Hydrolysis of a fat or an oil in a basic solution forms **soap.**

Section 25.5

- **Phospholipids** are lipids that contain a phosphate group.
- **Phosphoglycerides** differ from triglycerides in that a terminal OH group of glycerol is esterified with phosphoric acid instead of a fatty acid.
- Phosphoglycerides form **membranes** by arranging themselves in a **lipid bilayer.**
- **Sphingolipids** are like phosphoglycerides except they contain **sphingosine** instead of glycerol.

Section 25.6

- **Prostaglandins,** responsible for regulating a variety of physiological responses, are synthesized from arachidonic acid, a 20-carbon fatty acid.

Section 25.7

- **Terpenes** are made by joining five-carbon units, usually in a head-to-tail fashion.
- **Monoterpenes**—terpenes with two **isoprene units**—have 10 carbons, **sesquiterpenes** have 15, **diterpenes** have 20, **triterpenes** have 30, and **tetraterpenes** have 40.
- **Isopentenyl pyrophosphate** is the five-carbon compound used for the biosynthesis of terpenes.

Section 25.8

■ The reaction of **dimethylallyl pyrophosphate** (formed from isopentenyl pyrophosphate) with isopentenyl pyrophosphate forms **geranyl pyrophosphate,** a 10-carbon compound.

■ Reaction of geranyl pyrophosphate with another molecule of isopentenyl pyrophosphate forms **farnesyl pyrophosphate,** a 15-carbon compound.

■ Reaction of farnesyl pyrophosphate with another molecule of isopentenyl pyrophosphate forms **geranylgeranyl pyrophosphate,** a 20-carbon compound.

■ Two molecules of farnesyl pyrophosphate form **squalene,** a 30-carbon compound.

■ Two molecules of geranylgeranyl pyrophosphate form **phytoene,** a 40-carbon compound.

Section 25.9

■ Squalene is the precursor of **lanosterol,** which is the precursor of **cholesterol.**

Section 25.10

■ **Hormones** are chemical messengers; many hormones are steroids.

■ All steroids contain a tetracyclic ring system; β-substituents are on the same side of the ring system as the **angular methyl groups;** α-substituents are on the opposite side.

■ Cholesterol is the precursor of all the **steroid hormones.**

PROBLEMS

31. Dipalmitoylphosphatidylcholine is a surfactant found in the lining of the lungs. It prevents the lungs from collapsing when the lung volume is low and protects the lungs from injuries caused by inhaled particles. Draw its structure.

32. An optically active fat, when completely hydrolyzed, yields twice as much stearic acid as palmitic acid. Draw the structure of the fat.

33. Do all triacylglycerols have the same number of asymmetric centers?

34. a. How many different triacylglycerols are there in which one of the fatty acid components is lauric acid and two are myristic acid?
  b. How many different triacylglycerols are there in which one of the fatty acid components is lauric acid, one is myristic acid, and one is palmitic acid?

35. Cardiolipins are found in heart muscles. Draw the products formed when a cardiolipin undergoes complete acid-catalyzed hydrolysis.

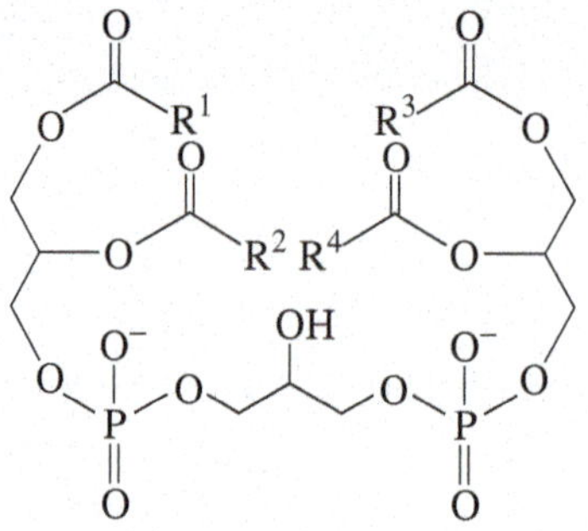

a cardiolipin

36. Nutmeg contains a simple, fully saturated triacylglycerol with a molecular weight of 722. Draw its structure.

37. Draw the product that is obtained from the reaction of cholesterol with the following reagents:
  a. H_2, Pd/C   **b.** acetyl chloride   **c.** H_2SO_4, Δ **d.** H_2O, H^+ **e.** a peroxyacid

38. A chemist synthesized the following samples of mevalonic acid and fed them to a group of lemon trees:

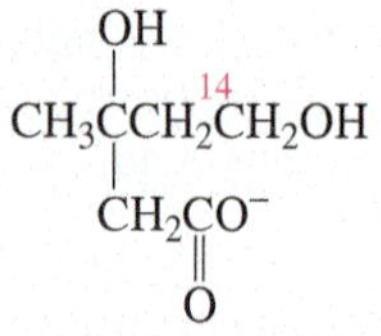

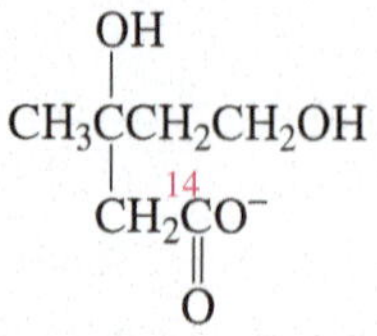

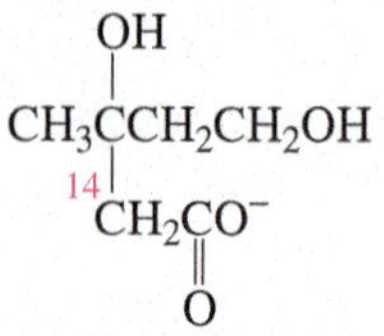

Which carbons in citronellal (page 1144), a compound isolated from lemon oil, will be labeled in trees that are fed the following?
  a. sample A   **b.** sample B   **c.** sample C

39. What compounds are formed when limonene reacts with ozone followed by treatment with dimethyl sulfide?

40. What compound is formed when limonene reacts with excess HBr? Is the compound chiral or achiral?

41. a. Propose a mechanism for the following reaction:

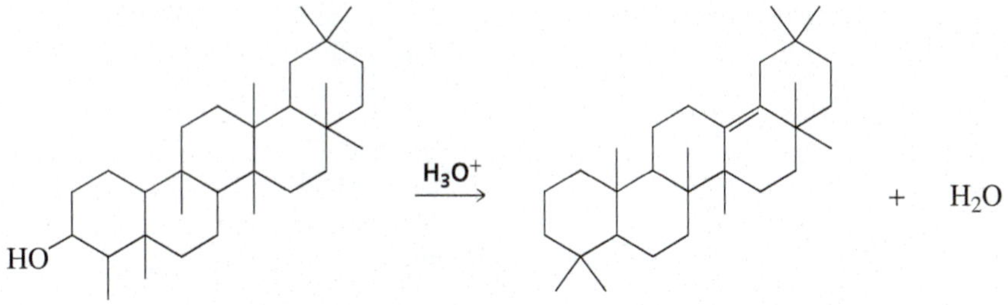
HO
H_3O^+
+ H_2O

b. To what class of terpene does the starting material belong? Mark off the isoprene units in the starting material.

42. Propose a mechanism for the biosynthesis of α-pinene from farnesyl pyrophosphate.

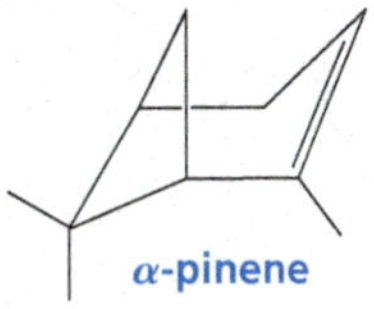

α-pinene

43. 5-Androstene-3,17-dione is isomerized to 4-androstene-3,17-dione by hydroxide ion. Propose a mechanism for this reaction.

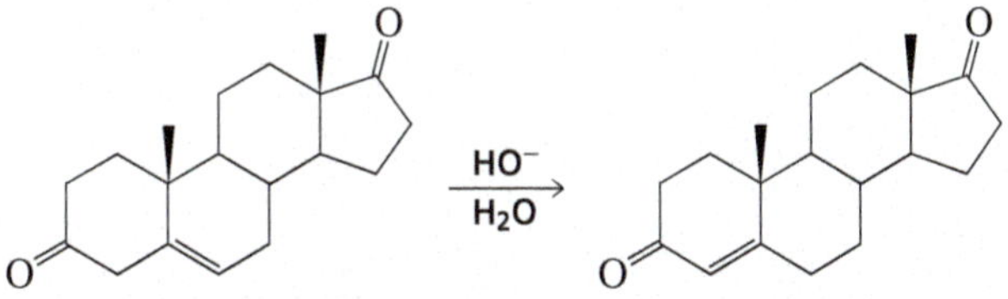
O
O
HO^-
H_2O
O
O

5-androstene-3,17-dione 4-androstene-3,17-dione

44. Both OH groups of the β,β-diol react with excess ethyl chloroformate, but only one OH group of the β,α-diol reacts under the same conditions. Explain the difference in reactivity.

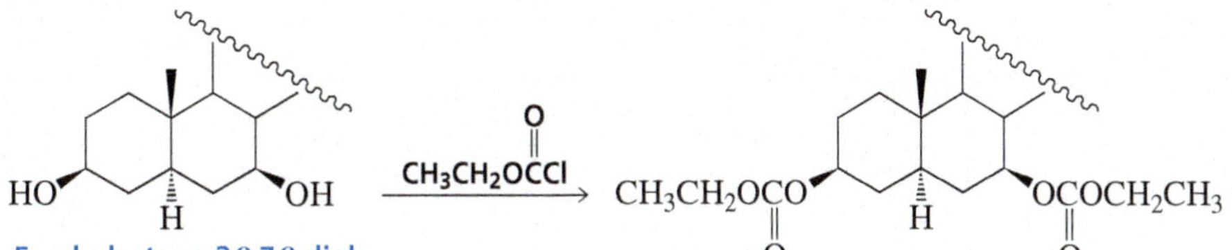
HO
H
OH
O
CH_3CH_2OCCl
CH_3CH_2OCO
O
H
$OCOCH_2CH_3$
O

5α-cholestane-3β,7β-diol

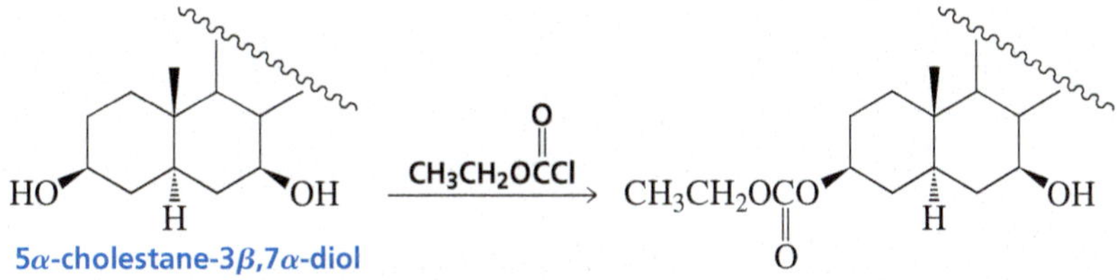
HO
H
OH
O
CH_3CH_2OCCl
CH_3CH_2OCO
O
H
OH

5α-cholestane-3β,7α-diol

45. Eudesmol is a sesquiterpene found in eucalyptus. Propose a mechanism for its biosynthesis.

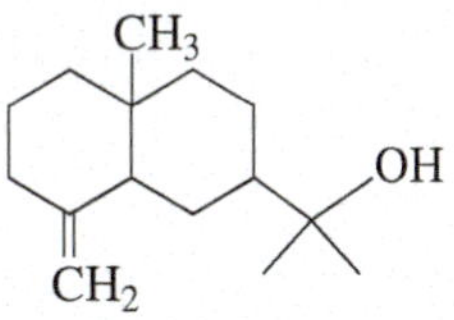
CH_3
OH
CH_2

eudesmol

46. What compounds are obtained when (+)-limonene (its structure is on page 1144) reacts with O_3 followed by dimethyl sulfide?

47. An optically active monoterpene (compound **A**) with molecular formula $C_{10}H_{18}O$ undergoes catalytic hydrogenation to form an optically inactive compound with molecular formula $C_{10}H_{20}O$ (compound **B**). When compound **B** is heated with acid, followed by reaction with O_3 and then with dimethyl sulfide, one of the products obtained is 4-methylcyclohexanone. Give possible structures for compounds **A** and **B**.

48. Diethylstilbestrol (DES) was given to pregnant women to prevent miscarriage, until it was found that the drug caused cancer in both the mothers and their female children. DES has estradiol activity, even though it is not a steroid. Draw DES in a way that shows that it is structurally similar to estradiol.

diethylstilbestrol
DES

49. If the following sesquiterpene is synthesized in a medium containing acetate with a ^{14}C-labeled carbonyl carbon, which carbons will be labeled?

26 The Chemistry of the Nucleic Acids

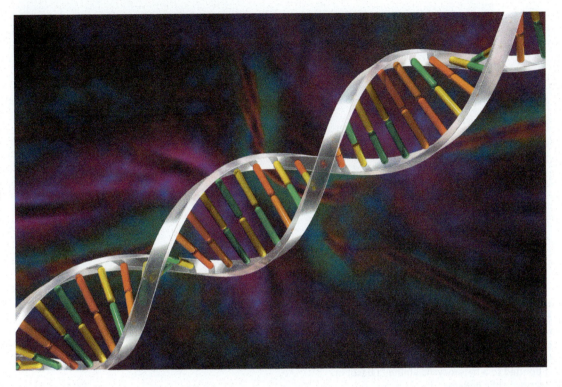

a double helix

We have studied two of the three major kinds of biopolymers: polysaccharides in Chapter 20 and proteins in Chapter 21. Now we will look at the third kind of biopolymer—nucleic acids. There are two types of nucleic acids: deoxyribonucleic acid (DNA) and ribonucleic acid (RNA). DNA encodes an organism's entire hereditary information and controls the growth and division of cells. In all organisms (except certain viruses), the genetic information stored in DNA is transcribed into RNA. This information can then be translated for the synthesis of all the proteins needed for cellular structure and function.

DNA was first isolated in 1869 from the nuclei of white blood cells. Because it was found in the nucleus and was acidic, it was called *nucleic acid*. Eventually, scientists found that the nuclei of all cells contain DNA. The fact that DNA is the carrier of genetic information was not known until 1944, when it was found that DNA could be transferred from one species to another, along with inheritable traits. In 1953, James Watson and Francis Crick described the three-dimensional structure of DNA—the famed double helix.

26.1 NUCLEOSIDES AND NUCLEOTIDES

Nucleic acids are chains of five-membered-ring sugars linked by phosphate groups. Notice that the linkages are **phosphodiesters** (Figure 26.1 on the next page).

- In RNA, the five-membered-ring sugar is D-ribose.
- In DNA, the five-membered-ring sugar is 2′-deoxy-D-ribose (D-ribose without an OH group in the 2′-position).

The Bases in DNA and RNA

The anomeric carbon of each sugar is bonded to a nitrogen of a heterocyclic compound in a β-glycosidic linkage. (Recall from Section 20.10 that a β-linkage is one in which the substituents at C-1 and C-4 are on the same side of the furanose ring.) Because the heterocyclic compounds are amines, they are commonly referred to as **bases**.

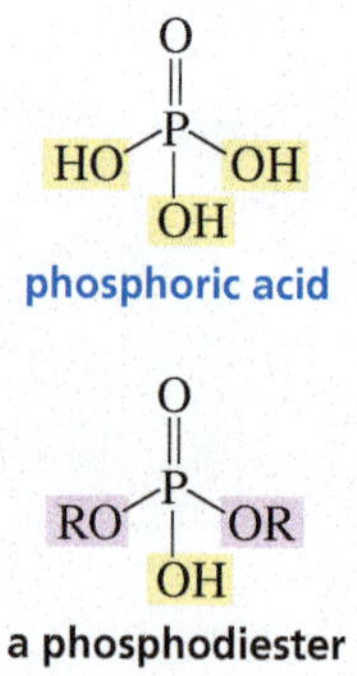

phosphoric acid

a phosphodiester

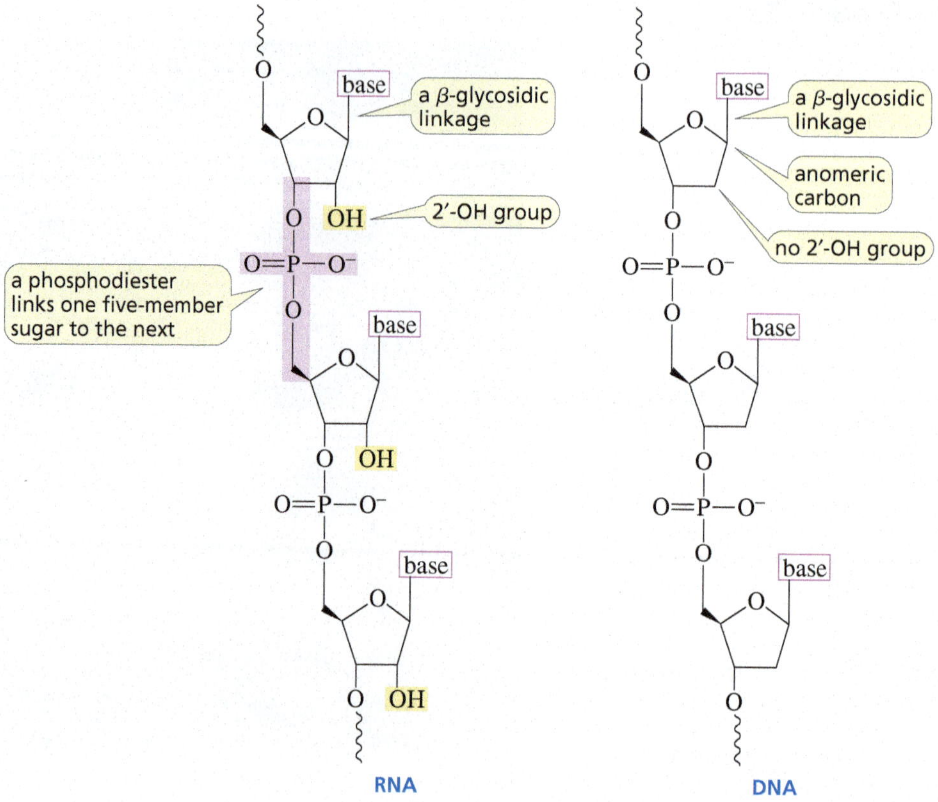

▶ **Figure 26.1**
Nucleic acids consist of a chain of five-membered-ring sugars linked by phosphate groups. The anomeric carbon of each sugar (D-ribose in RNA, 2′-deoxy-D-ribose in DNA) is bonded to a heterocyclic amine (a base) in a β-glycosidic linkage.

a β-glycosidic linkage

2′-OH group

a phosphodiester links one five-member sugar to the next

base

base

RNA

a β-glycosidic linkage

anomeric carbon

no 2′-OH group

base

base

DNA

The vast differences in heredity between different species and between different members of the same species are determined by the sequence of the bases in DNA. Surprisingly, there are only four bases in DNA: two are substituted purines (adenine and guanine), and two are substituted pyrimidines (cytosine and thymine).

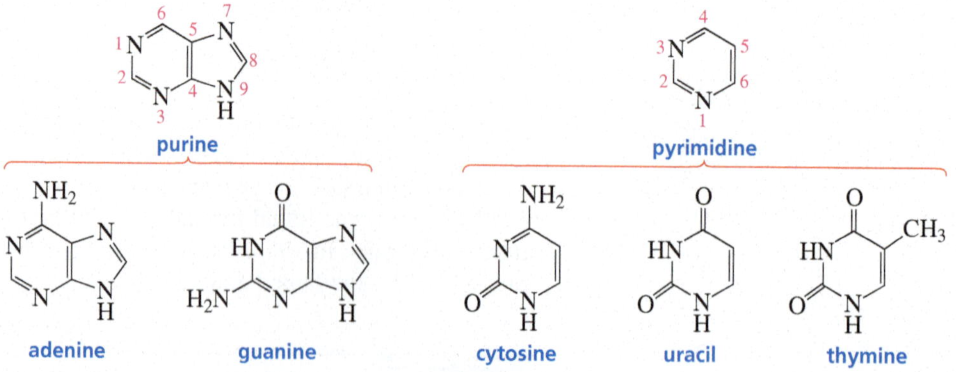

purine

pyrimidine

adenine

guanine

cytosine

uracil

thymine

RNA also contains only four bases. Three (adenine, guanine, and cytosine) are the same as those in DNA, but the fourth base in RNA is uracil instead of thymine. Notice that thymine and uracil differ only by a methyl group. (Thymine is 5-methyluracil.) The reason DNA contains thymine instead of uracil is explained in Section 26.10.

Nucleosides

The anomeric carbon of the furanose ring is bonded to purines at N-9 and to pyrimidines at N-1. A compound containing a base bonded to D-ribose or to 2′-deoxy-D-ribose is called a **nucleoside.** The ring positions of the sugar component of a nucleoside are indicated by primed numbers to distinguish them from the ring positions of the base. This is why the sugar component of DNA is referred to as 2′-deoxy-D-ribose. The nucleosides of RNA—where the sugar is D-ribose— are more precisely called ribonucleosides, whereas the nucleosides of DNA—where the sugar is 2′-deoxy-D-ribose—are called deoxyribonucleosides.

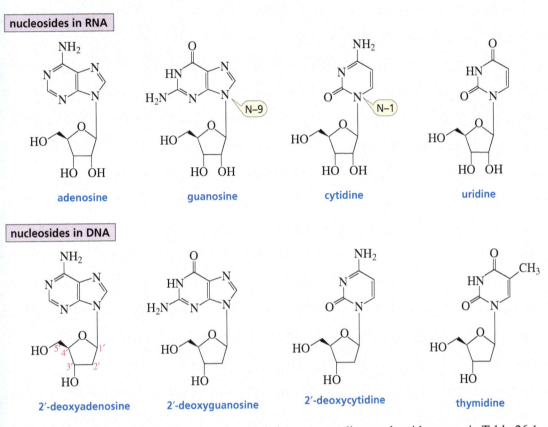

Nucleotides

Notice the difference in the base names and their corresponding nucleoside names in Table 26.1. For example, adenine is the base, whereas adenosine is the nucleoside; similarly, cytosine is the base, whereas cytidine is the nucleoside, and so forth. Because uracil is found only in RNA, it is shown attached to D-ribose but not to 2′-deoxy-D-ribose; because thymine is found only in DNA, it is shown attached to 2′-deoxy-D-ribose but not to D-ribose.

Table 26.1 The Names of the Bases, the Nucleosides, and the Nucleotides

Base	Ribonucleoside	Deoxyribonucleoside	Ribonucleotide	Deoxyribonucleotide
Adenine	Adenosine	2′-Deoxyadenosine	Adenosine 5′-phosphate	2′-Deoxyadenosine 5′-phosphate
Guanine	Guanosine	2′-Deoxyguanosine	Guanosine 5′-phosphate	2′-Deoxyguanosine 5′-phosphate
Cytosine	Cytidine	2′-Deoxycytidine	Cytidine 5′-phosphate	2′-Deoxycytidine 5′-phosphate
Thymine	—	Thymidine	—	Thymidine 5′-phosphate
Uracil	Uridine	—	Uridine 5′-phosphate	

Nucleotides

A **nucleotide** is a nucleoside with an OH group of the sugar bonded in an ester linkage to phosphoric acid. The nucleotides of RNA are more precisely called **ribonucleotides,** and those of DNA are called **deoxyribonucleotides.** The base names in nucleotides are the same as those in nucleosides.

Nucleoside = base + sugar
Nucleotide = base + sugar + phosphate

Because phosphoric acid can form an anhydride, nucleotides can exist as monophosphates, diphosphates, and triphosphates (Section 24.1). They are named by adding *monophosphate* or *diphosphate* or *triphosphate* to the name of the nucleoside.

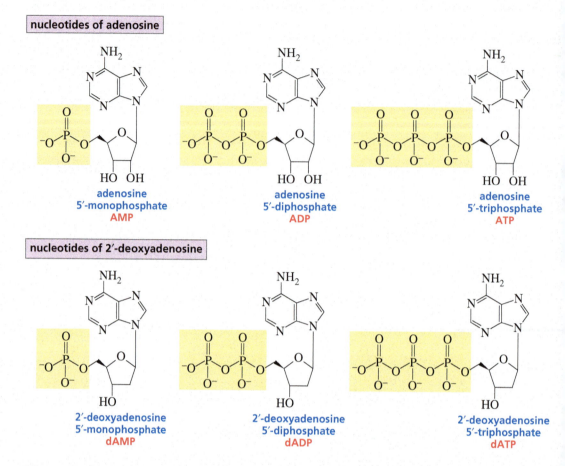

nucleotides of adenosine

adenosine
5'-monophosphate
AMP

adenosine
5'-diphosphate
ADP

adenosine
5'-triphosphate
ATP

nucleotides of 2'-deoxyadenosine

2'-deoxyadenosine
5'-monophosphate
dAMP

2'-deoxyadenosine
5'-diphosphate
dADP

2'-deoxyadenosine
5'-triphosphate
dATP

The names of the nucleotides are abbreviated (A, G, C, T, U—followed by MP, DP, or TP, depending on whether it is a monophosphate, diphosphate, or triphosphate—with a d in front if it contains 2'-deoxy-D-ribose instead of D-ribose: for example, ATP, dATP).

The Structure of DNA: Watson, Crick, Franklin, and Wilkins

James D. Watson was born in Chicago in 1928. He graduated from the University of Chicago at the age of 19 and received a Ph.D. three years later from Indiana University. In 1951, as a postdoctoral fellow at Cambridge University, Watson worked on determining the three-dimensional structure of DNA.

Francis H. C. Crick (1916–2004) was born in Northampton, England. Originally trained as a physicist, Crick did research on radar during World War II. After the war, deciding that the most interesting problem in science was the physical basis of life, he entered Cambridge University to study the structure of biological molecules by X-ray analysis. He was a graduate student when he carried out his portion of the work that led to the proposal of the double helical structure of DNA. He received a Ph.D. in chemistry in 1953.

Rosalind Franklin (1920–1958) was born in London. She graduated from Cambridge University and studied X-ray diffraction techniques in Paris. In 1951 she returned to England and accepted a position to develop an X-ray diffraction unit in the biophysics department at King's College. Her X-ray studies showed that DNA was a helix with the sugars and phosphate groups on the outside of the molecule. Tragically, Franklin never protected herself from her X-ray source and died without knowing the role her work had played in determining the structure of DNA, and without being recognized for her contribution.

Francis Crick (*left*) and James Watson (*right*)

Rosalind Franklin

Watson and Crick shared the 1962 Nobel Prize in Physiology or Medicine with Maurice Wilkins for determining the double-helical structure of DNA. Wilkins (1916–2004), who contributed X-ray studies that confirmed the double-helical structure, was born in New Zealand to Irish immigrants and moved to England six years later with his parents. He received a Ph.D. from Birmingham University. During World War II he joined other British scientists who were working with American scientists on the development of the atomic bomb. He returned to England in 1945 and, having lost interest in physics, turned his attention to biology.

Cyclic AMP

Cyclic AMP (cAMP), which is present in all life forms, controls a wide variety of biological processes. This cyclic nucleotide is called a "second messenger" because it serves as a link between at least 20 different hormones (the first messengers) and enzymes that regulate cellular function.

For example, secretion of a hormone, such as adrenaline, activates adenylate cyclase, the enzyme responsible for the synthesis of cyclic AMP from ATP. Cyclic AMP then activates a regulatory enzyme (Section 24.13), generally by phosphorylating it.

PROBLEM 1

In acidic solutions, nucleosides are hydrolyzed to a sugar and a heterocyclic base. Propose a mechanism for this reaction.

PROBLEM 2

Draw the structure for each of the following:

a. dCDP   c. dUMP   e. guanosine triphosphate

b. dTTP   d. UDP   f. adenosine 3'-monophosphate

26.2 NUCLEIC ACIDS ARE COMPOSED OF NUCLEOTIDE SUBUNITS

Nucleic acids are composed of long strands of nucleotide subunits (Figure 26.1).

- A **dinucleotide** contains two nucleotide subunits.
- An **oligonucleotide** contains 3 to 10 nucleotide subunits.
- A **polynucleotide** contains many nucleotide subunits.

DNA and RNA are polynucleotides.

Biosynthesis of Nucleic Acids

Nucleic acids are biosynthesized from nucleoside triphosphates, using enzymes called *DNA polymerases* (for the synthesis of DNA) or *RNA polymerases* (for the synthesis of RNA). The nucleotides are linked as a result of nucleophilic attack by a 3'-OH group of one nucleoside triphosphate on the α-phosphorus of another nucleoside triphosphate, breaking a phosphoanhydride bond and eliminating pyrophosphate (Figure 26.2). Thus, the phosphodiester joins the 3′-OH group of one nucleotide and the 5′-OH group of the next nucleotide, and the growing polymer is synthesized in the 5′ → 3′ direction. In other words, new nucleotides are added to the 3′-end.

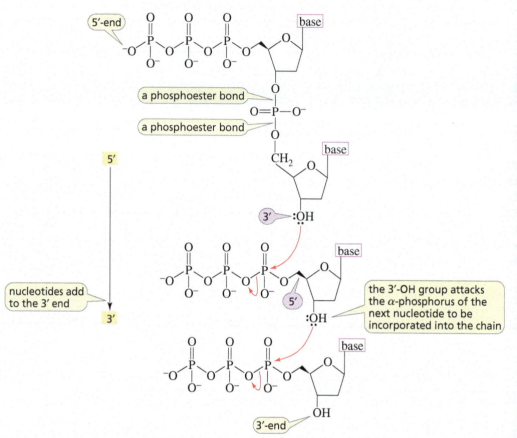

▶ Figure 26.2
Addition of nucleotides to a growing strand of DNA. Biosynthesis occurs in the 5′ → 3′ direction.

The pyrophosphate product is subsequently hydrolyzed, which makes the reaction that joins the nucleotides irreversible. Irreversibility is important if the genetic information in DNA is to be preserved (Section 24.1).

RNA strands are biosynthesized in the same way, using ribonucleoside triphosphates instead of 2′-deoxyribonucleoside triphosphates.

The Primary Structure of a Nucleic Acid

The **primary structure** of a nucleic acid is the sequence of bases in the strand. By convention, the sequence of bases is written in the 5′ → 3′ direction (the 5′-end is on the left). Remember that the nucleotide at the 5′-end of the strand has an unlinked 5′-triphosphate group, and the nucleotide at the 3′-end has an unlinked 3′-hydroxyl group.

ATGAGCCATGTAGCCTAATCGGC

5′-end   3′-end

DNA is synthesized in the 5′ → 3′ direction.

26.3 THE SECONDARY STRUCTURE OF DNA

Watson and Crick concluded, with the aid of Rosalind Franklin's X-ray data, that

- DNA consists of two strands of nucleotides, with the sugar–phosphate backbone on the outside and the bases on the inside.
- the strands are antiparallel (they run in opposite directions).
- the strands are held together by hydrogen bonds between the bases on one strand and the bases on the other strand (Figure 26.3).

The DNA Strands are Complementary

Experiments carried out by Erwin Chargaff were critical to Watson and Crick's proposal for the structure of DNA. These experiments showed that the number of adenines in DNA equals the number of thymines, and the number of guanines equals the number of cytosines. Chargaff also noted that the number of adenines and thymines relative to the number of guanines and cytosines is characteristic of a given species but varies from species to species. In human DNA, for example, 60.4% of the bases are adenines and thymines, whereas 74.2% of the bases are adenines and thymines in the DNA of the bacterium *Sarcina lutea*.

Chargaff's data showing that [adenine] = [thymine] and [guanine] = [cytosine] could be explained if adenine (A) always paired with thymine (T), and guanine (G) always paired with cytosine (C). This means the two strands are *complementary:* where there is an A in one strand, there is a T in the opposing strand; and where there is a G in one strand, there is a C in the other strand (Figure 26.3).

> *Thus, if you know the sequence of bases in one strand,*
> *you can figure out the sequence of bases in the other strand.*

Hydrogen Bonding Dictates Base Pairing

Why does A pair with T? Why does G pair with C? First of all, the width of the double-stranded molecule is relatively constant, so a purine must pair with a pyrimidine. If the larger purines paired, the strands would bulge; if the smaller pyrimidines paired, the strands would have to pull in to bring the two pyrimidines close enough to form hydrogen bonds. But what causes A to pair with T rather than with C (the other pyrimidine)?

The base pairing is dictated by hydrogen bonding. Learning that the bases exist in the keto form and not the enol form (Section 17.2) allowed Watson to explain the pairing.* Adenine forms two hydrogen bonds with thymine but would form only one hydrogen bond with cytosine. Guanine forms three hydrogen bonds with cytosine but would form only one hydrogen bond with thymine (Figure 26.4).

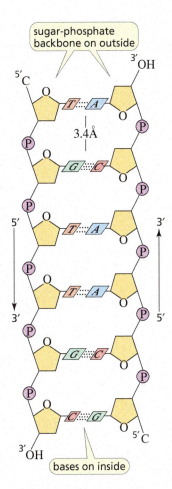

▲ **Figure 26.3**
The sugar–phosphate backbone of DNA is on the outside and the bases are on the inside. As pair with Ts and Gs pair with Cs. The two strands are antiparallel—that is, they run in opposite directions.

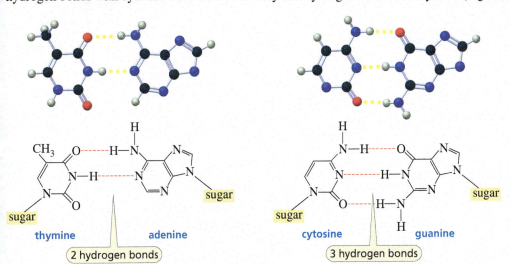

LEARN THE STRATEGY

◀ **Figure 26.4**
Base pairing in DNA: adenine and thymine form two hydrogen bonds; cytosine and guanine form three hydrogen bonds.

---

* Watson was having difficulty understanding the base pairing in DNA because he thought the bases existed in the enol form (see Problem 4). When Jerry Donohue, an American crystallographer, informed him that the bases more likely existed in the keto form, Chargaff's data could easily be explained by hydrogen bonding between adenine and thymine and between guanine and cytosine.

USE THE STRATEGY

PROBLEM 3

Indicate whether each functional group of the five heterocyclic bases in nucleic acids is a hydrogen bond acceptor (A), a hydrogen bond donor (D), or both (D/A).

PROBLEM 4

Using the D, A, and D/A designations in Problem 3, indicate how base pairing would be affected if the bases existed in the enol form.

PROBLEM 5 ♦

If one of the strands of DNA has the following sequence of bases running in the 5′ → 3′ direction,

5′—G—G—A—C—A—A—T—C—T—G—C—3′

a. what is the sequence of bases in the complementary strand?
b. what base is closest to the 5′-end in the complementary strand?

The Double Helix

The two antiparallel DNA strands are not linear but are twisted into a helix around a common axis (Figure 26.5a). The base pairs are planar and parallel to each other on the inside of the helix (Figure 26.5c). The secondary structure is, therefore, known as a **double helix.** The double helix resembles a circular staircase: the base pairs are the rungs, and the sugar–phosphate backbones are the handrails. The OH group of the phosphodiester linkages has a pK_a of about 2, so it is in its basic form (negatively charged) at physiological pH (Figure 26.2). The negatively charged phosphates repels nucleophiles, thereby preventing cleavage of the phosphodiester bonds.

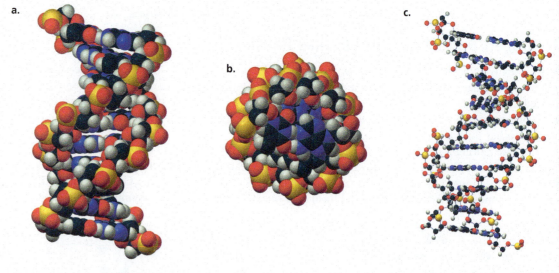

a.

b.

c.

▲ **Figure 26.5**
a. **The DNA double helix.**
b. **The view looking down the long axis of the helix. (The blue nitrogen atoms of the bases are on the inside; the yellow phosphorus atoms are on the outside.)**
c. **The bases are planar and parallel on the inside of the helix.**

Hydrogen bonding between base pairs is just one of the forces holding the two strands of the DNA double helix together. The bases are planar aromatic molecules that stack on top of one another, each pair slightly rotated with respect to the next pair, like a partially spread-out hand of cards. In this arrangement, there are favorable interactions between the mutually induced dipoles of adjacent pairs of bases. These interactions, known as **stacking interactions,** are weak attractive forces, but when added together they contribute significantly to the stability of the double helix.

Confinement of the bases to the inside of the helix has an additional stabilizing effect—it reduces the surface area of the relatively nonpolar residues that is exposed to water, which increases the entropy of the surrounding water molecules (Section 21.15). Stacking interactions are strongest between two purines and weakest between two pyrimidines.

There are two different alternating grooves in a DNA double helix; a **major groove** and a narrower **minor groove.** Proteins and other molecules can bind to the grooves. The hydrogen-bonding properties of the functional groups facing into each groove determine what kind of molecules will bind to the groove. For example, netropsin is an antibiotic that works by binding to the minor groove of DNA (Figure 26.6).

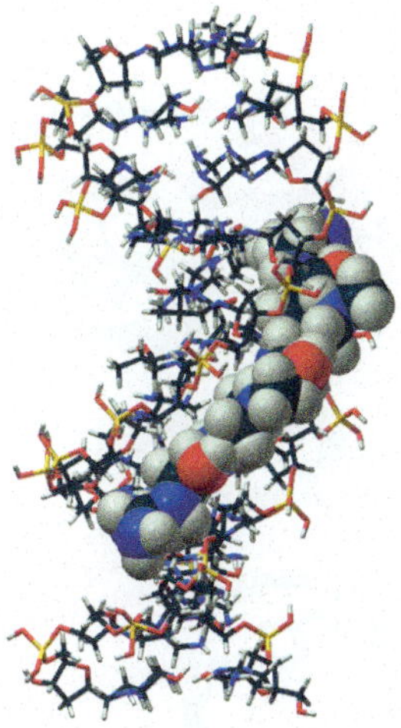

26.4 WHY DNA DOES NOT HAVE A 2′-OH GROUP

Unlike DNA, RNA is not stable because the 2′-OH group of ribose acts as a nucleophilic catalyst for the cleavage of RNA (Figure 26.7). This explains why the 2′-OH group is absent in DNA. DNA must remain intact throughout the life span of a cell in order to preserve the genetic information. Easy cleavage of DNA would have disastrous consequences for the cell and for life itself. RNA, in contrast, is synthesized as it is needed and is degraded once it has served its purpose.

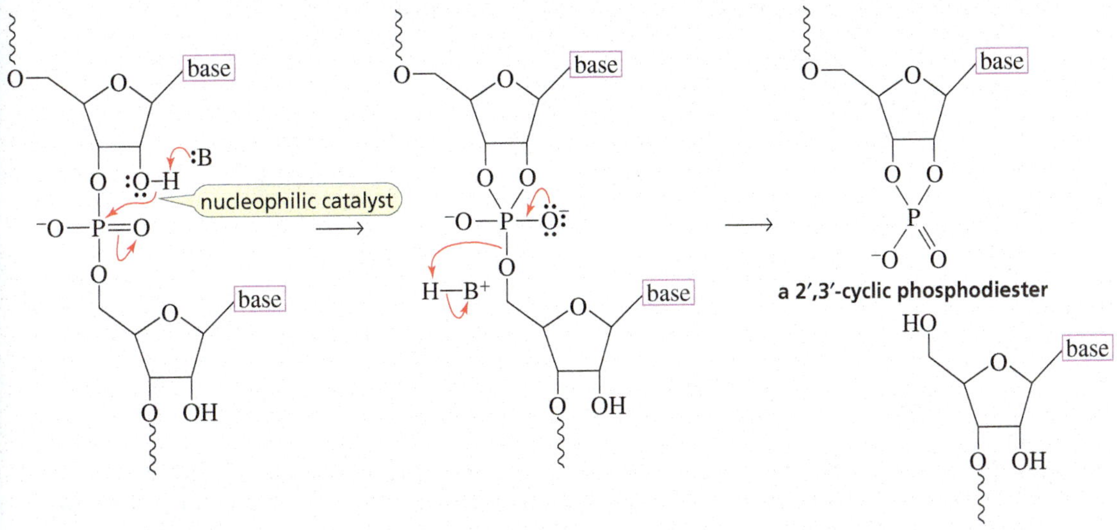

▲ Figure 26.6
The antibiotic netropsin bound in the minor groove of DNA.

▲ **Figure 26.7**
Catalysis of RNA cleavage by the 2′-OH group. RNA undergoes cleavage 3 billion times faster than DNA.

PROBLEM 6

The 2′,3′-cyclic phosphodiester that is formed (Figure 26.7) when RNA is cleaved, forms a mixture of nucleotide 2′- and 3′-phosphates when it reacts with water. Propose a mechanism for this reaction.

26.5 THE BIOSYNTHESIS OF DNA IS CALLED REPLICATION

The genetic information of a human cell is contained in 23 pairs of chromosomes. Each chromosome is composed of thousands of **genes** (segments of DNA). The total DNA from a human cell—the **human genome**—contains 3.1 billion base pairs.

Part of the excitement created by Watson and Crick's proposed structure for DNA was that the structure immediately suggested how DNA is able to pass on genetic information to succeeding generations. Because the two strands are complementary, both carry the same genetic information. Thus, when organisms reproduce, DNA molecules can be copied using the same base-pairing principle that is fundamental to their structure—that is, each strand can serve as the template for the synthesis of a complementary new strand (Figure 26.8 on the next page). The new (daughter) DNA molecules are identical to the original (parent) molecule, so they contain all the original genetic information. The synthesis of identical copies of DNA is called **replication.**

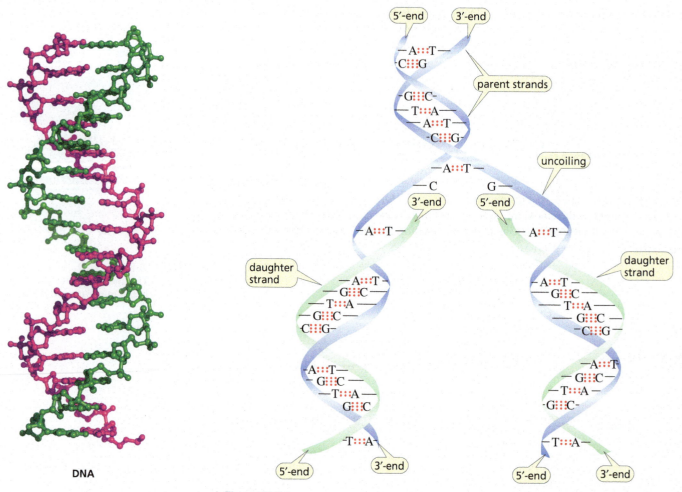

DNA

▲ **Figure 26.8**
Replication of DNA. The green daughter strand on the left is synthesized continuously in the 5′ → 3′ direction; the green daughter strand on the right is synthesized discontinuously in the 5′ → 3′ direction.

All the reactions involved in nucleic acid synthesis are catalyzed by enzymes. The synthesis of DNA takes place in a region of the molecule where the strands have started to separate. Because a nucleic acid can be synthesized only in the 5′ → 3′ direction, only the daughter strand on the left in Figure 26.8 is synthesized continuously in a single piece (because it is synthesized in the 5′ → 3′ direction).

LEARN THE STRATEGY

The other daughter strand needs to grow in a 3′ → 5′ direction, so it is synthesized discontinuously in small pieces. Each piece is synthesized in the 5′ → 3′ direction, and the fragments are joined together by an enzyme called DNA ligase (see Figure 21.10 on page 1021). Each of the two new molecules of DNA—called daughter molecules—contains one of the original parent strands (blue strand in Figure 26.8) plus a newly synthesized strand (green strand). This process is called **semiconservative replication.**

USE THE STRATEGY

PROBLEM 7

Using a dark line for the original parental DNA and a wavy line for DNA synthesized from parental DNA, show what the population of DNA molecules would look like in the fourth generation. (Parental DNA is the first generation.)

26.6 DNA AND HEREDITY

If DNA contains hereditary information, there must be a method to decode that information. The decoding occurs in two steps.

1. The sequence of bases in DNA provides a blueprint for the synthesis of RNA; the synthesis of RNA from a DNA blueprint is called **transcription** (Section 26.7).

2. The sequence of bases in RNA determines the sequence of amino acids in a protein; the synthesis of a protein from an RNA blueprint is called **translation** (Section 26.9).

Don't confuse transcription and translation: these words are used just as they are used in English. Transcription (DNA to RNA) is copying *within the same language*—in this case, the language of nucleotides. Translation (RNA to protein) is *changing to another language*—the language of amino acids. First we will look at transcription.

Transcription: DNA → RNA
Translation: mRNA → protein

Natural Products That Modify DNA

More than three-quarters of clinically approved anticancer drugs are natural products—compounds derived from plants, marine organisms, or microbes—that interact with DNA. Because cancer is characterized by the uncontrolled growth and proliferation of cells, compounds that interfere with the replication or transcription of DNA stop the growth of cancer cells. These drugs can interact with DNA by binding between the base pairs (called intercalation) or by binding to either its major or minor groove. The three anticancer drugs discussed here were isolated from *Streptomyces* bacteria found in soil.

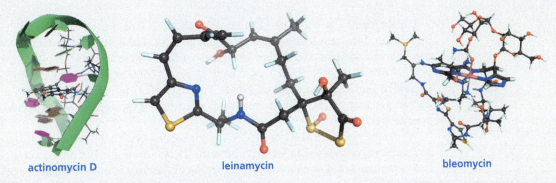

actinomycin D   leinamycin   bleomycin

Because intercalating compounds become sandwiched between the stacked bases in DNA, they are planar and often aromatic. Their binding to DNA is stabilized by stacking interactions with neighboring base pairs. Actinomycin D is an example of an intercalator. When this drug binds to DNA, it distorts the double helix, inhibiting both the replication and transcription of DNA. Actinomycin D has been used to treat a variety of cancers.

Drugs that bind to the major and minor grooves of DNA do so by a combination of hydrogen bonding, London dispersion forces, and electrostatic attractions—the same forces proteins use to bind their substrates. Leinamycin is an example of an anticancer drug that binds to the major groove. Once leinamycin is bound, it alkylates the N-7 position of a purine ring.

Bleomycin binds to the minor groove of DNA. Once in the minor groove, it uses a bound iron atom to remove a hydrogen atom from DNA, the first step in cleaving DNA. This drug has been approved for the treatment of Hodgkin lymphoma.

26.7 THE BIOSYNTHESIS OF RNA IS CALLED TRANSCRIPTION

Transcription starts when DNA unwinds at a particular site—called a *promoter site*—to form two single strands. One of the strands is called the **sense strand.** The complementary strand is called the **template strand.**

In order for RNA to be synthesized in the $5' \rightarrow 3'$ direction, the template strand is read in the $3' \rightarrow 5'$ direction (Figure 26.9 on the next page). The bases in the template strand specify the bases that need to be incorporated into RNA, following the same base-pairing principle used in the replication of DNA. For example, each guanine in the template strand specifies the incorporation of a cytosine into RNA, and each adenine in the template strand specifies the incorporation of a uracil into RNA. (Recall that in RNA, uracil is used instead of thymine.)

Because both RNA and the sense strand of DNA are complementary to the template strand, RNA and the sense strand of DNA have the same base sequence, except that RNA has a uracil wherever the sense strand has a thymine. Just as there are promoter sites in DNA that signal where to start RNA synthesis, there are sites signaling that no more bases should be added to the growing RNA chain.

RNA is synthesized in the 5′ → 3′ direction.

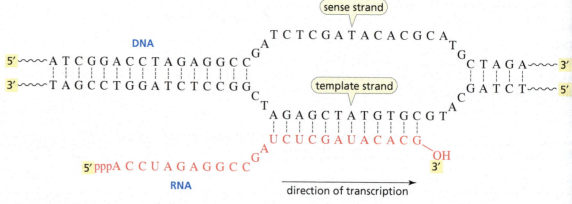

▲ **Figure 26.9**
Transcription: using DNA as a blueprint for the synthesis of RNA (ppp stands for the triphosphate group at the 5′ end of RNA).

Until recently, it was thought that only about 2% of the DNA in our cells was used to make proteins and the rest had no informational content. However, in the decade since the first human genome was sequenced (Section 26.12), more than 400 scientists have produced the *Encyclopedia of DNA Elements* (ENCODE), as a result of about 1600 experiments. This information has greatly expanded our knowledge about DNA. The biological purpose of about 80% of the DNA in the human genome has now been identified, and future experiments are expected to identify the purpose of the rest.

Apparently, a large amount of DNA is for regulation. There are about 150 types of human cells, and each one carries the DNA that codes for 21,000 proteins. But only a subset of these is activated in a particular cell. For example, the gene that makes hair is not activated in a cell that makes insulin and vice versa.

It is now known that there are about 30,000 additional genes that are a blueprint for RNA that is not subsequently translated to make proteins. Instead, the RNA is used for regulation. In other words, these RNA strands appear to be the switches that turn genes on and off. The enormous number of switches has surprised scientists. Now the problem is to find out how these switches work.

There Are More Than Four Bases in DNA

For some time it has been known that a fifth base is found in DNA—namely, 5-methylcytosine. This base silences genes so they are no longer transcribed. Recently, an enzyme was discovered that converts 5-methylcytosine to 5-hydroxymethylcytosine. It appears that 5-hydroxymethylcytosine also plays a role in turning genes on and off.

5-methylcytosine   5-hydroxymethylcytosine   5-formylcytosine

Now yet another base has been discovered. Again it is a 5-substituted cysteine—in this case, it is 5-formylcysteine. 5-Formylcysteine was found in embryonic stem cell DNA. Its function is yet to be determined but it is expected that it plays a role in transforming a fertilized egg into an embryonic stem cell.

PROBLEM 8 ◆

Why do both thymine and uracil specify the incorporation of adenine?

26.8 THE RNAs USED FOR PROTEIN BIOSYNTHESIS

RNA molecules are much shorter than DNA molecules and are generally single-stranded. Although DNA molecules have billions of base pairs, RNA molecules rarely have more than 10,000 nucleotides. There are several kinds of RNA. The RNAs used for protein biosynthesis are

- **messenger RNA (mRNA),** whose sequence of bases determines the sequence of amino acids in a protein
- **ribosomal RNA (rRNA),** a structural component of ribosomes—the particles on which the biosynthesis of proteins takes place
- **transfer RNA (tRNA),** the carrier of amino acids used for protein synthesis

tRNA molecules are much smaller than mRNA or rRNA molecules. A tRNA contains only 70 to 90 nucleotides. The single strand of tRNA is folded into a characteristic cloverleaf structure, with three loops and a little bulge next to the right-hand loop (Figure 26.10a). There are at least four regions with complementary base pairing. The three bases at the bottom of the loop directly opposite the 5′- and 3′-ends are called an **anticodon.** All tRNAs have a CCA sequence at the 3′-end (Figures 26.10a and 26.10b).

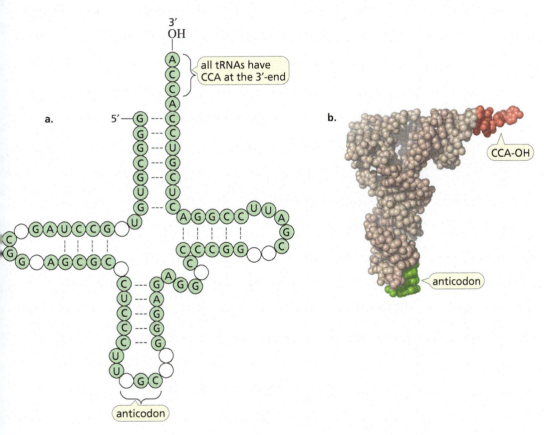

▲ **Figure 26.10**
  a. **A transfer RNA. Compared with other RNAs, tRNA contains a high percentage of unusual bases (shown as empty circles). These bases result from enzymatic modification of the four normal bases.**
  b. **A transfer RNA: the anticodon is green; the CCA at the 3′-end is red.**

An Amino Acid Is Attached to a tRNA at Its 3′-End

Each tRNA can carry an amino acid bound as an ester to its terminal 3′-OH group. The amino acid is inserted into a protein during protein biosynthesis. Each tRNA can carry only one particular amino acid. A tRNA that carries alanine is designated as tRNA^{Ala}.

The attachment of an amino acid to a tRNA is catalyzed by an enzyme called aminoacyl-tRNA synthetase. The mechanism for the reaction is shown below.

MECHANISM FOR THE ATTACHMENT OF AN AMINO ACID TO A tRNA

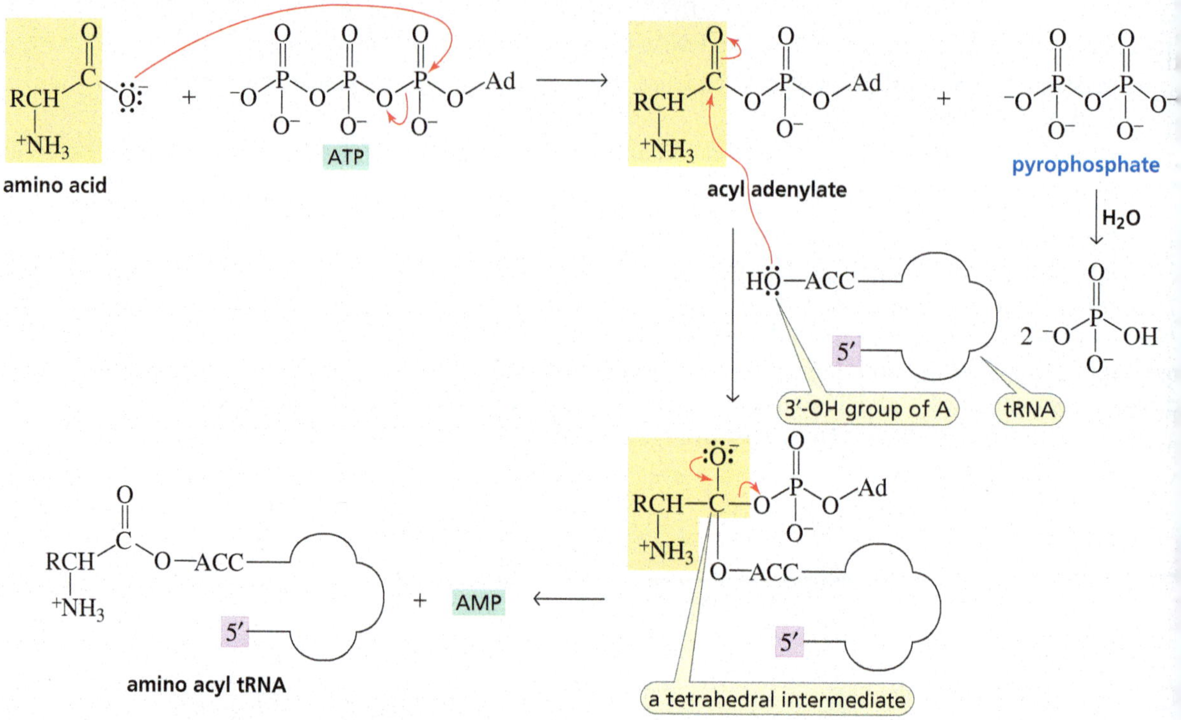

- The carboxylate group of the amino acid is activated by forming an acyl adenylate; now the amino acid has a good leaving group (Section 24.1).
- The pyrophosphate that is eliminated is subsequently hydrolyzed, ensuring the irreversibility of the phosphoryl transfer reaction (Section 24.1).
- The 3'-OH group of tRNA adds to the carbonyl carbon of the acyl adenylate, forming a tetrahedral intermediate.
- The aminoacyl tRNA is formed when AMP is eliminated from the tetrahedral intermediate.

Notice that the second and third steps are the two steps of a nucleophilic acyl substitution reaction.

All the steps for attaching the amino acid to its tRNA take place at the active site of the enzyme. Each amino acid has its own aminoacyl-tRNA synthetase. Each synthetase has two binding sites, one for the amino acid and one for the tRNA that will carry that amino acid (Figure 26.11).

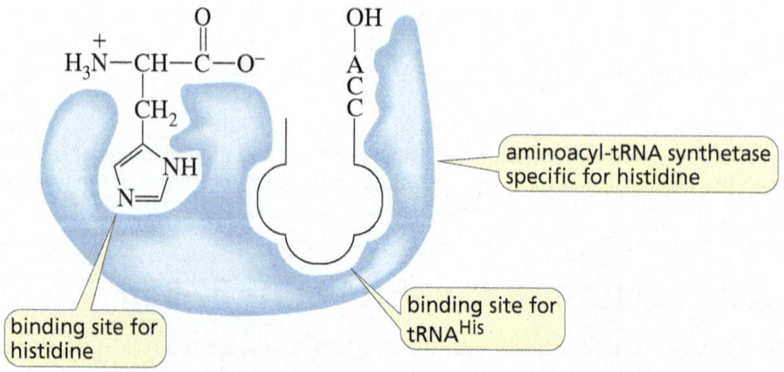

▲ Figure 26.11
An aminoacyl-tRNA synthetase has a binding site for the amino acid and a binding site for the tRNA that will carry that amino acid. In this example, histidine is the amino acid, and tRNA^His is the tRNA.

The Synthetases Correct Their Mistakes

It is critical that the correct amino acid be attached to the tRNA. Otherwise, the protein will not be synthesized with the desired amino acid sequence. Fortunately, the synthetases correct their own mistakes. For example, valine and threonine are approximately the same size, but threonine has an OH group in place of a CH_3 group of valine. Both amino acids, therefore, can bind at the amino acid–binding site of the aminoacyl-tRNA synthetase for valine, and both can then be activated by reacting with ATP to form an acyl adenylate. The aminoacyl-tRNA synthetase for valine has two adjacent catalytic sites. In addition to the site for attaching the acyl adenylate to tRNA, it has a site for hydrolyzing the acyl adenylate.

The site for attaching the acyl adenylate to tRNA is hydrophobic, so the valine acyl adenylate binds preferentially to that site. The site for hydrolyzing the acyl adenylate is polar, so the threonine acyl adenylate binds to that site. Thus, if threonine is activated by the aminoacyl-tRNA synthetase for valine, it will be hydrolyzed rather than transferred to the tRNA.

26.9 THE BIOSYNTHESIS OF PROTEINS IS CALLED TRANSLATION

Codons

A protein is biosynthesized from its N-terminal end to its C-terminal end by a process that reads the bases along the mRNA strand in the $5' \rightarrow 3'$ direction. The amino acid that is to be incorporated into a protein is specified by a three-base sequence called a **codon.** The bases are read consecutively and are never skipped. The three-base sequences and the amino acid that each sequence codes for are known as the **genetic code** (Table 26.2). A codon is written with the 5'-nucleotide on the left. For example, the codon UCA on mRNA codes for the amino acid serine, whereas CAG codes for glutamine.

Table 26.2 The Genetic Code

5'-Position		Middle position			3'-Position
	U	**C**	**A**	**G**	
U	Phe	Ser	Tyr	Cys	U
	Phe	Ser	Tyr	Cys	C
	Leu	Ser	Stop	Stop	A
	Leu	Ser	Stop	Trp	G
C	Leu	Pro	His	Arg	U
	Leu	Pro	His	Arg	C
	Leu	Pro	Gln	Arg	A
	Leu	Pro	Gln	Arg	G
A	Ile	Thr	Asn	Ser	U
	Ile	Thr	Asn	Ser	C
	Ile	Thr	Lys	Arg	A
	Met	Thr	Lys	Arg	G
G	Val	Ala	Asp	Gly	U
	Val	Ala	Asp	Gly	C
	Val	Ala	Glu	Gly	A
	Val	Ala	Glu	Gly	G

Yellow: nonpolar side chains
Green: AUG is part of the initiation signal (see Problem 26) and codes for internal methionine residues
Pink: neutral-polar side chains
Purple: charged polar side chains

Because there are four bases and the codons are triplets, 4^3 (or 64) different codons are possible. This is many more than are needed to specify the 20 different amino acids, so all the amino acids—except methionine and tryptophan—have more than one codon. It is not surprising, therefore, that methionine and tryptophan are the least abundant amino acids in proteins. Actually, 61 of the codons specify amino acids and three codons are stop codons. **Stop codons** tell the cell to "stop protein synthesis here."

How Translation Occurs

How the information in mRNA is translated into a polypeptide is shown in Figure 26.12. In this figure, serine, specified by the codon AGC, was the last amino acid incorporated into the growing polypeptide chain.

A protein is biosynthesized in the N-terminal → C-terminal direction

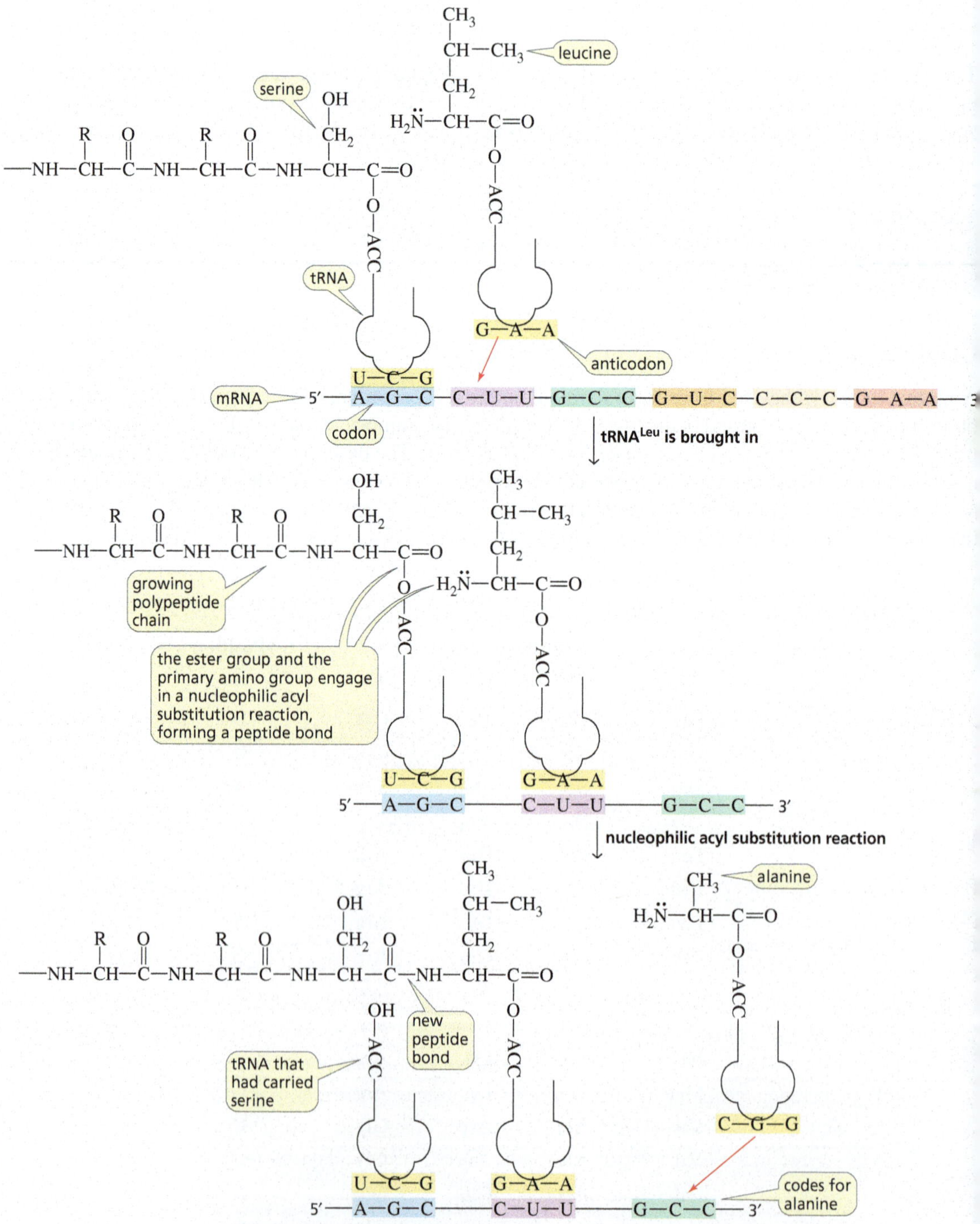

▶ Figure 26.12
Translation: the sequence of bases in mRNA determines the sequence of amino acids in a protein.

LEARN THE STRATEGY

- Serine was specified by the AGC codon in mRNA because the anticodon of the tRNA that carries serine is GCU (that is, 3′-UCG-5′). (Remember that a base sequence is read in the 5′ → 3′ direction, so the sequence of bases in an anticodon must be read from right to left.)
- The next codon, CUU, signals for a tRNA with an anticodon of AAG (3′-GAA-5′). That particular tRNA carries leucine. The amino group of leucine reacts in an enzyme-catalyzed nucleophilic acyl substitution reaction with the ester on the adjacent serine-carrying tRNA, displacing the tRNA that brought in serine.
- The next codon (GCC) specifies a tRNA that carries alanine. The amino group of alanine reacts in an enzyme-catalyzed nucleophilic acyl substitution reaction with the ester group on the adjacent leucine-carrying tRNA, displacing the tRNA that brought in leucine.

Subsequent amino acids are brought in one at a time in the same way, with the codon in mRNA specifying the amino acid to be incorporated by complementary base pairing with the anticodon of the tRNA that carries that amino acid.

PROBLEM 9 ♦

USE THE STRATEGY

If methionine is always the first amino acid incorporated into an oligopeptide, what oligopeptide is coded for by the following stretch of mRNA?

5′—G—C—A—U—G—G—A—C—C—C—C—G—U—U—A—U—U—A—A—A—C—A—C—3′

PROBLEM 10 ♦

Four Cs occur in a row in the segment of mRNA shown in Problem 9. What oligopeptide would be formed from the mRNA if one of the four Cs is cut out of the strand?

PROBLEM 11

UAA is a stop codon. Why does the UAA sequence in the segment of mRNA in Problem 9 not cause protein synthesis to stop?

Sickle Cell Anemia

Sickle cell anemia is an example of the damage that can be caused by a change in a single base of DNA (Problem 79 in Chapter 21). It is a hereditary disease caused when a GAG triplet becomes a GTG triplet in the sense strand of a section of DNA that codes for the β-subunit of hemoglobin (Section 21.16). As a consequence, the mRNA codon becomes GUG—which signals for incorporation of valine—rather than GAG, which would have signaled for incorporation of glutamate. The change from a polar glutamate to a nonpolar valine is sufficient to change the shape of the deoxyhemoglobin molecule. The change in shape stiffens the cells, making it difficult for them to squeeze through capillaries. Blockage of capillaries causes severe pain and can be fatal.

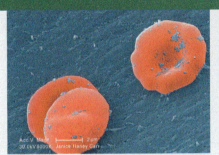
normal red blood cells

sickle red blood cell

Where Transcription and Translation Take Place

Protein synthesis takes place on a ribosome, which is composed of ribosomal RNA (rRNA) and protein (Figure 26.13 on the next page). The ribosome has three binding sites for RNA molecules. It binds the mRNA whose base sequence is to be read, the tRNA that carries the growing peptide chain, and the tRNA that carries the next amino acid to be incorporated into the protein.

1. Transcription of DNA occurs in the nucleus of the cell. The initial RNA transcript is the precursor of all RNA.
2. The initially formed RNA often must be chemically modified before it acquires biological activity. Modification can entail removing nucleotide segments, adding nucleotides to the 5′- or 3′-ends, or chemically altering certain nucleotides.
3. Proteins are added to rRNA to form ribosomes. tRNA, mRNA, and ribosomes leave the nucleus.
4. Each tRNA binds the appropriate amino acid.
5. tRNA, mRNA, and a ribosome work together to translate the information in mRNA into a protein.

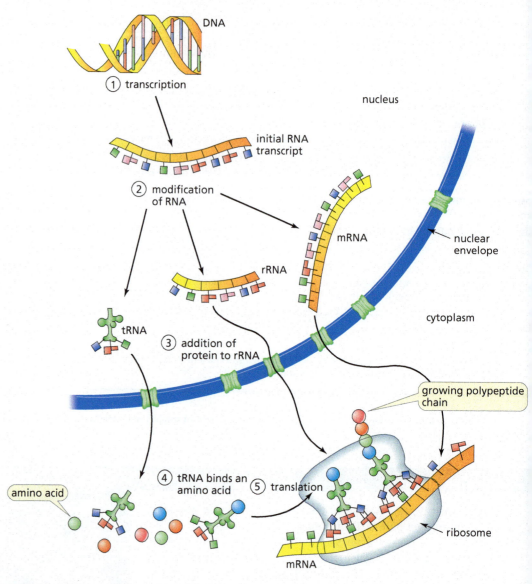

▲ Figure 26.13
Transcription and translation.

Antibiotics That Act by Inhibiting Translation

Puromycin is a naturally occurring antibiotic, one of several that acts by inhibiting translation. It does so by mimicking the 3′-CCA-aminoacyl portion of a tRNA, fooling the enzyme into transferring the growing peptide chain to the NH_2 group of puromycin rather than to the NH_2 group of the incoming 3′-CCA-aminoacyl tRNA (see the mechanism on page 1168). As a result, protein synthesis stops. Because puromycin blocks protein synthesis in eukaryotes as well as in prokaryotes, it is poisonous to humans and, therefore, is not a clinically useful antibiotic. To be clinically useful, an antibiotic must affect protein synthesis only in prokaryotic cells.

Clinically useful antibiotics	Mode of action
Tetracycline	prevents the aminoacyl-tRNA from binding to the ribosome
Erythromycin	prevents the incorporation of new amino acids into the protein
Streptomycin	inhibits the initiation of protein synthesis
Chloramphenicol	prevents the new peptide bond from being formed

puromycin

PROBLEM 12 ◆

A change in which base of a codon is least likely to damage a protein?

PROBLEM 13 ◆

Write the sequences of bases in the sense strand of DNA that resulted in the mRNA in Problem 9.

PROBLEM 14

List the possible codons on mRNA that specify each amino acid in Problem 9 and the anticodon on the tRNA that carries that amino acid.

26.10 WHY DNA CONTAINS THYMINE INSTEAD OF URACIL

In Section 23.7 we saw that dTMP is formed by methylating dUMP, with coenzyme N^5,N^{10}-methylenetetrahydrofolate supplying the methyl group.

R′ = 2′-deoxyribose-5-P

Because the incorporation of the methyl group into uracil oxidizes tetrahydrofolate to dihydrofolate, dihydrofolate must be reduced back to tetrahydrofolate to prepare the coenzyme for another catalytic reaction. The reducing agent is NADPH.

$$\text{dihydrofolate} \ + \ \text{NADPH} \ + \ \text{H}^+ \ \xrightarrow{\text{dihydrofolate reductase}} \ \text{tetrahydrofolate} \ + \ \text{NADP}^+$$

The NADP$^+$ formed in this reaction has to be reduced back to NADPH by NADH. Every NADH formed in a cell can result in the formation of 2.5 ATPs (Section 24.10). Therefore, reducing dihydrofolate comes at the expense of ATP production. This means that the synthesis of thymine is energetically expensive, so there must be a good reason for DNA to contain thymine instead of uracil.

The presence of thymine instead of uracil in DNA prevents potentially lethal mutations. Cytosine can tautomerize to form an imine (Section 17.2), which can be hydrolyzed to uracil (Section 16.8). The overall reaction is called a **deamination** because it removes an amino group.

If a C in DNA is deaminated to a U, the U will specify incorporation of an A into the daughter strand during replication instead of the G that would have been specified by C, and all the progeny of the daughter strand would have the same mutated chromosome. Fortunately, there is an enzyme that recognizes a U in DNA as a "mistake." It cuts it out and replaces it with a C before an incorrect base can be inserted into the daughter strand. If Us were normally found in DNA, the enzyme would not be able to distinguish between a normal U and a U formed by deamination of a cytosine. Having Ts in place of Us in DNA allows the Us that are found in DNA to be recognized as mistakes.

Unlike DNA that replicates itself, any mistake in RNA does not survive for long because RNA is continually degraded and then resynthesized from the DNA template. Therefore, changing a C to a U in RNA could lead to some copies of a defective protein, but most would not be defective. Thus, it is not worth incurring the loss of ATP to incorporate Ts into RNA.

Antibiotics Act by a Common Mechanism

Recently, it has been found that three different classes of antibiotics (a β-lactam, a quinolone, and an aminoglycoside) all kill bacteria in the same way. The antibiotics trigger the production of hydroxide radicals. The hydroxide radicals oxidize guanines to 8-oxoguanines. The cell is able to recognize 8-oxoguanines as mistakes and replace them with guanines. However, if there are too many 8-oxoguanines in DNA, the cell's repair mechanism becomes overwhelmed. Then, instead of cutting out the 8-oxoguanines, it breaks the DNA strand, which leads to cell death.

8-oxoguanine

PROBLEM 15 ◆

Adenine can be deaminated to hypoxanthine, and guanine can be deaminated to xanthine. Draw structures for hypoxanthine and xanthine.

PROBLEM 16

Explain why thymine cannot be deaminated.

26.11 ANTIVIRAL DRUGS

Relatively few clinically useful drugs have been developed for viral infections. The slow progress of this endeavor is due to the nature of viruses and the way they replicate. Viruses are smaller than bacteria and consist of nucleic acid—either DNA or RNA—surrounded by a coat of protein. Some viruses penetrate the host cell; others merely inject their nucleic acid into the cell. In either case, the viral nucleic acid is transcribed by the host and is integrated into the host genome.

Most **antiviral drugs** are analogues of nucleosides, interfering with the virus's nucleic acid synthesis. In this way, they prevent the virus from replicating.

Acyclovir

Acyclovir, the drug used against herpes viruses, has a three-dimensional shape similar to guanine. Therefore, acyclovir can fool the virus into incorporating the drug instead of guanine into its DNA. Once this happens, the DNA strand can no longer grow because acyclovir lacks a ribose with a 3′-OH group. The terminated DNA remains bound to DNA polymerase (the enzyme that catalyzes the synthesis of DNA), which irreversibly inactivates the enzyme (Section 26.2).

acyclovir
Aclovir®
used against
herpes simplex infections

cytarabine
Cytosar®
used against acute
myelocytic leukemia

idoxuridine
Herplex®
approved for topical
ophthalmic use

Cytarabine

Cytarabine, used for acute myelocytic leukemia, competes with cytosine for incorporation into viral DNA. Cytarabine contains an arabinose rather than a ribose (Table 20.1). Because the 2′-OH group is in the β-position (remember that the 2′-OH group of a natural ribonucleoside is in the α-position), the bases in cytarabine-modified DNA are not able to stack properly (Section 26.2).

Idoxuridine

Idoxuridine is approved (in the United States) only for the topical treatment of ocular infections, although it is used for herpes infections in other countries. Idoxuridine has an iodo group in place of the methyl group of thymine and is incorporated into DNA in place of thymine. Chain elongation can continue because idoxuridine has a 3′-OH group, but the resulting DNA is easily broken and is also not transcribed properly. (Also see AZT on page 1179.)

Influenza Pandemics

Every year we face an outbreak of influenza (the flu). Most of the time it is a virus that is already present in the population and, therefore, can be controlled by flu shots. But every once in awhile, a new influenza virus appears, which can cause a worldwide pandemic because it is not affected by any immunity a person may have to older strains of influenza and can, therefore, spread rapidly and infect a large number of people. Almost no effective antiviral drugs are available for the flu. (See Tamiflu on page 1055.)

The Russian flu of 1889–1890 was the first of the flu pandemics. It killed about 1 million people. The Spanish flu that broke out in 1918–1919 killed over 50 million people worldwide. The Asian flu of 1956–1958 killed about 2 million people before a vaccine was developed in 1957 to contain it. The Hong Kong flu of 1968–1969—so called because it affected 15% of the population of Hong Kong—had a much lower death rate—only about 750,000 people died—because people who had had the Asian flu had some immunity. Because this was the last worldwide pandemic, public health officials worry that another may occur soon.

Recent flu outbreaks that have been causes for concern are the avian flu (bird flu), discovered in 1997, and the swine flu, discovered in 2009. The avian flu was linked to chickens but it was subsequently transmitted to hundreds of people, 60% of whom died. The swine flu is a respiratory disease of pigs but it has been known to affect people. There are concerns that either of these flus could become a worldwide pandemic.

The carbohydrates attached to the surface of the viral protein account for the biggest difference in virus strains. The symptoms caused by viruses that bind primarily to sugars in the nose and throat are not as severe as those caused by viruses that bind to sugars deep in the lungs.

26.12   HOW THE BASE SEQUENCE OF DNA IS DETERMINED

In 2000, two teams of scientists (one from a private biotechnology company and one from the publicly funded Human Genome Project) announced that they had completed the first draft of the sequence of the 3.1 billion base pairs in human DNA; it cost ~ $2.7 billion and took about 10 years to complete. This was a huge achievement.

Since that time there has been an enormous advancement in the technology used for DNA sequencing and today a human genome can be sequenced in 3 days at a cost of less than $1,000. This has allowed the era of personalized medicine to begin. Soon we will be able to understand such things as what makes some people more susceptible to certain diseases, and why drugs work differently on different people. We are rapidly reaching the time when we will be given drugs that fit our genetic profile.

Restriction Endonucleases

Clearly, DNA molecules are too large to sequence as a unit. Therefore, DNA is first cleaved at specific base sequences, and the resulting DNA fragments are then sequenced individually.

The enzymes that cleave DNA at specific base sequences are called **restriction endonucleases,** and the DNA fragments they produce are called **restriction fragments.** Several hundred restriction endonucleases are now known; a few examples, the base sequence that each recognizes, and the point of cleavage in that base sequence are shown in the margin.

The base sequences that most restriction endonucleases recognize are *palindromes.* A palindrome is a word or a group of words that reads the same forward and backward. "Toot" and "race car" are examples of palindromes, as is "Was it a car or a cat I saw?" A restriction endonuclease recognizes

restriction endonuclease	recognition sequence
*Alu*I	AGCT TCGA
*Fnu*DI	GGCC CCGG
*Pst*I	CTGCAG GACGTC

a piece of DNA in which *the template strand is a palindrome of the sense strand.* In other words, the sequence of bases in the template strand (reading from right to left) is identical to the sequence of bases in the sense strand (reading from left to right).

PROBLEM 17 ◆

Which of the following base sequences is most likely to be recognized by a restriction endonuclease?

A ACGGGT   **C** ACGGCA   **E** ACATCGT

B ACGCGT   **D** ACACGT   **F** CCAACC

Pyrosequencing

Frederick Sanger (who received the 1953 Nobel Prize in Chemistry for being the first to determine the primary structure of a protein) and Walter Gilbert shared the 1980 Nobel Prize in Chemistry for their work on DNA sequencing. Over the years, their procedure has been replaced by other methods. A currently used technique is an automated procedure called **pyrosequencing.** In this method, a small piece of DNA primer is added to the restriction fragment whose sequence is to be determined. Nucleotides are then added to the primer by base pairing with the restriction fragment. This method—known as *sequencing-by-synthesis*—detects the identity of each base that adds to the primer.

Pyrosequencing requires DNA polymerase—the enzyme that adds nucleotides to a strand of DNA—and two additional enzymes that cause light to be emitted when pyrophosphate is detected.

DNA to be sequenced

3′—AGGCTCCAGTGATCCG —5′ $\xrightarrow{\text{DNA polymerase} \atop \text{2 additional enzymes}}$

P—TC

3′-protected
2′-deoxyribonucleoside
5′-triphosphate

primer

restriction fragment:
primer hybrid

Pyrosequencing also requires the four 2′deoxyribonucleoside 5′-triphosphates, each with a protected 3′-OH group.

a 3′-protected 2′-deoxyribonucleoside triphosphate

The restriction fragment:primer hybrid is attached to a solid support, similar to the solid support used in the automated synthesis of polypeptides (Section 21.11). The steps involved in pyrosequencing are:

- The enzymes and one of the four 3′-protected 2′-deoxyribonucleoside 5′-triphosphates (for example, 3′-protected dATP) are added to the column.
- The reagents are washed from the solid support.
- The process is repeated with a different 3′-protected 2′-deoxyribonucleoside 5′-triphosphate (for example, 3′-protected dGTP).
- The process is repeated with 3′-protected dCTP, and then repeated again with 3′-protected dTTP.
- The sequencer keeps track of which of the four nucleotides caused light to be observed—in other words, which nucleotide released pyrophosphate as a result of being added to the primer.
- The protecting group on the 3′-OH is removed.

The steps are repeated to determine the identity of the next nucleotide that adds to the primer. Pyrosequencing can determine the base sequence of a restriction fragment with as many as 500 nucleotides.

Once the sequence of bases in a restriction fragment is determined, the results can be checked by using the same process to obtain the base sequence of the fragment's complementary strand. The base sequence of the original piece of DNA can be determined by repeating the entire procedure with a different restriction endonuclease and noting overlapping fragments.

The DNA from a 40-million-year-old leaf preserved in amber has been sequenced.

26.13 GENETIC ENGINEERING

Genetic engineering (also called genetic modification) is the insertion of a segment of DNA into the DNA of a host cell so that the segment of DNA is replicated by the DNA-synthesizing machinery of the host cell. Genetic engineering has many practical applications. For example, replicating the DNA that codes for human insulin makes it possible to synthesize large amounts of the protein, eliminating the dependence on pigs for insulin and helping those who are allergic to pig insulin. Recall that pig insulin differs from human insulin by one amino acid (Section 21.8).

Agriculture is benefiting from genetic engineering. Crops are now being produced with new genes that increase their resistance to drought and insects. For example, genetically engineered cotton crops are resistant to the cotton bollworm, and genetically engineered corn is resistant to the corn rootworm. Genetically modified organisms (GMOs) have been responsible for a nearly 50% reduction in the use of chemicals for agricultural purposes in the United States. Recently, corn has been genetically modified to boost ethanol production, apples have been genetically modified to prevent them from turning brown when they are cut, and soybeans have been genetically modified to prevent trans fats from being formed when soybean oil is hydrogenated (Section 5.9).

Resisting Herbicides

Glyphosate, the active ingredient in a well-known herbicide called Roundup, kills weeds by inhibiting an enzyme that plants need to synthesize phenylalanine and tryptophan, amino acids they require for growth. Corn and cotton have been genetically engineered to tolerate the herbicide. Then, when fields are sprayed with glyphosate, the weeds are killed but not the crops.

These crops have been given a gene that produces an enzyme that uses acetyl-CoA to acetylate glyphosate in a nucleophilic acyl substitution reaction (Section 15.11). Unlike glyphosphate, N-acetylglyphosphate does not inhibit the enzyme that synthesizes phenylalanine and tryptophan.

corn genetically engineered to resist the herbicide glyphosate by acetylating it

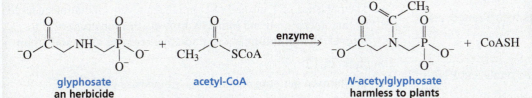

glyphosate
an herbicide

acetyl-CoA

N-acetylglyphosate
harmless to plants

Using Genetic Engineering to Treat the Ebola Virus

Plants have long been a source of drugs—morphine, ephedrine, and codeine are just a few examples (Section 10.9). Now scientists are attempting to obtain drugs from plants by biopharming. Biopharming uses genetic engineering techniques to produce drugs in crops such as corn, rice, tomatoes, and tobacco. To date, the only biopharmed drug approved by the Food and Drug Administration (FDA) is one that is manufactured in carrots and used to treat Gaucher's disease.

An experimental drug that was used to treat a handful of patients with Ebola, the virus that was spreading throughout West Africa, was obtained from genetically engineered tobacco plants. The tobacco plants were infected with three genetically engineered plant viruses that are harmless to humans and animals but have structures similar to that of the Ebola virus. As a result of being infected, the plants produced antibodies to the viruses. The antibodies were isolated from the plants, purified, and then used to treat the patients with Ebola.

The experimental drug had been tested in 18 monkeys who had been exposed to a lethal dose of Ebola. All 18 monkeys survived, whereas the three monkeys in the control group died. Typically, drugs go through rigorous testing on healthy humans prior to being administered to infected patients (see page 290). In the Ebola case, the FDA made an exception because it feared that the drug might be these patients' only hope. Five of the seven people given the drug survived. Currently, it takes about 50 kilograms of tobacco leaves and about 4 to 6 months to produce enough drug to treat one patient.

tobacco plants

ESSENTIAL CONCEPTS

Section 26.1

- **Nucleic acids** (DNA and RNA) are composed of long strands of nucleotide subunits linked by phosphodiester bonds.

- A **nucleoside** contains a base bonded to a sugar. A **nucleotide** is a nucleoside with an OH group of the sugar bonded to phosphoric acid by an ester linkage.

- The sugar in RNA is D-ribose; the sugar in DNA is 2'deoxy-D-ribose.

- The bases in DNA are A, G, C, and T; the bases in RNA are A, G, C, and U. The base is attached to the anomeric carbon of the sugar in a β-glycosidic linkage.

Section 26.2

- Nucleic acids are synthesized from nucleoside triphosphates. The nucleotides are linked as a result of nucleophilic attack by a 3'-OH group of one nucleotide on the α-phosphorus of another nucleotide.

- The **primary structure** of a nucleic acid is the sequence of bases in its strand.

Section 26.3

- DNA is double-stranded. The strands run in opposite directions and are twisted into a **double helix,** giving DNA a major groove and a minor groove.

- The bases are confined to the inside of the helix, and the sugar and phosphate groups are on the outside. The strands are held together by hydrogen bonds between bases of opposing strands and by **stacking interactions.**

- The two strands are **complementary:** A pairs with T and G pairs with C.

Section 26.4

- The difference in the sugars causes DNA to be stable and RNA to be easily cleaved.

Section 26.5

- DNA is synthesized in the $5' \rightarrow 3'$ direction by a process called **semiconservative replication.**

Section 26.6

- The sequence of bases in DNA provides the blueprint for the synthesis (**transcription**) of RNA.

- The sequence of bases in RNA determines the sequence of amino acids in a protein. The synthesis of a protein from an RNA a mRNA blueprint is called **translation.**

Section 26.7

- One strand of DNA is called the **sense strand;** the complementary strand is called the **template strand.**

- RNA is synthesized in the $5' \rightarrow 3'$ direction by reading the bases along the DNA template strand in the $3' \rightarrow 5'$ direction. The newly synthesized RNA strand and the DNA sense strand have the same sequence of bases (with a U in RNA where there is a T in DNA).

Section 26.8

- The RNAs used in protein biosynthesis are messenger RNA, ribosomal RNA, and transfer RNA.

- A tRNA carries an amino acid bound as an ester to its terminal 3'-position.

Section 26.9

- Each three-base sequence of mRNA—a codon—specifies the particular amino acid to be incorporated into a protein. The codons and the amino acids they specify are known as the genetic code.

- Protein synthesis proceeds from the N-terminal end to the C-terminal end by reading the bases along the mRNA strand in the $5' \rightarrow 3'$ direction.

Section 26.10

- The presence of T instead of U in DNA prevents mutations caused by tautomerization and imine hydrolysis (deamination) of C to form U. **Deamination** is a reaction that removes an amino group.

Section 26.12

- The **human genome** contains 3.1 billion base pairs.

- **Restriction endonucleases** cleave DNA at specific palindromes, forming **restriction fragments.**

- **Pyrosequencing** is a method used to determine the sequence of bases in the restriction fragments.

Section 26.13

- A large amount of a particular protein can be synthesized by **genetic engineering.**

PROBLEMS

18. What nonapeptide is coded for by the following fragment of mRNA?

$$5'—AAA—GUU—GGC—UAC—CCC—GGA—AUG—GUG—GUC—3'$$

19. What is the sequence of bases in the template strand of DNA that codes for the mRNA in Problem 18?

20. What is the sequence of bases in the sense strand of DNA that codes for the mRNA in Problem 18?

21. A segment of DNA has 18 base pairs. It has 7 cytosines in the segment.
   a. How many thymines are in the segment?   **b.** How many guanines are in the segment?

22. Name the following:

a. 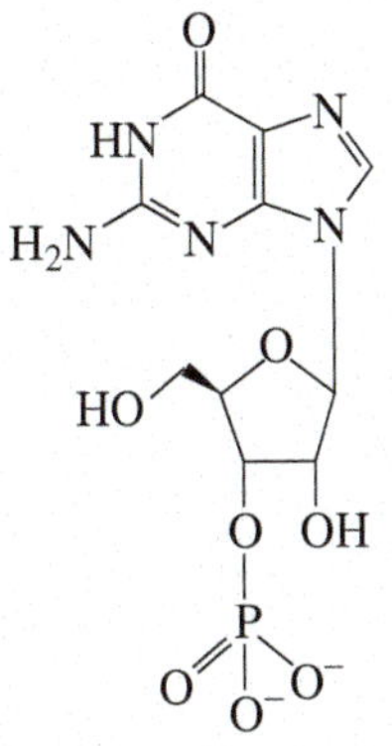   b. 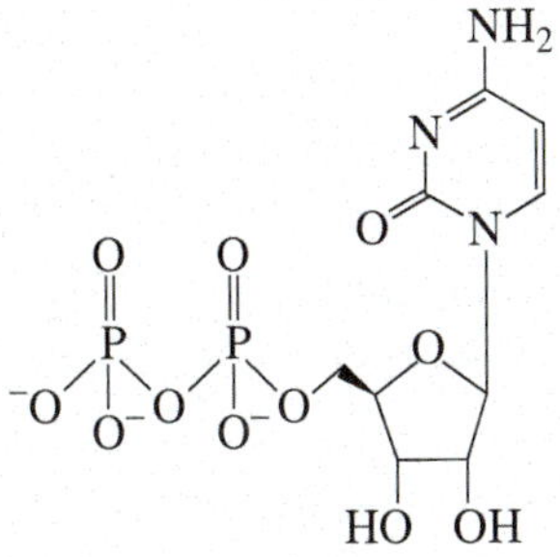   c. 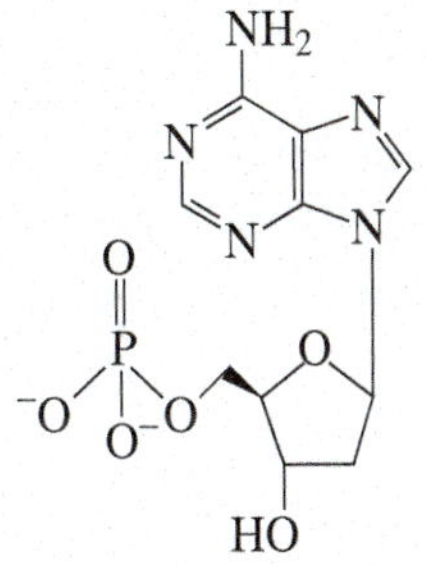   d.

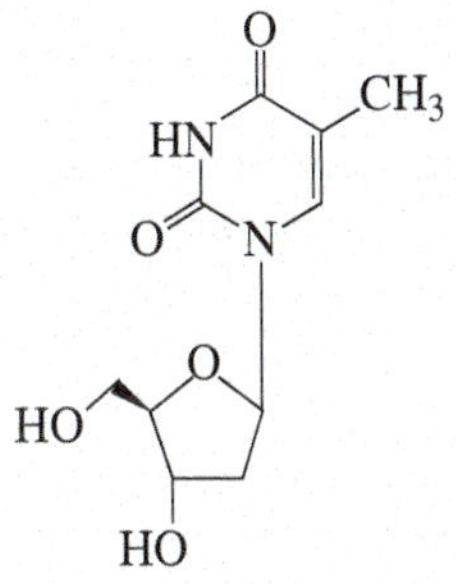

23. What is the base sequence of the segment of DNA that is responsible for the biosynthesis of the following hexapeptide?

Gly-Ser-Arg-Val-His-Glu

24. Propose a mechanism for the following reaction:

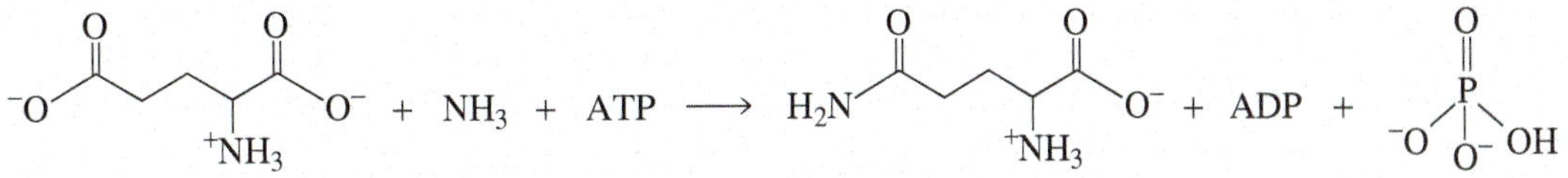

$$\text{(structure)} + NH_3 + ATP \longrightarrow \text{(structure)} + ADP + \text{(structure)}$$

25. What would be the C-terminal amino acid if the codon at the 3′-end of the mRNA in Problem 18 undergoes the following mutations?
   a. The first base is changed to A.
   b. The second base is changed to A.
   c. The third base is changed to A.
   d. The third base is changed to G.

26. The first amino acid incorporated into a polypeptide chain during its biosynthesis in prokaryotes is *N*-formylmethionine. Explain the purpose of the formyl group. (*Hint:* The ribosome has a binding site for the growing peptide chain and a binding site for the incoming amino acid.)

27. Match the codon with the anticodon:

Codon:	AAA	GCA	CUU	AGG	CCU	GGU	UCA	GAC
Anticodon:	ACC	CCU	UUU	AGG	UGA	AAG	GUC	UGC

28. a. Using the single-letter abbreviations for the amino acids in Table 21.2, write the sequence of amino acids in a tetrapeptide represented by the first four different letters in your first name. Do not use any letter twice. (Because not all letters are assigned to amino acids, you might have to use one or two letters in your last name.)
   b. Write one of the sequences of bases in mRNA that would result in the synthesis of that tetrapeptide.
   c. Write the sequence of bases in the sense strand of DNA that would result in formation of that fragment of mRNA.

29. Which of the following pairs of dinucleotides are present in equal amounts in DNA?
   A CC and GG
   B CG and GT
   C CA and TG
   D CG and AT
   E GT and CA
   F TA and AT

30. Human immunodeficiency virus (HIV) is the retrovirus that causes AIDS. AZT was one of the first drugs designed to interfere with retroviral DNA synthesis. When cells take up AZT, they convert it to AZT-triphosphate. Explain how AZT interferes with DNA synthesis.

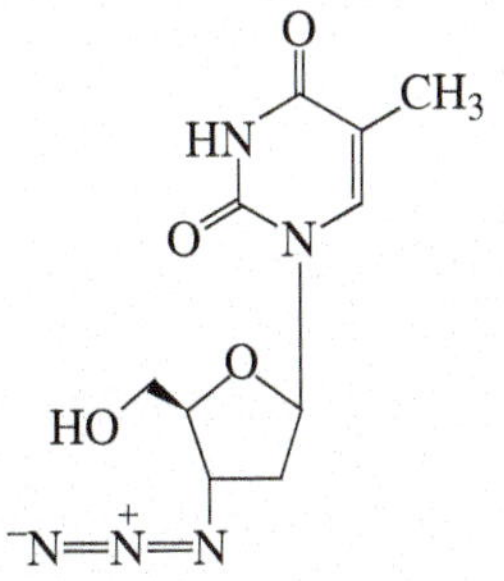

3′-azido-2′-deoxythymidine
AZT

31. If a mRNA contained only U and G in random sequence, what amino acids would be present in the protein when the mRNA is translated?

1179

32. Why is the codon a triplet rather than a doublet or a quartet?

33. RNAase, the enzyme that catalyzes the hydrolysis of RNA, has two catalytically active histidine residues at its active site. One of the histidine residues is catalytically active in its acidic form, and the other is catalytically active in its basic form. Propose a mechanism for RNAase.

34. The amino acid sequences of peptide fragments obtained from a normal protein were compared with those obtained from the same protein synthesized by a defective gene. They were found to differ in only one peptide fragment. Their amino acid sequences are shown here:

Normal: Gln-Tyr-Gly-Thr-Arg-Tyr-Val

Mutant: Gln-Ser-Glu-Pro-Gly-Thr

   a. What is the defect in DNA?
   b. It was later determined that the normal peptide fragment is an octapeptide with a C-terminal Val-Leu. What is the C-terminal amino acid of the mutant peptide?

35. Which cytosine in the following sense strand of DNA could cause the most damage to the organism if it is deaminated?

A—T—G—T—C—G—C—T—A—A—T—C

36. 5-Bromouracil, a highly mutagenic compound (that is, a compound that causes changes in DNA), is used in cancer chemotherapy. When administered to a patient, it is converted to the triphosphate and incorporated into DNA in place of thymine, which it resembles sterically. Why does it cause mutations? (*Hint:* The bromo substituent increases the stability of the enol tautomer.)

37. Sodium nitrite, a common food preservative (page 906), is capable of causing mutations in an acidic environment by converting cytosines to uracils. Explain how this occurs.

38. Why does DNA not unravel completely before replication begins?

39. *Staphylococcus* nuclease is an enzyme that catalyzes the hydrolysis of DNA. The reaction is catalyzed by Ca^{2+}, Glu 43, and Arg 87. Explain how the metal catalyst facilitates this reaction. Recall that the nucleotides in DNA have phosphodiester linkages.

PART EIGHT

Special Topics in Organic Chemistry

Chapter 27 Synthetic Polymers

Previous chapters discussed polymers synthesized by cells—proteins, carbohydrates, and nucleic acids. **Chapter 27** discusses polymers synthesized by chemists. These synthetic polymers have physical properties that make them useful components of thousands of materials that pervade and enhance our lives.

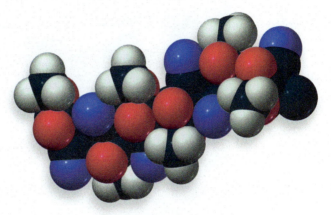

a few segments of Super Glue®

Chapter 28 Pericyclic Reactions

Chapter 28 discusses pericyclic reactions. These are reactions that occur as a result of a cyclic reorganization of electrons. In this chapter, you will learn how the conservation of orbital symmetry theory explains the relationships between the reactant, the product, and the reaction conditions in a pericyclic reaction. And you will see that the biosynthesis of vitamin D involves two sequential pericyclic reactions.

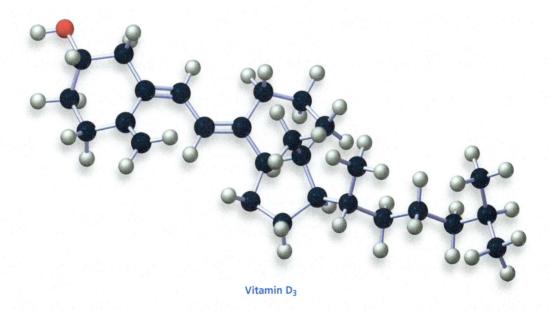

Vitamin D₃

27 Synthetic Polymers

boots made of synthetic rubber

Probably no group of synthetic compounds is more important to modern life than synthetic polymers. Unlike small organic molecules, which are of interest because of their chemical properties, these giant molecules—with molecular weights ranging from thousands to millions—are interesting primarily because of their physical properties that make them useful in everyday life. Some synthetic polymers resemble natural substances, but most are quite different from materials found in nature. Such diverse products as plastic bottles, DVDs, food wrap, artificial joints, Super Glue, toys, weather stripping, automobile body parts, and shoe soles are made of synthetic polymers.

A **polymer** is a large molecule made by linking together repeating units of small molecules called **monomers.** The process of linking them together is called **polymerization.**

$$n\,\text{M} \xrightarrow{\text{polymerization}} \text{—M—M—M—M—M—M—M—M—M—}$$

monomer   polymer

Polymers can be divided into two broad groups: **biopolymers** and **synthetic polymers.**

- Biopolymers are synthesized by cells. Examples of biopolymers are DNA, the storage molecule for genetic information; RNA and proteins, the molecules that facilitate biochemical transformations; and polysaccharides, compounds that store energy and also function as structural materials. The structures and properties of these biopolymers are presented in other chapters.
- Synthetic polymers are synthesized by scientists. This chapter explores synthetic polymers.

Humans first relied on *biopolymers* for clothing, wrapping themselves in animal skins and furs. Later, they learned to spin natural fibers into thread and to weave the thread into cloth. Today, much of our clothing is made of *synthetic polymers* (such as nylon, polyester, and polyacrylonitrile). It has been estimated that if synthetic polymers were not available, all the arable land in the United States would have to be used for the production of cotton and wool for clothing.

The first synthetic fiber was rayon. In 1865, the French silk industry was threatened by an epidemic that killed many silkworms. Louis Pasteur determined the source of the disease, but it was his assistant, Louis Chardonnet, who accidentally discovered the starting material for a synthetic fiber when, while wiping up some spilled nitrocellulose from a table, he noticed long silk-like strands adhering to both the cloth and the table.

A **plastic** is a polymer capable of being molded. The first commercial plastic, celluloid, was invented in 1856. It was used in the manufacture of billiard balls and piano keys, replacing scarce ivory and providing a reprieve for many elephants (only eight billiard balls could be obtained from an elephant tusk). Celluloid was also used for motion picture film until it was replaced by cellulose acetate, a more stable polymer.

The first synthetic rubber was synthesized by German chemists in 1917 in response to a severe shortage of raw materials as a result of blockading during World War I.

Polymer chemistry is part of the larger discipline of **materials science,** which involves the creation of new materials with improved properties to add to the metals, glass, fabrics, and woods we currently have. Approximately 30,000 polymer patents are currently in force. We can expect scientists to develop many more new materials in the years to come.

billiard balls

Synthetic polymers can be divided into two major classes.

Chain-Growth Polymers

Chain-growth polymers, also known as **addition polymers,** are made by **chain reactions**—the addition of monomers to the end of a growing chain. The end of a growing chain is reactive because it is a radical, a cation, or an anion. Polystyrene—used for hot drink cups, egg cartons, and insulation, among other things—is an example of a chain-growth polymer. Polystyrene is pumped full of air to produce the material used for insulation in house construction.

Chain-growth polymers are also called addition polymers.

Chain-growth polymers are made by chain reactions.

styrene   →   polystyrene
a chain-growth polymer

Step-Growth Polymers

Step-growth polymers, also called **condensation polymers,** are made by linking monomers as a result of removing (in most cases) a small molecule, generally water or an alcohol. The monomers have reactive functional groups at each end. Unlike chain-growth polymerization that requires the individual monomers to add to the end of a growing chain, step-growth polymerization allows any two reactive monomers or oligomers to be linked. Dacron is an example of a step-growth polymer. It was one of the first synthetic polymers to have a medical application—namely, for arterial grafts.

Step-growth polymers are also called condensation polymers.

Step-growth polymers are made by linking molecules with reactive functional groups at each end.

dimethyl terephthalate   +   1,2-ethanediol $\xrightarrow{\Delta}$ poly(ethylene terephthalate)   + $2n\ CH_3OH$
Dacron®
a step-growth polymer

27.2 AN INTRODUCTION TO CHAIN-GROWTH POLYMERS

The monomers used most commonly in chain-growth polymerization are ethylene (ethene) and substituted ethylenes (CH_2=CHR). Polymers formed from ethylene or substituted ethylenes are called **vinyl polymers.** Some of the many vinyl polymers synthesized by chain-growth polymerization are listed in Table 27.1.

Table 27.1 Some Important Chain-Growth Polymers and Their Uses

Monomer	Repeating unit	Polymer name	Uses
CH_2=CH_2	—CH_2—CH_2—	polyethylene	toys, water bottles, grocery bags
CH_2=CH—Cl	—CH_2—CH—Cl	poly(vinyl chloride) (PVC)	shampoo bottles, pipe, siding, flooring, clear food packaging
CH_2=CH—CH_3	—CH_2—CH—CH_3	polypropylene	molded caps, margarine tubs, indoor/outdoor carpeting, upholstery
CH_2=CH (phenyl)	—CH_2—CH— (phenyl)	polystyrene	egg cartons, hot drink cups, insulation
CF_2=CF_2	—CF_2—CF_2—	poly(tetrafluoroethylene) Teflon	nonstick surfaces, liners, cable insulation
CH_2=CH—C≡N	—CH_2—CH—C≡N	poly(acrylonitrile) Orlon, Acrilan	rugs, blankets, yarn, apparel, simulated fur
CH_2=C—CH_3, $COCH_3$, O	CH_3, —CH_2—C—, $COCH_3$, O	poly(methyl methacrylate) Plexiglas, Lucite	shatter-resistant alternative to glass
CH_2=CH—$OCCH_3$—O	—CH_2—CH—$OCCH_3$—O	poly(vinyl acetate) (PVA)	white glue, adhesives

Chain-growth polymerization proceeds by one of three mechanisms: **radical polymerization, cationic polymerization,** or **anionic polymerization.** Each mechanism has three distinct phases: an *initiation step* that starts the polymerization, *propagation steps* that allow the chain to grow, and *termination steps* that stop the growth of the polymer chain. We will see that the mechanism for a given chain-growth polymerization reaction depends on the structure of the monomer *and* on the initiator used to activate the monomer.

27.3 RADICAL POLYMERIZATION

The chain-initiating, chain-propagating, and chain-terminating steps that occur in *radical polymerization* are similar to the steps that take place in the radical reactions discussed in Section 12.2.

Radical polymerization requires a radical initiator. The radical initiator can be any compound with a weak bond that readily undergoes homolytic cleavage by heat or light to form radicals sufficiently energetic to convert an alkene into a radical.

MECHANISM FOR RADICAL POLYMERIZATION

initiation steps

$$RO{-}OR \xrightarrow[\text{or } h\nu]{\Delta} \underset{\text{radicals}}{2\,RO\cdot}$$

a radical initiator

$$RO\cdot \;+\; CH_2{=}\underset{\underset{Z}{|}}{CH} \longrightarrow RO{-}CH_2\underset{\underset{Z}{|}}{\overset{\bullet}{C}H}$$

the alkene monomer reacts with a radical

propagation steps

(propagating sites)

$$RO{-}CH_2\underset{\underset{Z}{|}}{\overset{\bullet}{C}H} \;+\; CH_2{=}\underset{\underset{Z}{|}}{CH} \longrightarrow RO{-}CH_2\underset{\underset{Z}{|}}{CH}CH_2\underset{\underset{Z}{|}}{\overset{\bullet}{C}H}$$

$$RO{-}CH_2\underset{\underset{Z}{|}}{CH}CH_2\underset{\underset{Z}{|}}{\overset{\bullet}{C}H} \;+\; CH_2{=}\underset{\underset{Z}{|}}{CH} \longrightarrow RO{-}CH_2\underset{\underset{Z}{|}}{CH}CH_2\underset{\underset{Z}{|}}{CH}CH_2\underset{\underset{Z}{|}}{\overset{\bullet}{C}H}$$

- The radical initiator breaks homolytically into radicals, and each of these radicals can react with a monomer, creating a monomer radical.
- In the first propagation step, the monomer radical reacts with another monomer, converting it to a radical.
- This radical reacts with another monomer, adding a new subunit to the chain. The unpaired electron is now at the end of the unit most recently added to the chain. This is called the **propagating site.**

Hundreds or even thousands of alkene monomers can add, one at a time, to the growing chain. Eventually, the chain reaction stops because the propagating sites are destroyed in a termination step. Propagating sites are destroyed when

- two chains combine at their propagating sites (chain combination).
- two chains undergo *disproportionation*: one chain is oxidized to an alkene and the other is reduced to an alkane as a result of hydrogen atom being transferred from one chain to another.
- a chain undergoes chain transfer.

termination steps

chain combination

$$2\,RO{\left[CH_2\underset{\underset{Z}{|}}{CH}\right]_n}CH_2\underset{\underset{Z}{|}}{\overset{\bullet}{C}H} \longrightarrow RO{\left[CH_2\underset{\underset{Z}{|}}{CH}\right]_n}CH_2\underset{\underset{Z}{|}}{CH}CH\underset{\underset{Z}{|}}{}CH_2{\left[CH\underset{\underset{Z}{|}}{}CH_2\right]_n}OR$$

disproportionation

$$2\,RO{\left[CH_2\underset{\underset{Z}{|}}{CH}\right]_n}CH_2\underset{\underset{Z}{|}}{\overset{\bullet}{C}H} \longrightarrow RO{\left[CH_2\underset{\underset{Z}{|}}{CH}\right]_n}CH{=}\underset{\underset{Z}{|}}{CH} \;+\; RO{\left[CH_2\underset{\underset{Z}{|}}{CH}\right]_n}CH_2\underset{\underset{Z}{|}}{CH_2}$$

chain transfer

$$-CH_2{\left[CH_2\underset{\underset{Z}{|}}{CH}\right]_n}CH_2\underset{\underset{Z}{|}}{\overset{\bullet}{C}H} \;+\; X{-}Y \longrightarrow -CH_2{\left[CH_2\underset{\underset{Z}{|}}{CH}\right]_n}CH_2\underset{\underset{Z}{|}}{CHX} \;+\; Y\cdot$$

In chain transfer, the growing chain reacts with a molecule XY in a manner that allows X· to terminate the chain, leaving behind Y· to initiate a new chain. Molecule XY can be a solvent, a radical initiator, or any molecule with a bond that can readily be cleaved homolytically. Chain transfer can be used to control the molecular weight of the polymer.

As long as the polymer has a high molecular weight, the groups (RO) at the ends of the polymer chains—arising from initiation and chain transfer—are relatively unimportant in determining its physical properties and are generally not even specified; it is the rest of the molecule that determines the properties of the polymer.

Head-to-Tail Addition

Chain-growth polymerization of substituted ethylenes exhibits a marked preference for **head-to-tail addition,** where the head of one monomer is attached to the tail of another, similar to what we saw in the biosynthesis of terpenes (Section 25.7). Notice that head-to-tail addition of a substituted ethylene results in a polymer in which every other carbon bears a substituent.

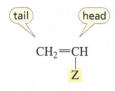

$$-CH_2CHCH_2CH-\qquad -CH_2CHCHCH_2-\qquad -CHCH_2CH_2CH-$$
$$\;\;\;\;\;\;\;\; \overset{|}{Z}\;\;\;\overset{|}{Z}\qquad\qquad\qquad \overset{|}{Z}\;\;\overset{|}{Z}\qquad\qquad\qquad \overset{|}{Z}\qquad\qquad\;\;\;\;\overset{|}{Z}$$

head-to-tail head-to-head tail-to-tail

Two factors favor head-to-tail addition. First, there is less steric hindrance at the unsubstituted sp^2 carbon of the alkene; therefore, the propagating site attacks it preferentially.

$$RO\cdot \;+\; CH_2{=}CH \longrightarrow -CH_2CHCH_2CHCH_2CHCH_2CHCH_2CHCH_2CH-$$

vinyl chloride   poly(vinyl chloride)

less sterically hindered sp^2 carbon

Second, radicals formed by addition to the unsubstituted sp^2 carbon can be stabilized by the substituent attached to the other sp^2 carbon. For example, when Z is a phenyl substituent, the benzene ring stabilizes the radical by electron delocalization.

In cases where Z is small (so that steric considerations are less important) and is not able to stabilize the growing end of the chain by electron delocalization, some head-to-head addition and some tail-to-tail addition occurs. This has been observed primarily when Z is fluorine. This abnormal mode of addition, however, has never been found to constitute more than 10% of the overall chain.

Monomers that Undergo Radical Polymerization

Monomers that most readily undergo chain-growth polymerization by a radical mechanism are those in which the substituent Z is a group able to stabilize the growing radical species by electron delocalization or by inductive electron withdrawal.

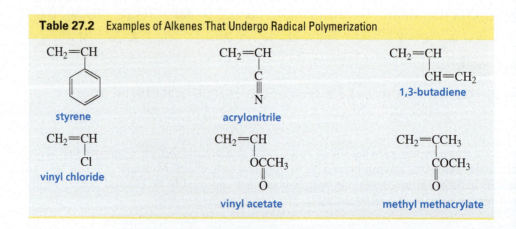

Table 27.2 Examples of Alkenes That Undergo Radical Polymerization

styrene	acrylonitrile	1,3-butadiene
vinyl chloride	vinyl acetate	methyl methacrylate

Radical polymerization of methyl methacrylate forms a clear plastic known as Plexiglas. The largest window in the world, made of a single piece of Plexiglas (54 ft long, 18 ft high, 13 in. thick), houses the sharks and barracudas in the Monterey Bay Aquarium.

Radical Initiators

The initiator for radical polymerization can be any compound with a weak bond that readily undergoes homolytic cleavage by heat or ultraviolet light to form radicals sufficiently energetic to convert an alkene into a radical. Several radical initiators are shown in Table 27.3. In all but one, the weak bond is an oxygen–oxygen bond.

Plexiglas windows

Table 27.3   Some Radical Initiators

$$\underset{\underset{CH_3}{|}}{\overset{\overset{CH_3}{|}}{CH_3C}}O-OH \longrightarrow \underset{\underset{CH_3}{|}}{\overset{\overset{CH_3}{|}}{CH_3C}}O\cdot \;+\; \cdot OH$$

$$K^+\,{}^-O\overset{O}{\underset{O}{\overset{\|}{\underset{\|}{S}}}}O-O\overset{O}{\underset{O}{\overset{\|}{\underset{\|}{S}}}}O^-K^+ \longrightarrow 2\,K^+\,{}^-O\overset{O}{\underset{O}{\overset{\|}{\underset{\|}{S}}}}O\cdot$$

$$\underset{\underset{CH_3}{|}}{\overset{\overset{CH_3}{|}}{CH_3C}}O-O\underset{\underset{CH_3}{|}}{\overset{\overset{CH_3}{|}}{C}}CH_3 \longrightarrow 2\,\underset{\underset{CH_3}{|}}{\overset{\overset{CH_3}{|}}{CH_3C}}O\cdot$$

$$\bigcirc-\overset{O}{\overset{\|}{C}}O-O\overset{O}{\overset{\|}{C}}-\bigcirc \longrightarrow 2\,\bigcirc-\overset{O}{\overset{\|}{C}}O\cdot \longrightarrow 2\,\bigcirc\cdot \;+\; 2\,CO_2$$

$$\underset{\underset{N}{\overset{\||}{\underset{C}{|}}}}{\overset{\overset{CH_3}{|}}{CH_3C}}-N{=}N-\underset{\underset{N}{\overset{\||}{\underset{C}{|}}}}{\overset{\overset{CH_3}{|}}{C}}CH_3 \longrightarrow 2\,\underset{\underset{N}{\overset{\||}{\underset{C}{|}}}}{\overset{\overset{CH_3}{|}}{CH_3C}}\cdot \;+\; N_2$$

Two factors enter into the choice of radical initiator for a particular chain-growth polymerization. The first is the solubility of the initiator. For example, potassium persulfate is often used if the initiator needs to be soluble in water, whereas an initiator with several carbons is chosen if the initiator needs to be soluble in a nonpolar solvent. The second factor is the temperature at which the polymerization reaction is to be carried out. For example, a *tert*-butoxy radical is relatively stable, so an initiator that forms a *tert*-butoxy radical is used for polymerizations carried out at relatively high temperatures.

Teflon: An Accidental Discovery

Teflon is a polymer of tetrafluoroethylene (Table 27.1). In 1938, a scientist needed some tetrafluoroethylene for the synthesis of what he hoped would be a new refrigerant. When he opened the cylinder of tetrafluoroethylene, no gas came out. He weighed the cylinder and found that it weighed more than an identical empty cylinder. In fact, it weighed the same as what a cylinder full of tetrafluoroethylene would weigh. Wondering what the cylinder contained, he cut it open and found a slippery polymer. Investigating the polymer further, he found that it was chemically inert to almost everything and could not be melted. In 1961, the first frying pan with a nonstick Teflon coating—"The Happy Pan"— was introduced to the public. Teflon is also used as a lubricant to reduce friction and in pipework that carries corrosive chemicals.

PROBLEM 1 ◆

What monomer would you use to form each of the following polymers?

a. 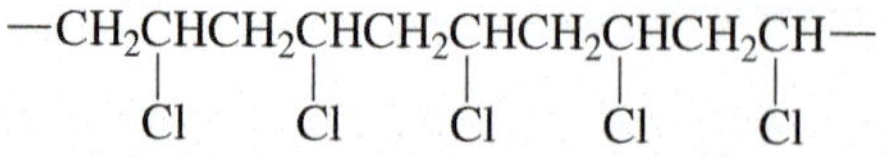 $-CH_2CHCH_2CHCH_2CHCH_2CHCH_2CH-$
            | | | | |
            Cl   Cl   Cl   Cl   Cl

b. 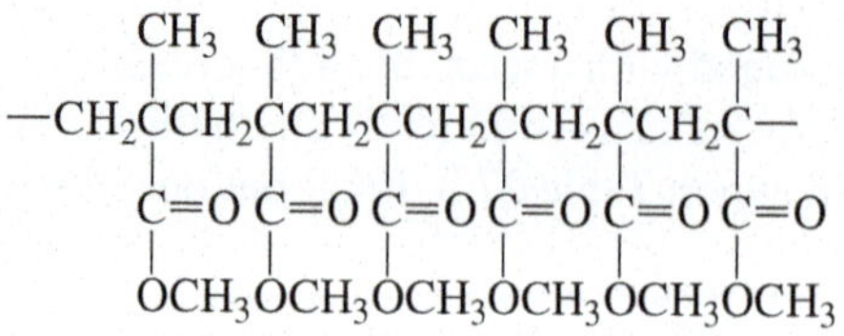
         CH_3 CH_3 CH_3 CH_3 CH_3 CH_3
          | | | | | |
$-CH_2CCH_2CCH_2CCH_2CCH_2C-$
          | | | | | |
         $C=O$ $C=O$ $C=O$ $C=O$ $C=O$ $C=O$
          | | | | | |
         OCH_3 OCH_3 OCH_3 OCH_3 OCH_3 OCH_3

c. $-CF_2CF_2CF_2CF_2CF_2CF_2CF_2CF_2CF_2CF_2-$

PROBLEM 2 ◆

Which polymer is more apt to contain abnormal head-to-head linkages: poly(vinyl chloride) or polystyrene?

PROBLEM 3

Draw a segment of polystyrene that contains two head-to-head, two tail-to-tail, and two head-to-tail linkages.

PROBLEM 4

Show the mechanism for the formation of a segment of poly(vinyl chloride) that contains three units of vinyl chloride and is initiated by hydrogen peroxide.

Branching of the Polymer Chain

If the propagating site removes a hydrogen atom from a chain, a branch can grow off the chain at that point. The propagating site can remove a hydrogen atom from a different polymer chain or from the same polymer chain.

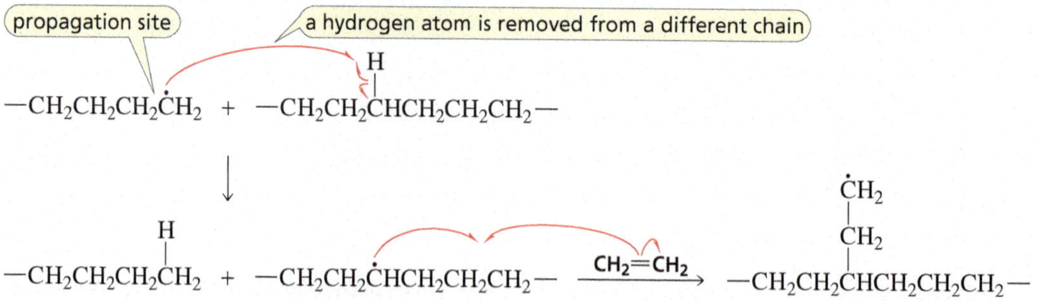

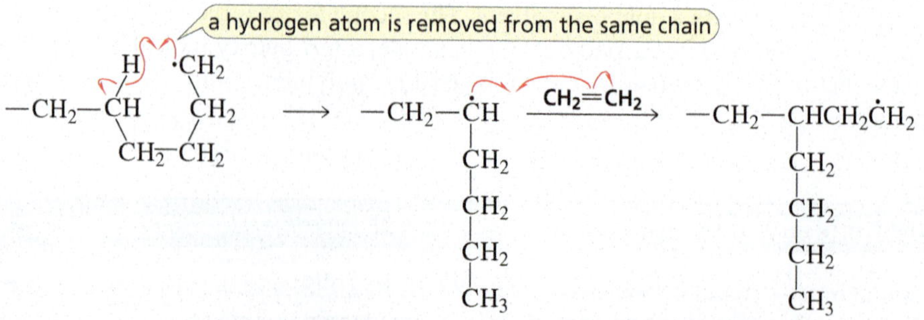

Removing a hydrogen atom from a carbon near the end of a chain leads to short branches, whereas removing a hydrogen atom from a carbon near the middle of a chain results in long branches.

chain with short branches   **chain with long branches**

Branched polymers are
more flexible.

Branching greatly affects the physical properties of the polymer. Unbranched chains can pack together more closely than branched chains can. Consequently, linear polyethylene (known as high-density polyethylene) is a relatively hard plastic, used for the production of such things as artificial hip joints, milk jugs, and toys; whereas branched polyethylene (low-density polyethylene) is a much more flexible polymer, used for trash bags and dry-cleaning bags.

Recycling Symbols

When plastics are recycled, the various types must be separated from one another. To aid in the separation, many states require manufacturers to place a recycling symbol on their products to indicate the type of plastic it is. You are probably familiar with these symbols, which are often embossed on the bottom of plastic containers. The symbols consist of three arrows around one of seven numbers; an abbreviation below the symbol indicates the type of polymer from which the container is made. The lower the number in the middle of the symbol, the greater is the ease with which the material can be recycled: 1 (PET) stands for poly(ethylene terephthalate), 2 (HDPE) for high-density polyethylene, 3 (V) for poly(vinyl chloride), 4 (LDPE) for low-density polyethylene, 5 (PP) for polypropylene, 6 (PS) for polystyrene, and 7 for all other plastics.

recycling symbols

PROBLEM 5 ♦

Polyethylene is used for the production of beach chairs as well as beach balls. Which of these items is made from more highly branched polyethylene?

PROBLEM 6

Draw a short segment of branched polystyrene.

Plasticizers

A plasticizer can be dissolved into a polymer to make it more flexible. The **plasticizer** decreases the attractions between the polymer chains, thereby allowing them to slide past one another. The most widely used plasticizer, di-2-ethylhexyl phthalate, is added to poly(vinyl chloride)—normally a brittle polymer—to make products such as vinyl raincoats, shower curtains, and garden hoses.

An important property to consider in choosing a plasticizer is its permanence—that is, how well the plasticizer remains in the polymer. The "new-car smell" appreciated by car owners is the odor of the plasticizer that has vaporized from the vinyl upholstery. When a significant amount of the plasticizer has evaporated, the upholstery becomes brittle and cracks.

di-2-ethylhexyl phthalate
a plasticizer

27.4 CATIONIC POLYMERIZATION

In cationic polymerization, the initiator is an electrophile (generally a proton) that adds to the monomer, causing it to become a carbocation. The initiator cannot be an acid such as HCl because its conjugate base (Cl⁻) will be able to react with the carbocation. Thus, the initiator most often used in cationic polymerization is a Lewis acid, such as BF_3, together with a proton-donating Lewis base, such as water. Notice that the reaction follows the rule that governs electrophilic addition reactions—that is, the electrophile (the initiator) adds to the sp^2 carbon bonded to the most hydrogens (Section 6.4).

MECHANISM FOR CATIONIC POLYMERIZATION

initiation steps

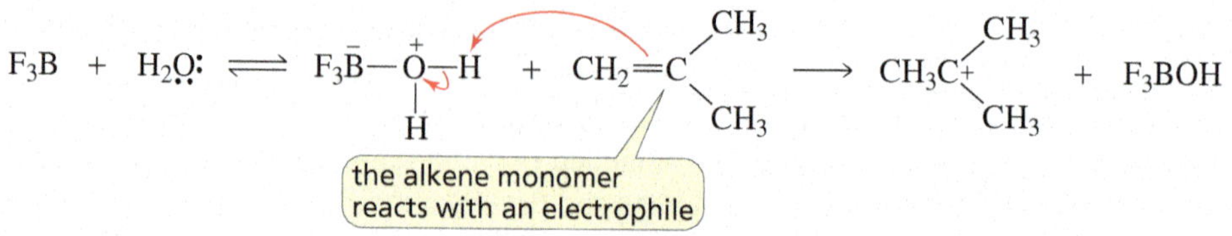

the alkene monomer reacts with an electrophile

propagating steps

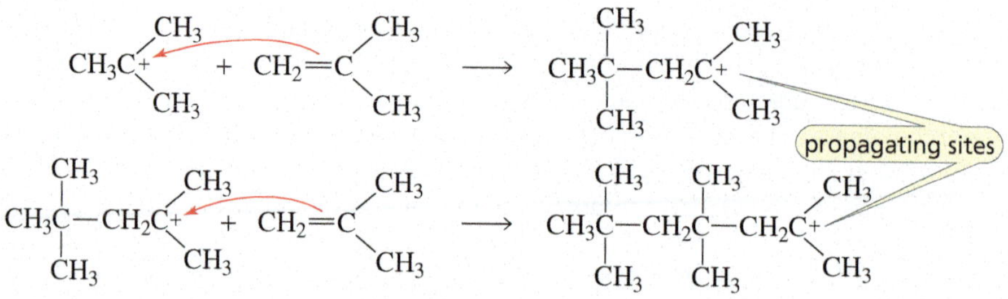

propagating sites

- The cation formed in the initiation step reacts with a second monomer, forming a new cation that reacts in turn with a third monomer. As each subsequent monomer adds to the chain, the new positively charged propagating site is at the end of the most recently added unit.

Cationic polymerization can be terminated by

- loss of a proton;
- addition of a nucleophile to the propagating site; or
- a chain-transfer reaction with the solvent (XY).

termination steps

loss of a proton

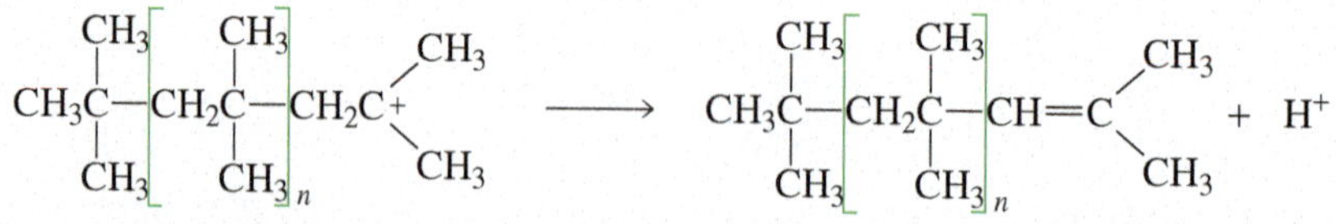

reaction with a nucleophile

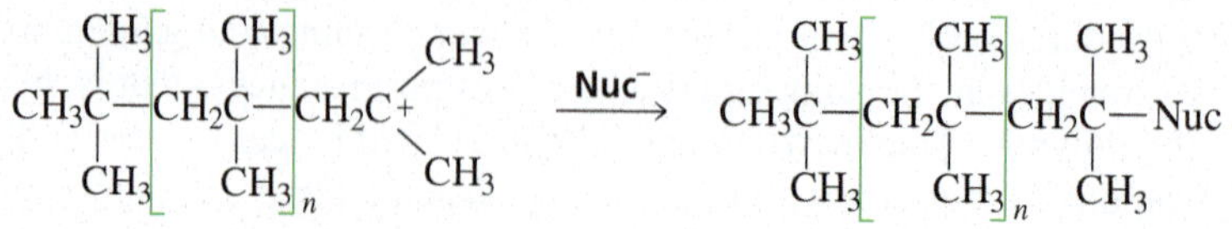

chain-transfer reaction with the solvent

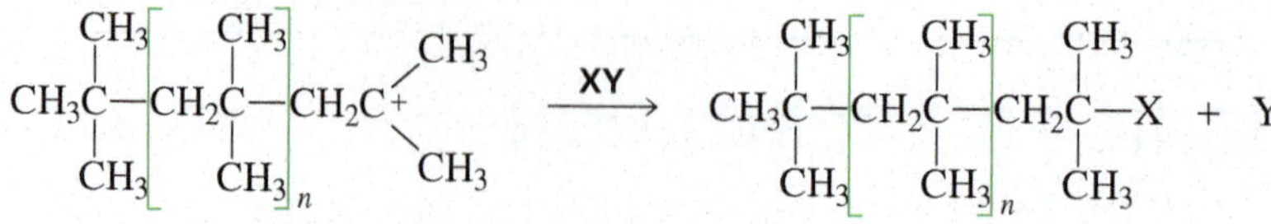

Rearrangement of the Carbocation

The carbocation intermediates formed during cationic polymerization, like any other carbocations, can undergo rearrangement by either a 1,2-hydride shift or a 1,2-methyl shift if rearrangement leads to a more stable carbocation (Section 6.7). For example, the polymer formed from the cationic polymerization of 3-methyl-1-butene contains both unrearranged and rearranged units.

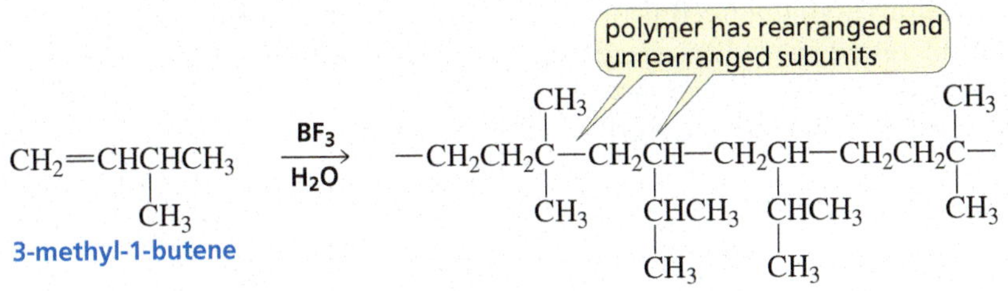

The unrearranged propagating site is a secondary carbocation, whereas the rearranged propagating site, obtained by a 1,2-hydride shift, is a more stable tertiary carbocation. The extent of rearrangement depends on the reaction temperature.

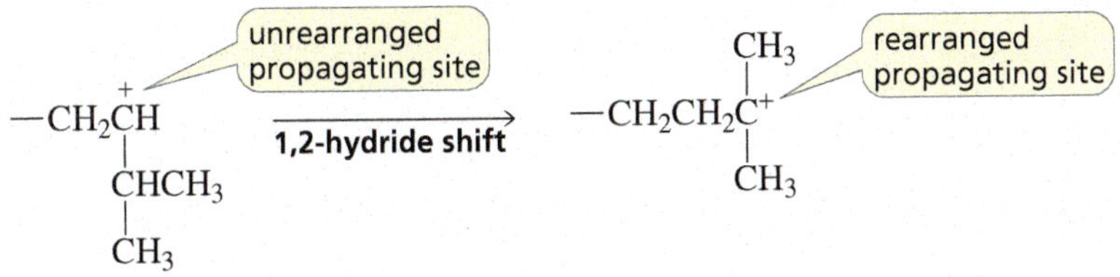

Monomers that are best able to undergo polymerization by a cationic mechanism are those with substituents that can stabilize the positive charge at the propagating site either by hyperconjugation (the first compound in Table 27.4; Section 6.2) or by donating electrons by resonance (the other two compounds in the table) (Section 8.10).

LEARN THE STRATEGY

Table 27.4 Examples of Alkenes That Undergo Cationic Polymerization

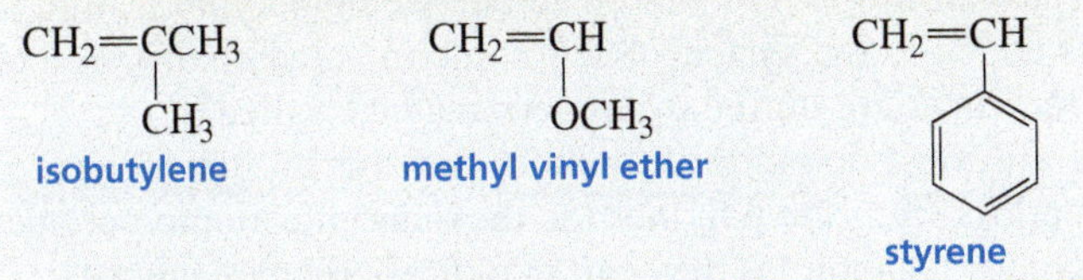

USE THE STRATEGY

PROBLEM 7 ◆

Rank the following groups of monomers from most able to least able to undergo cationic polymerization:

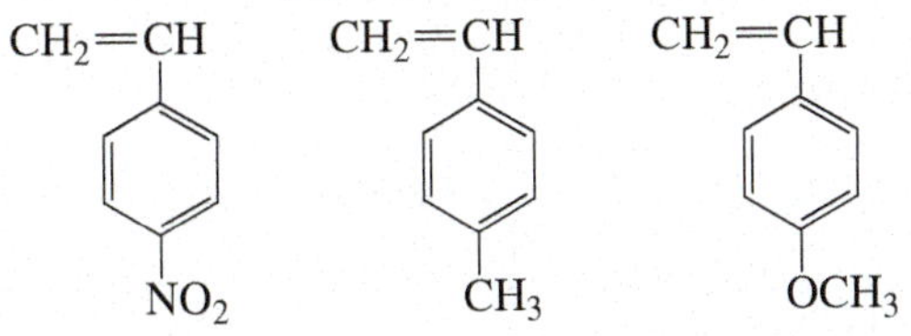

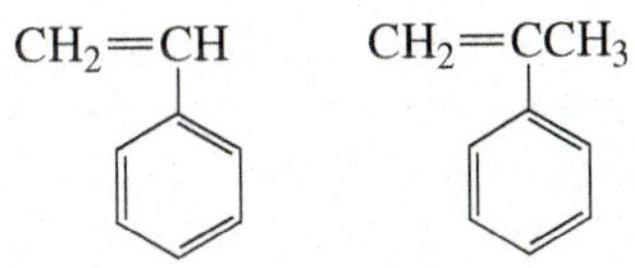

27.5   ANIONIC POLYMERIZATION

In anionic polymerization, the initiator is a nucleophile that reacts with the monomer to form a propagating site that is an anion. Nucleophilic attack on an alkene does not occur readily because alkenes are themselves electron rich. Therefore, the initiator must be a very good nucleophile, such as sodium amide or butyllithium (BuLi), and the alkene must contain a substituent that can withdraw electrons by resonance, which decreases the electron density of the double bond.

MECHANISM FOR ANIONIC POLYMERIZATION

initiation step

the alkene monomer reacts with a nucleophile

propagation steps

propagating sites

The chain can be terminated by reaction with an impurity in the reaction mixture. If all impurities are rigorously excluded, chain propagation will continue until all the monomer has been consumed. At this point, the propagating site will still be active, so the polymerization reaction will continue if more monomer is added to the system. Such nonterminated chains are called **living polymers** because the chains remain active until they are terminated ("killed").

Living polymers are most common in anionic polymerization because the chains cannot be terminated by proton loss from the polymer (as they can in cationic polymerization) or by chain combination or disproportionation (as they can in radical polymerization).

Alkenes that undergo polymerization by an anionic mechanism are those that can stabilize the negatively charged propagating site by resonance electron withdrawal (Table 27.5).

Table 27.5   Examples of Alkenes That Undergo Anionic Polymerization

acrolein   methyl methacrylate   styrene

Super Glue is a polymer made by anionic polymerization of methyl α-cyanoacrylate (see the top of the next page). Because the monomer has two electron-withdrawing groups, it requires only a moderately good nucleophile to initiate anionic polymerization, such as surface-absorbed water. You may well have experienced this reaction if you have ever spilled a drop of Super Glue on your fingers. A nucleophilic group on the surface of the skin initiates the polymerization reaction, with the result that two fingers can become firmly glued together.

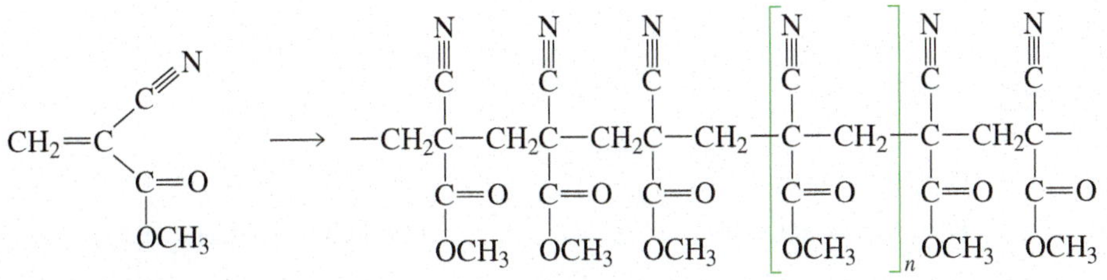

methyl α-cyanoacrylate   Super Glue®

The ability to form covalent bonds with groups on the surfaces of the objects to be glued together is what gives Super Glue its amazing strength. Polymers similar to Super Glue—namely, butyl, isobutyl, or octyl esters rather than methyl esters—are used by surgeons to close wounds.

PROBLEM 8 ◆

Rank the following groups of monomers from most able to least able to undergo anionic polymerization:

a.

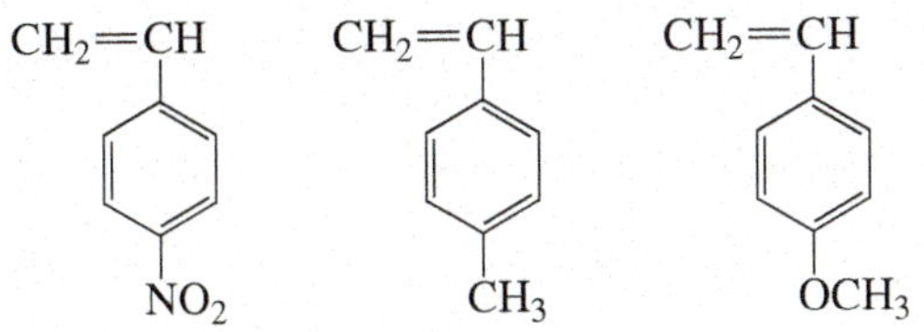

b. CH_2=$CHCH_3$ CH_2=$CHCl$ CH_2=CHC≡N

What Determines the Mechanism?

We saw that the substituent attached to the alkene determines the best mechanism for chain-growth polymerization. Alkenes with substituents that can stabilize radicals readily undergo radical polymerization, alkenes with electron-donating substituents that can stabilize cations undergo cationic polymerization, and alkenes with electron-withdrawing substituents that can stabilize anions undergo anionic polymerizations.

Some alkenes undergo polymerization by more than one mechanism. For example, styrene can undergo polymerization by radical, cationic, and anionic mechanisms because the phenyl group can stabilize benzylic radicals, benzylic cations, and benzylic anions. The mechanism followed for its polymerization depends on the nature of the initiator chosen to start the reaction.

PROBLEM 9 ◆

Why does methyl methacrylate not undergo cationic polymerization?

27.6 RING-OPENING POLYMERIZATIONS

Although ethylene and substituted ethylenes are the monomers most commonly used for the synthesis of chain-growth polymers, other compounds can form chain-growth polymers as well. For example, epoxides can undergo chain-growth polymerization via a ring-opening reaction. Polymerizations that involve ring-opening reactions are called **ring-opening polymerizations.**

If the initiator is a nucleophile, polymerization occurs by an anionic mechanism. From what you know about the reactions of epoxides, you should not be surprised that the nucleophile attacks the less sterically hindered carbon of the epoxide (Section 10.7).

LEARN THE STRATEGY

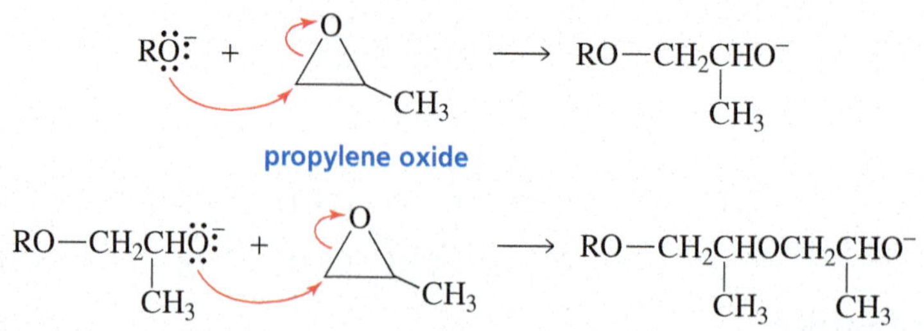

$$RO^- \; + \; \text{(propylene oxide epoxide)} \; \longrightarrow \; RO-CH_2CHO^-$$
with CH_3 on the central carbon.

propylene oxide

$$RO-CH_2\overset{|}{\underset{CH_3}{C}}HO^- \; + \; \text{(epoxide, }CH_3) \; \longrightarrow \; RO-CH_2CHOCH_2CHO^-$$
with CH_3 and CH_3 substituents.

If the initiator is an acid, epoxides are polymerized by a cationic mechanism. Notice that under acidic conditions the nucleophile attacks the more substituted carbon of the epoxide (Section 10.7).

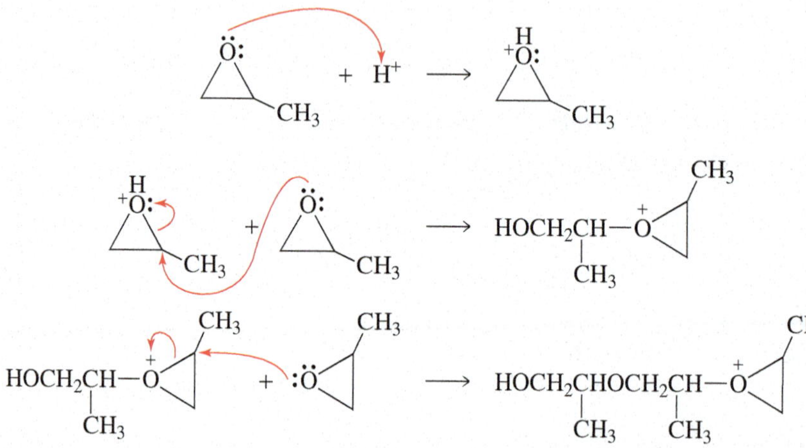

$$\text{(epoxide, }CH_3) \; + \; H^+ \; \longrightarrow \; \text{(protonated epoxide, }\overset{+}{O}H,\; CH_3)$$

$$\text{(protonated epoxide)} \; + \; \text{(epoxide)} \; \longrightarrow \; HOCH_2\overset{|}{\underset{CH_3}{C}}H-\overset{+}{O}\text{(ring, }CH_3)$$

$$HOCH_2\overset{|}{\underset{CH_3}{C}}H-\overset{+}{O}\text{(ring, }CH_3) \; + \; \text{(epoxide, }CH_3) \; \longrightarrow \; HOCH_2CHOCH_2\overset{|}{\underset{CH_3}{C}}H-\overset{+}{O}\text{(ring, }CH_3)$$

USE THE STRATEGY

PROBLEM 10
Describe the polymerization of 2,2-dimethyloxirane by

a. an anionic mechanism.

b. a cationic mechanism.

PROBLEM 11
Explain why, when propylene oxide undergoes anionic polymerization, nucleophilic attack occurs at the less substituted carbon of the epoxide, but when it undergoes cationic polymerization, nucleophilic attack occurs at the more substituted carbon.

PROBLEM 12 ◆
Which monomer and which type of initiator can you use to synthesize each of the following polymers?

a. 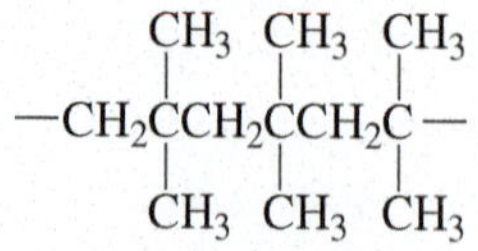
$$-CH_2\underset{CH_3}{\overset{CH_3}{C}}CH_2\underset{CH_3}{\overset{CH_3}{C}}CH_2\underset{CH_3}{\overset{CH_3}{C}}-$$

b. 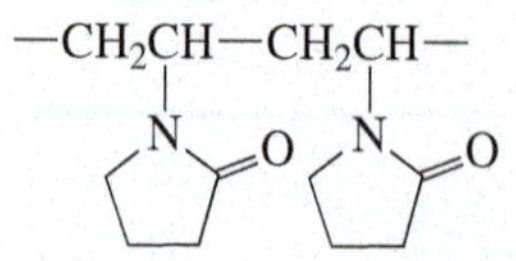
$$-CH_2CH-CH_2CH-$$
(each CH bearing an N-pyrrolidinone ring, $N{\text{—}}O$)

c. $-CH_2CH_2OCH_2CH_2O-$

d. 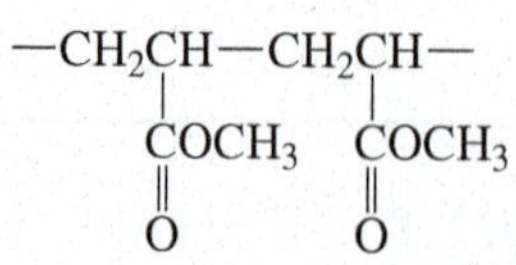
$$-CH_2CH-CH_2CH-$$
with $\underset{\overset{\|}{O}}{COCH_3}$ substituents on each CH.

PROBLEM 13 ◆
Draw a short segment of the polymer formed from cationic polymerization of 3,3-dimethyloxacyclobutane.

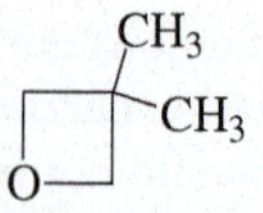

(structure of oxetane ring with two CH_3 groups at the 3-position and O in ring)

3,3-dimethyloxacyclobutane

27.7 STEREOCHEMISTRY OF POLYMERIZATION • ZIEGLER–NATTA CATALYSTS

Polymers formed from monosubstituted ethylenes can exist in three configurations: isotactic, syndiotactic, and atactic.

- An **isotactic polymer** has all of its substituents on the same side of the fully extended carbon chain. (*Iso* and *taxis* are Greek for "the same" and "order," respectively.)

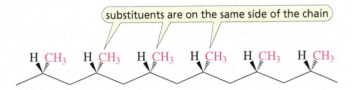

- In a **syndiotactic polymer,** the substituents regularly alternate on both sides of the carbon chain. (*Syndio* means "alternating.")

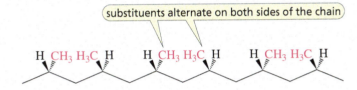

- The substituents in an **atactic polymer** are randomly oriented.

The configuration of the polymer affects its physical properties. Polymers in an isotactic or syndiotactic configuration are more likely to be crystalline solids (Section 27.12) because positioning the substituents in a regular order allows for a more regular packing arrangement. Polymers in an atactic configuration are more disordered and cannot pack together as well, so these polymers are less rigid and, therefore, softer.

Controlling the Configuration

The configuration of the polymer depends on the mechanism by which polymerization occurs. In general, radical polymerization leads primarily to branched polymers in the atactic configuration. Anionic polymerization can produce polymers with the most stereoregularity. The percentage of chains in the isotactic or syndiotactic configuration increases as the polymerization temperature decreases and the solvent polarity decreases.

In 1953, Karl Ziegler and Giulio Natta found that the configuration of a polymer could be controlled if the growing end of the chain and the incoming monomer were coordinated with an aluminum–titanium initiator. These initiators are now called **Ziegler–Natta catalysts.**

Long, unbranched polymers with either the isotactic or the syndiotactic configuration can be prepared using Ziegler–Natta catalysts. Whether the chain is isotactic or syndiotactic depends on the particular catalyst employed. These catalysts revolutionized the field of polymer chemistry because they allow the synthesis of stronger and stiffer polymers that have greater resistance to cracking and heat.

The Mechanism

A proposed mechanism for Ziegler–Natta-catalyzed polymerization of a substituted ethylene is shown on the top of the next page.

MECHANISM FOR ZIEGLER–NATTA–CATALYZED POLYMERIZATION OF A SUBSTITUTED ETHYLENE

- The monomer forms a complex with titanium (dotted arrow) at an open coordination site— that is, a site available to accept electrons.
- The coordinated alkene is inserted between the titanium and the growing polymer (R), thereby lengthening the polymer chain.
- Because a new coordination site opens up during insertion of the monomer, the process can be repeated over and over.

Polyacetylene is another polymer prepared by a Ziegler–Natta process. It can be converted to a **conducting polymer** because the conjugated double bonds in polyacetylene allow electron transport down its backbone after several electrons are removed from or added to the backbone (see page 333).

$$HC \equiv CH \xrightarrow{\text{a Ziegler–Natta catalyst}} -CH=CH-\left[CH=CH\right]_n CH=CH-$$

acetylene   polyacetylene

27.8 POLYMERIZATION OF DIENES

When the bark of a rubber tree is cut, a sticky white liquid oozes out. This same liquid is found inside the stalks of dandelions and milkweed. In fact, more than 400 plants produce this substance. The sticky material is *latex,* a suspension of rubber particles in water. Its biological function is to protect the plant after an injury by covering the wound like a bandage.

Natural Rubber

Natural rubber is a polymer of 2-methyl-1,3-butadiene, also called *isoprene*. On average, a molecule of rubber contains 5000 isoprene units. As in the case of other naturally occurring terpenes, the five-carbon compound actually used in the biosynthesis of rubber is isopentenylpyrophosphate (Section 25.8).

isoprene units   (*Z*)-poly(2-methyl-1,3-butadiene)
natural rubber

All the double bonds in natural rubber have the Z configuration. Rubber is a waterproof material because its tangled hydrocarbon chains have no affinity for water. Charles Macintosh was the first to use rubber as a waterproof coating for raincoats.

latex being collected from a rubber tree

Gutta-percha (from the Malaysian words *getah*, meaning "gum," and *percha*, meaning "tree") is a naturally occurring isomer of rubber in which all the double bonds have the *E* configuration. Like rubber, gutta-percha is exuded by certain trees, but it is much less common. It is also harder and more brittle than rubber. Gutta-percha is the filling material that dentists use in root canals. At one time, it was used for the casing of golf balls, but it becomes brittle in cold weather and tends to split on impact.

PROBLEM 14

Draw a short segment of gutta-percha.

Synthetic Rubber

By mimicking nature, scientists have learned to make synthetic rubbers with properties tailored to meet human needs. These materials have some of the properties of natural rubber, including being waterproof and elastic, but they have some improved properties as well; they are tougher, more flexible, and more durable than natural rubber.

One synthetic rubber, in which all the double bonds are cis, is formed by polymerizing 1,3-butadiene.

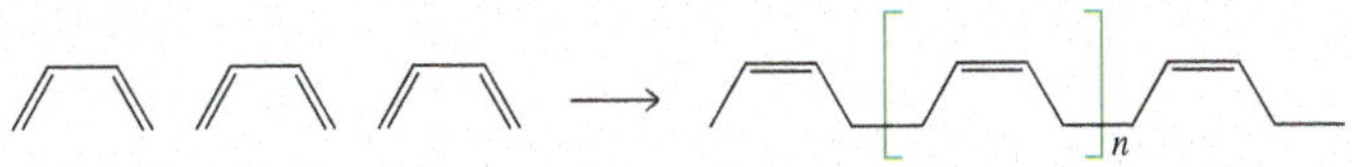

1,3-butadiene monomers → **(*Z*)-poly(1,3-butadiene)
a synthetic rubber**

Neoprene is a synthetic rubber made by polymerizing 2-chloro-1,3-butadiene. It is used to make wet suits, shoe soles, tires, hoses, and coated fabrics.

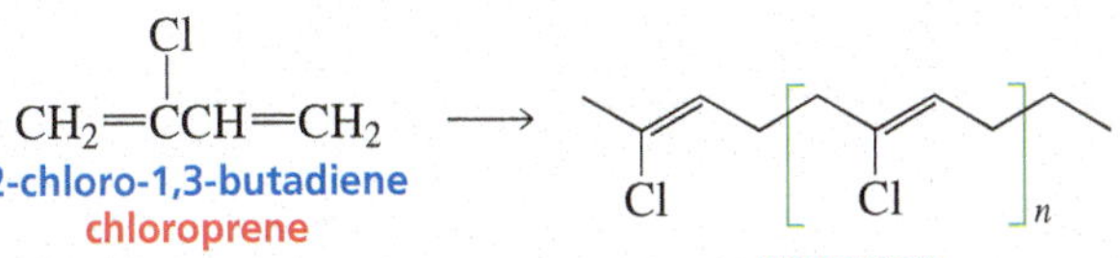

$CH_2{=}CCH{=}CH_2$
**2-chloro-1,3-butadiene
chloroprene**

neoprene

Vulcanization

A problem common to both natural and most synthetic rubbers is that the polymers are soft and sticky. However, they can be hardened by *vulcanization*. Charles Goodyear discovered this process while looking for ways to improve the properties of rubber. He accidentally spilled a mixture of rubber and sulfur on a hot stove. To his surprise, the mixture became hard but flexible. He called the heating of rubber with sulfur **vulcanization,** after Vulcan, the Roman god of fire.

Heating rubber with sulfur causes **cross-linking** of the separate polymer chains through disulfide bonds (Figure 27.1). Thus, the vulcanized chains are covalently bonded to each other in one giant molecule. Because the chains have double bonds, they have bends and kinks that allow them to stretch. When the rubber is stretched, the chains straighten out in the direction of the pull. The cross-linking prevents rubber from being torn when it is stretched; moreover, the cross-links provide a reference framework for the material to return to when the stretching force is removed.

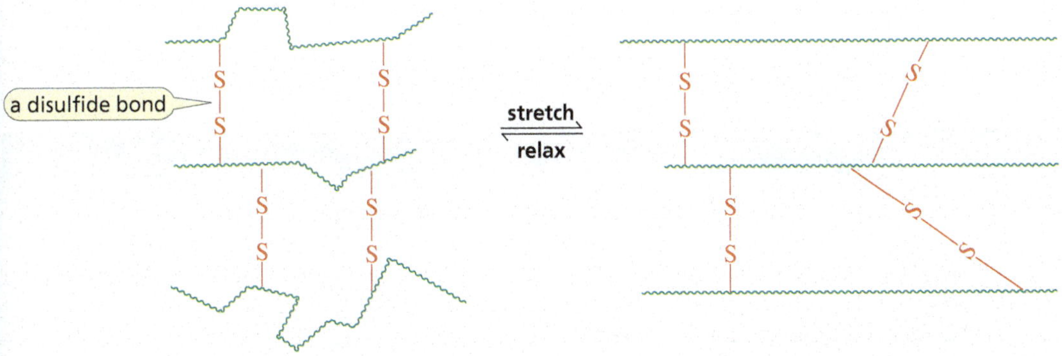

a disulfide bond

stretch ⇌ relax

◀ **Figure 27.1**
The rigidity of rubber is increased by cross-linking the polymer chains with disulfide bonds. When rubber is stretched, the randomly coiled chains straighten out and orient themselves along the direction of the stretch.

The physical properties of rubber can be controlled by regulating the amount of sulfur used in vulcanization. Rubber made with 1%–3% sulfur is soft and stretchy and is used to make rubber bands. Rubber made with 3%–10% sulfur is more rigid and is used in the manufacture of tires. Goodyear's name can be found on many tires sold today. The story of rubber is an example of a scientist taking a natural material and finding ways to improve its properties for useful applications.

PROBLEM 15

a. Draw three segments of the polymer that is formed from 1,4-polymerization of 1,3-butadiene in which all the double bonds are trans.

b. Draw three segments of the polymer that is formed from 1,2-polymerization of 1,3-butadiene.

27.9 COPOLYMERS

The polymers we have discussed so far are **homopolymers**—they are formed from only one type of monomer. Often, two or more different monomers are used to form a **copolymer.** Increasing the number of different monomers used to form a copolymer dramatically increases the number of different copolymers that can be formed. Even if only two kinds of monomers are used, copolymers with very different properties can be prepared by varying the amounts of each monomer. Both chain-growth polymers and step-growth polymers can be copolymers. Many of the synthetic polymers used today are copolymers. Table 27.6 shows some common chain-growth copolymers and the monomers from which they are synthesized.

Table 27.6   Some Examples of Copolymers and Their Uses

Monomers	Copolymer name	Uses
CH_2=CH–Cl (vinyl chloride) + CH_2=CCl–Cl (vinylidene chloride)	Saran	film for wrapping food
CH_2=CH–(phenyl) (styrene) + CH_2=CH–C≡N (acrylonitrile)	SAN	dishwasher-safe objects, vacuum cleaner parts
CH_2=CH–C≡N (acrylonitrile) + CH_2=CH–CH=CH_2 (1,3-butadiene) + CH_2=CH–(phenyl) (styrene)	ABS	bumpers, crash helmets, golf-club heads, luggage
CH_2=CCH$_3$–CH_3 (isobutylene) + CH_2=CHC=CH_2–CH_3 (isoprene)	butyl rubber	inner tubes, balls, inflatable sporting goods

---

Nanocontainers

Scientists have synthesized block copolymers that form micelles (Section 25.4). These spherical copolymers are currently being investigated for their possible use as nanocontainers (10–100 nanometers in diameter) for the delivery of non-water-soluble drugs to target cells. This strategy would allow a higher concentration of a drug to reach a cell than would occur in the natural aqueous milieu. In addition, targeting the drug to particular cells would lower the required dosage of the drug.

There are several types of copolymers. In an **alternating copolymer,** the two monomers alternate. A **block copolymer** consists of blocks of each kind of monomer. A **random copolymer** has a random distribution of monomers. A **graft copolymer** contains branches derived from one monomer grafted onto a backbone derived from another monomer. These structural differences extend the range of physical properties available to the scientist designing the copolymer.

an alternating copolymer	ABABABABABABABABABABABA
a block copolymer	AAAAABBBBBAAAAABBBBBAAA
a random copolymer	AABABABBABAABBABABBAAAB
a graft copolymer	AAAAAAAAAAAAAAAAAAAAAAAA

```
                          B        B        B
                          B        B        B
                          B        B        B
                          B        B        B
                          B        B        B
                          B        B        B
```

27.10 AN INTRODUCTION TO STEP-GROWTH POLYMERS

Step-growth polymers are formed by the intermolecular reaction of molecules with a functional group at each end. When the functional groups react, in most cases a small molecule such as H_2O, alcohol, or HCl is lost. This is why these polymers are also called *condensation polymers* (Section 17.11).

A step-growth polymer can be formed by the reaction of a single bifunctional compound with two different functional groups, A and B. Functional group A of one molecule reacts with functional group B of another molecule to form the compound (A—X—B) that undergoes polymerization.

$$A—B \quad A—B \quad \longrightarrow \quad A—X—B$$

A step-growth polymer can also be formed by the reaction of two different bifunctional compounds. One contains two A functional groups and the other contains two B functional groups. Functional group A reacts with functional group B to form the compound (A—X—B) that undergoes polymerization.

$$A—A \quad B—B \quad \longrightarrow \quad A—X—B$$

Step-growth polymers are made by combining molecules with reactive groups at each end.

Because the growing polymer will have the structure A—[polymer]—B, it can cyclize (an intramolecular reaction) instead of adding a new monomer (an intermolecular reaction), thereby terminating polymerization. We saw that an intramolecular reaction can be minimized by using a high concentration of the monomer (Section 9.16). Because large rings are harder to form than smaller ones, once the polymer chain has more than about 15 atoms, the tendency for cyclization decreases.

The formation of step-growth polymers, unlike the formation of chain-growth polymers, does not occur through chain reactions. Any two monomers (or short chains) can react. The progress of a typical step-growth polymerization is shown schematically in Figure 27.2 and explained on the next page.

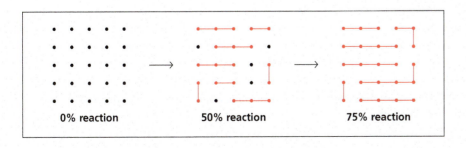

0% reaction	50% reaction	75% reaction

◀ **Figure 27.2**
Progress of a step-growth polymerization.

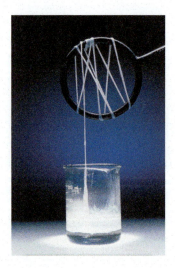

Nylon is pulled from a beaker of adipoyl chloride and 1,6-hexane-diamine.

When the reaction is 50% complete (12 bonds have formed between 25 monomers), the reaction products are primarily dimers and trimers. Even at 75% completion, no long chains have been formed. This means that if step-growth polymerization is to lead to long-chain polymers, very high yields must be achieved. We will see that the reactions involved in step-growth polymerizations are relatively simple (ester and amide formation). However, polymer chemists expend a great deal of effort to arrive at synthetic and processing methods that will result in high-molecular-weight polymers.

27.11 CLASSES OF STEP-GROWTH POLYMERS

Polyamides

Nylon 6 is an example of a step-growth polymer formed from a *monomer with two different functional groups.* The acyl chloride group of one monomer reacts with the amino group of another monomer to form an amide (Section 15.6). Thus, nylon is a **polyamide.** This particular nylon is called nylon 6 because it is formed from the polymerization of 6-aminohexanoyl chloride, a compound that contains six carbons.

nylon rope

$$H_2N(CH_2)_5\overset{\overset{\displaystyle O}{\|}}{C}Cl \xrightarrow[-HCl]{} -NH(CH_2)_5\overset{\overset{\displaystyle O}{\|}}{C}\!\!\left[\!NH(CH_2)_5\overset{\overset{\displaystyle O}{\|}}{C}\right]_n\!\!NH(CH_2)_5\overset{\overset{\displaystyle O}{\|}}{C}-$$

6-aminohexanoyl chloride nylon 6 **a polyamide**

A related polyamide, nylon 66, is an example of a step-growth polymer formed by *two different bifunctional monomers*—adipoyl chloride and 1,6-hexanediamine. It is called nylon 66 because the two starting materials each have 6 carbons.

$$Cl\overset{\overset{\displaystyle O}{\|}}{C}(CH_2)_4\overset{\overset{\displaystyle O}{\|}}{C}Cl \;+\; H_2N(CH_2)_6NH_2 \xrightarrow[-HCl]{} -\overset{\overset{\displaystyle O}{\|}}{C}(CH_2)_4\overset{\overset{\displaystyle O}{\|}}{C}\!\!\left[\!NH(CH_2)_6NH\overset{\overset{\displaystyle O}{\|}}{C}(CH_2)_4\overset{\overset{\displaystyle O}{\|}}{C}\right]_n\!\!NH(CH_2)_6NH$$

adipoyl chloride   1,6-hexanediamine nylon 66

Nylon first found wide use in textiles and carpets. Because it is resistant to stress, it is also used for such things as mountaineering ropes, tire cords, and fishing lines, and as a substitute for metal in bearings and gears. The usefulness of nylon precipitated a search for more new "super fibers" with super strength and super heat resistance.

PROBLEM 16 ♦

a. Draw a short segment of nylon 4.

b. Draw a short segment of nylon 44.

PROBLEM 17

Write an equation that explains what happens if a scientist working in the laboratory spills an aqueous solution of sulfuric acid on her nylon 66 hose.

Aramides

Kevlar is called a super fiber because of its strength; it is an aromatic polyamide. Aromatic polyamides are called **aramides.**

$$\text{1,4-benzenedicarboxylic acid} + \text{1,4-diaminobenzene} \xrightarrow[-H_2O]{\Delta}$$

Kevlar®
an aramide

The incorporation of aromatic rings into polymers results in polymers with great physical strength. Kevlar is five times stronger than steel on an equal weight basis. Army helmets are made of Kevlar, which is also used for lightweight bullet-resistant vests, automobile parts, high-performance skis, the ropes used on the Mars Pathfinder, and high-performance sails used in the America's Cup. Because it is stable at very high temperatures, it is used in the protective clothing worn by firefighters.

Kevlar owes its strength to the way the individual polymer chains interact with each other. The chains are hydrogen bonded, forming a sheet-like structure.

Mylar balloons

Polyesters

Dacron is the most common of the group of step-growth polymers known as **polyesters**—polymers containing many ester groups. Polyesters are used for clothing and are responsible for the wrinkle-resistant behavior of many fabrics. Dacron is made by the transesterification of dimethyl terephthalate with ethylene glycol (Section 15.7). The durability and moisture resistance of this polymer contribute to its "wash-and-wear" characteristics. Because PET is lightweight, it is also used for transparent soft drink bottles.

$$\text{dimethyl terephthalate} + \text{1,2-ethanediol (ethylene glycol)} \xrightarrow[-CH_3OH]{\Delta}$$

poly(ethylene terephthalate)
PET
Dacron®
a polyester

Poly(ethylene terephthlate) can also be processed to form a film known as Mylar. Mylar is tear resistant and, when processed, has a tensile strength nearly as great as that of steel. It is used in the manufacture of magnetic recording tape and sails. Aluminized Mylar was used to make the Echo satellite that was put into orbit around Earth as a giant reflector of microwave signals.

Kodel polyester also is formed by a transesterification reaction. The stiff polyester chain causes the fiber to have a harsh feel that can be softened by blending it with wool or cotton.

CH$_3$O—C—[benzene ring]—C—OCH$_3$ + HOCH$_2$—[cyclohexane]—CH$_2$OH $\xrightarrow[-CH_3OH]{\Delta}$

dimethyl terephthalate   **1,4-di(hydroxymethyl)cyclohexane (cis and trans)**

—C—[benzene ring]—C—[OCH$_2$—[cyclohexane]—CH$_2$O—C—[benzene ring]—C—]$_n$OCH$_2$—[cyclohexane]—CH$_2$—

Kodel®

> **PROBLEM 18**
>
> What happens to polyester slacks if aqueous NaOH is spilled on them?

Polycarbonates

Polyesters with two OR groups bonded to the same carbonyl carbon are known as **polycarbonates**. Lexan, a polycarbonate produced by the transesterification of diphenyl carbonate with bisphenol A, is a strong, transparent polymer used for bullet-proof windows and traffic-light lenses. In recent years, polycarbonates have become important in the automobile industry.

[benzene ring]—O—C—O—[benzene ring] + HO—[benzene ring]—C(CH$_3$)$_2$—[benzene ring]—OH $\xrightarrow[- \text{[benzene ring]—OH}]{\Delta}$

diphenyl carbonate   **bisphenol A**

[benzene ring]—C(CH$_3$)$_2$—[benzene ring]—O—C—O—[[benzene ring]—C(CH$_3$)$_2$—[benzene ring]—O—C—O—]$_n$[benzene ring]—C(CH$_3$)$_2$—[benzene ring]—O—C—

Lexan®
a polycarbonate

Lexan lens in an automobile

Health Concerns: Bisphenol A and Phthalates

Animal studies have raised concerns about human exposure to bisphenol A and phthalates. Pregnant rats exposed to bisphenol A have been found to have a 3 to 4 times higher incidence of precancerous lesions in their mammary ducts. Bisphenol A (BPA) is used in the manufacture of polycarbonates and epoxy resins (Section 27.11). Although there is no evidence that bisphenol A has an adverse impact on humans, most manufacturers of polycarbonates have stopped using the compound, and BPA-free water bottles are now found in stores.

Phthalates have been found to be endocrine disruptors—that is, they can alter the proper balance of hormones. Therefore, the primary risk they pose is to a developing fetus. It is difficult to avoid phthalates because of the numerous items (for example, the linings of most aluminum food and beverage cans) that contain them.

Epoxy Resins

Epoxy resins are the strongest adhesives known; they are extensively cross-linked systems. They can adhere to almost any surface and are resistant to solvents and to extremes of temperature. Epoxy cement is purchased as a kit consisting of a low-molecular-weight *prepolymer* (the most common is a copolymer of bisphenol A and epichlorohydrin) and a *hardener* that react when mixed to form a cross-linked polymer.

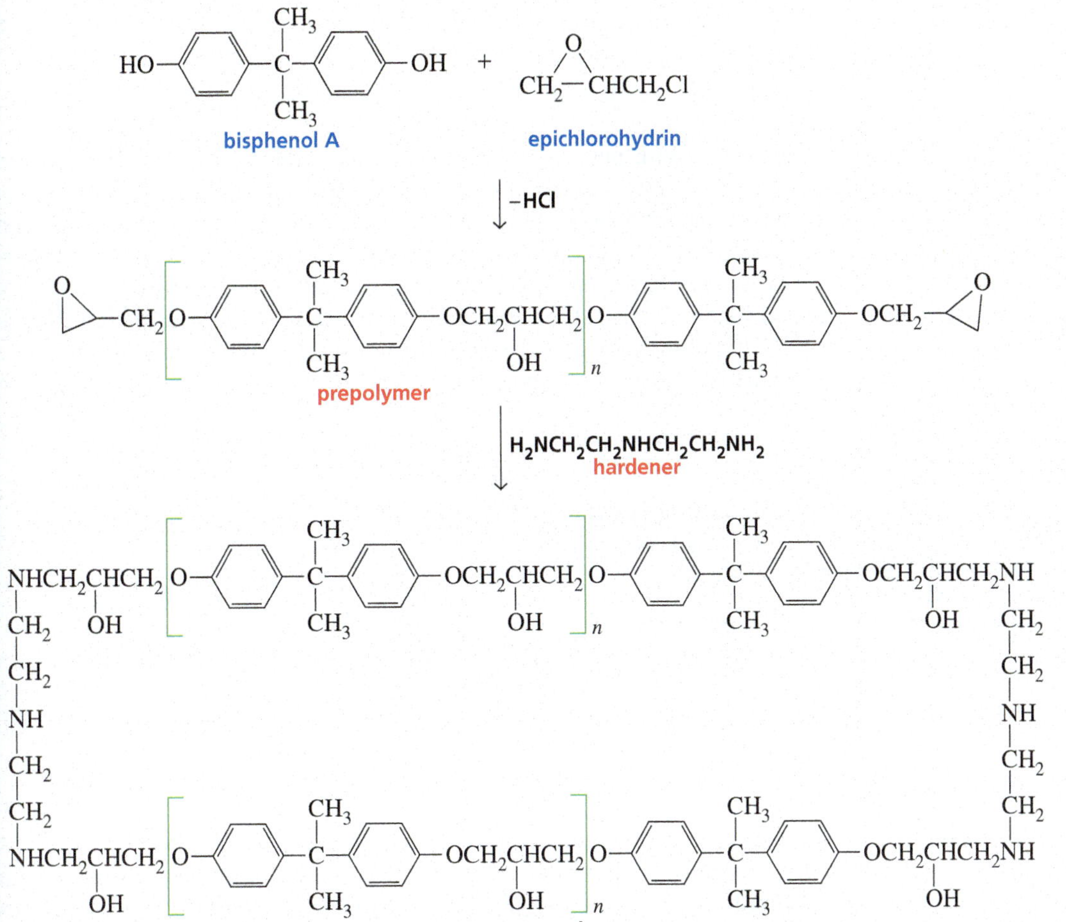

bisphenol A · epichlorohydrin

$-HCl$

prepolymer

$H_2NCH_2CH_2NHCH_2CH_2NH_2$
hardener

an epoxy resin

PROBLEM 19

a. Propose a mechanism for the formation of the prepolymer formed by bisphenol A and epichlorohydrin.
b. Propose a mechanism for the reaction of the prepolymer with the hardener.

Designing a Polymer

Today, polymers are being designed to meet ever more exacting and specific needs. A polymer used for making dental impressions, for example, must be soft enough initially to be molded around the teeth but must become hard enough later to maintain a fixed shape.

The polymer commonly used for dental impressions contains three-membered aziridine rings that react to form cross-links between the chains. Because aziridine rings are not very reactive, cross-linking occurs relatively slowly, so most of the hardening of the polymer does not occur until the polymer is removed from the patient's mouth.

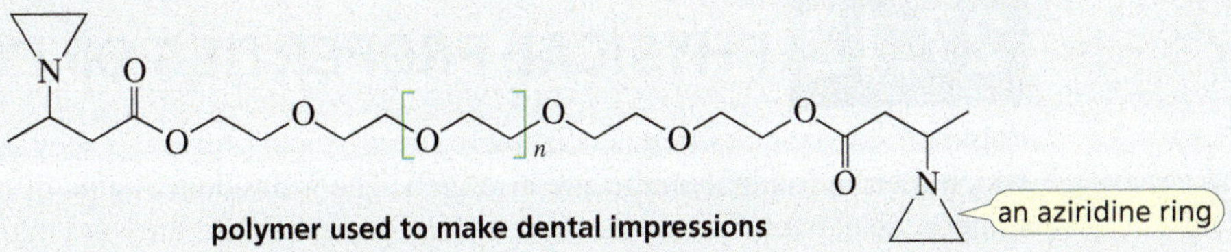

polymer used to make dental impressions · an aziridine ring

Polyurethanes

A **urethane**—also called a carbamate—is a compound that has an OR group and an NHR group bonded to the same carbonyl carbon. Urethanes can be prepared by treating an isocyanate with an alcohol, in the presence of a catalyst such as a tertiary amine (see the top of the next page).

$$RN=C=O \ + \ ROH \ \xrightarrow{} \ RNH-\overset{\displaystyle O}{\overset{\|}{C}}-OR$$

an isocyanate   an alcohol   a urethane

One of the most common **polyurethanes**—polymers that contain urethane groups—is prepared by the polymerization of toluene-2,6-diisocyanate and ethylene glycol. If the reaction is carried out in the presence of a blowing agent, the product is a polyurethane foam. Blowing agents are gases such as nitrogen or carbon dioxide. At one time, chlorofluorocarbons—low-boiling liquids that vaporize on heating—were used, but they have been banned for environmental reasons (Section 12.12). Polyurethane foams are used for furniture stuffing, bedding, carpet backings, and insulation. Notice that polyurethanes prepared from diisocyanates and diols are the only step-growth polymers we have seen in which a small molecule is *not* lost during polymerization.

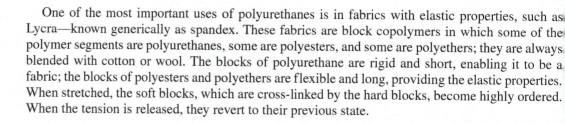

toluene-2,6-diisocyanate   + HOCH$_2$CH$_2$OH

ethylene glycol

a polyurethane

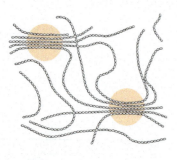

polyurethane foam

One of the most important uses of polyurethanes is in fabrics with elastic properties, such as Lycra—known generically as spandex. These fabrics are block copolymers in which some of the polymer segments are polyurethanes, some are polyesters, and some are polyethers; they are always blended with cotton or wool. The blocks of polyurethane are rigid and short, enabling it to be a fabric; the blocks of polyesters and polyethers are flexible and long, providing the elastic properties. When stretched, the soft blocks, which are cross-linked by the hard blocks, become highly ordered. When the tension is released, they revert to their previous state.

PROBLEM 20

Explain why, when a small amount of glycerol is added to the reaction mixture of toluene-2,6-diisocyanate and ethylene glycol during the synthesis of polyurethane foam, a much stiffer foam is obtained.

$$CH_2-CH-CH_2$$

$$\ \ OH \ \ \ OH \ \ \ OH$$

glycerol

27.12 PHYSICAL PROPERTIES OF POLYMERS

Polymers can be classified according to the physical properties they acquire as a result of the way in which their individual chains are arranged. The individual chains of a polymer are held together by London dispersion forces. Because these forces operate only at small distances, they are strongest if the polymer chains can line up in an ordered, closely packed array.

Regions of the polymer in which the chains are highly ordered with respect to one another are called **crystallites** (Figure 27.3). Between the crystallites are amorphous, noncrystalline regions in which the chains are randomly oriented. The more crystalline (the more ordered) the polymer is, the denser, harder, and more resistant it is to heat (Table 27.7). If the polymer chains possess substituents or have branches that prevent them from packing closely together, the density of the polymer is reduced.

▲ **Figure 27.3**

The polymer chains are highly ordered in the crystallites (indicated by circles). Between the crystallites are noncrystalline (amorphous) regions in which the polymer chains are randomly oriented.

Table 27.7 Properties of Polyethylene as a Function of Crystallinity					
Crystallinity (%)	55	62	70	77	85
Density (g/cm^3)	0.92	0.93	0.94	0.95	0.96
Melting point (°C)	109	116	125	130	133

Thermoplastic Polymers

Thermoplastic polymers have both ordered crystalline regions and amorphous noncrystalline regions. These plastics are hard at room temperature but soft enough to be molded when heated, because the individual chains can slip past one another at elevated temperatures. Thermoplastic polymers are the plastics we encounter most often in our daily lives—in combs, toys, and switch plates. They are the plastics that are easily cracked.

Thermosetting Polymers

Very rigid materials can be obtained if the polymer chains are cross-linked. The greater the degree of cross-linking, the more rigid is the polymer. Such cross-linked polymers are called **thermosetting polymers.** After they are hardened, they cannot be remelted by heating because the cross-links are covalent bonds, not London dispersion forces. Cross-linking reduces the mobility of the polymer chains, causing the polymer to be relatively brittle. Because thermosetting polymers do not have the wide range of properties characteristic of thermoplastic polymers, they are less widely used.

Melmac, a highly cross-linked thermosetting polymer of melamine and formaldehyde, is a hard, moisture-resistant material. Because it is colorless, Melmac can be made into materials with pastel colors. It is used to make counter surfaces and lightweight dishes.

PROBLEM 21

Propose a mechanism for the formation of Melmac.

PROBLEM 22

Bakelite, a durable plastic used for the casing of early radios and TVs, was the first of the thermosetting polymers. It is a highly cross-linked polymer formed from the acid-catalyzed polymerization of phenol and formaldehyde. Because it is much darker than Melmac, Bakelite's range of colors is limited. Propose a structure for Bakelite.

Melamine Poisoning

A few years ago, Chinese milk products were found to have been deliberately contaminated with melamine. Although melamine was never approved as a food additive, milk dealers added it to diluted milk to increase its apparent protein concentration. This milk was then used to make such things as powdered milk, infant formulas, cookies, and other foods. The contaminated milk sickened 300,000 Chinese children and killed at least 6 babies. The previous year, about 1000 dogs and cats died in the United Sates as a result of eating pet food with melamine-containing ingredients traced to Chinese suppliers. As a result of these tragedies, the Chinese legislature passed a food safety law and now requires producers to list all additives on food labels.

Elastomers

An **elastomer** is a polymer that stretches and then reverts to its original shape. It is a randomly oriented amorphous polymer, but it must have some cross-linking so that the chains do not slide past one another. When elastomers are pulled, the random chains stretch out and crystallize. The London dispersion forces are not strong enough to keep them in that arrangement, however, so when the stretching force is removed, the chains go back to their random shapes. Rubber and spandex are examples of elastomers.

Oriented Polymers

Polymer chains obtained by conventional polymerization can be stretched out and then packed together again in a more ordered, parallel arrangement than they had originally, resulting in polymers that are stronger than steel or that conduct electricity as well as copper. Such polymers are called **oriented polymers.** Converting conventional polymers into oriented polymers has been compared to "uncooking" spaghetti; the conventional polymer is analogous to disordered, cooked spaghetti, whereas the oriented polymer is like raw spaghetti.

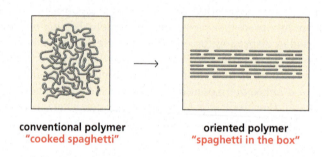

conventional polymer
"cooked spaghetti"

oriented polymer
"spaghetti in the box"

Dyneema, the strongest commercially available fabric, is an oriented polyethylene polymer with a molecular weight 100 times greater than that of high-density polyethylene. It is lighter than Kevlar and at least 40% stronger. A rope made of Dyneema can lift about nine times the weight as a steel rope of similar size! It is astounding that a chain of carbons can be stretched and reoriented to produce a material stronger than steel. Dyneema is used to make full-face crash helmets, protective fencing suits, and hang gliders.

compressed plastic bottles waiting to be recycled

27.13 RECYCLING POLYMERS

We saw in Section 27.3 that polymers are assigned a number from 1–6 that indicates the ease with which that kind of polymer can be recycled—the lower the number, the easier it can be recycled. Unfortunately, only polymers with the two lowest numbers PET (1)—the polymer used to make soft drink bottles—and HDPE (2)—the denser polymer used for juice and milk bottles—are recycled to any significant extent. This amounts to less than 25% of all polymers. The others are found in landfills.

PET is recycled by heating the polymer in an acidic solution of methanol. This transesterification reaction (Section 15.7) is the reverse of the transesterification reaction that formed the polymer (page 701). Because the products of recycling PET are the monomers used to make it, the products can be used to make more PET.

poly(ethylene terephthalate)
PET

dimethyl terephthalate **1,2-ethanediol**

27.14 BIODEGRADABLE POLYMERS

Biodegradable polymers are polymers that can be degraded by microorganisms such as bacteria, fungi, or algae. Polylactide (PLA), a biodegradable polymer of lactic acid, has found wide use. When lactic acid is polymerized, a molecule of water is lost that can hydrolyze the new ester bond, reforming lactic acid.

LEARN THE STRATEGY

However, if lactic acid is converted to a cyclic dimer, the dimer can form a polymer without the loss of water by ring-opening polymerization. (The red arrows show formation of the tetrahedral intermediate; the blue arrows show the subsequent elimination from the tetrahedral intermediate.)

cyclic dimer of lactic acid

Because lactic acid has an asymmetric center, there are several different forms of the polymer. The polymer's physical properties depend on the ratio of R and S enantiomers used in its synthesis. Polylactides are currently being used in nonwrinkle fabrics, microwavable trays, food packaging, and in several medical applications such as sutures, stents, and drug-delivery devices. They are also used for cold drink glasses. Unfortunately, hot drinks cause the polymer to liquify. Although polylactides are more expensive than nonbiodegradable polymers, their price is falling as their production increases.

glasses made of PLA

Polyhydroxyalkanoates (PHAs) are also biodegradable polymers. These are step-growth polymers of 3-hydroxycarboxylic acids. Thus, like PLA, they are polyesters. The most common PHA is PHB, a polymer of 3-hydroxybutyric acid; it can be used for many of the things that polypropylene is now used for. Unlike polypropylene, which floats, PHB sinks. PHBV, a PHA marketed under the trade name Biopol, is a copolymer of 3-hydroxybutyric acid and 3-hydroxyvaleric acid. It is being used for such things as wastepaper baskets, toothbrush holders, and soap dispensers. PHAs are degraded by bacteria to CO_2 and H_2O.

3-hydroxybutyric acid   **3-hydroxyvaleric acid**

USE THE STRATEGY

PROBLEM 23

a. Draw the structure of a short segment of PHB.
b. Draw the structure of a short segment of PHBV with alternating monomers.

ESSENTIAL CONCEPTS

Section 27.0

- A **polymer** is a giant molecule made by covalently linking repeating units of small molecules called **monomers.** The process of linking them is called **polymerization.**

- Polymers can be divided into two groups: **biopolymers,** which are synthesized by cells and **synthetic polymers,** which are synthesized by scientists.

Section 27.1

- Synthetic polymers can be divided into two classes: **chain-growth polymers,** also called **addition polymers;** and **step-growth polymers,** also known as **condensation polymers.**

Section 27.2

- **Chain-growth polymers** are made by **chain reactions,** which add monomers to the end of a growing chain.

- The chain reactions take place by one of three mechanisms: **radical polymerization, cationic polymerization,** or **anionic polymerization.**

- Each mechanism has an initiation step that starts the polymerization, propagation steps that allow the chain to grow at the **propagating site,** and termination steps that stop the growth of the chain.

- The choice of mechanism depends on the structure of the monomer and the initiator used to activate the monomer.

Sections 27.3–27.6

- In radical polymerization, the initiator is a radical; in cationic polymerization, it is an electrophile; and in anionic polymerization, it is a nucleophile.

- Chain-growth polymerization exhibits a preference for **head-to-tail addition.**

- Branching affects the physical properties of the polymer because unbranched chains can pack together more closely than branched chains can.

- A **plasticizer** is an organic compound that dissolves in the polymer and allows the chains to slide past one another.

- Nonterminated polymer chains are called **living polymers.**

- Epoxides can undergo **ring-opening polymerizations.**

Section 27.7

- The substituents are on the same side of the carbon chain in an **isotactic polymer,** alternate on both sides of the chain in a **syndiotactic polymer,** and are randomly oriented in an **atactic polymer.**

- The structure of a polymer can be controlled with **Ziegler–Natta catalysts.**

Section 27.8

- **Natural rubber** is a polymer of 2-methyl-1,3-butadiene. **Synthetic rubbers** have been made by polymerizing dienes other than 2-methyl-1,3-butadiene.

- Heating rubber with sulfur to cross-link the chains is called **vulcanization.**

Section 27.9

- **Homopolymers** are made of one kind of monomer; **copolymers** are made of more than one kind.

Section 27.10

- **Step-growth polymers** are formed by the reaction of two molecules with reactive functional groups at each end.

Section 27.11

- Nylon is a **polyamide. Aramides** are aromatic polyamides. Dacron is a **polyester.**

- **Polycarbonates** have two alkoxy groups bonded to the same carbonyl carbon.

- **Epoxy resins** are extensively cross-linked polymers, making them extremely strong adhesives.

- A **urethane** is a compound that has an OR and an NHR group bonded to the same carbonyl carbon.

Section 27.12

- **Crystallites** are highly ordered regions of a polymer. The more crystalline the polymer is, the denser, harder, and more resistant to heat it is.

- **Thermoplastic polymers** have crystalline and noncrystalline regions.

- **Thermosetting polymers** have cross-linked polymer chains. The greater the degree of cross-linking, the more rigid the polymer.

- An **elastomer** is a plastic that stretches and then reverts to its original shape.

Section 27.13

- PET can be recycled by heating it in methanol to convert it to the material from which it was synthesized.

Section 27.14

- **Biodegradable polymers** can be degraded by microorganisms.

PROBLEMS

24. Draw short segments of the polymers obtained from the following monomers. In each case, indicate whether the polymerization is a chain-growth or a step-growth polymerization.

a. CH_2=CHF

b. CH_2=CHCO$_2$H

c. HO(CH$_2$)$_5$COH (with C=O)

d. ClC(CH$_2$)$_5$CCl + H$_2$N(CH$_2$)$_5$NH$_2$ (with two C=O)

e. [aromatic ring with H$_3$C, —NCO, and OCN substituents] + HOCH$_2$CH$_2$OH

25. Draw the repeating unit of the step-growth polymer that is formed from the following pairs of monomers:

a. 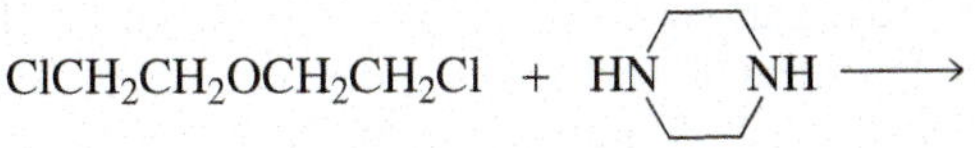

b. 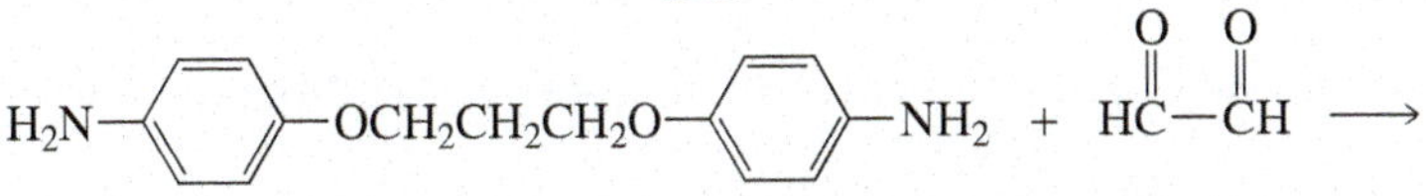

c. 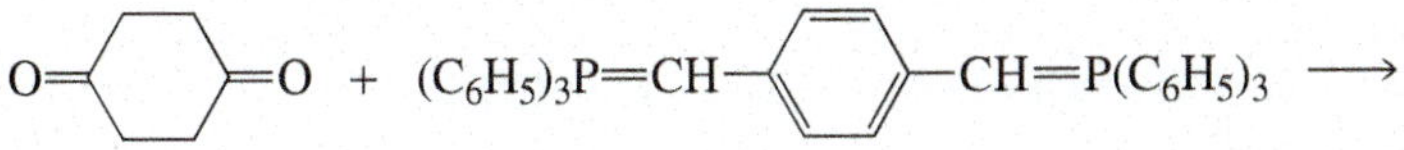

26. Use resonance structures to explain the stabilization of the intermediate formed in the cationic polymerization of styrene to form polystyrene.

27. Draw the structure of the monomer or monomers used to synthesize the following polymers, and indicate whether each is a chain-growth polymer or a step-growth polymer.

a. 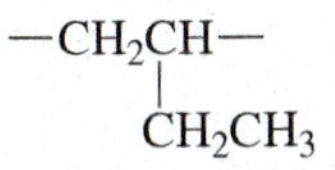

b. 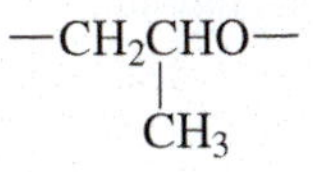

c. 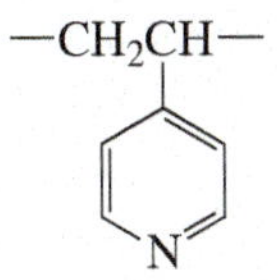

d. 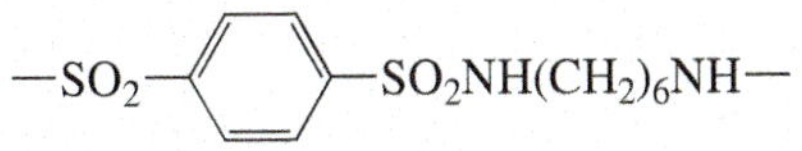

e. 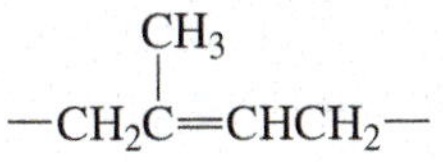

f. 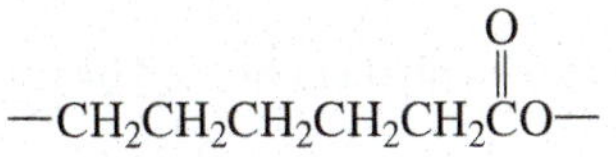

g. 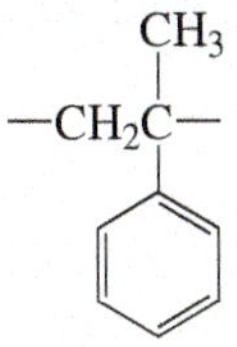

h. 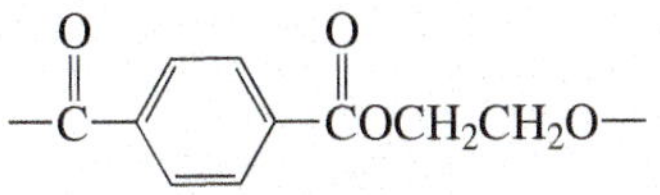

28. Explain why the configuration of a polymer of isobutylene is neither isotactic, syndiotactic, nor atactic.

29. Draw short segments of the polymers obtained from the following compounds under the given reaction conditions:

a. 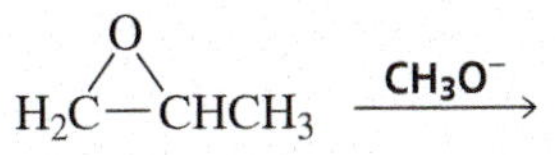

b. 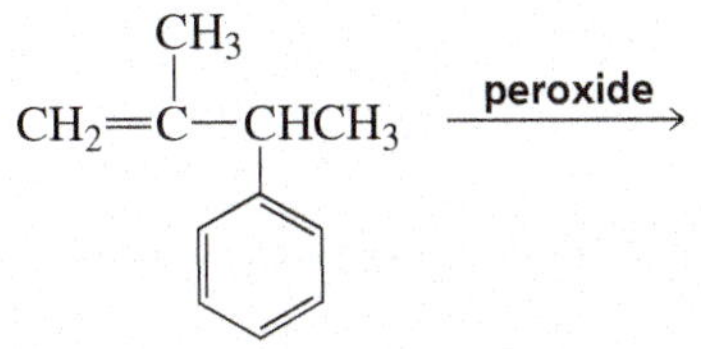

c. 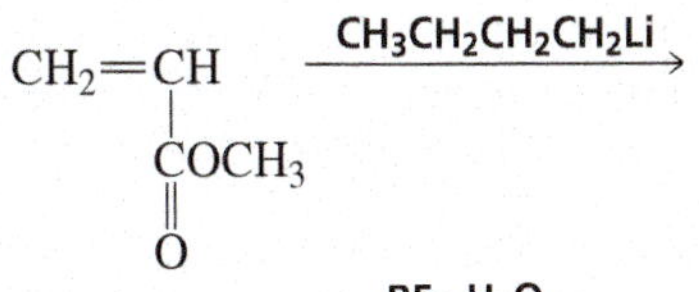

d. $CH_2{=}CHOCH_3 \xrightarrow{BF_3,\ H_2O}$

30. Quiana is a synthetic fabric that feels very much like silk.

a. Is Quiana a nylon or a polyester?

b. What monomers are used to synthesize it?

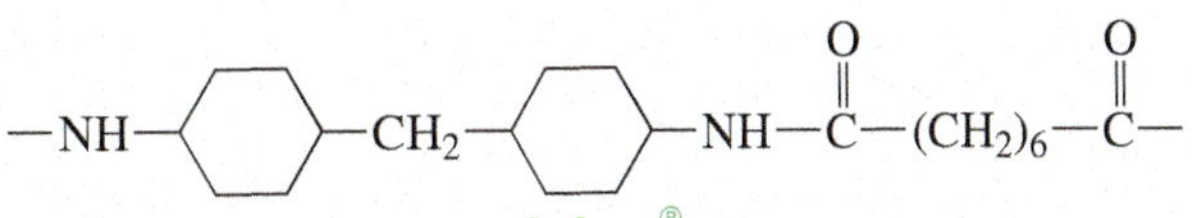

Quiana®

31. Show the mechanism for the transesterification reaction that recycles poly(ethylene terephthalate) to dimethyl terephthalate and 1,2-ethanediol when the polymer is heated in an acidic solution of methanol.

32. Explain why a random copolymer is obtained when 3,3-dimethyl-1-butene undergoes cationic polymerization.

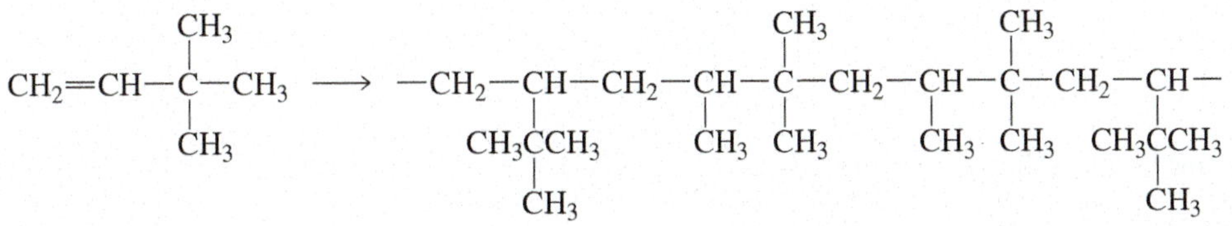

33. A chemist carried out two polymerization reactions. One flask contained a monomer that polymerizes by a chain-growth mechanism, and the other flask contained a monomer that polymerizes by a step-growth mechanism. When the reactions were terminated early in the process and the contents of the flasks analyzed, it was found that one flask contained a high-molecular-weight polymer and very little material of intermediate molecular weight. The other contained mainly material of intermediate molecular weight and very little high-molecular-weight material. Which flask contained which product? Explain.

34. Poly(vinyl alcohol) is a polymer used to make fibers and adhesives. It is synthesized by hydrolysis or alcoholysis of the polymer obtained from polymerization of vinyl acetate as shown below.

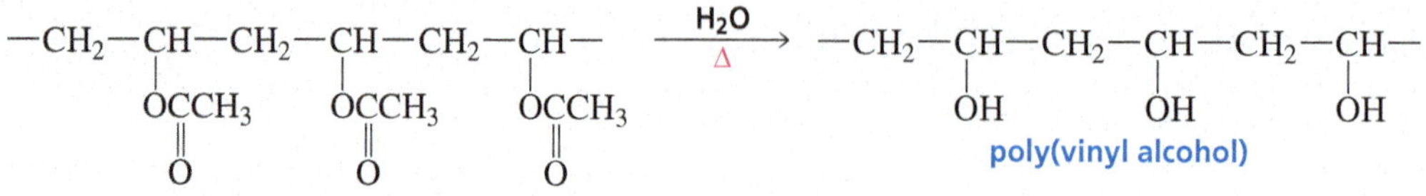

$$-CH_2-CH-CH_2-CH-CH_2-CH- \quad \xrightarrow[\Delta]{H_2O} \quad -CH_2-CH-CH_2-CH-CH_2-CH-$$

with $OCCH_3$ (C=O) substituents on the left (poly(vinyl acetate)) and OH substituents on the right (poly(vinyl alcohol))

a. Why is poly(vinyl alcohol) not prepared by polymerizing vinyl alcohol?

b. Is poly(vinyl acetate) a polyester?

35. Five different repeating units are found in the polymer obtained by cationic polymerization of 4-methyl-1-pentene. Identify these repeating units.

36. Draw a segment of a polymer (containing three monomers) that forms from anionic polymerization of the compound shown below.

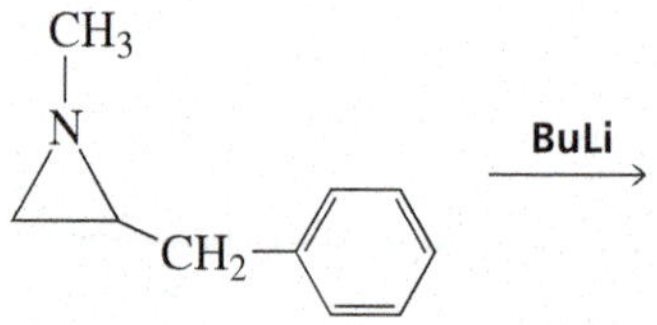

[structure: aziridine ring with N–CH_3, substituent CH_2–phenyl] $\xrightarrow{\text{BuLi}}$

37. If a peroxide is added to styrene, the polymer known as polystyrene is formed. If a small amount of 1,4-divinylbenzene is added to the reaction mixture, a stronger and more rigid polymer is formed. Draw a short section of this more rigid polymer.

$$CH_2=CH-\underset{\text{1,4-divinylbenzene}}{\text{(benzene ring)}}-CH=CH_2$$

38. A particularly strong and rigid polyester used for electronic parts is marketed under the trade name Glyptal. It is a polymer of terephthalic acid and glycerol. Draw a segment of the polymer and explain why it is so strong.

39. The following two compounds form a 1 : 1 alternating copolymer. No initiator is needed for the polymerization. Propose a mechanism for formation of the copolymer.

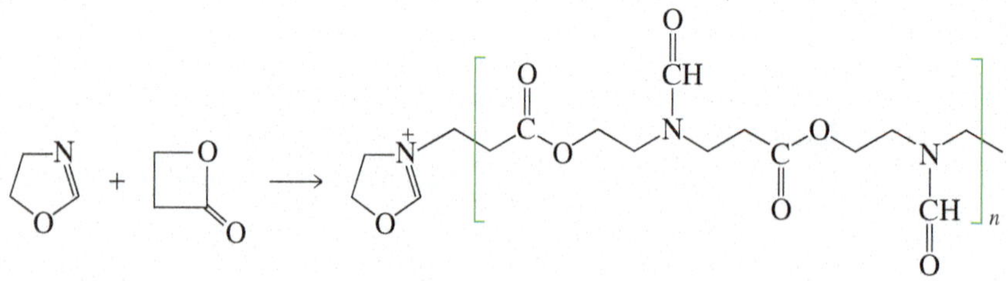

[structures: oxazoline ring + β-lactone (oxetanone) → copolymer segment]

40. Which monomer gives a greater yield of polymer, 5-hydroxypentanoic acid or 6-hydroxyhexanoic acid? Explain your choice.

41. When rubber balls and other objects made of natural rubber are exposed to air for long periods, they turn brittle and crack. Explain why this happens more slowly to objects made of polyethylene.

42. When acrolein undergoes anionic polymerization, a polymer with two types of repeating units is obtained. Draw the structures of the repeating units.

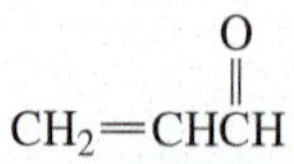

$$CH_2=CHCH \quad (C=O) $$
acrolein

43. Why do vinyl raincoats become brittle as they get old, even if they are not exposed to air or to any pollutants?

44. The polymer shown below is synthesized by hydroxide ion-promoted hydrolysis of a copolymer of *para*-nitrophenyl methacrylate and acrylate.

   a. Propose a mechanism for the formation of the copolymer.

   b. Explain why hydrolysis of the copolymer to form the polymer occurs much more rapidly than hydrolysis of *para*-nitrophenyl acetate.

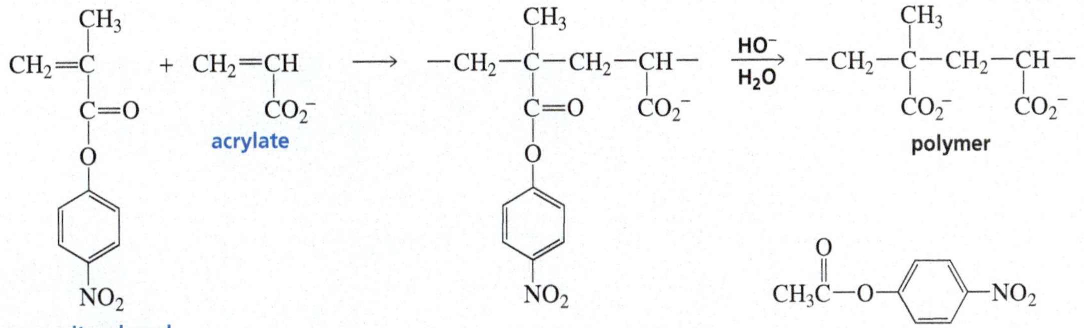

para-nitrophenyl
methacrylate   acrylate   polymer   *para*-nitrophenyl acetate

45. An alternating copolymer of styrene and vinyl acetate can be turned into a graft copolymer by hydrolyzing it and then adding ethylene oxide. Draw the structure of the graft copolymer.

46. How can head-to-head poly(vinyl bromide) be synthesized?

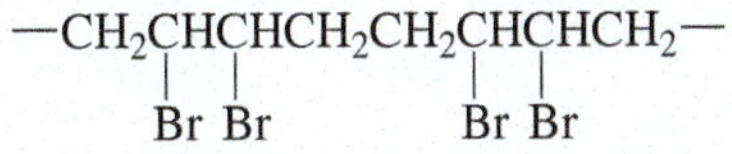

$$-CH_2CHCHCH_2CH_2CHCHCH_2-$$
$$Br\ \ BrBr\ \ Br$$

head-to-head poly(vinyl bromide)

47. Delrin (polyoxymethylene) is a tough self-lubricating polymer used in gear wheels. It is made by polymerizing formaldehyde in the presence of an acid catalyst.

   a. Propose a mechanism for formation of a segment of the polymer.

   b. Is Delrin a chain-growth polymer or a step-growth polymer?

28 Pericyclic Reactions

In this chapter, we will see why vitamin D is called the sunshine vitamin, how animals without sun-exposed skin get vitamin D, and why the reaction that causes skin cancer requires sunlight. We will also look at the reaction that causes fireflies to give off light.

Reactions of organic compounds can be divided into three classes—polar reactions, radical reactions, and pericyclic reactions. The most common are polar reactions.

- A **polar reaction** is one in which a nucleophile reacts with an electrophile. Both electrons in the new bond come from the nucleophile. Most of the reactions we saw in previous chapters are polar reactions.

  > **a polar reaction**

  $$H\text{—}\ddot{\underset{\cdot\cdot}{O}}{:}^- \; + \; \overset{\delta+}{CH_3}\text{—}\overset{\delta-}{Br} \; \longrightarrow \; CH_3\text{—}OH \; + \; Br^-$$

- A **radical reaction** is one in which a new bond is formed using one electron from each of the reactants.

  > **a radical reaction**

  $$CH_3\dot{C}H_2 \; + \; Cl\text{—}Cl \; \longrightarrow \; CH_3CH_2\text{—}Cl \; + \; \cdot Cl$$

- A **pericyclic reaction** is one in which the electrons in one or more reactants are reorganized in a cyclic manner. The only pericyclic reaction we have seen so far is the Diels–Alder reaction.

28.1   THERE ARE THREE KINDS OF PERICYCLIC REACTIONS

In this chapter we look at the three most common types of pericyclic reactions—*electrocyclic reactions, cycloaddition reactions,* and *sigmatropic rearrangements.*

Electrocyclic Reactions

An **electrocyclic reaction** is an intramolecular reaction in which a new σ (sigma) bond is formed between the ends of a conjugated π (pi) system. This reaction is easy to recognize—the product is a cyclic compound that has *one more ring* and *one less π bond* than the reactant.

an electrocyclic reaction

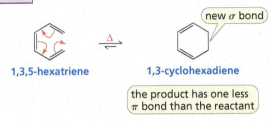

1,3,5-hexatriene   1,3-cyclohexadiene

new σ bond

the product has one less
π bond than the reactant

Electrocyclic reactions are reversible. In the reverse direction, an electrocyclic reaction is one in which a σ bond in a cyclic compound breaks, forming a conjugated system that has *one less ring* and *one more π bond* than the reactant.

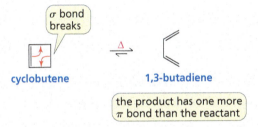

σ bond
breaks

cyclobutene   1,3-butadiene

the product has one more
π bond than the reactant

Cycloaddition Reactions

In a **cycloaddition reaction,** two different π-bond-containing molecules react to form a cyclic compound. Each of the reactants loses a π bond, and the resulting cyclic product has two new σ bonds. The Diels–Alder reaction, which you learned about in Section 8.14, is a familiar example of a cycloaddition reaction.

a cycloaddition reaction

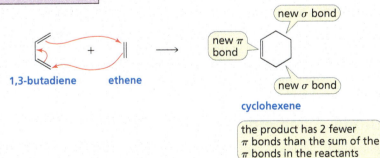

1,3-butadiene   ethene

new σ bond

new π
bond

new σ bond

cyclohexene

the product has 2 fewer
π bonds than the sum of the
π bonds in the reactants

Sigmatropic Rearrangements

In a **sigmatropic rearrangement,** a σ bond is broken in the reactant, a new σ bond is formed in the product, and the π bonds rearrange. The number of π bonds does not change, only their location changes. The σ bond that breaks can be in the middle of the π system or at the end of the π system.

sigmatropic rearrangements

σ bond is broken in the middle of the π system

Δ

σ bond is formed

product and reactant have the same number of π bonds, but their positions have changed

σ bond is broken at the end of the π system

σ bond is formed

Common Features of Pericyclic Reactions

Notice that electrocyclic reactions and sigmatropic rearrangements are *intra*molecular reactions. In contrast, cycloaddition reactions involve the interaction of two molecules—they are *inter*molecular reactions. The three kinds of pericyclic reactions share the following features:

- They are all concerted reactions—that is, all the electron reorganization takes place in a single step. Therefore, there is one cyclic transition state and no intermediate.
- They are highly stereoselective.
- They are generally not affected by catalysts or a change in solvent.

We will see that the configuration of the product formed in a pericyclic reaction depends on

- the configuration of the reactant;
- the number of conjugated double bonds or pairs of electrons in the reacting system; and
- whether the reaction is a thermal reaction or a photochemical reaction.

> A **thermal reaction** *takes place without the absorption of light.*

> A **photochemical reaction** *takes place when a reactant absorbs light.*

Despite its name, a thermal reaction does not necessarily require more heat than what is available at room temperature. Some thermal reactions do require additional heat in order to take place at a reasonable rate, but others readily occur at, or even below, room temperature.

For many years, pericyclic reactions puzzled chemists. Why did some pericyclic reactions take place only under thermal conditions, whereas others occurred only under photochemical conditions, and others were successfully carried out under both thermal and photochemical conditions?

The configurations of the products were also puzzling. After many pericyclic reactions had been investigated, chemists observed that if a pericyclic reaction could take place under both thermal and photochemical conditions, the configuration of the product formed under one set of conditions was different from the configuration of the product formed under the other set of conditions. For example, if the cis isomer was obtained under thermal conditions, the trans isomer was obtained under photochemical conditions and vice versa.

Conservation of Orbital Symmetry

It took two very talented chemists, each bringing his own expertise to the problem, to explain the puzzling behavior of pericyclic reactions. In 1965, R. B. Woodward, an experimentalist, and Roald Hoffmann, a theorist, developed the **conservation of orbital symmetry theory** to explain the relationships among the structure and configuration of the reactant, the conditions (thermal, photochemical, or both) under which the reaction takes place, and the configuration of the product. Because the behavior of pericyclic reactions is so precise, it is not surprising that everything about their behavior can be explained by one simple theory. The difficult part was having the insight that led to the theory.

The conservation of orbital symmetry theory states that *a pericyclic reaction requires the overlap of in-phase orbitals.* This theory was based on the **frontier orbital theory** put forth by Kenichi Fukui in 1954. Although Fukui's theory was more than 10 years old when the conservation of orbital symmetry theory was developed, it had been overlooked because of its mathematical complexity and Fukui's failure to apply it to stereoselective reactions.

PROBLEM 1 ♦

Examine the following pericyclic reactions. For each reaction, tell whether it is an electrocyclic reaction, a cycloaddition reaction, or a sigmatropic rearrangement.

28.2 **MOLECULAR ORBITALS AND ORBITAL SYMMETRY**

According to the conservation of orbital symmetry theory, the symmetry of a molecular orbital controls both the conditions under which a pericyclic reaction takes place and the configuration of the product that is formed. To understand pericyclic reactions, therefore, we must now review molecular orbital theory.

The overlap of *p* atomic orbitals to form molecular orbitals can be described mathematically using quantum mechanics. Fortunately, the result of the mathematical treatment can be described simply in nonmathematical terms by **molecular orbital (MO) theory.** You were introduced to MO theory in Sections 1.6 and 8.8. Take a few minutes to review the following key points raised in those sections.

NOTE TO THE STUDENT
• You will find additional information on MO theory in Special Topic II in the *Study Guide and Solutions Manual.*

- The two lobes of a *p* orbital have opposite phases. When two in-phase *p* atomic orbitals interact, a covalent bond is formed. The interaction of two out-of-phase *p* atomic orbitals subtracts from bonding because a node is created between the two nuclei.

- Electrons fill molecular orbitals according to the same rules that govern how they fill atomic orbitals (Section 1.2)—that is, an electron goes into the available MO with the lowest energy (the aufbau principle); only two electrons can occupy a particular MO and they must be of opposite spin (the Pauli exclusion principle); and an electron will occupy an empty degenerate orbital before it will pair up (Hund's rule).

- Each carbon that forms a π bond has a *p* atomic orbital, and the *p* atomic orbitals of the carbons combine to produce a set of MOs. Thus, a MO can be described by the **linear combination of atomic orbitals (LCAO).** In a MO, each electron that previously occupied a *p* atomic orbital belonging to an individual carbon now occupies the entire part of the molecule that is encompassed by the interacting *p* orbitals.

Molecular Orbital Description of Ethene

A molecular orbital description of ethene is shown in Figure 28.1 on the next page. (The opposite phases of the two lobes of a *p* orbital are indicated by different colors.) Because ethene has one π bond, it has two *p* atomic orbitals that combine to produce two molecular orbitals. (Recall that orbitals are conserved; Section 1.6.) The in-phase interaction of the two *p* atomic orbitals forms a π bonding MO, designated by ψ_1, and the out-of-phase interaction forms a π^* antibonding MO, designated by ψ_2. (ψ is the Greek letter *psi,* pronounced "sigh.") The energy of the π bonding MO is *lower* than that of the *p* atomic orbitals and the energy of the π^* antibonding MO is higher than that of the *p* atomic orbitals.

Recall that the bonding MO results from additive interaction of two atomic orbitals, whereas the antibonding MO results from subtractive interaction. In other words, the interaction of in-phase atomic

Orbitals are conserved: two atomic orbitals combine to produce two MOs, four atomic orbitals combine to produce four MOs, six atomic orbitals combine to produce six MOs, and so on.

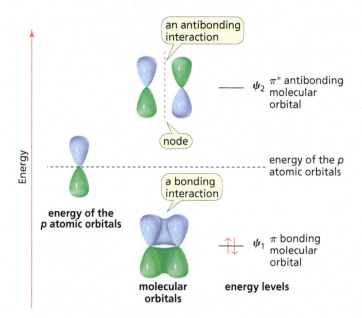

▶ **Figure 28.1**
The two *p* atomic orbitals of ethene combine to produce two MOs.

orbitals holds atoms together, whereas the interaction of out-of-phase atomic orbitals pushes atoms apart. Thus, in-phase overlap produces a bonding interaction and out-of-phase overlap creates a node. Recall that a node is a place in which there is zero probability of finding an electron (Section 1.5).

Because electrons reside in the available MOs with the lowest energy and two electrons can occupy a MO, the two π electrons of ethene reside in the π bonding MO. Figure 28.1 describes all molecules with one carbon–carbon double bond.

Molecular Orbital Description of 1,3-Butadiene

1,3-Butadiene has two conjugated π bonds, so it has four *p* atomic orbitals (Figure 28.2). Four atomic orbitals combine to produce four MOs: ψ_1, ψ_2, ψ_3, and ψ_4. Half are bonding MOs (ψ_1 and ψ_2), and the other half are antibonding MOs (ψ_3 and ψ_4). Like the MOs of ethene, the energy of the π bonding MOs are *lower* and the energy of the π^* antibonding MOs are higher than the energy of the *p* atomic orbitals. Because the four π electrons will reside in the available MOs with the lowest energy, two electrons are in ψ_1 and two are in ψ_2. Remember that although the MOs have different energies, they all coexist. Figure 28.2 describes all molecules with two conjugated carbon–carbon double bonds.

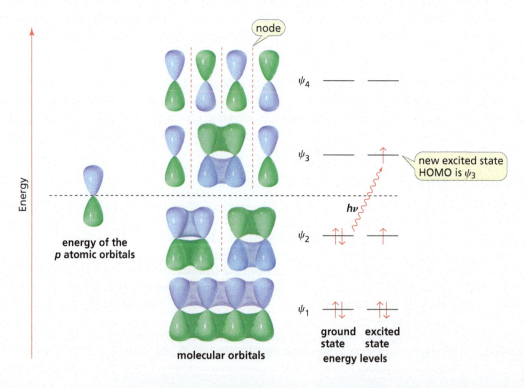

▶ **Figure 28.2**
The four *p* atomic orbitals of 1,3-butadiene combine to produce four MOs.

Analyzing Molecular Orbitals

Figure 28.2 shows that as the energy of the MO increases, the number of bonding interactions decreases and the number of nodes *between* the nuclei increases. For example, ψ_1 has three bonding interactions and zero nodes between the nuclei, ψ_2 has two bonding interactions and one node between the nuclei, ψ_3 has one bonding interaction and two nodes between the nuclei, and ψ_4 has zero bonding interactions and three nodes between the nuclei. *Overall, a MO is bonding if the number of bonding interactions is greater than the number of nodes between the nuclei, and a MO is antibonding if the number of bonding interactions is fewer than the number of nodes between the nuclei.*

The normal electronic configuration of a molecule is known as its **ground state.** In the ground state of 1,3-butadiene, the **highest occupied molecular orbital (HOMO)** is ψ_2, and the **lowest unoccupied molecular orbital (LUMO)** is ψ_3. If a molecule absorbs light of an appropriate wavelength, the light will promote an electron from its ground-state HOMO to its LUMO (from ψ_2 to ψ_3). The molecule is then in an **excited state.** In the excited state, the HOMO is ψ_3 and the LUMO is ψ_4.

LEARN THE STRATEGY

> *A molecule reacts from its ground state in a thermal reaction and from its excited state in a photochemical reaction.*

Some MOs are *symmetric*, and some are *antisymmetric*. Symmetric and antisymmetric MOs are easy to distinguish: if the lobes at the ends of the MO are in-phase (both have blue lobes on the top and green lobes on the bottom), then the MO is symmetric; if the two end lobes are out-of-phase, then the MO is antisymmetric. In Figure 28.2, ψ_1 and ψ_3 are **symmetric MOs,** whereas ψ_2 and ψ_4 are **antisymmetric MOs.**

Notice that as the MOs increase in energy, they alternate from being symmetric to being antisymmetric. Therefore,

A molecule reacts from its ground state in a thermal reaction and from its excited state in a photochemical reaction.

> *the ground-state HOMO and the excited-state HOMO always have opposite symmetries:*
> *if one is symmetric, the other is antisymmetric.*

Molecular Orbital Description of 1,3,5-Hexatriene

A molecular orbital description of 1,3,5-hexatriene, a compound with three conjugated double bonds, is shown in Figure 28.3. As a review, examine the figure and note

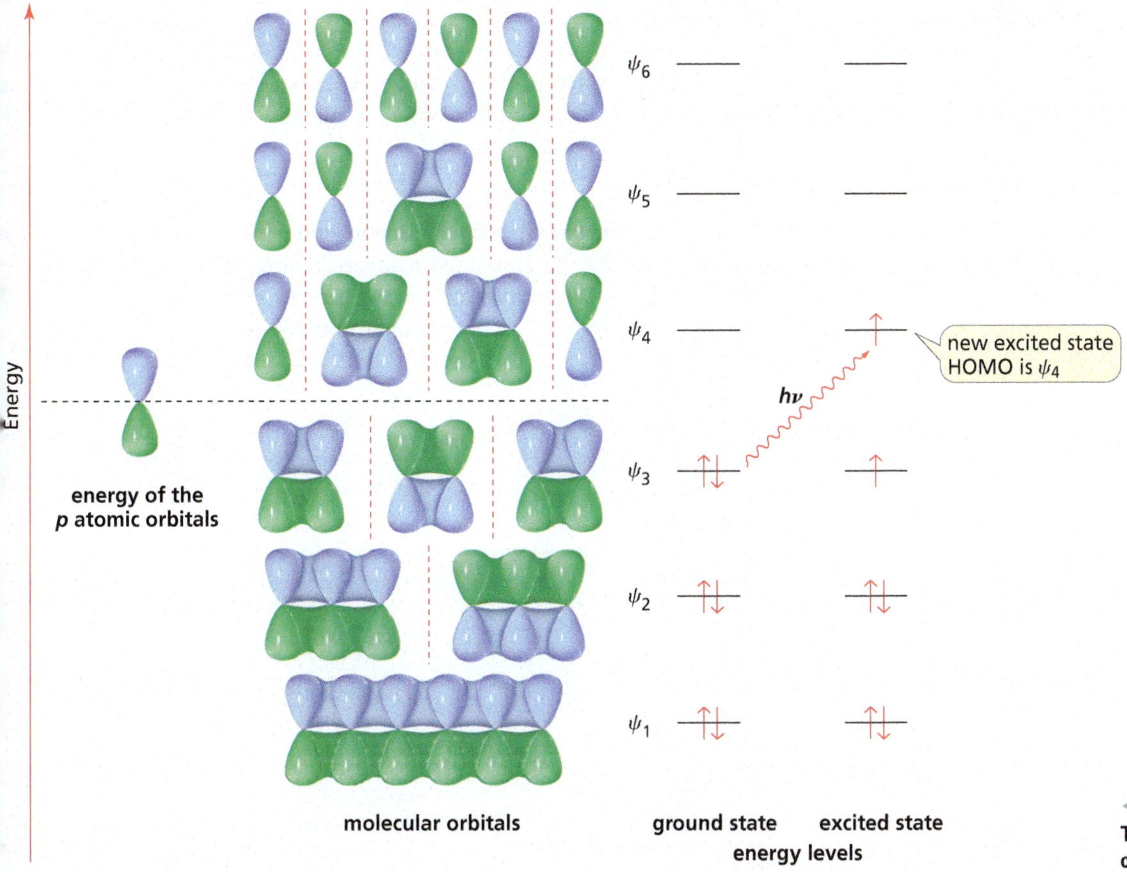

◀ **Figure 28.3**
The six *p* atomic orbitals of 1,3,5-hexatriene combine to produce six MOs.

- the distribution of electrons in the ground and excited states;
- that the number of bonding interactions decreases and the number of nodes increases as the MOs increase in energy;
- that as the MOs increase in energy, they alternate from symmetric to antisymmetric; and
- that compared to the ground state, the excited state has a new HOMO and a new LUMO.

The ground-state HOMO and the excited-state HOMO have opposite symmetries.

Although the chemistry of a compound is determined by all its MOs, we can learn a great deal about that chemistry by looking at only the **HOMO** and the **LUMO.** These MOs are known as the **frontier orbitals.** We will see that simply by evaluating *one* of the frontier orbitals of the reactant (or reactants) in a pericyclic reaction, we will be able to predict the conditions under which the reaction will occur (thermal or photochemical, or both) and the products that will form.

USE THE STRATEGY

PROBLEM 2 ◆

Answer the following questions for the MOs of 1,3,5-hexatriene:

a. Which are bonding MOs, and which are antibonding MOs?
b. Which MOs are the HOMO and the LUMO in the ground state?
c. Which MOs are the HOMO and the LUMO in the excited state?
d. Which MOs are symmetric, and which are antisymmetric?
e. What is the relationship between HOMO and LUMO and symmetric and antisymmetric MOs?

PROBLEM 3 ◆

a. How many MOs does 1,3,5,7-octatetraene have?
b. What is the designation of its HOMO (ψ_1, ψ_2, etc.)?
c. How many nodes does its highest energy MO have between the nuclei?

PROBLEM 4

Give a molecular orbital description for each of the following:

a. 1,3-pentadiene   b. 1,4-pentadiene   c. 1,3,5-heptatriene   d. 1,3,5,8-nonatetraene

28.3 ELECTROCYCLIC REACTIONS

An *electrocyclic reaction* is an intramolecular reaction in which the formation of a σ bond between the ends of the π system and the rearrangement of the π electrons leads to a cyclic product that has one less π bond than the reactant. An electrocyclic reaction is completely stereoselective; it is also stereospecific.

Example 1

NOTE TO THE STUDENT

- Recall that *E* means the high-priority groups are on opposite sides of the double bond, and *Z* means the high-priority groups are on the same side of the double bond (Section 4.2).

When (2E,4Z,6E)-octatriene undergoes an electrocyclic reaction under thermal conditions, only the cis product is formed. (Notice that the cis isomer is a meso compound.) In contrast, when (2E,4Z,6Z)-octatriene undergoes an electrocyclic reaction under thermal conditions, only the trans product is formed. (Notice that the trans product is a pair of enantiomers; Section 4.13.)

(2E,4Z,6E)-octatriene   cis-5,6-dimethyl-1,3-cyclohexadiene

(2E,4Z,6Z)-octatriene   trans-5,6-dimethyl-1,3-cyclohexadiene

However, when the reactions are carried out under photochemical conditions, the products have opposite configurations: the compound that forms the cis isomer under thermal conditions forms the trans isomer under photochemical conditions, and the compound that forms the trans isomer under thermal conditions forms the cis isomer under photochemical conditions.

(2E,4Z,6E)-octatriene   **trans-5,6-dimethyl-1,3-cyclohexadiene**

(2E,4Z,6Z)-octatriene   **cis-5,6-dimethyl-1,3-cyclohexadiene**

Δ = thermal conditions

hv = photochemical conditions

Example 2

Under thermal conditions, (2E,4Z)-hexadiene cyclizes to *cis*-3,4-dimethylcyclobutene (a meso compound), and (2E,4E)-hexadiene cyclizes to *trans*-3,4-dimethylcyclobutene.

(2E,4Z)-hexadiene   **cis-3,4-dimethylcyclobutene**

(2E,4E)-hexadiene   **trans-3,4-dimethylcyclobutene**

As we saw with the octatrienes, the configuration of the product changes if the reactions are carried out under photochemical conditions: (2E,4Z)-hexadiene forms the trans isomer, and (2E,4E)-hexadiene forms the cis isomer.

(2E,4Z)-hexadiene   **trans-3,4-dimethylcyclobutene**

(2E,4E)-hexadiene   **cis-3,4-dimethylcyclobutene**

Electrocyclic reactions are reversible. The cyclic compound is favored for electrocyclic reactions that form six-membered rings, whereas the open-chain compound is favored for electrocyclic reactions that form four-membered rings because of the angle strain associated with four-membered rings (Section 3.12).

Now we will use what we have learned about MOs to explain the configuration of the products obtained from the preceding reactions. Then we will be able to predict the configuration of the product(s) of any other electrocyclic reaction.

Conrotatory and Disrotatory Ring Closure

The product of an electrocyclic reaction results from the formation of a new σ bond. For this bond to form, the p orbitals at the ends of the conjugated system must rotate in order to overlap head-to-head as they rehybridize from sp^2 to sp^3. Rotation can occur in two ways. If both orbitals rotate in the same direction (both clockwise or both counterclockwise), ring closure is **conrotatory.**

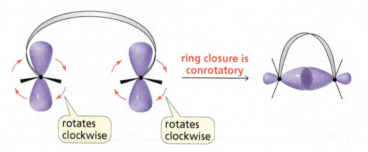

ring closure is conrotatory

rotates clockwise

rotates clockwise

If the orbitals rotate in opposite directions (one clockwise, the other counterclockwise), ring closure is **disrotatory.**

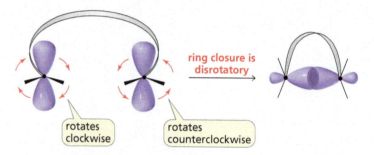

ring closure is disrotatory

rotates clockwise

rotates counterclockwise

Whether conrotatory or disrotatory ring closure occurs depends only on the symmetry of the HOMO of the compound undergoing ring closure. The symmetry of the HOMO determines the course of the reaction because this is where the highest energy electrons are. These are the most loosely held electrons and, therefore, the ones most easily moved during a reaction.

Symmetry-Allowed and Symmetry-Forbidden Pathways

To form the new σ bond, the orbitals must rotate so that in-phase p orbitals overlap. (Recall that in-phase overlap is a bonding interaction, whereas out-of-phase overlap would be an antibonding interaction.) If the HOMO is symmetric (the end orbitals are identical), rotation will have to be disrotatory to achieve in-phase overlap. In other words, disrotatory ring closure is symmetry-allowed, whereas conrotatory ring closure is symmetry-forbidden.

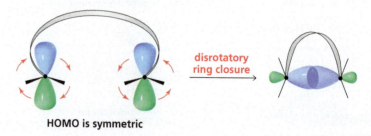

disrotatory ring closure

HOMO is symmetric

On the other hand, if the HOMO is antisymmetric, rotation has to be conrotatory in order to achieve in-phase overlap. In other words, conrotatory ring closure is symmetry-allowed, whereas disrotatory ring closure is symmetry-forbidden.

HOMO is antisymmetric

A **symmetry-allowed pathway** is one in which in-phase orbitals overlap, and a **symmetry-forbidden pathway** is one in which out-of-phase orbitals would overlap.

A symmetry-allowed pathway is one in which in-phase orbitals overlap.

- A symmetry-allowed reaction can take place under relatively mild conditions.
- A symmetry-forbidden reaction cannot take place by a concerted mechanism.

Explaining the Configuration of the Product(s) of an Electrocyclic Reaction

Now we are ready to learn why the electrocyclic reactions discussed at the beginning of this section form the products shown, and why the configuration of the product changes if the reaction conditions change from thermal to photochemical.

Example 1

Let's first look at the electrocyclic reaction of (2E,4Z,6E)-octatriene. The ground-state HOMO (ψ_3) of a compound with three conjugated π bonds is symmetric (Figure 28.3). This means that ring closure under *thermal conditions* is disrotatory. In disrotatory ring closure of (2E,4Z,6E)-octatriene, the methyl groups are both pushed up (or down), which results in formation of the cis product.

(2E,4Z,6E)-octatriene   ***cis*-5,6-dimethyl-1,3-cyclohexadiene**

Example 2

Now let's look at the electrocyclic reaction of (2E,4Z,6Z)-octatriene. Because its ground-state HOMO is also symmetric, it, too, undergoes disrotatory ring closure. In disrotatory ring closure, one methyl group is pushed up and the other is pushed down, which results in formation of the trans product. The enantiomer is obtained by reversing the groups that are pushed up and down.

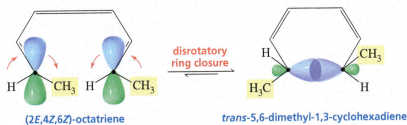

(2E,4Z,6Z)-octatriene   ***trans*-5,6-dimethyl-1,3-cyclohexadiene**

Example 3

If the electrocyclic reaction of the compound we just looked at in Example 2 takes place under *photochemical conditions*, we must consider the excited-state HOMO rather than the ground-state HOMO. The excited-state HOMO (ψ_4) of a compound with three π bonds is antisymmetric

(Figure 28.3). Therefore, under photochemical conditions, (2E,4Z,6Z)-octatriene undergoes conrotatory ring closure, so both methyl groups are pushed down (or up) and the cis product is formed.

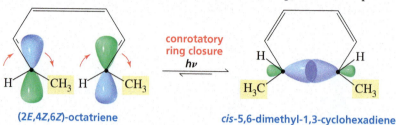

(2E,4Z,6Z)-octatriene   **cis-5,6-dimethyl-1,3-cyclohexadiene**

Similarly, (2E,4Z,6E)-octatriene, the compound in Example 1 that forms the cis isomer under thermal conditions, will form the trans enantiomers under photochemical conditions.

> **The symmetry of the HOMO of the compound undergoing ring closure controls the stereochemical outcome of an electrocyclic reaction.**

Now we can understand why the configuration of the product formed under photochemical conditions is the opposite of the configuration of the product formed under thermal conditions: the ground-state and excited-state HOMOs have opposite symmetries—that is, if one is symmetric, the other is antisymmetric. And the stereochemical outcome of an electrocyclic reaction depends only on the symmetry of the HOMO undergoing ring closure.

Example 4

Now let's see why ring closure of (2E,4Z)-hexadiene forms *cis*-3,4-dimethylcyclobutene. The compound undergoing ring closure has two conjugated π bonds. The ground-state HOMO of a compound with two conjugated π bonds is antisymmetric (Figure 28.2), so ring closure is conrotatory. Conrotatory ring closure of (2E,4Z)-hexadiene leads to the cis product.

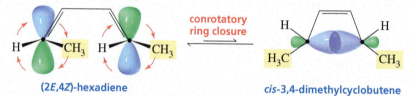

(2E,4Z)-hexadiene   **cis-3,4-dimethylcyclobutene**

Similarly, conrotatory ring closure of (2E,4E)-hexadiene leads to the trans product.

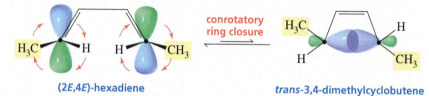

(2E,4E)-hexadiene   **trans-3,4-dimethylcyclobutene**

If the reaction is carried out under photochemical conditions, however, the excited-state HOMO of a compound with two conjugated π bonds is symmetric. Therefore, (2E,4Z)-hexadiene will undergo disrotatory ring closure and form the trans product, and (2E,4E)-hexadiene will undergo disrotatory ring closure and form the cis product.

Woodward–Hoffmann Rules for Electrocyclic Reactions

We saw that the ground-state HOMO of a compound with two conjugated double bonds is antisymmetric, whereas the ground-state HOMO of a compound with three conjugated double bonds is symmetric. If we examine the molecular orbital diagrams for compounds with four, five, six, and more conjugated double bonds, we will see that

> **The ground-state HOMO of a compound with an even number of conjugated double bonds is antisymmetric.**

> **The ground-state HOMO of a compound with an odd number of conjugated double bonds is symmetric.**

the ground-state HOMO of a compound with an even number of conjugated double bonds is antisymmetric, whereas the ground-state HOMO of a compound with an odd number of conjugated double bonds is symmetric.

Therefore, just knowing the number of conjugated double bonds in a compound, you can tell whether ring closure will be conrotatory (an even number of conjugated double bonds) or disrotatory (an odd number of conjugated double bonds) under thermal conditions. However, if the reaction takes place under photochemical conditions, everything is reversed because the ground-state and excited-state HOMOs have opposite symmetries; if the ground-state HOMO is symmetric, the excited-state HOMO is antisymmetric and vice versa.

What you have learned about electrocyclic reactions can be summarized by the **selection rules** listed in Table 28.1. These are also known as the **Woodward–Hoffmann rules** for electrocyclic reactions. The rules show that the mode of ring closure depends on the number of conjugated double bonds in the reactant *and* on whether the reaction is carried out under thermal or photochemical conditions. And once you know the mode of ring closure, you can determine the products of an electrocyclic reaction.

The selection rules in Table 28.1 tell us the symmetry-allowed mode of ring closure for electrocyclic reactions. We will see that there also are selection rules that tell us the symmetry-allowed mode of bond formation for cycloaddition reactions (Table 28.3 on page 1227) and the symmetry-allowed mode of rearrangement for sigmatropic rearrangements (Table 28.4 on page 1229). It can be rather burdensome to memorize these rules (and worrisome if they are forgotten during an exam), but they all can be summarized by the mnemonic TE-AC. How to use TE-AC is explained in Section 28.7.

Table 28.1 Woodward–Hoffmann Rules for Electrocyclic Reactions

Number of conjugated π bonds	Reaction conditions	Allowed mode of ring closure
Even number	Thermal	Conrotatory
	Photochemical	Disrotatory
Odd number	Thermal	Disrotatory
	Photochemical	Conrotatory

PROBLEM 5

a. For conjugated systems with two, three, four, five, six, and seven conjugated π bonds, construct quick MOs (just draw the lobes at the ends of the conjugated system as they are drawn on pages 1220 and 1221) to show whether the HOMO is symmetric or antisymmetric.
b. Using these drawings, convince yourself that the Woodward–Hoffmann rules in Table 28.1 are valid.

PROBLEM 6 ◆

a. Under thermal conditions, will ring closure of (2E,4Z,6Z,8E)-2,4,6,8-decatetraene be conrotatory or disrotatory?
b. Will the product have the cis or the trans configuration?
c. Under photochemical conditions, will ring closure be conrotatory or disrotatory?
d. Will the product have the cis or the trans configuration?

Using the Woodward–Hoffmann Rules for Electrocyclic Reactions

The series of reactions in Figure 28.4 illustrates how to determine the mode of ring closure and, therefore, the product of an electrocyclic reaction.

The first reaction has a reactant with three conjugated double bonds, which is undergoing ring closure under thermal conditions. Ring closure, therefore, is disrotatory (Table 28.1). Disrotatory ring closure of this reactant causes the substituents at the end of the π system (in this case hydrogens) to be cis in the ring-closed product. To determine the relative positions of the hydrogens, draw them in the reactant (A in Figure 28.4) and then draw arrows showing disrotatory ring closure.

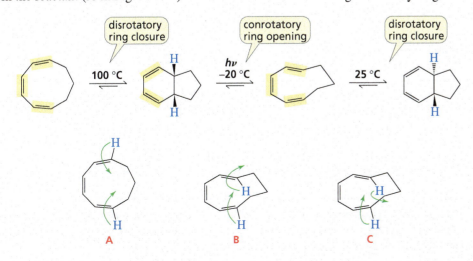

◀ **Figure 28.4**
Determining the stereochemistry of the product of an electrocyclic reaction.

The second reaction is a ring-opening electrocyclic reaction that takes place under photochemical conditions. The orbital symmetry rules used for a ring-closure reaction also apply to the reverse ring-opening reaction. The compound undergoing the reverse ring-closure reaction has three conjugated double bonds. The reaction occurs under photochemical conditions, so both ring opening and the reverse ring closure are conrotatory. (Notice that the number of conjugated double bonds we use to determine the mode of ring opening and ring closure is the number in the compound that is undergoing ring closure.) If conrotatory rotation is to result in a ring-closed product with cis hydrogens (shown by solid wedges), the hydrogens in the compound undergoing ring closure must point in the same direction (B in Figure 28.4).

The third reaction is a ring closure of a compound with three conjugated double bonds under thermal conditions, so ring closure is disrotatory. The hydrogens in the reactant point in the same direction (C in Figure 28.4), so disrotatory ring closure will cause them to be trans in the ring-closed product.

Notice that in all these electrocyclic reactions, if the bonds to the substituents (in this case, hydrogens) in the reactant point in *opposite directions* (as in A in Figure 28.4), the substituents will be cis in the product if ring closure is disrotatory and trans if ring closure is conrotatory. On the other hand, if the bonds point in the *same direction* (as in B or C in Figure 28.4), the substituents will be trans in the product if ring closure is disrotatory and cis if ring closure is conrotatory (Table 28.2).

Table 28.2 Configuration of the Product of an Electrocyclic Reaction

Substituents in the reactant	Mode of ring closure	Configuration of the product
Point in opposite directions	Disrotatory	cis
	Conrotatory	trans
Point in the same direction	Disrotatory	trans
	Conrotatory	cis

PROBLEM 7 ◆

Which of the following are correct? Correct any false statements.

a. A conjugated diene with an even number of double bonds undergoes conrotatory ring closure under thermal conditions.

b. A conjugated diene with an antisymmetric HOMO undergoes conrotatory ring closure under thermal conditions.

c. A conjugated diene with an odd number of double bonds has a symmetric HOMO.

PROBLEM 8 ◆

a. Identify the mode of ring closure for each of the following electrocyclic reactions.

b. Are the indicated hydrogens cis or trans?

28.4 CYCLOADDITION REACTIONS

In a *cycloaddition reaction,* two π-bond-containing reactants form a cyclic compound by rearranging their π electrons and forming two new σ bonds. The Diels–Alder reaction is the best-known example of a cycloaddition reaction (Section 8.14).

Classifying Cycloaddition Reactions

Cycloaddition reactions are classified according to the number of π electrons that interact to produce the product. The Diels–Alder reaction is a [4 + 2] cycloaddition reaction because one reactant has four interacting π electrons and the other reactant has two. Only the π electrons that participate in the electron rearrangement are counted.

[4 + 2] cycloaddition (a Diels–Alder reaction)

[2 + 2] cycloaddition

[8 + 2] cycloaddition

Suprafacial and Antarafacial Bond Formation

In a cycloaddition reaction, the new σ bonds in the product are formed by donation of electron density from one reactant to the other reactant. Because only an empty orbital can accept electrons, we must consider the HOMO of one of the reactants and the LUMO of the other. It does not matter which reactant's HOMO is used as long as electron donation occurs between the HOMO of one and the LUMO of the other.

There are two modes of orbital overlap for the simultaneous formation of two σ bonds, suprafacial and antarafacial. In **suprafacial** bond formation, both σ bonds form on the same side of the π system; in **antarafacial** bond formation, the two σ bonds form on opposite sides of the π system. Suprafacial bond formation is similar to syn addition, whereas antarafacial bond formation resembles anti addition (Section 6.13).

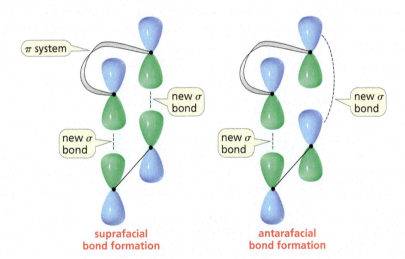

A cycloaddition reaction that forms a four-, five-, or six-membered ring *must* occur by suprafacial bond formation. The geometric constraints of these small rings make the antarafacial approach highly unlikely even if it is symmetry-allowed. (Remember that symmetry-allowed means the overlapping orbitals are in-phase.) Antarafacial bond formation is more likely in cycloaddition reactions that form larger rings.

[4 + 2] Cycloaddition Reactions

Figure 28.5 shows the required suprafacial orbital overlap for σ bond formation in a [4 + 2] cycloaddition reaction. Either the HOMO of the diene (a system with two conjugated double bonds) and the LUMO of the dienophile (a system with one double bond; shown on the left in Figure 28.5) or the HOMO of the dienophile and the LUMO of the diene (shown on the right in Figure 28.5) can be used to explain the reaction.

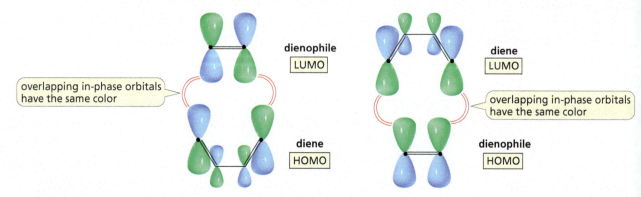

▲ **Figure 28.5**
Frontier molecular orbital analysis of a [4 + 2] cycloaddition reaction, which requires suprafacial overlap for bond formation. The HOMO of either of the reactants can be used with the LUMO of the other.

[2 + 2] Cycloaddition Reactions

A [2 + 2] cycloaddition reaction does not occur under thermal conditions but does take place under photochemical conditions.

The frontier molecular orbitals in Figure 28.6 show why this is so. Under thermal conditions, suprafacial overlap is not symmetry-allowed (the overlapping orbitals are out-of-phase). Antarafacial overlap is symmetry-allowed but is not possible because of the small size of the ring.

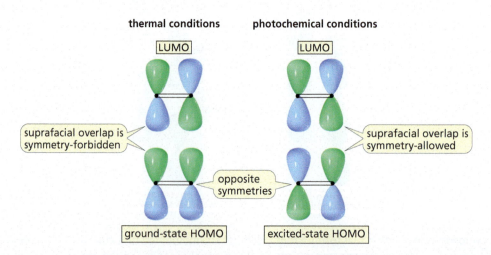

◀ **Figure 28.6**
Frontier molecular orbital analysis of a [2 + 2] cycloaddition reaction under thermal and photochemical conditions.

Under photochemical conditions, however, the reaction can take place because the symmetry of the excited-state HOMO is the opposite of the ground-state HOMO. Therefore, overlap of the excited-state HOMO of one alkene with the LUMO of the second alkene involves symmetry-allowed suprafacial bond formation.

Woodward–Hoffmann Rules for Cycloaddition Reactions

The selection rules for cycloaddition reactions are summarized in Table 28.3.

Table 28.3 Woodward–Hoffmann Rules for Cycloaddition Reactions

Sum of the number of bonds in the reacting systems of both reagents	Reaction conditions	Allowed mode of bond formation
Even number	Thermal	Antarafacial*
	Photochemical	Suprafacial
Odd number	Thermal	Suprafacial
	Photochemical	Antarafacial*

*Although antarafacial ring closure is symmetry-allowed for the indicated conditions, it can occur only with rings that have seven or more ring atoms.

PROBLEM 9 SOLVED

LEARN THE STRATEGY

Compare the reaction between 2,4,6-cycloheptatrienone and cyclopentadiene to the reaction between 2,4,6-cycloheptatrienone and ethene. Why does 2,4,6-cycloheptatrienone use two π electrons in one reaction and four π electrons in the other?

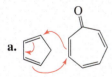

SOLUTION Both reactions are [4 + 2] cycloaddition reactions. When 2,4,6-cycloheptatrienone reacts with cyclopentadiene, it uses two of its π electrons because cyclopentadiene is the four-electron reactant. When 2,4,6-cycloheptatrienone reacts with ethene, it uses four of its π electrons because ethene is the two-electron reactant.

PROBLEM 10

USE THE STRATEGY

Explain why maleic anhydride reacts rapidly with 1,3-butadiene but does not react at all with ethene under thermal conditions.

maleic anhydride

PROBLEM 11 ♦

Will a concerted reaction take place between 1,3-butadiene and 2-cyclohexenone in the presence of ultraviolet light?

28.5 SIGMATROPIC REARRANGEMENTS

The third kind of pericyclic reaction we will consider is the group of reactions known as *sigmatropic rearrangements*. In a sigmatropic rearrangement, a σ bond in the reactant breaks, a new σ bond forms, and the π electrons rearrange.

Describing Sigmatropic Rearrangements

Some examples of sigmatropic rearrangements are shown below. Notice that the σ bond that breaks is a bond to an allylic carbon. It can be a σ bond between a carbon and a hydrogen, between a carbon and another carbon, or between a carbon and an oxygen, nitrogen, or sulfur. "Sigmatropic" comes from the Greek word *tropos,* which means "change." Thus, sigmatropic means "sigma-change."

LEARN THE STRATEGY
The numbering system used to describe a sigmatropic rearrangement differs from any numbering system you have seen previously. First, mentally break the σ bond to the allylic carbon in the reactant and label both atoms that were attached by the bond as 1. Then look at the new σ bond in the product. Count the number of atoms in each of the fragments that connect the broken σ bond and the new σ bond. The two numbers are put in brackets, with the smaller number written first. The following is a [2,3] sigmatropic rearrangement, because two atoms (N$=$N) connect the old and new σ bonds in one fragment and three atoms (C$-$C$=$C) connect the old and new σ bonds in the other fragment.

a [2,3] sigmatropic rearrangement

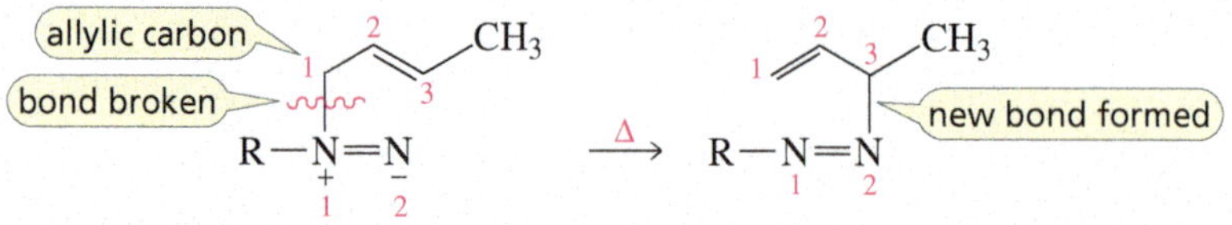

a [1,5] sigmatropic rearrangement

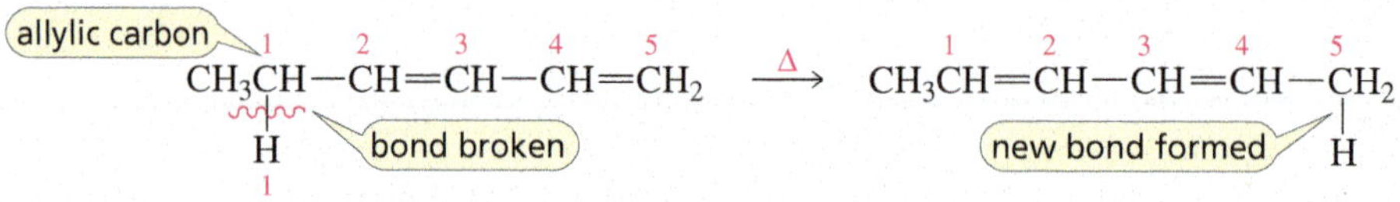

a [1,3] sigmatropic rearrangement

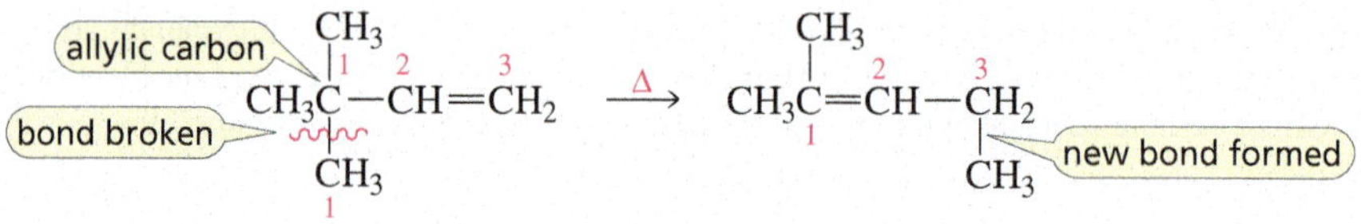

a [3,3] sigmatropic rearrangement

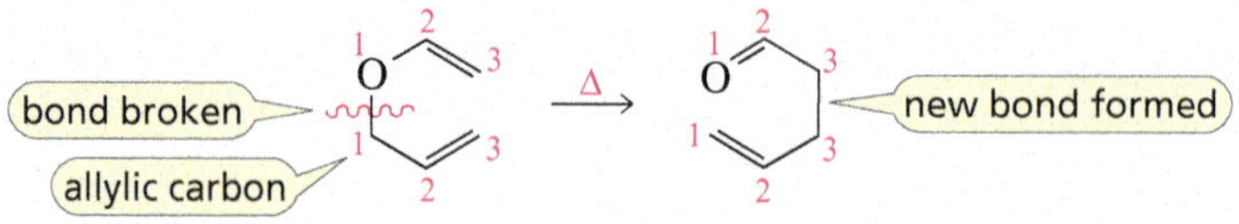

Notice that in the preceding examples, each reaction starts by breaking a bond to an allylic carbon.

USE THE STRATEGY

PROBLEM 12

a. Name the kind of sigmatropic rearrangement that occurs in each of the following reactions.
b. Using arrows, show the electron rearrangement that takes place in each reaction.

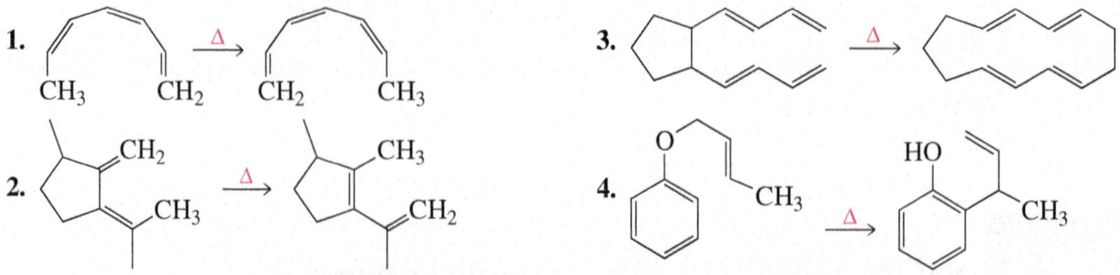

Suprafacial and Antarafacial Rearrangement

In the transition state of a sigmatropic rearrangement, the group that migrates is partially bonded to the migration origin and partially bonded to the migration terminus. There are two possible modes for rearrangement, analogous to those in cycloaddition reactions. The rearrangement is *suprafacial* f the migrating group remains on the same face of the π system; the rearrangement is *antarafacial* f the migrating group moves to the opposite face of the π system. If the transition state has six or fewer atoms in the ring, rearrangement must be suprafacial because of the geometric constraints of small rings.

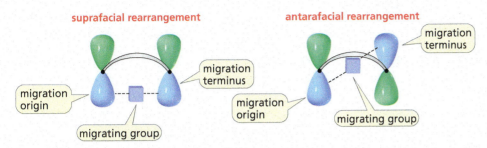

Notice that sigmatropic rearrangements have cyclic transition states.

Woodward–Hoffmann Rules for Sigmatropic Rearrangements

We can describe a [1,3] sigmatropic rearrangement as involving a π bond and a pair of σ electrons, or we can say that it involves two pairs of electrons. A [1,5] sigmatropic rearrangement involves two π bonds and a pair of σ electrons (three pairs of electrons), and a [1,7] sigmatropic rearrangement involves four pairs of electrons. The symmetry rules for sigmatropic rearrangements are nearly the same as those for cycloaddition reactions. The only difference is that we count the number of pairs of electrons rather than the number of π bonds. (Compare Tables 28.3 and 28.4.)

Table 28.4 Woodward–Hoffmann Rules for Sigmatropic Rearrangements

Number of pairs of electrons in the reacting system	Reaction conditions	Allowed mode of rearrangement
Even number	Thermal	Antarafacial*
	Photochemical	Suprafacial
Odd number	Thermal	Suprafacial
	Photochemical	Antarafacial*

*Although antarafacial rearrangement is symmetry-allowed for the indicated conditions, it can occur only with rings that have at least seven ring atoms.

Cope and Claisen Rearrangements

A **Cope rearrangement** is a [3,3] sigmatropic rearrangement of a 1,5-diene. A **Claisen rearrangement** is a [3,3] sigmatropic rearrangement of an allyl vinyl ether. Both rearrangements form six-membered-ring transition states. The reactions, therefore, must be able to take place by a suprafacial pathway. We saw that whether or not a suprafacial pathway is symmetry-allowed depends on the number of pairs of electrons involved in the rearrangement (Table 28.4). Because [3,3] sigmatropic rearrangements involve three pairs of electrons, they occur by a suprafacial pathway under thermal conditions. Therefore, both the Cope and Claisen rearrangements readily take place under thermal conditions.

The Ireland–Claisen rearrangement uses an allyl ester instead of the allyl vinyl ether used in the Claisen rearrangement. A base removes a proton from the α-carbon of the ester, and the enolate ion is trapped as a trimethylsilyl ether as a result of reacting with trimethylsilylchloride (Section 16.10). Mild heating results in a Claisen rearrangement.

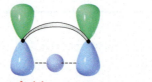

an allyl ester

PROBLEM 13 ◆

a. Draw the product of the following reaction:

b. If the terminal sp^2 carbon of the substituent attached to the benzene ring is labeled with ^{14}C, where will the label be in the product?

Migration of Hydrogen

When a hydrogen migrates in a sigmatropic rearrangement, the *s* orbital of the hydrogen is partially bonded to both the migration origin and the migration terminus in the transition state.

migration of hydrogen

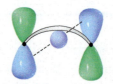

suprafacial rearrangement   **antarafacial rearrangement**

A [1,3] sigmatropic migration of hydrogen has a four-membered-ring transition state (see the reaction below). Because two pairs of electrons are involved, the selection rules require an antarafacial rearrangement under thermal conditions, since the HOMO is antisymmetric (Table 28.4). Consequently, 1,3-hydrogen shifts do not occur under thermal conditions because the four-membered-ring transition state does not allow antarafacial rearrangement.

1,3-Hydrogen shifts can take place if the reaction is carried out under photochemical conditions because the HOMO will be symmetric, so suprafacial rearrangement can occur (Table 28.4).

1,3-hydrogen shifts

Two products are obtained in the preceding reaction because the reactant has two different allylic hydrogens that can undergo a 1,3-hydrogen shift.

[1,5] Sigmatropic migrations of hydrogen are well known. They involve three pairs of electrons, so they take place by a suprafacial pathway under thermal conditions.

1,5-hydrogen shifts

[1,7] Sigmatropic hydrogen migrations involve four pairs of electrons. They can take place under thermal conditions because the eight-membered-ring transition state allows the required antarafacial rearrangement.

1,7-hydrogen shift

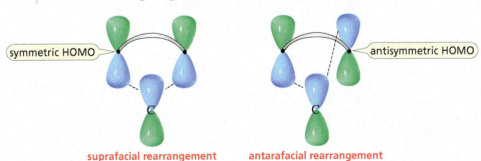

PROBLEM 14 ◆

Why was a deuterated compound used in the last reaction on the preceding page?

PROBLEM 15

Account for the difference in the products obtained under photochemical and thermal conditions:

PROBLEM 16 SOLVED

Show how 5-methyl-1,3-cyclopentadiene rearranges to form 1-methyl-1,3-cyclopentadiene and 2-methyl-1,3-cyclopentadiene.

SOLUTION Notice that both equilibria involve [1,5] sigmatropic rearrangements. Although a hydrogen moves from one carbon to an adjacent carbon, the rearrangements are not considered 1,2-shifts because these would not account for all the atoms involved in the rearranged π electron system.

Migration of Carbon

Unlike hydrogen that can migrate in only one way because of its spherical *s* orbital, carbon has two ways to migrate because it has a two-lobed *p* orbital. Carbon can simultaneously interact with the migration origin and the migration terminus using one lobe of its *p* orbital.

carbon migrating with one lobe of its p orbital interacting

symmetric HOMO   antisymmetric HOMO

suprafacial rearrangement   antarafacial rearrangement

Or, carbon can simultaneously interact with the migration source and the migration terminus using both lobes of its *p* orbital.

carbon migrating with both lobes of its *p* orbital interacting

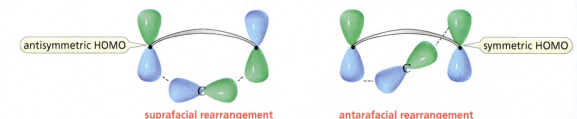

suprafacial rearrangement   antarafacial rearrangement

The drawings show that if the reaction requires a suprafacial rearrangement, carbon will migrate using one lobe of its *p* orbital if the HOMO is symmetric and will use both lobes if the HOMO is antisymmetric.

When carbon migrates using only one lobe of its *p* orbital, the migrating group retains its configuration because bonding is always to the same lobe. When carbon migrates using both lobes of its *p* orbital, bonding in the reactant and bonding in the product involve different lobes. Therefore, migration occurs with inversion of configuration.

The following [1,3] sigmatropic rearrangement has a four-membered-ring transition state that requires a suprafacial pathway. The reacting system has two pairs of electrons, so its HOMO is antisymmetric. Therefore, the migrating carbon uses both lobes of its *p* orbital, and, as a result, it undergoes inversion of configuration.

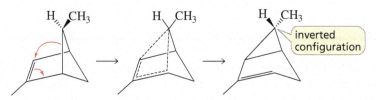

inverted configuration

PROBLEM 17

Explain why [1,3] sigmatropic migrations of hydrogen cannot occur under thermal conditions, but [1,3] sigmatropic migrations of carbon can.

PROBLEM 18 ◆

a. Will thermal 1,3-migrations of carbon occur with retention or inversion of configuration?
b. Will thermal 1,5-migrations of carbon occur with retention or inversion of configuration?

28.6 PERICYCLIC REACTIONS IN BIOLOGICAL SYSTEMS

Now we will look at some pericyclic reactions that occur in cells.

A Biological Cycloaddition Reaction

Exposure to ultraviolet light can cause skin cancer. This is one of the reasons many scientists have been concerned about the thinning ozone layer, because ozone protects organisms on the surface of Earth by absorbing ultraviolet radiation (Section 12.12).

One cause of skin cancer is the formation of *thymine dimers*. At any point in DNA where there are two adjacent thymines (Section 26.1), a [2 + 2] cycloaddition reaction can occur, which forms a thymine dimer. Because [2 + 2] cycloaddition reactions take place only under photochemical conditions, the reaction takes place only in the presence of ultraviolet light. Thymine dimers can cause cancer because they interfere with the structural integrity of DNA, which can lead to mutations and then to cancer.

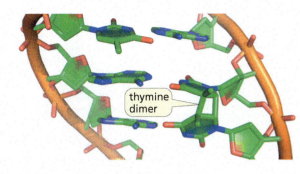

two adjacent thymine residues in DNA → mutation-causing thymine dimer

a segment of DNA

Fortunately, there is an enzyme (called DNA photolyase) that repairs damaged DNA. When the enzyme recognizes a thymine dimer, it reverses the [2 + 2] cycloaddition reaction to regenerate the original two thymines. A repair enzyme, however, is not perfect, and some damage always remains uncorrected. Rarely do people who do not have this repair enzyme live beyond the age of 20. Fortunately, this genetic defect is rare.

Bioluminescence

Bioluminescence is the emission of light by a living species. Fireflies are one of several species that luminesce (emit cold light) as a result of a reverse [2 + 2] cycloaddition reaction. Fireflies have an enzyme (luciferase) that catalyzes the reaction between luciferin, ATP, and molecular oxygen to form a compound with an unstable four-membered ring.

The purpose of ATP is to activate the carboxylate group by giving it a good leaving group (Section 24.1). A base removes a proton from an α-carbon, forming a nucleophile that can react with O_2. A nucleophilic acyl substitution reaction forms a compound with an unstable four-membered ring.

When the four-membered ring breaks in a reverse [2 + 2] cycloaddition reaction, the strain in the ring is relieved and a stable molecule of CO_2 is formed. The reaction releases so much energy that an electron in oxyluciferin is promoted to an excited state. When the electron in the excited state drops down to the ground state, a photon of light is released. Other species that luminesce (jellyfish, glow worms, some species of bacteria and fungi) do so by a similar mechanism.

fireflies

luciferin — ATP, luciferase → + pyrophosphate

an unstable four-membered ring

reverse 2 + 2 cycloaddition

has an electron in an excited state

oxyluciferin + CO_2 → + light

A Biological Electrocyclic Reaction and Sigmatropic Rearrangemen

Vitamin D is a general name for vitamins D_2 and D_3. Their only structural difference is the doubl bond in the hydrocarbon chain attached to the five-membered ring in vitamin D_2 that vitamin D_3 lacks

Vitamin D is formed from precursor molecules in the skin (7-dehydrocholesterol for D_3 and ergos terol for D_2) as a result of two pericyclic reactions that occur when skin is exposed to sunlight. Th first is an electrocyclic reaction that opens one of the six-membered rings to form provitamin D_3 (o provitamin D_2). This reaction occurs only under photochemical conditions (why sunlight is require for the reaction to take place).

The second pericyclic reaction is a [1,7] sigmatropic rearrangement of the provitamin that result in the formation of vitamin D_3 (or vitamin D_2). The sigmatropic rearrangement takes place unde thermal conditions and is slower than the electrocyclic reaction that takes place under photochemica conditions, so the vitamins continue to be synthesized for several days after exposure to sunlight. Th active form of the vitamin (calcitriol) requires two successive hydroxylations of vitamins D_3 and D_2 The first occurs in the liver and the second in the kidneys.

ergosterol has a double bond here

electrocyclic reaction
$h\nu$

7-dehydrocholesterol
ergosterol

provitamin D_2 has a double bond here

provitamin D_3
provitamin D_2

[1,7] sigmatropic rearrangement

vitamin D_2 has a double bond here

hydroxylations

calcitriol
active form of vitamin D

cholecalciferol (vitamin D_3)
ergocalciferol (vitamin D_2)

The Sunshine Vitamin

Vitamin D is not found in food, but a few foods contain the precursor molecules. For example, 7-dehydrocholesterol is present in dairy products and fatty fish, and ergosterol is present in some vegetables. For this reason, all milk sold in the United States is enriched with vitamin D_3, produced by shining ultraviolet light on the milk to convert 7-dehydrocholesterol to vitamin D_3.

We just saw that sunlight converts precursor molecules in the skin to vitamins D_3 and D_2, but many people wear sunscreen that blocks the UV light necessary to synthesize the vitamin. It has been estimated that 50% to 75% of the population has suboptimal levels of vitamin D because of the wide use of sun screens.

It has long been known that vitamin D improves calcium absorption in the intestines. A deficiency in vitamin D causes rickets—a disease characterized by deformed bones and stunted growth. Recent research suggests that a deficiency in vitamin D may increase the risk of cardiovascular disease, hypertension, and diabetes.

Too much vitamin D is harmful because it causes calcification of soft tissues. It is thought that skin pigmentation evolved to protect skin from the sun's UV rays in order to prevent the synthesis of too much vitamin D. This is consistent with the observation that peoples indigenous to coun- tries close to the equator have greater skin pigmentation.

Animals, Birds, Fish—And Vitamin D

Animals covered with fur and birds covered with feathers are not able to get vitamin D from sunlight because they do not have sun-exposed skin. They obtain vitamin D by excreting precursor molecules in their fur or feathers that sunlight converts to vitamin D. These are the same precursor molecules used to synthesize vitamin D in humans. The vitamin D stays on the fur or feathers, so when these animals groom themselves by licking their fur or plucking their feathers, they ingest the vitamin.

Obviously, fish also do not get vitamin D from sunlight. They get all their vitamin D from their diet (such as plankton) that they store in their livers, which makes them a good source of vitamin D for other species.

cat licking a paw

PROBLEM 19 ♦

Does the [1,7] sigmatropic rearrangement that converts provitamin D_3 to vitamin D_3 involve suprafacial or antarafacial rearrangement?

PROBLEM 20

Explain why the hydrogen and the methyl substituent are trans to one another after photochemical ring closure of provitamin D_3 to form 7-dehydrocholesterol.

PROBLEM 21 ♦

Chorismate mutase is an enzyme that promotes a pericyclic reaction by forcing the substrate to assume the conformation needed for the reaction. The product of the pericyclic reaction is prephenate that is subsequently converted into the amino acids phenylalanine and tyrosine. What kind of a pericyclic reaction does chorismate mutase catalyze?

chorismate → (chorismate mutase) → prephenate

28.7 SUMMARY OF THE SELECTION RULES FOR PERICYCLIC REACTIONS

The selection rules that determine the outcome of electrocyclic reactions, cycloaddition reactions, and sigmatropic rearrangements are summarized in Tables 28.1, 28.3, and 28.4, respectively. This is still a lot to remember. Fortunately, the selection rules for all pericyclic reactions can be summarized by TE-AC. How to use TE-AC is described below.

- If TE (Thermal/Even) describes the reaction, the outcome *is given* by AC (Antarafacial or Conrotatory).

- If *both* of the letters of TE are different (Photochemical/Odd), the outcome *is still given* by AC (Antarafacial or Conrotatory).

- If *one* of the letters of TE is different (the reaction is not Thermal/Even but is Thermal/Odd or Photochemical/Even), the outcome *is not given* by AC (that is, the outcome is Suprafacial or Disrotatory).

PROBLEM 22

Convince yourself that the TE-AC method for learning the information in Tables 28.1, 28.3, and 28.4 is valid.

ESSENTIAL CONCEPTS

Section 28.1

- A **pericyclic reaction** is one in which the electrons in the reactant(s) are reorganized in a cyclic manner.
- The three most common types of pericyclic reactions are *electrocyclic reactions*, *cycloaddition reactions*, and *sigmatropic rearrangements*.
- Pericyclic reactions are concerted, highly stereoselective reactions that are generally not affected by catalysts or by a change in solvent.
- The configuration of the product of a pericyclic reaction depends on the configuration of the reactant, the number of conjugated double bonds or pairs of electrons in the reacting system, and whether the reaction is **thermal or photochemical.**
- The **conservation of orbital symmetry theory** states that in-phase orbitals overlap during the course of a pericyclic reaction.

Section 28.2

- The two lobes of a *p* orbital have opposite phases. When two in-phase orbitals interact, a covalent bond is formed; two out-of-phase orbitals interact to create a node.
- If the lobes at the ends of the MO are in-phase, the MO is **symmetric;** if they are out-of-phase, the MO is **antisymmetric.**
- The **ground-state HOMO** of a compound with an even number of conjugated double bonds or an even number of pairs of electrons is antisymmetric; the ground-state HOMO of a compound with an odd number of conjugated double bonds or an odd number of pairs of electrons is symmetric.
- If a molecule absorbs light of an appropriate wavelength, the light will promote an electron from its ground state **HOMO** to its **LUMO.** The molecule is then in an excited state.
- The ground-state HOMO and the excited-state HOMO have opposite symmetries.
- A HOMO and LUMO are **frontier orbitals.**
- In a thermal reaction, the reactant is in its ground state; in a photochemical reaction the reactant is in an excited state.

Section 28.3

- An **electrocyclic reaction** is an intramolecular reaction in which a new σ bond is formed between the ends of a conjugated π system.

- To form the new σ bond, the orbitals at the ends of the conjugated system rotate so they can engage in in-phase overlap.
- If both orbitals rotate in the same direction, ring closure is **conrotatory;** if they rotate in opposite directions, it is **disrotatory.**
- A **symmetry-allowed pathway** is one in which in-phase orbitals overlap.
- If the HOMO is antisymmetric, conrotatory ring closure occurs; if it is symmetric, disrotatory ring closure occurs.

Section 28.4

- In a **cycloaddition reaction,** two different π-bond containing molecules react to form a cyclic compound by rearranging the π electrons and forming two new σ bonds.
- Bond formation is **suprafacial** if both σ bonds form on the same side of the π system; it is **antarafacial** if the two bonds form on opposite sides of the π system.
- Formation of rings with fewer than seven ring atoms requires suprafacial overlap.

Section 28.5

- In a **sigmatropic rearrangement,** a σ bond to an allylic carbon breaks in the reactant, a new σ bond forms in the product, and the π bonds rearrange.
- If the migrating group remains on the same face of the π system, the rearrangement is **suprafacial;** if it moves to the opposite face of the π system, it is **antarafacial.**

Section 28.6

- A [2 + 2] cycloaddition reaction between two adjacent thymines in DNA forms a mutation-causing **thymine dimer** that can lead to skin cancer.
- Vitamin D is made when skin is exposed to sunlight. The precursor molecules in the skin undergo an electrocyclic reaction under photochemical conditions followed by a sigmatropic rearrangement under thermal conditions and two hydroxylations.

Section 28.7

- The outcome of pericyclic reactions is given by a set of **selection rules,** which can be summarized by **TE-AC.**

PROBLEMS

23. Draw the product of each of the following reactions:

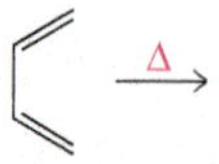

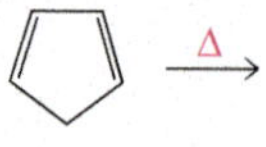

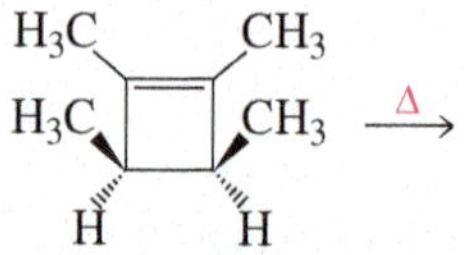

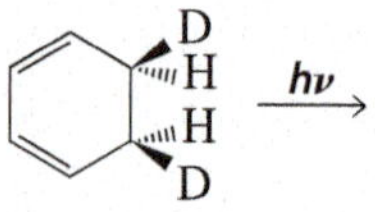

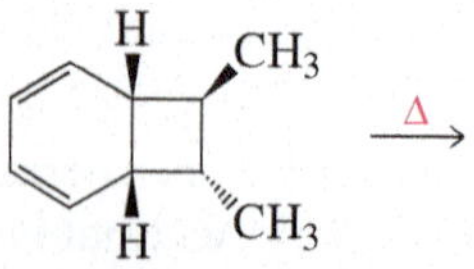

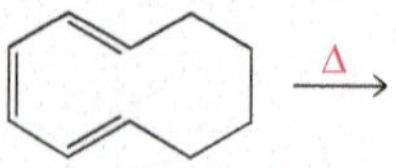

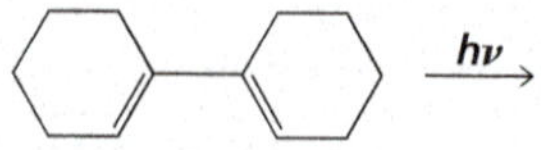

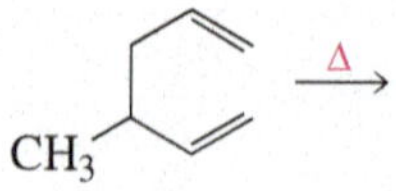

24. Draw the product of each of the following reactions:

a. 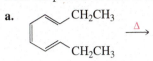 CH_2CH_3 / CH_2CH_3 $\xrightarrow{\Delta}$

c. CH_2CH_3 / CH_2CH_3 $\xrightarrow{h\nu}$

b. CH_2CH_3 / CH_2CH_3 $\xrightarrow{\Delta}$

d. CH_2CH_3 / CH_2CH_3 $\xrightarrow{h\nu}$

25. Account for the difference in the products of the following reactions:

26. Show how norbornane can be prepared from cyclopentadiene.

norbornane

27. Show how the reactant can be converted to the product in two steps.

28. Draw the product formed when each of the following compounds undergoes an electrocyclic reaction

a. under thermal conditions. **b.** under photochemical conditions.

1. CH_3 / CH_3 $\longrightarrow$

2. CH_3 / H_3C $\longrightarrow$

29. Draw the product of each of the following reactions:

a. $\xrightarrow{h\nu}$

c. OH $\xrightarrow{\Delta}$

e. CH_3 / O $\xrightarrow{\Delta}$

b. CH_3 / CH_3 $\xrightarrow{\Delta}$

d. O $\xrightarrow{\Delta}$

30. Could the reactions shown here take place by a concerted mechanism?

a.

b.

31. Dewar benzene is a highly strained isomer of benzene. In spite of its thermodynamic instability, it is very stable kinetically. It will slowly rearrange to benzene, but only if heated to a very high temperature. Why is it kinetically stable?

$\xrightarrow[\Delta]{\text{very slow}}$

Dewar benzene

32. What is the product of the following [1,3] sigmatropic rearrangement, **A** or **B**?

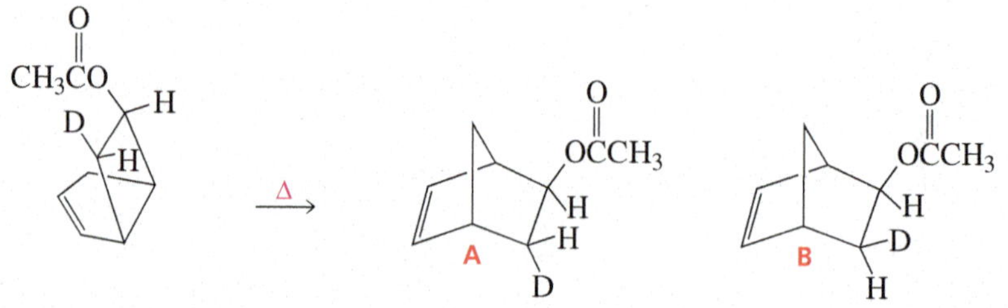

33. If the compounds shown here are heated, one will form one product from a [1,3] sigmatropic rearrangement and the other will form two products from two different [1,3] sigmatropic rearrangements. Draw the products of the reactions.

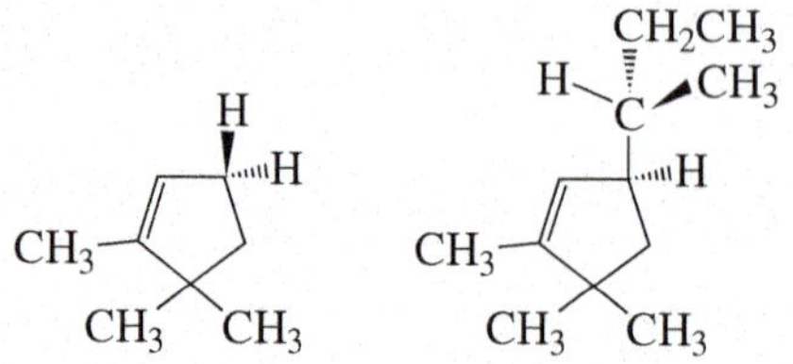

34. When the following compound is heated, a product is formed that shows an infrared absorption band at $1715\,\text{cm}^{-1}$. Draw the structure of the product.

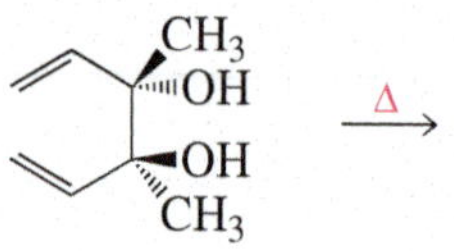

35. Two products are formed in the following [1,7] sigmatropic rearrangement, one due to hydrogen migration and the other to deuterium migration. Show the configuration of the products by replacing A and B with the appropriate atoms (H or D).

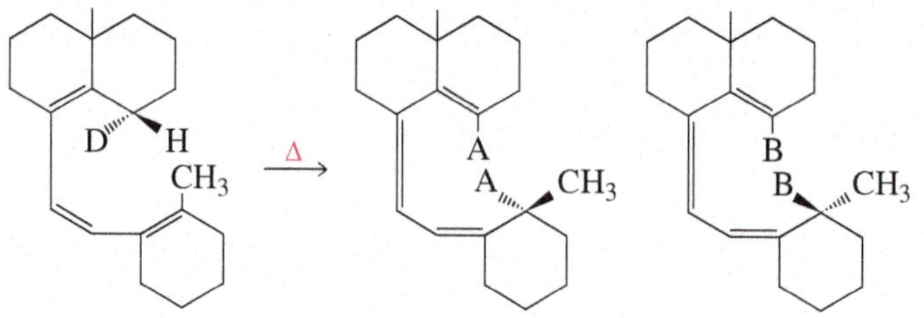

36. Propose a mechanism for the following reaction:

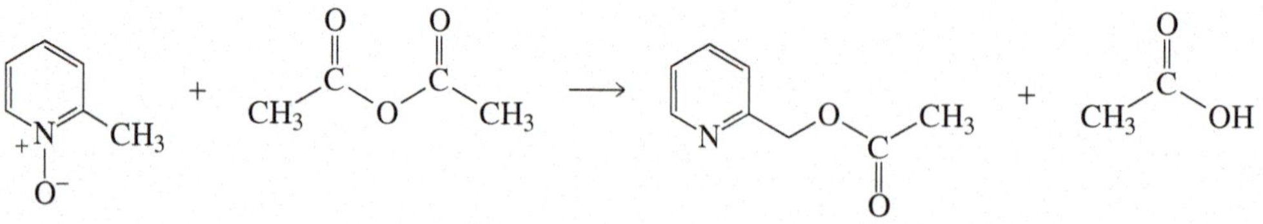

37. Draw the product of each of the following sigmatropic rearrangements:

a. [3,3] sigmatropic rearrangement Δ

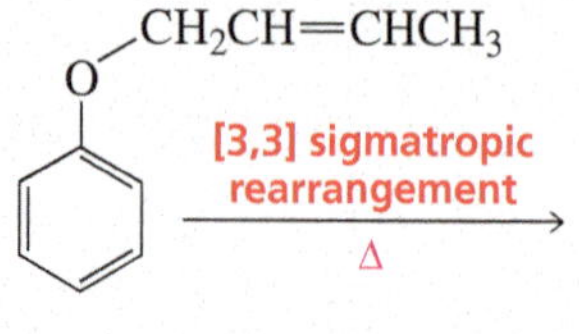

b. [3,3] sigmatropic rearrangement Δ

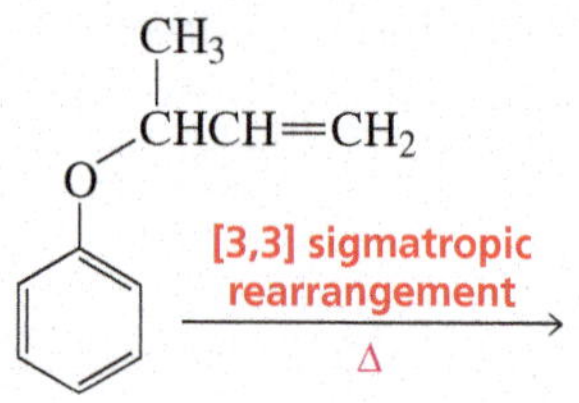

c. [5,5] sigmatropic rearrangement Δ

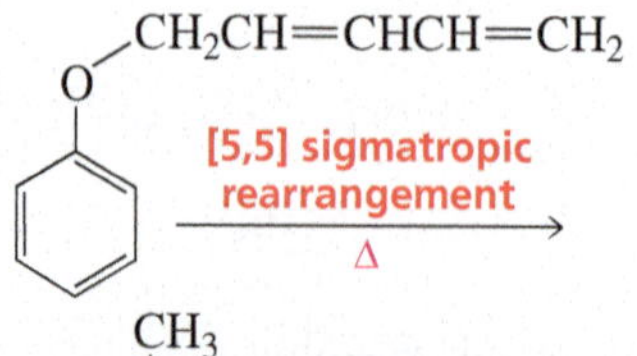

d. [5,5] sigmatropic rearrangement Δ

1238

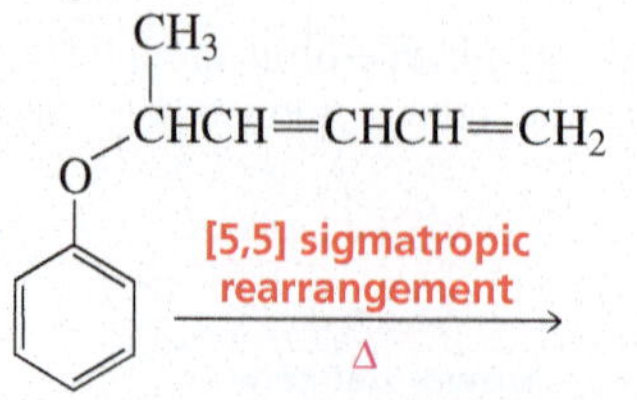

38. a. Propose a mechanism for the following reaction. (*Hint:* An electrocyclic reaction is followed by a Diels–Alder reaction.)
 b. What would be the product if *trans*-2-butene were used instead of ethene?

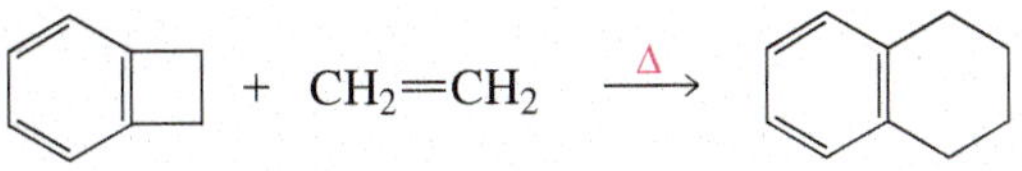

39. Explain why two different products are formed from disrotatory ring closure of (2*E*,4*Z*,6*Z*)-octatriene, but only one product is formed from disrotatory ring closure of (2*E*,4*Z*,6*E*)-octatriene.

40. *cis*-3,4-Dimethylcyclobutene undergoes thermal ring opening to form the two products shown. One of the products is formed in 99% yield, the other in 1% yield. Which is which?

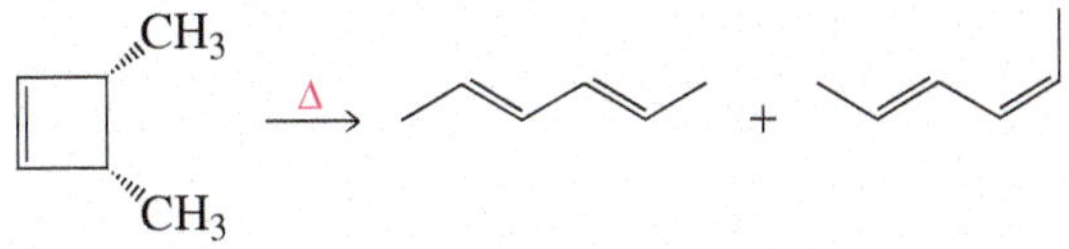

41. If isomer **A** is heated to about 100 °C, a mixture of isomers **A** and **B** is formed. Explain why there is no trace of isomer **C** or **D**.

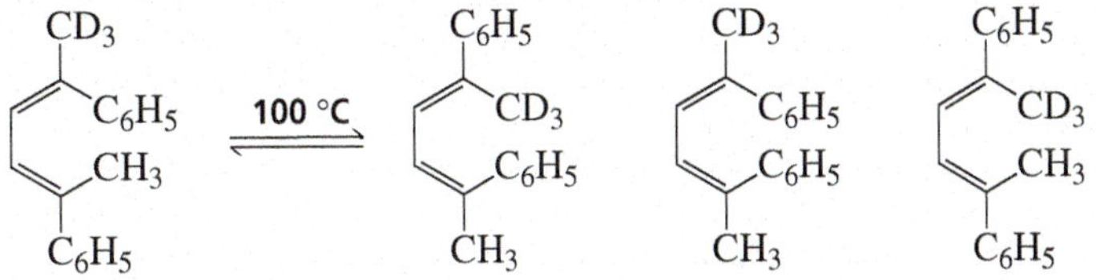

42. Propose a mechanism for the following reaction:

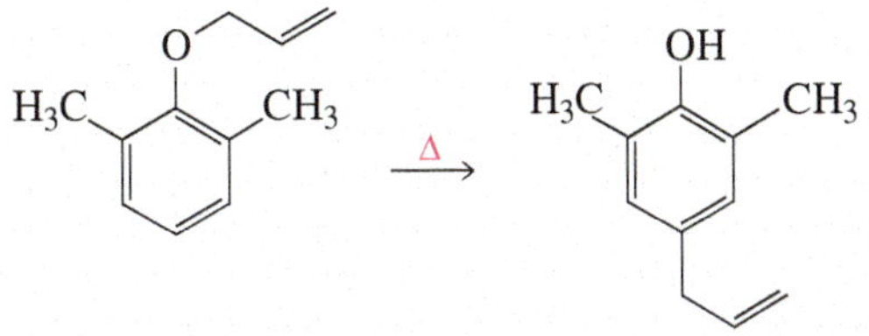

43. Explain why compound **A** will not undergo a ring-opening reaction under thermal conditions, but compound **B** will.

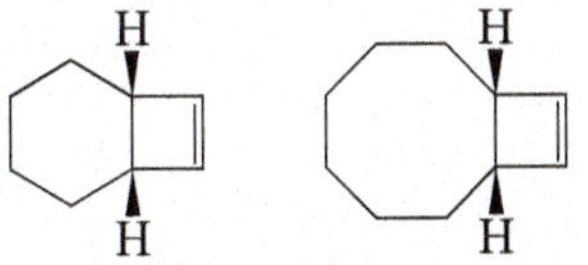

44. A student found that heating any one of the isomers shown here resulted in scrambling of the deuterium to all three positions on the five-membered ring. Propose a mechanism to account for this observation.

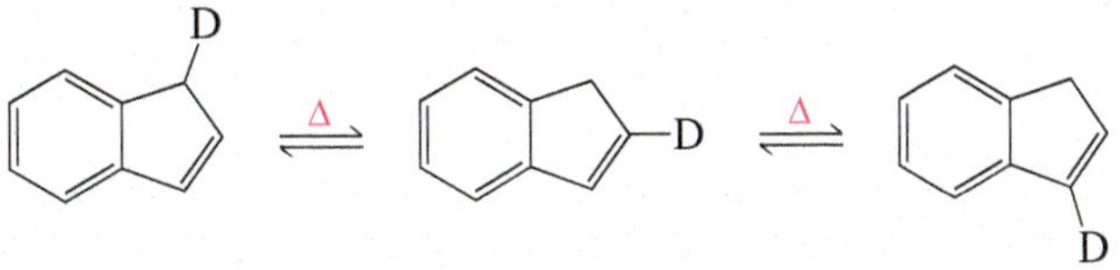

45. How can this transformation be carried out using only heat or light?

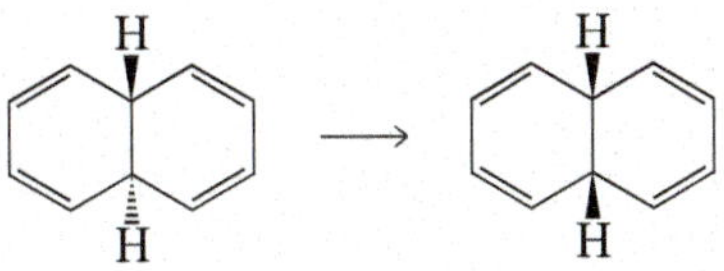

46. Show the steps involved in the following reaction:

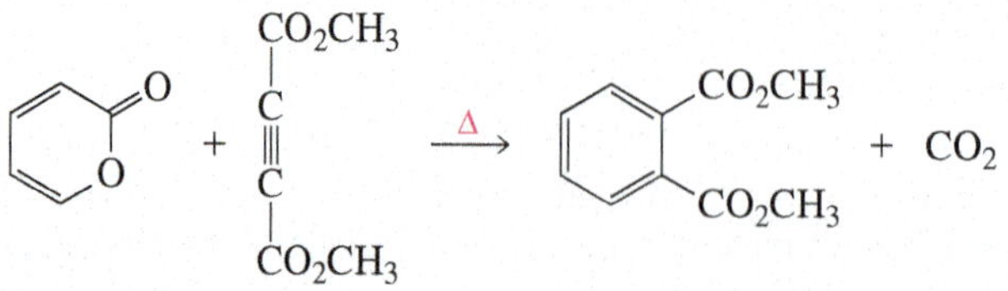

47. Propose a mechanism for the reaction shown below:

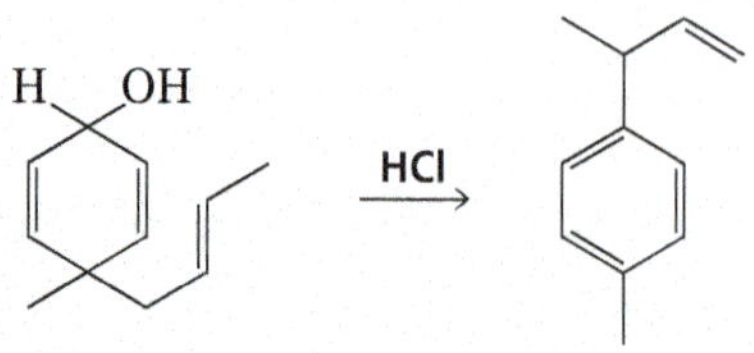

APPENDIX

pKa Values

Compound	pK_a	Compound	pK_a	Compound	pK_a
$CH_3C{\equiv}\overset{+}{N}H$	−10.1	O_2N—⟨benzene⟩—$\overset{+}{N}H_3$	1.0	CH_3—⟨benzene⟩—$\overset{O}{\overset{\|}{C}}OH$	4.3
HI	−10	pyrimidine $\overset{+}{N}H$	1.0	CH_3O—⟨benzene⟩—$\overset{O}{\overset{\|}{C}}OH$	4.5
HBr	−9				
$CH_3\overset{+OH}{\overset{\|}{C}}H$	−8	$HC{\overset{O}{\overset{\|}{}}}$—⟨benzene⟩—$\overset{+}{N}H_3$	1.2	⟨benzene⟩—$\overset{+}{N}H_3$	4.6
$CH_3\overset{+OH}{\overset{\|}{C}}CH_3$	−7.3	$Cl_2CH\overset{O}{\overset{\|}{C}}OH$	1.3	$CH_3\overset{O}{\overset{\|}{C}}OH$	4.8
HCl	−7	H_3PO_4	1.9		
⟨benzene⟩—SO_3H	−6.5	HSO_4^-	2.0	quinoline $\overset{+}{N}H$	4.9
$CH_3\overset{+OH}{\overset{\|}{C}}OCH_3$	−6.5	purine $H\overset{+}{N}$	2.5		
$CH_3\overset{+OH}{\overset{\|}{C}}OH$	−6.1	$FCH_2\overset{O}{\overset{\|}{C}}OH$	2.7	CH_3—⟨benzene⟩—$\overset{+}{N}H_3$	5.1
H_2SO_4	−5	$ClCH_2\overset{O}{\overset{\|}{C}}OH$	2.8	pyridine $\overset{+}{N}H$	5.2
pyrrolidinium $\overset{+}{N}H$	−3.8	$BrCH_2\overset{O}{\overset{\|}{C}}OH$	2.9	CH_3O—⟨benzene⟩—$\overset{+}{N}H_3$	5.3
$CH_3CH_2\overset{+}{\overset{H}{O}}CH_2CH_3$	−3.6	$ICH_2\overset{O}{\overset{\|}{C}}OH$	3.2	$CH_3\overset{+}{\underset{CH_3}{C}}{=}\overset{}{N}HCH_3$	5.5
$CH_3CH_2\overset{+}{\overset{H}{O}}H$	−2.4	HF	3.2		
$CH_3\overset{+}{\overset{H}{O}}H$	−2.5	HNO_2	3.4	$CH_3\overset{O}{\overset{\|}{C}}CH_2\overset{O}{\overset{\|}{C}}H$	5.9
H_3O^+	−1.7	O_2N—⟨benzene⟩—$\overset{O}{\overset{\|}{C}}OH$	3.4	$HO\overset{+}{N}H_3$	6.0
HNO_3	−1.3			H_2CO_3	6.4
CH_3SO_3H	−1.2	$CH_3\overset{O}{\overset{\|}{C}}$—⟨benzene⟩—$COOH$	3.7	$H_2PO_4^-$	6.7
$CH_3\overset{+OH}{\overset{\|}{C}}NH_2$	0.0	$HC\overset{O}{\overset{\|}{}}OH$	3.8	imidazolium $H\overset{+}{N}{\frown}NH$	6.8
$F_3C\overset{O}{\overset{\|}{C}}OH$	0.2	Br—⟨benzene⟩—$\overset{+}{N}H_3$	3.9	H_2S	7.0
$Cl_3C\overset{O}{\overset{\|}{C}}OH$	0.64	Br—⟨benzene⟩—$\overset{O}{\overset{\|}{C}}OH$	4.0	O_2N—⟨benzene⟩—OH	7.1
pyridinium $\overset{+}{N}$—OH	0.79	pyridine—$\overset{O}{\overset{\|}{C}}OH$	4.2	$HC\overset{O}{\overset{\|}{}}$—⟨benzene⟩—$OH$	7.7
				⟨benzene⟩—SH	7.8

[a] pK_a values are for the red H in each structure

(continued)

pK$_a$ Values (continued)

Compound	pK$_a$	Compound	pK$_a$	Compound	pK$_a$
aziridine (ring NH$_2^+$)	8.0	$CH_3CCH_2COCH_2CH_3$ (diketone)	10.7	pyrrole	~17
$H_2N\overset{+}{N}H_3$	8.1	$CH_3\overset{+}{N}H_3$	10.7	CH_3CH (acetaldehyde)	17
CH_3COOH	8.2	cyclohexyl–$\overset{+}{N}H_3$	10.7	$(CH_3)_3COH$	18
phthalimide (NH)	8.3	$(CH_3)_2\overset{+}{N}H_2$	10.7	CH_3CCH_3	20
$CH_3CH_2NO_2$	8.6	piperidine (protonated)	11.1	$CH_3COCH_2CH_3$	24.5
$CH_3CCH_2CCH_3$	8.9	$CH_3CH_2\overset{+}{N}H_3$	11.0	$HC{\equiv}CH$	25
$HC{\equiv}N$	9.1	pyrrolidine (protonated)	11.3	$CH_3C{\equiv}N$	25
morpholine (protonated)	9.3	$HOOH$	11.6	$CH_3CN(CH_3)_2$	30
Cl–C_6H_4–OH	9.4	HPO_4^{2-}	12.4	H_2	35
$\overset{+}{N}H_4$	9.4	CF_3CH_2OH	12.4	NH_3	36
$HOCH_2CH_2\overset{+}{N}H_3$	9.5	$CH_3CH_2OCCH_2COCH_2CH_3$	13.3	pyrrolidine	36
$H_3\overset{+}{N}CH_2CO^-$	9.8	$HC{\equiv}CCH_2OH$	13.5	CH_3NH_2	40
phenol (C_6H_5OH)	10.0	H_2NCNH_2	13.7	C_6H_5–CH_3	41
CH_3–C_6H_4–OH	10.1	$CH_3\overset{+}{N}(CH_3)CH_2CH_2OH$	13.9	benzene	43
CH_3O–C_6H_5	10.2	imidazole	14.4	$CH_2{=}CHCH_3$	43
HCO_3^-	10.2	CH_3OH	15.5	$CH_2{=}CH_2$	44
CH_3NO_2	10.2	H_2O	15.7	cyclopropene	46
H_2N–C_6H_4–OH	10.3	CH_3CH_2OH	16.0	CH_4	60
CH_3CH_2SH	10.5	CH_3CNH_2	16	CH_3CH_3	>60
$(CH_3)_3\overset{+}{N}H$	10.6	$C_6H_5CCH_3$	16.0		

APPENDIX

Kinetics

HOW TO DETERMINE RATE CONSTANTS

A **reaction mechanism** is a detailed analysis of how the chemical bonds (or the electrons) in the reactants rearrange to form the products. The mechanism for a given reaction must obey the observed rate law for the reaction. A **rate law** tells how the rate of a reaction depends on the concentration of the species involved in the reaction.

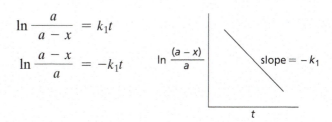

FIRST-ORDER REACTION

The rate is proportional to the concentration of one reactant:

$$A \xrightarrow{k_1} \text{products}$$

Rate law: rate $= k_1[A]$

To determine the first-order rate constant (k_1):

Change in the concentration of A with respect to time:

$$\frac{-d[A]}{dt} = k_1[A]$$

Let $a =$ the initial concentration of A;
Let $x =$ concentration of A that has reacted up to time t.
Therefore, the concentration of A left at time t is $(a - x)$.
Substituting into the previous equation gives

$$\frac{-d(a - x)}{dt} = k_1(a - x)$$

$$\frac{-da}{dt} + \frac{dx}{dt} = k_1(a - x)$$

$$0 + \frac{dx}{dt} = k_1(a - x)$$

$$\frac{dx}{(a - x)} = k_1 dt$$

Integrating the previous equation yields

$$-\ln(a - x) = k_1 t + \text{constant}$$

At $t = 0$, $x = 0$; therefore,

$$\text{constant} = -\ln a$$

$$-\ln(a - x) = k_1 t - \ln a$$

HALF-LIFE OF A FIRST-ORDER REACTION

The **half-life** ($t_{1/2}$) of a reaction is the time it takes for half the reactant to react (or for half the product to form). To derive the half-life of a reactant in a first-order reaction, we begin with the equation shown at the bottom of column 1.

$$\ln \frac{a}{(a - x)} = k_1 t$$

At $t_{1/2}$, $x = \dfrac{a}{2}$; therefore,

$$\ln \frac{a}{\left(a - \dfrac{a}{2}\right)} = k_1 t_{1/2}$$

$$\ln \frac{a}{\left(\dfrac{a}{2}\right)} = k_1 t_{1/2}$$

$$\ln 2 = k_1 t_{1/2}$$

$$0.693 = k_1 t_{1/2}$$

$$t_{1/2} = \frac{0.693}{k_1}$$

Notice that the half-life of a first-order reaction is independent of the concentration of the reactant.

SECOND-ORDER REACTION

The rate is proportional to the concentration of two reactants:

$$A + B \xrightarrow{k_2} \text{products}$$

Rate law: rate $= k_2[A][B]$

To determine the second-order rate constant (k_2):

Change in the concentration of A with respect to time:

$$\frac{-d[A]}{dt} = k_2[A][B]$$

Let a = the initial concentration of A;
Let b = the initial concentration of B;
Let x = the concentration of A that has reacted at time t.

Therefore, the concentration of A left at time $t = (a - x)$, and the concentration of B left at time $t = (b - x)$.
   Substitution gives

$$\frac{dx}{dt} = k_2(a - x)(b - x)$$

For the case where $a = b$ (this condition can be arranged experimentally),

$$\frac{dx}{dt} = k_2(a - x)^2$$

$$\frac{dx}{(a - x)^2} = k_2 \, dt$$

Integrating the equation gives

$$\frac{1}{(a - x)} = k_2t + \text{constant}$$

At $t = 0$, $x = 0$; therefore,

$$\text{constant} = \frac{1}{a}$$

$$\frac{1}{(a - x)} - \frac{1}{a} = k_2t$$

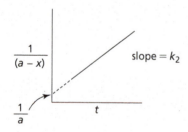

HALF-LIFE OF A SECOND-ORDER REACTION

$$\frac{1}{(a - x)} - \frac{1}{a} = k_2t$$

At $t_{1/2}$, $x = \dfrac{a}{2}$; therefore,

$$\frac{1}{a} = k_2t_{1/2}$$

$$t_{1/2} = \frac{1}{k_2a}$$

PSEUDO-FIRST-ORDER REACTION

It is easier to determine a first-order rate constant than a second-order rate constant because the kinetic behavior of a first-order reaction is independent of the initial concentration of the reactant. Therefore, a first-order rate constant can be determined without knowing the initial concentration of the reactant. The determination of a second-order rate constant requires not only that the initial concentration of the reactants be known but also that the initial concentrations of the two reactant be identical in order to simplify the kinetic equation.
   However, if the concentration of one of the reactants in a second-order reaction is much greater than the concentration of the other, the reaction can be treated as a first-order reaction. Such a reaction is known as a **pseudo-first-order reaction** and is given by

$$\frac{-d[A]}{dt} = k[A][B]$$

If $[B] \gg [A]$, then

$$\frac{-d[A]}{dt} = k'[A]$$

The rate constant obtained for a pseudo-first-order reaction (k', often called k_{obsd}) includes the concentration of B, but k can be determined by carrying out the reaction at several different concentrations of B and determining the slope of a plot of the observed rate versus [B].

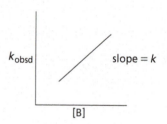

PROBLEMS

1. How long would it take for the reactant of a first-order reaction to decrease to one-half its initial concentration if the rate constant is $4.5 \times 10^{-3} \, s^{-1}$ and the initial concentration of the reactant is

   a. 1.0 M?   **b.** 0.50 M?

2. How long would it take for the reactants of a second-order reaction to decrease to one-half their initial concentration if the rate constant is $2.3 \times 10^{-2} \, M^{-1} \, s^{-1}$ and the initial concentration of both reactants is

   a. 1.0 M?   **b.** 0.50 M?

3. How many half-lives are required for a first-order reaction to reach >99% completion?

4. The initial concentration of a reactant undergoing a first-order reaction is 0.40 M. After five minutes, the concentration of the reactant is 0.27 M; after an additional five minutes, the concentration of the reactant is 0.18 M; and after an additional five minutes, the concentration of the reactant is 0.12 M.

   a. What is the average rate of the reaction during each five-minute interval?

   b. What is the rate constant of the reaction?

5. What percentage of a compound undergoing a first-order reaction with a rate constant of $2.70 \times 10^{-5} \, s^{-1}$ would have reacted at the end of two hours?

6. How long would it take for a first-order reaction with a rate constant of $5.30 \times 10^{-4} \text{ s}^{-1}$ to reach 70% completion?

7. The following data were obtained in a study of the rate of inversion of sucrose at 25 °C. The initial concentration of sucrose was 1.00 M.

Time (minutes)	0	30	60	90	130	180
Sucrose inverted (M)	0	0.100	0.195	0.277	0.373	0.468

   a. What is the order of the reaction? (Assume that the reaction is either first order or second order.)

   b. What is the rate constant of the reaction?

8. Calculate the activation energy of a first-order reaction that is 20% complete in 15 minutes at 40 °C and is 20% complete in 3 minutes at 60 °C.

9. Analysis of a hydrolysis reaction occurring in a dilute aqueous solution of sucrose shows that 80 grams of the original 100 grams of sucrose remain after 10 hours. At this rate, how much sucrose would be left after 24 hours? (Assume a pseudo-first-order reaction.)

Solutions to Problems

1. a. half-life of a first-order reaction $= t_{1/2} = \dfrac{\ln 2}{k_1}$

$$= \frac{0.693}{k_1}$$

$$= \frac{0.693}{4.5 \times 10^{-3} \text{ s}^{-1}}$$

$$= 154 \text{ seconds}$$

$$= 2 \text{ minutes, } 34 \text{ seconds}$$

  b. The half-life of a first-order reaction is independent of concentration, so the answer is the same as the answer for **a** (2 minutes, 34 seconds).

2. a. half-life of a second-order reaction $= t_{1/2} = \dfrac{1}{k_2 a}$

$$= \frac{1}{2.3 \times 10^{-2} \text{ M}^{-1} \text{ s}^{-1} (1.00 \text{ M})}$$

$$= 43 \text{ seconds}$$

  b. $t_{1/2} = \dfrac{1}{2.3 \times 10^{-2} \text{ M}^{-1} \text{ s}^{-1} (0.50 \text{ M})}$

$$= \frac{1}{1.5 \times 10^{-2} \text{ s}^{-1}}$$

$$= 87 \text{ seconds}$$

3. One-half of the compound reacts during the first half-life. One half of what is left after the first half-life reacts during the second half-life. One half of what is left after the second half-life reacts during the third half-life, etc. The following table shows that seven half-lives are required to reach >99% completion.

Number of half-lives	Percentage completion	
1	0.50 (100) = 50	50
2	0.50 (50) = 25	75
3	0.50 (25) = 12.5	87.5
4	0.50 (12.5) = 6.25	93.8
5	0.50 (6.25) = 3.125	96.9
6	0.50 (3.125) = 1.5625	98.4
7	0.50 (1.5625) = 0.78125	99.2

4. a. rate $= \dfrac{\text{change in concentration}}{\text{change in time}}$

1st interval:

$$\text{rate} = \frac{0.40 \text{ M} - 0.27 \text{ M}}{5 \text{ min}} = \frac{0.13 \text{ M}}{5 \text{ min}}$$

$$= 2.60 \times 10^{-2} \text{ M min}^{-1}$$

2nd interval:

$$\text{rate} = \frac{0.27 \text{ M} - 0.18 \text{ M}}{5 \text{ min}} = \frac{0.09 \text{ M}}{5 \text{ min}}$$

$$= 1.80 \times 10^{-2} \text{ M min}^{-1}$$

3rd interval:

$$\text{rate} = \frac{0.18 \text{ M} - 0.12 \text{ M}}{5 \text{ min}} = \frac{0.06 \text{ M}}{5 \text{ min}}$$

$$= 1.20 \times 10^{-2} \text{ M min}^{-1}$$

Thus, we see that the **rate** of the reaction is dependent on the concentration of the reactant; it decreases with decreasing concentration. Thus, the rate we have calculated is the average rate for each five-minute interval.

  b. rate $= k$ [reactant], using the average concentration of the reactant during the five-minute interval.

1st interval:

$$2.60 \times 10^{-2} \text{ M min}^{-1} = k(0.335 \text{ M})$$
$$k = 7.8 \times 10^{-2} \text{ min}^{-1}$$

2nd interval:

$$1.80 \times 10^{-2} \text{ M min}^{-1} = k(0.225 \text{ M})$$
$$k = 8.0 \times 10^{-2} \text{ min}^{-1}$$

3rd interval:

$$1.20 \times 10^{-2} \text{ M min}^{-1} = k(0.15 \text{ M})$$
$$k = 8.0 \times 10^{-2} \text{ min}^{-1}$$

The **rate constant**, as its name indicates, is constant during the course of the reaction.

5. $a =$ the initial concentration (set at 100%)

$x =$ the concentration that has reacted at time $= t$

$$\ln \frac{a}{a - x} = k_1 t$$

$$\ln \frac{100}{100 - x} = 2.70 \times 10^{-5} \text{ s}^{-1} (2 \text{ hours})$$

$$\ln \frac{100}{100 - x} = 2.70 \times 10^{-5} \text{ s}^{-1} (7200 \text{ seconds})$$

$$\ln \frac{100}{100 - x} = 0.194$$

$$\frac{100}{100 - x} = 1.21$$

$$100 = 121 - 1.21x$$

$$1.21\,x = 21$$

$$x = 17.4\%$$

6. $$\ln \frac{a}{a - x} = k_1 t$$

$$\ln \frac{100}{100 - 70} = 5.30 \times 10^{-4}\,\text{s}^{-1}\,t$$

$$\ln 3.33 = 5.30 \times 10^{-4}\,\text{s}^{-1}\,t$$

$$t = \frac{1.20}{5.30 \times 10^{-4}\,\text{s}^{-1}}$$

$$t = 2260 \text{ seconds}$$

$$t = 37.7 \text{ minutes}$$

7. To determine the order of the reaction, we will calculate the first-order and second-order rate constants based on the data provided. Since the rate constant for a reaction should be the same regardless of the set of data we use, the calculations will tell us the order of the reaction.

1st order

$$k_1 = \frac{-\ln \dfrac{a - x}{a}}{t}$$

$$k_1 = \frac{-\ln \dfrac{1.000 - 0.100}{1.000}}{30}$$

$$= 3.51 \times 10^{-3}$$

$$k_2 = \frac{-\ln \dfrac{1.000 - 0.195}{1.000}}{60}$$

$$= 3.62 \times 10^{-3}$$

$$k_1 = \frac{-\ln \dfrac{1.000 - 0.277}{1.000}}{90}$$

$$= 3.60 \times 10^{-3}$$

$$k_1 = \frac{-\ln \dfrac{1.000 - 0.373}{1.000}}{130}$$

$$= 3.59 \times 10^{-3}$$

$$k_1 = \frac{-\ln \dfrac{1.000 - 0.468}{1.000}}{180}$$

$$= 3.51 \times 10^{-3}$$

2nd order

$$k_2 = \frac{\dfrac{1}{a - x} - \dfrac{1}{a}}{t}$$

$$k_2 = \frac{\dfrac{1}{1.000 - 0.100} - \dfrac{1}{1.000}}{30}$$

$$= \frac{1.111 - 1.000}{30}$$

$$= 3.70 \times 10^{-3}$$

$$k_2 = \frac{\dfrac{1}{1.000 - 0.195} - 1.000}{60}$$

$$= 4.03 \times 10^{-3}$$

$$k_2 = \frac{\dfrac{1}{1.000 - 0.277} - 1.000}{90}$$

$$= 4.26 \times 10^{-3}$$

$$k_2 = \frac{\dfrac{1}{1.000 - 0.373} - 1.000}{130}$$

$$= 4.57 \times 10^{-3}$$

$$k_2 = \frac{\dfrac{1}{1.000 - 0.468} - 1.000}{180}$$

$$= 4.98 \times 10^{-3}$$

a. Because the calculated rate constants are relatively constant when the data are plugged into a first-order equation but vary considerably when the data are plugged into a second-order equation, one can conclude that the reaction is first-order.

b. $3.6 \times 10^{-3}\,\text{min}^{-1}$

8. Using the equation for calculating kinetic parameters on p. 224 of the text:

$$k \text{ at } 40\,°C = \frac{\ln \dfrac{1.00}{1.00 - 0.20}}{15 \text{ min}} = 1.49 \times 10^{-2} \text{ min}^{-1}$$

$$k \text{ at } 60\,°C = \frac{\ln \dfrac{1.00}{1.00 - 0.20}}{3 \text{ min}} = 7.44 \times 10^{-2} \text{ min}^{-1}$$

$$\ln k_2 - \ln k_1 = \frac{-E_a}{R}\left(\frac{1}{T_2} - \frac{1}{T_1}\right) \quad (40\,°C = 313 \text{ K}; 60\,°C = 333 \text{ K})$$

$$\ln (7.44 \times 10^{-2}) - \ln (1.49 \times 10^{-2}) = \frac{-E_a}{1.986 \times 10^{-3} \text{ kcal}}\left(\frac{1}{333} - \frac{1}{313}\right)$$

$$-2.60 - (-4.21) = \frac{-E_a}{1.986 \times 10^{-3}}(0.00300 - 0.00319)$$

$$1.61 = \frac{-E_a}{1.986 \times 10^{-3}}(-0.00019)$$

$$E_a = \frac{1.61 \times 1.986 \times 10^{-3}}{0.00019}$$

$$E_a = 16.8 \text{ kcal/mol}$$

9. Sucrose is hydrolyzed to form a mixture of glucose and fructose. Because there is excess water (it is the solvent), the reaction is a pseudo-first-order reaction.

First, the rate constant of the reaction must be determined:

$$\ln \frac{a}{a - x} = k_1 t$$

$$\ln \frac{100}{80} = k_1 \times 10 \text{ hours}$$

$$k_1 = 2.23 \times 10^{-2} \text{ hr}^{-1}$$

Now we can calculate the amount of sucrose that has reacted, and, therefore, the amount that would be left.

$$\ln \frac{a}{a - x} = k_1 t$$

$$\ln \frac{100}{100 - x} = 2.23 \times 10^{-2}\,(24 \text{ hr})$$

$$\ln \frac{100}{100 - x} = 0.535$$

$$\frac{100}{100 - x} = 1.71$$

$$100 = 171 - 1.71x$$

$$1.71x = 71$$

$$x = 41.5 \text{ g have reacted}$$

Therefore, 58.5 g would be left.

Summary of Methods Used to Synthesize a Particular Functional Group

SYNTHESIS OF ACETALS

1. Acid-catalyzed reaction of an aldehyde or a ketone with two equivalents of an alcohol (16.9).

SYNTHESIS OF ACID ANHYDRIDES

1. Reaction of an acyl halide with a carboxylate ion (15.16).
2. Heating a dicarboxylic acid (15.16).
3. Heating a dicarboxylic acid in the presence of acetic anhydride (15.16).

SYNTHESIS OF ACYL CHLORIDES OR ACYL BROMIDES

1. Reaction of a carboxylic acid with $SOCl_2$, PCl_3, or PBr_3 (15.18).

SYNTHESIS OF ALCOHOLS

1. Acid-catalyzed hydration of an alkene (6.5).
2. Hydroboration–oxidation of an alkene (6.8).
3. Reaction of an alkyl halide with HO^- (9.1, 9.3).
4. Reaction of an organocuprate with an epoxide (11.3).
5. Epoxidation of an alkene followed by reaction with NaH (10.7).
6. Reduction of an aldehyde, a ketone, an acyl chloride, an anhydride, an ester, or a carboxylic acid (16.5, 16.6).
7. Reaction of a Grignard reagent with an aldehyde, a ketone, an acyl chloride, or an ester (16.4).
8. Cleavage of an ether with HI or HBr (10.6).
9. Reaction of an organozinc reagent with an aldehyde or a ketone (p. 849).

SYNTHESIS OF ALDEHYDES

1. Hydroboration–oxidation of a terminal alkyne (7.8).
2. Oxidation of a primary alcohol with pyridinium chlorochromate or hypochlorous acid (10.5).
3. Swern oxidation of a primary alcohol with dimethyl sulfoxide, oxalyl chloride, and triethylamine (10.5).
4. Reaction of an acyl chloride with lithium tri(*tert*-butoxy) aluminum hydride (16.5).

5. Reaction of an ester with diisobutylaluminum hydride (DIBALH) (16.5).
6. Ozonolysis of an alkene, followed by reaction with dimethyl sulfide or zinc and acetic acid (6.11).

SYNTHESIS OF ALKANES

1. Catalytic hydrogenation of an alkene or an alkyne (5.9, 7.9, 16.7).
2. Reaction of a Grignard reagent with a source of protons (11.1).
3. Wolff–Kishner reduction of an aldehyde or a ketone (18.8).
4. Reduction of a phenone with H_2/Pd (18.8).
5. Reduction of a thioacetal or thioketal with H_2 and Raney nickel (16.11).
6. Reaction of an organocuprate with an alkylhalide (11.3).
7. Preparation of a cyclopropane by the reaction of an alkene with a carbene (p. 286).

SYNTHESIS OF ALKENES

1. Elimination of hydrogen halide from an alkyl halide (9.6, 9.7, 9.8).
2. Acid-catalyzed dehydration of an alcohol (10.4).
3. Hofmann elimination reaction: elimination of a proton and a tertiary amine from a quaternary ammonium hydroxide (10.10).
4. Exhaustive methylation of an amine, followed by a Hofmann elimination reaction (10.10).
5. Hydrogenation of an alkyne with Lindlar catalyst to form a cis alkene (7.6, 16.6).
6. Reduction of an alkyne with Na (or Li) and liquid ammonia to form a trans alkene (7.9, 16.7).
7. Formation of a cyclic alkene using a Diels–Alder reaction (8.14, 28.4).
8. Wittig reaction: reaction of an aldehyde or a ketone with a phosphonium ylide (16.13).
9. Reaction of an organocuprate with a halogenated alkene (11.3).
10. Heck reaction couples a vinyl halide with an alkene in a basic solution in the presence of PdL_2 (11.4).
11. Suzuki reaction couples a vinyl halide with an organoboron compound in the presence of (PdL_2) (11.4).
12. Alkene metathesis (11.5).

SYNTHESIS OF ALKYL HALIDES

1. Addition of hydrogen halide (HX) to an alkene (6.1).
2. Addition of HBr + a peroxide to an alkene (12.7).
3. Addition of hydrogen halide to an alkyne (7.6).
4. Radical halogenation of an alkane, an alkene, or an alkyl benzene (12.2, 12.9).
5. Reaction of an alcohol with a hydrogen halide, $SOCl_2$, PCl_3, or PBr_3 (10.1, 10.2).
6. Reaction of a sulfonate ester with halide ion (10.3).
7. Cleavage of an ether with a hydrogen halide (10.6).
8. Halogenation of an α-carbon of an aldehyde, a ketone, or a carboxylic acid (17.4, 17.5).

SYNTHESIS OF ALKYNES

1. Elimination of hydrogen halide from a vinyl halide (9.15).
2. Two successive eliminations of hydrogen halide from a vicinal dihalide or a geminal dihalide (9.15).
3. Reaction of an acetylide ion (formed by removing a proton from a terminal alkyne) with an alkyl halide (7.11).
4. Alkyne metathesis (12.5).

SYNTHESIS OF AMIDES

1. Reaction of an acyl chloride, an acid anhydride, or an ester with ammonia or with an amine (15.6, 15.7, 15.16).
2. Heating an ammonium carboxylate salt (15.10).
3. Reaction of a carboxylic acid and with dicyclohexylcarbodiimide followed by reaction with an amine (21.10).
4. Reaction of a nitrile with a secondary or tertiary alcohol (p. 736).

SYNTHESIS OF AMINES

1. Reaction of an alkyl halide with NH_3, RNH_2, or R_2NH (9.2).
2. Reaction of an alkyl halide with azide ion, followed by reduction of the alkyl azide (9.2).
3. Reduction of an imine, a nitrile, or an amide (15.15, 16.5, 16.8).
4. Reductive amination of an aldehyde or a ketone (16.8).
5. Gabriel synthesis of primary amines: reaction of a primary alkyl halide with potassium phthalimide (15.14).
6. Reduction of a nitro compound (18.10).
7. Condensation of a secondary amine and formaldehyde with a carbon acid (p. 851).

SYNTHESIS OF AMINO ACIDS

1. Hell–Volhard–Zelinsky reaction: halogenation of a carboxylic acid, followed by treatment with excess NH_3 (17.5).
2. Reductive amination of an α-keto acid (21.6).
3. The N-phthalimidomalonic ester synthesis (21.6).
4. The acetamidomalonic ester synthesis (21.6).
5. The Strecker synthesis: reaction of an aldehyde with ammonia, followed by addition of cyanide ion and hydrolysis (21.6).

SYNTHESIS OF CARBOXYLIC ACIDS

1. Oxidation of a primary alcohol (10.5).
2. Oxidation of an aldehyde (10.5, 16.12).
3. Oxidation of an alkyl benzene (18.10).
4. Hydrolysis of an acyl halide, an acid anhydride, an ester, an amide, or a nitrile (15.6, 15.7, 15.11, 15.12, 15.15, 15.16).
5. Haloform reaction: reaction of a methyl ketone with excess Br_2 (or Cl_2 or I_2) + HO^- (p. 847).
6. Reaction of a Grignard reagent with CO_2 (16.4).
7. Malonic ester synthesis (17.17).
8. Favorskii reaction: reaction of an α-haloketone with hydroxide ion (p. 851).

SYNTHESIS OF CYANOHYDRINS

1. Reaction of an aldehyde or a ketone with sodium cyanide and HCl (16.4).

SYNTHESIS OF DIHALIDES

1. Addition of Cl_2 or Br_2 to an alkene (6.9).
2. Addition of Cl_2 or Br_2 to an alkene (7.6).

SYNTHESIS OF 1,2-DIOLS

1. Reaction of an epoxide with hydroxide ion forms a trans 1,2-diol (10.7).
2. Reaction of an alkene with osmium tetroxide followed by hydrolysis with hydrogen peroxide forms a cis 1,2-diol (10.7).

SYNTHESIS OF DISULFIDES

1. Mild oxidation of a thiol (21.8).

SYNTHESIS OF ENAMINES

1. Reaction of an aldehyde or a ketone with a secondary amine (16.8).

SYNTHESIS OF EPOXIDES

1. Reaction of an alkene with a peroxyacid (6.10).
2. Reaction of a halohydrin with hydroxide ion (p. 480).
3. Reaction of an aldehyde or a ketone with a sulfonium ylide (p. 796).

SYNTHESIS OF ESTERS

1. Reaction of an acyl halide or an acid anhydride with an alcohol (15.6, 15.16).
2. Acid-catalyzed reaction of an ester or a carboxylic acid with an alcohol (15.8, 15.10).
3. Reaction of an alkyl halide with a carboxylate ion (9.15).
4. Reaction of a sulfonate ester with a carboxylate ion (10.3).
5. Oxidation of a ketone (16.12).
6. Preparation of a methyl ester by the reaction of a carboxylate ion with diazomethane (18.20).

SYNTHESIS OF ETHERS

1. Acid-catalyzed addition of an alcohol to an alkene (6.6).
2. Williamson ether synthesis: reaction of an alkoxide ion with an alkyl halide (9.15).
3. Formation of symmetrical ethers by heating an acidic solution of a primary alcohol (10.4).

SYNTHESIS OF HALOHYDRINS

1. Reaction of an alkene with Br_2 (or Cl_2) and H_2O (6.9).
2. Reaction of an epoxide with a hydrogen halide (10.7).

SYNTHESIS OF IMINES

1. Reaction of an aldehyde or a ketone with a primary amine (16.8).

SYNTHESIS OF KETONES

1. Addition of water to an alkyne (7.7).
2. Hydroboration–oxidation of an internal alkyne (7.8).
3. Oxidation of a secondary alcohol (10.5).
4. Ozonolysis of an alkene, followed by reaction with dimethyl sulfide or zinc and acetic acid (6.11).
5. Friedel–Crafts acylation of an aromatic ring (18.6).
6. Preparation of a methyl ketone by the acetoacetic ester synthesis (17.18).
7. Preparation of a cyclic ketone by the reaction of the next-size-smaller cyclic ketone with diazomethane (p. 798).

SYNTHESIS OF α,β-UNSATURATED KETONES

1. Selenenylation of a ketone, followed by oxidative elimination (p. 851).

SYNTHESIS OF NITRILES

1. Reaction of an alkyl halide with cyanide ion (9.2).
2. Reaction of an amide (with an NH_2 group) with $SOCl_2$ (15.15).

SYNTHESIS OF SUBSTITUTED BENZENES

1. Halogenation with Br_2 or Cl_2 and a Lewis acid (18.3).
2. Nitration with HNO_3 + H_2SO_4 (18.4).
3. Sulfonation: heating with H_2SO_4 (18.5).
4. Friedel–Crafts acylation (18.6).
5. Friedel–Crafts alkylation (18.7, 18.8).
6. Sandmeyer reaction: reaction of an arenediazonium salt with CuBr, CuCl, or CuCN (18.18).
7. Formation of a phenol by reaction of an arenediazonium salt with water (18.18).
8. Reaction of an organocuprate with an aryl halide (11.3).
9. Heck reaction: couples an aryl halide with an alkene in a basic solution in the presence of PdL_2 (11.4).
10. Suzuki reaction: couples an aryl halide with an organoborane in the presence of PdL_2 and triethylamine (11.4).

SYNTHESIS OF SULFIDES

1. Reaction of a thiol with an alkyl halide (10.11).
2. Reaction of a thiol with a sulfonate ester (10.3).

SYNTHESIS OF THIOLS

1. Reaction of an alkyl halide with hydrogen sulfide (9.2).
2. Catalytic hydrogenation of a disulfide (21.8).

Summary of Methods Employed to Form Carbon–Carbon Bonds

1. Alkene or alkyne metathesis (11.5).

2. Reaction of an acetylide ion with an alkyl halide or a sulfonate ester (7.11, 9.2, 10.3).

3. Diels–Alder and other cycloaddition reactions (8.14, 28.4).

4. Reaction of an organocuprate with an epoxide (11.3).

5. Friedel–Crafts alkylation and acylation (18.6, 18.7, 18.8).

6. Reaction of a cyanide ion with an alkyl halide or a sulfonate ester (9.2, 10.3).

7. Reaction of a cyanide ion with an aldehyde or a ketone (16.4).

8. Reaction of a Grignard reagent with an aldehyde, a ketone, an ester, an amide, or CO_2 (16.4).

9. Reaction of an organozinc reagent with an aldehyde or a ketone (p. 849).

10. Reaction of an alkene with a carbene (p. 286).

11. Reaction of an organocuprate with an α,β-unsaturated ketone or an α,β-unsaturated aldehyde (16.15).

12. Aldol addition (17.10, 17.12, 17.15).

13. Claisen condensation (17.13, 17.14).

14. Perkin condensation (p. 848).

15. Knoevenagel condensation (p. 848).

16. Reformatsky reaction (p. 849).

17. The benzoin condensation (p. 852).

18. Malonic ester synthesis and acetoacetic ester synthesis (17.17, 17.18).

19. Michael addition reaction (17.9).

20. Alkylation of an enamine (17.8).

21. Alkylation of the α-carbon of a carbonyl compound (17.7).

22. Reaction of an organocuprate with an aryl halide or a vinyl halide (11.3).

23. Heck reaction: couples a vinyl or an aryl halide with an alkene in a basic solution in the presence of PdL_2 (11.4).

24. Suzuki reaction: couples a vinyl or aryl halide with an organoborane in the presence of PdL_2 (11.4).

APPENDIX

Spectroscopy Tables

Mass Spectrometry	

Common fragment ions*			
m/z	Ion	m/z	Ion
14	CH_2	46	NO_2
15	CH_3	47	CH_2SH, CH_3S
16	O	48	$CH_3S + H$
17	OH	49	CH_2Cl
18	H_2O, NH_4	51	CHF_2
19	F, H_3O	53	C_4H_5
26	C≡N	54	CH_2CH_2C≡N
27	C_2H_3	55	C_4H_7, CH_2=CHC=O
28	C_2H_4, CO, N_2, CH=NH	56	C_4H_8
29	C_2H_5, CHO	57	C_4H_9, C_2H_5C=O
30	CH_2NH_2, NO	58	$\overset{O}{\overset{\|}{CH_3CCH_2}}$ + H, $C_2H_5CHNH_2$, $(CH_3)_2NCH_2$, $C_2H_5NHCH_2$, C_2H_2S
31	CH_2OH, OCH_3	59	$(CH_3)_2COH$, $CH_2OC_2H_5$, $\overset{O}{\overset{\|}{COCH_3}}$, $\underset{NH_2}{CH_2C}$=O + H, CH_3OCHCH_3
32	O_2 (air)	60	CH_3CHCH_2OH CH_2COOH + H, CH_2ONO
33	SH, CH_2F		
34	H_2S		
35	Cl		
36	HCl		
39	C_3H_3		
40	CH_2C≡N		
41	C_3H_5, CH_2C≡N + H, C_2H_2NH		
42	C_3H_6		
43	C_3H_7, CH_3C=O, C_2H_5N		
44	CH_2CH=O + H, CH_3CHNH_2, CO_2, NH_2C=O, $(CH_3)_2N$		
45	CH_3CHOH, CH_2CH_2OH, CH_2OCH_3, COOH, $CH_3CHO + H$		

*All of these ions have a single positive charge.

Mass Spectrometry

Common fragment lost			
Molecular ion minus	Fragment lost	Molecular ion minus	Fragment lost
1	H	41	CH_2=$CHCH_2$
15	CH_3	42	CH_2=$CHCH_3$, CH_2=C=O, CH_2—CH_2 (cyclopropane ring with CH_2), NCO
17	HO	43	C_3H_7, $CH_3\overset{O}{\overset{\|}{C}}$, CH_2=CHO, HCNO, CH_3 + CH_2=CH_2
18	H_2O	44	CH_2=CHOH, CO_2, N_2O, $CONH_2$, $NHCH_2CH_3$
19	F	45	CH_3CHOH, CH_3CH_2O, CO_2H, $CH_3CH_2NH_2$
20	HF	46	H_2O + CH_2=CH_2, CH_3CH_2OH, NO_2
26	CH≡CH, C≡N	47	CH_3S
27	CH_2=CH, HC≡N	48	CH_3SH, SO, O_3
28	CH_2=CH_2, CO, (HCN + H)	49	CH_2Cl
29	CH_3CH_2, CHO	51	CHF_2
30	NH_2CH_2, CH_2O, NO	52	C_4H_4, C_2N_2
31	OCH_3, CH_2OH, CH_3NH_2	53	C_4H_5
32	CH_3OH, S	54	CH_2=CHCH=CH_2
33	HS, (CH_3 and H_2O)	55	CH_2=$CHCHCH_3$
34	H_2S	56	CH_2=$CHCH_2CH_3$, CH_3CH=$CHCH_3$
35	Cl	57	C_4H_9
36	HCl, $2H_2O$	58	NCS, NO + CO, CH_3COCH_3
37	HCl + H	59	$CH_3O\overset{O}{\overset{\|}{C}}$, $CH_3\overset{O}{\overset{\|}{C}}NH_2$
38	C_3H_2, C_2N, F_2	60	C_3H_7OH
39	C_3H_3, HC_2N		
40	CH_3C≡CH		

^{1}H NMR Chemical Shifts

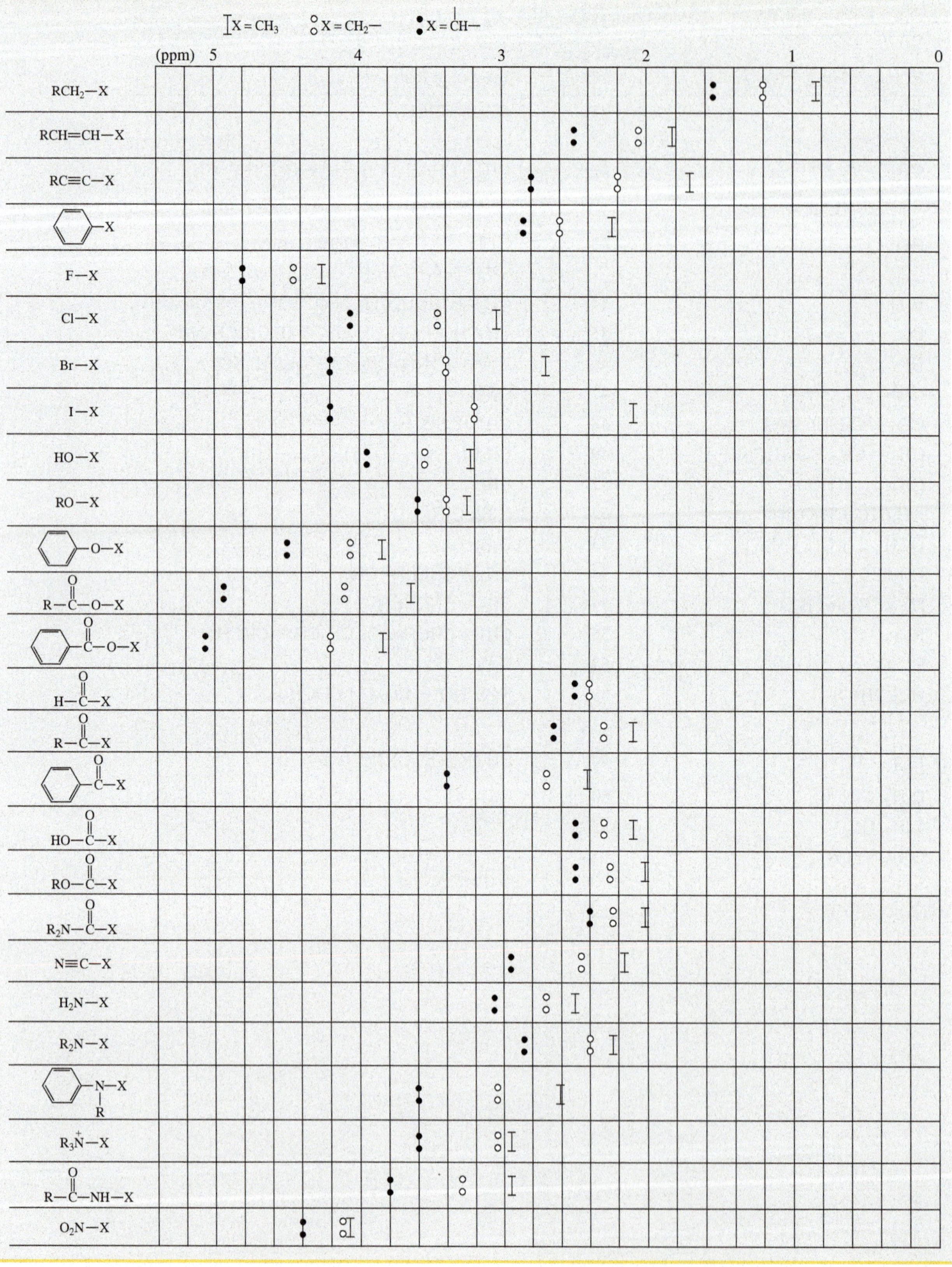

$| X = CH_3$ $\circ\, X = CH_2-$ $\bullet\, X = \overset{|}{C}H-$

Chart of ^{1}H NMR chemical shifts (ppm scale from 5 to 0) for various functional groups:

RCH$_2$—X

RCH=CH—X

RC≡C—X

⬡—X

F—X

Cl—X

Br—X

I—X

HO—X

RO—X

⬡—O—X

R—C(=O)—O—X

⬡—C(=O)—O—X

H—C(=O)—X

R—C(=O)—X

⬡—C(=O)—X

HO—C(=O)—X

RO—C(=O)—X

R$_2$N—C(=O)—X

N≡C—X

H$_2$N—X

R$_2$N—X

⬡—N(R)—X

R$_3\overset{+}{N}$—X

R—C(=O)—NH—X

O$_2$N—X

Characteristic Infrared Group Frequencies (S = strong, M = medium, W = weak). (Courtesy of N.B. Colthup, Stamford Research Laboratories, American Cyanamid Company, and the editor of the *Journal of the Optical Society*.) Overtone bands are marked **2ν**.

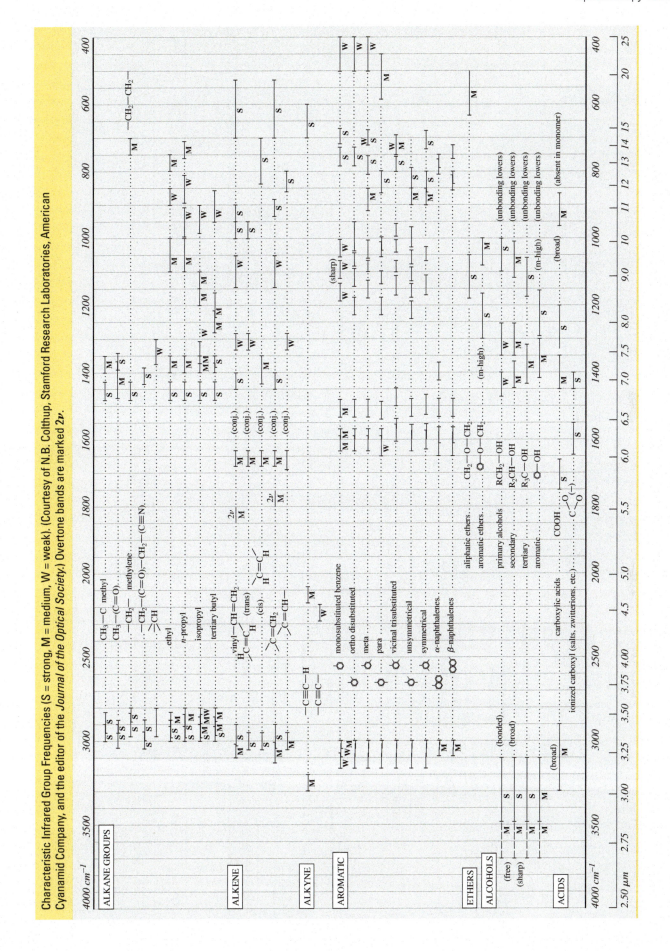

Characteristic Infrared Group Frequencies (continued)

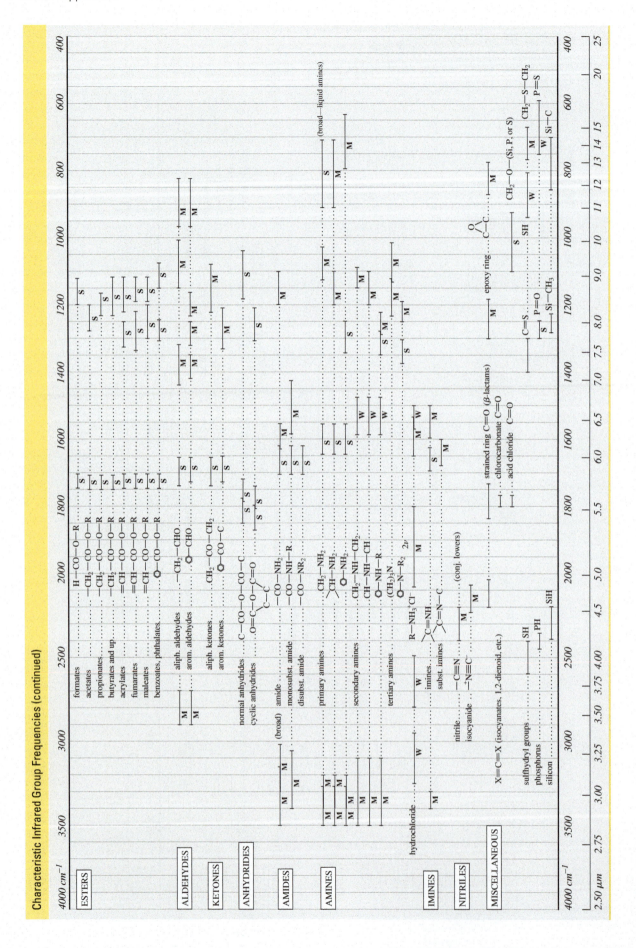

Characteristic Infrared Group Frequencies (continued)

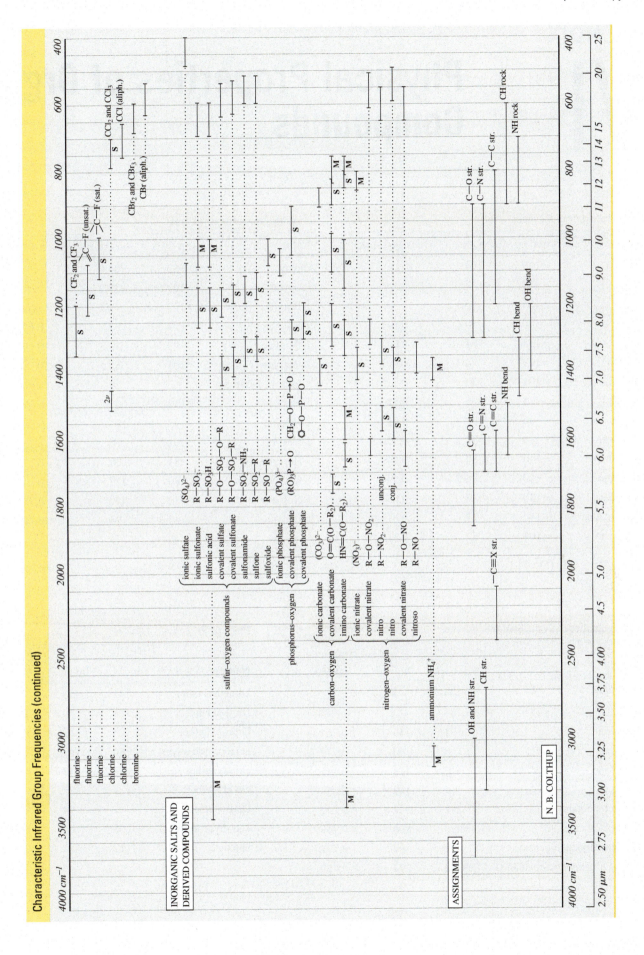

N. B. COLTHUP

VI Physical Properties of Organic Compounds

Physical Properties of Alkynes

Name	mp (°C)	bp (°C)	Density (g/mL)
Ethene	−169	−104	
Propene	−185	−47	
1-Butene	−185	−6.3	
1-Pentene		30	0.641
1-Hexene	−138	64	0.673
1-Heptene	−119	94	0.697
1-Octene	−101	122	0.715
1-Nonene	−81	146	0.730
1-Decene	−66	171	0.741
cis-2-Butene	−180	37	0.650
trans-2-Butene	−140	37	0.649
Methylpropene	−140	−6.9	0.594
cis-2-Pentene	−180	37	0.650
trans-2-Pentene	−140	37	0.649
Cyclohexene	−104	83	0.811

Physical Properties of Alkynes

Name	mp (°C)	bp (°C)	Density (g/mL)
Ethyne	−82	−84.0	
Propyne	−101.5	−23.2	
1-Butyne	−122	8.1	
2-Butyne	−24	27	0.694
1-Pentyne	−98	39.3	0.695
2-Pentyne	−101	55.5	0.714
3-Methyl-1-butyne		29	0.665
1-Hexyne	−132	71	0.715
2-Hexyne	−92	84	0.731
3-Hexyne	−101	81	0.725
1-Heptyne	−81	100	0.733
1-Octyne	−80	127	0.747
1-Nonyne	−50	151	0.757
1-Decyne	−44	174	0.766

Physical Properties of Cyclic Saturated Alkanes

Name	mp (°C)	bp (°C)	Density (g/mL)
Cyclopropane	−128	−33	
Cyclobutane	−80	−12	
Cyclopentane	−94	50	0.751
Cyclohexane	6.5	81	0.779
Cycloheptane	−12	118	0.811
Cyclooctane	14	149	0.834
Methylcyclopentane	−142	72	0.749
Methylcyclohexane	−126	100	0.769
cis-1,2-Dimethylcyclopentane	−62	99	0.772
trans-1,2-Dimethylcyclopentane	−120	92	0.750

Physical Properties of Ethers

Name	mp (°C)	bp (°C)	Density (g/mL)
Dimethyl ether	−141	−24.8	
Diethyl ether	−116	34.6	0.706
Dipropyl ether	−123	88	0.736
Diisopropyl ether	−86	69	0.725
Dibutyl ether	−98	142	0.764
Divinyl ether		35	
Diallyl ether		94	0.830
Tetrahydrofuran	−108	66	0.889
Dioxane	12	101	1.034

Physical Properties of Alcohols

Name	mp (°C)	bp (°C)	Solubility (g/100 g H$_2$O at 25 °C)
Methanol	−97.8	64	∞
Ethanol	−114.7	78	∞
1-Propanol	−127	97.4	∞
1-Butanol	−90	118	7.9
1-Pentanol	−78	138	2.3
1-Hexanol	−52	157	0.6
1-Heptanol	−36	176	0.2
1-Octanol	−15	196	0.05
2-Propanol	−89.5	82	∞
2-Butanol	−115	99.5	12.5
2-Methyl-1-propanol	−108	108	10.0
2-Methyl-2-propanol	25.5	83	∞
3-Methyl-1-butanol	−117	130	2
2-Methyl-2-butanol	−12	102	12.5
2,2-Dimethyl-1-propanol	55	114	∞
Allyl alcohol	−129	97	∞
Cyclopentanol	−19	140	s. sol.
Cyclohexanol	24	161	s. sol.
Benzyl alcohol	−15	205	4

Physical Properties of Amines

Name	mp (°C)	bp (°C)	Solubility (g/100 g H$_2$O at 25 °C)
Primary Amines			
Methylamine	−93	−6.3	v. sol.
Ethylamine	−81	17	∞
Propylamine	−83	48	∞
Isopropylamine	−95	33	∞
Butylamine	−49	78	v. sol.
Isobutylamine	−85	68	∞
sec-Butylamine	−72	63	∞
tert-Butylamine	−67	46	∞
Cyclohexylamine	−18	134	s. sol.
Secondary Amines			
Dimethylamine	−93	7.4	v. sol.
Diethylamine	−50	55	10.0
Dipropylamine	−63	110	10.0
Dibutylamine	−62	159	s. sol.
Tertiary Amines			
Trimethylamine	−115	2.9	91
Triethylamine	−114	89	14
Tripropylamine	−93	157	s. sol.

Physical Properties of Alkyl Halides

Name	bp (°C) Fluoride	Chloride	Bromide	Iodide
Methyl	−78.4	−24.2	3.6	42.4
Ethyl	−37.7	12.3	38.4	72.3
Propyl	−2.5	46.6	71.0	102.5
Isopropyl	−9.4	34.8	59.4	89.5
Butyl	32.5	78.4	100	130.5
Isobutyl		68.8	90	120
sec-Butyl		68.3	91.2	120.0
tert-Butyl		50.2	73.1	dec.
Pentyl	62.8	108	130	157.0
Hexyl	92	133	154	179

Physical Properties of Benzene and Substituted Benzenes

Name	mp (°C)	bp (°C)	Solubility (g/100 g H$_2$O at 25 °C)
Aniline	−6	184	3.7
Benzene	5.5	80.1	s. sol.
Benzaldehyde	−26	178	s. sol.
Benzamide	132	290	s. sol.
Benzoic acid	122	249	0.34
Bromobenzene	−30.8	156	insol.
Chlorobenzene	−45.6	132	insol.
Nitrobenzene	5.7	210.8	s. sol.
Phenol	43	182	s. sol.
Styrene	−30.6	145.2	insol.
Toluene	−95	110.6	insol.

Physical Properties of Carboxylic Acids

Name	mp (°C)	bp (°C)	Solubility (g/100 g H₂O at 25 °C)
Formic acid	8.4	101	∞
Acetic acid	16.6	118	∞
Propionic acid	−21	141	∞
Butanoic acid	−5	162	∞
Pentanoic acid	−34	186	4.97
Hexanoic acid	−4	202	0.97
Heptanoic acid	−8	223	0.24
Octanoic acid	17	237	0.068
Nonanoic acid	15	255	0.026
Decanoic acid	32	270	0.015

Physical Properties of Dicarboxylic Acids

Name	mp (°C)	Solubility (g/100 g H₂O at 25 °C)
Oxalic acid	189	S
Malonic acid	136	v. sol.
Succinic acid	185	s. sol.
Glutaric acid	98	v. sol.
Adipic acid	151	s. sol.
Pimelic acid	106	s. sol.
Phthalic acid	231	s. sol.
Maleic acid	130.5	v. sol.
Fumaric acid	302	s. sol.

Physical Properties of Acyl Chlorides and Acid Anhydrides

Name	mp (°C)	bp (°C)
Acetyl chloride	−112	51
Propionyl chloride	−94	80
Butyryl chloride	−89	102
Valeryl chloride	−110	128
Acetic anhydride	−73	140
Succinic anhydride		120

Physical Properties of Amides

Name	mp (°C)	bp (°C)
Formamide	3	200 d*
Acetamide	82	221
Propanamide	80	213
Butanamide	116	216
Pentanamide	106	232

*d means the substance decomposes.

Physical Properties of Esters

Name	mp (°C)	bp (°C)
Methyl formate	−100	32
Ethyl formate	−80	54
Methyl acetate	−98	57.5
Ethyl acetate	−84	77
Propyl acetate	−92	102
Methyl propionate	−87.5	80
Ethyl propionate	−74	99
Methyl butyrate	−84.8	102.3
Ethyl butyrate	−93	121

Physical Properties of Aldehydes

Name	mp (°C)	bp (°C)	Solubility (g/100 g H₂O at 25 °C)
Formaldehyde	−92	−21	v. sol.
Acetaldehyde	−121	21	∞
Propionaldehyde	−81	49	16
Butyraldehyde	−96	75	7
Pentanal	−92	103	s. sol.
Hexanal	−56	131	s. sol.
Heptanal	−43	153	0.1
Octanal		171	insol.
Nonanal		192	insol.
Decanal	−5	209	insol.
Benzaldehyde	−26	178	0.3

Physical Properties of Ketones

Name	mp (°C)	bp (°C)	Solubility (g/100 g H₂O at 25 °C)
Acetone	−95	56	∞
2-Butanone	−86	80	25.6
2-Pentanone	−78	102	5.5
2-Hexanone	−57	127	1.6
2-Heptanone	−36	151	0.4
2-Octanone	−16	173	insol.
2-Nonanone	−7	195	insol.
2-Decanone	14	210	insol.
3-Pentanone	−40	102	4.8
3-Hexanone		123	1.5
3-Heptanone	−39	149	0.3
Acetophenone	19	202	insol.
Propiophenone	18	218	insol.

Answers to Selected Problems

CHAPTER 1

REMEMBERING GENERAL CHEMISTRY: ELECTRONIC STRUCTURE AND BONDING

1. $8 + 8, 8 + 9, 8 + 10$ **2. a. 1.** 11 **2.** 18 **3.** 17   **b. 1.** 10 **2.** 18 **3.** 18   **3.** 35.453 amu

2. **a.** 3 **b.** 5 **c.** 6 **d.** 7

3. **a.** Cl $1s^2 2s^2 2p^6 3s^2 3p^5$; Br $1s^2 2s^2 2p^6 3s^2 3p^6 3d^{10} 4s^2 4p^5$; I $1s^2 2s^2 2p^6 3s^2 3p^6 3d^{10} 4s^2 4p^6 4d^{10} 5s^2 5p^5$ **b.** 7

4. **a.** 1 **b.** 4s

5. **a.** Cl—CH₃ **b.** H—OH **c.** H—F **d.** Cl—CH₃   **9. a.** KCl **b.** Cl₂

11. **a.** $\overset{\delta-}{HO}—\overset{\delta+}{H}$ **c.** $\overset{\delta+}{H_3C}—\overset{\delta-}{NH_2}$ **e.** $\overset{\delta-}{HO}—\overset{\delta+}{Br}$ **g.** $\overset{\delta+}{I}—\overset{\delta-}{Cl}$

   b. $\overset{\delta-}{F}—\overset{\delta+}{Br}$ **d.** $\overset{\delta+}{H_3C}—\overset{\delta-}{Cl}$ **f.** $\overset{\delta-}{H_3C}—\overset{\delta+}{Li}$ **h.** $\overset{\delta+}{H_2N}—\overset{\delta-}{OH}$

14. **a.** LiH and HF   **b.** Its hydrogen has the greatest electron density.   **c.** HF

15. **a.** oxygen **b.** oxygen **c.** oxygen **d.** hydrogen

18. **a.** CH₃CH₂OH   CH₃OCH₃

   b. CH₃CH₂CH₂OH   CH₃CHCH₃   CH₃CH₂OCH₃
                                  |
                                  OH

19. **a.** CH₃CH₂N̈H₂   **c.** CH₃CH₂ÖH   **e.** CH₃CH₂C̈l:

   b. CH₃N̈HCH₃   **d.** CH₃ÖCH₃   **f.** HÖN̈H₂

20. **a.** CH₃CH₂CH₂Cl   **c.** CH₃CH₂CNCH₂CH₃
                                        ‖ |
                                        O   CH₃

   b. CH₃COCH₂CH₃   **d.** CH₃CH₂C≡N
         ‖
         O

21. **a.** Cl **b.** O **c.** N **d.** C and H   **25.** yes   **26. a.** π* **b.** π **c.** σ* **d.** σ

27. The C—C bonds are formed by sp^3–sp^3 overlap; the C—H bonds are formed by sp^3–s overlap.   **33. a.** 120° **b.** 120° **c.** 107.3°   **34.** the nitrogen

35. most = water; least = methane   **37. a.** relative lengths: Br₂ > Cl₂ ; relative strengths: Cl₂ > Br₂   **b.** relative lengths; CH₃Br > CH₃Cl > CH₃F; relative strengths: CH₃F > CH₃Cl > CH₃Br

38. **a. 1.** C-I **2.** C-Cl **3.** H-Cl **b. 1.** C-Cl **2.** C-C **3.** H-F   **41.** σ

42. sp^2–sp^2 is stronger because the more s character, the stronger the bond

45. **a.** 109.5° **b.** 107.3° **c.** 109.5° **d.** 109.5°   **46.** a, e, g, h   **48.** hydrogen

CHAPTER 2

ACIDS AND BASES: CENTRAL TO UNDERSTANDING ORGANIC CHEMISTRY

1. CO₂ and CCl₄   **2. a.** HBr **b.** ⁻C≡N **c.** Br⁻ **d.** HC≡N **e.** HC≡N **f.** Br⁻ **g.** ⁻C≡N **h.** HBr   **3. a.** Cl⁻ + ⁺NH₄ **b.** HO⁻ + NH₃   **4. a. 1.** ⁺NH₄ **2.** HCl **3.** H₂O **4.** H₃O⁺ **b. 1.** ⁻NH₂ **2.** Br⁻ **3.** NO₃⁻ **4.** HO⁻   **5. a.** 5.2 **b.** 3.4 × 10⁻³   **6.** 8.16 × 10⁻⁸   **7.** $K_a = 1.51 \times 10^{-5}$; weaker

9. **a.** basic **b.** acidic **c.** basic

10. **a.** CH₃CH₂ÖH₂   **c.** CH₃   ⁺OH   **e.** CH₃CH₂   OH
                                     ‖C   ‖C
                                 | | | |
                                     OH   OH

   b. CH₃CH₂OH   **d.** CH₃CH₂N̈H₃

12. 40, 15, 5, 10   **13. a.** CH₃COO⁻ **b.** ⁻NH₂ **c.** H₂O

14. CH₃NH⁻ > CH₃O⁻ > CH₃NH₂ > CH₃CO⁻ > CH₃OH
                                          ‖
                                          O

15. acid   **18.** HO⁻ CH₃NH₂ HC≡C⁻

19. **a.** 2.0 × 10⁵ **b.** 3.2 × 10⁻⁷ **c.** 1.0 × 10⁻⁵ **d.** 4.0 × 10⁻¹³

20. ⁻CH₃ > ⁻NH₂ > HO⁻ > F⁻   **21.** CH₃CH₂⁻ > H₂C=CH⁻ > HC≡C⁻

22. the one on the right   **25.** F⁻ > Cl⁻ > Br⁻ > I⁻

26. **a.** oxygen **b.** H₂S **c.** CH₃SH   **27. a.** HBr **b.** CH₃CH₂CH₂ÖH₂

   c. the one on the right   **d.** CH₃CH₂CH₂SH   **28. a.** I⁻ **b.** F⁻

29. **a.** CH₃O⁻ **b.** HO⁻ **c.** NH₃ **d.** CH₃O⁻

30. **a.** CH₃OCH₂CH₂OH **b.** CH₃CH₂CF₂CH₂ÖH₂ **c.** CH₃CH₂OCH₂CH₂OH

   d. CH₃CH₂COH
                ‖
                O

31. CH₂CHCH₂COOH > CH₃CHCH₂COOH > CH₂CH₂CH₂COOH > CH₃CH₂CH₂COOH
        | | | |
        F F   F   F

32. **a.** CH₃CHCO⁻ **b.** CH₃CHCH₂CO⁻ **c.** CH₃CH₂CO⁻ **d.** CH₃CCH₂CH₂O⁻
            | ‖ | ‖   ‖   ‖
            Br O   Cl O   O   O

35. **a.** (resonance structures)

   b. (resonance structures)

36. CH₃S—OH   forms a more stable base because the electrons left behind when the proton is
        ‖   removed are shared by 3 oxygens.
        O

38. **a.** CH₃C≡⁺NH **b.** CH₃CH₃ **c.** F₃CCOH
                                            ‖
                                            O

   d. sp^2 CH₃   ⁺OH   CH₃   CH₃
                        ‖C   ⁺OCH₃
                    | | |
                                          H
              $pK_a = -7.3$ $pK_a = -3.6$

                              CH₃
                              |
   e. CH₃C≡⁺NH   CH₃C=⁺NHCH₃   CH₃CH₂N̈H₃

        $pK_a = -10.1$ $pK_a = 5.5$ $pK_a = 11.0$

39. **a.** CH₃COO⁻ **b.** CH₃CH₂⁺NH₃ **c.** H₂O **d.** Br⁻ **e.** ⁺NH₄ **f.** HC≡N **g.** NO₂⁻ **h.** NO₃⁻ **i.** HON̈H₃   **40.** 10.4   **42. a. 1.** charged **2.** charged **3.** charged **4.** charged **5.** neutral **6.** neutral **b. 1.** neutral **2.** neutral **3.** neutral **4.** neutral **5.** neutral **6.** neutral

45. **a.** 6.4 **b.** 7.3 **c.** 5.6

46. **a. 1.** 4.9 **2.** 10.7 **b. 1.** > 6.9 **2.** < 8.7

48. **a.** ≥ 12.7 **b.** ≤ 2.8

49. **a.** CH₃COO⁻ + H⁺ ⇌ CH₃COOH

   b. CH₃COOH + HO⁻ ⇌ CH₃COO⁻ + H₂O

CHAPTER 3

AN INTRODUCTION TO ORGANIC COMPOUNDS

1. **a.** 36 hydrogens **b.** 36 carbons   **3. a.** n-propyl alcohol or propyl alcohol

b. butyl methyl ether **c.** n-propylamine or propylamine

4. **a.** CH₃CHCH₂CH₃   **b.** CH₃CCH₃
            | |
            CH₃   CH₃
    2-methylbutane
                              2,2-dimethylpropane

5. CH₃CH₂CH₂CH₂Br   CH₃CHCH₂Br   CH₃CH₂CHCH₃   CH₃CBr
                         | | |
                         CH₃   Br   CH₃
                                                   CH₃
    butyl bromide   isobutyl bromide   sec-butyl bromide   tert-butyl bromide
        or
    n-butyl bromide

6. **c**

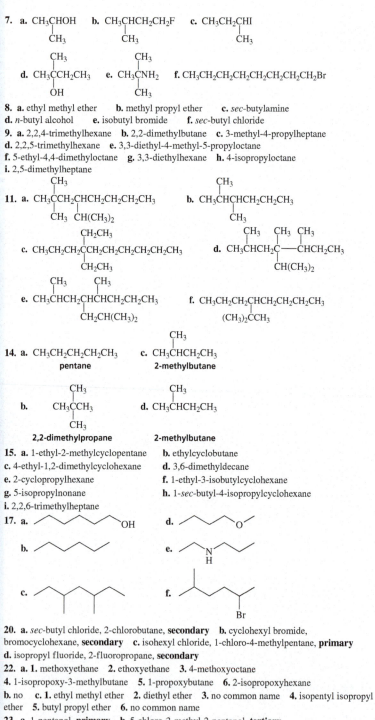

7. a. CH₃CHOH | **b.** CH₃CHCH₂CH₂F | **c.** CH₃CH₂CHI
with CH₃, CH₃, CH₃ substituents

d. CH₃CCH₂CH₃ (with CH₃ and OH) **e.** CH₃CNH₂ (with CH₃, CH₃) **f.** CH₃CH₂CH₂CH₂CH₂CH₂CH₂Br

8. a. ethyl methyl ether   **b.** methyl propyl ether   **c.** *sec*-butylamine
d. *n*-butyl alcohol   **e.** isobutyl bromide   **f.** *sec*-butyl chloride
9. a. 2,2,4-trimethylhexane **b.** 2,2-dimethylbutane **c.** 3-methyl-4-propylheptane
d. 2,2,5-trimethylhexane **e.** 3,3-diethyl-4-methyl-5-propyloctane
f. 5-ethyl-4,4-dimethyloctane **g.** 3,3-diethylhexane **h.** 4-isopropyloctane
i. 2,5-dimethylheptane

11. a. CH₃CCH₂CHCH₂CH₂CH₂CH₃ (CH₃; CH₃ CH(CH₃)₂)   **b.** CH₃CHCHCH₂CH₂CH₃ (CH₃; CH₃)
c. CH₃CH₂CH₂CCH₂CH₂CH₂CH₂CH₂CH₃ (CH₂CH₃; CH₂CH₃)   **d.** CH₃CHCHC—CHCH₂CH₃ (CH₃ CH₃ CH₃; CH(CH₃)₂)
e. CH₃CHCH₂CHCHCH₂CH₂CH₃ (CH₃ CH₃; CH₂CH(CH₃)₂)   **f.** CH₃CH₂CH₂CHCH₂CH₂CH₂CH₃ ((CH₃)₂CCH₃)

14. a. CH₃CH₂CH₂CH₂CH₃ pentane **c.** CH₃CHCH₂CH₃ (CH₃) 2-methylbutane
b. CH₃CCH₃ (CH₃ top, CH₃ bottom) 2,2-dimethylpropane **d.** CH₃CHCH₂CH₃ (CH₃) 2-methylbutane

15. a. 1-ethyl-2-methylcyclopentane   **b.** ethylcyclobutane
c. 4-ethyl-1,2-dimethylcyclohexane   **d.** 3,6-dimethyldecane
e. 2-cyclopropylhexane   **f.** 1-ethyl-3-isobutylcyclohexane
g. 5-isopropylnonane   **h.** 1-*sec*-butyl-4-isopropylcyclohexane
i. 2,2,6-trimethylheptane

17. a. (structure) OH   **d.** (structure) O
b. (structure)   **e.** (structure) N–H
c. (structure)   **f.** (structure) Br

20. a. *sec*-butyl chloride, 2-chlorobutane, **secondary** **b.** cyclohexyl bromide, bromocyclohexane, **secondary** **c.** isohexyl chloride, 1-chloro-4-methylpentane, **primary**
d. isopropyl fluoride, 2-fluoropropane, **secondary**
22. a. 1. methoxyethane **2.** ethoxyethane **3.** 4-methoxyoctane
4. 1-isopropoxy-3-methylbutane **5.** 1-propoxybutane **6.** 2-isopropoxyhexane
b. no   **c. 1.** ethyl methyl ether **2.** diethyl ether **3.** no common name **4.** isopentyl isopropyl ether **5.** butyl propyl ether **6.** no common name
23. a. 1-pentanol, **primary** **b.** 5-chloro-2-methyl-2-pentanol, **tertiary**
c. 5-methyl-3-hexanol, **secondary** **d.** 7-methyl-3,5-octanediol, both **secondary**

25. CH₃CCH₂CH₂CH₃ (CH₃; OH)   CH₃CH₂CCH₂CH₃ (CH₃; OH)   CH₃C—CHCH₃ (CH₃; OH CH₃)
(structures) 2-methyl-2-pentanol   3-methyl-3-pentanol   2,3-dimethyl-2-butanol

26. a. 4-chloro-3-ethylcyclohexanol, **secondary** **b.** 7,8-dimethyl-3-nonanol, **secondary**
c. 1-bromo-5,5-dimethyl-3-heptanol, **secondary** **d.** 4-methylcyclohexanol, **secondary**
27. a. hexylamine, 1-hexanamine, **primary** **b.** *sec*-butylisohexylamine, *N-sec*-butyl-4-methyl-1-pentanamine, **secondary** **c.** diethylmethylamine, *N*-ethyl-*N*-methylethanamine, **tertiary**
d. butylpropylamine, *N*-propyl-1-butanamine, **secondary** **e.** diethylpropylamine,
N,N-diethyl-1-propanamine, **tertiary** **f.** no common name, *N*-ethyl-3-methylcyclopentanamine,
secondary

28. a. tertiary **b.** primary **c.** secondary **d.** secondary
29. a. CH₃CHCH₂NHCH₂CH₂CH₃ (CH₃)   **b.** (structure N–H)
(structure, N–H, CH₃)
c. CH₃CHCH₂CH₂CH₂CH₂NH₂ (CH₃)   **d.** CH₃CH₂CH₂NCH₂CH₂CH₃ (CH₃)
(structure, NH₂)   (structure N)
e. CH₃CH₂CHNCH₃ (CH₂CH₃; CH₃)   **f.** (cyclohexyl NCH₂CH₃, CH₃)
(structure N)   (cyclohexyl N)
(structure N)

30. a. 6-methyl-1-heptanamine, isooctylamine, **primary**
b. cyclohexanamine, cyclohexylamine, **primary** **c.** 4-methyl-*N*-propyl-1- pentanamine,
isohexylpropylamine, **secondary** **d.** 2,5-dimethylcyclohexanamine, no common name, **primary**
31. a. 104.5° **b.** 107.3° **c.** 104.5° **d.** 109.5°
32. pentane
33. a. 1, 4, 5 **b.** 1, 2, 4, 5, 6
35. HO(structure)OH > (structure)OH > (structure)OH > (structure)NH₂ >
(structure) > (structure)
37. ethanol
38. a. (structure)OH, OH O > (structure)OH, O > (structure)OH > (structure)
b. HOCH₂CH₂CH₂OH > CH₃CH₂CH₂CH₂OH > CH₃CH₂CH₂CH₂Cl
39. hexethal
42. a. (Newman: CH₃, CH₃, H, H, H, H, CH₂CH₃) **b.** (Newman: CH₂CH₃, H, H, H, CH₃, CH₂CH₃) **c.** (Newman: CH₂CH₃, H, H, H₃C, CH₃, CH₂CH₃)
43. a. 135° **b.** 140° **46.** 84% **47.** 0.13 kcal/mol
48. a. cis **b.** cis **c.** cis **d.** trans **e.** trans **f.** trans
49. *cis*-1-*tert*-butyl-3-methylcyclohexane
51. a. one equatorial and one axial **b.** both equatorial and both axial
c. both equatorial and both axial **d.** one equatorial and one axial
e. one equatorial and one axial **f.** both equatorial and both axial
54. a. 3.6 kcal/mol **b.** 0

CHAPTER 4

ISOMERS: THE ARRANGEMENT OF ATOMS IN SPACE

1. a. CH₃CH₂CH₂OH   CH₃CHOH (CH₃)   CH₃CH₂OCH₃ **b.** 7

3. a. 1 and 3 **b. 1.** *cis*-2-hexene; *trans*-2-hexene **3.** *cis*-2-butene; *trans*-2-butene
5. CH₃CH₂CH₂CH=CH₂   CH₃CH=CCH₃ (CH₃)   CH₃CHCH=CH₂ (CH₃)   CH₃CH₂C=CH₂ (CH₃)

6. C **8. a.** ⁻I > ⁻Br > ⁻OH > ⁻CH₃
b. ⁻OH > ⁻CH₂Cl > ⁻CH=CH₂ > ⁻CH₂CH₂OH **9.** Z **11. a.** (E)-2-heptene
b. (Z)-3,4-dimethyl-2-pentene **c.** (Z)-1-chloro-3-ethyl-4-methyl-3-hexene
13. a, b, c, f, h **14.** a, c, and f **16.** a, c, and f **17. a.** 1 **b.** 3
20. a. —CH₂OH (1) —CH₃ (3) —H (4) —CH₂CH₂OH (2)
b. —CH₂Br (2) —OH (1) —CH₃ (4) —CH₂OH (3)
c. —CH(CH₃)₂ (2) —CH₂CH₂Br (3) —Cl (1) —CH₂CH₂CH₂Br (4)
d. —CH=CH₂ (2) —CH₂CH₃ (3) (phenyl) (1) —CH₃ (4)

22. a. *R* **b.** *R* **c.** *R*   **23. a.** *S* **b.** *R* **c.** *S* **d.** *S*

24. a. identical **b.** enantiomers **c.** enantiomers **d.** enantiomers

28. a. levorotatory **b.** dextrorotatory   **30. a.** *R* **b.** *R* **c.** *S*   **32.** *S*   **33.** +17

34. a. −24 **b.** 0

35. a. 0 **b.** +79 **c.** −79   **36. a.** do not know **b.** 98.5% dextrorotatory; 1.5% levorotatory   **39. a.** enantiomers **b.** identical **c.** diastereomers

40. a. 8 **b.** 2⁸ = 256   **43.** A, C, E   **46.** 1-chloro-1-methylcyclooctane, *cis*-1-chloro-5-methylcyclooctane, *trans*-1-chloro-5-methylcyclooctane

48. A = identical, B = enantiomer, C = diastereomer, D = identical

49. b, d, and f

54. asymmetric carbon on left = *R*; asymmetric carbon on right = *R*

56. a. (3*R*,4*S*)-3-chloro-4-methylhexane **b.** (2*S*,3*S*)-2-bromo-3-chloropentane **c.** (1*R*,3*S*)-3-bromocyclopentanol **d.** (2*R*,3*R*)-2,3-dichloropentane

57. the second one from the left   **61.**   **62. a.** A **b.** C

64. the one on the left

CHAPTER 5

ALKENES • THERMODYNAMICS AND KINETICS

2. a. C₄H₆ **b.** C₁₀H₁₆   **4. a.** 4 **b.** 1 **c.** 3 **d.** 13

6. 11

7. a. 4-methyl-2-pentene **b.** 2-chloro-3,4-dimethyl-3-hexene **c.** 1-bromocyclopentene **d.** 1-bromo-4-methyl-3-hexene **e.** 1,5-dimethylcyclohexene **f.** 1-butoxy-2-butene **g.** 1-bromo-2-methyl-1,3-pentadiene **h.** 8,8-dimethyl-1-nonene

9. a.   **c.** CH₃CH₂OCH=CH₂

b. BrCH₂CH₂CH₂C=CCH₃   **d.** CH₂=CHCH₂OH

11. a. 4 **b.** 4 **c.** 6

13. electrophiles: CH₃ĊHCH₃ ; nucleophiles: H⁻, CH₃O⁻, CH₃C≡CH, NH₃

18. a. all **b.** *tert*-butyl **c.** *tert*-butyl **d.** −1.7 kcal/mol or −7.1 kJ/mol

21. a. 1. A + B ⇌ C **d.** A ⇌ B + C   **2.** A ⇌ B + C

23. a. −21 kcal/mol **b.** −36 kcal/mol **c.** exothermic **d.** exergonic

26. A   **27. a.**   **b.**   **c.**

28. *cis*-3,4-dimethyl-3-hexene, *trans*-3-hexene, *cis*-3-hexene, *cis*-2,5-dimethyl-3-hexene

29. a. a and b **b.** b **c.** c

32. decreasing; increasing

33. a. it will decrease it **b.** it will increase it   **34. a.** the first reaction **b.** the first reaction   **36. a.** first step **b.** revert to reactants **c.** second step

37. a. 1 **b.** 2 **c.** see the *Solutions Manual* **d.** *k*₋₁ **e.** *k*₋₁ **f.** B ⟶ C

g. C ⟶ B   **38.** $\Delta H^{\ddagger}, E_a, \Delta G^{\ddagger}, \Delta S^{\ddagger}, k$

CHAPTER 6

THE REACTIONS OF ALKENES • THE STEREOCHEMISTRY OF ADDITION REACTIONS

2. a. 0 **b.** ethyl cation because of hyperconjugation   **3. a. 1.** 3 **2.** 3 **3.** 6 **b.** *sec*-butyl cation

4. a. CH₃CH₂ĊCH₃   >   CH₃CH₂ĊHCH₃   >   CH₃CH₂CH₂ĊH₂

b. CH₃CHCH₂ĊH₂   >   CH₃CHCH₂ĊH₂   >   CH₃CHCH₂ĊH₂
      | | |
      CH₃   Cl   F

5. a. products **b.** reactants **c.** reactants **d.** products

6. a. CH₃CH₂C=CH₂   **b.**
            |
            CH₃

7. a. CH₃CH₂CHCH₃   **c.**   **e.**
            |
            Br

b. CH₃CH₂ĊCH₃   **d.** CH₃ĊCH₂CH₂CH₃   **f.** CH₃CH₂CHCH₃
          | | |
          Br   Br   Br

8. a. CH₂=ĊCH₃   **b.** ⬡—CH₂CH=CH₂
          |
          CH₃

c.   **d.**   or

9. greater than −2.5 and less than 15   **10. a.** 3 **b.** 2 **c.** first step

11. a. CH₃CH₂CH₂CHCH₃   **c.** CH₃CH₂CH₂CHCH₂CH₃   **d.**
                | |
                OH   and OH

b.   CH₃CH₂CH₂CH₂CHCH₃
                         |
                         OH

17. a. CH₃ĊCH₂CH₃   **b.** CH₃CHCH₂CHCH₃   **c.**
          | | |
          Br   CH₃   Br

d.   **e.**   **f.**   and

18. 9-BBN

19. a. CH₃CHCHCH₃   **b.**   **20.** CH₃ĊCH₂CH₃
            | |
            OH   Br

25. a. CH₂CHCH₂CH₃   **c.** CH₂CHCH₂CH₃
          | | | |
          Br Br   Br OCH₂CH₃

b. CH₂CHCH₂CH₃   **d.** CH₂CHCH₂CH₃
        | | | |
        Br OH   Br OCH₃

27. a.   **c.**

b.   **d.**

28. a. 1-pentene **b.** cyclohexene **c.** 2,3-dimethyl-2-butene **d.** 2-methyl-2-pentene

31. a. 2,3-dimethyl-2-butene   **b.** *cis*-4-octene and *trans*-4-octene

32. whether the alkene has the *E* or the *Z* configuration

35. a double bond or an asymmetric center

36. a. no **b.** no **c.** no **d.** yes **e.** no **f.** no

41. a. 1. *trans*-3-heptene **2.** *cis*-heptene **b.** the enantiomer of each epoxide would also be formed **c. 1.** *R,R* **2.** *R,S*

42. a.   **b.**

c.   **d.**

50. a. 1-bromo-2-chloropropane **b.** equal amounts of *R* and *S*

51. a. (*R*)-malate and (*S*)-malate **b.** (*R*)-malate and (*S*)-malate

54. 3-Methylcyclohexanol would also be formed.

CHAPTER 7

THE REACTIONS OF ALKYNES: AN INTRODUCTION TO MULTISTEP SYNTHESIS

1. a. 5-bromo-2-pentyne **b.** 6-bromo-2-chloro-4-octyne
c. 1-methoxy-2-pentyne **d.** 3-ethyl-1-hexyne
2. a. 6-methyl-2-octyne **b.** 5-ethyl-4-methyl-1-heptyne **c.** 2-bromo-4-octyne
3. $C_{14}H_{20}$
4. a. $ClCH_2CH_2C\equiv CCH_2CH_3$ **d.** $CH_3CH_2CHC\equiv CCH_2CHCH_3$ (with CH₃, CH₃ substituents)

b. (cyclooctyne ring) **e.** $HC\equiv CCH_2CCH_3$ (with two CH₃)

c. $CH_3CHC\equiv CH$ (with CH₃) **f.** $CH_3C\equiv CCH_3$

6. a. 1-hepten-4-yne **b.** 4-methyl-1,4-hexadiene **c.** 5-vinyl-5-octen-1-yne
d. 3-butyn-1-ol **e.** 1,3,5-heptatriene **f.** 2,4-dimethyl-4-hexen-1-ol
7. a. 2-hepten-4-ol **b.** 1-hepten-5-yne **c.** 4-hepten-1-yne
8. a. sp^2-sp^2 **b.** sp^2-sp^3 **c.** $sp-sp^2$ **d.** $sp-sp^3$ **e.** $sp-sp$ **f.** sp^2-sp^2 **g.** sp^2-sp^3 **h.** $sp-sp^3$
i. sp^2-sp
9. pentane, 1-pentene, 1-pentyne
10. the cis isomer has a dipole moment; the trans isomer does not

12. a. $CH_2=CCH_3$ (with Br) **b.** CH_3CCH_3 (Br, Br) **c.** $CH_3C=CCH_3$ (Br, Br) **d.** $HC-CCH_3$ (Br Br, Br Br)

e. $CH_3CH_2CCH_3$ (Br, Br) **f.** $CH_3CCH_2CH_2CH_3$ (Br, Br) $+$ $CH_3CH_2CCH_2CH_3$ (Br, Br)

14. $CH_3CH_2CCH_2CH_2CH_2CH_3$ and $CH_3CH_2CH_2CCH_2CH_2CH_3$ (both with =O)

15. a. $CH_3C\equiv CH$ **b.** $CH_3CH_2C\equiv CCH_2CH_3$ **c.** $HC\equiv C-$(cyclohexyl)

16. a. $CH_2=CCH_3$ (OH) **b.** $CH_3CH=CCH_2CH_2CH_3$ (OH) and $CH_3CH_2C=CHCH_2CH_3$ (OH)
c. $CH_2=C-$(cyclohexyl) (OH) and $CH_3C=$(cyclohexyl) (OH)

18. ethyne (acetylene)
19. a. $CH_3CH_2CH_2C\equiv CH$ or $CH_3CH_2C\equiv CCH_3$ $\xrightarrow{H_2, Pd/C}$
b. $CH_3C\equiv CCH_3$ $\xrightarrow{H_2, \text{Lindlar catalyst}}$
c. $CH_3CH_2C\equiv CCH_3$ $\xrightarrow{Na, NH_3(liq)}$
d. $CH_3CH_2CH_2CH_2C\equiv CH$ $\xrightarrow{H_2, \text{Lindlar catalyst}}$ or $\xrightarrow{Na, NH_3(liq)}$

22. The carbanion that would be formed is a stronger base than the amide ion.
23. 25
24. a. $CH_3CH_2CH_2\bar{C}H_2 > CH_3CH_2CH=\bar{C}H > CH_3CH_2C\equiv\bar{C}$
b. $^-NH_2 > CH_3C\equiv\bar{C} > CH_3CH_2O^- > F^-$
26. a. $H_2C=\overset{+}{C}H$ **b.** $CH_3\overset{+}{C}H_2$

CHAPTER 8

DELOCALIZED ELECTRONS • AROMATICITY AND ELECTRONIC EFFECTS

1. a. 1. 4 **2.** 3 **b. 1.** 5 **2.** 5 **c. 1.** 7 **2.** 8

4. a. $CH_3CH_2CH=CH\overset{+}{C}H_2$ **c.** $CH_3\overset{+}{C}=CHCH_3$
b. $CH_3\overset{O}{C}CH=CHCH_3$ **d.** $CH_3-\overset{+NH_2}{\underset{||}{C}}-NH_2$

7. a. all are the same length **b.** 2/3 of a negative charge

10. (structure with O and methoxy group)
11. conjugated > isolated > cumulated
12. $CH_3\overset{CH_3}{C}=CHCH=\overset{CH_3}{C}CH_3 > CH_3CH=CHCH=CHCH_3 >$
2,5-dimethyl-2,4-hexadiene   2,4-hexadiene
$CH_3CH=CHCH=CH_2 > CH_2=CHCH_2CH=CH_2$
1,3-pentadiene   1,4-pentadiene

13. a. (cyclohexenyl with $\overset{+}{C}HCH_3$) **b.** $CH_3\overset{+}{NHCH_2}$ **c.** (phenyl isopropyl cation)
14. $\psi_3 = 3; \psi_4 = 4$
15. a. bonding $= \psi_1$ and ψ_2; antibonding $= \psi_3$ and ψ_4 **b.** symmetric $= \psi_1$ and ψ_3;
antisymmetric $= \psi_2$ and ψ_4 **c.** HOMO $= \psi_2$; LUMO $= \psi_3$
d. HOMO $= \psi_3$; LUMO $= \psi_4$ **e.** If the HOMO is symmetric, the LUMO is antisymmetric
and vice versa.
16. benzene has 6 bonding interactions; 1,3,5-hexatriene has 5 bonding interactions.
17. a. $CH_3\overset{O}{C}OH$ **b.** $CH_3CH=CH\overset{+}{N}H_2$ **c.** $CH_3CH=CHOH$
18. a. ethylamine **b.** ethoxide ion **c.** ethoxide ion **d.** phenolate ion
19. (phenyl)−COOH $>$ (phenyl)−OH $>$ (phenyl)−CH_2OH

20. a. donates electrons by resonance and withdraws electrons inductively **b.** donates electrons by
hyperconjugation **c.** withdraws electrons by resonance and withdraws electrons inductively
d. donates electrons by resonance and withdraws electrons inductively **e.** donates electrons by
resonance and withdraws electrons inductively **f.** withdraws electrons inductively
21. a. $ClCH_2\overset{O}{C}OH$ **b.** $O_2NCH_2\overset{O}{C}OH$ **c.** (phenyl with COOH) **d.** $H_3\overset{+}{N}CH_2\overset{O}{C}OH$
e. $H\overset{O}{C}OH$ **f.** (phenyl with COOH and Cl)

24. 1-bromo-1-phenylpropane
27. a. 6-bromo-6-methyl-1-heptene **b.** 5,6-dichloro-1-hexyne
c. 5-chloro-5-methylcycloheptene
28. a. $CH_3CHCHCH=CHCH_3$ (Cl, Cl) $+$ $CH_3CHCH=CHCHCH_3$ (Cl, Cl) **29.** (structure, tertiary allylic)
b. $CH_3CH_2C-C=CHCH_3$ (Br, CH_3CH_3) $+$ $CH_3CH_2C=CCHCH_3$ (Br, CH_3CH_3)
c. (cyclopentane with Br, Br) $+$ (cyclopentene with Br, Br) **d.** (cyclohexene with Cl)

32. a. 4,5-dichloro-4-methyl-2-pentene + 1,4-dichloro-2-methyl-2-pentene
b. 4-bromo-4-methyl-2-hexene + 2-bromo-4-methyl-3-hexene
33. a. Addition at C-1 forms the more stable carbocation.
b. To cause the 1,2- and 1,4-products to be different.
34. yes **35. a.** formation of the carbocation **b.** reaction of the carbocation with the nucleophile

39. a. (structure with two $\overset{O}{C}CH_3$ groups) **c.** (anhydride structure with CH₃ groups) **40.** (structure with CH₃O and CH=O)
b. (structure with C≡N) **d.** (decalin structure with CH₃ groups and O)

42. a. (structures with CH₃ and C≡N) **b.** (structures with C≡N and H₃C)

3. A and C 46. a. The product is a meso compound.
. The product is a racemic mixture.

7. a. [structure] CHC≡N / CH₂ b. [furan structure] + HC≡CH

c. [cyclopentadiene] + CH₃OCC≡CCOCH₃ d. [structures] + [cyclohexenone]

e. [structures with H, CH, C=C, CHO] f. [structures with H, COOH, C=C, HOOC, H]

8. none
9. a. 4 b. It will be aromatic if it is cyclic, if it is planar, and if every carbon in the ring has p orbital.
0. a. cyclopropenyl cation b. cycloheptatrienyl cation
2. a. Cyclopentadiene has the lower pK_a. b. Cyclopropane has the lower pK_a.
3. 3-bromocyclopropene
6. a. sp^3 b. sp^2 c. sp
7. quinoline = sp^2; indole = p; imidazole = 1 is sp^2 and 1 is p; purine = 3 are sp^2 and 1 is p; pyrimidine = both are sp^2

CHAPTER 9

SUBSTITUTION AND ELIMINATION REACTIONS OF ALKYL HALIDES

. a. tripled b. half as fast. c. no change
. decrease

. CH₃CH₂CH₂CH₂CH₂Br > CH₃CHCH₂CH₂Br (CH₃) > CH₃CH₂CHCH₂Br (CH₃) > CH₃CH₂CBr (CH₃)(CH₃)

. a. 2-methyoxybutane b. (S)-2-methoxybutane c. (R)-3-hexanol d. 3-pentanol
. a. 1-bromo-2-methylbutane b. 2-bromopropane c. 2-bromobutane
. 1-bromo-2-phenylethane 10. a. aprotic b. aprotic c. protic d. aprotic
1. a. RO⁻ b. RS⁻ c. RO⁻ 12. a. Br⁻ b. Cl⁻ c. CH₃O⁻ d. CH₃O⁻ e. ⁻NH₂ , ⁻NH₂ g. I⁻ h. Br⁻
4. a. CH₃CH₂Br + HO⁻ b. CH₃CHCH₂Br (CH₃) + HO⁻
. CH₃CH₂Cl + CH₃S⁻ d. CH₃CH₂Br + I⁻
5. a. CH₃CH₂OCH₂CH₂CH₃ b. CH₃CH₂C≡CCH₃
. CH₃CH₂N⁺(CH₃)₃Br⁻ d. CH₃CH₂SCH₂CH₃
8. a. 3-methoxy-3-methylpentane b. (R)-3-methoxy-3-methylhexane + (S)-3-methoxy--methylhexane
9. 2-iodo-2-methylpentane, 2-bromo-2-methylpentane, 2-chloro-2-methylpentane, 3-chloropentane
1. A and B
2. a. CH₃C=CHCH₃ (CH₃) b. CH₃C=CHCH₃ (CH₃) c. CH₃CH₂CCH=CH₂ (CH₃)
3. a. CH₃CH=CHCH₃ c. CH₃CH=CHCH=CH₂ e. [cyclohexene]
. CH₃C=CHCH₂CH₃ (CH₃) d. CH₂=CHCH₂CH₃ f. CH₃CHCH=CHCH₃ (CH₃)
4. a. [structure Br] b. [cyclohexane Br] c. [structure Br] d. [structure Br, Cl]
5. a. [structure Br] b. [cycloheptene Br] c. [structure Br] d. [structure Br phenyl]
6. CH₃C=CCH₂CH₃ (CH₃)(CH₃) > [structure H₃C, CH₃, CH₃CH, H] > [structure H₃C, H, CH₃CH, CH₃] > CH₃CHCCH₂CH₃ (CH₂)(CH₃)
7. most stable because elimination occurs through the carbocation

28. a. B b. B c. B d. A

33. a. 1. CH₃CH₂CH=CCH₃ (CH₃) 3. CH₃CH₂ (H) [C=C] b. no
2. CH₃CH₂ (H) [C=C] H / CH=CH₂ [structure phenyl H C=C H]

36. the eliminated atoms have to be in axial positions
37. the cis isomer; elimination only from the *cis* isomer occurs through the more stable conformer
39. a. 1-bromopropane b. 1-iodo-1-methylcyclohexane c. 2-bromo-2-methylbutane
41. it will increase
42. When Br is in an axial position, both adjacent Hs are in equatorial positions.
43. a. S_N2 difficult because of steric hindrance; no S_N1 because cannot form a primary carbocation b. No E2 because there are no hydrogens on a β-carbon; no E1 because cannot form a primary carbocation.
45. a. 1,3-cyclohexadiene b. 1-phenyl-1-propene c. 4-methyl-1,3-pentadiene
46. The answers are the same as those in 45.

48. a. [phenyl CH₂Br structure] Br b. [structure Br]

49. The answers are the same as those in 48.
54. it would increase
55. a. decrease b. decrease c. increase
56. a. CH₃Br + HO⁻ ⟶ CH₃OH + Br⁻
   b. CH₃I + HO⁻ ⟶ CH₃OH + I⁻
   c. CH₃Br + NH₃ ⟶ CH₃N⁺H₃ + Br⁻
   d. CH₃Br + HO⁻ →[DMSO] CH₃OH + Br⁻
   e. CH₃Br + NH₃ →[EtOH] CH₃N⁺H₃ + Br⁻
58. dimethyl sulfoxide
59. HO⁻ in 50% water/50% ethanol
60. a. CH₃CH=CH₂ b. CH₃CH₂CH=CH₂
62. CH₃C—OH (CH₃) + CH₃C—OCH₂CH₃ (CH₃) + CH₃C=CH₂ (CH₃)
64. a. HO⟋⟍Br c. HO⟋⟍Br
   b. HO⟋Br

CHAPTER 10

REACTIONS OF ALCOHOLS, ETHERS, EPOXIDES, AMINES, AND SULFUR-CONTAINING COMPOUNDS

1. no lone pair
6. a. CH₃CH₂CHCH₃ (Br)(Br) d. [structure Br]
   b. [cyclopentane CH₃ Cl] e. [cyclohexane CH₃ Cl]
   c. CH₃C—CHCH₃ (CH₃)(Br)(CH₃) f. [cyclohexane CH₃ CH₂CH₃ Cl]
8. [structure Cl] [structure Cl]
13. D 19. a. *trans*-2-pentene (major) + *cis*-2-pentene
   b. (E)-3,4-dimethyl-3-hexene (major) + (Z)-3,4-dimethyl-3-hexene
20. a. 3-ethyl-2,4-dimethyl-3-hexene b. (E)-3-ethyl-2,4-dimethyl-3-hexene
21. a. CH₃CH₂CHCH₂OH (CH₃) c. CH₃CH₂CHCH₂CH₃ (OH)
   b. [cyclohexane CH₃ OH] d. [cyclohexane CH₂OH]

22. b. or **c.**

22. 1. $CH_3CH_2\overset{O}{\overset{\|}{C}}CH_2CH_3$ **2.** $CH_3CH_2CH_2CH_2\overset{O}{\overset{\|}{C}}H$ **3.** no reaction

4. $CH_3\overset{O}{\overset{\|}{C}}CH_2\overset{O}{\overset{\|}{C}}CH_2CH_3$ **5.** (cyclohexanone structure)   **6.** $H\overset{O}{\overset{\|}{C}}CH_2CH_2\overset{O}{\overset{\|}{C}}H$

26. No, F^- and Cl^- are poor nucleophiles.

27. a. $HOCH_2\overset{CH_3}{\overset{|}{\underset{|}{\underset{OCH_3}{C}}}}CH_3$ **b.** $CH_3OCH_2\overset{OH}{\overset{|}{\underset{|}{\underset{CH_3}{C}}}}CH_3$ **c. and d.** $CH_3O\overset{CH_3OH}{\overset{|}{\underset{|}{\underset{CH_3CH_3}{C}}}}CCH_3$

28. noncyclic ether

31. a. $CH_3\overset{OH\ OH}{\overset{|\ \ \ |}{\underset{|}{\underset{CH_3}{C}}}}CHCH_2CH_3$ **b.** (cyclohexane with OH and CH_2OH)

40. (naphthalene epoxide structure)

44. $CH_3\overset{}{\underset{\underset{CH_3}{|}}{CH}}CH_2\overset{}{\underset{\underset{CH_3}{|}}{N}}CH_3$ + $CH_2=CHCH_3$

45. $CH_3\overset{}{\underset{\underset{CH_3}{|}}{C}}=CH_2$ + $\overset{CH_3}{\overset{|}{\underset{\underset{CH_3}{|}}{N}}}CH_2CH_2CH_3$

46. The only difference is the leaving group.

47. a. $CH_2=CHCH\overset{}{\underset{\underset{CH_3}{|}}{CH}}CH_2CH_2NCH_3$ **c.** $CH_3NCH_2\overset{}{\underset{\underset{CH_3}{|}}{CH}}CHCH_2CH=CH_2$

b. (methylenecyclohexane) + $N(CH_3)_3$

51. The first one is too insoluble; the second is too reactive; the third is too unreactive.

52. It has two sites that can be attacked by a nucleophile, and each site has an excellent leaving group.

CHAPTER 11

ORGANOMETALLIC COMPOUNDS

1. They all favor products.   **2.** an organosodium compound

3. $Ga(CH_3)_3$

8. a. $CH_3CH_2CH_2CH_2CH_2OH$

b. $CH_3CH=CHCH_2CH_2OH$ **c.** (benzene ring)—$CH_2CH_2CH_2OH$

14. a. (benzene with Br)   **b.** (styrene with Br)   **c.** (benzene with C=CH, Br)

15. 1-pentyne

17. a. (conjugated diene with C≡N)   **b.** (phenyl propenamide with NH_2)

c. (styrene with OCH_3 on ring)   **d.** (phenyl dienone with CH_3, O)

21. $CH_3\overset{O}{\overset{\|}{C}}$—(benzene)—$Br$ and $CH_2=CH$—(benzene)—OCH_3

$CH_3\overset{O}{\overset{\|}{C}}$—(benzene)—$CH=CH_2$ and Br—(benzene)—OCH_3

26. (benzene)—$C\equiv C$—(benzene) and CH_3—(benzene)—$C\equiv C$—(benzene)—CH_3

CHAPTER 12

RADICALS

3. a. the one on the third carbon from the left   **b.** 6   **4. a.** 3   **b.** 3   **c.** 5

d. 1   **e.** 5   **f.** 6   **g.** 2   **h.** 4   **i.** 4

8. a. 2-chloro-3-methylbutane   **b.** 35%   **9. a.** chlorination

b. bromination   **c.** no preference   **12. a.** D   **b.** B

14. a. $CH_3\overset{CH_3}{\overset{|}{\underset{|}{\underset{Br}{C}}}}CH_2CH_3$ **c.** $CH_3\overset{CH_3}{\overset{|}{\underset{|}{\underset{Br}{CH}}}}CHCHCH_3$

b. $CH_3\overset{CH_3}{\overset{|}{\underset{|}{\underset{Cl}{C}}}}CH_2CH_3$ **d.** $CH_3\overset{CH_3}{\overset{|}{\underset{|}{\underset{Cl}{C}}}}CH_2CH_3$

15. a. 2-methylpropane   **b.** butane   **19.** for 17: 2 (R and S) for 18: 3   **24.** 4

CHAPTER 13

MASS SPECTROMETRY • INFRARED SPECTROSCOPY • ULTRAVIOLET/VISIBLE SPECTROSCOPY

1. #2, #3, #5   **3.** $m/z = 57$ **5. a.** C_5H_{12} **b.** $C_6H_{12}O$ **c.** $C_5H_{10}O_2$ **d.** $C_6H_{12}NO$

6. a. C_6H_{14}, $C_5H_{10}O$, $C_4H_6O_2$ **b.** no   **8.** 2,6-dimethylheptane   **9.** 1 : 2 : 1

10. C_6H_{14} **11. a.** no   **b.** yes   **13. a.** 2-methoxy-2-methylpropane   **b.** 2-methoxybutane

c. 1-methoxybutane

14. a. $CH_2=\overset{+}{O}H$ **15. a.** 2-pentanone   **b.** 3-pentanone   **18. a.** 3-pentanol   **b.** 2-pentanol

19. a. $2000\ cm^{-1}$ **b.** 850 nm   **20. a.** IR   **b.** UV

21. a. 1. $C\equiv C$ **2.** $C-H$ stretch   **3.** $C\equiv N$ **4.** $C=O$ **b. 1.** $C-O$ **2.** $C-C$

22. a. carbon–oxygen stretch of a phenol   **b.** carbon–oxygen double-bond stretch of a ketone

c. $C-N$ stretch of aniline   **23.** sp^3

24. $C-O$ bond of pentanoic acid has partial double bond character

25. a. (δ-valerolactone) > (cyclohexanone) > (δ-valerolactam, NH)   **b.** (β-lactone) > (γ-butyrolactone) > (δ-valerolactone)

26. ethanol dissolved in carbon disulfide

27. $C-N$ stretch would be less intense   **28. a.** ketone   **b.** tertiary amine

31. 2-butyne, H_2, Cl_2, ethene   **32.** trans-2-hexene

33. methyl vinyl ketone

34. $4.1 \times 10^{-5}\,M$ **35.** $10{,}000\,M^{-1}cm^{-1}$ **36.** 219 nm

37. a. (stilbene) > (biphenyl) > (styrene) > (benzene)

b. (phenyl-N) > (phenyl-$\overset{+}{N}$) > (phenyl-N) > (cyclohexyl-N)

38. a. left = purple; right = blue   **b.** They would be the same color.

39. monitor increase in absorption at 340 nm   **40.** 5.0

CHAPTER 14

NMR SPECTROSCOPY

1. 43 MHz   **2. a.** 8.46 T   **b.** 11.75 T   **4. a.** 2   **b.** 1   **c.** 1   **d.** 4   **e.** 1   **f.** 4   **g.** 3   **h.** 3

i. 5   **j.** 3   **k.** 4   **l.** 3   **m.** 3   **n.** 2   **o.** 3   **7. a.** 600 Hz   **b.** 1000 Hz

8. a. 2.0 ppm   **b.** 2.0 ppm   **c.** 1000 Hz   **9. a.** 1.5 ppm   **b.** 150 Hz

10. to the right of the TMS signal   **11. a.** in each structure it is the proton(s) on the carbon on the right-hand side of the structure   **b.** in each structure it is the protons on the methyl group on the left-hand side of the structure

12. a. $CH_3\overset{}{\underset{\underset{Br\ Br}{|\ \ |}}{CHCH}}Br$ **c.** $CH_3CH_2\overset{}{\underset{\underset{Cl}{|}}{CH}}CH_3$ **e.** $CH_3CH_2CH=CH_2$

b. $CH_3\overset{}{\underset{\underset{CH_3}{|}}{CH}}OCH_3$ **d.** $CH_3\overset{O}{\overset{\|}{\underset{\underset{CH_3}{|}}{CHC}}}CH_2CH_3$ **f.** $CH_3OCH_2CH_2CH_3$

13. a. $CH_3CH_2CH_2Cl$ **b.** $CH_3CH_2\overset{}{\underset{\underset{Cl}{|}}{CH}}CH_3$ **c.** $CH_3CH_2\overset{O}{\overset{\|}{C}}H$

5. 9.25 ppm = hydrogens that protrude out; −2.88 ppm = hydrogens that point in

. The compounds have different integration ratios: 2 : 9, 1 : 3, and 2 : 1

. B   19. first spectrum = 1-iodopropane   21. a. 2-chloropropanoic acid

3-chloropropanoic acid   26. a. propyl benzene   b. 3-pentanone   c. ethyl benzoate

. They are separated by 4 sigma bonds.

. CH$_3$O—⬡—CH$_3$ 33. a. A   b. B and D   38. pure ethanol

).

CH$_3$CH$_2$—C(=O)—NH$_2$

4. a. CH$_3$(CH$_2$)$_4$C(=O)(CH$_2$)$_4$CH$_3$ b. Br—⬡—CH$_3$CH$_2$ c. ⬡=O

d. CH$_3$CH$_2$CH=CHCH$_2$CH$_3$

5. The hydrogens that produce a signal at ~1.0 ppm in the ^{1}H NMR spectrum are attached to
e carbon that produces a signal at ~19 ppm in the ^{13}C NMR spectrum.

CHAPTER 15

REACTIONS OF CARBOXYLIC ACIDS AND CARBOXYLIC ACID DERIVATIVES

. **a.** benzyl acetate   **b.** isopentyl acetate   **c.** methyl butyrate

. **a.** potassium butanoate, potassium butyrate   **b.** isobutyl butanoate, isobutyl butyrate

c. N,N-dimethylhexanamide, N,N-dimethylcaproamide   **d.** pentanoyl chloride, valeryl chloride

e. 5-methylhexanoic acid; δ-methylcaproic acid   **f.** propanamide; propionamide

g. 2-azacyclopentanone, γ-butyrolactam   **h.** cyclopentanecarboxylic acid

i. 5-methyl-2-oxacyclohexanone, β-methyl-δ-valerolactone

. In an alcohol because of no electron delocalization.

. The shortest bond has the highest frequency.

a. CH$_3$—C(=O)—O—CH$_3$ b. CH$_3$—C(=O)—O—CH$_3$
  (3, 2, 1)   (1, 2, 3)

1 = longest   1 = highest frequency
3 = shortest   3 = lowest frequency

. B

. **a.** acetic acid   **b.** no reaction

. **a.** new carboxylic acid derivative   **b.** no reaction   **c.** mixture of two products

. **a.** acetyl chloride   **b.** acetamide

0. **a.** no reaction   **b.** sodium acetate   **c.** no reaction   **d.** no reaction

1. true   15. **a.** ethyl alcohol   **b.** ammonia   **c.** phenol   **d.** benzyl alcohol

6. RO- is a stronger base than an amine   18. **a.** new   **b.** no reaction   **c.** mixture of two

9. **a.** phenyl acetate   **b.** phenyl acetate

0. **a.** carbonyl group of the ester is relatively unreactive, nucleophile is relatively unreactive,
aving group is a strong base   **b.** aminolysis, because an amine is a better nucleophile

1. **a.** protonated carboxylic acid, tetrahedral intermediate I, tetrahedral intermediate III,
$_3$O$^+$, CH$_3$O$^+$H$_2$ **b.** Cl$^-$, carboxylic acid, tetrahedral intermediate II, H$_2$O, CH$_3$OH

. H$_3$O$^+$ **d.** H$_3$O$^+$ if excess water is used; CH$_3$OH$_2$$^+$ if not.

3. **a.** benzoic acid and ethanol   **b.** butanoic acid and methanol   **c.** 5-hydroxypentanoic acid

5. **a.** isopropyl benzoate   **b.** ethyl acetate   28. **a.** propoxide ion   **b.** H$^+$ would destroy the
ucleophilicity of the amine; HO$^-$ and RO$^-$ would provide the wrong nucleophile

9. **a.** the alcohol   **b.** the carboxylic acid   31. **a.** butyrate ion and iodomethane

. acetate ion and 1-iodooctane

3. **a.** butanoyl chloride + ethylamine   **b.** benzoyl chloride + dimethylamine

4. B and E   36. B > C > A   37. −NH$_2$ is a stronger base than HO$^-$

8. **a.** pentyl bromide   **b.** isohexyl bromide   **c.** benzyl bromide   **d.** cyclohexyl bromide

0. **a.** butanenitrile, propyl cyanide   **b.** 4-methylpentanenitrile, isopentyl cyanide

1. **a.** 1-bromopropane   **b.** 1-bromo-2-methylpropane   **c.** bromocyclohexane

5. The reaction with amines because the carboxylate ion will be eliminated from the tetrahedral
ntermediate since it is a weaker base than the amine.

7. **a.** propanoic acid + thionyl chloride followed by phenol

. benzoic acid + thionyl chloride followed by ethylamine

CHAPTER 16

REACTIONS OF ALDEHYDES AND KETONES • MORE REACTIONS OF CARBOXYLIC ACID DERIVATIVES

1. **a.** 3-methylpentanal, β-methylvaleraldehyde   **b.** 4-heptanone, dipropyl ketone

c. 2-methyl-4-heptanone, isobutyl propyl ketone   **d.** 4-phenylbutanal, γ-phenylbutyraldehyde

e. 4-ethylhexanal, γ-ethylcaproaldehyde   **f.** 1-hepten-3-one, butyl vinyl ketone

2. if elsewhere would not be a ketone

4. **a.** 2-heptanone   **b.** chloromethyl phenyl ketone

5. **a.** 2-butanol   **b.** 2-methyl-2-pentanol   **c.** 1-methylcyclohexanol

6. CH$_3$CCH$_2$CH$_3$ (=O)   +   CH$_3$CH$_2$CH$_2$MgBr

CH$_3$CH$_2$CCH$_2$CH$_2$CH$_3$ (=O)   +   CH$_3$MgBr

7. **a.** two; (R)-3-methyl-3-hexanol and (S)-3-methyl-3-hexanol   **b.** one; 2-methyl-2-pentanol

9. B and D   11. A and C

14. RC≡C—C(OH)(R')—C≡CR

15. The proton will reside on cyanide ion because it is a stronger base than chloride ion.

16. no   17. The conjugate bases of the strong acids are weak and readily eliminated.

20. **a.** CH$_3$CHCH$_2$OH (CH$_3$)   **c.** (CH$_3$)$_3$C—⬡—OH

b. ⬡—OH   **d.** ⬡—CHCH$_3$ (OH)

21. **a.** 1-butanol + ethanol   **b.** benzyl alcohol   **c.** benzyl alcohol + methanol   **d.** 1-pentanol

22. **a.** ⬡—C(=O)—NHCH$_3$ **c.** CH$_3$C(=O)NHCH$_2$CH$_3$

b. CH$_3$C(=O)NH$_2$ **d.** CH$_3$C(=O)N(CH$_2$CH$_3$)(CH$_2$CH$_3$)

24. **a.** benzyl alcohol   **b.** 1-butanamine (butylamine)   **c.** cyclohexanol   **d.** no reaction

28. the OH group withdraws electrons inductively   29. ~8.5

30. **a.** 1 × 10^{-12} **b.** 1 × 10^{-9} **c.** 3.1 × 10^{-3} 34. a secondary amine and a tertiary amine

37. B   39. **a.** 1, 7, 8   **b.** 2, 3, 5   **c.** 4, 6

42. **a.** ⬡(OH)(CH$_2$OH)   **b.** NaBH$_4$ 45. **a.** 26%   **b.** 17%

47. **a.** phenyl propanoate   **b.** benzoic acid   **c.** 3-emthyl-2-oxacyclohexanone

d. *tert*-butyl 2-methypropanoate

CHAPTER 17

REACTIONS AT THE α-CARBON

1. The conjugate base of propene has delocalized electrons but they are delocalized onto a
carbon.

2. **a.** CH$_3$CCH$_2$C≡N (=O)   **b.** CH$_3$OCCH$_2$COCH$_3$ (=O)(=O)   **c.** CH$_3$CCH$_2$CH (=O)(=O)
   a β-keto nitrile   a β-diester   a β-keto aldehyde

3. The proton on the nitrogen is more acidic than the proton on the α-carbon.

4. Competing electron delocalization by the lone pair on N or O is more important for the amide
than for the ester.

5. **a.** CH$_3$CH (=O) > HC≡CH > CH$_2$=CH$_2$ > CH$_3$CH$_3$

b. CH$_3$CCH$_2$CCH$_3$ (=O)(=O) > CH$_3$CCH$_2$COCH$_3$ (=O)(=O) > CH$_3$OCCH$_2$COCH$_3$ (=O)(=O) > CH$_3$CCH$_3$ (=O)

c. [structures] > > NCH₃

7. a. CH₃CH=CCH₂CH₃ (OH)   **b.** [structure] C=CH₂ with OH

c. [structure OH]   **d.** [structure OH] and [structure OH] more stable

e. CH₃CH₂C=CHCCH₂CH₃ (OH, O) more stable   and   CH₃CH=CCH₂CCH₂CH₃ (OH, O)

f. [structure] —CH=CCH₃ (OH) and [structure] —CH₂C=CH₂ (OH) more stable

10. The rate-determining step must be removal of the proton from the α-carbon of the ketone.

12. [structure with D D D D and O]

15. a. 1   **b.** 2

19. a. [structure] + CH₃CCH₂CCH₃ (O O) + HO⁻

b. CH₃CCH=CH₂ (O) + CH₃CH₂OCCH₂COCH₂CH₃ (O O) + CH₃CH₂O⁻

c. [structure] + [structure N H] + CH₂=CH—C—CH₃ (O)

21. a. CH₃CH₂CH₂CH (O)   **c.** [structure] —CH₂CH (O)

b. CH₃CCH₃ (O)   **d.** CH₃CH₂CCH₂CH₃ (O)

22. [structure with O]

28. a. CH₃CH₂CH₂CCHCOCH₃ (O O) with CH₂ CH₃   **b.** CH₃CHCH₂CCHCOCH₂CH₃ (O O) with CH₃ CHCH₃ CH₃

29. A, B, and D

33. a. [structure OH CH₃ CCH₃ O]   **b.** [structure OH CH₃ O]

35. a. [structure OH O]   **c.** [structure OH CH O]

b. [structure HO O]   **d.** [structure OH CCH₃ O O]

---

38. A and D   **39. a.** methyl bromide   **b.** methyl bromide (twice)   **c.** benzyl bromide
d. isobutyl bromide   **41. a.** ethyl bromide   **b.** pentyl bromide   **c.** benzyl bromide
45. 7   **46. a.** 3   **b.** 7

CHAPTER 18

REACTIONS OF BENZENE AND SUBSTITUTED BENZENES

1. a. CH₃CHCH₂CH₂CH₂CH₃ [phenyl]   **c.** CH₃CH₂CHCH₂CH₃ [phenyl CH₂]   **d.** [phenyl CH₂Br]

b. [phenyl]—CH₂OH

2. Like Br₂, water is a Lewis base, so FeBr₃ reacts with water instead of Br₂.
5. a. ethylbenzene   **b.** isopropylbenzene   **c.** *sec*-butylbenzene   **d.** *tert*-butylbenzene
e. *tert*-butylbenzene   **f.** 3-phenylpropene
7. a. [phenyl]COOH   **c.** [phenyl]CH₂OCH₃

b. [phenyl]COOH with COOH   **d.** [phenyl]CH₂CH₂NH₂

9. a. *ortho*-ethylphenol   **b.** *meta*-bromochlorobenzene   **c.** *meta*-bromobenzaldehyde
d. *ortho*-ethylmethylbenzene
12. a. 1,3,5-tribromobenzene   **b.** 3-nitrophenol   **c.** *para*-bromomethylbenzene
d. 1,2-dichlorobenzene   **e.** 1-bromo-3-methylbenzene   **f.** 2-ethyl-4-iodobenzene
15. a. phenol > toluene > benzene > bromobenzene > nitrobenzene
   b. toluene > chloromethylbenzene > dichloromethylbenzene > difluoromethylbenzene
17. a. CH₂CH₂CH₃ [NO₂]   CH₂CH₂CH₃ [NO₂]   **c.** HC=O [NO₂]   **e.** SO₃H [NO₂]

b. Br [NO₂]   Br [NO₂]   **d.** C≡N [NO₂]   **f.** [cyclohexyl-phenyl] + [NO₂]

18. They are all meta directors.
21. a. no reaction   **b.** no reaction

b. NH₂ Br Br Br [structure]   **d.** CH₃ CH₃ + CH₃ CH₃ [structures]

22. a. no   **b.** yes
25. a. COOH [Br CH₃]   **c.** COOH [Br Cl]   **e.** HC=O [NO₂ OCH₃]

b. COOH COOH [Cl]   **d.** OCH₃ NO₂ [F]   **f.** CH₃ NO₂ [C(CH₃)₃]

26. yes
28. It would complex with the amino group, converting it into a meta director.
35. The amide bond to the left of the downward-pointing carbonyl group
39. a. 1-chloro-2,4-dinitrobenzene > *p*-chloronitrobenzene > chlorobenzene
b. chlorobenzene > *p*-chloronitrobenzene > 1-chloro-2,4-dinitrochlorobenzene

CHAPTER 19

MORE ABOUT AMINES • REACTIONS OF HETEROCYCLIC COMPOUNDS

1. a. 2,2-dimethylaziridine **b.** 4-ethylpiperidine **c.** 2-methylthiacyclopropane
d. 3-methylazacyclobutane **e.** 2,3-dimethyltetrahydrofuran
f. 2-ethyloxacyclobutane
3. The electron-withdrawing oxygen stabilizes the conjugate base.
4. a.   **b.** p$K_a \sim 8$ **c.** conjugate acid of 3-chloroquinuclidine

8. Both acidic and basic forms of pyrrole are aromatic; only the basic form of cyclopentadiene is aromatic.

10.   **13.**   >   >

14.   **15.**   >   >

16. Imidazole forms intramolecular hydrogen bonds that N-methylimidazole cannot form.
17. When a proton is removed from imidazole, the electrons left behind can be delocalized onto a nitrogen.
18. 20%

19.

20. Pyrimidine has a second nitrogen that withdraws electrons inductively.

CHAPTER 20

THE ORGANIC CHEMISTRY OF CARBOHYDRATES

1. D-Ribose is an aldopentose. D-Sedoheptulose is a ketoheptose. D-Mannose is an aldohexose.
3. a. L-glyceraldehyde **b.** L-glyceraldehyde **c.** D-glyceraldehyde
4. a. enantiomers **b.** diastereomers
5. a. D-ribose **b.** L-talose **c.** L-allose **d.** L-ribose
6. a. (2R,3S,4R,5R)-2,3,4,5,6-pentahydroxyhexanal **b.** (2S,3S,4R,5R)-2,3,4,5,6-pentahydroxyhexanal **c.** (2R,3S,4S,5R)-2,3,4,5,6-pentahydroxyhexanal
d. (2S,3R,4S,5S)-2,3,4,5,6-pentahydroxyhexanal **7.** D-psicose
8. a. A ketoheptose has four asymmetric centers ($2^4 = 16$ stereoisomers).
b. An aldoheptose has five asymmetric centers ($2^5 = 32$ stereoisomers).
c. A ketotriose has no asymmetric centers; therefore, it has no stereoisomers.
11. D-tagatose, D-galactose, D-talose, and D-sorbose
12. a. D-iditol **b.** D-iditol and D-gulitol **13. a. 1.** D-altrose **2.** L-gulose
3. L-galactose **2.** D-tagatose **2.** D-psicose **14. a.** L-gulose **b.** L-gularic acid
c. D-allose and L-allose, D-altrose and D-talose, L-altrose and L-talose, D-galactose and L-galactose **15. a.** D-gulose and D-idose **b.** L-xylose and L-lyxose **16. a.** D-allose and D-altrose **b.** D-glucose and D-mannose **c.** L-allose and L-altrose
18. A = D-glucose **B** = D-mannose **C** = D-arabinose **D** = D-erythrose

22.

23. the OH group at C-2 **24. a.** the OH group at C-2, C-3, and C-4
b. the OH group at C-3 and C-1

25.

26. A protonated amine is not a nucleophile.
28. a. α-D-talose (reducing) **b.** methyl α-D-galactoside (nonreducing)
c. ethyl β-D-psicoside (nonreducing) **29.** −74.2 **30. a.** Both sugars are glucoses.
b. β-1,6′ glycisidic linkage

31. a. amylose has α-1,4′-glycosidic linkages; cellulose has β-1,4′-glycosidic linkages
b. amylopectin has 1,6′-glycosidic linkages that create branches; amylose doesn't have branches
c. glycogen has more branches than amylopectin **d.** chitin has an N-acetylamino group instead of an OH group at the 2-position.
32. The electrons left behind when a proton is removed from the OH at C-3 are delocalized onto an oxygen; The electrons left behind when a proton is removed from the OH at C-2 are delocalized onto a carbon.
33. a. They cannot receive type A, B, or AB blood because these have sugar components that type O blood does not have. **b.** They cannot give blood to those with type A, B, or O blood because AB blood has sugar components that these other blood types do not have.

CHAPTER 21

AMINO ACIDS, PEPTIDES, AND PROTEINS

2. a. (R)-alanine **b.** (R)-aspartate **c.** The α-carbon of all the D-amino acids except cysteine has the R-configuration. **4.** Ile **5.** pH > 2.34 and pH < 9.69
6. because of the electron-withdrawing ammonium group
8. a.   **b.**

c.   **d.**

10. a. 5.43 **b.** 10.76 **c.** 5.68 **d.** 2.98 **11. a.** Asp **b.** Arg **c.** Asp **d.** Met
13. a. < 25% **b.** > 75% **15.** 2-methylpropanal **16.** Leucine and isoleucine have similar polarities and pI values, so they show up as one spot.
19. His > Val > Ser > Asp
20. One equivalent of ammonia will be protonated by the carboxylic acid.
21. a. L-Ala, L-Asp, L-Glu **b.** L-Ala and D-Ala, L-Asp and D-Asp, L-Glu and D-Glu
22. a. leucine **b.** methionine **23. a.** 4-bromo-1-butanamine **b.** benzylbromide
24. a. alanine **b.** isoleucine **c.** leucine
28. the bonds on either side of the α-carbon **30.** both have the S configuration
32. Leu-Val and VaL-Val **35. a.** 5.8% **b.** 4.4% **38.** Edman's reagent would release two amino acids in approximately equal amounts.
39. Gly-Arg-Trp-Ala-Glu-Leu-Met-Pro-Val-Asp
41. a. His-Lys Leu-Val-Glu-Pro-Arg Ala-Gly-Ala **b.** Leu-Gly-Ser-Met-Phe-Pro-Tyr Gly-Val
43. Leu-Tyr-Lys-Arg-Met-Phe-Arg-Ser
45. 110 Å in an α-helix and 260 Å in a straight chain
46. nonpolar groups on the outside and polar groups on the inside
47. a. cigar-shaped protein **b.** subunit of a hexamer

CHAPTER 22

CATALYSIS IN ORGANIC REACTIONS AND IN ENZYMATIC REACTIONS

1. $\Delta H^{\ddagger}, E_a, \Delta S^{\ddagger}, \Delta G^{\ddagger}, k_{rate}$ **3.** specific acid catalyzed
7. They do not have a negatively charged oxygen on one carbon and a carbonyl group on an adjacent carbon. **9.** close to one

10.

13. The nitro groups cause the phenolate ion to be a better leaving group than the carboxylate ion.
15. 2, 3, and 4 **16.** Ser-Ala-Leu; the terminal amino acid fits best in the hydrophobic pocket.
18. Arginine forms a direct hydrogen bond; lysine forms an indirect hydrogen bond.
20. NAM **23.** Enzyme-catalyzed reactions are highly stereoselective. **26.** 2
28. Putting a substituent on cysteine with iodoacetic acid could interfere with binding or catalysis of the substrate.

CHAPTER 23

THE ORGANIC CHEMISTRY OF THE COENZYMES, COMPOUNDS DERIVED FROM VITAMINS

1. It increases the susceptibility of the carbonyl carbon to nucleophilic attack, increases the nucleophilicity of water, and stabilizes the negative charge on the transition state.

2. [structure: tricarboxylic acid with ketone] + NADH + H⁺

3. [structure: pyruvate/lactate with OH] + NAD⁺

4. a. 7 **b.** 3 isolated from 2 others **5.** N-5 is a stronger base because the lone pair on N-5 unlike the lone pair on N-1 cannot be delocalized onto an oxygen. **13.** the one on the right **14. a.** alanine **b.** aspartate **15.** If the nitrogen is not protonated, it will be a poorer source for electron delocalization. **16.** The hydrogen bond formed by the OH group weakens the bond to the α-carbon.

20. CH_3CH_2CSCoA $\xrightarrow[\text{biotin}]{E}$ $CH_3CHCSCoA$ $\xrightarrow[\text{coenzyme B}_{12}]{E}$ CH_2CH_2CSCoA
    with COO⁻ groups

21. In THF, a carbonyl group is at C-4, and the bond between C-3 and C-4 is a single bond; in aminopterin, an amino group is at C-4, and the bond between C-3 and C-4 is a double bond. Intramolecular reactions that form 5-membered rings are faster than intermolecular reactions. **23.** Two thiols in the same molecule will react faster than two thiols in two separate molecules.

CHAPTER 24

THE ORGANIC CHEMISTRY OF THE METABOLIC PATHWAYS

5. The β-carbon has a partial positive charge. **6.** eight **7.** seven
10. two **11.** acetaldehyde reductase **12.** a ketone **13.** thiamine pyrophosphate
15. pyruvate **17.** a secondary alcohol **18.** citrate and isocitrate **19.** succinyl
21. a. succinyL-CoA synthetase **b.** aldolase, phosphoglycerate kinase, pyruvate kinase
22. a. 1 **b.** 1 + 5 = 6 **23. a.** glycerol kinase **b.** phophatidic acid phosphatase

CHAPTER 25

THE ORGANIC CHEMISTRY OF LIPIDS

2. glycerol palmaitate **6.** no **7.** Integral proteins will have more nonpolar amino acids than peripheral proteins. **8.** by synthesizing phospholidis that contain more saturated fatty acids **13.** The two halves are synthesized in a head-to-tail fashion and then joined together in a tail-to-tail linkage. **16.** one to reduce the ester to an aldehyde; the second to reduce the aldehyde to a primary alcohol **21.** The reaction of dimethylallyl pyrophosphate and isopentenyl pyrophosphate is an S_N1 reaction. **25.** There are two 1,2-hydride shifts and two 1,2-methyl shifts. The last step is elimination of a proton. **26.** A and B are cis fused; A and B are trans fused. **27.** β-substituent **30.** axial

CHAPTER 26

THE CHEMISTRY OF THE NUCLEIC ACIDS

5. a. 3′—C—C—T—G—T—T—A—G—A—C—G—5′ **b.** guanine
8. Thymine and uracil have the hydrogen bond donor and the hydrogen bond acceptor in the same place.
9. Met-Asp-Pro-Val-Ile-Lys-His **10.** Met-Asp-Pro-Leu-Leu-Asn
12. the third base
13. 5′—G-C-A-T-G-G-A-C-C-C-C-G-T-T-A-T-T-A-A-A-C-A-C—3′

15. [structures: xanthine, hypoxanthine] **17.** B

xanthine   hypoxanthine

CHAPTER 27

SYNTHETIC POLYMERS

1. a. $CH_2=CHCl$ **b.** $CH_2=CCH_3$ **c.** $CF_2=CF_2$
         with $COCH_3$, O group

2. poly(vinyl chloride) **5.** beach balls

7. a. [structures: three para-substituted styrenes with OCH₃, CH₃, NO₂]
CH₂=CH > CH₂=CH > CH₂=CH (OCH₃, CH₃, NO₂)

b. $CH_2=CHOCH_3$ > $CH_2=CHCH_3$ > $CH_2=CHCOCH_3$ (with O)

c. $CH_2=CCH_3$ $CH_2=CH$ [phenyl structures] >

8. a. [structures: three para-substituted styrenes NO₂, CH₃, OCH₃]
CH₂=CH > CH₂=CH > CH₂=CH (NO₂, CH₃, OCH₃)

b. $CH_2=CHC\equiv N$ > $CH_2=CHCl$ > $CH_2=CHCH_3$
9. The carbocation would be unstable because of the electron-withdrawing ester group.

12. a. $CH_2=CCH_3$ + BF_3 + H_2O (with CH₃) **c.** [cyclopropane oxide structure] + CH_3O^-

b. $CH_2=CH$ + BF_3 + H_2O (with pyrrolidinone structure) **d.** $CH_2=CH$ + BuLi (with COCH₃, O)

13. [structure: oxetane with O⁺—CH₂CCH₂OCH₂CCH₂OCH₂CCH₂OH chain, CH₃ groups]

14. [terpene structure]

16. a. —NHCH₂CH₂CH₂CNHCH₂CH₂CH₂C— (with O groups)

b. —NH(CH₂)₄NHCCH₂CH₂CNH(CH₂)₄NHCCH₂CH₂C— (with O groups)

23. a. [polyester structure] **b.** [polyester structure]

CHAPTER 28

PERICYCLIC REACTIONS

1. a. electrocyclic reaction **b.** sigmatropic rearrangement
c. cycloaddition reaction **d.** cycloaddition reaction
2. a. bonding orbitals = ψ_1, ψ_2, ψ_3; antibonding orbitals = ψ_4, ψ_5, ψ_6;
b. ground-state HOMO = ψ_3; ground-state LUMO = ψ_4
c. excited-state HOMO = ψ_4; excited-state LUMO = ψ_5
d. symmetric orbitals = ψ_1, ψ_3, ψ_5; antisymmetric orbitals = ψ_2, ψ_4, ψ_6
e. The HOMO and LUMO have opposite symmetries.
3. a. 8 **b.** ψ_4 **c.** 7 **6. a.** conrotatory **b.** trans **c.** disrotatory **d.** cis
7. a. correct **b.** correct **c.** correct **8. 1. a.** conrotatory **b.** trans
2. a. disrotatory **b.** cis
11. yes

13. [reaction structures with ¹⁴C labels] → → (with OH product)

14. If a nondeuterated reactant had been used, the products would have been identical to the reactant, so the rearrangement would not have been detectable.
18. a. inversion **b.** retention **19.** antarafacial
21. a [3,3] sigmatropic Claisen rearrangement

Glossary

absorption band   a peak in a spectrum that occurs as a result of the absorption of energy.

acetal   the product formed when a second equivalent of alcohol adds to a hemiacetal

$$R-\overset{\overset{\displaystyle OR}{|}}{\underset{\underset{\displaystyle OR}{|}}{C}}-H \quad \text{or} \quad R-\overset{\overset{\displaystyle OR}{|}}{\underset{\underset{\displaystyle OR}{|}}{C}}-R$$

acetamidomalonic ester synthesis   a method used to synthesize an amino acid that is a variation of the *N*-phthalimidomalonic ester synthesis.

acetoacetic ester synthesis   synthesis of a methyl ketone, using ethyl acetoacetate as the starting material.

achiral   an achiral molecule has a conformation identical to (superimposable upon) its mirror image.

acid (Brønsted)   a substance that loses a proton.

acid anhydride   loss of water from two molecules of a carboxylic acid.

$$\underset{R}{\overset{O}{\underset{}{\overset{\|}{C}}}}\underset{O}{\overset{}{}}\underset{R}{\overset{O}{\overset{\|}{C}}}$$

acid–base reaction   the reaction of an acid with a base.

acid catalyst   a catalyst that increases the rate of a reaction by donating a proton.

acid-catalyzed reaction   the hydration of a catalyst employed in the hydration of an alkene is an acid.

acid dissociation constant   a measure of the degree to which an acid dissociates in a dilute solution, so the concentration of water remains essentially constant.

activating substituent   a substituent that increases the reactivity of an aromatic ring. Electron-donating substituents activate aromatic rings toward electrophilic attack, and electron-withdrawing substituents activate aromatic rings toward nucleophilic attack.

active site   a pocket or cleft in an enzyme where the substrate is bound.

acyl adenylate   a carboxylic acid derivative with AMP as the leaving group.

acyl–enzyme intermediate   an intermediate formed when an amino acid residue of an enzyme is acetylated.

acyl group   a carbonyl group bonded to an alkyl group or to an aryl group.

acyl chloride
$$\underset{R}{\overset{O}{\overset{\|}{C}}}\underset{H}{}$$

acyl halide
$$\underset{R}{\overset{O}{\overset{\|}{C}}}\underset{Cl}{} \quad \text{or} \quad \underset{R}{\overset{O}{\overset{\|}{C}}}\underset{Br}{}$$

acyl phosphate   a carboxylic acid derivative with a phosphate leaving group.

1,2-addition product (direct addition)   the result of addition at the 1- and 2-positions.

1,4-addition product (conjugate addition)   addition to the 1- and 4-positions of a conjugated system.

addition polymer (chain-growth polymer)   a polymer made by adding monomers to the growing end of a chain.

addition reaction   a reaction in which atoms or groups are added to the reactant.

adrenal cortical steroids   glucocorticoids and mineralocorticoids.

alcohol   a compound in which a hydrogen of an alkane has been replaced by an OH group.

alcoholysis reaction   a reaction with an alcohol that converts one compound into two compounds.

aldaric acid   a dicarboxylic acid with an OH group bonded to each carbon. Obtained by oxidizing the aldehyde and primary alcohol groups of an aldose.

aldehyde   carbonyl carbon is bonded to a hydrogen and to an alkyl (or aryl) group (R).

$$\underset{R}{\overset{O}{\overset{\|}{C}}}\underset{H}{}$$

alditol   a compound with an OH group bonded to each carbon. Obtained by reducing an aldose or a ketose.

aldol addition   a reaction between two molecules of an aldehyde (or two molecules of a ketone) that connects the α-carbon of one with the carbonyl carbon of the other.

aldol condensation   an aldol addition followed by the elimination of water.

aldonic acid   a carboxylic acid with an OH group bonded to each carbon. Obtained by oxidizing the aldehyde group of an aldose.

aldose   a polyhydroxy aldehyde.

aliphatic   a nonaromatic organic compound.

alkaloid   a natural product, with one or more nitrogen heteroatoms, found in the leaves, bark, or seeds of plants.

alkane   a hydrocarbon that contains only single bonds.

alkene   a hydrocarbon that contains a double bond.

alkene metathesis (olefin metathesis)   breaks the double bond of an alkene (or the triple bond of an alkyne) and then rejoins the fragments.

alkylation reaction   a reaction that adds an alkyl group to a reactant.

alkyl group (alkyl substituent)   formed by removing a hydrogen from an alkane.

alkyl halide   a compound in which a hydrogen of an alkane has been replaced by a halogen.

alkyl substituent (alkyl group)   formed by removing a hydrogen from an alkane.

alkyl tosylate   a sulfonate ester formed from the reaction of TsCl and an alcohol.

alkyne   a hydrocarbon that contains a triple bond.

allene   a compound with two adjacent double bonds.

allosteric activator   a compound that activates an enzyme when it binds to a site on the enzyme (other than the active site).

allosteric inhibitor   a compound that inactivates an enzyme when it binds to a site on the enzyme (other than the active site).

allyl group $CH_2=CHCH_2-$

allylic carbon   an sp^3 carbon adjacent to a vinylic carbon.

allylic cation   a carbocation with a positive charge on an allylic carbon.

alternating copolymer   a copolymer in which two monomers alternate.

amide   has an NH_2, NHR, or NR_2 group in place of the OH group of a carboxylic acid. Amides are named by replacing "oic acid," "ic acid," or "ylic acid" of the acid name with "amide."

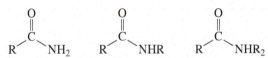

amine   an electron-withdrawing group bonded to an sp^3 carbon; they are not included in Group II because they do not undergo substitution and elimination reactions.

amine inversion   the configuration of an sp^3 hybridized nitrogen with a nonbonding pair of electrons that rapidly turns inside out.

amino acid   carboxylic acid with a protonated amino group on the α-carbon.

amino acid analyzer   an instrument that automates the ion-exchange separation of amino acids.

aminolysis reaction   a reaction with an amine that converts one compound into two compounds.

amino sugar   a sugar in which one of the OH groups is replaced by an NH_2 group.

anabolic steroids   steroids that aid in the development of muscle.

anabolism   reactions that living organisms carry out in order to synthesize complex molecules from simple precursor molecules.

androgens   male sex hormones.

angle strain   the strain introduced into a molecule as a result of its bond angles being distorted from their ideal values.

angstrom   unit of length; 100 picometers = 10^{-8} cm = 1 angstrom.

angular methyl group   a methyl substituent at the 10- or 13-position of a steroid ring system.

anion-exchange resin   a positively charged resin used in ion-exchange chromatography.

anionic polymerization   chain-growth polymerization in which the initiator is a nucleophile; the propagation site therefore is an anion.

annulation reaction   a ring-forming reaction.

anomeric carbon   the carbon in a cyclic sugar that is the carbonyl carbon in the open-chain form.

anomeric effect   the preference of certain substituents bonded to the anomeric carbon for the axial position.

anomers   two cyclic sugars that differ in configuration only at the carbon that is the carbonyl carbon in the open-chain form.

antarafacial bond formation   formation of two σ bonds on opposites sides of the π system.

antarafacial rearrangement   rearrangement in which the migrating group moves to the opposite face of the π system.

anti addition   an addition reaction in which two substituents are added at opposite sides of the molecule.

antiaromatic   a cyclic and planar compound with an uninterrupted ring of *p* orbital-bearing atoms containing an even number of pairs of π electrons.

antibiotic   a compound that interferes with the growth of a microorganism.

antibodies   compounds that recognize foreign particles in the body.

G-1

anticodon the three bases at the bottom of the middle loop in tRNA.

anti conformer the most stable of the staggered conformers.

anti elimination an elimination reaction in which the two substituents that are eliminated are removed from opposite sides of the molecule.

antisymmetric molecular orbital a molecular orbital in which the left (or top) half is not a mirror of the right (or bottom) half.

antiviral drug a drug that interferes with DNA or RNA synthesis in order to prevent a virus from replicating.

applied magnetic field the externally applied magnetic field.

aprotic polar solvent polar solvent molecules *do not have* a hydrogen bonded to an oxygen or to a nitrogen.

aramide an aromatic polyamide.

arene oxide an aromatic compound that has had one of its double bonds converted to an epoxide.

aromatic a cyclic and planar compound with an uninterrupted ring of *p* orbital-bearing atoms containing an odd number of pairs of π electrons.

Arrhenius equation relates the rate constant of a reaction to the experimental energy of activation and to the temperature at which the reaction is carried out ($k = Ae^{-E_a/RT}$).

aryl group a benzene or a substituted-benzene group.

asymmetric center (chiral center) an atom bonded to four different groups.

atactic polymer a polymer in which the substituents are randomly oriented on the extended carbon chain.

atomic number the number of protons (or electrons) in its nucleus. The atomic number is unique to a particular element.

atomic orbital a subshell associated with an atom.

atomic weight the average mass of the atoms in the naturally occurring element.

aufbau principle states that an electron will always go into that orbital with the lowest available energy.

automated solid-phase peptide synthesis an automated technique that synthesizes a peptide while its C-terminal amino acid is attached to a solid support.

auxochrome a substituent that when attached to a chromophore, alters the λ_{max} and intensity of absorption of UV/Vis radiation.

axial bond a bond of the chair conformation of cyclohexane that is perpendicular to the plane in which the chair is drawn (an up–down bond).

aziridine a three-membered-ring compound in which one of the ring atoms is a nitrogen.

azo linkage an —N=N— bond.

back-side attack nucleophilic attack on the side of the carbon opposite the side bonded to the leaving group.

Baeyer–Villiger oxidation oxidation of aldehydes or ketones with H_2O_2 to form carboxylic acids or esters, respectively.

basal metabolic rate the number of calories that would be burned if one stayed in bed all day.

base^1 a substance that gains a proton.

base^2 a heterocyclic compound (a purine or a pyrimidine) in DNA and RNA.

base catalyst a catalyst that increases the rate of a reaction by removing a proton.

base peak the tallest peak in a mass spectrum because it has the greatest relative abundance.

basicity a measure of how well a compound (a base) shares its lone pair with a proton.

Beer–Lambert law relationship among the absorbance of UV/Vis light, the concentration of the sample, the length of the light path, and the molar absorptivity ($A = cl\varepsilon$).

bending vibration a vibration that does not occur along the line of the bond. It results in changing bond angles.

benzyl group a benzene ring with a methylene group

benzylic carbon an *sp³* hybridized carbon bonded to a benzene ring.

benzylic cation a compound with a positive charge on a benzylic carbon.

bicyclic compound a compound containing two rings that share at least one carbon.

bifunctional molecule a molecule with two functional groups.

bile acids steroids that act as emulsifying agents so that water-insoluble compounds can be digested.

bimolecular reaction (second-order reaction) a reaction whose rate depends on the concentration of two reactants.

biochemistry (biological chemistry) the chemistry of living organisms.

biodegradable polymer a polymer that can be broken into small segments by an enzyme-catalyzed reaction.

bioorganic compound an organic compound found in biological systems.

biopolymer a polymer that is synthesized in nature.

biosynthesis synthesis in a biological system.

biotin (Vitamin H) the coenzyme required by enzymes that catalyze carboxylation of a carbon adjacent to an ester or a keto group.

block copolymer a copolymer in which there are regions (blocks) of each kind of monomer.

boat conformer the conformation of cyclohexane that roughly resembles a boat.

boiling point (bp) the temperature at which its liquid form becomes a gas (vaporizes).

bond length the internuclear distance between two atoms at minimum energy (maximum stability).

bond order the number of covalent bonds shared by two atoms.

bond strength the energy required to break a bond homolytically.

bridged bicyclic compound (a compound that contains two rings that share two nonadjacent carbons.

Brønsted acid a substance that donates a proton.

Brønsted base a substance that accepts a proton.

buffer solution a solution of a weak acid and its conjugate base.

carbanion a compound containing a negatively charged carbon.

carbocation a species containing a positively charged carbon.

carbocation rearrangement the rearrangement of a carbocation to a more stable carbocation

carbohydrate polyhydroxy aldehydes such as glucose, polyhydroxy ketones such as fructose, and compounds formed by linking polyhydroxy aldehydes and/or polyhydrox ketones together.

α-carbon a carbon bonded to a leaving group or adjacent to a carbonyl carbon.

β-carbon a carbon adjacent to an α-carbon.

carbon acid a compound containing a carbon that is bonded to a relatively acidi hydrogen.

carbonyl carbon the carbon of a carbonyl group.

carbonyl compound a compound that contains a carbonyl group.

carbonyl group a carbon doubly bonded to an oxygen.

carbonyl oxygen the oxygen of a carbonyl group.

carboxyl group COOH

carboxyl oxygen the single-bonded oxygen of a carboxylic acid or an ester.

carboxylic acid named by replacing the terminal "e" of the alkane name with "oic acid."

$$R-\overset{\overset{\displaystyle O}{\|}}{C}-OH$$

carboxylic acid derivative a compound that is hydrolyzed to a carboxylic acid.

carotenoid a class of compounds (a tetraterpene) responsible for the red and orange color of fruits, vegetables, and fall leaves.

catabolism reactions that living organisms carry out in order to break down comple molecules into simple molecules and energy.

catalyst decreases the energy barrier that must be overcome in the process of convertin the reactants to products.

catalytic hydrogenation the addition of hydrogen to a double or a triple bond with the ai of a metal catalyst.

cation-exchange resin a negatively charged resin used in ion-exchange chromatography.

cationic polymerization chain-growth polymerization in which the initiator is an electro phile; the propagation site therefore is a cation.

cephalin a phosphoacylglycerol in which the second OH group of phosphate has forme an ester with ethanolamine.

chain-growth polymer (addition polymer) a polymer made by adding monomers to th growing end of a chain.

chain reactions a growing polymer chain reacts with a molecule XY in a manner tha allows X to terminate the chain, leaving behind Y to initiate a new chain.

chair conformation the conformation of cyclohexane that roughly resembles a chair. It i the most stable conformation of cyclohexane.

chemically equivalent protons protons with the same connectivity relationship to the res of the molecule.

chemical shift the location of a signal in an NMR spectrum. It is measured downfield from a reference compound (most often, TMS).

chiral a chiral molecule has a nonsuperimposable mirror image.

chiral center a tetrahedral atom bonded to four different groups.

cholesterol a steroid that is the precursor of all other animal steroids.

chromatography a separation technique in which the mixture to be separated is dissolve in a solvent and the solvent is passed through a column packed with an absorber stationary phase.

chromophore the part of a molecule responsible for a UV or visible spectrum.

***s*-cis conformation** the conformation in which two double bonds are on the same side o a single bond.

cis-fused two cyclohexane rings fused together such that if the second ring were consid ered to be two substituents of the first ring, one substituent would be in an axial positio and the other would be in an equatorial position.

cis isomer the isomer with the hydrogens on the same side of the double bond or cyclic structure.

cis–trans isomers geometric isomers.

citric acid cycle (Krebs cycle) a series of reactions that converts the acetyl group o acetyl-CoA into two molecules of CO_2.

Claisen condensation   when two molecules of an *ester* undergo a condensation reaction.

Claisen-Schmidt condensation   the aldol addition product formed in this reaction loses water as soon as it is formed because the new double bond is conjugated with the carbonyl group and the benzene ring.

Claisen rearrangement   a [3,3] sigmatropic rearrangement of an allyl vinyl ether.

α-cleavage   homolytic cleavage of a group bonded to the α-carbon.

codon   a sequence of three bases in mRNA that specifies the amino acid to be incorporated into a protein.

coenzyme   a cofactor that is an organic molecule.

coenzyme A (CoASH)   reacts with the thioester in a transthioesterification reaction (one thioester is converted to another), substituting coenzyme A for dihydrolipoate.

coenzyme B₁₂   the coenzyme required by enzymes that catalyze certain rearrangement reactions.

cofactor   an organic molecule or a metal ion that certain enzymes need to catalyze a reaction.

coil conformation (loop conformation)   that part of a protein that is highly ordered, but not in an α-helix or a β-pleated sheet.

common intermediate   an intermediate that two compounds have in common.

common name   nonsystematic name.

competitive inhibitor   a compound that inhibits an enzyme by competing with the substrate for binding at the active site.

complete racemization   the formation of a pair of enantiomers in equal amounts.

complex carbohydrate   a carbohydrate containing two or more monosaccharides linked together.

concerted reaction   a reaction in which all the bond-making and bond-breaking processes occur in one step.

condensation polymer (step-growth polymer)   a polymer made by combining two molecules while removing a small molecule (usually water or an alcohol).

condensation reaction   a reaction combining two molecules while removing a small molecule (usually water or an alcohol).

conducting polymer   a polymer that can conduct electricity.

configuration   the three-dimensional structure of a particular atom in a compound. The configuration is designated by *R* or *S*.

configurational isomers   stereoisomers that cannot interconvert unless covalent bonds are broken. Because they cannot interconvert, configurational isomers can be separated.

conformation   the three-dimensional shape of a molecule at a given instant that can change as a result of rotations about σ bonds.

conformational isomers (conformers)   different spatial arrangements of the atoms that result from rotation about a single bond.

conformers (conformational isomers)   different spatial arrangements of the atoms that result from rotation about a single bond.

conjugate acid   when a base gains a proton, the resulting species with the proton is conjugate acid.

conjugate addition   1,4-addition to an α,β,-unsaturated carbonyl compound.

conjugate base   when an acid loses a proton, the resulting species without the proton is a conjugate base.

conjugated double bonds   double bonds separated by one single bond.

conrotatory ring closure   achieves head-to-head overlap of *p* orbitals by rotating the orbitals in the same direction.

conservation of orbital symmetry theory   a theory that explains the relationship between the structure and configuration of the reactant, the conditions under which a pericyclic reaction takes place, and the configuration of the product.

constitutional isomers   molecules that have the same atoms but differ in the way the atoms are connected.

contributing resonance structure (resonance contributor, resonance structure)   a structure with localized electrons that approximates the structure of a compound with delocalized electrons.

Cope rearrangement   a [3,3] sigmatropic rearrangement of a 1,5-diene.

copolymer   a polymer formed from two or more different monomers.

COSY spectrum   a 2-D NMR spectrum that shows coupling between sets of protons.

coupled protons   protons that split each other. Coupled protons have the same coupling constant.

coupled reactions   an endergonic reaction followed by an exergonic reaction, which drives the endergonic reaction toward products.

coupling constant   the distance (in hertz) between two adjacent peaks of a split NMR signal.

coupling reaction   a reaction that joins two CH-containing groups.

covalent bond   a bond formed as a result of sharing electrons between two nuclei.

cross-linking   connecting polymer chains by intermolecular bond formation.

crossed aldol addition   an aldol addition in which two different carbonyl compounds are used.

crown ether   a cyclic molecule that contains several ether linkages.

crystallites   regions of a polymer in which the chains are highly ordered.

C-terminal amino acid   the terminal amino acid of a peptide (or protein) that has a free carboxyl group.

cumulated double bonds   double bonds that are adjacent to one other.

cyanohydrin   product of the reaction of a carbon nucleophile that is added to an aldehyde

or a ketone
$$R - \overset{\overset{\displaystyle OH}{|}}{\underset{\underset{\displaystyle C \equiv N}{|}}{C}} - R(H)$$

cycloaddition reaction   a reaction in which two π-bond-containing molecules react to form a cyclic compound.

[4 + 2] cycloaddition reaction   a cycloaddition reaction in which four π electrons come from one reactant and two π electrons come from the other reactant.

cycloalkane   an alkane with its carbon atoms arranged in a ring.

deactivating substituent   a substituent that decreases the reactivity of an aromatic ring. Electron-withdrawing substituents deactivate aromatic rings toward electrophilic attack, and electron-donating substituents deactivate aromatic rings toward nucleophilic attack.

deamination   removes an amino group.

decarboxylation   loss of carbon dioxide from a molecule.

degenerate orbitals   orbitals that have the same energy.

dehydration   loss of water from a molecule.

dehydrogenase   an enzyme that carries out an oxidation reaction by removing hydrogen from the substrate.

dehydrohalogenation   elimination of a proton and a halide ion.

delocalization energy (resonance energy)   the extra stability a compound achieves as a result of having delocalized electrons.

delocalized electrons   electrons that are shared by three or more atoms.

denaturation   destruction of the highly organized tertiary structure of a protein.

deoxyribonucleic acid (DNA)   a polymer of deoxyribonucleotides.

deoxyribonucleotide   a nucleotide in which the sugar component is D -2′ deoxyribose.

deoxy sugar   sugars in which one of the OH groups is replaced by a hydrogen.

DEPT¹³C NMR spectrum   a series of four spectra that distinguishes among —CH₃, —CH₂, and —CH groups.

detergent   a salt of benzenesulfonic acid.

deuterium kinetic isotope effect   ratio of the rate constant obtained for a compound containing hydrogen and the rate constant obtained for an identical compound in which one or more of the hydrogens have been replaced by deuterium.

dextrorotatory   an optically active compound rotates the plane of polarization clockwise.

diastereomer   a configurational stereoisomer that is not an enantiomer.

diastereotopic hydrogens   two hydrogens bonded to a carbon that when replaced in turn with a deuterium, result in a pair of diastereomers.

1,3-diaxial interaction   the interaction between an axial substituent and the other two axial substituents on the same side of the cyclohexane ring.

diazonium ion $\text{Ar}\overset{+}{\text{N}} \equiv \text{N}$ or $\text{R}\overset{+}{\text{N}} \equiv \text{N}$

diazonium salt   a diazonium ion and an anion ($\text{Ar}\overset{+}{\text{N}} \equiv \text{NX}^-$).

Dieckmann condensation   an intramolecular Claisen condensation.

Diels–Alder reaction   a conjugated diene reacts with a compound containing a carbon–carbon double bond.

diene   a hydrocarbon with two double bonds.

dienophile   an alkene that reacts with a diene in a Diels–Alder reaction.

β-diketone   a ketone with a second carbonyl group at the β-position.

dimer   a molecule formed by the joining together of two identical molecules.

dinucleotide   two nucleotides linked by phosphodiester bonds.

dipeptide   two amino acids linked by an amide bond.

dipole–dipole interaction   an interaction between the dipole of one molecule and the dipole of another.

dipole moment   the measure of the magnitude of the charge on either atom (because the partial positive charge and the partial negative charge have the same magnitude) times the distance between the two charges.

direct addition   1,2-addition. to an α,β-unsaturated carbonyl compound (addition to the carbonyl compound)

disaccharide   a compound containing two monosaccharides linked together.

disconnection   breaking a bond to carbon to give a simpler species.

disproportionation   transfer of a hydrogen atom by a radical to another radical, forming an alkane and an alkene.

disrotatory ring closure   achieves head-to-head overlap of *p* orbitals by rotating the orbitals in opposite directions.

dissolving metal reduction   a reduction brought about by the use of sodium or lithium metal dissolved in liquid ammonia.

disulfide bridge   two cysteines in a protein can be oxidized to a disulfide, creating a bond.

DNA (deoxyribonucleic acid)   a polymer of deoxyribonucleotides.

doping   adding or removing electrons from a polymer with conjugated double bonds.

double bond   two bonds connecting two atoms.

doublet   an NMR signal split into two peaks.

doublet of doublets   an NMR signal split into four peaks of approximately equal height. Caused by splitting a signal into a doublet by one hydrogen and into another doublet by another (nonequivalent) hydrogen.

drug   a compound that reacts with a biological molecule, triggering a physiological effect.

drug resistance   biological resistance to a particular drug.

eclipsed conformation   a conformation in which the bonds on adjacent carbons are aligned as viewed looking down the carbon–carbon bond.

Edman's reagent   phenyl isothiocyanate. A reagent used to determine the N-terminal amino acid of a polypeptide.

effective magnetic field   the magnetic field that a proton "senses" through the surrounding cloud of electrons.

effective molarity   the concentration of the reagent that would be required in an inter-molecular reaction for it to have the same rate as an intramolecular reaction.

***E* isomer**   the isomer with the high-priority groups on opposite sides of the double bond.

elastomer   a polymer that can stretch and then revert to its original shape.

electrocyclic reaction   a reaction in which a π bond in the reactant is lost so that a cyclic compound with a new σ bond can be formed.

electromagnetic radiation   radiant energy that displays wave properties.

electronegative   elements that readily acquire an electron.

electronegativity   a measure of the ability of an atom to pull electrons toward itself.

electronic transition   promotion of an electron from its HOMO to its LUMO.

electron sink   site to which electrons can be delocalized.

electrophile   an electron-deficient atom or molecule.

electrophilic addition reaction   an addition reaction in which the first species that adds to the reactant is an electrophile.

electrophilic aromatic substitution reaction   a reaction in which an electrophile substi-tutes for a hydrogen of an aromatic ring.

electrophoresis   a technique that separates amino acids on the basis of their pI values.

electrostatic attraction   attractive force between opposite charges.

electrostatic catalysis   stabilization of a charge by an opposite charge.

elemental analysis   a determination of the relative proportions of the elements present in a compound.

β-elimination reaction (1,2-elimination reaction)   removal of two atoms or groups from adjacent carbons.

elimination reaction   a reaction that involves the elimination of atoms (or molecules) from the reactant.

enamine   an α,β-unsaturated tertiary amine.

enantiomerically pure   containing only one enantiomer.

enantiomeric excess (optical purity)   how much excess of one enantiomer is present in a mixture of a pair of enantiomers.

enantiomers   nonsuperimposable mirror-image molecules.

enantioselective reaction   a reaction that forms an excess of one enantiomer.

enantiotopic hydrogens   two hydrogens bonded to a carbon that is bonded to two other groups that are nonidentical.

endergonic reaction   the reaction will consume more energy than it releases; a reaction with a positive $\Delta G°$.

endo   a substituent is endo if it is closer to the longer or more unsaturated bridge.

endopeptidase   an enzyme that hydrolyzes a peptide bond that is not at the end of a peptide chain.

endothermic reaction   a reaction with a positive $\Delta H°$.

enediol rearrangement   interconversion of an aldose and one or more ketoses.

enkephalins   pentapeptides synthesized by the body to control pain.

enthalpy   the heat given off or the heat consumed during the course of a reaction.

entropy   a measure of the freedom of motion in a system.

enzyme   a protein that is a biological catalyst.

epimerization   changing the configuration of an asymmetric center by removing a proton from it and then reprotonating the molecule at the same site.

epimers   monosaccharides that differ in configuration at only one asymmetric center.

epoxidation   formation of an epoxide.

epoxide   an ether in which the oxygen is incorporated into a three-membered ring.

epoxy resin   substance formed by mixing a low-molecular-weight prepolymer with a com-pound that forms a cross-linked polymer.

equatorial bond   a bond of the chair conformer of cyclohexane that juts out from the ring in approximately the same plane that contains the chair.

equilibrium constant   the degree to which an acid (HA) dissociates in an aqueous solution is indicated by the reaction, K_{eq}.

E1 reaction   a first-order elimination reaction. rate = K[alkyl halide]

E2 reaction   a second-order elimination reaction. rate = K[alkyl halide][base]

erythro enantiomers   the pair of enantiomers with similar groups on the same side as drawn in a Fischer projection.

essential amino acid   an amino acid that humans must obtain from their diet because they cannot synthesize it at all or cannot synthesize it in adequate amounts.

essential oils   fragrances and flavorings isolated from plants that do not leave residues when they evaporate. Most are terpenes.

ester

estrogens   estrone and estradiol; female sex hormones.

ether   a compound containing an oxygen bonded to two carbons (ROR).

eukaryote   a unicellular or multicellular body whose cell or cells contain a nucleus.

excited-state electronic configuration   the electronic configuration that results when an electron in the ground-state electronic configuration has been moved to a higher energy orbital.

exergonic reaction   a reaction that will release more energy than it consumes; a reaction with a negative $\Delta G°$.

exhaustive methylation   reaction of an amine with excess methyl iodide to form a quater-nary ammonium iodide.

exo   a substituent is exo if it is closer to the shorter or more saturated bridge.

exopeptidase   an enzyme that hydrolyzes a peptide bond at the end of a peptide chain.

exothermic reaction   a reaction with a negative $\Delta H°$.

experimental energy of activation ($E_a = \Delta H^{\ddagger} - RT$)   a measure of the approximate energy barrier to a reaction. (It is approximate because it does not contain an entropy component.)

fat   a triester of glycerol that exists as a solid at room temperature.

fatty acid   carboxylic acids with long hydrocarbon chains that are found in nature.

Favorskii reaction   reaction of an α-haloketone with hydroxide ion.

feedback inhibitor   a compound that inhibits the enzyme that catalyzes its biosynthesis.

fibrous protein   a water-insoluble protein in which the polypeptide chains are arranged in bundles.

fingerprint region   the right-hand third of an IR spectrum where the absorption bands are characteristic of the compound as a whole.

first-order rate constant   the rate constant of a first-order reaction.

first-order reaction (unimolecular reaction)   a reaction whose rate depends on the con-centration of one reactant.

Fischer esterification reaction   reaction in which an ester can be prepared by treating a carboxylic acid with excess alcohol in the presence of an acid catalyst.

Fischer projection   a method of representing the spatial arrangement of groups bonded to an asymmetric center. The asymmetric center is the point of intersection of two perpen-dicular lines; the horizontal lines represent bonds that project out of the plane of the paper toward the viewer, and the vertical lines represent bonds that point back from the plane of the paper away from the viewer.

flagpole hydrogens   the two hydrogens in the boat conformation of cyclohexane that are closest to each other.

flavin adenine dinucleotide (FAD)   a coenzyme required in certain oxidation reactions. It is reduced to FADH$_2$, which forms 1.5 ATPs in oxidative phosphorylation when it is oxidized back to FAD.

formal charge   the number of valence electrons – (the number of nonbonding electrons + 1/2 the number of bonding electrons).

Fourier transform NMR (FT–NMR) spectrum   a technique in which all the nuclei are excited simultaneously by an rf pulse, their relaxation is monitored, and the data are math-ematically converted to a spectrum.

free energy of activation ($\Delta G^{\ddagger}$)   the energy barrier to a reaction.

free-induction decay   relaxation of excited nuclei.

frequency   the number of wave crests that pass by a given point in one second; frequency has units of hertz (Hz).

Friedel–Crafts acylation   an electrophilic substitution reaction that puts an acyl group on a benzene ring.

Friedel–Crafts alkylation   an electrophilic substitution reaction that puts an alkyl group on a benzene ring.

frontier orbitals   the HOMO and the LUMO.

frontier orbital theory   a theory that, like the conservation of orbital symmetry theory, explains the relationships among reactant, product, and reaction conditions in a pericyclic reaction.

functional group   determines the kinds of reactions a compound undergoes.

functional group interconversion   the conversion of one functional group into another functional group.

functional group region   the left-hand two-thirds of an IR spectrum where most functional groups show absorption bands.

furanose   a five-membered-ring sugar.

furanoside   a five-membered-ring glycoside.

Gabriel synthesis   conversion of an alkyl halide into a primary amine, using phthalimide as a starting material.

gauche X and Y are gauche to each other in this Newman projection:

gauche conformer a staggered conformer in which the largest substituents are gauche to each other.

gauche interaction the interaction between two atoms or groups that are gauche to each other.

gem-dialkyl effect two alkyl groups on a carbon, the effect of which is to increase the probability that the molecule will be in the proper conformation for ring closure.

gem-diol (hydrate) a compound with two OH groups on the same carbon.

geminal coupling the mutual splitting of two nonidentical protons bonded to the same carbon.

geminal dihalide a compound with two halogen atoms bonded to the same carbon.

gene a segment of DNA.

general-acid catalysis catalysis in which a proton is transferred to the reactant during the slow step of the reaction.

general-base catalysis catalysis in which a proton is removed from the reactant during the slow step of the reaction.

genetic code the amino acid specified by each three-base sequence of mRNA.

genetic engineering techniques synthesizing proteins.

geometric isomers have the same atoms, and the atoms are connected to each other in the same order, but they have different spatial arrangements.

Gibbs free-energy change ($\Delta G°$) the difference between the free energy of the products and the free energy of the reactants under standard conditions.

Gilman reagent (R_2CuLi) an organocuprate, prepared from the reaction of an organolithium reagent with cuprous iodide, used to replace a halogen with an alkyl group.

globular protein a water-soluble protein that tends to have a roughly spherical shape.

gluconeogenesis The series of reactions that convert two molecules of pyruvate into glucose.

glycol (diol) a compound containing two or more OH groups.

glycolysis the breakdown of glucose to two molecules of pyruvate.

glycoprotein a protein that is covalently bonded to a polysaccharide.

glycoside the acetal of a sugar.

N-glycoside a glycoside with a nitrogen instead of an oxygen at the glycosidic linkage.

glycosidic bond the bond between the anomeric carbon and the alkoxy oxygen.

α-1,4′-glycosidic linkage a linkage between the C-1 oxygen of one sugar and the C-4 of a second sugar with the oxygen atom of the glycosidic linkage in the axial position.

β-1,4′-glycosidic linkage a linkage between the C-1 oxygen of one sugar and the C-4 of a second sugar with the oxygen atom of the glycosidic linkage in the equatorial position.

graft copolymer a copolymer that contains branches of a polymer of one monomer grafted onto the backbone of a polymer made from another monomer.

Grignard reagent (Organomagnesium compound) prepared by adding an alkyl halide to magnesium metal shavings being stirred in an ether—usually diethyl ether or tetrahydrofuran (THF)—under anhydrous conditions. The reaction inserts magnesium between the carbon and the halogen.

ground-state electronic configuration a description of which orbitals the electrons of an atom or molecule occupy when all of the electrons of atoms are in their lowest-energy orbitals.

Hagemann's ester a compound prepared by treating a mixture of formaldehyde and ethylacetoacetate with base and then with acid and heat.

half-chair conformer the least stable conformation of cyclohexane.

haloform reaction the reaction of a halogen and HO— with a methyl ketone.

halogenation reaction take place only at high temperatures or in the presence of light with an appropriate wavelength.

halohydrin an organic molecule that contains a halogen and an OH group.

Hammond postulate states that the transition state will be more similar in structure to the species (reactants or products) that it is closer to energetically.

Haworth projection a way to show the structure of a sugar; the five- and six-membered rings are represented as being flat.

head-to-tail addition the head of one molecule is added to the tail of another.

heat of formation the heat given off when a compound is formed from its elements under standard conditions.

heat of hydrogenation the heat released in a hydrogenation reaction.

Heck reaction couples an aryl, benzyl, or vinyl halide or triflate with an alkene in a basic solution in the presence of $Pd(PPh_3)_4$.

Heisenberg uncertainty principle states that both the precise location and the momentum of an atomic particle cannot be simultaneously determined.

α-helix the backbone of a polypeptide coiled in a right-handed spiral with hydrogen bonding occurring within the helix.

Hell–Volhard–Zelinsky (HVZ) reaction heating a carboxylic acid with $Br_2 + P$ in order to convert it into an α-bromocarboxylic acid.

hemiacetal the product formed when one equivalent of an alcohol adds to an *aldehyde* or a *ketone*

hemiketal (hemiacetal is preferred)

Henderson–Hasselbalch equation $pK_a = pH + \log\dfrac{[HA]}{[A^-]}$

heptose a monosaccharide with seven carbons.

HETCOR spectra a 2-D NMR spectrum that shows coupling between protons and the carbons to which they are attached.

heteroatom an atom other than carbon or hydrogen.

heterocyclic compound (heterocycle) a cyclic compound in which one or more of the atoms of the ring are heteroatoms.

heterogeneous catalyst a catalyst that is insoluble in the reaction mixture.

heterolytic bond cleavage (heterolysis) breaking a bond with the result that both bonding electrons stay with one of the atoms.

hexose a monosaccharide with six carbons.

high-energy bond a bond that releases a great deal of energy when it is broken.

highest occupied molecular orbital (HOMO) the molecular orbital of highest energy that contains an electron.

high-resolution NMR spectroscopy NMR spectroscopy that uses a spectrometer with a high operating frequency.

Hofmann degradation exhaustive methylation of an amine, followed by reaction with Ag_2O, followed by heating to achieve a Hofmann elimination reaction.

Hofmann elimination (anti-Zaitsev elimination) a hydrogen is removed from the β-carbon bonded to the most hydrogens.

Hofmann elimination reaction elimination of a proton and a tertiary amine from a quaternary ammonium hydroxide.

Hofmann rearrangement conversion of an amide into an amine by using Br_2/HO.

homolog a member of a homologous series.

homologous series a family of compounds in which each member differs from the next by one before it in the series by one methylene (CH_2) group.

homolytic bond cleavage (homolysis) breaking a bond with the result that each of the atoms gets one of the bonding electrons.

homopolymer a polymer that contains only one kind of monomer.

Hooke's law an equation that describes the motion of a vibrating spring

$$v = \frac{1}{2\pi c}\left[\frac{f(m_1 + m_2)}{m_1 m_2}\right]^{1/2}$$

hormone an organic compound synthesized in a gland and delivered by the bloodstream to its target tissue in order to stimulate or inhibit some process.

Hückel's rule (4n + 2 rule) states that, for a compound to be aromatic, its cloud of electrons must contain $(4n + 2)\pi$ electrons, where n is an integer. This is the same as saying that the electron cloud must contain an odd number of pairs of π electrons.

human genome the total DNA of a human cell.

Hund's rule states that when there are degenerate orbitals, an electron will occupy an empty orbital before it will pair up with another electron.

hybrid orbital mixed orbitals that result from combining atomic orbitals

hydrate (gem-diol) the addition of water to an aldehyde or a ketone

hydrated water has been added to a compound.

hydration addition of water to a molecule.

hydrazone the imine derivative obtained from the reaction with hydrazine $R_2C=NNH_2$

hydride ion a negatively charged hydrogenion.

1,2-hydride shift the movement of a hydride ion from one carbon to an adjacent carbon.

hydroboration–oxidation the addition of borane to an alkene or an alkyne, followed by reaction with hydrogen peroxide and hydroxide ion.

hydrocarbon a compound that contains only carbon and hydrogen.

α-hydrogen a hydrogen bonded to an α-carbon.

hydrogenation addition of hydrogen.

hydrogen bond a special kind of dipole–dipole interaction that occurs between a hydrogen that is attached to an oxygen, a nitrogen, or a fluorine and a lone pair of an oxygen, a nitrogen, or a fluorine in another molecule.

hydrogen ion a positively charged hydrogen results from the loss of its sole electron.

hydrolysis reaction a reaction with water that converts one compound into two compounds.

hydrophobic interactions interactions between nonpolar groups. These interactions increase stability by decreasing the amount of structured water (increasing entropy).

hyperconjugation delocalization of electrons by overlap of a σ bond orbital with an empty orbital on an adjacent carbon.

imine (Schiff base) a compound with a carbon–nitrogen double bond $R_2C=NR$

inclusion compound a compound that specifically binds a metal ion or an organic molecule.

induced-dipole–induced-dipole interaction an interaction between a temporary dipole in one molecule and the dipole the temporary dipole induces in another molecule.

induced-fit model a model that describes the specificity of an enzyme for its substrate: the shape of the active site does not become completely complementary to the shape of the substrate until after the enzyme binds the substrate.

inductive electron donation donation of electrons through σ bonds.

inductive electron withdrawal withdrawal of electrons through sigma σ bonds.

informational strand (sense strand) the strand in DNA that is not read during transcription; it has the same sequence of bases as the synthesized mRNA strand (with a U, T difference).

infrared radiation electromagnetic radiation familiar to us as heat.

infrared (IR) spectroscopy identifies the kinds of bonds and, therefore, the kinds of functional groups in an organic compound.

infrared spectrum a plot of percent transmission versus wave number (or wavelength) of infrared radiation.

initiation step the step in which radicals are created, or the step in which the radical needed for the first propagation step is created.

interchain disulfide bridge a disulfide bridge between two cysteine residues in different peptide chains.

intermediate a chemical species that is a product of one step of a reaction and a reactant for the next step.

intermolecular reaction a reaction that takes place between two molecules.

internal alkyne an alkyne with the triple bond not at the end of the carbon chain.

intimate ion pair pair such that the covalent bond that joined the cation and the anion has broken, but the cation and anion are still next to each other.

intrachain disulfide bridge a disulfide bridge between two cysteine residues in the same peptide chain.

intramolecular catalysis catalysis in which the catalyst that facilitates the reaction is part of the molecule undergoing reaction.

intramolecular reaction a reaction that takes place within a molecule.

inversion of configuration turning the configuration of a carbon inside out like an umbrella in a windstorm, so that the resulting product has a configuration opposite that of the reactant.

iodoform test the addition of I_2/HO^- to a methyl ketone forms a yellow precipitate of triiodomethane.

ion–dipole interaction the interaction between an ion and the dipole of a molecule of the protic solvent.

ion-exchange chromatography a technique that uses a column packed with an insoluble resin to separate compounds on the basis of their charges and polarities.

isoelectric point (pI) the pH at which there is no net charge on an amino acid.

isolated double bonds double bonds separated by more than one single bond.

isomers compounds with the same molecular formula but different structures.

isotactic polymer a polymer in which all the substituents are on the same side of the fully extended carbon chain.

isotopes atoms with the same number of protons, but different numbers of neutrons.

IUPAC nomenclature systematic method of nomenclature of chemical compounds.

Kekulé structure a model that represents the bonds between atoms as lines.

ketal (acetal is preferred)
$$R-\overset{\overset{\displaystyle OR}{|}}{\underset{\underset{\displaystyle OR}{|}}{C}}-R$$

keto–enol interconversion (keto–enol tautomerization) interconversion of keto and enol tautomers.

keto–enol tautomers a ketone and its isomeric α,β-unsaturated alcohol.

β-keto ester an ester with a second carbonyl group at the β-position.

ketone carbonyl carbon bonded to two R groups
$$\overset{\overset{\displaystyle O}{\|}}{\underset{R \quad R}{C}}$$

ketose a polyhydroxy ketone.

Kiliani–Fischer synthesis a method used to increase the number of carbons in an aldose by one, resulting in the formation of a pair of C-2 epimers.

kinase an enzyme that puts a phosphate group on its substrate.

kinetic isotope effect a comparison of the rate of reaction of a compound with the rate of reaction of an identical compound in which one of the atoms has been replaced by an isotope.

kinetic product the more rapidly formed product produced from a reaction.

kinetic resolution separation of enantiomers on the basis of the difference in their rate of reaction with an enzyme.

kinetics the field of chemistry that deals with the rates of chemical reactions and the factors that affect those rates.

kinetic stability indicated by $\Delta G^{\ddagger}$. If $\Delta G^{\ddagger}$ is large, then the reactant is *kinetically stable* because it reacts slowly. If $\Delta G^{\ddagger}$ is small, then the reactant is *kinetically unstable*—it reacts rapidly. Similarly, if $\Delta G^{\ddagger}$ for the reverse reaction is large, then the product is kinetically stable, but if it is small, then the product is kinetically unstable.

kinetically controlled reactions that produce the kinetic product as the major product.

Knoevenagel condensation a condensation of an aldehyde or ketone with no α hydrogen and a compound with an α-carbon flanked by two electron-withdrawing groups.

Kolbe–Schmitt carboxylation reaction a reaction that uses CO_2 to carboxylate phenol.

lactam a cyclic amide.

lactone a cyclic ester.

λ_{max} the wavelength at which there is maximum UV/Vis absorbance.

lead compound the prototype in a search for other biologically active compounds.

leaning when a line drawn over the outside peaks of an NMR signal points in the direction of the signal given by the protons that cause the splitting.

leaving group the atom or group that is displaced in a nucleophilic substitution reaction.

Le Châtelier's principle states that if an equilibrium is disturbed, the components of the equilibrium will adjust in a way that will offset the disturbance.

lecithin a phosphoacylglycerol in which the second OH group of phosphate has formed an ester with choline.

levorotatory the compound that rotates polarized light in a counterclockwise direction.

Lewis acid non-proton-donating acids

Lewis base a base that has a pair of electrons that they can share.

Lewis structure a model that represents the bonds between atoms as lines or dots and the valence electrons as dots.

ligation sharing of lone-pair electrons with a metal ion.

linear combination of atomic orbitals (LCAO) the combination of atomic orbitals to produce a molecular orbital.

lipid a water-insoluble compound found in a living system.

lipid bilayer two layers of phosphoacylglycerols arranged so that their polar heads are on the outside and their nonpolar fatty acid chains are on the inside.

lipoate a coenzyme required in certain oxidation reactions.

living polymer a nonterminated chain-growth polymer that remains active. This means that the polymerization reaction can continue upon the addition of more monomer.

localized electrons electrons that are restricted to a particular region.

lock-and-key model a model that describes the specificity of an enzyme for its substrate: the substrate fits the enzyme as a key fits a lock.

London dispersion forces induced-dipole–induced-dipole interactions.

lone-pair electrons (nonbonding electrons) valence electrons not used in bonding.

long-range coupling splitting of a proton by a proton more than three σ bonds away.

loop conformation (coil conformation) that part of a protein that is highly ordered, but not in an α-helix or β-pleated sheet.

lowest unoccupied molecular orbital (LUMO) the molecular orbital of lowest energy that does not contain an electron.

Lucas test a test that determines whether an alcohol is primary, secondary, or tertiary.

magnetic resonance imaging (MRI) NMR used in medicine. The difference in the way water is bound in different tissues produces a variation in signal between organs as well as between healthy and diseased tissue.

major groove the wider and deeper of the two alternating grooves in DNA.

malonic ester synthesis the synthesis of a carboxylic acid, using diethyl malonate as the starting material.

Mannich reaction condensation of a secondary amine and formaldehyde with a carbon acid

Markovnikov's rule the actual rule is "When a hydrogen halide adds to an asymmetrical alkene, the addition occurs such that the halogen attaches itself to the sp^2 carbon of the alkene bearing the lowest number of hydrogen atoms." A more universal rule is "The electrophile adds to the sp^2 carbon that is bonded to the greater number of hydrogens."

mass number the sum of protons and neutrons in an atom.

mass spectrometry is used to find the molecular mass and the molecular formula of an organic compound; it is also used to identify certain structural features of the compound by identifying the fragments produced when the molecule breaks apart.

mass spectrum a graph of the relative abundance of each fragment plotted against its m/z values.

materials science the science of creating new materials to be used in place of known materials such as metal, glass, wood, cardboard, and paper.

McLafferty rearrangement rearrangement of the molecular ion of a ketone. The bond between the α- and β-carbons breaks, and a γ-hydrogen migrates to the oxygen.

mechanism-based inhibitor (suicide inhibitor) a compound that inactivates an enzyme by undergoing part of its normal catalytic mechanism.

mechanism of the reaction a description of the step-by-step process by which reactants are changed into products.

melting point (mp) the temperature at which its solid form is converted into a liquid.

membrane the material that surrounds a cell in order to isolate its contents.

mercaptan (thiol) the sulfur analog of an alcohol (RSH).

meso compound a compound that has asymmetric centers, it is achiral because it is superimposable on its mirror image.

metabolism reactions that living organisms carry out in order to obtain the energy and to synthesize the compounds they require.

meta-directing substituent a substituent that directs an incoming substituent meta to an existing substituent.

metal-ion catalysis catalysis in which the species that facilitates the reaction is a metal ion.

metalloenzyme an enzyme that has a tightly bound metal ion.

methylene (CH$_2$) group a CH$_2$ group.

2-methyl shift the movement of a methyl group with its bonding electrons from one carbon to an adjacent carbon.

micelle a spherical aggregation of molecules, each with a long hydrophobic tail and a polar head, arranged so that the polar head points to the outside of the sphere.

Michael reaction the addition of an α-carbanion to the β-carbon of an α,β-unsaturated carbonyl compound.

minor groove the narrower and more shallow of the two alternating grooves in DNA.

mixed anhydride an anhydride formed from two different acids.

mixed triacylglycerol a triacylglycerol in which the fatty-acid components are different.

molar absorptivity (ε) a constant that is characteristic of the compound at a particular wavelength.

molecular ion when the electron beam hits a molecule, it knocks out an electron.

molecular modification changing the structure of a lead compound.

molecular orbital the volume of space around a molecule where an electron is likely to be found.

molecular orbital (MO) theory combines the tendency of atoms to fill their octets by sharing electrons (the Lewis model) with their wave-like properties, assigning electrons to a volume of space called an orbital.

molecular recognition the ability of one molecule to recognize another molecule as a result of intermolecular interactions.

molozonide an unstable intermediate containing a five-membered ring with three oxygens in a row that is formed from the reaction of an alkene with ozone.

monomer a repeating unit in a polymer.

monosaccharide (simple carbohydrate) a single sugar molecule.

monoterpene a terpene that contains 10 carbons.

MRI scanner an NMR spectrometer used in medicine for whole-body NMR.

multiplet an NMR signal split into more than seven peaks.

multiplicity the number of peaks in an NMR signal.

multistep synthesis preparation of a compound by a route that requires several steps.

mutarotation a slow change in optical rotation to an equilibrium value.

mutase an enzyme that transfers a group from one position to another.

N + 1 rule an ^{1}H NMR signal for a hydrogen with N equivalent hydrogens bonded to an adjacent carbon is split into $N + 1$ peaks. A ^{13}C NMR signal for a carbon bonded to N hydrogens is split into $N + 1$ peaks.

neurotransmitter a compound that transmits nerve impulses.

nicotinamide adenine dinucleotide (NAD$^+$) a coenzyme required in certain oxidation reactions. It is reduced to NADH, which forms 2.5 ATPs in oxidative phosphorylation when it is oxidized back to NAD$^+$.

nicotinamide adenine dinucleotide phosphate (NADP$^+$) a coenzyme that is reduced to NADPH, which is used as a reducing agent in anabolic reactions.

NIH shift the 1,2-hydride shift of a carbocation (obtained from an arene oxide) that leads to an enone.

nitration substitution of a nitro group (NO$_2$) for a hydrogen of a benzene ring.

nitrile a compound that contains a carbon–nitrogen triple bond (RC≡N).

nitrosamine removal of a proton from nitrogen R$_2$NN=O

NMR spectroscopy the absorption of electromagnetic radiation to determine the structural features of an organic compound. In the case of NMR spectroscopy, it determines the carbon–hydrogen framework.

node that part of an orbital in which there is zero probability of finding an electron.

nonbonding electrons (lone-pair electrons) valence electrons not used in bonding.

nonbonding molecular orbital the p orbitals are too far apart to overlap significantly, so the molecular orbital that results neither favors nor disfavors bonding.

noncovalent interaction an interaction between molecules that is weaker than a covalent bond.

nonpolar covalent bond a bond formed between two atoms that share the bonding electrons equally.

nonreducing sugar a sugar that cannot be oxidized by reagents such as Ag$^+$ and Cu$^+$. Nonreducing sugars are not in equilibrium with the open-chain aldose or ketose.

N-phthalimidomalonic ester synthesis a method used to synthesize an amino acid that combines the malonic ester synthesis and the Gabriel synthesis.

N-terminal amino acid the terminal amino acid of a peptide (or protein) that has a free amino group.

nucleic acid the two kinds of nucleic acid are DNA and RNA.

nucleophile an electron-rich atom or molecule.

nucleophilic acyl substitution reaction (acyl transfer reaction) a reaction in which a group bonded to an acyl or aryl group is substituted by another group.

nucleophilic addition–elimination reaction a nucleophilic addition reaction that is followed by an elimination reaction. Imine formation is an example: an amine adds to the carbonyl carbon, and water is eliminated.

nucleophilic addition reaction a reaction that involves the addition of a nucleophile to a reagent.

nucleophilic aromatic substitution a reaction in which a nucleophile substitutes for a substituent of an aromatic ring.

nucleophilic catalysis catalysis that occurs as a result of a nucleophile forming a covalent bond with one of the reactants.

nucleophilic catalyst a catalyst that increases the rate of a reaction by acting as a nucleophile.

nucleophilicity a measure of how readily a compound is able to attack an electron-deficient atom.

nucleophilic substitution reaction a reaction in which a nucleophile substitutes for an atom or a group.

nucleoside a heterocyclic base (a purine or a pyrimidine) bonded to the anomeric carbon of a sugar (D-ribose or D-2'-deoxyribose).

nucleotide a heterocycle attached in the β-position to a phosphorylated ribose or deoxyribose.

observed rotation the amount the analyzer is rotated is observed in a polarimeter.

octet rule states that an atom will give up, accept, or share electrons in order to achieve a filled outer shell that contains eight electrons.

oil a triester of glycerol that exists as a liquid at room temperature.

olefin an alkene.

olefin metathesis (alkene metathesis) breaks the double bond of an alkene (or the triple bond of an alkyne) and then rejoins the fragments.

oligomer a protein with more than one peptide chain.

oligonucleotide 3 to 10 nucleotides linked by phosphodiester bonds.

oligopeptide 3 to 10 amino acids linked by amide bonds.

oligosaccharide 3 to 10 sugar molecules linked by glycosidic bonds.

open-chain compound an acyclic compound.

operating frequency the frequency at which an NMR spectrometer operates.

optical isomers stereoisomers that contain chirality centers.

optically active rotates the plane of polarized light.

optically inactive does not rotate the plane of polarized light.

optical purity (enantiomeric excess) how much excess of one enantiomer is present in a mixture of a pair of enantiomers.

orbital the volume of space around the nucleus in which an electron is most likely to be found.

organic compound compounds that are based on carbon.

organic synthesis preparation of organic compounds from other organic compounds.

organoboron compound a compound containing a C—B bond.

organocuprate (R$_2$CuLi) Gilman reagent, prepared from the reaction of an organolithium reagent with cuprous iodide, used to replace a halogen with an alkyl group.

Organomagnesium compound (Grignard reagent) prepared by adding an alkyl halide to magnesium metal shavings being stirred in an ether—usually diethyl ether or tetrahydrofuran (THF)—under anhydrous conditions. The reaction inserts magnesium between the carbon and the halogen.

organometallic compound a compound containing a carbon–metal bond.

oriented polymer a polymer obtained by stretching out polymer chains and putting them back together in a parallel fashion.

ortho-para-directing substituent a substituent that directs an incoming substituent ortho and para to an existing substituent.

oxidation loss of electrons by an atom or a molecule.

β-oxidation a series of four reactions that removes two carbons from a fatty acyl-CoA.

oxidation reaction a reaction in which the number of C—H bonds decreases or the number of C—O, C—N, or C—X (X = a halogen) increases.

oxidative addition the insertion of a metal between two atoms.

oxidative cleavage an oxidation reaction that cleaves the reactant into pieces.

oxidative phosphorylation a series of reactions that converts a molecule of NADH and a molecule of FADH$_2$ into 2.5 and 1.5 molecules of ATP, respectively.

oxime the imine derivative obtained from the reaction with hydroxylamine $R_2C{=}NOH$

oxyanion a compound with a negatively charged oxygen.

ozonide the five-membered-ring compound formed as a result of the rearrangement of a molozonide.

ozonolysis reaction of a carbon–carbon double or triple bond with ozone.

packing the fitting of individual molecules into a frozen crystal lattice.

paraffin an alkane.

parent hydrocarbon the longest continuous carbon chain in a molecule.

partial hydrolysis a technique that hydrolyzes only some of the peptide bonds in a polypeptide.

partial racemization formation of a pair of enantiomers in unequal amounts.

Pauli exclusion principle states that no more than two electrons can occupy an orbital and that the two electrons must have opposite spin.

pentose a monosaccharide with five carbons.

peptide polymer of amino acids linked together by amide bonds. A peptide contains fewer amino acid residues than a protein does.

peptide bond the amide bond that links the amino acids in a peptide or protein.

peptide nucleic acid (PNA) a polymer containing both an amino acid and a base designed to bind to specific residues on DNA or mRNA.

Perkin condensation a condensation of an aromatic aldehyde and acetic acid.

pericyclic reaction a concerted reaction that takes place as the result of a cyclic rearrangement of electrons.

peroxyacid a carboxylic acid with an extra oxygen.

perspective formula a method of representing the spatial arrangement of groups bonded to an asymmetric center. Two bonds are drawn in the plane of the paper; a solid wedge is used to depict a bond that projects out of the plane of the paper toward the viewer, and a hatched wedge is used to represent a bond that projects back from the plane of the paper away from the viewer.

pH the concentration of protons in a solution. This concentration is written as either $([H^+])$ or, because a proton in water is solvated, as $([H_3^+])$ $(pH = -\log[H^+])$.

pH-activity profile a plot of the activity of an enzyme as a function of the pH of the reaction mixture.

phenone

$$\underset{C_6H_5}{}\overset{O}{\underset{}{\overset{\|}{C}}}R$$

phenyl group a benzene ring is a substituent $C_6H_5{-}$

phenylhydrazone $R_2C{=}NNHC_6H_5$

pheromone a compound secreted by an animal that stimulates a physiological or behavioral response from a member of the same species.

phosphatase an enzyme that catalyzes the removal of a phosphoryl group from its substrate.

phosphatidic acid a phosphoacylglycerol in which only one of the OH groups of phosphate is in an ester linkage.

phosphoanhydride bond the bond holding two phosphoric acid molecules together.

phosphoglyceride (phosphoacylglycerol) a compound formed when two OH groups of glycerol form esters with fatty acids and the terminal OH group forms a phosphate ester.

phospholipid a lipid that contains a phosphate group.

phosphoryl transfer reaction the transfer of a phosphate group from one compound to another.

photochemical reaction a reaction that takes place when a reactant absorbs light.

photosynthesis the synthesis of glucose and O_2 from CO_2 and H_2O.

pH-rate profile a plot of the observed rate of a reaction as a function of the pH of the reaction mixture.

pi (π) bond a bond formed as a result of side-to-side overlap of p orbitals.

π bonding molecular orbital a molecular orbital that results when two in-phase atomic orbitals interact. Electrons in a bonding orbital increase bond strength.

π antibonding molecular orbital a molecular orbital that results when two atomic orbitals with opposite signs interact. Electrons in an antibonding orbital decrease bond strength.

(π) (pi)-complex a complex formed between an electrophile and a triple bond.

pinacol rearrangement rearrangement of a vicinal diol.

pK_a describes the tendency of a compound to lose a proton ($pK_a = -\log K_a$, where K_a is the acid dissociation constant).

plane of symmetry an imaginary plane that bisects a molecule into mirror images.

plane polarized light light that oscillates only in one plane.

plasticizer an organic molecule that dissolves in a polymer and allows the polymer chains to slide by each other.

β-pleated sheet the backbone of a polypeptide that is extended in a zigzag structure with hydrogen bonding between neighboring chains.

point of symmetry any line through a point of symmetry encounters identical environments at the same distance.

polar covalent bond a covalent bond between atoms of different electronegativites.

polarimeter an instrument that measures the direction and amount an optically active compound rotates the plane of polarization of plane-polarized light.

polarizability an indication of how readily an electron cloud can be distorted to create a strong induced dipole.

polar reaction the reaction between a nucleophile and an electrophile.

polyamide a polymer in which the monomers are amides.

polycarbonate a step-growth polymer in which the dicarboxylic acid is carbonic acid.

polyene a compound that has several double bonds.

polyester a polymer in which the monomers are esters.

polymer a large molecule made by linking together repeating units of small molecules.

polymer chemistry the field of chemistry that deals with synthetic polymers; part of the larger discipline known as materials science.

polymerization the process of linking up monomers to form a polymer.

polynucleotide many nucleotides linked by phosphodiester bonds.

polypeptide many amino acids linked by amide bonds.

polysaccharide a compound containing more than 10 sugar molecules linked together.

polyunsaturated fatty acid a fatty acid with more than one double bond.

polyurethane a polymer in which the monomers are urethanes.

porphyrin ring system consists of four pyrrole rings joined by one-carbon bridges.

primary alcohol an OH group attached to a primary carbon.

primary alkyl halide an alkyl halide in which the halogen is bonded to a primary carbon.

primary alkyl radical a radical with the unpaired electron on a primary carbon.

primary amine an amine with one alkyl group attached to the nitrogen.

primary carbocation a carbocation with the positive charge on a primary carbon.

primary carbon a carbon bonded to only one other carbon.

primary hydrogen a hydrogen bonded to a primary carbon.

primary structure (of a nucleic acid) the sequence of bases in a nucleic acid.

primary structure (of a protein) the sequence of amino acids in a protein.

prodrug a compound that does not become an effective drug until it undergoes a reaction in the body.

promoter site a short sequence of bases at the beginning of a gene.

propagating site the reactive end of a chain-growth polymer.

propagation step in the first of a pair of propagation steps, a radical (or an electrophile or a nucleophile) reacts to produce another radical (or an electrophile or a nucleophile) that reacts in the second to produce the radical (or the electrophile or the nucleophile) that was the reactant in the first propagation step.

pro-R hydrogen replacing this hydrogen with deuterium creates an asymmetric center with the R configuration.

pro-S hydrogen replacing this hydrogen with deuterium creates an asymmetric center with the S configuration.

prostacyclin a lipid, derived from arachidonic acid, that dilates blood vessels and inhibits platelet aggregation.

protecting group a reagent that protects a functional group from a synthetic operation that it would otherwise not survive.

protein a polymer containing 40 to 4000 amino acids linked by amide bonds.

protein prenylation attaching isopentenyl units to proteins.

protic polar solvent polar solvent molecules *have* a hydrogen bonded to an oxygen or to a nitrogen.

proton a positively charged hydrogen (H^+); a positively charged particle in an atomic nucleus.

proton-decoupled^{13}C NMR spectrum a ^{13}C NMR spectrum in which all the signals appear as singlets because there is no coupling between the nucleus and its bonded hydrogens.

proton transfer reaction the reaction of an acid with a base.

proximity effect an effect caused by one species being close to another.

pyranose a six-membered-ring sugar.

pyranoside a six-membered-ring glycoside.

pyridoxal phosphate (PLP) the coenzyme required by enzymes that catalyze certain transformations of amino acids.

pyrosequencing a technique used to determine the sequence of bases in a polynucleotide by detecting the identity of each base that adds to a primer.

quartet an NMR signal split into four peaks.

quaternary ammonium ion can undergo an elimination reaction with a strong base.

quaternary ammonium salt nitrogen compounds with four alkyl groups attached to the nitrogen.

quaternary structure a description of the way the individual polypeptide chains of a protein are arranged with respect to each other.

racemic mixture (racemate, racemic modification) a mixture of equal amounts of a pair of enantiomers.

radical (free radical) a species containing an atom with a single unpaired electron.

radical addition reaction an addition reaction in which the first species that adds is a radical.

radical anion a species with a negative charge and an unpaired electron.

radical cation a species with a positive charge and an unpaired electron.

radical chain reaction a reaction in which radicals are formed and react in repeating propagating steps.

radical inhibitor a compound that destroys reactive radicals by converting them to relatively stable radicals or to compounds with only paired electrons.

radical initiator a compound that creates radicals.

radical polymerization chain-growth polymerization in which the initiator is a radical; the propagation site is therefore a radical.

radical reaction a reaction in which a new bond is formed by using one electron from one reagent and one electron from another reagent.

radical substitution reaction a substitution reaction that has a radical intermediate.

random copolymer a copolymer with a random distribution of monomers.

rate constant a measure of how easy or difficult it is to reach the transition state of a reaction (to get over the energy barrier to the reaction).

rate-determining step (rate-limiting step) the step in a reaction that has the transition state with the highest energy.

rational drug design designing drugs with a particular structure to achieve a specific purpose.

R configuration after assigning relative priorities to the four groups bonded to an asymmetric center, if the lowest priority group is on a vertical axis in a Fischer projection (or pointing away from the viewer in a perspective formula), an arrow drawn from the highest priority group to the next-highest-priority group goes in a clockwise direction.

reaction coordinate diagram a reaction in which two or more constitutional isomers could be obtained as products but one of them predominates.

reactivity–selectivity principle states that the greater the reactivity of a species, the less selective it will be.

receptor site the site at which a drug binds in order to exert its physiological effect.

recombinant DNA DNA that has been incorporated into a host cell.

reduction reaction a reaction in which the number of C—H bonds increases or the number of C—O, C—N, or C—X (X = a halogen) decreases.

reducing sugar a sugar that can be oxidized by reagents such as Ag^+ or Br_2. Reducing sugars are in equilibrium with the open-chain aldose or ketose.

reduction gain of electrons by an atom or a molecule.

reductive amination the reaction of an aldehyde or a ketone with ammonia or with a primary amine in the presence of a reducing agent (H_2/Raney Ni).

reference compound a compound added to a sample whose NMR spectrum is to be taken. The positions of the signals in the NMR spectrum are measured from the position of the signal given by the reference compound.

reductive elimination the elimination of two groups attached to a metal.

Reformatsky reaction reaction of an organozinc reagent with an aldehyde or a ketone.

regioselective reaction a reaction that leads to the preferential formation of one constitutional isomer over another.

regulatory enzyme an enzyme that can be turned on and off.

relative configuration the configuration of a compound relative to the configuration of another compound.

relative rate obtained by dividing the actual rate constant by the rate constant of the slowest reaction in the group being compared.

replication the synthesis of identical copies of DNA.

resolution of a racemic mixture separation of a racemic mixture into the individual enantiomers.

resonance a compound with delocalized electrons is said to have resonance.

resonance contributor (resonance structure, contributing resonance structure) two structures shown for the conjugate base of the carboxylic acid.

resonance electron donation donation of electrons through p orbital overlap with neighboring π bonds.

resonance electron withdrawal withdrawal of electrons through p orbital overlap with neighboring π bonds.

resonance energy (delocalization energy) the extra stability associated with a compound as a result of its having delocalized electrons.

resonance hybrid a composite of the two resonance contributors. The double-headed arrow between the two resonance contributors is used to indicate that the actual structure is a hybrid.

restriction endonuclease an enzyme that cleaves DNA at a specific base sequence.

restriction fragment a fragment that is formed when DNA is cleaved by a restriction endonuclease.

retrosynthetic analysis (retrosynthesis) working backward (on paper) from the target molecule to available starting materials.

retrovirus a virus whose genetic information is stored in its RNA.

rf radiation radiation in the radiofrequency region of the electromagnetic spectrum.

ribonucleic acid (RNA) a polymer of ribonucleotides.

ribonucleotide a nucleotide in which the sugar component is D-ribose.

ribosome a particle composed of about 40% protein and 60% RNA on which protein biosynthesis takes place.

ring-expansion rearrangement rearrangement of a carbocation in which the positively charged carbon is bonded to a cyclic compound and, as a result of rearrangement, the size of the ring increases by one carbon.

ring-flip the conversion of the chair conformer of cyclohexane into the other chair conformer. Bonds that are axial in one chair conformer are equatorial in the other.

ring-opening polymerization a chain-growth polymerization that involves opening the ring of the monomer.

Ritter reaction reaction of a nitrile with a secondary or tertiary alcohol to form a secondary amide.

RNA (ribonucleic acid) a polymer of ribonucleotides.

RNA splicing the step in RNA processing that cuts out nonsense bases and splices informational pieces together.

Robinson annulation a reaction that puts two carbon–carbon bond-forming reactions together to form an a,b-unsaturated cyclic ketone.

Sandmeyer reaction the reaction of an arenediazonium salt with a copper(I) salt.

saponification hydrolysis of an ester (such as a fat) under basic conditions.

Schiemann reaction fluoro substitution occurs if the arenediazonium salt is heated with fluoroboric acid HBF_4.

Schiff base (imine) a compound with a carbon–nitrogen double bond $R_2C=NR$

S configuration after assigning relative priorities to the four groups bonded to an asymmetric center, if the lowest priority group is on a vertical axis in a Fischer projection (or pointing away from the viewer in a perspective formula), an arrow drawn from the highest priority group to the next-highest priority group goes in a counterclockwise direction.

secondary alcohol an OH group attached to a secondary carbon.

secondary alkyl halide an alkyl halide in which the halogen is bonded to a secondary carbon.

secondary alkyl radical a radical with the unpaired electron on a secondary carbon.

secondary amine an amine with two alkyl groups attached to the nitrogen.

secondary carbocation a carbocation with the positive charge on a secondary carbon.

secondary carbon a carbon bonded to two other carbons.

secondary hydrogen a hydrogen bonded to a secondary carbon.

secondary structure a description of the conformation of the backbone of a protein.

second-order rate constant the rate constant of a second-order reaction.

second-order reaction (bimolecular reaction) a reaction whose rate depends on the concentration of two reactants. rate = k [alkyl halide][nucleophile]

selection rules the rules that determine the outcome of a pericyclic reaction.

semiconservative replication the mode of replication that results in a daughter molecule of DNA having one of the original DNA strands plus a newly synthesized strand.

sense strand (informational strand) the strand in DNA that is not read during transcription; it has the same sequence of bases as the synthesized mRNA strand (with a U, T difference).

separated charges a positive and a negative charge that can be neutralized by the movement of electrons.

sesquiterpene a terpene that contains 15 carbons.

shielding phenomenon caused by electron donation to the environment of a proton. The electrons shield the proton from the full effect of the applied magnetic field. The more a proton is shielded, the farther to the right its signal appears in an NMR spectrum.

sigma (σ) bond a bond with a cylindrically symmetrical distribution of electrons—the electrons in the bond are symmetrically distributed about an imaginary line connecting the nuclei of the two atoms joined by the bond.

sigmatropic rearrangement a reaction in which a σ bond is broken in the reactant, a new σ bond is formed in the product, and the π bonds rearrange.

Simmons–Smith reaction formation of a cyclopropane using $CH_2I_2 + Zn(Cu)$.

simple carbohydrate (monosaccharide) a single sugar molecule.

simple triacylglycerol a triacylglycerol in which the fatty acid components are the same.

single bond one bond connecting two atoms.

singlet an unsplit NMR signal.

site-specific mutagenesis a technique that substitutes one amino acid of a protein for another.

skeletal structure show the carbon–carbon bonds as lines, but do not show the carbons or the hydrogens that are bonded to the carbons.

S_NAr reaction a nucleophilic aromatic substitution reaction.

S_N1 reaction a unimolecular nucleophilic substitution reaction.

S_N2 reaction a bimolecular nucleophilic substitution reaction.

soap a sodium or potassium salt of a fatty acid.

solid-phase synthesis a technique in which one end of the compound being synthesized is covalently attached to a solid support.

solvation the interaction between a solvent molecules and solute molecules (molecules dissolved in a solvent).

solvent-separated ion pair the cation and anion are separated by a solvent molecule.

solvolysis reaction with the solvent.

specific-acid catalysis catalysis in which the proton is fully transferred to the reactant before the slow step of the reaction takes place.

specific-base catalysis catalysis in which the proton is completely removed from the reactant before the slow step of the reaction takes place.

specific rotation a specified temperature and wavelength can be calculated from the observed rotation using the formula $[\alpha]_\lambda^T = \dfrac{\alpha}{l \times c}$

spectroscopy study of the interaction of matter and electromagnetic radiation.

sphingolipid a lipid that contains sphingosine.

sphingomyelin a sphingolipid in which the terminal OH group of sphingosine is bonded to phosphocholine or phosphoethanolamine.

spin–spin coupling the splitting of a signal in an NMR spectrum described by the $N + 1$ rule.

α-spin state nuclei in this spin state have their magnetic moments oriented in the same direction as the applied magnetic field.

β-spin state nuclei in this spin state have their magnetic moments oriented opposite the direction of the applied magnetic field.

splitting diagram (splitting tree) a diagram that shows the splitting of a set of protons one at a time.

squalene a triterpene that is a precursor of steroid molecules.

stacking interactions van der Waals interactions between the mutually induced dipoles of adjacent pairs of bases in DNA.

staggered conformation a conformation in which the bonds on one carbon bisect the bond angle on the adjacent carbon when viewed looking down the carbon–carbon bond.

step-growth polymer (condensation polymer) a polymer made by combining two molecules while removing a small molecule (usually of water or an alcohol).

stereochemistry the field of chemistry that deals with the structures of molecules in three dimensions.

stereocenter (stereogenic center) an atom at which the interchange of two substituents produces a stereoisomer.

stereoisomers isomers that differ in the way their atoms are arranged in space.

stereoselective reaction a reaction that leads to the preferential formation of one stereo-isomer over another.

stereospecific reaction a reaction in which the reactant can exist as stereoisomers and each stereoisomeric reactant leads to a different stereo isomeric product or set of products.

steric effects caused by the fact that groups occupy a certain volume of space.

steric hindrance refers to bulky groups at the site of a reaction that make it difficult for the reactants to approach each other.

steric strain the strain experienced by a molecule (that is, the additional energy it possesses) when atoms or groups are close enough for their electron clouds to repel each other.

steroid a class of compounds that contains a steroid ring system.

Stille reaction replaces the akenyl-organoboron compound of the Suzuki reaction with an alkenyl-organotin compound.

stop codon a codon at which protein synthesis is stopped.

straight-chain alkane an alkane in which the carbons form a continuous chain with no branches.

Strecker synthesis a method used to synthesize an amino acid: an aldehyde reacts with NH_3, forming an imine that is attacked by cyanide ion. Hydrolysis of the product gives an amino acid.

structural isomers (constitutional isomers) molecules that have the same molecular formula but differ in the way their atoms are connected.

structural protein a protein that gives strength to a biological structure.

α-substituent a substituent on the side of a steroid ring system opposite that of the angular methyl groups.

β-substituent a substituent on the same side of a steroid ring system as that of the angular methyl groups.

α-substitution reaction a reaction that puts a substituent on an α-carbon in place of an α-hydrogen.

substrate the reactant of an enzyme-catalyzed reaction.

subunit an individual chain of an oligomer.

suicide inhibitor (mechanism-based inhibitor) a compound that inactivates an enzyme by undergoing part of its normal catalytic mechanism.

sulfide (thioether) the sulfur analog of an ether.

sulfonate ester forms when an alcohol reacts with a sulfonyl chloride.

sulfonation substitution of a hydrogen of a benzene ring by a sulfonic acid group (SO_3H).

suprafacial bond formation the formation of two σ bonds on the same side of the π system.

suprafacial rearrangement rearrangement in which the migrating group remains on the same face of the π system.

Suzuki reaction couples an aryl, a benzyl, or a vinyl halide with an organo-borane in the presence of $Pd(PPh_3)_4$.

Swern oxidation oxidizes primary alcohols to aldehydes and secondary alcohols to ketones.

symmetrical anhydride an acid anhydride with identical R groups:

symmetrical ether an ether with two identical substituents bonded to the oxygen.

symmetric molecular orbital a molecular orbital in which the left half is a mirror image of the right half.

symmetry-allowed pathway a pathway that leads to overlap of in-phase orbitals.

symmetry-forbidden pathway a pathway that leads to overlap of out-of-phase orbitals.

syn addition an addition reaction in which two substituents are added to the same side of the molecule.

syndiotactic polymer a polymer in which the substituents regularly alternate on both sides of the fully extended carbon chain.

syn elimination an elimination reaction in which the two substituents that are eliminated are removed from the same side of the molecule.

synthetic equivalent the reagent that is actually used as the source of a synthon.

synthetic polymer a polymer that is not synthesized in nature.

synthon a fragment of a disconnection.

systematic nomenclature method of nomenclature based on structure.

target molecule desired end product of a synthesis.

tautomerism interconversion of tautomers.

tautomers rapidly equilibrating isomers that differ in the location of their bonding electrons.

template strand the strand in DNA that is read during transcription.

terminal alkyne an alkyne with the triple bond at the end of the carbon chain.

termination step when two radicals combine to produce a molecule in which all the electrons are paired.

terpene a lipid, isolated from a plant, that contains carbon atoms in multiples of five.

terpenoid a terpene that contains oxygen.

tertiary alcohol an OH group attached to a tertiary carbon.

tertiary alkyl halide an alkyl halide in which the halogen is bonded to a tertiary carbon.

tertiary alkyl radical a radical with the unpaired electron on a tertiary carbon.

tertiary amine an amine with three alkyl groups attached to the nitrogen.

tertiary carbocation a carbocation with the positive charge on a tertiary carbon.

tertiary carbon a carbon bonded to three other carbons.

tertiary hydrogen a hydrogen bonded to a tertiary carbon.

tertiary structure a description of the three-dimensional arrangement of all the atoms in a protein.

tetraene a hydrocarbon with four double bonds.

tetrahedral bond angle the angle between any two lines that point from the center to the corners of a tetrahedron is 109.5°

tetrahedral carbon a carbon, such as the one in methane, that forms covalent bonds using four equivalent sp^3 orbitals.

tetrahedral intermediate the intermediate formed in a nucleophilic acyl substitution reaction.

tetrahydrofolate (THF) the coenzyme required by enzymes that catalyze reactions that donate a group containing a single carbon to their substrates.

tetraterpene a terpene that contains 40 carbons.

tetrose a monosaccharide with four carbons.

thermal reaction a reaction that takes place without the reactant having to absorb light.

thermodynamic product the more stable product produced from a reaction.

thermodynamics the field of chemistry that describes the properties of a system at equilibrium.

thermodynamic stability is indicated by $\Delta G°$. If $\Delta G°$ is negative, then the product is *thermodynamically stable* compared with the reactant. If $\Delta G°$ is positive, then the product is *thermodynamically unstable* compared with the reactant.

thermodynamically controlled when a reaction is under thermodynamic control, the relative amounts of the products depend on their stabilities.

thermoplastic polymer a polymer that has both ordered crystalline regions and amorphous noncrystalline regions; it can be molded when heated.

thermosetting polymers cross-linked polymers that, after they are hardened, cannot be remelted by heating.

thiamine pyrophosphate (TPP) the coenzyme required by enzymes, which catalyze a reaction that transfers a two-carbon fragment to a substrate.

thiirane a three-membered-ring compound in which one of the ring atoms is a sulfur.

thin-layer chromatography (TLC) a technique that separates compounds on the basis of their polarity.

thioester an ester with a sulfur in place of the carboxyl oxygen

thioether (sulfide) the sulfur analog of an ether.

thiol (mercaptan) the sulfur analog of an alcohol.

threo enantiomers the pair of enantiomers with similar groups on opposite sides when drawn in a Fischer projection.

titration curve a plot of pH versus added equivalents of hydroxide ion.

Tollens test an aldehyde can be identified by observing the formation of a silver mirror in the presence of Tollens' reagent (Ag_2O/NH_3).

transamination a reaction in which an amino group is transferred from one compound to another.

s-trans conformer a conformation in which two double bonds are on opposite sides of a single bond.

transcription the synthesis of mRNA from a DNA blueprint.

transesterification reaction the reaction of an ester with an alcohol to form a different ester.

trans fused two cyclohexane rings fused together such that if the second ring were considered to be two substituents of the first ring, both substituents would be in equatorial positions.

transimination the reaction of a primary amine with an imine to form a new imine and a primary amine derived from the original imine.

transition metal catalyst a catalyst containing a transition metal, such as $Pd(PPh_3)_4$, that is used in coupling reactions.

trans isomer the isomer with the hydrogens on opposite sides of the double bond or cyclic structure. The isomer with identical substituents on opposite sides of the double bond.

transition state represents the highest-energy structures that are involved in the reaction.

transition state analog a compound that is structurally similar to the transition state of an enzyme-catalyzed reaction.

translation the synthesis of a protein from an mRNA blueprint.

transmetallation metal exchange.

triacylglycerol (triacylglycerol) the compound formed when the three OH groups of glycerol are esterified with fatty acids.

triene a hydrocarbon with three double bonds.

trigonal planar carbon an sp^2 hybridized carbon which is bonded to three atoms and three points define a plane.

triose a monosaccharide with three carbons.

tripeptide three amino acids linked by amide bonds.

triple bond three bonds connecting two atoms.

triplet an NMR signal split into three peaks.

triterpene a terpene that contains 30 carbons.

twist-boat conformation a conformation of cyclohexane.

ultraviolet light electromagnetic radiation with wavelengths ranging from 180 to 400 nm.

unsaturated hydrocarbon a hydrocarbon that contains one or more double or triple bonds.

unsymmetrical ether an ether with two different substituents bonded to the oxygen.

urethane a compound with a carbonyl group that is both an amide and an ester.

ultraviolet and visible (UV/Vis) spectroscopy the absorption of electromagnetic radiation in the ultraviolet and visible regions of the spectrum; used to determine information about conjugated systems.

valence electron an electron in an atom's outermost shell.

valence shell electron-pair repulsion (VSEPR) model provides information about compounds that have conjugated double bonds.

van der Waals forces (London dispersion forces) induced-dipole–induced-dipole interactions.

vector sum takes into account both the magnitudes and the directions of the bond dipoles.

vicinal dihalide a compound with halogens bonded to adjacent carbons.

vicinal diol (vicinal glycol) a compound with OH groups bonded to adjacent carbons.

vinyl group $CH_2{=}CH{-}$

vinylic carbon a carbon in a carbon–carbon double bond.

vinylic cation a compound with a positive charge on a vinylic carbon.

vinylic radical a compound with an unpaired electron on a vinylic carbon.

vinyl polymer a polymer in which the monomers are ethylene or a substituted ethylene.

visible light electromagnetic radiation with wavelengths ranging from 400 to 780 nm.

vitamin a substance needed in small amounts for normal body function that the body cannot synthesize at all or cannot synthesize in adequate amounts.

vitamin KH$_2$ the coenzyme required by the enzyme that catalyzes the carboxylation of glutamate side chains.

vulcanization heating rubber with sulfur to increase its hardness while maintaining its flexibility.

wavelength (λ) distance from any point on one wave to the corresponding point on the next wave; wavelength is generally measured in nanometers, (nm).

wavenumber ($\tilde{\nu}$) the number of waves in 1 cm.

wax an ester formed from a long-chain carboxylic acid and a long-chain alcohol.

wedge-and-dash structure a method of representing the spatial arrangement of groups. Wedges are used to represent bonds that point out of the plane of the paper toward the viewer, and dashed lines are used to represent bonds that point back from the plane of the paper away from the viewer.

Williamson ether synthesis formation of an ether from the reaction of an alkoxide ion with an alkyl halide.

Wittig reaction the reaction of an aldehyde or a ketone with a phosphonium ylide, resulting in the formation of an alkene.

Wohl degradation a method used to shorten an aldose by one carbon.

Wolff–Kishner reduction converts the carbonyl group of a ketone to a methylene group. This method reduces all ketone carbonyl groups, not just those adjacent to benzene rings.

Woodward–Hoffmann rules a series of selection rules for pericyclic reactions.

ylide a compound with opposite charges on adjacent covalently bonded atoms with complete octets.

X-ray crystallography a technique used to determine the structures of large molecules.

X-ray diffraction a technique used to obtain images used to determine the electron density within a crystal.

Zaitsev's rule the more substituted alkene product is obtained by removing a proton from the β-carbon that is bonded to the fewest hydrogens.

Ziegler–atta catalyst an aluminum–titanium initiator that controls the stereochemistry of a polymer.

Z isomer the isomer with the high-priority groups on the same side of the double bond.

zwitterion a compound with a negative charge and a positive charge on nonadjacent atoms.

Credits

Index